ANDERSONIAN LIBRARY
WITHDRAWN
FROM
LIBRARY STOCK
UNIVERSITY OF STRATHCLYDE

# BICYCLIC DIAZEPINES

*This is the Fiftieth Volume in the Series*

**THE CHEMISTRY OF HETEROCYCLIC COMPOUNDS**

**THE CHEMISTRY OF HETEROCYCLIC COMPOUNDS**

A SERIES OF MONOGRAPHS

**EDWARD C. TAYLOR,** *Editor*

**ARNOLD WEISSBERGER,** *Founding Editor*

# BICYCLIC DIAZEPINES

## Diazepines with an Additional Ring

*Edited by*

**R. Ian Fryer**

Department of Chemistry,
Rutgers, State University of New Jersey, Newark, New Jersey

AN INTERSCIENCE® PUBLICATION
John Wiley & Sons, Inc.
NEW YORK / CHICHESTER / BRISBANE / TORONTO / SINGAPORE

An Interscience® Publication



***Library of Congress Cataloging-in-Publication Data***

Bicyclic diazepines: diazepines with an additional ring / edited by R. Ian Fryer.
p. cm.—(Chemistry of heterocyclic compounds, ISSN 0069–3154; v. 50)
"An Interscience publication."
Includes bibliographical references.
ISBN 0-471-52148-5
1. Bicyclic diazepines. I. Fryer, R. Ian. II. Series.
RS431.B49B53 1991
547'.59—dc20 89-24943
CIP

Printed in United States of America
10 9 8 7 6 5 4 3 2 1

# Contributions

R. Ian Fryer, Department of Chemistry, Rutgers University, Newark, New Jersey 07102

A. Walser, Chemical Research Department, Hoffmann-La Roche, Inc., Nutley, New Jersey 07110

# The Chemistry of Heterocyclic Compounds
## Introduction to the Series

The chemistry of heterocyclic compounds constitutes one of the broadest and most complex branches of chemistry. The diversity of synthetic methods utilized in this field, coupled with the immense physiological and industrial significance of heterocycles, combine to make the general heterocyclic arena of central importance to organic chemistry.

*The Chemistry of Heterocyclic Compounds*, published since 1950 under the initial editorship of Arnold Weissberger, and later, until Dr. Weissberger's death in 1984, under our joint editorship, has attempted to make the extraordinarily complex and diverse field of heterocyclic chemistry as organized and readily accessible as possible. Each volume has dealt with syntheses, reactions, properties, structure, physical chemistry, and utility of compounds belonging to a specific ring system or class (e.g., pyridines, thiophenes, pyrimidines, three-membered ring systems). This series has become the basic reference collection for information on heterocyclic compounds.

Many broader aspects of heterocyclic chemistry are recognized as disciplines of general significance which impinge on almost all aspects of modern organic and medicinal chemistry, and for this reason we initiated several years ago a parallel series entitled *General Heterocyclic Chemistry*, which treated such topics as nuclear magnetic resonance, mass spectra, and photochemistry of heterocyclic compounds, the utility of heterocyclic compounds in organic synthesis, and the synthesis of heterocyclic compounds by means of 1,3-dipolar cycloaddition reactions. These volumes are of interest to all organic and medicinal chemists, as well as to those whose particular concern is heterocyclic chemistry.

It has become increasingly clear that this arbitrary distinction created as many problems as it solved, and we have therefore elected to discontinue the more recently initiated series *General Heterocyclic Chemistry* and to publish all forthcoming volumes in the general area of heterocyclic chemistry in *The Chemistry of Heterocyclic Compounds* series.

EDWARD C. TAYLOR

*Department of Chemistry*
*Princeton University*
*Princeton, New Jersey*

# Preface

The completion of this book has been greatly facilitated by the assistance of many people, too numerous to list by name, and the authors acknowledge their generous contributions of time and helpful advice. Special thanks are due to Andre Rosowsky, who, because of the unfortunate delays that occurred in the writing of the book, had to withdraw as editor. His editorial assistance in the early versions of the first few chapters was of inestimable value and provided a guide for the remainder of the work. Special thanks are also due Norman W. Gilman for his critical reading of the manuscript, his numerous helpful suggestions, and his ability to catch typographical mistakes. After completion of the last chapter, it became necessary to update the first four chapters with new material published through 1985. This work was incorporated into the body of these chapters, and the additional references have been included in the numbered reference lists. Our thanks are due to Zi-Qiang Gu and to Julia C. Pinto, who carried out a great portion of these revisions and additions.

The format generally followed for the presentation of these bicyclic diazepine compounds has been to discuss each system, first in order of ring fusion (i.e., [*a*]-fused before [*b*]-fused, etc.) and then by increasing size of the fused ring, least degree of saturation first, with discussions of carbocyclic fused rings preceding those of heterocyclic fused rings. In instances where the benzene annelated ring system was the most important, this ring system was given preference over other fused ring systems. The volume of material to be covered for the [*e*]-fused[1,4]-benzodiazepines was too great to allow use of this format and still retain some sense of clarity and ease of reading. It was therefore decided to divide these structures into five chapters. The first four of these, Chapters V–VIII, cover benzodiazepines, dihydrobenzodiazepines, dihydrobenzodiazepinones and -thiones, and tetrahydrobenzodiazepines. The literature was surveyed through 1983 for Chapters V and VI, and through 1984 for Chapters VII and VIII. Chapter IX covers other fused[*e*] [1,4]-diazepine ring systems, the literature being surveyed through 1985. Previous reviews of bicyclic diazepines have been noted and the material covered in such reviews has been incorporated in related chapters. 1,5-Benzodiazepine compounds are discussed in Chapter IV, under 1,4-diazepines with [*b*]-fused rings.

The literature subsequent to the cutoff dates for the bicyclic benzodiazepines has been largely due to the finding that, by appropriately substituting 1,4-benzodiazepine derivatives in the 3-position, it is possible to prepare compounds that are biologically active as cholecystokinin (CCK) receptor antagonists [see *e.g.*, B. E. Evans et al., *J. Med. Chem.* **30**, 1229 (1987)]. No attempt has been made to summarize this work. Most other 1,4-benzodiazepine chemistry currently reported relates to either tricyclic or tetracyclic ring systems and is beyond the scope of this review. Bridged diazepines are also considered to be beyond the scope of this review and are therefore excluded from discussion.

Finally, the authors express their gratitude to Professor E. C. Taylor, whose patience and understanding has been very much appreciated.

R. Ian Fryer
A. Walser

*Newark, New Jersey*
*Nutley, New Jersey*
*June* 1989

# Contents

# CHAPTER I

# Bicyclic 1,2-Diazepines

**R. Ian Fryer**

*Department of Chemistry, Rutgers, State University of New Jersey, Newark, New Jersey*

and

**A. Walser**

*Chemical Research Department, Hoffmann-La Roche Inc., Nutley, New Jersey*

# INTRODUCTION

This chapter reviews the synthesis and chemistry of 1,2-diazepines of structure **1**, with an additional carbocyclic or heterocyclic ring, of any size, fused to bonds *a–d*. Sections A to D of the chapter correspond to the site of annelation. These sections are subdivided according to the nature of the added ring. When necessary, compounds are arranged in groups according to the degree of saturation, with fully unsaturated species taking precedence.

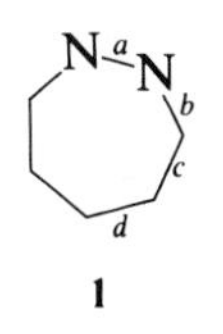


**1**

For both the [*c*]- and [*d*]-fused ring systems, benzannelated compounds represent the largest class and are given preference over other systems. In some instances, the small amount of published data does not warrant a separate section for the reactions of the compounds under discussion. Bridged seven-membered rings are not covered. All compounds are named and numbered according to current *Chemical Abstracts* practice.

1,2-Diazepines have been reviewed previously by M. Nastasi [*Heterocycles*, **4**, 1509 (1976)]; F. D. Popp and A. C. Noble (in *Advances in Heterocyclic Chemistry*, Vol. 8, A. R. Katritzky and A. J. Boulton, Eds., Academic Press, New York and London, 1967, p. 21); J. A. Moore and E. Mitchell (in *Heterocyclic Compounds*, Vol. 9, R. C. Elderfield, Ed., Wiley, New York, 1967, p. 224); G. A. Archer and L. H. Sternbach [*Chem. Rev.*, **68**, 747 (1968)]; and A. Nawojski [*Wiadomości Chemi*, **12**, 673 (1964)]. A review of 1,2-diazepines [V. Snieckus and J. Streith, *Acc. Chem. Res.*, **14**, 348 (1981)] also includes some bicyclic derivatives. One additional review article, covering 1,2-benzodiazepines and related compounds by T. Tsuchiya, appeared in *Yuki Gosei Kagaku Kyokashi* [**39**, 99 (1981); in Japanese].

## A. [*a*]-FUSED [1,2]DIAZEPINES

This section is devoted to bicyclic systems of general structure **2**. The few representatives of this system which are described in the literature are discussed in alphabetical order.

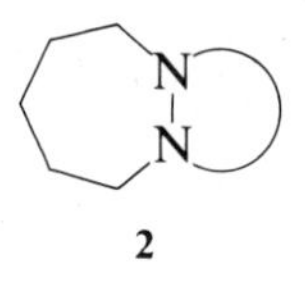

**2**

### 1. PYRAZOLO[1,2-*a*] [1,2]DIAZEPINES

None of the three possible tautomeric forms **3a**–**3c** of the unsaturated ring system has been reported in the literature. Only the synthesis of either partially or fully saturated compounds, which are named and numbered as derivatives of 1,7-diazabicyclo[5.3.0]decane (**4**), has been described.

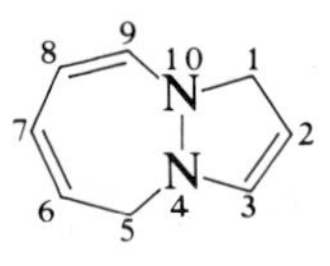

**3a**

1*H*,5*H*-Pyrazolo-[1,2-*a*] [1,2]diazepine

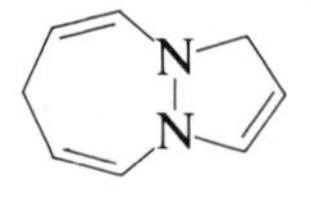

**3b**

1*H*,7*H*-Pyrazolo-[1,2-*a*] [1,2]diazepine

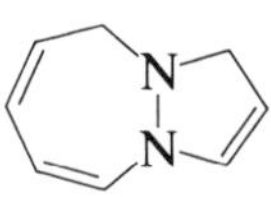

**3c**

1*H*,9*H*-Pyrazolo-[1,2-*a*] [1,2]diazepine

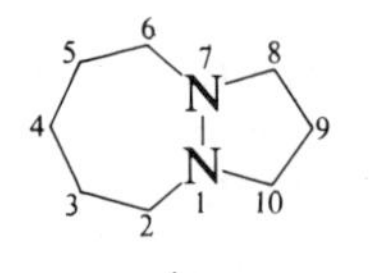

**4**

1,7-Diazabicyclo[5.3.0]decane

The highly substituted 2,3-dihydro-1*H*,5*H*-pyrazolo[1,2-*a*] [1,2]diazepines **8** were synthesized by Ulbrich and Kisch.[1] Photoinduced addition of iron pentacarbonyl to the pyrazolines **5** led to the complexes **6**, which were then allowed to react with diphenylacetylene (Eq. 1) to form new complexes formulated as **7**. Oxidative cleavage of the latter with bromine gave the diazepines **8**. The structures of **8** were assigned on the basis of analytical and spectral data and were supported by a few reactions. Thus 2 mol of bromine added to **8** (R = *i*-Pr) to yield the tetrabromo derivative **12**, which was not fully characterized but could be reconverted to the starting material by reduction with zinc in dimethyl sulfoxide. Thermolysis of **8** (R = Ph) gave the fragments **9**, **10**, and **11**, which can be derived from the assigned structure.

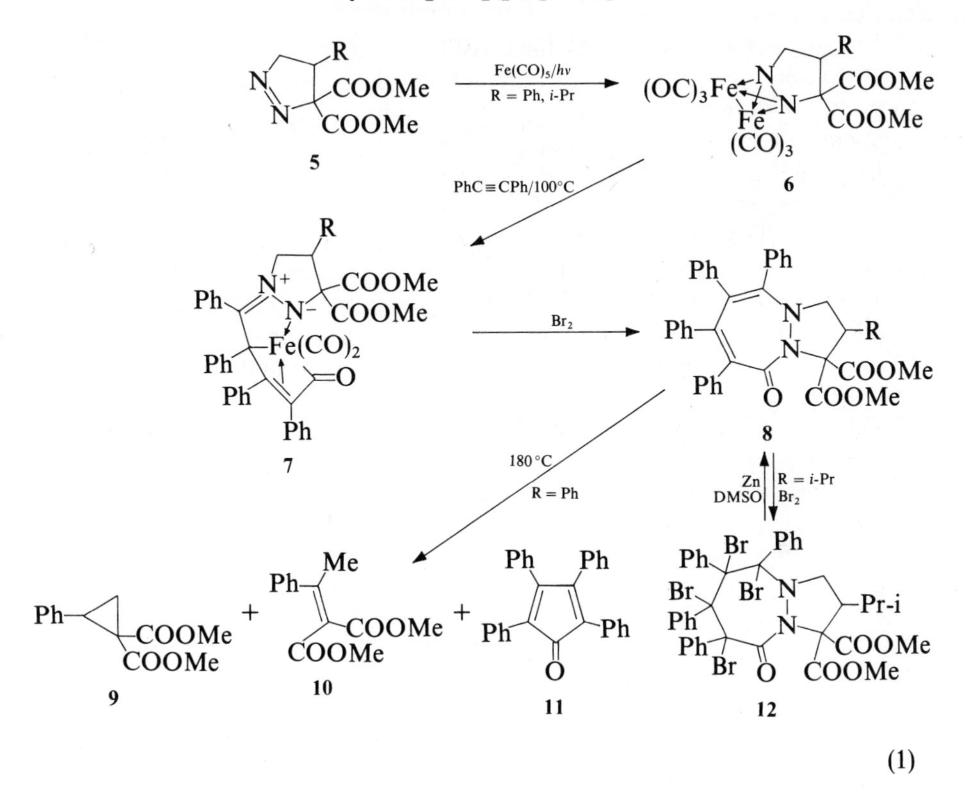


(1)

A pyrazolodiazepine with a saturated seven-membered ring, compound **16**, was obtained[2] by thermolysis of the pyrazolium chloride **14** via the chloropentylpyrazolone **15**. The pyrazolium salt was accessible by the reaction of *N*-aminopiperidine with methyl propiolate (Eq. 2) followed, by treatment of the resultant spirane **13** with hydrochloric acid. The structure assigned for **16** is supported by physical data, including $^{13}$C-nmr spectra.

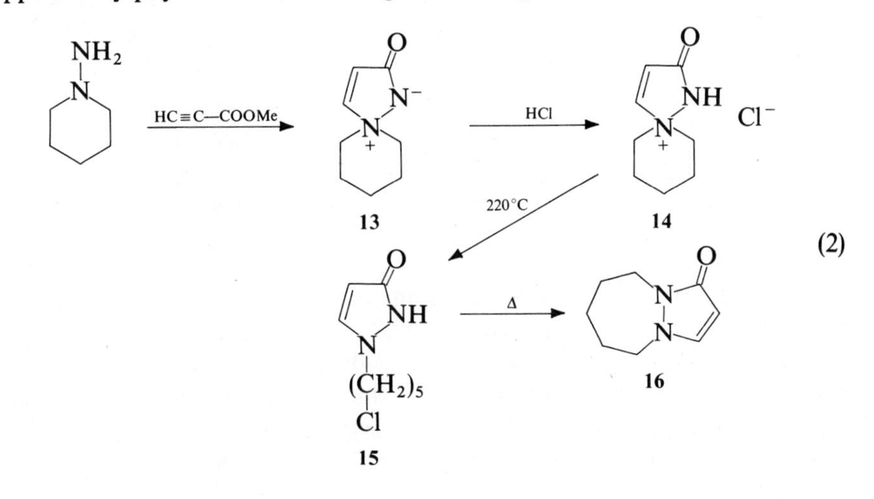


(2)

The one-pot preparation of the fully saturated parent compound **17** by condensation of pyrazolidine with glutaraldehyde, followed by reduction with sodium cyanoborohydride (Eq. 3) was reported by Nelsen and Weisman.[3] This compound and the homologous pyridazino[1,2-*a*] [1,2]diazepine **19**[4] were used with other tetrasubstituted hydrazines in physicochemical studies. In an attempt to determine the conformations of mono and bicyclic hydrazines, Nelsen and Buschek[5] applied photoelectron spectroscopy and estimated the dihedral angle of the lone pairs of electrons on the two nitrogens in the predominant conformer of compound **17** to be 138°. This value lies between the angles found for the two possible trans conformations of the monocyclic pyrazolidines. In a later study[6] $^{13}$C-nmr spectroscopy was used to determine the conformations of **17** and **19** and the equatorial–equatorial confirmation was found to predominate, although the spectra were not too informative due to a broadening at low temperatures (down to − 127°C). This was attributed to the flexibility of the seven-membered ring. The same compounds were compared with other tetra-alkylhydrazines in an investigation[4] of single-electron oxidation equilibria. On the basis of these experiments, it was concluded that the formation of the radical cation occurs preferentially in the seven-membered ring rather than in either the five- or six-membered ring.

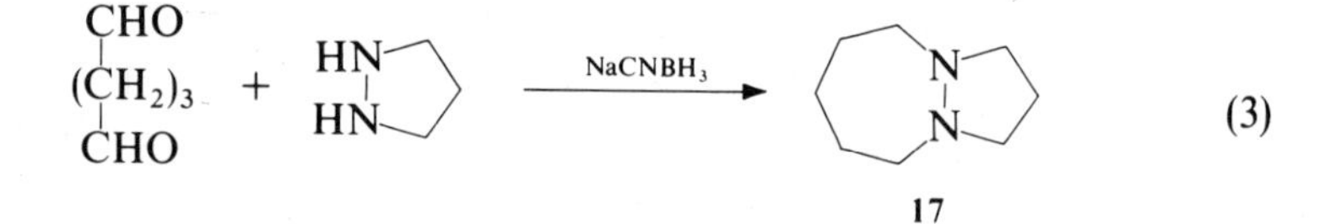


## 2. PYRIDAZINO[1,2-*a*] [1,2]DIAZEPINES

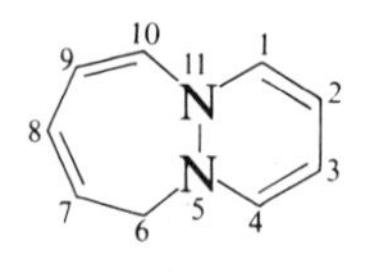


**18a**

6*H*-Pyridazino[1,2-*a*]-[1,2]diazepine

**18b**

8*H*-Pyridazino[1,2-*a*]-[1,2]diazepine

The parent ring system may exist formally as either of the tautomeric forms **18a** or **18b**. The only reported compounds belonging to this class are the fully saturated octahydropyridazino[1,2-*a*] [1,2]diazepines related to **19**. The unsubstituted derivative was prepared by reductive condensation of hexahydropyridazine with glutaraldehyde in the presence of sodium cyanoborohydride (Eq. 4).[4] Compound **19** may also be named 1,7-diazabicyclo[5.4.0]undecane.

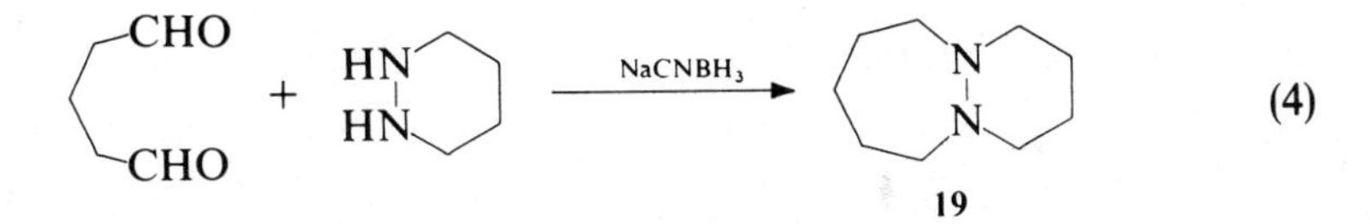


(4)

The synthesis of other *a*-fused derivatives has been reported in a U.S. patent.[76] In general, the synthesis starts from the known benzyloxycarbonyl-protected hexahydropyridazine (**20**), which is acylated by a 2-substituted 1,5-pentandioic acid chloride ester. Deprotection of the amine and ester **21** followed by cyclization affords the diazepine **22**. Removal of the *t*-Bu ester group at C-1 followed by conversion of the acid, Q = Cl to the corresponding Q = phthalimido derivative and finally to the amine **23** completes the synthetic scheme (Eq. 5). The patent further describes substitution of the primary amine.

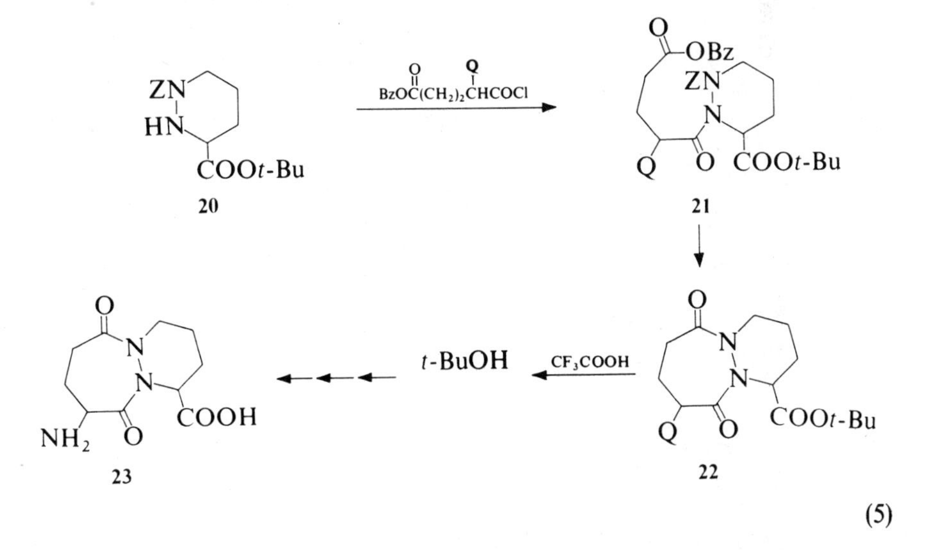


(5)

## 3. [1,2,4]TRIAZOLO[1,2-*a*] [1,2]DIAZEPINES

Compounds belonging to the fully saturated parent ring system were first prepared by Zinner and Deucker.[7] A double alkylation of the dipotassium salt **25** with 1,5-dibromopentane in dimethylformamide (Eq. 6) gave the triazolodiazepines **26**. More recently, a number of compounds of general structure **26**, where R represents a substituted phenyl moiety, appeared in the patent literature.[8] These compounds were prepared by the same alkylation process or by the cyclization of the carbamates **27** (X = O).

The dithione **28** (Y = S and R = 4-ClC$_6$H$_4$) was accessible either by reaction of the dione **26** with phosphorus pentasulfide or by condensation of the

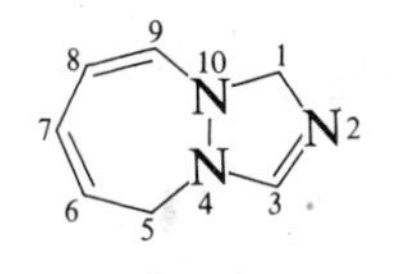

**24a**

1*H*,5*H*-[1,2,4]Triazolo-[1,2-*a*] [1,2]diazepine

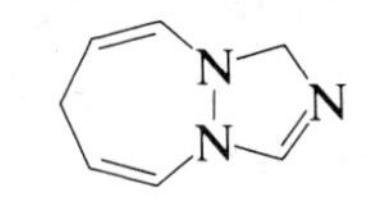

**24b**

1*H*,7*H*-[1,2,4]Triazolo-[1,2-*a*] [1,2]diazepine

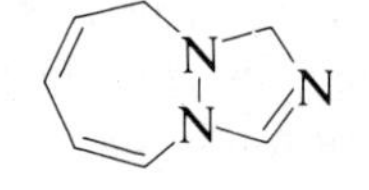

**24c**

1*H*,9*H*-[1,2,4]Triazolo-[1,2-*a*] [1,2]diazepine

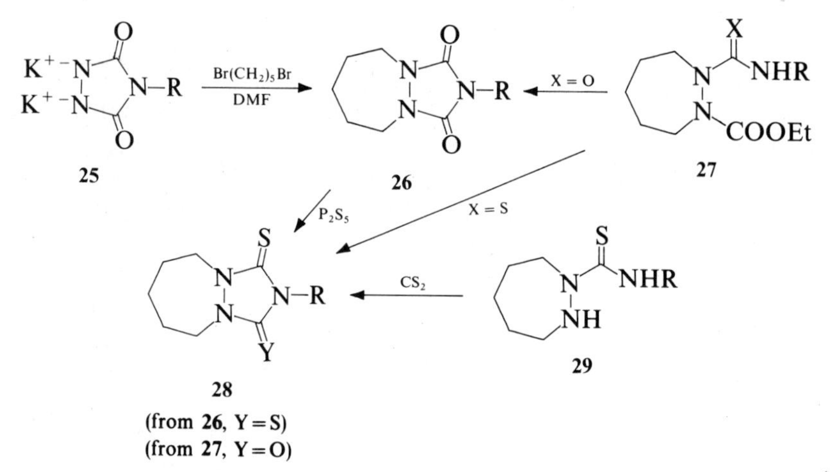


(6)

thiocarbazide **29** with carbon disulfide.[8b,c] Various monothiones were prepared by thermal cyclization of the urethanes **27** (X = S) in boiling cumene.[8a,b]

In what appears to be a typographical error, the structures of the 1,2,6,7,8,9-hexahydro-5*H*-[1,2,4]triazolo[1,2-*a*] [1,2]diazepinium salts, **30**, have been assigned to compounds derived from the treatment of an azomethine with a diazonium salt.[9] These compounds probably should be redrawn as the more likely [1,2,3]triazolo[3,4-*a*]azepinium salts **31**.

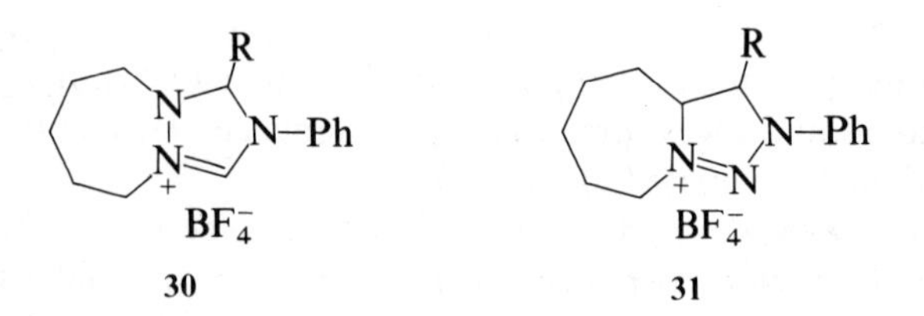

## B. [*b*]-FUSED [1,2]DIAZEPINES

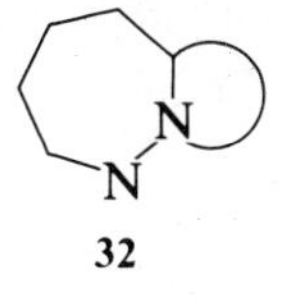


**32**

Only four known bicyclic systems with the general structure **32** have been reported in the literature; these are discussed in alphabetical order in this section.

### 1. AZETO[1,2-*b*] [1,2]DIAZEPINES

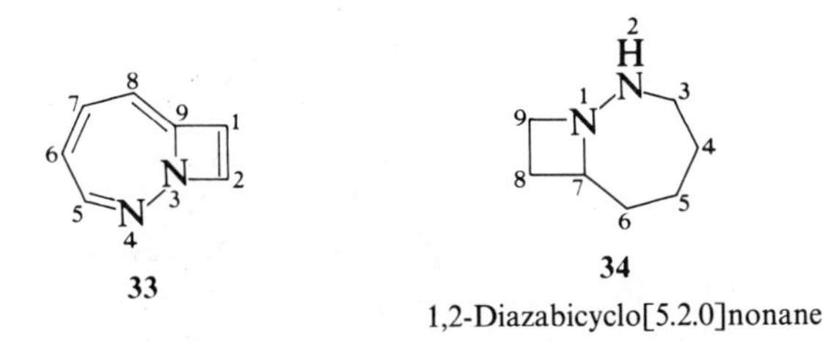


**33**

**34**
1,2-Diazabicyclo[5.2.0]nonane

Representatives of ring system **33** were synthesized by Streith and coworkers[10, 11] by the cycloaddition of ketenes to 1-acyl-1*H*-1,2-diazepines (Eq. 7). The compounds were named as derivatives of 1,2-diazabicyclo[5.2.0]nonane **34**. Thus, the reaction of a chloroketene generated in situ with **35** ($R_1$ = Ph) gave the adduct **36** ($R_2$ = H, $R_3$ = Cl) with a trans configuration. The cycloaddition of methylchloroketene led to a mixture of two diastereomers whose configurations were assigned on the basis of spectral data including nuclear Overhauser effects.

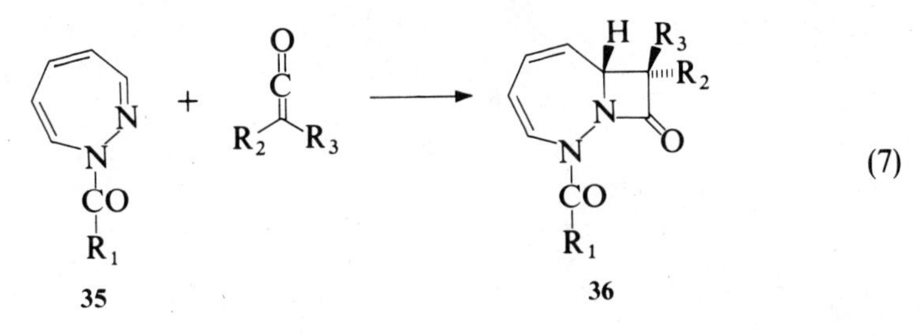


**35** **36** (7)

Although yields were given for several analogs, the initial publication[10] gave the spectral data and melting point only for **36** ($R_1$ = Ph, $R_2$ = H, $R_3$ = Cl).

A later publication[11] reported the addition of phthalimidoacetyl chloride to **35** ($R_1$ = EtO), which led to **36** ($R_1$ = EtO, $R_2$ = H, $R_3$ = 2-phthalimido) in 91% yield. Removal of the phthalimido protecting group afforded the corresponding amine **36** ($R_1$ = EtO, $R_2$ = H, $R_3$ = $NH_2$), which was reacylated with various aromatic carboxylic acids in the presence of dicyclohexylcarbodiimide.

Evidence for the assignment of the diene structure as shown in **36** included the formation of a Diels–Alder adduct between **36** and tetracyanoethylene. X-Ray analysis[12] of **36** furnished the final proof of structure.

Additional work by Streith *et al.* has expanded the use of these 2 + 2 cycloadditions.[104]

## 2. 1,2,4-OXADIAZOLO[4,5-*b*] [1,2]DIAZEPINES

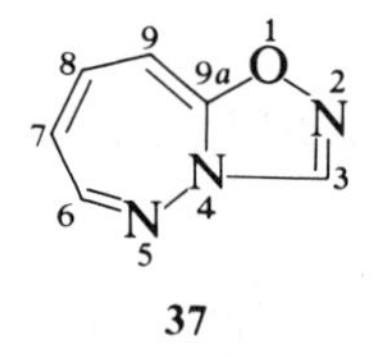

**37**

Although the parent ring system **37** is still unknown, substituted derivatives of the 5,9*a*-dihydro analog **39** were prepared[13] in good yields by a 1,3-dipolar addition of nitrile oxides to the acylated 1,2-diazepines **38** (Eq. 8). The site of cycloaddition was derived from spectroscopic data, including $^{13}$C-nmr spectra, and is in agreement with Huisgen's rule for 1,3-dipolar additions. The addition of mesitylnitrile oxide to **38** usually gave crystalline adducts, while the products obtained by the cycloaddition of benzonitrile oxide were described as unstable oils and were characterized only by spectroscopic data. In one case (starting with **38**, $R_1$ = EtO; $R_2$ = Me) a 1:2 adduct with benzonitrile oxide was isolated

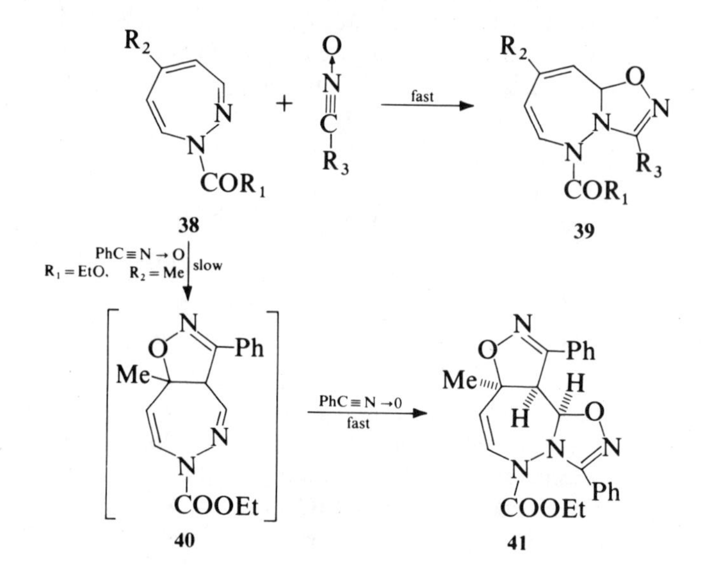


(8)

in low yield and assigned structure **41**. Since this compound could not be prepared from **39** ($R_1$ = EtO; $R_2$ = Me; $R_3$ = Ph) under the same cycloaddition conditions, the authors concluded that **41** probably is formed from an initial slow addition of benzonitrile oxide to the *d*-bond of **38**, leading to the intermediate **40**, followed by a fast reaction of **40** with a second molecule of nitrile oxide to yield the isolated product, **41**.

## 3. PYRROLO[1,2-*b*] [1,2]DIAZEPINES

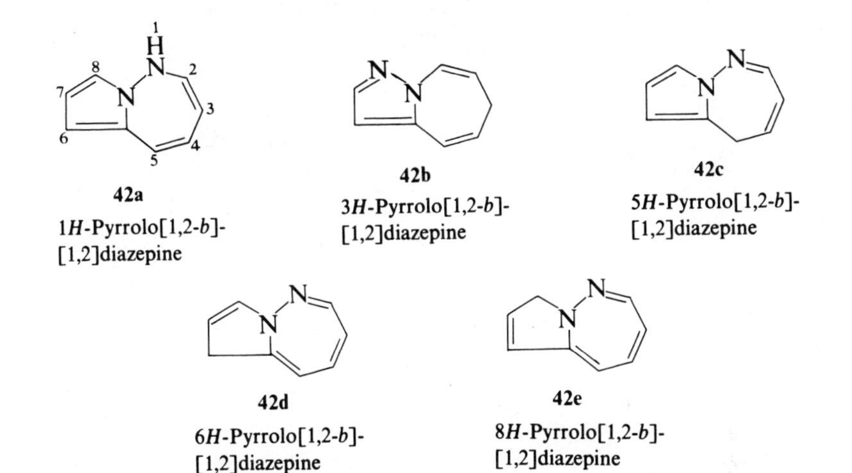


None of the five possible tautomers **42a**–**42e** of the parent ring system has been described in the literature. 2,5-Disubstituted 3*H*-pyrrolo[1,2-*b*] [1,2]diazepines (**44**) were obtained by Flitsch and coworkers[14] by the acid-catalyzed condensation of 1-aminopyrrole **43** with 1,4-diketones (Eq. 9). The 1,1-bipyrryls **45** were also isolated as major by-products.

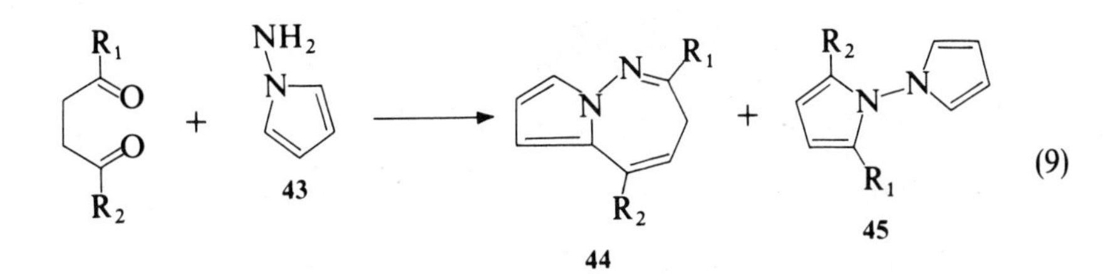


The structures of compounds **44** were compatible with spectroscopic data. Other, energetically less favored tautomers, were excluded on the basis of nmr data.

## 4. THIAZOLO[3,2-*b*] [1,2]DIAZEPINES

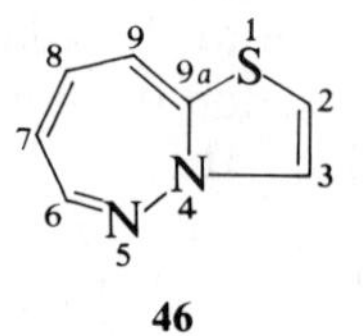


**46**

Representatives of the parent ring system **46** have been claimed incorrectly to be accessible by addition of dimethylacetylene dicarboxylate to *N*-aminothiazolium salts **47** (Eq. 10).[15] Potts and Choudhury[16] reinvestigated this reaction and found that the products were not thiazolodiazepines, but rather the pyrazoles **49**.

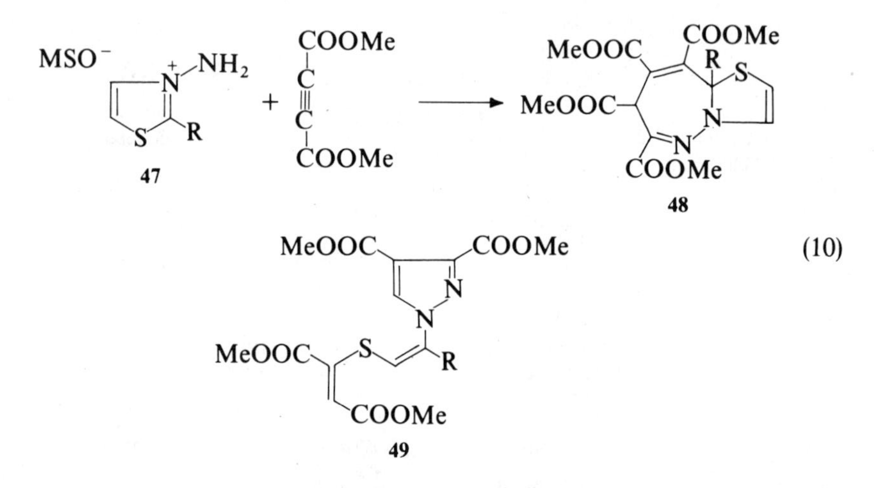


## C. [*c*]-FUSED [1,2]-DIAZEPINES

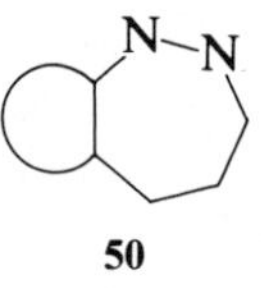


**50**

This section is concerned with bicyclic systems of general structure **50**. The benzo[*c*] [1,2]diazepines, which constitute the largest group of compounds within this structural class, are discussed first. Other [*c*]-fused [1,2]diazepines ring systems follow in alphabetical order in Sections 2–4.

# 1. BENZO[*c*] [1,2]DIAZEPINES (1,2-BENZODIAZEPINES)

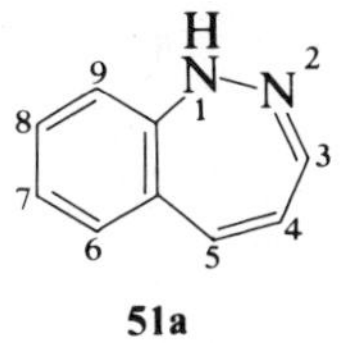

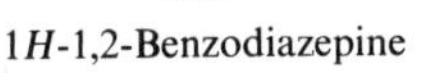

**51a**
1*H*-1,2-Benzodiazepine

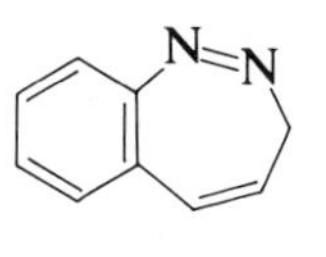


**51b**
3*H*-1,2-Benzodiazepine

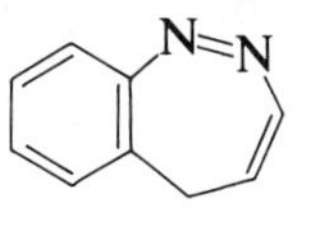


**51c**
5*H*-1,2-Benzodiazepine

If the energetically less favored quinoidal tautomers are disregarded, 1,2-benzodiazepines can occur in the three possible tautomeric forms **51a**–**51c**. No report of the synthesis of the parent ring system for 5*H*-1,2-benzodiazepine (**51c**) was found during our literature search.

## 1.1. 1*H*-1,2-Benzodiazepines

### *1.1.1. Synthesis*

The parent ring system **51a** together with substituted analogs were first synthesized in 1977 by Tsuchiya and coworkers.[17] Photolysis of *N*-iminoquinolinium dimers of type **54** in a mixture of methylene chloride and acetic acid (Eq. 11) led to the diazepines **55** in yields of 5–80%. The parent quinolines **52** and 2-aminoquinolines **57** were minor by-products of this reaction. The dimers were synthesized by carbonate treatment of the intermediate *N*-aminoquinolinium mesitylenesulfonates **53**, which were prepared by *N*-amination of the corresponding quinolines **52** with *O*-mesitylenesulfonylhydroxylamine.

The authors reported that nmr spectra of the dimers **54** recorded in the solvents used for photolysis indicated the presence of an equilibrium mixture of both **54** and the monomeric ylide **56**. A mechanism for the ring expansion of the ylide via the corresponding diaziridine followed by valence tautomerization was proposed in analogy with previously reported ring expansions of *N*-amino heterocycles (See Section 1.4.1 and Ref. 24).

The structures of the benzodiazepines were established both by spectroscopic data and by chemical transformations. While the synthetic utility of this method was not extensively investigated, the approach appears to be limited by the accessibility of the ylide dimers **54**. Thus, although a number of substituted *N*-aminoquinolinium salts **53** were prepared, not all these compounds formed the dimers **54**. Attempts to obtain benzodiazepines by treatment of the salts **53** with aqueous sodium carbonate and methylene chloride followed by direct irradiation of the organic phase were not successful.

Another synthesis of 1*H*-1,2-benzodiazepines, reported by Garanti and coworkers,[18] involves the intramolecular addition of a nitrile imide to an olefin

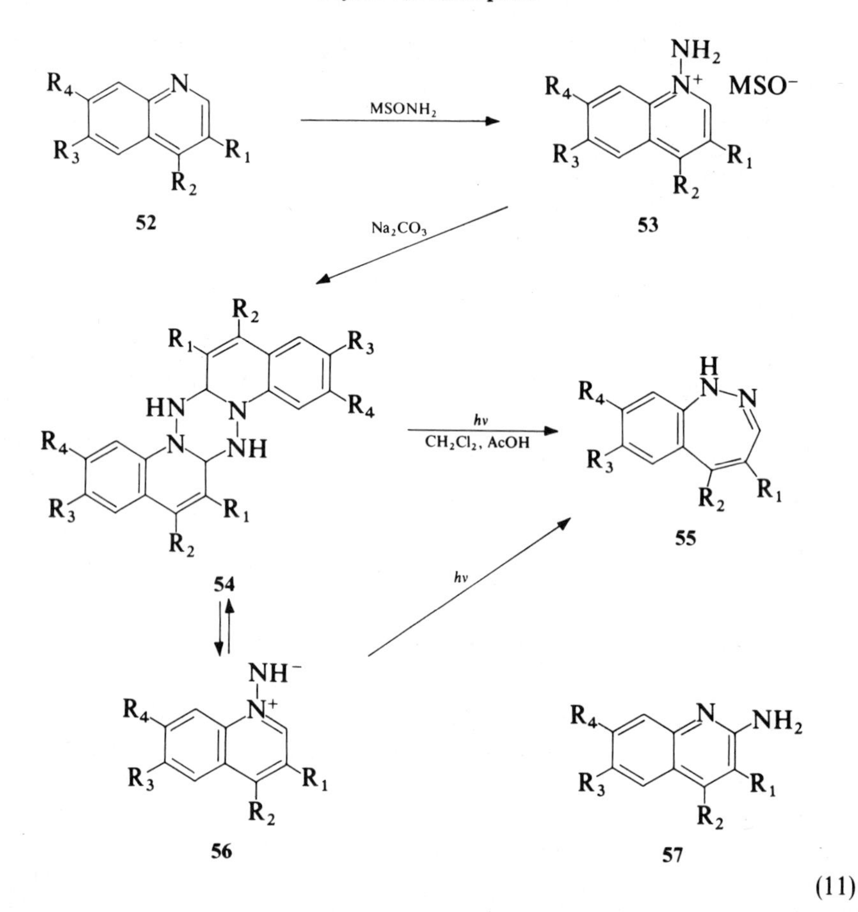


(11)

and is especially useful for the preparation of 1$H$-1,2-benzodiazepines with substituents in the 3- and 4-positions. Treatment of the phenylhydrazoyl chlorides **60** ($R_3$ = H) with triethylamine in boiling benzene led to HCl elimination and generated a transient nitrile imide, which cyclized immediately to the benzodiazepines **61** in good yields (Eq. 12). Compounds **60** were prepared by coupling of the diazonium salt **58** with ethyl 2-chloroacetoacetate or, in some instances, by dehydration of the carbinol **59** in the presence of catalytic amounts of *p*-toluenesulfonic acid. By introducing the olefinic bond after the coupling reaction, the cyclization of the diazonium salt **58** to a cinnoline by-product was suppressed. Compound **60** ($R_1$ = Me; $R_2$, $R_3$ = H) gave, in addition to the corresponding benzodiazepine **61**, the isomer **62** ($R_2$, $R_3$ = H) with the exocyclic double bond. The latter did not isomerize to **61** under the reaction conditions, implying that **62** was formed by a competing reaction pathway. Compounds **60** with a terminally disubstituted styrene ($R_2$, $R_3 \neq$ H) yielded the cyclopropa-[*c*]cinnolines **63** under the same conditions. The tetrasubstituted olefin **60** ($R_1$, $R_2$, $R_3$ = Me) gave the cinnoline **63** together with a 10% yield of the benzodiazepine **62** ($R_2$, $R_3$ = Me).

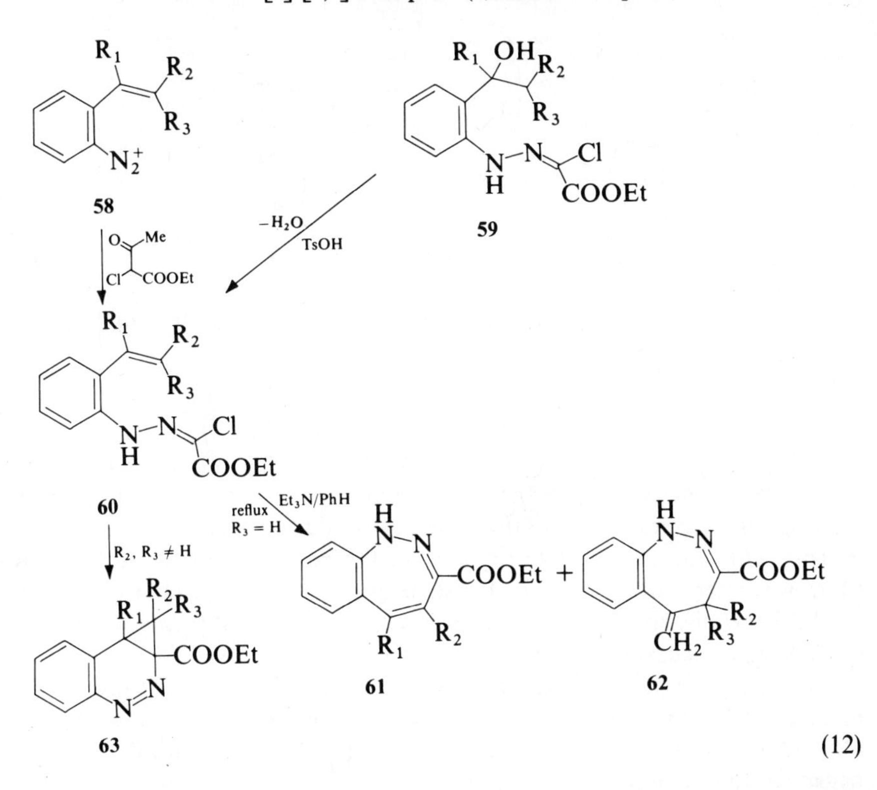


It was found that 3-carbethoxy compounds related to **61** (e.g., **64**) could be hydrolyzed with ethanolic sodium hydroxide to give the corresponding acids **65** in high yield. Thermal decomposition of these acids did not lead to the characterized decarboxylated products (Eq. 13).[77]

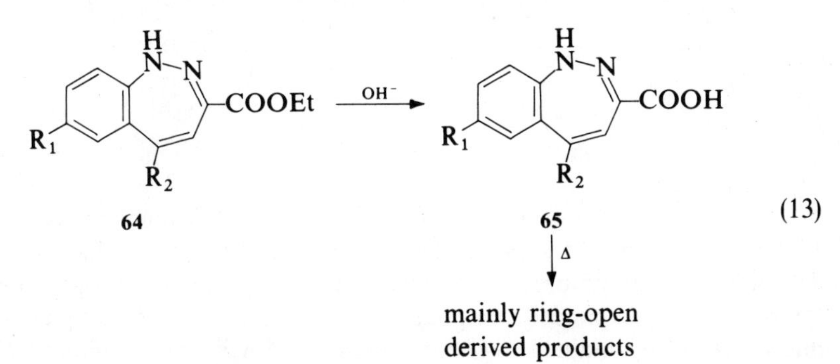


The authors explained the mechanism of formation of the products as proceeding by an intramolecular nucleophilic attack of the double bond on the electron-deficient carbon atom of the generated nitrile imide (Eq. 14). This would result in the formation of the dipole **66**, which is a valence tautomer of the tricyclic compound **67** and also of the 4*H*-1,2-benzodiazepine **68**. These high

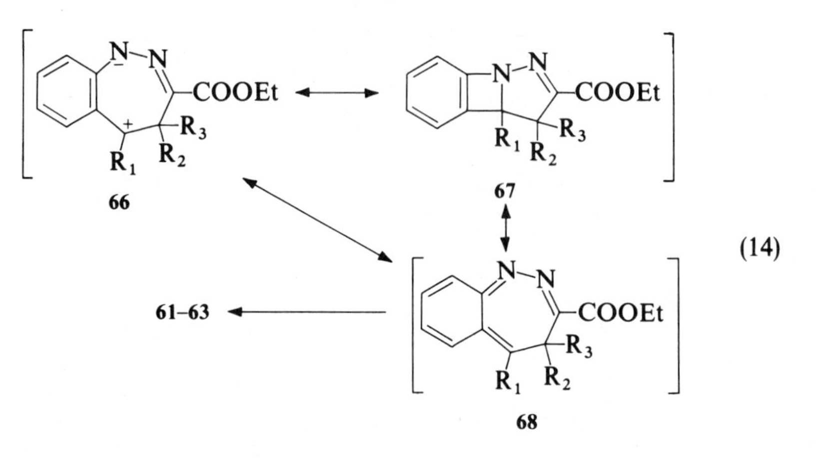


(14)

energy intermediates would then be stabilized by prototropic shifts or by electrocyclic reactions to form the observed products **61–63**.

This synthetic method is not limited to the benzodiazepine-3-carboxylates **61**, as was later demonstrated by the same group.[19] Several 7-chloro-5-phenyl-1*H*-1,2-benzodiazepines with electron-withdrawing substituents other than carbethoxy in the 3-position were prepared in good yields.

Also reported by the Garanti group was the use of 1-nitrohydrazones **69** as precursors rather than the 1-chloro compounds **60**. These intermediates allowed the preparation of 3-unsubstituted (or 3-alkyl-substituted) 1*H*-1,2-benzodiazepines **70** (Eq. 15).[78]

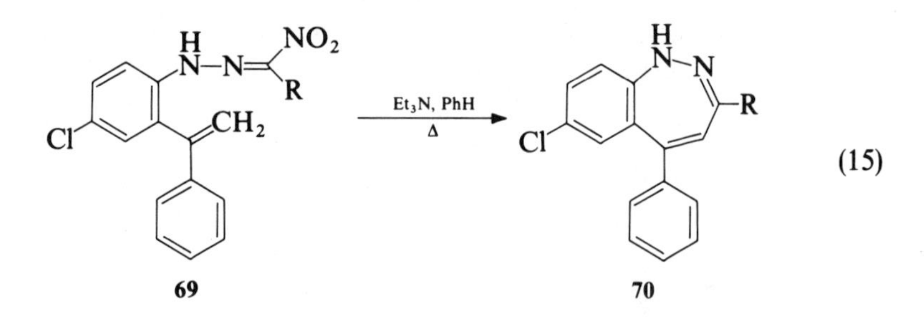


(15)

The acid-catalyzed rearrangement of cyclopropa[*c*]cinnolines (**63**) to give 1,2-benzodiazepines **71** seems to be of limited utility, since yields vary widely depending on the nature of the substitution of the fused cyclopropane ring. Besides the ring-expanded compound, varying amounts of fragmentation products were also obtained.[79] Oxidation of the 5-hydroxy compound **72** with activated $MnO_2$ led to the ketone **73** (Eq. 16).

In closely related work, Padwa and Nahm treated the *N*-substituted *o*-vinyl phenylhydrazone **60** ($R_1$, $R_2$, $R_3$ = H) with base at 80°C and obtained the corresponding diazepine **63** in 91% yield.[80] If the reaction was carried out at room temperature in the presence of silver carbonate, the only product obtained was the cinnoline **63** ($R_1$, $R_2$, $R_3$ = H) in 92% yield. Compound **63** could be

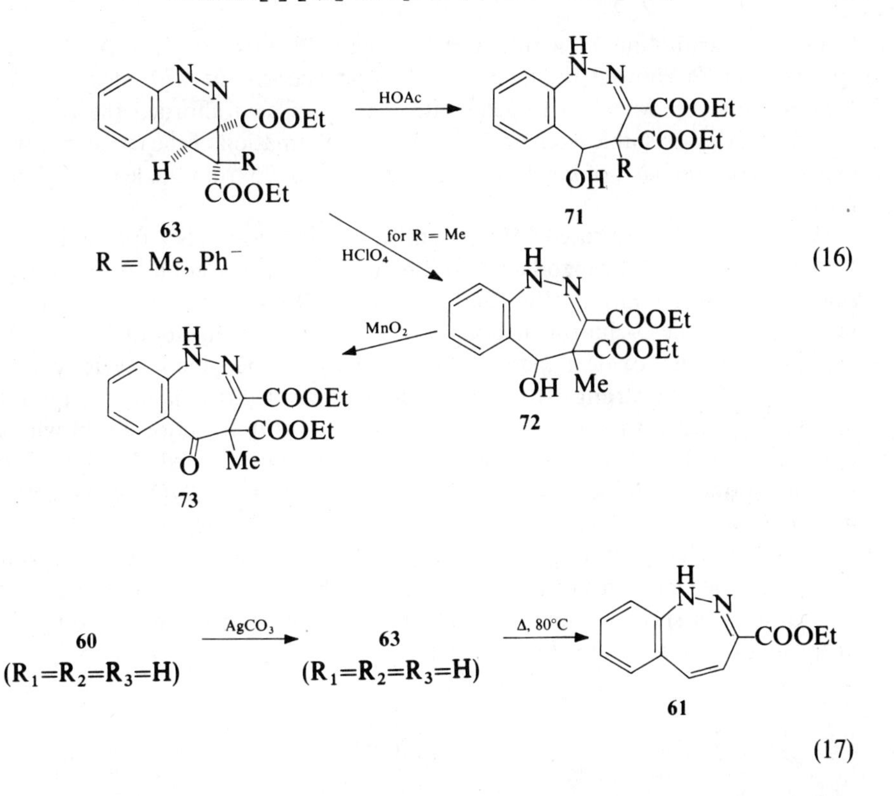


thermally converted at 80°C to the 1,2-benzodiazepine **61**. This ring expansion is readily explained in terms of an initial electrocyclic ring opening of the cyclopropane ring, followed by a 1,5-sigmatropic shift to give the more stable 1*H* tautomer.

Further mechanistic studies on these cycloadditions and rearrangements[81] were carried out in which the proposed *o*-quinoidal intermediate **75** was captured as the acetate **76** (Eq. 18). Thus treatment of either the endo or exo

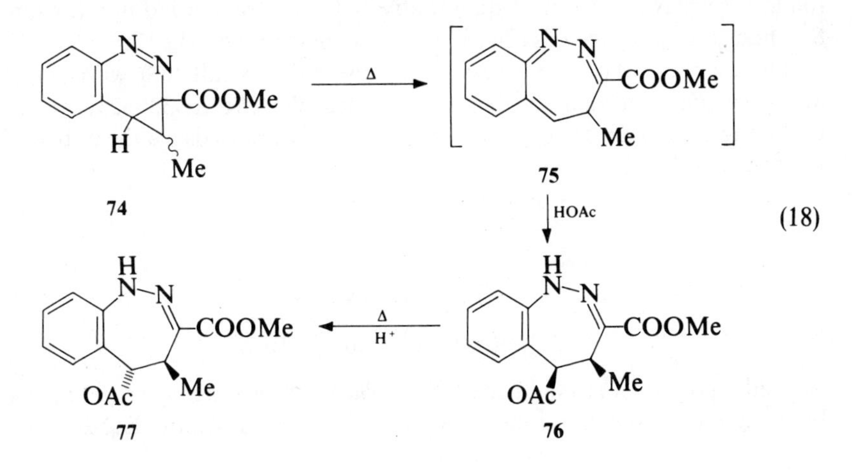

form of the cinnoline **74** with acetic acid gave **76** in high yield. A $^{13}$C-nmr spectrum of **76** showed a signal at 142.6 ppm, consistent with structure **76**. Heating this compound in toluene with a trace of acid afforded the thermodynamically more stable isomer **77** (Eq. 18). The formation of the cis isomer was explained on the basis of kinetic attack of acetate on **75** from the least hindered position.

A variety of 3-substituted 1*H*-1,2-benzodiazepines were also accessible[20] by the reaction of 3*H*-1,2-benzodiazepine 2-oxides **78** with nucleophiles under both acidic and basic conditions (Eq. 19). Treatment of **78** with dry hydrogen chloride in ether gave the 3-chlorobenzodiazepines **79** (X = Cl) in about 90% yield. While sodium methoxide in methanol converted **78** to the 3-methoxy derivatives **79** (X = MeO), the stronger base, ethoxide in ethanol, gave mainly the quinolines **80** together with small amounts of **79** (X = EtO). Reaction of **78** with a cyanide in methanol resulted in mixtures containing the nitrile **79** (X = CN), the corresponding amide (X = $CONH_2$) and the ester (X = COOMe). Dimethyl malonate anion converted **78** to the malonyl derivatives **79** [X = $CH(COOMe)_2$], although these compounds were not obtained in a crystalline state. The structures of these 3-substituted benzodiazepines were assigned on the basis of spectroscopic data. The authors rationalized their findings by proposing mechanisms for both the acid- and base-promoted reactions.

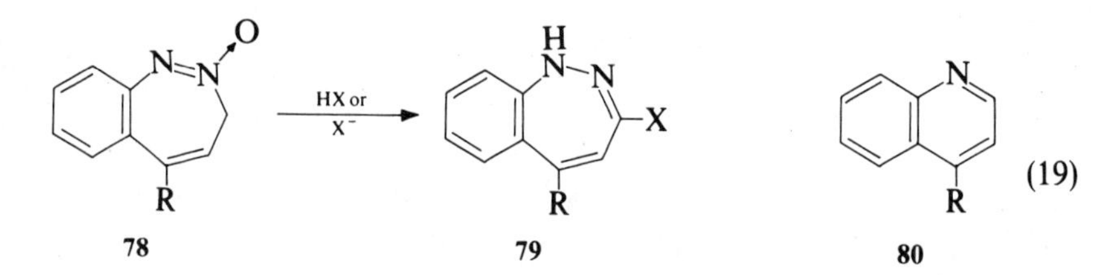


Since the 1*H*-1,2-benzodiazepine appears to be the thermodynamically most stable tautomer, this compound can also be prepared by isomerization of the corresponding 3*H* tautomer. The 3*H*-1,2-benzodiazepines **81** (R = H, Me) were found[21] to rearrange almost quantitatively to the corresponding 1*H* tautomer **82** under the influence of either basic or acidic catalysts (Eq. 20).

The reaction of tropone tosylhydrazone sodium salt (**85**) with acetylenes (e.g., **86a**–**86c**) afforded the benzodiazepines **89a**–**89c** in 33–52% yield. The cyclo-addition and rearrangement most probably proceeds via the norcaradiene intermediate **88**.[82]

### *1.1.2. Reactions*

#### 1.1.2.1. Reactions with Electrophiles

Methylation of **82** to the 1-methyl derivatives **83** was achieved in high yield[22] by using butyllithium and methyl iodide at low temperature (Eq. 20).

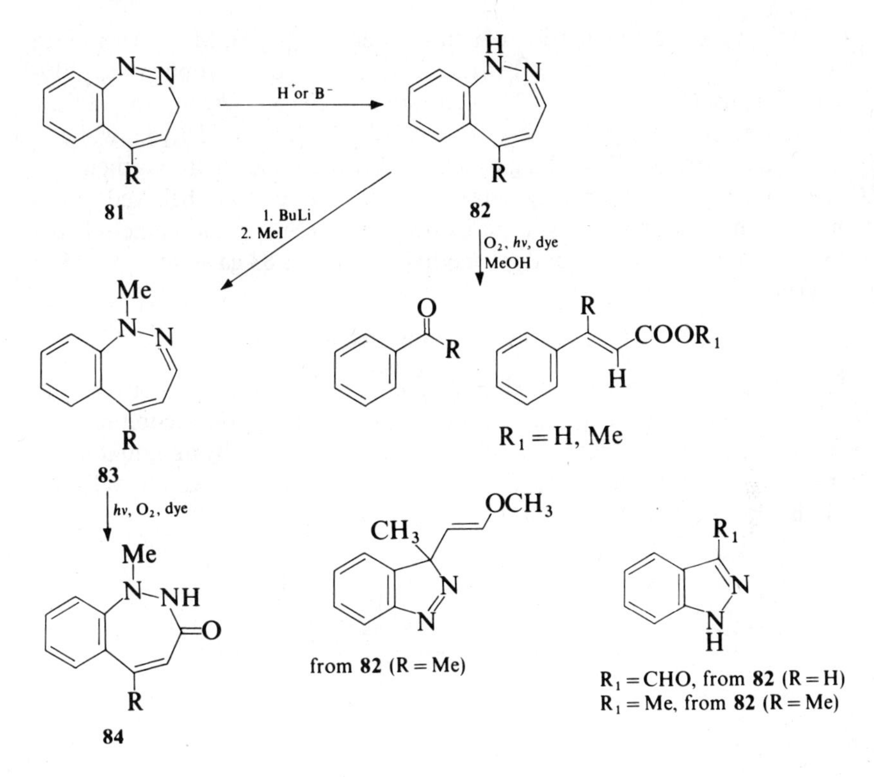
H⁺or B⁻
1. BuLi
2. MeI
O₂, hν, dye
MeOH
hν, O₂, dye
81
82
83
84
R
Me
COOR₁
R₁ = H, Me
OCH₃
CH₃
from 82 (R = Me)
R₁ = CHO, from 82 (R = H)
R₁ = Me, from 82 (R = Me)

(20)

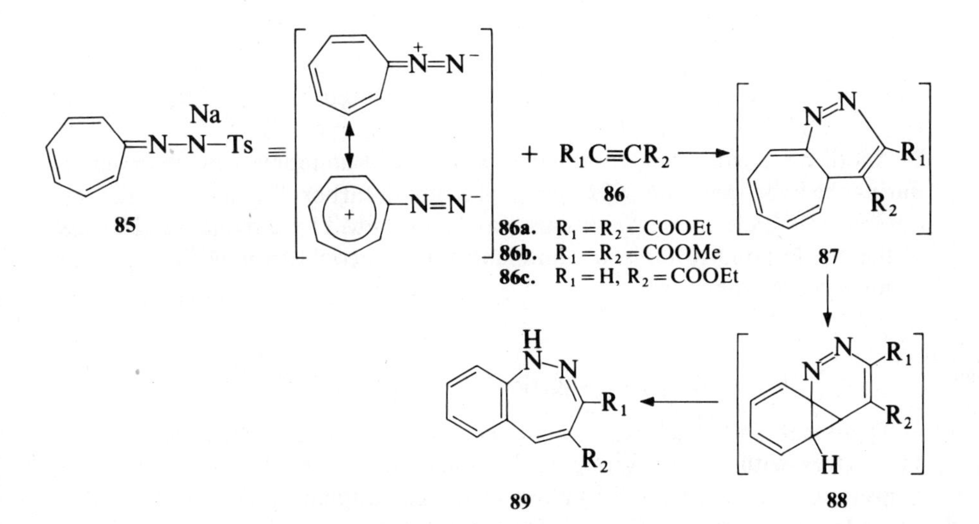
Na
=N–N–Ts
85
+ R₁C≡CR₂
86
86a. R₁ = R₂ = COOEt
86b. R₁ = R₂ = COOMe
86c. R₁ = H, R₂ = COOEt
87
88
89

(21)

Dye-sensitized photooxidation of compounds **82** (R = H, Me) gave an array of products[83] containing either benzaldehyde (27%) or acetophenone (30%), cinnamic acids (10%), and their corresponding esters (20%). In addition, compound **82** (R = H) yielded 2–3% of indazole-3-carboxaldehyde, while the 5-methyl derivative **82** (R = Me) gave 6% of 3-methyl-3-(2-methoxyethen-1-yl)-3*H*-indazole and 3% of 3-methylindazole. These products are believed to arise from 3*H*-benzodiazepine-3-hydroperoxides and 5*H*-benzodiazepine-5-hydroperoxides. Similar oxidation of the 1-methyl derivatives **83** gave the 3-ones **84** in good yield.

#### 1.1.2.2. Reactions with Nucleophiles

Treatment of **82** with sodium ethoxide in ethanol[17b] resulted in conversion to the 2-aminoquinolines **92** in high yield (Eq. 22). This ring contraction has been postulated to proceed via the ring-opened nitrile **90** formed by an initial proton abstraction from the 3-position followed by N—N bond cleavage. Recyclization of the nitrile intermediate would then lead to **92**.

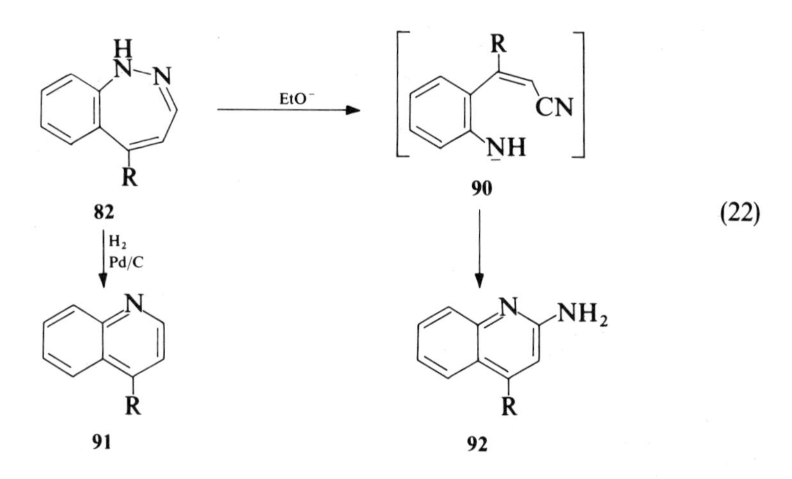


Another reductive ring contraction with loss of ammonia was observed[17b] during the hydrogenation of **82** over palladium on carbon. The formation of the quinoline **91** under these conditions may be explained by a reductive cleavage of the N—N bond to form an aminoimine, which could then cyclize with the elimination of ammonia.

#### 1.1.2.3. Oxidation Reactions

Treatment with lead tetraacetate of 1*H*-1,2,-benzodiazepines **93**, having an electron-withdrawing group such as methoxycarbonyl or cyano in the 3-position, resulted in the formation of the corresponding 5-acetoxy-5*H*-1,2-benzodiazepines **94** in 68% yield. These compounds were the first examples of 5*H*-1,2-benzodiazepines reported by Tsuchiya and Kurita.[102]

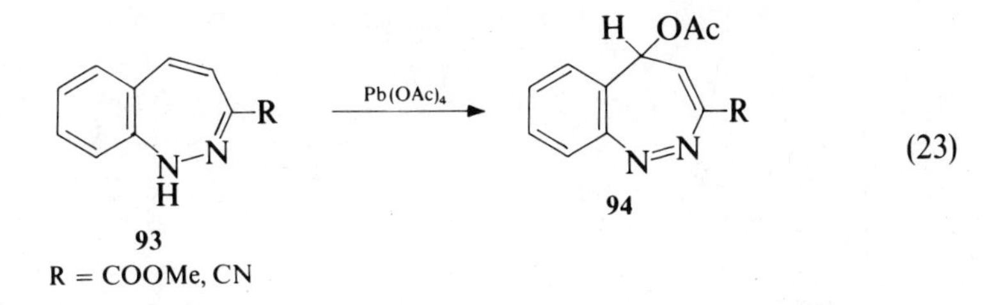


(23)

### 1.1.2.4. Acylation Reactions

The reaction of 3-methyl-1*H*-1,2-benzodiazepine (**95**; R = H) with ethyl chloroformate in benzene at room temperature, gave the *exo*-methylene compound **96** and 2-methylquinoline *N*-ethoxycarbonylimide (**97**) as products. The formation of a 1,3-diazepine (**98**) was not observed. However, when the R group in the 7-position of **95** was an electron-donating group such as a methyl or methoxy group, the formation of 3-ethoxycarbonyl-1*H*-1,3-benzo[*d*]diazepines (**98**) was obtained in addition to the *exo*-methylene compound **96** and the quinoline *N*-imides **97** (Eq. 24).[88, 103]

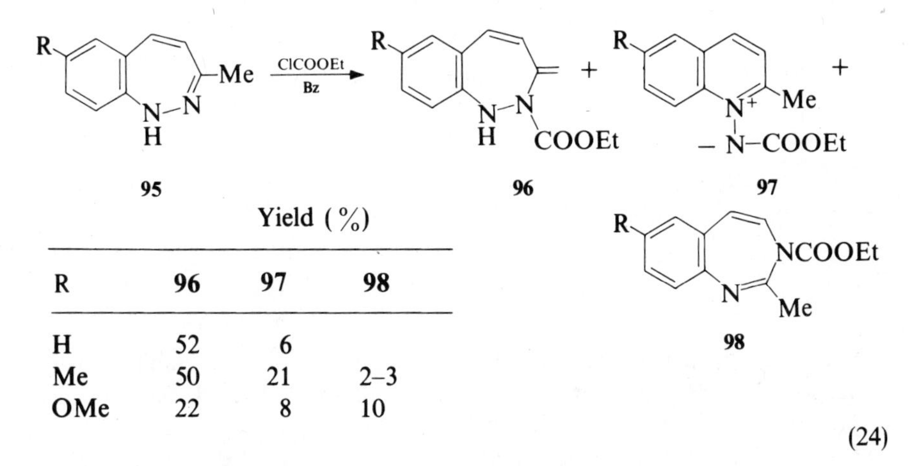


Yield (%)

| R | 96 | 97 | 98 |
|---|---|---|---|
| H | 52 | 6 | |
| Me | 50 | 21 | 2–3 |
| OMe | 22 | 8 | 10 |

(24)

## 1.2. 3*H*-1,2-Benzodiazepines

### *1.2.1. Synthesis*

The parent compound **81** (R = H) and its 5-methyl derivative (R = Me) were prepared in almost quantitative yield by Tsuchiya and Kurita[21,23] by dehydrogenation of the corresponding 2,3-dihydro-1*H*-1,2-benzodiazepines **99** with 4-phenyl-1,2,4-triazoline-3,5-dione (Eq. 26).

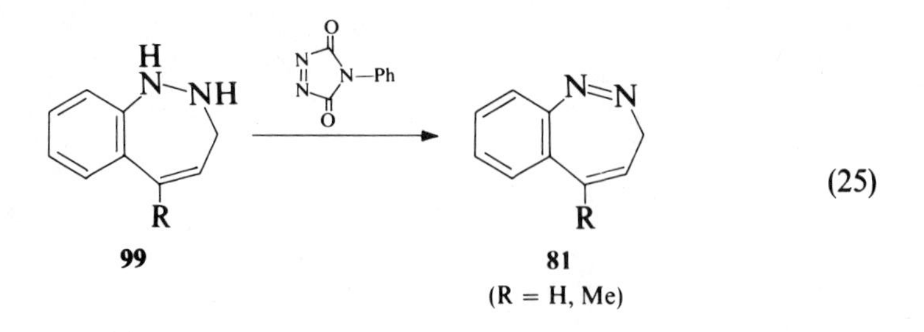


(25)

The 3-acetoxy-3*H*-1,2-benzodiazepines **100** ($R_1$ = Ac) were obtained by oxidation of **82** with lead tetraacetate in methylene chloride[21] or by treatment of the 2-oxide **78** with acetic acid.[20b] Compounds **100** ($R_1$ = Me) were prepared[21] from **82** by an oxidative addition of methanol using cupric nitrate in methanol (Eq. 26).

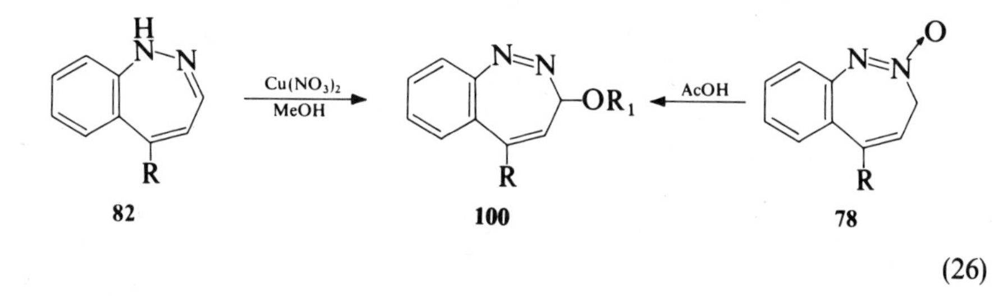


(26)

## *1.2.2. Reactions*

Since the 3*H*-benzodiazepines are thermodynamically less stable than the corresponding 1*H* tautomers, the treatment of **81** with acid, base, or heat resulted in conversion to **82** (Section 1.1.1).

Oxidation of **81** with *m*-chloroperbenzoic acid (Eq. 27) led to a 3:1 mixture of the 2-oxides **78** and the 1-oxides **101**. The products were separated by chromatography and the structures assigned on the basis of $^1$proton-nmr spectroscopy.[20a,b]

Irradiation of the diazepine **81** (R = H) in methylene chloride solution with a high pressure mercury lamp gave 90% of 3-vinylindazole **103** and 1–2% of the indene **104** (R = H).[17a] The 5-methyl analog **81** (R = Me) yielded no indazole under similar conditions, but gave instead a 70% yield of the 3-methylindene **104** (R = Me). This apparent discrepancy may be rationalized by the following mechanistic considerations. Thus, when R = H, the intermediate **102** can be stabilized by a 1,3-proton shift to form the indazole **103**. When R = Me, however, the intermediate can eliminate nitrogen and recyclize to the observed indene **104**.

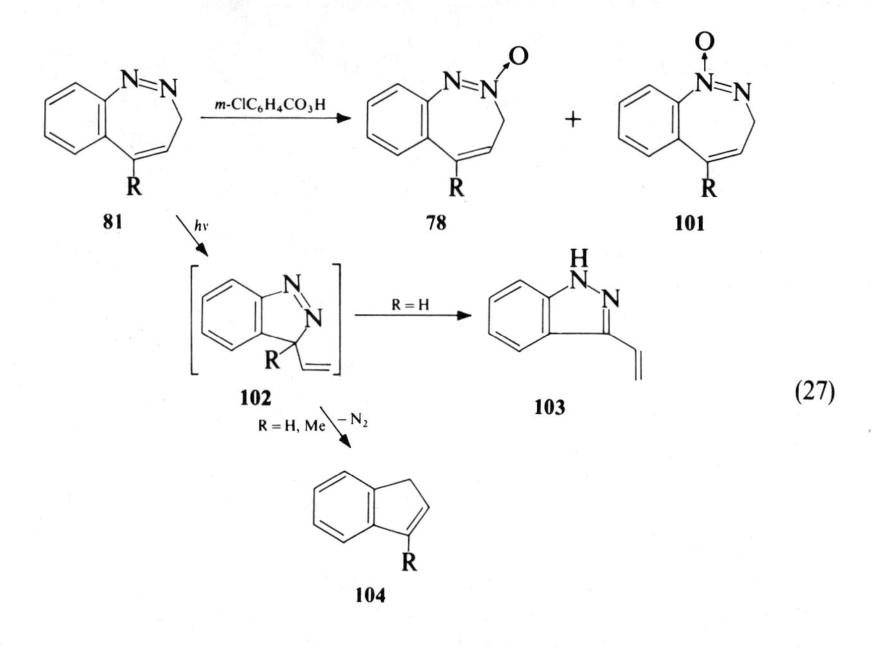


(27)

### 1.2.3. *Spectral Data*

References to characteristic spectral features are given in the tables of compounds (Section E). Kurita and Tsuchiya[17a] recorded the temperature-dependent proton-nmr spectra of compounds **81** (R = H, Me) and determined the energies of activation for the inversion of the seven-membered ring. At the temperature of coalescence, the energy of activation was calculated to be 11.7 kcal/mol for **81** (R = H) and 13.8 kcal/mol for the 5-methyl analog. The higher rigidity of the latter may be due to interaction of the 5-methyl group with the 6-proton.

## 1.3. 5*H*-1,2-Benzodiazepines

### 1.3.1. *Synthesis*

Treatment of the 1*H*-1,2-benzodiazepines **93** with $Pb(OAc)_4$ to give the 5*H*-1,2-benzodiazepines **94**,[84] was discussed under oxidation reactions of the 1*H* compounds (Section 1.1.2.3.). Also, isolated from this type of oxidation was the 3-vinyl indazole **106**. Since 3*H*-1,2-benzodiazepines are known to be susceptible to heat- and light-induced rearrangements,[85] the diazepines **105** are believed to be intermediates (Eq. 28). Oxidation of compounds **93** (R = H, Cl, or OMe) gave only the corresponding 3*H* tautomers **105** and/or the indazoles **106**. None of the 5*H* compounds were found.

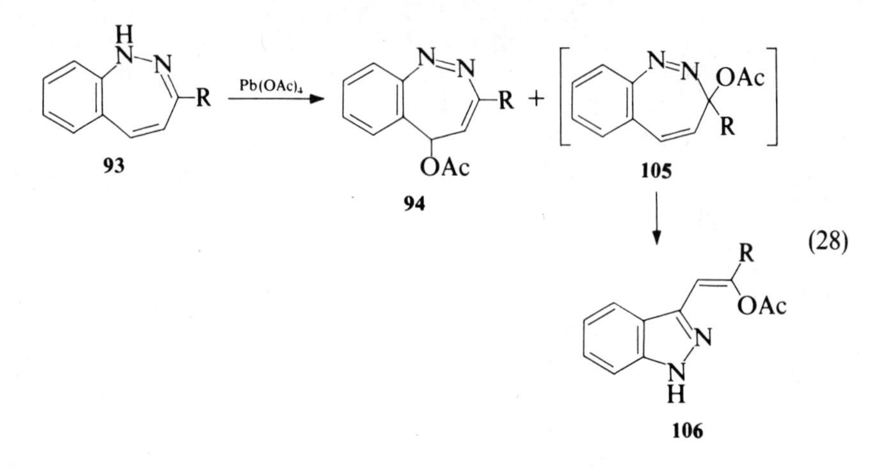


### *1.3.2. Reactions*

In analogy to the 3*H*-benzodiazepines, the 5*H* compounds **94** readily tautomerized into the corresponding 1*H* derivatives **103** by treatment with base ($Et_3N$).[85] Treatment of **94** with either acetic acid or methanol gave the adducts **108** by 1,4-additions. Irradiation of **94** gave the indole derivative **110** (75%), believed to have been formed via the intermediate valence tautomer **109** (Eq. 29).

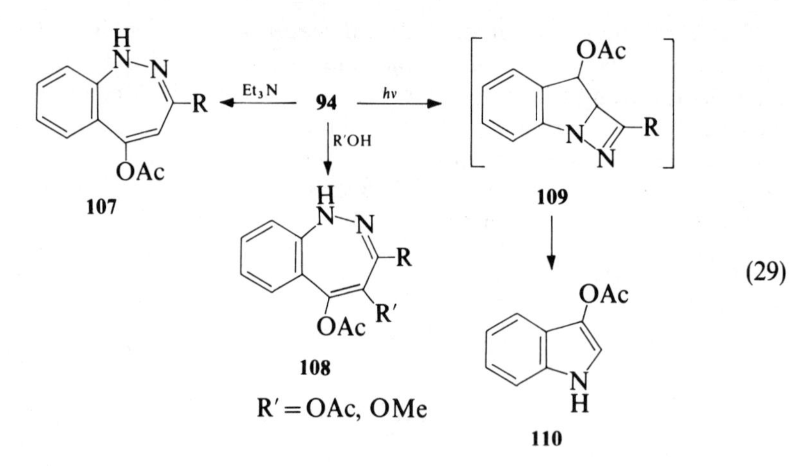


## 1.4. Dihydro-1,2-benzodiazepines

### *1.4.1. Synthesis*

Except for one 4,5-dihydro-1*H*-1,2-benzodiazepine substituted by a 3-carbethoxy and a 5-methylene group (viz., **62**), all the reported dihydro-1,2-

benzodiazepines are 2,3-dihydro-1*H* derivatives. No examples of either the 1,2-dihydro-5*H*- or the 4,5-dihydro-3*H*-1,2-benzodiazepine ring system were found. 2,3-Dihydro-1*H*-1,2-benzodiazepines **99** were prepared by reduction of compounds of type **82** with lithium aluminum hydride.[17a,b]

Reductive acylation of **82** with sodium borohydride in the presence of methyl chloroformate (Eq. 30) led to the 2-methoxycarbonyl derivatives **111** ($R_1$ = MeO).[17b] The same compounds were also formed, but in better yield, by the reaction of **82** with methyl chloroformate and sodium borohydride. Acetic anhydride at room temperature converted **99** to the 2-acetyl derivatives **111** ($R_1$ = Me), which could be further acetylated under more vigorous conditions to the diacetyl derivative **112**.

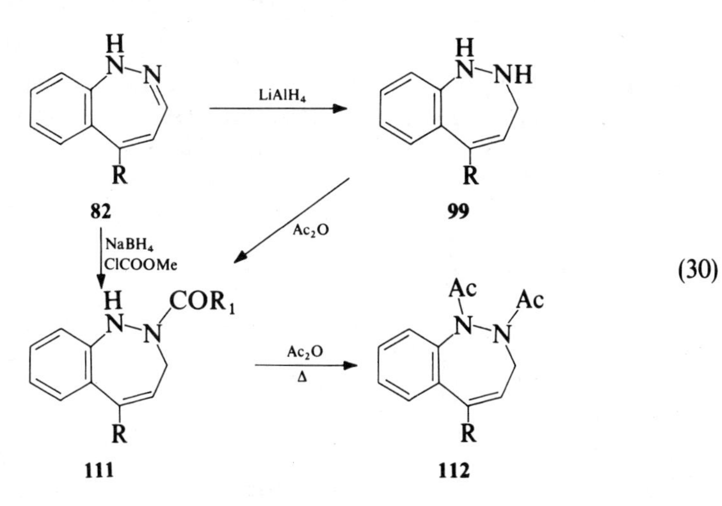


(30)

2-Acyl-2,3-dihydro-1,2-benzodiazepines bearing an alkoxy substituent in the 3-position **114**, were formed by photolysis of solutions of *N*-acylimino-quinolinium ylides **113** in alcohols (Eq. 31). While **114** ($R_1$, $R_2$ = Me) was isolated only in low yield and was not fully characterized,[24] the carbamate **114** ($R_1$ = EtO, $R_2$ = Et) was accessible in 60% yield.[25]

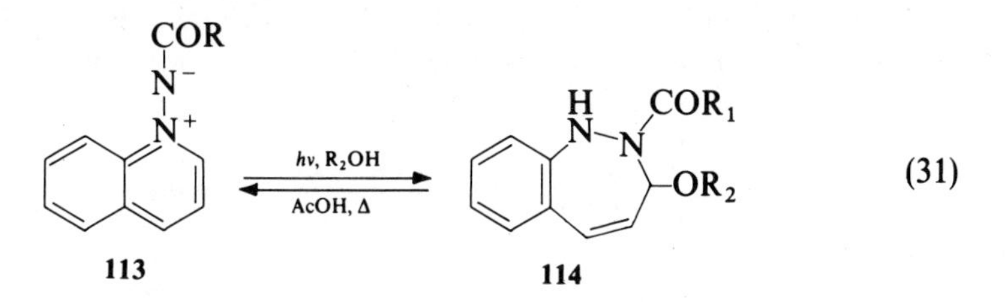


(31)

Two 1-methyl-1,2-dihydro-1,2-benzodiazepin-3(3*H*)-ones (**84**) were prepared by dye-sensitized photooxidation of the corresponding 1*H*-1,2-benzodiazepine (see Section 1.1.2.1).[23]

### 1.4.2. Reactions

Compound **114** ($R_1$ = EtO, $R_2$ = Et) was reconverted to the acetylimino quinolinium ylide **113** ($R_1$ = Me) upon heating in acetic acid.[25] Examples of the acylation of 2,3-dihydro-1*H*-1,2-benzodiazepines and of oxidations leading to 3*H*-1,2-benzodiazepines were mentioned above.

## 1.5. Tetrahydro-1,2-benzodiazepines

### 1.5.1. Synthesis

The 2,3,4,5-tetrahydro-1*H*-1,2-benzodiazepines **116** were synthesized[17b] from the corresponding 2,3-dihydro-1*H*-derivatives **115**, by catalytic hydrogenation over palladium on carbon (Eq. 32). Fission of the N—N bond was not observed under these conditions.

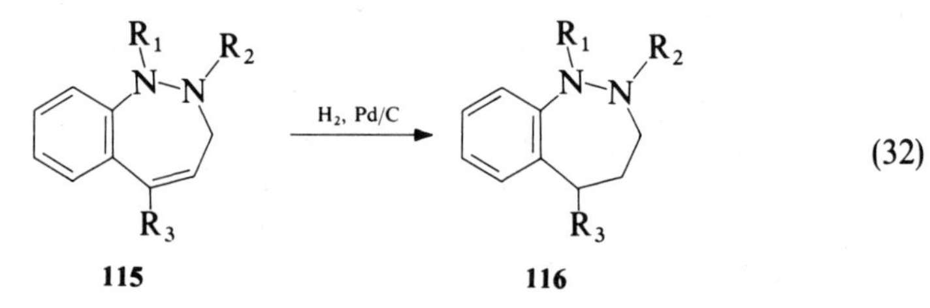


(32)

Selective acetylation of the more basic nitrogen of **116** ($R_1$, $R_2$ = H) with acetic anhydride at room temperature led in high yields to the 2-acetyl derivatives. The same compounds were obtained almost quantitatively by hydrogenation of the 2-acetyl-2,3-dihydro-1*H*-1,2-benzodiazepines **115** ($R_2$ = Ac). The diacetyl derivatives **116** ($R_1$, $R_2$ = Ac) were prepared analogously by hydrogenation of the appropriately substituted **115** or by acetylation of **116** ($R_1$, $R_2$ = H) with acetic anhydride under more vigorous conditions. Streith and coworkers[26] prepared the 2-acyl-6,7,8,9-tetrahydro-2*H*-1,2-benzodiazepines **119** by photolytic ring expansion (Eq. 33) of the *N*-acyliminotetrahydroquinolinium ylides **117**. The high regioselectivity of this reaction may be due to the

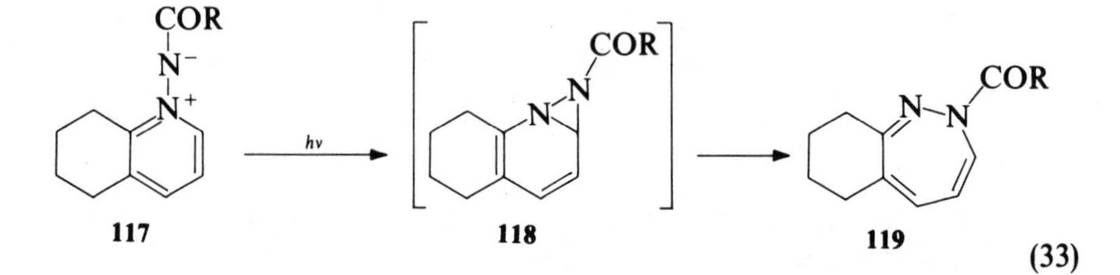


(33)

preferential formation of the intermediate diaziridine **118**, which can rearrange thermally to the diazepine.

The 1,2-diazepine **119** (R = EtO) was obtained as an oil in almost quantitative yield, whereas the corresponding acetyl derivative (R = Me), accessible in much lower yield, was crystalline. The latter compound formed an orange, crystalline iron tricarbonyl complex upon treatment with iron pentacarbonyl.

Although Fischer and Kuzel[27] synthesized **121** in 1883, most reported 1,2-benzodiazepines are of very recent origin. 1-Ethyl-1,2,4,5-tetrahydro-1,2-benzodiazepine-3(3*H*)-one **121**, which appears to be the first benzodiazepine ever prepared, was obtained by evaporating an aqueous acetic acid solution of the *o*-hydrazinophenylpropionic acid (**120**). Compound **120** was prepared by reduction of the corresponding nitrosoaniline with zinc and acetic acid.

This ring closure is limited to the synthesis of monosubstituted compounds, since the preferred cyclization is that which leads to the 1-aminoquinolone derivative **122**. Acid hydrolysis of **121** resulted[27] in ring cleavage to the starting hydrazino acid **120**.

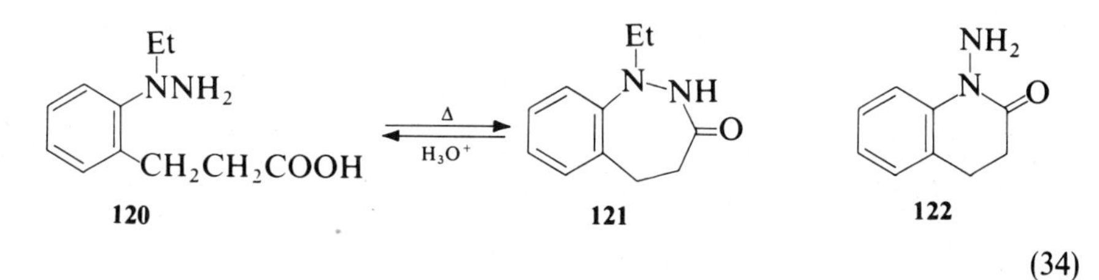


(34)

## 1.6. Hexahydro-1,2-benzodiazepines

### *1.6.1. Synthesis*

The structures **124** were assigned[28,29] to compounds obtained by reaction of the 1,5-diketones **123** with hydrazine in ethanol/acetic acid (Eq. 35). Since the compounds were apparently characterized by microanalysis alone, the structural assignment must be considered to be tentative.

Structure **125** was similarly assigned to the reaction product obtained by reaction of **123** ($R_1$ = H, $R_2$ = COOH) with methylhydrazine.[30] No spectral data for **125** were reported. Based on the paucity of the available evidence, these two compounds (**124** and **125**) might equally well be assigned as *N*-aminoquinoline derivatives—that is, structures represented by the proposed intermediate **126**.

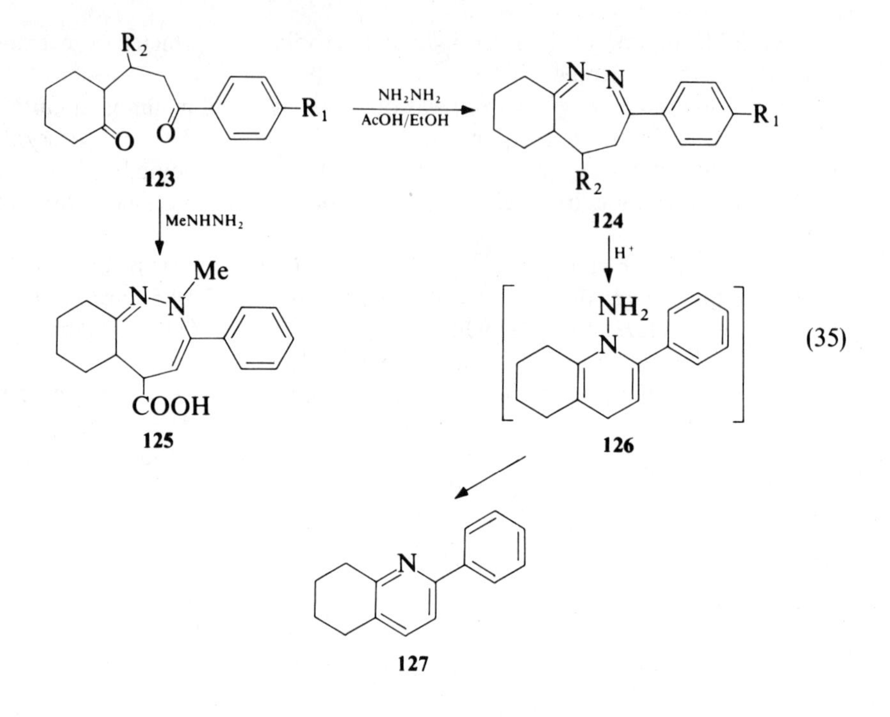


### 1.6.2. Reactions

It has been reported that treatment of **124** ($R_1$, $R_2$ = H) with strong acid, in the presence or absence of solvent, results in ring contraction to the tetrahydroquinoline **127**.[28] The authors postulate that this reaction is most likely to proceed via the intermediate **126**.

## 2. CYCLOPENTA[*c*] [1,2]DIAZEPINES

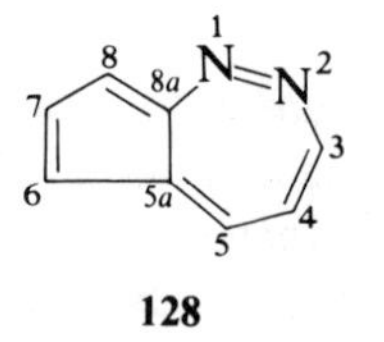


The parent compound **128** is as yet unknown although a few hexahydro derivatives **130**[28] and **131**[30] reportedly have been obtained by the reaction of the cyclopentanones **129** with hydrazine and methylhydrazine, respectively (Eq. 36). The correct assignment of these structures is doubtful (see Sections 1.6.1. and 1.6.2.).

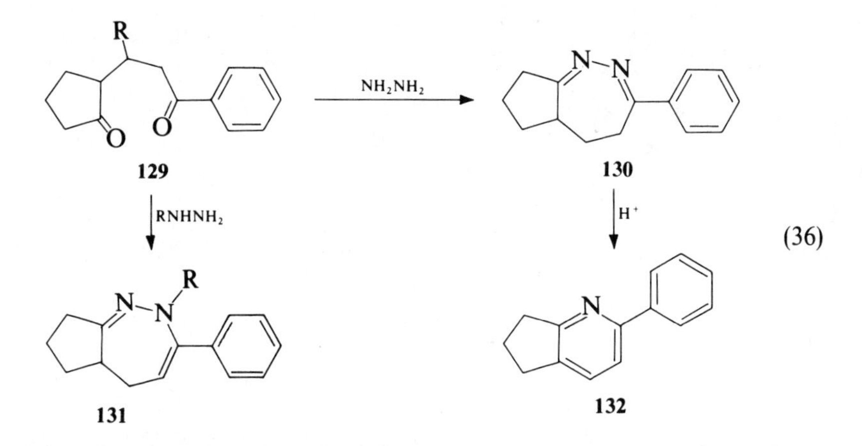


The structure of **130** is based solely on microanalytical data and on its ability to be converted to **132** under acidic conditions. The assignment of structure **131** (R = H, Me) was based on infrared spectral data.

## 3. CYCLOPROPA[*c*] [1,2]DIAZEPINES

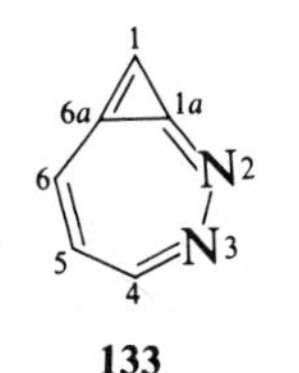


**133**
Cyclopropa[*c*] [1,2]diazepine

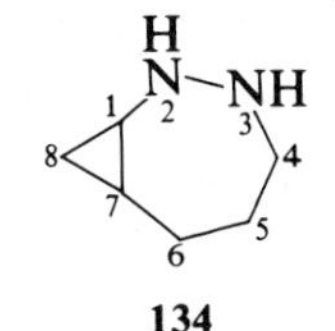


**134**
2,3-Diazabicyclo[5.1.0]octane

### 3.1. Synthesis

The only known example of the bicyclic system **133** or the corresponding reduced system **134** is the highly substituted compound **138**, which was prepared by Sasaki and coworkers[31] in 25% yield by a Diels–Alder type of addition (Eq. 37) of the cyclopropenone **136** to the 4*H*-pyrazole **138**. The postulated intermediate adduct **137** is believed to rearrange to the diazepine **138** as indicated. The structure of the product is supported by spectral data.

## 4. HETERO[*c*] [1,2]DIAZEPINES

Tsuchiya and coworkers extended their synthesis of 1,2-benzodiazepines[17] to the preparation of several heterocyclo[*c*] [1,2]diazepines.[32,86] Thus, the 1*H*-

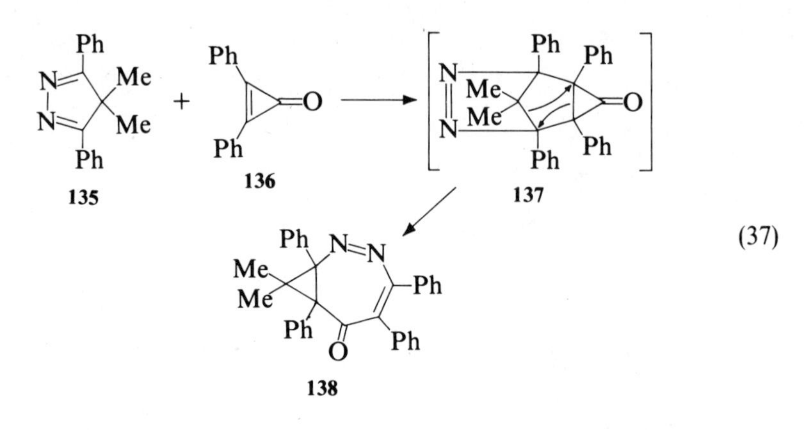


(37)

pyrido[3,2-*c*] [1,2]diazepine **141** was accessible in 25% yield by photolysis of the ylide dimer **140**, believed to exist in equilibrium with the monomer **139** under the reaction conditions (Eq. 38). The isomeric 1*H*-pyrido[2,3-*c*] [1,2]diazepine **143** was obtained analogously (Eq. 39) from the 1-amino-1,8-naphthyridinium ylide **142**.

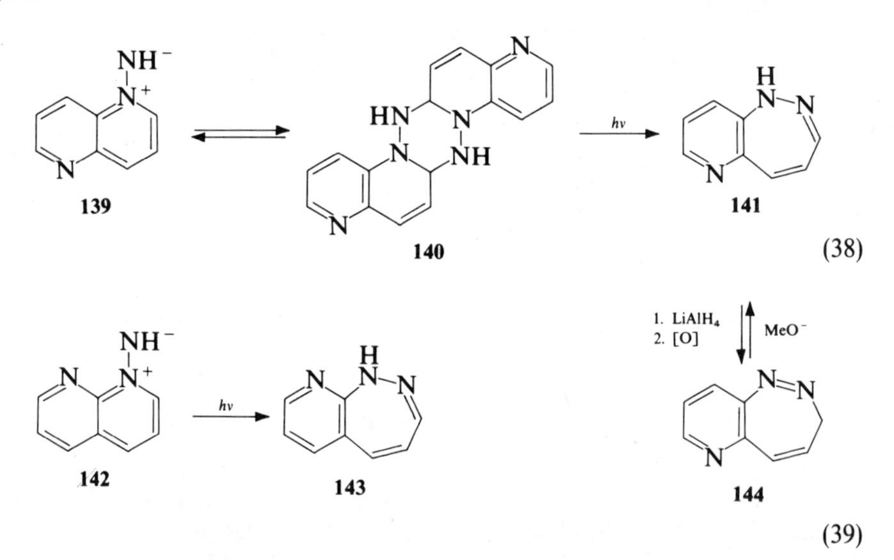


(38)

(39)

As in the benzodiazepine series, compound **141** was converted to the 3*H* tautomer **144** by reduction with lithium aluminum hydride followed by dehydrogenation with 4-phenyl-1,2,4-triazolin-3,5-dione. Treatment of **144** with sodium methoxide in methanol resulted in isomerization back to **141**, the thermodynamically more stable tautomer.

The same type of photochemical synthesis was used for the preparation of the 1*H*-pyrrolo[2,3-*c*] [1,2]diazepine.[87] Irradiation of the 2-methylpyridine *N*-acylimides **145** that were condensed with a thiophene, furan, or pyrrole ring on the *b*-side of the pyridine ring gave the corresponding fused 1*H*-1,2 and

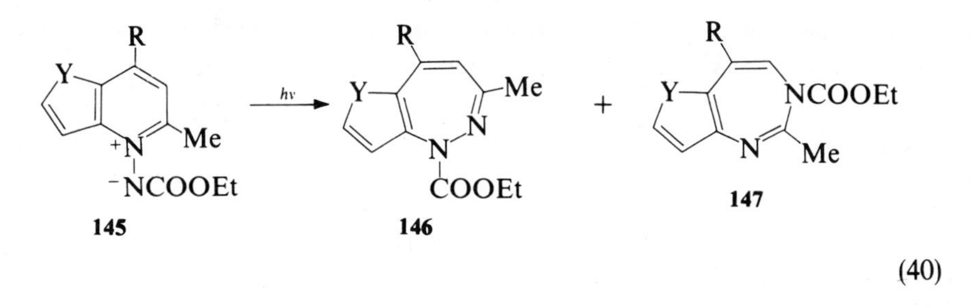


(40)

3*H*-1,3-diazepines **146** and **147**, whereas the *N*-unsubstituted *N*-imide gave only the 1*H*-1,2-diazepine **146** and no 1,3-diazepine (Eq. 40).

In this ring expansion reaction, the initial photoinduced rearrangement may take place on either side of the pyridine nitrogen to give two kinds of diaziridine intermediates, **148** and **149**; compound **148** may give 1,2-diazepines directly, whereas **149** may further rearrange to the aziridine intermediate **150**, followed by ring expansion to give the 1,3-diazepines **147** (Eq. 41).

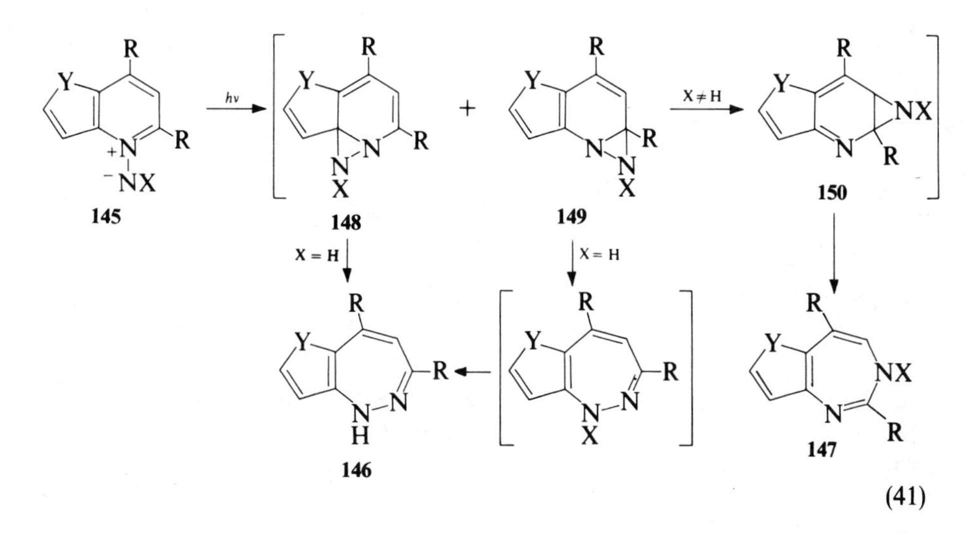


(41)

Similarly 1*H*-thieno[3,2-*c*] [1,2]diazepine **152** was prepared by photochemical synthesis from 4,6-dimethylthieno[3,2-*c*]pyridine *N*-imide **151** (Eq. 42).

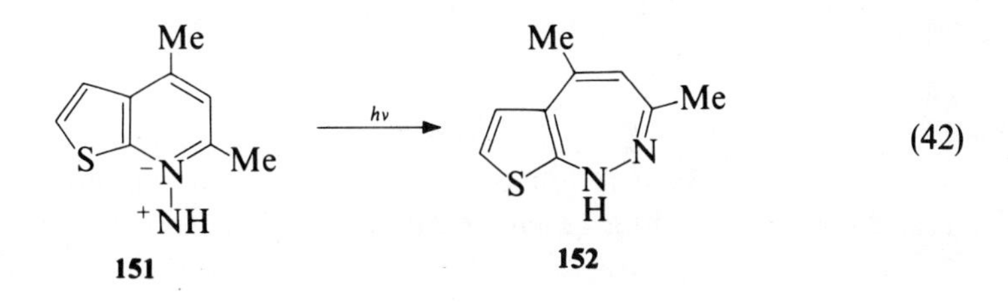


(42)

Treatment of *N*-unsubstituted compound **152** with ethyl chloroformate in anhydrous benzene resulted in a rearrangement with ring conversion to give the 3-ethoxycarbonyl-3*H*-1,3-thieno[3,2-*c*]diazepine **153** as the only product (Eq. 43).

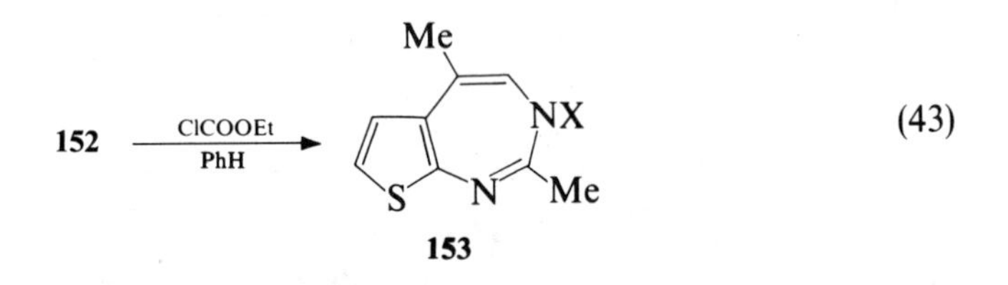


(43)

But treatment of the *N*-unsubstituted compound with ethyl chloroformate in pyridine instead of benzene afforded only the 1-ethoxycarbonyl-1*H*-1,2-thieno[3,2-*c*]diazepine **154**. Conversely, treatment of the *N*-substituted compound **154** with ethyl chloroformate in benzene resulted in the formation of the 1,2-diethoxycarbonyl-3-*exo*-methylene compound **155** instead of **153** (Eq. 44).

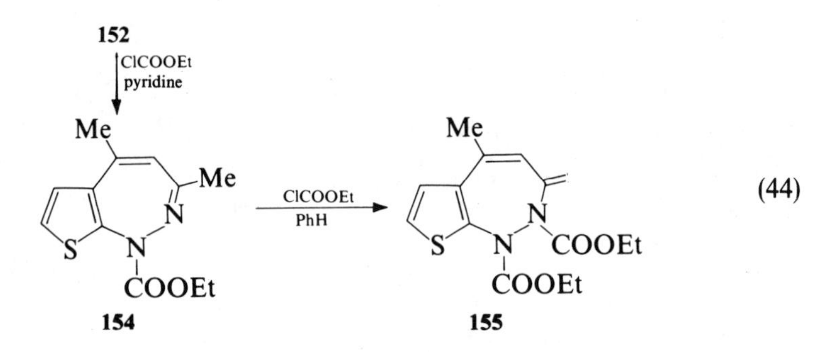


(44)

Morrison and coworkers reported the synthesis (Eq. 45) of the substituted 1*H*-pyrimido[4,5-*c*] [1,2]diazepines **158** by condensation of the hydrazine **156** with 1,3-diketones **157**.[33a,b,89]

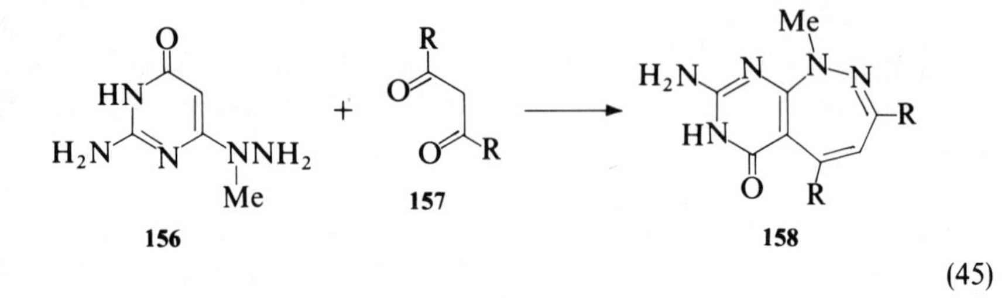


(45)

Compounds **158** were found to be susceptible to ring opening and closing rearrangements leading to either pyrido[2,3-*d*]pyrimidines, pyrimido[4,5-*c*]-pyridazines, or pyrazolo[2,3-*d*]pyrimidines.

# D. [*d*]FUSED [1,2]DIAZEPINES

This section describes the synthesis and properties of bicyclic systems of general formula **159**, beginning with the best known members of this class, the benzo[*d*] [1,2]diazepines.

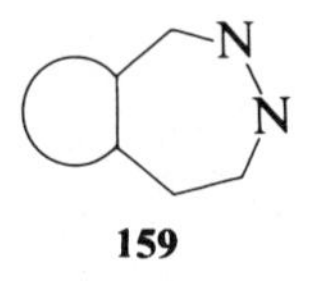

**159**

# 1. BENZO[*d*] [1,2]DIAZEPINES (2,3-BENZODIAZEPINES)

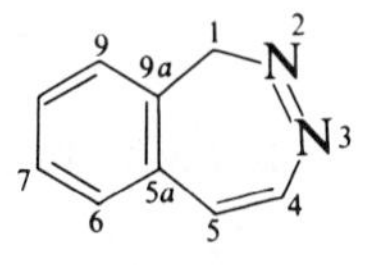

**160a**
1*H*-2,3-
Benzodiazepine

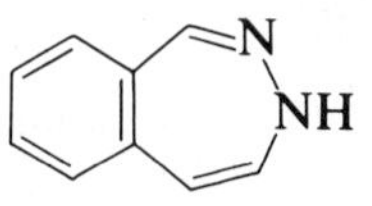

**160b**
3*H*-2,3-
Benzodiazepine

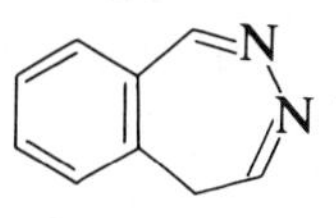

**160c**
5*H*-2,3-
Benzodiazepine

The 2,3-benzodiazepine ring system may theoretically exist in any of the three tautomeric forms **160a**–**160c** (the energetically less favored quinonoidal forms being disregarded).

## 1.1. 1*H*-2,3-Benzodiazepines

### *1.1.1. Synthesis*

Sharp and Thorogood[34] found that the thermal decomposition of the sodium salts of tosylhydrazones in aprotic solvents (Eq. 46) led in good yields to variously substituted 1*H*-2,3-benzodiazepines.

The preliminary account of the work by these authors was followed by a full paper[35] that gave a number of additional examples, including compounds with substituents in the benzene ring. Compounds **164** are most likely formed by an electrocyclic ring closure of the intermediate diazo compound **162** to yield the 4*H* tautomer **163**. A [1,5]-sigmatropic proton shift would then lead to the observed products. It would appear that the 1*H* compounds must form by a kinetically controlled route since, in general, the thermodynamically more stable 5*H* tautomers were obtained only in the presence of excess base. The structures of **164** were firmly established by spectral data and by an X-ray crystallographic

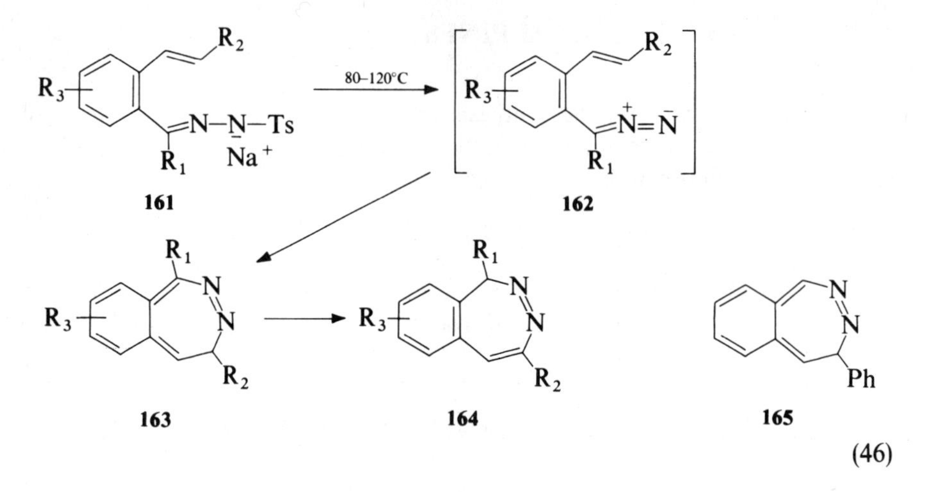


analysis of 1-methyl-4-phenyl-1*H*-2,3-benzodiazepine (**164**: $R_1$ = Me, $R_2$ = Ph, $R_3$ = H).[36]

This synthesis was later applied by Bendall,[37] who claimed to have obtained the 4*H* tautomer **165**, which rearranged to the 1*H* isomer upon standing in methanol solution for 2 days. Sharp and coworkers[35] showed that the compound claimed to be **165** was in fact the 1*H* tautomer **164** ($R_2$ = Ph; $R_1$, $R_3$ = H). The rearranged product Bendall believed to be the 1*H* tautomer was undoubtedly a photoproduct related to **166** as discussed below.

## 1.1.2. Reactions

Base-catalyzed or thermal isomerization (Eq. 47) of the 1*H*-2,3-benzodiazepines **164** leads to the thermodynamically more stable 5*H* tautomers **167**.[35] The sensitivity to light of **164** prompted the study of their photochemical transformation.[35,38,39] These compounds were found to undergo a rapid and virtually quantitative isomerization to the [1,2]diazeto[4,1-*a*]isoindoles **166**.

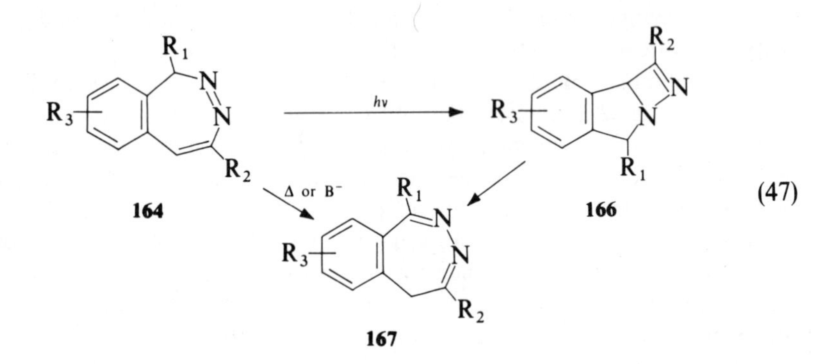

## 1.2. 3H-2,3-Benzodiazepines

### 1.2.1. Synthesis

Lida and Mukai were able to induce the diazepine **168** to undergo a (4 + 2) cycloaddition reaction with a pyrone activated by a methoxycarbonyl group.[90] Thus upon heating the pyrone **169** with the diazepine **168** at 80–110°C for 7 days, the benzodiazepines **171** were obtained in low yield. The reaction was postulated to proceed via the intermediate **170**, which would then decarboxylate and dehydrogenate to give **171** (Eq. 48).

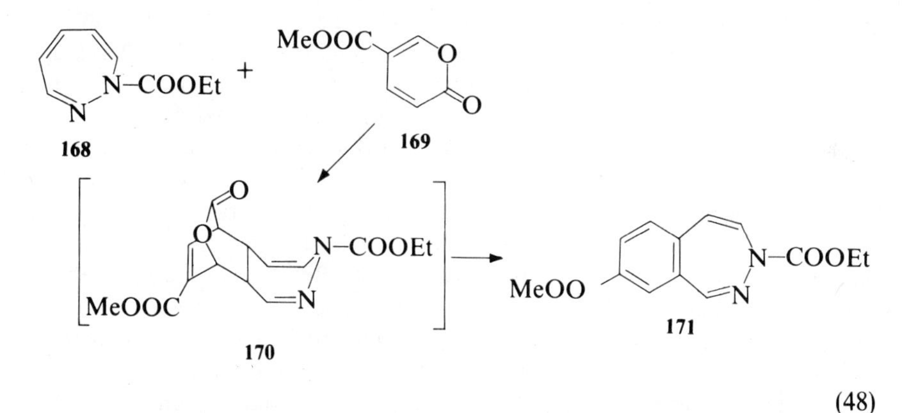


(48)

Kurita et al. reported that acetylating the benzodiazepines **172** (R = H or Me) afforded the 3*H* derivatives **173**. The acetate could be removed in either acid or base and original starting material recovered (Eq. 49).[91]

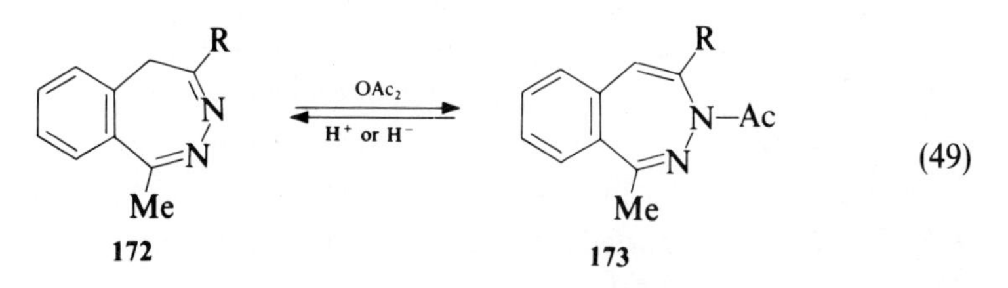


(49)

Photolysis of isoquinoline *N*-imides **174** under basic conditions gave the ring-expanded 5*H*-2,3-benzodiazepines **175** in yields of between 25 and 60%.[91,92] A reasonable mechanism was given, which proceeds via the photoinduced

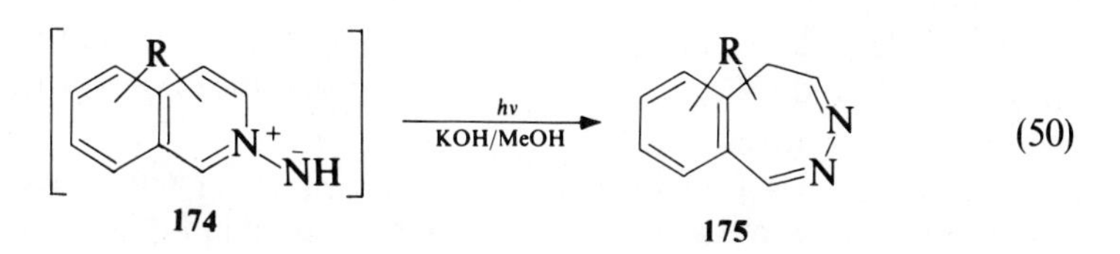


(50)

formation of a diaziridine followed by ring expansion and tautomerisation to the 5*H* product through the 1*H* intermediate (Eq. 50).

## 1.3. 5*H*-2,3-Benzodiazepines

### 1.3.1. Synthesis

As mentioned above, the tautomers **167** may be prepared by treatment of the 1*H*-2,3-benzodiazepines **164** with base or heat. If excess base is used during thermal decomposition of the sodium salts **161**, the 5*H* tautomers are obtained directly.[35] Small amounts of 5*H*-2,3-benzodiazepines were detected during pyrolysis of the photoproducts **166**.[39] The rather obvious route to these compounds (i.e., condensation of a 1,5-diketone with hydrazine) has not been widely used and seems to be exemplified[40] only by the synthesis of analogs of the 5*H*-2,3-benzodiazepine **180** by treatment of the 1,5-diketone **176** with hydrazine under acid catalysis (Eq. 51). If **176** was condensed with hydrazine hydrate below 100°C, the intermediate hydrazine **178** could be isolated in high yield and subsequently ring closed under acidic conditions. Reaction of the pyrylium salt **177** with hydrazine led to the alternative hydrazone **179**, which was found to be more labile than **178** and could be characterized only by spectroscopic methods. It was converted in good yield to **180** by heating in solution.

Compound **178** was initially thought to be the *N*-iminoisoquinolinium ylide **181**,[41] but detailed spectroscopic analysis[42] including $^{13}$C-nmr spectroscopy established the correct structure. The same synthetic method was later applied[43] for the preparation of $^{14}$C derivatives of **180** with the label in the ethyl and 8-methoxy groups.

### 1.3.2. Reactions

Acylation of **180** with *p*-nitrobenzoyl chloride or with acetic anhydride (Eq. 52) has been reported[44] to give the exocyclic double-bonded compounds **182** in 38 and 67% yield.

The products obtained from the acylation of 5*H*-2,3-benzodiazepines are strongly dependent on the nature of the acylating agent and the reaction conditions. The reaction of **183** with acetic anhydride in benzene was rapid at room temperature and gave reactive intermediate **184**. Quenching of **184** with water gave **185**, and when **185** was reacted with a range of O or S nucleophiles, products **186** and **187** were formed. It is noted that the diazepine ring of **186** and **187** was retained (Eq. 53).[97,98]

However, reaction of **183** with acyl halides in benzene or toluene induced a ring transformation that gave either 3-phenylisoquinoline *N*-imine salts **189** or

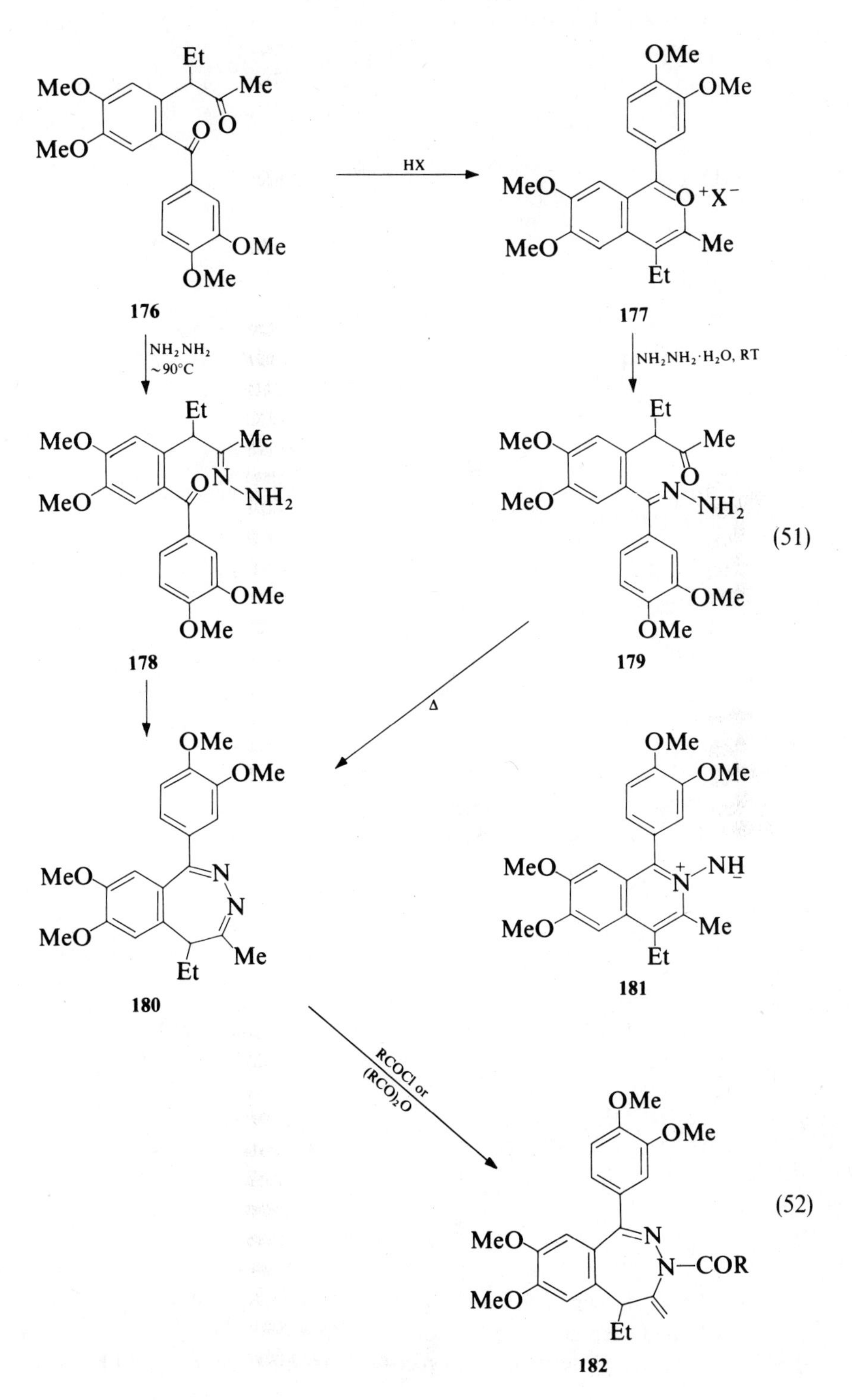
176
HX
177
$NH_2NH_2$
~90°C
$NH_2NH_2 \cdot H_2O$, RT
178
179
Δ
180
181
(51)
RCOCl or
$(RCO)_2O$
182
(52)

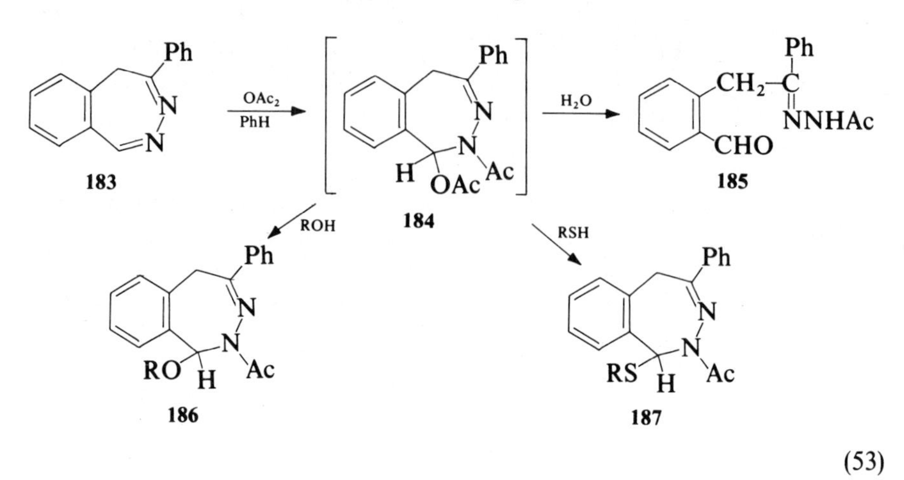


(53)

the formation of the acylated dimers **191** and **192** via a dehydrochlorinated intermediate **190** (Eq. 54).[99]

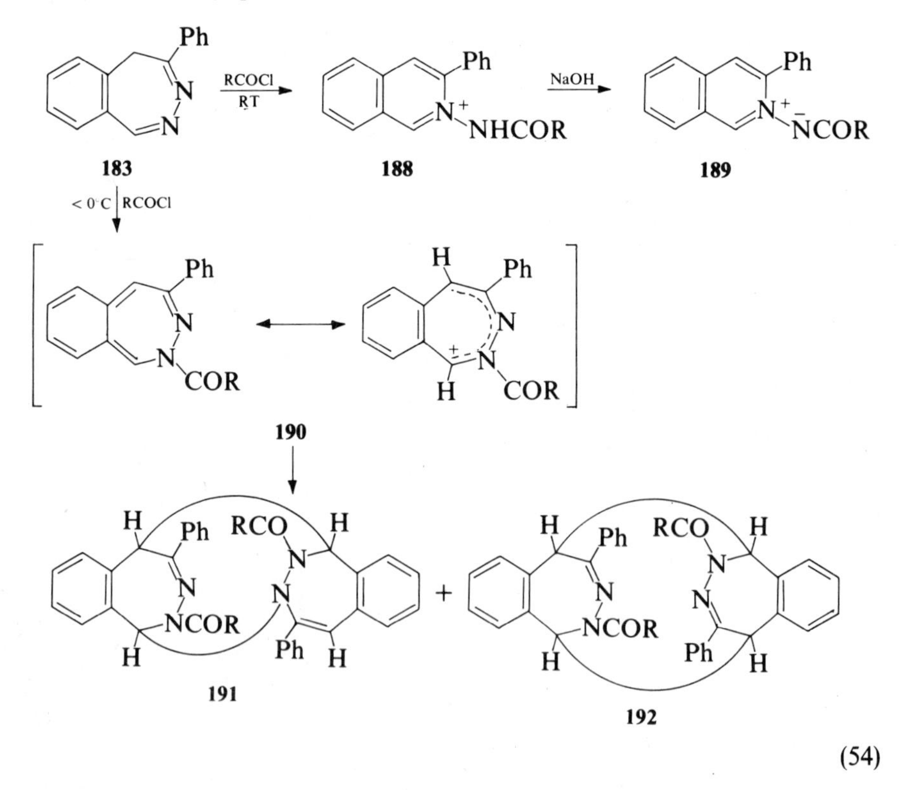


(54)

Similarly, as shown in Eq. 49, acylation of the 1-methyl and 1,4-dimethyl-(5*H*)-2,3-benzodiazepines gave the corresponding 3-acetyl derivatives **173**. Hydrolysis of the acetyl derivatives to recover starting material could be effected with either acid or base (Eq. 49).[91] For these compounds, however, 3*H* structures were assigned. Analogs without a 1-substituent gave only complex mixtures upon acylation.

Stepwise reduction of the seven-membered ring with sodium borohydride could be effected by the choice of solvent. Thus, use of ethanol at room temperature afforded the 3,4-dihydro compound **193**, while treatment in acetic acid gave the tetrahydro derivative **194**. 1,3-Dipolar cycloaddition of imines with nitrite oxides is well known, but it was interesting to observe that only a single product was obtained [viz., **195**: (R = H) 60–65%] when the unsubstituted analog **172** (R = H) was treated with mesitylnitrile oxide. This would indicate a difference in reactivity for the two imines (Eq. 55).[91]

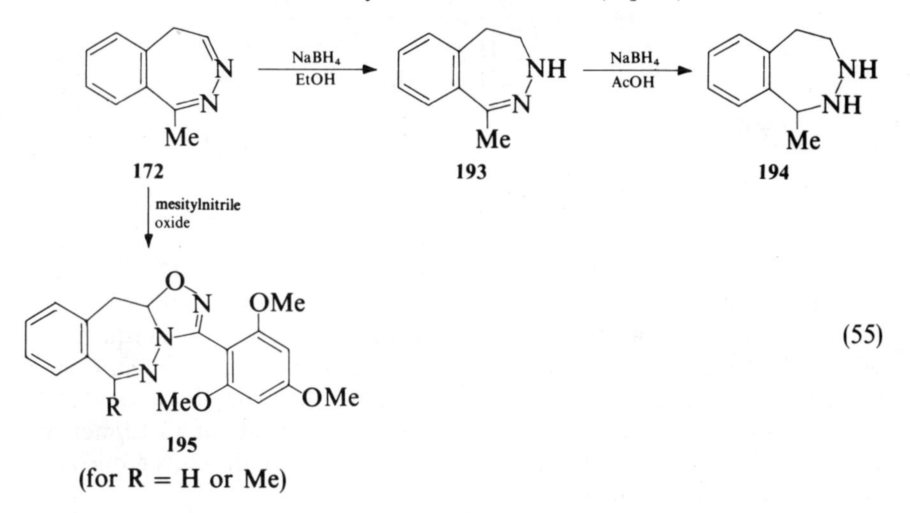


(55)

Similar borohydride reductions have been reported by Kovosi et al. in the patent literature (Eq. 56).[93]

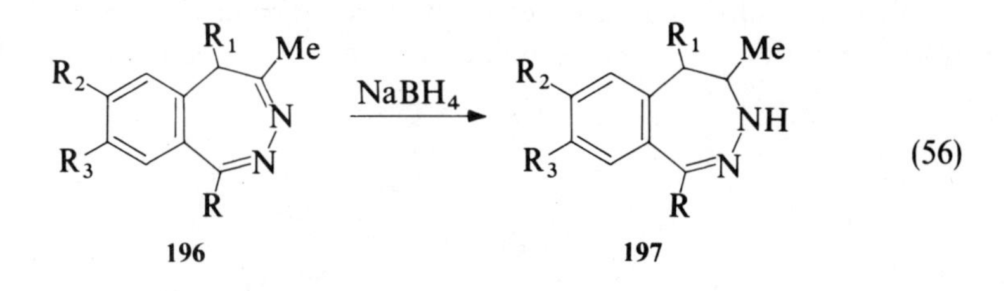


(56)

### 1.3.3. Spectral Data

The flexibility of the seven-membered ring was studied in 1*H*- and the 5*H*-2,3-benzodiazepines by means of temperature-dependent nmr spectroscopy.[35]

The protons at the 1-position of 1*H*-2,3-benzodiazepine give rise to an AB system with $J = 9$ Hz, which coalesces at $60 \pm 20°C$. The free energy of activation for ring inversion at the coalescence temperature was determined to be $15 \pm 1$ kcal/mol. If the carbon atom at the 1-position is substituted, no temperature dependence is observed, indicating a preference for one conformation.

The free energy of activation for ring inversion was also determined[35] for four differently substituted 5*H*-2,3-benzodiazepines **198** in which the C-5-protons appear as an AB system with $J = 12.5$ Hz. The coalescence temperature $T_c$ and free energies of activation for ring inversion at the coalescence temperature $\Delta G\ddagger$ are listed below.

**198**

| $R_1$ | $R_2$ | $R_3$ | $T_c$(°C) | $\Delta G\ddagger$ (kcal/mol) |
|---|---|---|---|---|
| Ph | H | H | $120 \pm 5$ | $19.5 \pm 0.3$ |
| Ph | H | MeO | $110 \pm 5$ | $19.0 \pm 0.3$ |
| Me | Ph | H | $137 \pm 5$ | $19.9 \pm 0.3$ |
| 4-MeC$_6$H$_4$ | Ph | H | $180 \pm 5$ | $22.1 \pm 0.3$ |

Proton and $^{13}$C-nmr spectra[42] of the 5*H*-2,3-benzodiazepine **175** showed the presence of two conformational isomers in solution, whereas only one conformation was observed in the solid state. In solution, conformational equilibrium was reached with a half-time of about 10 hours at room temperature. From the temperature dependence of this equilibrium, the difference between the free energies of the two conformations was estimated to be about 1 kcal/mol. The major conformer was assigned the configuration in which the ethyl group at the 5-position is pseudo equatorial.

Several salts of the 1-(3,4-dimethoxyphenyl)-7,8-dimethyl-5-ethyl-5*H*-2,3-benzodiazepin-3-ium iodide hydrate (1%) were prepared for pmr investigation,[94] to establish which of the two ring nitrogens was protonated. The authors concluded that only the N-3 atom could be protonated by strong acids. Quarternization with methyl iodide gave only the 3-methyl analog **199**.

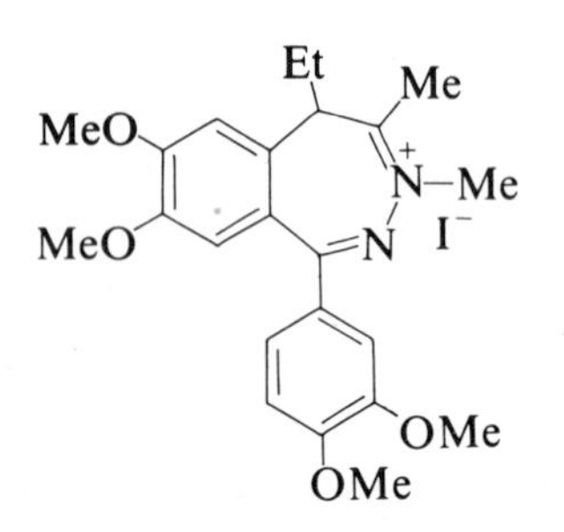


**199**

## 1.4. Dihydro-2,3-benzodiazepines

### *1.4.1. 4,5-Dihydro-1H-2,3-benzodiazepines*

The only representative of these cyclic azo compounds reported in the literature appears to be compound **201**, which was prepared by Schmitz and Ohme[45] in 20% yield by oxidation of the tetrahydro derivative **200** with hydrogen peroxide (Eq. 57).

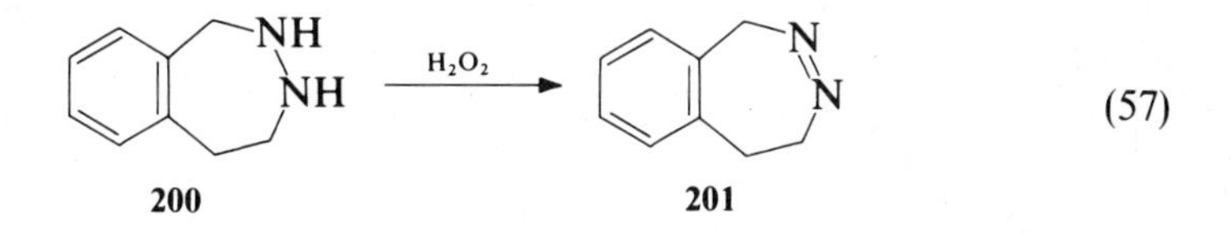


(57)

### *1.4.2. 4,5-Dihydro-3H-2,3-benzodiazepines*

The same investigators[45] obtained the dihydrobenzodiazepine **203** in 40% yield by thermal decomposition of the dimer **202** of the *N*-imino-3,4-dihydroisoquinolinium ylide (Eq. 58). Since this pyrolysis also led to the formation of isoquinoline, the latter was used as the solvent. The structure of **203** was confirmed by chemical transformations (see below). The cyclic hydrazone **203** was also accessible by acid-catalyzed rearrangement of the azo compound **201**.

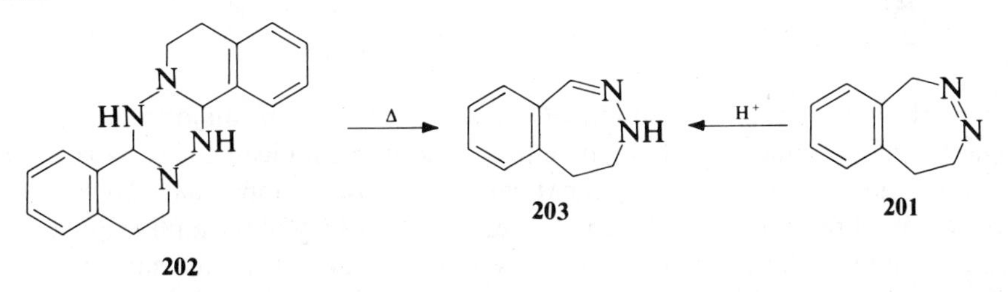


(58)

Tamura and coworkers[46] prepared 3-aryl-4,5-dihydro-3*H*-2,3-benzodiazepines **212** by reaction of *N*-aryl-*o*-formylphenethylamines **204** with hydroxylamine-*O*-sulfonic acid (Eq. 59). Compounds **204**, which may exist in equilibrium with the cyclic form **205**, are obtained by treatment of the *N*-phenyl-3,4-dihydroisoquinolinium salts **206** with hydroxide. The authors assumed that *N*-amination of **204** would lead to the hydrazone **211**, which then cyclizes to **212**. Streith and Fizet[47] showed, however, that *N*-amination of the phenethylamine is not involved; rather, a nucleophilic attack of the hydroxylamine, most likely on compound **206**, better explains the observed reaction products. The intermediate **208** formed in this manner may either lead to the diaziridinium salt **210**,

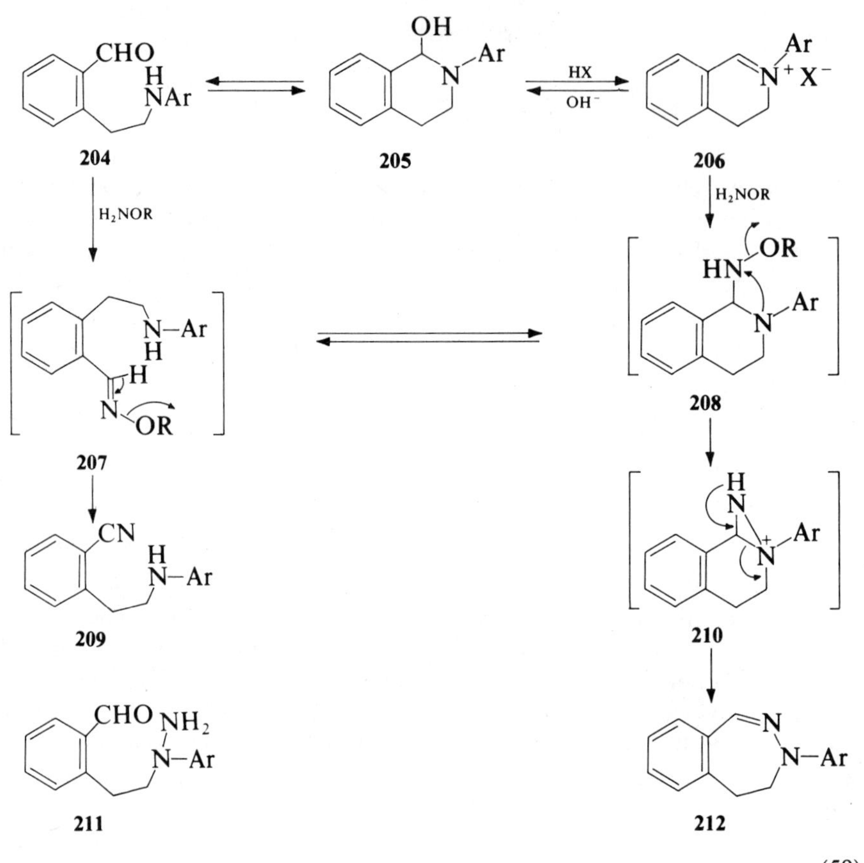


(59)

or via the ring-opened compound **207**, to the nitrile **209** (the major by-product). The benzodiazepine **212** would arise from **210** by the indicated ring expansion.

This mechanism is in agreement with the observations that the ease of formation of the benzodiazepine is related to the basicity of the aniline functionality in intermediate **208**. As the nucleophilicity of the aniline nitrogen decreased, the relative yields of benzodiazepines also decreased and the proportion of the nitrile by-products increased.

The 1-phenyl-substituted analogs **214** were reportedly synthesized (Eq. 60) by the reaction of the 2-(2-bromoethyl)benzophenones **213** with hydrazines at elevated temperature.[48] The structures of these compounds were supported by microanalytical data only. Dimers of a type related to compound **202** cannot be excluded from consideration.

The 4-phenyl-4,5-dihydro-3*H*-2,3-benzodiazepine **216** was reported to be accessible (Eq. 61) by partial catalytic hydrogenation of **215**.[37] This compound was characterized only by its melting point and by its infrared absorption band in the region of N—H bonds. Other structures, especially the tautomer **217**, cannot be excluded on the basis of the reported data.

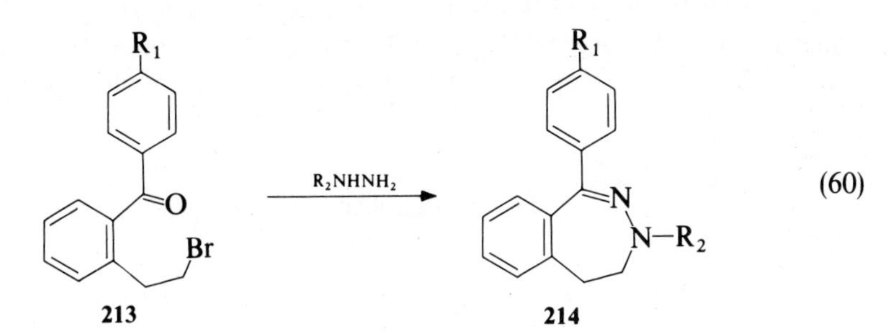


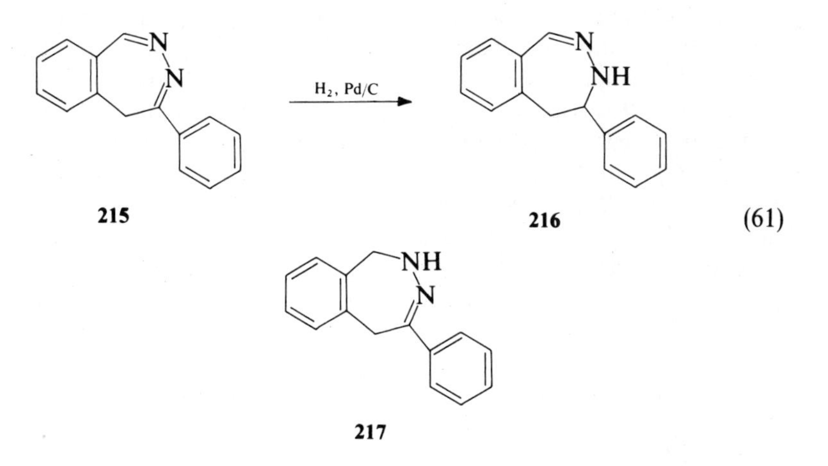


## 1.4.3. Reactions

The acid-catalyzed rearrangement of the azo compound **201** to the hydrazone **203** was mentioned above. Treatment of **203** with aqueous sulfuric acid[45] followed by alkaline workup (Eq. 62) led in almost quantitative yield to the

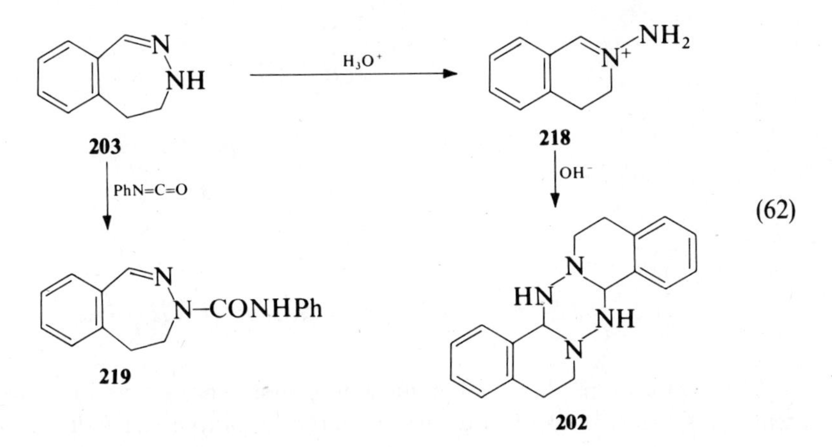

dimer **202**, possibly via the *N*-amino-3,4-dihydroisoquinolinium salt **218**. Acylation of **203** with phenylisocyanate[45] afforded the 3-(*N*-phenyl)carboxamide **219**.

## 1.5. Dihydro-2,3-benzodiazepinones

### *1.5.1. 2,3-Dihydro-2,3-benzodiazepin-1(1H-)-ones*

Treatment of the isocoumarin **220** with hydrazine in boiling ethanol (Eq. 63) led to the formation of the benzodiazepinone **221** as the major product and the *N*-aminoisoquinolone **222** as the minor product.[49] If the condensation was carried out in the presence of glacial acetic acid, the isoquinolone was the only product isolated. The structural assignment for **221** was based on its infrared spectrum (broad N—H) and its inability to react with aldehydes and isocyanates. This lack of reactivity was attributed to the weak basicity of the enamine nitrogen. No evidence to exclude the possibility that **221** is in fact a 2,5-dihydro derivative was given (see Section 1.5.2).

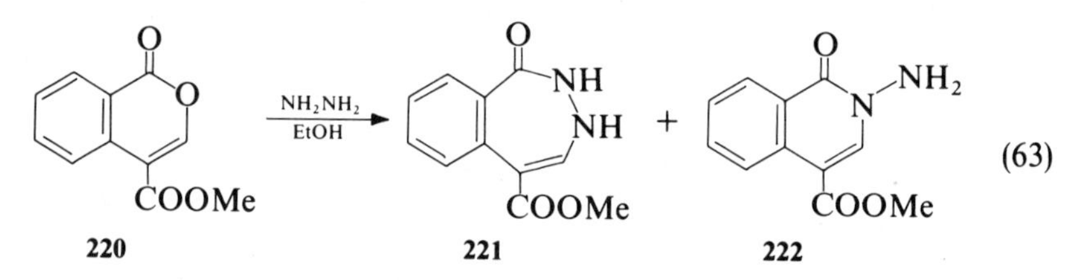


Among the many 4-aryl-substituted 2,5-dihydro analogs prepared in the same fashion, only the nitro-substituted compound **223** was observed to exist in the 2,3-dihydro tautomeric form, which has the double bond in conjugation with the two benzene rings.[54]

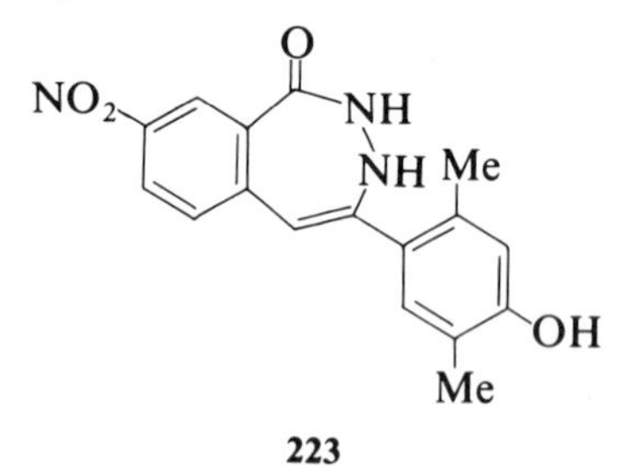


### *1.5.2. 2,5-Dihydro-2,3-benzodiazepin-1(1H-)-ones*

The synthesis of these compounds was first reported in the nineteenth century. Gottlieb[50] described the preparation of the benzodiazepine **225**

($R_1$ = Ph, $R_2$ = Me) by reaction of the isocoumarin **226** (X = O, $R_2$ = Me) with phenylhydrazine (Eq. 64). The same compound was obtained also from the condensation of the corresponding ketoacid **224** with phenylhydrazine. The analogous reaction of 2-(benzoylmethyl)benzoic acid **224** ($R_2$ = Ph) or the corresponding isocoumarin with hydrazine was subsequently studied by Wölbling,[51] Lieck,[52] and Buu-Hoi.[53] A number of 4-aryl-2,5-dihydro-2,3-benzodiazepin-1(1*H*)-ones were more recently prepared in high yield from the corresponding isocoumarins with hydrazine[54] and with monosubstituted hydrazines.[55] The 2-thioisocoumarins **226** (X = S, $R_2$ = Ar) reacted in the same fashion[56] with hydrazine to yield the benzodiazepines **225**. The extension of this method to the synthesis of the 5-phenyl analog **230** failed,[57] and the reaction of the isocoumarin **227** with methylhydrazine (Eq. 65) led exclusively to the isoquinolone **228**. Compound **230** was, however, synthesized[57] by displacement of bromide in **229** with phenylmagnesium bromide (Eq. 66).

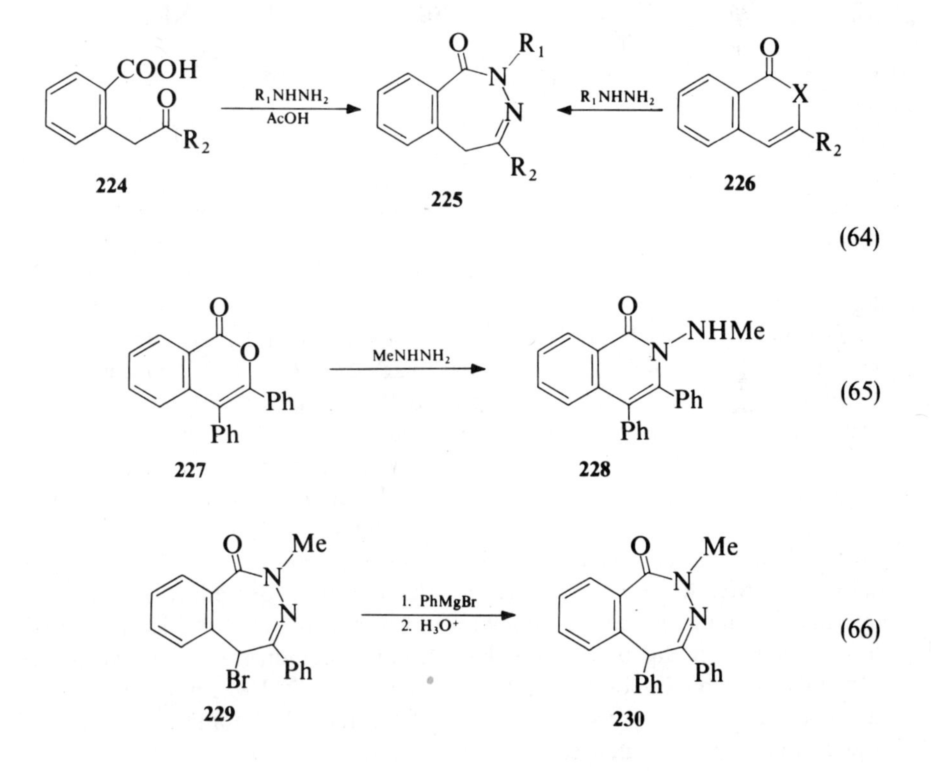


### 1.5.3. 3,5-Dihydro-2,3-benzodiazepin-4(4*H*)-ones

Halford and coworkers[58] obtained the first compounds of this type (Eq. 67) by thermal dehydration of the phenylhydrazone of *o*-acetylphenylacetic acid (**231**, $R_1$ = Me, $R_2$ = Ph). The parent compound **233** ($R_1$, $R_2$, $R_3$ = H) and the

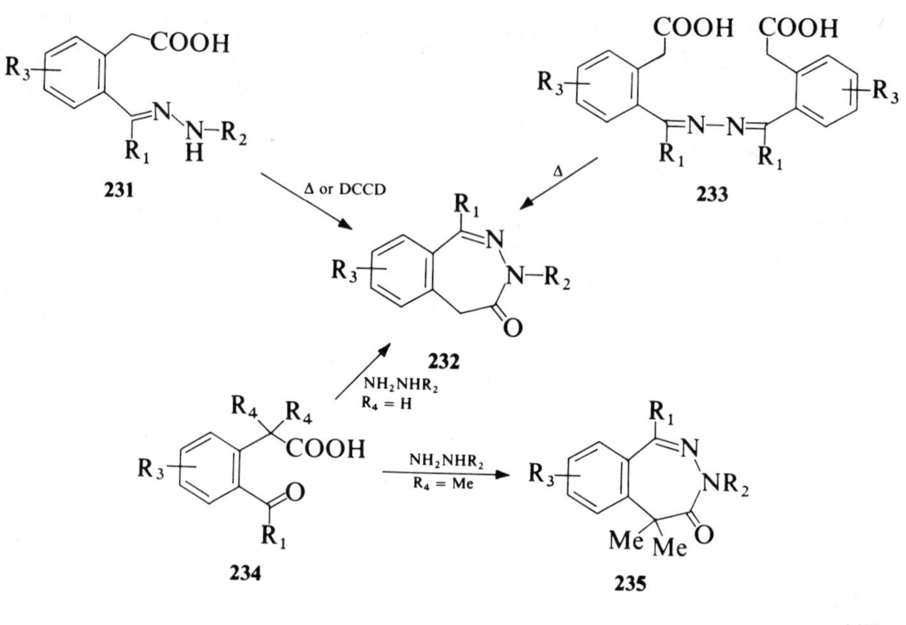


(67)

1-methyl analog **232** ($R_1$ = Me, $R_2$, $R_3$ = H) were accessible by pyrolysis of the semicarbazones **231** ($R_1$ = H, Me, $R_2$ = $CONH_2$; $R_3$ = H). Pyrolysis of the azine **233** ($R_1$ = Me; $R_3$ = H) also led to the benzodiazepine **232** ($R_1$ = Me; $R_2$, $R_3$ = H) in low yield.

Thermal cyclodehydration was later used successfully also for the preparation of 1-aryl-substituted benzodiazepines **232** ($R_1$ = Ph).[59] Wermuth and Flammang[60, 61] simplified and improved this synthesis and obtained good yields of benzodiazepines by heating under reflux a solution of the ketoacids **234** and a hydrazine in an inert solvent. The water formed in the reaction was removed azeotropically. The superiority of this modified method was evident from the preparation[61] of the 5,5-dimethyl derivative **235** ($R_1$ = Ph, $R_2$ = 2-morpholinoethyl, $R_3$ = H). Previous attempt to prepare a 5,5-dimethyl analog (**235**, $R_1$ = Me, $R_2$ = Ph, $R_3$ = H) by the pyrolysis procedure failed.[58] An alternate method for the cyclodehydration of arylhydrazones **231** ($R_2$ = aryl) to the corresponding benzodiazepines **232** was demonstrated by the use of dicyclohexylcarbodiimide.[62] The 1-benzyl derivative **237** was not accessible by this approach but could be synthesized[63] from the 1-bromomethyl compound **236** by reaction with phenylmagnesium bromide (Eq. 68).

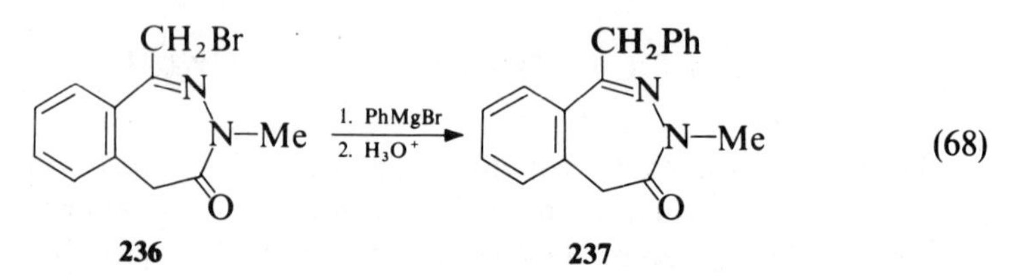


(68)

### *1.5.4. Reactions*

#### 1.5.4.1. Reactions with Electrophiles

Ring contraction of dihydro-2,3-benzodiazepinones to *N*-aminoisoquinolones **238** and **239** ($R_3$ = H) occurs with both the 1-ones **225**[49,51,55,56] and the 4-ones **232** ($R_3$ = H)[58,60] under the influence of acid catalysts (Eqs. 69 and 70). The same isoquinolones are, in general, by-products in the preparation of the benzodiazepines and may be the predominant or exclusive products, particularly under acidic conditions.[49, 55, 57]

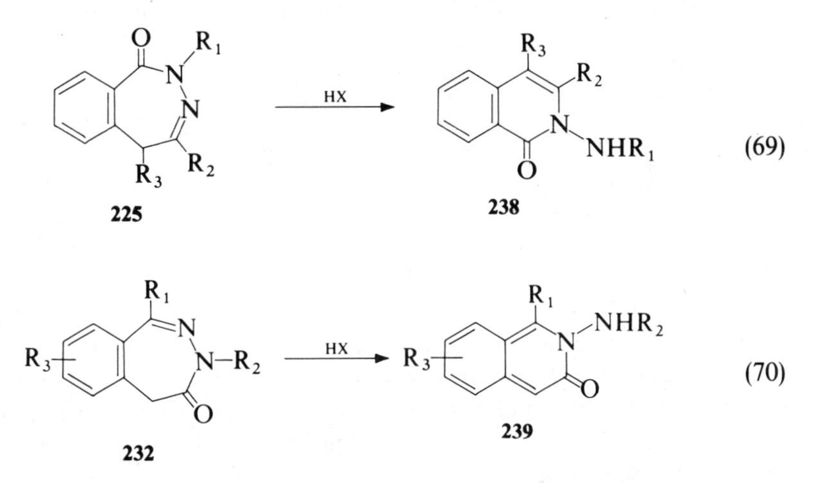


Reaction of compound **240** (R = Me) with *N*-bromosuccinimide in the presence of a peroxide catalyst (Eq. 71) afforded the 5-bromo derivative **229**.[57]

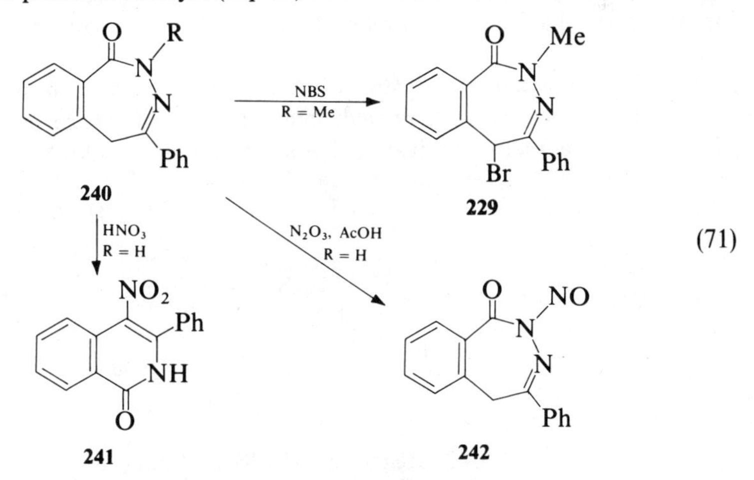


Flammang has studied the utility of a number of 1,3-disubstituted 4-oxo-3,5-dihydro-(4*H*)-2,3-benzodiazepines for the synthesis of 2-aminoisoquinolin-3-ones.[95] The starting materials were prepared by reacting the appropriate *o*-acyl or aroylphenylacetic acid with a substituted hydrazine. Heating the 2,3-benzodiazepine **243** thus obtained in $H_2SO_4$/HOAc at 100°C for 12 hours afforded the ring contraction product **244** in 0–30% yields. Best yields were obtained when R and $R_1$ are alkyl and aryl (Eq. 72). Most if not all of the compounds used in this study have been reported elsewhere. No spectra or physical characteristics were reported. A similar study was also reported for the ring contraction of 2,3-benzodiazepin-1-ones **243**.[96] These compounds underwent ring contraction in much higher yield (70–90%) when the starting diazepine was a secondary amide. Again, all the compounds used have been described elsewhere, therefore neither spectra nor physical characteristics are reported.

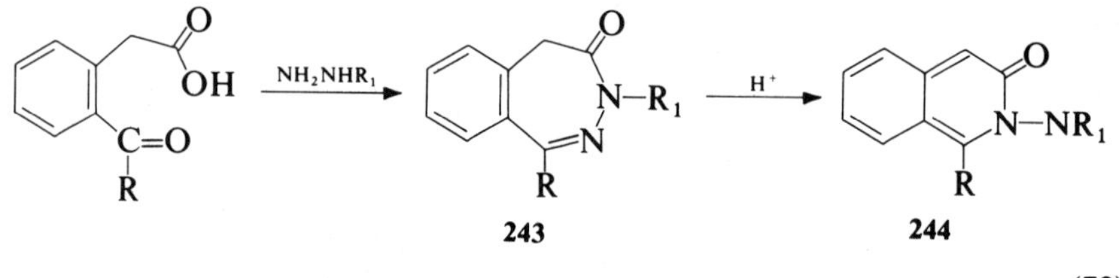


(72)

Wölbling[51] reported the conversion of the diazepinone **240** (R = H) to the 4-nitroisoquinolone **241** by treatment of **240** with nitric acid (Eq. 71). He also described the formation of an unstable nitroso derivative, to which he assigned structure **242**, as a result of the action of nitrogen oxides on a suspension of **240** (R = H) in acetic acid.

Alkylation of **240** (R = H) with methyl iodide and ethyl iodide led to the respective 2-alkyl derivatives.[51] Reaction of the 4-ones **232** ($R_2$ = H) with various alkyl halides was reported to yield the corresponding 3-alkyl derivatives.[59, 61]

Bromination of compound **245** with *N*-bromosuccinimide in the presence of a peroxide catalyst or with phenyltrimethylammonium perbromide (Eq. 73) gave the 1-bromomethyl derivative **236** in 15 and 40% yields.[63]

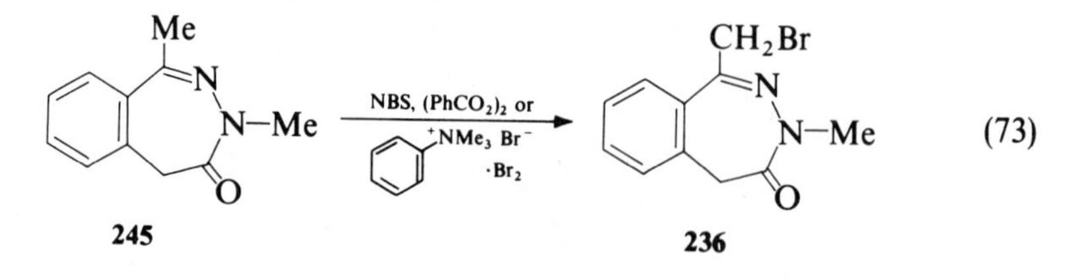


(73)

### 1.5.4.2. Reactions with Nucleophiles

Reduction of **246** (R = H) with zinc and hydrochloric acid afforded the isoquinolone **247**.[51] Reaction of the benzodiazepin-1-one **246** (R = Me) with phosphorus pentasulfide in boiling pyridine (Eq. 74) gave the thione **248**.[56] In

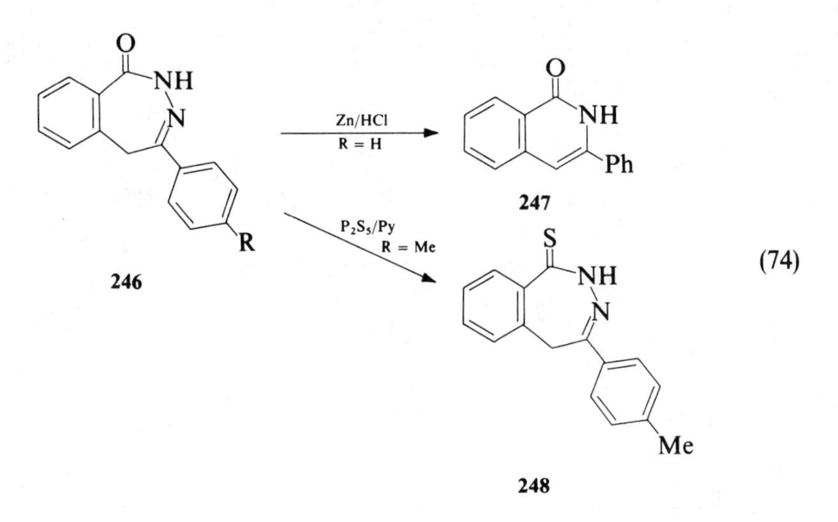


contrast to the corresponding ketone, this thione was found[56] to be resistant to ring contraction.

Cleavage of the hydrazide bond in **249** to give the open phenylhydrazone **250** could be effected (Eq. 75) by treatment with sodium hydroxide in boiling ethylene glycol.[58]

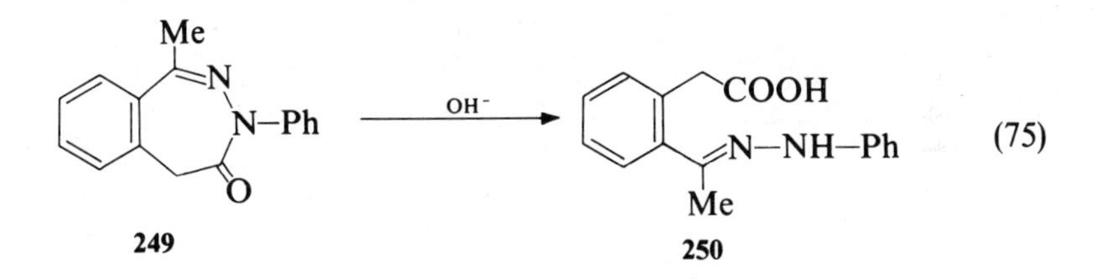


### *1.5.5. Tetrahydro-2,3-benzodiazepines*

The tetrahydrobenzodiazepine **200** (R = H) was prepared[45] by catalytic hydrogenation of the dihydrobenzodiazepine **203** over palladium on carbon (Eq. 76) and was characterized as a hydrochloride salt. The structure of **200** (R = H) was confirmed by an alternative synthesis via the alkaline hydrolysis of the phthaloyl derivative **253**. The latter was obtained by doubly alkylating phthalhydrazide with the dichloride **252**. Tetrahydrobenzodiazepines are also accessible by reduction of tetrahydrobenzodiazepinones with lithium aluminum hydride, exemplified by the conversion of **251** to **200** (R = Me), which formed a crystalline picrate.[64]

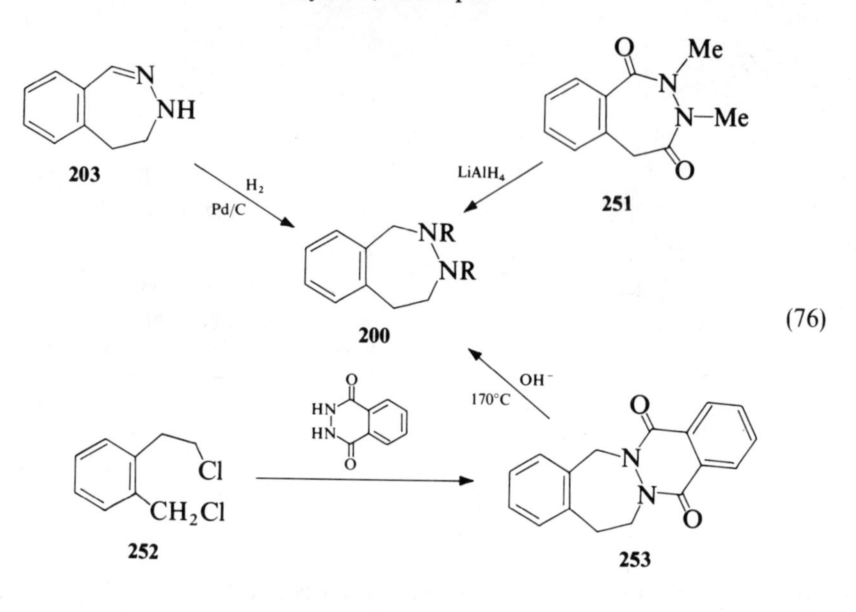


### 1.5.6. *Tetrahydro-2,3-benzodiazepinones*

The tetrahydro-2,3-benzodiazepin-4-ones **256** (R = Me, Ph) were recently reported by Cignarella and coworkers.[65] Condensation of the bromide **254** with a monosubstituted hydrazine (Eq. 77) led to the intermediates **255**, which were isolated and characterized. Compounds **255** slowly cyclized to the benzodiazepinones **256** at room temperature, or were more readily converted in high yield to compounds **256** by refluxing in acetic acid.

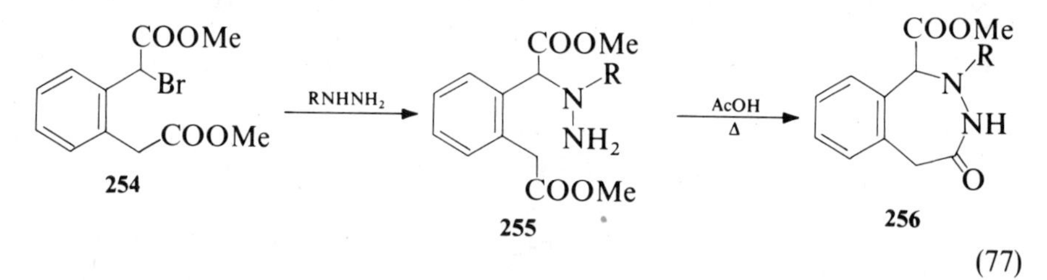


(77)

According to Rosen,[66] tetrahydrobenzodiazepinones of structure **258** could be obtained by the reaction of phenylacetic acid hydrazides **257** with formaldehyde under acidic conditions (Eq. 78). The structures assigned to these products

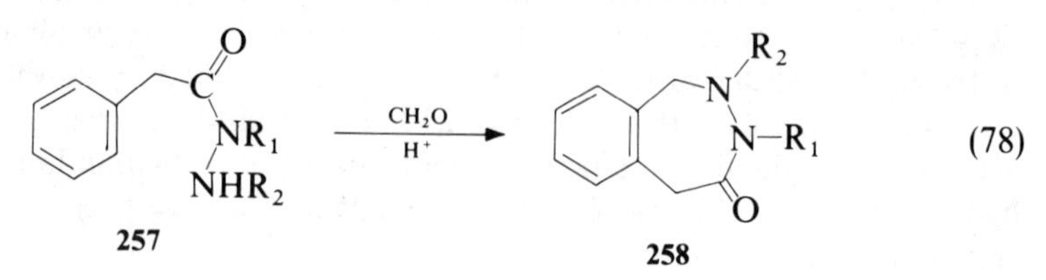


(78)

are doubtful, since the chemistry performed to establish the structures gave conflicting results.

Rosen and Popp[64] reinvestigated the reaction of homophthalic anhydride **259** with hydrazine (Eq. 79) and disproved much earlier reports[67] that this reaction leads to the tetrahydrobenzodiazepin-1,4-dione **260** (R = H). Only in the case of 1,2-dimethylhydrazine did this condensation lead to the benzodiazepine **260** (R = Me).

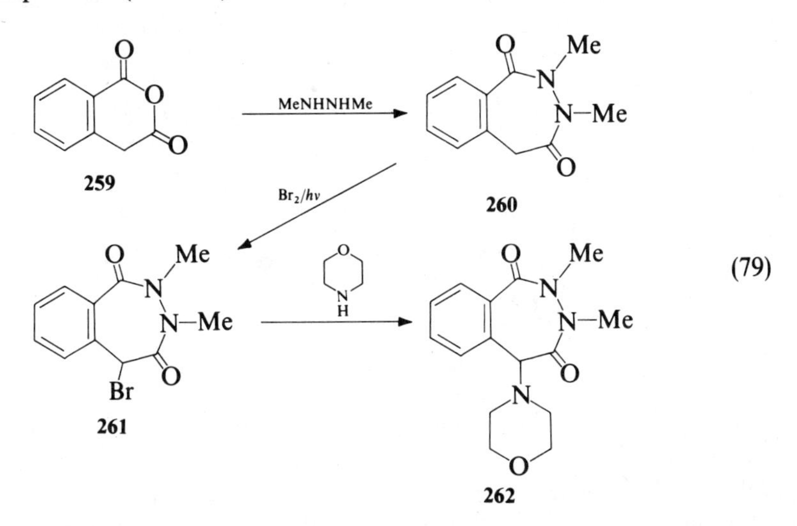


Bromination of **260** (R = Me) with bromine in the presence of light afforded[66] the 5-bromo derivative **261**, which reacted with morpholine to give what appears to be the 5-morpholino compound **262**. According to the analytical data, it seems that the latter compound was not properly purified.[66a]

## 2. CYCLOPENTA[*d*] [1,2]DIAZEPINES

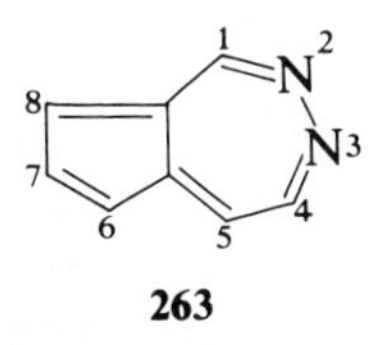


The only reported representative of ring system **263** is the 1,4-diphenyl derivative **267**.[68] This compound was formed in 40% yield (Eq. 80) by cycloaddition of the fulvene **264** and the 1,2,4,5-tetrazine **265** to give, presumably, the intermediate **266**. Loss of nitrogen and dimethylamine from this bridged intermediate would then lead to **267**. The structure of **267** is based on microanalysis, the mass spectrum and a characteristic ultraviolet absorption spectrum that resembles that of another diazaazulene.

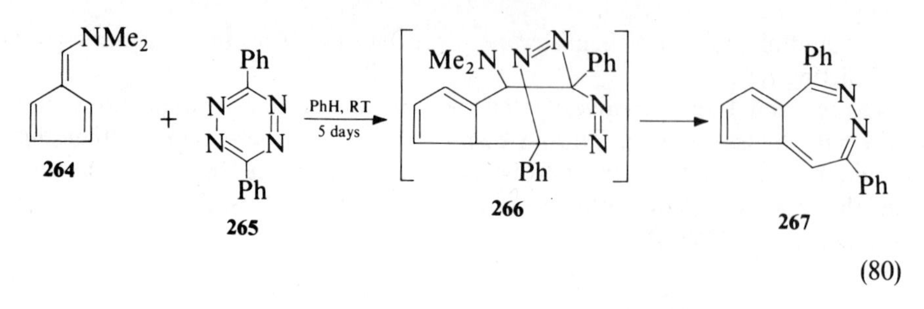


(80)

## 3. CYCLOPROPA[*d*] [1,2]DIAZEPINES

Cyclopropa[*d*] [1,2]diazepine

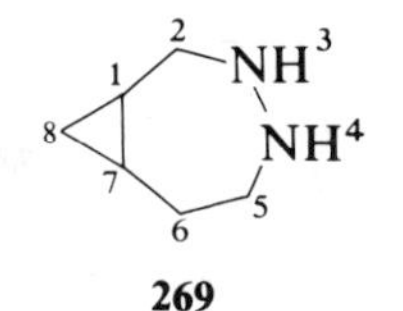


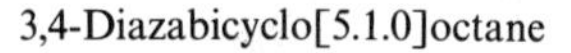
3,4-Diazabicyclo[5.1.0]octane

Compounds derived from the ring parent **268** have usually been named as 3,4-diazabicyclo[5.1.0]octanes **269** or trivially as homodiazepines. Streith and coworkers[69] obtained tetrahydrocyclopropa[*d*] [1,2]diazepines **272** by thermal or, in much lower yield, by photochemical extrusion of nitrogen (Eq. 81) from the pyrazolo[3,4-*d*] [1,2]diazepines **270** and **271**.

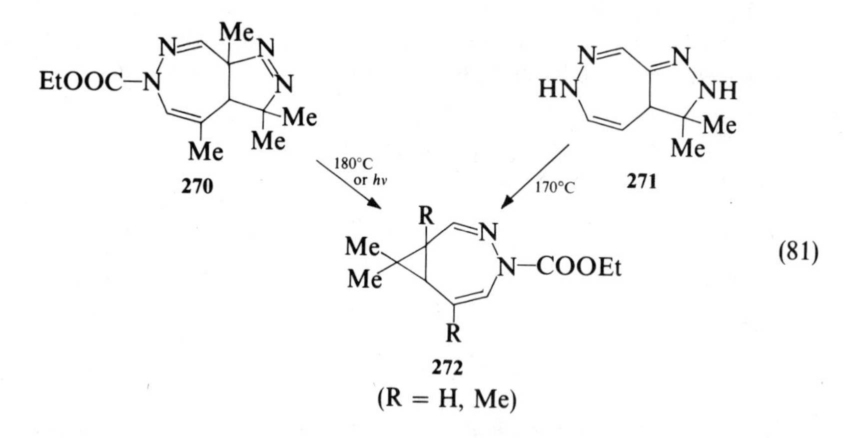


(81)

## 4. OXIRENO[*d*] [1,2]DIAZEPINES

1,2-Diazepine epoxides of structure **277**, which are derivatives of the parent ring systems **273** or **274**, were synthesized by Tsuchiya and coworkers[70] by

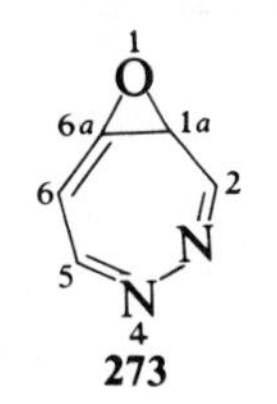


1a*H*-Oxireno[*d*] [1,2]diazepine

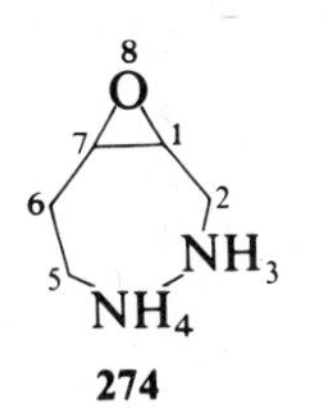


3,4-Diaza-8-oxabicyclo[5.1.0]octane

rearrangement of the endoperoxide **276** with potassium hydroxide in methanol at 0°C (Eq. 82). The endoperoxides **276** were prepared in 30–60% yields by dye-sensitized photooxygenation of the corresponding ethoxycarbonyl-1,2-diazepines **275**.

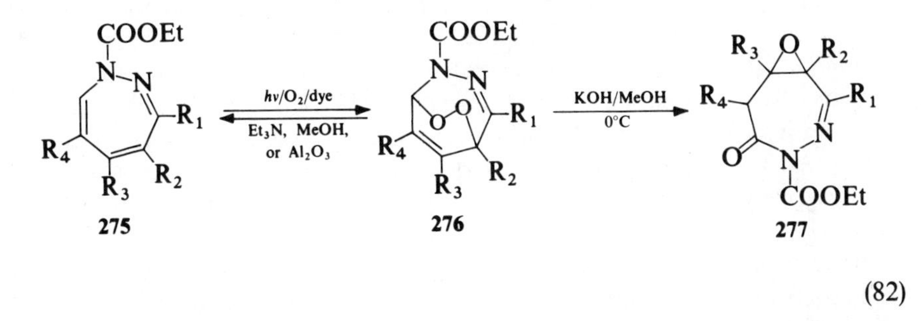


(82)

The same sequence of reactions carried out on the 2,3-dihydrodiazepine **278** led to the 2,3,4,5,6,6*a*-hexahydro-6*a*-methyl-1*aH*-oxireno[*d*] [1,2]diazepine **281**. In this case the intermediate **280** was isolated in 80% yield after treatment of the endoperoxide **279** with aluminum oxide in methylene chloride (Eq. 83). Interestingly, the attempted isolation of a corresponding intermediate in the conversion of **276** to **277** failed because treatment of **276** with either alumina or methanolic triethylamine converted the endoperoxide back to starting material **275**. The reactivity of these endoperoxides and the course of their rearrangement appear to depend on the nature of their substituents.

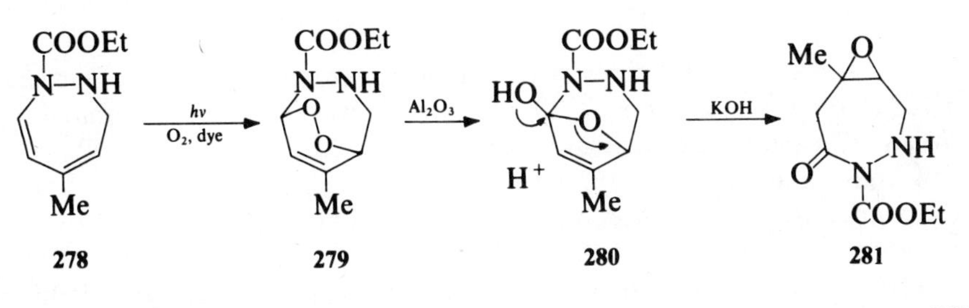


(83)

# 5. PYRAZOLO[3,4-*d*] [1,2]DIAZEPINES

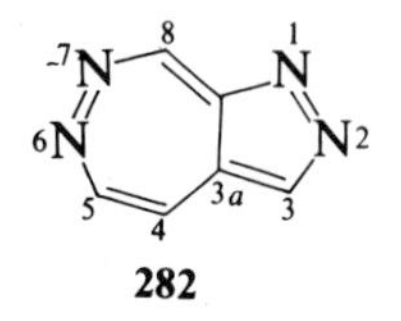

**282**

## 5.1. Synthesis

The parent compound **282** is still unknown. Tetrahydro derivatives with various substituents were synthesized by Streith and coworkers[69] by the cycloaddition of diazoalkanes to 1-acyl-1,2-diazepines. Diazomethane was inert toward most of these 1,2-diazepines and reacted only with compounds having a strongly electron-withdrawing substituent at the 1-position, such as the 1-benzenesulfonyl-1,2-diazepine **283** ($R_1 = SO_2Ph$, $R_2 = H$). 2-Diazopropane was found to be much more reactive, and it added regioselectively (Eq. 84) to yield the adducts **284** ($R_3 = Me$). The azo function undergoes rearrangement to the hydrazone form **285** if a 1,3-hydrogen shift is possible ($R_2 = H$).

Because of the limited stability of some of these adducts, they were characterized spectroscopically only. Detailed nmr studies[71] revealed an unusually large coupling between the vinylic proton at the 5-position with the allylic proton in the 3*a*-position. The coupling constant ranged from 2.5 to 3.5 Hz, depending on the substitution in **284** and **285**. This phenomenon was attributed to the conformational rigidity of this ring system, with the allylic proton being nearly perpendicular to the plane of the 4,5-olefinic bond.

The structure of the pyrazolo[3,4-*d*] [1,2]diazepine **287** was assigned to the product obtained[72] in moderate yield by condensation of the pyrone **286** with hydrazine (Eq. 85). The reported infrared absorption data would not exclude possible alternative structures.

## 5.2. Reactions

The thermal or photolytic extrusion of nitrogen from compounds of type **284** or **285** leading to cyclopropa[*d*] [1,2]diazepines was discussed for compounds **270** and **271** (Section 3). Oxidation of **288** ($R_1 = Me$, $R_2 = COOEt$, $R_3 = H$) with lead tetraacetate (Eq. 86) gave a 40% yield of an acetoxy derivative to which structure **289** (R = Me) was assigned and confirmed by X-ray analysis of the *p*-bromobenzoate ($R = 4\text{-}BrC_6H_4$).[73]

A more recent paper[74] described the acylation of the labile pyrazolodiazepines **288** ($R_1 = H$) to various stable 2-acyl derivatives **290**. The

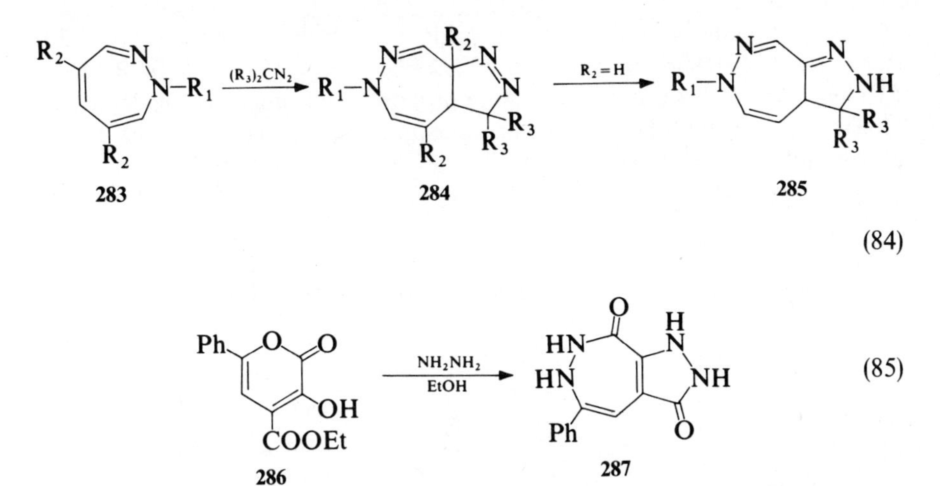


2-methoxymethyl derivative **291** was obtained by reaction of **288** ($R_1$ = H, $R_2$ = Ts, $R_3$ = H) with methanolic formaldehyde followed by workup with sodium borohydride (Eq. 86). Depending on the nature of the R, $R_2$, and $R_3$ groups, compounds **290** underwent conversions to **292** by shift of the double bond, to **293** by elimination of RH, and to **294** by elimination of RH accompanied by ring opening. Treatment of the methoxymethyl derivative **291** with

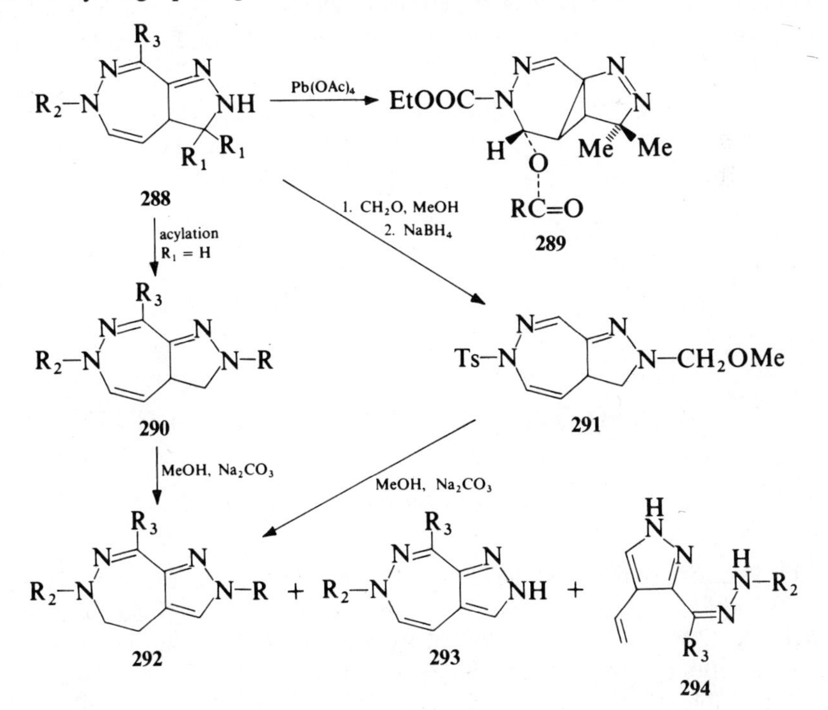


(86)

sodium carbonate in boiling methanol, for example, gave a 53% yield of **294** ($R_2$ = Ts, $R_3$ = H) and a 21% yield of **292** (R = $CH_2OMe$, $R_2$ = Ts, $R_3$ = H). Mechanisms for these conversions were proposed and supported by deuterium labeling experiments. The initial step appears to be the removal of the doubly allylic proton at the 3*a*-position.

## 6. PYRIDO[3,2-*d*] [1,2]DIAZEPINES

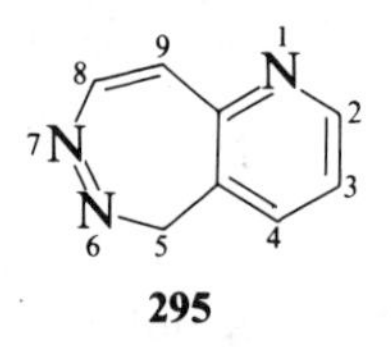


Of the four possible pyrido[*d*] [1,2]diazepine ring systems, only derivatives of the pyrido[3,2-*d*] [1,2]diazepine **295** are known in the literature.[75] In a manner analogous to the synthesis of 2,3-benzodiazepines (Section 1.5.2), compounds **297** were prepared by condensation of the ketoacids **296** with methylhydrazine in boiling ethanol (Eq. 87). The structures of **297** were supported by spectral data. In particular, nmr spectroscopy allowed differentiation from other tautomeric forms.

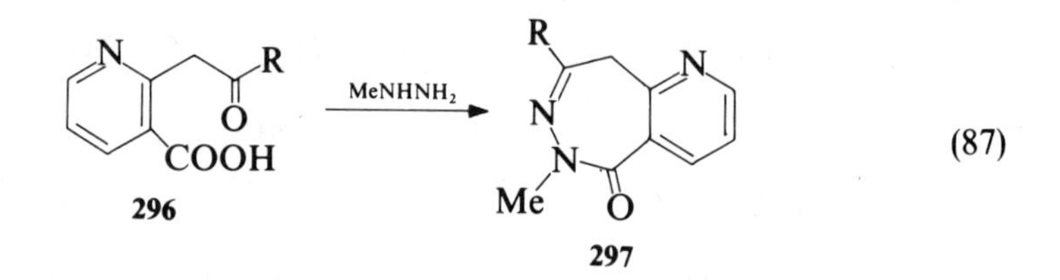


## 7. THIENO AND OTHER HETERO[*d*] [1,2]DIAZEPINES

### 7.1. Synthesis

Munro and Sharp reported the first route to synthesize thieno[3,2-*d*] **300** and thieno[2,3-*d*][1,2]diazepines **303** by cyclization of the α-(2-alkenylthienyl) diazoalkanes **298** and **301** (Eq. 88).[100] In contrast, 3-diazomethyl-4-(*trans*-2-phenylethenyl)thiophene **304** did not cyclize but gave carbene-derived products. The 1*H*-thieno[*d*][1,2]diazepines **300** and **303** were readily isomerized to the 5*H*-thieno[*d*][1,2]diazepines **305** and **306** by sodium ethoxide in ethanol.

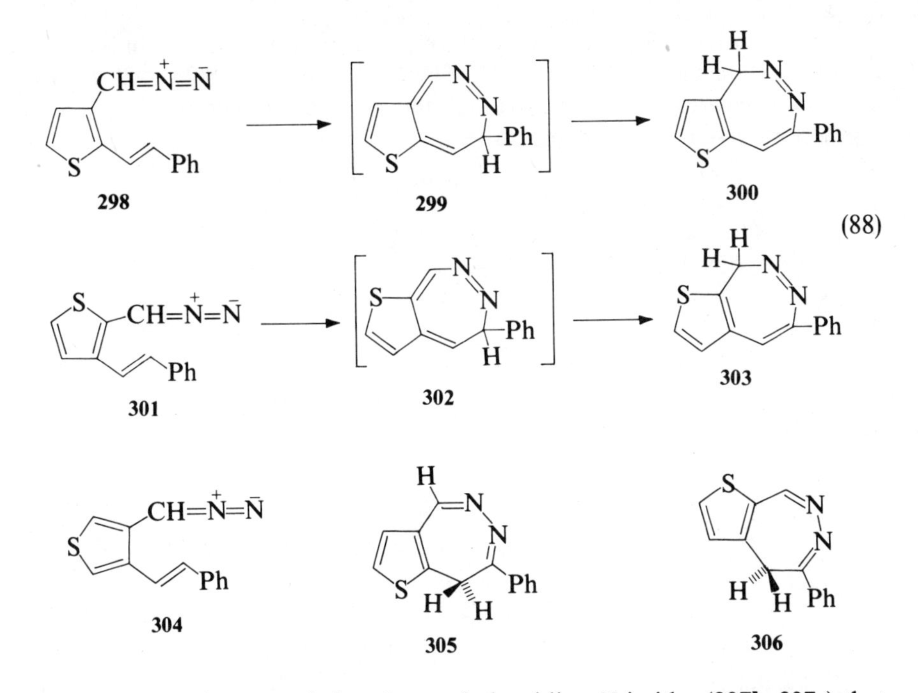


Tsuchiya et al. reported that the methylpyridine *N*-imides (**307b–307e**) that were condensed with a thiophene, furan, or pyrrole ring on the *c*-side of the pyridine ring gave, upon irradiation, the corresponding novel fused 3*H*-[1,2]diazepines (**308b–308e**) and products **309** and **310** as well (Eq. 89).[161]

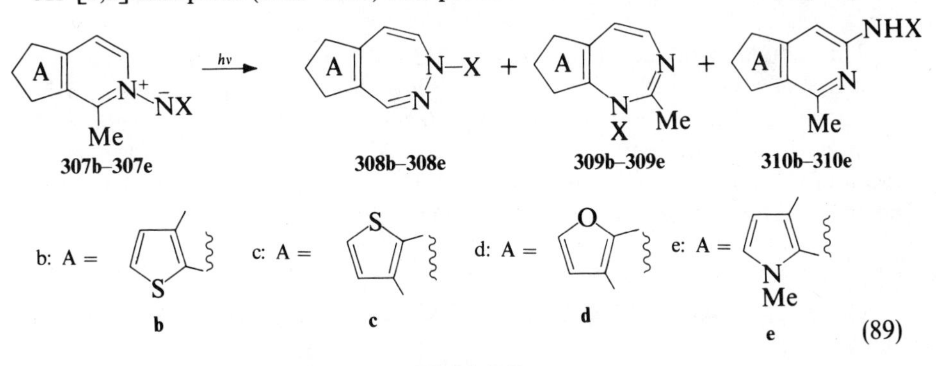


Yield (%)

| 308 b–e | 309 b–e | 310 b–e |
|---|---|---|
| 35% | 10% | 6% |
| 45% | 12% | 5% |
| 67% | — | 6% |
| 17% | 15% | 4% |

This photolysis can proceed by rearrangement to form two kinds of diaziridine intermediates, **311** and **312**. The latter can give **308b–308e** by ring expansion, whereas the former can further rearrange, followed by ring expansion, to give other products (Eq. 90).

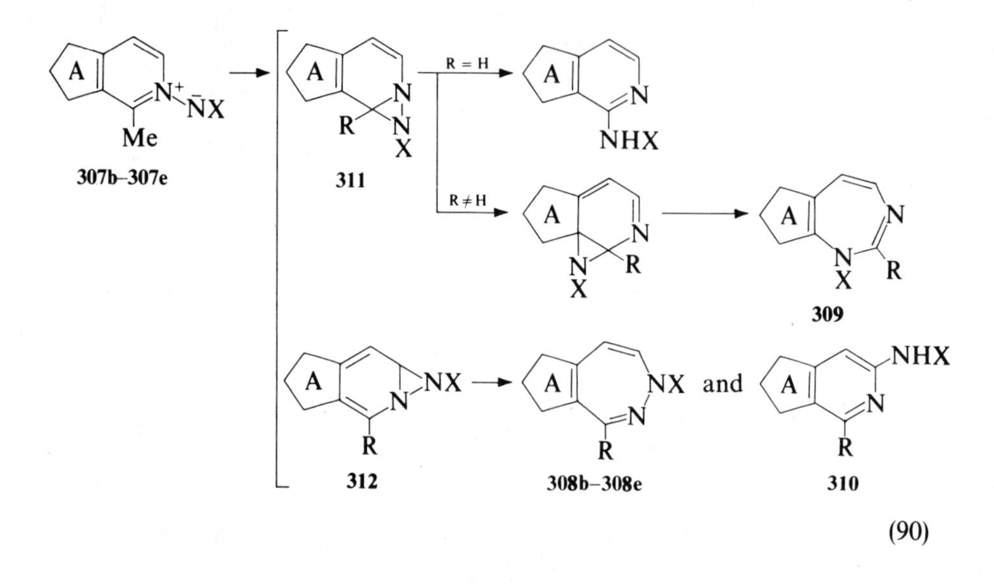


(90)

## 7.2. Reactions

The reduction of the [1,2]diazepine **313** with sodium borohydride in the presence of aqueous alkali gave the 1,2-dihydro (**314**) and 4,5-dihydro (**315**) compounds, whereas treatment of **313** with sodium borohydride in acetic acid gave the 1,2-dihydro compound **314** as the sole product in high yield. Acetylation of the dihydro compounds **314** and **315** with acetic anhydride gave the corresponding acetates **316** and **317** in high yields (Eq. 91).[101]

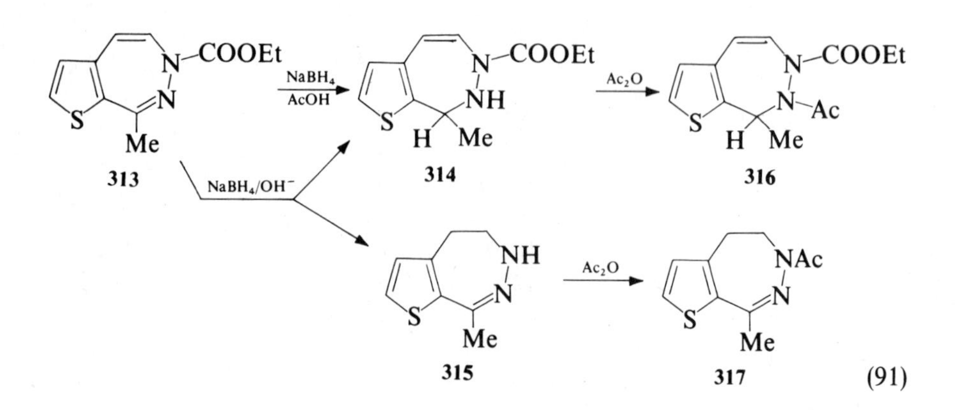


(91)

# E. TABLES OF COMPOUNDS

TABLE I-1. [*a*]-FUSED [1,2]DIAZEPINES

| Substituent | mp (°C); [bp (°C/torr)] | Solvent of Crystallization | Yield (%) | Spectra | Refs. |
|---|---|---|---|---|---|
| | | *Pyrazolo[1,2-a] [1,2]diazepines* | | | |
| ***2,3-Dihydro-1H-pyrazolo[1,2-a] [1,2]diazepin-5-(5H)-ones*** | | | | | |
| 2-(2-Pr)-3,3-$(COOMe)_2$-6,7,8,9-$(Ph)_4$ | 110–116 | $CH_2Cl_2$/MeOH | 57 | ir, ms, pmr | 1 |
| 3,3-$(COOMe)_2$-2,6,7,8,9-$(Ph)_5$ | 207–215d | | 60 | | 1 |
| ***6,7,8,9-Tetrahydro-5H-pyrazolo[1,2-a] [1,2]diazepin-1-(1H)-one*** | | | | | |
| None | 84 | $CH_2Cl_2/CCl_4$ | 42 | ir, $^{13}C$-nmr, pmr, uv | 2a, b |
| Hydrochloride | 175d | MeOH | 55 | pmr, uv | |
| ***Hexahydro-1H,5H-pyrazolo[1,2-a] [1,2]diazepine*** | | | | | |

TABLE I-1. —(contd.)

| Substituent | mp (°C); [bp (° C/torr)] | Solvent of Crystallization | Yield (%) | Spectra | Refs. |
|---|---|---|---|---|---|
| None | | | 32 | | 3 |
| | | *Pyridazino[1,2-a] [1,2]diazepines* | | | |
| ***Octahydro-6H-pyridazino[1,2-a] [1,2] diazepine*** | | | 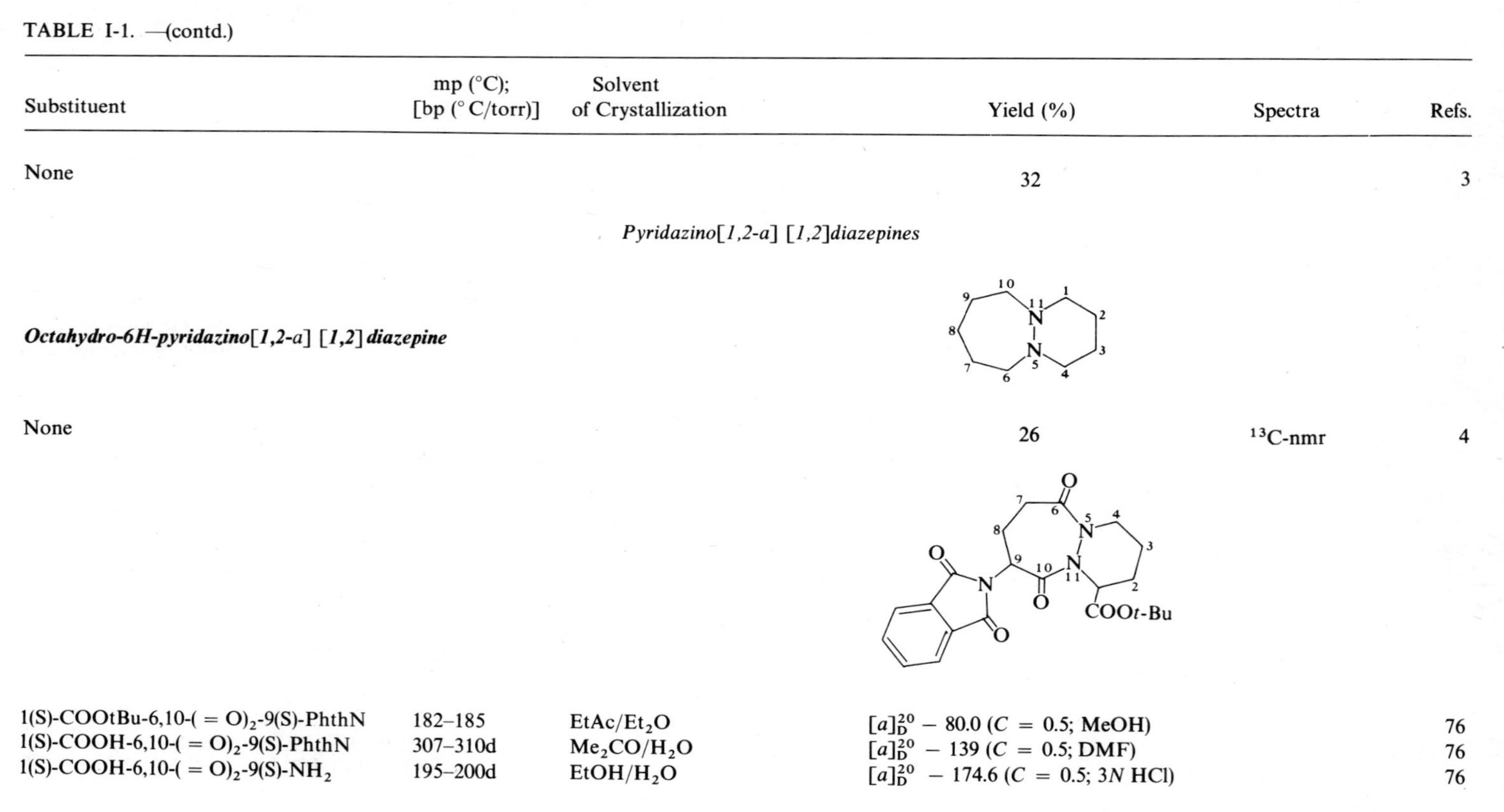 | | |
| None | | | 26 | $^{13}$C-nmr | 4 |
| 1(S)-COOtBu-6,10-( = O)$_2$-9(S)-PhthN | 182–185 | EtAc/$Et_2O$ | $[a]_D^{20}$ − 80.0 ($C$ = 0.5; MeOH) | | 76 |
| 1(S)-COOH-6,10-( = O)$_2$-9(S)-PhthN | 307–310d | $Me_2CO/H_2O$ | $[a]_D^{20}$ − 139 ($C$ = 0.5; DMF) | | 76 |
| 1(S)-COOH-6,10-( = O)$_2$-9(S)-$NH_2$ | 195–200d | EtOH/$H_2O$ | $[a]_D^{20}$ − 174.6 ($C$ = 0.5; 3$N$ HCl) | | 76 |

***Tetrahydro[1,2,4]triazolo[1,2-a] [1,2]diazepin-1,3-(2H,5H)-diones***

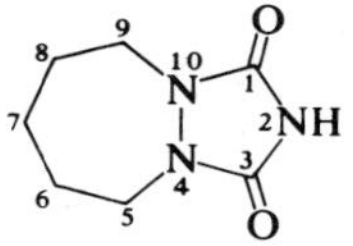

| | | | | |
|---|---|---|---|---|
| 2-Bu | [100–105/0.5] | | 20 | 7 |
| 2-Ph | 148 | | 21 | 7 |
| 2-(4-$BrC_6H_4$) | 134–135 | | 70 | 8b |
| 2-(4-$ClC_6H_4$) | 118–119.5 | | 74 | 8b |
| 2-[4-(4-$ClC_6H_4CH_2O$)$C_6H_4$] | 168–170 | | | 8b |
| 2-(4-$MeOC_6H_4$) | 134–135 | | 70 | 8b |
| 2-(4-$NO_2C_6H_4$) | 168–169 | | 72 | 8b |
| 2-(3-$CF_3C_6H_4$) | 103–104 | | | 8b |
| 2-(3,4-$Cl_2C_6H_3$) | 133–134 | | 75 | 8b |
| 2-(3,5-$Cl_2C_6H_3$) | 124–125 | | 74 | 8a, b |

***Tetrahydro[1,2,4]triazolo[1,2-a] [1,2]diazepin-1(2H,5H)-one-3-thiones***

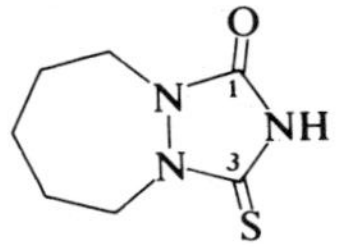

| | | | | |
|---|---|---|---|---|
| 2-(4-$BrC_6H_4$) | 150–152 | EtOH | 95 | 8b |
| 2-(4-$ClC_6H_4$) | 154.5–155 | EtOH | 95 | 8b |
| 2-(4-$MeOC_6H_4$) | 185–187 | EtOH | 92 | 8b |
| 2-(4-$MeC_6H_4$) | 208–210 | EtOH | 84 | 8b |
| 2-(3,5-$ClC_6H_3$) | 168–170 | EtOH | 77 | 8a, b |

***Tetrahydro[1,2,4]triazolo[1,2-a] [1,2]diazepin-1,3(2H,5H)-dithione***

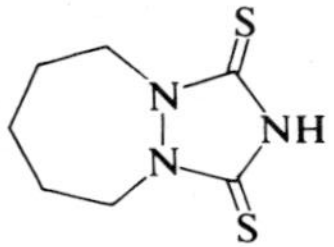

| | | | | |
|---|---|---|---|---|
| 2-(4-$ClC_6H_4$) | 200–201 | | 75 | 8b, c |

TABLE I-2. [*b*]-FUSED [1,2]DIAZEPINES

| Substituent | mp (°C); [bp (°C/torr)] | Solvent of Crystallization | Yield (%) | Spectra | Refs. |
|---|---|---|---|---|---|
| | | *Azeto[1,2-b] [1,2]diazepines* | | | |
| ***1,4-Dihydroazeto[1,2-b] [1,2]diazepin-2(2H)-ones*** | | | 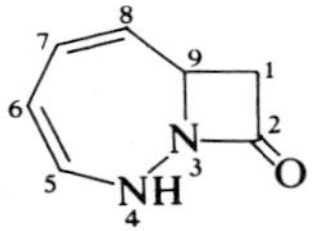 | | |
| 1-Cl-4-COPh, *trans* | 144–145 | | 70 | ir, pmr, uv | 10 |
| 1-Cl-4-COOEt, *trans* | | | 70 | | 10 |
| 1-Cl-4-COOPr-*i*, *trans* | | | 80 | | 10 |
| 1-Ph-4-COOEt | 110 | EtOH | 56 | | 10 |
| 1-$NH_2$-4-COOEt-7-Me, *trans* | | | 84 | ir, uv | 11 |
| 1,1-$Cl_2$-4-COPh | | | 80 | | 10 |
| 1-Cl-1-Me-COOEt, *cis* | | | 43 | | 10 |
| 1-Cl-1-Me-4-COOEt, *trans* | | | 6 | | 10 |
| 1-NHCOPh-4-COOEt-7-Me, *trans* | 50 | | 93 | ir, uv | 11 |
| 1-(4-$NO_2C_6H_4$CONH)-4-COOEt-7-Me, *trans* | | | 77 | ir, uv | 11 |
| 1-NHCO(2-COOH-$C_6H_4$)-4-COOEt-7-Me, *trans* | 158 | | 90 | ir, pmr, uv | 11 |
| 1-NHCOOPh-4-COOEt-7-Me, *trans* | 151 | | 92 | ir, pmr, $^{13}C$-nmr, uv | 11 |
| 1-Phthalimido-4-COOEt-7-Me, *trans* | 145 | | 91 | ir, pmr, $^{13}C$-nmr, uv | 11 |
| 1-N-[3-iminoisobenzofuran-1-one]-4-COOEt-7-Me, *trans* | 165 | | 90 | ir, pmr, uv | 11 |
| 1-NHCO(2-thienyl)-4-COOEt-7-Me, *trans* | 137 | | 92 | ir, uv | 11 |
| 1-Cl-4-COOEt-6-Me | Oil | | 61 | ir, uv, ms, pmr | 104 |
| 1-Br-4-COOEt-6-Me | Oil | | 56 | ir, uv, ms, pmr | 104 |
| 1-Me-4-COOEt | Oil | | 15 | ir, uv, ms, pmr | 104 |
| 1-*p*-ClTs-4-COOEt-6-Me | Oil | | 68 | ir, uv, ms, pmr | 104 |
| 1-MeCOO-4-COOEt-6-Me | Oil | | 95 | ir, uv, ms, pmr | 104 |

| Substituents | mp (°C) | Solvent | Yield (%) | Spectra | Ref. |
|---|---|---|---|---|---|
| 1-$PhCH_2OCONH$-4-COOEt-6-Me | 134 | $Et_2O$/Petr ether | 40 | ir, uv, pmr | 104 |
| 1-$N_3$-4-COOEt-6-Me | Oil | | 86 | ir, uv, ms, pmr | 104 |
| 1-Br-4-COPh | Oil | | 33 | ir, uv, ms, pmr | 104 |
| 1-*p*-ClTs-5-COPh | 133–134 | $CHCl_3/Et_2O$ | 81 | ir, uv, ms, pmr | 104 |
| 1-MeCOO-4-COPh-6-Me | 120 | $CHCl_3$/Petr ether | 79 | ir, uv, pmr | 104 |
| 1-$N_3$-4-COPh | 114 | EtOH | 55 | ir, uv, ms, pmr | 104 |
| 4-COOEt-6-Me | 77–78 | Hexane | 44 | ir, uv, ms, pmr | 104 |
| 4-$CONH_2$-6-Me | 175 | EtOAc | 69 | ir, uv, ms, pmr, | 104 |
| 1-Cl-4-COOEt-8-COOEt | Oil | | 60 | ir, uv, ms, pmr | 104 |
| 1,1-$Cl_2$-4-COPh | Oil | | 80 | ir, uv, ms, pmr | 104 |
| 1-Me-1-Cl-4-COPh | Oil | | 31 | ir, uv, ms, pmr | 104 |
| 1,1-$Me_2$-5-COOEt | 79 | $Et_2O$/Petr ether | 29 | ir, uv, pmr | 104 |
| 4-COOEt-1,1,6-$Me_3$ | Oil | | | ir, uv, ms, pmr | 104 |

***5-Carbethoxy-7-dimethylammonium-8,8-dimethylazeto[1,2-b] [1,2]diazepines***

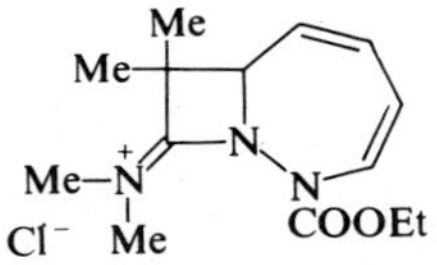

| Substituents | mp (°C) | Solvent | Yield (%) | Spectra | Ref. |
|---|---|---|---|---|---|
| None | | | 84 | ir, uv, pmr | 104 |
| 2,8,8-$Me_3$ | | | 85 | ir, uv, pmr | 104 |

*1,2,4-Oxadiazolo[4,5-b] [1,2]diazepines*

***5,9a-Dihydro-1,2,4-oxadiazolo[4,5-b] [1,2]diazepines***

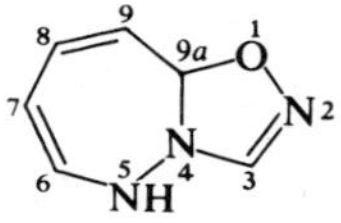

TABLE I-2. —(contd.)

| Substituent | mp (°C); [bp (°C/torr)] | Solvent of Crystallization | Yield (%) | Spectra | Refs. |
|---|---|---|---|---|---|
| 3-(2,4,6-$Me_3C_6H_2$)-5-COOEt | [70/0.01] | | 50 | ir, pmr, uv | 13 |
| 3-Ph-5-COPh-8-Me | | | 58 | ir, pmr, uv | 13 |
| 3-Ph-5-COOEt-8-Me | | | 60 | ir, pmr, uv | 13 |
| 3,8-$(Ph)_2$-5-COOEt | | | 62 | ir, pmr, uv | 13 |
| 3-(2,4,6-$Me_3C_6H_2$)-5-COPh-8-Me | 159 | $Et_2O$/Petr ether | 61 | ir, pmr, uv | 13 |
| 3-(2,4,6-$Me_3C_6H_2$)-5-COOEt-8-Me | 122 | $Et_2O$/Petr ether | 11.5 | ir, pmr, uv | 13 |
| 3-(2,4,6-$Me_3C_6H_2$)-5-COOEt-8-Ph | 157 | $Et_2O$/Petr ether | 64 | ir, pmr, uv | 13 |
| | | *Pyrrolo[1,2-b] [1,2]diazepines* | | | |
| ***3H-Pyrrolo[1,2-b] [1,2]diazepines*** | | | | | |
| 2,5-$(Me)_2$ | [79/0.4] | | 30–40 | pmr, uv | 14 |
| 2-Me-5-Ph | 93 | Petr ether | 5 | pmr, uv | 14 |
| 2,5-$(Ph)_2$ | 118 | EtOH/$H_2O$ | 10 | pmr, uv | 14 |
| 2-Ph-5-Me | [170–180/0.5] | | 9 | pmr, uv | 14 |

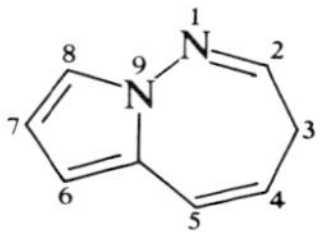

TABLE I-3. [*c*]-FUSED [1,2]DIAZEPINES

| Substituent | mp (°C); [bp (°C/torr)] | Solvent of Crystallization | Yield (%) | Spectra | Refs. |
|---|---|---|---|---|---|
| | | *1,2-Benzodiazepines* | | | |
| *1H-1,2-Benzodiazepines* | | | | | |
| None | 63–64 | $i$-$Pr_2O$ | 61 | ir, ms, pmr, uv | 17a, b |
| 4-Me | 87–88 | $i$-$Pr_2O$ | 60 | ir, ms, pmr, uv | 17b |
| 5-Me | 63.5–64 | $i$-$Pr_2O$ | 79 | ir, ms, pmr, uv | 17a, b |
| 7-COOMe | 114–115 | $i$-$Pr_2O$/PhH | 38 | pmr | 17b |
| 7-Cl | 73–74 | $i$-$Pr_2O$ | 50 | pmr | 17b, 18 |
| 7-MeO | 94.5–95.5 | $i$-$Pr_2O$ | 5 | pmr | 17b |
| 7-Me | 94.5–95.5 | $i$-$Pr_2O$ | 47 | pmr | 17b |
| 8-MeO | 103–104.5 | $i$-$Pr_2O$ | 62 | pmr | 17b |
| 8-Me | 93.5–95 | $i$-$Pr_2O$ | 70 | pmr | 17b |
| 3-COOEt | 105 | $Et_2O$/$i$-$Pr_2O$ | 17 | ir, pmr | 18b |
| 3-COOMe | 128–130 | $i$-$Pr_2O$/PhH | 33 | ir, ms, pmr | 20b |
| 3-$CONH_2$ | 189–191 | $i$-$Pr_2O$/PhH | 33 | ir, ms, pmr | 20b |
| 3-Cl | 68–69 | $i$-$Pr_2O$ | 86 | ir, pmr | 20a, b |
| 3-CN | 70–71 | $i$-$Pr_2O$ | 13 | ir, ms, pmr | 20b |
| 3-EtO | 63.5–65 | $i$-$Pr_2O$ | 10 | ir, ms, pmr | 20b |

TABLE I-3. —(contd.)

| Substituent | mp (°C); [bp (°C/torr)] | Solvent of Crystallization | Yield (%) | Spectra | Refs. |
|---|---|---|---|---|---|
| 3-MeO | 85–86 | $i$-$Pr_2O$ | 66 | ir, ms, pmr | 20a, b |
| 3-$CH(COOMe)_2$ | Oil | | 59 | ir, ms, pmr | 20a, b |
| 3,4-$(COOEt)_2$ | 114 | $Et_2O$/$i$-$Pr_2O$ | 68 | ir, pmr | 18a, b |
| 3-COOEt-4-CN | 168 | $Et_2O$/$i$-$Pr_2O$ | 65 | ir, pmr | 18a, b |
| 3-COOEt-4-Me | 119 | $i$-$Pr_2O$ | 66 | ir, pmr | 18b |
| 3-COOEt-4-Ph | 131 | $Et_2O$/$i$-$Pr_2O$ | 54 | ir, pmr | 18a, b |
| 3-COOEt-5-Me | 105 | $i$-$Pr_2O$ | 47 | ir, pmr | 18b |
| 3-COOEt-5-Ph | 180 | $Et_2O$/$i$-$Pr_2O$ | 75 | ir, pmr | 18b |
| 3-COOMe-5-Me | 159–160 | $i$-$Pr_2O$/PhH | 26 | ir, ms, pmr | 20b |
| 3-$CONH_2$-5-Me | 228.5–230 | EtOH | 34 | ir, ms, pmr | 20b |
| 3-Cl-5-Me | 86–87 | $i$-$Pr_2O$ | 89 | ir, pmr | 20a, b |
| 3-CN-5-Me | 67–68 | $i$-$Pr_2O$ | 12 | ir, ms, pmr | 20b |
| 3-EtO-5-Me | 91–93 | $i$-$Pr_2O$ | 9 | ir, ms, pmr | 20b |
| 3-MeO-5-Me | 94–95 | $i$-$Pr_2O$ | 55 | ir, ms, pmr | 20a b |
| 3-$CH(COOMe)_2$-5-Me | Oil | | 47 | ir, ms, pmr | 20a, b |
| 3-Ac-5-Ph-7-Cl | 140 | $i$-$Pr_2O$ | 62 | ir, pmr | 19 |
| 3-COPh-5-Ph-7-Cl | 168 | $i$-$Pr_2O$ | 81 | ir, pmr | 19 |
| 3-COOBu-$t$-5-Ph-7-Cl | 195d | $i$-$Pr_2O$ | 79 | ir, pmr | 19 |
| 3-(4-$NO_2C_6H_4$)-5-Ph-7-Cl | 158 | $i$-$Pr_2O$ | 55 | ir, pmr | 19 |
| 3-$SO_2$Ph-5-Ph-7-Cl | 135 | $i$-$Pr_2O$ | 54 | ir, pmr | 19 |
| 1-Me | 61–62 | | 80–90 | | 22 |
| 1,5-$(Me)_2$ | Oil | | 80–90 | | 22 |
| 3-COOH | 115d | | 89 | ir | 77 |
| 4-COOEt | 93–94 | | 52 | ir, uv, ms, pmr | 82 |
| 3-COOH-5-Ph-7-Cl | 126d | | 90 | ir | 77 |
| 3-COOH-5-Me | 120d | | 86 | ir | 77 |
| 3-COOH-5-$i$-Pr | 130d | | 87 | ir | 77 |
| 3-Me-7-Cl | 184 | $i$-$Pr_2O$ | 36 | ir, pmr | 78 |
| 3-Et-7-Cl | 140 | $i$-$Pr_2O$ | 45 | ir, pmr | 78 |

| | | | | | |
|---|---|---|---|---|---|
| 3-*n*-Pr-7-Cl | 143 | *i*-$Pr_2O$ | 38 | ir, pmr | |
| 3,4-COOMe | 154–155 | | 35 | ir, uv, ms, pmr | 82 |
| 3-COOMe-4-Me | 87–88 | | 88 | ir, uv, ms, pmr | 81 |
| 3-COOMe-4-Me-5-OAc, *cis* | 199–200 | | 100 | ir, uv, ms pmr, $^{13}C$-nmr | 81 |
| 3-COOMe-4-Me-5-OAc, *trans* | 150–151 | | | ir, uv, ms | 81 |
| ***3H-1,2-Benzodiazepines*** | | | | | |
| None | [125–128/1] | | 95 | ms, pmr, uv | 21, 23 |
| 3-AcO | 84-85 | *i*-$Pr_2O$ | 88 | ir, ms, pmr | 20b, 21 |
| 3-Meo | 53–54 | *i*-$Pr_2O$ | 74 | ms, pmr | 21 |
| 5-Me | [132–134/1] | | 93 | ms, pmr, uv | 21, 23 |
| 3-AcO-5-Me | 62–63 | *i*-$Pr_2O$ | 83 | ir, ms, pmr | 20b, 21 |
| 3-MeO-5-Me | Oil | | 44 | ms, pmr | 21 |
| 1-Oxide | 93–94 | PhH/*i*-$Pr_2O$ | 22 | ms, pmr, uv | 20a, b |
| 1-Oxide-5-Me | 93–94 | PhH/*i*-$Pr_2O$ | 20 | ms, pmr, uv | 20b |
| 2-Oxide | 85–86 | PhH/$Pr_2O$ | 55 | ms, pmr, uv | 20a, b |
| 2-Oxide-5-Me | 68-69 | PhH/*i*-$Pr_2O$ | 63 | ms, pmr, uv | 20 |
| ***5H-1,2-Benzodiazepines*** | | | | | |
| 3-$CO_2$Me-5-OAc | Oil | | 60 | pmr | 84 |
| 3-CN-5-OAc-1-[3,4-$(MeO)_2$-$C_6H_3$]-3,4-$Me_2$-5-Et-7,8-$(MeO)_2$-3-ium Iodide | Unstable | | 60 | | 84 |
| Hydrate | 150–152d | $H_2O$ | | uv, pmr | 94 |

TABLE I-3. —(contd.)

| Substituent | mp (°C); [bp (°C/torr)] | Solvent of Crystallization | Yield (%) | Spectra | Refs. |
|---|---|---|---|---|---|
| ***2,3-Dihydro-1H-1,2-benzodiazepines*** | | | | | |
| None | 56–58 | *i*-$Pr_2O$ | 96 | ir, ms, pmr | 17b |
| 2-Ac | 108–109 | *i*-$Pr_2O$ | 89 | ir, ms, pmr | 17b |
| 2-COOMe | 109–110 | *i*-$Pr_2O$ | 67 | ir, ms, pmr | 17b |
| 5-Me | 75–76 | *i*-$Pr_2O$ | 94 | ir, ms, pmr | 17b |
| 1,2-$(Ac)_2$ | 78–80 | PhH/*i*-$Pr_2O$ | 91 | ir, ms, pmr | 17b |
| 2-Ac-3-OMe | | | 6 | ir, pmr | 24 |
| 2-Ac-5-Me | 89–90 | PhH/*i*-$Pr_2O$ | 93 | ir, ms, pmr | 17b |
| 2-COOEt-3-EtO | 86–87 | Petr ether | 59 | ir, ms, pmr, uv | 25 |
| 2-COOMe-5-Me | 100–101 | *i*-$Pr_2O$ | 79 | ir, ms, pmr | 17b |
| 1,2-$(Ac)_2$-5-Me | 104.5–105.5 | PhH/*i*-$Pr_2O$ | 94 | ir, ms, pmr | 17b |
| ***4,5-Dihydro-1H-1,2-benzodiazepines*** | | | | | |
| 3,4-$(COOEt)_2$-4-Me-5-OH | 164–165 | *i*-$Pr_2O$ | 40 | ir, pms, ms | 79 |
| 3,4-$(COOEt)_2$-4-Me-5-OAc | 152–153 | *i*-$Pr_2O$ | 72 | ir, pmr, $^{13}C$-nmr | 79 |
| 3,4-$(COOEt)_2$-4-Ph-5-OAc | 188–190 | MeOH | 14 | $^{13}C$-nmr, ms ir, pmr | 79 |
| 3,4-$(COOEt)_2$-4-Me-5=O | 88–89 | *i*-$Pr_2O$/ *n*-Pentane | 52 | ir, pmr | 79 |

| | | | | | |
|---|---|---|---|---|---|
| 3-COOEt-5-OMe | 125–126 | Hexane/PhH | 97 | ir, uv, pmr | 81 |
| 3-COOEt-5-OEt | 121–122 | EtOH | | ir, uv, pmr | 81 |
| 3-COOEt-5-OAc | 155–156 | | 96 | ir, uv, pmr | 81 |
| 3-COOEt-5-Cl | 114–115 | | 99 | ir, uv, pmr | 81 |
| 3-COOEt-5-OH | 160–161 | | | ir, uv, pmr | 81 |
| ***1,2-Dihydro-1,2-benzodiazepin-3(3H)-ones*** | | | | | |
| 1-Me | 170–172 | | 60–65 | ir, pmr | 22 |
| 1,5-$(Me)_2$ | 201–203 | | 60–65 | ir, pmr | 22 |
| ***2,3,4,5-Tetrahydro-1H-1,2-benzodiazepines*** | | | | | |
| None | 56–57 | *i*-$Pr_2O$/Hexane | 90 | ir, ms, pmr | 17b |
| 2-Ac | 100–102 | *i*-$Pr_2O$/PhH | 97 | ir, ms, pmr | 17b |
| 5-Me | 45–46 | *i*-$Pr_2O$/Hexane | 86 | ir, ms, pmr | 17b |
| 1,2-$(Ac)_2$ | 91–93 | *i*-$Pr_2O$/Hexane | 97 | ir, ms, pmr | 17b |
| 2-Ac-5-Me | 113.5–115 | *i*-$Pr_2O$/PhH | 96 | ir, ms, pmr | 17b |
| 1,2-$(Ac)_2$-5-Me | 115–117 | *i*-$Pr_2O$/PhH | 97 | ir, ms, pmr | 17b |
| ***1,2,4,5-Tetrahydro-1,2-benzodiazepin-3(3H)-ones*** | | | | | |
| 1-Et | 165.5 | $H_2O$ | 60–70 | | 27 |

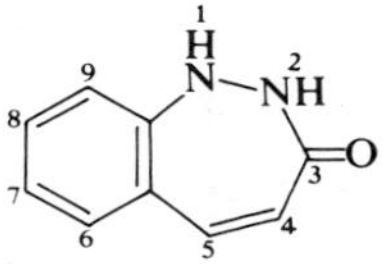

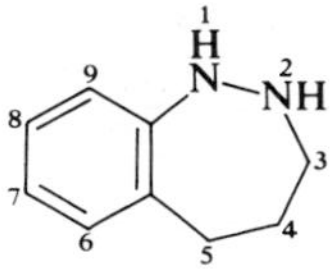

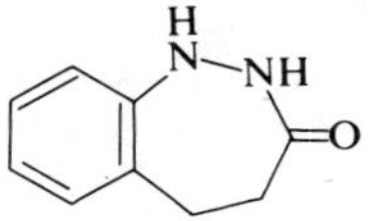

TABLE I-3. —(contd.)

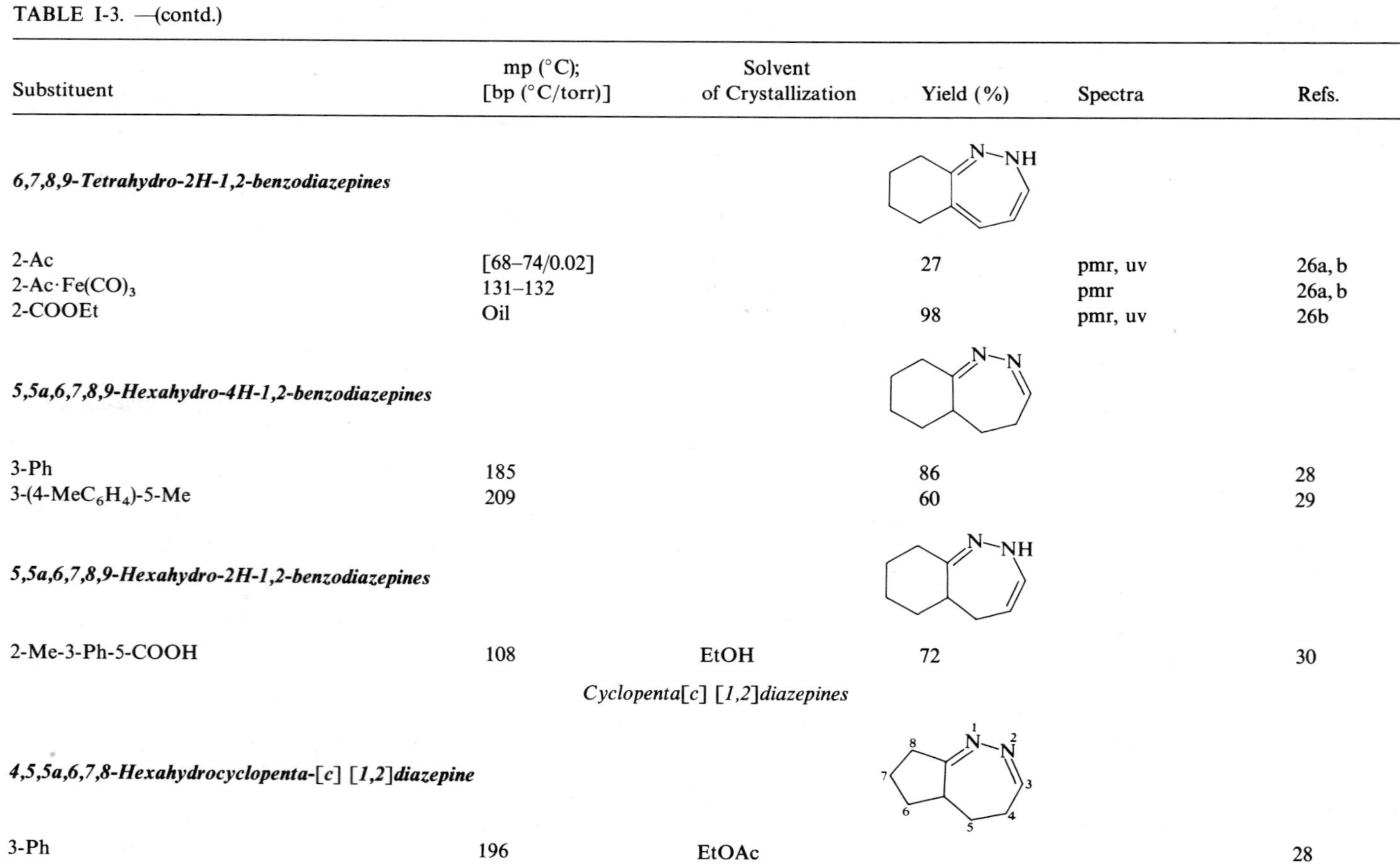

| Substituent | mp (°C); [bp (°C/torr)] | Solvent of Crystallization | Yield (%) | Spectra | Refs. |
|---|---|---|---|---|---|
| ***6,7,8,9-Tetrahydro-2H-1,2-benzodiazepines*** | | | | | |
| 2-Ac | [68–74/0.02] | | 27 | pmr, uv | 26a, b |
| 2-Ac·$Fe(CO)_3$ | 131–132 | | | pmr | 26a, b |
| 2-COOEt | Oil | | 98 | pmr, uv | 26b |
| ***5,5a,6,7,8,9-Hexahydro-4H-1,2-benzodiazepines*** | | | | | |
| 3-Ph | 185 | | 86 | | 28 |
| 3-(4-$MeC_6H_4$)-5-Me | 209 | | 60 | | 29 |
| ***5,5a,6,7,8,9-Hexahydro-2H-1,2-benzodiazepines*** | | | | | |
| 2-Me-3-Ph-5-COOH | 108 | EtOH | 72 | | 30 |
| | | *Cyclopenta[c] [1,2]diazepines* | | | |
| ***4,5,5a,6,7,8-Hexahydrocyclopenta-[c] [1,2]diazepine*** | | | | | |
| 3-Ph | 196 | EtOAc | | | 28 |

| | | | | | |
|---|---|---|---|---|---|
| ***2,5,5a,6,7,8-Hexahydrocyclopenta*[*c*] [*1,2*]*diazepines*** | | | 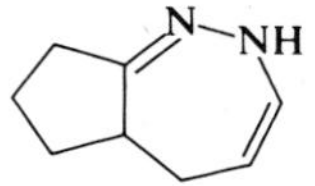 | | |
| 5-COOH | 150 | Petr ether | 56 | ir | 30 |
| 2-Me-5-COOH | 119 | Petr ether | 68 | ir | 30 |
| | | *Cyclopropa*[*c*] [*1,2*]*diazepines* | | | |
| ***1,1a-Dihydrocyclopropa*[*c*] [*1,2*]*diazepine-6(6H)-one*** | | | 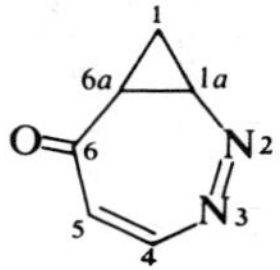 | | |
| 1,1-$(Me)_2$-1a,4,5,6a-$(Ph)_4$ | 186–187 | | 25 | ir, ms, pmr, uv | 31 |
| | | *Furo*[*c*] [*1,2*]*diazepines* | | | |
| ***1H-Furo*[*2,3-c*][*1,2*]*diazepine*** | | | 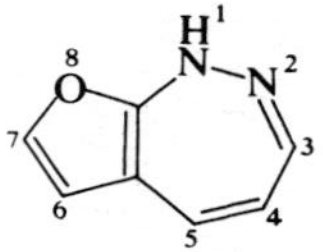 | | |
| None | Oil | | 25 | | 32 |
| ***1H-Furo*[*3,2-c*] [*1,2*]*diazepine*** | | | 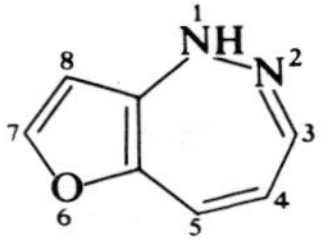 | | |
| None | Oil | | 25 | | 32 |

TABLE I-3. —(contd.)

| Substituent | mp (°C); [bp (°C/torr)] | Solvent of Crystallization | Yield (%) | Spectra | Refs. |
|---|---|---|---|---|---|
| | | *Pyrido*[*c*] [*1,2diazepines* | | | |
| ***1H-Pyrido*[2,3-*c*] [*1,2*]*diazepine*** | | | 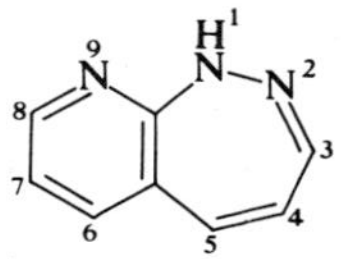 | | |
| None | 113–114 | | 20 | | 32 |
| ***1H-Pyrido*[2,3-*c*] [*1,2*]*diazepine*** | | | 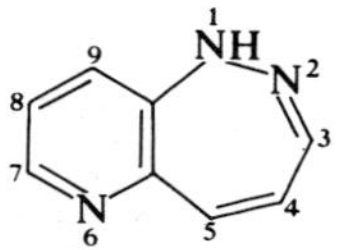 | | |
| None | 119–120 | | 25 | pmr | 32 |
| ***3H-Pyrido*[3,2-*c*] [*1,2*]*diazepine*** | | | 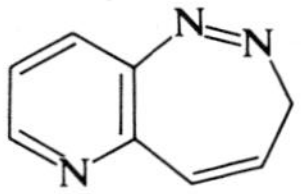 | | |
| None | 35–37 | | | pmr | 32 |

## *Pyrimido[4,5-c] [1,2]diazepines*

### ***1,7-Dihydropyrimido[4,5-c] [1,2]diazepin-6(6H)-ones***

| | | | | | |
|---|---|---|---|---|---|
| 1,3,5-$(Me)_3$-8-$NH_2$ | 250–253d | MeOH | 41 | ms, pmr, uv p$K_a$ | 33b |
| 1,3-$(Me)_2$-5-COOEt-8-$NH_2$ | 196d | $CCl_4$ | 11 | ms, pmr, uv | 33b |
| 1,5-$(Me)_2$-3-COOEt-8-$NH_2$ | 244–245d | EtOAc | 30 | ms, pmr, uv | 33b |
| 1-Me-3,5-$(Ph)_2$-8-$NH_2$ | 205–220d | $CCl_4$ | 29 | ms, pmr, uv | 33b |
| 1-Me-3-COOEt-5-Ph-8-$NH_2$ | 240.5–242.5d | PhH | 9 | ms, pmr, uv | 33b |
| 1-Me-3-COOEt-5-[2,4-$(MeO)_2$-$C_6H_3$]-8-$NH_2$ | 187–196d | PhH | 28 | pmr, uv | 33b |
| 1-Me-3-COOEt-5-[3,4-$(MeO)_2$-$C_6H_3$]-8-$NH_2$ | 162d | EtOAc/hexane | 7 | pmr, uv | 33b |
| 1-Me-3-COOMe-5-(3-OH-$C_6H_4$)-8-$NH_2$ | > 275 | MeoH | 49 | pmr, uv | 33b |
| 1-Me-3-COOMe-5-[3,4,5-$(MeO)_3C_6H_2$]-8-$NH_2$ | 258–260d | MeOH | 45 | ms, pmr, uv | 33b |
| 1-Me-3-COOMe-5-(3-Pyridyl)-8-$NH_2$ | 258–272d | MeOH | 56 | pmr, uv | 33b |

## *Thieno[c] [1,2]diazepines*

### ***1H-Thieno[2,3-c] [1,2]diazepines***

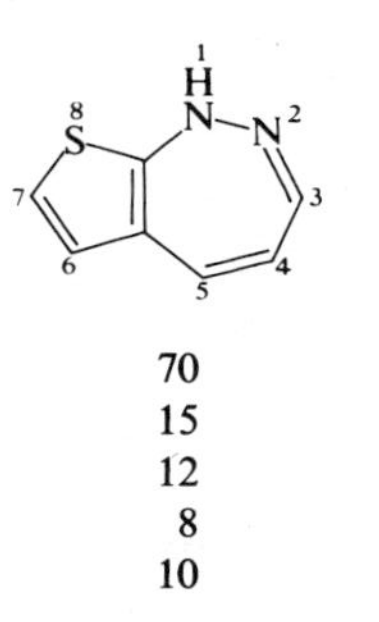

| | | | | | |
|---|---|---|---|---|---|
| None | 81–83 | | 70 | | 32 |
| 3,5-$Me_2$ | 96–97 | | 15 | ms, ir, uv, pmr | 87 |
| 1-COOEt-3,5-$Me_2$ | 134–135 | | 12 | ms, ir, uv, pmr | 87 |
| 1-Ac-3,5-$Me_2$ | 102–104 | | 8 | ms, ir, uv, pmr | 87 |
| 1-COPh-3,5-$Me_2$ | Oil | | 10 | ms, ir, uv, pmr | 87 |

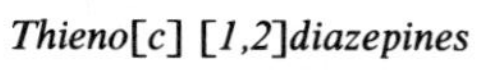
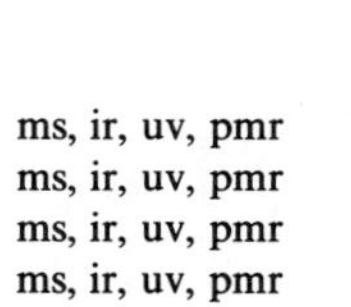

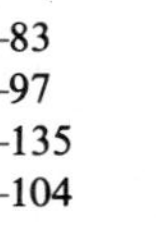
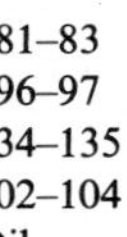

TABLE I-3. —(contd.)

| Substituent | mp (°C); [bp (°C/torr)] | Solvent of Crystallization | Yield (%) | Spectra | Refs. |
|---|---|---|---|---|---|
| ***1H-Thieno[3,2-c] [1,2]diazepine*** | | | 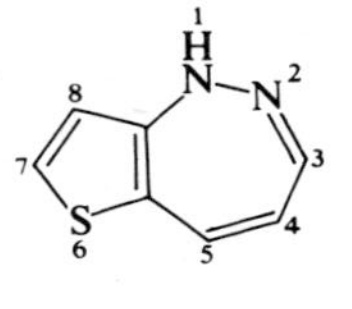 | | |
| None | 94–95 | | 65 | | 32 |
| | | *2,3-Benzodiazepines* | | | |
| ***1H-2,3-Benzodiazepines*** | | | 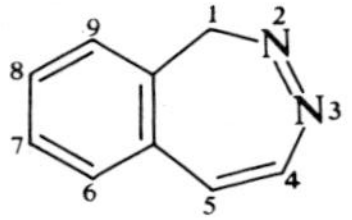 | | |
| None | 49–50 | | 41 | ms, pmr | 35 |
| 1-Et | 40–41 | | | ms, pmr | 35 |
| 1-Me | 47 | | 64 | ms, pmr | 34,35 |
| 1-Ph | 101–102 | | 62 | ms, pmr | 35 |
| 4-Ph | 132–133 | | 71 | ms, pmr | 35, 37 |
| 1-$PhCH_2$-4-Ph | 126–127 | | 65 | | 39 |
| 7,8-$(MeO)_2$ | 89–91 | | 70 | ms, pmr | 35 |
| 1-Me-4-Ph | 92–93 | | 88 | ms, pmr | 34, 35 |
| 1-(4-$MeC_6H_4$)-4-Ph | 146–147 | | 67 | ms, pmr | 34, 35 |
| 1-Ph-7,8-$(MeO)_2$ | 112–113 | | 84 | ms, pmr | 35 |

| Substituent | mp (°C) | Solvent | Yield (%) | Characterization | Ref. |
|---|---|---|---|---|---|
| ***3H-2,3-Benzodiazepines*** | | | 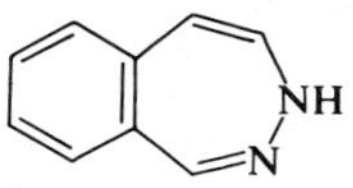 | | |
| 3-COOEt-8-COOMe | 84–86 | | 12 | pmr, ir, uv, ms | 90 |
| 1-Me-3-Ac | 110–111 | *i*-$Pr_2O$/Hexane | 59 | ir, pmr, ms, anal | 91 |
| 1,4-$Me_2$-3-Ac | Oil | | 50 | ir, pmr, ms, anal | 91 |
| ***5H-2,3-Benzodiazepines*** | | | 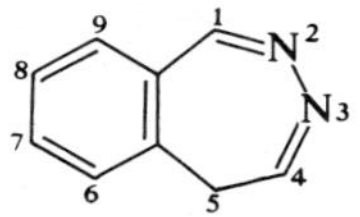 | | |
| 1-Ph | 152–153 | (EtOH) | 69; 46 | ms, pmr | 35 |
| 4-Ph | 108–110 | | 92 | | 37 |
| 1-$PhCH_2$-4-Ph | 148–149 | EtOH | 83 | | 39 |
| 1-Me-4-Ph | | | | ms, pmr | 35 |
| 1-(4-$MeC_6H_4$)-4-Ph | 269–270 | PhH/EtOH | 84 | ms, pmr | 35 |
| 1-Ph-7,8-$(MeO)_2$ | 162–163 | EtOH | 57; 82 | ms, pmr | 35 |
| 1-[3,4-$(MeO)_2C_6H_3$]-4- Me-5-Et-7,8-$(MeO)_2$ | 156–157 | *i*-PrOH | 71; 73 | ir, ms, $^{13}C$ nmr, pmr, uv | 40, 42 |
| 1-Me | Oil | | 35 | pmr | 91 |
| 4-Me | 106–108 | phH/Hexane | 45 | pmr | 91 |
| 7-Me | Oil | | 50 | pmr | 91 |
| 9-Me | Oil | | 55 | pmr | 91 |
| 1,4-di-Me | Oil | | 30 | pmr | 91 |
| 8-MeO | Oil | | 45 | pmr | 91 |
| 1-Me-3-Ac | 110–111 | *i*-$Pr_2O$/Hexane | 59 | ir, pmr, ms | 91 |
| 1,4-$Me_2$-3-Ac | Oil | | 50 | ir, pmr, ms | 91 |
| Hydrochloride | 220–221d | MeOH | | ir, pmr, uv | 40 |
| Hydrobromide | 218–219d | *i*-PrOH/$H_2O$ | | | 40 |
| Perchlorate | 200–202d | *i*-PrOH/$H_2O$ | | uv | 40 |
| Picrate | 206–207d | EtOH | | | 40 |

TABLE I-3. —(contd.)

| Substituent | mp (°C); [bp (°C/torr)] | Solvent of Crystallization | Yield (%) | Spectra | Refs. |
|---|---|---|---|---|---|
| 1-Et-4-[3,4-$(MeO)_2C_6H_3$]-6,7-$(MeO)_2$-2-Naphthol adduct | 168 | EtOH | 87.5 | | 40 |
| ***4,5-Dihydro-1H-2,3-benzodiazepine*** | | | | | |
| None | 63 | Pentane | 20 | ir | 45 |
| ***4,5-Dihydro-3H-2,3-benzodiazepines*** | | | | | |
| None | 72 | Petr ether | 42 | | 45 |
| Picrate | 180–190d | | | | |
| 1-Ph | 76–77 | Ligroin | 40 | ir, ms, pmr, | 48 |
| 3-Ph | 114–115 | MeOH | 57 | ir, ms, pmr, uv | 46 |
| 3-(4-$ClC_6H_4$) | 102–104 | MeOH | 50 | | 46 |
| 3-(4-$MeC_6H_4$) | 86–87 | MeOH | 47 | | 46 |
| 3-(2-$NO_2C_6H_4$) | | | 20 | | 47 |
| 3-(4-$NO_2C_6H_4$) | 154–155 | MeOH | 54 | | 46 |
| 3-CONHPh | 133–135 | $Et_2O$ | 55 | | 45 |
| 4-Ph | 113–115 | | | ir | 37 |
| 5-COOMe | 260–263 | $DMF/H_2O$ | 59.5 | | 49 |
| 1-Ph-3-$CH_2CH_2OH$ | 155–156 | EtOH | 42 | | 48 |
| 1-(4-$MeOC_6H_4$)-3-$CH_2CH_2OH$ | 163–165 | EtOH | 60 | | 48 |
| 4-(4-HO-2,5-$Me_2C_6H_2$]-8-$NO_2$ | 303d | MeOH/EtOH | 70–80 | | 54 |

*1,2-Dihydro-5H-2,3-benzodiazepines*

| | | | | | |
|---|---|---|---|---|---|
| 1-MeO-2-Ac-4-Ph | Oil [140/0.15 mm Hg] | | 88 | ir, pmr, $^{13}$C-nmr | 98 |
| 1-EtO-2-Ac-4-Ph | Oil [150/0.2 mm Hg] | | 81 | ir, pmr, $^{13}$C-nmr | 98 |
| 1-PhO-2-Ac-4-Ph | 116–117 | EtOH | 92 | ir, pmr, $^{13}$C-nmr | 98 |
| 1-EtS-2-Ac-4-Ph | Oil [190/0.4 mm Hg] | | 62 | ir, pmr, $^{13}$C-nmr | 98 |
| 1-PhS-2-Ac-4-Ph | Oil [150/0.15 mm Hg] | | 74 | ir, pmr, $^{13}$C-nmr | 98 |

*2,5-Dihydro-2,3-benzodiazepin-1(1H)-ones*

| | | | | | |
|---|---|---|---|---|---|
| 4-Ph | 202 | EtOH | 80–85 | pmr | 50, 55, 56 |
| 4-(4-$HOC_6H_4$) | 243d | | | | 53 |
| 4-(4-$MeOC_6H_4$) | 208 | EtOH | 80/88 | pmr | 56 |
| 4-(3-$MeC_6H_4$) | 190–191 | EtOH | 50 | | 52 |
| 4-(4-$MeC_6H_4$) | 233 | EtOH | 87/92 | pmr | 56 |
| 4-(2-Me-4-HO-$C_6H_3$) | 211d | MeOH/EtOH | 70–80 | pmr | 54 |
| 4-[2,3-$(Me)_2$-4-HO-$C_6H_2$] | 250d | MeOH/EtOH | 70–80 | pmr | 54 |
| 4-[2,6-$(Me)_2$-4-HO-$C_6H_2$] | 268d | MeOH/EtOH | 70–80 | pmr | 54 |
| 4-[3,5-$(Me)_2$-4-HO-$C_6H_2$] | 256d | MeOH/EtOH | 70–80 | pmr | 54 |
| 4-[4,5-$(Me)_2$-2-HO-$C_6H_2$] | 269d | MeOH/EtOH | 70–80 | pmr | 54 |
| 4-[4,6-$(Me)_2$-3-HO-$C_6H_2$] | 211d | MeOH/EtOH | 70–80 | pmr | 54 |
| 2-Et-4-Ph | 142 | EtOH/$H_2O$ | | | 51 |
| 2-(2-Diethylaminoethyl)-4-Ph Maleate | 112 | *i*-PrOH | 60 | | 55 |

TABLE I-3. —(contd.)

| Substituent | mp (°C); [bp (°C/torr)] | Solvent of Crystallization | Yield (%) | Spectra | Refs. |
|---|---|---|---|---|---|
| 2-(2-Hydroxyethyl)-4-Ph | 141 | EtOH | 33 | | 55 |
| 2-Me-4-Ph | 131, 133 | PhH | 60, 75 | ir, pmr, uv | 51, 55 |
| 2-(2-Morpholinoethyl)-4-Ph Maleate | 163 | *i*-PrOH | 76 | | 55 |
| 2-(2-Morpholinoethyl)-4-(4-$MeOC_6H_4$) Hydrochloride | 208 | EtOH | 80 | | 55 |
| 2-(2-Morpholinoethyl)-4-(4-$MeC_6H_4$) Maleate | 148 | PhH | 65 | | 55 |
| 2-Morpholinomethyl-4-Ph | 164 | EtOH | 30 | ir, pmr | 55 |
| 2-(3-Morpholinopropyl)-4-Ph Hydrochloride | 218 | *i*-PrOH | 62 | | 55 |
| 2-NO-4-Ph | 110 | $EtOH/H_2O$ | | | 51 |
| 2-Ph-4-Me | 198–199 | EtOH | | | 50 |
| 2-(2-Pyrrolindinoethyl)-4-Ph Maleate | 157 | *i*-PrOH | 50 | | 55 |
| 2-(3-Pyrrolidinopropyl)-4-Ph Maleate | 165 | EtOH | 50 | | 55 |
| 4-Ph-8-MeO | 147 | EtOH | 50 | | 55 |
| 4-[2,3-$(Me)_2$-4-$HOC_6H_2$]-8-HO | 259d | MeOH/EtOH | 70–80 | pmr | 54 |
| 4-[2,5-$(Me)_2$-4-$HOC_6H_2$]-8-$NH_2$ | 277d | MeOH/EtOH | 70–80 | pmr | 54 |
| 4-[2,5-$(Me)_2$-4-$HOC_6H_2$]-8-MeO | 259d | MeOH/EtOH | 70–80 | pmr | 54 |
| 4-[3,5-$(Me)_2$-4-$HOC_6H_2$]-8-MeO | 288d | MeOH/EtOH | 70–80 | pmr | 54 |
| 4-[4,5-$(Me)_2$-2-$HOC_6H_2$]-8-HO | 295d | MeOH/EtOH | 70–80 | pmr | 54 |
| 2-Me-4,5-$Ph_2$ | 145 | EtOH | 55 | ir, pmr, uv | 57 |
| 2-Me-4-Ph-5-Br | 164 | | | ir, pmr | 57 |
| ***2,5-Dihydro-2,3-benzodiazepin-1(1H)-thione*** | | 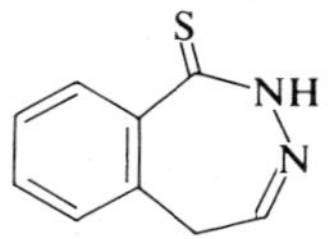 | | | |
| 4-(4-$MeC_6H_4$) | 240 | PhH/EtOH | | pmr | 56 |

***3,5-Dihydro-2,3-benzodiazepin-4(4H)-ones***

| | | | | | |
|---|---|---|---|---|---|
| None | | Analysis only | | | 58 |
| 1-Me | 184.5–185.5 | EtOH | | | 58 |
| 1-Ph | 208; 212 | *i*-PrOH | 60 | | 60, 61 |
| 1-$PhCH_2$-3-Me | 139 | EtOH | 60 | ir, pmr, uv | 63 |
| 1-$BrCH_2$-3-Me | 148 | | 15; 40 | ir, pmr | 63 |
| 1,3-$Me_2$ | | | | | 63 |
| 1-Me-3-Ph | 158–158.5 | | 43 | | 58 |
| 1-Ph-3-Me | 131 | | 70; 62 | ir, pmr, uv | 60, 61c |
| 1-Ph-3-(2-Morpholinoethyl) Maleate | 72 | $DMF/H_2O$ | 50; 86.7 | | 61 |
| | 159 | $DMF/H_2O$ | 85 | | 61 |
| 1-Ph-3-(3-Morpholinopropyl) Maleate | | | | | |
| | 169 | EtOH | 40 | | 61 |
| 1-Ph-3-(2-Pyrrolidinoethyl) Maleate | 167 | *i*-PrOH | 50 | | 61 |
| 1-Ph-8-Cl | 229–230 | $CHCl_3$/EtOH | 65 | | 59 |
| 1-(4-$ClC_6H_4$)-3-2-Morpholinoethyl | | | | | |
| Maleate | 185 | EtOH | 35 | | 61 |
| 1-(4-$MeOC_6H_4$-3-(2-Morpholinoethyl) | | | | | |
| Maleate | 189; 190 | EtOH | 40; 35 | | 60, 61 |
| 1-Ph-3-(2-Dimethylaminoethyl)-8-Cl | 83–86 | Hexane | 34 | | 59 |
| 1-Ph-3-(2-Morpholinoethyl)-8-Cl | | | | | |
| Maleate | 147 | *i*-PrOH | 77; 53 | | 60, 61 |
| 1-Ph-3-Me-8-Cl | 127; 137–138 | EtOH | 60; 48; 62 | ir, pmr, uv | 59–61 |
| 1-Ph-3-(2-Pyrrolidinoethyl)-8-Cl | | | | | |
| Hydrochloride hydrate | 215 | EtOH | 50 | | 61 |
| 1-Me-7,8-$(MeO)_2$ | 210–212 | EtOH | 45 | | 59 |
| 1-Me-3-(2-Dimethylaminoethyl)- | | | | | |
| 7,8-$(MeO)_2$ | 155–157 | EtOH/$Et_2O$ | 85 | | 59 |

TABLE I-3. [*c*]-FUSED [1,2]DIAZEPINES

| Substituent | mp (°C); [bp (°C/torr)] | Solvent of Crystallization | Yield (%) | Spectra | Refs. |
|---|---|---|---|---|---|
| 1-Me-3-Ph-7,8-$(MeO)_2$ | 194 | EtOH | 82 | ir, pmr | 62 |
| 1-Me-3-(4-$BrC_6H_4$)-7,8-$(MeO)_2$ | 189 | EtOH | 72 | ir, pmr | 62 |
| 1-Me-3-(4-$ClC_6H_4$)-7,8-$(MeO)_2$ | 185 | EtOH | 64 | ir, pmr | 62 |
| 1-Ph-3-(2-Morpholinoethyl)-5,5-$Me_2$ | 169 | *i*-PrOH | 20 | | 61 |
| ***2,3,4,5-Tetrahydro-1H-2,3-benzodiazepines*** | | | | | |
| None | | | | | |
| Hydrochloride | 139–143 | $H_2O$, HCl | 40 | | 45 |
| 2,3-$Me_2$ | | | | | |
| Picrate | 129–130 | | 75 | | 64 |
| ***1,2,3,5-Tetrahydro-2,3-benzodiazepin-4(4H)-ones*** | | | | | |
| 1-COOMe-2-Me | 121–122 | $H_2O$ | 82 | ir, pmr | 65 |
| 1-COOMe-2-Ph | 149–151 | EtOH | 80 | ir, pmr | 65 |
| ***3,5-Dihydro-2H-2,3-benzodiazepin-1,4-diones*** | | | | | |
| 2,3-$Me_2$ | 118–119 | Hexane | 84 | ir, ms, pmr | 64 |
| 2,3-$Me_2$-5-Br | 162–163 | PhH/Hexane | 76 | ir, pmr | 66 |

| Substituent | mp (°C) or bp (°C/torr) | Solvent | Yield (%) | Spectra | Ref. |
|---|---|---|---|---|---|
| ***Cyclopenta[d] [1,2]diazepine*** | | | 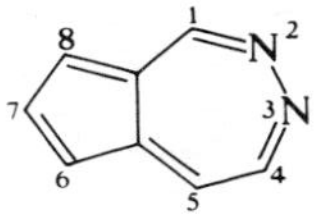 | | |
| 1,4-$Ph_2$ | 289–292 | $CHCl_3$ | 40 | ms, uv | 68 |
| *Cyclopropa[d] [1,2]diazepines* | | | | | |
| ***3,5a,6,6a-Tetrahydrocyclopropa[d] [1,2]diazepines*** | | | 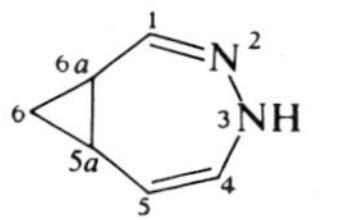 | | |
| 3-COOEt-6,6-$Me_2$ | Oil | | 35 | ir, ms, pmr, uv | 69 |
| 3-COOEt-5,6,6,6*a*-$Me_4$ | 65–67 | | 78 | ir, ms, pmr, uv | 69 |
| *Oxireno[d] [1,2]diazepines* | | | | | |
| ***1a,4,6,6a-Tetrahydrooxireno[d] [1,2]diazepin-5(5H)-ones*** | | | 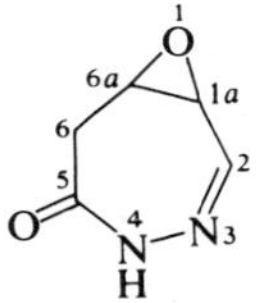 | | |
| 4-COOEt | 177–178 | Acetone/$Et_2O$ | 35 | ir, ms, pmr, uv | 70a,b |
| 2-Me-4-COOEt | 114–115 | Acetone/$Et_2O$ | 45 | ir, ms, pmr, uv | 70b |
| 4-COOEt-6*a*-Me | 78–79 | Acetone/$Et_2O$ | 40 | ir, ms, pmr, uv | 70b |
| 1*a*,6-$Me_2$-4-COOEt | 102–103 | Acetone/$Et_2O$ | 55 | ir, ms, pmr, uv | 70b |
| ***1a,2,3,4,6,6a-Hexahydrooxireno[d] [1,2]diazepin-5(5H)-one*** | | | 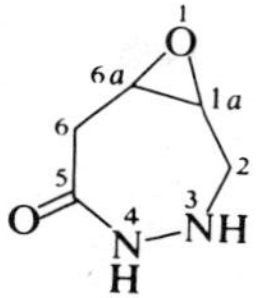 | | |
| 4-COOEt-6*a*-Me | Oil | | 35; 75 | ir, ms, pmr, uv | 70b |

TABLE I-3. —(contd.)

| Substituent | mp (°C); [bp (°C/torr)] | Solvent of Crystallization | Yield (%) | Spectra | Refs. |
|---|---|---|---|---|---|
| | | *Pyrazolo[3,4-d] [1,2]diazepines* | | | |
| ***1,6-Dihydropyrazolo[3,4-d] [1,2]diazepine*** | | | 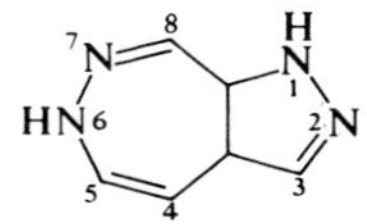 | | |
| 6-COPh-8-Me | 179–180 | PhH | 78 | ir, pmr, uv | 74 |
| ***2,3,3a,6-Tetrahydropyrazolo[3,4-d] [1,2]diazepines*** | | | 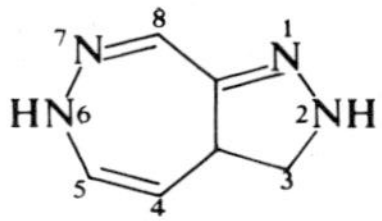 | | |
| 6-$PhSO_2$ | 148–149 | | 76 | ir, ms, pmr, uv | 69a, b |
| 2-Ac-6-$PhSO_2$ | 179–180 | $EtOH/H_2O$ | 56 | ir, pmr, uv | 74 |
| 2-Ac-6-(4-$MeC_6H_4SO_2$) | 178 | $EtOH/H_2O$ | 60 | ir, pmr, uv | 74 |
| 2-$MeOCH_2$-6-(4-Me-$C_6H_4SO_2$) | 144–145 | EtOH | 39 | ir, pmr, uv | 74 |
| 2-Ac-6-COPh-8-Me | 167–168 | EtOH | 57 | ir, pmr, uv | 74 |
| 2-Ac-6-$PhSO_2$-8-Me | 147 | MeOH | 49 | | 74 |
| 2-(4-$MeC_6H_4SO_2$)-6-COPh-8-Me | 161 | EtOH | 52 | ir, pmr, uv | 74 |
| 3,3-$Me_2$-6-COPh | [50/0.01 mm] | | | ir, pmr, uv | 69a |
| 3,3-$Me_2$-6-COOEt | 135–137 | PhH | 80 | ir, ms, pmr, uv | 69a |
| 3,3-$Me_2$-6-COOPr-*i* | 136–138 | PhH | 61 | ir, ms, pmr, uv | 69a |
| 3,3-$Me_2$-6-$PhSO_2$ | [40/0.01 mm] | | 50 | ir, ms, pmr, uv | 69a |
| 3,3-$(Me)_2$-6-(4-Me-$C_6H_4SO_2$) | 148–149 | | 49 | ir, ms, pmr, uv | 69a |
| 2-Ac-3,3-$Me_2$-6-COOEt | | | | pmr | 71 |

| Substituents | mp (°C) | Solvent | Yield (%) | Spectra | Ref. |
|---|---|---|---|---|---|
| ***2,4,5,6-Tetrahydropyrazolo[3,4-d] [1,2]diazepines*** | | | | | |
| 6-COPh | 248 | MeOH | 24 | ir, pmr, uv | 74 |
| 2-$MeOCH_2$-6-(4-Me-$C_6H_4SO_2$) | 154–155 | EtOH | 21 | ir, pmr | 74 |
| ***3,3a,6,8a-Tetrahydropyrazolo[3,4-d] [1,2]diazepines*** | | | | | |
| 6-COPh | 86–89 | | 44 | ir, ms | 69a |
| 4,8a-$Me_2$-6-COOEt | 81–83 | PhH | 18 | ir, pmr, uv | 69a |
| ***1,2,6,7-Tetrahydropyrazolo[3,4-d] [1,2]diazepin-3,8(3H,8H)-dione*** | | | | | |
| 5-Ph | > 310 | | 29 | ir | 72 |

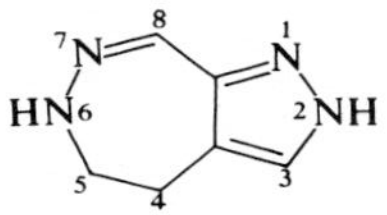

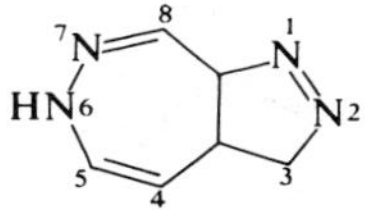

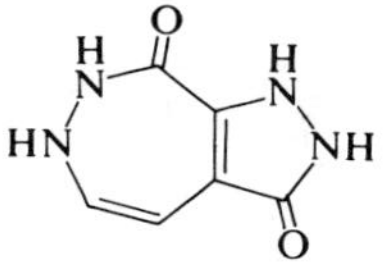

*Pyrido[3,2-d] [1,2]diazepines*

| Substituents | mp (°C) | Solvent | Yield (%) | Spectra | Ref. |
|---|---|---|---|---|---|
| ***6,9-Dihydropyrido[3,2-d] [1,2]diazepin-5(5H)-ones*** | | | | | |
| 6,8-$Me_2$ | 84 | Petr ether | 74 | ir, pmr, uv | 75 |
| 6-Me-8-Ph | 114.5–115 | Petr ether | 60 | ir, pmr | 75 |

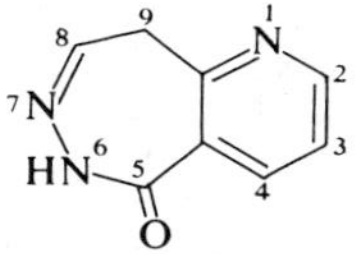

TABLE I-3. —(contd.)

| Substituent | mp (°C); [bp (°C/torr)] | Solvent of Crystallization | Yield (%) | Spectra | Refs. |
|---|---|---|---|---|---|
| ***1H-Thieno*[3,2-*d*][*1,2*]*diazepine*** | | | 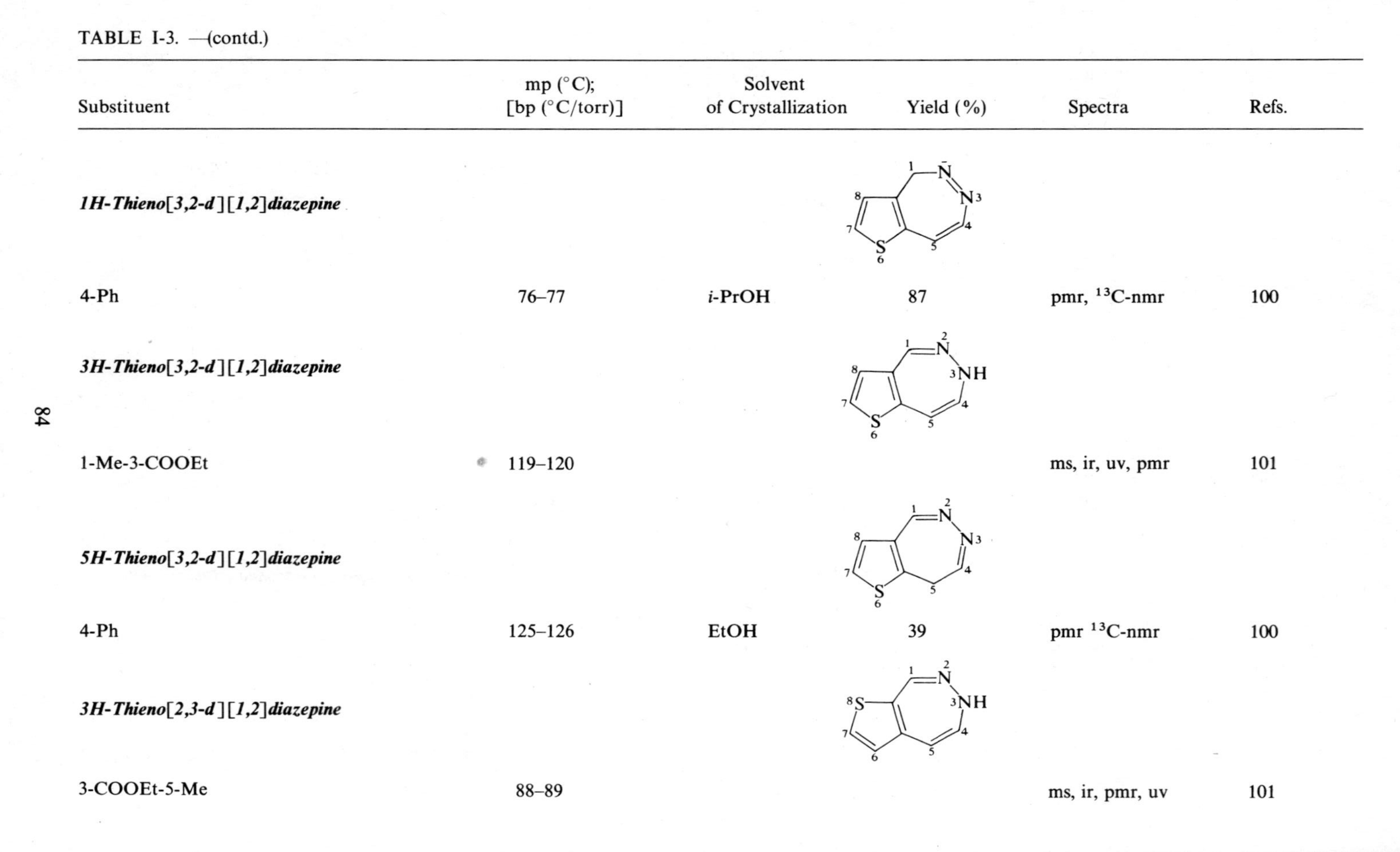 | | |
| 4-Ph | 76–77 | *i*-PrOH | 87 | pmr, $^{13}$C-nmr | 100 |
| ***3H-Thieno*[3,2-*d*][*1,2*]*diazepine*** | | | | | |
| 1-Me-3-COOEt | 119–120 | | | ms, ir, uv, pmr | 101 |
| ***5H-Thieno*[3,2-*d*][*1,2*]*diazepine*** | | | | | |
| 4-Ph | 125–126 | EtOH | 39 | pmr $^{13}$C-nmr | 100 |
| ***3H-Thieno*[2,3-*d*][*1,2*]*diazepine*** | | | | | |
| 3-COOEt-5-Me | 88–89 | | | ms, ir, pmr, uv | 101 |

| | | | | | |
|---|---|---|---|---|---|
| ***5H-Thieno[2,3-d][1,2]diazepine*** | | | 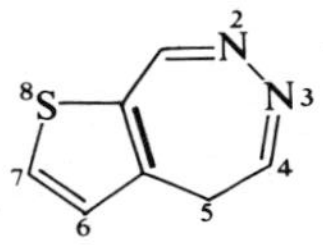 | | |
| 4-Ph | 133–134 | EtOH | 74 | pmr $^{13}$C-nmr | 100 |
| ***1,2-Dihydro-3H-thieno[2,3-d][1,2]diazepines*** | | | 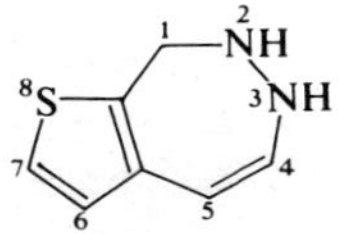 | | |
| 1-Me-3-COOEt | Oil | | 27 | ms, ir, pmr, anal | 101 |
| 1-Me-2-Ac-3-COOEt | Oil | | 91 | ms, ir, pmr, anal | 101 |
| 4-Ph | 117–119 | *i*-PrOH | 65 | pmr, $^{13}$C-nmr | 100 |
| ***4,5-Dihydro-3H-thieno[2,3-d][1,2]diazepines*** | | | 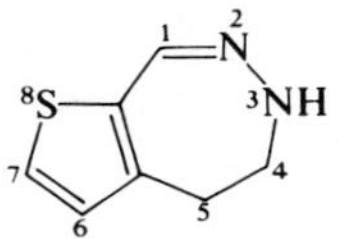 | | |
| 1-Me | Oil | | 34 | ms, ir, pmr, anal | 101 |
| 1-Me-3-Ac | 92–94 | *n*-Hexane/*i*-$Pr_2O$ | 94 | ms, ir, pmr, anal | 101 |

## F. REFERENCES

1. B. Ulbrich and H. Kisch, *Angew. Chem.*, **90**, 388 (1978).
2. (a) W. Sucrow, M. Slopianka, and V. Bardakos, *Angew. Chem.*, **87**, 551 (1975). (b) W. Sucrow and M. Slopianka, *Chem. Ber.*, **111**, 780 (1978).
3. S. F. Nelsen and G. R. Weisman, *Tetrahedron Lett.*, 2321 (1973).
4. S. F. Nelsen, V. Peacock, and G. R. Weisman, *J. Am. Chem. Soc.*, **98**, 5269 (1976).
5. S. F. Nelsen and J. M. Buschek, *J. Am. Chem. Soc.*, **96**, 6987 (1964).
6. S. F. Nelsen and G. R. Weisman, *J. Am. Chem. Soc.*, **98**, 7007 (1976).
7. G. Zinner and W. Deucker, *Arch. Pharm.*, **296**, 14 (1963).
8. (a) Ger. Offen. 2, 554,866, June 1976. (Mitsubishi Chemical Industries Ltd.): Tokyo or Jap. Kokai, 76 70820, June 1976; (b) Dutch Patent 7,507,233 (Mitsubishi Chemical Industries, Ltd.), Tokyo, December 1975 or Jap. Kokai 77 83687, July 1977; (c) Jap. Kokai 76 86489, July 1976.
9. S. S. Mathur and H. Suschitzky, *J. Chem. Soc., Perkin Trans. I*, 2474 (1975).
10. J. P. Luttringer and J. Streith, *Tetrahedron Lett.*, 4163 (1973).
11. J. Streith and G. Wolff, *Heterocycles*, **5**, 471 (1976).
12. R. Allmann and T. Debaerdemaeker, *Cryst. Struct. Commun.*, **3**, 365 (1974).
13. J. Streith, G. Wolff, and H. Fritz, *Tetrahedron*, **33**, 1349 (1977).
14. W. Flitsch, U. Krämer, and H. Zimmermann, *Chem. Ber.*, **102**, 3268 (1969).
15. H. Koga, M. Hirobe, and T. Okamoto, *Chem. Pharm. Bull.*, **22**, 482 (1974).
16. K. T. Potts and D. R. Choudhury, *J. Org. Chem.*, **42**, 1648 (1977).
17. (a) T. Tsuchiya, J. Kurita, H. Igeta, and V. Snieckus, *J. Chem. Soc., Chem. Commun.*, 640, 1974; (b) T. Tsuchiya, J. Kurita, and V. Snieckus, *J. Org. Chem.*, **42**, 1856 (1977).
18. (a) L. Garanti, A. Scandroglio, and G. Zecchi, *Tetrahedron Lett.*, 3349 (1975); (b) L. Garanti and G. Zecchi, *J. Chem. Soc., Perkin Trans. I*, 2092 (1977).
19. L. Chiodini, L. Garanti, and G. Zecchi, *Synthesis*, 603 (1978).
20. (a) T. Tsuchiya and J. Kurita, *J. Chem. Soc., Chem. Commun.*, 419 (1976); (b) T. Tsuchiya and J. Kurita, *Chem. Pharm. Bull.*, **26**, 1896 (1978).
21. T. Tsuchiya and J. Kurita, *Chem. Pharm. Bull.*, **26**, 1890 (1978).
22. T. Tsuchiya, J. Kurita, and K. Takayama, *Heterocycles* , **9**, 1549 (1978).
23. T. Tsuchiya and J. Kurita, *J. Chem. Soc., Chem. Commun.*, 936 (1974).
24. T. Shiba, K. Yamane, and H. Kato, *J. Chem. Soc., Chem. Commun.*, 1592 (1970).
25. Y. Tamura, S. Matsugashita, H. Ishibashi, and M. Ikeda, *Tetrahedron*, **29,** 2359 (1973).
26. (a) A. Frankowski, J. Streith, and H. Normant, *C. R. Acad. Sci., Paris* (C), **276**, 959 (1973); (b) A. Frankowski and J. Streith, *Tetrahedron*, **33**, 427 (1977).
27. E. Fischer and H. Kuzel, *Anal. Chem.*, **221**, 261 (1883).
28. N. S. Gill, K. B. James, F. Lions, and K. T. Potts, *J. Am. Chem. Soc.*, **74**, 4923 (1952).
29. A. Sammour, A. Raouf, M. Elkasaby, and M. A. Ibrahim, *Acta. Chim. Acad. Sci. Hung.*, **78**, 399 (1973).
30. A. Sammour and M. El-Hashash, *Egypt. J. Chem.*, **16**, 381 (1973).
31. T. Sasaki, K. Kanematsu, Y. Yukimoto, and E. Kato, *Synth. Commun.*, **3**, 249 (1973).
32. T. Tsuchiya, M. Enkaku, and H. Sawanishi, *Heterocycles*, **9**, 621 (1978).
33. (a) R. W. Morrison, Jr., W. R. Mallory, and V. L. Styles, Presentation at the 174th National Meeting of the American Chemical Society, Chicago, Aug. 28–Sept. 2, 1977, ORGN 9; (b) European Patent Appl. 78,100,435.3 February 1979 (Wellcome Foundation, Ltd.).
34. J. T. Sharp and P. B. Thorogood, *J. Chem. Soc., Chem. Commun.*, 1197 (1970).
35. A. A. Reid, J. T. Sharp, H. R. Sood, and P. B. Thorogood *J. Chem. Soc., Perkin Trans. I*, 2543 (1973).
36. R. O. Gould and S. E. B. Gould, *J. Chem. Soc., Perkin Trans. II*, 1075 (1974).
37. V. I. Bendall, *J. Chem. Soc., Chem. Commun.*, 823 (1972).
38. A. A. Reid, J. T. Sharp, and S. J. Murray, *J. Chem. Soc., Chem. Commun.*, 827 (1972).
39. A. A. Reid, H. R. Sood, and J. T. Sharp, *J. Chem. Soc., Perkin Trans. I*, 362 (1976).

40. (a) J. Körösi and T. Láng, *Chem. Ber.*, **107**, 3883 (1974); (b) J. Körösi, T. Láng, E. Komlos, and L. Erdelyi, U.S. Patent 3,736,315, May 1973 (Egyesult Gyogyszeres Tapszergyar, Budapest, Hungary).
41. Hungarian Patent 155,572, Dez. 1966 (Gyógyszerkutató Intézet); *Chem. Abstr.*, **70**, 115026a (1969).
42. A. Neszmélyi, E. Gáez-Baitz, G. Horváth, T. Láng, and J. Körösi, *Chem. Ber.*, **107**, 3894 (1974).
43. G. Zólyomi, O. Bánfi, T. Láng, and J. Körösi, *Chem. Ber.*, **107**, 3904 (1974).
44. M. Lampert-Sreter, *Acta Chim. Acad. Sci. Hung.*, **83**, 115 (1974).
45. E. Schmitz and R. Ohme, *Chem. Ber.*, **95**, 2012 (1962).
46. Y. Tamura, J. Minamikawa, H. Matsushima, and M. Ikeda, *Synthesis*, 159 (1973).
47. J. Streith and C. Fizet, *Tetrahedron Lett.*, 3297 (1977).
48. C. Van der Stelt, P. S. Hofman, and W. T. Nauta, *Recl. Trav. Chim.*, **84**, 633 (1965).
49. M. D. Nair and P. A. Malik, *Indian J. Chem.*, **10**, 341 (1972).
50. J. Gottlieb, *Chem. Ber.*, **32**, 958 (1899).
51. H. Wölbling, *Chem. Ber.*, **38**, 3845 (1905).
52. A. Lieck, *Chem. Ber.*, **38**, 3853 (1905).
53. M. Buu-Hoi, *Compt. Rend.*, **209**, 321 (1939).
54. A. Rose and N. P. Buu-Hoi, *J. Chem. Soc.*, 2205 (1968).
55. M. Flammang and C. C. Wermuth, *Eur. J. Med. Chem.* (*Chim. Ther.*), **12**, 121 (1977).
56. L. Legrand and N. Lozach, *Bull. Soc. Chim. Fr.*, 2237 (1970).
57. M. Flammang, *C. R. Acad. Sci., Paris* (*C*), **286**, 671 (1978).
58. J. O. Halford, R. W. Raiford, Jr., and B. Weissmann, *J. Org. Chem.*, **26**, 1898 (1961).
59. K. Nagarajan, J. David, and R. K. Shah, *J. Med. Chem.*, **15**, 1091 (1972).
60. C. C. Wermuth and M. Flammang, *Tetrahedron Lett.*, 4293 (1971).
61. (a) M. Flammang and C. C. Wermuth, *Eur. J. Med. Chem.* (*Chim. Ther.*), **11**, 83 (1976); (b) Fr. Patent 2,104,998, April 1972 (Synthelabo S.A.); (c) Fr. Patent 2,085,645, December 1971 (Synthelabo S.A.).
62. A. Sotiriadis, P. Catsoulacos, and D. Theodoropoulos, *J. Heterocycl. Chem.*, **11**, 401 (1974).
63. M. Flammang, *C. R. Acad. Sci, Paris* (*C*), **283**, 593 (1976).
64. G. Rosen and F. D. Popp, *J. Heterocycl. Chem.*, **6**, 9 (1969).
65. G. Cignarella, R. Cerri, F. Savelli, and A. Maselli, *J. Heterocycl. Chem.*, **14**, 465 (1977).
66. (a) G. M. Rosen, Dissertation "A study of 2,3-benzodiazepines." b) *Diss. Abstr.*, **32**, 1463 (1971).
67. W. F. Whitmore and R. C. Cooney, *J. Am. Chem. Soc.*, **66**, 1237 (1944).
68. T. Sasaki, K. Kanematsu, and T. Kataoka, *J. Org. Chem.*, **40**, 1201 (1975).
69. (a) G. Kiehl, J. Streith, and G. Taurand, *Tetrahedron*, **30**, 2851 (1974); (b) G. Taurand and J. Streith, *Tetrahedron Lett.*, 3575 (1972).
70. (a) T. Tsuchiya, H. Arai, H. Hasegawa, and H. Igeta, *Tetrahedron Lett.*, 4103 (1974); (b) T. Tsuchiya, H. Arai, H. Hasegawa, and H. Igeta, *Chem. Pharm. Bull.*, **25**, 2749 (1977).
71. Y. L. Chow, J. Streith, and G. Taurand, *Org. Magn. Resonance*, **5**, 155 (1973).
72. M. Hieda, K. Omura, and S. Yurugi, *J. Pharm. Soc. Japan*, **92**, 1327 (1972).
73. J. Streith, G. Kiehl, and H. Fritz, *Tetrahedron Lett.*, 631 (1974).
74. J. R. Frost and J. Streith, *J. Chem. Soc., Perkin Trans. I*, 1297 (1978).
75. D. E. Ames and W. D. Doddis, *J. Chem. Soc., Perkin Trans. I*, 705 (1972).
76. M. R. Attwood, C. H. Hassall, R. W. Lambert, G. Lawton, and S. Redhaw, U.S. Patent 4,512,924, April 1985.
77. L. Garanti and G. Zecchi, *J. Heterocycle. Chem.*, **16**, 1061 (1979).
78. L. Garanti, G. Testoni, and G. Zecchi, *Synthesis*, 380 (1979).
79. L. Garanti and G. Zecchi, *J. Chem. Soc., Perkin Trans. I*, 1195 (1979).
80. A. Padwa and S. Nahm, *J. Org. Chem.*, **44**, 4746 (1979).
81. A. Padwa and S. Nahm, *J. Org. Chem.*, **46**, 1402 (1981).
82. K. Saito, *Chem. Lett.* (*Japan*), 463 (1983).
83. T. Tsuchiya, K. Takayama, and J. Kurita, *Chem. Pharm. Bull.*, **27**: 10, 2476 (1979).
84. T. Tsuchiya and J. Kurita, *J. Chem. Soc., Chem. Commun.*, 803 (1979).

85. T. Tsuchiya and J. Kurita, *Chem. Pharm. Bull.*, **86**, 1890 (1978).
86. T. Tsuchiya, M. Enkaku, and H. Sawanishi, *Heterocycles*, **12**, 11 (1979).
87. T. Tsuchiya, M. Enkaku, and S. Okajima, *Chem. Pharm. Bull.*, **29**, 3173 (1981).
88. J. Kurita, M. Enkaku, and T. Tsuchiya, *Chem. Pharm. Bull.*, **31**, 3684 (1983).
89. R. W. Morrison, Jr., W. R. Mallory, and V. L. Styles, U.S. Patent 4,235,905, November 1980.
90. S. Lida and T. Mukai, *Heterocycles*, **11**, 401 (1978).
91. J. Kurita, M. Enkaku, and T. Tsuchiya, *Chem. Pharm. Bull.*, **30**, 3764 (1982).
92. M. Enkaku, J. Kurita, and T. Tsuchiya, *Heterocycles*, **16**, 11 (1981).
93. J. Kovosi et al., U.S. Patent 4,423,044, December 1983.
94. A. Neszmelyi, T. Láng, and J. Kövösi, *Acta Chim. Hung.*, **114**, 293 (1983).
95. M. Flammang, *C. R. Acad. Sci. Paris (C)*, **290**, 349 (1980).
96. M. Flammang and C.-G. Wermuch, *C. R. Acad. Sci. Paris (C)*, **290**, 361 (1980).
97. D. P. Munro and J. T. Sharp, *Tetrahedron Lett.*, **23**, 345 (1982).
98. D. P. Munro and J. T. Sharp, *J. Chem. Soc., Perkin Trans. I*, 1133 (1984).
99. K. R. Motion, D. P. Munro, J. T. Sharp, and M. D. Walkinshaw, *J. Chem. Soc., Perkin Trans. I*, 2027 (1984).
100. D. P. Munro and J. T. Sharp, *J. Chem. Soc., Perkin Trans. I*, 1718 (1980).
101. T. Tsuchiya, H. Sawanish, M. Enkaku, and T. Hirai, *Chem. Pharm. Bull.*, **29**, 1539 (1981).
102. T. Tsuchiya and J. Kurita, *Chem. Pharm. Bull.*, **28**, 1842 (1980).
103. J. Kurita, M. Enkaku, and T. Tsuchiya, *J. Chem. Soc., Chem. Commun.*, 990 (1982).
104. R. Allman, T. Debaerdemaeker, G. Kiehl, J. P. Luttringer, T. Tschamber, A. Wolff, and J. Streith, *Justus Liebigs Ann. Chem.*, 1361 (1983).

CHAPTER II

# Bicyclic 1,3-Diazepines

**R. Ian Fryer**

*Department of Chemistry,*
*Rutgers, State University of New Jersey, Newark,*
*New Jersey*

and

**A. Walser**

*Chemical Research Department,*
*Hoffmann-La Roche Inc., Nutley,*
*New Jersey*

## INTRODUCTION

This chapter reviews the synthesis and chemistry of 1,3-diazepines of structure **1** with an additional ring of any size fused to side *a*, *c*, *d*, or *e*. The added rings can be carbocyclic or heterocyclic.

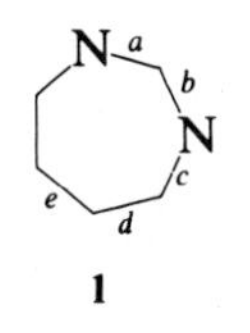

**1**

Sections A through D of the chapter deal with [*a*]-, [*c*]-, [*d*]-, and [*e*]-fused 1,3-diazepines. Each section is subdivided according to the nature of the added ring and the subsections, in general, are listed in alphabetical order. An exception is made for the [*d*]- and [*e*]-fused systems in which the benzodiazepines constitute the largest and most important class of compounds. These derivatives are given preference over the other ring systems and are discussed first in Sections C and D.

When necessary, individual classes of compounds are further divided into groups according to the degree of saturation, with the fully unsaturated species being given precedence. In some instances, the paucity of published data does not warrant separate subsections for the reactions of the compounds under discussion. Bridged seven-membered rings are not covered. All compounds are named and numbered according to current *Chemical Abstract* rules.

1,3-Diazepines have been reviewed by G. Hornyak, K. Lampert, and G. Simig [*Kem. Kozlem.*, **33**, 81 (1970)]; G. A. Archer and L. H. Sternbach [*Chem. Rev.*, **68**, 747 (1968)]; F. D. Popp and A. J. Noble (in *Advances in Heterocyclic Chemistry*, Vol. 8, A. R. Katritzky and A. J. Boulton, Eds., Academic Press, New York and London, 1967, p. 21); J. A. Moore and E. Mitchell (in *Heterocyclic Compounds*, Vol. 9, R. C. Elderfield, Ed., Wiley, New York, 1967, p. 224); A. Nawojski [*Wiadomości Chem.*, **23**, 673 (1964)]; and G. De Stevens [*Rec. Chem. Progr.*, **23**, 105 (1962)].

## A. [*a*]-FUSED [1,3]DIAZEPINES

Only seven bicyclic ring systems with the general structure **A** have been reported in the literature. These are discussed in alphabetical order in the subsections that follow.

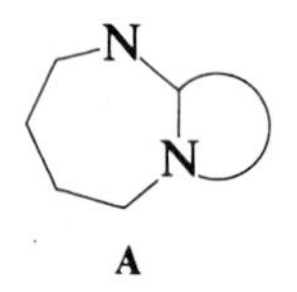


**A**

### 1. 1,3-DIAZETO[1,2-*a*] [1,3]DIAZEPINES

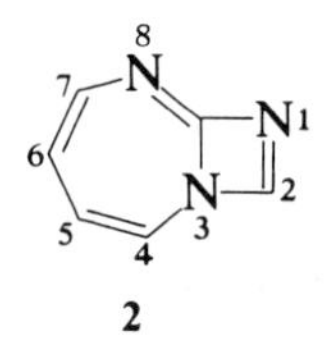


**2**

The only member of this ring system that is described in the literature is the disubstituted hexahydro derivative **5**.[1] This diazeto diazepine was prepared in good yield (Eq. 1) by reaction of 1,4-diaminobutane with *N*-phenyldichloromethanimine (**3**). Compound **5** is believed to form by condensation of the diazepine intermediate **4** with a second molecule of dichloroimine. The structural assignment for **5** was based on spectroscopic data.

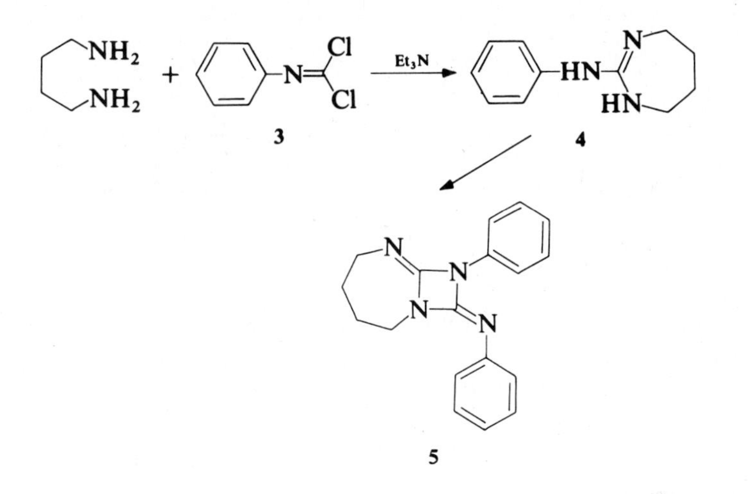

(1)

## 2. IMIDAZO[1,2-*a*] [1,3]DIAZEPINES

### 2.1. Synthesis

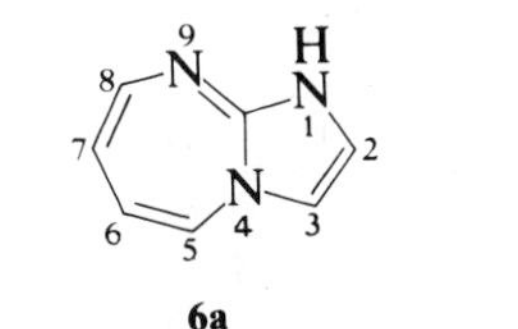

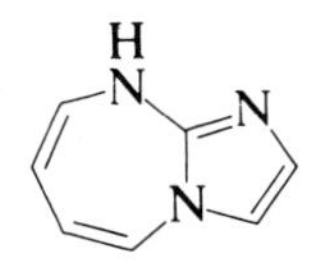

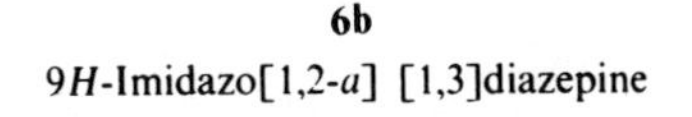

**6a**
1*H*-Imidazo[1,2-*a*] [1,3]diazepine

**6b**
9*H*-Imidazo[1,2-*a*] [1,3]diazepine

Partially reduced derivatives of both the 1*H* (**6a**) and 9*H* (**6b**) imidazo[1,2,-*a*] [1,3]diazepines are known, but none of the five possible tautomers of the parent ring system has been reported in the literature.

The 2,3,5,6,7,8-hexahydro-1*H*-imidazo[1,2,-*a*] [1,3]diazepine **8** was prepared in 63% yield (Eq. 2) by dehydrohalogenation of the 2-chloroethylamidine **7**.[2] The structural assignment was based on both analytical data and on analogous cyclizations.

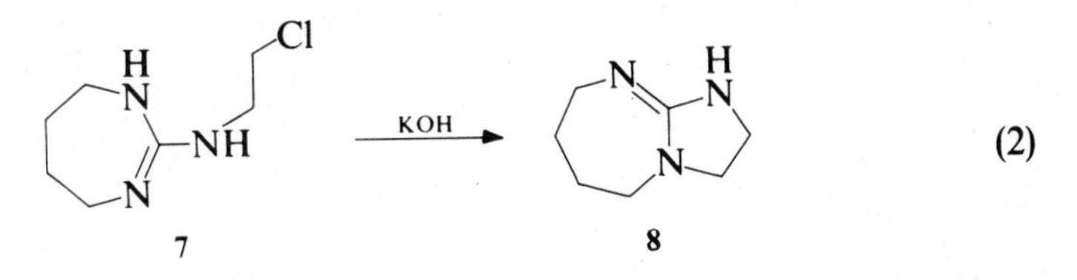

(2)

1-Substituted analogs **10** of the same ring system were shown to be accessible (Eq. 3) by intramolecular alkylation of the imidazolidine **9**.[3,93]

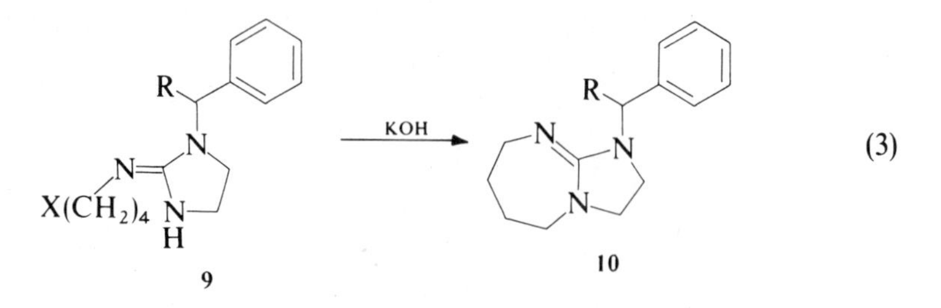


Compound **13**, the 2,2-diphenyl-substituted analog of **6a**, was prepared in low yield[94] by treatment of **11** with Meerwein reagent in methylene chloride, followed by heating with ethanolic ammonia in a sealed tube.

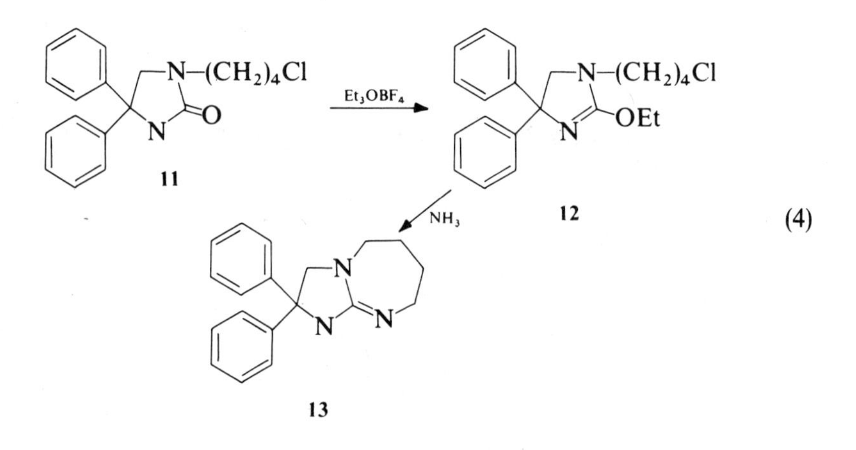


Compounds **16**, bearing a phenyl substituent at position 9, were obtained in low yields[4] by a double alkylation of the 2-anilinoimidazolines **14** with the 1,4-dibromobutanes **15** (Eq. 5). The stereochemistry of the 6,7-disubstituted derivatives was not discussed.

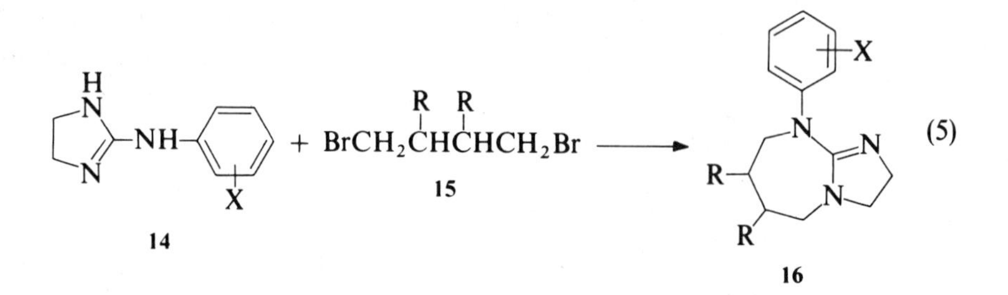


The imidazo[1,2-*a*] [1,3]diazepine **18** was synthesized[5] in low yield (Eq. 6) as a mixture of two diastereomers by cyclization of the anhydrodipeptide **17** in an aqueous medium at pH 5 and 100°C.

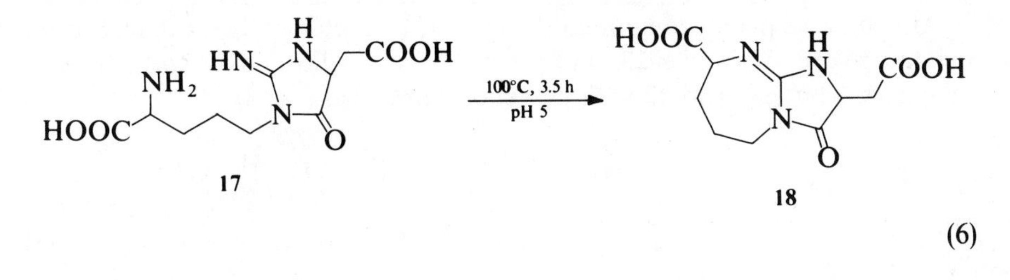


(6)

## 2.2. Reactions

Nitration of 2,3,5,6,7,8-hexahydro-1*H*-imidazo[1,2-*a*] [1,3]diazepine **8** with nitric acid in acetic anhydride (Eq. 7) afforded a quantitative yield of the nitrate salt of an *N*-nitro compound to which structure **19** was assigned.[2a]

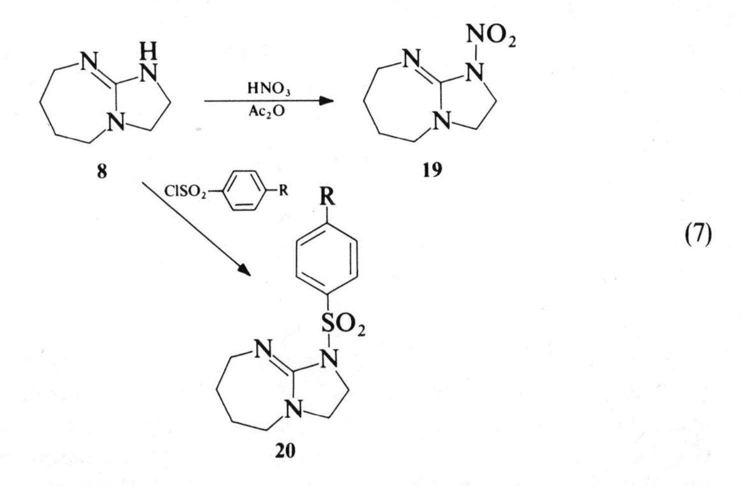


(7)

Reaction of **8** with *p*-substituted benzenesulfonyl chlorides led to the *N*-sulfonyl derivatives **20**.[6] The possibility that the nitro group in **19** and the sulfonyl group in **20** are in fact attached to the nitrogen at the 9-position instead of the 1-position has not been excluded.

## 2.3. Imidazo[1,5-*a*] [1,3]diazepines

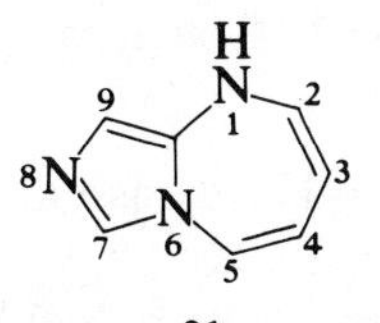


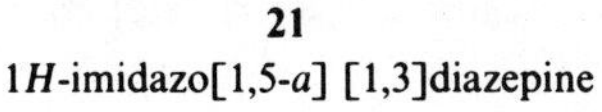
**21**
1*H*-imidazo[1,5-*a*] [1,3]diazepine

Although the parent compound represented by tautomer **21** is still unknown, the tetrahydro derivative **23** was prepared[7] by cyclization of the 5-amino-imidazole-1-butyric acid **22** with polyphosphoric acid (Eq. 8).

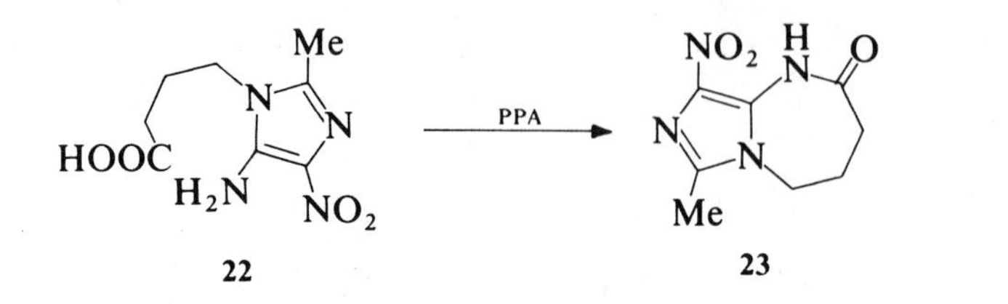


(8)

## 3. PYRIDO[1,2-*a*] [1,3]DIAZEPINES

### 3.1. Synthesis

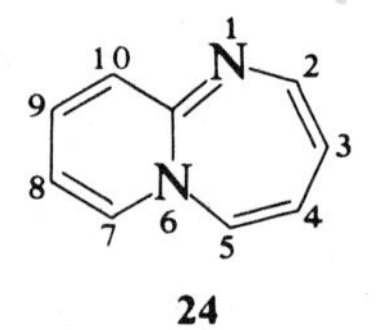


A few substituted analogs of the parent ring system **24** have been described in the patent literature.[8] Compounds **27** were isolated as minor products from the cyclization of the enamines **25** with phosphorylchloride–polyphosphoric acid (Eq. 9). In general, the pyridopyrimidines **26** were the major products of the reaction, and only in the case of $R_1$ = Me and $R_2$ = H was the pyridodiazepine obtained in good yield. Dehalogenation of **27** ($R_1$, $R_2$ = H) by catalytic hydrogenation over palladium on carbon led to compound **28** ($R_1$, $R_2$ = H). Compound **28** ($R_1$ = Me, $R_2$ = H) has been claimed[8] to be formed by a mysterious reductive cyclization of the enamine **25** ($R_1$ = Me, $R_2$ = H) upon treatment with phosphorylbromide–polyphosphoric acid.

### 3.2. Reactions

Nitration of **28** ($R_1$ = Me, $R_2$ = H) was assumed to occur at the 5-position to give **29** as the product.

Alkaline hydrolysis of the ester **27** ($R_1$ = Me, $R_2$ = H) gave the corresponding acid **30**.

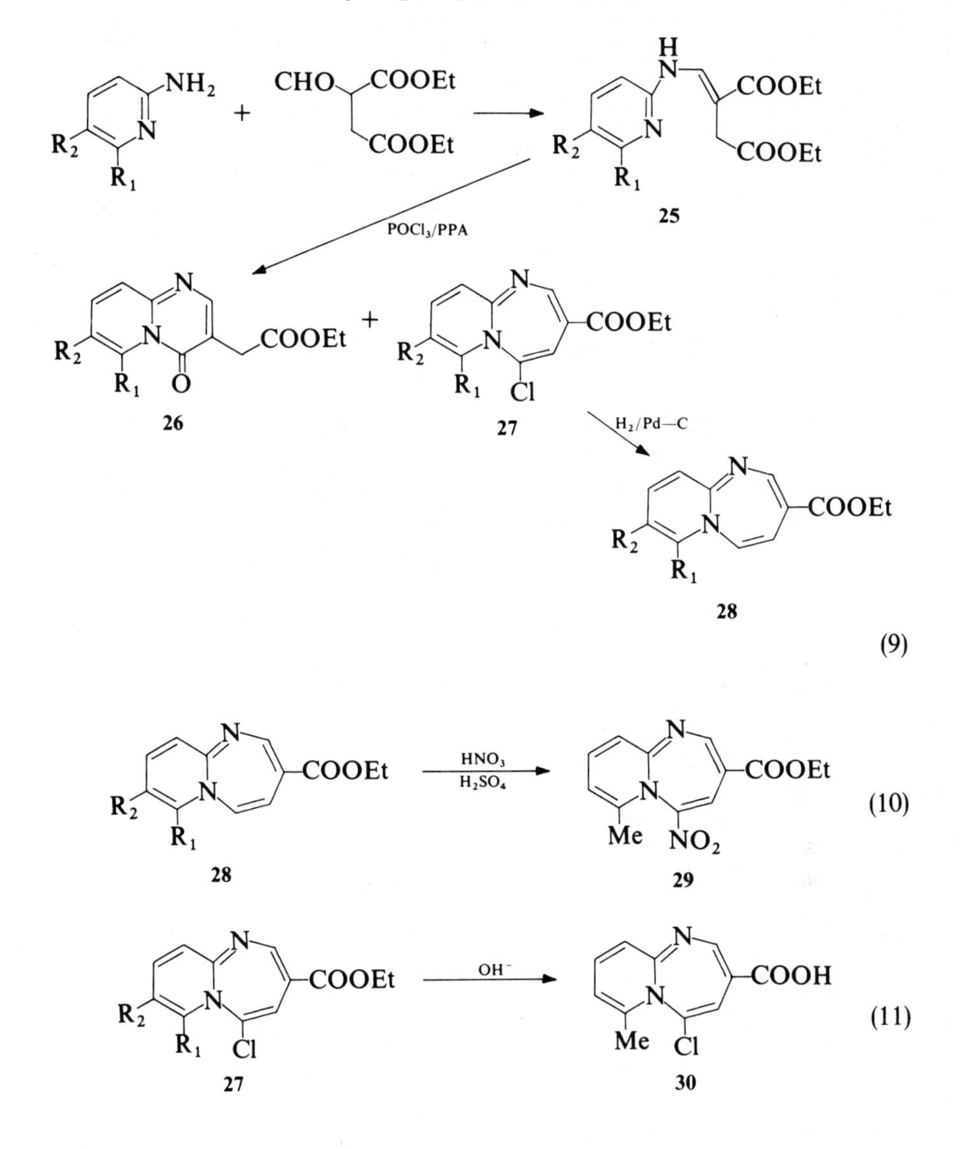


### 3.3. Dihydropyrido[1,2-*a*] [1,3]diazepines

Thermal cyclization of the enamines **25** led to the isolation of both the substituted 4,5-dihydropyrido[1,2-*a*] [1,3]diazepin-5-ones **31** and the pyridopyrimidines **24** (Eq. 12).

Other tautomeric forms of **31** have not been ruled out as alternative structures. It is quite likely that the 5-chloro derivatives **27** described in Section 3.1 arise by subsequent reaction of the initially formed lactams **31** with phosphorylchloride.

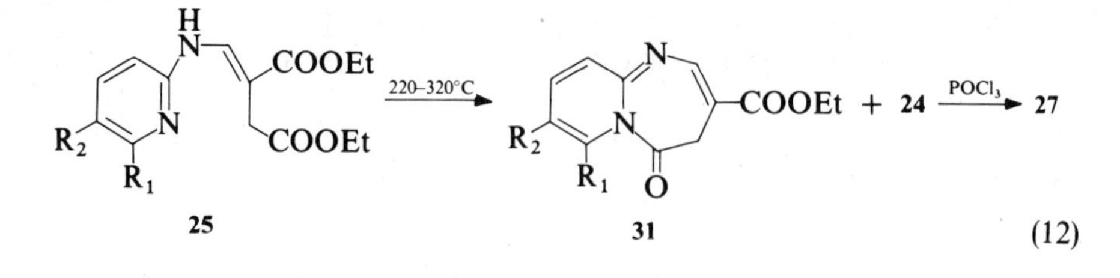


### 3.4. Tetrahydropyrido[1,2-*a*] [1,3]diazepines

Fozard and Jones[9] reported the preparation of the quaternary salt **34** by means of a thermal ring closure of the 4-bromobutyramide **33** formed upon acylation of **32** (R = H) with 4-bromobutyryl bromide (Eq. 13). The structure of **34** was corroborated by ring opening to **36**, which was characterized as a picrate salt. The same picrate was obtained from the product of alkylation of 2-pyridinamine with 4-bromobutanoic acid ethyl ester.

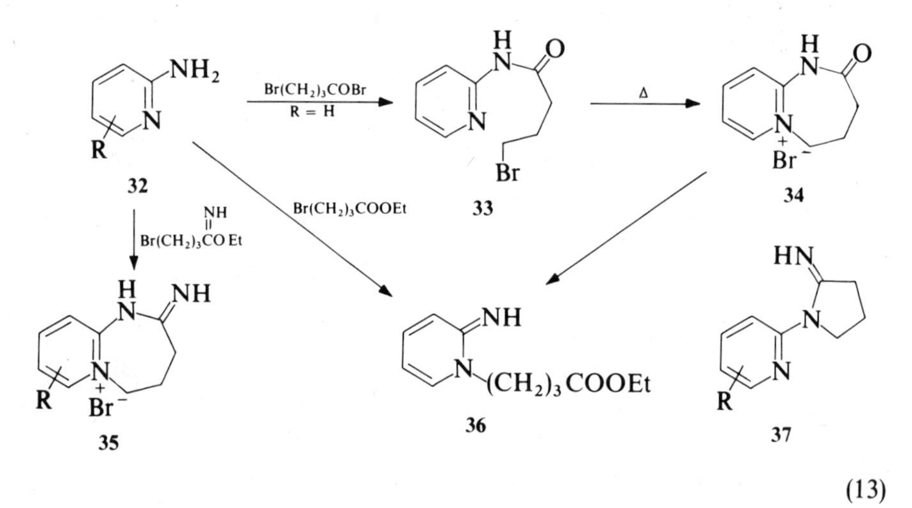


The corresponding imides **35** have been claimed[10] to be formed in moderate yields during the condensation of 4-bromobutyrimidate with the 2-pyridinamines **32**. Although nmr spectroscopy was used to support the structures **35**, the alternative 1-(2-pyridyl)-2-pyrrolidinimine structures **37** have not been rigorously excluded. Preference for the formation of a five-membered over a seven-membered ring has been observed by Ott and Hess[11] in the reaction of 2-pyridinamines with tetraalkyl succinic acid anhydride at high temperature. They considered the pyrido[1,2-*a*] [1,3]diazepine structures for the products of this condensation, but were able to show that these compounds were in fact *N*-(2-pyridyl) succinimides.

According to a patent,[12] the 9-phenyl derivative **39** was obtained by reaction of the *N*-(4-chlorophenyl)pyridone **38** with 1,4-butanediamine in boiling 2-ethoxyethanol (Eq. 14).

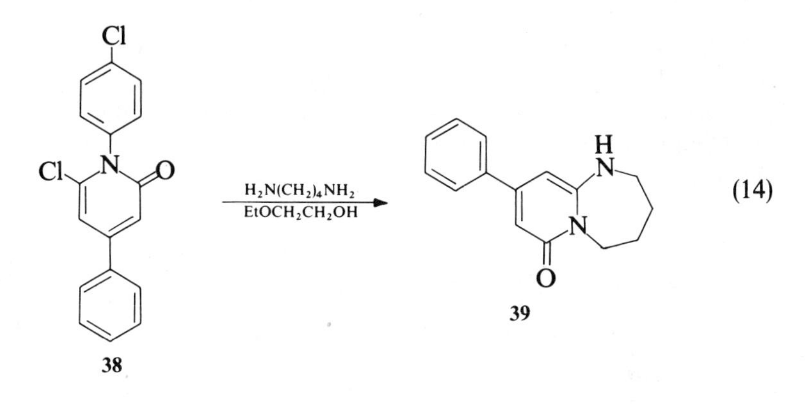


# 4. PYRIMIDO[1,2-*a*] [1,3]DIAZEPINES

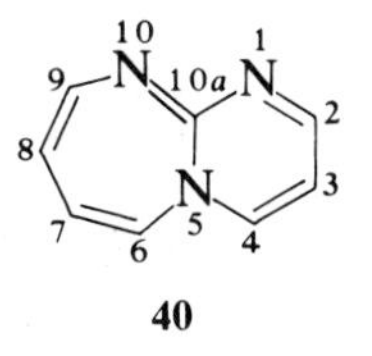


The parent ring system **40** has not been described, and the hexahydro derivative **42** appears to be the only representative of this class of compounds so far prepared. Compound **42** was isolated in low yield[13] from the alkylation of the sodium salt of 2-amino-4-hydroxy-6-methylpyrimidine **41** with 1,4-dibromobutane (Eq. 15).

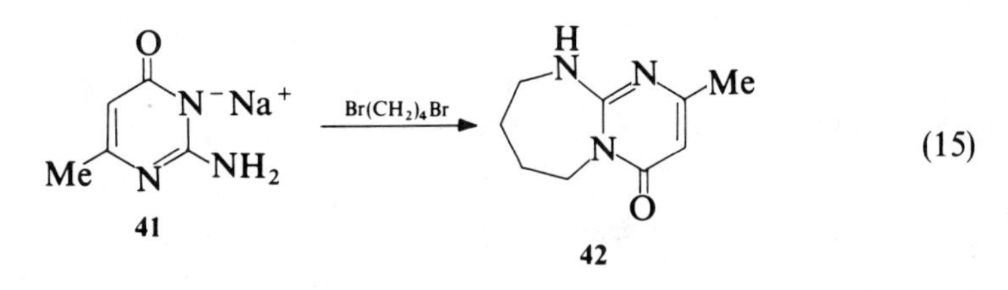


## 4.1. Pyrimido[1,6-*a*] [1,3]diazepines

Again, fully unsaturated compounds of the parent ring system **43** are unknown. The structures of the 1,10*a*-dihydro derivatives **45** (X = O, $CH_2$) were

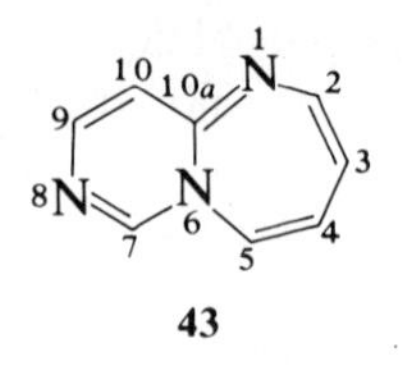


**43**

assigned to the products obtained[14] in very low yield by reaction of 4-aminopyrimidines **44** (X = O, $CH_2$) with dimethyl acetylenedicarboxylate (Eq. 16).

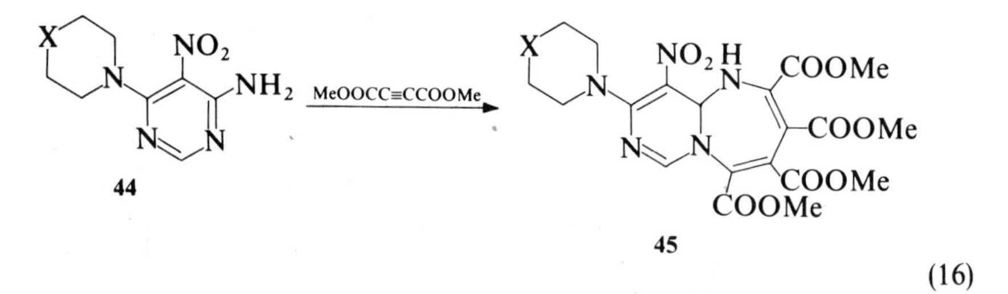


(16)

Cyclization of compound **46** in boiling acetic anhydride (Eq. 17) has been reported to yield the trione **47**.[15a,b]

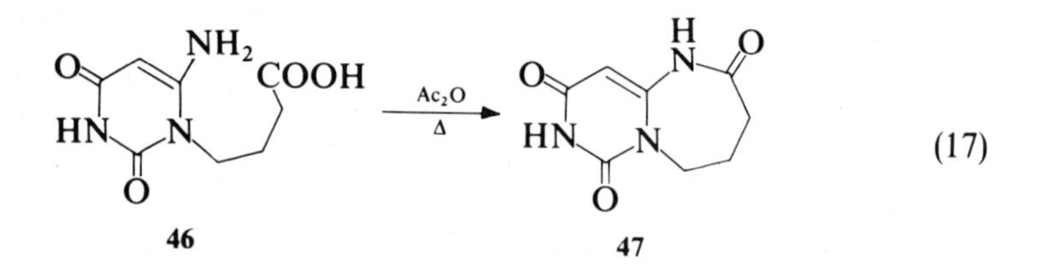


(17)

## 5. PYRROLO[1,2-*a*] [1,3]DIAZEPINES

No derivatives of the five possible tautomers, **48a–48e** of the parent ring system have been described in the literature.

### 5.1. Tetrahydropyrrolo[1,2-a] [1,3]diazepines

The unsubstituted tetrahydropyrrolodiazepine **50** ($R_1 = R_2 = R_3 =$ H) was reportedly obtained by Beckmann rearrangement of the oxime **49** (Eq. 18).[16]

The substituted analog **50** ($R_1$, $R_2$ = Me, $R_3$ = CN), was prepared by ring closure of the pyrrole derivative **51**, via intramolecular alkylation (Eq. 18).[17] The formation of a five-membered pyrolidine ring appears to be precluded in this

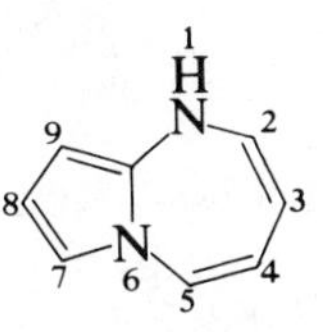

**48a**
1*H*-Pyrrolo[1,2-*a*] [1,3]diazepines

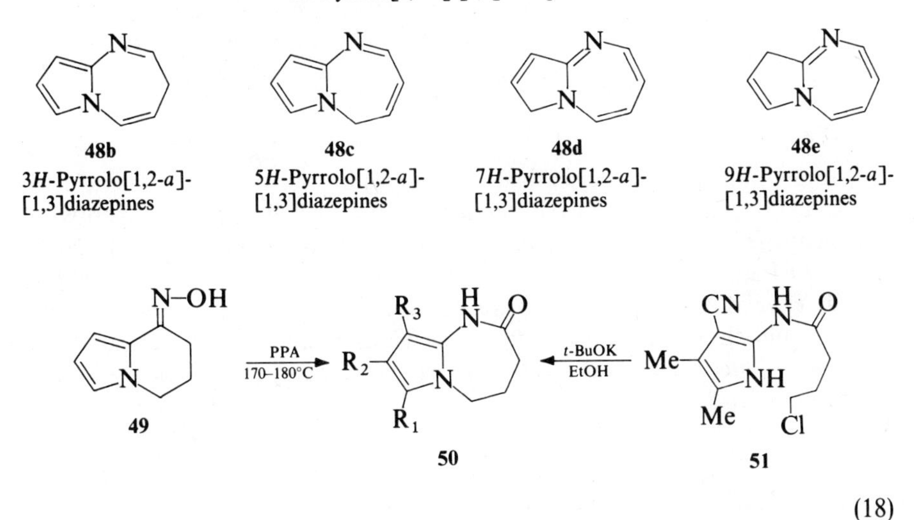


(18)

alkylation because anion formation occurs preferentially on the pyrrole nitrogen.

The substituted dihydropyrrolodiazepine **54**,[96] was obtained by the catalytic hydrogenation of the cyclopropene ring of **53**, which in turn was prepared by treatment of **52** with an excess of diazomethane at room temperature (Eq. 19).

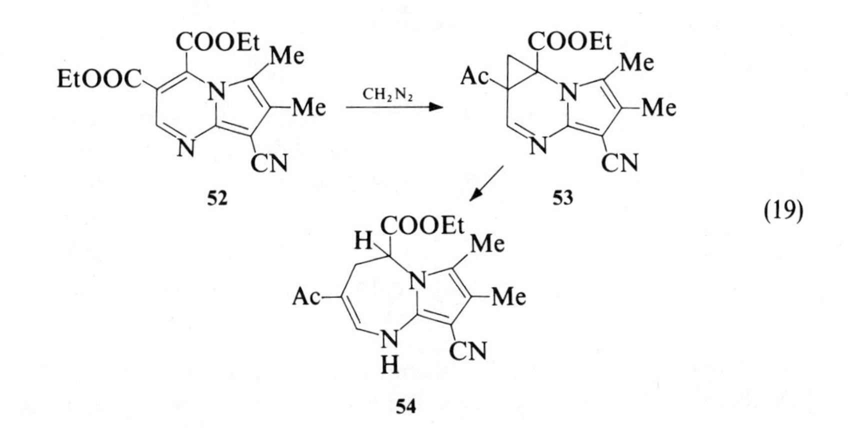


(19)

### 5.2. Hexahydropyrrolo[1,2-*a*] [1,3]diazepines

All the reported hexahydro derivatives of this ring system contain a saturated pyrrole ring. Thus, the hexahydro-2*H*-pyrrolo[1,2-*a*] [1,3]diazepine **56** was prepared[18] in 80% yield by acid-catalyzed ring closure of 1-(4-aminobutyl)-2-pyrrolidinone (**55**) at elevated temperature (Eq. 20).

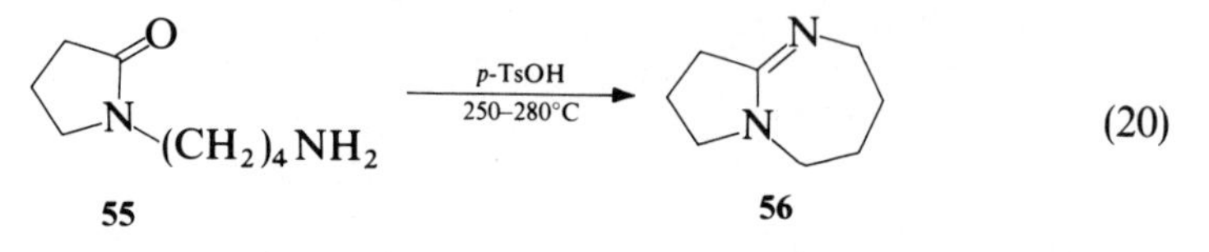


Ducker and Gunter[19] described the interesting formation of the related hexahydropyrrolodiazepine **61** by dimerization and hydration of 4-methyl-3-pentenenitrile (**57**). Treatment of **57** with concentrated sulfuric acid at 0–20°C afforded **61** in 23% yield. These authors proposed that dimerization of the initially formed carbonium ion **58** would lead to the 1,6-diazacyclodecane **59**, which would then undergo hydration and transannular ring closure to form the final product (Eq. 21). Hydrolysis of **61** catalyzed by silver oxide gave the 1,6-diazecin-2,7-dione **60**.

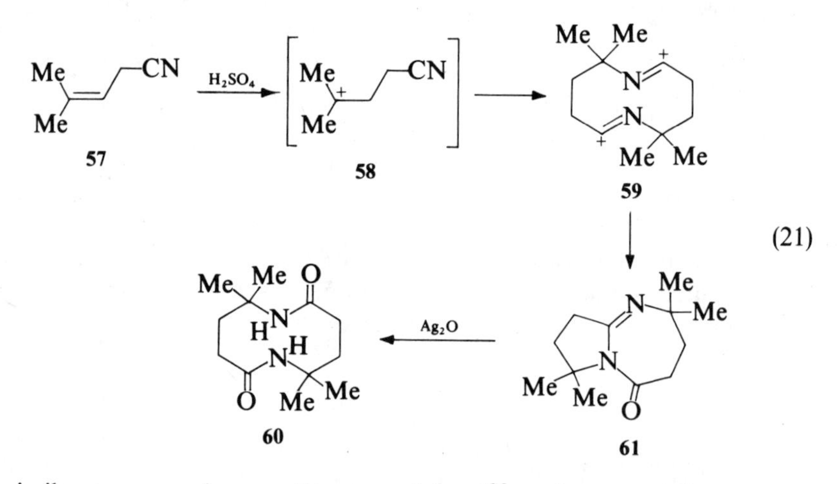


A similar transannular reaction was claimed[20] to be responsible for the interconversion (Eq. 22) of the 1,6-diazecin-2,7-dione **62** and the pyrrolidinone **64**. Neither the intermediate pyrrolodiazepine **63** nor its dehydration product **65** were isolated, however.

### 5.3. Octahydropyrrolo[1,2-*a*] [1,3]diazepines

Octahydropyrrolodiazepines bearing a carbonyl group at the 5- or 7-position have been prepared by two routes. The first method involves catalytic reduction

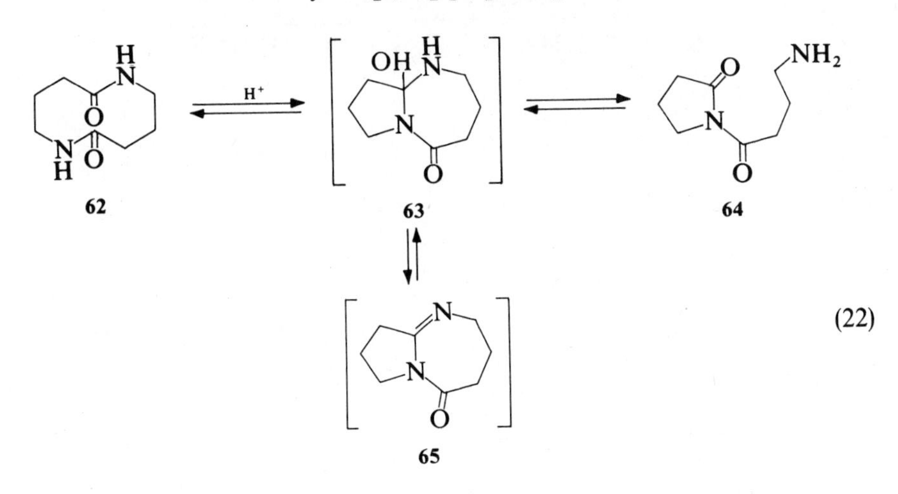


of the corresponding tetrahydro derivative, as exemplified by the hydrogenation of amidine **61** over platinum[19] to give **66** in 65% yield (Eq. 23). In the second method, a 4-ketocarboxylic acid is condensed with 1,4-diaminobutane. Wollweber[21] used this route to obtain **68** ($R_2$ = Me) from the reaction of ethyl levulinate **67** ($R_1$ = Et, $R_2$ = Me) with 1,4-diaminobutane at 100°–150°C (Eq. 24). The same principle was also used for the synthesis of compounds **68** ($R_2$ = aryl).[22] In this instance a mixture of the ketoacid **67** ($R_1$ = H, $R_2$ = aryl) and 1,4-diaminobutane was heated in refluxing chlorobenzene.

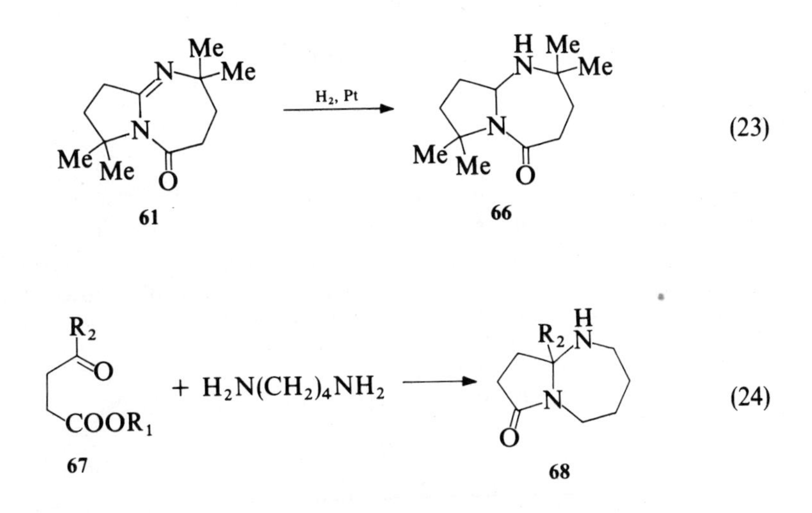


Other substituted analogs of **68** are claimed in the patent literature,[23] but no physical data are given.

# 6. THIAZOLO[3,2-*a*] [1,3]DIAZEPINES

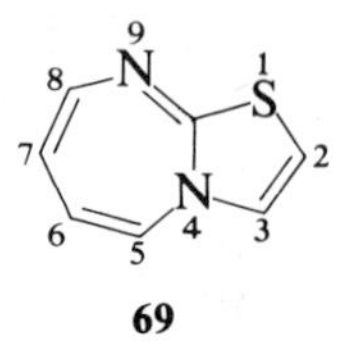


**69**

Although the fully unsaturated parent compound **69** has not been described, a number of reduced derivatives have been prepared.

## 6.1. Tetrahydrothiazolo[3,2-*a*] [1,3]diazepines

Chadha and coworkers[24] have applied the standard thiazole synthesis (i.e., the condensation of a 2-haloketone with a thioamide) in converting the 1,3-diazepine **70** into a series of 5,6,7,8-tetrahydrothiazolo[3,2-*a*][1,3]diazepines **72**. A later publication[25] reported modification of this synthesis by forming the α-bromoketone in situ with *N*-bromosuccinimide in benzene. If the condensation of the bromoketone with the thiourea **70** was performed under milder conditions (in acetone at room temperature), the intermediate hydroxythiazolidines **71** could be isolated[26] and subsequently dehydrated to **72** under more vigorous, acid-catalyzed conditions (Eq. 25).

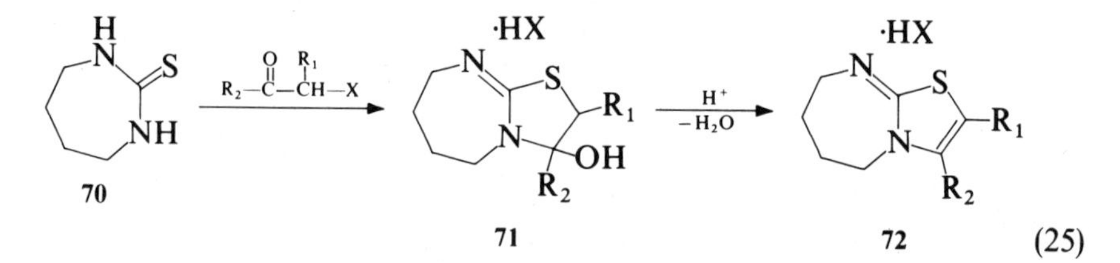


(25)

The same reaction scheme was used by Wei and Bell[27a,b] for the preparation of the hydroxythiazolidine **71** ($R_1 = CH_2COOH$, $R_2 = 4\text{-ClC}_6H_4$). The thiourea **70** was heated under reflux with the appropriate bromoketone in acetic acid. In this instance dehydration required prolonged boiling in acetic acid. The thiazolodiazepines **72** were generally isolated and characterized as salts.

## 6.2. Hexahydrothiazolo[3,2-*a*] [1,3]diazepines

### *6.2.1. Synthesis*

The 2,3,5,6,7,8-hexahydro derivative **73** was prepared[28] by reaction of the 1,3-diazepine-2-thione derivative **70** with 1,2-dibromoethane in boiling ethanol

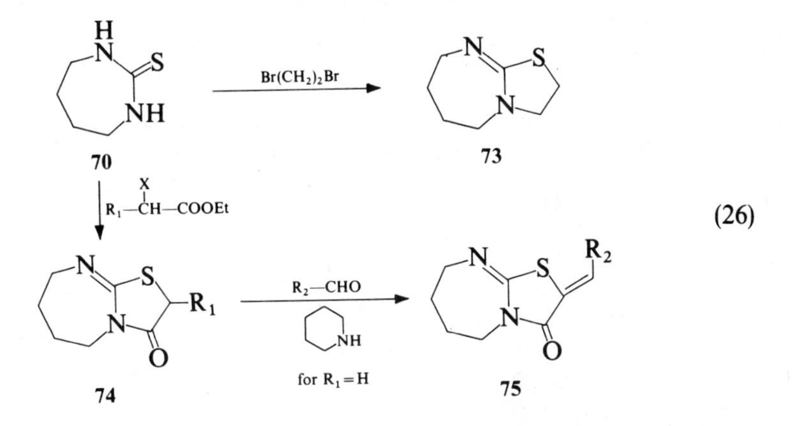


(26)

(Eq. 26). Compound **73** was characterized as the hydrobromide salt, and the structure assignment was based on analytical and infrared absorption data.

Condensation of the cyclic thiourea **70** with α-haloesters was found to give the 3-ones **74** in good yield.[24] Treatment of **74** ($R_1 = H$) with aromatic aldehydes in the presence of piperidine gave the 2-(arylmethylene) derivatives **75**, which could also be prepared directly, by a one-pot procedure, from **70**. The stereochemistry of the exocyclic double bond in **75** was not discussed. For steric reasons, the configuration depicted may be preferred. The synthesis of a series of substituted 2,3-diphenyl-3-ols has been reported in the patent literature.[97]

### 6.2.2. *Reactions*

Treatment[27a] of the hydroxyacid **76** with acetic anhydride under reflux (Eq. 27) led to the lactone **77**. The stereochemistry of the ring fusion was not studied.

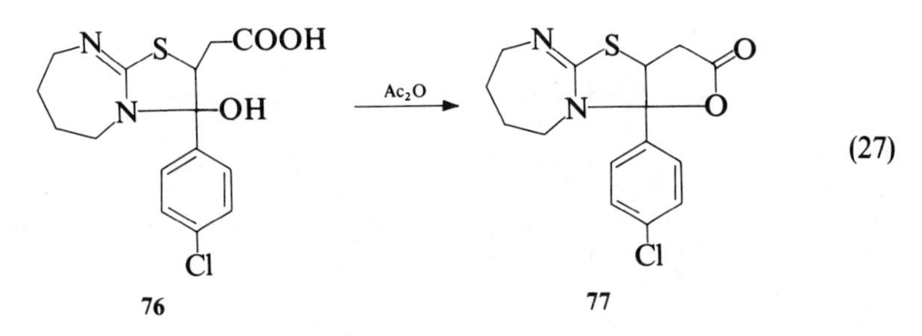


(27)

The addition of bromine to **78** led to the dibromo derivative **80**, the stereochemistry of which was not specified. Acid hydrolysis of the 2-benzylidene derivative **78** opened the seven-membered ring to give the thiazolidin-2,4-dione **79**.[24] This compound was also synthesized (Eq. 28) by condensation of the related hydrolysis product **81** ($R_1 = H$) with benzaldehyde.

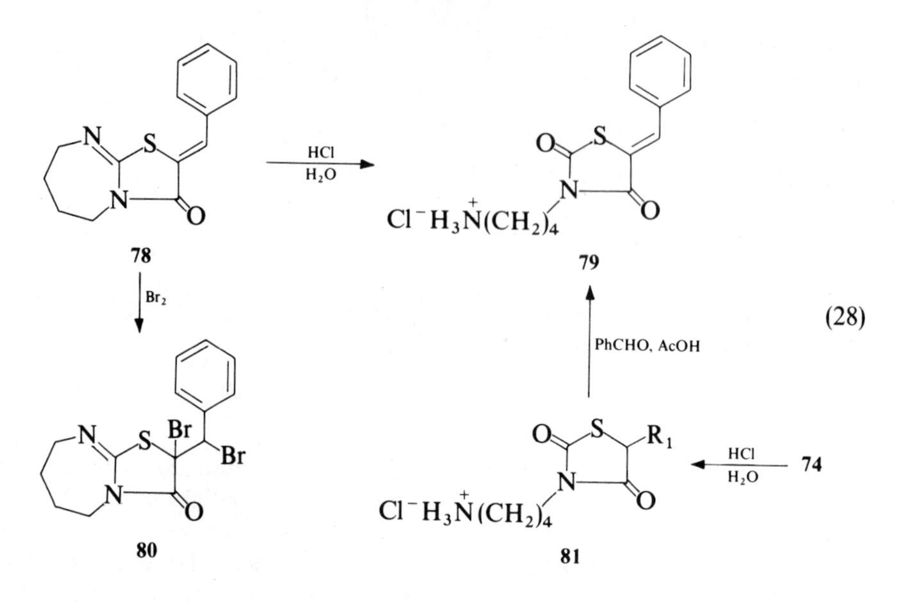


## 6.3. Octahydrothiazolo[3,2-*a*] [1,3]diazepines

Structure **83** has been assigned[29] to the product obtained by fusion of the isothiuronium salt **82** with succinic anhydride. The structure of **83** is questionable, since it is based solely on poor microanalytical data.

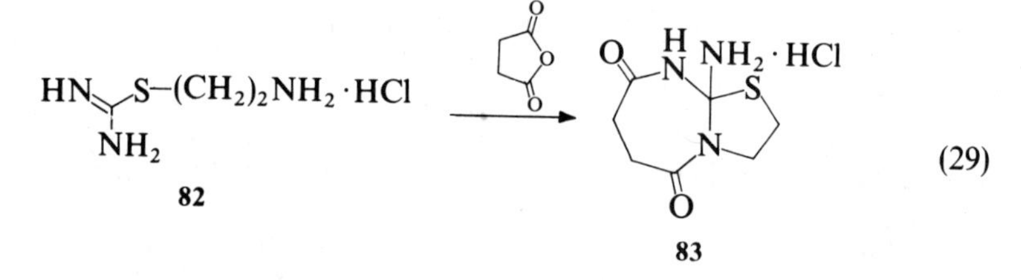


## 6.4. Pyrazolo[1,5-*a*] [1,3]diazepines

3,7,8-Trisubstituted-6*H*-pyrazolo[1,5-a] [1,3]diazepin-6-ones (**85**) were prepared by heating a solution of the corresponding 3,6,7-trisubstituted-pyrazolo[1,5-*a*]pyrimidines (**84**) in acetic acid and water at 70°C.[95]

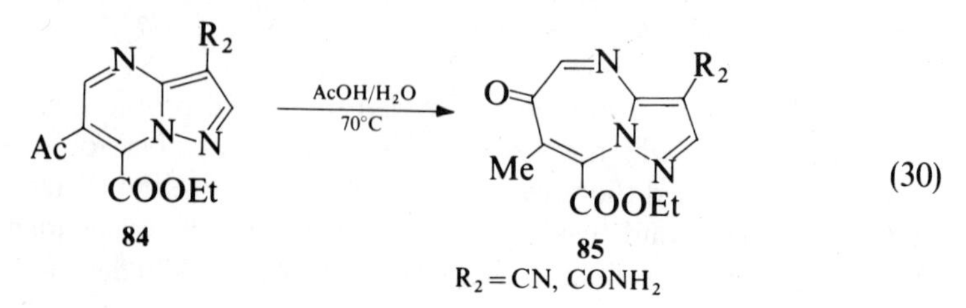

Structure assignments of compounds **85** were based on analytical and spectroscopic data.

## 7. 1,3,5-TRIAZINO[1,2-*a*] [1,3]DIAZEPINES

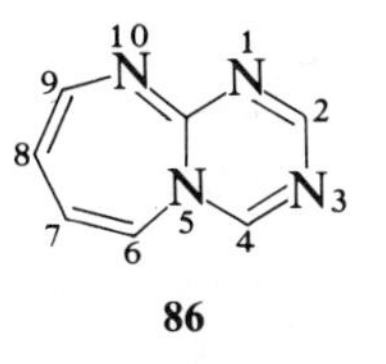


**86**

A search of the chemical literature has not revealed the existence of any derivatives of the unsaturated ring system **86**. Octahydro derivatives of structure **88** were obtained[30] by reaction of the 2-benzylthio-1,3-diazepine **87** with aryl isothiocyanates (Eq. 31). The success of this synthesis appears to depend on the reactivity of the isothiocyanate. Thus, methyl isothiocyanate only added to the nitrogen of the diazepine to form the 1:1 adduct **89**, whereas (4-dimethylaminophenyl)isothiocyanate failed to react at all.

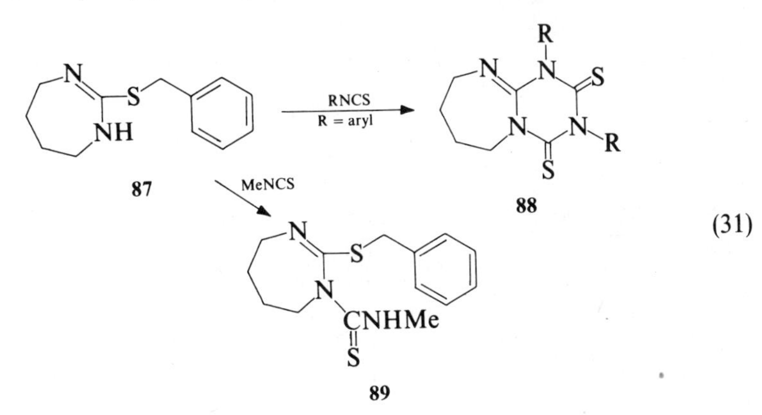


(31)

### 7.1. [1,2,4]-Triazino[4,3-*a*] [1,3]diazepines

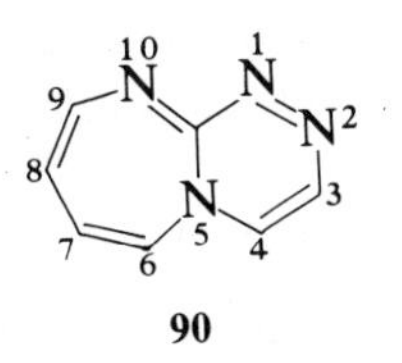


**90**

Once again, only reduced derivatives of the parent ring system **90** have been reported in the literature. Compounds **93** were synthesized[31] by conversion of

2-ethoxy-4,5,6,7-tetrahydro-1*H*-1,3-diazepine (**91**) to the intermediate hydrazine **92** followed by condensation with α-ketoesters (Eq. 32). Infrared spectroscopic studies led to the conclusion that **93** is the predominant tautomeric form of the product in solution, whereas **94** is favored in the solid state.[32]

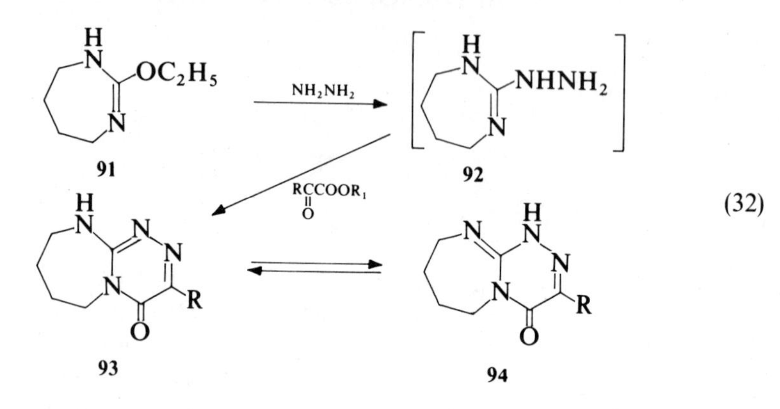


## B. [*c*]-FUSED [1,3]DIAZEPINES

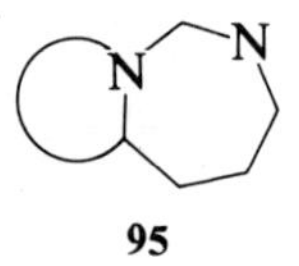


A search of the literature revealed only one (patent) citation describing [*c*]-fused [1,3]diazepines of general structure **95**.[98] The two compounds reported were synthesized from 3-(4-imidazolyl)propylamine (**96**) either by treatment with diphenylcyanoiminocarbonate (**97**: R = CN) or by treatment with diphenylbenzoyliminocarbonate (**97**: R = COPh). The initially formed intermediates **98** were cyclized in situ to give the corresponding 8-substituted [*c*]-fused [1,3]diazepines **99** (Eq. 33).[98]

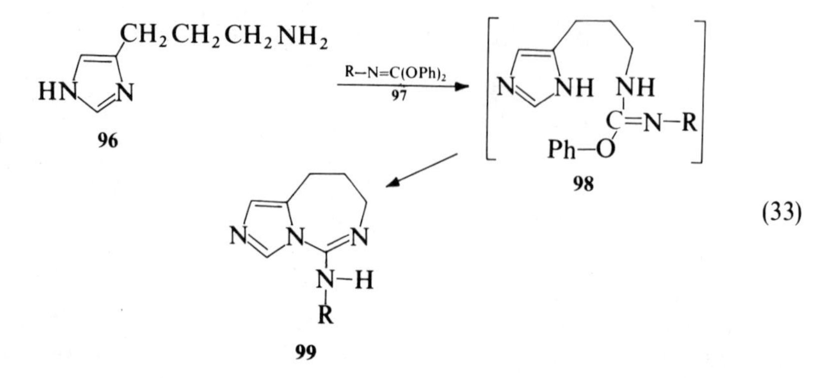

## C. [*d*]-FUSED [1,3]DIAZEPINES

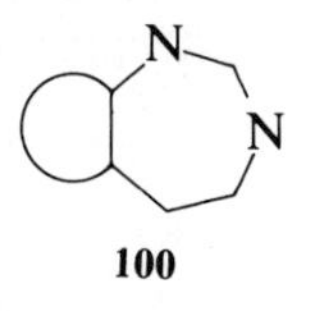


**100**

The most important class of compounds of general structure **100** consists of the 1,3-benzodiazepines. These derivatives are, therefore, given precedence over other fused systems in this section.

## 1. BENZO[*d*]-1,3-DIAZEPINES (1,3-BENZODIAZEPINES)

### 1.1. Synthesis

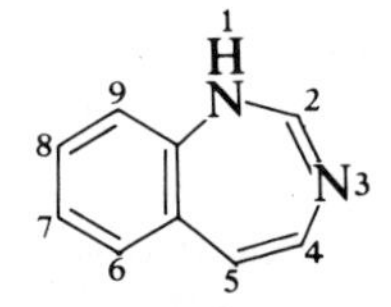


**100a**
1*H*-1,3-Benzodiazepines

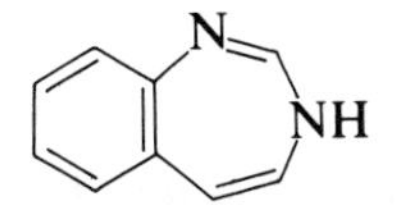


**100b**
3*H*-1,3-Benzodiazepines

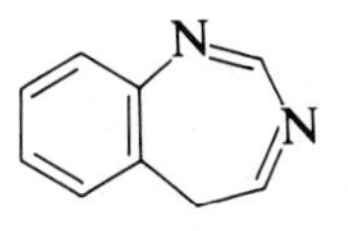


**100c**
5*H*-1,3-Benzodiazepines

If the energetically less favored quinonoidal structures are disregarded, the parent 1,3-benzodiazepine, derived from the generic **100**, may occur in any of the three tautomeric forms **100a**–**100c**. Although none of the parent tautomers has been synthesized, derivatives of all these systems have been described. The substituted 1*H*-1,3-benzodiazepine, **102** appears to be the first reported derivative of **100a**. This compound was reportedly obtained[33] by oximation of ketone **101** (Eq. 34). The structure was supported by spectral data, but solid proof is lacking.

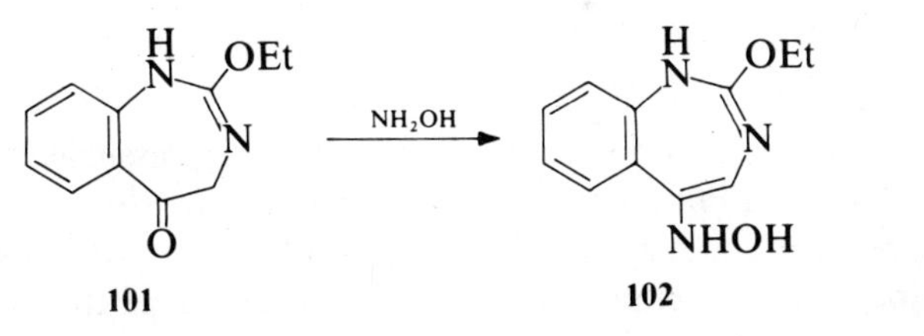


(34)

More recently, some substituted 1*H*-1,3-benzodiazepine derivatives (**104**), were prepared by a photoinduced rearrangement of the isoquinoline *N*-imides **103** (Eq. 35).[99,101] If the R in the 5-position is an electron-donating group, the ring expansion is favored and benzodiazepines can be isolated in relatively good yields.

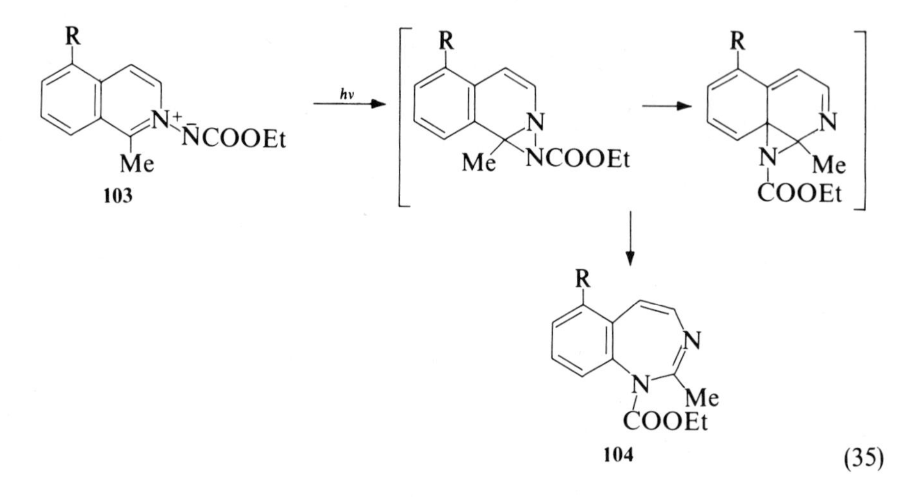


(35)

In a similar manner, the 3*H*-1,3-benzodiazepine derivatives **106** were obtained by irradiation of the 2-methylquinoline *N*-imides **105** with an electron-donating group, such as OMe or NMe at either the 6- or the 8-position, (Eq. 36).[99]

The 3-substituted 3*H*-1,3-benzodiazepine **108** was isolated[34] in 25% yield from the reaction mixture obtained by dehydration of **107** with molecular sieves in boiling benzene (Eq. 37). The indoline **109** was the major product of cyclization. The indoline was probably formed by transfer of the acetyl group to the primary aromatic amino group, followed by isomerization of the enamine to the imine, which would then cyclize to **109**. Compound **107** was prepared by reduction of the corresponding nitro compound, which in turn was made by the addition of the anion of *N*-acetyl-*o*-toluidine to 2-nitrophenylacetylene.

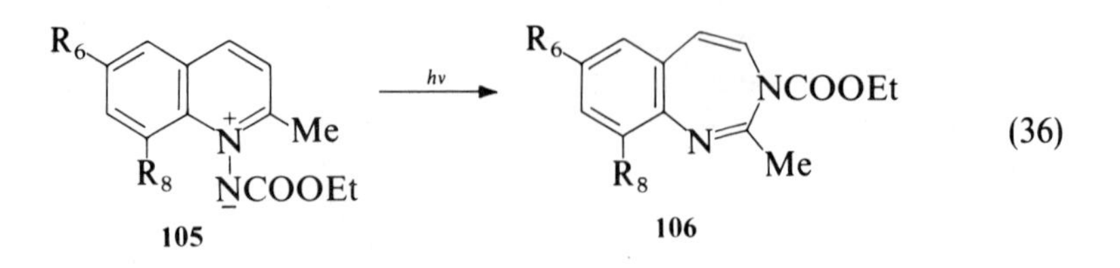


(36)

The structures of 5*H*-1,3-benzodiazepines **111** were assigned[35] to the minor products formed by addition of acetylene dicarboxylates (Eq. 38) to the quinazoline $N^3$-oxide **110**. The major components isolated from the reaction mixture were the ring-opened compounds **112**. The structures of the benzodiazepines

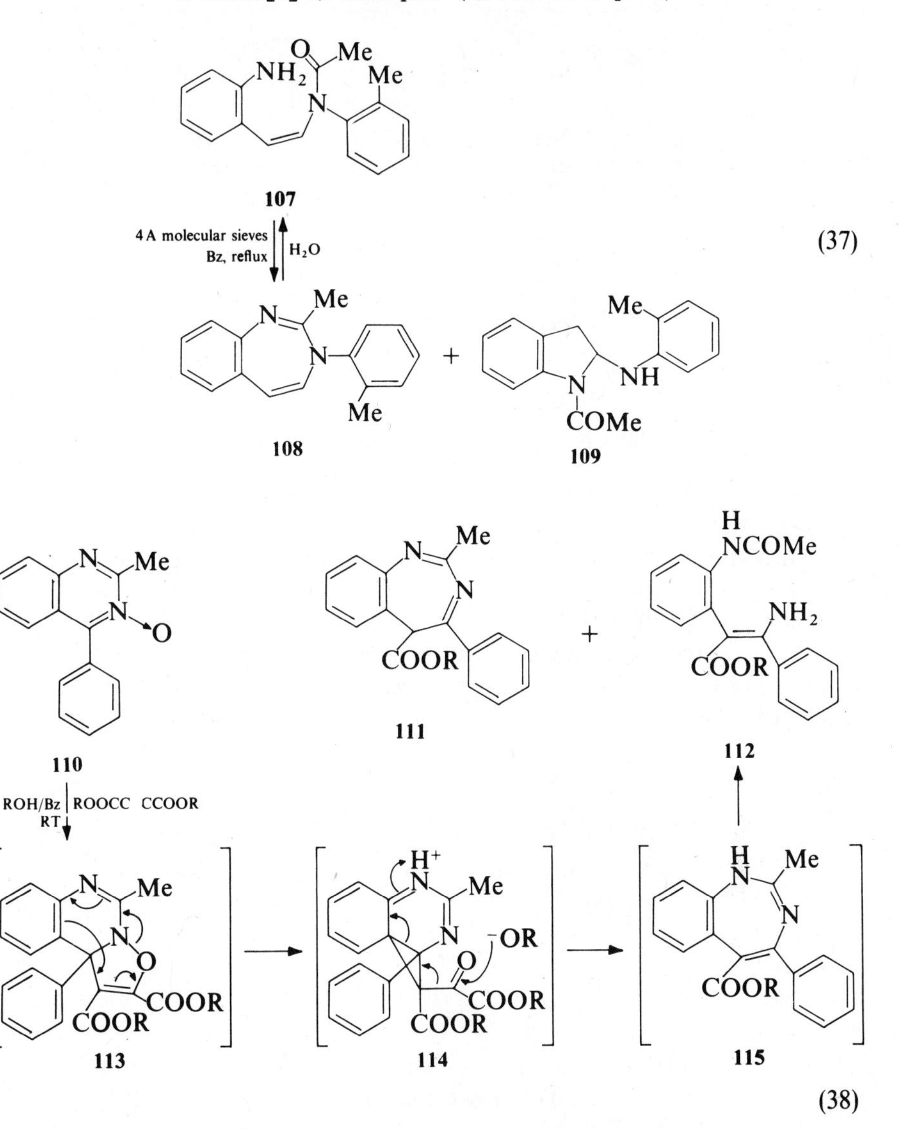


were elucidated with the help of spectral and analytical data and were supported by the formation of the alkaline hydrolysis products discussed in Section 1.2. Mechanistically, the formation of **111** was postulated to occur via the intermediates **113** and **114** and the 1*H* tautomer **115**.

### 1.2. Reactions

Treatment of **116** with methanol or ethanol and acetic acid resulted in the formation of solvent adducts **117**, which decomposed during isolation to give

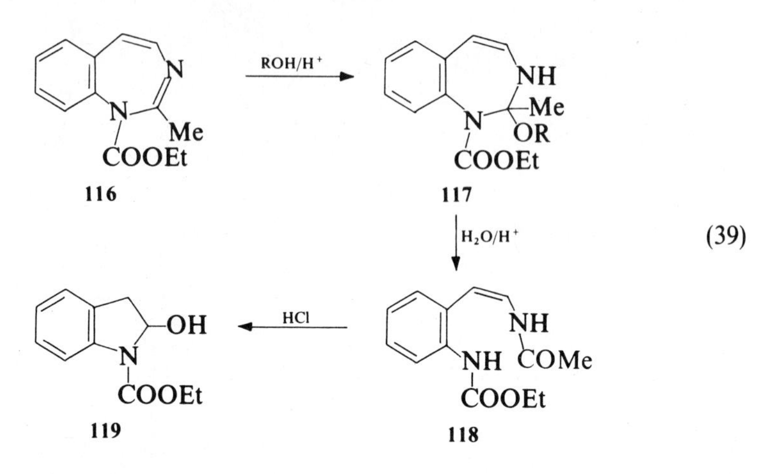


the ring-opened product **118** and the 2-hydroxy-2,3-dihydroindole derivative **119** (Eq. 39).[99]

Hydrolytic ring cleavage of **108** to **107** was reported[34] to occur upon contact with water. Alkaline hydrolysis of the ester **111** was accompanied by ring contraction (Eq. 40), with formation of the indole **120**. The authors[35] suggest that rearrangement and ring contraction could occur by a transannular attack at the 2-position by the carbanion that was initially formed at the 5-position. Decarboxylation may occur at any stage of this reaction sequence.

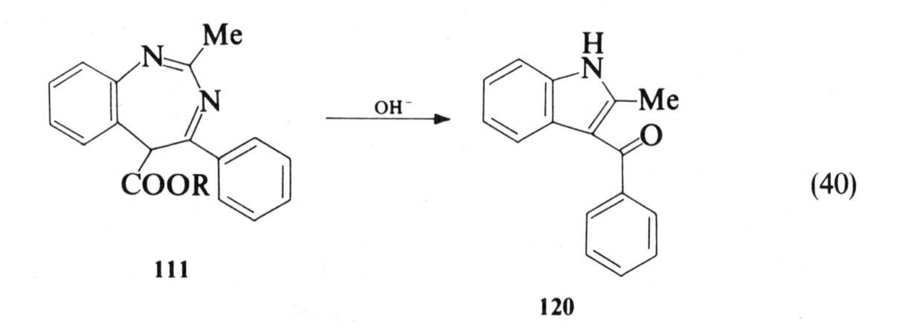


## 1.3. Dihydro-1,3-benzodiazepines

### *1.3.1. Synthesis*

A general synthetic method for the preparation of 4,5-dihydro-3*H*-1,3-benzodiazepines **122** is the condensation of *o*-aminophenethylamines **121** with imidates[36] or amidines (Eq. 41).[37,38,100,103,104] The first method was investigated extensively,[36] and the yield of benzodiazepines was found to be dependent on the nature of the $R_2$ substituent in the imidate. Good results were obtained with $R_2$ representing alkyl or haloalkyl groups, or phenyl groups with electron-withdrawing substituents. A modification of this bimolecular reaction

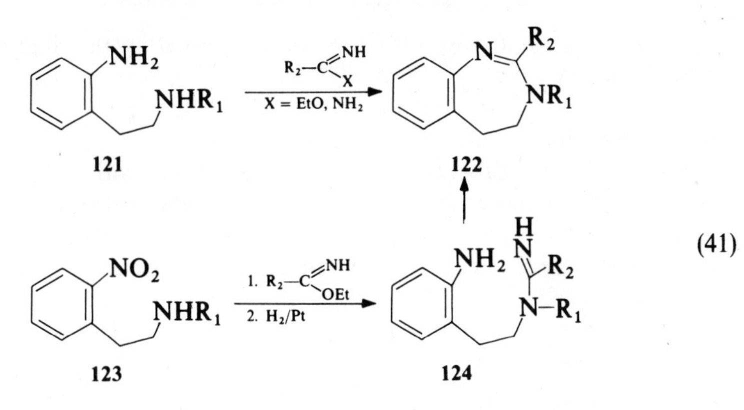


(41)

involving intramolecular amine exchange on the preformed amidine **124** often gave better results.[36] Compound **124** was prepared by reaction of the *o*-nitrophenethylamine **123** with the imidate, followed by catalytic reduction of the nitro group.

In the single instance reported,[34] ring closure by dehydration gave a good yield of a dihydrobenzodiazepine. Thus heating the amide **125** under reflux in a mixture of methylene chloride and phosphorylchloride for 16 hours (Eq. 42) gave a 40% yield of the diazepine **126**.

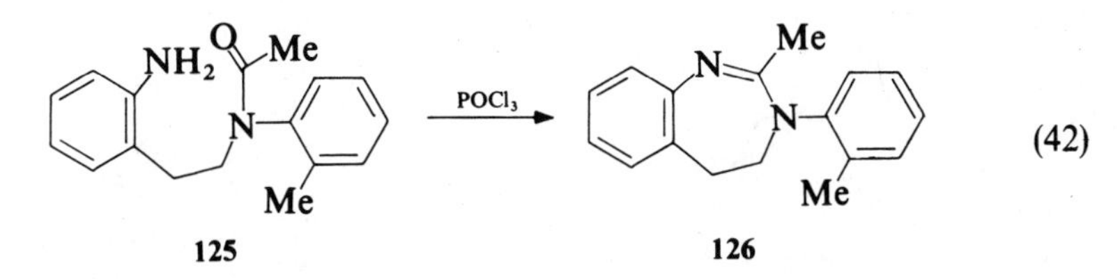


(42)

A variety of 2-amino-4,5-dihydro-3*H*-benzodiazepines have been described in the patent literature. Compounds **132** can be prepared by displacement of the thiomethyl group in **127** with amines in refluxing acetonitrile,[39,40] or by reaction of the urea **128** (X = O) with phosphorylchloride and an amine.[41] In addition (Eq. 43), compounds **132** may be synthesized by first converting the

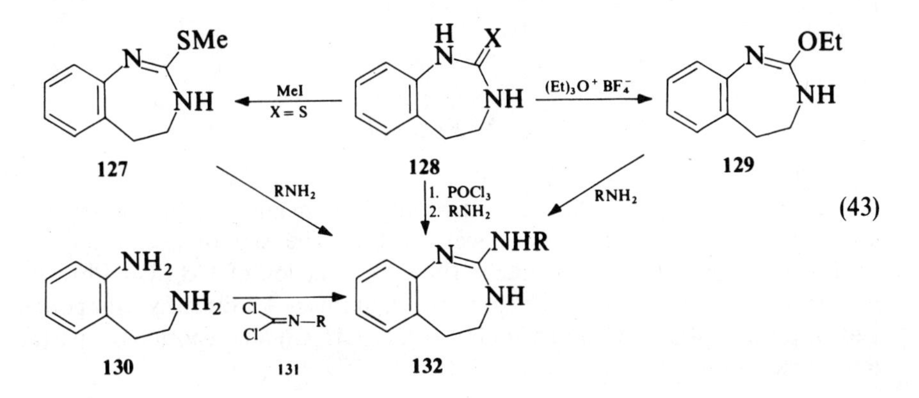


(43)

urea **128** to the *O*-ethylisourea **129** and then treating this imino ether with amines.[42] In an example of still another method,[43] reaction of *o*-aminophenethylamine **130** with dichloroaryl isocyanides **131** led to compounds **132** bearing substituted anilino groups (R = aryl) at the 2-position.

2-Ethoxy-4,5-dihydro-1*H*-1,3-benzodiazepines **134** with substituents at the 5-position were prepared by Taylor and Tully,[33] who added hydride, Grignard, and aryllithium reagents to the carbonyl group of **133**. The resulting tertiary alcohols **134** were generally isolated and characterized as salts (Eq. 44).

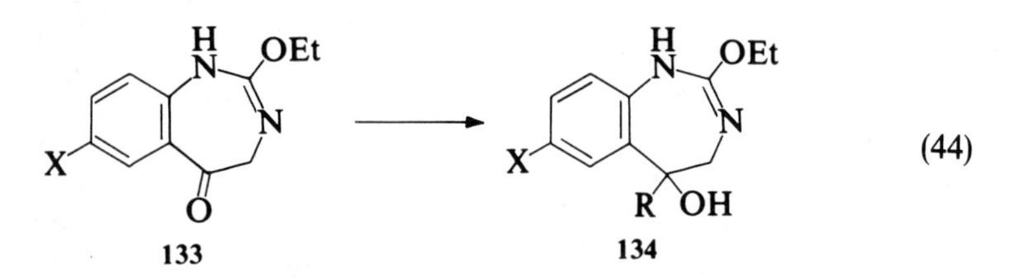


An unusual pathway leading to the 3-substituted derivative **137** was published by Elslager and coworkers,[44] who obtained this compound in low yield by desulfurization of the thiazolo[2,3-*b*] [1,3]benzodazepine **136** with Raney nickel (Eq. 45). Compound **136** was obtained from the reaction of the 2-thione **135** with α-bromoacetophenone.

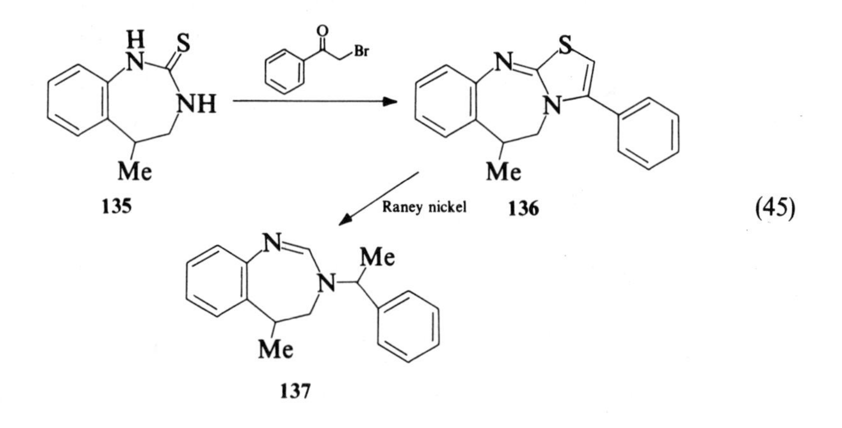


## *1.3.2. Reactions*

Alkylation of **138** (R = Ph, Me) with *n*-butyllithium and methyl *p*-toluenesulfonate yielded[36] the 3-methyl derivatives **139**. The site of alkylation was confirmed by unambiguous synthesis (Eq. 46). Reaction of **138** (R = Ph) with methyl iodide in boiling benzene led to the quaternary 1,3-dimethyl compound **140**. Intramolecular alkylation of **138** (R = $CH_2CH_2Cl$) was reported to give the imidazobenzodiazepine **141**.

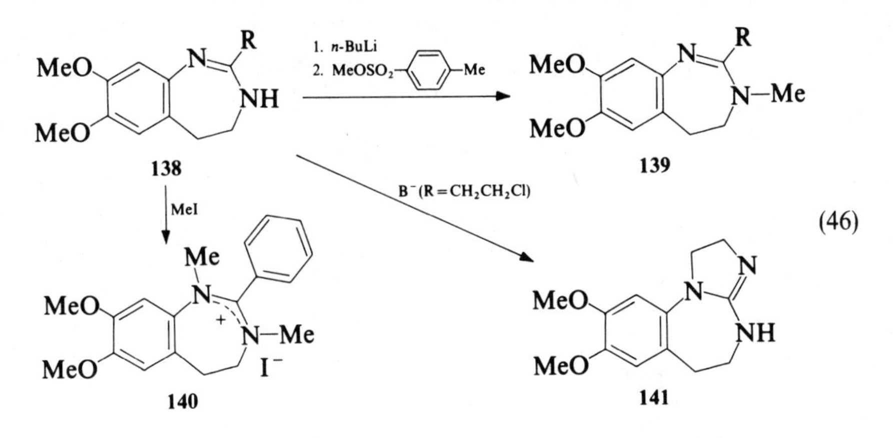


(46)

Hydrolysis and methoxyaminolysis of the benzodiazepine **142** was studied in detail.[38] Methoxyaminolysis of **142** at pH 7.11 was shown to exhibit three distinct phases (Eq. 47). This was interpreted in terms of the initial formation of the kinetically favored product **143**, followed by isomerization to the thermodynamically favored product **144**, and finally, cleavage to a mixture of the diamine **145** and methoxyformamide.

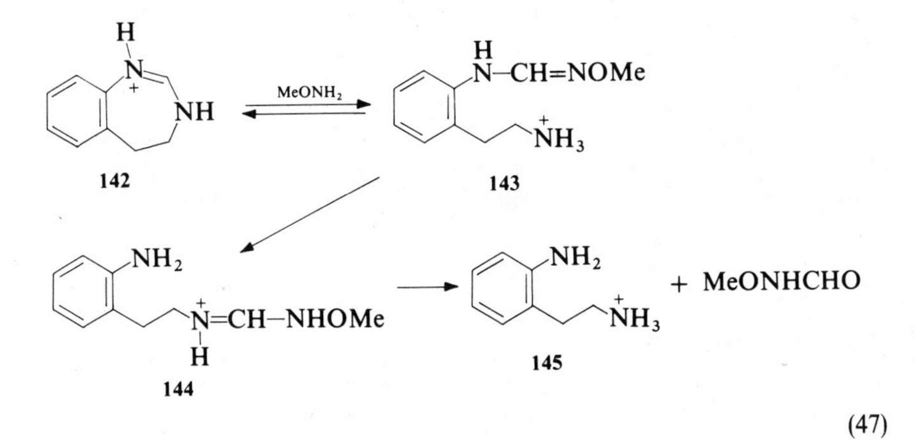


(47)

Treatment of **138** ($R = CH_2Cl$) with amines $R_1R_2NH$ led to chloride displacement and formation of **138** ($R = CH_2NR_1R_2$).[36] Taylor and Tully[33] investigated the dehydration of the tertiary alcohol **146**. When the hydrochloride salt of **146** was heated at 100°C for 20 minutes in dimethylformamide and the resulting mixture was poured into water (Eq. 48), the products **147**, **148**, and **149** were isolated in 38, 23, and 22% yield. When the free base of **146** was heated under reflux in benzene solution for 48 hours, only the ring-opened carbamate **148** was isolated (52% yield).

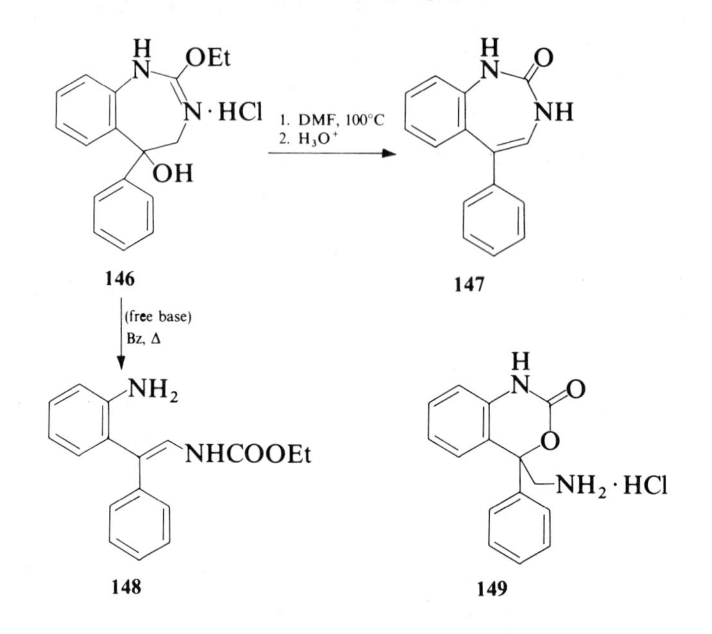


(48)

## 1.4. Dihydro-1,3-benzodiazepinones

### *1.4.1. Synthesis*

An attempt to prepare the 1,3-dihydro-2*H*-1,3-benzodiazepin-2-one **147** from the carbinol **150** by standard methods of dehydration failed.[33] As mentioned above however, this compound was formed as the major product by treatment of the hydrochloride **146** with dimethylformamide at 100°C followed by aqueous workup (Eq. 49).

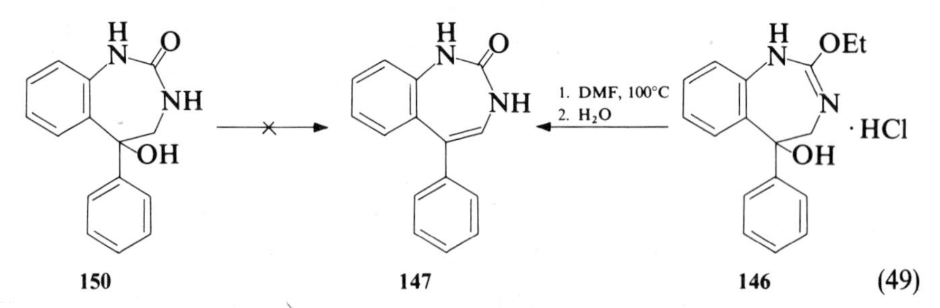


(49)

An interesting synthesis by Woerner and coworkers[45] involved reaction of the 2*H*-azirine **151** with the ketenimine **152** to give diazepinone **154** in 15% yield (Eq. 50). The intermediate aziridine **153** also was obtained in comparable amounts and could be converted almost quantitatively to **154** by heating in trichlorobenzene. The reaction probably proceeds via oxidative elimination of benzophenone, which is also isolated from the reaction mixture.

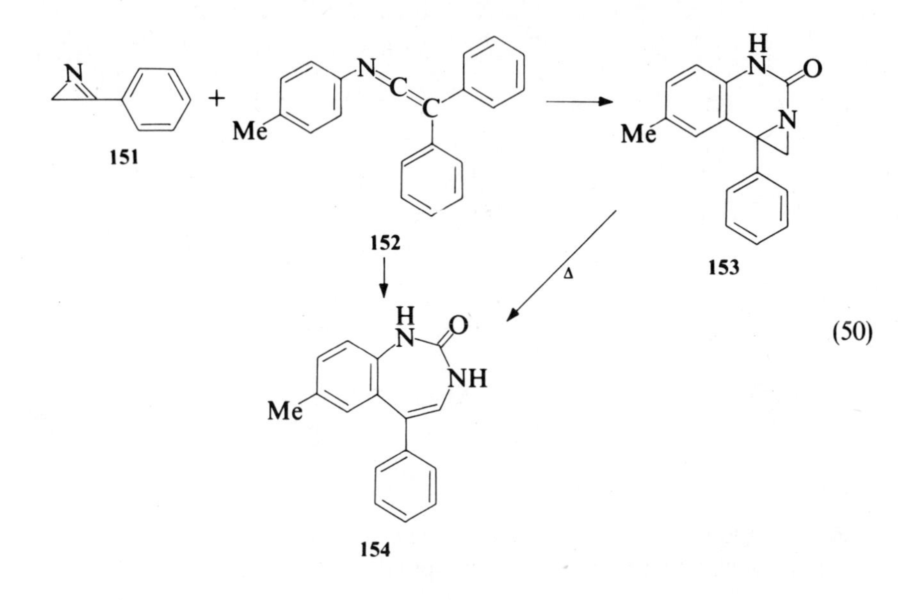


(50)

A fairly general synthesis of the dihydrobenzodiazepin-4-ones **156** has been disclosed in the patent literature.[46] Condensation of *o*-aminophenylacetic acids **155** with imidates in boiling *n*-butanol (Eq. 51) was claimed to give the benzodiazepines **156** in moderate yields. Similar approaches to these compounds reported in the literature have been less rewarding. For example, according to Rodriguez and coworkers,[36a] reaction of the phenylacetamide **157** (R = H) with imidates failed to yield benzodiazepines. Attempts by Golik and Taub[47] to prepare **156** (R = Me) by cyclodehydration of the diamide **157** (R = Ac) were also unsuccessful.

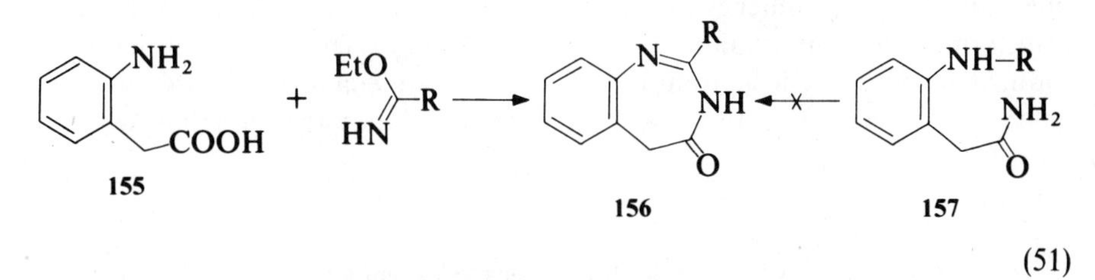


(51)

Dihydro-1,3-benzodiazepin-5-ones were found to be accessible via modification of the urea carbonyl in dione **158**. Triethyloxonium tetrafluoroborate converted **158** to the isourea **159**,[33] which could be converted to the 2-amino derivatives **160** by treatment with amines in boiling toluene (Eq. 52).

1,3-Benzodiazepin-5-one **161** was isolated unexpectedly in about 1% yield by Farrar,[48] from the reaction of *p*-anisidine with formaldehyde in hydrochloric acid (Eq. 53). The structure of **161** was derived from spectroscopic data. The presence of a ketone function was confirmed by the formation of a semicarbazone and a phenylhydrazone. The compound was found to undergo

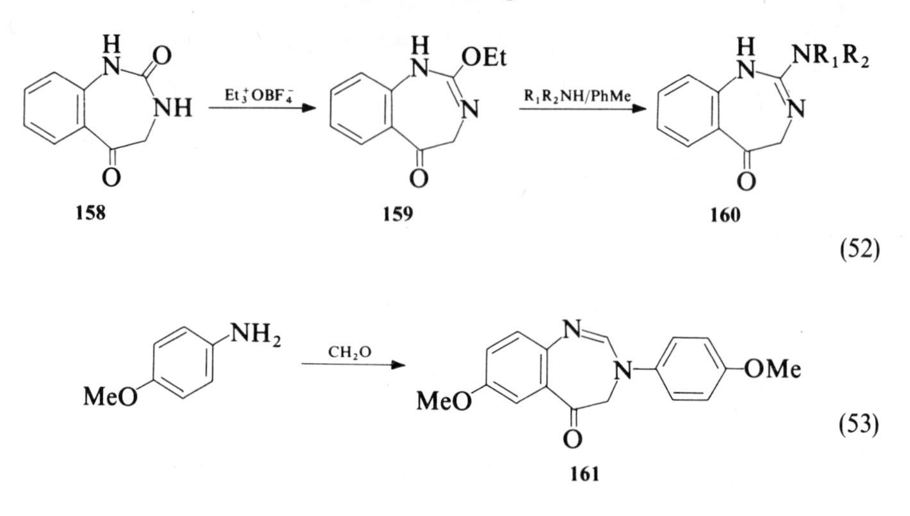


oxidation upon melting or upon treatment with permanganate in acetone to give a product which may be the corresponding 4,5-dione.

### *1.4.2. Reactions*

Reactions of the 2-ethoxy-1,3-benzodiazepin-5-one **159**,[33] such as reduction with sodium borohydride, addition of organometallic reagents, oxime formation, and displacement of the ethoxy group by amines have been discussed above. When **159** was subjected to prolonged heating under reflux with morpholine in toluene (Eq. 54), or if the reaction was carried out in the presence of one equivalent of *p*-toluenesulfonic acid, the initially formed guanidine **162** was transformed into the quinoline **165** in 30% yield.[33] This ring contraction is thought to occur by formation of the intermediate enamine **163**, followed by valence tautomerization to the aziridine **164** and finally the quinoline **165**.

## 1.5. Tetrahydro-1,3-benzodiazepines

2,3,4,5-Tetrahydro-1*H*-1,3-benzodiazepines are cyclic aminals and therefore would be expected to be sensitive compounds. These compounds can be obtained by reduction of the appropriate tetrahydrobenzodiazepinone with lithium aluminum hydride, as exemplified by the work of de Stevens and Dughi.[49] These authors prepared the 3-methyl derivative **167** (R = H) by reduction of **166** (R = H), or in two steps via the intermediate 1-benzyl derivative **167** (R = $CH_2Ph$) (Eq. 55). The product was characterized as the maleate salt.

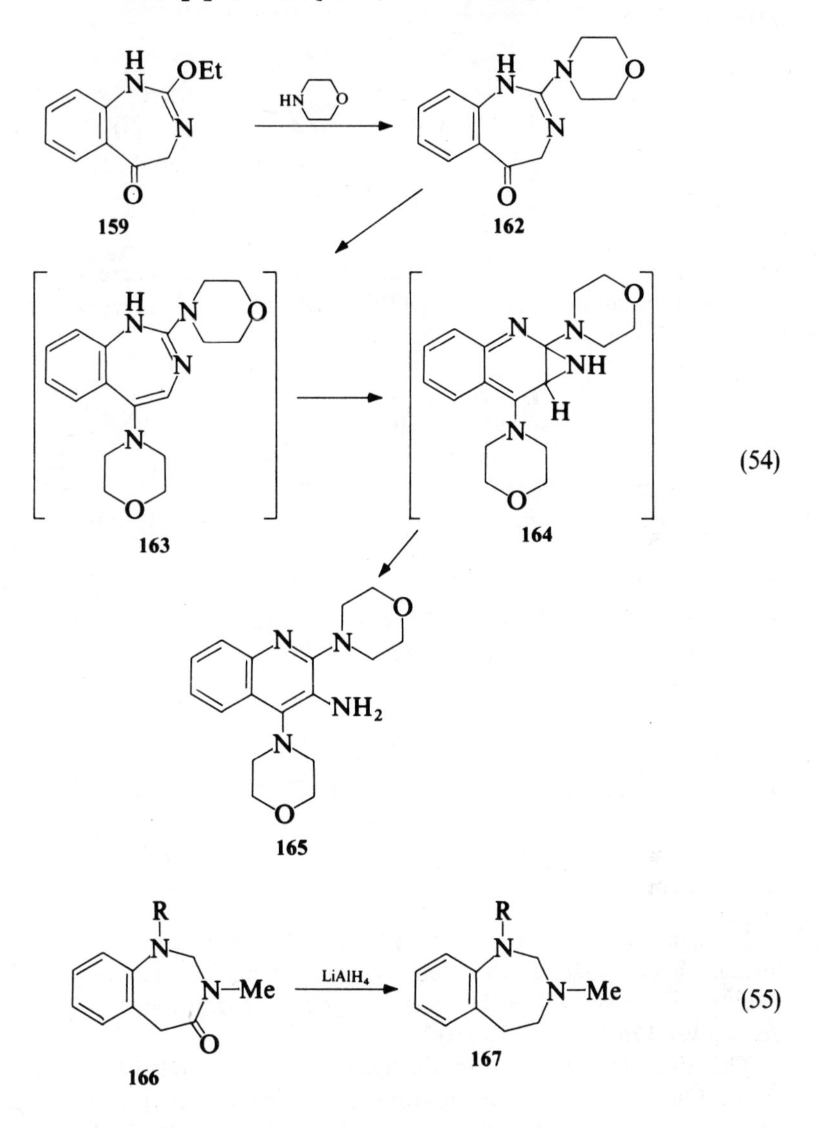


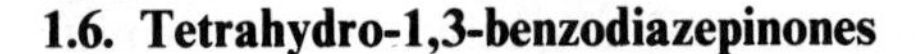

## 1.6. Tetrahydro-1,3-benzodiazepinones

### *1.6.1. Synthesis*

Tetrahydro-1,3-benzodiazepin-2-ones of structure **169** have been prepared by reaction of aminophenethylamines **168** with phosgene[50] or *N,N′*-carbonyldiimidazole[36, 42, 51] (Eq. 56).

Derivatives with substituents at the 5-position were accessible[33] via addition reactions to the carbonyl group of the 2,5-diones **170**. Thus, reduction with

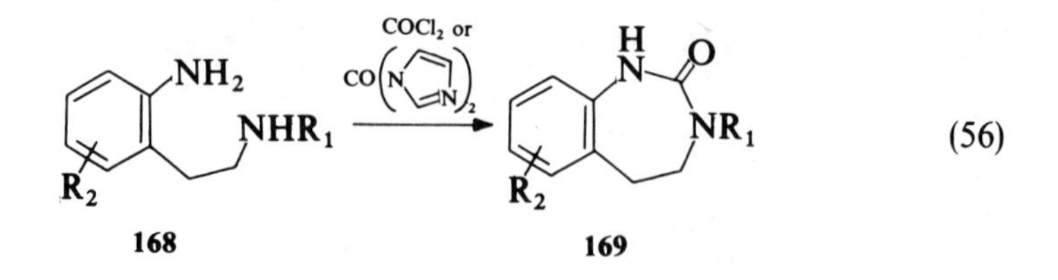


sodium borohydride led to the alcohol **171** ($R_2$ = H), whereas reaction with methylmagnesium iodide or aryllithium afforded the corresponding tertiary carbinols (Eq. 57).

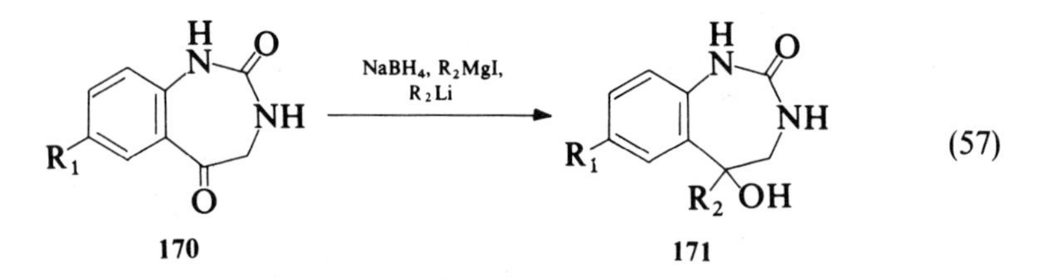


The 5-hydroxy derivative **173** was synthesized[52] in 90% yield by acid-catalyzed cyclization of the urea **172** (Eq. 58).

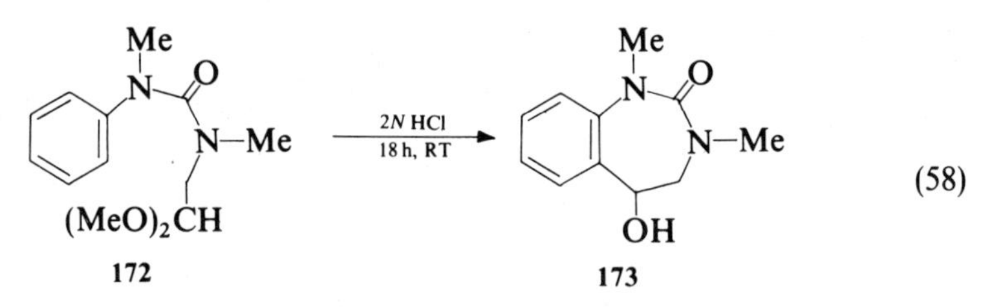


Condensation of *o*-aminophenylacetamides of general formula **174** with formaldehyde yielded the tetrahydro-1,3-benzodiazepin-4-ones **175** (Eq. 59).[46, 50, 52, 53] With $R_2$ = H, formaldehyde reacted further to form 3-methylol derivatives **175** ($R_2 = CH_2OH$).[54]

This ring closure with formaldehyde was also extended to the unsaturated amide **176**, obtained by condensation of the appropriate phenylacetamide **174** ($R_1$, $R_3$ = H; $R_2$ = Me) with α-methylphenylacetaldehyde (Eq. 59). In this manner, **177**, a compound of unspecified stereochemistry, was prepared in 20% yield.

A quite unusual synthesis of a tetrahydro-1,3-benzodiazepin-4-one has been reported by Tsuge and Watanabe,[55] who heated a benzene solution of a mixture of the azaspiropentane **178** and the nitrone **179** under reflux and isolated the crystalline products **180**, **182**, and **184** (11.4, 8.7, and 7.2% yield, respectively). The benzodiazepine **182** was found to undergo ring opening (Eq. 60) to **184** upon heating in methanol. Since **184** was also prepared in 60% yield by addition of the nitrone **179** to the ketenimine **181**, it was concluded that the spirane **178** underwent thermal fragmentation to the ketenimine **181** and ethylene. Reaction

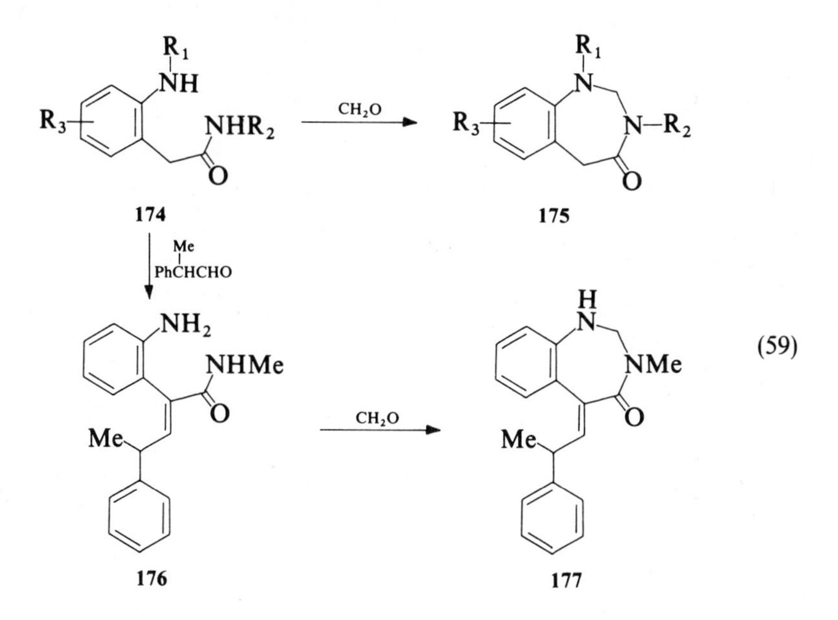


(59)

of the ketenimine with the nitrone is believed to proceed via the intermediates **183** and **185**.

## *1.6.2. Reactions*

The conversion of the urea carbonyl in tetrahydro-1,3-benzodiazepin-2-ones to 2-ethoxy[33,42] and 2-amino[36] groups was presented above. Similarly, the reduction of the carbonyl group in tetrahydro-1,3-benzodiazepin-4-ones with lithium aluminum hydride to form tetrahydro-1,3-benzodiazepines[49] has been discussed.

In attempts to dehydrate the alcohols **186** (see above), Taylor and Tully[33] observed that heating **186** ($R_1$, $R_2$, $R_3$ = H) under reflux in acetic acid led to the formation of the acetate **187**. On the other hand, when the tertiary alcohol **186** ($R_1$, $R_2$ = H, $R_3$ = Ph) was heated in formic acid for 24 hours, only the oxindole **188** was isolated (Eq. 60). The mechanism of this interesting ring contraction was not discussed. It is likely that the oxindole was formed by a transannular reaction after initial dehydration.

The 1,3-dimethyl derivative **186** ($R_1$, $R_2$ = Me; $R_3$ = H) was found to rearrange[52] to the oxazolidinone **189** under mild basic conditions, such as heating in the presence of bicarbonate or aluminum oxide (Eq. 61). More vigorous hydrolytic conditions, acidic or basic, gave the expected phenethylamine **190**.

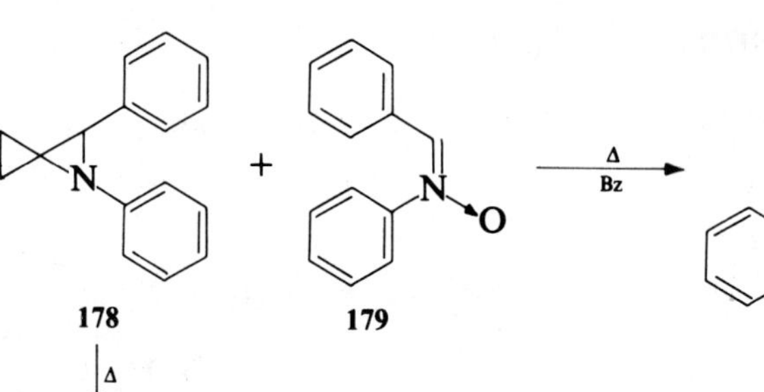


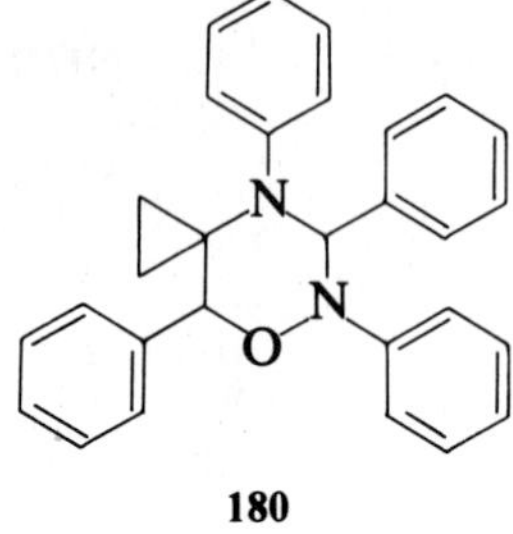


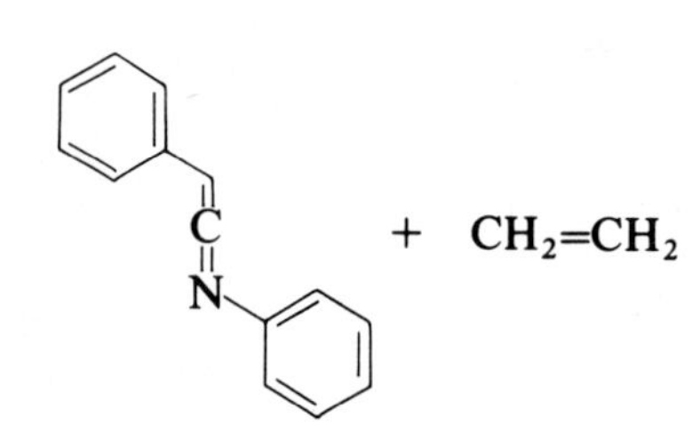


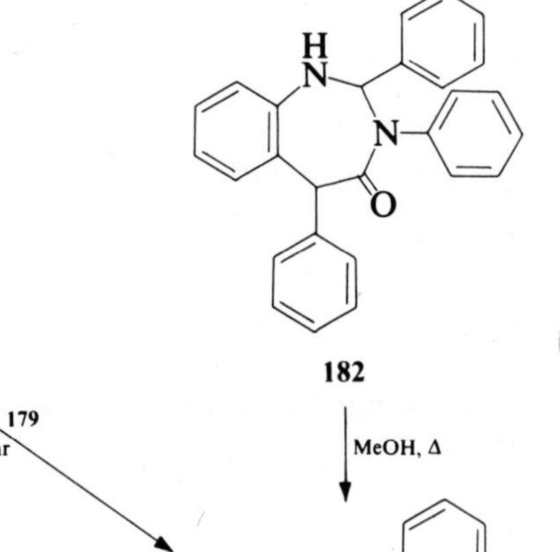


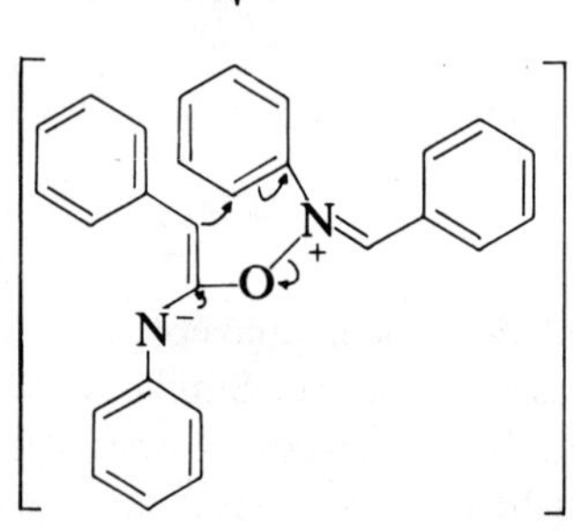

185

(60)

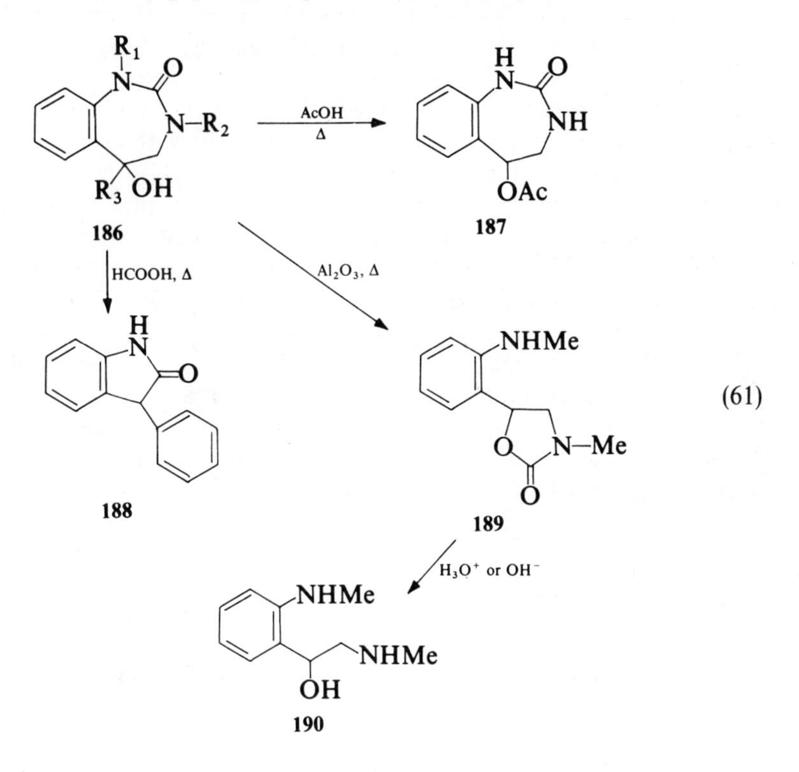


(61)

## 1.7. Tetrahydro-1,3-benzodiazepinediones

### 1.7.1. Synthesis

The synthesis of tetrahydro-1,3-benzodiazepin-2,4-diones was achieved[56] by cyclization of the 2-ureidophenylacetic acids **192** using acetyl chloride for activation of the carboxyl group (Eq. 62). Other attempts to cyclize **192** by

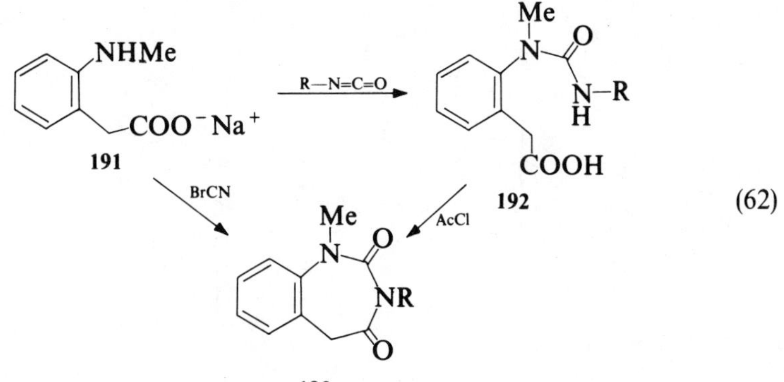


(62)

dehydration with acetic anhydride or dicyclohexylcarbodiimide failed and led to other products. The ureas **192** were prepared by reacting the sodium salt of methylaminophenylacetic acid **191** with isocyanates. Compound **193** (R = H) was also obtained in lower yield by reaction of **191** with cyanogen bromide.

The reaction of sodium phenylacetylide with phenyl isocyanate was reported by Bird[57] to lead to **194** in 4% yield (Eq. 63). The structure of this product was assigned on the basis of spectral data and its conversion to the dihydro derivative **198** by catalytic hydrogenation. The formation of **194** can be rationalised on the basis of an electrophilic attack by the alkyne substituent at the position ortho to the nitrogen of the aromatic ring in **196**. The hydantoin **197**, which is the major product of this reaction, could be formed by electrophilic attack of the alkyne on nitrogen as indicated in **195**.

Compound **194** was also obtained by Ohshiro and coworkers[58] in 12% yield by heating a mixture of phenyl isocyanate, diphenylcarbodiimide, and phenylacetylene at 140–190°C in the presence of iron pentacarbonyl.

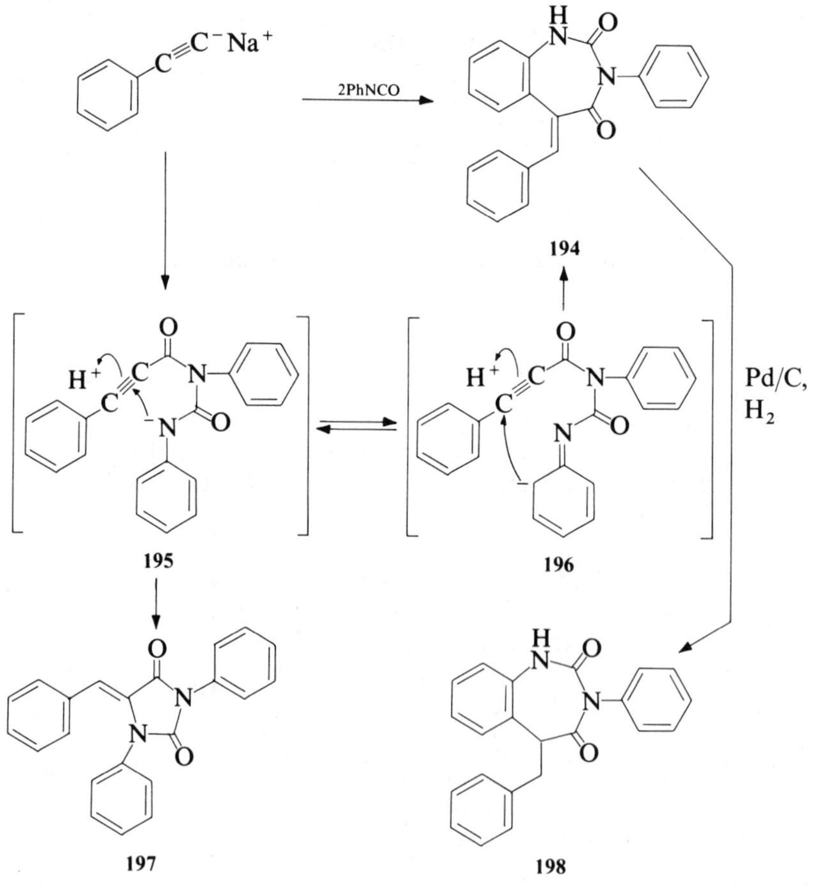


(63)

Other compounds also related to **194**, the 2-imino derivatives **199**, were obtained by a three-component reaction involving an arylcarbodiimide, phenylbromoacetylene, and iron pentacarbonyl.[59] Coupling of phenylbromoacetylene to give 1,4-diphenylbutadiyne was also observed, and in some instances it was the major reaction product. *N*,*N*-Bis(*o*-tolyl)carbodiimide, for example, gave a 17% yield of the benzodiazepine **199** (R = 2-$MeC_6H_4$, X = 9-Me) and 36% of the butadiyne. This may be due to steric hindrance of the arylation step (i.e., cyclization of intermediate **202**). The authors presented the following mechanistic explanation for the formation of **199**. Successive insertion of 2 mol of the carbodiimide into the initially generated acetylide complex **200** would lead to intermediate **201**. Fission of the iron–nitrogen bond and transfer

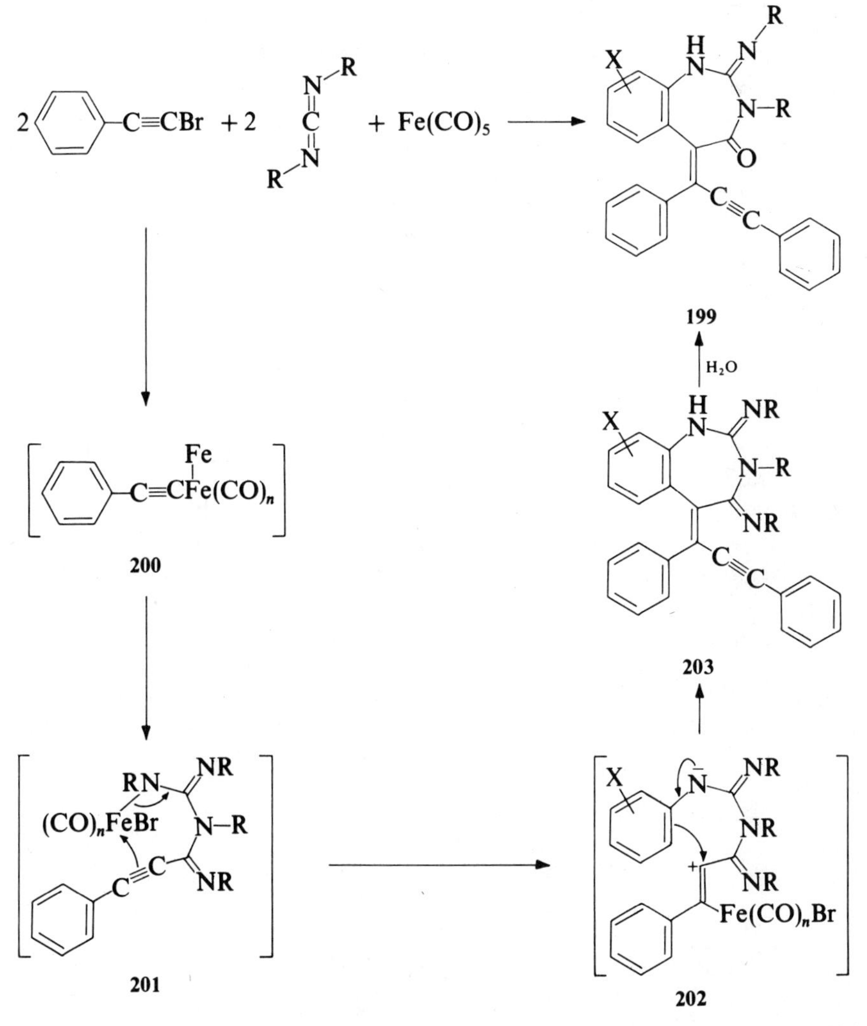


(64)

to the triple bond would result in **202**, which can cyclize and rearomatize by proton shift. Introduction of another phenylacetylene by coupling would lead to the 4-imino derivative **203**. This compound was not isolated but underwent hydrolysis during workup to yield the observed product **199** (Eq. 64).

The tetrahydro-1,3-benzodiazepin-2,5-dione **206** was prepared (Eq. 65) by reaction of the diaminoacetophenone **204** with carbonyldiimidazole.[33] The initially formed imidazolcarboxamide **205** could be isolated and subsequently cyclized to the benzodiazepine in 88% yield by heating in water at 90°C.

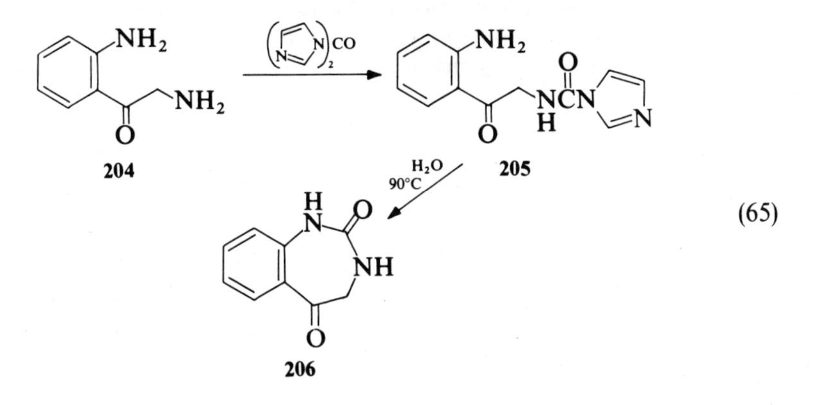


(65)

## *1.7.2. Reactions*

Electrophilic substitution reactions were used[33] successfully to prepare the 7-halo and 7-nitro derivatives **208**. The 7-nitro compound was reduced to the corresponding amine **209** with stannous chloride (Eq. 66). Chloroacetylation of the amine yielded the chloroacetate **210** (R = Cl), which was converted to basic derivatives by displacement of chloride with morpholine or piperidine.

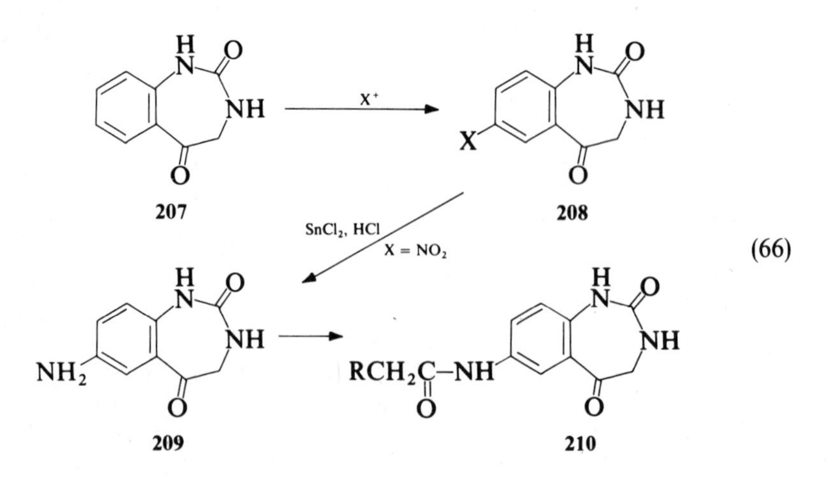


(66)

Reaction of **207** and **208** (X = $NO_2$) with hydroxylamine (Eq. 67) gave the corresponding oximes **211** (X = H, $NO_2$). When **207** was treated with pyrrolidine and a catalytic amount of *p*-toluenesulfonic acid in boiling toluene, the quinoline **212** was obtained. This would indicate that the anticipated enamine is formed, but is so reactive that it undergoes ring contraction in situ.

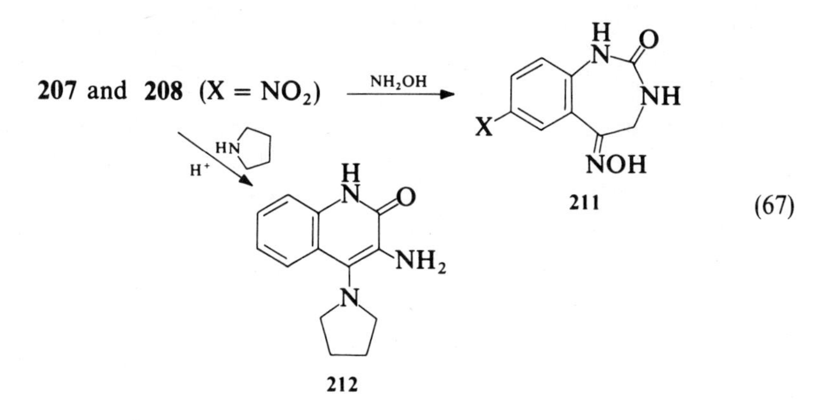


(67)

The expected reactivity shown by the 5-oxo group during the addition of hydride or metallorganic reagents has been discussed. The conversion of the urea carbonyl at the 2-position to 2-ethoxy and 2-amino groups also was mentioned above.

## 1.8. Tetrahydro-1,3-benzodiazepin-2-thiones

### *1.8.1. Synthesis*

The tetrahydro-1,3-benzodiazepin-2-thiones **214** have been prepared (Eq. 68) by reaction of the phenethylamines **213** with thiophosgene in the presence of imidazole[36] or by condensation with carbon disulfide.[39, 40, 44]

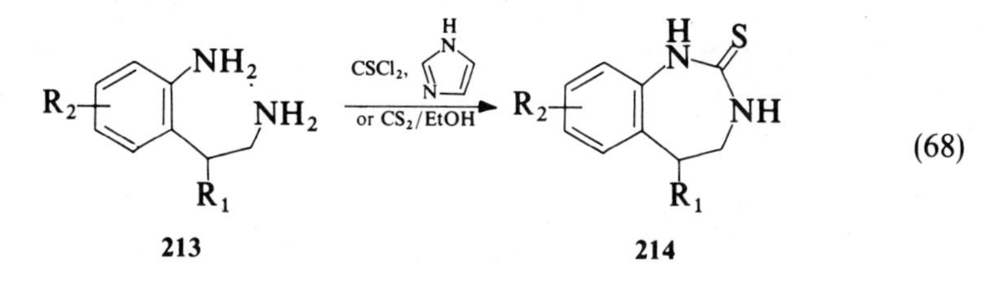


(68)

Compound **216** was obtained in 78% yield[56] by cyclodehydration of the thiourea **215**, which was accessible by reaction of the *o*-methylaminophenylacetic acid sodium salt **191** with methyl isothiocyanate (Eq. 69).

Ghosh[60] assigned the structures of 1,3-benzodiazepin-2-thioxo-4,5-diones **219** to the compounds obtained by dehydration–cyclization of the thioureas **218**

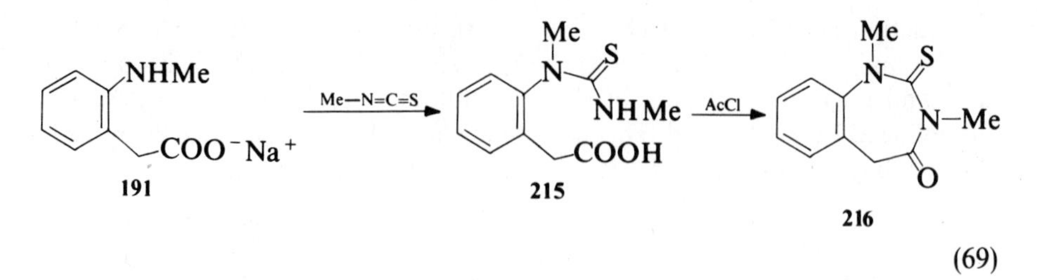


(69)

in boiling acetic anhydride (Eq. 70). The latter were formed by reaction of the sodium salt of the acid **217** with aryl isothiocyanates. The alternative isatin structure was excluded because the compound was found to be stable to hot hydrochloric acid.

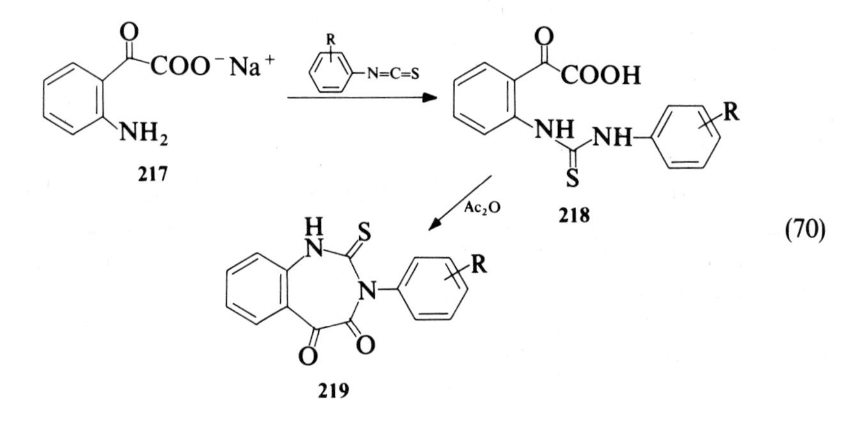


(70)

### *1.8.2. Reactions*

The conversion of the 2-thiones **214** to 2-thiomethyl derivatives by alkylation[36b, 39, 40] has been discussed. Reaction of **220** with a variety of $\alpha$-haloketones and $\alpha$-haloesters led to the thiazolobenzodiazepines **221**, **222**, and **223** (Eq. 71).[44]

## 1.9. Tetrahydro-2-cyanoimino-1,3-benzodiazepine

The 2-cyanoimino-1,3-benzodiazepine **226** was prepared by cyclization of *N*-2-(2-aminophenyl)ethylbenzylamine (**225**) with dimethyl cyanoimidothiocarbonate. Compound **225** was prepared by hydrogenation of the corresponding nitro derivative **224**, using a platinum catalyst. Reaction of **226** with ethyl bromoacetate gave the 1-ethoxycarbonylmethyl derivative **227** (Eq. 72).[102]

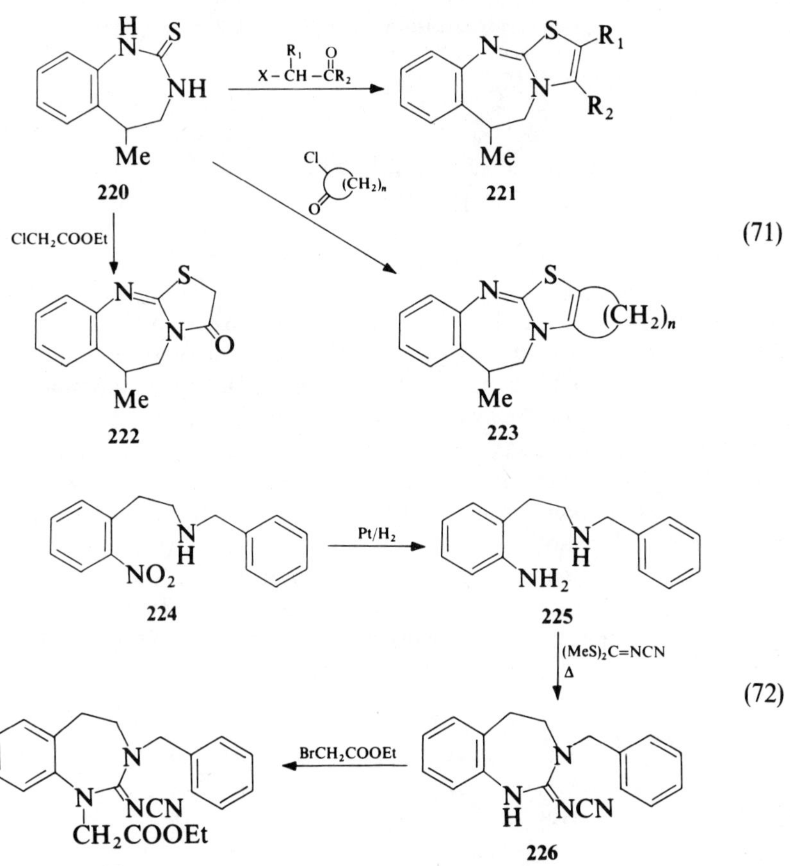


## 2. IMIDAZO[4,5-*d*] [1,3]DIAZEPINES

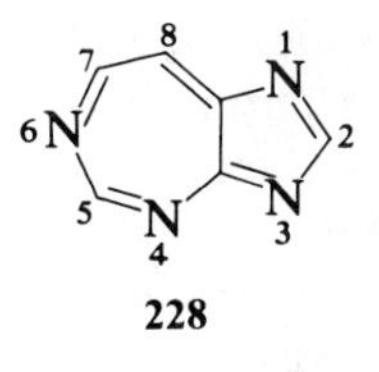


Of the large number of possible bicyclic systems consisting of a 1,3-diazepine with a [*d*]-fused heterocycle, representatives of the imidazo[4,5-*d*] [1,3]-diazepine ring system appear to be the only known compounds of the type. The parent ring system, **228**, is still unreported, but tetrahydro derivatives have been isolated from natural sources.

### 2.1. Tetrahydroimidazo[4,5-*d*] [1,3]diazepines

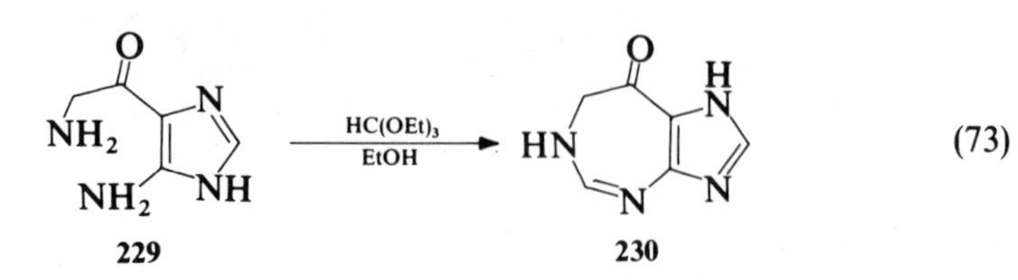


According to a U.S. patent,[61] the 1,6,7,8-tetrahydroimidazo [4,5-*d*] [1,3]diazepin-8-one **230** was prepared by reaction of the diamine **229** with thiethyl orthoformate in boiling ethanol (Eq. 73). This compound is the key intermediate used for the synthesis of the naturally occurring glycosides coformycin (**231**) and 2′-desoxycoformycin (covidarabine, **232**). Both compounds were isolated from bacterial fermentations, coformycin[62] from culture filtrates of *Streptomyces kamikawaensis* SF-557 covidarabine[63] from *Streptomyces antibioticus* NRRL 3238. Both compounds are potent inhibitors of adenosine deaminase and therefore enhance the activity of nucleosides such as formycin and vidarabine. The structures of both coformycin[64] and covidarabine[65] have been elucidated by spectroscopic means and confirmed by X-ray crystallographic analyses.

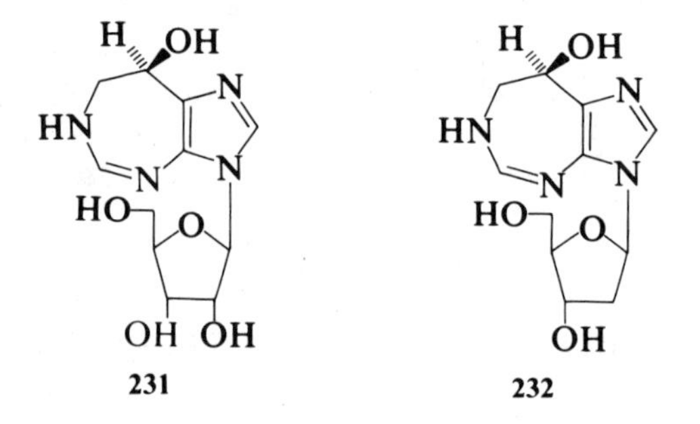


A synthesis of coformycin, which may mimic its biological formation, was achieved by ring expansion of a purine riboside.[66] Photochemical addition of methanol to the acetylated $\beta$-D-ribofuranosyl purine **233** (R = Ac) led stereospecifically to **234** (R = Ac) in 96% yield. The alcohol was converted to the mesylate, which was treated with potassium *t*-butoxide to form the intermediate aziridine **235**. This intermediate was not isolated but subjected to hydrolysis to yield coformycin **231** in 35% overall yield from **234** (Eq. 74). A Belgian patent[67] claims the synthesis of isocoformycin **236** in 21% yield by treatment of the mesylate of **234** with aqueous hydroxide.

Baker and Putt[61] have described a total synthesis of **232**. The imidazodiazepinone **230** was alkylated with 3,5-di-*O*-(*p*-toluoyl)-$\alpha$-D-erythropentofuranosyl chloride **237** after protection with bis(trimethylsilyl) acetamide.

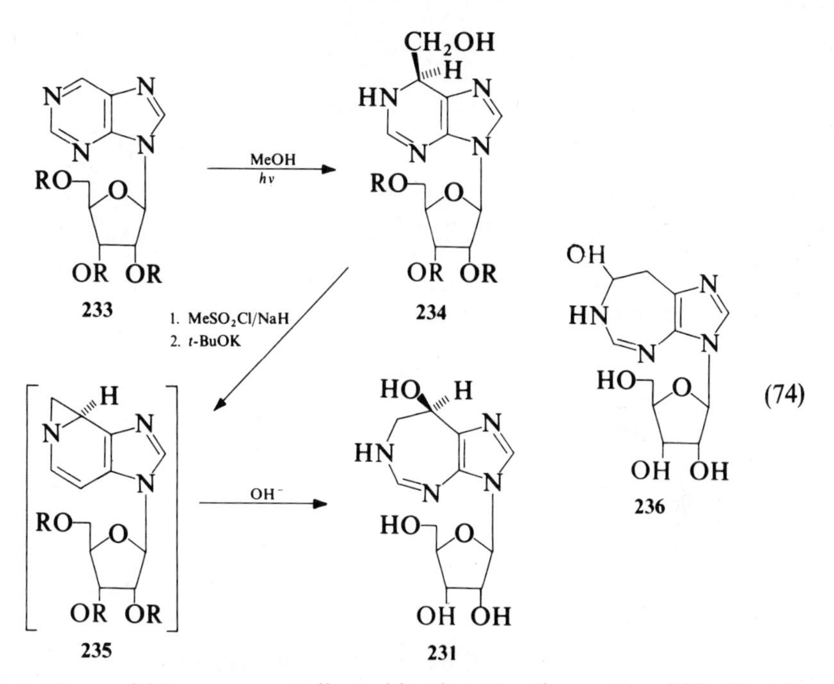


Deprotection with aqueous sodium bicarbonate then gave **238** (R = 4-$MeC_6H_4CO$). Removal of the *p*-toluoyl groups by transesterification followed by reduction of the carbonyl group with sodium borohydride led to a mixture of **232** and its 8*S* isomer (Eq. 75). The sequence of reduction and transesterification could also be reversed.

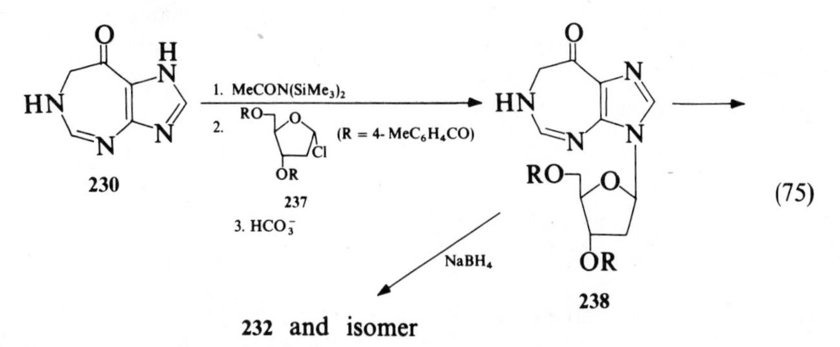


More recently, the synthesis of the related 8-desoxy analogs have been reported.[105]

## D. [*e*]-FUSED [1,3]DIAZEPINES

This section reviews diazepines of general formula **239**. The benzannelated system is given priority.

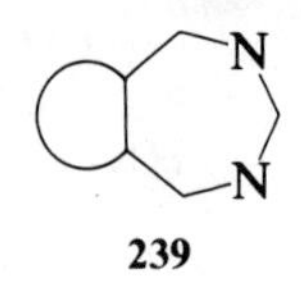

239

# 1. BENZO[*e*] [1,3]DIAZEPINES (2,4-BENZODIAZEPINES)

Neither of the two possible parent 2,4-benzodiazepine tautomers **240a** or **240b** has been described.

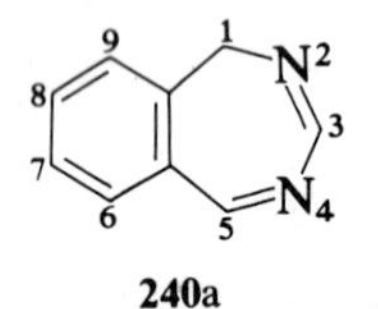

**240a**

1*H*-2,4-Benzodiazepines

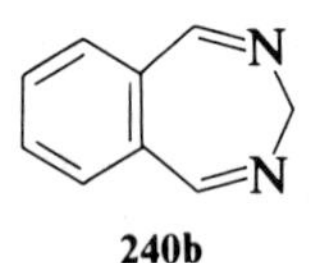

**240b**

3*H*-2,4-Benzodiazepines

## 1.1. 4,5-Dihydro-2,4-benzodiazepines

### *1.1.1. Synthesis*

Some of the methods used for the synthesis of 1,3-benzodiazepines were also found to be useful for the preparation of 2,4-benzodiazepines of the cyclic amidine type. Thus, condensation of the diamine **241** with imidates,[36a, 68] amidines,[37] or nitriles[69] led to 2,5-dihydro-1*H*-2,4-benzodiazepines **242**, generally characterized as salts (Eq. 76).

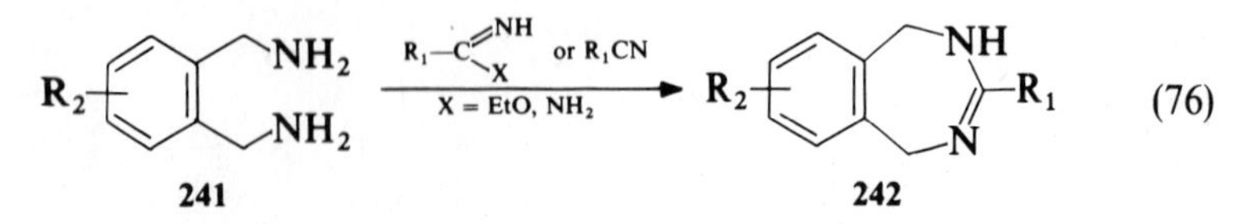

(76)

3-Thioalkyl derivatives **244** were obtained[37–39, 70–72] by alkylation of the thione **243** (Y = S). 3-Amino derivatives **245** were accessible (Eq. 77) either by treatment of the urea **243** (Y = O) with phosphoryl chloride and an amine[36a] or by displacement of the methylthio group of **244** (R = Me) with an amine.[73]

### *1.1.2. Reactions*

Several unsuccessful attempts were made to *N*-alkylate compounds **242**.[36a] The displacement of the 3-methylthio group of **244** (R = Me) by amines was cited in Section 1.1.1.

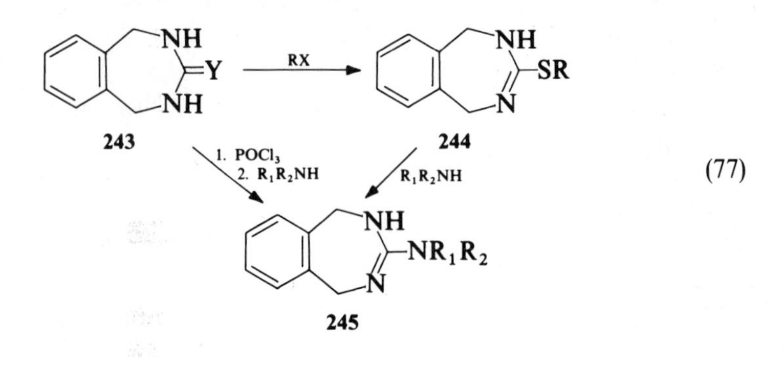


(77)

## 1.2. Dihydro-2,4-benzodiazepinones

### 1.2.1. Synthesis

Syntheses of 2,3-dihydro-5-phenyl-1*H*-2,4-benzodiazepin-1-ones **253** have been published[74,75] and patented.[76] Treatment of the chloromethyl derivative **248** ($R_1$ = H, $R_2$ = OH) with aqueous ammonia in dioxane (Eq. 78) gave, after chromatography, a 12% yield of the benzodiazepine **253** ($R_1$ = H). This low yield conversion of the chloride **248** to the benzodiazepine **253** was improved considerably by preparing **250** ($R_2$ = OH) via the azide **249** and subjecting **250** ($R_2$ = OH) to acid-catalyzed dehydration. Boiling a solution of **250** ($R_1$ = Cl, $R_2$ = OH) in benzene in the presence of a catalytic amount of para-toluenesulfonic acid afforded **253** ($R_1$ = Cl). The formation of **253** from **250** ($R_2$ = OH) may proceed via the benzophenone **251** (X = O) or the diazetidine **252**. Studies in our laboratories[77] showed that the reaction of **247** ($R_1$ = Cl) with thionyl chloride also led to the dichloride **248** ($R_1$, $R_2$ = Cl) which could be converted to the diamine **250** ($R_1$ = Cl, $R_2$ = $NH_2$). This diamine was also transformed to the benzodiazepine upon heating in toluene solution containing acetic acid, possibly via the intermediate imine **251** (X = NH).

Attempts[47] to prepare the 2,5-dihydro-3-methyl-1*H*[2,4]benzodiazepin-1-one **255** by cyclodehydration of the benzamide **254** were unsuccessful (Eq. 79).

### 1.2.2. Reactions

Oxidation of the benzodiazepinones **253** with *m*-chloroperbenzoic acid (Eq. 80) afforded[74,75] a mixture of the 4-oxide **256** and the oxaziridine **257**. Separation of these compounds was achieved by chromatography.

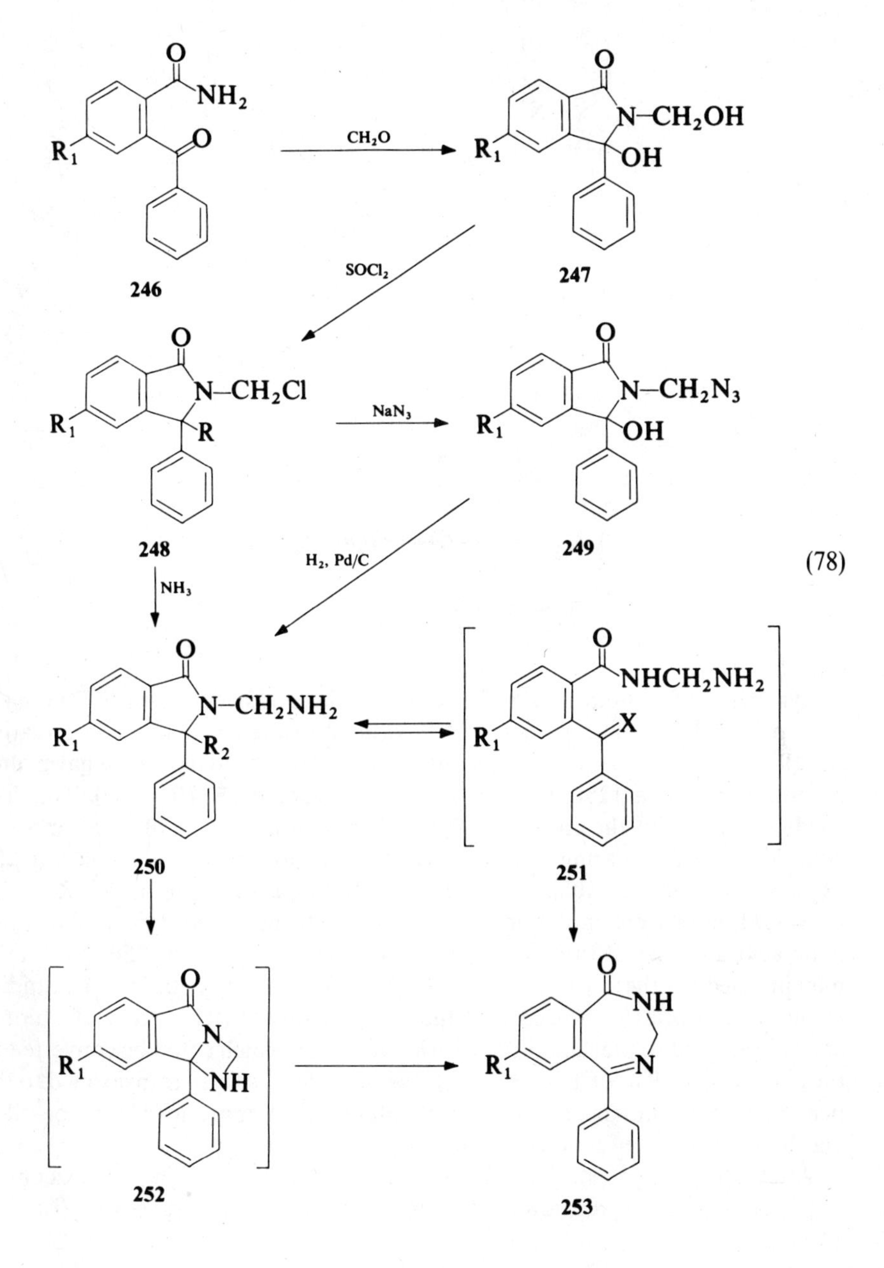

NH2
O
R1
CH2O
N–CH2OH
OH
246
247
SOCl2
N–CH2Cl
R
NaN3
N–CH2N3
248
249
H2, Pd/C
NH3
N–CH2NH2
R2
NHCH2NH2
X
250
251
NH
N
252
253


(78)

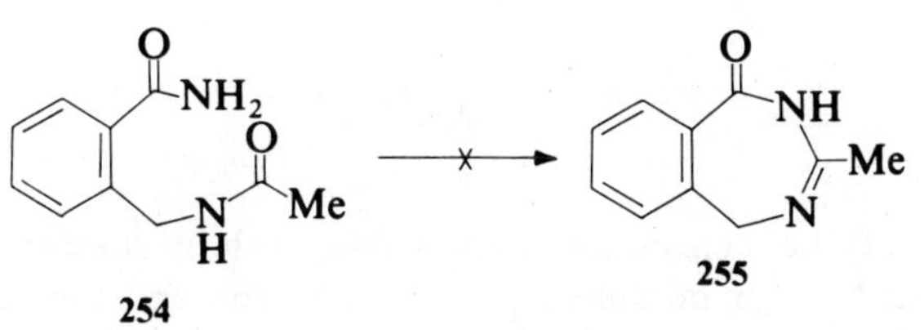

NH2
O
N
H
Me
NH
Me
N
254
255


(79)

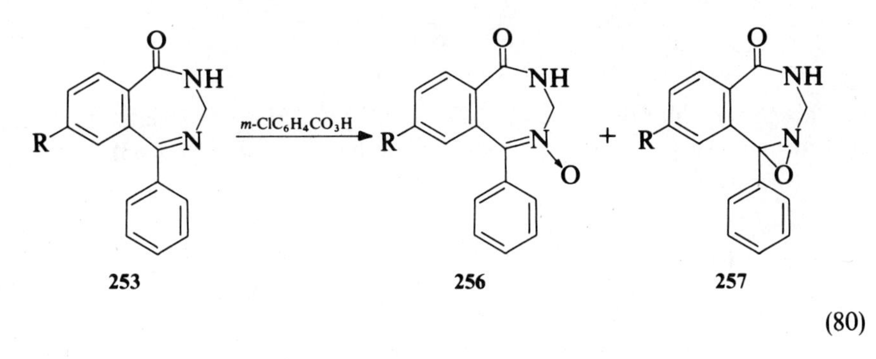


(80)

## 1.3. Tetrahydro-2,4-benzodiazepines

The synthesis of 2,4-diaryl-2,3,4,5-tetrahydro-1*H*-2,4-benzodiazepines **259** was described in the early 1900s by Scholtz and coworkers,[78,79] who obtained these compounds by reaction of the diamines **258** with aldehydes in boiling ethanol (Eq. 81). With $R_1$ = 2-Me, cyclization had to be effected in concentrated hydrochloric acid.[79]

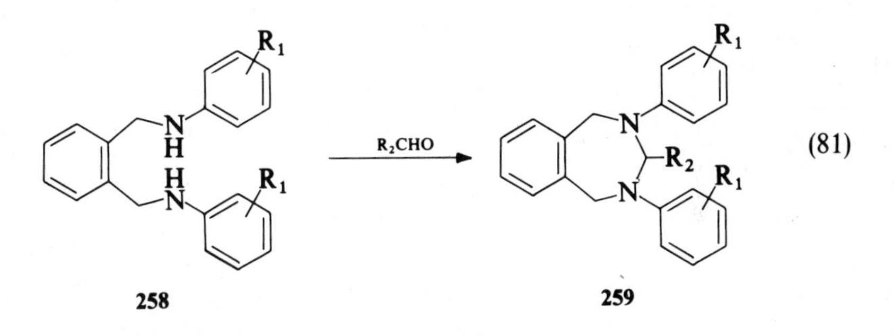


(81)

## 1.4. Tetrahydro-2,4-benzodiazepinones

### 1.4.1. Synthesis

Reaction of the diamines **260** with carbonyldiimidazole (Eq. 82) gave the tetrahydro-2,4-benzodiazepin-3-ones **261** in high yield.[36a,80]

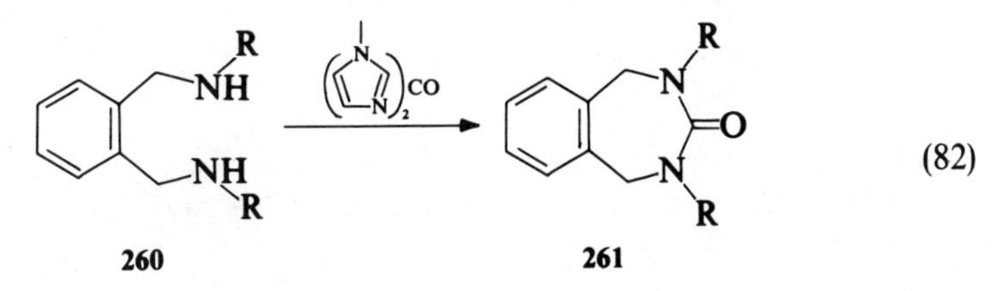


(82)

### *1.4.2. Reactions*

Phosphoryl chloride converted **262** into the *O*-dichlorophosphate **263**,[36a] an intermediate used in the preparation of 3-amino derivatives (Eq. 83).

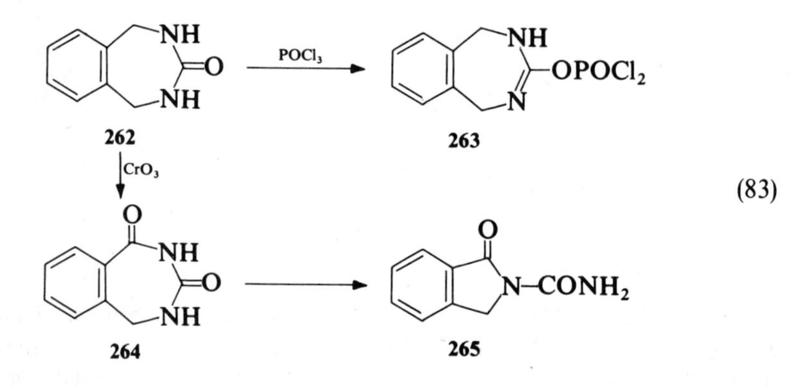


Oxidation of **262** with chromium trioxide was shown[80] to give the isoxindole **265** rather than the expected dione **264**. It was postulated that **264** is in fact the primary oxidation product, but that it undergoes spontaneous rearrangement and ring contraction to give **265**.

## 1.5. Tetrahydro-2,4-benzodiazepinediones

### *1.5.1. Synthesis*

Felix and Fryer[80] obtained the tetrahydro-2,4-benzodiazepin-1,3-dione **267** by oxidation of the 2,4-dibenzyl derivative **266** with chromium trioxide (Eq. 84). They reported that the minor product from the reaction of benzaldehyde azine with carbon monoxide, once believed to be **264**,[81] actually has the ring-contracted structure **265**.

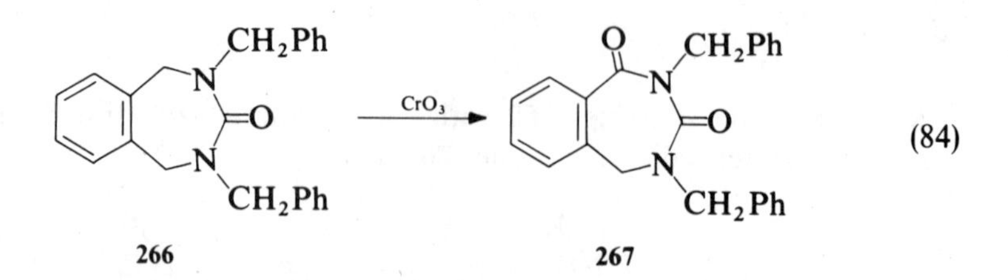


A facile synthesis of the 1,5-dione **269** was described,[82,106] in which condensation of phthaloyl chloride with the amidine **268** gave the benzodiazepine in 80% yield (Eq. 85).

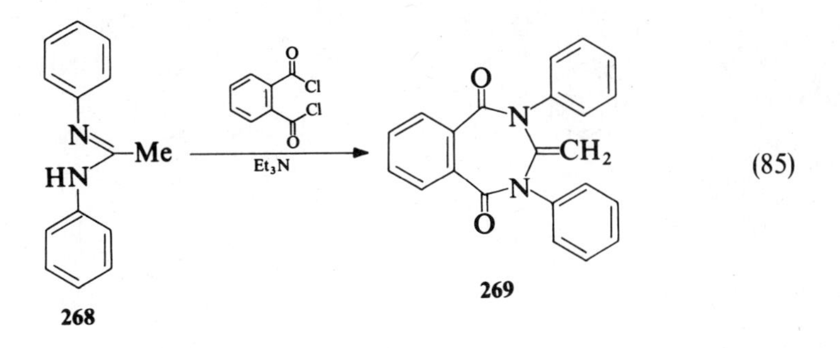


(85)

### 1.5.2. Reactions

Compound **269** was cleaved by hot polyphosphoric acid to give a mixture of *N*-phenylphthalimide and acetanilide in 76 and 48% yield. The phthalimide was also isolated from the reaction of **269** with chromic acid or bromine.

## 1.6. Tetrahydro-2,4-benzodiazepin-1,3,5-triones

Syntheses of the trione **271** ($R_1$, $R_2$ = H)[83,84] and ($R_1$ = H; $R_2$ = alkyl, aryl)[85] have been reported, but the assigned structures were shown to be wrong.[86] More recently,[87] the 2,4-dialkyl derivative **271** ($R_1$, $R_2$ = $CH_2CH_2Cl$) was prepared in 95% yield by reaction of the oxazoline **270** with phthaloyl chloride (Eq. 86).

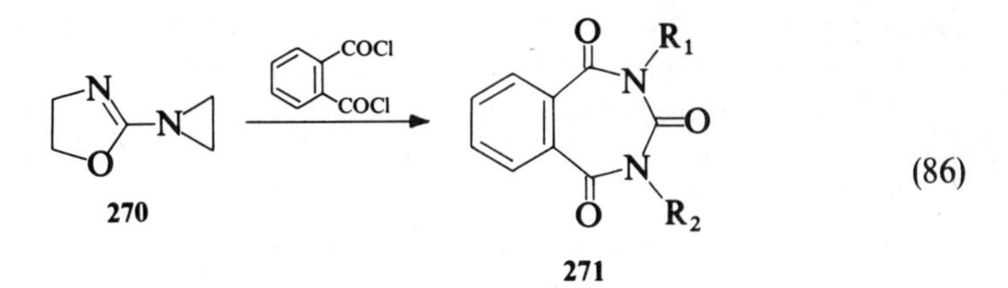


(86)

## 1.7. 2,4-Benzodiazepinthiones

### 1.7.1. Synthesis

Treatment of **272** with phosphorus pentasulfide in pyridine gave the corresponding thione **273** (Eq. 87).[88]

Tetrahydro-2,4-benzodiazepin-3-thiones **275** were prepared by reaction of the diamine **274** with thiophosgene in the presence of imidazole,[68] or by ring closure of the dithiocarbamic acid adduct **276** obtained from carbon disulfide and **274** (Eq. 88).[70]

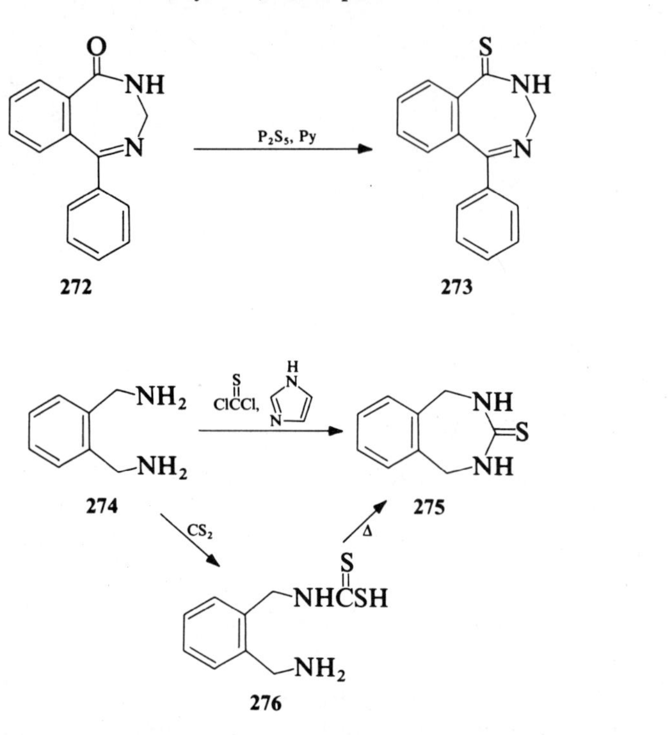


Fujita and Sato[89] synthesized a variety of tetrahydro-3-thioxo-2,4-benzodiazepin-1-ones **278** by base-catalyzed reaction of 2-chloromethylbenzoyl chloride with the disubstituted thioureas **277** (Eq. 89). The isomeric 2,4-benzothiazepines **279** were usually the predominant products, and the ratio of the two products **278** and **279** was found to be dependent on the substituents $R_1$, $R_2$ of the thiourea and also on the choice of solvent and base. A high proportion

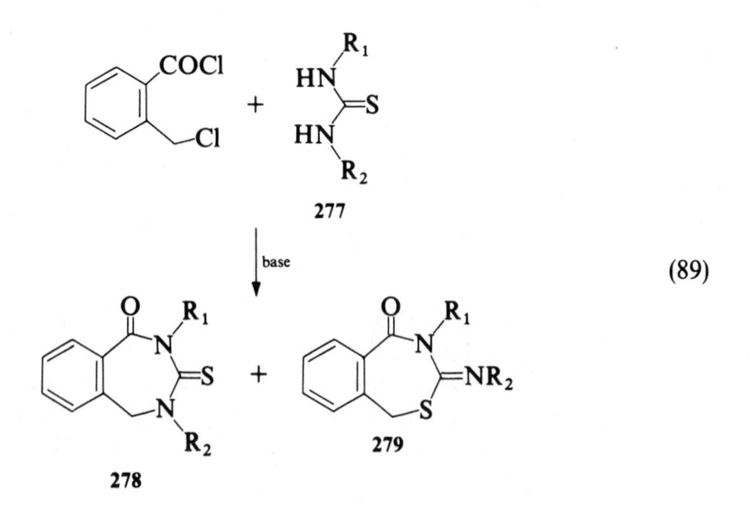

of the benzodiazepine **278** was obtained when $R_1$ and $R_2$ were methyl or benzyl groups. The structures of the products **278** obtained from unsymmetrically substituted thioureas were assigned on the basis of nmr spectroscopic comparison with the compounds prepared from symmetrically substituted thioureas. The 5-methylene protons in **278** appear as an AB system with $J = 15$ Hz, indicating slow ring inversion at room temperature.

### 1.7.2. Reactions

Heating the thione **273** with acetylhydrazine in *n*-butanol gave the triazolophthalazine **280** in 60% yield (Eq. 90).[88]

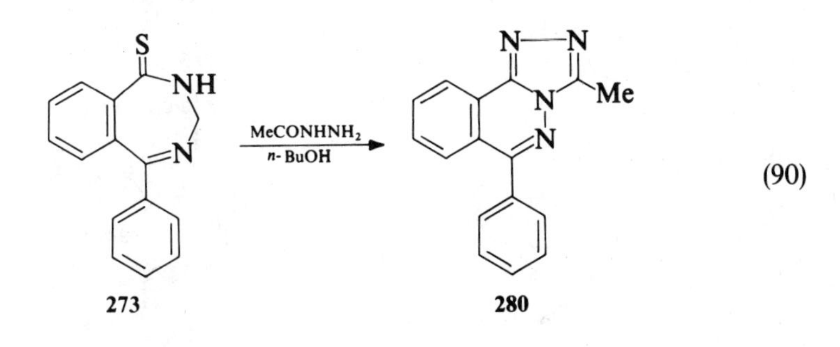


(90)

Alkylations of the 3-thione **275** leading to 2,5-dihydro-3-alkylthio-1*H*-2,4-benzodiazepines[70,72] were cited earlier. A double alkylation of **275** with 1,2-dibromoethane (Eq. 91) led to the tricyclic compound **281**.[70] Reaction with α-haloesters afforded **282**,[70] while α-haloketones yielded the thiazolo derivatives **283**.[70,72]

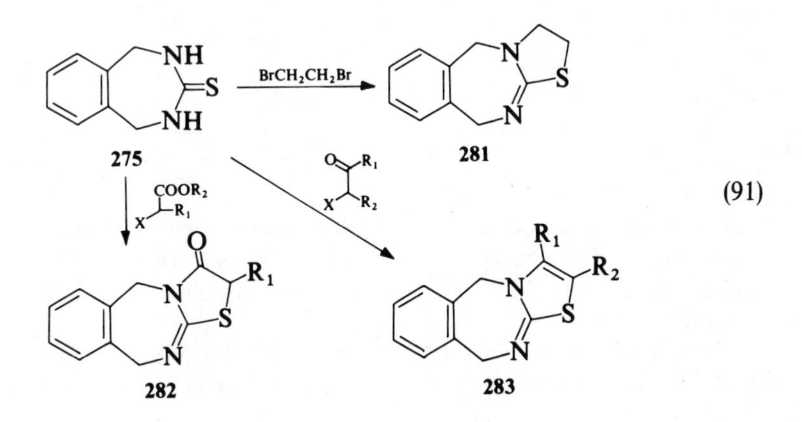


(91)

### 1.8. 3-Cyanoimino-2,4-benzodiazepine

#### *1.8.1. Synthesis*

Ishikawa and Watanabe[102] reported the synthesis of the 3-cyanoimino-2,4-benzodiazepine (**285**) by reaction of *o*-xylene diamine (**284**) with dimethyl cyanoimidodithiocarbonate. Treatment of **285** with ethyl bromoacetate gave *N*-substituted compound **286**, which upon heating in *t*-butanol and hydrochloric acid, gave the 2,3,5,10-tetrahydro-1*H*-imidazo[2,1-*b*] [2,4]benzodiazepin-2-one (**287**) (Eq. 92).

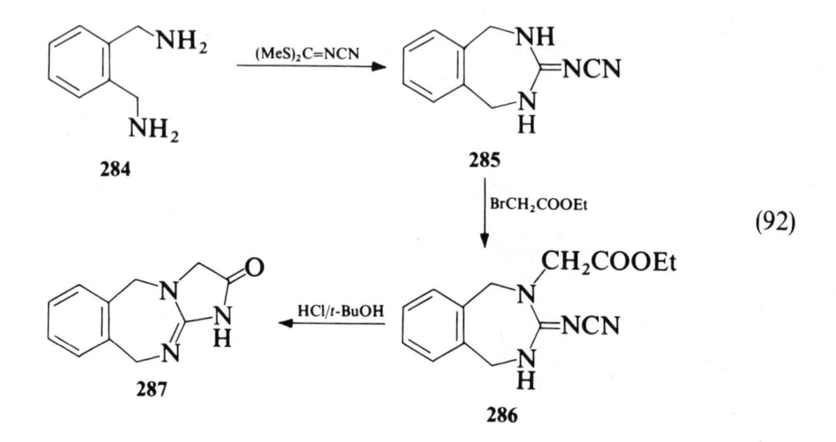


(92)

## 2. CYCLOPENTA[*e*] [1,3]DIAZEPINES

### 2.1. Synthesis

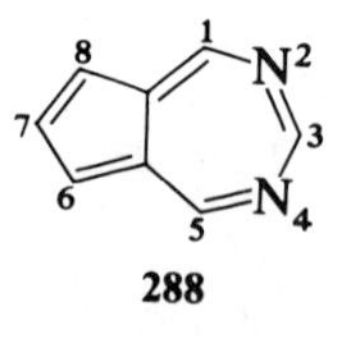


Although the parent ring system **288** (also named 5,7-diazaazulene) is still unknown, the 3-dimethylamino derivative **290** was synthesized[90] in 50% yield by condensation of the cyclopentadiene **289** with dimethylguanidine in boiling ethanol (Eq. 93).

The analogous reaction of **289** with acetamidine did not lead to the anticipated 3-methylcyclopenta[*e*] [1,3]diazepine, but instead gave a very low yield of

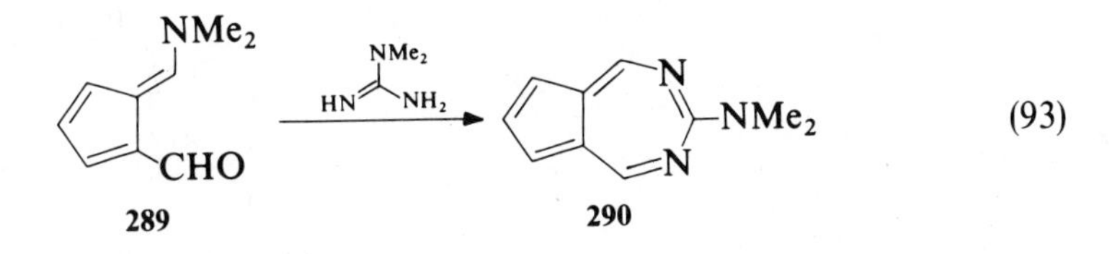


**290**. The structure of **290** was derived from spectral data and confirmed by X-ray crystallographic analysis.[91]

### 2.2. Reactions

According to ultraviolet and nmr spectra, protonation of **290** occurs on the ring nitrogens. The action of acylating agents on **290** led to polymerization. Alkaline hydrolysis (Eq. 94) converted **290** to the sodium salt **291** in 70% yield.[90a]

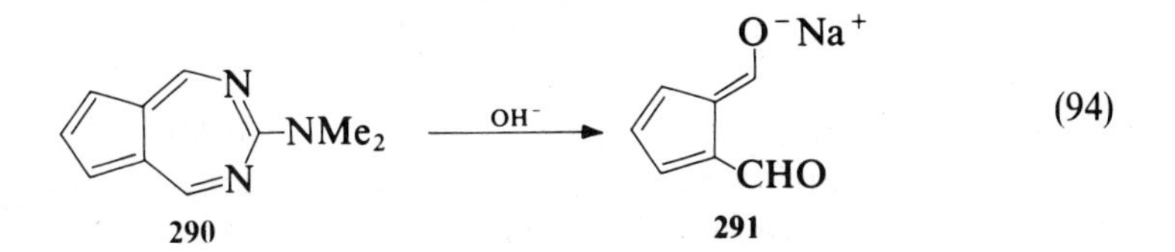


## 3. THIENO[3,4-*e*] [1,3]DIAZEPINES

### 3.1. Synthesis

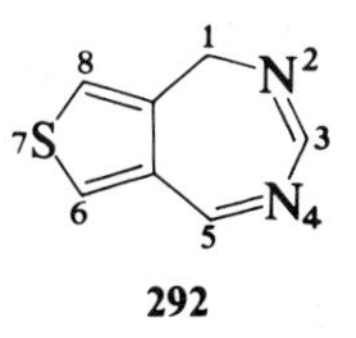


Thieno [3,4-*e*] [1,3]diazepines represented by the 1*H* tautomer **292** have not been prepared, although tetrahydro derivatives are disclosed in the literature[92] and are derived from the reaction of the thiophene **293** with hexamethylenetetramine (Eq. 95).

Steam treatment of the quaternary hexamine salt of **293** gave the formyl derivative **294** in about 30% yield. The formation of **294** requires a redox reaction of a bis(hydroxymethyl) intermediate. The structure of **294** was supported both by alkaline hydrolysis to **295** and acid hydrolysis to the diamine **296**. Reaction of **296** with formaldehyde and formic acid or condensation of **295** with methyl formate led to a recovery of **294**.

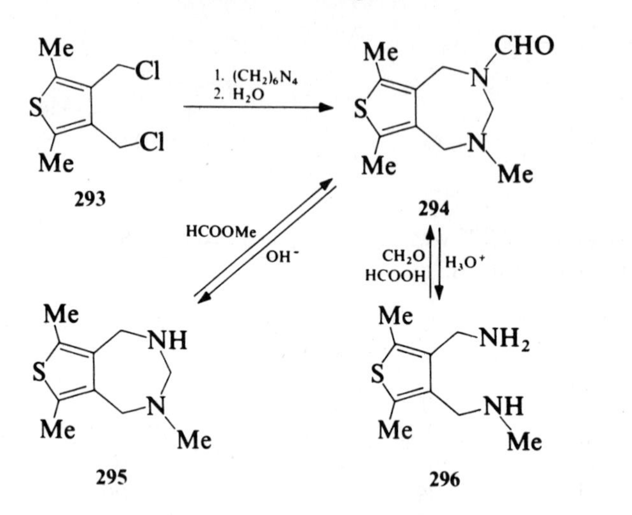


(95)

### 3.2. Reactions

Quaternization of **294** with methyl iodide led to **297**, which was cleaved readily to the diamine **298** under hydrolytic conditions (Eq. 96). Alkaline and acid hydrolysis of **294** was discussed in the preceding section.

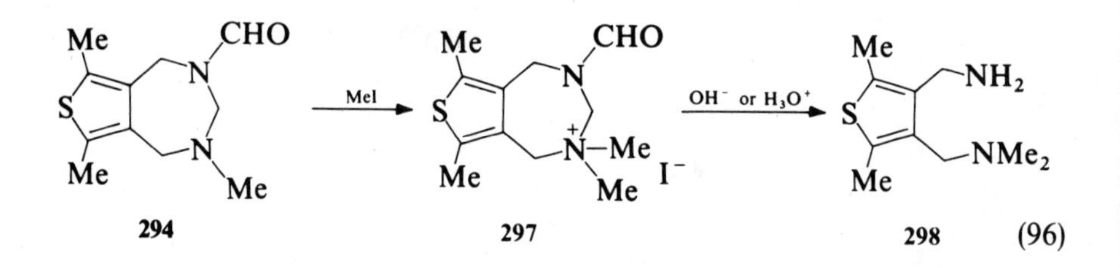


(96)

## 4. FURO[3,4-*e*] [1,3]DIAZEPINE AND PYRROLO[3,4-*e*] [1,3]DIAZEPINE

[1,3]Diazepin-1,5-(2*H*)-dione **303** and [1,3]diazepin-1,5-(2*H*,7*H*)-dione **304** were prepared by treatment of the appropriate heterocyclic dicarboxylic acid

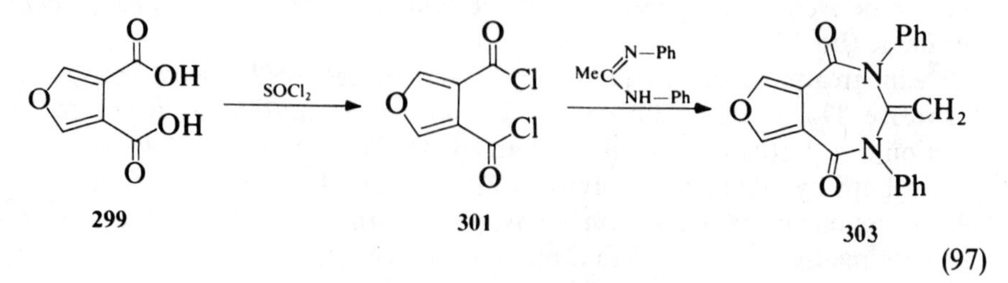


(97)

chloride **301** (Eq. 97) and **302** (Eq. 98) with *N*,*N*′-diphenylacetamidine in the presence of triethylamine.[106]

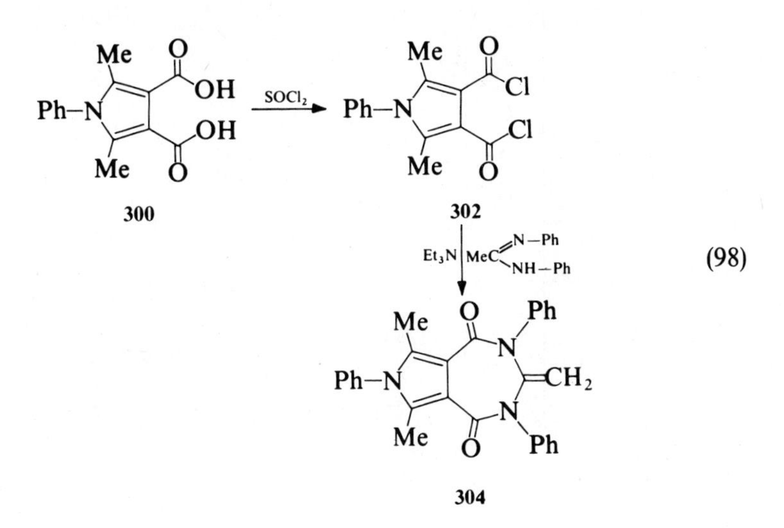


(98)

# E. TABLES OF COMPOUNDS

TABLE II-1. [*a*]-FUSED [1,3]DIAZEPINES

| Substituent | mp (°C); [bp (°C/torr)] | Solvent of Crystallization | Yield (%) | Spectra | Refs. |
|---|---|---|---|---|---|
| | *Diazeto[1,2-a] [1,3]diazepines* | | | | |
| ***1,2,4,5,6,7-Hexahydro-1,3-diazeto[1,2-a] [1,3]diazepine*** | | | 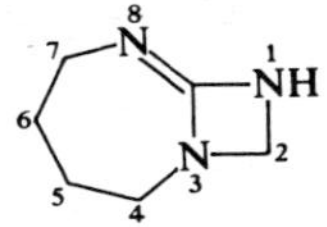 | | |
| 1-Ph-2-(Ph)imino | 105–106 | MeOH | 66 | ir, pmr | 1 |
| | *Imidazo[1,2-a] [1,3]diazepines* | | | | |
| ***2,3,5,6,7,8-Hexahydro-1H-imidazo[1,2-a] [1,3]diazepines*** | | | 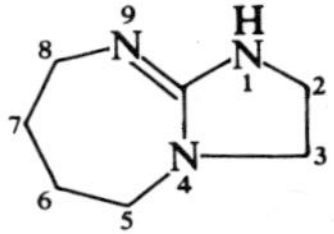 | | |
| None | 109–111 | Ac | 63 | | 2a |
| | [118–122/0.26] | | | $pK_a$ | 2c |
| Picrate | 216–218 | MeOH | | | 2a |
| 1-$NO_2$ | 166.5–167.5 | $Et_2O$ | 100 | | 2a |
| 1-(4-$NH_2C_6H_4SO_2$) | 207–208.5 | EtOAc | 63 | | 6 |
| 1-(4-$AcNHC_6H_4SO_2$) | 236–238 | EtOAc | 72 | | 6 |
| 1-(4-$MeC_6H_4SO_2$) | 110.5–111.5 | EtOAc | 86.5 | | 6 |

| | | | | | |
|---|---|---|---|---|---|
| ***2,3,5,6,7,8-Hexahydro-1H-imidazo[1,2-a] [1,3]diazepines*** | | | | | |
| 1-$CH_2$-Ph | 200–203 | Acetone | 30 | pmr, ir, anal | 93 |
| 2,2-$(Ph)_2$ | 163–164 | AcOEt | 35 | ir, anal | 94 |
| ***2,3,5,6,7,8-Hexahydro-9H-imidazo[1,2-a] [1,3]diazepines*** 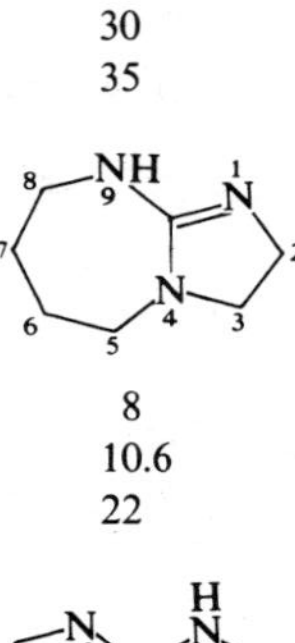 | | | | | |
| 9-(3-$ClC_6H_4$) | Oil | | 8 | | 4 |
| 6,7-$(OH)_2$-9-(2,4-$Cl_2C_6H_3$) | 213–214 | | 10.6 | | 4 |
| 6,7-$(OH)_2$-9-(2,6-$Cl_2C_6H_3$) | 214–215 | Ac | 22 | | 4 |
| ***1,2,5,6,7,8-Hexahydroimidazo[1,2-a] [1,3]diazepin-3-(3H)-one*** 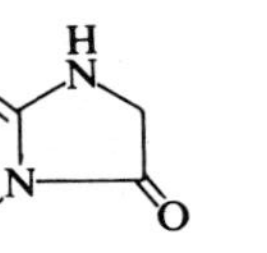 | | | | | |
| 2(RS)-$CH_2COOH$-8(S)-COOH | 295d | | | pmr | 5 |
| *Imidazo[1,5-a] [1,3]diazepines* | | | | | |
| ***1,3,4,5-Tetrahydroimidazo[1,5-a] [1,3]diazepin-2(2H)-one*** | | | | | |
| 7-Me-9-$NO_2$ | 164–165 | EtOH | 52 | ir, uv | 6 |

TABLE II-1. —(contd.)

| Substituent | mp (°C); [bp (°C/torr)] | Solvent of Crystallization | Yield (%) | Spectra | Refs. |
|---|---|---|---|---|---|
| | | *Pyrido[1,2-a] [1,3]diazepines* | | | |
| | | | 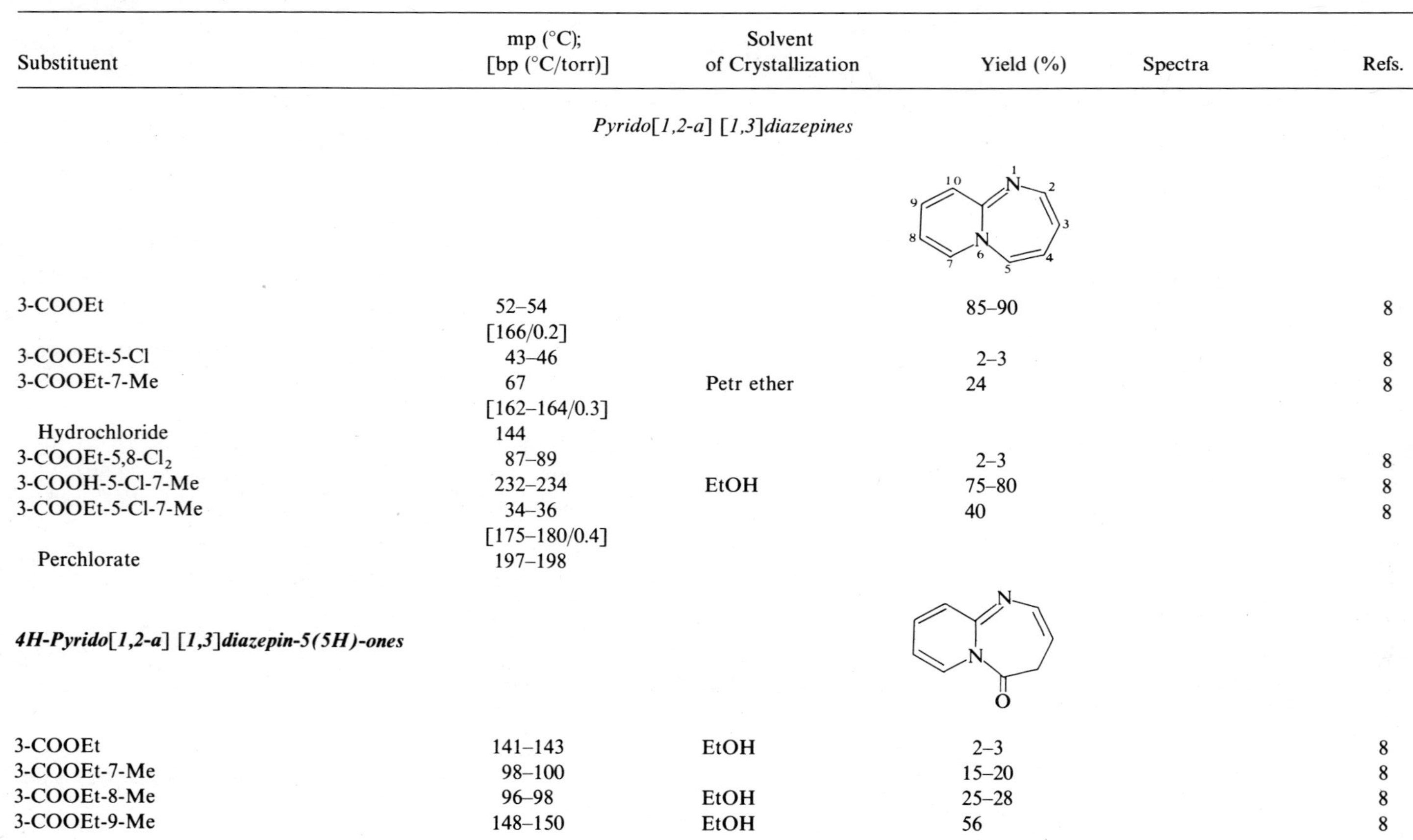 | | |
| 3-COOEt | 52–54 | | 85–90 | | 8 |
| | [166/0.2] | | | | |
| 3-COOEt-5-Cl | 43–46 | | 2–3 | | 8 |
| 3-COOEt-7-Me | 67 | Petr ether | 24 | | 8 |
| | [162–164/0.3] | | | | |
| Hydrochloride | 144 | | | | |
| 3-COOEt-5,8-$Cl_2$ | 87–89 | | 2–3 | | 8 |
| 3-COOH-5-Cl-7-Me | 232–234 | EtOH | 75–80 | | 8 |
| 3-COOEt-5-Cl-7-Me | 34–36 | | 40 | | 8 |
| | [175–180/0.4] | | | | |
| Perchlorate | 197–198 | | | | |
| ***4H-Pyrido[1,2-a] [1,3]diazepin-5(5H)-ones*** | | | | | |
| 3-COOEt | 141–143 | EtOH | 2–3 | | 8 |
| 3-COOEt-7-Me | 98–100 | | 15–20 | | 8 |
| 3-COOEt-8-Me | 96–98 | EtOH | 25–28 | | 8 |
| 3-COOEt-9-Me | 148–150 | EtOH | 56 | | 8 |

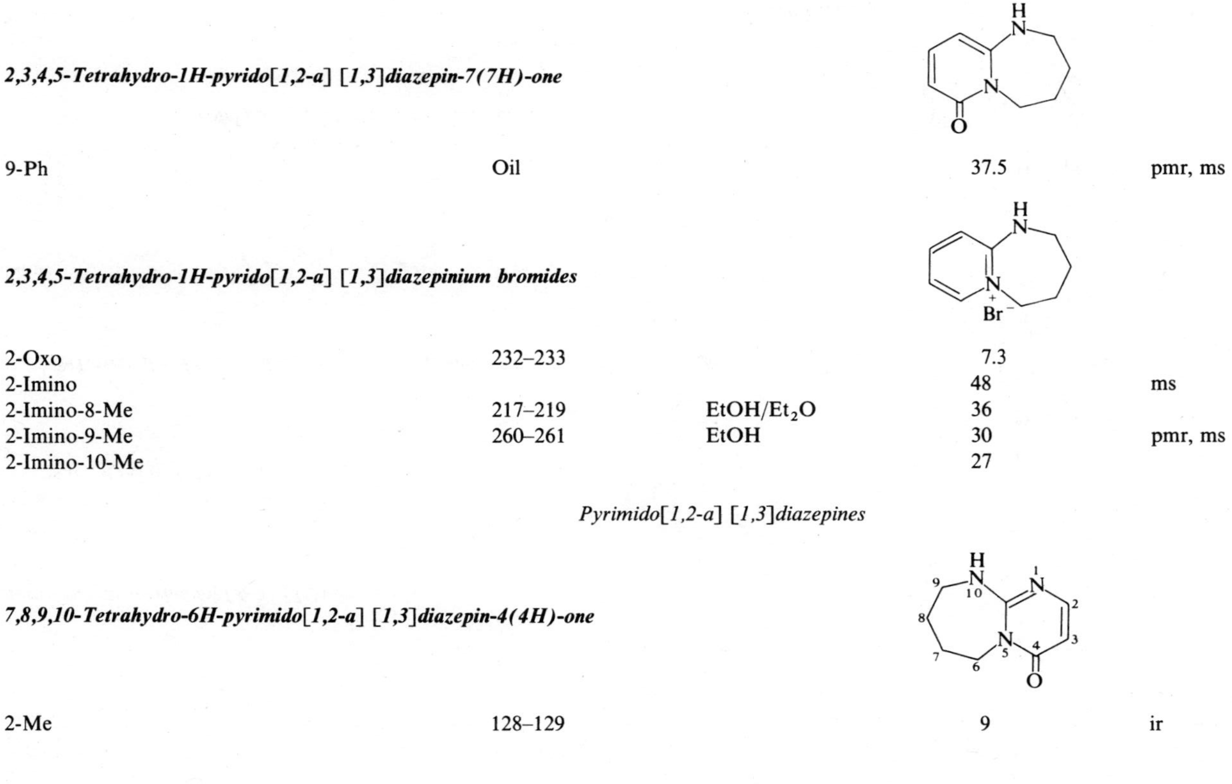

***2,3,4,5-Tetrahydro-1H-pyrido[1,2-a] [1,3]diazepin-7(7H)-one***

| | | | | | |
|---|---|---|---|---|---|
| 9-Ph | Oil | | 37.5 | pmr, ms | 12 |

***2,3,4,5-Tetrahydro-1H-pyrido[1,2-a] [1,3]diazepinium bromides***

| | | | | | |
|---|---|---|---|---|---|
| 2-Oxo | 232–233 | | 7.3 | | 9 |
| 2-Imino | | | 48 | ms | 10 |
| 2-Imino-8-Me | 217–219 | EtOH/$Et_2O$ | 36 | | 10 |
| 2-Imino-9-Me | 260–261 | EtOH | 30 | pmr, ms | 10 |
| 2-Imino-10-Me | | | 27 | | 10 |

*Pyrimido[1,2-a] [1,3]diazepines*

***7,8,9,10-Tetrahydro-6H-pyrimido[1,2-a] [1,3]diazepin-4(4H)-one***

| | | | | | |
|---|---|---|---|---|---|
| 2-Me | 128–129 | | 9 | ir | 13 |

TABLE II-1. —(contd.)

| Substituent | mp (°C); [bp (°C/torr)] | Solvent of Crystallization | Yield (%) | Spectra | Refs. |
|---|---|---|---|---|---|
| | | *Pyrimido[1,6-a] [1,3]diazepines* | | | |
| ***1,10a-Dihydropyrimido[1,6-a] [1,3]diazepines*** | | | 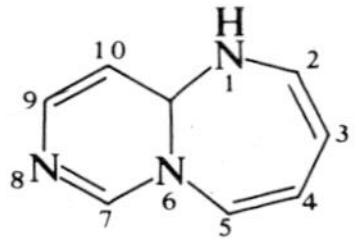 | | |
| 2,3,4,5-$(COOMe)_4$-9-Piperidino-10-$NO_2$ | 235–237 | | 6 | | 14 |
| 2,3,4,5-$(COOMe)_4$-9-Morpholino-10-$NO_2$ | 250–251 | | 6 | ir, ms | 14 |
| ***3,4,5,8-Tetrahydro-1H-pyrimido[1,6-a] [1,3]diazepine-2,7,9(2H,7H,9H)-trione*** | | | 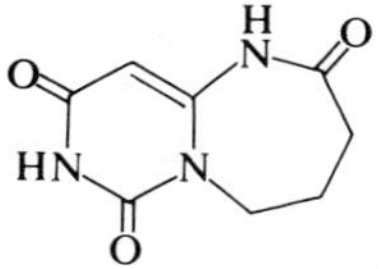 | | |
| None | 319–321 | | 76 | uv | 15 |
| | | *Pyrrolo[1,2-a] [1,3]diazepines* | | | |
| ***1,3,4,5-Tetrahydropyrrolo[1,2-a] [1,3]diazepin-2(2H)-ones*** | | | 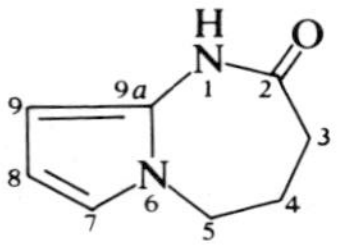 | | |
| None | 115–116 | | | ir, uv | 16 |
| 7,8-$(Me)_2$-9-CN | 181–182 | PhH/Cyclohexane | 76 | ir, pmr | 17 |

| | | | | | |
|---|---|---|---|---|---|
| ***2,3,4,5,8,9-Hexahydro-7H-pyrrolo[1,2-a] [1,3]diazepine*** | | | 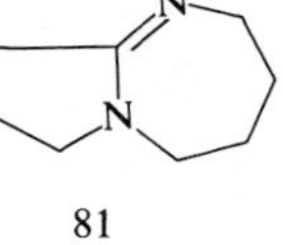 | | |
| None | [108–110/14] | | 81 | | 18 |
| ***2,3,4,7,8,9-Hexahydropyrrolo[1,2-a] [1,3]diazepin-5(5H)-one*** | | | 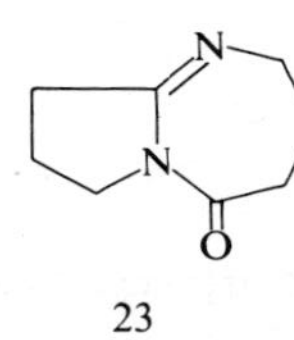 | | |
| 2,2,7,7-$(Me)_4$ | 44–45 | Pentane | 23 | ir, pmr, uv | 19 |
| ***Perhydropyrrolo[1,2-a] [1,3]diazepin-5(5H)-one*** | | | 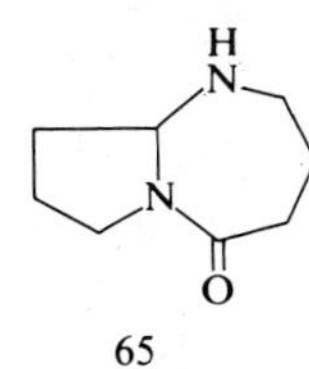 | | |
| 2,2,7,7-$(Me)_4$ | 85 | Pentane | 65 | ir, pmr | 19 |
| ***Perhydropyrrolo[1,2-a] [1,3]diazepin-7(7H)-ones*** | | | 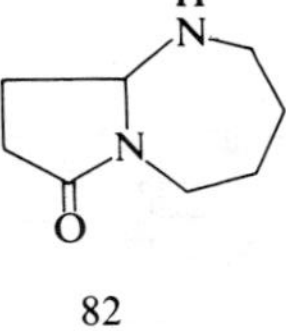 | | |
| 9*a*-Me | 118–119 [150/0.5] | | 82 | ir, pmr | 21 |
| 9*a*-Ph | 109 | | | | 22 |

TABLE II-1. —(contd.)

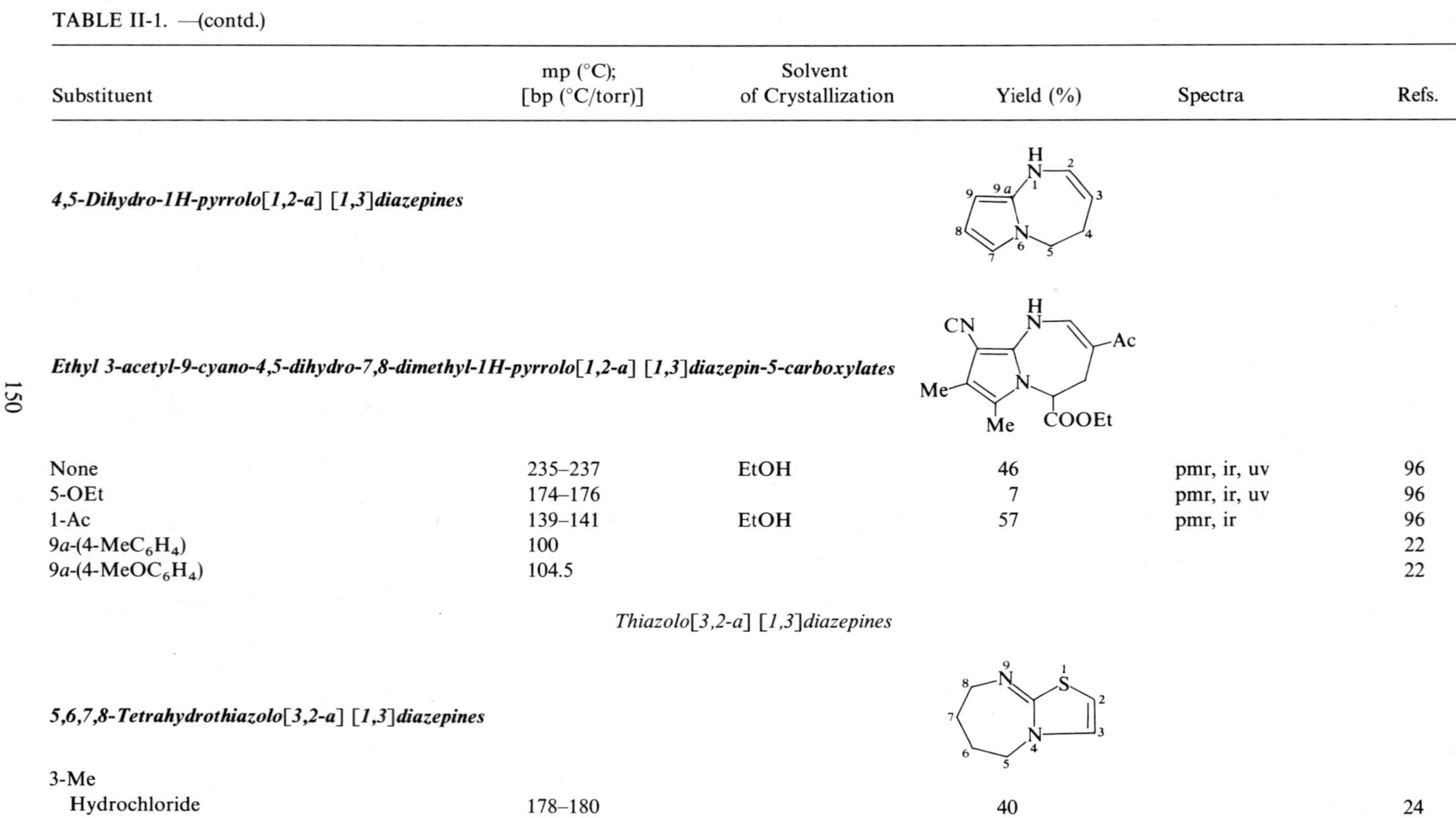

| Substituent | mp (°C); [bp (°C/torr)] | Solvent of Crystallization | Yield (%) | Spectra | Refs. |
|---|---|---|---|---|---|
| ***4,5-Dihydro-1H-pyrrolo[1,2-a] [1,3]diazepines*** | | | | | |
| ***Ethyl 3-acetyl-9-cyano-4,5-dihydro-7,8-dimethyl-1H-pyrrolo[1,2-a] [1,3]diazepin-5-carboxylates*** | | | | | |
| None | 235–237 | EtOH | 46 | pmr, ir, uv | 96 |
| 5-OEt | 174–176 | | 7 | pmr, ir, uv | 96 |
| 1-Ac | 139–141 | EtOH | 57 | pmr, ir | 96 |
| 9*a*-(4-$MeC_6H_4$) | 100 | | | | 22 |
| 9*a*-(4-$MeOC_6H_4$) | 104.5 | | | | 22 |
| *Thiazolo[3,2-a] [1,3]diazepines* | | | | | |
| ***5,6,7,8-Tetrahydrothiazolo[3,2-a] [1,3]diazepines*** | | | | | |
| 3-Me | | | | | |
| Hydrochloride | 178–180 | | 40 | | 24 |

| | | | | | |
|---|---|---|---|---|---|
| 3-Ph | 94–96 | | | | 26 |
| Hydrobromide | 216 | | 64 | | 24 |
| 3-(4-$BrC_6H_4$) | 125–126 | $CHCl_3$ | 55 | ir | 24 |
| Hydrobromide | 212–214 | | 55 | | 25 |
| 3-(4-$ClC_6H_4$) | | | | | |
| Hydrobromide | 219–220 | EtOH | 49 | ir | 25 |
| Hydrochloride | 206–207 | Acetone | | | 26 |
| 3-(4-$HOC_6H_4$) | | | | | |
| Hydrobromide | 210–211 | | 35 | | 25 |
| 3-[3,4-$(HO)_2C_6H_3$] | | | | | |
| Hydrochloride | 264 | EtOH | 58 | ir | 24 |
| 3-[2,3,4-$(HO)_3C_6H_2$] | | | | | |
| Hydrochloride | 295d | | 73 | | 24 |
| 3-(4-$MeC_6H_4$) | | | | | |
| Hydrobromide | 211–212 | | 39 | | 25 |
| 3-(4-$NO_2C_6H_4$) | 140d | | 70 | | 24 |
| Hydrobromide | 210–211d | | 84 | | 24 |
| 3-(4-$PhC_6H_4$) | 133 | | 72 | | 24 |
| Hydrobromide | 235 | | 77 | | 24 |
| Hydrochloride | 253–254d | | | | |
| 3-(3-$CF_3C_6H_4$) | | | | | |
| Hydrochloride | 198–199 | | | | 26 |
| 2-Me-3-Ph | | | | | |
| Hydrobromide | 238–239 | | 54 | | 25 |
| 2,3-$(Ph)_2$ | 102–105 | | 65 | | 24 |
| ***2,3,5,6,7,8-Hexahydrothiazolo[3,2-a] [1,3]diazepines*** | | | | | |
| None | | | | | |
| Hydrobromide | 223–224 | $EtOH/Et_2O$ | 59 | ir | 28 |
| 3-HO-3-Ph | | | | | |
| Hydrobromide | 180–182 | Acetone | 95 | | 26 |

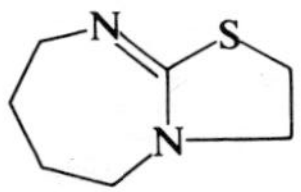

TABLE II-1. —(contd.)

| Substituent | mp (°C); [bp (°C/torr)] | Solvent of Crystallization | Yield (%) | Spectra | Refs. |
|---|---|---|---|---|---|
| 3-HO-3-(4-$ClC_6H_4$) | | | | | |
| Hydrobromide | 193–194 | Acetone | 92 | | 26 |
| 3-HO-3-(4-$MeOC_6H_4$) | | | | | |
| Hydrobromide | 164–165 | Acetone | 28 | | 26 |
| 3-HO-3-(4-$PhC_6H_4$) | | | | | |
| Hydrobromide | 237–238 | Acetone | 92 | | 26 |
| 3-HO-3-(3-$CF_3C_6H_4$) | | | | | |
| Hydrobromide | 186–188 | | 42 | | 26 |
| 2-$CH_2COOH$-3-HO-3-(4-$ClC_6H_4$) | | | | | |
| Hydrobromide | 163–165 | MeCN | 44 | ir | 27a |
| | 165–167 | MeCN | 82 | | 27b |
| ***5,6,7,8-Tetrahydrothiazolo[3,2-a] [1,3]diazepin-3(2H)-ones*** | | | | | |
| None | | | | | |
| Hydrochloride | 235–236d | EtOAc/EtOH | 75.5 | ir | 24 |
| 2-Me | | | | | |
| Hydrobromide | 226–228 | EtOH | 75 | ir | 24 |
| 2-Benzylidene | 107–108 | PhH/Petr ether | 71, 62 | ir | 24 |
| 2-[4-Dimethylaminobenzylidene] | 153–154 | EtOAc/Petr ether | 73 | | 24 |
| 2-Br-2-CHBrPh | 217–218d | | 65 | | 24 |
| 2-Br-2-[CHBr(4-$Me_2NC_6H_4$)] | 271–272d | | 60 | | 24 |

***2,3,5,6,7,8-Hexahydrothiazolo*[3,2-*a*] [*1,3*]*diazepine-3-ols***

| | | | |
|---|---|---|---|
| 2-(*o*-Chlorophenyl)-3-(3,4-dimethoxyphenyl) | 98–100 | EtOH | 97 |
| Hydrochloride | 197–199 | EtOH | 97 |
| 2-Phenyl-3-(5-methyl-2-thienyl) | 70–90 | EtOH | 97 |
| 2-phenyl-3-phenyl | 100–103 | | |
| Hydrochloride | 197–199 | EtOH/Ether | 97 |
| 2-(*p*-Chlorophenyl)-3-(*p*-Chlorophenyl) | | | |
| Hydrobromide | 216–218 | Acetone | 97 |
| 2-(*p*-Chlorophenyl)-3-phenyl | | | |
| Hydrobromide | 190–192 | Acetone | 97 |
| 2-Phenyl-3-(*p*-tolyl) | | | |
| Hydrobromide | 208–210 | Acetone | 97 |
| 2-Phenyl-3-(5-methyl-2-furyl) | | | |
| Hydrochloride | 193–195 | Ether | 97 |
| 2-Phenyl-3-(3,4-xylyl) | | | |
| Hydrochloride | 184–186 | Acetone | 97 |
| 2-(*p*-Chlorophenyl)-3-[3,4-(methylenedioxy)phenyl] | | | |
| Hydrochloride | 184–186 | EtOH | 97 |
| 2-Phenyl-3-(3-methyl-2-thienyl) | | | |
| Hydrochloride | 149–155 | Acetone | 97 |
| 2-Phenyl-3-(2,4-dimethoxyphenyl) | | | |
| Hydrochloride | 202–203 | EtOH | 97 |
| 2-(*o*-Chlorophenyl)-3-(*o*-Chlorophenyl) | | | |
| Hydrochloride | 224–226 | EtOH | 97 |
| 2-(*o*-Chlorophenyl)-3-[3,4-(methylenedioxy)-phenyl | | | |
| Hydrochloride | 180–182 | Acetone | 97 |

TABLE II-1. —(contd.)

| Substituent | mp (°C); [bp (°C/torr)] | Solvent of Crystallization | Yield (%) | Spectra | Refs. |
|---|---|---|---|---|---|
| 2-(*o*-Fluorophenyl)-3-[3,4-(methylenedioxy)-phenyl] | | | | | |
| Hydrochloride | 184–186 | | | | 97 |
| 2-(*o*-Fluorophenyl)-3-(2-thienyl) | | | | | |
| Hydrochloride | 192–195 | | | | 97 |
| ***2,3,7,9a-Tetrahydrothiazolo[3,2,-a] [1,3]diazepin-5,8(6H,9H)dione*** 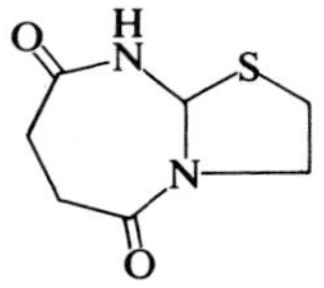 | | | | | |
| 9*a*-$NH_2$ | | | | | |
| Hydrochloride | 209–210 | EtOH/$H_2O$ | 34.5 | | 29 |
| *Pyrazolo[1,5-a] [1,3]diazepines* | | | | | |
| ***8-Carbethoxy-7-methyl-6H-pyrazolo[1,5-a] [1,3]diazepin-6-ones*** 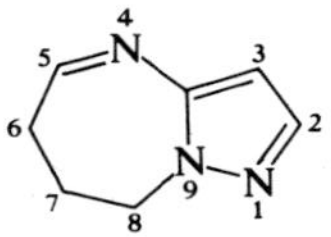 | | | | | |
| 3-CN | 112–113 | EtOH | 74.4 | pmr, ir, ms, $^{13}C$-nmr, anal | 95 |
| 3-$CONH_2$ | 195–196 | EtOH | 83.3 | pmr, ir, ms, anal | 95 |

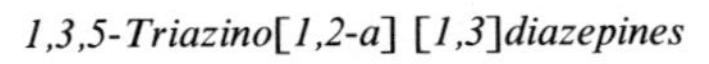

*1,3,5-Triazino[1,2-a] [1,3]diazepines*

***6,7,8,9-Tetrahydro-1,3,5-triazino[1,2-a] [1,3]diazepin-2,4(1H,3H)-dithione***

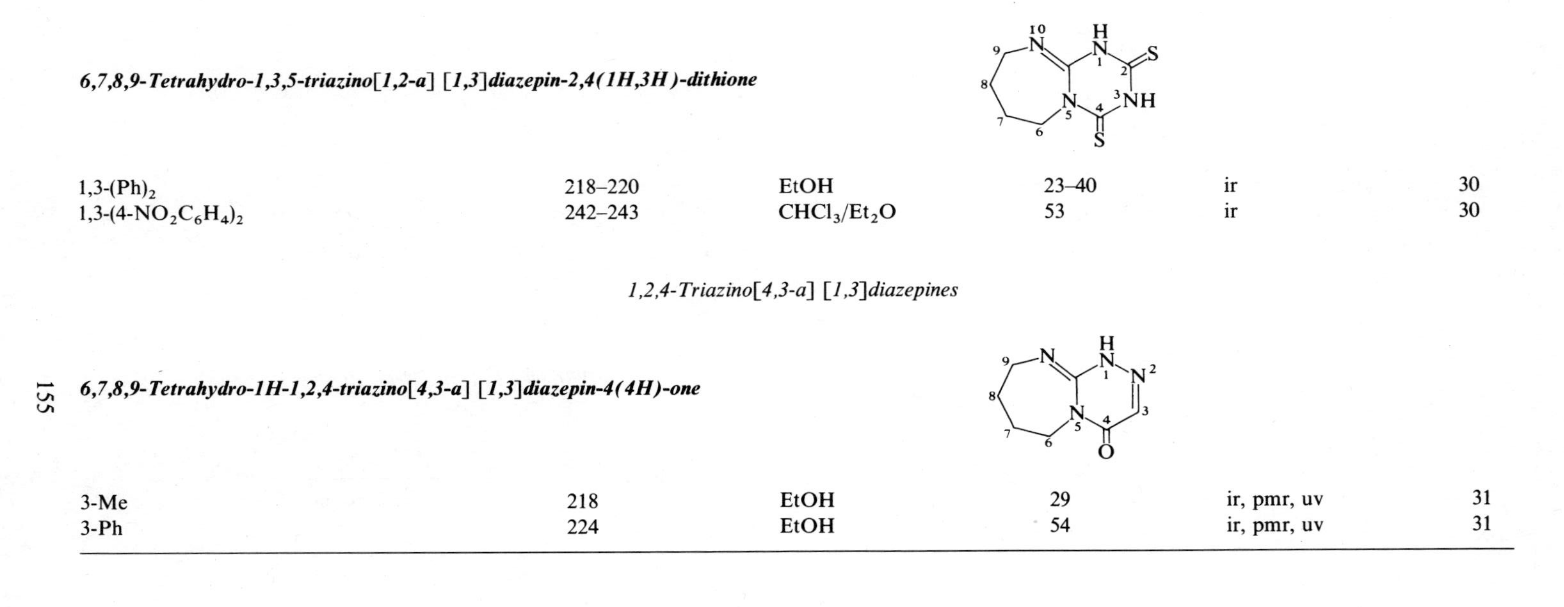

| | | | | | |
|---|---|---|---|---|---|
| 1,3-$(Ph)_2$ | 218–220 | EtOH | 23–40 | ir | 30 |
| 1,3-$(4\text{-}NO_2C_6H_4)_2$ | 242–243 | $CHCl_3/Et_2O$ | 53 | ir | 30 |

*1,2,4-Triazino[4,3-a] [1,3]diazepines*

***6,7,8,9-Tetrahydro-1H-1,2,4-triazino[4,3-a] [1,3]diazepin-4(4H)-one***

| | | | | | |
|---|---|---|---|---|---|
| 3-Me | 218 | EtOH | 29 | ir, pmr, uv | 31 |
| 3-Ph | 224 | EtOH | 54 | ir, pmr, uv | 31 |

TABLE II-2. [*c*]-FUSED [1,3]-DIAZEPINES

| Substituent | mp (°C); [bp (°C/torr)] | Solvent of Crystallization | Yield (%) | Spectra | Refs. |
|---|---|---|---|---|---|
| ***4,5-Dihydro**[**3,4-c**] [**1,3diazepines**]* | | | 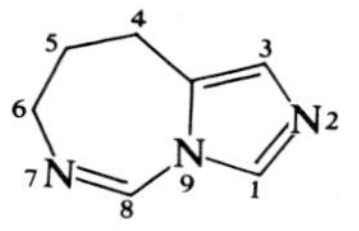 | | |
| 8-(NH-CO-Ph) | 158–160 | EtOH | | | 98 |
| 8-(NH-CN) | 210–215 | | | | 98 |

TABLE II-3. [*d*]-FUSED [1,3]DIAZEPINES

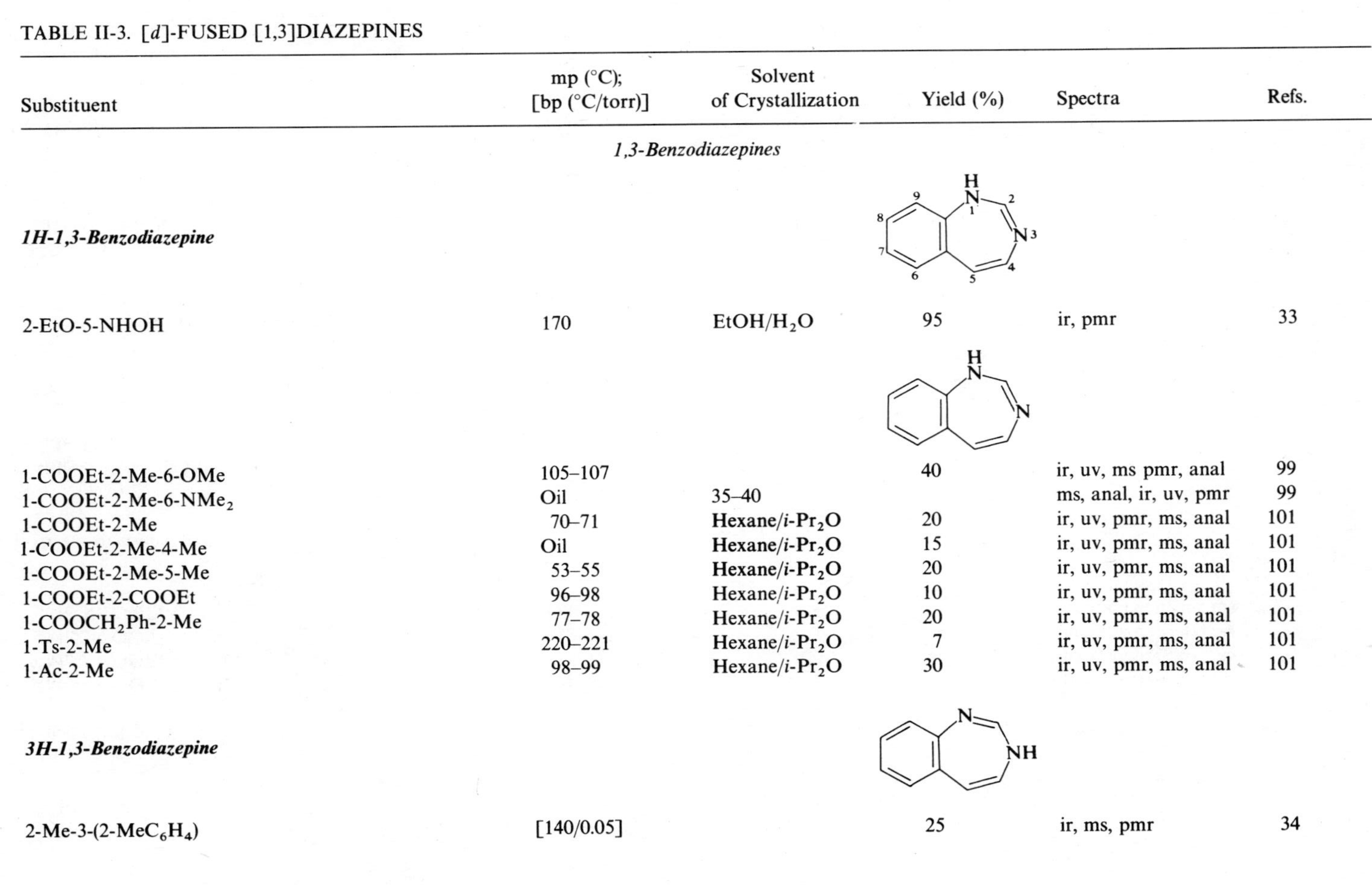

| Substituent | mp (°C); [bp (°C/torr)] | Solvent of Crystallization | Yield (%) | Spectra | Refs. |
|---|---|---|---|---|---|
| | | *1,3-Benzodiazepines* | | | |
| ***1H-1,3-Benzodiazepine*** | | | | | |
| 2-EtO-5-NHOH | 170 | EtOH/$H_2O$ | 95 | ir, pmr | 33 |
| 1-COOEt-2-Me-6-OMe | 105–107 | | 40 | ir, uv, ms pmr, anal | 99 |
| 1-COOEt-2-Me-6-$NMe_2$ | Oil | 35–40 | | ms, anal, ir, uv, pmr | 99 |
| 1-COOEt-2-Me | 70–71 | Hexane/*i*-$Pr_2O$ | 20 | ir, uv, pmr, ms, anal | 101 |
| 1-COOEt-2-Me-4-Me | Oil | Hexane/*i*-$Pr_2O$ | 15 | ir, uv, pmr, ms, anal | 101 |
| 1-COOEt-2-Me-5-Me | 53–55 | Hexane/*i*-$Pr_2O$ | 20 | ir, uv, pmr, ms, anal | 101 |
| 1-COOEt-2-COOEt | 96–98 | Hexane/*i*-$Pr_2O$ | 10 | ir, uv, pmr, ms, anal | 101 |
| 1-$COOCH_2$Ph-2-Me | 77–78 | Hexane/*i*-$Pr_2O$ | 20 | ir, uv, pmr, ms, anal | 101 |
| 1-Ts-2-Me | 220–221 | Hexane/*i*-$Pr_2O$ | 7 | ir, uv, pmr, ms, anal | 101 |
| 1-Ac-2-Me | 98–99 | Hexane/*i*-$Pr_2O$ | 30 | ir, uv, pmr, ms, anal | 101 |
| ***3H-1,3-Benzodiazepine*** | | | | | |
| 2-Me-3-(2-$MeC_6H_4$) | [140/0.05] | | 25 | ir, ms, pmr | 34 |

TABLE II-3. —(contd.)

| Substituent | mp (°C); [bp (°C/torr)] | Solvent of Crystallization | Yield (%) | Spectra | Refs. |
|---|---|---|---|---|---|
| 2-Me-3-COOEt-7-OMe | 62–64 | | 48 | pmr, anal, ir, uv, ms | 99 |
| 2-Me-3-COOEt-7-$NMe_2$ | 115–117 | | 26 | pmr, anal, ir, uv, ms | 99 |
| 2-Me-3-COOEt-7-Me | 110–111 | | 8 | pmr, anal, ir, uv, ms | 99 |
| 2-Me-3-COOEt-9-Me | Oil | | | pmr, anal, ir, uv, ms | 99 |
| ***5H-1,3-Benzodiazepines*** 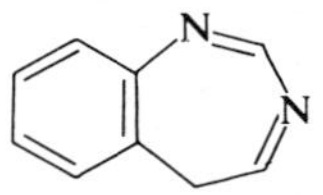 | | | | | |
| 2-Me-4-Ph-5-COOMe | 168–170.5 | Acetone/$Et_2O$/Hexane | | ir, ms, pmr, uv | 35 |
| 2-Me-4-Ph-5-COOEt | 112–113 [130/0.05] | $Et_2O$/Hexane | | ir, ms, pmr, uv | 35 |
| ***4,5-Dihydro-1H-1,3-benzodiazepines*** 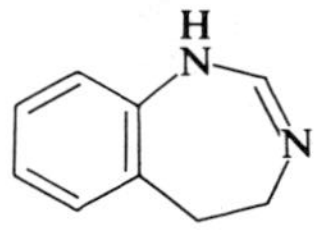 | | | | | |
| 2-EtO-5-OH | | | | | |
| Hydrochloride | 155d | EtOAc/MeOH | 63 | ir, pmr | 33 |
| 2-EtO-5-$CH_2$Ph-5-OH | | | | | |
| Hydrochloride | 158d | EtOAc/MeOH | 52 | ir, pmr | 33 |
| 2-EtO-5-Et-5-OH | | | | | |
| Hydrochloride | 153d | EtOAc/MeOH | 58 | ir, pmr | 33 |
| 2-EtO-5-Me-5-OH | | | | | |
| Hydrochloride | 155d | EtOAc/MeOH | 55 | ir, pmr | 33 |
| 2-EtO-5-Ph-5-OH | | | | | |
| Hydrochloride | 163, 215 | EtOAc/MeOH | 70 | ir, pmr | 33 |

| | | | | | |
|---|---|---|---|---|---|
| 2-EtO-5-(2-$ClC_6H_4$)-5-HO | | | | | |
| Sulfate | 145d | EtOAc/MeOH | 72 | ir, pmr | 33 |
| 2-EtO-5-(3-$ClC_6H_4$)-5-HO | | | | | |
| Hydrochloride | 153–155 | EtOAc/MeOH | 58 | ir, pmr | 33 |
| 2-EtO-5-(4-$ClC_6H_4$)-5-HO | | | | | |
| Hydrochloride | 155 | EtOAc/MeOH | 66 | ir, pmr | 33 |
| 2-EtO-5-(4-$EtC_6H_4$)-5-HO | | | | | |
| Hydrochloride | 158–160 | EtOAc/MeOH | 56 | ir, pmr | 33 |
| 2-EtO-5-(3-$FC_6H_4$)-5-HO | | | | | |
| Hydrochloride | 162–164 | EtOAc/MeOH | 63 | ir, pmr | 33 |
| 2-EtO-5-(4-$FC_6H_4$)-5-HO | | | | | |
| Hydrochloride | 159 | EtOAc/MeOH | 63 | ir, pmr | 33 |
| 2-EtO-5-(2-$MeOC_6H_4$)-5-HO | | | | | |
| Hydrochloride | 156, > 200 | EtOAc/MeOH | 52 | ir, pmr | 33 |
| 2-EtO-5-(3-$MeOC_6H_4$)-5-HO | | | | | |
| Hydrochloride | 145d | EtOAc/MeOH | 52 | ir, pmr | 33 |
| 2-EtO-5-(4-$MeOC_6H_4$)-5-HO | | | | | |
| Hydrochloride | > 210d | EtOAc/MeOH | 50 | ir, pmr | 33 |
| 2-EtO-5-(2-$MeC_6H_4$)-5-HO | | | | | |
| Hydrochloride | 173, > 200 | EtOAc/MeOH | 49 | ir, pmr | 33 |
| 2-EtO-5-(3-$MeC_6H_4$)-5-HO | | | | | |
| Hydrochloride | 156–158d | EtOAc/MeOH | 54 | ir, pmr | 33 |
| 2-EtO-5-(4-$MeC_6H_4$)-5-HO | | | | | |
| Hydrochloride | 147d | EtOAc/MeOH | 46 | ir, pmr | 33 |
| 2-EtO-5-(2-Thienyl)-5-HO | | | | | |
| Hydrochloride | 153, > 200 | EtOAc/MeOH | 45 | ir, pmr | 33 |
| 2-EtO-5-(3-$CF_3C_6H_4$)-5-HO | | | | | |
| Hydrochloride | 156–157d | EtOAc/MeOH | 47.5 | ir, pmr | 33 |
| 2-EtO-5-HO-7-$NO_2$ | 153 | EtOH/$H_2O$ | 66 | ir, pmr | 33 |

TABLE II-3. —(contd.)

| Substituent | mp (°C); [bp (°C/torr)] | Solvent of Crystallization | Yield (%) | Spectra | Refs. |
|---|---|---|---|---|---|
| ***4,5-Dihydro-3H-1,3-benzodiazepines*** | | | | | |
| None | 114–116 | $CH_2Cl_2$ | | pmr | 38 |
| Acetate | 112–114 | PhH | 71 | ir, pmr, uv | 38 |
| $2\text{-}NHCH_2Ph$ | | | | | |
| Hydrochloride | 190–192 | EtOH | | | 39 |
| $2\text{-}NH(CH_2)_2Cl$ | | | | | |
| Hydrochloride | 178–180 | *i*-PrOH | | | 42 |
| $2\text{-}NH(CH_2)_2$ [3,4-$(MeO)_2$-$C_6H_3$] | | | | | |
| Hydrochloride | 187–189.5 | *i*-PrOH | | | 39 |
| $2\text{-}NH(CH_2)_2NMe_2$ | | | | | |
| Dihydrochloride | 203.5–205.5 | *i*-PrOH | | | 41 |
| $2\text{-}NH(CH_2)_3NMe_2$ | | | | | |
| Dihydrochloride | 234–236 | *i*-PrOH | | | 41 |
| 2-EtO | | | | | |
| Tetrafluoborate | 118–121 | $CH_2Cl_2/Et_2O$ | | | 39, 41 |
| 2-[4-(2-Furoyl)-1-piperazino] | | | | | |
| Hydroiodide | 204–206 | *i*-PrOH | | | 40 |
| $2\text{-}NH(CH_2)_2OH$ | | | | | |
| Hydrochloride | 199–200 | *i*-PrOH/$Et_2O$ | | | 42 |
| 2-Me | | | | | |
| Hydrochloride | 240–242 | *i*-PrOH | | ir, ms, pmr, uv | 37 |
| Tetrachloromercurate | 211–213 | *i*-PrOH | 25 | ir, uv | 37 |
| 2-[4-Me-1-Piperazino] | | | | | |
| Hydroiodide | 249–250 | EtOH | | | 40 |

| | | | | | |
|---|---|---|---|---|---|
| 2-[4-(1,3-Benzodioxol-5-yl)methyl]-1-piperazino | | | | | |
| Hydroiodide | 273–275 | MeOH | | | 40 |
| 2-Morpholino | | | | | |
| Hydroiodide | 184–186 | $MeCN/Et_2O$ | | | 40 |
| 2-$NH(CH_2)_2Ph$ | | | | | |
| Hydroiodide | 203.5–206 | EtOH | | | 39 |
| 2-[4-(2-Thienoyl)-1-piperazino | | | | | |
| Hydroiodide | 232–234.5 | MeOH | | | 40 |
| 2-MeS | | | | | |
| Hydroiodide | 175 | $EtOH/MeOH/Et_2O$ | | | 39, 41 |
| 2-Me-3-(2-$MeC_6H_4$) | [160/0.1] | | 70 | ir, ms, pmr | 34 |
| 3-(CHMePh)-5-Me | | | | | |
| Hydrochloride | 269–272 | Acetone | ~ 10 | | 44 |
| 2-$NH_2$-7,8-$(MeO)_2$ | | | | | |
| Hydrochloride | 200–203 | EtOH | 34 | ir, pmr, uv | 36a |
| | 233 | EtOH | | | 36b |
| 2-$CH_2Ph$-7,8-$(MeO)_2$ | | | | | |
| Hydrochloride | 194–196 | EtOH | 48 | | 36a |
| 2-(3-COOEt-$C_6H_4$)-7,8-$(MeO)_2$ | | | | | |
| Hydrochloride | 134–137 | EtOH | | | 36b |
| 2-(4-COOEt-$C_6H_4$)-7,8-$(MeO)_2$ | | | | | |
| Hydrochloride | 251–253 | EtOH | | | 36b |
| 2-(3-COOH-$C_6H_4$)-7,8-$(MeO)_2$ | | | | | |
| Hydrochloride | 258–288 | EtOH | | | 36b |
| 2-$CH_2Cl$-7,8-$(MeO)_2$ | 126 | PhH | | | |
| Hydrochloride | 282 | EtOH | 86 | | 36a |
| 2-(3-$ClC_6H_4$)-7,8-$(MeO)_2$ | | | | | |
| Hydrochloride, hydrate | 157–163 | EtOH | | | 36b |
| 2-(4-$ClC_6H_4$)-7,8-$(MeO)_2$ | | | | | |
| Hydrochloride | 273–276 | EtOH | 34 | | 36a |
| 2-[3,4-$(MeO)_2C_6H_3$]$CH_2$-7,8-$(MeO)_2$ | | | | | |
| Hydrochloride | 192–194 | EtOH | | | 36b |
| 2-[3,4-$(MeO)_2C_6H_3$]-7,8-$(MeO)_2$ | | | | | |
| Hydrochloride, hydrate | 224–225 | EtOH | | | 36b |

TABLE II-3. —(contd.)

| Substituent | mp (°C); [bp (°C/torr)] | Solvent of Crystallization | Yield (%) | Spectra | Refs. |
|---|---|---|---|---|---|
| 2-NH$(CH_2)_2$NMe$_2$-7,8-(MeO)$_2$ | | | | | |
| Dihydrochloride | 227 | MeOH | | | 41 |
| 2-CH$_2$NMe$_2$-7,8-(MeO)$_2$ | | | | | |
| Dihydrochloride, hydrate | 248–250 | EtOH | 63 | | 36a |
| 2-NH$(CH_2)_3$NMe$_2$-7,8-(MeO)$_2$ | | | | | |
| Dihydrochloride | 264–265 | *i*-PrOH/MeOH | | | 41 |
| 2-[4-(Me)$_2$NSO$_2$C$_6$H$_4$]-7,8-(MeO)$_2$ | | | | | |
| Hydrochloride | 250–252 | EtOH | | | 36b |
| 2-(4-EtOC$_6$H$_4$)-7,8-(MeO)$_2$ | | | | | |
| Hydrochloride | 225–227 | EtOH | | | 36b |
| 2-(3-FC$_6$H$_4$)-7,8-(MeO)$_2$ | | | | | |
| Hydrochloride | 257–259 | EtOH | | | 36b |
| 2-(4-FC$_6$H$_4$)-7,8-(MeO)$_2$ | | | | | |
| Hydrochloride | 281–286 | EtOH | | | 36b |
| 2-[4-(2-Furoyl)-1-piperazino]-7,8-(MeO)$_2$ | | | | | |
| Hydroiodide, hydrate | 235 | MeOH | | | 40 |
| 2-(4-HOC$_6$H$_4$)-7,8-(MeO)$_2$ | | | | | |
| Hydrochloride | 300–302 | EtOH | | | 36b |
| 2-(1-Imidazolyl)CH$_2$-7,8-(MeO)$_2$ | | | | | |
| Dihydrochloride | 269–271 | EtOH | 65 | | 36a |
| 2-(4-MeOC$_6$H$_4$)CH$_2$-7,8-(MeO)$_2$ | | | | | |
| Hydrochloride | 204–205 | EtOH | | | 36b |
| 2-CH$_2$OMe-7,8-(MeO)$_2$ | | | | | |
| Hydrochloride | 214–216 | EtOH | 64 | | 36a |
| 2-(3-MeOC$_6$H$_4$)-7,8-(MeO)$_2$ | | | | | |
| Hydrochloride | 140–144 | EtOH | | | 36b |
| 2-(4-MeOC$_6$H$_4$)-7,8-(MeO)$_2$ | | | | | |
| Hydrochloride | 243 | EtOH | 38 | | 36a, b |

| | | | | | |
|---|---|---|---|---|---|
| 2-Me-7,8-$(MeO)_2$ | 110–112 | | | | 36b |
| Hydrochloride | 300 | EtOH | 51 | | 36a |
| 2-(4-Me-Piperazino)-7,8-$(MeO)_2$ | | | | | |
| Hydroiodide | 278–280 | EtOH | | | 40 |
| 2-Morpholino-7,8-$(MeO)_2$ | | | | | |
| Hydroiodide | 257–259 | $CH_3CN$ | | | 40 |
| 2-(3-$NO_2C_6H_4$)-7,8-$(MeO)_2$ | | | | | |
| Hydrochloride | 262–264 | EtOH | 36 | | 36a |
| 2-(4-$NO_2$-$C_6H_4$)-7,8-$(MeO)_2$ | | | | | |
| Hydrochloride | 200 | EtOH | 41 | | 36a |
| 2-Ph-7,8-$(MeO)_2$ | 107–109 | EtOAc | 88 | | 36a |
| Hydrochloride | 276–278 | EtOH | | | 36a |
| 2-(4-Ph-1-Piperazino)-$CH_2$-7,8-$(MeO)_2$ | | | | | |
| Dihydrochloride | 228 | | | | 36b |
| 2-(Piperidino)$CH_2$-7,8-$(MeO)_2$ | | | | | |
| Dihydrochloride | 273–275 | EtOH | 62 | | 36a |
| 2-Pyrrolidino-7,8-$(MeO)_2$ | | | | | |
| Hydroiodide | 268–270 | EtOH | | | 40 |
| 2-MeS-7,8-$(MeO)_2$ | | | | | |
| Hydroiodide | 197–200 | MeOH/$Et_2O$ | | | 39, 41 |
| | 209–211 | EtOH | | | 36b |
| 2-(3-$CF_3C_6H_4$)-7,8-$(MeO)_2$ | | | | | |
| Hydrochloride | 237–239 | EtOH | | | 36a |
| 2-(4-$CF_3C_6H_4$)-7,8-$(MeO)_2$ | | | | | |
| Hydrochloride | 281–282 | EtOH | | | 36a |
| 2-[3,4,5-$(MeO)_3C_6H_2$]-7,8-$(MeO)_2$ | | | | | |
| Hydrochloride, hydrate | 165–167 | EtOH | | | 36b |
| 2,3-$Me_2$-7,8-$(MeO)_2$ | | | | | |
| Hydrochloride | 267–269 | EtOH | 56 | | 36a |
| 2-Ph-3-Me-7,8-$(MeO)_2$ | 150–152 | EtOH | | | 36b |
| Hydrochloride | 267–268 | EtOH | 52 | | 36a |
| Methiodide | 187 | PhH | 59 | | 36a |
| 2-Ph-3-Me-4-Ph | | | | | |
| Hydrochloride | 258–259 | MeCN | 29 | ir, pmr, anal | 100 |

TABLE II-3. —(contd.)

| Substituent | mp (°C); [bp (°C/torr)] | Solvent of Crystallization | Yield (%) | Spectra | Refs. |
|---|---|---|---|---|---|
| 2-Ph-4-Ph | | | | | |
| Hydrochloride | 244–247 | MeCN | 40 | ir, pmr, anal | 100 |
| 2-Ph-3-Me-4-[4-Cl-$C_6H_4$] | | | | | |
| Hydrochloride | 269–270 | EtOH/$Et_2O$ | 51 | ir, pmr, anal | 100 |
| 2-Ph-3-Me-4-[4-MeO-$C_6H_4$] | | | | | |
| Hydrochloride | 247–249 | MeCN | 79 | ir, pmr, anal | 100 |
| 2-Ph-3-Me-4-[4-Me-$C_6H_4$] | | | | | |
| Hydrochloride | 268–270 | EtOH/$Et_2O$ | 57 | ir, pmr, anal | 100 |
| 2-Ph-3-Me-4-[4-F-$C_6H_4$] | | | | | |
| Hydrochloride | 270–272 | EtOH/$Et_2O$ | 66 | ir, pmr, anal | 100 |
| 2-Ph-3-Me-4-[3,4-$(OMe)_2$-$C_6H_3$] | | | | | |
| Hydrochloride | 175–180 | EtOH/$Et_2O$ | 42 | ir, pmr, anal | 100 |
| 2-SH-4-Ph | 210–212 | *i*-BuOH | 58 | ir, pmr, anal | 100 |
| 2-MeS-4-Ph | | | | | |
| Hydroiodide | 212–216 | MeOH/$Et_2O$ | 79 | ir, pmr, anal | 100 |
| 2-C-N($CH_2$-$CH_2$)$_2$N-Me-4-Ph | 93–98 | Cyclohexane | 63 | ir, pmr, anal | 100 |
| 2-C-N($CH_2CH_2$)$_2$O-4-Ph | 128–129 | Toluene | 61 | ir, pmr, anal | 100 |
| 2-SH-3-Me-4-Ph | 174–178 | *i*-BuOH | 52 | ir, pmr, anal | 100 |
| 2-MeS-3-Me-4-Ph | 175-183 | MeCN/$Et_2O$ | 92 | ir, pmr, anal | 100 |
| 2,3-$Me_2$ | | | | | |
| Hydrochloride | 240–241 | *i*-PrOH | 83 | | 103 |
| 2,3-$Me_2$-4-Ph | 145–146 | *i*-PrOH/Hexane | 62 | | 103 |
| 2,3-$Me_2$-4-Ph | | | | | |
| Hydrochloride | 241–242 | *i*-PrOH | | | 103 |
| 4-Ph | | | | | |
| Hydrochloride | 185–187 | MeOH/$Et_2O$ | | anal | 104 |
| 2-Me-4-Ph | | | | | |
| Hydrochloride | 194–197 | EtOH/$Et_2O$ | | anal | 104 |

| | | | | | |
|---|---|---|---|---|---|
| 3-Me-4-Ph | | | | | |
| Hydrochloride | 251–254 | MeOH/$Et_2O$ | | anal | 104 |
| 2-Me-3-Propyl-4-Ph | | | | | |
| Hydrochloride | 272–275 | $Et_2O$ | | anal | 104 |
| 2-Me-3-Cyclohexylmethyl-4-Ph | | | | | |
| Hydrochloride | 256–258 | MeOH/Hexane | | anal | 104 |
| 2-Et-3-Me-4-Ph | | | | | |
| Hydrochloride | 250–255 | Acetone | | anal | 104 |
| 2-Me-3-Benzyl-4-Ph | | | | | |
| Hydrochloride | 245–247 | $Et_2O$ | | anal | 104 |
| 2,3-$Me_2$-4-F-$C_6H_4$ | | | | | |
| Hydrochloride | 253–255 | EtOH/$Et_2O$ | | anal | 104 |
| 2,3-$Me_2$-4-(4-Me-$C_6H_4$) | | | | | |
| Hydrochloride | 285–288 | EtOH | | anal | 104 |
| 2,3-$Me_2$-4-(4-MeO-$C_6H_4$) | 130–132 | MeCN | | anal | 104 |
| 2-Et-3-Me-4-(4-MeO-$C_6H_4$) | | | | | |
| Hydrochloride | 240–242 | *i*-PrOH | | anal | 104 |
| 2-Et-4-Ph | | | | | |
| Hydrochloride | 242–245 | *i*-PrOH | | anal | 104 |
| 2,3-$Me_2$-4-(3,4-$Me_2$-$C_6H_3$) | | | | | |
| Hydrochloride | 222–226 | EtOH/$Et_2O$ | | anal | 104 |
| 2-Et-3-Me-4-[3,4-$(MeO)_2$-$C_6H_3$] | 214–216 | EtOH/$Et_2O$ | | anal | 104 |

*Dihydro-1,3-benzodiazepinones*

***1,3-Dihydro-1,3-benzodiazepin-2(2H)-ones***

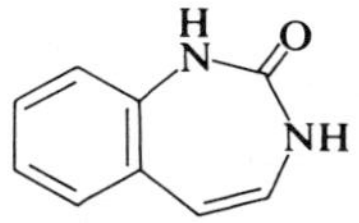

| | | | | | |
|---|---|---|---|---|---|
| 5-Ph | 250 | MeOH | 38 | ir, pmr | 33 |
| 5-Ph-7-Me | 256–260 | PhH | 15 | ir, pmr | 45 |

TABLE II-3. —(contd.)

| Substituent | mp (°C); [bp (°C/torr)] | Solvent of Crystallization | Yield (%) | Spectra | Refs. |
| --- | --- | --- | --- | --- | --- |
| ***3,5-Dihydro-1,3-benzodiazepin-4(4H)-ones*** | | | 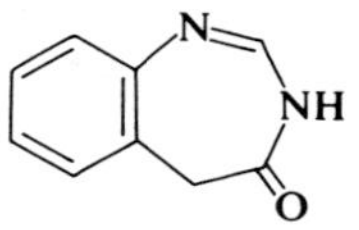 | | |
| 2-(4-$ClC_6H_4$) | 239–242 | MeOH | 35 | | 46 |
| 2-Et | 196–198 | *i*-PrOH | 33 | | 46 |
| 2-(4-$MeOC_6H_4$) | 231–232 | *i*-PrOH | 18 | | 46 |
| 2-Me | 223–224 | *i*-PrOH | 29 | | 46 |
| 2-(4-$NO_2C_6H_4$) | 294 | Dioxane | 54 | | 46 |
| 2-Ph | 242–243 | Dioxane | 41 | | 46 |
| ***1,4-Dihydro-1,3-benzodiazepin-5(5H)-ones*** | | | 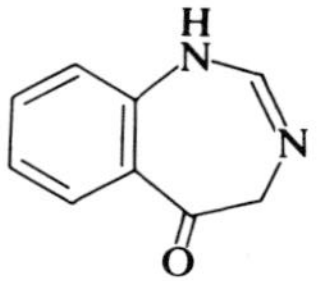 | | |
| 2-EtO | 88 | Petr ether | 84 | ir, pmr | 33 |
| 2-Morpholino | 150, 170 | EtOAc | 41.5 | ir, pmr | 33 |
| 2-Pyrrolidino | 152–154d | EtOAc | 40 | ir, pmr | 33 |
| 2-EtO-7-$NO_2$ | 110–11 | MeOH/$H_2O$ | 70 | ir, pmr | 33 |

| | | | | | |
|---|---|---|---|---|---|
| ***3,4-Dihydro-1,3-benzodiazepin-5(5H)-ones*** | | | | | |
| 3-(4-$MeOC_6H_4$)-7-MeO | 188, 225 | Ac | 1 | ir, pmr | 48 |
| Phenylhydrazone, hydrate | 136–138 | | | | 48 |
| Semicarbazone | 218–219d | | | | 48 |
| ***2,3,4,5-Tetrahydro-1H-1,3-benzodiazepines*** | | | | | |
| 3-Me | | | | | |
| Maleate | 99–101 | EtOAc | | uv | 49 |
| 1-$CH_2$Ph-3-Me | Oil | | | | 49 |
| ***1,3,4,5-Tetrahydro-1,3-benzodiazepin-2(2H)-ones*** | | | | | |
| None | 169–171 | $CHCl_3$/Hexane | ~ 40 | | 36b, 42 |
| 3-$(CH_2)_2NMe_2$ | 114–116 | EtOAc | | | 51 |
| Hydrochloride | 231–233 | EtOH | | | 51 |
| 3-Et | 148–150 | MeOH | 31 | | 50 |
| 3-Me | 129–131 | MeOH | 35 | | 50 |
| 3-$(CH_2)_2$-Pyrrolidino | | | | | |
| Hydrochloride | 215–217 | | | | 51 |
| 5-OAc | 206d | MeOH/$Et_2O$ | 32 | ir, pmr | 33 |
| 5-OH | 181 | MeOH/$Et_2O$ | 51 | ir, pmr | 33 |
| 3-$(CH_2)_2NMe_2$-6-Cl | 173–175 | | | | 51 |

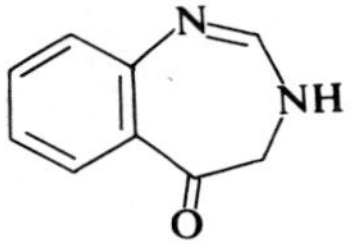

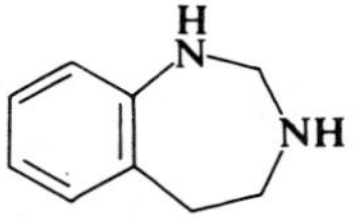

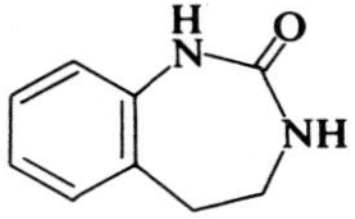

TABLE II-3. —(contd.)

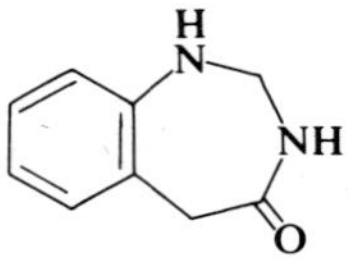

| Substituent | mp (°C); [bp (°C/torr)] | Solvent of Crystallization | Yield (%) | Spectra | Refs. |
|---|---|---|---|---|---|
| 3-$(CH_2)_2NMe_2$-8-Cl | 148–150 | | | | 51 |
| Hydrochloride | 256–258 | | | | 51 |
| 3-$(CH_2)_2NMe_2$-8-(*i*-Pr) | | | | | |
| Hydrochloride | 215–217 | | | | 51 |
| 5-HO-5-Me | 210d | EtOAc | 29 | ir, pmr | 33 |
| 5-HO-5-(2-Me-$C_6H_4$) | 244d | EtOH | 70 | ir, pmr | 33 |
| 5-HO-5-Ph | 224 | EtOAc | 38 | ir, pmr | 33 |
| 5-HO-7-$NH_2$ | 195d | EtOH | 69 | ir, pmr | 33 |
| 5-HO-7-$NO_2$ | 207 | DMF/$CHCl_3$ | 76 | ir, pmr | 33 |
| 7,8-$(MeO)_2$ | 244–247 | $H_2O$ | 79 | | 36a |
| 7,8-(OPr-*i*$)_2$ | 280–285 | | | | 36b |
| 1,3-$Me_2$-5-HO | [160–162/0.7] | | 90 | ir, ms, pmr | 52 |
| 3-Me-7,8-$(MeO)_2$ | 197–199 | | | | 36b |
| ***1,2,3,5-Tetrahydro-1,3-benzodiazepin-4(4H)-ones*** | | | | | |
| None | | | 25 | | 54 |
| 3-$CH_2$Ph | 192 | | | | 49 |
| 3-Et | 144–145 | | | | 49 |
| 3-$CH_2$OH | | | 6 | | 54 |
| 3-Me | 150 | $H_2O$ | 78 | ir, uv | 49 |
| 3-Ph | 197–199 | EtOH/$H_2O$ | 71.5 | | 50 |
| 3-Pr | 128.5–130 | | | | 49 |
| 1-$CH_2$Ph-3-Me | 118–120 | EtOH/$H_2O$ | 87 | ir, uv | 49 |
| 3-Et-7-Cl | 140–142 | $H_2O$ | 71.5 | | 50 |

| | | | | | |
|---|---|---|---|---|---|
| 3-Me-7-Br | 172–174 | $H_2O$ | 65 | | 50 |
| 3-Me-7-Cl | 154–156 | $H_2O$ | 71 | | 50 |
| 3-Ph-7-Br | 217–218 | $EtOH/H_2O$ | 55 | | 50 |
| 3-Ph-7-Cl | 217–219 | $EtOH/H_2O$ | 68 | | 50 |
| 2,3,5-$(Ph)_3$ | 156–157 | | 9 | ir, ms, pmr, $^{13}C$-nmr | 55 |
| 2-Phenylimino-3-Ph-5-[1,3-$Ph_2$-2-propyn-1-ylidene] | 150–151 | PhH/Hexane | 14–41 | ir, ms, pmr | 59 |
| 1-$CH_2$Ph-3-Me-7,8-$(OMe)_2$ | 154–155 | | 73 | | 53 |
| 1-(4-$ClC_6H_4CH_2$)-3-Me-7,8-$(MeO)_2$ | 140–142 | EtOH | 57 | | 53 |
| 1-(4-$MeOC_6H_4$)-3-Me-7,8-$(MeO)_2$ | 130–132 | | 72.5 | | 53 |
| 2-(*p*-Tolylimino)-3-(4-Me-$C_6H_4$)-5-[1,3-$Ph_2$-2-propyn-1-ylidene]-7-Me | 188–190 | PhH/Hexane | 33 | ir, ms, pmr | 59 |
| 2-(*o*-Tolylimino)-3-(2-Me-$C_6H_4$)-5-[1,3-$Ph_2$-2-propyn-1-ylidene]-9-Me | 181–182 | PhH/Hexane | 17 | ir, ms, pmr | 59 |
| ***3,5-Dihydro-1H-1,3-benzodiazepine-2,4-diones*** | | | | | |
| 1-Me | 203 | | 18 | ir, uv | 56 |
| 1,3-$(Me)_2$ | 136 | | 34 | ir, uv | 56 |
| 1-Me-3-Ph | 186 | EtOH | 45 | ir, uv | 56 |
| 3-Ph-5-$CH_2$Ph | 122–124 | $MeOH/H_2O$ | | ir | 57 |
| 3-Ph-5-Benzylidene | 202–203 | | 4; 12 | ir, uv | 57, 58 |
| ***3,4-Dihydro-1H-1,3-benzodiazepine-2,5(5H)-diones*** | | | | | |
| None | 220d | $DMF/CHCl_3$ | 88 | ir, pmr | 33 |

TABLE II-3. —(contd.)

| Substituent | mp (°C); [bp (°C/torr)] | Solvent of Crystallization | Yield (%) | Spectra | Refs. |
|---|---|---|---|---|---|
| 5-Oxime | 255d | $CHCl_3$/MeO $CH_2CH_2OH$ | 77.5 | ir | 33 |
| 7-$NH_2$ | 255d | EtOH | 90 | ir, pmr | 33 |
| 7-Br | 254 | DMF | | ir, pmr | 33 |
| 7-Cl | 240 | DMF/MeOH | 57.5 | ir, pmr | 33 |
| 7-$NHCOCH_2Cl$ | 250d | DMF/MeOH | 89 | ir, pmr | 33 |
| 7-(Morpholinoacetyl)amino | 255 | DMF/$H_2O$ | 89 | ir, pmr | 33 |
| 7-$NO_2$ | 260d | DMF/MeOH | 60 | ir, pmr | 33 |
| 5-Oxime | 250d | EtOH | 70 | ir | 33 |
| 7-(Piperidinoacetyl)amino | 255 | DMF/$H_2O$ | 84 | ir, pmr | 33 |
| ***1,3,4,5-Tetrahydro-1,3-benzodiazpine-2(2H)-thiones*** | | | 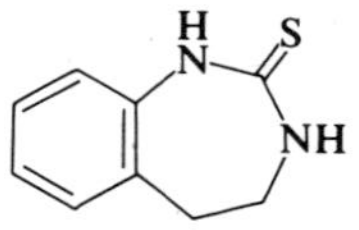 | | |
| None | 195 | | | | 39, 40 |
| 5-Me | 180–182 | | 65 | ir, pmr | 44 |
| 7,8-$(MeO)_2$ | 251–253 | | | | 36b, 39, 40 |
| ***3,5-Dihydro-2-thioxo-1H-1,3-benzodiazepin-4(4H)-one*** | | | 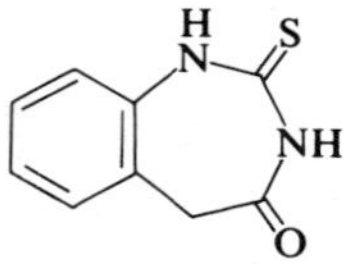 | | |
| 1,3-$(Me)_2$ | 136 | EtOH | 78 | ir, uv | 56 |

| | | | | | |
|---|---|---|---|---|---|
| ***1,3-Dihydro-2-thioxo-1,3-benzodiazepine-4,5(4H, 5H)-diones*** | | | | | |
| 3-Ph | 155–157 | Acetone | | | 60 |
| 3-(2-Me$C_6H_4$) | 205–206 | EtOH | | | 60 |
| ***1,3,4,5,-Tetrahydro-2-cyanoimino-1,3-benzodiazepines*** | | | | | |
| 3-$CH_2$Ph | 177–179 | $Me_2CO$ | 16 | ir, pmr, anal | 102 |
| 1-$CH_2$COOEt-3-$CH_2$-Ph | 131–133 | $Et_2O$ | 72 | ir, pmr, anal | 102 |
| ***3,6,7,8-Tetrahydroimidazo[4,5-d] [1,3]diazepines*** | | | | | |
| 3-(β-D-Ribofuranosyl)-8(*R*)-OH (coformycin) | 182–184 | $H_2O$ | | ms, uv, ORD, $pK_a$, X-ray | 62, 64 |
| 3-(2-Deoxy-β-D-ribofuranosyl)-8(*R*)-OH (covidarabine) | 220–225 | | | ms, pmr, $^{13}C$ nmr, uv, X-ray, $pK_a$ | 65 |
| Tetraacetate | 68–78 | | | ms, pmr | 65 |

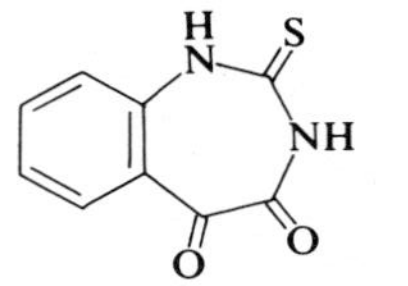

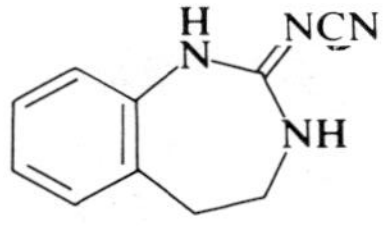

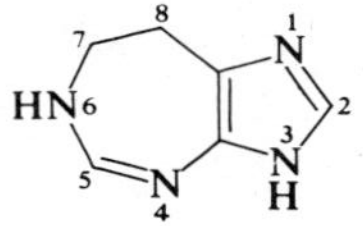

TABLE II-3. —(contd.)

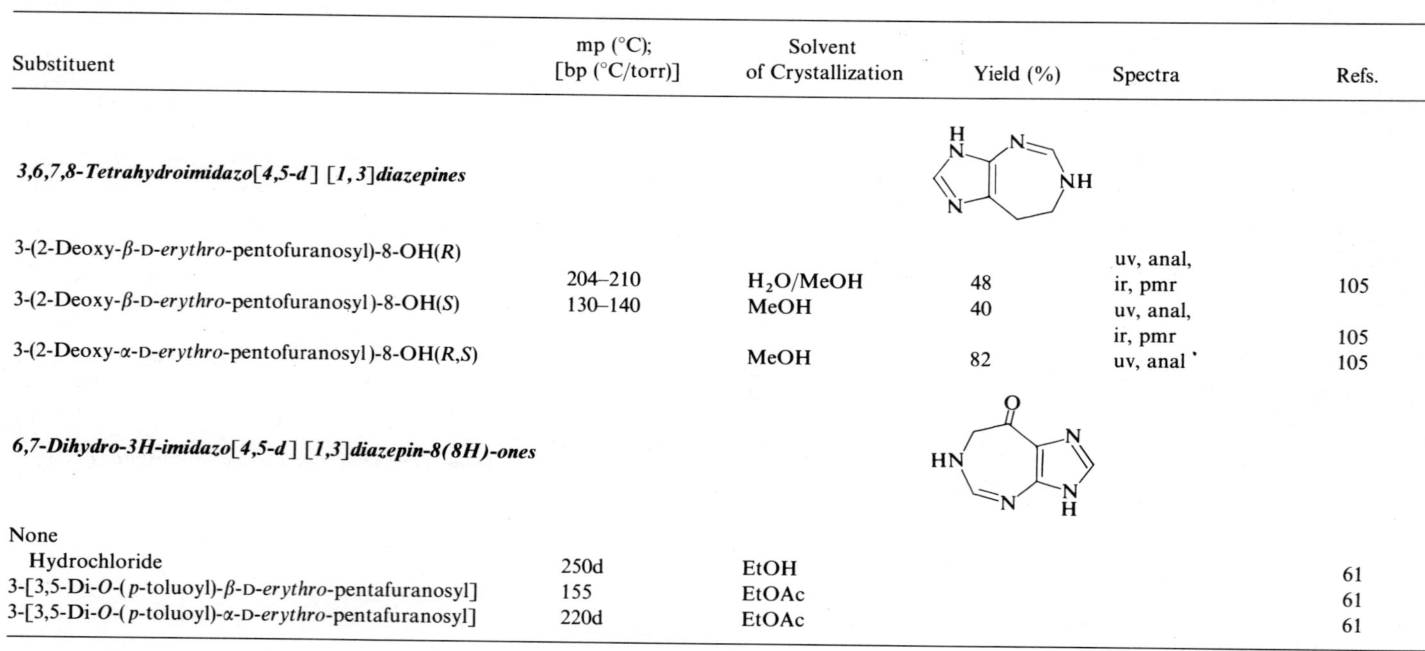

| Substituent | mp (°C); [bp (°C/torr)] | Solvent of Crystallization | Yield (%) | Spectra | Refs. |
|---|---|---|---|---|---|
| ***3,6,7,8-Tetrahydroimidazo[4,5-d] [1,3]diazepines*** | | | | | |
| 3-(2-Deoxy-β-D-*erythro*-pentofuranosyl)-8-OH(*R*) | 204–210 | $H_2O$/MeOH | 48 | uv, anal, ir, pmr | 105 |
| 3-(2-Deoxy-β-D-*erythro*-pentofuranosyl)-8-OH(*S*) | 130–140 | MeOH | 40 | uv, anal, ir, pmr | 105 |
| 3-(2-Deoxy-α-D-*erythro*-pentofuranosyl)-8-OH(*R*,*S*) | | MeOH | 82 | uv, anal | 105 |
| ***6,7-Dihydro-3H-imidazo[4,5-d] [1,3]diazepin-8(8H)-ones*** | | | | | |
| None | | | | | |
| Hydrochloride | 250d | EtOH | | | 61 |
| 3-[3,5-Di-*O*-(*p*-toluoyl)-β-D-*erythro*-pentafuranosyl] | 155 | EtOAc | | | 61 |
| 3-[3,5-Di-*O*-(*p*-toluoyl)-α-D-*erythro*-pentafuranosyl] | 220d | EtOAc | | | 61 |

TABLE II-4. [*e*]-FUSED [1,3]DIAZEPINES

| Substituent | mp (°C); [bp (°C/torr)] | Solvent of Crystallization | Yield (%) | Spectra | Refs. |
|---|---|---|---|---|---|
| | | *2,4-Benzoidazepines* | | | |
| ***4,5-Dihydro-1H-2,4-benzodiazepines*** | | | 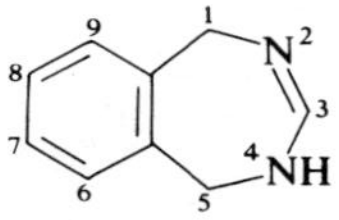 | | |
| 3-$NH_2$ | | | | | |
| Hydrochloride | 279–281 | EtOH | 76 | ir, uv | 36a |
| 3-$PhCH_2$ | | | | | |
| Hydrochloride | 256–259 | EtOH | 67 | | 36a |
| 3-(4-COOEt-piperazino) | | | | | |
| Hydroiodide | 200–202 | *i*-PrOH | | | 73a, b |
| 3-$CH_2Cl$ | | | | | |
| Hydrochloride | 263–264 | EtOH | 53 | | 36a |
| 3-(4-$ClC_6H_4$) | | | | | |
| Hydrochloride | 266–267 | EtOH | 46 | | 36a |
| 3-$OPOCl_2$ | | | | | |
| Hydrochloride | 215 | EtOH | 72 | | 36a |
| 3-$NMe_2$ | | | | | |
| Hydrochloride | 253–255 | EtOH | 69 | | 36a |
| 3-$CH_2NMe_2$ | | | | | |
| Dihydrochloride | 239–241 | EtOH | 78 | | 36a |
| 3-$CH_2OMe$ | | | | | |
| Hydrochloride | 175–178 | EtOH | 28 | | 36a |
| 3-(4-$MeOC_6H_4$) | | | | | |
| Hydrochloride | 262–263 | EtOH | | | 68 |
| 3-Me | 148–149 | PhH | | ir, ms, pmr | 37 |
| Hydrochloride | 283–285 | EtOH | 59 | | 36a |
| Trichloromercurate | 173–174 | *i*-PrOH | 58 | ir, pmr | 37 |

TABLE II-4. —(contd.)

| Substituent | mp (°C); [bp (°C/torr)] | Solvent of Crystallization | Yield (%) | Spectra | Refs. |
|---|---|---|---|---|---|
| 3-NHMe | | | | | |
| Hydrochloride | 210–211 | EtOH | | | 68 |
| 3-[4-(1,3-Benzodioxol-5-yl)piperazino] | | | | | |
| Hydroiodide, hydrate | 179–181.5 | EtOH | | | 73a, b |
| 3-[4-(1,3-Benzodioxol-5-yl)methylpiperazino] | | | | | |
| Hydroiodide | 202–204 | MeOH | | | 73a, b |
| 3-(3-$NO_2C_6H_4$) | | | | | |
| Hydrochloride | 260 | EtOH | 55 | | 36a |
| 3-(4-$NO_2C_6H_4$) | | | | | |
| Hydrochloride | 278–280 | EtOH | 36 | | 36a |
| 3-Ph | | | | | |
| Hydrochloride | 232–234 | EtOH/$Et_2O$ | | | 68 |
| Maleate | 191–193 | EtOH | | | 68 |
| 3-Piperidinomethyl | | | | | |
| Dihydrochloride | 288–289 | EtOH | | | 68 |
| 3-[4-(2-Thienyl)-1-piperazino] | | | | | |
| Hydroiodide | 192–193.5 | EtOH | | | 73a |
| 3-[4-(2-Thienylmethyl)-1-piperazino] | | | | | |
| Hydrochloride | 194–196 | *i*-PrOH | | | 73a, b |
| 3-$PhCH_2S$ | | | | | |
| Hydrochloride | 162–164 | *i*-PrOH/$Et_2O$ | 70 | | 72 |
| 3-MeS | | | | | |
| Hydroiodide | 226d | MeOH/$Et_2O$ | 100 | | 70 |
| 3-(2-$ClC_6H_4CH_2S$) | | | | | |
| Hydrochloride | 209–211 | *i*-PrOH | 61 | | 72 |
| 3-(2,5-$Me_2C_6H_2SCH_2$)$C_6H_3$] | | | | | |
| Hydrochloride | 216–218 | *i*-PrOH | 62 | | 72 |
| 3-(2-Pyridylmethylthio) | | | | | |
| Hydrochloride | > 250 | | | | 71 |

| | | | | | |
|---|---|---|---|---|---|
| 3-(4-$CF_3C_6H_4$) | | | | | |
| Hydrochloride | > 300 | EtOH | 58 | | 36a |
| 3-$CCl_3$-7-Me | 110.5–112.5 | PhH | | | 69 |
| ***2,3-Dihydro-2,4-benzodiazepin-1(1H)-ones*** | | | | | |
| 5-Ph | 214 | EtOAc | 12; 60 | ms, pmr, $^{13}C$ nmr | 74, 75 |
| 4-Oxide | 215 | EtOAc/$CHCl_3$ | 40 | ms, pmr | 74, 75 |
| 5-Ph-7-Cl | 139 | MeOH | 8; 55 | ms, pmr | 74, 75 |
| 4-Oxide | 232–233 | $CHCl_3$/$CCl_4$ | 5 | ms, pmr | 75 |
| ***2,3,4,5-Tetrahydro-1H-2,4-benzodiazepines*** | | | | | |
| 2,4-$(Ph)_2$ | 196 | EtOH | | | 78 |
| 2,4-(2-$MeC_6H_4)_2$ | 139 | EtOH/Py | | | 79 |
| 2,4-(4-$MeC_6H_4)_2$ | 159–160 | EtOH | | | 78 |
| 2,4-(2-$MeC_6H_4$)-3-Ph | 180 | | | | 79 |
| ***1,2,4,5-Tetrahydro-2,4-benzodiazepin-3(3H)-ones*** | | | | | |
| None | 256–260 | $CHCl_3$/EtOH | 92 | | 80 |
| | > 300 | EtOH | 84 | | 36a |
| 2,4-$(PhCH_2)_2$ | Oil | | 92 | ms | 80 |

TABLE II-4. —(contd.)

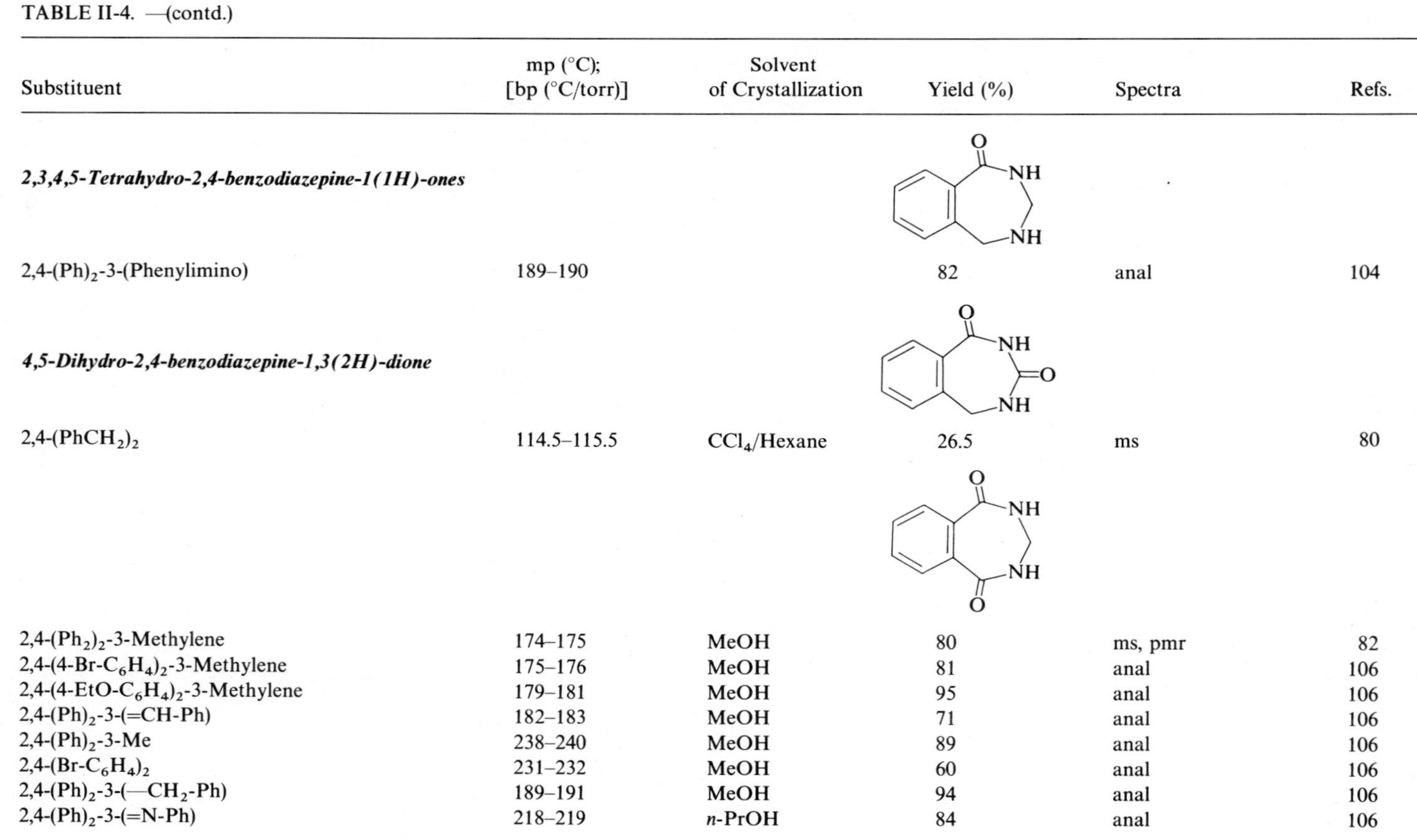


| Substituent | mp (°C); [bp (°C/torr)] | Solvent of Crystallization | Yield (%) | Spectra | Refs. |
|---|---|---|---|---|---|
| ***2,3,4,5-Tetrahydro-2,4-benzodiazepine-1(1H)-ones*** | | | | | |
| 2,4-$(Ph)_2$-3-(Phenylimino) | 189–190 | | 82 | anal | 104 |
| ***4,5-Dihydro-2,4-benzodiazepine-1,3(2H)-dione*** | | | | | |
| 2,4-$(PhCH_2)_2$ | 114.5–115.5 | $CCl_4$/Hexane | 26.5 | ms | 80 |
| 2,4-$(Ph_2)_2$-3-Methylene | 174–175 | MeOH | 80 | ms, pmr | 82 |
| 2,4-(4-Br-$C_6H_4)_2$-3-Methylene | 175–176 | MeOH | 81 | anal | 106 |
| 2,4-(4-EtO-$C_6H_4)_2$-3-Methylene | 179–181 | MeOH | 95 | anal | 106 |
| 2,4-$(Ph)_2$-3-(=CH-Ph) | 182–183 | MeOH | 71 | anal | 106 |
| 2,4-$(Ph)_2$-3-Me | 238–240 | MeOH | 89 | anal | 106 |
| 2,4-(Br-$C_6H_4)_2$ | 231–232 | MeOH | 60 | anal | 106 |
| 2,4-$(Ph)_2$-3-(—$CH_2$-Ph) | 189–191 | MeOH | 94 | anal | 106 |
| 2,4-$(Ph)_2$-3-(=N-Ph) | 218–219 | *n*-PrOH | 84 | anal | 106 |

| | | | | | |
|---|---|---|---|---|---|
| ***2,4-Benzodiazepine-1,3,5-(2H,4H)-trione*** | | | 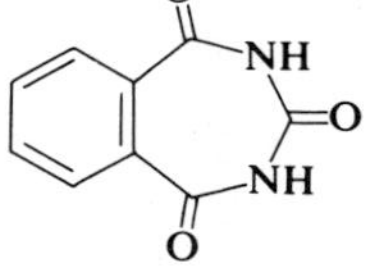 | | |
| 2,4-$(CH_2CH_2Cl)_2$ | 87.5–89 | *i*-PrOH | 95; 57 | ir, pmr | 87 |
| ***2,3-Dihydro-2,4-benzodiazepine-1(1H)-thione*** | | | 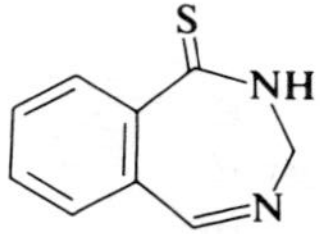 | | |
| 5-Ph | 203 | | 38 | ms, pmr | 88 |
| ***1,2,4,5-Tetrahydro-2,4-benzodiazepine-3(3H)-thione*** | | | 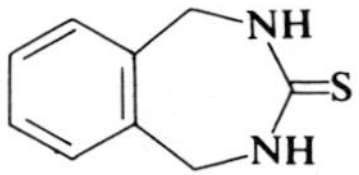 | | |
| None | 283–284 | | | | 68 |
| | 290d | | 72 | | 70 |
| ***2,3,4,5-Tetrahydro-3-thioxo-1H-2,4-benzodiazepin-1-ones*** | | | 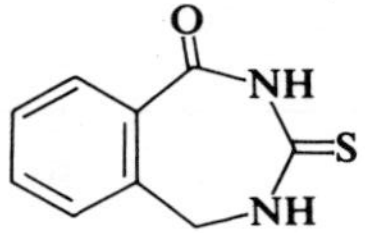 | | |
| 2,4-$(CH_2CH{=}CH_2)_2$ | 44–45 | *i*-$Pr_2O$ | 9.5 | ir, uv | 89a |
| 2-$(CH_2CH{=}CH_2)$-4-Ph | 137–138 | *i*-PrOH | 13.5 | ir, uv | 89a |
| 2,4-$(PhCH_2)_2$ | 135–136 | *i*-PrOH | 50 | ir, pmr, uv | 89a |
| 2-$PhCH_2$-4-(2-Me)Pr | 124–125 | *i*-PrOH | 13.5 | ir, pmr, uv | 89a |
| 2-$PhCH_2$-4-Ph | 155–157 | *i*-PrOH | 13.8 | ir, pmr, uv | 89a |
| 2,4-$(4-ClC_6H_4CH_2)_2$ | 163–164 | EtOH | 23 | ir, pmr, uv | 89a |

TABLE II-4. —(contd.)

| Substituent | mp (°C); [bp (°C/torr)] | Solvent of Crystallization | Yield (%) | Spectra | Refs. |
|---|---|---|---|---|---|
| 2,4-(4-$MeOC_6H_4CH_2)_2$ | 117.5 | EtOH | 18.3 | ir, pmr, uv | 89a |
| 2,4-$Me_2$ | 170–171.5 | *i*-PrOH | 45.5 | ir, pmr, uv | 89a |
| 2-Me-4-Pyrrolidinomethyl | 136–137 | *i*-PrOH | 8.3 | ir, pmr, uv | 89a |
| 2,4-(4-$MeC_6H_4CH_2)_2$ | 130–131 | EtOH | 25.5 | ir, pmr, uv | 89a |
| 2-Isobutyl-4-$PhCH_2$ | 91–93.5 | *i*-PrOH | 7.5 | ir, pmr, uv | 89a |
| 2-(3-Morpholinopropyl)-4-Me | 117–118 | *i*-PrOH | 9 | ir, pmr, uv | 89a |
| 2-(2-Pyrrolidinoethyl)-4-Ph | 168–170 | *i*-PrOH | 6 | ir, pmr | 89a |
| ***1,2,4,5-Tetrahydro-3-cyanoimino-2,4-benzodiazepines*** | | | NH, =NCN, NH | | |
| None | 249–251 | EtOH | 78 | ir, anal | 102 |
| 2-$CH_2COOEt$ | 189–191 | $CHCl_3/Et_2O$ | 65 | ir, pmr, anal | 102 |
| ***Cyclopenta[e] [1,3]diazepines*** | | | 1, $N^2$, 3, $N_4$, 5, 6, 7, 8 | | |
| 3-$NMe_2$ | 103 | Petr ether | 40–50 | ir, pmr, uv | |
| Picrate | 170d | | | X-ray | 90 |

| | | | | | |
|---|---|---|---|---|---|
| ***2,3,4,5-Tetrahydro-1H-thieno*[3,4-*e*] [*1,3*]*diazepines*** | | | 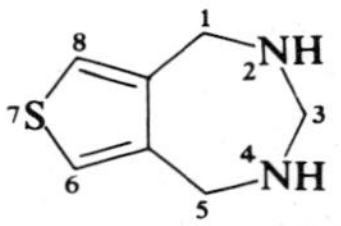 | | |
| 4,6,8-$(Me)_3$ | [125–128/5] | | 40 | | 92 |
| Dipicrate | 183d | $EtOH/Et_2O$ | | | 92 |
| 2-CHO-4,6,8-$(Me)_3$ | [132–132/5] | Heptane | 30 | | 92 |
| Methiodide | 216–218 | $EtOH/H_2O$ | | | 92 |
| ***3,4-Dihydro-Furo*[*3,4-e*] [*1,3*]*diazepine-1,5(2H)-dione*** | | | 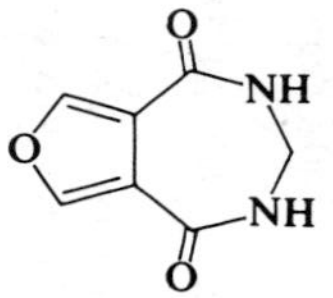 | | |
| 2,4-$(Ph)_2$-3-Methylene | 186–188 | MeOH | 76 | anal | 106 |
| ***3,4-Dihydropyrrolo*[*3,4-e*] [*1,3*]*diazepine-1,5(2H,7H)-dione*** | | | 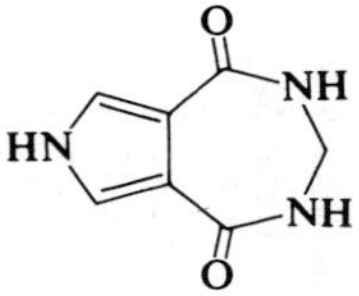 | | |
| 6,8-$(Me)_2$-3-Methylene-2,4,7-triphenyl | 235–237 | MeOH | 76 | anal | 106 |

# F. REFERENCES


1. J. Burkhardt and K. Hamann, *Chem. Ber.*, **101**, 3428 (1968).
2. (a) A. F. McKay and M.-E. Kreling, *Can. J. Chem.*, **35**, 1438 (1957). (b) A. F. McKay and M.-E. Kreling, Brit. Patent 826,837, January 1960. (c) A. F. McKay and M.-E. Kreling, *Can. J. Chem.*, **40**, 1160 (1962).
3. Jap. Patent J5-3079890, July 1978 (Daiichi Phar.).
4. (a) Ger. Offen. 2,118,261, April 1971 (C. H. Boehringer Sohn, Ingelheim). (b) Dutch Patent 71,743, October 1972 (C. H. Boehringer Sohn, Ingelheim), *Chem. Abstr.*, **78**, 29773.
5. C. R. Lee and R. J. Pollitt, *Biochem. J.*, **126**, 79 (1972).
6. (a) M.-E. Kreling and A. F. McKay, *Can. J. Chem.*, **36**, 775 (1958). (b) A. F. McKay and M.-E. Kreling, U.S. Patent 2,865,913, December 1958.
7. V. Sunjic, T. Fajdiga, and M. Japelj, *J. Heterocycl. Chem.*, **7**, 211 (1970).
8. Ger. Offen. 2,315,422, October 1973, (Chinoin Gyogyszeres Vegyeszeti Termekek Gyara RT).
9. A. Fozard and G. Jones, *J. Chem. Soc.*, 2763 (1964).
10. Y. Okamoto, A. Takada, and T. Ueda, *Chem. Pharm. Bull.*, **20**, 725 (1972).
11. E. Ott and F. Hess, *Arch. Pharm.*, **276**, 181 (1938).
12. Ger. Offen. 2,731,982, January 1978 (Yamanouchi Pharm. Co. Ltd, Tokyo).
13. V. S. Reznik, I Sh. Salikhov, Yu. Shvetsov, A. N. Shirshov, V. S. Bakulin, and B. E. Ivanov, *Izv. Akad. Nauk SSSR, Ser. Khim.*, 880 (1977).
14. L. P. Prikazchikova, L. K. Kurilenko, and V. M. Cherkasov, *Ukr. Khim. Zh.*, **42**, 518 (1976).
15. (a) I. B. Toperman and O. Yu. Magidson, *Khim.-Farm. Zh.*, **2**, 35 (1968). (b) O. Yu. Magidson and I. B. Toperman, U.S.S.R 224,522, August 1968.
16. Japan Patent 74, 27,877, July 1974 (Fujisawa Pharm. Co. Ltd.); *Chem. Abstr.*, **82**, 156, 398g.
17. J. W. Sowell, and C. De Witt, Blanton, *J. Pharm. Sci.*, **65**, 908 (1976).
18. Fr. Patent 1,491,791, August 1967 (Farbenfarbriken Bayer).
19. J. W. Ducker and M. J. Gunter, *Aust. J. Chem.*, **21**, 2809 (1968).
20. G. I. Glover, R. B. Smith, and H. Rapoport, *J. Am. Chem. Soc.*, **87**, 2003 (1965).
21. (a) H. Wollweber, *Angew. Chem., Int. Edit., Engl.*, **8**, 69 (1969). (b) Ger. Offen. 1,802,468, May 1970 (Farbenfabriken Bayer).
22. Belg. Patent 659,530, August 1965 (J. R. Geigy).
23. W. J. Houlihan, U.S. Patent 3,334,099, August 1967 (Sandoz.)
24. V. K. Chadha, H. S. Chandhary, and H. K. Pujari, *Aust. J. Chem.*, **22**, 2697 (1969).
25. K. S. Dhaka, V. K. Chadha, and H. K. Pujari, *Indian J. Chem.*, **11**, 554 (1973).
26. R. E. Manning, U.S. Patent 3,763,142, October 1973 (Sandoz-Wander); Belg. Patent 777,241, June 1972 (Sandoz S.A.).
27. (a) S. C. Bell and P. H. L. Wei, *J. Med. Chem.*, **19**, 524 (1976). (b) P. H. L. Wei and S. C. Bell, U.S. Patent 3,853,872, December 1974 (American Home Products Corp.).
28. V. K. Chadha, K. S. Sharma, and H. K. Pujari, *Indian J. Chem.*, **9**, 1216 (1971).
29. V. Wolf and W. Braun, *Arzneimittel-Forsch.*, **10**, 304 (1960).
30. F. D'Angeli, C. Di Bello, and V. Giormani, *Gazz. Chim. Ital.*, **95**, 735 (1965).
31. M. Brugger and F. Korte, *Justus Liebigs Ann. Chem.*, **764**, 112 (1972).
32. D. J. Le Count and P. J. Taylor, *Tetrahedron*, **31**, 433 (1975).
33. J. B. Taylor and W. R. Tully, *J. Chem. Soc., Perkin Trans.*, 1331 (1976).
34. F. M. F. Chen and T. P. Forrest, *Can. J. Chem.*, **51**, 881 (1973).
35. U. Stauss, H. P. Härter, M. Neuenschwander, and O. Schindler, *Helv. Chim. Acta*, **55**, 771 (1972).
36. (a) H. R. Rodriguez, B. Zitko, and G. de Stevens, *J. Org. Chem.*, **33**, 670 (1968). (b) H. R. Rodriguez and G. de Stevens, U.S. Patent 3,681,340, August 1972.
37. J. M. Desmarchelier, N. A. Evans, R. F. Evans, and R. B. Johns, *Aust. J. Chem.*, **21**, 257 (1968).
38. B. A. Burdick, P. A. Benkovic, and S. J. Benkovic, *J. Am. Chem. Soc.*, **99**, 5716 (1977).
39. J. T. Suh and R. A. Schnettler, U.S. Patent 3,780,023, December 1973.
40. J. T. Suh and R. A. Schnettler, U.S. Patent 3,838,122, September 1974.

41. J. T. Suh and R. A. Schnettler, U.S. Patent 3,780,024, December 1973.
42. T. Jen, P. Bender, H. Van Hoeven, B. Dienel, and B. Loer, *J. Med. Chem.*, **16**, 407 (1973).
43. J. T. Suh and R. A. Schnettler, U.S. Patent 3,849,400, November 1974.
44. E. F. Elslager, D. F. Worth, and S. C. Perricone, *J. Heterocycl. Chem.*, **6**, 491 (1969).
45. F. P. Woerner, H. Reimlinger, and R. Merenyi, *Chem. Ber.*, **104**, 2768 (1971).
46. Ger. Offen. 1,947,062, March 1970 (Yeda Research and Development Co. Ltd.).
47. U. Golik and W. Taub, *J. Heterocycl. Chem.*, **12**, 1155 (1975).
48. W. V. Farrar, *Chem. Ind.* (London) 1808 (1968).
49. G. de Stevens and M. Dughi, *J. Am. Chem. Soc.*, **83**, 3087 (1961).
50. O. Hromatka, M. Knollmüller, and H. Deschler, *Monatsh. Chem.*, **100**, 469 (1969).
51. W. B. Wright, Jr., U.S. Patent 3,474,090, October 1969 (American Cyanamid Co.).
52. T. P. Forrest, G. A. Dauphinee, and F. M. F. Chen, *Can. J. Chem.*, **52**, 2725 (1974).
53. G. de Stevens, U.S. Patents 3,157,642, November 1964, and 3,310,582, March 1967.
54. G. de Stevens, *Rec. Chem. Prog.*, **23**, 105 (1962).
55. O. Tsuge and H. Watanabe, *Heterocycles*, **7**, 907 (1977).
56. P. Gyulai and K. Lempert, *Magyar Kem. Foly.*, **76**, 96 (1970).
57. C. W. Bird, *J. Chem. Soc.*, 5762 (1965).
58. Y. Ohshiro, K. Kinugasa, T. Manami, and T. Agawa, *J. Org. Chem.*, **35**, 2136 (1970).
59. (a) A Baba, Y. Ohshiro, and T. Agawa, *J. Organomet. Chem.*, **87**, 247 (1975). (b) Japan. Kokai, 75,121,290, September 1975 (Mitsubishi Chemical Co. Ltd.).
60. T. N. Ghosh, *J. Indian Chem. Soc.*, **10**, 583 (1933).
61. D. C. Baker and S. R. Putt, U.S. Patent 4,117,229, September 1978 (Warner-Lambert Co.).
62. T. Tsurnoka, N. Ezaki, S. Amano, C. Uchida, and T. Niida, *Meiji Seika Kenkyu Nempo* 17 (1967); *Chem. Abstr.*, **69**, 8514r.
63. A. Ryder, H. W. Dion, P. W. K. Woo, and J. D. Howells, U.S. Patent 3,923,785, 1975 (Parke, Davis & Co.).
64. H. Nakamura, G. Koyama, Y. Iitaka, M. Ohno, N. Yagisawa, S. Kondo, K. Maeda, and H. Umezawa, *J. Am. Chem. Soc.*, **96**, 4327 (1974).
65. P. W. K. Woo, H. W. Dion, S. M. Lange, L. F. Dahl, and L. J. Durham, *J. Heterocycl. Chem.*, **11**, 641 (1974).
66. (a) M. Ohno, N. Yagisawa, S. Shibahara, S. Kondo, K. Maeda, and H. Umezawa, *J. Am. Chem. Soc.*, **96**, 4326 (1974). (b) H. Umezawa, K. Maeda and S. Kondo, U.S. Patent 3,959,257, May 1976.
67. Belg. Patent 864,711, November 1978 (Zaidan Hojin Biseibutsu Kagaku Kenkyu Kai).
68. Brit. Patent 1,183,135, March 1970 (Ciba Ltd.).
69. Ger. Offen., 2,601,137, July 1976 (Merck & Co.).
70. E. F. Elslager, D. F. Worth, N. F. Haley, and S. C. Perricone, *J. Heterocycl. Chem.*, **5**, 609 (1968).
71. Ger. Offen. 2,504,252, August 1975 (Aktiebolaget Hässle, Mölndal, Sweden).
72. E. F. Elslager, J. R. McLean, S. C. Perricone, D. Potoczak, H. Veloso, D. F. Worth, and R. H. Wheelock, *J. Med. Chem.*, **14**, 397 (1971).
73. (a) R. A. Schnettler and J. T. Suh, U.S. Patent 3,867,388, February 1975. (b) U.S. Patent 3,905,980, September 1975 (Colgate-Palmolive Co.).
74. U. Golik, *Tetrahedron Lett.*, 1327 (1975).
75. U. Golik, *J. Heterocycl. Chem.*, **12**, 903 (1975).
76. W. Taub and U. Golik, U.S. Patent 3,939,152, February 1976 (Yeda Research & Development Co., Ltd.).
77. E. Reeder, unpublished results.
78. M. Scholtz and K. Jaross, *Berichte*, **34**, 1504 (1901).
79. M. Scholtz and R. Wolfrum, *Berichte*, **43**, 2304 (1910).
80. A. M. Felix and R. I. Fryer, *J. Heterocycl. Chem.*, **5**, 291 (1968).
81. A. Rosenthal and S. Millward, *Can. J. Chem.*, **42**, 956 (1964).
82. H. W. Heine and C. Tintel, *Tetrahedron Lett.*, 23 (1978).
83. A. Piutti, *Justus Liebigs Ann. Chem.*, **214**, 17 (1882).

84. T. W. Evans and W. M. Dehn, *J. Am. Chem. Soc.*, **51**, 3651 (1929).
85. C. S. Smith and C. J. Cavallito, *J. Am. Chem. Soc.*, **61**, 2218 (1939).
86. D. Grdenić and A. Bezjak, *Arch. Kem.*, **25**, 101 (1953).
87. D. A. Tomalia, N. D. Ojha, and B. P. Thill, *J. Org. Chem.*, **34**, 1400 (1969).
88. U. Golik, *J. Heterocycl. Chem.*, **13**, 613 (1976).
89. (a) H. Fujita and Y. Sato, *Chem. Pharm. Bull.*, **23**, 1764 (1975). (b) Japan Kokai, 75,117,790, September 1975 (Sankyo Co., Ltd.); *Chem. Abstr.*, **84**, 105670.
90. (a) U. Müller-Westerhoff and K. Hafner, *Tetrahedron Lett.*, 4341 (1967). (b) K. Hafner, *J. Heterocycl. Chem.*, **13**, 33 (1976). (c) Brit. Patent 1,226,179, March 1971 (Studiengesellschaft Kohle m.b.H.).
91. H. J. Lindner, *Chem. Ber.*, **103**, 1828 (1970).
92. M. S. Kondakova and I. L. Goldfarb, *Bull. Acad. Sci. USSR*, 570 (1958).
93. A. Kosasayama, T, Konno, K. Higashi, and F. Ishikawa, *Chem. Pharm. Bull.*, **27**: 4, 841 (1979).
94. A. Kosasayama, T. Konno, K. Higashi, and F. Ishikawa, *Chem. Pharm. Bull.*, **27**: 4, 848 (1979).
95. T. Kurihara, K. Nasu, F. Ishimori, and T. Tani, *J. Heterocycl. Chem.*, **18**, 163 (1981).
96. T. Kurihara, K. Nasu, and Y. Adachi, *J. Heterocycl. Chem.*, **20**, 81 (1983).
97. A. S. Tomcufcik, W. B. Wright, Jr., and J. W. Marsico, Jr., U.S. Patent 4,344,954, August 1982.
98. D. W. Hills and G. R. W. Harpenden, U.S. Patent 4,375,435, March 1983.
99. T. Tsuchiya, S. Okajima, M. Enkaku, and J. Kurita, *Chem. Pharm. Bull.*, **30**: 10, 3757 (1982).
100. L. L. Setescak, F. W. Dekow, J. M. Kitzen, and L. L. Martin, *J. Med. Chem.*, **27**, 401 (1984).
101. T. Tsuchiya, M. Enkaku, and S. Okajima, *Chem. Pharm. Bull.*, **28**: 9, 2602 (1980).
102. F. Ishikawa and Y. Watanabe, *Chem. Pharm. Bull.*, **28**: 4, 1307 (1980).
103. T. B. K. Lee and G. E. Lee, U.S. Patent, 4,374,067, February 1983.
104. L. L. Martin, M. Worm and C. A. Crichlow, U.S. Patent, 4,409,145, October 1983.
105. E. Chan, S. R. Putt, and H. D. H. Showalter, *J. Org. Chem.*, **47**, 3457 (1982).
106. H. W. Heine, D. W. Ludovici, J. A. Pardoen, R. C. Weber II, E. Bonsall, and K. R. Osterhout, *J. Org. Chem.*, **44**: 22, 3843 (1979).

CHAPTER III

# 1,4-Diazepines with [*a*]- or [*d*]-Fused Rings

**R. Ian Fryer**

*Department of Chemistry, Rutgers, State University of New Jersey, Newark, New Jersey*

and

**A. Walser**

*Chemical Research Department, Hoffmann-La Roche Inc., Nutley, New Jersey*

## INTRODUCTION

The ring systems reviewed in this chapter contains an additional ring fused to sides *a* or *d* of the 1,4-diazepine **1**.

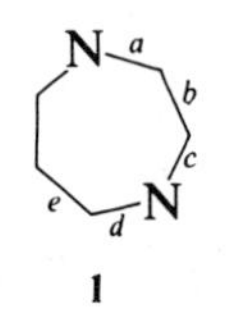


**1**

The extensive work carried out in the area of [*b*]-fused[1,4]diazepines (e.g., 1,5-benzodiazepines) is the subject of a separate chapter while the more voluminous work carried out in the area of [*e*]-fused[1,4]diazepines (e.g., 1,4-benzodiazepines) is further subdivided into individual chapters depending on the position of the double bonds in the seven-membered ring.

1,4-Dazepines with either [*a*]- or [*d*]-fused rings have one nitrogen common to both rings and have been reviewed by F. D. Popp and A. C. Noble (in *Advances in Heterocyclic Chemistry* Vol. 8, A. R. Katritzky and A. J. Boulton, Eds., Academic Press, New York and London, 1967, p. 21).

## A. [*a*]-FUSED [1,4]DIAZEPINES

This section embraces all bicyclic systems of general structure **2**. The ring systems are discussed in alphabetical order.

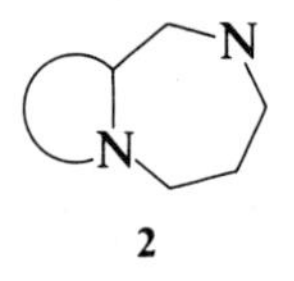


**2**

## 1. AZIRINO[1,2-*a*] [1,4]DIAZEPINES

### 1.1. Synthesis

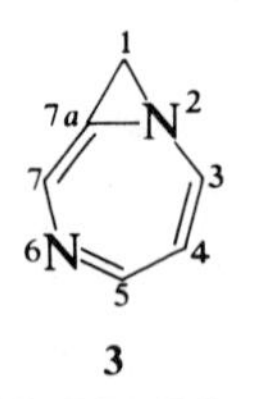


**3**

1*H*-Azirino[1,2-*a*] [1,4]diazepine

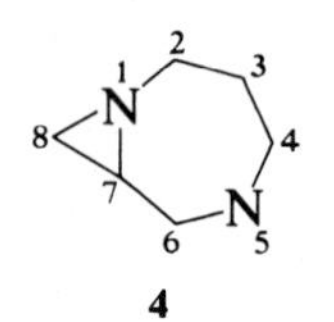


**4**

1,5-Diazabicyclo[5.1.0]octane

The parent ring system represented by the 1*H* tautomer **3** is still unknown. A 3,7*a*-dihydro derivative, **6**, was synthesized in about 25% yield by Padwa and Gehrlein[1] by the reaction of *trans*-3-benzoyl-2-phenylaziridine (**5**) with cinnamaldehyde in the presence of ammonia and ammonium bromide in ethanol at room temperature (Eq. 1). The indicated relative configuration was assigned on the basis of nuclear Overhauser effects, which suggested a proximal orientation of the hydrogens in the 1- and 3-positions.

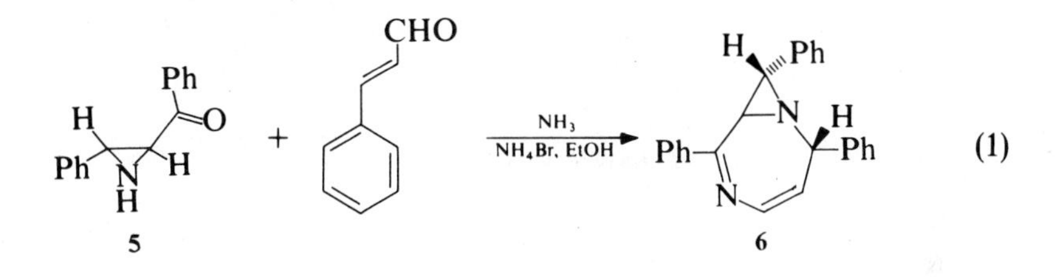


## 1.2. Reactions

Sodium methoxide in methanol converted **6** in nearly quantitative yield to the dihydropyrimidine **7** (Eq. 2). A mechanism for this base-catalyzed rearrangement was proposed. Irradiation of a benzene solution of **6** gave the acyclic compound **8** in 80% yield. Thermal rearrangement of **8** was shown to give **10** and **11**. Compounds **10** and **11** were also formed by thermolysis of **6** in boiling xylene. The major product of this reaction was the bicyclic compound **9**.

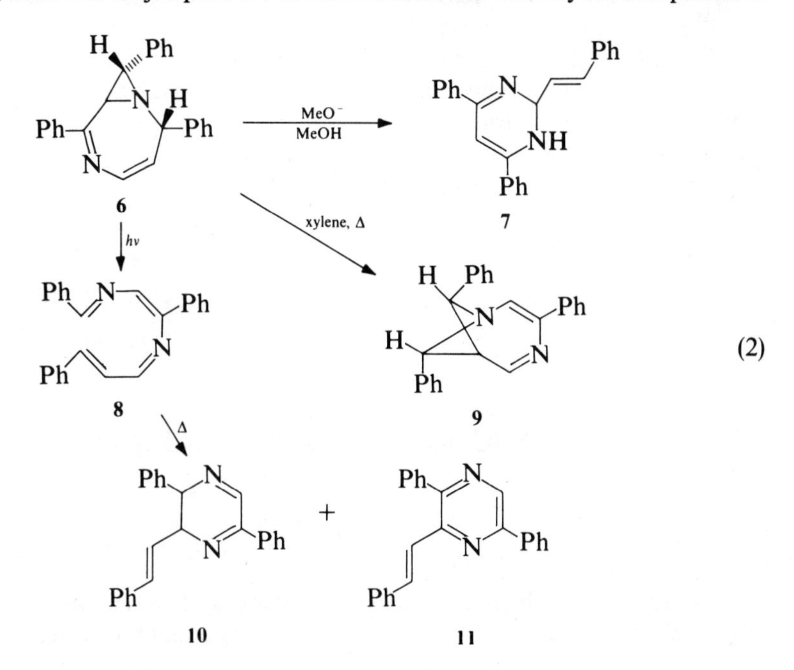

An additional reaction of **6** with fumaronitrile is reported under pyrrolo-[1,2-*a*] [1,4]diazepines. (Section 4.1).

## 2. IMIDAZO[1,5-*a*] [1,4]DIAZEPINES

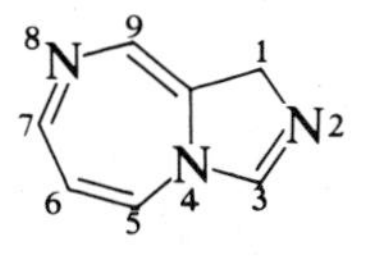

**12a**
1*H*-Imidazo[1,5-*a*] [1,4]diazepine

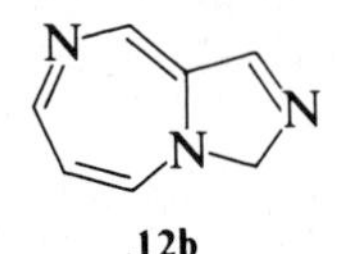

**12b**
3*H*-Imidazo[1,5-*a*] [1,4]diazepine

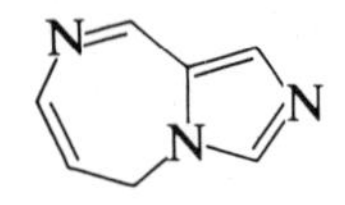

**12c**
5*H*-Imidazo[1,5-*a*]-[1,4]diazepine

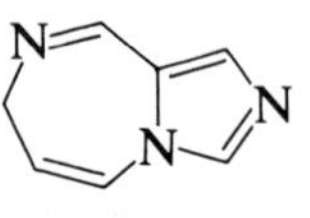

**12d**
7*H*-Imidazo[1,5-*a*]-[1,4]diazepine

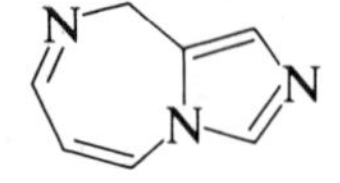

**12e**
9*H*-Imidazo[1,5-*a*]-[1,4]diazepine

This parent ring system may exist in any of the five tautomeric forms **12a**–**12e**. The tautomers with the aromatic five-membered ring should be thermodynamically favored. None of the fully unsaturated systems are known.

The synthesis of substituted tetrahydro derivatives with an intact imidazole ring has been reported.[2] Reaction of the 1-(3-aminopropyl)imidazoles **13** with formaldehyde in a buffered medium (Eq. 3) led to the imidazodiazepines **14** in 30–60% yields. Compounds **13** were accessible by cyanoethylation of the parent imidazole followed by reduction of the nitrile, either catalytically or by means of lithium aluminium hydride.

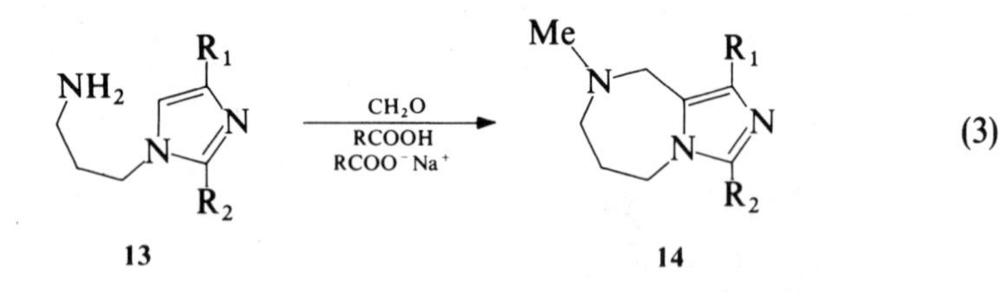


(3)

## 3. PYRIDO[1,2-*a*] [1,4]DIAZEPINES

### 3.1. Synthesis

A tetrahydro derivative of the parent ring system **15** with the pyridine nitrogen quaternized was prepared by Fozar and Jones.[3] Thermal ring closure

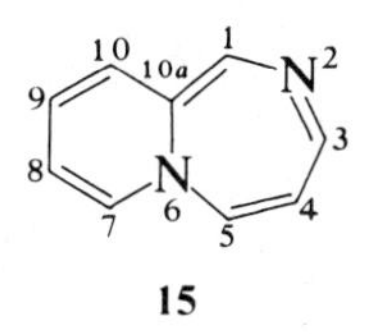


and quaternization of the bromide **16** gave **17** (Eq. 4). Compound **16** was obtained by treatment of the corresponding alcohol with phosphorus tribromide.

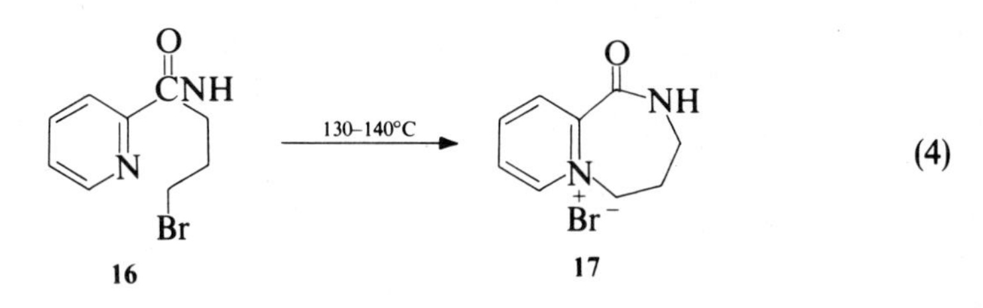


Perhydro derivatives of this ring system have received more synthetic attention. Paquette and Scott[4] obtained the 3-one **19** as the minor product of the Schmidt reaction of the ketone **18**. The major product from this rearrangement was the [*d*]-fused [1,4]diazepine **20** (Eq. 5).

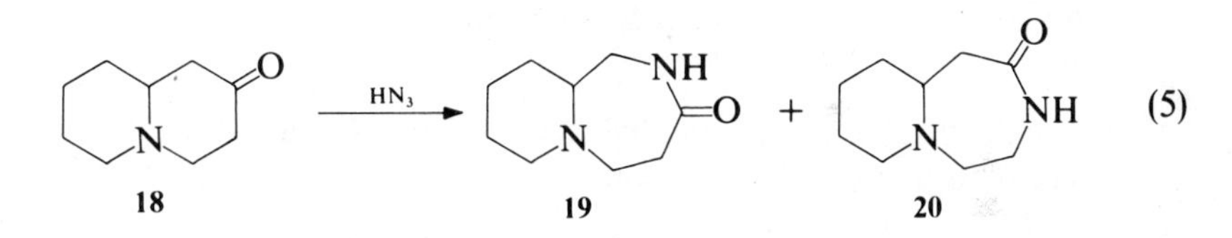


Reaction of the piperidine derivative **21** with benzylamines **22** has been claimed[5] to yield the decahydropyrido[1,2-*a*] [1,4]diazepines **23** (Eq. 6).

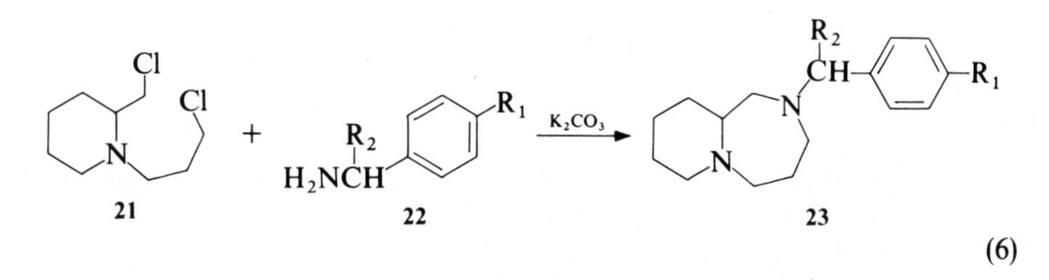


(6)

A variety of perhydro derivatives **25** bearing an aroyl group at the 4-position (Eq. 7) were prepared by condensation of the piperidine **24** with formaldehyde and an aryl methyl ketone in boiling acetic acid.[6] The reported yields were 25–77%. It is likely that the initially formed $\alpha,\beta$-unsaturated ketone undergoes Michael addition by one amino group and that the ring is formed by condensation of a second molecule of formaldehyde with the remaining amino group and

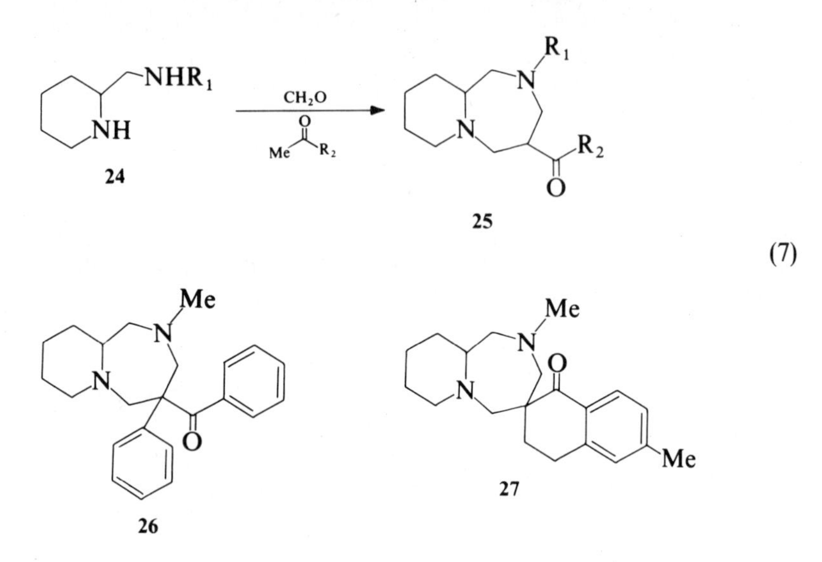


the α-carbon of the ketone. This condensation was also carried out with benzyl phenyl ketone and 6-methoxy-1-tetralone to give **26** and **27**, respectively.

## 3.2. Reactions

Quaternization of **19** with methyl iodide afforded[4] the salt **28**, which upon treatment with hydroxide was reconverted to **19** by elimination of the methyl group (Eq. 8).

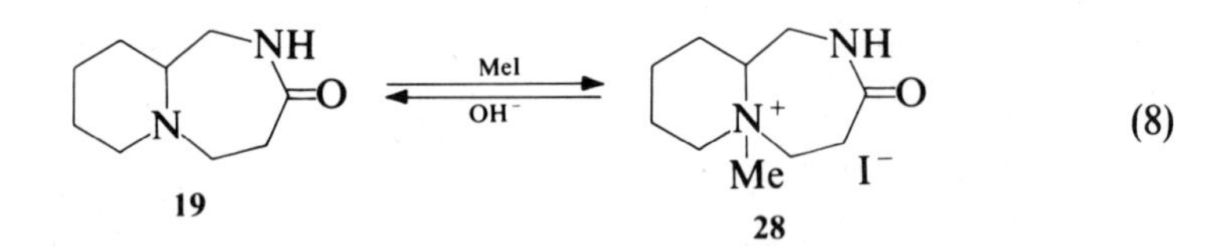


Compounds **29** ($R_1 = Me$) were reduced with lithium aluminum hydride or sodium borohydride to give a mixture of diastereomeric alcohols **30**[6] (Eq. 9). Reaction of **29** with Grignard reagents led to the carbinols **31**, of which generally only one isomer was isolated, often in high yield. Treatment of the tertiary carbinol **31** ($R_2$, $R_3 = Ph$) with sulfuric acid led to the diphenylmethylene derivative **32**. Replacement of the hydroxyl group in **30** ($R_2 = Ph$) with a morpholino moiety to give **33** was achieved via the chloride.

Two 2-substituted perhydro pyrido[1,2-*a*] [1,4]diazepines **35**, which were disclosed in the patent literature,[7] were prepared by alkylation of the 2-hydroxyethyl derivative **34** with a diphenylmethylhalide (Eq. 10).

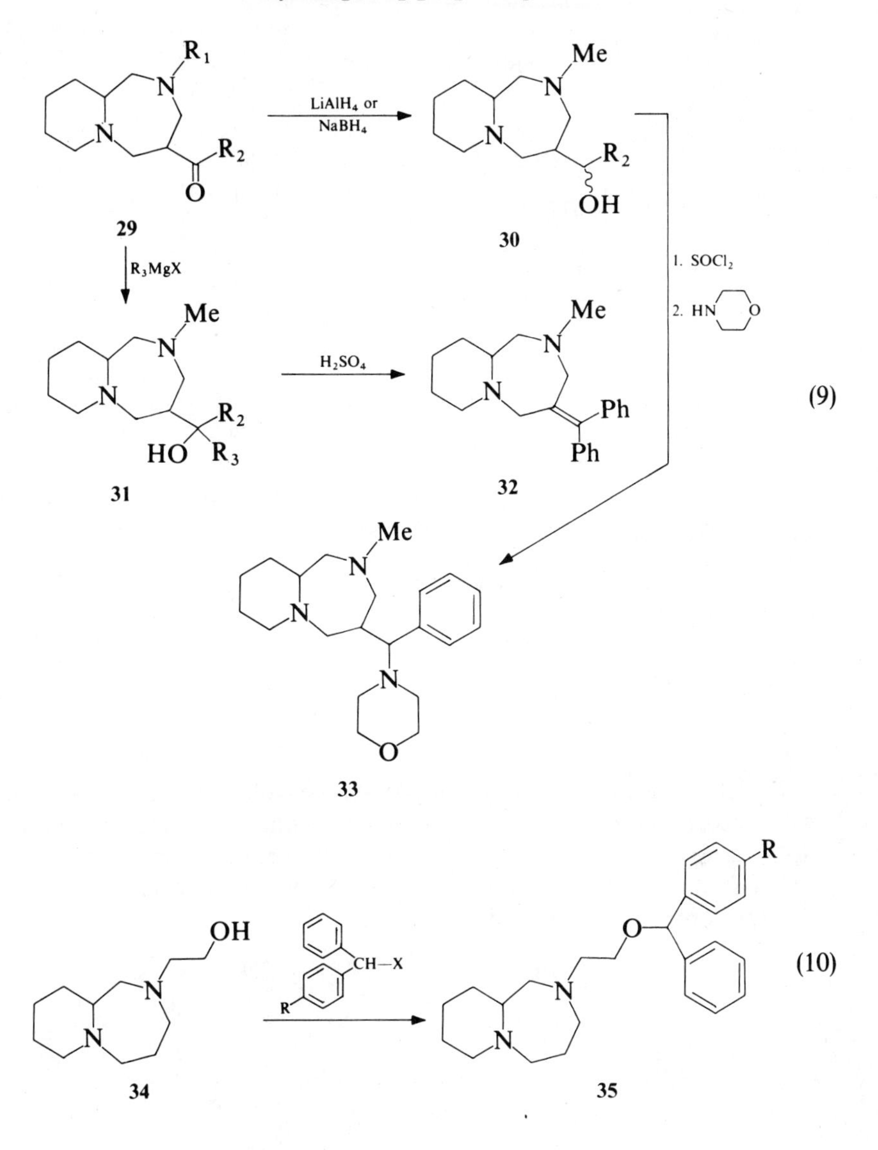




# 4. PYRROLO[1,2-*a*] [1,4]DIAZEPINES

## 4.1. Synthesis

None of the tautomeric forms **36a–36e** of the parent ring system are known. The tetrahydro derivative **38** with an intact pyrrole ring was described in the

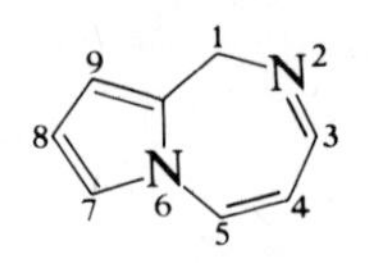

**36a**
1*H*-Pyrrolo[1,2-*a*] [1,4]diazepines

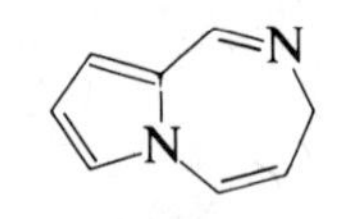

**36b**
3*H*-Pyrrolo[1,2-*a*] [1,4]diazepines

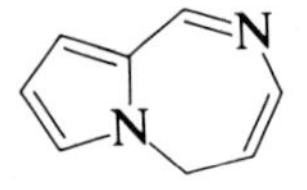

**36c**
5*H*-Pyrrolo[1,2-*a*]-[1,4]diazepines

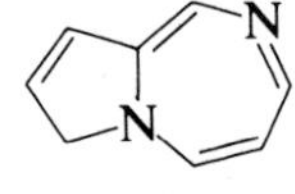

**36d**
7*H*-Pyrrolo[1,2-*a*]-[1,4]diazepines

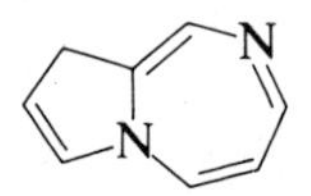

**36e**
9*H*-Pyrrolo[1,2-*a*]-[1,4]diazepines

patent literature.[8] This compound was reportedly obtained by Beckmann rearrangement from the oxime **37** (Eq. 11).

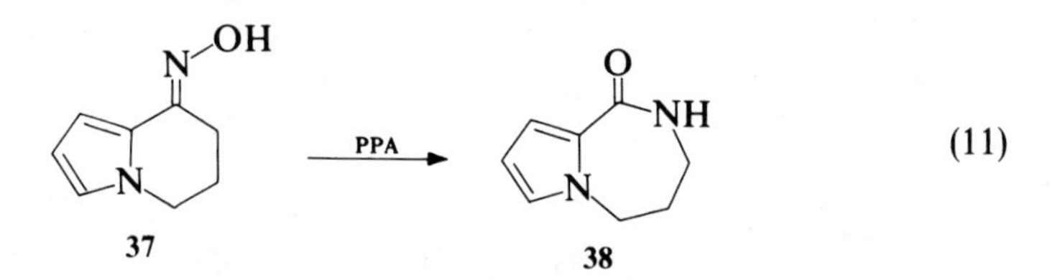


(11)

The polysubstituted 7,8,9,9*a*-tetrahydro-5*H*-pyrrolo[1,2-*a*] [1,4]diazepine **39** was formed in 76% yield by the 1,3-dipolar addition of fumaronitrile to the azirinodiazepine **6** in boiling xylene[1] (Eq. 12). The relative stereochemistry was derived from nmr data and was supported by the stereochemistry of the hydrolysis product **40** obtained in 70% yield.

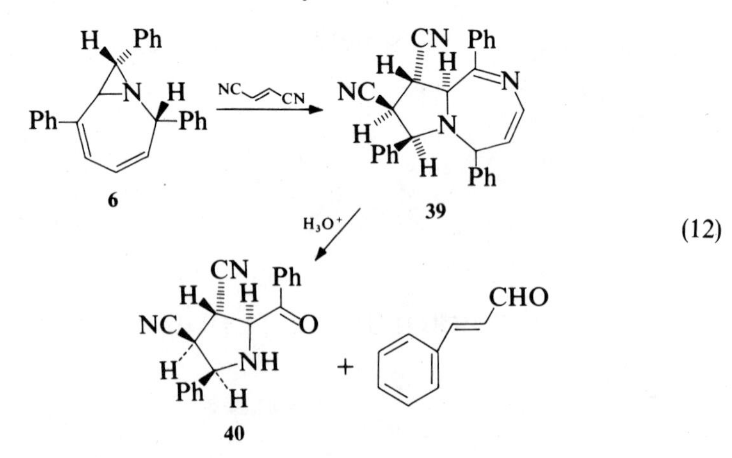


(12)

The saturated pyrrolodiazepin-1-one, **43** was reportedly synthesised from L-proline ethyl ester as shown in Eq. 13.[12] No experimental details were given.

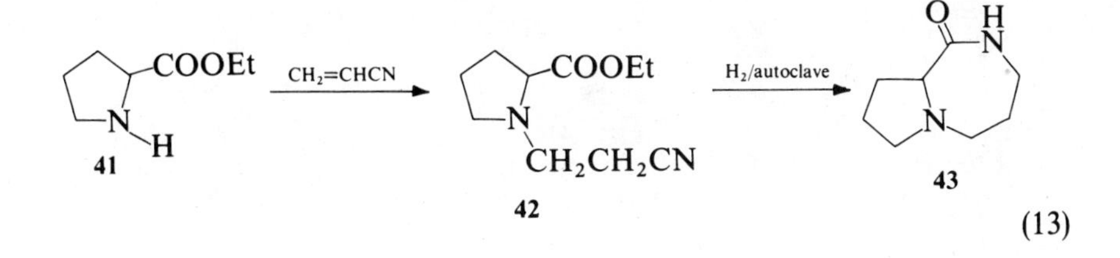


(13)

### 4.2. Reactions

Reduction of **43** using lithium aluminum hydride was reported to give the fully saturated pyrrolo[1,2-*a*] [1,4]diazepine **44**. Treatment of **44** with either substituted or unsubstituted benzhydryl bromides gave the *N*-substituted derivative **45** (Eq. 14).[12]

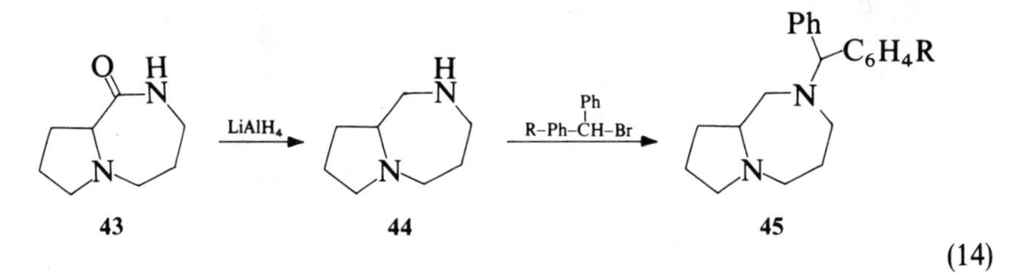


(14)

## B. [d]-FUSED [1,4]DIAZEPINES

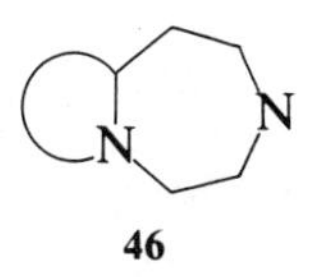


Ring systems of general structure **46**, which are the subject of this section, are discussed in alphabetical order.

## 1. IMIDAZO[1,2-*d*] [1,4]DIAZEPINES

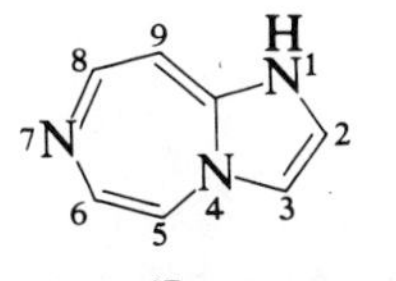


**47a**
1*H*-Imidazo[1,2-*d*] [1,4]diazepines

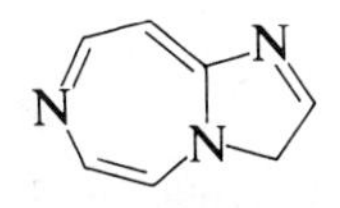


**47b**
3*H*-Imidazo[1,2-*d*] [1,4]diazepines

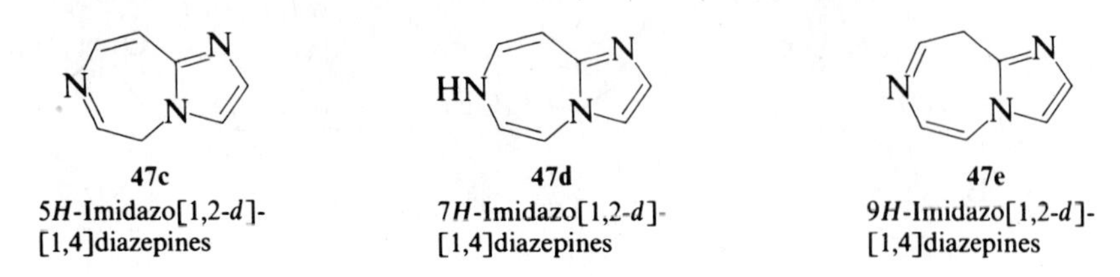


**47c** 5*H*-Imidazo[1,2-*d*]-[1,4]diazepines

**47d** 7*H*-Imidazo[1,2-*d*]-[1,4]diazepines

**47e** 9*H*-Imidazo[1,2-*d*]-[1,4]diazepines

## 1.1. Synthesis

None of the tautomeric forms **47a-47e** of the parent ring system have been described in the literature. Hexahydro derivatives were disclosed in the patent literature.[9, 10] These compounds were synthesized by fusion of the imidazolone ring to the preformed diazepine (Eq. 15). The *N*-protected diazepinone **48** (R = benzyloxycarbonyl) was converted to the imino ether **49**, which reacted with glycine to yield the amidine **50**. Cyclodehydration of **50** by boiling in 2-methoxyethanol led to the imidazo[1,2-*d*] [1,4]diazepine **52**. The protecting group was removed by treatment with hydrogen bromide in glacial acetic acid.

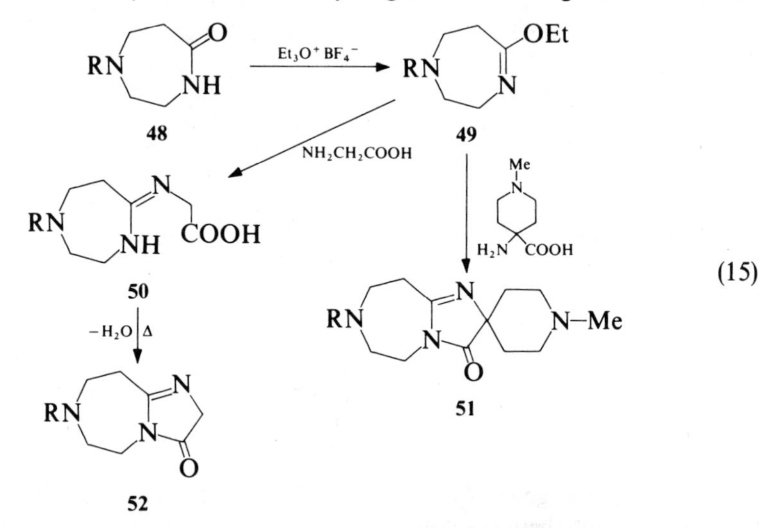


(15)

Condensation of **49** with 4-amino-*N*-methylpiperidine-4-carboxylic acid in boiling methanol gave the spiro compound **51** (R = benzyloxycarbonyl) directly in 45% yield. Removal of the protecting group followed by Eschweiler–Clarke methylation afforded **51** (R = Me).

## 1.2. Reactions

Although hydrogenation and quaternization reactions of **52** have been mentioned,[9,10] experimental details are sparse and the products were not completely characterized.

# 2. PYRIDO[1,2-*d*] [1,4]DIAZEPINES

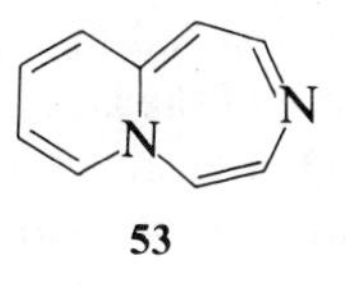


The parent ring system **53** has not yet been synthesized. Compounds with a quaternized pyridine ring were prepared by Blicke and Hughes[11] by cyclization of **55** with sodium iodide in acetone to give **56**. Compound **55** was obtained by chloroacetylation of the pyridine **54**.

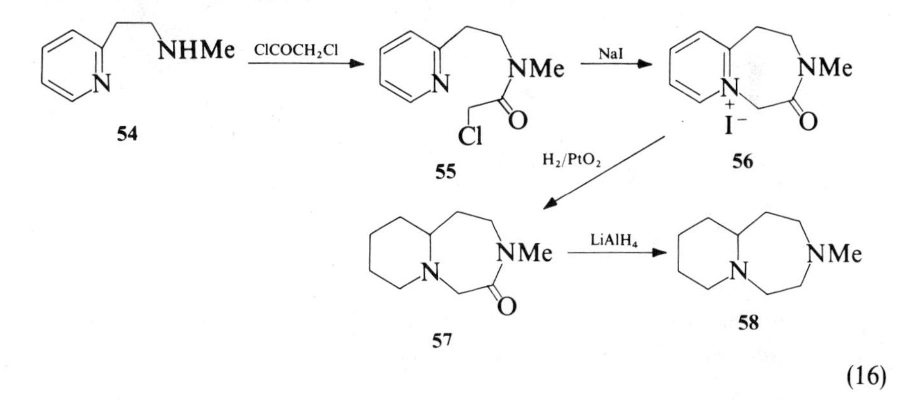


(16)

The perhydro derivative **57** was accessible in good yield by hydrogenation of the quaternary salt **56** over platinum. Further reduction of **57** with lithium aluminum hydride (Eq. 16) gave **58**. Compound **58** was also reported[11] to be formed upon reduction of **61** (R = H) with lithium aluminum hydride. The intermediate **61** was synthesized by quaternization of the chloroacetylpiperidine **59** to give **60**, followed by thermal demethylation (Eq. 17). The phenyl analog **59**

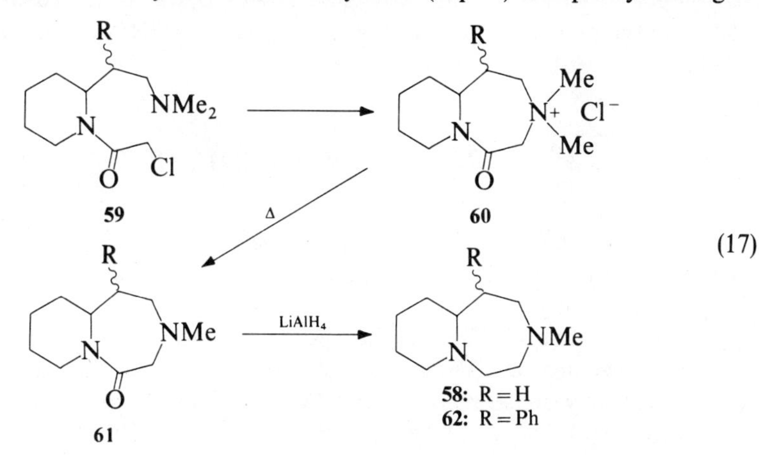


(17)

(R = Ph) was obtained as a mixture of two diastereomers, which were separated by fractional crystallization. Both isomers were converted to their corresponding perhydro derivatives **62**.

As previously mentioned [Section A.3.1], Schmidt reaction on the quinolizidine derivative **18** led to a mixture of the lactams **19** and **20**. The major product **20** was obtained in 20% yield.[4] Quaternization with methyl iodide (Eq. 18) gave the 6-methiodide **63**, which underwent Hoffmann elimination to give the *cis*-olefin **64**. The quaternary salt **63** could be regenerated in 29% yield by treatment of **64** with hydrogen iodide.

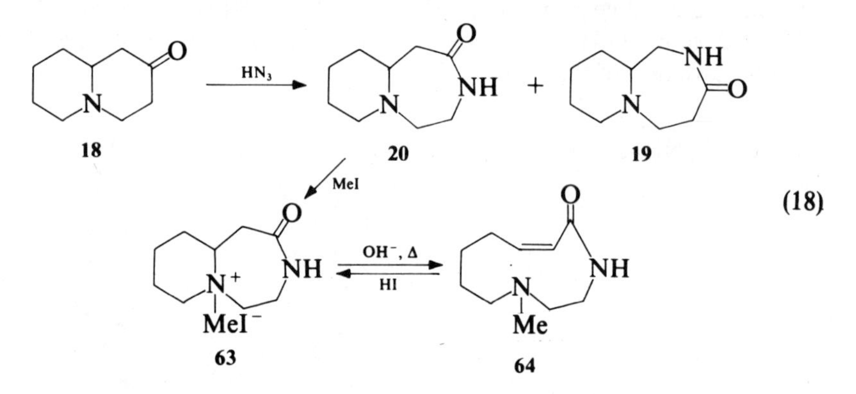


Various 3-substituted perhydropyrido[1,2-*d*] [1,4]diazepines **66** were prepared by reaction of the dihalide **65** with amines (Eq. 19).[5]

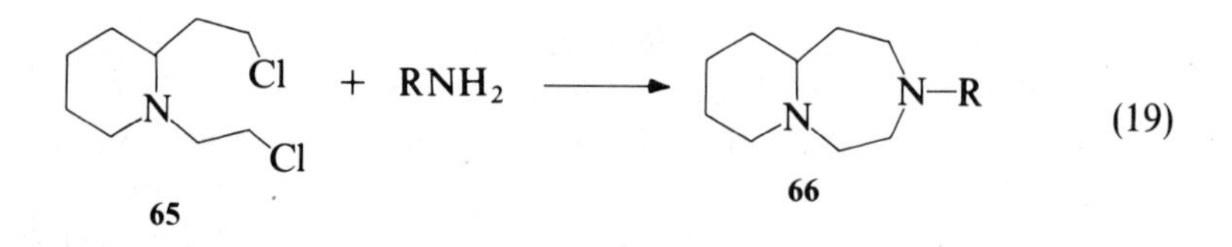


Compound **66** [R = 2-(diphenylmethoxy)ethyl] was disclosed in a patent[7] and was prepared by alkylation of **66** (R = $CH_2CH_2OH$) with diphenylmethyl halide.

## 3. PYRROLO[1,2-*d*] [1,4]DIAZEPINES

This ring system may exist in any of the tautomeric forms **67a**–**67e**. The tautomers with an aromatic pyrrole ring **67a**–**67c** are considered to be thermodynamically favored.

A representative of this bicyclic system was prepared containing a fully saturated diazepine ring. Thus, cleavage of the phthalimido protecting group of an appropriately substituted pyrrole **68** with aqueous hydrazine led to the tetrahydropyrrolo[1,2-*d*] [1,4]diazepines **69** in yields of 27–37% (Eq. 20).[13]

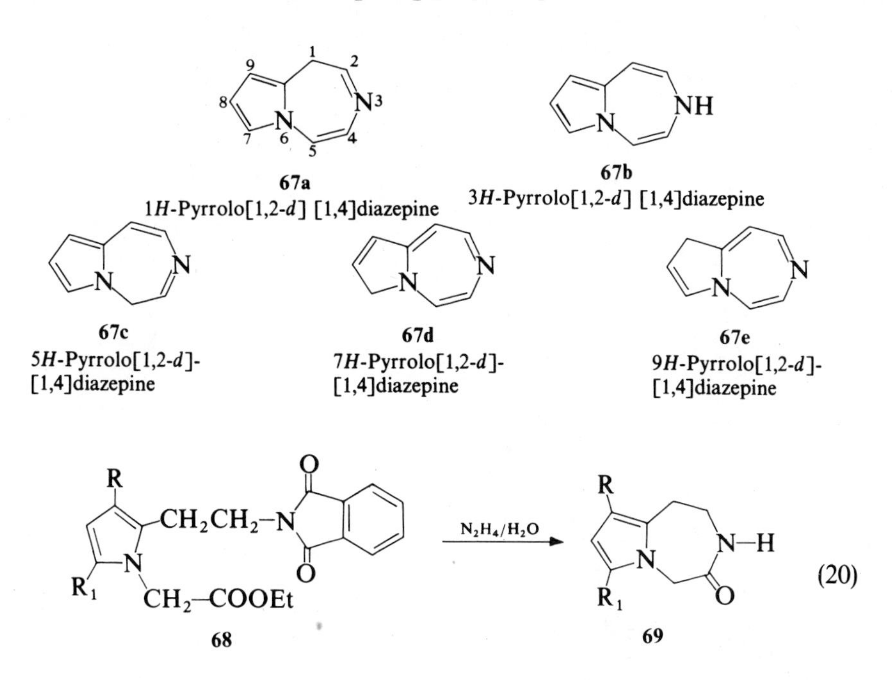


Previously, an octahydro analog had been prepared by Paquette and Scott,[4] who carried out a Schmidt reaction on the indolizine **70** (Eq. 21). This procedure gave **71** in 22% yield.

As already described for the homologous pyrido[1,2-*d*] [1,4]diazepine, **71** could be quaternized with methyl iodide to give **72**. Hoffmann elimination gave the *cis*-$\alpha,\beta$-unsaturated compound **73** in 20% yield. Again, treatment of **73** with hydrogen iodide re-formed the quaternary **72** via transannular Michael addition.

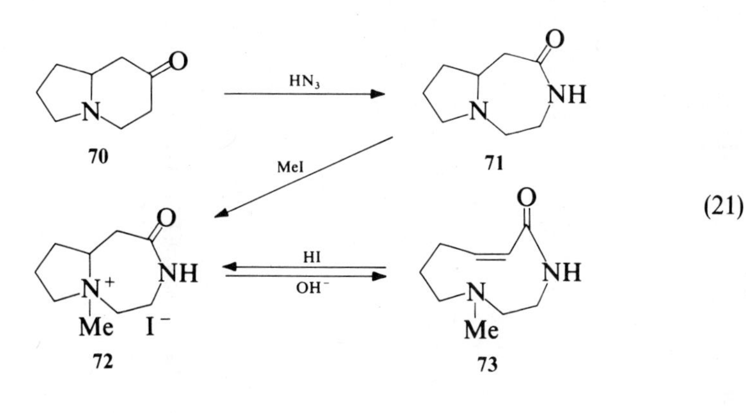

## C. TABLES OF COMPOUNDS

TABLE III-1. [*a*]-FUSED [1,4]DIAZEPINES

| Substituent | mp (°C); [bp (° C/torr)] | Solvent of Crystallization | Yield (%) | Spectra | Refs. |
|---|---|---|---|---|---|
| | | *Azirino[1,2-a] [1,4]diazepines* | | | |
| ***3,7a-Dihydro-1H-azirino[1,2-a] [1,4]diazepine*** | | | | | |
| $1,3,7-Ph_3$ | 157–158 | EtOH | 22 | ir, ms, pmr, uv | 1 |
| | | *Imidazo[1,5-a] [1,4]diazepines* | | | |
| ***6,7,8,9-Tetrahydro-5H-imidazo[1,5-a] [1,4]diazepines*** | | | | | |
| 8-Me | [110/1.0] | | 45 | pmr | 2a |
| Dipicrate | 220d | | | | |
| $1\text{-}i\text{-}Pr\text{-}3,8\text{-}(Me)_2$ | [105/0.04] | | 33 | pmr | 2a |
| Dipicrate | 216d | | | | |
| 1-Et-3-*n*-Pr-8-Me | [108/0.04] | | 38 | pmr | 2a |
| Dipicrate | 196d | | | | |
| 1-Et-3-*i*-Pr-8-Me | [98/0.07] | | 59 | pmr | 2a |
| Dipicrate | 225d | | | | |

| | | | | | |
|---|---|---|---|---|---|
| 1-Pr-3-*i*-Pr-8-Me | [105/0.04] | | 43 | pmr | 2a |
| Dipicrate | 208d | | | | |
| 1,3-(*i*-Pr)$_2$-8-Me | [109/0.4] | | 47 | pmr | 2a |
| Dipicrate | 220d | | | | |

*Pyrido[1,2-a] [1,4]diazepines*

***2,3,4,5-Tetrahydro-1-oxo-1H-pyrido[1,2-a] [1,4]diazepiniums***

| | | | | | |
|---|---|---|---|---|---|
| Bromide | 218–219 | EtOH | 22 | ir, uv | 3 |
| Picrate | 206.5–209 | EtOH/Ac | | | 3 |

***1,2,3,4,5,6,7,9,10,10a-Decahydropyrido[1,2-a] [1,4]diazepines***

| | | | | | |
|---|---|---|---|---|---|
| 2-PhCH$_2$ | [148–150/1.0] | | | | 5 |
| Dihydrochloride | 253 | 5 | | | |
| 2-(4-FC$_6$H$_4$CH$_2$) | [155–160/1.0] | | | | 5 |
| Dihydrochloride | 207–209 | EtOH/Et$_2$O | | | 5 |
| 2-[1,3-Benzodioxol-5-yl]methyl | [153–156/2.0] | | | | 5 |
| Dihydrochloride | 222 | | | | 5 |
| 2-CHPh$_2$ | [196–198/1.5] | | | | 5 |
| Dihydrochloride | 244–246 | EtOH/Et$_2$O | | | 5 |
| 2-(4-ClC$_6$H$_4$)CH(Ph) | [194–196/2.0] | | | | 5 |
| Dihydrochloride | 242 | | | | 5 |
| 2-4-FC$_6$H$_4$CH$_2$CH$_2$ | [166–169/2.0] | | | | 5 |
| Dihydrochloride | 118 | | | | 5 |

TABLE III-1. —(contd.)

| Substituent | mp (° C); [bp (° C/torr)] | Solvent of Crystallization | Yield (%) | Spectra | Refs. |
| --- | --- | --- | --- | --- | --- |
| 2-[3,4-$(MeO)_2C_6H_3CH_2CH_2$] | [170–176/0.5] | | | | 5 |
| Dihydrochloride | 204 | | | | 5 |
| 2-[CH(Ph)$CH_2$(Ph)] | [206–208/2.0] | | | | 5 |
| Dihydrochloride | 178 | | | | 5 |
| 2-($Ph_2CHOCH_2CH_2$) | [103–107/2.0] | | | | 7 |
| Dihydrochloride | 199–202 | | | | 7 |
| 2-(4-$ClC_6H_4$)(Ph)$CHOCH_2CH_2$ | [130–132/1.0] | | | | 7 |
| Dihydrobromide | 217–219 | | | | 7 |
| 2-($Ph_2CHCH_2CH_2$) | [188–191/1.0] | | | | 5 |
| Dimaleate | 197 | | | | 5 |
| 2-[(4-$ClC_6H_4$)(Ph)$CHCH_2CH_2$] | [200–202/1.0] | | | | 5 |
| Dimaleate | 189 | | | | 5 |
| 2-$PhCH_2$-4-PhCO | | | | | |
| Dihydrochloride | 210 | MeOH/EtOH | 35 | | 6 |
| 2-$Me_2N(CH_2)_3$ | | | | | |
| Trihydrochloride | 197 | MeOH/$Et_2O$ | 25 | | 6 |
| 2-Me-4-PhCO | 67 | Pentane | | | 6 |
| Dihydrochloride | 217 | | 72 | | 6 |
| 2-Me-4-(4-$ClC_6H_4$CO) | 75–76 | *i*-$Pr_2O$ | | | 6 |
| Dihydrochloride | 170–171 | EtOH | 50 | | 6 |
| 2-Me-4-(CHClPh) | | | | | |
| Dihydrochloride | 195–197 | MeOH | 80 | | 6 |

| | | | | |
|---|---|---|---|---|
| 2-Me-4-(4-$FC_6H_4CO$) | 63–65 | $Et_2O$ | | 6 |
| Dihydrochloride | 225–226 | MeOH | 71 | 6 |
| 2-Me-4-(2-Furoyl) | | | | |
| Monohydrate | 49–50 | *i*-$Pr_2O$ | | 6 |
| Dihydrochloride | 228–230 | MeOH | 63 | 6 |
| 2-Me-4-[CH(OH)Ph] | | | | |
| Isomer A | 164–166 | Acetone | 46 | 6 |
| Dihydrochloride | 248–250 | MeOH | | 6 |
| Isomer B | 109–111 | Acetone | 42 | 6 |
| Dihydrochloride | 267–269 | MeOH | | 6 |
| 2-Me-4-[CH(OH)(4-$ClC_6H_4$)] | | | | |
| Isomer A | 146–148 | *i*-$Pr_2O$ | 47 | 6 |
| Isomer B | 126–127 | *i*-$Pr_2O$ | 46 | 6 |
| 2-Me-4-[CH(OH)(4-$MeOC_6H_4$)] | | | | |
| Isomer A | 145–146 | *i*-$Pr_2O$ | | 6 |
| Dihydrochloride | 254–255 | MeOH | 82 | 6 |
| Isomer B | 121–124 | *i*-$Pr_2O$ | 82 | 6 |
| Dihydrochloride | 243–244 | MeOH | | 6 |
| 2-Me-4-[CH(OH)(2-Naphthyl)] | | | | |
| Isomers A and B | 87–113 | | 97.5 | 6 |
| 2-Me-4-[CH(OH)(2-Thienyl)] | | | | |
| Isomer A | 140–142 | | 40 | 6 |
| Dihydrochloride | 240–241 | MeOH | | 6 |
| Isomer B | 96–98 | | | 6 |
| Dihydrochloride | 259–260 | MeOH | 41 | 6 |
| 2-Me-4-[C(OH)$(Ph)_2$] | 124–126 | IPE | 90 | 6 |
| Dihydrochloride | 272–273 | | | 6 |
| 2-Me-4-[C(OH)(Ph)(4-$FC_6H_4$)] | 134–135 | MeOH | 69 | 6 |
| Dihydrochloride | 281–283 | MeOH | | 6 |
| 2-Me-4-[C(OH)(Ph)(3-Indolyl)] | 65–67 | | 90 | 6 |

TABLE III-1. —(contd.)

| Substituent | mp (° C); [bp(° C/torr)] | Solvent of Crystallization | Yield (%) | Spectra | Refs. |
|---|---|---|---|---|---|
| 2-Me-4-[C(OH)(Ph)(4-$MeOC_6H_4$)] | | | | | |
| Dihydrochloride | 248–249 | MeOH | 65 | | 6 |
| 2-Me-4-[C(OH)(Ph)(3-$CF_3C_6H_4$)] | | | | | |
| Dihydrochloride | 297–298 | DMSO/$H_2O$ | 81 | | 6 |
| 2-Me-4-[C(OH)(4-F-$C_6H_4$)$_2$] | | | | | |
| Dihydrochloride | 270–272 | EtOH | 77 | | 6 |
| 2-Me-4-[C(OH)(4-$MeOC_6H_4$)$_2$] | | | | | |
| Dihydrochloride | 214–216 | EtOH | 90 | | 6 |
| 2-Me-4-[C(OH)(4-$MeOC_6H_4$)(2-Thienyl)] | | | | | |
| Dihydrochloride | 224–225 | MeOH | 72 | | 6 |
| 2-Me-4-[CH(Ph)(Morpholino)] | 138–139 | Petr ether | 76 | | 6 |
| 2-Me-4-(2-Naphthoyl) | 92 | Petr ether | | | 6 |
| Dihydrochloride | 198 | MeOH/$Et_2O$ | 50 | | 6 |
| 2-Me-4-(4-$MeOC_6H_4CO$) | | | | | |
| Dihydrochloride | 214–216 | EtOH | 52 | | 6 |
| 2-Me-4-(3-$NO_2C_6H_4CO$) | | | | | |
| Dihydrochloride | 220–221 | MeOH | 56 | | 6 |
| 2-Me-4-(4-$NO_2C_6H_4CO$) | | | | | |
| Dihydrochloride | 185–186 | MeOH | 64 | | 6 |
| 2-Me-4-(2-Thienoyl) | | | | | |
| Monohydrate | 66–67 | *i*-$Pr_2O$ | | | 6 |
| Dihydrochloride | 236–238 | MeOH | 77 | | 6 |

| | | | | | |
|---|---|---|---|---|---|
| 2-Me-4-[$(Ph)_2$-Methylene] | 116–117 | Hex | 81 | | 6 |
| Dihydrochloride | 250–252 | MeCN | | | 6 |
| 2-(4-$MeC_6H_4SO_2$)-4-PhCO | 163–164 | MeOH | | | 6 |
| 2-Me-4-PhCO-4-Ph Dihydrochloride | 206 | *i*-PrOH | | | 6 |
| ***2,4,5,7,8,9,10,10a-Octahydro-1H-pyrido[1,2-a] [1,4]diazepin-3(3H)-ones*** | | | 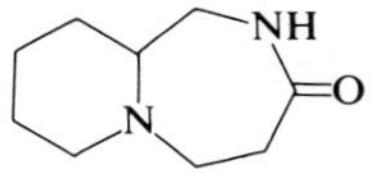 | | |
| None | 95–105 | EtOAc/Hexane | 8 | | 4 |
| 6-Methiodide | 234–236 | MeOH | | ir | 4 |
| *Pyrrolo[1,2-a] [1,4]diazepines* | | | | | |
| ***7,8,9,9a-Tetrahydro-5H-pyrrolo[1,2-a] [1,4]diazepine*** | | | 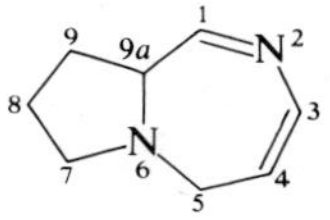 | | |
| 1,5,7-$Ph_3$-8,9-$(CN)_2$ | 144–145 | EtOH | 76 | ir, ms, pmr, uv | 1 |

TABLE III-1. —(contd.)

| Substituent | mp (°C); [bp (°C/torr)] | Solvent of Crystallization | Yield (%) | Spectra | Refs. |
|---|---|---|---|---|---|
| ***2,3,4,5-Tetrahydropyrrolo[1,2-a][1,4]diazepin-1(1H)-one*** | | | 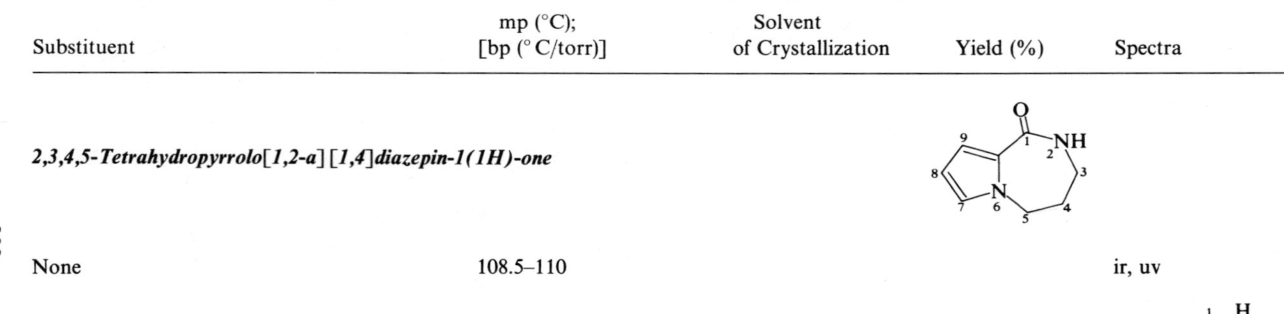 | | |
| None | 108.5–110 | | | ir, uv | 8 |
| ***2,3,4,5,7,8,9,9a-Octahydro-1H-pyrrolo[1,2-a][1,4]diazepines*** | | | | | |
| 2-Benzhydryl Dihydrochloride | 183–185 | *i*-PrOH | | $[\alpha]_D - 15.9(C = 1, H_2O)$ | 12 |
| 2-(*p*-chlorobenzhydryl) | [190–191/15 mmHg] | | | $[\alpha]_D - 16.7(C = 1, CHCl_3)$ | 12 |

TABLE III-2. [*d*]-FUSED [1,4]DIAZEPINES

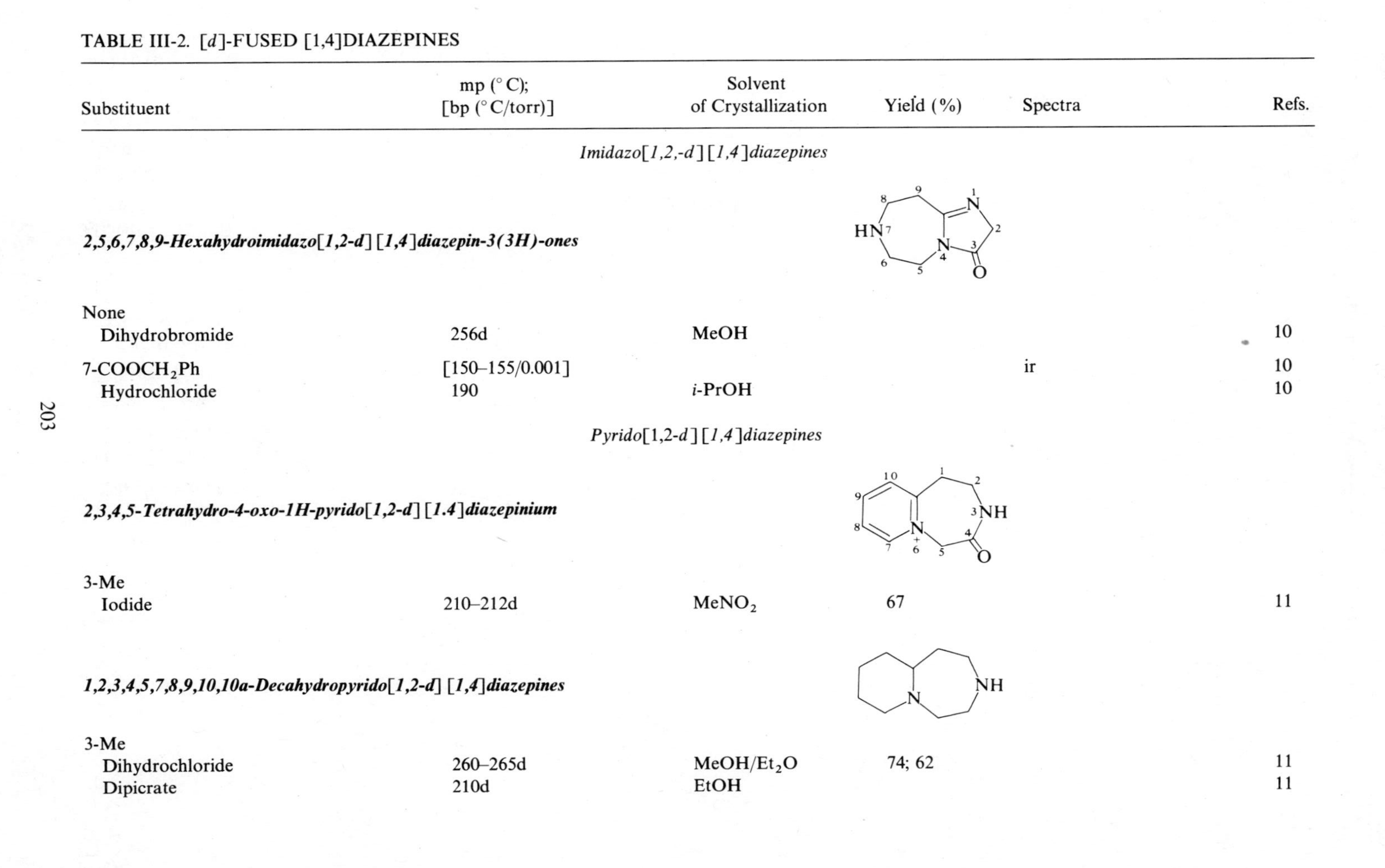

| Substituent | mp (°C); [bp (°C/torr)] | Solvent of Crystallization | Yield (%) | Spectra | Refs. |
|---|---|---|---|---|---|
| *Imidazo[1,2,-d][1,4]diazepines* | | | | | |
| ***2,5,6,7,8,9-Hexahydroimidazo[1,2-d][1,4]diazepin-3(3H)-ones*** | | | | | |
| None | | | | | |
| Dihydrobromide | 256d | MeOH | | | 10 |
| 7-$COOCH_2Ph$ | [150–155/0.001] | | | ir | 10 |
| Hydrochloride | 190 | *i*-PrOH | | | 10 |
| *Pyrido[1,2-d][1,4]diazepines* | | | | | |
| ***2,3,4,5-Tetrahydro-4-oxo-1H-pyrido[1,2-d][1.4]diazepinium*** | | | | | |
| 3-Me | | | | | |
| Iodide | 210–212d | $MeNO_2$ | 67 | | 11 |
| ***1,2,3,4,5,7,8,9,10,10a-Decahydropyrido[1,2-d][1,4]diazepines*** | | | | | |
| 3-Me | | | | | |
| Dihydrochloride | 260–265d | $MeOH/Et_2O$ | 74; 62 | | 11 |
| Dipicrate | 210d | EtOH | | | 11 |

TABLE III-2. —(contd.)

| Substituent | mp(° C); [bp (° C/torr)] | Solvent of Crystallization | Yield (%) | Spectra | Refs. |
|---|---|---|---|---|---|
| 3-$PhCH_2$ | [150–153/1.5] | | | | 5 |
| Dihydrochloride | 251 | | | | 5 |
| 3-[1,3-Benzodioxol-5-yl]methyl | [145–146/2.0] | | | | 5 |
| Dihydrochloride | 199 | | | | 5 |
| 3-($4-FC_6H_4CH_2$) | [160–163/1.0] | | | | 5 |
| Dihydrochloride | 142 | | | | 5 |
| 3-$CHPh_2$ | [196–198/1.0] | | | | 5 |
| Dihydrochloride | 235 | | | | 5 |
| 3-($4-ClC_6H_4$)(Ph)CH | [198–201/1.0] | | | | 5 |
| Dihydrochloride | 230 | | | | 5 |
| 3-[3,4-$(MeO)_2C_6H_3CH_2CH_2$] | [185–187/2.0] | | | | 5 |
| Dihydrochloride | 204 | | | | 5 |
| 3-[CH(Ph)$CH_2$Ph] | [185–186/1.0] | | | | 5 |
| Dihydrochloride | 234–236 | EtOH/$Et_2O$ | | | 5 |
| 3-[3,3-$Ph_2CCH_2CH_2$] | [210–213/3.0] | | | | 5 |
| Dimaleate | 150 | | | | 5 |
| 3-[Ph($4-ClC_6H_4$)$CHCH_2CH_2$] | [207–210/1.5] | | | | 5 |
| Dimaleate | 131 | | | | 5 |
| 3-($Ph_2CHOCH_2CH_2$) | [115–116/1.5] | | | | 7 |
| Dihydrobromide | 218–220 | | | | 7 |

| | | | | |
|---|---|---|---|---|
| 1-Ph-3-Me | | | | |
| Isomer A | 56–58 | | | 11 |
| | [123–125/0.3] | | 82 | 11 |
| Dihydrochloride | 269–272d | EtOH | | 11 |
| Isomer B | 106–108 | MeOH/$H_2O$ | 90 | 11 |
| Dihydrochloride | 278–280d | EtOH | | 11 |
| ***3,4,5,7,8,9,10,10a-Octahydropyrido[1,2-d] [1,4]diazepin-2(1H)-ones*** | | 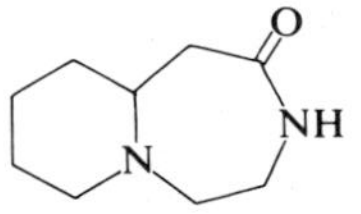 | | |
| None | 135–136 | EtOAc | 20 | 4 |
| 6-Methiodide | 263–264 | MeOH | 73 | 4 |
| ***1,2,5,7,8,9,10,10a-Octahydropyrido[1,2-d] [1,4]diazepin-4(3H)-ones*** | | 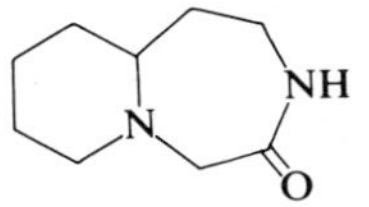 | | |
| 3-Me | 38–40 | | 82 | 11 |
| | [119–122/1.0] | | | 11 |
| Hydrochloride | 220–222 | EtOH/$Et_2O$ | | 11 |
| Picrate | 158–161 | EtOH | | 11 |
| ***1,2,3,7,8,9,10,10a-Octahydropyrido[1,2-d] [1,4]diazepin-5(4H)-ones*** | | 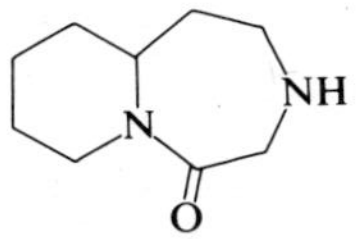 | | |
| 3-Me | [106–108/0.3] | | 48 | 11 |
| Picrate | 207–211 | Acetone | | 11 |
| Chloroaurate | 153–156 | $H_2O$/HCl | | 11 |
| 3-Methochloride | 215 | $CHCl_3$ | 90 | 11 |

TABLE III-2. —(contd.)

| Substituent | mp (° C); [bp (° C/torr)] | Solvent of Crystallization | Yield (%) | Spectra | Refs. |
|---|---|---|---|---|---|
| 1-Ph-3-Me | [158–162/0.1] | | 64 | | 11 |
| Isomer A | 175–177 | EtOH | | | 11 |
| Chloroaurate | 223–225d | $EtOH/Et_2O$ | 95 | | 11 |
| 3-Methochloride | 114–116 | $MeOH/H_2O$ | | | 11 |
| Isomer B | [170–175/0.3] | | 81 | | 11 |
| Chloroaurate | 181–184 | EtOH | | | 11 |
| 3-Methochloride | 220–222d | $CHCl_3$/Acetone | 56 | | 11 |
| *Pyrrolo[1,2-d] [1,4]diazepines* | | | | | |
| ***1,3,4,5,7,8,9,9a-Octahydropyrrolo[1,2-d] [1,4]diazepin-2(2H)-ones*** | | | 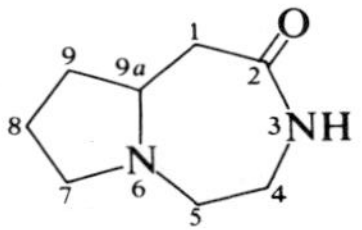 | | |
| None | 65–70 | Hexane | 22 | | 4 |
| 6-Methiodide | 234.5–236 | EtOH/MeOH | | | 4 |
| ***1,2,3,5-Tetrahydropyrrolo[1,2-d] [1,4]diazepine-4(4H)-ones*** | | | 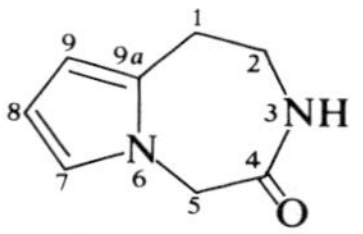 | | |
| 7-Me | 137–138 | EtOH | 27 | ir, pmr | 13 |
| 7,9-$Ph_2$ | 218–219 | EtOH | 37 | ir, pmr | 13 |

## D. REFERENCES


1. A. Padwa and L. Gehrlein, *J. Am. Chem. Soc.*, **94**, 4933 (1972).
2. (a) M. Yamauchi and M. Masui, *Chem. Pharm. Bull.*, **24**, 1480 (1976). (b) M. Yamauchi and M. Masui, *Chem. Ind.*, 31 (1976).
3. A. Fozard and G. Jones, *J. Chem. Soc.*, 2763 (1974).
4. L. A. Paquette and M. K. Scott, *J. Org. Chem.*, **33**, 2379 (1968).
5. H. Kato and T. Mori, Ger. Offen., 2,141,464, March 1972 (Hokuriku Seiyaku Co., Ltd.).
6. H. Ulbrich, H. Biere, G. Paschelke, H. Wachtel, D. Palenschat, R. Horowski, and W. Kehr, Ger. Offen. 2,347,390, April 1975.
7. Japan Patent 74, 28,755, July 1974 (Hokuriku Pharm. Co. Ltd.); *Chem. Abstr.*, **82**, 156,396.
8. A. Morimoto and T. Watanabe, Japan Patent 74, 27,877, July 1974 (Fujisawa Pharm. Co. Ltd.).
9. R. G. Groit, U.S. Patent 3,313,819, April 1967 (Sandoz, Hanover, NJ).
10. R. G. Groit, U.S. Patent 3,609,140, September 1971 (Sandoz, Hanover, NJ).
11. F. F. Blicke and J. L. Hughes, *J. Org. Chem.*, **26,** 3257 (1961).
12. H. Kato, T. Hishikana, and E. Koshinaka, U.S. Patent 4,093,630, June 1978 (Hokurika Pharm. Co. Ltd.).
13. H. Stetter and P. Lappe, *Ann. Chem.*, 703 (1980).

CHAPTER IV

# [1,4]Diazepines with [*b*]-Fused Rings

**R. Ian Fryer**

*Department of Chemistry, Rutgers,*
*State University of New Jersey, Newark,*
*New Jersey*

and

**A. Walser**

*Chemical Research Department,*
*Hoffmann-La Roche Inc., Nutley,*
*New Jersey*

## INTRODUCTION

This chapter reviews bicyclic systems of general structure **1**. The extensive work carried out in the area of 1,5-benzodiazepines is given precedence over other ring systems, which are subsequently described in alphabetical order. The synthesis and reactions of these [*b*]-fused [1,4]diazepines are divided into subsections based on the degree of saturation of the ring system.

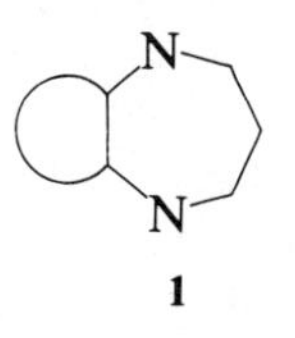


**1**

Previous reviews of this general area have been carried out by D. Lloyd and H. P. Cleghorn (in *Advances in Heterocyclic Chemistry*, Vol. 17, A. R. Katritzky and A. J. Boulton, Eds., Academic Press, New York and London, 1974, p. 27); G. A. Archer and L. H. Sternbach [*Chem. Rev.*, **68**, 747 (1968)]; F. D. Popp and A. C. Noble (in *Advances in Heterocyclic Chemistry*, Vol. 8, A. R. Katritzky and

A. J. Boulton, Eds., Academic Press, New York and London, 1967, p. 21); J. A. Moore and E. Mitchell (in *Heterocyclic Compounds*, Vol. 8, R. C. Elderfield, Ed., Wiley, New York, 1967, p. 224); and E. Schulte [*Dtsch. Apothek. Z.*, **115**, 1253 (1975)].

## 1. 1,5-BENZODIAZEPINES

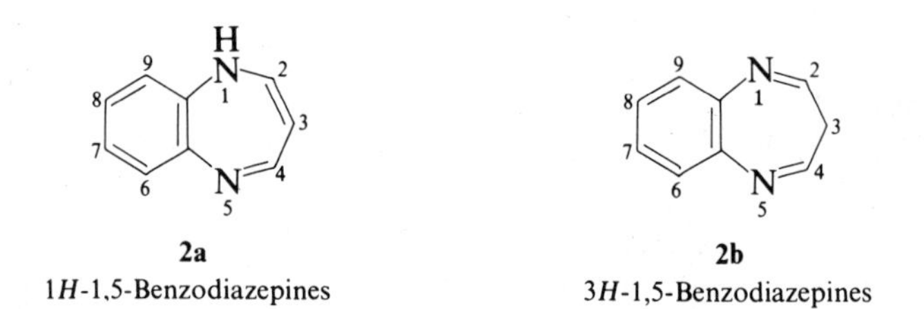


**2a**
1*H*-1,5-Benzodiazepines

**2b**
3*H*-1,5-Benzodiazepines

1,5-Benzodiazepines may exist in either of the tautomeric forms **2a** or **2b**. While the 3*H* tautomer is, in general, thermodynamically preferred, monoprotonation renders the 1*H* tautomer energetically more favorable, and most of the salts of 1,5-benzodiazepines occur in this form. For this reason the tautomers are not disscussed separately.

### 1.1. Synthesis

The most common method of synthesis is the condensation of *ortho*-phenylenediamines with 1,3-dicarbonyl compounds.

#### *1.1.1. From o-Phenylenediamines and 1,3-Dialdehydes*

The parent compound **2a** was prepared[1] and isolated in the form of various salts by the reaction of *o*-phenylenediamine **3** with 1-ethoxy-1,3,3-trimethoxypropane in ethanol–acetic acid followed by treatment of the mixture with a strong acid to form the salt.

Reaction of the diamine **3** with malondialdehyde **4** (R = H) (Eq. 1) under neutral conditions at room temperature led only to the monoimine **5** (R = H). Phenylmalondialdehyde gave, according to Rupe and Huber,[2] a red benzodiazepine (**7**, R = Ph) incorrectly assigned the 1*H*-tautomeric structure. Ruske and Hüfner[3] obtained the same compound **7** (R = Ph) and demonstrated with the help of infrared spectra that the red benzodiazepine was actually the 3*H* tautomer. The same tautomer was also isolated in the synthesis of the 3-bromo derivative **7** (R = Br), prepared in 33% yield[3] from the diamine and bromomalondialdehyde **4** (R = Br). Both the 3-phenyl- and the 3-bromobenzodiazepine gave lighter colored hydrochloride salts and apparently undergo protonation

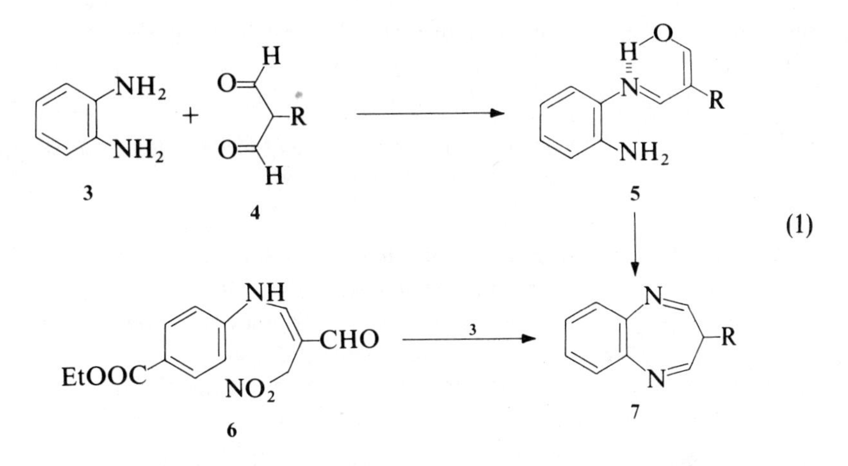


without tautomerization. A blue-black hydrochloride salt of the 3-phenyl derivative, which is most probably the protonated 1$H$ tautomer, was observed[3] when a different synthesis was used. Reaction of the phenylenediamine with nitromalondialdehyde **4** ($R = NO_2$) in hot, dilute hydrochloric acid gave the highly insoluble red benzodiazepine **7** ($R = NO_2$).[4,5] The 7-chloro derivative was obtained[6] in an analogous manner from 4-chloro-1,2-phenylenediamine.

Another example involving a substituted malondialdehyde was published more recently. *o*-Phenylenediamine was condensed with the dialdehyde **4** (R = 6-chloro-2-benzoxazolyl) in propanol while maintaining the pH between 3 and 5 by addition of formic acid. The corresponding benzodiazepine **7** was isolated in 70% yield, again as a red compound.[7]

Derivatives of malondialdehydes such as the enamine **6** have also been employed[5] in the preparation of the 3-nitrobenzodiazepine **7** ($R = NO_2$). Thus, condensation of **6** with the diamine **3** gave **7** and 4-aminobenzoic acid ethyl ester (Eq. 1).

A stepwise approach to the synthesis of 1,5-benzodiazepines was demonstrated by Rupe and Huber,[2] who obtained the 3-phenylbenzodiazepine **7** (R = Ph) by a reductive cyclization of the nitro compound **8** using either iron in acetic acid or tin in hydrochloric acid as the reducing agent (Eq. 2). The latter

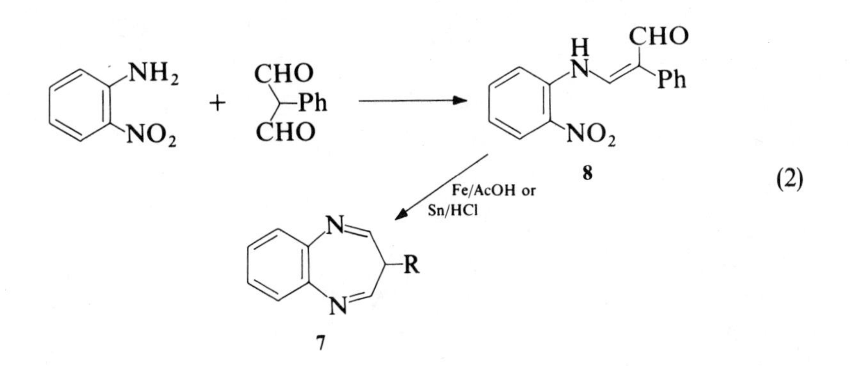

method led to a blue-black hydrochloride salt, most likely the protonated 1*H* tautomer.

### *1.1.2. From o-Phenylenediamines and 3-Ketoaldehydes (a-Hydroxymethyleneketones)*

Weissenfels and coworkers[8,9] studied the condensation of *o*-phenylenediamine with various α-hydroxymethyleneketones. Initially, these authors prepared tricyclic 1,5-benzodiazepines from α-formylcycloalkanones,[8] and later they applied[9,318] this reaction to the aliphatic compounds **9**. The primary products **10**, were isolated and cyclized to **11** by means of perchloric acid in alcohol (Eq. 3).

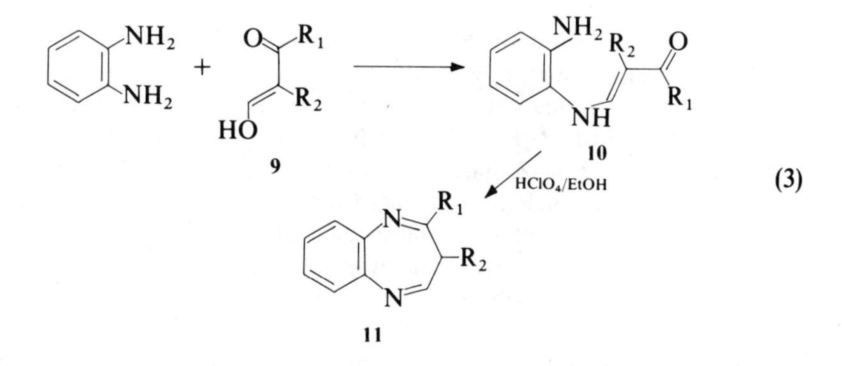


(3)

### *1.1.3. From o-Phenylenediamines and β-Chlorovinylaldehydes or Ketones*

It has been found that 1,5-benzodiazepines are also accessible, and can be synthesized in good yield, by the condensation of *o*-phenylenediamine with β-chloro-α,β-unsaturated aldehydes **12** (Eq. 4). This method was first applied to

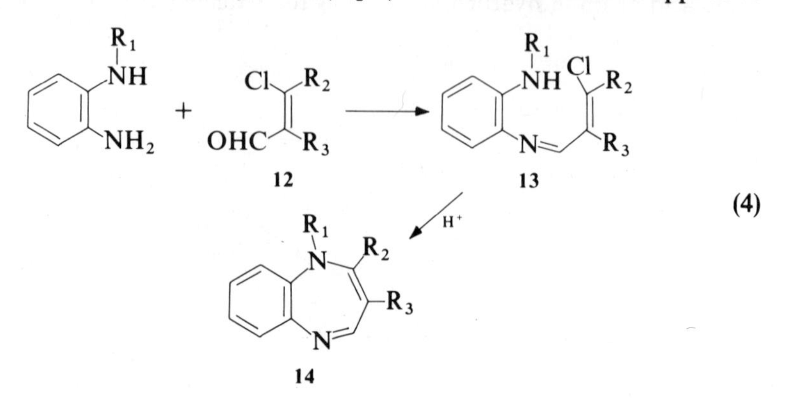


(4)

1-formyl-2-chlorocycloalkenes[10] and later extended to the $\beta$-chlorocinnamaldehydes **12** ($R_2$ = Ph).[11] The same route was also used successfully to prepare 1-methyl-1,5-benzodiazepines **14** ($R_1$ = Me)[12] from *N*-methyl-*o*-phenylenediamine. The imines **13** were shown[11] to be intermediates in this synthesis.

Ruske and Grimm[13] have also demonstrated that 2-methyl-1*H*-1,5-benzodiazepine hydrochloride (**14**, $R_1 = R_3$ = H, $R_2$ = Me) can be obtained by the condensation of *o*-phenylenediamine with $\beta$-chlorovinyl methyl ketone in ethanolic hydrogen chloride. Reaction of the dimethyldiamine **15** with the $\beta$-chloroaldehyde **16** afforded the quaternary salt **17** (Eq. 5).[12]

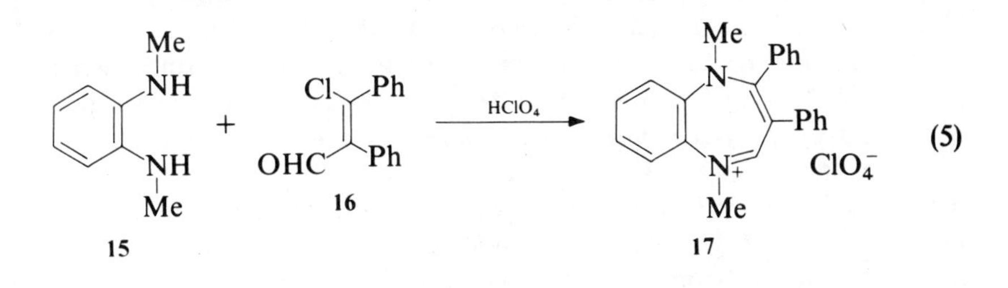


### *1.1.4. From o-Phenylenediamines and 1,3-Diketones*

The prototype of this reaction was the condensation of *o*-phenylenediamine **18** ($R_1$ = H) with pentane-2,4-dione **19** ($R_2 = R_3$ = Me), first described by Thiele and Steimmig[14] in 1907. This method has been widely used for the synthesis of variously substituted 1,5-benzodiazepines.[304] The reaction was shown to be pH-dependent,[1,15] and the best yields[16,17] were obtained under slightly acidic conditions. The benzodiazepines were generally isolated as the deeply colored salts of the 1*H*-tautomeric form **20**. According to Schwarzenbach and Lutz,[18] liberation of the base leads to the short-lived, yellow 1*H*-benzodiazepine **21**, which spontaneously tautomerizes to the more stable 3*H* isomer **22** (Eq. 6).

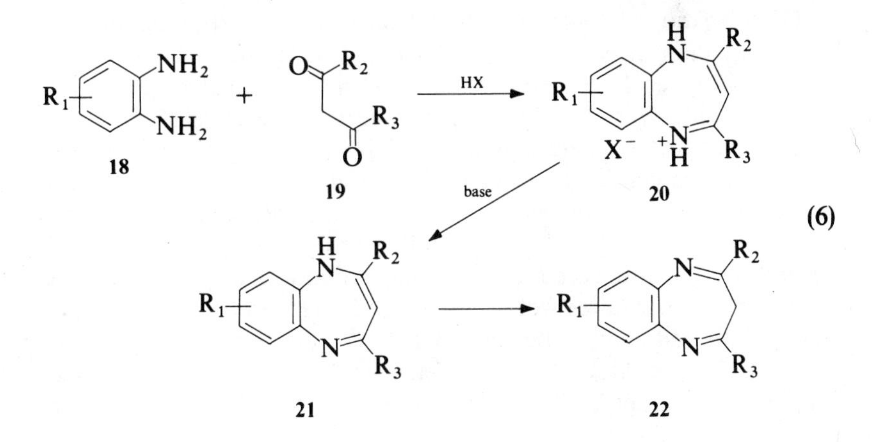

Substituted phenyl derivatives of 2,4-dimethyl-1,5-benzodiazepines have been prepared by the same procedure.[1,19–25] A 2-methoxymethyl derivative[1] and the 2,4-di(bromomethyl)-1,5-benzodiazepine[26] were obtained by the condensation of *o*-phenylenediamine with an appropriately functionalized pentane-2,4-dione. Thc synthesis of 2-phenyl-4-methyl-1,5-benzodiazepine described long ago by Thiele and Steimmig[14] was repeated.[16,27,28] Matsumoto and coworkers[28] claimed to have obtained the 1*H* tautomer as a hydrate by condensation of benzoylacetone with *o*-phenylenediamine in boiling xylene containing a catalytic amount of *p*-toluenesulfonic acid. It is more likely that this hydrate is identical to the uncyclized primary adduct **23** ($R_2$ = Me, $R_3$ = Ph) (Eq. 7) described by Sulca and coworkers.[29]

Several analogs with substituents in the 2-phenyl group were prepared[16,30,304] using Thiele's procedure or by heating under reflux a propanol solution of *o*-phenylenediamine with the appropriate 1,3-dione in the presence of molecular sieves.[31] Thus, treatment of the diamine with the appropriate heteroaroylacetones has been shown to give 2-(2-selenophenyl)-4-methyl-1,5-benzodiazepine,[32] 2-(3-ethoxy-2-quinoxalyl)-4-methyl-1,5-benzodiazepine,[33] and 2-(4-hydroxy-3-coumarinyl)-4-methyl-1,5-benzodiazepine.[34]

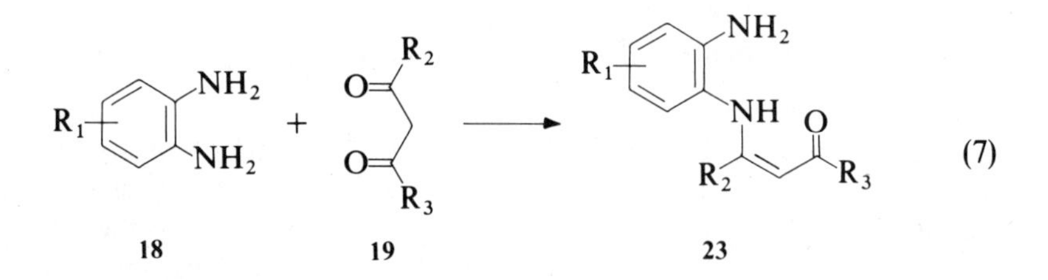


1,5-Benzodiazepines bearing polyfluoroalkyl substituents in the 2- and/or 4-position were also prepared in good yield,[35a,b] provided the second substituent in the 2- or 4-position was either phenyl or perfluoroalkyl. The proportion of uncyclized primary product **23** recovered from the reaction was found to increase as the polyfluoroalkyl chain was extended in length.

The formation of 2,4-diaryl-3*H*-1,5-benzodiazepines required more vigorous conditions than those chosen by Thiele and Steimmig.[14] Thus, Finnar[36] obtained 2,4-diphenyl-3*H*-1,5-benzodiazepine in 50% yield by condensation of dibenzoylmethane with *o*-phenylenediamine in boiling ethanol containing acetic acid. A comparable yield was reported by Barltrop and coworkers,[16] who performed this condensation in boiling xylene in the presence of *p*-toluenesulfonic acid with azeotropic removal of water. Eiden and Heja[33,37] as well as Kulkarni and Thakar,[38,39] synthesized substituted 2,4-diphenyl-3*H*-1,5-benzodiazepines using acetic acid in combination with other solvents at elevated temperatures. The latter workers favored the 1*H*-tautomeric structure on the basis of infrared spectra, but nmr data[31,37] indicate that the 3*H*-tautomer is preferred. The reactions of *o*-phenylenediamine with 1,3,4-triones **24** afforded the 1,5-benzodiazepines **25**. The formation of quinoxalines **26** was not observed.

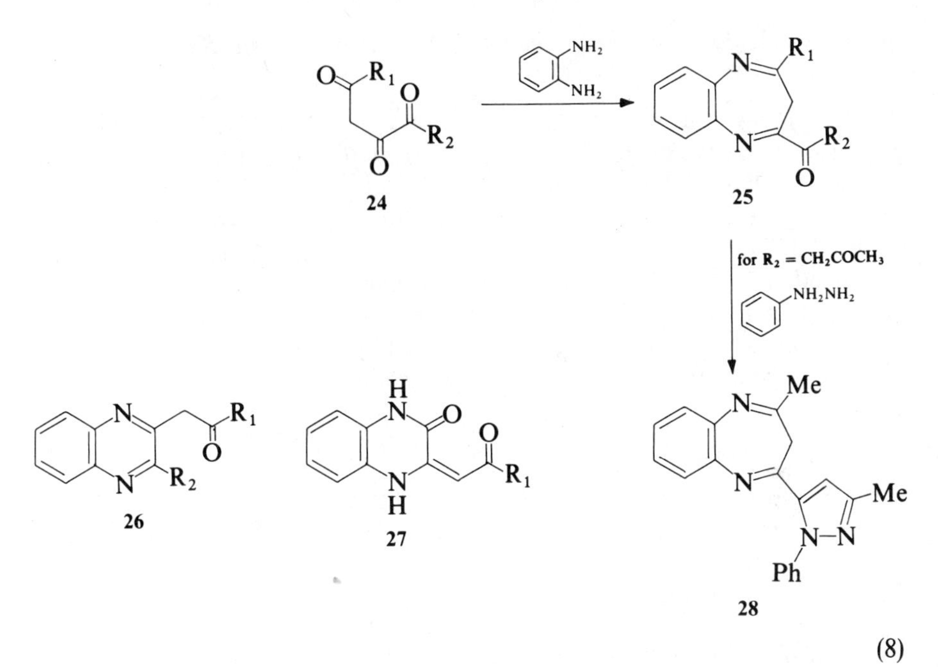


(8)

Thus, condensation of the diamine with hexane-2,3,5-trione **24** ($R_1 = R_2 = Me$) was reported to yield 2-acetyl-4-methyl-3*H*-1,5-benzodiazepine (**25**; $R_1 = Ac$, $R_2 = Me$),[40] while reaction with 1,6-diphenylhexane-1,3,4,6-tetraone yielded 83% of the corresponding benzodiazepine **25** ($R_1 = Ph$, $R_2 = CH_2COPh$).[36] A similar compound was obtained from the corresponding octane-2,4,5,7-tetraone. In this instance the remaining 1,3-dicarbonyl system reacted with phenylhydrazine to yield as the final product the pyrazolyl derivative **28**.[36]

The structure of 4-methyl-1,5-benzodiazepine-2-carboxylic acid was tentatively assigned to the product obtained by condensation of *o*-phenylenediamine with acetylpyruvic acid.[41] More recently,[42,43] anilides of 4-aryl-1,5-benzodiazepine-2-carboxylic acids were prepared from the diamine and aroylpyruvic acid anilides. In this case, the formation of the quinoxalones **27** was also observed. Interestingly, the electronic character of the aroyl group was found to be the determining factor in the formation of either the quinoxalone **27** or the benzodiazepine **25**. Thus, electron-donating groups in the para position of the benzoyl group promoted benzodiazepine formation, while electron withdrawing substituents favored the formation of a quinoxalone.

Using the appropriate 1,3-diketone, condensation with *o*-phenylenediamine allowed the preparation of 3-substituted 1,5-benzodiazepines. Among the compounds obtained in this manner were the 2,3,4-trimethyl derivate,[44,45] various 3-arylsulfonyl derivatives,[46] the 3-benzoyl-2,4-diphenyl-1,5-benzodiazepine,[29] the 3-oximino-2,4-dimethyl-1,5-benzodiazepine,[14,47,48] and 3-diphenylmethylene-1,5-benzodiazepine.[49]

### 1.1.5. From o-Phenylenediamines and Acetylenic Ketones

The synthesis of 1,5-benzodiazepines of type **22** by the reaction of *o*-phenylenediamines **18** with acetylenic ketones **29** (Eq. 9) has been reported both by Ried and Koenig[50] and by Andreichikov and coworkers.[43,51] In particular this method has been applied to the preparation of variously substituted 2,4-diphenyl-3*H*-1,4-benzodiazepines.[52–54] The reaction has been carried out in boiling ethanol,[52] methanol, or mixtures of ethanol and acetic acid.[53,54] The melting points reported for 2,4-diphenyl-7-methyl-3*H*-1,4-benzodiazepine[53,54] differ widely. The lower melting point reported by Amey and Heindel[53] would be more in line with those of closely related compounds.

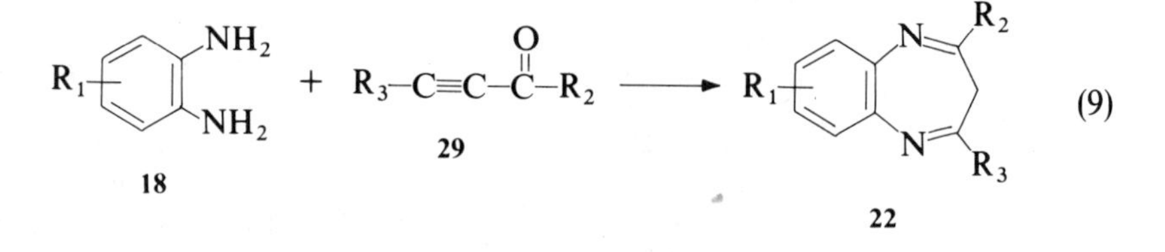


### 1.1.6. From o-Phenylenediamines and Masked 1,3-Diketones

Van Allan and Chie Chang[55] obtained 2-phenacyl-4-phenyl-1*H*-1,5-benzodiazepines **33** by rearrangement of the pyrone derivatives **30** (Eq.10). The rearrangement was effected using 1,1-dimethylhydrazine in boiling acetonitrile. The starting imino-4*H*-pyranes **30** were synthesized by the addition of an appropriate *o*-phenylenediamine to the corresponding pyrylium salt, generated with phosphorus oxychloride. It was postulated[55] that the pyrone is opened by dimethylhydrazine via **31** to form intermediate **32**, which, for steric reasons, would cyclize exclusively to the benzodiazepine rather than to the pyridine. The location of the substituent X in **33** has not been unequivocally established. On the basis of the nmr spectrum, the 1*H*-tautomeric form was assigned to **33** (X = H). The 3*H* tautomer **34** with an enolized carbonyl would fit the spectral data also.

2,4-Diphenyl-3*H*-1,5-benzodiazepines **37** were accessible in 70–80% yields by hydrogenation of the isoxazolones **35** over palladium on carbon.[56] Reductive cleavage of the N—O bond would lead, as shown, to the intermediate **36**, which would then cyclize with concomitant decarboxylation to yield the observed product **37** (Eq. 11).

### 1.1.7. Other Syntheses

Bindra and Le Goff[57] incorrectly reported the unexpected formation of 2-benzoyl-4-phenyl-1*H*-1,5-benzodiazepine **39** by reaction of *o*-phenylenediamine

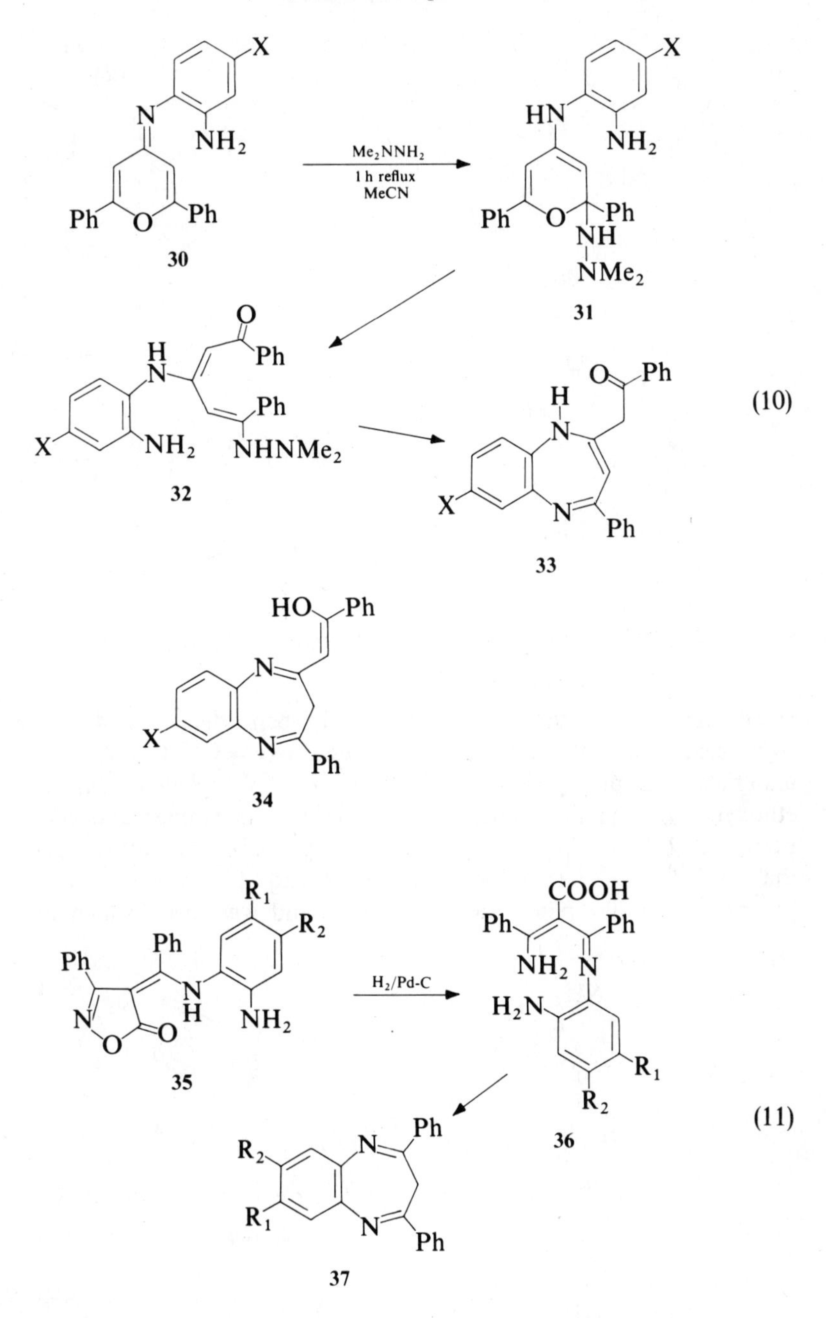


with 1,2-dibenzoylethylene in boiling acetic acid. Bass and coworkers[58] reinvestigated this reaction and were able to show that the product was actually the quinoxaline **40**, formed by dehydrogenation of the intermediate dihydro derivative **38** (Eq. 12).

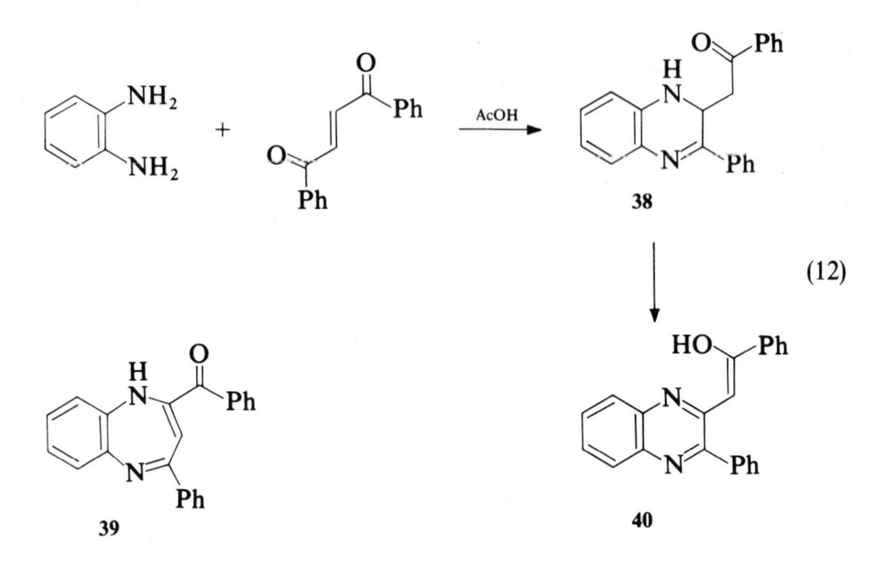


### *1.1.8. Synthesis of 2(4)-Amino-1,5-benzodiazepines*

Okamoto and Ueda[59] obtained the 3*H*-1,5-benzodiazepines **42** in high yields by cyclization of the *β*-anilinoacrylonitriles **41** ($R_1$ = CN, COOEt). The starting materials were prepared by condensation of *o*-phenylenediamine with either ethoxymethylenemalononitrile or ethyl ethoxymethylenecyanoacetate. Reaction of **42** ($R_1$ = CN, $R_2$ = H) with aqueous guanidine or methylguanidine gave the 2,4-diamino-3*H*-1,5-benzodiazepine **44** (Eq. 13) instead of the expected pyrimidobenzodiazepine. The same compound was also formed by alkaline

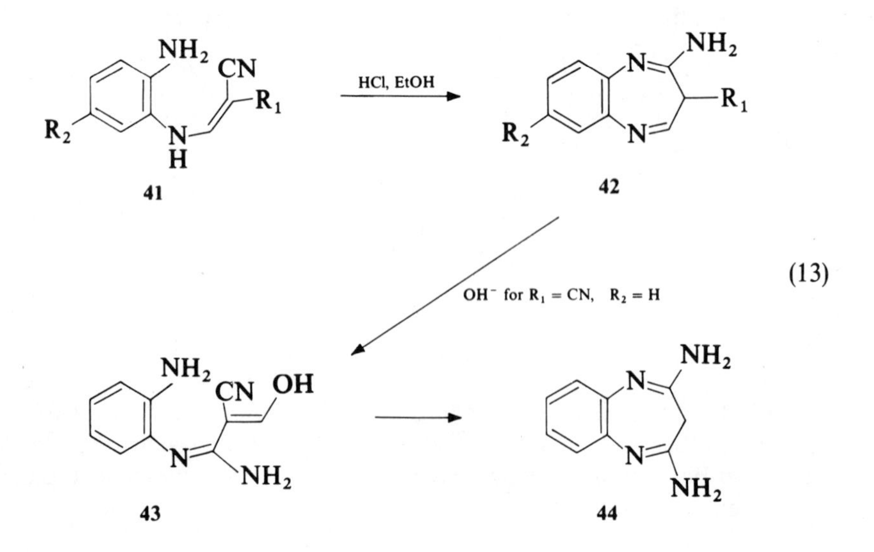

hydrolysis of the nitrile **42** ($R_1$ = CN, $R_2$ = H). This conversion was mechanistically explained by an opening of the ring to form the intermediate hydroxymethylene derivative **43**, which could then recyclize and decarbonylate to yield the diamine **44**. In a later paper,[60] Okamoto and Ueda described syntheses of some 8-chloro derivatives **42** ($R_2$ = Cl).

A high yield synthesis[61,311] of 2-amino-4-(2-chlorophenyl)-3*H*-1,5-benzodiazepine **46** involved the condensation of the acetylenic imidate **45** with *o*-phenylenediamine (Eq. 14).

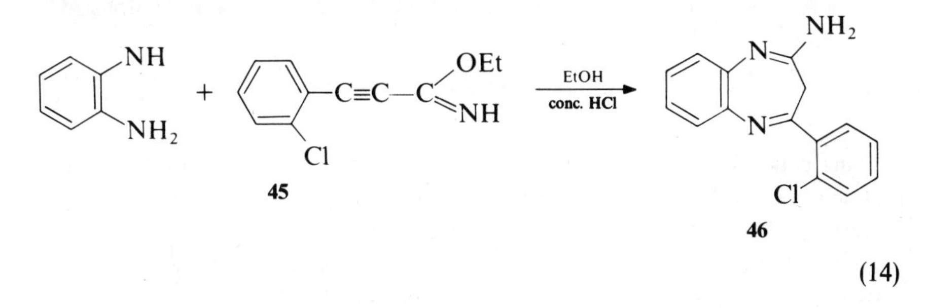


(14)

Reaction of *o*-phenylenediamines with *N*-alkyl-*N*-phenylethoxycarbonylacetamides (**47**) and phosphorus oxychloride at elevated temperature gave the 2,4-disubstituted 3*H*-1,5-benzodiazepines **49**[62] in low yield, most likely via a 2-amino-4-chloro intermediate **48** (Eq. 15).

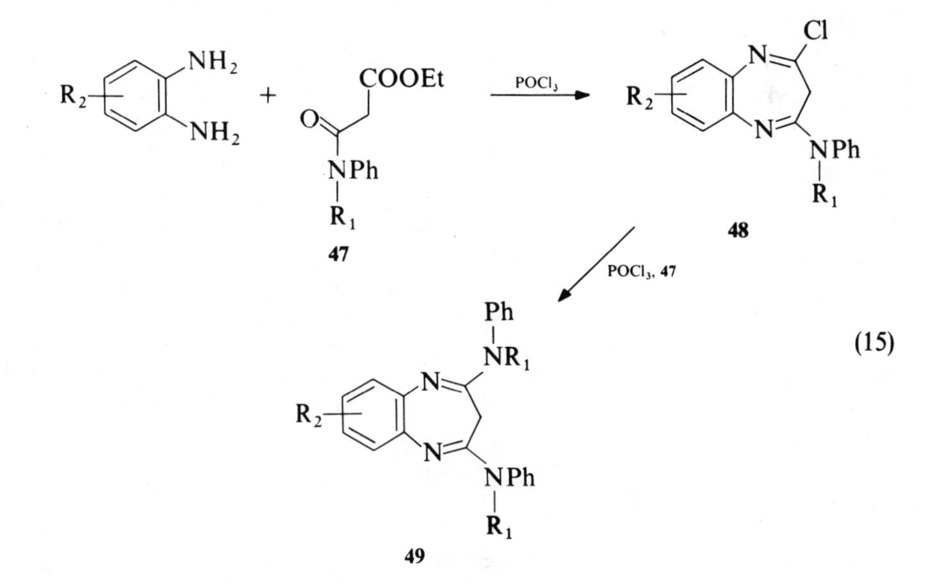


(15)

The most widely used method[316,319] for the preparation of 2- and or 4-amino-1,5-benzodiazepines **51** is the displacement of the methylthio group in **50** with an amine (Eq. 16).

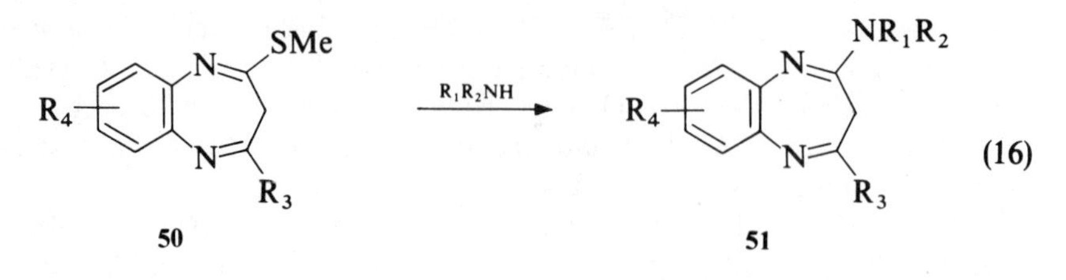


(16)

Using this displacement reaction, a variety of 2-cycloalkylamino-4-phenyl-3*H*-1,5-benzodiazepines[63,64] and 2,4-diamino derivatives[65] were synthesized.

### *1.1.9. Synthesis of 2-Alkoxy-1,5-benzodiazepines*

This class of iminoethers has not been extensively investigated. Stachel[66] claimed to have obtained the 2-ethoxy-1*H*-1,5-benzodiazepine **53** in about 20% yield by the condensation of *o*-phenylenediamine with the ketene acetal **52** (X = S) in methanol solution and at room temperature (Eq. 17). While the different reactivity of the *O*- and *S*-ketene acetals allowed the isolation of the mono-adduct **53**, the analogous reaction of the bisketene acetal **52** (X = O) led to the bis(1,5-benzodiazepine) **54** in better yield. The structures assigned to **53** and **54** lack spectroscopic support.

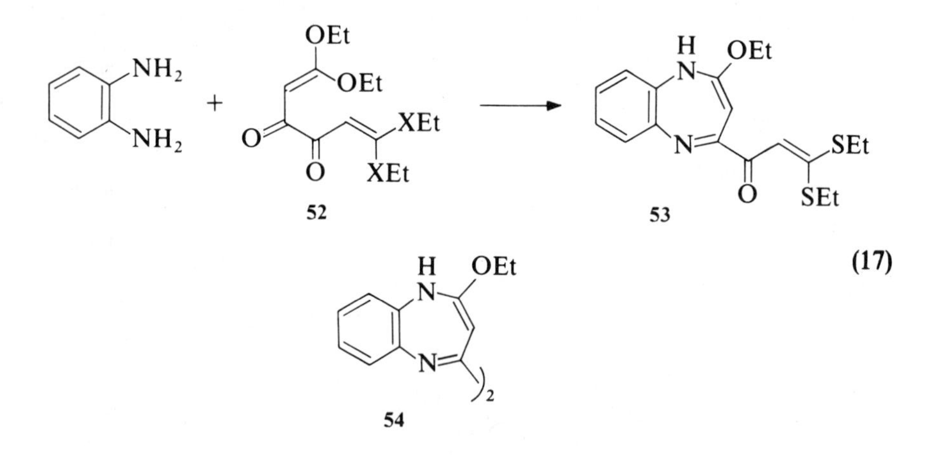


(17)

### *1.1.10. Synthesis of 2-Alkylthio-1,5-benzodiazepines*

The title compounds **56** have been exclusively prepared by the standard method, alkylation of the corresponding thiones **55** (Eq. 18).

Following their initial publication,[67] Nardi and coworkers prepared a series of thioethers **56** ($R_1$ = substituted phenyl, $R_3$ = basic side chains).[68,70,71] In connection with a mass spectrometric study,[70] analogs in which $R_3$ was —Ac or

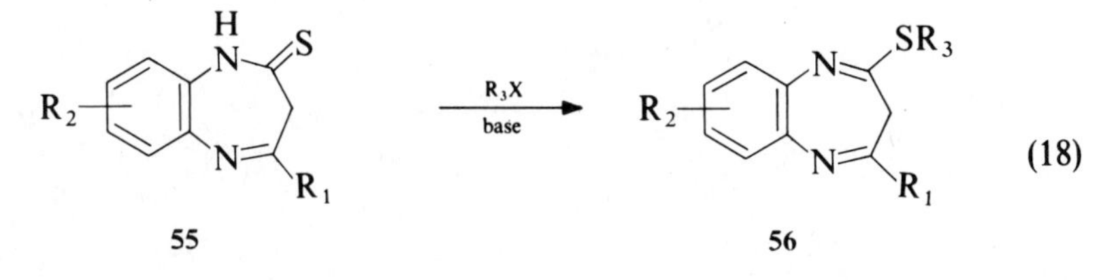


(18)

—$CD_3$ were synthesized. Utilizing the procedure above, several 2-methylthio compounds were prepared[63–65] as intermediates for the synthesis of the corresponding 2-amino derivatives. Nardi and coworkers[72] also investigated the alkylation of the thiones **57** ($R_1$ = Me or $CH_2Ph$), but isolated only the alkylation product **58** ($R_2$ = $CH_2CH_2NEt_2$) in a crude state. Acylation of **57** ($R_1$ = Me or $CH_2Ph$) with acetic anhydride, on the other hand, led to the crystalline acetylthio derivatives **58** ($R_2$ = Ac) (Eq. 19).

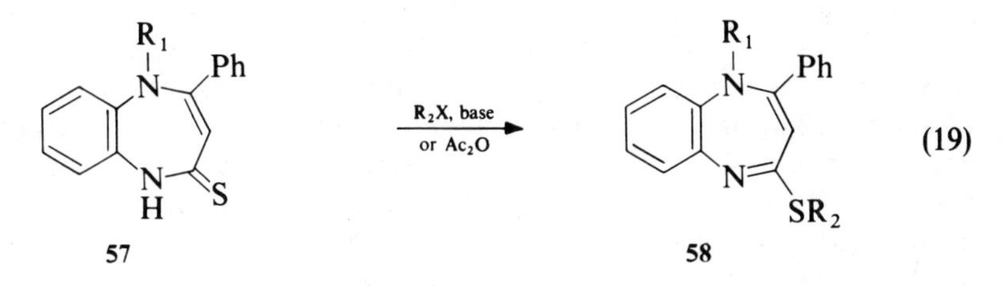


(19)

Peseke described[73] and patented[74] the preparation of the 2-hydroxy-4-methylthio-1*H*-1,5-benzodiazepine-3-carboxylate **60** by mild acid hydrolysis of the amidine **59** (Eq. 20).

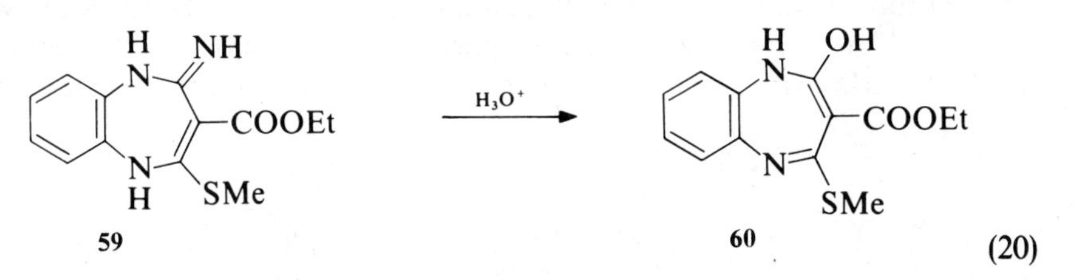


(20)

In view of the insufficient spectral evidence presented, other tautomeric structures for **60** might also be considered.

## 1.2. Reactions

### *1.2.1. Reactions with Electrophiles*

#### 1.2.1.1. Protonation

1,5-Benzodiazepines (**61**) are weak bases and are protonated by strong acids to form deeply colored monocations. The color of the cations is attributed to the protonated 1*H*-tautomeric form **62**, in which the charge is delocalized (Eq. 21).

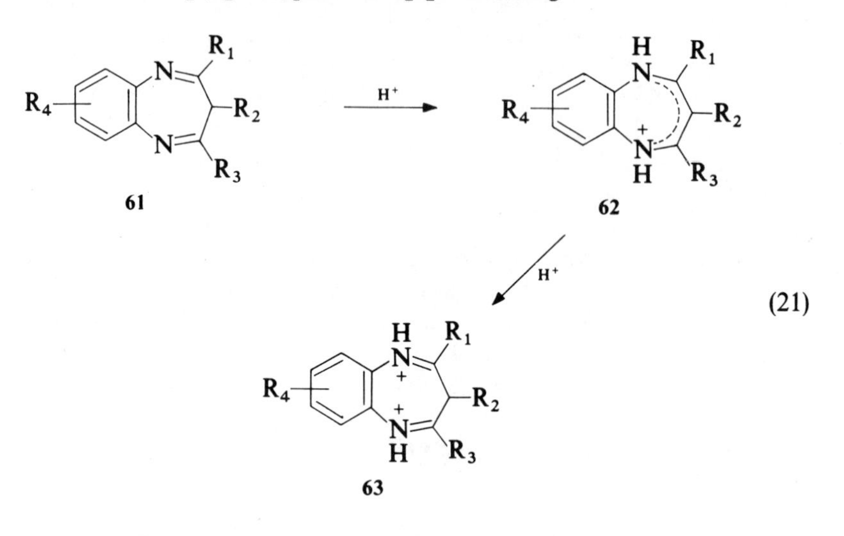


Strong acids in nonaqueous systems convert the colored cation to a colorless dication **63**. The p$K_a$ for the first protonation of the 2,4-dimethyl derivative was determined to be 4.5 by potentiometric methods[75,76] and 5.76 by spectroscopic means.[77] The p$K_a$ for the monocation–dication equilibrium appears to be approximately $-1.0$.[1,78]

### 1.2.1.2. Halogenation

Bromination of 2,4-dimethyl-3*H*-1,5-benzodiazepine (**64**, R = Me) with bromine in a mixture of acetic acid and nitromethane, was reported[33] to yield the 3-bromo derivative **65** as the hydrobromide salt (Eq. 22). This method was also used[79] for the synthesis of the 2,4-diphenyl derivative **65** (R = Ph). Reaction of **64** (R = Me) with bromine in chloroform led to the hydrobromide of the

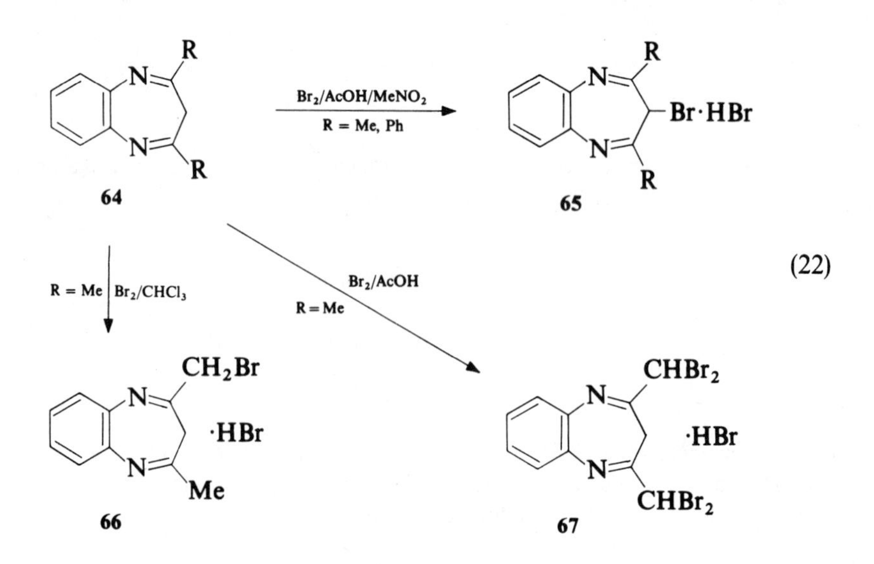

bromomethyl derivative **66**.[19] Treatment of the hydrogen sulfate salt of **64** (R = Me) with bromine in acetic acid afforded the tetrabromo derivative **67**.[19] Lloyd and coworkers[1] had prepared the same tetrabromo compound previously but had assigned an incorrect structure.

Further bromination of **67** in acetic acid at room temperature led to a nonabromide, which was assigned structure **70** (Eq. 23). This compound was also obtained by bromination of **68** via the hexabromide **69**. Hydrogenation of **70** over platinum on carbon[19] led to the hexabromide **69**.

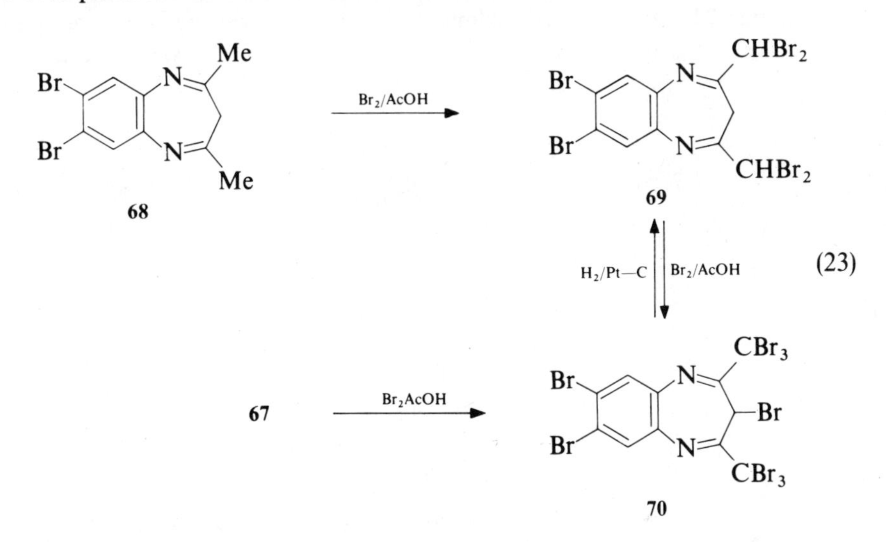


### 1.2.1.3. Nitration

The direct introduction of a nitro substituent into the 1,5-benzodiazepine nucleus has not been reported. While an initial attempt to nitrate the benzodiazepine **71** led only to tars, Levshina and coworkers,[80] using a nitration mixture of nitric and sulfuric acids (Eq. 24), were able to isolate compounds **72**, **73**, and **74** in 40, 26, and 8% yields, after fractional crystallization.

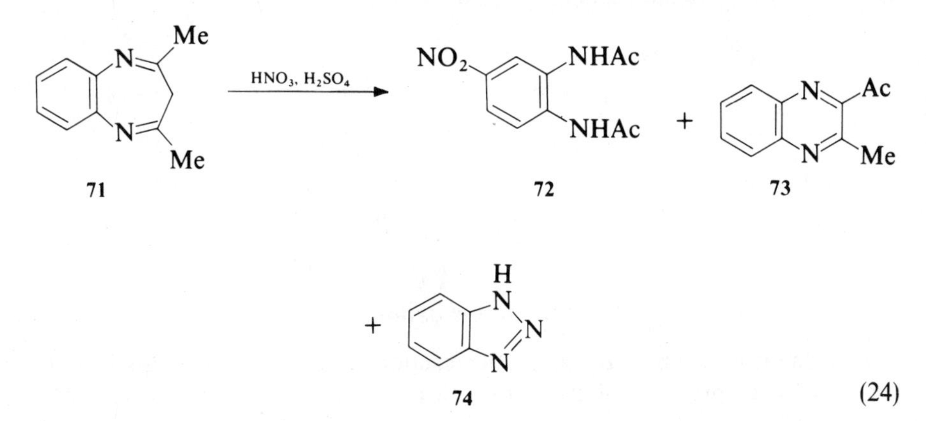

The products may be rationalized as being formed by a combination of electrophilic attack at the 3-position, acid hydrolysis, and oxidative ring cleavage. The benzotriazole **74** probably arose from nitrosation of *o*-phenylenediamine.

### 1.2.1.4. Nitrosation

Barltrop and coworkers,[16] who studied the action of nitrous acid on the benzodiazepine **71**, found that the major products of this reaction were the quinoxaline **73** and 2-methylbenzimidazole (**75**). A nitroso derivative assigned the 1,5-benzodiazepine structure **76** was also isolated in low yield (Eq. 25).

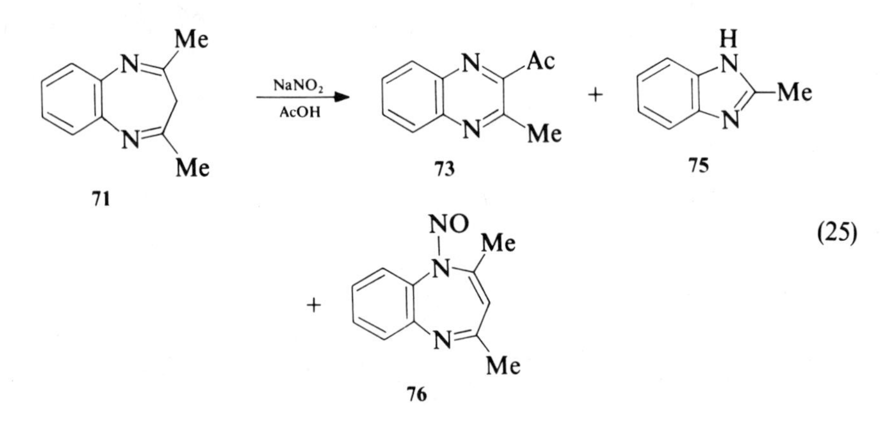


This reaction is believed to proceed via initial formation of the *N*-nitroso compound **76**, which then undergoes further transformations to give the major reaction products **73** and **75**.

### 1.2.1.5. Reactions with Other Nitrogen Electrophiles

Diazonium cations have also been reported to react at the 3-position of 1,5-benzodiazepines. Thus, coupling of 2,4-diphenyl-3*H*-1,5-benzodiazepine (**77**) with 4-nitrobenzenediazonium cation led to a product which was assigned the phenylhydrazone structure **78** (Eq. 26).[16]

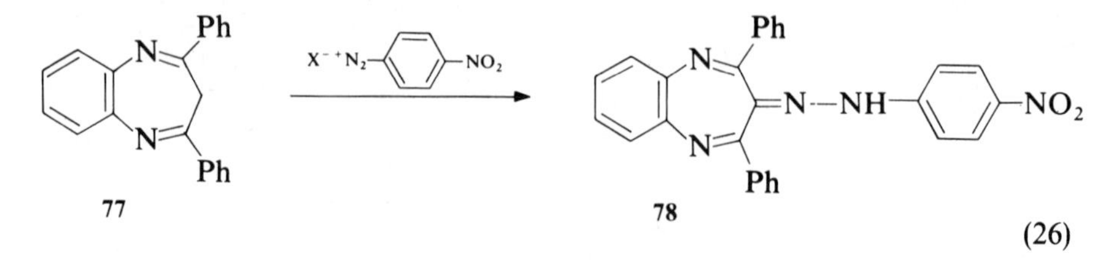


### 1.2.1.6. Oxidation

As with nitration and nitrosation, the reaction of 1,5-benzodiazepines **79** with peracids led to products of either ring cleavage or ring contraction (Eq. 27).

Thus, treatment of **79** with monopersulfuric acid, or preferably peracetic acid, gave the 2-acetylquinoxalines **80** ($R_1$ = Me, $R_2$ = Me, Ph).[16] On the other hand, oxidation of the 2,4-diphenyl derivative **79** ($R_1 = R_2 =$ Ph) with aqueous peracetic acid gave the diamide **81** together with the minor products **82**–**84**.[81] It was speculated[81] that **80** and **81** arose from different reaction paths, but it is more likely that the aqueous reaction medium is responsible for the predominance of ring cleavage over ring contraction in the case of the 2,4-diphenyl derivative.

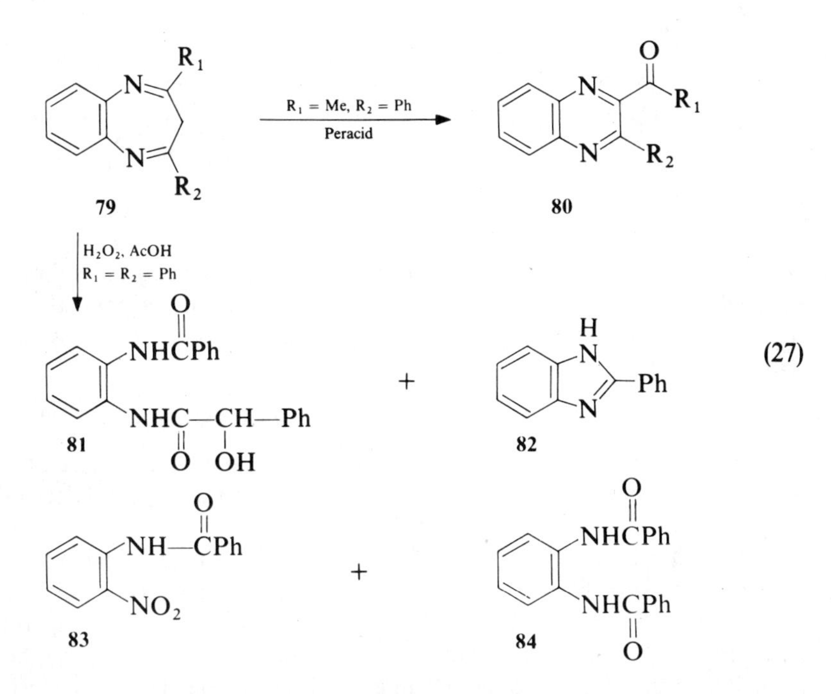


Photooxidation of the benzodiazepines **79** also led to the ring-contracted products **80** (Eq. 28). Interesting solvent effects were observed in this reaction.[82] Thus irradiation of a benzene solution of 2-methyl-4-phenyl-3*H*-1,5-benzodiazepine with a high pressure mercury arc under an oxygen atmosphere led to 2-benzoyl-3-methylquinoxaline (**80**; $R_1$ = Ph, $R_2$ = Me) in 25% yield, while in acetic acid solution, 2-acetyl-3-phenylquinoxaline was obtained, as in the peracid oxidation.

Photooxidation of 2,4-dimethylbenzodiazepine in 0.1 *N* sulfuric acid gave 2-methylbenzimidazole as the major product and only 5% of the quinoxaline **80** ($R_1 = R_2 =$ Me).

The quinoxalin-2(1*H*)-one **85** that was isolated as a by-product was shown to be a photooxidation product of the 2-acyl-3-methylquinoxaline **80** ($R_2$ = Me). In the case of the 2,4-diphenylbenzodiazepine, photooxidation in benzene gave low yields of both the 3-one **86** ($R_1 = R_2$ = Ph) and the ring-open diamide **84**.[82] The 3-ones **86** are believed to be precursors of the 2-acylquinoxalines in both

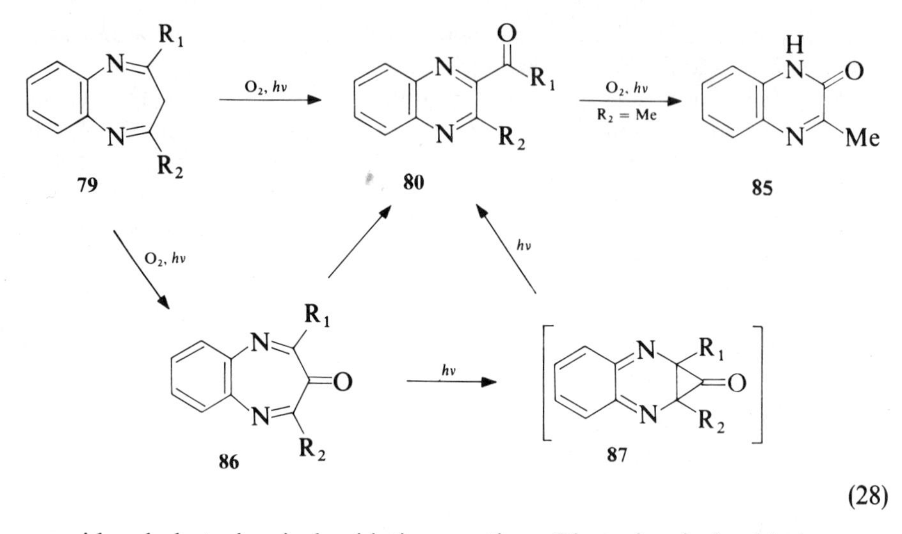


(28)

peracid and photochemical oxidation reactions. Photochemical oxidation may proceed via the intermediate valence tautomer **87**, while in the peracid oxidation it is probable that **86** undergoes solvolytic ring opening followed by reclosure with ring contraction.

### 1.2.1.7. Alkylation, Reactions with Aldehydes

Treatment of 2,4-dimethyl-3*H*-1,5-benzodiazepine with a mixture of sodium amide and methyl iodide in tetrahydrofuran[16] led to the 3-methyl derivative **88** (Eq. 29).

Base-catalyzed condensation of **71** with benzaldehyde in the presence of aqueous potassium hydroxide led to a mixture of **89** and **91**, while only the latter compound[16] was obtained using ethoxide in ethanol as the base. Piperonaldehyde, on the other hand, gave only the 3-substituted derivative **90**, together with a small amount of the diadduct **92**, in either base system. The distyryl **91** could be partially hydrolyzed to **89** upon treatment with aqueous alkali.

### 1.2.1.8. Acylation

Reaction of the 7-amino-1,5-benzodiazepine **93** ($R = NH_2$) with an equimolar amount of acetic anhydride in benzene solution led to the corresponding acetate **93** (R = NHAc).[20] An excess of reagent caused ring cleavage with formation of the enamine **94** and the diamide **95** (Eq. 30). The unsubstituted benzodiazepine **93** (R = H) could be similarly ring opened.

Benzodiazepines with 2-phenyl substituents have been shown to be less prone to ring cleavage. Boiling acetic anhydride converted **96** to the diacetates **97**.[31] Partial hydrolysis of the phenolic acetate afforded **99**. Selective acylation of the phenol to yield **98** could be carried out by the low temperature reaction of **96** with an acyl halide in the presence of triethylamine (Eq. 31).

The condensation of 2-methyl-3*H*-1,5-benzodiazepines **100** with diethyl oxalate was reported[27] to give the tricyclic compounds **101**. A later paper revised

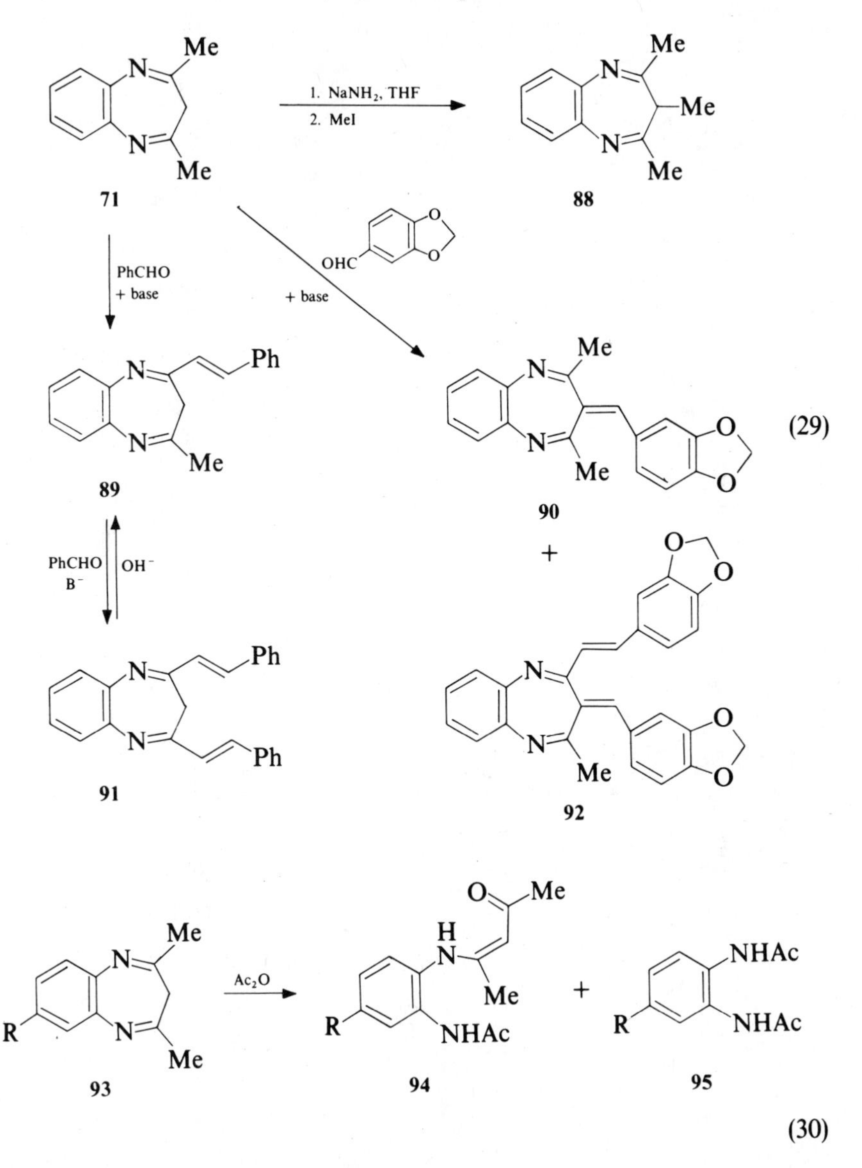


these structures on the basis of spectral data and assigned[83] the isomers **102** as the correct structures (Eq. 32).

Acylation of **96** on carbon (3-position) could be carried out readily by using amide acetals such as dimethylformamide dimethylacetal to give the enamines **103**, which subsequently underwent ring closure to the benzopyranobenzodiazepines **104** (Eq. 33).[31,37]

Paterson and Proctor[84] reported the formation of the sulfonamide **105** by reaction of 2,4-dimethyl-3*H*-1,5-benzodiazepine with tosyl chloride in pyridine. The product was described as a deep red, amorphous material (mp > 360°C),

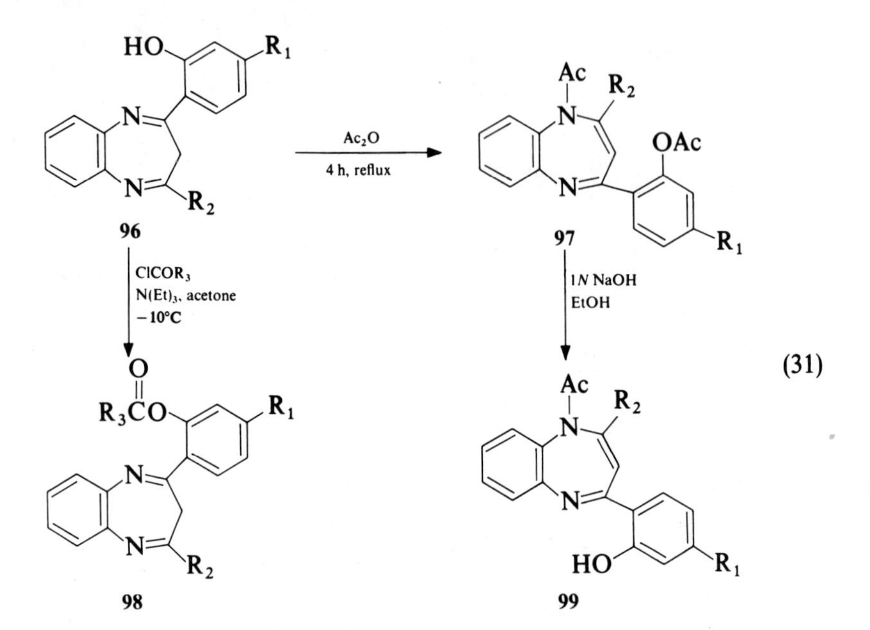
HO
R1
N
N
R2
96
Ac2O
4 h, reflux
Ac
R2
OAc
97
R1
ClCOR3
N(Et)3, acetone
−10°C
1N NaOH
EtOH
O
R3CO
R1
R2
98
Ac
R2
HO
R1
99

(31)

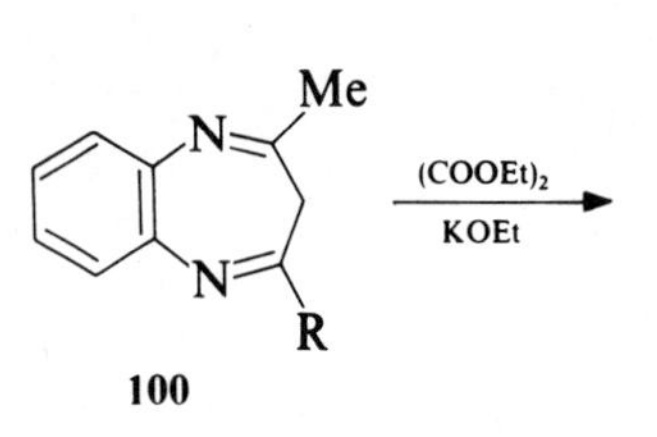
Me
N
N
R
100
(COOEt)2
KOEt

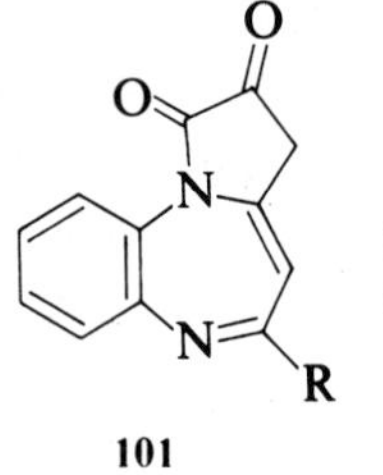
O
O
N
N
R
101

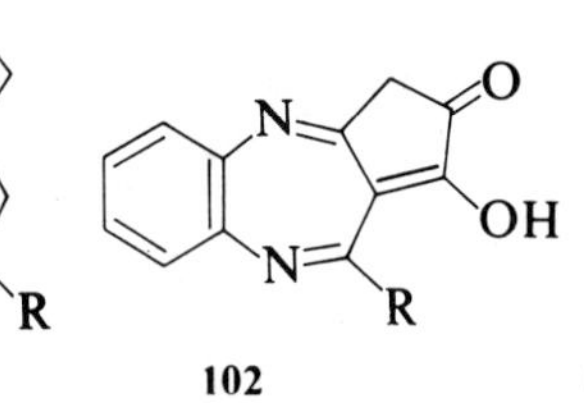
N
O
OH
N
R
102

(32)

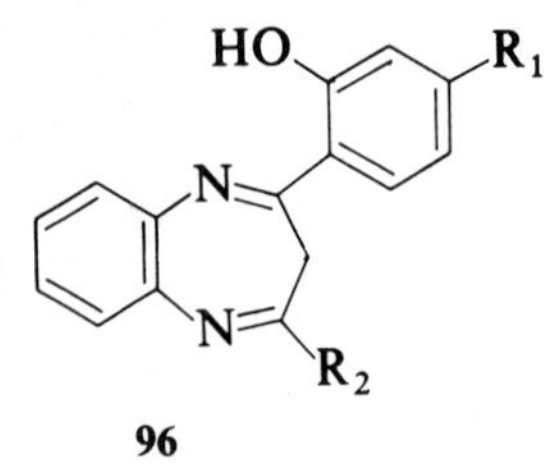
HO
R1
N
N
R2
96
R3
Me2NC(OMe)2

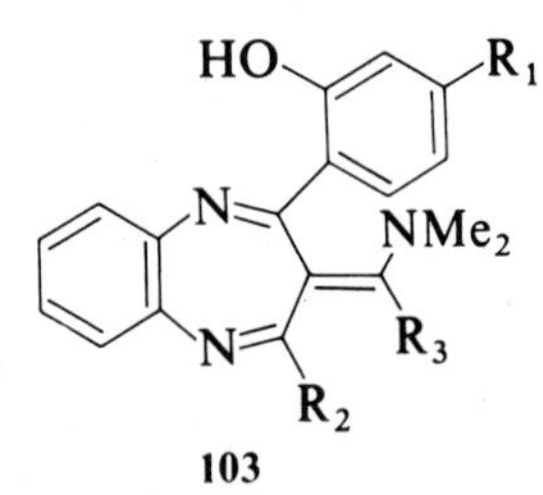
HO
R1
N
NMe2
N
R3
R2
103

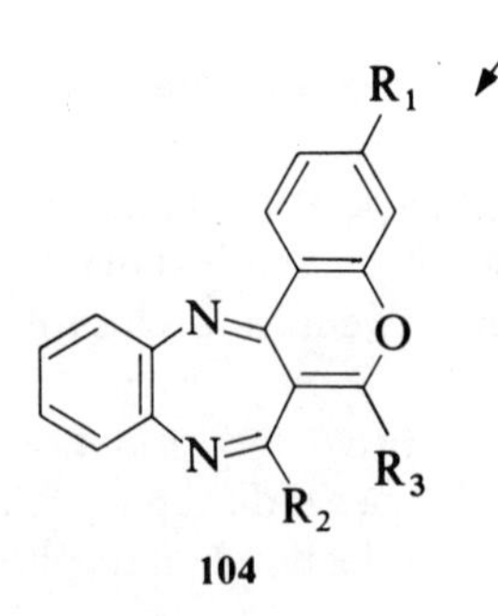
R1
N
O
N
R3
R2
104

(33)

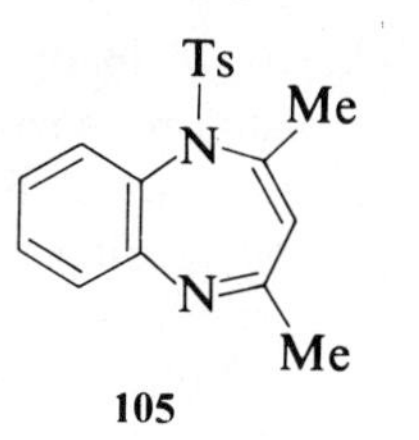


105

unaffected by refluxing with acids and alkalies. These observations suggest that the assigned structure may be incorrect.

## *1.2.2. Reactions with Nucleophiles*

### 1.2.2.1. Hydrolysis

The imine bonds in the benzodiazepines **79** are hydrolyzed readily by aqueous acid to give the enamine **106** and, upon further hydrolysis, *o*-phenylenediamine and a 1,3-diketone. These components may recombine to form the enamine **106**, which is in equilibrium with both the dihydrobenzimidazole **107** and the benzodiazepine **79**. Aromatization of **107** by the indicated mechanism would lead irreversibly to a mixture of the corresponding benzimidazole **108** and a methyl ketone (Eq. 34).

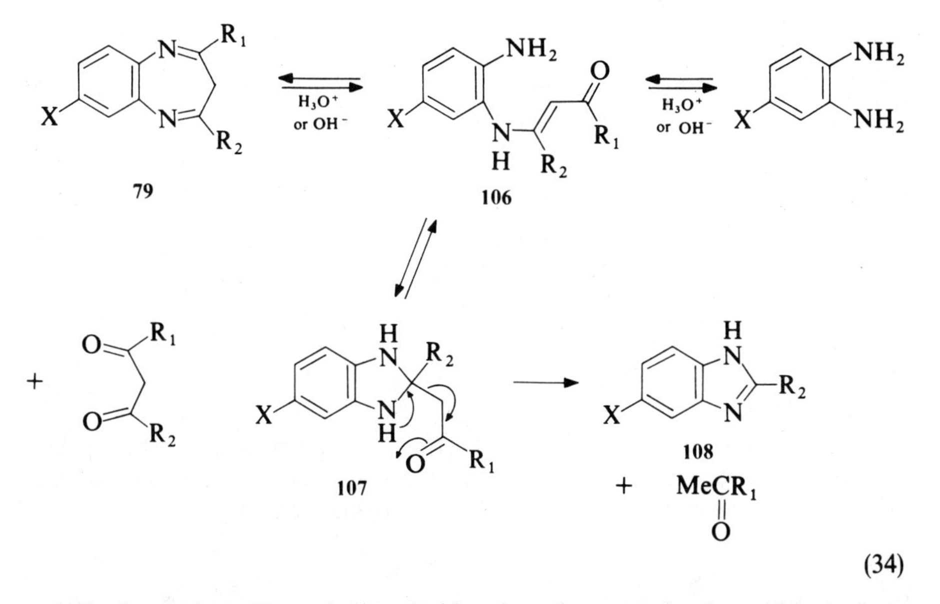


(34)

The formation of 2-methylbenzimidazole and acetone by the acid hydrolysis of **79** ($R_1 = R_2$ = Me) was reported by Thiele and Steimmig.[14] The unsymmetrical 2-methyl-4-phenyl-3*H*-1,5-benzodiazepine **79** (X = H, $R_1$ = Me, $R_2$ = Ph) yielded, as expected, 2-methylbenzimidazole and 2-phenylbenzimidazole together with acetone and acetophenone.[14] The hydrolysis of the selenyl derivative **79** ($R_1$ = Me, $R_2$ = 2-selenyl) yielded only 2-methylbenzimidazole.[32] This would suggest that the dihydrobenzimidazole **107** ($R_1$ = 2-selenyl, $R_2$ = Me) is both formed and fragmented preferentially.

Hydrolytic cleavage of 1,5-benzodiazepines to *o*-phenylenediamines and 1,3-diketones would also explain the formation of 2,3-dimethylquinoxaline from **79** ($R_1$ = $R_2$ = Me) in the presence of butane-2,3-dione and ferric chloride[47] or aqueous acid.[16]

The imine functions of 1,5-benzodiazepines are also cleaved by alkaline hydrolysis. This cleavage is more facile if electron-withdrawing substituents are present in the aromatic ring. Thus, treatment of the 7-nitro derivative of **79** with aqueous hydroxide at room temperature led to the enamine **106** (X = $NO_2$).[1,21] More vigorous conditions afforded the diamine.[1]

The hydrolytic behavior of 3-arylidene derivatives is very similar to that of the 3-unsubstituted compounds. Treatment of **90** with hydrochloric acid in boiling ethanol gave the benzimidazole **109**, 2-methylbenzimidazole, and the methylketone **110** (Eq. 35).[16]

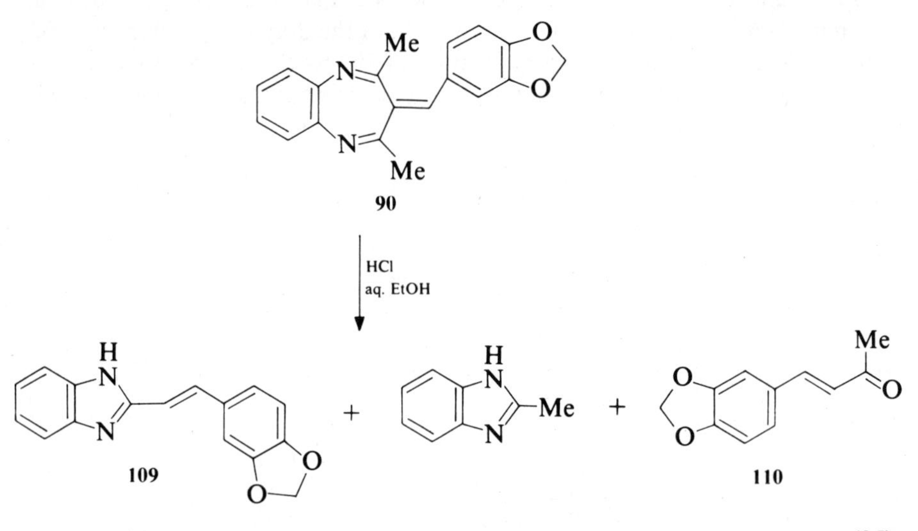


(35)

Reaction of the 3-benzoyl derivative **111** with concentrated hydrochloric acid resulted in cleavage to give the 2,4-diphenylbenzodiazepine **77** and benzoic acid, along with 2-phenylbenzimidazole and dibenzoylmethane (Eq. 36).[29]

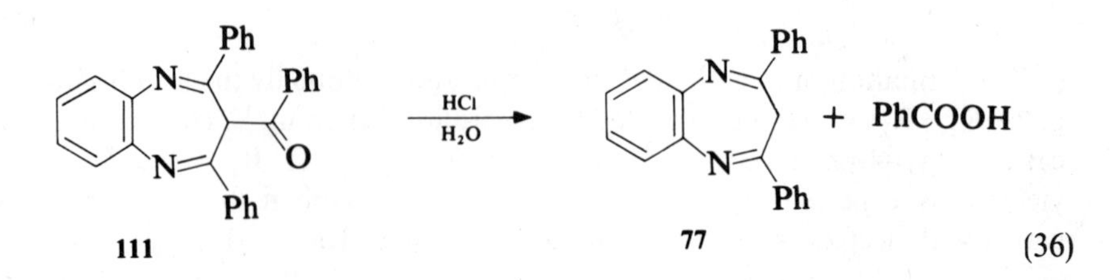


(36)

The same type of ketone cleavage would also explain the formation of 2,4-dimethyl-3*H*-1,5-benzodiazepine (**71**) from *o*-phenylenediamine and triacetylmethane (Eq. 37).[16]

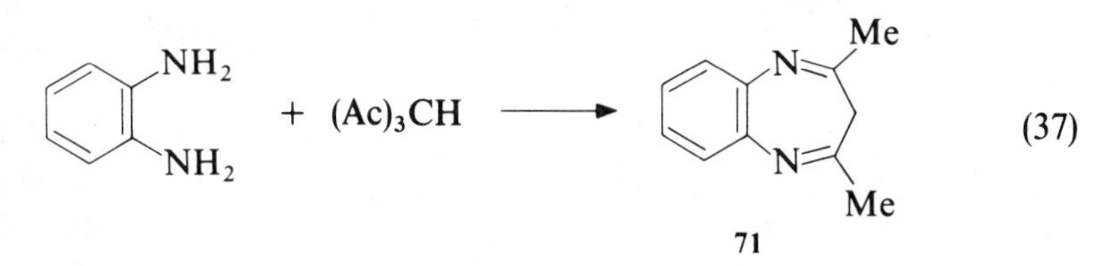


Hydrolysis of the 3-oximino derivative **112** was shown to lead to the formation of quinoxalines.[16,47] Thus, acid hydrolysis of **112** with 10% sulfuric acid in the presence of ferric chloride led to a mixture of 2-acetyl-3-methylquinoxaline (**73**) and the corresponding oxime **113** (Eq. 38). When **112** was treated with an acetic acid–ethanol mixture at room temperature for 30 hours, the oxime **113** and 2-methylbenzimidazole were isolated. The hydrolysis of **112** with hot water led to 2-methylbenzimidazole.[48]

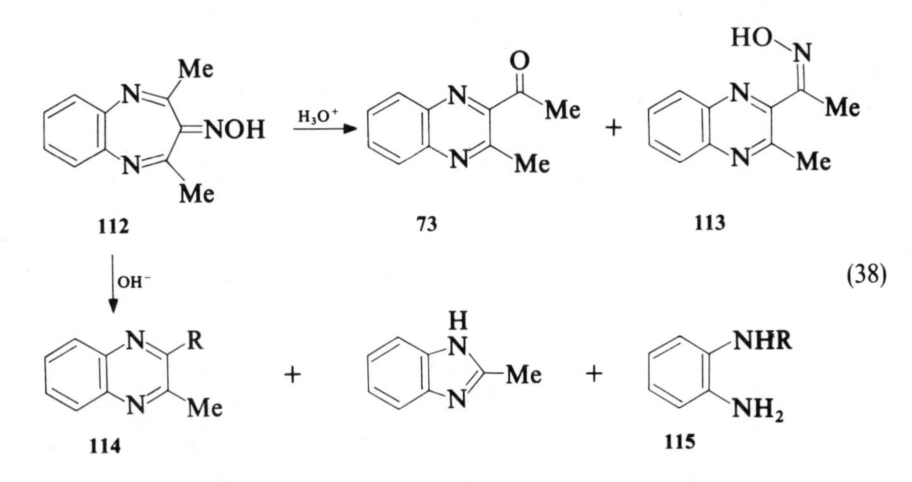


Alkaline hydrolysis of **112** was even more complex and gave an array of products, of which the quinoxaline **114** (R = OH), 2-methylbenzimidazole, and *o*-phenylenediamine **115** (R = H) were isolated (Eq. 38).[16] When the reaction was carried out with sodium carbonate in aqueous ethanol, the 2-aminoquinoxaline **114** ($R = NH_2$) and the acetate **115** (R = Ac) were formed in addition to the three products mentioned above.

Mild acid hydrolysis of the 2-alkylthiobenzodiazepines **116** yielded the lactam **117** as the major product (Eq. 39).[67] Further degradation was observed with the methylthio derivative **116** (R = Me), to yield *o*-phenylenediamine and benzimidazoles.[67]

Alkaline hydrolysis of the 2-aminobenzodiazepine **118** gave different products depending on the reaction conditions.[60] The conversion of **118** to the 2,4-

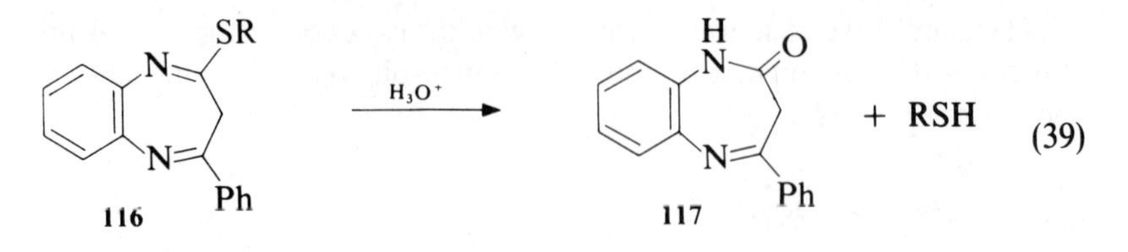


diaminobenzodiazepine was mentioned in connection with the synthesis of the latter (see Section 1.1.8). Aqueous sodium hydroxide under controlled conditions transformed **118** to the 1,3,5-benzotriazepine **122** in 38% yield. This reaction may proceed via the intermediates **119** and **120** (Eq. 40).

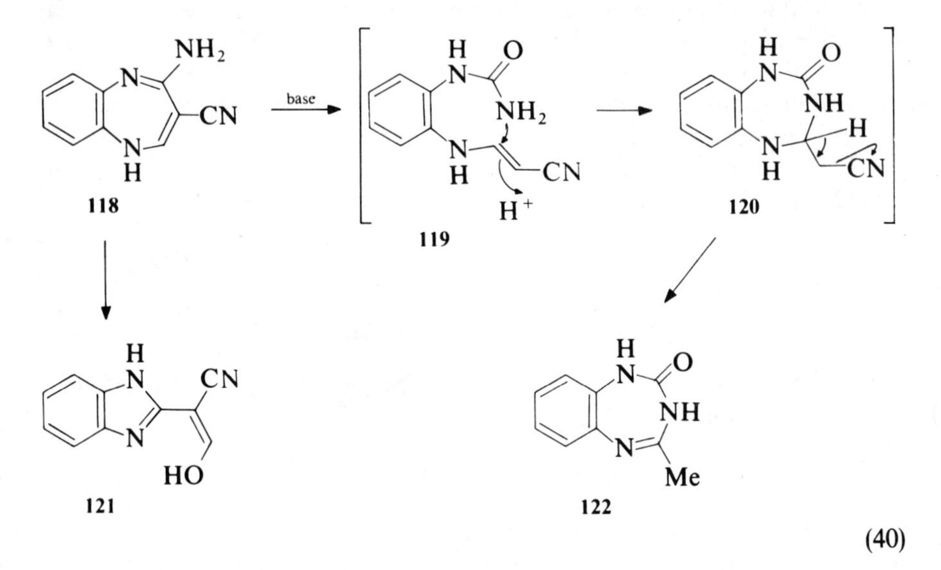


(40)

Treatment of **118** with a weak aqueous base such as 2-aminopyrimidine or ammonia gave the benzimidazole **121**, apparently by hydrolytic cleavage of the 4,5-bond followed by a recyclization to the benzimidazole.

1.2.2.2. Reaction with Nitrogen Nucleophiles

Cleavage of the benzodiazepine **123** ($R_1 = R_3 = Me$, $R_2 = H$) by phenylhydrazine to give *o*-phenylenediamine and 3,5-dimethyl-1-phenylpyrazole **124** ($R_1 = R_3 = Me$, $R_2 = H$) was observed by Thiele and Steimmig (Eq. 41).[14] Finnar[36] obtained high yields of pyrazoles in the reaction of symmetrically

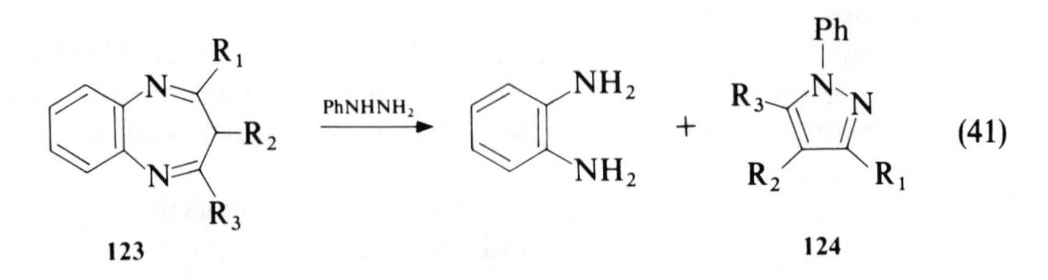

substituted 2,4-disubstituted 3*H*-1,5-benzodiazepines with phenylhydrazine in acetic acid.

The unsymmetrical 2-methyl-4-phenyl-3*H*-1,5-benzodiazepine **123** ($R_1$ = Me, $R_2$ = H, $R_3$ = Ph) gave 3-methyl-1,5-diphenylpyrazole in 68% yield, suggesting preferential attack at the less hindered 2-position.

Compound **123** ($R_1$ = Ph, $R_2$ = H, $R_3$ = benzoylacetyl) yielded the dipyrazole **124** ($R_1$ = Ph, $R_2$ = H, $R_3$ = 1,3-diphenylpyrazol-5-yl).[36] Barltrop and coworkers[16] applied the same reaction to the product obtained by methylation of 2,4-dimethyl-1,5-benzodiazepine and obtained the pyrazole **124** ($R_1 = R_2 = R_3$ = Me). This showed that methylation had occurred at the 3-position in **123** ($R_1$ = $R_2$ = $R_3$ = Me).

Reaction of **123** ($R_1$ = $R_3$ = H, $R_2$ = 6-chlorobenzoxazol-2-yl) with boiling hydrazine or phenylhydrazine was reported to give the pyrazoles **125** (R = H, Ph) in low yields.[7]

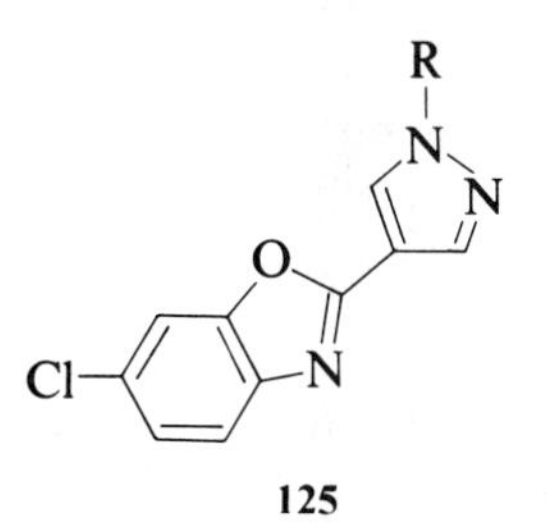

**125**

The action of several nitrogen nucleophiles on the 3-hydroxyiminobenzodiazepine **112** has been studied by O'Callaghan and Twomey.[48] For example, substituted hydrazines in methanol containing hydrogen chloride converted **112** to hydrazones of 2-acetyl-3-methylquinoxalines **126** (Eq. 42), not to the expected

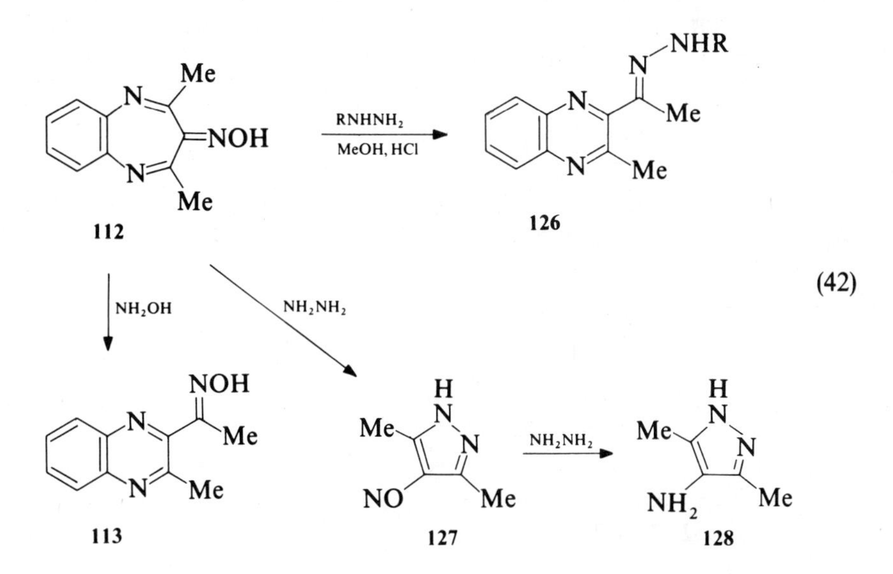


(42)

benzodiazepine-3-hydrazones. The reaction of **112** with either semicarbazide or thiosemicarbazide hydrochloride was initially reported to lead to benzodiazepin-3-carbazones,[85] but the assigned structures were later corrected[48] to quinoxalines **126** (R = $CONH_2$, $CSNH_2$).

The reaction of **112** with hydroxylamine hydrochloride afforded the oxime **113**. Depending on the reaction temperature either the *syn*- or *anti*-oxime was obtained. Hydrazine cleaved **112** to *o*-phenylenediamine and the 4-nitrosopyrazole **127**. In the presence of excess hydrazine, the 4-nitrosopyrazole was reduced to the 4-aminopyrazole **128**. The reaction of 2-alkylthiobenzodiazepines with amines to produce the corresponding 2-amino-substituted derivatives was discussed above (see Section 1.1.8).

#### 1.2.2.3. Reductions

Reduction of the imine double bonds in 1,5-benzodiazepines is discussed in connection with the synthesis of dihydro- and tetrahydro- derivatives (see Sections 2.1.6. and 4.1.4). The nitro group in 2,4-dimethyl-7-nitro-3*H*-1,5-benzodiazepine has been catalytically reduced with Raney nickel to an amino group without affecting the imine bonds.[21]

### *1.2.3. Miscellaneous Reactions*

The hydrochloride of the *tert*-butyl ester **129** was reported[43] to rearrange to quinoxaline **130** upon dissolution in organic solvents or on storage (Eq. 43).

Since water has to be involved in this transformation, the starting material may be in the form of a hydrate. The same type of reaction may apply to the pyrolysis of benzodiazepinium salts, which, in the case of 2,4-dimethyl-1*H*-1,5-benzodiazepine hydrochloride, was reported[1] to yield 2-methylbenzimidazole and acetone.

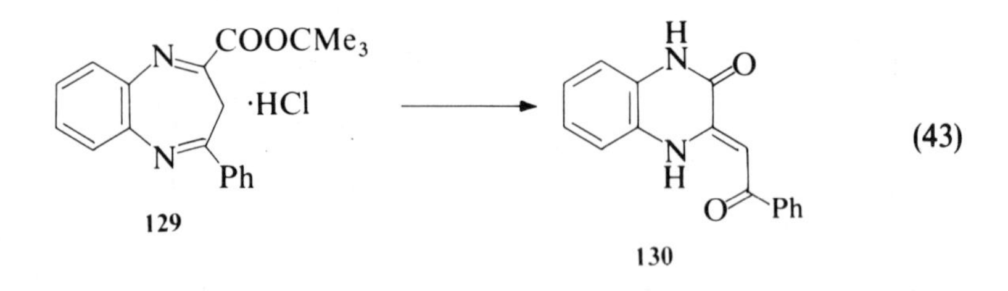


### *1.2.4. Metal Complexes*

Ouchi and coworkers[86,87] and Hunter and Webb[88,89] have prepared a variety of metal halide and metal sulfate complexes of structure **131** ($R_1 = R_2$ = Me) and **132** ($R_1$ = Me, $R_2$ = Ph) (Eq. 44). The latter compounds generally

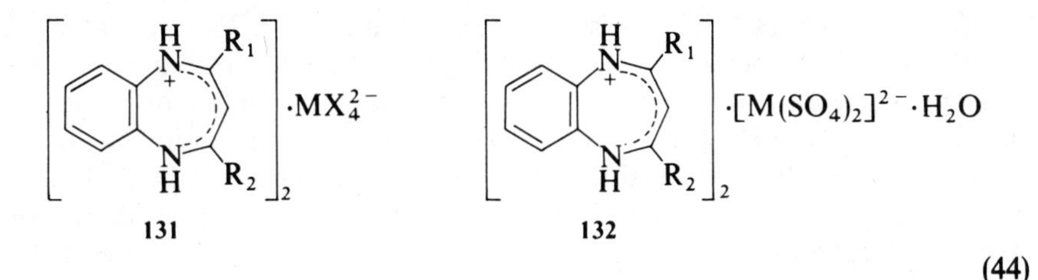


(44)

contained water of crystallization. The complex with ferrosulfate had been discovered earlier[90] but was assigned the wrong stoichiometry. Both spectroscopic and magnetic properties of these complexes were reported.

The structure of the tetraiodozincate **133** was determined by X-ray crystallographic analysis.[91] The X-ray data confirmed the spectroscopic evidence that the benzodiazepinium cation does not coordinate with the metal in these complexes.

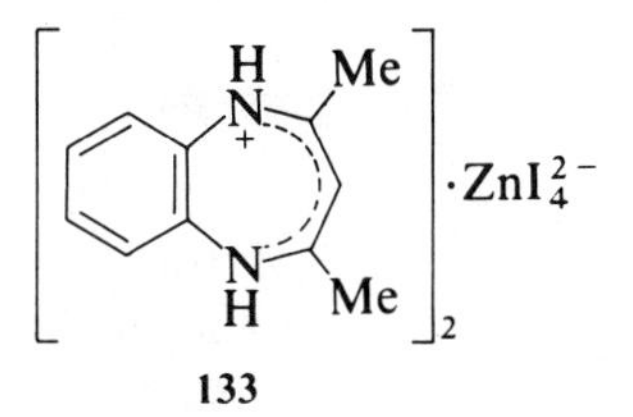


## 1.3. Spectra and Physical Data

The structures of the 1,5-benzodiazepines and the corresponding mono- and dications were, in large part, assigned on the basis of spectroscopic data. Nuclear magnetic resonance spectra were particularly informative.[92–94] For example, Mannschreck and coworkers[95] studied the temperature-dependent proton-nmr spectra of 2,4-disubstituted 3*H*-1,5-benzodiazepines. At normal operating temperatures the protons at C-3 generally appear as a singlet, but at low temperatures an AB system is observed. This result demonstrates the nonequivalence of the protons at C-3 and indicates that the 3*H*-1,5-benzodiazepine ring is nonplanar. The conformation of the seven-membered ring can probably be represented best by the equilibrium between A and B (Eq. 45).

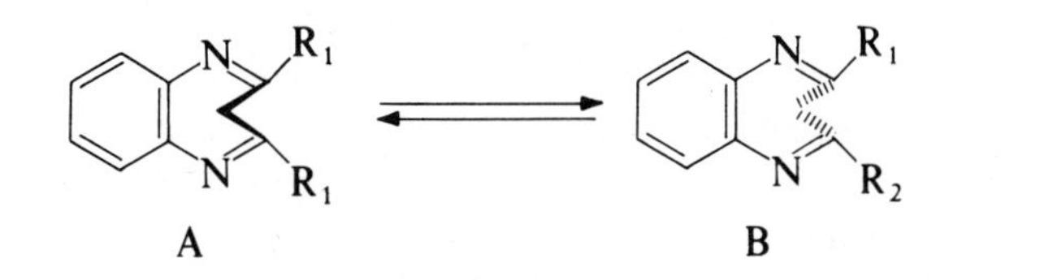


(45)

When the A ⇆ B interconversion is very rapid (as is the case at room temperature), the protons at C-3 become magnetically equivalent and appear as a singlet. The activation energy $\Delta G_c$ at the coalescence temperature was determined to vary between 11 and 13 kcal/mol for 2,4-disubstituted 3*H*-1,5-benzodiazepines. Phenyl-substituted derivatives usually gave higher values for $\Delta G_c^{\#}$, possibly as a result of greater steric hindrance in the transition state for the A ⇆ B interconversion.

The structures of 2,4-dimethyl-1,5-benzodiazepinium chloride dihydrate (and the isomorphous bromide) were analyzed by X-ray crystallography, which showed that the seven-membered ring was almost planar.[96] A polarographic study of several 2-phenyl-4-pyridyl-3*H*-1,5-benzodiazepines revealed two main waves corresponding to a consecutive reduction of the 4,5- and the 1,2-imine double bonds.[97]

## 2. DIHYDRO-1,5-BENZODIAZEPINES

### 2.1. Synthesis

2,3-Dihydro-1*H*-1,5-benzodiazepines of structure **135** are generally accessible by the condensation of *o*-phenylenediamine with either an α,β-unsaturated carbonyl compound **134**[310,313] or an appropriate precursor thereof, such as a β-halo- or a β-aminoketone (Eq. 46).

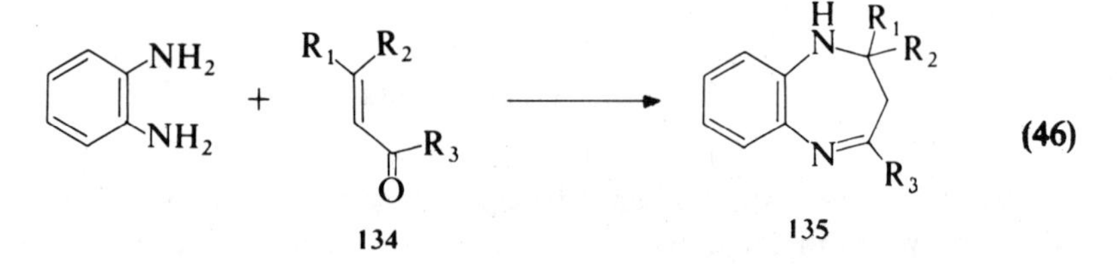


#### 2.1.1. *From o-Phenylenediamines and α,β-Unsaturated Aldehydes*

Ried and Stahlhofen[98] reported the synthesis of **135** ($R_1 = Me$, $R_2 = R_3 = H$) by the condensation of *o*-phenylenediamine with crotonaldehyde. The same reaction has also been applied[99] to the diphenylmethane derivative **136** to prepare **137** (Eq. 47). The structure of **137** was assumed to be symmetrical, with the more reactive amino group forming the imine.

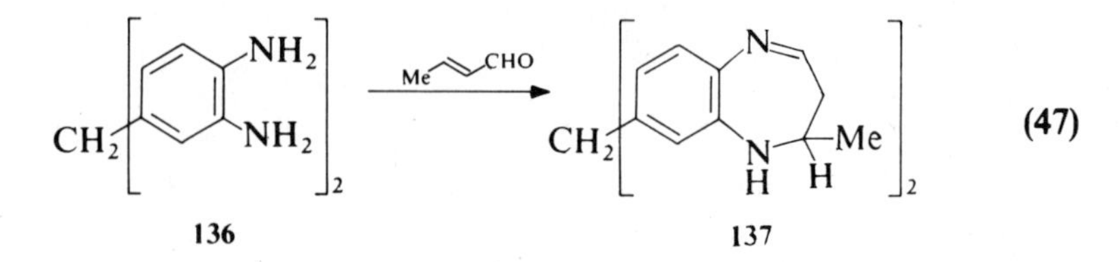

### 2.1.2. *From o-Phenylenediamines and α,β-Unsaturated Ketones*

The synthesis of 2,3-dihydro-2,2,4-trimethyl-1*H*-1,5-benzodiazepine **135** ($R_1 = R_2 = R_3$ = Me) from *o*-phenylenediamine and mesityl oxide was first described by Ried and Stahlhofen[98] and Mushkalo.[100] Several later papers[101–105] report the preparation of the same compound and the corresponding 7,8-dichloro derivative[106] by generating mesityl oxide in situ from acetone in the presence of an acid catalyst. The condensation was facilitated by using either boron trifluoride etherate[103] or 2-naphthylsulfonic acid[105] as a catalyst.

Fukushima and coworkers[107] have shown that this method can also be used for the preparation of 2,3-dihydro-2,4-diphenyl-1*H*-1,5-benzodiazepine. Reid and Stahlhofen[98] were not able to obtain 1,5-benzodiazepines from the reaction of *o*-phenylenediamine with cinnamaldehyde, benzylidene acetophenone, or benzylidene acetone. The latter investigators isolated the open-chain adducts **138** (Eq. 48) resulting from Michael addition of the amine to the double bond. By increasing the nucleophilic character of the carbonyl group, as in 1,3-dipyridylpropenones, Samula and Jurkowska-Kowalczyk[108] were able to obtain 2,4-diaryl derivatives **135** ($R_1 = R_3$ = Aryl, $R_2$ = H) in good yield. These authors also isolated the intermediate **139**, indicating that addition to the reactive carbonyl group may be the first step in the reaction pathway.

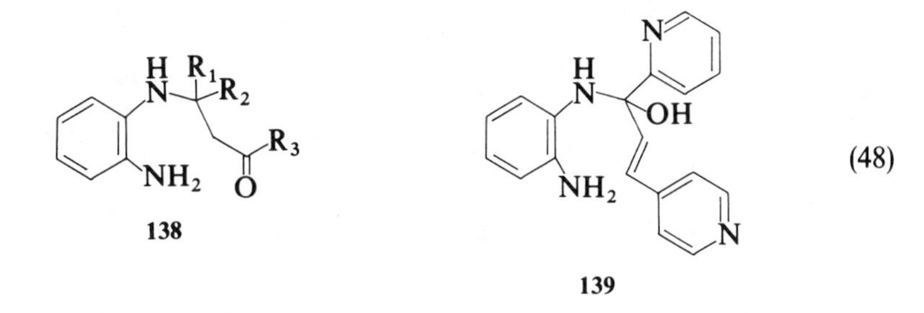


Herbert and Suschitzky[103] were able to obtain the benzodiazepine **135** ($R_1$ = Me, $R_2 = R_3$ = Ph) as a hydrate in 80% yield by the condensation of *o*-phenylenediamine with acetophenone in the presence of boron trifluoride etherate.

### 2.1.3. *Condensation of o-Phenylenediamines with β-Haloketones*

Both Mushkalo[100] and Ried and Torinus[109] have demonstrated the equivalence of 2-bromo-2-methylpentan-4-one and mesityl oxide for the synthesis of the 2,3-dihydro-2,2,4-trimethyl-1*H*-1,5-benzodiazepine (**135**, $R_1 = R_2 = R_3$ = Me). This same bromoketone has also been used for the synthesis of the corresponding nitro- and carboxy-substituted analogs.[110] The position of the aromatic substituent in these products was not determined.

Mushkalo[100] showed that this method may also be utilized in the preparation of *N*-substituted derivatives. Treatment of 2-bromo-2-methylpentan-4-one with 2-aminodiphenylamine afforded 2,3-dihydro-2,2,4-trimethyl-1-phenyl-1*H*-1,5-benzodiazepine. This compound was not fully characterized but was used for further reactions. Hideg and Hidcg-Hankovzky[111] prepared 2,3-dihydro-4-phenyl-7,8-dimethyl-1*H*-1,5-benzodiazepine by reaction of the appropriate diamine with *β*-chloropropiophenone.

### *2.1.4. From o-Phenylenediamines and β-Aminoketones*

The accessibility of Mannich bases of type **140** makes this synthesis attractive. Hideg and Hankovzky[112] used this method for the preparation of numerous compounds of structure **141** (Eq. 49). The same reaction was utilized by Werner and coworkers[113,114] and also by Curtze and Thomas.[115]

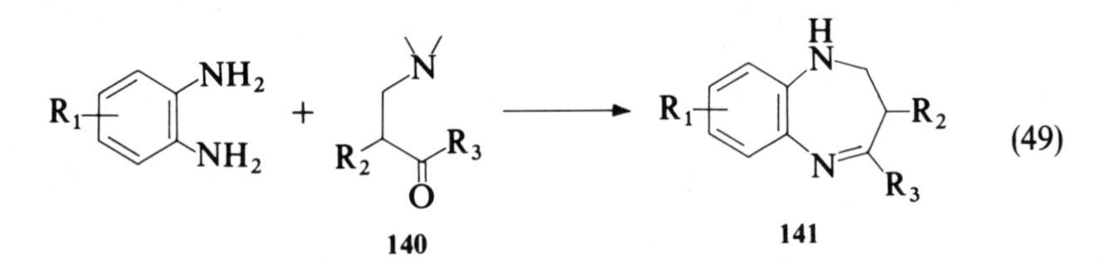


(49)

### *2.1.5. From o-Phenylenediamines and β-Hydroxyketones*

Hideg and Hideg-Hankovzky[111] have reported that *β*-hydroxyketones can be successfully utilized for the synthesis of dihydro-1,5-benzodiazepines. Condensation of 4,5-dimethyl *o*-phenylenediamine with the hydroxyketone **142** in refluxing xylene with azeotropic removal of water gave the benzodiazepine **143** in 78% yield (Eq. 50).

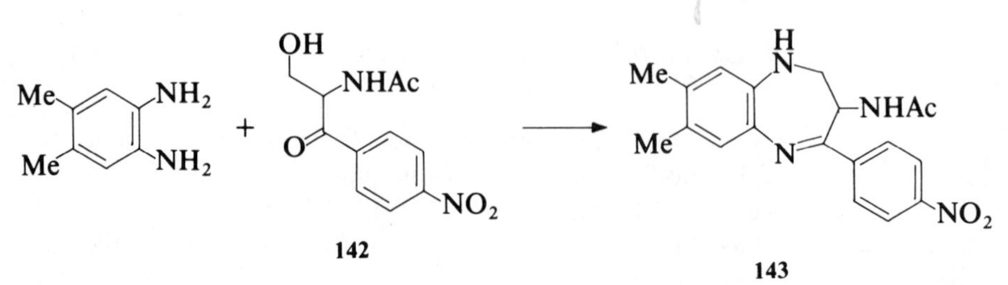


(50)

### 2.1.6. *By Partial Reduction of 1,5-Benzodiazepines*

Reduction of 3*H*-1,5-benzodiazepines **64** with sodium borohydride to 2,3-dihydro-1*H*-1,5-benzodiazepines **144** has been reported for both the 2,4-dimethyl[116] and the 2,4-diphenyl derivatives (Eq. 51).[117]

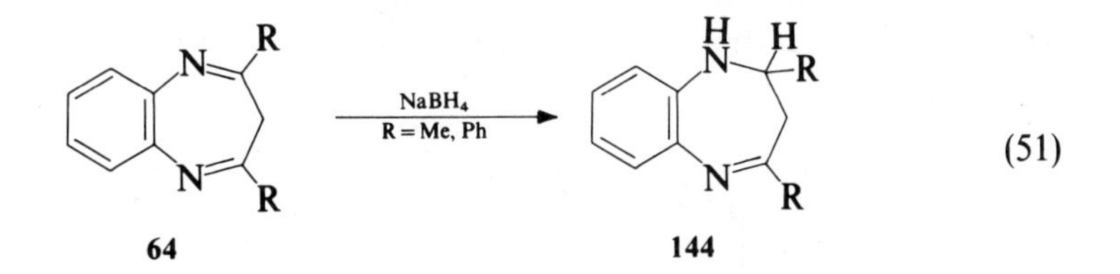


### 2.1.7. *Other Syntheses*

The 4-amino-2,3-dihydro-1*H*-1,5-benzodiazepine **146** was obtained[118] by cyclization of the acrylonitrile adduct **145** using dry hydrogen chloride in tetrahydrofuran (Eq. 52).

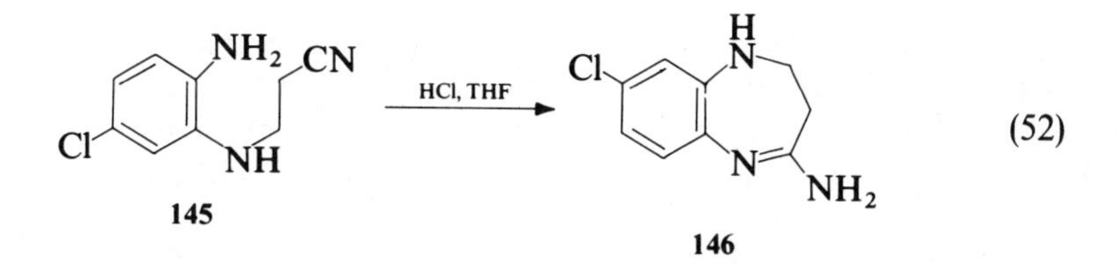


The formation of the dihydrobenzodiazepine-2-thiocarboxamide **148** by treatment of **147**, the adduct of hydrogen cyanide with 2,4-dimethyl-1,5-benzodiazepine, with ammonium hydrogen sulfide was reported by Bodforss (Eq. 53).[119]

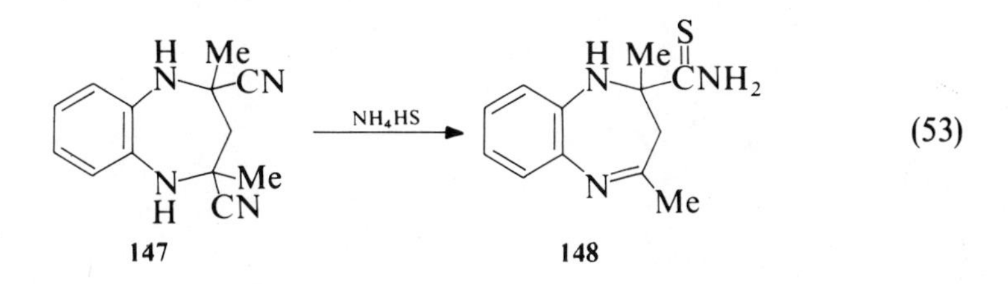

## 2.2. Reactions

### *2.2.1. Reactions with Electrophiles*

#### 2.2.1.1. Nitrosation

Ried and Stahlhofen[98] obtained the dinitroso derivative assigned structure **150** by reaction of the dihydrobenzodiazepine **149** with sodium nitrite in acetic acid (Eq. 54).

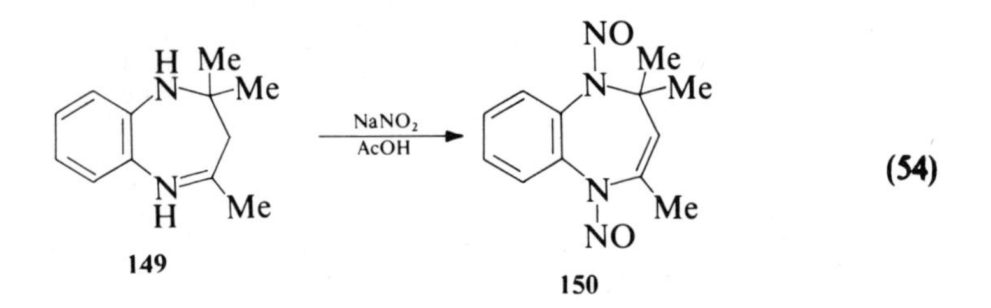


(54)

#### 2.2.1.2. Oxidation

The oxidation of 8-chloro-4-amino-2,3-dihydro-1*H*-1,5-benzodiazepine to the corresponding 2-one[118] is discussed in Section 3.1.5.

#### 2.2.1.3. Alkylation and Reactions with Aldehydes

Treatment of the dihydrobenzodiazepine **151** with methyl iodide in benzene led to quaternization of the imine nitrogen, yielding **152**.[100] The salts of **151** as well as **152** were condensed with aldehydes to form cyanine dyes **153** ($R_3$ = H, Me). Similar condensations were performed with 4-dimethylamino-

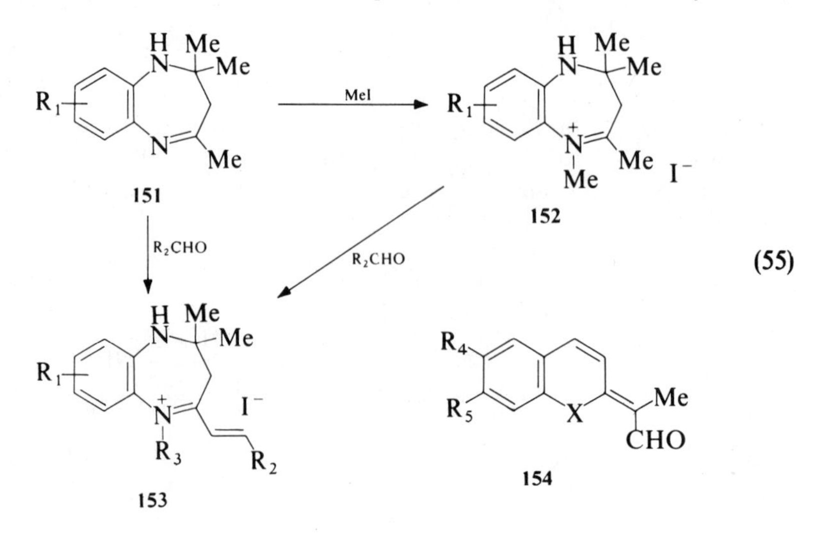


(55)

benzaldehyde[100,110] and with the benzopyran or benzothiopyran derivatives **154** (X = O, S) in the presence of acetic anhydride (Eq. 55).[120]

Aldehyde equivalents such as the enamides **155**[100] and **156**[110] (Eq. 56) were also successfully condensed with **151** to give the corresponding dyes.

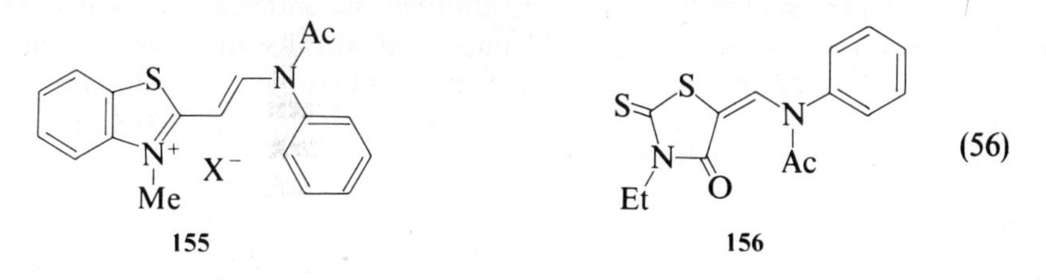


Condensation of **151** ($R_1$ = 8-$NO_2$) with the dianil of glutaconic aldehyde afforded the diadduct **157** (Eq. 57).[110]

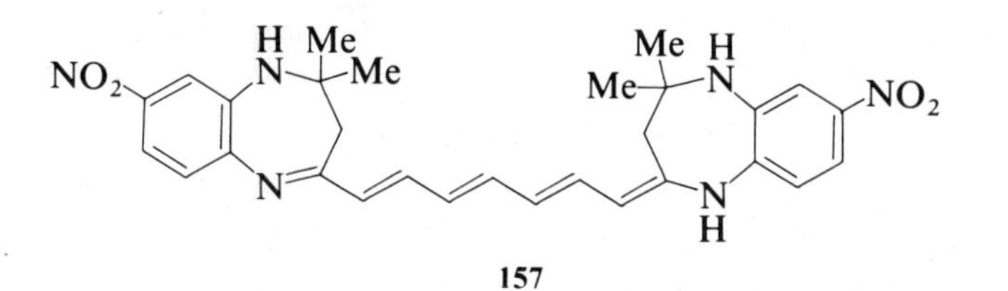


(57)

**158**

Treatment of **151** with triethyl *o*-formate in pyridine[100] or in acetic anhydride at 100°C afforded the dyes **158**.[110] These dyes were most likely obtained via the intermediate 4-(2-ethoxyethylene)-1,5-benzodiazepine **153** ($R_2$ = OEt, $R_3$ = H).

### 2.2.1.4. Acylation

According to the patent literature,[105] reaction of **149** with furan-2-carboxylic acid chloride in the presence of triethylamine gave the 4-furoyl derivative **159**

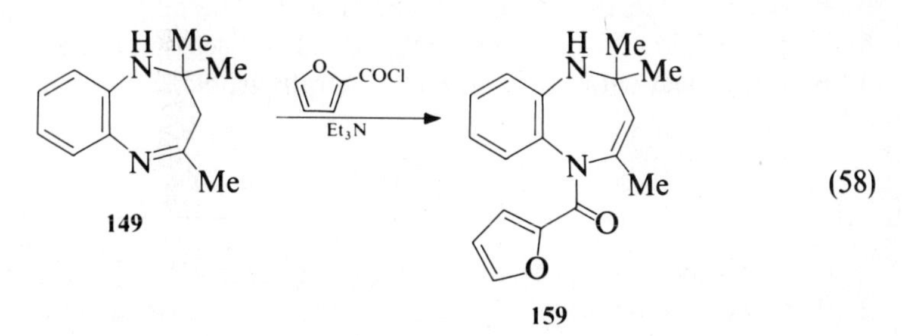

(Eq. 58). The 1-substituted product was not reported and probably is not formed because of steric hindrance.

Compounds lacking geminal methyl groups in the 2-position undergo acylation on the 1-nitrogen. Thus, treatment of **160** with methyl chloroformate or benzoyl chloride in pyridine (Eq. 59) afforded the anticipated products **161**.[115] Similar acylations were performed on 2,3-dihydro-7,8-dimethyl-4-phenyl-1*H*-1,5-benzodiazepines using chloroacetic acid anhydride and sodium acetate in acetic acid.[111] Treatment of **160** with acetic anhydride at 100°C yielded the tetracyclic compound **162** (R = Me).

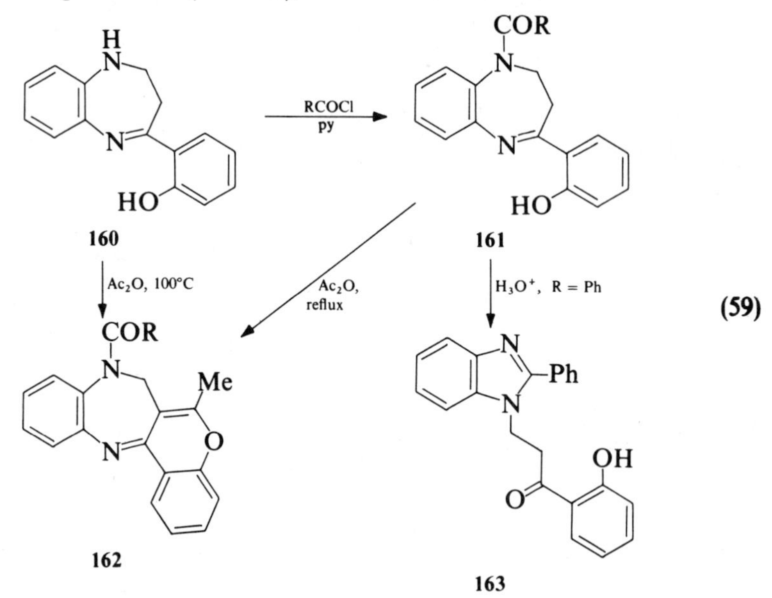


(59)

The urethane **161** (R = MeO) required the more vigorous conditions of heating under reflux for conversion to the benzochroman **162** (R = MeO).

### *2.2.2. Reactions with Nucleophiles*

Hydrolysis of **161** (R = Ph) with 20% hydrochloric acid gave the benzimidazole **163** via an initial cleavage of the imine double bond followed by recyclization to the five-membered ring.[115] The addition of hydrogen and hydrogen cyanide to the imine double bond is discussed in Sections 4.1.5. and 4.1.6., which deal with the synthesis of tetrahydrobenzodiazepines.

## 2.3. Spectral Data

Room temperature proton- and $^{13}$C-nmr spectra[117,121] of the 2,3-dihydro-2,2,4-trimethyl-1*H*-1,5-benzodiazepine **149** show a single resonance for the protons in the 3-position and for the geminal methyl groups. This would indicate a rapid conformational change at ambient temperature. By measuring the broadening of the methyl signal at lower temperatures, Hunter and Webb,[122] have determined that the free energy of activation for the conformational change is about 9 kcal/mol. According to the proton-nmr analysis, the 2,3-dihydro-2,4-diphenyl-1*H*-1,5-benzodiazepine showed a fixed conformation and the ABX pattern assigned to the C-3 protons did not collapse to a singlet up to a temperature of 140°. The mass spectral fragmentation of these compounds has also been studied.[122]

## 3. DIHYDRO-1,5-BENZODIAZEPIN-2-ONES

Dihydro-1,5-benzodiazepin-2-ones can exist either as the 3*H* tautomer **164** or the 5*H* tautomer **165**. Since the latter has rarely been observed, the thermodynamically more stable tautomer is probably the 3*H* compound.

### 3.1. Synthesis

#### *3.1.1. From o-Phenylenediamines and β-Ketoesters*

The condensation of *o*-phenylenediamine with *β*-ketoesters may be expected to lead to a 4-substituted 1,5-benzodiazepin-2-one, a 2-substituted benzimidazole, or a mixture of both products.[306] Sexton[123] studied the reaction of *o*-phenylenediamine with ethyl acetoacetate and obtained two products, a benzodiazepine and a benzimidazole, to which he assigned structures **167** (R = Me) and **171** (R = Me). The structure assignment for **171** (R = Me) was later corrected by Davoll[124] and by Rossi and coworkers,[125] who recognized that the imidazole derivative was the 1-(2-propenyl)benzimidazolone **169** (R = Me) arising from a thermal rearrangement of **167** (Eq. 60).

It has been well established that the initial step in the reaction of *o*-phenylenediamine with a *β*-ketoester is the formation of the enamine **166**. Cyclization of **166** to the benzodiazepine **167** predominates under neutral or alkaline conditions at moderate temperatures, whereas acid catalysis favors the formation of the imidazoles **170**. These compounds are most likely formed by cyclization of **166** to the intermediate **168**, which may then aromatize with loss

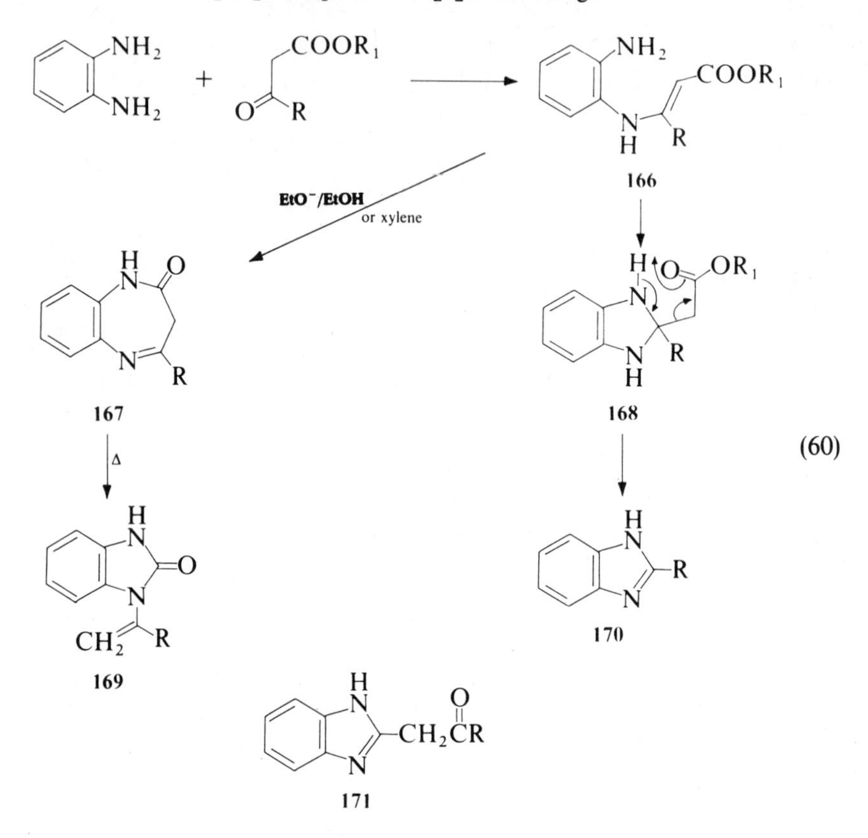


of acetate. Acetate elimination should be facilitated by the indicated intramolecular proton shift (Eq. 60).

The cyclization of the enamines **166** with alkoxide in alcohol has evolved as the procedure of choice for the synthesis of benzodiazepin-2-ones.[106,124] Condensation of *o*-phenylenediamine with a variety of $\beta$-ketoesters in refluxing toluene or xylene has been successfully used for the selective preparation of the benzodiazepines **167** (e.g., with R = Me,[126] $CH_2Ph$,[127] $CF_3$,[128] Ph,[129] $CCl_3$,[130] 1-adamantyl,[131] and 3-pyridyl[132]).

When $\beta$-ketoesters are condensed with unsymmetrically substituted *o*-phenylenediamines, both the expected regioisomers **172** and **174** are formed (Eq. 61). Such mixtures were obtained with X = Br,[133] Cl,[134] Me,[132,134,135] MeO,[134] and (1,3-dihydro-4-methyl-1,5-benzodiazepin-2(2*H*)-one-7-(or8)yl)methyl.[136] The composition of this mixture was found to depend more on the nature of substituent X than on R.[134] In some instances a remarkable regioselectivity was reported. Thus, Kost and coworkers[137] obtained the benzodiazepine **172** (X = R = Me) in 87% yield by carrying out the reaction of the diamine with ethyl acetoacetate in boiling xylene at high dilution. It is feasible that under these conditions, the initially formed enamine **173** rearranges to **176**, freeing the more basic amino group for ring closure with the ester carbonyl. If the two enamines

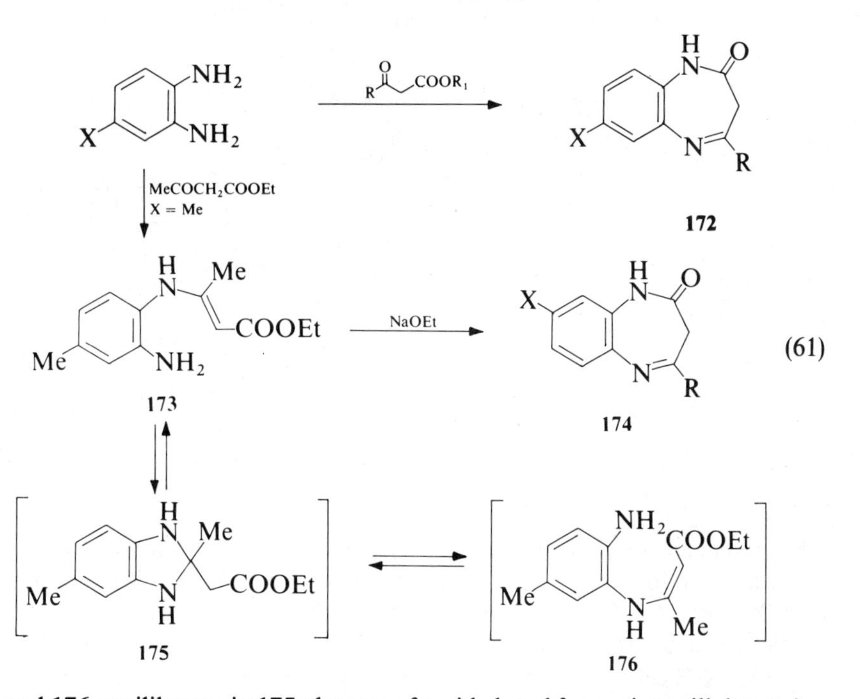


**173** and **176** equilibrate via **175**, the rate of amide bond formation will determine the product, and this amide bond would be expected to form more readily with the more basic amino group. The regioisomer **174** (X = R = Me), on the other hand, was obtained in 66% yield by cyclization of the enamine **173** with sodium ethoxide in ethanol.[138] Compound **173** resulted from the preferential reaction of the keto group of ethyl acetoacetate with the more basic nitrogen of toluidine under mild conditions.

The situation for 4-nitrophenylenediamine is reversed.[139–141] The amino group meta to the nitro substituent is the more nucleophilic one and would favor formation of the enamine **177**. Cyclization of **177** with sodium ethoxide in ethanol led to the benzodiazepine **172** (X = $NO_2$, R = Me). The regioisomer **174** (X = $NO_2$, R = Me) was however accessible in 88% yield, by utilizing the high dilution technique in boiling xylene. In this case, a small amount of tautomer **178** was also observed.[139]

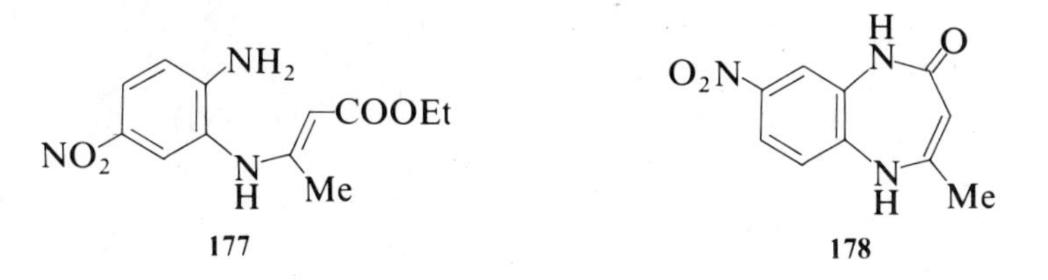


The same considerations were also shown to apply in the preparation of the dichloro derivatives **179** (X = Cl) and **180** (X = Cl).[142] Compound **179** (X = Cl) was synthesized via the enamine in 89% yield, while the isomer **180** (X = Cl) was

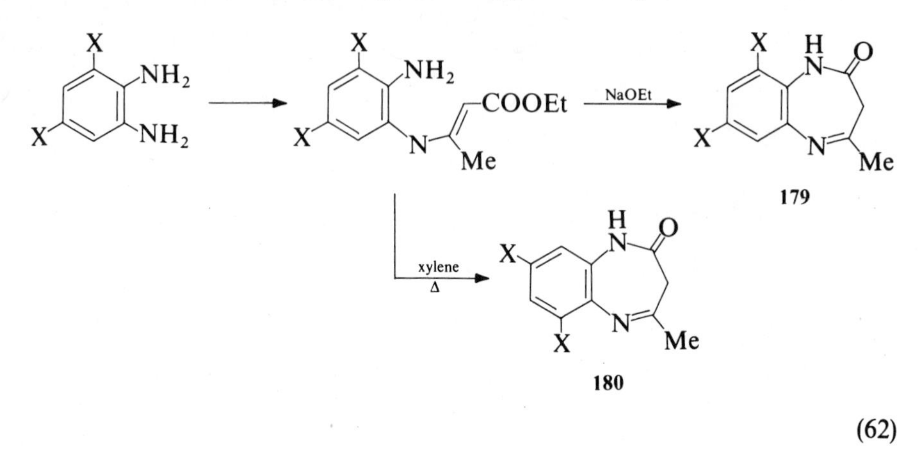


(62)

obtained in 80% yield by refluxing in xylene. The dibromo analogs **179** (X = Br) and **180** (X = Br)[143] as well as the bromo derivatives **172** (X = Br, R = Me) and **174** (X = Br, R = Me)[144] were accessible in the same manner (Eq. 62).

Pennini and coworkers[145] studied the condensation of the diamines **115** (R = Me, $CH_2Ph$) with ethyl benzoylacetate in boiling xylene and obtained the corresponding benzodiazepinones **184**, together with some benzimidazole derivatives. These authors isolated the benzoylacetate **181** (R = $CH_2Ph$), which upon further heating in xylene formed the benzodiazepine **184** (R = $CH_2Ph$). This conversion most likely proceeds by an acyl migration via **182** to give intermediate **183**, which closes to the seven-membered ring (Eq. 63).

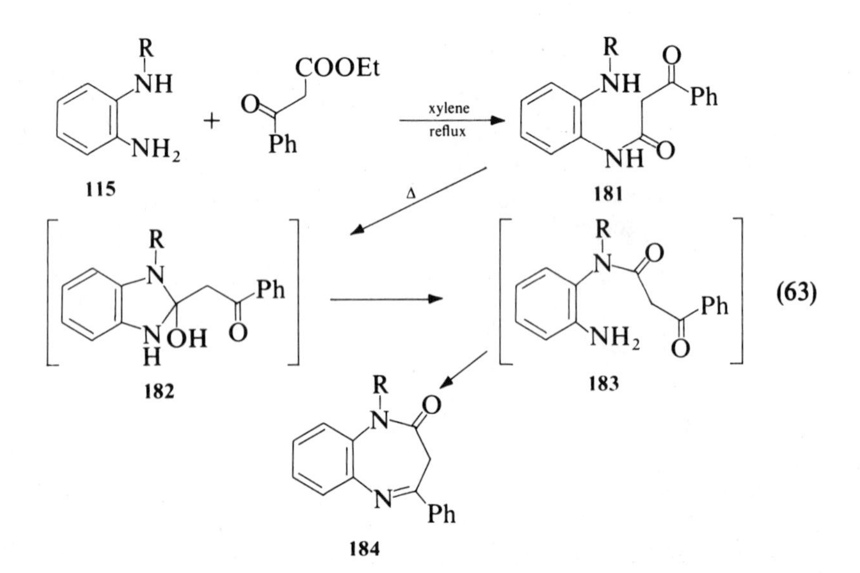


(63)

### 3.1.2. *From o-Phenylenediamines and Diketene*

Ried and Stahlhofen[146] described the exothermic reaction of *o*-phenylenediamine with diketene leading to the benzodiazepine **185** (R = H) (Eq. 64). The same reaction was later extended to the preparation of 4-, 7-, and 8-substituted analogs.[49, 315]

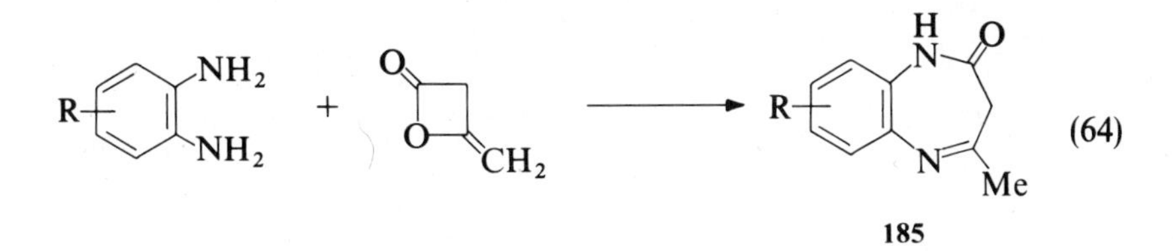


### 3.1.3. *From o-Phenylenediamines and Cyclobutenediones*

The first synthesis of this type was reported by Ried and Isenbruck[147] who isolated the benzodiazepine **189** from the reaction of *o*-phenylenediamine with the cyclobutenedione **186** (Eq. 65). The initial product of this condensation is the pyrrolobenzimidazole **187**, which must be formed by air oxidation of an intermediate imidazoline. Further addition of phenylenediamine to the reactive amide bond in **187** could then lead to intermediate **188** which would cyclize to the observed product **189**.

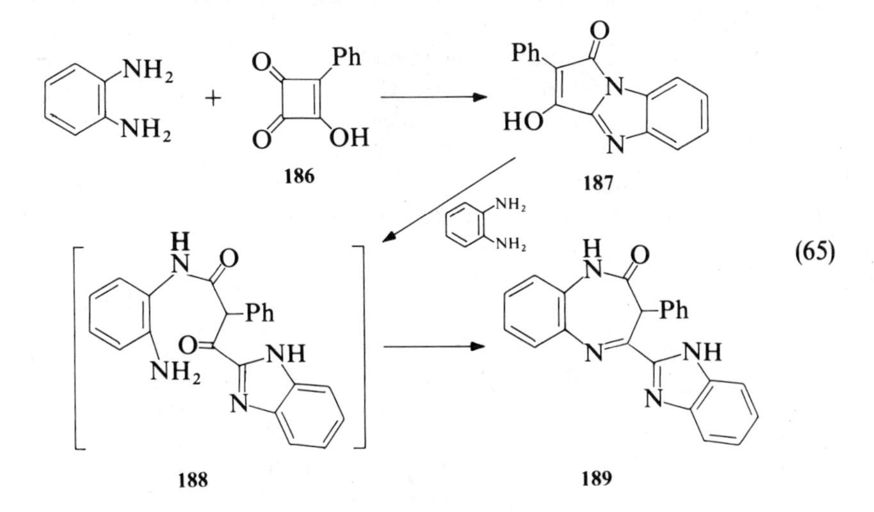


3-Hydroxy-2,4-diphenylcyclobutenone **190** was reported[148] to form a salt with *o*-phenylenediamine that, upon pyrolysis in boiling toluene, gave the benzodiazepine **193** (R = H) in 57% yield (Eq. 66). The 5*H*-tautomeric structure was assigned on the basis of spectral data. The same compound was also

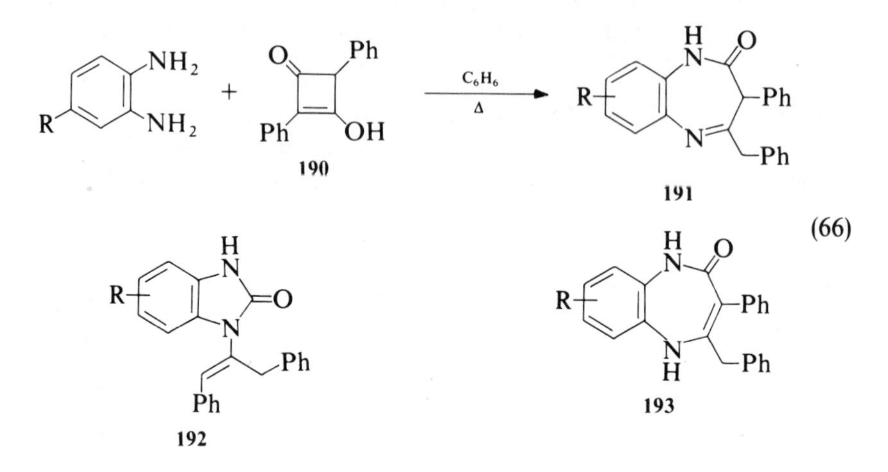


isolated from the condensation of the diamine with ethyl 2,4-diphenyl-3-oxobutyrate. Substituted diamines gave lower yields of **193**, possibly because of the formation of regioisomers, only one of which was isolated. The relative position of the substituent (7 or 8) was unassigned. We believe that the structures assigned by these investigators are probably incorrect and that the products thought to be the benzodiazepines **193** are actually the benzimidazolones **192**. Evidence for this is given by Zubovics and coworkers[149] who studied the same reaction in boiling benzene and obtained the benzodiazepinones **191** with different melting points. These authors confirmed their structural assignment for compounds **191**, including the location of the substitutent R, by an alternate synthesis.

Zubovics and coworkers[149] condensed the substituted nitroanilines **194** with the cyclobutenone **190** and obtained the anilides **195**, which were then reductively cyclized to the benzodiazepines **196** ($R_1$ = Cl, $R_2$ = H; $R_1$ = MeO, $R_2$ = H; $R_1$ = Me, $R_2$ = H; and $R_1$ = H, $R_2$ = Me) (Eq. 67). Comparison of the derivatives prepared by reductive cyclization with the products obtained by direct condensation from substituted phenylenediamines indicated that the

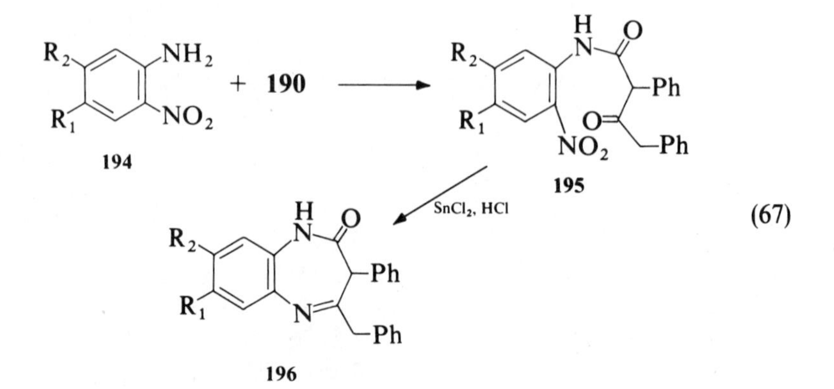

more basic amino group of the diamine preferentially attacks the cyclobutenone to form the amide bond.

Similar reductive cyclizations have been used for the preparation of both 7-chloro-1,3-dihydro-4-phenyl-1,5-benzodiazepin-2(2*H*)-one[134] and 7-amino-1,3-dihydro-4-phenyl-1,5-benzodiazepin-2(2*H*)-one.[150]

### *3.1.4. Other Methods*

The utility of an isoxazole-4-carboxylic acid as a 1,3-dicarbonyl equivalent was demonstrated by Ajello and coworkers[151] in the synthesis of the 1,5-benzodiazepinone **199**. Thus, reduction of the anilide **197** over Raney nickel gave **198**, which was cyclized to the benzodiazepine **199** in 50% yield by heating in toluene in the presence of a catalytic amount of *p*-toluenesulfonic acid. The acetyl group in **199** was easily removed by acid hydrolysis to give **200** (Eq. 68).

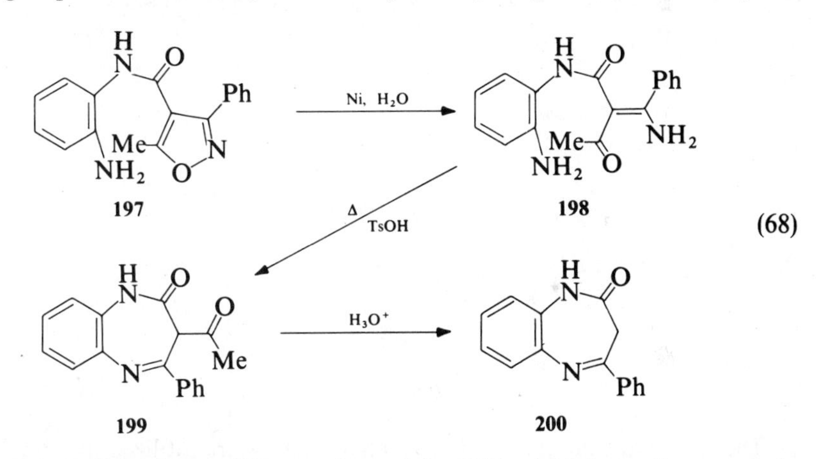


The benzodiazepinone **202** was formed in 50% yield from the coumarin derivative **201** and 2 equivalents of *o*-phenylenediamine in refluxing xylene[152] (Eq. 69). The benzimidazole **203** isolated as a second product resulted from the

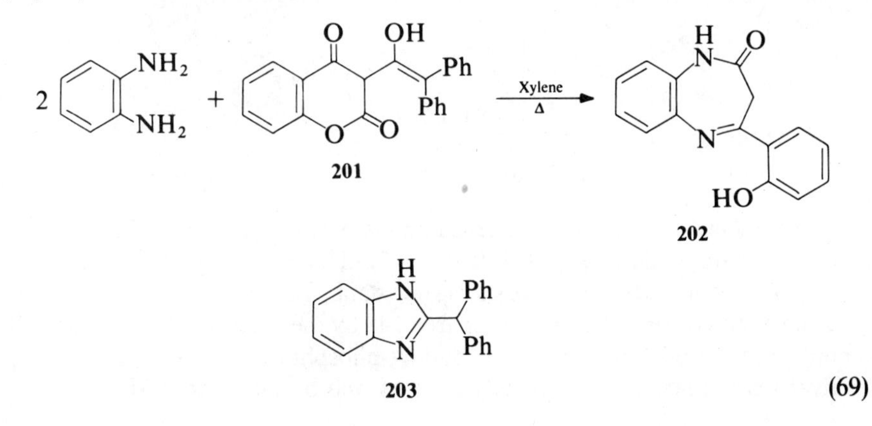

reaction of *o*-phenylenediamine with diphenylacetaldehyde formed from **201** by retroaldol cleavage.

### *3.1.5. Dihydrobenzodiazepin-2-ones with Heteroatom Substituents in Position 4*

4-Amino-1,3-dihydro-1-phenyl-1,5-benzodiazepin-2(2*H*)-ones **206** were prepared[153] from cyanoacetanilides **204** by reduction of the nitro group with zinc and hydrochloric acid followed by ring closure of **205** by hydrogen chloride (Eq. 70).

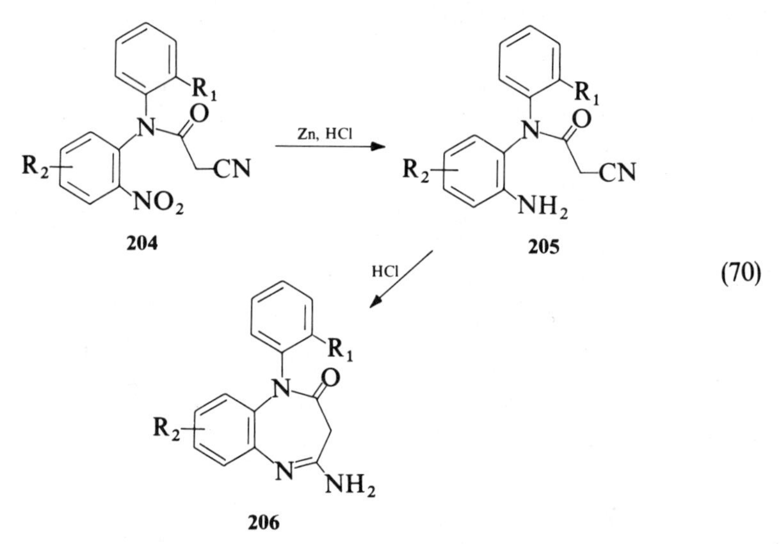


(70)

The reactive malonic acid derivatives **207** were utilized by Buyle and Viehe[154,155] for the synthesis of the 4-aminobenzodiazepines **208** (Eq. 71). The location of the substituent $R_3$ (7- or 8-position) in **208** has not been determined.

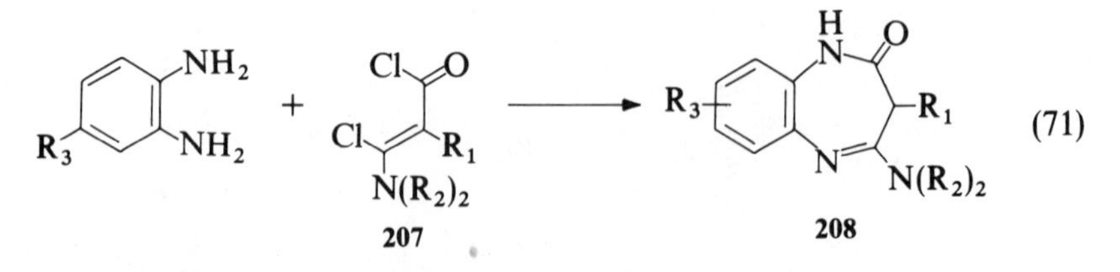


(71)

*N*-Phenyl-substituted *o*-phenylenediamine yielded a mixture of the two isomeric compounds **209** and **210** (Eq. 72). The predominant product was assigned structure **210** on the basis of nmr data.

Roma and coworkers[156] also prepared **208** by the reaction of *o*-phenylenediamine with malonic ester amide in boiling phosphorus oxychloride. 1-Phenyl derivatives similar to **209** were accessible as well by this procedure.

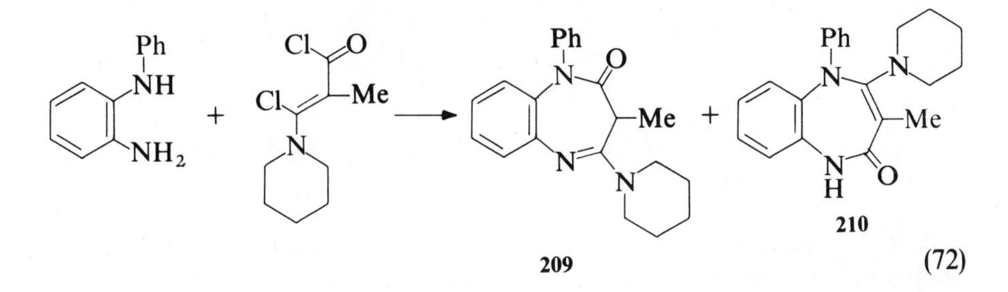


Conversion of a lactam to an amidine via an iminohalide has been used as a general method of preparation of many 4-aminobenzodiazepinones. Thus, the substituted 1-phenyl derivatives **212** ($R_2$ = aryl) were obtained by sequential reaction of the tetrahydrodiones **211** with a phosphorus pentahalide and an amine.[118,157] Another synthesis employed titanium tetrachloride and methyl amine to form the 4-methylamino analog.[158] The thiones **213** and the imino thio ethers **214** (X = S) were also used as precursors for the amidines **212**.[118,157] 4-Hydrazine analogs **212** ($R_3 = NH_2$, $R_4$ = H) were accessible by these methods,[159,160] as well as by reaction of the nitrosoamidine **212** ($R_3$ = NO, $R_4$ = Me) with hydrazine (Eq. 73).[158]

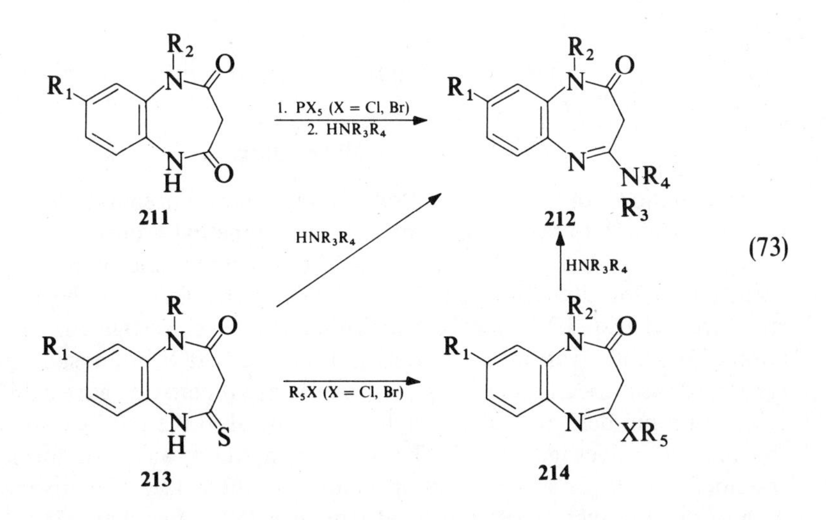


The imino ethers **214** (X = O) were prepared by reaction of the dione **211** with Meerwein salt[161] or via the iminochloride. Treatment of the nitrosoamidines **212** ($R_2$ = Ph, $R_3$ = NO, $R_4$ = Me) with alcohol and base also led to the imino ethers.[158] Thio ethers **214** (X = S) were obtained[161] by alkylation of the thiones **213** (Eq. 73).

The 1,5-benzodiazepin-2-ones may also be obtained by oxidation of **215** with chromium trioxide in sulfuric acid[118] to give the 4-amino compound **216** (Eq. 74).

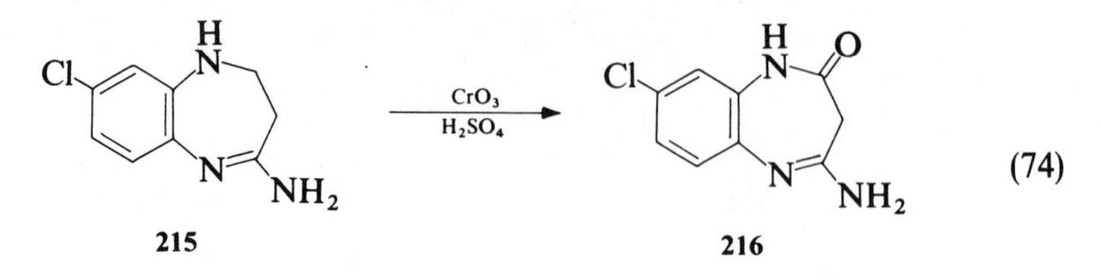


The 3-quinoxalinyl-1,5-benzodiazepine **218** was prepared via a ring transformation mechanism that occurs by reacting 3-(*N*,*N*-dimethylcarbamoyl)-furo-[2,3-*b*]quinoxaline hydrochloride **217** with *o*-phenylenediamine (Eq. 75).[305]

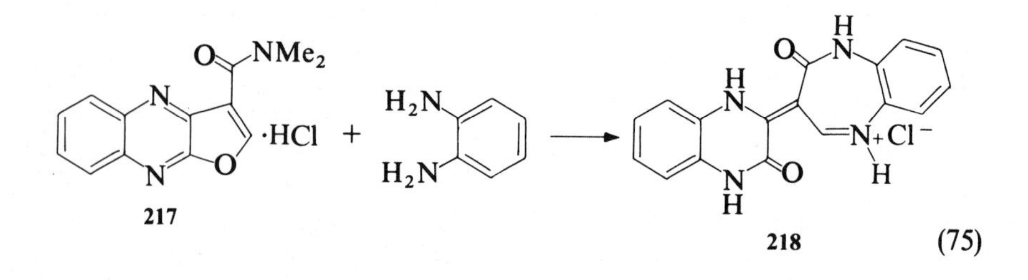


## 3.2. Reactions

### 3.2.1. Reactions with Electrophiles

#### 3.2.1.1. Halogenation

Bromination of 4-methyl-1,3-dihydro-1,5-benzodiazepin-2-(2*H*)-one has been reported[162] to take place primarily on the methyl group.

Barchet and Merz[132] obtained a monobromo compound by reaction of the 4-phenyl analog **200** with bromine and incorrectly assigned the tautomeric structure **221** (Eq. 76). Solomko and coworkers[133] reinvestigated the bromination of **200** under various conditions and showed that *N*-bromosuccinimide in carbon tetrachloride and in the presence of benzoyl peroxide gave a 76% yield of the 3-bromo derivative **219**, which was identical to the product obtained by the previous investigators.[132] The same compound was also formed with bromine in acetic or sulfuric acid or mixtures of these acids. The spectroscopic data of this product clearly indicated structure **219** rather than **221**.

When bromination was carried out with bromine in sulfuric acid containing silver sulfate, a mixture of the 4-(4-bromophenyl) derivative **220** (R = H) and the 3-bromo-4-(4-bromophenyl) compound **220** (R = Br) was obtained (Eq. 76).

#### 3.2.1.2. Nitration

Nitration of the 4-methyl derivative **222** ($R_1$ = Me, $R_2$ = H) with potassium nitrate in sulfuric acid at −10° to −5°C gave the 7-nitro compound **223**

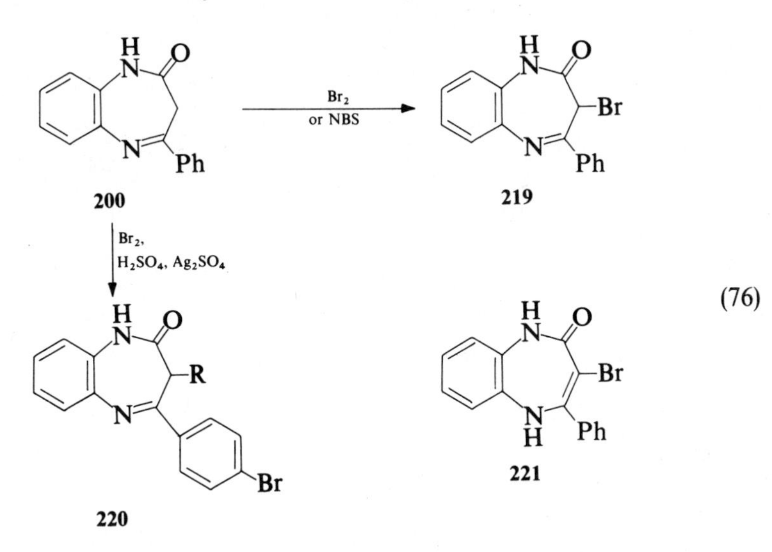


($R_1$ = Me, $R_2$ = H)[141] in good yield (Eq. 77). The 4-phenyl-1,5-benzodiazepinone **222** ($R_1$ = Ph, $R_2$ = H)[150] as well as the 8-methyl derivatives **222** ($R_1$ = Me, Ph, 4-$MeOC_6H_4$, $R_2$ = Me) were similarly converted to the corresponding 7-nitro derivatives **223** in high yields.[163] Compounds **222** ($R_1$ = Me, Ph, 4-$MeOC_6H_4$, $R_2$ = Cl, Br) were also successfully nitrated at the 7-position.[164] Under similar conditions of nitration the 8-methoxy-4-methyl benzodiazepinone **222** ($R_1$ = Me, $R_2$ = MeO) gave the 7-nitro compound **223** ($R_1$ = Me, $R_2$ = MeO) and the 9-nitro analog **224** in approximately equal amounts.[163]

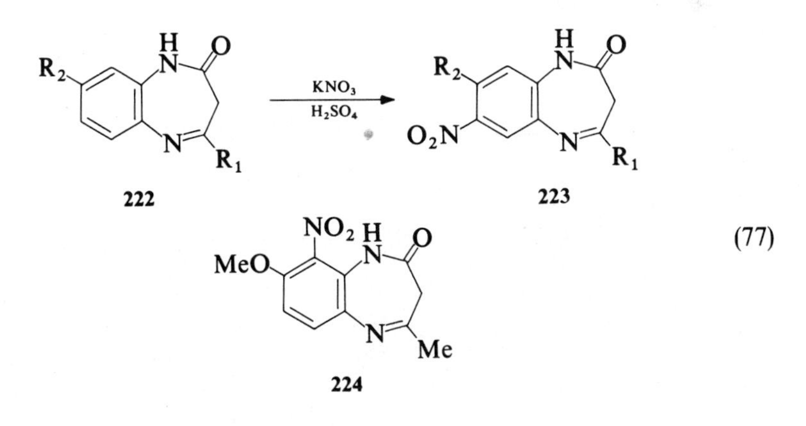


In the case of the 7-methyl-substituted benzodiazepin-2-ones **225** (R = Me, Ph, 4-$MeOC_6H_4$) the primary product of nitration was shown to be the 1-nitro compound **226**, which subsequently rearranged to the 9-nitro derivative **228** (Eq. 78). To a lesser extent, the 8-nitro analog **227** was also formed by a slower reaction pathway.[163]

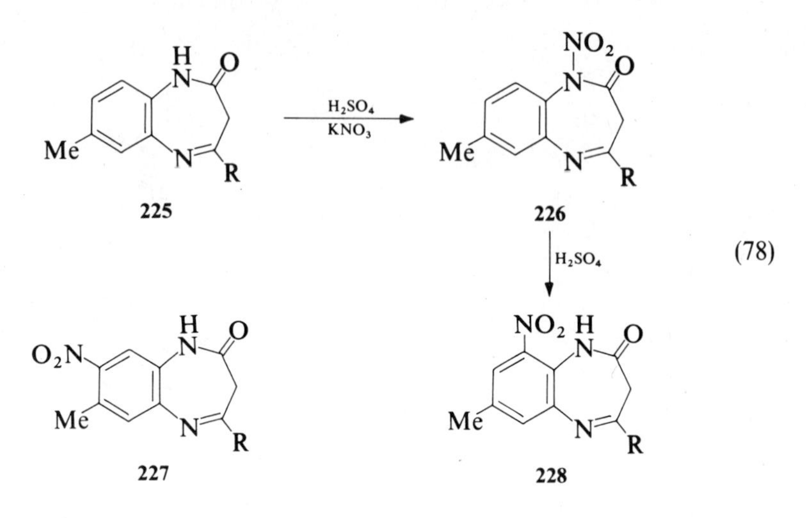


### 3.2.1.3. Nitrosation

According to Davoll,[124] the reaction of 4-methyl-1,3-dihydro-1,5-benzodiazepin-2(2*H*)-one with nitrous acid in 2 *N* hydrochloric acid led to an unstable nitroso compound, which was assigned structure **229** (Eq. 79).

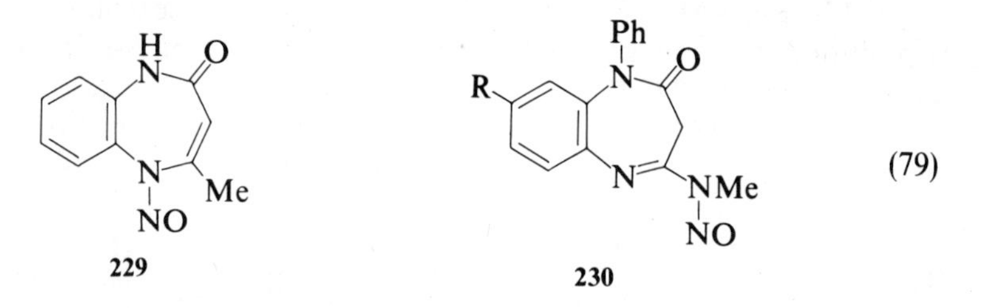


The nitrosoamidines **230** were obtained[158] by treatment of the corresponding amidines with sodium nitrite in glacial acetic acid.

### 3.2.1.4. Reaction with Other Nitrogen Electrophiles

Benzenediazonium chloride reacted with 4-methyl-1,3-dihydro-1,5-benzodiazepin-2(2*H*)-one in 2 *N* hydrochloric acid to give a red crystalline compound that was assigned the hydrazone structure **231**.[124]

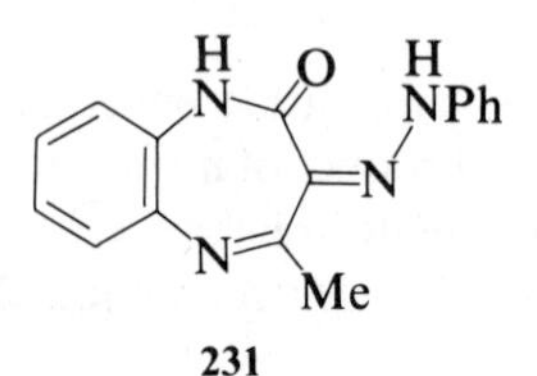

#### 3.2.1.5. Oxidation

The benzodiazepinone **191** was found to undergo air oxidation in boiling benzene to yield the benzoyl derivative **232** as the major product, in addition to a small amount of the dione **233** (Eq. 80).[149] Compound **232** was also isolated from the reaction of *o*-phenylenediamine with α,γ-diphenylacetoacetate in boiling xylene.

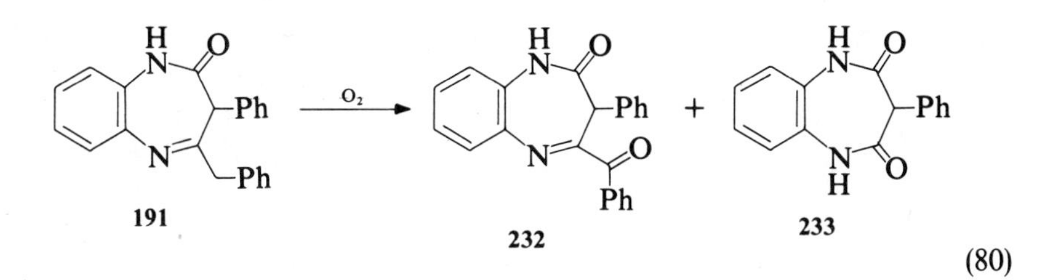


(80)

#### 3.2.1.6. Alkylation and Reaction with Aldehydes

The lactam nitrogen of the benzodiazepinones **234** could be alkylated by using an alkyl halide and base (e.g. methyl iodide and potassium carbonate,[126] 2-dimethylaminoethyl bromide and sodium amide,[165] or 2-chloromethyl-3,4-dihydroimidazole, and sodium amide[166]) to give in each instance the appropriate 1-substituted derivatives **235** (Eq. 81).

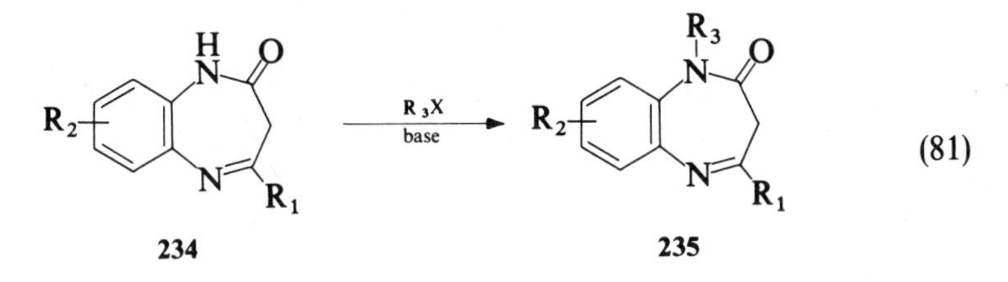


(81)

The alkylation of 4-amino-1-phenylbenzodiazepin-2-ones **236** (X = Cl, $NO_2$) was studied by Bauer and coworkers,[167] who showed that the course of the reaction was affected by the choice of solvent and reagent. Treatment of **236** (X = $NO_2$) with methyl iodide and sodium methoxide in dimethylacetamide gave a yield of the dimethylamino derivative **237** (X = $NO_2$, R = Me) that exceeded 80%, while the same reagent in tetrahydrofuran gave mainly the monomethylated product **237** (X = $NO_2$, R = H), together with the benzimidazolone **239** (X = $NO_2$). The latter compound was the major product if the methylation was carried out in methanol. This ring contraction was shown to proceed via the 2,4-dione and will be discussed under reactions of the 2,4-diones. Methylation with fluorosulfonic acid methyl ester, on the other hand, gave high yields of the 5-methylated imino compounds **238** (X = $NO_2$, Cl) (Eq. 82).[167]

Cyclization of the chloroethyl derivative **240** by an intramolecular alkylation using potassium carbonate in benzene–dimethylacetamide was reported[167] to

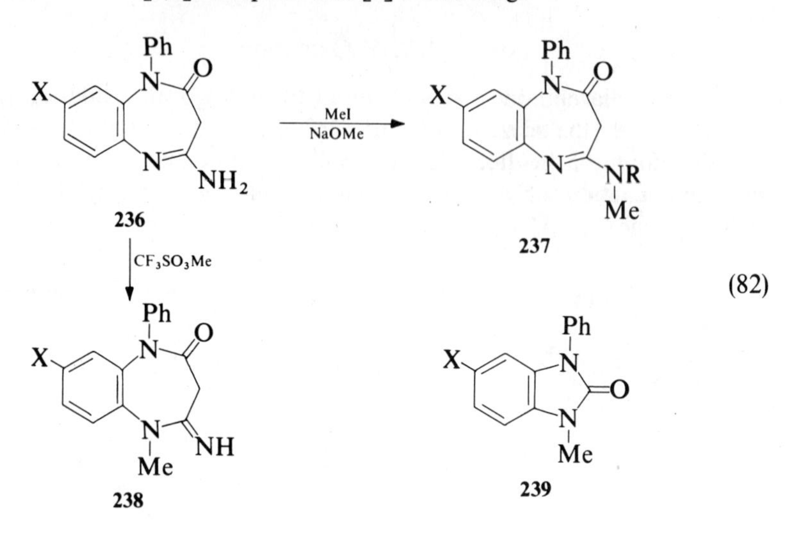


give the triazepinobenzodiazepine **241** and the imidazolidinone **242**, the latter being the major product (Eq. 83).

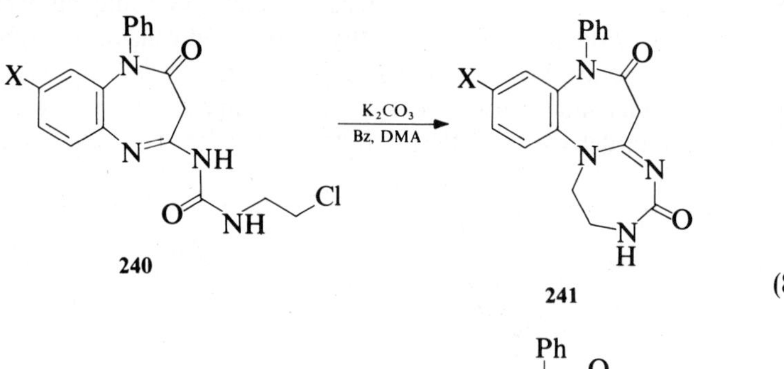


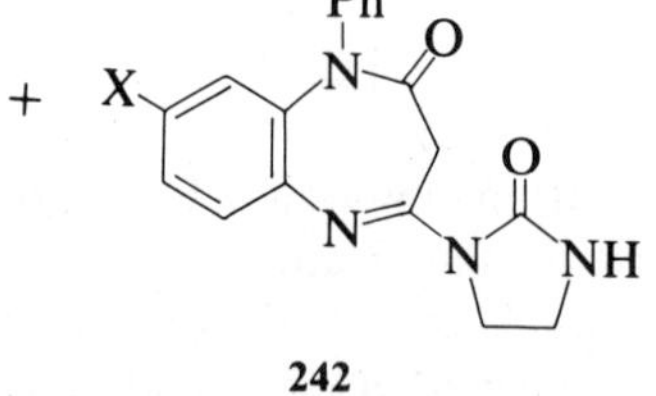


Condensation of 4-methyl-1,3-dihydro-1,5-benzodiazepin-2(2*H*)-one with benzaldehyde and piperidine in boiling benzene led to the styrene derivative **243** in 20% yield.[49]

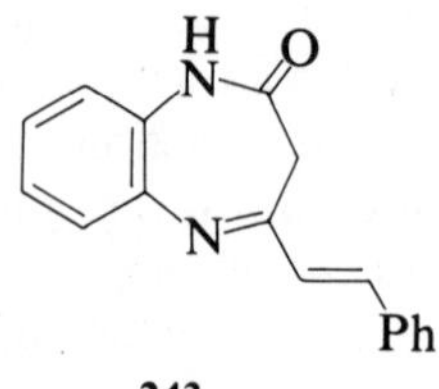

### 3.2.1.7. Acylation

It is surprising that no *N*-acylations of the benzodiazepin-2-ones **234** have been reported. According to the patent literature,[168] reaction of **234** ($R_1$ = Me) with substituted phenylisocyanates afforded the 3-carboxamides **244** in almost 90% yield (Eq. 84).

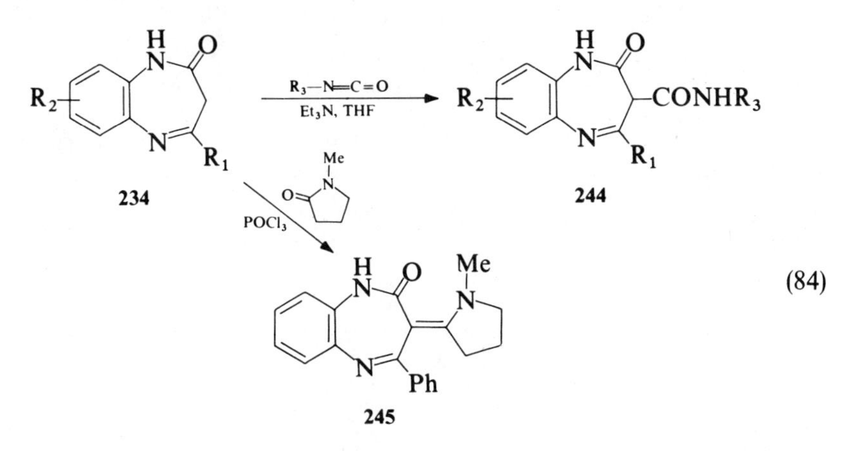


(84)

Acylation on the 3-position was also observed under Vilsmeier conditions. Thus, a mixture of *N*-methylpyrrolidinone and phosphorus oxychloride converted **234** ($R_1$ = Ph, $R_2$ = H) to the enamine **245**.[169] The stereochemistry of **245** is unknown and the structure lacks rigorous proof.

The 4-amino derivatives **246** ($R_1$ = H, Me) were converted to the mono-acylated products **247** by reaction with acid chlorides, chloroformates, or isocyanates (Eq. 85).[167,170] Under more vigorous conditions, the diacylated

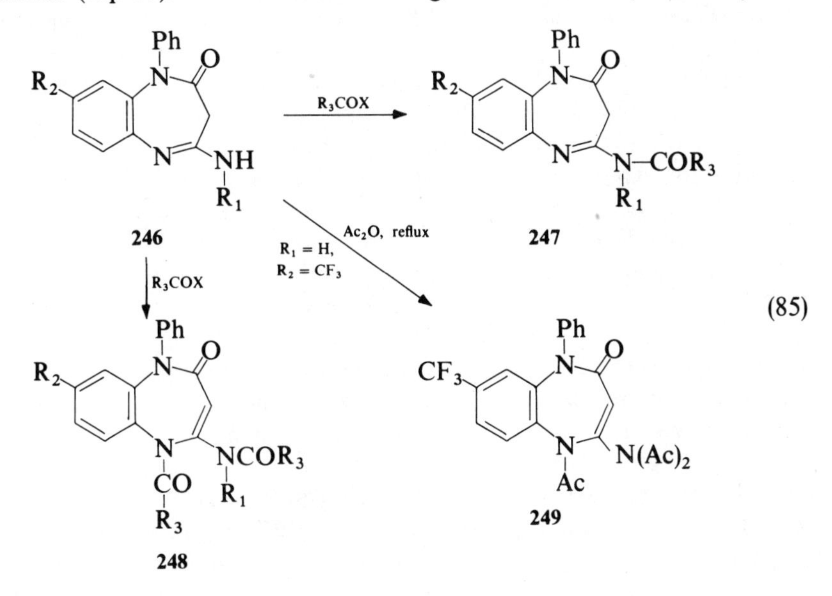


(85)

analogs **248** were obtained. Prolonged reflux of **246** ($R_1 = H$, $R_2 = CF_3$) in acetic anhydride gave the triacetyl derivative **249**.[167]

Acylation of the 4-hydrazinobenzodiazepin-2-one **250** with chloroacetyl chloride gave the chloroacetate **251** in high yield. This compound was converted to the triazolobenzodiazepine **252** by heating to 140°C in acetic acid (Eq. 86).[159]

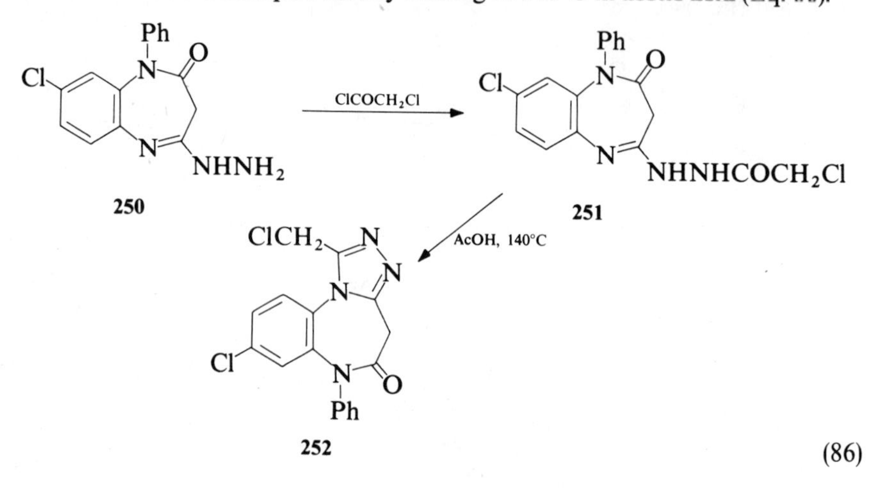


(86)

### 3.2.2. *Reactions with Nucleophiles*

#### 3.2.2.1. Hydrolysis

Acid hydrolysis of the benzodiazepin-2-ones **234** ($R_1 = Me$, $R_2 = H$) with dilute sulfuric acid at reflux gave 2-methylbenzimidazole and acetone (Eq. 87).[124] The nitro analogs **234** ($R_1 = Me$, $R_2 =$ 7- and 8-nitro) afforded the analogous reaction products,[140] although complete hydrolytic cleavage of **234** ($R_1 = Me$, $R_2 = 7\text{-}NO_2$) to 4-nitrophenylenediamine was also observed.[141] Sexton,[123] on the other hand, reported the formation of 2-hydroxybenzimidazole and acetone by heating of **234** ($R_1 = Me$, $R_2 = H$) in 20% aqueous sulfuric acid. The 4-trifluoromethyl derivative **234** ($R_1 = CF_3$, $R_2 = H$) was hydrolyzed in boiling 10% aqueous sulfuric acid to give a mixture of the benzimidazole **254** ($R_1 = CF_3$) and *o*-phenylenediamine, while treatment with 10% aqueous sodium hydroxide at reflux led to 2-methylbenzimidazole.[128] Phenacylbromide was isolated from the acid hydrolysis of the 3-bromo derivative **255** ($R_1 = Br$, $R_2 = Ph$) (Eq. 87), while *p*-bromoacetophenone was recovered from the hydrolysis of **254** ($R_1 = Br$, $R_2 = 4\text{-}BrC_6H_4$),[133] thus confirming the site of the attachment of the bromine. The cleavage of the acetyl group of **255** ($R_1 = Ac$, $R_2 = Ph$) by heating with hydrochloric acid in ethanol was mentioned above (Section 3.1.4).

Both acidic and basic hydrolyses are best rationalized by an initial cleavage of the imine bond followed by recyclization to the benzimidazole intermediate **253**. This intermediate may then either dehydrate to give **254** or fragment as indicated to form the 2-hydroxybenzimidazole. 2-Methylbenzimidazole would result from cleavage of the ketone **254** as shown in Eq. 87.

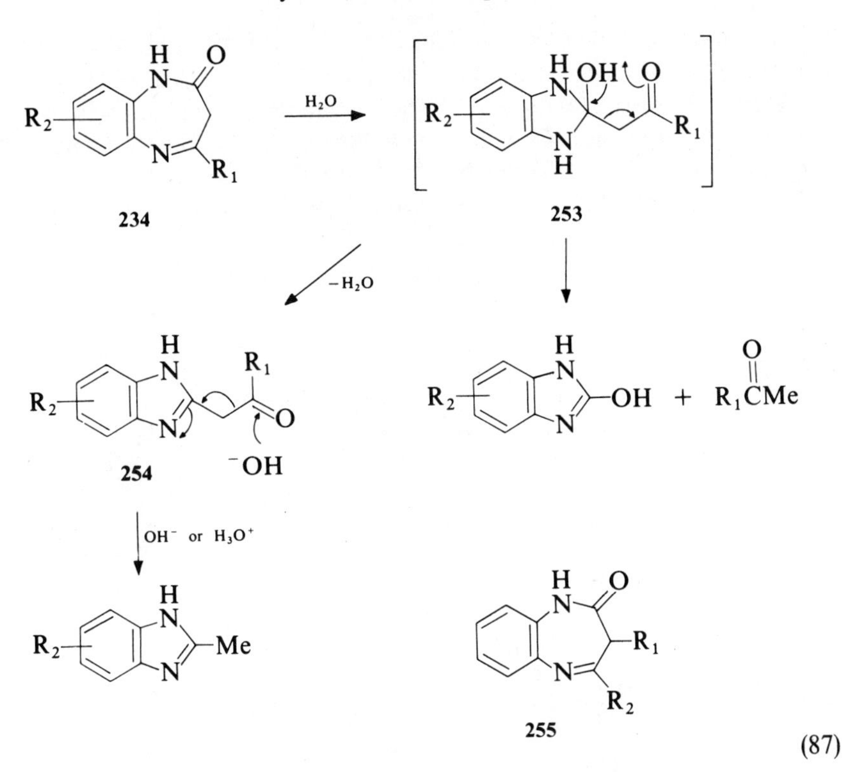


### 3.2.2.2. Reactions with Nitrogen Nucleophiles

Treatment of the 1,5-benzodiazepinone **234** ($R_1 = CF_3$, $R_2 = H$) with hydroxylamine in ethanol at room temperature was reported[128] to open the amide bond rather than the imine to yield the hydroxamic acid **256**, although the alternate structure **257** has not been ruled out (Eq. 88). Both compounds could undergo the observed recyclization to the benzodiazepine **234** ($R_1 = CF_3$, $R_2 = H$) upon treatment with acid.

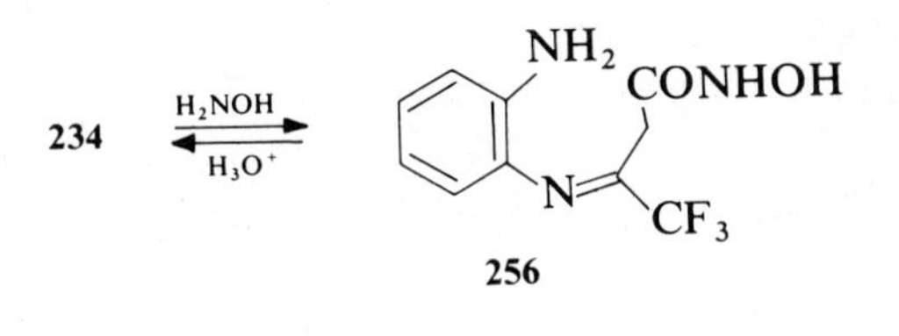


(88)

257

The 4-amino group of the benzodiazepin-2-ones **258** has been described to undergo acid-catalyzed exchange with acetals of aminoacetaldehyde to yield the substituted amidines **259** (Eq. 89).[153] These compounds could be cyclized to the imidazobenzodiazepines **261** by heating in boiling formic acid. Compounds **261** were also accessible by the reaction of **258** with α-bromoketones.[153] Amine exchange with propargylamine was accompanied by in situ cyclization to the 1-methylimidazobenzodiazepine **260**.[153]

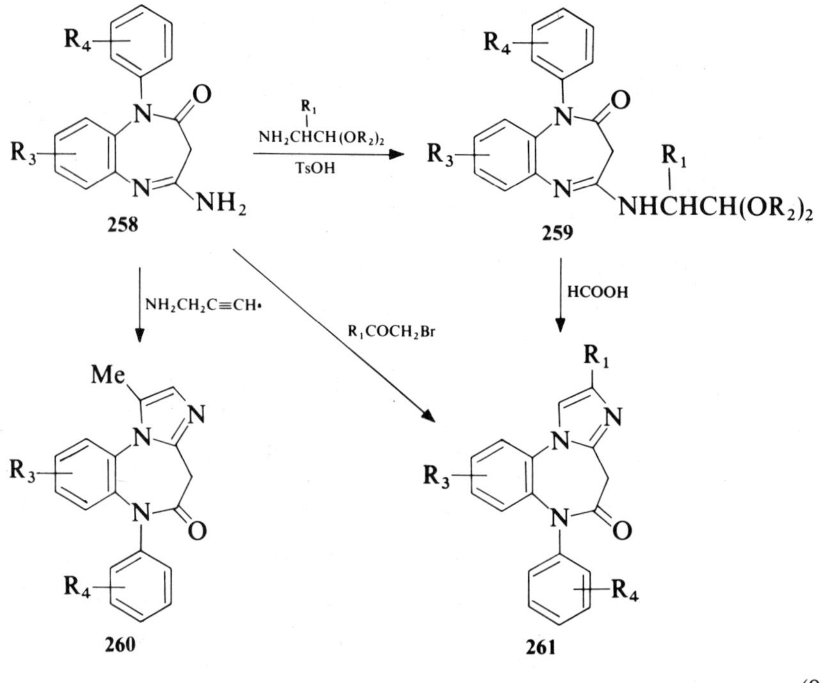


(89)

Displacement of a 4-alkoxy or 4-alkylthio group by amines[118] was mentioned during the discussion of the synthesis of 4-amino compounds in Section 3.1.5. Also discussed was the utilization of an *N*-nitrosomethylamine function as a leaving group to prepare 4-hydrazino derivatives.[158] Reaction of these nitrosoamidines **230** with acetylhydrazine and triethylamine in boiling butanol led to the formation of the triazolobenzodiazepines **262** (Eq. 90).[158]

#### 3.2.2.3. Reactions with Carbon Nucleophiles (Carbanions)

The nitrosoamidine **230** reacted with the carbanions of nitromethane and dimethyl malonate to give the 4-methylene derivatives **263** ($R_1 = NO_2$, $R_2 = H$, $R_1$, $R_2$ = COOMe).[171]

Condensation of **230** with the anion of ethyl isocyanoacetate gave the imidazobenzodiazepine **264** in one step.[172]

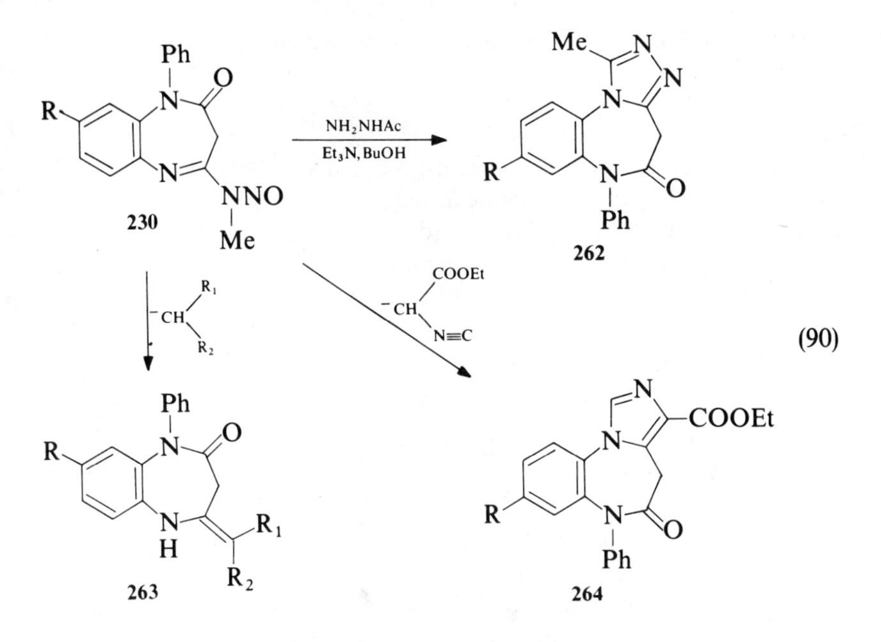


### 3.2.2.4. Reactions with Sulfur Nucleophiles

4-(Dialkylamino)-1-phenyl-1,3-dihydro-1,5-benzodiazepin-2-one (**265**), treated with phosphorus pentasulfide, afforded the corresponding benzodiazepin-2-thiones **266** (Eq. 91). A reaction of **266** with sodium hydride and methyl iodide gave the corresponding 2-methylthio derivatives **267**. Treatment of **267** with dialkylamines led to the formation of 2,4-bis-(dialkylamino)-1-phenyl-1$H$-1,5-

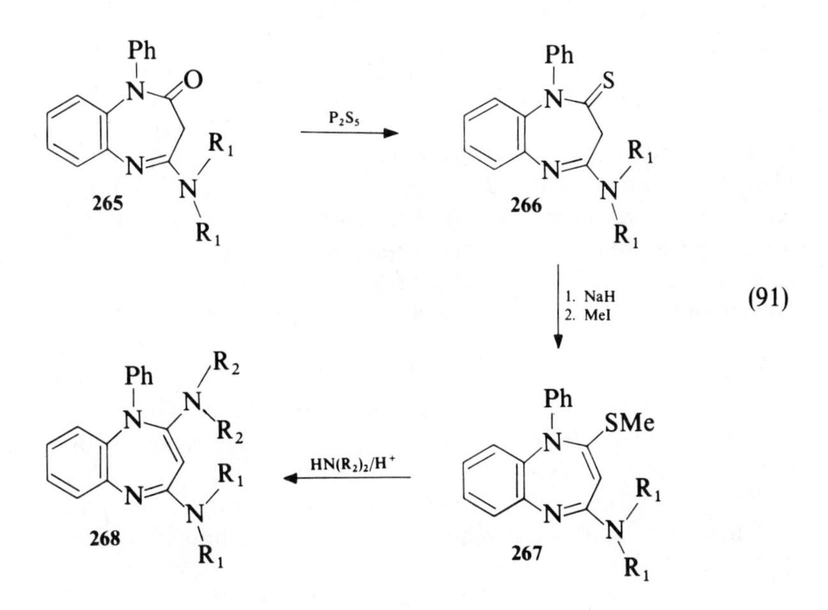

benzodiazepines **268**. The bulkiness of $R_2$ had a great effect on the yields of these compounds (**268**).

### 3.2.2.5. Other Reactions

The reaction of 4-(dialkylamino)-1,3-dihydro-2*H*-1,5-benzodiazepin-2-ones **269** with *N*,*N*-dimethylformamide in the presence of phosphorus pentachloride at room temperature produced 4-(dialkylamino)-3-[(dimethylamino)methylene]-1,3-dihydro-2*H*-1,5-benzodiazepin-2-one (**270**).[309] Compounds **270** (Eq. 92) are useful starting materials for the synthesis of tricyclic 1,5-benzodiazepine derivatives.

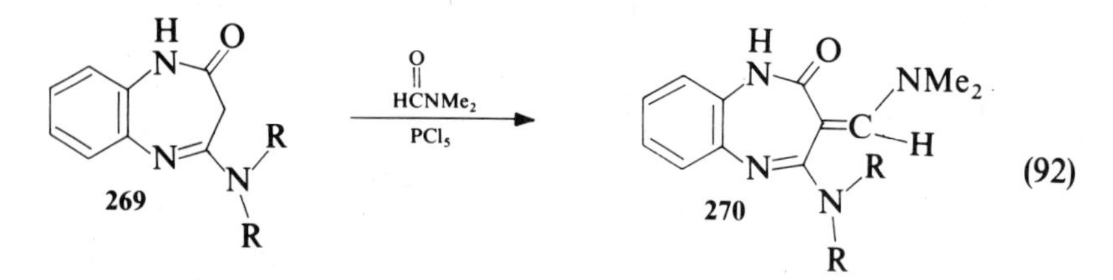


Reaction with hydrazine of **270**, which has the least steric hindrance of the 4-dialkylamino substituents (R = Me), afforded the pyrazolo[3,4-*b*] [1,5]benzodiazepine derivatives **271** (Eq. 93). However reaction of **270** with guanidine or amidines gave 5*H*-pyrimido-[4,5-*b*] [1,5]benzodiazepine derivatives **272**.

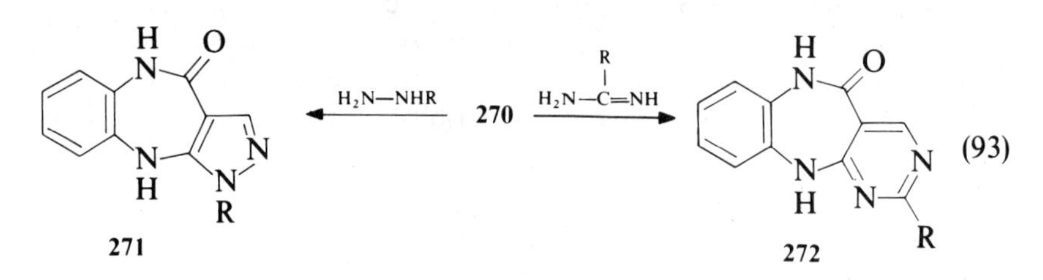


Treatment of **273** with a diazonium salt leads to **274**, azo coupling ocurring at the exocyclic methyl group. Tosyl azide, on the other hand, reacts at the 3-position to give compound **275**. Further reaction of **275** with benzoylphenylketene gives the diazooxazinobenzodiazepine **276** without loss of the diazo nitrogen. However. interaction of **275** with phenyl isocyanate, yields the pyrazolo[4,3-*b*] [1,5]benzodiazepine **277** (Eq. 94).[312]

### 3.2.2.6. Reduction

The reductions of the imine and the carbonyl functions of 1,3-dihydro-1,5-benzodiazepin-2(2*H*)-ones are discussed in Section 4, dealing with the synthesis of tetrahydrobenzodiazepines.

The nitro group in 1,3-dihydro-7-nitro-4-phenyl-1,5-benzodiazepin-2(2*H*)-one has been catalytically reduced to an amino group with Raney nickel and hydrogen.[150]

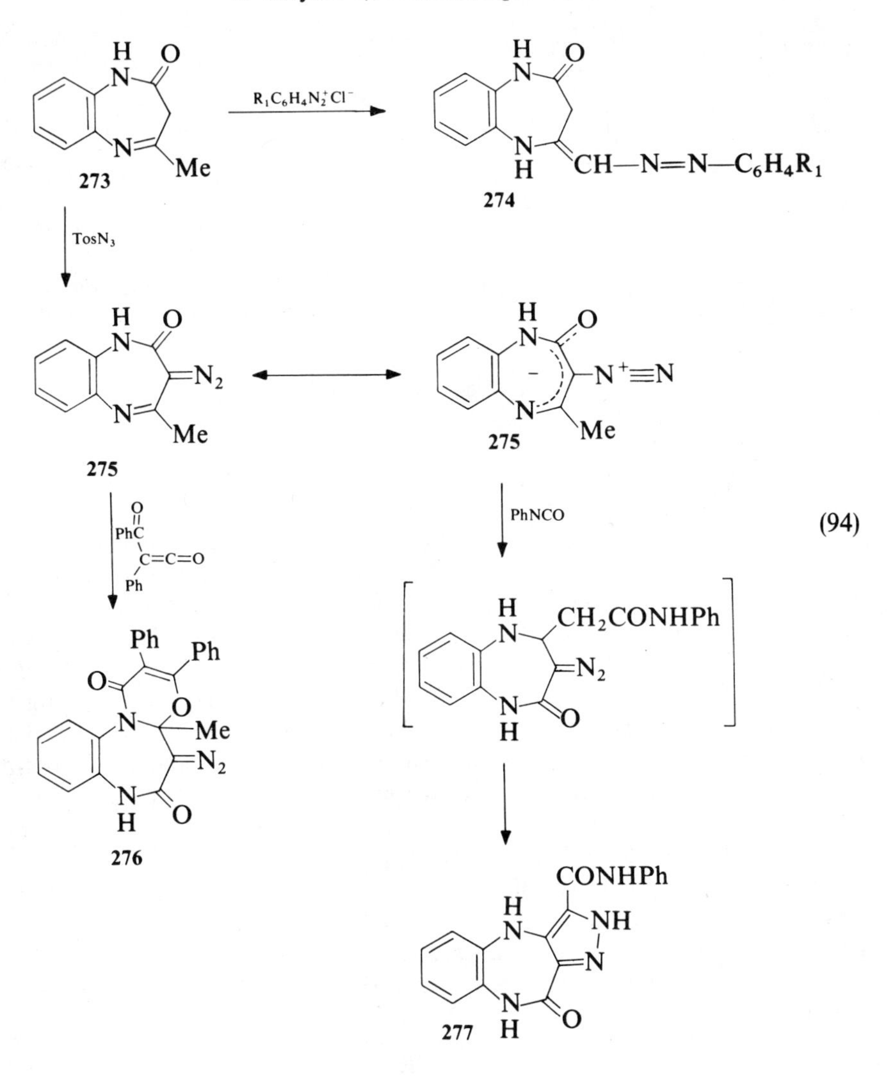


### *3.2.3. Pyrolysis*

Thermal ring contraction of 1,3-dihydro-1,5-benzodiazepin-2(3*H*)-ones **234** to the benzimidazoles **278** appears to be quite general and has been used as a diagnostic tool in establishing the location of the $R_2$ substituent.[135–137,140–144] As indicated below (Eq. 95), the rearrangement can be formulated as a transannular attack of the imine nitrogen on the amide carbonyl followed by cleavage of the 2,3-bond.

Compounds **234**, with substituents in the 3-position (e.g., 3-phenyl[127,149] and 3-acetyl[151] derivatives) have been reported to undergo the same ring contrac-

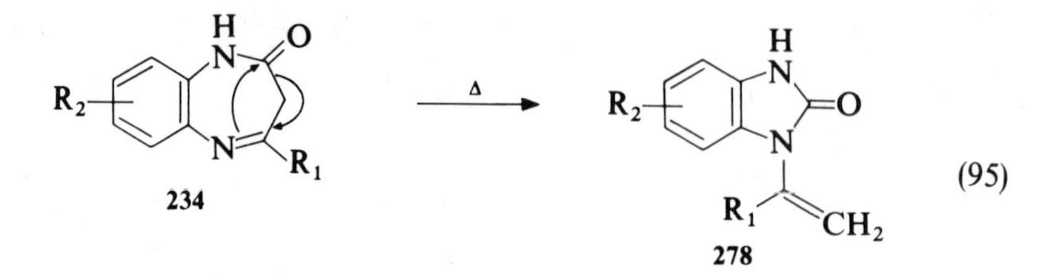


tion. The imidazolones **278** were also formed by heating the benzodiazepines in boiling 2-ethoxyethanol in the presence of sodium 2-ethoxyethoxide.[124,141]

### 3.3. Spectral Data

Nmr spectra of the benzodiazepinones **235** unsubstituted at position 1 ($R_3$ = H) exhibit a singlet for the two methylene protons at C-3. This is indicative of a rapid flipping of the seven-membered ring between two boat-type conformations. From a study of the signal broadening at low temperature, Mannschreck and coworkers[95] estimated the activation energy for the ring inversion to be about 9.5 kcal/mol for the 4-methyl derivative **235** ($R_1$ = Me, $R_2$=$R_3$ = H). Benassi and coworkers[173] reported a value of 10 kcal/mol for the 4-phenyl-substituted compound **235** ($R_1$ = Ph, $R_2$=$R_3$ = H). Ring flipping is somewhat inhibited by the introduction of a substituent at the 1-position, which gives rise to an AB system for the methylene protons with a coupling constant at room temperature of 12 Hz. The free energy of activation at 298°K was determined to be 15.6 kcal/mol for **235** ($R_1$ = Ph, $R_2$ = H, $R_3$ = Me) and 16 kcal/mol for **235** ($R_1$ = Ph, $R_2$ = H, $R_3$ = $CH_2Ph$).[173]

Low energy barriers to ring flipping are also observed with 4-amino derivatives **235** ($R_1$ = $NR_4R_5$, $R_2$ = 8-Cl, $NO_2$, $R_3$ = Ph), in which the methylene protons at C-3 appear as a broad singlet or broad AB system.[167] The C-3 protons readily undergo exchange under acid catalysis (DCl in DMSO).[167]

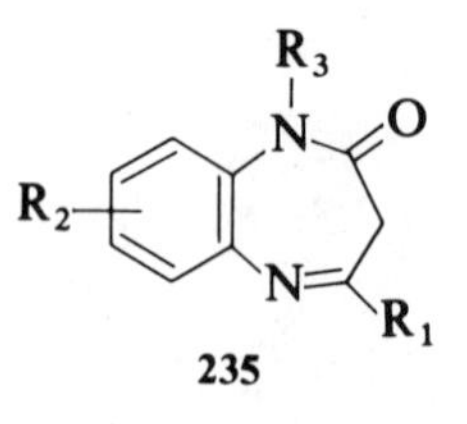


The mass spectra of compounds **235** ($R_1$ = Me; $R_2$ = 7- or 8-Br, Cl, Me, $NO_2$; $R_3$ = H) were reported.[174] The principal path of fragmentation under ion impact was rupture of the 1,2-bond leading to ejection of ketene. The 7-substituted isomers may be distinguished from the 8-substituted analogs on the basis of the relative intensities of the peaks corresponding to the molecular ions and its fragmentation species.[174]

# 4. TETRAHYDRO-1*H*-1,5-BENZODIAZEPINES

## 4.1. Synthesis

### 4.1.1. *By Alkylation of o-Phenylenediamine Derivatives*

The synthesis of the parent compound **281**, dates back to the nineteenth century. Hinsberg and Strupler[175] prepared the 1,5-dibenzenesulfonate **280** (R = Ph) by alkylation of **279** (R = Ph) with 1,3-dibromopropane (Eq. 96). Subsequent cleavage of the sulfonamides with acid led to **281**. The synthesis of **281** was eventually improved by Stetter,[176] who obtained the 1,5-ditosylamide **280** (R = 4-$MeC_6H_4$). Hydrolysis of this compound with 90% sulfuric acid for several days at room temperature led to the monotosylamide **282**, while **281** was obtained in 70% yield under more vigorous conditions. More recently,[177] this hydrolysis was reported to give **281** in 75% yield. This method was also applied to the synthesis of 1-phenyltetrahydro-1*H*-1,5-benzodiazepine[178] and 1-(2-methoxyphenyl)tetrahydro-1*H*-1,5-benzodiazepine.[179] Alkylation of **279** (R = 4-$MeC_6H_4$) with 1,3-dibromobutane gave the corresponding 2-methyl analog of **280**.[124] The utility of this synthetic path was enhanced by the finding[180] that the *N*-tosyl group can be conveniently removed photochemically in the presence of sodium borohydride.

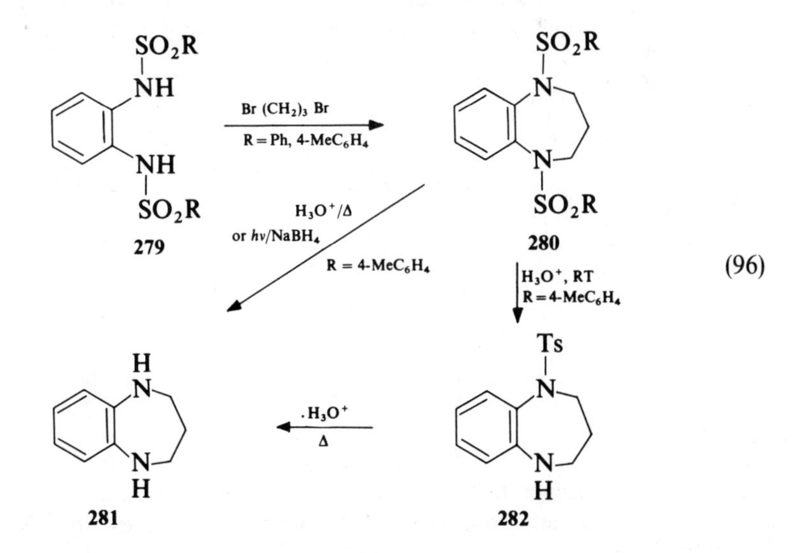


(96)

Alkylation of the ditosylamide **279** (R = 4-$MeC_6H_4$) with 2,3-dibromopropanol unexpectedly led to the 3-hydroxy-1,5-benzodiazepine **284**.[181] The same product was also obtained with epibromohydrin, indicating that the epoxide **283** is probably an intermediate in this synthesis (Eq. 97). Hydrolysis of the

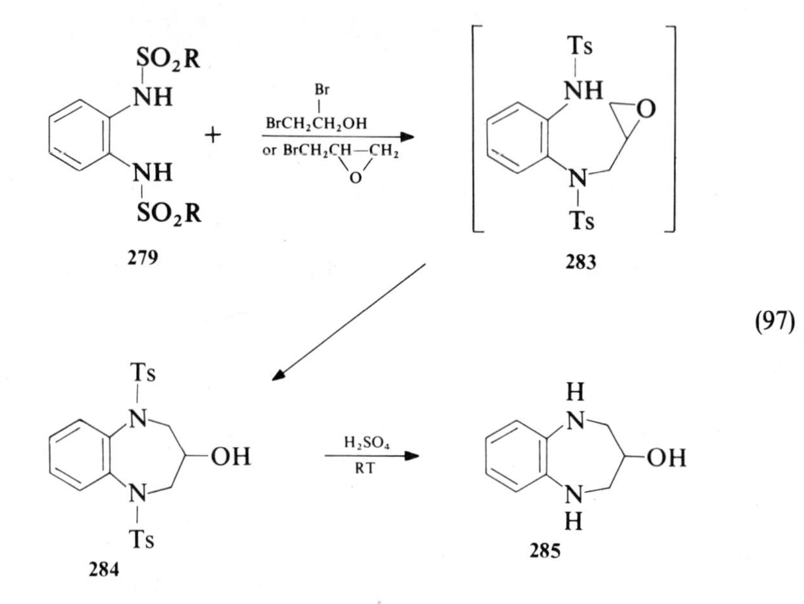


ditosylamide **284** was carried out with sulfuric acid at room temperature to give **285**.

Another example of diazepine formation by double alkylation was the reaction of *N*,*N*′-dimethyl *o*-phenylenediamine with 3-chloro-2-chloromethylpropene. Heating of these reactants under reflux in methanol in the presence of triethanolamine gave the benzodiazepine **286** in moderate yield (Eq. 98).[182]

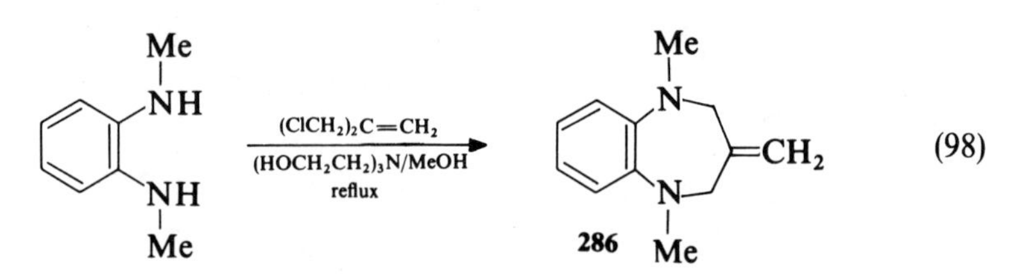


### 4.1.2. *By Aromatic Substitution or via Benzynes*

Reaction of 1,3-di(methylamino)propane with 1,3-dichloro- and 1,2,3- or 1,2,4-trichlorobenzenes in the presence of sodium amide has been shown[183] to lead to 1,5-dimethyltetrahydro-1*H*-1,5-benzodiazepines **289** (X = H, Cl) (Eq. 99).

Compound **289** (X = H) was formed in 40% yield from 1,3-dichlorobenzene. Both 1,2,3-trichlorobenzene and 1,2,4-trichlorobenzene gave mixtures of the 6-chloro and 7-chlorobenzodiazepines **289** (X = 6-Cl, 7-Cl), with the latter being the major component. The dichloroanilines **287** were isolated as intermediates

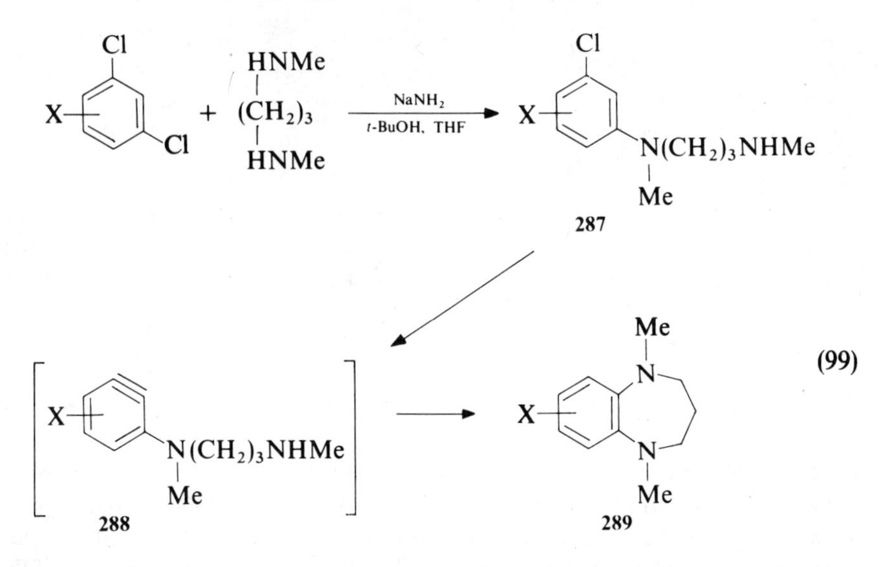


of the reaction of the trichlorobenzenes with the diamine. These intermediates are most likely formed by a nucleophilic substitution reaction. If the chloro substituent X in **287** were ortho to the aniline nitrogen, it would be converted to the benzodiazepine **289** (X = 6-Cl or 7-Cl) by another substitution reaction. In the case of **287** (X = H or Cl), the aryne **288** (X = H or Cl) is a likely intermediate (Eq. 99).

### *4.1.3. By Cyclodehydration*

This method has been used successfully for the synthesis of the 6-amino derivative **291**,[184] obtained in high yield by heating the alcohol **290** in boiling aqueous hydrogen bromide (Eq. 100). The compound was characterized as a dihydrochloride.

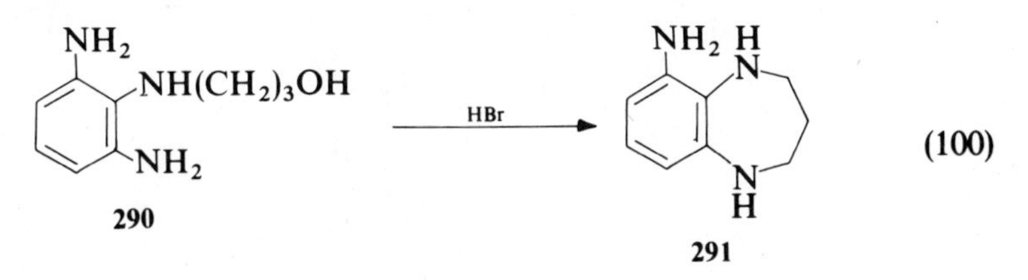


### *4.1.4. By Reduction of 1,5-Benzodiazepines and 2,3-Dihydro-1H-1,5-Benzodiazepines*

The most widely applied method for the synthesis of tetrahydrobenzodiazepines has been the reduction of the imine double bonds in **292** or **293** (Eq. 101).

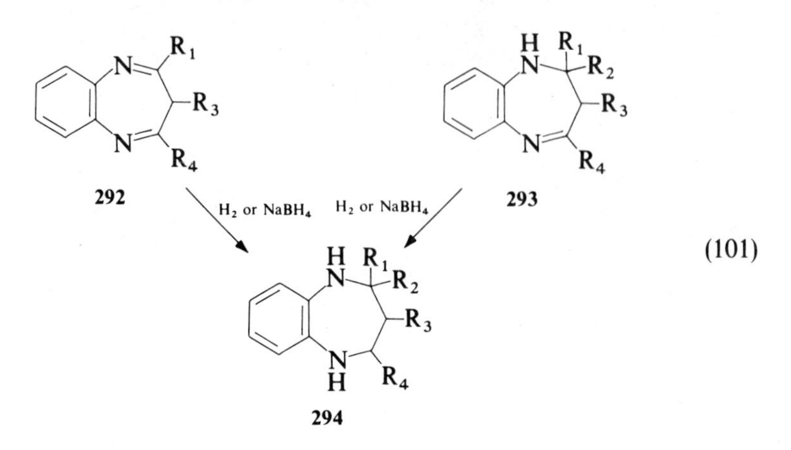


(101)

Reduction was achieved either by catalytic hydrogenation[98,107,111,113] or by reaction with sodium borohydride.[111,112,116,117,185,186] While catalytic hydrogenation led to mixtures of cis and trans isomers,[107] reduction with sodium borohydride was more stereoselective, leading to the cis isomer in the case of the 2,4-dimethyl derivative **294** ($R_1$ = Me; $R_2$, $R_3$ = H; $R_4$ = Me).[117,185,186]

### *4.1.5. By Reduction of Benzodiazepinones*

Several syntheses of tetrahydrobenzodiazepines **297** by lithium aluminum hydride reduction of dihydrobenzodiazepinones **295** or tetrahydrobenzodiazepinones **296** have been reported (Eq. 102).[107,111,117,124,187]

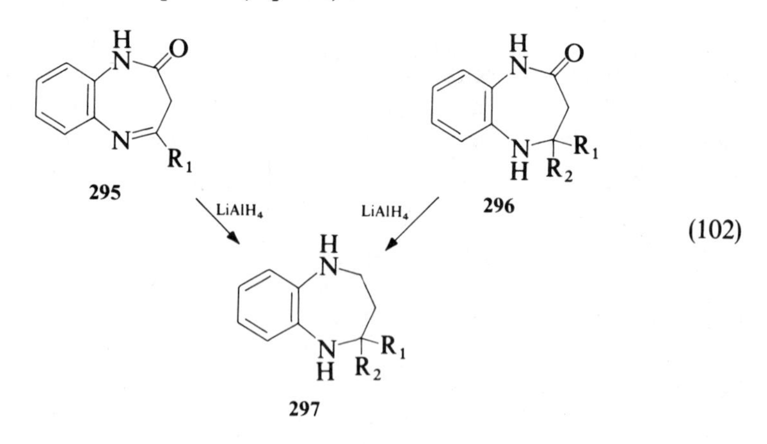


(102)

### *4.1.6. By Addition of Hydrogen Cyanide to Imine Bonds*

Hydrogen cyanide adds to the imine double bonds of 1,5-benzodiazepines and 2,3-dihydro-1*H*-1,5-benzodiazepines to form the 2,4-dicyano derivatives

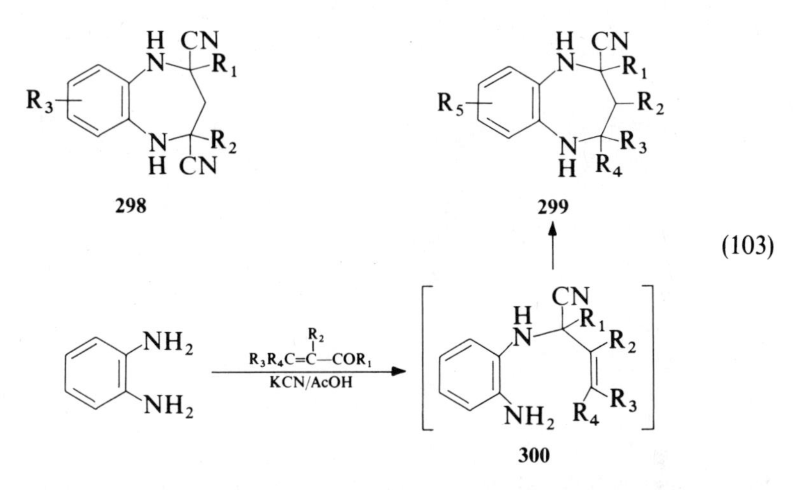


**298**[119] and 2-cyano derivatives **299**[119] (Eq. 103). Compounds **299** were also formed by condensation of *o*-phenylenediamine with $\alpha,\beta$-unsaturated carbonyl compounds in the presence of hydrogen cyanide.[99,188] When $R_1 = R_2 = R_3 = H$ and $R_4$ = Ph or 2-furyl, the intermediate adducts **300** can be isolated.[188,189] The stereochemistry of the hydrogen cyanide adducts **298** and **299** has not been established.

Corresponding acid derivatives **302** could be obtained from **301** in very low yields by alkaline hydrolysis (Eq. 104). Esters **303** were also prepared from **301** by alcoholysis.[320]

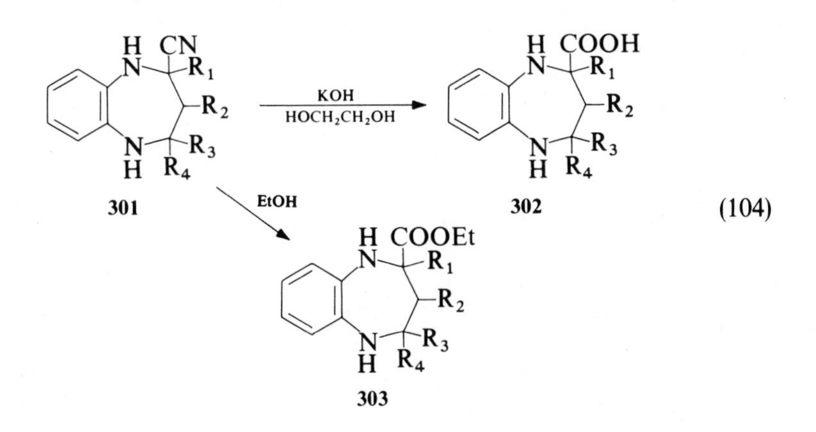


### 4.1.7. *Other Syntheses*

Reductive desulfurization of the diazepinophenothiazine **304** with Raney nickel afforded the 1-phenyltetrahydro-1,5-benzodiazepine **306** (R = Ph) in 62% yield (Eq. 105).[178] This method was later extended[190] to the diazepinodi-

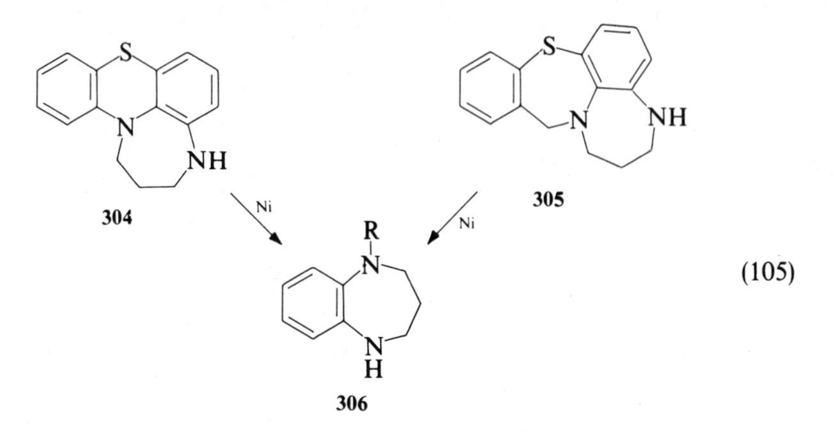


(105)

benzthiazepine **305**, which was desulfurized to the 1-benzyltetrahydro-1,5-benzodiazepine **306** (R = $CH_2Ph$).

An unusual reductive cleavage of the phenoxazine **307** (R = Ts or H) upon irradiation in presence of sodium borohydride was reported[179] to form a 55% yield of the 1-(2-hydroxyphenyl)tetrahydro-1,5-benzodiazepine **308** (Eq. 106).

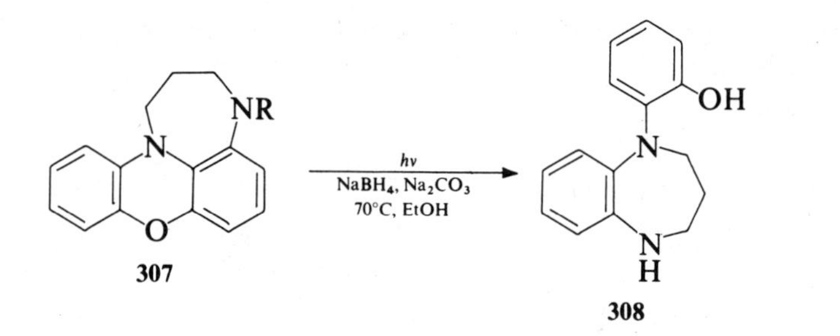


(106)

## 4.2. Reactions

### 4.2.1. *Reactions with Electrophiles*

#### 4.2.1.1. Nitrosation

The formation of a 1,5-dinitroso derivative (**309**) upon treatment of the parent tetrahydro-1*H*-1,5-benzodiazepine **281** with nitrous acid was first observed by Hinsberg and Strupler.[175] The preparation of 1,5-dinitroso compounds was also described for the 3-hydroxy analog[181] **285**, and also for the 2,2,4-trimethyl[98] and 2,4-dicyano-2,4-dimethyl analogs.[119]

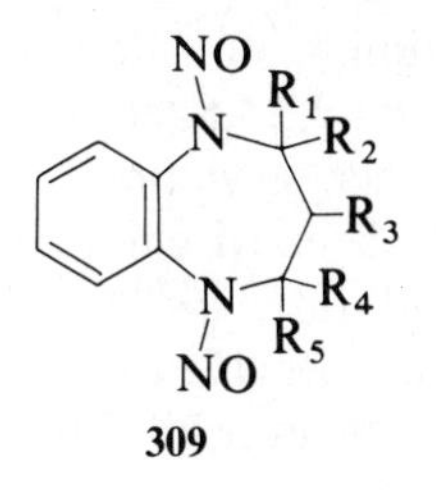


### 4.2.1.2. Oxidation

Oxidation of the 3-hydroxytetrahydrobenzodiazepine **284** with chromium trioxide and sulfuric acid was reported to give the corresponding ketone **310** (Eq. 107).[181]

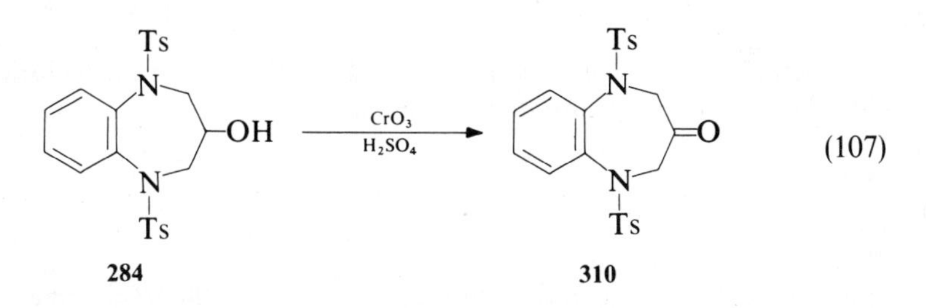


### 4.2.1.3. Alkylation

The monotosylamide **282** was benzylated using sodium hydride and benzyl bromide.[190] Phenylation of the same compound with iodobenzene and potassium carbonate in the presence of copper afforded 1-phenyl-5-tosyl-2,3,4,5-tetrahydro-1*H*-1,5-benzodiazepine.[178] Reaction of **311** (R = Me) or the corresponding phenol (**311**, R = H) with methyl iodide in ethanolic potassium hydroxide led to the same 5-methyl derivative **312** (Eq. 108).

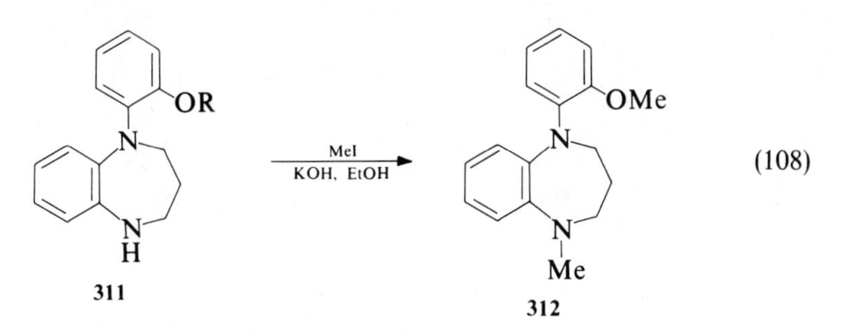


### 4.2.1.4. Acylation

Reaction of 2,3,4,5-tetrahydro-1*H*-1,5-benzodiazepine with *p*-toluenesulfonyl chloride has been reported to give either a monotosylamide[178] or a 1,5-

ditosylamide.[124] *N*-Tosylation of 2-methyl-2,3,4,5-tetrahydro-1*H*-1,5-benzodiazepine[124] and 1-benzyl-2,3,4,5-tetrahydro-1*H*-1,5-benzodiazepine[190] has also been described. Other acylations have been carried out with 3-bromopropionyl chloride,[191] benzoyl chloride,[177] acetic anhydride,[181] and (2-naphthyloxy)thiocarbonyl chloride.[192,308]

Treatment of the 6-amino derivative **291** with boiling formic acid yielded the tricyclic compound **313** in 60% yield (Eq. 109).[184]

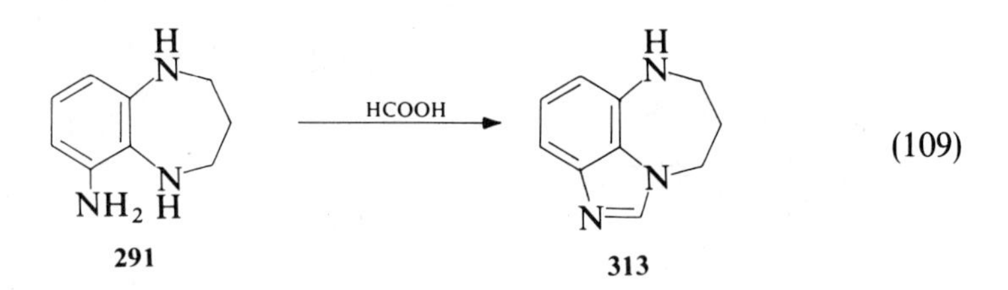


Reaction of the 3-hydroxy derivative **284** with either mesyl chloride or tosyl chloride gave the appropriate sulfonamides.[181]

### 4.2.2. Reactions with Nucleophiles

Hydrolytic cleavage of the sulfonamides **280** and **284** to give the tetrahydrobenzodiazepine **281** and the 3-hydroxy derivative **285** was mentioned above.

The conversion of 2,4-dicyano-2,4-dimethyl-2,3,4,5-tetrahydro-1*H*-1,5-benzodiazepine (**147**) to the thiocarboxamide **148**[119] by reaction with ammonium hydrosulfide has also been described (Eq. 53).

Selective removal of the benzoyl group in compound **314** was achieved with lithium aluminum hydride in ether (Eq. 110).[177] Sodium borohydride in pyridine effected the same transformation, but in poor yield.

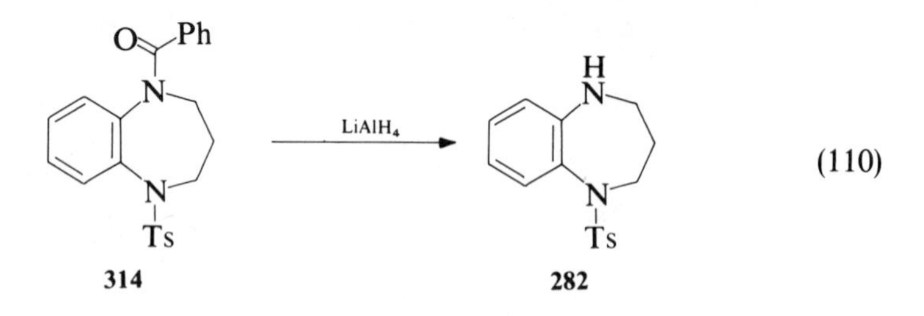


### 4.2.3. Photoreactions

Photolysis of 1-tosyl-2,3,4,5-tetrahydro-1*H*-1,5-benzodiazepine (**282**) in 70% ethanol containing sodium borohydride and sodium carbonate was reported to give the parent tetrahydrobenzodiazepine **281** in quantitative yield.[180] Irradiation in benzene solution gave only a 12% yield of **281** and a small amount of a compound assigned structure **315**. Using the conditions above **280** (R = 4-Me-$C_6H_4$) and **316** were cleaved to **281** and **311** (R = Me).[179]

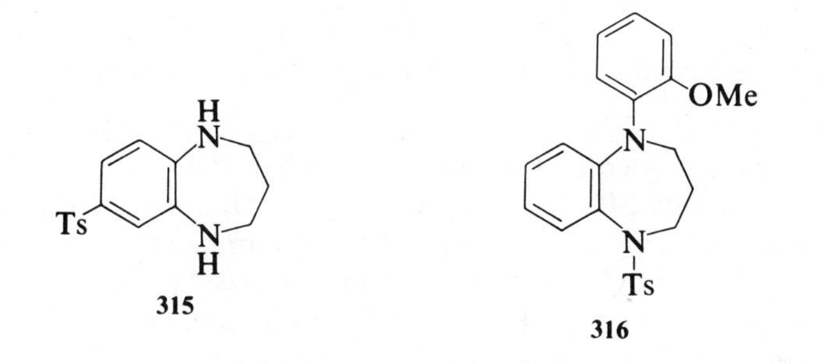


## 4.3. Spectral Data

Proton-nmr spectra were employed to assign the cis configuration to tetrahydrobenzodiazepines (**317**) obtained by reduction of benzodiazepinium salts with sodium borohydride.[117,185,186] A detailed analysis of the nmr spectrum of **317** ($R_1 = R_2$ = Me) indicated the preference of conformation **318**.[185]

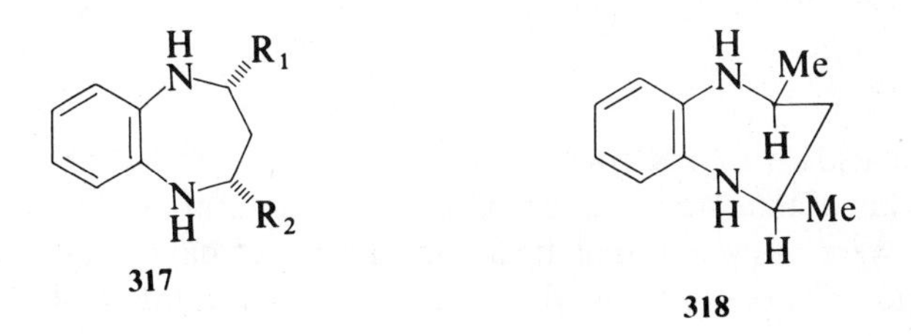


Variable-temperature nmr studies indicate[122] that the unsubstituted tetrahydrobenzodiazepine and the 2,2-dimethyl derivative undergo rapid conformational changes at room temperature. The 2-methyl, 2,4-dimethyl, and 2,2,4-trimethyl derivatives, on the other hand, exist in stable conformations that do not undergo interconversion at temperatures up to 140°C.

The mass spectra of several tetrahydrobenzodiazepines were recorded, and the nature of the fragmentation ions has been elucidated.[117]

# 5. TETRAHYDRO-1,5-BENZODIAZEPINONES

## 5.1. 1,3,4,5-Tetrahydro-1,5-benzodiazepin-2(2*H*)-ones

### 5.1.1. *Synthesis*

#### 5.1.1.1. *o*-Phenylenediamine and α,β-Unsaturated Carboxylic Acids

The first synthesis of this type, the condensation of *o*-phenylenediamine with acrylic acid, was reported by Bachman and Heisey.[193] These authors obtained the parent compound **319** ($R_1=R_2=R_3 = R_4 =$ H) in 67% yield by heating the diamine and 60% aqueous acrylic acid in concentrated hydrochloric acid on the steam bath for 3 hours (Eq. 111). Ried and Urlass[194] obtained the 4-methyl analog **319** ($R_2 =$ Me, $R_1=R_3=R_4 =$ H) by fusion of the diamine with crotonic acid. The same compound was later prepared[195–197] using the conditions of Bachman and Heisey.[193] 4,4-Dimethyl derivatives **319** ($R_1 =$ H, $R_2=R_3 =$ Me, $R_4 =$ H, Cl, Me) were similarly synthesized by condensation of the appropriate diamine with 3-methylcrotonic acid.[198] Condensation of 2-aminodiphenylamine with acrylic acid afforded the 1-phenyl-benzodiazepinone **319** ($R_1 =$ Ph, $R_2=R_3=R_4 =$ H) in low yield.[199] Dandegaonker and Desai[200] prepared a series of 4-phenyl derivatives **319** ($R_2 =$ substituted Ph, $R_1=R_3=R_4 =$ H) by fusion of the diamine with cinnamic acids.

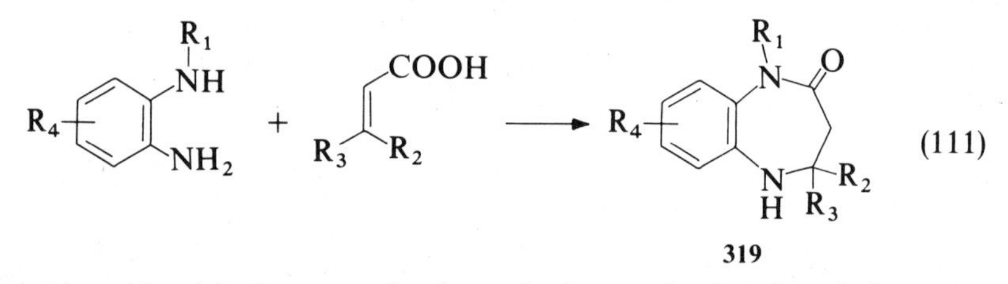


Cinnamic acids with electron-releasing substituents in the phenyl ring gave excellent yields of the condensation products. Structural assignments for these compounds were based on analytical and ultraviolet data only. According to Solomko and coworkers,[201] acrylic acid may be advantageously replaced by acrylamide.

#### 5.1.1.2. *o*-Phenylenediamine and β-Halocarboxylic Acids

β-Halocarboxylic acids, which are well-known substitutes for α,β-unsaturated carboxylic acids, have also been used for the preparation of the benzodiazepinones **319**. Thus, reaction of *o*-phenylenediamine with 3-bromobutyric acid led to the formation of **319** ($R_1=R_2=R_4 =$ H, $R_3 =$ Me) in reasonable yield.[109]

While 2,2-dialkyl-3-halopropionic acids **320** cannot be converted to α,β-unsaturated acids, they could still be utilized for the preparation of the benzodiazepines **321**, although the yields obtained were low (Eq. 112).[202,203]

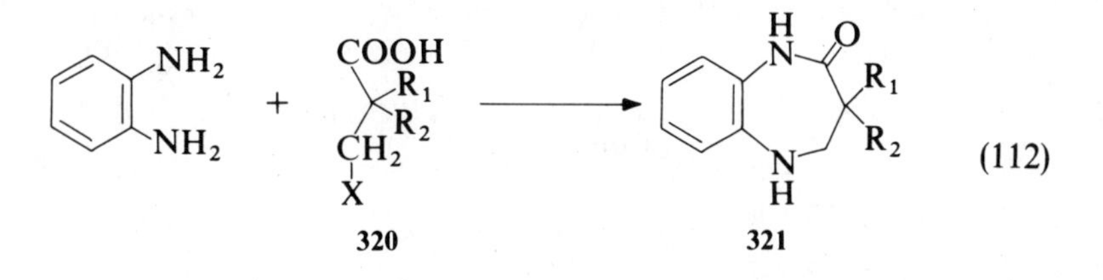


### 5.1.1.3. By Ring Closure of *o*-Phenylenediamine Derivatives

3-(2-Aminoanilino)propionic acids **323** ($R_2$ = OH), prepared by reduction of the corresponding nitro derivatives **322**, were cyclized in high yield to the benzodiazepinones **324** either by heating in boiling acetic acid in the presence of *p*-toluenesulfonic acid[204] or by heating with polyphosphoric acid (Eq. 113).[205,206] Reduction and ring closure could be carried out in one step by means of zinc and phosphoric acid in dioxane.[204] Dicyclohexylcarbodiimide has been used for a similar ring closure.[203]

The cyclization or reductive ring closure of esters and amides of the propionic acids **322** and **323** ($R_2$ = MeO or $NH_2$) has been shown to give the benzodiazepinones **324** as well.[201] Even the hydrazide **323** ($R_1 = R_3$ = H, $R_2$ = $NHNH_2$) was found to undergo ring closure to **324** ($R_1 = R_3$ = H).[206]

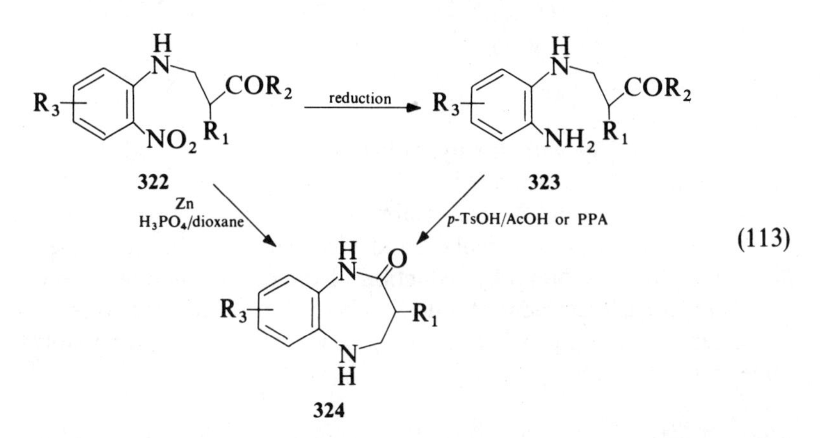


The 1-phenyl-substituted compounds **328** were obtained by cyclization of the propionic acids **325** using either thionyl chloride or acetic anhydride (for $R_1$ = Me) (Eq. 114).[207] Other ring closures have been effected by treatment of the 3-bromopropionanilides **326** with sodium bicarbonate and sodium iodide in boiling methanol.[207] In addition, compounds **328** ($R_3$ = H) were obtained in good yield by treatment of the 3-chloropropionanilides **327** (X = Cl) with potassium carbonate in dimethylformamide[199,208,209] or with sodium amide in liquid ammonia.[199,203,208,323] The conversion of **327** to **328** requires an acyl migration that can be rationalized by the formation of an intermediate $\beta$-lactam such as **329**. Further intramolecular attack as illustrated would yield the benzodiazepinone **330** (Eq. 115).

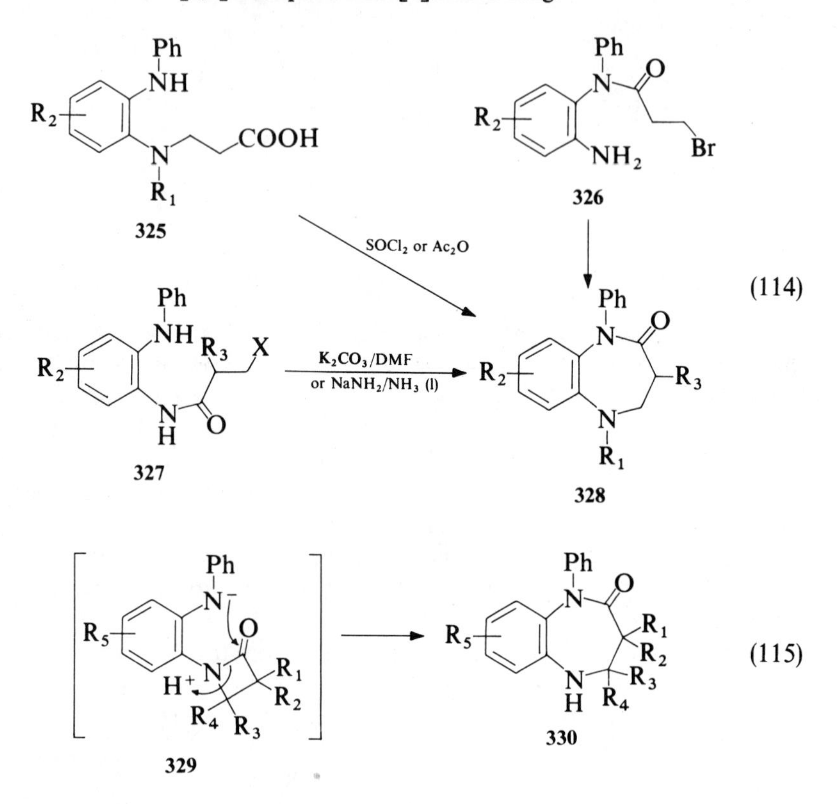


Compounds **330** with methyl substituents in the 3- and 4-positions were synthesized by this method.[210]

The possibility that $\beta$-lactams are intermediates in this reaction was supported by earlier work of Nicolaus and coworkers,[203] who were able to prepare $\beta$-lactams **331** ($X = NH_2$) by reduction of the corresponding nitro compound **331** ($X = NO_2$). Treatment of **331** ($X = NH_2$) with acid effected rearrangement to the benzodiazepinones **332** (Eq. 116). Base treatment might possibly effect the same transformation.

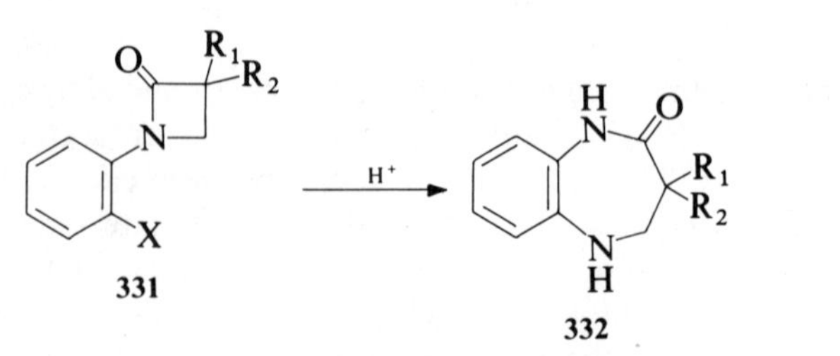


### 5.1.1.4. By Ring Expansion (Schmidt Reaction)

Misiti and coworkers[187] have carried out the Schmidt reaction on 1,2,3,4-tetrahydroquinolin-4-ones **(333)** and obtained various ratios of 1,3,4,5-

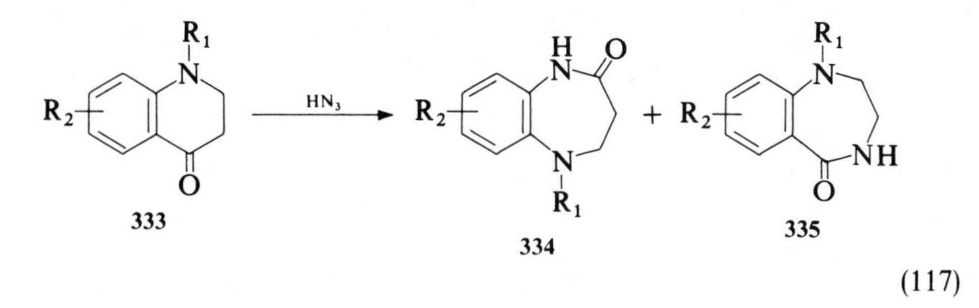


(117)

tetrahydro-1,5-benzodiazepin-2(2*H*)-ones (**334**) and 1,2,3,4-tetrahydro-1,4-benzodiazepin-5(5*H*)-ones (**335**) (Eq. 117). These authors showed that earlier investigators[211] were incorrect in their structural assignments. The best yields of 1,5-benzodiazepin-2-ones **334** were obtained by using 1-acetyl derivatives (**333**, $R_1$ = Ac) while 1-methyl- or 1-phenyltetrahydroquinolin-4-ones led mainly to the 1,4-benzodiazepin-5-ones **335**.

#### 5.1.1.5. By Reduction of 1,3-Dihydro-1,5-benzodiazepin-2(2*H*)-ones

Reduction of the imine double bond in 1,3-dihydro-1,5-benzodiazepin-2(2*H*)-ones (**336**) has been widely utilized for the preparation of the tetrahydro derivatives **337** (Eq. 118). Catalytic hydrogenation over Raney nickel was employed to prepare **337** ($R_1 = R_3$ = H, $R_2$ = Me;[146] $R_1$ = H, $R_2$ = Me, $R_3$ = 8-Cl;[106] $R_1$ = H, $R_2$ = Me, $R_3$ = 8-Me;[137,138] $R_1$ = H, $R_2$ = Me, $R_3$ = 8-Br;[144] and $R_1 = R_3$ = H, $R_2$ = Ph or 2-furyl).[129] The same reduction was also achieved by using palladium as the catalyst.[125,127,136,141,149,212] Under these conditions a nitro group in the 7-position was simultaneously reduced to the amino group to give **337** ($R_1$ = H, $R_2$ = Me, $R_3$ = 7-$NH_2$).[141] It was possible to reduce the imine double bond and retain the nitro function by using sodium borohydride, as shown by the preparation of **337** ($R_1$ = H, $R_2$ = Me, $R_3$ = 8-$NO_2$).[139] When compounds of type **336** are synthesized by the reductive cyclization of a nitro ketone, excessive reduction may lead directly to the tetrahydrobenzodiazepinone **337**.[134,149]

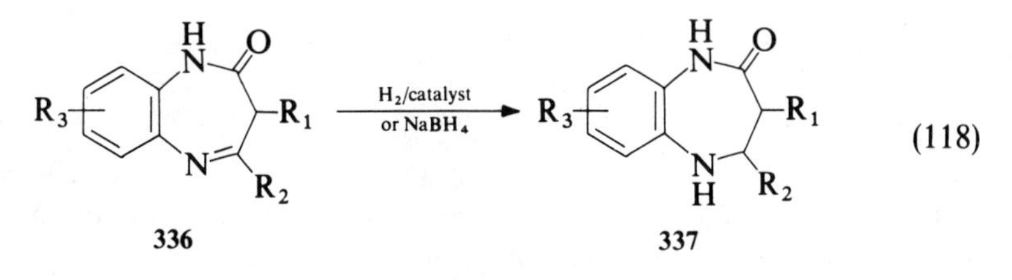


(118)

#### 5.1.1.6. Other Syntheses

4-(Alkoxycarbonyl)methylene-substituted 1,5-benzodiazepin-2-ones (**339**) were obtained by Merz and coworkers[213,214] by condensation of *o*-phenylenediamine with dialkyl acetone dicarboxylates or their equivalents, the piperidin-4-one derivatives **338**, ($R_1$ = Me, Et; $R_2$ = Ph, 2-pyridyl, 3-pyridyl, 2-quinolinyl; $R_3$ = H, Me, $PhCH_2$) in boiling xylene (Eq. 119).

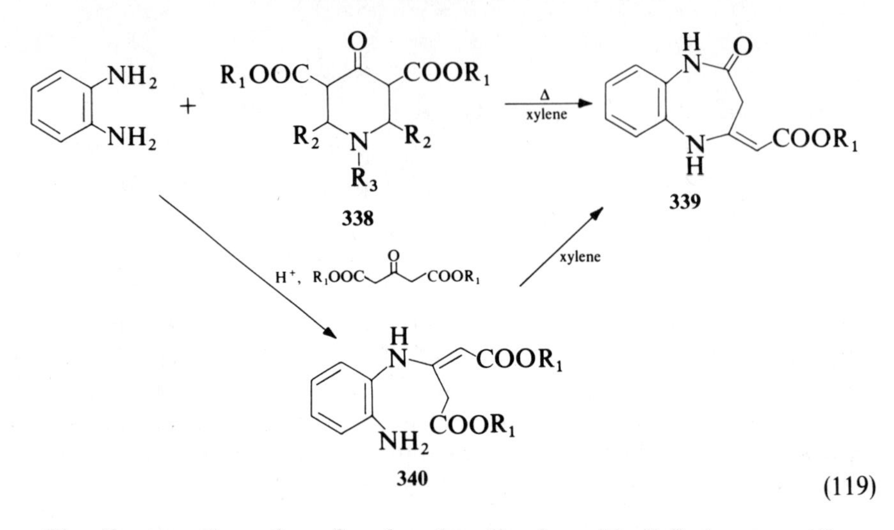


(119)

The direct condensation of *o*-phenylenediamine with dialkyl acetone dicarboxylates and the cyclization of the intermediate enamine **340** in boiling xylene gave superior yields. A different approach was used for the synthesis of the 1-phenyl analog **342**, which was prepared by hydrolysis and decarboxylation of the malonylidene derivative (Eq. 120).[171] The formation of **341** was mentioned in Section 3.2.2.3.

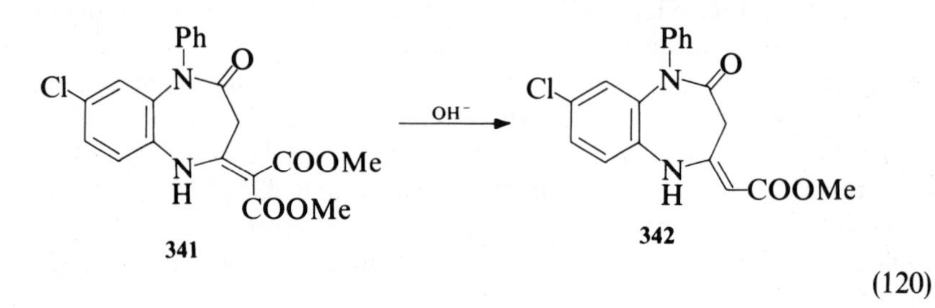


(120)

## *5.1.2. Reactions*

### 5.1.2.1. Reactions with Electrophiles

A. Halogenation. Bromination of the 7,8-dichloro derivative **343** using either bromine in chloroform or a mixture of *N*-bromosuccinimide and benzoyl peroxide yielded the 6-bromo compound **344** (Eq. 121).[106]

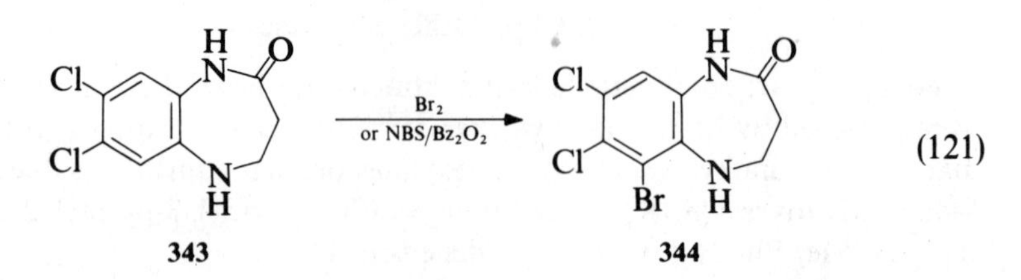


(121)

B. Nitrosation. Reaction of the benzodiazepines **345** with nitrous acid afforded the 5-nitroso compounds **346** (Eq. 122). Nitrosations were carried out for $R_1, R_2, R_3 = H$;[194] $R_1, R_2 = H, R_3 = Me$;[146,194] $R_1, R_2 = H, R_3 = Ph$;[129] and $R_1 = Ph, R_2 = Et, R_3 = H$.[203]

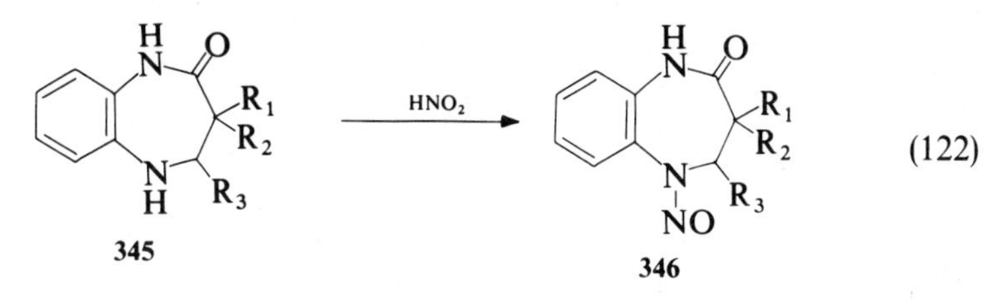


Nitrosation of the methoxycarbonylmethylene derivative **342** occurred exclusively on carbon to give the oxime **347** in high yield (Eq. 123).[171] The indicated stereochemistry of the oxime appears to be favored because of hydrogen bonding to the imine nitrogen.

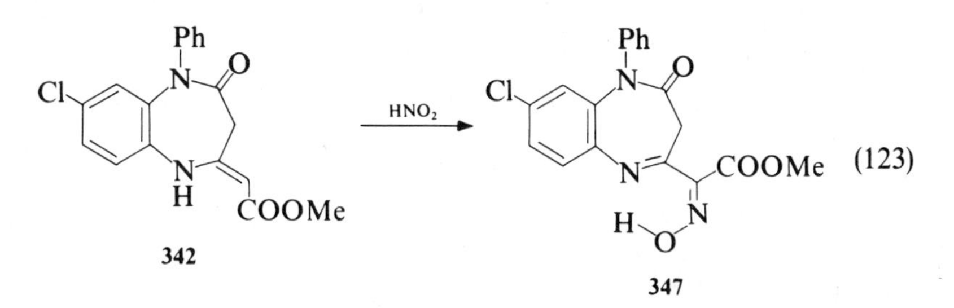


C. Oxidation. Nicolaus and coworkers[203] oxidized the 3,3-disubstituted tetrahydrobenzodiazepinone **348** to the dihydro compound **349** by means of ferric chloride in boiling ethanol (Eq. 124).

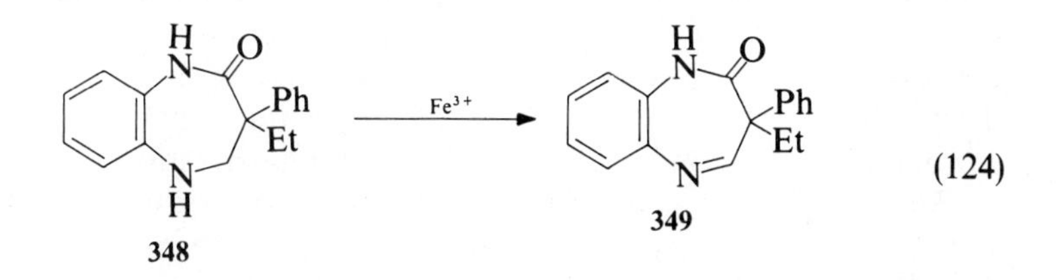


The chromic acid oxidation of tetrahydrobenzodiazepinones to the corresponding 2,4-diones is discussed in Section 6.1.1, which describes the synthesis of these compounds.

D. Alkylation. Reaction of **348** with methyl iodide or ethyl iodide and sodium bicarbonate in boiling methanol led to the corresponding 5-alkyl derivatives.[203] However, arylation of the related 8-substituted compounds **350** took place on the lactam nitrogen to give the 1-aryl analogs **351** ($R_3 = H$) (Eq. 125).[204, 215] These arylations were carried out by a modified Ullmann reaction

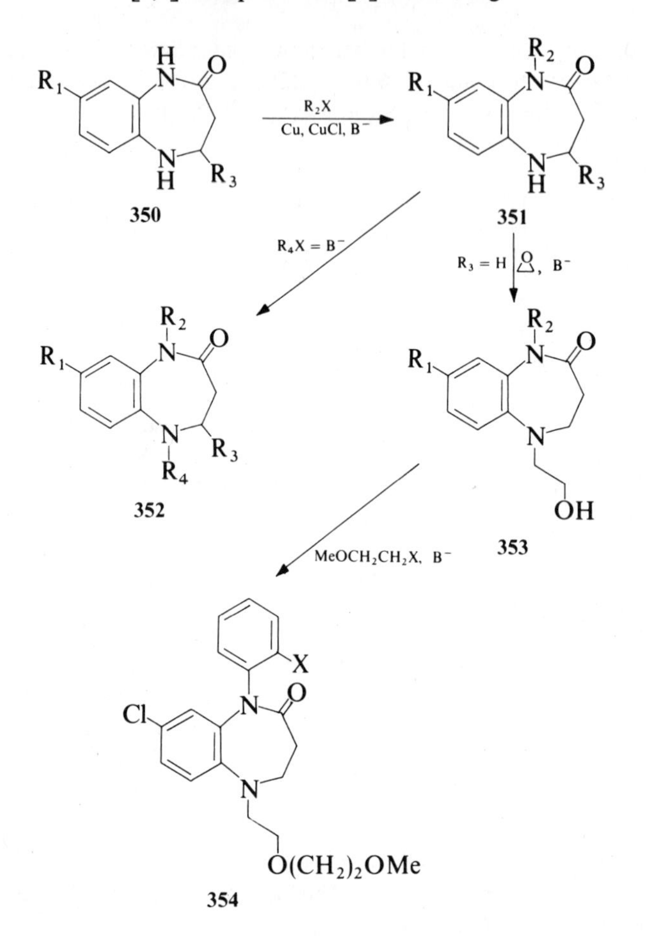


(125)

using an aryl halide, copper powder, cuprous chloride, and an acid receptor such as potassium acetate or pyridine in a polar nonprotic solvent (e.g., *N*,*N*-dimethylacetamide). The reaction temperature ranged from 140 to 160°C and the yields were 45–95%.[204] In general, 2-substituted aryl halides, including 2-bromopyridine, gave lower yields of the *N*-aryl derivatives.

Compounds **351** could be further alkylated at the 5-position by a variety of reagents such as alkyl halides,[199, 204, 207–210] alkenyl halides,[199, 204, 207, 208, 210] alkynyl halides,[215] and hydroxyalkyl halides[207, 215] to give **352**, or with ethylene oxide[107, 215] to give **353**. Alkylation of the hydroxy group in **353** led to the ether **354**. Basic side chains were attached to the 1-position of **355** ($R_1$ = Me, Et) by alkylation with 2-(*N*,*N*-dimethylamino)ethyl bromide[212] to give **356** ($R_1$, $R_2$ = Me) or by reaction with *N*-(2-bromoethyl)-*N*-methyl-1-adamantylamine and sodium amide in toluene[208] to give **356** ($R_1$ = Me, $R_2$ = 1-adamantyl) (Eq. 126).

E. Acylation, Reaction with Aldehydes. Acylation agents preferentially attack the weakly basic nitrogen at the 5-position as shown by the acetylation of

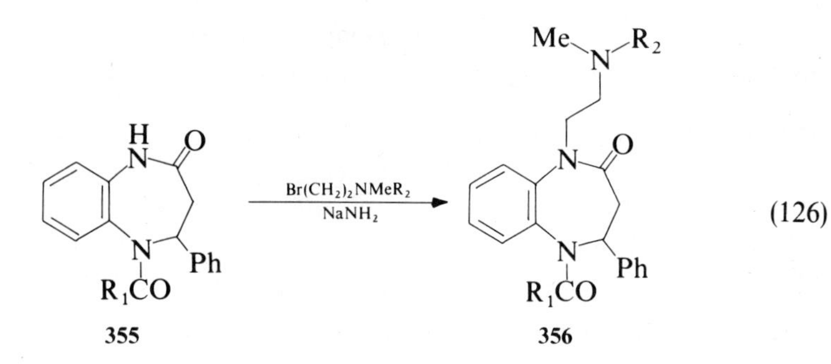


(126)

the 4-methyl-[146, 194] and 4-phenyl-1,3,4,5-tetrahydro-1,5-benzodiazepin-2-(2*H*)-ones[212, 216] by acetic anhydride. The same type of acylation was also achieved with acetyl chloride[141, 203] and propionyl chloride.[212] A variety of 5-acyl derivatives of 1-phenyl-2,3,4,5-tetrahydro-1,5-benzodiazepin-2(1*H*)-ones were prepared by acylations using either formic acid,[207, 215, 217] various carboxylic acid chlorides,[204, 217] sulfonyl chlorides,[204] chloroformates,[204, 217] or isocyanates.[204] 5-Carboxamides were obtained by reaction of the tetrahydrobenzodiazepinone with sodium cyanate and acid[203, 217] or by reacting this compound with phosgene and subsequently with an amine.[217] The 5-hydroxyalkyl derivatives **357** were acylated with either acetic or succinic anhydrides in pyridine to yield the corresponding esters **358** (R = Me, $CH_2CH_2COOH$) (Eq. 127).[217]

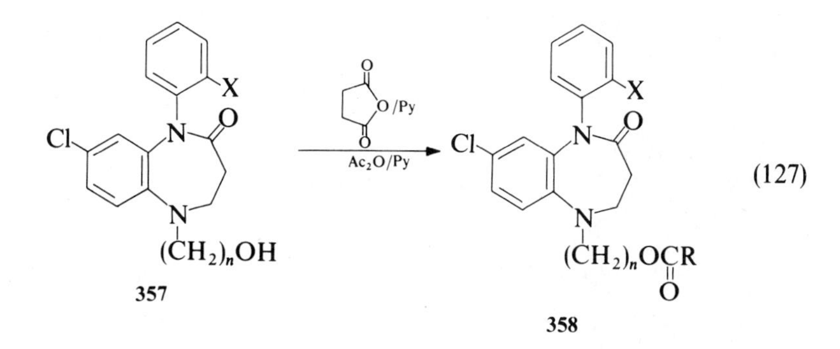


(127)

Condensation of the 4-(aminomethyl)benzodiazepinone **359**, obtained by catalytic reduction of the 4-(nitromethylene) derivative in the presence of triethyl *ortho*-acetate, afforded the imidazoline **360** (Eq. 128).[171]

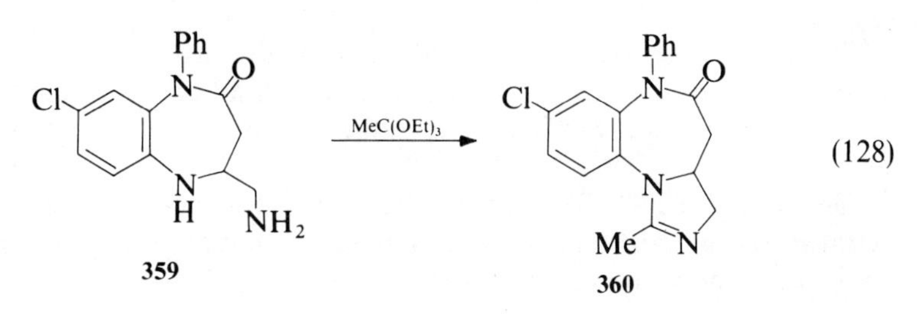


(128)

The 5-aminotetrahydrobenzodiazepinone **361** was acetylated with acetic anhydride to give a diacetate, formulated as either **362** or **363** (Eq. 129).[194]

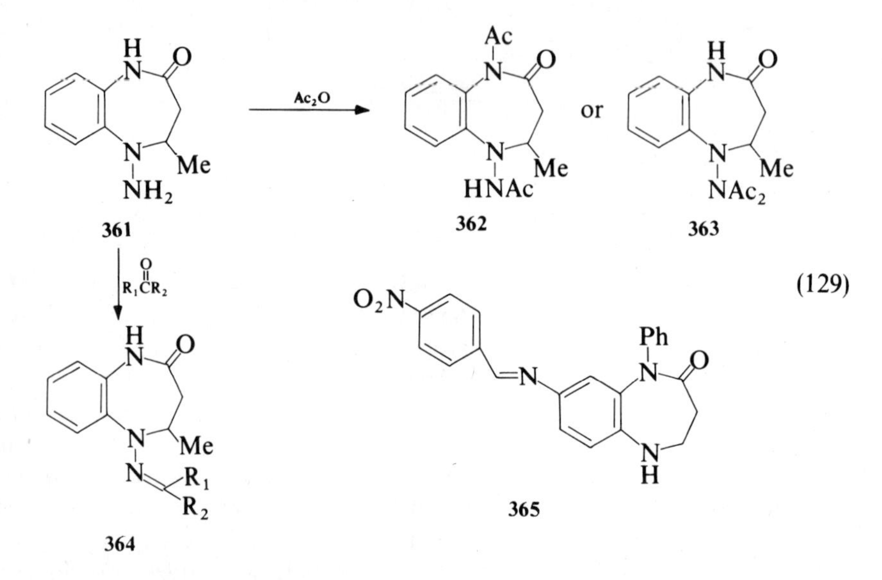


Compound **361** formed the hydrazones **364** in good yields by reaction with benzaldehyde, 2-nitrobenzaldehyde, and benzophenone.[194] The imine **365** was obtained by condensation of the appropriate 8-amino derivative with 4-nitrobenzaldehyde.[204]

### 5.1.2.2. Reactions with Nucleophiles

A. Hydrolysis, Alcoholysis. Hydrolytic cleavage of the amide bond in **348** to give the acid **366** was effected in boiling concentrated hydrochloric acid (Eq. 130).[203]

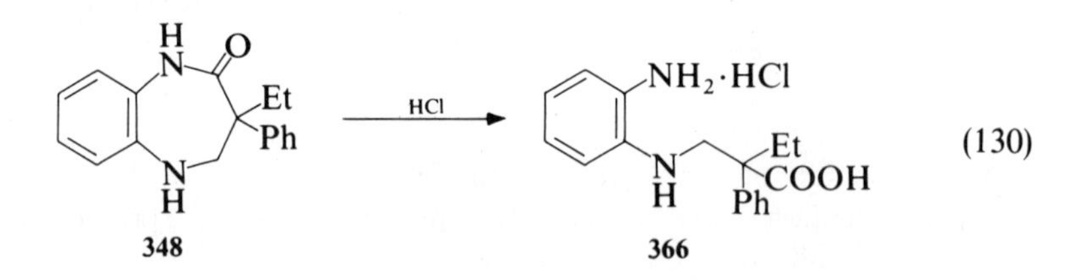


While vigorous alkaline or acid hydrolysis cleaved the alkoxycarbonylmethylene derivative **339** to *o*-phenylenediamine and acetone, a more controlled alkaline hydrolysis gave a high yield of the decarboxylated product **367** with an endocyclic double bond (Eq. 131).[214]

Treatment of **339** (R = Me, Et) with benzyl alcohol, 2-ethoxyethanol, or tetrahydrofuran-2-methanol and a catalytic amount of sodium gave good yields of the corresponding transesterified products.[214]

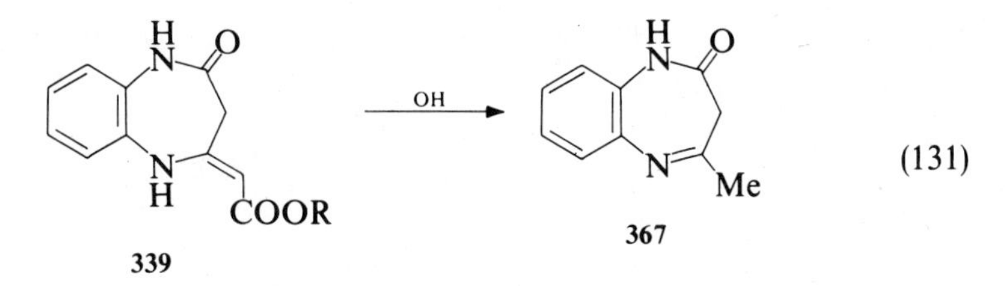


(131)

B. Substitution Reactions. The 5-(2-hydroxyethyl) derivative **368** was converted to the 2-chloroethyl analog **369** by treatment with thionyl chloride and pyridine (Eq. 132).[217]

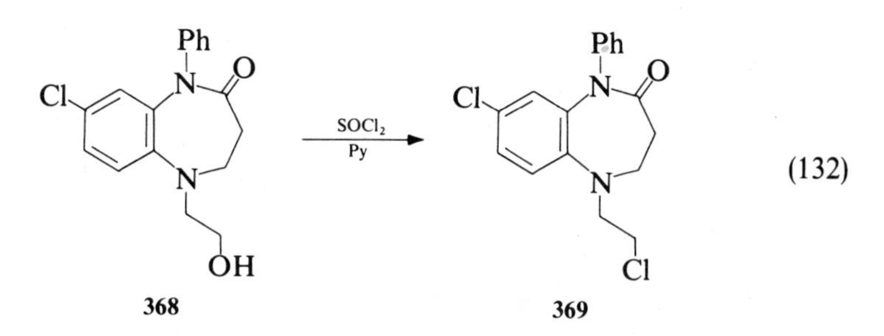


(132)

The Sandmeyer reaction was employed to transform the 8-amino compounds **370** to the 8-cyano and 8-hydroxy derivatives **372** (X = CN, OH; R = H). To avoid nitrosation of the 5-position during diazotization, the nitrogen was protected by trifluoroacetylation. The protecting group was easily removed, after reaction of the diazonium salt **371** (R = $COCF_3$) with the nucleophile, by heating with aqueous ammonia.[204] Reaction of the diazonium salts **371** (R = $COCF_3$, NO) with 4-(*N*,*N*-diethylamino)aniline afforded the corresponding azo compounds **372** (X = $Et_2NC_6H_4N{=}N$—; R = $COCF_3$, NO) (Eq. 133).[204]

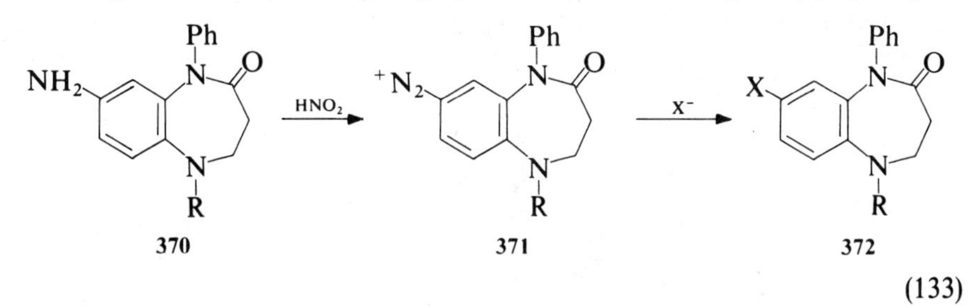


(133)

C. Reductions. The reduction of the 2-carbonyl group by lithium aluminum hydride was discussed above (Section 4.1). 8-Nitro derivatives were catalytically reduced to the 8-amino analogs using palladium[141] or Raney nickel.[204] The latter catalyst was also effective in the reduction of the nitromethylene compound **373** to the amine **359** without dehalogenation (Eq. 134).[171]

The 5-nitroso derivatives **346** were reduced to the corresponding 5-amino analogs by zinc and acetic acid.[194, 203]

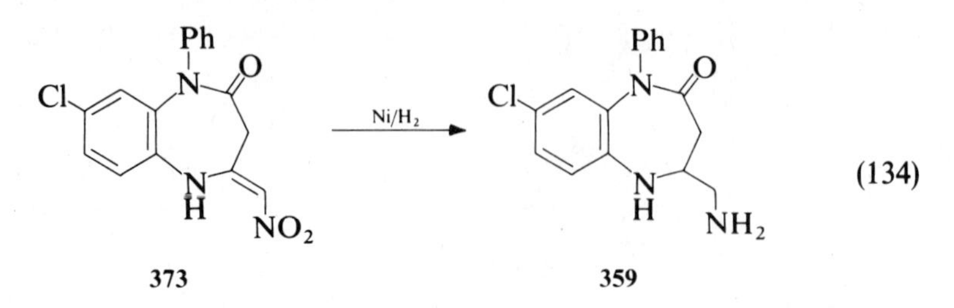


(134)

Reduction of the 1-(2-nitrophenyl) derivatives **374** with stannous chloride, zinc and phosphoric acid, or iron and hydrochloric acid led directly to the benzimidazo[1,2-*a*] [1,5]benzodiazepines **375** (Eq. 135).[218]

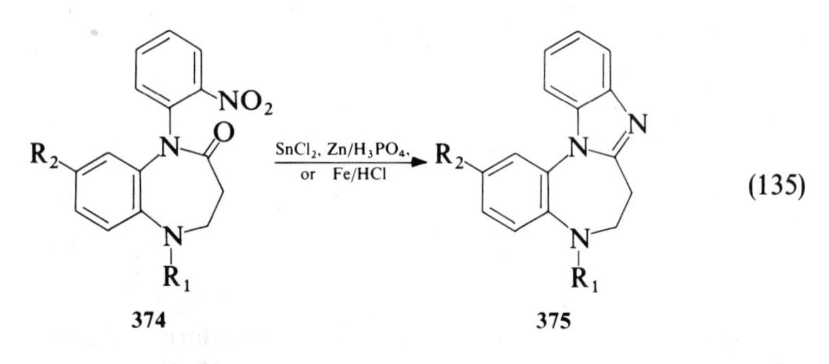


(135)

## 5.2. 1,2,4,5-Tetrahydro-1,5-benzodiazepin-3(3*H*)-ones

### *5.2.1. Synthesis*

Only the di-*N*-tosyl derivative **310** of the parent ring system has been described. Paterson and Proctor[84] prepared **310** by condensing the ditosylamide **376** with dibromoacetone in toluene in the presence of sodium carbonate (Eq. 136).

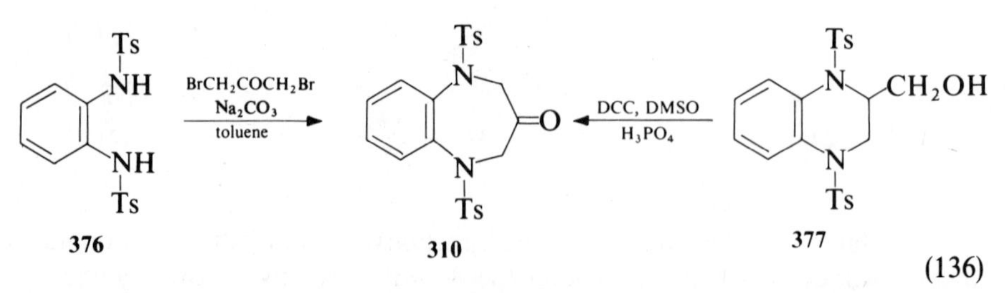


(136)

The same compound was later[219] isolated from the oxidation of the tetrahydroquinoxaline-2-methanol **377** with dicyclohexylcarbodiimide (DCC) and dimethyl sulfoxide in the presence of phosphoric acid. A third synthesis of **310** by Jones oxidation of the corresponding alcohol **284** was noted above (see Section 4.2.1.2).

### 5.2.2. *Reactions*

The ketone **310** formed typical derivatives including the oxime,[181] hydrazone,[181] tosylhydrazone,[181] and 2,4-dinitrophenylhydrazone.[84, 181] Sodium borohydride reduced the carbonyl group to the corresponding alcohol.[181]

According to Paterson and Proctor,[84] elimination of sulfinic acid from **310** led to a red monotosylamide, which on the basis of spectral data was assigned structure **378** (Eq. 137). Depending on the base and solvent combination, the yield varied between 50 and 78%, with potassium *t*-butoxide in dimethyl sulfoxide giving the best results.

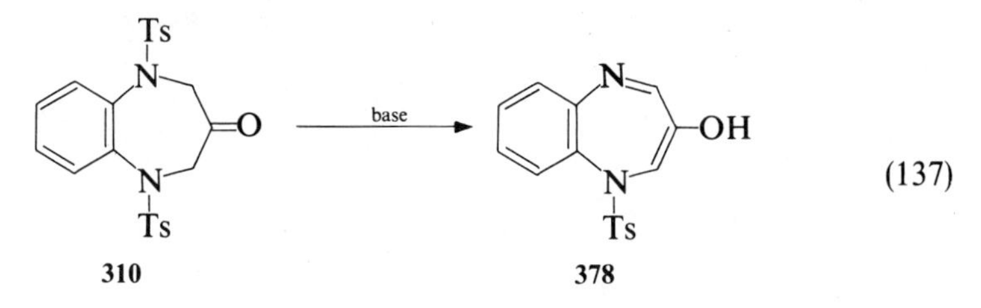


(137)

## 6. TETRAHYDRO-1,5-BENZODIAZEPINEDIONES AND TRIONES

This section reviews the synthesis and chemistry of compounds related to structures **379**, **380**, **381** and of variously substituted analogs. 3,3-Dihydroxy-2,4-ones are considered to be the hydrated form of the triones **381**, and as such are discussed below (Section 6.2). No representatives of the 2,3-dione structure **380** have been described in the literature.

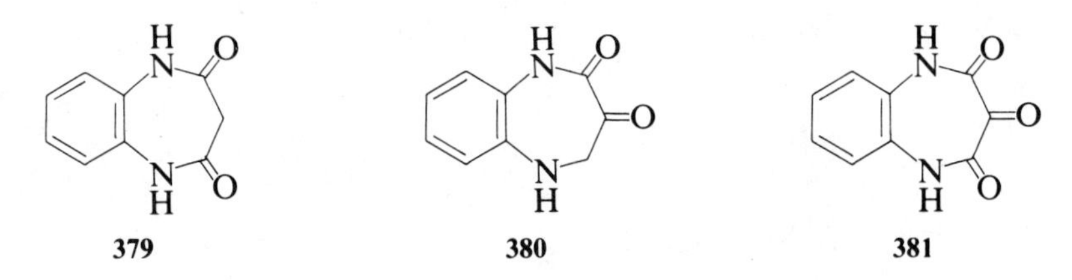


### 6.1. 3,5-Dihydro-1*H*-1,5-Benzodiazepine-2,4(2*H*,4*H*)-Diones

### 6.1.1. *Synthesis*

#### 6.1.1.1. *o*-Phenylenediamine and Malonic Acid or Esters

The condensation of *o*-phenylenediamine with malonic acid constitutes the most fundamental synthesis of the parent compound **379** and dates back to studies carried out by Meyer[220–222] early in this century and later by Phillips.[223] Phillips obtained both the dione **379** and the amide **382** (R = H) by heating a mixture of *o*-phenylenediamine and malonic acid in boiling 4 *N* hydrochloric acid. The amide **382** (R = H) was quantitatively cyclized to **383** (R = H) upon further heating in dilute hydrochloric acid (Eq. 138). Following Phillip's procedure, Shriner and Boermans[224] obtained the parent compound **379** in 62% yield. Several analogs with substituents at the 7-position were similarly prepared from the substituted *o*-phenylenediamines.[225–226]

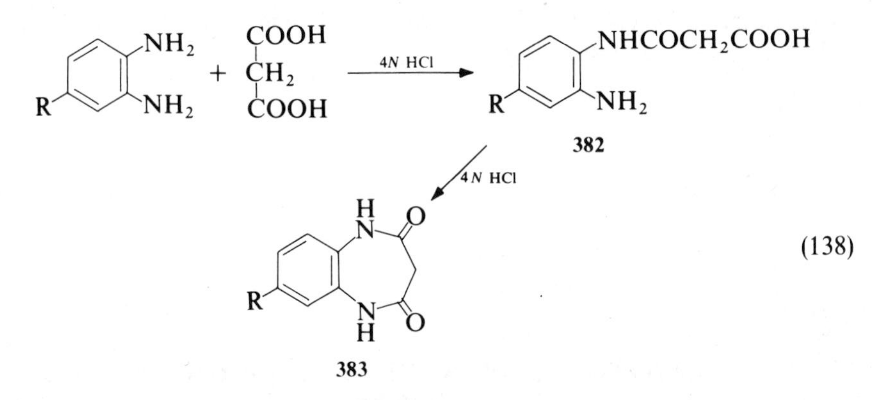


(138)

Meyer[220–222] also showed that malonic acid could be replaced by malonic esters. A modification of this method using base instead of acid catalysis was found to be preparatively useful for the synthesis of many 3-substituted derivatives.[227] The 3-allyl derivatives have been extensively studied by Brobanski and Wagner,[228–230] who used sodium ethoxide and toluene to obtain improved yields of the benzodiazepinediones **384**. This procedure was successfully extended to the preparation of **384** ($R_1$ = alkyl) from N-alkyl *o*-phenylenediamines (Eq. 139).[229]

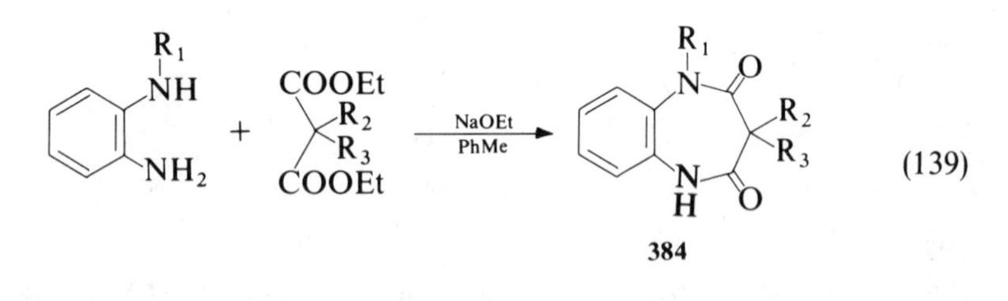


(139)

### 6.1.1.2. *o*-Phenylenediamine and Malonyl Chlorides

The reaction of malonyl chlorides with *o*-phenylenediamines has been used for the synthesis of benzodiazepin-2,4-diones. Particularly successful were the preparations of 7-nitro,[231] 3,3-dially[228–230] and 1-phenyl-5-acyl derivatives.[232] Compounds **385** ($R_1$ = aryl or heteroaryl, $R_4$ = H or alkyl) have also been claimed to be accessible by this process (Eq. 140).[233]

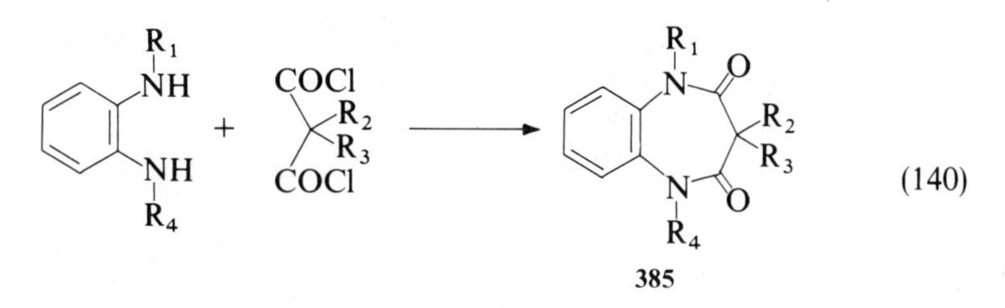


### 6.1.1.3. By Cyclization of *o*-aminoanilides

The most widely used syntheses of benzodiazepin-2,4-diones involve formation of the two amide bonds in separate operations. Thus, 2-nitroanilines (**386**) are reacted with the chloride ester of malonic acid to yield the corresponding 2-nitroanilides **387**. Reduction of the nitro group leads to the amines **388**, which can cyclize with loss of alcohol ($R_3$ = OH) to form the benzodiazepin-2,4-diones **389** (Eq. 141).

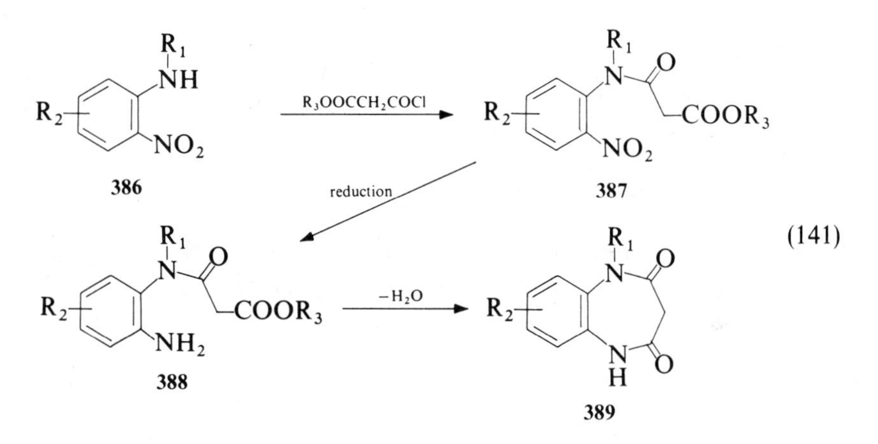


This approach was chosen by Rossi and coworkers[234, 235] and by Weber et al.[236] for the preparation of compounds **389** ($R_1$ = Me or Ph). They found that it was not necessary to isolate the intermediate amines **388**. Cyclization of **388** was effected by acid catalysis and occurred spontaneously during the reduction step, particularly when zinc in hydrochloric acid was used as the reducing agent.[234] Weber and coworkers[237−239] describe the cyclization of the amines **388** with various substituents under basic conditions using sodium ethoxide in ethanol at room temperature. Cyclization of the diacylated 2-aminodiphenylamines **390** to compounds **391** by means of thionyl chloride was claimed in a Canadian patent (Eq. 142).[240]

### 6.1.1.4. By Oxidation

According to the patent literature,[241] 1-phenyl-8-nitro-substituted benzodiazepin-2,4-diones (**393**) were prepared in high yield by oxidation of the

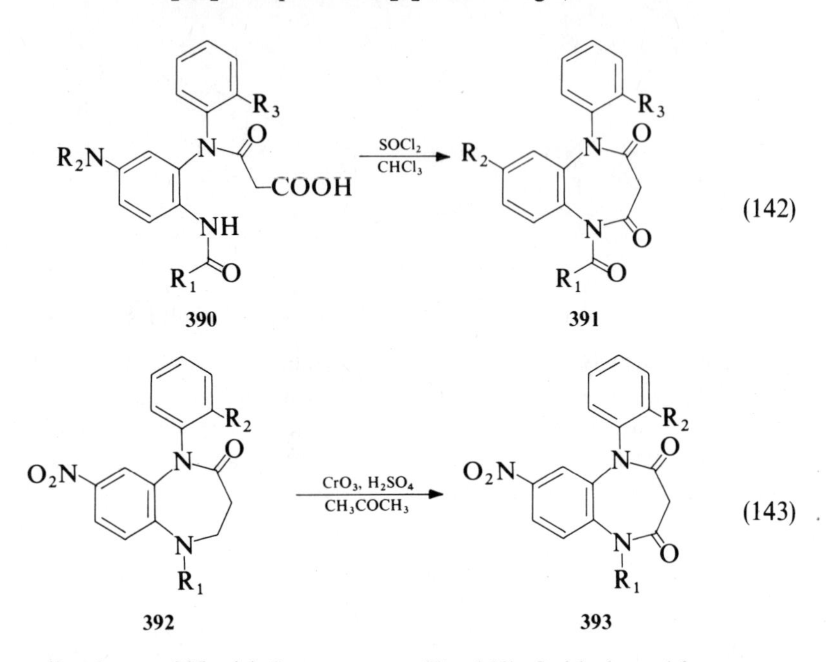


corresponding 2-ones **392** with Jones reagent (Eq. 143). Oxidation with manganese dioxide reportedly gave inferior yields.

### 6.1.1.5. Other Syntheses

Van Alphen[242] described the formation of the parent benzodiazepinedione **379** by the reaction of *o*-phenylenediamine with carbon suboxide (Eq. 144).

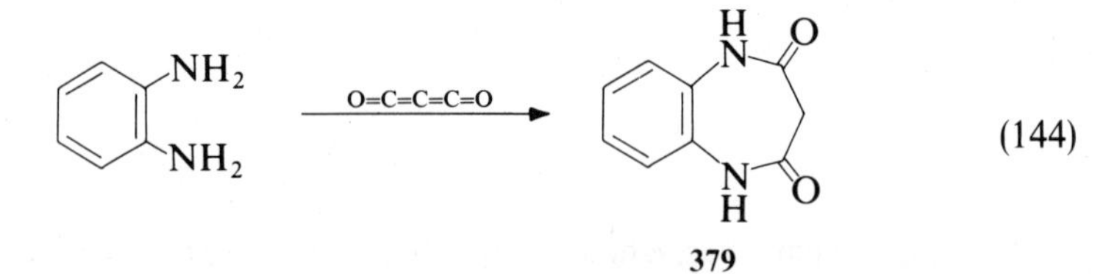


The 3-substituted 1,5-benzodiazepin-2,4-diones **395** were prepared by the reaction of *o*-phenylenediamine with diethyl α-(*N*-phenylbenzenesulfonamido)-malonate **394** (Eq. 145).[317]

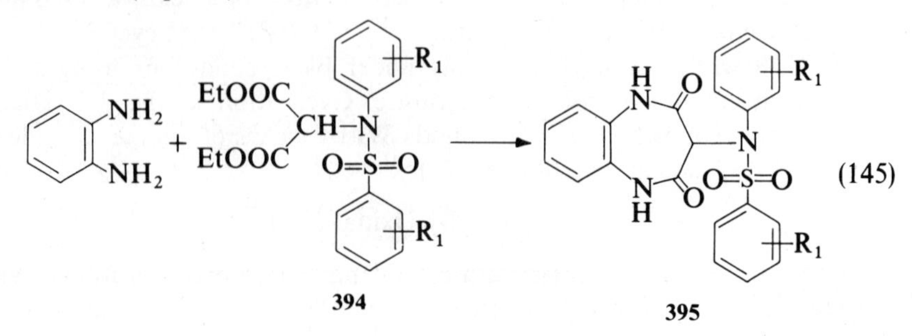

### 6.1.2. Reactions

#### 6.1.2.1. Reactions with Electrophiles

A. Halogenation. The malonic acid moeity in the tetrahydro-1,5-benzodiazepin-2,4-diones **396** retains some of its expected reactivity towards electrophilic reagents. Thus, both bromine and chlorine, either in glacial acetic acid or in chloroform, halogenate **396** ($R_1$ = alkyl) at the 3-position to give **397** (Eq. 146).[243] Good yields of the 3-bromo derivatives were also obtained by bromination of the 3-carbanion generated with sodium hydride in tetrahydrofuran.

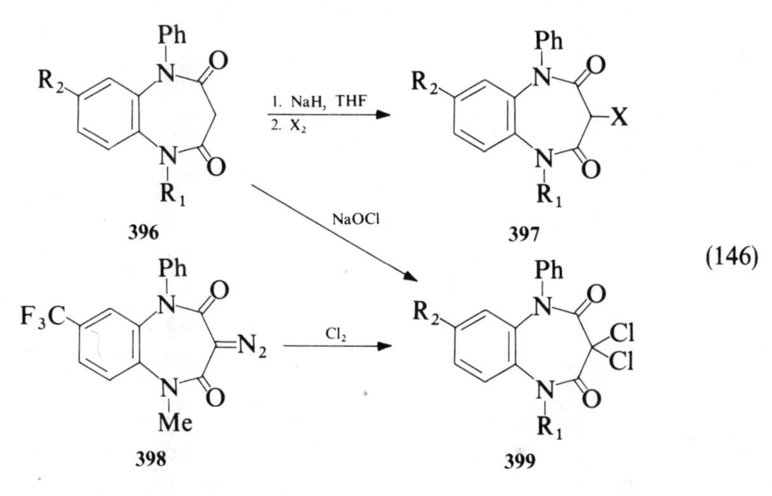


(146)

Chlorination of **396** ($R_1$ = H) with sodium hypochlorite led to the dichloro derivative **399**. Compound **399** ($R_1$ = Me, $R_2$ = $CF_3$) was accessible by treatment of the diazo compound **398** with chlorine in the presence of a copper catalyst.[243]

B. Nitration. According to the patent literature,[226] nitration of the tetrahydrobenzodiazepin-2,4-dione **383** (R = H) with fuming nitric acid at – 10°C gives the aromatic nitro derivative **400**. At higher temperatures (0°C to room temperature), nitration also occurs at the 3-position to give **401** (R = H, Cl, F, COOH) (Eq. 147).

C. Nitrosation. Although malonic acid derivatives are generally nitrosated with ease, Weber and Bauer[243] were unable to obtain any nitrosated product from the treatment of **396** with an alkylnitrite. Nitrous acid converted the 3-amino derivative **403** (R = Me, X = Cl) to the diazo compound **404** (R = Me, X = Cl) (Eq. 148).[243]

D. Reaction with Other Nitrogen Electrophiles. The 3-carbanions **402** reacted with electrophilic nitrogen compounds. Chloramine yielded the 3-amino derivatives **403**,[243] while tosylazide formed the quite stable 3-diazo compounds **404** (Eq. 148).[243,244]

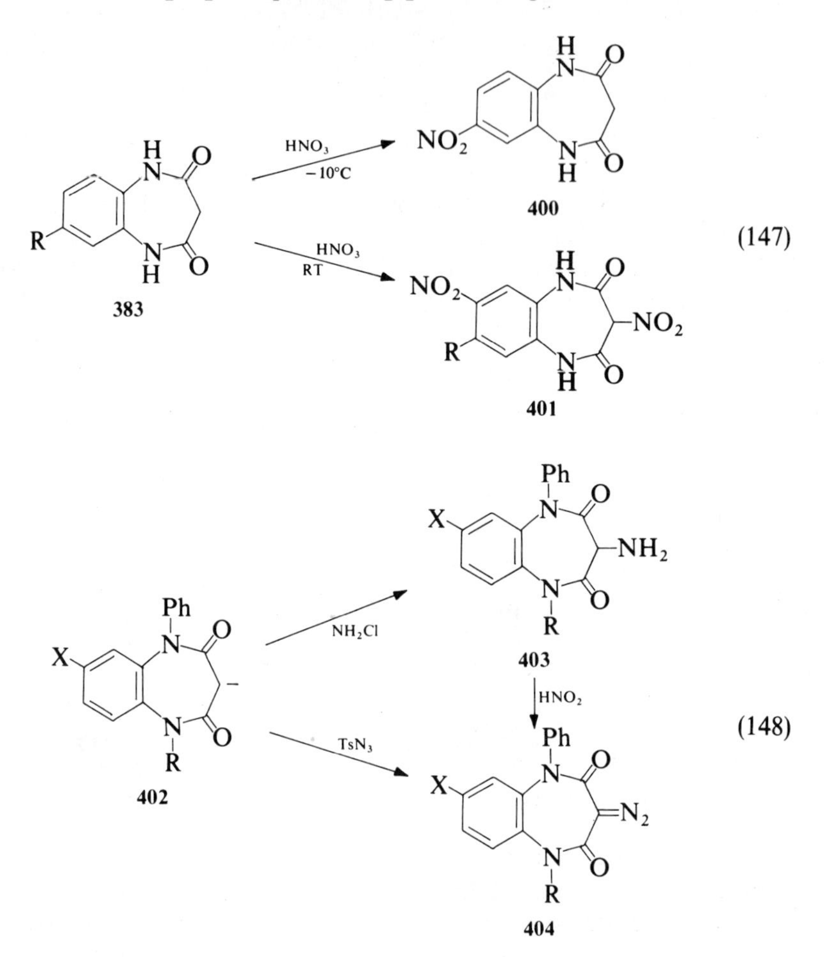


E. Oxidation. Bauer and Weber[245–248] investigated the oxidation of the 3-aminomethylene derivatives **405** with various reagents. Oxidation with potassium permanganate or chromic acid at room temperature gave the formyl derivatives **406** in about 50% yield (Eq. 149). If the oxidation was performed with permanganate in acetone at −26 to −30°C followed by warming to room temperature, the 3-hydroxy derivatives **407** were obtained as the major products. The formyl compounds **406**, formed as by-products under these conditions, were separated by conversion to the hemiacetals **408**. Under more vigorous reaction conditions, permanganate in acetone and sulfuric acid, the enamine **405** or the 3-hydroxy compounds **406** and **407** were oxidized to the 3,3-dihydroxy derivatives **409**. Manganese dioxide has also been claimed to effect the conversion of **407** to **409**.[247]

F. Alkylation. Alkylations and arylations of the nitrogens at both the 1- and 5-positions have been extensively studied and a large variety of substituted derivatives have been prepared.

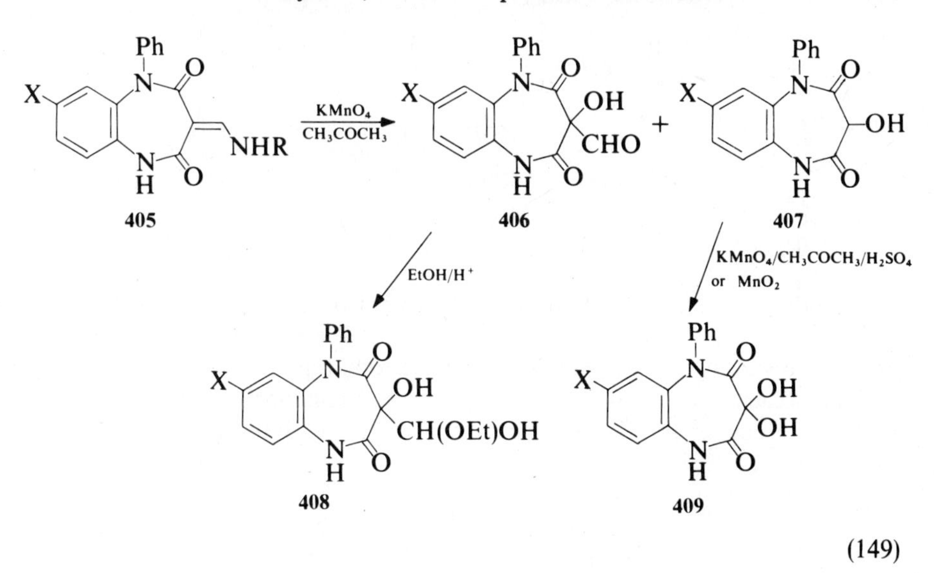


(149)

Shriner and Boermans[224] demonstrated that methylation of **410** ($R_1$ = H) with sodium ethoxide and methyl iodide in ethanol led to the 1,5-dimethyl derivative **411** ($R_1$ = H, $R_2$ = Me). Büchi and coworkers[227] also obtained exclusive *N*-alkylation upon reaction of **410** ($R_1$ = Bu) with *n*-butyl bromide in ethanolic potassium hydroxide (Eq. 150). The acidity of the amide protons allows the use of relatively weak bases such as hydroxide in protic solvents. Under these conditions C-alkylation at the 3-position does not occur.

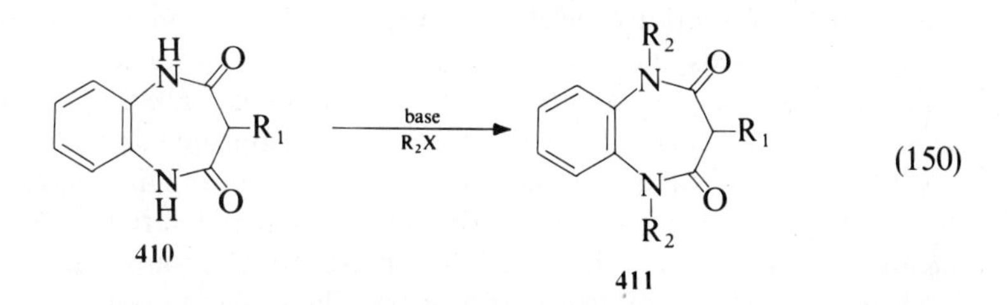


(150)

Rossi and coworkers[234,235] describe the alkylation of 1-methyl and 1-phenyl derivatives using a variety of halides and base–solvent combinations. Similar substitution reactions were carried out by Wagner[228] on 3,3-diallyl derivatives. Alkylations of 1-phenyl derivatives have been extensively described in the patent literature.[236,249,250] Methylation with methyl iodide and base has also been performed on a 3,3-dichloro compound[243] and a 3-dimethylaminomethylene derivative.[247]

Strong bases in aprotic solvents generate the 3-carbanion of 1,5-disubstituted compounds such as **412**. These carbanions react with alkyl halides to form the 3-substituted derivatives **413** (Eq. 151).[239] 3,3-Disubstituted compounds have not been prepared by this method.

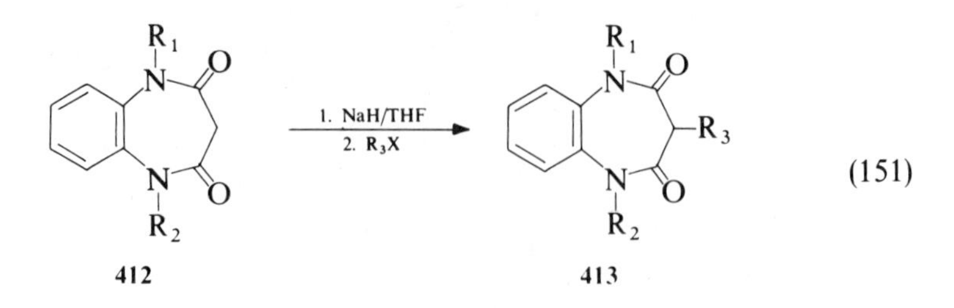


(151)

A modified Ullmann reaction was found to be suitable for the introduction of aryl and heteroaryl substituents at the 1-position. Weber and coworkers[237–239] prepared a variety of 1,5-disubstituted 1,5-benzodiazepin-2,4-diones **415** by this method. Reaction of **414** with an aryl or heteroaryl halide was performed in the presence of copper powder and potassium acetate at 100–160°C in dimethylformamide (Eq. 152).

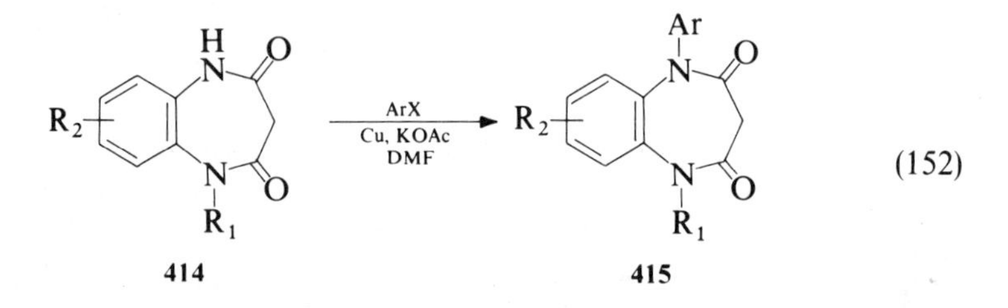


(152)

Weber et al. also carried out kinetic studies comparing the rate of *N*-phenylation of the benzodiazepin-2,4-dione with that of acetanilide. A higher rate of arylation was observed for the cyclic amide, which was attributed to its fixed spatial orientation. The effect on the rate of arylation of substituents at the 7- and 8-positions indicated that electron-withdrawing substituents at the 8-position enhanced reactivity. The influence of the 5-alkyl substituent on the arylation rate appears to be mainly steric. The 5-isopropyl derivative exhibited a distinctly greater reactivity than the compounds bearing a straight-chain alkyl group. This phenomenon was attributed to the greater rigidity of the seven-membered ring when substituted by an isopropyl group. This rigidity was evident by analysis of the nmr spectral data. An unusual intramolecular alkylation on oxygen was observed by Wagner and coworkers[228,230,307] during a study of the hydration of 3-allyl derivatives. Brief treatment of compounds **416** (R = *i*-Pr, Ph, cyclohexyl) with concentrated sulfuric acid or 85% phosphoric acid yielded the corresponding furobenzodiazepines **417** (Eq. 153).

When R represented a less bulky substituent such as allyl, the hydrated form **418** was obtained.[228] The addition of hydroxide to the 2-position carbon was thus dependent on the steric bulk of the substituent R on the neighboring carbon atom.

G. Acylation. Acylation of the amide functions in 1,5-benzodiazepin-2,4-diones **414** occurs on nitrogen to give **419** (Eq. 154). Generally, the amide is deprotonated by strong bases prior to reaction with an acyl halide or anhy-

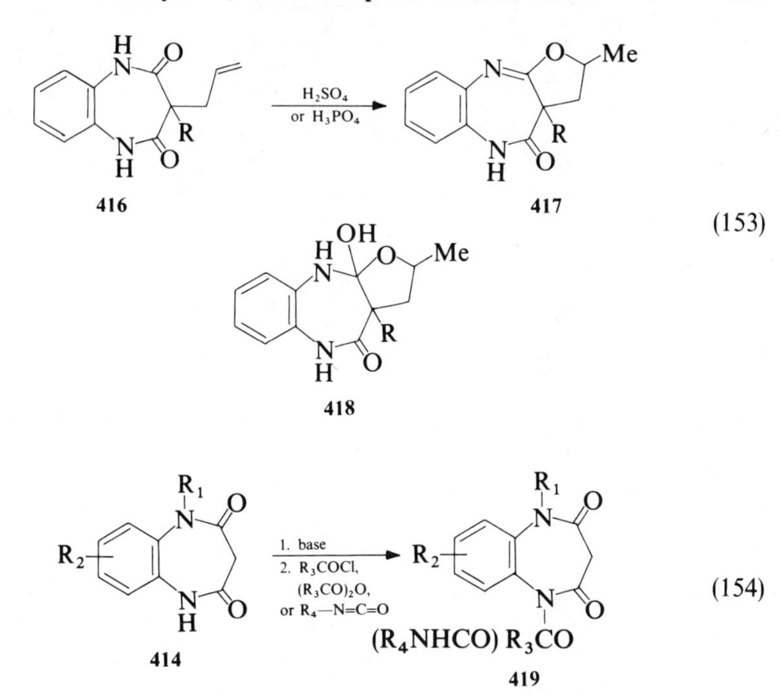


dride.[234,251] The reaction with methylisocyanate[251] was carried out with a catalytic amount of triethylamine, while acylation with acetic or propionic anhydride[251] required heating neat or in the presence of pyridine. Thus, acetylation of the 3-hydroxy derivative **420** with acetic anhydride for 3 hours at 85°C gave the 3-acetoxy compound **421** (R = H) in 73% yield, while under the more vigorous conditions of reflux for 4–5 hours the *O,N*-diacetyl derivative **421** (R = Ac) was obtained (Eq. 155).[246]

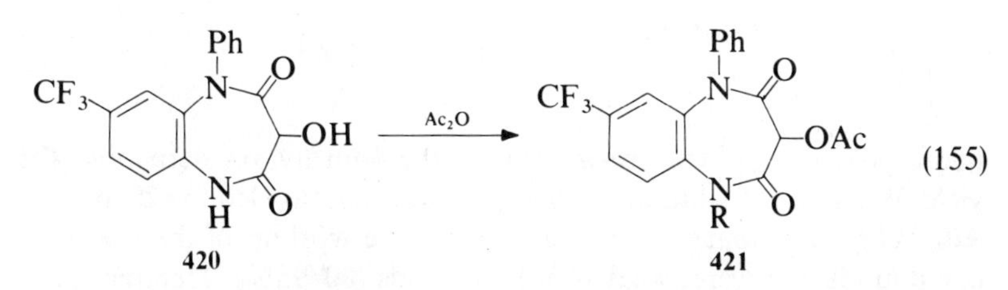


The 3-benzoyloxy analog of **421** (R = H) was prepared in moderate yield by treating **420** with sodium hydride and benzoyl chloride.[246] Benzoylation of the 3-carbanion of **422** led in low yield to the *O*-benzoylated enolate **423** of the 3-benzoyl derivative (Eq. 156).[243]

The enamines of 3-formyl derivatives were found to be readily accessible by reaction of compounds **424** with phosphorus pentachloride and dimethylformamide.[118,248,252] Studies of this reaction revealed that the first step may be the formation of an O—P bond involving the free amide function followed by reaction with immonium chloride to yield the intermediate **425** (Eq. 157). Mild

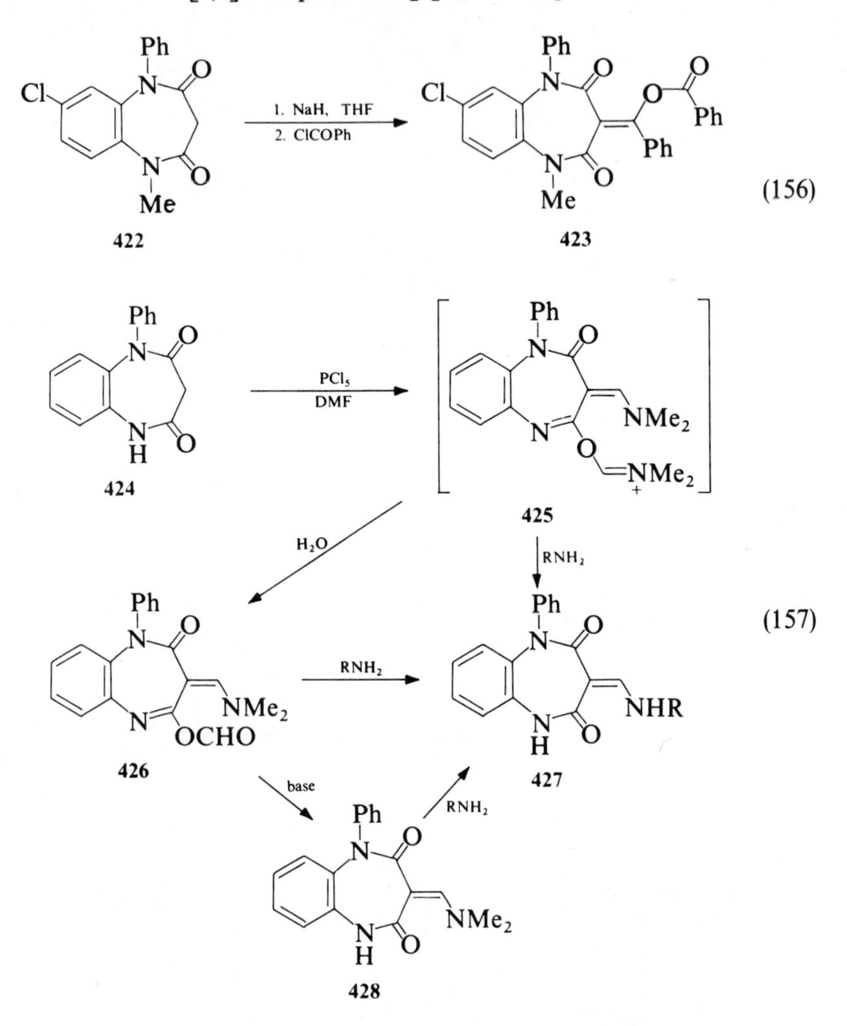


hydrolytic workup with ice water gave the 4-formyloxy derivative **426** in 72% yield. Workup with dilute sodium hydroxide solution led to **428**, presumably via **426**. When a primary amine was used in the workup of the reaction mixture, compounds **427** were isolated in high yields (80–95%). Treatment of **428** with *n*-butylamine at room temperature resulted in the expected amine exchange to give **427** (R = *n*-Bu).

### 6.1.2.2. Reactions with Nucleophiles

A. Hydrolysis and Alcoholysis. Shriner and Boermans[224] subjected the 1,5-dimethyl derivative **429** to acid hydrolysis and isolated the benzimidazolidine **430** in 61% yield. This compound most likely was formed by cleavage of an amide bond and recylization to the imidazolium salt **431**, which could then undergo decarboxylation and hydration to give the observed product (Eq. 158).

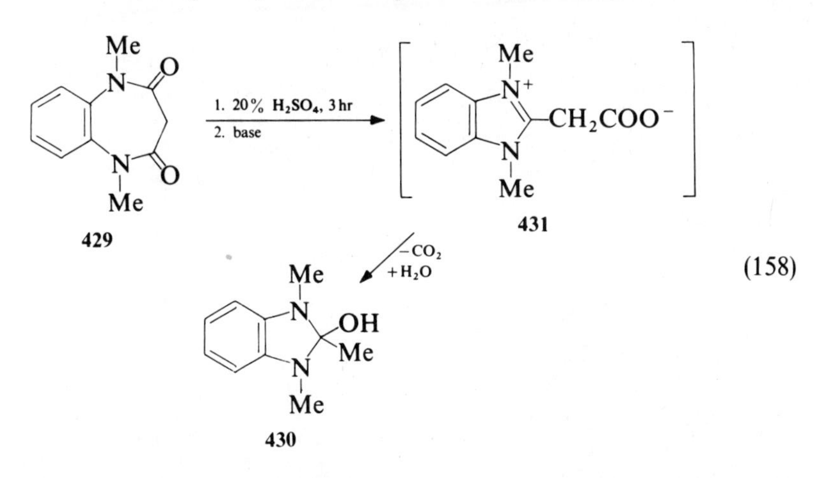


(158)

Weber and coworkers[243] attempted to prepare the 3-benzoyl derivative **432** by mild alkaline hydrolysis of the benzoylated enolate **423** but observed only cleavage to the 3-unsubstituted compound **422** (Eq. 159).

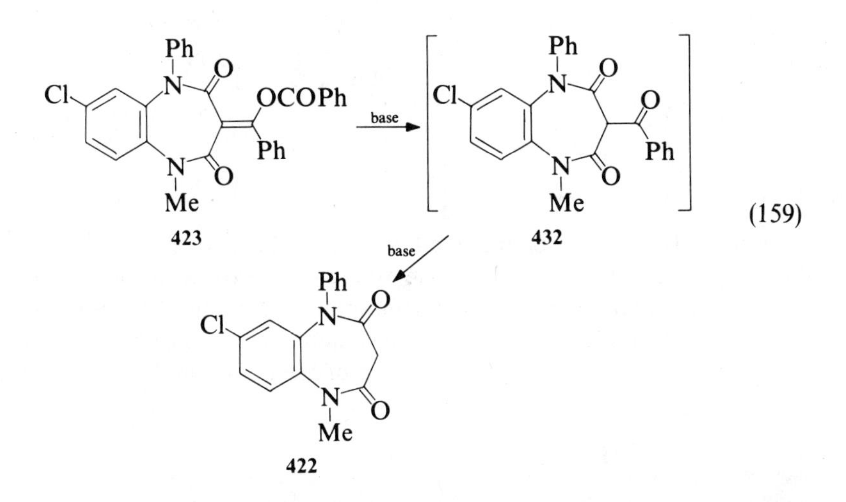


(159)

Hydration of the double bond in 3-allyl-1,5-benzodiazepin-2,4-diones (**433**) was extensively studied by Wagner and coworkers.[228–230] Using 85% phosphoric acid at 100°C they obtained the benzimidazoles **437** as the major products. Under milder conditions, such as concentrated sulfuric acid at room temperature, compounds **435** and their ring-opened equivalents **436** could be isolated. Since **436** was converted to the benzimidazole **437** by heating in phosphoric acid, it is likely that **435** and **436** are the initial products of the reaction leading to **437** (Eq. 160).

The initial step is believed to be protonation of the double bond with formation of a carbonium ion, which is intramolecularly trapped by the amide oxygen to form the cyclic carbonium ion **434**. If $R_1$ is hydrogen, loss of this

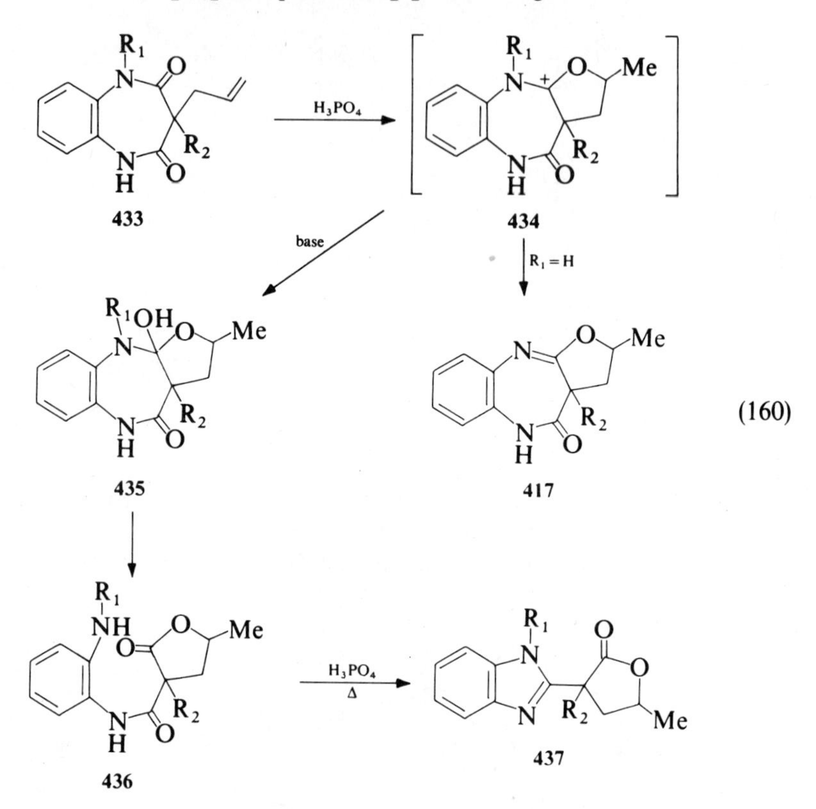


(160)

proton leads to **417**. Such imino ethers were stable enough to be isolated when $R_2$ was sufficiently bulky to impede attack of hydroxide at the 2-position.[238]

The quantitative conversion of compounds **385** to the benzimidazolones **440** by treatment with catalytic amounts of sodium ethoxide in ethanol was reported by Weber and coworkers.[239] Ethanolysis of one of the amide bonds followed by

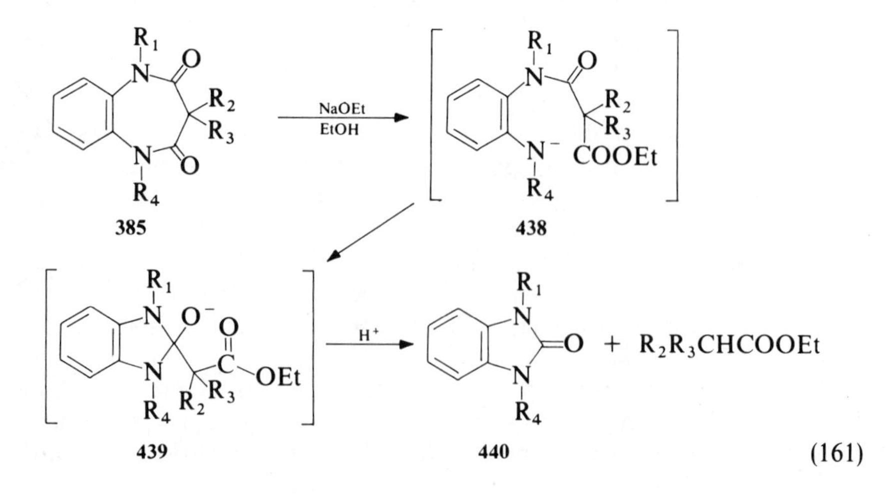


(161)

recyclization of intermediates **438** would give the intermediate hydroxybenzimidazolidines **439**. Retroaldol-type cleavage would then lead to the benzimidazolones **440** and ethyl acetate ($R_2$, $R_3$ = H) (Eq. 161).

B. Substitution Reactions. The attempt to displace halogen in **397** (X = Cl, Br) with nucleophiles was not successful.[243] The reaction of **397** ($R_1$ = Me) with methoxide gave only the benzimidazolone **441**, while treatment with dimethylamine in the presence of copper effected dehalogenation to yield **396** (Eq. 162).

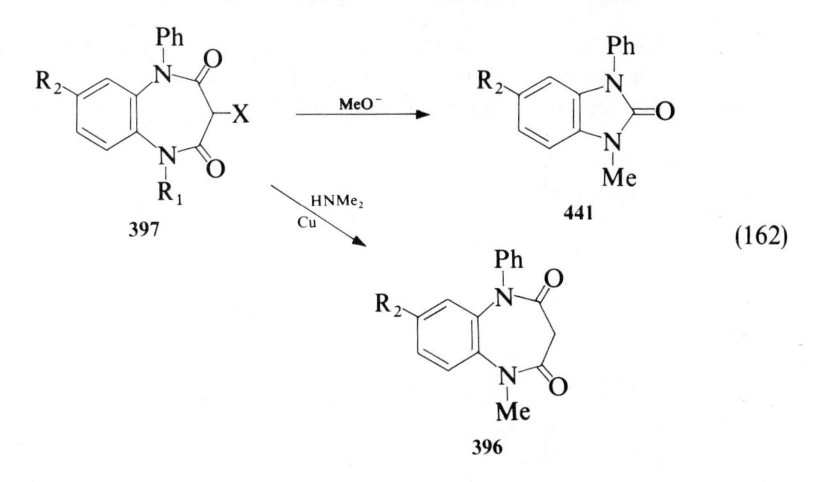


(162)

The 3-diazo derivatives **404**, on the other hand, proved to be useful as substitutes for nucleophilic displacement reactions. Protonation to the diazonium salts **442** followed by loss of nitrogen generates the carbonium ions **443**. Nucleophiles represented by $Y^-$ may then add to give the 3-substituted derivatives **444** (Eq. 163).

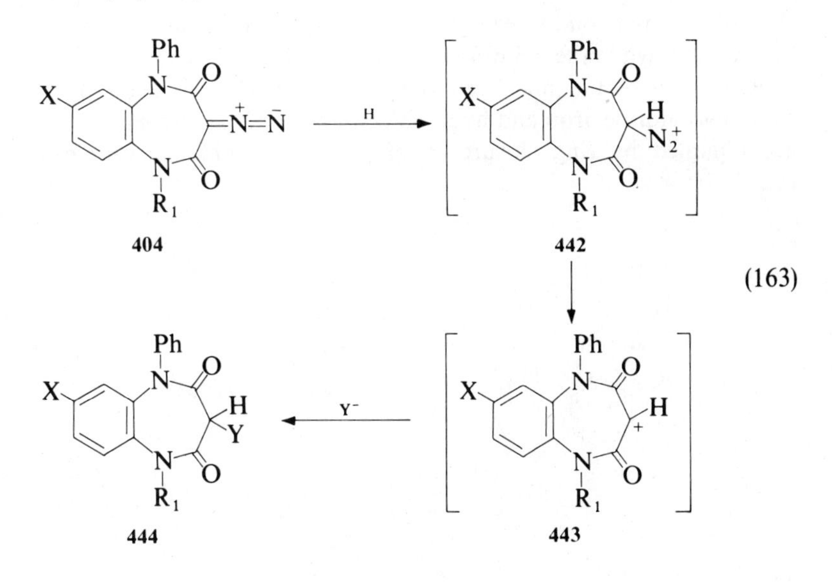


(163)

Reaction of **404** with water in the presence of copper or copper salts afforded the 3-hydroxy derivatives **444** (Y = OH) in high yields. Decomposition of the diazo compound in alcohols in the presence of catalytic amounts of boron trifluoride and copper powder led to the 3-alkoxy analogs **444** (Y = alkoxy). Boiling **404** in carboxylic acids under similar conditions produced 3-acyloxy compounds (**444**, Y = acyloxy). The 3-chloro compound **444** (Y = Cl) was obtained by treatment of the diazo derivative with hydrogen chloride, potassium chloride, and copper powder in acetonitrile.[243]

C. Reduction. Lithium aluminum hydride in tetrahydrofuran was capable of reducing both carbonyl groups in **410** ($R_1$ = H, Et) to give good yields of the tetrahydro-1,5-benzodiazepines **445** (Eq. 164).[231]

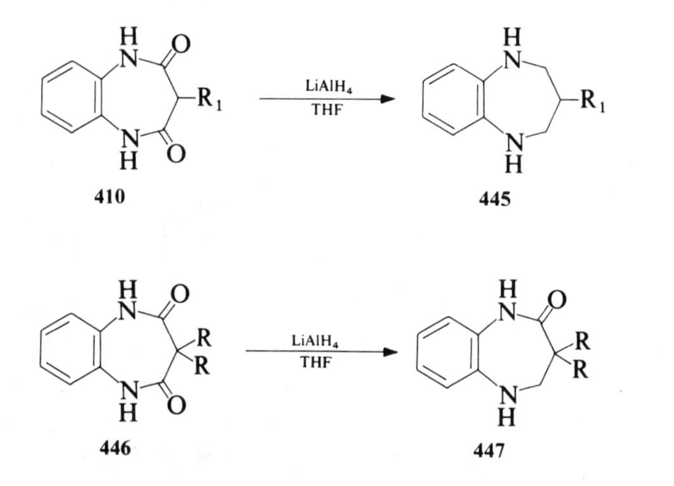


(164)

Under similar conditions, the 3,3-diethyl derivative **446** (R = Et) was reduced only to **447**, while the 3,3-dipropyl analog did not undergo any reduction.[231]

Reduction of the nitro group in the 2-nitrophenyl derivatives **448** with stannous chloride, iron and hydrochloric acid, or zinc and phosphoric acid was accompanied by ring closure to give the benzimidazobenzodiazepines **449** (Eq. 165).[218]

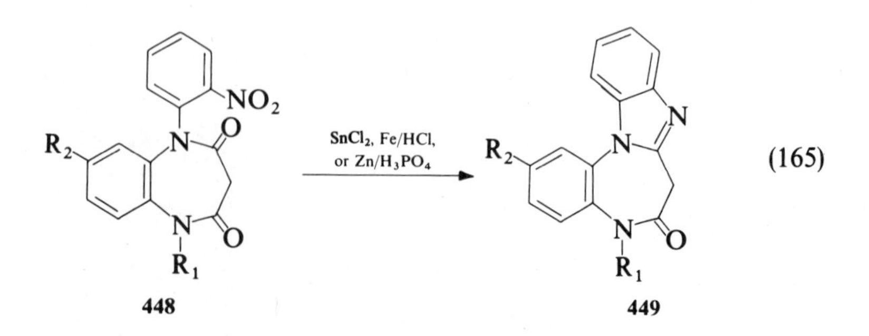


(165)

## 6.2. 1,5-Dihydro-1,5-benzodiazepin-2,3,4(2*H*,3*H*,4*H*)-triones

### 6.2.1. *Synthesis*

The triones **450** were prepared by thermal dehydration of the corresponding 3,3-dihydroxy derivatives **409** (Eq. 166).[246,247]

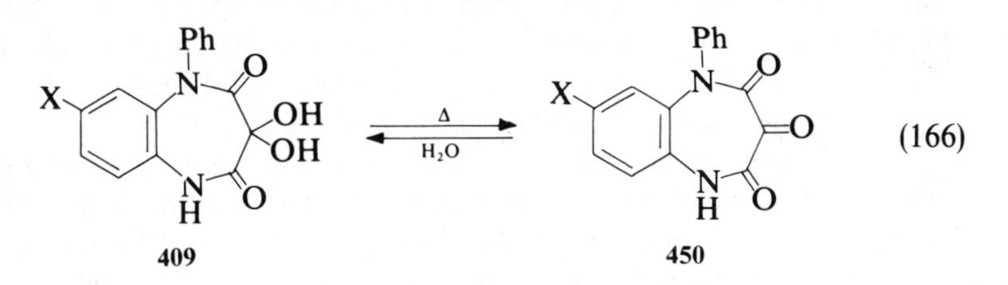


### 6.2.2. *Reactions*

Heating the trione **450** ($X = CF_3$) under reflux in wet xylene for one hour gave the quinoxaline **451** and carbon dioxide (Eq. 167).[246] The 3-hydroxy compound **453** ($X = CF_3$), which was detected in the reaction mixture, may have been formed by hydride abstraction from the solvent. The 3-hydroxy derivatives **453** are accessible in high yields by reduction of **450** with zinc and acetic acid.[246] Other reducing agents such as tin and hydrochloric acid or

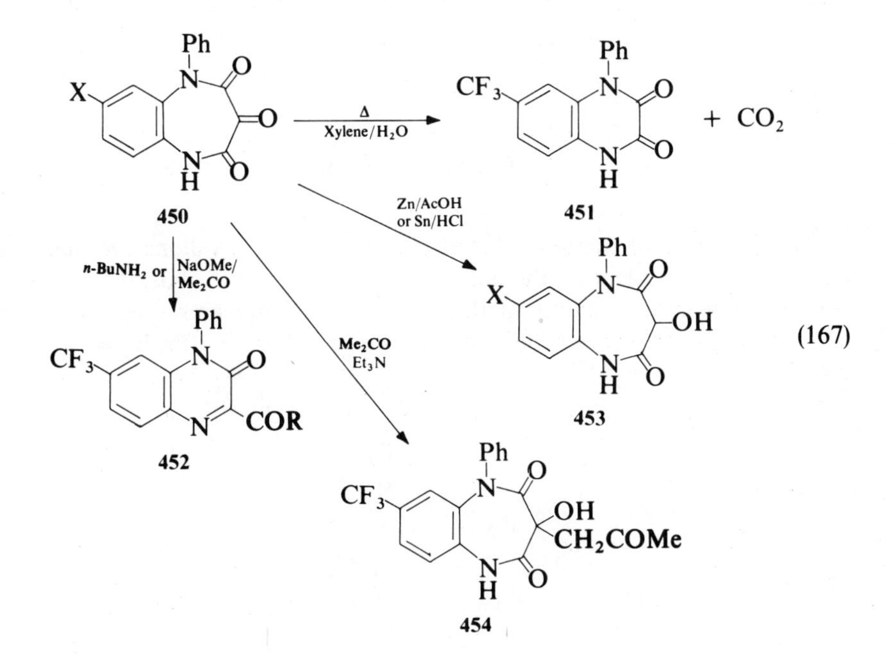

sodium borohydride, or catalytic hydrogenation have also been claimed to effect the same transformation.[245]

Bauer and Weber[246] describe the addition of nucleophiles to the reactive 3-keto group of the trione **450** (X = $CF_3$). The high reactivity of this group was evident from the easy formation of the hydrate **409** (Eq. 166). Reaction of **450** (X = $CF_3$) with *n*-butylamine in boiling benzene gave the quinoxaline derivative **452** (R = NHBu-*n*). The corresponding methylester **452** (R = OMe) was isolated from the reaction of **450** with sodium methoxide in acetone. Also isolated from this reaction was the quinoxalinedione **451** and the aldol product **454**. The latter compound was also prepared in high yield by heating an acetone solution of **450** (X = $CF_3$) under reflux in the presence of triethylamine.

The formation of the quinoxalines **452** may best be explained by a benzilic acid type rearrangement as illustrated below. The reaction is initiated by the addition of a nucleophile $X^-$ to the 3-keto group to give the anion **455**. Ring contraction as indicated would then lead to intermediate **456**, which can either undergo dehydration to give **452** or be cleaved to **451** (Eq. 168).

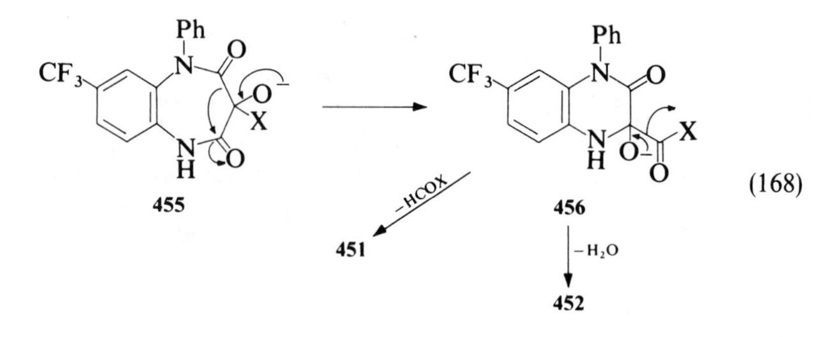


## 6.3. Spectral Data

The flexibility of the seven-membered ring of the 2,4-dione **389** has been shown to be dependent on the bulk of the substituent $R_1$. Thus, the methylene protons appear as a singlet in the room temperature nmr spectrum of **389** ($R_1$ = Et, $R_2$ = 8-Cl) while the same protons in **389** ($R_1$ = *i*-Pr, $R_2$ = 8-Cl) give rise to an AB system.[239]

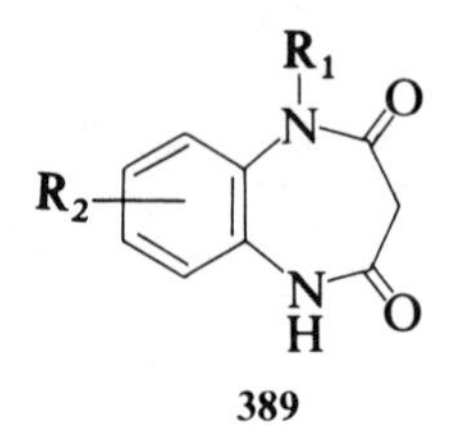

The large coupling constant of 14.5 Hz between the enamine proton and the NH in compounds **427** was attributed to the transoid orientation of these protons, which are fixed by hydrogen bonding.[252]

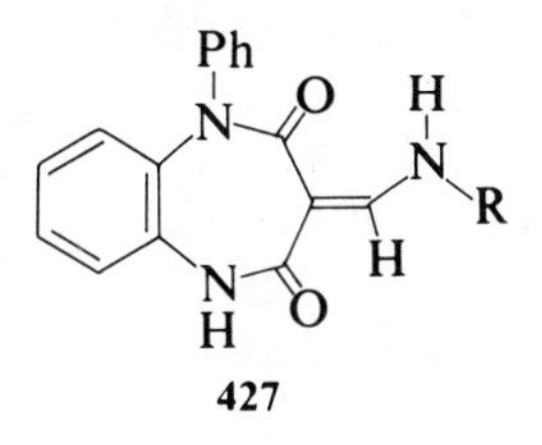


Mass spectrometry was essential in studying the biotransformations of triflubazam, **415** ($R_1$ = Me, $R_2$ = 8-$CF_3$, Ar = Ph).[253,254] High pressure liquid chromatography was used to separate the metabolites of this compound.[255]

# 7. 1,5-BENZODIAZEPINETHIONES

## 7.1. Synthesis

### *7.1.1. o-Phenylenediamine and Thioacid Derivatives*

1,3-Dihydro-2*H*-1,5-benzodiazepin-2-thiones **457** with a substituted phenyl group at the 4-position were prepared[256] in good yield by condensation of *o*-phenylenediamine with an appropriate benzoyl dithioacetic acid (Eq. 169).

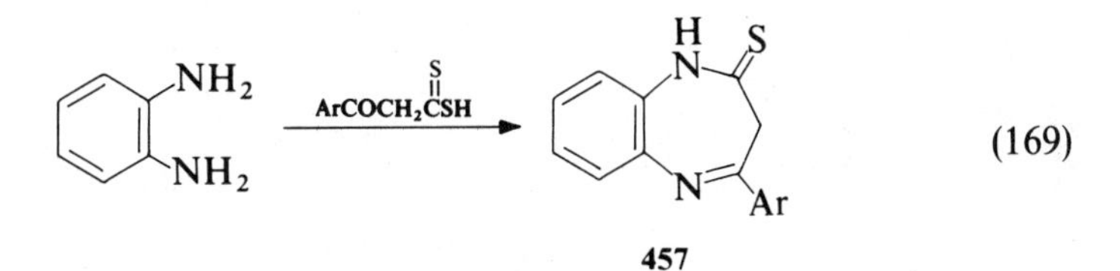


(169)

3-Carboxylic acid esters of structure **459** were found to be accessible from *o*-phenylenediamine and the dithiolane **458** ($n$ = 1)[257] or the trithiolane **458** ($n$ = 2) (Eq. 170).[258]

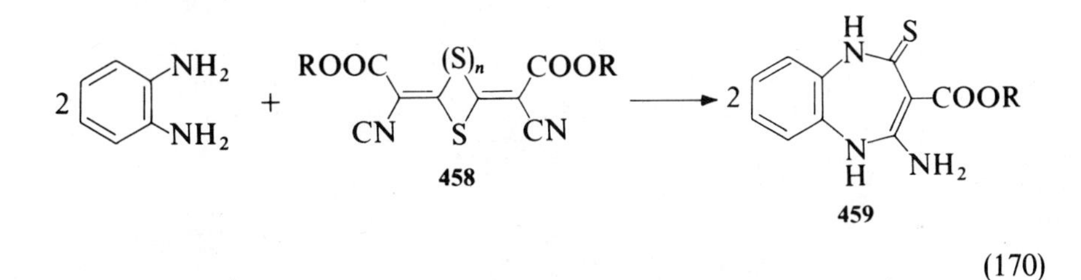


(170)

### *7.1.2. By Thiation of Lactam Oxygen*

Thiation of the corresponding 1,5-benzodiazepin-2-ones or 2,4-diones has been the most widely used method of synthesizing thiones.

Szarvasi and coworkers[205] extended the synthesis of the 1,3,4,5-tetrahydro-1,5-benzodiazepin-2(2*H*)-thione **460** ($R_1 = R_2$ = H) described by Kiprianov and Khilya[259] to several substituted derivatives and obtained compounds **461** ($R_1$ = H) in 66–84% yield by allowing the lactams **460** ($R_1$ = H) to react with phosphorus pentasulfide in boiling pyridine (Eq. 171).

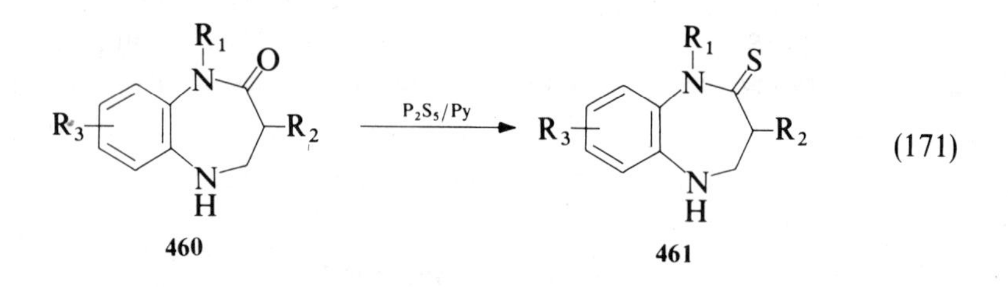


The same reagents were likewise used for the preparation of the 1-aryl analogs **461** ($R_1$ = aryl, $R_2$ = H) and for the thiation of **462** to give **463** (Eq. 172).[145]

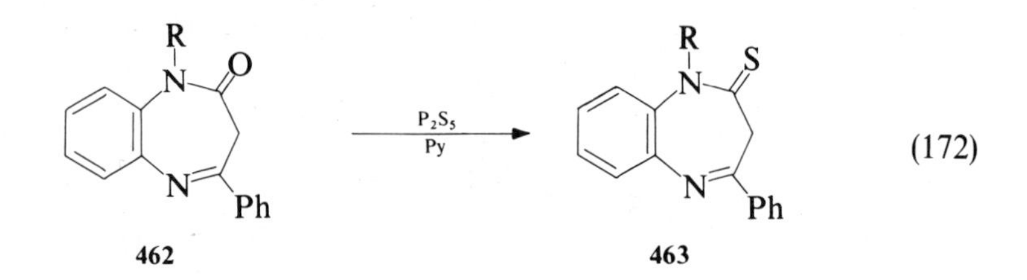


Good regioselectivity was observed during the thiation of the 2,4-diones **464**, affording mainly the 4-thione **465** (X = O) and only small amounts of the 2,4-dithione **465** (X = S) (Eq. 173).[159,161,260]

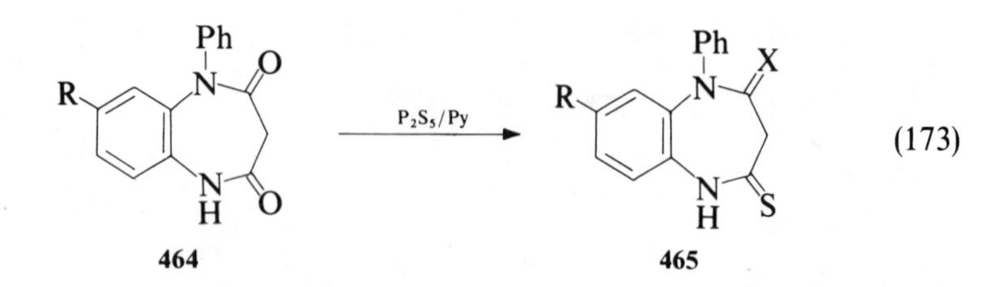

## 7.2. Reactions

### 7.2.1. *Reactions with Electrophiles*

Alkylation and acylation of the thiones was discussed above in connection with the synthesis of dihydrobenzodiazepin-2-ones containing alkylthio substituents at the 4-position. Reaction of the thione **461** ($R_1$, $R_2$, $R_3$ = H) with bromoacetone in benzene has been reported[259] to yield the thiazolium salt **466**. The thiazolobenzodiazepine **467** was obtained in low yield by reaction of the thione **465** (R = $CF_3$, X = O) with 1-bromo-2,2-diethoxyethane.[260] Structure **468** was assigned[261] to the product obtained by alkylation of **459** (R = Et) with methyl 2-bromopropionate in the presence of sodium ethoxide.

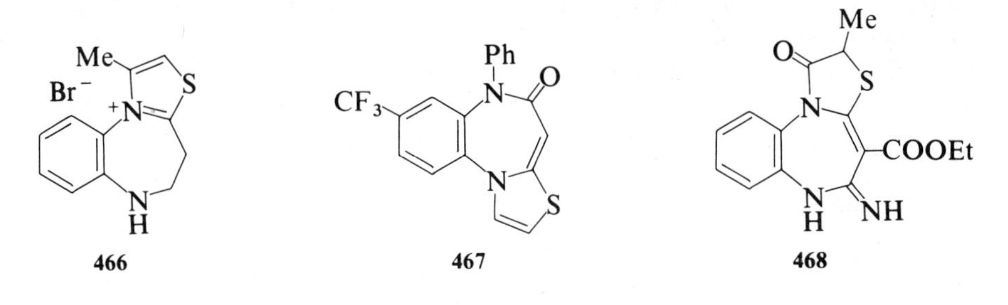


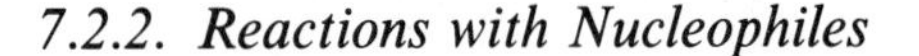

### 7.2.2. *Reactions with Nucleophiles*

Thiation has been used to activate the lactam carbonyl to prepare the 4-hydrazino derivatives **470** (Eq. 174).[159,160] The triazolobenzodiazepines **472**

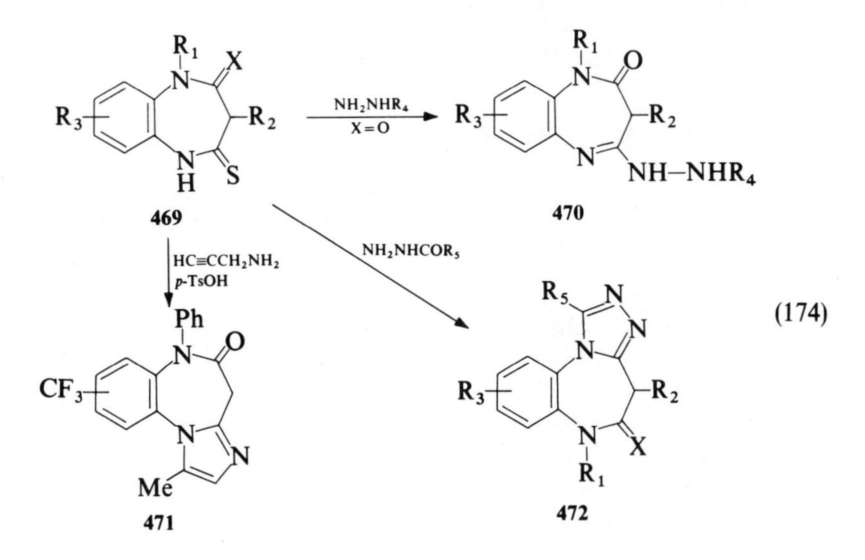


(174)

($R_1$ = Ph; X = O, S, $H_2$) were obtained in one step by condensing the appropriate thiones **469** either with acetylhydrazine in refluxing butanol[159,260] or with various aroylhydrazines by heating neat or in refluxing trimethylbenzene or *n*-butanol.[205] The imidazobenzodiazepine **471** was accessible by reaction of the appropriate thione with propargylamine in *n*-butanol containing *p*-toluenesulfonic acid.[260]

### 7.3. Spectral Data

Benassi and coworkers[173] carried out an nmr study at various temperatures with the thiones **463**. Compound **463** (R = H) showed a singlet for the C-3 protons at room temperature, as did the corresponding 2-one. At lower temperatures this singlet changed into an AB system. The 1-methyl- and 1-benzyl-substituted analogs **463** (R = Me, $CH_2Ph$) exhibited an AB system for the methylene protons even at room temperature. The coalescence temperature and free energy of activation for ring inversion were determined and are given below.

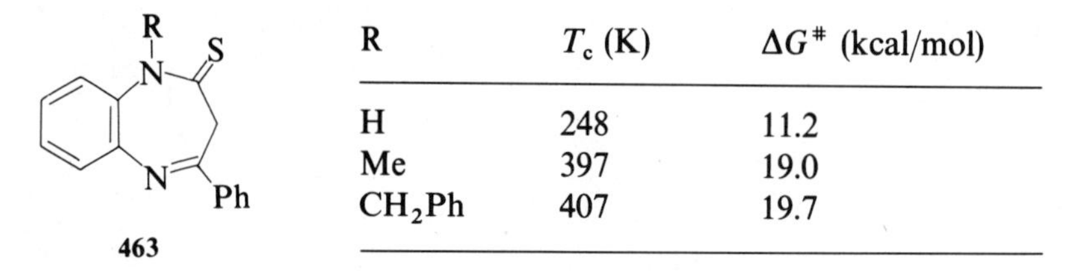


| R | $T_c$ (K) | $\Delta G^{\#}$ (kcal/mol) |
|---|---|---|
| H | 248 | 11.2 |
| Me | 397 | 19.0 |
| $CH_2Ph$ | 407 | 19.7 |

The mass spectral fragmentation of thiones **463** (R = H) with substituents on the 4-phenyl ring has been described.[262]

## 8. HEXAHYDRO-1,5-BENZODIAZEPINES

### 8.1. 5*a*,6,7,8,9,9*a*-Hexahydro-1*H*-1,5-benzodiazepines

#### *8.1.1. Synthesis and Spectral Data*

The title compounds were accessible by condensation of 1,2-diaminocyclohexane **473** with 1,3-diketones (Eq. 175). The 2,4-dimethyl derivative **474** (R = Me) was first prepared and characterized by Lloyd and coworkers.[1,263] The mass spectral data of this compound and of the 2,4-diphenyl analog were also reported.[264] Potter and coworkers[265] prepared the cis and trans isomers of **474** (R = Me) and described the proton-nmr spectra of these compounds.

$^{13}$C-Nmr spectra for of **474** (R = Me) were also obtained and compared with those of other 1,4-diazepines.[266]

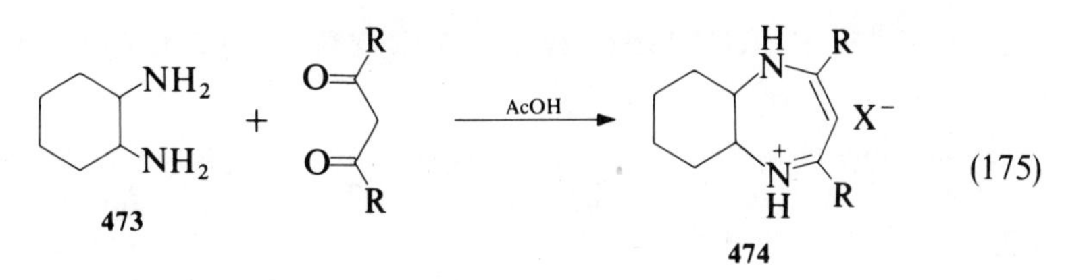


## 8.2. 2,3,4,6,7,8-Hexahydro-1*H*-1,5-benzodiazepines

### 8.2.1. *Synthesis*

McDougall and Malik[267] studied the reaction of 1,2-diketones with 1,3-diamines and obtained the diazepine **475** by condensation of cyclohexan-1,2-dione with 1,3-diaminopropane (Eq. 176). The structure of **475** was supported by infrared and ultraviolet data.

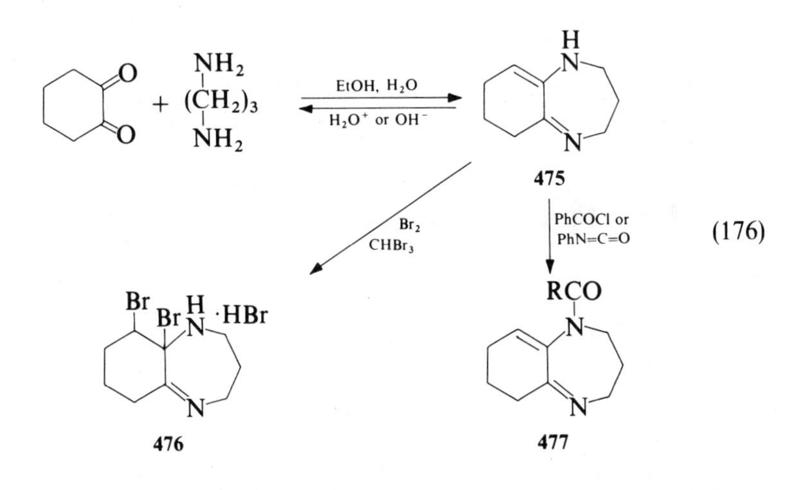


### 8.2.2. *Reactions*

Alkaline or acid hydrolysis of **475** resulted in cleavage to the starting diketone and diamine. Isomerization attempts with ethoxide and dehydrogenation experiments with chloranil were unsuccessful. Reaction of **475** with bromine in bromoform gave the hydrobromide of a dibromo derivative, to which structure **476** was assigned. Acylation with benzoyl chloride or phenylisocyanate led to unstable and ill-defined products assigned structures **477** (R = Ph, R = NHPh).

## 9. PERHYDRO-1*H*-1,5-BENZODIAZEPINES

Settimj and coworkers[268] studied the reaction of diketones with diamines in the presence of cyanide (Strecker's synthesis). The 5*a*,9*a*-dicyanoperhydro-1*H*-1,5-benzodiazepine **478** was prepared in 87% yield by condensation of cyclohexan-1,2-dione with 1,3-diaminopropane in the presence of hydrogen cyanide (Eq. 177). The dimer **479** was isolated as a by-product.

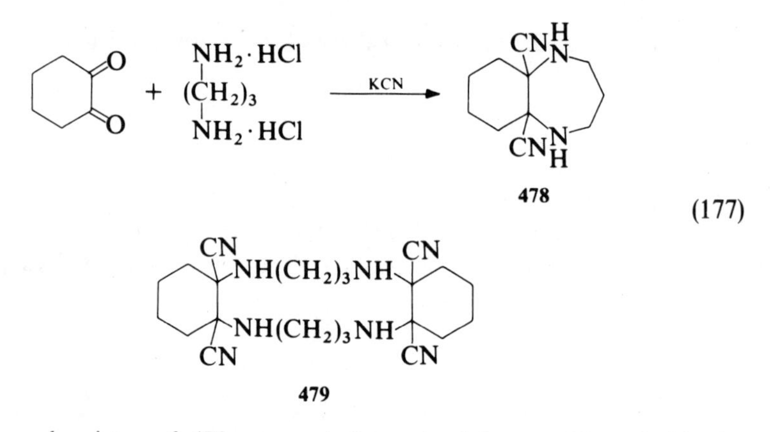


The stereochemistry of **478** was not determined but could probably be assigned the thermodynamically more stable trans configuration.

## 10. CYCLOBUTA[*b*] [1,4]DIAZEPINES

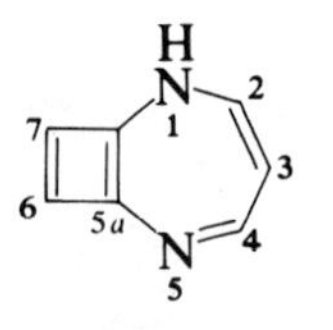


**480a**
1*H*-Cyclobuta[*b*][1,4]diazepines

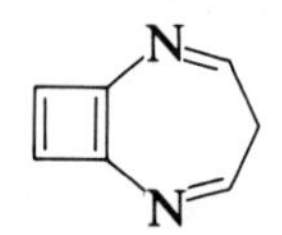


**480b**
3*H*-Cyclobuta[*b*][1,4]diazepines

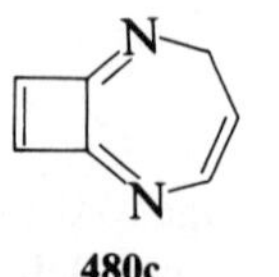


**480c**
2*H*-Cyclobuta[*b*][1,4]diazepines

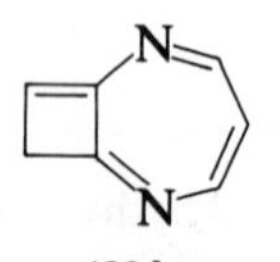


**480d**
6*H*-Cyclobuta[*b*][1,4]diazepines

The only derivatives belonging to this ring system, represented by the four possible tautomers **480a**–**480d**, were synthesized by Seitz and Morck.[269] These authors reported the synthesis of the diazepines **483** by condensation of the diaminocyclobutenediones **481** with diethyl malonate at elevated temperature. Reaction of **481** (R = Me) led directly to the diazepine **483** (R = Me). The same condensation with **481** (R = H), on the other hand, gave the intermediate **482** (R = H) and required the use of sodium ethoxide to effect ring closure to the diazepine **483** (R = H) (Eq. 178).

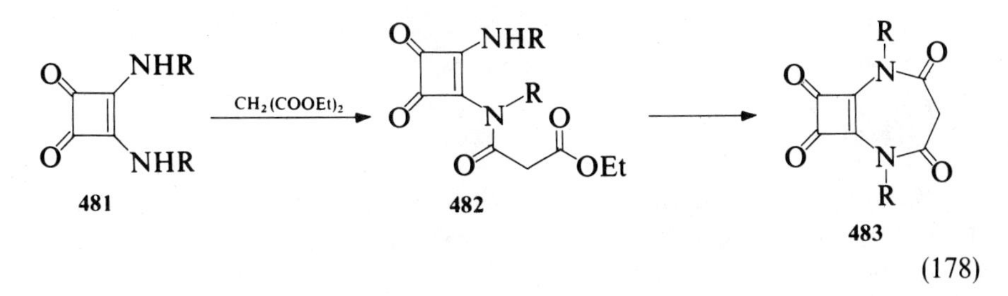


(178)

This difference of reactivity of methylated versus unmethylated **481** was attributed to a stabilization of the intermediate **482** (R = H) by intramolecular hydrogen bonding.

## 11. CYCLOPENTA[*b*] [1,4]DIAZEPINES

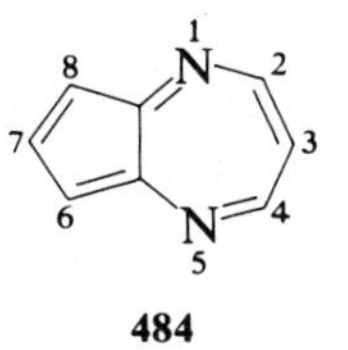


The aromatic diazaazulene **484** itself has not been reported in the literature. The hexahydro derivative **485** was accessible by condensation of 1,2-diaminocyclopentane with acetylacetone (Eq. 179). The effects of pH and temperature on the reaction of trans-1,2-diaminocyclopentane with acetylacetone were studied by Lloyd and Marshall.[270]

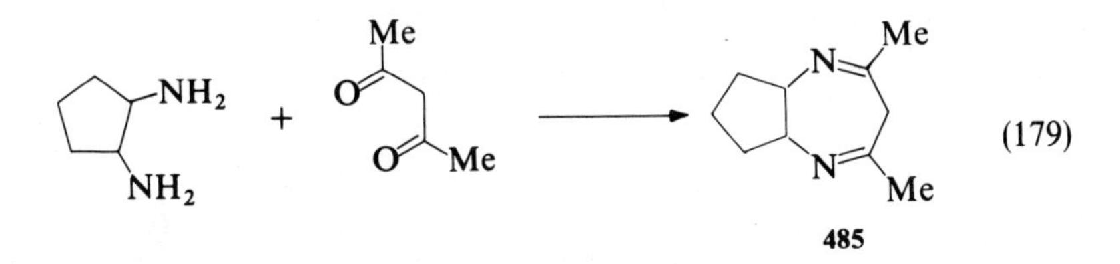


(179)

Yields of 80% were obtained at 59 °C over a pH range of 4–10. At room temperature a minimum yield was observed at pH 8, while maximum yields were obtained at both pH 4 and pH 11. The trans as well as cis isomers of **485** were later prepared for proton-nmr investigations.[265]

## 12. ISOXAZOLO[4,5-*b*] [1,4]DIAZEPINES

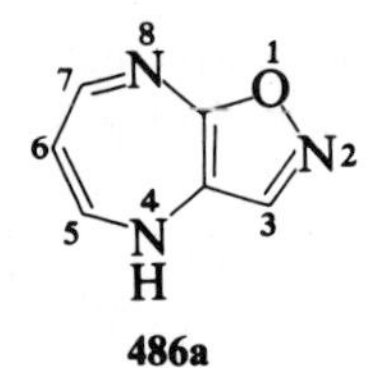

**486a**

4*H*-Isoxazolo[4,5-*b*]-[1,4]diazepines

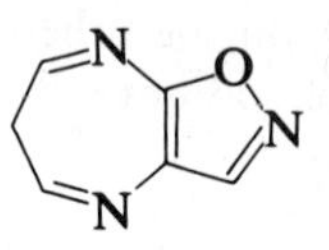

**486b**

6*H*-Isoxazolo[4,5-*b*]-[1,4]diazepines

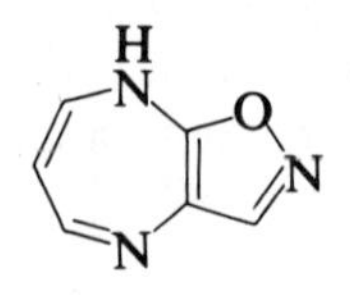

**486c**

8*H*-Isoxazolo[4,5-*b*]-[1,4]diazepines

This ring parent may theoretically exist in any of eight tautomeric forms. Three of these, **486a**–**486c**, have an intact isoxazole ring and are considered to be thermodynamically favored.

### 12.1. Synthesis

Only the 3,5,7-trisubstituted derivatives **489** of the 6*H* tautomers **486b** have been reported.[271,272] These compounds were synthesized by condensation of the 4,5-diaminoisoxazoles **487** with pentane-2,4-dione (Eq. 180). Treatment of **487** (R = Ph) with aqueous pentane-2,4-dione at room temperature resulted in the formation of the enamine that has been assigned structure **488**. Ring closure of **488** to **489** (R = Ph) was achieved in refluxing ethanol in the presence of triethylamine.

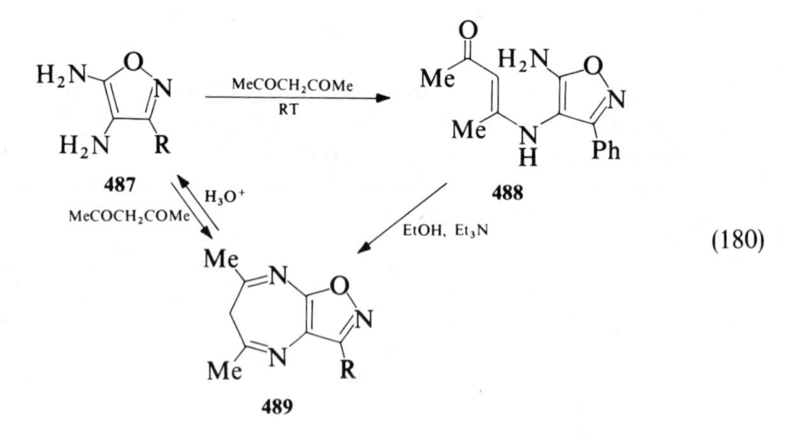


(180)

### 12.2. Reactions

Protonation of **489** led to a deep violet-blue diazepinium cation believed to be derived from the 4*H* tautomer **486a** or the 8*H* compound **486c**. The formation of a dication was not observed. Acid hydrolysis resulted in cleavage of **489** to **487**.

## 13. 1,2,5-OXADIAZOLO[3,4-*b*] [1,4]DIAZEPINES

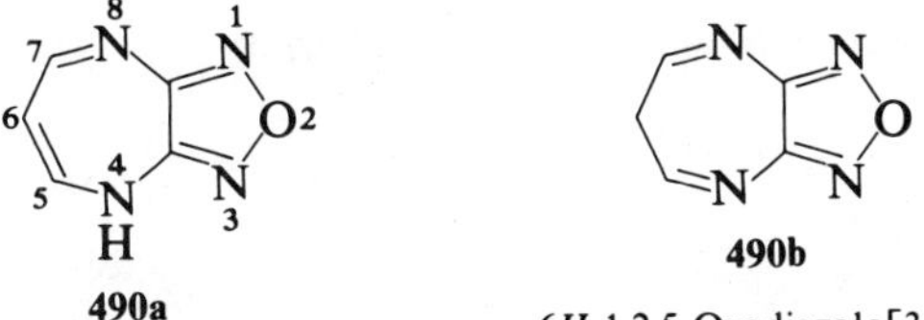

**490a**

4*H*-1,2,5-Oxadiazolo[3,4-*b*] [1,4]diazepines

**490b**

6*H*-1,2,5-Oxadiazolo[3,4-*b*] [1,4]diazepines

5,7-Disubstituted derivatives **492** of the parent ring system **490** were prepared by Gasco and coworkers[273] in the usual manner: that is, by condensation of the diamine **491** with 1,3-dicarbonyl compounds (Eq. 181).

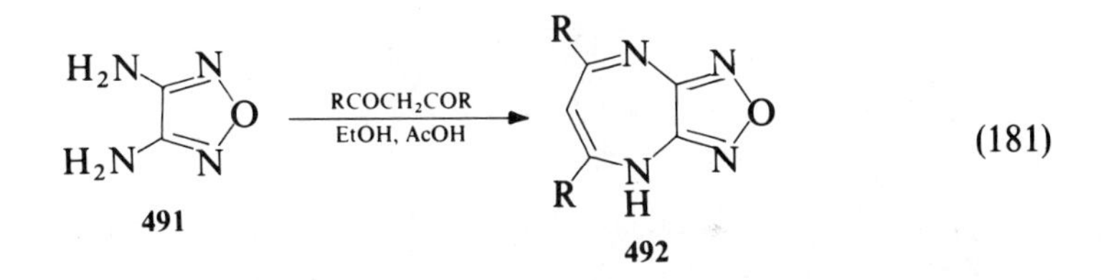


Unlike the corresponding benzodiazepine derivatives, these compounds exist in the 4*H* tautomeric form **490a**. Thus, the 5,7-dimethyl derivative showed two methyl signals in the nmr spectra at low temperature, with coalescence at – 29°C indicating localization of the exchangeable proton at the 4-position. At higher temperature the proton is rapidly exchanged between the 4- and 8-positions, rendering the methyl groups attached to the 5- and 7-positions magnetically equivalent. The unsymmetrical 5-methyl-7-phenyl analog showed a preference for only one tautomer, the proton being attached to either the 4- or 8-nitrogen.

## 14. PYRAZOLO[3,4-*b*] [1,4]DIAZEPINES

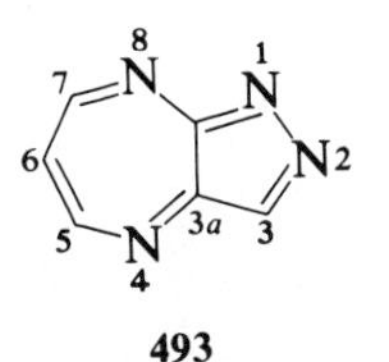

**493**

## 14.1. Synthesis

The polysubstituted 1,6-dihydro derivative **495** of the parent ring system **493** was synthesized by Affane-Nguema and coworkers.[274] The diaminopyrazole **494** was condensed with acetylacetone to give an 85–90% yield of the diazepine **495** (Eq. 182). Under milder conditions the intermediate enamine **496** was formed in quantitative yield and could be cyclized (R = H) by refluxing in butanol.

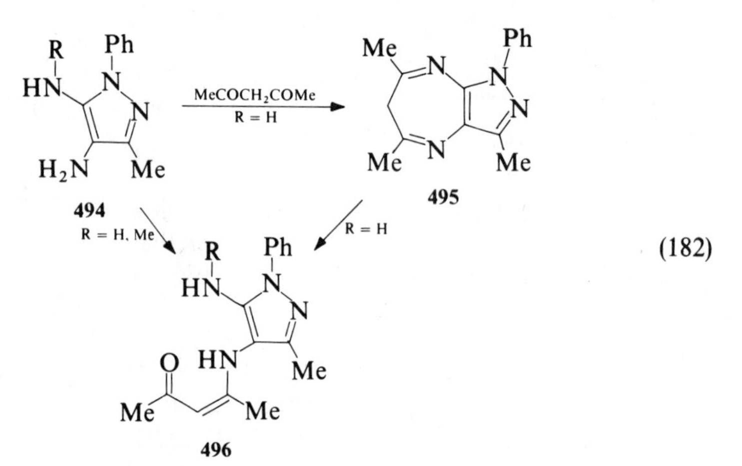


(182)

Cyclization could not be effected with the corresponding 3-methylamino-pyrazole **496** (R = Me).

The condensation of the diaminopyrazoles **494** ($R_1$ = H, Me) with ethyl acetoacetate and ethyl 2-acetopropionate was also studied.[274] Again the corresponding enamine intermediate **497** could be isolated in high yield and subsequently cyclized with sodium ethoxide in ethanol to the diazepinones **498** ($R_1$ = H; $R_2$ = H, Me) (Eq. 183).

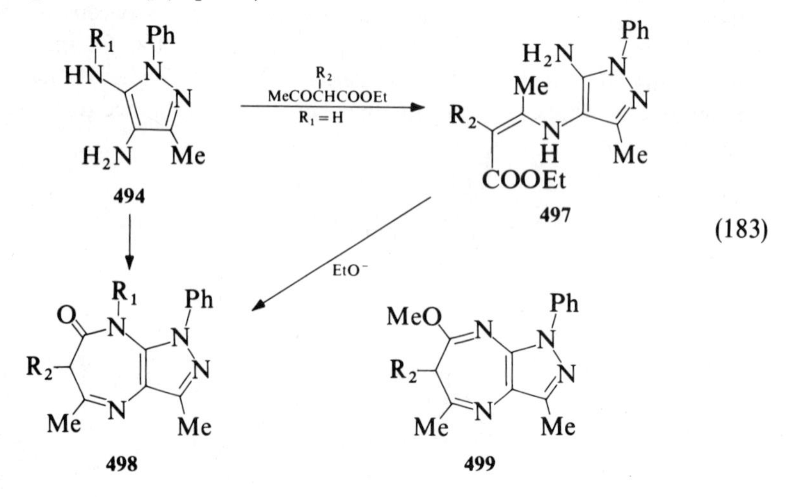


(183)

Reaction of the diamines **494** ($R_1$ = H, Me) with the ketoester in boiling xylene gave the diazepinones **498** in much lower yield. Since the enamine **497** was not cyclized under thermal conditions, a different reaction path must account for the formation of **498** in refluxing xylene. In this instance, the amide bond may be established before the imine is formed.

### 14.2. Reactions

Alkylation of the pyrazolodiazepinones **498** ($R_1$ = H) with diazomethane led to the *N*- and *O*-methylated products **498** ($R_1$ = Me) and **499**, in 75 and 20% yield.[274] However, methylation with sodium methoxide and methyl iodide gave only **498** ($R_1$ = Me).[275] Reaction of **498** ($R_1$ = Me, $R_2$ = H) with methyl iodide in a sodium–liquid ammonia system yielded 50% of the 6,6-dimethyl compound **500** and a lesser amount of the monomethylated product **498** ($R_1 = R_2$ = Me). When the 8-position nitrogen carried a hydrogen substituent, reductive alkylation of the imine bond was observed as well. Thus, **498** ($R_1$ = H, $R_2$ = H) reacted with methyl iodide and sodium in liquid ammonia to give **501** ($R_1$ = H, $R_2$ = Me) and **501** ($R_1 = R_2$ = Me) in 40 and 30% yield. Diazomethane reacted with **501** ($R_1 = R_2$ = H) to form the 8-methyl derivative **501** ($R_1$ = Me, $R_2$ = H) as the major product. Small amounts of the dimethylated compound **501** ($R_1$ = $R_2$ = Me) were also obtained. The methylthio derivative **503** resulted from the alkylation of the thione **502** with diazomethane (Eq. 174). This thione was prepared by treatment of the corresponding lactam **498** ($R_1$ = $R_2$ = H) with phosphorus pentasulfide and pyridine.

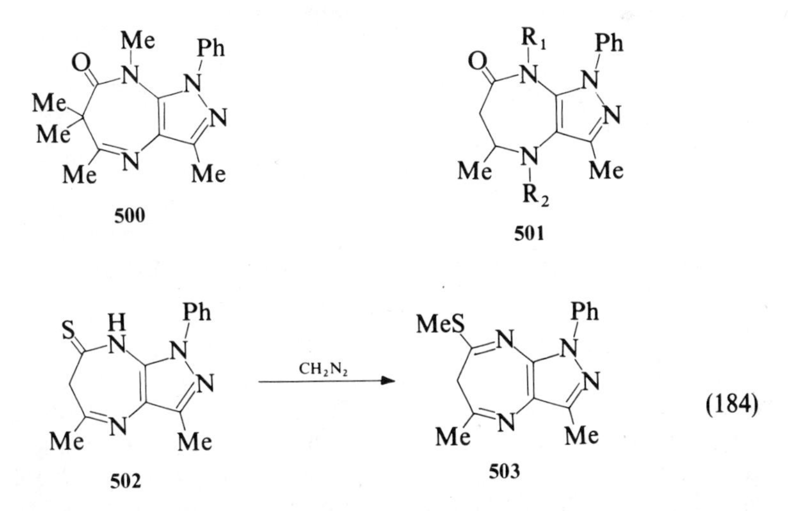


Hydrogenation of the imine double bond in **498** ($R_1 = R_2$ = H) over palladium on carbon gave high yields of **501** ($R_1 = R_2$ = H). The same transformation was also achieved in lower yield by reduction with sodium in liquid ammonia.

Reduction of **498** ($R_1 = R_2 = H$) with lithium aluminum hydride in boiling ether for 6 hours afforded a mixture of **505** (R = H) and **501** ($R_1 = R_2 = H$) together with **504** (R = H) (Eq. 175). More vigorous conditions, 24 hours of reflux, led to the single product **505** (R = H) in 90% yield. Similar conditions converted **495** to **505** (R = Me), as well as **504** (R = H, Me) to **505** (R = H, Me). A highly selective reduction of the 7,8-imine leading to **504** (R = Me) was observed during the hydrogenation of **495** over palladium on carbon.

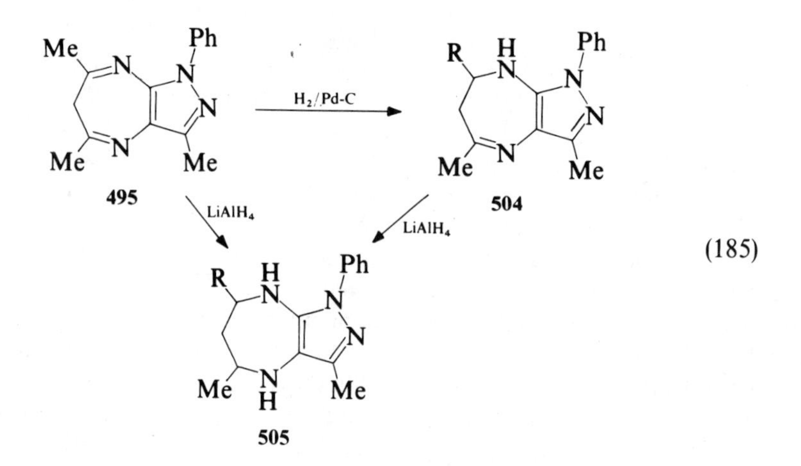


(185)

## 14.3. Spectral Data

The mass spectral fragmentation of many of these pyrazolodiazepines has been reported.[276]

## 14.4. 4,8-Dihydropyrazolo[3,4-*b*] [1,4]diazepin-5,7(1*H*,6*H*)-diones

A pyrazolone compound was reacted at 0–5°C with benzenediazonium chloride in glacial acetic acid to form the phenylhydrazone **506**, which upon treatment with boiling phosphorus oxychloride can be chlorinated at the 5-position to give the 4-phenylazo-5-chloropyrazole **507**. Subsequently, the chlorine was substituted with an appropriate amine ($R_5NH_2$) at 100–160°C to form **508**. Reaction of **508** with malonic acid derivative yielded **509** as a product. The azo group of **509** can be split by catalytic hydrogenation (with Pd, Pt, or Raney nickel) to form **510**, cyclization of which gave **511** (Eq. 186).[321, 322]

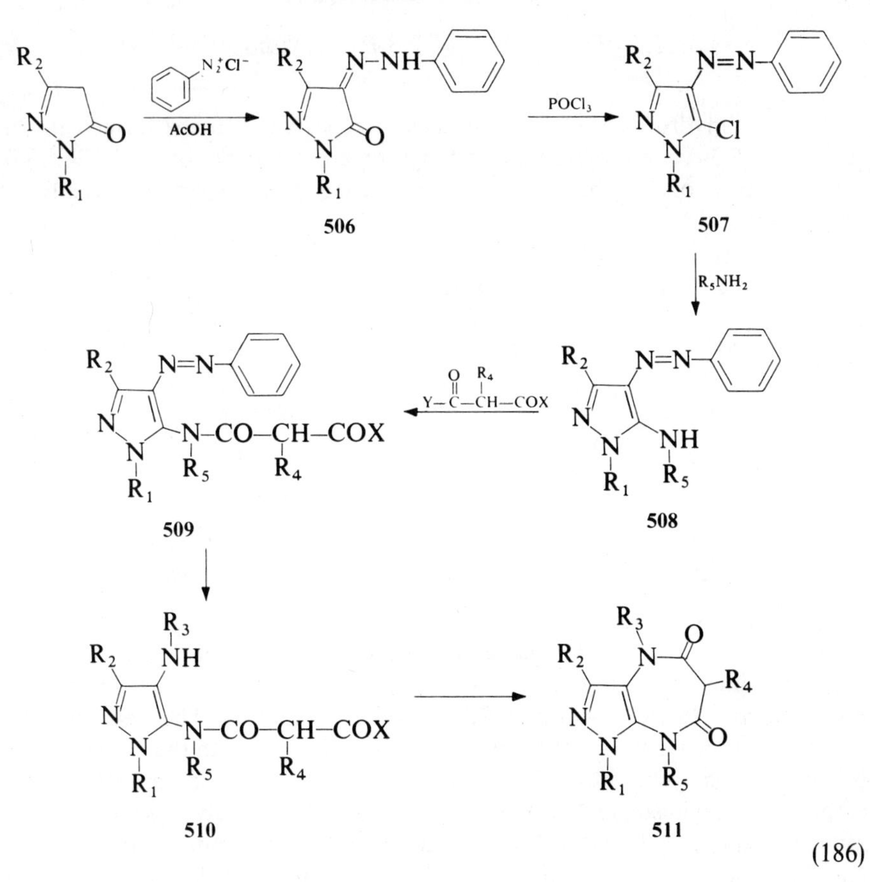


(186)

# 15. PYRIDO[*b*] [1,4]DIAZEPINES

## 15.1. Pyrido[2,3-*b*] [1,4]Diazepines

Examples of this parent ring system represented by the tautomers **512a–512c** have not been described in the literature. Compounds with higher degree of saturation are discussed below.

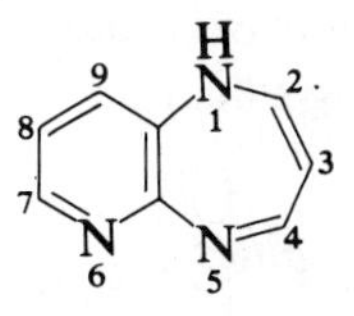

**512a**
1*H*-Pyrido[2;3-*b*]-[1,4]diazepines

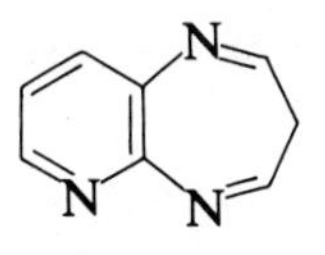

**512b**
3*H*-Pyrido[2,3-*b*]-[1,4]diazepines

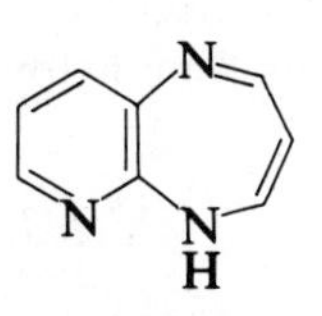

**512c**
5*H*-Pyrido[2,3-*b*]-[1,4]diazepines

### 15.1.1. *Dihydropyrido[2,3-b] [1,4]diazepines*

The 4,5-dihydro derivative **514** has been reported[111] to be formed in the reaction of 2,3-diaminopyridine with the *β*-dimethylaminoketone **513** (Eq. 187). The structure of **514** has not been confirmed either spectroscopically or chemically.

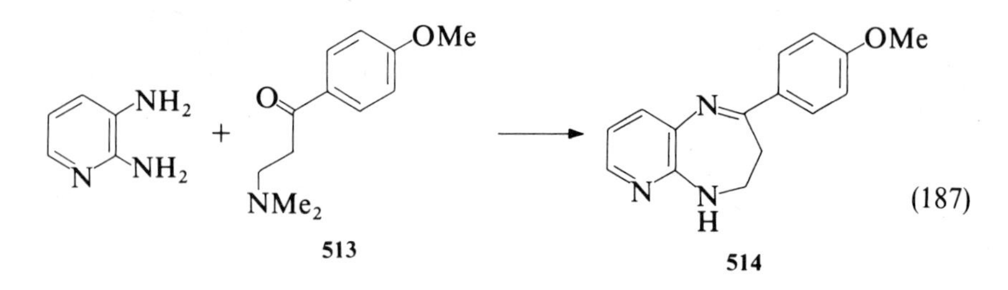


### 15.1.2. *Dihydropyrido[2,3-b]* [1,4]*diazepinones*

Israel and coworkers[277] have investigated the condensation of 2,3-diaminopyridine with ethyl acetoacetate and reported the formation of the regioisomers **516** (R = Me) and **517** (R = Me) (Eq. 188). The enamine **515**, arising from the reaction of the more basic amino group with the ketone, could be isolated and cyclized to **517** (R = Me) by treatment with sodium ethoxide in ethanol. The regioisomer **516** (R = Me) was formed thermally together with the imidazopyridines **520** and **521**. According to spectroscopic data **516** (R = Me) and **517** also exist in the tautomeric forms **518** and **519**. Structure **519** was assigned to the colorless crystalline form and structure **517** to the yellow tautomer in solution. On the basis of nmr data, **516** (R = Me) were present in solution together with the tautomer **518** in a ratio of about 1:3.

The reaction of 2,3-diaminopyridine with ethyl acetoacetate in refluxing xylene was also investigated by Nawojski[278] and later by Lavergne and coworkers[279] who reported, respectively, yields of a 8.5 and 60% of the diazepinone **517** (R = Me). The analogous reaction with ethyl benzoylacetate[280] gave 65% of the diazepine **517** (R = Ph). Condensation products with 4-methoxybenzoyl acetate and 3,4,5,-trimethoxybenzoylacetate were also described.[111] Two regioisomers corresponding to analogs of **516** (R = Me) and **517** (R = Me) were similarly obtained by the condensation of 5-bromo-2,3-diamino-4-methylpyridine with ethyl acetoacetate in boiling toluene.[281]

### 15.1.3. *Tetrahydropyrido[2,3-b] [1,4]diazepines*

The tetrahydro derivative **523** was prepared[111] by catalytic hydrogenation of the corresponding dihydro compound **522** (Eq. 189). Desulfurization of the

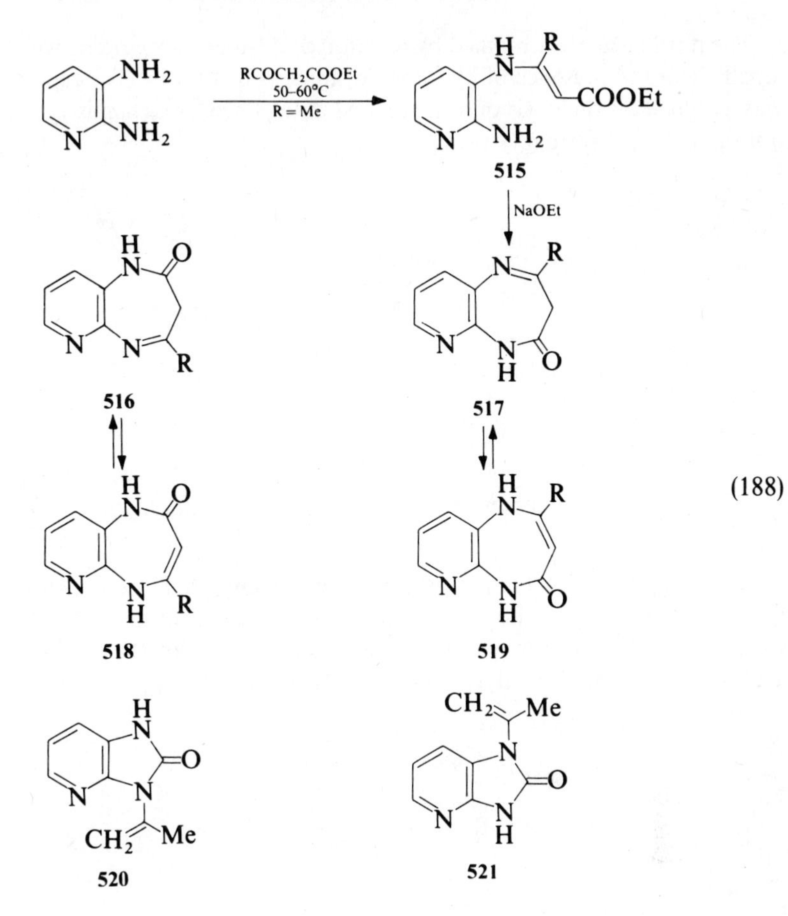


thiones **524** (R = H, Br) gave good yields of the reduced product **525** (Eq. 190).[282]

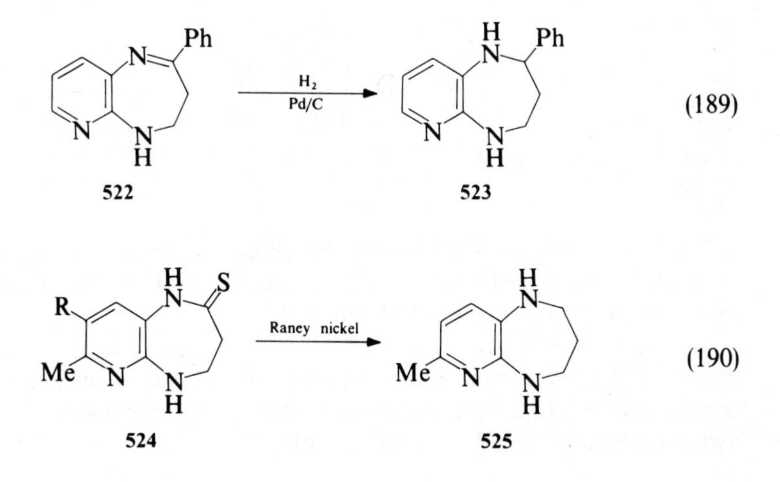

The nitrile **526** was obtained by reaction of 2,3-diaminopyridine with cinnamaldehyde in the presence of hydrogen cyanide (Eq. 191).[283] The regiochemistry was assigned on the basis of nmr spectral data. The stereochemistry, which was not discussed, is probably trans.

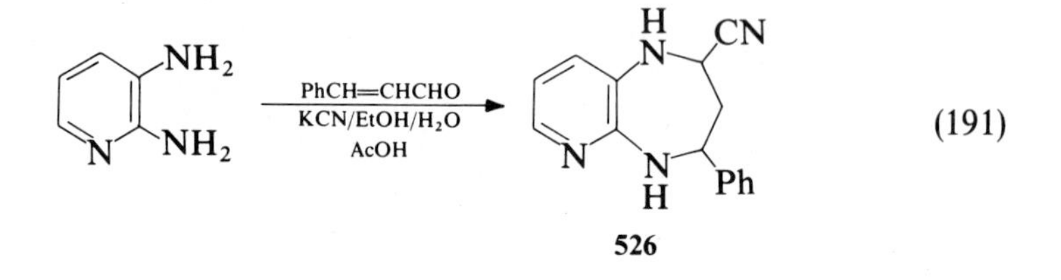


(191)

### 15.1.4. *Tetrahydropyrido[2,3-b] [1,4]diazepinones*

According to the patent literature[284] the 5-phenyl-7-chloro-substituted compounds **530** can be synthesized by thermal cyclization of the amino acids **529**, which are obtained by reaction of the pyridine **527** with the anilinopropionic acid **528** followed by catalytic reduction of the nitro group (Eq. 192).

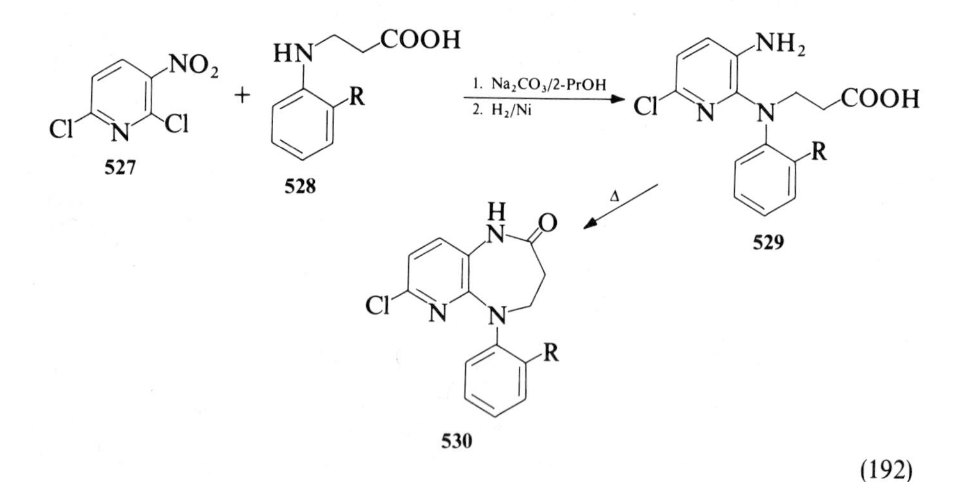


(192)

The tetrahydropyrido[2,3-*b*] [1,4]diazepin-2-ones **532** were isolated as minor products of the Schmidt reaction on the ketone **531**.[282] The major products were the isomeric diazepines **533** (Eq. 193).

Compounds **532** were converted in high yield to the corresponding thiones **534** by treatment with phosphorus pentasulfide in pyridine.[282] The tetrahydropyrido[2,3-*b*] [1,4]diazepin-4-one **535** was prepared[285] by catalytic hydrogenation of **517** (R = Me) (Eq. 194).

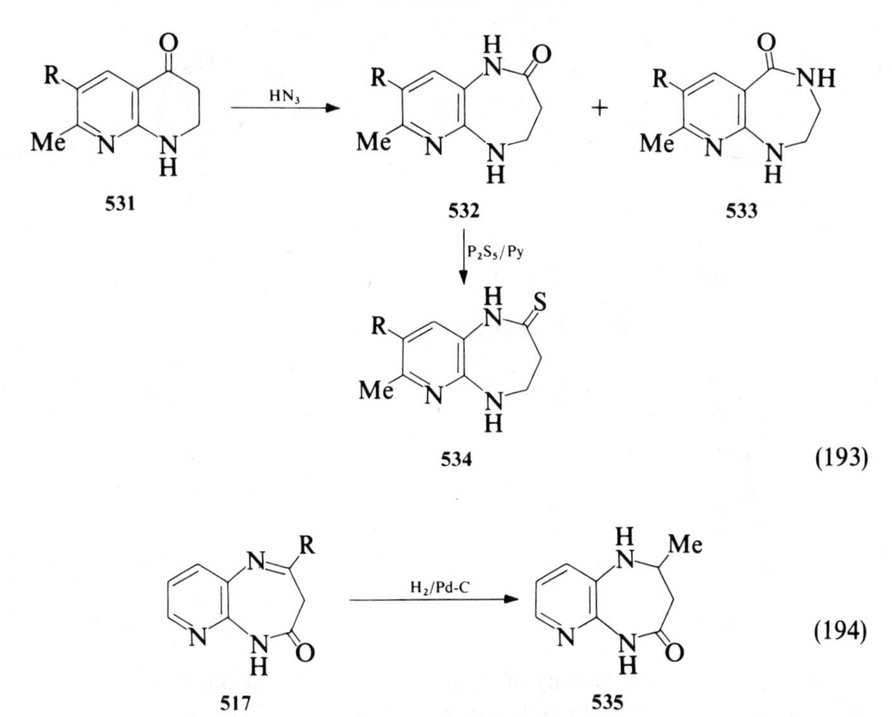


### 15.1.5. *Tetrahydropyrido[2,3-b] [1,4]diazepinediones*

The diones **536** ($R_1$ = H) were accessible[286] in 12–20% yield by condensation of substituted malonic acid esters with 2,3-diaminopyridine (Eq. 195). A better yield was reported[287] for the preparation of the 3,3-diallyl derivative by using sodium ethoxide in boiling xylene.

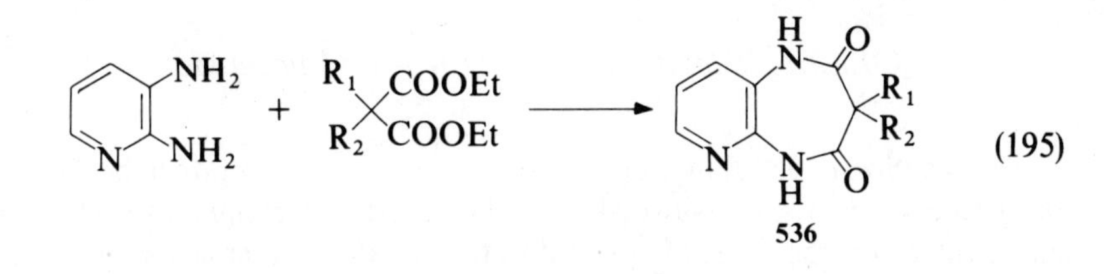


Ring closure of the malonic ester amides **537** (Eq. 196) with sodium ethoxide in ethanol has been utilized for the synthesis of chloro-substituted pyridine derivatives including the 5-phenyl compounds **538** ($R_1$ = Ph).[284] The starting materials **537** were prepared by acylation of the appropriate 2-amino-3-nitropyridine followed by reduction of the nitro group.

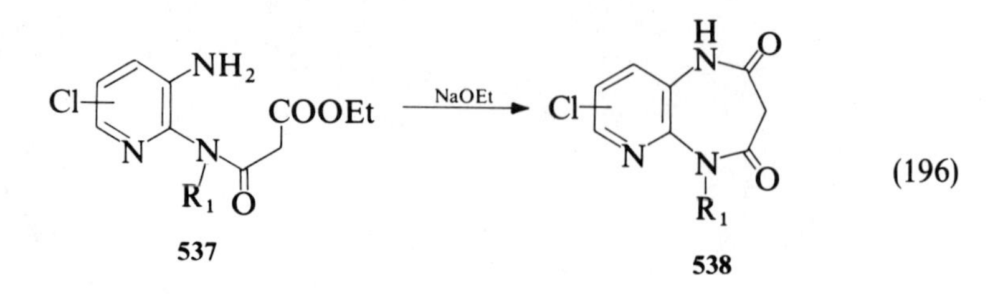


(196)

## 15.2. Pyrido[3,4-*b*] [1,4]diazepines

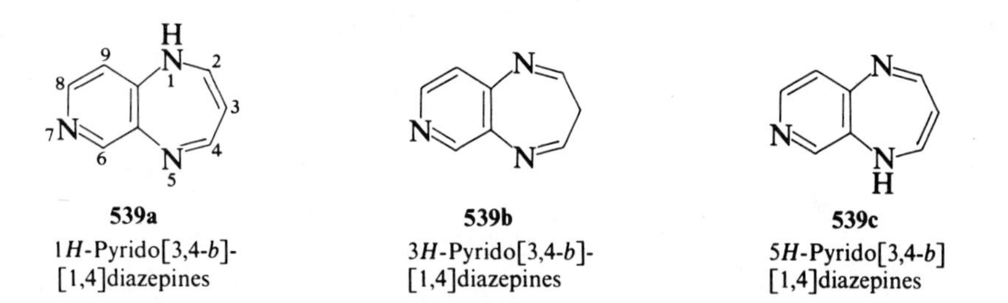


Maintaining the aromaticity of the pyridine ring would allow this ring system to exist as any of the three tautomers **539a–539c**. Derivatives of the 3*H*-tautomeric form, compounds **541**, have been synthesized by the condensation of the diaminopyridines **540** with acetylacetone (Eq. 197).[288]

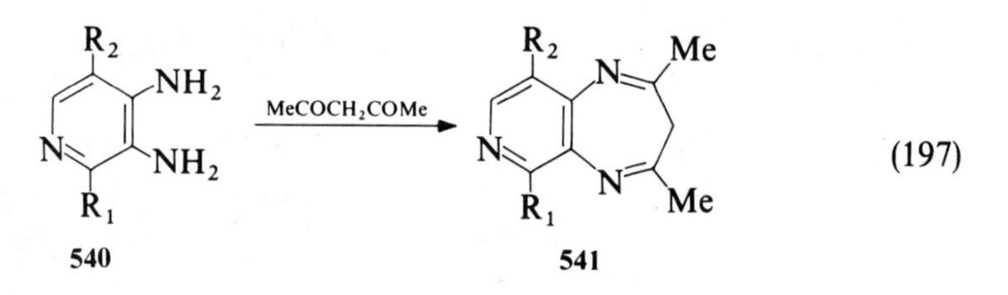


(197)

### 15.2.1. *Dihydropyrido[3,4-b] [1,4]diazepinones*

The reaction of 3,4-diaminopyridine with ethyl acetoacetate in boiling xylene was first investigated by Nawojski,[289] who incorrectly assigned structure **544** to this product on the basis of spectral data (Eq. 198). Later work by Israel and Jones[290] showed that this assignment was incorrect. The latter authors obtained **542** in 60% yield as a mixture of the 1*H* and 3*H* tautomers by condensation of the reactants in boiling toluene. The regioisomer **544**, on the other hand, was synthesized via the enamine **543** by ring closure with sodium ethoxide in ethanol. Compound **544** existed only as the 3*H* tautomer in solution. Fragmentation of these compounds upon electron impact was discussed.[279]

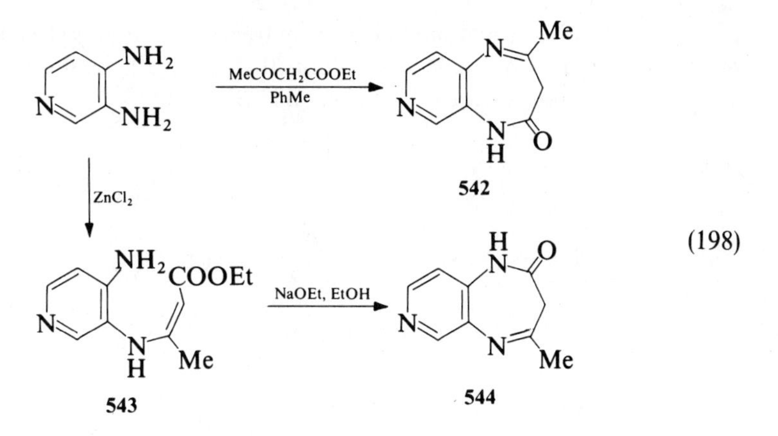


(198)

### 15.2.2. *Tetrahydropyrido[3,4-b] [1,4]diazepines and Diones*

Structure **545** (undefined stereochemistry) was assigned to the product obtained by the reaction of 3,4-diaminopyridine with cinnamaldehyde in the presence of hydrogen cyanide (Eq. 199).[283]

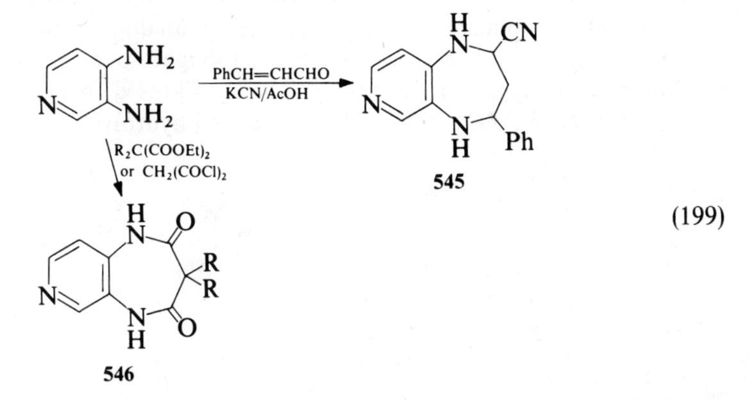


(199)

Compound **546** (R = H) was prepared in 58% yield by reaction of 3,4-diaminopyridine with malonyl chloride.[231] The 3,3-diallyl derivative **546** (R = $CH_2CH{=}CH_2$) was accessible by condensation of the same diamine with diethyl diallylmalonate in boiling xylene in the presence of ethoxide.[287]

## 15.3. Reactions

### 15.3.1. *Reactions with Electrophiles*

Both tetrahydropyrido[2,3-*b*] [1,4]diazepin-2-ones **547** (X = $H_2$, $R_1$ = Ph) and the corresponding 2,4-diones (X = O) were alkylated at the 1-position by reaction with sodium hydride followed by treatment with alkyl halides such as methyl iodide, allyl bromide, and 2-(dimethylamino)ethyl chloride to give the corresponding 1-substituted derivatives **548** (Eq. 200).[284]

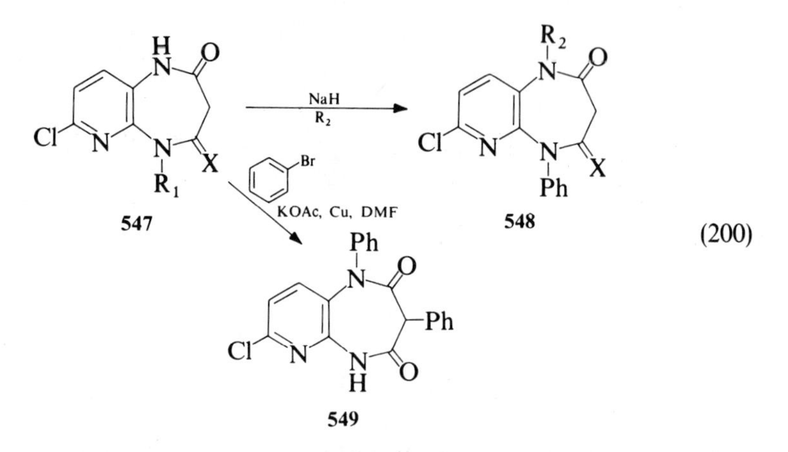


Arylation of **547** ($X = O$, $R_1 = H$) with bromobenzene, potassium acetate, and copper in dimethylformamide gave a low yield of the 1,3-diphenyl derivative **549**.[284] Conversion of the 3,3-diallyl derivative **550** to the lactone **551** (Eq. 201) by treatment with phosphoric acid proceeded in analogy to the corresponding benzodiazepine.[228, 230] The initial step is thought to be an intramolecular alkylation on oxygen. Interestingly, the [3,4]-fused pyridodiazepine **546** (R = allyl) behaved quite differently and yielded hydrolysis products only.

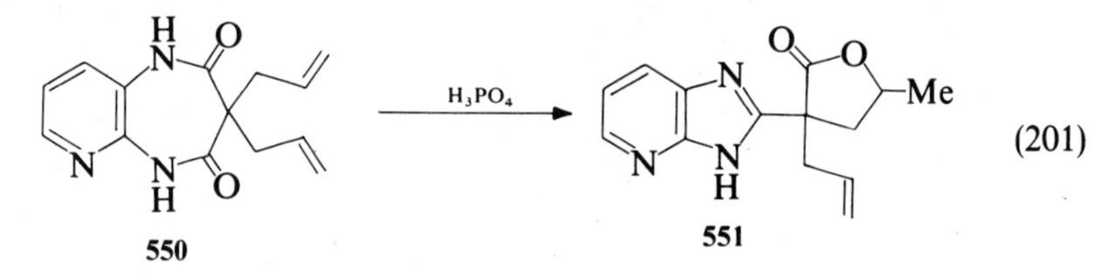


### 15.3.2. Reactions with Nucleophiles

Transformations of the lactam carbonyl to the thiolactam by treatment with phosphorus pentasulfide in pyridine were also successful in this series of compounds.[282, 291] Further conversion of the thione **552** to the amidine **553** by reaction with methylamine has been claimed (Eq. 202).[291]

### 15.3.3. Thermal Reactions

Thermal ring contraction of the dihydropyridodiazepinones **554** to the imidazopyridines **555**[292] appears to be general for compounds bearing the

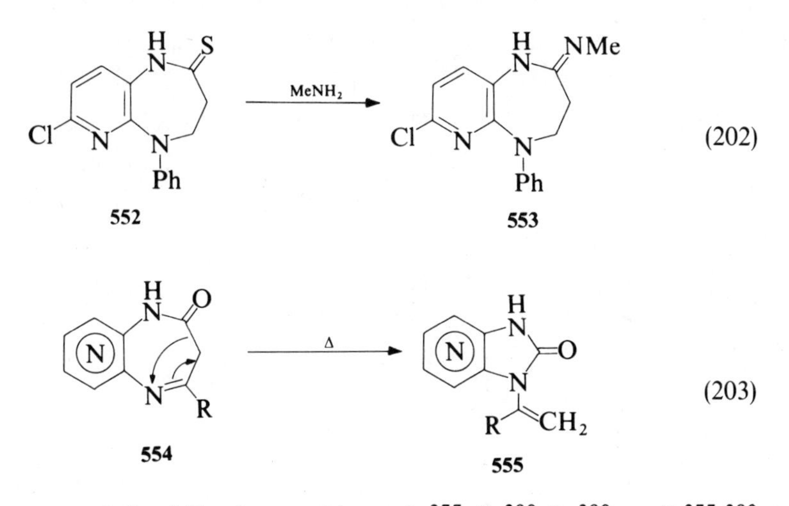


nitrogen at any of the following positions: 6-,[277] 7-,[290] 8-,[290] or 9-[277,280] (Eq. 203).

Compounds **555** were isolated as the major products during the preparation of **554** in boiling xylene.[278,281,290] This general reaction is a useful diagnostic tool for the determination or confirmation of the structures of dihydropyrido-diazepinones. Israel and coworkers[292] suggested a concerted mechanism for this reaction involving a [1,3]sigmatropic shift from carbon to nitrogen.

# 16. PYRIMIDO[4,5-*b*] [1,4]DIAZEPINES

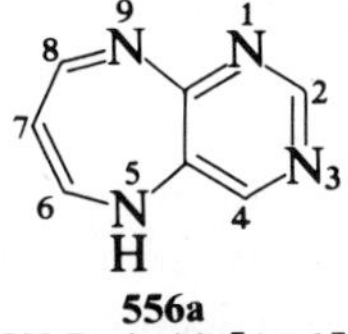

**556a**
5*H*-Pyrimido[4,5-*b*]-[1,4]diazepines

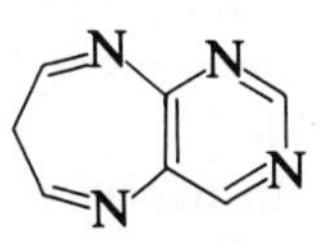

**556b**
7*H*-Pyrimmdo[4,5-*b*]-[1,4]diazepines

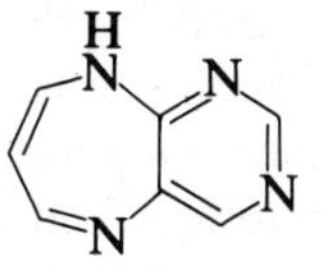

**556c**
9*H*-Pyrimido[4,5-*b*]-[1,4]diazepines

Compounds with this parent ring system represented by the tautomers **556a–556c** have not been reported in the literature.

## 16.1. Dihydropyrimido[4,5-*b*] [1,4]diazepinones

The first synthesis of a pyrimido[4,5-*b*] [1,4]diazepine was reported by Nyberg and coworkers,[293] who, on the basis of spectral data, assigned structure **557** (R = Me) to the product isolated in 38% yield from the condensation of ethyl acetoacetate with 4,5-diaminopyrimidine in boiling xylene (Eq. 204). The

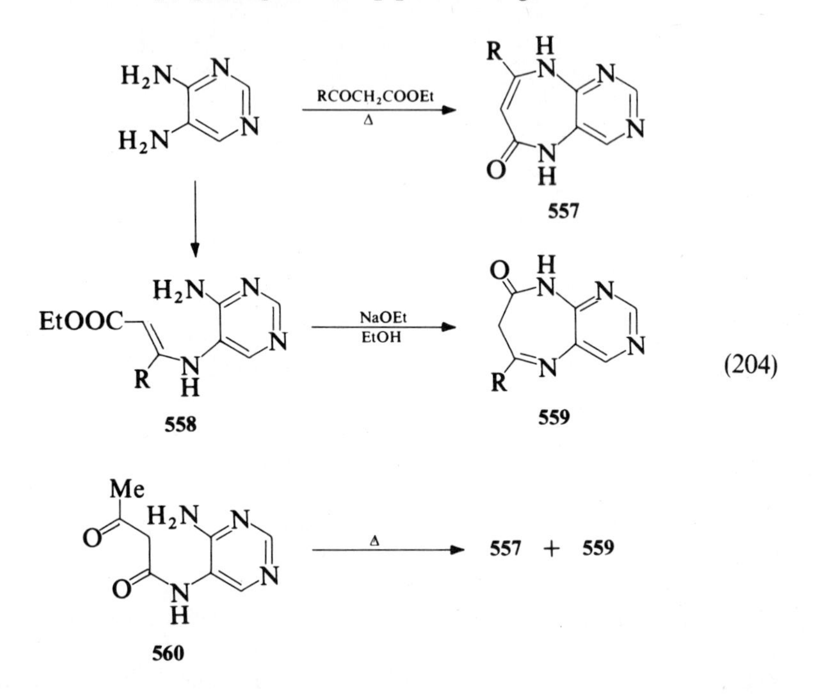


same reaction was later reinvestigated by Israel and coworkers,[294] who confirmed the structure of **557** and also prepared the regioisomer **559** (R = Me) by cyclization of the enamine **558** (R = Me) with sodium ethoxide in ethanol. Thermal ring closure of the acetoacetamide **560**, thought to be the intermediate in the formation of **557**, led to a mixture of **557** (R = Me) and **559** (R = Me), with the former being the predominant product. The formation of **559** (R = Me) from **560** may be due to an intramolecular transacylation. Some evidence of tautomerism was reported for **559** (R = Me).[294] A sublimed sample of this compound was colorless and exhibited an ultraviolet spectrum different from that of material recrystallized from ethanol. The structures **557** and **559** were confirmed by the base-catalyzed rearrangement to isopropenyl purinones (see Section 16.6, Reactions).

A similar condensation of 4,5-diaminopyrimidine with ethyl (3,4,5-trimethoxybenzoyl)acetate in refluxing xylene was reported[111] to give an 82% yield of the 5*H* tautomer of **559** (R = 3,4,5-$(MeO)_3C_6H_2$).

## 16.2. Tetrahydropyrimido[4,5-*b*] [1,4]diazepinones

The synthesis of **562** by condensation of the diamine **561** with 1,3-diketones in polyphosphoric acid has been described by Fukushima and coworkers (Eq. 205).[295]

Structure **563** was assigned to the intermediate enamines obtained by heating the diamine and the diketone in boiling butanol. Ring closure of **563** was effected

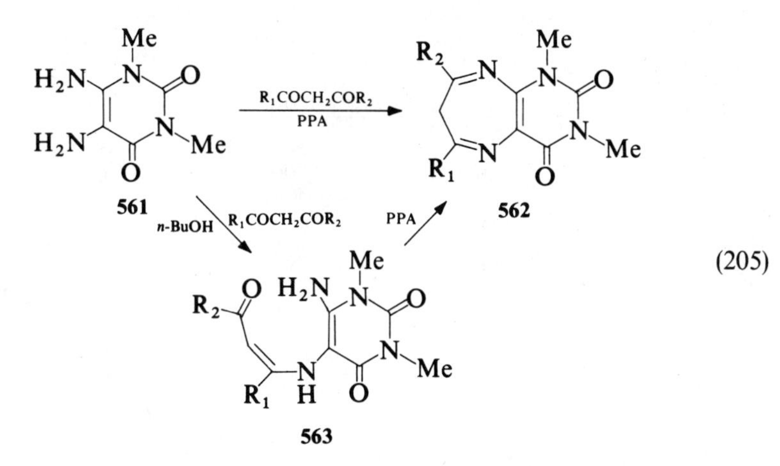


(205)

by treatment with polyphosphoric acid. While the symmetrical diketones gave acceptable yields of product, the reaction of **561** with benzoylacetone gave the expected mixture of regioisomers **562** ($R_1$ = Me, $R_2$ = Ph) and **562** ($R_1$ = Ph, $R_2$ = Me) in low yield.

## 16.3. Hexahydropyrimido[4,5-*b*] [1,4]diazepinones

The diamine **561** reacted smoothly with acetone or mesityl oxide (4-methylpent-3-en-2-one) to form a diazepine, which was assigned structure **564** on the basis of spectral data.[296] In particular, the infrared spectrum showed an absorption band indicating a free NH. The regioisomer would be expected to exhibit a chelated NH absorption band. According to the patent literature,[297] condensation of **561** with *β*-ketoesters led to the enamines **565** which could be cyclized in good yield to the diazepines **567** by treatment with sodium alkoxides or sodium hydroxide in methanol or ethanol (Eq. 206).

The diazepines **567** were originally formulated as the 5*H* tautomers, but the later work of Fukushima and coworkers[298] indicates clearly that the phenyl-substituted derivative **567** ($R_1$ = Ph, $R_2$ = H) has the structure shown in Eq. 206. These authors also synthesized the regioisomer **568** by thermal ring closure of the *β*-ketoamide **566**, which was prepared by heating the diamine **561** with ethyl benzoylacetate in xylene. Again, the tautomeric form shown was preferred on the basis of nmr spectroscopy.

An interesting condensation product was isolated[299] from the reaction of **561** with the unsaturated 1,4-diketone **569** in boiling acetic acid. On the basis of spectral data, the structure of the product was assigned as **570** (Eq. 207). A strong argument was the lack of the NH absorption bands in the infrared spectrum recorded in Nujol. However, it is not unlikely that **570** is in fact the 7*H* tautomer in the solid state and a mixture of the 7*H* and 5*H* tautomers in solution. In view of Bass's work on related benzodiazepines, structure **571** also has to be considered as a viable alternative.

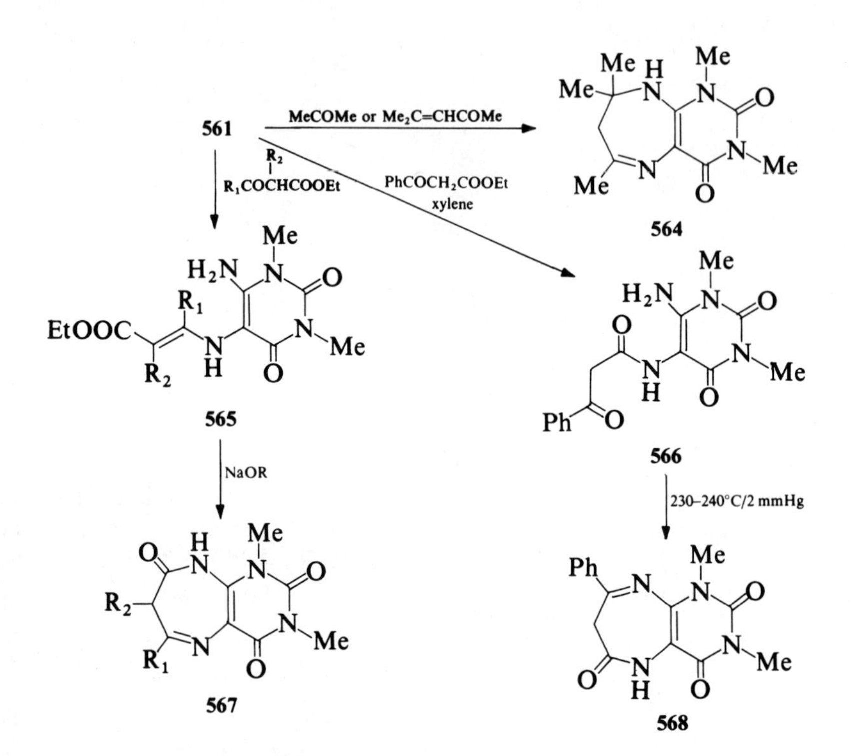
561
MeCOMe or $Me_2C{=}CHCOMe$
$R_1COCH(R_2)COOEt$
$PhCOCH_2COOEt$
xylene
564
565
566
NaOR
230–240°C/2 mmHg
567
568

(206)

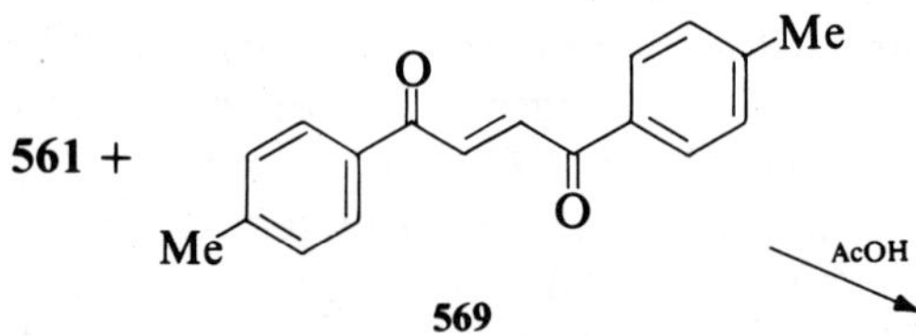
561 +
569
AcOH

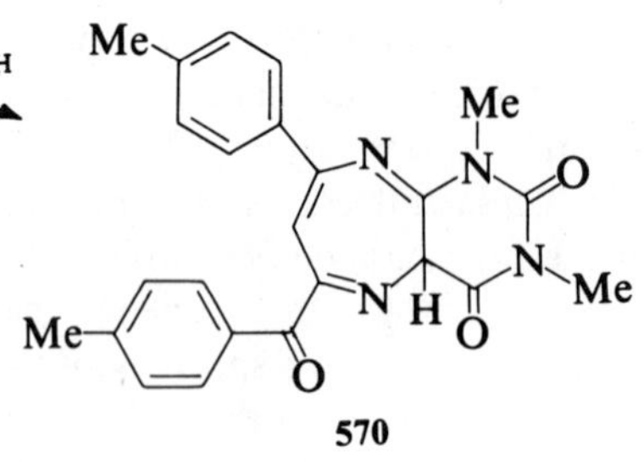
570

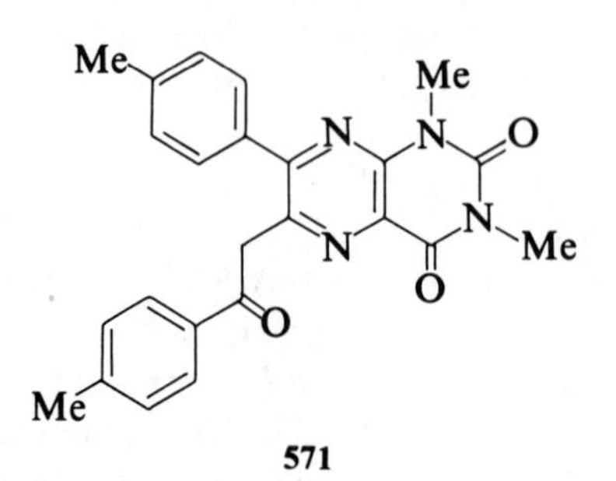
571

(207)

## 16.4. Octahydropyrimido[4,5-*b*] [1,4]diazepinones

The octahydrotriones **573** were prepared in good yield[297] by hydrogenation of the imine bond in **567** or by cyclization of the esters **572** with sodium ethoxide in ethanol. Compounds **572** were accessible by hydrogenation of the enamines **565** over platinum catalyst (Eq. 208).

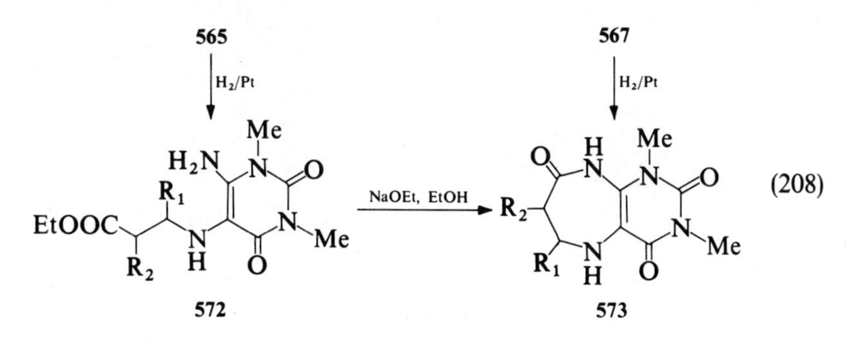


The stereochemistry of the product **573** ($R_1$ = Me, $R_2$ = Et), obtained by hydrogenation of the corresponding compound **567**, was not established.

## 16.5. Hexahydropyrimido[4,5-*b*] [1,4]diazepine-2-thiones

While the condensation of the diaminopyrimidinedione **561** with ethyl acetoacetate in boiling xylene led only to an acetoacetamide of type **566**, the corresponding 2-thione **574** gave a high yield of the diazepine **575** (Eq. 209).[300]

The reaction of **574** with diethyl ethoxymethylenemalonate was reported to lead to a diazepine of possible structure **576**.[301]

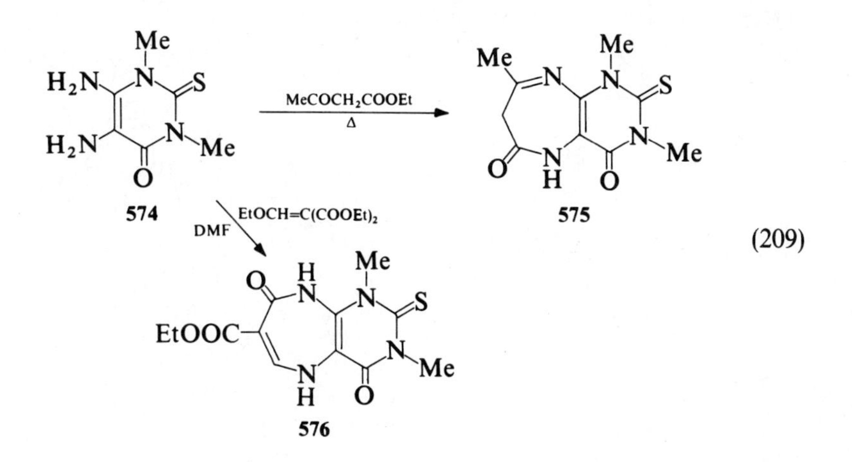

## 16.6. Reactions

### *16.6.1. Reactions with Electrophiles*

Nitrosation of **564** afforded a dinitroso compound that was assigned structure **577** (Eq. 210).[296]

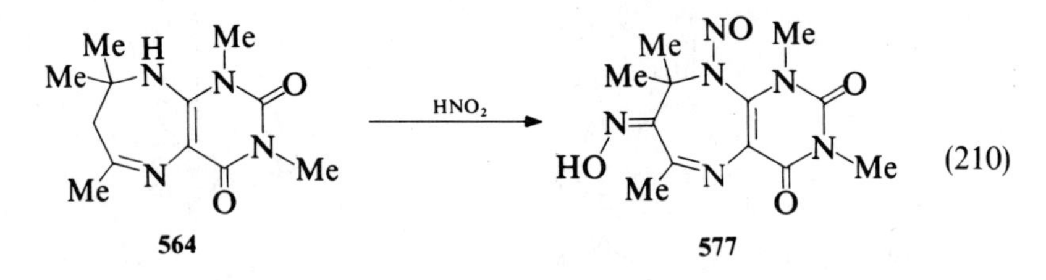


### *16.6.2. Reactions with Nucleophiles*

Compound **557** (R = Me) was found to be very susceptible to hydrolysis. Cleavage of the enamine occurred during recrystallization from 95% ethanol to give the acetoacetamide **560** (Eq. 211).[294]

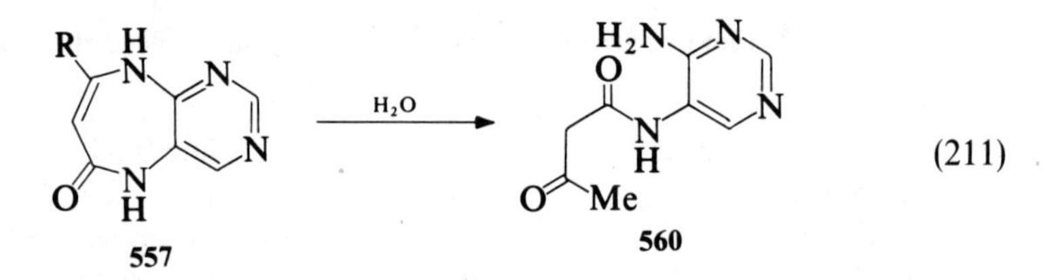


A similar ring opening took place during the reaction of **575** with 2,4-dinitrophenylhydrazine in aqueous ethanolic–sulfuric acid, leading to the crystalline hydrazone **578** (Eq. 212).[300]

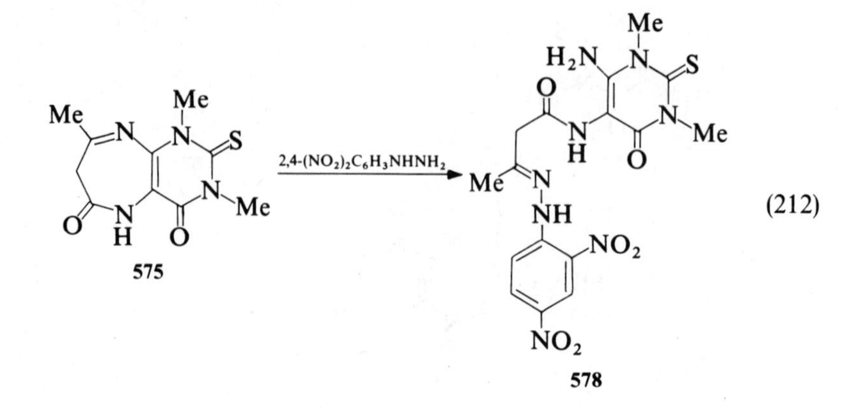

Reaction of **559** (R = Me) with sodium 2-ethoxyethoxide in hot 2-ethoxyethanol resulted in ring contraction to **579** (Eq. 213).[294] This rearrangement helped in the structure determination of **559** (R = Me).

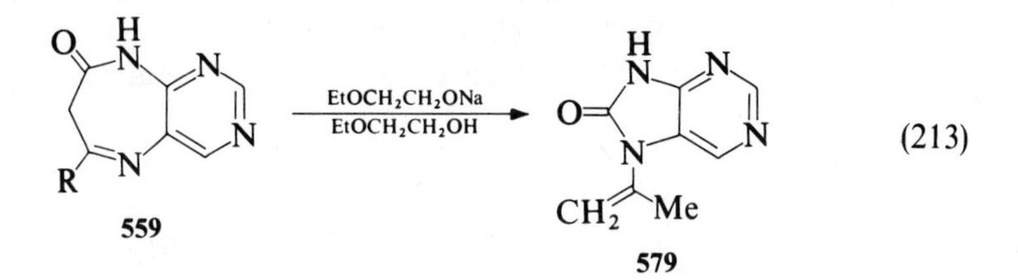


(213)

Hydrogenation of **564** with palladium on carbon was reported[296] to give a perhydro derivative of structure **580** of unknown stereochemistry (Eq. 214).

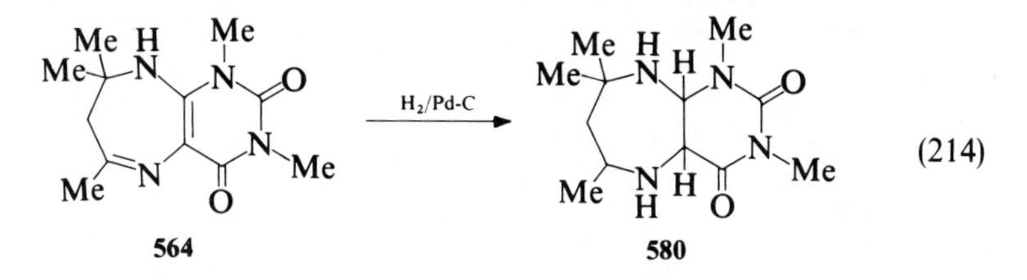


(214)

Reduction of the imine functionality in **567** was mentioned above.

### *16.6.3. Pyrolysis*

Pyrolysis of either **557** (R = Me) or **559** (R = Me) at 250°C gave the purinone **581** (Eq. 215),[294] not the expected isopropylene-substituted analogs.

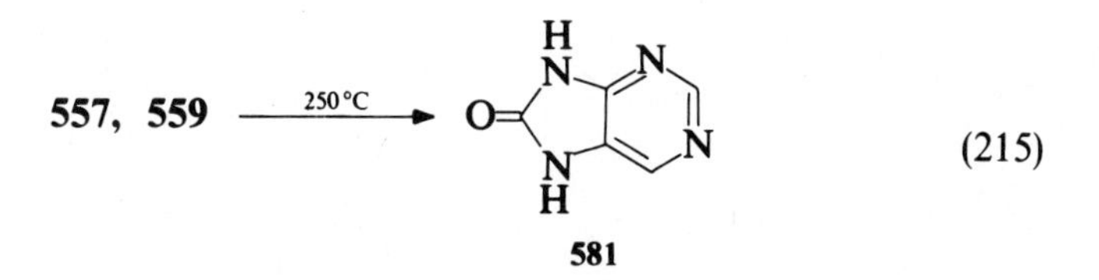


(215)

## 17. PYRROLO[3,4-*b*] [1,4]DIAZEPINES

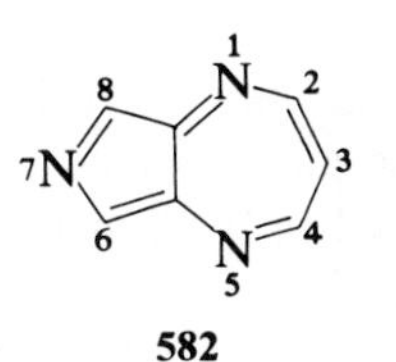

The only representatives of the pyrrolo[3,4-*b*] [1,4]diazepine ring system **582** reported in the literature appear to be the hexahydrodiones **585**.[302] These compounds were obtained by condensation of the diaminopyrrolinone **583** with ethyl acetoacetate and cyclization of the resulting enamine **584** using sodium ethoxide in ethanol (Eq. 216). Nmr spectra in dimethyl sulfoxide indicate the presence of a mixture of the tautomers **585** and **586** in a ratio of approximately 1:2.

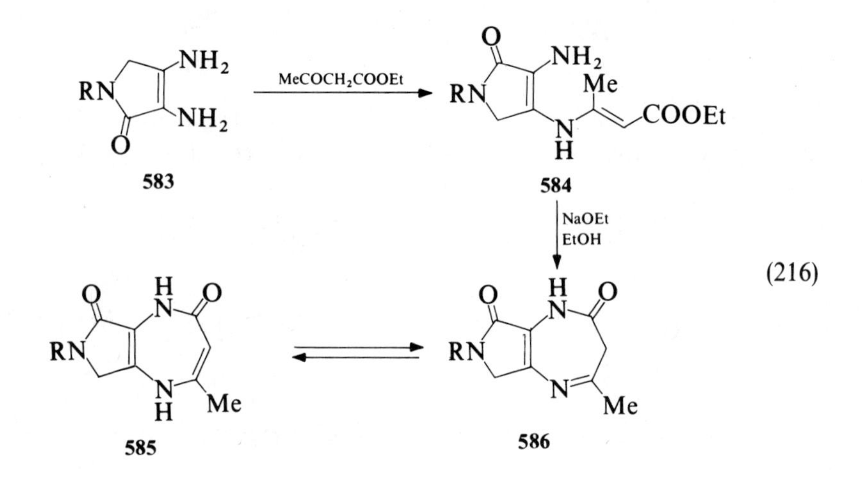


## 18. TRIAZOLO[4,5-*b*] [1,4]DIAZEPINES

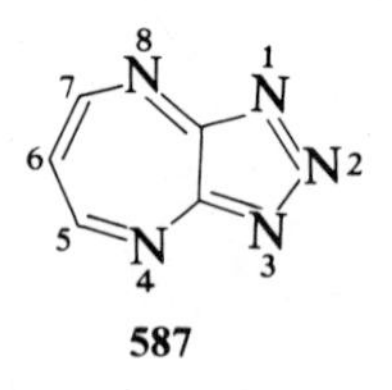


Compounds derived from the parent ring system **587** were described by Lovelette and Long.[303] These authors reacted the diaminotriazole **588** with ethyl acetoacetate and obtained in 90% yield the enamine that was assigned structure **589** (Eq. 217). This assignment was based on the assumption that the carbonyl group would react preferentially with the more reactive 4-amino function of **588**. Compound **589** was then cyclized as usual to the diazepine **590** by heating in ethanolic sodium ethoxide under reflux. An nmr spectrum in deuterated pyridine solution indicated the presence of tautomer **590**.

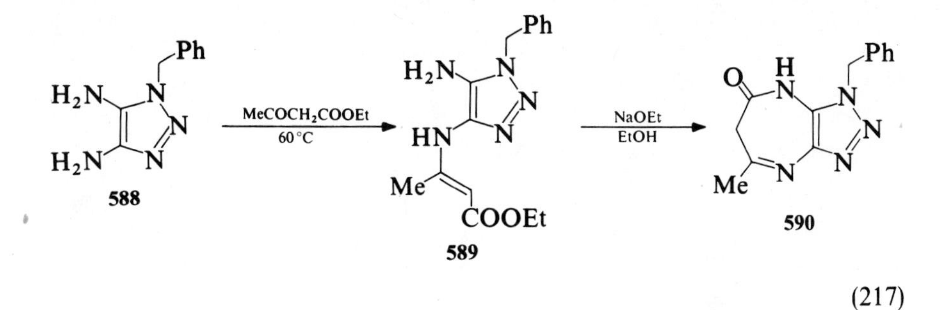


(217)

# 19. TABLES OF COMPOUNDS

TABLE IV-1. [*b*]-FUSED[1,4]DIAZEPINES

| Substituent | mp (°C); [bp (°C/torr)] | Solvent of Crystallization | Yield (%) | Spectra | Refs. |
|---|---|---|---|---|---|
| | | *1,5-Benzodiazepines* | | | |
| ***1H-1,5-Benzodiazepines*** | | 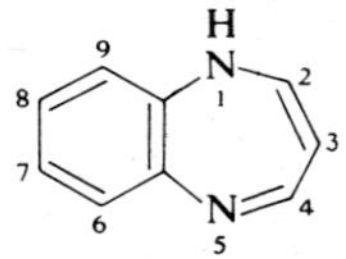 | | | |
| 1-Me-2-(4-Br$C_6H_4$) | | | | | |
| Perchlorate | 160–180d | | 69 | uv | 12 |
| 1-Me-2-(4-Cl$C_6H_4$) | | | | | |
| Perchlorate | 179–181d | | 77 | uv | 12 |
| 1-Me-2-(4-$NO_2C_6H_4$) | | | | | |
| Perchlorate | 196d | | 34 | uv | 12 |
| 2-PhCO-4-Ph | 169 | | 36 | pmr, uv, ms | 57 |
| 2-(PhCO$CH_2$)-4-Ph | 185 | PhMe | 88 | pmr | 55 |
| 2-EtO-4-(3,3-Diethylthio-2-propen-1-on-1-yl) | 162–165d | MeCN | 19 | | 66 |
| 2-EtO-4-(2-Ethoxy-1*H*-1,5-benzodiazepin-4-yl) | 166–167.5 | MeCN/Dioxane | 35 | | 66 |
| 3-COOEt-4-$NH_2$ | | | | | |
| Hydrochloride | 230d | EtOH | 80 | | 59, 60 |
| 3-CN-4-$NH_2$ | 203d | EtOH | | | 60 |
| Hydrochloride | 280d | | 92 | ir, pmr, ms | 59, 60 |
| 1-Me-2,3-$(Ph)_2$ | | | | | |
| Perchlorate | 210–211d | | 72 | uv | 12 |
| 5-Methoperchlorate | 115–116d | | 64 | uv | 12 |

| | | | | | |
|---|---|---|---|---|---|
| 1-Me-2-Ph-4-AcS | 166–169 | EtOAc | 82 | | 72 |
| 1-Me-2-Ph-4-$SCH_2CH_2NEt_2$ | Oil | | | | 72 |
| 1-Me-2-Ph-4-C(MeO)$(Ph)_2$ | 178 | $CHCl_3$/MeOH | 42 | ir, pmr | 50 |
| 1,2-$(Ph)_2$-4-C(MeO)$(Ph)_2$ | 180 | PhH/MeOH | 57 | ir, pmr | 50 |
| 1-Ac-2-Me-4-(2-AcO-5-Cl-$C_6H_3$) | 153–154 | EtOH | 56 | ir, ms | 31 |
| 1-Ac-2-Ph-4-(2-$AcOC_6H_4$) | 121–123 | EtOH | 43 | ir, ms | 31 |
| 1-Ac-2-Ph-4-(2-AcO-5-$ClC_6H_3$) | 135–137 | Cyclohexane | 70 | ir, ms | 31 |
| 1-Ac-2-Ph-4-(2-$HOC_6H_4$) | 163–164 | Cyclohexane | 28 | ir, ms | 31 |
| 1-Ac-2-Ph-4-(2-HO-5-$ClC_6H_3$) | 171–172 | Cyclohexane | | ir, ms | 31 |
| 1-$PhCH_2$-2-Ph-4-AcS | 179–182 | PhH/petr ether | 30 | pmr | 72 |
| 1-$PhCH_2$-2-Ph-4-$SCH_2CH_2NEt_2$ | Oil | | | | 72 |
| 1-(4-$MeC_6H_4SO_2$)-2,4-$Me_2$ | 360 | | | | 84 |
| Perchlorate | 185 | $H_2O$ | | | 84 |
| 1-NO-2,4-$Me_2$ | 80 | Petr ether | 7.5 | | 16 |
| 2-HO-3-COOEt-4-MeS | 202–204 | EtOH | 50 | ir, pmr | 73, 74 |
| 3-COOEt-4-$NH_2$-8-Cl | | | | | |
| Hydrochloride | 223d | $H_2O$ | 59 | ms | 60 |
| 3-CN-4-$NH_2$-8-Cl | 205d | EtOH | 60 | ir, ms | 60 |
| 1-Ph-2-SMe-4-$NMe_2$ | 116–117 | EtOH | 67.7 | ir, pmr | 314 |
| 1-Ph-2-SMe-4-$NEt_2$ | 84–85 | EtOH | 69.6 | ir, pmr | 314 |
| 1-Ph-2-SMe-4-$N(CH_2)_4$ | 182–183 | EtOH | 73.8 | ir, pmr | 314 |
| 1-Ph-2-$NMe_2$-4-$NMe_2$ | 148–149 | EtOH | 68.4 | ir, pmr | 314 |
| 1-Ph-2-$NEt_2$-4-$NMe_2$ | 152–153 | EtOH | 8.7 | ir, pmr | 314 |
| 1-Ph-2-$N(CH_2)_5$-4-$NMe_2$ | 151–152 | EtOH | 72.2 | ir, pmr | 314 |
| 1-Ph-2-$NMe_2$-4-$NEt_2$ | 95–96 | EtOH | 59.5 | ir, pmr | 314 |
| 1-Ph-2-$NEt_2$-4-$N(CH_2)_5$ | 113–114 | Petr ether | 61.8 | ir, pmr | 314 |
| 1-Ph-2-$N(CH_2)_5$-4-$NMe_2$ | 190–191 | EtOAc | 57 | ir, pmr | 314 |
| 1-Ph-2,4-$(N(CH_2)_4)_2$ | 202–203 | EtOAc | 90.7 | ir, pmr | 314 |
| 1-Ph-2-$N(CH_2)_4$-4-$N(CH_2)_5$ | 213–214 | Acetone | 84.3 | ir, pmr | 314 |
| 1-Ph-2-$N(CH_2)_4$-4-Morpholino | 179–180 | *n*-BuOH | 62.9 | ir, pmr | 314 |
| 1-Ph-2-$N(CH_2)_4$-4-($N(CH_2)_4$N-Me) | 185 | *n*-BuOH | 60.9 | ir, pmr | 314 |
| 2,4-$Me_2$-7-$CF_3$ | 214–215 | EtOH/$Et_2O$ | 56 | ir, pmr | 318 |
| 2-Ph-4-(4-$FC_6H_4$)-7-$CF_3$ | 228–230 | EtOH/$Et_2O$ | 38 | ir, pmr | 318 |

TABLE IV-1. —(contd.)

| Substituent | mp (°C); [bp (°C/torr)] | Solvent of Crystallization | Yield (%) | Spectra | Refs. |
|---|---|---|---|---|---|
| 2-(4-$FC_6H_4$)-4-Me-7-$CF_3$ | 140 | EtOH/$Et_2O$ | 35 | ir, pmr | 318 |
| 2-(4-$FC_6H_4$)-4-Me-8-MeO | 256 | EtOH/$Et_2O$ | 40 | ir, pmr | 318 |
| 2-(4-$FC_6H_4$)-4-Me-7-$NO_2$ | 205 | EtOH/$Et_2O$ | 42 | ir, pmr | 318 |
| 2-Ph-4-(4-$FC_6H_4$)-7-$NO_2$ | 223–225 | EtOH/$Et_2O$ | 45 | ir, pmr | 318 |
| 2,4-(4-$FC_6H_4$)$_2$-7-$NO_2$ | 155 | EtOH/$Et_2O$ | 39 | ir, pmr | 318 |
| 2-(4-$FC_6H_4$)-4-Me-8-Cl | 205 | EtOH/$Et_2O$ | 35 | ir, pmr | 318 |
| 2-Ph-4-(4-$FC_6H_4$)-8-Cl | 233 | EtOH/$Et_2O$ | 35 | ir, pmr | 318 |
| 2,4-(4-$FC_6H_4$)$_2$-8-Cl | 232 | EtOH/$Et_2O$ | 38 | ir, pmr | 318 |
| ***3H-1,5-Benzodiazepines***[‡] | | | | | |
| None | | | | | |
| Hydrochloride·$H_2O$ | 196.5 | | 80 | | 1 |
| Hydrobromide·$2H_2O$ | 187–189 | | | | 1 |
| Perchlorate·$H_2O$ | 192–196 | MeOH/$H_2O$ | 82 | | 1 |
| ***Monosubstituted*** | | | | | |
| 2-(4-$BrC_6H_4$) | | | | | |
| Perchlorate | 235–236 | MeOH | 74 | uv | 11 |
| Hydrochloride | 180–182 | MeOH | | | 11 |
| 2-(4-$ClC_6H_4$) | | | | | |
| Perchlorate | 221–222 | MeOH | 78 | uv | 11 |
| Hydrochloride | 162–163 | MeOH | | | 11 |

[‡] 2-Amino-, 2,4-diamino-, and 2-thio-substituted derivatives follow polysubstituted.

| | | | | | |
|---|---|---|---|---|---|
| 2-(4-$Me_2NC_6H_4$) | | | | | |
| Dihydrochloride | 179–181 | MeOH | 70 | uv | 11 |
| 2-(4-$MeOC_6H_4$) | | | | | |
| Perchlorate | 180–181 | MeOH | 78 | uv | 11 |
| Hydrochloride | 225–226d | MeOH | | | 11 |
| 2-Me | | | | | |
| Hydrochloride | 145d | EtOH | | | 13 |
| 2-Ph | | | | | |
| Hydrochloride | 183d | MeOH | 82 | uv | 11 |
| Perchlorate | 212–213d | MeOH | | | 11 |
| 2-(4-$NO_2C_6H_4$) | | | | | |
| Perchlorate | 196–198 | MeOH | 75 | uv | 11 |
| Hydrochloride | 196–197 | MeOH | | | 11 |
| 2-(4-$PhC_6H_4$) | | | | | |
| Hydrochloride | 142d | MeOH | 65 | uv | 11 |
| 3-Br | 184d | | 33 | ir | 3 |
| 3-(6-Cl-Benzoxazol-2-yl) | $> 300$ | | 70 | ir, pmr | 7 |
| 3-$NO_2$ | $> 360$ | | 57–81 | | 4–6 |
| 3-Ph | 268d | | 23 | | 2, 3 |
| ***2,3 Disubstituted*** | | | | | |
| 2-$PhCH_2$-3-Ph | | | | | |
| Perchlorate | 161–162 | MeOH | 78 | uv | 9 |
| 2,3-$Me_2$ | | | | | |
| Perchlorate | 167–168 | MeOH | 39 | uv | 9 |
| 2,3-$Ph_2$ | | | | | |
| Hydrochloride | 215–216d | MeOH | 74 | uv | 11 |
| Perchlorate | 259–260d | MeOH | | | |
| ***2,4 Disubstituted*** | | | | | |
| 2-($AcCH_2CO$)-4-Me | 186–188d | EtOH | 58 | ir | 36 |
| 2-(2-$AcOC_6H_4$)-4-Ph | 110–113 | Cyclohexane | 30 | ir, ms | 31 |
| 2-(2-AcO-5-$ClC_6H_3$)-4-Ph | 146–147 | Cyclohexane | 85 | ir, ms | 31 |
| 2-(2-AcO-5-$ClC_6H_3$)-4-Me | 131–134 | Cyclohexane | 38 | ir, ms | 31 |

TABLE IV-1. —(contd.)

| Substituent | mp (°C); [bp (°C/torr)] | Solvent of Crystallization | Yield (%) | Spectra | Refs. |
|---|---|---|---|---|---|
| 2-Ac-4-Me | 121–123 | MeOH | | | 40 |
| 2-(2-PhCOOC$_6$H$_4$)-4-Ph | 117–118 | Petr ether | 40 | ir, ms | 31 |
| 2-(2-PhCOO-5-ClC$_6$H$_3$)-4-Ph | 148–153 | EtOH | 60 | ir, ms | 31 |
| 2-(PhCOCH$_2$CO)-4-Ph | 210 | CHCl$_3$ | 83 | ir | 36 |
| 2,4-(BrCH$_2$)$_2$ | ca. 110d | Acetone/H$_2$O | 65 | | 26 |
| Picrate | 138–141 | PhH | | | 26 |
| Picrolonate | 149–151 | PhH | | | 26 |
| 2-BrCH$_2$-4-Me | | | | | |
| Hydrobromide | 201 | MeOH | | ir, pmr | 19 |
| 2-(3-BrC$_6$H$_4$)-4-Ph | 154 | EtOH | 30 | | 38 |
| 2-(4-BrC$_6$H$_4$)-4-Ph | 159–160 | EtOH/H$_2$O | 81 | | 52 |
| 2,4-(Br$_2$CH)$_2$ | | | | | |
| Hydrobromide | 360; 400 | AcOH | 71; 80 | ir, pmr | 1, 19 |
| 2-COOH-4-Me | 258 | | | | 41 |
| 2-COO-*t*-Bu-4-Ph | | | | | |
| Hydrochloride | | | 50 | | 43 |
| 2-COOEt-4-Ph | 87–88 | EtOH | 26 | | 51 |
| 2-(COO-*i*-Pr)-4-Ph | 115–116 | *i*-PrOH | 35 | | 51 |
| Hydrochloride | 191–192 | | | | 51 |
| 2-(COO-*i*-Pr)-4-(4-MeC$_6$H$_4$) | 97–98 | EtOH | 19 | | 51 |
| 2-CONHPh-4-Ph | 201–202 | MeCN | 67; 84 | ir | 42, 43 |
| 2-CONHPh-4-(4-MeOC$_6$H$_4$) | 228–229 | MeCN | 76 | ir | 43 |
| 2-CONHPh-4-(4-MeC$_6$H$_4$) | 237–238 | MeCN | 87 | ir | 43 |
| 2-CONH(4-BrC$_6$H$_4$)-4-Ph | 202–203 | MeCN | 64 | ir | 43 |
| 2-CONH(4-BrC$_6$H$_4$)-4-(4-MeC$_6$H$_4$) | 242–243 | MeCN | 99 | ir | 43 |
| 2-CONH(4-MeOC$_6$H$_4$)-4-Ph | 179–180 | MeCN | 42 | ir | 43 |
| 2-CONH(4-MeOC$_6$H$_4$)-4-(4-MeC$_6$H$_4$) | 201–202 | MeCN | 96 | ir | 43 |
| 2-CONH(4-MeOC$_6$H$_4$)-4-(4-BrC$_6$H$_4$) | 207–208 | MeCN | 44 | ir | 43 |

| | | | | | |
|---|---|---|---|---|---|
| 2-CONH(4-$MeC_6H_4$)-4-Ph | 201–202 | MeCN | 63 | ir | 43 |
| 2-(3-$ClC_6H_4$)-4-Ph | 171 | EtOH | 51 | | 38 |
| 2-(4-$ClC_6H_4$)-4-Ph | 149–150 | EtOH/$H_2O$ | 81 | | 52 |
| 2-(2-OCOOEt-5-$ClC_6H_3$)-4-Ph | 116–118 | EtOH | 83 | ir, ms | 31 |
| 2-(3-EtO-Quinoxalin-2-yl)-4-Me | 135–136 | $Et_2O$ | 61 | pmr, uv | 33 |
| Perchlorate | 220d | EtOH | 93 | | |
| 2-$CF_3$-4-Ph | 81 | | 89 | | 35 |
| 2-$C_3F_7$-4-Ph | 98 | | 75 | | 35 |
| 2-$C_5F_{11}$-4-Ph | 86 | $CCl_4$ | 60, 70 | | 35 |
| 2-$C_7F_{15}$-4-Ph | 102 | | 46 | | 35 |
| 2,4-$(C_7F_{15})_2$ | 66 | | 76 | | 35 |
| 2-(2-$HOC_6H_4$)-4-Me | 125–128 | Cyclohexane | 51 | | 31 |
| 2-(2-$HOC_6H_4$)-4-Ph | 184–189 | | 43–67 | ms, pmr, uv | 37, 38 |
| 2-(2-$HOC_6H_4$)-4-(3-Pyridyl) | 218–221 | | 26 | | 39a, 97 |
| 2-(2-$HOC_6H_4$)-4-(4-Pyridyl) | 205–206 | | 24 | | 39a, 97 |
| 2-(2-$HOC_6H_4$)-4-(2-Thienyl) | 206 | EtOH | 35 | | 39a |
| 2-(2,4-$(HO)_2C_6H_3$)-4-(3-Pyridyl) | 280–281 | | | | 97 |
| 2-(2,4-$(HO)_2C_6H_3$)-4-(4-Pyridyl) | 256–257 | | | | 97 |
| 2-(2-HO-3-$BrC_6H_3$)-4-Ph | 170 | | 27 | | 38 |
| 2-(2-HO-5-$BrC_6H_3$)-4-Ph | 197 | | 35 | ir | 38 |
| 2-(2-HO-5-$BrC_6H_3$)-4-(3-Pyridyl) | 202 | | 31 | | 39a |
| 2-(2-HO-5-$BrC_6H_3$)-4-(4-Pyridyl) | 165 | | 26 | | 39a |
| 2-(2-HO-5-$BrC_6H_3$)-4-(2-Thienyl) | 193 | | 27 | | 39a |
| 2-(2-HO-3-$ClC_6H_3$)-4-Ph | 155 | | 34 | | 38 |
| 2-(2-HO-3-$ClC_6H_3$)-4-(3-Pyridyl) | 169 | | 40 | | 39a |
| 2-(2-HO-3-$ClC_6H_3$)-4-(4-Pyridyl) | 169 | | 23 | | 39a |
| 2-(2-HO-3-$ClC_6H_3$)-4-(2-Thienyl) | 185 | | 30 | | 39a |
| 2-(2-HO-3,5-$Cl_2C_6H_2$)-4-Ph | 199 | | 20 | | 38 |
| 2-(2-HO-3,5-$Cl_2C_6H_2$)-4- (3-Pyridyl) | 193 | | 33 | | 39a |
| 2-(2-HO-3,5-$Cl_2C_6H_2$)-4-(4-Pyridyl) | 186 | | 27 | | 39a |
| 2-(2-HO-3,5-$Cl_2C_6H_2$)-4-(2-Thienyl) | 195 | | 44 | | 39a |
| 2-(2-HO-3-Cl-5-$EtC_6H_2$)-4-Ph | 203 | | 27 | | 38 |
| 2-(2-HO-4-$ClC_6H_3$)-4-Ph | 183–184 | | 30 | | 38 |

TABLE IV-1. —(contd.)

| Substituent | mp (°C); [bp (°C/torr)] | Solvent of Crystallization | Yield (%) | Spectra | Refs. |
|---|---|---|---|---|---|
| 2-(2-HO-4-Cl$C_6H_3$)-4-(3-Pyridyl) | 235 | | 25 | | 39a |
| 2-(2-HO-4-Cl$C_6H_3$)-4-(4-Pyridyl) | 208 | | 20 | | 39a |
| 2-(2-HO-4-Cl$C_6H_3$)-4-(2-Thienyl) | 146 | | 30 | | 39a |
| 2-(2-HO-5-Cl$C_6H_3$)-4-Me | 156–157 | MeOH | 60 | | 31 |
| 2-(2-HO-5-Cl$C_6H_3$)-4-Ph | 183–185 | EtOH | 36; 39 | ir | 31, 38 |
| 2-(2-HO-5-Cl$C_6H_3$)-4-(3-Pyridyl) | 190 | | 31 | | 39a |
| 2-(2-HO-5-Cl$C_6H_3$)-4-(4-Pyridyl) | 192 | | 23 | | 39a |
| 2-(2-HO-5-Cl$C_6H_3$)-4-(2-Thienyl) | 176 | | 25 | | 39a |
| 2-(2-HO-5-F$C_6H_3$)-4-Ph | 194 | | 29 | | 38 |
| 2-(2-HO-4-MeO$C_6H_3$)-4-(3-Pyridyl) | 209–210 | | | | 97 |
| 2-(2-HO-4-MeO$C_6H_3$)-4-(4-Pyridyl) | 187–189 | | | | 97 |
| 2-(2-HO-3-Me$C_6H_3$)-4-Ph | 151 | | 37 | ir | 38 |
| 2-(2-HO-3-Me$C_6H_3$)-4-(3-Pyridyl) | 208–209 | | 30 | | 39a |
| 2-(2-HO-3-Me$C_6H_3$)-4-(4-Pyridyl) | 145 | | 30 | | 39a |
| 2-(2-HO-3-Me$C_6H_3$)-4-(2-Thienyl) | 165 | | 33 | | 39a |
| 2-(2-HO-3,4-$Me_2C_6H_2$)-4-Ph | 193 | | 28 | | 38 |
| 2-(2-HO-3,4-$Me_2C_6H_2$)-4-(3-Pyridyl) | 221 | | 40 | | 39a |
| 2-(2-HO-3,4-$Me_2C_6H_2$)-4-(4-Pyridyl) | 205 | | 26 | | 39a |

| | | | | |
|---|---|---|---|---|
| 2-(2-HO-3,4-$Me_2C_6H_2$)-4-(2-Thienyl) | 208 | | 30 | | 39a |
| 2-(2-HO-3,5-$Me_2C_6H_2$)-4-Ph | 213; 168 | | 35; 26 | | 38 |
| 2-(2-HO-3,5-$Me_2C_6H_2$)-4-(3-Pyridyl) | 161 | | 25 | | 39a |
| 2-(2-HO-3,5-$Me_2C_6H_2$)-4-(4-Pyridyl) | 150 | | 20 | | 39a |
| 2-(2-HO-3,5-$Me_2C_6H_2$)-4-(2-Thienyl) | 190 | | 36 | | 39a |
| 2-(2-HO-4-$MeC_6H_3$)-4-Ph | 147 | | 37 | ir | 38 |
| 2-(2-HO-4-$MeC_6H_3$)-4-(3-Pyridyl) | 225 | | 30 | | 39a |
| 2-(2-HO-4-$MeC_6H_3$)-4-(4-Pyridyl) | 150 | | 34 | | 39a |
| 2-(2-HO-4-$MeC_6H_3$)-4-(2-Thienyl) | 206–207 | | 30 | | 39a |
| 2-(2-HO-4-Me-5-$ClC_6H_2$)-4-Ph | 215 | | 45 | | 38 |
| 2-(2-HO-4-Me-5-$ClC_6H_2$)-4-(3-Pyridyl) | 203 | | 44 | | 39a |
| 2-(2-HO-$C_6H_4$)-4-(2-MeO-$C_6H_4$) | 169–170 | EtOH | 58 | ir, pmr | 304 |
| 2-(2-HO-$C_6H_4$)-4-(3-MeO-$C_6H_4$) | 173–175 | EtOH/petr. $Et_2O$ | 56 | ir, pmr | 304 |
| 2-(2-HO-$C_6H_4$)-4-(2-$NO_2$-$C_6H_4$) | 139–141 | EtOH/EtOAc | 62 | gc/ms | 304 |
| 2-(2-HO-$C_6H_4$)-4-(4-$NO_2$-$C_6H_4$) | 268–270 | EtOH/$CHCl_3$ | 40 | gc/ms | 304 |
| 2-(2-$ClC_6H_4$)-4-$NH_2$ | | | 51 | anal. | 311 |
| 2-(2-$ClC_6H_4$)-4-$NH_2$ | 192–196 | Bz/Hexane | 41 | ir, pmr | 311 |
| 2-Ph-4-$NH_2$ | | | | | 311 |
| 2-Ph-4-$N(CH_2)_5$ | 110–113 | EtOH/$H_2O$ | 49 | ir, pmr, uv | 311 |
| 2-Ph-4-$N(CH_2)_4NMe$ | 151–154 | EtOH/$Et_2O$ | 68 | ir, pmr | 311 |
| 2-NHMe-Ph | 166 | Ligroin | 56 | | 316 |
| 2-$NHCHMe_2$-4-Ph | 131–132 | Ligroin | 45 | | 316 |
| 2-NHCHMePh-4-Ph | 83–95 | EtOH | 68 | | 316 |
| 2-$NHCHMeCH_2Ph$-4-Ph | 140 | Hexane | 32 | | 316 |
| 2-$NHCH_2CH_2Ph$-4-Ph | 124 | Ligroin | 61 | | 316 |
| 2-$NHCH_2CH_2OH$-4-Ph | 164 | EtOAc | 53 | | 316 |
| 2-$NHCH_2CH_2NEt_2$-4-Ph | 99–100 | EtOH/$H_2O$ | 70 | | 316 |
| 2-$NH(CH_2)_3NEt_2$-4-Ph | 120–121 | Ligroin | 56 | | 316 |
| 2-$NHC_6H_{11}$-4-Ph | 126 | Hexane | 38 | | 316 |

TABLE IV-1. —(contd.)

| Substituent | mp (°C); [bp (°C/torr)] | Solvent of Crystallization | Yield (%) | Spectra | Refs. |
|---|---|---|---|---|---|
| 2-$NMe_2$-4-Ph | 110–111 | *i*-PrOH | 30 | | 316 |
| 2-$N(CH_2)_4$-4-Ph | 159 | Ligroin | 85 | | 316 |
| 2-$N(CH_2)_5$-4-Ph | 119–121 | Ligroin | 71 | | 316 |
| 2-$N(CH_2)_4O$-4-Ph | 151 | Ligroin | 81 | | 316 |
| 2-$N(CH_2)_4NMe$-4-Ph | 158 | Hexane | 77 | | 316 |
| 2-(2-HO-4-Me-5-$Clc_6H_2$)-4-(4-Pyridyl) | 180 | | 48 | | 39a |
| 2-(2-HO-4-Me-5-$Clc_6H_2$)-4-(2-Thienyl) | 182–183 | | 32 | | 39a |
| 2-(2-HO-4,5-$Me_2C_6H_2$-(3-Pyridyl) | 198 | | 30 | | 39a |
| 2-(2-HO-4,5-$Me_2C_6H_2$-(4-Pyridyl) | 208 | | 22 | | 39a |
| 2-(2-HO-4,5-$Me_2C_6H_2$-(2-Thienyl) | 160–162 | | 26 | | 39a |
| 2-(2-HO-5-$MeC_6H_3$)-4-Ph | | | | ir | 38 |
| 2-(2-HO-5-$MeC_6H_3$)-4-(3-Pyridyl) | 184 | | 42 | | 39a |
| 2-(2-HO-5-$MeC_6H_3$)-4-(4-Pyridyl) | 135 | | 141 | | 39a |
| 2-(2-HO-5-$MeC_6H_3$)-4-2-Thienyl) | 165 | | 34 | | 39a |
| 2-(4-HO-Coumarin-3-yl)-4-Me | 277–278 | EtOH | 35.5 | ir | 34 |
| 2,4-$(ICH_2)_2$ | 97–99d | Acetone/$H_2O$ | | | 26 |
| Hydrochloride | 126–128 | | | | 26 |
| Picrate | 159–161 | | | | 26 |
| Picrolonate | 132–136 | | | | 26 |
| 2-$MeOCH_2$-4-Me | | | | | |
| Hydrochloride | 185–188 | EtOH/$Et_2O$ | 30 | | 1 |
| 2-(3-$MeOC_6H_4$)-4-Me | | | | | |
| Hydrogen sulfate | 163 | | 47 | uv | 16 |
| 2-(4-$MeOC_6H_4$)-4-Me | | | | | |
| Hydrogen sulfate | 191 | | 86 | uv | 16 |
| 2-($Ph_2CMeO$)-4-Ph | 151 | $CH_2Cl_2$/MeOH | 75 | ir, ms, pmr | 50 |
| 2-(4-$MeOC_6H_4$)-4-Ph | 105–106 | EtOH/$H_2O$ | 85 | | 52 |

| | | | | | |
|---|---|---|---|---|---|
| 2,4-$Me_2$ | 131–132 | $Et_2O$ | 75–80 | ir, pmr, uv | 14–17, 19 |
| Hydrobromide · 2 $H_2O$ | 216 | | | | 1 |
| Hydrochloride · 2 $H_2O$ | 204–206 | $H_2O$ | 70; 87 | uv | 1, 14–16 |
| Hydrogen sulfate | 225–227 | | 74; 90 | uv | 16, 17 |
| Perchlorate | 204–205 | | | | 1 |
| 2-Me-4-Ph | 87–88 | | | | 16, 27, 28 |
| Hydrochloride | 190–192 | | | | 29 |
| Hydrogen sulfate · $H_2O$ | 173–175 | | 75 | uv | 16, 29 |
| 2-Me-4-(4-$EtSeC_6H_4$) | 102–103 | Acetone/petr ether | ~ 40 | | 30 |
| 2-Me-4-(4-$MeSeC_6H_4$) | 107–108 | Acetone/petr ether | ~ 40 | | 30 |
| 2-Me-4-(4-$PrSeC_6H_4$) | 92–93 | Acetone/petr ether | ~ 40 | | 30 |
| 2-Me-4-(Selenen-2-yl) | 98–99 | $EtOH/H_2O$ | 38 | | 32 |
| 2-Me-4-(1-Ph-3-Me-Pyrazol-5-yl) | 166–167 | MeOH | 57 | | 36 |
| 2-Me-4-(PhCH=CH) | 128–129 | PhH/petr ether | 23 | uv | 16 |
| 2-(3-$MeC_6H_4$)-4-Ph | 166 | | 30 | | 38 |
| 2-(4-$MeC_6H_4$)-4-Ph | 160–161 | $EtOH/H_2O$ | 79 | | 52 |
| 2-(4-$NO_2C_6H_4$)-4-Ph | 204–205 | MeCN | 52 | | 54 |
| 2,4-$Ph_2$ | 140–141 | EtOH | 45–86 | uv | 16, 36, 50, 52, 56 |
| Hydrobromide · 0.5$H_2O$ | 221 | $EtOH/H_2O$ | | | 16 |
| Hydrochloride · $H_2O$ | 242–244d | | | uv | 16 |
| Hydrochloride · 3$H_2O$ | 242–244d | | | | 16 |
| Picrate | 216–217 | MeOH | | | 16 |
| 2,4-$(PhCH=CH)_2$ | 164–165 | | 45 | uv | 16 |
| 3-$NO_2$-7-Cl | > 360 | | 70 | ir | 7 |
| ***2,3,4-Trisubstituted*** | | | | | |
| 2,4-$Me_2$-3-Br | | | | | |
| Hydrobromide | 207; 225 | $MeNO_2/Et_2O$ | | ms | 3, 79 |
| 2,3,4-$Me_3$ | 85 | Petr ether, | 84 | uv | 16 |
| Hydrochloride | 214 | subl | 85 | uv | 16, 45 |
| 2,4-$Me_2$-3-$PhSO_2$ | 206 | EtOH | | pmr | 46 |
| 2,4-$Me_2$-3-(4-$MeC_6H_4SO_2$) | 213 | EtOH | | pmr | 46 |
| 2,4-$Me_2$-3-(3-COOH-4-$HOC_6H_3SO_2$) | 216 | EtOH | | pmr | 46 |
| 2,4-$Me_2$-3-(3-COOMe-4-$HOC_6H_3SO_2$) | 194 | EtOH | | pmr | 46 |

TABLE IV-1. —(contd.)

| Substituent | mp (°C); [bp (°C/torr)] | Solvent of Crystallization | Yield (%) | Spectra | Refs. |
|---|---|---|---|---|---|
| 2,4-$Me_2$-3-(3-COOEt-4-$HOC_6H_3SO_2$) | 174 | EtOH | | pmr | 46 |
| 2,4-$Me_2$-3-(3-COOH-4-HO-Naphthyl-1-sulfonyl) | 233 | EtOH | | pmr | 46 |
| 2,4-$Me_2$-3-(3-COOMe-4-HO-Naphthyl-1-sulfonyl) | 252 | EtOH | | pmr | 46 |
| 2,4-$Me_2$-3-(3-COOEt-4-HO-Naphthyl-1-sulfonyl) | 214 | EtOH | | pmr | 46 |
| 2,4-$Me_2$-3-(3-COOPr-4-HO-Naphthyl-1-sulfonyl) | 209 | EtOH | | pmr | 46 |
| 2,4-$Me_2$-3-(3-COOBu-4-HO-Naphthyl-1-sulfonyl) | 206 | EtOH | | pmr | 46 |
| 2-Me-3-$PhSO_2$-4-Ph | 184 | EtOH | | pmr | 46 |
| 2-Me-3-(4-$MeC_6H_4SO_2$)-4-Ph | 182 | EtOH | | pmr | 46 |
| 2-Me-3-(3-COOH-4-$HOC_6H_3SO_2$)-4-Ph | 217 | EtOH | | pmr | 46 |
| 2-Me-3-(3-COOMe-4-$HOC_6H_3SO_2$)-4-Ph | 241 | EtOH | | pmr | 46 |
| 2-Me-3-(3-COOEt-4-$HOC_6H_3SO_2$)-4-Ph | 226 | EtOH | | pmr | 46 |
| 2-Me-3-(3-COOMe-4-HO-Naphthyl-1-sulfonyl)-4-Ph | 230 | EtOH | | pmr | 46 |
| 2-Me-3-(3-COOEt-4-HO-Naphthyl-1-sulfonyl)-4-Ph | 228 | EtOH | | pmr | 46 |
| 2-Me-3-(3-COOPr-4-HO-Naphthyl-1-sulfonyl)-4-Ph | 220 | EtOH | | pmr | 46 |
| 2-Me-3-(3-COOBu-4-HO-Naphthyl-1-sulfonyl)-4-Ph | 207 | EtOH | | pmr | 46 |
| 2,4-$Ph_2$-3-PhCO | 169–170 | EtOH | 30 | ir, uv | 29 |
| 2,4-$Ph_2$-3-Br Hydrobromide·0.5$H_2O$ | 234 | | | | 79 |

| *2,4,7(8)-Trisubstituted* | | | | |
|---|---|---|---|---|
| 2-($PhCOCH_2$)-4-Ph-7-Cl | 193 | PhMe | 91 | | 55 |
| 2-($PhCOCH_2$)-4-Ph-7-$NO_2$ | 230 | Py/MeOH | 84 | | 55 |
| 2,4-($C_7F_{15}$)$_2$-7-Me | 63 | | 60 | | 35 |
| 2-$C_3F_7$-4-Ph-7(8)-Me | 97 | | 50 | | 35 |
| 2-$C_5F_{11}$-4-Ph-7(8)-Me | 80–81 | | 58 | | 35 |
| 2-$C_7F_{15}$-4-Ph-7(8)-Me | 85–86 | | 38 | | 35 |
| 2-(2-$HOC_6H_4$)-4-Ph-7-COOH | 130 | EtOH | 22 | | 39b |
| 2-(2-$HOC_6H_4$)-4-Ph-7-Cl | 189 | EtOH | 42 | | 39b |
| 2-(2-$HOC_6H_4$)-4-(3-Pyridyl)-7-Cl | 180 | EtOH | 25 | | 39b |
| 2-(2-$HOC_6H_4$)-4-(2-Thienyl)-7-Cl | 109 | EtOH | 29 | | 39b |
| 2-(2-HO-3-$MeC_6H_3$)-4-Ph-7-COOH | 250 | EtOH | 30 | | 39b |
| 2-(2-HO-3-$MeC_6H_3$)-4-Ph-7-Cl | 178 | EtOH | 44 | | 39b |
| 2-(2-HO-3,5-(Me)$_2C_6H_2$)-4-Ph-7-COOH | 280–282 | EtOH | 31 | | 39b |
| 2-(2-HO-3,5-(Me)$_2C_6H_2$)-4-Ph-7-Cl | 198 | EtOH | 31 | | 39b |
| 2-(2-HO-5-$ClC_6H_3$)-4-Ph-7-COOH | 193 | EtOH | 33 | | 39b |
| 2-(2-HO-5-$ClC_6H_3$)-4-Ph-7-Cl | 176 | EtOH | 38 | | 39b |
| 2-(2-HO-5-$ClC_6H_3$)-4-2-Thienyl)-7-Cl | 150 | EtOH | 29 | | 39b |
| 2-(2-HO-4,5-(Me)$_2C_6H_2$)-4-Ph-7-Cl | 204 | EtOH | 29 | | 39b |
| 2-(2-HO-5-$MeC_6H_3$)-4-Ph-7-COOH | 120 | EtOH | 25 | | 39b |
| 2-(2-HO-5-$MeC_6H_3$)-4-Ph-7-Cl | 204–205 | EtOH | 32 | | 39b |
| 2-(2-HO-5-$MeC_6H_3$)-4-(2-Thienyl)-7-COOH | 259 | EtOH | 22 | | 39b |
| 2-(2-HO-5-$MeC_6H_3$)-4-(2-Thienyl)-7-Cl | 208–210 | EtOH | 27 | | 39b |
| 2,4-$Me_2$-7-$NH_2$ | 179–180 | PhH | 76 | ir, uv | 20 |
| Hydrochloride | 210–213 | EtOH | | uv | 20, 21 |
| Hydrochloride·0.5$H_2O$ | 203–206 | EtOH | 62 | | 21 |
| 2,4-$Me_2$-7-NHAc | | | | | |
| Hydrate | 139–142 | PhH/EtOH | 30 | ir, uv | 20 |
| Hydrochloride·$H_2O$ | 226 | | | | 20 |
| 2,4-$Me_2$-7-COOH | | | | | |
| Perchlorate | 243–244 | EtOH | 55 | uv | 1 |
| 2,4-$Me_2$-7-Cl | | | | | |
| Perchlorate | 187–188 | EtOH | 41 | uv | 1 |
| Chloroplatinate | 180–185 | | 54.5 | | |

TABLE IV-1. —(contd.)

| Substituent | mp (°C); [bp (°C/torr)] | Solvent of Crystallization | Yield (%) | Spectra | Refs. |
|---|---|---|---|---|---|
| 2,4-$Me_2$-7-$CH_2N(CH_2CH_2Cl)_2$ | 118–122 | PhH/petr ether | | | 21 |
| Dihydrochloride·1.6$H_2O$ | 62–66d | EtOH/$Et_2O$ | | | 22 |
| Dihydrochloride·0.87$H_2O$ | 147–149d | EtOH/$Et_2O$ | | | 22 |
| Dihydrochloride·0.32$H_2O$ | 154–155d | EtOH/$Et_2O$ | | ir, uv | 22 |
| 2,4,7-$Me_3$ | | | | | |
| Perchlorate | 197–198 | EtOH | | uv | 1 |
| 2,4-$Me_2$-7-MeO | | | | | |
| Hydrogen sulfate | | | | uv | 1 |
| 2,4-$Me_2$-7-[(2,4-Dimethyl-3*H*-1,5-benzodiazepin-7-yl) methyl] | | | | | |
| Dihydrochloride | 159–163 | EtOH/$Et_2O$ | | ir, pmr, uv | 24 |
| 2,4-$Me_2$-7-[2-(2,4-Dimethyl-3*H*-1,5-benzodiazepin-7-yloxy) ethoxy] | | | | | |
| Hydrogen sulfate | | | | uv | 1 |
| 2,4-$Me_2$-7-($CH_2NHCH_2COOH$) | | | | | |
| Hydrochloride·$H_2O$ | 83–85 | | 44 | | 25 |
| 2,4-$Me_2$-7-($CH_2NHCH_2COOEt$) | | | | | |
| Hydrochloride·$H_2O$ | 149–151 | | 26 | | 25 |
| 2,4-$Me_2$-7-[$CH_2NH(4\text{-}COOHC_6H_4)$] | | | | | |
| Hydrochloride·$H_2O$ | 199–202 | | 42 | | 25 |
| 2,4-$Me_2$-7-[$CH_2NH(4\text{-}COOEt\text{-}C_6H_4)$] | | | | | |
| Hydrochloride·$H_2O$ | 137–139 | | 77.5 | | 25 |
| 2,4-$Me_2$-7-[$CH_2NHCH(PhCH_2)COOH$] | | | | | |
| Hydrochloride·$H_2O$ | 189–191 | | 45 | | 25 |
| 2,4-$Me_2$-7-[$CH_2NHCH(PhCH_2)COOEt$] | | | | | |
| Hydrochloride·$H_2O$ | 170–172 | | 34 | | 25 |
| 2,4-$Me_2$-7-[$CH_2NHCH(Ph)CH_2COOH$] | | | | | |
| Hydrochloride·$H_2O$ | 188–192 | | 47 | | 25 |

| | | | | | |
|---|---|---|---|---|---|
| 2,4-$Me_2$-7-[$CH_2NHCH(Ph)CH_2COOEt$] | | | | | |
| Hydrochloride·$H_2O$ | 222–223 | | 87 | | 25 |
| 2,4-$Me_2$-7-$NO_2$ | | | | | |
| Hydrochloride | 200–202 | EtOH | 83.5 | uv | 1 |
| Nitrate | 147 | | | | |
| 2,4-$Me_2$-7-MeS | | | | | |
| Hydrochloride | 199–200 | EtOH | 51 | uv | 1 |
| 2-Me-4-Ph-7-[(2-Me-4-Ph-3*H*-1,5-Benzodiazepin-7-yl)methyl] | 108–111 | $CHCl_3$/petr ether | | ir, pmr, uv | 24 |
| Hydrogen sulfate | 196–199 | EtOH/$Et_2O$ | | | 24 |
| 2-(4-$NO_2C_6H_4$)-4-Ph-7-COOH | 292d | MeCN | 37 | ir, pmr | 54 |
| 2-(4-$NO_2C_6H_4$)-4-Ph-7-Cl | 190–191 | DMF/$H_2O$ | 47 | ir, pmr | 54 |
| 2-(4-$NO_2C_6H_4$)-4-Ph-7-Me | 178–180 | MeCN | 38 | ir, pmr | 54 |
| 2,4-$Ph_2$-7-COOH | 260 | | 53 | | 39b |
| 2,4-$Ph_2$-7-Cl | 162–163 | MeCN | 3; 46 | ir, pmr | 54, 39b |
| | 165–167 | EtOH | 37 | pmr | 53 |
| 2,4-$Ph_2$-7-Me | 270–271 | MeCN | 40 | ir, pmr | 54 |
| | 117–118 | EtOH | 25 | pmr | 53 |
| | 111 | EtOH | 70–80 | | 56 |
| 2,4-$Ph_2$-7-$NO_2$ | 243–244 | EtOH | 50; 11 | ir, pmr | 53, 54 |
| | 236 | EtOH | 36 | | 39b |
| 2-Ph-4-$NH_2$-7-Cl | | | | | |
| Hydrochloride | 295 | | 34 | ir, pmr, uv | 311 |
| 2-Ph-4-$NH_2$-7-$NO_2$ | | | | | 311 |
| 2-Ph-4-SMe-7-$NO_2$ | 157 | | 55 | ir, pmr, uv | 319 |
| 2-(4-$ClC_6H_4$)-4-SMe-7-$NO_2$ | 177 | | 73 | ir, pmr, uv | 319 |
| 2-(4-$BrC_6H_4$)-4-SMe-7-$NO_2$ | 182 | | 78 | ir, pmr, uv | 319 |
| 2-(4-$ClC_6H_4$)-4-SMe-7-Cl | 141 | | 56 | ir, pmr, uv | 319 |
| 2-(2-Thienyl)-4-SMe-7-$NO_2$ | 222 | | 97 | ir, pmr, uv | 319 |
| 2-(4-$ClC_6H_4$)-4-$NH_2CH_2$-Ph-7-$NO_2$ | 223 | | 43 | | 319 |
| 2-(4-$ClC_6H_4$)-4-Morpholino-7-$NO_2$ | 233 | | 90 | | 319 |
| 2-(4-$ClC_6H_4$)-4-$N(CH_2)_4NMe$-7-$NO_2$ | 211 | | 58 | | 319 |

TABLE IV-1. —(contd.)

| Substituent | mp (°C); [bp (°C/torr)] | Solvent of Crystallization | Yield (%) | Spectra | Refs. |
|---|---|---|---|---|---|
| ***Polysubstituted*** | | | | | |
| 2,4-$(Br_2CH)_2$-7,8-$Br_2$ | | | | | |
| Hydrobromide | 285 | AcOH | | ir | 19 |
| 2,4-$(Br_3C)_2$-3,7,8-$Br_3$ | 195–196 | THF | 68 | ir, pmr | 19 |
| 2,4-$(C_7F_{15})_2$-7,8-$Me_2$ | 89 | | 85 | | 35 |
| 2-$C_3F_7$-4-Ph-7,8-$Me_2$ | 120 | | 82 | | 35 |
| 2-$C_5F_{11}$-4-Ph-7,8-$Me_2$ | 115 | | 60 | | 35 |
| 2-$C_7F_{15}$-4-Ph-7,8-$Me_2$ | 107 | | 85 | | 35 |
| 2,4-$Me_2$-7,8-$(COOMe)_2$ | | | | | |
| Hydrogen sulfate | 236–237d | EtOH | 86 | ir, pmr | 23 |
| 2,4-$Me_2$-7,8-$Br_2$ | 130–132 | | | ir, pmr | 19 |
| Hydrogen sulfate | 220–221 | MeOH/$Et_2O$ | 62 | ir, pmr | 19 |
| 2-Me-4-Ph-7,8-$Me_2$ | 115–117 | MeCN | 58 | ir, pmr | 54 |
| 2-(4-$NO_2C_6H_4$)-4-Ph-7,8-$Me_2$ | 208–209 | MeCN | 40 | ir, pmr | 54 |
| 2,4-$Ph_2$-6-$NO_2$-8-Cl | 230–234 | MeCN | 31 | ir, pmr | 54 |
| 2,4-$Ph_2$-7,8-$Me_2$ | 163–164 | MeCN | 21; 70–80 | ir, pmr | 54, 56 |
| ***2-Amino-substituted 1,5-benzodiazepines*** | | | | | |
| 2-$NH_2$-4-(2-$ClC_6H_4$) | 167–169 | $CH_2Cl_2$ | 51 | | 61 |
| Hydrochloride | 285d | EtOH | 95 | | 61 |
| 2-[$HO(CH_2)_2NMe$]-4-Ph | 111–112.5 | *i*-PrOH | 72 | uv | 63 |
| 2-$NMe_2$-4-Ph | 108–110.5 | EtOH | 63 | uv | 63 |
| 2-(4-Me-Piperazino)-4-Ph | 155–157 | MeOH | 50 | uv | 63, 64 |
| 2-(4-Me-Piperazino)-4-(4-$ClC_6H_4$) | 155.5–157.5 | Cyclohexane | 42 | uv | 63, 64 |

| | | | | | |
|---|---|---|---|---|---|
| 2-(4-Me-Piperazino)-4-(4-$FC_6H_4$) | 157–158 | EtOAc/Hexane | 46 | uv | 63, 64 |
| 2-(4-Me-Piperazino)-4-(4-$MeOC_6H_4$) | 148.5–150 | EtOAc/Hexane | 56 | uv | 63, 64 |
| 2-(4-Me-Piperazino)-4-(4-$MeC_6H_4$) | 129.5–132 | EtOAc/Hexane | 36 | uv | 63, 64 |
| 2-(4-Me-Piperazino)-4-Ph-8-Cl | 125.5–126.5 | MeOH | 34 | uv | 63, 64 |
| 2-(4-Me-Piperazino)-4-Ph-7,8-$Me_2$ | 131.5–134.5 | Hexane | 64 | uv | 63, 64 |
| 2-(4-Me-Piperazino)-4-(4-$FC_6H_4$)-7,8-$Me_2$ | 169–170 | EtOAc/Hexane | 22 | uv | 63, 64 |
| 2-Morpholino-4-Ph | 150–152 | EtOH | 70 | uv | 63 |
| 2-(4-Ph-Piperazino)-4-Ph | 172–172.5 | EtOAc/Hexane | 52 | uv | 63, 64 |
| 2-Piperidino-4-Ph | 115.5–118 | MeOH | 64 | uv | 63 |
| 2-[4-($PhCH_2$)-Piperidino]-4-Ph | 97–99.5 | Hexane | 29 | uv | 63 |
| 2-[4-($Ph_2CH$)-Piperidino]-4-Ph | 170–172 | EtOAc/Hexane | 58 | uv | 63 |
| 2-(4-Ph-Piperidino)-4-Ph | 205–207.5 | $CHCl_3$/Hexane | 70 | uv | 63 |
| 2-(4-Ph-Piperidino)-4-(4-$ClC_6H_4$) | 170–171 | EtOAc/Hexane | 52 | uv | 63 |
| 2-(4-Ph-Piperidino)-4-(4-$FC_6H_4$) | 173–175 | $CHCl_3$/Hexane | 43 | uv | 63 |
| 2-(4-Ph-Piperidino)-4-(4-$MeOC_6H_4$) | 158–160 | $CHCl_3$/Hexane | 53 | uv | 63 |
| 2-(4-Ph-Piperidino)-4-(4-$MeC_6H_4$) | 158–160 | $CHCl_3$/Hexane | 43 | uv | 63 |
| 2-(4-Ph-Piperidino)-4-Ph-8-Cl | 134.5–136 | EtOAc/Hexane | 48 | uv | 63 |
| 2-(4-Ph-Piperidino)-4-Ph-7,8-$Me_2$ | 174.5–176.5 | EtOAc/Hexane | 63 | uv | 63 |
| 2-[4-(4-$ClC_6H_4$)-Piperidino]-4-Ph | 198.5–200 | EtOAc | 51 | uv | 63 |
| 2-[4-(3-$CF_3C_6H_4$)-Piperidino]-4-Ph | 161.5–162.5 | EtOAc/Hexane | 74 | uv | 63 |
| 2-[4-(4-$CF_3C_6H_4$)-Piperidino]-4-Ph | 157–158.5 | EtOAc/Hexane | 79 | uv | 63 |
| 2-[4-(4-$MeOC_6H_4$)-Piperidino]-4-Ph | 158.5–160 | EtOAc/Hexane | 51 | uv | 63 |
| 2-[4-(1-Piperidinyl) Piperidino]4-Ph | 155–157 | MeOH | 34 | uv | 63 |
| 2-Pyrrolidino-4-Ph | 155–156 | EtOH | 81 | uv | 63 |
| 2-(4-Ph-1,2,3,6-Tetrahydropyridin-1-yl)-4-Ph | 180.5–182 | EtOAc | 68 | uv | 63 |
| 2-[4-(4-$ClC_6H_4$)-1,2,3,6-Tetrahydropyridin-1-yl]-4-Ph | 199–202 | EtOAc | 46 | uv | 63 |
| 2-[4-(4-$CF_3C_6H_4$)-1,2,3,6-Tetrahydropyridin-1-yl]-4-Ph | 168–170 | $CHCl_3$/Hexane | 29 | uv | 63 |
| 2-[4-(4-$MeC_6H_4$)-1,2,3,6-Tetrahydropyridin-1-yl]-4-Ph | 208–211 | $CHCl_3$/Hexane | 24 | uv | 63 |
| ***2,4-Diamino-substituted*** | | | | | |
| 2,4-$(NH_2)_2$ | > 260 | $H_2O$ | 50.5 | ms, pmr | 59, 60 |
| Dihydrochloride | > 260 | EtOH | | ms | 60 |
| 2-$NEt_2$-4-$NMe_2$ | 95–96 | *i*-PrOH | 94.5 | ir, pmr | 65 |
| 2-$NEt_2$-4-Morpholino | 119–120 | *i*-PrOH | 72 | ir, pmr | 65 |

TABLE IV-1. —(contd.)

| Substituent | mp (°C); [bp (°C/torr)] | Solvent of Crystallization | Yield (%) | Spectra | Refs. |
|---|---|---|---|---|---|
| 2-$NEt_2$-4-Pyrrolidino | 69–71 | Petr ether | 44.5 | ir, pmr | 65 |
| 2,4-$(EtNPh)_2$ | 117–118 | EtOH | 11 | ir, pmr | 62 |
| 2,4-$(EtNPh)_2$-7-Cl | 137–138 | EtOH | 18.5 | ir, pmr | 62 |
| 2,4-$(EtNPh)_2$-7,8-$Me_2$ | 152–153 | Cyclohexane | 11 | ir, pmr | 62 |
| 2,4-$(NMe_2)_2$ | 147–148 | Petr ether | 65 | ir, pmr | 65 |
| 2-$NMe_2$-4-Pyrrolidino | 175–177 | *i*-PrOH | 84.5 | ir, pmr | 65 |
| 2,4-$(MeNPh)_2$ | 164–165 | Cyclohexane | 10.5 | ir, pmr | 62 |
| 2,4-$(MeNPh)_2$-7-Cl | 161–162 | EtOH | 14 | ir, pmr | 62 |
| 2,4-$(MeNPh)_2$-7,8-$Me_2$ | 198–199 | Cyclohexane | 23 | ir, pmr | 62 |
| 2-Morpholino-4-pyrrolidino | 206–207 | *i*-PrOH | 67 | ir, pmr | 65 |
| 2-Piperidino-4-pyrrolidino | 149–150 | *i*-PrOH | 79 | ir, pmr | 65 |
| 2,4-$(Pyrrolidino)_2$ | 198–199 | Cyclohexane | 61.5 | ir, pmr | 65 |
| ***2-Thio-substituted*** | | | | | |
| 2-$(NH_2COCH_2S)$-4-Ph | 172 | PhH | 85 | ms, pmr | 71 |
| 2-$(EtOOCCH_2S)$-4-Ph | 59 | Petr ether | 71 | ms, pmr | 71 |
| 2-$(NCCH_2S)$-4-Ph | 153 | Petr ether | 76 | ms, pmr | 71 |
| 2-$(PhCH_2S)$-4-Ph | 88 | Petr ether | 52 | pmr | 67 |
| 2-$[Et_2N(CH_2)_2S]$-4-Ph | | | | | |
| Hydrochloride | 153 | *i*-PrOH | 64 | | 67, 68 |
| Methiodide | 180 | EtOH | 85 | | 68, 70 |
| 2-$[Et_2N(CH_2)_2S]$-4-$(4-BrC_6H_4)$ | | | | | |
| Hydrochloride | 154 | EtOAc | 81 | | 68, 69 |
| Methiodide | 185 | EtOH | 73 | | 68, 69 |
| 2-$[Et_2N(CH_2)_2S]$-4-$(2-ClC_6H_4)$ | | | | | |
| Hydrochloride | 149–150 | EtOAc | 70 | | 69 |
| | 175 | *i*-PrOH | 61.5 | | 68 |
| Methiodide | 192–193 | EtOH | 67 | | 68, 69 |

| | | | | |
|---|---|---|---|---|
| 2-[$Et_2N(CH_2)_2S$]-4-(3-$ClC_6H_4$) | | | | |
| Hydrochloride | 175 | *i*-PrOH | 62 | 69 |
| | 149–150 | EtOAc | 61 | 68 |
| Methiodide | 177–178 | EtOH | 68 | 68, 69 |
| 2-[$Et_2N(CH_2)_2S$]-4-(4-$ClC_6H_4$) | | | | |
| Hydrochloride | 159 | EtOAc | 67; 83 | 68, 70 |
| Methiodide | 179 | Acetone | 85 | 68 |
| 2-[$Et_2N(CH_2)_2S$]-4-(2,4-$Cl_2C_6H_3$) | | | | |
| Hydrochloride | 188 | EtOAc | 90 | 68, 69 |
| Methiodide | 176 | EtOH | 70 | 68, 69 |
| 2-[$Et_2N(CH_2)_2S$]-4-(3,4-$Cl_2C_6H_3$) | | | | |
| Hydrochloride | 169–170 | EtOAc | 80 | 68, 69 |
| Methiodide | 186 | EtOH | 60 | 68, 69 |
| 2-[$Et_2N(CH_2)_2S$]-4-(2-Furyl) | | | | |
| Citrate | 119 | EtOAc | 55 | 69 |
| Methiodide | 139–140 | MeOH | 62 | 69 |
| 2-[$Et_2N(CH_2)_2S$]-4-(4-$MeOC_6H_4$) | | | | |
| Hydrochloride | 167 | *i*-PrOH/$Et_2O$ | 60 | 68, 70 |
| Methiodide | 184 | EtOH | 88 | 68, 70 |
| 2-[$Et_2N(CH_2)_2S$]-4-[2,4-$(MeO)_2C_6H_3$] | 81–82 | Petr ether | | 68 |
| Methiodide | 167 | EtOH | | 68 |
| 2-[$Et_2N(CH_2)_2S$]-4-(4-$MeC_6H_4$) | 62 | Petr ether | 68 | 68, 70 |
| Hydrochloride | 156 | EtOAc | | |
| Methiodide | 183 | EtOH | 85 | 68, 70 |
| 2-[$Et_2N(CH)_2S$]-4-(4-$PhOC_6H_4$) | | | | |
| Hydrochloride | 150–152 | *i*-PrOH/$Et_2O$ | 95 | 68, 70 |
| Methiodide | 180 | EtOH | 77 | 68, 70 |
| 2-[$Et_2N(CH_2)_2S$]-4-(4-$PhC_6H_4$) | | | | |
| Hydrochloride | 168–169 | EtOAc | 63; 77 | 68, 70 |
| Methiodide | 206 | MeOH | 83 | 68, 70 |
| 2-[$Et_2N(CH_2)_2S$]-4-(2-Thienyl) | | | | |
| Hydrochloride | 192 | EtOAc | 75 | 69 |
| Citrate | 136 | EtOAc | | 69 |
| Methiodide | 124 | EtOH | 72 | 69 |

TABLE IV-1. —(contd.)

| Substituent | mp (°C); [bp (°C/torr)] | Solvent of Crystallization | Yield (%) | Spectra | Refs. |
|---|---|---|---|---|---|
| 2-[$Et_2N(CH_2)_2S$]-4-(4-$PhCH_2SC_6H_4$) | | | | | |
| Hydrochloride | 140 | *i*-PrOH | 90 | | 68, 69 |
| Methiodide | 173 | EtOH | 80 | | 68, 69 |
| 2-[$Et_2N(CH_2)_2S$]-4-[4-(Cyclohexylthio)]$C_6H_4$ | | | | | |
| Hydrochloride | 172 | EtOH | 90 | | 68, 69 |
| Methiodide | 188 | EtOH | 78 | | 68, 69 |
| 2-[$Et_2N(CH_2)_2S$]-4-(4-$EtSC_6H_4$) | | | | | |
| Citrate | 113 | EtOAc | 52 | | 68, 69 |
| Methiodide | 192 | EtOH | 60 | | 68, 69 |
| 2-[$Et_2N(CH_2)_2S$]-4-(4-$MeSC_6H_4$) | | | | | |
| Hydrochloride | 164 | EtOAc | 67 | | 68, 69 |
| Methiodide | 195 | MeOH | 80 | | 68, 69 |
| 2-[$Et_2N(CH_2)_2S$]-4-(4-$PrSC_6H_4$) | | | | | |
| Hydrochloride | 136 | EtOAc | 63 | | 68, 69 |
| Methiodide | 186 | EtOH | 76 | | 68, 69 |
| 2-[$Et_2N(CH_2)_2S$]-4-[4-*i*-$PrSC_6H_4$] | | | | | |
| Citrate | 116 | EtOAc | 40 | | 68, 69 |
| Methiodide | 199–200 | EtOH | 71 | | 68, 69 |
| 2-[$Et_2N(CH_2)_2S$]-4-(4-$BuSC_6H_4$) | 42 | Petr ether | 68 | | 68, 69 |
| Methiodide | 177 | EtOH | 73 | | 68, 69 |
| 2-[$Et_2N(CH_2)_2S$]-4-[4-(3-Me-Bu)$C_6H_4$] | | | | | |
| Citrate | 100 | EtOAc | 70 | | 68, 69 |
| Methiodide | 172 | EtOH | 60 | | 68, 69 |
| 2-[$Et_2N(CH_2)_2S$]-4-(2-$PhSC_6H_4$) | | | | | |
| Hydrochloride | 184 | *i*-PrOH | 85 | | 68, 69 |
| Methiodide | 196 | EtOH | 68 | | 68, 69 |

| | | | | | |
|---|---|---|---|---|---|
| 2-[$Et_2N(CH_2)_2S$]-4-(3-$PhSC_6H_4$) | | | | | |
| Hydrochloride | 166 | EtOH | 93 | | 68, 69 |
| Methiodide·$H_2O$ | 105 | EtOH/$H_2O$ | 60 | | 68, 69 |
| 2-[$Et_2N(CH_2)_2S$]-4-(4-$PhSC_6H_4$) | | | | | |
| Hydrochloride | 150 | EtOAc | 90 | | 68, 70 |
| Methiodide | 162 | *i*-PrOH | 86 | | 68, 70 |
| 2-[$Et_2N(CH_2)_2S$]-4-[4-Dodecylthio-$C_6H_4$] | | | | | |
| Citrate | 117 | EtOH | 65 | | 68, 69 |
| Methiodide | 142 | EtOH | 63 | | 68, 69 |
| 2-[$Et_2N(CH_2)_2S$]-4-Ph | | | | | |
| Hydrochloride·$H_2O$ | 87 | EtOH/$Et_2O$ | 50 | | 68, 70 |
| Methiodide | 158 | Acetone | 73 | | 68, 70 |
| 2-[$Et_2N(CH_2)_2S$]-4-(4-$ClC_6H_4$) | | | | | |
| Hydrochloride | 162 | EtOAc | 56; 61 | | 68, 70 |
| Methiodide | 163 | EtOH | 47 | | 68, 70 |
| 2-[$Et_2N(CH_2)_2S$]-4-(4-$PhSC_6H_4$) | | | | | |
| Citrate | 114–115 | EtOAc | 65 | | 68, 70 |
| Methiodide | 174 | EtOH | 70 | | 68, 70 |
| 2-[$Et_2N(CH_2)_2S$]-4-(4-$PhSC_6H_4$) | | | | | |
| Tartrate·$H_2O$ | 58–62 | *i*-PrOH | 64 | | 68 |
| 2-[$Et_2N(CH_2)_2S$]-4-(4-$PhSC_6H_4$) | | | | | |
| Citrate | 138 | EtOAc | 52.5 | | 68 |
| 2-MeS-4-$NEt_2$ | 93–94 | *i*-PrOH | 89 | ir, pmr | 65 |
| 2-MeS-4-$NMe_2$ | 61–62 | Petr ether | 88 | ir, pmr | 65 |
| 2-MeS-4-Ph | 87–88 | Petr ether | 70 | pmr | 67 |
| 2-$CD_3S$-4-Ph | 84–85 | Hexane | | ms | 71 |
| 2-MeS-4-Pyrrolidino | 123–124 | *i*-PrOH | 88.5 | ir, pmr | 65 |
| 2-[$Me_2N(CH_2)_2S$]-4-Ph | | | | | |
| Hydrochloride | 165–166 | EtOAc | 79 | | 68, 70 |
| Methiodide | 182 | MeOH | 75 | | 68, 70 |
| 2-[$Me_2N(CH_2)_2S$]-4-(4-$PhSC_6H_4$) | | | | | |
| Citrate | 78 | *i*-PrOH | 53 | | 68, 70 |
| Methiodide | 201 | EtOH | 80 | | 68, 70 |

TABLE IV-1. —(contd.)

| Substituent | mp (°C); [bp (°C/torr)] | Solvent of Crystallization | Yield (%) | Spectra | Refs. |
|---|---|---|---|---|---|
| 2-[$Me_2N(CH_2)_3S$]-4-Ph | | | | | |
| Hydrochloride · 2-PrOH | 80 | *i*-PrOH | 43; 60 | | 68, 70 |
| Methiodide | 169 | *i*-PrOH | 71 | | 68, 70 |
| 2-[$Me_2N(CH_2)_3S$]-4-(4-$PhSC_6H_4$) | | | | | |
| Hydrochloride | 111–112 | *i*-PrOH | 53 | | 68, 70 |
| Methiodide | 201 | EtOH | 87 | | 68, 70 |
| 2-[2-(4-Me-Piperazinoethyl)thio]-4-Ph | | | | | |
| Dihydrochloride | 172 | *i*-PrOH/$Et_2O$ | 80 | | 68, 70 |
| Dimethiodide | 170 | MeOH | 78 | | 68, 70 |
| Methiodide | 121–122 | MeOH/$Et_2O$ | | | 68, 70 |
| 2-[2-(4-Me-Piperazinoethyl)thio]-4-(4-$PhSC_6H_4$) | 86 | Petr ether | 64 | | 68, 70 |
| Citrate | 175 | EtOH | | | 68, 70 |
| Dimethiodide | 177 | EtOH/$H_2O$ | 65 | | 68, 70 |
| Methiodide | 185–188 | EtOH | 92.5 | | 68, 70 |
| 2-[3-(4-Me-Piperazinopropyl)thio]-4-Ph | | | | | |
| Dihydrochloride | 228 | *i*-PrOH | 28; 41 | | 68, 70 |
| Dimethiodide | 226 | MeOH | 72 | | 68, 70 |
| Methiodide | 167 | MeOH | 65 | | 68, 70 |
| 2-[2-(Morpholinoethyl)thio]-4-Ph | | | | | |
| Hydrochloride | 208 | *i*-PrOH/$Et_2O$ | 68 | | 68, 70 |
| Methiodide | 174 | MeOH | 53 | | 68, 70 |
| 2-[2-Morpholinoethyl)thio]-4-(4-$PhSC_6H_4$) | 90 | *i*-PrOH | 72 | | 70 |
| Hydrochloride | 215 | *i*-PrOH | 72.5 | | 68, 70 |
| Methiodide | 173 | EtOH | 82 | | 68, 70 |
| 2-[2-(Piperidinoethyl)thio]-4-Ph | | | | | |
| Hydrochloride | 181 | *i*-PrOH/$Et_2O$ | 66 | | 68, 70 |
| Methiodide | 188 | *i*-PrOH | 79 | | 68, 70 |

| | | | | | |
|---|---|---|---|---|---|
| 2-[2-(Piperidinoethyl)thio]-4-(4-$PhSC_6H_4$) | 78 | Ligroin | 72 | | 70 |
| Hydrochloride·$H_2O$ | 144–145 | *i*-PrOH | | | 68, 70 |
| Methiodide | 180 | EtOH | 77 | | 68, 70 |

*1,5-Benzodiazepin-3-ones*

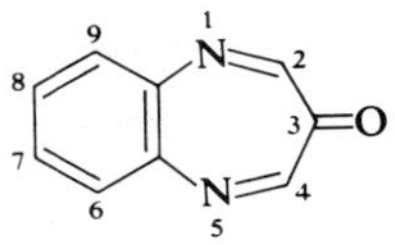

| | | | | | |
|---|---|---|---|---|---|
| 2,4-$Ph_2$ | 120–121 | | 1.4 | | 82 |
| 2,4-$Me_2$ | | | | | |
| Oxime | 215d | MeOH | 80–100 | | 16, 47, 48 |
| 4-$NO_2$-Phenylhydrozone | 252–253 | EtOH/$H_2O$ | 19.5 | | 16 |
| Semicarbazone | 257–260 | Dioxane | | | 85 |
| Thiosemicarbazone | 229–230 | | | | 85 |
| *N*-Me-Thiosemicarbazone | 240–242 | | | | 85 |

***3-Methylene-1,5-benzodiazepines***

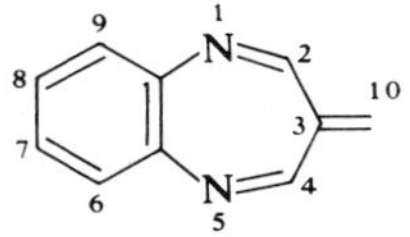

| | | | | | |
|---|---|---|---|---|---|
| 2-(2-$HOC_6H_4$)-4-Me-10-$NMe_2$ | 72–80 | EtOH/$H_2O$ | 94 | | 31 |
| 2-(2-$HOC_6H_4$)-4-Ph-10-$NMe_2$ | 75–80 | MeOH/$H_2O$ | ~ 100 | | 37 |
| 2-(2-HO-5-$ClC_6H_3$)-4-Ph-10-$NMe_2$ | 146–148 | EtOH | 61 | | 31 |
| 2-(2-HO-5-$ClC_6H_3$)-4-Me-10-Me-10-$N(Me)_2$ | 80–90 | EtOH/$H_2O$ | 85 | | 31 |
| 2-(2-HO-5-$ClC_6H_3$)-4-Ph-10-Me-10-$NMe_2$ | 60–65 | MeOH/$H_2O$ | 81 | | 31 |
| 2-(2-$HOC_6H_4$)-4-Ph-10-Me-10-$NMe_2$ | 207 | Ligroin | 42 | | 37 |
| 2,4-$Me_2$-10-[1,3-Benzodioxol-5-yl] | 192–193 | PhH/Petr ether | 51 | uv | 16 |
| 2,4-$Me_2$-10,10-$Ph_2$ | 156–157 | MeOH | 62.5 | ir, pmr | 49 |
| 2,4-$Me_2$-7-Cl-10,10-$Ph_2$ | 156 | MeOH | 50 | | 49 |
| 2,4-$Me_2$-7-$NO_2$-10,10-$Ph_2$ | 205–206 | MeOH | 7 | | 49 |

TABLE IV-1. —(contd.)

| Substituent | mp (°C); [bp (°C/torr)] | Solvent of Crystallization | Yield (%) | Spectra | Refs. |
|---|---|---|---|---|---|
| 2,4,7-$Me_3$-10,10-$Ph_2$ | 141–142 | MeOH | 40 | | 49 |
| 2-Me-4-[2-(1,3-Benzodioxol-5-yl) ethylene] 10-(1,3-Benzodioxol-5-yl) | 257–258 | PhH | 1.5 | uv | 16 |

*Dihydro-1,5-benzodiazepines*

***2,3-Dihydro-1H-1,5-benzodiazepines***

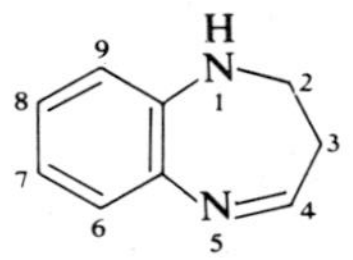

| Substituent | mp (°C); [bp (°C/torr)] | Solvent of Crystallization | Yield (%) | Spectra | Refs. |
|---|---|---|---|---|---|
| ***Monosubstituted*** | | | | | |
| 2-Me | 234 | EtOH | 21 | | 98 |
| Picrate | 185, 204 | EtOH | | | 98, 110 |
| 4-Ph | 42–43 | | 60 | | 112 |
| 4-(3-$NH_2C_6H_4$) | 128–130 | EtOAc | 94 | | 113, 114 |
| 4-(4-$FC_6H_4$) | 154–156 | | | | 111 |
| 4-(2-$HOC_6H_4$) | 119–120 | | 90 | | 112–114 |
| 4-(2-HO-4-$MeOC_6H_3$) | 141–142 | EtOAc | 74.5 | | 113 |
| 4-[2,4-$(MeO)_2C_6H_3$] | 125–126 | EtOH | 92 | | 113 |
| 4-[3,4,5-$(MeO)_3C_6H_2$] | 103–104 | MeOH | 92 | | 113, 114 |
| 4-(3-$NO_2C_6H_4$) | 115.5–116 | EtOH | 82 | | 113, 114 |
| ***Disubstituted*** | | | | | |
| 1-COPh-4-(2-$HOC_6H_4$) | 192 | EtOH | 41 | | 115 |
| 1-COOMe-4-(2-$HOC_6H_4$) | 192 | EtOH | 32 | | 115 |

| | | | | | |
|---|---|---|---|---|---|
| 2,4-$Me_2$ | 64–65 | | 85 | ir | 116 |
| Picrate | 182 d | PhH | | | 116 |
| 2,4-$Ph_2$ | 127–129 | Petr ether | 25 | pmr | 107, 117, 122 |
| 2-Me-4-[2-(4-Dimethylaminophenyl)ethenyl] | 129–130 | MeOH | 85 | ir, uv | 310 |
| Hydrobromide | 165 | EtOH | 60 | uv | 100 |
| 2-(3-Pyridyl)-4-(2-$HOC_6H_4$) | 165–166 | EtOH | 54 | ir, pmr | 108 |
| 2-(3-Pyridyl)-4-(2-HO-4-$MeOC_6H_3$) | 163–164 | EtOH | 58 | | 108 |
| 2-(4-Pyridyl)-4-(2-$HOC_6H_4$) | 162–164 [145–147] | EtOH($Et_2O$) | 64 | | 108 |
| 2-(4-Pyridyl)-4-(2-HO-4-$MeOC_6H_3$) | 190 d | EtOH | 64 | | 108 |
| 2-(4-Pyridyl)-4-(2-pyridyl) | 165–167 | EtOH | 9 | | 108 |
| 2-Me-8-(2-Methyl-2,3-dihydro-1*H*-1,5-benzo-diazepin-8-yl) methyl | 264 d | DMSO, $H_2O$ | 97; 60 | ir, pmr, uv | 99 |
| 2-(4-$BrC_6H_4$)-4-Ph | 142 | MeOH | 63 | ir, uv | 310 |
| 2-(4-$MeOC_6H_4$)-4-Ph | 146–147 | MeOH | 54 | ir, uv | 310 |
| 2-(4-$NO_2C_6H_4$)-4-Ph | 189–190 | | 77 | ir, uv | 310 |
| 2-Ph-4-(4-$MeC_6H_4$) | 127 | MeOH | 57 | ir, uv | 310 |
| 2-Ph-4-(4-$OMeC_6H_4$) | 139–140 | | 46 | ir, uv | 310 |
| 2-Ph-4-(4-$PhC_6H_4$) | 179–180 | | 45 | ir, uv | 310 |
| 2-Ph-4-(4-$BrC_6H_4$ | 142–143 | MeOH | 77 | ir, uv | 310 |
| 2-Ph-4-(4-$ClC_6H_4$) | 129–130 | MeOH | 60 | ir, uv | 310 |
| 2-Ph-4-(4-$NO_2C_6H_4$) | 104–105 | | 76 | ir, uv | 310 |
| 2-$CCl_3$-4-Ph | 125–129 | EtOH/Heptane | 76 | | 313 |
| 2-$CCl_3$-(4-$ClC_6H_4$) | 112–115 | EtOH/Heptane | 70 | | 313 |
| 2-$CCl_3$-(4-$BrC_6H_4$) | 118–121 | EtOH/Heptane | 78 | | 313 |
| 2-$CCl_3$-(4-$MeC_6H_4$) | 119–122 | EtOH/Heptane | 72 | | 313 |
| 2-$CCl_3$-(4-$MeOC_6H_4$) | 93–96 | EtOH/Heptane | 70 | | 313 |
| 2-$CCl_3$-(4-$NO_2C_6H_4$) | 150–152 | EtOH/Heptane | 52 | | 313 |
| 3-Me-4-(4-$MeC_6H_4$) | 36–39 | | 76 | | 112 |
| 4-Ph-7-COOH | 360 | | | | 111 |
| 4-Ph-7(8)-Cl | 45–46 | | 48 | | 112 |
| 4-(4-$MeOC_6H_4$)-7-Me | 194–195 | | | | 111 |
| 4-[2,4-$(MeO)_2C_6H_3$]-7-Me | 125–126 | | | | 111 |

TABLE IV-1. —(contd.)

| Substituent | mp (°C); [bp (°C/torr)] | Solvent of Crystallization | Yield (%) | Spectra | Refs. |
|---|---|---|---|---|---|
| 4-(2-Thienyl)-7-Me | 43–44 | | | | 111 |
| 4-$NH_2$-8-Cl | 143–144 | $CH_2Cl_2$/*i*-$Pr_2O$ | | | |
| | | | 65 | | 118 |
| ***Trisubstituted*** | | | | | |
| 2,2,4-$Me_3$ | 127 | Ligroin | 77,85 | ir, pmr, uv | 98, 100–105, 109, 122 |
| | | | | $^{13}C$-nmr | 104, 121 |
| | [125–127/2] | | | $n_D$ | 110 |
| Methiodide | 212 | Acetone | 35 | | 110 |
| Picrate | 161 d | EtOH | | | 106 |
| 2,2-$Me_2$-4-[2-(4-Dimethylaminophenyl)ethenyl] | 122–132 | | | uv | 100 |
| Hydrobromide | 222–223 | EtOH | 66 | uv | 100 |
| 5-Methoperchlorate | 175–176 | MeOH | 79 | | 110 |
| 2,2-$Me_2$-4-[3-(4,4-Dimethyl-1,3,4,5-tetrahydro-2*H*-1,5-benzodiazepin-2-ylidene)propen-1-yl] | | | | | |
| Hydrobromide | 204 | EtOH | 40 | uv | 100 |
| 2,2-$Me_2$-4-[3-(1,4,4-Trimethyl-1,3,4,5-tetrahydro-2*H*-1,5-benzodiazepin-2-ylidene)propen-1-yl] | | | | | |
| 5-Methobromide | 142 | MeOH | 31 | uv | 110 |
| 2,2-$Me_2$-4-[3-(2*H*-Benzochroman-2-ylidene)but-1-en-1-yl] | | | | | |
| Hydrochloride | 206 | $Ac_2O$ | 51 | uv | 120 |
| 2,2-$Me_2$-4-[3-(6,7-Dimethyl-2*H*-benzothiochroman-2-ylidene)but-1-en-1-yl] | | | | | |
| Hydrochloride | 137 | $Ac_2O$/AcOH | 60 | uv | 120 |
| 2,2-$Me_2$-4-[3-(6,7-Diphenyl-2*H*-benzothiochroman-2-ylidene)but-1-en-1-yl] | | | | | |
| Hydrochloride | 187 | $Ac_2O$/AcOH | 5 | uv | 120 |

| Substituents | mp (°C) | Solvent | Yield (%) | Spectra | Ref. |
|---|---|---|---|---|---|
| 2,2-$Me_2$-4-[2-(4-Hydroxy-2,3-dihydro-2-thioxothiazol-5-yl)ethenyl] | 205–206 | EtOH | 45 | uv | 110 |
| 2,2-$Me_2$-4-[2-(3-Ethyl-4-Hydroxy-2,3-dihydro-2-thioxothiazol-5-yl)ethenyl | 230–231 | MeOH | 75 | uv | 110 |
| 2,2-$Me_2$-4-[3-(3-Methyl-2,3-dihydrobenzothiazol-2-ylidene)propen-1-yl] | | | | | |
| Hydroiodide | 215 | EtOH | 58 | uv | 100 |
| 2-Me-2,4-$Ph_2$ | 102 | Petr ether | 80 | ir, pmr | 103 |
| 2,4-$Me_2$-2-$CSNH_2$ | 240 d | $CHCl_3$ | | | 119 |
| 2,3-$Me_2$-4-[4-Dimethylaminophenyl)ethenyl] | | | | | |
| Hydrobromide | 198 | | 68 | uv | 110 |
| 3-NHAc-4-(4-$NO_2C_6H_4$)-7-Cl | 241–242 | | | | 111 |
| 4-(1-Naphthyl)-7,8-$Me_2$ | 115–117 | | | | 111 |
| 4-(2-Naphthyl)-7,8-$Me_2$ | 176–179 | | | | 111 |
| 4-Ph-7,8-Me | 137–138 | EtOH/$H_2O$ | 85 | ir, pmr | 111 |
| 4-(4-$EtOC_6H_4$)-7,8-$Me_2$ | 192–193 | | | | 111 |
| 4-(2-$HOC_6H_4$)-7,8-$Me_2$ | 134–136 | | 87 | | 112 |
| 4-(3-$HOC_6H_4$)-7,8-$Me_2$ | 300 | | | | 111 |
| 4-(3-$MeOC_6H_4$)-7,8-$Me_2$ | 47–48 | | 65 | | 112 |
| 4-(4-$MeOC_6H_4$)-7,8-$Me_2$ | 173–174 | Xylene | 92 | uv | 112 |
| 4-(4-$NO_2C_6H_4$)-7,8-$Me_2$ | 279–282 | | 28 | | 112 |
| ***Tetrasubstituted*** | | | | | |
| 1-Ac-4-(4-$MeOC_6H_4$)-7,8-$Me_2$ | 81–83 | | | | 111 |
| 1-$COCH_2Cl$-4-(4-$EtOC_6H_4$)-7,8-$Me_2$ | 138–139 | | 78 | | 111 |
| 1-Ph-2,2-$Me_2$-4-[2-(4-Dimethylaminophenyl)ethenyl] | | | | | |
| Hydrobromide | 208 | EtOH | 36 | uv | 100 |
| 2,2,4-$Me_3$-6-COOH | | | | | |
| Hydrobromide | 192–193 | MeOH | 60 | | 110 |
| 2,2,4-$Me_3$-6-$NO_2$ | | | | | |
| Hydrobromide | 104–106 | MeOH | 60 | | 110 |
| 2,2,4-$Me_3$-8-COOH | | | | | |
| Hydrobromide | 171–172 | MeOH | 60 | | 110 |

TABLE IV-1. —(contd.)

| Substituent | mp (°C); [bp (°C/torr)] | Solvent of Crystallization | Yield (%) | Spectra | Refs. |
|---|---|---|---|---|---|
| 2,2,4-$Me_3$-8-$NO_2$ | 144.5 | MeOH/$H_2O$ | | | 110 |
| Hydrobromide | 204–205 | MeOH | 50 | | 110 |
| 2,2-$Me_2$-4-[2-(4-Dimethylaminophenyl)ethenyl]-6-COOH | | | | | |
| Hydrobromide | 216–217 | MeOH | 84 | uv | 110 |
| 2,2-$Me_2$-4-[2-(4-Dimethylaminophenyl)ethenyl]-6-$NO_2$ | 158–159 | | | uv | 110 |
| Hydrobromide | | | 78 | uv | 110 |
| 2,2-$Me_2$-4-[2-(4-Dimethylaminophenyl)ethenyl-8-COOH | 168–169 | MeOH | | uv | 110 |
| Hydrobromide | 203–205 | MeOH | 89 | uv | 110 |
| 2,2-$Me_2$-4-[2-(4-Dimethylaminophenyl)ethenyl]-8-$NO_2$ | 144–145 | MeOH | | uv | 110 |
| Hydrobromide | 213–215 | MeOH | 71 | uv | 110 |
| 2,2-$Me_2$-4-[3-(9-Nitro-4,4-dimethyl-1,3,4,5-tetrahydro-2*H*-1,5-benzodiazepin-2-ylidene)propen-1-yl]-6-$NO_2$ | | | | | |
| Hydrobromide | 220–221 | MeOH | 40 | uv | 110 |
| 2,2-$Me_2$-4-[3-(7-Nitro-4,4-dimethyl-1,3,4,5-tetrahydro-2*H*-1,5-benzodiazepin-2-ylidene)propen-1-yl]-8-$NO_2$ | | | | | |
| Hydrobromide | 235 | MeOH | 44 | uv | 110 |
| 2,2-$Me_2$-4-[7-(7-Nitro-4,4-dimethyl-1,3,4,5-tetrahydro-2*H*-1,5-benzodiazepin-2-ylidene)heptatrien-1-yl]-8-$NO_2$ | 153 | EtOH | | | |
| Hydrobromide | | | 79 | uv | 110 |
| 3-NHAc-4-(4-$NO_2C_6H_4$)-7,8-$Me_2$ | 205–207 | EtOH | 78 | ir | 111 |
| 3-Me-4-(4-$MeOC_6H_4$)-7,8-$Me_2$ | 118–120 | | | | 111 |
| 3-Ph-4-[2,4-$(HO)_2C_6H_3$]-7,8-$Me_2$ | 350 | | | | 111 |
| ***Pentasubstituted*** | | | | | |
| 1-$COCH_2Cl$-3-NHAc-4-(4-$NO_2C_6H_4$)-7,8-$Me_2$ | 198–200 | | | | 111 |
| 2,2,4-$Me_3$-7,8-$Cl_2$ | 113 | Hexane | | ir, ms, pmr, uv | 106 |
| 2,2,4-$Me_3$-6,8-$(NO_2)_2$ | | | | | |
| Hydrobromide | 169–171 | MeOH | 72 | | 110 |

| Substituents | mp (°C) | Solvent | Yield (%) | Spectra | Ref. |
|---|---|---|---|---|---|
| 2,2-$Me_2$-4-[2-(4-Dimethylaminophenyl)ethenyl)-6,8-(7,9)-$Me_2$ | | | | | |
| Hydrochloride | 178 | EtOH/Acetone | 34 | uv | 100 |
| 2,2-$Me_2$-4-[2-(4-Dimethylaminophenyl)ethenyl-6,8-$(NO_2)_2$ | | | | | |
| Hydrobromide | 213–215 | MeOH | 77 | uv | 110 |
| ***2,5-Dihydro-1H-1,5-benzodiazepines*** | | | | | |
| 2,2,4-$Me_3$-5-(2-Furoyl) | 120 | | 73 | | 105 |
| 2,2-$Me_2$-4-[2-(4-Dimethylaminophenyl)ethenyl]-5-Ph | | | | | |
| Hydrobromide | 185 | | | uv | 100 |
| 1,5-$(NO)_2$-2,2,4-$Me_3$ | 170 | | | | 98 |

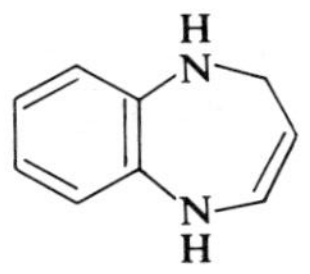

*Dihydro-1,5-benzodiazepinones*

| Substituents | mp (°C) | Solvent | Yield (%) | Spectra | Ref. |
|---|---|---|---|---|---|
| ***1,3-Dihydro-1,5-benzodiazepin-2(2H)-ones*** | | | | | |
| ***Monosubstituted*** | | | | | |
| 4-(1-Adamantyl) | 251–252 | Acetone | 54 | ir, uv | 131 |
| 4-$CH_2Ph$ | 147–149 | EtOH/Petr ether | 88 | ir, uv | 127 |
| 4-$Et_2N$ | 151–152 | EtOH | 23 | pmr | 156 |
| 4-EtNMe | 135–136 | EtOH | 12 | pmr | 156 |
| 4-EtNPh | 181–182 | EtOH | 10 | pmr | 156 |
| 4-$Me_2N$ | 201–202 | EtOH | 21.5 | pmr | 156 |
| 4-(2-Furyl) | 250 | Dioxan | | | 129 |
| Picrate | 195d | EtOH | | | 129 |

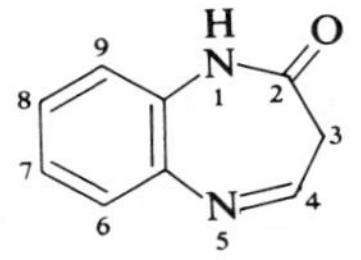

TABLE IV-1. —(contd.)

| Substituent | mp (°C); [bp (°C/torr)] | Solvent of Crystallization | Yield (%) | Spectra | Refs. |
|---|---|---|---|---|---|
| 4-Me | 147–151 | PhH/Petr ether | 35–84 | ir, uv | 123–125 |
| Picrate | 177d | EtOH | | | 146 |
| 4-Ph | 209–210 | EtOH | 87 | ir, uv | 129, 132, 134 |
| Picrate | 168 | EtOH | | | 129 |
| 4-(4-$BrC_6H_4$) | 207–208 | EtOH/$H_2O$ | 85.5 | pmr | 133 |
| 4-(2-$HOC_6H_4$) | 256–257 | EtOH | 50 | uv | 152 |
| 4-(3-Pyridyl) | 219–220 | | | ir, uv | 132 |
| 4-Pyrrolidino | 214–215 | PhH | 45 | pmr | 156 |
| 4-Styryl | 218–219 | EtOH | 21 | | 143 |
| 4-$CCl_3$ | 193–194 | PhH | 75 | ir, pmr | 130 |
| 4-$CF_3$ | 183–185 | PhH | 80–88 | uv | 128 |
| 3-(3-Oxo-1,2,3,4-tetrahydroquinoxalin-2-ylidene) hydrochloride | | | | | 305 |
| 3-(3-Oxo-3,4-dihydroquinoxalin-2-ylidene) | | | 91 | | 305 |
| 3-(3-Chloroquinoxalin-2-ylidene) | | | 26 | | 305 |
| 4-NHMe | 230–231 | EtOH | 59 | ir, pmr | 306 |
| 4-NHEt | 222 | EtOH | 61 | ir, pmr | 306 |
| 4-$(CH_2)_2COOEt$ | 143 | Bz | 66 | ir, pmr | 315 |
| ***Disubstituted*** | | | | | |
| 1-$CH_2Ph$-4-Ph | 128–129 | Petr ether | 19 | | 145 |
| 1-$Me_2N(CH_2)_2$-4-Ph | | | | | |
| Hydrochloride | 191–193 | | 10 | | 165 |
| 1-[2-(4,5-Dihydroimidazolyl)methyl]-4-Ph | | | | | |
| Hydrochloride | 281d | | | | 166 |
| 1-Me-4-Ph | 69–73 | Petr ether | 54 | | 145 |
| 1-Ph-4-$NH_2$ | 251–252 | PhH Acetone | 43 | ir, pmr | 153 |
| 1-Ph-4-$Et_2N$ | 164–165 | EtOH | 13.5 | pmr | 156 |

| | | | | | |
|---|---|---|---|---|---|
| 1-Ph-4-$(EtO)_2CHCH_2NH$ | 150–151 | $CH_2Cl_2$/Hexane | 90 | | 153 |
| 1-Ph-4-$Me_2N$ | 184–185 | EtOH | 23 | pmr | 156 |
| 1-Ph-4-$(MeO)_2CHCH_2NH$ | 171–172 | $CH_2Cl_2$/Hexane | 90 | | 153 |
| 1-Ph-4-MeNH | 181–183 | $CH_2Cl_2$/Hexane | | | 158 |
| 1-Ph-4-MeNNO | 139–140 | EtOAc/Hexane | | | 158 |
| 1-Ph-4-Pyrrolidino | 163–164 | EtOH | 16 | pmr | 156 |
| 3-Ac-4-Ph | 195–196 | EtOH | 50 | ir, ms, pmr, uv | 151 |
| 3-Br-4-Ph | 188 | EtOH | 55 | ir, pmr, uv | 133 |
| 3-Br-4-(4-$BrC_6H_4$) | 207–208 | EtOH | 98 | pmr | 133 |
| 3-Et-4-$Et_2N$ | 154–155 [157] | EtOH | 6; 65 | ir, pmr | 154, 156 |
| 3-Me-4-$Et_2N$ | 150–151 | EtOH | 7 | pmr | 156 |
| 3-Me-4-(1-Piperidino) | 156 | Petr ether | 53 | ir, pmr | 154 |
| 3-Ph-4-(2-Benzimidazolyl) | 275d | DMF/EtOH | 19 | ir, ms, pmr | 147 |
| 3-Ph-4-$CH_2Ph$ | 128–130 | *i*-PrOH | 42; 43 | ir, pmr | 149 |
| | 190 | MeCN | 51; 57 | ir, pmr | 148 |
| 3-Ph-4-COPh | 187–190 | *i*-PrOH | 50 | ir, ms, pmr | 149 |
| 3-Ph-4-$Et_2N$ | 192 | EtOAc/Hexane | 45 | ir, pmr | 154 |
| 3-(4-$ClC_6H_4NHCO$)-4-Me | 162–163 | | 88 | | 168 |
| 3-(4-$MeOC_6H_4NHCO$)-4-Me | 131–132 | *i*-$Pr_2O$ | 89 | | 168 |
| 3-(4-$CF_3OC_6H_4NHCO$)-4-Me | 125.5–126.5 | *i*-$Pr_2O$ | | | 168 |
| 3-(=$CHNMe_2$)-4-NHMe | 202–203 | *i*-Pr | 41 | ir, pmr | 306 |
| 3-(=$CHNMe_2$)-4-NHEt | 201–203 | *i*-Pr | 43 | ir, pmr | 306 |
| 3-(=$CHNMe_2$)-4-$NMe_2$ | 243–244 | Acetone | 39 | ir, pmr | 309 |
| 3-(=$CHNMe_2$)-4-$NEt_2$ | 183–184 | Toluene | 49 | ir, pmr | 309 |
| 3-(=$CHNMe_2$)-4-Pyrrolidine | 234–235 | Acetone | 65 | ir, pmr | 309 |
| 3-$N_2$-4-Me | 243–245 | | 75 | anal | 312 |
| 3-$N_2$-4-Ph | 157 | | 41 | anal | 312 |
| 4-$NH_2$-8-Cl | 274–277 | Acetone | 31.5 | | 118 |
| 4-Me-7-Br | 186 | PhH/Hexane | 37 | | 144 |
| 4-Me-8-Br | 202 | PhH | 77 | ir, pmr | 143, 144 |
| 4-Me-7-Cl | 183 | EtOH | 32, 89 | ir, pmr, uv | 134 |
| | 140–141 | PhH | | | 143 |

TABLE IV-1. —(contd.)

| Substituent | mp (°C); [bp (°C/torr)] | Solvent of Crystallization | Yield (%) | Spectra | Refs. |
|---|---|---|---|---|---|
| 4-Me-8-Cl | 178 | EtOH | 93 | ir, uv | 106, 143 |
| 4-Me-7-MeO | 127 | EtOH | 56 | uv | 134 |
| 4-Me-8-MeO | 185 | MePh | 82 | uv | 134 |
| 4,7-$Me_2$ | 184 | EtOH | 38 | uv | 134 |
| | 164–165 | PhH/Petr ether | 87 | | 137 |
| 4,8-$Me_2$ | 174–175 | PhH/Petr ether | 66 | pmr | 137, 138 |
| 4-Me-7-[4-Methyl-1,3-dihydro-2-oxo-2*H*-1,5-benzodiazepin-7-yl)methyl] | 133–136 | PhH | 70 | ir, pmr, uv | 136 |
| 4-Me-8-[(4-Methyl-1,3-dihydro-2-oxo-2*H*-1,5-benzodiazepin-8-yl)Methyl] | 165–168 | PhMe | | ir, pmr, uv | 136 |
| 4-Me-7-$NO_2$ | 241–243 | DMF | 78 | pmr, uv | 139, 141 |
| | 227–228 | DMF | 71 | | 140 |
| 4-Me-8-$NO_2$ | 204–205 | Acetone | 91 | ir, pmr, uv | 139 |
| | 197–197.5 | Acetone | 88 | | 140 |
| 4-Ph-7-$NH_2$ | 217 | EtOH | 70 | | 150 |
| 4-Ph-7-Br | 240–241 | | | uv | 133 |
| 4-Ph-8-Br | 215–216 | EtOH | 83 | uv | 133 |
| 4-Ph-7-Cl | 230 | EtOH | 65 | uv | 134 |
| 4-Ph-8-Cl | 217 | EtOH | 85 | uv | 134 |
| 4-Ph-7-Me | 225 | EtOH | 60 | uv | 134 |
| 4-Ph-8-Me | 213 | EtOH | 81 | uv | 134 |
| 4-Ph-7-$NO_2$ | 254 | DMF | 71 | | 150 |
| 4-(3-Pyridyl)-7(8)-Me | 213–217 | | | ir, uv | 132 |
| 4-$CF_3$-7-Me | 202–203 | $CCl_4$ | 56 | ir, pmr, uv | 135 |
| 4-$CF_3$-8-Me | 238–240 | $CCl_4$ | 25 | ir, pmr, uv | 135 |
| 4-Ph-8-$NO_2$ | 239 | | 44 | | 319 |
| 4-(4-$ClC_6H_4$)-8-$NO_2$ | 250 | | 44 | | 319 |

| | | | | | |
|---|---|---|---|---|---|
| ***Trisubstituted*** | | | | | |
| 1-$NO_2$-4,7-$Me_2$ | 122 | $Et_2O$/Hexane | 26 | ir | 163 |
| 1-Ph-3-Me-4-(1-Piperidino) | 250 | MeOH | 7 | ir, pmr | 154 |
| 1-Ph-4-(Allylamino)-8-Cl | 173–176 | THF/$i$-$Pr_2O$ | 48 | | 157a |
| 1-Ph-4-$NH_2$-8-Br | 248–249 | $CH_2Cl_2$/$Et_2O$ | 49; 63 | | 157 |
| Methanesulfonate | 276–278 | | | | 157a |
| 1-Ph-4-$NH_2$-8-Cl | 242–243 | $i$-PrOH/MeOH | 39, 51 | | 157 |
| | 247–249 | $CH_2Cl_2$/$i$-$Pr_2O$ | 32 | | 153 |
| 1-Ph-4-$NH_2$-8-F | 222–224 | $CH_2Cl_2$/$Et_2O$ | 43 | | 157a |
| 1-Ph-4-$NH_2$-8-$NO_2$ | 219–220 | | | | 157b |
| Methanesulfonate | 238–240 | DMF | 68 | | 157a |
| 1-Ph-4-$NH_2$-8-$CF_3$ | 227–228 | $CH_2Cl_2$/$i$-$Pr_2O$ | 54; 58 | | 157a |
| Hydrochloride | 282–283 | $CH_2Cl_2$/$Et_2O$ | | | 157b |
| Methanesulfonate | 247–249 | | | | 157b |
| 1-Ph-4-AcNH-8-Br | 248–250 | | | | 167, 170 |
| 1-Ph-4-AcNH-8-Cl | 210–211 [226–227] | $i$-$Pr_2O$ | 85 | | 170 |
| 1-Ph-4-AcNH-8-$NO_2$ | 231–237d | | | | 167, 170 |
| 1-Ph-4-AcNH-8-$CF_3$ | 184–186 | $i$-$Pr_2O$ | 82; 90 | | 170 |
| 1-Ph-4-AcNMe-8-$NO_2$ | 174–176 | | | | 167, 170 |
| 1-Ph-4-PhCONH-8-$NO_2$ | 228–230 | EtOAc/$i$-$Pr_2O$ | 50 | | 167, 170 |
| 1-Ph-4-$PhCH_2OCONH$-8-$NO_2$ | 178–180 | | | | 167, 170 |
| 1-Ph-4-BuO-8-$CF_3$ | 98–99 | | | | 161 |
| 1-Ph-4-BuNH-8-Cl | 147–148 | | | | 157 |
| 1-Ph-4-$t$-BuNH-8-Cl | 275–276 | MeOH | 80; 88 | | 157 |
| 1-Ph-4-BuNEt-8-Cl | 185–188 | | | | 157b |
| 1-Ph-4-BuNHCONMe-8-$NO_2$ | 128–130 | EtOAc/$i$-$Pr_2O$ | 44 | | 170 |
| 1-Ph-4-$ClCH_2CONHNH$-8-Cl | 210–215 | | 86 | | 159 |
| 1-Ph-4-$ClCH_2CH_2NHCONH$-8-Br | 200–201 | | | | 167, 170 |
| 1-Ph-4-$ClCH_2CH_2NHCONH$-8-$NO_2$ | 196–198 | | | | 167, 170 |
| 1-Ph-4-$ClCH_2CH_2NHCONH$-8-$CF_3$ | 188–190 | | | | 167, 170 |
| 1-Ph-4-$ClCH_2CH_2NHCO$ NMe-8-Br | 145–147 | | | | 167, 170 |

TABLE IV-1. —(contd.)

| Substituent | mp (°C); [bp (°C/torr)] | Solvent of Crystallization | Yield (%) | Spectra | Refs. |
|---|---|---|---|---|---|
| 1-Ph-4-(Cyclohexylcarbonylamino)-8-$NO_2$ | 188–192 | | | | 167, 170 |
| 1-Ph-4-(Cyclohexylcarbonylamino)-8-$CF_3$ | 253–255 | | | | 167, 170 |
| 1-Ph-4-(Diallylamino)-8-Cl | | | | | |
| Hydrochloride | 200d | | | | 157b |
| 1-Ph-4-$(EtO)_2CHCH(Me)$ | | | | | |
| NH-8-Cl | 171–173 | $CH_2Cl_2$/Hexane | 73 | | 153 |
| 1-Ph-4-$Me_2N$-8-Cl | 142–145 | | | | 157b |
| 1-Ph-4-$Me_2N$-8-$NO_2$ | 219–220 | $CH_2Cl_2$/$i$-$Pr_2O$ | 80; 87 | | 157 |
| 1-Ph-4-$Me_2N$-8-$CF_3$ | 154–155 | $i$-$Pr_2O$ | 63 | | 157 |
| 1-Ph-4-$Me_2N(CH_2)_2NH$-8-Cl | 278–280 | | | | 157b |
| 1-Ph-4-$Me_2N(CH_2)_2NH$-8-$NO_2$ | 148–153 | | | | 157b |
| 1-Ph-4-$Me_2N(CH_2)_2NH$-8-$CF_3$ | 171–173 | | | | 157b |
| 1-Ph-4-EtO-8-Br | 142–143 | | | | 161 |
| 1-Ph-4-EtO-8-Cl | 97–100 | MeOH | 52 | | 161 |
| 1-Ph-4-EtO-8-$NO_2$ | 182–184 | | | | 161 |
| 1-Ph-4-EtO-8-$CF_3$ | 145–147 | $i$-$Pr_2O$ | 50.5 | | 161 |
| 1-Ph-4-EtOOCNH-8-Cl | 187–189 | EtOAc/$i$-$Pr_2O$ | 68 | | 170 |
| 1-Ph-4-EtOOCNH-8-$NO_2$ | 212–213 | EtOAc/$i$-$Pr_2O$ | 62 | | 167, 170 |
| 1-Ph-4-$EtO(CH_2)_3NH$-8-Cl | 169–170 | $Et_2O$ | 76 | | 157a |
| 1-Ph-4-EtNH-8-Cl | 183–184 | | | | 157b |
| 1-Ph-4-EtNH-8-$NO_2$ | 243–245 | EtOH | 91 | | 157b |
| 1-Ph-4-EtNH-8-$CF_3$ | 181–183 | $i$-$Pr_2O$ | 43 | | 157a |
| 1-Ph-4-EtNHCONH-8-Cl | 226–228 | | | | 170 |
| 1-Ph-4-$NHNH_2$-8-Cl | | | | | |
| Sesquihydrate | 102–103 | | 47 | | 158, 159 |
| 1-Ph-4-$HO(CH_2)_2NH$-8-$NO_2$ | 191–193 | MeOH | 67 | | 157 |
| 1-Ph-4-$HO(CH_2)_2NH$-8-$CF_3$ | 188–189 | | | | 157 |
| 1-Ph-4-MeO-8-Cl | 136–138 | $Et_2O$/Hexane | | | 158a |

| | | | | |
|---|---|---|---|---|
| 1-Ph-4-MeO-8-$CF_3$ | 174–175 | | | 161 |
| 1-Ph-4-[MeOOCC(=NOH)]-8-Cl | 271–273 | THF | | 171, 158b |
| 1-Ph-4-MeNH-8-Cl | 217–220 | $CH_2Cl_2$/*i*-$Pr_2O$ | 68 | 157a |
| 1-Ph-4-MeNH-8-$NO_2$ | 217–219 | $CH_2Cl_2$/*i*-$Pr_2O$ | 75 | 157a |
| 1-Ph-4-MeNH-8-$CF_3$ | 201–203 | $CH_2Cl_2$/*i*-$Pr_2O$ | 92; 95.5 | 157a |
| 1-Ph-4-MeNNO-8-Cl | 123–126 | $CH_2Cl_2$/Hexane | | 158 |
| 1-Ph-4-(4-Methylpiperazino)-8-Cl | 217–219 | | | 157b |
| 1-Ph-4-[4-(2-Methylphenyl)piperazino]-8-$NO_2$ | 223–225 | | | 157 |
| 1-Ph-4-(2-Methylpropanoylamino)-8-Cl | 173–175 | | | 170 |
| | 123–126 | $CH_2Cl_2$/Hexane | | 158 |
| 1-Ph-4-MeS-8-Cl | 113–115 | | | 161 |
| 1-Ph-4-MeS-8-$NO_2$ | 160–161 | | | 161 |
| 1-Ph-4-MeS-8-$CF_3$ | 177–179 | *i*-$Pr_2O$ | 48 | 161 |
| 1-Ph-4-(2-Oxo-1-imidazolinyl)-8-Br | 267–270 | | | 167, 170 |
| 1-Ph-4-(2-Oxo-1-imidazolinyl)-8-$NO_2$ | 268–270 | | | 167, 170 |
| 1-Ph-4-(2-Oxo-1-imidazolinyl)-8-$CF_3$ | 253–255 | | | 167 |
| 1-Ph-4-Piperidino-8-$NO_2$ | 175–176 | | | 157b |
| 1-Ph-4-Piperidino-8-$CF_3$ | 144–146 | | | 157b |
| 1-Ph-4-*i*-PrNH-8-Cl | 240–241 | | | 157b |
| 1-Ph-4-*i*-PrNH-8-$NO_2$ | 251–253 | | | 157b |
| 1-Ph-4-*i*-PrNH-8-$CF_3$ | 215–217 | *i*-$Pr_2O$ | 42 | 157a |
| 1-Ph-4-*i*-PrO-8-$CF_3$ | 192–193 | *i*-$Pr_2O$ | 45 | 161 |
| 1-(2-$BrC_6H_4$)-4-EtO-8-Cl | 194–196 | | | 161 |
| 1-(2-$ClC_6H_4$)-4-AcNMe-8-Cl | 176–178 | | | 167 |
| 1-(2-$ClC_6H_4$)-4-$NH_2$-8-Cl | 260–262 | $CH_2Cl_2$/$Et_2O$ | 38 | 157 |
| 1-(2-$ClC_6H_4$)-4-EtO-8-Cl | 191–192 | | | 161 |
| 1-(2-$ClC_6H_4$)-4-MeNH-8-Cl | 224–225 | | | 157 |
| 1-(2-$FC_6H_4$)-4-$NH_2$-8-Cl | 258–259 | EtOAc | 35 | 153 |
| 1-(2-$FC_6H_4$)-4-BuNH-8-Cl | 200–202 | | | 157 |
| 1-(2-$FC_6H_4$)-4-EtO-8-Cl | 128–130 | | | 161 |
| 1-(2-$NO_2C_6H_4$)-4-AcNMe-8-$CF_3$ | 178–180 | | | 167, 170 |
| 1-(2-$NO_2C_6H_4$)-4-EtO-8-Cl | 161–162 | | | 161 |
| 1-(2-$NO_2C_6H_4$)-4-MeNH-8-Cl | 260–262 | | | 157 |

TABLE IV-1. —(contd.)

| Substituent | mp (°C); [bp (°C/torr)] | Solvent of Crystallization | Yield (%) | Spectra | Refs. |
|---|---|---|---|---|---|
| 1-(2-$CF_3C_6H_4$)-4-BuNH-8-Cl | 184–185 | | | | 157 |
| 1-(2-$CF_3C_6H_4$)-4-EtO-8-Cl | 123–125 | | | | 161 |
| 3-Et-4-$Et_2$N-7(8)-Cl | 136 | EtOAc | 56 | | 154 |
| 3-Ph-4-$PhCH_2$-7-Cl | 174–176 | *i*-PrOH | 48 | ir, pmr | 149 |
| 3-Ph-4-$PhCH_2$-8-Cl | 148–151 | *i*-PrOH | 22 | ir, pmr | 149 |
| 3-Ph-4-$PhCH_2$-7-MeO | 152–154 | MeCN | 35; 30 | ir, pmr | 149 |
| 3-Ph-4-$PhCH_2$-7(8)-MeO | 172–174 | | 34 | | 148 |
| 3-Ph-4-$PhCH_2$-7-Me | 158–160 | MeCN | 48 | ir, pmr | 149 |
| 3-Ph-4-$PhCH_2$-8-Me | 188–192 | MeCN | 35 | ir, pmr | 149 |
| 3-Ph-4-$PhCH_2$-7(8)$NO_2$ | 200–202 | | 30 | | 148 |
| 4-Me-6,8-$Br_2$ | 203 | PhH | 80 | ir, pmr | 143 |
| 4-Me-7,9-$Br_2$ | 184 | PhH | 90 | ir, pmr | 143 |
| 4-Me-6,8-$Cl_2$ | 213–214 | PhH | 80 | pmr | 142 |
| 4-Me-7,8-$Cl_2$ | 219d | EtOH/$H_2O$ | 80 | ir, uv | 106 |
| 4-Me-7,9-$Cl_2$ | 189 | PhH | 89 | pmr | 142 |
| 4-Me-8-MeO-9-$NO_2$ | 195 | EtOH | 34 | ms, pmr, uv | 164 |
| 4,7,8-$Me_3$ | 186 | PhH | 88.5 | ir, pmr | 143 |
| 4,7-$Me_2$-8-$NO_2$ | 218 | EtOH | 5.5 | ms, pmr | 163 |
| 4,7-$Me_2$-9-$NO_2$ | 136 | EtOH | 33.5 | ms, pmr | 163 |
| 4,8-$Me_2$-7-$NO_2$ | 235 | DMF | 87 | ms, pmr | 163 |
| 4-Me-7-$NO_2$-8-Br | 248 | DMF | 62 | ms, pmr, uv | 164 |
| 4-Me-7-$NO_2$-8-Cl | 248 | DMF | 92 | ms, pmr, uv | 164 |
| 4-Me-7-$NO_2$-8-MeO | 258 | DMF | 40 | ms, pmr, uv | 164 |
| 4-Ph-7-Me-8-$NO_2$ | 246 | MePH | 24 | ms, pmr | 163 |
| 4-Ph-7-Me-9-$NO_2$ | 223 | EtOH | 30 | ms, pmr | 163 |
| 4-Ph-7-$NO_2$-8-Br | 267 | DMF | 70 | ms, pmr, uv | 164 |
| 4-Ph-7-$NO_2$-8-Cl | 265 | DMF | 82 | ms, pmr, uv | 164 |
| 4-Ph-7-$NO_2$-8-Me | 268 | DMF | 93 | ms, pmr | 163 |

| | | | | | |
|---|---|---|---|---|---|
| 4-(4-$MeOC_6H_4$)-7-Me-8-$NO_2$ | 261 | PhMe | 77 | ms, pmr | 163 |
| 4-(4-$MeOC_6H_4$)-7-$NO_2$-8-Br | 277 | DMF | 60 | uv | 164 |
| 4-(4-$MeOC_6H_4$)-7-$NO_2$-8-Cl | 272 | DMF | 96 | uv | 164 |
| 4-(4-$MeOC_6H_4$)-7-$NO_2$-8-Me | 259 | DMF | 78 | ms, pmr | 163 |
| ***Polysubstituted*** | | | | | |
| 3-PhNHCO-4-Me-7,8-$Me_2$ | 255 | THF | | | 168 |
| 4-Me-6,7,8,9-$Cl_4$ | 260 | Dioxan | 50 | ir | 126 |
| 1,4-$Me_2$-6,7,8,9-$Cl_4$ | 247–249 | PhH | | ir, pmr | 126 |
| 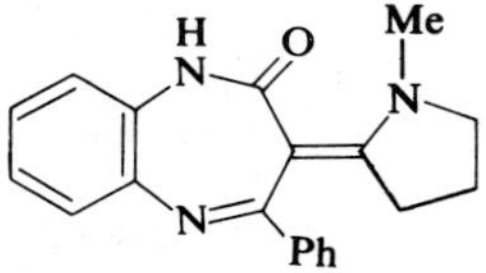 | 210–212 | PhH | 77 | | 169 |

***1,5-Dihydro-1,5-benzodiazepin-2(2H)-ones***

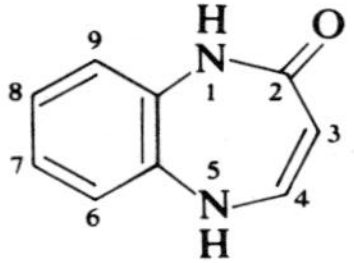

| | | | | | |
|---|---|---|---|---|---|
| 3-(3-Oxo-3,4-dihydroquinoxalin-2-ylidene) hydrochloride | | | 72 | | 305 |
| 1-Ph-4-$Ac_2N$-5-Ac-8-$CF_3$ | 204–206 | *i*-$Pr_2O$ | 27 | ir, pmr | 167 |
| 1-Ph-4-PhCONH-5-PhCO-8-$NO_2$ | 270–273 | | | | 167 |
| 1-Ph-4-PhCONH-5-PhCO-8-$CF_3$ | 252–254 | | | | 167 |
| 1-Ph-4-(Cyclohexylcarbonylamino)-5-(cyclohexyl-carbonyl)-8-$NO_2$ | 240–242 | | | | 167 |
| 1-Ph-4-(Cyclohexylcarbonylamino)-5-(cyclohexyl-carbonyl)-8-$CF_3$ | 242–244 | | | | 167 |
| 1-Ph-4-(Propenoylamino)-5-propenoyl-8-Cl | 211–213 | | | | 167 |
| 1-Ph-4-($CF_3$CONH)-5-$CF_3$CO-8-$NO_2$·$Me_2$NAc | 147–149 | | | | 167 |

TABLE IV-1. —(contd.)

| Substituent | mp (°C); [bp (°C/torr)] | Solvent of Crystallization | Yield (%) | Spectra | Refs. |
|---|---|---|---|---|---|
| 3-Me-4-Piperidino-5-Ph | 155 | Hexane | 42 | ir, pmr | 154 |
| 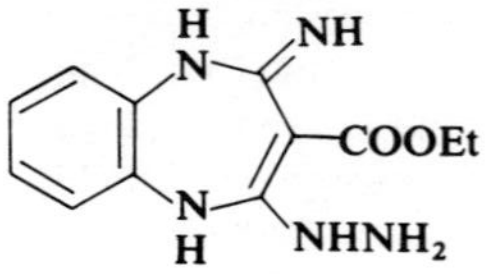 (structure: 1H-1,5-benzodiazepine, 2-NH, 3-COOEt, 4-NHNH$_2$) | 127–134 | EtOH | 96 | | 160 |
| ***1H-1,5-benzodiazepin-2,3(2H,3H)-diones*** | | 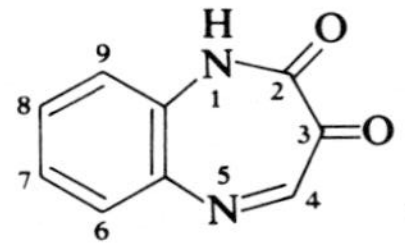 | | | |
| 4-Me | | | | | |
| 3-Phenylhydrazone | 252d | | 68 | | 124 |
| *Tetrahydro-1,5-benzodiazepines* | | | | | |
| ***2,3,4,5-Tetrahydro-1H-1,5-benzodiazepine*** | | 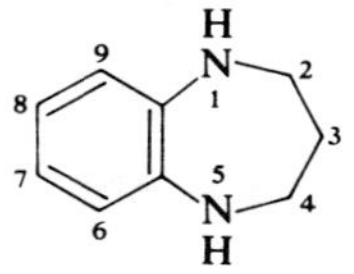 | | | |
| None | 102–104 [125–130/3 mm] | Ligroin | 69, 99 | | 175–178, 180, 187 178, 231 |
| ***Monosubstituted*** | | | | | |
| 1-PhCH$_2$ | | | 50.5 | ir | 190 |
| 1-Me | [125–130/0.005] | | | ir, pmr | 187 |

| | | | | | |
|---|---|---|---|---|---|
| 1-(4-$MeC_6H_4SO_2$) | 166–167 | EtOH | 72; 68 | | 176–178 |
| 1-Ph | [175/3.5 mm] | | 62 | | 178 |
| 1-(2-$HOC_6H_4$) | 185–186 | EtOH | 55 | ir, pmr | 179 |
| 1-(2-$MeOC_6H_4$) | 90–91 | EtOH | 63 | ir, pmr | 179 |
| 2-Me | 97–98 | | 33 | | 124 |
| 2-Ph | 106–107 | MeOH | | pmr, uv | 107 |
| 2-(3-$H_2NC_6H_4$) | 118–119 | EtOH | | | 113 |
| 2-[2,4-$(MeO)_2C_6H_3$] | 120–122 | EtOH | 72 | | 113 |
| 3-Et | 82–83 | | 67 | | 231 |
| 3-HO | 139–141 | EtOH | 81 | ir, pmr, uv | 181 |
| 6-$NH_2$ | | | | | |
| Dihydrochloride | 231–237 | | 89 | | 184 |
| 7-MeO | [115–120/0.005 mm] | | | | |
| | | | ir, pmr | 187 | |
| ***Disubstituted*** | | | | | |
| 1-Et-2-Me | 51 | | | | 192 |
| 1-PhCO-5-(4-$MeC_6H_4SO_2$) | 183–184 | EtOH | 93 | ir | 177 |
| 1-$PhCH_2$-5-(4-$MeC_6H_4SO_2$) | 124–125 | | 40; 55 | ir, pmr | 190 |
| 1,5-$(BrCH_2CH_2CO)_2$ | 127–129 | PhH | 80 | | 191 |
| 1,5-$Me_2$ | | | 41 | pmr | 183 |
| 1-Me-5-(2-$MeOC_6H_4$) | 193–194d | EtOH | 31; 60 | pmr | 179 |
| 1,5-(4-$MeC_6H_4SO_2)_2$ | 193 | AcOH | 75 | | 176 |
| 1-(4-$MeC_6H_4SO_2$)-5-(2-$MeOC_6H_4$) | 182–183 | EtOH | 75 | ir | 179 |
| 1-(4-$MeC_6H_4SO_2$)-5-Ph | 138 | | 45; 28 | | 178 |
| 1,5-$(NO)_2$ | 120 | | | | 175 |
| 1,5-$(PhSO_2)_2$ | 204–205 | | | | 175 |
| 2-CN-4-(2-Furyl) | 122–123.5 | | 68 | | 188 |
| 2-CN-4-Me | 159–161 | | 86 | | 188 |
| 2-CN-4-Ph | 134–135 | | 99 | | 188 |
| 2-CN-4-(2-Thienyl) | 155–156 | | 63 | | 188 |
| 2,4-$Me_2$, *cis* | 59–60 | Petr ether | 44; 90 | ir, ms, pmr | 116, 117, 186 |
| Picrate | 186; 175d | PhH | | | 116, 117 |
| Perchlorate | 200d | EtOH/$Et_2O$ | | | 116 |

TABLE IV-1. —(contd.)

| Substituent | mp (°C); [bp (°C/torr)] | Solvent of Crystallization | Yield (%) | Spectra | Refs. |
|---|---|---|---|---|---|
| 2-Me-4-Ph, *cis*, *trans* mixture | 64–65 | Petr ether | 26 | pmr, uv | 107, 116 |
| *cis*-Hydrochloride | 220d | EtOH/$H_2O$ | 100 | | 186 |
| 2-Me-4-(4-$BrC_6H_4$), *cis* | | | | | |
| Hydrochloride | 245–246 | EtOH/$H_2O$ | 100 | | 116, 186 |
| 2-Me-4-(4-$ClC_6H_4$), *cis* | | | | | |
| Hydrochloride | 227–229; 275 | EtOH/$H_2O$ | 100 | | 116, 186 |
| 2-Me-4-(4-$EtOC_6H_4$), *cis* | | | | | |
| Hydrochloride | 205d | EtOH/$H_2O$ | 100 | | 116, 186 |
| 2-Me-4-(4-$MeOC_6H_4$), *cis* | | | | | |
| Hydrochloride | 215d | EtOH/$H_2O$ | 100 | | 116, 186 |
| 2-Me-4-(4-$MeC_6H_4$), *cis* | | | | | |
| Hydrochloride | 230d | EtOH/$H_2O$ | 100 | | 116, 186 |
| 2-Me-4-(3-$NO_2C_6H_4$) | | | | | |
| Hydrochloride | 115–117d | EtOH/$H_2O$ | 100 | | 116, 186 |
| 2,4-$Ph_2$ | | | | | |
| *cis* | 136–137 | | 7; 5.5 | pmr, uv | 107 |
| *trans* | 140–141 | MeOH | 13 | pmr, uv | 107 |
| 2-Ph-7-Cl | 183–184 | | | | 111 |
| 2-[3,4,5-$(MeO)_3C_6H_2$]-7-Cl | 58–59 | | | | 111 |
| 2-$CH_2NH_2$-4-Ph | 183–184 | Bz | 25 | ir, pmr, uv | 320 |
| 2-$CH_2NHCOCH{=}CH$(4-$PhC_6H_4$) | 151–153 | MeOH | 62 | ir, uv | 320 |
| 2-$CH_2N{=}CHCH{=}CH$(4-$PhC_6H_4$) | 123–127 | MeOH | 64 | ir, uv | 320 |
| 2-$CH_2NHCO(C_6H_4$-4-Me) | 147–149 | PhH/Hexane | 57 | ir, uv | 320 |
| 2-$CH_2NHCO$(4-$PhC_6H_4$) | 220–221 | MeOH | 64 | ir, uv | 320 |
| 2-COOH-4-Ph | 207–210 | $CHCl_3$ | 8 | ir, uv | 320 |
| 2-COOH-4-Me | 170–171 | $CHCl_3$ | 5 | ir, uv | 320 |
| 2-COOMe-4-Ph | 158–162 | EtOH | 57 | ir, uv | 320 |
| 2-COOEt-4-Ph | 146–157 | EtOH | 35 | ir, uv | 320 |
| Hydrochloride | 155–160 | EtOH | | | 320 |

| | | | | | |
|---|---|---|---|---|---|
| 2-COOEt-4-Me | 154–162 | EtOH | 32 | ir, uv | 320 |
| Hydrochloride | 147–151 | EtOH | | | |
| ***Trisubstituted*** | | | | | |
| 1-Et-2-Me-5-(2-Naphthyloxy-thiocarbonyl | 112 | EtOH | | | 192 |
| 1,5-$(Ac)_2$-2-(2-$HOC_6H_4$) | 205–206 | | | | 111 |
| 1,5-$(Ac)_2$-3-HO | 222–223 | EtOH | 93.5 | ir, pmr, uv | 181 |
| 1,5-$(4\text{-}MeC_6H_4SO_2)_2$-2-Me | 202–204 | | 54 | | 124 |
| 1,5-$(4\text{-}MeC_6H_4SO_2)_2$-3-HO | 194–195 | EtOH | 41 | ir, pmr, uv | 181 |
| 1,5-$(4\text{-}MeC_6H_4SO_2)_2$-3-(4-$MeC_6H_4SO_3$) | 192–193 | $CHCl_3$/Petr ether | 97.5 | ir, pmr, uv | 181 |
| 1,5-$(4\text{-}MeC_6H_4SO_2)_2$-3-$MeSO_3$ | 203–204 | $CHCl_3$/Petr ether | 74 | ir, pmr, uv | 181 |
| 1,5-$(NO)_2$-3-HO | 126–127 | PhH | 93 | ir, pmr, uv | 181 |
| 1,5-$Me_2$-6-Cl | | | 8.5; 6 | pmr | 183 |
| 1,5-$Me_2$-7-Cl | | | 31; 37 | pmr | 183 |
| 2,2,4-$Me_2$ | 69 | Petr ether | | | 98 |
| 2-CN-2-Me-4-Ph | 155–157 | | 90 | | 188 |
| 2-CN-4-Me-7-(2-Cyano-4-methyl-2,3,4,5-tetrahydro-1*H*-1,5-benzodiazepin-7-yl)methyl | 132–135 | PhH | 88 | ir, pmr, uv | 99 |
| 2-CN-4-Ph-7-(2-Cyano-4-phenyl-2,3,4,5-tetrahydro-1*H*-1,5-benzodiazepin-7-yl)methyl | 142–144 | PhH | 72 | ir, pmr, uv | 99 |
| 2-(1-Naphthyl)-7,8-$Me_2$ | 77–79 | | | | 111 |
| Dihydrochloride | 188–190 | | | | 111 |
| 2-(2-Naphthyl)-7,8-$Me_2$ | 119–121 | | | | 111 |
| Dihydrochloride | 181–183 | | | | 111 |
| 2-Ph-7,8-$Me_2$ | 86–88 | EtOH/$H_2O$ | 67; 75 | | 111 |
| Dihydrochloride | 196–198 | EtOH/$Et_2O$ | | | 111 |
| 2-(4-$EtOC_6H_4$)-7,8-$Me_2$ | 70–72 | | | | 111 |
| 2-(4-$FC_6H_4$)-7,8-$Me_2$ | 66–68 | | | | 111 |
| Dihydrochloride | 179–181 | | | | 111 |
| 2-(2-$HOC_6H_4$)-7,8-$Me_2$ | 128–129 | | 85 | | 112 |
| 2-(4-$MeOC_6H_4$)-7,8-$Me_2$ | 121–123 | | 85 | | 111 |
| | 127–128 | EtOH/$H_2O$ | 70 | uv | 112 |
| 2-Me-2-COOEt-4-Ph | 90–93 | EtOH | 48 | ir, uv | 320 |

TABLE IV-1. —(contd.)

| Substituent | mp (°C); [bp (°C/torr)] | Solvent of Crystallization | Yield (%) | Spectra | Refs. |
|---|---|---|---|---|---|
| ***Tetrasubstituted*** | | | | | |
| 1,5-$Ac_2$-2-CN-4-Me | 148–152 | EtOH | 37 | ir, pmr | 308 |
| 1,5-$(COCH_2Cl)_2$-2-CN-4-Me | 140–145 | EtOH | 36 | ir, pmr | 308 |
| 1,5-$(COEt)_2$-2-CN-4-Me | 147–149 | EtOH | 41 | ir, pmr | 308 |
| 1,5-$(COPh)_2$-2-CN-4-Me | 203–205 | EtOH | 39 | ir, pmr | 308 |
| 1,5-$(CO\text{-}4\text{-}ClC_6H_4)_2$-2-CN-4-Me | 204–206 | EtOH | 30 | ir, pmr | 308 |
| ***Polysubstituted*** | | | | | |
| 1,5-$Ac_2$-2-$(2\text{-}HOC_6H_4)$-7,8-$Me_2$ | 251–252 | | | | 111 |
| 1,5-$Ac_2$-2-$(4\text{-}MeOC_6H_4)$-7,8-$Me_2$ | 187.5 | EtOH/$H_2O$ | 90 | | 111 |
| 1,5-$(BrCH_2CH_2CO)_2$-2,4-$Me_2$ | 134–136 | PhH | 60 | | 191 |
| 1,5-$(BrCH_2CH_2CO)_2$-2-Me-4-Ph | 155–157 | PhH | 55 | | 191 |
| 1,5-$(NO_2)$-2,2,4-$Me_3$ | 157 | EtOH/$H_2O$ | | | 98 |
| 1,5-$(NO_2)$-2,4-$(CN)_2$-2,4-$Me_2$ | 183d | Acetone | | | 119 |
| 1,5-$Ac_2$-2-Me-2-COOEt-4-Ph | 173–175 | EtOH | 57 | ir, pmr | 308 |
| 1,5-$(COCH_2Cl)_2$-2-Me-2-COOEt-4-Ph | 131–132 | EtOH | 40 | ir, pmr | 308 |
| 1,5-$(COEt)_2$-2-Me-2-COOEt-4-Ph | 150–152 | EtOH | 57 | ir, pmr | 308 |
| 1,5-$(COPh)_2$-2-Me-2-COOEt-4-Ph | 173–175 | EtOH | 68 | ir, pmr | 308 |
| 1,5$(CO\text{-}4\text{-}ClC_6H_4)_2$-2-Me-2-COOEt-4-Ph | 202–214 | EtOH | 85 | ir, pmr | 308 |
| 2-CN-2,4,4-$Me_3$ | 141–144 | EtOH | 80; 35 | | 119, 188 |
| 2,4-$(CN)_2$-2,4-$Me_2$ | 160d | EtOH | | | 119 |
| 2-CN-2,3,4-$Me_3$ | 165–168 | | 74 | | 188 |
| 2-$[3,4\text{-}(HO)_2C_6H_3]$-3-Ph-7,8-$Me_2$ | 208–209 | | | | 111 |

| | | | | |
|---|---|---|---|---|
| 2-(4-$MeOC_6H_4$)-3,7,8-$Me_3$ | 35–36 | | | 111 |
| Dihydrochloride | 238–240 | | | 111 |
| 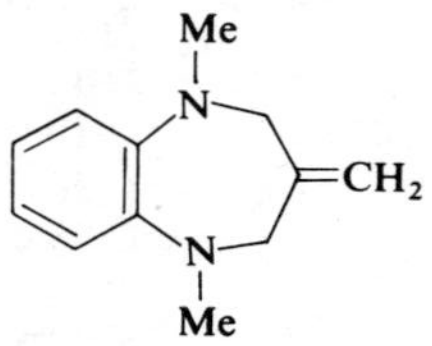 1,5-Dimethyl-3-methylene-2,3,4,5-tetrahydro-1H-1,5-benzodiazepine (N(Me), N(Me), =$CH_2$) | [98–100/1 mm] | | 34 | 182 |

*1,3,4,5-Tetrahydro-1,5-benzodiazepin-2(2H)-ones*

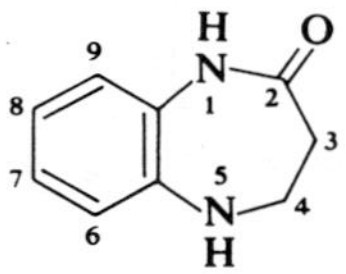

| | | | | |
|---|---|---|---|---|
| None | 141–142 | PhH/Hexane | 17–67 | 187, 193 |
| | | | | 201, 205, 206 |
| Hydrochloride | 191–192 | | | 201 |
| Picrate | 186–187 | EtOH | | 201 |
| ***Monosubstituted*** | | | | |
| 1-Ph | 170–171 | EtOAc | 76; 85 | 199, 323 |
| 1-(2-$ClC_6H_4$) | 149–150 | EtOAc | | 199, 323 |
| 1-(3-$ClC_6H_4$) | 183–184 | *i*-PrOH | | 199. 323 |
| 1-(2,5-$Cl_2C_6H_3$) | 197–198 | EtOAc | | 199, 323 |
| 1-(2,6-$Cl_2C_6H_3$) | 211–212 | $CHCl_3$ | | 199, 323 |
| 1-(4-$ClC_6H_4$) | 233–234 | DMF | | 199, 323 |
| 1-(2-$MeOC_6H_4$) | 127–128 | EtOAc | | 199, 323 |
| 1-(4-$MeOC_6H_4$) | 188–189 | EtOH | | 199, 323 |
| 1-(4-$MeC_6H_4$) | 198–199 | *i*-PrOH | 81 | 199, 323 |
| 1-(3-$CF_3C_6H_4$) | 147–148 | *i*-$Pr_2O$ | | 199, 323 |
| 3-Me | 201–202 | EtOH | 84 | 205 |
| 4-Me | 185–186 | MeOH | 44–100 | 109, 146, 141, 194–197 |
| Picrate·MeOH | 203 | MeOH | | 194 |

TABLE IV-1. —(contd.)

| Substituent | mp (°C); [bp (°C/torr)] | Solvent of Crystallization | Yield (%) | Spectra | Refs. |
|---|---|---|---|---|---|
| 4-Ph | 163–167 | EtOAC | | | 129, 212 |
| | 195–196 | | 46 | | 200 |
| 4-(4-Ac-3-$MeOC_6H_3$) | 155–156 | | 49.5 | | 200 |
| 4-(2-$H_2NC_6H_4$) | | | | | |
| Hydrochloride | 230–231 | $EtOH/Et_2O$ | 49.5 | | 200 |
| 4-(4-$H_2NC_6H_4$) | | | | | |
| Hydrochloride | 255–257 | $EtOH/Et_2O$ | 47 | | 200 |
| 4-(2-$HOC_6H_4$) | 148–149 | | 54 | | 200 |
| 4-(3-$HOC_6H_4$) | 169–170 | | 64 | | 200 |
| 4-(4-HO-3-$MeOC_6H_3$) | 141–142 | | 50.5 | | 200 |
| 4-(4-($MeOC_6H_4$) | 129–130 | | 77 | | 200 |
| 4-(2-$NO_2C_6H_4$) | 80–81 | | 40 | | 200 |
| 4-(4-$NO_2C_6H_4$) | 184–185 | | 30.5 | | 200 |
| 5-Ac | 159–161 | EtOAc/Hexane | 80 | | 187 |
| 5-Me | 138–140 | PhH/Hexane | 20.5 | | 187 |
| 5-NO | 195–196 | EtOH | 90 | | 194 |
| 5-Ph | [221] 181.5–182.5 | (PhH) | 4.5 | | (211) |
| | | EtOH | (91) | | 187 |
| 7-MeO | 158–159 | EtOAc | 28.5 | | 187 |
| 8-Br | 180–181 | EtOAc/$i$-$Pr_2O$ | 76 | | 204 |
| 8-Cl | 183–184 | EtOAc/Hexane | 32; 65; 87 | | 106, 187, 204 |
| 8-MeO | 146.5–147 | PhH | 51 | | 201 |
| 8-$NO_2$ | 274.5–275 | EtOH | 79 | | 201 |
| | 254–255 | MeCN | 84 | | 204 |
| 8-$CF_3$ | 233d | | 55 | | 204 |
| 4-(2-Furyl) | 169 | PhH | | | 129 |

| *Disubstituted* | | | | |
|---|---|---|---|---|
| 1-Ph-3-Me | 146–147 | EtOAc | 83 | 210, 323 |
| 1-Ph-4-Me | 130–131 | EtOAc | | 210, 323 |
| 1-Ph-5-Allyl | 71–72 | Petr ether | | 199, 323 |
| 1-Ph-5-Et | 92–93 | *i*-$Pr_2O$ | | 199, 323 |
| 1-Ph-5-Me | 103–104 | *i*-$Pr_2O$ | | 199, 323 |
| 1-Ph-7-Cl | 203–204 | MeCOEt | | 199, 323 |
| 1-Ph-7-MeO | 216–217 | AcOH | | 199, 323 |
| 1-Ph-7-Me | 164–165 | EtOAc | | 199, 323 |
| 1-Ph-7-$CF_3$ | 168–169 | EtOAc/Petr ether | | 199, 209, 323 |
| 1-Ph-8-$NH_2$ | 284–285 | *i*-$Pr_2O$ | 88 | 204 |
| 1-Ph-8-Br | 192–193 | EtOH | 60 | 209 |
| | 183–185 | | 82 | 204 |
| 1-Ph-8-Cl | 179–180 | EtOH | | 199, 204, 323 |
| Hydrochloride | 158–161 | | | 207 |
| 1-Ph-8-CN | 182–185 | EtOAc/$Et_2O$ | 48 | 204 |
| 1-Ph-8-(4-Diethylaminophenyl)azo | 260 | EtOH/$H_2O$ | 83.5 | 204 |
| 1-Ph-8-HO | 192–193 | $Et_2O$ | 38 | 204 |
| 1-Ph-8-$NO_2$ | 241 | | | 204 |
| 1-Ph-8-(4-Nitrophenyl)-methanimino | 188 | PhMe | 58 | 204 |
| 1-Ph-8-$CF_3$ | 164–165 | *i*-PrOH | | 209 |
| 1-Ph-9-Cl | 213–214 | EtOAc | | 199, 323 |
| 1-(2-Br$C_6H_4$)-8-Cl | 186–188 | | | 207 |
| 1-(4-Br$C_6H_4$)-8-Cl | 216–217 | MeCOEt | | 199, 323 |
| 1-(2-Cl$C_6H_4$)-5-Me | 128–129 | *i*-PrOH | | 199, 323 |
| 1-(2-Cl$C_6H_4$)-7-$CF_3$ | 176–177 | *i*-PrOH | | 199, 209, 323 |
| 1-(2-Cl$C_6H_4$)-8-Cl | 188–189 | *i*-PrOH | | 199, 323 |
| 1-(2-Cl$C_6H_4$)-9-Cl | 212–213 | EtOH | | 199, 323 |
| 1-(2-Cl$C_6H_4$)-8-$NO_2$ | 232–233 | | | 207 |
| 1-(2-Cl$C_6H_4$)-8-$CF_3$ | 134 | | | 207 |
| 1-(2,3-$Cl_2C_6H_3$)-5-Me | 134–135 | *i*-PrOH | | 199, 323 |
| 1-(2,3-$Cl_2C_6H_3$)-8-Cl | 144–145 | *i*-PrOH | | 208 |
| 1-(2,5-$Cl_2C_6H_3$)-5-Me | 133–134 | *i*-PrOH | | 199, 323 |

TABLE IV-1. —(contd.)

| Substituent | mp (°C); [bp (°C/torr)] | Solvent of Crystallization | Yield (%) | Spectra | Refs. |
|---|---|---|---|---|---|
| 1-(2,6-$Cl_2C_6H_3$)-8-Cl | 250–251 | AcOH | | | 208 |
| 1-(3-$ClC_6H_4$)-5-Me | 87–88 | *i*-$Pr_2O$ | | | 199, 323 |
| 1-(4-$ClC_6H_4$)-4-Me | 159–160 | *i*-PrOH | | | 210, 323 |
| 1-(4-$ClC_6H_4$)-5-Me | 89–90 | *i*-$Pr_2O$ | | | 199, 323 |
| 1-(4-$ClC_6H_4$)-8-Cl | 203–204 | EtOAc | | | 199, 323 |
| 1-(2-$FC_6H_4$)8-Cl | 168–169 | *i*-PrOH | | | 199, 323 |
| 1-(2-$MeOC_6H_4$)-5-Me | 129–130 | EtOAc | | | 199, 208, 323 |
| 1-(2-$MeOC_6H_4$)-8-Cl | 147–148 | EtOAc | | | 199 |
| 1-(3-$MeOC_6H_4$)-8-Cl | 178–179 | | | | 199, 323 |
| 1-(4-$MeOC_6H_4$)-5-Allyl | [189–192/0.1] | *i*-PrOH | | | 199, 323 |
| 1-(4-$MeOC_6H_4$)-5-Me | 78–79 | *i*-$Pr_2O$ | | | 199, 323 |
| 1-(4-$MeOC_6H_4$)-8-Cl | 198–199 | EtOH | 73 | | 199, 323 |
| 1-(2-$MeC_6H_4$)-8-Cl | 150–151 | *i*-PrOH | | | 199, 323 |
| 1-(3-$MeC_6H_4$)-8-Cl | 172–173 | EtOH | | | 199, 323 |
| 1-(4-$MeC_6H_4$)-8-Cl | 208–209 | MeCOEt | | | 199, 323 |
| 1-(2-$NO_2C_6H_4$)-8-Cl | 171–172 | $CH_2Cl_2$/*i*-$Pr_2O$ | 61 | | 207 |
| 1-(2-$NO_2C_6H_4$)-8-$NO_2$ | 233–235 | | | | 207 |
| 1-(2-$NO_2C_6H_4$)-8-$CF_3$ | 156 | | | | 207 |
| 1-(2-$CF_3C_6H_4$)-8-Cl | 132–133 | | | | 207 |
| | 147–148 | EtOAc/Petr ether | | | 208 |
| 1-(3-$CF_3C_6H_4$)-3-Me | 168–169 | *i*-PrOH | | | 199, 323 |
| 1-(3-$CF_3C_6H_4$)-5-Me | 104–105 | *i*-$Pr_2O$ | | | 208, 323 |
| 1-(3-$CF_3C_6H_4$)-8-Cl | 148–149 | *i*-$Pr_2O$ | | | 208, 323 |
| 1-(2-Pyridyl)-8-Br | 174–176 | | | | 204 |
| 1-(2-Pyridyl)-8-$CF_3$ | 150–152 | | | | 207 |
| 3,3-$Et_2$ | 211–213 | EtOH | 58 | ir | 203 |
| 3-Et-3-Ph | 143–145 | Ligroin | 88 | ir | 203 |
| Hydrochloride | 218–219 | EtOH | | | |

| | | | | | |
|---|---|---|---|---|---|
| 3,3-$Me_2$ | 191 | PhH | 22 | ir, pmr | 202 |
| 3-Me-3-Ph | | | | | |
| Hydrochloride | 228–231 | *i*-PrOH | 74 | ir | 203 |
| 3-Me-7-Cl | 165–167 | EtOH | 94 | pmr | 205 |
| 3-Me-8-Cl | 195–196 | EtOH | 86 | | 205 |
| 3-Me-8-MeO | 142–143 | EtOH | 34; 50 | | 205 |
| 3,3-$Ph_2$ | 206–207 | EtOH | 11 | ir | 203 |
| 3,3-$Pr_2$ | 171–173 | *i*-PrOH | 74 | ir, pmr | 203 |
| 4,4-$Me_2$ | 248–250 | $CHCl_3$/Hexane | 46.5 | | 198 |
| 4-Me-5-Ac | 177 | $H_2O$ | 97 | | 194 |
| | 190 | Ligroin | | | 146 |
| 4-Me-5-$NH_2$ | 176 | PhH | 86 | | 194 |
| Oxalate | 173d | EtOAc | | | 194 |
| 4-Me-5-(*N*-Diphenylmethanimino) | 204 | MeOH | 67 | | 194 |
| 4-Me-5-[*N*-(2-Nitrophenyl)-methanimino] | 201 | | 88 | | 194 |
| 4-Me-5-NO | 185 | MeOH | 77 | | 146, 194 |
| 4-Me-5-(*N*-Phenylmethanimino) | 184 | MeOH | 72 | | 194 |
| 4-Me-7-$NH_2$ | | | | | |
| Hemithydrate | 156–158 | | | | 141 |
| Dihydrochloride | 243–245d | | 85 | | 141 |
| Picrate | 154–156 | | | | 141 |
| 4-Me-7-(Methyl-1,3,4,5-tetrahydro-2-oxo-1,5-benzo-diazepin-7-yl)-methyl | 116–119 | EtOH | | | 136 |
| 4-Me-8-$NH_2$ | 175–176.5 | EtOAc/Petr ether | 70 | | 141 |
| Picrate | 174–176.5 | | | | 141 |
| 4-Me-8-Br | 213–214 | PhH/Hexane | 84; 86.5 | | 144 |
| 4-Me-8-Cl | 198 | MeOH/$H_2O$ | 31 | pmr, uv | 106 |
| 4,8-$Me_2$ | 196–197 | | 70; 87 | ir | 137, 138 |
| 4-Me-8-$NO_2$ | 247–250 | MeOH | 72; 94 | ir, uv | 139, 141 |
| 4-Ph-5-Ac | 200–202 | MeCN | | | 212, 216 |
| 4-Ph-5-EtCO | 224–226 | MeCN | | | 212 |
| 4-Ph-5-NO | 181d | MeOH | | | 129 |
| 4-Ph-7-Cl | 161 | PhH | 5 | | 134 |

TABLE IV-1. —(contd.)

| Substituent | mp (°C); [bp (°C/torr)] | Solvent of Crystallization | Yield (%) | Spectra | Refs. |
|---|---|---|---|---|---|
| 5-Ac-7-MeO | 219–221 | EtOAc/Hexane | 81 | ir, pmr | 187 |
| 5-Ac-8-Cl | 240–241 | EtOH | 72 | ir, pmr | 187 |
| 5-Me-7-MeO | 150–152 | EtOAc/Hexane | 32 | ir, pmr | 187 |
| 5-Me-8-Cl | 140–142 | EtOAc/Hexane | 23 | ir, pmr | 187 |
| 7,8-$Cl_2$ | 280 | $MeOCH_2CH_2OH/H_2O$ | 40; 70 | | 106 |
| 7,8-$Me_2$ | 176–176.5 | PhH | 56 | pmr, uv | 201 |
| ***Trisubstituted*** | | | | | |
| 1-Ac-4-Me-5-AcNH | 170 | $CHCl_3/Et_2O$ | 76 | | 194 |
| 1-$Me_2N(CH_2)_2$-4-Ph-5-Ac | 132–134 | Hexane | | | 212 |
| Hydrochloride | 238–240 | *i*-PrOH | | | 212 |
| 1-$Me_2N(CH_2)_2$-4-Ph-5-COEt | 122–125 | PhH/Hexane | | | 212 |
| Hydrochloride | 226–228 | *i*-PrOH | | | 212 |
| 1-2(*N*-Methyl-1-adamantylamino)ethyl-4-Ph-5-Ac | | | | | |
| Hydrochloride | | | | | 216 |
| 1-Ph-3,3-$Me_2$ | 151–152 | *i*-PrOH | | | 199, 210, 323 |
| 1-Ph-3,4-$Me_2$ | 193–194 | EtOH | | | 210, 323 |
| 1-Ph-3,5-$Me_2$ | 101–102 | *i*-$Pr_2O$ | | | 199, 210, 323 |
| 1-Ph-3-Me-7-Cl | 172–173 | EtOAc | | | 199, 210, 323 |
| 1-Ph-3-Me-8-Cl | 128–129 | *i*-PrOH | | | 199, 210, 323 |
| 1-Ph-4,4-$Me_2$ | 146–147 | *i*-$Pr_2O$ | | | 199, 210, 323 |
| 1-Ph-4,5-$Me_2$ | 97–98 | $Et_2O$ | | | 199, 210, 323 |
| 1-Ph-4-Me-5-Allyl | 108–109 | *i*-$Pr_2O$ | | | 210, 323 |
| 1-Ph-4-Me-7-Cl | 170–172 | EtOH | | | 199, 210, 323 |
| 1-Ph-4-Me-8-Cl | 167–168 | *i*-PrOH | 78 | | 210, 323 |
| 1-Ph-5-Ac-8-Cl | 209–210 | MeOH | 88 | | 217 |
| | 188–190 | | | | 204 |
| 1-Ph-5-Ac-8-$NO_2$ | 178–179 | | | | 204 |
| 1-Ph-5-($AcOCH_2CH_2$)-8-Cl | 118–119 | *i*-PrOH | 92 | | 217 |

| | | | | |
|---|---|---|---|---|
| 1-Ph-5-(Adamantyl) carbonyl-8-Cl | 160–168 | | | 204 |
| 1-Ph-5-Allyl-8-Br | 158–160 | | 86.5 | 204, 207 |
| 1-Ph-5-Allyl-8-Cl | 135–136 | EtOH | 83 | 199b, 207, 323 |
| 1-Ph-5-Allyl-8-$NO_2$ | 126–127 | | | 207 |
| 1-Ph-5-Allyl-8-$CF_3$ | 144–145 | | | 207 |
| 1-Ph-5-$H_2NCO$-8-Cl | 197–198 | EtOH | 85 | 217 |
| 1-Ph-5-PhCO-8-$NO_2$ | 235–237 | | | 204 |
| 1-Ph-5-$PhCH_2$-8-$NO_2$ | Oil | | | 207 |
| 1-Ph-5-(2-Butenoyl)-8-Cl | 108–109 | | | 204 |
| 1-Ph-5-(But-2-en-1-yl)-8-Br | 129–130 | | | 207 |
| 1-Ph-5-(But-2-en-1-yl)-8-$NO_2$ | 135–136 | *i*-$Pr_2O$ | 63 | 207 |
| 1-Ph-5-BuOCO-8-Cl | 108–109 | | | 204 |
| 1-Ph-5-Bu-8-Br | 130–131 | | | 204 |
| 1-Ph-5-Bu-8-$NO_2$ | 114–115 | | | 204 |
| 1-Ph-5-Bu-8-$CF_3$ | 88–89 | | | 204 |
| 1-Ph-5-$ClCH_2CO$-8-$CF_3$ | 185–187 | | | 204 |
| 1-Ph-5-(4-Chloro-*cis*-but-2-en-l-yl)-8-$NO_2$ | 167–168 | | | 204 |
| 1-Ph-5-(2-Chloroethyl)-8-Cl | 180–181 | MeCOEt | 77 | 217 |
| 1-Ph-5-(Cyclohexylmethyl)-8-$NO_2$ | Oil | | | 215 |
| 1-Ph-5-(Cyclopropyl) methyl-8-$NO_2$ | 127 | | | 215, 217 |
| 1-Ph-5-$Et_2NCO$-8-$CF_3$ | 126–128 | | | 204 |
| 1-Ph-5-$Ph_2CHCO$-8-Cl | 233–234 | | | 204 |
| 1-Ph-5-EtOCO-8-Cl | 172–173 | EtOAc | 84 | 217 |
| 1-Ph-5-EtOCO-8-$NO_2$ | 127–129 | | 87 | 204 |
| 1-Ph-5-Et-8-Cl | 135–136 | EtOH | | 199, 323 |
| 1-Ph-5-Et-8-$NO_2$ | 160–162 | | | 207 |
| 1-Ph-5-EtNHCO-8-Cl | 214–216 | | | 204 |
| 1-Ph-5-EtOOCCO-8-Cl | 186–188 | | | 204 |
| 1-Ph-5-EtOOCCO-8-$NO_2$ | 152–154 | | | 204 |
| 1-Ph-5-EtOOCCO-8-$CF_3$ | 158–160 | | | 204 |
| 1-Ph-5-CHO-8-Br | 158–159 | *i*-$Pr_2O$/Acetone | 78 | 215 |
| 1-Ph-5-CHO-8-Cl | 150–151 | *i*-PrOH | 89 | 217 |

TABLE IV-1. —(contd.)

| Substituent | mp (°C); [bp (°C/torr)] | Solvent of Crystallization | Yield (%) | Spectra | Refs. |
|---|---|---|---|---|---|
| 1-Ph-5-CHO-8-$NO_2$ | 190–191 | | | | 215 |
| 1-Ph-5-CHO-8-$CF_3$ | 141–142 | | | | 215 |
| 1-Ph-5-*n*-Hexyl-8-Br | 144–145 | | | | 215 |
| 1-Ph-5-*n*-Hexyl-8-$NO_2$ | Oil | | | | 215 |
| 1-Ph-5-($HOCH_2CH_2$)-8-Br | 154–155 | MeOH | 51; 64 | | 215 |
| 1-Ph-5-($HOCH_2CH_2$)-8-Cl | 160–161 | *i*-PrOH | 82 | | 217 |
| 1-Ph-5-($HOCH_2CH_2$)-8-$NO_2$ | 188–193 | | | | 215 |
| 1-Ph-5-($HOCH_2CH_2$)-8-$CF_3$ | 91–92 | $Et_2O$ | 27 | | 215 |
| 1-Ph-5-[$HO(CH_2)_3$]-8-Br | 161–163 | | 16 | | 215 |
| 1-Ph-5-($MeOCH_2CH_2$)-8-Cl | 108–109 | EtOH | 67 | | 217 |
| 1-Ph-5-Me-8-Br | 122–124 | | | | 204 |
| 1-Ph-5-Me-8-Cl | 143–144 | EtOH | 93 | | 199b, 323 |
| 1-Ph-5-Me-8-$NO_2$ | 123–125 | | 63 | | 204 |
| 1-Ph-5-Me-8-$CF_3$ | 91–92 | | | | 204 |
| 1-Ph-5-Me-7-Cl | 105–106 | *i*-$Pr_2O$ | | | 199, 323 |
| 1-Ph-5-Me-9-Cl | 160–161 | *i*-PrOH | | | 199, 323 |
| 1-Ph-5-(2-$MeC_6H_4CO$)-8-$NO_2$ | 172–179 | | | | 204 |
| 1-Ph-5-(4-$MeC_6H_4SO_2$)-8-Cl | 212–213 | | | | 204 |
| 1-Ph-5-(1-Naphthylsulfonyl)-8-Cl | 181–182 | | | | 204 |
| 1-Ph-5-(4-$NO_2C_6H_4OCO$)-8-$NO_2$ | 152–154 | | | | 204 |
| 1-Ph-5-($PhCH_2CO$)-8-Cl | 176–178 | | | | 204 |
| 1-Ph-5-PhNHCO-8-Cl | 250–253 | | 93 | | 204 |
| 1-Ph-5-(*trans*-3-Phenylpropenoyl)-8-$NO_2$ | 168–170 | | | | 204 |
| 1-Ph-5-(Propenoyl)-8-Cl | 176–178 | | | | 204 |
| 1-Ph-5-Pr-8-Br | 141–142 | | | | 204, 207 |
| 1-Ph-5-Pr-8-Cl | 125–126 | | | | 207 |
| 1-Ph-5-Pr-8-$NO_2$ | 125–126 | | | | 207 |
| 1-Ph-5-Pr-8-$CF_3$ | 143–145 | | | | 207 |

| | | | | |
|---|---|---|---|---|
| 1-Ph-5-*i*-Pr-8-Br | 123–124 | | | 207 |
| 1-Ph-5-*i*-Pr-8-$NO_2$ | 170–171 | | | 207, 215 |
| 1-Ph-5-Stearyl-8-Cl | 72–74 | | | 204 |
| 1-Ph-5-[2-(Succinyloxy)-ethyl]-8-Cl | 146–147 | *i*-PrOH | 88 | 217 |
| 1-Ph-5-[3-(Succinyloxy)-propyl]-8-Cl | 141–142 | *i*-PrOH | 86 | 217 |
| 1-Ph-5-$CF_3$CO-8-$NH_2$ | 206–207 | *i*-$Pr_2O$ | 85 | 204 |
| 1-Ph-5-$CF_3$CO-8-(4-Diethylaminophenyl)azo | 183–186 | EtOH/$H_2O$ | 42 | 204 |
| 1-Ph-5-$CF_3$CO-8-$NO_2$ | 189 | | | 204 |
| 1-Ph-7,8-$Cl_2$ | 193–194 | MeOH | | 323 |
| 1-(2-$ClC_6H_4$)-5-Ac-8-Cl | 222–223 | MeCOEt | 85 | 217 |
| 1-(2-$ClC_6H_4$)-5-$H_2$NCO-8-Cl | 235–236 | EtOH | 92 | 217 |
| 1-(2-$ClC_6H_4$)-5-PhCO-8-$CF_3$ | 187–188 | | | 204 |
| 1-(2-$ClC_6H_4$)-5-EtOCO-8-Cl | 159–160 | *i*-PrOH | 77 | 217 |
| 1-(2-$ClC_6H_4$)-5-$Et_2$NCO-8-Cl | 171–172 | *i*-PrOH | 69 | 217 |
| 1-(2-$ClC_6H_4$)-5-CHO-8-Cl | 182–183 | *i*-PrOH | 84 | 217 |
| 1-(2-$ClC_6H_4$)-5-CHO-8-$CF_3$ | 143–144 | | | |
| 1-(2-$ClC_6H_4$)-5-$HOCH_2CH_2$-8-Cl | 148–149 | EtOAc | 70 | 217 |
| 1-(2-$ClC_6H_4$)-5-$MeOCH_2CH_2$-8-Cl | 98–99 | *i*-PrOH | 58 | 217 |
| 1-(2-$ClC_6H_4$)-5-Me-7-$CF_3$ | 116–117 | *i*-PrOH | 82 | 209 |
| 1-(2-$ClC_6H_4$)-5-Me-8-Cl | 187–188 | *i*-PrOH | | 199, 323 |
| 1-(2-$ClC_6H_4$)-5-Me-9-Cl | 175–176 | *i*-PrOH | | 199, 323 |
| 1-(2-$ClC_6H_4$)-5-MeNHCO-8-Cl | 223–224 | EtOH | 75 | 217 |
| 1-(2-$ClC_6H_4$)-5-(Prop-2-ynyl)-8-$NO_2$ | 168–170 | | | 207 |
| 1-(2-$ClC_6H_4$)-5-[2-(Succinyloxy)-ethyl]-8-Cl | 149–150 | EtOAc | 81 | 217 |
| 1-(4-$ClC_6H_4$)-4,5-$Me_2$ | 80–82 | Petr ether | | 199, 323 |
| 1-(4-$ClC_6H_4$)-5-Me-8-Cl | 135–136 | *i*-PrOH | | 199, 323 |
| 1-(2-$FC_6H_4$)-5-Ac-8-Cl | 187–189 | | | 204 |
| 1-(2-$FC_6H_4$)-5-CHO-8-Cl | 146–147 | | | 207 |
| 1-(2-$FC_6H_4$)-5-Me-8-Cl | 116–117 | EtOH/$H_2O$ | | 199, 323 |
| 1-(2-$MeOC_6H_4$)-5-Me-8-Cl | 150–151 | EtOAc/Petr ether | | 199, 323 |
| 1-(4-$MeOC_6H_4$)-5-Me-8-Cl | 144–145 | *i*-PrOH | 79 | 199, 323 |
| 1-(2-$MeC_6H_4$)-5-Allyl-8-Cl | 154–155 | *i*-PrOH | | 208 |
| 1-(2-$MeC_6H_4$)-5-Me-8-Cl | 148–149 | *i*-PrOH | | 199, 323 |

TABLE IV-1. —(contd.)

| Substituent | mp (°C); [bp (°C/torr)] | Solvent of Crystallization | Yield (%) | Spectra | Refs. |
|---|---|---|---|---|---|
| 1-(3-$MeC_6H_4$)-5-Me-8-Cl | 135–136 | *i*-PrOH | | | 199, 323 |
| 1-(2-$NO_2C_6H_4$)-5-CHO-8-Cl | 211–214 | | | | 207 |
| 1-(2-$NO_2C_6H_4$)-5-Me-8-Cl | 176–178 | | | | 207 |
| 1-(2-Pyridyl)-5-Ac-8-Cl | 184 | | | | 204 |
| 1-(2-Pyridyl)-5-Allyl-8-Br | 144–146 | | | | 204 |
| 3,5-$Et_2$-3-Ph | 131–133 | $Et_2O$ | | ir | 203 |
| 3-Et-3-Ph-5-Ac | 175–176 | EtOH | 69 | ir | 203 |
| 3-Et-3-Ph-5-$NH_2$ | 139–141 | $Et_2O$ | 68 | ir | 203 |
| 3-Et-3-Ph-5-$H_2NCO$ | 226–228 | PhH | 37 | ir | 203 |
| 3-Et-3-Ph-5-Me | 138–140 | $Et_2O$ | 63 | ir | 203 |
| 3-Et-3-Ph-5-NO | 177–179 | EtOH | 94 | ir | 203 |
| 3-Et-3-Ph-7-Br | 167–170 | PhH/Petr ether | 28 | ir | 203 |
| 3-Et-3-Ph-8-Br | 162–164 | AcOH | 36 | ir | 203 |
| 3-Me-7,8-$Cl_2$ | 179–181 | EtOH | 71 | | 205 |
| 4,4,8-$Me_3$ | 230–232 | $CHCl_3$/Hexane | 47 | uv | 198 |
| 4-$Me_2$-8-Cl | 239–240 | $CHCl_3$/Hexane | 20 | uv | 198 |
| 4-Me-7,8-$Cl_2$ | 238 | DMSO/$H_2O$ | 60 | pmr, uv | 106 |
| 4-PH-7,9-$Cl_2$ | 115–116 | | 49 | | 200 |
| 4-(2-$H_2NC_6H_4$)-7,9-$Cl_2$ | 148–149 | | 33.5 | | 200 |
| 4-(4-$H_2NC_6H_4$)-7,9-$Cl_2$ | 188–190 | | 33 | | 200 |
| 4-(2-$HOC_6H_4$)-7,9-$Cl_2$ | 138–139 | | 25 | | 200 |
| 4-(3-$HOC_6H_4$)-7,9-$Cl_2$ | 158–159 | | 32 | | 200 |
| 4-(4-HO-3-$MeOC_6H_3$)-7,9-$Cl_2$ | 150–152 | | 27.5 | | 200 |
| 4-(4-$MeOC_6H_4$)-7,9-$Cl_2$ | 123–124 | | 26.5 | | 200 |
| 4-(2-$NO_2C_6H_4$)-7,9-$Cl_2$ | 161–162 | | 26 | | 200 |
| 4-(4-$NO_2C_6H_4$)-7,9-$Cl_2$ | 195–196 | | 23.5 | | 200 |
| 6-Br-7,8-$Cl_2$ | 260 | EtOH | 75 | pmr, uv | 106 |

***Polysubstituted***

| | | | | | |
|---|---|---|---|---|---|
| 1-Ph-3,3-$Me_2$-8-Cl | 143–133 | *i*-PrOH | | | 199, 323 |
| 1-Ph-3,4-$Me_2$-8-Cl | 194–195 | EtOH | | | 199, 323 |
| 1-Ph-3,5-$Me_2$-7-Cl | 116–117 | *i*-$Pr_2O$ | | | 199, 323 |
| 1-Ph-3,5-$Me_2$-8-Cl | 151–152 | *i*-PrOH | | | 199, 323 |
| 1-Ph-4,4-$Me_2$-8-Cl | 180–181 | *i*-$Pr_2O$ | | | 199, 323 |
| 1-Ph-4,5-$Me_2$-7-Cl | 122–123 | *i*-$Pr_2O$ | | | 199 |
| 1-PH-4,5-$Me_2$-8-Cl | 113–114 | EtOAc | 82 | | 199, 323 |
| 1-Ph-5-Me-7,8-$Cl_2$ | 109–110 | *i*-PrOH | | | 199, 323 |
| 1-Ph-3,4,5-$Me_3$-8-Cl | 116–117 | *i*-PrOH | | | 199, 323 |
| 1-Ph-3,3,5-$Me_2$-8-Cl | 191–192 | *i*-PrOH | | | 199, 323 |

***4-Methylene-1,3,4,5-tetrahydro-1,5-benzodiazepin-2(2H)-ones***

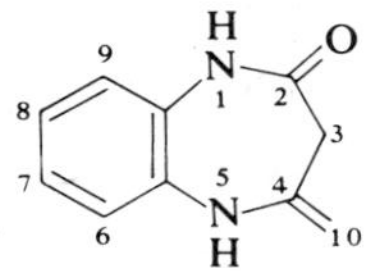

| | | | | | |
|---|---|---|---|---|---|
| 10-$PhCH_2OCO$ | 197 | $PhCH_2OH$ | 87 | | 214 |
| 10-EtOCO | 228 | EtOH | 30–95 | ir, pmr | 213, 214 |
| 10-$EtOCH_2CH_2OCO$ | 160 | EtOH | 90 | | 214 |
| 10-MeOCO | 255 | EtOH | 30–95 | | 213–214 |
| 10-(Tetrahydrofuran-2-yl)-methoxycarbonyl | 173 | EtOH | 55 | | 214 |
| 1-Ph-10-$NO_2$ | 233–235 | $CH_2Cl_2$/EtOH | | | 171 |
| 1-Ph-8-Cl-10-MeOCO | 199–201 | $CH_2Cl_2$/MeOH | | | 171 |
| 1-Ph-8-Cl-10-$NO_2$ | 260–262 | $CH_2Cl_2$/MeOH | | | 160 |
| 1-Ph-8-Cl-10,10-$(MeOCO)_2$ | 155–158 | MeOH | | | 171 |

***1,2,4,5-Tetrahydro-1,5-benzodiazepin-3(3H)-ones***

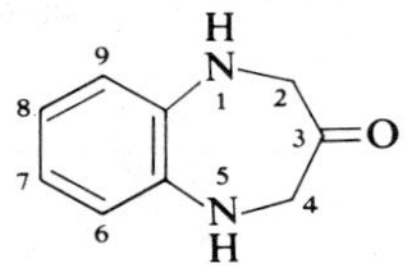

TABLE IV-1. —(contd.)

| Substituent | mp (°C); [bp (°C/torr)] | Solvent of Crystallization | Yield (%) | Spectra | Refs. |
|---|---|---|---|---|---|
| 1,5-(4-$MeC_6H_4SO_2)_2$ | 176–178 | EtOAc | 25 | ir, pmr | 219 |
| | 179–180 | $CHCl_3$/EtOH | 89 | ir, pmr | 181 |
| Oxime | 198–198.5 | EtOH/$H_2O$ | | | 181 |
| Hydrazone | 184–185 | $CHCl_3$/EtOH | | | 181 |
| Tosylhydrazone | 187–188 | EtOH | | | 181 |
| 2,4-Dinitrophenylhydrazone | 200–201 | $CHCl_3$/EtOH | | | 181 |
| ***3,5-Dihydro-1H-1,5-benzodiazepin-2,4(2H,4H)diones*** | | 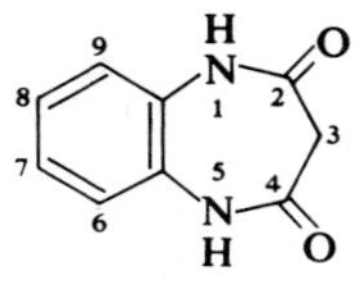 | | | |
| None | > 360 | | 62 | | 220, 221, 223–227 |
| ***Monosubstituted*** | | | | | |
| 1-$PhCH_2$ | 214–216 | EtOH/Dioxane | | | 234 |
| 1-Bu | 168–169 | EtOH | | | 234 |
| 1-Cyclohexyl | 184–185 | EtOH/$H_2O$ | | | 234 |
| 1-Me | 240–242 | $H_2O$ | | | 234, 235 |
| 1-Ph | 271–272 | EtOH | | | 234 |
| 1-(4-$ClC_6H_4$) | 248–250 | AcOH | | | 234 |
| 1-(2-$MeOC_6H_4$) | 295–297 | | | | 236 |
| 1-(3-$MeOC_6H_4$) | 213 | | | | 236 |
| 1-(4-$MeOC_6H_4$) | 286–290 | $H_2O$ | 54 | | 236 |
| 1-(2-Me-4-$ClC_6H_3$) | 245 | | | | 234, 236 |
| 3-Allyl | 282 | PhN/EtOH | 49 | | 227 |
| 3-*n*-Bu | 312–315 | PhN/EtOH | 47 | | 227 |

| | | | | |
|---|---|---|---|---|
| 3-Et | 300d | PhH | 66.5 | 227 |
| 3-Me | 326d | PhN | 44 | 220–222, 227 |
| 3-$n$-Pr | 285 | PhN | 54 | 227 |
| 3-NH-(2-$PhSO_2C_6H_4$) | | | | |
| hydrochloride | 271d | | 31 | 317 |
| 3-N(Ph)($SO_2Ph$) | 250d | | 49 | 317 |
| 3-N(Ph)(4-$MeSO_2C_6H_4$) | 268d | | 46 | 317 |
| 3-N(4-$MeOC_6H_4$)($SO_2Ph$) | 263d | | 48 | 317 |
| 3-N(4-$MeOC_6H_4$)(4-$MeSO_2C_6H_4$) | 278d | | 49 | 317 |
| 3-N(2-$MeC_6H_4$)($SO_2Ph$) | 270d | | 50 | 317 |
| 3-N(2-$MeC_6H_4$)(4-$MeSO_2C_6H_4$) | 282d | | 49 | 317 |
| 7-COOH | > 330 | | | 226 |
| 7-Cl | 335–340d | AcOH | | 226 |
| 7-F | > 310 | | | 226 |
| 7-Me | > 300 | | | 221 |
| 7-$NO_2$ | 295d | EtOH | 73 | 226 |
| ***Disubstituted*** | | | | |
| 1-(4-$AcOC_6H_4CO$)-5-Me | 192–195 | EtOH | | 234 |
| 1-PhCO-5-Me | 187–189 | EtOH/$H_2O$ | 32.5 | 234 |
| 1-$PhCH_2$-5-Me | 193–194 | AcOH/$H_2O$ | | 234 |
| 1,3-$n$-$Bu_2$ | 136–137 | Ligroin | | 227 |
| 1-$n$-Bu-5-Me | 141–143 | EtOH/$H_2O$ | | 234 |
| 1-(4-$ClC_6H_4CO$)-5-Me | 190–193 | EtOH | | 234 |
| 1-Cyclopentyl-7-Cl | 146–148 | | | 239 |
| 1-Et-7-Cl | 168–170 | | | 239 |
| 1,5-$Me_2$ | 255 | EtOH | 37 | 224, 234 |
| 1-Me-5-$i$-Pr | 158–160 | $i$-PrOH | | 234 |
| 1-Me-7-Br | 221–222 | | | 239 |
| 1-Me-7-Cl | 204–206 | EtOH/$H_2O$ | 79, 82.5 | 234, 237 |
| 1-Me-7-F | 227–230 | | | 239 |
| 1-Me-7-MeO | 201–203 | $i$-PrOH/$H_2O$ | | 234, 235 |
| 1-Me-8-MeO | 213–215 | | | 239 |
| 1-Me-7-$NO_2$ | 230–231 | | | 241 |

TABLE IV-1. —(contd.)

| Substituent | mp (°C); [bp (°C/torr)] | Solvent of Crystallization | Yield (%) | Spectra | Refs. |
|---|---|---|---|---|---|
| 1-Me-7-$CF_3$ | 192–194 | $H_2O$ | 91 | | 237–239 |
| 1-Me-8-$CF_3$ | 193–194 | | | | 239 |
| 1-Ph-5-*n*-Bu | 122–124 | *i*-PrOH | | | 234, 235 |
| 1-Ph-5-Me | 144–146 | *i*-PrOH | | | 234 |
| 1-Ph-5-($Me_2NCH_2CH_2$) | 131–132 | $EtOH/H_2O$ | | | 234 |
| 1-Ph-5-[$Me_2N(CH_2)_3$] | 111–113 | $PhH/Et_2O$ | | | 234 |
| 1-Ph-5-(4-$MeOC_6H_4$) | 192–194 | | | | 237 |
| 1-Ph-5-$Me_2POCH_2$ | 253–255 | MeOH/PhMe | 76 | | 250 |
| 1-Ph-7-$CF_3$ | 224–225 | *i*-PrOH | 87 | | 239 |
| 1-Ph-8-Br | 281–284 | $AcOH/H_2O$ | | | 234 |
| 1-Ph-8-Cl | 290–292 | AcOH | 42 | | 234 |
| 1-Ph-8-CN | 270–272 | | | | 233 |
| 1-Ph-8-MeO | 239–241 | $EtOH/H_2O$ | | | 234 |
| 1-Ph-8-$NO_2$ | 272–274 | MeCN | 92 | | 241 |
| 1-(2-$BrC_6H_4$)-8-Cl | 255–256 | | | | 236 |
| 1-(2-$ClC_6H_4$)-8-Cl | 263 | | | | 236, 237 |
| 1-(2-$ClC_6H_4$)-8-$NO_2$ | 250–252 | | | | 241 |
| 1-(2,4-$Cl_2C_6H_3$)-8-Cl | 200–202 | | | | 237 |
| 1-(4-$ClC_6H_4$)-5-Me | 192–194 | $EtOH/H_2O$ | | | 234, 237 |
| 1-(4-$ClC_6H_4$)-8-Cl | 264–266 | EtOH | | | 234, 237 |
| 1-(2-$FC_6H_4$)-5-Me | 162–163 | | | | 236 |
| 1-(2-$FC_6H_4$)-8-Cl | 255–257 | | | | 236 |
| 1-(2-$FC_6H_4$)-8-$NO_2$ | 260–263 | | | | 241 |
| 1-(2-$MeOC_6H_4$)-5-Et | 194–195 | | | | 236, 237 |
| 1-(2-$MeOC_6H_4$)-5-Me | 205–207 | | | | 236, 237 |
| 1-(3-$MeOC_6H_4$)-5-Me | 127 | | | | 236, 237 |
| 1-(4-$MeOC_6H_4$)-5-PhCO | 104–106 | | | | 251 |
| 1-(4-$MeOC_6H_4$)-5-Me | 175 | EtOAc | 65 | | 236, 237 |

| | | | | |
|---|---|---|---|---|
| 1-(3-$MeC_6H_4$)-5-Me | 163–164 | | | 237 |
| 1-(2,3-$Me_2C_6H_3$)-5-Me | 222–224 | | | 237 |
| 1-(2-$NO_2C_6H_4$)-8-$NO_2$ | 285–286 | | | 241 |
| 1-(2-$CF_3C_6H_4$)-8-$NO_2$ | | | | |
| Ethanolate | 118d | EtOH | | 241 |
| 1,3-*n*-$Pr_2$ | 131–133 | EtOH | | 227 |
| 1-*n*-Pr-7-Cl | 169–171 | | | 239 |
| 1-*i*-Pr-7-Cl | 174–176 | | | 239 |
| 1-(2-Pyridyl)-8-Cl | 268–270 | $CH_2CH_2$/*i*-$Pr_2O$ | | 233 |
| 3,3-Diallyl | 240–242 | EtOH | 35; 76 | 227, 228, 307 |
| 3-Allyl-3-$PhCH_2$ | 257 | EtOH | 26 | 230 |
| 3-Allyl-3-Bu | 212–213 | EtOH | 37 | 228b, 230 |
| 3-Allyl-3-cyclohexyl | 302 | EtOH | 17 | 230 |
| 3-Allyl-3-($Et_2NCH_2CH_2$) | 216 | EtOH | 40 | 230 |
| 3-Allyl-3-[$Me_2N(CH_2)_3$] | 223 | EtOH | 40 | 230 |
| 3-Allyl-3-Et | 250–254 | EtOH | 31; 34 | 228b, 230 |
| 3-Allyl-3-Me | 248–249 | EtOH | 42; 79.5 | 228b, 230 |
| 3-Allyl-3-(3-methylbut-1-yl) | 242 | EtOH | 27 | 230 |
| 3-Allyl-3-Ph | 291 | EtOH | 60 | 230 |
| 3-Allyl-3-*i*-Pr | 280 | EtOH | 15; 23.5 | 228b, 230 |
| 3,3-*n*-$Bu_2$ | 236 | EtOH | 36 | 227 |
| 3-*n*-Bu-7-$NO_2$ | 270 | PhH | 78 | 231 |
| 3,3-$Et_2$ | 262–264 | EtOH | 43 | 227 |
| 3-Et-3-(3-Ethylbut-1-yl) | 254–255 | EtOH | 41 | 227 |
| 3-Et-3-*n*-Pentyl | 243.5 | EtOH | 44 | 227 |
| 3-Et-3-*i*-Pr | 269 | EtOH | 35 | 227 |
| 3-Et-7-$NO_2$ | 310–312 | PhMe | 65 | 231 |
| 3,3-$Me_2$ | 313–315 | Py | 54 | 227 |
| 3,7-$(NO_2)_2$ | 212–213 | EtOH/$H_2O$ | | 226 |
| 3,3-*n*-$Pr_2$ | 252.5 | EtOH | 43 | 227 |
| 3-Pr-7-$NO_2$ | 305–308 | PhMe | 70 | 231 |

TABLE IV-1. —(contd.)

| Substituent | mp (°C); [bp (°C/torr)] | Solvent of Crystallization | Yield (%) | Spectra | Refs. |
|---|---|---|---|---|---|
| ***Trisubstituted*** | | | | | |
| 1-PhCO-5-Me-8-Cl | 178–179 | AcOH/$H_2O$ | | | 234 |
| 1-$PhCH_2$-3,3-Diallyl | 155 | PhH/Petr ether | 71 | | 237 |
| 1-$PhCH_2$-5-Me-8-Cl | 145–148 | EtOH/Ligroin | | | 234 |
| 1,3,5-*n*-$Bu_3$ | 96–97 | Ligroin | | | 227 |
| 1-*n*-Bu-3,3-$Et_2$ | 144–146 | Ligroin | | | 227 |
| 1-(2-Carboxyethyl)-3,3-Diallyl | 114 | Acetone/$H_2O$ | 15 | | 237 |
| 1-(Carboxymethyl)-3,3-Diallyl | 246 | Acetone | 22 | | 237 |
| 1-($Et_2NCH_2CH_2$)-3,3-Diallyl | Oil | | 35 | | 237 |
| 1-[$Me_2N(CH_2)_3$]-3,3-Diallyl | Oil | | 30 | | 237 |
| 1-Et-3,3-Diallyl | 172 | EtOH | 34 | | 230 |
| 1-Me-3,3-Diallyl | 126 | EtOH | 40 | | 230 |
| 1-Naphtyl-5-Me-8-Cl | 209–211 | | | | 237 |
| 1-Ph-3-AcO-8-$CF_3$ | 274–276 | EtOAc | 73 | ir, pmr | 246 |
| 1-Ph-3-PhCOO-8-$CF_3$ | 267–269 | $Et_2O$ | 37.5 | | 246 |
| 1-Ph-3-HO-8-Br | 264–266 | EtOH | 94 | | 245, 246 |
| 1-Ph-3-HO-8-Cl | 278–280 | THF/$Et_2O$ | | ir, pmr | 245, 246 |
| 1-Ph-3-HO-8-$CF_3$ | 260–264 | THF | 72 | | 245 |
| | 243d | | 58 | | 246 |
| 1,5-$Ph_2$-7-Cl | 255–256 | $CH_2Cl_2$/Petr ether | | | 237, 239 |
| 1-Ph-5-Ac-8-Cl | 201–203 | PhH/Ligroin | | | 234, 240, 251 |
| 1-Ph-5-Allyl-8-Cl | 203–206 | EtOH/$H_2O$ | 40 | | 234, 235, 237 |
| 1-Ph-5-Allyl-8-$NO_2$ | 237–239 | | 75 | | 241 |
| 1-Ph-5-PhCO-8-Br | 200–201 | | | | 240, 251 |
| 1-Ph-5-PhCO-8-Cl | 215–216 | Acetone | 55; 57 | | 240, 251 |
| 1-Ph-5-PhCO-7-Me | 214–216 | | | | 251 |
| 1-Ph-5-PhCo-7-$CF_3$ | 213–215 | | | | 251 |
| 1-Ph-5-PhCO-8-$CF_3$ | 176–178 | | | | 240, 251 |

| | | | | |
|---|---|---|---|---|
| 1-Ph-5-$PhCH_2$-8-Cl | 181–182 | | | 237 |
| | 173–175 | *i*-PrOH | | 234, 235 |
| 1-Ph-5-*n*-Bu-8-Cl | 159–161 | EtOH/$H_2O$ | | 234, 235 |
| 1-Ph-5-$ClCH_2CO$-8-Cl | 208–209 | | | 251 |
| 1-Ph-5-(2-$ClC_6H_4CO$)-8-Cl | 222–224 | | | 240, 251 |
| 1-Ph-5-(2,4-$Cl_2C_6H_3CO$)-8-Cl | 204–205 | | | 251 |
| 1-Ph-5-(2-$ClC_6H_4$)-7-Cl | 243–245 | | | 237 |
| 1-Ph-5-(*trans*-3-Chloroprop-2-en-1-yl)-8-Cl | 153–154 | | | 237 |
| 1-Ph-5-(3-Chloroprop-1-yl)-8-Cl | 156–158 | | | 237 |
| 1-Ph-5-Cyclohexyl-8-Cl | 231–233 | | | 237 |
| 1-Ph-5-(Cyclohexylcarbonyl)-8-Cl | 157–159 | | | 240, 251 |
| 1-Ph-5-(Cyclopentyl)-8-Cl | 158–160 | | | 239 |
| 1-Ph-5-(Cyclopropyl)methyl-8-Cl | 215–218 | EtOH | | 234, 235, 237 |
| 1-Ph-5-($Et_2NCH_2CH_2$)-8-Cl | 146–148 | EtOH/Ligroin | | 234, 237 |
| 1-Ph-5-($Me_2NCH_2CH_2$)-8-Cl | 156–158 | EtOH/$Et_2O$ | 58 | 234, 237 |
| Methiodide | 256–259 | EtOH/$H_2O$ | | 234 |
| 1-Ph-5-[$Me_2N(CH_2)_3$]-8-Cl | 123–125 | EtOH/$H_2O$ | | 234 |
| 1-Ph-5-$Me_2POCH_2$-8-Cl | 257–258 | | 71 | 250 |
| 1-Ph-5-*n*-Dodecanoyl-8-Cl | 89–91 | | | 251 |
| 1-Ph-5-EtOCO-8-Cl | 180–183 | EtOH | | 234, 251 |
| 1-Ph-5-$EtOCH_2CH_2$-8-Cl | 135–137 | | | 237 |
| 1-Ph-5-$EtOOCCH_2$-8-Cl | 156–158 | PhH/Ligroin | | 234 |
| 1-Ph-5-Et-8-Br | 201–203 | | | 237 |
| 1-Ph-5-Et-8-Cl | 227–228 | EtOH | | 234, 235, 237 |
| 1-Ph-5-Et-8-$NO_2$ | 257–259 | | 90 | 241 |
| 1-Ph-5-Et-7-$CF_3$ | 176–178 | | | 237 |
| 1-Ph-5-(2-$FC_6H_4CO$)-8-Cl | 176–178 | | | 251 |
| 1-Ph-5-CHO-8-Cl | 210–211 | | | 240, 251 |
| 1-Ph-5-(2-Furoyl)-8-Cl | 228–230 | | | 251 |
| 1-Ph-5-*n*-Hexadecanoyl-8-Cl | 78–80 | | | 251 |
| 1-Ph-5-$HOCH_2CH_2$-8-Cl | 208–210 | | | 237 |
| 1-Ph-5-$HOCH_2CH_2$-8-$NO_2$ | 187–188 | EtOH | | 241 |
| 1-Ph-5-$HOCH_2CH_2$-8-$CF_3$ | 153–154 | | | 237 |

TABLE IV-1. —(contd.)

| Substituent | mp (°C); [bp (°C/torr)] | Solvent of Crystallization | Yield (%) | Spectra | Refs. |
|---|---|---|---|---|---|
| 1-Ph-5-HO($CH_2$)$_3$-8-Cl | 211–213 | | | | 237 |
| 1-Ph-5-HO($CH_2$)$_3$-8-$CF_3$ | 157–159 | | | | 237 |
| 1-Ph-5-(1-Hydroxy-2-propyl)-8-Cl | 192–194 | | | | 237 |
| 1-Ph-5-$MeOCH_2CH_2$-8-Cl | 175–178 | | | | 237 |
| 1-Ph-5-$MeOCH_2$-8-Cl | 164–165 | | | | 237 |
| 1-Ph-5-[3,4-$(MeO)_2C_6H_3CO$]-8-Cl | 134–137 | | | | 240, 251 |
| 1-Ph-5-Me-8-Ac | 134–137 | | | | 237 |
| 1-Ph-5-Me-8-Br | 201–203 | EtOH | | | 234 |
| 1-Ph-5-Me-7-Cl | 161–162 | | | | 237 |
| 1-Ph-5-Me-8-Cl | 166–168 | EtOH/$H_2O$ | | | 234 |
| | 182–184 | *i*-PrOH | | | 237 |
| 1-Ph-5-Me-9-Cl | 172–174 | | | | 237 |
| 1-Ph-5-Me-8-CN | 260–262 | | | | 237 |
| 1-Ph-5-Me-8-F | 185–187 | | | | 237 |
| 1-Ph-5-Me-7-MeO | 162–164 | | | | 237, 239, 249 |
| 1-Ph-5-Me-8-MeO | 130–132 | EtOH/$H_2O$ | | | 234, 237 |
| 1-Ph-5,7-$Me_2$ | 154–156 | | | | 237 |
| 1-Ph-5,8-$Me_2$ | 194–195 | | | | 237, 249 |
| 1-Ph-5-Me-8-MeOCO | 145–147 | | | | 237 |
| 1-Ph-5-Me-8-$NO_2$ | 179–181 | $CH_2Cl_2$/*i*-$Pr_2O$ | 88; 40 | | 241 |
| 1-Ph-5-Me-7-$CF_3$ | 130–131 | | 89 | | 237, 239 |
| 1-Ph-5-Me-8-$CF_3$ | 204–205 | $CH_2Cl_2$/*i*-$Pr_2O$ | 75; 45 | | 237, 239 |
| 1-Ph-5-MeNHCO-8-Cl | 306–308 | | | | 251 |
| 1-Ph-5-(3-Methyl-1-butyl)-8-Cl | 136–138 | PhH/Ligroin | | | 234 |
| 1-Ph-5(3-Methylbut-2-en-1-yl)-8-Cl | 154–156 | | | | 237 |
| 1-Ph-5-(2-$MeC_6H_4CO$)-8-Cl | 197–200 | | | | 240, 251 |
| 1-Ph-5-(4-$MeC_6H_4CO$)-8-Cl | 194–196 | | | | 251 |
| 1-Ph-5(2,4-$Me_2C_6H_3$)-7-Cl | 244–245 | | | | 237 |

| | | | | |
|---|---|---|---|---|
| 1-Ph-5-(2-Methylpropanoyl)-8-Cl | 116–118 | | | 240 |
| 1-Ph-5-(2-Methylpropenoyl)-8-Cl | 149–151 | EtOH/$H_2O$ | | 234, 235 |
| 1-Ph-5-(2-Methylpropyl)-8-Cl | 198–200 | EtOH/$H_2O$ | | 234 |
| 1-Ph-5-(4-$NO_2C_6H_4CO$)-8-Cl | 216–218 | | | 240, 251 |
| 1-Ph-5-(Piperidino)propyl-8-Cl | 142–144 | | | 237 |
| 1-Ph-5-$PhCH_2CO$-8-Cl | 127–129 | | | 240, 251 |
| 1-Ph-5-(*trans*-3-phenyl-propenoyl)-8-Cl | 205–206 | | | 240, 251 |
| 1-Ph-5-Propanoyl-8-Cl | 194–196 | PhH | 80–90 | 234, 240, 251 |
| 1-Ph-5-*n*-Pr-8-Cl | 199–200 | | | 234, 235, 239 |
| 1-Ph-5-*i*-Pr-8-Cl | 143–145 | PhH/Ligroin | | 234, 239 |
| 1-Ph-5-*n*-Pr-8-$NO_2$ | 239–241 | | | 241 |
| 1-Ph-5-*i*-Pr-8-$NO_2$ | 212–213 | | 82 | 241 |
| 1-Ph-5-(2-Thienoyl)-8-Cl | 210–212 | | | 251 |
| 1-Ph-5-$CF_3CO$-8-Cl | 173–175 | | | 240, 251 |
| 1-(2-$AcC_6H_4$)-5-Me-8-Cl | 205–206 | | | 237, 239 |
| 1-(2-$BrC_6H_4$)-5-PhCO-8-Cl | 216–220 | | | 240 |
| 1-(2-$BrC_6H_4$)-5-Me-8-Br | 205–206 | | | 237 |
| 1-(2-$BrC_6H_4$)-5-Me-8-Cl | 210–212 | | | 236, 237, 249 |
| 1-(2-$BrC_6H_4$)-5-Me-8-F | 190–192 | | | 237 |
| 1-(2-$BrC_6H_4$)-5-Me-8-$CF_3$ | 194–195 | | | 237 |
| 1-(2-$ClC_6H_4$)-5-PhCO-8-Cl | 105–107 | | | 240 |
| 1-(2-$ClC_6H_4$)-5-Et-8-Cl | 207–209 | | | 236, 237 |
| 1-(2-$ClC_6H_4$)-5-$HOCH_2CH_2$-8-Cl | 197–199 | | | 236, 237 |
| 1-(2-$ClC_6H_4$)-5-(1-Hydroxy-2-propyl)-8-Cl | 156–158 | | | 236, 237 |
| 1-(2-$ClC_6H_4$)-5-Me-8-Cl | 222–224 | | | 236 |
| 1-(2-$ClC_6H_4$)-5-Me-8-F | 195–197 | | | 236, 239 |
| 1-(2-$ClC_6H_4$)-5-Me-8-$CF_3$ | 175–177 | | | 237 |
| 1-(2-$ClC_6H_4$)-5-*i*-Pr-8-Cl | 215–217 | | | 236, 237 |
| 1-(3-$ClC_6H_4$)-5-Me-8-Cl | 191–192 | | | 237 |
| 1-(4-$ClC_6H_4$)-5-PhCO-8-Cl | 243–245 | | | 251 |
| 1-(4-$ClC_6H_4$)-5-Me-8-Cl | 227–229 | EtOH | | 234, 237 |
| 1-(2-$CNC_6H_4$)-5-Me-8-Cl | 209–210 | | | 237 |
| 1-(2-$EtC_6H_4$)-5-Me-8-Cl | 179–180 | | | 237 |
| 1-(2-$FC_6H_4$)-5-PhCO-8-Cl | 140–141 | | | 240 |

TABLE IV-1. —(contd.)

| Substituent | mp (°C); [bp (°C/torr)] | Solvent of Crystallization | Yield (%) | Spectra | Refs. |
|---|---|---|---|---|---|
| 1-(2-$FC_6H_4$)-5-Bu-8-$NO_2$ | 164–166 | | | | 241 |
| 1-(2-$FC_6H_4$)-5-($Me_2NCH_2CH_2$)-8-Cl | 134–136 | | | | 236, 237 |
| 1-(2-$FC_6H_4$)-5-Me-8-Cl | 153–154 | | | | 237 |
| 1-(2-$FC_6H_4$)-5-Me-8-$NO_2$ | 195–196 | | | | 236 |
| Methanolate | 113–115 | MeOH | | | 241 |
| 1-(2-$FC_6H_4$)-5-Me-8-$CF_3$ | 184–186 | | | | 236, 237, 239 |
| 1-(4-$HOC_6H_4$)-5-Me-8-$CF_3$ | 268–270 | | | | 239 |
| 1-(2-$MeOC_6H_4$)-5-Me-8-Cl | 221–222 | | | | 237 |
| 1-(2-$MeOC_6H_4$)-5-Me-8-$CF_3$ | 199–201 | | | | 239 |
| 1-(2-$MeC_6H_4$)-5-Me-8-Cl | 201–203 | | | | 237 |
| 1-(4-$MeC_6H_4$)-5-PhCO-8-Cl | 165–168 | | | | 251 |
| 1-(4-$MeC_6H_4$)-5-Me-8-Cl | 203–204 | | | | 237 |
| 1-(2,3-$Me_2C_6H_3$)-5-Et-8-Cl | 201–203 | | | | 237 |
| 1-(2,3-$Me_2C_6H_3$)-5-Me-8-Cl | 200–202 | | | | 237 |
| 1-(2,4-$Me_2C_6H_3$)-5-Me-8-Cl | 190–192 | | | | 237 |
| 1-(2-Me-4-$ClC_6H_3$)-5-Me-8-Cl | 202–204 | | | | 236, 237 |
| 1-(2-$NO_2C_6H_4$)-5-Allyl-8-$NO_2$ | 158–160 | | | | 241 |
| 1-(2-$NO_2C_6H_4$)-5-PHCO-8-Cl | 192 | | | | 240 |
| 1-(2-$NO_2C_6H_4$)-5-Cyclohexyl-8-Cl | 182–183 | | | | 237, 238 |
| 1-(2-$NO_2C_6H_4$)-5-Et-8-$NO_2$ | 226–228 | | | | 241 |
| 1-(2-$NO_2C_6H_4$)-5-$HO(CH_2)_3$-8-Cl | 162–163 | | | | 237, 238 |
| 1-(2-$NO_2C_6H_4$)-5-Me-8-Cl | 206–208 | | | | 237, 238 |
| 1-(2-$NO_2C_6H_4$)-5-Me-8-$CF_3$ | 230–232 | $CH_2Cl_2$/Petr ether | 80 | | 237, 239 |
| 1-(2-$NO_2C_6H_4$)-5-*i*-Pr-8-$NO_2$ | 240–242 | | | | 241 |
| 1-(3-$NO_2C_6H_4$)-5-Me-8-$CF_3$ | 185–186 | | | | 238 |
| 1-(4-$NO_2C_6H_4$)-5-Me-8-$CF_3$ | 150–153 | | | | 238 |
| 1-(2-$CF_3C_6H_4$)-3-OH-8-Cl | 281–282 | EtOH | 89 | | 245, 246 |
| 1-(2-$CF_3C_6H_4$)-5-PhCO-8-Cl | 165 | | | | 240 |

| | | | | | |
|---|---|---|---|---|---|
| 1-(2-$CF_3C_6H_4$)-5-Me-8-Cl | 204–205 | | | | 236, 237 |
| 1-(2-$CF_3C_6H_4$)-5-Me-8-$CF_3$ | 164–165 | $i$-$Pr_2O$ | 78 | | 237, 249 |
| 1-(3-$CF_3C_6H_4$)-5-Me-8-Cl | 192–193 | | | | 236, 237, 249 |
| 1-(2-Pyridyl)-5-$AcOCH_2CH_2$-8-Cl | 196–198 | | | | 237 |
| 1-(2-Pyridyl)-5-$PhCH_2$-8-Cl | 216–218 | | | | 237 |
| 1-(2-Pyridyl)-5-Bu-8-Cl | 148–149 | | | | 237 |
| 1-(2-Pyridyl)-5-Cyclohexyl-8-Cl | 190 | | | | 237 |
| 1-(2-Pyridyl)-5-Et-7-Cl | 194–196 | | | | 237 |
| 1-(2-Pyridyl)-5-Et-8-Cl | 194–196 | | | | 237 |
| 1-(2-Pyridyl)-5-Et-8-$CF_3$ | 153–155 | | | | 237 |
| 1-(2-Pyridyl)-5-$HOCH_2CH_2$-8-Cl | 176–178 | | | | 237 |
| 1-(2-Pyridyl)-5-$HOCH_2CH_2$-8-$CF_3$ | 149–151 | | | | 237 |
| 1-(2-Pyridyl)-5-Me-8-Br | 242–243 | | | | 237 |
| 1-(2-Pyridyl)-5-Me-8-Cl | 231–233 | $CH_2Cl_2$/Petr ether | 50–55 | | 237, 239 |
| 1-(2-Pyridyl)-5-Me-8-$NO_2$ | 176–177 | | | | 241 |
| 1-(2-Pyridyl)-5-Me-8-$CF_3$ | 164–168 | | | | 237 |
| 1-(2-Pyridyl)-5-Ph-7-Cl | 203–204 | | | | 237 |
| 1-(2-Pyridyl)-5-Pr-8-Cl | 177–178 | | | | 237 |
| 1-(2-Pyridyl)-5-*i*-Pr-8-Cl | 165–167 | | | | 237 |
| 1-(4-Cl-2-Pyridyl)-5-Me-8-Cl | 216–217 | | | | 237, 239 |
| 1-(5-Me-2-Pyridyl)-5-Me-8-Cl | 225–227 | | | | 237, 239 |
| 1-(3-Pyridyl)-5-Et-8-Cl | 196–198 | | | | 237, 239 |
| 1-(3-Pyridyl)-5-Me-8-Cl | 164–166 | | | | 237 |
| 1-(2-Pyrimidyl)-5-Me-8-Cl | 243–245 | | | | 237, 239 |
| 1-(2-Thienyl)-5-Me-8-Cl | 173–174 | | | | 237, 239 |
| 3,8-$(NO_2)_2$-7-COOH | | | | | 226 |
| 3,8-$(NO_2)_2$-7-Cl | 168–169 | $H_2O$ | | | 226 |
| 3,8-$(NO_2)_2$-7-F | 187–188d | | | | 226 |
| 3-Et-3-Allyl-7-Cl | 242–244 | EtOH | 30 | ir, pmr | 307 |
| 3,3-Diallyl-7-Cl | 229–230 | EtOH | 34 | ir, pmr | 307 |
| ***Tetrasubstituted*** | | | | | |
| 1,5-$(PhCH_2)_2$-3,3-$(Bu)_2$ | 122–123 | EtOH | | | 227 |
| 1,5-$(PhCH_2)_2$-3,3-$Et_2$ | 133–134 | EtOH | | | 227 |

TABLE IV-1. —(contd.)

| Substituent | mp (°C); [bp (°C/torr)] | Solvent of Crystallization | Yield (%) | Spectra | Refs. |
|---|---|---|---|---|---|
| 1,5-$n$-$Bu_2$-3,3-$Et_2$ | 90–92 | Ligroin | | | 227 |
| 1,5-$(HOOCCH_2)_2$-3,3-Diallyl | 193 | Acetone/$H_2O$ | 11 | | 229 |
| 1,5-$(Et_2NCH_2CH_2)_2$-3,3-Diallyl | Oil | | 20 | | 229 |
| 1,5-$(Me_2NCH_2CH_2)_2$-3,3-Diallyl | 28 | PhH/Petr ether | 17 | | 229 |
| 1,5-$Me_2$-3,3-$Et_2$ | | EtOH | | | 227 |
| 1,5-(COOMe)2-3-Et-3-Allyl | 112–114 | PhH | 72 | ir, pmr | 307 |
| 1,5-(COOEt)2-3-Et-3-Allyl | 76–78 | PhH | 70 | ir, pmr | 307 |
| 1,5-(COOMe)-3,3-diallyl | 106–107 | PhH | 75 | ir, pmr | 307 |
| 1,5-$(COOMe)_2$-3-Et-3-(2-OH-Allyl) | 136–138 | EtOH | 90 | ir, pmr | 307 |
| 1,5-$(COOEt)_2$-3-Et-3-(2-OH-Allyl) | Oil | MeOH | 84 | ir, pmr | 307 |
| 1,5-$(COOMe)_2$-3-Allyl-3-(2-OH-Allyl) | 112–114 | MeOH | 88 | ir, pmr | 307 |
| 1-Ph-3-AcO-5-Ac-8-$CF_3$ | 176–178 | EtOAc | 60.5 | | 246 |
| 1-Ph-3-AcO-5-Me-8-Cl | 240–244 | MeCN | 57 | ir, pmr | 243 |
| 1-Ph-3-$AcCH_2$-3-HO-8-$CF_3$ | 206 | $CH_2Cl_2$ | 79 | ir, pmr | 246 |
| 1-Ph-3-$NH_2$-5-Me-8-Cl | 219–222 | $CH_2Cl_2$/$i$-$Pr_2O$ | 17 | | 243 |
| 1-Ph-3-PhCOO-5-Me-8-Cl | 219–220 | | 60 | | 243 |
| 1-Ph-3-Br-5-Me-8-Cl | 278–279 | MeCN | 95 | ir, pmr | 243 |
| 1-Ph-3-Br-5-Me-8-$CF_3$ | 276–278 | MeCN | 80 | | 243 |
| 1-Ph-BuO-5-Me-8-$CF_3$ | 195–196 | | 52 | | 243 |
| 1-Ph-3,3-$Cl_2$-8-$CF_3$ | 237–238 | MeCN | 48 | ir, pmr | 243 |
| 1-Ph-2,8-$Cl_2$-5-Me | 283–285 | MeCN | 35 | | 243 |
| 1-Ph-3-$(EtO)_2$CH-3-HO-8-Cl | 188–189 | EtOH | 69 | | 246 |
| 1-Ph-3-EtO-5-Me-8-$CF_3$ | 226–227 | | 40 | | 243 |
| 1-Ph-3-Et-3-Me-8-Cl | 208–210 | THF | | | 237 |
| 1-Ph-3-CHO-3-HO-8-Cl | | | 45 | ir, pmr | 246 |
| 1-Ph-3-CHO-3-HO-8-$NO_2$ | 208–210 | EtOAc/MeCN | 51 | | 246 |
| 1-Ph-3-HCOO-5-Me-8-$CF_3$ | 220–222 | | 30 | | 243 |
| 1-Ph-3,3-$(HO)_2$-8-Br | 225d | $CH_2Cl_2$ | 88.5: 71 | | 246, 247 |
| 1-Ph-3,3-$(HO)_2$-8-Cl | 190d | | | | 247 |

| | | | | | |
|---|---|---|---|---|---|
| 1-Ph-3,3-$(HO)_2$-8-$NO_2$ | 190d | $CH_2Cl_2$ | 63; 66 | | 246, 247 |
| 1-Ph-3,3-$(HO)_2$-8-$CF_3$ | 245d | $CH_2Cl_2$ | 72 | | 246 |
| 1-Ph-3-HO-5-Allyl-8-$CF_3$ | 178–180 | | | | 245 |
| 1-Ph-3-HO-5-Et-8-Br | 256–257 | | 51 | | 243, 245 |
| 1-Ph-3-HO-5-Et-8-$CF_3$ | 235–236 | | 48 | | 243, 245 |
| 1-Ph-3-HO-5-$HOCH_2CH_2$-8-$CF_3$ | 217–219 | | | | 245 |
| 1-Ph-3-HO-5-Me-8-Br | 269–271 | | | | 245 |
| 1-Ph-3-HO-5-Me-8-Cl | 262–264 | MeCN | 82 | | 243, 245 |
| 1-Ph-3-HO-5-Me-8-F | 255–258 | | | | 245 |
| 1-Ph-3-HO-5-Me-8-$NO_2$ | 252–254 | | 40 | | 243, 245 |
| 1-Ph-3-HO-5-Me-8-$CF_3$ | 245–248 | Dioxan | 80 | ir, pmr | 243, 245 |
| 1-Ph-3-HO-5-*i*-Pr-8-$CF_3$ | 201–203 | | | | 245 |
| 1-Ph-3-MeO-5-Me-8-Cl | 258–260 | | 61 | | 243 |
| 1-Ph-3-MeO-5-Me-8-$CF_3$ | 254–256 | MeOH | 88 | ir, pmr | 243 |
| 1-Ph-3,5-$Me_2$-8-Cl | 218–220 | MeOH | 64 | | 237 |
| 1-Ph-3-Me-5-Pr-8-Cl | 155–157 | | | | 237 |
| 1-Ph-3-Me-5-*i*-Pr-8-Cl | 116 | | | | 237 |
| 1-Ph-3-*i*-PrO-5-Me-8-$CF_3$ | 157–158 | | 25 | | 243 |
| 1-(2-$BrC_6H_4$)-3-HO-5-Me-8-Cl | 253 | | | | 245 |
| 1-(2-$ClC_6H_4$)-3-HO-5-Me-8-$CF_3$ | 188–190 | | | | 245 |
| 1-(4-$ClC_6H_4$)-3-HO-5-Me-8-$CF_3$ | 239–241 | | | | 245, 247 |
| 1-(2-$FC_6H_4$)-3-HO-5-Me-8-$CF_3$ | 195–197 | | | | 245 |
| 1-(3-$FC_6H_4$)-3-HO-5-Me-8-$CF_3$ | 188–189 | | | | 245 |
| 1-(2-$MeC_6H_4$)-3,5-$Me_2$-8-Cl | 195–197 | | | | 237 |
| 1-(2-$MeC_6H_4$)-3-Me-5-Et-8-Cl | 173–174 | | | | 237 |
| 1-(2--$NO_2C_6H_4$)-3-HO-5-Me-8-$CF_3$ | 189–190 | | | | 245 |
| 1-(3-$NO_2C_6H_4$)-3-HO-5-Me-8-$CF_3$ | 230–233 | | | | 245 |
| 1-(2-$CF_3C_6H_4$)-3,3-$(HO)_2$-8-Cl | 177d | | 37 | | 246 |
| 1-(2-$CF_3C_6H_4$)-3-HO-5-Me-8-Cl | 260–262 | | | | 245 |
| 1-(3-$CF_3C_6H_4$)-3-HO-5-Me-8-$CF_3$ | 187–188 | | | | 245 |
| 1,5-*n*-$Pr_2$-3,3-$Et_2$ | 89–90 | EtOH | | | 227 |
| 1-(2-Pyridyl)-3-HO-5-Me-8-Br | 218–220 | | | | 245 |
| 1-(2-Pyridyl)-3-HO-5-Me-8-Cl | 204–206 | MeOH | 75 | | 245 |

TABLE IV-1. —(contd.)

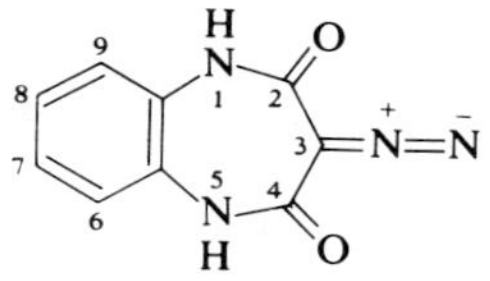

| Substituent | mp (°C); [bp (°C/torr)] | Solvent of Crystallization | Yield (%) | Spectra | Refs. |
|---|---|---|---|---|---|
| ***Pentasubstituted*** | | | | | |
| 1-Ph-3,3-$Cl_2$-5-Me-8-$CF_3$ | 228–230 | $CH_2Cl_2$/MeOH | 50 | | 243 |
| 1-Ph-3,3-$(HO)_2$-5-Allyl-8-$CF_3$ | 174–176 | | | | 247 |
| 1-Ph-3,3-$(HO)_2$-5-Me-8-Br | 188–190 | THF | 60 | | 247 |
| 1-Ph-3,3-$(HO)_2$-5-Me-8-$CF_3$ | | | | | |
| Hydrate | 176d | Acetone/$H_2O$ | 77 | | 247 |
| 1-Ph-3,3-$(HO)_2$-5-*i*-Pr-8-$CF_3$ | 177–178 | | | | 247 |
| 1-(2-$ClC_6H_4$)-3,3-$(HO)_2$-5-Me-8-$CF_3$ | 163d | | | | 247 |
| 1-(4-$ClC_6H_4$)-3,3-$(HO)_2$-5-Me-8-$CF_3$ | 193d | | | | 247 |
| 1-(2-$FC_6H_4$)-3,3-$(HO)_2$-5-Me-8-Cl | 169d | | | | 247 |
| 1-(4-$HOC_6H_4$)-3,3-$(HO)_2$-5-Me-8-$CF_3$ | 193d | | | | 247 |
| 1-(2-$NO_2C_6H_4$)-3,3-$(HO)_2$-5-Me-8-$CF_3$ | 177d | | | | 247 |
| 1-(3-$NO_2C_6H_4$)-3,3-$(HO)_2$-5-Me-8-$CF_3$ | 230d | | | | 245 |
| 1-(2-$CF_3C_6H_4$)-3,3-$(HO)_2$-5-Me-8-$CF_3$ | 187–188 | | | | 245 |
| 1-(3-$CF_3C_6H_4$)-3,3-$(HO)_2$-5-Me-8-$CF_3$ | 130d | | | | 247 |
| 1-(2-Pyridyl)-3,3-$(HO)_2$-5-Me-8-Cl | 165d | | | | 247 |
| ***3-Diazo-1,5-dihydro-3H-1,5-benzodiazepin-2,4(2H,4H)-diones*** | | | | | |
| ***Trisubstituted*** | | | | | |
| 1-Ph-5-Allyl-8-$CF_3$ | 138–140d | | | | 244 |
| 1-Ph-5-Et-8-Br | 145-146d | | | | 244 |
| 1-Ph-5-Et-8-$CF_3$ | 147–148d | | | | 244 |
| 1-Ph-5-$HOCH_2CH_2$-8-$CF_3$ | 135–140d | | | | 244 |
| 1-Ph-5-Me-8-Br | 160d | | | | 244 |

| | | | | | |
|---|---|---|---|---|---|
| 1-Ph-5-Me-8-Cl | 174–176d | EtOAc | 70; 54 | ir | 243, 244 |
| 1-Ph-5-Me-8-$NO_2$ | 156–158d | MeOH | 46 | | 243, 244 |
| 1-Ph-5-Me-8-$CF_3$ | 157d | EtOH | 75; 44 | | 243, 244 |
| 1-Ph-5-Me-8-F | 163–165d | | | | 244 |
| 1-Ph-5-*i*-Pr-8-$CF_3$ | 126–128d | | | | 244 |
| 1-(2-$ClC_6H_4$)-5-Me-8-$CF_3$ | 134–135d | | | | 244 |
| 1-(2-$FC_6H_4$)-5-Me-8-$CF_3$ | 107–110d | | | | 244 |
| 1-(3-$FC_6H_4$)-5-Me-8-$CF_3$ | 95–98d | | | | 244 |
| 1-(2-$NO_2C_6H_4$)-5-Me-8-$CF_3$ | 156–157d | | | | 244 |
| 1-(2-$CF_3C_6H_4$)-5-Me-8-Cl | 172d | | | | 244 |
| 1-(2-Pyridyl)-5-Me-8-Br | 158–160d | | | | 244 |
| 1-(2-Pyridyl)-5-Me-8-Cl | 162–163d | | | | 244 |

***1,5-Dihydro-3-methylene-3H-1,5-benzodiazepin-2,4(2H,4H)-diones***

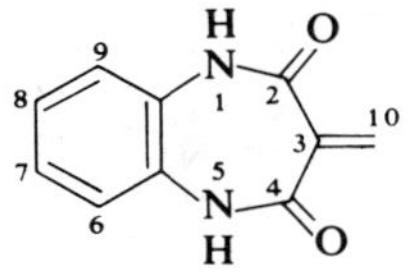

*Trisubstituted*

| | | | | | |
|---|---|---|---|---|---|
| 1-Ph-8-Br-10-BuNH | 235d | DMF/$H_2O$ | 85.5 | | 245, 248, 252 |
| 1-Ph-8-Cl-10-Allylamino | 215–218 | | | | 245, 248, 252 |
| 1-Ph-8-Cl-10-$NH_2$ | 282–285 | | 80–95 | | 245, 248, 252 |
| 1-Ph-8-Cl-10-BuNH | 181–185 | | | | 248, 252 |
| 1-Ph-8-Cl-10-$Me_3CNH$ | 255–256 | | | | 245, 248, 252 |
| 1-Ph-8-Cl-10-$Et_2NCH_2CH_2NH$ | 282–285 | | | | 245, 248, 252 |
| 1-Ph-8-Cl-10-$Me_2N$ | 234–235 | EtOAc | 70.5 | | 245, 248, 252 |
| 1-Ph-8-Cl-10-MeNH | 253–254 | | | | 245, 248, 252 |
| 1-Ph-8-Cl-10-(2-Methylpropyl)amino | 198–199 | | | | 245, 248, 252 |
| 1-Ph-8-$NO_2$-10-$NH_2$ | 250–251 | EtOAc | 77 | | 245, 248, 252 |
| 1-Ph-8-$NO_2$-10-BuNH | 226–227 | | | | 245, 248, 252 |
| 1-Ph-8-$NO_2$-10-$Me_2N$ | 277–278 | | 82–90 | | 245, 248, 252 |
| 1-Ph-8-$CF_3$-10-*n*-BuNH | 155–156 | | | | 245, 248, 252 |
| 1-Ph-8-$CF_3$-10-$Me_2N$ | 219–220 | | 82–90 | | 252 |

TABLE IV-1. —(contd.)

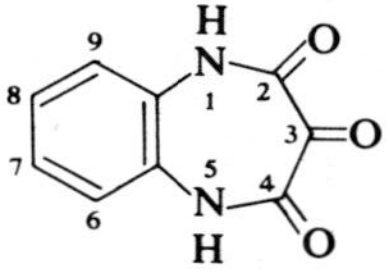

| Substituent | mp (°C); [bp (°C/torr)] | Solvent of Crystallization | Yield (%) | Spectra | Refs. |
|---|---|---|---|---|---|
| 1-(2-Br$C_6H_4$)-8-Cl-10*n*-BuNH | 203 | | | | 245, 248, 252 |
| 1-(2-Cl$C_6H_4$)-8-Cl-10*n*-BuNH | 208–210 | | | | 245, 248, 252 |
| 1-(2-F$C_6H_4$)-8-Cl-10-*n*-BuNH | 208 | | | | |
| 1-(2-$NO_2C_6H_4$)-8-Cl-10-*n*-BuNH | 183 | | | | 248, 252 |
| 1-(2-$CF_3C_6H_4$)-8-Cl-10-*n*-BuNH | 208–210 | | | | 248, 252 |
| ***Polysubstituted*** | | | | | |
| 1-Ph-5-Me-8-Cl-10-PhCOO-10-Ph | 222–223 | MeOH | 20 | ir, pmr | 243 |
| 1-Ph-5-Me-8-$CF_3$-10-$Me_2N$ | 164–166 | Acetone/$Et_2O$ | 76 | | 245, 247 |
| ***1,5-Dihydro-3H-1,5-benzodiazepin-2,3,4(2H,4H)-triones*** | | | | | |
| 1-Ph-8-Br | 243d | | 69; 88.5 | | 246, 247 |
| 1-Ph-8-Cl | 239d | | 72 | | 246, 247 |
| 1-Ph-8-$CF_3$ | 254d | | 84 | ir, pmr | 246, 247 |

*1,5-Benzodiazepin-thiones*

***1,3-Dihydro-1,5-benzodiazepin-2(2H)-thiones***

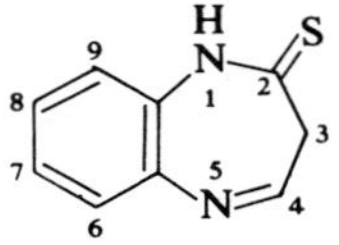

| | | | | | |
|---|---|---|---|---|---|
| ***Monosubstituted*** | | | | | |
| 4-Ph | 228–230 | EtOAc | 79 | ms | 256, 262 |
| 4-[4-($PhCH_2S$)$C_6H_4$] | 197 | EtOAc | | | 256 |
| 4-(4-$BrC_6H_4$) | 238 | EtOAc | | | 256 |
| 4-(4-*n*-$BuSC_6H_4$) | 184 | EtOAc | | | 256 |
| 4-(2-$ClC_6H_4$) | 199 | EtOH | | ms | 256, 262 |
| 4-(3-$ClC_6H_4$) | 224–225 | EtOAc | | ms | 256, 262 |
| 4-(4-$ClC_6H_4$) | 243–245 | EtOAc | 73 | ms | 256, 262 |
| 4-(2,4-$Cl_2C_6H_3$) | 178 | EtOH | | | 256 |
| 4-(3,4-$Cl_2C_6H_3$) | 228–229 | EtOAc | | | 256 |
| 4-(4-*n*-Dodecylthiophenyl) | 163 | EtOAc | | | 256 |
| 4-(4-$EtSC_6H_4$) | 206 | EtOAc | | | 256 |
| 4-(2-$MeOC_6H_4$) | 195 | MeOH | | ms | 256, 262 |
| 4-(3-$MeOC_6H_4$) | 191 | EtOAc | | ms | 256, 262 |
| 4-(4-$MeOC_6H_4$) | 233d | EtOAC | | ms | 256, 262 |
| 4-[2,4$(MeO)_2C_6H_3$] | 179 | PhH/Petr ether | | | 256 |
| 4-(2-$MeC_6H_4$) | 186 | PhH | | ms | 256, 262 |
| 4-(3-$MeC_6H_4$) | 204 | EtOAc | | ms | 256, 262 |
| 4-(4-$MeC_6H_4$) | 243–245 | EtOAc | | ms | 256, 262 |
| 4-[1-Methylcyclohex-1-yl)methylthio]-phenyl | 189–190 | EtOAc | | | 256 |
| 4-(4-$MeSC_6H_4$) | 214 | EtOAc | | | 256 |
| 4-[4-(3-*n*-Pentylthio)phenyl] | 171 | EtOAc | | | 256 |
| 4-(4-$PhOC_6H_4$) | 225 | EtOAc | 50 | | 256 |
| 4-(4-$PhC_6H_4$) | 229–230 | EtOAc | 73 | | 256 |
| 4-(2-$PhSC_6H_4$) | 214 | EtOAc | | | 256 |
| 4-(4-$PhSC_6H_4$) | 229–230 | EtOAc | 50 | | 256 |
| 4-(4-$PrSC_6H_4$) | 196 | *i*-PrOH | | | 256 |
| 4-(4-*i*-$PrSC_6H_4$) | 201 | EtOAc | | | 256 |
| ***Disubstituted*** | | | | | |
| 1-$PhCH_2$-4-Ph | 127–128 | EtOH | 70 | | 145 |
| 1-Me-4-Ph | 109–110 | EtOH | 82 | | 145 |

TABLE IV-1. —(contd.)

| Substituent | mp (°C); [bp (°C/torr)] | Solvent of Crystallization | Yield (%) | Spectra | Refs. |
|---|---|---|---|---|---|
| ***1,5-Dihydro-1,5-benzodiazepin-2(2H)-thiones*** | | 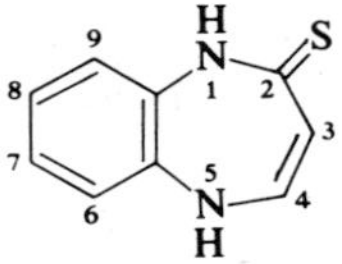 | | | |
| 3-COOEt-4-$NH_2$ | 165–167 | EtOH | 85 | | 257 |
| 3-COOMe-4-$NH_2$ | 190–193 | EtOH | 80 | | 258 |
| ***1,3,4,5-Tetrahydro-1,5-benzodiazepin-2(2H)-thiones*** | | 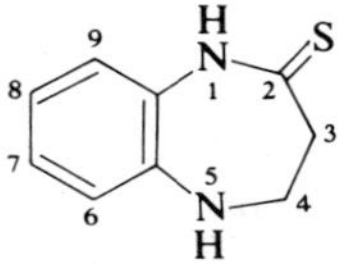 | | | |
| None | 159–160 | MePh | 47 | | 205, 259 |
| 3-Me | 194–195 | EtOH | 72 | | 205 |
| 3-Me-7-Cl | 160–162 | EtOH | 74 | | 205 |
| 3-Me-8-Cl | 196–197 | EtOH | 84 | | 205 |
| 3-Me-8-MeO | 144–145 | EtOH | 80 | | 205 |
| 3-Me-7,8-$Cl_2$ | 193–195 | EtOH | 66 | | 205 |
| ***3,5-Dihydro-4-oxo-1H-1,5-benzodiazepin-2(2H,4H)-thiones*** | | 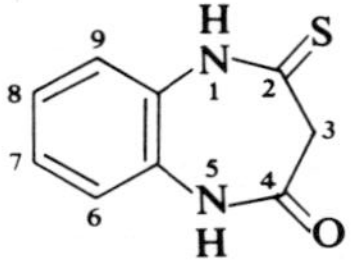 | | | |
| 5-Ph-7-Cl | | | | | 159 |
| 5-Ph-7-$CF_3$ | 247.5–248.5 | MeCN | 55 | | 260 |
| | 251–253 | | | | 161 |

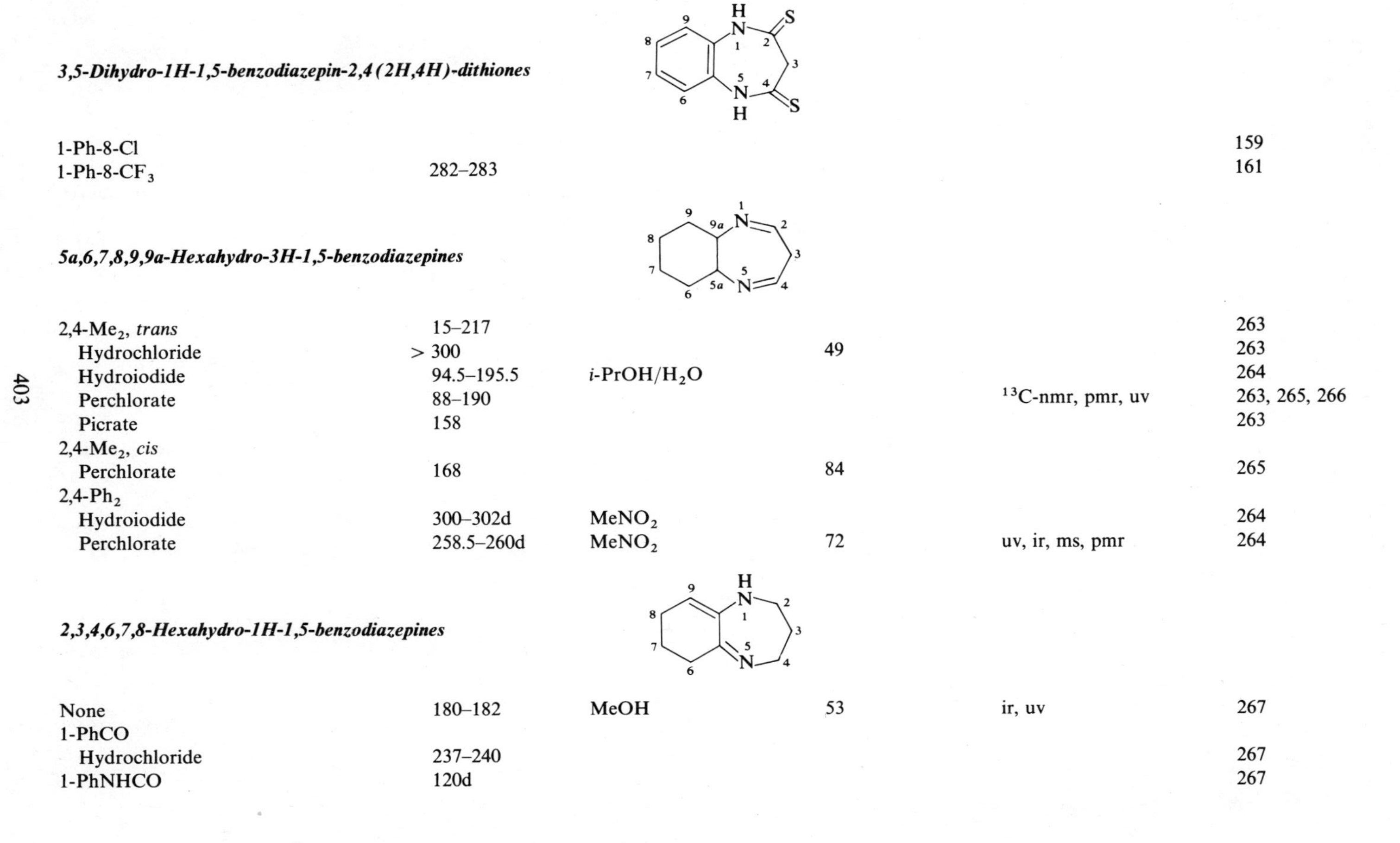

| | | | | | |
|---|---|---|---|---|---|
| ***3,5-Dihydro-1H-1,5-benzodiazepin-2,4(2H,4H)-dithiones*** | | | | | |
| 1-Ph-8-Cl | | | | | 159 |
| 1-Ph-8-$CF_3$ | 282–283 | | | | 161 |
| ***5a,6,7,8,9,9a-Hexahydro-3H-1,5-benzodiazepines*** | | | | | |
| 2,4-$Me_2$, *trans* | 15–217 | | | | 263 |
| Hydrochloride | > 300 | | 49 | | 263 |
| Hydroiodide | 94.5–195.5 | *i*-PrOH/$H_2O$ | | | 264 |
| Perchlorate | 88–190 | | | $^{13}C$-nmr, pmr, uv | 263, 265, 266 |
| Picrate | 158 | | | | 263 |
| 2,4-$Me_2$, *cis* | | | | | |
| Perchlorate | 168 | | 84 | | 265 |
| 2,4-$Ph_2$ | | | | | |
| Hydroiodide | 300–302d | $MeNO_2$ | | | 264 |
| Perchlorate | 258.5–260d | $MeNO_2$ | 72 | uv, ir, ms, pmr | 264 |
| ***2,3,4,6,7,8-Hexahydro-1H-1,5-benzodiazepines*** | | | | | |
| None | 180–182 | MeOH | 53 | ir, uv | 267 |
| 1-PhCO | | | | | |
| Hydrochloride | 237–240 | | | | 267 |
| 1-PhNHCO | 120d | | | | 267 |

TABLE IV-1. —(contd.)

| Substituent | mp (°C); [bp (°C/torr)] | Solvent of Crystallization | Yield (%) | Spectra | Refs. |
|---|---|---|---|---|---|
| ***2,3,4,6,7,8,9,9a-Octahydro-1H-1,5-benzodiazepines*** | | 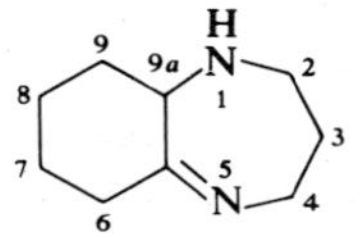 | | | |
| 9,9a-$Br_2$ Hydrobromide | 170–173 | $CHBr_3$ | 33 | | 267 |
| ***Perhydro-1H-1,5-benzodiazepines*** | | 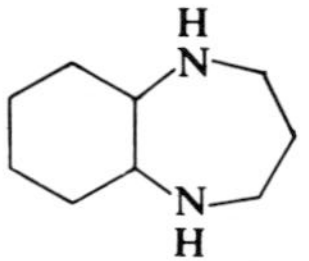 | | | |
| 6*a*, 9*a*-$(CN)_2$ | 193–194 | EtOH | 87 | | 268 |
| *Cyclobuta[b] [1,4]diazepines* | | | | | |
| ***3,5Dihydro-1H-cyclobuta[b] [1,4]diazepin-2,4,6,7(2H, 4H,6H,7H)-tetraones*** | | | | | |
| None | > 250d | 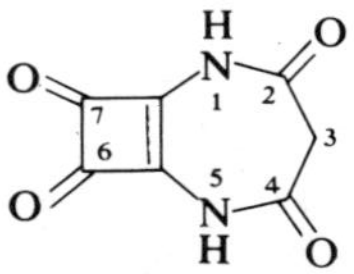 EtOH | 60 | ir, pmr | 269 |
| 1,5-$Me_2$ | > 250d | EtOH | 30 | ir, pmr | 269 |

*Cyclopenta[b] [1,4]diazepines*

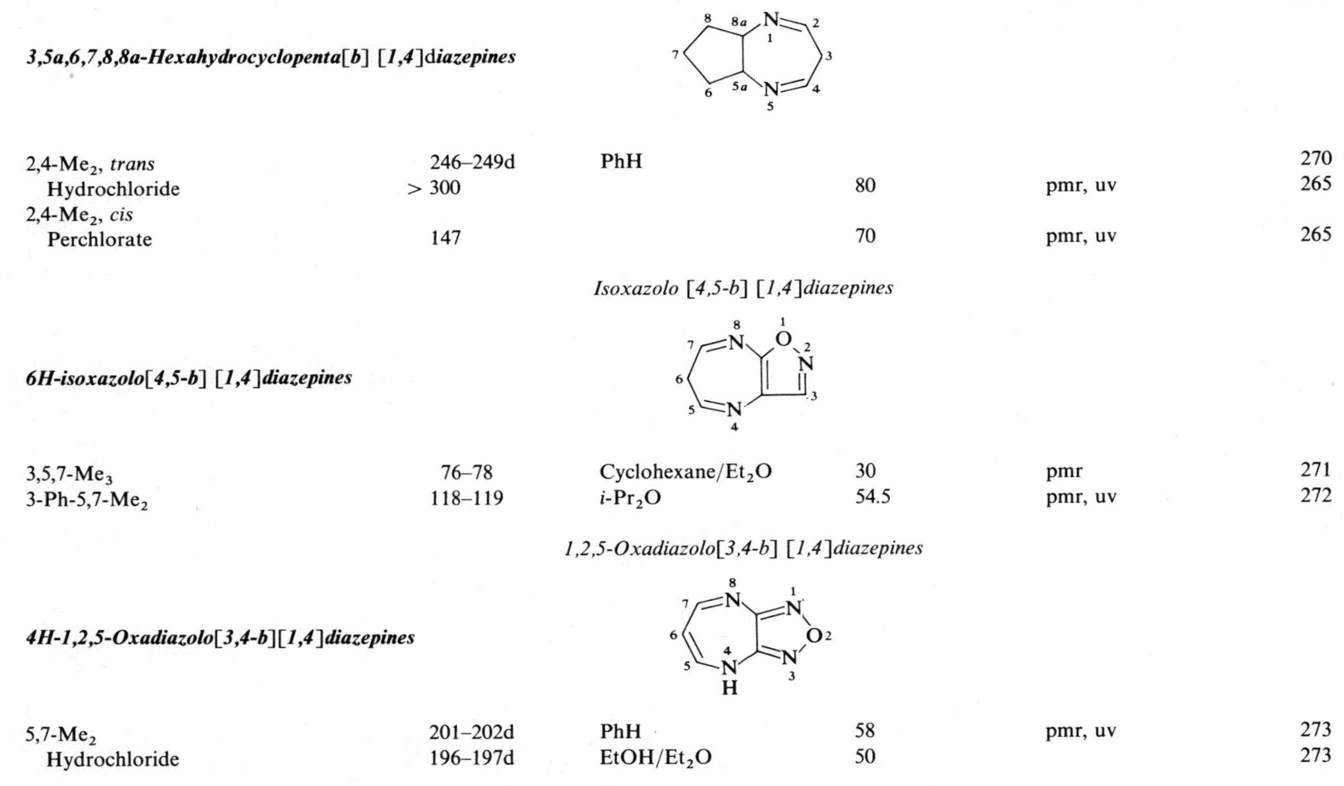


**3,5a,6,7,8,8a-Hexahydrocyclopenta[b] [1,4]diazepines**

| | | | | | |
|---|---|---|---|---|---|
| 2,4-$Me_2$, *trans* | 246–249d | PhH | | | 270 |
| Hydrochloride | > 300 | | 80 | pmr, uv | 265 |
| 2,4-$Me_2$, *cis* | | | | | |
| Perchlorate | 147 | | 70 | pmr, uv | 265 |

*Isoxazolo [4,5-b] [1,4]diazepines*

**6H-isoxazolo[4,5-b] [1,4]diazepines**

| | | | | | |
|---|---|---|---|---|---|
| 3,5,7-$Me_3$ | 76–78 | Cyclohexane/$Et_2O$ | 30 | pmr | 271 |
| 3-Ph-5,7-$Me_2$ | 118–119 | *i*-$Pr_2O$ | 54.5 | pmr, uv | 272 |

*1,2,5-Oxadiazolo[3,4-b] [1,4]diazepines*

**4H-1,2,5-Oxadiazolo[3,4-b][1,4]diazepines**

| | | | | | |
|---|---|---|---|---|---|
| 5,7-$Me_2$ | 201–202d | PhH | 58 | pmr, uv | 273 |
| Hydrochloride | 196–197d | EtOH/$Et_2O$ | 50 | | 273 |

TABLE IV-1. —(contd.)

| Substituent | mp (°C); [bp (°C/torr)] | Solvent of Crystallization | Yield (%) | Spectra | Refs. |
|---|---|---|---|---|---|
| 5-Me-7-Ph | 154–155d | PhH | 75 | pmr, uv | 273 |
| Hydrochloride | 181–182d | $EtOH/Et_2O$ | 86 | | 273 |
| 5,7-$Ph_2$ | 221–222 | PhH | | pmr, uv | 273 |
| | | *Pyrazolo[3,4-b][1,4]diazepines* | | | |
| ***1,6-Dihydropyrazolo[3,4-b] [1,4]diazepines*** | | | | | |
| 1-Ph3,5,7-$Me_3$ | 100–102 | PhH | 85; 90 | ms, pmr | 274 |
| 1-Ph-3,5-$Me_2$-7-MeO | | | 20 | ms, pmr | 274 |
| 1-Ph-3,5-$Me_2$-7-MeS | 135–136 | $C_6H_6$ | 90 | ms, pmr | 275 |
| 1-Ph-3,5,6-$Me_3$-7-MeO | | | 15 | ms, pmr | 274 |
| ***1,4,5,6-Tetrahydropyrazolo[3,4-b][1,4]diazepines*** | | | | | |
| 1-Ph-3,5-$Me_2$-7-MeO | 40–42 | Pentane | 10 | ms, pmr | 275 |
| 1-Ph-3,4,5-$Me_3$-7-MeO | | | 5 | ms, pmr | 275 |

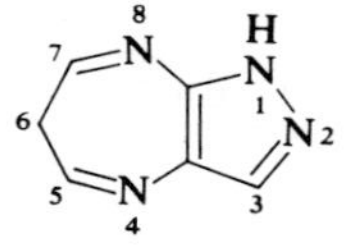

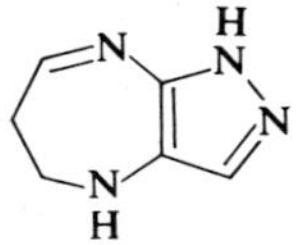

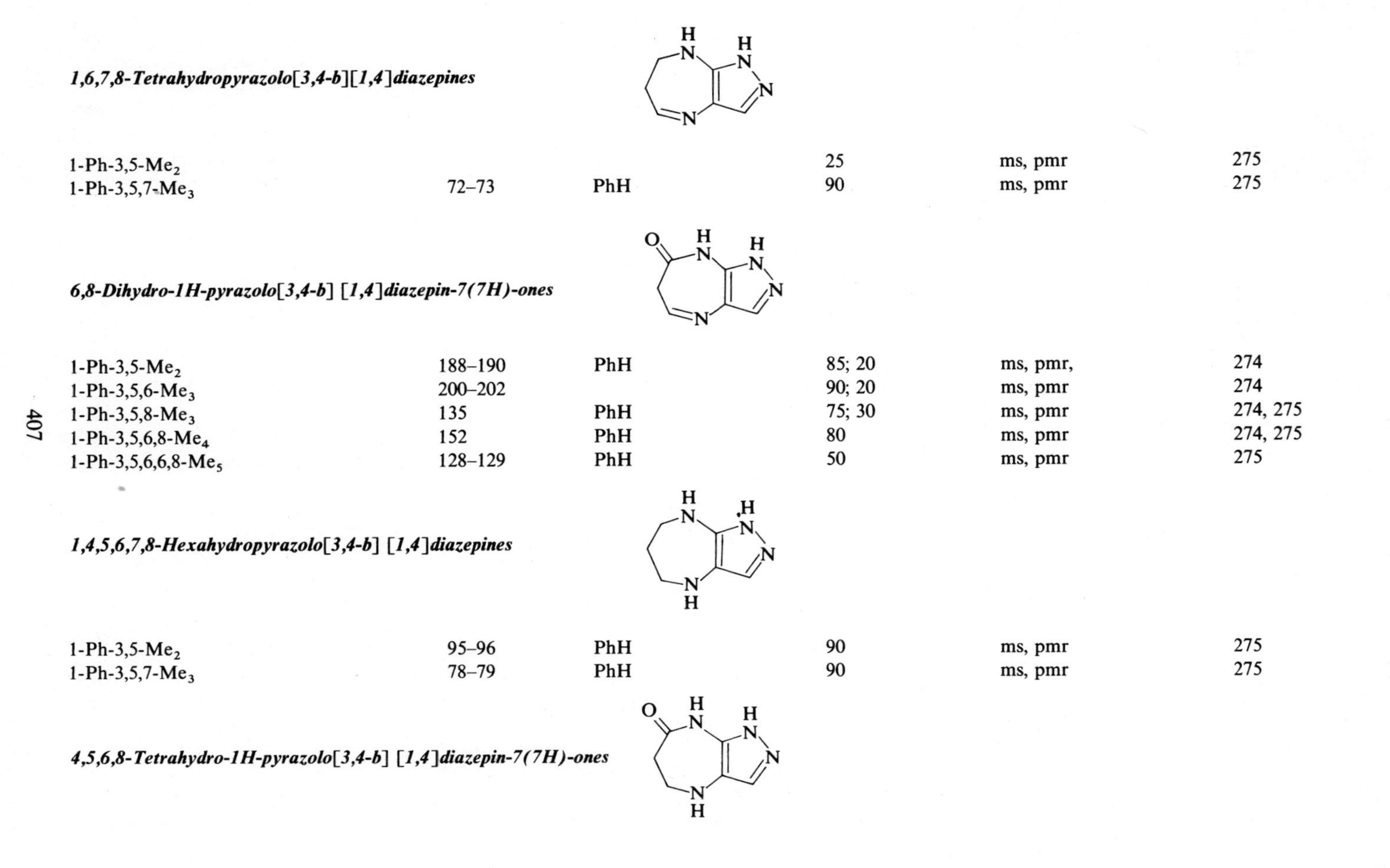

| | | | | | |
|---|---|---|---|---|---|
| ***1,6,7,8-Tetrahydropyrazolo[3,4-b][1,4]diazepines*** | | | | | |
| 1-Ph-3,5-$Me_2$ | | | 25 | ms, pmr | 275 |
| 1-Ph-3,5,7-$Me_3$ | 72–73 | PhH | 90 | ms, pmr | 275 |
| ***6,8-Dihydro-1H-pyrazolo[3,4-b] [1,4]diazepin-7(7H)-ones*** | | | | | |
| 1-Ph-3,5-$Me_2$ | 188–190 | PhH | 85; 20 | ms, pmr, | 274 |
| 1-Ph-3,5,6-$Me_3$ | 200–202 | | 90; 20 | ms, pmr | 274 |
| 1-Ph-3,5,8-$Me_3$ | 135 | PhH | 75; 30 | ms, pmr | 274, 275 |
| 1-Ph-3,5,6,8-$Me_4$ | 152 | PhH | 80 | ms, pmr | 274, 275 |
| 1-Ph-3,5,6,6,8-$Me_5$ | 128–129 | PhH | 50 | ms, pmr | 275 |
| ***1,4,5,6,7,8-Hexahydropyrazolo[3,4-b] [1,4]diazepines*** | | | | | |
| 1-Ph-3,5-$Me_2$ | 95–96 | PhH | 90 | ms, pmr | 275 |
| 1-Ph-3,5,7-$Me_3$ | 78–79 | PhH | 90 | ms, pmr | 275 |
| ***4,5,6,8-Tetrahydro-1H-pyrazolo[3,4-b] [1,4]diazepin-7(7H)-ones*** | | | | | |

TABLE IV-1. —(contd.)

| Substituent | mp (°C); [bp (°C/torr)] | Solvent of Crystallization | Yield (%) | Spectra | Refs. |
|---|---|---|---|---|---|
| 1-Ph-3,5-$Me_2$ | 120–121 | PhH | 90; 20 | ms, pmr | 275 |
| 1-Ph-3,4,5-$Me_3$ | 114–115 | $Et_2O$ | 40 | ms, pmr | 275 |
| 1-Ph-3,5,6-$Me_3$ | 169–170 | PhH | 90; 40 | ms, pmr | 275 |
| 1-Ph-3,5,8-$Me_3$ | 108–109 | PhH | 90; 5 | ms, pmr | 275 |
| 1-Ph-3,4,5,6-$Me_4$ | 165–166 | PhH | 35 | ms, pmr | 275 |
| 1-Ph-3,4,5,8-$Me_4$ | 88–90 | PhH | 30 | ms, pmr | 275 |
| 1-Ph-3,5,8-$Me_4$ | 144–145 | PhH | 90; 5 | ms, pmr | 275 |
| ***4,8-Dihydropyrazolo[3,4-b] [1,4]diazepin-5,7(1H,6H)-diones*** | | | | | |
| 1-Et-3-Me-8-(2-$MeC_6H_4$) | 221 | | | | 321 |
| 1-Et-3-Me-8-(2-$ClC_6H_4$) | 206–207 | | | | 321 |
| 1,3-$Me_2$-8-(4-$MeC_6H_4$) | 251 | | | | 321 |
| 1-Et-3-Me-8-(4-$ClC_6H_4$) | 186 | | | | 321 |
| 1-Et-3-Me-8-(4-$MeC_6H_4$) | 183–184 | | | | 321 |
| 1,3-$Me_2$-8-(4-$ClC_6H_4$) | 261–262 | | | | 321 |
| 1,3-$Me_2$-8-(2-$MeC_6H_4$) | 235 | | | | 321 |
| 1,3-$Me_2$-8-(2-$ClC_6H_4$) | 228 | | | | 321 |
| 1-Ph-2-Me-8-(4-$MeC_6H_4$) | 245–246 | | | | 321 |
| 1,3,4-$Me_3$-8-(4-$MeC_6H_4$) | 165 | | | | 321 |
| 1,3,4-$Me_3$-8-(2-$MeC_6H_4$) | 163 | | | | 321 |
| 1,3,4-$Me_3$-8-(4-$ClC_6H_4$) | 165–166 | | | | 321 |
| 1-Et-3,4-$Me_2$-8-(2-$ClC_6H_4$) | 181–182 | | | | 321 |
| 1-Et-3,4-$Me_2$-8-(4-$MeC_6H_4$) | 157 | | | | 321 |
| 1-Et-3,4,6-$Me_3$-8-(4-$MeC_6H_4$) | 152 | | | | 321 |

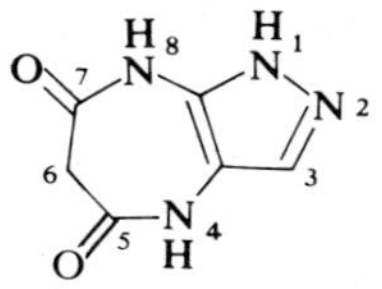

| | | |
|---|---|---|
| 1-Et-3,4-$Me_2$-8-(4-$ClC_6H_4$) | 185–186 | 321 |
| 1,3,4-$Me_3$-8-(2-$ClC_6H_4$) | 188 | 321 |
| 1,3-$Me_2$-4-($CH_2CCH$) | 161 | 321 |
| 1-Et-3,4-$Me_2$-8-(2-$MeC_6H_4$) | 161–162 | 321 |
| 1,3,8-$Me_3$-4-(2-$ClC_6H_4$) | 181 | 322 |
| 1,3,8-$Me_3$-4-(4-$MeC_6H_4$) | 220 | 322 |
| 1,3,8-$Me_3$-4-(4-$FC_6H_4$) | 219 | 322 |
| 1,3,8-$Me_3$-4-(4-$ClC_6H_4$) | 246 | 322 |
| 1,3,8-$Me_3$-4-(3-$FC_6H_4$) | 207–208 | 322 |
| 1-Et-3,8-$Me_2$-4-(2-$ClC_6H_4$) | 163 | 322 |
| 1-Et-3,8-$Me_2$-4-(4-$FC_6H_4$) | 155 | 322 |
| 1-Et-3,8-$Me_2$-4-(4-$ClC_6H_4$) | 153 | 322 |
| 1-Et-3,8-$Me_2$-4-(4-$MeC_6H_4$) | 156 | 322 |
| 1-Et-3,8-$Me_2$-4-(3-$ClC_6H_4$) | 194 | 322 |
| 1-Et-3,8-$Me_2$-4-(3-$MeC_6H_4$) | 194 | 322 |
| 1-Et-3,8-$Me_2$-4-(3-$FC_6H_4$) | 167 | 322 |
| 1-*i*-Pr-3,8-$Me_2$-4-(2-$ClC_6H_4$) | 229 | 322 |
| 1-*i*-Pr-3,8-$Me_2$-4-(2-$MeC_6H_4$) | 154 | 322 |
| 1-*i*-Pr-3,8-$Me_2$-4-(4-$FC_6H_4$) | 198 | 322 |
| 1-*i*-Pr-3,8-$Me_2$-4-(3-$ClC_6H_4$) | 160 | 322 |
| 1-*i*-Pr-3,8-$Me_2$-4-(4-Cl-$C_6H_4$) | 196 | 322 |
| 1-*i*-Pr-3,8-$Me_2$-4-(2-$FC_6H_4$) | 182 | 322 |
| 1-*i*-Pr-3-Me-8-($CH_2CH_2OMe$) | 136 | 322 |
| 1-*i*-Pr-3,8-$Me_2$-4-(3-$FC_6H_4$) | 155 | 322 |
| 1,3,8-$Me_3$-4-(2-$FC_6H_4$) | 181 | 322 |
| 1-Et-3,8-$Me_2$-4-(2-$FC_6H_4$) | 117 | 322 |
| 1,8-$Me_2$-3-Et | 181 | 322 |
| 1,8-$Me_2$-3-Cyclohexyl | 223 | 322 |
| 1-Et-3-Cyclohexyl-8-Me | 163 | 322 |
| 3,8-$Me_2$ | 262 | 322 |
| 1-Et-3-*n*-Pr-8-Me | 129 | 322 |
| 1,8-$Me_2$-3-*n*-Pr | 142 | 322 |
| 1-$CH_2Ph$-3,8-$Me_2$ | 158 | 322 |
| 1,3-$Me_2$-8-$CH_2Ph$ | 134 | 322 |

TABLE IV-1. —(contd.)

| Substituent | mp (°C); [bp (°C/torr)] | Solvent of Crystallization | Yield (%) | Spectra | Refs. |
|---|---|---|---|---|---|
| 1,8-$Me_2$ | 145 | | | | 322 |
| 1,3-$Me_2$ | 211 | | | | 322 |
| 1,3-$Me_2$-8-Et | 191 | | | | 322 |
| 1,3-$Me_2$-8-*n*-Pr | 188 | | | | 322 |
| 1,3-$Me_2$-8-Cyclopropane | 172 | | | | 322 |
| 1,3-$Me_2$-8-$CH_2CH_2NEt_2$ | 121 | | | | 322 |
| 1,3-$Me_2$-8-$CH_2CCH$ | 177 | | | | 322 |
| 1,8-$Me_2$-3-$CF_3$ | 163 | | | | 322 |
| *Pyrido[2,3-b][1,4]diazepines* | | | | | |
| ***4,5-Dihydro-3H-Pyrido[2,3-b][1,4]diazepines*** 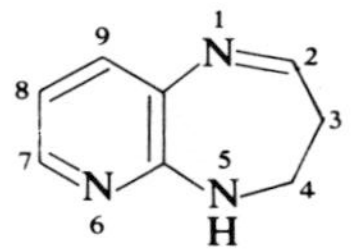 | | | | | |
| 2-(4-$MeOC_6H_4$) | 125–126 | | | | 111 |
| Dihydrochloride | 207–208 | | | | 111 |
| ***1,3-Dihydropyrido[2,3-b][1,4]diazepin-2(2H)-ones*** 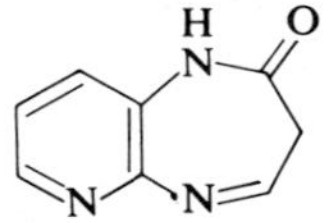 | | | | | |
| 4-Me | 195–197 | | | uv | 277 |
| 4,9-$Me_2$-8-Br | 289–291 | PhH | 39 | | 281 |

| | | | | | |
|---|---|---|---|---|---|
| ***3,5-Dihydropyrido[2,3-b][1,4]diazepin-4-(4H)-ones*** 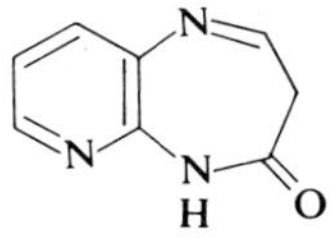 | | | | | |
| 2-Me | 189–190 | EtOH | 60 | ir, pmr, uv | 277–279 |
| 2-Ph | 258 | Xylene | 65 | uv | 280 |
| 2-(4-$MeOC_6H_4$) | 267–268 | | | | 111 |
| 2-[3,4,5-$(MeO)_3C_6H_2$] | 180–181 | | | | 111 |
| 2,9-$Me_2$-8-Br | 252–255 | PhH | 82; 20 | | 281 |
| ***2,3,4,5-Tetrahydro-1H-pyrido[2,3-b][1,4]diazepines*** 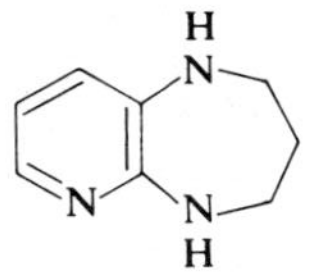 | | | | | |
| 2-Ph | 95–98 | | | | 111 |
| 7-Me | 82–85 | | | pmr | 282 |
| Hydrochloride | 206–213 | MeOH/EtOAc | | | 283 |
| 2-Methylimino-5-Ph-7-Cl | 180–182 | *i*-PrOH | 83 | | 291 |
| ***1,3,4,5-Tetrahydropyrido[2,3-b][1,4]diazepin-2(2H)-ones*** 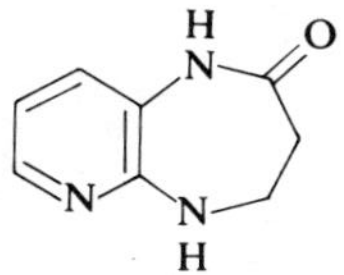 | | | | | |
| 7-Me | 194–196 | PhH | 2 | pmr | 282 |
| 7-Me-8-Br | 250–252 | PhH | 12 | pmr | 282 |
| 5-Ph-7-Cl | 270–275 | DMF | 50 | | 284 |
| 5-(2-$FC_6H_4$)-7-Cl | 264–267 | | 43 | | 284 |
| 1-Allyl-5-Ph-7-Cl | 128–130 | EtOH | 60 | | 284 |

TABLE IV-1. —(contd.)

| Substituent | mp (°C); [bp (°C/torr)] | Solvent of Crystallization | Yield (%) | Spectra | Refs. |
|---|---|---|---|---|---|
| 1-($Me_2NCH_2CH_2$)-5-Ph-7-Cl | 128–130 | $MeOH/H_2O$ | 50 | | 284, 291 |
| 1-Me-5-Ph-7-Cl | 189–191 | EtOH | 45 | | 284 |
| | 160–162 | EtOH | | | 291 |
| 1-Me-5-(2-$FC_6H_4$)-7-Cl | 144–146 | | 55 | | 284, 291 |
| ***1,2,3,5-Tetrhydropyrido[2,3-b] [1,4]diazepin-4(4H)-ones*** | | 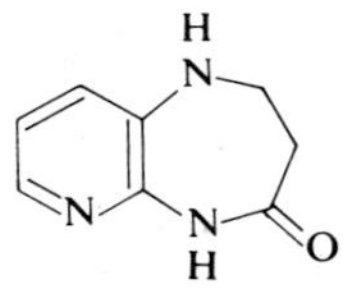 | | | |
| 2-Me | 172–173 | $MeOH/Et_2O$ | | ir, pmr, uv | 285 |
| ***1,3,4,5-Tetrahydropyrido[2,3-b] [1,4]diazepin-2(2H)-thiones*** | | 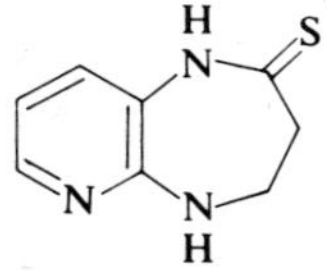 | | | |
| 7-Me | 248–249 | EtOH | 80 | pmr | 282 |
| 5-Ph-7-Cl | 220–222 | PhH | 23 | | 291 |
| 7-Me-8-Br | 225–227 | EtOH | 94 | pmr | 282 |
| ***3,5-Dihydro-1H-pyrido[2,3-b] [1,4]diazepin-2,4(2H,4H)-diones*** | | 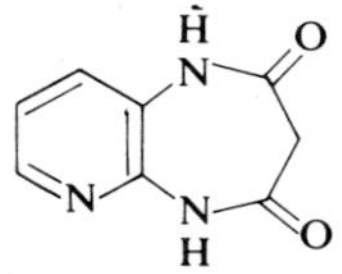 | | | |

| | | | | | |
|---|---|---|---|---|---|
| 3-$PhCH_2$ | 303–304d | $H_2O$ | 13 | | 286 |
| 3-*n*-Bu | 267–268 | EtOH | 17 | | 286 |
| 3-Et | 298 | MeOH | 12 | | 286 |
| 3-*n*-Pr | 330d | EtOH | 20 | | 286 |
| 7-Cl | 300 | $H_2O$ | 60 | | 284 |
| 8-Cl | 300d | DMF/$H_2O$ | 50 | | 284 |
| 3,3-Diallyl | 194 | AcOH/$H_2O$ | 49.5 | ir, pmr | 287 |
| 1,3-$Ph_2$-7-Cl | 238–242 | EtOH | 15 | | 284 |
| | *Pyrido[3,4-b] [1,4]diazepines* | | | | |
| ***3H-Pyrido[3,4-b] [1,4]diazepines*** | | | | | |
| 2,4-$Me_2$ | 120–121 | | 48.5 | pmr | 288 |
| 2,4-$Me_2$-6-Cl | 168–169 | | 60 | pmr | 288 |
| 2,4-$Me_2$-9-$NO_2$ | 201–202 | | 83 | pmr | 288 |
| ***1,3-Dihydropyrido[3,4-b] [1,4]diazepin-2(2H)-ones*** | | | | | |
| 4-Me | 205–207 | EtOH | 67 | ir, pmr | 290 |
| | 174–177 | MeOH/$Et_2O$ | 34.5 | ir, pmr, uv | 289 |
| ***3,5-Dihydropyrido[3,4-b][1,4]diazepin-4(4H)-ones*** | | | | | |
| 2-Me | 180–182 | EtOAc/Cyclohexane | | pmr, uv | 290 |

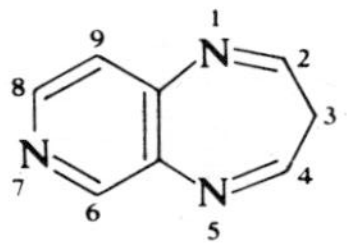

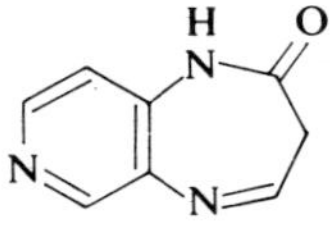

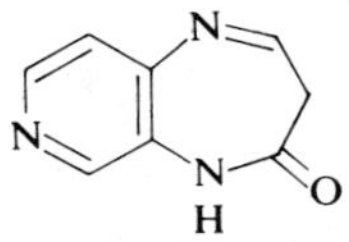

TABLE IV-1. —(contd.)

| Substituent | mp (°C); [bp (°C/torr)] | Solvent of Crystallization | Yield (%) | Spectra | Refs. |
|---|---|---|---|---|---|
| ***1,5-Dihydropyrido[3,4-b] [1,4]diazepin-4(4H)-ones*** | | 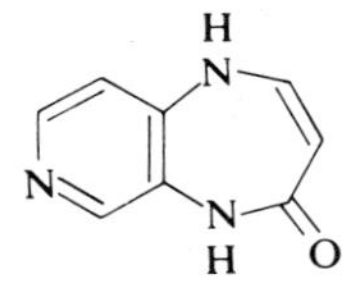 | | | |
| 2-Me | 168–171 | | | pmr, uv | 290 |
| ***2,3,4,5-Tetrahydro-1H-pyrido**[3,4-b][1,4]diazepines* | | 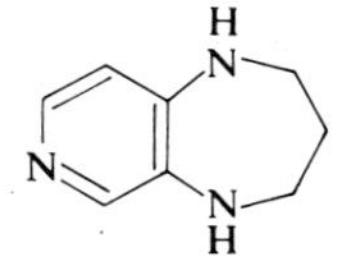 | | | |
| 2-CN-4-Ph | 209–212 | $CHCl_3$ | | ir, ms, pmr, uv | 283 |
| ***3,5-Dihydro-1H-pyrido[3,4-b][1,4]diazepin-2,4(2H,4H)-diones*** | | 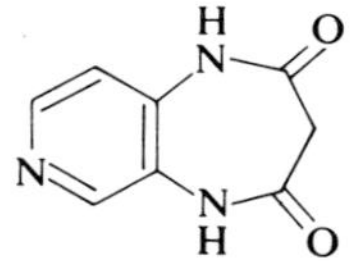 | | | |
| None | > 360 | Acetone/Petr ether | 58 | | 231 |
| 3,3-Diallyl | 236 | $AcOH/H_2O$ | 21 | ir | 287 |

*Pyrimido[4,5-b][1,4]diazepines*

***5,9-Dihydropyrimido[4,5-b][1,4]diazepin-6(6H)-ones***

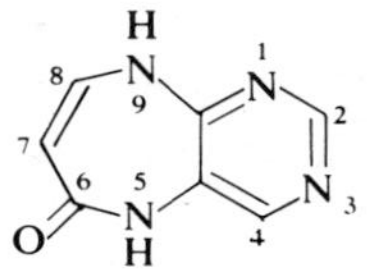

| | | | | | |
|---|---|---|---|---|---|
| 8-Me | 250–252d | Xylene | 38; 58 | pmr, uv | 293, 294 |

***7,9-Dihydropyrimido[4,5-b][1,4]diazepin-8(8H)-ones***

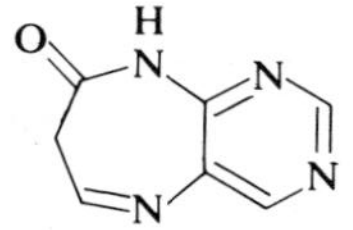

| | | | | | |
|---|---|---|---|---|---|
| 6-Me | 240–241 | EtOH | 89 | ir, uv | 294 |
| 6-(4-MeOC$_6$H$_4$) | 196–197 | | | | 111 |
| 6-[3,4,5-(MeO)$_3$C$_6$H$_2$] | 206–207 | Xylene/EtOH | 82 | ir | 111 |

***3,7-Dihydro-1H-pyrimido[4,5-b] [1,4]diazepin-2,4(2H,4H)-diones***

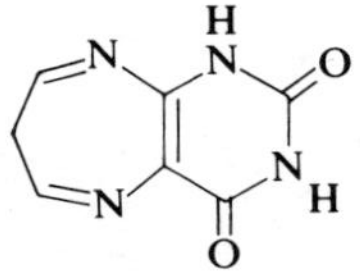

| | | | | | |
|---|---|---|---|---|---|
| 1,3,6,8-Me$_4$ | 185–186 | | 48; 54 | ir, ms, pmr, uv | 295 |
| 1,3,6-Me$_3$-8-Ph | 228d | | 10; 29 | ir, ms, pmr, uv | 295 |
| 1,3,8-Me$_3$-6-Ph | 190d | | 1.3 | ir, ms, pmr, uv | 295 |
| 1,3-Me$_2$-6,8-Ph$_2$ | 230–232d | EtOH | 40 | ir, ms, pmr, uv | 295 |

TABLE IV-1. —(contd.)

| Substituent | mp (°C); [bp (°C/torr)] | Solvent of Crystallization | Yield (%) | Spectra | Refs. |
|---|---|---|---|---|---|
| ***3,4a-Dihydro-1H-pyrimido[4,5-b] [1,4]diazepin-2,4(2H,4H)-diones*** | | 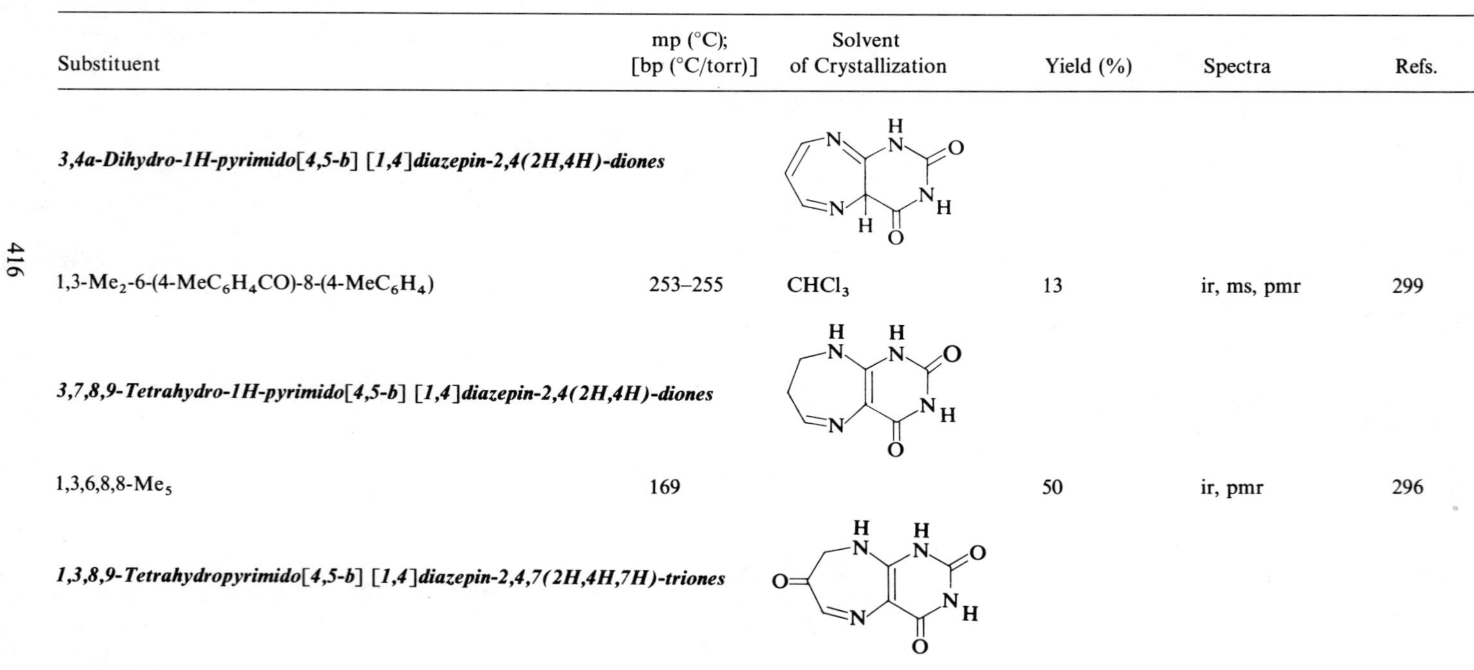 | | | |
| 1,3-$Me_2$-6-(4-$MeC_6H_4CO$)-8-(4-$MeC_6H_4$) | 253–255 | $CHCl_3$ | 13 | ir, ms, pmr | 299 |
| ***3,7,8,9-Tetrahydro-1H-pyrimido[4,5-b] [1,4]diazepin-2,4(2H,4H)-diones*** | | | | | |
| 1,3,6,8,8-$Me_5$ | 169 | | 50 | ir, pmr | 296 |
| ***1,3,8,9-Tetrahydropyrimido[4,5-b] [1,4]diazepin-2,4,7(2H,4H,7H)-triones*** | | | | | |
| 1,3,6,8,8-$Me_5$-9-NO-7-oxime | 190d | | | ir | 296 |

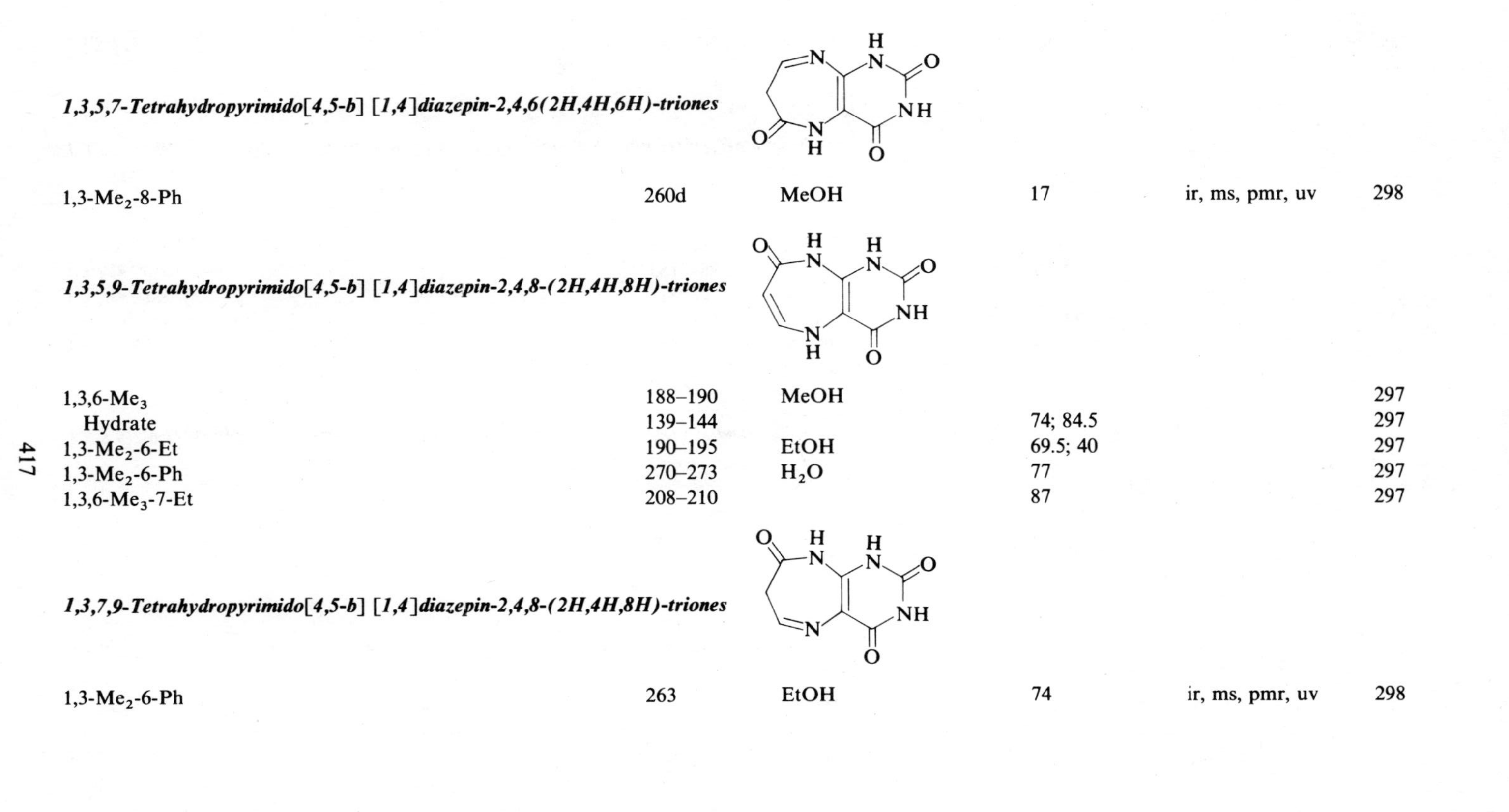

| | | | | | |
|---|---|---|---|---|---|
| ***1,3,5,7-Tetrahydropyrimido[4,5-b] [1,4]diazepin-2,4,6(2H,4H,6H)-triones*** | | | | | |
| 1,3-$Me_2$-8-Ph | 260d | MeOH | 17 | ir, ms, pmr, uv | 298 |
| ***1,3,5,9-Tetrahydropyrimido[4,5-b] [1,4]diazepin-2,4,8-(2H,4H,8H)-triones*** | | | | | |
| 1,3,6-$Me_3$ | 188–190 | MeOH | | | 297 |
| Hydrate | 139–144 | | 74; 84.5 | | 297 |
| 1,3-$Me_2$-6-Et | 190–195 | EtOH | 69.5; 40 | | 297 |
| 1,3-$Me_2$-6-Ph | 270–273 | $H_2O$ | 77 | | 297 |
| 1,3,6-$Me_3$-7-Et | 208–210 | | 87 | | 297 |
| ***1,3,7,9-Tetrahydropyrimido[4,5-b] [1,4]diazepin-2,4,8-(2H,4H,8H)-triones*** | | | | | |
| 1,3-$Me_2$-6-Ph | 263 | EtOH | 74 | ir, ms, pmr, uv | 298 |

TABLE IV-1. —(contd.)

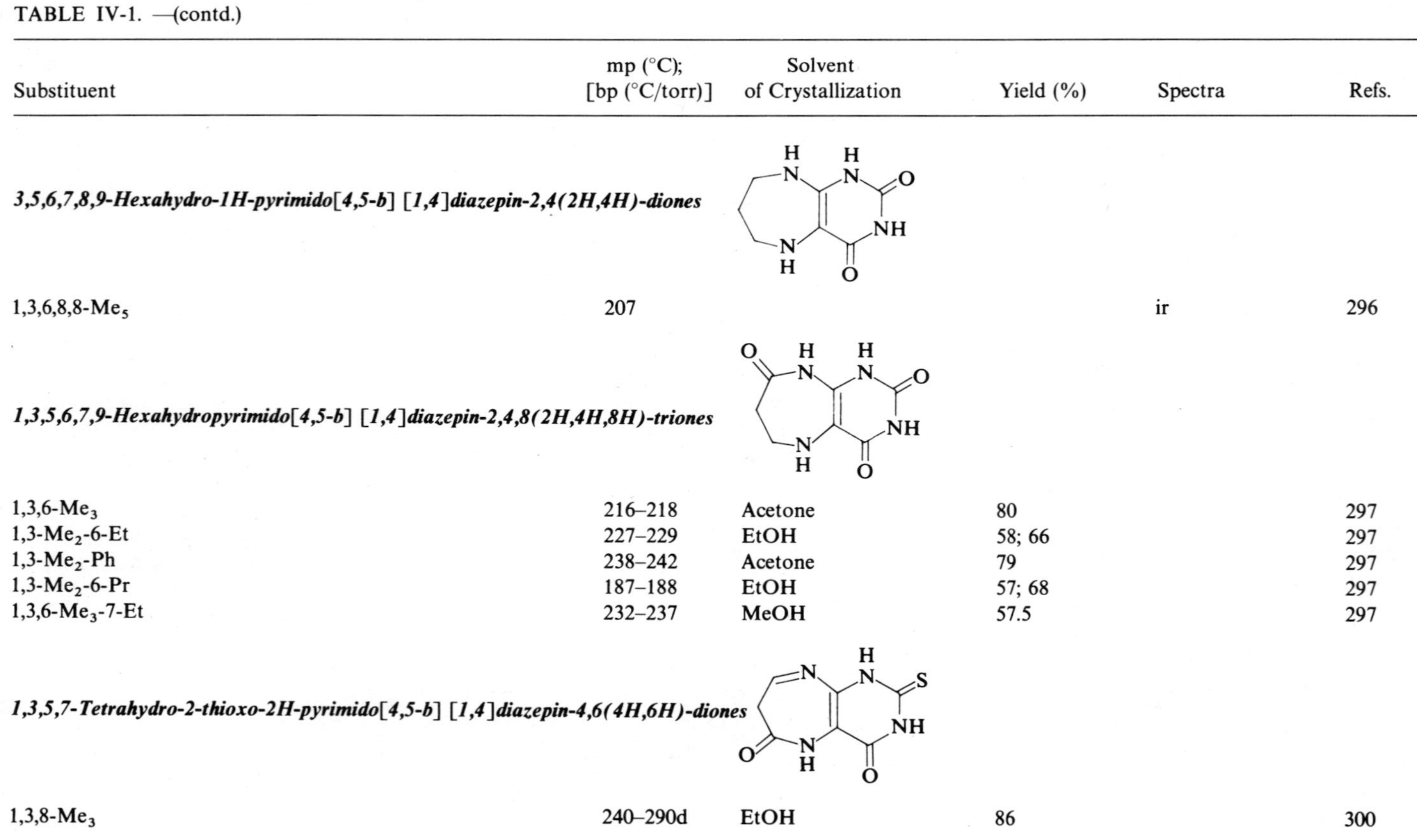

| Substituent | mp (°C); [bp (°C/torr)] | Solvent of Crystallization | Yield (%) | Spectra | Refs. |
|---|---|---|---|---|---|
| ***3,5,6,7,8,9-Hexahydro-1H-pyrimido[4,5-b] [1,4]diazepin-2,4(2H,4H)-diones*** | | | | | |
| 1,3,6,8,8-$Me_5$ | 207 | | | ir | 296 |
| ***1,3,5,6,7,9-Hexahydropyrimido[4,5-b] [1,4]diazepin-2,4,8(2H,4H,8H)-triones*** | | | | | |
| 1,3,6-$Me_3$ | 216–218 | Acetone | 80 | | 297 |
| 1,3-$Me_2$-6-Et | 227–229 | EtOH | 58; 66 | | 297 |
| 1,3-$Me_2$-Ph | 238–242 | Acetone | 79 | | 297 |
| 1,3-$Me_2$-6-Pr | 187–188 | EtOH | 57; 68 | | 297 |
| 1,3,6-$Me_3$-7-Et | 232–237 | MeOH | 57.5 | | 297 |
| ***1,3,5,7-Tetrahydro-2-thioxo-2H-pyrimido[4,5-b] [1,4]diazepin-4,6(4H,6H)-diones*** | | | | | |
| 1,3,8-$Me_3$ | 240–290d | EtOH | 86 | | 300 |

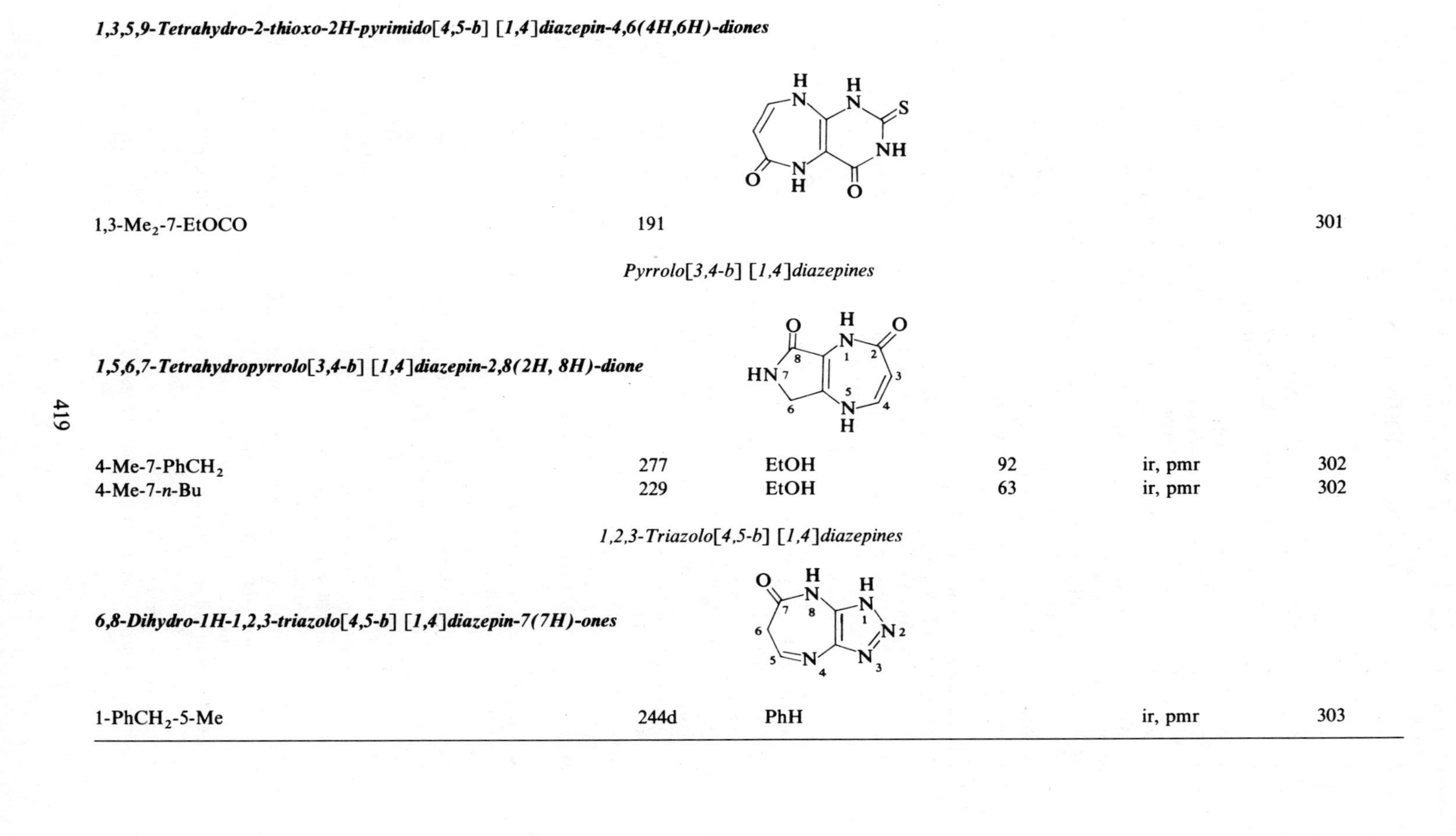

***1,3,5,9-Tetrahydro-2-thioxo-2H-pyrimido[4,5-b] [1,4]diazepin-4,6(4H,6H)-diones***

| | | | | | |
|---|---|---|---|---|---|
| 1,3-$Me_2$-7-EtOCO | 191 | | | | 301 |

*Pyrrolo[3,4-b] [1,4]diazepines*

***1,5,6,7-Tetrahydropyrrolo[3,4-b] [1,4]diazepin-2,8(2H, 8H)-dione***

| | | | | | |
|---|---|---|---|---|---|
| 4-Me-7-$PhCH_2$ | 277 | EtOH | 92 | ir, pmr | 302 |
| 4-Me-7-*n*-Bu | 229 | EtOH | 63 | ir, pmr | 302 |

*1,2,3-Triazolo[4,5-b] [1,4]diazepines*

***6,8-Dihydro-1H-1,2,3-triazolo[4,5-b] [1,4]diazepin-7(7H)-ones***

| | | | | | |
|---|---|---|---|---|---|
| 1-$PhCH_2$-5-Me | 244d | PhH | | ir, pmr | 303 |

## 20. REFERENCES


1. D. Lloyd, R. H. McDougall, and D. R. Marshall, *J. Chem. Soc.*, 3785 (1965).
2. H. Rupe and A. Huber, *Helv. Chim. Acta*, **10**, 846 (1927).
3. W. Ruske and E. Hüfner, *J. Prakt. Chem.*, **18**, 156 (1962).
4. F. E. King and P. C. Spensley, *J. Chem. Soc.*, 2144 (1952).
5. R. M. Acheson, *J. Chem. Soc.*, 4731 (1956).
6. D. P. Clifford, D. Jackson, R. V. Edwards, and P. Jeffrey, *Pestic. Sci.*, **7**, 453 (1976).
7. S. M. Jain and R. A. Pawar, *Indian J. Chem.*, **13**, 304 (1975).
8. M. Weissenfels, U. Thust, and M. Mühlstädt, *J. Prakt. Chem.*, **13**, 304 (1975).
9. M. Weissenfels, R. Kache, and W. Kräuter, *J. Prakt. Chem.*, **35**, 166 (1967).
10. M. Weissenfels, *Z. Chem.*, **4**, 458 (1964).
11. M. Weissenfels, H. Schurig, and Huhsam, *Chem. Ber.*, **100**, 584 (1967).
12. M. Weissenfels and G. Dill, *Z. Chem.*, **7**, 456 (1967).
13. W. Ruske and G. Grimm, *J. Prakt. Chem.*, **18**, 163 (1962).
14. J. Thiele and G. Steimmig, *Chem. Ber.*, **40**, 955 (1907).
15. C. A. C. Haley and P. Maitland, *J. Chem. Soc.*, 3155 (1951).
16. J. A. Barltrop, C. G. Richards, D. M. Russell, and G. Ryback, *J. Chem. Soc.*, 1132 (1959).
17. P. Neumann and F. Vogtle, *Chem. Exp. Didakt.*, **2**, 267 (1976).
18. G. Schwarzenbach and G. Lutz, *Helv. Chim. Acta*, **23**, 1147 (1940).
19. R. L. Williams, W. Peaston, and J. Kern, *J. Chem. Soc., Perkin Trans. I*, 2353 (1976).
20. K. V. Levshina, L. P. Glazyrina, and T. S. Safonova, *Khim. Geterotsikl. Soedin.*, **6**, 1135 (1970); *Chem. Heterocycl. Compd.*, **6**, 1061 (1970).
21. K. V. Levshina, T. A. Andrianova, and T. S. Safonova, *Zh. Org. Khim.*, **5**, 162 (1969); *J. Org. Chem.* (USSR), **5**, 158 (1969).
22. K. V. Levshina, E. I. Yumasheva, and T. S. Safonova, *Khim. Geterotsikl. Soedin.*, **7**, 556 (1971); *Chem. Heterocycl. Compd.*, **7**, 519 (1971).
23. R. L. Williams and S. W. Shalaby, *J. Heterocycl. Chem.*, **10**, 891 (1973).
24. A. Nawojski and W. Nawrocka, *Rocz. Chem.*, **48**, 1073 (1974).
25. K. V. Levshina, E. I. Yumasheva, T. S. Safonova, A. I. Kravchenko, and V. A. Chernov, *Khim.-Farm. Zh.*, **5**, 18 (1971).
26. A. Becker, *Helv. Chim. Acta*, **32**, 1584 (1949).
27. S. Veibel and S. F. Hromadko, *Chem. Ber.*, **93**, 2752 (1960).
28. M. Matsumoto, Y. Matsumura, A. Lio, and T. Yonezawa, *Bull. Chem. Soc. Japan*, **43**, 1496 (1970).
29. M. Sulca, A. Ya. Strakov, and I. Hirsbergs, *Latv. PSR Zinat. Akad. Vestis, Kim. Ser.*, **186** (1971).
30. E. Hannig and H. Ziebandt, *Pharmazie*, **23**, 688 (1968).
31. F. Eiden and G. Heja, *Arch. Pharm.*, **310**, 964 (1977).
32. Yu. K. Yur'ev, N. M. Magdesieva, and V. V. Titov, *Zh. Org. Khim.*, **1**, 163 (1965).
33. F. Eiden and G. Bachmann, *Arch. Pharm.*, **306**, 401 (1973).
34. M. Trkovnik, M. Kules, M. Lacan, and B. Babarevic, *Z. Naturforsch*, **29**, 580 (1974).
35. (a) R. E. Pastor, C. A. Giovannoni, and A. R. Cambon, *Chim. Ther.* (*Eur J. Med. Chem.*) **9**, 175 (1974). (b) A. Cambon, C. Giovannoni, R. Paspor, and J. Riess, Ger. Offen. 2,424,652, December 1974; *Chem. Abstr.*, **84**, 44179d (1976).
36. I. L. Finnar, *J. Chem. Soc.*, 4094 (1958).
37. (a) F. Eiden and G. Heja, *Synthesis*, 148 (1973). (b) F. Eiden and G. Heja, Ger. Offen. 2,343,528; March 1975; *Chem. Abstr.*, **83**, 43395p (1975).
38. S. U. Kulkarni and K. Z. Thakar, *J. Indian Chem. Soc.*, **52**, 849 (1975).
39. (a) S. U. Kulkarni and K. Z. Thakar, *J. Indian Chem. Soc.*, **53**, 279 (1976); (b) S. U. Kulkarni and K. Z. Thakar, *J. Indian Chem. Soc.*, **53**, 283 (1976).
40. H. Stetter and F. von Prann, *Chem. Ber.*, **102**, 1643 (1969).
41. J. Schmitt, *Justus Liebigs Ann. Chem.*, **569**, 17 (1950).

42. Yu. S. Andreichikov, S. P. Tendryakova, Yu. A. Nalimova, S. G. Pitirimova, and L. A. Voronova, *J. Org. Chem.* (USSR) **13**, 483 (1977).
43. Yu. S. Andreichikov, S. G. Pitirimova, S. P. Tendryakova, R. F. Saraeva, and T. N. Tokmakova, *Zh. Org. Khim.*, **14**, 169 (1978); *J. Org. Chem.* (USSR), **14**, 156 (1978).
44. S. B. Vaisman, *Trans. Inst. Chem. Kharkov Univ.*, **4**, 157 (1938); *Chem. Abstr.*, **34**, 5847 (1940).
45. J. O. Halford and R. M. Fitch, *J. Am. Chem. Soc.*, **85**, 3354 (1963).
46. R. K. Singhal, S. Kumar, U. C. Pant, and B. C. Joshi, *Natl. Appl. Sci. Bull.*, **26**, 1 (1974).
47. J. A. Barltrop and C. G. Richards, *Chem. Ind.* (London) 466 (1957).
48. C. N. O'Callaghan and D. Twomey, *J. Chem. Soc.*, 600 (1969).
49. S. Motoki, C. Urakawa, A. Kano, Y. Fushimi, T. Hirano, and K. Murata, *Bull. Chem. Soc. Japan*, **43**, 809 (1970).
50. W. Ried and E. Koenig, *Justus Liebigs Ann. Chem.*, **755**, 24 (1972).
51. Yu. S. Andreichikov and R. F. Saraeva, *Khim. Geterotsikl. Soedin.*, **8**, 1702 (1972); *Chem. Heterocyclic. Compd.*, **8**, 1544 (1972).
52. S. P. Korshunov, V. M. Kazantseva, L. A. Vopilina, V. S. Pisareva, and N. V. Utekhina, *Khim. Geterotsikl. Soedin.*, **9**, 1421 (1973); *Chem. Heterocycl. Compd.*, **9**, 1287 (1973).
53. R. L. Amey and N. D. Heindel, *Org. Prep. Proc. Int.*, **8**, 306 (1976).
54. W. Ried and R. Teubner, *Justus Liebigs Ann. Chem.*, 741 (1978).
55. J. A. Van Allan and S. Chie Chang, *J. Heterocycl. Chem.*, **11**, 1065 (1974).
56. W. Müller, U. Kraatz, and F. Korte, *Chem. Ber.*, **106**, 332 (1973).
57. A. P. Bindra and E. Le Goff, *Tetrahedron Lett.*, 1523 (1974).
58. R. G. Bass, D. D. Crichton, H. K. Meetz, and A. F. Johnson, Jr., *Tetrahedron Lett.*, 2073 (1975).
59. Y. Okamoto and T. Ueda, *J. Chem. Soc. Chem. Commun.*, 367 (1973).
60. Y. Okamoto and T. Ueda, *Chem. Pharm. Bull.*, **23**, 1391 (1975).
61. P. C. Unangst and P. L. Southwick, *J. Heterocycl. Chem.*, **10**, 399 (1973).
62. G. Roma, A. Balbi, and A. Ermili, *Farmaco, Ed. Sci.*, **32**, 393 (1977).
63. C. R. Ellefson, C. M. Woo, A. Miller, and J. R. Kehr, *J. Med. Chem.*, **21**, 952 (1978).
64. C. R. Ellefson and F. M. Hershenson, U.S. Patent 4,123,430, October 1978; *Chem. Abstr.*, **90**, 87528h (1979).
65. G. Roma, E. Vigevani, A. Balbi, and A. Ermili, *Farmaco, Ed. Sci.*, **34**, 62 (1979).
66. H. D. Stachel, *Chem. Ber.*, **95**, 2171 (1962).
67. D. Nardi, A. Tajana, and S. Rossi, *J. Heterocycl. Chem.*, **10**, 815 (1973).
68. D. Nardi, E. Massarani, and L. Degen, Swiss Patent 555,347, October 1974; *Chem. Abstr.*, **82**, 43480s (1974).
69. D. Nardi, E. Massarani, A. Tajana, R. Cappelletti, and M. Veronese, *Farmaco, Ed. Sci.*, **30**, 727 (1975).
70. D. Nardi, E. Massarani, A. Tajana, R. Cappelletti, and M. Salvaterra, *Farmaco, Ed. Sci.*, **30**, 248 (1975).
71. A. Trka, A. Frigerio, D. Nardi, A. Tajana, and U. Rapp, *Farmaco, Ed. Sci.*, **33**, 885 (1978).
72. D. Nardi, R. Pennini, and A. Tajana, *J. Heterocycl. Chem.*, **12**, 825 (1975).
73. K. Peseke, *Tetrahedron*, **32**, 483 (1976).
74. East German Patent 117,219, January 1976; *Chem. Abstr.*, **85**, 21499q (1976).
75. G. Schwarzenbach and K. Lutz, *Helv. Chim. Acta*, **23**, 1162 (1940).
76. D. Lloyd and D. R. Marshall, *Chem. Ind.* (London), 1760 (1964).
77. S. Veibel and J. I. Nielsen, *Mat. Fys. Medd. Dan. Vid. Selsk.*, **35**, 6 (1966).
78. D. Lloyd and H. P. Cleghorn, in *Advances in Heterocyclic Chemistry*, Vol. **17**, A. R. Katritzky and A. J. Boulton, Eds., Academic Press, New York and London, 1974, p. 35.
79. P. W. W. Hunter and G. A. Webb, *J. Inorg. Nucl. Chem.*, **35**, 1457 (1973).
80. K. V. Levshina, L. P. Glazyrina, and T. S. Safonova, *Khim. Geterotsikl. Soedin.*, **6**, 1133 (1970); *Chem. Heterocycl. Compd.*, **6**, 1059 (1970).
81. M. Matsumoto, A. Iio, and T. Yonezawa, *Bull. Chem. Soc. Japan*, **43**, 281 (1970).
82. T. Yonezawa, M. Matsumoto, and H. Kato, *Bull. Chem. Soc. Japan*, **41**, 2543 (1968).
83. S. Veibel and J. I. Nielsen, *Chem. Ber.*, **99**, 2709 (1966).

84. W. Paterson and G. R. Proctor, *J. Chem. Soc.*, 485 (1965).
85. V. C. Barry, M. L. Conalty, C. N. O'Callaghan, and D. Twomey, *Proc. Roy. Irish Acad., Sect. B*, **65**, 309 (1967).
86. A. Ouchi, T. Takeuchi, M. Nakatani, and Y. Takahashi, *Bull. Chem. Soc. Japan*, **44**, 434 (1971).
87. A. Furuhashi, S. Komatsu, and A. Ouchi, *Bull. Chem. Soc. Japan*, **45**, 2942 (1972).
88. P. W. W. Hunter and G. A. Webb, *J. Inorg. Nucl. Chem.*, **34**, 1511 (1972).
89. P. W. W. Hunter and G. A. Webb, *Inorg. Nucl. Chem. Lett.*, **9**, 271 (1973).
90. B. Emmert and H. Gsottschneider, *Chem. Ber.*, **66**, 1871 (1933).
91. P. L. Orioli and H. C. Lip, *Cryst. Struct. Commun.*, **3**, 477 (1974).
92. H. A. Staab and F. Vögtle, *Chem. Ber.*, **98**, 2701 (1965).
93. W. J. Barry, I. L. Finar, and E. F. Mooney, *Spectrochim. Acta*, **21**, 1095 (1965).
94. K. F. Turchin, *Khim. Geterotsikl. Soedin.*, **10**, 828 (1974).
95. A. Mannschreck, G. Rissmann, F. Vogtle, and D. Wild, *Chem. Ber.*, **100**, 335 (1967).
96. J. C. Speakman and F. B. Wilson, *Acta Crystallogr.*, **32**, 622 (1976).
97. K. Butkiewi, *J. Electroanal. Chem.*, **90**, 271 (1978).
98. W. Ried and P. Stahlhofen, *Chem. Ber.*, **90**, 815 (1957).
99. A. Nawojski and W. Nawrocka, *Rocz. Chem.*, **49**, 1915 (1975).
100. L. K. Mushkalo, *Nauk Zapiski Kiiv. Derzhav. Univ.*, **16**; 15, Zbirnik Khim. Fak. No. 8, 133 (1957); *Chem. Abstr.*, **53**, 18057 (1959).
101. A. N. Kost, S. F. Solomko, L. N. Polovina, and L. G. Gergel, *Khim. Geterotsikl. Soedin.*, **7**, 553 (1971); *Chem. Heterocycl. Compd.*, **7**, 516 (1971).
102. Q. Q. Dang, R. Caujolle, and T. D. Thi Bang, *C.R. Acad. Sci. Paris (C)*, **272**, 1518 (1971).
103. J. A. L. Herbert and H. Suschitzky, *J. Chem. Soc., Perkin Trans. I*, 2657 (1974).
104. W. Jehn and R. Radeglia, *J. Prakt. Chem.*, **317**, 1035 (1975).
105. J. R. DeBaun, F. M. Pallos, and D. R. Baker, U.S. Patent 3,978,227; August 1976; *Chem. Abstr.*, **86**, 5498d (1977).
106. R. M. Acheson and W. R. Tully, *J. Chem. Soc.*, 1117 (1970).
107. S. Fukushima, Y. Akahori, I. Sakamoto, K. Noro, and S. Kazama, *Yakugaku Zasshi*, **90**, 1076 (1970).
108. K. Samula and E. Jurkowska-Kowalczyk, *Rocz. Chem.*, **48**, 2287 (1974).
109. W. Ried and E. Torinus, *Chem. Ber.*, **92**, 2902 (1959).
110. L. K. Mushkalo and V. A. Chuiguk, *Ukr. Khim. Zh.*, **35**, 740 (1969).
111. K. Hideg and O. Hideg-Hankovzky, *Acta Chim.* (Budapest), **75**, 137 (1973).
112. K. Hideg and O. H. Hankovzky, *Acta. Chim.* (Budapest), **57**, 213 (1968).
113. W. Werner, W. Jungstand, W. Gutsche, and K. Wohlrabe, East German Patent 122,247; September 1976; *Chem. Abstr.*, **87**, 39550a (1977).
114. W. Werner, W. Zschiesche, J. Guttner, and H. Heineke, *Pharmazie*, **31**, 282 (1976).
115. J. Curtze and K. Thomas, *Justus Liebigs Ann. Chem.*, 328 (1974).
116. N. M. Omar, *Indian J. Chem.*, **12**, 498 (1974).
117. P. W. W. Hunter and G. A. Webb, *Tetrahedron*, **28**, 5573 (1972).
118. A. Bauer, P. Danneberg, K. H. Weber, and K. Minck, *J. Med. Chem.*, **16**, 1011 (1973).
119. S. Bodforss, *Justus Liebigs Ann. Chem.*, **745**, 99 (1971).
120. L. K. Mushkalo, M. Chabuby, M. Weissenfels, M. Pulst, and H.-J. Hense, *Z. Chem.*, **14**, 187 (1974).
121. R. Radeglia and W. Jehn, *J. Prakt. Chem.*, **318**, 1049 (1976).
122. P. W. W. Hunter and G. A. Webb, *Tetrahedron*, **29**, 147 (1973).
123. W. A. Sexton, *J. Chem. Soc.*, 303 (1942).
124. J. Davoll, *J. Chem. Soc.*, 308 (1960).
125. A. Rossi, A. Hunger, J. Kebrle, and K. Hoffmann, *Helv. Chim. Acta*, **43**, 1298 (1960).
126. D. E. Burton, A. J. Lambie, D. W. J. Lane, G. T. Newbold, and A. Percival, *J. Chem. Soc.*, 1268 (1968).
127. A. Rossi, A. Hunger, J. Kebrle, and K. Hoffmann, *Helv. Chim. Acta*, **43**, 1046 (1960).
128. F. B. Wigton and M. M. Joullié, *J. Am. Chem. Soc.*, **81**, 5212 (1959).
129. W. Ried and P. Stahlhofen, *Chem. Ber.*, **90**, 828 (1957).

130. D. K. Wald and M. M. Joullié, *J. Org. Chem.*, **31**, 3369 (1966).
131. R. T. Blickenstaff and N. Wells, *Org. Prep. Proc. Int.*, **6**, 197 (1974).
132. R. Barchet and K. W. Merz, *Tetrahedron Lett.*, 2239 (1964).
133. Z. F. Solomko, V. I. Avramenko, and L. V. Pribega, *Khim. Geterotsikl. Soedin.*, **14**, 411 (1978); *Chem. Heterocycl. Compd.*, **14**, 340 (1978).
134. T. S. Chmilenko and Z. F. Solomko, *Khim. Geterotsikl. Soedin.*, **13**, 834 (1977); *Chem. Heterocycl. Compd.*, **13**, 681 (1977).
135. M. Israel, L. C. Jones, and M. M. Joullié, *J. Heterocycl. Chem.*, **8**, 1015 (1971).
136. A. Nawojski and W. Nawrocka, *Rocz. Chem.*, **49**, 203 (1975).
137. A. N. Kost, Z. F. Solomko, N. M. Prikhod'ko, and S. S. Teteryuk, *Khim. Geterotsikl. Soedin.*, **7**, 1556 (1971); *Chem. Heterocycl. Compd.*, **7**, 1447 (1971).
138. A. M. Kost, Z. F. Solomko, V. G. Vinokurov, and V. S. Tkachenko, *J. Org. Chem.* (USSR), **8**, 2080 (1972).
139. Z. F. Solomko, V. S. Tkachenko, A. N. Kost, V. A. Budylin, and V. L. Pikalov, *Khim. Geterotsikl. Soedin.*, **11**, 533 (1975); *Chem. Heterocycl. Compd.*, **11**, 470 (1975).
140. A. N. Kost, Z. F. Solomko, V. A. Budylin, and T. S. Semenova, *Khim. Geterotsikl. Soedin.*, **8**, 696 (1972); *Chem. Heterocycl. Compd.*, **8**, 632 (1972).
141. B. A. Puodzhyunaite and Z. A. Talaikite, *Khim. Geterotsikl. Soedin.*, **10**, 833 (1974); *Chem. Heterocycl. Compd.*, **10**, 724 (1974).
142. Z. F. Solomko, V. L. Pikalov, and L. V. Pribega, *Ukr. Khim. Zh.*, **40**, 768 (1974).
143. A. A. Stolyarchuk, Yu. N. Furman, V. L. Pikalov, Z. F. Solomko, and V. S. Tkachenko, *Chim.-Farm. Zh.*, **9**, 19 (1975).
144. Z. F. Solomko and V. L. Pikalov, *Vopr. Khim. Khim. Tekhnol.*, **39**, 93 (1975).
145. R. Pennini, A. Tajana, and D. Nardi, *Farmaco, Ed. Sci.*, **31**, 120 (1976).
146. W. Ried and P. Stahlhofen, *Chem. Ber.*, **90**, 825 (1957).
147. W. Ried and G. Isenbruck, *Chem. Ber.*, **105**, 337 (1972).
148. S. Linke, J. Kurz, and C. Wünsche, *Justus Liebigs Ann. Chem.*, 936 (1973).
149. Z. Zubovics, G. Fehér, and L. Toldy, *Acta Chim. Acad. Sci. Hung.*, **92**, 293 (1977).
150. T. S. Chmilenko, Z. F. Solomko, and A. N. Kost, *Khim. Geterotsikl. Soedin.*, **13**, 525 (1977); *Chem. Heterocycl. Compd.*, **13**, 423 (1977).
151. E. Ajello, O. Migliara, L. Ceraulo, and S. Petruso, *J. Heterocycl. Chem.*, **11**, 339 (1974).
152. F. Eiden and E. Zimmermann, *Arch. Pharm.*, **309**, 619 (1976).
153. T. Hara, H. Fujimori, Y. Kayama, T. Mori, K. Itoh, and Y. Hashimoto, *Chem. Pharm. Bull.*, **25**, 2584 (1977).
154. R. Buyle and H. G. Viehe, *Tetrahedron*, **25**, 3453 (1969).
155. R. Buyle and H. G. Viehe, U.S. patent 3,644,374; February 1972; *Chem. Abstr.*, **76**, 140798k (1972).
156. G. Roma, A. Ermili, and A. Balbi, *Farmaco, Ed. Sci.*, **32**, 81 (1977).
157. (a) A. Bauer, K.-H. Weber, K. Minck, and P. Danneberg, U.S. patent 3,944,579; March 1976; *Chem. Abstr.*, **85**, 21491f (1976). (b) A. Bauer, K.-H. Weber, K. Minck, and P. Danneberg, U.S. patent 3,862,136; January 1975 (Boehringer Ingelheim GmbH).
158. R. I. Fryer, L. H. Sternbach, and A. Walser, U.S. patent 4,111,934; September 1978; *Chem. Abstr.*, **90**, 152253f (1979).
159. R. B. Moffett, B. V. Kamdar, and P. F. Voigtlander, *J. Med. Chem.*, **19**, 192 (1976).
160. K. Peseke, East Ger. Patent 116,826; December 1975; *Chem. Abstr.*, **83**, 5703w (1975).
161. K.-H. Weber, A. Bauer, P. Danneberg, and K. Minck, U.S. patent 3,711,467; January 1973; *Chem. Abstr.*, **78**, 97729n (1973).
162. Z. F. Solomko, V. L. Pikalov, and V. I. Avramenko, All-Union Institute of Scientific and Technical Information Deposited Paper 1992–75; Ref. *Zh. Khim.*, **24**, 284 (1975).
163. Z. F. Solomko, T. S. Chmilenko, P. A. Sharbatyan, N. I. Shtemenko, and S. I. Khimyuk, *Khim. Geterotsikl. Soedin.*, **14**, 122 (1978); *Chem. Heterocycl. Compd.*, **14**, 100 (1978).
164. Z. F. Solomko, T. S. Chmilenko, P. A. Sharbatyan, L. V. Shevchenko, and N. Ya. Bozhanova, *Khim. Geterotsikl. Soedin.*, **14**, 551 (1978); *Chem. Heterocycl. Compd.*, **14**, 455 (1978).
165. J. Krapcho and C. F. Turk, *J. Med. Chem.*, **9**, 191 (1966).

166. L. H. Werner, S. Ricca, A. Rossi, and G. de Stevens, *J. Med. Chem.*, **10**, 575 (1967).
167. A. Bauer, K.-H. Weber, K.-H. Pook, and H. Daniel, *Justus Liebigs Ann. Chem.*, 969 (1973).
168. J. M. McManus, U.S. patent 3,595,858; July 1971; *Chem. Abstr.*, **75**, 110345x (1971).
169. M. R. Chandramohan and S. Seshadri, *Indian J. Chem.*, **12**, 940 (1974).
170. A. Bauer, K.-H. Weber, and P. Danneberg, German patent 2,306,770; September 1974; *Chem. Abstr.*, **82**, 4331z (1975).
171. A. Walser and R. I. Fryer, U.S. patent 4,111,931; September 1978 (Hoffman-La Roche Inc.).
172. A. Walser, U.S. patent 4,118,386; October 1978; *Chem. Abstr.*, **89**, 59906n (1978).
173. R. Benassi, P. Lazzeretti, F. Taddei, D. Nardi, and A. Tajana, *Org. Magn. Resonance*, **8**, 387 (1976).
174. A. N. Kost, P. A. Sharbatyan, P. B. Terent'ev, Z. F. Solomko, V. S. Tkachenko, and L. G. Gergel, *Zh. Org. Khim.*, **8**, 2113 (1972); *J. Org. Chem.* (USSR), **8**, 2160 (1972).
175. O. Hinsberg and A. Strupler, *Justus Liebigs Ann. Chem.*, **287**, 220 (1895).
176. H. Stetter, *Chem. Ber.*, **86**, 197 (1953).
177. H. Shirai, T. Hayazaki, and A. Maki, *Nagoya Shiritsu Daigaku Yakugakubu Kenkyu Nempo*, **19**, 53 (1971).
178. T. Ichii, *J. Pharm. Soc. Japan* **82**, 992 (1962).
179. H. Shirai and T. Hayazaki, *Yakugaku Zasshi*, **91**, 1109 (1971).
180. H. Shirai, T. Hayazaki, and M. Kawai, *Nagoya Shiritsu Daigaku Yakugakubu Kenkyu Nempo*, **20**, 45 (1972).
181. G. H. Fisher and H. P. Schultz, *J. Org. Chem.*, **39**, 631 (1974).
182. H. Richter, K. Schulze, and M. Mühlstädt, *Z. Chem.*, **8**, 220 (1968).
183. L. Lallaz and P. Caubère, *Synthesis*, **10**, 657 (1975).
184. W. Knobloch and G. Lietz, *J. Prakt. Chem.*, **36**, 113 (1967).
185. N. M. Omar, A. F. Youssef, M. A. El-Gency, and M. M. Kassab, *Can. J. Pharm. Sci.*, **11**, 89 (1976).
186. N. M. Omar, A. F. Youssef, and M. A. El-Gency, *Arch. Pharm., Chem. Sci. Ed.*, **3**, 89 (1975).
187. D. Misiti, F. Gatta, and R. Landi-Vittory, *J. Heterocycl. Chem.*, **8**, 231 (1971).
188. J. S. Walia, L. A. Heindl, A. S. Walia, and P. S. Walia, *J. Chem. Soc., Chem. Commun.*, 962 (1972).
189. J. S. Walia, P. S. Walia, L. A. Heindl, and P. Zbylot, *J. Chem. Soc., Chem. Commun.*, 108 (1972).
190. H. Shirai, T. Hayazaki, and A. Maki, *Yakugaku Zasshi*, **91**, 1228 (1971).
191. T. S. Safonova, K. W. Lewschina, W. A. Tschernow, T. A. Andrianowa, N. A. Grinewa, and S. M. Minakowa, Germ. Offen. 2,328,870; January 1975; *Chem. Abstr.*, **82**, 156397p (1975).
192. H. Böshagen and M. Plempel, U.S. patent 3,911,126; October 1975; *Chem. Abstr.*, **84**, 155700m (1976).
193. G. B. Bachman and L. V. Heisey, *J. Am. Chem. Soc.*, **71**, 1985 (1949).
194. W. Ried and G. Urlass, *Chem. Ber.*, **86**, 1101 (1953).
195. H. Wahl and M. T. Le Bris, *Ind. Chim. Belge*, **32**, 145 (1967).
196. M. T. Le Bris, *Bull. Soc. Chim. Fr.*, 3411 (1967).
197. S. Raines and C. A. Kovacs, *J. Heterocycl. Chem.*, **4**, 305 (1967).
198. N. K. Khakimova, Ch. Sh. Kadyrov, and A. A. Shazhenov, *Uzb. Khim. Zh.*, **19**, 53 (1975); *Chem. Abstr.*, **83**, 79210p (1975).
199. (a) O. Bub., U.S. patent 4,108,852; August 1978; (b) O. Bub, German Offen. 1,913,536; October 1970; *Chem. Abstr.*, **73**, 120691e (1970).
200. S. H. Dandegaonker and G. B. Desai, *Indian J. Chem.*, **1**, 298 (1963).
201. Z. F. Solomko, A. N. Kost, L. N. Polovina, and M. A. Salimov, *Khim. Geterotsikl. Soedin.*, **7**, 987 (1971); *Chem. Heterocycl. Compd.*, **7**, 922 (1971).
202. G. Holan, J. J. Evans, and M. Linton, *J. Chem. Soc., Perkin Trans. I*, 1200 (1977).
203. B. J. R. Nicolaus, E. Bellasio, G. Pagani, L. Mariani, and E. Testa, *Helv. Chim. Acta*, **48**, 1867 (1965).
204. A. Bauer, K.-H. Weber, and M. Unruh, *Arch. Pharm.*, **305**, 557 (1972).
205. E. Szarvasi, M. Grand, J.-C. Depin, and A. Betbeder-Matibet, *Eur. J. Med. Chem. (Chim. Ther.)*, **13**, 113 (1978).

206. Z. F. Solomko, V. T. Braichenko, and M. S. Malinovskii, *Khim. Geterotsikl. Soedin.*, **8**, 428 (1972); *Chem. Heterocycl. Compd.*, **8**, 395 (1972).
207. Belg. Patent 751,834; December 1970 (C.H. Boehringer Sohn, Ingelheim).
208. O. Bub, Ger. Offen. 2,062,226; September 1971; *Chem. Abstr.*, **76**, 14602d (1972).
209. O. Bub, Ger. Offen. 2,062,237; September 1971; *Chem. Abstr.*, **76**, 3913e (1972).
210. O. Bub, Ger. Offen. 1,953,647; May 1971; *Chem. Abstr.*, **75**, 49153q (1971).
211. P. I. Ittyerah and F. G. Mann, *J. Chem. Soc.*, 467 (1958)
212. J. Krapcho and C. Turk, U.S. patent 3,321,468; May 1967; *Chem. Abstr.*, **68**, 21970k (1968).
213. K. W. Merz, R. Haller, and E. Muller, *Naturwissenschaften*, **50**, 663 (1963).
214. E. Muller, R. Haller, and K. W. Merz, *Justus Liebigs Ann. Chem.*, **697**, 193 (1966).
215. A. Bauer, K.-H. Weber, H. Merz, K. Zeile, R. Giesemann, and P. Danneberg, U.S. patent 3,816,409; June 1974; *Chem. Abstr.*, **81**, 105588y (1974).
216. J. Bernstein, U.S. patent 3,341,521; September 1967; *Chem. Abstr.*, **68**, 95875e (1968).
217. O. Bub, H.-P. Hofmann, H. Kreiskott, and F. Zimmermann, U.S. patent 3,847,905; November 1974 (Knoll AG).
218. A. Bauer, K.-H., Weber, P. Danneberg, and F.-J. Kuhn, German Offen. 2,231,560; January 1974; *Chem. Abstr.*, **80**, 96043s (1974).
219. M. P. Mertes and A. J. Lin, *J. Med. Chem.*, **13**, 77 (1970).
220. R. Meyer, *Justus Liebigs Ann. Chem.*, **327**, 1 (1903).
221. R. Meyer, *Justus Liebigs Ann. Chem.*, **347**, 17 (1906).
222. R. Meyer and H. Luders, *Justus Liebigs Ann. Chem.*, **415**, 29 (1918).
223. M. A. Phillips, *J. Chem. Soc.*, 2393 (1928).
224. R. L. Shriner and P. G. Boermans, *J. Am. Chem. Soc.*, **66**, 1810 (1944).
225. G. Glotz, *Bull. Soc. Chim. Fr.*, **3**, 511 (1936).
226. D. R. Buckle, B. C. C. Cantello, and N. J. Morgan, Brit. patent 1,460,936; January 1977; *Chem. Abstr.*, **87**, 68438a (1977).
227. J. Büchi, H. Dietrich, and E. Eichberger, *Helv. Chim. Acta*, **39**, 957 (1956).
228. (a) B. Brobanski, W. Roman, and E. Wagner, *Farmaco, Ed. Sci.*, **26**, 3 (1971); (b) B. Bobranski, E. Wagner, and W. Roman, Polish patent 71,544; October 1974; *Chem. Abstr.*, **83**, 79299z (1975).
229. E. Wagner, *Disst. Pharm. Pharmacol.*, **24**, 391 (1972).
230. E. Wagner, *Rocz. Chem.*, **48**, 1289 (1974).
231. M. M. El-Enany, K. M. Choneim, and M. Khalifa, *Pharmazie*, **32**, 79 (1977).
232. Dutch patent 7,204,008; September 1972 (C.H. Boehringer Sohn, Ingelheim).
233. K.-H Weber, H. Merz, K. Zeile, R. Giesemann, and P. Danneberg, Ger. Offen. 1,918,073; October 1970; *Chem. Abstr.*, **74**, 3673k (1971).
234. S. Rossi, O. Pirola, and R. Maggi, *Chim. Ind.* (Milan) **51**, 479 (1969).
235. S. Rossi, U.S. patent 3,984,398; October 1976 (Roussel-UCLAF).
236. K.-H. Weber, K. Zeile, P. Danneberg, R. Giesemann, and K. H. Hauptmann, U.S. patent 3,836,653; September 1974; *Chem. Abstr.*, **86**, 29894f (1977).
237. K.-H. Weber, H. Merz, and K. Zeile, Ger. Offen. 1,934,607; January 1970; *Chem. Abstr.*, **72**, 100771g (1970).
238. K.-H. Weber, H. Merz, R. Giesemann, and P. Danneberg, Ger. Offen. 1,966,128; August 1971; *Chem. Abstr.*, **75**, 110348a (1971).
239. K.-H. Weber, A. Bauer, and K.-H. Hauptmann, *Justus Liebigs Ann. Chem.*, **756**, 128 (1972).
240. K.-H. Weber, and A. Bauer, Canadian patent 992,540; July 1976; *Chem. Abstr.*, **86**, 55499a (1977).
241. K.-H. Weber, A. Bauer, H. Merz, and K. Minck, U.S. patent 3,678,033; July 1972 (Boehringer Ingelheim GmbH).
242. J. Van Alphen, *Recl. Trav. Chim.*, **43**, 823 (1924).
243. K.-H. Weber and A. Bauer, *Justus Liebigs Ann. Chem.*, **763**, 66 (1972).
244. K.-H. Weber, A. Bauer, and K.-H. Pook, Ger. Offen. 2,103,746; August 1972; *Chem. Abstr.*, **77**, 140176b (1972).

245. K.-H. Weber, A. Bauer, P. Danneberg, K. Minck, K.-H. Pook, U.S. patent 3,707,538; December 1972 (Boehringer Ingelheim GmbH).
246. A. Bauer and K.-H. Weber, *Justus Liebigs Ann. Chem.*, **762**, 73 (1972).
247. A. Bauer, K.-H. Weber, P. Danneberg, and K. Minck, U.S. patent 3,711,468; January 1973 (Boehringer Ingelheim GmbH).
248. A. Bauer, K.-H. Weber, and K.-H. Pook, U.S. patent 3,766,169; October 1973 (Boehringer Ingelheim GmbH).
249. K.-H. Hauptmann, K.-H. Weber, K. Zeile, P. Danneberg, and R. Giesemann, Brit. patent 1,217,217; December 1970; *Chem. Abstr.*, **74**, 141893h (1971).
250. H. Kuch and I. Hoffman, U.S. patent 3,718,645; February 1973 (Farbwerke Hoechst AG).
251. K.-H. Weber, K. Zeile, R. Giesemann, P. Danneberg, U.S. patent 3,624,076; November 1971 (Boehringer Ingelheim GmbH).
252. A. Bauer, K.-H. Pook, and K.-H. Weber, *Justus Liebigs Ann. Chem.*, **757**, 87 (1972).
253. K. B. Alton, J. E. Patrick, C. Shaw, and J. L. McGuire, *Drug Metab. Dispos.*, **3**, 445 (1975).
254. K. B. Alton, R. M. Grimes, C. Shaw, J. E. Patrick, and J. L. McGuire, *Drug Metab. Dispos.*, **3**, 352 (1975).
255. R. E. Huettemann and A. P. Shroff, *J. Pharm. Sci.*, **64**, 1339 (1975).
256. D. Nardi, E. Massarani, and L. Degen, Swiss patent 532,592; February, 1973; *Chem. Abstr.*, **78**, 136346k (1973).
257. K. Peseke, East Ger. patent 99,796; August 1973; *Chem. Abstr.*, **80**, 83081c (1974).
258. K. Peseke, East Ger. patent 106,041; August 1974; *Chem. Abstr.*, **81**, 169569a (1974).
259. A. I. Kiprianov and V. P. Khilya, *Zh. Org. Khim.*, **3**, 1091 (1976); *J. Org. Chem.* (USSR), **3**, 1051 (1967).
260. A. W. Chow, F. J. Gyurik, and R. C. Parish, *J. Heterocycl. Chem.*, **13**, 163 (1976).
261. K. Peseke, East Ger. patent 105,235, April 1974; *Chem. Abstr.*, **81**, 169568z (1974).
262. G. Belvedere, A. Frigerio, A. Malorni, and D. Nardi, *Boll. Chim. Farm.*, **114**, 151 (1975).
263. D. Lloyd, R. H. McDougall, and D. R. Marshall, *J. Chem. Soc. C*, **780**, (1966).
264. D. Lloyd, H. McNab, and D. R. Marshall, *Aust. J. Chem.*, **30**, 365 (1977).
265. G. W. H. Potter, M. W. Coleman, and A. M. Monro, *J. Heterocycl. Chem.*, **12**, 611 (1975).
266. D. Lloyd, R. K. Mackie, H. McNab, and K. S. Tucker, *Tetrahedron*, **32**, 2339 (1976).
267. R. H. McDougall and S. H. Malik, *J. Chem. Soc. C*, 2044 (1969).
268. G. Settimj, F. Fraschetti, and S. Chiavarelli, *Gazz. Chim. Ital.*, **96**, 749 (1966).
269. G. Seitz and H. Morck, *Arch. Pharm.*, **307**, 113 (1974).
270. D. Lloyd and D. R. Marshall, *J. Chem. Soc.*, 2597 (1956).
271. E. Abushanab, D. Y. Lee, and L. Goodman, *J. Heterocycl. Chem.*, **10**, 181 (1973).
272. G. Desimoni and G. Minoli, *Tetrahedron*, **26**, 1393 (1970).
273. A. Gasco, G. Rua, E. Menziani, G. M. Nano, and G. Tappi, *J. Heterocycl. Chem.*, **7**, 131 (1970).
274. J.-P. Affane-Nguema, J.-P. Lavergne, and P. Viallefont, *J. Heterocycl. Chem.*, **14**, 391 (1977).
275. J.-P. Affane-Nguema, J.-P. Lavergne, and P. Viallefont, *J. Heterocycl. Chem.*, **14**, 1013 (1977).
276. J.-P. Affane-Nguema, J.-P. Lavergne, and P. Viallefont, *Org. Mass Spectrom.*, **12**, 136 (1977).
277. M. Israel, L. C. Jones, and E. J. Modest, *J. Heterocycl. Chem.*, **4**, 659 (1967).
278. A. Nawojski, *Rocz. Chem.*, **42**, 1641 (1968).
279. J.-P. Lavergne, P. Viallefont, and J. Daunis, *Org. Mass. Spectrom.*, **11**, 680 (1976).
280. M. Israel and L. C. Jones, *J. Heterocycl. Chem.*, **6**, 735 (1969).
281. M. Israel and L. C. Jones, *J. Heterocycl. Chem.*, **10**, 201 (1973).
282. S. Carboni, A. DaSettimo, D. Bertini, P. L. Ferrarini, O. Livi, and I. Tonetti, *Farmaco, Ed. Sci.*, **30**, 237 (1975).
283. A. Nawojski and W. Nawrocka, *Rocz. Chem.*, **48**, 2275 (1974).
284. W. Bebenburg, Ger. Offen. 2,360,852; June 1974; *Chem. Abstr.*, **81**, 105591u (1974).
285. A. Nawojski, *Rocz. Chem.*, **43**, 573 (1969).
286. K. Winterfeld and M. Wildersohn, *Pharm. Acta. Helv.*, **45**, 323 (1970).
287. B. Brobanski and J. Stankiewicz, *Diss. Pharm. Pharmacol.*, **24**, 301 (1972).
288. N. S. Miroshnichenko and A. V. Stetsenko, *Ukr. Khim. Zh.*, **41**, 105 (1975).
289. A. Nawojski, *Rocz. Chem.*, **43**, 979 (1969).

290. M. Israel and L. C. Jones, *J. Heterocycl. Chem.*, **8**, 797 (1971).
291. Aust. Patent 325,054; October 1975; *Chem. Abstr.*, **84**, 44200d (1976).
292. M. Israel, L. C. Jones, and E. J. Modest, *Tetrahedron Lett.*, 4811 (1968).
293. W. H. Nyberg, C. W. Noell, and C. C. Cheng, *J. Heterocycl. Chem.*, **2**, 110 (1965).
294. M. Israel, S. K. Tinter, D. H. Trites, and E. J. Modest, *J. Heterocycl. Chem.*, **7**, 1029 (1970).
295. S. Fukushima, A. Ueno, K. Noro, K. Iwagaya, T. Noro, K. Morinaga, Y. Akahori, H. Ishihara, and Y. Saiki, *Yakugaku Zasshi*, **97**, 52 (1977).
296. Q. Q. Dang, R. Caujolle, and T. B. T. Dang, *C.R. Acad. Sci. Paris, Ser. C*, **274**, 885 (1972).
297. French Patent 1,497,861; September 1967; *Chem. Abstr.*, **69**, 96792h (1968).
298. S. Fukushima, A. Ueno, K. Noro, T. Noro, K. Morinaga, Y. Akahori, H. Ishihara, and Y. Saiki, *Yakugaku Zasshi*, **96**, 1453 (1976).
299. S. Wawzonek, *J. Org. Chem.*, **41**, 310 (1976).
300. P. H. Stahl and K. W. Merz, *Pharmazie* **22**, 630 (1967).
301. P. H. Stahl, R. Barchet, and K. W. Merz, *Arzneimittel-Forsch.*, **18**, 1214 (1968).
302. H. v. Dobeneck, A. Uhl, and L. Forster, *Justus Liebigs Ann. Chem.*, 476 (1976).
303. C. A. Lovelette and L. Long, Jr., *J. Org. Chem.*, **37**, 4124 (1972).
304. A. L. Linand and J. M. Hoch, *Arzneimittel-Forsch.*, **34**, 640 (1984).
305. Y. Kurasawa, J. Satoh, M. Ogura, Y. Okamato, and A. Takada, *Heterocycles*, **22**, (1984).
306. G. Roma, M. Di Braccio, M. Mazzei, and A. Ermili, *Farmaco, Ed. Sci.*, **39**, 477 (1984).
307. E. Wagner, *Pol. J. Chem.*, **56**, 131 (1982).
308. A. Nawojski, W. Nawrocka, and H. Liszkiewicz, *Pol. J. Pharmacol. Pharm.*, **35**, 531 (1983).
309. G. Roma, A. Balbi, A. Ermili, and E. Vigevani, *Farmaco, Ed. Sci.*, **38**, 546 (1983).
310. V. D. Orlov, N. N. Kolos, F. G. Yaremenko, and V. F. Lavrushin, *Khim. Geterotsikl. Soedin.*, **16**, 697 (1980).
311. P. C. Unangst, *J. Heterocycl. Chem.*, **18**, 1257 (1981).
312. L. Capuano and K. Gartner, *J. Heterocycl. Chem.*, **18**, 1341 (1981).
313. H. Voigt, *Z. Chem.*, **21**, 103 (1981).
314. G. Roma, M. Di Braccio, M. Mazzei, and A. Ermili, *Farmaco, Ed. Sci.*, **35**, 997 (1980).
315. T. Kato, N. Katagiri, and R. Sato, *J. Chem. Soc., Perkin Trans. I*, 525 (1979).
316. A. Tajana, R. Pennini and D. Nardi, *Farmaco, Ed, Sci.*, **35**, 181 (1980).
317. R. Grover and B. C. Joshi, *J. Indian Chem. Soc.*, **56**, 1220 (1979).
318. K. C. Joshi, V. N. Pathak, P. Arya, and P. Chand, *Pharmazie*, **34**, 718 (1979).
319. A. Ushirogochi, Y. Tominaga, Y. Matsuda, and G. Kobayashi, *Heterocycles*, **14**, 7 (1980).
320. A. Nawojski, W. Nawrocka, and H. Liszkiewicz, *Pol. J. Pharmacol. Pharm.*, **34**, 423 (1982).
321. G. Rackur and I. Hoffmann, U.S. Patent 4,305,952; December 1981.
322. G. Rackur and I. Hoffman, U.S. Patent 4,302,468; November 1981.
323. O. Bub, U.S. Patent 4,239,684; December 1980, (Knoll AG).

# Chapters V–IX

# [*e*]-Fused[1,4]Diazepines

Chapters V–IX are devoted to the synthesis and chemistry of [*e*]-fused [1,4]diazepines, represented by the general structure **1**.

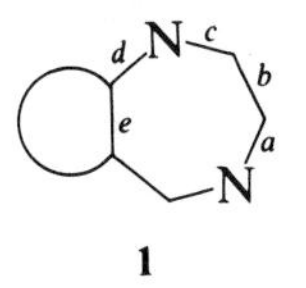


**1**

Because of the large volume of published material describing members of this general ring system, a division into several chapters seemed appropriate.

The 1,4-benzodiazepines constitute the largest class of compounds falling into this category. They will be discussed in Chapters V–VIII, separated according to the degree of saturation of the diazepine ring.

The related hetero[*e*][1,4]diazepines will be treated after the benzo-fused system and will be presented in alphabetical order in Chapter IX. This review includes previously reviewed material.

Earlier reviews were published by:


G. A. Archer and L. H. Sternbach, *Chem. Rev.*, **68**, 747 (1968).

S. J. Childress and M. I. Gluckman, *J. Pharm. Sci.*, **53**, 577 (1964).

F. D. Popp and A. C. Noble, in *Advances in Heterocyclic Chemistry*, Vol. 8, A. R. Katritzky and A. J. Boulton, Academic Press, New York and London, 1967, p. 21.

J. A. Moore and E. Mitchell, in *Heterocyclic Compounds*, Vol. 9, K. C. Elderfield, Ed., Wiley, New York, 1967, p. 227.

L. H. Sternbach, L. O. Randall, and S. Gustafson, in *Psychopharmacological Agents*, Vol. 1, M. Gordon, Ed., Academic Press, New York, 1964, p. 137.

L. H. Sternbach, L. O. Randall, R. Banzinger and H. Lehr, in *Medicinal Research Series*, Vol. 25, A. Burger, Ed., Dekker, New York, 1968, p. 237.

A. V. Bogatskii and S. A. Andronati, *Russ. Chem. Rev.*, **39**, 1064 (1970).

R. I. Fryer, *J. Heterocycl. Chem.*, **9**, 747 (1972).

L. H. Sternbach in *The Benzodiazepines*, S. Garattini, E. Mussini, and L. O. Randall, Eds., Raven Press, New York, 1973, p. 1.

L. O. Randall, W. Schallek, L. H. Sternbach, and R. Y. Ning, in *Psychopharmacological Agents*, Vol. 3, M. Gordon, Ed., Academic Press, New York, 1974, p. 175.

E. Schulte, *Dtsch. Apoth. Z.*, **115**, 1253 (1975).

L. H. Sternbach, in *Progress in Drug Research*, Vol. 22, E. Tucker, Ed., Birkhäuser Verlag, Basel and Stuttgart, 1978, p. 229.

L. H. Sternbach, *Angew. Chem., Int. Ed. Engl.*, **10**, 34 (1971).

H. Schütz, *Benzodiazepines, A Handbook*, Springer-Verlag, New York, 1982.

CHAPTER V

# 1,4-Benzodiazepines

**A. Walser**

*Chemical Research Department,*
*Hoffmann-La Roche Inc.,*
*Nutley, New Jersey*

and

**R. Ian Fryer**

*Department of Chemistry, Rutgers,*
*State University of New Jersey,*
*Newark, New Jersey*

# INTRODUCTION

This chapter reviews the synthesis and reactions of 1,4-benzodiazepines represented by the three tautomers **2a**–**2c**. Representatives of all three tautomers **2a**–**2c** have been reported in the literature and are discussed separately in Sections 1–3.

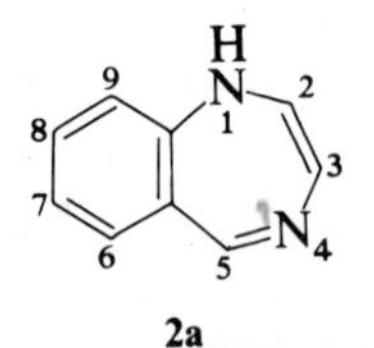

**2a**
1*H*-1,4-Benzodiazepine

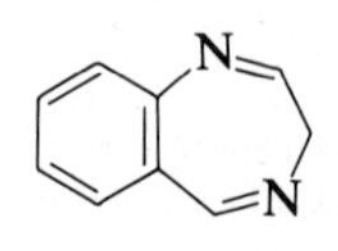

**2b**
3*H*-1,4-Benzodiazepine

**2c**
5*H*-1,4-Benzodiazepine

# 1. 1*H*-1,4-BENZODIAZEPINES

## 1.1. Synthesis

The parent ring system, **2a**, has not yet been reported. The 5-methoxy-8-chloro analog **4** (R = MeO) and several 5-amino derivatives **4** (R = $NHR_2$) were obtained[1a] in 15–50% yields by photolysis of the 4-azidoquinoline **3** in the presence of methoxide or an amine, respectively (Eq. 1). The enamine structure **4** (R = MeO) was assigned on the basis of proton- and $^{13}C$-nmr spectroscopy.[1b] The preliminary account of this work[1a] contains no analytical or spectroscopic data for the product **4**.

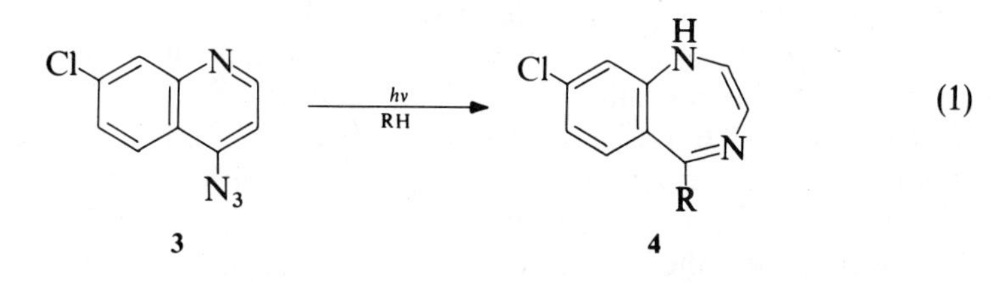


The trisubstituted derivatives **8** (X = H, F) were prepared by desulfurization of the 2-thioalkylbenzodiazepines **6** ($R_1$ = Me; $R_2$ = H; $R_3$ = Cl; $R_4$ = Me, Et; X = H, F) with Raney nickel in the presence of diethylamine[2,4] or, in better yield, by dehydration of the 2-hydroxy compounds **7** (X = H, F) with mesyl chloride and pyridine (Eq. 2).[2–4]

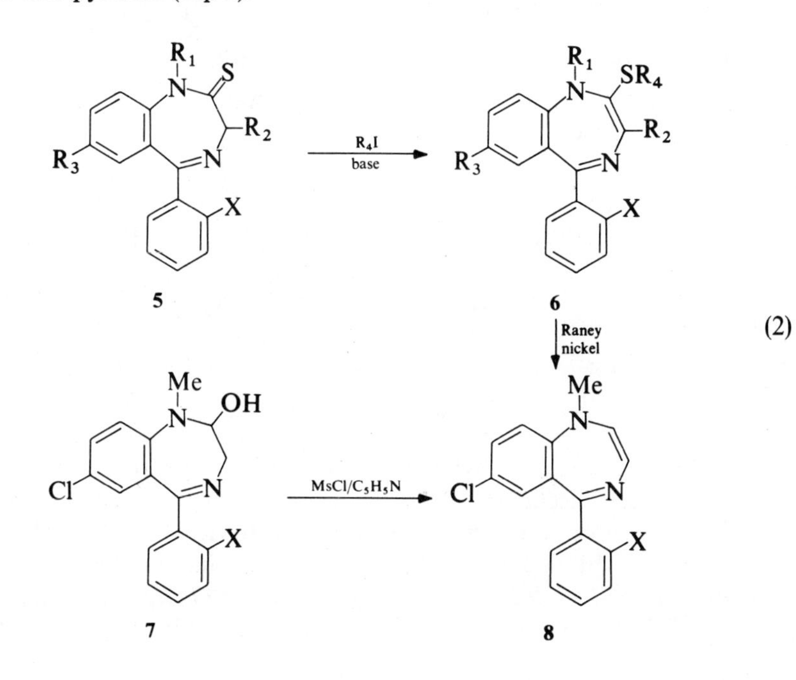

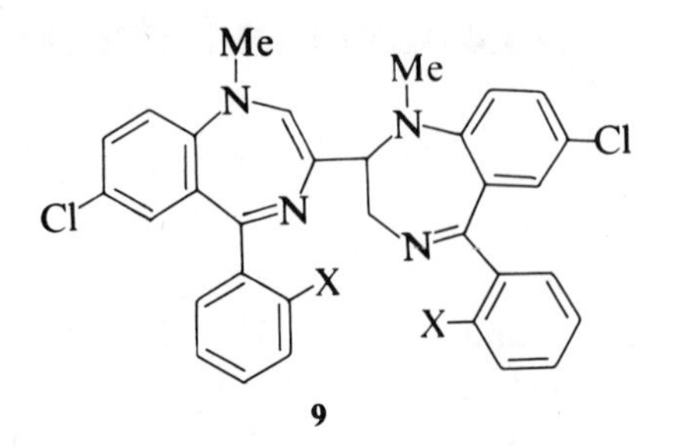


The latter reaction gave as by-products orange dimers to which structure **9** was assigned on the basis of spectral and analytical data.[5a] The 2-thioalkyl compounds **6** were prepared[2-4,6] by alkylation of the appropriate 2-thiones, **5**, the preparation of which is discussed in Chapter VI, covering dihydro-1,4-benzodiazepines. Oxidation of compounds **6** with *m*-chloroperbenzoic acid led to the corresponding sulfoxides[2,3,6] and sulfones.[2,3]

The 2-cyano-1*H*-1,4-benzodiazepine **11** was obtained in 66% yield by treatment of the corresponding 3*H* tautomer **10** with triethylamine in boiling tetrahydrofuran (Eq. 3).[7] In this case, the enamine is stabilized by the cyano group, rendering the 1*H* tautomer the thermodynamically preferred form. A similar stabilization was encountered with the 3-carboxamide **13**, which was synthesized[8] by cyclization of the formamidine **12** in boiling toluene (Eq. 4). Compound **12** was accessible by cleavage of nitrazepam with methylamine followed by reaction with dimethylformamide dimethylacetal.

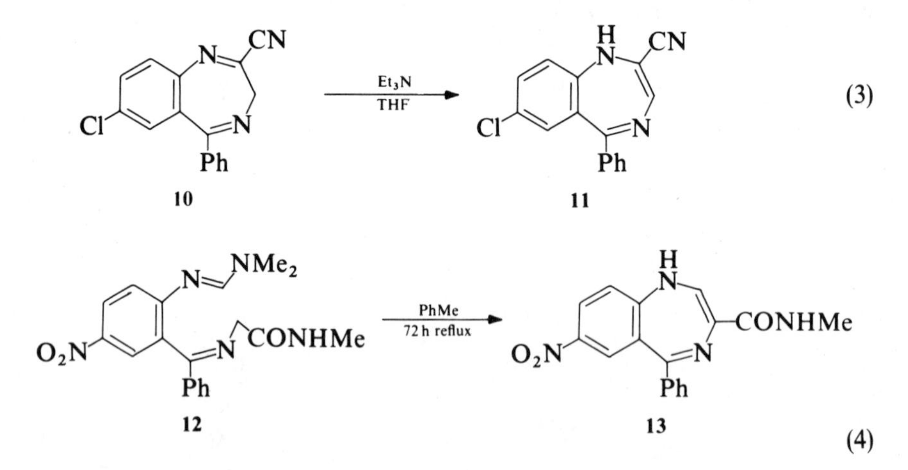


Acetylation of the 2-methylaminobenzodiazepine **14** with acetyl chloride or with a mixture of acetic anhydride and sodium acetate gave the diacetate **15** (Eq. 5).[9] Reaction of the benzodiazepin-2-ones **16** ($R_1$ = H; $R_2$ = H, COOEt; X = Cl, $NO_2$) with sodium methoxide and ethyl chloroformate in dimethylformamide at − 30°C led to the *N,O*-dicarboxylates **17** ($R_1$ = COOEt; $R_2$ = H, COOEt).[5b] The 1-methyl compound **16** ($R_1$ = Me; $R_2$ = COOEt; X = Cl)

under similar conditions yielded the *O*-acylated derivative **17** ($R_1$ = Me, $R_2$ = COOEt, X = Cl) (Eq. 6).[5b]

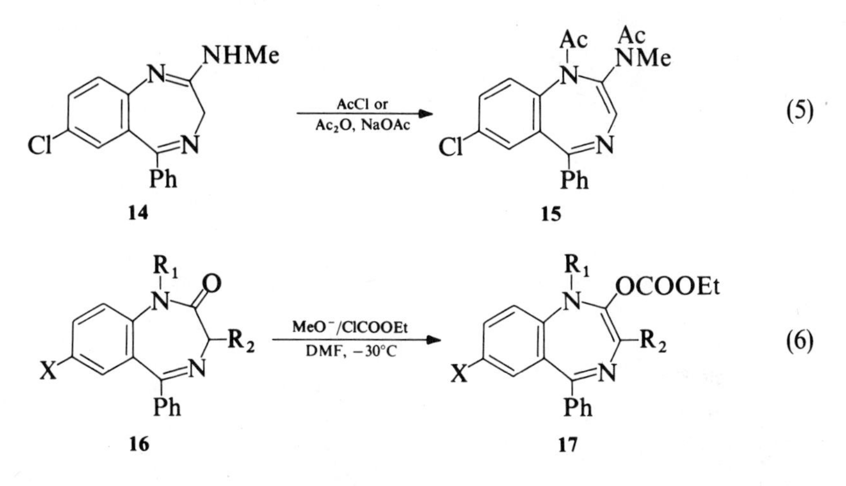


In the patent literature, structure **18** has been assigned to the product obtained from the reaction of chlordiazepoxide (4-oxide of **14**) with formaldehyde and hydrochloric acid.[10]

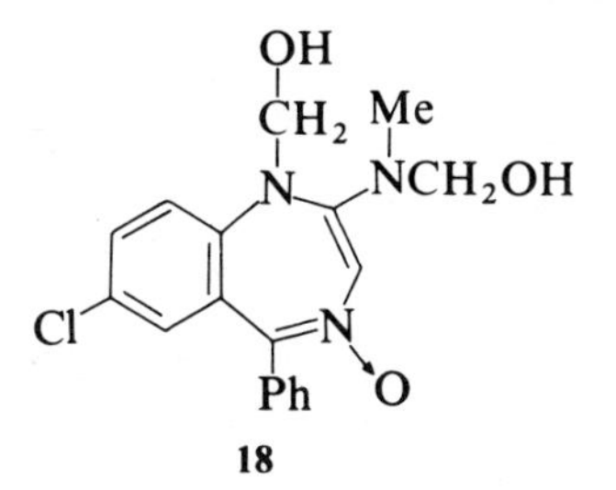


## 1.2. Reactions

### 1.2.1. *Reactions with Electrophiles*

The sulfones **19** were found to undergo rearrangement upon oxidation with *m*-chloroperbenzoic acid to give the benzodiazepin-2-ones **21** (Eq. 7).[2] This transformation was assumed to proceed via an initial formation of epoxide **20**, which then could rearrange with migration of the methylsulfonyl group as indicated.[2]

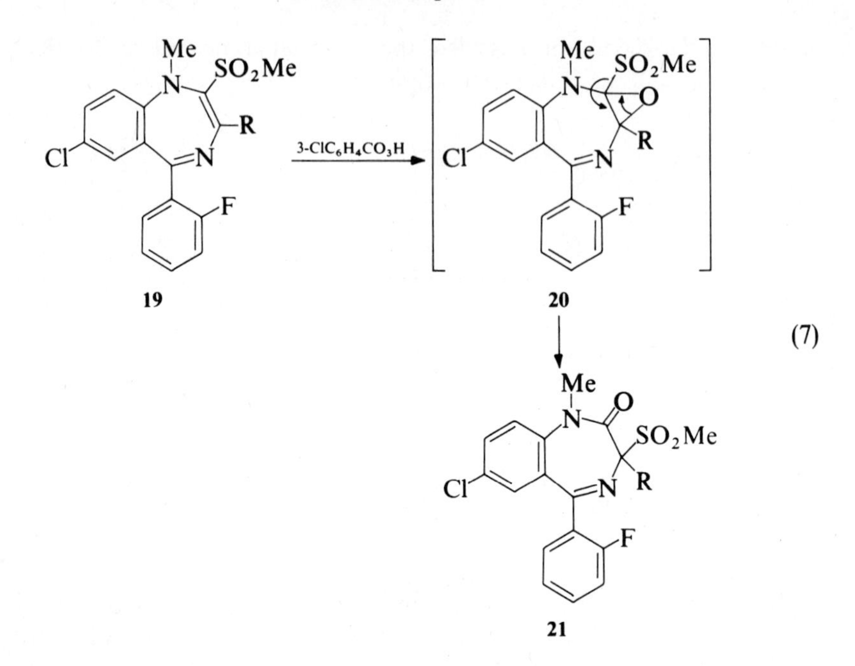


Oxidation of **13** with chromium trioxide in acetic acid led to the quinazoline-2-carboxamide **23**, most likely via the proposed intermediate **22** (Eq. 8).[8]

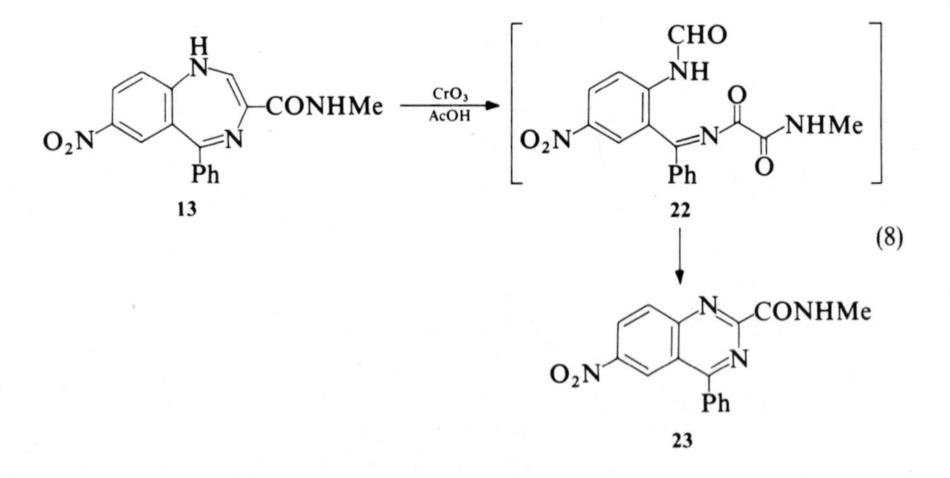


The 2-thioethyl compound **24** reacted with diethyl azodicarboxylate in boiling dioxane to form the adduct **25** (Eq. 9).[2] Dimethyl acetylene dicarboxylate under similar conditions added to **24** to yield the dibenzodiazepine **28**.[2] Cyclobutadienes were proposed as possible intermediates,[2] however, it is equally plausible that the reagent adds first to the 3-position of **24** to form

**26**. Diels–Alder reaction with a second molecule of acetylene dicarboxylate would then lead to **27**, which could aromatize to **28** by loss of ethanethiol.

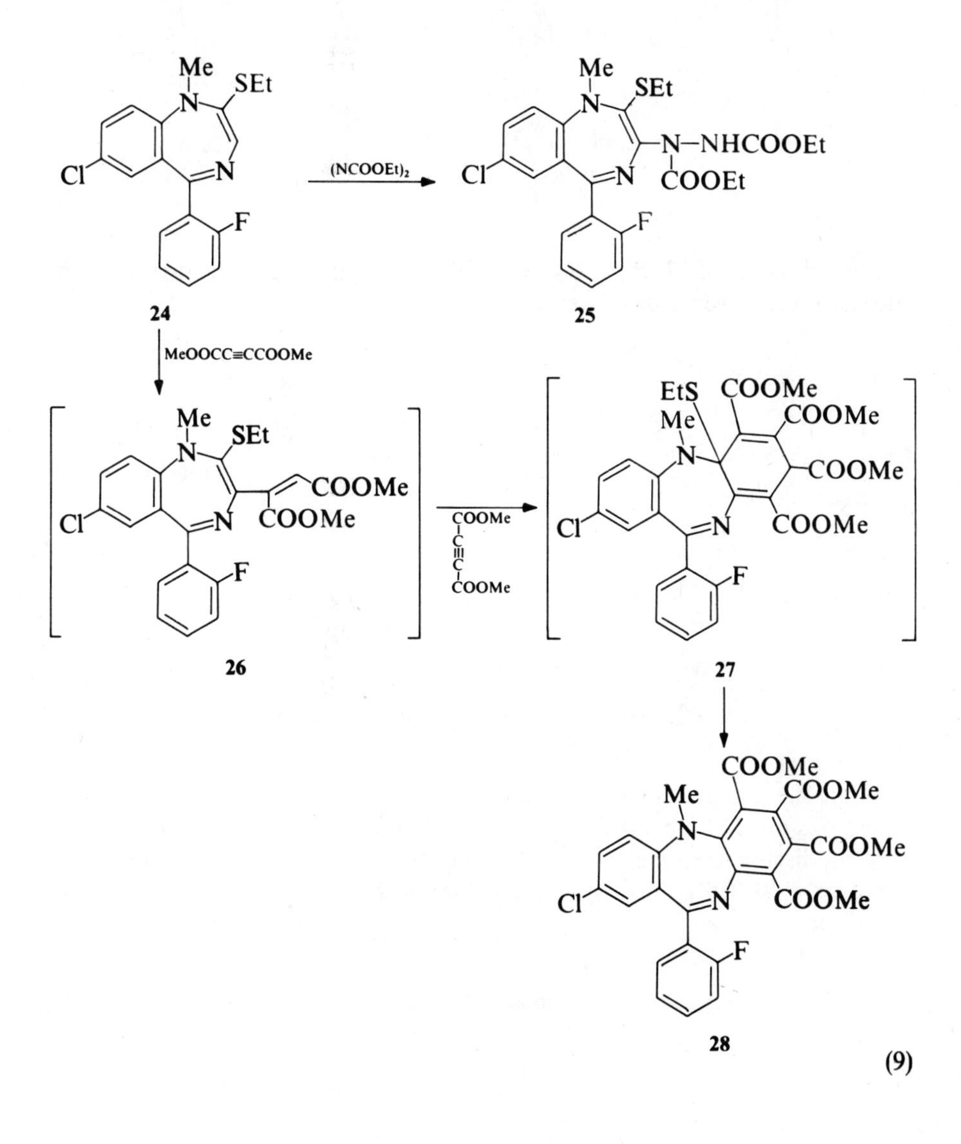


(9)

### *1.2.2. Reactions with Nucleophiles*

Acid hydrolysis of the 3-carboxamide **13** gave the 2-substituted indole **30** (R = COCONHMe) as the major product plus a minor amount of **30** (R = H).[8] This hydrolytic ring contraction was formulated to proceed via the benzophenone intermediate **29** (Eq. 10).[8]

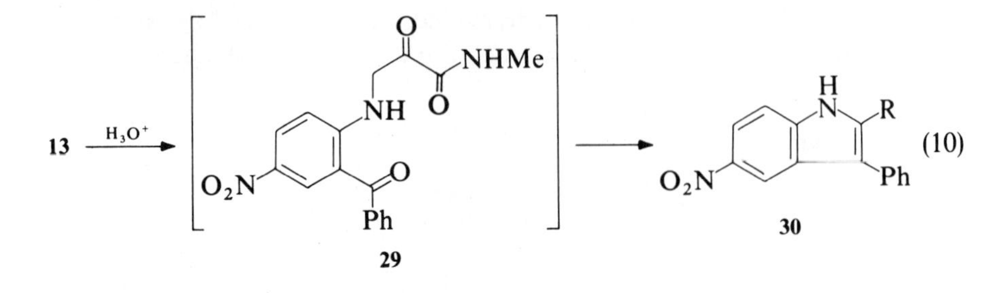


Mild cleavage of **17** ($R_1 = R_2 =$ COOEt, X = Cl) with methoxide in methanol gave the 1,3-dicarboxylate **31** (Eq. 11).[5b]

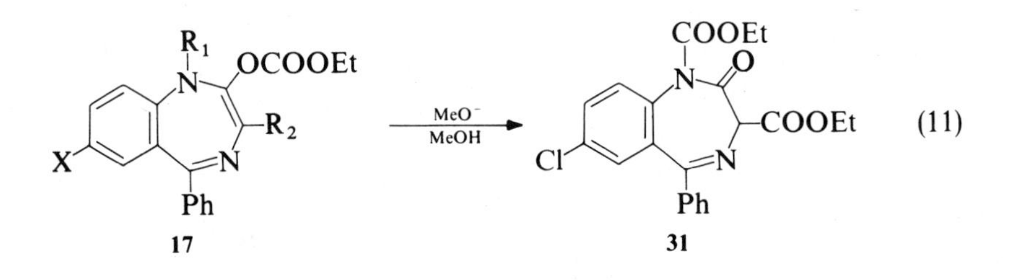


## 2. 3*H*-1,4-BENZODIAZEPINES

While the parent compound is still unknown, the abundance of substituted derivatives of 3*H*-1,4-benzodiazepines reported in the literature made it necessary to discuss this material in several subsections. The compounds are, therefore, classified according to the nature of the substituent attached to the 2-position.

### 2.1. 2-Hydrogen- and 2-Carbon-Substituted 3*H*-1,4-Benzodiazepines

#### *2.1.1. Synthesis*

The 5-phenyl-7-chloro derivative **33** (R = H) and its 4-oxide appear to be the only 2-unsubstituted 3*H*-1,4-benzodiazepines described in the literature.[7,11] These compounds were prepared by oxidation of the corresponding 2,3-dihydro derivatives **32** (R = H) with ordinary manganese dioxide in boiling benzene. Good yields of **33** (R = H) were found to be contingent on prior azeotropic removal of water from the manganese dioxide and by addition of a small amount of acetic acid. The dimer **34** was isolated as a by-product (Eq. 12).

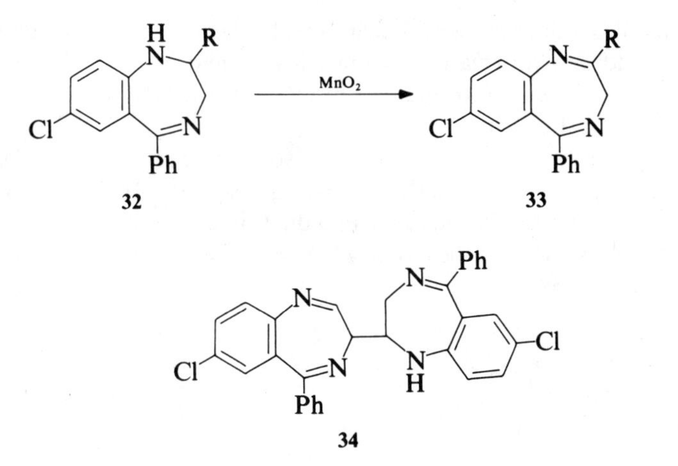


The 2-methyl analog **33** (R = Me)[12,13] and its corresponding 4-oxide[14] were also accessible by oxidation of the appropriate 2,3-dihydro derivatives with manganese dioxide. The same procedure was used for the preparation of the 2-cyanobenzodiazepine **33** (R = CN) and the 2-carboxamides **33** (R = $CONH_2$)[7,15] and **33** (R = $CONMe_2$).[7b]

The 2-methyl 4-oxides **36** could be obtained in high yields by the ring expansion of 2-chloromethyl-2-methyl-1,2-dihydroquinazoline 3-oxides **35** using sodium hydroxide in boiling ethanol. This method was applied for the synthesis of 5-phenyl derivatives **36** ($R_1$ = Ph; $R_2$ = H, Cl, $NO_2$; $R_3$ = H),[12-14,16,17] 5-(2-fluorophenyl) compounds,[18] and a 5-methyl analog **36** ($R_1$ = Me, $R_2$ = $R_3$ = MeO).[18]

This ring expansion was postulated to proceed[14] by initial abstraction of the proton in the 1-position followed by ring opening to the oxime anion **37**. This intermediate would then undergo an intramolecular alkylation on nitrogen to form the seven-membered ring (Eq. 13).

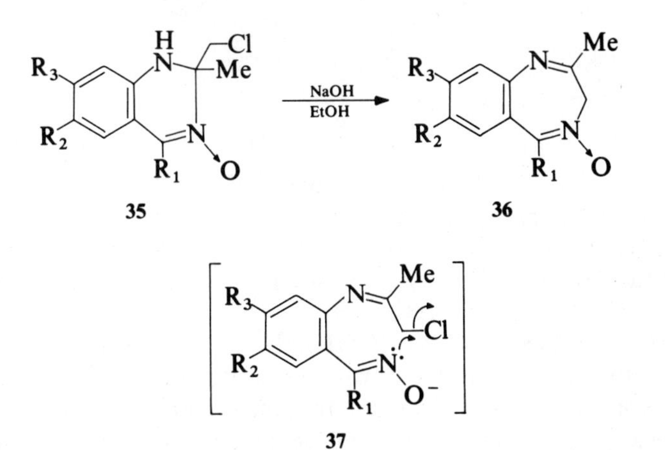

2-Methyl benzodiazepines and their 4-oxides can also be obtained by alkaline hydrolysis and decarboxylation of the 2-acetic acid ester **42** ($R_1$ = COOMe) or by cleavage of the corresponding *t*-butyl ester **42** ($R_1$ = COO*t*-Bu) with trifluoroacetic acid (Eq. 15).[5]

The 2-cyanobenzodiazepine **39** was isolated from the reaction of the 2-nitromethylene derivative **38** (X = F) with phosphorus trichloride in pyridine (Eq. 14).[19] This transformation involves reduction of the nitro group to a nitroso derivative tautomeric with the oxime **41** (X = F). Dehydration of **41** would then lead to the nitrile **39**.

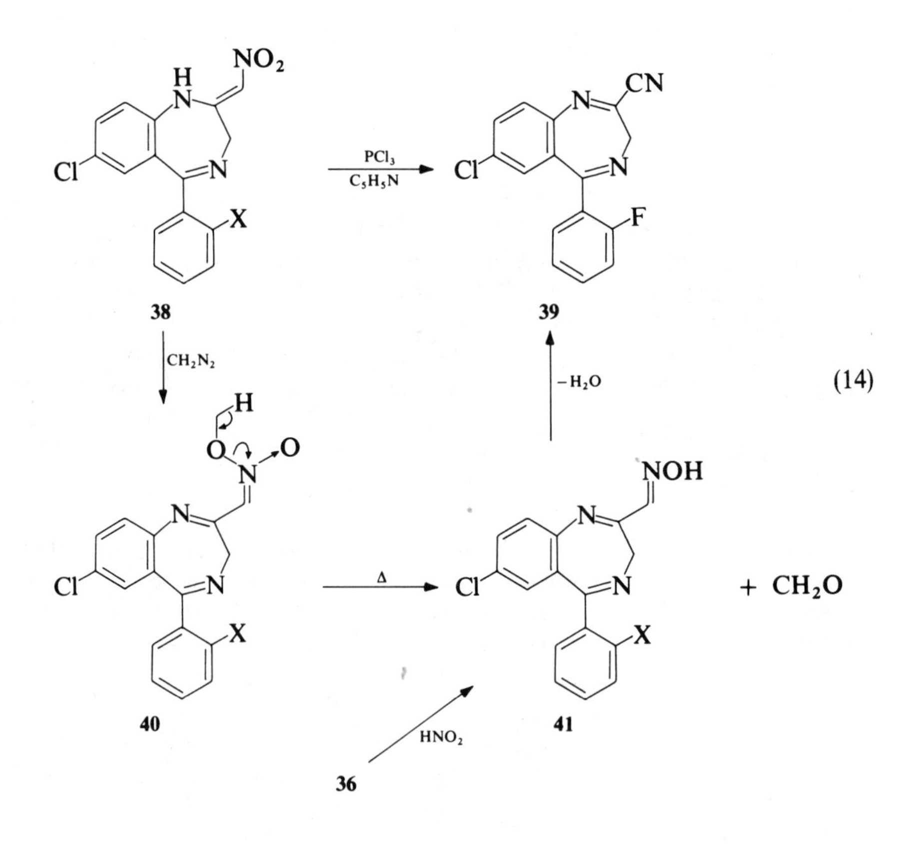


The oximes **41** (X = F) and the 4-oxide of **41** (X = H) were formed by methylation of the appropriate nitromethylene derivative **38** with diazomethane and subsequent thermolysis in boiling benzene or toluene.[19] The 4-oxide of **41** (X = F) was also obtained by nitrosation of the 2-methyl benzodiazepine **36** ($R_1$ = 2-$FC_6H_4$, $R_2$ = Cl, $R_3$ = H).[18]

The related oximes **43** ($R_1$ = $COOR_2$,[20–22] Ac,[20] $CONR_4R_5$,[5] 2-pyridyl,[5] $SO_2Me$,[5] $SO_2NMe_2$[5]) and the nitroximes **43** ($R_1$ = $NO_2$)[23,24] were analogously prepared in high yields by nitrosation of the appropriate 2-methylene benzodiazepines **42** with sodium nitrite in glacial acetic acid

(Eq. 15). The nitro group in the nitroximes **43** ($R_1 = NO_2$) was susceptible to displacement with nucleophiles, and these compounds were converted to amidoximes **45** ($R_2 = NR_4R_5$, $R_3 = H$) by reaction with amines at room temperature.[24] Methylthiolate displaced the nitro group to give the thiomethyl oxime **45** ($R_2$ = MeS, $R_3$ = H).[24] Methylation of the nitroxime **43** ($R_1 = NO_2$) with diazomethane yielded **44** ($R_3$ = Me).[23,24] When this methylation was carried out in tetrahydrofuran, the 4-methoxybutyl derivatives **44** [$R_3 = MeO(CH_2)_3$] were formed as by-products,[24e] apparently by generation of a tetrahydrofuran oxonium ion. Compound **44** ($R_3$ = Me) reacted also with nucleophiles such as amines[24] and alkoxides[24e] to give the *N*-methoxyamidines **45** ($R_2 = NR_4R_5$, $R_3$ = Me) and the *N*-methoxyimidates **45** ($R_2$ = MeO, $R_3$ = Me).

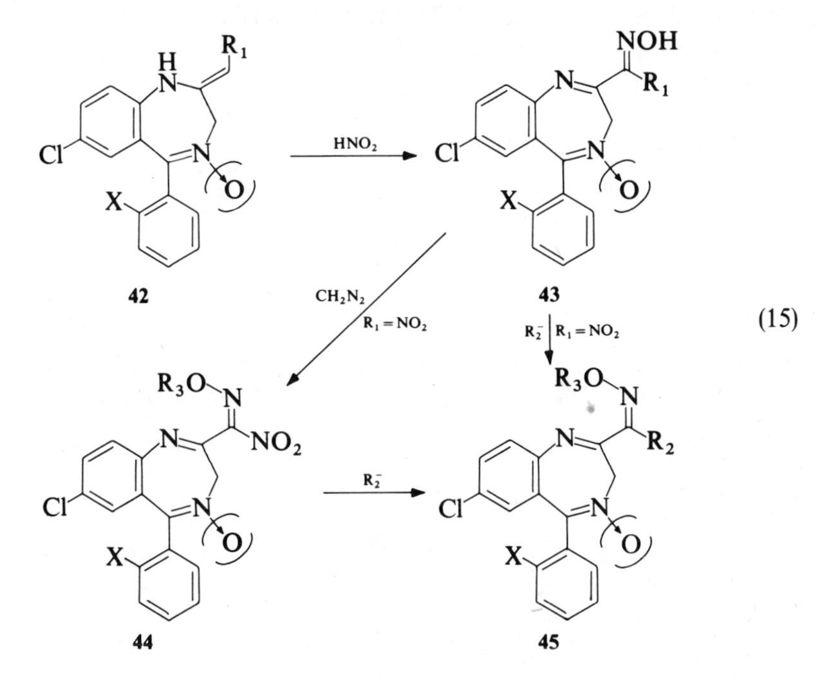

(15)

The 8-chloro-2-phenyl benzodiazepine **47** (R = MeO) resulted from the reaction of the 5-one **46** with Meerwein reagent.[25] The imino ether was converted to the 5-hydrazino compound **47** ($R = NHNH_2$) by treatment with hydrazine at room temperature (Eq. 16).[25]

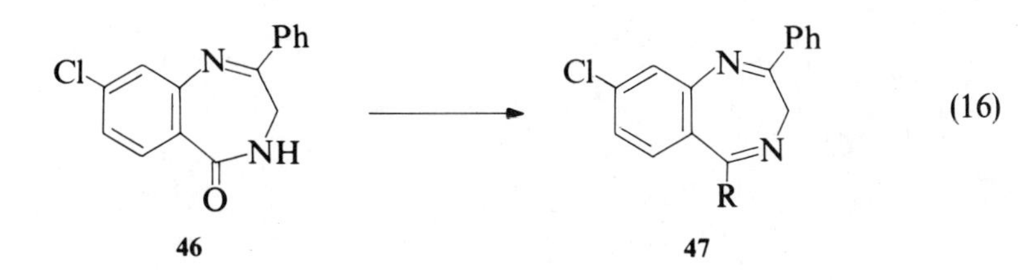

(16)

The 4-oxide **49** was prepared by addition of phenylmagnesium bromide to the lactam **48** (Eq. 17).[26]

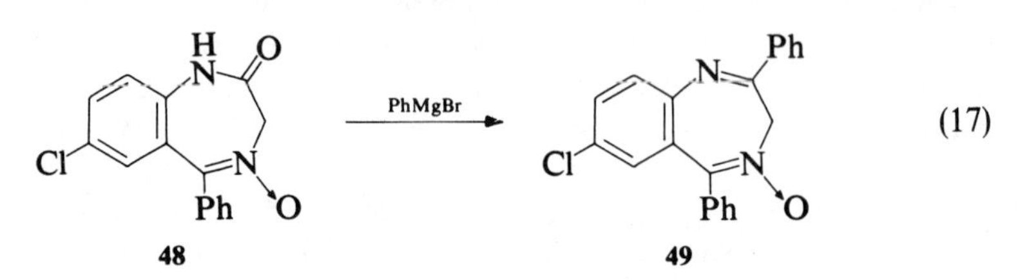


Formation of a C—C bond at the 2-position has been achieved by reaction of the phosphoryl imidate **50** with the anion of acetaminomalonic ester yielding the 3*H*-benzodiazepines **51** (Eq. 18). This synthesis was described for R = 2-pyridyl, 2-halophenyl and X = Cl, Br, CN.[21]

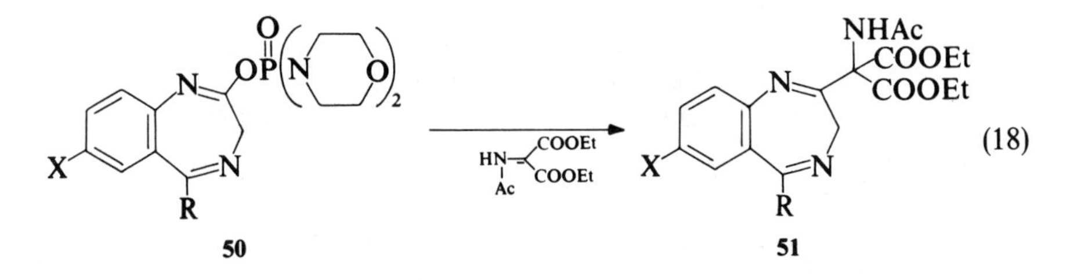


Structure **53** was assigned[27] to the product isolated in 5% yield from the reaction of the 2-amino derivative **52** with acetylacetone (Eq. 19).

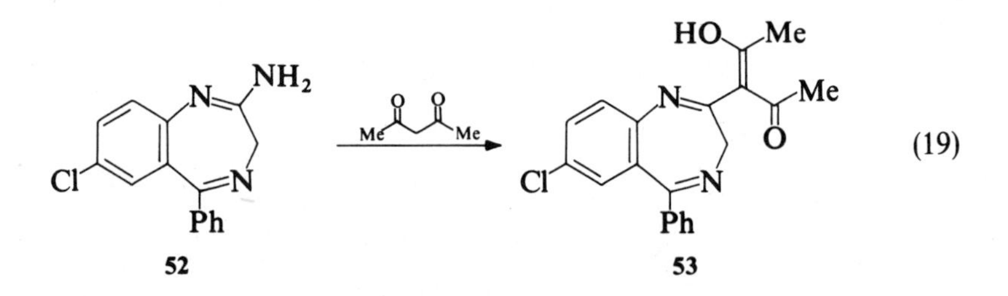


3*H*-1,4-Benzodiazepines can also be prepared by the isomerization of the 5*H*-tautomers in the presence of methoxide,[18] as demonstrated by the conversion of **54** to **55** (Eq. 20).

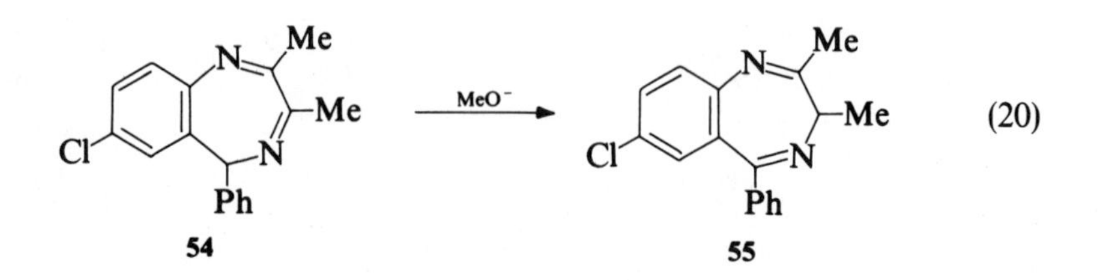

The 2-substituted 3*H*-1,4-benzodiazepine **57** resulted from an intramolecular alkylation on carbon of the 3-chloropropionyl derivative **56**, using triethylamine in boiling dimethylformamide (Eq. 21).[28]

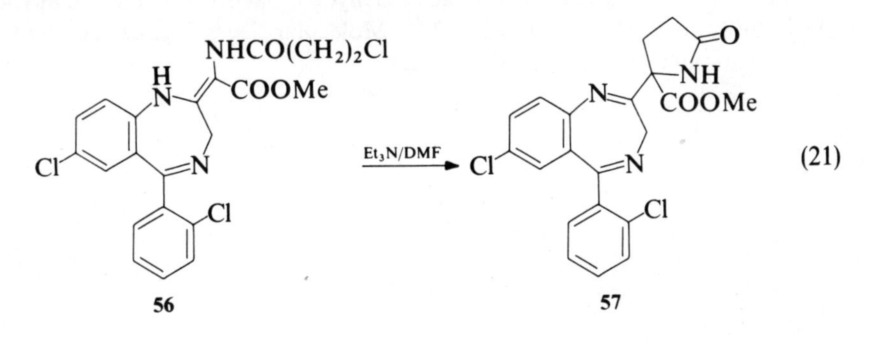


## 2.1.2. Reactions

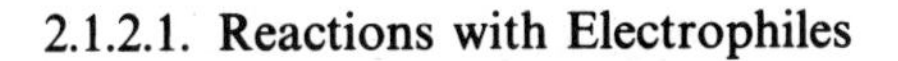

### 2.1.2.1. Reactions with Electrophiles

Treatment of the 4-oxides **58** (X = H, F) with *t*-butyl hypochlorite led to the 2-chloromethyl derivatives **59** (Eq. 22).[18]

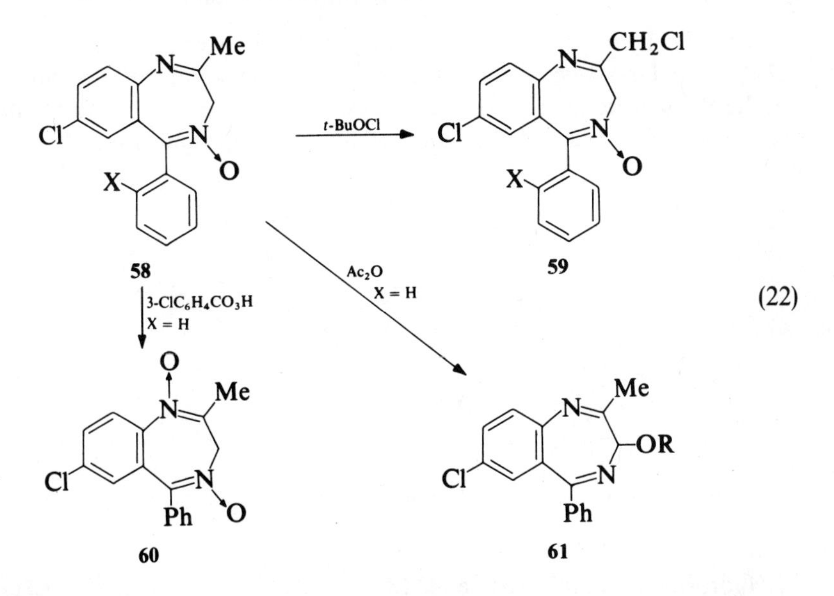


Oxidation of **58** (X = H) with 3-chloroperbenzoic acid gave the 1,4-dioxide **60**.[18] The nitrosation of **58** (X = F) to the 2-carboxaldoxime was mentioned above.

Methylation of the nitroximes **43** ($R_1 = NO_2$) to the *O*-methyl oximes **44** ($R_3$ = Me) has also been discussed.

The 3-acetoxy compound **61** (R = Ac) was prepared by rearrangement of the nitrone **58** (X = H) in acetic anhydride. The acetate was hydrolyzed to the alcohol **61** (R = H) using aqueous hydroxide in tetrahydrofuran.[17]

The amidoxime **62** reacted with acetaldehyde in methylene chloride at room temperature to give the oxadiazole **63**.[24h] Molecular sieves were used to bind the water eliminated in this reaction (Eq. 23).

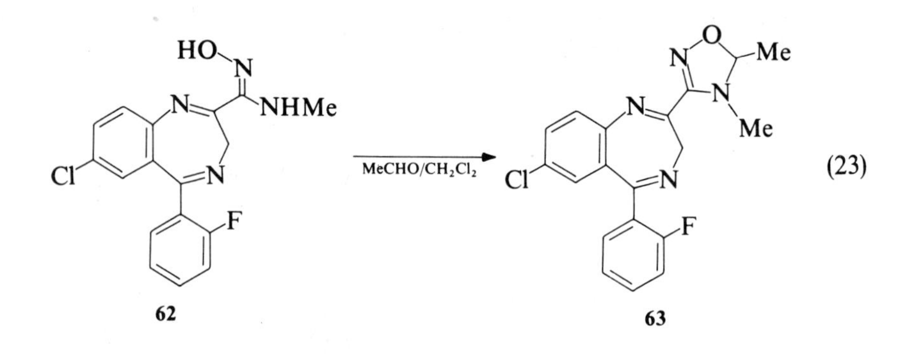


### 2.1.2.2. Reactions with Nucleophiles

Catalytic hydrogenation over Raney nickel converted **58** (X = H) to the 2,3-dihydro-1*H*-1,4-benzodiazepine **32** (R = Me) in which both the nitrone and the 1,2-imine were reduced (Eq. 24). Selective reduction of the nitrone function was reported for the 7-deschloro compound by employing the same reagent.[12,13,17]

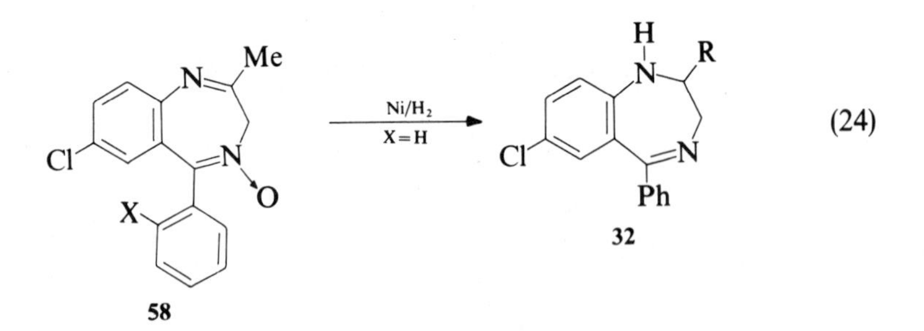


Hydrogenation of the oximes **43** ($R_1$ = COOR; Ac) over Raney nickel led to the enediamines **64** (Eq. 25).[20,21] With the exception of **64** ($R_1$ = COOMe, X = Cl),[20] these enediamines were not characterized but were converted further to tricyclic systems such as imidazo[1,5-*a*][1,4]benzodiazepines. In the presence of methanolic ammonia, this catalytic hydrogenation proceeded with retention of the nitrone functionality.[5b]

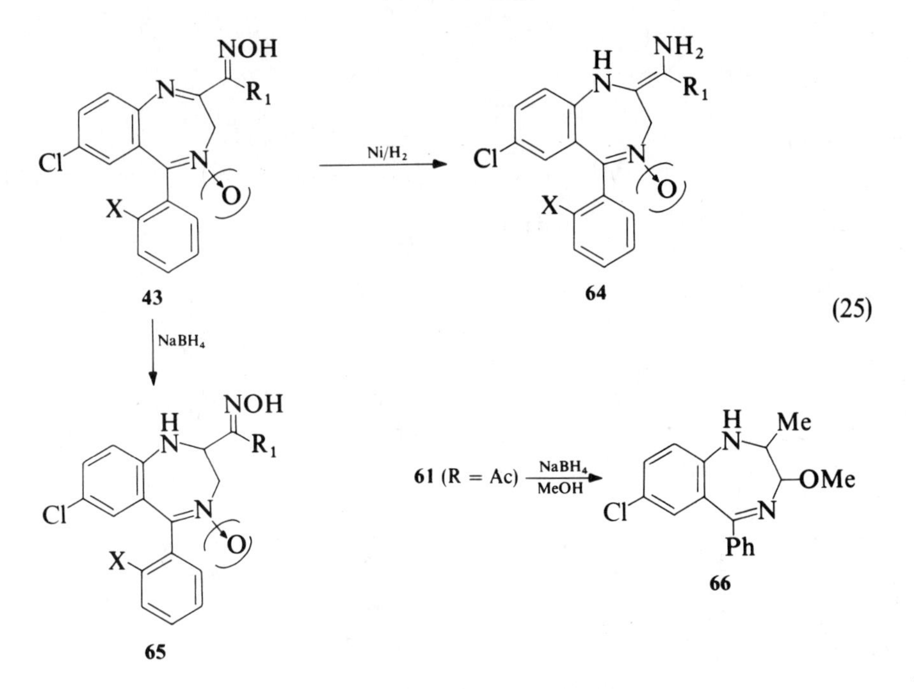


Selective reductions of the 1,2-imine were carried out by using borohydride in ethanol both for the 2-methyl-3*H*-1,4-benzodiazepines **58**[12-14,17] and for oximes of structure **43**. Thus compounds **43** [($R_1$ = H),[19] ($R_1$ = COO*t*-Bu, $CONMe_2$, 2-pyridyl),[5b] ($R_1$ = $NR_4R_5$),[23,24] ($R_1$ = MeS)[24e]] and the corresponding oxime *O*-methyl ethers [($R_1$ = morpholino)[23] and ($R_1$ = MeO)[24e]] were reduced to the corresponding 1,2-dihydro derivatives **65**. Under the same conditions the nitroximes **43** ($R_1$ = $NO_2$) and the corresponding *O*-methyl ethers underwent initial displacement of the nitro group by hydride followed by a reduction of the 1,2-imine to yield **65** ($R_1$ = H) and the corresponding *O*-methyl derivative, respectively.[24]

Reduction of the 3-acetoxy compound **61** (R = Ac) with sodium borohydride in methanol yielded the 3-methoxy-2,3-dihydrobenzodiazepine **66** of undetermined stereochemistry.[12,13,17]

The benzodiazepine **67** reacted with a variety of nucleophiles to give the 2-substituted dihydro compounds **32** (Eq. 26). Thus, there exist reports of addition of water,[29] methanol,[7,11] hydrogen cyanide,[7,11,15] isopropylamine,[11] diethylamine,[11] aniline,[11] piperidine,[6,11] 2-mercaptoethanol,[6,30,31] and benzylthiol.[11] Addition of hydrogen sulfide yielded the 2,5-bridged sulfide **68**,[7,30] and mercaptoacetic acid formed the cycloaddition products **69**[7,30] and **70**.[7]

The cyano group in **10** was readily displaced by other nucleophiles.[7] Refluxing in methanol led to the 2-methoxy derivative **33** (R = MeO), while reaction with pyrrolidine at room temperature gave **33** (R = pyrrolidino).[7] Condensation with acetylhydrazine in boiling butanol led to the triazolobenzodiazepine **71** in 60% yield (Eq. 27).[7,32]

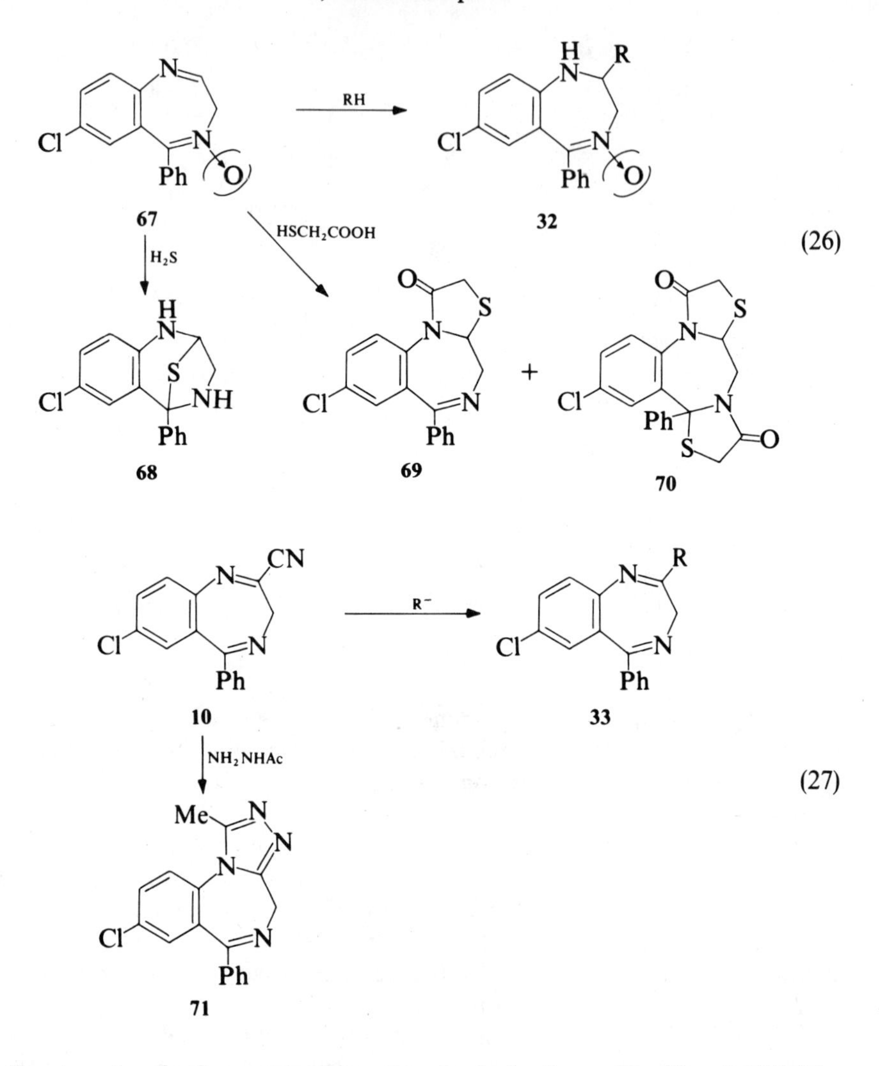


Treatment of the acetaminomalonyl derivatives **51** (R = 2-$ClC_6H_4$, 2-$FC_6H_4$; X = Cl) with sodium ethoxide in ethanol yielded the decarboxylated compounds **72** (X = Cl, F) (Eq. 28).[21]

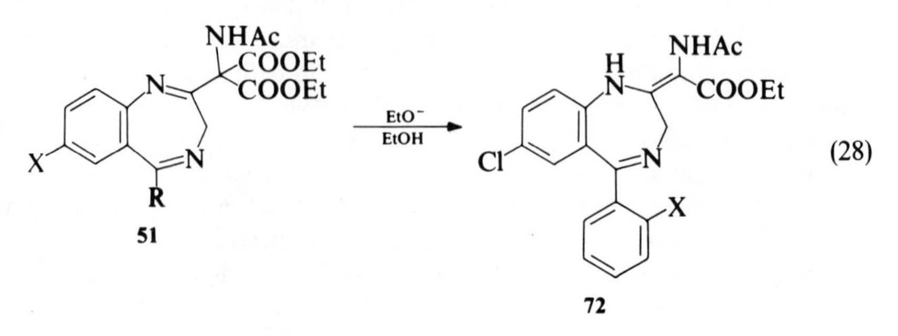

Refluxing ethanolic hydrogen chloride converted the 3*H*-benzodiazepine **67** to the 3-aminoquinoline **73**.[29] A similar ring contraction was observed with the 2-substituted compound **53**, which led to the pyrroloquinoline **75**.[27] The aziridine **74** was postulated as an intermediate in this ring contraction (Eq. 29).

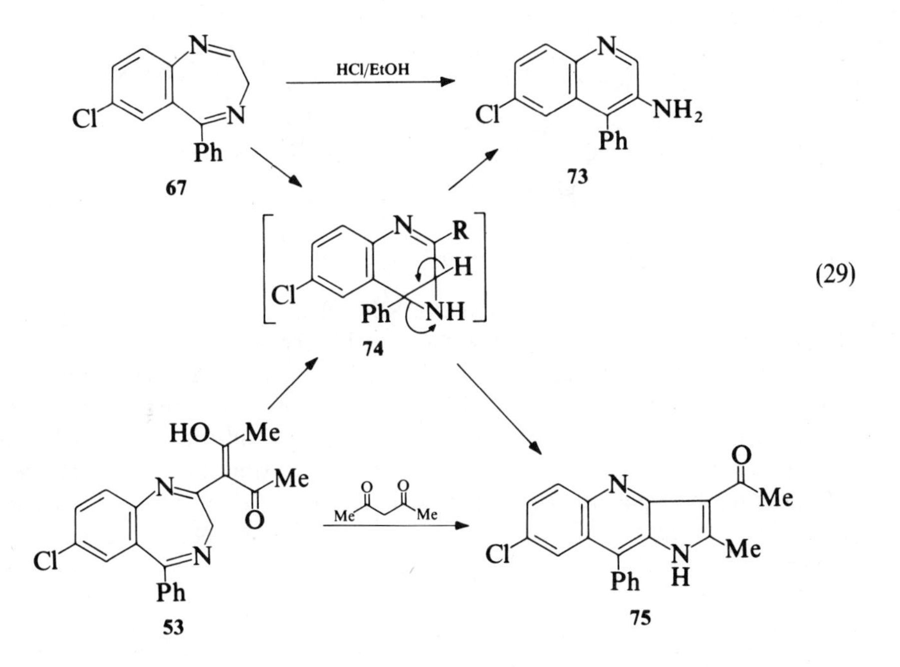


## 2.2. 2-Amino-substituted 3*H*-1,4-Benzodiazepines

### *2.2.1. Synthesis*

#### 2.2.1.1. Synthesis by Ring Expansion

Following Sternbach's pioneering discovery of chlordiazepoxide (**79**) by the reaction of 6-chloro-2-chloromethyl-5-phenylquinazoline-3-oxide (**76**) with methylamine,[33] this approach to the synthesis of 2-amino-1,4-benzodiazepines was applied for the preparation of a great variety of compounds. The initial step of the ring enlargement is now thought to involve the addition of the nucleophile (e.g., methylamine) at the electrophilic 2-position of the quinazoline 3-oxide to form the intermediate adduct **77**. Cleavage of the 2,3-bond as indicated would then lead to the oxime anion **78**, which is set up for an intramolecular alkylation on nitrogen to give nitrone **79** (Eq. 30).

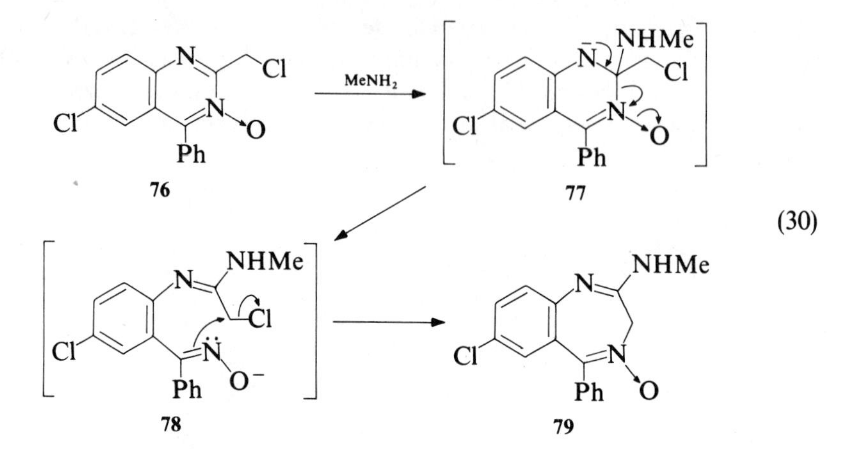


(30)

It has been shown that the reaction of a quinazoline 3-oxide of general structure **80** with a nucleophile may lead to either the ring expansion product **81** or the displacement product **82** (Eq. 31). The course of the reaction, hence the structure of the product, depend on both the nucleophile and the substituents on the quinazoline 3-oxide. Thus, electron-withdrawing substituents attached at the 6-position of the quinazoline 3-oxide enhance the ring enlargement reaction by rendering the 2-position more electrophilic. An electron-releasing group such as methyl at the 6- or 8-positions made the displacement reaction more competitive.[34,35]

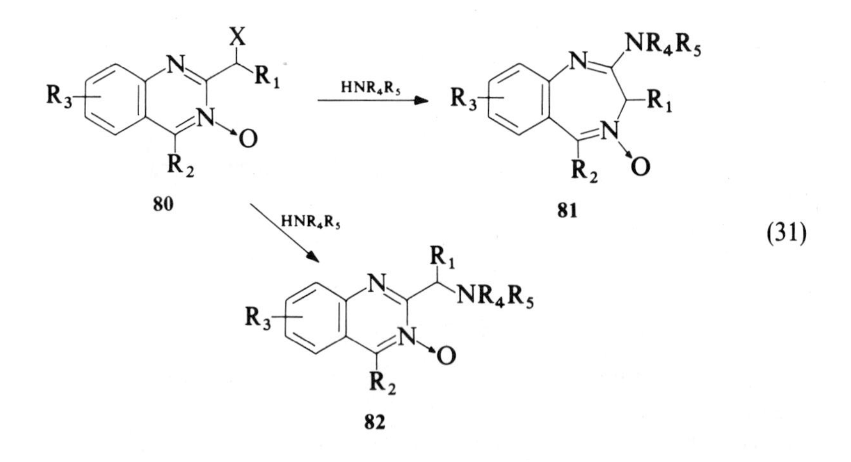


(31)

Successful ring expansions were carried out with ammonia and a large variety of primary aliphatic amines. The quinazoline 3-oxides used were most typically substituted by X = Cl, Br, $R_1$ = H, Me, Et, and $R_2$ = Ph[33–46] or substituted Ph,[34,41,44,46–51] $R_2$ = H,[52] Me,[53,54] cycloalkyl,[55] benzocycloheptyl,[56] 2-thienyl,[41] 2-pyridyl,[57] pentadeuterophenyl.[58] The substituent $R_3$ on the benzene moiety of

**80** was chosen from halogen,[33,34] nitro,[38] cyano, carbomethoxy,[37] trifluoromethyl,[36,39,42] methylthio,[40] and others. Although the leaving group X employed has been mostly chloride or bromide, the methanesulfonyloxy moiety was reported to be equivalent in this ring expansion reaction.[59] Mixtures of products resulting from both ring expansion and displacement of chloride were observed when **80** (X = Cl) was treated with other primary amines such as allylamine,[33] 2-methoxyethylamine,[33] 2-ethanolamine,[33] propargylamine,[43] 2,2-dimethoxyethylamine,[43] and with secondary amines such as dimethylamine[60] and pyrrolidine.[60-62] Treatment with piperazine was reported to give only the corresponding displacement product.[33]

Guanidine converted **80** ($R_1$ = H, $R_2$ = Ph, $R_3$ = 6-Cl, X = Cl) to the corresponding benzodiazepine **81** [$R_4$ = H, $R_5$ = $C(NH)NH_2$].[41] A similar ring expansion with hydrazine was reported[63] to lead to 2-hydrazinobenzodiazepines, but these structures were later found to be incorrect.

Chlordiazepoxide **79** was also obtained by treatment of the 1,2-dihydro-2-dichloromethylquinazoline 3-oxide **83** with methylamine.[17,64] It is likely that elimination of hydrogen chloride from **83** generated the 2-chloromethylquinazoline 3-oxide **76**, which would then have undergone the usual ring expansion in situ (Eq. 32).

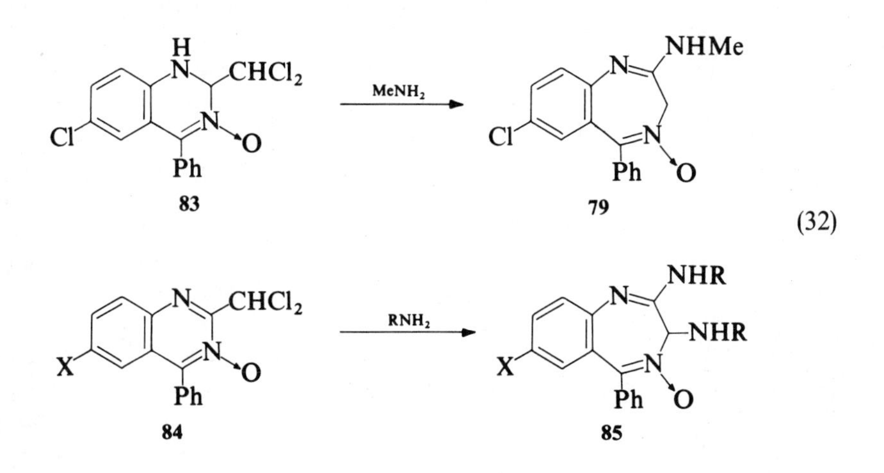


The 2-dichloromethylquinazolines **84** (X = Cl, $NO_2$, $CF_3$) were reported to react with a variety of primary amines to give the 3-aminobenzodiazepines **85**.[65] The structures of compounds **85** were not verified.

### 2.2.1.2. Synthesis from 2-Thio Compounds

Many 2-amino-3*H*-1,4-benzodiazepines **88** were prepared by reaction of the 2-thiones **86** or the 2-methylthio compounds **87** with amines. This procedure was extensively utilized for the general synthesis of amidines (Eq. 33).

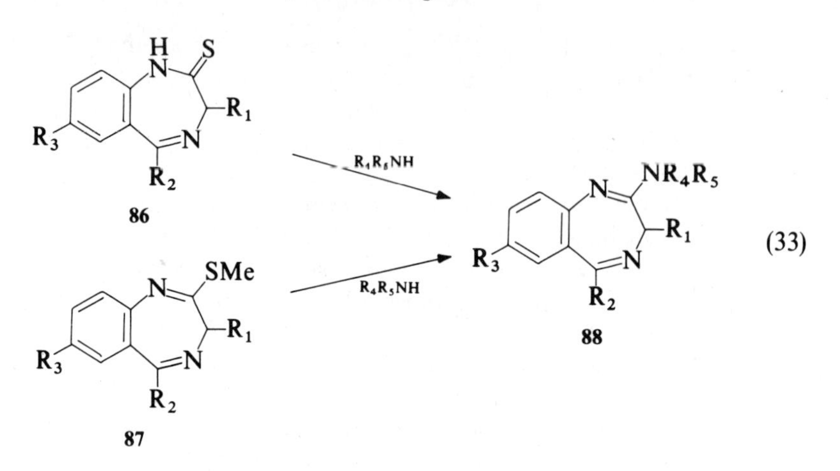


The following amines were successfully employed in this reaction: ammonia,[66–70] alkylamines,[70–72] dialkylamines,[71,72b] pyrrolidine,[70] piperidine,[71,72] 2-diethylaminoethylamine,[71,72b] 2-morpholinoethylamine,[71,72b] 2-methoxyethylamine, propargylamine,[74,75] 3-methoxypropylamine,[73] 2,2-dimethoxyethylamine,[74–77] glycine,[78–80] 1-amino-4-piperidino-2-butyne,[81] 1-amino-4-hydroxy-2-butyne,[82a] and 2-substituted 2-aminomethyl-1,3-dioxolanes.[82] These transformations were generally carried out with **86** or **87** ($R_1$ = H, Me; $R_2$ = Ph, substituted Ph, or 2-pyridyl;[76] $R_3$ = H, halogen, or nitro).

### 2.2.1.3. Other Syntheses

2-Amino-3*H*-1,4-benzodiazepines **88** and their 4-oxides were prepared in good yields by amine exchange under acid catalysis in boiling ethanol[83] or in dimethyl sulfoxide.[75,81] This displacement of ammonia with other amines appears to be quite general and was carried out with a variety of primary amines[83,84] including propargylamine,[75] 2,2-dimethoxyethylamine,[74] and 1-amino-4-diethylamino-2-butyne.[81]

The reaction of the lactams **89** with titanium tetrachloride and ammonia or a primary or secondary amine provided another general access to compounds **88** including their 4-oxides.[85] According to the more recent patent literature, this method was applied to prepare 2-(1-imidazolyl)-3*H*-1,4-benzodiazepines.[86] Conversion of the lactams **89** to phosphorylimidates **90** by reaction of the lactam anion with a phosphoryl halide followed by displacement of the phosphate by ammonia or an amine was an alternate approach to compounds **88** (Eq. 34). The crystalline dimorpholinochlorophosphate ($R_6$ = morpholino) received particular attention because it led to crystalline phosphorimidates **90** ($R_6$ = morpholino).[70,36–88] The compounds **90** do not have to be isolated but can be further reacted in situ, as demonstrated by the conversion of the lactam **89** ($R_1$ = H, $R_2$ = Ph, $R_3$ = Cl) to the 2-methylaminobenzodiazepine by means of diphenylchlorophosphate ($R_6$ = PhO)[87] or diethylchlorophosphate ($R_6$ = EtO).[5a] More recently the cyclic chlorophosphate [$(R_6)_2 = OCH_2CH_2O$]

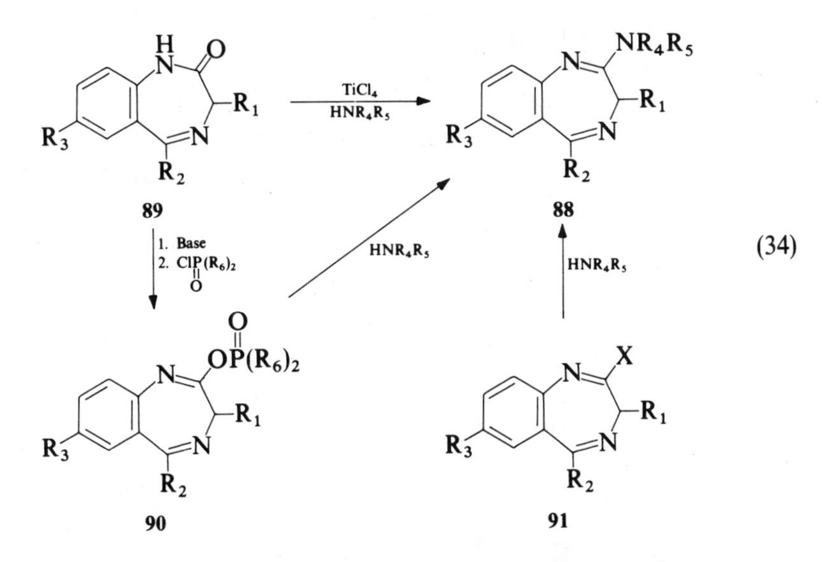


was successfully applied for the preparation of various compounds **88**.[88] Compounds **90** ($R_1$ = AcO) were treated with secondary amines to yield **88** ($R_1$ = AcO), hydrolysis of which led to the corresponding 3-hydroxy derivatives.[88]

Benzodiazepines with other leaving groups in the 2-position were transformed to 2-aminobenzodiazepines. Thus the 2-chloro compound **91** (X = Cl, $R_1$ = MeO, $R_2$ = 2-$ClC_6H_4$, $R_3$ = Cl) reacted with ammonia to give the corresponding 2-amino derivative.[89] A 2-(1-imidazolyl) derivative was synthesized by reaction of **91** (X, $R_3$ = Cl; $R_1$ = H; $R_2$ = 2-$ClC_6H_4$) with imidazole in boiling tetrahydrofuran.[86] Again, isolation of the 2-chloro compound, the synthesis of which is discussed in Section 2.6.1, is optional. Chlorides generated in situ were reacted with anilines to prepare 2-(phenylamino) derivatives.[90] 2-Cyano-[7] and 2-alkoxybenzodiazepines[91] **91** (X = CN or MeO) underwent similar nucleophilic displacements with amines. The nitrosomethylamino group in **91** [X = N(NO)Me] was also described to be a suitable leaving group for the preparation of 2-aminobenzodiazepines.[92] Reaction of a 2-phosphonate **91** [X = $PO(OEt)_2$, $R_1$ = MeO, $R_2$ = 2-$ClC_6H_4$, $R_3$ = Cl] with ammonia at 50°C for 6 hours was reported to give a 75% yield of the corresponding 2-aminobenzodiazepine.[89]

A direct conversion of the lactam **89** ($R_1$ = H, $R_2$ = Ph, $R_3$ = Cl) to the 2-pyrrolidinobenzodiazepine by heating with pyrrolidine in boiling toluene in the presence of *p*-toluenesulfonic acid was described in a patent.[85a]

While the successive treatment with phosphorus oxychloride and an amine was rarely applied[70] to the benzodiazepin-2-ones, this procedure was used to prepare the 5-aminobenzodiazepines **95** from the 5-one **92**.[18] The 5-chloro compound **94** was isolated and characterized.[18] Reaction of this chloroimine with phenylmagnesium bromide led to the 5-phenyl-3*H*-1,4-benzodiazepine **96** (Eq. 35).[93]

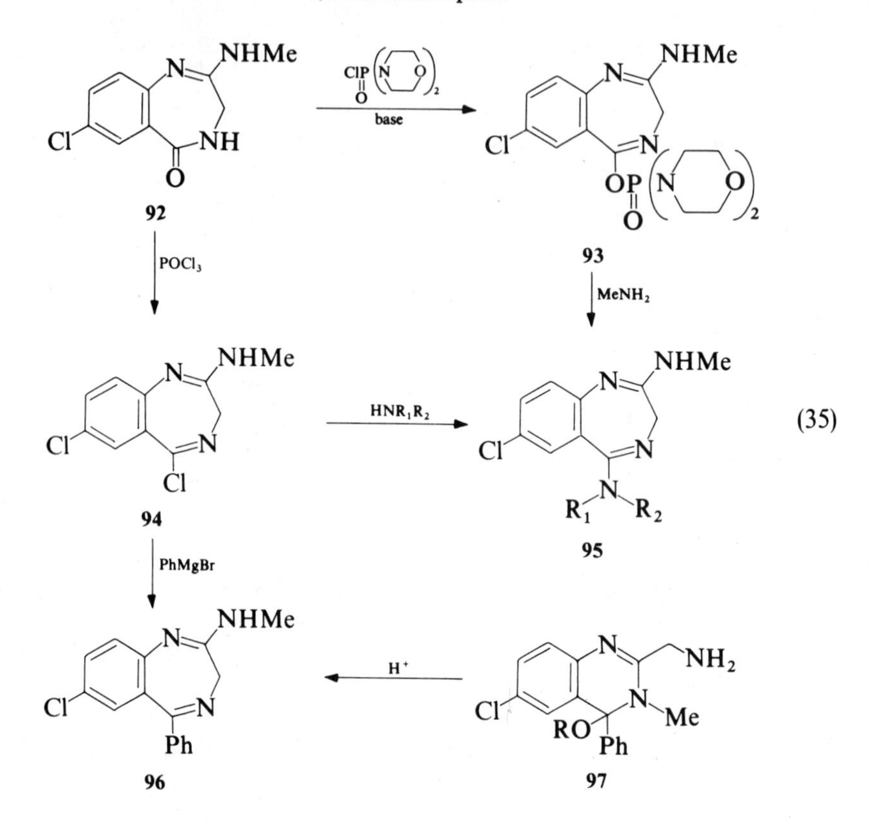


The 2,5-diamino derivative **95** ($R_1$ = H, $R_2$ = Me) was prepared by reaction of the 5-phosphate **93** with methylamine.[87a,b] Compound **93** resulted from the phosphorylation of **92** with sodium hydride and dimorpholinophosphinic chloride in tetrahydrofuran.[87b]

The benzodiazepine **96** was formed in 45–70% yields by short heating of the quinazoline derivatives **97** (R = H, Me, Et) with acetic acid.[94] This rearrangement involves cleavage of the six-membered ring and recyclization to the seven-membered ring.

2-Amino derivatives **100** were synthesized in good yields by cyclization of the cyanomethylimines **99** with hydroxide or with hydrogen chloride in methanol (Eq. 36).[95,96] The required nitriles **99** were accessible by acid-catalyzed amine exchange from the 2-hydroxyethylimines **98**.

A process for the preparation of chlordiazepoxide **79** made use of the addition of phenylmagnesium bromide to the nitrone **101** (Eq. 37).[52] The resulting hydroxyamine **102** was then oxidized with ferricyanide to yield 74% of the desired product.[97] Mercuric oxide also effected this transformation.[33] Chlordiazepoxide was formed also by thermal- or acid-catalyzed rearrangement of its photoproduct, the oxaziridine **103**.[98] Dehydration of **102** by treat-

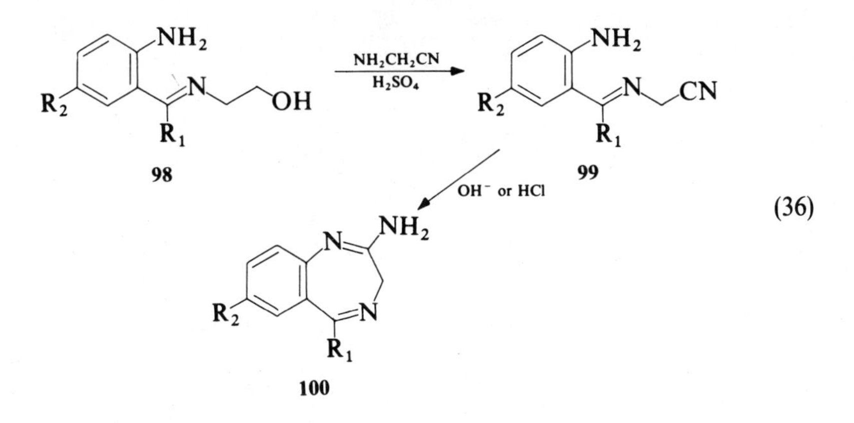


(36)

ment with thionyl chloride in boiling chloroform led to the imine, 4-desoxychlordiazepoxide.[33]

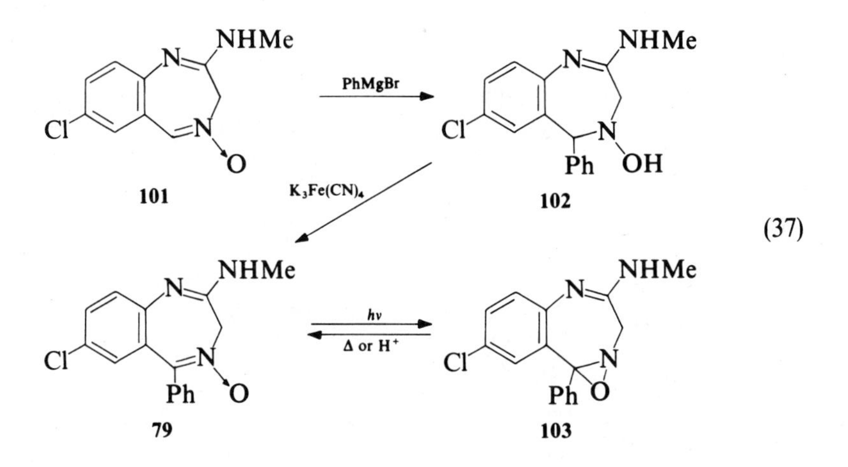


(37)

## 2.2.2. *Reactions*

### 2.2.2.1. Reactions with Electrophiles

A. Oxidation. The 2-aminobenzodiazepines **104** were oxidized with *m*-chloroperbenzoic acid to give the 1-oxides **105**[99] or the 1,4-dioxide **106** ($R_1$ = H, $R_2$ = Cl) (Eq. 38).[100] Peracetic acid converted the acetylderivative **107** to the corresponding 4-oxide.[101]

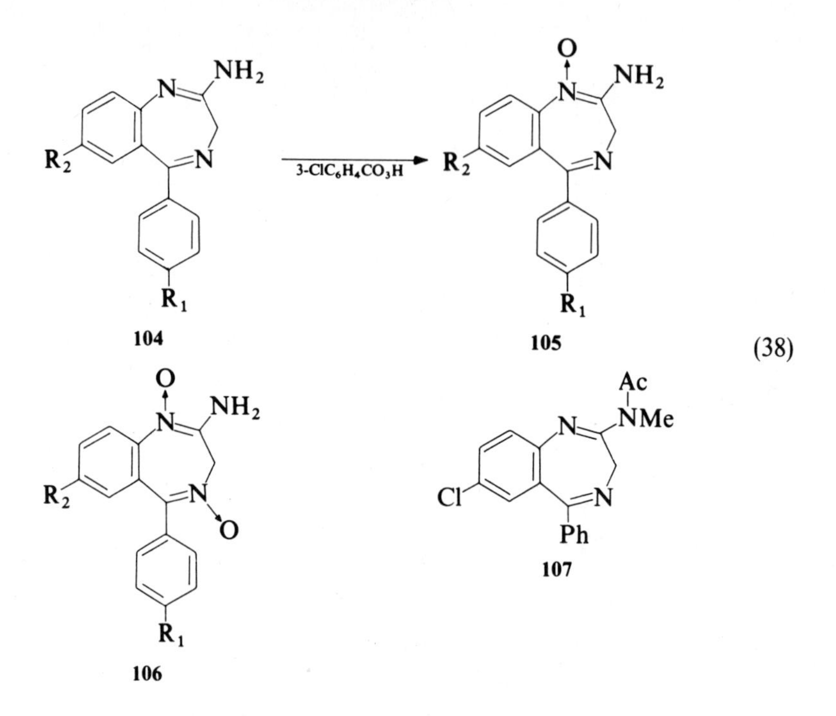


Oxidation of the 2-amidino compounds **108** with hypochlorite or with lead tetraacetate afforded the triazolobenzodiazepines **109** (Eq. 39).[102,103]

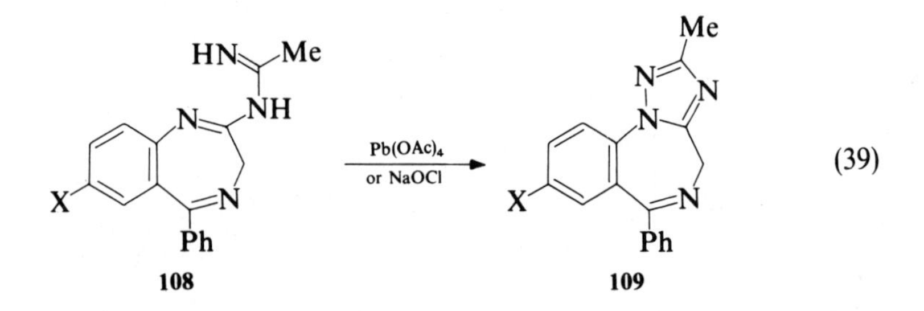


B. Reactions with Nitrogen Electrophiles. Nitrosation of the 2-methylaminobenzodiazepines **110** with sodium nitrite in glacial acetic acid[21,92,94] or with nitrosylchloride in pyridine[21] gave the *N*-nitroso compounds **111** (Eq. 40).

Treatment of chlordiazepoxide **110** ($R_1$ = Ph, $R_2$ = Cl, 4-oxide) with sodium nitrite in 3N-hydrochloric acid led in high yield to the quinazoline-2-carboxaldoxime **113**.[94] The same compound was obtained in 95% yield by action of 3N-hydrochloric acid on the nitrosoamidine **111** ($R_1$ = Ph; $R_2$ = Cl, 4-oxide). The formation of **113** may be rationalized by a reversible hydrolytic cleavage of the nitrone followed by ring closure to the intermediate **112**, which then undergoes irreversible oxidation to the oxime, possibly via an *N*-nitroso hydroxyamine.[94]

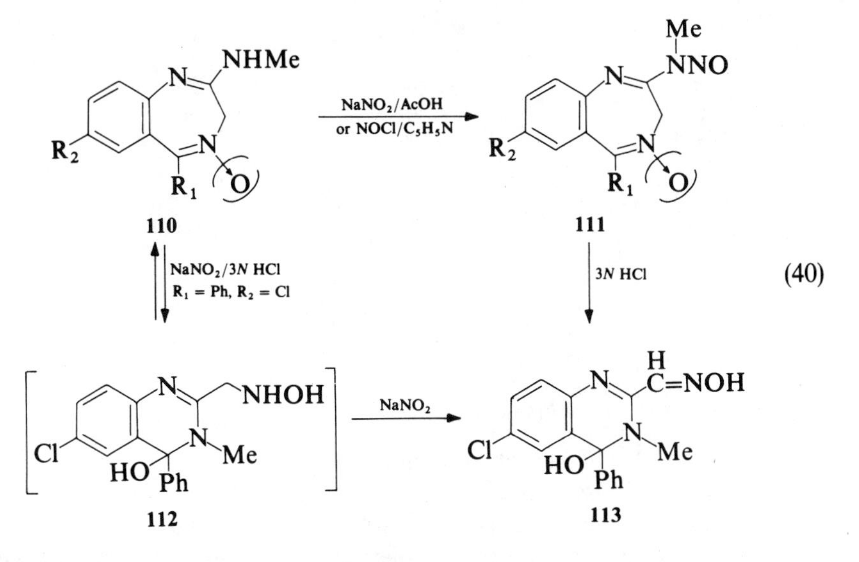


## C. Reactions with Carbon Electrophiles

C.1. *Alkylation.* Chlordiazepoxide **79** was alkylated at the nitrogen attached to the 2-position by sodium hydride in dimethylformamide and the following alkyl halides: methyl iodide,[104] benzyl chloride,[105] allyl bromide,[105] and methoxymethyl chloride[105] to give the appropriate product **114** (Eq. 41). Similar alkylations were more recently reported by Matsuo and coworkers[73] on the amidine **110** ($R_1$ = 2-$ClC_6H_4$, $R_2$ = Cl). Michael addition of acrylonitrile and ethyl acrylate to chlordiazepoxide yielded compounds **114** with R = $CH_2CH_2CN$ and R = $CH_2CH_2COOEt$, respectively.

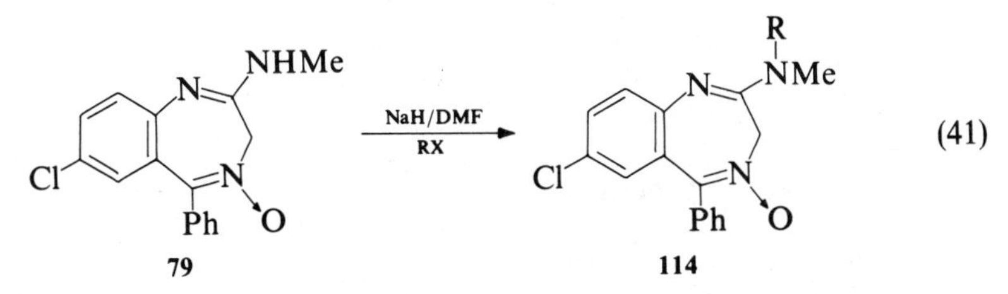


Attempted further methylation of **115** on carbon by the anion of dimethyl sulfoxide and dimethyl sulfate led to ring contraction with formation of five products **116, 117**, and **118** in the amounts indicated (Eq. 42).[104]

Mechanisms accounting for these rearranged products were proposed.[104] The initial step was assumed to be abstraction of a proton from the 3-position followed by a transannular attack of the tautomeric 5-carbanion onto the 2-position to give the intermediate **119**. Opening of the four-membered ring as indicated would lead to the oxime **120**, which can suffer elimination of dimethylamine to give **121**. The conversion of **121** to **122** involves a reduction, possibly by the anion of dimethyl sulfoxide used as a solvent. Methylation of **122** may then

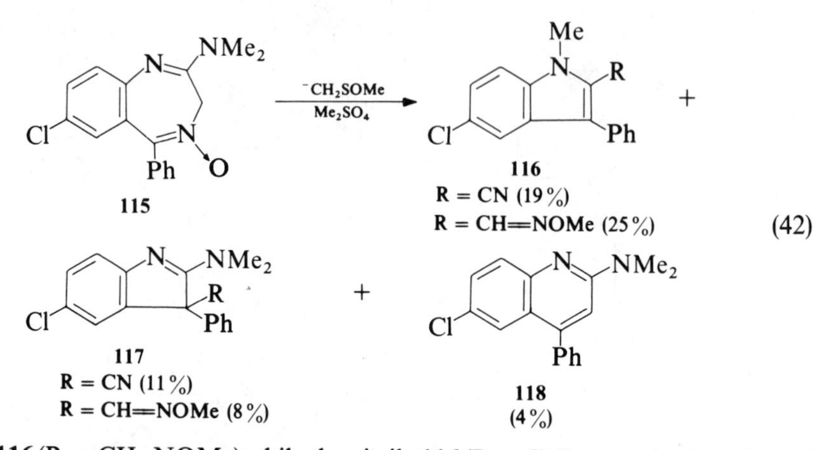


give **116** (R = CH=NOMe) while the nitrile **116** (R = CN) may arise from loss of water from **122** or loss of methanol from the oxime methyl ether.

The formation of compounds **117** from intermediate **120** was explained by the indicated migration of the aldoxime moiety from the 2-position to the 3-position of the indole.

The quinoline **118** was thought to result from elimination of HNO from an intermediate dihydroquinoline **124**, being formed by protonation of the postulated intermediate **123**, which would arise from **120** by the indicated mechanism. It is possible that the anion of dimethyl sulfoxide may be involved in the formation of **118**.

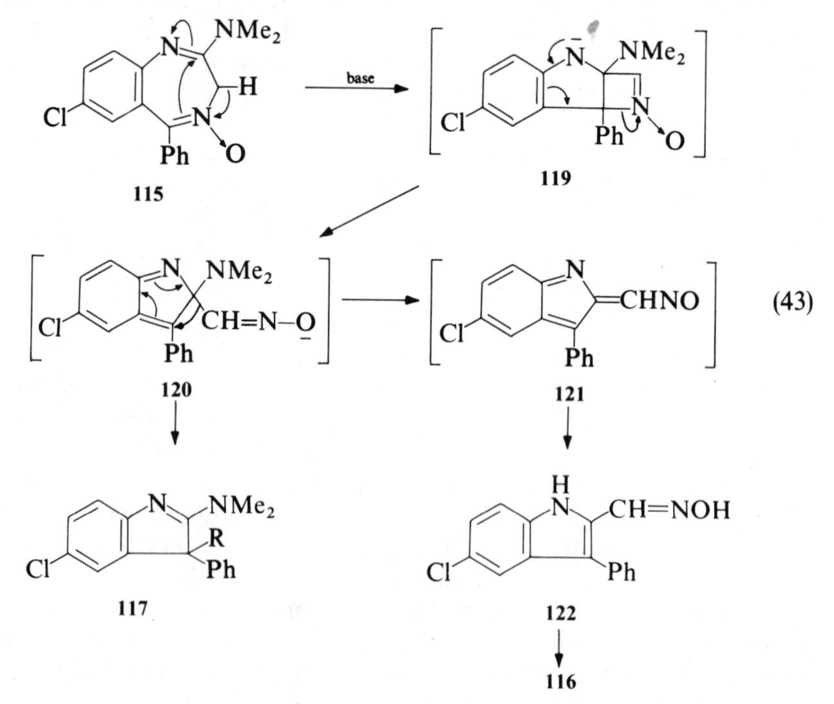

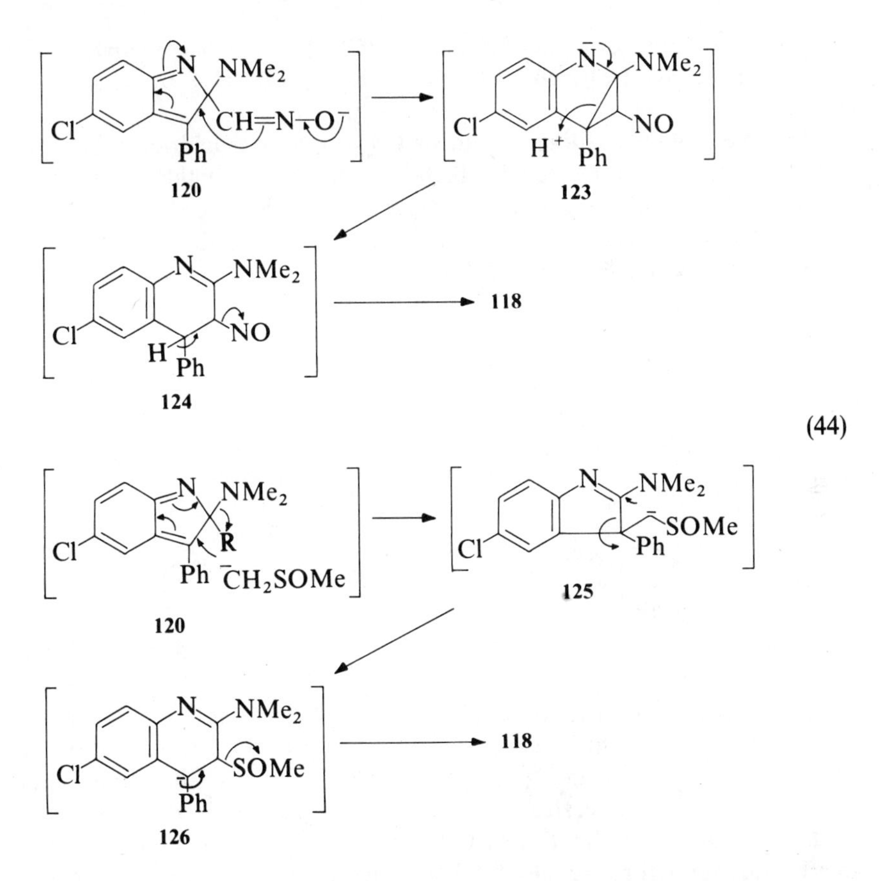


For example, intermediate **120** (R = CN, CH=NOMe) may be attacked by the dimethyl sulfoxide anion to give the 3*H*-indole **125**, which could undergo ring expansion as indicated to the dihydroquinoline **126**. Elimination of HSOMe instead of HNO would then account for the formation of **118**.

Alkylations of **127** were carried out selectively on the phenolic group with 1-benzyloxycarbonylamino-3-bromopropane and with 2-bromoacetic acid methyl ester to give **128** with $R = (CH_2)_3NHCOOCH_2Ph$ and $R = CH_2COOMe$, respectively (Eq. 45).[49] Cleavage of the carbobenzoxy group

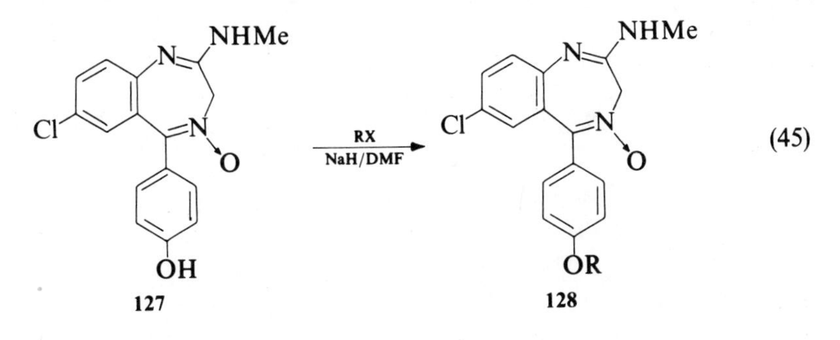

led to the corresponding amine **128** [R = $(CH_2)_3NH_2$]. The methyl ester was converted to the hydrazide **128** (R = $CH_2CONHNH_2$) by reaction with hydrazine.[49]

Treatment of the 2-chloroethyl ureas **129** with sodium hydride in tetrahydrofuran caused an intramolecular alkylation yielding the imidazolidinones **130** (R = H; X = H, Cl) (Eq. 46).[67] Compound **130** (R = H, X = H) was then methylated on nitrogen using methyl iodide and ethoxythallium in dimethylformamide to give **130** (R = Me, X = H).[67]

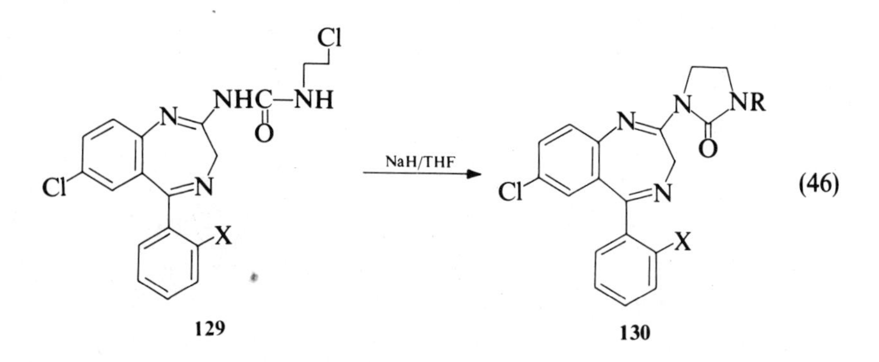


C.2. *Reactions with Aldehydes, Ketones, and Equivalents.* According to the patent literature,[106] chlordiazepoxide hydrochloride **79** formed addition products with acetaldehyde, propionaldehyde, butyraldehyde, and furan-2-carboxaldehyde to which structures **131** were assigned (Eq. 47). The aminoacetal **132** was obtained in about 10% yield by treatment of **131** (R = 2-furyl) with methanolic sodium hydroxide.[106c] The structures **133** assigned to the products formed by treatment of **131** (R = Et, Pr) with hydroxide[106a,b] are most likely incorrect.

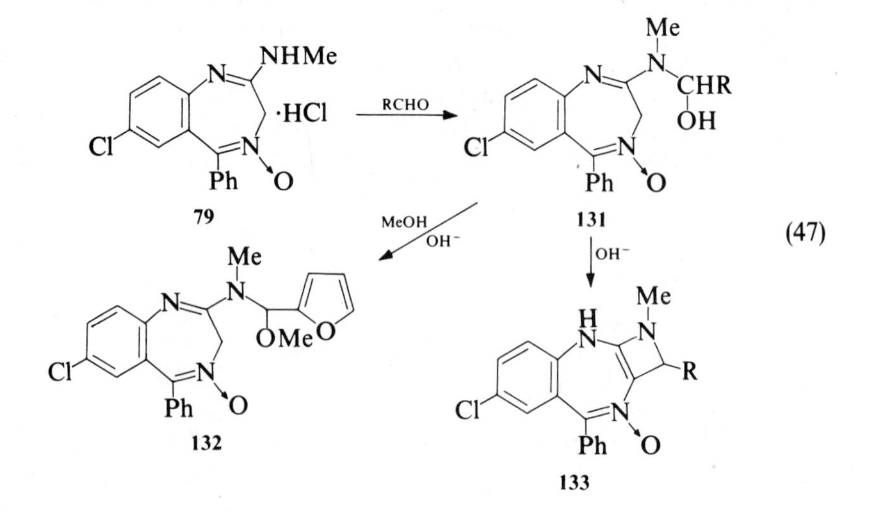

Intramolecular condensations with aldehydes and ketones were applied for the synthesis of imidazo[1,2-*a*][1,4]benzodiazepines. Thus, treatment of the acetals or ketals **134** with Lewis acids[77a] or with concentrated sulfuric acid[74–76,82] yielded the imidazobenzodiazepines **135** ($R_2$ = H). This ring system was also formed by reaction of the acetylenes **136** with concentrated sulfuric acid[74–76,82] or with *p*-toluenesulfonic acid in boiling butanol.[75]

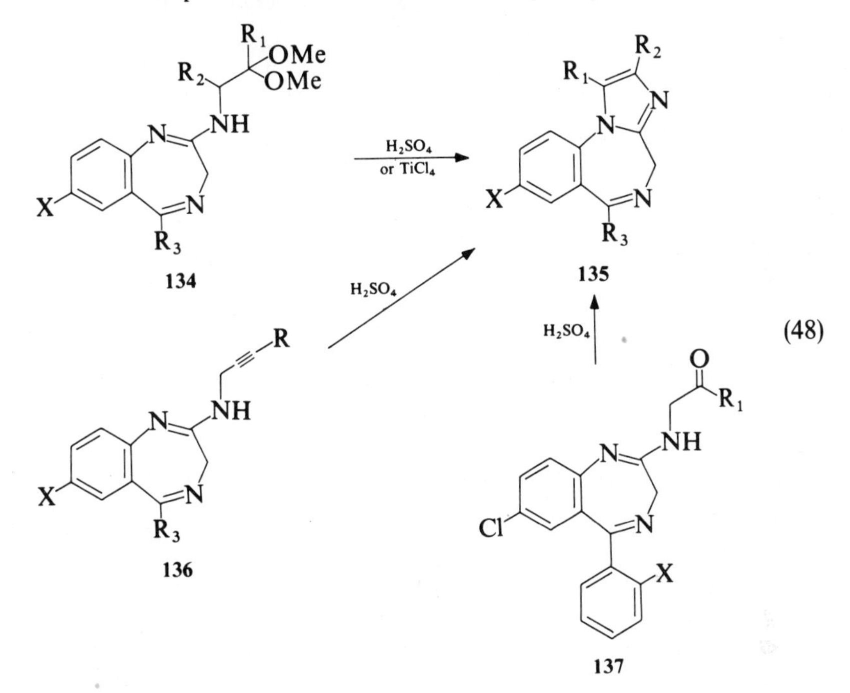


Cyclodehydration of **137** ($R_1$ = phthalimidomethyl; X = H, Cl) with concentrated sulfuric acid gave **135** ($R_1$ = phthalimidomethyl, $R_2$ = H, $R_3$ = phenyl or 2-$ClC_6H_4$, X = Cl).[82] The condensation of **52** with α-bromoketones provided still another access to compounds **135** ($R_3$ = Ph, X = Cl).[107]

The product resulting from the reaction of **52** with bromoacetone was **135** ($R_1$ = H, $R_2$ = Me, $R_3$ = Ph, X = Cl) (Eq. 49). This means that the 2-amino group of **52** attacked the carbonyl carbon of the bromoketone to form an intermediate imine **138** ($R_3$ = Me, $R_4$ = H), which would convert to **135** by formal elimination of hydrogen bromide via an intramolecular alkylation on the 1-position nitrogen and subsequent tautomerization. Condensation of **52** with 3-bromo-2-butanone ($R_3$ = $R_4$ = Me) and with 3-bromo-2-pentanone ($R_3$ = Me, $R_4$ = Et) led to the expected 1,2-disubstituted imidazo[1,2-*a*][1,4]benzodiazepines, although the 2-ethyl- and the 2-propyl-substituted compounds were interestingly formed as the major products in these condensations. The enamine **139** was proposed as an intermediate. The indicated Michael-type cyclization would violate Baldwin's rule for ring closure. The cyclopropanone imine **140** may be an alternate possible intermediate (Eq. 49).

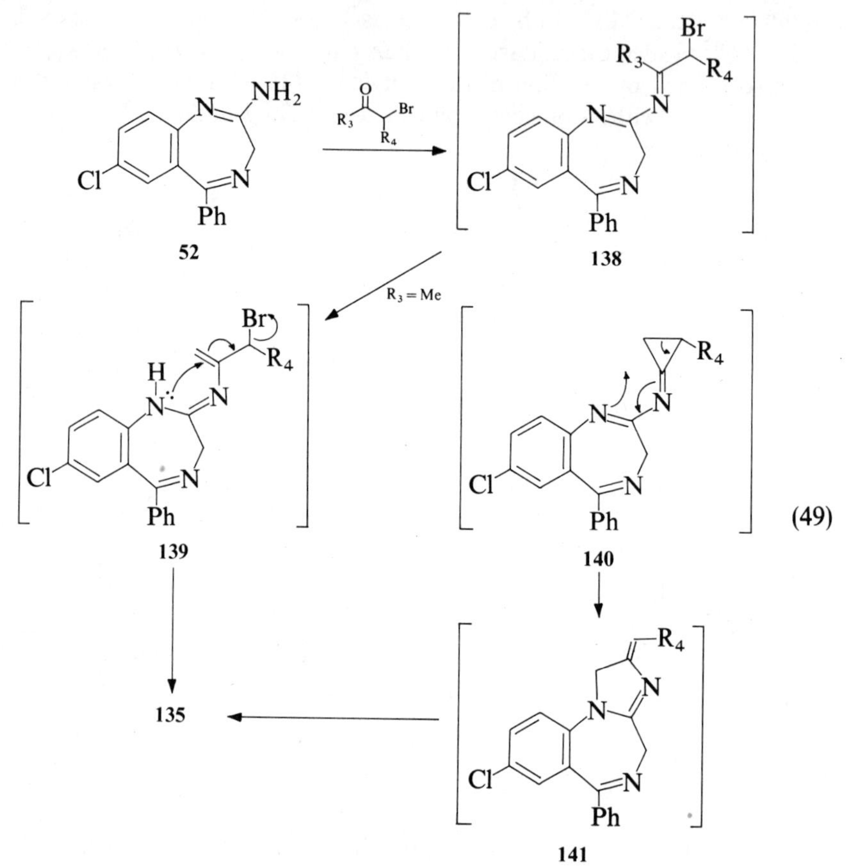


Reaction of the 2-aminobenzodiazepine **52** with ethoxymethylenemalonic acid diethyl ester afforded **142**, which was cyclized thermally to the pyrimidobenzodiazepine **143** (Eq. 50).[108]

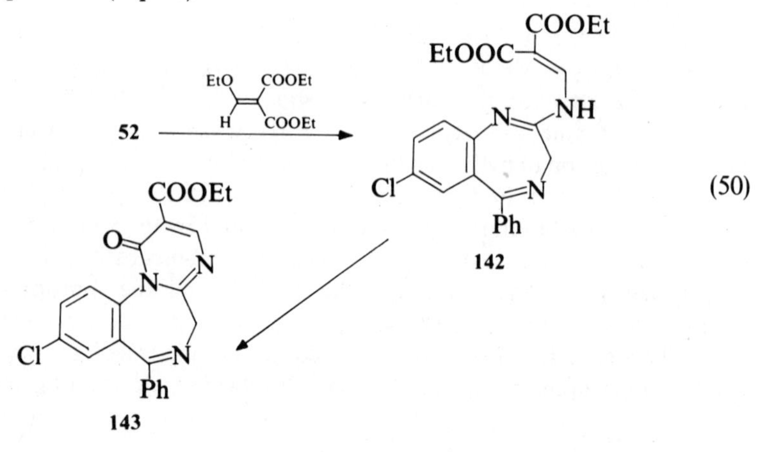

An interesting transformation was observed during the reaction of the 1-oxides **105** with acetylene dicarboxylate in methanol at room temperature (Eq. 51). The products **146** were obtained in good yields and were converted to compounds **147** by heating in ethanol. A plausible mechanism for this transformation is a 3,3-sigmatropic rearrangement of the intermediate adduct **144** leading to **145**. The latter would then aromatize and enolize to the product **146** by proton shift.

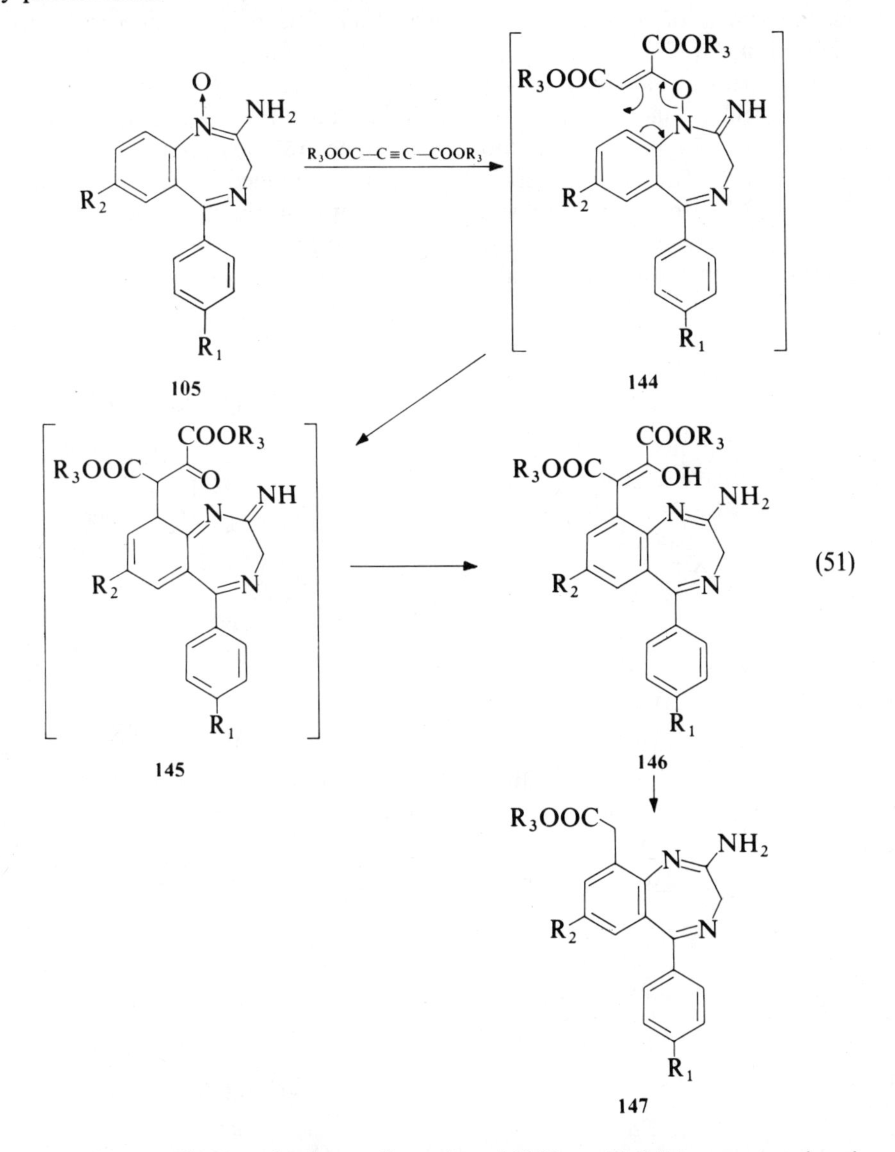


C.3. *Acylation.* The 2-aminobenzodiazepines **148** ($R_1$ = H, Me) were acetylated with acetic anhydride in pyridine or with acetyl chloride in pyridine at room temperature to the acetyl derivatives **149** ($R_2$ = Me).[33,34,48,51,110] With the

4-oxides **148**, the rearrangement of the nitrone to the 3-acetoxy derivative **151** ($R_2$ = Me) occurs to a minor extent under these conditions. This "Polonovski type" of rearrangement becomes prevalent under more vigorous conditions.[109] The same observations were made for the reaction of chlordiazepoxide with propionic anhydride or butyryl chloride in pyridine.[110] Short heating of chlordiazepoxide in acetic anhydride at 100°C[111] or reaction with acyl halides in dimethylformamide[110] or tetrahydrofuran at reflux[112] caused "Polonovski" rearrangement of the nitrone without acylation of the methylamino group, thus giving **150** (Eq. 52).

Treatment of the 4-oxide **148** ($R_1$ = H, X = H, Y = 7-Cl) with acetic anhydride yielded in addition to the diacetate **151** ($R_1$ = H, $R_2$ = Me, X = H, Y = 7-Cl) the oxazolobenzodiazepine **152** and, unexpectedly, the 3-one **153**.[111] The same diacetate **151** was also obtained by acetylation of the 2-amino-3-hydroxy-3*H*-1,4-benzodiazepine with acetic anhydride.[111] Compound **152** results from cyclization of the appropriate diacetate **151** with acetylation on the 1-position nitrogen.[111]

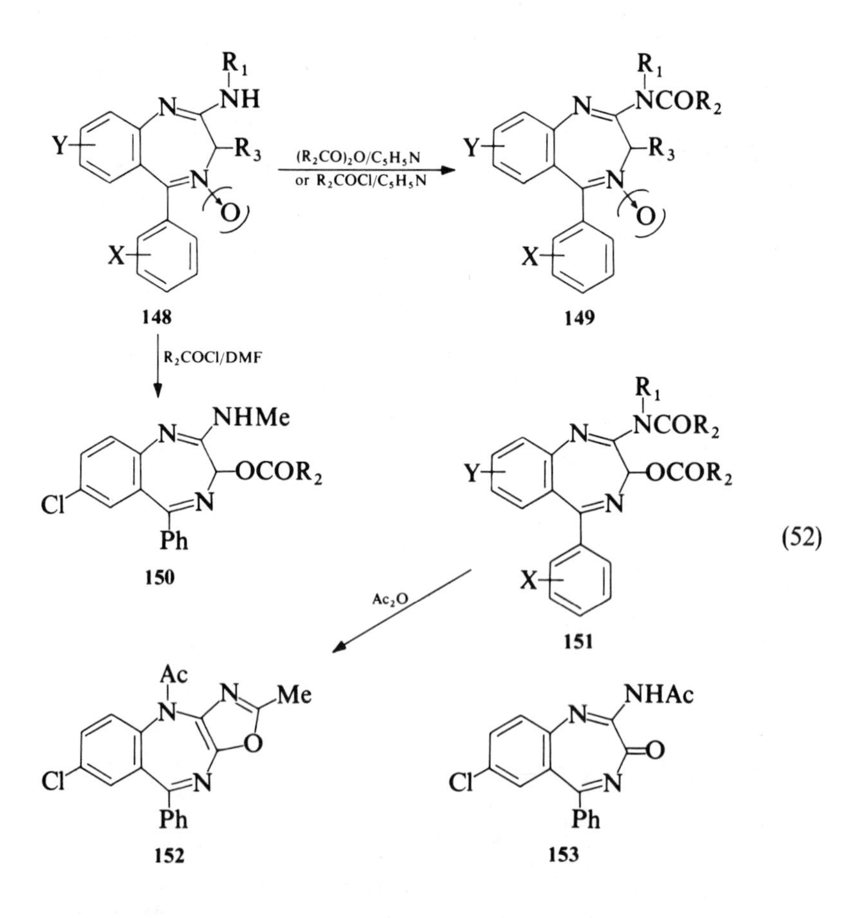


(52)

The formation of the 3-one **153** involves an oxidation. Bell and coworkers suggested no mechanism.[111] It is possible that **153** is formed by an oxygen-to-carbon migration of the acetyl cation in the tautomer **154** to give intermediate **155**, which would convert to **153** by elimination of acetaldehyde (Eq. 53). Compound **153** was obtained in 37% yield by heating the 1-oxide **105** ($R_1$ = H, $R_2$ = Cl) with acetic anhydride at 100°C for 3.5 hours.[99] While acetylation of **148** ($R_1$ = H, $R_3$ = 2-methylimidazolyl, X = H, Y = 7-Cl) with acetic anhydride at 95°C for a few minutes led to the *N*-acetyl derivative **149**, longer heating in acetic anhydride gave 38% yield of **153**.[99]

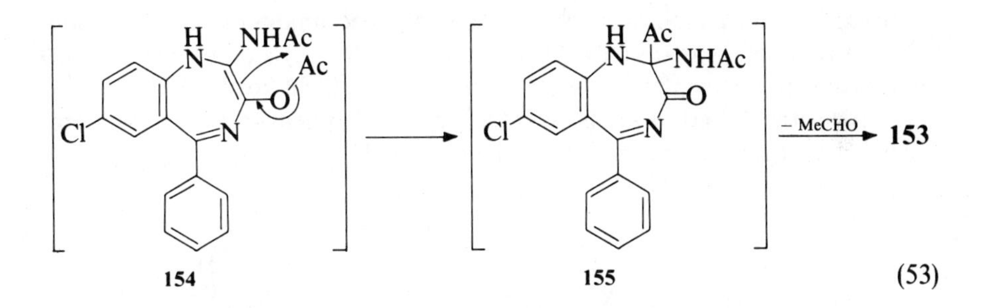


(53)

Acylation of the 2-aminobenzodiazepines **148** ($R_1$ = H) with diketene at room temperature gave high yields of the acetoacetates **156**.[113–115] These compounds were cyclized with dehydration to the pyrimidobenzodiazepines **157**

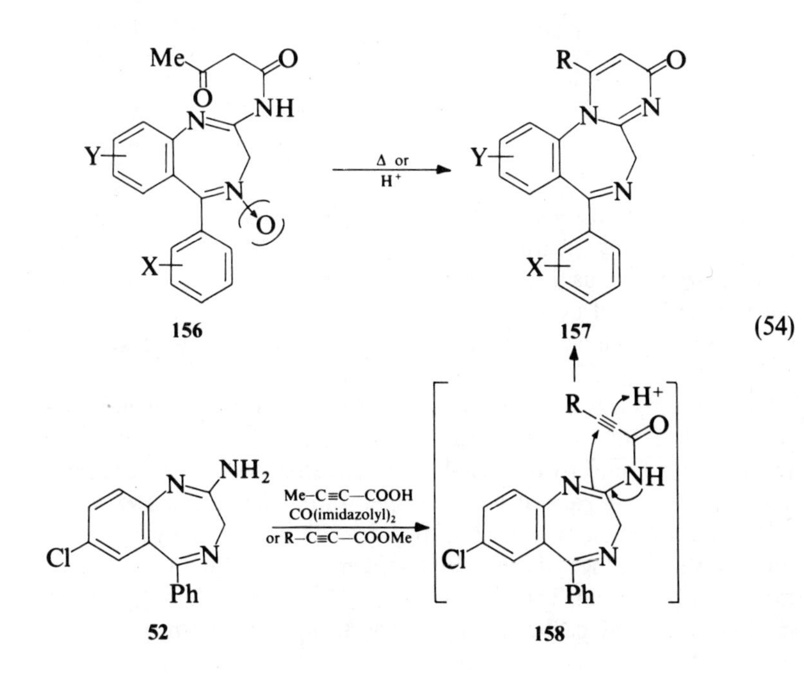


(54)

by heating or in higher yields by treatment with methanolic hydrogen chloride at 0°C (Eq. 54).[113] The ring system **157** was also synthesized by condensation of the 2-aminobenzodiazepine **52** with acetylenic carboxylic acids or esters. Thus, reaction of **52** with acetylene carboxylic acid methyl ester or with acetylene dicarboxylic acid dimethyl ester in boiling methanol gave the pyrimidobenzodiazepines **157** (R = H and R = COOMe, respectively).[116] The compound **157** (R = Me, X = H, Y = 9-Cl) was obtained by condensation of **52** with 2-butynoic acid and carbonyldiimidazole in dimethylformamide.[116] The formation of compounds **157** by this process implies that the intermediate acyl derivatives **158** are the primary products and undergo the indicated cyclization.

Reaction of the 2-aminobenzodiazepines with isocyanates or isothiocyanates led to a variety of ureas **159** (X = O) and thioureas (X = S) (Eq. 55). The reagents employed were methyl,[66,117] ethyl,[117] cyclopropyl,[66] 2-chloroethyl,[66,67] (ethoxycarbonyl)methyl,[66] ethoxycarbonyl,[66,118] acetyl,[66] phenyl,[117] and 4-chlorophenyl[117] isocyanate, and methyl isothiocyanate.[117]

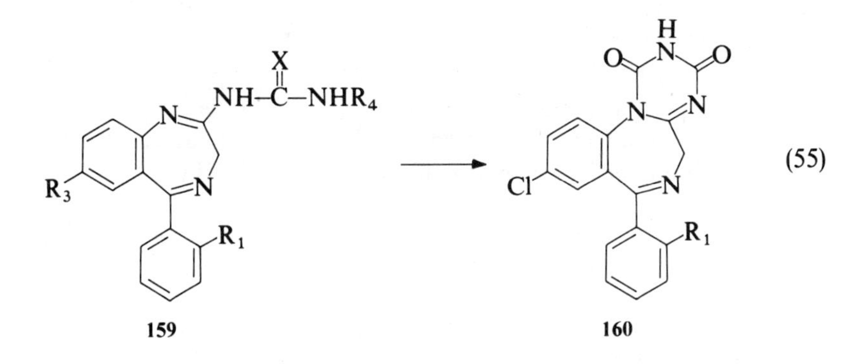


The urea **159** ($R_4$ = COOEt) was cyclized to the triazinobenzodiazepine **160** in 40% yield by heating in xylene.[66,118] If $R_3$ in the starting 2-aminobenzodiazepine was an amino group, the reaction with methyl isocyanate led to acylation of both amino groups.[70]

The reaction of chlordiazepoxide with methyl isocyanate was more complex because of the participation of the nitrone in the reaction. Thus, treatment of chlordiazepoxide with methyl isocyanate in tetrahydrofuran yielded 8.5% of **161** (R = H) and 60% of the urea **161** (R = CONHMe) (Eq. 56).[119] This major product was found to be unstable: it converted to the imidazobenzodiazepine **164** upon long storage at room temperature. Compound **164** was prepared in 27% yield by reacting chlordiazepoxide with methyl isocyanate in tetrahydrofuran in the presence of triethylamine. Under these conditions the urea **162** was formed in 11% yield. The conversion of **161** (R = CONHMe) to **164** with loss of methylamine and carbon dioxide was formulated to proceed via the intermediate **163**.

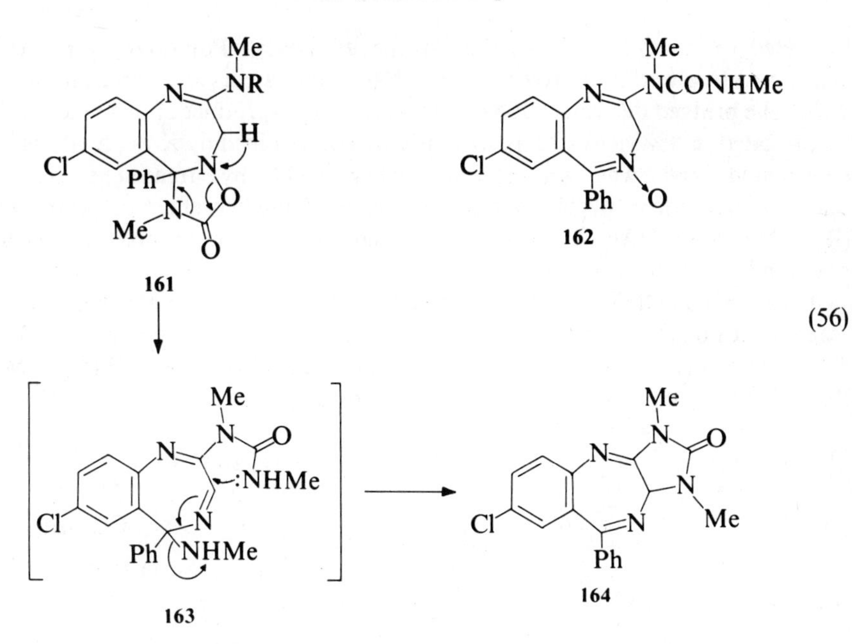


Reaction of the 2-aminobenzodiazepines **165** with oxalyl chloride and pyridine in tetrahydrofuran at low temperature afforded the imidazo[1,2-*a*]-[1,4]benzodiazepines **166** (Eq. 57).[68,69] A 61% yield was obtained when $R_1$ represented methyl.[68] This reaction was also carried out with oxalyl chloride in boiling benzene[120] for **165** ($R_1$ = Me, X = H). When chlordiazepoxide was

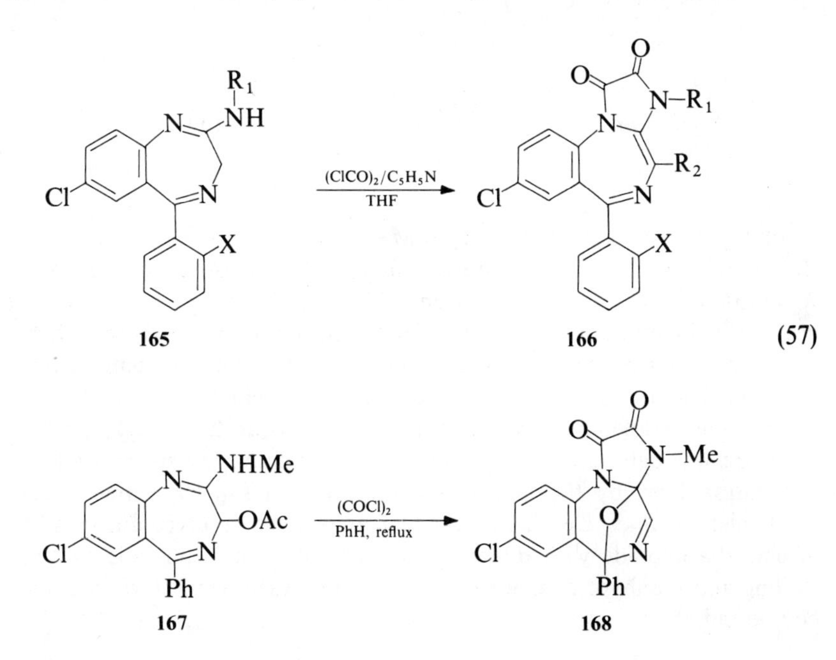

subjected to these conditions, the nitrone underwent Polonovski-type rearrangement to give the chloro compound **166** ($R_1$ = Me, $R_2$ = Cl) in nearly 80% yield. The bridged derivative **168** was formed as a by-product of this reaction. It was isolated in 8% yield and its structure was determined by X-ray analysis.[120] Compound **168** was formed in higher yield by treatment of the 3-acetoxy derivative **167** with oxalyl chloride in boiling benzene. Compound **166** ($R_1$ = Me, $R_2$ = OAc, X = H) was isolated as a minor by-product of this reaction.[120]

The amino alcohols **169** ($R_1$ = H, Me; $R_2$ = OH) as well as chlordiazepoxide reacted with oxalic ester chloride in boiling benzene to give the oxazolobenzodiazepine **170** (Eq. 58).[120] In the case of chlordiazepoxide and **169** ($R_1$ = Me, $R_2$ = OH), this conversion involves a demethylation.

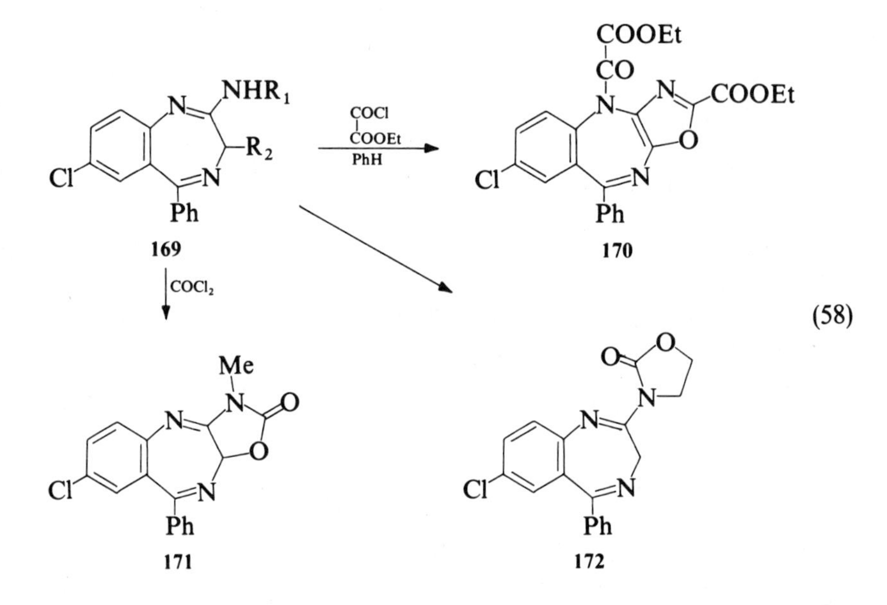


Phosgene reacted with **169** ($R_1$ = Me, $R_2$ = OH) to yield more than 70% of the oxazolone **171**.[120] The corresponding 3-acetoxy compound **169** ($R_1$ = Me, $R_2$ = OAc) formed the chlorocarbonyl derivative instead.[120b] The same reagent converted the ethanolamine **169** ($R_1$ = $CH_2CH_2OH$, $R_2$ = H) to **172**.[7b]

The reaction of chlordiazepoxide **79** with ethyl chloroformate in dimethylformamide in the presence of sodium hydride yielded 50% of the urethane **173**.[121] The indolenine **174** was isolated as a by-product in 5% yield and its structure was determined by X-ray crystallographic analysis.[121] Its formation was rationalized by the mechanism indicated in Eq. 59, which follows the mechanisms proposed earlier for the rearrangements observed during alkylation of chlordiazepoxide with dimethyl sulfate in dimethyl sulfoxide and base.[104] Boiling acetic anhydride converted **173** to the oxazolone **171** via the 3-acetoxy compound.[121]

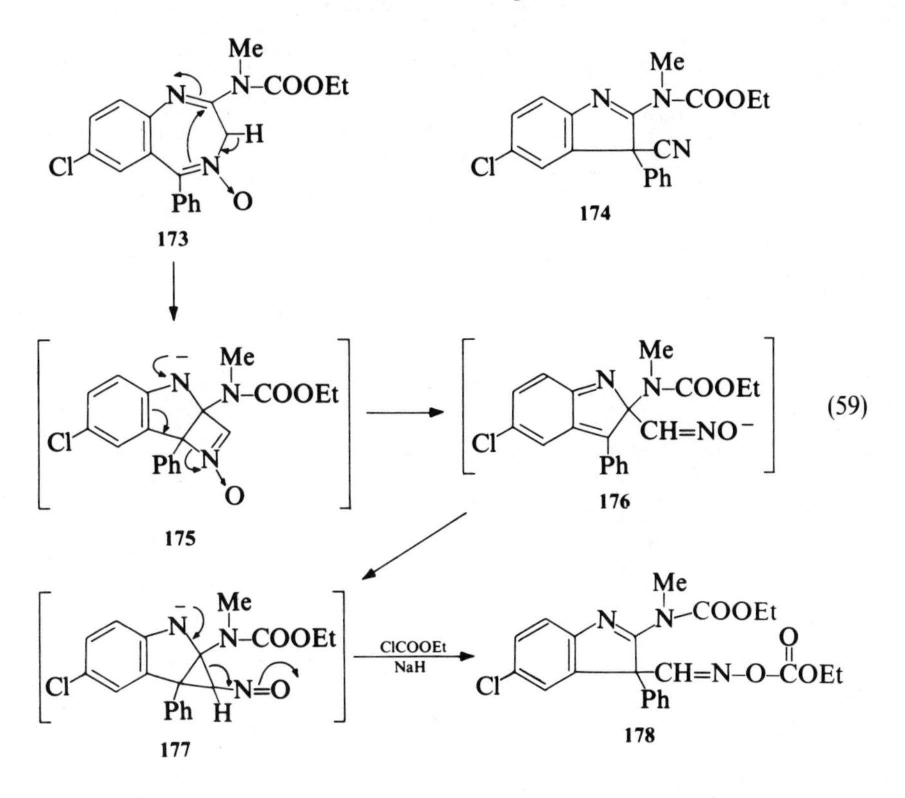


Acylation of the 2-aminobenzodiazepines with dimethylmalonyl chloride was studied by Moffett.[122] Compound **165** ($R_1$ = H, X = H, Y = Cl) was thus transformed into the tricyclic compound **179** (Eq. 60). Under more vigorous conditions, **179** underwent further acylation with dimethylmalonyl chloride and led to the dimer **181b**. Reaction of the 2-methylamino analog **165** ($R_1$ = Me, X = H, Y = Cl) with this reagent cleanly gave **181a** in 80% yield.[122]

Intramolecular acylation of **165** ($R_1$ = $CH_2COOH$) by means of dicyclohexylcarbodiimide led to the imidazobenzodiazepines **180**.[78-80] A 5-oxide of **180** (X = H, Y = Cl) was obtained in 10% yield by heating the 4-oxide of **165** ($R_1$ = $CH_2COOEt$; X = H, Y = Cl) in refluxing acetic acid in the presence of sodium acetate.[80]

Triethyl orthoacetate or propionate at 150°C reacted with 2-aminobenzodiazepine hydrochlorides[104] to give the imidates **182**, which afforded the amidines **183** upon further treatment with ammonia in methanol (Eq. 61).[102,103]

The tricyclic compound **184** was formed in 25% yield by reacting the 2-aminobenzodiazepine **138** with formamide and phosphoryl chloride at 110°C.[123]

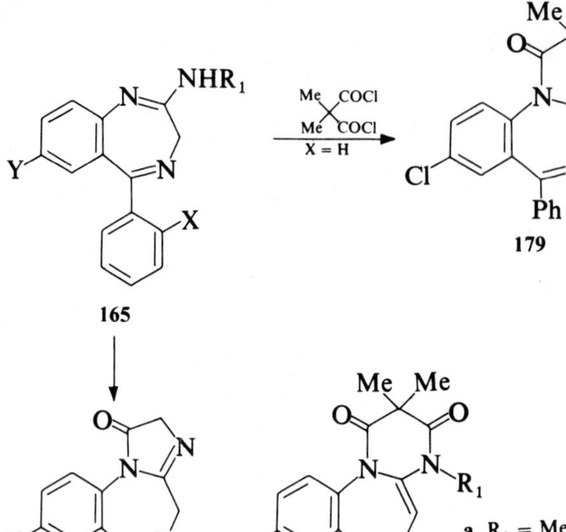
NHR$_1$
Y
X
165
Me COCl
Me COCl
X = H
Me Me
O O
N N
Cl N
Ph
179

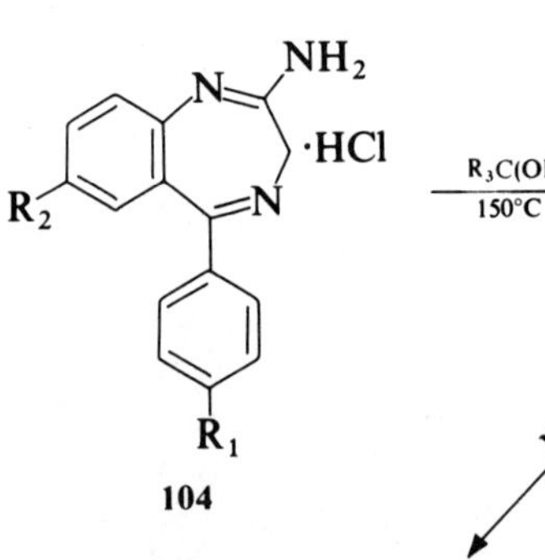
O
N
N
Y N
X
180

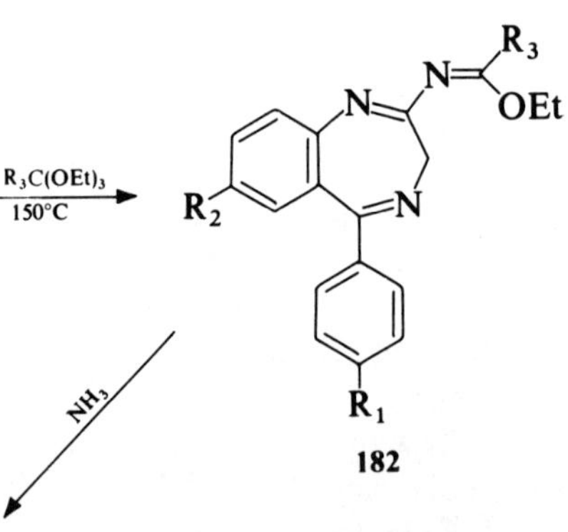
Me Me
O O
N N R$_1$
Cl N
Ph
181

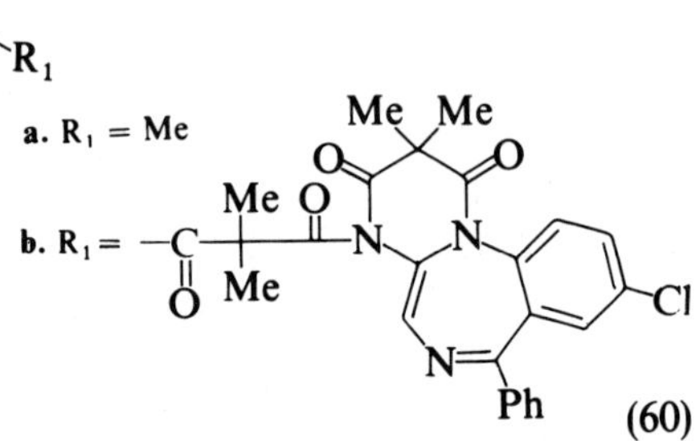
a. R$_1$ = Me
b. R$_1$ =
Me Me
O O
Me O
N N
O Me
Cl
N
Ph

(60)

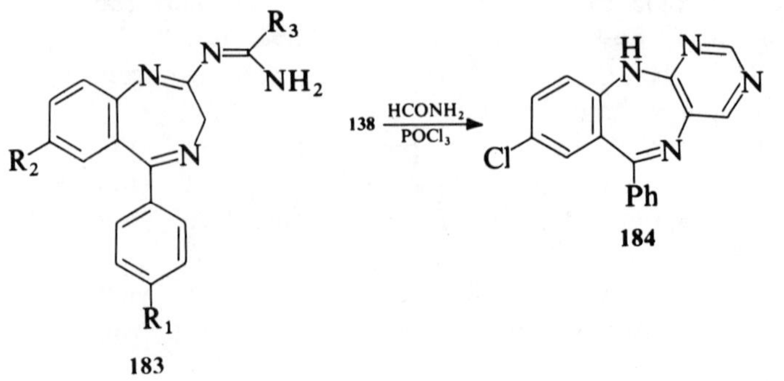
NH$_2$
·HCl
R$_2$ N
R$_1$
104
R$_3$C(OEt)$_3$
150°C
R$_3$
N OEt
R$_2$ N
R$_1$
182
NH$_3$
R$_3$
N NH$_2$
R$_2$ N
R$_1$
183
138 HCONH$_2$
POCl$_3$
H
N N
N
Cl N
Ph
184

(61)

### 2.2.2.2. Reactions with Nucleophiles

A. Reduction. The reduction of the 4-oxides to the corresponding imines was generally carried out with phosphorus trichloride in a chlorinated hydrocarbon solvent.[33,34,40,41,51,60] The same reagent was also used for the reduction of the 1-oxide.[99] Raney nickel[33] or palladium on carbon and hydrogen[41] also converted 4-oxides to the 4-desoxy compounds.

Hydrogenation of the nitrones **185** (R = H, Me) or the 4-desoxy compound (R = H) over platinum led to the 4,5-dihydro derivatives **186** (Eq. 62).[33,60] Using palladium as catalyst the hydrogenation of chlordiazepoxide **185** (R = H) gave in 70% yield the 7-deschloro derivative of **186** (R = H).[33]

Lithium aluminum hydride in tetrahydrofuran reduced chlordiazepoxide to the hydroxylamine **102**.[33]

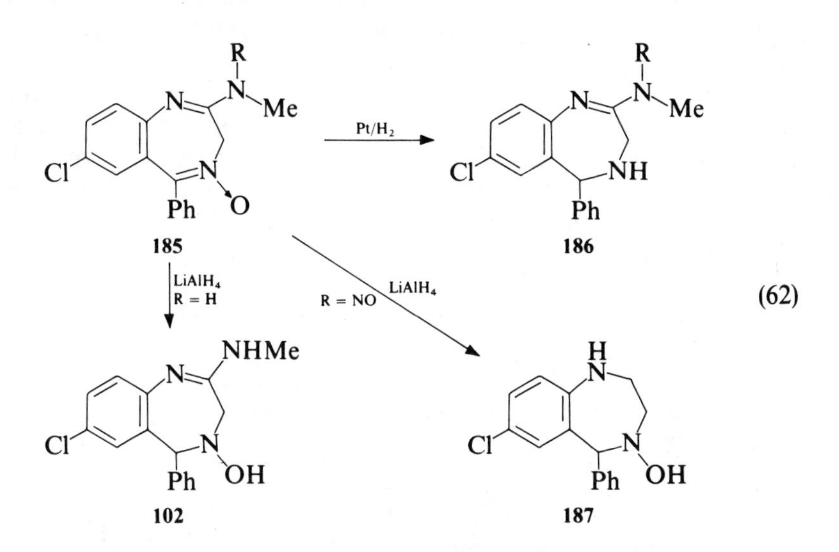


The same reagent converted the nitrosoamidine **185** (R = NO) in 80% yield to the tetrahydro derivative **187**.[92] The nitrosoamidine apparently suffered displacement by hydride to give the intermediate 1,2-imine, which then reduced to **187**. Reduction of the 3-one **153** with lithium aluminum hydride gave about 55% of the deacetylated alcohol **188** (Eq. 63).[111]

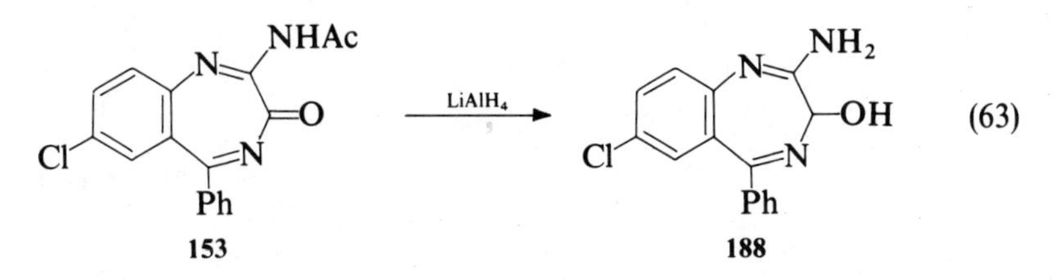


7-Nitro compounds were reduced to the 7-amino analogs by means of stannous chloride in hydrochloric acid.[70]

B. Reactions with Oxygen Nucleophiles. Acid salts of 2-aminobenzodiazepines **189** ($R_1 = R_2 = R_3 = H$; X = Cl, $NO_2$) were converted to the corresponding 2-ones **190** by boiling in methanol (Eq. 64).[95] If this hydrolysis was carried out with deuterium chloride in deuterium oxide–deuteromethanol, the lactam obtained contained 87% of deuterium in the 3-position.[124] Exchange of the protons in the 3-position of 2-amino-7-chloro-5-phenyl-3*H*-1,4-benzodiazepine by deuterium was achieved by heating with sodium deuteromethoxide in deuterated methanol.[124] Acid hydrolysis of the acetyl derivatives **189** ($R_1$ = Me, $R_2$ = Ac, $R_3$ = H) or their 4-oxides also gave the appropriate lactams **190**.[51,125] Methylacetamide was isolated in this case, indicating cleavage of the bond between C-2 and N after attack of hydroxide at the 2-position.[125] This type of hydrolysis converted the diacetate **189** ($R_1$ = Me, $R_2$ = Ac, $R_3$ = OAc, X = Cl) to the 3-acetoxy lactam **190** ($R_3$ = OAc, X = Cl). This transformation was carried out with either ethanolic hydrogen chloride[111] or with 1 *N* hydrochloric acid in dioxane.[110]

Alkaline hydrolysis of the same diacetate with one equivalent of hydroxide at room temperature gave 34% yield of the monoacetate **189** ($R_1$ = Me, $R_2$ = H, $R_3$ = OAc, X = Cl) by preferential cleavage of the *N*-acetyl group.[110] Further hydrolysis with 1 *N* sodium hydroxide led in high yield to the 2-methylamino-3-hydroxybenzodiazepine.[110]

Cleavage of the *N*-acetyl group with formation of chlordiazepoxide from its *N*-acetyl derivative was observed during chromatography over basic alumina.[101] Treatment of the 3-acetoxy compound **189** ($R_1$ = H, $R_2$ = Me, $R_3$ = OAc, X = Cl) with 1 *N* hydrochloric acid at 85°C or of the corresponding 3-hydroxy derivative at room temperature gave the quinazoline-2-carboxyaldehyde **192**.[110] This aldehyde and 2-amino-5-chlorobenzophenone

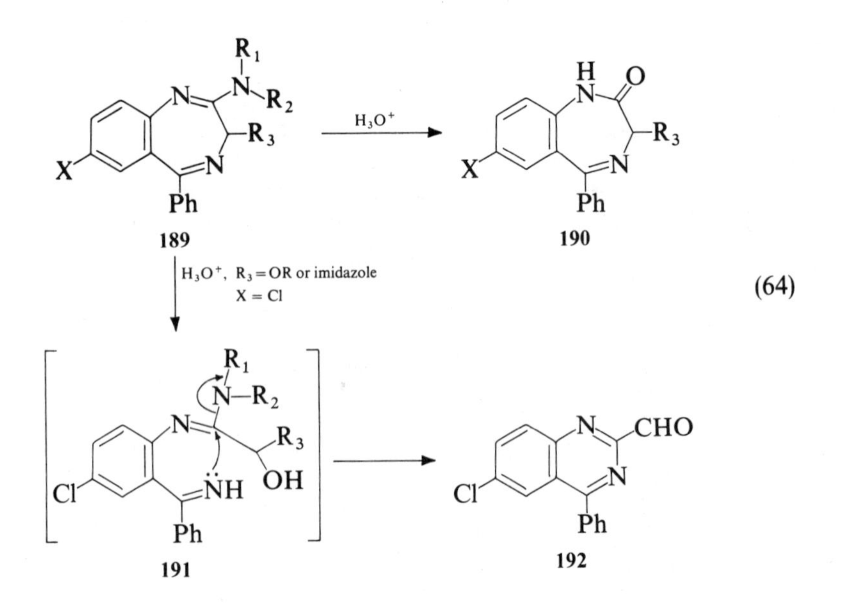


(64)

were isolated also from the acid hydrolysis of the 3-imidazolyl compound **189** ($R_1 = R_2 = H$, $R_3$ = 2-methylimidazolyl, X = Cl).[99] The hemiacetal of **192** was obtained in 86% yield from the reaction of the 3-hydroxy compound with ethanolic hydrogen chloride at room temperature.[110] The formation of the quinazoline-2-carboxaldehyde by these reactions may be the result of hydrolytic cleavage of the 3,4-bond to give **191** and subsequent recyclization and elimination of the 2-amino function.

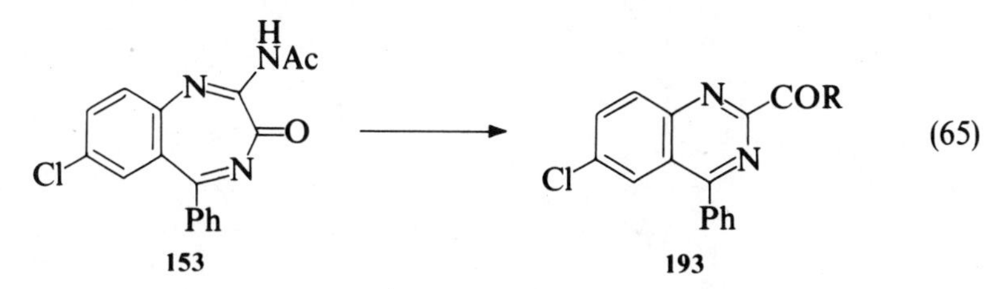


(65)

The 3-one **153** suffered similar ring contractions. Thus the ethyl ester **193** (R = EtO) was formed during the reaction of **153** with ethanolic hydrogen chloride (Eq. 65).[111] Acetic acid converted **153** to the *N*-acetylcarboxamide **193** (R = NHAc), while reaction of **153** with hydroxide afforded the carboxylic acid **193** (R = OH).[111] These rearrangements are again initiated by fission of the 3,4-bond in **153**, which is a reactive acylimine in this case.

Treatment of the nitrosoamidine **194** with 1 *N* hydrochloric acid in tetrahydrofuran yielded 32% of the lactam **195**, 32% of chlordiazepoxide **79**, and 14% of the quinazoline **113** (Eq. 66).[94] The latter was formed in 95% yield by reacting **194** with 3 *N* hydrochloric acid.[94] Hydroxide converted **194** in high yield to the lactam **195** with liberation of diazomethane.[92,94]

Accordingly, alkoxides reacted smoothly with **194** to form the 2-alkoxybenzodiazepines **196**.[92]

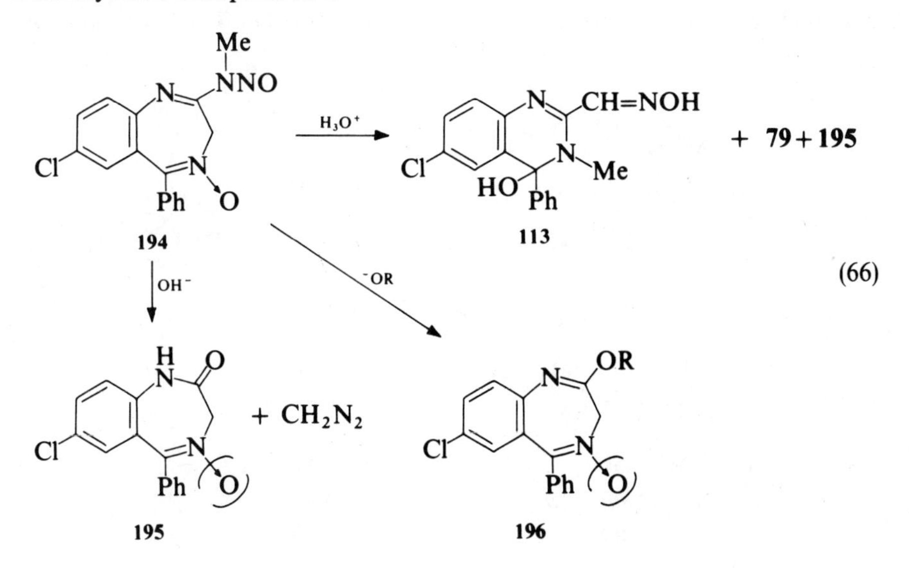


(66)

Acid hydrolysis of the 1-oxide **197** led to the benzisoxazole **198** (Eq. 67).[99] This compound was also formed in about 50% yield by reacting the 4-oxide of **197** with hydrogen sulfide.[100] Mild treatment of the 4-oxide of **197** with aqueous acetic acid gave 88% yield of the 1-hydroxy compound **199**.[100]

2-Amino-3-acyloxybenzodiazepines **150** underwent alcoholysis with ethanol or methanol to form the 3-alkoxy derivatives.[9,53]

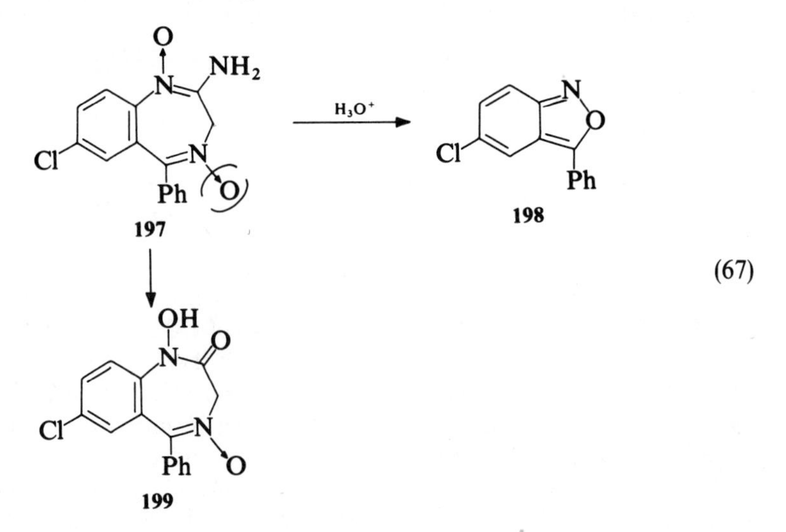


C. Reactions with Nitrogen Nucleophiles. The acid-catalyzed exchange of the 2-amino group by other amines and the displacement of the 2-nitrosomethylamino function by amines was discussed above (Section 2.2.1.3).

Various 2-aminobenzodiazepines **200** were claimed to react with acylhydrazines to form the triazolobenzodiazepines **201**,[126] in particular by heating in hexamethyl phosphoric triamide[126c] at 160–170°C (Eq. 68). The nitrosoamidine

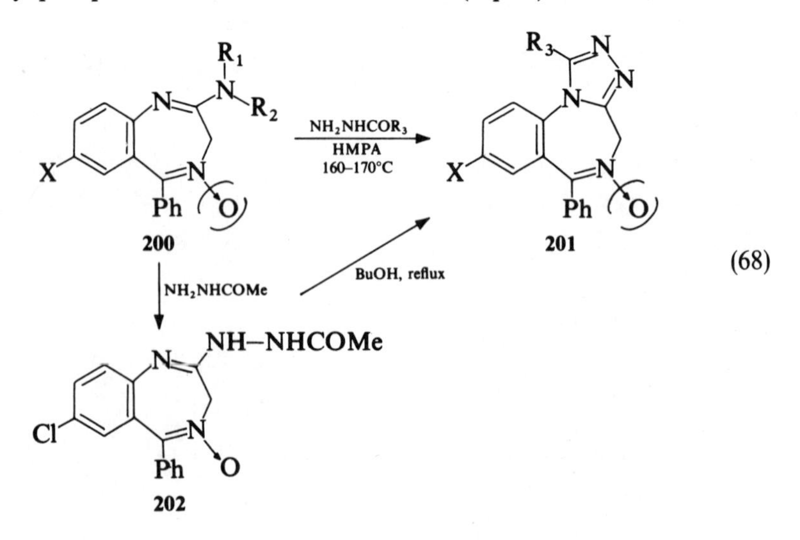

**200** ($R_1$ = Me; $R_2$ = NO; X = Cl, 4-oxide) reacted with acetylhydrazine to give initially the hydrazide **202**, which cyclized with loss of water to the triazole **201**.[92]

3-Hydroxybenzodiazepines **203** were converted to 3-amino analogs **205** by treatment with thionyl chloride and subsequently reacting the crude chloride with an amine.[9] More recently, 2,3-diaminobenzodiazepines **205** were prepared in good yields by reacting the 3-hydroxy compounds with 2-dialkylamino-4,5-dihydro-1,3-dioxa-2-phosphole **204** (Eq. 69).[45]

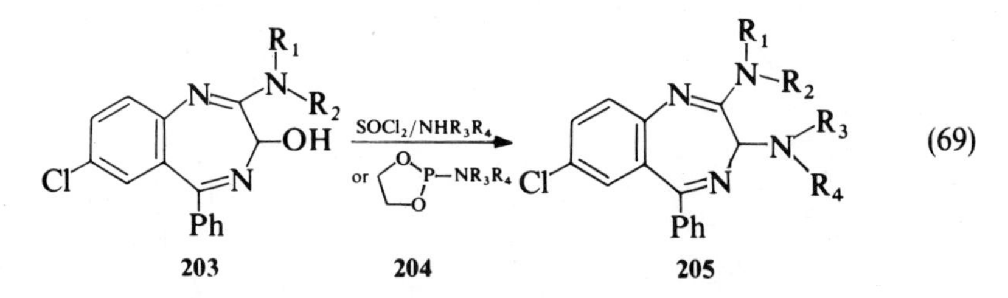


Ammonia converted the carbonyl chloride **206** to the urea **207** (Eq. 70).[120b]

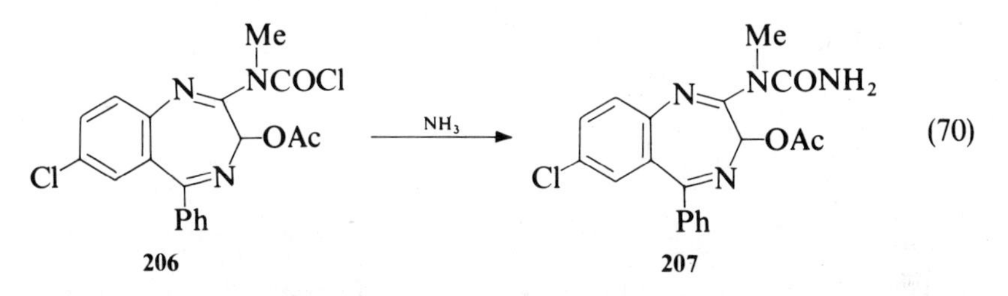


Ammonolysis of the imidazobenzodiazepine **208** was accompanied by ring opening and yielded the amide **209** (R = $NH_2$).[120a] Reaction with ethanol and triethylamine accordingly gave the 3-ethoxy derivative **209** (R = EtO) (Eq. 71).[120b]

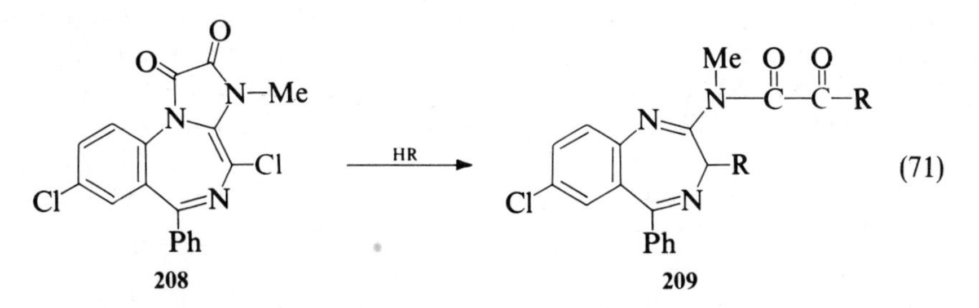


The reaction of the 1-oxide **105** with carbonyldiimidazole or a phosgene–imidazole combination interestingly yielded the 3-imidazolyl compounds **211**.[99] The oxadiazolone **210** was postulated as an intermediate.[99] The addition of the nucleophile to the 4-position of **210** with expulsion of carbon dioxide would lead to the observed products (Eq. 72).

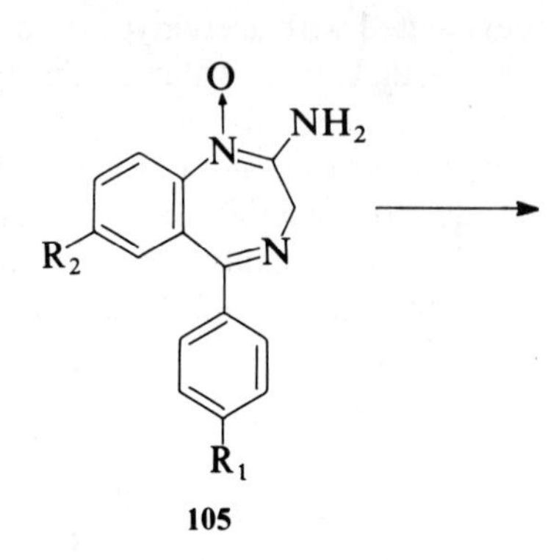


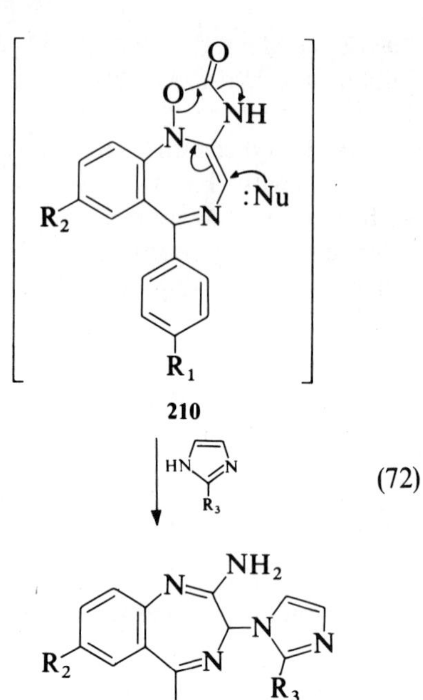


This reaction was used for the synthesis of a variety of compounds bearing 2-methylimidazole, 2-ethylimidazole, and benzimidazole substituents in the 3-position.[99]

D. Reactions with Carbon Nucleophiles. The reaction of the 2-aminobenzodiazepine **52** with 1,3-dicarbonyl compounds at 145–160°C afforded the pyrroloquinolines **212** (R = Me, EtO, *t*-BuO) (Eq. 73).[27] Boiling **52** with acetylacetone gave, besides the pyrroloquinoline **75** (Eq. 29), 5% yield of the benzodiazepine **53**, which was proposed to be the initial product of the overall conversion.[27] For discussion of mechanism see Section 2.1.2.2.

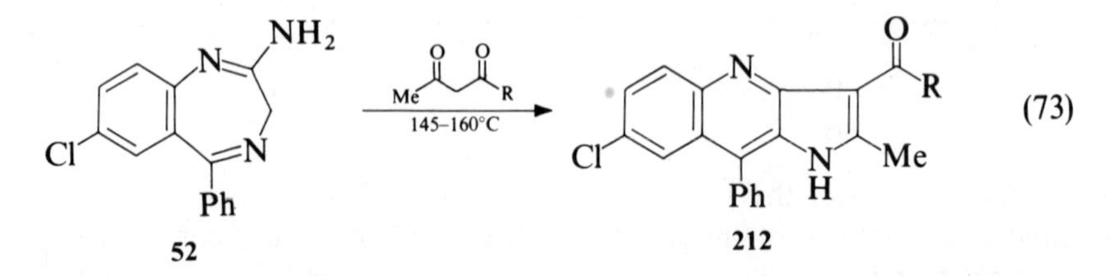


The nitrosoamidines **111** reacted with stabilized carbanions such as the anions of nitroalkanes,[19] malonic esters,[92] and malononitrile[92] to give the

2-methylene derivatives **213** ($R_3 = NO_2$; $R_4$ = H, Me; $R_3 = R_4 = COOR_5$, CN) (Eq. 74).

The anion of ethyl isocyanoacetate underwent a similar displacement reaction with nitrosochlordiazepoxide to give the imidazobenzodiazepine **215**, most likely via the intermediate isonitrile **214**.[127] Protonation of **214** on the isonitrile carbon followed by cyclization as indicated would lead to the imidazole **215**.

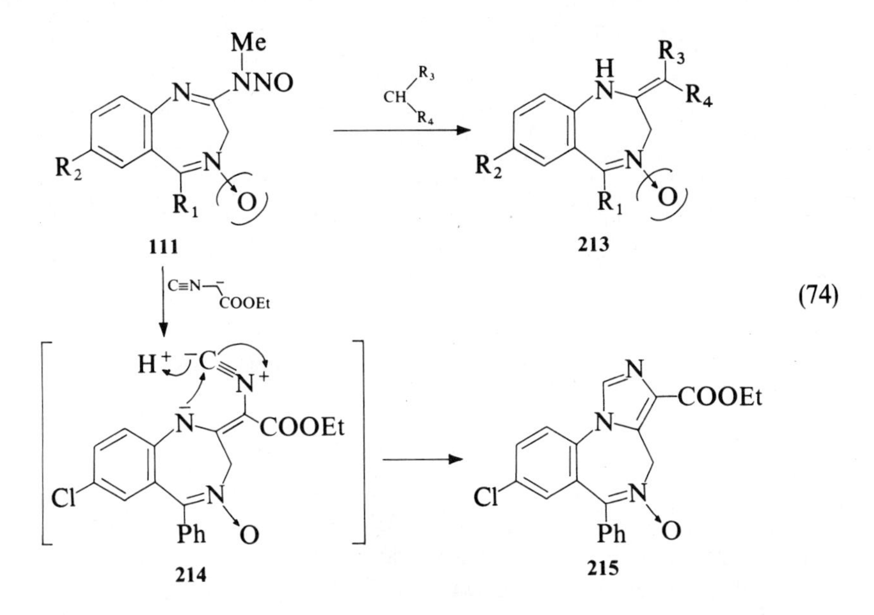


(74)

### 2.2.2.3. Thermal and Photoreactions

Heating of the 5-phosphoryl derivative **93** in boiling 1,3,5-trimethylbenzene for 2 hours yielded the 5-one **216**, resulting from migration of the phosphoryl group from oxygen to nitrogen (Eq. 75).[87b]

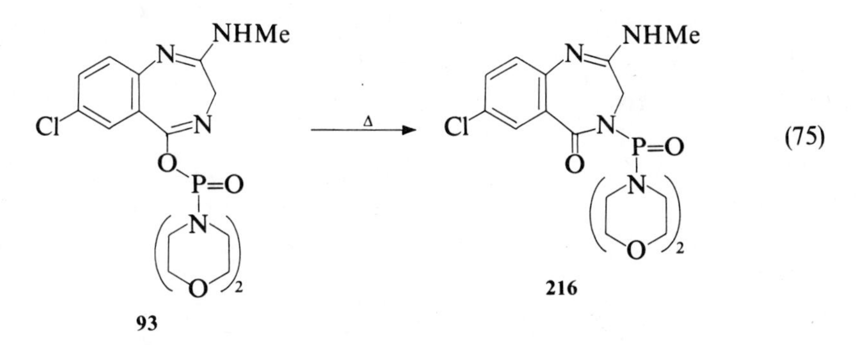


(75)

The photochemistry of chlordiazepoxide was investigated in detail.[98,128–130] Irradiation at 350 nm converted the nitrone to the oxaziridine **103**, which reverts to chlordiazepoxide thermally or by acid catalysis (Eq. 76).[98, 128] Further irradiation of the oxaziridine at 300 nm led to the photoproducts **217** and **218**, which can still suffer further photo rearrangement.[128]

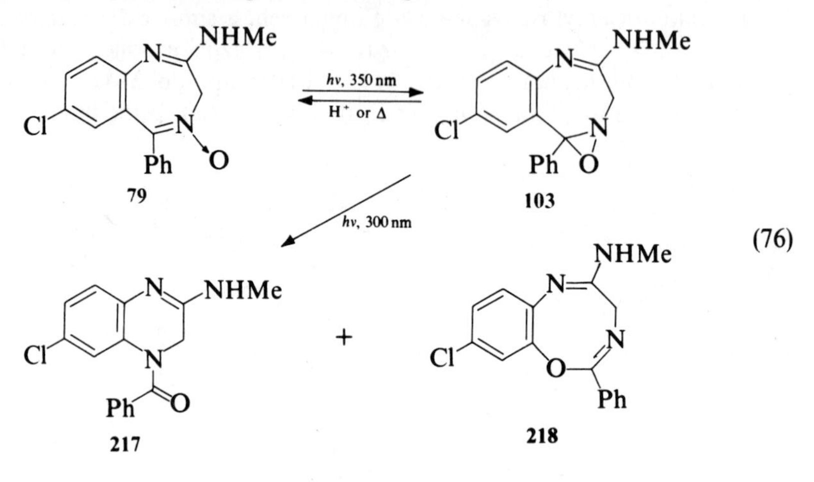


(76)

## 2.3. 2-Hydroxyamino-3*H*-1,4-benzodiazepines

### *2.3.1. Synthesis*

The 2-hydroxyamino compounds **220** were prepared by reacting the 2-thiones **219** with hydroxy- or alkoxyamines[112,131–135] or by displacing a suitable leaving group $R_3$ in **221** (Eq. 77). Such leaving groups were the

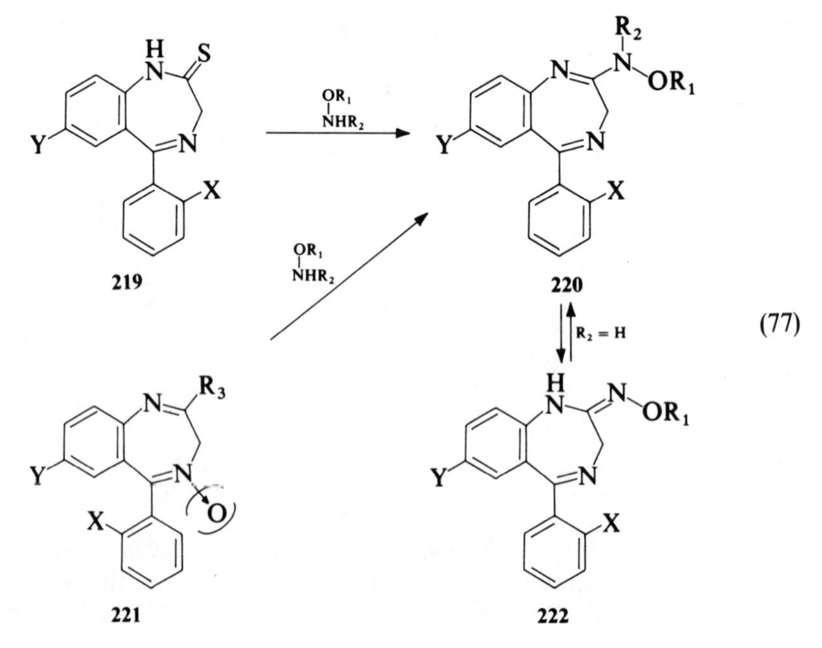


(77)

methylthio[135] and the nitrosomethylamino moieties.[92] The products were formulated with either an endo- or exocyclic double bond. The tautomers **222** appear to be more compatible with the spectroscopic data[92] than **220** ($R_2$ = H).

### 2.3.2. *Reactions*

Alkylation of the 4-oxide of **223** (R = H) with ethyl bromoacetate and potassium *t*-butoxide as base led to the *O*-alkylated product **224** ($R_1$ = $CH_2COOEt$, 4-oxide), which was reduced to **224** ($R_1$ = $CH_2COOEt$) by means of phosphorus trichloride.[5a]

Addition of vinyl ether to the 4-oxide of **223** (R = H) yielded the acetal **224** [$R_1$ = CH(Me)OEt, 4-oxide] (Eq. 78).[5a]

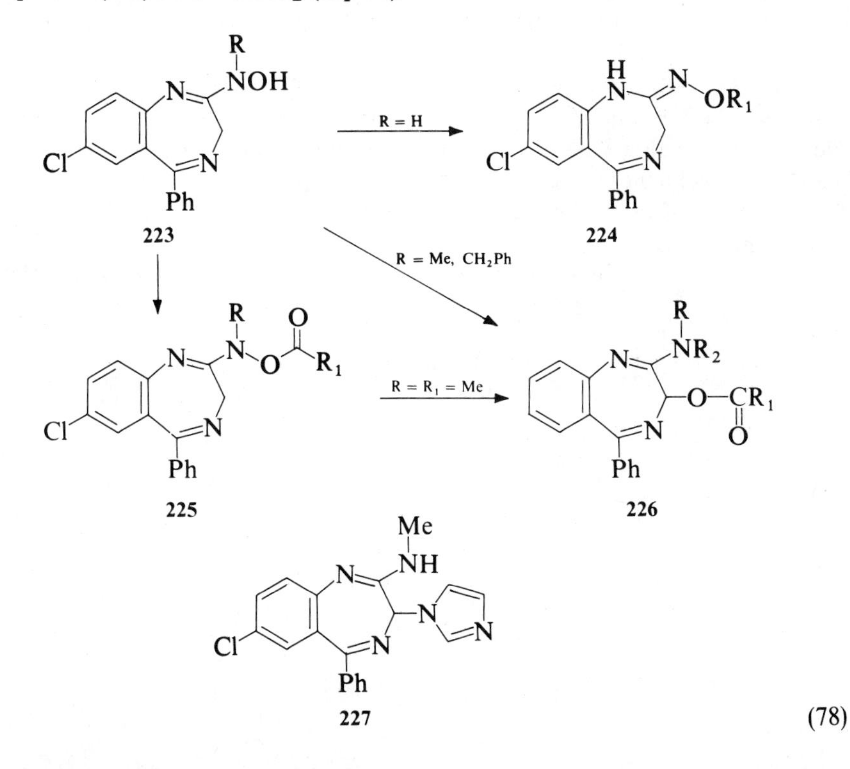


(78)

Acylations of 2-hydroxyaminobenzodiazepines occurred on oxygen as shown by the conversion of **223** (R = H) to the *O*-acetate **225** (R = H, $R_1$ = Me) by acetic anhydride in pyridine,[135] by reaction of **223** (R = Me) with acetyl chloride or benzoyl chloride in the cold,[112a] and by treatment of **223** (R = Me) with isocyanates.[112]

While the reaction of **223** (R = H) with carbonyldiimidazole yielded **225** (R = H, $R_1$ = imidazolyl), the corresponding *N*-methyl compound **223** (R = Me) was converted to the 3-(1-imidazolyl)benzodiazepine **227**[112a] under comparable conditions. This transformation involves formally a reduction of the nitrogen and oxidation of the carbon in the 3-position. This type of rearrangement was also observed when **223** (R = Me or $CH_2Ph$) was heated at 100°C with acetic anhydride, leading to **226** (R = Me or $CH_2Ph$, $R_1$ = Me, $R_2$ = H) in high yields.[112a] Compound **226** (R = $R_1$ = Me, $R_2$ = H) was also formed by heating the *O*-acetate **225** (R = $R_1$ = Me) with acetic acid in dimethylformamide.[112a] The corresponding propionate **226** (R = Me, $R_1$ = Et, $R_2$ = H) was similarly obtained by heating **225** (R = $R_1$ = Me) with propionic acid. Reaction of **223** (R = Me) with a mixture of formic acid and acetic anhydride at room temperature led to the *N*-formyl derivative **226** (R = $R_1$ = Me, $R_2$ = CHO).[112a] These experiments indicate that the rearrangement of **223** or **225** to **226** is not intramolecular. The authors[112a] propose the following mechanism: a nitronium ion is generated by cleavage of the N—O bond, which can then isomerize to a carbonium ion (3-position) and is trapped by the nucleophile.

Phosgene in the presence of triethylamine converted **223** (R = H) to the oxadiazolone **229** (Eq. 79).[132, 134, 135] When the same reagents were applied to the *N*-methyl analog **223** (R = Me), the oxazolone **171** was formed in 20% yield with rearrangement.[112a] Compound **171** was prepared in higher yield by thermal cyclization of the carbonate **226** (R = Me, $R_1$ = EtO, $R_2$ = H).[112a] Treatment of **223** (R = Me) with ethyl chloroformate in tetrahydrofuran resulted in rearrangement with formation of the 3-ethoxy derivative **228**.[112a]

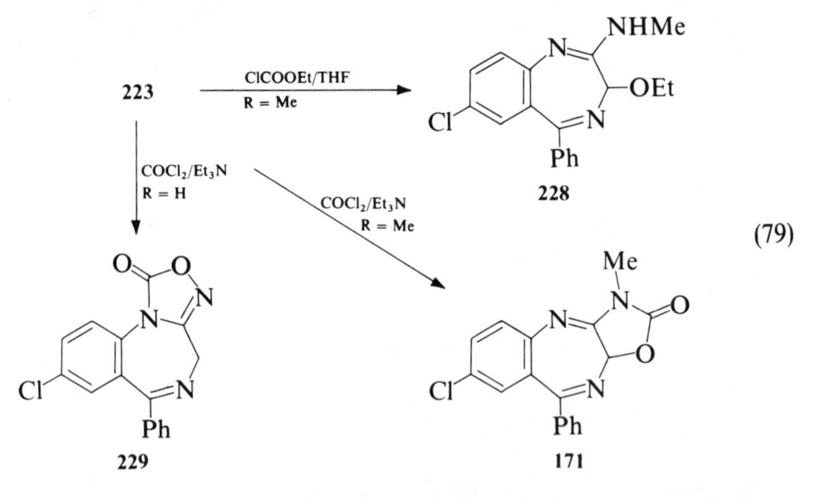


## 2.4. 2-Hydrazino-3*H*-1,4-benzodiazepines

### *2.4.1. Synthesis*

The first reported synthesis[63] of a 2-hydrazinobenzodiazepine by ring expansion of 2-chloromethyl-6-chloro-4-phenylquinazoline 3-oxide with hydrazine was later shown to be in error.[136, 137] Compounds **230** were first prepared by reacting the 2-thiones **86** with hydrazine[134, 136–141] or by exchanging the amino group in **231** with hydrazine in the presence of an acid catalyst (Eq. 80).[47, 136–140] The latter method was also suitable to prepare the 4-oxides of **230**.[47, 136–140]

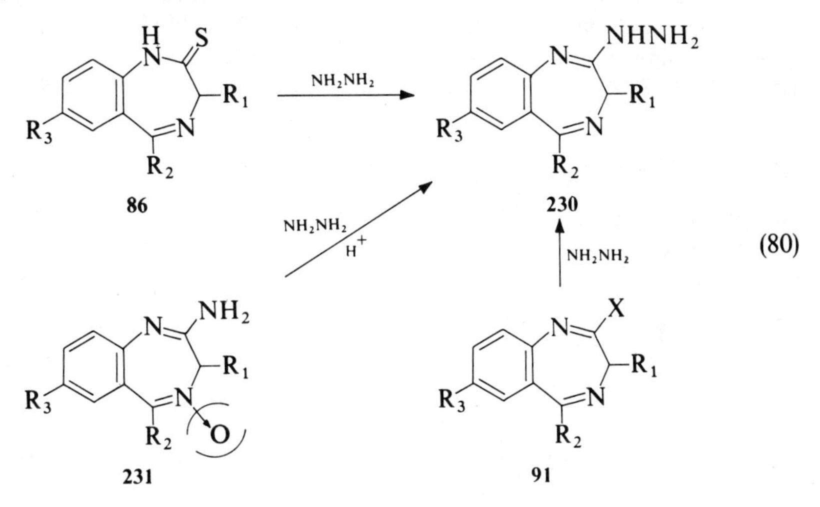


(80)

The benzodiazepines **91** with a leaving group X in the 2-position were employed as well for the synthesis of 2-hydrazino compounds. Suitable groups displaced with hydrazine were benzylthio,[140] methylthio,[136–138] ethoxy,[142] nitrosomethylamino,[92, 143, 144] and dimorpholinophosphoryloxy.[87]

Substituted hydrazines reacted in a similar fashion. Treatment of the thione **86** ($R_1$ = H, $R_2$ = Ph, $R_3$ = Cl) with 1,2-dimethylhydrazine gave **232** ($R_1$ = Me, $R_2$ = Me, $R_3$ = H) in 95% yield.[145] Reaction of the same thione with methylhydrazine led to a mixture of hydrazinobenzodiazepines **232** ($R_1$ = Me, $R_2$ = $R_3$ = H) and ($R_1$ = $R_3$ = H, $R_2$ = Me).[146–149] The former compound, resulting from the attack of the thione by the more nucleophilic methylated nitrogen, could be crystallized from the mixture.[148] The corresponding 4-oxide was obtained by treatment of *N*-nitrosochlordiazepoxide with methylhydrazine.[92] The structure was confirmed by conversion to a formaldehyde hydrazone[92] and by an X-ray analysis of the acetaldehyde hydrazone of **232** ($R_1$ = Me,

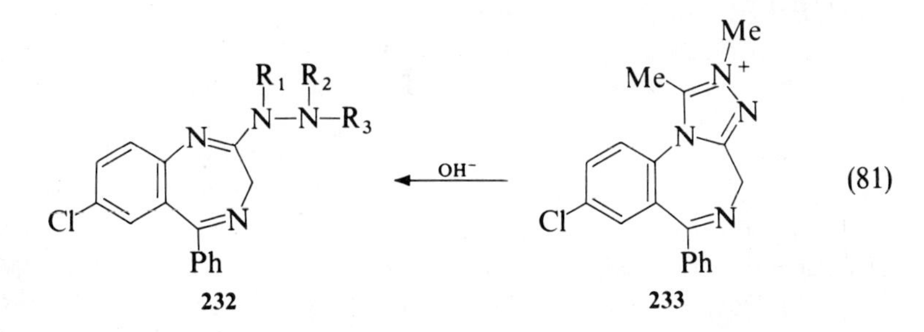


(81)

$R_2 = R_3 = H$). Acid-catalyzed displacement of the methylamino group in chlordiazepoxide with methylhydrazine led to the 4-oxide of **232** ($R_1 = R_2 = H$, $R_3 = Me$). This compound was not characterized but was directly converted to triazolobenzodiazepines.[5a]

Acylated hydrazines such as acetylhydrazine or ethoxycarbonylhydrazine underwent reactions with 2-thiones,[150, 152] *N*-nitrosochlordiazepoxide,[143, 144] and the 2-(dimorpholinophosphoryloxy) derivative[87] to give the hydrazides **232** ($R_1 = R_2 = H$, $R_3 =$ Ac or COOEt) or the corresponding 4-oxides. Since these hydrazides may be cyclodehydrated in situ to triazolobenzodiazepines (see Section 2.4.2, Reactions), they were often not isolated.

The acetylhydrazine derivative **232** ($R_1 = H$, $R_2 = Me$, $R_3 = Ac$) was formed by hydrolytic ring opening of the quaternized triazolobenzodiazepine **233** (Eq. 81).[148]

## *2.4.2. Reactions*

### 2.4.2.1. Reactions with Electrophiles

A. Nitrosation. Nitrosation of the 2-hydrazinobenzodiazepine **232** ($R_1 = R_2 = R_3 = H$) gave the tetrazolobenzodiazepine **234** in 80% yield.[134, 153]

B. Reactions with Aldehydes and Ketones. The hydrazones **235** resulted from condensation of the appropriate hydrazine with aldehydes or ketones.[136, 145, 148, 154, 155] If $R_3$ and $R_5$ in the hydrazones **235** are both hydrogen, these compounds may be in equilibrium with the cyclic form **237**. The latter may be removed from the equilibrium by oxidation to the triazole **239** (Eq. 82).

Oxidizing agents used to convert **235** to **239** include air,[155] manganese dioxide,[154] and diethyl azodicarboxylate.[154] Reaction of the hydrazine **232** ($R_1 = R_2 = H$, $R_3 = Me$) or its 4-oxide with aliphatic aldehydes yielded the dihydrotriazolobenzodiazepines **236**[146-148] or their 5-oxides,[5a] respectively. Compound **236** (R = Me) suffered oxidation and hydrolytic cleavage to the hydrazide **238** in the presence of air and water.[148] Since **238** was also obtained by hydrolysis of the quaternary triazolobenzodiazepine **233**, the latter is most likely the intermediate in this conversion.

Hydrazones of functionalized aldehydes and ketones were used for the synthesis of other heterofused benzodiazepines as well. Thus the α-chlorohydrazones **240** underwent an intramolecular alkylation to form the triazinobenzodiazepines **241**.[156, 157] Examples for $R_1 = H$,[156] Me,[156] and $CH_2Cl$[151] were reported (Eq. 83).

Treatment of the dimethylacetal of hydrazones **242** ($R_1 = Me$, $R_2 = H$) with concentrated sulfuric acid yielded more than 50% of the triazino derivatives **241** ($R_1 = Me$, $R_2 = OH$).[158] Hydrogen fluoride at $-80°C$ cyclized the hydrazone of butane-2,3-dione **242** ($R_1 = R_2 = Me$, $X = H$) to the methylene derivative **243** ($R_1 = Me$, $X = H$, $Y = CH_2$) in 37% yield.[159] Hydrazones of α-ketoacids were converted to the triazinobenzodiazepines **243** ($Y = O$) by treatment with

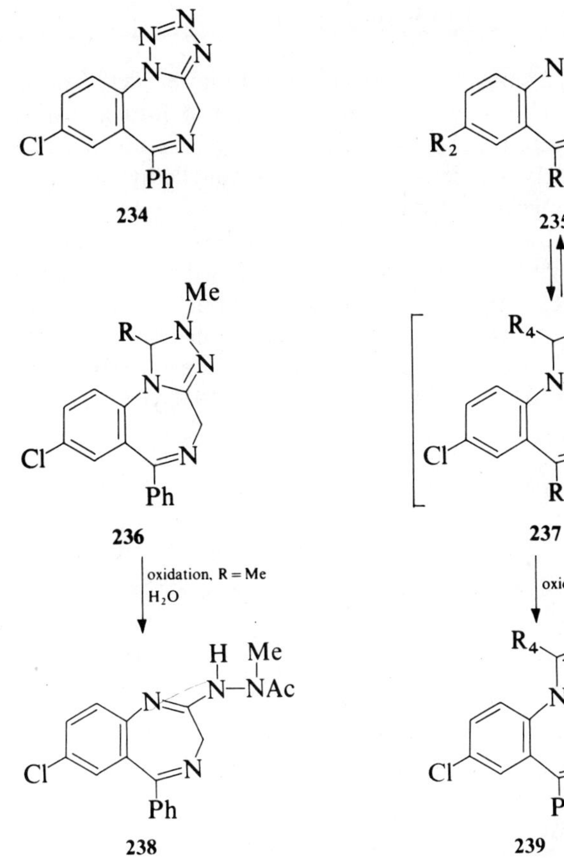
234
235
$R_3 = R_5 = H$
236
237
oxidation, R = Me
$H_2O$
oxidation
238

(82)

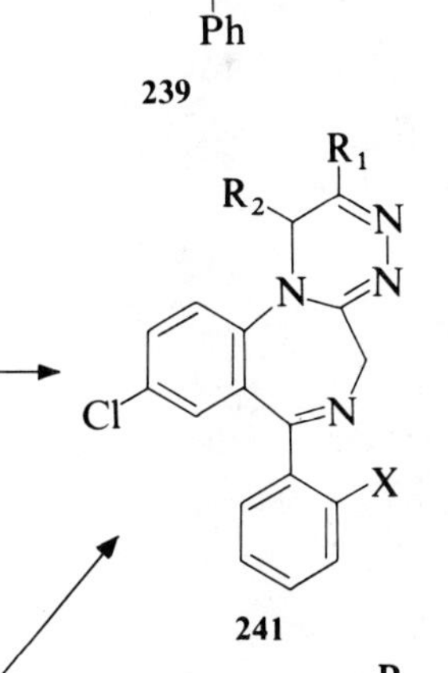
239
241

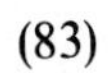
(83)

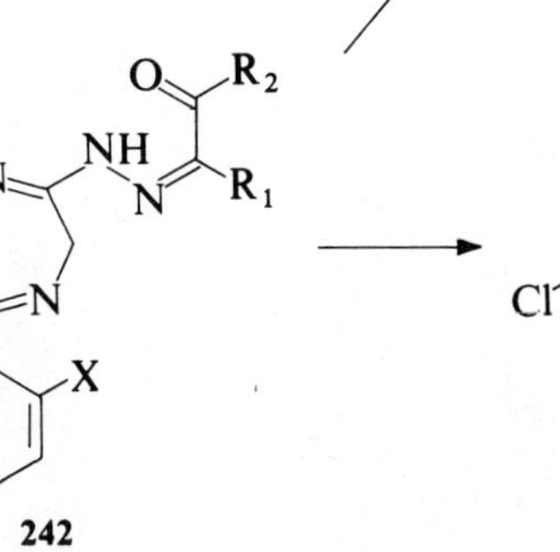
242

243

carbonyldiimidazole in boiling tetrahydrofuran or in lower yield by heating in glacial acetic acid.[160] The same compounds were also formed by refluxing the esters **242** ($R_2$ = RO) in 1,2,4-trichlorobenzene.[160] Compounds **243** with $R_1$ = H, Me, (ethoxycarbonyl)methyl, 2-(ethoxycarbonyl)ethyl, and 3-(diethylamino)propyl were synthesized in this fashion.[160]

The reaction of 2-hydrazinobenzodiazepines with 1,3-dicarbonyl compounds was studied also.[161-163] Pentane-2,4-dione condensed with the hydrazinobenzodiazepine **230** ($R_1$ = H, $R_2$ = Ph, $R_3$ = Cl) or its 4-oxide to give the pyrazoline **244**, which dehydrated to the pyrazole **245** upon treatment with trifluoroacetic acid.[161] Boiling the 4-oxide of **244** in xylene gave 40% yield of the triazolobenzodiazepine **239** ($R_4$ = Me, 5-oxide) beside 19% yield of the pyrazole **245** (4-oxide).

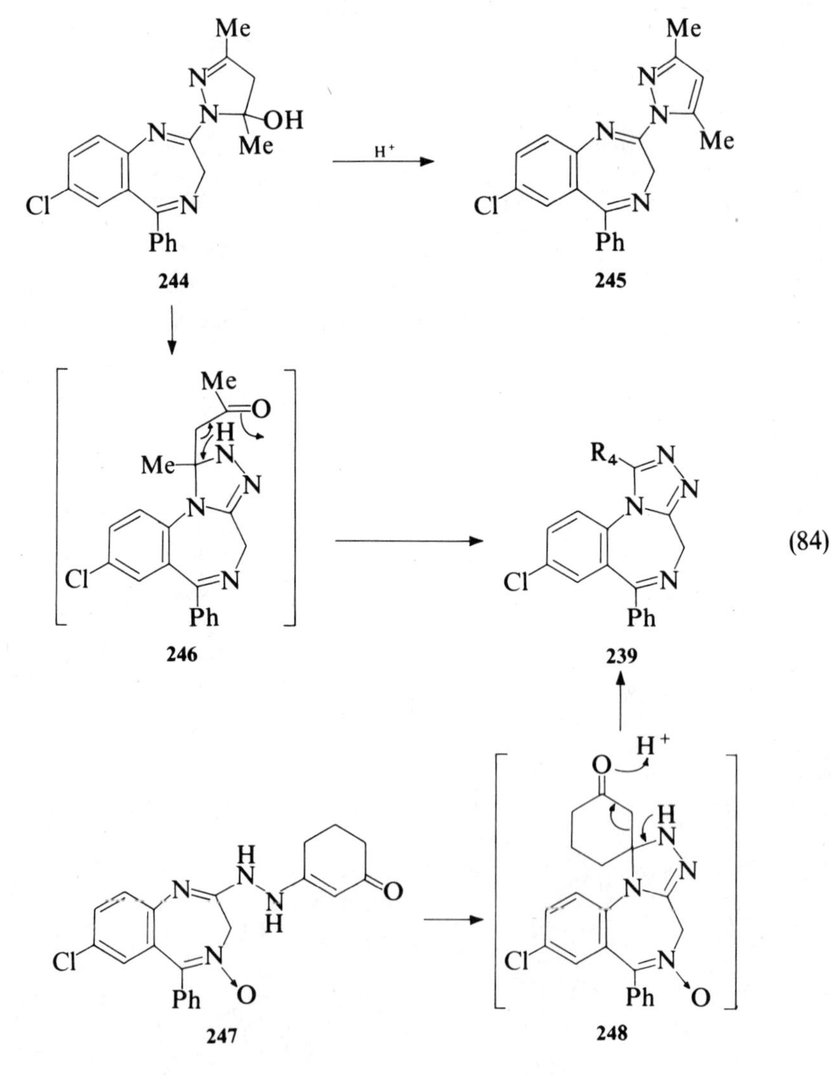

The triazolobenzodiazepine **239** ($R_4$ = Me) was the exclusive product when **244** was heated to reflux in butanol in the presence of triethylamine overnight (Eq. 84).[161–163] The hydrazone of the cyclohexane-1,3-dione **247** was converted to the triazolobenzodiazepine **239** ($R_4$ = 4-oxobutyl, 5-oxide) under similar conditions, although in much lower yield. The spiro compound **248** may be an intermediate in this reaction.[161]

Methyl acetoacetate formed the hydrazone **249**, which partially cyclized to the pyrazolone **250** upon boiling in butanol–triethylamine. The major product formed was the 2-butoxybenzodiazepine **251** (Eq. 85).[161]

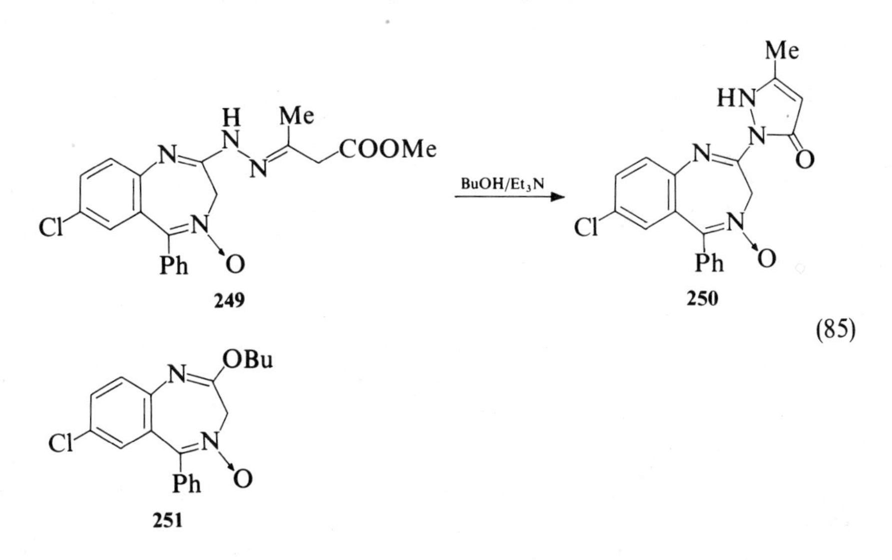


(85)

Since the pyrazolone **250** did not convert to **251** under the reaction conditions, the latter must have been formed by displacement of the 2-substituent from **249** or of some other intermediate species.

C. Acylation. 2-Hydrazinobenzodiazepines generally acylate at the terminal nitrogen to give **252**. Such acylations have been carried out with acid anhydrides,[47,139,143,144,148] acid chlorides,[47,145,164] isocyanates,[138,165] and isothiocyanates.[138,165] Acylations with concomitant ring closure by dehydration led to the triazolobenzodiazepines **253**.[140,141,166–172] These compounds were also obtained in high yields by reaction of 2-hydrazinobenzodiazepines with orthoesters.[47,136,137,139,143,144,172,173] Heating the 2-hydrazinobenzodiazepine with trifluoroacetic acid at 100°C for 2 hours yielded 19% of the 1-trifluoromethyltriazolobenzodiazepine **253** ($R_1$ = Ph, $R_2$ = Cl, $R_3$ = $CF_3$).[172] Formic acid reacted in a similar fashion to give **253** ($R_1$ = Ph, $R_2$ = Cl, $R_3$ = H).

The cyclization–dehydration of the hydrazides **252** (X = O) to the triazoles **253** was carried out by heating,[134,140,146,152] by refluxing in butanol,[134,171] pyridine,[139,140,146] or acetic acid[166–171] and by heating with polyphosphoric acid (Eq. 86).[140]

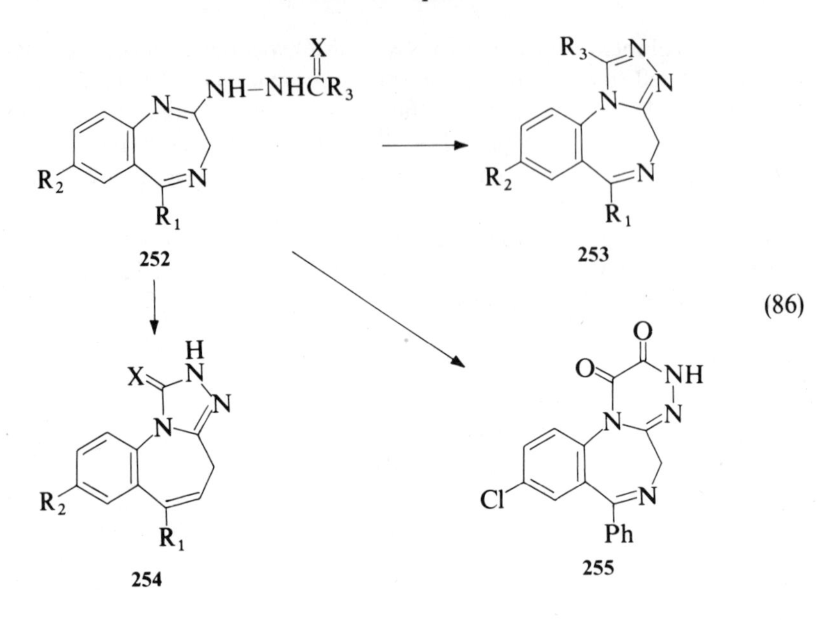


Compounds **252** (X = NH, $R_1$ = Ph, $R_2$ = Cl, $R_3$ = Me), which were accessible by reacting the 2-hydrazinobenzodiazepine with acetimidate or acetamidine, were thermally cyclized to the triazolo compounds.[140] The triazolobenzodiazepine **253** ($R_1$ = Ph, $R_2$ = Cl, $R_3$ = H) was also formed by treatment of the 2-hydrazino derivative with formamide and sulfuric acid at 100°C, or by heating with formamidine hydrochloride in the presence of 2-methylimidazole at 160°C.[140]

Heating the carbethoxyhydrazine derivatives **252** ($R_1$ = Ph or 2-Cl$C_6H_4$, $R_2$ = Cl, $R_3$ = EtO, X = O) at about 200–230°C effected cyclization to the triazolone **254** (X = O).[134,152,174] The same compounds were obtained by thermolysis or long refluxing in pyridine of the semicarbazones **252** ($R_3$ = NHR, X = O).[138,165] The thiosemicarbazones **252** ($R_3$ = NHR, X = S) underwent ring closure under the same conditions to the triazolethione **254** (X = S).[138,165] This compound was also formed by treating the 2-hydrazinobenzodiazepine with thiophosgene.[172,175,176] Cyanogen bromide reacted with the 2-hydrazinobenzodiazepine to give the 1-aminotriazolo compound **253** ($R_1$ = Ph, $R_2$ = Cl, $R_3$ = $NH_2$).[172,175,176]

Cyclization of the oxalyl derivative **252** ($R_1$ = Ph, $R_2$ = Cl, $R_3$ = COOEt, X = O) by boiling in pyridine for 2 hours yielded 80% of the triazinobenzodiazepine **255**.[164] The 4-oxide of **256** ($R_1$ = H, $R_2$ = Me) reacted with phosgene to form the triazolone **257**.[5a] The thiatriazolobenzodiazepine **258** was synthesized by treating the hydrazide **256** ($R_1$ = H, $R_2$ = Ac) with thionyl chloride in pyridine (Eq. 87).[177]

Reaction of the methylhydrazine **256** ($R_1$ = Me, $R_2$ = H) with orthoesters under acid catalysis yielded the quaternary triazolium salts **259** (R = H, Me).[148,149]

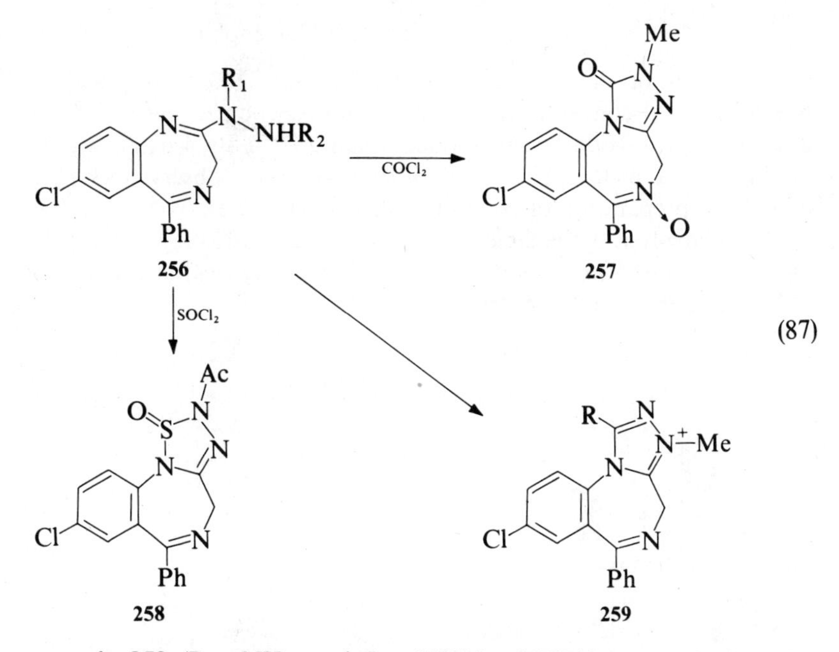


Compounds **259** ($R = NH_2$ and $R = NHMe$, NHPh) were obtained by condensation of the same hydrazine with cyanogen bromide and dichloroimines, respectively.[148,149] Phosgene and thiophosgene converted this hydrazine to zwitterionic compounds **259** ($R = O^-$, $S^-$).[148,149]

The 2-hydrazinobenzodiazepines **260**, which are disubstituted on the terminal nitrogen, were cyclized to the triazinobenzodiazepines **261** (R = H, Me) by treatment with glacial acetic acid (Eq. 88).[145] The structure **262** (R = Ac) was assigned to the product formed in 30% yield by treating compound **262** (R = H) with acetic anhydride in acetic acid.[145]

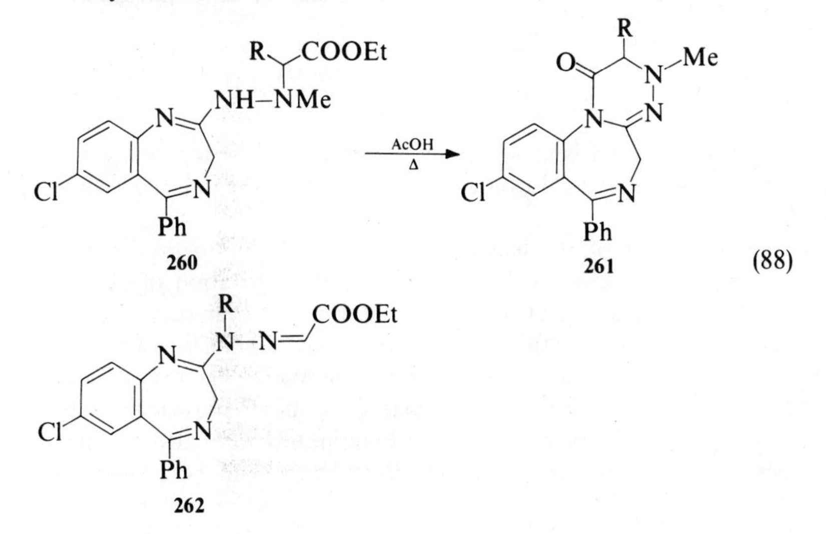

#### 2.4.2.2. Reactions with Nucleophiles

Treatment of the 2-hydrazinobenzodiazepine **263** with Raney nickel in boiling ethanol converted it to the 2-aminobenzodiazepine **52**.[137] The dimer **264**[137] or its 4-oxide[5a] were formed when **263** or its 4-oxide was heated in boiling methanol (Eq. 89). These dimers were also observed as by-products during the preparation of 2-hydrazinobenzodiazepines.

Acid hydrolysis of the dimer **264** or the hydrazine **263** regenerated the lactam **265**.[137] The latter was also obtained in 40% yield by treatment of the hydrazine **263** with methanolic hydrogen chloride.[136]

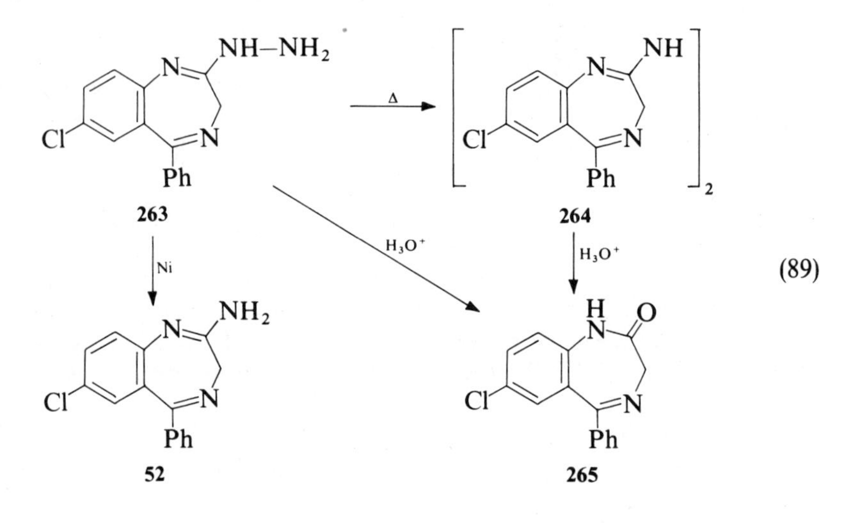


(89)

## 2.5. 2-*O*- and 2-*S*- Substituted 3*H*-1,4-Benzodiazepines

### 2.5.1. *Synthesis*

2-Alkoxy-3*H*-1,4-benzodiazepine 4-oxides **267** were obtained by ring expansion of the 2-chloromethylquinazoline 3-oxides **80** with alkoxides (Eq. 90).[5a,91,178]

Displacement of the leaving group X in 3*H*-1,4-benzodiazepines **268** by alkoxides or phenoxides[90] was used in the preparation of these imidates as well. Suitable leaving groups were chloride,[179] cyanide,[7] methylthio,[180,181] (2-dimethylamino)ethylthio,[181] nitrosomethylamino,[92] dimorpholinophosphoryloxy,[87] and (*N*-hydroxyimino)nitromethyl.[24a] The reaction of the lactams **269** with diazomethane[182-184] or diazopropane[185] provided another access to 2-alkoxybenzodiazepines. The latter compounds were also obtained by alkylation of the 2-ones **269** using an alkyl halide and potassium carbonate in acetone.[70]

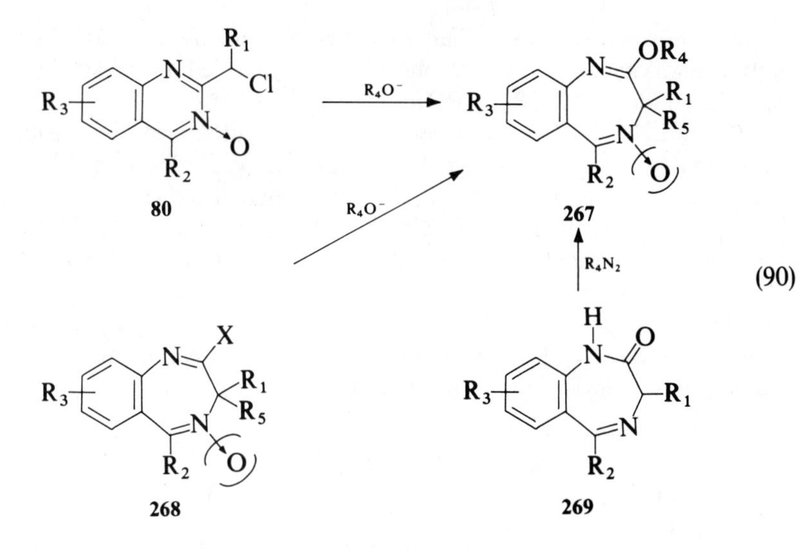


Reductive ring closure of the azides **270** ($X = N_3$) with Raney nickel in ethanol or zinc and ammonium chloride led to the 2-ethoxybenzodiazepines **271**. Triphenylphosphine as reducing agent allowed the conversion of **270** ($X = N_3$, $R = NO_2$) to the corresponding benzodiazepine.[142] Compounds **271** were also prepared by treatment of the bromide **270** (X = Br) with liquid ammonia in the presence of sodium iodide (Eq. 91).[142]

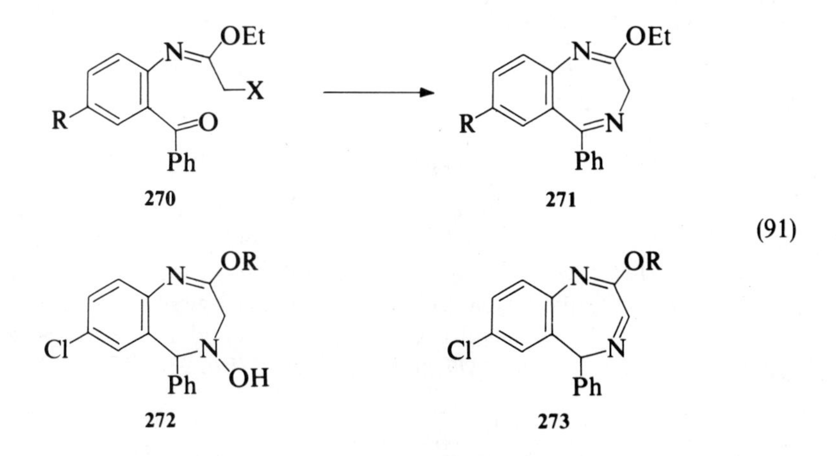


The dehydration of the hydroxylamines **272** (R = Me, Pr) by refluxing in 1,4-dimethylpiperazine or in methanol containing methoxide gave the corresponding 3*H*-1,4-benzodiazepines.[185] The 5-*H*-benzodiazepines **273** were also converted to the 3*H* tautomers under these conditions.[185]

Phosphorylation of the lactams **269** on oxygen with chlorophosphates[20,87,88] and base gave compounds **267** [$R_4 = PO(OR_6)_2$], which were generally not isolated but further reacted with nucleophiles in situ. Dimorpholinophosphorylchloride as phosphorylating agent led to crystalline isolable phosphorimidates **267** [$R_4$ = PO(morpholino)$_2$].[21,87]

2-Thioalkyl-3*H*-1,4-benzodiazepines were synthesized by reacting the 2-thiones **86** with alkylating agents. Alkylations were carried out with a variety of reagents such as dimethyl sulfate,[71,72,73] methyl iodide,[72,73] alkyl halides containing tertiary amino groups,[181,186] and bromoacetic acid.[187] Reaction of *N*-nitrosochlordiazepoxide with ethanethiol and base gave the 2-ethylthio derivative in moderate yield.[92]

Direct conversion of the lactam **269** to the 2-thioalkyl-3*H*-1,4-benzodiazepines **268** (X = SR) was achieved by means of a combination of thiol and titanium tetrachloride[82a,188] or tetrachlorosilane.[188]

### *2.5.2. Reactions*

#### 2.5.2.1. Reactions with Electrophiles

The acetic acids **274** (R = H, Me) were cyclized to the thiazolobenzodiazepines **275** by treatment with acetic anhydride and triethylamine (Eq. 92).[187]

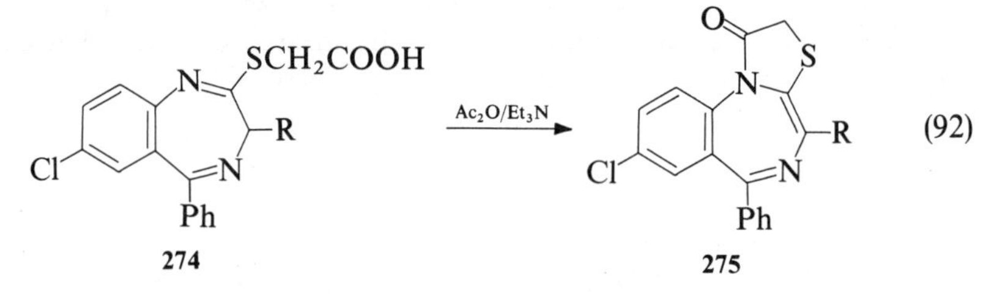


#### 2.5.2.2. Reactions with Nucleophiles

The utility of 2-alkylthio, 2-alkoxy, and 2-phosphoryloxy benzodiazepines for the synthesis of 2-amino, 2-hydroxyamino, and 2-hydrazino derivatives was discussed earlier during the review of the synthesis of these compounds.

Hydrolyses of 2-alkoxy and 2-thiomethyl benzodiazepines led to the lactams. Low temperature hydrolysis cleaved the 1,2-imine bond in 2-alkoxy benzodiazepines **276** to give the esters **277** in 45–50% yield (Eq. 93).[181]

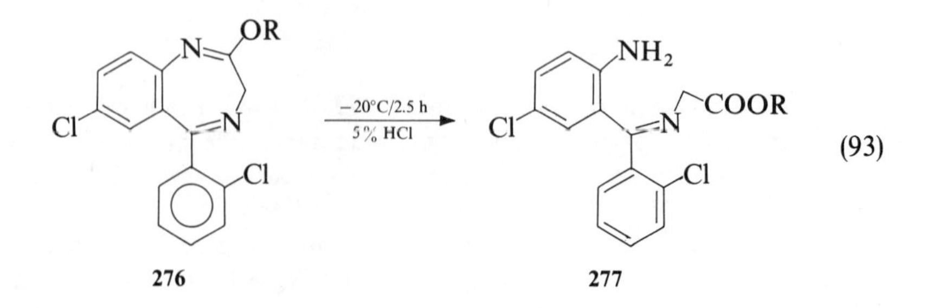

The phosphoryl derivative **278** hydrolyzed to the lactam **265** in 52% yield upon standing in water–tetrahydrofuran for 7 days at room temperature.[87a] The 2-morpholino compound **279** was the major by-product of this hydrolysis (Eq. 94).[87a]

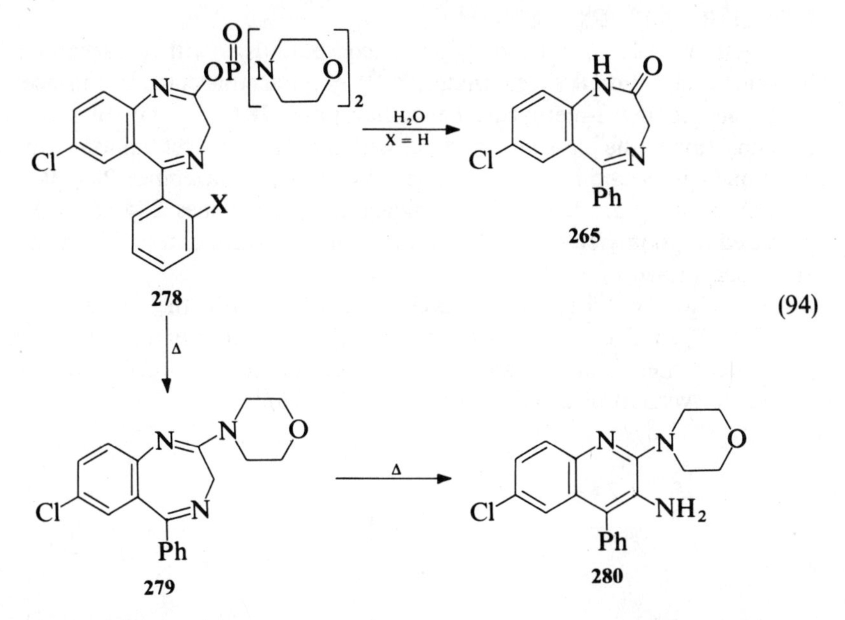


(94)

2-Methylthio benzodiazepines **281** ($R_1$ = SMe) were converted to triazolobenzodiazepines **282** in one step by heating with acylhydrazines in hexamethyl phosphoric triamide at 100–140°C (Eq. 95).[125,189–191]

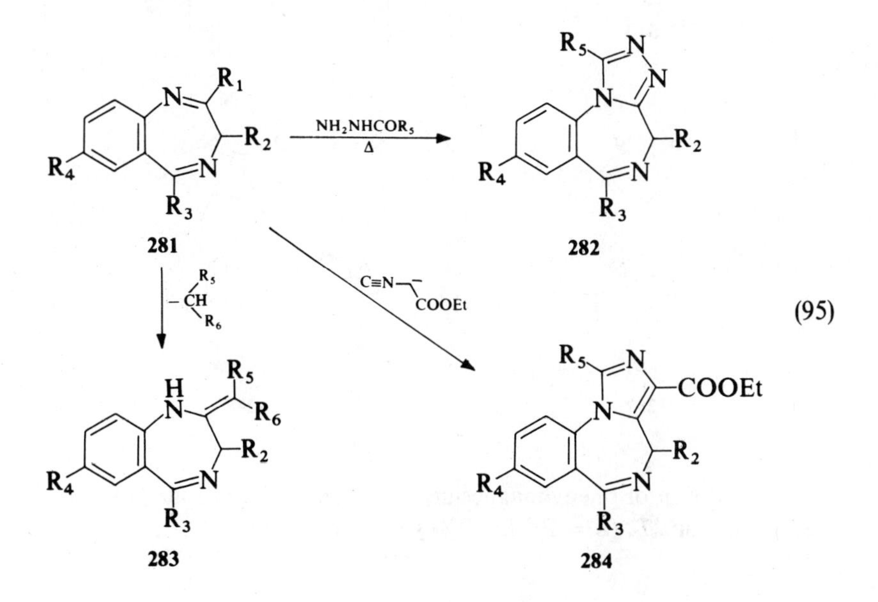


(95)

Reaction of **278** (X = H) with acetylhydrazine in boiling butanol also gave the triazolobenzodiazepine **282** ($R_2$ = H, $R_3$ = Ph, $R_4$ = Cl, $R_5$ = Me) in one step and in 80% yield.[87a] The same compound was claimed to be formed under similar conditions from the 2-methoxy benzodiazepine **281** ($R_1$ = MeO, $R_2$ = H, $R_3$ = Ph, $R_4$ = CN).[184]

Reaction of 2-phosphoryloxy compounds with carbanions of nitromethane,[87a] dialkyl malonates,[20,87a] ethyl acetoacetate,[20] and acetylacetone[20] led to the 2-methylene benzodiazepines **283** (Eq. 95). In some cases, condensation of the phosphorimidates with the anion of acetaminomalonic ester mentioned in Section 2.1.1 gave the imidazobenzodiazepines **284** ($R_5$ = Me) directly as a by-product.[21] The imidazobenzodiazepines **284** ($R_5$ = H) were prepared in good yields by reaction of the phosphorimidates with the anion of ethyl isocyanoacetate.[127]

The anion of the nitrone **285** condensed with the phosphorimidate **278** (X = F) to give the imidazobenzodiazepine **288** in a one-pot reaction. The postulated intermediates **286** and **287** indicate the proposed course of this overall conversion of **278** (X = F) to **288** (Eq. 96).[192]

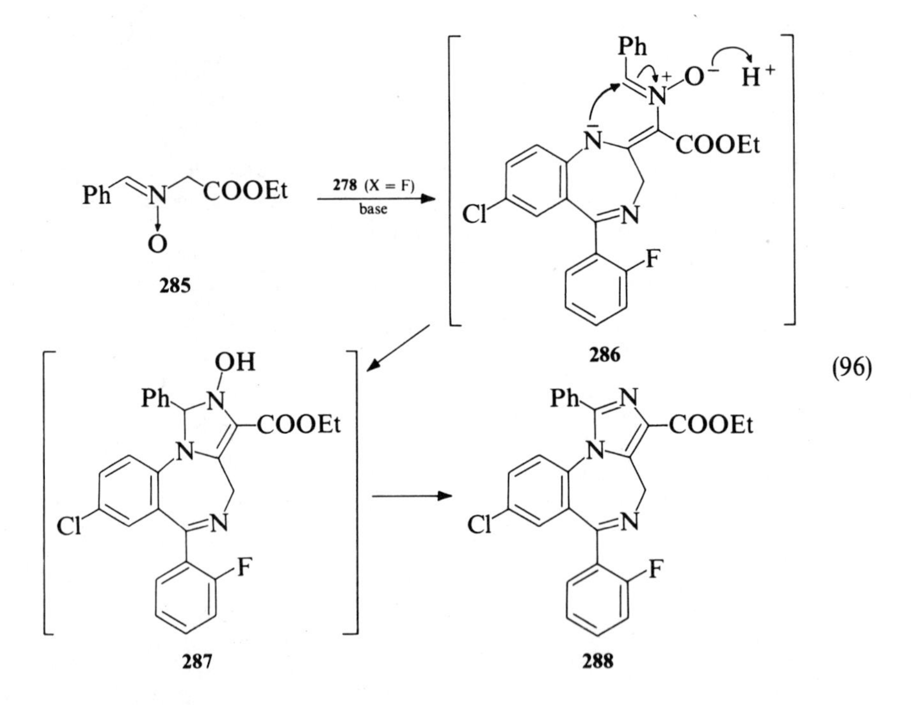


(96)

The addition of phenylmagnesium bromide to the nitrone **289** afforded the hydroxyamine **272** (R = Pr) in 70% yield (Eq. 97).[185]

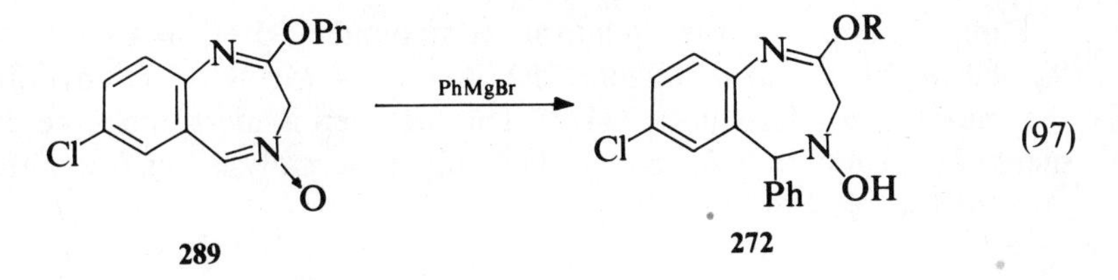


### 2.5.2.3. Thermal Reactions

Thermolysis of the phosphorimidate **278** (X = H) in boiling 1,2,4-trichlorobenzene gave the 2-morpholinobenzodiazepine **279** in 26% yield, in addition to 17% of the 3-aminoquinoline **280** (Eq. 94).[87a] The latter was also formed in 35% yield by heating **279** in this solvent at reflux for 4 hours.[87a]

## 2.6. 2-Halo-3*H*-1,4-benzodiazepines

### *2.6.1. Synthesis*

2-Halobenzodiazepines **291** (X = F, Cl, Br) were claimed to be accessible by reaction of the corresponding lactam **290** with carbonyl dihalides in a mixture of benzene and pyridine at room temperature.[193] The yields appear to be generally low. An exception is the 3-methoxy derivative **291** ($R_1$ = MeO, $R_2$ = Cl, $R_3$ = Cl), which was obtained in 50% yield by treatment of the corresponding lactam with phosphorus pentachloride in boiling methylene chloride (Eq. 98).[194]

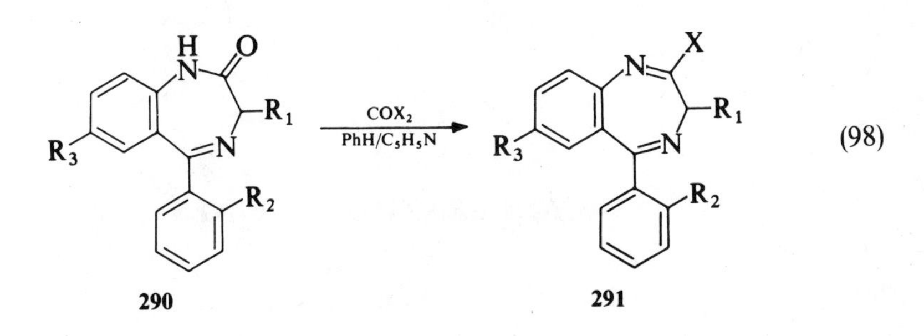


### *2.6.2. Reactions*

2-Halobenzodiazepines were subjected to nucleophilic displacement reactions with ammonia,[89] amines,[193] imidazole,[86] and ethoxide.[179] Trialkylphosphite reacted with **291** (X = Cl, $R_1$ = MeO, $R_2$ = $R_3$ = Cl) to give the 2-phosphonates **291** [X = $PO(OR)_2$] in 50–75% yield.[195] Ammonia at 50°C converted the latter to the 2-aminobenzodiazepine.[89]

Short boiling of the 2-chloro compound **291** ($X = Cl$, $R_1 = MeO$, $R_2 = R_3 = Cl$) with the tetrazoles **292** ($R$ = Me, 4-$MeOC_6H_4$) in pyridine gave the triazolobenzodiazepines **294**.[196] The first step is most likely the displacement of the chloride to form **293** which, upon thermolysis, can lose nitrogen to give the triazole **294** (Eq. 99).

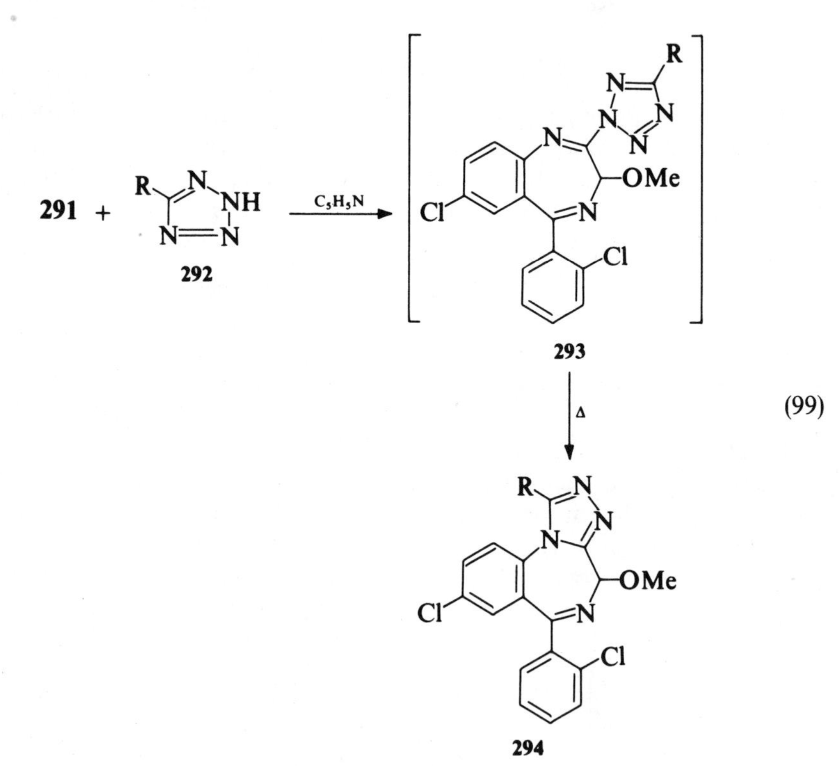


# 3. 5*H*-1,4-BENZODIAZEPINES

## 3.1. Synthesis

5-Phenyl-5*H*-1,4-benzodiazepines **298** ($R_1$ = H, Me, $ClCH_2$; $R_2$ = Cl, $NO_2$) were obtained by thermal rearrangement of the azirinoquinazolines **296**, which were accessible by treating the appropriate 2-chloromethyl-1,2-dihydroquinazolines **295** with potassium *t*-butoxide in tetrahydrofuran (Eq. 100).[11–14,16,17]

Compound **298** ($R_1$ = Me, $R_2$ = Cl) was also prepared in 40% yield by oxidation of the 2,5-dihydro-1*H*-1,4-benzodiazepine **297** with manganese dioxide.[12–14,17] Reaction of **298** ($R_1$ = $ClCH_2$, $R_2$ = Cl) with dimethylamine yielded the corresponding amine **298** ($R_1$ = $Me_2NCH_2$, $R_2$ = Cl).[18]

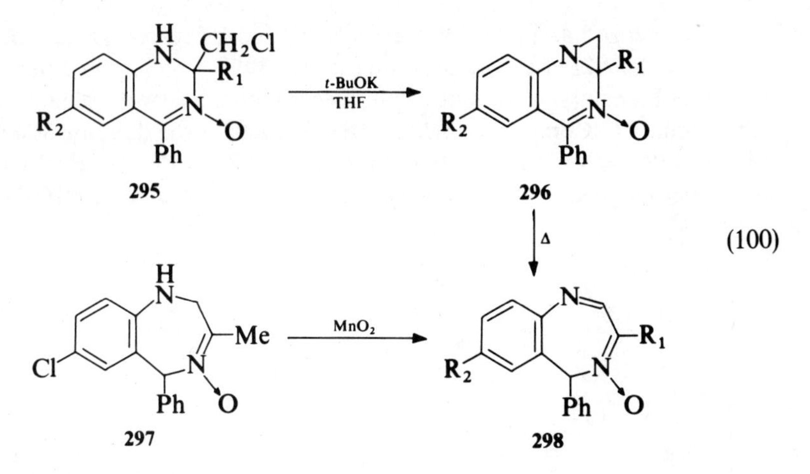


The structure **300** was assigned to the products obtained by treatment of the quinazolines **299** with methoxide (Eq. 101).[18]

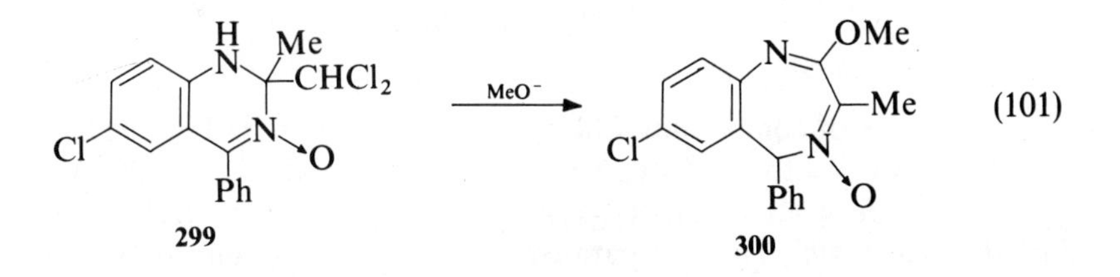


The 5*H* tautomer **302** of chlordiazepoxide was formed in 70% yield by oxidation of the hydroxyamine **301** (R = NHMe) with potassium ferricyanide[197] or hypochlorite.[197] Dehydration of **301** (R = PrO) by refluxing with phenyl isocyanate in a mixture of isobutyl acetate and 1,4-dimethylpiperazine for 15 minutes gave a 96% yield of the 5*H*-benzodiazepine **303** (Eq. 102).[185]

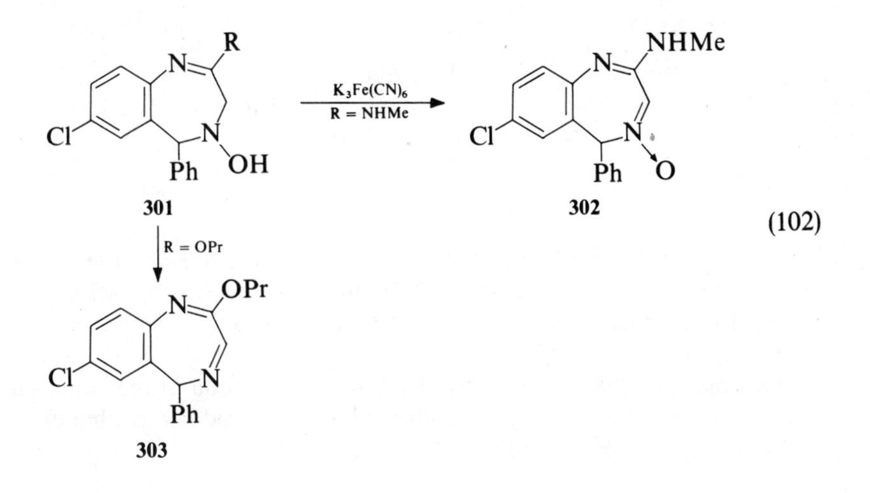

5*H*-1,4-Benzodiazepines **306** were synthesized by rearrangement and dehydration of the 1,2,3,4-tetrahydroquinazolines **305** by boiling in benzene in the presence of a catalytic amount of *p*-toluenesulfonic acid with separation of the water formed.[18] Compounds **305** are the adducts of the diamine **304** with 1,2-diketones ($R_3$ = Me, Ph) (Eq. 103). In the case of $R_3$ = Ph, the relative positions of the methyl group and the phenyl moiety in **306** were not established.[18]

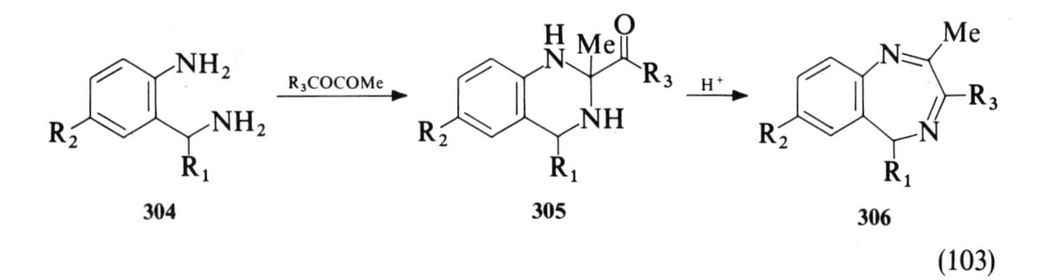


(103)

## 3.2. Reactions

Reductions of compounds **298** with lithium aluminum hydride or borohydride reagents led to the 4-hydroxy tetrahydrobenzodiazepines **307**.[11–14,16,17] The reduction product **307** ($R_1$ = Me, $R_2$ = Cl) was shown to be a mixture of diastereomers, which were separated by fractional crystallization. The stereochemistry was not assigned.[12,14,17] Reduction of **298** ($R_1$ = $CH_2Cl$, $R_2$ = Cl) with lithium aluminum hydride led to the 3-methyl compound **307** ($R_1$ = Me, $R_2$ = Cl) (Eq. 104).[12–14,17] The tetrahydro derivatives **308**, of unknown stereochemistry, resulted from the treatment of compounds **306** with sodium borohydride in methanol.[18]

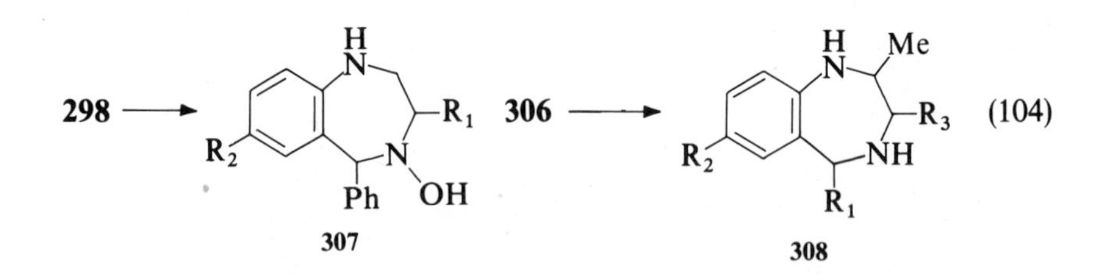


(104)

The 5*H* tautomer **302** was converted to chlordiazepoxide in 80% yield by treatment with alkoxide in ethanol.[197] Similar tautomerization of 2-alkoxy-5*H*-1,4-benzodiazepines such as **303** to the 3*H* tautomers was mentioned earlier in Section 2.5.1.[185]

Treatment of **298** ($R_1$ = H, $R_2$ = Cl) with methoxide in methanol effected tautomerization to the 3*H*-benzodiazepine which added methanol to the 1,2-imine bond to give **309** (Eq. 105).[11]

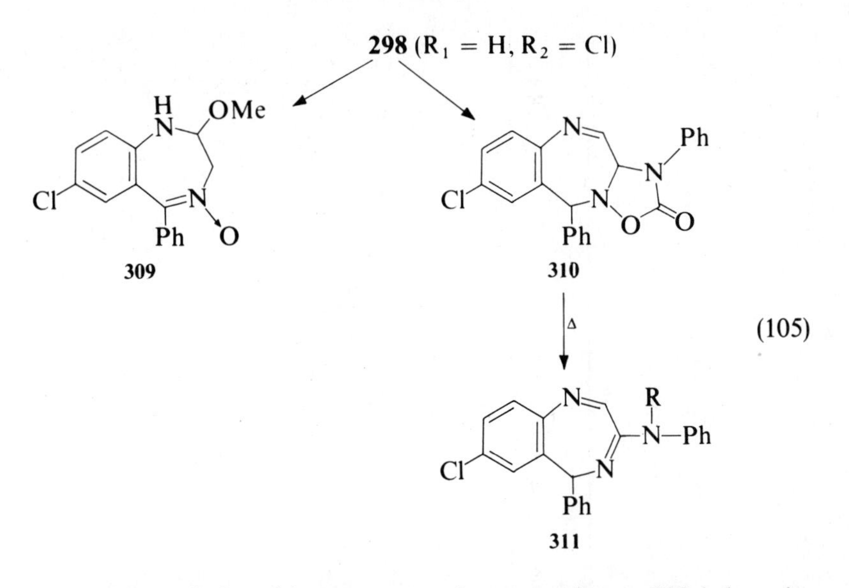


Dipolar addition of phenyl isocyanate to the same nitrone **298** led to the tricyclic compound **310**.[18] Treatment of **310** with triethylamine in benzene effected elimination of carbon dioxide to give a product assigned structure **311** (R = H) on the basis of spectral and analytical data. Acetylation of **311** (R = H) yielded an acetyl derivative formulated as **311** (R = Ac) (Eq. 105).[18]

# 4. TABLES OF COMPOUNDS

TABLE V-1. 1*H*-1,4-BENZODIAZEPINES

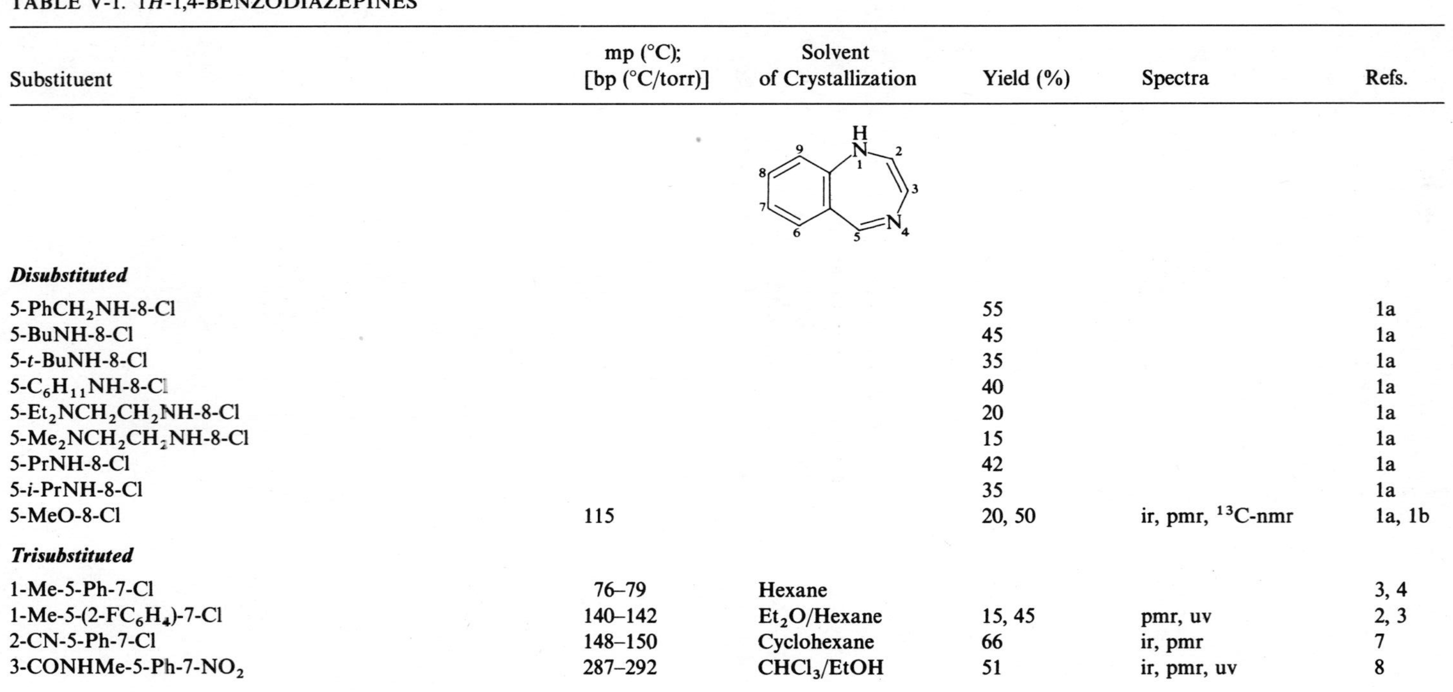


| Substituent | mp (°C); [bp (°C/torr)] | Solvent of Crystallization | Yield (%) | Spectra | Refs. |
|---|---|---|---|---|---|
| ***Disubstituted*** | | | | | |
| 5-$PhCH_2NH$-8-Cl | | | 55 | | 1a |
| 5-BuNH-8-Cl | | | 45 | | 1a |
| 5-*t*-BuNH-8-Cl | | | 35 | | 1a |
| 5-$C_6H_{11}NH$-8-Cl | | | 40 | | 1a |
| 5-$Et_2NCH_2CH_2NH$-8-Cl | | | 20 | | 1a |
| 5-$Me_2NCH_2CH_2NH$-8-Cl | | | 15 | | 1a |
| 5-PrNH-8-Cl | | | 42 | | 1a |
| 5-*i*-PrNH-8-Cl | | | 35 | | 1a |
| 5-MeO-8-Cl | 115 | | 20, 50 | ir, pmr, $^{13}C$-nmr | 1a, 1b |
| ***Trisubstituted*** | | | | | |
| 1-Me-5-Ph-7-Cl | 76–79 | Hexane | | | 3, 4 |
| 1-Me-5-(2-$FC_6H_4$)-7-Cl | 140–142 | $Et_2O$/Hexane | 15, 45 | pmr, uv | 2, 3 |
| 2-CN-5-Ph-7-Cl | 148–150 | Cyclohexane | 66 | ir, pmr | 7 |
| 3-CONHMe-5-Ph-7-$NO_2$ | 287–292 | $CHCl_3$/EtOH | 51 | ir, pmr, uv | 8 |

| | | | | | |
|---|---|---|---|---|---|
| ***Tetrasubstituted*** | | | | | |
| 1-Ac-2-AcNMe-5-Ph-7-Cl | 200–203 | $CH_2Cl_2$/Petr ether | | ir, pmr | 9 |
| 1-EtOOC-2-EtOOCO-5-Ph-7-Cl | 104–107 | EtOAc/Petr ether | | ir, pmr | 5b |
| 1-EtOOC-2-EtOOCO-5-Ph-7-$NO_2$ | 106–110 | EtOAc/Petr ether | | pmr | 5b |
| 1-[$Et_2NCH_2CH_2$]-2-MeS-5-(2-$FC_6H_4$)-7-Cl | Oil | | | | 3 |
| 1-$HOCH_2$-2-$MeNCH_2OH$-5-Ph-7-Cl Hydrochloride | 189–190d | | | | 10 |
| 1-Me-2-EtS-5-(2-$FC_6H_4$)-7-Cl | 103–104 | $CH_2Cl_2$/MeOH | 79 | | 2, 6 |
| 1-Me-2-MeSO-5-Ph-7-Cl | 148–151 | $CH_2Cl_2$/Petr ether | | | 3, 6 |
| 1-Me-2-MeSO-5-(2-$FC_6H_4$)-7-Cl | 130–144 | MeOH | 48 | pmr | 2 |
| 1-Me-2-$MeSO_2$-5-Ph-7-Cl | 184–188 | $CH_2Cl_2$/MeOH | | | 3 |
| 1-Me-2-$MeSO_2$-5-(2-$FC_6H_4$)-7-Cl | 145–151 | $CH_2Cl_2$/MeOH | 55 | pmr | 2 |
| 1-Me-2-MeS-5-Ph-7-Cl | 77–81 | $Et_2O$/Petr ether | | | 4, 6 |
| 1-Me-2-MeS-5-(2-$ClC_6H_4$)-7-$NO_2$ | 135–140 | $CH_2Cl_2$/Petr ether | | | 3, 6 |
| 1-Me-2-MeS-5-(2-$FC_6H_4$)-7-Cl | 72–76 | MeOH | 83 | pmr | 2 |
| 1-Me-3-[7-Cl-2,3-Dihydro-1-Me-5-Ph-1*H*-1,4-benzodiazepin-2-yl]-5-Ph-7-Cl | 175–178 | $Et_2O$ | | ir, ms, pmr, uv | 5a |
| 1-Me-3-[7-Cl-2,3-Dihydro-5-(2-$FC_6H_4$)-1-Me-1*H*-1,4-benzodiazepin-2-yl]-5-(2-$FC_6H_4$)-7-Cl | 225–230 | $CH_2Cl_2/Et_2O$ | | ir, ms, pmr, uv | 5a |
| ***Pentasubstituted*** | | | | | |
| 1,3-$(EtOOC)_2$-2-EtOOCO-5-Ph-7-Cl | 110–116 | $Et_2O$/Hexane | | ir, pmr | 5b |
| 1-Me-2-EtOOCO-3-EtOOC-5-Ph-7-Cl | 158–161 | $Et_2O$/Hexane | | ir, pmr | 5b |
| 1-Me-2-EtS-3-N(COOEt)NHCOOEt-5-(2-$FC_6H_4$)-7-Cl | 118–120 | $Et_2O$/Hexane | | ir, ms, pmr | 2 |
| 1,3-$Me_2$-2-$MeSO_2$-5-(2-$FC_6H_4$)-7-Cl | 138–140 | MeOH | 83 | | 2, 3 |
| 1,3-$Me_2$-2-MeS-5-(2-$FC_6H_4$)-7-Cl | 149–152 | MeOH | 74 | pmr | 2, 3 |

TABLE V-2. 3*H*-1,4-BENZODIAZEPINES

| Substituent | mp (°C); [bp (°C/torr)] | Solvent of Crystallization 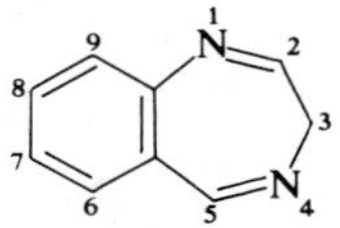 | Yield (%) | Spectra | Refs. |
|---|---|---|---|---|---|
| ***Disubstituted*** | | | | | |
| 2-Me-5-Ph | 108–110 | Cyclohexane | | | 12, 13, 17 |
| 2-Me-5-Ph, 4-Oxide | 127–138 | EtOAc | | | 12, 13, 17 |
| 2-C(NOH)COO*t*-Bu-5-(2-ClC$_6$H$_4$) | 218–219 | CH$_2$Cl$_2$/EtOH | | | 20b |
| 5-Ph-7-Cl | 101–104 | EtOAc | 64 | ir, pmr | 7, 11 |
| 5-Ph-7-Cl, 4-Oxide | 159–161 | EtOAc | | | 11 |
| 5-(2-FC$_6$H$_4$)-7-I | 124–126 | EtOAc | | | 18 |
| ***Trisubstituted*** | | | | | |
| 2-C(NHAc) (COOEt)$_2$-5-(2-ClC$_6$H$_4$)-7-Cl | 153–155 | EtOH | | | 21 |
| 2-C(NHAc) (COOEt)$_2$-5-(2-FC$_6$H$_4$)-7-Cl | 185–195 | EtOH | | | 21 |
| 2-C(NHAc) (COOEt)$_2$-5-(2-FC$_6$H$_4$)-7-CN | 138–140 | CH$_2$Cl$_2$/Et$_2$O | | | 21 |
| 2-C(NHAc) (COOEt)$_2$-5-(2-Pyridyl)-7-Br | 178–180 | *i*-PrOH | | | 21 |
| 2-CONH$_2$-5-Ph-7-Cl | 219–221 | CHCl$_3$/Et$_2$O | 85 | ir, ms, pmr | 7, 15 |
| 2-CONH$_2$-5-(2-FC$_6$H$_4$)-7-Cl | 178–180 | CH$_2$Cl$_2$/Petr ether | | | 15 |
| 2-CONMe$_2$-5-Ph-7-Cl | 124–127 | CH$_2$Cl$_2$/Petr ether | | | 7b |
| 2-ClCH$_2$-5-Ph-7-Cl, 4-Oxide | 122d | CH$_2$Cl$_2$/Petr ether | | ir, pmr, uv | 18 |
| 2-ClCH$_2$-5-(2-FC$_6$H$_4$)-7-Cl, 4-Oxide | 135d | EtOAc/Hexane | | ir, uv | 18 |
| 2-CN-5-Ph-7-Cl | 151–154 | Et$_2$O | 80 | ir, ms, pmr | 7, 15 |
| 2-CN-5-(2-FC$_6$H$_4$)-7-Cl | 106–110 | Et$_2$O/Hexane | 31 | pmr, uv | 19 |
| 2-CHNOH-5-Ph-7-Cl | 217–221 | EtOH | | ir, pmr, uv | 19b |
| 2-CHNOH-5-Ph-7-Cl, 4-Oxide | 226–231 | CH$_2$Cl$_2$/Et$_2$O | 56 | pmr, uv | 19 |
| 2-CHNOH-5-(2-FC$_6$H$_4$)-7-Cl | 250–251d | CH$_2$Cl$_2$/MeOH/EtOAc | 27 | pmr, uv | 19 |
| 2-CHNOH-5-(2-FC$_6$H$_4$)-7-Cl, 4-Oxide | 203–215d | EtOH | | ir, pmr, uv | 18 |

| | | | | | |
|---|---|---|---|---|---|
| 2-C(NOH)Ac-5-(2-ClC$_6$H$_4$)-7-Cl | 189–192 | CH$_2$Cl$_2$/Hexane | 88.5 | ir, pmr, uv | 20a |
| 2-C(NOH)NH$_2$-5-(2-FC$_6$H$_4$)-7-Cl | 248–249 | THF/EtOH | | | 23 |
| 2-C(NOH)NH$_2$-5-(2-FC$_6$H$_4$)-7-Cl, 4-Oxide | 258–260 | MeOH/EtOH/THF | | | 23 |
| 2-C(NOH)NHMe-5-(2-FC$_6$H$_4$)-7-Cl | 223–225d | CH$_2$Cl$_2$/EtOAc/Hexane | | | 23 |
| 2-C(NOH)NMe$_2$-5-(2-FC$_6$H$_4$)-7-Cl | 160–164 | MeOH/EtOAc | | | 23 |
| 2-C(NOH)NMe$_2$-5-(2-FC$_6$H$_4$)-7-Cl, 4-Oxide | 190–192d | MeOH/EtOAc | | | 23 |
| 2-C(NOH)COOMe-5-Ph-7-Cl | 235–237d | THF/MeOH | | | 21 |
| 2-C(NOH)COOMe-5-Ph-7-Cl, 4-Oxide | 237–239d | DMF/MeOH | | | 21 |
| 2-C(NOH)COOMe-5-(2-ClC$_6$H$_4$)-7-Cl | 223–225d | MeOH/THF | 97 | ir, pmr, uv | 20a, 21 |
| 2-C(NOH)COOMe-5-(2-FC$_6$H$_4$)-7-Cl | 238–241d | AcOH/H$_2$O | | | 21 |
| 2-C(NOH)COOMe-5-(2-Pyridyl)-7-Br | 203–204d | THF/PhH | | ir, pmr, uv | 20b |
| 2-C(NOH)COOEt-5-(2-ClC$_6$H$_4$)-7-Cl | 219–221 | CH$_2$Cl$_2$/EtOH | | | 20b |
| 2-C(NOH)COOEt-5-(2-ClC$_6$H$_4$)-7-Cl, 4-Oxide | 244–246 | THF/EtOH | | | 20b |
| 2-C(NOH)COOEt-5-(2-FC$_6$H$_4$)-7-Cl | 230–233 | THF/EtOH | | pmr | 20b |
| 2-C(NOH)COOEt-5-(2-FC$_6$H$_4$)-7-Cl, 4-Oxide | 225–226 | THF/EtOH | | | 20b |
| 2-C(NOH)COO *i*-Pr-5-(2-FC$_6$H$_4$)-7-Cl | 244–245d | THF/*i*-PrOH | | | 20b |
| 2-C(NOH)COO*i*-Pr-5-(2-FC$_6$H$_4$)-7-Cl, 4-Oxide | 231–233 | CH$_2$Cl$_2$/EtOH | | | 20b |
| 2-C(NOH)COO*t*-Bu-5-(2-ClC$_6$H$_4$)-7-Cl, 4-Oxide | 233–234 | *i*-PrOH | | | 20b |
| 2-C(NOH)COO*t*-Bu-5-(2-FC$_6$H$_4$)-7-Cl | 212–214 | CH$_2$Cl$_2$/Hexane | | | 20b |
| 2-C(NOH)COO*t*-Bu-5-(2-FC$_6$H$_4$)-7-Cl, 4-Oxide | 191–192 | CH$_2$Cl$_2$/EtOH | | | 20b |
| 2-C(NOH)COO*t*-Bu-5-(2-F-4-MeOC$_6$H$_3$)-7-Cl | 215–216 | Et$_2$O | | | 20b |
| 2-C(NOH)CN-5-(2-FC$_6$H$_4$)-7-Cl | 248–250 | THF/EtOH | | | 20b |
| 2-C(NOH)CONMe$_2$-5-Ph-7-Cl | 260–262d | THF/EtOH | | | 20b |
| 2-C(NOH)CONMe$_2$-5-Ph-7-Cl, 4-Oxide | 254–255 | THF/EtOH | | pmr | 20b |
| 2-C(NOH)CONMe$_2$-5-(2-FC$_6$H$_4$)-7-Cl | 245–255 | CH$_2$Cl$_2$/EtOH | | | 20b |
| 2-C(NOH)CONEt$_2$-5-(2-FC$_6$H$_4$)-7-Cl | 251–253 | CH$_2$Cl$_2$/EtOAc | | ir, uv | 20b |
| 2-C(NOH)SO$_2$NMe$_2$-5-(2-ClC$_6$H$_4$)-7-Cl | 168–170 | THF/EtOH | | | 20b |
| 2-C(NOH)SO$_2$NMe$_2$-5-(2-FC$_6$H$_4$)-7-Cl | 149–150 | MeOH/Et$_2$O | | pmr | 20b |
| 2-C(NOH)SMe-5-(2-FC$_6$H$_4$)-7-Cl, 4-Oxide | 220–222d | THF/EtOAc | | pmr | 24 |
| 2-C(NOH)NO$_2$-5-(2-FC$_6$H$_4$)-7-Cl | 220–230d | Et$_2$O | | | 22 |
| 2-C(NOH)NO$_2$-5-(2-FC$_6$H$_4$)-7-Cl, 4-Oxide | d | MeOH/EtOAc | | | 22 |
| 2-C(NOH)Morpholino-5-(2-FC$_6$H$_4$)-7-Cl | 191–193 | MeOH/EtOAc | | | 23 |
| 2-C(NOH)Morpholino-5-(2-FC$_6$H$_4$)-7-Cl, 4-Oxide | 211–213d | MeOH/THF/EtOAc | | | 23 |
| 2-C(NOH) (4-Me-Piperazino)-5-(2-FC$_6$H$_4$)-7-Cl | 198–200 | EtOH | | | 23 |

TABLE V-2. —(contd.)

| Substituent | mp (°C); [bp (°C/torr)] | Solvent of Crystallization | Yield (%) | Spectra | Refs. |
|---|---|---|---|---|---|
| 2-C(NOH) (4-Me-Piperazino)-5-(2-$FC_6H_4$)-7-Cl, 4-Oxide | 184–186d | EtOH | | | 23 |
| 2-C(NOH) (2-Pyridyl)-5-(2-$FC_6H_4$)-7-Cl | 220–222 | MeOH | | pmr | 20b |
| 2-C(NOH) (2-Pyridyl)-5-(2-$FC_6H_4$)-7-Cl, 4-Oxide·0.66EtOH | 191–193d | $CH_2Cl_2$/EtOH | | pmr | 20b |
| 2-C(NOH)Pyrrolidino-5-(2-$FC_6H_4$)-7-Cl | 175–178 | THF/EtOH | | | 23 |
| 2-C(NOH)Pyrrolidino-5-(2-$FC_6H_4$)-7-Cl, 4-Oxide | 168–172 | MeOH/EtOAc | | | 23 |
| 2-C(NOMe)$NH_2$-5-(2-$FC_6H_4$)-7-Cl | 120–124 | $Et_2O$/Hexane | | | 23 |
| 2-C(NOMe)NHMe-5-(2-$FC_6H_4$)-7-Cl | 140–142 | $Et_2O$/Hexane | | | 23 |
| 2-C(NOMe)OMe-5-(2-$FC_6H_4$)-7-Cl | 165–166 | EtOAc/Hexane | | pmr | 24 |
| 2-C(NOMe)OMe-5-(2-$FC_6H_4$)-7-Cl, 4-Oxide·0.5$Et_2O$ | 80–82 | $Et_2O$ | | pmr | 24 |
| 2-C(NOMe)Morpholino-5-(2-$FC_6H_4$)-7-Cl | 141–143 | $CH_2Cl_2$/$Et_2O$/Hexane | | | 24 |
| 2-C(NOMe)Morpholino-5-(2-$FC_6H_4$)-7-Cl, 4-Oxide | 170–173 | $Et_2O$/Hexane | | | 23 |
| 2-C(NOMe)$NO_2$-5-(2-$FC_6H_4$)-7-Cl | 130–133 | $Et_2O$/Hexane | | | 23 |
| 2-C(NOMe)$NO_2$-5-(2-$FC_6H_4$)-7-Cl, 4-Oxide | 207–209 | $Et_2O$ | | pmr | 22 |
| 2-C(NOMe) (2-Pyridyl)-5-(2-$FC_6H_4$)-7-Cl | 171–173 | *i*-PrOH | | pmr | 20b |
| 2-C[NO$(CH_2)_4$OMe]-5-(2-$FC_6H_4$)-7-Cl | 98–100 | $Et_2O$/Hexane | | ir, pmr, uv | 24 |
| 2-C[NO$(CH_2)_4$OMe]-5-(2-$FC_6H_4$)-7-Cl, 4-Oxide | 125–128 | $Et_2O$ | | pmr | 24 |
| 2-Me-5-Ph-7-Cl | 132–135 | Hexane | | | 12, 13 |
| 2-Me-5-Ph-7-Cl, 4-Oxide | 167–169d | EtOH | 78; 34 | pmr, uv | 12–14, 16, 17 |

| | | | | | |
|---|---|---|---|---|---|
| 2-Me-5-Ph-7-Cl, 1,4-Dioxide | 138–140 | EtOAc/Hexane | | ir, pmr, uv | 18 |
| 2-Me-5-Ph-7-$NO_2$, 4-Oxide | 196–200d | EtOAc | | | 12, 13, 17 |
| 2-Me-5-(2-$FC_6H_4$)-7-Cl, 4-Oxide | 171–173 | EtOH | | ir, uv | 18 |
| 2,5-$(Ph)_2$-7-Cl, 4-Oxide | 204–206 | MeOH | | ir, pmr | 26 |
| 2-Ph-5-MeO-8-Cl | 105–106 | Petr ether | 38 | | 25 |
| 2-Ph-5-$NHNH_2$-8-Cl | 162–164 | EtOAc/Hexane | 35 | | 25 |
| 2-(Phthalimido)$CH_2$-5-Ph-7-Cl, 4-Oxide | 186–188d | EtOAc/$Et_2O$ | | ir, pmr, uv | 18 |
| 2-(5-COOMe-Pyrrolidin-2-one-5-yl)-5-(2-$ClC_6H_4$)-7-Cl | 210–212d | EtOAc/Hexane | | ir, pmr, uv | 28 |
| 2-(4,5-Dihydro-4,5-$Me_2$-1,2,4-Oxadiazol-3-yl)-5-(2-$FC_6H_4$)-7-Cl | 133–135 | $Et_2O$/Hexane | | ir, pmr, uv | 24b |
| ***Tetrasubstituted*** | | | | | |
| 2,3-$(Me)_2$-5-Ph-7-Cl | 142–144 | Hexane | | pmr | 18 |
| 2-Me-3-AcO-5-Ph-7-Cl | 130–131.5 | EtOH | | | 17 |
| 2-Me-3-HO-5-Ph-7-Cl | 120–125d | EtOH/$H_2O$ | | | 17 |
| 2,5-$(Me)_2$-7,8-$(MeO)_2$, 4-Oxide | 146–149d | EtOAc | | ir, pmr, uv | 18 |

TABLE V-3. 2-AMINO-3*H*-1,4-BENZODIAZEPINES

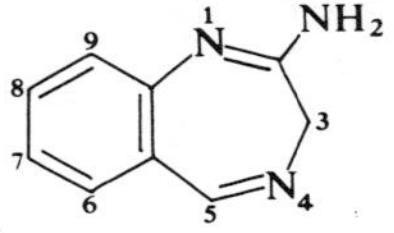

| Substituent | mp (°C); [bp (°C/torr)] | Solvent of Crystallization | Yield (%) | Spectra | Refs. |
|---|---|---|---|---|---|
| ***Monosubstituted*** | | | | | |
| 5-Ph | 217–218 | $Me_2CO$ | 70; 76 | | 95, 96 |
| 1-Oxide, hemihydrate | 145–147 | MeOH, $Me_2CO$ | 74 | | 99 |
| 5-(2-$ClC_6H_4$) | 240–242 | $CH_2Cl_2$/MeOH | 92 | | 66 |
| ***Disubstituted*** | | | | | |
| 3-(1-Imidazolo)-5-Ph | 258–261 | MeOH | 73 | | 99 |
| 5-Me-7-Cl | 238–238.5d | | 87 | pmr | 95 |
| 5-Ph-7-Br, 4-Oxide | 261–262 | MeOH | 66.7 | | 34 |
| Hydrochloride | 243–244 | MeOH | | | 34, 46 |
| 5-Ph-7-Cl | 236–237 | EtOH | 61; 75 | | 34, 41, 66, 68, 96 |
| Hydrochloride | 264–265d | | | | 41 |
| Dihydrochloride | 245–246d | | | | 95 |
| 1-Oxide, hemihydrate | 159–160 | MeOH/EtOAc | 75 | | 99 |
| 4-Oxide | 255–256; 278–283 | MeOH | 60 | | 33, 46, 85a |
| Hydrochloride | 245–246 | MeOH/$Et_2O$ | | | 33, 46 |
| 1,4-Dioxide | 221 | EtOH | 47 | ir, ms, pmr, uv | 100 |
| | 210–211 | MeOH/EtOAc | 53 | | 99 |
| 5-Ph-7-MeO | 184–185 | $CHCl_3$ | 80; 85 | | 95, 96 |
| 1-Oxide, hemihydrate | 158–161 | MeOH/$Me_2CO$ | 70 | | 99 |
| 5-Ph-7-Me | 222–223d | EtOH | 63 | | 95, 96 |
| 1-Oxide·0.33$H_2O$ | 161–163 | MeOH/$Me_2CO$ | 75 | | 99 |
| 5-Ph-7-$NO_2$ | 227–228d; 214–216 | THF | 93; 70 | | 95, 96 |

| | | | | | |
|---|---|---|---|---|---|
| Dihydrochloride | 234–235d; 245–246 | | | | 95 |
| 4-Oxide | 243d | EtOH | 97 | | 38 |
| 5-Ph-7-$CF_3$ | 190–193 | $Me_2CO$/Hexane | 58 | | 95, 96 |
| Dihydrochloride | 215–223d | | | | 96 |
| 4-Oxide | 240–242 | MeOH | 81.5 | uv | 36, 39 |
| 5-(2-$ClC_6H_4$)-7-Cl | 228–230 | $CH_2Cl_2$/MeOH | 87.5 | | 66, 118 |
| | 218–220 | $CHCl_3$/Hexane | 6 | | 95 |
| 5-(2-$ClC_6H_4$)-7-$NO_2$ | 250–254 | $CH_2Cl_2$/EtOH | | | 70 |
| 5-(2-$FC_6H_4$)-7-$NH_2$ | 248–254 | $CH_2Cl_2$ | | | 70 |
| 5-(2-$FC_6H_4$)-7-$NO_2$ | 240–242 | $CH_2Cl_2$/EtOAc | | | 70 |
| 5-(4-$ClC_6H_4$)-7-Cl, 4-Oxide | 252–254 | DMF/$H_2O$ | 72 | | 47 |
| 5-(4-$MeOC_6H_4$)-7-Cl | 237–238 | MeOH | 86 | | 95 |
| 1-Oxide·0.66$H_2O$ | 158–160 | MeOH/$Me_2CO$ | 67 | | 99 |
| 4-Oxide | 237–238 | $Me_2CO$ | 54 | | 47 |
| 5-(4-$HOC_6H_4$)-7-Cl, 4-Oxide | 302–303d | DMF/$H_2O$ | | | 38b |
| ***Trisubstituted*** | | | | | |
| 3-(1-Benzimidazolo)-5-Ph-7-Cl | 285–286 | MeOH/$CHCl_3$ | 70 | | 99 |
| 3-(1-Benzimidazolo)-5-Ph-7-Me | 270–273 | MeOH/$CH_2Cl_2$ | 92 | | 99 |
| 3-(1-Benzimidazolo)-5-(4-$MeOC_6H_4$)-7-Cl | 302–303 | MeOH/$CHCl_3$ | 82 | | 99 |
| 3-(2-Et-1-Imidazolo)-5-Ph-7-Me | 253–254 | MeOH | 59 | | 99 |
| 3-(2-Et-1-Imidazolo)-5-(4-$MeOC_6H_4$)-7-Cl | 250–252 | MeOH | 55 | | 99 |
| 3-HO-5-Ph-7-Cl | 176–181d | $CH_2Cl_2$/MeOH | | | 85b |
| 3-(1-Imidazolo)-5-Ph-7-Cl | 265–266 | MeOH/$CHCl_3$ | 78; 39 | | 99 |
| 3-MeO-5-(2-$ClC_6H_4$)-7-Cl | 216–219 | MeCN | 60; 75 | | 70 |
| 3-Me-5-(2-$ClC_6H_4$)-7-$NO_2$ ( + ) | 236–240 | EtOH | | $[\alpha]_D$ + 728.8 (DMSO 1%) | 70 |
| 3-(2-Me-1-Imidazolo)-5-Ph-7-Cl | 270–271 | MeOH | 60; 62 | | 99 |
| 3-(2-Me-1-Imidazolo)-5-Ph-7-MeO, hemihydrate | 250–252 | MeOH | 79 | | 99 |
| 5-Ph-7-Cl-9-$CH_2$COOEt | 228–230d | MeOH | 80 | ir, pmr | 109 |
| 5-Ph-7-Cl-9-C(OH)COOEt, hemihydrate | 208–209 | MeOH | 79 | ir, pmr | 109 |
| 5-(4-$MeOC_6H_4$)-7-Cl-9-$CH_2$COOEt | 206–207d | $Me_2CO$/Hexane | 84 | | 109 |
| 5-(4-$MeOC_6H_4$)-7-Cl-9-C (COOEt)C(OH)COOEt | 200–202 | MeOH | 54 | ir, pmr | 109 |

TABLE V-4. 2-METHYLAMINO-3*H*-1,4-BENZODIAZEPINES

| Substituent | mp (°C); [bp(°C/torr)] | Solvent of Crystallization | Yield (%) | Spectra | Refs. |
|---|---|---|---|---|---|
| ***Monosubstituted*** | | | | | |
| 5-Me, 4-Oxide | 191–192 | $Me_2CO$ | | | 38b |
| 5-Ph | 219–221 | | | | 41 |
| | 227–229 | DMF | | | 51 |
| 4-Oxide | 216–218 | $Me_2CO$ | | | 34, 41 |
| | 190–191 | | | | 46 |
| 4-Oxide, hydrochloride | 225–226 | $Me_2CO$/MeOH $Et_2O$/Petr ether | | | 34, 46 |
| 5-(4-$MeOC_6H_4$), 4-Oxide | 251–252 | EtOH/Petr ether | 70 | ir, uv | 34, 46, 48 |
| Hydrochloride | 218–219 | MeOH/$Me_2CO$ Petr ether | | | 34, 46 |
| 5-[3,4-$(MeO)_2C_6H_3$], 4-Oxide | 193–194 | EtOH/Petr ether | 63 | | 34 |
| Hydrochloride | 223–224 | MeOH/$Me_2CO$ Petr ether | | | 34 |
| 7-Cl, 4-Oxide | 245d | EtOH/$Et_2O$ | | | 38b, 52 |
| ***Disubstituted*** | | | | | |
| 5-(Benzocycloheptan-7-yl)-7-Cl, 4-Oxide | | | | | |
| Hemiethanolate | 212–213 | EtOH | 87 | ir, pmr, uv | 56 |
| 5,7-$Cl_2$ | 197–200 | MeOH | | | 18 |
| 5-Cyclohexyl-7-Cl | 218–220 | | | | 41 |
| 4-Oxide | 230–231 | MeOH | | | 55 |
| | 239–241 | | | | 41 |

| | | | | | |
|---|---|---|---|---|---|
| 5-MeNH-7-Cl | 248–250 | EtOH | 48 | ir, ms, uv | 87a |
| 5-(4-Me-Piperazino)-7-Cl | 202–204d | *i*-PrOH/EtOAc | | | 18 |
| 5-(Morpholino)$_2$PO$_2$-7-Cl | 210–212 | EtOAc | 47 | ir, ms | 87b |
| 5-(1-Piperidino)-7-Cl | 226–228 | *i*-PrOH | | | 18 |
| 5-Ph-7-Br, 4-Oxide | 242–243 | $Me_2CO$ | 68 | | 34, 46 |
| Hydrochloride | 239–240d | $MeOH/Et_2O$ | | | 34, 46 |
| 5-Ph-7-Cl | 242–245 | $Me_2CO$ | 95 | | 33, 41, 73, 85a, 93b |
| Hydrochloride | 260–261 | $MeOH/Et_2O$ | | | 33 |
| 4-Oxide | 238.5–240.5 | EtOH | 82 | | 33, 46, 59, 74 |
| | | | | X-ray | 198 |
| | | | | $^{13}$C-nmr, pmr | 199 |
| Hydrochloride | 215–216 | MeOH | | | 33 |
| Hydrogen sulfate | 214–215 | $MeOH/Me_2CO$ | | | 33 |
| Dihydrogen phosphate | 206–207 | $H_2O/Me_2CO$ | | | 33 |
| 5-Ph-7-MeO, 4-Oxide | 231–233 | PhH/Hexane | | | 38b |
| 5-Ph-7-MeOOC, 4-Oxide | 259–260 | MeOH | | uv | 37 |
| 5-Ph-7-Me | 218–220 | $CH_2Cl_2$ | 60 | | 34, 41 |
| 4-Oxide | 214–215 | $Me_2CO$ | 50 | | 34, 46 |
| Hydrochloride | 224–225 | $MeOH/Me_2CO$ $Et_2O$ | | | 46 |
| 5-Ph-7-(2-Me-1,3-Dioxolan-2-yl)·0.35EtOH | 194–197 | EtOH | | | 85c |
| 5-Ph-7-MeS, 4-Oxide | 245–246 | MeCN | | | 40 |
| 5-Ph-7-$NO_2$ | 225–228 | MeOH | | | 85a |
| 4-Oxide | 260–261d | $EtOH/Et_2O$ | 59 | | 38 |
| 5-Ph-7-$CF_3$, 4-Oxide | 264–265 | $Et_2O$/Hexane | 85 | uv | 36, 39, 42 |
| | 267–268 | MeOH | | | |
| Hydrochloride | 222d | $MeOH/Et_2O$ | | | 39 |
| 5-$C_6D_5$-7-Cl, 4-Oxide | 229–231 | | 61 | | 58 |
| 5-PhNH-7-Cl | 200–202 | EtOH | | | 18 |
| 5-(4-$H_2NC_6H_4$)-7-Cl, 4-Oxide | 180–185 and 288–290 | $CH_2Cl_2$/MeOH | 50 | ir, ms | 49 |
| 5-(2-$ClC_6H_4$)-7-Cl, 4-Oxide | 247–248 | PhH | 80 | | 34 |
| Hydrochloride | 243–246d | $MeOH/Et_2O$ | | | 34 |

TABLE V-4. —(contd.)

| Substituent | mp (°C); [(bp °C/torr)] | Solvent of Crystallization | Yield (%) | Spectra | Refs. |
|---|---|---|---|---|---|
| 5-(2-ClC$_6$H$_4$)-7-NO$_2$ | 219–221 | $CH_2Cl_2$/EtOH | | | 51 |
| 5-(4-ClC$_6$H$_4$)-7-Cl | 241–242 | $CHCl_3$ | 82 | | 34 |
| 4-Oxide | 254–255 | EtOH | 68 | | 34, 44, 51 |
| Hydrochloride | 245d | EtOH | | | 34 |
| 5-(2-FC$_6$H$_4$)-7-Ac | 181–183 | | | | 85c |
| 4-Oxide | > 203d | MeCN | | | 85c |
| 5-(2-FC$_6$H$_4$)-7-Cl | 204–206 | THF/$Et_2O$ | 98 | | 178 |
| 5-(2-FC$_6$H$_4$)-7-Et | 172–174 | MeCN | | | 51 |
| 4-Oxide | 201–203 | MeCN | | | 85c |
| 5-(2-FC$_6$H$_4$)-7-(MeCHOH) | 216–218 | MeCN | | | 85c |
| 5-(2-FC$_6$H$_4$)-7-I, 4-Oxide | 218–219d | EtOH | | | 18 |
| 5-(2-FC$_6$H$_4$)-7-MeNHCONH | 185–186 | $CH_2Cl_2$ | | | 70 |
| 5-(2-FC$_6$H$_4$)-7-(MeN=CMe) | 212–215 | MeCN | | | 85c |
| 4-Oxide | > 180d | MeCN | | | 85c |
| 5-(2-FC$_6$H$_4$)-7-NO$_2$ | 214–216 | EtOAc | | | 70 |
| 5-(2,6-F$_2$C$_6$H$_3$)-7-Cl, 4-Oxide | 230–234d | MeOH/$Et_2O$/Petr ether | | | 9 |
| 5-(4-HOC$_6$H$_4$)-7-Cl, 4-Oxide | 278–279d | EtOAc | | | 49, 50 |
| 5-(4-MeC$_6$H$_4$)-7-Br | 258–259 | $CHCl_3$ | 54 | | 34 |
| 4-Oxide | 255–256 | EtOH | 41.5 | | 34, 44, 51 |
| 5-[4-(H$_2$N(CH$_2$)$_3$O)C$_6$H$_4$]-7-Cl, 4-Oxide, dihydrochloride | 211–215d | *i*-PrOH/MeOH | 80 | ir | 49 |
| 5-[4-PhCH$_2$OCONH(CH$_2$)$_3$OC$_6$H$_4$]-7-Cl, 4-Oxide | 103–106 | $CH_2Cl_2$/$Et_2O$ | 19 | ir, ms | 49 |
| 5-[4-H$_2$NNHCOCH$_2$OC$_6$H$_4$]-7-Cl, 4-Oxide | 254–256 | DMF/MeOH/$H_2O$ | 90 | ir | 49 |
| 5-[4-MeOOCCH$_2$OC$_6$H$_4$]-7-Cl, 4-Oxide | 106–110 | $CH_2Cl_2$/Hexane | 59 | ir | 49 |
| 5-(2-pyridyl)-7-Br | 208–214 | $CH_2Cl_2$/Hexane | | | 143, 144 |
| 4-Oxide | 231–233d | $Me_2CO$ | | | 38b |
| 5-(2-thienyl)-7-Cl, 4-Oxide hydrochloride | 256–257 | | | | 41 |

| Compound | mp (°C) | Solvent | Yield (%) | Data | Ref. |
| --- | --- | --- | --- | --- | --- |
| ***Trisubstituted*** | | | | | |
| 3-AcO-5-Ph-7-Cl | 202–203 | $CH_2Cl_2/Et_2O$ | | ir | 110, 112a |
| 3-AcO-5-Ph-7-$NO_2$ | 222–222.5 | THF/Hexane | | | 38b |
| 3-AcO-5-Ph-7-$CF_3$ hydrochloride | 206–207 | MeCN | | | 38b |
| 3-AcO-5-(4-$MeOC_6H_4$)-7-Cl | 202–203 | THF/Hexane | | | 110 |
| 3-$H_2N$-5-Ph-7-Cl | 168–170 | $CH_2Cl_2$/Hexane | | | 9 |
| 3-$C_6H_4COO$-5-Ph-7-Cl | 215–216 | $Me_2CO$ | | ir | 110 |
| 3-$BrCH_2COO$-5-Ph-7-Cl | 175–177 | $CH_2Cl_2$/Hexane | | | 9 |
| 3-PrCOO-5-Ph-7-Cl | 171–172 | | | ir | 110 |
| 3-$NCCH_2CH_2$-5-(2-$FC_6H_4$)-7-Cl | 162–165 | $Et_2O$/Petr ether | | | 70 |
| 3-$(EtO)_2CHCH_2NH$-5-Ph-7-Cl | 95–97 | $Et_2O$/Petr ether | | | 9 |
| 3-$Et_2N$-5-Ph-7-Cl | 170–172 | EtOH | 62 | | 45 |
| 3-$Me_2N$-5-Ph-7-Cl | 167–169 | EtOH | 68 | | 45 |
| 3-EtO-5-Ph-7-Cl Hydrochloride | 215–216 | THF/$Et_2O$ | 63 | ir | 112a |
| 3-EtO-5-Ph-7-$NO_2$ | 222–223 | THF/Hexane | | | 38b |
| 3-$EtO_2CO$-5-Ph-7-Cl hydrochloride | 163–164 | $Et_2O$ | | ir, ms, pmr | 112a |
| | 178–180 | THF/$Et_2O$ | | ir | 112a |
| 3-$EtO_2CCONH$-5-Ph-7-Cl | 198–201 | EtOAc/Hexane | | | 9 |
| 3-HO-5-Ph-7-Cl | 184–186 | $CH_2Cl_2$/Petr ether | 87 | | 110 |
| | 191–192d | DMF/$H_2O$ | | | 110 |
| 3-HO-5-Ph-7-$NO_2$ | 161–163 | MeCN | | | 38b |
| 3-HO-5-Ph-7-$CF_3$ | 177–178d | PhH/Hexane | | | 38b |
| 3-(1-Imidazolo)-5-Ph-7-Cl | 252–253 | PhH | 60 | ir, ms, pmr | 112a |
| 3-MeO-5-Ph-7-Cl | 155–160 | $CH_2Cl_2$/Petr ether | | | 9 |
| 3-Me-5-Ph-7-Cl-, 4-Oxide | 246–247 | $Me_2CO$ | | | 46 |
| Hydrochloride | 190–191 | $Me_2CO$/MeOH/$Et_2O$ | | | 46 |
| 3-Me-5-Ph-7-$CF_3$, 4-Oxide | 257–258d | MeCN | | | 38b |
| 3-Me-5-(2-$ClC_6H_4$)-7-$NO_2$ ( + ) | 207–209 | $CH_2Cl_2$/Hexane | | $[\alpha]_D$ + 497.2 1% in $CH_2Cl_2$ | 70 |
| 3-Me-5-(2-$FC_6H_4$)-7-Cl | 203–205 | EtOH | | | 70 |
| 3-MeNH-5-Ph-7-Cl, 4-Oxide | 150–151 | MeCN | | | 65 |
| 3-MeNH-5-Ph-7-$NO_2$, 4-Oxide | 161–162 | MeCN | | | 65 |
| 3-MeNH-5-Ph-7-$F_3C$, 4-Oxide | 162–163 | MeOH/$H_2O$ | | | 65 |

TABLE V-4. —(contd.)

| Substituent | mp (°C); [(bp °C/torr)] | Solvent of Crystallization | Yield (%) | Spectra | Refs. |
|---|---|---|---|---|---|
| 3-MeNCH$_2$Ph-5-Ph-7-Cl | 169–171 | EtOH | 77 | | 45 |
| 3-(3-Me-Butanoyloxy)-5-Ph-7-Cl | 180–181 | Et$_2$O | | | 38b |
| 3-MeNPh-5-Ph-7-Cl | 156–158 | EtOH | 81 | | 45 |
| 3-Morpholino-5-Ph-7-Cl | 212–214 | EtOH | 80 | | 45 |
| 3-(3-Oxobut-1-yl)-5-(2-ClC$_6$H$_4$)-7-NO$_2$ | 198 | EtOAc/Hexane | | | 70 |
| 3-(3-Oxobut-1-yl)-5-(2-FC$_6$H$_4$)-7-NO$_2$ | 181–183 | Et$_2$O | | | 70 |
| 3-Piperidino-5-Ph-7-Cl | 157–159 | EtOH | 74 | | 45 |
| 3-EtCOO-5-Ph-7-Cl | 197–198 | Et$_2$O | 25 | ir | 110 |
| 3-Pr-5-Ph-7-Cl, 4-Oxide | 221–222 | Me$_2$CO/Petr ether | | | 33 |
| Hydrochloride | 187–189 | MeOH/Et$_2$O | | | 33 |
| 3-Pyrrolidino-5-Ph-7-Cl | 163–165 | EtOH | 85 | | 45 |
| 5-Ph-7,8-Cl$_2$, 4-Oxide | 233–234 | MeOH | | | 34, 46 |
| Hydrochloride | 231–232 | MeOH/Et$_2$O/Petr ether | | | 34, 46 |
| 5-Ph-7,9-Cl$_2$, 4-Oxide | 251–252 | MeOH | 71 | | 34, 44 |
| Hydrochloride | 204–207 | EtOH | | | 34 |
| 5-Ph-7-Cl-9-Me, 4-Oxide | 219–220 | Me$_2$CO | | | 38b |
| 5-Ph-7,8-Me$_2$, 4-Oxide | 259–261 | Me$_2$CO | | | 34, 46 |
| Hydrochloride | 230–231 | MeOH/Me$_2$CO | | | 34, 46 |
| 5-Ph-7,9-Me$_2$, 4-Oxide | 215–216 | CH$_2$Cl$_2$/Petr ether | | | 38b |
| Hydrochloride | 225–226 | MeOH/Et$_2$O | | | 38b |
| 5-(4-ClC$_6$H$_4$)-7,8-Me$_2$, 4-Oxide | 258–259 | MeOH | 50 | | 34, 46 |
| Hydrochloride | 247–248 | H$_2$O/EtOH/Et$_2$O | | | 34, 46 |
| 5-(4-NO$_2$C$_6$H$_4$)-7,8-Me$_2$, 4-Oxide | | | | | |
| Hydrochloride | 264–265 | MeOH | | | 38b |
| ***Tetrasubstituted*** | | | | | |
| 3,3-[NCCH$_2$CH$_2$]-5-(2-FC$_6$H$_4$)-7-Cl | 194–196 | Et$_2$O | | | 70 |

TABLE V-5. *N*-MONOSUBSTITUTED-2-AMINO-3*H*-1,4-BENZODIAZEPINES

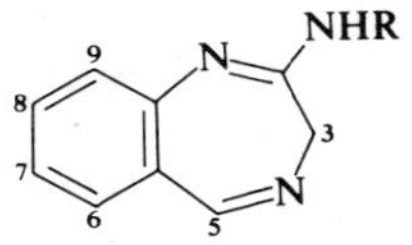

| Substituent | mp (°C); [bp (°C/torr)] | Solvent of Crystallization | Yield (%) | Spectra | Refs. |
|---|---|---|---|---|---|
| ***R; Other Substituents*** | | | | | |
| ***Monosubstituted*** | | | | | |
| Bu; 5-Ph | 130–131 | $Et_2O$/*i*-$Pr_2O$ | | | 84 |
| $(EtO)_2CHCH_2$; 5-Ph | 102–103 | Hexane | | | 83 |
| $(MeO)_2CHCH_2$; 5-(2-$ClC_6H_4$) | 181–182 | EtOAc | | | 77a |
| Et; 3-PhNMe | 133–135 | EtOH | 85 | | 45 |
| ***Disubstituted*** | | | | | |
| 1-Adamantyl; 5-Ph-7-Cl·0.5EtOAc | 224–225 | EtOAc | 16.5 | | 25 |
| $H_2NCH_2CH_2$; 5-Ph-7-Cl, 4-Oxide | 170–171 | $CH_2Cl_2/Et_2O$/Petr ether | 43 | | 33 |
| Dihydrochloride | 219–220 | $MeOH/Et_2O$ | | | 33 |
| $N_3CH_2COCH_2$; 5-Ph-7-Cl | 181.5–187.5 | EtOAc/Hexane | | | 82b |
| (1-Aziridino)$CH_2CH_2$-5-Ph-7-Cl, 4-Oxide | 163–165 | $CH_2Cl_2/Et_2O$ | | pmr, uv | 92 |
| $PhCH_2$; 5-Ph-7-Cl, 4-Oxide | 223–225 | | | | 41 |
| $BrCH_2COCH_2$; 5-Ph-7-Cl | 206–209 | EtOAc/Hexane | 12 | | 82b |
| Bu; 5-Ph-7-Cl | 167–169 | $MeOH/H_2O$ | | | 61 |
| 4-Oxide | 202–203 | $Me_2CO$ | 50 | | 33, 46 |
| Hydrochloride | 171–173 | *i*-PrOH/$Me_2CO/Et_2O$ | | | 33, 46 |
| $HOOCCH_2$; 5-Ph-7-Cl | 215–220 | EtOH | 77 | ir, pmr | 78–80 |
| $HOOCCH_2$; 5-Ph-7-$NO_2$ | 154–155 | | 66 | | 78–80 |
| $HOOCCH_2$; 5-(2-$ClC_6H_4$)-7-Cl | 136–139 | $MeOH/Et_2O$ | 75; 87 | | 78–80 |
| $HOOCCH_2$; 5-(2-$ClC_6H_4$)-7-$NO_2$ | 158–161 | $Et_2O$ | 83 | | 78–80 |
| $HOOCCH_2$; 5-(2-$FC_6H_4$)-7-Cl | 147–150 | | 88 | | 78, 79 |

TABLE V-5. —(contd.)

| Substituent | mp (°C); [bp (°C/torr)] | Solvent of Crystallization | Yield (%) | Spectra | Refs. |
|---|---|---|---|---|---|
| $HOOCCH_2$; 5-(2-$FC_6H_4$)-7-$NO_2$ | 144–147 | | 45 | | 78 |
| 2-$HOOCC_6H_4$; 5-Ph-7-Cl | 215–219 | | | | 81 |
| (1-Cyclohexene)$CH_2CH_2$; 5-Ph-7-Cl, 4-Oxide | 204–205 | | | | 38b |
| Cyclohexyl; 5-Ph-7-Cl, 4-Oxide | 257–258 | $CH_2Cl_2/Et_2O$ | | | 38b |
| Cyclopropyl; 5-Ph-7-Cl, 4-Oxide | 257–259d | MeOH | | | 38b |
| (Cyclopropyl)$CH_2$; 5-Ph-7-Cl, 4-Oxide | 234–236 | $CH_2Cl_2$/Hexane | | | 38b |
| $(EtO)_2CHCH_2$; 5-Ph-7-Cl | 131–132 | Cyclohexane | 47 | | 25, 83 |
| $(EtO)_2CHCH_2$; 5-Ph-7-$NO_2$ | 114–115 | $Et_2O$/Hexane | | | 83 |
| $(EtOOC)_2C{=}CH$; 5-Ph-7-Cl | 105–107 | $Et_2O$ | | | 108 |
| 4-$Et_2N$-2-butyn-1-yl; 5-Ph-7-Cl, 4-Oxide | 177 | | 20 | | 81 |
| $Et_2NCH_2CH_2$; 5-Ph-7-Cl | 105–107 | Pentane | | | 61 |
| Hydrochloride | 252–254 | $EtOH/Et_2O$ | | | 61 |
| Dihydrochloride | 249–250 | *i*-PrOH | | | 61 |
| Dipicrate | 203–204 | $THF/Et_2O$ | | | 83, 84 |
| 4-Oxide | 154–156 | Cyclohexane | 60 | | 45 |
| Dihydrochloride·$0.5H_2O$ | 237–238 | | | | 41 |
| $(MeO)_2CHCH_2$; 5-Ph-7-Cl | 162 | EtOH | 56; 75 | | 43, 74, 75, 77a |
| 4-Oxide | 196–198 | PhH/Cyclohexane | 76 | | 43, 74, 75 |
| $(MeO)_2CHCH_2$; 5-(2-$ClC_6H_4$)-7-Cl | 175.5–176.5 | EtOAc/Hexane | 85 | | 77a |
| $(MeO)_2CHCH_2$; 5-(2-Pyridyl)-7-Br | 155–157 | EtOAc/Hexane | | | 77b |
| 3,4-$(MeO)_2C_6H_3CH_2CH_2$; 5-Ph-7-Cl | 197–200 | $CH_2Cl_2$/Hexane | | | 9 |
| $(MeO)_2CHCHMe$; 5-(2-$ClC_6H_4$)-7-Cl | 157 | EtOAc/Hexane | | | 77a |
| $Me_2NCH_2CH_2$; 5-Ph-7-Cl, dihydrochloride | 269–270d | AcOH | | | 25 |
| 4-Oxide, dihydrochloride | 262–263 | | | | 41 |
| $Me_2N(CH_2)_3$; 5-Ph-7-Cl | 159–160 | *i*-$Pr_2O$ | | | 83, 84 |
| 4-Oxide, dihydrochloride | 242–243d | | | | 41 |
| $Me_2N(CH_2)_3$; 5-Ph-7-$NO_2$ | 130–131 | MeOH | | | 83, 84 |
| 2-$EtOOCC_6H_4$; 5-Ph-7-Cl | 163–166 | | | | 81 |

| | | | | |
|---|---|---|---|---|
| EtOOCCH$_2$; 5-Ph-7-Cl | 97–98 | $i$-Pr$_2$O | | 83 |
| 4-Oxide | 208 | $i$-PrOH | 22 | 75 |
| EtOOCCH$_2$; 5-Ph-7-NO$_2$ | 194–195 | Me$_2$CO/Hexane | | 83 |
| Et; 5-Ph-7-Br | 224 | MeOH | 78 | 34 |
| 4-Oxide | 246–248 | Me$_2$CO | 86 | 34 |
| Hydrochloride | 232–233 | MeOH/Et$_2$O | | 34 |
| Et; 5-Ph-7-Cl, 4-Oxide | 231–233 | Me$_2$CO | 69 | 33, 46 |
| Hydrochloride | 208–209 | EtOH/Et$_2$O | | 33, 46 |
| (2-Furyl)CH$_2$; 5-Ph-7-Cl | 150–151 | $i$-Pr$_2$O | | 83 |
| 4-Oxide | 225–227 | PhH | 86 | 45 |
| Hexyl; 5-Ph-7-Cl | 149–150 | EtOH/H$_2$O | | 61 |
| HOCH$_2$CH$_2$; 5-Ph-7-Cl | 172–173 | Et$_2$O | | 83, 84 |
| 4-Oxide | 216–218 | MeOH | 45 | 33, 46 |
| Hydrochloride | 210–211d | MeOH/Et$_2$O | | 33, 46 |
| HO(CH$_2$)$_3$; 5-Ph-7-Cl | 203–205 | EtOAc | | 83, 84 |
| MeOCH$_2$CH$_2$; 5-Ph-7-Cl, 4-Oxide | 225–226 | Me$_2$CO | | 33, 46 |
| Hydrochloride | 207–209 | MeOH/Et$_2$O | | 33, 46 |
| MeO(CH$_2$)$_2$; 5-(2-ClC$_6$H$_4$)-7-Cl | 185–189 | EtOH | 78.6 | 73 |
| MeO(CH$_2$)$_3$; 5-(2-ClC$_6$H$_4$)-7-Cl | 175.5–179 | MeO(CH$_2$)$_3$NH$_2$ | 89.1 | 73 |
| (2-Me-4,5-Dihydro-1-imidazolo)CH$_2$CH$_2$; | | | | |
| 5-Ph-7-Cl, 4-Oxide | | | | |
| Dihydrochloride·0.5H$_2$O | 245–246 | | | 41 |
| (4-Me-1-Piperazino) (CH$_2$)$_3$; 5-Ph-7-Cl, 4-Oxide | 216–218 | | | 41 |
| (Morpholino)CH$_2$CH$_2$; 5-Ph-7-Cl | 196–198 | CH$_2$Cl$_2$/Pentane | | 61 |
| 4-Oxide, dihydrochloride·H$_2$O | 277–278d | | | 41 |
| (Morpholino) (CH$_2$)$_3$; 5-Ph-7-Cl | | | | |
| 4-Oxide, dihydrochloride | 231–232d | | | 41 |
| Ph; 5-Ph-7-Cl | 205–208 | PhH/Petr ether | | 81 |
| PhCH$_2$CH$_2$; 5-Ph-7-Cl | 137–138 | EtOH/H$_2$O | | 61 |
| 4-Oxide | 203–205 | EtOH | 79 | 45 |
| (Phthalimido)CH$_2$COCH$_2$; 5-Ph-7-Cl | 142–145 | Me$_2$CO | | 82b |
| (Phthalimido)CH$_2$COCH$_2$; 5-(2-ClC$_6$H$_4$)-7-Cl | 200–202 | MeOH/EtOAc | | 82b |
| 4-(1-Piperidino)-2-butyn-1-yl; 5-Ph-7-Cl | 168 | EtOAc | 23 | 81 |
| 4-Oxide | | | 27 | 81 |

TABLE V-5. —(contd.)

| Substituent | mp (°C); [bp (°C/torr)] | Solvent of Crystallization | Yield (%) | Spectra | Refs. |
|---|---|---|---|---|---|
| Propen-3-yl; 5-Ph-7-Cl, 4-Oxide | 202–204 | MeOH | 35 | | 33, 46 |
| Hydrochloride | 221–227d | MeOH/$Me_2CO$/$Et_2O$ | | | 33, 46 |
| Propen-3-yl; 5-(2-$ClC_6H_4$)-7-$NO_2$ | 160 | $Et_2O$ | | | 70 |
| *i*-Pr; 5-Ph-7-Cl, 4-Oxide | 248–250 | EtOH | | | 85a |
| Propyn-3-yl; 5-Ph-7-Cl | 203.5–205 | MeOH | | | 76 |
| | 220 | *i*-PrOH | 78 | | 43, 74, 75 |
| 4-Oxide | 222 | BuOH | 70 | | 43, 74 |
| ***Trisubstituted*** | | | | | |
| $AcNHCH_2CH_2$; 3-AcO-5-Ph-7-Cl | 175–180 | $CH_2Cl_2$/$Et_2O$/Petr ether | | | 9 |
| $AcNHCH_2CH_2$; 3-OH-5-Ph-7-Cl | 195–198 | $CH_2Cl_2$/$Et_2O$ | | | 9 |
| $PhCH_2$; 3-AcO-5-Ph-7-Cl | 208–210 | $Et_2O$ | 98 | ir, ms, pmr | 112a |
| $PhCH_2$; 3-$PhCH_2NH$-5-Ph-7-Cl, 4-Oxide | 135–138 | MeCN | | | 65 |
| Bu; 3-BuNH-5-Ph-7-Cl, 4-Oxide | 142–144 | MeCN | | | 65 |
| Cyclohexyl; 3-(cyclohexyl)NH-5-Ph-7-Cl | | | | | |
| 4-Oxide | 149–151 | EtOAc | | | 65 |
| Cyclopentyl; 3-(cyclopentyl)NH-5-Ph-7-Cl | | | | | |
| 4-Oxide | 175–176d | EtOAc | | | 65 |
| (Cyclopropyl)$CH_2$; 3-(cyclopropyl)$CH_2NH$-5-Ph-7-Cl, 4-Oxide | 155–156 | PhH/Hexane | | | 65 |

| | | | | | |
|---|---|---|---|---|---|
| $Et_2NCH_2CH_2$; 3-AcO-5-Ph-7-Cl | | | 62 | | 45 |
| $Et_2NCH_2CH_2$; 3-HO-5-Ph-7-Cl | 169–171 | Cyclohexane | 91 | | 45 |
| Et; 3-AcO-5-Ph-7-Cl | 201–203 | Cyclohexane | 78 | | 45 |
| Et; 3-$PhCH_2NMe$-5-Ph-7-Cl | 158–160 | EtOH | 72 | | 45 |
| Et; 3-$Et_2N$-5-Ph-7-Cl | 121–123 | EtOH | 58 | | 45 |
| Et; 3-$Me_2N$-5-Ph-7-Cl | 106–108 | EtOH | 77 | | 45 |
| Et; 3-EtNH-5-Ph-7-Cl, 4-Oxide | 152–153 | MeCN | | | 65 |
| Et; 3-HO-5-Ph-7-Cl | 201–203 | EtOH | 92 | | 45 |
| Et; 3-Morpholino-5-Ph-7-Cl | 190–192 | EtOH | 71 | | 45 |
| Et; 3-Piperidino-5-Ph-7-Cl | 190–192 | EtOH | 57 | | 45 |
| Et; 3-Pyrrolidino-5-Ph-7-Cl | 138–140 | EtOH | 71 | | 45 |
| (2-Furyl)$CH_2$; 3-AcO-5-Ph-7-Cl | 167–168 | PhH | 68 | | 45 |
| (2-Furyl)$CH_2$; 3-HO-5-Ph-7-Cl | 153–155 | PhH | 87 | | 45 |
| (2-Furyl)$CH_2$; 3-morpholino-5-Ph-7-Cl | 174–176 | EtOH | 80 | | 45 |
| $PhCH_2CH_2$; 3-AcO-5-Ph-7-Cl | 195–197 | PhH | 75 | | 45 |
| $PhCH_2CH_2$; 3-HO-5-Ph-7-Cl | 200–202 | PhH | 89 | | 45 |
| *i*-Pr; 3-*i*-PrNH-5-Ph-7-Cl, 4-Oxide | 167–168.5 | Hexane | | | 65 |
| ***Tetrasubstituted*** | | | | | |
| Bu; 3,3-$Me_2$-5-(2-$FC_6H_4$)-7-$NH_2$ | | | | ir, ms, pmr | 85d |

TABLE V-6. *N,N*-DISUBSTITUTED-2-AMINO-3*H*-1,4-BENZODIAZEPINES

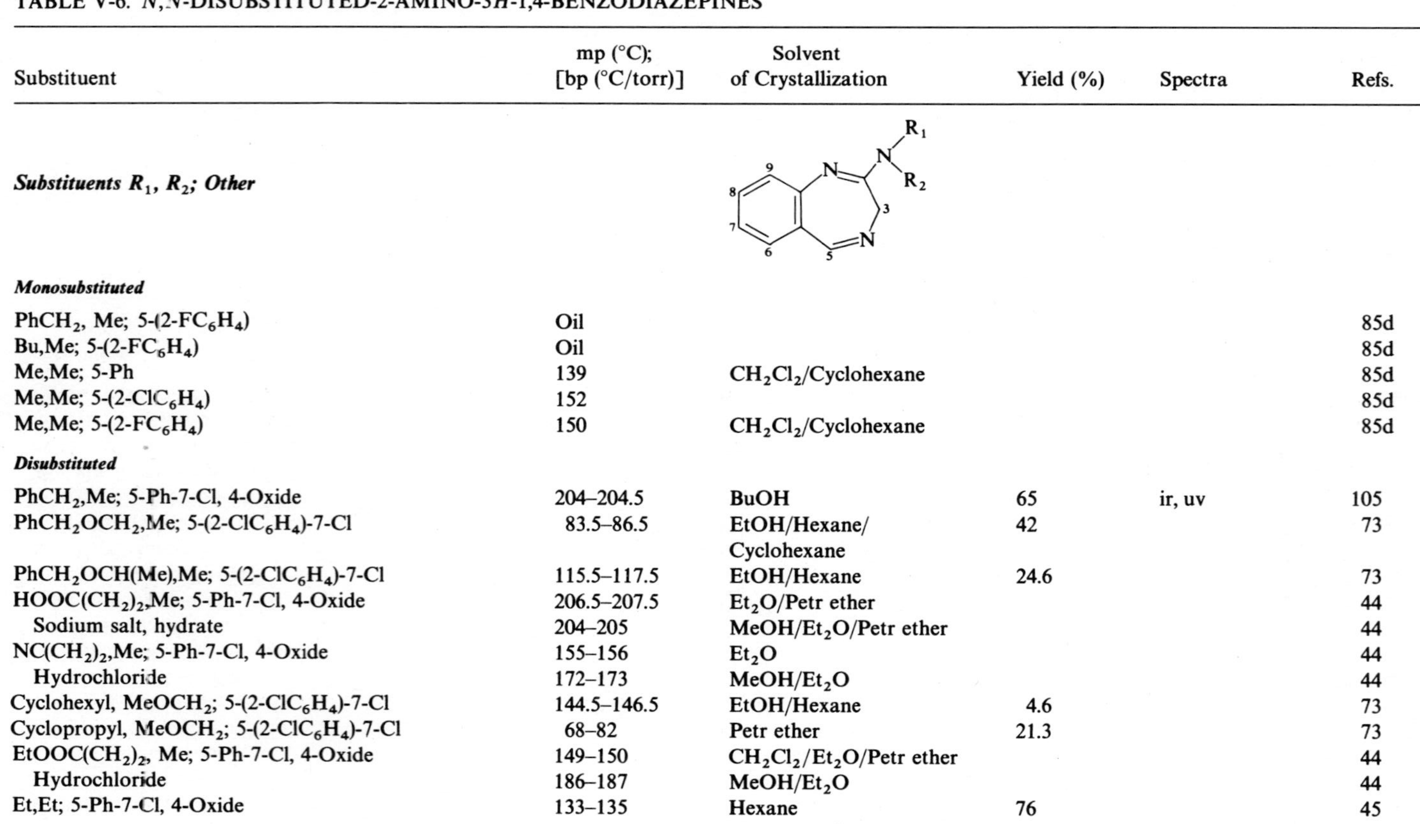

| Substituent | mp (°C); [bp (°C/torr)] | Solvent of Crystallization | Yield (%) | Spectra | Refs. |
|---|---|---|---|---|---|
| ***Substituents $R_1$, $R_2$; Other*** | | | | | |
| ***Monosubstituted*** | | | | | |
| $PhCH_2$, Me; 5-(2-$FC_6H_4$) | Oil | | | | 85d |
| Bu,Me; 5-(2-$FC_6H_4$) | Oil | | | | 85d |
| Me,Me; 5-Ph | 139 | $CH_2Cl_2$/Cyclohexane | | | 85d |
| Me,Me; 5-(2-$ClC_6H_4$) | 152 | | | | 85d |
| Me,Me; 5-(2-$FC_6H_4$) | 150 | $CH_2Cl_2$/Cyclohexane | | | 85d |
| ***Disubstituted*** | | | | | |
| $PhCH_2$,Me; 5-Ph-7-Cl, 4-Oxide | 204–204.5 | BuOH | 65 | ir, uv | 105 |
| $PhCH_2OCH_2$,Me; 5-(2-$ClC_6H_4$)-7-Cl | 83.5–86.5 | EtOH/Hexane/ Cyclohexane | 42 | | 73 |
| $PhCH_2OCH(Me)$,Me; 5-(2-$ClC_6H_4$)-7-Cl | 115.5–117.5 | EtOH/Hexane | 24.6 | | 73 |
| $HOOC(CH_2)_2$,Me; 5-Ph-7-Cl, 4-Oxide | 206.5–207.5 | $Et_2O$/Petr ether | | | 44 |
|   Sodium salt, hydrate | 204–205 | MeOH/$Et_2O$/Petr ether | | | 44 |
| $NC(CH_2)_2$,Me; 5-Ph-7-Cl, 4-Oxide | 155–156 | $Et_2O$ | | | 44 |
|   Hydrochloride | 172–173 | MeOH/$Et_2O$ | | | 44 |
| Cyclohexyl, $MeOCH_2$; 5-(2-$ClC_6H_4$)-7-Cl | 144.5–146.5 | EtOH/Hexane | 4.6 | | 73 |
| Cyclopropyl, $MeOCH_2$; 5-(2-$ClC_6H_4$)-7-Cl | 68–82 | Petr ether | 21.3 | | 73 |
| $EtOOC(CH_2)_2$, Me; 5-Ph-7-Cl, 4-Oxide | 149–150 | $CH_2Cl_2$/$Et_2O$/Petr ether | | | 44 |
|   Hydrochloride | 186–187 | MeOH/$Et_2O$ | | | 44 |
| Et,Et; 5-Ph-7-Cl, 4-Oxide | 133–135 | Hexane | 76 | | 45 |

| | | | | | |
|---|---|---|---|---|---|
| Et,$MeOCH_2$; 5-(2-$ClC_6H_4$)-7-Cl | 98.5–101 | EtOH/Hexane | 23.8 | | 73 |
| (2-Furyl)CHOH,Me; 5-Ph-7-Cl, | | | | | |
| 4-Oxide, hydrochloride | 148–150d | MeOH/$Et_2O$/Petr ether | | | 106c,d |
| (2-Furyl)CHOMe,Me; 5-Ph-7-Cl, 4-Oxide | 141–142 | MeOH/$Et_2O$/Petr ether | 10 | | 106c |
| (1-Hydroxybut-1-yl),Me; 5-Ph-7-Cl, | | | | | |
| 4-Oxide, hydrochloride | 194–195d | MeOH/$Et_2O$ | | | 106a,b,d |
| Butanal adduct | 193d | MeOH/$Et_2O$/Petr ether | | | 106a |
| (1-Hydroxyeth-1-yl), Me; 5-Ph-7-Cl, | | | | | |
| 4-Oxide, hydrochloride | 170–171d | MeOH/$Et_2O$ | | | 106d |
| (1-Hydroxprop-1-yl),Me; 5-Ph-7-Cl, | | | | | |
| 4-Oxide, hydrochloride | 192–194d | MeOH/$Et_2O$ | | | |
| Propanal adduct | 194–198d | MeOH/$Et_2O$/Petr ether | 15 | | 106 |
| $MeO(CH_2)_2$,Me; 5-(2-$ClC_6H_4$)-7-Cl | 112–115 | EtOH/Hexane | 70.3 | | 73 |
| $MeOCH_2$,Me; 5-Ph-7-Cl, 4-Oxide | 139–140 | EtOH | | ir, uv | 105 |
| $MeOCH_2$, Me; 5-(2-$ClC_6H_4$)-7-Cl | 111–112.5 | EtOH/Hexane | 38.4 | | 73 |
| $MeOCH_2$, *i*-Pr; 5-(2-$ClC_6H_4$)-7-Cl | 159.5–161 | EtOH | 23.1 | | 73 |
| $MeO(CH_2)_3$, Me; 5-(2-$ClC_6H_4$)-7-Cl | Oil | | 99.4 | | 73 |
| Me,Me; 5-Ph-7-Cl | 178–180 | $Me_2CO$ | | | 60, 61 |
| | 205–209 | EtOH | | | 85a |
| 4-Oxide | 200–203 | EtOH | | | 60 |
| Me,Me; 5-(2-$FC_6H_4$)-7-MeNHCONH | 235–236 | EtOAc/$Et_2O$ | | | 70 |
| Me,Me; 5-(2-$FC_6H_4$)-7-$NO_2$ | 275–276 | EtOAc | | | 70 |
| Me,Propen-3-yl; 5-Ph-7-Cl, 4-Oxide | 126–126.5 | *i*-PrOH | | ir,uv | 105 |
| Me,*i*-$PrO(CH_2)_2$; 5-(2-$ClC_6H_4$)-7-Cl | 82.5–85 | Hexane/Petr ether | 67.8 | | 73 |
| Me, 3-(10,11-Dihydro-5*H*-dibenzo[*a*,*e*]cyclo- | | | | | |
| hepten-5-ylidene)propyl; 5-Ph-7-Cl, hydrochloride | ~ 110 | $Et_2O$ | | ir, ms | 85c |

TABLE V-6. —(contd.)

| Substituent | mp (°C); [bp (°C/torr)] | Solvent of Crystallization | Yield (%) | Spectra | Refs. |
|---|---|---|---|---|---|
| ***Trisubstituted*** | | | | | |
| Et,Et; 3-AcO-5-Ph-7-Cl | 140–142 | PhH | 75 | | 45 |
| Et,Et; 3-HO-5-Ph-7-Cl | 150–152 | PhH | 85 | | 45 |
| Et,Me; 3-Me-5-(2-$ClC_6H_4$)-7-$NO_2$ (+) | 180–185 | EtOH | | $[\alpha]_D$ + 48.6° $C$: 1% $CH_2Cl_2$ | 70 |
| Me,Me; 3-AcO-5-Ph-7-Cl | 132–134 | EtOH | 82 | | 45 |
| Me,Me; 3-$PhCH_2NMe$-5-Ph-7-Cl | 144–146 | EtOH | 59 | | 45 |
| Me,Me; 3-$Et_2N$-5-Ph-7-Cl | 120–122 | EtOH | 65 | | 45 |
| Me,Me; 3-$Me_2N$-5-Ph-7-Cl | 161–163 | EtOH | 71 | | 45 |
| Me,Me; 3-HO-5-Ph-7-Cl | 142–144 | PhH | 85 | | 45 |
| Me,Me; 3-Morpholino-5-Ph-7-Cl | 160–162 | EtOH | 66 | | 45 |
| Me,Me; 3-Piperidino-5-Ph-7-Cl | 167–169 | EtOH | 78 | | 45 |
| Me,Me; 3-Pyrrolidino-5-Ph-7-Cl | 173–175 | EtOH | 75 | | 45 |
| ***Tetrasubstituted*** | | | | | |
| Me,Me; 3,3-$Me_2$-5-(2-$FC_6H_4$)-7-$NO_2$ | | | | ir, ms, pmr | 85d |

TABLE V-7. 3*H*-1,4-BENZODIAZEPINES WITH A NITROGEN HETEROCYCLE ATTACHED AT POSITION 2

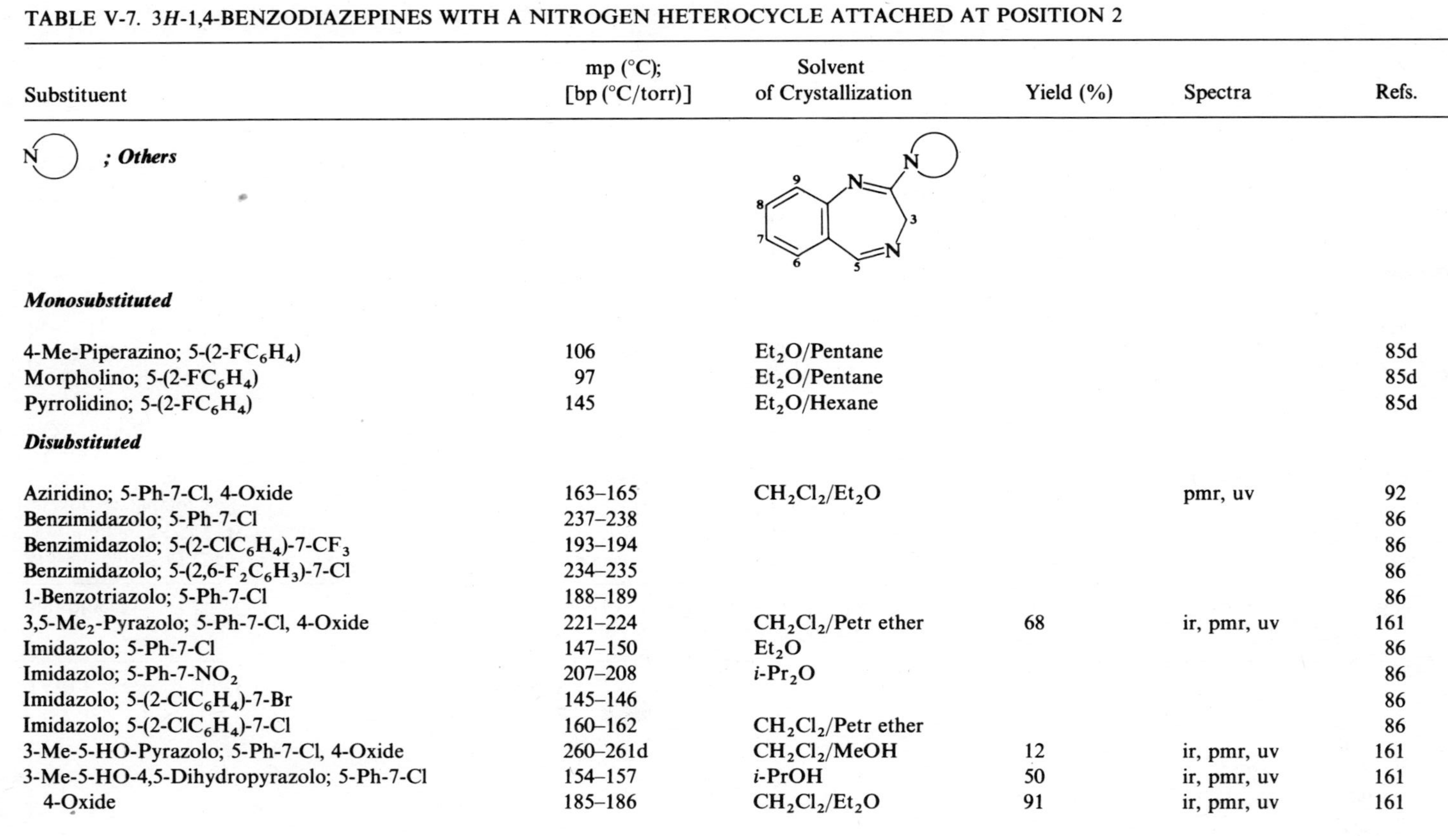

| Substituent | mp (°C); [bp (°C/torr)] | Solvent of Crystallization | Yield (%) | Spectra | Refs. |
|---|---|---|---|---|---|
| N( ) ; ***Others*** | | | | | |
| ***Monosubstituted*** | | | | | |
| 4-Me-Piperazino; 5-(2-$FC_6H_4$) | 106 | $Et_2O$/Pentane | | | 85d |
| Morpholino; 5-(2-$FC_6H_4$) | 97 | $Et_2O$/Pentane | | | 85d |
| Pyrrolidino; 5-(2-$FC_6H_4$) | 145 | $Et_2O$/Hexane | | | 85d |
| ***Disubstituted*** | | | | | |
| Aziridino; 5-Ph-7-Cl, 4-Oxide | 163–165 | $CH_2Cl_2$/$Et_2O$ | | pmr, uv | 92 |
| Benzimidazolo; 5-Ph-7-Cl | 237–238 | | | | 86 |
| Benzimidazolo; 5-(2-$ClC_6H_4$)-7-$CF_3$ | 193–194 | | | | 86 |
| Benzimidazolo; 5-(2,6-$F_2C_6H_3$)-7-Cl | 234–235 | | | | 86 |
| 1-Benzotriazolo; 5-Ph-7-Cl | 188–189 | | | | 86 |
| 3,5-$Me_2$-Pyrazolo; 5-Ph-7-Cl, 4-Oxide | 221–224 | $CH_2Cl_2$/Petr ether | 68 | ir, pmr, uv | 161 |
| Imidazolo; 5-Ph-7-Cl | 147–150 | $Et_2O$ | | | 86 |
| Imidazolo; 5-Ph-7-$NO_2$ | 207–208 | *i*-$Pr_2O$ | | | 86 |
| Imidazolo; 5-(2-$ClC_6H_4$)-7-Br | 145–146 | | | | 86 |
| Imidazolo; 5-(2-$ClC_6H_4$)-7-Cl | 160–162 | $CH_2Cl_2$/Petr ether | | | 86 |
| 3-Me-5-HO-Pyrazolo; 5-Ph-7-Cl, 4-Oxide | 260–261d | $CH_2Cl_2$/MeOH | 12 | ir, pmr, uv | 161 |
| 3-Me-5-HO-4,5-Dihydropyrazolo; 5-Ph-7-Cl | 154–157 | *i*-PrOH | 50 | ir, pmr, uv | 161 |
| 4-Oxide | 185–186 | $CH_2Cl_2$/$Et_2O$ | 91 | ir, pmr, uv | 161 |

TABLE V-7. —(contd.)

| Substituent | mp (°C); [bp (°C/torr)] | Solvent of Crystallization | Yield (%) | Spectra | Refs. |
|---|---|---|---|---|---|
| Morpholino; 5-Ph-7-Cl | 80–90 | Hexane | 74 | ir, ms, uv | 87a |
| Morpholino; 5-(2-$FC_6H_4$)-7-CN | 191–195 | $CH_2Cl_2$/Hexane | | | 9 |
| 2-Oxooxazolidino; 5-Ph-7-Cl | 223–225 | PhH/$Et_2O$ | | | 7b |
| 2-Oxopyrrolidino; 5-Ph-7-Cl | 247–247.5 | *i*-PrOH | 66 | ir, ms, pmr | 66, 67 |
| 2-Oxopyrrolidino; 5-(2-$ClC_6H_4$)-7-Cl | 197–199 | EtOAc/$Et_2O$ | 31 | | 66, 67 |
| 2-Oxo-3-($Me_2NCH_2CH_2$)-pyrrolidino; 5-Ph-7-Cl | 129–130.5 | Cyclohexane | 52 | | 66 |
| 2-Oxo-3-Me-pyrrolidino; 5-Ph-7-Cl | 212–215 | *i*-PrOH | 82 | | 66, 67 |
| Piperidino; 5-Ph-7-Cl | 115–116 | EtOH | 60; 89 | | 61, 72a |
| 4-Oxide | 164–170 | $CH_2Cl_2$/Hexane | | | 85a |
| Pyrazolo; 5-Ph-7-Cl | 172–173 | | | | 86 |
| Pyrazolo; 5-(2-$FC_6H_4$)-7-Cl | 184–185 | | | | 86 |
| Pyrrolidino; 5-Ph-7-Cl | 139–141 | *i*-PrOH | | | 60, 85a |
| | 135–138 | MeOH | | | |
| 4-Oxide | 197–198 | EtOH | | | 60, 61 |
| Pyrrolidino; 5-(2-$ClC_6H_4$)-7-$NO_2$ | 197–200 | $CH_2Cl_2$/$Et_2O$/Petr ether | | | 70 |
| 1,2,4-Triazolo; 5-Ph-7-Cl | 201–202 | | | | 86 |

TABLE V-8. *N*-METHYL-*N*-NITROSO-2-AMINO-3*H*-1,4-BENZODIAZEPINES

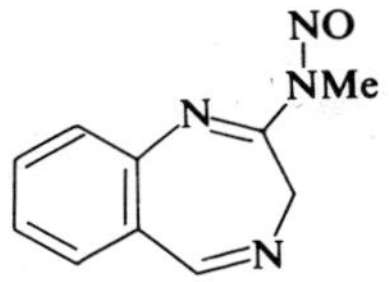

| Substituent | mp (°C); [bp (°C/torr)] | Solvent of Crystallization | Yield (%) | Spectra | Refs. |
|---|---|---|---|---|---|
| 5-Ph | 192–199d | | | | 21 |
| 5-Ph-7-Cl, 4-Oxide | 158–160d | $Et_2O$/Hexane | | | 143, 144 |
| 5-(2-$ClC_6H_4$)-7-$NO_2$ | 167–169 | $CH_2Cl_2$/EtOH | | | 21 |
| 5-(2-$FC_6H_4$)-7-Cl | 110–112 | $Et_2O$ | 60 | pmr, uv | 19a |
| 5-(2-Pyridyl)-7-Br | 102–106 | *i*-PrOH | | | 143, 144 |

TABLE V-9. *N*-ACYL 2-AMINO-3*H*-1,4-BENZODIAZEPINES

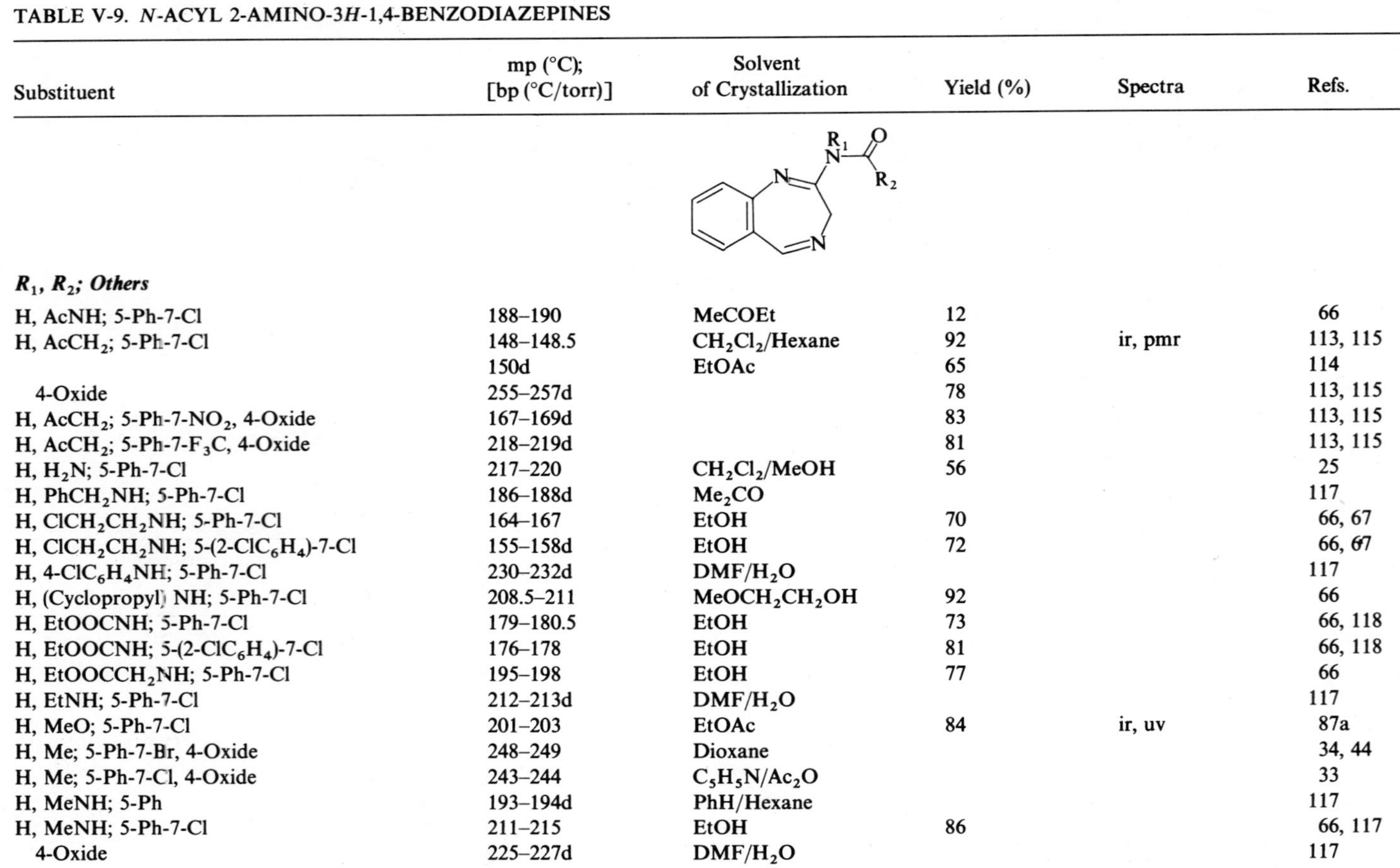


| Substituent | mp (°C); [bp (°C/torr)] | Solvent of Crystallization | Yield (%) | Spectra | Refs. |
|---|---|---|---|---|---|
| ***$R_1$, $R_2$; Others*** | | | | | |
| H, AcNH; 5-Ph-7-Cl | 188–190 | MeCOEt | 12 | | 66 |
| H, $AcCH_2$; 5-Ph-7-Cl | 148–148.5 | $CH_2Cl_2$/Hexane | 92 | ir, pmr | 113, 115 |
| | 150d | EtOAc | 65 | | 114 |
| 4-Oxide | 255–257d | | 78 | | 113, 115 |
| H, $AcCH_2$; 5-Ph-7-$NO_2$, 4-Oxide | 167–169d | | 83 | | 113, 115 |
| H, $AcCH_2$; 5-Ph-7-$F_3C$, 4-Oxide | 218–219d | | 81 | | 113, 115 |
| H, $H_2N$; 5-Ph-7-Cl | 217–220 | $CH_2Cl_2$/MeOH | 56 | | 25 |
| H, $PhCH_2NH$; 5-Ph-7-Cl | 186–188d | $Me_2CO$ | | | 117 |
| H, $ClCH_2CH_2NH$; 5-Ph-7-Cl | 164–167 | EtOH | 70 | | 66, 67 |
| H, $ClCH_2CH_2NH$; 5-(2-$ClC_6H_4$)-7-Cl | 155–158d | EtOH | 72 | | 66, 67 |
| H, 4-$ClC_6H_4NH$; 5-Ph-7-Cl | 230–232d | DMF/$H_2O$ | | | 117 |
| H, (Cyclopropyl) NH; 5-Ph-7-Cl | 208.5–211 | $MeOCH_2CH_2OH$ | 92 | | 66 |
| H, EtOOCNH; 5-Ph-7-Cl | 179–180.5 | EtOH | 73 | | 66, 118 |
| H, EtOOCNH; 5-(2-$ClC_6H_4$)-7-Cl | 176–178 | EtOH | 81 | | 66, 118 |
| H, $EtOOCCH_2NH$; 5-Ph-7-Cl | 195–198 | EtOH | 77 | | 66 |
| H, EtNH; 5-Ph-7-Cl | 212–213d | DMF/$H_2O$ | | | 117 |
| H, MeO; 5-Ph-7-Cl | 201–203 | EtOAc | 84 | ir, uv | 87a |
| H, Me; 5-Ph-7-Br, 4-Oxide | 248–249 | Dioxane | | | 34, 44 |
| H, Me; 5-Ph-7-Cl, 4-Oxide | 243–244 | $C_5H_5N/Ac_2O$ | | | 33 |
| H, MeNH; 5-Ph | 193–194d | PhH/Hexane | | | 117 |
| H, MeNH; 5-Ph-7-Cl | 211–215 | EtOH | 86 | | 66, 117 |
| 4-Oxide | 225–227d | DMF/$H_2O$ | | | 117 |

| | | | | | |
|---|---|---|---|---|---|
| H, MeNH; 5-Ph-7-MeO | 197–199 | EtOH | | | 117 |
| H, MeNH; 5-Ph-7-Me | 212–213d | EtOH | | | 117 |
| H, MeNH; 5-Ph-7-$NO_2$ | 214–215d | DMF/$H_2O$ | | | 117 |
| H, MeNH; 5-Ph-7-$F_3C$ | 207–210 | MeOH | | | 117 |
| H, MeNH; 5-(2-$FC_6H_4$)-7-MeNHCONH | 255–257d | EtOAc | | | 70 |
| H, MeNH; 5-(4-$MeOC_6H_4$)-7-Cl | 210–211 | MeOH | | | 117 |
| H, PhNH; 5-Ph-7-Cl | 228–229d | DMF/$H_2O$ | | | 117 |
| H, PhNH; 5-Ph-7-$NO_2$ | 230–231d | DMF/$H_2O$ | | | 117 |
| Bu, Me; 5-Ph-7-Cl | 87–88 | Hexane | | | 38b |
| 4-Oxide | 130–132 | $Me_2CO$/Hexane | | | 38b |
| Me, $H_2N$; 3-AcO-5-Ph-7-Cl | 137–140 | $CH_2Cl_2$/$Et_2O$ | | | 9 |
| | 195–204 | | | | |
| Me, $H_2NCO$; 3-$H_2N$-5-Ph-7-Cl | 280–284 | THF/EtOH | | | 9 |
| Me, Cl; 3-AcO-5-Ph-7-Cl | 138–143 | $CH_2Cl_2$/Hexane | | | 9 |
| Me, $ClCH_2CH_2NH$; 5-Ph-7-Cl | 112–116 | *i*-PrOH | 72 | | 66 |
| Me, EtOOC; 3-EtO-5-Ph-7-Cl | Oil | | | ms, pmr | 9 |
| Me, Et; 5-Ph-7-Cl, 4-Oxide | 213–214 | $CH_2Cl_2$/$Me_2CO$ | | ir | 44, 110 |
| Me, H; 3-AcO-5-Ph-7-Cl | 192–193 | $Et_2O$/Petr ether | 32 | ir, ms, pmr | 112a |
| Me, Me; 3-AcO-5-Ph-7-Cl | 145–146 | $Et_2O$ | 34 | ir | 110 |
| | 159–160 | | | | |
| Me, Me; 5-Ph-7-Cl | 162 | $Et_2O$/Petr ether | | | 51, 125 |
| 4-Oxide | 186–187 | $Et_2O$/Petr ether | 82 | | 33, 44, 51 |
| Me, Me; 5-Ph-7-Me, 4-Oxide | 205–206 | $Me_2CO$ | 54 | | 34, 44, 110 |
| Me, Me; 5-Ph-7,8-$Me_2$, 4-Oxide | 193–194 | $Me_2CO$/$Et_2O$ | | | 34, 44, 51 |
| Me, Me; 5-(4-$ClC_6H_4$)-7-Cl, 4-Oxide | 191–192 | $Me_2CO$ | | | 34, 44 |
| Me, Me; 5-(4-$MeOC_6H_4$), 4-Oxide | 181–182 | $Me_2CO$ | | | 34, 44 |
| Me, Me; 5-(4-$MeOC_6H_4$)-7-Cl, 4-Oxide | 218.5–219.5 | $CH_2Cl_2$/Hexane | | ir | 48 |
| | 188–190 | | | | 110 |
| Me, Me; 5-(4-$MeC_6H_4$)-7-Br, 4-Oxide | 209–210 | $Me_2CO$/Petr ether | | | 34, 44, 51 |
| Me, MeNH; 5-Ph-7-Cl | 235–243 | *i*-PrOH | 94 | | 66 |
| 4-Oxide | 175–177 | $CHCl_3$/EtOAc | 11 | ir, ms, pmr, uv | 119 |

TABLE V-9. —(contd.)

| Substituent | mp (°C); [bp (°C/torr)] | Solvent of Crystallization | Yield (%) | Spectra | Refs. |
|---|---|---|---|---|---|
| Me, Pr; 5-Ph-7-Cl, 4-Oxide | 169–170 | $Et_2O$ | | ir | 44, 110 |
| | | [structure: 2-(N–C(=X)R)amino-3H-1,4-benzodiazepine, H N–C(=X)R] | | | |
| ***X, R; Others*** | | | | | |
| NH, $NH_2$; 5-Ph-7-Cl, 4-Oxide | | | | | |
| Hydrochloride | 255–256d | | | | 41 |
| NH, Et; 5-Ph-7-Cl | 146–147 | $i$-$Pr_2O$ | | | 102, 103 |
| NH, Me; 5-Ph-7-Cl | 174-174.5 | EtOAc | | | 102, 103 |
| 4-Oxide | 181–182 | EtOAc | | | 102, 103 |
| NH, Me; 5-Ph-7-$NO_2$ | 179–180 | EtOAc | | | 102, 103 |
| NH, Me; 5-(4-$MeOC_6H_4$)-7-Cl | 136–138 | $Me_2CO$ | | | 102, 103 |
| S, MeNH; 5-Ph-7-Cl | 211–212d | $DMF/H_2O$ | | | 117 |
| S, MeNH; 5-Ph-7-$NO_2$ | 191–192d | EtOH | | | 117 |
| | | [structure: 2-(N=C(Me)OEt)-3H-1,4-benzodiazepine] | | | |
| 5-(4-$MeOC_6H_4$)-7-Cl | 142–143 | $i$-$Pr_2O$ | | | 102, 103 |

TABLE V-10. *N*-HYDROXY 2-AMINO-3*H*-1,4-BENZODIAZEPINES

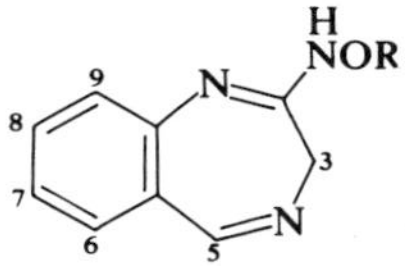

| Substituent | mp (°C); [bp (°C/torr)] | Solvent of Crystallization | Yield (%) | Spectra | Refs. |
|---|---|---|---|---|---|
| ***R; Others*** | | | | | |
| H; 5-Ph-7-Cl | 126–130 | EtOAc | 58 | uv | 131, 132, 135 |
| 4-Oxide | 250–255d | $CH_2Cl_2$/EtOH | 73 | ir, pmr, uv | 92 |
| H; 5-(2-$ClC_6H_4$)-7-$NO_2$ | 155–160 | $Et_2O$/Hexane | | | 70 |
| Ac; 5-Ph-7-Cl | 212.5–213.5 | MeOH/EtOAc | 56 | ir, pmr, uv | 135 |
| Ac; 5-(2-$ClC_6H_4$)-7-$NO_2$ | 126–127d | $CH_2Cl_2$ | | | 70 |
| $PhCH_2$; 5-Ph-7-Cl | 180–181.5 | MeOH | 54 | uv | 131, 135 |
| *t*-Bu; 5-Ph-7-Cl | 251.5–252.5 | EtOAc | 14 | uv | 131, 135 |
| $Et_2NCH_2CH_2$; 5-Ph-7-Cl·0.5$H_2O$ | 99–100 | EtOAc/Hexane | 91 | uv | 131, 135 |
| $EtOOCCH_2$; 5-Ph-7-Cl | 113–114 | $Et_2O$/Hexane | 31 | uv | 131, 135 |
| | 121–123 | | | | 19b |
| 4-Oxide | 181–183 | $Et_2O$ | | ir, pmr, uv | 19b |
| EtO(Me)CH; 5-Ph-7-Cl, 4-Oxide | 214–216d | $CH_2Cl_2$/$Et_2O$ | | ir, pmr, uv | 19b |
| (1-Imidazolo)CO; 5-Ph-7-Cl | 105.5–106.5 | EtOAc/Petr ether | 52 | ir, pmr, uv | 135 |
| Me; 5-Ph-7-Cl | 185–186 | MeOH | 62.5 | uv | 131, 135 |
| 4-Oxide | 232–234 | $CH_2Cl_2$/EtOH | 79 | ir, pmr, uv | 92 |
| Me; 5-(2-$ClC_6H_4$)-7-Cl | 158.5–159.5 | EtOAc | 70 | uv | 131, 135 |
| PhNHCO; 3-$PhCH_2OCOO$-5-Ph-7-Cl | 124–128d | EtOH | | | 112b |
| PhNHCO: 3-$Me_2NCOO$-5-Ph-7-Cl | 99d | | | | 112b |
| PhNHCO; 3-$Me_2NCOO$-5-(2-$ClC_6H_4$)-7-Cl | 117d | | | | 112b |
| PhNHCO; 5-Ph-7-Cl | 212–215 | DMF/$H_2O$ | | | 112b |
| PhNHCO; 5-Ph-7-$NO_2$ | 182–185d | EtOH | | | 112b |
| PhNHCO; 5-(2-$ClC_6H_4$)-7-Cl | 224–226 | EtOH | | | 112b |

TABLE V-10. —(contd.)

| Substituent | mp (°C); [bp (°C/torr)] | Solvent of Crystallization | Yield (%) | Spectra | Refs. |
|---|---|---|---|---|---|
| PhNHCO; 5-(2-$ClC_6H_4$)-7-$NO_2$ | 201–203 | EtOH | | | 112b |
| PhNHCO; 5-(2-$FC_6H_4$)-7-Cl | 126–128 | *i*-PrOH | | | 112b |
| PhNHCO; 5-(2-pyridyl)-7-Br | 190–192 | EtOH | | | 112b |
| (4-*t*-$BuC_6H_4$)NHCO; 5-Ph-7-Cl | 221–223 | EtOH | | | 112b |
| (3-$ClC_6H_4$)NHCO; 3-$Me_2$NCOO-5-Ph-7-Cl | 105d | | | | 112b |
| (3-$ClC_6H_4$)NHCO; 3-$Me_2$NCOO-5-(2-$ClC_6H_4$)-7-Cl | 121d | | | | 112b |
| (3-$ClC_6H_4$)NHCO; 3-Me-5-(2-$ClC_6H_4$)-7-$NO_2$ (+) | 137–139 | | | | 112b |
| (3-$ClC_6H_4$)NHCO; 5-Ph-7-Cl | 157–159 | EtOH | | | 112b |
| Hydrochloride | 210 | MeOH/DMF | | | 112b |
| (3-$ClC_6H_4$)NHCO; 5-Ph-7-$NO_2$ | 123–125 | EtOH | | | 112b |
| (3-$ClC_6H_4$)NHCO; 5-(2-$ClC_6H_4$)-7-Cl | 125–137 | EtOH | | | 112b |
| (3-$ClC_6H_4$)NHCO; 5-(2-$ClC_6H_4$)-7-$NO_2$ | 130–135 | EtOH | | | 112b |
| (3-$ClC_6H_4$)NHCO; 5-(2-$FC_6H_4$)-7-Cl | 194–196 | EtOH | | | 112b |
| (3-$ClC_6H_4$)NHCO; 5-(2-$FC_6H_4$)-7-$NO_2$ | 126–130 | MeOH/EtOH | | | 112b |
| (3-$ClC_6H_4$)NHCO; 5-(2-Pyridyl)-7-Br | 190–191 | | | | 112b |
| (3-Cl-4-$MeC_6H_3$)NHCO-5-Ph-7-Cl | 218–220 | MeOH | | | 112b |
| (4-$ClC_6H_4$)NHCO; 5-Ph-7-Cl | 200–203 | MeOH | | | 112b |
| (3,4-$Cl_2C_6H_3$)NHCO; 5-Ph-7-Cl | 218–219 | EtOH | | | 112b |
| (3,4-$Cl_2C_6H_3$)NHCO; 5-(2-$ClC_6H_4$)-7-$NO_2$ | 190–192d | EtOH | | | 112b |
| (4-$EtOOCC_6H_4$)NHCO; 5-Ph-7-Cl | 212–214 | EtOH | | | 112b |
| (2-$FC_6H_4$)NHCO; 5-Ph-7-Cl | 193–195 | EtOH | | | 112b |
| (3-$MeC_6H_4$)NHCO; 5-Ph-7-Cl | 156 | EtOH | | | 112b |
| (4-$MeC_6H_4$)NHCO; 5-Ph-7-Cl | 220–221 | EtOH | | | 112b |
| (2,5-$Me_2C_6H_3$)NHCO; 5-Ph-7-Cl | 181–183 | EtOH | | | 112b |

| | | | | | |
|---|---|---|---|---|---|
| (2,6-$Me_2C_6H_3$)NHCO; 5-Ph-7-Cl | 198–200 | EtOH | | | 112b |
| (2-$F_3CC_6H_4$)NHCO; 5-Ph-7-Cl | 190 | MeOH/DMF | | | 112b |
| (3-$F_3CC_6H_4$)NHCO; 5-Ph-7-Cl | 95 | EtOH | | | 112b |
| (4-$F_3CC_6H_4$)NHCO; 5-Ph-7-Cl | 215–218 | *i*-PrOH | | | 112b |
| Propen-3-yl; 5-Ph-7-Cl | 134–135.5 | EtOAc | 36.5 | uv | 131, 135 |
| Propen-3-yl; 5-Ph-7-$NO_2$ | 178–180 | EtOAc/Hexane | | | 131 |
| Propen-3-yl; 5-(2-$ClC_6H_4$)-7-Cl | 130–131 | EtOAc/Hexane | 38 | uv | 131, 135 |
| 2-$^{14}C$ | 129–131 | | 84 | | 133 |
| (1-Pyrrolidino)$CH_2CH_2$; 5-Ph-7-Cl hydrate | 137.5–138.5 | EtOAc | 42; 44 | uv | 131, 135 |
| ***$R_1$, $R_2$; Others*** | | | | | |
| $PhCH_2$, H; 5-Ph-7-Cl | 181–183 | $Et_2O$ | 48 | ir, ms, pmr | 112a |
| $PhCH_2$, (Cyclohexyl)NHCO; 5-Ph-7-Cl | 162–165 | $Et_2O$/Petr ether | 97 | ir, pmr | 112a |
| $PhCH_2$, PhNHCO; 5-Ph-7-Cl | 158–161 | $Et_2O$/Petr ether | 95 | ir, pmr | 112a |
| Me, H; 5-Ph-7-Cl | 167–168 | EtOAc | 87 | ir, ms, pmr | 112a |
| 4-Oxide | 213–215d | EtOH | | | 85c |
| Me, Ac; 5-Ph-7-Cl | 126–128 | $Et_2O$/Petr ether | 98 | ir, ms, pmr | 112a |
| Me, PhCO; 5-Ph-7-Cl | 113–115 | $Et_2O$/Petr ether | 75 | ir, ms, pmr | 112a |
| Me, (Cyclohexyl)NHCO; 5-Ph-7-Cl | 147–149 | $Et_2O$/Petr ether | 94 | ir, pmr | 112a |
| Me, MeNHCO; 5-Ph-7-Cl | 126–128 | $Et_2O$/Petr ether | 98 | ir, ms, pmr | 112a |

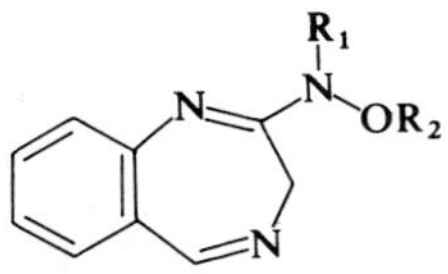

TABLE V-11. 2-HYDRAZINO-3*H*-1,4-BENZODIAZEPINES

| Substituent | mp (°C); [bp/(°C/torr)] | Solvent of Crystallization | Yield (%) | Spectra | Refs. |
|---|---|---|---|---|---|
| ***R; Others*** | | 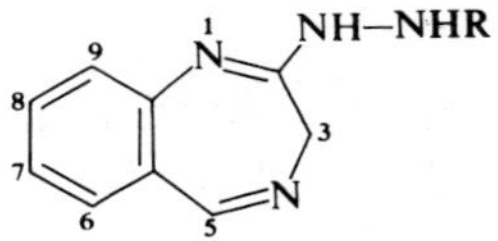 | | | |
| H; 3-(2-Me-Prop-1-yl)-5-Ph-7-Cl | 168–169 | $CHCl_3$/Hexane | 65 | | 47, 140 |
| H; 5-Ph·0.33PhH | 116–118d | $CH_2Cl_2$/PhH | 74 | | 47, 140 |
| H; 5-Ph-7-Cl | 217.5–219 | EtOAc | | | 134, 137, 175 |
| | 203–205d | $CH_2Cl_2$/PhH | 63–81 | pmr, uv | 47, 136, 146 |
| 4-Oxide | 262–263d | $CHCl_3/Et_2O$ | 94 | pmr | 47, 137, 143, 144 |
| | 288–290 | $CH_2Cl_2/Et_2O$ | 83 | ms, pmr, uv | 92 |
| H; 5-Ph-7-MeO | 110–120 | $CHCl_3/Et_2O$ | 77 | | 47, 140 |
| H; 5-Ph-7-Me | 240–241 | $CHCl_3/Et_2O$ | 95 | | 47, 140 |
| | 213–215 | | | | 142 |
| H; 5-Ph-7-$NO_2$ | 226d | EtOH | | | 139a, 140 |
| 4-Oxide | 266d | | 87 | | 47 |
| H; 5-Ph-7-$F_3C$ | 133–135d | $CHCl_3$/Hexane | 94 | | 47, 140 |
| 4-Oxide | 285–287d | $CHCl_3$/PhH | 98 | | 47 |
| H; 5-(2-$ClC_6H_4$)-7-Cl | 220–223d | $CHCl_3$/Hexane | 91 | | 47 |
| | 230–233 | THF/MeOH | | | 143, 144 |
| H; 5-(4-$ClC_6H_4$)-7-Cl, 4-Oxide | > 300 | $CHCl_3$/PhH | 90 | | 47 |
| H; 5-(2-$FC_6H_4$)-7-Et | 126–129 | MeOH | | | 85c |
| H; 5-(4-$MeOC_6H_4$)-7-Cl | 214–220 | PhH | 80 | | 140 |
| 4-Oxide | > 300 | $CHCl_3$/PhH | 90 | | 47 |
| H; 5-(2-pyridyl)-7-Br | 218–228d | MeOH | | | 141 |
| | 224–226 | $CHCl_3$/MeOH | | | 173 |

| | | | | | |
|---|---|---|---|---|---|
| Ac; 5-Ph-7-Cl | 202–204d | $CHCl_3$/MeOH | 81 | | 47, 150 |
| | 209–210 | MeOH | | | |
| 4-Oxide | 256–258 | DMF/$H_2O$ | 71; 73 | | 47, 140 |
| | 272–275d | DMF | | | 143, 144 |
| Ac; 5-Ph-7-$NO_2$·0.5$H_2O$ | 184–185 | DMF/$H_2O$ | 82 | | 47 |
| Ac; 5-(2,6-$F_2C_6H_3$)-7-Cl | 274–277 | EtOH | | | 150 |
| PhCO; 5-Ph-7-Cl | 207–208d | $CHCl_3$/MeOH | 79; 89 | | 47, 140 |
| (7-Cl-5-Ph-3H-1,4-Benzodiazepin-2-yl); 5-Ph-7-Cl | 253–254 | $CHCl_3$/Hexane | 26 | ms | 136, 137 |
| (Cyclohexyl)CO; 5-Ph-7-Cl | 224–225 | DMF/$H_2O$ | 93; 97 | | 47, 140 |
| | 205–206d | | | | |
| EtOOC; 5-Ph-7-Cl | 198–199d | $CH_2Cl_2$/EtOAc | 77 | uv | 134, 152 |
| EtOOC; 5-(2-$ClC_6H_4$) | 226–228 | EtOH | | | |
| | 214–215 | DMF | | | 25 |
| EtOOC; 5-(3-$MeOC_6H_4$)-7-Cl | 177.5–178.5 | EtOH/$Et_2O$ | | | 25 |
| EtOOCCO; 5-Ph-7-Cl | 173–175d | EtOH | 53 | | 145, 164 |
| HCO; 5-Ph-7-Cl·0.5MeOH | 161–162 | MeOH | 54 | | 47, 140 |
| MeOOC; 5-Ph-7-Cl | 201–203d | EtOAc | | | 85c |
| Me; 5-Ph-7-Cl·0.25EtOAc | 209d | EtOAc | | | 146 |
| MeNHCO; 5-Ph-7-Cl | 247d | DMF/$H_2O$ | | | 146, 165 |
| 4-Oxide | 251–252d | DMF/$H_2O$ | | | 146, 165 |
| MeNHCO; 5-Ph-7-$NO_2$ | 239–240d | DMF/$H_2O$ | | | 146, 165 |
| MeNHCO; 5-(4-$MeOC_6H_4$)-7-Cl | 234–235d | DMF/$H_2O$ | | | 146, 165 |
| (3-Oxocyclohexen-1-yl)-5-Ph-7-Cl, 4-Oxide | 223–224d | $CH_2Cl_2$/MeOH/$Et_2O$ | 85 | ir, pmr, uv | 161 |
| (4-Oxo-2-penten-2-yl)-5-Ph-7-Cl | 154–157 | *i*-PrOH | | | 162, 163 |
| 4-Oxide | 185–186 | $CH_2Cl_2$/$Et_2O$ | | | 162, 163 |
| PhNHCO; 5-Ph-7-Cl | 220–221d | DMF/$H_2O$ | | | 146, 165 |
| PhNHCS; 5-Ph-7-Cl | 199–205d | $Me_2CO$ | | | 146, 165 |
| $PhCH_2CO$: 5-Ph-7-Cl | 224–225 | DMF/$H_2O$ | 90 | | 47 |
| EtCO; 5-Ph-7-Cl | 186–187d | $CHCl_3$/MeOH | 94 | | 47 |

TABLE V-11. —(contd.)

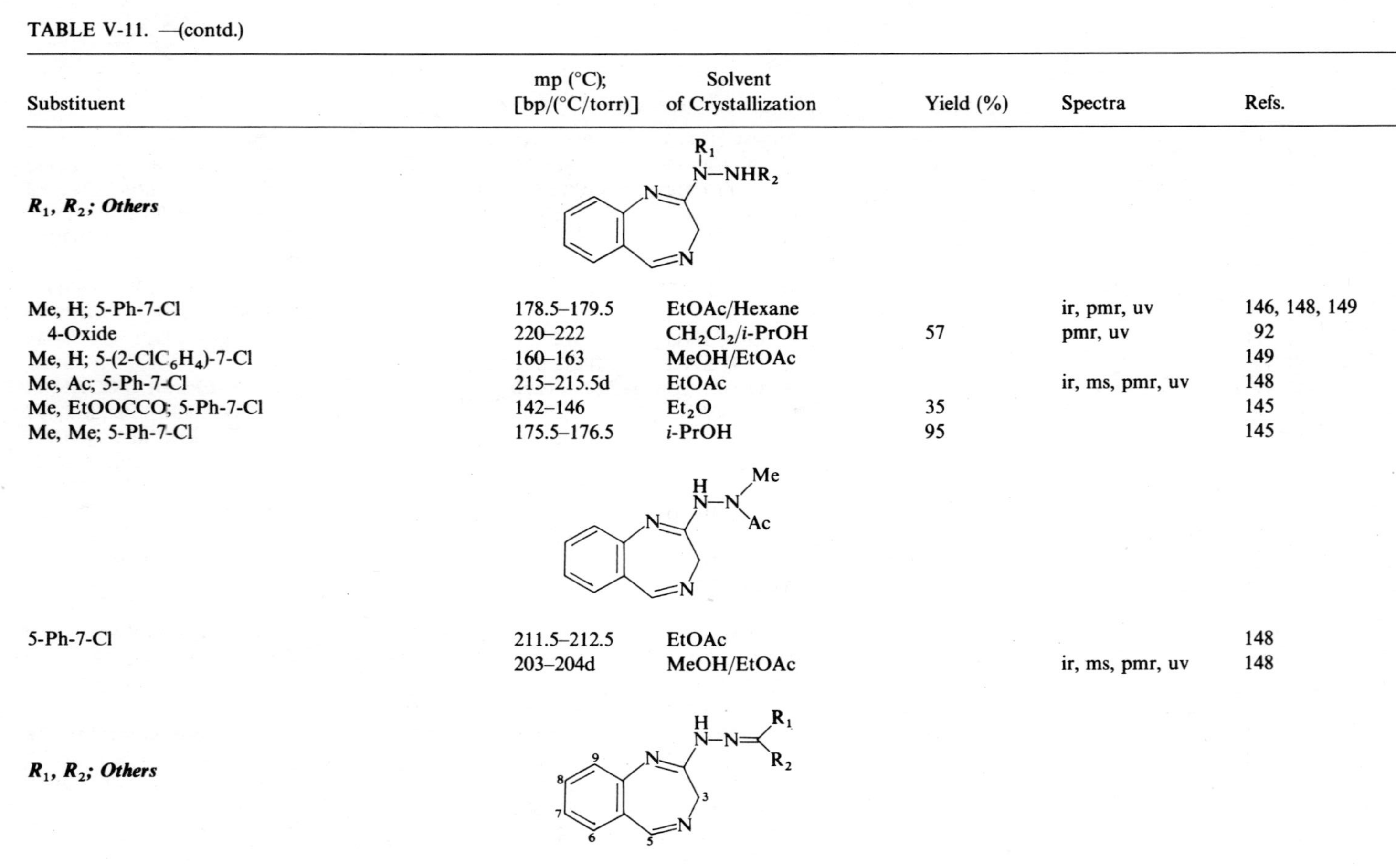

| Substituent | mp (°C); [bp/(°C/torr)] | Solvent of Crystallization | Yield (%) | Spectra | Refs. |
|---|---|---|---|---|---|
| ***R₁, R₂; Others*** | | | | | |
| Me, H; 5-Ph-7-Cl | 178.5–179.5 | EtOAc/Hexane | | ir, pmr, uv | 146, 148, 149 |
| 4-Oxide | 220–222 | $CH_2Cl_2$/*i*-PrOH | 57 | pmr, uv | 92 |
| Me, H; 5-(2-$ClC_6H_4$)-7-Cl | 160–163 | MeOH/EtOAc | | | 149 |
| Me, Ac; 5-Ph-7-Cl | 215–215.5d | EtOAc | | ir, ms, pmr, uv | 148 |
| Me, EtOOCCO; 5-Ph-7-Cl | 142–146 | $Et_2O$ | 35 | | 145 |
| Me, Me; 5-Ph-7-Cl | 175.5–176.5 | *i*-PrOH | 95 | | 145 |
| 5-Ph-7-Cl | 211.5–212.5 | EtOAc | | | 148 |
| | 203–204d | MeOH/EtOAc | | ir, ms, pmr, uv | 148 |
| ***R₁, R₂; Others*** | | | | | |

| | | | | | |
|---|---|---|---|---|---|
| H, HOOC; 5-Ph-7-Cl methanolate | 164–166d | MeOH | 77 | ir, pmr, uv | 145, 160 |
| H, $ClCH_2$; 5-Ph-7-Cl | 210–235d | $Et_2O$ | 52 | | 145, 156 |
| H, Et; 5-Ph-7-Cl | 149–150 | $Me_2CO$ | | | 155 |
| H, $CH{=}CH_2$; 5-(2-$ClC_6H_4$)-7-Cl | 157–159 | $CH_2Cl_2$/Petr ether | 22 | | 154 |
| H, $CH(OH)CH_2OH$; 5-Ph-7-Cl, 4-Oxide | 201–203d | $CH_2Cl_2/Et_2O$/MeOH | 72 | ir, uv | 154 |
| H, MeOOC; 5-Ph-7-Cl, 4-Oxide | 190–192d | $CH_2Cl_2$/*i*-PrOH | 73 | pmr, uv | 154 |
| H, Me; 5-Ph-7-Cl | 151–152 | $Et_2O$ | 85 | ir, ms, pmr, uv | 154, 155 |
| | 162–164 | | | | 148 |
| 4-Oxide | 198–200d | MeOH | 95 | pmr, uv | 154 |
| H, Ph; 5-Ph-7-Cl | 168–169 | | 60 | | 155 |
| $H_2N$, Me, 5-Ph-7-Cl | 203–205d | $Me_2CO$ | | | 140 |
| HOOC, $HOOCCH_2CH_2$; 5-Ph-7-Cl | 212d | MeOH | 94.5 | | 145, 160 |
| HOOC, $Et_2N(CH_2)_3$; 5-Ph-7-Cl | | | | pmr | 160 |
| HOOC, Me; 5-Ph-7-Cl | 183–185d | EtOAc | 76 | | 145, 160 |
| Methanolate | 145–165d | MeOH | 97 | | 160 |
| HOOC, Me; 5-(2-$ClC_6H_4$)-7-Cl, ethanolate | 147–151d | EtOH | 94 | | 160 |
| $ClCH_2$, $ClCH_2$; 5-Ph-7-Cl | 165–250d | $CH_2Cl_2$/MeOH | 64.5 | | 145, 157 |
| $ClCH_2$, Me; 5-Ph-7-Cl | 210–227d | EtOAc | 79 | | 145, 156 |
| $ClCH_2$, Me; 5-(2-$ClC_6H_4$)-7-Cl | 202–217d | *i*-PrOH | 97 | | 145, 156 |
| $(MeO)_2CH$, Me; 5-Ph-7-Cl | 182–183 | EtOAc/Hexane | 60 | | 145, 158 |
| $(MeO)_2CH$, Me; 5-(2-$ClC_6H_4$)-7-Cl | 157–158 | EtOAc/Hexane | 81 | | 145, 158 |
| EtOOC, $EtOOCCH_2$; 5-Ph-7-Cl | | | | pmr | 160 |
| EtOOC, Me; 5-Ph-7-Cl | | | | pmr | 160 |
| EtOOC, Me; 5-(2-Pyridyl)-7-Br | 230–233d | $CH_2Cl_2$/MeOH/$Et_2O$ | 72 | | 145 |
| MeOOC, $MeOOC(CH_2)_2$; 5-Ph-7-Cl | | | | pmr | 160 |
| MeOOC, Me; 5-Ph-7-Cl | 156–157 | $CHCl_3/Et_2O$ | 67 | | 145, 160 |
| | 171.5–176 | EtOAc | 53 | | 145 |
| MeOOC, Me; 5-(2-$ClC_6H_4$)-7-Cl methanolate | 109d | MeOH | 77.5 | | 145, 160 |
| $MeOOCCH_2$, Me; 5-Ph-7-Cl, 4-Oxide | 165–167 | $CH_2Cl_2/Et_2O$/Petr ether | 30 | ir, pmr, uv | 161 |
| Me, Me; 3-($Me_2CHCH_2$)-5-Ph-7-Cl | 189–190 | | | | 47 |
| Me, Me; 5-Ph-7-Cl | 184.5–185.5 | $Me_2CO$ | | pmr, uv | 47, 136, 137 |
| 4-Oxide | 223–224 | $Me_2CO$ | | pmr, uv | 47, 136, 137 |
| Me, Me; 5-Ph-7-Me | 194–195 | | | | 47 |

TABLE V-11. —(contd.)

| Substituent | mp (°C); [bp/(°C/torr)] | Solvent of Crystallization | Yield (%) | Spectra | Refs. |
|---|---|---|---|---|---|
| Me, Me; 5-Ph-7-$NO_2$ | 203–205 | | | | 47 |
| 4-Oxide | 244–245 | | | | 47 |
| Me, Me; 5-Ph-7-$F_3C$ | 224–225 | | | | 47 |
| Me, Me; 5-(2-$ClC_6H_4$)-7-Cl | 167–168 | $Me_2CO$/Hexane | 97 | | 47, 145 |
| Me, Me; 5-(4-$ClC_6H_4$)-7-Cl, 4-Oxide | 211–213 | | | | 47 |
| ***$R_1$, $R_2$; Others*** | 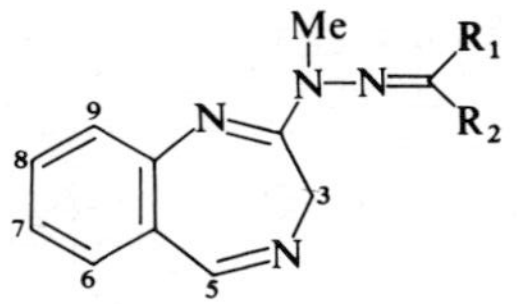 | | | | |
| H, H; 5-Ph-7-Cl | 137–138 | $CH_2Cl_2/Et_2O$ | | | 7b |
| H, Me; 5-Ph-7-Cl | 199–200 | EtOAc | | ir, ms, pmr, uv | 148 |
| | | | | X-ray | 148 |
| ***R; Others*** | 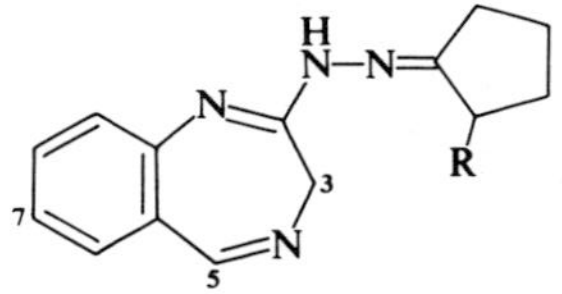 | | | | |
| Cl; 5-Ph-7-Cl | | | 40 | | 145 |
| MeO; 5-Ph-7-Cl | 174–177d | | 47 | | 145 |

TABLE V-12. 2-*OR*-3*H*-1,4-BENZODIAZEPINES

| Substituent | mp (° C); [bp (° C/torr)] | Solvent of Crystallization | Yield (%) | Spectra | Refs. |
|---|---|---|---|---|---|
| ***R; Others*** | | | | | |
| Bu; 3,3-$Me_2$-5-(2-$ClC_6H_4$)-7-$NO_2$ | 82 | Petr ether | | | 70 |
| Bu; 3,3-$Me_2$-5-(2-$FC_6H_4$)-7-$NO_2$-9-Cl | 113 | Hexane/Petr ether | | | 70 |
| Bu; 5-Ph-7-Cl, 4-Oxide | 91–93 | $CH_2Cl_2$/Petr ether | 30 | ir, pmr, uv | 161 |
| $(EtO)_2PO_2$; 3-MeO-5-(2-$ClC_6H_4$)-7-Cl | 132–134 | PhMe/Hexane | 50–75 | | 195 |
| $Et_2N(CH_2)_2$; 3,3-$Me_2$-5-(2-$FC_6H_4$)-7-$H_2N$-9-Cl | 106–107 | EtOH/Petr ether | | | 70 |
| $Et_2N(CH_2)_2$; 3,3-$Me_2$-5-(2-$FC_6H_4$)-7-$NO_2$-9-Cl | 107–108 | $Et_2O$/Hexane | | | 70 |
| $(MeO)_2PO_2$; 3-MeO-5-(2-$ClC_6H_4$)-7-Cl | 123–125 | PhMe | 50–75 | | 195 |
| $Me_2N(CH_2)_2$; 5-Ph-7-Cl, 4-Oxide | 105–110 | $Et_2O$ | 42 | pmr, uv | 92 |
| $(Morpholino)_2PO_2$; 3-$(EtO)_2CH$-5-(2-$ClC_6H_4$)-7-$NO_2$ | 216 | $Et_2O$ | | | 70 |
| $(Morpholino)_2PO_2$; 3-Me-5-(2-$ClC_6H_4$)-7-$NO_2$ (+) | 230–234 | PhH | | $[\alpha]_D$ + 375.5° (1% in $CH_2Cl_2$) | 70 |
| $(Morpholino)_2PO_2$; 3-Me-5-(2-$ClC_6H_4$)-7-$NO_2$ (−) | 228–232 | PhH | | $[\alpha]_D$ − 375.5° (1% in $CH_2Cl_2$) | 70 |
| $(Morpholino)_2PO_2$; 3-Me-5-(2-$FC_6H_4$)-7-Cl | 90–95 | $Et_2O$ | | | 21 |
| $(Morpholino)_2PO_2$; 5-Ph-7-Cl | 189–191 | EtOAc | | ir, ms, uv | 87b |
| 4-Oxide | 160–166 | PhH/Hexane | | | 200 |
| $(Morpholino)_2PO_2$; 5-Ph-7-(2-Me-1,3-dioxolan-2-yl | 178–182 | $CH_2Cl_2$/$Et_2O$ | | | 9 |
| $(Morpholino)_2PO_2$; 5-Ph-7-$NO_2$ | 208–209 | EtOAc | | | 21 |
| $(Morpholino)_2PO_2$; 5-(2-$ClC_6H_4$)-7-Cl | 185–187 | $CH_2Cl_2$/$Et_2O$ | | ir, ms, uv | 87b |
| $(Morpholino)_2PO_2$; 5-(2-$ClC_6H_4$)-7-$NO_2$ | 214–216 | $CH_2Cl_2$/$Et_2O$ | | ir, ms, uv | 87b |

TABLE V-12. —(contd.)

| Substituent | mp (° C); [bp (° C/torr)] | Solvent of Crystallization | Yield (%) | Spectra | Refs. |
|---|---|---|---|---|---|
| $(Morpholino)_2PO_2$; 5-(2-$FC_6H_4$)-7-Cl | 140–142 | PhH/Hexane | | | 7b |
| $(Morpholino)_2PO_2$; 5-(2-$FC_6H_4$)-7-CN | 194–197 | $CH_2Cl_2/Et_2O$ | | | 9 |
| $(Morpholino)_2PO_2$; 5-(2-$FC_6H_4$)-7-I | 104–112 | $CH_2Cl_2/Et_2O$ | | | 21 |
| $(Morpholino)_2PO_2$; 5-(2-$FC_6H_4$)-7-$NO_2$ | 169–172 | EtOAc | | | 21 |
| $(Morpholino)_2PO_2$; 5-(2-pyridyl)-7-Br | 180–182 | $CH_2Cl_2$/EtOAc | | | 21 |
| Et; 3-MeO-5-(2-$ClC_6H_4$)-7-Cl | 110–113 | $Et_2O$ | 55 | | 179 |
| Et; 3-Me-5-Ph-7-Cl, 4-Oxide | 156–157 | Hexane | 22 | | 178 |
| Et; 3,3-$Me_2$-5-(2-$ClC_6H_4$)-7-$NO_2$ | 151 | Petr ether | | ir, pmr | 70 |
| Et; 3,3-$Me_2$-5-(2-$FC_6H_4$)-7-$NO_2$-9-Cl | | $Et_2O$/Hexane | | ir, pmr | 70 |
| Et; 5-Ph-7-Cl | Oil | | | | 142 |
| Et; 5-Ph-7-$NO_2$ | 143–145 | $Et_2O$/Hexane | | | 142 |
| Et; 5-(4-$MeOC_6H_4$)-7-Cl | 119–120 | $i$-$Pr_2O$ | | | 142 |
| $HO(CH_2)_2$; 3,3-$Me_2$-5-(2-$FC_6H_4$)-7-$NO_2$-9-Cl | 142–145 | $Et_2O$ | | | 70 |
| $HO(CH_2)_2$; 5-Ph-7-Cl | 142–144 | $Et_2O$/Hexane | 82 | ir, pmr, uv | 87a |
| $HO(CH_2)_2$; 5-(2-$FC_6H_4$)-7-Cl | 116–118 | $Et_2O/CH_2Cl_2$/Petr ether | | | 9 |
| Me; 3-Me-5-Ph-7-Cl | 105–107.5 | Hexane | | | 18 |
| 4-Oxide | 194–196 | PhH | | | 18 |
| Me; 3,3-$Me_2$-5-(2-$FC_6H_4$)-7-$NO_2$-9-Br | 210–211 | $CH_2Cl_2$/Petr ether | | | 70 |
| Me; 5-Ph-7-Cl | 88–94 | $Et_2O$/Petr ether | | | 180 |
| | 94–97 | Petr ether | 87 | | 87a |
| | [141–148/0.1] | | | pmr | 185 |
| 4-Oxide | 186–188 | MeOH | 94 | | 92 |
| Me; 5-Ph-7-CN | 142–142.5 | EtOAC/Hexane | | | 184 |
| Me; 5-(2-$FC_6H_4$)-7-Cl | 79–81 | $Et_2O$/Hexane | | | 9 |
| 4-Oxide | 192–195 | $Et_2O$ | | | 24e |

| | | | | | |
|---|---|---|---|---|---|
| Me; 5-(2-$FC_6H_4$)-7-$NO_2$ | 155–157 | | | | 201 |
| Ph; 5-Ph-7-Cl | 131–133 | $Et_2O$/Petr ether | | | 90 |
| Ph; 5-Ph-7-$NO_2$ | 178–181 | | | | 90 |
| Ph; 5-(2-$ClC_6H_4$)-7-Br | 155–158 | | | | 90 |
| Ph; 5-(2-$ClC_6H_4$)-7-Cl | 151–152 | $Et_2O$/Petr ether | | | 90 |
| Ph; 5-(2-$ClC_6H_4$)-7-F | 110–112 | | | | 90 |
| 4-(AcNH)$C_6H_4$; 5-Ph-7-Cl | 168–169 | | | | 90 |
| 4-(AcNH)$C_6H_4$; 5-(2-$ClC_6H_4$)-7-Cl | 188–191 | | | | 90 |
| 2-($H_2NCO$)$C_6H_4$; 5-Ph-7-Cl | 186–187 | | | | 90 |
| 2-($PhCH_2O$)$C_6H_4$; 5-Ph-7-Cl | 119–122 | | | | 90 |
| 4-$BrC_6H_4$; 5-Ph-7-Cl | 105–110 | | | | 90 |
| 4-$ClC_6H_4$; 5-Ph-7-Cl | 105–110 | | | | 90 |
| 3-($Me_2N$)$C_6H_4$; 5-Ph-7-Cl | 117–118 | | | | 90 |
| 4-($Me_2N$)$C_6H_4$; 5-Ph-7-Cl | 129–130 | | | | 90 |
| 4-($Me_2N$)$C_6H_4$; 5-(2-$ClC_6H_4$)-7-Cl | 120–122 | | | | 90 |
| 4-($Me_2NCH_2$)$C_6H_4$; 5-Ph-7-Cl | 171–172 | $Me_2CO/Et_2O$ | | | 90 |
| 4-($Me_2NCH_2$)$C_6H_4$; 5-(2-$ClC_6H_4$)-7-Cl | 112–115 | | | | 90 |
| 4-($Me_2NCH_2$)$C_6H_4$; 5-(2-$FC_6H_4$)-7-Cl | 102–106 | | | | 90 |
| 3-$MeC_6H_4$; 5-Ph-7-Cl | 93–94 | | | | 90 |
| 4-$NO_2C_6H_4$; 5-Ph-7-Cl | 133–135 | | | | 90 |
| 4-(PrOOC)$C_6H_4$; 5-Ph-7-Cl | 111 | | | | 90 |
| Propen-3-yl; 5-Ph-7-Cl, 4-Oxide | 120–122 | $Me_2CO$/Hexane | 49 | pmr, uv | 92 |
| Pr; 3,3-$Me_2$-5-(2-$FC_6H_4$)-7-$NO_2$-9-Cl | 117–118 | $Et_2O$/Hexane/Petr ether | | | 70 |
| Pr; 5-Ph-7-Cl | [164–170/0.25] | | | pmr | 185 |
| Pr; 7-Cl, 4-Oxide | 148–149 | $Me_2CO$ | 36 | | 185 |
| *i*-Pr; 3,3-$Me_2$-5-(2-$ClC_6H_4$)-7-$NO_2$ | 110–112 | $Et_2O$/Petr ether | | | 70 |
| *i*-Pr; 3,3-$Me_2$-5-(2-$FC_6H_4$)-7-$H_2N$-9-Cl | 92–93 | Hexane | | | 70 |
| *i*-Pr; 3,3-$Me_2$-5-(2-$FC_6H_4$)-7-$NO_2$-9-Cl | 102 | Hexane/Petr ether | | | 70 |

TABLE V-13. 2-*SR*-3*H*-1,4-BENZODIAZEPINES

| Substituent | mp (° C); [bp (° C/torr)] | Solvent of Crystallization | Yield (%) | Spectra | Refs. |
|---|---|---|---|---|---|
| ***R; Others*** | | | | | |
| $H_2NCOCH_2$; 5-(2-$ClC_6H_4$)-7-Cl | 73–75 | PhH | 85 | | 181 |
| Bu; 5-(2-$ClC_6H_4$)-7-Cl | Oil | | 80 | | 181 |
| $PhCOOCH_2CH_2$; 5-(2-$ClC_6H_4$)-7-Cl | 123–125 | PhH/Hexane | 84 | | 181 |
| $HOOCCH_2$; 3-Me-5-Ph-7-Cl | 182–184d | $CH_2Cl_2$/EtOH | | | 187 |
| $HOOCCH_2$; 5-Ph-7-Cl | 107–111 | EtOH/$H_2O$ | | uv | 187 |
| $HOOCCH_2$; 5-(2-$ClC_6H_4$)-7-Cl | 191d | $Et_2O$ | 17.2 | | 181 |
| [5-(2-$ClC_6H_4$)-7-Cl-3H-1,4-Benzodiazepin-2-yl]-$S(CH_2)_2$; 5-(2-$ClC_6H_4$)-7-Cl | 173–175 | EtOH | 11.3 | ir, pmr | 181 |
| $Bu_2N(CH_2)_2$; 5-(2-$ClC_6H_4$)-7-Cl fumarate | 169–171 | $Me_2CHCH_2Ac$ | 58 | | 181, 186 |
| $Et_2N(CH_2)_2$; 5-(2-$ClC_6H_4$)-7-Cl fumarate | 158–160 | $Me_2CHCH_2Ac$ | 44 | | 181, 186 |
| $Hex_2N$, $(CH_2)_2$; 5-(2-$ClC_6H_4$)-7-Cl fumarate | 113–115 | $Et_2O$ | 59 | | 181, 186 |
| $Me_2N(CH_2)_{10}$; 5-(2-$ClC_6H_4$)-7-Cl fumarate | 53–58 | EtOAc/$Et_2O$ | 48 | pmr | 181, 186 |
| $Me_2N(CH_2)_2$; 5-Ph-7-Cl fumarate | 197–199 | MeOH | 55 | | 181, 186 |
| $Me_2N(CH_2)_2$; 5-Ph-7-$NO_2$ maleate | 131–134 | EtOH/$Et_2O$ | 48.5 | | 181, 186 |
| $Me_2N(CH_2)_2$; 5-(2-$ClC_6H_4$)7-Cl fumarate | 66–72 | | | ir, pmr | 181 |
| | 158–161 | *i*-PrOH | 72 | | 181 |
| | | EtOAc | 48; 62; 75 | | 181, 187 |
| Maleate | 140–142 | *i*-PrOH/$Me_2CO$ | | | 186 |
| $Me_2N(CH_2)_2$; 5-(2-$ClC_6H_4$)-7-$NO_2$ maleate | 196–198 | MeOH | 22.7 | | 181, 186 |
| $Me_2N(CH_2)_2$; 5-(2-$FC_6H_4$)-7-Br | 105–108 | $Et_2O$/Hexane | 76.4 | | 181, 186 |

| | | | | | |
|---|---|---|---|---|---|
| $Me_2N(CH_2)_2$; 5-(2-$FC_6H_4$)-7-Cl fumarate | 164–166 | *i*-PrOH | 68 | | 181, 186 |
| $Me_2N(CH_2)_2$; 5-(2-$FC_6H_4$)-7-$NO_2$ fumarate | 153–157 | EtOH | 73.2 | | 181, 186 |
| $Me_2N(CH_2)_2$; 5-(2-$HOC_6H_4$)-7-Cl | | | | ms | 186 |
| $Me_2N(CH_2)_2$; 5-(2-$MeOC_6H_4$)-7-Cl maleate | 142–145 | $Me_2CO$/EtOAc | 70 | | 181, 186 |
| $Me_2N(CH_2)_2$; 5-(2-$MeOC_6H_4$)-8-Cl fumarate | 172–174 | | | | 186 |
| $Me_2N(CH_2)_6$; 5-(2-$ClC_6H_4$)-7-Cl maleate | 108–110 | EtOAc | 65 | | 181, 186 |
| $Me_2N(CH_2)_3$; 5-Ph-7-Cl maleate | 91; 130–133 | $Me_2CO/Et_2O$ | | | 202 |
| $Me_2N(CH_2)_3$; 5-(2-$ClC_6H_4$)-7-Cl fumarate | 154–156 | EtOH/$Me_2CO$ | | | 186 |
| | | *i*-PrOH | 41; 71 | | 181 |
| $Me_2NCH_2CHMe$; 5-(2-$ClC_6H_4$)-7-Cl fumarate | 178–180 | *i*-PrOH/$Et_2O$ | 34.6; 52 | | 181, 186 |
| *i*-$Pr_2N(CH_2)_2$; 5-(2-$ClC_6H_4$)-7-Cl fumarate | 156–158 | *i*-PrOH | 56 | | 181 |
| $EtO(CH_2)_2$; 5-(2-$ClC_6H_4$)-7-Cl | Oil | | 37 | | 181 |
| Et; 5-Ph-7-Cl, 4-Oxide | 142–144 | $CH_2Cl_2/Et_2O$/Hexane | 28 | pmr, uv | 92 |
| Et; 5-(2-$ClC_6H_4$)-7-Cl | 114–118 | EtOH | 73.5 | | 181 |
| $EtS(CH_2)_2$; 5-(2-$ClC_6H_4$)-7-Cl | 76–77 | Hexane | 63.5 | | 181 |
| (Hexahydroazepino)$(CH_2)_2$; 5-(2-$ClC_6H_4$)-7-Cl fumarate | 170–171 | *i*-PrOH | 66.4 | | 181, 186 |
| (Hexahydroazepino)$(CH_2)_3$; 5-(2-$ClC_6H_4$)-7-Cl fumarate | 164–167 | THF/$Et_2O$ | 39 | | 186 |
| [4-(2-$MeOC_6H_4$)Piperazino]-$(CH_2)_3$; 5-(2-$ClC_6H_4$)-7-Cl-fumarate | 180–181 | *i*-PrOH | | | 186 |
| Me; 3-Me-5-(2-$ClC_6H_4$)-7-$NO_2$ | 160 | EtOH/$Et_2O$ | | | 70 |
| Me; 3-Me-5-(2-$FC_6H_4$)-7-Cl | 145–147 | EtOH | | | 70 |
| Me; 3-(3-Hydroxyiminobut-1-yl)-5-(2-$ClC_6H_4$)-7-$NO_2$ | 165–167 | $CH_2Cl_2/Et_2O$/Petr ether | | | 70 |
| Me; 3-(3-Methyliminobut-1-yl)-5-(2-$ClC_6H_4$)-7-$NO_2$ | 165–167 | $Et_2O$/Petr ether | | | 70 |
| Me; 3-(3-Oxobut-1-yl)-5-(2-$ClC_6H_4$)-7-$NO_2$ | 135 | EtOH/Petr ether | | | 70 |
| Me; 3-(3-Oxobut-1-yl)-5-(2-$FC_6H_4$)-7-$NO_2$ | 163 | EtOH/$Et_2O$ | | | 70 |
| Me; 5-Ph-7-Cl | 132–134 | EtOH | 76 | | 72, 73 |
| Me; 5-(2-$ClC_6H_4$) | 109–111 | EtOAc/Petr ether | | | 126a,b |
| Me; 5-(2-$ClC_6H_4$)-7-Cl | 118–120 | EtOAc/cyclohexane | 45 | | 131, 135 |
| | 123–125 | EtOH | 84 | | 181 |
| Me; 5-(2-$ClC_6H_4$)-7-$NO_2$ | 164–166 | EtOH | | | 70 |
| Me; 5-(2-$FC_6H_4$)-7-Cl | 72–77 | $Et_2O$/Petr ether | | | 9 |

TABLE V-13. —(contd.)

| Substituent | mp (° C); [bp (° C/torr)] | Solvent of Crystallization | Yield (%) | Spectra | Refs. |
|---|---|---|---|---|---|
| Me; 5-(2-$FC_6H_4$)-7-MeNHCONH | 211–212 | EtOAc | | | 70 |
| Me; 5-(2-$FC_6H_4$)-7-$NO_2$ | 161–163 | EtOH | | | 70 |
| (4-Me-Piperazino)$(CH_2)_2$; 5-Ph-7-Cl difumarate | 193–194d | THF | 42 | | 186 |
| (Morpholino)$(CH_2)_2$; 5-(2-$ClC_6H_4$)-7-Cl fumarate | 191–193 | EtOH | 53.5 | | 181, 186 |
| (Morpholino)$(CH_2)_3$; 5-(2-$ClC_6H_4$)-7-Cl fumarate | 169–171 | *i*-PrOH | 59.5 | | 181, 186 |
| Ph; 5-Ph-7-Cl | 105–108 | | | | 90 |
| (Phthalimido)$(CH_2)_3$; 5-(2-$ClC_6H_4$)-7-Cl | 143–146 | $Me_2CHCH_2Ac/Et_2O$ | 77 | | 181 |
| (Piperidino)$(CH_2)_2$; 5-Ph-7-$F_3C$ fumarate | 132–134 | $Me_2CO/Et_2O$ | 38.5 | | 181, 186 |
| (Piperidino)$(CH_2)_2$; 5-(2-$ClC_6H_4$)-7-Cl fumarate | 198–199 | MeOH | 67 | | 181, 186 |
| (Piperidino)$(CH_2)_3$; 5-(2-$ClC_6H_4$)-7-Cl fumarate | 110–112 | $THF/Et_2O$ | 76.8 | | 181, 186 |
| Pr; 5-(2-$ClC_6H_4$)-7-Cl | 71–72 | Hexane | 75.5 | | 181 |
| (Pyrrolidino)$(CH_2)_2$; 5-(2-$ClC_6H_4$)-7-Cl fumarate | 184–188 | *i*-PrOH | 55 | | 181, 186 |
| (Pyrrolidino) $(CH_2)_3$; 5-(2-$ClC_6H_4$)-7-Cl fumarate | 148–151 | *i*-PrOH | 54.8 | | 181, 186 |

TABLE V-14. 2-HALO-3*H*-1,4-BENZODIAZEPINES

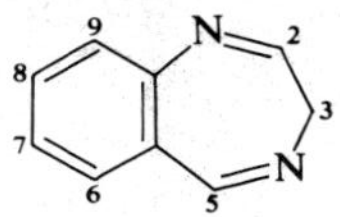

| Substituent | mp (° C); [bp (° C/torr)] | Solvent of Crystallization | Yield (%) | Spectra | Refs. |
|---|---|---|---|---|---|
| 2-Br-5-Ph-7-Cl | 137–139 | | | | 193 |
| 2,7-$Cl_2$-3-MeO-5-(2-$ClC_6H_4$) | 137–141 | MeCN/Cyclohexane | 50 | | 194 |
| 2,7-$Cl_2$-5-Ph | 139–142 | Ligroin | | | 193 |
| 2,7-$Cl_2$-5-(2-$ClC_6H_4$) | 125–126 | | | | 193 |
| 2,7-$Cl_2$-5-(2-$FC_6H_4$) | 88–90 | $Et_2O$/Petr ether | | | 193 |
| 2,7-$Cl_2$-5-(2,6-$F_2C_6H_3$) | 135–139 | | | | 193 |
| 2-Cl-5-(2-$ClC_6H_4$)-7-Br | 134–137 | $Et_2O$/Petr ether | | | 193 |
| 2-Cl-5-(2-$ClC_6H_4$)-7-$F_3C$ | 112–115 | | | | 193 |
| 2-F-5-Ph-7-Cl | 121–124 | | | | 193 |

TABLE V-15. 5*H*-1,4-BENZODIAZEPINES

| Substituent | mp (° C); [bp (° C/torr)] | Solvent of Crystallization | Yield (%) | Spectra | Refs. |
|---|---|---|---|---|---|
| 2-MeO-3-Me-5-Ph-7-Cl | 139–141 | Hexane | | | 18 |
| 4-Oxide | 191–194 | EtOAc | | | 18 |
| 2-Me-3-Ph | 122–124 | Hexane | | | 18 |
| 2,3-$Me_2$-5-Ph-7-Cl | 151–154 | MeOH | | | 18 |
| 2-MeNH-5-Ph-7-Cl, 4-Oxide | 210–212 | EtOH | | | 197 |
| 2-PrO-5-Ph-7-Cl | 79–80 [154/0.1] | Petr ether | 96 | | 185 |
| 3-$ClCH_2$-5-Ph-7-Cl, 4-Oxide | 125–128d | EtOAc | | pmr, uv | 12–14 |
| 3-$ClCH_2$-5-Ph-7-$NO_2$, 4-Oxide | 139–141 | EtOH | | | 18 |
| 3-Me-5-Ph, 4-Oxide | 182–184 | $Me_2CO$/Hexane | | | 12, 13 |
| | 196–200 | $Et_2O$ | | | 12, 13 |
| 3-Me-5-Ph-7-Cl, 4-Oxide | 169–170d | $CH_2Cl_2$/Hexane | | pmr, uv | 14 |
| | 165–175d | EtOAc | | | 12, 13 |
| 3-Me-5-Ph-7-$NO_2$, 4-Oxide | 224–226d | PhH/Hexane | | | 12, 13, 17 |
| 3-Ph | 117–120 | EtOH | | | 18 |
| 3-Ph-7-Br | 127–129 | EtOH | | | 18 |
| 3-PhNH-5-Ph-7-Cl | 107–109d | $Et_2O$/Hexane | | | 18 |
| 3-PhNAc-5-Ph-7-Cl | 167–170d | EtOAc/Petr ether | | | 18 |
| 5-Ph-7-Cl, 4-Oxide | 157–158.5 | EtOH | | pmr, uv | 11, 14 |

# 5. REFERENCES


1. (a) F. Hollywood, E. F. V. Scriven, H. Suschitzky, D. R. Thomas, and R. Hull, *J. Chem. Soc., Chem. Commun.*, 806 (1978). (b) F. Hollywood, Z. U. Khan, E. F. V. Scriven, R. K. Smalley, H. Suschitzky, and D. R. Thomas, *J. Chem. Soc., Perkin Trans. I*, 431 (1982).
2. R. I. Fryer, D. L. Coffen, J. V. Earley, and A. Walser, *J. Heterocycl. Chem.*, **10**, 473 (1973).
3. J. V. Earley, R. I. Fryer, and A. Walser, U.S. Patent 3,836,521, September 1974.
4. J. V. Earley, R. I. Fryer, and A. Walser, U.S. Patent 3,838,116, September 1974.
5. (a) A. Walser and R. I. Fryer, Hoffmann-La Roche, Nutley, NJ, unpublished data. (b) A. Walser and J. Hellerbach, Hoffmann-La Roche, Nutley, NJ, unpublished data.
6. J. V. Earley, R. I. Fryer, and A. Walser, U.S. Patent 3,869,448, March 1975.
7. (a) D. L. Coffen, J. P. DeNoble, E. L. Evans, G. F. Field, R. I. Fryer, D. A. Katonak, B. J. Mandel, L. H. Sternbach, and W. J. Zally, *J. Org. Chem.*, **39**, 167 (1974). (b) D. L. Coffen et al., Hoffmann-La Roche, Nutley, NJ, unpublished data.
8. R. I. Fryer, J. V. Earley, and L. H. Sternbach, *J. Org. Chem.*, **32**, 3798 (1967).
9. R. I. Fryer et al., Hoffmann-La Roche, Nutley, NJ, unpublished data.
10. U. D. Shenoy, U.S. Patent 4,006,135, February 1977.
11. Neth. Patent 6,614,923, April 1967 (Hoffmann-La Roche & Co., Switzerland).
12. G. F. Field and L. H. Sternbach, U.S. Patent 3,594,365, July 1971.
13. G. F. Field and L. H. Sternbach, U.S. Patent 3,594,364, July 1971.
14. G. F. Field, W. J. Zally, and L. H. Sternbach, *J. Am. Chem. Soc.*, **89**, 332 (1967).
15. D. L. Coffen and R. I. Fryer, U.S. Patent 3,849,399, November 1974.
16. G. F. Field, W. J. Zally, and L. H. Sternbach, *Tetrahedron Lett.*, 2609 (1966).
17. G. F. Field and L. H. Sternbach, U.S. Patent 3,625,959, December 1971.
18. G. F. Field et al., Hoffmann-La Roche, Nutley, NJ, unpublished data.
19. (a) A. Walser, L. E. Benjamin Sr., T. Flynn, C. Mason, R. Schwartz, and R. I. Fryer, *J. Org. Chem.*, **43**, 936 (1978). (b) A. Walser et al., Hoffmann-La Roche, Nutley, NJ, unpublished data.
20. (a) A. Walser, T. Flynn, and R. I. Fryer, *J. Heterocycl. Chem.*, **15**, 577 (1978). (b) A. Walser et al., Hoffmann-La Roche, Nutley, NJ, unpublished data.
21. (a) Belg. Patent 839,364, September 1976 (Hoffmann-La Roche & Co., Switzerland). (b) A. Walser and R. I. Fryer, U.S. Patent 4,280,957, July 1981.
22. G. F. Field and W. J. Zally, U.S. Patent 4,238,610, December 1980.
23. (a) A. Walser, U.S. Patent 4,226,771, October 1980. (b) A. Walser, U.S. Patent 4,240,962, December 1980. (c) A. Walser, U.S. Patent 4,247,463, January 1981.
24. (a) A. Walser, U.S. Patent 4,226,768, October 1980. (b) A. Walser, U.S. Patent 4,257,946, March 1981. (c) A. Walser, U.S. Patent 4,244,868, January 1981. (d) A. Walser, U.S. Patent 4,244,867, January 1981. (e) A. Walser, Hoffmann-La Roche, Nutley, NJ, unpublished data.
25. R. B. Moffett and B. V. Kamdar, *J. Heterocycl. Chem.*, **16**, 793 (1979).
26. A. Szente and J. Hellerbach, Hoffmann-La Roche and Co., Basel, Switzerland, unpublished data.
27. J. Szmuszkovicz, L. Baczynskyi, C. C. Chidester, and D. Duchamp, *J. Org. Chem.*, **41**, 1743 (1976).
28. A. Walser, T. Flynn, and R. I. Fryer, *J. Heterocycl. Chem.*, **20**, 791 (1983).
29. K. Meguro and Y. Kuwada, *Yakugaku Zasshi*, **93**, 1263 (1973).
30. D. L. Coffen and R. I. Fryer: (a) U.S. Patent 3,850,948, November 1974, (b) U.S. Patent 3,906,001, September 1975.
31. D. L. Coffen and R. I. Fryer, U.S. Patent 3,932,399, January 1976.
32. D. L. Coffen and R. I. Fryer, U.S. Patent 3,849,434, November 1974.
33. L. H. Sternbach and E. Reeder, *J. Org. Chem.*, **26**, 1111 (1961).
34. L. H. Sternbach, E. Reeder, O. Keller, and W. Metlesics, *J. Org. Chem.*, **26**, 4488 (1961).
35. M. C. J. Kuchar, Ph.D. Thesis, Brigham Young University, University Microfilms Order 64-6643; *Diss. Abstr.*, **25**, 1572 (1964).

36. G. Saucy and L. H. Sternbach, *Helv. Chim. Acta*, **45**, 2226 (1962).
37. L. H. Sternbach, G. Saucy, F. A. Smith, M. Müller, and J. Lee, *Helv. Chim. Acta*, **46**, 1720 (1963).
38. (a) L. H. Sternbach, R. I. Fryer, O. Keller, W. Metlesics, G. Sach, and N. Steiger, *J. Med. Chem.* **6**, 261 (1963). (b) L. H. Sternbach et al., Hoffmann-La Roche, Nutley, NJ, unpublished data.
39. L. H. Sternbach and G. Saucy, U.S. Patent 3,341,592, September 1967.
40. O. Keller, N. Steiger, and L. H. Sternbach, U.S. Patent 3,121,103, February 1964.
41. S. C. Bell, C. Gochman, and S. J. Childress, *J. Med. Pharm. Chem.*, **5**, 63 (1962).
42. M. Gordon, I. J. Pachter, and J. W. Wilson, *Arzneim-Forsch.*, **13**, 802 (1963).
43. J.-P. Maffrand, G. Ferrand, and F. Eloy, *Chim. Ther.*, **9**, 539 (1974).
44. E. Reeder and L. H. Sternbach, U.S. Patent 3,051,701, August 1962.
45. F. Gatta, M. R. Del Giudice, L. Di Simone, and G. Settimj, *J. Heterocycl. Chem.*, **17**, 865 (1980).
46. L. H. Sternbach, U.S. Patent 2,893,992, July 1959.
47. K. Meguro, H. Tawada, H. Miyano, Y. Sato, and Y. Kuwada, *Chem. Pharm. Bull.*, **21**, 2382 (1973).
48. G. N. Walker, *J. Org. Chem.*, **27**, 1929 (1962).
49. J. V. Earley, R. I. Fryer, and R. Y. Ning, *J. Pharm. Sci.*, **68**, 845 (1979).
50. R. V. Davis and R. I. Fryer, U.S. Patent 4,083,948, April 1978.
51. E. Reeder and L. H. Sternbach: (a) U.S. Patent 3,371,085, February 1968, (b) U.S. Patent 3,270,053, August 1966.
52. Neth. Patent 6,608,039, December 1966 (Grindstedvaerket, Denmark).
53. L. H. Sternbach et al., unpublished data.
54. H. S. Broadbent, R. C. Anderson, M. C. J. Kuchar, and P. D. Ziemer; First International Congress of Heterocyclic Chemistry, Albuquerque, NM, 1967; *Abstracts.*
55. L. Berger and L. H. Sternbach, U.S. Patent 3,179,656, April 1965.
56. Z. Vejdelek, M. Rajsner, E. Svatek, J. Holubek, and M. Protiva, *Collect. Czech. Chem. Commun.*, **44**, 3604 (1979).
57. R. I. Fryer, R. A Schmidt, and L. H. Sternbach, *J. Pharm. Sci.*, **53**, 624 (1964).
58. A. F. Fentiman, Jr. and R. L. Foltz, *J. Labelled Compd. Radiopharm.*, **13**, 579 (1977).
59. H. M. Wuest, U.S. Patent 3,189,602, June 1965.
60. J. L. Spencer, U.S. Patent 3,462,419, August 1969.
61. G. A. Archer and L. H. Sternbach, U.S. Patent 3,678,036, July 1972.
62. Neth. Patent 6,413,180, May 1965 (Hoffmann-La Roche & Co., Switzerland).
63. M. E. Derieg, R. I. Fryer, and L. H. Sternbach, *J. Chem. Soc., C*, 1103 (1968).
64. Neth. Patent 6,512,614, March 1966 (Hoffmann-La Roche & Co., Switzerland).
65. Neth. Patent 6,603,736, September 1966 (Hoffmann-La Roche & Co., Switzerland).
66. R. B. Moffett and A. D. Rudzik, *J. Med. Chem.*, **16**, 1256 (1973).
67. R. B. Moffett, U.S. Patent 3,847,935, November 1974.
68. Ger. Offen. 2,252,079, May 1973 (Upjohn Co., U.S.).
69. R. B. Moffett, U.S. Patent 3,846,443, November 1974.
70. A. Szente and A. Fischli, Hoffmann-La Roche & Co., Basel, Switzerland, unpublished data.
71. Belg. Patent 634,438, January 1964 (Hoffmann-La Roche & Co., Switzerland).
72. (a) G. A. Archer and L. H. Sternbach, *J. Org. Chem.*, **29**, 231 (1964). (b) G. A. Archer and L. H. Sternbach, U.S. Patent 3,422,091, January 1969.
73. M. Matsuo, K. Taniguchi, and I. Ueda, *Chem. Pharm. Bull.*, **30**, 1481 (1982).
74. J. P. Maffrand, G. Ferrand, and F. Eloy, *Tetrahedron Lett.*, 3449 (1973).
75. Belg. Patent 798,677, August 1973 (Centre d'Etudes pour l'Industrie Pharmaceutique, France).
76. J. B. Hester and A. R. Hanze: (a) U.S. Patent 3,933,794, January 1976, (b) U.S. Patent 3,927,016, December 1975, (c) U.S. Patent 3,917,627, November 1975.
77. M. Gall: (a) U.S. Patent 3,763,179, October 1973, (b) U.S. Patent 3,992,393, November 1976.
78. I. R. Ager, G. W. Danswan, D. R. Harrison, D. P. Kay, P. D. Kennewell, and J. B. Taylor, *J. Med. Chem.*, **20**, 1035 (1977).
79. J. B. Taylor and D. R. Harrison: (a) U.S. Patent 4,134,976, January 1979, (b) D. R. Harrison, U.S. Patent 4,044,142, August 1977.
80. J. B. Taylor and D. R. Harrison, U.S. Patent 4,185,102, January 1980.

81. French Patent 2,244,525, April 1975 (Centre d'Etudes pour l'Industrie Pharmaceutique, France).
82. (a) M. Gall and B. V. Kamdar, *J. Org. Chem.*, **46**, 1575 (1981). (b) M. Gall, U.S. Patent 3,910,946, October 1975.
83. K. Meguro, H. Tawada, and Y. Kuwada, U.S. Patent 3,795,673, March 1974.
84. K. Meguro, H. Natsugari, H. Tawada, and Y. Kuwada, *Chem. Pharm. Bull.*, **21**, 2366 (1973).
85. (a) J. V. Earley, R. I. Fryer, and L. H. Sternbach, U.S. Patent 3,644,335, February 1972. (b) J. V. Earley and R. I. Fryer, Hoffmann-La Roche, Nutley, NJ, unpublished data. (c) R. Y. Ning and L. H. Sternbach, Hoffmann-La Roche, Nutley, NJ, unpublished data. (d) Q. Branca and A. Fischli, Hoffmann-La Roche & Co., Basel, Switzerland, unpublished data.
86. Ger. Offen. 2,947,076, June 1981 (Asta-Werke AG).
87. (a) R. Y. Ning, R. I. Fryer, P. B. Madan, and B. C. Sluboski, *J. Org. Chem.*, **41**, 2724 (1976). (b) R. Y. Ning, R. I. Fryer, P. B. Madan, and B. C. Sluboski, *J. Org. Chem.*, **41**, 2720 (1976).
88. M. R. Del Giudice, F. Gatta, C. Pandolfi, and G. Settimj, *Farmaco, Ed. Sci.*, **37**, 343 (1982).
89. J. H. Sellstedt, U.S. Patent 3,897,416, July 1975.
90. Ger. Offen. 2,947,075, June 1981 (Asta-Werke AG).
91. Neth. Patent 6,412,484, May 1965 (Hoffmann-La Roche & Co., Switzerland).
92. A. Walser and R. I. Fryer, *J. Org. Chem.*, **40**, 153 (1975).
93. (a) G. F. Field, L. H. Sternbach, and W. J. Zally, U.S. Patent 3,678,038, July 1972. (b) G. F. Field, L. H. Sternbach, and W. J. Zally, U.S. Patent 3,624,073, November 1971.
94. A. Walser, R. I. Fryer, L. H. Sternbach, and M. C. Archer, *J. Heterocycl. Chem.*, **11**, 619 (1974).
95. K. Meguro, H. Tawada, and Y. Kuwada, *Yakugaku Zasshi*, **93**, 1253 (1973).
96. K. Meguro, Y. Kuwada, and T. Masuda, U.S. Patent 3,687,941, August 1972.
97. P. Nedenskov and M. Mandrup, *Acta Chem. Scand., B*, **31**, 701 (1977).
98. L. H. Sternbach, B. A. Koechlin, and E. Reeder, *J. Org. Chem.*, **27**, 4671 (1962).
99. H. Natsugari, K. Meguro, and Y. Kuwada, *Chem. Pharm. Bull.*, **27**, 2608 (1979).
100. R. Y. Ning, R. I. Fryer, and B. C. Sluboski, *J. Org. Chem.*, **42**, 3301 (1977).
101. Neth. Patent 6,514,541, May 1966 (Hoffmann-La Roche & Co., Switzerland).
102. H. Tawada, K. Meguro, and Y. Kuwada, U.S. Patent 3,887,541, June 1975.
103. H. Tawada, K. Meguro, and Y. Kuwada, U.S. Patent 3,703,525, November 1972.
104. N. W. Gilman, J. F. Blount, and L. H. Sternbach, *J. Org. Chem.*, **37**, 3201 (1972).
105. S. Farber, H. M. Wuest, and R. I. Meltzer, *J. Med. Chem.*, **7**, 235 (1964).
106. U. D. Shenoy: (a) U.S. Patent 4,145,417, March 1979, (b) U.S. Patent 4,061,745, December 1977, (c) U.S. Patent 4,065,474, December 1977, (d) U.S. Patent 4,060,608, November 1977.
107. T. Hara, K. Itoh, and N. Itoh, *J. Heterocycl. Chem.*, **13**, 1233 (1976).
108. J. Szmuszkovicz, U.S. Patent 3,842,080, October 1974.
109. H. Natsugari and Y. Kuwada, *Chem. Pharm. Bull.*, **27**, 2618 (1979).
110. L. H. Sternbach, E. Reeder, A. Stempel, and A. I. Rachlin, *J. Org. Chem.*, **29**, 332 (1964).
111. S. C. Bell, C. Gochman, and S. J. Childress, *J. Org. Chem.*, **28**, 3010 (1963).
112. (a) H.-G. Schecker and G. Zinner, *Arch. Pharm.*, **313**, 926 (1980). (b) M. Forsch and H. Gerhards, EP 0,041,242, December 1981.
113. H. Natsugari, K. Meguro, and Y. Kuwada, *Chem. Pharm. Bull.*, **27**, 2927 (1979).
114. A. I. Hanze, U.S. Patent 3,734,912, May 1973.
115. Y. Kuwada, H. Natsugari, and K. Meguro, U.S. Patent 4,175,079, November 1979.
116. Ger. Offen. 2,400,425, July 1974 (Upjohn Co., U.S.).
117. K. Meguro, Y. Kuwada, Y. Nagawa, and T. Masuda, U.S. Patent 3,652,754, March 1972.
118. R. B. Moffett, U.S. Patent 3,773,765, November 1973.
119. R. B. Moffett, *J. Org. Chem.*, **39**, 568 (1974).
120. (a) R. I. Fryer, J. V. Earley, and J. F. Blount, *J. Org. Chem.*, **42**, 2212 (1977). (b) R. I. Fryer and J. V. Earley, Hoffmann-La Roche, Nutley, NJ, unpublished data.
121. T. Miyadera, T. Hata, C. Tamura, and R. Tachikawa, *Chem. Pharm. Bull.*, **29**, 2193 (1981).
122. R. B. Moffett, U.S. Patent 3,822,259, July 1974.
123. S. Kobayashi, *Chem. Lett.*, 967 (1974).
124. H. Natsugari, K. Meguro, and Y. Kuwada, *Chem. Pharm. Bull.*, **27**, 2589 (1979).

125. L. H. Sternbach and E. Reeder, *J. Org. Chem.*, **26**, 4936 (1961).
126. (a) H. Allgeier and A. Gagneux, U.S. Patent 3,867,536, February 1975. (b) H. Allgeier and A. Gagneux, U.S. Patent, 3,946,032, March 1976. (c) Swiss Patent 561,722, May 1975 (Ciba-Geigy AG, Switzerland).
127. A. Walser, U.S. Patent 4,118,386, October 1978.
128. P. J. G. Cornelissen, G. M. J. Beijersbergen van Henegouwen, and K. W. Gerritsma, *Int. J. Pharm.*, **3**, 205 (1979).
129. G. F. Field and L. H. Sternbach, U.S. Patent 3,697,545, October 1972.
130. G. F. Field and L. H. Sternbach, *J. Org. Chem.*, **33**, 4438 (1968).
131. J. B. Hester, U.S. Patent 3,649,617, March 1972.
132. J. B. Hester, U.S. Patent 3,857,854, December 1974.
133. R. S. P. Hsi and T. D. Johnson, *J. Labelled Compds. Radiopharm.*, **12**, 613 (1976).
134. J. B. Hester, Jr., D. J. Duchamp, and C. G. Chidester, *Tetrahedron Lett.*, 4039 (1970).
135. J. B. Hester, Jr. and A. D. Rudzik, *J. Med. Chem.*, **17**, 293 (1974).
136. K. Meguro and Y. Kuwada, *Chem. Pharm. Bull.*, **21**, 2375 (1973).
137. K. Meguro and Y. Kuwada, *Tetrahedron Lett.*, 4039 (1970).
138. K. Meguro and Y. Kuwada, U.S. Patent 3,864,356, February 1975.
139. (a) K. Meguro and Y. Kuwada, U.S. Patent 3,907,820, September 1975. (b) U.S. Patent 4,235,775, November 1980.
140. K. Meguro and Y. Kuwada, U.S. Patent 4,116,956, September 1978.
141. J. B. Hester, Jr., U.S. Patent 3,995,043, November 1976.
142. H. Tawada, H. Natsugari, K. Meguro, and Y. Kuwada, U.S. Patent 4,102,881, July 1978.
143. L. H. Sternbach and A. Walser, U.S. Patent 3,970,664, July 1976.
144. L. H. Sternbach and A. Walser: (a) U.S. Patent 3,879,406, April 1975, (b) U.S. Patent 4,044,016, August 1977.
145. R. B. Moffett, G. N. Evenson, and P. F. Von Voigtlander, *J. Heterocycl. Chem.*, **14**, 1231 (1977).
146. J. B. Hester and J. Szmuszkovicz, U.S. Patent 3,862,956, January 1975.
147. Ger. Offen. 2,242,059, March 1973 (Upjohn Co., U.S.).
148. J. B. Hester, Jr., C. G. Chidester, and J. Szmuszkovicz, *J. Org. Chem.*, **44**, 2688 (1979).
149. J. B. Hester, Jr., U.S. Patent 4,082,761, April 1978.
150. J. B. Hester, Jr., U.S. Patent 3,741,957, June 1973.
151. J. B. Hester, Jr., U.S. Patent 3,701,782, October 1972.
152. J. B. Hester, Jr., U.S. Patent 3,646,055, February 1972.
153. J. B. Hester, Jr., U.S. Patent 3,717,653, February 1973.
154. A. Walser and G. Zenchoff, *J. Med. Chem.*, **20**, 1694 (1977).
155. Ger. Offen. 2,242,938, March 1973 (Upjohn Co., U.S.).
156. R. B. Moffett, U.S. Patent 4,016,165, April 1977.
157. R. B. Moffett, U.S. Patent 4,028,356, June 1977.
158. G. N. Evenson: (a) U.S. Patent 4,107,159, August 1978, (b) U.S. Patent 4,086,230, April 1978.
159. G. N. Evenson, U.S. Patent 4,073,785, February 1978.
160. R. B. Moffett, U.S. Patent 4,017,492, April 1977.
161. A. Walser and G. Zenchoff, *J. Heterocycl. Chem.*, **15**, 161 (1978).
162. R. I. Fryer and A. Walser, U.S. Patent 3,901,907, August 1975.
163. R. I. Fryer and A. Walser, U.S. Patent 3,864,328, February 1975.
164. R. B. Moffett, U.S. Patent 4,073,784, February 1978.
165. K. Meguro and Y. Kuwada, U.S. Patent 3,865,811, February 1975.
166. J. B. Hester, Jr., U.S. Patent 4,039,551, August 1977.
167. J. B. Hester, Jr., U.S. Patent 4,081,452, March 1978.
168. J. B. Hester, Jr., U.S. Patent 4,009,175, February 1977.
169. J. B. Hester, Jr., U.S. Patent 4,141,902, February 1979.
170. J. B. Hester, Jr., U.S. Patent 4,021,441, May 1977.
171. M. Gall, U.S. Patent 4,010,177, March 1977.
172. J. B. Hester, Jr. and P. Von Voigtlander, *J. Med. Chem.*, **22**, 1390 (1979).
173. J. B. Hester, Jr., U.S. Patent 3,996,230, December 1976.

174. J. B. Hester, Jr., U.S. Patent 3,708,592, January 1973.
175. J. B. Hester, Jr., U.S. Patent 3,751,426, August 1973.
176. Neth. Patent 7,205,629, October 1972 (Upjohn Co., U.S.).
177. J. B. Hester, Jr., U.S. Patent 3,737,434, June 1973.
178. S. C. Bell, T. S. Sulkowski, C. Gochman, and S. J. Childress, *J. Org. Chem.*, **27**, 562 (1962).
179. J. H. Sellstedt and D. M. Teller, U.S. Patent 4,056,525, November 1977.
180. J. V. Earley, R. I. Fryer, R. Y. Ning, and L. H. Sternbach, U.S. Patent 3,681,341, August 1972.
181. M. Matsuo, K. Taniguchi, and I. Ueda, *Chem. Pharm. Bull.*, **30**, 1141 (1982).
182. F. M. Vane and W. Benz, *Org. Mass. Spectrom.*, **14**, 233 (1979).
183. Neth. Patent 6,412,300, April 1965 (Hoffmann-La Roche & Co., Switzerland).
184. Ger. Offen. 2,302,525, August 1973 (Upjohn Co., U.S.).
185. Belg. Patent 777,138, January 1972 (Grindstedvaerket, Denmark).
186. I. Ueda and M. Matsuo, U.S. Patent 4,094,870, June 1978.
187. J. B. Hester, Jr., U.S. Patent 3,897,446, July 1975.
188. T. Watanabe, M. Matsuo, K. Taniguchi, and I. Ueda, *Chem. Pharm. Bull.*, **30**, 1473 (1982).
189. H. Allgeier and A. Gagneux, U.S. Patent 4,178,378, December 1979.
190. Swiss Patent 562,822, June 1975 (Ciba-Geigy AG, Switzerland).
191. Swiss Patent 551,723, May 1975 (Ciba-Geigy AG, Switzerland).
192. A. Walser, R. F. Lauer, and R. I. Fryer, *J. Heterocycl. Chem.*, **15**, 855 (1978).
193. Belg. Patent 823,613, April 1975 (Asta-Werke AG, West Germany).
194. J. H. Sellstedt, U.S. Patent 3,882,101, May 1975.
195. J. H. Sellstedt, U.S. Patent 3,880,835, April 1975.
196. J. H. Sellstedt and D. M. Teller, U.S. Patent 3,880,877, April 1975.
197. T. T. Moller, P. Nedenkov, and H. B. Rasmussen, U.S. Patent 3,635,949, January 1972.
198. V. Bertolasi, M. Sacerdoti, and G. Gilli, *Acta Crystallogr.*, **38**, 1768 (1982).
199. R. Haran and J. P. Tuchagues, *J. Heterocycl. Chem.*, **17**, 1483 (1980).
200. C. W. Perry et al., Hoffmann-La Roche, Nutley, NJ, unpublished data.
201. W. Zwahlen et al., Hoffmann La Roche & Co., Basel, Switzerland, unpublished data.
202. E. Kyburz et al., Hoffmann-La Roche & Co., Basel, Switzerland, unpublished data.

# CHAPTER VI

# Dihydro-1,4-Benzodiazepines

**A. Walser**

*Chemical Research Department,*
*Hoffmann-La Roche Inc.,*
*Nutley, New Jersey*

and

**R. Ian Fryer**

*Department of Chemistry, Rutgers,*
*State University of New Jersey, Newark,*
*New Jersey*

## INTRODUCTION

This chapter covers the syntheses and reactions of dihydro-1,4-benzodiazepines.

Retaining the aromaticity of the benzene ring, the dihydro-1,4-benzodiazepines **1–4** may be formulated. Representatives of all four tautomers, which have been described in the literature, are discussed in this chapter.

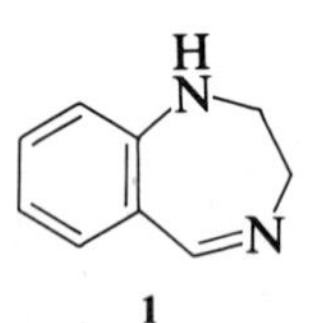

**1**

2,3-Dihydro-1*H*-1,4-benzodiazepine

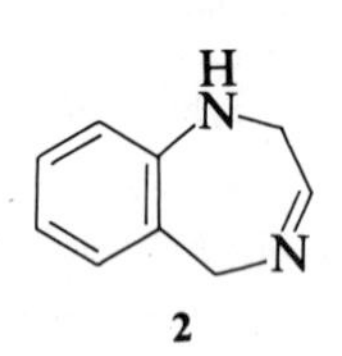

**2**

2,5-Dihydro-1*H*-1,4-benzodiazepine

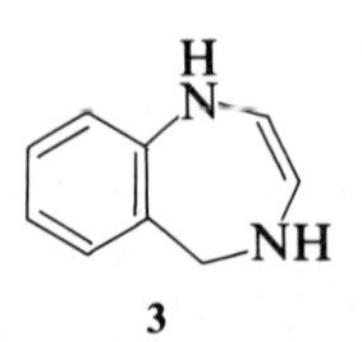

**3**

4,5-Dihydro-1*H*-1,4-benzodiazepine

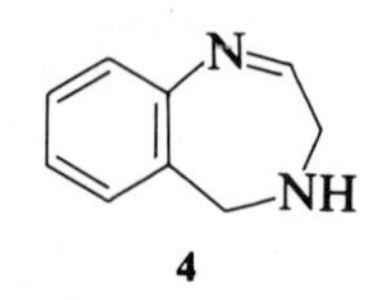

**4**

4,5-Dihydro-3*H*-1,4-benzodiazepine

# 1. 2,3-DIHYDRO-1*H*-1,4-BENZODIAZEPINES

## 1.1. Synthesis of 2,3-Dihydro-1*H*-1,4-benzodiazepines

### *1.1.1. From (2-Halophenyl)carbonyl Compounds*

The 2,3-dihydro-1*H*-1,4-benzodiazepines **7** were prepared by reacting the (2-halophenyl)carbonyl compounds **5** (X = halogen) with the diamine **6** ($R_3$ = H, Me) at elevated temperatures (Eq. 1).

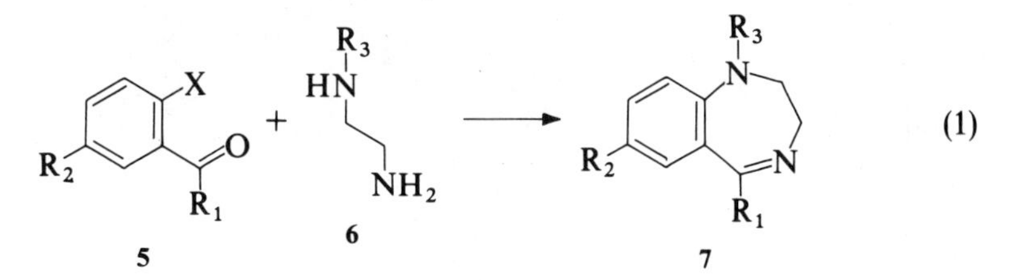


This synthesis worked particularly well when $R_2$ was *para* to the halogen X and was a strongly electron-withdrawing substituent such as a nitro or a trifluoromethyl group. The following compounds **7** were successfully made by this procedure:

| Refs. | $R_1$ | $R_2$ | $R_3$ |
|---|---|---|---|
| 1, 2 | Ph | $NO_2$ | H |
| 2 | Ph | $F_3C$ | H |
| 3, 4 | 2-Pyridyl<br>3-Pyridyl<br>4-Pyridyl | $NO_2$, $F_3C$ | H |
| 5, 6 | 2-Pyrimidyl | $NO_2$ | H |
| 7 | 2-Thiazolyl<br>2-Me-3-pyrazolyl<br>1-Me-2-imidazolyl<br>1,2-thiazol-5-yl | $NO_2$ | H, Me |

Less activated halides could be employed if the reaction of the diamine **6** was carried out in boiling nitrobenzene in the presence of cupric acetate and potassium carbonate as shown by the preparation of compounds **7** [$R_1$ = Ph; $R_2$ = H, Cl; $R_3$ = H, Me, (cyclopropyl)methyl] from the imines of the benzophenones **5** ($R_1$ = Ph; $R_2$ = H, Cl; X = Cl).[8] According to a Belgian patent,[9] medazepam (**12**) was obtained by treating **8** under similar conditions with the diamine **6** ($R_3$ = Me) in the presence of hydrazine hydrate. Syntheses of **7**

($R_1$ = alkenyl; $R_2$ = Cl, $F_3C$; $R_3$ = H) from ketone **5** (X = Cl) and ethylenediamine were claimed in another patent,[10] but the compounds were not properly characterized.

Medazepam (**12**) was reported[11] to be synthesized by a stepwise modification shown in Eq. 2. The benzophenone **8** was reacted with the protected diamine **9** at 180–200°C in the presence of cupric acetate and potassium carbonate to form **10**, hydrazinolysis of which led to **12**.

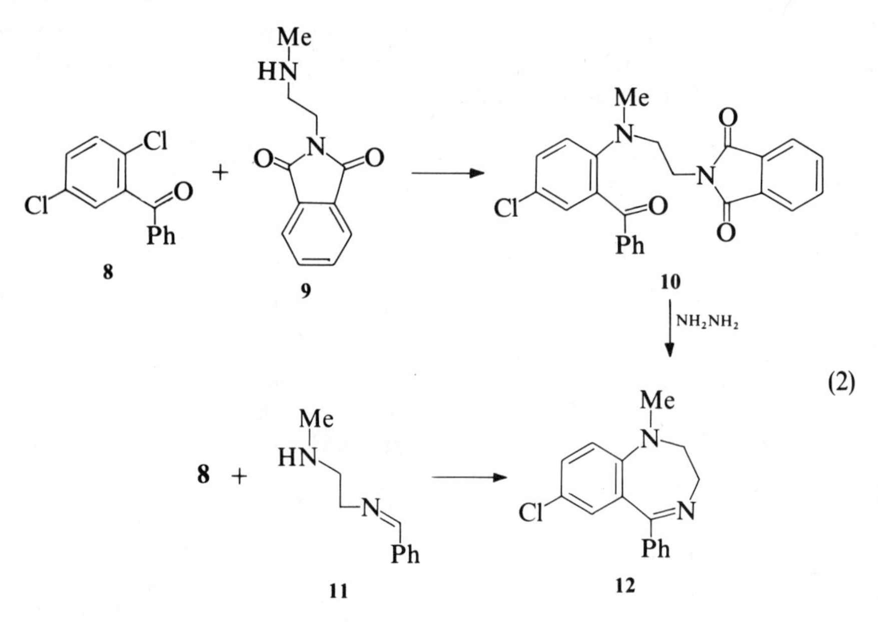


Compound **12** was also claimed[12] to be accessible by reaction of **8** with the benzaldimine **11** in boiling nitrobenzene.

An example of activation of the 2-chloro substituent by a *p*-sulfonamide functionality was provided by the conversion of **5** ($R_1$ = 2-amino-5-chlorophenyl; $R_2$ = $SO_2NH_2$, X = Cl) to the corresponding benzodiazepine by heating in ethylenediamine.[13]

The susceptibility of fluorine on aromatic rings toward displacement by nucleophiles made the ketones **5** (X = F, $R_1$ = 2-pyridyl, $R_2$ = H)[3] and (X = F, $R_1$ = indol-3-yl, $R_2$ = H)[14] suitable starting materials for the preparation of the corresponding benzodiazepines.

### *1.1.2. From 2-Aminobenzophenones*

Alkylation of the 2-aminobenzophenone **13** ($R_1$ = H, $R_2$ = Cl) with *N*-protected 2-aminoethyl bromides in hot dimethylformamide gave moderate yields of the products **14** ($R_3$ = phthalimido or benzamido). Removal of the protecting groups by acid hydrolysis led to the benzodiazepine **15** (Eq. 3).[15]

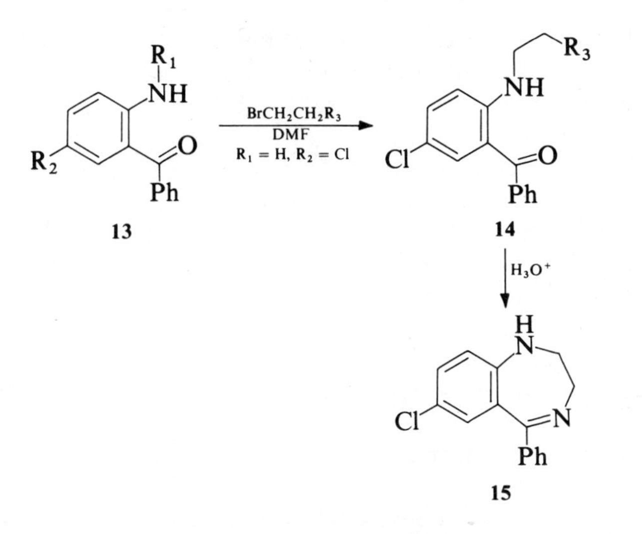


The preparation of **10** by alkylation of the methylaminobenzophenone **13** ($R_1$ = Me, $R_2$ = Cl) with *N*-(2-haloethyl)phthalimide and subsequent hydrazinolysis to **12** was claimed in a Belgian patent.[16]

*N*-(2-Haloethyl)-aminobenzophenones **16** ($R_1$ = tosyl) were obtained by alkylation of the tosylated aminobenzophenone **13** ($R_1$ = tosyl) with 1,2-dibromoethane or 1-bromo-2-chloroethane. The tosyl group was removed by treatment with sulfuric acid to give **16** ($R_1$ = H), which was further alkylated with methyl iodide in dimethylformamide in the presence of barium oxide to give **16** ($R_1$ = Me).[17]

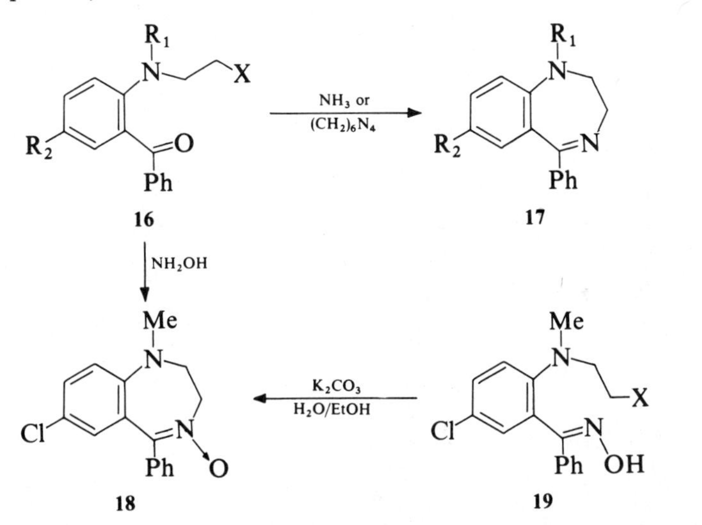


The halides **16** (X = Br, Cl) were converted to the benzodiazepines **17** ($R_1$ = H, Me; $R_2$ = Cl) by treatment with methanolic ammonia at 120–130°C or in better yields by reaction with hexamethylenetetramine in refluxing ethanol.[17,18] In the latter case, the initial formation of a quaternary adduct, which yielded the benzodiazepine upon reflux in ethanol, was observed.[18] The "hexamine" procedure was also applied to the preparation of **17** ($R_1 = F_3CCH_2$, $R_2 = Cl$) (Eq. 4).[19]

Reaction of the chloride **16** ($R_1$ = Me, $R_2 = X = Cl$) with hydroxylamine in ethanol–water at reflux led to the nitrone **18**.[20–23] Since the same compound was obtained by treating the oxime **19** (X = Cl) with potassium carbonate,[21–23] it is likely that **19** is an intermediate in the conversion of **16** to **18**. The oxime **19** (X = Cl) resulted from the reaction of **19** (X = OH) with thionyl chloride in pyridine.[21–23]

Mihalic and coworkers[24–26] studied the transformation of the halides **16** to benzodiazepines in detail and established the intermediacy of aziridinium ions **21** (Eq. 5). Thus, when the dideuterated chloride **20** ($R_1$ = Me, $R_2 = X = Cl$, $R_3 = R_4 = D$, $R_5 = H$) was treated with ethanolic ammonia at 130°C a mixture of the benzodiazepines **22** (45%) and **24** (55%) was obtained. A similar ratio of the 2-methyl- and 3-methyl-benzodiazepines resulted when the bromide **20** ($R_1 = R_5$ = Me, $R_2$ = Cl, $R_3 = R_4$ = H, X = Br) was subjected to ammonia. This demonstrates that the aziridinium ion **21** is formed and opened by the

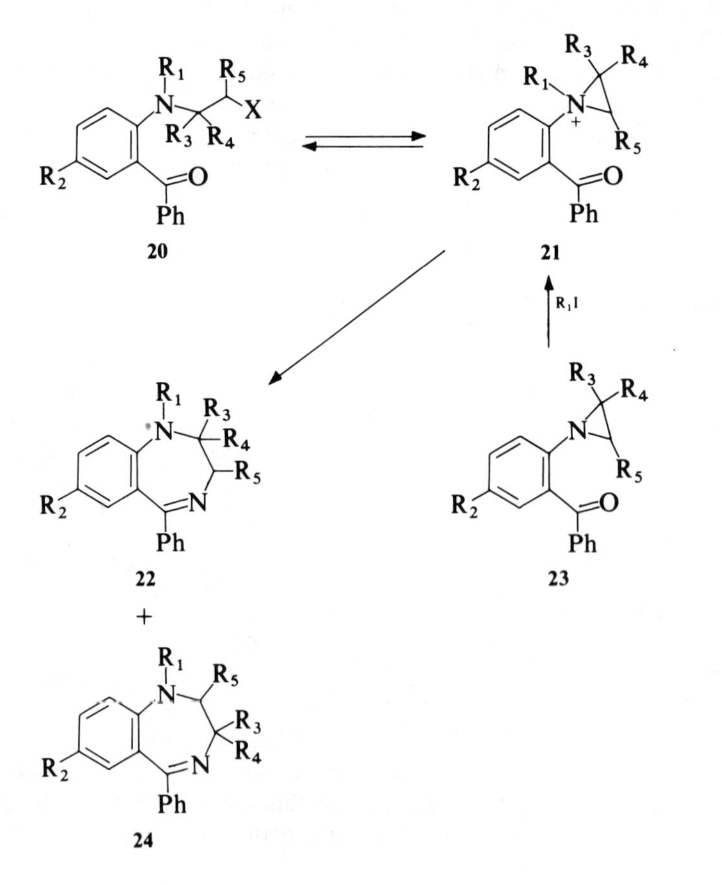


(5)

nucleophile, ammonia in this case, in both possible ways with little regioselectivity. The stereoselectivity of this reaction was investigated[24,25] using the enantiomeric benzophenone **20** ($R_1 = R_4 = R_5 = H$, $R_2 = NO_2$, $R_3 = Et$, $X = Br$) and aziridine **23** ($R_2 = NO_2$, $R_3 = Et$, $R_4 = R_5 = H$) both with the *S* absolute configuration. Both the halide and the aziridine were converted to a mixture of 2-ethyl and 3-ethyl benzodiazepines. The benzodiazepines from the halide and from the aziridine were in the ratios of 92:8 and 3:1. The 2-ethyl benzodiazepine retained the absolute stereochemistry whether the halide or the aziridine was used as starting material. The 3-ethyl benzodiazepine was formed with retention of configuration from the halide but with a high degree of inversion from the aziridine. The authors[24] gave mechanistic explanations for the results observed. Based on these studies, an efficient synthesis of medazepam (**12**) was developed.[26] The aziridine **23** ($R_2 = Cl$, $R_3 = R_4 = R_5 = H$) was quaternized with methyl iodide to give the iodide **20** ($R_1 = Me$, $R_2 = Cl$, $R_3 = R_4 = R_5 = H$, $X = I$) via the unstable aziridinium salt **21**. Treatment of this iodide with hexamethylenetetramine or ammonia led to medazepam in 92% overall yield from the aziridine.

Compound **12** was also formed by a double alkylation of the benzophenone imine **25** with 1,2-dibromoethane (Eq. 6).[27] The 1-desmethyl analog **15** was obtained in low yield by reacting 2-amino-5-chlorobenzophenone with aziridine and aluminum chloride.[15]

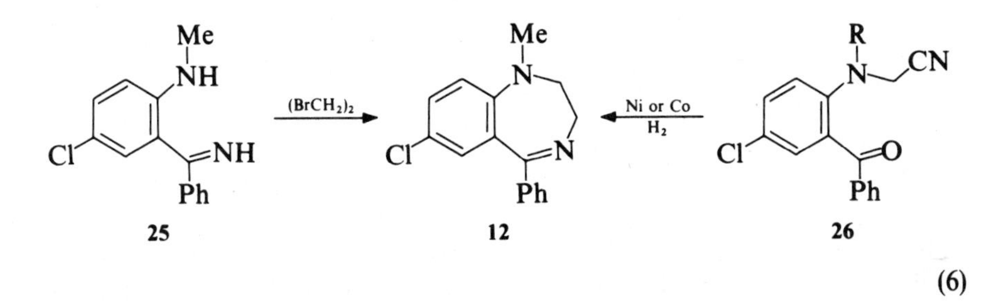


(6)

A process leading to **12** in good yield involved catalytic hydrogenation of the nitrile **26** (R = Me) using Raney nickel[28,29] or cobalt[28] as a catalyst (Eq. 6). The nitrile **26** (R = Me) was prepared either by reacting 2-methylamino-5-chlorobenzophenone with formaldehyde and hydrogen cyanide[29] or by alkylation with chloroacetonitrile.[28]

According to the patent literature,[30,31] 2,3-dihydro-1*H*-1,4-benzodiazepines **28** were obtained by cyclization and dehydration of the ethanolimines **27** in polyphosphoric acid alkyl esters at elevated temperature (Eq. 7). Thus, medazepam and its 1-desmethyl analog were claimed to be accessible in 90% yield by heating the appropriate aminobenzophenone hydroxyethylimine with phosphorus pentoxide and ethanol at 90°C for 3 hours.[30] At higher temperatures (140°–160°C) alkylation of the 1-position nitrogen by the alkylphosphate was observed.[31]

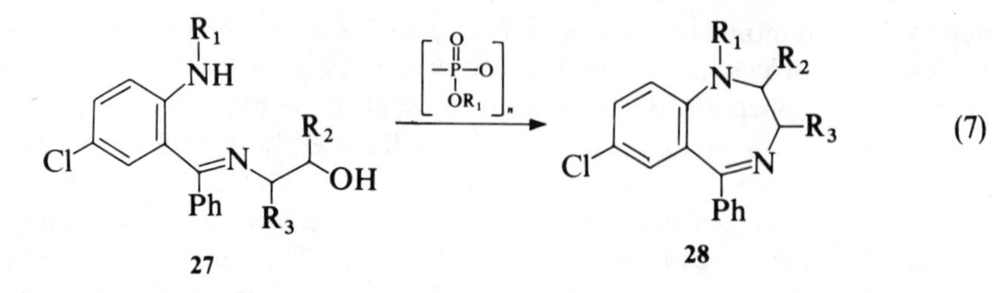


Giacconi and coworkers[32] described the synthesis of the 2,3-diamino-substituted compounds **31** by reacting the benzophenone imines **29** (R = H, Me; Y = Cl, $NO_2$) with the 1,2-ethanediiminium dibromides **30** (X = O, NMe, $CH_2$) (Eq. 8). In the case of R = H, the benzodiazepines **31** could be isolated only when Y was a nitro group. If Y was chlorine or hydrogen, elimination of the 2-aminofunction with formation of the 1,2-imine bond occurred readily, giving 3-substituted amino-3*H*-1,4-benzodiazepines.

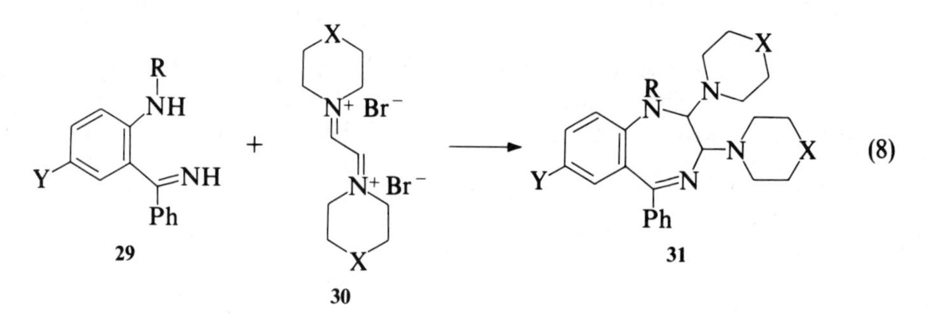


### *1.1.3. By Oxidation of Indoles*

The oxidative cleavage of the 2,3-bond in 3-substituted indoles leading to derivatives of (2-aminophenyl)ketones has been widely used for the synthesis of benzodiazepines. Thus, 1-acylated-2,3-dihydro-1*H*-1,4-benzodiazepines **33** were obtained from the oxidation of 3-phenyl-indoles **32** bearing a 2-aminoethyl moiety on the indole nitrogen (Eq. 9).

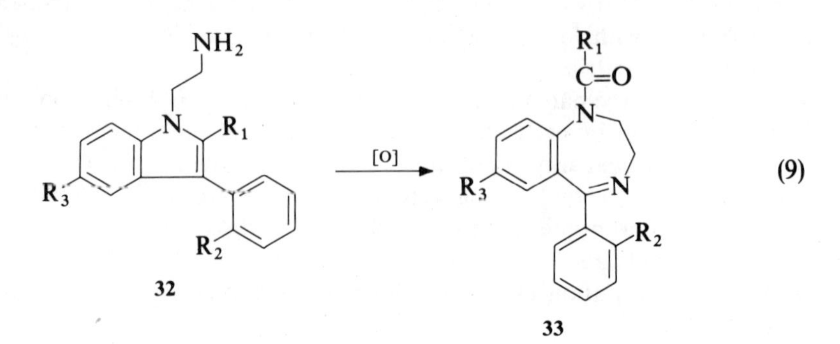

The reaction of **32** ($R_1 = R_2 = H, R_3 = Cl$) with chromium trioxide in glacial acetic acid–water was reported[33] to give the 1-formyl derivative **33** ($R_1 = R_2 = H, R_3 = Cl$). In our hands this procedure did not lead to the expected product. Instead we obtained an unidentified compound with the reported melting point[33] which, however, was quite different from that of an authentic sample of 7-chloro-1-formyl-2,3-dihydro-5-phenyl-1*H*-1,4-benzodiazepine prepared by formylation of the parent benzodiazepine.[34]

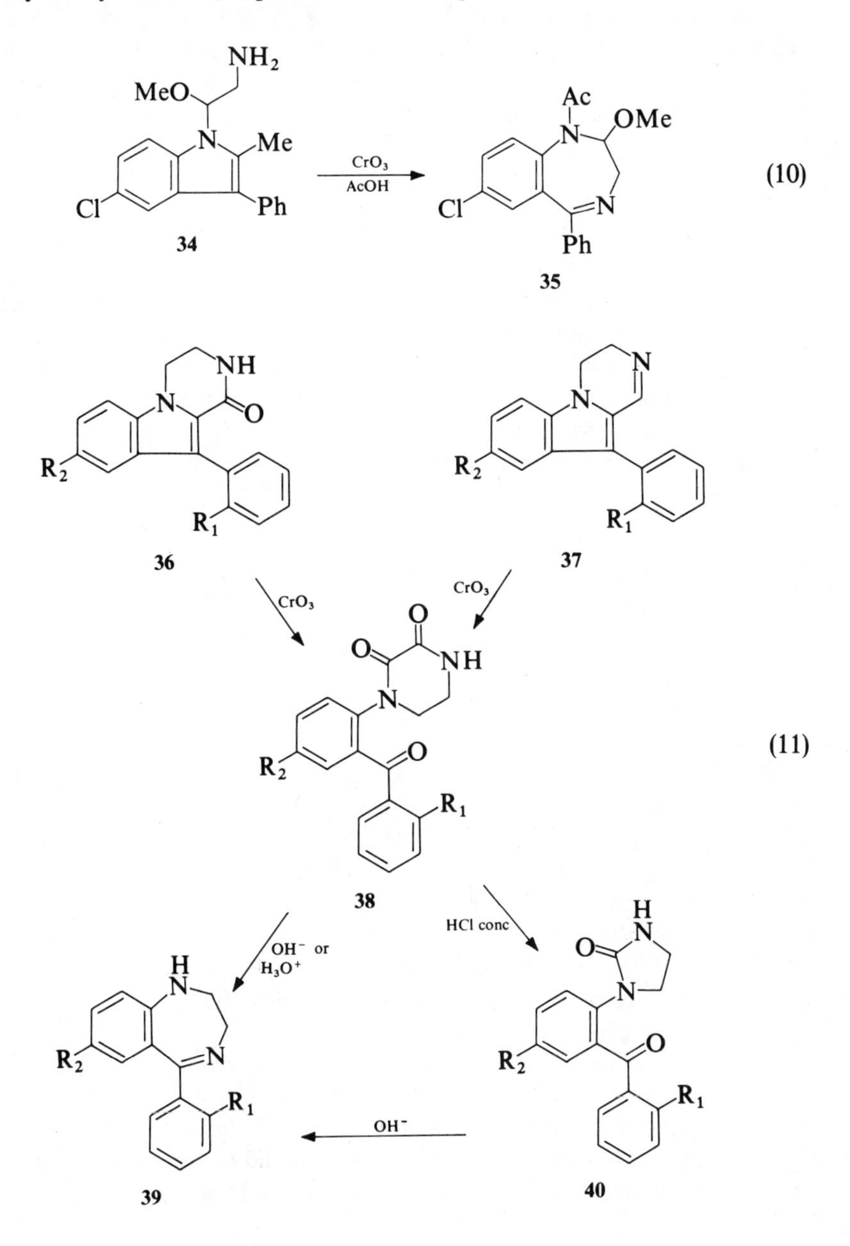

If the indole **32** carried a substituent at the 2-position, the oxidation to the benzodiazepine **33** worked satisfactorily as shown for compounds with $R_1$ = Me,[34,35] $CH_2OMe$,[34] $CH_2NMe_2$,[36] and $CONMe_2$.[36] In addition to chromium trioxide, ozone and periodate[35] were successfully used to perform this oxidation. The indole **34** with a modified side chain at the 1-position allowed the preparation of the 2-methoxy analog **35** (Eq. 10).[34]

Oxidation of the pyrazinoindoles **36**[34,37–39] and **37**[34] with chromium trioxide yielded the piperazin-2,3-diones **38**, which were converted to the benzodiazepines **39** by alkaline[37–39] or acid hydrolysis (Eq. 11).[34] Heating of **38** with concentrated hydrochloric acid was reported to give the imidazolidinone **40**, which was subsequently converted to the benzodiazepine by alkaline hydrolysis.[40]

### *1.1.4. By Bischler–Napieralski Type Reactions*

Kaegi and coworkers were the first to employ the Bischler–Napieralski reaction for the preparation of 1-alkyl-2,3-dihydro-1*H*-1,4-benzodiazepines.[41] Treating the benzamide **41** ($R_1$ = Me, $R_2$ = Ph, $R_3$ = Cl), radiolabeled on the carbonyl carbon, with a mixture of phosphorus oxychloride and phosphorus pentoxide at 110°C for 16 hours led to $^{14}C_5$-labeled medazepam (**42**: $R_1$ = Me, $R_2$ = Ph, $R_3$ = Cl) in high yield (Eq. 12).

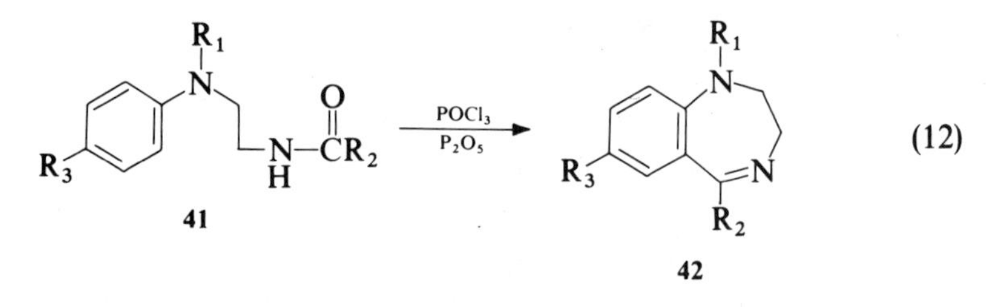


This procedure was applied to the synthesis of the $^{14}C_5$-labeled analog ($R_2$ = 2-$FC_6H_4$),[42] the nitro-substituted compound **42** ($R_1$ = Me, $R_2$ = 2-$NO_2C_6H_4$; $R_3$ = Cl),[43] the 1-trifluoroethyl-substituted derivative **42** ($R_1$ = $F_3CCH_2$, $R_2$ = 2-$FC_6H_4$, $R_3$ = Cl),[44] and several other analogs.[45a] Compounds with aliphatic substituents at the 5-position ($R_1$ = Me; $R_2$ = $ClCH_2$, $Cl_2CH$, trichloroethylene; $R_3$ = Cl) were also obtained by this method.[45b]

The cyclization of the benzamides **41** ($R_1$ = Me; $R_2$ = Ph; $R_3$ = H, Me) to the corresponding benzodiazepines was also effected by treatment with polyphosphoric acid at higher temperatures (~155°C).[46]

Liepmann and collaborators[47] extended this reaction to the benzamides **43** and observed some interesting results. Refluxing **43** ($R_1$ = Ph, $R_2$ = Cl) with phosphorus oxychloride gave a mixture of the diazocine **44** and the benzodiazepine **48**. The authors showed that **44** and **48** did not interconvert under the reaction conditions employed for their formation and that **48** was formed via the aziridinium ion **45** and the chloride **47**. Under more vigorous conditions, by heating in tetrachloroethane at 140°C,[47] the diazocines thermally rearranged to the 2-chloromethyl-1,4-benzodiazepines **48**, most likely via the aziridinium salt **46** (Eq. 13).

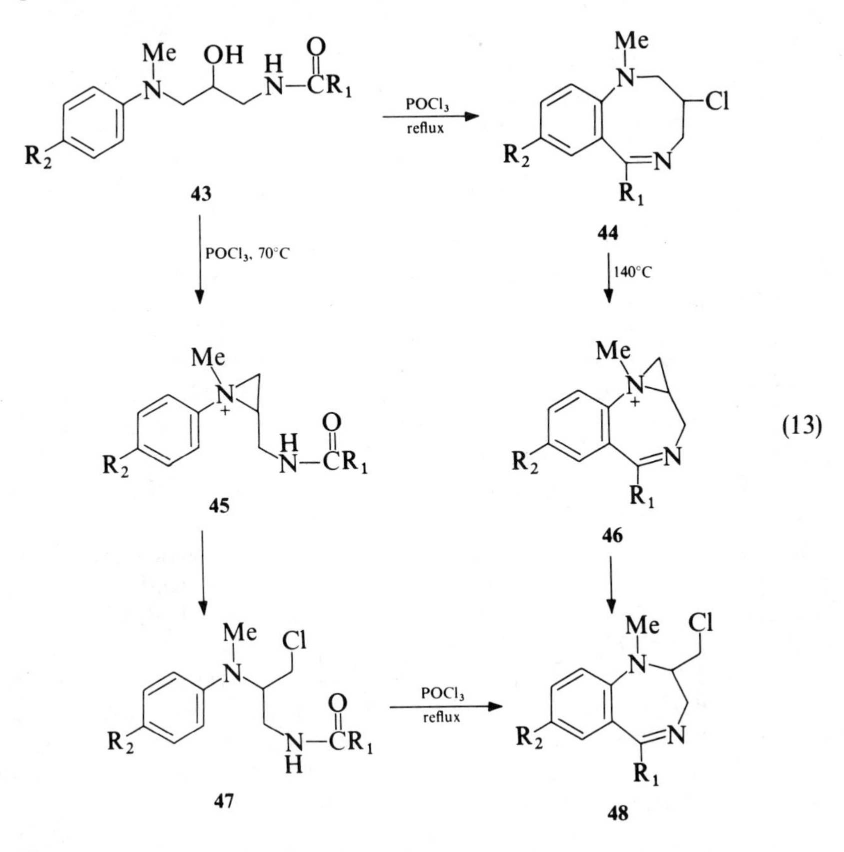


(13)

The usefulness of this synthesis of 2-carbon-substituted benzodiazepines was enhanced by the finding that the diazocines **44** undergo a ring contraction when they are reacted with various nucleophiles. Thus, treatment of **44** with alkoxides,[47–49,54] cyanide,[47,48] azide,[50,51] potassium phthalimide,[47,50,51] hydroxide,[54] acetate,[47] phenoxide,[52] or amines[47,48,50] yielded the corresponding benzodiazepines **49** (Eq. 14).

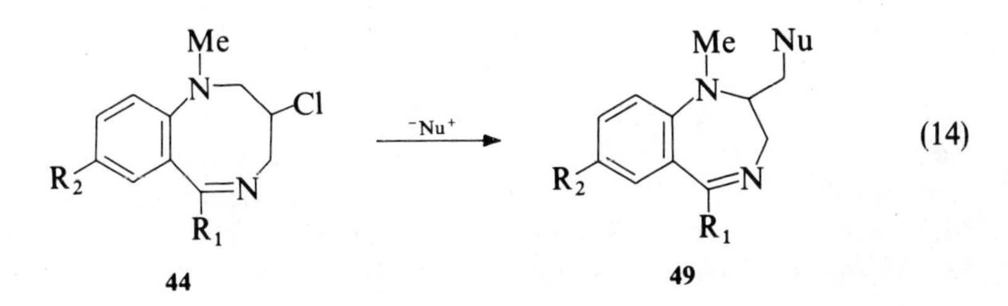


(14)

Reaction of the 3-hydroxydiazocine **50** with thionyl chloride in boiling benzene resulted in ring contraction and afforded the benzodiazepine **51** in 80% yield (Eq. 15).[47]

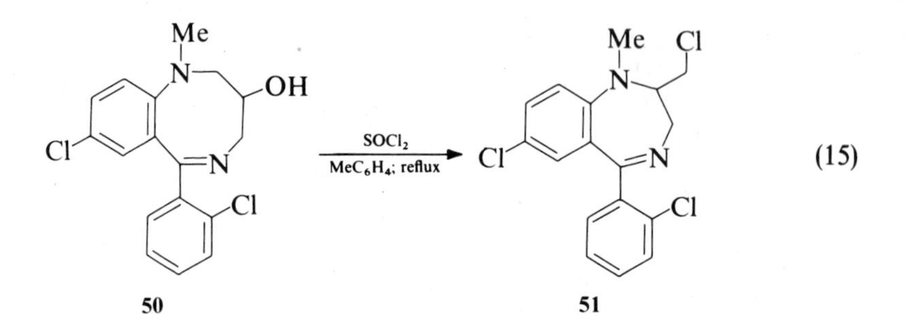


A modification of the Bischler–Napieralski conditions in which the *o*-substituted benzamides **43** ($R_1$ = 2-ClC$_6$H$_4$, 2-FC$_6$H$_4$; $R_2$ = Cl) were reacted with phosphorus pentachloride and subsequently with aluminum chloride or stannic chloride gave improved yields.[47,53]

The methods described above were not limited to the synthesis of 5-phenyl-substituted benzodiazepines but were also applied to the preparation of compounds **49** with $R_1$ = 2-thienyl,[51,54] 2-furyl,[51,54] 3-thienyl,[51] and 3-furyl.[51] Compounds **44** and **48** with $R_1$ = 2-pyridyl, 3-pyridyl, and 4-pyridyl were claimed to be obtained by reaction of the corresponding 3-hydroxybenzazocine with carbon tetrachloride and triphenylphosphine.[51]

1-Alkyl-2,3-dihydro-1*H*-1,4-benzodiazepines were also accessible via nitrilium salts. Reaction of the anilines **52** ($R_1$ = H, Cl; X = Cl) with nitriles **53** ($R_2$ = Ph, 2-ClC$_6$H$_4$, 1-cyclohexenyl, 2-chloro-1-cyclohexenyl) in the presence of aluminum chloride or stannic chloride at 110–130°C gave moderate yields of the corresponding benzodiazepines **54** (Eq. 16).[34]

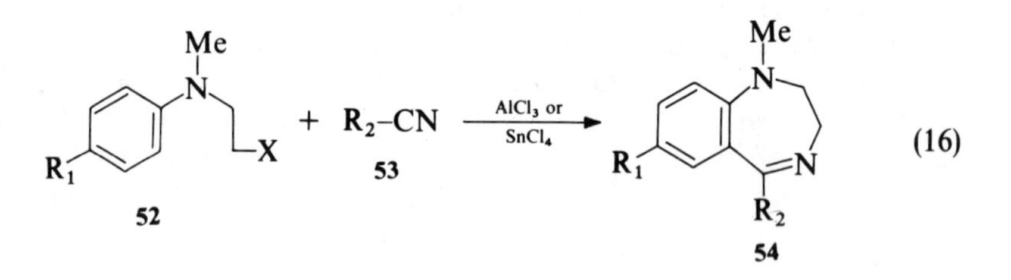


The alcohol **52** (X = OH) could be used as the starting material as well, but the reaction with the nitrile required aluminum chloride and the higher temperature of 150–160°C.[34] The 5-thioethyl and 5-thiophenyl substituted compounds **54** ($R_1$ = Cl; $R_2$ = EtS, PhS) were obtained in low yields by treating the chloride **52** with the respective thiocyanate and stannic chloride.[55]

### 1.1.5. By Reduction

Benzodiazepines of higher oxidation level were reduced to 2,3-dihydro-1*H*-1,4-benzodiazepines by various methods. The parent compound **56** was isolated from the reduction of the 2,5-dione **55** with lithium aluminum hydride in refluxing tetrahydrofuran (Eq. 17).[56]

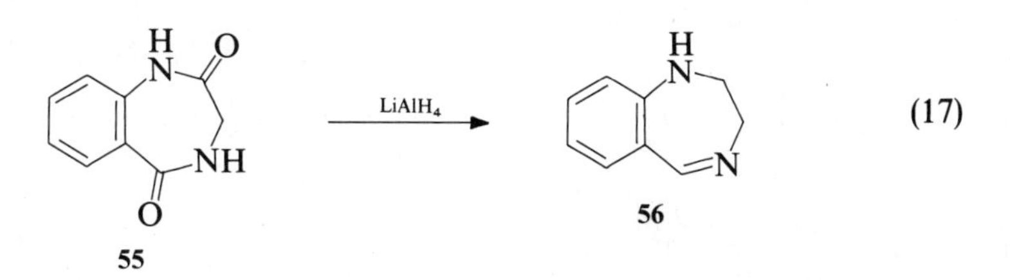


The same reagent in boiling ether[57] or tetrahydrofuran[2,15] and diborane in tetrahydrofuran at $-15$ to $-10$°C,[58,59] effectively converted the 2-ones **57** to **59**. Reduction of the 4-oxides **58** (R = H, Me) with lithium aluminum hydride in ether led to the hydroxyamines **60**,[57] while in boiling tetrahydrofuran compounds **59** ($R_1$ = H, Me; $R_2$ = H; $R_3$ = Ph; X = 7-Cl) were formed as major products, apparently by base-induced dehydration of **60** (Eq. 18).[34]

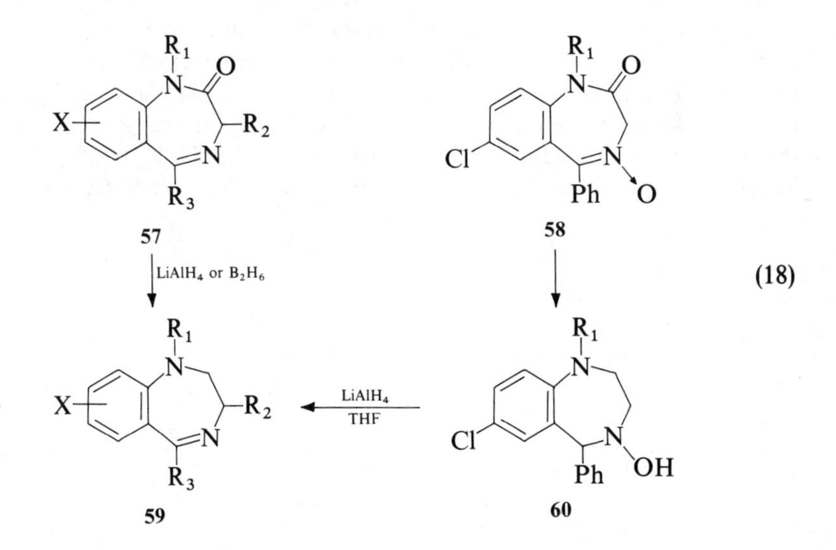


Lithium aluminum hydride at $-40$ to $-50$°C reduced the 5-phenyl-1,4-benzodiazepine-2-ones **61** (Y = H, F) to the 2-hydroxy compounds **62** (Y = H, F) (Eq. 19).[58,60] When diazepam (**61**: Y = H) was exposed to this reagent at room temperature, a dimer was isolated in 23% yield; it was assigned structure **63** (Eq. 19).

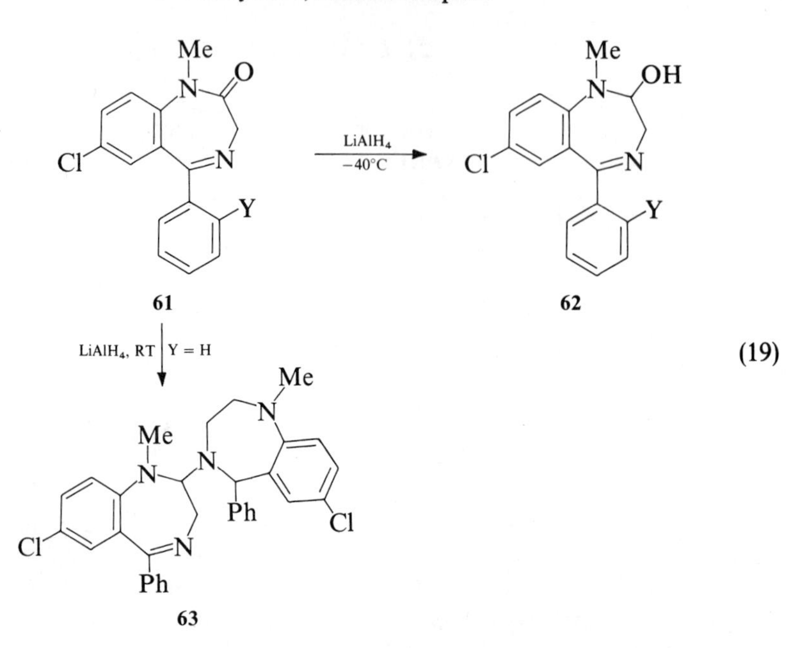


A variety of 2,3-dihydro-1$H$-1,4-benzodiazepines **59** were obtained by other reduction procedures, such as desulfurization of the 2-thiones **64**[61] or the 2-thioalkyl derivatives **65** ($R_3$ = EtS) with Raney nickel[62]; reduction of the 3$H$-1,4-benzodiazepines **66** with lithium aluminum hydride for $R_3$ = ($N$-nitroso)methylamino,[63] $R_3$ = dimorpholinophosphoryloxy;[63] and treatment of **65** ($R_3$ = H) with Raney nickel and hydrogen (Eq. 20).[62] Catalytic hydrogenation of the 4-oxide of **66** ($R_2$ = Ph, $R_3$ = Me, X = 7-Cl) over Raney nickel gave the corresponding 2,3-dihydro-1$H$-1,4-benzodiazepine **67** ($R_3$ = Me).[64] The 1,2-imine bond in several 2-substituted 3$H$-1,4-benzodiazepines **66** was reduced

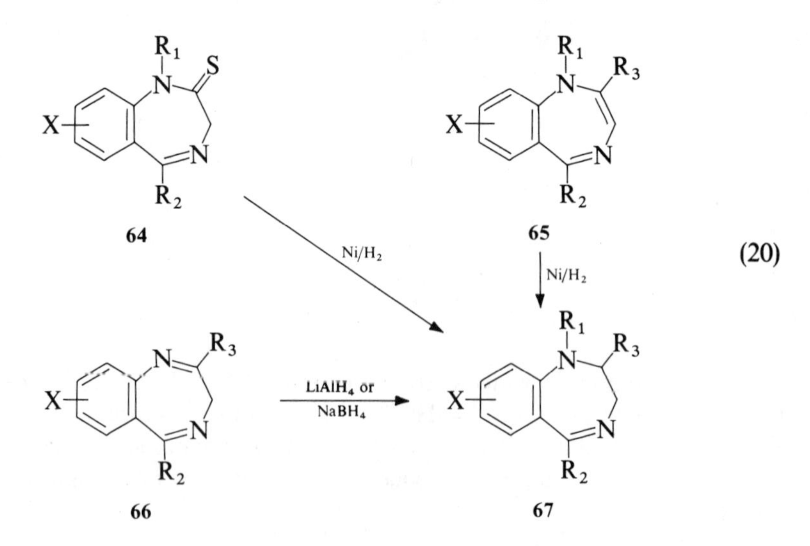

selectively with sodium borohydride. Such reductions were reported for **66** with $R_3$ = Me,[64] hydroxyiminomethyl,[65] (hydroxyimino)nitromethyl,[66] (hydroxyimino)aminomethyl and related compounds,[66] (hydroxyimino)methylthiomethyl,[66] and (methoxyimino)methoxymethyl.[66] The nitrone functionality in the corresponding 4-oxides was not touched by sodium borohydride under the conditions used. This reagent also converted the 3-amino derivatives **68** effectively to the 2,3-dihydro derivatives **69** (Eq. 21).[32]

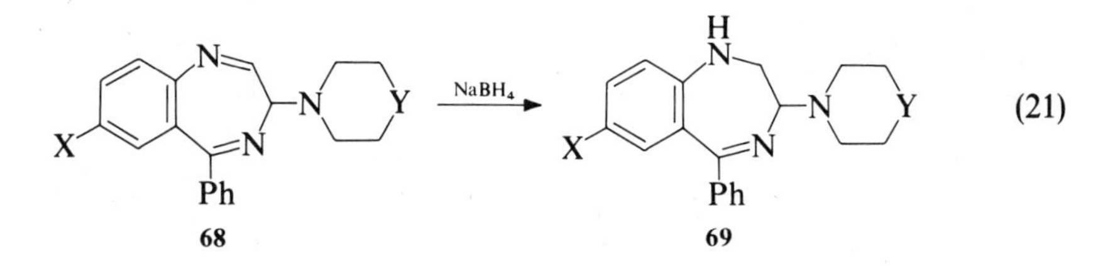


(21)

## 1.1.6. *By Addition and Substitution*

2-Substituted 2,3-dihydro-1*H*-1,4-benzodiazepines **71** were accessible by addition of nucleophiles to the 3*H*-1,4-benzodiazepines **70**. Such additions were successfully carried out with amines,[67,68] alcohols,[67,68] hydrogen cyanide,[67,68] ethanethiol,[68] benzylthiol,[67] and water.[69a] Base-catalyzed addition of dimethyl malonate to **70** (R = Ph, X = 7-Cl) led to **71** (Nu = $CH(COOMe)_2$, R = Ph, X = 7-Cl) (Eq. 22).[63]

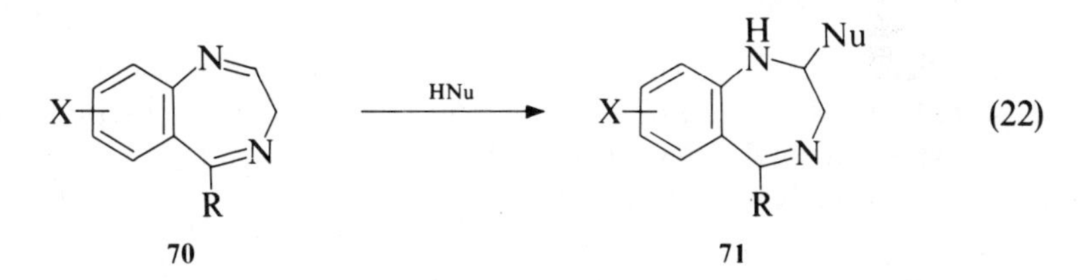


(22)

Displacement of the 2-hydroxy group in 1-substituted benzodiazepines **72** ($R_1$ = Me, $CH_3CH_2$) provided another method for the introduction of substituents at the 2-position. This approach was employed to prepare 2-cyano derivatives **73** (Nu = CN) by use of sodium cyanide in acetic acid or acetone cyanohydrin (Eq. 23).[70,71]

Reaction of **72** ($R_1$ = $CF_3CH_2$, $R_2$ = Ph, X = Cl), or its cyclic isomer **74**, with ethanethiol yielded the 2-ethylthio compound **73** ($R_1$ = $CF_3CH_2$, $R_2$ = Ph, X = Cl, Nu = EtS).[72] The bridged sulfide **75** was similarly cleaved by 2-hydroxyethanethiol to afford **73** ($R_1$ = H, $R_2$ = Ph, X = Cl, Nu = $HOCH_2CH_2S$).[68c] Introduction of a substituent at the 2-position was accomplished by taking advantage of the reactivity of carbons activated by a nitrosamino function. Compound **76** was reacted with benzaldehyde in the presence of potassium *t*-butoxide to give a mixture of the diastereomeric

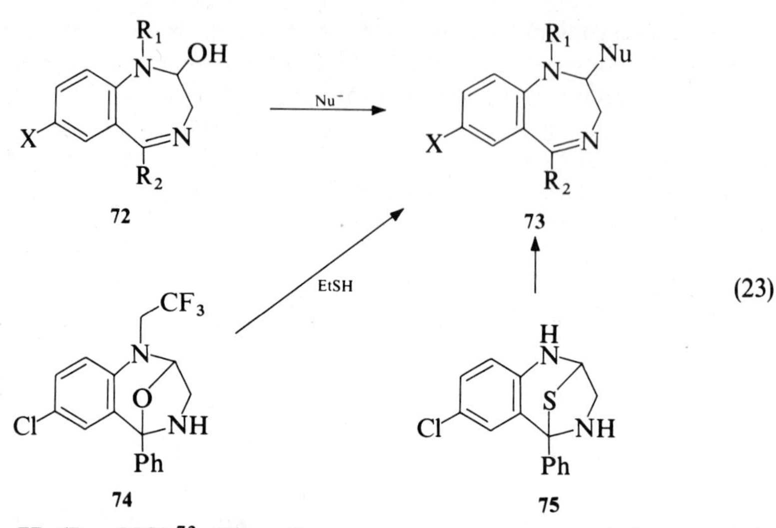


(23)

alcohols **77** (R = NO).[73] The nitroso group was removed by catalytic hydrogenation over Raney nickel, yielding **77** (R = H) (Eq. 24).

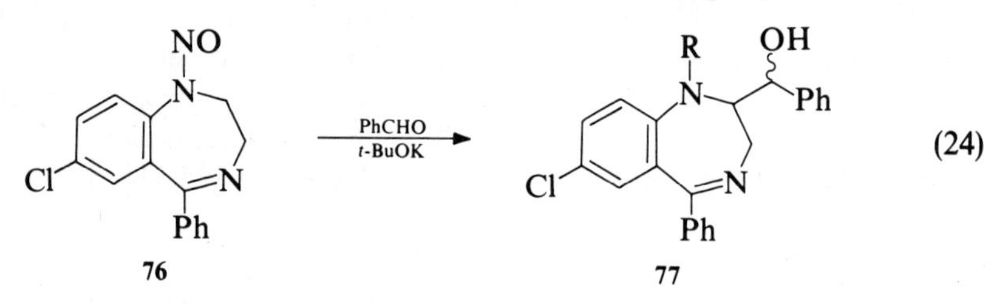


(24)

The benzodiazepines bearing a heteroatom attached at the 5-position were prepared from the 5-ones **78** (X = O) or 5-thiones **78** (X = S). Methylation of the latter with dimethyl sulfate led to the 5-methylthio compound **79** ($R_1 = R_4 = Me$, $R_2 = H$, X = S) (Eq. 25).[55]

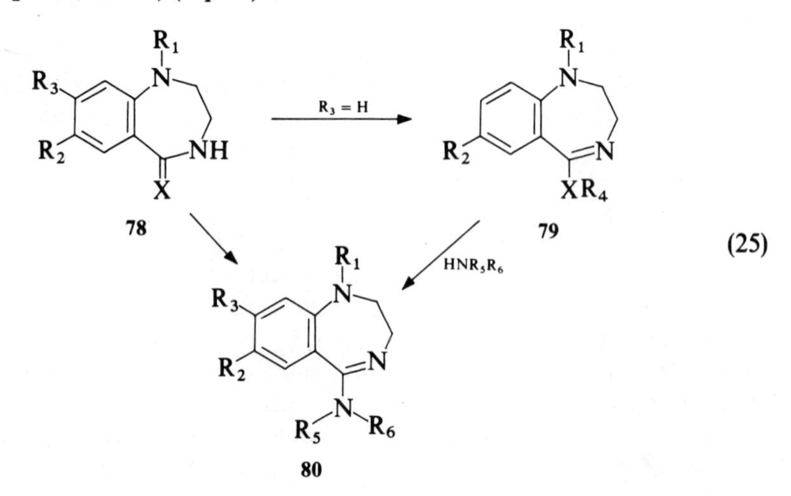


(25)

The corresponding iminoether **79** ($R_1$ = Me, $R_2$ = Cl, $R_4X$ = EtO) was obtained by treatment of the appropriate 5-one with boron trifluoride etherate in the presence of 2-chloromethyloxirane.[74,75] Reaction of **79** ($R_4X$ = MeS, EtO) with amines,[55,74] hydroxyamines,[74] and hydrazines[76] led to 5-amino-substituted benzodiazepines **80**. Compounds **80** ($R_1$ = aryl; $R_2$ = H; $R_3$ = Cl, $CF_3$) were prepared by reacting the corresponding 5-one **78** (X = O) with phosphorus pentachloride and an amine.[77]

### *1.1.7. By Oxidation*

2,3-Dihydro-1*H*-1,4-benzodiazepines **82** ($R_2$ = H, Ph, substituted Ph) were also prepared by oxidation of the corresponding tetrahydrobenzodiazepines **81**. The required amine-to-imine oxidation was effected by a variety of reagents such as manganese dioxide,[78] diethyl azodicarboxylate,[78] DDQ,[78,79] ruthenium dioxide,[80] chloranil,[79] hypoiodite,[79] and sulfur in boiling dimethylformamide.[81] A low yield photochemical oxidation involving irradiation in dimethyl sulfoxide was also reported (Eq. 26).[82]

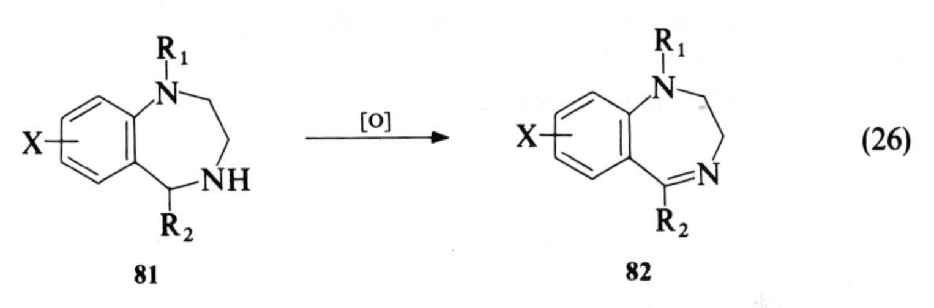


The nitrones **84** resulted from the oxidation of the hydroxyamines **83** ($R_1$ = H, Ac; $R_2$ = H, Me) with mercuric oxide (Eq. 27).[64,83,84]

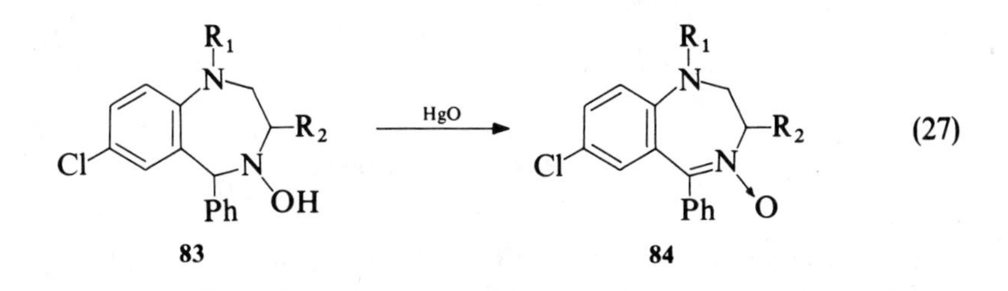


### *1.1.8. By Elimination*

2,3,4,5-Tetrahydro-1*H*-1,4-benzodiazepines **85**, bearing a leaving group, $R_2$, at the 4-position, were converted to 2,3-dihydro-1*H*-1,4-benzodiazepines **86**, most likely by initial deprotonation at C-5 followed by loss of the leaving group (Eq. 28).

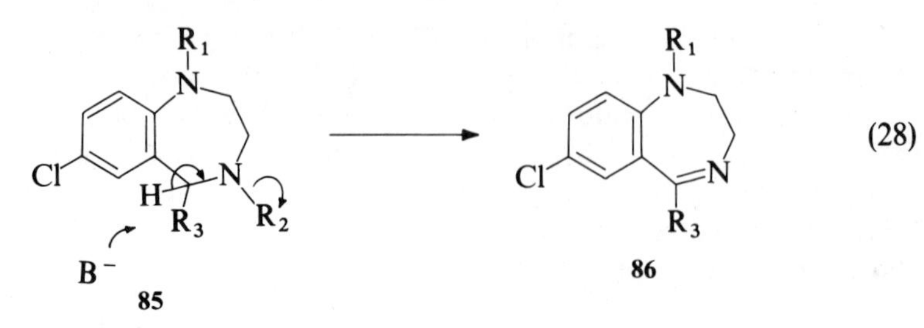


(28)

Thus the 4-sulfonates **85** ($R_1$ = Me, $R_2$ = $MeSO_2$, 4-$MeC_6H_4SO_2$; $R_3$ = H, Ph) were transformed into the corresponding compounds **86** by treatment with sodium hydride in dimethylformamide.[78,85] The 4-acyl derivatives **85** ($R_1$ = Me; $R_2$ = CHO, Ac; $R_3$ = Ph) similarly underwent eliminations of formaldehyde or acetaldehyde, respectively, to give the imines **86**. Elimination of water from the 4-hydroxyderivatives **85** ($R_1$ = H, Me; $R_2$ = OH; $R_3$ = Ph) was achieved in high yield by heating in strong base such as sodium hydroxide in boiling ethanol.[34] Medazepam (**86**: $R_1$ = Me, $R_3$ = Ph), was also obtained when the 4-acetoxy compound **85** ($R_1$ = Me, $R_2$ = AcO, $R_3$ = Ph) was reacted with strong bases.[34] Refluxing of the 4-hydroxy compound with dicyclohexylcarbodiimide in toluene effected elimination of water as well.[34] In this case, the intermediate isourea **87**, which could undergo the indicated elimination to medazepam (**12**), may be formed (Eq. 29).

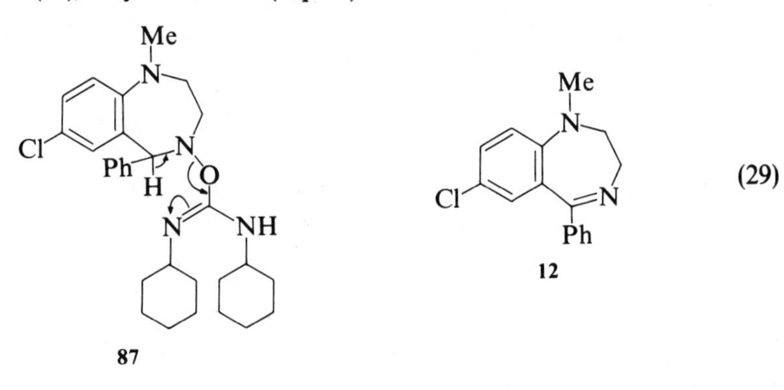


(29)

## *1.1.9. Other Syntheses*

2-Hydroxybenzodiazepines **72** ($R_1$ = H, Me; $R_2$ = Ph; X = Cl, $NO_2$) were synthesized by treatment of the acetals **88** with aqueous acid (Eq. 30).[69]

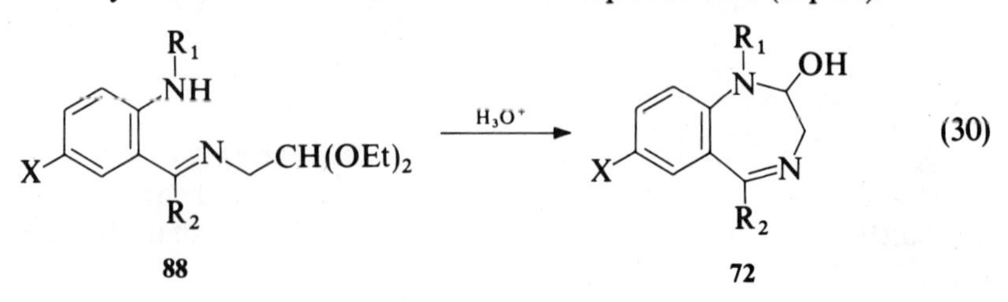


(30)

A synthesis reported in the patent literature[86] involves oxidation of the benzhydrol **89** with manganese dioxide or chromium trioxide in acetic acid to the corresponding benzophenone, which undergoes in situ cyclization to the diazepine **42** (Eq. 31).

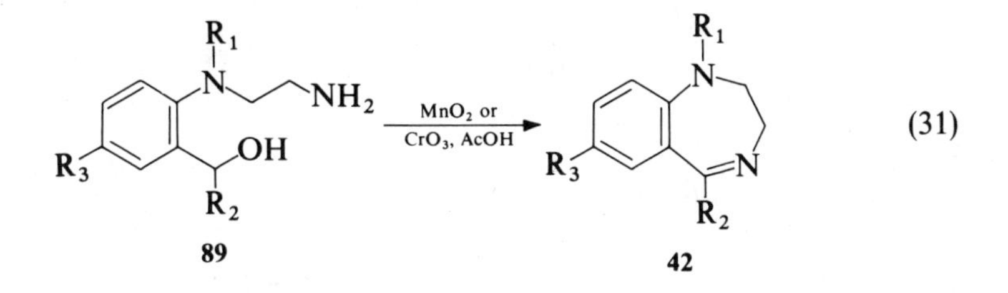


Thermal rearrangement of the oxaziridines **90** ($R_1$ = CHO, Ac) afforded the corresponding nitrones **91** (Eq. 32).[83,84]

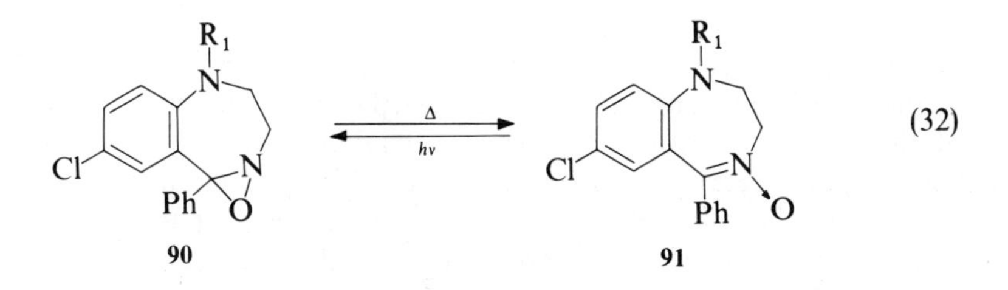


### 1.2. Synthesis of 2-Imino-2,3-dihydro-1*H*-1,4-benzodiazepines

The title compounds bearing a hydrogen at the 1-position exist as 2-amino-3*H*-1,4-benzodiazepines and were discussed in Chapter V.

The 2-imino derivatives **93** ($R_1$ = Me, $R_2$ = Ph) were prepared in 80% yield by cyclization of the nitriles **92** with ethanolic hydrogen chloride.[87] A 1-benzyl-7-nitro analog was similarly synthesized.[87a] The imines **92** were obtained by an amine exchange of the ethanolamine Schiff base with 2-aminoacetonitrile hydrogen sulfate in the presence of 2-methyl imidazole.[87]

Compound **93** ($R_1$ = Me, $R_2$ = $Me_2N$, $R_3$ = Ph, X = Cl) resulted from the reaction of the 2-thione **94** ($R_3$ = Ph) with 1,1-dimethylhydrazine.[88] The acetylhydrazine **93** ($R_1$ = Me, $R_2$ = AcNH, $R_3$ = 2-$FC_6H_4$, X = Cl) was likewise formed from the appropriate 2-thione and acetylhydrazine.[89]

The 1-cyano-2-methylimine **93** ($R_1$ = CN, $R_2$ = Me, $R_3$ = Ph, X = Cl) was isolated from the reaction of the quaternary triazolobenzodiazepine **95** with aqueous alkali (Eq. 33).[90]

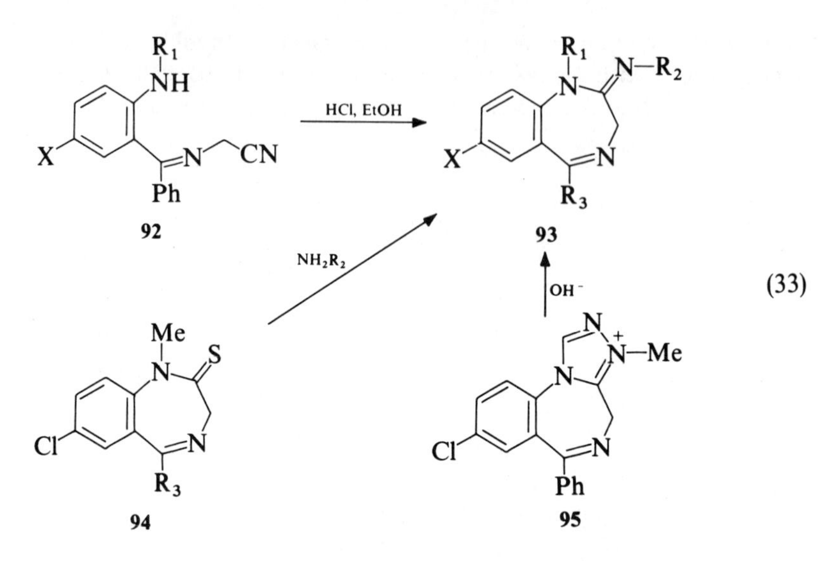


### 1.3. Synthesis of 2-Methylene-1,3-dihydro-2*H*-1,4-benzodiazepines

The benzodiazepines **97** with an exocyclic double bond at the 2-position were synthesized by reacting 3*H*-1,4-benzodiazepines (**96**) having a leaving group X at the 2-position with a stabilized carbanion, $^{-}CHR_4R_5$. Suitable leaving groups X were *N*-nitrosomethylamino,[91,92] dimorpholinophosphoryloxy,[92,93] diethoxyphosphoryloxy,[63,92,93] and diphenoxyphosphoryloxy.[63] The carbanions employed were derived from dialkyl malonates,[91–93] malononitrile,[91] malonic acid ethyl ester dimethylamide,[63] alkyl cyanoacetate,[63] alkyl acetoacetate,[63,93] acetylacetone,[93] diethyl acetonedicarboxylate,[63] and nitroalkanes (Eq. 34).[65,92] The 4-oxides of **97** could be obtained in the same fashion.

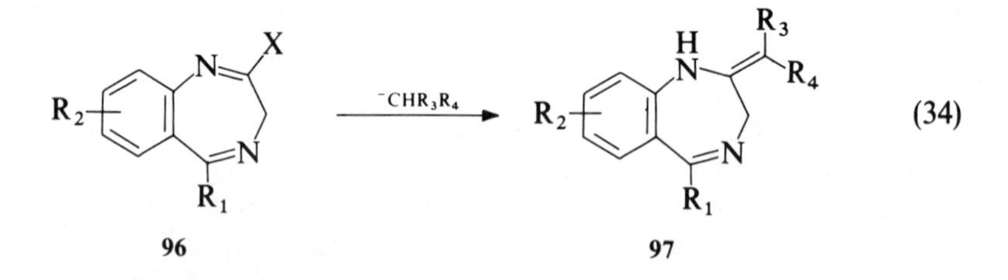


Compounds **99** ($R_3$ = COOR) were accessible by alkaline hydrolysis of the corresponding malonyl derivatives **97** ($R_3 = R_4$ = COOR) followed by decarboxylation, which occurred usually under the conditions of hydrolysis (refluxing in alcohol–water in the presence of hydroxide). Under these conditions, compounds **97** with $R_3$ = COOMe, $R_4$ = Ac or with $R_3 = R_4$ = Ac lost an acetyl group to form the ester **99** ($R_3$ = COOMe),[93] and the methylketone **99**

($R_3$ = Ac),[93] respectively. The tertiary butyl ester **97** ($R_3$ = COO*t*-Bu, $R_4$ = Ac) was cleaved by trifluoroacetic acid and spontaneously decarboxylated to give **99** ($R_3$ = Ac, 4-desoxy).[63] The same reagent converted **97** ($R_3$ = COO*t*-Bu, $R_4$ = CN) to the corresponding acid, which was thermally decarboxylated to **99** ($R_3$ = CN, 4-desoxy).[63]

A more direct synthesis of the nitrones **99** was provided by the ring expansion of the quinazolines **98** with carbanions. The carbanions successfully employed in this ring expansion include those generated from nitromethane,[94] ethyl acetate, *i*-propyl acetate, *t*-butyl acetate, *N,N*-dimethylacetamide, 1-acetyl-4-methylpiperazine, acetonitrile, dimethyl sulfone, *N,N*-dimethylmethylsulfonamide and 2-methylpyridine.[95] The anions were formed with a strong, sterically hindered base, preferably lithium diisopropylamide.

In the case of acetonitrile anion, the primary adduct **100** could be isolated and further converted to **99** ($R_3$ = CN) by treatment with a strong base like potassium *t*-butoxide (Eq. 35).[95]

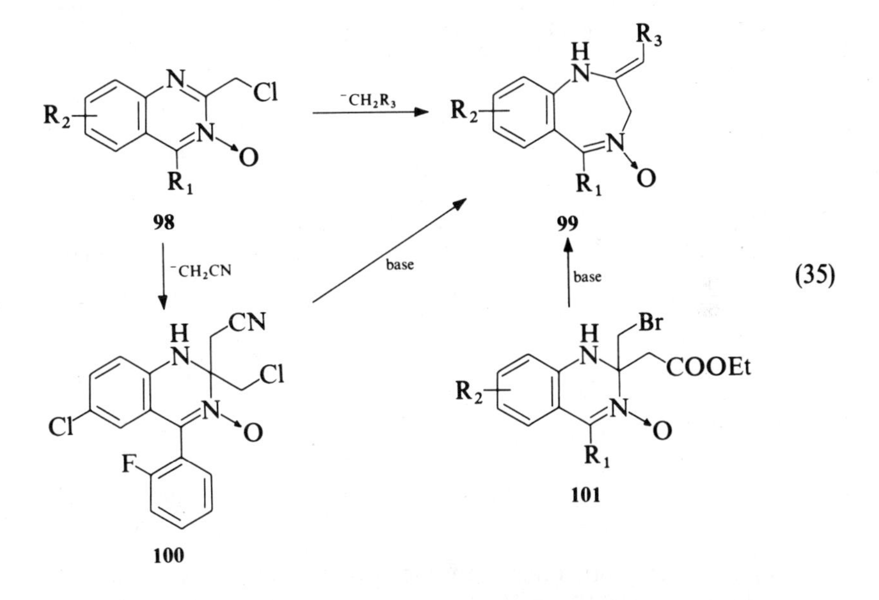


Analogously, reaction of the dihydroquinazolines **101** with alkoxide led to the benzodiazepines **99** ($R_3$ = COOEt).[96] As shown by the preparation of **103** (R = H, Me), the 2-dichloromethyl-1,2-dihydroquinazoline **102** (R = H, Me) may serve as the starting material in place of **98**. However, the yield in this variation was considerably lower (Eq. 36).[94] The anion of nitromethane was conveniently formed with lithium amide in dimethyl sulfoxide.[94]

The malonyl derivatives **107** were also found to be formed in good yields by reaction of the lactams **104** with the phosphorylated malonic ester of possible structure **105**.[97] The phosphorylated malonic ester may react with the lactam,

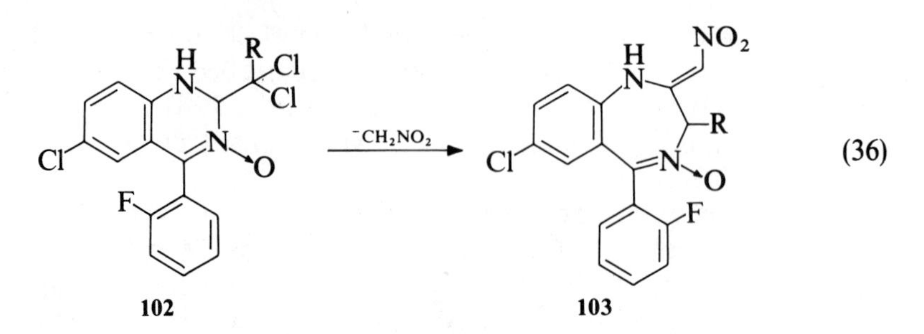


generating the iminophosphate **106** with loss of ethanol; **106** could then rearrange in the indicated fashion to form **107** (Eq. 37).

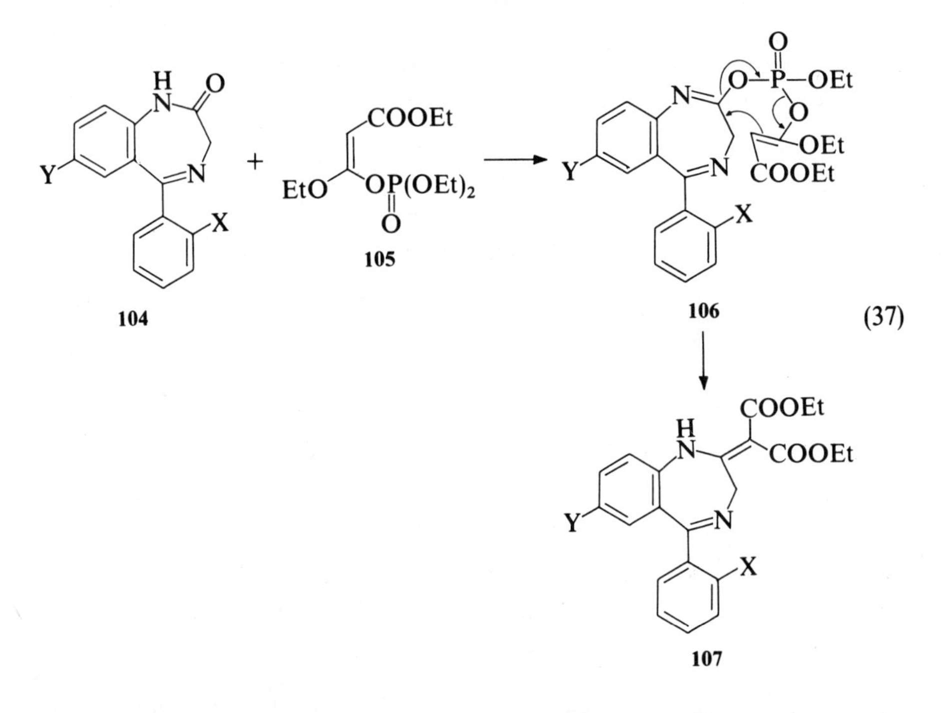


Oxidation was another means for the introduction of an exocyclic methylene into compounds **108** ($R_1$ = H, Me; $R_2$ = CN, $CONH_2$).[63,98] Manganese dioxide and chromium trioxide were the reagents of choice.

When the tricyclic compound **110** was subjected to activated manganese dioxide, the formyl derivative **109** ($R_1$ = H, $R_2$ = CON(CHO)Me, X = F, Y = Cl) was obtained as the major product (Eq. 38).[63]

The $\alpha$-amino derivatives **112** ($R_1$ = ROOC, Ac; $R_4$ = H) resulted from the catalytic reduction of the oximes **111** over Raney nickel.[92,93] In the presence of ammonia, the nitrone function of the corresponding 4-oxides survived the catalytic hydrogenation.[95] With few exceptions, the unstable amines **112** were not characterized but used directly in further conversions (see Eq. 62, below).

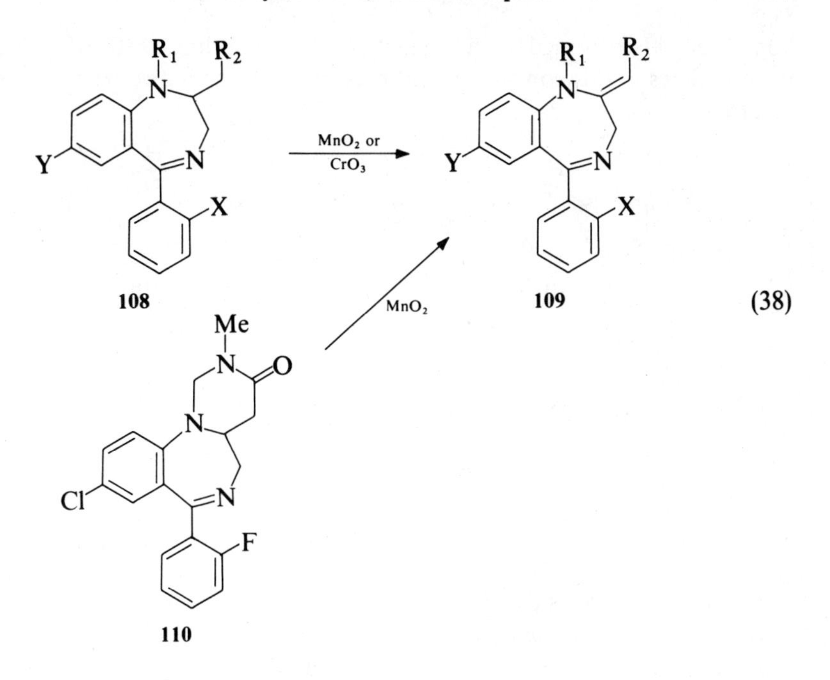


*N*-Acetyl derivatives **112** ($R_1$ = EtOOC, $R_4$ = Ac) were also prepared by decarboxylation of the malonates **113** (Eq. 39).[92]

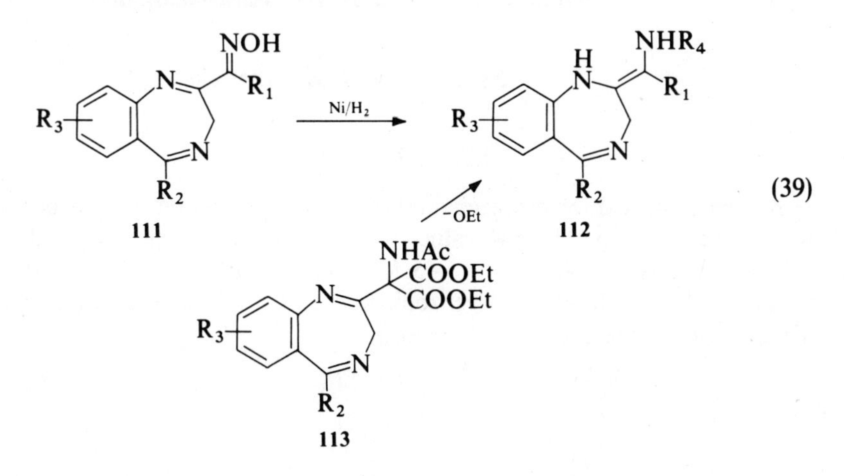


### 1.4. Synthesis of 3-Methylene-1,2-dihydro-3*H*-1,4-benzodiazepines

These compounds were prepared exclusively by ring expansion of quinazolines. Compounds **116** ($R_1$ = H; $R_2$ = H, Cl) and their 4-oxides were obtained by treatment of the corresponding quinazolines **114** with potassium *t*-butoxide in ether or tetrahydrofuran (Eq. 40).[99] The isomers with an endocyclic double bond were also formed in this reaction. The aziridines **115** are most likely intermediates in this conversion and may undergo the indicated rearrangement to **116**.

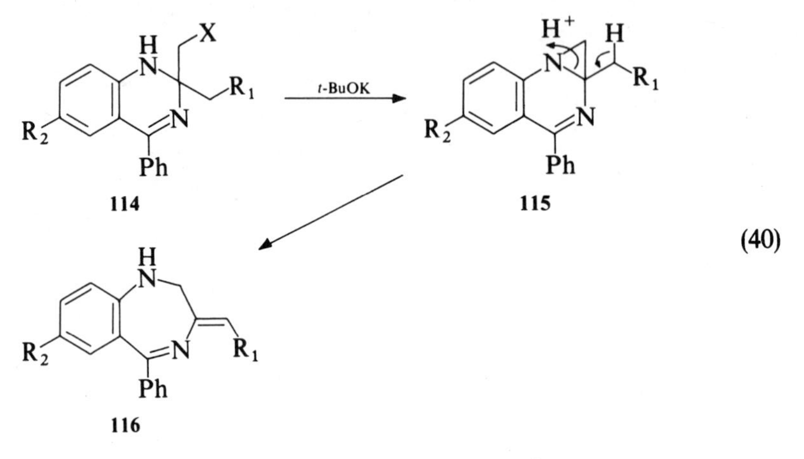


Compound **116** ($R_1$ = EtOOC; $R_2$ = Cl, 4-oxide) was isolated from the reaction of the quinazoline 3-oxide **114** ($R_1$ = EtOOC; $R_2$ = Cl; X = Br, 3-oxide) with potassium *t*-butoxide in ethanol.[96]

## 1.5. Reactions of 2,3-Dihydro-1*H*-1,4-benzodiazepines

### *1.5.1. Reactions with Electrophiles*

#### 1.5.1.1. Protonation

As demonstrated for medazepam (**12**), the imine nitrogen is more basic than the aniline nitrogen. The p$K$ values were determined spectroscopically to be 6.25 for the 4-position and −1.45 for the 1-position.[100] The 4-protonated species of 2,3-dihydro-1*H*-1,4-benzodiazepines are yellow to orange, while the diprotonated forms are virtually colorless.

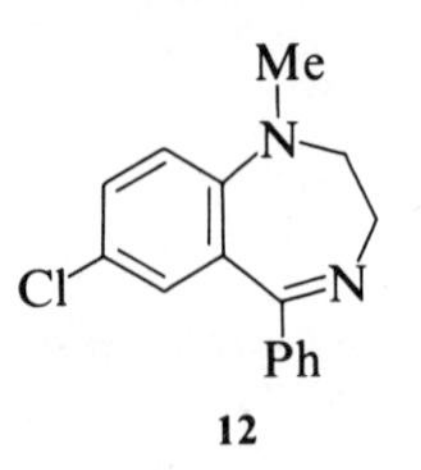


#### 1.5.1.2. Halogenation

Introduction of a halogen atom at the 7-position was reported for a variety of 5-phenyl-substituted 2,3-dihydro-1*H*-1,4-benzodiazepines. Medazepam (**12**) was prepared by chlorination of **117** ($R_1$ = Me, $R_2$ = H, $R_3$ = Ph) with *N*-chlorosuccinimide, *t*-butyl hypochlorite, or chlorine (Eq. 41).[101] *N*-Chloro- and *N*-bromosuccinimide were used to halogenate compounds **117** ($R_1$ = Me; $R_2$ = $MeOCH_2$, $ClCH_2$; $R_3$ = Ph, 2-$ClC_6H_4$) at the 7-position.[48,49,102] Reaction of **117** ($R_1$ = H, Me; $R_2$ = H; $R_3$ = Ph, 2-$FC_6H_4$) with iodine monochloride in acetic acid mixed with sulfuric acid gave high yields of the 7-iodo derivatives.[103,104]

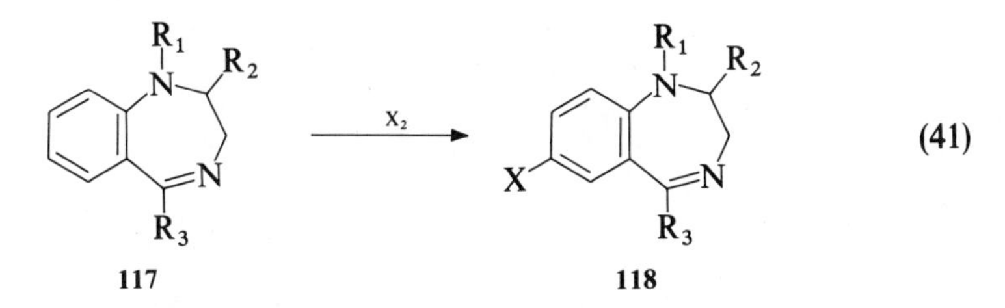


(41)

The chlorination of the 2-acetylidene analog **119** with *t*-butyl hypochlorite led to a mixture of two olefin isomers **120**, which were separated by chromatography (Eq. 42).[63]

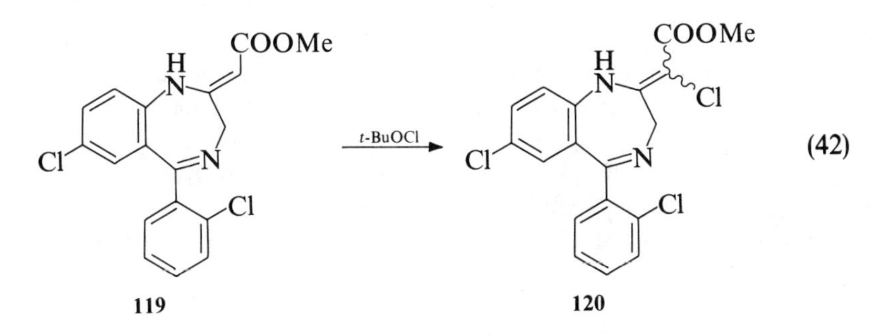


(42)

### 1.5.1.3. Oxidation

2,3-Dihydro-1*H*-1,4-benzodiazepines **121** ($R_1$ = H, Me; $R_2$ = H; $R_3$ = Ph) were oxidized to the corresponding 2-ones **122** by chromium trioxide in acetic acid[46] or in sulfuric acid–acetone.[105] The same reagent oxidized the 2-hydroxy compounds to the 2-ones as well.[69] Ruthenium tetroxide allowed the introduction of a carbonyl group into the 2-position more effectively. Medazepam was thus converted to diazepam in 55% yield using this reagent in chloroform at 0°C.[80] This procedure was applied also to **121** ($R_1$ = $CF_3CH_2$; $R_2$ = H; $R_3$ = 2-$ClC_6H_4$, 2-$FC_6H_4$; X = Cl).[106] *N*-Bromosuccinimide in tetrahydrofuran and aqueous sodium bicarbonate solution was reported to oxidize **121** ($R_1$ = Me, $R_2$ = H, $R_3$ = 2-$FC_6H_4$, X = Cl) to the corresponding 2-one.[42]

Milkowski and coworkers[107] studied the oxidation of **121** ($R_1$ = Me, $R_2$ = $ClCH_2$, $R_3$ = Ph, X = Cl) under various conditions. Potassium permanganate in dilute hydrochloric acid yielded 57% of diazepam and 10% of the anthranil **123** as the major products. Diazepam was also the major product of

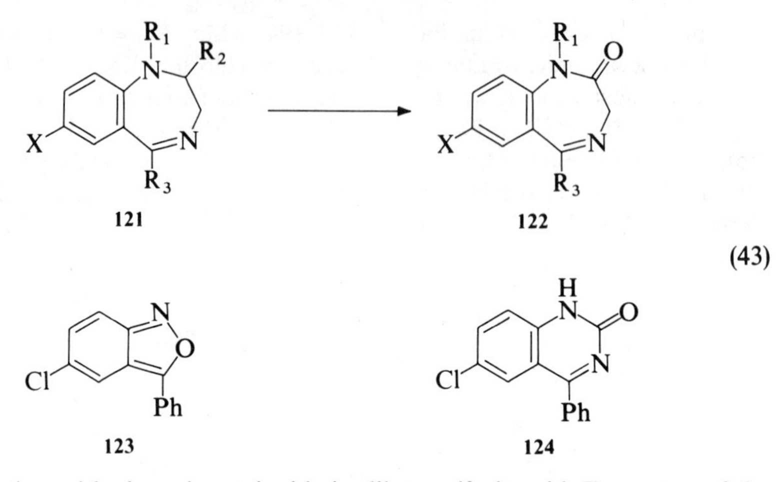


the oxidation with chromium trioxide in dilute sulfuric acid. Treatment of the same compound with chromium trioxide in pyridine gave the quinazolinone **124** (30%), the 1-formyl derivative **121** ($R_1$ = HCO, $R_2$ = $ClCH_2$, $R_3$ = Ph, X = Cl) (18%), desmethyldiazepam (14%), and diazepam (4%). The 2-chloromethyl and the 2-hydroxymethyl compounds **121** ($R_1$ = Me; $R_2$ = $ClCH_2$, $HOCH_2$; $R_3$ = 2-$ClC_6H_4$; X = Cl) were oxidized to the corresponding 2-ones in 20% yield using potassium permanganate in dilute hydrochloric acid (Eq. 43).[47]

The oxidation of 2,3-dihydro-1*H*-1,4-benzodiazepines to 3*H*-1,4-benzodiazepines[67,68] by manganese dioxide was discussed in Chapter V, dealing with the synthesis of 3*H*-1,4-benzodiazepines. When the 5-unsubstituted benzodiazepine **125** was subjected to this reagent in refluxing benzene in the presence of a catalytic amount of acetic acid, the dimer **126** was isolated in 12% yield (Eq. 44).[68a]

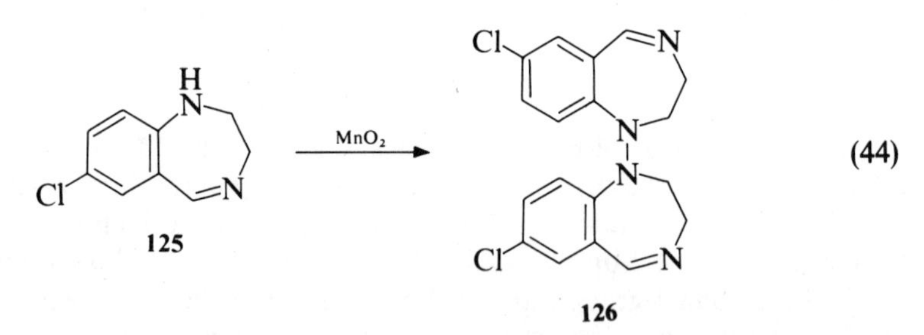


Manganese dioxide and chromium trioxide were used also for the introduction of an exocyclic double bond into the 2-acetonitriles[98] and a 2-acetamide[63] (see Section 1.3).

1-Acylated-2,3-dihydro-1*H*-1,4-benzodiazepines **127** ($R_1$ = Ac, HCO; $R_2$ = Ph) were oxidized by peracetic or *m*-chloroperoxybenzoic acid to the oxaziridines **90**, which were thermally rearranged to the corresponding nitrones.[83,84] A direct conversion of the imine **127** ($R_1$ = $F_3CCH_2$, $R_2$ = 2-$FC_6H_4$) to the *N*-oxide **128** by *m*-chloroperoxyzoic acid was also reported (Eq. 45).[44]

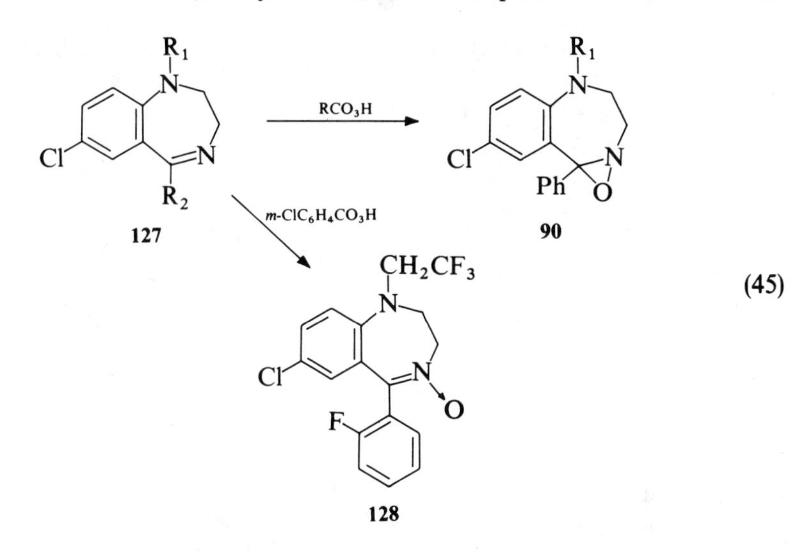


(45)

### 1.5.1.4. Reactions with Nitrogen Electrophiles

A. Nitrosation. The dihydrobenzodiazepine **129** ($R_1$ = H, $R_2$ = Ph) was nitrosated by sodium nitrite in glacial acetic acid[83c] or by nitrosyl chloride in pyridine[73] to the 1-nitroso derivative **130** ($R_1$ = H, $R_2$ = Ph) (Eq. 46). Similar nitrosations were carried out on the nitrile **129** ($R_1$ = CN, $R_2$ = Ph)[68d] and the 2-acetic acid methyl ester **129** ($R_1$ = $MeOOCCH_2$, $R_2$ = 2-$FC_6H_4$)[62b] to yield the corresponding 1-nitroso compounds.

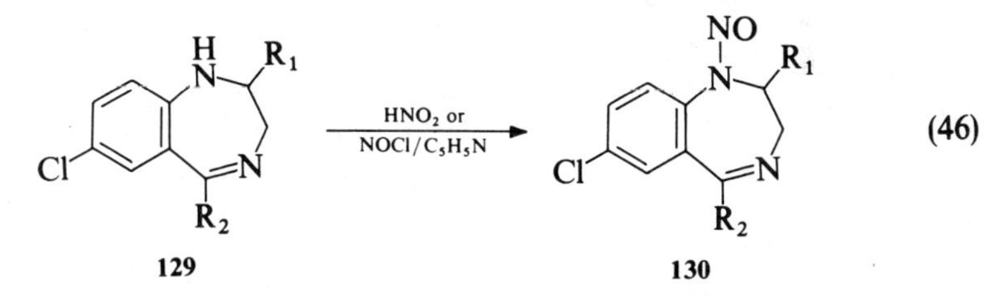


(46)

Reaction of the 2-methylene derivatives **131** with sodium nitrite in glacial acetic acid led in high yields to the oximes **111** (Eq. 47). This conversion was described for a variety of substituents $R_1$ and also for the corresponding 4-oxides. Applicable substituents were $R_1$ = ROOC,[92,93,95,96] Ac,[93] CN,[95] $NO_2$,[66] $Me_2NOC$,[95] $Me_2NSO_2$,[95] and 2-pyridyl.[95]

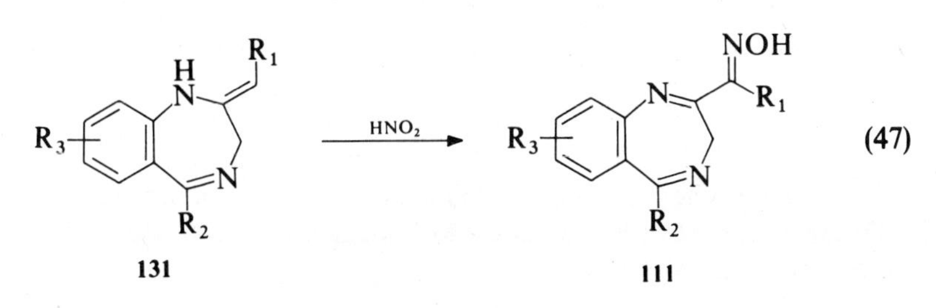


(47)

Nitrosation of the α-amino derivatives **132** provided a high yield synthesis of the triazolobenzodiazepines **133** (Eq. 48).[63] Transformations of this type were carried out for R = MeO, EtO, *t*-BuO, and Me.[63]

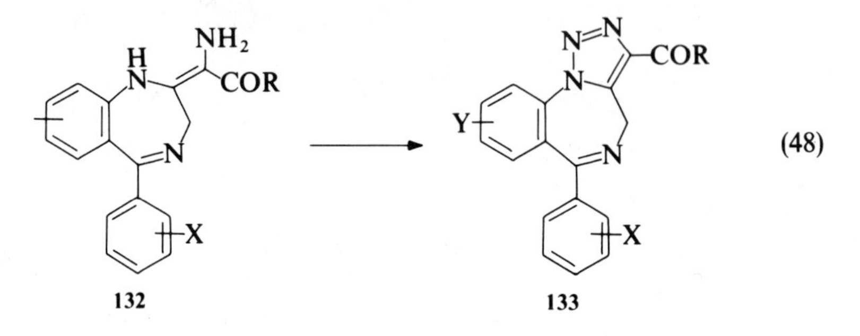


Several 7-amino-substituted 2,3-dihydro-1*H*-1,4-benzodiazepines[134] were subjected to the Sandmeyer reaction (e.g., diazotization to **135** and reaction of the diazonium salts with nucleophiles) (Eq. 49).[2–5,7,112,113]

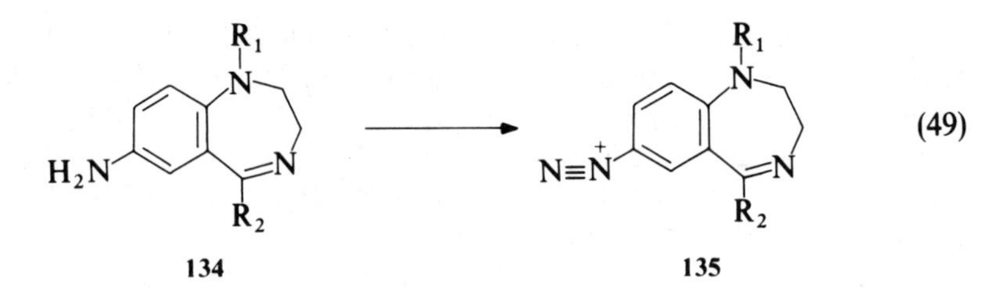


B. Nitration. Nitration of the benzodiazepine **136** led to the 7-nitro compound **137** (R = H) and the 7,9-dinitro derivative **137** ($R = NO_2$). If the 7-position was occupied by a chlorine, the nitro group entered the 9-position (Eq. 50).[106]

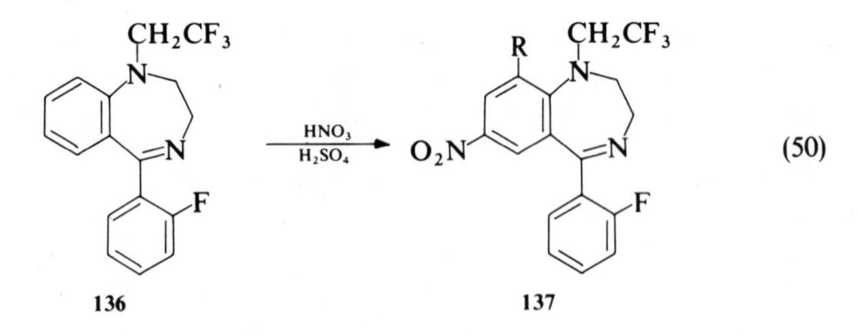


Nitration at the 7-position of **138** with concomitant demethylation (yielding **139**) was observed during reaction with nitric acid in a mixture of sulfuric and acetic acids (Eq. 51).[48]

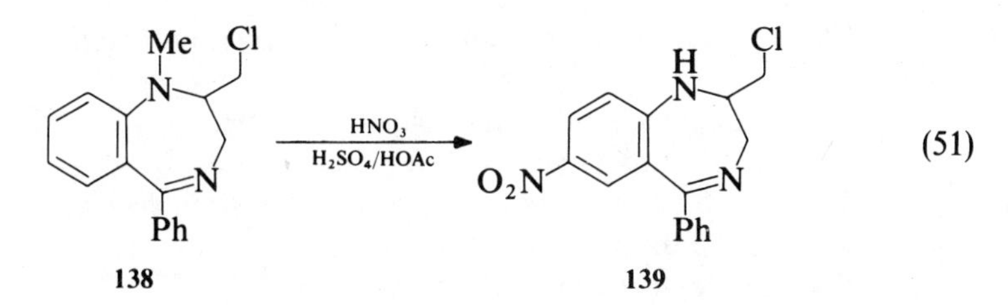


C. Reactions with other Nitrogen Electrophiles. Diethyl azodicarboxylate added to the 2-methylene compounds **140** (R = MeO, Me; X = F, Cl) to give the adducts **141** of undetermined stereochemistry (Eq. 52).[63, 108]

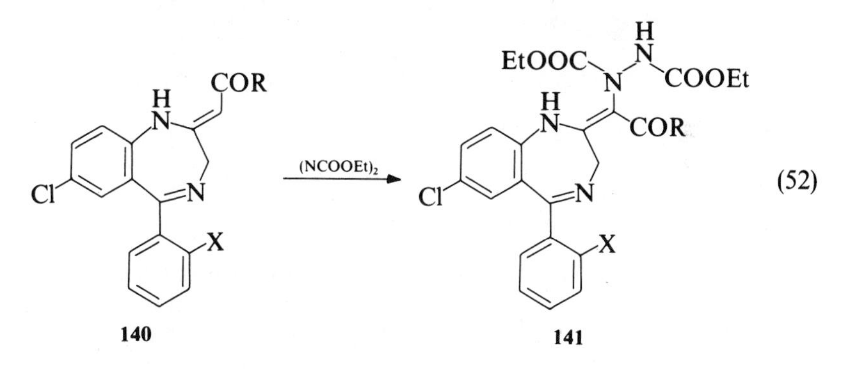


### 1.5.1.5. Reactions with Carbon Electrophiles

A. Alkylation. Methylations at the 1-position were carried out with dimethyl sulfate–sodium methoxide in dimethylformamide[3−7] and with methyl iodide using sodium hydride[15, 109, 110] or phenyllithium[37, 111] as base. A high yield methylation by means of formaldehyde and sodium cyanoborohydride was also described.[32] Similar alkylations were performed with cyclopropylmethyl bromide,[111] 2-diethylaminoethyl chloride,[111, 112] 2-phthalimido-1-bromoethane,[111] and 2-bromo-*N*-methylacetamide.[113] Quaternization of the 4-position nitrogen was achieved with dimethyl sulfate in refluxing benzene[114a] or with ethyl iodide.[114b]

Alkylations of various functional groups on 2,3-dihydro-1*H*-1,4-benzodiazepines were described. The alcohol **108** ($R_1$ = Me; $R_2$ = OH; X = Y = Cl) was transformed into the propargyl ether **108** ($R_1$ = Me; $R_2$ = HC≡C–$CH_2$; X

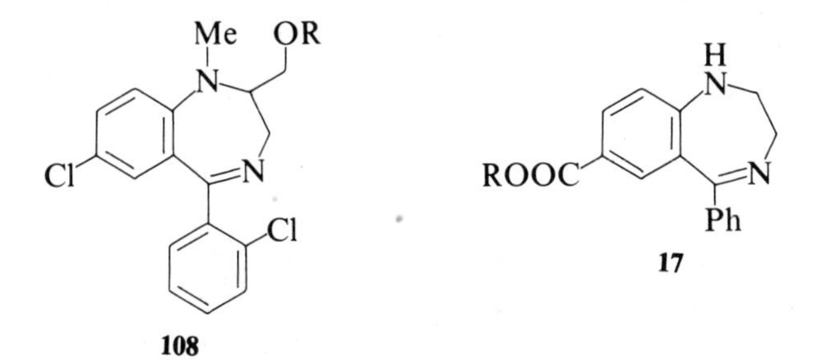

= Y = Cl) by reaction with 1-bromo-2-propyne and sodium hybride in benzene.[52]

The benzodiazepine-7-carboxylic acid **17** ($R_1$ = H; $R_2$ = COOH) was esterified by diazomethane to **17** ($R_1$ = H; $R_2$ = COOMe).[2]

The 2-chloroethylamides **142** underwent an intramolecular alkylation on oxygen to form the oxazolines **143** when treated with potassium carbonate in dimethylformamide in the presence of iodide (Eq. 53).[115]

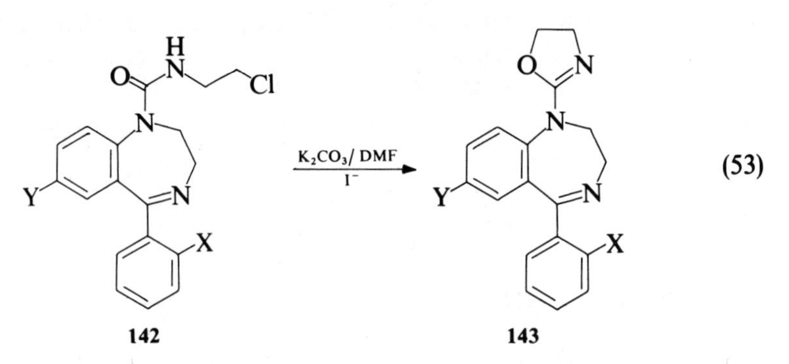


Similarly, potassium carbonate in acetonitrile effected conversion of the chloroacylamides **144** ($n$ = 1–3) to the cyclic compounds **145** (Eq. 54).[115]

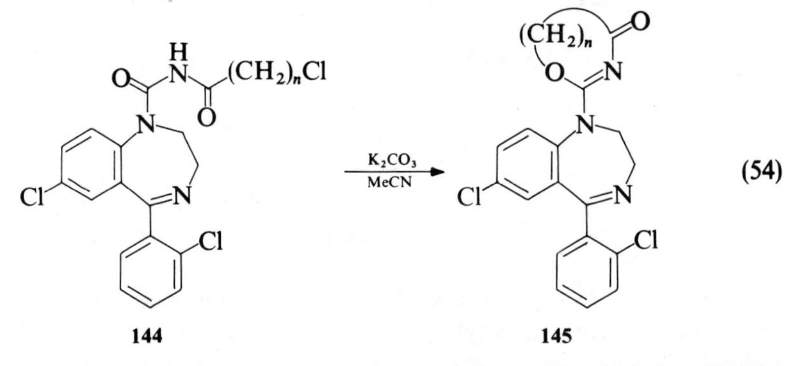


Intramolecular alkylations of the α-chloroacyl derivatives **146** (R = H, Me) by treatment with triethylamine in hot dimethylformamide yielded the pyrazinodiazepines **147** (R = H, Me) (Eq. 55).[116]

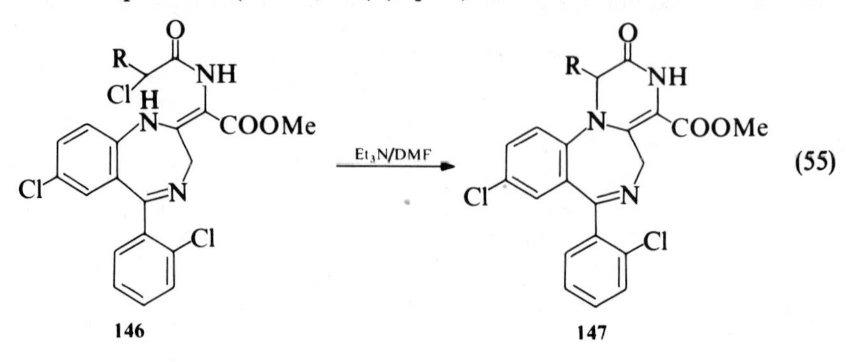

The 3-chloropropanoyl analog **148** gave under the same reaction conditions the pyrrolinone **149** (Eq. 56).[116]

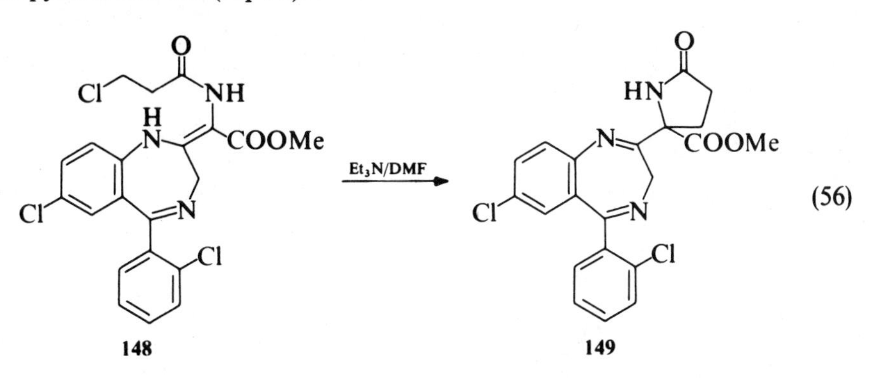


The 1-malonyl derivative **150** of unknown stereochemistry was subjected to intramolecular alkylation, which resulted in a mixture of two diastereomeric pyrrolidinones **151**, separated by chromatography (Eq. 57).[63]

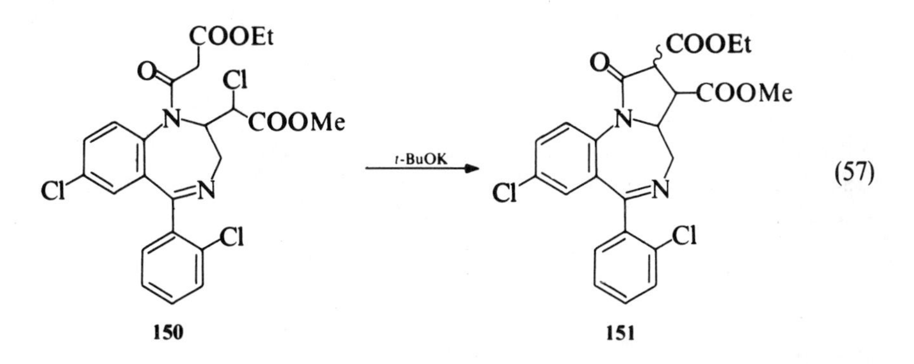


B. Reactions with Aldehydes, Ketones, and Epoxides. The oximes **152** (R = H; $R_1$ = H, $Me_2N$, pyrrolidino, morpholino, MeO, MeS, 2-pyridyl) reacted with aldehydes in hot glacial acetic acid to give moderate yields of the imidazobenzodiazepines **153**.[66] This reaction was also applied to a few 4-oxides. Under milder reaction conditions (e.g., refluxing in dichloroethane in the presence of pivalic acid), formaldehyde converted **152** ($R = R_1 = H$) to a mixture of the oxadiazine **154** ($R_1 = R_2 = H$) and the imidazoline *N*-oxide **155** (Eq. 58.)[66]

If $R_1$ in **152** was an electron-withdrawing moiety such as an ester (*t*-BuOOC) or amide ($Me_2NCO$), the oxadiazines **154** ($R_1$ = *t*-BuOOC, $Me_2NCO$; $R_2$ = H, Me) were formed exclusively.[63] Methylation of the oxime oxygen in **152** (R = Me) improved the yields of the imidazobenzodiazepines **153** by blocking the formation of the oxadiazines.[66]

Treatment of the morpholino derivative **156** with formaldehyde and pivalic acid in refluxing 1,2-dichloroethane gave the spiro compound **159** as the major product.[66] Its formation was rationalized by hydroxymethylation of the oxime

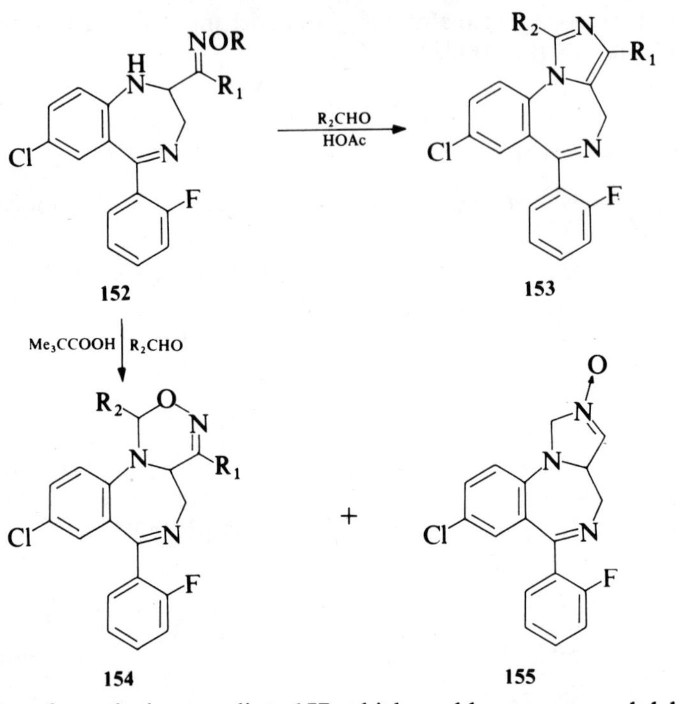


nitrogen to form the intermediate **157**, which could rearrange and dehydrate to **158**. The latter would then undergo ring closure by addition to the 1,2-imine (Eq. 59).[66]

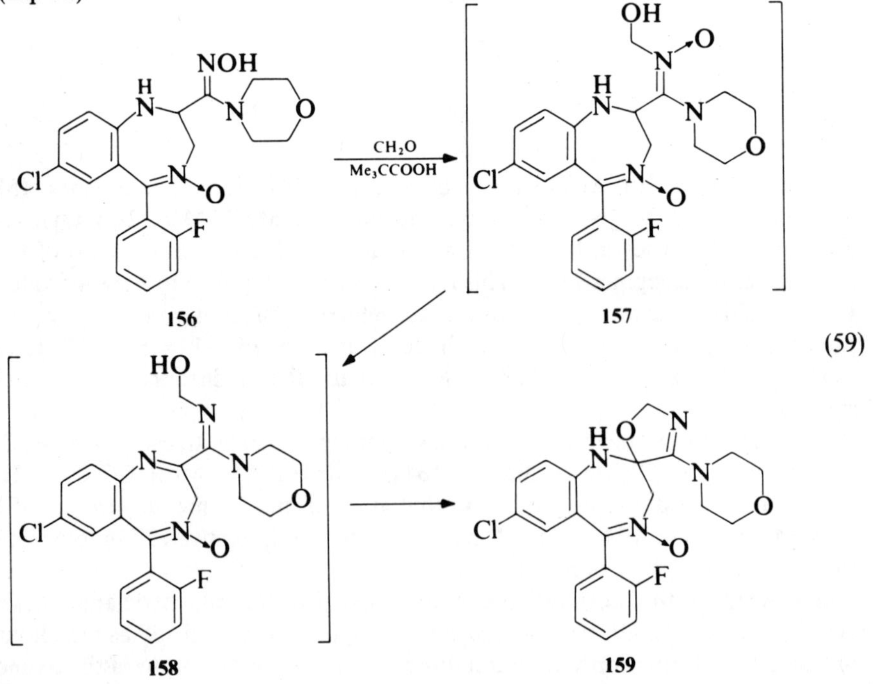

The structure of **159** was determined by single-crystal X-ray analysis.

Ring formations with formaldehyde were carried out with other functionalized 2,3-dihydro-1,4-benzodiazepines. The 2-hydroxyethylthio derivative **160** was thus converted to the tricyclic compound **161** in 78% yield.[68]

The alcohol **162** and the methylamide **164** were similarly transformed into the 1,3-oxazine **163** and the pyrimidine **110** (Eq. 60).

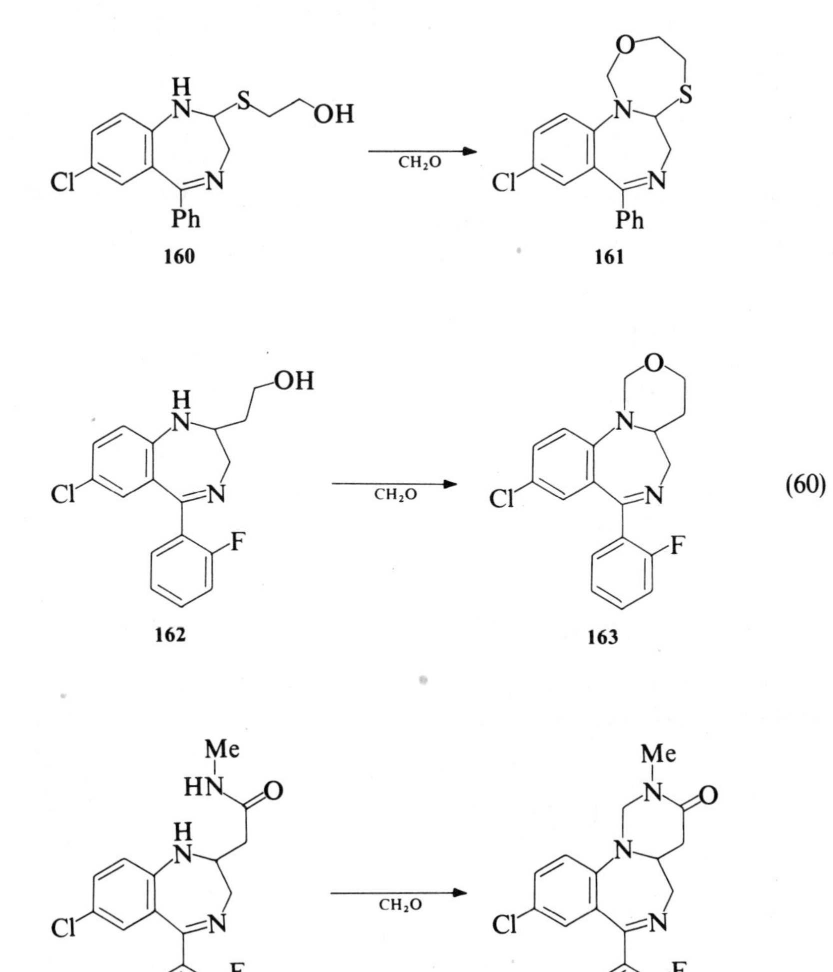


The amines **132** (R = MeO, EtO, *t*-BuO, Me) reacted with aliphatic and aromatic aldehydes to form unstable adducts, which oxidized in the presence of air or manganese dioxide to the imidazo[1,5-*a*][1,4]benzodiazepines **165** (Eq. 61).[92,93]

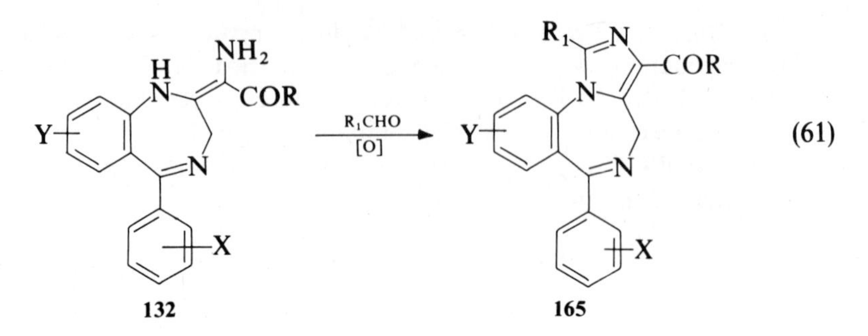


Acetylacetone added to **132** (R = MeO, X = 2-Cl, Y = 7-Cl) to form the enamine **166**, which was thermally converted to the imidazobenzodiazepine with elimination of acetone. This transformation was postulated to proceed through the dihydroimidazole **167** as indicated in Eq. 62.[93]

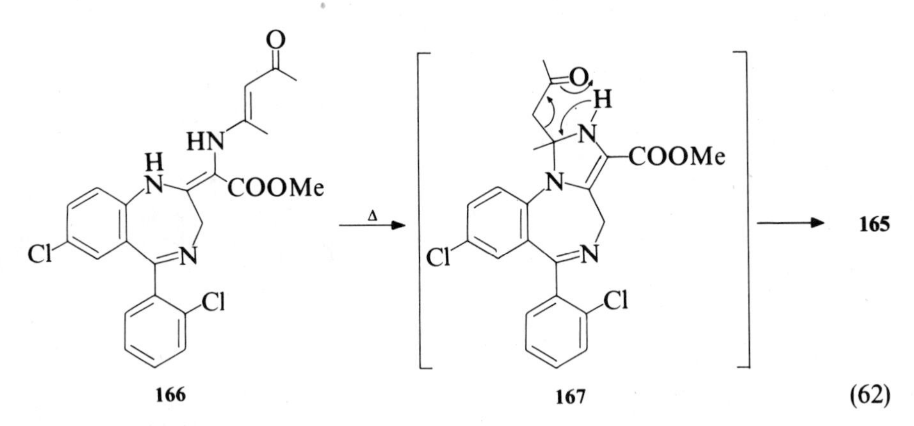


Lewis acid catalyzed addition of ethylene oxide to the imine function in medazepam (**12**) yielded the oxazolidine **168** (Eq. 63).[117]

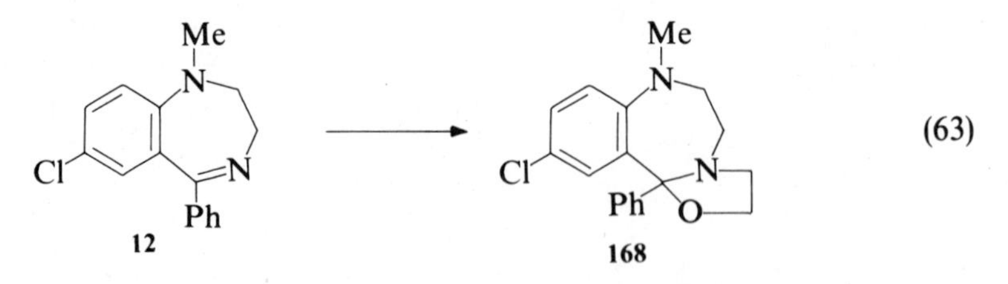


C. Acylation. Additions to Imine. Acylations at the 1-position nitrogen were carried out for many 2,3-dihydro-1*H*-1,4-benzodiazepines using acid anhydrides, acid chlorides, and isocyanates. The reagents employed were formic acid–acetic anhydride,[15,83] acetic anhydride[57] in combination with pyridine[15,83] or potassium carbonate,[68] acetyl chloride in dimethylformamide,[83] malonic acid chloride ethyl ester,[63] and the following isocyanates: methyl,[68] 2-chloroacetyl, 2-chloroethyl, 3-bromopropanoyl, 4-chlorobutanoyl, and 2-chloroacetylthio.[115]

The 2-aminomethyl derivatives **169** ($R_1$ = Me) were reacted with various acid anhydrides,[50,51,118] acid chlorides,[50,51,118] isocyanates,[118] and ethyl chloroformate[118] to yield the derivatized amines **170** (Eq. 64).

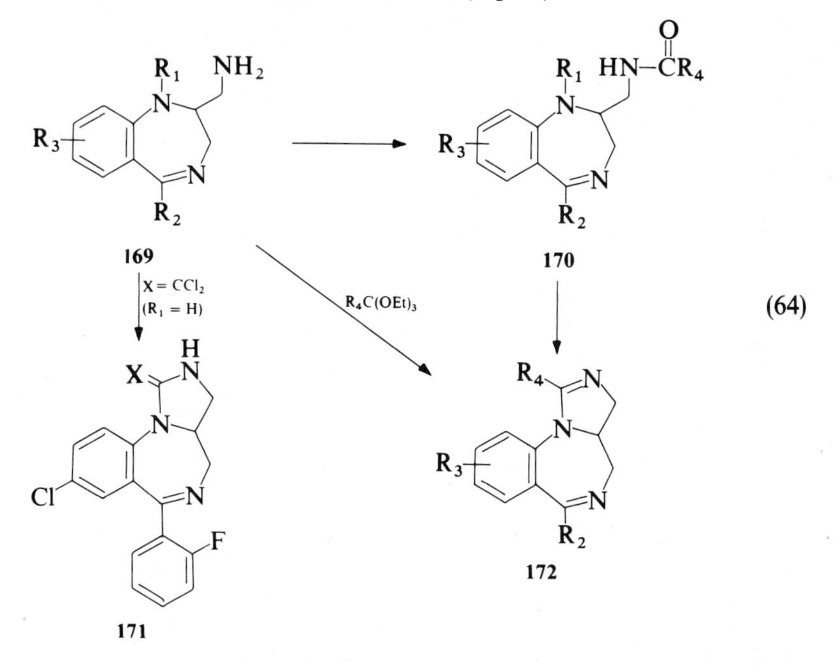


Compound **169** ($R_1$ = H, $R_2$ = 2-$FC_6H_4$, $R_3$ = 7-Cl) was monoacetylated on the primary amino group by acetic anhydride in methanol[65] and diacetylated in pyridine–acetic anhydride to give **170** ($R_1$ = Ac, $R_2$ = 2-$FC_6H_4$, $R_3$ = 7-Cl, $R_4$ = Me).[65]

Phosgene and thiophosgene reacted with **169** ($R_1$ = H, $R_2$ = 2-$FC_6H_4$, $R_3$ = 7-Cl) to give the imidazolidinone **171** (X = O) and the thione **171** (X = S) (Eq. 64).[119] Treatment of **169** ($R_1$ = H) with orthoesters afforded the imidazolines **172**.[65] These imidazolines were also obtained by cyclization of the acylated derivatives **170** with polyphosphoric acid at elevated temperature.[65]

Reaction of the amidine **171a** with dimethylformamide dimethylacetal in refluxing toluene led to the imidazole **172a** (Eq. 65).[66a]

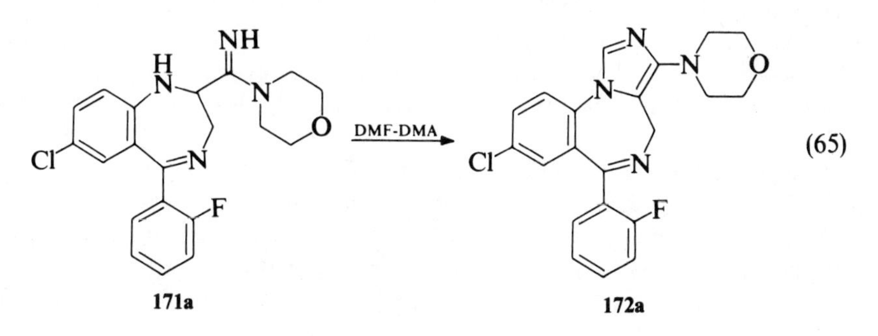

The diastereomeric 2-(hydroxybenzyl) derivatives **173** reacted with phosgene to form the corresponding oxazolidinones **174** (Eq. 66).[73] The stereochemistry of the compounds was not determined.

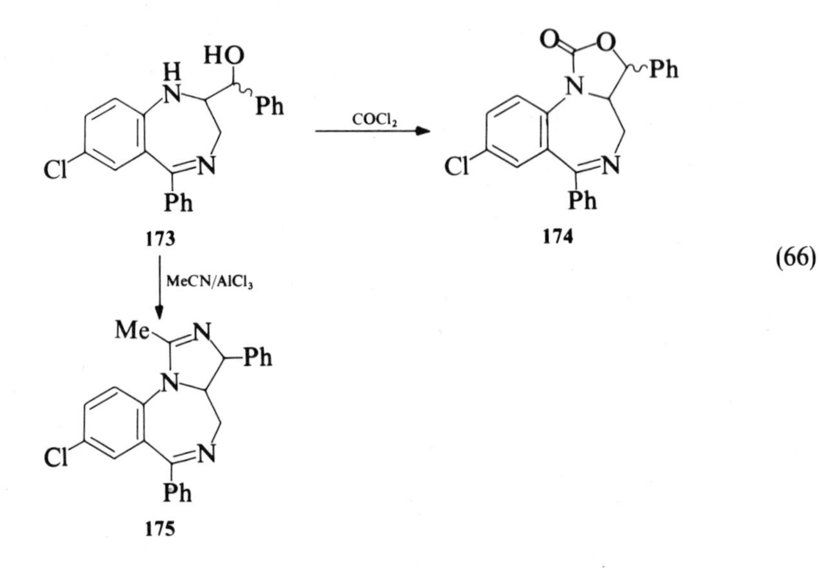


(66)

One of the diastereomers **173** was converted to the imidazoline **175** by treatment with acetonitrile and aluminum chloride.[73]

The enamines of structure **132** were transformed into imidazo[1,5-*a*] [1,4]-benzodiazepines **165** by acylation and subsequent ring closure or, better, by treatment with orthoesters or amide acetals (Eq. 67).[92,93]

Reaction of **132** (R = MeO, X = 2-Cl, Y = 7-Cl) with phosgene or thiophosgene led to the imidazole derivatives **176** (X = Cl, Z = O, S).[119] The 1-one **176** (X = Cl, Z = O) was also obtained by treatment of the appropriate **132** with carbonyldiimidazole.[108] Base-catalyzed cyclization of the urethane **177**, resulting from the acylation of **132** with ethyl chloroformate, provided still another access to the 1-ones **176** (X = F, Z = O) (Eq. 67).[119]

An intramolecular acylation on carbon with dehydration (Dieckmann-type cyclization) allowed the conversion of the malonyl derivative **178** to the tricyclic compound **179** (Eq. 68).[63]

The 2-hydroxymethyl substituted 7-chloro-2,3-dihydro-1-methyl-5-phenyl-1*H*-1,4-benzodiazepine was reacted with ethyl isocyanate to give the corresponding urethane.[52] Methyl isocyanate acylated the 2-imino compound **180** to the urea **181** (Eq. 69).[120]

Acylations at the 4-position nitrogen with subsequent addition of a nucleophile to the 5-position were observed. Medazepam (**12**) reacted with ethyl chloroformate in the presence of aqueous sodium carbonate solution to give benzophenone **184** via the 5-hydroxy compound **182**.[121] Intramolecular trapping by the nucleophile afforded tricyclic compounds. Thus, reaction of **12** with

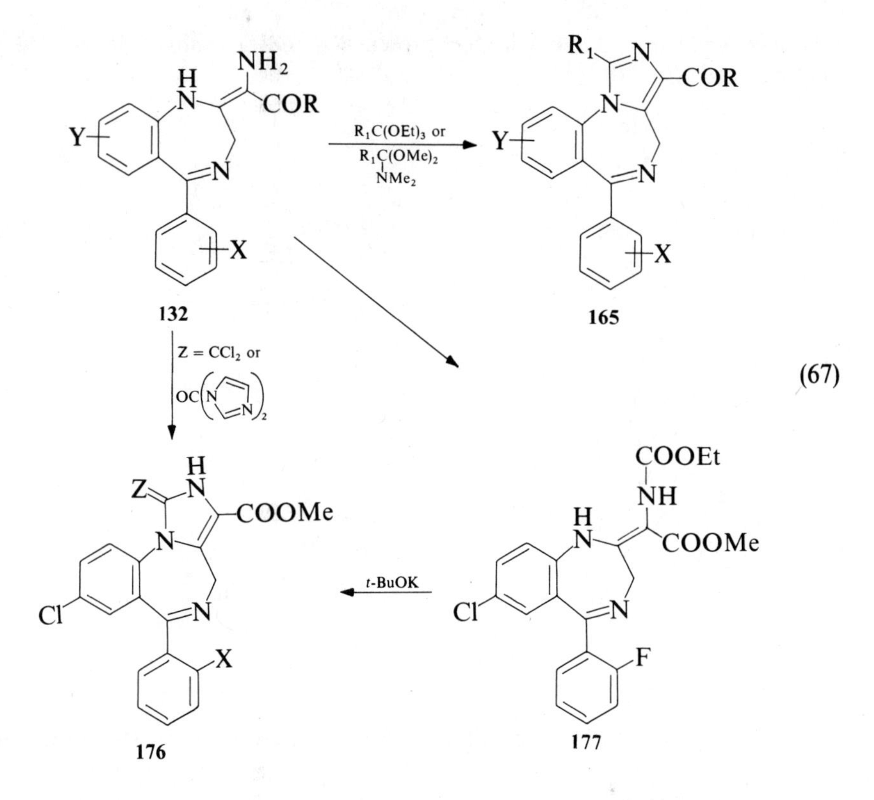

NH2
H
N
COR
Y
N
X
132
R1C(OEt)3 or
R1C(OMe)2
NMe2
R1
N
COR
N
Y
N
X
165
Z = CCl2 or
OC N N 2


(67)

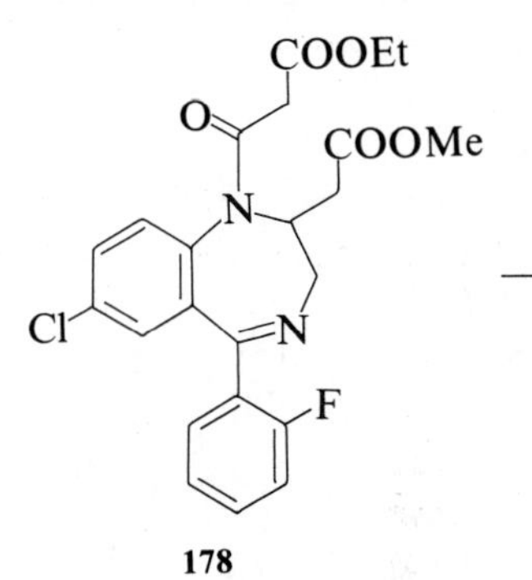

H
Z
N
COOMe
N
Cl
N
X
176


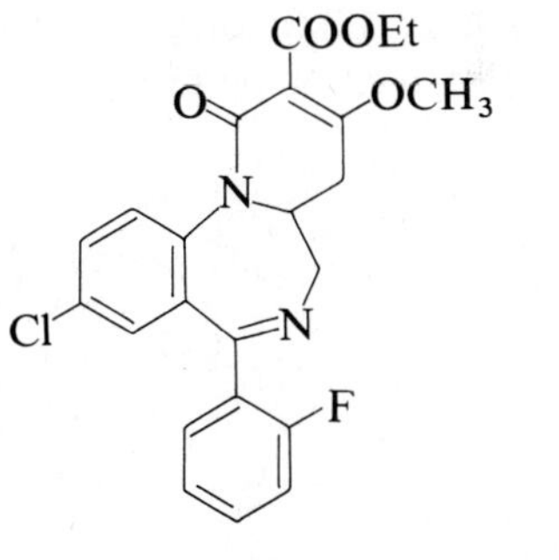

t-BuOK
COOEt
NH
H
N
COOMe
Cl
N
F
177


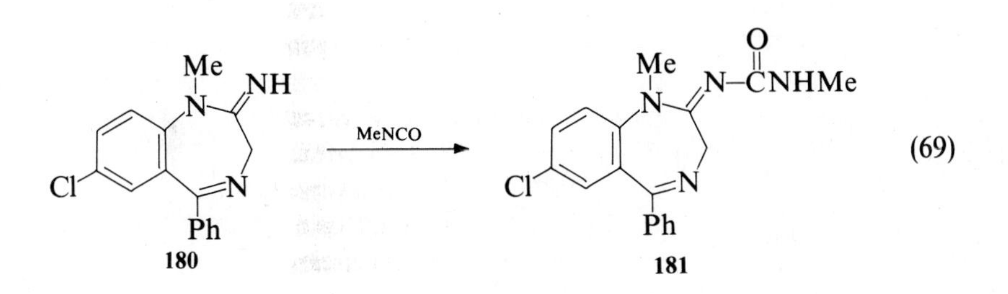

COOEt
O
COOMe
N
Cl
N
F
178
t-BuOK
COOEt
O
OCH3
N
Cl
N
F
179


(68)


Me
N
NH
Cl
N
Ph
180
MeNCO
O
Me
N
N–CNHMe
Cl
N
Ph
181


(69)

malonic acids (R = H, Et) in the presence of acetic anhydride yielded the lactones **183** (Eq. 70).[122]

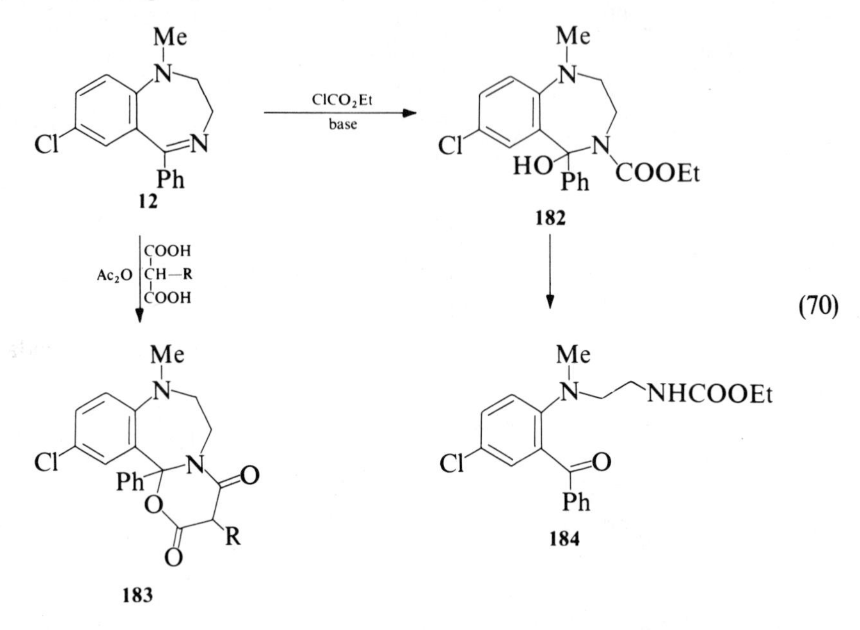


Related reactions are the addition of ketenes and nitrile oxides to the imine bond.

Gunda and Enebäck[123] reported the addition of the *N*-protected glycine **185** to medazepam (**12**) to form the *β*-lactam **186**. Phosphorus oxychloride was used to activate the carboxyl group of **185** (Eq. 71).

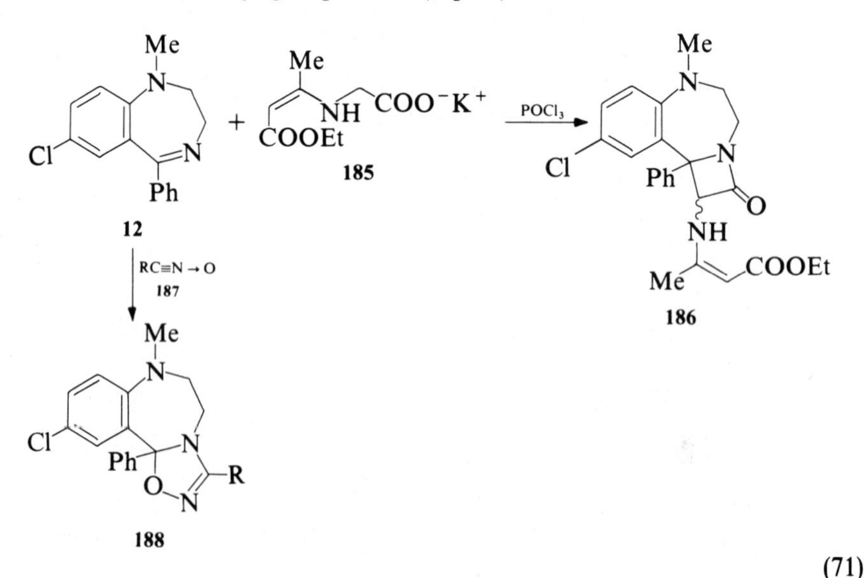

Dipolar additions of the nitrile oxides **187** (R = Ph, COPh) to medazepam yielding **188** were studied by Jaunin and coworkers (Eq. 71).[124]

The 4-oxides **189** (R = CHO, Ac) were converted to the 3-acetoxy derivatives **190** by treatment with refluxing acetic anhydride (Eq. 72).[83]

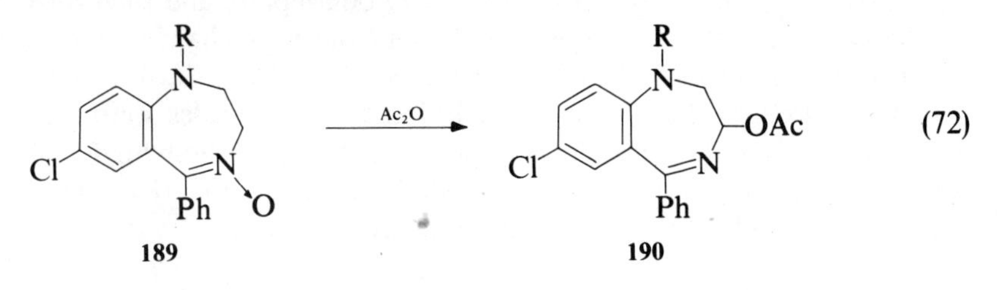


The 5-hydroxylamino compound **191** was reacted with carbonyldiimidazole to form the tricyclic derivative **192** (Eq. 73).[74a]

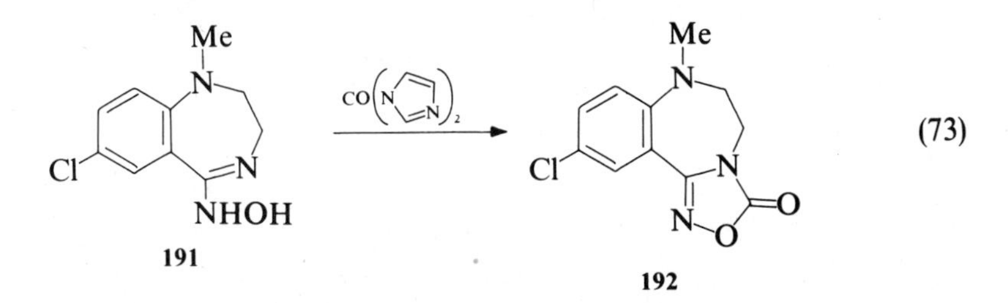


Acylation of the aromatic amino group in 5-(2-aminophenyl)-substituted and 7-amino-substituted compounds was carried out by standard methods.[7,43]

A 1-amino-substituted benzodiazepine was acylated by acetic anhydride.

### 1.5.1.6. Sulfonation

Sulfonations at the 1-position were effected by tosyl and mesyl chloride.[125] Selective tosylation of the 2-aminomethyl group was reported for **193** ($R_1$ = H, $R_2$ = $H_2NCH_2$, X = F).[126]

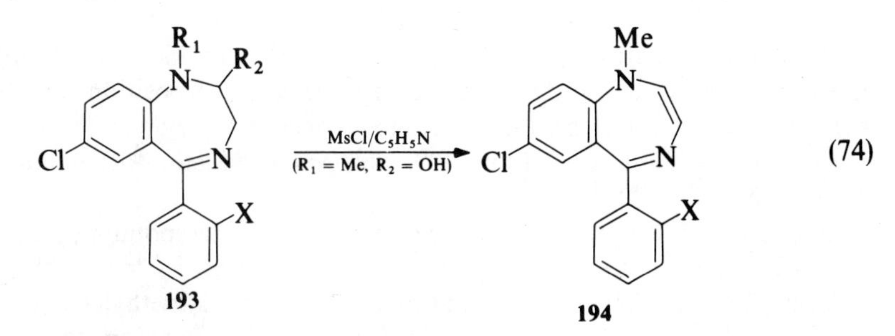


The 2-hydroxy derivatives **193** ($R_1$ = Me; $R_2$ = HO; X = H, F) were treated with mesyl chloride and pyridine to form the 1*H*-benzodiazepines **194** (Eq. 74).[60,127]

## *1.5.2. Reactions with Nucleophiles*

### 1.5.2.1. Reduction

The reduction of 2,3-dihydro-1*H*-1,4-benzodiazepines and their 4-oxides to 2,3,4,5-tetrahydro derivatives is discussed in the next chapter in the section describing the synthesis of such compounds.

The 2-nitromethylene compounds **195** and their 4-oxides were reduced by hydrogen with Raney nickel as catalyst to the 2-aminomethyl derivatives **196**.[65,92] Sodium borohydride in ethanol converted **195** ($R_1 = H$, $R_2 = 2\text{-}FC_6H_4$, $X = Cl$) to the bridged nitro compound **197**.[65]

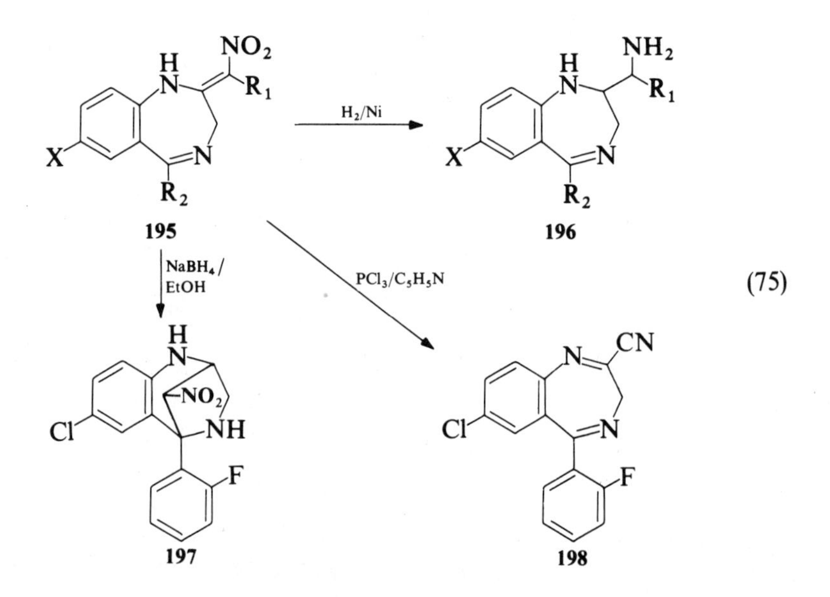


This transformation involves reduction of the double bond followed by intramolecular attack of the α-nitro carbanion at the 5-position. Treatment of the same nitro compound **195** with phosphorus trichloride in the presence of pyridine led to the nitrile **198** in 30% yield via a reduction accompanied by a dehydration (Eq. 75).[65]

Reductions of aromatic nitro functions to the corresponding amines were generally carried out with Raney nickel and hydrogen[2–5,7,42] or with stannous chloride. This catalyst was also used to hydrogenate a 3-methylene benzodiazepine to the corresponding 3-methyl derivative[99] and a 2-azidomethyl derivative to the corresponding amine,[50] to remove a 1-nitroso moiety,[73] and to reduce the *N*-hydroxyamidine **199** to the corresponding amidine **171a** (Eq. 76).[66a]

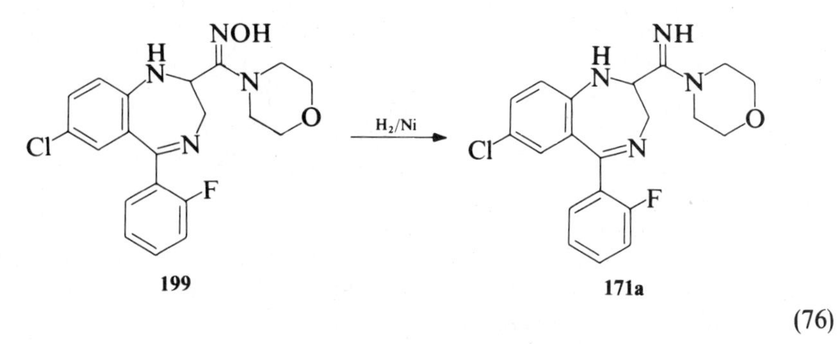


(76)

Phosphorus trichloride was most commonly used to reduce 4-oxides to the corresponding parent compounds.[21-23,91,96,99] Hexachlorodisilane was used as well.[66b] The exocyclic double bond in the esters **200** (R = H, Me; X = F, Cl) was reduced by triethylsilane in trifluoroacetic acid to the 2-acetic acid esters **201** (R = H, $CH_3$; X = F, Cl) (Eq. 77).

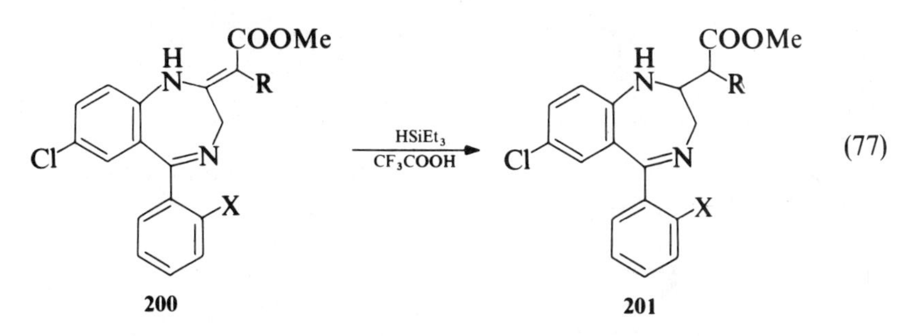


(77)

Lithium aluminum hydride converted the ester **201** (R = H, X = F) to the corresponding alcohol[63] and reduced a 2-carboxamide to the aminomethyl derivative **196** ($R_1$ = H, R = 2-$FC_6H_4$, X = Cl).[92] This reagent reduced the 1-nitroso compound to the 1-amino analog.[83c] A 7-acetyl group was reduced by sodium borohydride to the hydroxyethyl moiety.[104]

### 1.5.2.2. Reactions with Oxygen and Sulfur Nucleophiles

A. Hydrolysis. The imine function in 2,3-dihydro-1*H*-1,4-benzodiazepines **202** was cleaved by aqueous hydrochloric acid to give the benzophenones **203** as hydrochloride salts (Eq. 78).[2,15] The free base of **203** (R = H, Me; X = $NO_2$) was stable enough to be isolated,[2] while **203** (R = H, X = Cl) cyclized back to the benzodiazepine spontaneously upon liberation of the base.[2] The 7-trifluoromethyl analog **202** (R = H, X = $F_3C$) yielded the carboxylic acid **203** (X = HOOC) together with the decarboxylated compound. These were recyclized to the corresponding benzodiazepines **202**.[2] An analogous hydrolytic ring opening occurred when the 4-quaternary salt was subjected to aqueous base.[114] The acetyl group at the 1-position was removed by aqueous sulfuric acid.[34]

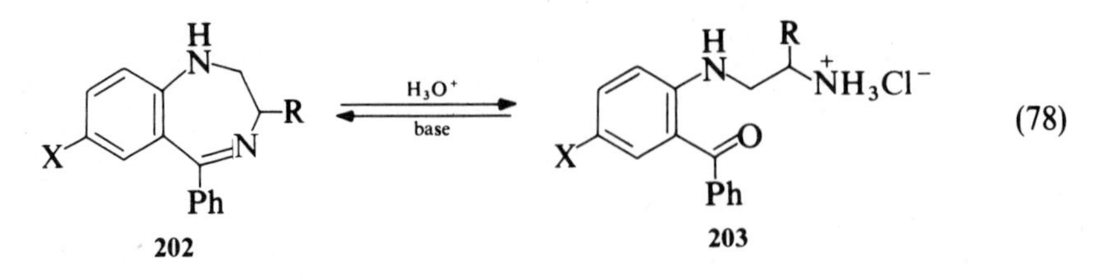


(78)

Acid hydrolysis converted 2-cyano derivatives to the carboxamides.[68,70] Treatment of the 2-cyanomethylene compounds **204** with concentrated sulfuric acid yielded the corresponding amides **205** (Eq. 79).[62,98]

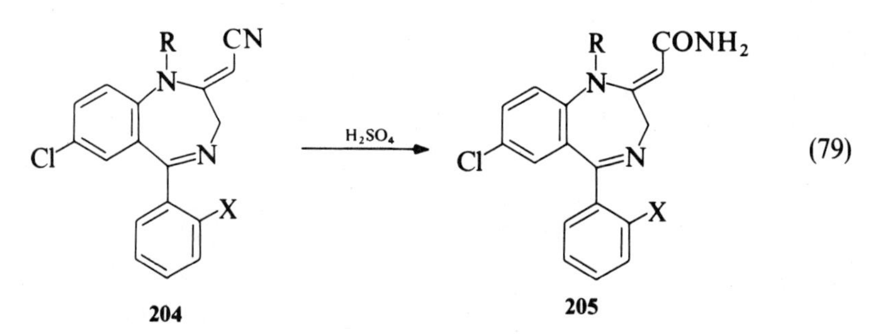


(79)

Mild acid hydrolysis (20% HCl at room temperature) converted the 3-amino derivatives **69** to the 3-hydroxy analogs **206** (Eq. 80).[32]

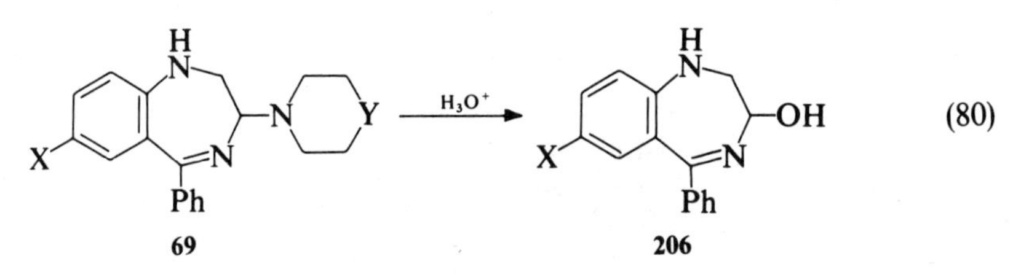


(80)

The 5-(3-indolyl)benzodiazepine **207** underwent rearrangement during acid hydrolysis to give the quinoline **210**. The reaction, as outlined in Eq. 81, was proposed to proceed via intermediates **208** and **209**.[128]

Alkaline hydrolysis of the 2-malonylidene compounds **211** ($R_2$ = $COOR_1$, Ac) to the corresponding acetylidene derivatives **211** ($R_2$ = H) was mentioned in Section 1.1.3. Further hydrolysis and decarboxylation led to the 2-methyl-3*H*-1,4-benzodiazepine **212** (Eq. 82).[63]

The benzodiazepine 2-acetic acid methyl ester **201** (R = H, X = F), a similar ethyl ester,[118] and an iminoether[67] were hydrolyzed by hydroxide in methanol to the corresponding carboxylic acids. Mild treatment of the 3-acetoxy derivatives **213** (R = Ac, Me) with hydroxide in methanol or with aluminum oxide gave the corresponding alcohols,[83] while more vigorous conditions (refluxing in methanolic sodium hydroxide for 3 hours) led to the indole 214.[110] The indole-2-carboxaldehyde **216** was formed from the 3-hydroxy compound **215** under

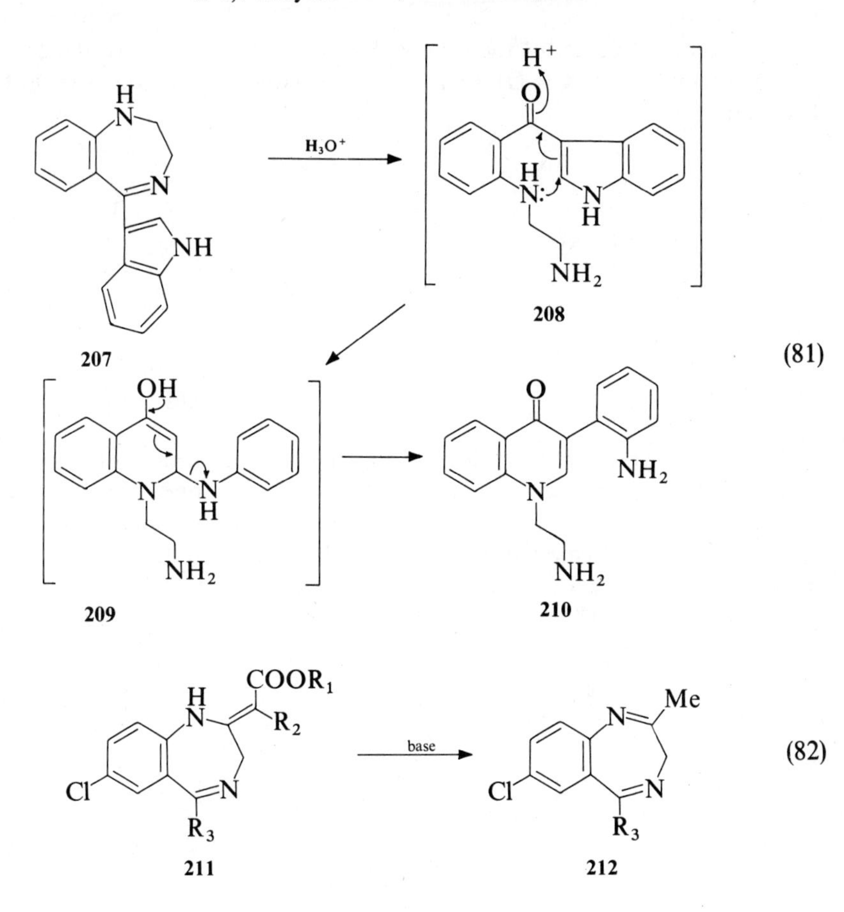


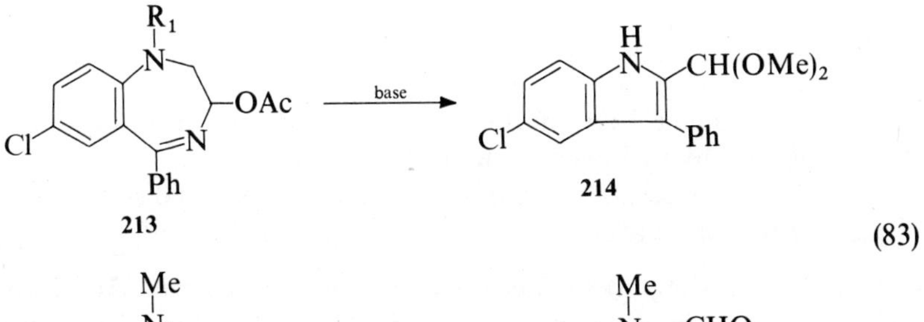


(83)

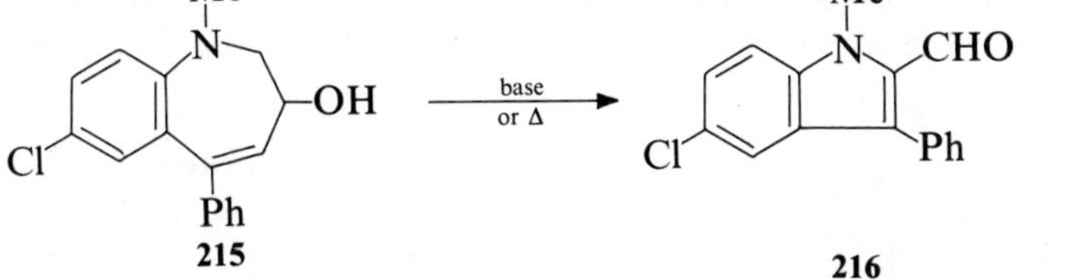


comparable conditions (Eq. 83).[110] The 4-oxide **189** (R = Ac) was deacetylated by refluxing in ethanolic sodium hydroxide.[84] Displacement of the chloride in 2-chloromethyl benzodiazepines by hydroxide was reported.[47–49] According to the patent literature,[129] the dihalides or ditosylates **217** gave the oxazinobenzodiazepines **218** (X = O) upon treatment with sodium hydroxide in tetrahydrofuran (Eq. 84).

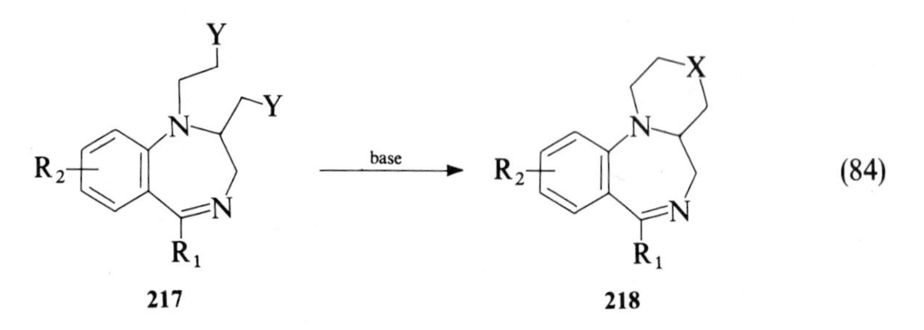


B. Other Oxygen Nucleophiles. 2-Cyanobenzodiazepines **219** ($R_1$ = H, Me; X = H, F) were converted to the iminoesters **220** by treatment with methanolic hydrogen chloride[67] or methanol in the presence of cyanide (Eq. 85).[70,71b] Further reaction with methanolic hydrogen chloride led to the 2-carboxylic acid methyl ester,[67] which was also prepared by the esterification of the corresponding 2-carboxylic acid.[71a]

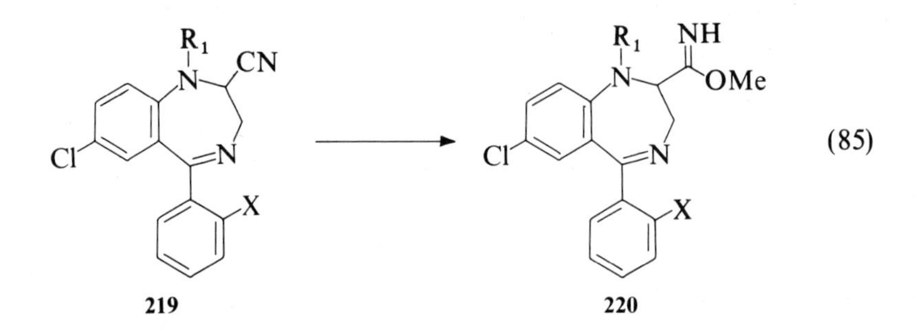


A benzodiazepine 2-acetonitrile was similarly transformed into the 2-acetic acid ethyl ester by hydrogen chloride in ethanol.[118]

The chloride in several 2-chloromethyl derivatives was displaced by oxygen nucleophiles such as alkoxides,[47–49,54,102] phenoxide,[52] and acetate.[47]

C. Sulfur Nucleophiles. The 2-hydroxy benzodiazepine **221** reacted with hydrogen sulfide in acetic acid to yield the bridged sulfide **222** (Eq. 86).[68]

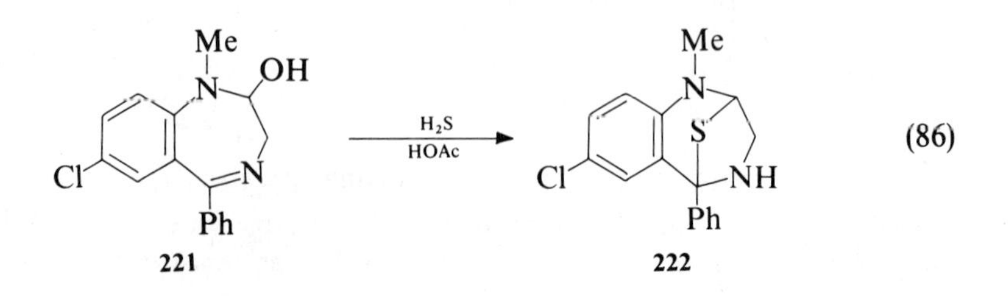

The thiazinobenzodiazepines **218** (X = S) were reported to be synthesized from the dihalides or ditosylates **217** by treatment with sodium hydrogen sulfide (Eq. 84).[129]

### 1.5.2.3. Reactions with Nitrogen Nucleophiles

The halogen of 2-halomethyl derivatives was displaced by nitrogen nucleophiles such as sodium azide,[50,51] potassium phthalimide,[47,50] morpholine,[47] piperidine,[48] and other amines.[50] Reaction of the dihalide **217** with ammonia or methylamine led to the pyrazinobenzodiazepines **218** (X = NH, NMe).[129] The 2-methoxy functionality was displaced by piperidine.[67]

The 5-ethoxy moiety in **223** (X = EtO, Y = Cl) was replaced by various *N*-nucleophiles: hydroxylamine,[74a,b] methoxyamine,[74b] hydroxylamine *O*-acetic acid ethyl ester,[74b] ethanolamine,[74b] 2-(diethylamino)ethylamine,[74a,b] 4-phenylpiperazine,[74b] and hydrazine derivatives.[74a,75,76,130] In the last case, cyclizations to **224**[74a,130] and **225** were observed (Eq. 87).[74a,75] Similar reactions were carried out with the 5-methylthio compound **223** (X = MeS, Y = H).[55]

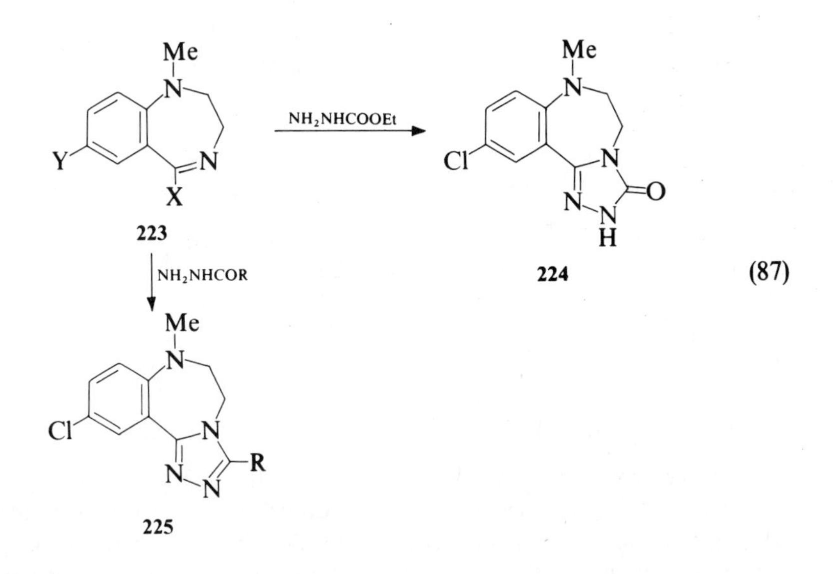


(87)

Benzodiazepine-2-acetic acid esters or acids were converted to the corresponding amides either by direct aminolysis[63] or via activation of the carboxyl group.[118] Reaction of the 2-acetylidene derivatives **226** (X = H, F) with methanolic ammonia in the presence of ammonium chloride at 100°C yielded the spiro compounds **227** and the pyrroloquinolines **228** as the major products (Eq. 88).[63]

Cyclization of the nitrile **229** by refluxing in methanol and concentrated aqueous ammonia led to a mixture of **230** (R = H) and **230** (R = OH) in 11.5 and 33% yield.[68a] A better conversion of **229** to **230** (R = H) was achieved by

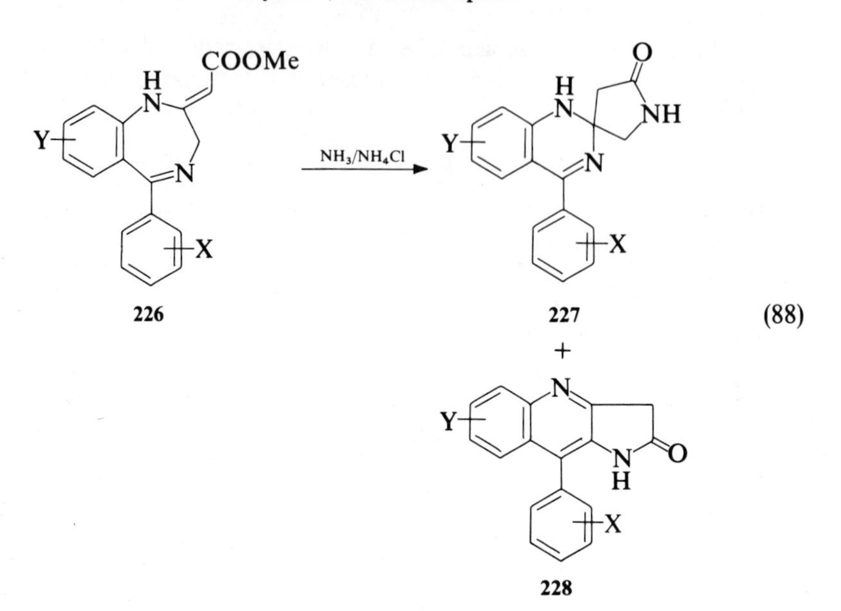


treating first with aqueous hydrochloric acid and subsequently with ammonia (Eq. 89).

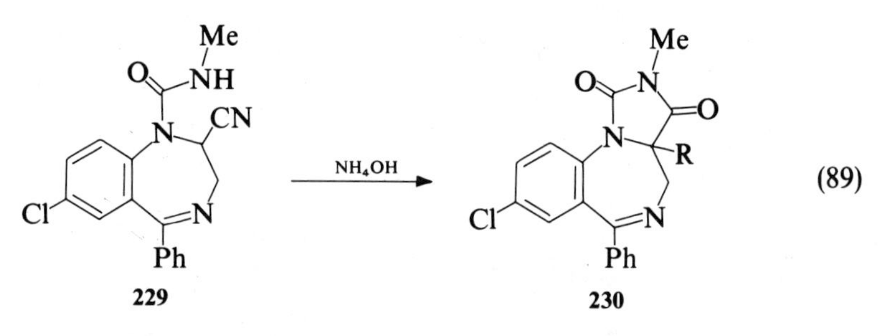


### 1.5.2.4. Reactions with Carbon Nucleophiles

Displacement of the chloride in 2-chloromethyl-2,3-dihydro-1*H*-1,4-benzodiazepines by cyanide led to the 2-acetonitriles.[47] Acetone cyanohydrin or sodium cyanide in acetic acid reacted with 2-hydroxy benzodiazepines to give the 2-cyano analogs.[70,71] The 7-cyano derivative **232** was obtained from the

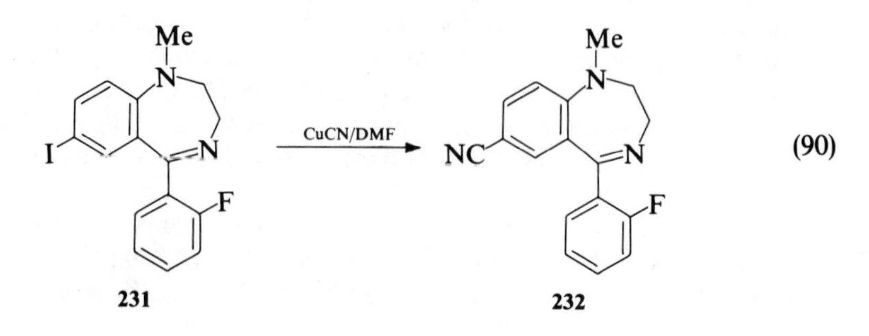

7-iodo compound **231** by means of cuprous cyanide in dimethylformamide (Eq. 90).[104] The nitrile was further reacted with methyllithium to yield the 7-acetyl compound.[104,131] A 7-acetyl derivative was prepared also by reacting the 7-diazonium salt with acetaldehyde semicarbazone.[132]

#### 1.5.2.5. Miscellaneous Reactions

Attempted cyclization of the acetyl derivative **233** by base led to the 2-cyanoindole **234** (Eq. 91).[68a]

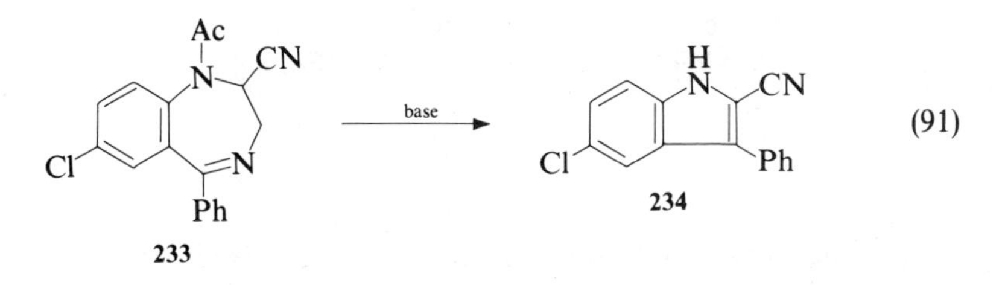


Treatment of the 1-sulfonates **235** (R = Me, 4-$MeC_6H_4$) with sodium hydride in dimethylformamide resulted in ring opening to the vinylimines **236** (Eq. 92).[125]

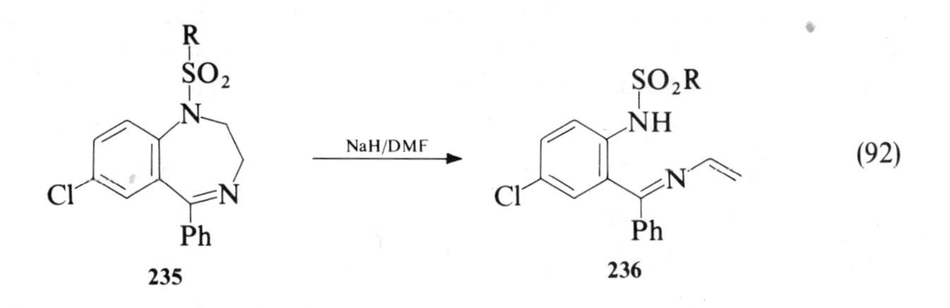


### *1.5.3. Photo and Thermal Reactions*

The photochemical rearrangement of the nitrones **91** ($R_1$ = Ac, CHO) to the oxaziridines **90** was described by Metlesics and coworkers (Eq. 32).[83,84] The reverse reaction was accomplished by heat.[83,84]

Irradiation of the 1-alkyl derivatives **237** led to the indoles **238** in a preparatively useful reaction (Eq. 93).[133] The 1-alkyl group is required, since compounds **237** with R = H or Ac were photostable. A *p*-nitro group on the 5-phenyl substituent also made the benzodiazepine inert to this transformation.[133]

The photolysis proceeded well in methanol, glacial acetic acid, or acetic anhydride but poorly in benzene and not at all in dry acetonitrile.

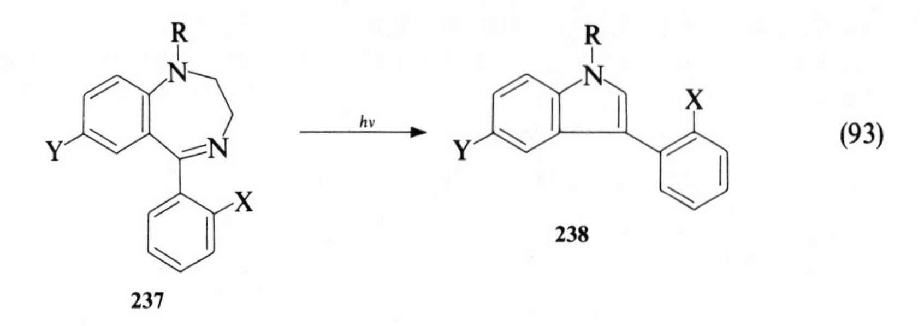


The 3-hydroxy compound **215** was rearranged thermally in nearly quantitative yield to the indole-2-carboxyaldehyde **216** (Eq. 83).

### 1.6. Spectral Data

References to spectral data are given in the tables (Section 5). The ultraviolet, infrared and proton-nmr spectra of 2-chloromethyl and 2-hydroxymethyl benzodiazepines were thoroughly compared with those of the isomeric diazocines.[134]

Romeo and coworkers[135] studied the conformation of medazepam and 1-desmethyl medazepam by proton-nmr spectroscopy using a lanthanide shift reagent. Computer-simulated lanthanide shifts were found to be consistent with complexation of the imine nitrogen. Two rapidly interconverting pseudoboat conformers **a** and **b** were observed in deuterochloroform at room temperature,

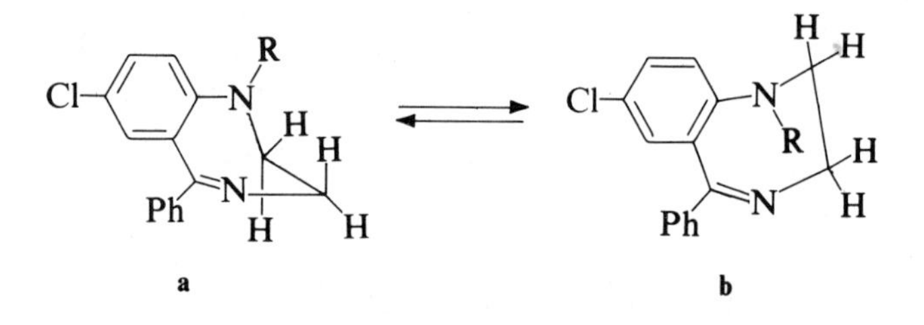


An X-ray analysis of medazepam[136] revealed that the seven-membered ring is in a twist–boat conformation similar to that of the 2-ones.

## 2. 2,5-DIHYDRO-1*H*-1,4-BENZODIAZEPINES

The parent compound **239** is still unknown. The 4-desoxy analog of **242** (R = H, X = Cl) was isolated from the electrochemical reduction of oxazepam in basic medium.[137] The 3-methyl-substituted 4-oxides **242** were prepared by oxidation of the hydroxyamines **241** (X = H, Cl) with mercuric oxide or by manganese dioxide (Eq. 94).[64,99]

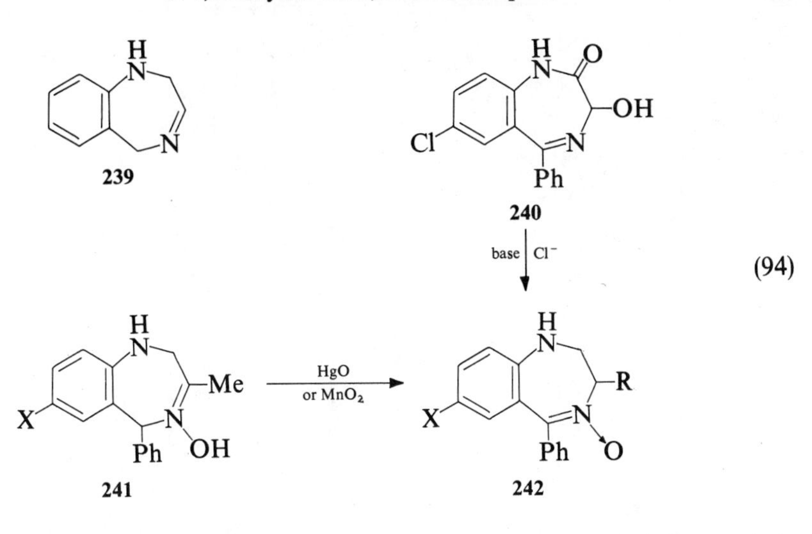


Two amidines (**244**: R = Ph, 4-$MeC_6H_4SO_2$) were prepared by catalytic hydrogenation of the 5*H*-benzodiazepines **243** over platinum (Eq. 95).[99c]

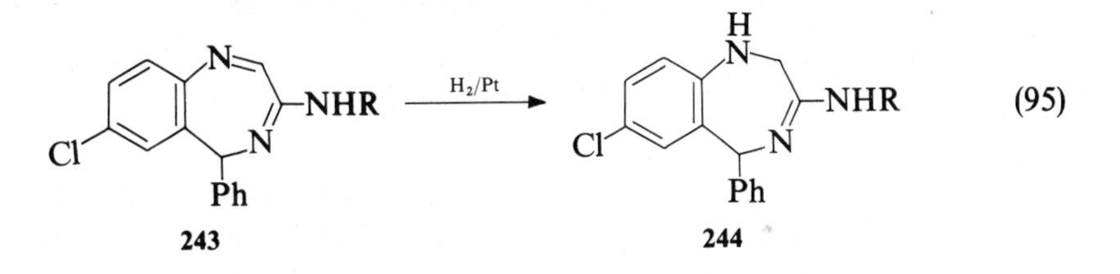


## 3. 4,5-DIHYDRO-1*H*-1,4-BENZODIAZEPINES

Only one substituted representative of the parent compound **245** has been reported in the literature.

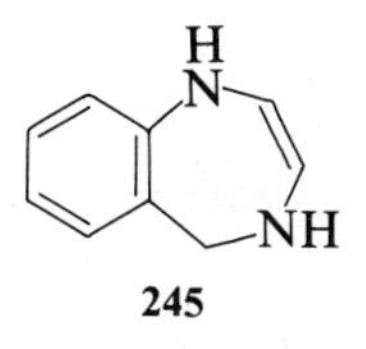


Deyrup and Gill[138] obtained **247** in 62% yield by reaction of **246** with a catalytic amount of potassium *t*-butoxide in refluxing diglyme for 10 hours. The structure was assigned on the basis of spectral data and chemical evidence, which consisted in the facile oxidation of **247** to **249** by methanolic hypochlorite. The chloride **248** was postulated as an intermediate (Eq. 96).

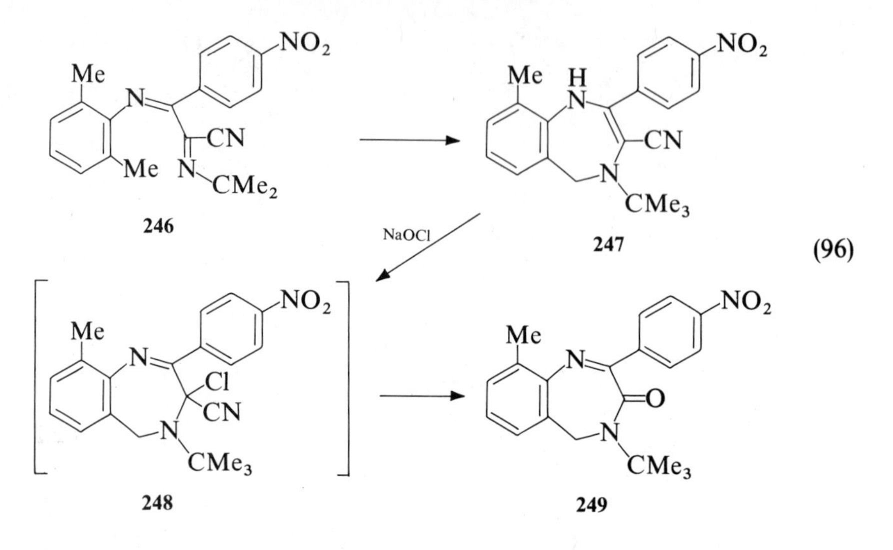


(96)

# 4. 4,5-DIHYDRO-3*H*-1,4-BENZODIAZEPINES

## 4.1. Synthesis

While the parent compound **250** has not been synthesized, several 2-amino derivatives **252** have been prepared by catalytic hydrogenation of the 4,5-double bond in 3*H*-1,4-benzodiazepines **251** using platinum and palladium as catalysts.[139] In the latter case dechlorination at the 7-position was a concomitant reaction and gave **252** (X = H) (Eq. 97).[139]

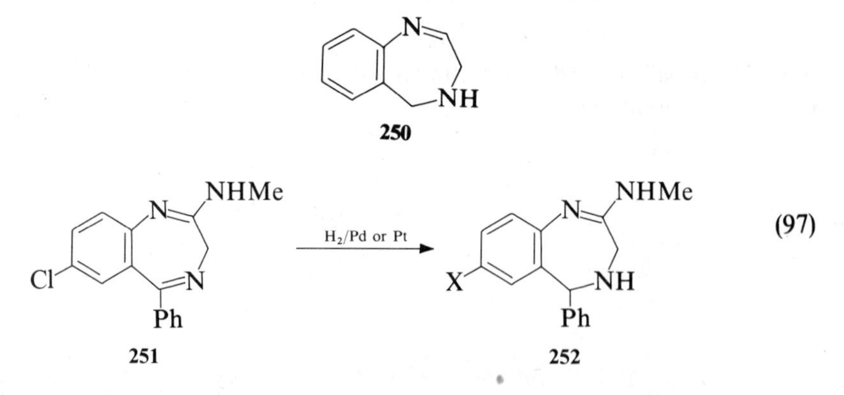


(97)

The addition of phenylmagnesium bromide to the nitrones **253** (R = NHMe, alkoxy) opened a route to 4-hydroxy analogs **254** (Eq. 98).[140–142]

Compound **252** (X = Cl) was also prepared by reaction of the 2-one with methylamine and titanium tetrachloride.[143]

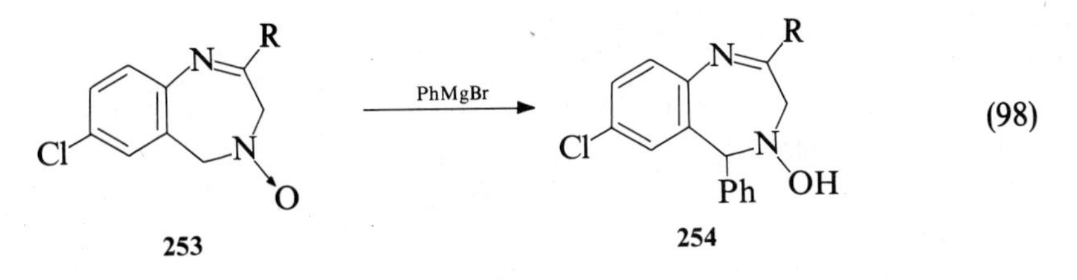


## 4.2. Reactions

For a discussion of the oxidation of the hydroxyamine **254** (R = NHMe) to chlordiazepoxide and its 5*H* tautomer,[141,144] and the dehydration of **254** (R = alkoxy) to the 5*H*-benzodiazepines,[141,142] see Chapter V (Sections 2.2.1.3, 3.1, and 2.5.1).

An attempt to catalytically hydrogenate the 1,2-imine bond in **252** (X = Cl) failed.[145] Electrochemical reduction gave the dihydroquinazoline **258** in 80% yield.[145] Its formation may be rationalized by the intermediates **255–257**, as indicated in Eq. 99.

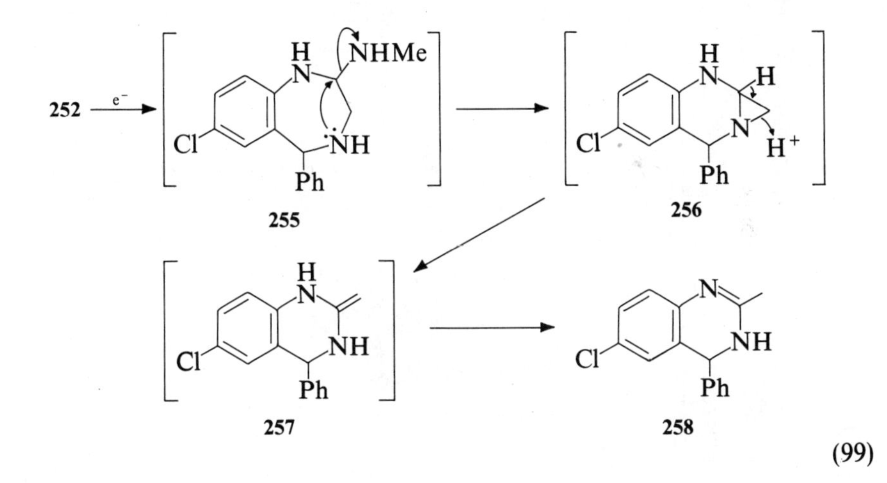

# 5. TABLES OF COMPOUNDS

TABLE VI-1. 2,3-DIHYDRO-1$H$-1,4-BENZODIAZEPINES

| Substituent | mp (°C); [bp (°C/torr)] | Solvent of Crystallization | Yield (%) | Spectra | Refs. |
|---|---|---|---|---|---|
| | | 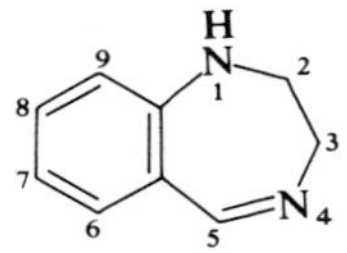 | | | |
| None | 244–246 | MeOH | | ir, uv | 56 |
| | *Monosubstituted* | | | | |
| 5-(2-Benzofuryl) hydrochloride | 225–226d | MeOH/$Et_2O$ | | | 128b |
| 5-(5-Br-3-Indolyl) | 217–219 | EtOH/$H_2O$ | | | 128b |
| 1-(2,3-Dihydro-1$H$-1,4-benzodiazepin-1-yl) | 178–180 | Sublimed | 12 | ir, ms, pmr | 68a |
| 5-(3-Indolyl) | 217–219 | EtOH/$H_2O$ | | | 14 |
| Hydrochloride | 180–185d | | | | 128b |
| Methiodide | 259–262 | MeOH/$Et_2O$ | | | 128b |
| 5-(1-Me-3-Indolyl) | 144–149 | $CH_2Cl_2$/Petr ether | | | 128b |
| 5-Ph | 145–147 | EtOH | 77 | | 2, 8, 35 |
| 5-(2-$H_2N$-5-$ClC_6H_3$) hydrochloride | 230–260 | $Et_2O$/MeOH | | | 14c |
| 5-(2-$ClC_6H_4$) | 165–167 | PhH/Petr ether | | | 25 |
| 5-(2-$FC_6H_4$) | 146–150 | $Et_2O$ | | | 103 |
| 5-(2-Pyridyl) | 152–154 | $CH_2Cl_2$/Hexane | 6, 11 | | 3a |
| 5-(2-Pyrimidyl) | 171–173 | PhH/Hexane | | | 3b |
| 5-(2-Thiazolyl) | 158–160 | $Me_2CO$/Hexane | | | 101b |

***1,5-Disubstituted***

| | | | | | |
|---|---|---|---|---|---|
| 1-Ac-5-Ph | 149–150 | PhH/Cyclohexane | 81 | | 35 |
| 1-Et-5-Ph | [140–150/0.5] | | 49 | | 35 |
| 1-Me-5-$NH_2$ hydroiodide | 157 | $CHCl_3$/PhMe | 80 | | 55 |
| 1-Me-5-(4-Benzylpiperazino) dimaleate | 173–174 | MeOH | 40 | | 55 |
| 1-Me-5-$ClCH_2$ hydrochloride | 170–180d | *i*-PrOH/$Me_2CO$ | | | 45b |
| 1-Me-5-$Cl_2CH$ | 52–54 | $Et_2O$/Hexane | | | 45b |
| 1-Me-5-($Et_2NCH_2CH_2NH$) dipicrate | 174 | MeCN | 73 | | 55 |
| 1-Me-5-($Et_2N(CH_2)_3NH$) dipicrate | 171 | MeCN | 58 | | 55 |
| 1-Me-5-(4-EtOCO-Piperazino) | 96–97 | Heptane | 50 | | 55 |
| 1-Me-5-$H_2NNH$ | 148 | MeOH | 72 | ir, pmr | 76 |
| 1-Me-5-(4-HO$(CH_2)_2$-Piperazino) dimaleate | 166 | MeOH/$Et_2O$ | 73 | | 55 |
| 1-Me-5-(2-Indolyl) | 158–160 | $Et_2O$ | | | 128b |
| 1-Me-5-(4-Me-Piperazino) dipicrate | 238 | MeCN | 46 | | 55 |
| 1-Me-5-(3-Me-Piperidino) picrate | 131 | EtOH | 78 | | 55 |
| 1-Me-5-(4-Me-Piperidino) picrate | 197–198 | EtOH | 83 | | 55 |
| 1-Me-5-[2-(1-Me-2,3-Dihydro-1*H*-1,4-benzodiazepin-5-yl)hydrazino] | 221 | $Me_2CO$ | | ir, ms, pmr | 76 |
| 1-Me-5-MeS | 66 | Petr ether | 69 | ir, pmr | 55 |
| 1-Me-5-Morpholino | 88 | Petr ether | 85 | | 55 |
| 1-Me-5-Ph | 111–113 | | | | |
| | [150/0.5] | Hexane | 39 | | 8, 46 |
| Picrate | [157] | MeOH | | | 46 |
| 1-Me-5-(2-$ClC_6H_4$) | 97–99 | Hexane | | | 99c |
| 1-Me-5-(4-$ClC_6H_4$) | 151–153 | MeOH | | | 109 |
| 1-Me-5-(2-$FC_6H_4$) | 114–117 | EtOAc/Petr ether | | | 104 |
| 1-Me-5-(2-$F_3CC_6H_4$) | 83–85 | Hexane | | | 109 |
| Hydrochloride | 250–252 | EtOH/$Et_2O$ | | | 86, 109 |
| 1-Me-5-PhNHNH | 147–149 | EtOH | 70 | ir, pmr | 76 |
| 1-Me-5-(4-Ph-Piperazino) | 108–109 | EtOAc | 45 | | 55 |
| 1-Me-5-Piperidino | 101 | Petr ether | 56 | | 55 |
| 1-Me-5-Pyrrolidino | [120–122/0.2] | | 70 | | 55 |

TABLE VI-1. —(contd.)

| Substituent | mp (°C); [bp (°C/torr)] | Solvent of Crystallization | Yield (%) | Spectra | Refs. |
|---|---|---|---|---|---|
| 1-Me-5-Trichloroethylene | 57–59 | Hexane | | | 45b |
| Hydrochloride | 185–190d | i-PrOH/$Et_2O$ | | | 45b |
| 1-Ph-5-Me | 91–92 | Hexane | | | 128b |
| 1-$F_3CCH_2$-5-(2-$FC_6H_4$) | 91–93 | $Et_2O$/Hexane | 66 | | 106 |
| ***1,7-Disubstituted*** | | | | | |
| 1-Me-7-Cl hydrochloride | 227–230 | MeOH/$Et_2O$ | | | 78 |
| ***2,5-Disubstituted*** | | | | | |
| 2-$H_2NCH_2$-5-Ph | Oil | | | | 49 |
| 2-$H_2NCH_2$-5-(2-$FC_6H_4$) | Oil | | | | 14c |
| Dimaleate hemihydrate | 147–150 | MeOH/$Et_2O$ | | | 14c |
| 2-Me-5-Ph | 109–110 | $Et_2O$/Petr ether | 85–90 | | 25 |
| Hydrochloride | 249–252 | MeOH/$Et_2O$ | | | 99c |
| 4-Oxide | 148–151 | PhH | | | 99a, b |
| ***3,3-Disubstituted*** | | | | | |
| 3-MeO-3-Ph | 144–145 | Cyclohexane | | | 99c |
| ***3,5-Disubstituted*** | | | | | |
| 3-Me-5-Ph | 165–167 | | 10–15 | | 25 |
| ***3,7-Disubstituted*** | | | | | |
| 3-EtO-7-Br | 138–140 | $Et_2O$ | | | 99c |
| 3-(2-Propynyloxy)-7-Br | 120–122d | $CH_2Cl_2$/Petr ether | | | 99c |
| ***5,7-Disubstituted*** | | | | | |
| 5-(3-Indolyl)-7-Br | 225–228 | $CHCl_3$/Hexane | | | 128b |
| 5-(3-Indolyl)-7-Cl | 243–245 | $CHCl_3$/$CH_2Cl_2$ | | | 128b |

| | | | | | |
|---|---|---|---|---|---|
| 5-(3-Indolyl)-7-$F_3C$ | 231–233 | $EtOH/H_2O$ | | | 128b |
| 5-(5-Isothiazolyl)-7-$NO_2$ | 198–200 | $CH_2Cl_2/MeOH/Hexane$ | | | 7 |
| 5-(1-Me-2-Imidazolyl)-7-$NO_2$ | 278–280d | MeOH | | | 7 |
| 5-(1-Me-5-Pyrazolyl)-7-$NO_2$ | 199–201 | $CH_2Cl_2/EtOAc$ | | | 7 |
| 5-Ph-7-$H_2N$ dihydrochloride | > 250 | | | | 2 |
| 5-Ph-7-Br | 173–175 | | | | 38 |
| 5-Ph-7-HOOC hydrochloride | | | | | |
| Hydrate | 315–316d | $H_2O$ | 62 | | 2 |
| 5-Ph-7-Cl | 174–176 | $CH_2Cl_2$/Petr ether | 50, 71, 89 | | 15, 37, 57 |
| Hydrochloride | 245–247 | $MeOH/Et_2O$ | | | 15 |
| 4-Oxide | 247–248 | EtOH | | | 57 |
| | 240–243 | MeOH | 87 | uv | 83, 84 |
| Ethiodide hydrate | 200–204 | $MeOH/H_2O$ | | | 14c |
| 5-Ph-7-MeO dihydrochloride·MeOH | 188d | $MeOH/Et_2O$ | | | 89 |
| 5-Ph-7-MeOOC | 191–193 | $CH_2Cl_2$/Petr ether | | | 2 |
| Hydrochloride | 251–252d | $MeOH/Et_2O$ | | | 2 |
| 5-Ph-7-Me | 130–132 | $EtOH/H_2O$ | | | 3b |
| 5-Ph-7-$NO_2$ | 210–211 | $CHCl_3/MeOH$ | 75–85 | | 1 |
| 5-Ph-7-$F_3C$ | 116–118 | Hexane | 61 | | 2 |
| Hydrochloride | 283–285 | $MeOH/Et_2O$ | | | 2 |
| 5-(2-$H_2N$-5-$ClC_6H_3$)-7-$H_2NSO_2$ | 225–226 | $EtOH/H_2O$ | | | 13 |
| 5-(2-$ClC_6H_4$)-7-Cl | 175–177 | PhH/Petr ether | | | 38, 59 |
| 4-Oxide | 215–217 | EtOH | 45 | | 61 |
| 5-(2-$ClC_6H_4$)-7-$NO_2$ | 180–183 | $CH_2Cl_2$/Hexane | | | 89 |
| 5-(2,6-$F_2C_6H_3$)-7-Cl | 180–186 | $CH_2Cl_2$/Hexane | | | 14c |
| 5-(2-$FC_6H_4$)-7-Cl | 163–164.5 | | | | 38, 59 |
| Ethiodide | 244–247 | $CH_2Cl_2$/$i$-PrOH/$Et_2O$ | | | 14c |
| 5-(2-$FC_6H_4$)-7-I | 157–159 | $CHCl_3$/Hexane | | | 103 |
| 5-(2-$FC_6H_4$)-7-$NO_2$ | 220 | $Me_2CO/Et_2O$ | | | 89 |
| 5-(2-Pyridyl)-7-$H_2N$ | 148–150 | $CH_2Cl_2$/Hexane | 79 | | 3a |
| 5-(2-Pyridyl)-7-Br | 197–198 | $CH_2Cl_2$/Hexane | 70 | | 3a, 25, 37 |
| Hydrochloride | 147–150 | $MeOH/Et_2O$ | | | 3a |
| 5-(2-Pyridyl)-7-Cl | 187–188 | EtOH | 64 | | 37, 25 |

TABLE VI-1. —(contd.)

| Substituent | mp (°C); [bp (°C/torr)] | Solvent of Crystallization | Yield (%) | Spectra | Refs. |
|---|---|---|---|---|---|
| 5-(2-Pyridyl)-7-$NO_2$ | 261 | EtOH | 73 | | 3, 4 |
| | 209–211 | $Me_2CO$ | 19 | | 37 |
| Hydrochloride | >310 | $MeOH/Et_2O$ | | | 3a |
| 5-(2-Pyridyl)-7-$F_3C$ | 183–184 | PhH/Hexane | 17 | | 3a, 25 |
| 5-(3-Pyridyl)-7-$NO_2$ | 212–215 | $Me_2CO$ | 65 | | 3a |
| 5-(4-Pyridyl)-7-$NO_2$ | 281–283 | EtOH | 56 | | 3a |
| 5-(2-Pyrimidyl)-7-$NO_2$ | 222–224 | $EtOH/CH_2Cl_2$/Hexane | 43 | ir | 5, 6 |
| 5-(2-Thiazolyl)-7-$NO_2$ | 229–230d | $THF/Et_2O$ | | | 7 |
| 5-(2-Thienyl)-7-$NO_2$ | 148–149 | EtOH | | | 45a |
| | | *Trisubstituted* | | | |
| ***1,2,5-Trisubstituted*** | | | | | |
| 1-Me-2-$H_2NCH_2$-5-Ph dihydrochloride | 209–213 | *i*-$PrOH/Me_2CO/Et_2O$ | | | 50 |
| 1-Me-2-$N_3CH_2$-5-Ph hydrochloride | 181–183 | $Me_2CO$/*i*-PrOH | | | 50 |
| 1-Me-2-$ClCH_2$-5-Ph hydrochloride hemihydrate | 195–198 | *i*-PrOH | | | 48a |
| 1-Me-2-$ClCH_2$-5-(2-$ClC_6H_4$) hydrochloride | 198–200 | | | | 48a |
| 1-Me-2-$ClCH_2$-5-(2-$FC_6H_4$) | Oil | | | ir | 53 |
| 1-Me-2-$HOCH_2$-5-(2-$ClC_6H_4$) | 133–134 | | | | 48a |
| 1-Me-2-$MeOCH_2$-5-(2-$ClC_6H_4$) | Oil | | | | 102 |
| 1-Me-2-(4-Me-Piperazino)$CH_2$-5-Ph | Oil | | | | 48a |
| 1-Me-2-$PrNHCH_2$-5-Ph | Oil | | | | 50 |
| 1-Me-2-(2-Propynyl)$NHCH_2$-5-Ph | Oil | | | | 50 |
| 1-Me-2-(2-Thienoyl)$NHCH_2$-5-Ph, *dl*-tartrate·0.4EtOH | 110–125 | $EtOH/Et_2O$ | | | 50 |
| 1-Me-2-(2-Thienoyl)$NHCH_2$-5-(2-$FC_6H_4$) hydrochloride | | | | | 51a, b |
| 1-Me-2-(2-Thienoyl)N(Me)$CH_2$-5-Ph, *dl*-tartrate | 104–115d | | | | 50 |
| 1-Me-2-(3-Thienoyl)$NHCH_2$-5-Ph hydrochloride | 234–237.5 | $EtOAc/Me_2CO/Et_2O$ | | | 50 |
| ***1,5,6-Trisubstituted*** | | | | | |
| 1-Me-5-Ph-6-MeO | 176.5–177.5 | $CHCl_3/Et_2O$ | | | 14c |

***1,5,7-Trisubstituted***

| Compound | mp (°C) | Solvent | Yield (%) | Spectra | Refs. |
|---|---|---|---|---|---|
| 1-Ac-5-Ph-7-Cl | 165–166 | $CH_2Cl_2/Et_2O$/Petr ether | 92 | | 15, 83a |
| 4-Oxide | 222–224 | EtOH | 35, 76 | uv | 15, 83, 57 |
| 1-Ac-5-Ph-7-$NO_2$ | 160–160.5 | *i*-PrOH | 39 | | 58 |
| 1-Ac-5-(2-Pyridyl)-7-$H_2N$ | | $CH_2Cl_2$/Hexane | | | 3b |
| 1-Ac-5-(2-Pyridyl)-7-Br | 166–168 | $CH_2Cl_2$/Hexane | | | 3b |
| 1-AcNH-5-Ph-7-Cl | 127–129 | $CH_2Cl_2$/Petr ether | | | 83c |
| 1-$H_2N$-5-Ph-7-Cl hydrochloride | 204–207 | EtOH | | | 83c |
| 1-BuNHCO-5-Ph-7-Cl | 154–156 | $CH_2Cl_2/Et_2O$ | | | 83c |
| 1-$ClCH_2CONHCO$-5-(2-$ClC_6H_4$)-7-Cl | 152–156 | $Et_2O$ | | | 115 |
| 1-$Cl(CH_2)_2NHCO$-5-Ph-7-Cl | 160–161 | $CH_2Cl_2$/Hexane | | | 115 |
| 1-$Cl(CH_2)_2NHCO$-5-(2-$ClC_6H_4$)-7-Cl | 150–153 | $Et_2O$ | | | 115 |
| 1-$Cl(CH_2)_2NHCO$-5-(2-$FC_6H_4$)-7-Cl | 138–140 | $Et_2O$ | | | 115 |
| 1-$Cl(CH_2)_2NHCO$-5-(2-$FC_6H_4$)-7-F | 160–162 | $Et_2O$ | | | 115 |
| 1-$Cl(CH_2)_3$-5-Ph-7-$NO_2$ hydrochloride | | $MeOH/Et_2O$ | | | 3b |
| 1-(Cyclopropyl)$CH_2$-5-Ph-7-Cl | 78–80 | *i*-$Pr_2O$ | | | 111 |
| Hydrochloride | 232–233d | | | | 8, 86 |
| 1-(Cyclopropyl)$CH_2$-5-Ph-7-$F_3C$ hydrochloride | 244–246d | | | | 59 |
| 1-(Cyclopropyl)$CH_2$-5-(2-$ClC_6H_4$)-7-Cl | 88–89 | | | | 59, 111 |
| 1-$Et_2NCH_2CO$-5-Ph-7-Cl | 185–186 | *i*-PrOH | | | 36 |
| 1-$Et_2NCH_2CO$-5-(2-$FC_6H_4$)-7-Cl dihydrochloride | 199–201d | *i*-PrOH/$Et_2O$ | | | 36 |
| 1-$Et_2N(CH_2)_2$-5-Ph-7-Cl dihydrochloride | 234–236 | i-PrOH | | | 112, 59, 111 |
| 1-$Et_2N(CH_2)_2$-5-Ph-7-$NO_2$ dihydrochloride | 234–236 | $MeOH/Et_2O$ | | | 111, 112 |
| 1-$Et_2NCOCO$-5-Ph-7-Cl | 171–173 | PhH/Petr ether | | | 36 |
| 1-(4,5-Dihydrooxazol-2-yl)-5-Ph-7-Cl | 125–126 | $CH_2Cl_2$/Hexane | | | 115 |
| 1-(4,5-Dihydrooxazol-2-yl)-5-(2-$ClC_6H_4$)-7-Cl | 110–111 | $Et_2O$/Hexane | | | 115 |
| 1-(4,5-Dihydrooxazol-2-yl)-5-(2-$FC_6H_4$)-7-Cl | 180–181 | $CH_2Cl_2$/Hexane | | | 115 |
| 1-(4,5-Dihydro-4-oxooxazol-2-yl) 5-(2-$ClC_6H_4$)-7-Cl | 224–228 | $Et_2O$ | | | 115 |
| 1-(4,5-Dihydrothiazol-2-yl)-5-(2-$ClC_6H_4$)-7-Cl | 123–124 | $CH_2Cl_2$/Hexane | | | 115 |
| 1-(4,5-Dihydro 4-oxothiazol-2-yl)-5-(2-$ClC_6H_4$)-7-Cl | 217–218 | $CH_2Cl_2$/Hexane | | | 115 |
| 1-(5,6-Dihydro-4-oxooxazin-2-yl)-5-(2-$ClC_6H_4$)-7-Cl | 124–135 | $Et_2O$/Hexane | | | 115 |
| 1-$Me_2NCOCH_2$-5-Ph-7-$H_2N$ | 187–189 | $CH_2Cl_2$/Hexane | | | 3b |
| 1-$Me_2NCOCH_2$-5-Ph-7-Cl hydrochloride | 243–245 | $MeOH/Et_2O$ | | | 3b |

TABLE VI-1. —(contd.)

| Substituent | mp (°C); [bp (°C/torr)] | Solvent of Crystallization | Yield (%) | Spectra | Refs. |
|---|---|---|---|---|---|
| 1-$Me_2NCOCH_2$-5-Ph-7-$NO_2$ hydrochloride | 242–245 | $MeOH/Et_2O$ | | | 3b |
| 1-$Me_2N(CH_2)_3$-5-Ph-7-Cl dihydrochloride | 255–257 | $MeOH/Et_2O$ | | | 3b |
| 1-$Me_2N(CH_2)_3$-5-Ph-7-$NO_2$ dihydrochloride, hemihydrate | 260–262 | $MeOH/Et_2O$ | | | 3b |
| 1-$EtOCH_2$-5-Ph-7-Cl citrate | 136–140 | $EtOH/Et_2O$ | | | 59 |
| 1-Et-5-Ph-7-Cl | 163–165 | | | | 18, 31 |
| 1-HCO-5-Ph-7-Cl | 116–119 | $CH_2Cl_2$/Petr ether | 34 | | 83, 15 |
| 4-Oxide | 150–153 | $CH_2Cl_2/Et_2O$ | 50 | uv | 83, 84 |
| 1-$MeOCH_2CO$-5-Ph-7-Cl | 140–141 | $Et_2O$/Hexane | | | 34 |
| 1-Me-5-(1-Adamantyl)-7-Cl | 74–77 | Petr ether | | | 117 |
| 1-Me-5-(2-Cl-Cyclohexen-1-yl)-7-Cl | 86–88 | Hexane | | | 34 |
| 1-Me-5-(1-Cyclohexen-1-yl)-7-Cl | 86 | Hexane | | | 34 |
| 1-Me-5-[$Et_2N(CH_2)_2NH$)]-7-Cl dihydrochloride | 253–255 | MeOH/EtOAc | 71 | | 74b, a |
| 1-Me-5-EtO-7-Cl | 38.5–39 | Petr ether | 62 | | 75, 74a, b |
| Hydrochloride | 116.5 | $CH_2Cl_2$/EtOAc | | | 75 |
| 1-Me-5-$EtOOCCH_2ONH$-7-Cl | 130–130.5 | EtOAc/Hexane | 28 | | 74b |
| 1-Me-5-EtS-7-Cl | 81–82 | *i*-PrOH | 15 | ir, pmr | 55 |
| 1-Me-5-HONH-7-Cl | 186.5–189 | $CH_2Cl_2$/EtOAc | 77 | | 74a, b |
| 1-Me-5-$HO(CH_2)_2NH$-7-Cl | 115–116.5 | EtOAc | 60 | | 74b |
| 1-Me-5-(Isothiazol-5-yl)-7-$NO_2$ | 173–175 | $CHCl_3$/Hexane | | | 7 |
| 1-Me-5-MeONH-7-Cl | 118–119 | Hexane | 22 | | 74b |
| 1-Me-5-(1-Me-Imidazol-2-yl)-7-$H_2N$ | 170–172 | $MeOH/Et_2O$ | | | 7 |
| 1-Me-5-(1-Me-Imidazol-2-yl)-7-Cl | 157–159 | EtOAc/Hexane | | | 7 |
| 1-Me-5-(1-Me-Imidazol-2-yl)-7-$NO_2$ | 175–178 | $MeOH/Et_2O$ | | | 7 |
| 1-Me-5-(1-Me-Pyrazol-5-yl)-7-$NO_2$ | 155–157 | EtOAc/Hexane | | | 7 |
| 1-Me-5-Ph-7-Ac | 109–113 | $CH_2Cl_2$/Petr ether | | | 131, 132 |
| 1-Me-5-Ph-7-AcNH | 176–177 | $CH_2Cl_2$/Petr ether | | | 2 |
| 1-Me-5-Ph-7-$H_2N$ | 158–159 | $Et_2O$ | | | 2 |
| 1-Me-5-Ph-7-$N_3$ | 80–82 | Petr ether | | | 101b |

| | | | | | |
|---|---|---|---|---|---|
| 1-Me-5-Ph-7-Br | 104–105 | Hexane | | | 2, 59 |
| Hydrochloride | 257–258d | MeOH/$Et_2O$ | | | 2 |
| 1-Me-5-Ph-7-(2-$HOOCC_6H_4CO$) | 270–290 | $CH_2Cl_2$/MeOH | | | 88 |
| 1-Me-5-Ph-7-Cl | 102–103 | Hexane | 58, 75 | | 2, 37, 61a |
| Hydrochloride | 248–250 | MeOH/$Et_2O$ | | | 21 |
| 4-Oxide | 165–167 | $Et_2O$/Hexane | | uv | 21, 83b, 110 |
| | 139–142 | $CH_2Cl_2$/Petr ether | 53 | | 83a |
| 1-Me-5-Ph-7-CN | 149–150 | $Et_2O$ | 22 | | 2 |
| 1-Me-5-Ph-7-$Me_2N$ | 115–117 | EtOH/$H_2O$ | | | 109 |
| Dihydrochloride | 252–254 | EtOH/$Et_2O$ | | | 109 |
| 1-Me-5-Ph-7-I | 153–155 | EtOH | | | 103 |
| 1-Me-5-Ph-7-Me | 138 | Hexane | 32 | | 46 |
| 1-Me-5-Ph-7-$NO_2$ | 186–188 | *i*-PrOH | 76 | | 2, 59, 109 |
| 1-Me-5-Ph-7-$N_2SO_3H$ | 234–235d | $H_2O$ | | | 132b |
| 1-Me-5-Ph-7-$F_3C$ | 150.5–152 | PhH/Hexane | 51 | | 2, 59 |
| Hydrochloride | 261–262 | MeOH/$Et_2O$ | | | 2 |
| 1-Me-5-(2-$AcNHC_6H_4$)-7-Cl | 200–201 | $Me_2CO$/Petr ether | | | 45a |
| 1-Me-5-(2-$H_2NC_6H_4$- | | | 83 | | 43 |
| 1-Me-5-(2-$ClC_6H_4$)-7-Cl | 93–94.5 | $Et_2O$/Petr ether | | | 34 |
| 1-Me-5-(2-$ClC_6H_4$)-7-I | 106–109 | Hexane | | | 99c |
| 1-Me-5-(4-$ClC_6H_4$)-7-Cl | 106–108 | $CH_2Cl_2$/$Et_2O$ | | | 126b |
| 1-Me-5-(2,3-$Cl_2C_6H_3$)-7-Cl | 116–117 | Petr ether | | | 45a |
| 1-Me-5-(2,6-$Cl_2C_6H_3$)-7-Cl | 143.5–144.5 | $Me_2CO$ | | | 45a |
| 1-Me-5-(2-$FC_6H_4$)-7-Ac | 112–114 | $Et_2O$/Hexane | | | 104 |
| 1-Me-5-(2-$FC_6H_4$)-7-$H_2N$ | 144 | $Et_2O$ | | | 89 |
| 1-Me-5-(2-$FC_6H_4$)-7-Cl | 100–103 | EtOAc | | pmr | 42 |
| Hydrochloride | 246–247d | $Me_2CO$/MeOH | | | 59 |
| 1-Me-5-(2-$FC_6H_4$)-7-CN | 135–137 | $Et_2O$ | | | 104 |
| 1-Me-5-(2-$FC_6H_4$)-7-CH(OH)Me | 125–127 | $Et_2O$/Pentane | | | 104 |
| 1-Me-5-(2-$FC_6H_4$)-7-I | 102–105 | Hexane | | | 104 |
| 1-Me-5-(2-$FC_6H_4$)-7-MeNHCONH | 188 | $CH_2Cl_2$/$Et_2O$ | | | 89 |
| 1-Me-5-(2-$MeSC_6H_4$)-7-Cl | 100–101 | $Et_2O$ | | | 45a |
| 1-Me-5-[2-MeS(O) $C_6H_4$]-7-Cl | 166–167 | $CH_2Cl_2$/$Et_2O$ | | | 45a |
| 1-Me-5-(2-$NO_2C_6H_4$)-7-Cl hydrochloride | 197–208 | $Me_2CO$ | 37 | | 43 |

TABLE VI-1. —(contd.)

| Substituent | mp (°C); [bp (°C/torr)] | Solvent of Crystallization | Yield (%) | Spectra | Refs. |
|---|---|---|---|---|---|
| 1-Me-5-(PhNH)-7-Cl | 162–164 | EtOH | | | 99c |
| 1-Me-5-(4-Ph-Piperazino)-7-Cl | 130–131 | $Et_2O$/Petr ether | 23 | | 74b |
| 1-Me-5-PhS-7-Cl | 152–153 | Hexane | 7 | ir, pmr | 55 |
| 1-Me-5-(2-Pyridyl)-7-$H_2N$ | 165–167 | $CH_2Cl_2$/Petr ether | 64 | | 3a, 4 |
| 1-Me-5-(2-Pyridyl)-7-Cl | 152–154 | EtOH | 31 | | 3a, 4 |
| 1-Me-5-(2-Pyridyl)-7-$Me_2N$ | 120.5–122 | $CH_2Cl_2$/Heptane | | | 101b |
| 1-Me-5-(2-Pyridyl)-7-$NO_2$ | 181–184 | PhH/EtOH | 84 | | 3a, 4 |
| Hydrochloride | 200–202d | MeOH/$Et_2O$ | | | 3a |
| 1-Me-5-(2-Pyridyl)-7-$F_3C$ | 127–129 | Hexane | 19 | | 3a, 4 |
| 1-Me-5-(3-Pyridyl)-7-$H_2N$ | 170–173 | $CH_2Cl_2$/Hexane | 66 | | 3a |
| 1-Me-5-(3-Pyridyl)-7-Cl | 88–90 | | 23 | | 3a |
| 1-Me-5-(3-Pyridyl)-7-$NO_2$ | 266–269d | MeOH | 66 | | 3a |
| 1-Me-5-(4-Pyridyl)-7-$H_2N$ | 174–176 | EtOH | 49 | | 3a |
| 1-Me-5-(4-Pyridyl)-7-Cl dihydrochloride | 223–225 | MeOH/$Et_2O$ | 56 | | 3a |
| 1-Me-5-(4-Pyridyl)-7-$NO_2$ hydrochloride | 250–251d | MeOH/$Et_2O$ | 49 | | 3a |
| 1-Me-5-(2-Pyrimidyl)-7-Cl | 107–109 | $CH_2Cl_2$/Petr ether | 34 | ir, ms | 6 |
| 1-Me-5-(2-Pyrimidyl)-7-$NO_2$ | 181–183d | $CH_2Cl_2$/Hexane | 61 | | 6 |
| 1-Me-5-(2-Thiazolyl)-7-(1-adamantoyl)NH | 262–263 | EtOH | | | 7 |
| 1-Me-5-(2-Thiazolyl)-7-$H_2N$ | 164–166 | PhH/Petr ether | | | 7 |
| 1-Me-5-(2-Thiazolyl)-7-Cl | 138–140 | $CHCl_3$/Hexane | | | 7 |
| 1-Me-5-(2-Thiazolyl)-7-$NO_2$ | 190–192 | $CHCl_3$/Petr ether | | | 7 |
| 1-Me-5-(2-Thienyl)-7-$NO_2$ | 174–175 | $Me_2CO$ | | | 45a |
| 1-Me-5-(Trichloroethenyl)-7-Cl | 102–104 | $Et_2O$/Hexane | | | 45b |
| 1-MeNHCO-5-Ph-7-Cl | 180–182 | $CH_2Cl_2$/$Et_2O$ | | | 83c |
| 1-MeNHCS-5-Ph-7-Cl | 179–180 | $CH_2Cl_2$/Petr ether | | | 83c |
| 1-$MeNHCOCH_2$-5-Ph-7-$H_2N$ | 190–192 | $CH_2Cl_2$/Hexane | | | 113 |
| 1-$MeNHCOCH_2$-5-Ph-7-Cl | 187–189 | $CH_2Cl_2$/Hexane | | | 113 |
| Hydrochloride | 252–254 | MeOH/$Et_2O$ | | | 113 |

| | | | | | |
|---|---|---|---|---|---|
| 1-MeNHCOCH$_2$-5-Ph-7-NO$_2$ | 223–225 | Me$_2$CO | | | 113 |
| Hydrochloride | 258–260 | Me$_2$CO/MeOH | | | 113 |
| 1-MeNHCOCH$_2$-5-(2-Pyridyl)-7-Br | 230–232 | Me$_2$CO/Petr ether | | | 113 |
| 1-MeNHCOCH$_2$-5-(2-Pyridyl)-7-NO$_2$ | 183–85 | PhH/Hexane | | | 113 |
| 1-(4-MeC$_6$H$_4$)SO$_2$-5-Ph-7-Cl | 154–156 | CHCl$_3$/EtOH | 70 | | 125 |
| 1-MeSCH$_2$-5-Ph-7-Cl maleate | 158–159 | *i*-PrOH | | | 59 |
| 1-MeSO$_2$-5-Ph-7-Cl | 170–173 | CHCl$_3$/EtOH | 61 | | 125 |
| 1-NO-5-Ph-7-Cl | 119–121 | MeOH | 94 | | 83c, 73 |
| 1-(Phthalimido)-(CH$_2$)$_2$-5-Ph-7-Cl | 175–176 | *i*-PrOH/Cyclohexane | | | 111 |
| 1-HSO$_3$NHCO-5-Ph-7-Cl | 170d | H$_2$O | | | 124b |
| 1-(4,5,6,7-Tetrahydro-4-oxo-1,3-oxazepin-2-yl)-5-(2-ClC$_6$H$_4$)-7-Cl | 187–189.5 | Et$_2$O | | | 115 |
| 1-F$_3$CCH$_2$-5-Ph-7-Cl | 66–67.5 | Petr ether | 40 | | 106 |
| 1-F$_3$CCH$_2$-5-(2-ClC$_6$H$_4$)-7-Cl | 188–198 | EtOH/Et$_2$O | 42 | | 106 |
| 4-Oxide | 137–138 | CH$_2$Cl$_2$/Hexane | 66 | | 106 |
| 1-F$_3$CCH$_2$-5-(2-FC$_6$H$_4$)-7-Br | 97–98 | Et$_2$O/Hexane | 82 | | 106 |
| 1-F$_3$CCH$_2$-5-(2-FC$_6$H$_4$)-7-Cl | 83–85 | CH$_2$Cl$_2$/Hexane | 75 | | 44b, 106 |
| 4-Oxide | 163–165 | CH$_2$Cl$_2$/Hexane | 83 | | 44b, 106 |
| 1-F$_3$CCH$_2$-5-(2-FC$_6$H$_4$)-7-NO$_2$ | 123–125 | CH$_2$Cl$_2$/Hexane | 14 | | 106 |
| ***1,5,8-Trisubstituted*** | | | | | |
| 1-Me-5-Ph-8-MeO | 120–121 | CH$_2$Cl$_2$/Et$_2$O/Petr ether | | | 14c |
| 1-Ph-5-H$_2$N-8-Cl | 216–217 | Et$_2$O | 53 | | 77 |
| 1-Ph-5-MeNH-8-Cl | 195–196 | Et$_2$O | 52 | ir, pmr | 77 |
| ***1,5,9-Trisubstituted*** | | | | | |
| 1-Me-5-Ph-9-Cl | | | | | |
| Hydrochloride | 235–243 | MeOH/Et$_2$O | | | 14c |
| 1-Me-5-(2-FC$_6$H$_4$)-9-Cl | 138–143 | Et$_2$O/Petr ether | | | 14c |
| ***2,5,7-Trisubstituted*** | | | | | |
| 2-AcNHCH$_2$-5-(2-FC$_6$H$_4$)-7-Cl maleate | 169–171 | *i*-PrOH/Et$_2$O | | | 63 |
| 2-H$_2$NCO-5-Ph-7-Cl | 210–213 | CH$_2$Cl$_2$/Et$_2$O | 60 | ir, ms, pmr | 68a, b |

TABLE VI-1. —(contd.)

| Substituent | mp (°C); [bp (°C/torr)] | Solvent of Crystallization | Yield (%) | Spectra | Refs. |
|---|---|---|---|---|---|
| 2-$H_2NCO$-5-(2-$FC_6H_4$)-7-Cl | 192–193 | $CH_2Cl_2$/Petr ether | | | 68b |
| 2-$H_2NCOCH_2$-5-(2-$FC_6H_4$)-7-Cl | 228–230 | THF/EtOAc | | | 63 |
| 2-$H_2N$(HON)C-5-(2-$FC_6H_4$)-7-Cl | 216–218 | EtOAc/Hexane | | | 66b |
| 2-$H_2NCH_2$-5-Ph-7-Cl, 4-Oxide | 165–167 | EtOAc/Hexane | | | 99c |
| 2-$H_2NCH_2$-5-(2-$FC_6H_4$)-7-Cl dimaleate | 196–198 | MeOH/*i*-PrOH/$H_2O$ | 83 | | 65 |
| 2-Aziridino(MeON)C-5-(2-$FC_6H_4$)-7-Cl | 161–164 | EtOAc/Hexane | | | 66b |
| 2-*t*-BuOOC(HON)C-5-(2-$FC_6H_4$)-7-Cl | 203–204d | EtOAc | | | 95 |
| 2-HOOC-5-Ph-7-Cl hydrochloride | 218–221 | $H_2O$/HCl | | | 68d |
| 4-Oxide | 205–208d | MeOH | | | 67 |
| 2-$HOOCCH_2$-5-(2-$FC_6H_4$)-7-Cl | 126–130 | MeOH/$H_2O$ | | | 63 |
| 2-$HOOCCH_2S$-5-Ph-7-Cl | 161–165 | $CH_2Cl_2$/MeOH | | | 68d |
| 2-$HOOCCH_2NHCO$-5-Ph-7-Cl | 268–270 | DMF/$Et_2O$ | | | 68d |
| 2-$ClCH_2$-5-Ph-7-$NO_2$ | 148–149 | | | | 48a |
| Hydrochloride | 213–214 | *i*-PrOH/$Et_2O$ | | | 48a |
| 2-Cl(MeOOC)CH-5-(2-$FC_6H_4$)-7-Cl | 126–127 | $Et_2O$/Hexane | | | 45b |
| 2-CN-5-Ph-7-Cl | 181–183d | $CH_2Cl_2$/Hexane | 60 | ir, ms, pmr | 68a, b |
| 2-CN-5-(2-$FC_6H_4$)-7-Cl | 163–164 | $CH_2Cl_2$/Cyclohexane | | | 68b |
| 2-$CNCH_2$-5-(2-$FC_6H_4$)-7-Cl | 199–201 | EtOAc/Hexane | | | 45b |
| 2-[(Cyclohexyl)$NHCOCH_2NHCO$]-5-Ph-7-Cl | 238 | $CH_2Cl_2$/$Et_2O$/EtOH | | | 68d |
| 2-$Et_2N$-5-Ph-7-Cl, 4-Oxide | 142–144d | EtOAc | | | 67 |
| 2-$Et_2N(CH_2)_2NHCO$-5-Ph-7-Cl | 157–158 | $CH_2Cl_2$/$Et_2O$/ Petr ether | | | 68d |
| 2-(4,5-Dihydroimidazol-1-yl)-5-Ph-7-Cl, 4-oxide | 187–188 | MeOH | 50 | | 67 |
| | 173–176 | $CH_2Cl_2$/Hexane | | | |
| 2-$Me_2NCO$-5-Ph-7-Cl, hydrate | 143–145 | $Et_2O$ | | | 68d |
| 2-$Me_2NCO$(HON)C-5-(2-$FC_6H_4$)-7-Cl | 214–215 | $CH_2Cl_2$/EtOH | 50 | | 95 |
| 2-$Me_2N$(HON)C-5-(2-$FC_6H_4$)-7-Cl | 178–180 | MeOH/EtOAc | | | 66b |
| 4-Oxide·0.166$Et_2O$ | 145–150d | EtOAc/$Et_2O$ | | | 66b |
| 2-$Me_2NSO_2$(HON)C-5-(2-$FC_6H_4$)-7-Cl | 242–244 | THF/EtOH | 66 | | 95 |

| | | | | | |
|---|---|---|---|---|---|
| 2-$Me_2NSO_2CH_2$-5-(2-$FC_6H_4$)-7-Cl | 181–183 | EtOH | | | 45b |
| 2-EtO-5-Ph-7-Cl, 4-Oxide | 132–133 | EtOH | | | 67 |
| 2-$EtOOCCH_2NHCO$-5-Ph-7-Cl | 142–144 | $CH_2Cl_2$/Petr ether | | | 68d |
| 2-Et-5-Ph-7-$NO_2$ | 146–148 | $CCl_4$/Hexane | 65 | ir, pmr | 25 |
| 2-*S* Configuration | | | | $[\alpha]_D$ | 24, 25 |
| 2-HO-5-Ph-7-Cl hydrochloride | 125d | $Me_2CO/H_2O$ | | pmr | 69b |
| Hydrate | 125d | $EtOH/H_2O$/HCl | 79.5 | pmr | 69a |
| 2-HO-5-Ph-7-$NO_2$ hydrochloride·$0.5H_2O$ | 178d | $EtOH/H_2O$/HCl | 80 | pmr | 69a, b |
| 2-$HOCH_2CH_2$-5-(2-$FC_6H_4$)-7-Cl | 139–141 | $CH_2Cl_2/Et_2O$ | | | 63 |
| 2-HO$(CH_2)_2$S-5-Ph-7-Cl | 135–140d | $Et_2O$ | 78 | ir | 68a |
| 2-(HON)CH-5-Ph-7-Cl, 4-Oxide | 195–198d | EtOH | | | 99c |
| 2-(HON)CH-5-(2-$FC_6H_4$)-7-Cl | 195–197 | $CH_2Cl_2$/EtOAc | 58 | uv, pmr | 65 |
| 4-Oxide | 184–186 | MeOH/EtOAc | | | 66c |
| 2-HOCH(Ph)-5-Ph-7-Cl | 196–197 | EtOAc/Hexane | 93 | pmr | 73 |
| 2-MeO-5-Ph-7-Cl | 179–181d | MeOH/$Et_2O$ | | | |
| Hydrochloride | 185–193d | MeOH/$Et_2O$ | 85 | | 68a, 67 |
| 4-Oxide | 127–131 | MeOH | | | 67 |
| 2-MeO-5-Ph-7-$F_3C$, 4-Oxide | 179–181d | MeOH/$Et_2O$ | | | 67 |
| 2-MeOOC-5-Ph-7-Cl, 4-Oxide | 154–156 | *i*-PrOH/$H_2O$ | | | 67 |
| 2-$MeOOCCH_2$-5-(2-$FC_6H_4$)-7-Cl | 122–124 | $CH_2Cl_2$/Petr ether | | | 14c |
| 2-MeO(HN)C-5-Ph-7-Cl, 4-Oxide | 168–171d | EtOAc | | | 67 |
| 2-Me-5-Ph-7-$H_2N$ | 190–194d | EtOAc/Hexane | 92 | | 64 |
| 2-Me-5-Ph-7-Cl | 146–147.5 | EtOAc | 50 | uv | 99a, b, 25 |
| Hydrochloride | 260–265d | MeOH | | | 99a, b |
| 4-Oxide | 200–202.5 | EtOH | 75 | pmr, uv | 99a, b |
| Hydrochloride | 185–193d | EtOH/$Et_2O$ | | | 99a, b |
| 2-Me-5-Ph-7-$NO_2$ | 152–153 | $Et_2O$ | 2 | | 64 |
| Hydrochloride | 276–277 | MeOH | | | 64 |
| 2-$MeNHCOCH_2$-5-(2-$FC_6H_4$)-7-Cl | 167–170 | $CH_2Cl_2$/Hexane | | | 63 |
| 2-$MeNHCOCH_2$-5-(5-Br-2-$FC_6H_3$)-7-Cl | 144–147 | EtOAc/Hexane | | | 45b |
| 2-MeNH(HON)C-5(2-$FC_6H_4$)-7-Cl | 215–217d | MeOH/EtOAc | | | 66 |
| 2-(4-$MeC_6H_4$)$SO_2NHCH_2$-5-(2-$FC_6H_4$)-7-Cl | 145–147 | $Et_2O$ | 61 | ir, ms, pmr | 126 |
| 2-$MeSO_2CH_2$-5-(2-$F_6H_4$)-7-Cl | 143–145 | EtOAc/Hexane | | | 45b |
| 4-Oxide | 186–188 | MeOH | | | 45b |

TABLE VI-1. —(contd.)

| Substituent | mp (°C); [bp (°C/torr)] | Solvent of Crystallization | Yield (%) | Spectra | Refs. |
|---|---|---|---|---|---|
| 2-MeS(HON)C-5-(2-$FC_6H_4$)-7-Cl, 4-Oxide | 215–217d | MeOH/EtOAc | | | 66a |
| 2-[Morpholino(HON)C]-5-(2-$FC_6H_4$)-7-Cl | 133–137 | MeOH/EtOAc/Hexane | | | 66 |
| 4-Oxide | 192–193d | MeOH/EtOAc | | | 66 |
| 2-[Morpholino(HN)C]-5-(2-$FC_6H_4$)-7-Cl | 175–178 | THF/Hexane | | | 66a |
| 2-$O_2NCH_2$-5-(2-$FC_6H_4$)-7-Cl | | | | | |
| 4-Oxide | 142–143d | EtOAc/Hexane | | | 99c |
| 2-PhNH-5-Ph-7-Cl, 4-Oxide | 157–159 | EtOAc | | | 67 |
| 2-$PhCH_2S$-5-Ph-7-Cl, 4-Oxide | 116–123d | EtOAc | | | 67 |
| 2-(Phthalimido)$CH_2$-5-Ph-7-Cl, 4-Oxide | 226–228 | EtOH | | | 99c |
| 2-Piperidino-5-Ph-7-Cl | 130–135d | EtOAc | 94 | ir | 67 |
| 4-Oxide | 177–181d | $CH_2Cl_2$/Hexane | | | |
| 2-*i*-PrNH-5-Ph-7-Cl, 4-Oxide | 142–145d | EtOAc | | | 67 |
| 2-*i*-$PrOCH_2$-5-Ph-7-$NO_2$, hydrochloride | 233–235 | | | | 52 |
| 2-(2-Pyridyl)(HON)C-5-(2-$FC_6H_4$)-7-Cl | 211–213 | MeOH | | | 95 |
| 2-Pyrrolidino(HON)C-5-(2-$FC_6H_4$)-7-Cl | 200–202d | THF/EtOH | | | 66 |
| 4-Oxide | 182–183d | EtOAc/MeOH | | | 66 |
| 2-$F_3CCONHCH_2$-5-(2-$FC_6H_4$)-7-Cl | 110–112 | | | | |
| | 140–143 | $CH_2Cl_2$/Hexane | | | 14c |
| ***2,5,9-Trisubstituted*** | | | | | |
| 2-$ClCH_2$-5-Ph-9-$NO_2$ | 123–125 | | | | 48a |
| ***3,3,7-Trisubstituted*** | | | | | |
| 3-CN-3-Me-7-Br | 143–145 | $CH_2Cl_2$/Hexane | | | 99c |
| 3-MeO-3-Me-7-Br | 162–163 | *i*-PrOH/THF/Petr ether | | | 99c |
| ***3,5,7-Trisubstituted*** | | | | | |
| 3-HOOC-5-Ph-7-Cl | 159–160 | $H_2O$ | | | 99c |
| 3-CN-5-Ph-7-Cl | 144–146 | $CHCl_3$/Hexane | | | 99c |

| | | | | | |
|---|---|---|---|---|---|
| 3-EtO-5-Ph-7-Cl | 144–148 | $EtOH/H_2O$ | | | 99c |
| 4-Oxide | 196–199d | EtOH | | | 99c |
| 3-EtOOC-5-Ph-7-Cl hydrochloride | 216–218d | $EtOH/Et_2O$ | | | 99c |
| 3-Et-5-Ph-7-$NO_2$, 3*S* | | | $[a]_D$ | | 24, 25 |
| 3-Et-5-Ph-7-$NO_2$, 3*R* | | | $[\alpha]_D$ | ir, pmr | 25 |
| 3-Et-5-Ph-7-$NO_2$ | 233–235 | EtOAc/Petr ether | 8 | | 25 |
| 3-Me-5-Ph-7-$H_2N$ dihydrochloride | 277–280d | $EtOH/Et_2O$ | | | 2 |
| 3-Me-5-Ph-7-Cl | 127–128 | $Et_2O$/Petr ether | 70 | | 2, 99a, b, 59 |
| 4-Oxide | 190–195d | EtOAc | | | |
| | 200–203 | EtOAc | 10 | uv, pmr | 99a, b, |
| 3-Me-5-Ph-7-$NO_2$ | | | 80 | | 2 |
| ***5,7,8-Trisubstituted*** | | | | | |
| 5-Ph-7-$H_2NSO_2$-8-Cl | 255–256 | $MeOH/H_2O$ | | | 117 |
| ***5,7,9-Trisubstituted*** | | | | | |
| 5-Ph-7-Cl-9-Br hydrochloride | 247–249d | $MeOH/Et_2O$ | | | 83c |
| | | *Tetrasubstituted* | | | |
| ***1,2,5,7-Tetrasubstituted*** | | | | | |
| 1-Ac-2-$AcNHCH_2$-5-(2-$FC_6H_4$)-7-Cl | 213–215 | $CH_2Cl_2/Et_2O$ | 43 | ir, pmr, uv | 65 |
| 1-Ac-2-CN-5-Ph-7-Cl | 208–210 | $EtOAc/Et_2O$ | 45 | ir, ms, pmr | 68a |
| 1-Ac-2-MeO-5-Ph-7-Cl | 195–197 | MeOH | | | 34 |
| 1-$ClCH_2CH_2$-2-$ClCH_2$-5-Ph-7-Cl | 114–116 | | | | 48a |
| 1-$EtOOCCH_2CO$-2-$MeOOCCH_2$-5-(2-$FC_6H_4$)-7-Cl | 97–99 | $Et_2O$/Hexane | | | 63 |
| 1-$EtOOCCH_2CO$-2-MeOOCCHCl-5-(2-$ClC_6H_4$)-7-Cl | | | | | |
| Isomer A | 141–143 | $CH_2Cl_2$/Hexane | | | 45b |
| Isomer B | 143–145 | $CH_2Cl_2/Et_2O$ | | | 45b |
| 1-Et-2-$HOCH_2$-5-Ph-7-Cl hydrochloride | 196–202 | | | | 48a |
| 1-Et-2-Me-5-Ph-7-Cl | 205–210 | | 46 | | 31 |
| 1-CHO-2-$ClCH_2$-5-Ph-7-Cl | 151–152 | *i*-PrOH | 13 | ir, ms, pmr, uv | 107 |
| 1-$MeOOCCH_2CO$-2-(2-HO-1-Propyl)-5-(2-$ClC_6H_4$)-7-Cl | 158–160 | $CH_2Cl_2$/Hexane | | | 45b |

TABLE VI-1. —(contd.)

| Substituent | mp (°C); [bp (°C/torr)] | Solvent of Crystallization | Yield (%) | Spectra | Refs. |
|---|---|---|---|---|---|
| 1-Me-2-AcNHCH$_2$-5-Ph-7-Cl hydrochloride | 184–186 | *i*-PrOH/Et$_2$O | | | 48a, 118 |
| 1-Me-2-AcNHCH$_2$-5-(2-ClC$_6$H$_4$)-7-Cl | 113–121 | | | | 118 |
| 1-Me-2-H$_2$NCO-5-Ph-7-Cl | 204–206 | MeOH | | | 70, 71 |
| 1-Me-2-H$_2$NCONHCH$_2$-5-Ph-7-Cl hydrochloride | 231–232 | *i*-PrOH/Et$_2$O | | | 48a, 118 |
| 1-Me-2-H$_2$NCONHCH$_2$-5-(2-ClC$_6$H$_4$)-7-Cl hydrochloride | 200–205 | | | | 118 |
| 1-Me-2-H$_2$NCOCH$_2$-5-Ph-7-Cl hydrochloride | 224–226 | | | | 48a, 118 |
| 1-Me-2-H$_2$NCH$_2$-5-Ph-7-Cl dihydrochloride·0.5 *i*-PrOH | 206–209 | | | | 48a |
| 1-Me-2-PhCOOCH$_2$-5-Ph-7-Cl hydrochloride | 175 | *i*-PrOH/Et$_2$O | | | 48a |
| 1-Me-2-BrCH$_2$-5-(2-ClC$_6$H$_4$)-7-Cl hydrochloride | 186–188 | | | | 53 |
| 1-Me-2-HOOC-5-Ph-7-Cl hydrochloride | 247–249 | H$_2$O/HCl | | | 71a |
| 1-Me-2-HOOCCH$_2$-5-Ph-Cl hydrochloride | 229d | MeOH/Et$_2$O | | | 48a, 118 |
| 1-Me-2-HOOCCH$_2$-5-(2-(ClC$_6$H$_4$)-7-Cl hydrochloride | 221–222 | | | | 48a |
| 1-Me-2-ClCH$_2$-5-(2-Furyl)-7-Cl hydrochloride | 198–199 | Me$_2$CO/EtOH | | | 54 |
| 1-Me-2-ClCH$_2$-5-Ph-7-Br hydrochloride | 95–98 | | | | 48a |
| 1-Me-2-ClCH$_2$-5-Ph-7-Cl hydrochloride·*i*-PrOH | 110–112 | *i*-PrOH | 75 | | 48a |
| | 178–180 | | | | 50 |
| 1-Me-2-ClCH$_2$-5-Ph-7-F hydrochloride | 180–184 | | | | 48a, 53 |
| 1-Me-2-ClCH$_2$-5-Ph-7-MeO hydrochloride | 191–193 | | | | 48a |
| 1-Me-2-ClCH$_2$-5-Ph-7-Me hydrochloride·*i*-PrOH | 130–133 | | | | 48a |
| 1-Me-2-ClCH$_2$-5-Ph-7-NO$_2$ hydrochloride | 213–214 | | | | 50 |
| 1-Me-2-ClCH$_2$-5-(2-ClC$_6$H$_4$)-7-Cl | 116–119 | Et$_2$O | 7.5 | ir, pmr, uv | 47 |
| Hydrochloride | 181–183 | EtOH/Me$_2$CO | | ir, uv | 47, 48a, 53 |
| 1-Me-2-ClCH$_2$-5-(3,4-Cl$_2$C$_6$H$_3$)-7-Cl hydrochloride | 139–141 | | | | 48a |
| 1-Me-2-ClCH$_2$-5-(2-FC$_6$H$_4$)-7-Cl hydrochloride | 227–229 | EtOAc/Me$_2$CO | 43 | ir, pmr, uv | 47, 48a, 53 |
| 1-Me-2-ClCH$_2$-5-(2-F$_3$CC$_6$H$_4$)-7-Cl | Oil | | | ir | 53 |
| 1-Me-2-ClCH$_2$-5-(2-Thienyl)-7-Cl hydrochloride | 184–186 | | | | 54 |

| | | | | | |
|---|---|---|---|---|---|
| 1-Me-2-(7-Cl-1-Me-5-Ph-2,3,4,5-Tetrahydro-1*H*-1,4-benzodiazepin-4-yl)-5-Ph-7-Cl | 281–285d | THF | 23 | | 58 |
| 1-Me-2-(3-Cl-2-MeC$_6$H$_3$)NHCOCH$_2$-5-Ph-7-Cl hydrochloride | 219–222 | | | | 118 |
| 1-Me-2-(4-ClC$_6$H$_4$)OCH$_2$-5-Ph-7-Cl hydrochloride | 192–200 | *i*-PrOH/Et$_2$O | | | 48a, 52 |
| 1-Me-2-(Cinnamoyl)NHCH$_2$-5-(2-ClC$_6$H$_4$)-7-Cl hydrochloride | 220–223 | | | | 118 |
| 1-Me-2-CN-5-Ph-7-Cl | 120–122 | MeOH | | | 70, 71 |
| 1-Me-2-CN-5-(2-FC$_6$H$_4$)-7-Cl | 124–125 | MeOH | | | 70, 71 |
| 1-Me-2-(Cyclohexyl)NHCOCH$_2$-5-(2-ClC$_6$H$_4$)-7-Cl hydrochloride | 238–240 | | | | 118 |
| 1-Me-2-CNCH$_2$-5-Ph-7-Cl hydrochloride | 213–215 | *i*-PrOH/Et$_2$O | | | 48a, 118 |
| 1-Me-2-CNCH$_2$-5-(2-ClC$_6$H$_4$)-7-Cl | 105–107 | Et$_2$O | 87.5 | ir, pmr, uv | 47 |
| Hydrochloride | 198–202 | Me$_2$CO | | | 47 |
| | 171–174 | | | | 48a |
| 1-Me-2-Me$_2$NCOCH$_2$-5-Ph-7-Cl hydrochloride | 102–105 | | | | 118 |
| 1-Me-2-Me$_2$NCOCH$_2$-5-(2-ClC$_6$H$_4$)-7-Cl hydrochloride | 172–177 | | | | 118 |
| 1-Me-2-EtOOCNHCH$_2$-5-Ph-7-Cl hydrochloride | 196–197 | *i*-PrOH/Et$_2$O | | | 118 |
| 1-Me-2-EtOOCNHCH$_2$-5-(2-ClC$_6$H$_4$)-7-Cl hydrochloride | 205 | *i*-PrOH/Et$_2$O | | | 118 |
| 1-Me-2-EtOOCCH$_2$-5-Ph-7-Cl hydrochloride | 206–208 | EtOH/Et$_2$O | | | 48a, 118 |
| 1-Me-2-EtOCH$_2$-5-(2-BrC$_6$H$_4$)-7-Br hydrochloride | 154–156 | EtOH | | | 49 |
| 1-Me-2-EtOCH$_2$-5-(2-ClC$_6$H$_4$)-7-Br hydrochloride | 191–194 | EtOH | | | 49 |
| 1-Me-2-EtOCH$_2$-5-(2-ClC$_6$H$_4$)-7-Cl hydrochloride | 158–161 | *i*-PrOH/Et$_2$O | | | 48a, 52 |
| 1-Me-2-EtOCH$_2$-5-(2,6-Cl$_2$C$_6$H$_3$)-7-Cl hydrochloride | 211–212 | | | | 52 |
| 1-Me-2-EtOCH$_2$-5-(2-IC$_6$H$_4$)-7-Br hydrochloride | 204–207 | EtOH | | | 49 |
| 1-Me-2-EtOCH$_2$-5-(2-F$_3$CC$_6$H$_4$)-7-Br | 102–104 | EtOH | | | 49 |
| 1-Me-2-EtNHCOOCH$_2$-5-Ph-7-Cl | 175–177 | Me$_2$CO | | | 48a, 52 |
| 1-Me-2-H$_2$NNHCOCH$_2$-5-Ph-7-Cl hydrochloride | 200 | | | | 118 |
| 1-Me-2-HO-5-Ph-7-Cl | 148–151d | Et$_2$O | 97 | ir, ms | 58 |
| Hydrochloride·0.66H$_2$O | 108–110d | EtOH/H$_2$O/HCl | 90 | pmr | 69a, b |
| 1-Me-2-HO-5-(2-FC$_6$H$_4$)-7-Cl | 143–146 | THF/Et$_2$O | 37 | pmr | 60 |

TABLE VI-1. —(contd.)

| Substituent | mp (°C); [bp (°C/torr)] | Solvent of Crystallization | Yield (%) | Spectra | Refs. |
|---|---|---|---|---|---|
| 1-Me-2-HO$(CH_2)_2$NHCO$CH_2$-5-Ph-7-Cl hydrochloride | 208–210 | | | | 118 |
| 1-Me-2-HO$CH_2$-5-(2-Furyl)-7-Cl hydrochloride | 226–227 | $Me_2CO$/EtOH | | | 54 |
| 1-Me-2-HO$CH_2$-5-Ph-7-Br hydrochloride | 241–242 | *i*-PrOH/$Et_2O$ | | | 48a |
| 1-Me-2-HO$CH_2$-5-Ph-7-Cl hydrochloride | 227–235 | *i*-PrOH/$Et_2O$ | | | 48a |
| 1-Me-2-HO$CH_2$-5-Ph-7-F hydrochloride·*i*-PrOH | 99–101 | *i*-PrOH/$Et_2O$ | | | 48a |
| 1-Me-2-HO$CH_2$-5-Ph-7-MeO hydrochloride | 186–189 | *i*-PrOH/$Et_2O$ | | | 48a |
| 1-Me-2-HO$CH_2$-5-Ph-7-Me hydrochloride | 192–195 | *i*-PrOH/$Et_2O$ | | | 48a |
| 1-Me-2-HO$CH_2$-5-Ph-7-MeS hydrochloride | 196–201 | *i*-PrOH/$Et_2O$ | | | 48a |
| 1-Me-2-HO$CH_2$-5-(2-Br$C_6H_4$)-7-Cl hydrochloride | 205–206 | *i*-PrOH/$Et_2O$ | | | 48a |
| 1-Me-2-HO$CH_2$-5-(2-Cl$C_6H_4$)-7-Cl | 173–175 | *i*-PrOH | 25 | ir, pmr, uv | 47, 48a |
| Hydrochloride | 218–220 | EtOH | | | 47 |
| Maleate | 115–119 | EtOH | | | 47 |
| 1-Me-2-HO$CH_2$-5-(2,3-$Cl_2C_6H_3$)-7-Cl hydrochloride | 226–229 | *i*-PrOH/$Et_2O$ | | | 48a |
| 1-Me-2-HO$CH_2$-5-(2,4-$Cl_2C_6H_3$)-7-Cl hydrochloride | 225 | | | | 48a |
| 1-Me-2-HO$CH_2$-5-(2,6-$Cl_2C_6H_3$)-7-Cl hydrochloride | 216–220 | *i*-PrOH/$Et_2O$ | | | 48a |
| 1-Me-2-HO$CH_2$-5-(3,4-$Cl_2C_6H_3$)-7-Cl hydrochloride | 242–245 | *i*-PrOH/$Et_2O$ | | | 48a |
| 1-Me-2-HO$CH_2$-5-(2-F$C_6H_4$)-7-Cl | 173–175 | $Et_2O$ | 33 | ir, pmr, uv | 47, 48a |
| Hydrochloride | 224–227 | EtOH/$Me_2CO$ | | | |
| 1-Me-2-HO$CH_2$-5-(2-Me$C_6H_4$)-7-Cl hydrochloride | 186–189 | *i*-PrOH/$Et_2O$ | | | 48a |
| 1-Me-2-HO$CH_2$-5-(2-$F_3CC_6H_4$)-7-Cl hydrochloride | 196–201 | *i*-PrOH/$Et_2O$ | | | 48a |
| 1-Me-2-HO$CH_2$-5-(3-$F_3CC_6H_4$)-7-Cl hydrochloride | 226–228 | *i*-PrOH/$Et_2O$ | | | 48a |
| 1-Me-2-HO$CH_2$-5-(2-Thienyl)-7-Cl hydrochloride | 214–218 | | | | 54 |
| 1-Me-2-(1-HO-2-propyl)NHCO$CH_2$-5-(2-Cl$C_6H_4$)-7-Cl hydrochloride | 208–210 | | | | 118 |
| 1-Me-2-MeOOC-5-Ph-7-Cl hydrochloride | 212–214 | MeOH/$Et_2O$ | | | 71a |
| 1-Me-2-MeO$(CH_2)_2$O$CH_2$-5-(2-Cl$C_6H_4$)-7-Cl hydrochloride | 167–172 | | | | 52 |

| | | | | | |
|---|---|---|---|---|---|
| 1-Me-2-MeO(HN)C-5-Ph-7-Cl | 142–144 | $Et_2O$ | | | 70, 71b |
| 1-Me-2-$MeOCH_2$-5-(2-Furyl)-7-Cl | | | | | |
| hydrochloride | 219–221 | *i*-PrOH | | | 54 |
| 1-Me-2-$MeOCH_2$-5-Ph-7-Cl | | | | | |
| hydrochloride | 198–210 | *i*-PrOH/$Et_2O$ | | | 48a, 52 |
| 1-Me-2-$MeOCH_2$-5-(2-$BrC_6H_4$)-7-Br | | | | | |
| hydrochloride | 185–187 | EtOH | | | 49 |
| 1-Me-2-$MeOCH_2$-5-(2-$BrC_6H_4$)-7-Cl | | | | | |
| hydrochloride | 181–182 | | | | 52 |
| 1-Me-2-$MeOCH_2$-5-(2-$ClC_6H_4$)-7Br | | | | | |
| hydrochloride | 193–196 | EtOH | | | 49 |
| 1-Me-2-$MeOCH_2$-5-(2-$ClC_6H_4$)-7-Cl | | | | | |
| hydrochloride | 192–195 | *i*-PrOH/$Me_2CO$ | 73 | ir, pmr | 47, 52 |
| 1-Me-2-$MeOCH_2$-5-(2,6-$Cl_2C_6H_3$)-7-Cl | | | | | |
| hydrochloride | 199–200 | | | | 52 |
| 1-Me-2-$MeOCH_2$-5-(2-$FC_6H_4$)-7-Br | | | | | |
| hydrochloride | 183–185 | EtOH | | | 49 |
| 1-Me-2-$MeOCH_2$-5-(2-$FC_6H_4$)-7-Cl | | | | pmr | 47 |
| hydrochloride | 179–181 | EtOH/$Me_2CO$ | | ir, uv | 47, 52 |
| 1-Me-2-$MeOCH_2$-5-(2-$IC_6H_4$)-7-Br | | | | | |
| hydrochloride | 223–225 | EtOH | | | 49 |
| 1-Me-2-$MeOCH_2$-5-(2-$F_3CC_6H_4$)-7-Br | | | | | |
| hydrochloride·0.5$H_2O$ | 128–130 | EtOH | | | 49 |
| 1-Me-2-$MeOCH_2$-5-(2-$F_3CC_6H_4$)-7-Cl | 95–97 | | | | 52 |
| 1-Me-2-$MeOCH_2$-5-(2-Thienyl)-7-Cl | | | | | |
| hydrochloride | 226–229 | $Me_2CO$/$Et_2O$ | | | 54 |
| 1-Me-2-Me-5-Ph-7-Cl | 94–95.5 | | 13.5 | | 31, 24 |
| 1-Me-2-$MeNHCOCH_2$-5-Ph-7-Cl | | | | | |
| hydrochloride | 183–186 | | | | 118 |
| 1-Me-2-$MeNHCOCH_2$-5-(2-$ClC_6H_4$)-7-Cl | | | | | |
| hydrochloride | 222–226 | | | | 118 |
| 1-Me-2-$MeNHCONHCH_2$-5-Ph-7-Cl | | | | | |
| dihydrochloride | 140–150 | *i*-PrOH/$Et_2O$ | | | 48a |

TABLE VI-1. —(contd.)

| Substituent | mp (°C); [bp (°C/torr)] | Solvent of Crystallization | Yield (%) | Spectra | Refs. |
|---|---|---|---|---|---|
| 1-Me-2-MeNHCONHCH$_2$-5-(2-ClC$_6$H$_4$)-7-Cl hydrochloride | 195 | | | | 118 |
| 1-Me-2-MeNHCH$_2$CONHCH$_2$-5-Ph-7-Cl | 132–133 | | | | 118 |
| 1-Me-2-MeNHCOOCH$_2$-5-Ph-7-Cl | 157–158 | | | | 48a |
| 1-Me-2-(3-Me-butanoyl)NHCH$_2$-5-(2-ClC$_6$H$_4$)-7-Cl hydrochloride | 230–231 | | | | 118 |
| 1-Me-2-(4-Me-Piperazino)-5-Ph-7-Cl trihydrochloride·EtOH | 214–216 | EtOH | | | 48a |
| 1-Me-2-(4-MeC$_6$H$_4$)SCH$_2$-5-Ph-7-Cl hydrochloride | 185–187 | *i*-PrOH/Et$_2$O | | | 48a |
| 1-Me-2-(Morpholino)CH$_2$-5-Ph-7-Cl dihydrochloride | 237–245 | *i*-PrOH/Et$_2$O | | | 48a |
| 1-Me-2-(Morpholino)CH$_2$-5-(2-ClC$_6$H$_4$)-7Cl | 136–138 | EtOH | 72 | ir, pmr, uv | 47 |
| 1-Me-2-PhOCH$_2$-5-Ph-7-Cl hydrochloride | 180 | *i*-PrOH | | | 48a, 52 |
| 1-Me-2-PhNHCONHCH$_2$-5-Ph-7-Cl hydrochloride | 198–207 | *i*-PrOH/Et$_2$O | | | 48a, 118 |
| 1-Me-2-PhNHCOOCH$_2$-5-Ph-7-Cl | 140–145 | | | | 48a, 52 |
| 1-Me-2-PhCH$_2$NHCH$_2$-5-Ph-7-Cl dihydrochloride | 165–168 | *i*-PrOH/Et$_2$O | | | 48a |
| 1-Me-2-(Phthalimido)CH$_2$-5-Ph-7-Cl | 151–152 | MeOH | | | 48a, 118 |
| 1-Me-2-(Phthalimido)CH$_2$-5-(2-ClC$_6$H$_4$)-7-Cl | 167–167.5 | MeOH/Me$_2$CO | 60 | ir, pmr, uv | 47 |
| 1-Me-2-(Piperidino)CH$_2$-5-Ph-7-Cl | 143–145 | Et$_2$O | 65 | | 48a |
| 1-Me-2-(2-Propenyloxy)CH$_2$-5-(2-ClC$_6$H$_4$)-7-Cl | 184–185 | *i*-PrOH/Et$_2$O | | | 52 |
| 1-Me-2-(*i*-PrNHCOCH$_2$)-5-Ph-7-Cl hydrochloride | 220–223 | *i*-PrOH/Et$_2$O | | | 48a, 118 |
| 1-Me-2-(*i*-PrNHCOCH$_2$)-5-(2-ClC$_6$H$_4$)-7-Cl hydrochloride | 240–241 | | | | 48a |

| | | | | | |
|---|---|---|---|---|---|
| 1-Me-2-(*i*-PrNHCONHCH$_2$)-5-Ph-7-Cl | | | | | |
| dihydrochloride | 180–182 | *i*-PrOH/Et$_2$O | | | 48a |
| 1-Me-2-PrOCH$_2$-5-(2-BrC$_6$H$_4$)-7-Br | | | | | |
| hydrochloride | 143–146 | EtOH | | | 49 |
| 1-Me-2-(*i*-PrOCH$_2$)-5-(2-ClC$_6$H$_4$)-7-Br | | | | | |
| hydrochloride | 189–191.5 | EtOH | | | 49 |
| 1-Me-2-PrOCH$_2$-5-(2-ClC$_6$H$_4$)-7-Br | | | | | |
| hydrochloride | 152–154 | EtOH | | | 49 |
| 1-Me-2-PrOCH$_2$-5-(2-ClC$_6$H$_4$)-7-Cl | | | | | |
| hydrochloride | 144–146 | | | | 52 |
| 1-Me-2-(*i*-PrOCH$_2$)-5-(2-ClC$_6$H$_4$)-7-Cl | | | | | |
| hydrochloride | 193–196 | *i*-PrOH/Et$_2$O | | | 48a, 52 |
| 1-Me-2-[3,4,5-(MeO)$_3$C$_6$H$_2$]CONHCH$_2$-5-Ph-7-Cl | | | | | |
| hydrochloride | 180 | *i*-PrOH/Et$_2$O | | | 48a |
| 1-MeNHCO-2-CN-5-Ph-7-Cl | 229–230 | CH$_2$Cl$_2$/Et$_2$O | 71 | ir, ms, pmr | 68a |
| 1-NO-2-CN-5-Ph-7-Cl | 146–149 | CH$_2$Cl$_2$/Hexane | | | 68d |
| 1-NO-2-MeOOCCH$_2$-5-(2-FC$_6$H$_4$)-7-Cl | 86–90 | Et$_2$O/Petr ether | | | 62b |
| 1-NO-2-Ph(OH)CH-5-Ph-7-Cl | | | | | |
| Isomer A | 170–172d | CH$_2$Cl$_2$/Et$_2$O/Hexane | 30.5 | ir, pmr, uv | 73 |
| Isomer B | 180–183d | EtOAc/Hexane | | pmr, uv | 73 |
| 1-PhCH$_2$-2-HOCH$_2$-5-Ph-7-Cl | | | | | |
| hydrochloride | 196–202 | | | | 48a |
| 1-F$_3$CCH$_2$-2-EtS-5-Ph-7-Cl | 110–111 | EtOH/H$_2$O | | | 72 |
| ***1,3,5,7-Tetrasubstituted*** | | | | | |
| 1-Ac-3-AcO-5-Ph-7-Cl | 177–179 | CH$_2$Cl$_2$/Hexane | 59 | uv | 83, 110 |
| 1-Ac-3-HO-5-Ph-7-Cl | 165–167d | CH$_2$Cl$_2$/Et$_2$O | | | 83a |
| 1-HCO-3-AcO-5-Ph-7-Cl | 165–167 | CH$_2$Cl$_2$/Et$_2$O | | | 83 |
| 1-Me-3-EtO-5-Ph-7-Cl, 4-Oxide | 106–109 | Et$_2$O/Petr ether | | | 99c |
| 1-Me-3-Et-5-Ph-7-Cl | 114–115 | | 55 | | 31 |
| 1-Me-3-Et-5-(2-(FC$_6$H$_4$)-7-Cl | 120–121 | | 24.5 | | 31 |
| 1-Me-3-HO-5-Ph-7-Cl | 155–156d | CH$_2$Cl$_2$/Petr ether | 34 | uv | 83, 110 |
| 1-Me-3-Me-5-Ph-7-Cl | 102–104 | CH$_2$Cl$_2$/Petr ether | | | 24, 45b |
| 1-F$_3$CCH$_2$-3-AcO-5-(2-ClC$_6$H$_4$)-7-Cl | 127–131 | Et$_2$O/Hexane | 63 | | 106 |

TABLE VI-1. —(contd.)

| Substituent | mp (°C); [bp (°C/torr)] | Solvent of Crystallization | Yield (%) | Spectra | Refs. |
|---|---|---|---|---|---|
| ***Other Tetrasubstituted*** | | | | | |
| 1-$F_3CCH_2$-5-(2-$FC_6H_4$)-7-Cl-9-$NO_2$ | 138–139 | PhH/Hexane | 41 | | 106 |
| 1-$F_3CCH_2$-5-(2-$FC_6H_4$)-7,9-$(NO_2)_2$ | 188–190 | $CH_2Cl_2$/Hexane | 11 | | 106 |
| 2-$ClCH_2$-5-Ph-7-Cl-9-$NO_2$ | | | | | |
| hydrochloride | 232–235 | | | | 48a |
| 2-$ClCH_2$-5-Ph-7,9-$(NO_2)_2$ | 170–174 | | | | 48a |
| 2,2-$Me_2$-5-Ph-7-Cl | 131–133 | Hexane | | | 99c |
| 2-Me-3-MeO-5-Ph-7-Cl | 174–176 | MeOH | | | 99a, b |
| 2-$H_2NCH_2$-3-Me-5-(2-$FC_6H_4$)-7-Cl | | | | | |
| dimaleate | 188–189 | *i*-PrOH | | | 45b |
| 5-Ph-6,7,8-$(MeO)_3$ | | | | | |
| hydrochloride | 226–227 | EtOH/$Et_2O$ | | | 28c |
| 5-Ph-7,8,9-$(MeO)_3$ | | | | | |
| hydrochloride | 169d | EtOH/$Et_2O$ | | | 28c |
| ***Pentasubstituted*** | | | | | |
| 1-Me-2-$CH_2OH$-5-[3,4-$(MeO)_2C_6H_3$]-7,8-$(MeO)_2$ | | | | | |
| hydrochloride | 111–115 | | | | 48a |
| 1-Me-5-Ph-6,7,8-$(MeO)_3$ hydrochloride | 207d | EtOH/$Et_2O$ | | | 28c |
| 1-Me-5-Ph-6,8-$(MeO)_2$-7-OH | 203–205 | EtOH | | | 28c |

TABLE VI-2. 2-IMINO-2,3-DIHYDRO-1$H$-1,4-BENZODIAZEPINES

| Substituent | mp (°C); [bp (°C/torr)] | Solvent of Crystallization | Yield (%) | Spectra | Refs. |
|---|---|---|---|---|---|
| ***R; Other*** | | 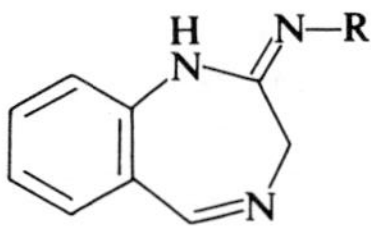 | | | |
| AcNH; 1-Me-5-(2-$FC_6H_4$)-7-Cl | 230–232 | $CH_2Cl_2/Et_2O$ Petr ether | | | 89 |
| $Me_2N$; 1-Me-5-Ph-7-Cl | 148–153 | $Et_2O$/Petr ether | | | 14c |
| H; 1-Me-5-Ph-7-Cl dihydrochloride | 198–200d | EtOH | 79 | | 87a, b |
| Me; 1-Me-5-Ph-7-Cl | 186–188d | EtOAc/Hexane | | uv | 90 |
| MeNHCO; 1-Me-5-Ph-7-Cl | 167–169 | MeOH/$i$-$Pr_2O$ | | | 120 |

TABLE VI-3. 2-METHYLENE-2,3-DIHYDRO-1*H*-1,4-BENZODIAZEPINES

| Substituent | mp (°C); [bp (°C/torr)] | Solvent of Crystallization | Yield (%) | Spectra | Refs. |
|---|---|---|---|---|---|
| ***R; Other*** | | 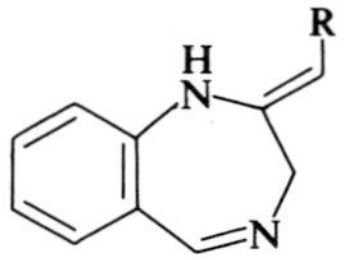 | | | |
| Ac; 5-(2-$ClC_6H_4$)-7-Cl | 143–145; 118–120 | $Et_2O$/Hexane | 62 | ir, pmr, uv | 93 |
| Ac; 5-(2-$FC_6H_4$)-7-Cl | 157–159 | $CH_2Cl_2$/Hexane | | | 63 |
| $H_2NCO$;1-Me-5-Ph-7-Cl·0.4EtOH | 179–181 | EtOH | 75 | | 98 |
| $H_2NCO$; 1-Me-5-(2-$ClC_6H_4$)-7-Br·0.75$Me_2CO$ | 110–115 | $Me_2CO$ | | | 98 |
| $H_2NCO$; 1-Me-5-(2-$ClC_6H_4$)-7-Cl | 212–217 | EtOH | | | 98 |
| $H_2NCO$; 1-Me-5-(2-$ClC_6H_4$)-7-$NO_2$ | 211–214 | | | | 98 |
| $H_2NCO$; 1-Me-5-(2-$FC_6H_4$)-7-Br·0.5$Me_2CO$ | 154–155 | $Me_2CO$ | | | 98 |
| $H_2NCO$; 1-Me-5-(2-$FC_6H_4$)-7-Cl·0.5$H_2O$·0.33$Me_2CO$ | 162–163 | $Me_2CO$ | | | 98 |
| $H_2NCO$; 1-Me-5-(2-$F_3CC_6H_4$)-7-Br | 115–120 | | | | 98 |
| $H_2NCO$; 1-Me-5-(2-$F_3CC_6H_4$)-7-Cl | 101–105 | | | | 98 |
| $H_2NCO$; 5-(2-$FC_6H_4$)-7-Cl | 243–245 | THF/EtOAc | | | 63 |
| PhCONH; 5-Ph-7-Cl, 4-Oxide | 207–210d | THF/$H_2O$ | | | 99c |
| *t*-BuOOC; 5-(2-$ClC_6H_4$) | 170–173 | $CH_2Cl_2$/EtOH | | | 63 |
| 4-Oxide | 219–220 | $CH_2Cl_2$/EtOH | | | 63 |
| *t*-BuOOC; 5-(2-$ClC_6H_4$)-7-Cl, 4-Oxide | 168–170 | $Et_2O$ | 87 | | 95 |
| *t*-BuOOC; 5-(2-$ClC_6H_4$)-7-F, 4-Oxide | 220–221 | $Et_2O$ | | | 63 |
| *t*-BuOOC; 5-(2-$ClC_6H_4$)-8-Cl | 158–160 | $CH_2Cl_2$/EtOH | | | 63 |
| 4-Oxide | 197–199 | $Et_2O$ | | | 63 |
| *t*-BuOOC; 5-(4-$ClC_6H_4$), 4-Oxide | 160–162 | Cyclohexane | | | 63 |

| | | | | | |
|---|---|---|---|---|---|
| $t$-BuOOC; 5-(4-$ClC_6H_4$)-7-Cl, 4-Oxide | 195–196 | Hexane | | | 63 |
| $t$-BuOOC; 5-(2-$FC_6H_4$) | 155–157 | MeOH | | | 63 |
| 4-Oxide | 186–188d | EtOH | | | 63 |
| $t$-BuOOC; 5-(2-$FC_6H_4$)-7-Cl | 158–160 | Hexane | | | 95 |
| 4-Oxide | 182–183 | $CH_2Cl_2$/Hexane | 98 | | 95 |
| $t$-BuOOC; 5-(2-F-4-$MeOC_6H_3$)-7-Cl | 178–180 | $Et_2O$/Hexane | | | 63 |
| 4-Oxide | 128–131 | $Et_2O$/Hexane | | | 63 |
| CN; 1-Me-5-Ph-7-Cl | 225–227 | EtOH | 60 | ir | 98 |
| CN; 1-Me-5-(2-$ClC_6H_4$)-7-Br | 178–179 | | | | 98 |
| CN; 1-Me-5-(2-$ClC_6H_4$)-7-Cl | 162–165 | $i$-PrOH | | ir | 98 |
| CN; 1-Me-5-(2-$ClC_6H_4$)-7-$NO_2$ | 83–90 | | | | 98 |
| CN; 1-Me-5-(2-$FC_6H_4$)-7-Br | 167–168 | | | | 98 |
| CN; 1-Me-5-(2-$FC_6H_4$)-7-Cl | 160–162 | | | | 98 |
| CN; 1-Me-5-(2-$F_3CC_6H_4$)-7-Br | 128–130 | | | | 98 |
| CN; 1-Me-5-(2-$F_3CC_6H_4$)-7-Cl | 125–130 | | | | 98 |
| CN; 5-(2-$FC_6H_4$)-7-Cl | 189–191 | EtOH | | | 95 |
| 4-Oxide | 247–248 | THF/EtOH | | | 95 |
| $Me_2NCO$; 3-AcO-5-(2-$FC_6H_4$)-7-Cl | 245–246 | $CH_2Cl_2/Et_2O$ | | | 63 |
| $Me_2NCO$; 5-Ph-7-Cl | 182–184 | $CH_2Cl_2/Et_2O$ | | | 95 |
| 4-Oxide | 221–223 | $CH_2Cl_2$/EtOH | | | 95 |
| $Me_2NCO$; 5-(2-$ClC_6H_4$)-7-Cl | 176–178 | EtOAc/Hexane | 71 | | 95 |
| 4-Oxide | 221–225 | EtOAc | 66 | | 95 |
| $Me_2NCO$; 5-(2-$FC_6H_4$)-7-Cl | 170–172 | EtOH | | | 95 |
| 4-Oxide | 159–160 | EtOH | 67 | | 95 |
| $Me_2NSO_2$; 5-(2-$ClC_6H_4$)-7-Cl | 161–163 | EtOH | 60 | | 95 |
| 4-Oxide | 182–184 | EtOAc/Hexane | 72.5 | | 95 |
| $Me_2NSO_2$; 5-(2-$FC_6H_4$)-7-Cl | 150–153 | EtOAc/Hexane | 65 | | 95 |
| 4-Oxide | 168–170 | EtOAc/$Et_2O$ | 61 | | 95 |
| EtOOC; 5-(2-$ClC_6H_4$), 4-Oxide | 172–174 | $CH_2Cl_2/Et_2O$ | | | 63 |
| EtOOC; 5-Ph-7-Cl, 4-Oxide | 138–140 | EtOAc/Hexane | | | 96 |
| EtOOC; 5-(2-$ClC_6H_4$)-7-Cl | 103–105 | $CH_2Cl_2$/EtOH | | | 63 |
| 4-Oxide | 207–209 | EtOH | | | 99c |
| EtOOC; 5-(2-$FC_6H_4$)-7-Cl | 109–111 | EtOH | | | 95 |
| 4-Oxide | 148–150 | EtOAc/Hexane | | | 95, 96 |

TABLE VI-3. —(contd.)

| Substituent | mp (°C); [bp (°C/torr)] | Solvent of Crystallization | Yield (%) | Spectra | Refs. |
|---|---|---|---|---|---|
| CHO(Me)NCO; 5-(2-$FC_6H_4$)-7-Cl | 180–183 | $Et_2O$ | | | 63 |
| MeOOC; 5-Ph | 130–132 | Cyclohexane | | | 63 |
| MeOOC; 5-(4-$ClC_6H_4$) | 129–131 | EtOAc/Hexane | | | 63 |
| MeOOC; 5-Ph-7-Cl | 171–173 | $Et_2O$ | | | 92, 96 |
| 4-Oxide | 215–216 | MeOH | | | 92, 96 |
| MeOOC; 5-(2-$ClC_6H_4$)-7-Cl | 158–159 | $CH_2Cl_2$/MeOH | 90, 96 | ir, pmr, uv | 93 |
| MeOOC; 5-(2-$ClC_6H_4$)-7,9-$Cl_2$ | 177–179 | EtOH | | | 63 |
| MeOOC; 5-(2-$FC_6H_4$)-7-Cl | 161–162 | $CH_2Cl_2$/Hexane | | | 92, 96 |
| 4-Oxide | 192–193 | MeOH/$Et_2O$ | | | 96 |
| MeOOC; 5-(2-Pyridyl)-7-Br | 178–180 | MeOH | | | 63 |
| MeNHCO; 5-(2-$FC_6H_4$)-7-Cl | 203–205 | $CH_2Cl_2$/Hexane | | | 63 |
| (4-Me-Piperazino)CO; 5-(2-$FC_6H_4$)-7-Cl, 4-Oxide | 202–204 | EtOH | 30 | | 95 |
| $MeSO_2$; 5-(2-$FC_6H_4$)-7-Cl | 131–134 | $CH_2Cl_2$/EtOH | 44 | | 95 |
| 4-Oxide | 185–187d | THF/EtOH | 80 | | 95 |
| $NO_2$; 3-Me-5-(2-$ClC_6H_4$)-7-$NO_2$ | 215–218 | $Et_2O$/Petr ether | | | 89 |
| $NO_2$; 3-Me-5-(2-$FC_6H_4$)-7-Cl | 219–221 | $CH_2Cl_2$/MeOH | | | 92 |
| 4-Oxide | 216–218 | $CH_2Cl_2$/ETOAc | 87 | | 94, 92 |
| $NO_2$; 5-Ph | 141–142 | EtOH | | | 92 |
| $NO_2$; 5-Ph-7-Cl | 184–186 | $CH_2Cl_2$/$Et_2O$/Hexane | | | 92 |
| 4-Oxide | 253–254d | $CH_2Cl_2$ | 57 | | 94 |
| $NO_2$; 5-Ph-7-(2-Me-1, 3-Dioxolan-2-yl) | 158–161 | $CH_2Cl_2$/Hexane | | | 92 |
| $NO_2$; 5-(2-$ClC_6H_4$)-7-Cl | 182–185 | EtOH | | | 92 |
| $NO_2$; 5-(2-$ClC_6H_4$)-7-$NO_2$ | 240–243d | $CH_2Cl_2$/EtOH | | | 92 |
| $NO_2$; 5-(2-$FC_6H_4$), 4-Oxide | 209–212d | THF/Hexane | 84 | | 94 |
| $NO_2$; 5-(2-$FC_6H_4$)-7-Cl | 174–176 | $CH_2Cl_2$/EtOH | | | 92 |
| 4-Oxide | 240–243 | DMF/$H_2O$ | 93 | | 94 |
| $NO_2$; 5-(2-$FC_6H_4$)-7-Et | 138–141 | $Et_2O$/Hexane | | | 92 |
| $NO_2$; 5-(2-$FC_6H_4$)-7-I | 214–216 | $CH_2Cl_2$/$Et_2O$ | | | 14c |
| $NO_2$; 5-(2-$FC_6H_4$)-7-$NO_2$, 4-Oxide | 216–220 | THF/Hexane | 36 | | 94 |

| | | | | | |
|---|---|---|---|---|---|
| $NO_2$; 5-(2-Pyridyl)-7-Br | 240–245d | THF/EtOH | | | 92 |
| *i*-PrOOC; 5-(2-$FC_6H_4$)-7-Cl | 144–145 | *i*-PrOH | | | 95 |
| 4-Oxide | 175–177 | $CH_2Cl_2$/EtOH | 94 | | 95 |
| 2-Pyridyl; 5-(2-$FC_6H_4$)-7-Cl | 169–172 | $CH_2Cl_2$/Hexane | 79 | | 95 |
| 4-Oxide | 212–214d | EtOAc | 38.5 | | 95 |
| ***$R_1$, $R_2$; Other*** | | | | | |
| Ac, Ac; 5-(2-$ClC_6H_4$)-7-Cl | 205–207 | $CH_2Cl_2$/*i*-PrOH | 27.5 | ir, pmr, uv | 93 |
| Ac, *t*-BuOOC; 5-(2-$FC_6H_4$)-7-Cl | 160–162 | Cyclohexane | | | 63 |
| Ac, EtOOCNH(EtOOC)N; 5-(2-$ClC_6H_4$)-7-Cl | 184–186 | $CH_2Cl_2$/Hexane | | | 63 |
| Ac, MeOOC; 5-(2-$ClC_6H_4$)-7-Cl | 142–144 | *i*-PrOH | 25 | | 93 |
| AcNH, EtOOC; 5-(2-$ClC_6H_4$)-7-Cl maleate | 139–142d | EtOAc | | | 92 |
| AcNH, EtOOC; 5-(2-$FC_6H_4$)-7-Cl maleate | 149–151 | EtOAc | | | 92 |
| AcNH, MeOOC; 5-(2-$ClC_6H_4$)-7-Cl | 177–179 | EtOAc/Hexane | | | 63 |
| $H_2N$, *t*-BuOOC; 5-(4-$ClC_6H_4$) | 190–195 | $CH_2Cl_2$/EtOH | | | 63 |
| $H_2N$, EtOOC; 5-(2-$ClC_6H_4$)-7-Cl | 145–150d | $Et_2O$/Hexane | | ir, pmr, uv | 92, 93 |
| ·EtOH | 119–121 | THF/EtOH | 72.5 | pmr | 92, 93 |
| $H_2N$, MeOOC; 5-(2-Pyridyl)-7-Br | 193–196 | $CH_2Cl_2$/EtOH | | | 63 |
| $H_2NCO$,CN; 5-Ph-7-Cl, 4-Oxide | 246–248d | $Et_2O$ | | | 63 |
| PhCONH, MeOOC; 5-(2-$FC_6H_4$)-7-Cl | 217–219 | EtOAc/Hexane | | | 92 |
| *t*-BuOOC, *t*-BuOOC; 5-Ph-7-Cl, 4-Oxide | 209–214d | $CH_2Cl_2$/$Et_2O$ | | | 63 |
| *t*-BuOOC, CN; 5-(2-$FC_6H_4$)-7-Cl | 174–177 | $Et_2O$/Hexane | | | 63 |
| HOOC, CN; 5-(2-$FC_6H_4$)-7-Cl | 182–184d | THF/Hexane | | | 63 |
| Cl, MeOOC; 5-(2-$ClC_6H_4$)-7-Cl | | | | | |
| Isomer A | 179–180 | EtOAc/Hexane | | | 63 |
| Isomer B | 152–154 | $Et_2O$/Hexane | | | 63 |
| $ClCH_2CONH$, MeOOC; 5-(2-$ClC_6H_4$)-7-Cl | 192–194d | $CH_2Cl_2$/$Et_2O$ | 98 | pmr | 116 |
| (2-Cl-Propanoyl)NH, MeOOC; 5-(2-$ClC_6H_4$)-7-Cl | 128–130 | $Et_2O$ | | | 63 |
| (3-Cl-Propanoyl)NH, MeOOC; 5-(2-$ClC_6H_4$)-7-Cl | 176–179d | $CH_2Cl_2$/$Et_2O$ | 54 | pmr | 116 |
| CN, CN; 5-Ph-7-Cl | 274–276 | THF/EtOAc | 58 | | 91 |
| 4-Oxide·0.5 dioxane | 240–242d | Dioxane | 79 | uv. pmr | 91 |

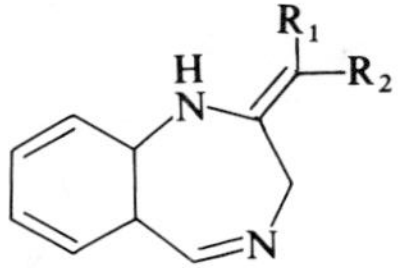

TABLE VI-3. —(contd.)

| Substituent | mp (°C); [bp (°C/torr)] | Solvent of Crystallization | Yield (%) | Spectra | Refs. |
|---|---|---|---|---|---|
| CN, EtOOC; 5-Ph-7-Cl, 4-Oxide | 207–208 | $CH_2Cl_2$/EtOH | | | 63 |
| CN, EtOOC; 5-(2-$FC_6H_4$)-7-Cl | 236–238.5 | | | | 97 |
| $Me_2NCO$, EtOOC; 5-Ph-7-Cl | 198–200 | $CH_2Cl_2$/$Et_2O$ | | | 63 |
| 4-Oxide | 251–253 | $CH_2Cl_2$/Hexane | | | 63 |
| EtOOC, EtOOC; 5-(2-$ClC_6H_4$) | 112–115 | $Et_2O$ | | | 63 |
| EtOOC, EtOOC; 5-Ph-7-Cl | 141–144 | | | | 97 |
| EtOOC, EtOOC; 5-Ph-7-$NO_2$ | 127.5–130 | | | | 97 |
| EtOOC, EtOOC; 5-(2-$ClC_6H_4$)-7-Cl | 191.5–194.5 | | | | 97 |
| EtOOC, EtOOC; 5-(2-$ClC_6H_4$)-7-$NO_2$ | 176–179 | | | | 97 |
| EtOOC, EtOOC; 5-(2-$FC_6H_4$)-7-Cl | 197–200 | $CH_2Cl_2$/$Et_2O$ | | | 63, 97 |
| EtOOC, EtOOC; 5-(2-$FC_6H_4$)-7-$NO_2$ | 182–184 | | | | 97 |
| EtOOC, EtOOC; 5-(2-Pyridyl)-7-Br | 190.5–193.5 | | | | 97 |
| EtOOC, $EtOOCCH_2CO$; 5-(2-$ClC_6H_4$)-7-Cl | 151–154 | $CH_2Cl_2$/$Et_2O$ | | | 63 |
| EtOOCNH, MeOOC; 5-(2-$FC_6H_4$)-7-Cl | 188–191 | $CH_2Cl_2$/EtOAc/Hexane | | | 119a, b |
| EtOOCNH(EtOOC)N, MeOOC; 5-(2-$ClC_6H_4$)-7-Cl | 211–214 | EtOAc/Hexane | 95 | ir, pmr, uv | 108 |
| EtOOCNH(EtOOC)N, MeOOC; 5-(2-$FC_6H_4$)-7-Cl | 195–200 | EtOAc/Hexane | | | 63 |
| MeOOC, MeOOC; 5-Ph | 156–158 | *i*-PrOH | | | 63 |
| MeOOC, MeOOC; 5-(2-$ClC_6H_4$) | 135–138 | $Et_2O$ | | | 63 |
| MeOOC, MeOOC; 5-(4-$ClC_6H_4$) | 135–138 | EtOH | | | 63 |
| MeOOC, MeOOC; 5-Ph-7-Cl | 165–166 | *i*-PrOH | 83 | ir, uv, pmr | 91, 92 |
| | 138–140 | | | | |
| 4-Oxide | 194–195 | $CH_2Cl_2$/Hexane | 69 | ir, pmr, uv | 91, 92 |
| MeOOC, MeOOC; 5-(2-$ClC_6H_4$)-7-Cl | 205–207 | EtOAc | | | 92 |
| MeOOC, MeOOC; 5-(2-$ClC_6H_4$)-7-$NO_2$ | 172–175 | $CH_2Cl_2$/$Et_2O$/Hexane | | | 63 |
| MeOOC, MeOOC; 5-(2-$ClC_6H_4$)-7,9-$Cl_2$ | 128–130 | EtOH | | | 63 |
| MeOOC, MeOOC; 5-(2-$FC_6H_4$)-7-Cl | 170–172 | $CH_2Cl_2$/EtOH | | | 92 |
| MeOOC, MeOOC; 5-(2-Pyridyl)-7-Br | 171–173 | EtOAc | | | 63 |
| Me, $NO_2$; 5-(2-$FC_6H_4$)-7-Cl | 153–155 | $CH_2Cl_2$/EtOH | | | 14c |
| (2-Oxo-3-penten-4-yl)NH, MeOOC; 5-(2-$ClC_6H_4$)-7-Cl | 218–220d | $CH_2Cl_2$/EtOAc | 62 | ir, pmr, uv | 93 |
| (Propenoyl)NH, MeOOC; 5-(2-$ClC_6H_4$)-7-Cl | 203–206 | EtOAc/$Et_2O$ | 90 | pmr | 116 |

TABLE VI-4. 3-METHYLENE-2,3-DIHYDRO-2*H*-1,4-BENZODIAZEPINES

| Substituent | mp (°C); [bp (°C/torr)] | Solvent of Crystallization | Yield (%) | Spectra | Refs. |
|---|---|---|---|---|---|
| ***R; Others*** | | 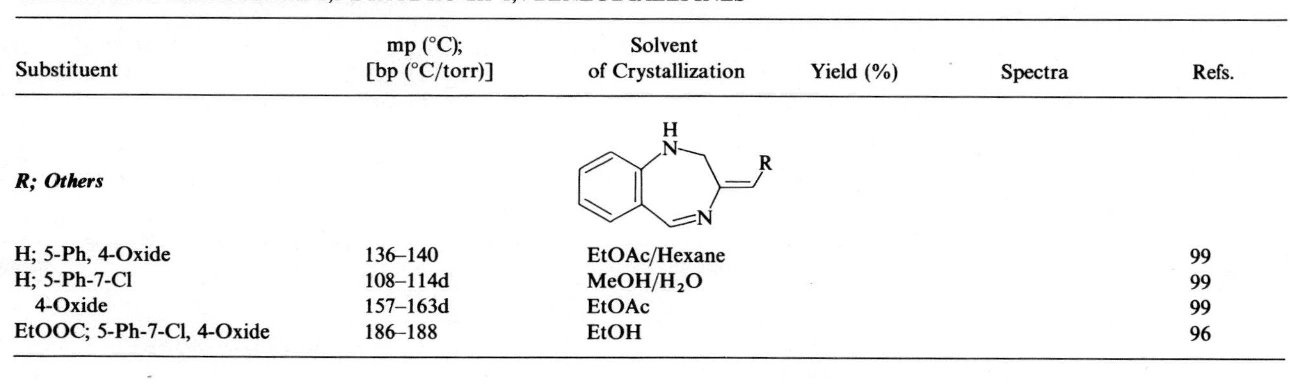 | | | |
| H; 5-Ph, 4-Oxide | 136–140 | EtOAc/Hexane | | | 99 |
| H; 5-Ph-7-Cl | 108–114d | $MeOH/H_2O$ | | | 99 |
| 4-Oxide | 157–163d | EtOAc | | | 99 |
| EtOOC; 5-Ph-7-Cl, 4-Oxide | 186–188 | EtOH | | | 96 |

TABLE VI-5. 2,5-DIHYDRO-1*H*-1,4-BENZODIAZEPINES

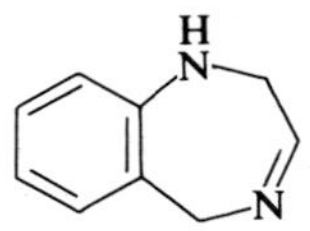

| Substituent | mp (°C); [bp (°C/torr)] | Solvent of Crystallization | Yield (%) | Spectra | Refs. |
|---|---|---|---|---|---|
| 5-Ph-7-Cl | 191–192 | | | ir, ms, uv | 137 |
| 3-Me-5-Ph-7-Cl, 4-Oxide | 170–176d | MeOH | 70 | pmr, uv | 64, 99a, b |
| 3-(4-$MeC_6H_4SO_2$)NH-5-Ph-7-Cl | 228–230d | $Me_2CO$ | | ir, uv | 99c |
| 3-PhNH-5-Ph-7-Cl | 202–208d | EtOAc/Hexane | | ir, uv | 99c |

TABLE VI-6. 4,5-DIHYDRO-1,4-BENZODIAZEPINES

| Substituent | mp (°C); [bp (°C/torr)] | Solvent of Crystallization | Yield (%) | Spectra | Refs. |
|---|---|---|---|---|---|
| 4,5-Dihydro-1*H*-1,4-benzodiazepines | | | | | |
| 2-(4-$NO_2C_6H_4$)-3-CN-4-*t*-Bu-9-Me | | | 62 | ir, pmr | 138 |
| 4,5-Dihydro-3*H*-1,4-benzodiazepines | | | | | |
| 2-MeNH-5-Ph | 153–155 | $Et_2O$/Petr ether | | | 139 |
| Hydrochloride | 240–242 | EtOH/$Et_2O$ | 70 | | 139 |
| Dihydrochloride | 240–242 | MeOH/$Et_2O$/Petr ether | | | 139 |
| 2-MeNH-5-Ph-7-Cl | 179–180 | $Et_2O$/Petr ether | 63 | | 139, 143 |
| Dihydrochloride | 226–238 | MeOH/$Et_2O$ | | | 139 |
| 2-MeNH-4-HO-5-Ph-7-Cl | | | | | |
| Isomer A | 181–183 | $Et_2O$ | 87 | | 140, 141 |
| Isomer B | 196–198 | | 38 | | 141 |
| 2-MeNAc-4-AcO-5-Ph-7-Cl | 133–134 | $CH_2Cl_2$/Petr ether | | | 45a |
| 2-PrO-4-HO-5-Ph-7-Cl | 126–127 | Cyclohexane | 69 | | 142 |
| 2-MeO-4-HO-5-Ph-7-Cl | | | | | 141, 142 |

## 6. REFERENCES


1. J. A. Hill, A. W. Johnson, and T. J. King, *J. Chem. Soc.*, 4430 (1961).
2. L. H. Sternbach, G. A. Archer, and E. Reeder, *J. Org. Chem.*, **28**, 3013 (1963).
3. (a) G. A. Archer, A. Stempel, S. S. Ho, and L. H. Sternbach, *J. Chem. Soc., C.*, 1031 (1966). (b) G. A. Archer, unpublished data, Hoffmann-La Roche, Nutley, NJ.
4. R. I. Fryer, R. A. Schmidt, and L. H. Sternbach, U.S. Patent 3,403,161, September 1968.
5. G. A. Archer, R. I. Kalish, R. Y. Ning, B. C. Sluboski, A. Stempel, T. V. Steppe, and L. H. Sternbach, *J. Med. Chem.*, **20**, 1312 (1977).
6. A. M. Felix, R. I. Fryer, and L. H. Sternbach, U.S. Patent 3,546,212, December 1970.
7. Ger. Offen. 2,223,648, November 1972 (Hoffmann-La Roche & Co., Switzerland).
8. H. Moriyama, H. Yamamoto, S. Inaba, and H. Nagata, U.S. Patent 3,817,984, June 1974.
9. Belg. Patent 770,749, August 1971 (Lab. Pharmedical S.A., Luxembourg).
10. S. T. Ross, U.S. Patent 3,555,010, January 1971.
11. Belg. Patent 772,935, October 1971 (Lab. Pharmedical S.A., Luxembourg).
12. Belg. Patent 794,358, May 1973 (Lab. Pharmedical S.A., Luxembourg).
13. S. C. Bell and C. Gochman, U.S. Patent 3,538,082, November 1970.
14. (a) R. I. Fryer and L. H. Sternbach, U.S. Patent 3,391,158, July 1968. (b) R. I. Fryer and L. H. Sternbach, U.S. Patent 3,391,159, July 1968. (c) R. I. Fryer, unpublished data, Hoffmann-La Roche, Nutley, NJ.
15. L. H. Sternbach, E. Reeder, and G. A. Archer, *J. Org. Chem.*, **28**, 2456 (1963).
16. Belg. Patent 772,825, October 1971 (Lab. Pharmedical S.A., Luxembourg).
17. N. Blasevic and F. Kajfez, *J. Heterocycl. Chem.*, **8**, 845 (1971).
18. Neth. Patent 7,108,245, December 1971 (Compagnia di Ricerca Chimica S.A., Switzerland).
19. W. Schlesinger, U.S. Patent 4,155,904, May 1979.
20. G. A. Archer and L. H. Sternbach, U.S. Patent 3,583,978, June 1971.
21. G. A. Archer and L. H. Sternbach, U.S. Patent 3,803,233, April 1974.
22. G. A. Archer and L. H. Sternbach, U.S. Patent 3,887,604, June 1975.
23. G. A. Archer and L. H. Sternbach, U.S. Patent 3,646,011, February 1972.
24. M. Mihalic, V. Sunjic, and F. Kajfez, *Tetrahedron Lett.*, 1011 (1975).
25. Swiss Patent 610,311, April 1979 (Compagnia di Ricerca Chimica S.A., Switzerland).
26. M. Mihalic, V. Sunjic, F. Kajfez, and M. Zinic, *J. Heterocycl. Chem.*, **14**, 941 (1977).
27. G. A. Archer and L. H. Sternbach, U.S. Patent 3,553,199, January 1971.
28. (a) J. Hellerbach and G. Zanetti, U.S. Patent 3,906,025, September 1975. (b) Neth. Patent 7,205,955, December 1972 (Hoffmann-La Roche & Co., Switzerland). (c) G. Zanetti, unpublished data, Hoffmann-La Roche & Co., Basel, Switzerland.
29. G. N. Walker, A. R. Engle, and R. J. Kempton, *J. Org. Chem.*, **37**, 3755 (1972).
30. Belg. Patent 775,710, May 1972 (VEB Arzneimittelwerk, Germany).
31. F. Kajfez, V. Sunjic, and V. Caplar, U.S. Patent 4,226,766, October 1980.
32. P. Giacconi, E. Rossi, R. Stradi, and R. Eccel, *Synthesis*, 789 (1982).
33. H. Yamamoto, S. Inaba, T. Okamoto, T. Hirohashi, K. Ishizumi, M. Yamamoto, I. Maruyama, K. Mori, and T. Kobayashi, U.S. Patent 4,002,611, January 1977.
34. A. Walser and J. Hellerbach, Hoffmann-La Roche & Co., Basel, Switzerland, unpublished data.
35. F. Gatta and S. Chiavarelli, *Farmaco, Ed. Sci.*, **32**, 33 (1977).
36. H. Yamamoto, S. Inaba, T. Okamoto, T. Hirohashi, K. Ishizumi, M. Yamamoto, I. Murayama, K. Mori, and T. Kobayashi, U.S. Patent 3,663,534, May 1972.
37. S. Inaba, K. Ishizumi, T. Okamoto, and H. Yamamoto, *Chem. Pharm. Bull.*, **20**, 1628 (1972).
38. H. Yamamoto, S. Inaba, T. Okamoto, T. Hirohashi, K. Ishizumi, M. Yamamoto, I. Maruyama, K. Mori, T. Kobayashi, and T. Izumi, U.S. Patent 3,702,321, November 1972.
39. Neth. Patent 7,112,297, March 1972 (Sumitomo Chemical Co., Japan).
40. H. Yamamoto, S. Inaba, T. Okamoto, T. Hirohashi, K. Ishizumi, M. Yamamoto, I. Maruyama, T. Kobayashi, and T. Izumi, U.S. Patent 3,875,142, April 1975.

41. H. H. Kaegi, *J. Labelled Compd.*, **4**, 363 (1968).
42. I. Nakatsuka, K. Kawahara, T. Kamada, F. Shono, and A. Yoshitake, *J. Labelled Compd.*, **13**, 453 (1977).
43. M. E. Derieg, R. I. Fryer, and L. H. Sternbach, U.S. Patent 3,651,046, March 1972.
44. M. Steinman, U.S. Patent 3,723,414, March 1973.
45. (a) L. H. Sternbach et al., unpublished data, Hoffmann-La Roche, Nutley, NJ. (b) A. Walser and R. I. Fryer, unpublished results, Hoffmann-La Roche, Nutley, NJ.
46. K.-H. Wünsch, H. Dettmann, and S. Schönberg, *Chem. Ber.*, **102**, 3891 (1969).
47. H. Liepmann, W. Milkowski, and H. Zeugner, *Eur. J. Med. Chem.*, **11**, 501 (1976).
48. W. Milkowski, S. Funke, R. Hüschens, H.-G. Liepmann, W. Stühmer, and H. Zeugner: (a) U.S. Patent 3,998,809, December 1976, (b) U.S. Patent 4,096,141, June 1978.
49. W. Milkowski, S. Funke, R. Hüschens, H.-G. Liepmann, W. Stühmer, and H. Zeugner, U.S. Patent 4,098,786, July 1978.
50. Ger. Offen. 2,952,279, June 1981 (Kali-Chemie Pharma GmbH, W. Germany).
51. H. Zeugner, D. Roemer, H. Liepmann, and W. Milkowski: (a) EP 0,068,240, January 1983 (Kali-Chemie Pharma GmbH, W. Germany), (b) U.S. Patent 4,325,957, April 1982.
52. Ger. Offen. 2,353,160 November 1974 (Kali-Chemie AG, W. Germany).
53. Ger. Offen. 2,448,259, April 1976 (Kali-Chemie Pharma GmbH, W. Germany).
54. Ger. Offen. 2,314,993, October 1974 (Kali-Chemie Pharma GmbH, W. Germany).
55. C. Corral, R. Madronero, and S. Vega, *J. Heterocycl. Chem.*, **14**, 985 (1977).
56. (a) M. Uskokovic, J. Iacobelli, and W. Wenner, *J. Org. Chem.*, **27**, 3606 (1962). (b) M. Uskokovic and W. Wenner, U.S. Patent 3,261,828, July 1966.
57. T. S. Sulkowski and S. J. Childress, *J. Org. Chem.*, **28**, 2150 (1963).
58. K. Ishizumi, S. Inaba, and H. Yamamoto, *J. Org. Chem.*, **37**, 4111 (1972).
59. K. Ishizumi, K. Mori, T. Okamoto, T. Akase, T. Izumi, M. Akatsu, Y. Kume, S. Inaba, and H. Yamamoto, U.S. Patent 4,044,003, August 1977.
60. R. Ian Fryer, D. L. Coffen, J. V. Earley, and A. Walser, *J. Heterocycl. Chem.*, **10**, 473 (1973).
61. (a) G. A. Archer and L. H. Sternbach, *J. Org. Chem.*, **29**, 231 (1964). (b) G. A. Archer and L. H. Sternbach, U.S. Patent 3,678,036, July 1972. (c) G. A. Archer and L. H. Sternbach, U.S. Patent 3,422,091, January 1969.
62. (a) J. V. Earley, R. I. Fryer, and A. Walser, U.S. Patent 3,838,116, September 1974. (b) J. V. Earley, R. I. Fryer, and A. Walser, unpublished data.
63. A. Walser, T. Flynn, C. Mason, and R. I. Fryer, unpublished data, Hoffmann-La Roche, Nutley, N.J.
64. G. F. Field, W. J. Zally, and L. H. Sternbach, *J. Am. Chem. Soc.*, **89**, 332 (1967).
65. A. Walser, L. E. Benjamin, Sr., T. Flynn, C. Mason, R. Schwartz, and R. I. Fryer, *J. Org. Chem.*, **43**, 936 (1978).
66. (a) A. Walser and R. I. Fryer, *J. Heterocycl. Chem.*, **20**, 551 (1983). (b) A. Walser, U.S. Patents 4,226,768, October 1980; 4,240,962, December 1980; 4,244,868, January 1981. (c) A. Walser, U.S. Patents 4,226,771, October 1980; 4,247,463, January 1981.
67. Neth. Patent 6,614,923, April 1967 (Hoffmann-La Roche & Co., Switzerland).
68. (a) D. L. Coffen, J. P. DeNoble, E. L. Evans, G. F. Field, R. Ian Fryer, D. A. Katonak, B. J. Mandel, L. H. Sternbach, and W. J. Zally, *J. Org. Chem.*, **39**, 167 (1974). (b) D. L. Coffen and R. Ian Fryer, U.S. Patent 3,849,399, November 1974. (c) D. L. Coffen and R. Ian Fryer, U.S. Patents 3,932,399, January 1976; 3,906,001, September 1975; 3,850,948, November 1974. (d) D. L. Coffen et al., unpublished results, Hoffmann-La Roche, Nutley, NJ.
69. (a) K. Meguro and Y. Kuwada, *Yakugaku Zasshi*, **93**, 1263 (1973). (b) K. Meguro, H. Tawada, Y. Kuwada, and T. Masuda, U.S. Patent 3,692,772, September 1972.
70. R. Y. Ning and M. A. Schwartz, U.S. Patent 3,925,358, December 1975.
71. (a) L. H. Sternbach and R. Y. Ning, U.S. Patent 3,873,525, March 1975. (b) Belg. Patent 818,132, January 1975 (Hoffmann-La Roche & Co., Switzerland).
72. M. Steinman, U.S. Patent 3,856,787, December 1974.
73. A. Walser, R. F. Lauer, and R. Ian Fryer, *J. Heterocycl. Chemistry*, **15**, 855 (1978).

74. (a) J. B. Hester, D. J. Duchamp, and C. G. Chidester, *Tetrahedron Lett.*, 1609 (1971). (b) J. B. Hester, U.S. Patent 3,896,109, July 1975.
75. J. B. Hester, U.S. Patent 3,714,178. January 1973.
76. R. Madronero and S. Vega, *J. Heterocycl. Chem.*, **15**, 1127 (1978).
77. A. Bauer and K.-H. Weber, *Z. Naturforsch.*, **29**, 670 (1974).
78. G. A. Archer and L. H. Sternbach, U.S. Patent 3,671,517, June (1972).
79. E. Reeder and L. H. Sternbach, U.S. Patent 3,624,071, November 1971.
80. A. M. Felix, J. V. Earley, R. I. Fryer, and L. H. Sternbach, *J. Heterocycl. Chem.*, **5**, 731 (1968).
81. Neth. Patent 7,404,131, October 1974 (Dumex Ltd., Denmark).
82. K. Ishizumi, K. Mori, S. Inaba, and H. Yamamoto, *Chem. Pharm. Bull.*, **23**, 2169 (1975).
83. W. Metlesics and L. H. Sternbach: (a) U.S. Patent 3,644,336, February 1972, (b) U.S. Patent 2,498,973, March 1970, (c) unpublished data, Hoffmann-La Roche, Nutley, NJ.
84. W. Metlesics, G. Silverman, and L. H. Sternbach, *J. Org. Chem.*, **28**, 2459 (1963).
85. R. I. Fryer and L. H. Sternbach: (a) U.S. Patent 3,625,957, December 1971, (b) U.S. Patent 3,706,734, December 1972.
86. Swiss Patent 544,762, January 1974 (Sumitomo Chemical Co., Japan).
87. (a) K. Meguro, H. Tawada, and Y. Kuwada, *Yakugaku Zasshi*, **93**, 1253 (1973). (b) Ger. Offen. 1,966,616, May 1973 (Takeda Chemical Industries Ltd., Japan).
88. J. Earley and R. I. Fryer, unpublished data, Hoffmann-La Roche, Nutley, NJ.
89. A. Szente and J. Hellerbach, unpublished data, Hoffmann-La Roche & Co., Basel, Switzerland.
90. J. B. Hester, Jr., U.S. Patent 4,082,761, April 1978.
91. A. Walser and R. Ian Fryer, *J. Org. Chem.*, **40**, 153 (1975).
92. A. Walser and R. I. Fryer, U.S. Patent 4,280,957, July 1981.
93. A. Walser, T. Flynn, and R. I. Fryer, *J. Heterocycl. Chem.*, **15**, 577 (1978).
94. R. I. Fryer, J. V. Earley, N. W. Gilman, and W. Zally, *J. Heterocycl. Chem.*, **13**, 433 (1986).
95. A. Walser, T. Flynn, C. Mason, and R. I. Fryer, *J. Heterocycl. Chem.*, **23**, 1303 (1984).
96. G. F. Field and W. J. Zally, U.S. Patent 4,238,610, December 1980.
97. H. Niemczyk, U.S. Patent 4,335,042, June 1982.
98. H. Liepmann, R. Hüschens, W. Milkowski, H. Zeugner, R. Budden, and J. Bahlsen, U.S. Patent 4,170,649, October 1979.
99. (a) G. F. Field and L. H. Sternbach, U.S. Patent 3,594,364, July 1971. (b) G. F. Field and L. H. Sternbach, U.S. Patent 3,594,365, July 1971. (c) G. F. Field, unpublished data, Hoffmann-La Roche, Nutley, NJ.
100. R. Maupas and M. B. Fleury, *Analysis*, **10**, 187 (1982).
101. R. Y. Ning and L. H. Sternbach, U.S. Patent 3,635,948, January 1972.
102. W. Milkowski, R. Budden, S. Funke, R. Hüschens, H.-G. Liepmann, W. Stühmer, and H. Zeugner, U.S. Patent 4,244,869, January 1981.
103. G. F. Field and L. H. Sternbach, U.S. Patent 3,651,047, March 1972.
104. R. Y. Ning and L. H. Sternbach, U.S. Patent 3,682,892, August 1972.
105. R. I. Fryer and L. H. Sternbach, U.S. Patent 3,322,753, May 1967.
106. M. Steinman, J. G. Topliss, R. Alekel, Y. Wong, and E. E. York, *J. Med. Chem.*, **16**, 1354 (1973).
107. W. Milkowski, R. Hüschens, and H. Kuchenbecker, *J. Heterocycl. Chem.*, **17**, 373 (1980).
108. A. Walser and T. Flynn, *J. Heterocycl. Chem.*, **17**, 1697 (1980).
109. Swiss Patent 510,033, August 1971 (Hoffmann-La Roche & Co., Switzerland).
110. W. Metlesics, G. Silverman, and L. H. Sternbach, *J. Org. Chem.*, **29**, 1621 (1964).
111. T. Okamoto, T. Akase, T. Izumi, M. Akatsu, Y. Kume, S. Inaba, and H. Yamamoto, U.S. Patent 3,882,244, August 1974.
112. G. A. Archer, R. I. Fryer, and L. H. Sternbach, U.S. Patent 3,299,053, January 1967.
113. G. A. Archer and L. H. Sternbach, U.S. Patent 3,236,838, February 1966.
114. S. Afr. Patent 66/5349 (Hoffmann-La Roche Inc., 1967).
115. P. K. Yonan, U.S. Patent 4,208,327, June 1980.
116. A. Walser, T. Flynn, and R. I. Fryer, *J. Heterocycl. Chem.*, **20**, 791 (1983).
117. M. E. Derieg and R. I. Fryer, unpublished results, Hoffmann-La Roche, Nutley, NJ.

118. Ger. Offen. 2,353,187, November 1974 (Kali-Chemie AG, W. Germany).
119. A. Walser, R. I. Fryer, and L. Benjamin: (a) U.S. Patent 4,125,726, November 1978, (b) U.S. Patent 4,147,875, April 1979.
120. K. Meguro, Y. Kuwada, Y. Nagawa, and T. Masuda, U.S. Patent 3,652,754, March 1972.
121. A. Szente, J. Hellerbach, and A. Walser, Hoffmann-La Roche & Co., Basel, Switzerland, unpublished data.
122. J. Hellerbach and A. Szente, U.S. Patent 3,696,095, October 1972.
123. T. E. Gunda and C. Enebäck, *Acta Chem. Scand.*, **37**, 75 (1983).
124. (a) R. Jaunin, W. E. Oberhänsli, and J. Hellerbach, *Helv. Chim. Acta*, **55**, 2975 (1972). (b) R. Jaunin, unpublished data, Hoffmann-La Roche & Co., Basel, Switzerland.
125. R. I. Fryer, D. Winter, and L. H. Sternbach, *J. Heterocycl. Chem.*, **4**, 355 (1967).
126. (a) R. I. Fryer, J. Blount, E. Reeder, E. J. Trybulski, and A. Walser, *J. Org. Chem.*, **43**, 4480 (1978). (b) E. J. Trybulski, unpublished data, Hoffmann-La Roche, Nutley, NJ.
127. J. V. Earley, R. I. Fryer, and A. Walser, U.S. Patent 3,836,521, September 1974.
128. (a) E. E. Garcia, J. G. Riley, and R. I. Fryer, First International Congress of Heterocyclic Chemistry, Albuquerque, NM, 1967; *Abstracts.* (b) E. E. Garcia, unpublished data, Hoffmann-La Roche, Nutley, NJ.
129. (a) H. Liepmann, R. Hüschens, W. Milkowski, H. Zeugner, I. Hell, and K.-U. Wolf, U.S. Patent 4,338,314, July 1982. (b). EP $A_1$0,008,045.
130. J. B. Hester, U.S. Patent 3,717,654, February 1973.
131. R. Y. Ning and L. H. Sternbach, U.S. Patent 3,746,702, July 1973.
132. (a) P. A. Wehrli, R. I. Fryer, and L. H. Sternbach, U.S. Patent 3,553,206, January 1971. (b) P. A. Wehrli, unpublished data, Hoffmann-La Roche, Nutley, NJ.
133. M. Steinman and Y.-S. Wong, *Tetrahedron Lett.*, 2087 (1974).
134. E. Finner, F. Rosskopf, and W. Milkowski, *Eur. J. Med. Chem.-Chim. Thep.*, **11**, 508 (1976).
135. G. Romeo, M. C. Aversa, P. Giannetto, P. Ficarra, and M. G. Vigorita, *Org. Magn. Resonance*, **15**, 33 (1981).
136. G. Gilli, V. Bortolasi, and M. Sacerdoti, *Acta Crystallogr.*, **34**, 3793 (1978).
137. M. M. Ellaithy, J. Volke, and J. Hlavaty, *Collect. Czech. Chem. Commun.*, **41**, 3014 (1976).
138. J. A. Deyrup and J. C. Gill, *Tetrahedron Lett.*, 4845 (1973).
139. L. H. Sternbach and E. Reeder, *J. Org. Chem.*, **26**, 1111 (1961).
140. Neth. Patent 6,608,039, December 1966 (Grindstedvaerket, Denmark).
141. P. Nedenskov and M. Mandrup, *Acta Chim. Scand.*, **31**, 701 (1977).
142. Belg. Patent 777,138, January 1972 (Grindstedvaerket, Denmark).
143. J. V. Earley, R. I. Fryer, and L. H. Sternbach, U.S. Patent 3,644,335, February 1972.
144. T. T. Moller, P. Nedenskov, and H. B. Rasmussen, U.S. Patent 3,635,949, January 1972.
145. H. Oelschläger and H. Hoffmann, *Arch. Pharm.*, **300**, 817 (1967).

CHAPTER VII

# Dihydro-1,4-Benzodiazepinones and Thiones


**A. Walser**

*Chemical Research Department,*
*Hoffmann-La Roche Inc.,*
*Nutley, New Jersey*

and

**R. Ian Fryer**

*Department of Chemistry, Rutgers,*
*State University of New Jersey,*
*Newark, New Jersey*

# INTRODUCTION

The subject of this chapter is the chemistry of dihydro-1,4-benzodiazepinones and thiones. Leaving the aromaticity of the benzene ring undisturbed, the seven structures shown in Eq. 1 may be formulated. Representatives of the two tautomeric 2-ones **1** and **2**, the 3-ones **3** and **4**, and the 5-ones **5** and **6** (all with X = O) have been reported in the literature. Of the corresponding thiones (X = S) the derivatives of the 2-thione **1** (X = S) have been described. The chemistry of these compounds is discussed in Sections 1–8, followed by the syntheses and reactions of the diones in Section 9.

The 2-ones **1** were extensively explored and are by far the most numerous; several thousand derivatives have been reported in the literature and are listed in Tables VII.1.

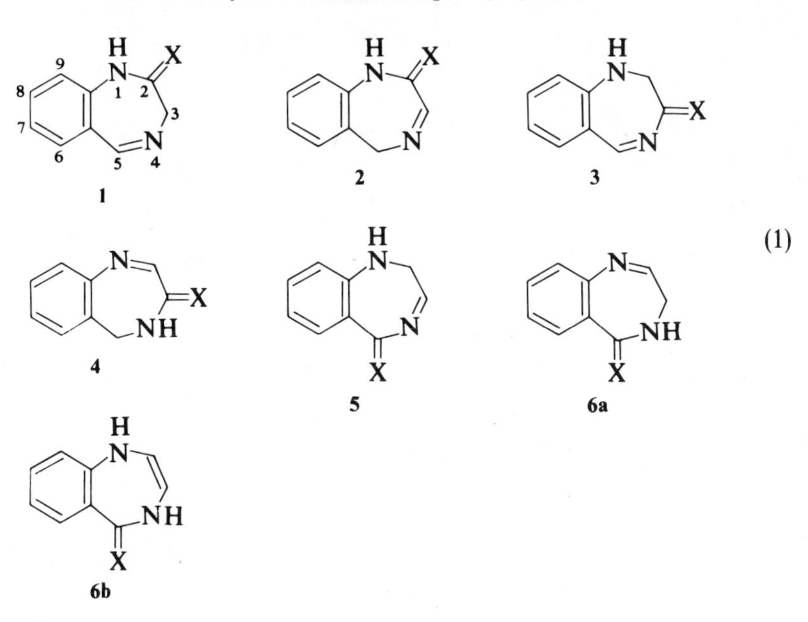


## 1. 1,3-DIHYDRO-1,4-BENZODIAZEPIN-2(2*H*)-ONES

### 1.1. Synthesis

#### *1.1.1. From (2-Aminophenyl)carbonyl Compounds*

Most syntheses of 1,3-dihydro-1,4-benzodiazepin-2(2*H*)-ones (**10**) start from the (2-aminophenyl)carbonyl compounds **7** and incorporate the α-amino acid moiety by first establishing the amide bond to give **8** and then closing the seven-membered ring by forming the 4,5-imine bond to give the benzodiazepine **10**. Procedures using the reversed sequence i.e., first forming the imine to yield **9** and subsequently establishing the amide bond to form **10** were also widely used. The various methods involving these two pathways (Eq. 2) are discussed in the subsections that follow.

1.1.1.1. Via 2-Haloacetanilides

The 2-haloacylanilides **12** (Eq. 3) were generally prepared by acylation of the anilines **11** with the halides of α-halo alkanoic acids. The halogen X in **12** was then displaced directly by ammonia or by a nitrogen nucleophile, which could be converted to an amino group. For the direct displacement by ammonia, the bromides **12** (X = Br) were most often used. A large variety of benzodiazepin-2-ones **14** were prepared by treating the 2-bromoacetyl derivatives **12** (X = Br, $R_4$ = H) with liquid ammonia and a cosolvent such as methylene chloride or tetrahydrofuran.[1–29]

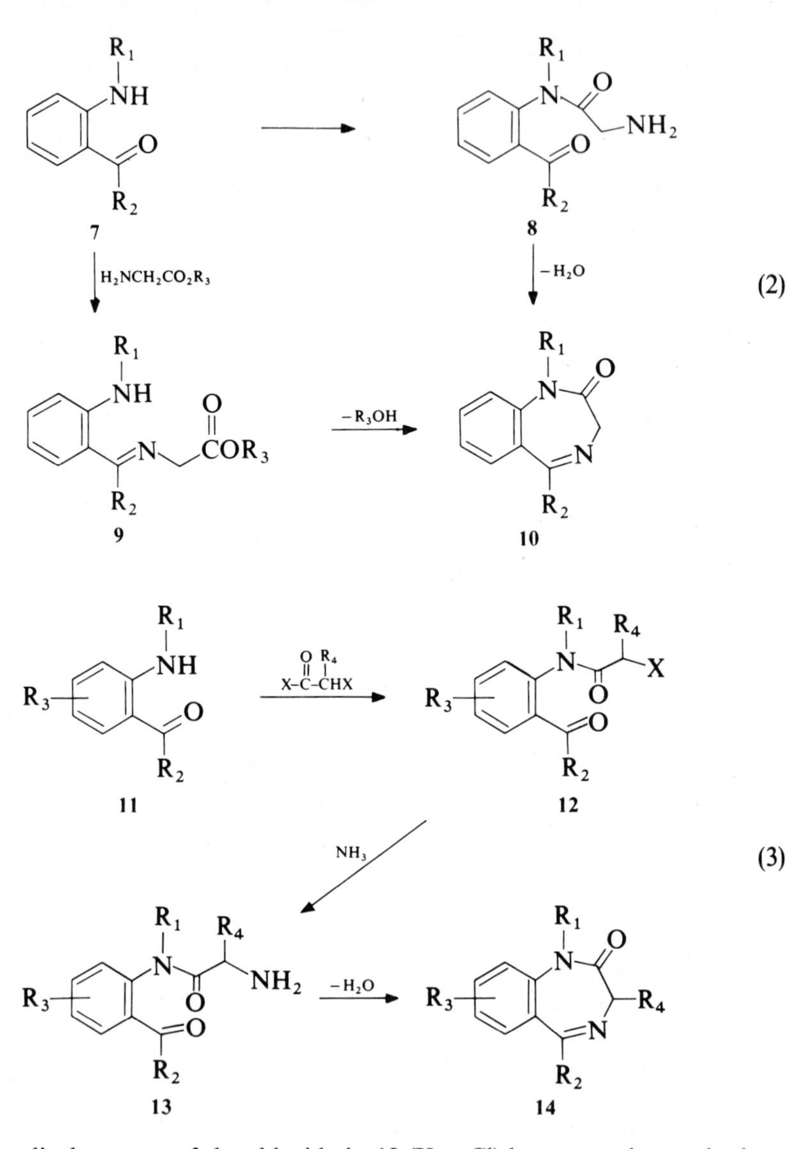


Direct displacement of the chloride in **12** (X = Cl) by ammonia required an elevated temperature. Diazepam **14** ($R_1$ = Me, $R_2$ = Ph, $R_3$ = 7-Cl, $R_4$ = H) was obtained in 65% yield by heating the appropriate chloride **12** with ethanolic ammonia at 80°C in a sealed tube.[30] A somewhat higher yield was achieved by substituting ammonia with ammonium carbonate.[30] According to the patent literature, **14** ($R_1$ = Me, $R_2$ = Ph, $R_3$ = $R_4$ = H) was accessible in 63% yield by heating the corresponding chloride **12** in dimethyl sulfoxide at 100°C in the presence of ammonia.[31] The iodide **12** (X = I) reacts with ammonia under mild conditions, such as liquid ammonia in tetrahydrofuran.[32,33] The conversion, in situ, of the chloride to the iodide, by addition of catalytic amounts of potassium iodide, has been proved to be advantageous.[34–36] Acetanilides **12** with X = 4-Me-phenylsulfonyloxy were successfully converted to the benzodiazepinones by ammonolysis in a variety of solvents.[37]

The amino compounds **13**, resulting from the displacement of the leaving group X in **12** with ammonia, were isolated in many cases when $R_1$ represented hydrogen. When $R_1$ was alkyl, the amino derivatives **13** cyclized spontaneously to the benzodiazepinones **14** (Eq. 3). The rapid cyclization could be suppressed by protonation of the amino group. The salts of **13** ($R_1$ = alkyl) have been obtained in crystalline form. The higher stability of the amines **13** ($R_1$ = H) was attributed to hydrogen bonding of the aniline proton to the orthocarbonyl oxygen as shown for **15**. The existence of such hydrogen bonding was evident from spectroscopic data, in particular from proton-nmr and infrared spectra.[38–40] The hydrogen bonding puts the molecule into a conformation which cannot undergo ring closure, and the hydrogen bond must be destroyed to allow the reactants to approach each other. This may be achieved by heating and is facilitated by employing polar protic and aprotic solvents. Heating **13** ($R_1$ = H) in ethanol or toluene in the presence of acetic acid is the preferred method for ring closure to **14**.[39] According to a Dutch patent,[41] ring closure was achieved by heating in dimethyl sulfoxide at 70–75°C. Due to steric crowding of the carbonyl group in the 2,6-dichlorobenzoyl derivative, **13** ($R_1 = R_4$ = H, $R_2$ = 2,6-$Cl_2C_6H_3$, $R_3$ = 4-Cl) could not be cyclized to the benzodiazepinone.[42] The corresponding 1-benzyl-substituted analog was induced to undergo ring closure by heating in toluene in the presence of pivalic acid, demonstrating again the more facile cyclization of **13** bearing a substituent on the aniline nitrogen.

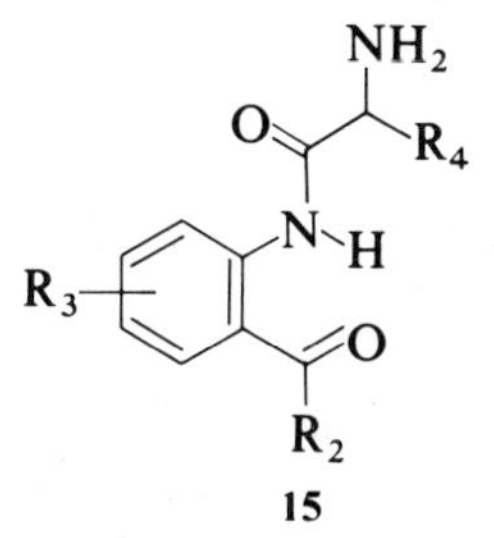


**15**

By employing weak acidic media for ring closure of the aminoacetanilides, the Smiles type of rearrangement,[14,43] which is routinely observed when compounds **13** ($R_1 \neq$ H, $R_3$ = electron-withdrawing group at the 4-position) are treated with ammonia, can be largely suppressed. Compounds **13**, which have the propensity to undergo the Smiles-type rearrangement, are not accessible by direct amination of the halides and are, therefore, better prepared via the azide. The bromide or chloride in **12** (X = Br, Cl) is displaced by azide in dimethylformamide at room temperature, and the resulting azidoacetanilide is reduced catalytically[11,42,44,45] to the amine, which can cyclize in situ or in a separate step to the benzodiazepinone.

The reduction of the azidoacetanilides to the corresponding amines was also achieved with triphenylphosphine[46,47] and triethyl phosphite.[47] High yields of the benzodiazepinones were obtained when the azide was heated with triphenylphosphine in toluene. In a first reaction step, the azide is converted to an iminophosphorane, which cyclizes with elimination of triphenylphosphine oxide. Triphenylphosphine appears to function as a catalyst in this second step.[46] Nitro-substituted benzodiazepinones were also accessible by this method.

Hexamethylenetetramine has been used as an alternate source of ammonia in the preparation of benzodiazepines.[48-54] The haloacetanilides **12** (X = Br, Cl; $R_1 \neq H$) were converted to the corresponding benzodiazepinones **14**, via the intermediates **16**, by heating to reflux in ethanol in the presence of hexamethylenetetramine. When $R_1$ represented hydrogen, the imidazolidinones **17** and **18** were formed almost exclusively (Eq. 4).[49-51] A notable exception was the hydrobromide of **12** ($R_1 = R_4 = H$, $R_2$ = 2-pyridyl, $R_3$ = 4-Br), which could be converted to bromazepam **14** ($R_1 = R_4 = H$, $R_2$ = 2-pyridyl, $R_3$ = 7-Br), under conditions which transformed the corresponding phenyl analog to the imidazolinone **17**.[52] The formation of the imidazolidinones **17** and **18** could be avoided either by reacting the halide with hexamine in the presence of ammonia,[53] or by treating the halide with a mixture of ammonia and formaldehyde.[53] A slightly modified process using hexamethylenetetramine in combination with an aqueous solution of an ammonium salt such as ammonium bromide was also patented.[54]

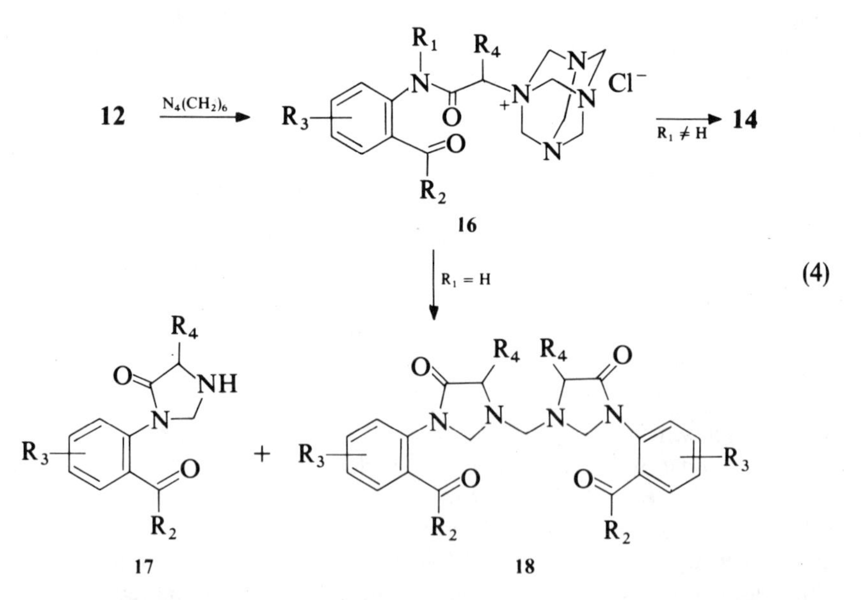


The fate of hexamine in the conversion of **12** to **14** was studied in detail by Clarke and coworkers.[51] The quaternary salt **16** (Eq. 4) is formed in the first step and undergoes cleavage by an internal redox reaction to give methyl formate and hexamine methochloride as the main decomposition products. According to a Belgian patent,[55] hexamethylenetetramine can be replaced by dinitroso-pentamethylenetetramine to prepare 1-substituted benzodiazepinones and bromazepam, 7-bromo-1,3-dihydro-5-(2-pyridyl)-1,4-benzodiazepin-2(2*H*)-one. Phthalimide was also used to displace halogen from the haloacetamido compounds (**12**) (see Section 1.1.1.2).

Displacing the leaving group in **12** (X = I or sulfonyloxy) by hydroxylamine led to the hydroxyamine **19**, which was cyclized preferably with acid catalysis to the nitrone **20** (Eq. 5).[56–58]

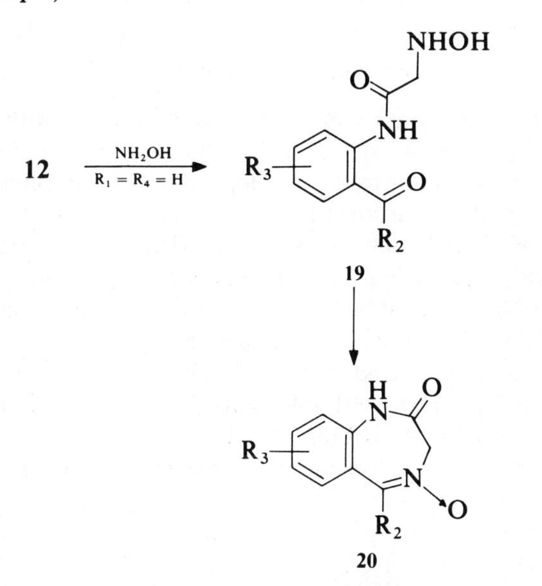


(5)

According to the patent literature,[59] the 3-acetamino derivative **22** was obtained by ammonolysis of the chloride **21**. In the first step, the chloride is displaced by ammonia in boiling methanol and the intermediate obtained was cyclized by refluxing in methylene chloride–acetic acid (Eq. 6).

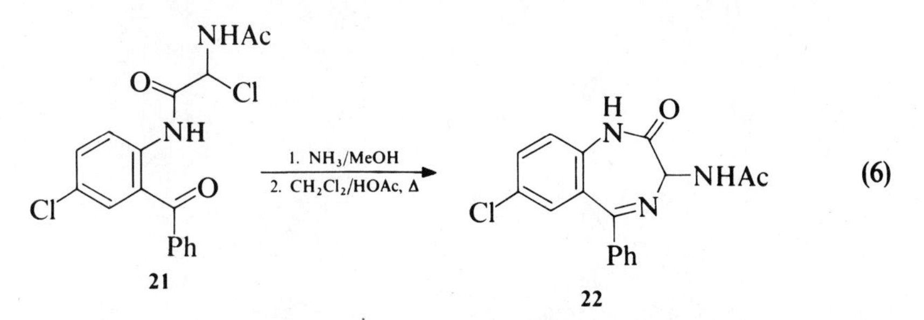


(6)

The formation of the 3-chloronitrone **24** by reaction of the dichloroacetanilide **23** with hydroxylamine in methanol was patented as well (Eq. 7).[60] In this case, displacement of a chloride by hydroxylamine is unlikely, and **24** is probably formed by intermediacy of the benzophenone oxime.

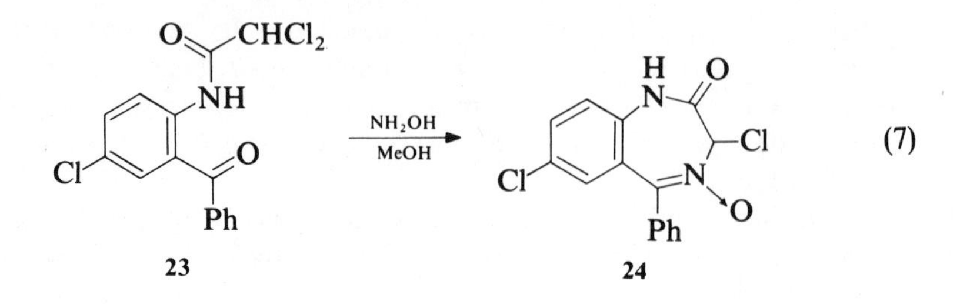


## 1.1.1.2. By Reaction with *N*-protected α-Amino Acids

(2-Aminophenyl)carbonyl compounds were coupled with *N*-protected α-amino acids to give the anilides **25** and **26** (Eq. 8). Because of the weakly basic character of the aniline nitrogen, the formation of the amide bond requires strong activation of the carboxyl function of the amino acid, preferentially the acid chloride. High yield acylations were carried out on various 2-aminobenzophenones with 2-phthalimidoacetyl chloride.[61–73] The phthalimido group in **25** was removed by standard hydrazinolysis and the amine obtained, if required, was cyclized to the benzodiazepinone as mentioned above. In many cases the substituent $R_1$ could be introduced into **25** ($R_1$ = H) by alkylation.[71–73] Reaction of **25** ($R_1 = R_4$ = H, $R_2$ = Ph, $R_3$ = 4-Cl) with methyl isocyanate gave the corresponding urea **25** ($R_1$ = MeNHCO), which upon hydrazinolysis yielded the benzodiazepinone with loss of the methylaminocarbonyl group.[74]

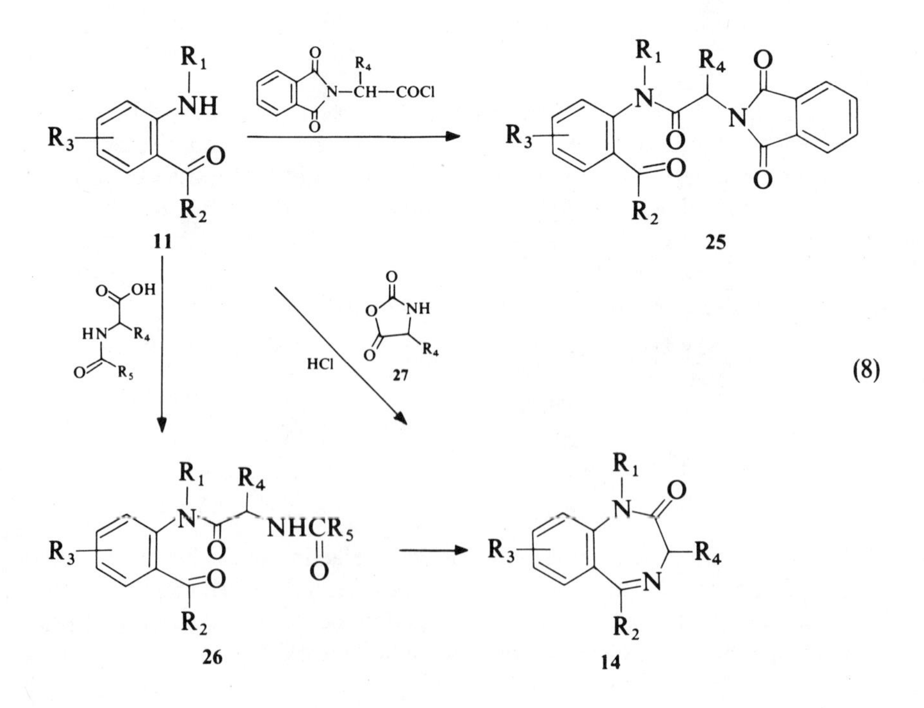

The benzyloxycarbonyl moiety represents another widely used *N*-protecting group. *N*-Benzyloxycarbonyl amino acids were coupled with the anilines **11** by activation of the carboxyl group via mixed anhydrides, dicyclohexylcarbodiimide, or acid chlorides to form **26** ($R_5$ = $PhCH_2O$).[3,11,20,33,39,40,75,76] A convenient method for activation of the carboxyl group occurs in situ by the generation of the acid chloride by reacting the *N*-protected amino acid with phosphorus pentachloride at low temperature in methylene chloride or tetrahydrofuran.[39,77–79] Phosphorus pentachloride was successfully replaced by thionyl chloride.[80] The benzyloxycarbonyl group was removed by hydrogenolysis or hydrogen bromide in acetic acid.

The *t*-butoxycarbonyl protecting group was applied with equal success, in particular for the preparation of 3-substituted optically active benzodiazepinones.[40,81] In this case, dicyclohexylcarbodiimide served as a carboxyl activating reagent and hydrogen bromide in acetic acid was used for deprotection. The amines corresponding to **26** were thus obtained as the hydrobromide salts and were cyclized as previously mentioned after conversion to the free bases. In a rare instance, the methoxycarbonyl group served as a protecting group[82] and was cleaved by hydrogen bromide in boiling ethyl acetate. An application of the trifluoroacetyl group for the same purpose was described more recently.[83] *N*-Trifluoroacetylglycine was activated by conversion to the acid chloride and reacted with the aminobenzophenone to yield **26** ($R_5$ = $CF_3$). Treatment with alkali at room temperature cleaved the trifluoroacetyl group to give the corresponding amine, which was cyclized in the usual manner.

Compounds **26** ($R_1$ = H) could be selectively alkylated on the aniline nitrogen as shown by conversion of the benzyloxycarbonyl derivatives **26** ($R_1$ = H) to the corresponding compounds with $R_1$ = $CH_2OMe$[72] and with $R_1$ = $CH_2NHCOOEt$.[84]

The activation and the protection of glycine are present simultaneously in the reagent, 2-isocyanatoacetyl chloride. The application of this reagent for the preparation of benzodiazepinones has been described in the patent literature.[85] The intermediate isocyanate was converted to the benzodiazepine by stirring in solution. A variant of this procedure takes advantage of the Leuch's anhydride, the 1,3-oxazolidin-2,5-dione, which can be looked at as an internally protected and activated amino acid. The reaction of **27** with the aniline **11** was generally carried out in methylene chloride or ether, catalyzed by hydrogen chloride (Eq. 8). A wide variety of benzodiazepinones were claimed to be accessible in high yields by this method.[86–89] The 3-carboxylic acid ethyl ester **14** ($R_4$ = COOEt) was similarly obtained by reaction of the appropriate aminobenzophenone with **27** ($R_4$ = COOEt) and hydrogen chloride in methylene chloride followed by ring closure in boiling acetic acid.[82]

### 1.1.1.3. By Reaction with α-Amino Acid Esters

A general synthesis of benzodiazepin-2-ones involves the condensation of the (2-aminophenyl)carbonyl compounds **11** with esters of amino acids. It is likely that in this case the amino group of the amino acid ester reacts first with the carbonyl function of **11** to form the imine **9** (Eq. 2), which then cyclizes with elimination of the alcohol $R_3OH$. This method was first reported by Walker[90] and by Sternbach and coworkers.[2] It was subsequently applied for the preparation of many analogs.[91–102]

In a standard procedure, the hydrochloride salt of the amino acid ester was heated with the aminobenzophenone in pyridine. The salt of the amino acid ester was present in large excess, because it undergoes conversion to diketopiperazine under the reaction conditions applied. The yields were better for compounds with $R_1 = H$. The application of glycine ester attached to a polymer in the synthesis of benzodiazepin-2-ones was reported with yields of 30–60%.[103]

The synthesis employing amino acid esters was considerably improved by facilitating the first step, the formation of the imine. This was achieved by using the imine **28** instead of the corresponding ketone and exchanging the amine of **28** with the amino acid ester. The acid-catalyzed ring closure of the product **29** to the benzodiazepin-2-one **14** proceeds in high yield (Eq. 9).

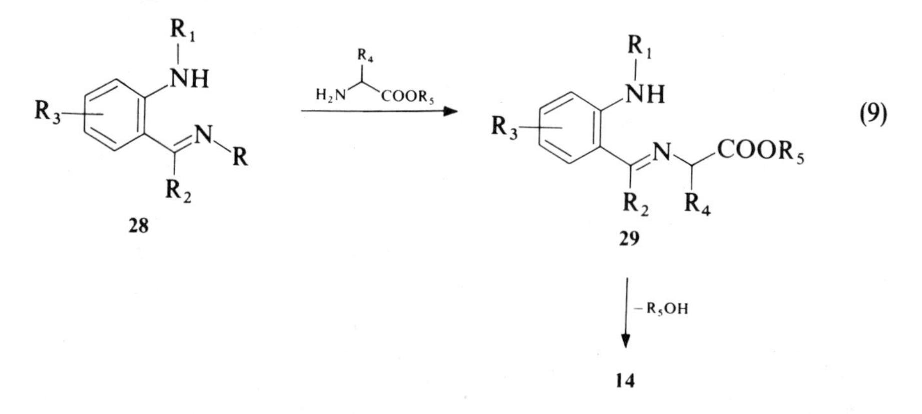


This method was used with particular success for the preparation of benzodiazepin-2-one-3-carboxylic acid esters **14** ($R_4 = COOR$) starting from the imines **28** ($R = H$). The imines were accessible by the reaction of the appropriate nitrile with a Grignard reagent $R_2MgX$.[104–106] In addition to the imine, substituted imines **28** with $R = Me$,[108] $R = CH_2CH_2OH$,[109] and $R = CH_2CH_2NH_2$[110] were also applied for this purpose.

### 1.1.1.4. By Reaction with α-Amino Acids

The condensation of a (2-aminophenyl)carbonyl compound with an α-amino acid by elimination of two molecules of water constitutes the most formally direct synthesis of 1,4-benzodiazepin-2-ones. Reeder and Sternbach[91a] reported the formation of the anticipated benzodiazepinone by reaction of 2-amino-5-chlorobenzophenone with glycine in boiling pyridine in the presence of hydrogen chloride. The yield of this reaction was improved considerably by carrying out the reaction in a refluxing mixture of pivalic acid and toluene and azeotropically removing water. Diazepam, **31**, was thus obtained in about 60% yield from the benzophenone **30** (Eq. 10).[108]

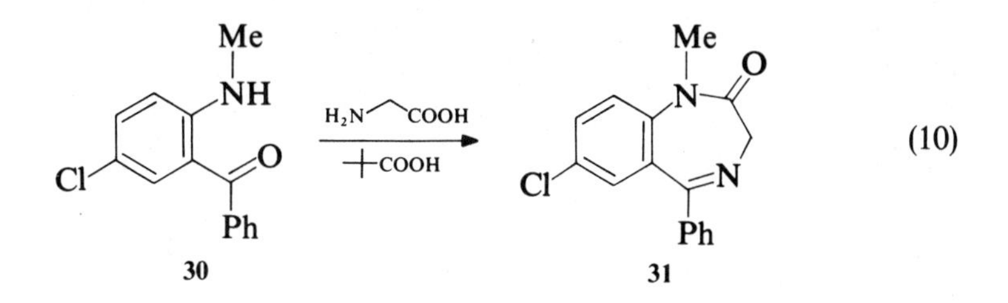


Other procedures resorted to in situ activation of glycine by using phosphorus oxychloride,[111,112] phosphorus trichloride,[113] or thionyl chloride.[114] Another equivalent method employs the acid chloride hydrochloride of the amino acid.[115]

### *1.1.2. By Oxidation of Indoles*

The oxidative cleavage of the 2,3-double bond in substituted indoles to form the acylated (2-aminophenyl)carbonyl compounds was also studied for the preparation of 1,4-benzodiazepin-2-ones. While the synthesis of the required indoles may be cumbersome, this method can be useful to obtain certain substituted benzodiazepinones which are difficult to prepare by other routes.

The most commonly used oxidizing agents were chromium trioxide in acetic acid–water, hydrogen peroxide with ammonium molybdate, and ozone. As pointed out in Chapter V, for the chromic acid oxidation to proceed smoothly to the desired product, the indole has to be substituted at both the 2- and the 3-positions. Treatment of the indole **32** ($R_1$ = Me, $R_2 = R_4$ = H, $R_3$ = 5-Cl) with chromium trioxide yielded exclusively isatin.[116] When ozone in acetic acid was used, a low yield (6.5%) of the expected benzodiazepin-2-one was obtained.[116] A great number of variously substituted 1,4-benzodiazepin-2-ones were prepared by this method.[116–133] Based on our experience with this procedure,[108] the oxidation of the indole **32** ($R_2$ = Ph) with chromium trioxide leads to the carbonyl compound **13**, which can cyclize in situ to the benzodiazepin-2-one **14** (Eq. 11). By keeping the amino group in **13** protonated by a strong acid, the cyclization was slowed down enough to allow the isolation of the salt of **13**.

Oxidation of the 1-(2-phthalimidoacetyl)indoles **33** (R = H, Me) with chromium trioxide led to the benzophenones **35** ($R_1$ = H) and the corresponding acyl derivatives **35** ($R_1$ = Ac, CHO), which were converted to the benzodiazepinone **36** by hydrazinolysis. Ozonization of **33** gave the ozonides **34**, which were transformed directly to **36** by hydrazinolysis or to the benzophenones **35** by heating in ethanol (Eq. 12).[131,132] In the case of the oxidation of

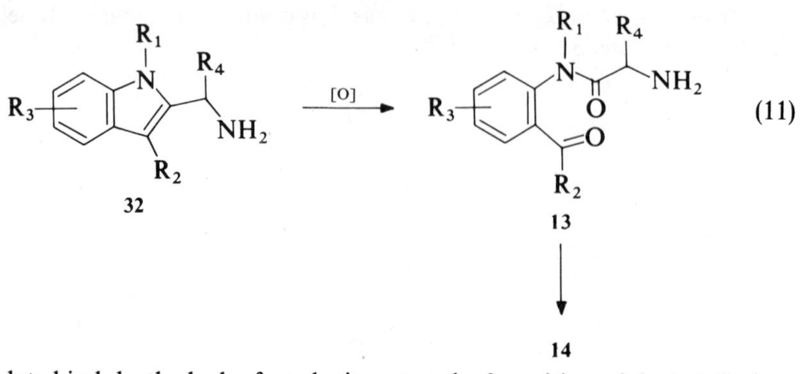


a 1-acylated indole, the lack of a substituent at the 2-position of the indole does not seem to be detrimental to the course of the reaction.

According to the patent literature,[125] this method was applied to the preparation of alkyl- and alkoxy-substituted 5-phenyl-1,4-benzodiazepin-2-ones. We were, on the other hand, not successful in using the chromic acid oxidation for the synthesis of 7-methoxy-substituted compounds.[134] The 7-amino derivatives **38** were obtained by oxidation of the phthalimido compounds **37**, followed by cleavage of the protecting groups with hydrazine (Eq. 13).[133]

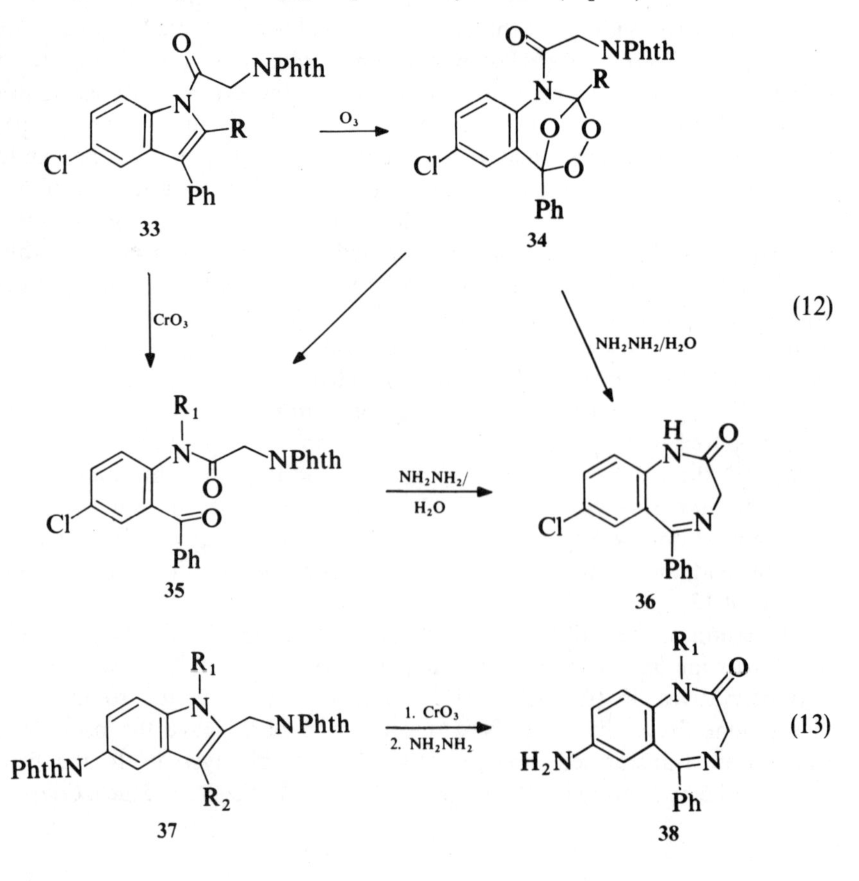

### *1.1.3. By Ring Expansion*

The 4-oxides of 1,3-dihydro-1,4-benzodiazepin-2(2*H*)-ones, **40**, were obtained in good yields by reaction of the 2-chloromethylquinazolines **39** with hydroxide.[91,93,97,135–137] Following Sternbach's discovery of the ring expansion of **39** by primary amines to give 2-amino-3*H*-1,4-benzodiazepines (see Chapter V, Section 2.2.1.1), a variety of other nucleophiles were found to effect the same type of reaction. In the present case of ring expansion with hydroxide, the same mechanism probably applies. The hydroxide ion is thought to attack the 2-position of the quinazoline, generating the anion **41**, which can undergo ring opening to form the oxime anion **42**. Intramolecular alkylation on nitrogen would then convert **42** to the benzodiazepinone **40** (Eq. 14).

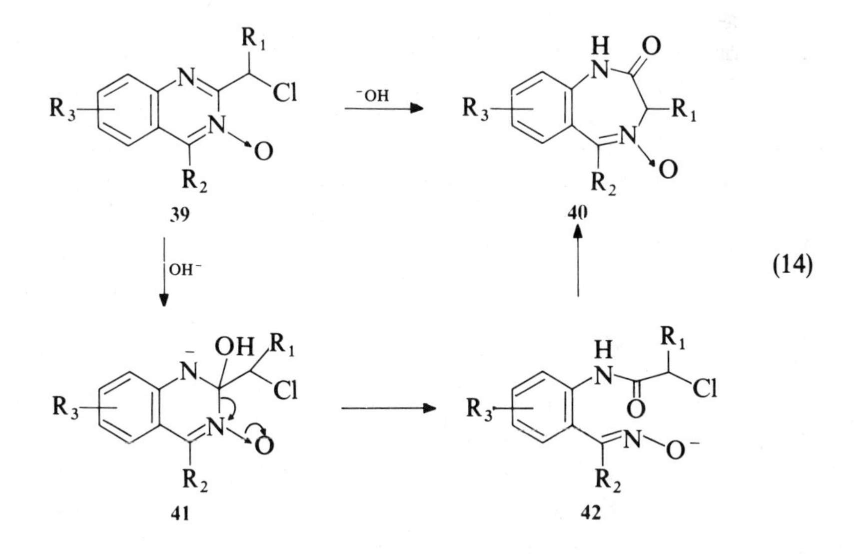


The 1,2-dihydro-2-dichloromethylquinazoline-3-oxides **43** were also found to undergo the same reaction. It is likely that an initial elimination of hydrogen chloride from **43** would lead to the chloromethylquinazoline **39** ($R_1$ = H) (Eq. 15).[138–141] This process was applied to the preparation of 3-fluoro-[142] and 3-chloro-[138,139,143] substituted benzodiazepinones. Thus the respective 2-dihaloquinazolines **39** ($R_1$ = F, Cl) were reacted with hydroxide at 0–5°C to give the 3-halo derivatives **40** ($R_1$ = F, Cl; $R_2$ = Ph; $R_3$ = 7-Cl, $NO_2$, $CF_3$) (Eq. 14).

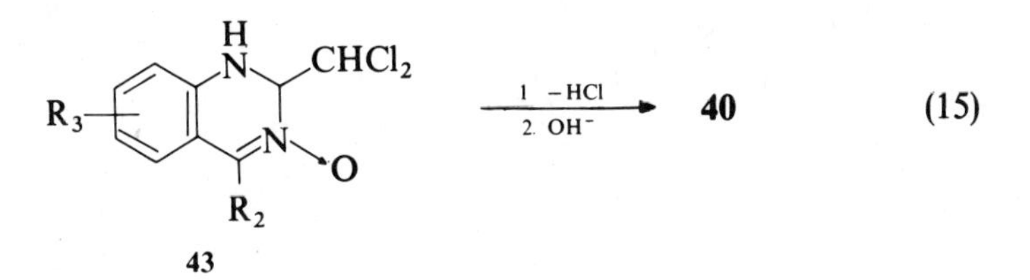

Treatment of the 1,2-dihydro-2-hydroxyquinazoline **44** with sodium hydride in tetrahydrofuran at 0°C was reported to result in a low yield ring expansion to the 3-fluoro compound **45** (Eq. 16).[144]

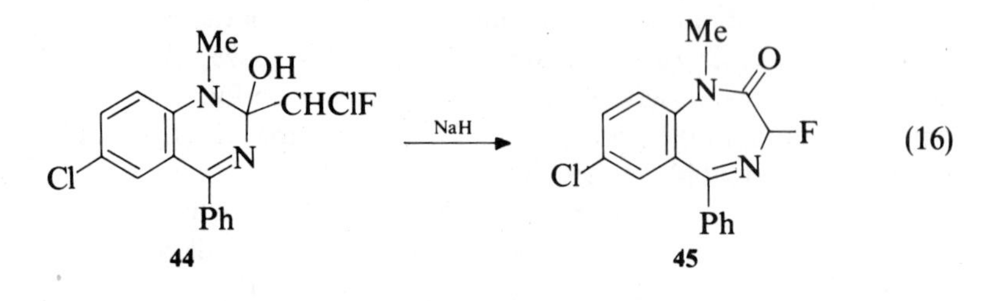

(16)

The tetrahydroquinolones **46** ($R_2$ = COOR) were transformed to the corresponding benzodiazepin-2-ones by heating in refluxing toluene or benzene in the presence of acetic acid.[39,145–147] The 3-carboxylates **47** ($R_2$ = COOR) are most likely formed by a reversible ring opening of **46** to the benzophenone **48**, which can dehydrate to the benzodiazepine **47** (Eq. 17). It is possible for the aziridine **49** to be an intermediate, since this compound could also be converted to the benzodiazepine **47**.[39] When the quinolone **46** ($R_1$ = Me, $R_2$ = COOMe, X = H, Y = Cl) was heated to reflux in 80% acetic acid for 20 hours, it rearranged and decarboxylated to form diazepam (**31**) directly.[145]

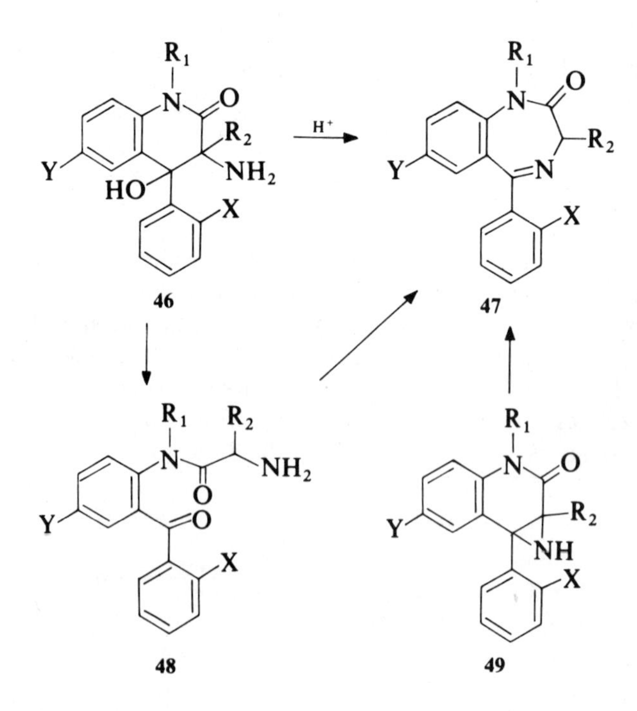

(17)

### *1.1.4. By Ring Contraction*

The 1,2-benzoxazocines **51**, which were formed by treatment of the *syn*-oximes **50** with base, were found to rearrange to the 3-hydroxybenzodiazepines **53** upon reaction with hydroxide or methoxide in methanol at room temperature (Eq. 18).[148]

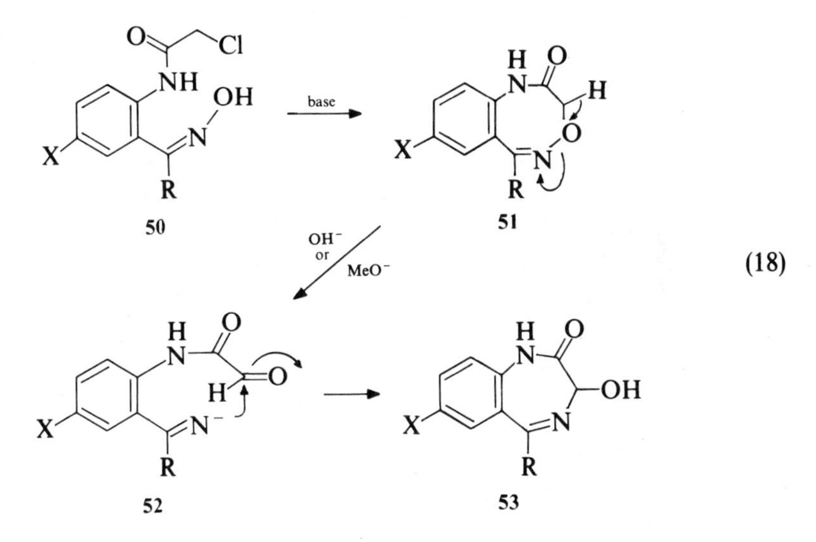


(18)

The mechanism of this ring contraction was proposed to involve abstraction of a proton from the methylene group followed by cleavage of the N—O bond to lead to the imine anion **52**. This intermediate would then undergo ring closure as indicated to form the 3-hydroxybenzodiazepinone **53**.

A related ring contraction was reported for the 1,4,5-benzotriazocinium salts **55**, which were obtained by intramolecular quaternization of the chloroacetamido hydrazones **54** (Eq. 19).[149] Treatment of the triazocinium salts **55** with methoxide in methanol resulted in a rearrangement to the 3-aminobenzodiazepinones **57** in high yield. Fusion of the hydrazones **54** at 190°C was accompanied by evolution of an unidentified gas, and the benzodiazepinone **58** was isolated from the tarry residue. The formation of the benzodiazepinone was explained by intermediacy of the diazepinium salt **56**, a compound isolated from a slightly less vigorous fusion of **55**.

### *1.1.5. By Oxidation*

1,3-Dihydro-1,4-benzodiazepin-2(2*H*)-ones were prepared by the oxidation of less saturated benzodiazepines by means of a variety of oxidizing agents. The oxidation of the 4,5-bond in the tetrahydro-2-ones **59** to form **60** was achieved by the following methods: bromine and sodium hydroxide,[72,150] chromium

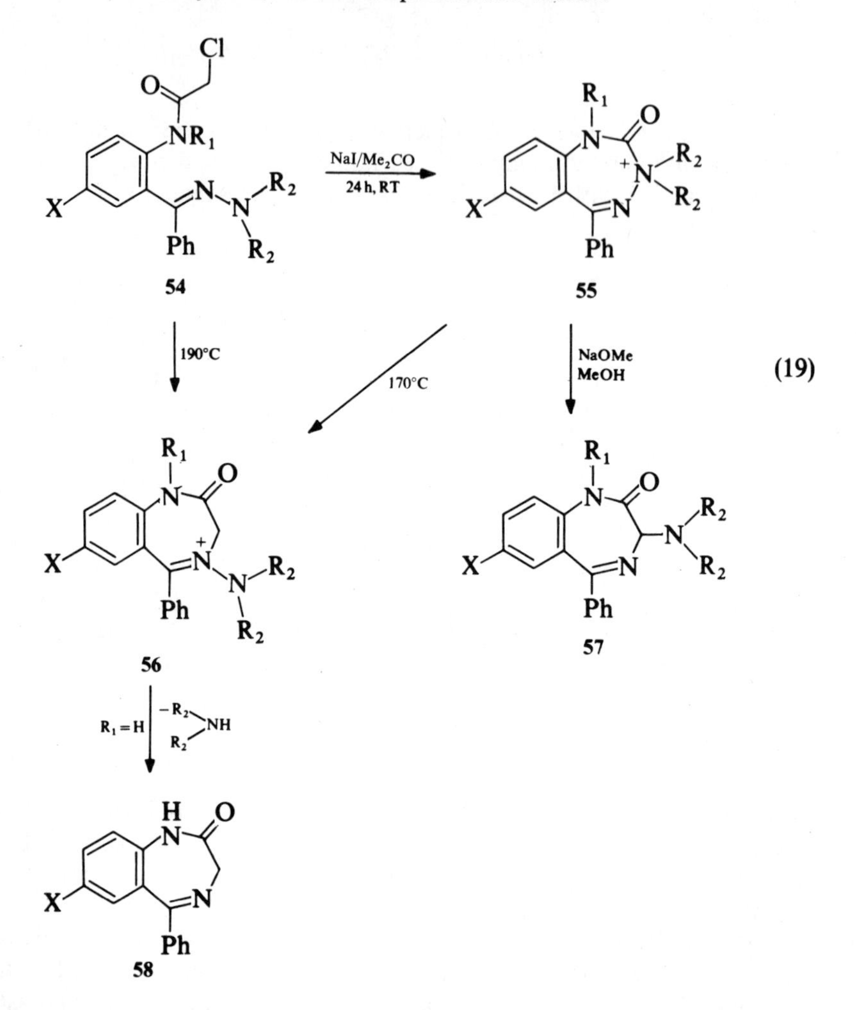


trioxide,[151,152] selenium dioxide,[151] silver oxide,[151] potassium permanganate,[151,153] dichlorodicyanobenzoquinone,[154] and dimethyl sulfoxide in combination with ultraviolet light (Eq. 20).[155,156]

Ruthenium tetroxide,[157,158] manganese dioxide, diethyl azodicarboxylate,[134] and iodine in combination with lead tetraacetate[134,152c] were also used to oxidize the amine **59** to the imine **60**. A Dutch patent[159] claims the preparation of diazepam in 40% yield by oxidation of the 4-formyl derivative **61** ($R_1$ = Me, $R_2$ = Ph; $R_3$ = formyl, X = 7-Cl) with chromium trioxide in acetic acid. Another publication[156] reported a 26% yield for this transformation. The carbonyl group at the 2-position was also introduced by oxidation of 2,3-dihydro-1*H*-1,4-benzodiazepines **62** ($R_3$ = H) with various oxidizing reagents. Thus, compounds **62** ($R_3$ = H) were converted to the 2-ones **60** by means of ruthenium tetroxide,[28,157,158,160] chromium trioxide,[28,161,162] *N*-bromosuc-

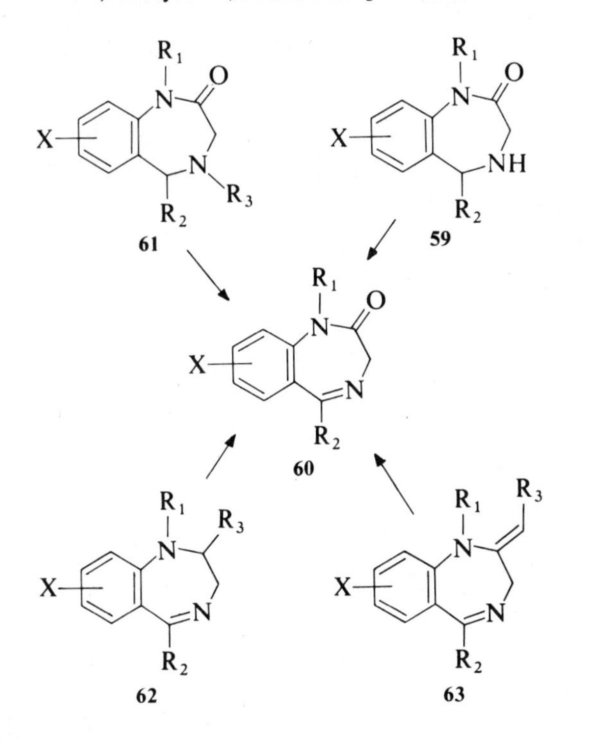


(20)

cinimide in tetrahydrofuran, containing aqueous sodium bicarbonate solution,[163] or with manganese dioxide.[138]

The 2-ones were also obtained in high yield by oxidation of the 2-hydroxy derivatives **62** ($R_3$ = OH).[164,165] Various 2-substituted 2,3-dihydro-1*H*-1,4-benzodiazepines **62** ($R_3 = CH_2X$) were oxidized down to the 2-ones **60** by potassium permanganate[166-168] or by chromic acid.[167] Such oxidations were described for **62** with $R_3 = CH_2Cl$,[166-168] $CH_2OH$,[166,168] $CH_2OMe$,[168] piperidinomethyl,[168] and COOH.[169] These reagents also cleaved the double bond in the 2-methylene benzodiazepines **63**.[134]

### *1.1.6. By Elimination and Rearrangement*

Elimination of the moiety $R_3H$ from the 4-substituted compounds **61**, constitutes another synthesis of the benzodiazepin-2-ones **60**. Eliminations of this type were successfully carried out with the 4-sulfonyl derivatives **61** ($R_3 = MeSO_2$, $4\text{-}MeC_6H_4SO_2$), using strong base in a variety of solvents (Eq. 21).[72,154,170-173] Aprotic solvents favored the abstraction of a proton from the 3-position and gave predominantly the 5*H* tautomers **64**. The 5*H* tautomer appears to be the kinetic product and can be equilibrated to the thermodynamically more favored 3*H* compound **60** by treatment with strong base in a protic solvent such as ethoxide in ethanol.[72] By carrying out the elimination in a protic solvent, the 3*H* tautomers were obtained directly as the major products.[170]

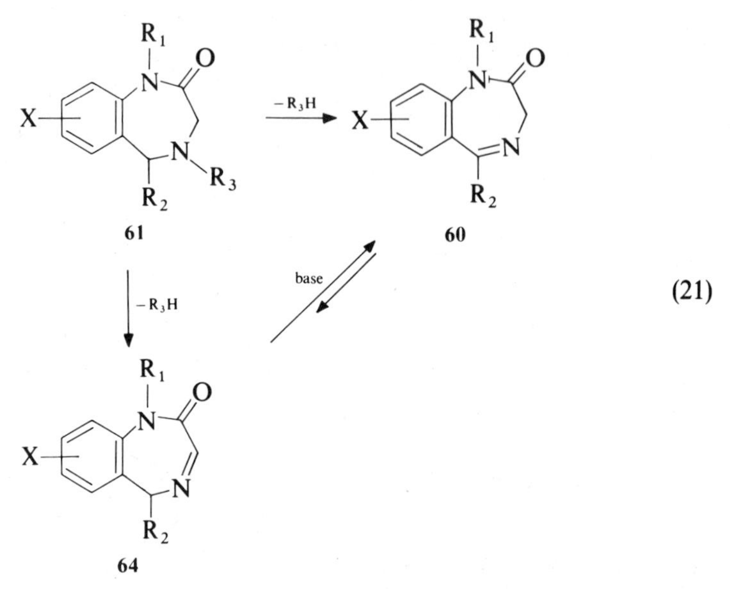


The 4-acetoxy derivatives **61** ($R_3$ = AcO) underwent a similar transformation. Heating the 4-acetoxy compound with ethanol and potassium *t*-butoxide to reflux led to the 3*H* tautomer **60**, while the use of diethylamine or triethylamine in boiling ethanol afforded the 5*H* tautomer **64**.[72,174,175] Elimination of water from the 4-hydroxy compounds **61** ($R_3$ = OH) was achieved by means of phenyl isocyanate in combination with a weak base such as pyridine or 1,4-dimethylpiperazine.[174] This method gave the 3*H* tautomers in 75–90% yield. Elimination of water was also observed with dicyclohexylcarbodiimide in pyridine, thionylchloride in chloroform, concentrated sulfuric acid[177] and phosphorus oxychloride in pyridine at reflux.[141] Dicyclohexylcarbodiimide in boiling toluene also dehydrated the 4-hydroxy compound efficiently.[72] According to the patent literature,[177b] reaction of the 5*H* tautomer **64** ($R_1$ = H, $R_2$ = Ph, X = 7-Cl) with *N*-bromosuccinimide and subsequently with aqueous dimethylformamide gave a low yield of the 3-hydroxy-3*H*-1,4-benzodiazepine, oxazepam.

### *1.1.7. By Hydrolysis*

The benzodiazepin-2-ones **66** were frequently obtained by hydrolyses of the 3*H*-1,4-benzodiazepines **65** bearing a leaving group at the 2-position (Eq. 22). The leaving groups X, which were easily displaced by hydroxide, include the *N*-nitrosomethylamino moiety,[178] an alkylthio group,[179] and the α-methoxyiminonitromethyl residue.[180] Other groups, X, were replaced by hydroxide under acid catalysis such as the 2-amino,[181,182] *N*-acetylmethylamino,[135,183,184] hydrazino,[185,186a] methylthio,[186b] and alkoxy moieties.[187,188]

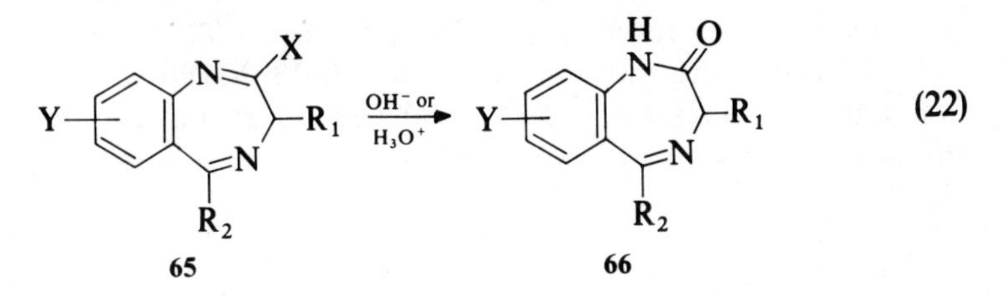


The 1-hydroxy compound **68** resulted from the acid hydrolysis of the 1-oxide **67** (Eq. 23).[189] Related solvolytic reactions leading to 2-ones were described by Fryer and coworkers.[190] Treatment of the imidazolindiones **69** and **70** with liquid ammonia led to the 2-ones **71** with R = EtO and R = MeNHCOCONH, respectively (Eq. 24).

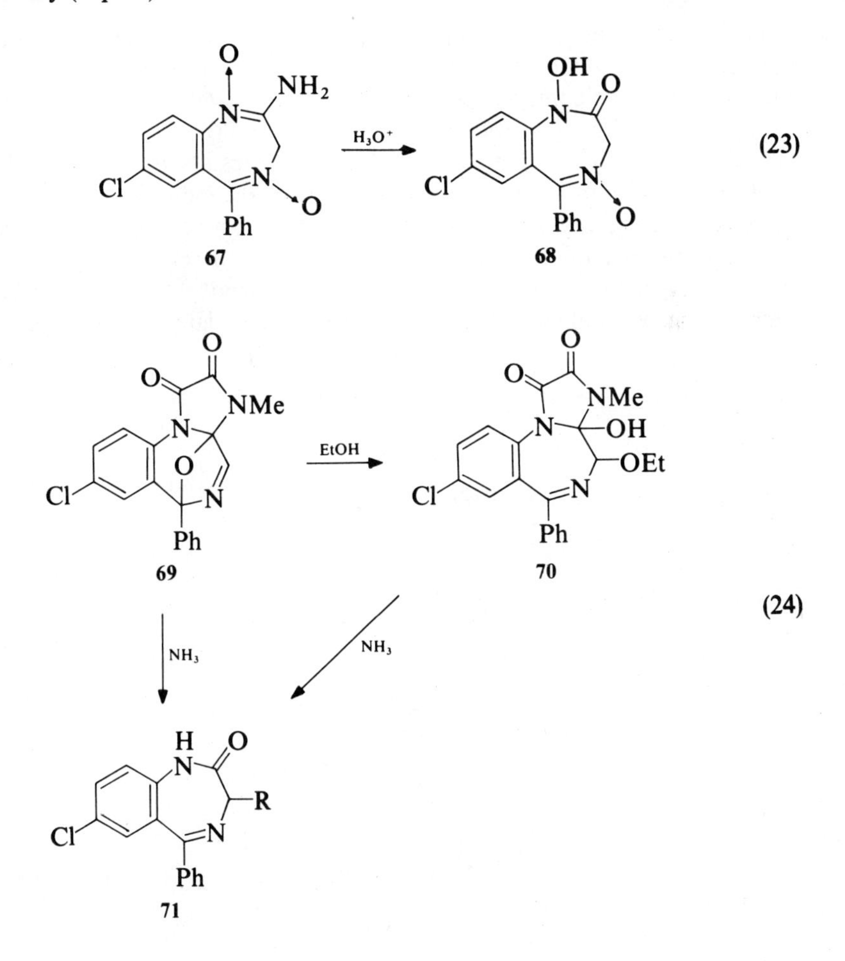

### *1.1.8. Other Syntheses*

The 5-phenyl derivatives **58** (X = H, Cl, $NO_2$) were reported to be formed in moderate yields by reaction of the benzisoxazoles **72** with glycine ethyl ester hydrochloride **73** in 2-methylimidazole at 120–130°C (Eq. 25).[74,191] The conversion of the benzisoxazole to the benzodiazepin-2-one involves a reduction step, and it is not apparent which ingredient in the reaction mixture serves as the reducing agent.

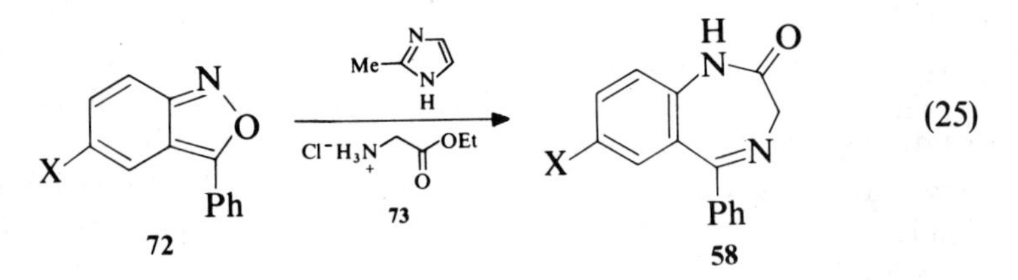


(25)

The 1-chloroacetylisatins **74** (X = H, Br) reacted with ammonia or primary and secondary aliphatic amines in ethanol to give the chloroacetanilides **75** (Eq. 26).[192] These compounds were then converted to the benzodiazepine-5-carboxamides by conventional procedures. Although further reaction of **75** with ammonia led to the benzodiazepine,[76] better results were achieved by first displacing the chloride with iodide. Amination of the iodide with liquid ammonia gave the corresponding aminoacetanilides, which were then cyclized under acid catalysis. Heating the isatins **74** (X = H, Cl) with hexamethylenetetramine in alcohol, led directly to the benzodiazepin-5-carboxylic acid esters **76**. The intermediate acetanilides **75** (R = OR′) were not isolated in this instance.[193]

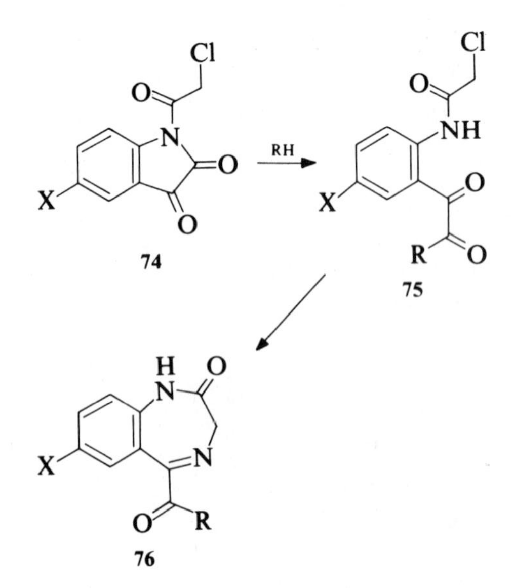


(26)

Both the 5-carboxamides and the corresponding esters were subjected to further derivatization. Hydrogenolysis of the allyl ester afforded the corresponding carboxylic acid, which was decarboxylated by heating in tetrahydrofuran to yield the parent compound.[193] Other functional group transformations are discussed in Sections 1.2.

The 4-acyl derivatives of benzodiazepin-2,5-diones were found to react with phenylmagnesium chloride at the 5-carbonyl group.[194] The benzophenone **78** (R = Me) was thus prepared in 86% yield from the 4-acetyl-2,5-dione **77** (R = Me) (Eq. 27). The transformation of **78** into diazepam, **31**, was achieved by methanolysis or by treatment with hydroxylamine followed by hydrolysis of the oxime **79** with sodium bisulfite in aqueous ethanol at reflux temperature. The 4-trifluoroacetyl derivative **77** (R = $CF_3$) was converted to diazepam, **31**, in 30% yield by reaction with phenylmagnesium chloride followed by acid hydrolysis.[194]

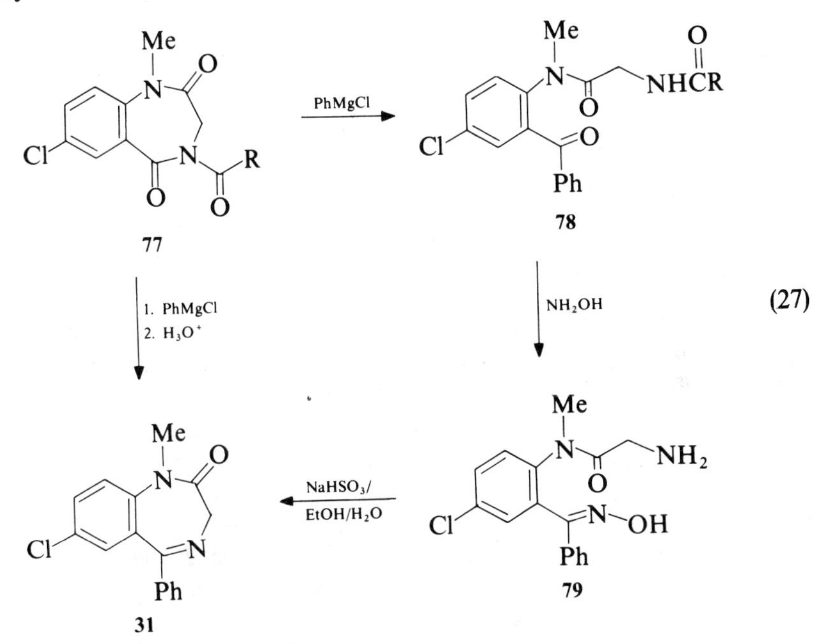


1,3-Dihydro-1,4-benzodiazepin-2(2*H*)-ones, bearing a heteroatom attached at the 5-position, were synthesized from the 2,5-diones **80**. The 5-alkoxy derivatives **81** ($R_1 \neq H$) were obtained by reaction of the 2,5-diones with triethyloxonium tetrafluoroborate in methylene chloride at room temperature or by displacement of the 5-chloride in **82** by alkoxides (Eq. 28).[195,196] The iminochloride **82** resulted from the treatment of the 2,5-dione **80** with phosphorus pentachloride in carbon tetrachloride and chloroform at reflux.[195,196] This chlorine atom (**82**) was also readily displaced by a variety of primary and secondary amines and phenyl thiol to give **83**.[196] The reaction of **82** with hydrazine derivatives[197,198] is discussed below (Section 1.2.2.3).

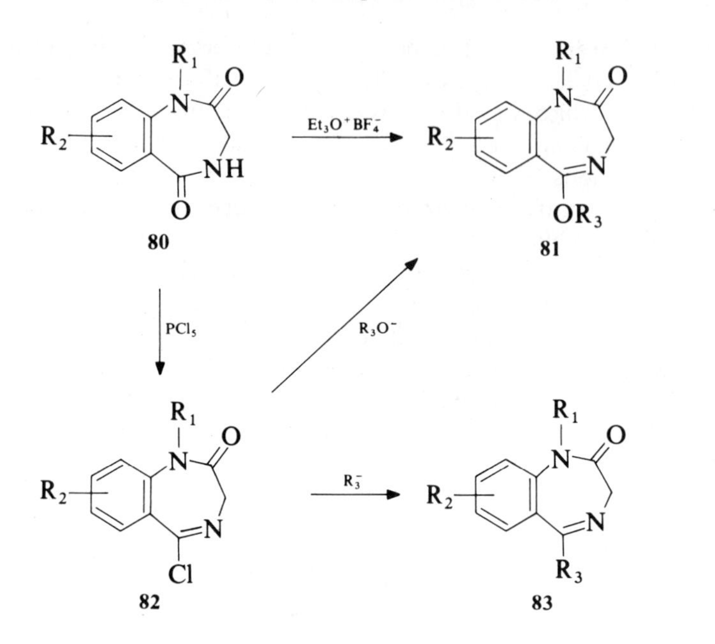


Bell and coworkers[199–201] obtained the 3-acetamino derivative **87** (R = Ac) by treatment of the *N*-acetoxyacetamide **84** with ammonia. The authors proposed the following mechanism: elimination of acetic acid from **84** would lead to the *N*-acetylimine **85**, which could then add ammonia to give intermediate **86**. Cyclization with dehydration would lead to the observed product **87** (Eq. 29). Methanolic hydrogen chloride at room temperature cleaved the acetyl group in **87** to give the 3-amino derivative. This method was later applied for the preparation of several 3-amino-1,3-dihydro-1,4-benzodiazepin-2(2*H*)-ones.[202]

A related transformation was reported in the patent literature.[203] Reaction of the hydroxylamine **88** with ethyl chloroformate gave the *N,O*-diacylated compound **89**, which was converted to the carbamate **87** (R = COOEt) by treatment with ammonia. Compound **89** was also formed from the nitrone **90** by treatment with ethyl chloroformate. The oxazolidine **91** was considered to be a likely intermediate in the conversion of **90** to **89**.

The hydroxylamine **88** could be prepared by reducing the nitroacetanilide **92** ($R_1$, $R_2$ = H) with zinc and acetic acid (Eq. 30).[204] Cyclization of the hydroxylamine **88** to the nitrone was carried out by acid catalysis such as treatment with ethanolic hydrogen chloride. When the nitroacetanilide **92** ($R_1$ = Me, $R_2$ = H) was subjected to zinc–acetic acid reduction, diazepam was isolated.[204] It is likely that in this case the nitrone **93** was formed during the reduction and was further reduced to the imine **95** ($R_1$ = Me, $R_2$ = H) by excess reagent.

Reduction of the nitro derivative **92** ($R_1$ = Me, $R_2$ = COOEt) or the oxime **94** with zinc and acetic acid followed by treatment with acetic acid in boiling benzene gave the 3-carboxylates **95** ($R_1$ = Me, $R_2$ = COOEt) in moderate

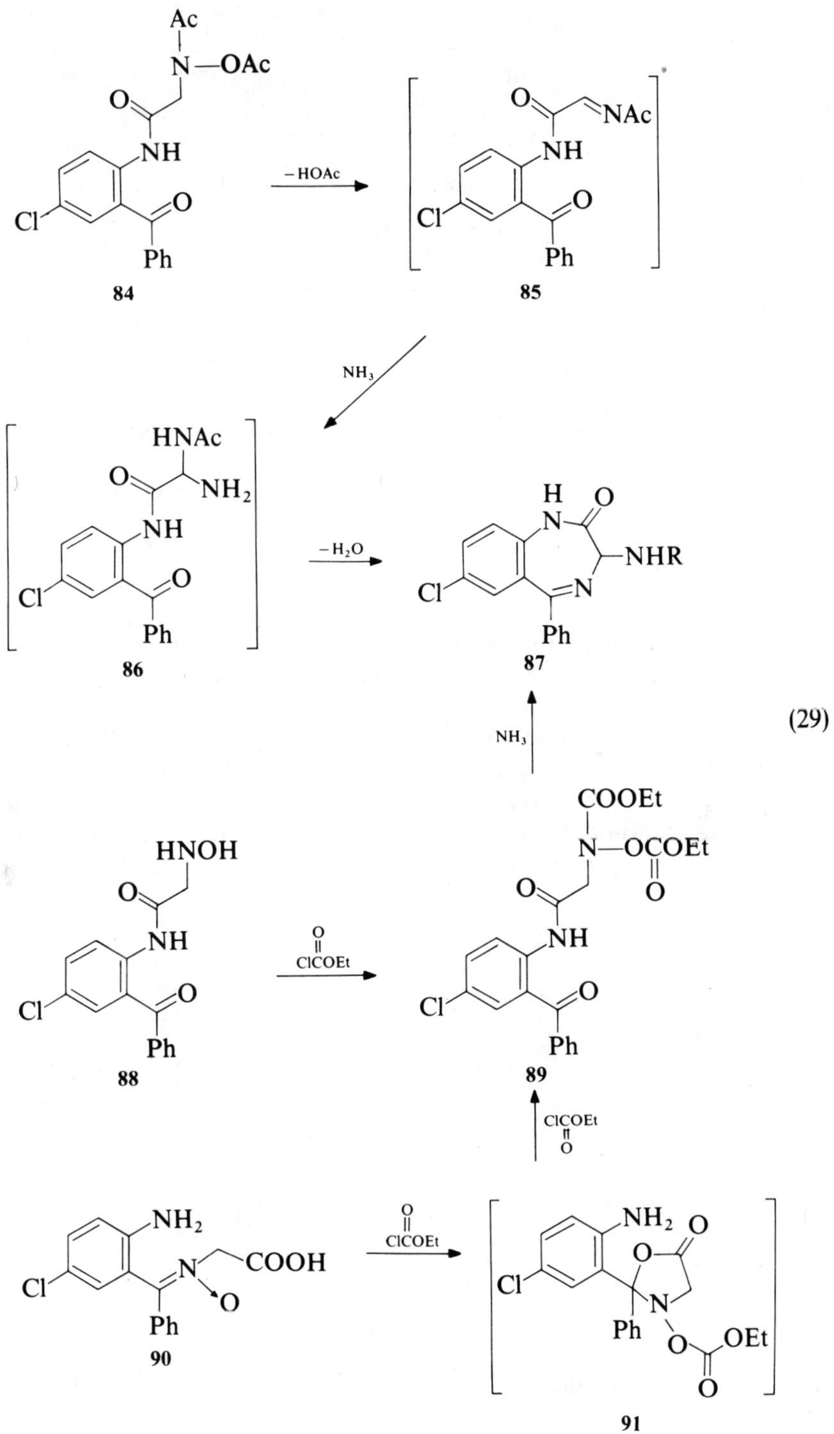

Ac
N—OAc
O
NH
Cl
O
Ph
84
– HOAc
O
NAc
NH
Cl
O
Ph
85
$NH_3$
HNAc
O
$NH_2$
NH
Cl
O
Ph
86
$-H_2O$
H
O
N
NHR
Cl
N
Ph
87
$NH_3$
COOEt
N—OCOEt
O
O
HNOH
O
NH
Cl
O
Ph
88
ClCOEt
NH
Cl
O
Ph
89
ClCOEt
$NH_2$
Cl
N
COOH
O
Ph
90
ClCOEt
$NH_2$
O
O
Cl
N
Ph
O
OEt
O
91


(29)

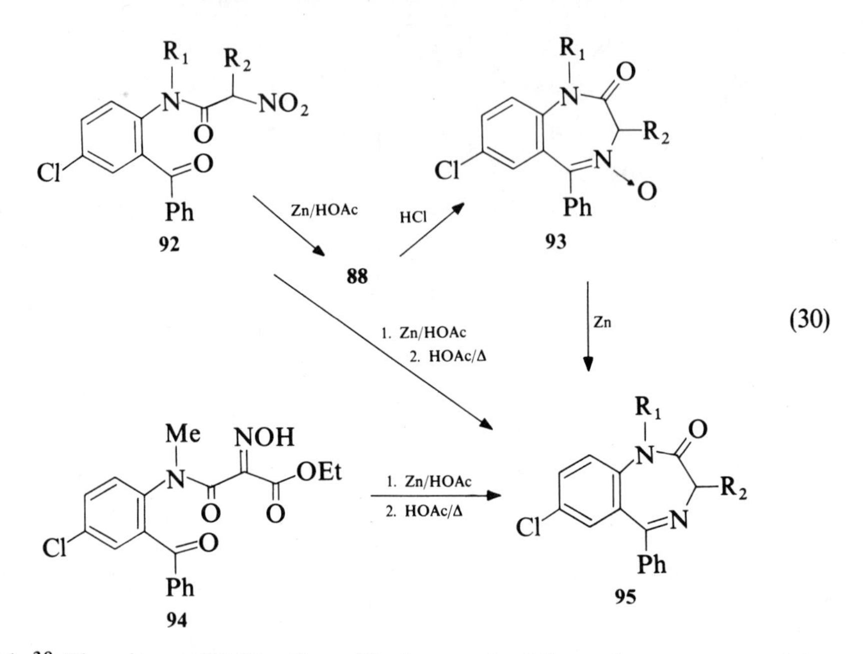


yields.[39] The nitrone **93** ($R_1$, $R_2$ = H) also resulted from the treatment of the diacylated hydroxylamine **84** with ethanolic hydrogen chloride.[56,205]

A synthesis of the 3-hydroxy-4-oxide, **97**, was reported in the patent literature.[206,207] The 2,2-diacetoxyacetanilide **96** was ring closed either by treatment with trifluoroacetic acid (Eq. 31) or by reacting **96** with hydroxide to form the corresponding 2,2-dihydroxy derivative and subjecting this intermediate to trifluoroacetic acid or triethoxythallium. The nitrone **97** was converted to the 3-chloro derivative **98** by means of phosphorus trichloride. Reductive ring

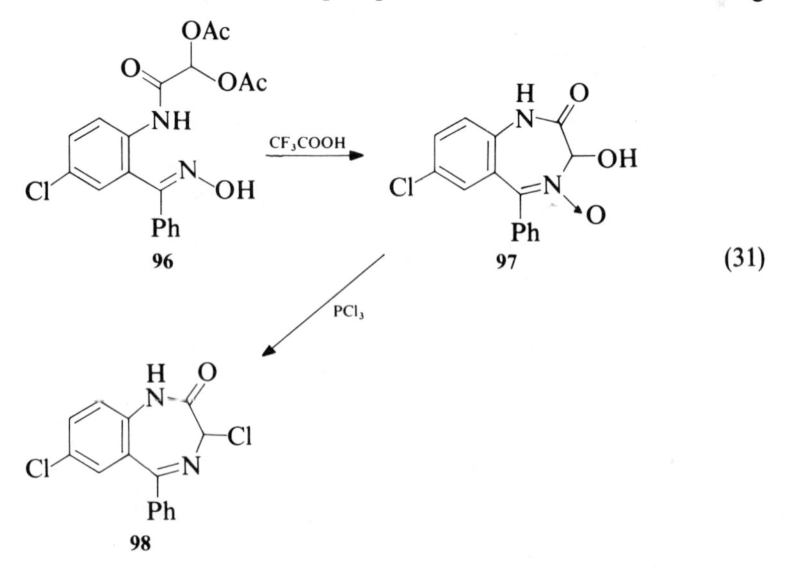

closure of the oxime **96** using palladium on carbon in the presence of trifluoroacetic acid was claimed to give the 3-hydroxy benzodiazepine **97** (4-desoxy).

A synthesis of nitrazepam (**101**) involves the reaction of the imino ether **99** (X = Br, I) with liquid ammonia, which leads to the glycine ester imine **100** (Eq. 32).[208] The latter was then cyclized to **101** by treatment with strong base such as methoxide or by heating with 2-methylimidazole at 140°C. In the case of the iodide **99** (X = I), the reaction with liquid ammonia also yielded nitrazepam directly.

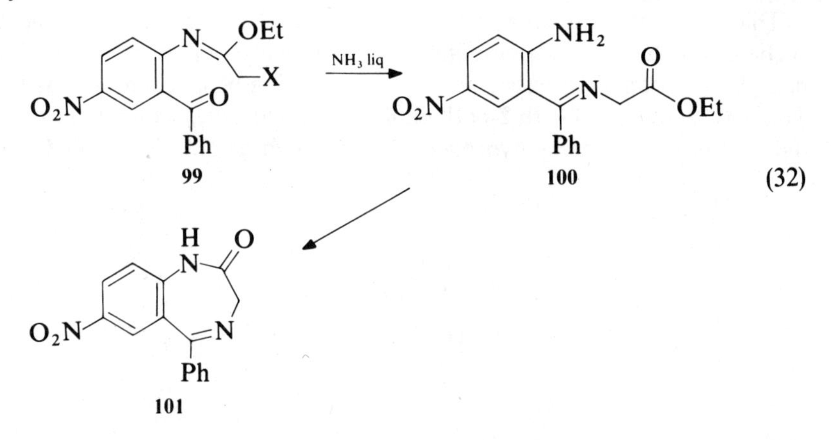


Diazepam was prepared by reacting the imine **102** (R = Me, X = Cl) with bromoacetyl bromide in a two-phase system of benzene and aqueous hydroxide (Eq. 33).[209] Reaction of the imine **102** with 2-bromo-2-fluoroacetyl chloride or 2-chloro-2-fluoroacetyl chloride and sodium hydride in tetrahydrofuran or

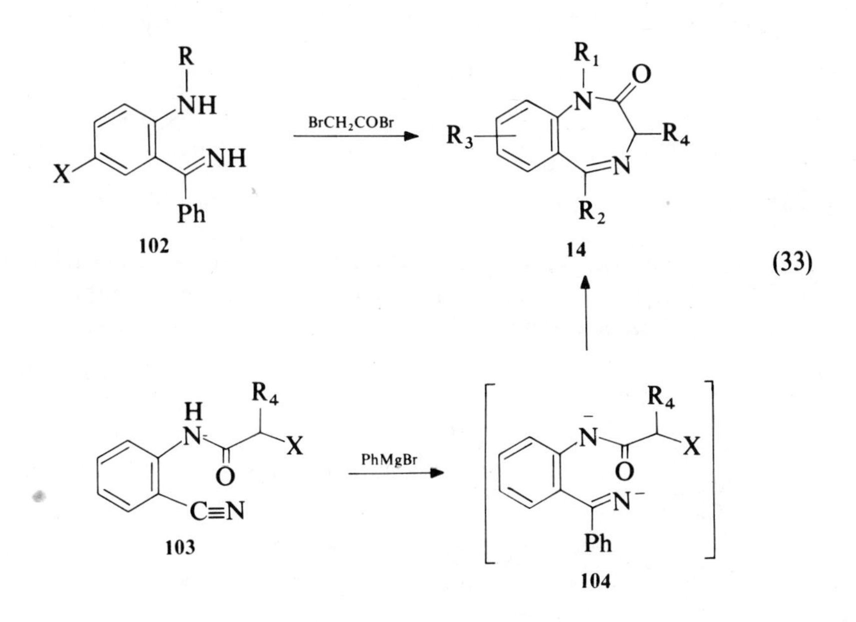

dimethylformamide yielded similarly the 3-fluorodiazepam.[144] Nitrazepam was obtained by reacting the urea **102** (R = MeNHCO, X = $NO_2$) with bromoacetyl bromide and potassium carbonate in dimethylformamide at 60–65°C.[74] Bergman and coworkers[210,211] reported the syntheses of the benzodiazepines **14** ($R_1$, $R_3$ = H; $R_2$ = Ph; $R_4$ = H, Ph) by reacting the nitriles **103** (X = Cl; $R_4$ = H, Ph) with phenylmagnesium bromide. The imine anion **104** is believed to be the intermediate leading to the benzodiazepines and the quinazoline by-products.

Diazepam (**31**), was obtained in high yield by heating the hydrazone **105** in acetic acid (Eq. 34).[212] Compound **105** was prepared in several steps; first, the benzyloxycarbonylhydrazone of the corresponding benzophenone was formed. This was then reacted with 2-phthalimidoacetyl chloride. Sequential cleavage of the protecting groups by hydrogenolysis and hydrazinolysis afforded **105**.

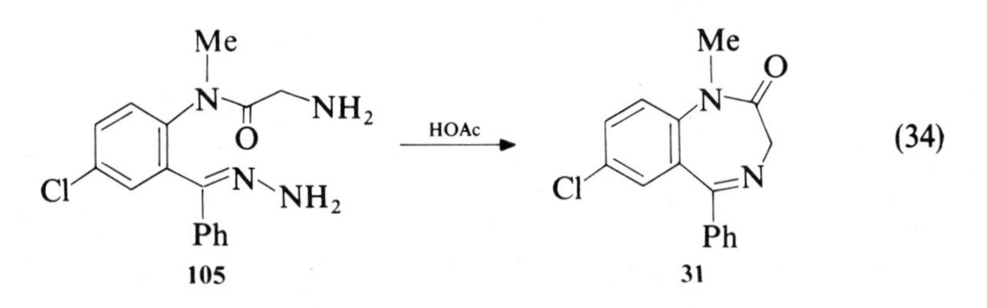


The imidazolidines **106** were converted to the benzodiazepines **107** by heating to reflux in ethanolic hydrogen chloride (Eq. 35).[213] Compounds **106** resulted from the hexamethylenetetramine process as discussed in Section 1.1.1.1.

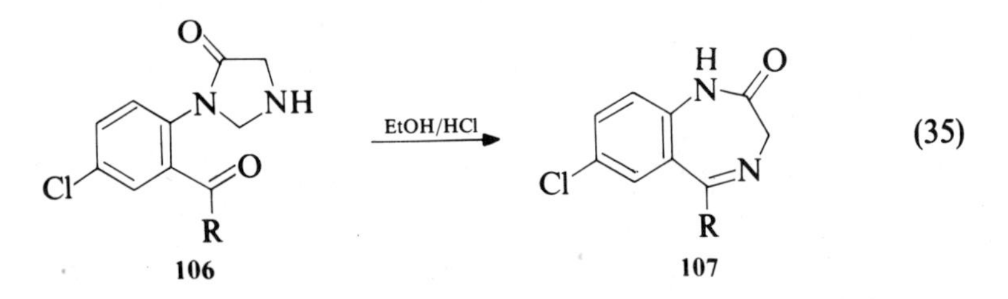


A Hoffmann degradation was used for the preparation of the 1,3-dimethyl derivative **109**. Thus, bromine and sodium hydroxide converted the amide **108** to the benzodiazepine **109** (Eq. 36).[214]

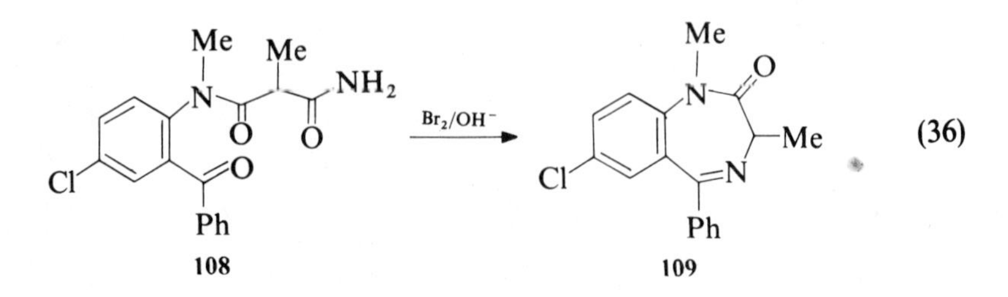

The oxaziridines **110** were rearranged to the 3-hydroxy or 3-alkoxy derivatives **111** (R = OH, alkoxy) in the presence of ferrous sulfate in water or in alcohol, respectively (Eq. 37).[215] Reaction of **110** (R = Me) with ethylamine in tetrahydrofuran at room temperature yielded 12% of the 3-ethylamino derivative **111** ($R_2$ = EtNH).[215b] The rearrangement of the oxaziridine to the nitrone, thermally or by acid catalysis, was described earlier.[216–218] A reductive transformation of the oxaziridines **110** (R = H, Me) to the corresponding imines was achieved with hydrogen iodide or with hydroxylamine.[137]

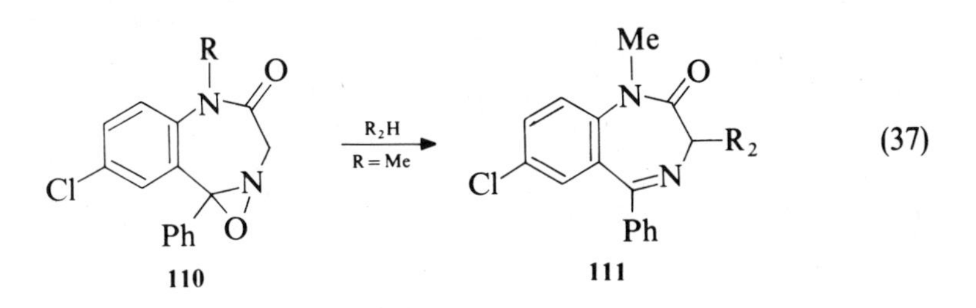


(37)

The tetracyclic compounds **112**, which resulted from the treatment of the bromoacetanilide with ammonia in methanol, were reacted with methanolic hydrogen chloride to give the benzodiazepines **113** (Eq. 38).[123]

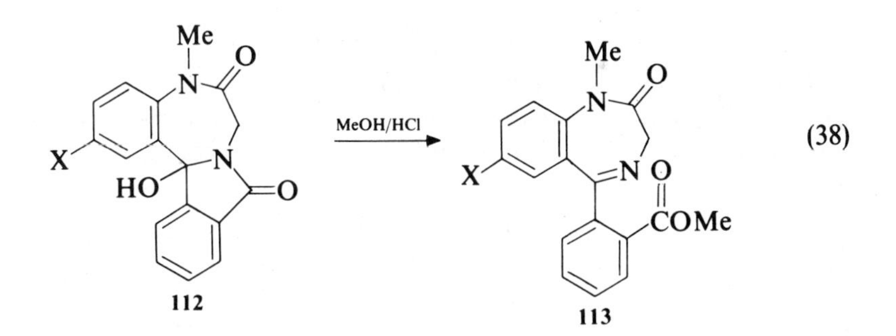


(38)

Migration of methyl from oxygen to nitrogen occurred when 2-methoxy-3*H*-1,4-benzodiazepine **114** was heated to 240–260°C, leading to the formation of **31** (Eq. 39).[219]

According to a Dutch patent,[220] a synthesis of diazepam, **31**, was achieved in moderate yields by condensation of the chloroacetanilide **115** with benzonitrile in titanium tetrachloride at 186–188°C. While a similar process involving nitrilium ions worked for the synthesis of 1-methyl-2,3-dihydro-1*H*-1,4-benzodiazepines (see Chapter VI), we were not able to prepare 2-ones such as diazepam by this method.[108]

Dealkylation of the tetracyclic derivatives **116** ($n$ = 2, 3) was reported[221] to occur by the use of acetic anhydride in combination with sodium acetate or boron trifluoride, leading to diazepam, **31**.

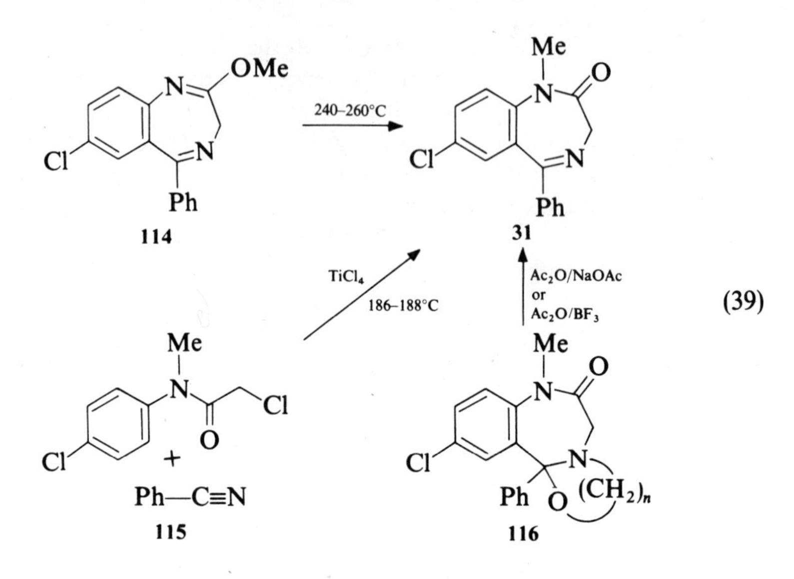


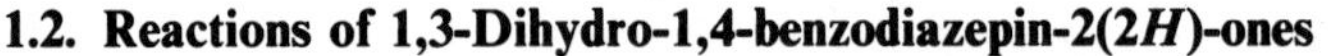

## 1.2. Reactions of 1,3-Dihydro-1,4-benzodiazepin-2(2*H*)-ones

### *1.2.1. Reaction with Electrophiles*

#### 1.2.1.1. Halogenation

Reaction of the 5-substituted benzodiazepin-2(2*H*)-ones **117** ($R_2 \neq H$) with hypochlorite led to the 1-chloro compounds **118**, which, in many instances, were stable enough to be characterized (Eq. 40).[222–224] By heating in a solvent in the

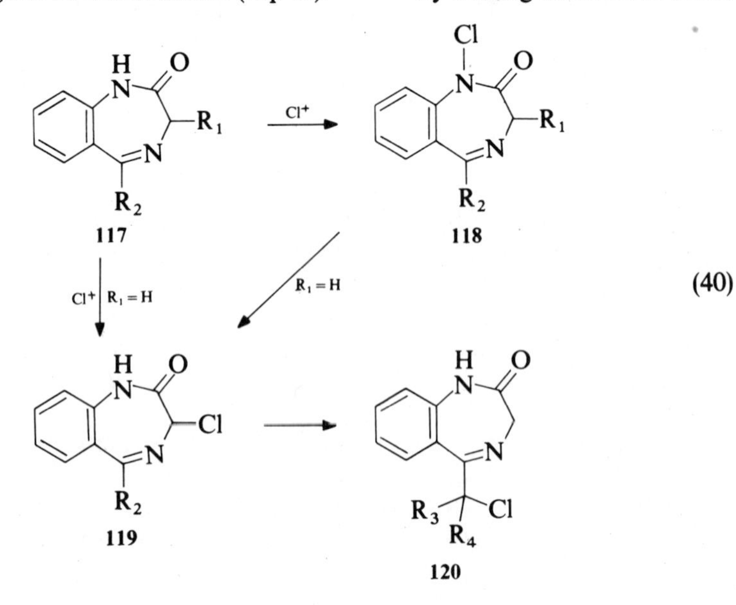

presence of a radical initiator, the 1-chloro derivatives rearranged to the 3-chloro analogs **119**. When $R_2$, in **119**, represented an alkyl or cycloalkyl residue, the chlorine migrated further onto this group to form **120**.[222,224] Chlorination of **117** ($R_2$ = Ph, 2-pyridyl) with *N*-chlorosuccinimide in benzene in the presence of azodiisobutyronitrile afforded the 3-chloro derivatives **119** directly.[225]

Reaction of the 3-cyano analog **95** ($R_1$ = Me, $R_2$ = CN) with *N*-chlorosuccinimide was reported to lead to the 3-chlorinated product **121** (X = Cl) (Eq. 41).[226] This compound was not properly characterized but converted to the 3-methoxy analog **121** (X = MeO). The bromination of the 3-carboxylic acid ethyl ester **95** ($R_1$ = H, $R_2$ = COOEt), by treatment with bromine in methylene chloride at room temperature in the presence of benzoyl peroxide, similarly afforded the 3-bromo derivative, which was converted in situ to the 3-hydroxy and 3-methoxy analogs.[101,227]

Fluorination at the 3-position was achieved by reacting the 3-carbanion, generated with lithium diisopropylamide or potassium *t*-butoxide, with perchloroyl fluoride or trifluoromethoxy fluoride.[228]

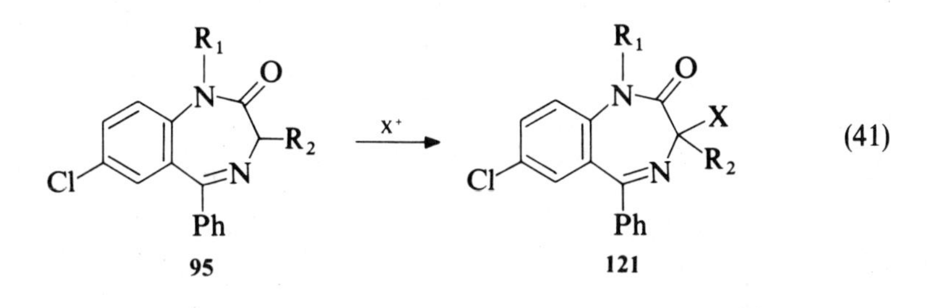


Chlorinations at the 7-position were described for **117** ($R_1$ = H, $R_2$ = Ph) by using a combination of chlorine, ferric chloride and nitrobenzene[31] and, under similar conditions, for **117** ($R_1$ = AcO, $R_2$ = Ph).[229] 7-Nitro compounds **122** ($R_1 = R_2$ = Me; $R_3$ = 2-Cl$C_6H_4$, 2-F$C_6H_4$; Y = $NO_2$) were chlorinated at the 9-position by treatment with chlorine in 1,2-dichloroethane in the presence of formic acid and hydrogen chloride to give **123** (X = Cl) (Eq. 42).[230] These conditions also allowed the introduction of a chlorine into the 9-position of **122** ($R_1$, $R_2$ = H; $R_3$ = 2-Cl$C_6H_4$; Y = Cl).[134] Bromination at the 9-position was similarly carried out with *N*-bromosuccinimide in a mixture of formic acid and hydrochloric acid.[230]

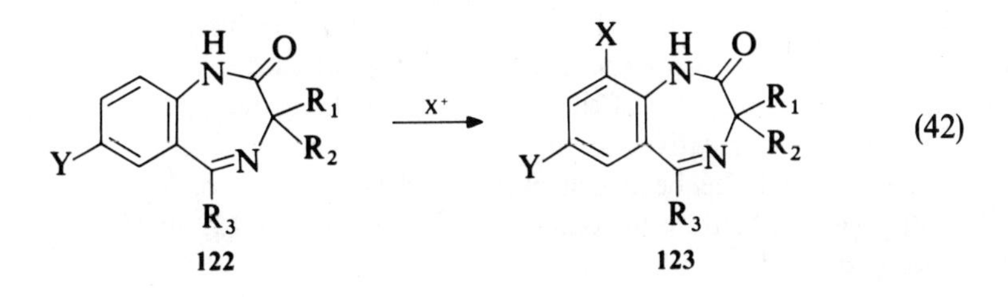

Halogenations of the 7-amino derivatives **124** ($R_1$ = Me; $R_2$ = H, Me, Et; $R_3$ = H, Me; $R_4$ = 2-$ClC_6H_4$, 2-$FC_6H_4$; Y = H) were carried through by reactions with chlorine in hydrochloric acid or with bromine in acetic acid to give the corresponding 6-halogenated analogs (Eq. 43).[230,231] The 6,8-dichloro derivatives **125** (X = Cl) were a result of the chlorination of the 7-amino compounds **124** (Y = H) with *N*-chlorosuccinimide at room temperature,[80,230] while the 6,8-dibromo analogs **125** (X = Br) were prepared similarly with *N*-bromosuccinimide in methylene chloride.[230] Further treatment of the 7-amino-9-chloro compounds **124** (Y = Cl) with *N*-chlorosuccinimide afforded the 6,8,9-trichloro compound **125** (X = Y = Cl) along with the dichloro compounds **126** and **127**.[230]

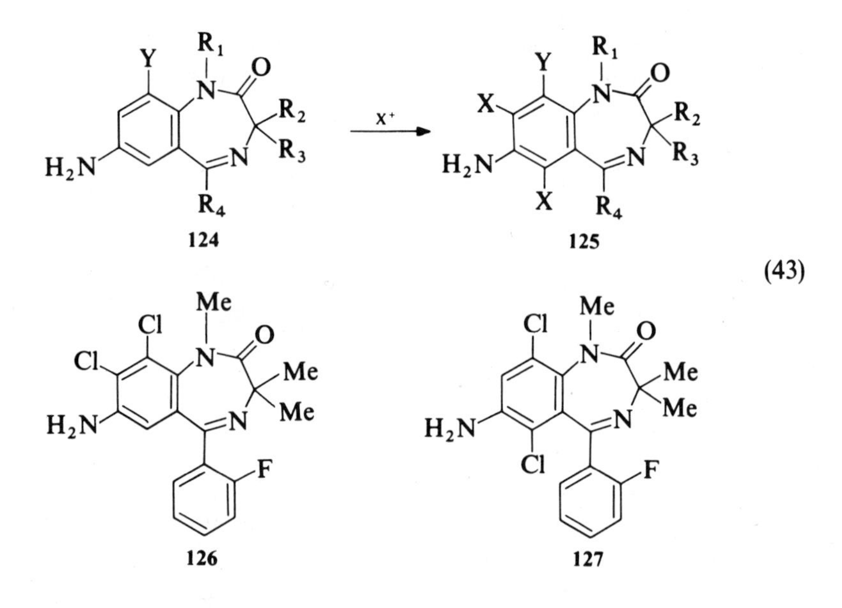


## 1.2.1.2. Oxidation

Oxidations by peracids have been widely used to convert benzodiazepin-2(2*H*)-ones to the corresponding 4-oxides.[8,9,34,44,67,69,72,232–236] Compounds with dialkylaminoalkyl residues attached at the 1-position were oxidized with *m*-chloroperbenzoic acid at the nitrogen in the side chain first and then at the 4-position.[238,239] Similarly, 1-methylthioalkyl derivatives were oxidized by peracetic or *m*-chloroperbenzoic acid to the sulfoxides, to the sulfones and then further to the 4-oxides of the sulfones.[72,234] Hydrogen peroxide converted the 7-methylthio derivatives to the corresponding sulfoxides.[7,93]

Reaction of the benzodiazepines **128** ($R_1$ = H, Me; $R_2$ = H, OH; X = H, Cl, F) with ruthenium tetroxide afforded the 3-ones **129** in approximately 50% yield (Eq. 44).[157,158]

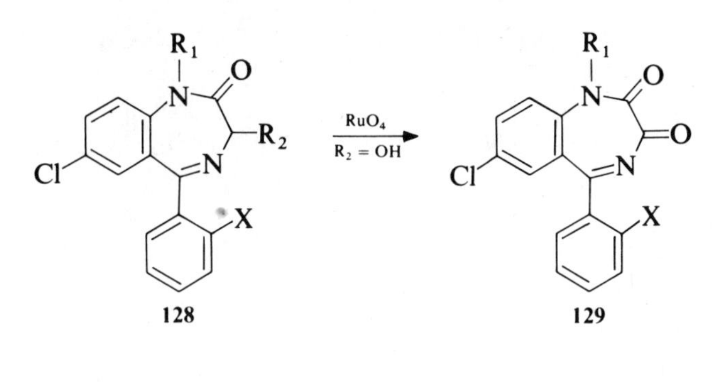


(44)

130

The major metabolic pathway of benzodiazepin-2-ones has been shown to involve hydroxylation at the 3-position. The microsomal oxidation of the racemic **128** ($R_1$, $R_2$ = Me; X = H) was investigated and reported[240] to yield 5% of the 3-hydroxy derivative **130** ($R_1 = R_2$ = Me; X = H). Microbiological oxidation of a variety of 1-methyl- and 1-cyclopropylmethyl-substituted benzodiazepin-2-ones led to optically active 3-hydroxy compounds with partial loss of the substituent at the 1-position.[241] The 3-carboxylic acid esters **128** ($R_1$ = H, Me, $MeOCH_2$; $R_2$ = COOR) were found to be readily oxidized by molecular oxygen in the presence of a strong base.[242] This reaction most likely proceeds via the 3-hydroperoxide, which may undergo reductive cleavage to the product. The 3-phosphonates **128** ($R_1$ = Me, $R_2$ = $(EtO)_2OP$ or $(MeO)_2OP$, X = Cl) underwent a similar type of oxidation when treated with sodium hydride and oxygen.[243] The corresponding 3-one **129** was thus obtained in 21% yield. The analog lacking the 1-methyl substituent could be prepared by avoiding an acid workup. The intermediate anion of **129** ($R_1$ = H) was reacted with trimethylsilyl chloride and the silylated amide was hydrolyzed under neutral conditions to give the 3-one **129** ($R_1$ = H, X = Cl) in 43% yield.

Ceric ammonium nitrate was described as a suitable reagent for the transformation of the 7-ethyl derivative **131** into the 7-acetyl analog **132** (Eq. 45).[244]

The olefinic side chain attached to the 1-position of **133** (R = H, $AcOCH_2$) and also the 4-oxides thereof were converted to the corresponding diols **134** (R = H, $HOCH_2$) by means of potassium permanganate (Eq. 46).[237]

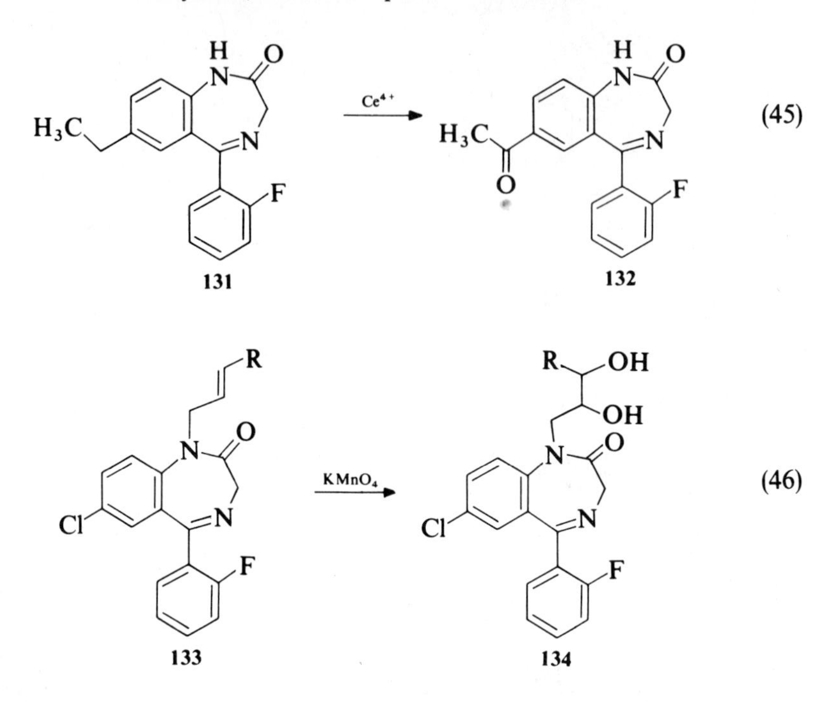


Oxidation of the 7-hydroxyaminobenzodiazepinones **135** ($R_1$ = Me, $MeOCH_2$) with manganese dioxide afforded the 7-nitroso derivatives **136** (Eq. 47).[245,247] If this oxidation is allowed to proceed slowly, the nitroso compound condenses with the hydroxyamine to yield the azoxy compound **137**.[245]

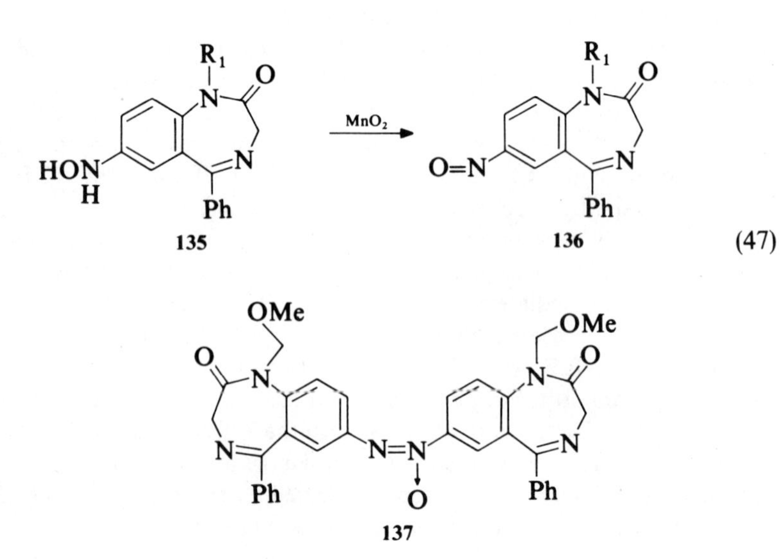

The aminophenol **138** was oxidized with Fremy's salt to yield the aminoquinone **139** (Eq. 48).[245]

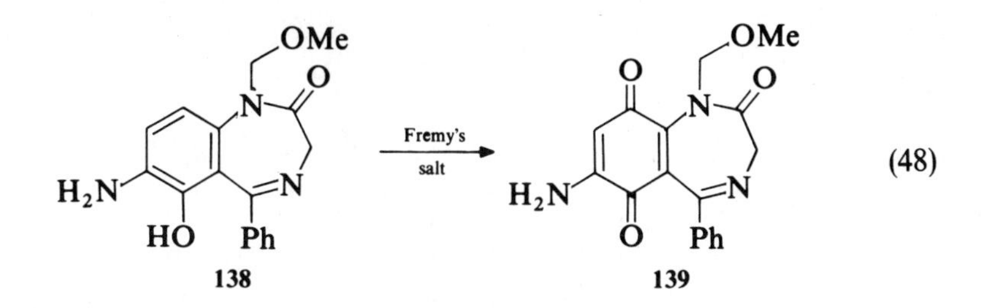


(48)

A U.S. patent [248] outlines the preparation of a mixture of 5-(trifluoromethoxy)phenyl-substituted compounds **140** by treatment of **36** with trifluoromethoxy fluoride at $-78°C$ in hydrogen fluoride in the presence of light (Eq. 49).

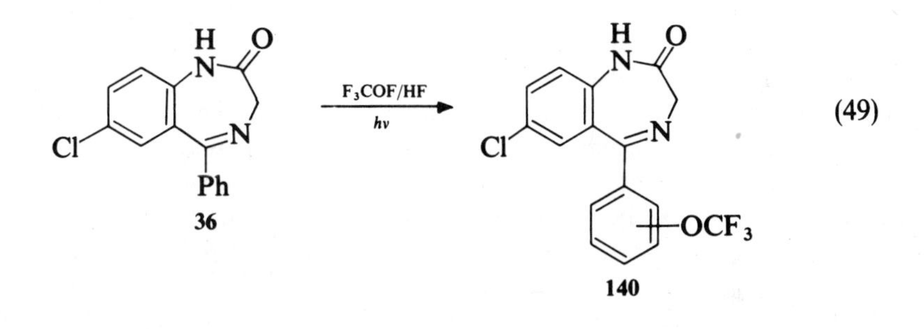


(49)

### 1.2.1.3. Reactions with Nitrogen Electrophiles

A. Nitrosation. 3-Aminobenzodiazepin-2-ones were nitrosated in aqueous medium to give the 3-hydroxy derivatives.[199,200,202] If the nitrosation was carried out in acetic acid, the 3-acetoxy benzodiazepin-2-one was obtained.[199] Reaction of the 3-hydrazino compound with nitrous acid and subsequent treatment with hydroxide also led to the 3-hydroxybenzodiazepin-2-one, most likely by intermediacy of the 3-azido analog.[249] Nitrosation of the 1-aminobenzodiazepin-2-one was reported to cause deamination, leading to the corresponding lactam.[250] Several 7-amino-substituted compounds were converted to the diazonium salts, which were isolated in some instances as tetrafluoroborate salts.[251] Reactions of diazonium salts in situ[134,252,253] are further discussed in Section 1.2.2.

B. Nitration. Reaction of 7-unsubstituted benzodiazepin-2-ones **141** (X = H) with potassium nitrate in concentrated sulfuric acid at 0–5°C led to the 7-nitro derivatives **142** (Eq. 50). Such nitrations were carried out with compounds **141** as follows:

| $R_1$ | $R_2$ | $R_3$ | Refs. |
|---|---|---|---|
| H | H | Ph | 136, 254, 255 |
| H | H | 2-$ClC_6H_4$, 4-$ClC_6H_4$, 2-$FC_6H_4$, 2-$O_2NC_6H_4$ | 136 |
| | | 2-$F_3CC_6H_4$ | 6 |
| H | H | Cyclohexyl | 94 |
| $F_3CCH_2$ | H | Ph | 69 |
| H | H | 2-Pyridyl, 4-Pyridyl | 3, 22 |
| H | H | 2-Thienyl | 22 |
| H | COOEt | Ph | 104, 107 |

In the 5-(2-thienyl) case,[22] the location of the nitro group introduced was not firmly established, and the possibility of primary nitration of the thienyl moiety is not ruled out. An 8-methyl-substituted compound was nitrated under these conditions at the 7-position.[136]

Under slightly more vigorous nitration conditions (i.e., potassium nitrate in concentrated sulfuric acid at room temperature), 7-nitro- and 7-halo-substituted benzodiazepin-2-ones were nitrated in the meta position of the 5-phenyl group.[69,256] When the 5-phenyl ring was substituted by a halogen (Cl or F) in the *ortho* position, nitration was directed to the *para* position relative to this halogen to give 2′-halo-5′-nitro derivatives **144**.[134,152,257–259] Further nitration of **141** (X = Cl, $R_1 = R_2 =$ H, $R_3$ = 3-$O_2NC_6H_4$) occurred at the 9-position, leading to **143**.[152]

In the absence of a strong acid, the imine is not protonated and the nitronium ion attacks the 4-position nitrogen. Thus treatment of the 1-substituted benzodiazepin-2-ones **145** (R = Me, Cl; X = Cl, $NO_2$) with fuming nitric acid in acetic anhydride yielded the addition products **146** (Eq. 51).[260] While the 1-unsubstituted 7-nitro compound underwent the same addition reaction, the corresponding 7-chloro analog **145** (X = Cl, R = H) did not. The reason for this divergent reactivity is not clear.

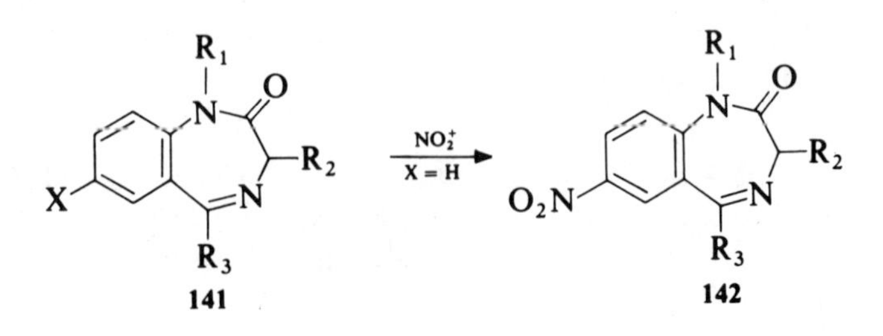

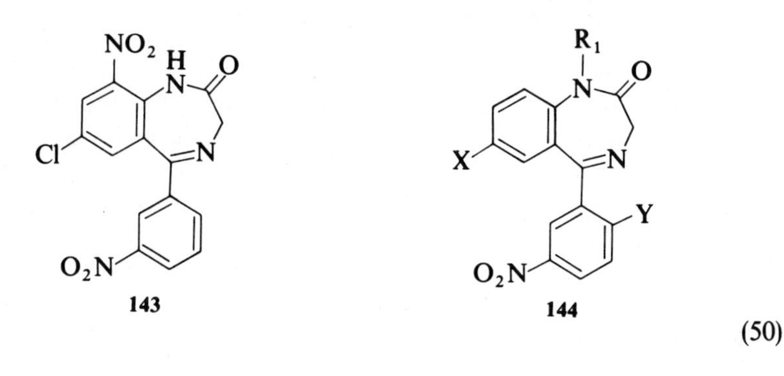


(50)

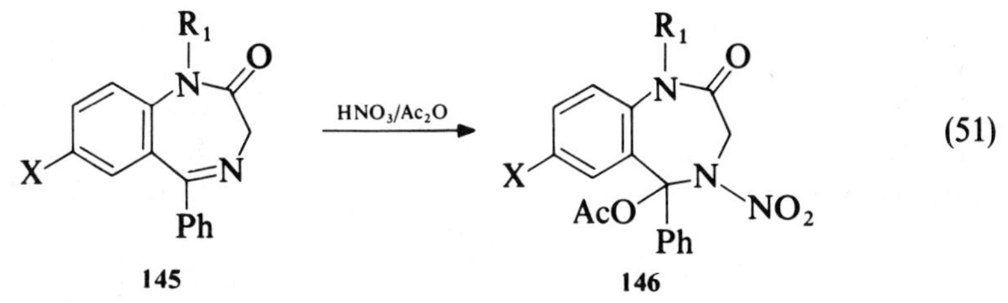


(51)

C. Reaction with Other Nitrogen Electrophiles. Aminations of benzodiazepin-2-ones in the 1-position were carried out by means of chloramine[250] or *O*-(2,4-dinitrophenyl)hydroxylamine.[261] The benzodiazepinone was first deprotonated by a strong base to form the anion at the 1-position, which was then reacted with the electrophilic amine.

The 4′-hydroxy derivative **147** was coupled with the diazonium salt **148** to afford the azo compound **149** (Eq. 52).[8]

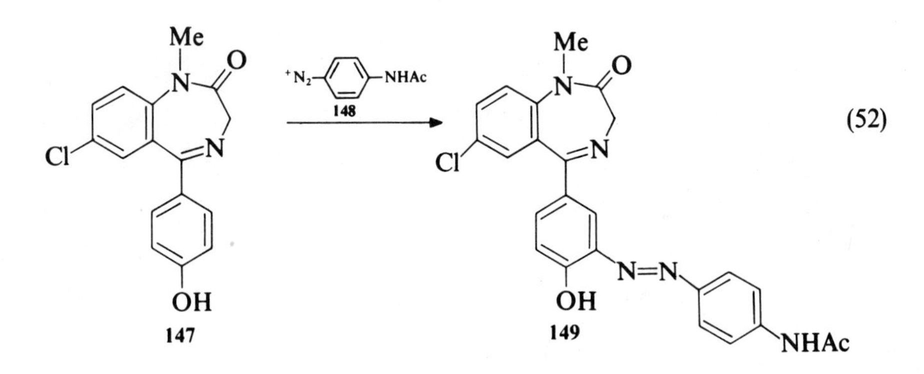


(52)

### 1.2.1.4. Reactions with Carbon Electrophiles

A. Alkylation. A great variety of 1-substituted 1,4-benzodiazepin-2-ones **151** have been prepared by alkylation of the lactams **150**. Generally, the proton at the 1-position is abstracted by a base to form the anion, which is reacted with the alkylating agent.

Methylations at the 1-position were carried out in several ways: with dimethyl sulfate in aqueous sodium hydroxide,[34,142,262,263] in a two-phase system of methylene chloride and sodium hydroxide,[54] with sodium methoxide as base in methanol,[135] in benzene,[10] in toluene,[2,16] and in dioxane.[5] Using methyl iodide as the alkylating reagent, the following base–solvent combinations were applied: sodium methoxide in dimethylformamide,[2,7,18,91,93,94,222,264] in methanol,[15] in a mixture of methanol and dimethylformamide,[9] with potassium hydroxide in tetrahydrofuran,[265] with potassium *t*-butoxide in dimethylformamide,[77,141] and in tetrahydrofuran,[26] with sodium hydride in dimethylformamide,[2,13,32,95] in tetrahydrofuran,[266,267] and in a mixture of dimethylformamide, 1,2-dimethoxyethane, and 2-methoxyethanol;[27] with sodium amide in a mixture of dimethylformamide and tetrahydrofuran,[162] with solid potassium carbonate in acetone,[62,80,230,231,251] and barium oxide in dimethylformamide.[40,81,134] The last combination was especially useful for methylation without racemization of 3-substituted optically active compounds. The use of phenyllithium as base has been described in the patent literature.[268] Radiolabeled methyl iodide was reacted with **150** ($R_2$ = H, $R_3$ = Ph, $R_4$ = 7-Cl) and **150** ($R_2$ = H, $R_3$ = 2-$FC_6H_4$, $R_4$ = 7-$O_2N$) in acetone and sodium hydroxide to give the corresponding 1-methyl analogs with a carbon-11 label.[269]

Other methylations occurred by treatment with trimethylsulfonium iodide or trimethylsulfoxonium iodide in dimethyl sulfoxide in the presence of sodium hydride or butyllithium.[270]

Diazomethane was reported to methylate both the 1-position nitrogen[91] and the 2-position oxygen yielding the 2-methoxy-3*H*-1,4-benzodiazepines **153** (Eq. 53).[271,272]

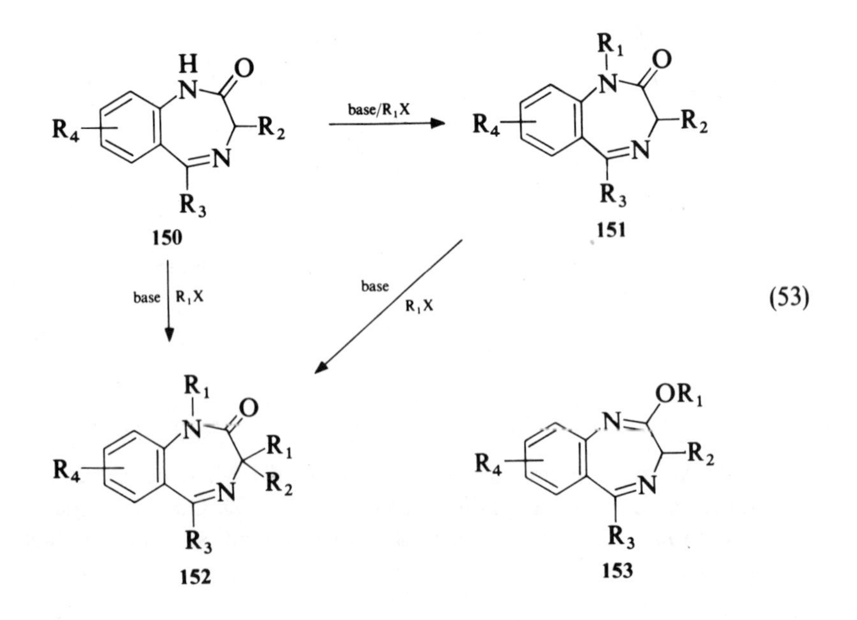

Higher, branched and unbranched alkyl groups were introduced into the 1-position in the same fashion.[2,18,91,135,136,251] Similarly, reaction of the anion of **150** with allyl and propargyl halides afforded compounds with an unsaturated alkyl residue at the 1-position.[2,18,91,135,273–275] Compounds bearing an alkyl group with an aromatic or aliphatic ring were also accessible by this procedure. Examples are 1-benzyl[2,91] and 1-(cyclopropyl)methyl derivatives.[268,276]

The high reactivity of the 1-chloro-1-alkoxyalkyl derivatives (α-chloro ethers) allows the easy preparation of a number of 1-substituted benzodiazepin-2-ones by alkylation. The anions of **150** were again generated with methoxide in dimethylformamide[72,276,277] or with sodium hydride in tetrahydrofuran[71,72] or in dimethylformamide.[278]

The use of such α-chlorothioethers as chloromethylmethylsulfide led in analogy to 1-methylthiomethyl derivatives **151** ($R_1$ = $MeSCH_2$).[72,134]

A large number of differently functionalized alkyl moieties were attached to the 1-position nitrogen of **150** in a similar manner. Such alkylations were carried out with the following reagents: 2-alkoxyethyl chlorides,[66,89,128,234,276,279,280] 2-alkoxyethyl sulfates,[281] 2-acetoxyethyl bromide,[236] 1-chloro-2,3-diacetoxypropane,[236] 1-bromo-2,3,4-triacetoxybutane,[236] 2-bromoethanol,[233,283,284] 3-bromopropanol,[284] 1-chloro-2-hydroxy-3-methoxypropane,[285] 2-methylthioethyl chloride[128,234] and the corresponding sulfoxide,[86c,286] alkyl halides functionalized with amino groups,[17,65,150,238,254,268,287,288] 3-chloro-1-ethylpiperidine,[289] 1-(4-chlorobut-1-yl)-4-(4-fluorobenzoyl)piperidine,[290] 1-benzyloxycarbonylamino-2-bromoethane,[233] and 1-chloromethyl pyrrolidin-2-one.[134] Other alkylations at the 1-position were described for compounds having two halogens of different reactivity. Thus a 2,2,2-trifluoroethyl group was introduced by reaction of 1,1,1-trifluoro-2-iodoethane with the anion generated by sodium methoxide in dimethylformamide.[69] Reaction of the anion with 1-bromo-3-chloropropane yielded the 1-(3-chloropropyl) derivative **151** ($R_1$ = $Cl(CH_2)_3$).[291] The higher reactivity of the chloride in the 1-position of 1,2-dichloro-1-methoxyethane facilitated the synthesis of **151** ($R_1$ = 2-chloro-1-methoxyeth-1-yl; $R_2$ = H, COOEt; $R_3$ = Ph; $R_4$ = 7-Cl, $NO_2$).[72,278] Compounds with a (2-chloroethoxy)methyl group in the 1-position were also accessible.[72] Alkylation reactions were also successful with α-haloketones,[65,284] α-haloesters,[66,292] and α-halocarboxamides.[293] Longer chain fatty acids were introduced by reaction of the anion of **150** with ω-haloalkanoic acids or esters thereof.[294] Examples of alkylations with 2-bromobut-2-enoic acid methyl ester (trans isomer) were also reported.[294] A butyrophenone side chain was similarly attached to the 1-position of **150** ($R_2$ = H; $R_3$ = Ph, 2-$FC_6H_4$; $R_4$ = 7-Cl) in the form of its 1,3-dioxolane derivative followed by hydrolytic liberation of the carbonyl group to yield the corresponding benzodiazepines **151** [$R_1$ = 3-(4-fluorobenzoyl)prop-1-yl].[295] Benzodiazepin-2-one-1-acetaldehydes have not been described, but the dimethyl acetal **150** ($R_1$ = 2,2-dimethoxy-1-ethyl, $R_2$ = H, $R_3$ = Ph, $R_4$ = 7-Cl) was obtained by alkylation of the appropriate anion **150** with 1-bromo-2,2-dimethoxyethane.[296]

Phosphorus-containing alkyl residues were attached to the amide nitrogen by means of chloroalkyl dialkylphosphine oxides to give **151** [$R_1$ = $(R_5)_2PO(CH_2)_x$ where $x$ = 1, 2, 3 and $R_5$ = Me, Et, Pr].[25] Another type of alkylation reaction at the 1-position is a Michael addition of the anion to acrylonitrile, leading to **151** ($R_1$ = $NCCH_2CH_2$; $R_2$ = H, OH; $R_3$ = Ph, 2-$FC_6H_4$, 2-$ClC_6H_4$; $R_4$ = 7-Cl, $NO_2$).[91,280] When $R_2$ = OH, a selective reaction at the 1-position occurred with benzyltriethyl ammonium hydroxide as a base.[280] Other examples of addition reactions include the addition of ethylvinyl ether to **150** ($R_2$ = H, $R_3$ = Ph, $R_4$ = 7-$NO_2$) by heating in acetic acid[72] and addition of enamides generated in situ from *N*-tosyloxycarbamates and triethylamine.[84]

If the benzodiazepine **150** is treated with 2 equivalents of a strong base in an aprotic solvent, a 1,3-dianion is generated. Reactions of the 1,3-dianion with alkylating agents have resulted in the formation of 1,3-disubstituted derivatives **152**. 1-Substituted compounds were similarly deprotonated at the 3-position and reacted with carbon electrophiles. Thus the 4-oxide of diazepam was alkylated at the 3-position with methyl iodide and benzyl chloride in dimethylformamide at low temperatures, using potassium *t*-butoxide for deprotonation, to yield the 3-methyl and the 3-benzyl derivatives, respectively.[297] The same method was applied to introduce 3-allyl substituents[298] onto this compound and the 5-methyl analog as well as to methylate the 1-methoxymethyl derivative **151** ($R_1$ = $MeOCH_2$; $R_2$ = H, $R_3$ = Ph; $R_4$ = 7-Cl, 4-oxide) at the 3-position.[277] The 3-carbanion of diazepam was generated with lithium diisopropyl amide in tetrahydrofuran at $-60°C$ and subsequently reacted with alkyl iodides and benzyl chlorides.[299] Double alkylations at the 3-position were observed with butyl iodide and propyl iodide, while the 3,3-dimethyl compound was not formed under the same conditions.[299] Methylation of the 3-cyano compound **151** ($R_1$ = Me, $R_2$ = CN, $R_3$ = Ph, $R_4$ = 7-Cl) was reported to yield the 3-methyl derivative.[226] This reaction was carried out with methyl iodide and potassium carbonate in dimethyl sulfoxide.

Reaction of the 4-oxide of diazepam, **154**, with 1,4-dibromobutane was found to give the spiro compound **155** (Eq. 54).[134]

Base-catalyzed addition of methyl acrylate to **154** yielded the 3-substituted derivative **156**.[298]

The nitrogen in the 4-position of 1,3-dihydro-1,4-benzodiazepin-2(2*H*)-ones **151** is weakly basic ($pK_a \approx 2.8$) and forms quaternary salts with reactive alkyl halides or sulfates. The quaternary salts described in the literature are listed as salts of the parent base in Tables VII.1. Quaternizations with methyl iodide[34,35,179] and dimethyl sulfate[179,293,300] were most common (Eq. 55).

Benzodiazepin-2-ones with amino groups in side chains attached at the 1- or 3-position were also converted to quaternary ammonium salts. Thus, basic esters of structure **158** were reacted with methyl bromide in benzene at room temperature to give the ammonium salts **159** (Eq. 56).[147,301] This reagent was also used to quaternize the 7-dimethylamino group of **151** ($R_1$ = Me, $R_2$ = H,

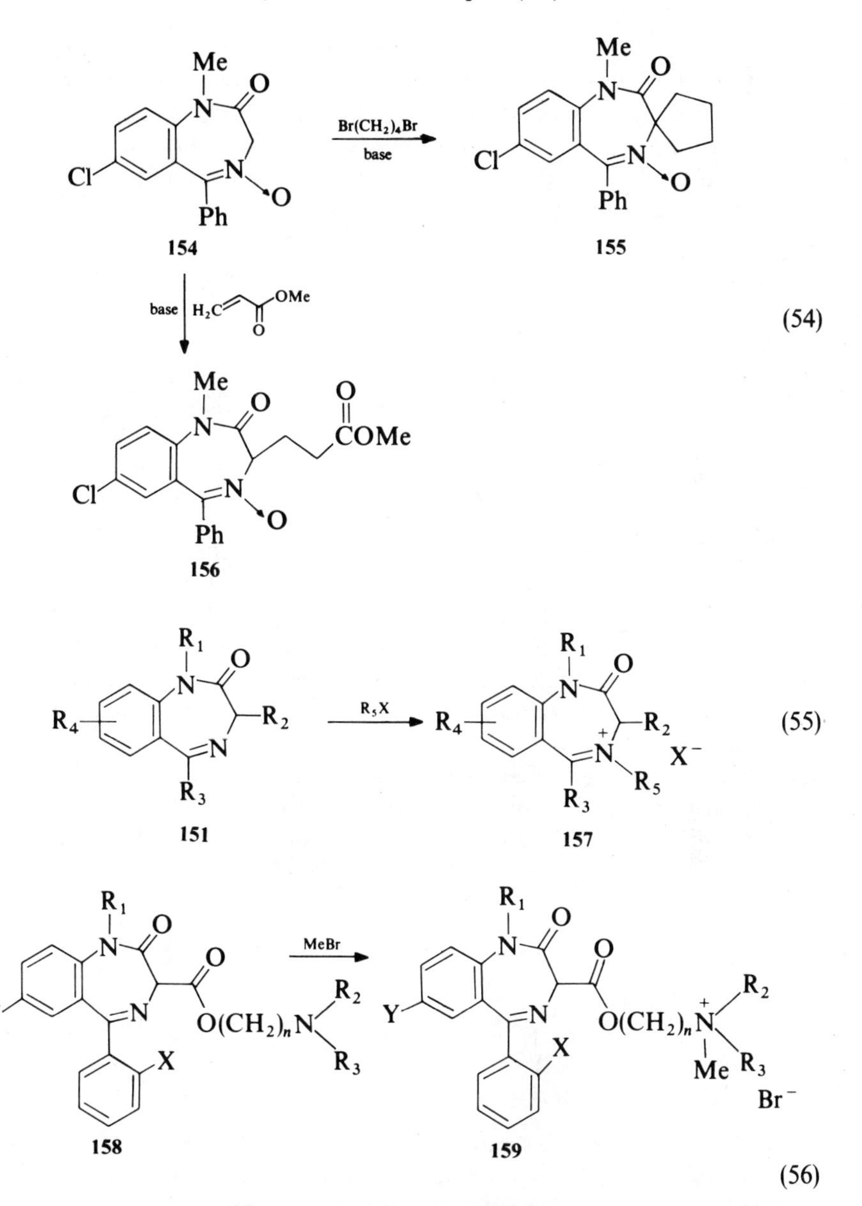


$R_3$ = Ph, $R_4$ = 7-$Me_2N$).[95] A 3-diethylamino derivative was also converted to its quaternary salt by reaction with methyl bromide.[302]

Functional groups in various positions have been subjected to alkylation reactions. For example, 1-(2-hydroxyethyl) derivatives were reacted with ethyl bromoacetate,[150c,150d] 1-(2-methylaminoethyl) analogs were alkylated on nitrogen with this same reagent and with *N*-methyl bromoacetamide,[150c,d] and 3-hydroxy-1,4-benzodiazepin-2-ones were reacted with acrylonitrile to yield the corresponding 3-(2-cyanoethoxy) derivatives.[280] Diazomethane was used to methylate the 1-hydroxy compound **68**[189] and to convert the *N*-hydroxyamide **160** (R = $F_3CO$) into the methoxy amide **161** (R = $F_3CO$).[245] The 7-hydroxyamino group in **160** (R = H) was also *N,O*-dimethylated by means of methyl iodide and potassium *t*-butoxide in dimethylformamide.[245] This combination was likewise applied to convert **160** (R = Ac) into **161** (R = Ac) (Eq. 57).[245]

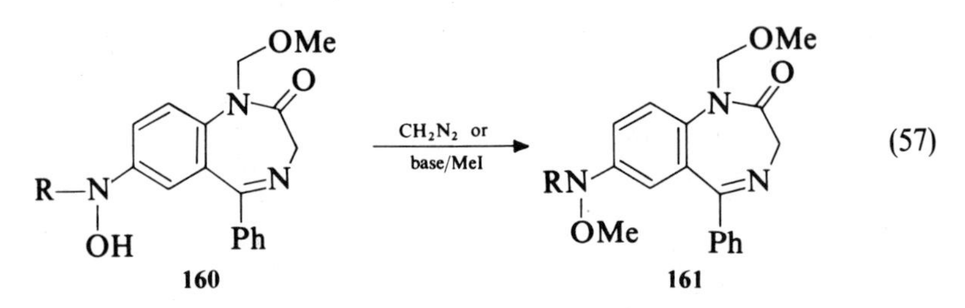


Chloromethylmethyl ether was reacted with the hydroxy group of **151** ($R_1$ = Et, $R_2$ = H, $R_3$ = 2-$FC_6H_4$, $R_4$ = 7-(1-hydroxyethyl)) using *N,N*-dimethyl aniline as an acid acceptor to give the corresponding 7-(1-methoxy-1-ethyl) derivative.[251] Reaction of a 7-aminobenzodiazepine with chloromethylmethylsulfide afforded the 7-methylthiomethylamino derivative.[134] Reductive methylations of the 7-nitro derivatives, by a reaction with formaldehyde, Raney nickel, and hydrogen, led to the 7-dimethylamino analogs.[95,255] Introduction of aryl substituents into the 1-position was possible by a modified Ullmann reaction. The lactams **162** (R = Me, Et; X = Cl, $F_3C$) were reacted with aryl bromides in dimethylacetamide at 140°C in the presence of copper powder and potassium acetate (Eq. 58).[19]

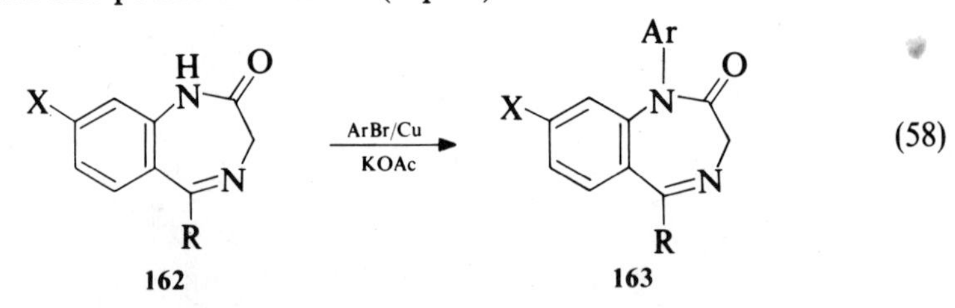


2,4-Dinitrobenzene was reported to react with diazepam and tetramethylammonium hydroxide to form a blue adduct to which structure **164** was assigned on the basis of spectral data (Eq. 59).[303]

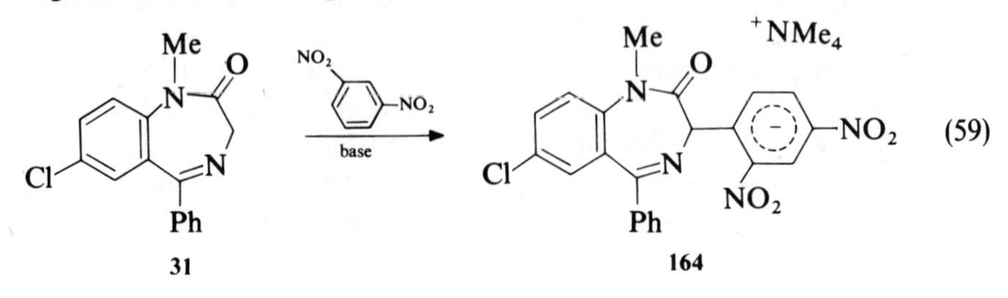

1,4-Benzodiazepin-2-ones, which bear at the 1-position a side chain containing a leaving group, underwent intramolecular alkylation on oxygen, affording **166**. The chlorides **165** ($R_1$ = H, MeO; $R_2$ = H, COOEt; X = Cl, I) were treated with sodium hydride in dimethylformamide at 0–10°C (Eq. 60). The products, in particular **166** ($R_2$ = H), are sensitive to acid. The ethoxycarbonyl function of the keteneacetal-type structure **166** ($R_2$ = COOEt) exerts a stabilizing effect on the compound.[278]

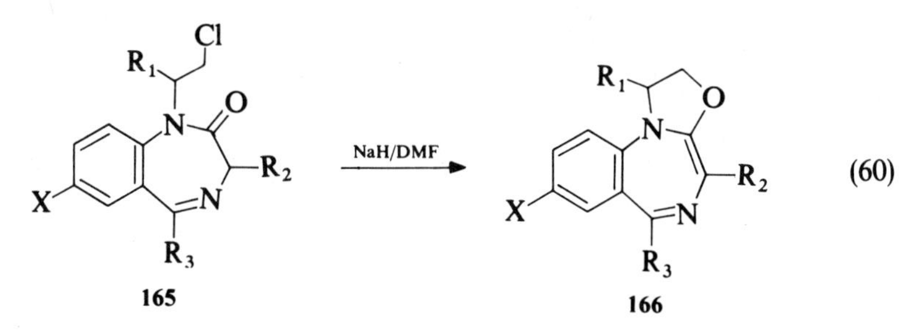


(60)

B. Reaction with Aldehydes, Ketones, and Epoxides. The formation of 3-hydroxymethyl derivatives **168** ($R_5 = R_6$ = H) by condensation of **167** with formaldehyde and base was first described by Sternbach and coworkers.[91a] However, it was initially thought that the product of the condensation was the 1-hydroxymethyl derivative instead of the 3-hydroxymethyl derivative. This procedure was later applied to prepare other 3-hydroxymethyl analogs.[264] Broger[192] studied the reaction of diazepam with aromatic aldehydes in the presence of base and isolated the carbinols **168** ($R_5$ = H, $R_6$ = aryl), which underwent dehydration to the 3-arylidenes **170**. A Soviet group[304] prepared several 3-methylene derivatives **170** by reaction of the lactams **167** ($R_1$ = H; $R_2$ = Ph; X = Br, Cl, Me) with substituted benzaldehydes, heterocyclic aldehydes, and isatins. They carried out these condensations in acetic anhydride with sodium acetate at 140°C. Generation of the 3-carbanion with lithium diisopropyl amide and the addition of aldehydes and ketones was an effective way to prepare these compounds.[299] Some of the ketones that successfully added to the 3-position of diazepam were acetone, acetophenone, cyclohexanone, and benzophenone.

The 7-amino-substituted compounds were condensed with several aromatic aldehydes to give the appropriate aldimines.[202] The aldimines were also formed from 3-amino derivatives.[202] The 7-hydrazino benzodiazepine was converted by reaction with ethyl pyruvate to the hydrazone.[134] Ethylene oxide was found to react in the presence of a Lewis acid, such as aluminum chloride, to give the oxazolino derivatives **169** (Eq. 61).[305,306] If $R_1$ of the starting material **167** was hydrogen, alkylation of the nitrogen with formation of the 1-(2-hydroxyethyl) derivative **169** ($R_1$ = $HOCH_2CH_2$) was also observed. Reaction with propylene oxide in the presence of tin tetrachloride yielded a mixture of diastereoisomers in a ratio of 3:2. The major isomer had the $R_3$ methyl group in the trans configuration with respect to the $R_2$ phenyl moiety.[306]

The oxazolines **169** were reported to rearrange to quaternary salts **171** by treatment with hydrogen chloride gas or by heating in ethanol with *p*-toluenesulfonic acid.[307] The salts **171** could rearrange back to the oxazolines **169** by reaction with sodium carbonate, pyridine, or just water.

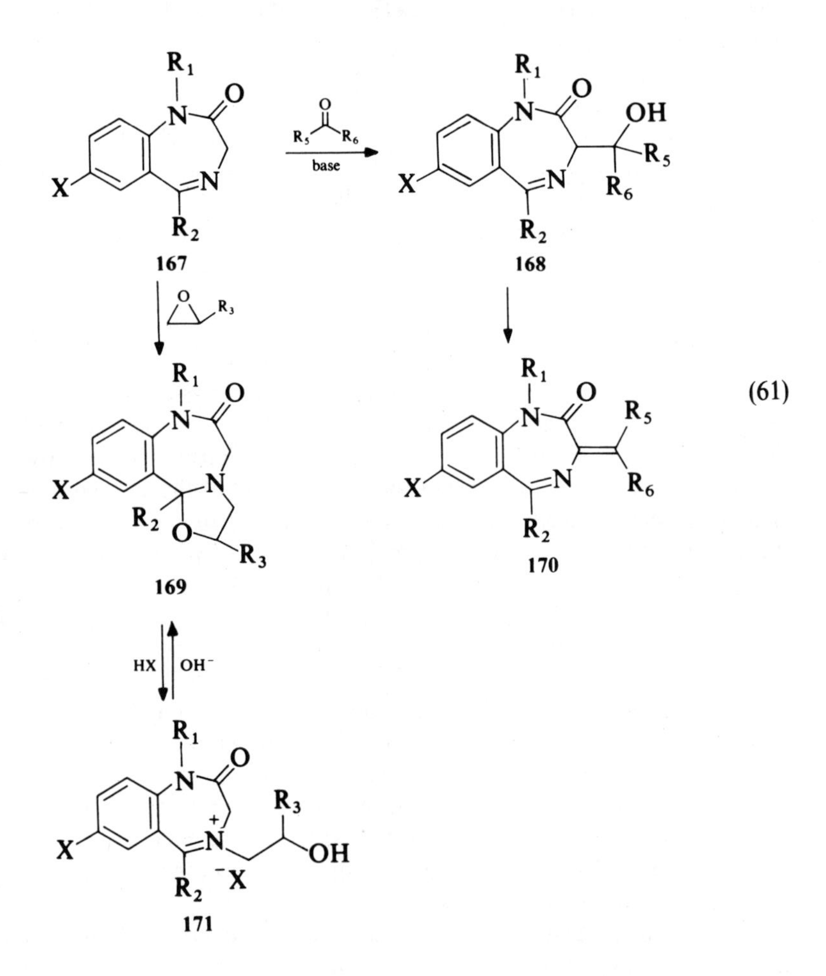


(61)

C. Acylations. A great variety of acylations were carried out with 1,3-dihydro-1,4-benzodiazepin-2(2*H*)-ones and in particular with compounds bearing functional groups amenable to derivatization by acylation (Eq. 62). Acylations at the 1-position nitrogen of **172** to give the 1-acyl derivatives **173** were possible with acetic anhydride at reflux,[44,134,148,227,301,308,309] with 2-chloroacetic anhydride at 100°C,[308] with ethyl and methyl chloroformate and sodium hydride at low temperature,[134] or with ethyl chloroformate and pyridine.[227]

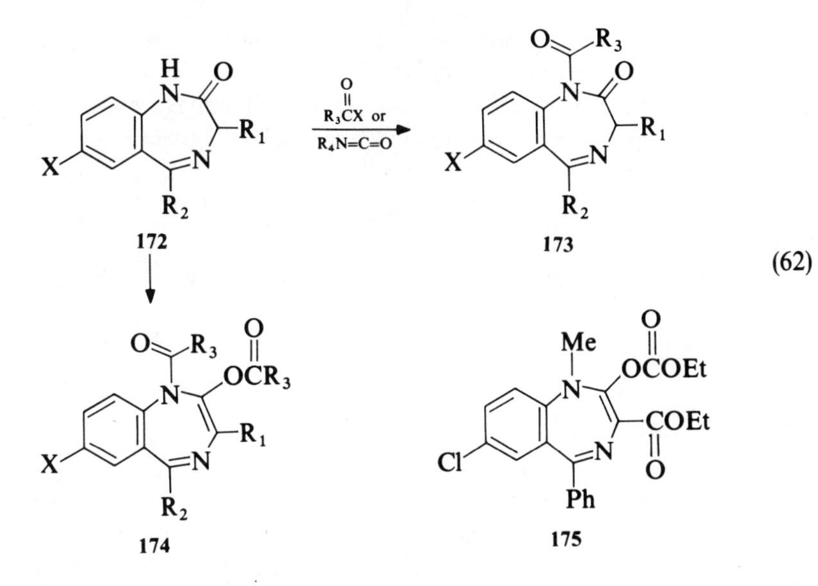


The 3-carboxylates **172** ($R_1$ = COOEt; $R_2$ = Ph; X = Cl, $NO_2$) were reacted with ethyl chloroformate and sodium hydride in dimethylformamide at low temperature to yield the *N,O*-diacylated compounds **174** ($R_1$ = COOEt, $R_2$ = Ph, $R_3$ = EtO, X = Cl).[134] The 1-methyl analog gave, under comparable conditions, the *O*-acylated compound **175**.[134] A related double acylation on the 1-nitrogen and the 2-oxygen of the lactams **176** (X = H, Cl) was described in the patent literature.[310] Dimethylmalonyl chloride reacted with the 5-phenyl-1,4-benzodiazepin-2-ones **176** in the presence of triethylamine to form the tricyclic compounds **177** in approximately 75% yield (Eq. 63).

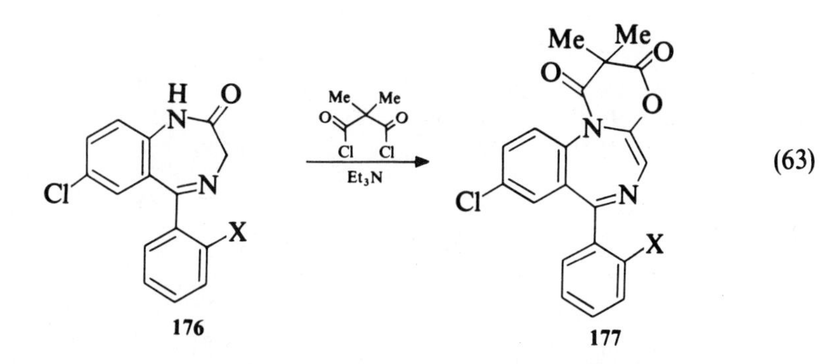


"Acylation" at the 1-position nitrogen was also possible with isocyanates. Several 5-phenyl-substituted lactams **172** were reacted with a variety of alkyl and alkenyl isocyanates to give the urea derivatives **173** ($R_3$ = $R_4$NH).[275,311,312] 3-Fluoro-substituted benzodiazepinones were reacted with methyl and ethyl isocyanate to give the 1-alkylaminocarbonyl derivatives.[266,267]

Introduction of acyl groups into the 3-position of diazepam (**31**) was described by Reitter and coworkers.[299] They generated the anion of diazepam with lithium diisopropylamide (LDA) and added ethyl acetate and methyl benzoate to prepare the 3-acetyl and the 3-benzoyl derivatives **178** (R = Me, Ph) in 12 and 20% yield (Eq. 64).

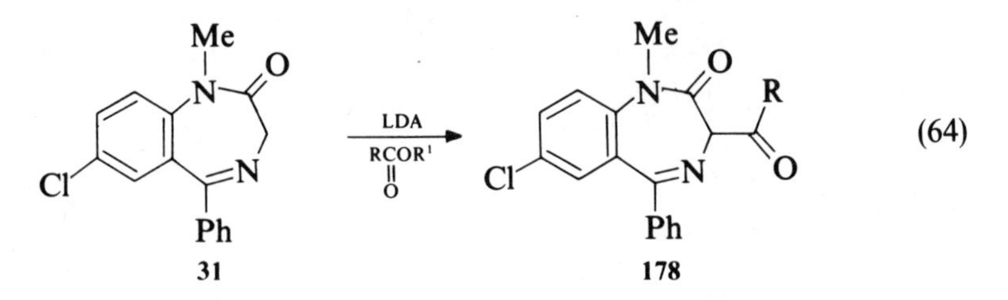


The 3-phosphonate **179** ($R_1$ = $PO(OEt)_2$, $R_2$ = 2-$ClC_6H_4$) was acetylated at the 3-position by reacting the anion generated with sodium hydride in dimethoxyethane with acetyl chloride to yield **180** [$R_1$ = $PO(OEt)_2$, $R_2$ = 2-$ClC_6H_4$, $R_3$ = Me].[243] The anion of the 3-cyano derivative **179** ($R_1$ = CN, $R_2$ = Ph), generated with sodium hydride in 1,2-dimethoxyethane, reacted with ethyl chloroformate to yield the 3-carboxylic acid ethyl ester **180** ($R_1$ = CN, $R_2$ = Ph, $R_3$ = OEt) (Eq. 65).[226]

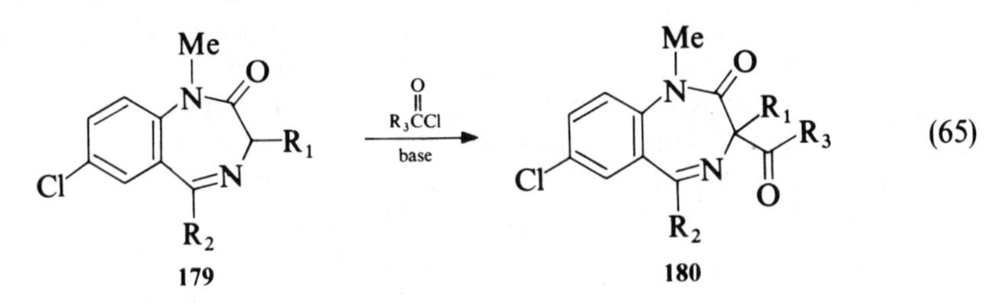


Trichloroacetyl isocyanate in tetrahydrofuran at room temperature was reported to acylate the 3-position of **36** to give the 3-trichloroacetylaminocarbonyl derivative **181** (Eq. 66).[313]

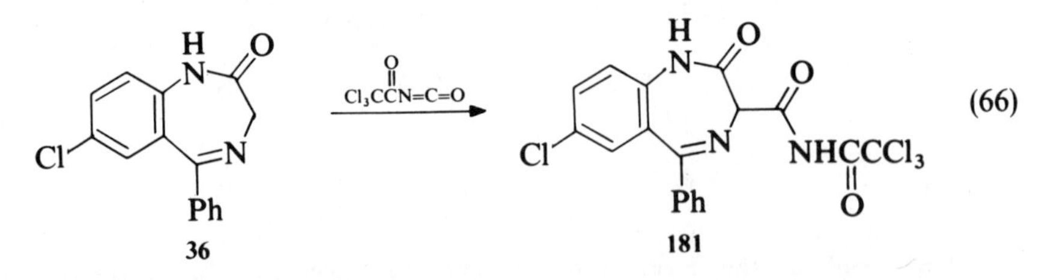


As described in the patent literature,[314] the reaction of various lactams **167** with ethyl formate and sodium hydride afforded the enolized 3-formyl derivatives **182** [$R_1$ = H, Me, $F_3CCH_2$, $Et_2N(CH_2)_2$; $R_2$ = Ph, 2-$ClC_6H_4$; 2-$FC_6H_4$, X = Cl, Br, I, $NO_2$, F; Y = OH].[314] The same compounds were also obtained by hydrolysis of the corresponding enamines **182** (Y = $Me_2N$), which were a product of the reaction of the lactams **167** with dimethylformamide diethylacetal.[315]

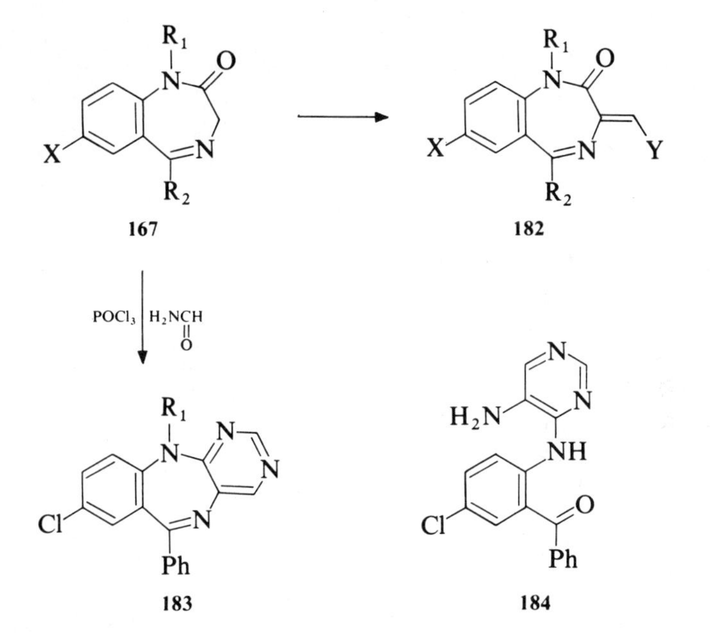


A similar acylation at the 3-position of **167** ($R_1$ = H, Me) using formamide in phosphorus oxychloride led to the pyrimidine **183** and its hydrolysis product **184**.[316]

Acylations of the benzodiazepines **167** ($R_1$ = H, $R_2$ = Ph, X = Cl) with chloroformate esters in the presence of potassium carbonate afforded the imine addition products **185** ($R_2$ = Ph; $R_3$ = Me, Et).[108]

Trapping of the intermediate acyliminium ion by a nucleophile delivered intramolecularly led to the cyclic derivatives **186** and **187** (Eq. 68). The 1,3-oxazinone **186** resulted from the reaction of diazepam, **167** ($R_1$ = Me, $R_2$ = Ph, X = Cl), with diketene.[317–319] Reaction of **167** ($R_1$ = Me; $R_2$ = Ph, 2-$FC_6H_4$; X = Cl, $NO_2$) with malonic acids ($R_3$ = H, Et) and acetic anhydride afforded the oxazinediones **187**.[320]

Coffen and coworkers[321] reported the addition of mercaptoacetic acid to the imine bond of diazepam **167** ($R_1$ = Me, $R_2$ = Ph, X = Cl). The adduct **188** was obtained in 38% yield by heating diazepam and mercaptoacetic acid in benzene for 5 days at reflux. Intramolecular trapping of the acyliminium ion by the α-carbon of the acyl group led to the β-lactams **189**. β-Lactam formations were observed with chloroacetyl chloride and triethylamine[301] and with the enamine obtained from glycine and ethyl acetoacetate in the presence of phosphorus oxychloride.[322] Jaunin and coworkers[323] studied the dipolar addition of nitrile oxides to the imine bond of **167** ($R_1$ = H, Me; $R_2$ = Ph) and obtained the expected adducts **190** ($R_1$ = H, Me; $R_2$ = Ph; $R_3$ = Ph, benzoyl, COOEt) (Eq. 69).

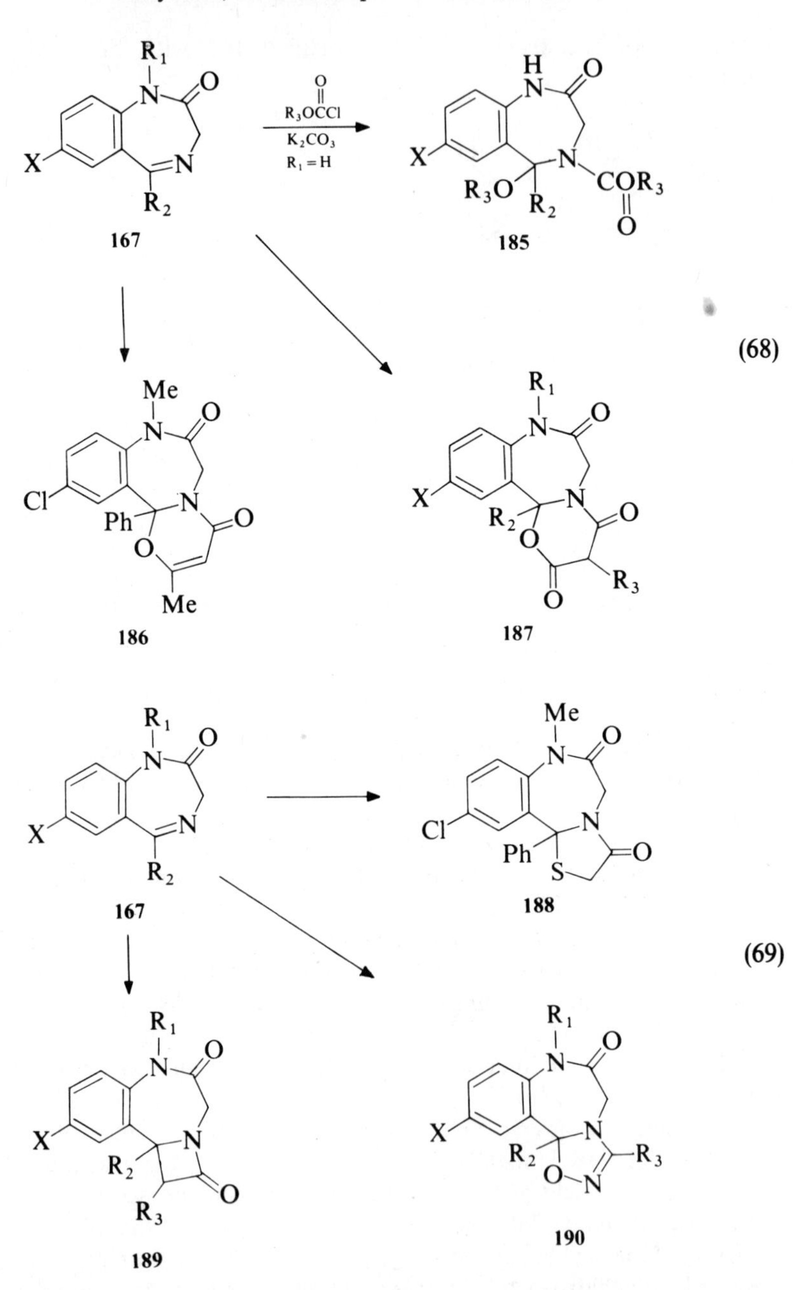


They also described the transformation of the 3-carboxamide **191** to the tricyclic derivatives **193**.[226] Treatment of **191** with oxalyl chloride led to the intermediate acyliminium ion **192**, which when reacted with such nucleophiles as ethanol, water, and mercaptoacetic acid ethyl ester, formed the adducts **193** (R = EtO, HO, $EtOOCCH_2S$) (Eq. 70).

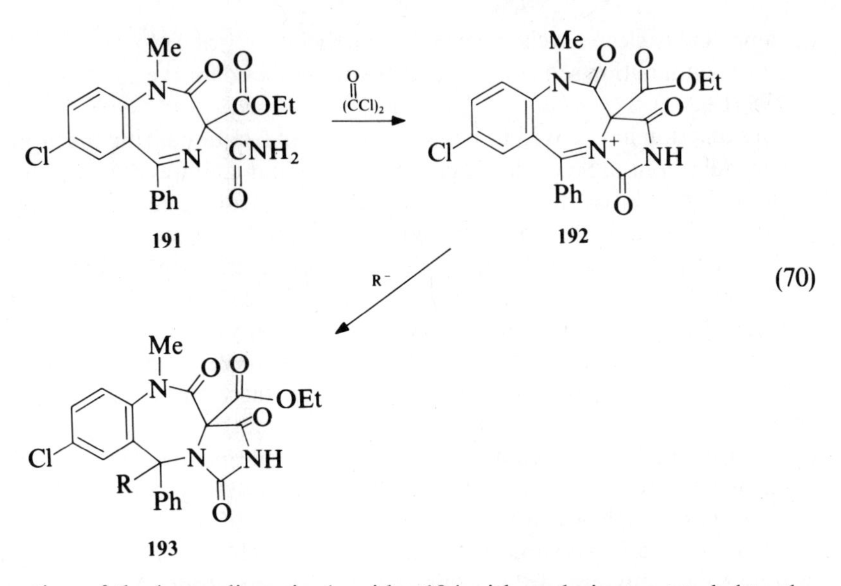


Reaction of the benzodiazepin-4-oxides **194** with acylating agents led to the 3-acyloxy compounds **195**. This "Polonovski" rearrangement is initiated by acylation of the nitrone oxygen to give the acyloxyiminium ion **196**, which will then add an acyloxy anion at the 5-position. Elimination of the carboxylic acid $R_4COOH$ from this adduct should then lead to the intermediate **197**. Migration of the acyloxy group from the 5-position to the 3-position may then occur by an

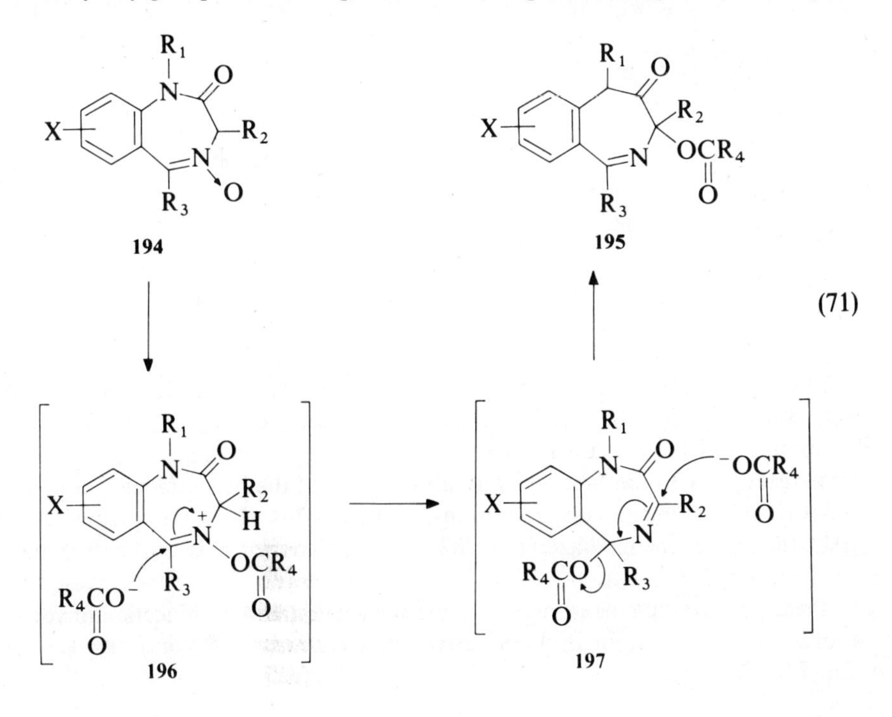

intramolecular electrocyclic mechanism to yield the isolated products **195**. An addition–elimination sequence may also be considered for the conversion of **197** to **195** (Eq. 71).

This reaction has proved to be as useful way of functionalizing the 3-position of benzodiazepines. Since the discovery by Bell and Childress[262] of a variation of the Polonovski rearrangement, many 3-acyloxy compounds have been prepared by this method. The most commonly used reagent was acetic anhydride at temperatures exceeding 80°C, preferably at reflux (see References 1, 8, 9, 44, 58, 69, 72, 97, 148, 184, 189, 227, 232–234, 254, 262, 263, 277, 282, 288, 309, 324 and 325). Other anhydrides used were 2-propylpentanoic anhydride,[326] chloroacetic anhydride,[263,265,327,328] pivalic anhydride,[329] and trifluoroacetic anhydride.[266,267] Acid chlorides could also be successfully used in place of anhydrides.[329,263] Reaction of the 4-oxide with ethyl chloroformate was reported to lead to the 3-chloro derivative,[263] while methyl chloroformate at reflux temperature also gave the 3-methoxycarbonyloxy compound.[265] The 3-chloro compound was also formed by treatment of the 4-oxide with oxalyl chloride.[134] The 3-acetylthio derivative was a product of the reaction of the nitrone with a mixture of thioacetic acid and its anhydride.[227]

The tricyclic compound **199**[330] was a product of an acylation of the nitrone **198** ($R_1$ = H, Me) by methyl isocyanate at the 4-position oxygen and an intramolecular trapping of the iminium ion. When $R_1$ = H, a simultaneous acylation at the 1-position occurs to give **199** ($R_1$ = MeNHCO) as a product (Eq. 72).

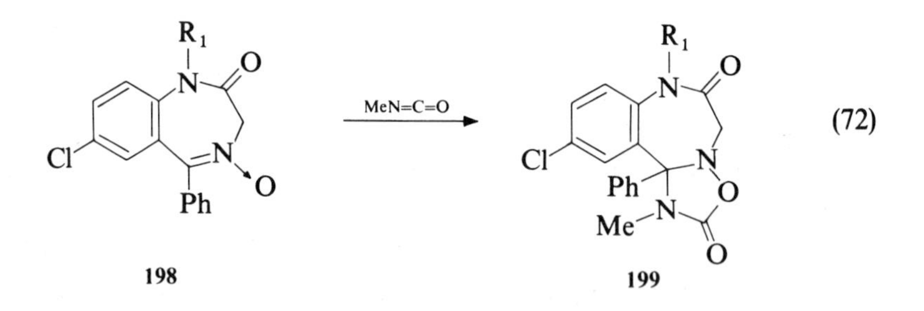


The 3-allyl derivatives **200** ($R_1$ = Me, $MeOCH_2$; $R_2$ = H, Me; $R_3$ = H, Me) were converted to the dienes **201** by heating in acetic anhydride. The 3-acetoxy compound which is initially formed undergoes elimination of acetic acid.[298] The same reaction with the 3-benzyl derivative, afforded the benzylidene compound **203** (Eq. 73). Under the same conditions, the dienes **201** ($R_2$ = H) were reacted to yield the pyrrolobenzodiazepines **202**. Mechanisms for this cyclization have been proposed.

Treatment of the corresponding 5-methyl analog **204** with acetic anhydride afforded the 5-acetoxymethyl derivative **205** by a normal Polonovski reaction (Eq. 74).[298]

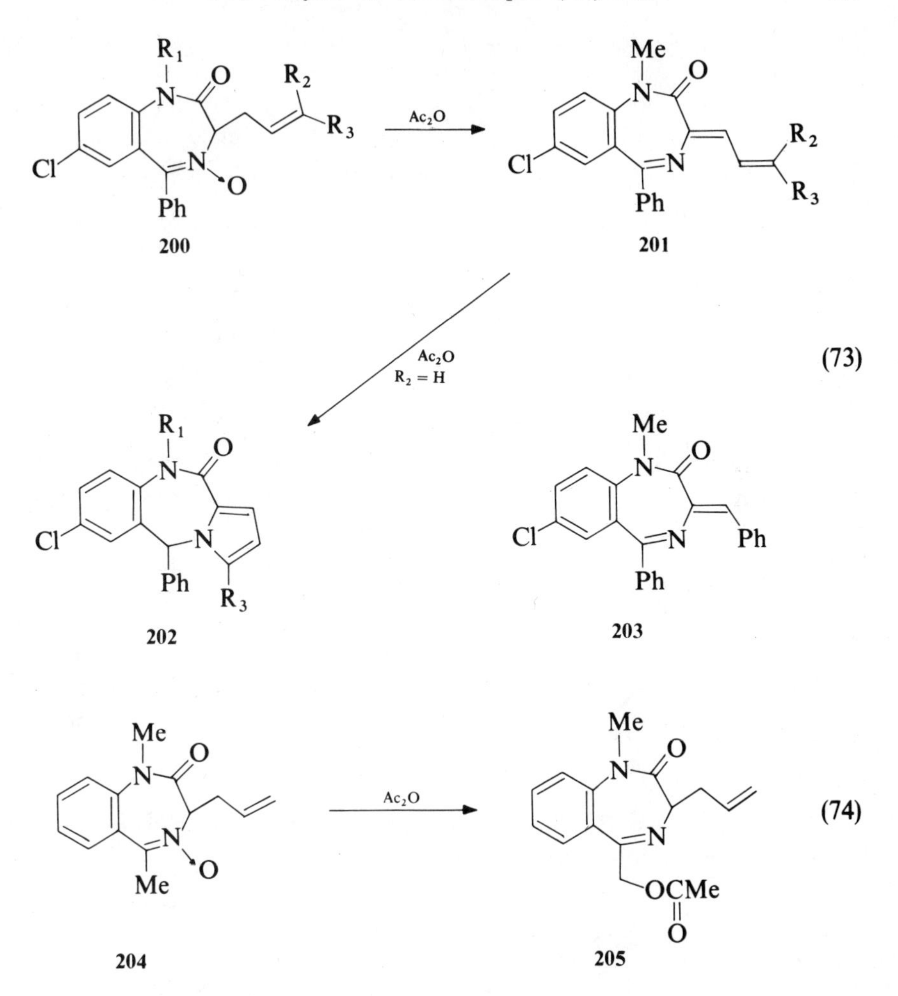


The 5-hydrazino derivatives **206** ($R_1$ = H) were obtained in situ by cleavage of the *t*-butoxycarbonyls **206** ($R_1$ = *t*-BuOOC). These derivatives were converted to the tricyclic systems **207** and **208** by acylation of the 4-position nitrogen. Reaction of the intermediate hydrazine with trifluoroacetic anhydride led to **207** ($R_2 = F_3C$) in 70% yield.[198] The analog **207** ($R_2 = ClCH_2$) was similarly obtained by treating the hydrazine **206** ($R_1$ = *t*-BuOOC) with chloroacetic acid and subsequently with chloroacetic acid anhydride.[198b] Heating the same compound with a mixture of oxalic acid and diethyl oxalate led to both the triazole **207** ($R_2$ = EtOOC) and the triazine **208** (Eq. 75).

In addition to the acylations involving atoms of the benzodiazepine ring system, numerous acylations of functional groups attached to various positions were carried out. Thus 1-amino[250] and 1-hydroxy[189] benzodiazepin-2-ones were acetylated by reaction with acetic anhydride. Reaction of the 1-amino

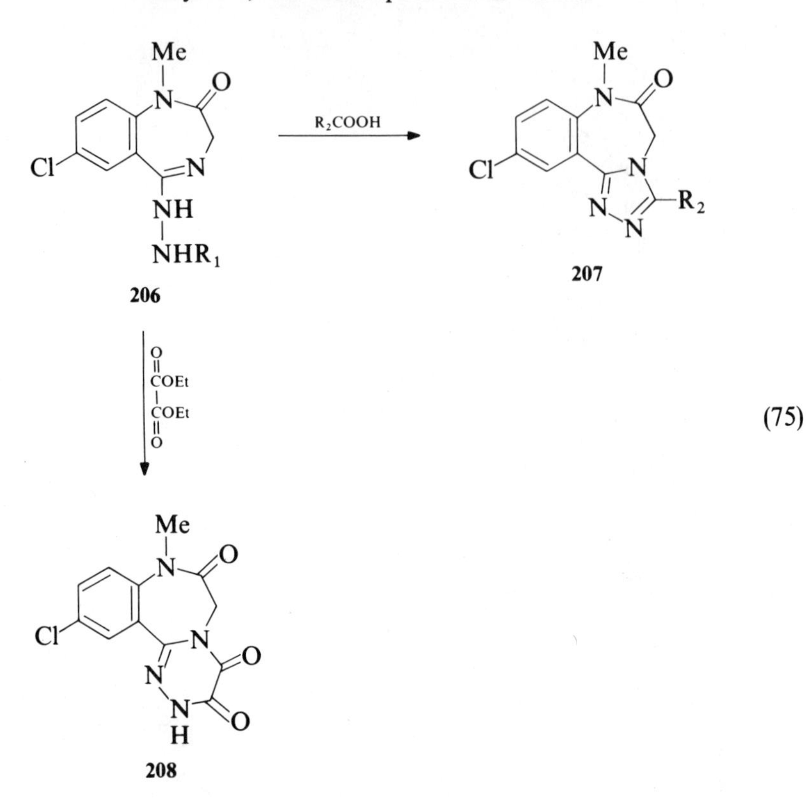


compounds **209** (X = H, Cl) with carboxamides in hot polyphosphoric acid (PPA) gave the triazolobenzodiazepines **210** (Eq. 76).[261]

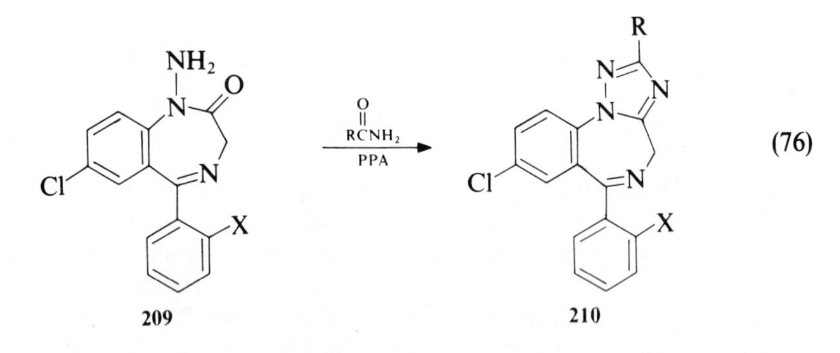


A 1-(2-aminoethyl) derivative was acylated with bromoacetyl bromide[331] and was also converted to the thioisocyanate derivative.[332] Both compounds served as possible irreversible ligands for the benzodiazepine binding site. 1-(2-Hydroxyethyl) compounds were acylated with a variety of anhydrides, acid chlorides, and isocyanates.[72,283] The 3-amino,[199,202,227] and in particular, the 3-hydroxybenzodiazepin-2-ones, were converted to many different acyl derivatives. Reactions described included those with anhydrides,[333] anhydrides and pyridine,[60,234] acid chlorides and chloroformates or chlorocarbamates in the presence of pyridine[334–337] or 4-dimethylaminopyridine,[265,327,328,338] and reactions with carboxylic acids activated by carbonyldiimidazole or dicyclohexylcarbodiimide.[329,339,340]

The sodium salt of the 3-hydroxymethylene derivative **182** ($R_1$ = Me, $R_2$ = 2-$ClC_6H_4$, X = Br, Y = OH), shown in Eq. 67, afforded the 3-acetoxymethylene analog upon treatment with acetic anhydride in methanol.[314]

The 5-carboxylic acid hydrazide **211** ($R_1$ = H) was acetylated by acetic anhydride to the corresponding acetyl derivative **211** ($R_1$ = Ac). Compound **211** ($R_1$ = Ac) was then cyclized with dehydration by polyphosporic acid to give the oxadiazolyl analog **212** (Eq. 77).[193]

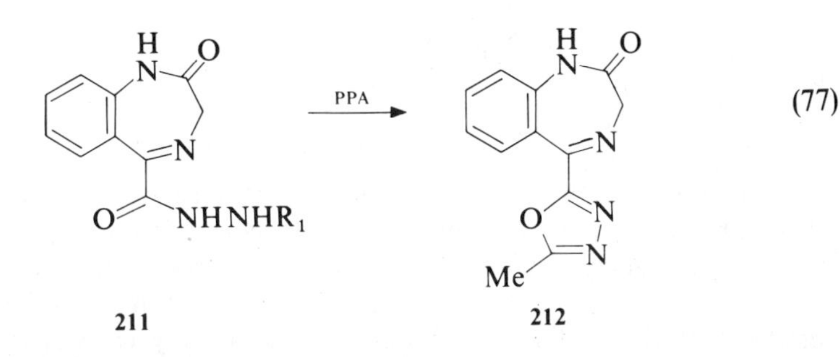


Many acylations were carried out on 3-hydroxymethyl compounds giving 3-acyloxymethyl benzodiazepin-2-ones.[264] The 7-amino derivatives were converted to acetyl,[134,251,301] trifluoroacetyl,[80] and a variety of urea derivatives by reaction of the intermediate isocyanate with amines.[80,230,231] 7-Aminomethyl and 7-(1-amino-1-ethyl) analogs were likewise converted to the corresponding ureas by the same method.[251] Reaction of 7-(1-hydroxy-1-ethyl) compounds with isocyanates led to the appropriate urethanes.[251] The 7-hydroxyamino derivative **213** reacted with acetic anhydride in the presence of pyridine to yield the *N*,*O*-diacetyl analog, which was selectively hydrolized to the *N*-acetyl derivative **214** (R = Me). Trifluoroacetic anhydride in pyridine, at −50 to −30°C, reacted in the same fashion, while the same reagents in boiling methylene chloride rearranged the hydroxyamine to the aminophenol **215** (Eq. 78).[245]

The amino group of 5-(2-aminophenyl)-7-chloro-1-methyl-1,4-benzodiazepin-2(2*H*)-one was acetylated with acetic anhydride.[162]

The nitrogen in the side chain of flurazepam was attacked by cyanogen bromide, leading to dealkylation and formation of the *N*-cyano compound.[233] Benzyl chloroformate[150c] and ethyl chloroformate were also used for similar dealkylation–acylation reactions.

Fryer and Sternbach[341,342] described the reaction of diazepam, **216** (R = Me), and nordiazepam, **216** (R = H), with acetic anhydride. Depending on the reaction conditions the formation of different products was observed. Diazepam was heated to reflux for 2 hours in the presence of sodium acetate

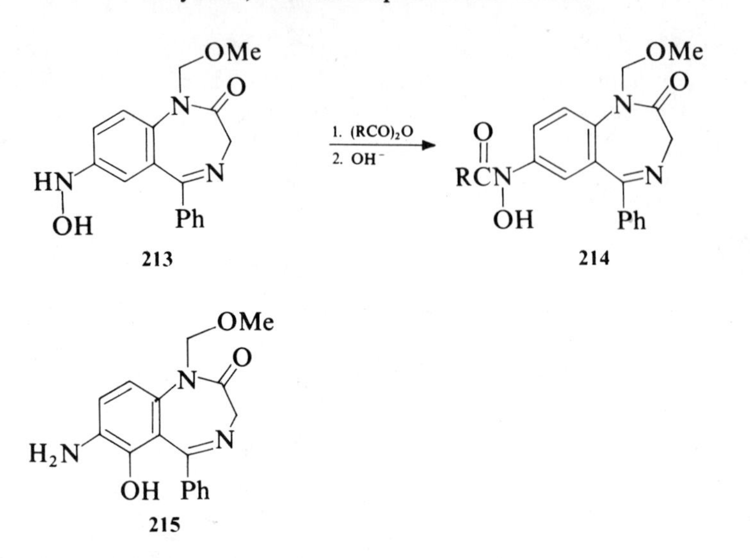


trihydrate to yield 23% of the quinolone **217**. Its formation was proposed to be initiated by acetylation of the imine nitrogen, generating an acetyliminium ion, which could undergo ring contraction to an aziridinoquinoline. Rearrangement involving a proton shift would then lead to the isolated product. It is also possible that the acetyliminium ion is hydrolytically cleaved to the benzophenone **218**, which would undergo cyclization and dehydration to form the quinolone **217** (Eq. 79). When nordiazepam was heated to reflux for 3 hours in acetic anhydride in the presence of catalytic amounts of concentrated sulfuric acid, the oxazoloquinoline **219** was obtained in 23% yield, along with a small amount (2%) of the benzophenone **218** (R = H).

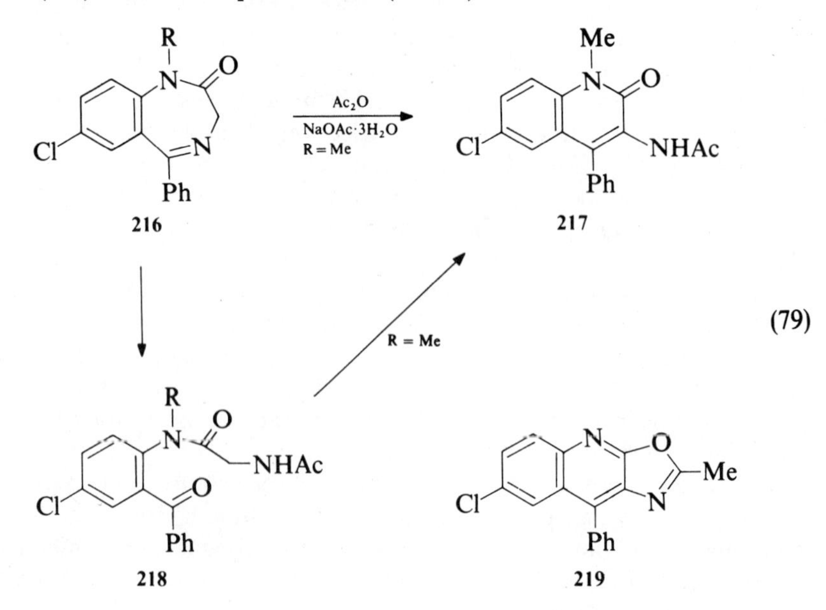

Prolonged heating (19 hours) of nordiazepam, **58** (X = Cl), or the 7-deschloro analog **58** (X = H), in a mixture of acetic anhydride and pyridine, afforded a 10% yield of the isoindoles **220** (X = Cl and H, respectively) (Eq. 80).[342] The possible mechanisms involved in the formation of these isoindoles are discussed in Section 1.2.2.7.

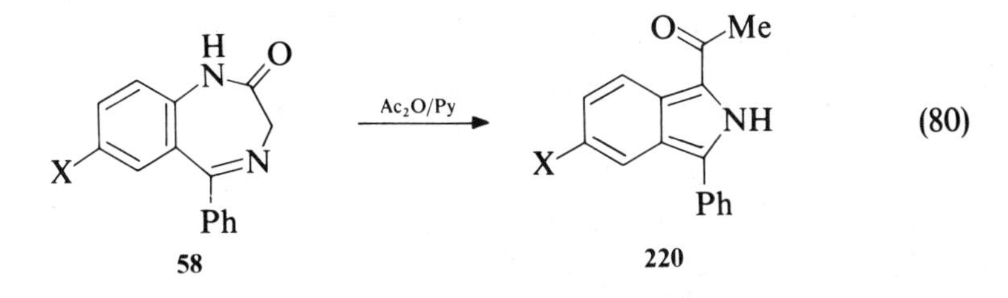


D. Sulfonation and Phosphorylation. The methylsulfonyl group was introduced into the 3-position of **167** ($R_1$ = Me, $R_2$ = 2-$FC_6H_4$, X = Cl) by means of sodium hydride and methanesulfonyl chloride[343] in dimethylformamide. Chlorosulfonic acid reacted with the 7-aminobenzodiazepine to yield the corresponding sulfamic acid.[134] Szente[301] prepared the 1-methylsulfonyl derivatives of benzodiazepin-2-ones by a reaction of the lactams with methanesulfonyl chloride and base.

Phosphorylation of benzodiazepin-2-ones on the oxygen of the lactam led to 2-phosphoryloxy derivatives, the syntheses and reaction of which were described in Chapter V. Phosphorylations were carried out with dimorpholinophosphinic chloride,[114,344,345] diphenylchlorophosphate,[134,346] and diethylchlorophosphate.[134,345] The hydroxy group of 1-(2-hydroxyethyl) derivatives was also phosphorylated.[347] The 3-hydroxy function of **221**, when reacted with the cyclic chlorophosphate **222** and sodium hydride, gave the phosphate **223** (Eq. 81).[348]

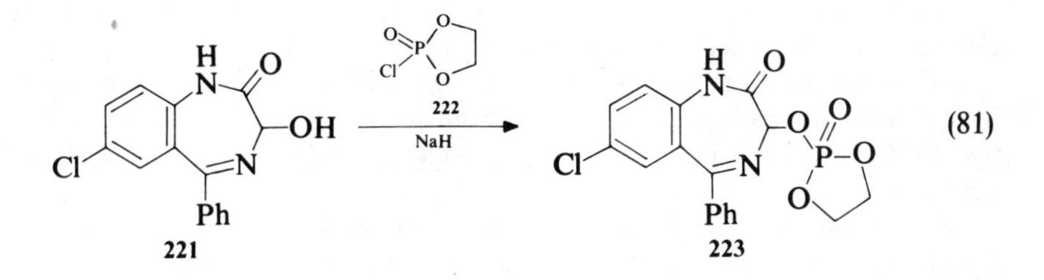


The rearrangement of the nitrone **224** to the quinoxaline **225** by treatment with phosphorus oxychloride at reflux may be rationlized in the following manner. The initial phosphorylation of the nitrone oxygen gives the postulated cyclic intermediate **226**. The indicated ring contraction would generate the carbonium ion **227**, which could then react with chloride ion as shown, to form the isolated product **225** (Eq. 82).[349]

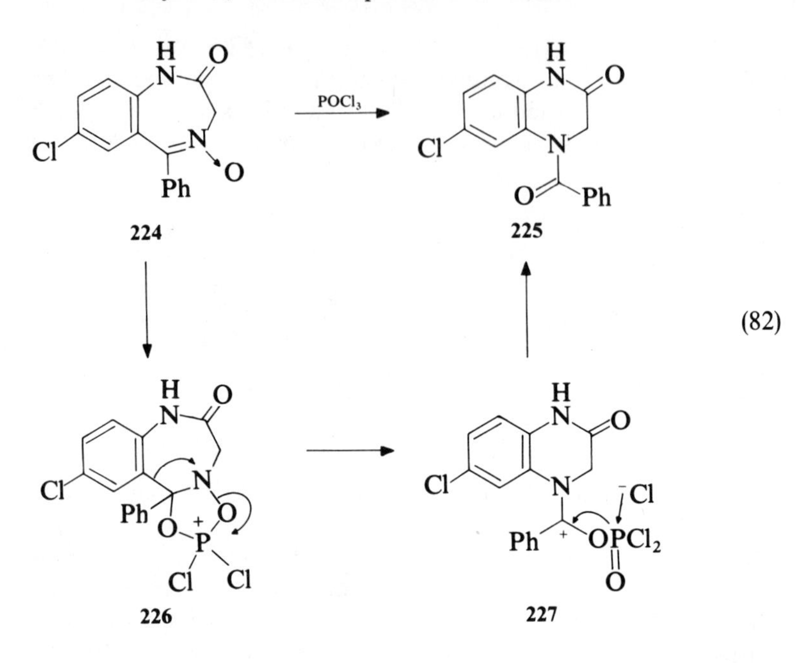


### *1.2.2. Reactions with Nucleophiles*

#### 1.2.2.1. Reduction

The reduction of the 2-carbonyl function to a hydroxy function and further to a hydrocarbon was described in Chapter VI. The reduction of the 4, 5-imine bond to the amine and the hydroxamine will be discussed in Chapter VIII, dealing with the tetrahydrobenzodiazepines.

Electrochemical reduction of the 4,5-imine bond at the dropping mercury electrode has been used widely in analytical work.[350,351]

3-Chlorobenzodiazepinones were dehalogenated to the parent compounds by hydrogen and palladium on carbon[262] or by Raney nickel in the presence of hydroxide.[143] In the latter case, the 4-oxide partially survived the dehalogenation.

Sodium borohydride reduced the 3-hydroxymethylene derivative to the 3-hydroxymethyl analog.[264] Hydrogenation of the 3-methylene derivative over palladium on carbon afforded the 3-methyl compound.[243] Reduction of the 4-oxides to the correponding imines was accomplished by a variety of reducing agents. Phosphorus trichloride was most commonly used. (See References 6, 7, 72, 91, 93, 97, 135, 138, 142, 143, 147, 150, 206, and 207).

Other trivalent phosphorus compounds such as trimethylphosphite[142] and triethylphosphite[189] were equally useful. Zinc–acetic acid was also used, although further reduction of the imine could occur.[150b] Hydrogen in combination with a catalyst, such as Raney nickel,[91,135,254] platinum,[142] or palladium on carbon[34,35] was also used as a reducing agent.

Nitro groups attached at the 2-,[162] 3-,[134] and 4-positions[134] of the 5-phenyl moiety were reduced to the corresponding amino compounds by Raney nickel and hydrogen. 7-Nitro derivatives were similarly reduced to the amino compounds using hydrogen and Raney nickel,[6,8,36,105,107,136,202,293] platinum,[252] and palladium on carbon.[252] Stannous chloride and hydrochloric acid were also successful reducing agents.[80,104,230,251–253] Stannous chloride in the presence of sodium acetate reduced the 7-nitro compounds to the corresponding hydroxyamino derivatives.[245,246]

The reduction, in situ, of the 7-diazonium salt with stannous chloride afforded the 7-hydrazinobenzodiazepine.[134] The 7-hydroxyiminoalkyl and the 7-cyano derivatives were reduced to the corresponding amines by Raney nickel and hydrogen.[152c,251] Desulfurization of the 7-methylthiomethylamino compound with Raney nickel led to the 7-methylamino analog.[134] Reductive amination of the 7-acetyl compounds with sodium cyanoborohydride and aminoethanol or dimethylamine were described by Branca, Fischli, and Szente.[251] The sodium borohydride reduction of 7-acetyl and 7-pentanoyl derivatives yielded the 7-hydroxylakyl benzodiazepines.[33,251] The 1-(acetyl) methyl and 1-(benzoyl)methyl analogs were similarly reduced to the corresponding alcohols by sodium borohydride.[284]

The allyl ester of the benzodiazepine 5-carboxylic acid was cleaved by hydrogenolysis using platinum as a catalyst.[193]

#### 1.2.2.2. Reactions with Halogen Nucleophiles

The reaction of 1,3-dihydro-1,4-benzodiazepin-2(2*H*)-ones with such reagents as phosgene,[352–354] phosphorus pentachloride,[355] or a combination of carbon tetrachloride and triphenylphosphine[353] has been discussed in Chapter V, Section 25 dealing with 3*H*-1,4-benzodiazepines.

The combination of carbon tetrachloride and triphenylphosphine was used to dehydrate a 3-carboxamide to the corresponding nitrile.[266] The 3-hydroxy compounds were readily converted to the 3-chloro derivatives by treatment with thionyl chloride.[215b,227,262,263,265,324,356–358] The reaction of 3-hydroxy-4-oxide with phosphorus trichloride gave the reduced 3-chloro compound, which could be further reacted in situ.[207] Partial exchange of fluorine by chlorine was observed when the 3-fluoro-4-oxide was reduced to the corresponding imine with phosphorus trichloride.[142] The 3-fluorobenzodiazepines were obtained by halogen exchange, using silver fluoride with the 3-bromo- or the 3-iodo-compounds.[228] The 3-fluoro compound was also prepared by reaction of the 4-oxide with antimony pentachloride in hydrogen fluoride[228] or by treatment of the 3-hydroxy compound with hydrogen fluoride and potassium fluoride in pyridine.[228] Other methods used to form the 3-fluoro compounds were the reaction of the 3-chloro compound with antimony pentachloride in hydrogen fluoride, treatment of the 3-amino derivative with an alkylnitrite in hydrogen fluoride or diazotization of the 3-amino compound with sodium nitrite in hydrogen fluoride–pyridine, thallation at the 3-position and subsequent reaction with borontrifluoride, and finally electrolytic oxidation in hydrogen fluoride.[228] It was also found[266,267] that the 3-fluoro analogs were a product of the reaction of 3-hydroxy compounds with diethylaminosulfur trifluoride.

7-Diazonium salts were reacted with iodide[251,253] or chloride[293] to yield the corresponding 7-halo derivatives. A modified version of the Sandmeyer reaction was used to prepare a radioactive 7-bromo compound by incorporating bromine-75.[359] The diazonium bromide was converted to a triazene with piperidine and then heated in carbon tetrachloride with methanesulfonic acid. The 7-acetyl compound was converted to the 7-(1,1-difluoro-1-ethyl) derivative by reaction with molybdenum hexafluoride.[360]

Compounds substituted by a 7-methylsulfinyl group were reacted with thionyl chloride to give the corresponding 7-chloromethylthio analogs.[7,93] The same reaction was used to prepare the 7-(1-chloro-1-ethylthio) compounds.

Boron trifluoride in combination with acetonitrile was reacted with the 4-oxide **224** to give the 3-hydroxy compound **221** (Eq. 83).[235] Boron tribromide has been used to convert the 3-methoxy derivative to the 3-hydroxy analog.[101]

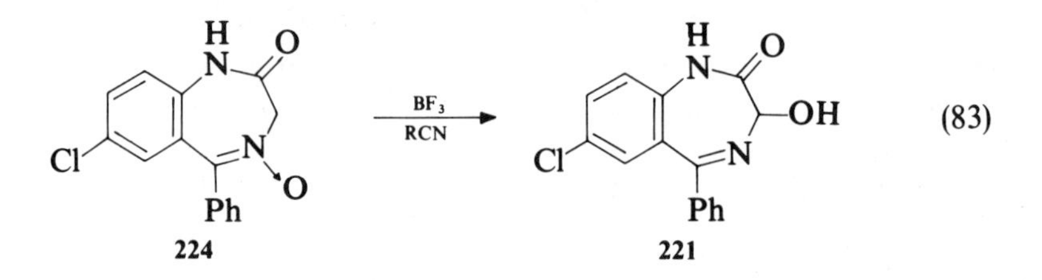


### 1.2.2.3. Reactions with Oxygen and Sulfur Nucleophiles

A. Hydrolysis. The 4,5-imine bond in **167** is most sensitive to acid hydrolysis. Since the hydrolysis of the imine to the ring-opened form **228** is reversible, benzodiazepin-2-ones can tolerate low pH media at room temperature. The reversible ring opening was studied spectroscopically and the pH dependence of the equilibrium between protonated diazepam and its ring-opened form was determined.[361] A similar study of the hydrolytic opening of flunitrazepam, flurazepam, and fludiazepam at body temperature was conducted.[362] The 4-quaternary salts **167** ($R_3$ = alkyl) are readily hydrolyzed to the ring-opened benzophenones **229**.[363] Under more vigorous conditions of acid hydrolysis, benzodiazepin-2-ones were cleaved to the corresponding benzophenones **230** (Eq. 84).[6,69,94,256]

The 3,4-bond of the 3-hydroxybenzodiazepin-2-ones appears to be most sensitive to acid hydrolysis. Treatment of oxazepam (**221**) for 10 minutes in boiling acetic acid led to the quinazoline-2-carboxaldehyde **231** in about 60% yield.[183,349,364] Hydrolysis of oxazepam in a mixture of hydrochloric acid and ethanol or methanol led also to the 2,2′-bis-quinazoline **232** and the indolylquinazoline **233** (Eq. 85)[364] The 4-oxide of oxazepam could also exist in a ring-opened form.[206,207]

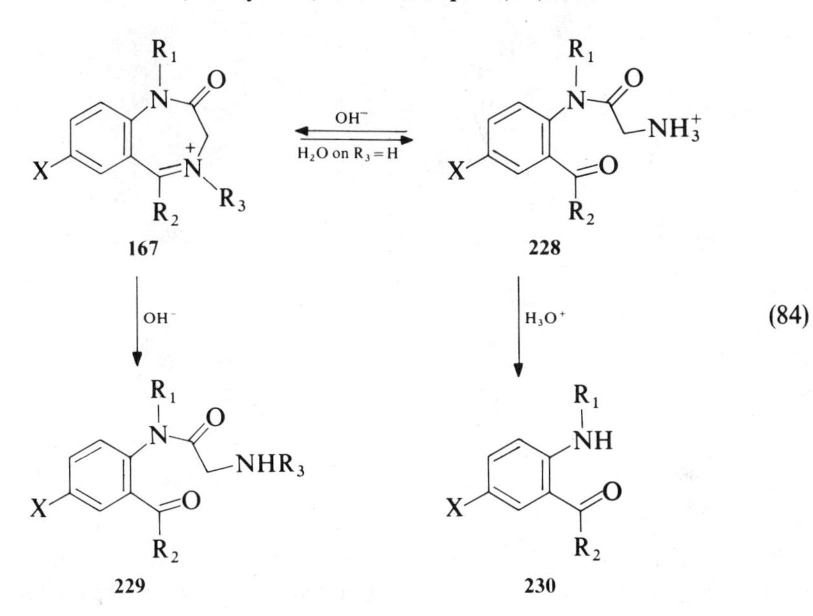


(84)

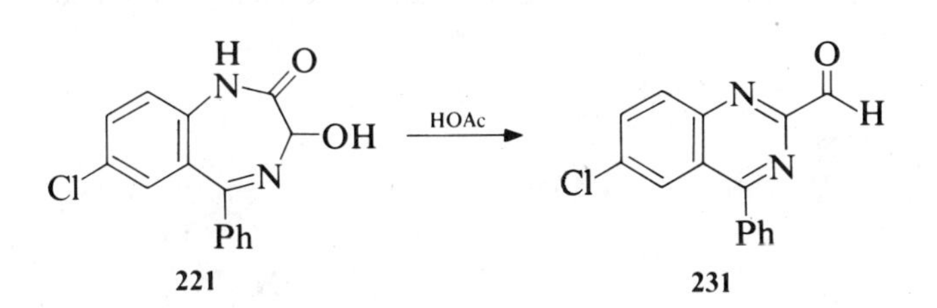


(85)

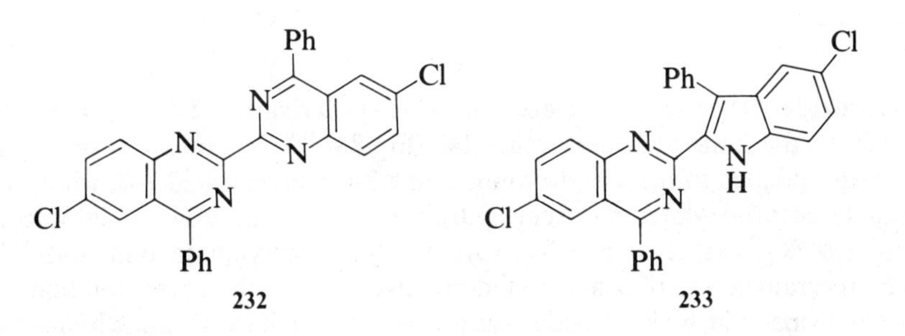


Acid hydrolysis of the 1-amino derivatives **234** or the 4-desoxy analogs led to the indazoles **235** (Eq. 86).[250] Acid or base-catalyzed hydrolysis of the 1-hydroxy compounds **236** or the corresponding 4-desoxy derivatives yielded the benzisoxazoles **237**.[188]

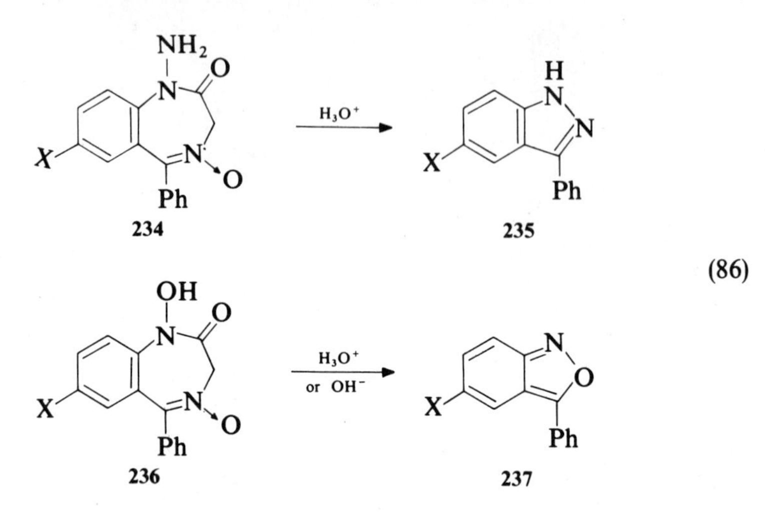


While 1-unsubstituted benzodiazepin-2-ones **167** ($R_1$ = H) are quite stable to alkali, due to formation of the amide anion, hydroxide induces ring opening in 1-alkyl and 1-acyl derivatives to give the salt of the carboxylic acids **238** (Eq. 87).[42,134] The sterically hindered imine **238** ($R_1$ = Me, $R_2$ = 2,6-$Cl_2C_6H_3$) recyclized upon treatment with acid rather than hydrolyzing to the benzophenone.[42]

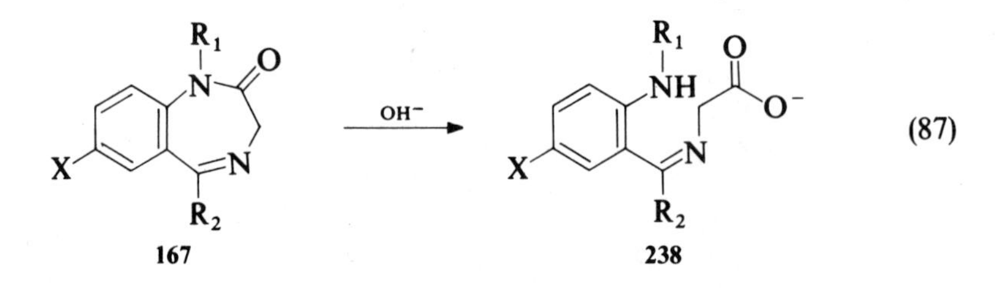


Hydroxide converted the 1-acetyl-3-acetoxy derivative **239** ($R_1$ = Ac; $R_2$, $R_3$ = H) to the 2-methyl quinazoline **240** (Eq. 88).[309] Treatment of oxazepam with hydroxide led to the dihydroquinazoline-2-carboxylic acid **242**, which was thermally decarboxylated.[262] Acid hydrolysis of the 3-acetoxy-3-methyl compound **239** ($R_1$ = H, $R_2$ = Me, $R_3$ = Ac) yielded 2-acetylquinazoline **241**.[227]

The rearrangement of 3-hydroxy derivatives **243** to the corresponding 4,5-dihydro-3-ones **244**, by hydroxide, was a side reaction of the alkaline hydrolyses of 3-acyloxy compounds. Several cases of this protic rearrangement were reported (Eq. 89).[148,233,262,265,288,365] 3-Amino derivatives **245** were analogously rearranged to the amidines **246**.[215b] To avoid this rearrangement, hydrolysis of the 3-acyloxy group must be carried out under mild conditions, such as hydroxide in methanol at room temperature.[8,44,58,72,148,232–234,263,277,282,288,293] The 3-trifluoroacetoxy function was easily cleaved.[266,267]

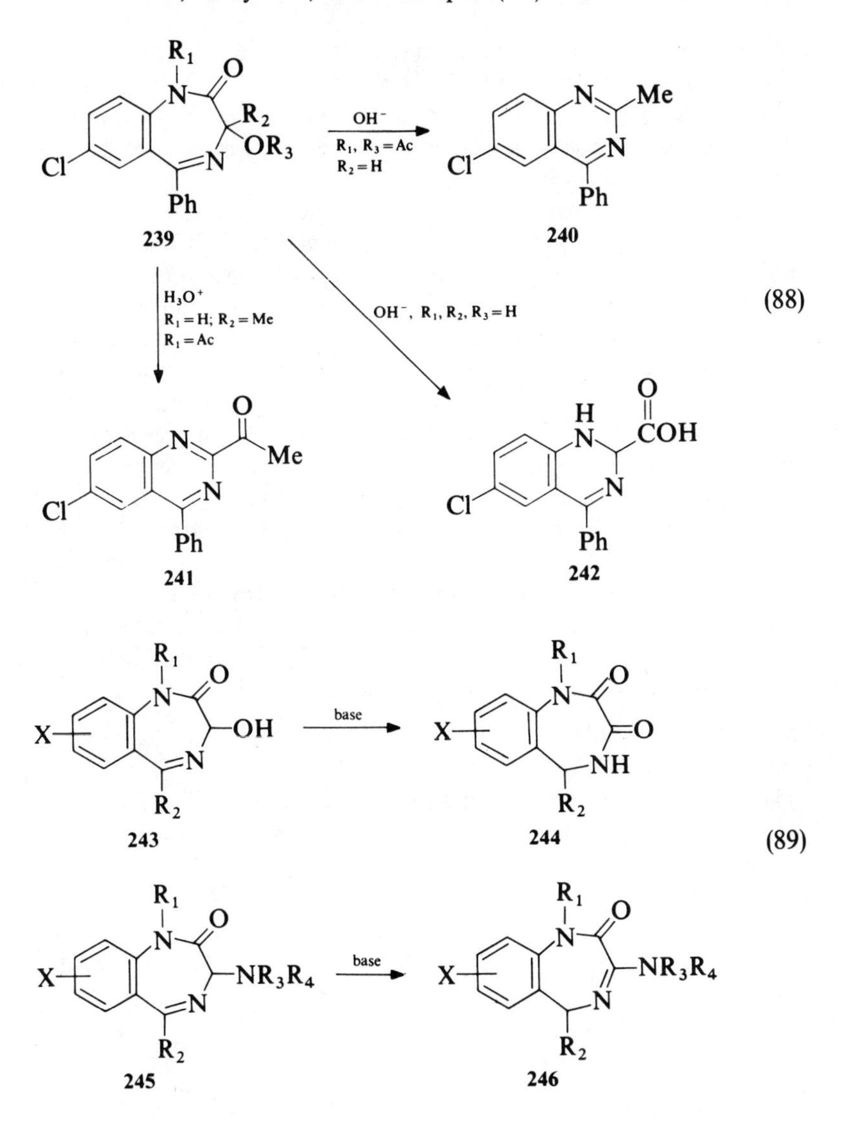


The 3-thiol was prepared by alkaline hydrolysis of the 3-acetylthio analog.[227] Acetoxyalkyl groups attached at the 1-position[72] and at the 5-position[298] were hydrolyzed to the alcohols.

Benzodiazepin-2-one-3-carboxylic acid esters were converted by alkaline hydrolyses to the corresponding acids,[72,77,101,102,105,145,242] which are stable as alkali salts but decarboxylate readily when in the free acid form. Diazepam-3-carboxylic acid, liberated from its alkali salt at low temperature, could be partially converted to the methyl ester by treatment with diazomethane, indicating that the decarboxylation at low temperature is not instantaneous.[134] 1-Alkanoic acid ethyl ester was hydrolyzed to the corresponding acid by alkali.[294]

Hydroxide was reacted with the imide **181** (Eq. 66) to form the corresponding 3-carboxamide by selective hydrolysis of the trichloroacetyl group.[313] Hydroxide was also used to cleave the trifluoroacetamide in the 4′-position to give the 4′-amino compounds.[8,9] A 3-hydroxy compound was isolated from the reaction of the 3-azido derivative with hydroxide.[249]

Acid hydrolysis was used in the preparation of 3-amino derivatives from the corresponding acetamide[199,202] or thioacetamide.[202] It was also used to cleave an acetone ketal to form the diol,[334] to ring open an aziridine to form an amino alcohol,[231] to convert a *t*-butyl ether to the corresponding alcohol,[231] a t-butyl ester to the corresponding acid,[294] and finally to convert an epoxide to the corresponding diol.[237] Concentrated sulfuric acid was able to hydrate nitriles to the amides[150c,226,233] and to debenzylate 1-benzyl derivatives.[134,42] Hydrolysis of a cyanamide with sulfuric acid led to the corresponding amine.[233] A 3-(dimethylamino)methylene derivative was converted to the 3-(hydroxymethylene) analog by treatment with aqueous oxalic acid in tetrahydrofuran.[314] An amino group, protected by a benzyloxycarbonyl group, was deprotected by reaction with hydrogen bromide in acetic acid.[8,147,150c] Hydrolysis of the 3-diazonium salt led to the 3-hydroxy compound.[199,202] The reaction of nordiazepam in deuterium oxide and deuteromethanol, at reflux temperature, in the presence of deuterium chloride, resulted in a 50% exchange of the protons at the 3-position of nordiazepam.[182]

**B. Reaction with Alkoxides and Acyloxides.** The 1,2-bond, in the 1-(methoxycarbonyl) derivative **247** (X = Cl, $NO_2$), was selectively cleaved by methanol and triethylamine to give the ring-opened imines **248** (Eq. 90).[134] The stereochemistry of the imine was retained in this reaction.

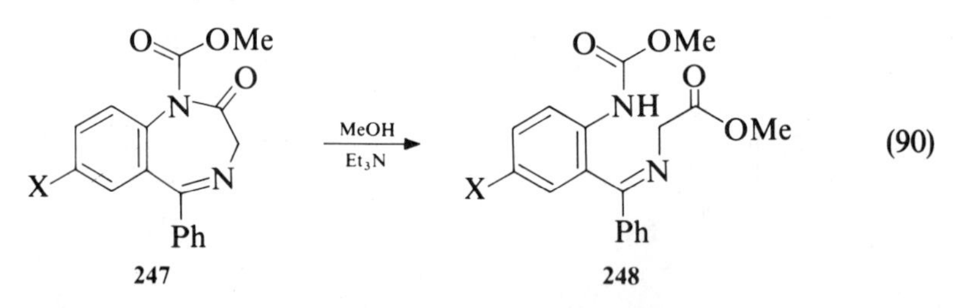


Methanolyses have been carried out on several acetoxy derivatives using methanol and methoxide or triethylamine. Thus 1-acetoxyalkyl-substituted benzodiazepin-2-ones were converted to the corresponding 1-hydroxyalkyl derivatives.[236] A 3-acetoxy group[72] and a 3-acetamino function[201] were similarly converted to the 3-hydroxy and the 3-amino derivatives. The formation of 3-alkoxy compounds from 3-acyloxy derivatives was also observed.[263,366] Optically active alkoxy derivatives were thus obtained by treating the enantiomeric 3-hemisuccinates with thionyl chloride and an alcohol.[366]

When the 3-acetoxy-3-methyl compounds **249** ($R_1$ = Me, $MeOCH_2$; $R_2$ = Ac) were treated with methanol and methoxide, a rearrangement to the bridged benzodiazepine **250** was observed (Eq. 91).[277] Mechanisms for this transformation were discussed. The 3-hydroxy-3-methyl compound **249** ($R_2$ = H) was obtained by treatment of the acetoxy derivative with concentrated sulfuric acid.[277]

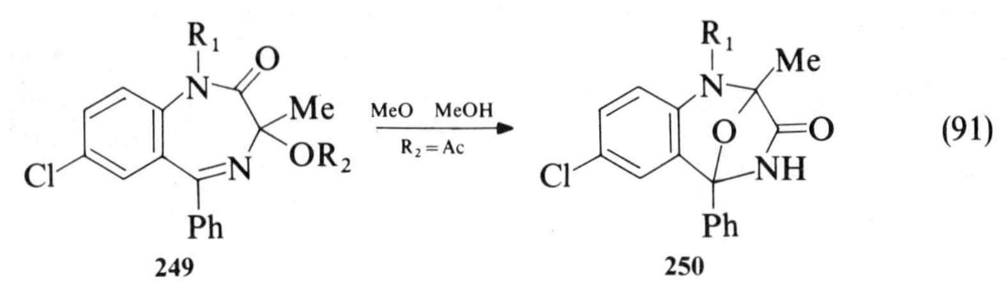


Methanolysis served to cleave a trifluoroacetyl group from the 7-trifluoroacetyl-*N*-methoxyamino derivative[245] and to convert an *N*-(2-acetoxyethyl) urea to the corresponding hydroxyethylurea.[231]

3-Carboxylic acid aminoalkyl esters are the products of a base-catalyzed transesterification.[147] The reactive 3-chlorobenzodiazepin-2-ones were treated with many alkoxides to form the 3-alkoxy derivatives.[101,143,215,223,225–227,288,324,334,357] The latter were also obtained by etherification of the 3-hydroxy compounds (e.g., by treatment with methanol and sulfuric acid).[227] The 3-acetoxy compounds were similarly obtained by displacing the chloride with acetate.[143,207] Carboxylic acids were reacted with the 3-chloro compounds in the presence of triethylamine to form the 3-acyloxy derivatives.[339] The 3-diazonium salt could also be reacted to form the 3-acetoxy compound.[199] Exchange of chloride with acetate occurred at the side chains attached at the 1-[72] and 7-positions.[231] Acetalization of the 3-hydroxymethylene derivative with trialkyl orthoformates in the appropriate alcohol led to the 3-(dialkoxy)methyl analogs, which were thermalized in the presence of sodium acetate to afford the enol ethers.[314]

5-Chlorobenzodiazepin-2-ones were reacted with alkoxides and phenoxides to yield the 5-alkoxy and 5-phenoxy derivatives.[195] The aromatic fluorines in the 5-(2-fluorophenyl) and the 5-(2,6-difluorophenyl) derivatives were displaced by alkoxides.[152c]

The 5-chloroalkyl derivatives were dehalogenated to the corresponding alkenes by lithium carbonate in dimethylformamide.[222,224]

C. Reaction with Sulfur Nucleophiles. The conversion of 2-ones to 2-thiones will be discussed in Section 2, below.

Reaction of 3-chlorobenzodiazepin-2-ones with thiourea or thioacetamide gave the 3-thiol compound.[367] The 3-thioalkyl derivatives were similarly obtained by treating the 3-chloro compound with thiols.[227,367] Treatment of the 3-acetoxy analog with thioacetic acid at 100°C for 50 minutes resulted in the formation of the 3-acetylthio compound.[227]

A 5-phenylthio compound was prepared by reacting the 5-chlorobenzodiazepine with phenylthiol.[196] A 3-acetamino compound was converted to the corresponding 3-thioacetamide by treatment with phosphorus pentasulfide.[202]

### 1.2.2.4. Reactions with Nitrogen Nucleophiles

The transformation of 1,3-dihydro-1,4-benzodiazepin-2(2*H*)-ones to 2-amino-3*H*-1,4-benzodiazepines was discussed in Chapter V.

Methylamine in dimethylformamide was reacted with nitrazepam (**101**) at room temperature to cleave the 1,2-bond and give the ring-opened amide **251** in 77% yield.[368] Piperidine at reflux temperature effected the same type of transformation (Eq. 92).

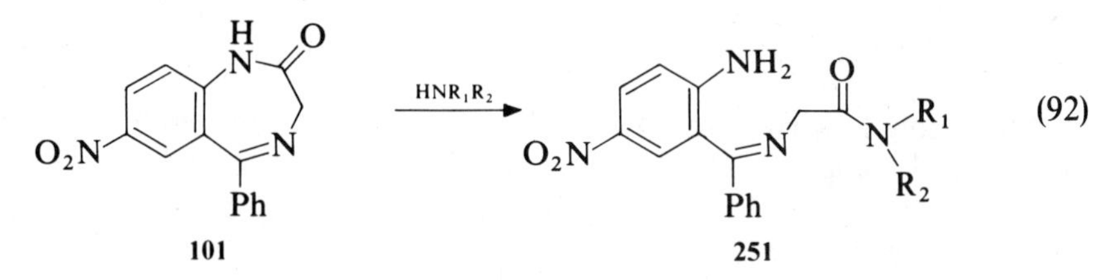


Intramolecular attack of the amino group in **252** ($X = NH_2$), at the 2-carbonyl group, led to the tricyclic amidines **253**.[233,291] These compounds were also obtained directly by displacement of the chloride in **252** (X = Cl) with ammonia in hot ethanol containing sodium iodide (Eq. 93).[291]

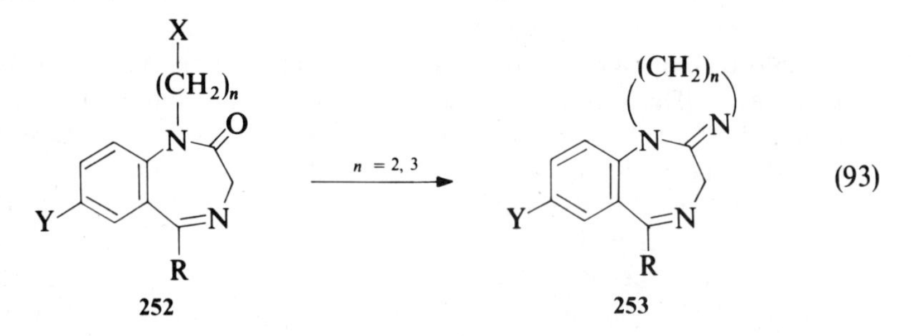


3-Chlorobenzodiazepin-2-ones were reacted with a variety of amines to form the 3-amino derivatives. The amines included ammonia,[199,225,358] primary[227,358] and secondary[358] amines, and hydrazine.[249] Tertiary amines formed the quaternary salts when treated with the 3-chlorobenzodiazepin-2-ones.[227,334,356] The 3-amino compounds were also formed by substitution of a 3-phosphoryloxy group[348] or by treating the 3-hydroxy derivatives with a phosphorylamide.[369] The 3-acetoxy group of the 3-acetoxy-3-methyl derivative was directly displaced by diethylamine.[227,325] Amines have been demonstrated to be strong enough bases to deprotonate the 3-position and to rearrange the 3-amino compounds to the tautomeric amidines (see Eq. 89).[215b,260]

5-Chlorobenzodiazepin-2-ones were converted to the 5-amino compounds by treatment with various amines.[196] The reaction of the chloride **254** with hydrazine and hydrazine derivatives was studied by Wade and coworkers.[197,198] Hydrazine in ethanol at room temperature converted the 5-chloro compound to the 5-hydrazino analog **255** (R = H) in 32% yield,[198b] while the same reagent, neat or in boiling benzene, led to ring cleavage and gave the triazine **256**. Heating the chloro compound **254** with formylhydrazine in refluxing 1,2-dimethoxyethane afforded the triazolobenzodiazepine **207** ($R_2 = H$) in 84% yield. Other triazoles **207** ($R_2$ = Ph, Me, $CH_2CN$) were similarly obtained. The triazolone **257** was prepared in 94% yield by heating to reflux for 4 hours, the 5-chloro compound with ethoxycarbonylhydrazine in toluene.[198b] Refluxing the chloride **254** with oxamic hydrazide afforded the triazole carboxamide **207** ($R_2 = CONH_2$). The morpholino derivative of oxamic hydrazide yielded 84% of the triazine **208** and only 7% of the triazole **207** ($R_2$ = morpholinocarbonyl) after heating in dimethylformamide at 100°C for 30 minutes (Eq. 94).

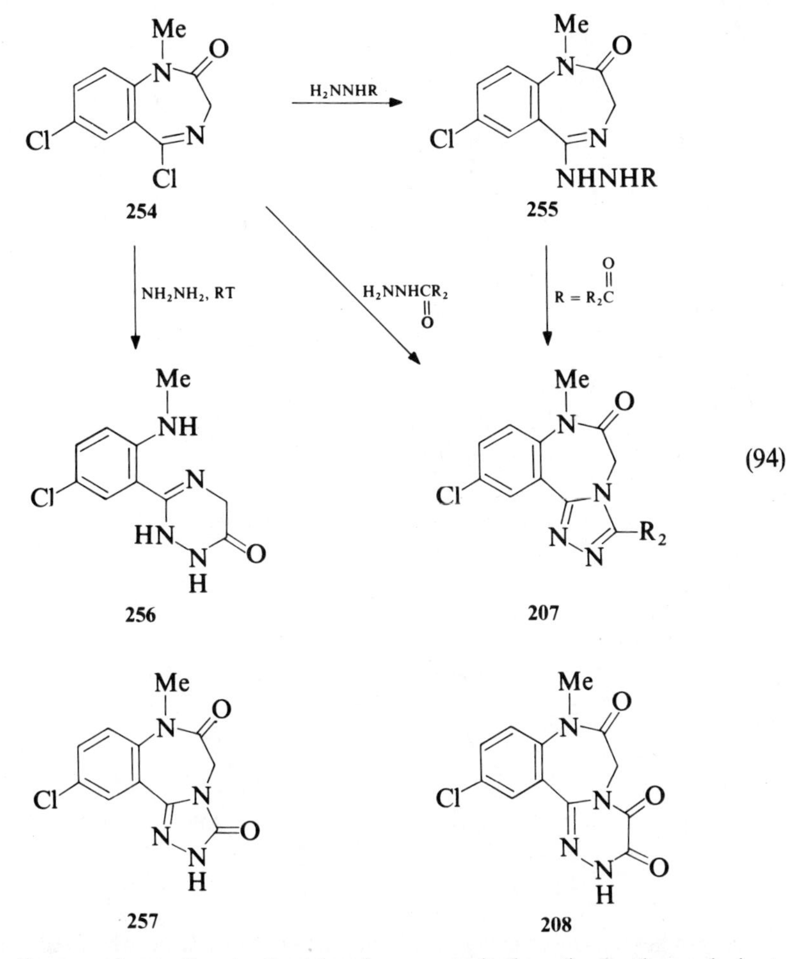


A similar transformation to the triazole occurred when the 5-ethoxy derivative was heated to reflux with 2-chlorobenzoyl hydrazide in diglyme.[370]

Side chain chlorides at the 1-, 3-, and 5- positions were displaced by nitrogen nucleophiles. 1-Chloroalkyl derivatives were reacted with ammonia,[291] with potassium phthalimide,[72] and with amines.[150c,254,287] Chloroacetyl functions attached at the 1- and 3-positions were similarly converted to the basic derivatives by treatment with a variety of amines.[263,265,283,328] The 5-(haloalkyl) derivatives were reacted with amines to form the 5-(aminoalkyl) analogs.[224] Reaction of the 5-(1-chloro-1-cyclohexyl) compound **258** with diethylamine led to elimination of hydrogen chloride with formation of the 5-cyclohexylidene derivative **259** (Eq. 95).[224] Other amines such as dimethylamine and *N*-methylpiperazine yielded the displacement products **260**. The enamine **262** was obtained when the chloride **261** (R = Pr) was treated with pyrrolidine.[224] Reaction of the dichloromethyl derivative **261** (R = Cl) with dimethylamine led in low yield to the aminal **263**.[224]

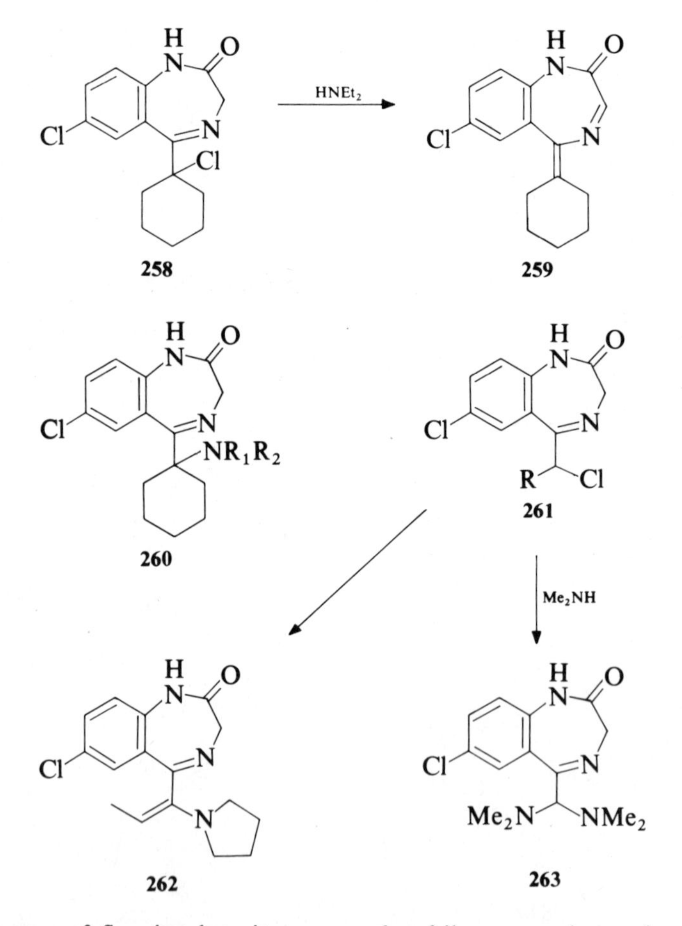


(95)

Exchanges of fluorine by nitrogen nucleophiles were observed with 5-(2-fluorophenyl)- and 5-(2,6-difluorophenyl)-substituted compounds.[134,152c]

The 7-diazonium salts were converted to the 7-azido analogs[252] and, in one instance, to a triazene.[359] Reaction of a 7-nitroso derivative with methylhydroxylamine led to the methylazoxy compound.[245,246]

Ester functions at the 3-, 5-, and 7- positions have been converted to amides and hydrazides by direct amination.[5,15,105,107,193,226] Carboxylic acid and esters not directly attached to the ring system were also converted to amides and hydrazides.[8,102,105d,357]

3-Phenoxycarbonyloxy derivatives were reacted with ammonia and primary and secondary amines to yield urethanes.[336] Several 3-aminomethylene derivatives were accessible by an amine exchange of the 3-(dimethylaminomethylene) compound.[315]

The cyclization (aqueous potassium carbonate) of the thiourea **264** to the triazole **265** involves intramolecular attack of the carbonyl group by a urea nitrogen (Eq. 96).[193]

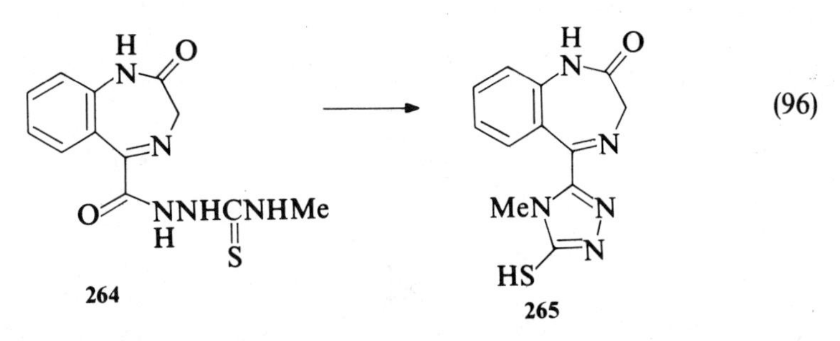


(96)

The reaction of oxazepam (**221**) with methylamine in aqueous ethanol led to the methylimine of the quinazoline-2-carboxaldehyde **266**. Treatment with hydrazine hydrate in boiling ethanol afforded the corresponding hydrazone **266** (R = $NH_2$) (Eq. 97).[349]

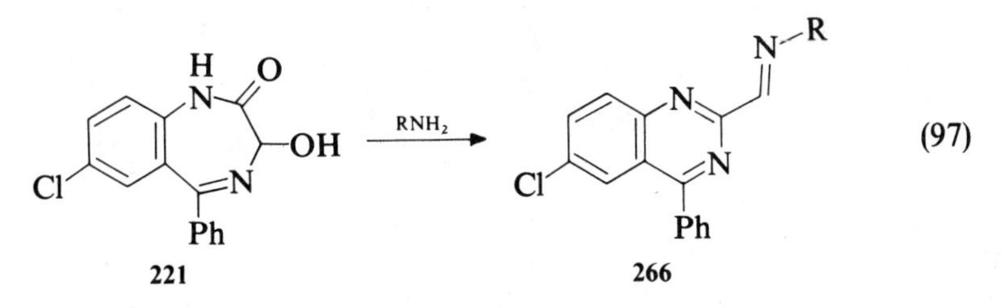


(97)

### 1.2.2.5. Reactions with Carbon Nucleophiles

Methyllithium was reacted with the 7-cyano derivatives to yield the 7-acetyl compounds via the imines.[11] These acetyl compounds were further converted, by this reagent, to the 2-hydroxy-2-propyl analogs.[33] The 3-carboxylic acid ethyl ester reacted with excess methylmagnesium iodide to form the same type of carbinol,[134] the 3-(2-hydroxy-2-propyl) derivative. Phenylmagnesium bromide added to the nitrone **267** (R = H) to give the 4-hydroxy compounds **268** (R = H).[176,177] The 5-phenyl nitrone **267** (R = Ph, X = Cl) underwent addition of phenylmagnesium bromide at the amide carbonyl, leading to the 2-phenyl-3*H*-1,4-benzodiazepine **269** (Eq. 98).[301] Methylmagnesium iodide added in the same fashion at the 5-position[301] of the 5-phenyl derivative, giving **268** (R = Me, X = Cl). Phenyllithium reacted, at 30°C, with the carbonyl group of diazepam (**31**) to form the ring-opened phenylketone **270**.[301] Ethylmagnesium bromide underwent the same type of reaction. According to Field,[371] treatment of the 5-chloro compound **254** with phenylmagnesium bromide resulted in the formation of the oxazole **271** by attack at the carbonyl group followed by ring opening and cyclization of the enolized ketone onto the chloride (Eq. 99).

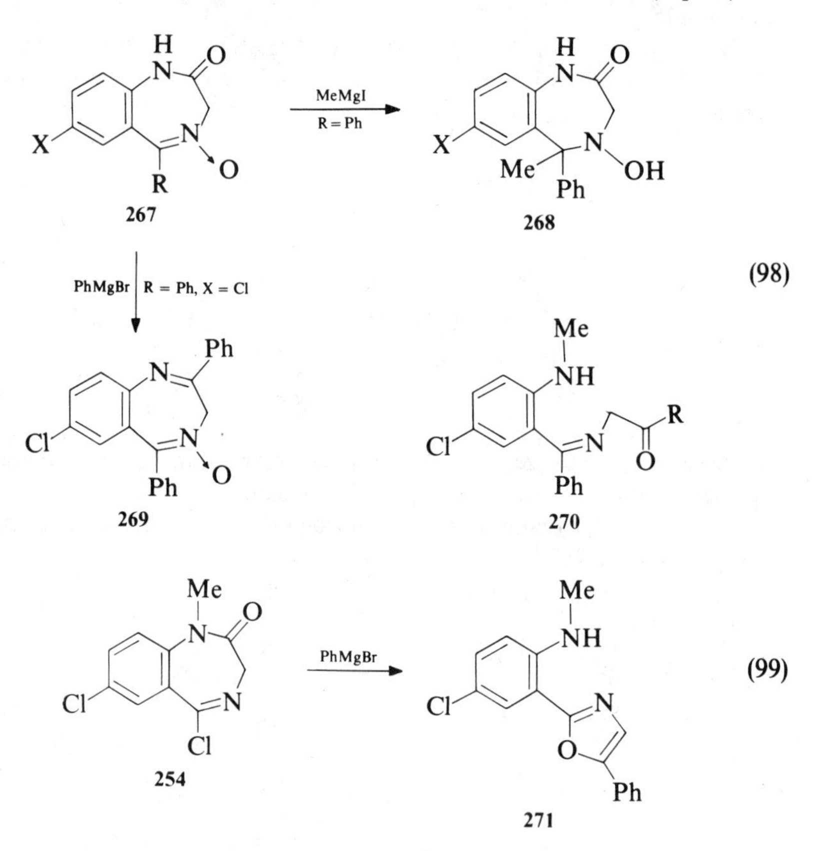


Cuprous cyanide in dimethylformamide was used to displace a 7-iodo substituent with a cyano group.[251] Displacement of the chloride in 5-(1-chloro-1-cyclohexyl) derivative **258** (Eq. 95) by cyanide led to the corresponding nitrile.[224]

#### 1.2.2.6. Photo and Thermal Reactions

Irradiation of the nitrones **272** with ultraviolet light led to the oxaziridines **273** (Eq. 100). Compound **273** (X = Cl) was obtained in 78% yield by irradiating a tetrahydrofuran solution with a Hanovia medium pressure mercury lamp and a Pyrex filter.[218,372] While the oxaziridine **273** (X = Cl) was relatively stable to further photolysis, the corresponding methylthio analog **273** (X = MeS) was further reacted to yield a mixture of the quinoxaline **275** and the diazoxecine **274**. The photochemistry of nitrazepam was studied in detail by Roth and Adomeit[373] and by Cornelissen and van Henegouwen.[374]

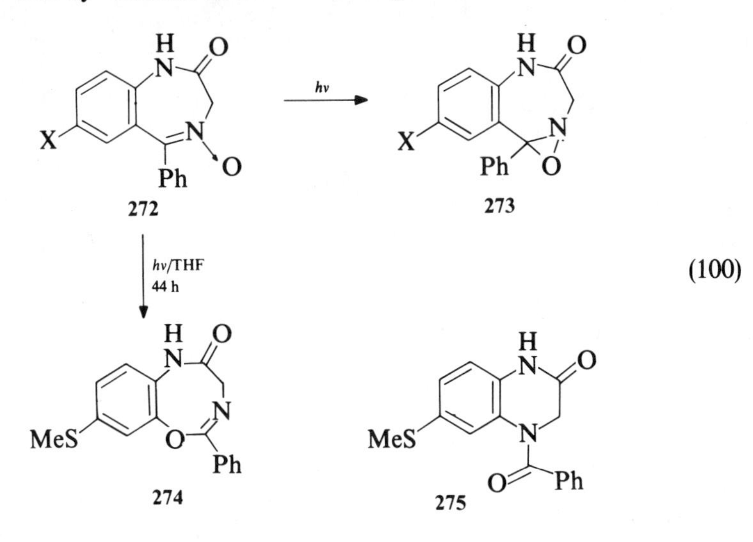


Irradiation with ultraviolet or sunlight in protic solvents led to photoreduction of the nitro group to the amino moiety. The 7-nitroso and 7-hydroxylamino compounds, which were formed at the intermediate stage, underwent further reaction in situ to give the azoxy and azo compounds. If the irradiation was carried out in aqueous acidic medium, the compounds were hydrolyzed to the corresponding benzophenones. Flunitrazepam and clonazepam behave similarly.[374a] The photochemical degradation of diazepam with ultraviolet light at 254 nm yielded benzophenones, 4-phenylquinazolinones, 4-phenylquinazolines, and glycine.[375]

Among the few thermal reactions carried out with benzodiazepin-2-ones, the formation of the 1-vinyl derivatives **227** by pyrrolysis of the *N*-oxides of 1-(2-dialkylaminoethyl) compounds **276** (R = Me) was preparatively useful.[238] While the 2-(2-dimethylamino)ethyl *N*-oxide **276** (R = Me) led exclusively to the alkene, the corresponding diethylamino derivative **276** (R = Et) yielded the hydroxyamine **278** in addition to the vinyl compound **277** (Eq. 101).

Heating the 1,3-dicarboxylic acid ester **279** in boiling ethanol resulted in ring contraction to the isoindole **280** (Eq. 102).[134]

### 1.2.2.7. Other Reactions

The ring contraction of benzodiazepin-2-ones to isoindoles, which has been shown to proceed thermally in Section, 1.2.2.6, is generally a base-catalyzed reaction.[300,376] The anion generated at the 3-position is believed to attack the electrophilic carbon at the 9*a*-position intramolecularly, leading to cleavage of the 9*a*, 1-bond. Tautomerization of the 1*H*-isoindole **281** ($R_2$ = H) would then produce the isoindole **283** (Eq. 103). This rearrangement was found to proceed

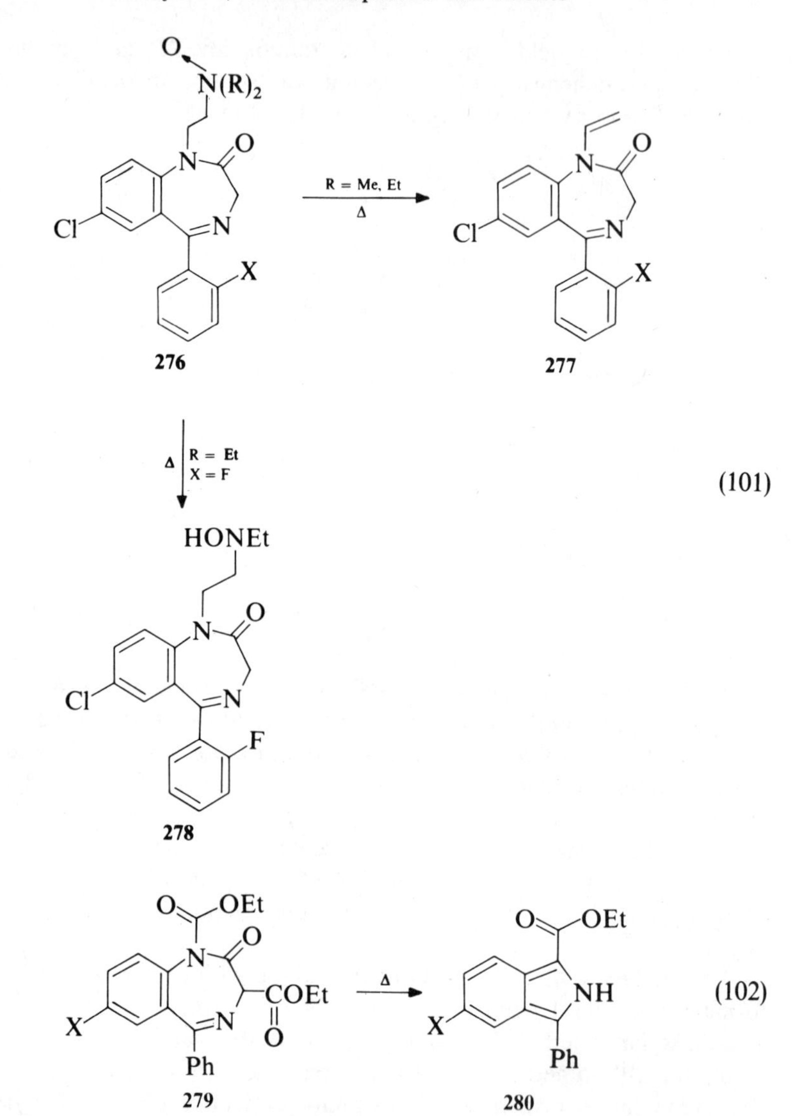


more readily with the 3-carboxylic acid ethyl esters **141** ($R_2$ = COOR), which is consistent with the easier formation of the carbanion at the 3-position. 7-Nitro groups and electron-withdrawing substituents at the 1-position also facilitated this rearrangement.[134] The intermediates **281** ($R_2$ = COOR) preferentially lose the carboxamide function to form the isoindole-1-carboxylic acid esters **282**.

The quaternary salt of **141** ($R_1$ = Me, $R_2$ = H, $R_3$ = Ph, X = H) was shown to undergo the same ring contraction to the isoindole **283** ($R_1$ = Me, $R_3$ = Ph, X = H).[300]

Ethyl propiolate, in refluxing tetrahydrofuran, was added to the nitrone function of **154** to form the isoxazoline **284** in approximately 40% yield. The

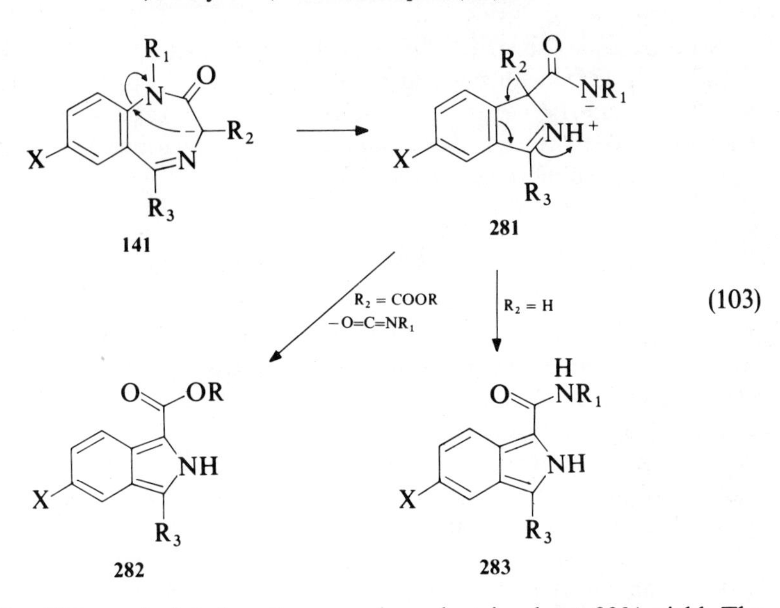


quinoxaline **286** was isolated as the second product in about 33% yield. The structures were determined by X-ray crystallography (Eq. 104).[377]

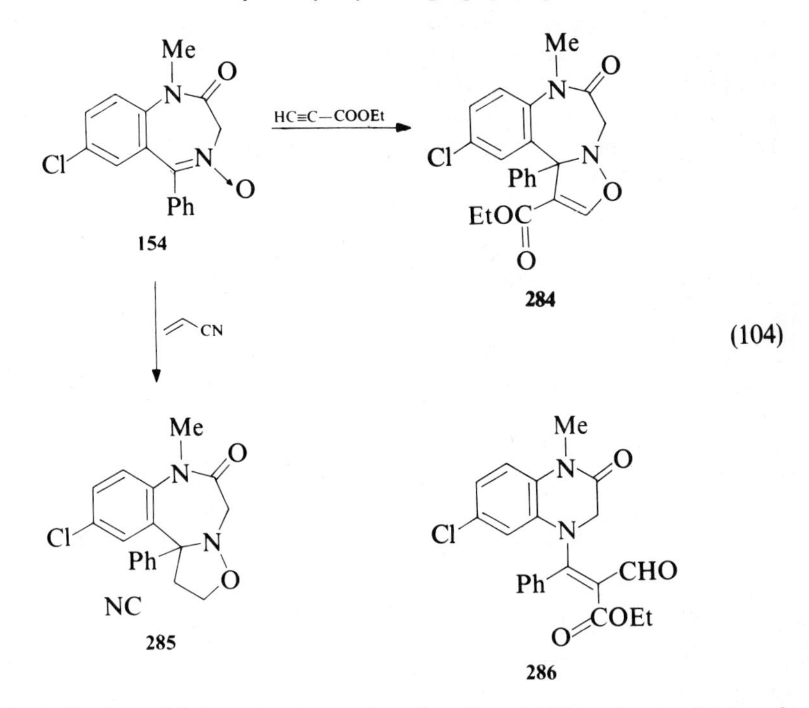


Acrylonitrile also added to the nitrone functionality of **154** to give a mixture of two isomers to which the structure **285** was assigned on the basis of spectral data.[134]

Benzodiazepin-2-ones can serve as ligands to metal ions. Complexes with cobalt chloride,[378] with zinc and cadmium halides,[379] with styphnates,[380] and with palladium(II)[381] were characterized. The complexes formed with bromazepam and chlorides, sulfates, and perchlorates of various divalent metal ions were investigated more recently.[382,383]

### 1.3. Spectral Data

The pH dependence of the ultraviolet spectra of benzodiazepin-2-ones was studied to determine the $pK_a$ values of several compounds.[350,351,384] Ultraviolet and circular dichroic spectra of optically active 3-substituted benzodiazepin-2-ones were analyzed.[385]

Proton-nmr spectroscopy allowed the investigation of the conformational mobility of the seven-membered ring. 1,3-Unsubstituted benzodiazepin-2-ones generally show the 3-protons as a broad singlet at room temperature. 1-Substituted derivatives are more rigid and the 3-protons appear as a distinct AB system at ambient temperature. The coalescence temperature and the activation energy for the interconversion of the two enantiomeric conformations **287** and **288** (Eq. 105) were determined for diazepam,[386,387] nordiazepam,[386,387] nitrazepam,[388] and several other analogs.[386-390] While the free energy of activation for the ring inversion is about 10–12 kcal/mol for the 1-unsubstituted compounds, the values for compounds carrying a substituent at the 1-position are considerably higher, between 16 and 19 kcal/mol. The energy barrier has been shown to be sufficiently large in the 1-*t*-butyl analogs to allow resolution of the two enantiomers.[391]

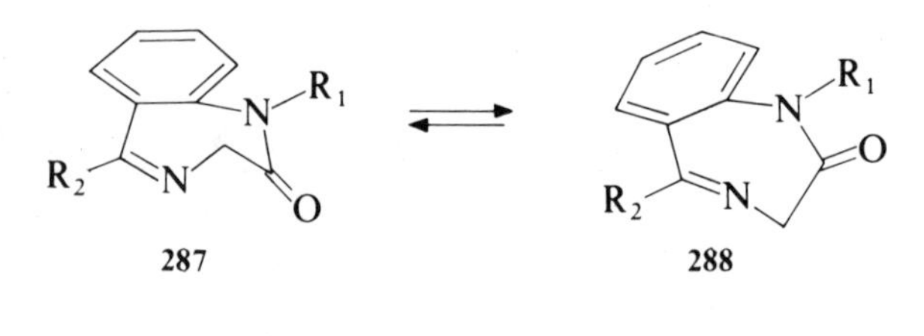


(105)

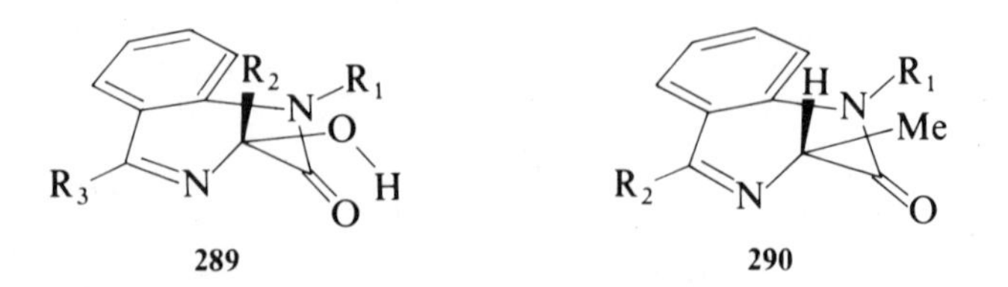


The conformation **290** with pseudoequatorial orientation of the 3-substituent is energetically favored with 3-monosubstituted compounds. Thus 3-methyl and 3-hydroxy derivatives were found to exist in only one conformation at room temperature.[392,393] This has been confirmed by X-ray crystallography for both *R*- and *S*-3-methyl-1,4-benzodiazepin-2-ones (**290**[$R_1$ = H], is the *S*-alanine derived enantiomer). The methyl groups for both enantiomers were found to be in pseudoequatorial orientation in the crystal, *R* and *S* stereochemistry being maintained by inversion of the 7-membered ring.[501] In the 3,3-disubstituted compound **289** ($R_2$ = Me), two conformations were observed. The orientation of the 3-methyl group in the pseudoaxial arrangement was preferred, possibly due in part to hydrogen bonding of the 3-hydroxy group to the 2-carbonyl oxygen.[393] Paramagnetic shift reagents were used for both proton[390,393] and carbon-13 nmr studies.[390,394] Carbon-13 nmr spectroscopy[390,394–398] confirmed the conformational data derived from proton-nmr data. More nitrogen-15 nmr data have been reported.[399,400] Although mass spectra of many benzodiazepin-2-ones were recorded (see Table VII.1), few studies about the fragmentation mechanism were carried out.[401]

The solid state structures of several benzodiazepin-2-ones were determined by X-ray crystallography.[402–414]

## 2. 1,3-DIHYDRO-1,4-BENZODIAZEPIN-2(2*H*)-THIONES

### 2.1. Synthesis

The 2-thiones **292** were, in most instances, prepared by treatment of the corresponding 2-ones **291** with phosphorus pentasulfide in pyridine (Eq. 106).[24,79,100,109,179,255,415–421] Tetrahydrofuran in combination with

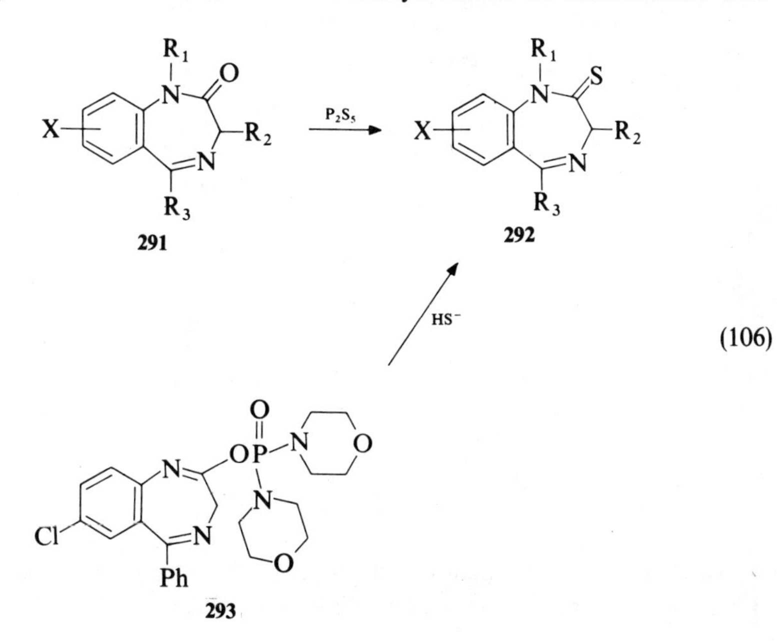


(106)

triethylamine,[79] dioxane,[422] and xylene[423] was also used as a reaction medium. The conversion of the 4-oxide of **291** ($R_1$, $R_2$ = H; $R_3$ = Ph; X = 7-Cl) to the thione by means of phosphorus pentasulfide and pyridine was accompanied by a reduction to yield **292** ($R_1$, $R_2$ = H; $R_3$ = Ph; X = 7-Cl).[424] The same thione was obtained in 78% yield by treatment of the iminophosphate **293** with hydrogen sulfide and triethylamine.[346] A European patent application describes the preparation of a 3-benzyloxycarbonyloxy-2-thione by thiation of the corresponding 2-one using dimeric 4-methoxyphenyl thionophosphine sulfide.[425]

## 2.2. Reactions

### 2.2.1. *Reactions with Electrophiles*

Alkylation of 2-thiones **292** takes place on sulfur, leading to the 2-thioalkyl-3*H*-1,4-benzodiazepines **294** (Eq. 107), or in case of 1-substituted compounds to the 2-thioalkyl-1*H*-1,4-benzodiazepines **295**. (See Chapter V for the synthesis and reactions of these compounds.) Alkylations were carried out with methyl or ethyl iodide and sodium hydride in dimethyl or diethylformamide,[343,415] with dimethyl sulfate and sodium hydroxide,[255,426] and with amino-substituted alkylchlorides and various bases.[179,427] Bromoacetic acid in methanolic sodium hydroxide alkylated the thione in the same fashion.[428]

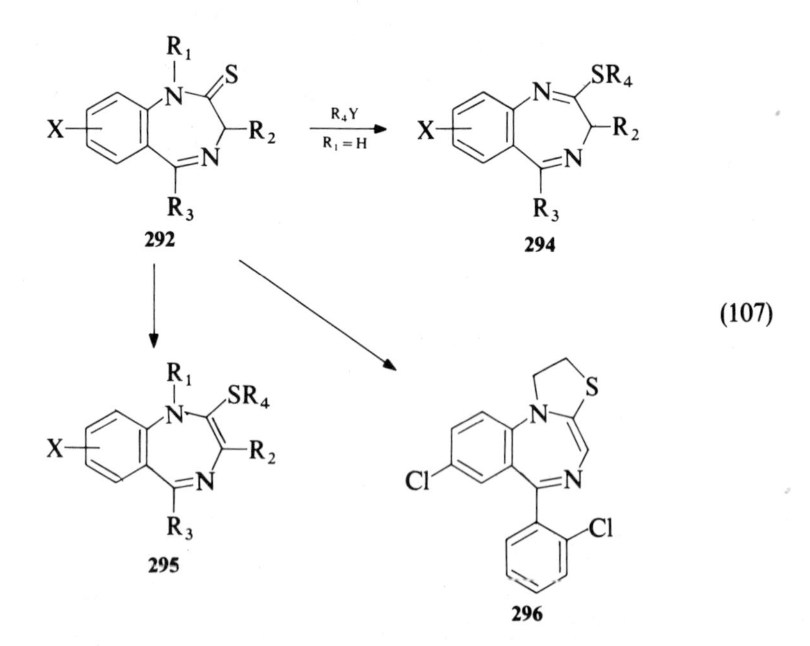


(107)

Alkylation of the 2-thione **292** ($R_1$, $R_2$ = H; $R_3$ = 2-$ClC_6H_4$; X = 7-Cl) with 1,2-dibromoethane gave 11% of the dimer and 4.3% of the tricyclic compound

**296.**[179] Similar thiazolines (**300**) were formed by reacting the 2-thiones **297** (X = H, Cl) with 1,2-dichloro-1-methoxyethane and potassium *t*-butoxide in dimethylformamide and subsequently with sodium hydride.[134] Initial alkylation of the sulfur by the more reactive chloride formed the intermediate **298**, which cyclizes, in the presence of sodium hydride, to the observed product **300** (Eq. 108).

The only acylation of a 2-thione described was the reaction of **297** (X = H) with dimethylmalonyl dichloride, producing the thiazine derivative **299** in 28% yield.[310]

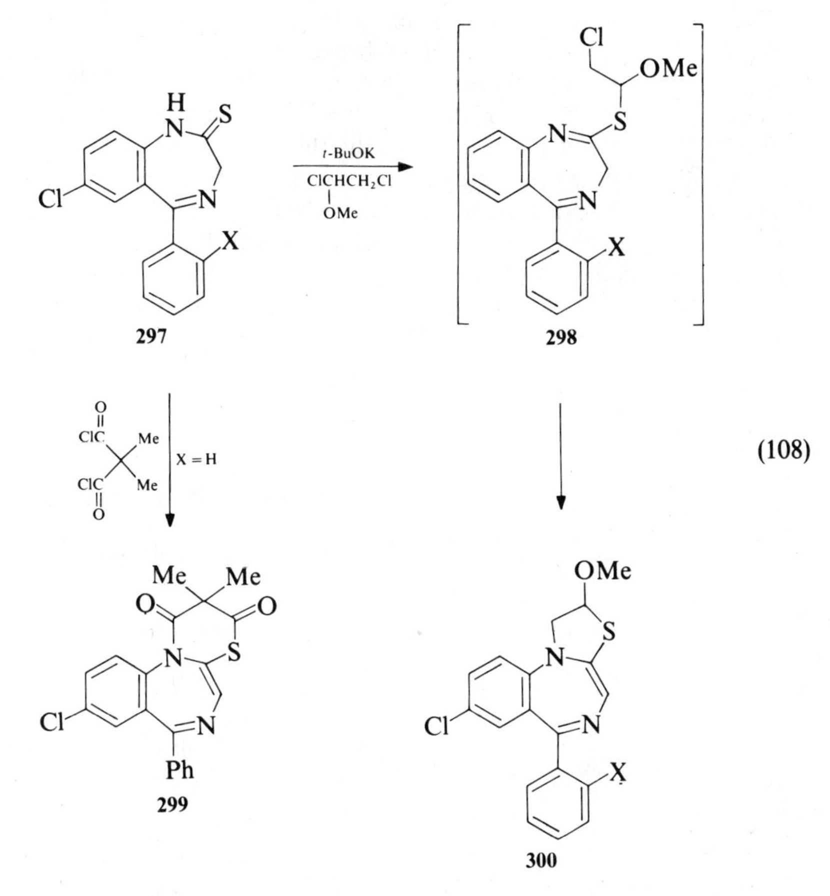


(108)

## 2.2.2. *Reactions with Nucleophiles*

Reduction, by desulfurization, of the 2-thiones with Raney nickel led to the 2,3-dihydro-1*H*-1,4-benzodiazepines.[255] (See in Chapter VI, Section 1.1.5.).

Lithium aluminum hydride in tetrahydrofuran at 0°C reduced the 1-substituted 2-thiones **301** (R = Me, $F_3CCH_2$) to the corresponding thiols, which in turn cyclized to the bridged derivatives **302**.[429]

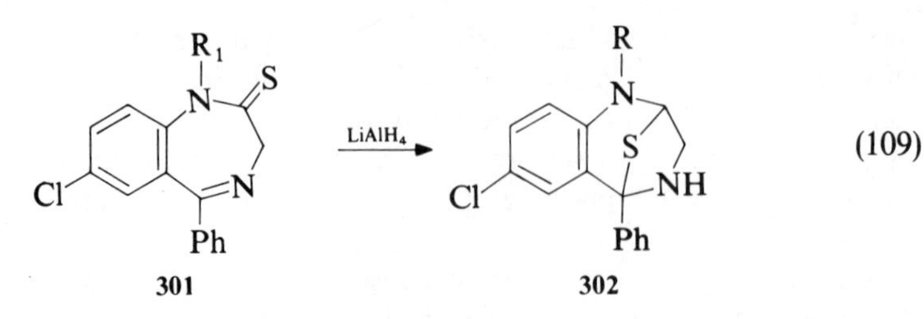


The 2-thiones **303** were reacted with a variety of nitrogen nucleophiles to form the 2-amino-3*H*-1,4-benzodiazepines **304** (Eq. 110). The nitrogen nucleophiles included various amines,[24b,313,417,430–443] hydroxyamines,[418,425,426,444–446] hydrazines[186,292,296,447] and acylated hydrazines.[24a,313,448–454] If the reaction of the thione with the acylhydrazine was carried out at high temperatures, such as boiling in butanol for several hours, the intermediate 2-acylhydrazino derivatives **304** ($R_3$ = H, $R_4$ = NH-acyl) were cyclized in situ to the triazolobenzodiazepines **305** (Eq. 110). Other suitably functionalized amines **304** ($R_3$ = H, $R_4$ = 2-alkynyl, 2-acylmethyl) attached at the 2-position were further converted to imidazobenzodiazepines **306** by acid-catalyzed ring closure.[430,431,437–442]

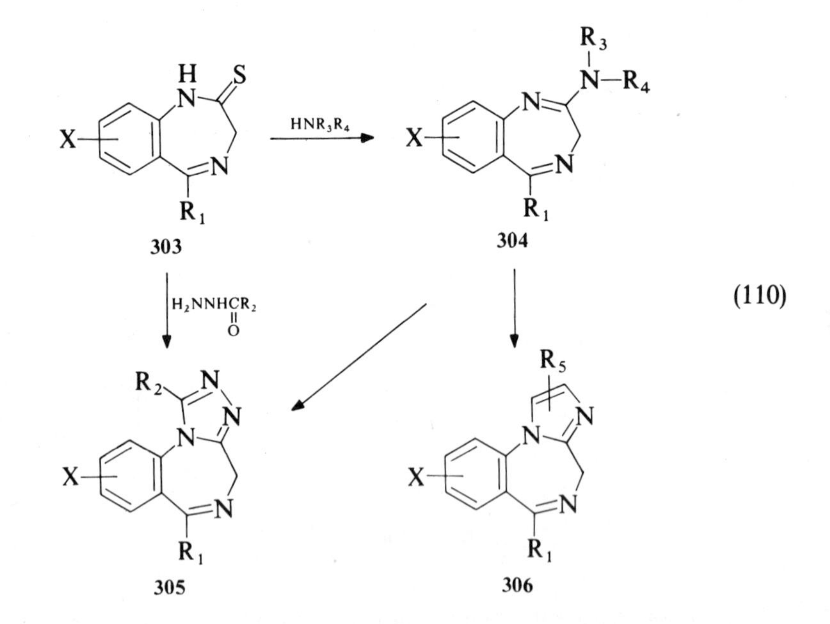


The thione **307** underwent the same rearrangement to the isoindole **308** as the corresponding 2-one (Eq. 111). Heating of **307** with sodium hydride in dimethylformamide yielded 80% of the thiocarboxamide **308**.[300]

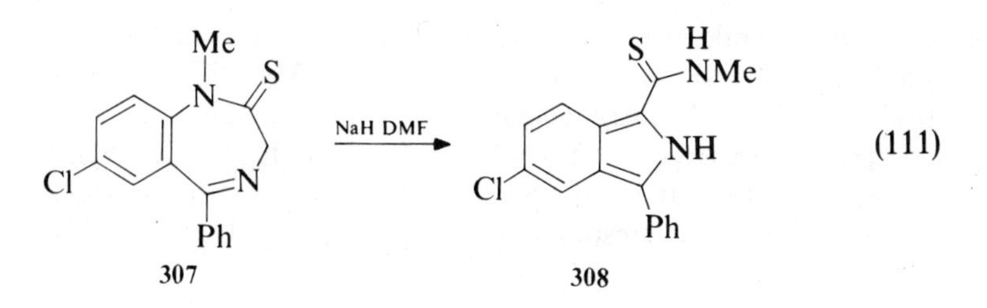


## 3. 1,5-DIHYDRO-1,4-BENZODIAZEPIN-2(2*H*)-ONES

### 3.1. Synthesis

The 1,5-dihydro-1,4-benzodiazepin-2(2*H*)-ones **310** were obtained by elimination of leaving groups $R_2$ from the 4-position nitrogen of **309**. Acetic acid was satisfactorily eliminated from 4-acetoxy compounds **309** ($R_2$ = AcO, $R_3$ = Ph, $R_4$ = H) by treatment with triethylamine[32,455] or diethylamine in boiling ethanol.[175] Dehydration of the 4-hydroxy derivative **309** ($R_2$ = OH, $R_3$ = Ph, $R_4$ = H), by heating in pyridine in the presence of phosphorus oxychloride, yielded the 1,5-dihydro tautomer in addition to the 1,3-dihydro compound.[141] Elimination of equivalents of sulfinic acid from the 4-tosyl derivative **309** ($R_2$ = 4-$MeC_6H_4SO_2$) by means of sodium hydride in benzene gave the 1,5-dihydro compound in 35% yield.[170–172] 5-Methoxy compounds **310** ($R_3$ = Ph, $R_4$ = MeO) were prepared by reaction of the 4-nitro derivatives **309** ($R_2$ = $NO_2$, $R_3$ = Ph, $R_4$ = MeO) with sodium hydride in dimethylformamide at −10°C (Eq. 112).[456]

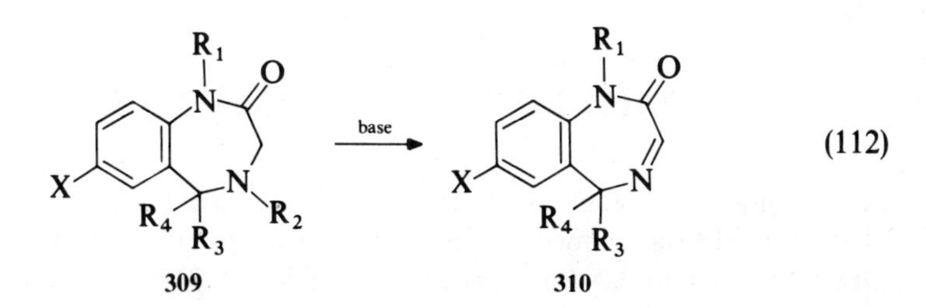


An efficient synthesis of the 1,5-dihydro tautomers would be possible if the anions of 1,3-dihydro-1,4-benzodiazepin-2(2*H*)-ones **141**, generated by abstraction of a proton from the 3-position, could be exclusively reprotonated at the 5-position. Kinetic protonation does not seem to be selective, however. Under equilibrium conditions, the 1,3-dihydro compounds are energetically favored if the 3-position is unsubstituted. The 1,5-dihydro compounds become the more stable tautomers in the 3-hydroxy, 3-methoxy, and 3-amino derivatives. It is thus possible to rearrange the 3-hydroxy compounds quantitatively to the

2,3-diones, while the 3-methoxy and the 3-amino analogs rearrange to the imidates **311** ($R_2 = OR$) and amidines **311** ($R_2 = NRR'$), respectively (Eq. 113).[149,457] A rearrangement occurred during alkylation of the 3-methoxy compound **141** ($R_1$ = H, $R_2$ = MeO, $R_3$ = 2-$ClC_6H_4$, X = Cl) with 1-chloro-2-diethylaminoethane and potassium hydroxide in the presence of potassium iodide, giving the corresponding imidate **311** (Eq. 113).[457]

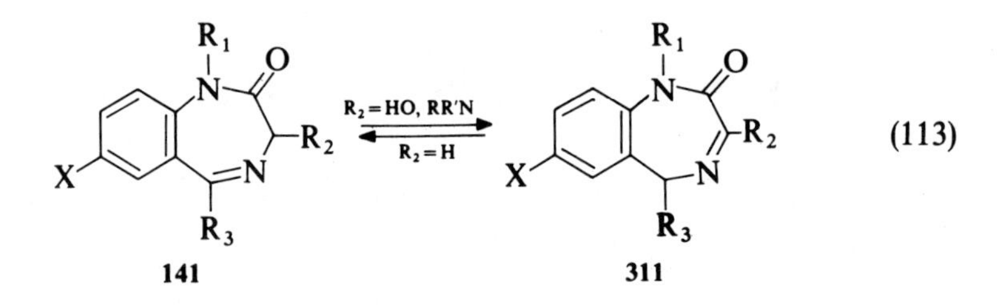


(113)

### 3.2. Reactions

The rearrangement, by based-catalyzed equilibration, of 1,5-dihydro-1,4-benzodiazepin-2-(2*H*)-ones **311** ($R_2$ = H) to the corresponding 1,3-dihydro compounds **141** was discussed in Section 1.1.6.

The 5-methoxy analogs **310** ($R_1$ = H, $R_3$ = Ph, $R_4$ = MeO) were alkylated at the 1-position by methyl iodide and sodium hydride in dimethylformamide.[456]

From the reaction of **311** ($R_1$, $R_2$ = H, $R_3$ = Ph, X = Cl) with *N*-bromosuccinimide in benzene and subsequent treatment with water–dimethylformamide, oxazepam, **141** ($R_1$ = H, $R_2$ = OH, $R_3$ = Ph, X = Cl) was isolated in 3% yield.[458]

## 4. 1,2-DIHYDRO-1,4-BENZODIAZEPIN-3(3*H*)-ONES

Although the parent compound is still unknown, the 2,2-dimethyl-5-phenyl derivative **313** was formed by reaction of the magnesium salt of the imine anion **312** with 2-bromo-2,2-dimethylacetyl bromide (Eq. 114).[459] The structure of this product was established by X-ray analysis.

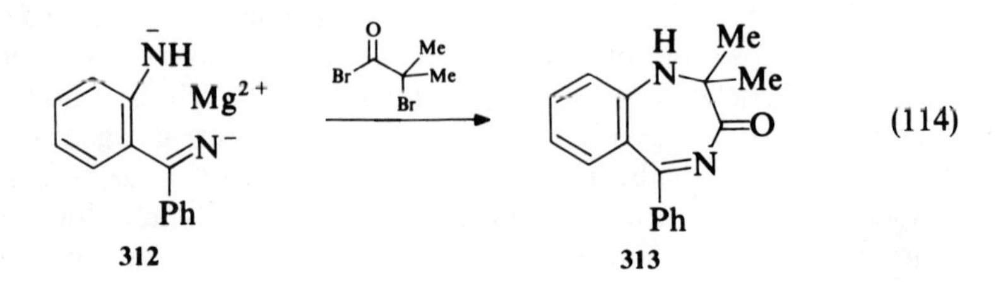


(114)

## 5. 4,5-DIHYDRO-1,4-BENZODIAZEPIN-3(3*H*)-ONES

The parent ring **317** was prepared in 43% yield by photolysis of the 3-azidoquinoline **314** in methanol containing potassium methoxide (Eq. 115).[460,461] This ring expansion was formulated to proceed via the imidate **315**. The aminoquinoline **316** was formed as a by-product.

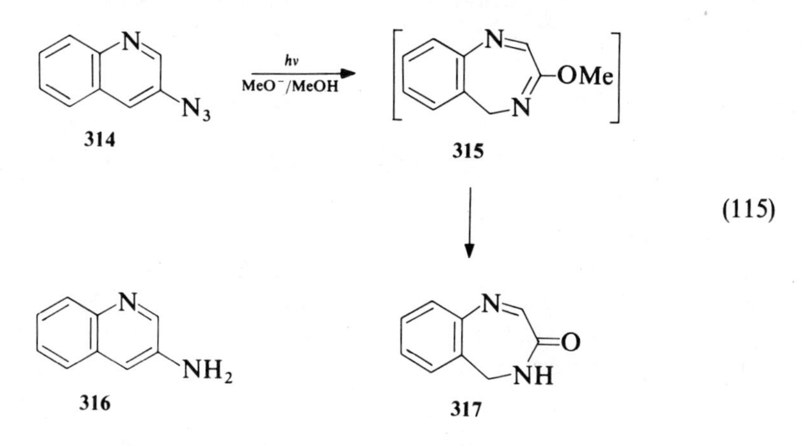


The 2-ethyl-5-phenyl derivative **320** was isolated from the reaction of the acylated anthranilonitrile **318** with phenylmagnesium bromide.[462] Compound **320** and the addition product **321**, with phenylmagnesium bromide, were assumed to form via the aziridinone **319** as indicated in Eq. 116.

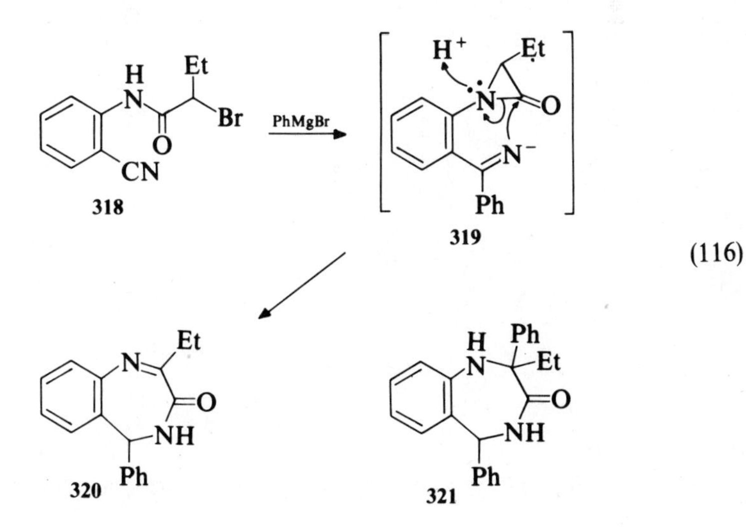


Compound **323** was the product formed when the emamine **322** was treated with sodium hypochlorite in methanol (Eq. 117).[463]

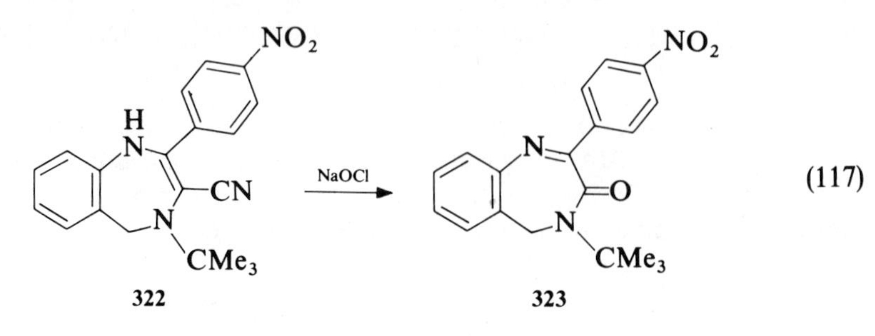


## 6. 1,2-DIHYDRO-1,4-BENZODIAZEPIN-5(5*H*)-ONES

The only representative of this class of compounds was prepared by Field and coworkers.[464] The tetrahydroquinazolinone **324** was reacted with potassium *t*-butoxide in tetrahydrofuran, at room temperature, to give **325** in approximately 50% yield (Eq. 118). Catalytic hydrogenation over platinum converted **325** to the tetrahydro derivative **326**.

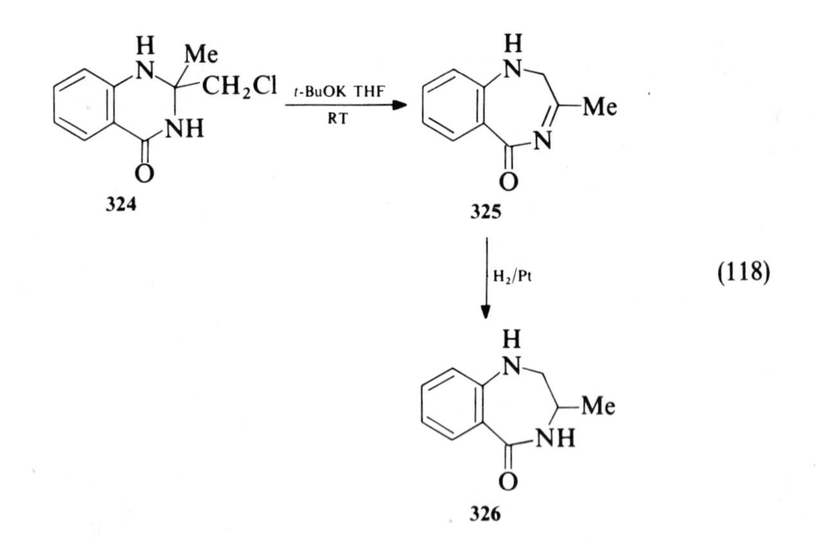


## 7. 1,4-DIHYDRO-1,4-BENZODIAZEPIN-5(5*H*)-ONES

The 8-chloro compound **328** was formed in 20% yield by photolysis of the 4-azidoquinoline **327** (Eq. 119).[461]

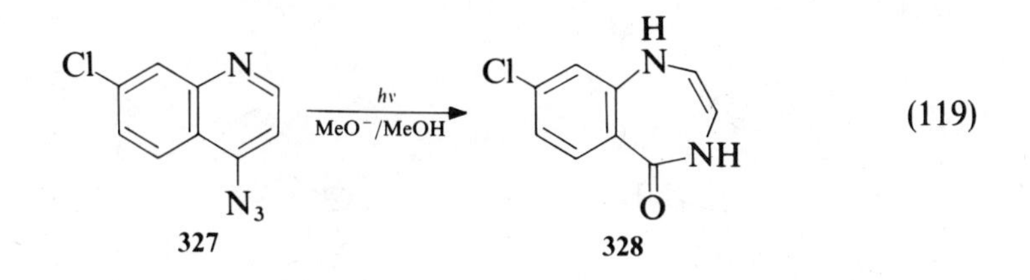


## 8. 3,4-DIHYDRO-1,4-BENZODIAZEPIN-5(5*H*)-ONES

While the parent compound **330** ($R_1$, $R_2$, X = H) appears to be still unknown, the 2-phenyl derivatives **330** ($R_1$ = Ph, 2-$ClC_6H_4$) were synthesized by cyclization of the anthranil amides **329**.[465-468] These anthranil amides were prepared either by reaction of isatoic anhydride with α-aminoacetophenones or by reduction of the nitro compounds with iron and hydrochloric acid (Eq. 120).[468]

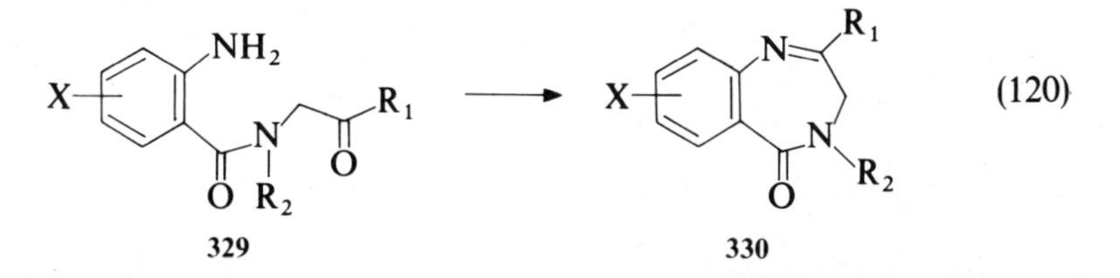


2-Amino derivatives **332** ($XR_1$ = NRR′) were obtained by amination of the 2,5-diones **331** (X = O) with an amine and titanium tetrachloride.[344,469-471] Treatment of the 2-thiones **331** (X = S) with amines HNRR′ led also to the amidines **332** ($XR_1$ = NRR′).[471] 2-Thiomethyl derivatives **332** ($XR_1$ = MeS) were obtained by alkylation of the 2-thiones (Eq. 121). The thiomethyl analogs reacted with amines to form the amidines.[471]

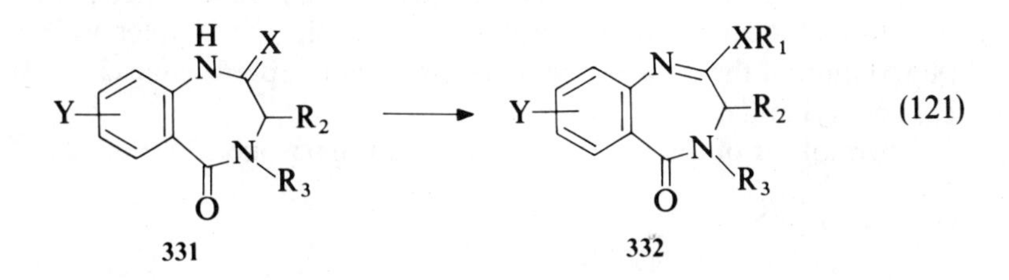


Nitrosation of the 2-methylamino compound **333** (R = H) led to the nitrosoamidine **333** (R = NO), which reacted with carbanions of nitromethane and dimethylmalonate to form other 2-substituted compounds **334** with liberation of diazomethane (Eq. 122).[134] The malonyl derivative **334** ($R_1$, $R_2$ = COOMe) was hydrolyzed to the acetylidene compound **335**. Nitrosation of the latter led to the oxime **336**, which was transformed into the imidazobenzodiazepine **337** by reduction and subsequent condensation with triethyl orthoacetate (Eq. 122).[134]

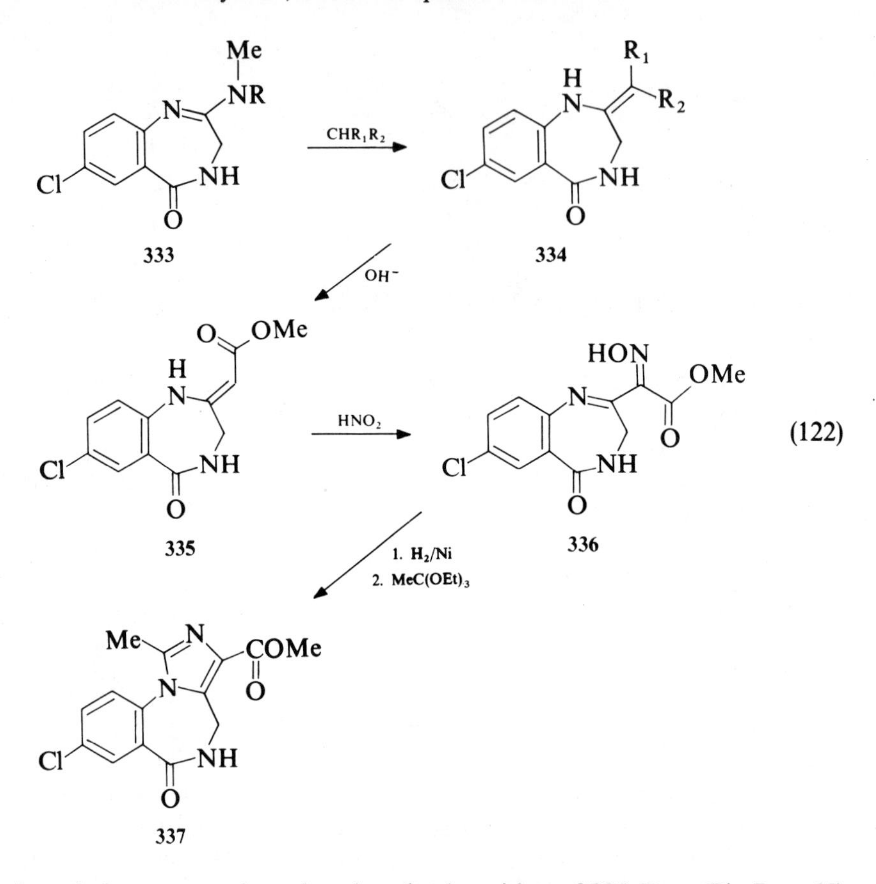


A methyl group was introduced at the 4-position of **330** ($R_1$ = Ph, $R_2$ = H) by alkylation of **330** ($R_1$ = Ph, $R_2$ = H) with methyl iodide and sodium amide in toluene.[468] Treatment of the 5-one **333** (R = H) with phosphorus oxychloride afforded the 5-chloro-3*H*-1,4-benzodiazepine, an intermediate, where further variation of the 5-position substituent is possible (see Chapter V).[469,470] Phosphorylation of the same 5-one with dimorpholinophosphinic chloride gave the phosphorylimidate.[338,344]

Thermolysis of the imidate **338** afforded the 5-one **339** (Eq. 123).[344]

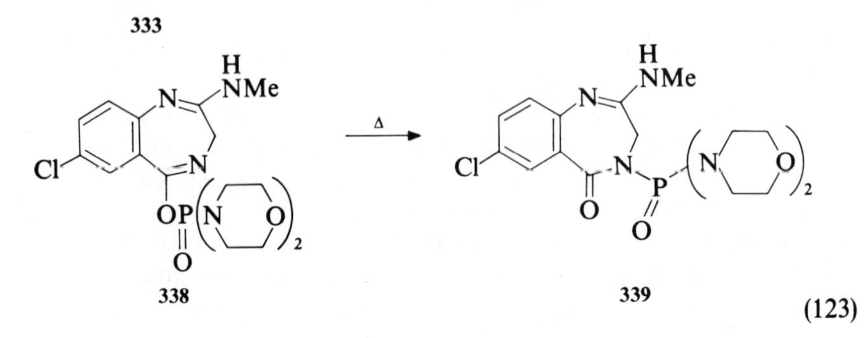

For reductions of the 3,4-dihydro-1,4-benzodiazepin-5(5*H*)-ones to tetrahydro derivatives, see Chapter VIII, which deals with the synthesis and reactions of the tetrahydro-1,4-benzodiazepines.

## 9. DIHYDRO-1,4-BENZODIAZEPINEDIONES

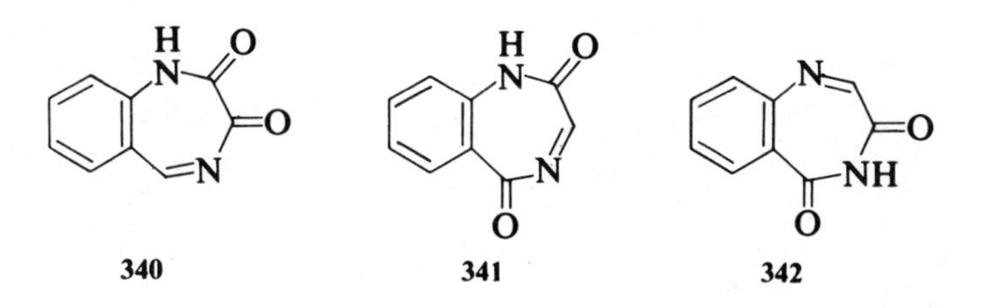


Of the three possible diones **340–342**, representatives of only the 2,3-diones **340** have been synthesized. The 2,3-diones **343** were obtained by oxidation of the 2-ones **141** ($R_1$ = H, Me; $R_2$ = H, OH, $R_3$ = Ph) with ruthenium tetroxide or by treatment of the 3-hydroxy compounds **141** ($R_2$ = OH) with activated manganese dioxide (Eq. 124).[157,158] Another method involves the oxidation of the 3-phosphoranes obtained by treatment of the 3-triphenylphosphonium salts **141** ($R_2$ = $PPh_3$) with molecular oxygen.[243]

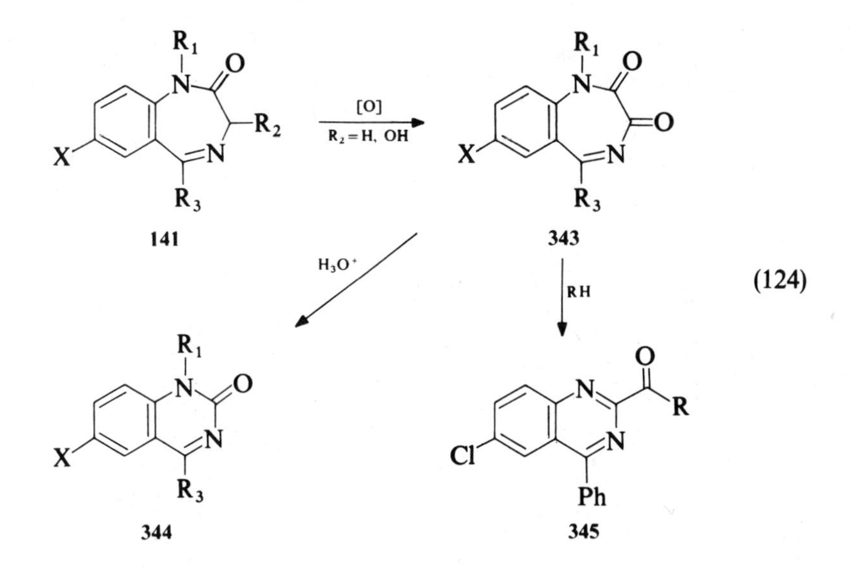


The 2,3-diones **343** are readily attacked by nucleophiles. Hydrolysis led to the quinazolinones **344**, which were also formed as by-products during the preparation of the diones by oxidation.[157] Reaction of **343** ($R_1$ = H, $R_3$ = Ph, X = Cl) with aqueous potassium carbonate afforded 40% of the quinazoline-2-carboxylic acid **345** (R = OH). The amide **345** (R = $NH_2$) was similarly formed and was isolated in 7% yield from the oxidation of oxazepam with ruthenium tetroxide.[157]

An early report in the literature,[472] describing the synthesis of the benzodiazepin-3,5-dione **347**, was later corrected by Gilman and coworkers,[473] who showed that the reported benzodiazepine **347** was actually the benzoxazinone **348**. Compound **348** was obtained by heating the hydrazide **346** in acetic anhydride for several hours at reflux (Eq. 125).

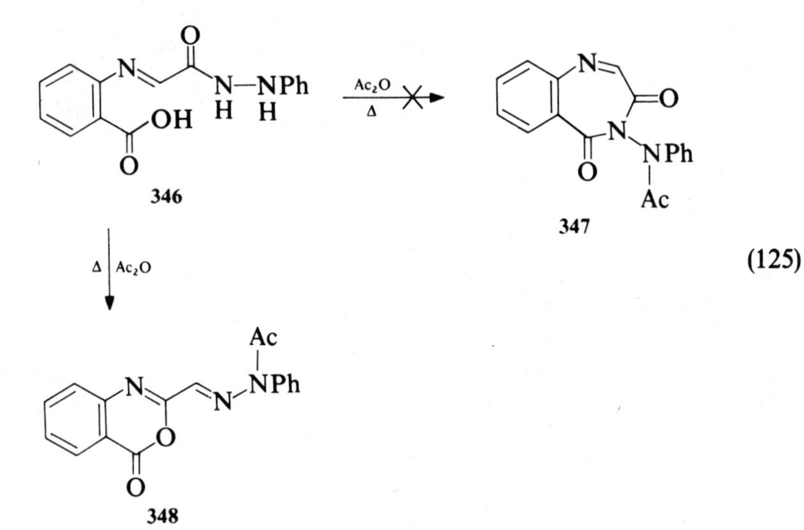


(125)

# 10. TABLE OF COMPOUNDS

TABLE VII-1. DIHYDRO-1,4-BENZODIAZEPINONES AND DIHYDRO-1,4-BENZODIAZEPINTHIONES

| Substituent | mp (°C) or; [bp (°C/torr)] | Solvent of Crystallization | Yield (%) | Spectra | Refs. |
|---|---|---|---|---|---|
| | *1,3-Dihydro-1,4-benzodiazepin-2(2H)-ones* | | | | |
| ***Unsubstituted*** | | | | | |
| None | 225–227 | MeCN | 83 | ir, pmr, uv | 1 |
| | 215–217d | | 86 | ir, pmr | 193 |
| 4-Oxide | 259–260d | $EtOH/H_2O$ | 70 | ir, ms, pmr | 1 |
| ***Monosubstituted*** | | | | | |
| 3-AcO | 197–198d | THF | 51 | ir, pmr, uv | 1 |
| 5-AcNHNHCO | 233–234 | | 90 | | 193 |
| 5-(Allyloxy)CO | 128–129 | | 86 | | 193 |
| 5-$H_2$NCO | 262–264 | MeOH/Acetone | | | 192 |
| 5-(2-Benzofuryl) | 196–197d | MeCOEt | | | 67b |
| 5-Bu | 75–76 | | | | 474 |
| 5-HOOC | 139–140 | | 80 | | 193 |
| 5-Cycloheptyl | 158–160 | | | | 86b |
| 5-(Cyclohex-1-en-1-yl)-7-Cl | 207–208 | EtOAc | 80 | | 224 |
| 5-Cyclohexyl | 199–201 | | | | 70 |
| 5-Et | 123–124 | EtOAc | 80 | ir, pmr | 21 |
| 5-Ferrocenyl | 224–227d | $CH_2Cl_2$ | 91 | | 13 |
| 5-($NH_2$NHCO) | 213–215d | | 81 | | 193 |
| 5-(3-$H_2$NNHCO-Thien-2-yl) | 235–237 | MeOH | | | 86b |
| 5-(2-Indolyl) | 223–225 | EtOH | | | 67b |

TABLE VII-1. —(contd.)

| Substituent | mp (°C) or; [bp (°C/torr)] | Solvent of Crystallization | Yield (%) | Spectra | Refs. |
|---|---|---|---|---|---|
| 5-Me | | | | | |
| Hydrochloride | 285–286 | | | | 34 |
| 4-Oxide | 235–236 | $H_2O$ | | | 34 |
| | 235–240 | Acetone/MeOH | 32 | | 298 |
| 5-(1-Me-3-Indolyl) | 268–269 | EtOH | | | 67b |
| 5-(5-Me-1,3,4-Oxadiazol-2-yl) | 200 | | 38 | | 193 |
| 5-(4-Me-3-HS-1,2,4-Triazol-5-yl) | | | | | 193 |
| 5-(3,5-$Me_2$-Isoxazol-4yl | 276–277d | EtOH | | | 13b |
| 5-(3,6-$Me_2$-Pyrazin-2-yl) | 269–270d | EtOH | | | 13b |
| 5-(2,2-$Me_2$-Propyl) | 138–139 | PhH/Hexane | | | 1b |
| 5-(MeNHCSNHNHCO) | 214–215 | | 68 | | 193 |
| 5-MeOOC | 173–175 | | 51 | ir, pmr | 193 |
| 5-(3-MeOOC-Thien-2-yl) | 193–194 | MeOH | | | 86b |
| 5-(Naphth-1-yl)-7-Cl | 192–194 | PhH/Hexane | | | 476 |
| 5-Octyl-7-Cl | 98–99 | PhH | | | 1b |
| 5-Ph | 178–179 | EtOH | 85.5 | | 2, 35 |
| Hydrochloride | 251–253 | MeCN | | | 34 |
| 4-Oxide | 250 | | | | 34 |
| 5-(2-$BrC_6H_4$) | 220–222 | $CH_2Cl_2$/EtOAc/Hexane | | | 134 |
| 5-(2-$ClC_6H_4$) | 212–213 | $CH_2Cl_2$/$Et_2O$ | | | 91 |
| 4-Oxide | 230–232 | $CH_2Cl_2$/EtOH | | | 134 |
| 5-(2,4-$Cl_2C_6H_3$) | 218–220 | EtOAc/$Et_2O$ | | | 134 |
| 5-(4-$ClC_6H_4$) | 262–263 | EtOH | | | 91 |
| 5-(2-$HOOCC_6H_4$)·$H_2O$ | 215–230 | | | | 15 |
| | 243–244d | $H_2O$/AcOH | | | 2b |
| 5-(2-$FC_6H_4$) | 180–181 | Acetone/Hexane | 77 | | 2 |
| 5-(2,6-$F_2C_6H_3$) | 242–244 | MeOH | | | 152c |
| 5-(2-F-4-$ClC_6H_3$) | 232–233 | EtOH | | | 152b |

| | | | | | |
|---|---|---|---|---|---|
| 5-(2-F-5-$O_2NC_6H_3$) | 219–222 | | 38 | | 152a |
| 5-(4-$FC_6H_4$) | 188–190 | Acetone/Hexane | | | 152c |
| 5-(2-$F_3CC_6H_4$) | 187–188 | $Et_2O$/Hexane | 67 | uv | 4 |
| 5-[2,5-$(F_3C)_2C_6H_3$] | 280 | | | | 17 |
| 5-(3-$F_3CC_6H_4$) | 204–205 | PhH/Acetone | 37 | uv | 4 |
| 5-(4-$F_3CC_6H_4$) | 219–220 | $Et_2O$/Hexane | 70 | uv | 4 |
| 5-(2-$O_2NC_6H_4$) | 206–208 | PhH | | | 41, 136 |
| 5-(3-$O_2NC_6H_4$) | 224–227 | MeOH | 27 | | 152a |
| 5-(4-$O_2NC_6H_4$) | 279–281 | MeOH | 25 | | 152a |
| 5-(2-Ph-Ethyl) | 152–153 | PhH/Petr ether | | | 2b |
| 5-(2-Pyridyl) | 232–234d | Acetone | 69 | | 3 |
| 5-(4-Pyridyl) | 206–207 | PhH | 75 | | 3 |
| 5-(2-Thiazolyl) | 248–249 | $CHCl_3$/Petr ether | | | 27 |
| 4-Oxide | 255–257 | HOAC | | | 33b |
| 5-(2-Thienyl) | 197–198 | PhH/Hexane | | | 22 |
| 7-Br | 251–254 | Acetone | | | 247 |
| 7-Cl | 247–252d | Acetone | | | 247 |
| 4-Oxide | 255–257 | PhH | | | 247 |
| ***Disubstituted*** | | | | | |
| ***1,5-Disubstituted*** | | | | | |
| 1-(2-AcO-Ethyl)-5-(2-$FC_6H_4$) | 135–137 | | | | 86b |
| 1-[2-(N-$H_2NCOCH_2$-N-Me-Amino)ethyl]-5-Ph | 129–131 | EtOH | | | 302 |
| 1-Benzyl-5-(2-$BrC_6H_4$) | 135–138 | *i*-PrOH | | | 134 |
| 1-*t*-Bu-5-Ph | 83–85 | $Et_2O$/Hexane | | | 391b |
| 1-*t*-Bu-5-(2-$ClC_6H_4$) | 151–153 | $Et_2O$/Hexane | | | 391b |
| 1-(Cyclopropyl)$CH_2$-5-Ph | | | | | |
| Hydrochloride | 204d | MeOH/$Et_2O$ | | | 86b |
| 1-[2-(EtO-Acetoxy)ethyl]-5-(2-$FC_6H_4$) | 154–158 | | | | 86b |
| 1-(2-$Et_2N$-Ethyl)-5-Me | 64–65 | Pentane | | | 255c |
| 1-(2-$Et_2N$-Ethyl)-5-Ph | 80–81 | Hexane | | | 302 |
| 1-(2-$Et_2N$-Ethyl)-5-(4-$ClC_6H_4$) | 90–94 | Hexane | | | 302 |
| 1-(2-$Et_2N$-Ethyl)-5-(3-$F_3CC_6H_4$) | 58–60 | Hexane | | | 302 |

TABLE VII-1. —(contd.)

| Substituent | mp (°C) or; [bp (°C/torr)] | Solvent of Crystallization | Yield (%) | Spectra | Refs. |
|---|---|---|---|---|---|
| 1-$F_3CCH_2$-5-Ph | 138–139 | EtOH | 65 | | 69 |
| 1,5-$Me_2$ | 155–156 | Hexane | | | 2b |
| 4-Oxide | 268–271 | EtOAc/MeOH | 61 | pmr | 298 |
| 1-Me-5-$H_2NCO$ | 203–205 | MeOH | | | 192 |
| 1-Me-5-CN | 107–109 | THF/Petr ether | | | 192 |
| 1-Me-5-Ferrocenyl | 163–164d | $CH_2Cl_2$/Hexane | 91 | | 13 |
| 1-Me-5-(3,5-$Me_2$-Isoxazol-4-yl) | 209–210 | $CH_2Cl_2$/Hexane | | | 13b |
| 1-Me-5-Ph | 154–156 | EtOH | 44 | | 2 |
| | | | | X-ray | 403 |
| Methiodide | 176–180 | MeOH/$Et_2O$ | 91 | | 300 |
| 1-Me-5-(4-$H_2NC_6H_4$) | 276–279 | EtOAc | | | 134 |
| 1-Me-5-(2-$ClC_6H_4$) | 135–137 | PhH/Petr ether | | | 91 |
| 1-Me-5-(2,4-$Cl_2C_6H_3$) | 119–121 | $Et_2O$/Petr ether | | | 134 |
| 1-Me-5-(4-$ClC_6H_4$) | 160–162 | MeOH | | | 255a |
| 1-Me-5-(2-$FC_6H_4$) | 113–114 | $Et_2O$ | 44 | | 2 |
| 1-Me-5-(2-F-5-$H_2NSO_2C_6H_3$) | 225–226 | EtOAc | | | 301 |
| 1-Me-5-(4-$FC_6H_4$) | 169–170 | Acetone/Peter ether | | | 152c |
| 1-Me-5-(2-$F_3CC_6H_4$) | 135–137 | EtOH/$H_2O$ | | | 255a |
| 1-Me-5-(3-$F_3CC_6H_4$) | 121–122 | EtOH | | | 302 |
| 1-Me-5-(4-$O_2NC_6H_4$) | 178–180 | EtOAc/Hexane | | | 134 |
| 1-Me-5-(3,6-$Me_2$-Pyrazin-2-yl) | 123–124d | $CH_2Cl_2$/Hexane | | | 13b |
| 1-Me-5-MeOOC | 122–124 | EtOAc/Petr ether | | | 477 |
| 1-Me-5-(2-Pyridyl) | 199–200 | | | | 86b |
| 1-Me-5-(2-Thiazolyl) | 185–188 | $CH_2Cl_2$/$Et_2O$ | | | 27 |
| 1-Me-5-(2-Thienyl) | 107–109 | PhH/Hexane | | | 22 |
| 1-$MeNHCOCH_2$-5-Me | 212–213 | *i*-PrOH | | | 255c |
| 1-$MeNHCOCH_2$-5-Ph | 215 | Acetone | | | 293 |
| 1-(2-$Me_2N$-Ethyl)-5-(2-$ClC_6H_4$) | 113–114 | $CH_2Cl_2$/Petr ether | | | 134 |

| | | | | | |
|---|---|---|---|---|---|
| 1-(2-$Me_2$N-Ethyl)-5-[2,5-$(F_3C)_2C_6H_3$] | | | | | |
| Dihydrochloride | ~ 100 | $Et_2O$ | | | 17 |
| 1-(3-$Me_2$N-Propyl)-5-Ph | 145–146 | $CH_2Cl_2/Et_2O$/Petr ether | | | 2b |
| 1-(3-$Me_2$N-Propyl)-5-[2,5-$(F_3C)_2C_6H_3$] | | | | | |
| Dihydrochloride | ~ 100 | | | | 17 |
| 1-(3-$Me_2$N-Propyl)-5-Me | 66–67 | $Et_2O$/Pentane | | | 255c |
| 1-$MeOCH_2$-5-Ph | 72–74 | $Et_2O$ | | | 72c |
| 1-$MeOOCCH_2$-5-(2-$ClC_6H_4$) | 166–168 | MeOH | | | 292 |
| 1-(2-Oxopyrrolidin-1-yl)$CH_2$-5-Ph | 120 | | | | 478 |
| ***1,7-Disubstituted*** | | | | | |
| 1-(2-$Et_2$N-Ethyl)-7-Cl | | | | | |
| 4-Oxide | 88–90 | $Et_2O$ | 61 | | 176a |
| 1-Me-7-Br | 117–120 | $Et_2O$ | | | 152c |
| 1-Me-7-Cl | 105–107 | EtOH/Cyclohexane | 6.5 | ir, ms, pmr | 116 |
| ***3,5-Disubstituted*** | | | | | |
| 3-AcNH-5-Ph | 222–224 | | 42 | | 202 |
| 3-AcO-5-Ph | 229–231 | | | | 263 |
| 3-AcO-5-(2-$ClC_6H_4$) | 290–291 | $CH_2Cl_2$/EtOAc | | | 134 |
| 3-$H_2$N-5-Ph | 178–179 | | 62 | | 202 |
| 3-(4-*t*-BuOOCNH-Butanoyloxy)-5-Ph · 0.5*i*-PrOH | 124–126 | *i*-PrOH | | | 478 |
| 3-(N-*t*-BuOOC-D-Phenylalanine ester)-5-Ph | | | | | |
| Isomer A | 116–119 | Cyclohexane | | | 478 |
| Isomer B | 204–205 | | | | 478 |
| 3-(N-*t*-BuOOC-L-Phenylalanine ester)-5-Ph | | | | | |
| Isomer A | 136 | | | | 478 |
| Isomer B | 131 | | | | 478 |
| 3-HOOC-5-Ph | | | | | |
| Dipotassium salt | | | | ir, uv | 104 |
| Potassium salt | | | | ir, uv | 104 |
| 3-Et-5-Ph | 169 | | 20 | | 100 |
| 3-EtOOC-5-Ph | 226 | EtOAc | 70 | uv | 104 |
| 4-Oxide | 192–194 | EtOAc/EtOH | | | 108 |

TABLE VII-1. —(contd.)

| Substituent | mp (°C) or; [bp (°C/torr)] | Solvent of Crystallization | Yield (%) | Spectra | Refs. |
|---|---|---|---|---|---|
| 3-EtOOC-5-(4-$ClC_6H_4$) | 222–227d | MeCN | | | 302 |
| 3-EtOOC-5-(2-$F_3CC_6H_4$) | 183 | | | | 82 |
| 3-EtOOC-5-(4-$MeC_6H_4$) | 189–191 | MeCN | | | 302 |
| 3-EtOOC-5-(4-$MeOC_6H_4$) | 187–192 | EtOH | | | 302 |
| 3-HO-5-Ph | 194–195d | EtOH | 44 | | 148 |
| 3-Me-5-Ph | 203–204 | PhH | 28 | | 100 |
| (+)-Enantiomer | 162–163 | EtOH | | | 301 |
| 3-*R*-Me-5-(2-$ClC_6H_4$)·0.5EtOAc | 82–85d | EtOAc/Hexane | | | 134 |
| 3-*S*-Me-5-(2-$ClC_6H_4$)·0.5EtOAc | 82–85d | EtOAc/Hexane | | | 134 |
| 3-R-Me-5-(4-$ClC_6H_4$) | 233–235 | $CH_2Cl_2$/EtOH | | | 134 |
| 3-S-Me-5-(4-$ClC_6H_4$) | 234–235 | $CH_2Cl_2$/EtOH | | | 134 |
| 3-Me-5-(2-$FC_6H_4$) | 174 | Cyclohexane | | | 251b |
| 3-$Me_2N$-5-Ph | 236–238 | | 84 | | 149 |
| 3-[2-(2-Oxopyrrolidin-1-yl)acetoxy]-5-Ph | 245 | | | | 478 |
| 3-*i*-Pr-5-Ph | 233 | | 10 | | 100 |
| ***5,6-Disubstituted*** | | | | | |
| 5-Ph-6-Cl | 244–245 | PhH/Hexane | 18 | | 2 |
| ***5,7-Disubstituted*** | | | | | |
| 5-$H_2NCO$-7-Br | 273–274d | MeOH/Acetone | | | 192 |
| 5-(Benzocyclohept-2-yl)-7-Cl, complex with 2-(2-$H_2N$-5-Cl-benzoyl)benzocycloheptane | 159–160 | EtOH/PhH | 28 | pmr | 49 |
| 5-(1,3-Benzodioxolan-5-yl)-7-Cl | 207–210 | $CH_2Cl_2$/PhH | | | 92 |
| 5-Benzyl-7-Cl | 168 | EtOAc | 33 | | 18 |
| | 155–159 | $CH_2Cl_2$/*i*-$Pr_2O$ | | | 124 |
| 5-Bu-7-Cl | 188 | EtOAc | 79 | | 18 |
| 5-*t*-Bu-7-Cl | 192–193 | PhH/Hexane | | | 1b |
| | 169–179 | PhH/Hexane | | | 1b |

| | | | | | |
|---|---|---|---|---|---|
| 5-$C_6D_5$-7-Cl | 215–216 | | 41 | | 97 |
| 4-Oxide | 233–234d | | 78 | | 97 |
| 5-(1-Cl-Cyclohexan-1-yl)-7-Cl | 196–198 | EtOAc | 77 | ir, pmr, uv | 222 |
| 5-(1-Cl-Cyclohexan-1-yl)-7-$NO_2$ | 247 | Xylene | 87 | | 224 |
| 5-(1-Cl-Cyclopentan-1-yl)-7-Cl | 191d | *i*-$Pr_2O$ | 70 | | 224 |
| 5-(1-Cl-eth-1-yl)-7-Cl | 197d | EtOAC | 61 | | 224 |
| 5-(1,1-$Cl_2$-eth-1-yl)-7-Cl | 160; 190d | MeOH | 73 | | 224 |
| 5-$Cl_2CH$-7-Cl | 210d | EtOH | 56 | | 224 |
| 5-$Cl_3C$-7-Cl | 185d | MeOH | 40 | | 224 |
| 5-(1-Cl-but-1-yl)-7-Cl | 128–129 | *i*-$Pr_2O$ | 81 | | 224 |
| 5-(1,1-$Cl_2$-but-1-yl)-7-Cl | 208 | EtOAc | 78 | | 224 |
| 5-(1-Cl-1-Me-prop-1-yl)-7-Cl | 141 | $Et_2O$ | 50 | | 224 |
| 5-(1-CN-Cyclohexan-1-yl)-7-Cl | 236 | MeOH | 50 | | 224 |
| 5-(Cyclohexen-1-yl)-7-Cl | 207–208 | EtOAc | 80 | ir, pmr, uv | 222 |
| 5-Cyclohexyl-7-Cl | 212 | EtOAc | 90 | | 18 |
| | 200–202 | | | | 34 |
| 5-Cyclohexyl-7-$F_3C$ | 179–183 | Hexane | | | 1b |
| 5-Cyclohexyl-7-Me | 165–166 | EtOAc | 66 | | 18 |
| 5-Cycloheptyl-7-Cl | 159–161 | $MeOH/H_2O$ | | | 70 |
| 5-Cyclohexyl-7-$NO_2$ | 232–233 | Acetone | | | 70 |
| 5-(Cylcopent-en-1-yl)-7-Cl | 204–205 | EtOAc | 38 | | 224 |
| 5-Cyclopentyl-7-Cl | 182 | | 62 | | 18 |
| | 170–171 | $MeOH/H_2O$ | | | 70 |
| 5-Cyclopropyl-7-Cl | 198–200 | Acetone | | | 1b |
| 5-(1,3-Dithian-2-yl)-7-Cl | 157–159 | $CH_2Cl_2$/Hexane | | | 391b |
| 5-Et-7-Cl | 134, 156 | EtOAc | 51; 75 | | 18, 21 |
| 5-(3-EtO-Propyl)-7-Cl | 112–113 | PhH/Hexane | | | 1b |
| 5-Ferrocenyl-7-I | 168–171d | $CH_2Cl_2$/Hexane | 91 | | 13 |
| 5-(2-Furyl)-7-Cl | 245–246 | Acetone | | | 22 |
| 5-(2-HO-Ethyl)NHCO-7-Br | 196–198 | $CH_2Cl_2$/MeOH | | | 192 |
| 5-(2-Indolyl)-7-Cl | 297–300d | Acetone | | | 67b |
| 5-(Isothiazol-1-yl)-7-Cl | 247–249 | EtOH | | | 13b |
| 5-Me-7-Br | 223–224 | MeOH | | | 2b |
| 5-Me-7-Cl | 226–228 | EtOAc | 66 | | 18 |

TABLE VII-1. —(contd.)

| Substituent | mp (°C) or; [bp (°C/torr)] | Solvent of Crystallization | Yield (%) | Spectra | Refs. |
|---|---|---|---|---|---|
| 5-Me-7-$NO_2$ | 221–223 | EtOH | | | 2b |
| 5-(3-Me-Butyl)-7-Cl | 185–188 | EtOH | | | 1b |
| 5-(1-Me-Imidazol-2-yl)-7-Cl | 243–245d | Acetone/Petr ether | | | 27 |
| 5-(3-Me-Isoxazol-5-yl)-7-Cl | 243–245 | MeOH | | | 479 |
| 5-(5-Me-Isoxazol-3-yl)-7-Cl | 241–242 | MeOH | | | 479 |
| 5-(3,5-$Me_2$-Isoxazol-4-yl)-7-I | 253–254d | $CH_2Cl_2$/Hexane | | | 13b |
| 5-(4-Me-Piperazin-1-yl)-7-Cl | 241–243d | EtOAc | | | 134 |
| 5-[(1-(4-Me-Piperazin-1-yl)cyclohex-1-yl]-7-Cl | 240 | EtOAc | 83 | | 224 |
| 5-(1-Me-prop-1-en-1-yl)-7-Cl | 168 | *i*-$Pr_2O$ | 30 | | 224 |
| | 221–222 | | | | 86b |
| 5-(3,6-$Me_2$-Pyrazin-2-yl)-7-Br | 264–266 | EtOH | 47 | | 28 |
| 5-(2-Me-Pyrazol-3-yl)-7-Cl | 196–197 | $CHCl_3$/Hexane | | | 27 |
| | 216–218 | EtOAc/Petr ether | | | 27 |
| 5-(4-Me-3-SH-1,2,4-Triazol-5-yl)-7-Cl | > 300 | | 61 | | 193 |
| 5-(1-$Me_2$N-Cyclohex-1-yl)-7-Cl | 222 | EtOH | 58 | | 224 |
| 5-(3-$Me_2$N-Propyl)-7-Cl | 209–211 | EtOH/$Et_2O$ | | | 1b |
| 5-$(Me_2N)_2$CH-7-Cl | 188d | EtOAc | 10 | | 224 |
| 5-MeOOC-7-Cl | 173–174 | | 32 | | 193 |
| 5-(4-$MeOOCCH_2O$-$C_6H_4$)-7-Cl | 235–243 | MeOH | | | 152c |
| 5-(6-MeS-2-Pyridyl)-7-Br | 238 | EtOH | | | 480 |
| 5-(6-MeOS-2-Pyridyl)-7-Br | 255 | EtOH | | | 480 |
| 5-(6-$MeO_2S$-2-Pyridyl)-7-Br | 248 | EtOH | | | 480 |
| 5-Pentyl-7-Cl | 104 | Cyclohexane | 75 | | 18 |
| 5,7-$Ph_2$ | 234–235 | PhH/Petr ether | 20 | | 2 |
| 5-Ph-7-Ac | 184–186 | PhH/Petr ether | | | 11 |
| | 192–193 | *i*-PrOH | | | 44 |
| 4-Oxide | 208–209 | Acetone | | | 32, 44 |
| 5-Ph-7-AcNH | | | | | |
| 4-Oxide | > 215 | MeOH/EtOH | | | 251c |

| | | | | | |
|---|---|---|---|---|---|
| 5-Ph-7-(Adamant-1-yl)CONH | 327–329d | MeCN | | | 33b |
| 5-Ph-7-$H_2N$ | 228–231 | EtOH | | | 136 |
| 4-Oxide | 274–275d | DMF/EtOH | 69 | | 252 |
| 5-Ph-7-$H_2NCO$ | 268–271 | | | uv | 5 |
| 5-Ph-7-($H_2N$-Hydroxyiminomethyl) | 273–275d | DMF/EtOH | | | 33b |
| 5-Ph-7-$H_2NSO_2$ | 287–288d | | 32 | | 98 |
| 5-Ph-7-$N_3$ | 174–175d | $CH_2Cl_2$/Hexane | 50 | | 252 |
| 4-Oxide | 186–188d | THF/Hexane | 83 | | 252 |
| 5-Ph-7-(Benzoyl)NH | 261–262 | MeOH | | | 251c |
| 5-Ph-7-Br | 220–221 | Acetone | 76 | | 2 |
| 4-Oxide | 230–231 | $CH_2Cl_2$/Petr ether | | | 91a |
| 5-Ph-7-(4-$BrC_6H_4CHN$) | 196–198 | | 50 | | 202 |
| 5-Ph-7-Bu | 145–147 | PhH/Petr ether | | | 33b |
| 5-Ph-7-*t*-Bu | 244–245 | EtOH | | | 479 |
| 5-Ph-7-BuS | | | | | |
| Hydrochloride | 247–249 | EtOH/MeCN | | | 7 |
| 5-Ph-7-$HOOCCH_2S$ | 230 | EtOH/$H_2O$ | | | 2b |
| 5-Ph-7-(1-Carboxy-1-ethyl) | 252–255 | $Et_2O$/Petr ether | 9 | ir, pmr | 20 |
| 5-Ph-7-Cl | 216–217 | Acetone | 52 | | 20 |
| 4-Oxide | 238–239 | EtOH | | | 30 |
| Hydrochloride | 251–252 | EtOH | | | 30 |
| Methiodide | 250–251 | Acetone | | | 30, 31 |
| Tosylate | 280–281 | PhH | 99 | | 498 |
| $BF_3 \cdot Et_2O$ Adduct of 4-oxide | 160–165d | | | | 235 |
| 5-Ph-7-(4-$ClC_6H_4$) | 271 | PhH | | | 481 |
| 5-Ph-7-(4-$ClC_6H_4CHN$) | 192–193 | | 72 | | 202 |
| 5-Ph-7-(5-Cl-2-$HOC_6H_3$)-$N_2$ | 288–290 | PhH | 54 | | 202 |
| 5-Ph-7-Diazonium | | | | | |
| Tetrafluoroborate | 227d | $HBF_4$ | | | 251b |
| 5-Ph-7-(1,3-Dioxolan-2-yl) | 153–154 | EtOH | | | 32 |
| 5-Ph-7-(1-Cl-Ethyl)S | 195–196 | EtOH | | | 7 |
| Hydrochloride | 236–238 | | | | 7 |
| 5-Ph-7-$ClCH_2S$ | | | | | |
| Hydrochloride | 258–260d | MeOH | | | 7 |

TABLE VII-1. —(contd.)

| Substituent | mp (°C) or; [bp (°C/torr)] | Solvent of Crystallization | Yield (%) | Spectra | Refs. |
|---|---|---|---|---|---|
| 5-Ph-7-CN | 256–257 | $MeNO_2$ | | | 5, 15 |
| 5-Ph-7-(Cyclohexyl)CONH | 188–190 | MeOH/Acetone/Hexane | | | 251c |
| 5-Ph-7-(1,3-Dihydro-2-oxo-5-Ph-2*H*-1,4-benzodiazepin-7-yl) | 348d | THF | | | 481 |
| | 238d | | | | 86b |
| 5-Ph-7-Et | 194–195 | PhH/Petr ether | | | 33b |
| 5-Ph-7-EtO | 174–176 | $CH_2Cl_2$/EtOH | | | 134 |
| 5-Ph-7-(2-EtOOC-*trans*-Ethen-1-yl) | 190–192 | EtOH | | | 33b |
| 5-Ph-7-EtS | | | | | |
| Hydrochloride | 273 | EtOH/MeCN | | | 7 |
| 5-Ph-7-(Et)OS | 195–196 | Acetone | | | 7 |
| 5-Ph-7-(4-$Et_2NC_6H_4$CHN) | 255–257 | | 40 | | 202 |
| 5-Ph-7-OCH | 155–158 | MeCN | | | 33b |
| Hydrazone | 219–221 | DMF/$H_2O$ | | | 33b |
| Oxime · 0.5EtOH | 240–242 | EtOH/$H_2O$ | | | 33b |
| Thiosemicarbazone | 266–268d | DMF/$H_2O$ | | | 33b |
| 5-Ph-7-F | 197–198 | Acetone/Petr ether | 76 | | 2 |
| 5-Ph-7-$F_2$CHO | 182–183 | PhMe | 70 | ir, ms, pmr, uv | 10 |
| 5-Ph-7-$F_2$CHS | 174–175 | PhMe | 84 | ir, ms, pmr, uv | 10 |
| 5-Ph-7-$F_2CHSO_2$ | 247–248 | EtOH | 46 | ir, ms, pmr, uv | 10 |
| 5-Ph-7-$F_3$C | 204–205 | PhH/Hexane | 25 | | 4 |
| 4-Oxide | 217–218 | PhH/Hexane | 84 | | 4 |
| 5-Ph-7-(2-$F_3$C-1,3-Dithiolan-2-yl) | 215–216 | $CH_2Cl_2$ | | | 257 |
| 5-Ph-7-(Formyl)NH | 227–229 | MeOH/Acetone/Hexane | | | 251c |
| 5-Ph-7-(2-Furyl)CHN | 168–169 | | 39 | | 202 |
| 5-Ph-7-(Hexafluoro-2-HO-prop-2-yl) | 271–273 | EtOAc/$Et_2O$ | | | 482 |
| 5-Ph-7-HO | 287–290 | MeCN | 65 | | 2 |
| 5-Ph-7-HONH | 186–187 | $CH_2Cl_2$/EtOH | | | 246 |
| 5-Ph-7-(1-HO-Butyl) | 172–174 | $Et_2O$/Pentane | | | 33a |

| | | | | | |
|---|---|---|---|---|---|
| 5-Ph-7-(1-HO-Ethyl) | 214–216 | $Et_2O$ | | | 33a |
| 4-Oxide | 193–195 | Acetone/Hexane | | | 33b |
| 5-Ph-7-(2-HO-Ethyl)S | | | | | |
| Hydrochloride | 252–253d | EtOH/*i*-PrOH | | | 7 |
| 5-Ph-7-(1-HO-Pentyl) | 171–174 | $Et_2O$/Pentane | | | 33b |
| 5-Ph-7-(4-$HOC_6H_4CHN$) | 276–277 | | 70 | | 202 |
| 5-Ph-7-(3-HO-Propanoyl) | 165–185 | $CH_2Cl_2$/Hexane | | | 33b |
| 5-Ph-7-(1-Propyl) | 179–181 | $Et_2O$ | | | 33a |
| 5-Ph-7-(2-HO-Prop-2-yl) | 227–229 | $Et_2O$ | | | 33a |
| 5-Ph-7-(3-HO-Propyn-1-yl) | 205–207 | $CHCl_3$/Hexane | | | 33b |
| 5-Ph-7-I | 226–228 | *i*-PrOH | | | 371 |
| 4-Oxide | 247–248 | EtOH | | | 371 |
| 5-Ph-7-Me | 208–209 | EtOH/Petr ether | 82 | | 2 |
| 4-Oxide | 235–236 | | | | 34 |
| | 226–227 | $CH_2Cl_2$/Petr ether | | | 91a |
| 5-Ph-7-(2-Me-1,3-Dioxolan-2-yl) | 250–252 | EtOH | | | 32 |
| 4-Oxide | 206–208 | Acetone/Hexane | | | 33b |
| 5-Ph-7-(2-Me-Propanoyl)NH | 246 | EtOAc/Hexane | | | 251c |
| 5-Ph-7-(2,2-$Me_2$-Propanoyl)NH | 177–180 | EtOAc/Hexane | | | 251c |
| 5-Ph-7-$MeNHSO_2$ | 293–296 | EtOH/DMF | | | 2b |
| 5-Ph-7-MeO | 217–218 | PhH | 42 | | 2 |
| 4-Oxide | 189–190 | Acetone/Hexane | | | 1b |
| 5-Ph-7-(4-MeO-Benzoyl)NH | 304–307 | MeOH | | | 251c |
| 5-Ph-7-MeOOC | 219–220 | MeOH | | uv | 5 |
| 5-Ph-7-MeONC | 193–195 | MeCN | | | 33b |
| 5-Ph-7-MeS | 216–218 | Acetone | | | 7 |
| 4-Oxide | 191–193 | EtOH/$H_2O$ | | | 7 |
| | 193–194 | Acetone | | uv | 218 |
| 5-Ph-7-(Me)OS | 254d | EtOH/MeCN | | | 7 |
| 5-Ph-7-(Me)$O_2$S | 256–258 | Acetone/ | | | 7 |
| 4-Oxide | 256–257d | DMF/EtOH | | | 33b |
| 5-Ph-7-$Me_2N$ | 245–247 | EtOAc | | | 95 |
| N-Oxide | 196–197d | EtOH/$Et_2O$ | | | 72c |
| 5-Ph-7-$Me_2NCH_2$ | 175–179 | $CH_2Cl_2$/Petr ether | | | 38b |

TABLE VII-1. —(contd.)

| Substituent | mp (°C) or; [bp (°C/torr)] | Solvent of Crystallization | Yield (%) | Spectra | Refs. |
|---|---|---|---|---|---|
| 5-Ph-7-$Me_2NCHN$ | | | | | |
| Dihydrochloride | 260–262d | EtOH | | | 2b |
| 5-Ph-7-(4-$Me_2NC_6H_4CHN$) | 240–243 | | 50 | | 202 |
| 5-Ph-7-$Me_2NSO_2$ | 226–227 | *i*-PrOH | 30 | | 98 |
| 5-Ph-7-(Morpholino)$CH_2$ | 219–221 | Acetone | | | 38b |
| 5-Ph-7-$NO_2$ | 224–226 | EtOH | | | 136 |
| 4-Oxide | 208–218d | EtOH | | | 139 |
| | 218–220d | EtOH/Petr ether | | | 136 |
| 5-Ph-7-(2-$O_2NC_6H_4CHN$) | 154–156 | | 55 | | 202 |
| 5-Ph-7-(3-$O_2NC_6H_4CHN$) | 237–238 | | 37 | | 202 |
| 5-Ph-7-(4-$O_2NC_6H_4CHN$) | 278–280 | | 50 | | 202 |
| 5-Ph-7-(4-$O_2N$-Benzoyl)NH | 279–281 | MeOH | | | 251c |
| 5-Ph-7-(Oxiran-2-yl) | 186d | $Et_2O$ | | | 33b |
| 5-Ph-7-(Pentanoyl | 63–66 | $Et_2O$/Pentane | | | 33b |
| 5-Ph-7-(PhCHN) | 168–170 | EtOH | 61 | ir | 202 |
| 5-Ph-7-(Piperidin-1-yl) | 250–252 | EtOH/Petr ether | | | 95 |
| 5-Ph-7-Propanoyl | 172–175 | $Et_2O$ | | | 33b |
| 5-Ph-7-(Propanoyl)NH | 251–252 | MeOH/Acetone/Hexane | | | 251c |
| 5-Ph-7-$Pr_2$(O)P | 214–215d | *i*-PrOH/$Et_2O$ | | | 483 |
| 5-Ph-7-(Pyrrol-1-yl) | 250–258 | MeOH | | | 67b |
| 5-Ph-7-(3,4,5-Triazatricyclo[5.2.1.0]dec-4-en-3-yl) | 197–198d | EtOAc | | | 33b |
| 5-Ph-7-Vinyl | 213–215 | THF | | | 33b |
| 5-(4-$AcOC_6H_4$)-7-Cl | 225–228 | $CH_2Cl_2$/Hexane | | | 152c |
| 5-(4-$H_2NC_6H_4$)-7-Cl | 262–266 | Acetone/$H_2O$ | | | 302 |
| 4-Oxide | 256–258 | THF/MeOH | 39 | ir, ms | 8 |
| 5-(4-$H_2NCOCH_2OC_6H_4$)-7-Cl | 235–243 | | 19 | ir | 8 |
| 5-(2-$BrC_6H_4$)-7-Cl | 208–209 | $CH_2Cl_2$/$Et_2O$ Petr ether | | | 2b |
| 5-(2-$BrC_6H_4$)-7-F | 194–196 | THF/EtOH | | | 134 |

| | | | | | |
|---|---|---|---|---|---|
| 5-(4-$BrC_6H_4$)-7-Br | | | | | |
| 4-Oxide | 260–261d | | | | 358 |
| 5-(4-$BrC_6H_4$)-7-Cl | 260–264 | $CHCl_3$/EtOH | | | 152c |
| 5-(2-$HOOCC_6H_4$)-7-Cl | 207–208d | THF/PhH | | | 2b |
| 5-(2-$ClC_6H_4$)-7-AcNH | 292–300 | THF/MeOH Petr ether | | | 152c |
| 5-(2-$ClC_6H_4$)-7-$H_2N$ | 230–232 | EtOH | | | 136 |
| 5-(2-$ClC_6H_4$)-7-$N_3$ | 186–187d | $CH_2Cl_2$/Hexane | 70 | | 252 |
| 5-(2-$ClC_6H_4$)-7-*t*-Bu | 274–275 | $Et_2O$ | | | 134 |
| 5-(2-$ClC_6H_4$)-7-Cl | 199–201 | EtOH | 75 | | 2 |
| 4-Oxide | 248–249 | PhH/Petr ether | | | 91a |
| Methiodide | 198–200d | MeCN | 93 | ir, pmr | 179 |
| 5-(2-$ClC_6H_4$)-7-CN | 232–233 | EtOH | | uv | 5, 15 |
| 5-(2-$ClC_6H_4$)-7-EtO | 228–230 | $CH_2Cl_2$/EtOAc | | | 134 |
| 5-(2-$ClC_6H_4$)-7-F | 197–199 | EtOH | | | 134 |
| 5-(2-$ClC_6H_4$)-7-HONH | 182–183 | $CH_2Cl_2$/MeOH/$Et_2O$ | | | 134 |
| 5-(2-$ClC_6H_4$)-7-Me | 223–224 | MeOH | | uv | 5, 91 |
| 5-(2-$ClC_6H_4$)-7-MeO | 224–228 | EtOAc/$Et_2O$ | | | 134 |
| 5-(2-$ClC_6H_4$)-7-MeS | 221–223 | EtOH | | | 7 |
| 5-(2-$ClC_6H_4$)-7-$Me_2N$ | 245–248 | EtOAc | | | 95 |
| 5-(2-$ClC_6H_4$)-7-$NO_2$ | 237–239 | $CH_2Cl_2$/EtOH | | | 12 |
| 4-Oxide | 254d | EtOH | 64 | | 8 |
| 5-(2-$ClC_6H_4$)-7-$HO_3SNH$ | | | | | |
| Ammonium salt | 284–288d | $H_2O$ | | | 134 |
| 5-(3-$ClC_6H_4$)-7-$NO_2$ | 182 | EtOH/Hexane | | | 301 |
| 5-(2,3-$Cl_2C_6H_3$)-7-Cl | 225–227 | $CH_2Cl_2$/Petr ether | | | 2b |
| 5-(2,4-$Cl_2C_6H_3$)-7-Cl | 231–233 | Acetone | 83 | | 500 |
| 5-(2,5-$Cl_2C_6H_3$)-7-Cl | 270–271 | Acetone/Petr ether | | | 2b |
| 5-(2,6-$Cl_2C_6H_3$)-7-Cl | 231–233 | $CH_2Cl_2$/$Et_2O$ | 51 | | 42 |
| 5-(2-Cl-6-$FC_6H_3$)-7-Cl | 234–235 | EtOH | | | 2b |
| 5-(2-Cl-4-$HOC_6H_3$)-7-Cl | 222–225 | | | | |
| | 280–290 | THF/Hexane | | | 152c |
| 5-(2-Cl-5-$HOC_6H_3$)-7-Cl | 292–294 | MeOH/EtOAc | | | 134 |
| 5-(2-Cl-4-$MeOC_6H_3$)-7-Cl | 229–230 | $CH_2Cl_2$/EtOH | | | 134 |
| 5-(2-Cl-5-$MeOC_6H_3$)-7-Cl | 221–223 | EtOAc/Hexane | | | 134 |

TABLE VII-1. —(contd.)

| Substituent | mp (°C) or; [bp (°C/torr)] | Solvent of Crystallization | Yield (%) | Spectra | Refs. |
|---|---|---|---|---|---|
| 5-(2-Cl-5-$O_2NC_6H_3$)-7-$NO_2$ | 330–331d | $H_2SO_4/H_2O$ | | | 258 |
| 5-(3-$ClC_6H_4$)-7-Cl | 247–248 | EtOH | 70 | | 2 |
| 5-(3,4-$Cl_2C_6H_3$)-7-Cl | 245–247 | $CH_2Cl_2$/MeOH | | | 92 |
| 5-(3,5-$Cl_2C_6H_4$)-7-Cl | 237–238 | Acetone | | | 2b |
| 5-(4-$ClC_6H_4$)-7-Br | | | | | |
| 4-Oxide | 260–261d | | | | 34 |
| 5-(4-$ClC_6H_4$)-7-Cl | 247–248 | EtOH | | | 91a |
| 5-(4-$ClC_6H_4$)-7-F | 230–232 | Acetone | 92 | | 500 |
| 5-(4-$ClC_6H_4$)-7-MeO | 219–221 | $CH_2Cl_2$/EtOH | | | 134 |
| 5-(4-$ClC_6H_4$)-7-$NO_2$ | 253–254 | $CH_2Cl_2$ | | | 136, 86b |
| 4-Oxide | 250–252 | MeOH | | | 91a |
| 5-(3-Cl-2-Pyridyl)-7-Cl | 221–222d | $CH_2Cl_2$ | | | 484 |
| 5-(2-$EtOC_6H_4$)-7-EtO | 184–187 | EtOAc/Hexane | | | 134 |
| 5-(2-$FC_6H_4$)-7-Ac | 211–213 | $Et_2O$/Petr ether | | | 11 |
| 4-Oxide | 215–216d | Acetone | | | 2b |
| 5-(2-$FC_6H_4$)-7-AcNH | | | | | |
| 4-Oxide | 280–300 | $CH_2Cl_2$/EtOH | | | 301 |
| 5-(2-$FC_6H_4$)-7-AcS | 168–170 | $Et_2O$ | | | 301 |
| 5-(2-$FC_6H_4$)-7-$H_2N$ | 264–266 | EtOH | | | 136 |
| 5-(2-$FC_6H_4$)-7-($H_2N$-Hydroxyiminomethyl) | 261–263d | DMF | | | 33b |
| 5-(2-$FC_6H_4$)-7-$H_2NCH_2$ | | | | | |
| Picrate | 194–198 | THF/MeOH | | | 152c |
| 5-(2-$FC_6H_4$)-7-$N_3$ | 172–173 | PhH | | | 485 |
| 5-(2-$FC_6H_4$)-7-Br | 186–187 | Acetone/Petr ether | 84 | | 2 |
| 5-(2-$FC_6H_4$)-7-(Butanoyl)NH | 200–204 | EtOAc/Petr ether | | | 301 |
| 5-(2-$FC_6H_4$)-7-Cl | 205–206 | PhH/Hexane | 90 | | 2 |
| 4-Oxide | 220–223 | Acetone/MeOH Petr ether | 67 | | 233 |
| 5-(2-$FC_6H_4$)-7-CN | 239–240 | MeOH | | uv | 5 |
| 2-(2-$FC_6H_4$)-7-(Cyclopentyl)CONH | 165–170 | EtOAc/$Et_2O$ | | | 301 |

| | | | | |
|---|---|---|---|---|
| 5-(2-$FC_6H_4$)-7-(Cyclopropyl)CONH | 278–280 | EtOAc | | 301 |
| 5-(2-$FC_6H_4$)-7-Et | 199–201 | PhH/Petr ether | | 33b |
| 4-Oxide | 173–175 | $CH_2Cl_2$/$Et_2O$ | | 33b |
| 5-(2-$FC_6H_4$)-7-EtNH | 176–178 | $Et_2O$ | | 301 |
| 5-(2-$FC_6H_4$)-7-EtO | 187–190 | EtOAc/Hexane | | 134 |
| 5-(2-$FC_6H_4$)-7-$EtOCS_2$ | 182–184 | $Et_2O$ | | 301 |
| 5-(2-$FC_6H_4$)-7-F | 197–200 | Acetone | 93 | 500 |
| 5-(2-$FC_6H_4$)-7-(1,1-$F_2$-Ethyl) | 210–212 | EtOH | | 360 |
| 5-(2-$FC_6H_4$)-7-[5-(2-$FC_6H_4$)-1,3-Dihydro-2-oxo-2*H*-1,4-benzodiazepin-7-yl]thio | 250–256d | EtOH | | 301 |
| 5-(2-$FC_6H_4$)-7-OCHNH | 250–254 | EtOAc | | 301 |
| 5-(2-$FC_6H_4$)-7-HO | 310–311d | EtOAc/$CH_2Cl_2$ | | 301 |
| 5-(2-$FC_6H_4$)-7-(1-HO-Ethyl) | 222–224 | $Et_2O$ | | 33a |
| 4-Oxide·0.6 $H_2O$ | 125–140 | $Et_2O$ | | 33b |
| 5-(2-$FC_6H_4$)-7-(2-HO-Ethyl)NHCONH | Amorphous | | | 301 |
| 5-(2-$FC_6H_4$)-7-(1-HO-Iminoethyl) | 232–233 | EtOAc | | 251b |
| 5-(2-$FC_6H_4$)-7-(2-HO-prop-2-yl) | 230–232 | $Et_2O$ | | 33a |
| 5-(2-$FC_6H_4$)-7-HS | 194–198d | $Et_2O$ | | 301 |
| 5-(2-$FC_6H_4$)-7-I | 222–224 | EtOH | | 26 |
| 5-(2-$FC_6H_4$)-7-MeNHCONH | 183–190d | EtOAc/$Et_2O$ | | 301 |
| 5-(2-$FC_6H_4$)-7-[1-(4-$MeC_6H_4SO_2$-Hydrazono-ethyl]·$2H_2O$ | 240–242 | EtOH | | 33b |
| 4-Oxide | 192–196 | EtOH | | 141 |
| 5-(2-$FC_6H_4$)-7-MeO | 195–198 | $CH_2Cl_2$/EtOAc | | 134 |
| 5-(2-$FC_6H_4$)-7-MeS | 193–195 | MeOH/$Et_2O$ | | 192 |
| 4-Oxide | 186–188d | $CH_2Cl_2$/Hexane | | 192 |
| 5-(2-$FC_6H_4$)-7-MeS(O) | 240–242d | $CH_2Cl_2$/$Et_2O$ | | 301 |
| 5-(2-$FC_6H_4$)-7-$MeSO_2$ | | | | |
| 4-Oxide | 240–245d | EtOAc | | 301 |
| 5-(2-$FC_6H_4$)-7-$NO_2$ | 222–224 | EtOH | | 12 |
| | 210–211 | Acetone | | 136 |
| 4-Oxide | 232–233 | THF/$Et_2O$ | | 2b |
| 5-(2-$FC_6H_4$)-7-(Octadecanoyl)NH | 116–118 | $Et_2O$ | | 301 |

TABLE VII-1. —(contd.)

| Substituent | mp (°C) or; [bp (°C/torr)] | Solvent of Crystallization | Yield (%) | Spectra | Refs. |
|---|---|---|---|---|---|
| 5-(2,6-$F_2C_6H_3$)-7-Cl | 251–253 | EtOH | | | 23 |
| 4-Oxide | 290–300d | THF/ETOH | | | 152c |
| 5-(2,6-$F_2C_6H_3$)-7-$NO_2$·MeOH | 237–243 | MeOH | | | 152c |
| 5-(2-F-5-$H_2NC_6H_3$)-7-Cl | 212 | Acetone | | | 259 |
| 5-(2-F-4-$HOC_6H_3$)-7-Cl | 298–303 | $CH_2Cl_2$/MeOH | | | 152c |
| 5-(2-F-6-$HOC_6H_3$)-7-Cl | 210–220 | | | | |
| | 239–244 | $CH_2Cl_2$/MeOH | | | 152c |
| 5-(2-F-5-$IC_6H_3$)-7-Cl | 180 | MeOH/$Et_2O$/Hexane | | | 259 |
| 5-(2-F-6-$Me_2NC_6H_3$)-7-Cl | 204–208 | $CH_2Cl_2$/Hexane | | | 152c |
| 5-(2-F-4-$MeOC_6H_3$)-7-Cl | 208–211 | EtOAc/MeOH | | | 134 |
| 5-(2-F-6-$MeOC_6H_3$)-7-Cl | 207–213 | $CH_2Cl_2$/Hexane | | | 152c |
| 5-(2-F-6-$MeSC_6H_3$)-7-Cl | 205–209 | $CH_2Cl_2$/Petr ether | | | 152c |
| 5-(2-F-5-$O_2NC_6H_3$)-7-Cl | 275 | MeOH/$C_6H_6$ | | | 259 |
| 5-(2-F-5-$O_2NC_6H_3$)-7-$NO_2$ | 295–298 | Acetone | 57; 81 | | 152a, 256 |
| 5-(3-$FC_4H_4$)-7-Cl | 200–201 | Acetone | 47 | | 2 |
| 5-(2-$F_3CC_6H_4$)-7-$H_2N$ | 197–198 | PhH | | | 6 |
| | 221–222 | | | | 6 |
| 5-(2-$F_3CC_6H_4$)-7-$N_3$ | 178–179d | $CH_2Cl_2$/Hexane | 67 | | 252 |
| 5-(4-$FC_6H_4$)-7-Cl | 223–224 | Acetone/Hexane | 75 | | 2 |
| 5-(4-$FC_6H_4$)-7-F | 184–185 | Acetone/Hexane | | | 152c |
| 5-(3-F-2-Pyridyl)-7-Cl | 220–221 | $CH_2Cl_2$/Hexane | | | 484 |
| 5-(4-$F_3CONHC_6H_4$)-7-Cl | 273–276 | $CH_2Cl_2$/$Et_2O$ | 43 | | 8 |
| 4-Oxide | 295–297d | $CH_2Cl_2$/MeOH | 66 | ir | 8 |
| 5-(2-$F_3CC_6H_4$)-7-Br | 183–185 | Acetone/Hexane | | | 6 |
| 5-(2-$F_3CC_6H_4$)-7-Cl | 190–192 | PhH/Heptane | | | 6 |
| 5-(2-$F_3CC_6H_4$)-7-$F_3C$ | 226–227 | PhH/Hexane | 43 | | 4 |
| 5-(2-$F_3CC_6H_4$)-7-MeS | 199–200 | PhH | | | 7 |
| 5-(2-$F_3CC_6H_4$)-7-$Me_2N$ | 254–256 | MeOH | | | 95 |
| 5-(2-$F_3CC_6H_4$)-7-$NO_2$ | 233–234 | Acetone/MeOH | | | 6 |

| | | | | | |
|---|---|---|---|---|---|
| 5-(4-$H_2NNHCOCH_2O$)$C_6H_4$-7-Cl·0.5 mol THF | 180–185 | | | | |
| | 210–220 | THF/Hexane | 59 | ir | 8 |
| 5-(2-$HOC_6H_4$)-7-Cl | 285–287 | MeCN | 47 | | 2 |
| 5-(2-HO-4-$FC_6H_3$)-7-Cl | 290–295 | $CHCl_3$/EtOH | | | 152c |
| 5-(4-$HOC_6H_4$)-7-Cl | 271–272 | Acetone | 67 | ir | 8 |
| 4-Oxide | 280–281d | Dioxane/$H_2O$ | | | 1b |
| 5-(4-$MeOOCCH_2OC_6H_4$)-7-Cl | 272–275 | EtOH/$CHCl_3$ | 41 | ir | 8 |
| 5-(2-$MeOC_6H_4$)-7-Cl | 205–207 | EtOH/$H_2O$ | 45 | | 2 |
| 5-(2-$MeOC_6H_4$)-7-MeO | 189–192 | EtOAc/Hexane | | | 134 |
| 5-(3-$MeOC_6H_4$)-7-Cl | 219–220 | Acetone/Hexane | 54 | | 2 |
| 5-(4-$MeOC_6H_4$)-7-Cl | 212–214 | PhH/Hexane | 52 | | 2 |
| 5-(1-Me-1-Pr)-7-Cl | 136 | *i*-$Pr_2O$ | 71 | | 18 |
| 5-(2-$MeC_6H_4$)-7-Cl | 180–181 | $Et_2O$ | 88 | | 2 |
| 5-(2-$MeC_6H_4$)-7-F | 175–176 | | | | |
| | 185–186 | EtOAc/Hexane | | | 134 |
| 5-(2,4-$Me_2C_6H_3$)-7-Cl | 210–212 | $Et_2O$/Hexane | | | 92 |
| 5-(3,4-$Me_2C_6H_3$)-7-Cl | 240–242 | $CH_2Cl_2$/Hexane | | | 92 |
| 5-(3-$MeC_6H_4$)-7-Cl | 198–199 | PhH/Petr ether | 48 | | 2 |
| 5-(4-$MeC_6H_4$)-7-Br | 239–240 | Acetone | 66 | | 2 |
| 4-Oxide | 237–238 | $CH_2Cl_2$/Petr ether | | | 91a |
| 5-(2-$Me_2NC_6H_4$)-7-Cl | 239–240 | $CH_2Cl_2$/Hexane | | | 152c |
| 5-(2-$MeSC_6H_4$)-7-Cl | 184–185 | Acetone/Hexane | | | 7 |
| 5-(Morpholino)CO-7-Br | 237–238 | EtOAc | | | 192 |
| 5-(2-$O_2NC_6H_4$)-7-Br | 220–222 | PhH/Hexane | | | 162 |
| 5-(2-$O_2NC_6H_4$)-7-$NO_2$ | 226–228 | THF | | | 136, 86b |
| 5-(3-$O_2NC_6H_4$)-7-Cl | 234–238 | MeOH | 54 | | 256 |
| 5-(4-$O_2NC_6H_4$)-7-Cl | 248–250 | MeCN | | | 302 |
| 5-(3-$O_2NC_6H_4$)-7-$NO_2$ | 250–254 | Acetone/Hexane | 72 | | 256 |
| 4-Oxide | 243–244d | DMF/EtOH | | | 33b |
| 5-(4-$PhC_6H_4$)-7-Cl | 272–275 | EtOAc | | | 92 |
| 5-(2-Ph-Ethenyl)-7-Cl | 210–212 | MeOH/$H_2O$ | | | 479 |
| 5-$PhSO_2CH_2$-7-Cl | 236–238 | MeCN | | | 302 |
| 5-(Piperidin-1-yl)CO-7-Br | 183–186 | EtOAc | | | 192 |
| 5-(Piperidin-1-yl)CO-7-$NO_2$ | 225–228 | Acetone | | | 192 |

TABLE VII-1. —(contd.)

| Substituent | mp (°C) or; [bp (°C/torr)] | Solvent of Crystallization | Yield (%) | Spectra | Refs. |
|---|---|---|---|---|---|
| 5-[2-(Piperidin-1-yl)-$C_6H_4$]-7-Cl | 239–240 | EtOH | | | 152c |
| 5-Pr-7-Cl | 168 | EtOAc | 80 | | 18 |
| 5-*i*-Pr-7-Cl | 153–154 | $Et_2O$ | 81 | | 18 |
| 5-(2-Pyrazinyl)-7-Cl | 184–185d | MeOH | 84 | ir, ms, pmr | 28 |
| 5-(2-Pyridyl)-7-Br | 237–239 | Acetone | 20 | | 3 |
| 3:1 Complex with $FeCl_2$ | 291–295d | EtOH/AcOH | | | 33b |
| 2:1 Complex with $FeCl_3$ | 234d | AcOH | | | 33b |
| 2:1 Complex with $ZnCl_2$ | 280–282d | AcOH | | | 33b |
| 1′-Oxide | 256–258d | $CH_2Cl_2$/THF/EtOH | | | 152c |
| 4-Oxide | 263d | EtOH | | | 247 |
| 4,1′-Dioxide | 252–253d | DMF/EtOH | | | 33b |
| 5-(2-Pyridyl)-7-Cl | 225–226d | Acetone | 57 | | 3 |
| 5-(2-Pyridyl)-7-$F_3C$ | 242–244 | MeOH | | | 1b |
| 5-(2-Pyridyl)-7-I | 253–254d | EtOH | | | 371 |
| 5-(2-Pyridyl)-7-MeS | 203–205 | Acetone | | | 192 |
| 5-(2-Pyridyl)-7-$NO_2$ | 254–255 | Acetone | 45 | | 3 |
| 5-(2-Pyridyl)$CH_2$-7-Cl | 204–205 | PhH | | | 94c |
| 5-(4-Pyridyl)-7-Br | 228–229 | Acetone | 39 | | 3 |
| 5-(4-Pyridyl)-7-$NO_2$ | 242–243 | Acetone | 43 | | 3 |
| 5-(2-Pyrimidyl)-7-Cl | 241–242 | $CHCl_2$/Hexane | 36 | | 28 |
| 5-(4-Pyrimidyl)-7-Cl | 256–257d | EtOAc | 51 | | 28 |
| 5-[(1-Pyrrolidin-1-yl)-cyclopent-1-yl]-7-Cl | 241 | EtOH | 70 | | 224 |
| 5-(2-Pyrrolyl)-7-Cl | 262–263 | EtOH | | | 22 |
| 5-(2-Thiazolyl)-7-Cl | 259–260 | EtOAc/Petr ether | | | 27 |
| 5-(2-Thiazolyl)-7-I | 269–271d | $CH_2Cl_2$/MeOH | | | 27 |
| 5-(2-Thienyl)-7-Cl | 212–214 | PhH/Hexane | | | 22 |
| 4-Oxide | 252–254d | MeOH | | | 22, 34 |
| 5-(2-Thienyl)-7-MeS | 235–236 | Acetone | | | 2b |
| 5-(2-Thienyl)-7-$NO_2$ | 265–266 | MeCN | | | 22 |
| 5-(3-Thienyl)-7-Br | 238–241 | $CH_2Cl_2$/EtOH | | | 134 |

| | | | | |
|---|---|---|---|---|
| ***5,8-Disubstituted*** | | | | |
| 5-Me-8-Cl | 203–204 | EtOAc/*i*-$Pr_2O$ | 94 | 19 |
| 5-Ph-8-Cl | 214–215 | Acetone | 30 | 2 |
| 5-Ph-8-$F_3C$ | 184–186 | PhH/Petr ether | 47 | 4 |
| | 216–218 | | | 86b |
| 5-Ph-8-Me | 255–256 | MeOH/MeCN | | 91 |
| 5-Ph-8-MeO | 186–188 | Acetone/Hexane | 47 | 2 |
| | 190–192 | Acetone/Hexane | | 16 |
| 5-Ph-8-$NO_2$ | 252d | | | 86b |
| | 262d | EtOH | | 136 |
| 5-(2-$ClC_6H_4$)-8-F | 239–242 | EtOAc/Hexane | | 134 |
| 5-(2-$ClC_6H_4$)-8-$NO_2$ | 270d | EtOAc | | 258 |
| 5-(2-$FC_6H_4$)-8-$H_2N$ | 250–253 | $H_2O$ | | 257 |
| 5-(2-$FC_6H_4$)-8-F | 177–179 | EtOAc/Hexane | | 134 |
| 5-(2-$FC_6H_4$)-8-Me | 208–210 | $CH_2Cl_2$/EtOH | | 257 |
| 5-(2,6-$F_2C_6H_3$)-8-Cl | 200–202 | MeOH | | 152c |
| 5-(2-F-5-$O_2NC_6H_3$)-8-AcNH | 285–287 | THF/MeOH/$Et_2O$ | | 257 |
| 5-(2-Thienyl)-8-MeO | 213–214 | PhH/Hexane | | 22 |
| ***5,9-Disubstituted*** | | | | |
| 5-Ph-9-Cl | 175–176 | Hexane/Acetone | 25 | 2 |
| 5-Ph-9-Me | 184–185 | $CH_2Cl_2$/Petr ether | | 91 |
| 5-Ph-9-$NO_2$ | 146–147 | | | 86b |
| | 144–145 | EtOH | | 136 |
| 5-(2-$ClC_6H_4$)-9-$NO_2$ | 198–200 | $CH_2Cl_2$/$Et_2O$/Petr ether | | 301 |
| ***Trisubstituted*** | | | | |
| ***1,3,5-Trisubstituted*** | | | | |
| 1-Ac-3-AcO-5-(2-$ClC_6H_4$) | 153–156 | EtOAc/Hexane | | 134 |
| 1-Allyl-3-R-Me-5-(4-$ClC_6H_4$) Hydrochloride | 193–195d | EtOAc/EtOH | | 134 |
| 1-Allyl-3-S-Me-5-(4-$ClC_6H_4$) Hydrochloride | 190–194d | EtOAc/EtOH | | 134 |
| 1-*t*-Bu-3-Me-5-(2-$ClC_6H_4$) | 115–117 | $Et_2O$/Petr ether | | 391b |

TABLE VII-1. —(contd.)

| Substituent | mp (°C) or; [bp (°C/torr)] | Solvent of Crystallization | Yield (%) | Spectra | Refs. |
|---|---|---|---|---|---|
| 1-Me-3-Allyl-5-Me | 69–71 | $Et_2O$/Hexane | | | 134 |
| 4-Oxide | 138–141 | EtOAc/Hexane | 57 | pmr | 298 |
| 1-Me-3-Allyl-5-$AcOCH_2$ | 82–84 | $Et_2O$/Hexane | 60 | ir, pmr, uv | 298 |
| 1-Me-3-Allyl-5-CHO | 68–71 | $Et_2O$/Hexane | 55 | ir, pmr, uv | 298 |
| 1-Me-3-Allyl-5-$HOCH_2$ | 100–106 | $CH_2Cl_2$/$Et_2O$/Hexane | 61 | ir, pmr, uv | 298 |
| 1-Me-3-S-Benzyl-5-Ph | 135–137 | Acetone/$H_2O$ | | [α] | 81 |
| 1-Me-3-EtOOC-5-Ph | | | | | |
| 4-Oxide | 188–190 | EtOH | | | 108 |
| 1-Me-3-S-Me-5-(4-$ClC_6H_4$) | 115–117 | $Et_2O$/Hexane | | | 134 |
| 1,3-$Me_2$-5-(2,4-$Cl_2C_6H_3$) | 151–152 | $CH_2Cl_2$/EtOH | | | 134 |
| ***1,5,6-Trisubstituted*** | | | | | |
| 1-Me-5-(2-$FC_6H_4$)-6-Cl | 142–143 | $Et_2O$ | | | 301 |
| 1-Me-5-(2-F-5-$H_2NSO_2C_6H_3$)-6-Cl | 246–247 | EtOAc/$Et_2O$ | | | 301 |
| ***1,5,7-Trisubstituted*** | | | | | |
| 1-Ac-5-Ph-7-$NO_2$ | 150 | $Et_2O$ | | | 301 |
| 1-Ac-5-(2-$ClC_6H_4$)-7-Cl | 95 | $Et_2O$/Petr ether | | | 301 |
| 1-Ac-5-(2-$ClC_6H_4$)-7-$NO_2$ | 142 | $Et_2O$/Petr ether | | | 301 |
| 1-Ac-5-(2-$FC_6H_4$)-7-$Ac_2N$ | 184–186 | $Et_2O$ | | | 301 |
| 1-($Ac_4$-β-D-glucopyranosyl)-5-Ph-7-Cl | 90d | Cyclohexane | | | 301 |
| 1-AcNH-5-(2-$ClC_6H_4$)-7-Cl | 221–223 | $Et_2O$ | 65 | | 250 |
| 1-(2-AcNH-Ethyl)-5-(2-$FC_6H_4$)-7-Cl | 223–225 | THF | | | 152c |
| 1-$AcCH_2$-5-Ph-7-Cl | 169–171 | Acetone | | | 293, 284 |
| 1-$AcCH_2$-5-(2-Pyridyl)-7-Br | 156–158 | $CH_2Cl_2$/Hexane | | | 284 |
| 1-(2-AcO-Ethoxy)$CH_2$-5-Ph-7-Cl | 92–94 | $Et_2O$/Hexane | | | 72 |
| 1-(2-AcO-1-MeO-Ethyl)-5-Ph-7-AcNH | 166–167 | PhH | | | 486 |
| 1-(2-AcO-1-MeO-Ethyl)-5-Ph-7-Cl | 110–111 | EtOH | | | 72 |
| 1-(2-AcO-Ethyl)-5-Ph-7-Cl | 102–103 | | | | 86b |
| Hydrochloride | 188–195d | | | | 88 |

| | | | | | |
|---|---|---|---|---|---|
| 1-(2-AcO-Ethyl)-5-(2-$ClC_6H_4$)-7-Cl | | | | | |
| 4-Oxide | 161–163 | *i*-PrOH/*i*-$Pr_2O$ | 62 | | 365 |
| 1-(2-AcO-Ethyl)-5-(2-$FC_6H_4$)-7-Cl | 103–105 | $CH_2Cl_2$/Hexane | | | 283, 86b |
| 4-Oxide | 161–163 | PrOH | 58 | | 282, 365 |
| 1-[2,3-$(AcO)_2$-Propyl]-5-Ph-7-Cl | 139–141 | | | | 285 |
| 1-(2-AcO-3-MeO-Propyl)-5-Ph-7-Cl | 121–123 | | | | 285 |
| 1-(4-AcO-2-Me-But-2-en-1-yl)-5-Ph-7-Cl | 121–122 | MeOH | | | 486 |
| 1-(4-AcO-*trans*-But-2-en-1-yl)-5-(2-$FC_6H_4$)-7-Cl | | | | ir, pmr | 236 |
| 1-Allyl-5-Ph-7-Cl | 105–106 | Hexane | 57 | | 2 |
| 4-Oxide | 150–151 | Acetone/Petr ether | | | 91a |
| 1-Allyl-5-Ph-7-$F_3C$ | 127–128 | $Et_2O$/Petr ether | | | 2b |
| 1-Allyl-5-Ph-7-$NO_2$ | 124–125 | $Et_2O$ | | | 2b |
| 1-Allyl-5-(2-$ClC_6H_4$)-7-Cl | 128–130 | EtOH | 40 | | 2 |
| 1-Allyl-5-(4-$ClC_6H_4$)-7-Cl | 145–146 | $CH_2Cl_2$/Petr ether | 85 | | 2 |
| 1-Allyl-5-(2-$FC_6H_4$)-7-Cl | 126–127 | *i*-PrOH | | | 236 |
| 4-Oxide | 183 | EtOH | | | 236 |
| 1-Allyl-5-(2-$FC_6H_4$)-7-(cyclopentyl)CONH | 176–178 | $Et_2O$/Hexane | | | 301 |
| 1-Allyl-5-(2-$FC_6H_4$)-7-$NO_2$ | 150–151 | EtOAc | | | 251c |
| 1-Allyl-5-(4-$MeOC_6H_4$)-7-Cl | 125–126 | EtOAc/Hexane | | | 134 |
| 1-Allyl-5-Cyclohexyl-7-Cl | 111–112 | Hexane | 75 | | 18 |
| 1-(Allyl)NHCO-5-Ph-7-Cl | 102–105 | *i*-PrOH | 67 | | 312 |
| | | | 89 | | 311 |
| 1-(Allyl)NHCO-5-Ph-CN | 137–139d | *i*-PrOH | 75 | | 311, 312 |
| 1-(Allyl)$NHCOCH_2$-5-Ph-7-Cl | 185–186 | EtOH | | | 293 |
| 1-(2-Allyloxy)ethyl-5-Ph-7-Cl | 91–93 | | | | 86d, 128 |
| 4-Oxide | 138–139 | | | | 234 |
| 1-(2-Allyloxy)ethyl-5-(2-$FC_6H_4$)-7-Cl | 88–90 | | | | 86d, 128 |
| 4-Oxide | 108–110 | | | | 234 |
| 1-$H_2N$-5-Ph-7-Cl | 70–75 | | | | 261 |
| 4-Oxide | 225 | $CH_2Cl_2$/PhH | 64 | ir | 250 |
| 1-$H_2N$-5-Ph-7-$NO_2$ | 155–157d | PhH | 44 | ir | 250 |
| 4-Oxide | 200–203d | PhH | 20 | ir | 250 |
| 1-$H_2N$-5-(2-$ClC_6H_4$)-7-Cl | 202–204 | MeOH | 78 | ir | 250, 261 |

TABLE VII-1. —(contd)

| Substituent | mp (°C) or; [bp (°C/torr)] | Solvent of Crystallization | Yield (%) | Spectra | Refs. |
|---|---|---|---|---|---|
| 1-$H_2N$-5-(2-$ClC_6H_4$)-7-$NO_2$ | 207–212 | $CH_2Cl_2$/Petr ether | | | 152c |
| 1-$H_2N$-5-(2-Pyridyl)-7-Br | 149–151 | MeOH | | | 247 |
| 1-[2-($H_2N$-Acetoxy)-1-MeO-ethyl]-5-Ph-7-Cl | 115–116 | $Et_2O$/Cyclohexane | | | 486 |
| 1-(4-$H_2NCO$-Butyl)-5-Ph-7-Cl | 143–145 | $CH_2Cl_2$/$Et_2O$ | | | 302 |
| 1-(4-$H_2NCO$-Butyl)-5-(2-$FC_6H_4$)-7-Cl | 102–104 | $CH_2Cl_2$/$Et_2O$ | | | 302 |
| 1-[2-(N-$H_2NCO$-N-Et-Amino)ethyl]-5-(2-$FC_6H_4$)-7-Cl | 185–188 | $CH_2Cl_2$/$Et_2O$ | 90 | | 233 |
| 1-$H_2NCOCH_2$-5-Ph-7-Cl | 233–235 | Acetone | | | 293, 86b |
| 1-$H_2NCOCH_2$-5-Ph-7-CN | 239–242 | Acetone | | | 38b |
| 1-$H_2NCOCH_2$-5-(2-$FC_6H_4$)-7-Cl | 198–201 | MeOH/$Et_2O$/Petr ether | | | 152c |
| 1-[2-(N-[1-$H_2NCO$-1-Ethyl]-N-Me-amino)-ethyl]-5-Ph-7-Cl | 143–145 | EtOH | | | 150c |
| 1-[2-(N-2-$H_2NCO$-Ethyl-N-Me-amino)ethyl]-5-Ph-7-Cl | 140–143 | EtOH | | | 150d |
| 1-[2-($H_2NCO$-Methoxy)ethyl]-5-Ph-7-Cl | 97–100 | MeCN | | | 150d |
| 1-[2-($H_2NCO$-Methoxy)ethyl]-5-(2-$ClC_6H_4$)-7-Cl | 198–200 | EtOH | | | 150d |
| 1-[2-($H_2NCO$-Methoxy)ethyl]-5-(2-$FC_6H_4$)-7-Cl | 187–189 | EtOH | | | 150c |
| 1-[2-($H_2NCO$-Methoxy)ethyl]-5-(2-pyridyl)-7-Br | 143–144 | EtOH | | | 150c, d |
| 1-[2-(N-$H_2NCOCH_2$-N-Et-Amino)ethyl]-5-Ph-7-Cl | 116–118 | MeCN | | | 150c, d |
| 1-[2-(N-$H_2NCOCH_2$-N-Me-Amino)ethyl]-5-Ph-7-$H_2N \cdot$EtOH | 110–112 | EtOH | | | 302 |
| 1-[2-(N-$H_2NCOCH_2$-N-Me-Amino)ethyl]-5-Ph-7-Cl | 146–148 | Acetone | | | 150c, d |
| 4-Oxide | 162–164d | MeCN | | | 150c |
| 1-[2-(N-$H_2NCOCH_2$-N-Me-Amino)ethyl)-5-Ph-7-$NO_2$ | 215–218d | MeCN | | | 150c, d |
| 1-[2-(N-$H_2NCOCH_2$-N-Me-Amino)ethyl]-5-(2-$ClC_6H_4$)-7-Cl | 152–154 | $CH_2Cl_2$/$Et_2O$ | | | 150c, d |
| | 130–132 | EtOH | | | 302 |
| 4-Oxide | 162–164d | | | | 150d |
| 1-[2-(N-$H_2NCOCH_2$-N-Me-Amino)ethyl]-5-(2-$FC_6H_4$)-7-Cl | 130–132 | MeCN | | | 150c, d |

| | | | | |
|---|---|---|---|---|
| 1-[2-(N-$H_2NCOCH_2$-N-Me-Amino)ethyl]-5-(2-$FC_6H_4$)-7-$NO_2$ | 187–189 | EtOH | | 150c, d |
| 1-[2-(N-$H_2NCOCH_2$-N-Me-Amino)ethyl]-5-(2-pyridyl)-7-Br | 176–178 | EtOH | | 150c, d |
| 1-[3-(N-$H_2NCOCH_2$-N-Me-Amino)propyl]-5-Ph-7-Cl | 136–138 | $CH_2Cl_2/Et_2O$ | | 150c, d |
| 1-[3-($H_2NCOCH_2S$)Propyl]-5-Ph-7-Cl | 135–137 | EtOH | | 150d, 263 |
| 1-(2-$H_2N$-Ethoxy)$CH_2$-5-Ph-7-Cl | | | | |
| Hydrochloride | 147–150 | EtOAc/MeOH | | 72 |
| 1-(2-$H_2N$-Ethyl)-5-(2-$FC_6H_4$)-7-Cl | | | | |
| Dihydrochloride | 218–221d | EtOH | | 233 |
| 1-(2-$H_2N$-Ethyl)-5-(2-F-5-$IC_6H_3$)-7-Cl | | | | |
| Hydrobromide | | | | 259 |
| 1-(2-$H_2NCO$-Ethyl)-5-Ph-7-Cl | 200–201 | Acetone | | 293 |
| 1-(2-$H_2NCOCH_2O$-Ethyl)-5-Ph-7-Cl | 90–110 | $Et_2O/CH_2Cl_2$/Petr ether | | 302 |
| 1-[2-$H_2NCO$(Me)N-Ethyl]-5-Ph-7-$F_3C$ | 175–176 | $CH_2Cl_2/Et_2O$ | | 152c |
| 1-[3-$H_2NCO$(Me)N-Propyl]-5-Ph-7-$NO_2$ | 169–172 | $CH_2Cl_2/Et_2O$ | | 152c |
| 1-[3-$H_2NCO$(Me)N-Propyl]-5-(2-pyridyl)-7-Br | 196–200 | $CH_2Cl_2$/MeOH Ether | | 152c |
| 1-(Benzoyl)$CH_2$-5-Ph-7-Cl | 174–175 | EtOH | | 293, 284 |
| 1-Benzyl-5-Ph-7-Cl | 105–106 | Hexane | 57 | 2 |
| | 174–175 | $CH_2Cl_2/Et_2O$/Peter ether | | 91a |
| 4-Oxide | 151–152 | $CH_2Cl_2/Et_2O$/Peter ether | 40 | 91a, 109 |
| 1-Benzyl-5-(2-$ClC_6H_4$)-7-MeO | 115–118 | $Et_2O$/Hexane | | 134 |
| 1-Benzyl-5-(2,6-$Cl_2$-$C_6H_3$)-7-Cl | 206–207 | $CH_2Cl_2/Et_2O$ | 78 | 42 |
| 1-Benzyl-5-(2-$FC_6H_4$)-7-Cl | 128–130 | $Et_2O$/Hexane | | 152c |
| 1-Benzyl-5-(3-thienyl)-7-Br | 174–177 | EtOAc/Hexane | | 134 |
| 1-[2-(N-Benzyl-N-$CNCH_2$-amino)ethyl]-5-Ph-7-Cl | 128–130 | $CH_2Cl_2/Et_2O$ | | 302 |
| 1-[2-(N-Benzylmethyl amino)ethyl]5-Ph-7-Cl | Oil | | | 302 |
| 1-[2-(N-Benzyl-N-$MeOOCCH_2$-amino)ethyl]-5-Ph-7-Cl | 145–147 | MeOH | | 150c |
| 1-[2-(Benzyloxy-CONH)ethyl]-5-(2-$FC_6H_4$)-7-Cl | 142–145 | $Et_2O$ | 45 | 233 |
| 1-[2-(Benzyloxy-CO-N-Me-amino)ethyl]-5-Ph-7-Cl | | | | |
| Hydrochloride | 176–180 | Acetone | | 150c |
| 1-(Benzyloxy)$CH_2$-5-Ph-7-Cl | 103–108 | EtOH | | 71 |
| 1-(Benzyloxy)$CH_2$-5-Ph-7-$NO_2$ | 156–158 | EtOH | 62 | 56 |

TABLE VII-1. —(contd)

| Substituent | mp (°C) or; [bp (°C/torr)] | Solvent of Crystallization | Yield (%) | Spectra | Refs. |
|---|---|---|---|---|---|
| 1-(Benzyloxy)$CH_2$-5-(2-$ClC_6H_4$)-7-Cl | 104–106 | $Et_2O$/Petr ether | 61 | | 56 |
| 1-(Benzyloxy)$CH_2$-5-(2-$ClC_6H_4$)-7-$NO_2$ | 128–130 | EtOH | | | 71 |
| 1-(1-Benzyloxy-1-ethyl)-5-(2-$ClC_6H_4$)-7-Cl | 120–122 | EtOH/Hexane | 53 | | 71, 56 |
| 1-(2-Benzyloxy)ethyl-5-Ph-7-Cl | 133–135 | | | | 89 |
| 4-Oxide | 155–156 | | | | 89 |
| 1-(2-Benzyloxy)ethyl-5-(2-$ClC_6H_4$)-7-Cl | 104–106 | $Et_2O$/Peter ether | | | 71 |
| 1-(2-Benzyloxy)ethyl-5-(2-$FC_6H_4$)-7-Cl | 114–116 | | | | 89 |
| 4-Oxide | 123–124 | | | | 89 |
| 1-[2-(2-Br-Acetamino)ethyl]-5-(2-$FC_6H_4$)-7-Cl | 169–171 | | | ir, ms, pmr | 331 |
| Hydrochloride | 223–225d | EtOH | 79 | ir, ms, pmr | 331 |
| 1-(4-Br-Butyl)-5-Ph-7-Cl | 107–109 | MeCOEt/Hexane | | | 38b |
| 1-(4-Br-Butyl)-5-(2-pyridyl)-7-Br | 119–122 | $Et_2O$ | | | 67b |
| 1-(3-Br-Propyl)-5-Ph-7-Cl | 89–93 | $Et_2O$/Hexane | | | 38b |
| 4-Oxide | 163–165 | $CH_2Cl_2$/MeOH | | | 134 |
| 1-(3-Br-Propyl)-5-(2-$ClC_6H_4$)-7-Cl | 106–112 | $CH_2Cl_2$/Hexane | | | 192 |
| 1-(3-Br-Propyl)-5-(2-$FC_6H_4$)-7-Cl | 95–98 | $Et_2O$ | | | 38b |
| 4-Oxide | 172–175 | $CH_2Cl_2$/$Et_2O$/Peter ether | | | 152c |
| 1-[2-(5-Br-3-Pyridincarbonyloxy)ethyl]-5-(2-$FC_6H_4$)-7-Cl | 140–143 | $CH_2Cl_2$/Hexane | | | 283 |
| 1-(*trans*-But-2-en-1-yl)NHCO-5-Ph-7-Cl | 114–116 | Cyclohexane | 23 | | 311, 312 |
| 1-Bu-5-(2-$FC_6H_4$)-7-$NO_2$ | 134–136 | $CH_2Cl_2$/Hexane | | | 251c |
| 1-*t*-Bu-5-Ph-7-Cl | 100–103 | Pentane | 67 | | 14 |
| 1-*t*-Bu-5-(2-$ClC_6H_4$)-7-$H_2N$ | | | | | |
| Picrate | 223–225d | EtOH/$H_2O$ | | | 391b |
| 1-*t*-Bu-5-(2-$ClC_6H_4$)-7-(2-MeOOC-Benzoyl)(Me)N | 198–199 | $CH_2Cl_2$/Hexane | | | 391b |
| 1-*t*-Bu-5-(2-$ClC_6H_4$)-7-Phthalimido | 185–187 | Acetone/Hexane | | | 391b |
| 1-*t*-Bu-5-(2-$ClC_6H_4$)-7-$NO_2$·0.5 $CH_2Cl_2$ | 210–212 | EtOAc/Hexane | | | 391b |
| | 208–210d | $CH_2Cl_2$/Hexane | | | 391b |

| | | | | |
|---|---|---|---|---|
| 1-{N-[6-(4-{3-*t*-BuNH-2-HO-Propoxy}phenoxy)hexyl]NHCOCH$_2$}-5-Ph-7-Cl·0.5H$_2$O | 113–118 | Et$_2$O | | 487 |
| 1-[3-Bu (Me)N-Propyl]-5-Ph-7-Cl | Oil | | | 302 |
| 1-[2-(*trans*-But-2-enoyloxy)-1-MeO-ethyl]-5-Ph-7-Cl | 130–131 | EtOH | | 302 |
| 1-[2-(1-*t*-BuOOC-L-Prolylamino)ethyl]-5-(2-FC$_6$H$_4$)-7-Cl | 88–92 | Et$_2$O/Petr ether | | 157b |
| 1-(3-*t*-BuOOC-*trans*-Prop-2-en-l-yl)-5-Ph-7-Cl | 163–165 | MeCN | | 302 |
| 1-(1-*t*-BuOOC-*trans*-Prop-l-en-1-yl)-5-Ph-7-Cl | 164–166 | MeCN | | 294 |
| 1-(2-BuO-Ethyl)-5-Ph-7-Cl | | | | |
| 4-Oxide | 163–164 | | | 234 |
| 1-(2-BuO-Ethyl)-5-(2-FC$_6$H$_4$)-7-Cl | | | | |
| 4-Oxide | 111–112 | | | 234 |
| 1-[2-(HCONH-Acetoxy)-1-MeO-ethyl]-5-Ph-7-Cl | 110–112 | EtOAc/Petr ether | | 486 |
| 1-(4-HOOC-Butyl)-5-(2-FC$_6$H$_4$)-7-Cl | | | | |
| Hydrosulfate | 205–209 | Acetone/Et$_2$O | 65 | 294 |
| 1-(2-HOOC-Ethyl)-5-Ph-7-NO$_2$ | 198–206 | MeOH | 30 | 294 |
| 1-(2-HOOC-Ethyl)-5-(2-ClC$_6$H$_4$)-7-H$_2$N | 262–264 | THF/Hexane | | 152c |
| 1-(2-HOOC-Ethyl)-5-(2-ClC$_6$H$_4$)-7-NO$_2$ | 188–191 | MeOH | 10 | 294 |
| 1-(2-HOOC-Ethyl)-5-(2-FC$_6$H$_4$)-7-Cl | 184–188 | CH$_2$Cl$_2$/Et$_2$O | 44 | 294 |
| 1-(2-HOOC-Ethyl)-5-(2-FC$_6$H$_4$)-7-NO$_2$ | 188–192 | MeOH | 19 | 294 |
| 1-(2-HOOC-Ethyl)-5-(2-pyridyl)-7-Br | 102–109 | MeOH/THF | 8 | 294 |
| 1-HOOCCH$_2$-5-Ph-7-Cl | 194–196 | Acetone/Et$_2$O | | 134 |
| 1-HOOCCH$_2$-5-Ph-7-NO$_2$·0.5 H$_2$O | 189–191d | MeOH/*i*-PrOH | | 2b |
| 1-HOOCCH$_2$-5-(2-FC$_6$H$_4$)-7-Cl | 216–225d | MeOH/Et$_2$O/Petr ether | | 152c |
| 1-HOOCCH$_2$-5-(2,6-F$_2$C$_6$H$_3$)-7-Cl·H$_2$O | 172–176 | CH$_2$Cl$_2$/MeOH | | 152c |
| 1-HOOC(MeO)CH-5-Ph-7-Cl | | | | |
| Ammonium salt | 212–215d | MeOH/H$_2$O | | 108 |
| 1-[2-(HOOC-Methoxy)ethyl]-5-Ph-7-Cl | | | | |
| Tosylate | 179–181d | | | 150c |
| 1-[2-(HOOCCH$_2$O)-Ethyl]-5-(2-FC$_6$H$_4$)-7-Cl | 174–180 | Acetone/Et$_2$O | | 152c |
| 1-(9-HOOC-Nonyl)-5-(2-FC$_6$H$_4$)-7-Cl | Oil | | | 152c |
| 1-{2-[N-(3-HOOC-Propanoyl)-N-Et-amino]ethyl}-5-(2-FC$_6$H$_4$)-7-Cl | 120–130 | Amorphous | | 152c |

TABLE VII-1. —(contd.)

| Substituent | mp (°C) or; [bp (°C/torr)] | Solvent of Crystallization | Yield (%) | Spectra | Refs. |
|---|---|---|---|---|---|
| 1-[2-(3-HOOC-Propanoyloxy)ethyl]-5-Ph-7-Cl | | | | | |
| 4-Oxide | 198d | MeOH | 48 | | 365 |
| 1-[2-(3-HOOC-Propanoyloxy)ethyl]-5-(2-$FC_6H_4$)-7-Cl | 152–156 | MeOH/$Et_2O$ | | | 283 |
| 1-[2-(3-HOOC-Propanoyloxy)-1-MeO-ethyl]-5-Ph-7-Cl | 167–169 | Acetone | | | 302 |
| 4-Oxide | 191–192 | $CH_2Cl_2$/$Et_2O$ | | | 302 |
| 1-[2-(3-HOOC-Propanoyloxy)-1-MeO-ethyl]-5-(2-$FC_6H_4$)-7-Cl | 110–112 | Acetone | | | 302 |
| 1-[2-(3-HOOC-*cis*-Propenoyloxy)ethyl]-5-(2-$FC_6H_4$)-7-Cl | 153–157 | $CH_2Cl_2$/Hexane | | | 283 |
| Sodium salt·MeOH | 143–147 | | | | 283 |
| 1-(3-HOOC-*trans*-Prop-2-en-1-yl)-5-Ph-7-Cl | | | | | |
| Tosylate | 268–270d | MeOH/$Et_2O$ | | | 294 |
| Sodium salt | 220–222 | MeOH/Acetone | | | 283 |
| 1-(3-HOOC-Propyl)-5-Ph-7-Cl | 166–168 | MeOH/$Et_2O$ | | | 488 |
| 1-(3-HOOC-Propyl)-5-(2-$FC_6H_4$)-7-Cl | 173–178 | MeOH/$Et_2O$/Petr ether | 87 | | 294 |
| 1-(3-HOOC-Propyl)-5-(2-$FC_6H_4$)-7-I | 190–192 | $CHCl_3$/Hexane | 55 | | 294 |
| 1-(3-HOOC-3,3-$Me_2$-Propyl)-5-Ph-7-Cl | 176–178 | | | | 294 |
| 1-Cl-5-(1-Cl-Cyclohex-1-yl)-7-Cl | 102 | *i*-$Pr_2O$ | 62 | | 224 |
| 1-Cl-5-(Cyclohex-1-en-1-yl)-7-Cl | 134 | *i*-$Pr_2O$ | 56 | | 224 |
| 1-Cl-5-Cyclohexyl-7-Cl | 163–164 | *i*-$Pr_2O$ | 91 | | 222 |
| 1-Cl-5-Ph-7-Cl | 143–144d | EtOAc | 64; 91 | | 11, 224 |
| 1-[2-(Cl-Acetoxy)ethyl-5-(2-$FC_6H_4$)]-7-Cl | 135–137 | $CH_2Cl_2$/Petr ether | | | 283 |
| 1-[2-(Cl-Acetoxy)-1-MeO-ethyl]-5-Ph-7-Cl | 124–126 | $Et_2O$/Petr ether | | | 302 |
| 1-[2-(Cl-Acetoxy)-1-MeO-ethyl]-5-Ph-7-$NO_2$ | 117–120 | $CH_2Cl_2$/$Et_2O$ | | | 302 |
| 1-[2-($Cl_2$-Acetoxy)-1-MeO-ethyl]-5-Ph-7-Cl | 116–118 | $Et_2O$/Petr ether | | | 302 |
| 1-[2-(4-Cl-Benzoyloxy)-1-MeO-ethyl]-5-Ph-7-Cl | 150–153 | EtOH | | | 302 |
| 1-(4-Cl-Butyl)-5-(2-$FC_6H_4$)-7-Cl | 90–93 | $Et_2O$/Hexane | | | 254 |
| 1-{2-[2-(7-Cl-1,3-Dihydro-2-oxo-5-Ph-2*H*-1,4-benzodiazepin-1-yl)ethoxy]ethyl}-5-Ph-7-Cl | 164–167 | MeOH/$Et_2O$ | | | 487 |

| | | | | |
|---|---|---|---|---|
| 1-[3-(7-Cl-1,3-Dihydro-2-oxo-5-Ph-2*H*-1,4-benzodiazepin-1-yl)-propyl]-5-Ph-7-Cl | 239–241 | $CH_2Cl_2$/MeOH | | 487 |
| 1-[2-(2-Cl-Ethoxy)ethyl]-5-(2-$FC_6H_4$)-7-Cl | | | | |
| Hydrochloride | 176–183d | MeOH/$Et_2O$ | | 152c |
| 1-[2-Cl-1-(2-Cl-Ethoxy)ethyl]-5-Ph-7-Cl | 147–148 | $CH_2Cl_2$/EtOH | | 108 |
| 1-[2-Cl-1-(2-Cl-Ethoxy)ethyl]-5-Ph-7-$NO_2$ | 163–166 | $CH_2Cl_2$/EtOH | | 108 |
| 1-(2-Cl-Ethoxy)$CH_2$-5-Ph-7-$H_2N$ | 190–191 | $CH_2Cl_2$/Hexane | | 108 |
| 1-(2-Cl-Ethoxy)$CH_2$-5-Ph-7-Cl | 96–98 | | | |
| | 128–130 | MeOH | | 72 |
| 1-(2-Cl-Ethoxy)$CH_2$-5-Ph-7-$NO_2$ | 97–99 | EtOH | | 72 |
| 1-(2-Cl-Ethoxy)$CH_2$-5-(2-$ClC_6H_4$)-7-Cl | 92–93 | | | |
| | 131–133 | $Et_2O$ | | 72 |
| 1-(2-Cl-Ethoxy)$CH_2$-5-(2-$FC_6H_4$)-7-I | 135–137 | EtOH | | 108 |
| 1-(2-Cl-Ethyl)-5-Ph-7-Cl | 168–169 | Acetone/Hexane | | 254 |
| 1-(2-Cl-Ethyl)-5-(2-$FC_6H_4$)-7-Cl | 111–114 | $Et_2O$ | | 278 |
| 1-(2-Cl-Ethyl)NHCO-5-Ph-7-Cl | 110–112 | *i*-PrOH | 33 | 311, 312 |
| 1-(2-Cl-1-EtO-Ethyl)-5-Ph-7-Cl | 129–132 | EtOH | | 302 |
| 1-(2-Cl-1-MeO-Ethyl)-5-Ph-7-$H_2N$ | 210–220d | EtOH | | 108 |
| 1-(2-Cl-1-MeO-Ethyl)-5-Ph-7-Cl | 152–172 | $CH_2Cl_2$/MeOH | | 72 |
| 4-Oxide | 214–215 | EtOAc/EtOH | | 72 |
| 1-(2-Cl-1-MeO-Ethyl)-5-Ph-7-$NO_2$ | 162–164 | $CH_2Cl_2$/MeOH | | 72 |
| 1-(2-Cl-1-MeO-Ethyl)-5-(2-$ClC_6H_4$)-7-$NO_2$ | 196–198 | EtOH | | 302 |
| 1-(2-Cl-1-MeO-Ethyl)-5-(2-$FC_6H_4$)-7-Cl | 170–171 | MeOH | | 302 |
| 1-(2-Cl-1-MeO-Ethyl)-5-(2-$FC_6H_4$)-7-I | 224–226d | $CH_2Cl_2$/EtOH | | 278 |
| 1-(2-Cl-1-MeO-Ethyl)-5-(2-pyridyl)-7-Br | 158–160 | $CH_2Cl_2$/$Et_2O$/Petr ether | | 302 |
| 1-(2,2,2-$Cl_3$-Ethoxy)$CH_2$-5-Ph-7-$NO_2$ | 113–118 | EtOH | | 72 |
| 1-(3-Cl-Propyl)-5-Ph-7-Cl | 86–89 | $Et_2O$/Hexane | | 125h |
| | 87–90 | $Et_2O$/Hexane | 86 | 287, 291 |
| 1-(3-Cl-Propyl)-5-Ph-7-$F_3C$ | 118–123 | $Et_2O$/Hexane | 50 | 291 |
| 1-(3-Cl-Propyl)-5-Ph-7-$NO_2$ | | | | |
| Hydrochloride | | | | 38b |
| 1-(3-Cl-Propyl)-5-(2-$FC_6H_4$)-7-Cl | 86–89 | $Et_2O$/Hexane | 48 | 287, 291 |
| 1-(3-Cl-Propyl)-5-(2-pyridyl)-7-Br | 103–106 | $Et_2O$/Hexane | 42 | 291 |
| 1-[(4-Cl-Benzyloxy)methyl]-5-Ph-7-Cl | 92–93 | $Et_2O$/Petr ether | | 71 |

TABLE VII-1. —(contd.)

| Substituent | mp (°C) or; [bp (°C/torr)] | Solvent of Crystallization | Yield (%) | Spectra | Refs. |
|---|---|---|---|---|---|
| 1-[(4-Cl-Benzyloxy)methyl]-5-Ph-7-$NO_2$ | 146–147 | EtOH | 16 | | 71, 56 |
| 1-{4-[7-Br-1,3-Dihydro-2-oxo-5-(2-pyridyl)-2*H*-1,4-benzodiazepin-1-yl]butyl}-5-(2-pyridyl)-7-Br | 257–260 | $CH_2Cl_2$/MeOH | | | 152c |
| 1-(7-Cl-1,3-Dihydro-2-oxo-5-Ph-2*H*-1,4-benzodiazepin-1-yl)$CH_2$-5-Ph-7-Cl | 255–257 | DMF/MeCN | | | 33b |
| 1-{4-[7-Cl-1,3-Dihydro-2-oxo-5-Ph-2*H*-1,4-benzodiazepin-1-yl]butyl}-5-Ph-7-Cl | 265–286 | $CHCl_3$/EtOH | | | 152c |
| 1-{4-[7-Cl-5-(2-$FC_6H_4$)-1,3-Dihydro-2-oxo-2*H*-1,4-benzodiazepin-1-yl]butyl}-5-(2-$FC_6H_4$)-7-Cl | 209–213 | $CH_2Cl_2$/MeOH | | | 152c |
| 1-{4-[5-(2-$ClC_6H_4$)-1,3-Dihydro-7-$NO_2$-2-oxo-2*H*-1,4-benzodiazepin-1-yl]butyl}-5-(2-$ClC_6H_4$)-7-$NO_2$ | 272–275 | $CHCl_3$/MeOH | | | 152c |
| 1-(6-Cl-4-Ph-Quinazolin-2-yl)$CH_2$-5-Ph-7-Cl | 181–182 | $Et_2O$ | | | 2b |
| 4,3′-Dioxide | 273–274 | $CHCl_3$/EtOH | | | 2b |
| 1-(2-CN-Ethyl)-5-Ph-7-Cl | 117–118 | $Et_2O$/Petr ether | | | 91 |
| 1-(2-CN-Ethyl)-5-(2-$FC_6H_4$)-7-Cl | 150 | $Et_2O$ | | | 301 |
| 1-[2-(N-CN-N-Et-Amino)ethyl]-5-(2-$FC_6H_4$)-7-Cl | 132–134 | $CH_2Cl_2$/MeOH | 76 | | 233 |
| 1-[2-(N-CN-Ethyl-N-Me-amino)ethyl]-5-Ph-7-Cl | 82–83 | | | | 150c |
| Dihydrochloride | > 200d | EtOH/$Et_2O$ | | | 150c |
| 1-[2-CN(Me)N-Ethyl]-5-Ph-7-$F_3C$ | 50–54 | $Et_2O$/Petr ether | | | 152c |
| 1-[3-CN(Me)N-Propyl]-5-Ph-7-$NO_2$ | 154–156 | $CH_2Cl_2$/$Et_2O$ | | | 152c |
| 1-[3-CN(Me)N-Propyl]-5-(2-pyridyl)-7-Br | 100–104 | $CH_2Cl_2$/$Et_2O$/Petr ether | | | 152c |
| 1-(2-$NCCH_2O$-Ethyl)-5-Ph-7-Cl | | | | | |
| Hydrochloride | > 220d | MeOH/$Et_2O$ | | | 150c |
| 1-[2-$NCCH_2O$-Ethyl]-5-(2-$FC_6H_4$)-7-Cl | 100–101 | $Et_2O$/Heptane | | | 150c |
| 1-[2-$NCCH_2O$-Ethyl]-5-(2-pyridyl)-7-Br | 126–128 | EtOH | | | 150c |
| 1-$NCCH_2$-5-Ph-7-Cl | | | | | |
| Hydrochloride | 219–221d | | | | 86b |
| 1-$CNCH_2$-5-Ph-7-$NO_2$ | 207–208 | EtOH | | | 86b |
| 1-[2-(N-$CNCH_2$-N-Me-Amino)ethyl]-5-Ph-7-Cl | 108–110 | $Et_2O$ | | | 150c |

| | | | | |
|---|---|---|---|---|
| 1-[2-(N-CNCH$_2$-N-Me-Amino)ethyl]-5-(2-ClC$_6$H$_4$)-7-Cl | 115–117 | EtOH | | 302 |
| 1-(2-CNCH$_2$O-Ethyl)-5-(2-FC$_6$H$_4$)-7-Cl | 101–102 | Et$_2$O/Hexane | | 302 |
| 1-(Cyclobutyl)CH$_2$-5-Ph-7-Cl | 165–167 | | | 113 |
| 1-(2-Cyclobutylmethoxy)ethyl-5-(2-ClC$_6$H$_4$)-7-Cl | | | | |
| Hydrochloride | 177 | MeOH/Et$_2$O | | 281 |
| 1-(Cyclohexyl)NHCO-5-Ph-7-Cl | 134–136 | *i*-PrOH | 46 | 311, 312 |
| 1-[3-(Cyclohexyl)NH-propyl]-5-Ph-7-Cl | | | | |
| Dihydrochloride | 265d | MeOH/Et$_2$O | | 302 |
| 1-Cyclopentyl-5-(2-FC$_6$H$_4$)-7-NO$_2$ | 164–166 | EtOAc | | 251c |
| 1-(2-Cyclopentylmethoxy)ethyl-5-(2-ClC$_6$H$_4$)-7-Cl | | | | |
| Hydrochloride | 165–166 | MeOH/Et$_2$O | | 281 |
| 1-[2-(Cyclopropyl)COO-1-MeO-ethyl]-5-Ph-7-Cl | 113–115 | EtOH | | 72 |
| 1-(Cyclopropyl)NHCO-5-Ph-7-Cl | 142–144d | *i*-PrOH | 32 | 311, 312 |
| | 213–215 | | | 311, 312 |
| 1-(Cyclopropyl)CH$_2$-5-benzyl-7-Cl | | | | |
| Hydrochloride | 195d | MeOH/PhH | | 124 |
| 1-(Cyclopropyl)CH$_2$-5-Ph-7-Cl | 145–146 | EtOH | 65 | 62 |
| 4-Oxide | 153–155 | | 76 | 232 |
| 1-(Cyclopropyl)CH$_2$-5-(2-ClC$_6$H$_4$)-7-Cl | 128–129 | | | 113 |
| 1-(Cyclopropyl)CH$_2$-5-(2-FC$_6$H$_4$)-7-Cl | 86–88 | | | 89 |
| Hydrochloride | 195d | EtOH/*i*-Pr$_2$O | | 86b |
| 1-(Cyclopropyl)CH$_2$-5-(2-MeC$_6$H$_4$)-7-Cl | 144–145 | | | 276 |
| 1-[2-(Cyclopropyl)methoxyethyl]-5-Ph-7-Cl | 98–100 | Et$_2$O/Petr ether | 51 | 66 |
| 1-[2-(Cyclopropyl)methoxyethyl]-5-(2-FC$_6$H$_4$)-7-Cl | 85 | | | |
| Hydrochloride | 173 | Acetone/Et$_2$O | 52.5 | 66 |
| 1-[(Cyclopropylmethoxycarbonyl)CH$_2$]-5-Ph-7-Cl | 138–140 | Pentane | 49 | 66 |
| 1-[(Cyclopropylmethoxycarbonyl)CH$_2$]-5-(2-FC$_6$H$_4$)-7-Cl | | | | |
| 1-[2-(Cyclopropyl methylthio)ethyl]-5-Ph-7-Cl | 82 | | 45 | 66 |
| 1-[2-(Cyclopropyl methylthio)ethyl]-5-(2-FC$_6$H$_4$)-7-Cl | 95 | | 54 | 66 |
| 1-Decyl-5-(2-FC$_6$H$_4$)-7-NO$_2$ | Oil | | | 251c |
| 1-{2-[3-(10,11-Dihydro-5*H*-dibenzo[*a,d*]cyclohepten-5-ylidene)propyl]MeNCOO}-5-(2-FC$_6$H$_4$)-7-Cl | 85–95 | Et$_2$O/Petr ether | | 33b |

TABLE VII-1. —(contd.)

| Substituent | mp (°C) or; [bp (°C/torr)] | Solvent of Crystallization | Yield (%) | Spectra | Refs. |
|---|---|---|---|---|---|
| 1-{2-[3-(10,11-Dihydro-5*H*-dibenzo[*a,d*]cyclohepten-5-ylidene)propyl]MeNCOON(Et)}-5-(2-$FC_6H_4$)-7-Cl | 95–110 | $Et_2O$/Petr ether | | | 33b |
| 1-(1,4-Dioxan-2-yl)$CH_2$-5-Ph-7-Cl | 142–144 | *i*-PrOH/*i*-$Pr_2O$ | | | 285 |
| 1-(1,4-Dioxan-2-yl)$CH_2$-5-(2-$FC_6H_4$)-7-Cl | 129–131 | | | | 285 |
| 1-Et-5-(4-$ClC_6H_4$)-7-Cl | 128–130 | EtOH | 74 | | 2 |
| 1-Et-5-(2-$FC_6H_4$)-7-$H_2N$ | 207–208 | EtOH | | | 72c |
| 1-Et-5-(2-$FC_6H_4$)-7-(1-$H_2N$-Ethyl) | 53–55 | | | | 251b |
| 1-Et-5-(2-$FC_6H_4$)-7-Cl | 103–105 | Acetone/Petr ether | | | 152c |
| 1-Et-5-(2-$FC_6H_4$)-7-$NO_2$ | 105 | EtOAc | | | 72c |
| 1-Et-5-Cyclohexyl-7-Cl | 114–115 | Hexane | 75 | | 18 |
| 1-Et-5-EtO-7-Cl | 91–93 | EtOH | 46 | | 195 |
| 1-Et-5-Pentyl-7-Cl | [150–160/0.05] | | 81 | | 18 |
| 1-Et-5-Ph-7-$H_2N$ | 220 | EtOH | | | 72c |
| 1-Et-5-Ph-7-Cl | 127–128 | Acetone | | | 91a |
| 4-Oxide | 207–208 | Acetone/Petr ether | | | 91a |
| | 211–212 | | | | 358 |
| 1-Et-5-Ph-7-$NO_2$ | 165–166 | $CH_2Cl_2$/$Et_2O$/Petr ether | | | 2b |
| 1-Et-5-(2-$FC_6H_4$)-7-Ac | 154–156 | $CH_2Cl_2$/Cyclohexane | | | 251 |
| 1-Et-5-(2-$FC_6H_4$)-7-(1-AcO-Ethyl) | Oil | | | | 251 |
| 1-Et-5-(2-$FC_6H_4$)-7-(1-$H_2N$-Ethyl) | 53–55 | | | | 251 |
| 1-Et-5-(2-$FC_6H_4$)-7-[1-(Butylaminocarbonyloxy)ethyl] | 48–50 | Cyclohexane | | | 251 |
| 1-Me-5-(2-$FC_6H_4$)-7-(1-EtOOCO-Ethyl) | Oil | | | | 251b |
| 1-Et-5-(2-$FC_6H_4$)-7-(1-Hydroxyiminoethyl) | 236–238 | $CH_2Cl_2$/Hexane | | | 251 |
| 1-Et-5-(2-$FC_6H_4$)-7-(1-HO-Ethyl) | 55–57 | Cyclohexane | | | 251 |
| 1-Et-5-(2-$FC_6H_4$)-7-[1-($MeOCH_2O$)Ethyl] | Oil | | | | 251 |
| 1-Et-5-(2-$FC_6H_4$)-7-[1-(MeO-Methoxyimino)ethyl] | Oil | | | | 251b |
| 1-Et-5-(2-$FC_6H_4$)-7-[1-(Phenylaminocarbonyloxy)ethyl] | 83–85 | Cyclohexane | | | 251 |
| 1-Et-5-(2-Pyridyl)-7-Br | 143–145 | $CH_2Cl_2$/$Et_2O$/Hexane | | | 134 |

| | | | | |
|---|---|---|---|---|
| 1-[2-(N-Et-N-MeO(NH)C-amino)ethyl]-5-(2-$FC_6H_4$)-7-Cl | 91–100 | $Et_2O$ | | 152c |
| 1-(1-Et-Piperidin-3-yl)-5-Ph-7-Cl | | | | |
| Dihydrochloride | 220–223d | $EtOH/Et_2O$ | | 289 |
| 1-(1-Et-Piperidin-3-yl)-5-Ph-7-$NO_2$ | | | | |
| Dihydrochloride | 213–215 | EtOH | | 289 |
| 1-(1-Et-Piperidin-3-yl)-5-(2-$FC_6H_4$)-7-Cl | | | | |
| Dihydrochloride | 217–220d | EtOH | | 289 |
| 1-$EtNHCOCH_2$-5-Ph-7-Cl | 211–212 | Acetone | | 293, 86b |
| 1-[2-(EtNHCO-Methoxy)ethyl]-5-(2-$ClC_6H_4$)-7-Cl | 136–138 | $CH_2Cl_2/Et_2O$ | | 150c |
| 1-(2-EtNHCOO-1-MeO-Ethyl)-5-Ph-7-Cl | 117–119 | $Et_2O$/Petr ether | | 302 |
| 1-(2-EtNH-Ethyl)-5-(2-$FC_6H_4$)-7-Cl | 80–85 | $Et_2O$/Petr ether | | 233 |
| Dihydrochloride | 215–217 | $MeOH/Et_2O$ | 35 | 233 |
| 1-[2-Et(HO)N-Ethyl]-5-(2-$FC_6H_4$)-7-Cl | 132–133 | $Et_2O$ | 15 | 238 |
| 1-[2-(EtO-Acetoxy)ethyl]-5-(2-$FC_6H_4$)-7-Cl | 154–158 | Acetone/$Et_2O$ | | 283 |
| 1-EtOOC-5-Ph-7-$NO_2$ | 167–169d | EtOAc/Hexane | | 108 |
| 1-(1-EtOOCNH-1-Ethyl)-5-Ph-7-Cl | 155–157 | EtOH | | 84 |
| 1-(2-EtOOCNH-Ethyl)-5-Ph-7-Cl | 127–129 | $Et_2O$/Petr ether | | 302 |
| 1-(EtOOCNH)$CH_2$-5-Ph-7-$H_2N$ | 210 | $MeOH/Et_2O$ | | 72c |
| 1-(EtOOCNH)$CH_2$-5-Ph-7-Cl | 126–129 | $MeOH/H_2O$ | | 84 |
| 1-(EtOOCNH)$CH_2$-5-Ph-7-$NO_2$ | 157–160 | MeOH | | 84 |
| 1-(EtOOCNH)$CH_2$-5-(2-$ClC_6H_4$)-7-Cl | 147–149 | EtOH | | 84 |
| 1-(EtOOCNH)PhCH-5-Ph-7-Cl | 177–180 | EtOH | | 84 |
| 1-(4-EtOOC-Butyl)-5-(2-$FC_6H_4$)-7-Cl | | | | |
| Hydrochloride | 137–147 | $EtOH/Et_2O$ | 45 | 294 |
| 1-(4-EtOOC-Butyl)-5-(2-$FC_6H_4$)-7-I | | | | |
| Hydrochloride | 119–122 | $EtOH/Et_2O$ | | 294 |
| 1-$EtOOCCH_2$-5-Ph-7-Cl | 115–117 | | | 86b |
| | 127–129 | Acetone/Hexane | | 293 |
| 1-$EtOOCCH_2$-5-(2-$FC_6H_4$)-7-$NO_2$ | 139–142 | $CH_2Cl_2$/MeOH | | 152c |
| 1-$EtOOCCH_2$-5-(2,6-$F_2$-$C_6H_3$)-7-Cl | 159–161 | $CH_2Cl_2$/EtOH | | 152c |
| 1-$(EtOOC)_2$CH-5-Ph-7-Cl | 139–140 | Cyclohexane | | 486 |
| 1-$(EtOOC)_2$CH-5-(2-$FC_6H_4$)-7-Cl | 132–134 | $Et_2O$/Petr ether | | 72c |
| 1-($EtOOCCH_2$)NHCO-5-Ph-7-Cl | 125–128 | *i*-PrOH | 79 | 311 |

TABLE VII-1. —(contd.)

| Substituent | mp (°C) or; [bp (°C/torr)] | Solvent of Crystallization | Yield (%) | Spectra | Refs. |
|---|---|---|---|---|---|
| 1-[2-(EtOOC-Methoxy)ethyl]-5-Ph-7-Cl | Oil | | | | 150d |
| 1-[1,1-(EtOOC)$_2$-2-Phthalimidoethyl]-5-Ph-7-Cl | 194–195 | EtOAc | | | 486 |
| 1-(3-EtOOC-Propyl)-5-PH-7-$NO_2$ | | | | | |
| Hydrochloride | 175–180 | EtOH/$Et_2O$ | | | 152c |
| 1-(3-EtOOC-Propyl)-5-(2-$ClC_6H_4$)-7-$NO_2$ | | | | | |
| Hydrochloride | 170–174 | EtOH/$Et_2O$ | | | 152c |
| 1-(3-EtOOC-Propyl)-5-(2-$FC_6H_4$)-7-Cl | | | | | |
| Hydrochloride | 196–198 | EtOH/$Et_2O$ | 64 | | 294 |
| 1-(3-EtOOC-Propyl)-5-(2-$FC_6H_4$)-7-I | | | | | |
| Hydrochloride | 187–189 | EtOH/$Et_2O$ | 50 | | 294 |
| 1-[2-(2-EtO-Ethoxy)ethyl]-5-Ph-7-Cl | 78–79 | | | | 89 |
| 1-[2-(EtO-Ethoxy)ethyl]-5-(2-$FC_6H_4$)-7-Cl | 76–77 | | | | 89 |
| 1-(1-EtO-1-Ethyl)-5-Ph-7-$H_2N$ | 152–153 | EtOAc/Petr ether | | | 72c |
| 1-(1-EtO-1-Ethyl)-5-Ph-7-$NO_2$ | 184–187 | EtOH | | | 72 |
| 1-(2-EtO-Ethyl)-5-Ph-7-Cl | 156–158 | PhH/Hexane | | | 125h |
| Hydrochloride | 212–214d | MeOH/Acetone | | | 279 |
| 1-(2-EtO-Ethyl)-5-(2-$ClC_6H_4$)-7-Cl | 142–143 | | | | 125h |
| 1-(2-EtO-Ethyl)-5-(2-$FC_6H_4$)-7-Cl | 98–100 | *i*-PrOH | | | 125h |
| Hydrochloride | 201–203d | *i*-PrOH/$Et_2O$ | | | 125h |
| 1-(2-EtO-Ethyl)-5-(2-$MeC_6H_4$)-7-Cl | 141–142 | | | | 125h |
| 4-Oxide | 118–119 | | | | 234 |
| 1-[2,2-(EtO)$_2$-Ethyl-5-(2-$FC_6H_4$)-7-Cl | 110–113 | $Et_2O$/Petr ether | | | 152c |
| 1-(1-EtO-2-HO-Ethyl)-5-Ph-7-Cl | 186–189 | MeCN | | | 302 |
| 1-{3-[4-(2-EtO-Ethyl)piperazin-1-yl]propyl}-5-(2-$FC_6H_4$)-7-Cl | | | | | |
| Trimaleate | 125–133 | Acetone | 24 | | 287, 86b |
| 1-[1-EtO-2-(Pyridin-3-yl)COO-ethyl]-5-Ph-7-Cl | 116–117 | EtOH | | | 302 |
| 1-(2-EtOOC-Ethyl)-5-Ph-7-Cl | | | | | |
| Hydrochloride | 225–227 | MeOH/$Et_2O$ | | | 293 |

| | | | | | |
|---|---|---|---|---|---|
| 1-$EtOCH_2$-5-Ph-7-$H_2N$ | 171–172 | $CH_2Cl_2$/Hexane | | | 108 |
| 1-$EtOCH_2$-5-Ph-7-Cl | 131–132 | $CH_2Cl_2$/Hexane | | | 72 |
| 1-$EtOCH_2$-5-Ph-7-$NO_2$ | 105–107 | EtOH | | | 72 |
| 1-(1-EtO-1-Ethyl)-5-Ph-7-$NO_2$ | 172–174 | EtOH | | | 72 |
| 1-(2-$EtSO_2$-Ethyl)-5-(2-$FC_6H_4$)-7-Cl | 122–123 | | | | 127 |
| 1-[2-($Et_2N$-Acetoxy)ethyl]-5-(2-$FC_6H_4$)-7-Cl | 85–88 | $CH_2Cl_2$/$Et_2O$/Petr ether | | | 283 |
| 1-(4-$Et_2N$-Butyl)-5-(2-$FC_6H_4$)-7-Cl | | | | | |
| Hydrochloride·0.5$H_2O$ | 138–140 | Acetone/$Et_2O$ | | | 254 |
| 1-$Et_2NCOCH_2$-5-Ph-7-Cl | 148–149 | Acetone/Hexane | | | 293, 86b |
| 1-(2-$Et_2NCOO$-1-MeO-Ethyl)-5-Ph-7-Cl | 109–111 | $Et_2O$/Petr ether | | | 302 |
| 1-(2-$Et_2N$-Ethyl)-5-Ph-7-$H_2N$ | 137–138 | | | | 72c |
| 1-(2-$Et_2N$-Ethyl)-5-Ph-7-Br | 94–96 | Hexane | | | 302 |
| 1-(2-$Et_2N$-Ethyl)-5-Ph-7-Cl | 79–81 | Pentane | 46 | | 55, 287 |
| 4-Oxide | 121–122 | $Et_2O$ | 32 | | 287 |
| 1-(2-$Et_2N$-Ethyl)-5-Ph-7-F | 85–86 | $Et_2O$/Petr ether | | | 2b |
| 1-(2-$Et_2N$-Ethyl)-5-Ph-7-$F_3C$ | | | | | |
| Dihydrochloride | 218–221 | MeOH/$Et_2O$ | | | 86b |
| | 150–155d | $CH_2Cl_2$/$Et_2O$ | | | 302 |
| 1-(2-$Et_2N$-Ethyl)-5-Ph-7-Me | 84–86 | Hexane | | | 302 |
| 1-(2-$Et_2N$-Ethyl)-5-Ph-7-$NO_2$ | | | | | |
| Dihydrochloride | 232–233d | MeOH/$Et_2O$ | | | 86b |
| 1-(2-$Et_2N$-Ethyl)-5-(2-$H_2NC_6H_4$)-7-Cl | 135–137 | $CH_2Cl_2$/$Et_2O$ | | | 391b |
| 1-(2-$Et_2N$-Ethyl)-5-(2-$ClC_6H_4$)-7-$H_2N$ | 160–162 | EtOAc | | | 72c |
| 1-(2-$Et_2N$-Ethyl)-5-(2-$ClC_4H_4$)-7-Cl | 68–70 | Hexane | 35 | | 287, 86b |
| 4-Oxide | 164–166 | EtOH | 78 | pmr | 288, 365 |
| 1-(2-$Et_2N$-Ethyl)-5-(2-$ClC_6H_4$)-7-$NO_2$ | 87 | EtOAc | | | 72c |
| N-oxide·2$H_2O$ | 114–116d | PhH | | | 258 |
| 1-(2-$Et_2N$-Ethyl)-5-(3,4-$Cl_2C_6H_3$)-7-Cl | | | | | |
| Dihydrochloride | 205–207d | EtOH | | | 302 |
| 1-(2-$Et_2N$-Ethyl)-5-(4-$ClC_6H_4$)-7-Cl | 117–118 | EtOH | | | 302 |
| 1-(2-$Et_2N$-Ethyl)-5-(2-$FC_6H_4$)-7-$H_2N$ | 164–165 | $CH_2Cl_2$/Petr Ether | | | 2b |
| 1-(2-$Et_2N$-Ethyl)-5-(2-$FC_6H_4$)-7-Cl | | | | | |
| Dihydrochloride | 190–220d | MeOH/$Et_2O$ | 43; 25 | | 287, 55 |
| | 211–212d | *i*-PrOH | | | 86b |

TABLE VII-1. —(contd.)

| Substituent | mp (°C) or; [bp (°C/torr)] | Solvent of Crystallization | Yield (%) | Spectra | Refs. |
|---|---|---|---|---|---|
| 4-Oxide | 122–124 | $Et_2O$/Petr ether | 74 | | 233 |
| ω-Oxide·$H_2O$ | 90–95 | EtOAc | 71 | pmr | 238 |
| 4,ω-Dioxide | 141–142 | $CH_2Cl_2EtOAc/Et_2O$ | 50 | pmr | 238 |
| 1-(2-$Et_2$N-Ethyl)-5-(2-$FC_6H_4$)-7-I | 92–95 | $Et_2O$/Petr ether | | | 371 |
| 1-(2-$Et_2$N-Ethyl)-5-(2-$FC_6H_4$)-7-$NO_2$ | | | | | |
| Dihydrochloride | 196–197 | MeOH/$Et_2O$ | | | 2b |
| 1-(2-$Et_2$N-Ethyl)-5-(3-$FC_6H_4$)-17-Cl | 80–81 | *i*-prOH | | | 489 |
| 1-(2-$Et_2$N-Ethyl)-5-(4-$FC_6H_4$)-7-Cl | 102–103 | *i*-PrOH | | | 489 |
| 1-(2-$Et_2$N-Ethyl)-5-(4-$MeOC_6H_4$)-7-Cl | 109–111 | Hexane | | | 302 |
| 1-(2-$Et_2$N-Ethyl)-5-(2-pyridyl)-7-Br | | | | | |
| Hydrochloride | 176–180d | MeOH/$Et_2O$ | 50 | | 287 |
| 1-{2-[N-(2-$Et_2$N-Ethyl)-$NHCOCH_2O$]ethyl)}-5-Ph-7-Cl | 93–95 | $CH_2Cl_2/Et_2O$ | | | 302 |
| 1-{2-[N-(2-$Et_2$N-Ethyl)-$NHCOCH_2O$]ethyl}-5-(2-$FC_6H_4$)-7-Cl | 84–86 | *i*-$Pr_2O$ | | | 302 |
| 1-(3-$Et_2$N-Propyl)-5-Ph-7-Cl | 89–91 | Hexane | 61 | | 287 |
| Dihydrochloride | 208–210 | $EtOH/Et_2O$/Acetone | 85 | | 287 |
| 1(3-$Et_2$N-Propyl)-5-(2-$FC_6H_4$)-7-Cl | | | | | |
| Dihydrochloride | 182–192d | $EtOH/E_2O$ | 86 | | 287, 100 |
| 1-[$Et_2P(O)CH_2$]-5-Ph-7Cl | 178–180 | Xylene | 45 | | 25 |
| 1-[$Et_2P(O)CH_2$]-5-(4-*i*-$PrC_4H_4$)-7-Cl | Oil | | | | 25 |
| 1-(4-F-Benzoyl)propyl-5-Ph-7-Cl | Oil | | 60 | | 295 |
| 1-(4-F-Benzoyl)propyl-5-(2-$FC_6H_4$)-7-Cl | 149–159 | *i*-PrOH | | | 295 |
| 1-$F_3CCH_2$-5-Ph-7-Cl | 164–166 | Acetone/Hexane | 50 | | 69, 125h, 86b |
| Hydrochloride | 195–197d | | | | 125h |
| 4-Oxide | 192–194 | Acetone/$Et_2O$ | 70 | | 69 |
| 1-$F_3CCH_2$-5-Ph-7-$NO_2$ | 161–162 | Acetone/MeOH | 20 | | 69 |
| 1-$F_3CCH_2$-5-(2-$ClC_6H_4$)-7-Cl | 105–107 | $CH_2Cl_2$/Hexane | 57 | | 69 |
| 4-Oxide | 208–201 | $CH_2Cl_2$/Hexane | 19 | | 69 |

| | | | | | |
|---|---|---|---|---|---|
| 1-$F_3CCH_2$-5-(2-$FC_6H_4$)-7-Cl | 124–127 | $CH_2Cl_2$/Hexane | 65 | | 69 |
| 4-Oxide | 196–199 | $CH_2Cl_2$/Hexane | 55 | | 69 |
| 1-$F_3CCH_2$-5-(3-$O_2NC_6H_4$)-7-Cl | 230–231 | Acetone | 74 | | 69 |
| 1-$F_3CF_2CCH_2$-5-Ph-7-Cl | 196–197 | $CHCl_3$/Hexane | 64 | | 69 |
| 1-{2-[(2-Furoyl)oxy]-1-MeO-Ethyl}-5-Ph-7-Cl | 129–130 | EtOH | | | 302 |
| 1-(β-D-glucopyranosyl)-5-Ph-7-Cl | 160–162 | EtOAc/EtOH | | | 301 |
| 1-[2-(D-Gluconamino)ethyl]-5-Ph-7-Cl·0.9EtOH | 98–100 | EtOH | | | 258 |
| 1-(Hexyloxy)$CH_2$-5-Ph-7-$NO_2$ | 108–109 | $Et_2O$ | | | 72 |
| 1-(3-$H_2$NNHCO-Propyl)-5-Ph-7-Cl | | | | | |
| Dihydrochloride | 135d | *i*-PrOH | | | 488 |
| 1-HO-5-Ph-7-Cl | | | | | |
| 4-Oxide | 219–220 | EtOAc | 88 | ir, ms, pmr, uv | 189 |
| 1-[2,3,4-$(HO)_3$-Butyl]-5-(2-$FC_6H_4$)-7-Cl | | | | | |
| Hydrochloride | 143–145 | MeOH/THF | | | 236 |
| 1-(2-HO-Ethoxy)ethyl-5-(2-$FC_6H_6$)-7-Cl | | | | | |
| Hydrochloride | 180–185d | MeOH/$Et_2O$ | | | 283 |
| 1-(2-HO-Ethoxy)$CH_2$-5-Ph-7-Cl | 129–131 | $Et_2O$ | | | 72 |
| 1-(2-HO-Ethyl)-5-Ph-7-$N_3$ | 158–159d | PhH/Hexane | 20 | | 252 |
| 1-(2-HO-Ethyl)-5-Ph-7-Cl | 158–160 | EtOH | | | 86a, 125h |
| 4-Oxide | 235d | EtOH | 47 | | 365 |
| 1-(2-HO-Ethyl)-5-Ph-7-$F_3C$ | 115–116 | | | | 86a |
| 1-(2-HO-Ethyl)-5-Ph-7-$NO_2$ | 235–236 | EtOH/$CHCl_3$ | | | 86a, 125h |
| | 239–240d | Acetone | | | 279 |
| 4-Oxide | 178d | MeCN | 50 | | 365 |
| 1-(2-HO-Ethyl)-5-(2-$ClC_6H_4$)-7-Cl | 115–116 | | | | 86a |
| Hydrochloride | 195–200d | MeOH/$Et_2O$ | | | 152c |
| 1-(2-HO-Ethyl)-5-(2-$FC_6H_4$)-7-Cl | 114–116 | EtOH/*i*-$Pr_2O$ | | | 88, 125h |
| Hydrochloride | 194–196 | MeOH/$Et_2O$ | 46 | | 86a, 233 |
| 1-(2-HO-Ethyl)-5-(2-$FC_6H_4$)-7-$H_2N$·0.5MeOH | 128–135 | MeOH | | | 152c |
| 1-(2-HO-Ethyl)-5-(2-$FC_6H_4$)-7-I | | | | | |
| Hydrochloride | 206–208 | MeOH/$Et_2O$ | | | 371 |
| 1-(2-HO-Ethyl)-5-(2-$FC_6H_4$)-7-$NO_2$ | 199–203 | MeOH | | | 152c |
| 1-{3-[4-(2-HO-Ethyl)-piperazin-1-yl]propyl}-5-Ph-7-Cl | | | | | |
| Dimaleate | 121–123 | Acetone | 79 | | 287 |

TABLE VII-1. —(contd.)

| Substituent | mp (°C) or; [bp (°C/torr)] | Solvent of Crystallization | Yield (%) | Spectra | Refs. |
|---|---|---|---|---|---|
| 1-{3-[4-(2-HO-Ethyl)piperazin-1-yl]propyl}-5-($2-FC_6H_4$)-7-Cl Dimaleate | 120–123 | Acetone | 51 | | 287 |
| 1-(2-HO-1-MeO-Ethyl)-5-Ph-7-AcNH | 229–230 | EtOAc | | | 486 |
| 1-(2-HO-1-MeO-Ethyl)-5-Ph-7-$H_2N$ | 220–221 | EtOAc | | | 486 |
| 1-(2-HO-1-MeO-Ethyl)-5-Ph-7-Cl | 157–159 | EtOH | | | 72 |
| 4-Oxide | 216–217 | EtOH | | | 302 |
| 1-(2-HO-1-MeO-Ethyl)-5-Ph-7-$NO_2$ | 201–203 | MeCN | | | 302 |
| 1-(2-HO-1-MeO-Ethyl)-5-($2-ClC_6H_4$)-7-$NO_2$ | 188–190 | EtOH | | | 302 |
| 1-(2-HO-1-MeO-Ethyl-5-($2-FC_6H_4$)-7-Cl | 172–174 | MeCN | | | 302 |
| 1-(2-HO-1-MeO-Ethyl)-5-(2-Pyridyl)-7-Br | 201–203 | EtOH | | | 302 |
| 1-(4-HO-2-Me-But-2-en-1-yl)-5-Ph-7-Cl | 110–111 | EtOAc | | | 486 |
| 1-(2-HO-2-Ph-Ethyl)-5-Ph-7-Cl | | | | | |
| Hydrochloride | 216–217 | Acetone/MeOH | | | 284 |
| 1-(2-HO-Propyl)-5-Ph-7-Cl | | | | | |
| Hydrochloride | 203–205 | Acetone/MeOH/$Et_2O$ | | | 284 |
| 1-(2-HO-Propyl)-5-(2-pyridyl)-7-Br | 126–128 | $Et_2O$/Petr ether | | | 284 |
| 1-(2-HO-3-MeO-Propyl)-5-Ph-7-Cl | 116–118 | *i*-$Pr_2O$/$CH_2Cl_2$ | | | 285 |
| 1-[2,3-$(HO)_2$-Propyl]-5-Ph-7-$NO_2$ | 174–176 | MeOH/EtOAc | | | 486 |
| 1-[2,3-$(HO)_2$-Propyl]-5-Ph-7-Cl | 154–155 | | | | 285 |
| 4-Oxide | 187–189 | | | | 285 |
| 1-[2,3-$(HO)_2$-Propyl]-5-($2-ClC_6H_4$)-7-Cl | | | | | |
| Hydrochloride | 188–190 | MeOH/$Et_2O$ | | | 192 |
| 1-[2,3-$(HO)_2$-Propyl]-5-($2-ClC_6H_4$)-7-$NO_2$ | 135–137 | EtOAc | | | 258 |
| 1-[2,3-$(HO)_2$-Propyl]-5-($2-FC_6H_4$)-7-Cl | 120–122 | EtOH | | | 236 |
| Hydrochloride | 174–175 | EtOH/EtOAc | | | 236 |
| 4-Oxide | 165–166 | EtOAc | | | 236 |
| 1-[2,3-$(HO)_2$-Propyl]-5-($2-FC_6H_4$)-7-I | | | | | |
| Hydrochloride | 171 | EtOH/EtOAc | | | 236 |

| | | | | | |
|---|---|---|---|---|---|
| 1-(2-HO-3-*i*-PrNH-Propyl)-5-Ph-7-Cl | | | | | |
| Hydrochloride | 257–258d | $CH_2Cl_2$/*i*-PrOH | | | 285 |
| 1-(3-HO-Propyl)-5-Ph-7-Cl | 156–158 | Acetone/Petr ether | | | 279, 86a |
| 1-{3-[4-(2-HO-3-*i*-PrNH-Propoxy)phenoxy]propyl}-5-Ph-7-Cl (*S*)-Enantiomer | 110–111 | $CH_2Cl_2/Et_2O$ | | | 487 |
| 1-{3-[4-(2-{4-[2-HO-3-*i*-PrNH-Propoxy]phenoxy}ethyl)piperazin-1-yl]propyl}-5-Ph-7-Cl | | | | | |
| (*S*)-Enantiomer | | | | | |
| Trimaleate | 153–155 | MeOH/EtOAc | | | 487 |
| 1-(3-*l*-Propyl)-5-(2-$ClC_6H_4$)-7-Cl | 139–142 | $CH_2Cl_2$/Hexane | | | 192 |
| 1,5-$Me_2$-7-Cl | 143 | Cyclohexane | 43 | | 18 |
| 1-Me-5-Allyloxy-7-Cl | 64 | $Et_2O$/Pentane | 28 | | 195 |
| 1-Me-5-$H_2NCO$-7-Br | 200–203 | $CHCl_3$/Petr ether | | | 192 |
| 1-Me-5-(Aziridin-1-yl)-7-Cl | | | | | |
| Hemifumarate | 224–225d | EtOH | | | 196 |
| 1-Me-5-(1,3-Benzodioxolan-5-yl)-7-Cl | 145–147 | $CH_2Cl_2$/Hexane | | | 92 |
| 1-Me-5-Benzyl-7-Cl | 110 | *i*-$Pr_2O$ | 46 | | 18 |
| Hydrochloride | 214–216 | MeOH/$C_6H_6$ | 60 | ir, pmr | 116 |
| 1-Me-5-(2-Br-Ethoxy)-7-Cl | 110 | $Et_2O$/Pentane | 40 | | 195 |
| 1-Me-5-Bu-7-Cl | 55–56 | Hexane | 80 | | 18 |
| | [160/0.1] | | | | 124 |
| 1-Me-5-*t*-Bu-7-Cl | 80–81 | Hexane | | | 1b |
| 1-Me-5-$C_6D_5$-7-Cl | 131–132 | | 49 | | 97 |
| 4-Oxide | 184–186 | | 81 | | 97 |
| 1-Me-5,7-$Cl_2$ | 133–135 | PhH/$CCl_4$ | 55 | ir, ms, pmr | 198b |
| Hydrochloride | 70–80d | | | | 196 |
| 1-Me-5-(2,2-$Cl_2$-Ethoxy)-7-Cl | 123 | $Et_2O$/Pentane | 38 | | 195 |
| 1-Me-5-(2-Cl-ethyl)NHCO-7-Br | 208–211 | $Et_2O$/Hexane | | | 192 |
| 1-Me-5-CN-7-Br | 170–171 | $CH_2Cl_2$/Petr ether | | | 192 |
| 1-Me-5-(Cyclohexen-1-yl)-7-Cl | 144 | EtOAc | 52 | ir, pmr, uv | 222, 86b |
| 1-Me-5-(Cyclohexen-1-yl)-7-$NO_2$ | 163 | | | | 86b |
| 1-Me-5-Cyclohexyl-7-Cl | 126–127 | Hexane | 82 | | 18 |
| | 149–150 | | | | 86b |

TABLE VII-1. —(contd.)

| Substituent | mp (°C) or; [bp (°C/torr)] | Solvent of Crystallization | Yield (%) | Spectra | Refs. |
|---|---|---|---|---|---|
| 1-Me-5-(Cyclopent-1-en-1-yl)-7-Cl | 150 | $i$-$Pr_2O$ | 50 | | 224 |
| 1-Me-5-Cyclopentyl-7-Cl | 149–150 | EtOAc | | | 70 |
| | 157–158 | | | | 18 |
| 1-Me-5-Et-7-Cl | 97 | Cyclohexane | 70 | | 18 |
| 1-Me-5-EtO-7-Cl | 123 | $CCl_4$/Pentane | 75 | | 195 |
| 1-Me-5-EtO-7-Me | 87–89 | EtOAc | 46 | | 195 |
| 1-Me-5-EtO-7-$NO_2$ | 108–109 | $CCl_4$/Pentane | 57 | | 195 |
| 1-Me-5-[2,2-$(EtO)_2$-ethyl]NH-7-Cl | 112–114 | Acetone/Hexane | | | 391b |
| 1-Me-5-(2,2,2-$F_3$-Ethoxy)-7-Cl | 128 | $Et_2O$/Pentane | 32 | | 195 |
| 1-Me-5-(2,2,2-$F_3$-1-Me-Ethoxy)-7-Cl | 95 | $Et_2O$/Pentane | 38 | | 195 |
| 1-Me-5-Ferrocenyl-7-I | 154–156d | $CH_2Cl_2$/Hexane | 79 | | 13 |
| 1-Me-5-(2-HO-Ethyl)NHCO-7-Br | 185–188 | Acetone | | | 192 |
| 1-Me-5-(Isothiazol-5-yl)-7-Cl | 189–191 | $CH_2Cl_2$/Hexane | | | 13b |
| 1-Me-5-(1-Me-1-Pr)-7-Cl | [150–160/0.05] | | 81 | | 18 |
| 1-Me-5-(4-Me-1-Piperazinyl)-7-Cl | 171 | $CHCl_3$ | | | 196 |
| 1-Me-5-(MeNPh)-7-Cl | | | | | |
| Hydrochloride | 273–275 | MeOH | | | 196 |
| 1-Me-5-(3-$Me_2$N-Propylamino)-7-Cl | | | | | |
| Dihydrochloride | 268d | MeOH | | | 196 |
| 1-Me-5-MeO-7-Cl | 114–116 | $CCl_4$/Pentane | 60 | | 195 |
| 1-Me-5-(3-Morpholin-1-yl-propylamino)-7-Cl | | | | | |
| Dihydrochloride | 275d | EtOH/$Et_2O$ | | | 196 |
| 1-Me-5-Ph-7-Ac | 120–125 | $CH_2Cl_2$/$Et_2O$ | | | 160 |
| 4-Oxide | 158–160 | MeOH | | | 2b |
| 1-Me-5-Ph-7-Ac(Allyl)N | 65–67 | | | | 251c |
| 1-Me-5-Ph-7-AcNH | 196–199 | MeOH | | | 251c |
| 4-Oxide | 218–219 | EtOH/$Et_2O$ | | | 251c |

| | | | | | |
|---|---|---|---|---|---|
| 1-Me-5-Ph-7-Ac(Me)N | 226–229 | EtOH | | | 247 |
| 1-Me-5-Ph-7-$H_2N$ | 229–231 | | | | |
| | 238–240 | EtOH | | | 247 |
| 1-Me-5-Ph-7-($H_2N$-Hydroxyiminomethyl) | 253–255d | DMF/EtOH | | | 33b |
| 1-Me-5-Ph-7-($H_2N$-Methoxyiminomethyl) | 194–196 | PhH/Hexane | | | 33b |
| 1-Me-5-Ph-7-$H_2NCH_2$ | 230d | $Et_2O$ | | | 251 |
| Dihydrochloride | 212–215d | MeOH/$Et_2O$ | | | 38b |
| 1-Me-5-Ph-7-$N_3$ | 123–125d | Acetone/Hexane | 46 | | 252 |
| 1-Me-5-Ph-7-Br | 132–133 | $Et_2O$ | | | 2b |
| 1-Me-5-Ph-7-*t*-Bu | 122–124 | MeOH/$H_2O$ | | | 479 |
| 1-Me-5-Ph-7-Cl | 127–130 | $Et_2O$ | | | |
| 4-Oxide | 178–180 | PhMe | | | 30 |
| 1-Me-5-Ph-7-$ClCH_2S$ | 83–85 | $Et_2O$ | | | 2b |
| 1-Me-5-Ph-7-(4-$ClC_6H_4$) | 161–162 | EtOH | | | 481 |
| 1-Me-5-Ph-7-CN | 157–159 | PhH/Petr ether | | uv | 5 |
| 1-Me-5-Ph-7-(1,3-Dioxolan-2-yl) | 138–139 | $Et_2O$/Pentane | | | 32 |
| 1-Me-5-Ph-7-[N-(1-EtOOC)-1-Iminoethyl]NH | 179–180 | $CH_2Cl_2$/$Et_2O$/Hexane | | | 134 |
| 1-Me-5-Ph-7-OCH | 123–125 | $Et_2O$/Pentane | | | 33b |
| 1-Me-5-Ph-7-F | 109–110 | $Et_2O$/Petr ether | 60 | | 2 |
| 1-Me-5-Ph-7-$F_2CHS$ | 123–124 | $Et_2O$/Petr ether | 69 | ir, ms, pmr, uv | 10 |
| 1-Me-5-Ph-7-$F_2CHSO_2$ | 140 | $Et_2O$/Petr ether | 70 | ir, ms, pmr, uv | 10 |
| 1-Me-5-Ph-7-$F_3C$ | 63–65 | | | | |
| | [123/0.07] | | | | 490 |
| 4-Oxide | 184–186 | PhH/Hexane | 71 | uv | 4 |
| 1-Me-5-Ph-7-(Formyl)NH | 145–147 | MeOH/Acetone/Hexane | | | 251c |
| 1-Me-5-Ph-7-$H_2NNH$ | 178–180 | *i*-PrOH | | | 134 |
| 1-Me-5-Ph-7-HONH | 211–213 | $CH_2Cl_2$/EtOH | 49 | | 245 |
| 1-Me-5-Ph-7-(2-HO-Ethylamino)$CONHCH_2$ | 201 | EtOH | | | 251 |
| 1-Me-5-Ph-7-I | 136–138 | EtOH | | | 371 |
| 1-Me-5-Ph-7-Me | 131–133 | $Et_2O$/Hexane | | | 302 |
| 1-Me-5-Ph-7-(2-Me-1,3-dioxolan-2-yl) | 122–124 | Acetone/Hexane | | | 32 |
| 1-Me-5-Ph-7-(2-Me-Propanoyl)NH | 162–165 | EtOAc/Hexane | | | 251c |
| 1-Me-5-Ph-7-MeNH | 134–135 | $Et_2O$ | | | 134 |

TABLE VII-1. —(contd.)

| Substituent | mp (°C) or; [bp (°C/torr)] | Solvent of Crystallization | Yield (%) | Spectra | Refs. |
|---|---|---|---|---|---|
| 1-Me-5-Ph-7-$Me_2N$ | 141–143 | EtOH/$Et_2O$/Hexane | | | 95 |
| 7-Methobromide | 190d | MeOH/$Et_2O$ | | | 95 |
| 1-Me-5-Ph-7-$Me_2NCONH$ | 130 | Amorphous | | | 251b |
| 1-Me-5-Ph-7-$Me_2NCHN$ | 161–163 | $CH_2Cl_2$/Hexane | | | 247 |
| 1-Me-5-Ph-7-MeO | 109–110 | $Et_2O$ | | | 72c |
| 1-Me-5-Ph-7-MeS | 35–45 | Hexane | | | 7 |
| 4-Oxide | 156–158 | MeOH | | | 192 |
| 1-Me-5-Ph-7-Me(O)S | | | | | |
| Hydrochloride | 201–202 | MeOH/$Et_2O$ | | | 2b |
| 1-Me-5-Ph-7-$MeSO_2$ | 178–179 | $CH_2Cl_2$/$Et_2O$ | | | 2b |
| 1-Me-5-Ph-7-$MeSCH_2NH$ | 153–155 | $CH_2Cl_2$/$Et_2O$ Hexane | | | 134 |
| 1-Me-5-Ph-7-(1-Me-1*H*-Tetrazol-5-yl) | 212–214 | PhH/Hexane | 8 | | 252 |
| 1-Me-5-Ph-7-(2-Me-Tetrazol-5-yl) | 201–203 | PhH/Hexane | 46 | | 252 |
| 1-Me-5-Ph-7-$NO_2$ | 159–161 | $CH_2Cl_2$/MeOH | 71 | | 36 |
| 4-Oxide | 219–221d | Acetone/Hexane | | | 1b |
| 1-Me-5-Ph-7-NO | 126–129 | MeOH | | | 247 |
| 1-Me-5-Ph-7-(Propanoyl)NH | 215–217 | MeOH/EtOAc/Hexane | | | 251c |
| 1-Me-5-Ph-7-[(Pyrrolidin-1-yl)$CONHCH_2$] | ~ 106 | Amorphous | | | 251b |
| 1-Me-5-Ph-7-(1*H*-Tetrazol-5-yl) | 278–280d | DMF/$H_2O$ | 82 | | 252 |
| 1-Me-5-Ph-7-(3,4,5-Triazatricyclo[5.2.1.0]dec-4-en-3-yl) | 200–201d | EtOH | 87 | | 252 |
| 1-Me-5-(2-benzyl-$NHC_6H_4$)-7-Cl | 173–176 | $CH_2Cl_2$/$Et_2O$ | | | 152c |
| 1-Me-5-(2-$AcNHC_6H_4$)-7-Cl | 250–251 | MeCN | 100 | | 162 |
| 1-Me-5-[3-(4-$AcNHC_6H_4$-azo)-4-$HOC_6H_3$]-7-Cl | 276–278 | $CH_2Cl_2$/Petr ether | | | 152c |
| 1-Me-5-(2-$H_2NC_6H_3$)-7-Br | 200–205 | $CH_2Cl_2$/Hexane | 87 | | 162 |
| 1-Me-5-(2-$H_2NC_6H_3$)-7-Cl | 203–207 | $CH_2Cl_2$/Hexane | 57 | | 162 |
| 1-Me-5-[3-(4-$H_2NC_6H_4$-azo)-4-$HOC_6H_3$]-7-Cl | 262–268 | $CH_2Cl_2$/MeOH | 45 | ir | 8 |
| 1-Me-5-(2-$ClC_6H_4$)-7-AcNH | Foam | | | | 258 |
| 1-Me-5-(2-$ClC_6H_4$)-7-$H_2N$ | 239–240 | EtOH | | | 72c |

| | | | | |
|---|---|---|---|---|
| 1-Me-5-(2-ClC$_6$H$_4$)-7-Cl | 135–136 | EtOH | 78 | 2 |
| 4-Oxide | 218–220 | | | 358 |
| 1-Me-5-(2-ClC$_6$H$_4$)-7-(2-HO-ethyl)NHCONH | 150–165d | Acetone | | 301 |
| 1-Me-5-(2-ClC$_6$H$_4$)-7-MeNHCONH | 216–220 | Acetone/Petr ether | | 301 |
| 1-Me-5-(2-ClC$_6$H$_4$)-7-Me$_2$N | 110–115 | Et$_2$O/Hexane | | 95 |
| 1-Me-5-(2-ClC$_6$H$_4$)-7-MeS | 115–118 | Et$_2$O/CH$_2$Cl$_2$ | | 192 |
| 1-Me-5-(2-ClC$_6$H$_4$)-7-NO$_2$ | 194–195 | Et$_2$O | | 136 |
| 4-Oxide | 201–204 | MeCN | | 302 |
| 1-Me-5-(2,3-Cl$_2$C$_6$H$_3$)-7-Cl | 164–165 | Acetone/Petr ether | | 2b |
| 1-Me-5-(2,4-Cl$_2$C$_6$H$_3$)-7-Cl | 178–181 | Acetone/Hexane | 70 | 500 |
| 1-Me-5-(2,6-Cl$_2$C$_6$H$_3$)-7-Cl | 160–161 | Et$_2$O/Petr ether | | 42 |
| 1-Me-5-(2-Cl-5-H$_2$NC$_6$H$_3$)-7-Cl | 210–212 | EtOH | | 134 |
| 1-Me-5-(2-Cl-5-MeOC$_6$H$_3$)-7-Cl | 130–132 | Et$_2$O | | 134 |
| 1-Me-5-(2-Cl-5-O$_2$NC$_6$H$_3$)-7-Cl | 237–239 | CH$_2$Cl$_2$/EtOH | | 134 |
| 1-Me-5-(3,4-Cl$_2$C$_6$H$_3$)-7-Cl | 154–157 | MeOH/Petr ether | | 92 |
| 1-Me-5-(4-ClC$_6$H$_4$)-7-Cl | 154–156 | Et$_2$O | 74 | 2 |
| 4-Oxide | 238–240 | THF/Hexane | | 152c |
| 1-Me-5-(4-ClC$_6$H$_4$)-7-F | 160–163 | Acetone/Hexane | 73 | 500 |
| 1-Me-5-(4-ClC$_6$H$_4$)-7-NO$_2$ | 131–135 | Acetone/Hexane | | 136 |
| 1-Me-5-(4-Cl-2-Me-Phenoxy)-7-Cl | 179 | Et$_2$O/Pentane | 55 | 195 |
| 1-Me-5-[3-Cl-2-Pyridyl]-7-Cl | 125–126 | CH$_2$Cl$_2$/Hexane | | 484 |
| 1-Me-5-(2-FC$_6$H$_4$)-7-Ac | 117–120 | CH$_2$Cl$_2$/Petr ether | | 33b |
| 4-Oxide·0.5 acetone | 85–86 | Acetone | | 2b |
| 1-Me-5-(2-FC$_6$H$_4$)-7-AcNH | 172–175 | EtOH/Hexane | | 134 |
| 4-Oxide | 212 | CH$_2$Cl$_2$/Et$_2$O | | 301 |
| 1-Me-5-(2-FC$_6$H$_4$)-7-AcNHCH$_2$ | 199–200 | CH$_2$Cl$_2$/Hexane | | 251b |
| 1-Me-5-(2-FC$_6$H$_4$)-7-Ac(Me)N | | EtOAc | | 301 |
| 1-Me-5-(2-FC$_6$H$_4$)-7-(2-AcO-ethyl)NHCONH | 196–198 | EtOAc | | 231 |
| 1-Me-5-(2-FC$_6$H$_4$)-7(Allyl)NHCSNH | 168–170 | EtOAc | | 301 |
| 1-Me-5-(2-FC$_6$H$_4$)-7-(Allyloxy)CONH | 140–144 | Et$_2$O | | 301 |
| 1-Me-5-(2-FC$_6$H$_4$)-7-H$_2$N | 206–208d | THF/Hexane | 64 | 252, 253 |
| 1-Me-5-(2-FC$_6$H$_4$)-7-(2-H$_2$N-Acetyl)NH | 134–137 | EtOAc/Et$_2$O | | 301 |
| 1-Me-5-(2-FC$_6$H$_4$)-7-NHCONH | 216–217d | EtOH/Et$_2$O | | 301 |

TABLE VII-1. —(contd.)

| Substituent | mp (°C) or; [bp (°C/torr)] | Solvent of Crystallization | Yield (%) | Spectra | Refs. |
|---|---|---|---|---|---|
| 1-Me-5-(2-$FC_6H_4$)-7-(2-$H_2N$-Ethoxy)CONH | 127–130d | $MeOH/Et_2O$ | | | 301 |
| 1-Me-5-(2-$FC_6H_4$)-7-$H_2NC(NH)NH$ | 248–249d | EtOAc/EtOH | | | 301 |
| 1-Me-5-(2-$FC_6H_4$)-7-$H_2NCH_2$ | Amorphous | | | | 251 |
| 1-Me-5-(2-$FC_6H_4$)-7-$H_2NSO_2$ | 299–210 | $CH_2Cl_2$ | | | 301 |
| 1-Me-5-(2-$FC_6H_4$)-7-$N_3$ | 97–99d | EtOAc/Heptane | 44 | | 252 |
| 1-Me-5-(2-$FC_6H_4$)-7-(Aziridin-1-yl)CONH | 186–190d | $EtOAc/Et_2O$ | | | 231 |
| 1-Me-5-(2-$FC_6H_4$)-7-(Benzoyl)NH | 230–234 | $EtOAc/CH_2Cl_2$ | | | 301 |
| 1-Me-5-(2-$FC_6H_4$)-7-(Benzyl)NHCONH | 174–178 | $EtOAc/Et_2O$ | | | 231 |
| 1-Me-5-(2-$FC_6H_4$)-7-(Benzyloxy)CONH | 150d | $EtOAc/Et_2O$/Hexane | | | 301 |
| 1-Me-5-(2-$FC_6H_4$)-7-(Benzyloxy)CO(Me)N | 118–122 | $Et_2O$/Hexane | | | 301 |
| 1-Me-5-(2-$FC_6H_4$)-7-(Benzyloxy)CSNH | 192–193 | $Et_2O$ | | | 301 |
| 1-Me-5-(2-$FC_6H_4$)-7-Br | 132–133 | $Et_2O$ | 3 | | 2 |
| 1-Me-5-(2-$FC_6H_4$)-7-(Butanoyl)NH | 140–150 | $EtOAc/Et_2O$ | | | 301 |
| 1-Me-5-(2-$FC_6H_4$)-7-(Butanoyl)MeN | 105–107 | $Et_2O$/Petr ether | | | 301 |
| 1-Me-5-(2-$FC_6H_4$)-7-BuNHCONH | 213 | $CH_2Cl_2$/EtOAc | | | 301 |
| 1-Me-5-(2-$FC_6H_4$)-7-*t*-BuNHCONH | 210d | EtOH/Petr ether | | | 231 |
| 1-Me-5-(2-$FC_6H_4$)-7-(But-2-yl)OOCNH | 130–140d | $Et_2O$ | | | 301 |
| 1-Me-5-(2-$FC_6H_4$)-7-(But-2-yl)SCONH | 174 | $Et_2O$/Petr ether | | | 301 |
| 1-Me-5-(2-$FC_6H_4$)-7-(But-2-yl)SCSNH | 176–180d | $Et_2O$/Petr ether | | | 301 |
| 1-Me-5-(2-$FC_6H_4$)-7-BuOOCNH | 154–158 | $CH_2Cl_2/Et_2O$/Petr ether | | | 301 |
| 1-Me-5-(2-$FC_6H_4$)-7-*t*-$BuOOCNHCH_2$ | Amorphous | | | | 251b |
| 1-Me-5-(2-$FC_6H_4$)-7-(2-*t*-BuO-Ethyl)NHCONH | 177 | EtOAc | | | 231 |
| 1-Me-5-(2-$FC_6H_4$)-7-(2-*t*-BuO-Ethyl)(benzyl)NCONH | 82–86 | $CH_2Cl_2$/Hexane | | | 231 |
| | 160 | $CH_2Cl_2$/Hexane | | | 301 |
| 1-Me-5-(2-$FC_6H_4$)-7-($HOOCCH_2NH$)CONH | 170–175d | $EtOH/H_2O$ | | | 301 |
| 1-Me-5-(2-$FC_6H_4$)-7-Cl | 69–74 | $MeOH/H_2O$ | 81 | | 2 |
| Hydrochloride | 218–219 | EtOH | | | 86b |
| 4-Oxide | 177–178 | $CH_2/Cl_2/Et_2O$/Petr ether | | | 152c |
| 1-Me-5-(2-$FC_6H_4$)-7-(2,2,2-$Cl_3$-Acetyl)NHCONH | 146–148d | $EtOAc/Et_2O$ | | | 301 |

| | | | |
|---|---|---|---|
| 1-Me-5-(2-$FC_6H_4$)-7-(2-Cl-Ethyl)NHCONH | 240d | EtOAc | 231 |
| 1-Me-5-(2-$FC_6H_4$)-7-(4-$ClC_6H_4$)NHCONH | 223–225 | $CH_2Cl_2$/$Et_2O$ | 301 |
| 1-Me-5-(2-$FC_6H_4$)-7-bis-(4-$ClC_6H_4$NHCO)N | 188 | $CH_2Cl_2$/EtOH | 301 |
| 1-Me-5-(2-$FC_6H_4$)-7-CN | 177 | | 251 |
| 1-Me-5-(2-$FC_6H_4$)-7-(Cyclohexyl)NHCONH | 233–234 | EtOAc/$Et_2O$ | 301 |
| 1-Me-5-(2-$FC_6H_4$)-7-(Cyclohexyl)CONH | 245–250 | EtOAc/$CH_2Cl_2$ | 301 |
| 1-Me-5-(2-$FC_6H_4$)-7-(Cyclopentyl)CONH | 202–203 | EtOAc/$Et_2O$ | 301 |
| 1-Me-5-(2-$FC_6H_4$)-7-(Cyclopentyl)MeN | 122–124 | $Et_2O$/Hexane | 301 |
| 1-Me-5-(2-$FC_6H_4$)-7-(Cyclopropyl)NHCONH | 230 | EtOAc/EtOH/$Et_2O$ | 301 |
| 1-Me-5-(2-$FC_6H_4$)-7-(Cyclopropyl)CONH | 180–185d | EtOAc | 301 |
| 1-Me-5-(2-$FC_6H_4$)-7-Diazonium Tetrafluoroborate | 120–125 | | 251 |
| 1-Me-5-(2-$FC_6H_4$)-7-[1,3-Dihydro-5-(2-$FC_6H_4$)-1-Me-2-oxo-2*H*-1,4-benzodiazepin-7-yl]NHCONH | 230d | EtOH/$Et_2O$/Petr ether | 301 |
| 1-Me-5-(2-$FC_6H_4$)-7-[1,3-Dihydro-5-(2-$FC_6H_4$)-1-Me-2-oxo-2*H*-1,4-benzodiazepin-7-yl]NHCONHNHCONH | 230–245d | EtOAc | 301 |
| 1-Me-5-(2-$FC_6H_4$)-7-[1,3-Dihydro-5-(2-$FC_6H_4$)-1-Me-2-oxo-2*H*-1,4-benzodiazepin-7-yl]-NHCONH$CH_2CH_2$NHCONH | 220–232d | Acetone | 301 |
| 1-Me-5-(2-$FC_6H_4$)-7-{4-[1,3-Dihydro-5-(2-$FC_6H_4$)-1-Me-2-oxo-2*H*-1,4-benzodiazepin-7-yl]NHCO-piperazin-1-yl]NHCO-piperazin-1-yl}CONH | 221d | EtOH/$Et_2O$ | 301 |
| 1-Me-5-(2-$FC_6H_4$)-7-[1,3-Dihydro-5-(2-$FC_6H_4$)-1-Me-2-oxo-2*H*-1,4-benzodiazepin-7-yl]oxyazo | 240–242 | $CH_2Cl_2$/$Et_2O$ | 251b |
| 1-Me-5-(2-$FC_6H_4$)-7-[1,3-Dihydro-5-(2-$FC_6H_4$)-2-oxo-2*H*-1,4-benzodiazepin-7-yl]S | 120–140d | $Et_2O$/Hexane | 301 |
| 1-Me-5-(2-$FC_6H_4$)-7-Et | 105–107 | $Et_2O$/Petr ether | 33b |
| 1-Me-5-(2-$FC_6H_4$)-7-EtNHCONH | 168–170d | THF/$Et_2O$ | 301 |
| 1-Me-5-(2-$FC_6H_4$)-7-$Et_2$NCONH | 138–139 | Acetone/$Et_2O$ | 231 |
| 1-Me-5-(2-$FC_6H_4$)-7-Et(Me)NCONH | 165–168 | EtOAc/$Et_2O$ | 231 |
| 1-Me-5-(2-$FC_6H_4$)-7-(EtOOC$CH_2$NH)CONH | 212–214 | MeOH/EtOAc | 301 |
| 1-Me-5-(2-$FC_6H_4$)-7-(2-EtO-Ethyl)NHCONH | 180–182 | EtOAc/$Et_2O$ | 301 |
| 1-Me-5-(2-$FC_6H_4$)-7-EtOCSNH | 150–152 | EtOH/$Et_2O$ | 301 |
| 1-Me-5-(2-$FC_6H_4$)-7-EtO$CS_2$ | 80–85d | $Et_2O$/Hexane | 301 |

TABLE VII-1. —(contd.)

| Substituent | mp (°C) or; [bp (°C/torr)] | Solvent of Crystallization | Yield (%) | Spectra | Refs. |
|---|---|---|---|---|---|
| 1-Me-5-(2-$FC_6H_4$)-7-EtSCONH | 221–222d | EtOAc/$Et_2O$ | | | 301 |
| 1-Me-5-(2-$FC_6H_4$)-7-EtSCSNH | 150–154d | $CH_2Cl_2$/$Et_2O$ | | | 301 |
| 1-Me-5-(2-$FC_6H_4$)-7-F | 110–111 | Hexane | 82 | | 500 |
| 1-Me-5-(2-$FC_6H_4$)-7-$F_3$CONH | 173–175 | Acetone/Hexane | | | 134 |
| 1-Me-5-(2-$FC_6H_4$)-7-$F_3$CO(Me)N | 189–191 | EtOH | | | 134 |
| 1-Me-5-(2-$FC_6H_4$)-7-(Formyl)NH | 106–107 | MeOH/$Et_2O$ | | | 301 |
| 1-Me-5-(2-$FC_6H_4$)-7-(Formyl)MeN | 176–177 | EtOAc | | | 301 |
| 1-Me-5-(2-$FC_6H_4$)-7-(2-Furoyl)NH | 170–180 | EtOAc/$CH_2Cl_2$ | | | 301 |
| 1-Me-5-(2-$FC_6H_4$)-7-(2-Furoyl)MeN | 195–200 | EtOAc/Petr ether | | | 301 |
| 1-Me-5-(2-$FC_6H_4$)-7-HONH | 228–230d | THF/EtOH | 52 | | 245 |
| 1-Me-5-(2-$FC_6H_4$)-7-[2-(2-HO-Ethoxy)ethoxy]CONH | 146–148 | EtOAc | | | 301 |
| 1-Me-5-(2-$FC_6H_4$)-7-(1-HO-Ethyl) | 133–135 | $Et_2O$/Pentane | | | 33a |
| 1-Me-5-(2-$FC_6H_4$)-7-(2-HO-Ethyl)NHCONH | 156–160d | Acetone | | | 231 |
| 4-Oxide | 256–261 | Acetone | | | 301 |
| | 163–170d | EtOH/$Et_2O$ | | | 301 |
| 1-Me-5-(2-$FC_6H_4$)-7-(2-HO-Ethylaminocarbonyl-aminomethyl) | 125 | | | | 251 |
| 1-Me-5-(2-$FC_6H_4$)-7-(2-HO-Ethyl)(Me)NCONH | 196–198 | Acetone | | | 231 |
| 1-Me-5-(2-$FC_6H_4$)-7-[4-(2-HO-Ethyl)piperazin-1-yl]-CONH | 235 | EtOH | | | 301 |
| 1-Me-5-(2-$FC_6H_4$)-7-(2-HO-Ethoxy)CONH | 188–190 | $CH_2Cl_2$ | | | 301 |
| 1-Me-5-(2-$FC_6H_4$)-7-HO(Me)NCONH | 198–200d | EtOAc/EtOH | | | 301 |
| 1-Me-5-(2-$FC_6H_4$)-7-(2-HO-Prop-1-yl)NHCONH | 149–151d | EtOAc | | | 231 |
| 1-Me-5-(2-$FC_6H_4$)-7-(1-HO-Prop-2-yl)NHCONH | 165–168d | Acetone | | | 231 |
| 1-Me-5-(2-$FC_6H_4$)-7-(3-HO-Propyl)NHCONH | 118–119 | Acetone | | | 231 |
| 1-Me-5-(2-$FC_6H_4$)-7-(2-HO-Prop-1-yloxy)CONH | 116–122d | Acetone/$Et_2O$/Petr ether | | | 301 |
| 1-Me-5-(2-$FC_6H_4$)-7-HS | 125–126 | $Et_2O$/Hexane | | | 301 |
| 1-Me-5-(2-$FC_6H_4$)-7-(2-HS-Ethyl)NHCONH | 186–190d | $CH_2Cl_2$/$Et_2O$ | | | 301 |
| 1-Me-5-(2-$FC_6H_4$)-7-$HSO_3$ | 322d | MeOH/$H_2O$ | | | 301 |

| | | | | |
|---|---|---|---|---|
| 1-Me-5-(2-F$C_6H_4$)-7-I | 107–110 | $Et_2O$/Petr ether | | 26 |
| Hydrochloride | 226–229 | MeOH/$Et_2O$ | | 26 |
| 4-Oxide | 178–181 | EtOH | | 141 |
| 1-Me-5-(2-F$C_6H_4$)-7-(2,2-$Me_2$-Oxazolidin-3-yl)CONH | 184–188d | EtOAc/$Et_2O$ | | 301 |
| 1-Me-5-(2-F$C_6H_4$)-7-(4-Me$C_6H_4SO_2$)NHCONH | 230–234d | Acetone | | 301 |
| 1-Me-5-(2-F$C_6H_4$)-7-(4-Me-Piperazin-1-yl)CONH | 245–246 | EtOAC | | 301 |
| 1-Me-5-(2-F$C_6H_4$)-7-(2-Me-Propanoyl)NH | 142–150 | EtOAc/$Et_2O$ | | 301 |
| 1-Me-5-(2-F$C_6H_4$)-7-MeNH | 109–111 | $Et_2O$ | | 134 |
| 1-Me-5-(2-F$C_6H_4$)-7-MeNHCONH | 173 | EtOAc | | 231 |
| 1-Me-5-(2-F$C_6H_4$)-7-MeNHCO(HO)N·EtOH | 206–208d | EtOH | | 301 |
| 1-Me-5-(2-F$C_6H_4$)-7-MeNHCO(Me)N | 75–80 | EtOH/Petr ether | | 301 |
| 1-Me-5-(2-F$C_6H_4$)-7-MeNH$SO_2$ | 200 | EtOAc/$Et_2O$ | | 301 |
| 1-Me-5-(2-F$C_6H_4$)-7-MeNHCSNH | 210–212 | $CH_2Cl_2$ | | 301 |
| 1-Me-5-(2-F$C_6H_4$)-7-$Me_2$NCONH·0.5 EtOH | 97 | EtOH/Petr ether | | 231 |
| 1-Me-5-(2-F$C_6H_4$)-7-($Me_2$NCONH$CH_2$) | 105 | | | 251 |
| 1-Me-5-(2-F$C_6H_4$)-7-$Me_2$N$SO_2$ | 198–200 | EtOAc/$Et_2O$ | | 301 |
| 1-Me-5-(2-F$C_6H_4$)-7-MeONHCONH | 144–146d | EtOAc | | 301 |
| 1-Me-5-(2-F$C_6H_4$)-7-MeOOCNH | 143–145d | $Et_2O$ | | 301 |
| 1-Me-5-(2-F$C_6H_4$)-7-(2-MeO-Ethoxy)CONH | 140–142 | $CH_2Cl_2$/$Et_2O$ | | 301 |
| 1-Me-5-(2-F$C_6H_4$)-7-MeO(Me)NCONH | 220–222 | $CH_2Cl_2$/EtOAc | | 301 |
| 1-Me-5-(2-F$C_6H_4$)-7-(4-MeO$C_6H_4$)NHCONH | 240d | EtOAc | | 231 |
| 1-Me-5-(2-F$C_6H_4$)-7-MeS | Amorphous | | | 192 |
| 1-Me-5-(2-F$C_6H_4$)-7-Me(O)S | 178d | $CH_2Cl_2$/$Et_2O$ | | 301 |
| 1-Me-5-(2-F$C_6H_4$)-7-Me$SO_4$ | 173–174 | EtOAc/$Et_2O$ | | 301 |
| 4-Oxide | 198–200 | EtOAc | | 301 |
| 1-Me-5-(2-F$C_6H_4$)-7-Me$SO_2$NH | 166–168 | | | 301 |
| 1-Me-5-(2-F$C_6H_4$)-7-(Morpholino)CONH | 237–238d | EtOAc | | 301 |
| 1-Me-5-(2-F$C_6H_4$)-7-$NO_2$ | 165–167 | $Et_2O$/Petr ether | 35 | 55 |
| | 170–172 | $CH_2Cl_2$/Hexane | | 136 |
| 4-Oxide | 166–167 | $CH_2Cl_2$/$Et_2O$ | | 2b |
| 1-Me-5-(2-F$C_6H_4$)-7-(4-$NO_2$-Benzyloxy)CONH | 247–250d | $CH_2Cl_2$ | | 301 |
| 1-Me-5-(2-F$C_6H_4$)-7-NO | Oil | | | 251b |
| 1-Me-5-(2-F$C_6H_4$)-7-(Octadecanoyl)NH·EtOH | 75 | EtOH | | 301 |
| 1-Me-5-(2-F$C_6H_4$)-7-(2-Oxoimidazolidin-1-yl) | 129–132d | $Et_2O$/EtOH | | 301 |

TABLE VII-1. —(contd.)

| Substituent | mp (°C) or; [bp (°C/torr)] | Solvent of Crystallization | Yield (%) | Spectra | Refs. |
|---|---|---|---|---|---|
| 1-Me-5-(2-$FC_6H_4$)-7-(Pentanoyl)NH | 140–144 | EtOAc/$Et_2O$ | | | 301 |
| 1-Me-5-(2-$FC_6H_4$)-7-PhNHCONH | 209–210 | EtOAc/$Et_2O$ | | | 301 |
| 1-Me-5-(2-$FC_6H_4$)-7-(Piperazin-1-yl)CONH | 181d | EtOH/$Et_2O$ | | | 301 |
| 1-Me-5-(2-$FC_6H_4$)-7-(Piperidin-1-yl)CONH | 237–239 | EtOAc | | | 301 |
| 1-Me-5-(2-$FC_6H_4$)-7-*i*-PrNHCONH | 185–188d | EtOAc | | | 231 |
| 1-Me-5-(2-$FC_6H_4$)-7-*i*-PrNH$SO_2$ | 173–174 | EtOAc/$Et_2O$ | | | 301 |
| 1-Me-5-(2-$FC_6H_4$)-7-*i*-PrOOCNH | 164–166 | $CH_2Cl_2$/$Et_2O$ | | | 301 |
| 1-Me-5-(2-$FC_6H_4$)-7-(Pyrrolidin-1-yl)CONH | 159–160 | Acetone | | | 231 |
| 1-Me-5-(2-$FC_6H_4$)-7-(Pyrrolidin-1-yl)CONH$CH_2$ | 145 | | | | 251 |
| 1-Me-5-(2-$FC_6H_4$)-7-SCN | 150 | $Et_2O$ | | | 301 |
| 1-Me-5-(2-$FC_6H_4$)-7-(Thiazolidin-3-yl)CONH | 185d | $CH_2Cl_2$/EtOAc | | | 231 |
| 1-Me-5-(2-$FC_6H_4$)-7-(2-Thienoyl)NH | 162–163d | EtOAc | | | 301 |
| 1-Me-5-(2,6-$F_2C_6H_3$)-7-Cl | 158–162 | $CH_2Cl_2$/MeOH | | | 152c |
| 1-Me-5-(2,6-$F_2C_6H_3$)-7-$NO_2$ | 182–184 | $CH_2Cl_2$/Hexane | | | 152c |
| 1-Me-5-(2-F-5-$H_2NSO_2C_6H_3$)-7-$H_2NSO_2$ | 280d | Acetone/$CH_2Cl_2$ | | | 301 |
| 1-Me-5-(2-F-5-MeNH$SO_2C_6H_3$)-7-MeNH$SO_2$ | 190 | EtOAc | | | 301 |
| 1-Me-5-{2-F-6-[(2-HOOC-Ethyl)MeN]$C_6H_3$}-7-Cl | 188–197 | MeOH/$H_2O$ | | | 152c |
| 1-Me-5-[2-F-6-(HOOC$CH_2$)MeN]-7-Cl | 178–180 | $CH_2Cl_2$/$Et_2O$/MeOH | | | 152c |
| 1-Me-5-[2-F-6-(3-HOOC-Propanoyl)NH]-7-Cl | 214–223d | MeOH | | | 152c |
| 1-Me-5-[2-F-6-(EtOOC$CH_2$)MeN]$C_6H_3$-7-Cl | 174–178 | $CH_2Cl_2$/EtOH | | | 152c |
| 1-Me-5-(2-F-6-HO$C_6H_3$)-7-Cl | 180–188 | $CH_2Cl_2$/Petr ether | | | 152c |
| 1-Me-5-(2-F-6-HOOC$CH_2OC_6H_3$)-7-Cl | 235–238 | THF/$Et_2O$/Petr ether | | | 152c |
| 1-Me-5-(2-F-6-MeNH$C_6H_3$)-7-Cl | 172–177 | $CH_2Cl_2$/Hexane | | | 152c |
| 1-Me-5-(2-F-6-$Me_2NC_6H_3$)-7-Cl | 123–127 | $Et_2O$/Petr ether | | | 152c |
| 1-Me-5-(2-F-6-MeO$C_6H_3$)-7-Cl | 177–185 | PhH/Hexane | | | 152c |
| 1-Me-5-(2-F-6-MeOOC$CH_2OC_6H_3$)-7-Cl | 173–177 | $CH_2Cl_2$/Petr ether | | | 152c |
| 1-Me-5-(2-F-Phenoxy)-7-Cl | 128 | $Et_2O$/Pentane | 50 | | 195 |
| 1-Me-5-(3-$FC_6H_4$)-7-Cl | Oil | | | | 91a |
| 1-Me-5-(4-$FC_6H_4$)-7-Cl | 160–162 | $Et_2O$ | | X-ray | 403 |

| | | | | | |
|---|---|---|---|---|---|
| 1-Me-5-(3-F-2-Pyridyl)-7-Cl | 119–122 | $CH_2Cl_2$/Hexane | | | 484 |
| 1-Me-5-(2-$F_3CC_6H_4$)-7-$H_2N$ | 213–215 | MeOH | | | 247 |
| 1-Me-5-(2-$F_3CC_6H_4$)-7-$NO_2$ | 198–199 | Acetone | | | 6 |
| 1-Me-5-(2-$F_3CC_6H_4$)-7-$Me_2N$ | 110–115 | $Et_2O$/Hexane | | | 95 |
| 1-Me-5-[3-(4-$F_3CCONHC_6H_4$)azo-4-$HOC_6H_3$]-7-Cl | 276–278 | $CH_2Cl_2$/Petr ether | 70 | ir | 8 |
| 1-Me-5-$H_2NNH$-7-Cl | 139–141 | PhMe/Me-C-Hexane | 32 | ir, pmr | 198b |
| 1-Me-5-(4-$HOC_6H_4$)-7-Cl | 255–259 | MeOH | 6 | | 8 |
| 1-Me-5-[4-(2-HO-3-*i*-PrNH-Propoxy)$C_6H_4$]-7-Cl | | | | | |
| Maleate | 167–170 | Acetone/$Et_2O$ | | | 487 |
| 1-Me-5-(2-$MeC_6H_4$)-7-Cl | 137–139 | MeOH | 90 | | 2 |
| 1-Me-5-(2,4-$Me_2C_6H_3$)-7-Cl | 174–176 | $CH_2Cl_2$/Hexane | | | 92 |
| 1-Me-5-(2-$Me_2NC_6H_4$)-7-Cl | 157–158 | $Et_2O$ | | | 152c |
| 1-Me-5-(1-Me-Imidazol-2-yl)-7-Cl | 169–171 | $CHCl_3$/Hexane | | | 27 |
| 1-Me-5-(2-Me-Pyrazol-3-yl)-7-Cl | 144–147 | $CHCl_3$/Hexane | | | 27 |
| 1-Me-5-(2-$MeOC_6H_4$)-7-Cl | 161–162 | PhH/Hexane | 8 | | 2 |
| 1-Me-5-(4-$MeOC_6H_4$)-7-Cl | 112–115 | Hexane | | | 1b |
| 1-Me-5-[3,4,5-$(MeO)_3C_6H_2$]-7-Cl | 140–142 | | | | 86b |
| 1-Me-5-(2-$O_2NC_6H_4$)-7-Br | 189–190 | $CH_2Cl_2$/Hexane | 68 | | 162 |
| 1-Me-5-(2-$O_2NC_6H_4$)-7-Cl | 174–175 | $CH_2Cl_2$/Hexane | 40 | | 162 |
| 1-Me-5-(2-$O_2NC_6H_4$)-7-$NO_2$ | 209–212 | MeOH | | | 136, 86b |
| 1-Me-5-(3-$O_2NC_6H_4$)-7-$NO_2$ | 218–219 | Acetone | | | 152c |
| 1-Me-5-(4-$O_2NC_6H_4$)-7-Cl | 162–163 | EtOH | | | 302 |
| 1-Me-5-(4-$PhC_6H_4$)-7-Cl | 187–189 | $CH_2Cl_2$/MeOH | | | 92 |
| 1-Me-5-PhO-7-Cl | 111–113 | $Et_2O$/Pentane | 55 | | 195 |
| 1-Me-5-PhS-7-Cl | 147 | MeOH | | | 196 |
| 1-Me-5-[2-(Piperidin-1-yl)$C_6H_4$]-7-I | 182–184 | EtOAc/$Et_2O$ | | | 33b |
| 1-[4-(4-Me-Piperazin-1-yl)butyl]-5-(2-pyridyl)-7-Br | | | | | |
| Dihydrochloride | 150–163d | EtOH/$Et_2O$ | | | 67b |
| 1-[2-(4-Me-Piperazin-1-yl)ethyl]-5-Ph-7-Cl | 159–160 | Acetone/Hexane | 66 | | 287, 86b |
| Diamaleate | 158–160 | MeOH/Hexane | 59 | | 287 |
| 1-[2-(4-Me-Piperazin-1-yl)ethyl]-5-(2-$FC_6H_4$)-7-Cl | | | | | |
| Trihydrochloride | 225–234 | MeOH/$Et_2O$ | 37 | | 287 |
| 1-[3-(4-Me-Piperazin-1-yl)propyl]-5-Ph-7Cl | | | | | |
| Diamaleate | 180–182 | MeOH | 80 | | 287 |

TABLE VII-1. —(contd.)

| Substituent | mp (°C) or; [bp (°C/torr)] | Solvent of Crystallization | Yield (%) | Spectra | Refs. |
|---|---|---|---|---|---|
| 1-[3-(4-Me-Piperazin-1-yl)propyl]-5-Ph-7-Cl | | | | | |
| Dimaleate | 178–179 | MeOH/$Et_2O$ | | | 255c |
| 1-[3-(4-Me-Piperazin-1-yl)propyl]-5-(2-$FC_6H_4$)-7-Cl | | | | | |
| Dimaleate | 185–187 | MeOH | 78 | | 287 |
| 1-Me-5-Pr-7-Cl | 72; 87 | Hexane | 70 | | 18 |
| 1-Me-5-*i*-Pr-7-Cl | 108 | Hexane | 75 | | 18 |
| 1-Me-5-PrO-7-Cl | 70 | $Et_2O$/Pentane | 15 | | 195 |
| 1-Me-5-(Propyn-3-yloxy)-7-Cl | 120 | $Et_2O$/Pentane | 40 | | 195 |
| 1-Me-5-(2-Pyridyl)-7-$H_2N$ | 228–230 | $CH_2Cl_2$/PhH | | | 33b |
| 1-Me-5-(2-Pyridyl)-7-Br | 136–137 | EtOAc | 70 | | 3 |
| 4,1′-Dioxide | 245–248 | $CH_2Cl_2$/Hexane | | | 152c |
| 1-Me-5-(2-Pyridyl)-7-$Me_2N$ | 151–153 | $CH_2Cl_2$/Heptane | | | 33b |
| 1-Me-5-(2-Pyridyl)-7-MeS | 134–137 | $CH_2Cl_2$/$Et_2O$/Hexane | | | 192 |
| 1-Me-5-(2-Pyridyl)-7-$NO_2$ | 217–219d | PhH | | | 33b |
| 1-Me-5-(2-Pyrimidyl)-7-Cl | 157–159 | $CH_2Cl_2$/$Et_2O$/Hexane | 6 | ir | 28 |
| 1-Me-5-(2-Pyrimidyl)-7-$NO_2$ | 194–197d | $CH_2Cl_2$/Hexane | | | 158 |
| 1-Me-5-(2-Thiazolyl)-7-Cl | 204–206 | $CH_2Cl_2$/MeOH | | | 27 |
| 1-Me-5-(2-Thiazolyl)-7-I | 200–201 | $CH_2Cl_2$/EtOH | | | 27 |
| 1-(3-Me-But-1-yl)-5-Ph-7-Cl | 84–85 | $Et_2O$/Petr ether | | | 2b |
| 1-(3-Me-But-1-yl)-5-Ph-7-$NO_2$ | 121–122 | $Et_2O$ | | | 136 |
| 1-(2,2-$Me_2$-1,3-Dioxolan-4-yl)$CH_2$-5-Ph-7-Cl | 115–118 | | | | 285 |
| 1-(2-Me-1,3-Dioxolan-2-yl)propyl-5-(2-$FC_6H_4$)-7-Cl | 152–153 | EtOH | | | 302 |
| 1-[2-(5-Me-Isoxazol-3-yl)COO-1-MeO-ethyl]5-Ph-7-Cl | 145–147 | EtOH | | | 302 |
| 4-Oxide | 210–211 | $CH_2Cl_2$/$Et_2O$ | | | 302 |
| 1-[2-(4-Me-Oxazol-5-yl)COO-1-MeO-ethyl]-5-Ph-7-Cl | 151–154 | EtOH | | | 302 |
| 1-(4-$MeC_6H_4SO_2NHCO$)-5-Ph-7-$NO_2$ | 242–243 | Acetone | | | 491 |
| 1-[2-(2,2-$Me_2$-Propanoyloxy)-l-MeO-ethyl]-5-Ph-7-Cl | 101–103 | $Et_2O$/Petr ether | | | 302 |
| 1-(2-Me-Prop-1-yl)-5-Ph-7-Cl | 96–98 | $Et_2O$ | | | 2b |
| 1-(2,2-$Me_2$-Propyl)-5-Ph-7-Cl | 139–141 | Hexane | | | 1b |

| | | | | | |
|---|---|---|---|---|---|
| 1-[2-Me-4-(Pyridin-3-yl)COO-but-2-en-1-yl]-5-Ph-7-Cl | 142–143 | MeOH | | | 486 |
| 1-[2-$Me_3$SiO-1-MeO-Ethyl]5-Ph-7-Cl | 109–110 | MeCN | | | 302 |
| 1-MeNHCO-5-Ph-7-$H_2N \cdot Et_2O$ | 131–135 | $Et_2O$ | | | 2b |
| 1-MeNHCO-5-Ph-7-Cl | 148–149 | $Et_2O$ | 90 | | 311 |
| 1-MeNHCO-5-Ph-7-$F_3C$ | 159–160 | $CH_2Cl_2/Et_2O$/Petr ether | | | 2b |
| 1-MeNHCO-5-Ph-7-I | 154–155 | $CH_2Cl_2/Et_2O$/Petr ether | | | 2b |
| 1-MeNHCO-5-Ph-7-$NO_2$ | 160 | PhH/Acetone | | ir, pmr | 74 |
| 1-MeNHCO-5-(2-$ClC_6H_4$)-7-$NO_2$ | 141–142 | $CH_2Cl_2/Et_2O$ | | | 2b |
| 1-MeNHCO-5-(2-$FC_6H_4$)-7-$H_2N \cdot Et_2O$ | 127–130 | $Et_2O$ | | | 301 |
| 1-MeNHCO-5-(2-$FC_6H_4$)-7-Cl | 145–146 | $CH_2Cl_2/Et_2O$ | | | 2b |
| 1-MeNHCO-5-(2-$FC_6H_4$)-7-I | 158–159 | $CH_2Cl_2/Et_2O$/Petr ether | | | 2b |
| 1-MeNHCO-5-(2-$FC_4H_4$)-7-MeNHCONH | 186d | EtOAc/$Et_2O$/EtOH | | | 301 |
| 1-MeNHCO-5-(2-$FC_4H_4$)-7-$NO_2$ | 144 | $CH_2Cl_2/Et_2O$ | | | 301 |
| 1-MeNHCO-5-(2-Pyridyl)-7-Br | 158–160 | | | | |
| | 240–245 | $CH_2Cl_2/Et_2O$/Petr ether | | | 2b |
| 1-(1-MeNHCO-Ethyl)-5-Ph-7-Cl | 189–190 | Acetone | | | 293 |
| 1-$MeNHCOCH_2$-5-Ph-7-$H_2N$ | 176–178 | $CH_2Cl_2/Et_2O$ | | | 2b |
| 1-$MeNHCOCH_2$-5-Ph-7-Br | 257–259 | EtOH | | | 255c |
| 1-$MeNHCOCH_2$-5-Ph-7-Cl | 252–254d | Acetone | | | 293, 86b |
| Methohydrosulfate | 173–175 | MeOH/$Et_2O$ | | | 293 |
| 4-Oxide | 269 | Acetone | | | 293 |
| 1-$MeNHCOCH_2$-5-Ph-7-CN | 224–226 | Acetone | | | 38b |
| 1-$MeNHCOCH_2$-5-Ph-7-$F_3C$ | 260–261 | Acetone | | | 293 |
| 1-$MeNHCOCH_2$-5-Ph-7-$NO_2$ | 231–232 | Acetone | | | 293 |
| 1-$MeNHCOCH_2$-5-(2-$ClC_6H_4$)-7-Cl | 197–199 | $CH_2Cl_2$/Hexane | | | 293 |
| 1-$MeNHCOCH_2$-5-(2-$FC_6H_4$)-7-Cl | 212–214 | | | | 86b |
| 1-$MeNHCOCH_2$-5-(2-$FC_6H_4$)-7-$NO_2$ | 207–208 | Acetone | | | 293 |
| 1-$MeNHCOCH_2$-5-(2-Pyridyl)-7-Br | 230–232 | Acetone/Petr ether | | | 293 |
| 1-[2-(MeNHCO-Methoxy)ethyl]-5-Ph-7-Cl | 140–142 | $CH_2Cl_2/Et_2O$ | | | 150d |
| 1-[2-(MeNHCO-Methoxy)ethyl]-5-(2-$ClC_6H_4$)-7-Cl | 131–133 | $CH_2Cl_2/Et_2O$ | | | 302 |
| 1-[2-(N-$MeNHCOCH_2$-N-Me-Amino)ethyl]-5-Ph-7-Cl | 111–113 | MeCN/$Et_2O$ | | | 150c, d |
| 1-[2-(N-$MeNHCOCH_2$-N-Me-Amino)ethyl]-5-Ph-7-$NO_2$ | 158–160 | EtOH | | | 302 |
| 1-(2-MeNHCOO-Ethyl)-5-(2-$FC_6H_4$)-7-Cl | | | | | |
| Hydrochloride | 175–180 | MeOH/$Et_2O$ | | | 283 |

TABLE VII-1. —(contd.)

| Substituent | mp (°C) or; [bp (°C/torr)] | Solvent of Crystallization | Yield (%) | Spectra | Refs. |
|---|---|---|---|---|---|
| 1-(2-MeNHCOO-1-MeO-Ethyl)-5-Ph-7-Cl | 135–137 | $Et_2O$/Petr ether | | | 302 |
| 1-(2-MeHN-Ethoxy)$CH_2$-5-Ph-7-Cl | | | | | |
| Oxalate | 208–210 | EtOH | | | 72 |
| 1-(2-MeNH-Ethyl)-5-5-Ph-7-Cl | | | | | |
| Dihydrochloride | 220–223d | MeOH/$Et_2O$ | | | 150c |
| 1-[2-(2-Me-4-$NO_2$-Imidazol-1-yl)ethyl]-5-Ph-7-Cl | 230–232 | | | | 55 |
| 1-[2-(2-Me-5-$NO_2$-Imidazol-1-yl)ethyl]-5-Ph-7-Cl | 168–170 | PhH/$Et_2O$ | 47 | | 55 |
| 1-(3-MeNH-Propyl)-5-(2-$FC_6H_4$)-7-Cl | | | | | |
| Dihydrochloride | 193–196d | MeOH/$Et_2O$ | 86 | | 287, 86b |
| 4-Oxide | 134–139 | THF/$Et_2O$ | | | 152c |
| 1-MeO-5-Ph-7-Cl | | | | | |
| 4-Oxide | 195–197 | EtOAc | 34 | ir, ms, pmr | 189 |
| 1-[2-[3,4,5-$(MeO)_3$Benzoyloxy]ethyl]-5-(2-$FC_6H_4$)-7-Cl | 161–163 | $CH_2Cl_2$/Hexane | | | 283 |
| 1-MeOOC-5-Ph-7-Cl | 168–170 | EtOAc/Hexane | | | 108 |
| 1-MeOOC-5-Ph-7-$NO_2$ | 165–167d | EtOAc/Hexane | | | 108 |
| 1-MeOOC(MeO)CH-5-Ph-7-Cl | 195–196 | MeOH | | | 108 |
| 1-MeOOC(MeO)CH-5-Ph-7-$NO_2$ | 157–159 | MeOH | | | 108 |
| 1-(2-MeOOC-Methoxy)ethyl-5-Ph-7-Cl | | | | | |
| Hydrochloride | 179–182d | MeOH/$Et_2O$ | | | 150c |
| 1-MeOOC$CH_2$-5-Ph-7-Cl | 137–138 | MeOH | 42 | | 293, 292 |
| 1-MeOOC$CH_2$-5-Ph-7-$NO_2$ | 139–141 | MeOH | | | 255c |
| 1-MeOOC$CH_2$-5-(2-$ClC_6H_4$)-7-Cl | 193–194 | $CH_2Cl_2$/MeOH | | | 292 |
| 1-MeOOC$CH_2$-5-(2-F-6-MeOOC$CH_2$O-$C_6H_3$)-7-Cl | 147–150 | $CH_2Cl_2$/$Et_2O$ | | | 152c |
| 1-(MeOOC)MeOCH-5-Ph-7-Cl | 195–196 | $CH_2Cl_2$/MeOH | | | 72 |
| 1-(MeOOC)MeOCH-5-Ph-7-$NO_2$ | 157–159 | MeOH | | | 72 |
| 1-[2-(N-MeOOC$CH_2$-Amino)ethyl]-5-Ph-7-Cl | | | | | |
| Tosylate | 147–150d | Acetone/$Et_2O$ | | | 150c |
| 1-(3-MeOOC-3,3-$Me_2$-Propyl)-5-Ph-7-Cl | 130–132 | MeOH | | | 294 |
| 1-(1-MeOOC-Prop-1-en-1-yl)-5-Ph-7-Cl | 125–127 | $Et_2O$ | | | 294 |

| | | | | | |
|---|---|---|---|---|---|
| 1-(3-MeOOC-Prop-2-en-1-yl)-5-Ph-7-Cl, *trans* | 125–127 | MeOH | | | 302 |
| 1-[2-(MeO-Acetoxy)-1-MeO-ethyl]-5-Ph-7-Cl | 103–105 | $Et_2O$/Petr ether | | | 302 |
| 1-[2-(2-MeO-Ethoxy)ethyl]-5-Ph-7-Cl | 93–89 | *i*-$Pr_2O$/*i*-PrOH | | | 89 |
| 4-Oxide | 132–133 | | | | 89 |
| 1-[2-(2-MeO-Ethoxy)ethyl]-5-(2-$FC_6H_4$)-7-Cl | 70–72 | | | | 89 |
| 1-(1-MeO-1-Ethyl)-5-Ph-7-Cl | 131–132 | $CH_2Cl_2$/Hexane | | | 108 |
| 1-(1-MeO-1-Ethyl)-5-Ph-7-$NO_2$ | 189–190 | MeOH | | | 72 |
| 1-(2-MeO-Ethoxy)$CH_2$-5-Ph-7-$NO_2$ | 120–121 | MeOH | | | 72 |
| 1-(2-MeO-Ethyl)-5-Ph-7-Cl | 108–109 | | | | 45, 125h |
| 4-Oxide | 137–138 | *i*-PrOH | | | 234 |
| 1-(2-MeO-Ethyl)-5-(2-$FC_6H_4$)-7-Cl | 93–95 | | | | 125h |
| 1-(2-MeO-Ethyl)-5-(2-$MeC_6H_4$)-7-Cl | 115–117 | *i*-PrOH/*i*-$Pr_2O$ | | | 276 |
| 1-[2,2-$(MeO)_2$-Ethyl]-5-Ph-7-Cl | 117–119 | *i*-PrOH | 73 | | 296 |
| 1-[1-MeO-2-(1,5-$Me_2$-Pyrazol-3-yl)COO]-5-Ph-7-Cl | 197–199 | EtOH | | | 302 |
| 1-(1-MeO-2-PhNHCOO-ethyl)-5-Ph-7-Cl | 139–142d | $Et_2O$/Petr ether | | | 302 |
| 1-[1-MeO-2-(3-Ph-*trans*-Propenoyloxy)ethyl]-5-Ph-7-Cl | 144–146 | EtOH | | | 302 |
| 1-[1-MeO-2-(3-Phthalimidopropanoyloxy)ethyl]-5-Ph-7-Cl | 157–158 | EtOH | | | 486 |
| 1-[1-MeO-2-(Pyridin-3-yl)COO-ethyl]-5-Ph-7-AcNH | 203–205 | MeOH | | | 486 |
| 1-[1-MeO-2-(Pyridin-3-yl)COO-ethyl]-5-Ph-7-$NO_2$ | 150–152 | MeOH | | | 486 |
| 1-[1-MeO-2-(Pyridin-3-yl)COO-ethyl]-5-(2-$FC_6H_4$)-7-Cl | 155–156 | $Et_2O$ | | | 486 |
| 1-[1-MeO-2-(Pyridin-3-yl)COO-ethyl]-5-(2-pyridyl)-7-Br | 147–150d | *i*-PrOH | | | 302 |
| 1-$MeOCH_2$-5-Ph-7-AcNH | 150–151 | *i*-PrOH | | | 72c |
| 1-$MeOCH_2$-5-Ph-7-Ac(HO)N | 205–208d | $CH_2Cl_2$/*i*-PrOH | 45 | pmr, uv | 245 |
| 1-$MeOCH_2$-5-Ph-7-Ac(MeO)N | 125–128 | $CH_2Cl_2$/$Et_2O$/Hexane | 46 | pmr, uv | 245 |
| 1-$MeOCH_2$-5-Ph-7-$H_2N$ | 146 | EtOAc/Petr ether | | | 72 |
| 4-Oxide | 186–188 | *i*-PrOH/$Et_2O$ | | | 108 |
| 1-$MeOCH_2$-5-Ph-7-$H_2NCONH$ | 223–224 | EtOH/$Et_2O$ | | | 72c |
| 1-$MeOCH_2$-5-Ph-7-(Benzyloxy)CONH | 169–171 | $CH_2Cl_2$/Petr ether | | | 108 |
| 1-$MeOCH_2$-5-Ph-7-Cl | Oil | | | | 72 |
| 4-Oxide | 164–166 | MeOH | 68 | | 277, 72 |
| 1-$MeOCH_2$-5-Ph-7-$F_3C$ | 122–124 | Hexane | 53 | | 56 |
| 1-$MeOCH_2$-5-Ph-7-$F_3CONH$ | 172–174 | $CH_2Cl_2$/$Et_2O$ | | | 108 |
| 4-Oxide | 230–232d | $CH_2Cl_2$/$Et_2O$ | | | 108 |
| 1-$MeOCH_2$-5-Ph-7-$F_3CCO(HO)N$ | 175–178d | $CH_2Cl_2$/MeOH/$Et_2O$ | 75 | ir, pmr, uv | 245 |

TABLE VII-1. —(contd.)

| Substituent | mp (°C) or; [bp (°C/torr)] | Solvent of Crystallization | Yield (%) | Spectra | Refs. |
|---|---|---|---|---|---|
| 1-$MeOCH_2$-5-Ph-7-$F_3CCO(MeO)N$ | 110–112 | $Et_2O$/Hexane | 53 | pmr, uv | 245 |
| 1-$MeOCH_2$-5-Ph-7-HONH | 168–170 | $CH_2Cl_2/Et_2O$ | 70 | ir, pmr, uv | 245 |
| 1-$MeOCH_2$-5-Ph-7-MeNH | 116–119 | $CH_2Cl_2$/Petr ether | | | 108 |
| 1-$MeOCH_2$-5-Ph-7-$Me_2N$ | 116–117 | EtOH | | | 72c |
| 1-$MeOCH_2$-5-Ph-7-MeO | | | | | |
| Hydrochloride | 198–200 | MeOH/$Et_2O$ | | | 72c |
| 1-$MeOCH_2$-5-Ph-7-MeOOCNH | 201–203 | THF/Petr ether | | | 108 |
| 1-$MeOCH_2$-5-Ph-7-MeOOC(Me)N | 135–138 | $CH_2Cl_2$/Petr ether | | | 108 |
| 1-$MeOCH_2$-5-Ph-7-Me(MeO)N | | | 12 | pmr | 245 |
| 1-$MeOCH_2$-5-Ph-7-MeN(O)N | 110–115 | | 37.5 | | 245 |
| 1-$MeOCH_2$-5-Ph-7-MeONH | 147–149 | $Et_2O$ | 84 | pmr, uv | 245 |
| 1-$MeOCH_2$-5-Ph-7-MeS(O) | Amorphous | | 55 | | 56 |
| 1-$MeOCH_2$-5-Ph-7-$MeSO_2NH$ | 163–165 | EtOH | | | 72c |
| 1-$MeOCH_2$-5-Ph-7-$NO_2$ | 139–141 | PhH/EtOH | | | 72 |
| 4-Oxide | 213–215 | $CH_2Cl_2$/EtOAc | | | 134 |
| 1-$MeOCH_2$-5-Ph-7-NO | Oil | | 55 | pmr | 245 |
| 1-$MeOCH_2$-5-Ph-7-N(O)N-[1,3-Dihydro-1-$MeOCH_2$ 2-oxo-5-Ph-2*H*-1,4-benzodiazepin-7-yl] | 240–242 | $CH_2Cl_2$/Hexane | 53 | ms, pmr, uv | 245 |
| 4-Oxide | 212–215 | $CH_2Cl_2$/EtOAc | | | 72 |
| 1-$MeOCH_2$-5-(2-$ClC_6H_4$)-7-$H_2N$ | 200–202 | $CH_2Cl_2$/Hexane | | | 134 |
| 1-$MeOCH_2$-5-(2-$ClC_6H_4$)-7-Cl | 139–140 | MeOH | | | 72 |
| 1-$MeOCH_2$-5-(2-$ClC_6H_4$)-7-HONH | 205–208d | THF/*i*-PrOH | 90 | | 245 |
| 1-$MeOCH_2$-5-(2-$ClC_6H_4$)-7-$NO_2$ | 136–137 | MeOH | | | 72 |
| 1-$MeOCH_2$-5-(2-$FC_6H_4$)-7-$H_2N$ | 169–170 | EtOH/$Et_2O$ | | | 72c |
| 1-$MeOCH_2$-5-(2-$FC_6H_4$)-7-Cl | 113–114 | MeOH | | | 72 |
| 4-Oxide | 150–151 | | | | 234 |
| 1-$MeOCH_2$-5-(2-$FC_6H_4$)-7-I | 115–117 | Acetone/Heptane | | | 134 |
| 1-$MeOCH_2$-5-(2-$FC_6H_4$)-7-$Me_2N$ | 164–165 | EtOH | | | 72c |
| 1-$MeOCH_2$-5-(2-$FC_6H_4$)-7-$NO_2$ | 105–107 | EtOH | | | 134 |

| | | | | | |
|---|---|---|---|---|---|
| 1-MeOCH$_2$-5-(2-MeC$_6$H$_4$)-7-Cl | 130–131 | | | | 276 |
| 1-MeOCH$_2$-5-(2-Pyridyl)-7-H$_2$N | 162–164 | CH$_2$Cl$_2$/Hexane | | | 134 |
| 1-MeOCH$_2$-5-(2-Pyridyl)-7-NO$_2$ | 164–166 | CH$_2$Cl$_2$/Hexane | | | 134 |
| 1-(2-MeO-Benzoyl)CH$_2$-5-Ph-7-Cl | 183–185 | EtOH | 67 | ir, pmr, uv | 65 |
| 1-[2-(3,4,5-MeO$_3$-Benzoyloxy)ethyl]-5-(2-FC$_6$H$_4$)-7-Cl | 161–163 | | | | 86b |
| 1-[2-(3,4,5-MeO$_3$-Benzoyloxy)-1-MeO-ethyl]-5-Ph-7-Cl | 146–148 | CH$_2$Cl$_2$/Hexane | | | 72 |
| 1-[2-(4-MeOC$_6$H$_4$)Acetoxy-1-MeO-ethyl]-5-Ph-7-Cl | 95–97 | Et$_2$O/Hexane | | | 72 |
| 1-{3-[4-(2-MeOC$_6$H$_4$)-Piperazin-1-yl]propyl}-5-Ph-7-Cl | | | | | |
| Dihydrochloride | 245–250 | CH$_2$Cl$_2$/Acetone | | | 488 |
| 1-[2-(2-MeO-3-Pyridincarbonyloxy)ethyl]-5-(2-FC$_6$H$_4$)-7-Cl | | | | | |
| Dihydrochloride | 154–158 | MeOH/Et$_2$O | | | 283 |
| 1-[2-(N-Me$_2$NCO-N-Me-Amino)ethyl]-5-Ph-7-Cl | 128–131 | Et$_2$O | | | 302 |
| 1-Me$_2$NCOCH$_2$-5-Ph-7-Cl | 179–181 | Acetone/Hexane | | | 293, 86b |
| Hydrochloride | 243–245 | MeOH/Et$_2$O | | | 293 |
| 1-[2-(Me$_2$NCO-Methoxy)ethyl]-5-Ph-7-Cl | 120–122 | CH$_2$Cl$_2$/Et$_2$O | | | 150c, d |
| 1-(2-Me$_2$N-Ethoxy)CH$_2$-5-Ph-7-Cl | | | | | |
| Hydrochloride | 175–177 | Acetone/Et$_2$O | | | 72 |
| 1-(2-Me$_2$N-Ethyl)-5-Ph-7-Cl | 96–98 | Et$_2$O/Hexane | 23 | | 287 |
| Methochloride·H$_2$O | 185–187 | MeOH/Et$_2$O | | | 255c |
| 4-Oxide | 146–147 | Acetone/Petr ether | 64 | | 287 |
| | 211–212 | | | | 358 |
| Hydrochloride | 217–218 | EtOH/Et$_2$O | 58 | | 287 |
| 1-(2-Me$_2$N-Ethyl)-5-Ph-7-F$_3$C | | | | | |
| Dihydrochloride | 217–221 | MeOH/Et$_2$O | 52 | | 287 |
| 1-(2-Me$_2$N-Ethyl)-5-Ph-7-NO$_2$ | 121–122 | Et$_2$O/Petr ether | 44 | | 287 |
| Dihydrochloride | 232–233d | MeOH/Et$_2$O | 40 | | 287 |
| 1-(2-Me$_2$N-Ethyl)-5-(2-ClC$_6$H$_4$)-7-Cl | 178–180 | CH$_2$Cl$_2$/Et$_2$O/Hexane | 76 | | 238 |
| Dihydrochloride·*i*-PrOH | 168–170 | *i*-PrOH | | | 134 |
| 4-Oxide | 152–154 | Et$_2$O/Pentane | 18 | | 288 |
| $\omega$-Oxide·H$_2$O | 131–133 | CH$_2$Cl$_2$/Et$_2$O | 63 | | 238 |

TABLE VII-1. —(contd,)

| Substituent | mp (°C) or; [bp (°C/torr)] | Solvent of Crystallization | Yield (%) | Spectra | Refs. |
|---|---|---|---|---|---|
| 1-(2-$Me_2N$-Ethyl)-5-(2-$FC_6H_4$)-7-$NO_2$ | | | | | |
| Dihydrochloride | 232–233d | | | | 126 |
| 1-(2-$Me_2N$-Ethyl)-5-(2,6-$F_2C_6H_3$)-7-Cl | 156–159 | $CH_2Cl_2$/Hexane | | | 152c |
| 1-(2-$Me_2N$-1-Me-Ethyl)-5-Ph-7-Cl | 134–135 | Hexane | 55 | | 287 |
| Dihydrochloride | 165–168d | *i*-PrOH/$Et_2O$ | 50 | | 86b, 287 |
| 1-(3-$Me_2N$-Propyl)-5-Ph-7-Br | | | | | |
| Dinitrate | 154–155 | MeOH/$Et_2O$ | | | 2b |
| 4-Oxide | 121–122 | $Et_2O$ | | | 2b |
| 1-(3-$Me_2N$-Propyl)-5-Ph-7-Cl | 90–92 | Hexane | 64 | | 86b, 287 |
| 1-(2-$Me_2N$-Propyl)-5-Ph-7-$NO_2$ | | | | | |
| Hydrochloride | 192–193d | EtOH | 60 | | 287 |
| 1-(3-$Me_2N$-Propyl)-5-(2-$FC_6H_4$)-7-Cl | | | | | |
| Dihydrochloride | 202–207d | MeOH/$Et_2O$ | | | 86b |
| | 180–200d | MeOH/$Et_2O$ | 61 | | 287 |
| 4-Oxide | 149–151 | $CH_2Cl_2$/$Et_2O$/Hexane | | | 152c |
| 1-(3-$Me_2N$-Propyl)-5-(2-Pyridyl)-7-Br | | | | | |
| Dihydrobromide | 130–146d | MeOH/$Et_2O$ | | | 152c |
| Dihydrochloride | 181–183d | MeOH/$Et_2O$ | 48 | | 86b, 287 |
| 4-Oxide | | | | | |
| Hydrochloride | 135–150d | *i*-PrOH/$Et_2O$ | | | 67b |
| 1-(3-$Me_2N$-Propyl)-5-(4-Pyridyl)-7-Br | 97–99 | $Et_2O$/Cyclohexane | | | 38b |
| 1-$Me_2CN$-5-Ph-7-Cl | 170–172 | $Et_2O$ | | | 247 |
| 1-[$Me_2P(O)CH_2CH_2$]-5-Ph-7-Cl | 170–175 | Xylene/Ligroin | 42 | | 25 |
| 1-[$Me_2P(O)(CH_2)_3$]-5-Ph-7-Cl | 154 | PhH/Cycliohexane | 47 | | 25 |
| 1-[$Me_2P(O)CH_2$]-5-Ph-7-Cl | 174–175 | Xylene | 52; 80 | | 25 |
| | 193–195 | Xylene | | | 25 |
| 1-[$Me_2P(O)CH_2$]-5-(2-$ClC_6H_4$)-7-Cl | 240d | | 40 | | 25 |
| 1-[$Me_2P(O)CH_2$]-5-(4-*i*-$PrC_6H_4$)-7-Cl | 150–153 | | 45 | | 25 |
| 1-[2-($Me_3$-Acetoxy)-1-MeO-ethyl]-5-Ph-7-Cl | 101–103 | $Et_2O$/Hexane | | | 72 |

| | | | | |
|---|---|---|---|---|
| 1-(2-MeS-Ethyl)-5-Ph-7-Cl | 114–115 | | | 128 |
| Hydrochloride | 165–167d | | | 86b |
| 4-Oxide | 174–175 | | | 234 |
| 1-(2-MeS-Ethyl)-5-(2-$FC_6H_4$)-7-Cl | 77–79 | | | 128 |
| 1-$MeSCH_2$-5-Ph-7-$H_2N$·0.5 *i*-PrOH | 123–125d | *i*-PrOH | | 134 |
| 1-$MeSCH_2$-5-Ph-7-Cl | 115–117 | $Et_2O$/Hexane | | 72 |
| 1-$MeSCH_2$-5-Ph-7-$MeSCH_2NH$ | 145–146 | $CH_2Cl_2/Et_2O$ | | 134 |
| 1-$MeSCH_2$-5-Ph-7-$NO_2$ | 139–140 | EtOH | | 72 |
| 1-$MeSCH_2$-5-(2-$ClC_6H_4$)-7-Cl | 127–129 | EtOAc/Hexane | | 72 |
| 1-[2-MeS(O)-Ethyl]-5-Ph-7-Cl | 166–167 | EtOAc | | 86c, 92 |
| | 107–109 | *i*-PrOH/*i*-$Pr_2O$ | | 234 |
| 1-$MeSO_2$-5-Ph-7-$NO_2$ | 222d | $CH_2Cl_2$ | | 301 |
| 1-$MeSO_2$-5-(2-$ClC_6H_4$)-7-$NO_2$ | 215–217 | $CH_2Cl_2$ | | 301 |
| 1-[2-$MeSO_2$-Ethyl]-5-Ph-7-Cl | 160–161 | Acetone | | 87 |
| 4-Oxide | 214–215 | | | 234 |
| 1-[2-$MeSO_2$-Ethyl]-5-(2-$FC_6H_4$)-7-Cl | 155–156 | | | 87 |
| 1-MeS(O)$CH_2$-5-Ph-7-$H_2N$ | 189–191 | *i*-PrOH | | 134 |
| 1-MeS(O)$CH_2$-5-Ph-7-Cl | 158–159 | EtOAc/Hexane | | 72 |
| 1-MeS(O)$CH_2$-5-Ph-7-$NO_2$ | 190–192 | EtOAc | | 72 |
| 1-MeS(O)$CH_2$-5-(2-$ClC_6H_4$)7-Cl | 150–152 | EtOAc/$Et_2O$ | | 72 |
| 1-$MeSO_2CH_2$-5-Ph-7-$H_2N$ | 160–161 | $CH_2Cl_2/Et_2O$ | | 134 |
| 1-$MeSO_2CH_2$-5-Ph-7-Cl | 162–164 | EtOAc/$Et_2O$ | | 72 |
| 1-$MeSO_2CH_2$-5-Ph-7-$NO_2$ | 190–192 | $CH_2Cl_2$/EtOH | | 72 |
| 1-$MeSO_2CH_2$-5-(2-$ClC_6H_4$)-7-Cl·EtOH | 110–115d | EtOH | | 72 |
| 4-Oxide | 248–250d | $CH_2Cl_2$/EtOH | | 134 |
| 1-$MeSO_2CH_2$-5-(2-$FC_6H_4$)-7-Cl | | | | |
| 4-Oxide | 155–157 | | | 234 |
| 1-(2-Morpholinoethyl)-5-Ph-7-Cl | 144–146 | Acetone/Hexane | 83 | 287, 86b |
| Maleate | 156–157 | Acetone | 79 | 287 |
| 1-[2-(Morpholino)$_2PO_2$-ethyl]-5-(2-$FC_6H_4$)-7-Cl | 107–110 | $H_2O/Et_3PO_4$ | | 33b |
| 1-[3-(4-$NO_2$-Phenoxy)CO-propyl]-5-Ph-7-Cl | 123–125 | PhH/*i*-PrOH | | 488 |
| 1-(Oxiran-2-yl)$CH_2$-5-(2-$ClC_6H_4$)-7-Cl | 135–138 | $Et_2O$/Hexane | | 192 |
| 1-(3-Oxobutyl)-5-(2-$ClC_6H_4$)-7-$NO_2$ | 170–174 | EtOAc/Petr ether | | 301 |
| 1-(3-Oxobutyl)-5-(2-$FC_6H_4$)-7-Cl | 121–123 | $Et_2O$/Petr ether | | 301 |

TABLE VII-1. —(contd.)

| Substituent | mp (°C) or; [bp (°C/torr)] | Solvent of Crystallization | Yield (%) | Spectra | Refs. |
|---|---|---|---|---|---|
| 1-(2-Oxo-Pyrrolidin-1-yl)$CH_2$-5-Ph-7-$NO_2$ | 173–175 | EtOAc | | | 108 |
| 1-(2-Oxo-Pyrrolidin-1-yl)$CH_2$-5-(2-$ClC_6H_4$)7-$NO_2$ | 196 | EtOH | | | 478 |
| 1-(2-Oxo-Pyrrolidin-1-yl)$CH_2$-5-(2-pyridyl)-7-Br | 197–199 | EtOAc | | | 478 |
| 1,5-$Ph_2$-7-Cl | 197–199 | $CH_2Cl_2$/Petr ether | | | 247 |
| 1-Ph-5-(2-$ClC_6H_4$)-7-$NO_2$ | 198–200 | $CH_2Cl_2$/Petr ether | | | 391b |
| 1-Ph-5-EtO-7-Cl | 151–152 | EtOH | 38 | | 195 |
| 1-Ph-5-Me-7-Cl | | | | | |
| Hydrochloride | 202–204d | MeOH/Acetone | 60 | ir, ms, pmr | 116 |
| 1-(Ph-Methylimino)-5-(2-$ClC_6H_4$)-7-Cl | 169–171 | $Et_2O$ | 68 | | 250 |
| 1-(2-$FC_6H_4$)-5-Me-7-Cl | | | | | |
| Hydrochloride | 199–200d | | 55 | | 130 |
| 1-PhNHCO-5-Ph-7-Cl | 136–138d | $CH_2Cl_2$/Petr ether | | | 247 |
| 1-PhNHCO$CH_2$-5-Ph-7-Cl | 220–222 | Acetone | | | 293 |
| 1-(2-$Ph_2PO_4$-Ethyl)-5-(2-$FC_6H_4$)-7-Cl | Oil | | | | 33b |
| 1-(2-PhO-Ethyl)-5-Ph-7-Cl | 165–166 | $CHCl_3$/*i*-PrOH | | | 45, 89 |
| 4-Oxide | 238–240 | | | | 89 |
| 1-(2-PhO-Ethyl)-5-(2-$FC_6H_4$)-7-Cl | 131–132 | | | | 89 |
| 4-Oxide | 187–188 | | | | 234 |
| 1-[3-(4-Ph-Piperazin-1-yl)propyl]-5-Ph-7-Cl | | | | | |
| Hydrochloride | 254–257 | EtOH/$H_2O$ | 69 | | 65 |
| 1-(Phthalimido)acetyl-5-(2-$ClC_6H_4$)-7-$NO_2$ | 222–224 | $CH_2Cl_2$/$Et_2O$ | | | 301 |
| 1-(2-Phthalimidoethoxy)$CH_2$-5-Ph-7-Cl | 138–139 | $CH_2Cl_2$/MeOH | | | 72 |
| 1-(3-Phthalimidopropyl)-5-Ph-7-Cl | 167–170 | Cyclohexane | | | 38b |
| 1-[2-(Piperazin-1-yl)ethyl]-5-Ph-7-Cl | | | | | |
| Dimaleate | 172–178d | Acetone | 69 | | 287 |
| 1-[3-(Piperazin-1-yl)propyl]-5-Ph-7-Cl | | | | | |
| Dimaleate | 120–122 | Acetone | | | 254 |
| 1-[3-(Piperazin-1-yl)propyl]-5-(2-$FC_6H_4$)-7-Cl | | | | | |
| Hydrochloride | 235–255 | MeOH/$Et_2O$ | 78 | | 287 |

| | | | | |
|---|---|---|---|---|
| 1-[2-(Piperidin-1-yl)ethyl]-5-Ph-7-Cl | 90–92 | Hexane | 90 | 287 |
| Maleate | 172–173 | Acetone | 81 | 287 |
| Maleate | 172–173 | | | 86b |
| 1-Pr-5-Ph-7-$NO_2$ | 127–128 | $Et_2O$ | | 2b |
| 1-Pr-5-(2-$FC_6H_4$)-7-(Cyclopentyl)CONH | 200–204 | $Et_2O$ | | 301 |
| 1-Pr-5-(2-$FC_6H_4$)-7-$NO_2$ | 107–108 | EtOAc/Hexane | | 251c |
| 1-[2-(PrNH-Acetoxy)ethyl]-5-(2-$FC_6H_4$)-7-Cl | | | | |
| Hydrochloride | 219–222d | MeOH/$Et_2O$ | | 283 |
| 1-(2-PrO-Ethyl)-5-Ph-7-Cl | 99–100 | | | 125h |
| 4-Oxide | 169–170 | | | 234 |
| 1-(2-PrO-Ethyl)-5-(2-$FC_6H_4$)-7-Cl | 85–87 | | | 125h |
| 4-Oxide | 135–136 | | | 234 |
| 1-Pr$OCH_2$-5-Ph-7-$H_2N$ | 176–177 | $CH_2Cl_2$/Hexane | | 134 |
| 1-Pr$OCH_2$-5-Ph-7-$NO_2$ | 73–76 | EtOH | | 72 |
| 1-[$Pr_2P(O)CH_2$]-5-Ph-7-Cl | 157–160 | Xylene | 65 | 25 |
| 1-*i*-Pr-5-(2-$ClC_6H_4$)-7-Cl | 148–150 | $Et_2O$ | 17 | 2 |
| 1-{2-[4-(2-{4-[3-*i*-PrNH-2-HO-Propoxy]phenoxy}-ethyl)piperazin-1-yl]ethyl}-5-Ph-7-Cl | | | | |
| (*S*)-Enantiomer | | | | |
| Trimaleate | 137–141 | MeOH/EtOAC | | 487 |
| I-(2-*i*-Pr$SO_2$-Ethyl)-5-(2-$FC_6H_4$)-7-Cl | | | | |
| Hydrochloride | 186–187 | | | 127 |
| 1-(1-Propyn-3-yl)-5-Ph-7-Cl | 144–146 | EtOH | 61 | 55 |
| Hydrochloride | 229–231d | | | 273 |
| Hydrogen sulfate | 166–168d | | | 273 |
| 1-(1-Propyn-3-yl)-5-Ph-7-$NO_2$ | 156–158 | PhH/Hexane | | 273 |
| Hydrochloride | 216–218 | | | 273 |
| 1-(1-Propyn-3-yl)-5-(2-$ClC_6H_4$)-7-Cl | 140–142 | EtOH | | 91 |
| Hydrochloride | 180–182 | | | 273 |
| 1-(1-Propyn-3-yl)-5-(2-$FC_6H_4$)-7-$H_2N$ | 205–208 | $Et_2O$ | | 301 |
| 1-(1-Propyn-3-yl)-5-(2-$FC_6H_4$)-7-Cl | 70–72 | MeOH | | 273 |
| Hydrochloride | 183–185 | | | 273 |
| 1-(1-Propyn-3-yl)-5-(2-$FC_6H_4$)-7-(cyclopentyl)CONH | 170–174 | $Et_2O$/Hexane | | 301 |
| 1-(1-Propyn-3-yl)-5-(2-$FC_6H_4$)-7-$NO_2$ | 159–160 | EtOAc | | 251c |

TABLE VII-1. —(contd.)

| Substituent | mp (°C) or; [bp (°C/torr)] | Solvent of Crystallization | Yield (%) | Spectra | Refs. |
|---|---|---|---|---|---|
| 1-[2-(2-Pyrazinyl)COO-1-MeO-ethyl]-5-Ph-7-Cl | 113–116d | MeOH | | | 302 |
| 1-[2-(1-Pyridinium)-1-MeO-ethyl]-5-Ph-7-Cl | | | | | |
| Tosylate | 205–210d | $EtOH/Et_2O$ | | | 302 |
| 1-[2-(3-Pyridinecarbonyloxy)ethyl]-5-(2-$FC_6H_4$)-7-Cl | 138–140 | MeOH | | | 283, 86b |
| 1-[2-(4-Pyridyl)COO-ethoxy]$CH_2$-5-(2-$FC_6H_4$)-7-Cl | 139–142 | $CH_2Cl_2$/Petr ether | | | 283, 86b |
| 1-[2-(2-Pyridyl)COO-1-MeO-ethyl]-5-Ph-Cl | 152–154 | $EtOH/Et_2O$ | | | 302 |
| 1-[2-(4-Pyridyl)COO-1-MeO-ethyl]-5-Ph-7-Cl | 177–180 | EtOH | | | 302 |
| 1-(2-Pyridyl)$CH_2$-5-Ph-7-Cl | | | | | |
| Dihydrochloride | 244–246 | $MeOH/Et_2O$ | | | 255c |
| 1-[2-(Pyrrolidin-1-yl)ethoxy]$CH_2$-5-Ph-7-Cl | | | | | |
| Hydrochloride | 163–166 | Acetone/$Et_2O$ | | | 72 |
| 1-[2-(Pyrrolidin-1-yl)ethyl]-5-Ph-7-Cl | 106–107 | Hexane | 88 | | 287 |
| Maleate | 157–159 | Acetone | 79 | | 287, 86b |
| 1-(2-SCN-Ethyl)-5-(2-$FC_6H_4$)-7-Cl | 143–145 | EtOH | | | 331, 332 |
| Hydrochloride | 217–220d | $MeOH/Et_2O$ | | | 331, 332 |
| 1-(2-Tetrahydropyran-2-yloxy)ethyl-5-Ph-7-Cl | 126–127 | $i$-$Pr_2O$/$i$-PrOH | | | 89 |
| 1-(2-Tetrahydropyran-2-yloxy)ethyl-5-(2-$FC_6H_4$)-7-Cl | 102–104 | | | | 89 |
| 4-Oxide | 145–146 | | | | 89 |
| 1-[2-(2-Thienyl)COO-1-MeO-ethyl]-5-Ph-7-Cl | 134–15 | EtOH | | | 486 |
| 1-Vinyl-5-(2-$ClC_6H_4$)-7-Cl | 114–115 | Hexane | 92 | pmr | 238 |
| 1-Vinyl-5-(2-$FC_6H_4$)-7-Cl | 89–90 | $Et_2O$/Hexane | 57 | pmr | 238 |
| 4-Oxide | 183–184 | $CH_2Cl_2/Et_2O$ | 50 | | 238 |
| 1-(2-Vinyloxy)ethyl-5-Ph-7-Cl | 89–91 | $i$-PrOH/$i$-$Pr_2O$ | | | 128 |
| Hydrochloride | 216–218d | $CHCl_3$/$i$-PrOH | | | 86d, 128 |
| 1-(2-Vinyloxy)ethyl-5-(2-$ClC_6H_4$)-7-Cl | 118–120 | | | | 86d |
| 1-(2-Vinyloxy)ethyl-5-(2-$FC_6H_4$)-7-Cl | 82–84 | | | | 86d |
| 1-{3-[4-(2-Vinyloxyethyl)-piperazinyl]propyl}-5-(2-$FC_6H_4$)-7-Cl | | | | | |
| Dimaleate | 115–122 | Acetone | 15 | | 287 |

| | | | | | |
|---|---|---|---|---|---|
| ***1,5,8-Trisubstituted*** | | | | | |
| 1-(2-$Et_2N$-Ethyl)-5-Ph-8-Cl | 87–89 | Hexane | | | 302 |
| 1-Me-5-EtO-8-Cl | 95–97 | EtOH | 65 | | 195 |
| | 118–120 | EtOAc/Hexane | | | 391b |
| 1-Me-5-[2,2-$(EtO)_2$-Ethyl]NH-8-Cl | 128–130 | EtOAc/Hexane | | | 391b |
| 1-Me-5-Ph-8-Cl | 194–195 | Acetone/Hexane | | | 152c |
| 1-Ph-5-Et-8-Cl | 156–158 | | | | 19 |
| 1-Ph-5-Et-8-$F_3C$ | 111–112 | | | | 19 |
| 1-Ph-5-Me-8-Cl | 171–172 | EtOAc/$i$-$Pr_2O$ | 86 | | 19 |
| 1-(2-$FC_6H_4$)-5-Me-8-Cl | 162–167 | | | | 19 |
| 1-(2-$F_3CC_6H_4$)-5-Me-8-Cl | 138–141 | | | | 19 |
| 1-(2-$F_3CC_6H_4$)-5-Me-8-Cl | 138–141 | | | | 19 |
| 1-(2-$O_2NC_6H_4$)-5-Me-8-Cl | 176–178 | | | | 19 |
| 1-(2-Pyridyl)-5-Me-8-Cl | 146–148 | | | | 19 |
| 1-Ph-5-Me-8-$F_3C$ | 133–135 | | | | 19 |
| ***1,5,9-Trisubstituted*** | | | | | |
| 1-Me-5-Ph-9-Cl | 139–143 | $CH_2Cl_2$/Petr ether | | | 152c |
| 1-Me-5-(2-$FC_6H_4$)-9-Cl | 148 | $Et_2O$ | | | 301 |
| ***3,3,5-Trisubstituted*** | | | | | |
| 3,3-$Me_2$-5-(2-$FC_6H_4$) | 212 | Cyclohexane | | | 251b |
| 3,3-(*spiro*-Adamant-2-yl)-5-Ph | 265–276 | | | ir, uv | 115 |
| ***3,5,7-Trisubstituted*** | | | | | |
| 3-AcNH-5-Ph-7-Br | 293–294 | | 41 | | 202 |
| 3-AcNH-5-Ph-7-Cl | 268–269 | MeCN | | | 59b |
| | 274–275 | EtOH | | | 199 |
| 3-AcNH-5-Ph-7-Me | 214–216 | | | | 202 |
| 3-(AcNH-Acetoxy)-5-Ph-7-Cl | 210–211 | AcEt | | | 339 |
| 3-(4-AcNH-Butanoyloxy)-5-Ph-7-Cl | 199–201 | MeCN | | | 339 |
| 3-(4-AcNH-Butanoyloxy)-5-(2-$ClC_6H_4$)-7-Cl·0.5 AcEt | 120–121 | AcEt | | | 339 |
| 3-AcO-5-$C_6D_5$-7-Cl | 236–237d | | 83 | | 97 |
| 3-AcO-5-Ph-7-Ac | 238–239 | Acetone | | | 2b |
| | 227–229 | MeCN | 96 | | 44 |

TABLE VII-1. —(contd.)

| Substituent | mp (°C) or; [bp (°C/torr)] | Solvent of Crystallization | Yield (%) | Spectra | Refs. |
|---|---|---|---|---|---|
| 3-AcO-5-Ph-7-AcNH | 264–266 | $EtOH/CH_2Cl_2/EtOAc$ | | | 301 |
| 3-AcO-5-Ph-7-Cl | 239–241 | $EtOH/H_2O$ | | | 199, 225 |
| | 233–237 | $EtOH/H_2O$ | | | 223 |
| | 242–243 | $Ac_2O$ | | | 263 |
| 3-AcO-5-Ph-7-$F_3C$ | 230–231 | PhH/Hexane | | | 1b |
| 3-AcO-5-Ph-7-(1-HO-Ethyl) | 233–235 | Acetone/Hexane | | | 33b |
| 3-AcO-5-Ph-7-MeS | 253–254d | $CHCl_3$/Hexane | | | 1b |
| 3-AcO-5-Ph-7-$NO_2$ | 267–268 | THF/Hexane | | | 1b |
| 3-AcO-5-(2-$ClC_6H_4$)-7-AcNH | 231–233 | $CHCl_3$ | | | 134 |
| 3-AcO-5-(2-$ClC_6H_4$)-7-$H_2N$ | 233–237d | $CH_2Cl_2$/THF/Petr ether | | | 152c, 263 |
| 3-AcO-5-(2-$ClC_4H_4$)-7-Cl | 262–264 | | | | |
| 3-AcO-5-(2-$ClC_6H_4$)-7-$NO_2$ | 279–281 | PhH | | | 8 |
| 3-AcO-5-(4-$ClC_6H_4$)-7-Br | 256–257 | | | | 263 |
| 3-AcO-5-(2-$FC_5H_4$)-7-Ac | 267–268 | Acetone | | | 2b |
| 3-AcO-5-(2-$FC_6H_4$)-7-AcNH | 264–268 | EtOH/EtOAc | | | 301 |
| 3-AcO-5-(2-$FC_6H_4$)-7-Cl | 239–247 | MeOH | 76 | | 233 |
| 3-AcO-5-(2-$FC_6H_4$)-7-(1-HO-Ethyl) | 213–215 | Acetone/Hexane | | | 33b |
| 3-AcO-5-(2-$FC_6H_4$)-7-I | 215–218 | $CH_2Cl_2$/Hexane | | | 371 |
| 3-AcO-5-(2-$FC_6H_4$)-7-$NO_2$ | 245–246 | $CH_2Cl_2/Et_2O$/Petr ether | | | 2b |
| 3-AcO-5-(2,6-$F_2C_6H_3$)-7-Cl | 247–260d | THF/Hexane | | | 152c |
| 3-AcO-5-(4-$F_3$CONH)$C_6H_4$-7-Cl | 257–262d | $CH_2Cl_2$/THF/Hexane | 87 | ir, ms | 8 |
| 3-AcO-5-(2-Pyridyl)-7-Br | 237–238d | THF/Hexane | 50 | ir, pmr | 148 |
| 3-AcO-5-(2-Thienyl)-7-Cl | 269 | | | | 263 |
| 3-$AcOCH_2$-5-Ph-7-Cl | 172–173 | Acetone/Hexane | | | 1b |
| 3-$AcOCH_2$-5-(2-$ClC_6H_4$)-7-AcNH | 234–236 | EtOAc/Petr ether | | | 301 |
| 3-$AcOCH_2$-5-(2-$ClC_6H_4$)-7-$NO_2$ | 170d | $Et_2O$ | | | 301 |
| 3-$AcOCH_2$-5-(2-$FC_6H_4$)-7-AcNH | 215–217d | $CH_2Cl_2/EtOAc$ | | | 301 |

| | | | | | |
|---|---|---|---|---|---|
| 3-(AcO)Ph-Acetoxy-5-(2-ClC$_6$H$_4$)-7-Cl | 210–211 | | | pmr | 338 |
| Mixture of diastereoisomers | >200 | | | pmr | 338 |
| 3-AcS-5-Ph-7-Cl | 208–209 | | | pmr | 338 |
| 3-(Adamantyl)CONH-5-Ph-7-Cl | 264 | | 72 | | 202 |
| 3-(D-Alanyl)CO-5-Ph-7-Cl | 250–252d | EtOH | | | 301 |
| 3-(L-Alanyl)CO-5-Ph-7-Cl | 278–280d | EtOH | | | 301 |
| 3-(L-Alanyl)CO-5-Ph-7-NO$_2$·0.5 EtOH | 280d | EtOH | | | 301 |
| 3-(Allyl)NH-5-(2-ClC$_6$H$_4$)-7-NO$_2$ | 176d | CH$_2$Cl$_2$/Hexane | | | 301 |
| 3-(Allyl)$_2$N-5-(2-ClC$_6$H$_4$)-7-NO$_2$ | 192–194d | CH$_2$Cl$_2$/EtOH | | | 301 |
| 3-H$_2$N-5-Ph-7-Br | 210–211 | | 42 | | 202 |
| 3-H$_2$N-5-Ph-7-Cl | 205–206 | EtOH | 45 | | 199, 202 |
| | | CH$_2$Cl$_2$/Et$_2$O | 83 | | 260 |
| | 218–220d | MeCN | | | 225 |
| | 195–197d | MeCN | | | 143 |
| Hydrochloride·0.5EtOH | 225–226 | EtOH | | | 199, 358 |
| 3-H$_2$N-5-Ph-7-Me | 200–202 | | 60 | | 202 |
| 3-H$_2$N-5-(2-ClC$_6$H$_4$)-7-NO$_2$ | | | | | |
| Hydrochloride·acetone | 212–216d | MeOH/Acetone | | | 302 |
| 3-H$_2$NCO-5-Ph-7-Cl | 255–256 | MeOH | 74 | | 105 |
| 3-H$_2$NCO-5-(2-ClC$_6$H$_4$)-7-NO$_2$ | 290–295 | EtOH/Petr ether | | | 30 |
| 3-H$_2$NCOCH$_2$-5-Ph-7-Cl | 260–264d | EtOH | | | 302 |
| 3-H$_2$NCOO-5-Ph-7-Cl | 230–232 | | | | 336 |
| 3-(3-H$_2$NCO-1-Pyridinium)-5-Ph-7-Cl | | | | | |
| Chloride | 244–246d | MeOH/Et$_2$O | | | 334 |
| 3-(3-H$_2$NCO-1-Pyridinium)-5-(2-ClC$_6$H$_4$)-7-Cl | | | | | |
| Chloride | 261–263d | Acetone | 80 | | 356 |
| 3-(4-H$_2$NCO-1-Pyridinium)-5-(2-ClC$_6$H$_4$)-7-Cl | | | | | |
| Chloride | 222–224d | Acetone | 90 | | 356 |
| 3-N$_3$CO-5-Ph-7-Cl | 123d | CH$_2$Cl$_2$/MeOH | | | 488 |
| 3-(Aziridin-1-yl)-5-(2-ClC$_6$H$_4$)-7-NO$_2$ | 230d | CH$_2$Cl$_2$/Et$_2$O | | | 301 |
| 3-Benzoyloxy-5-Ph-7-Cl | 251–252 | | | | 263 |
| 3-Benzyl-5-Ph-7-Cl | | | | | |
| *R*-(+)-Enantiomer | 107–110 | Cyclohexane | | [α] | 112 |
| 3-Benzyl-5-(2-ClC$_6$H$_4$)-7-NO$_2$ | 202–204 | Et$_2$O | | | 301 |

TABLE VII-1. —(contd.)

| Substituent | mp (°C) or; [bp (°C/torr)] | Solvent of Crystallization | Yield (%) | Spectra | Refs. |
|---|---|---|---|---|---|
| 3-Benzyl(Me)N-5-(2-$ClC_6H_4$)-7-$NO_2$ | 199–201 | MeCN | | | 302 |
| 3-(2-Benzylidene-1-Me-hydrazino)-5-(2-$ClC_6H_4$)-7-$NO_2$ | 194 | MeOH | | | 301 |
| 3-(4-$BrC_6H_4$)CHN-5-Ph-7-Cl | 197–200 | | 40 | | 202 |
| 3-Bu-5-(2-$ClC_6H_4$)-7-$NO_2$ | 188–190 | $CH_2Cl_2$/$Et_2O$/Petr ether | | | 301 |
| 3-BuNH-5-(2-$ClC_6H_4$)-7-$NO_2$ | 180–182d | EtOH/$Et_2O$ | | | 301 |
| 3-(But-2-yl)NH-5-(2-$ClC_6H_4$)-7-$NO_2$ | 174–175 | EtOH/Hexane | | | 301 |
| 3-*t*-BuNH-5(2-$ClC_6H_4$)-7-$NO_2$ | 184–187 | EtOH | | | 302 |
| 3-BuO-5-Ph-7-Cl | | | | | |
| (+)-Enantiomer | 131 | EtOH | | [α] | 366 |
| 3-BuO-5-(2-$ClC_6H_4$)-7-$NO_2$ | 210–212 | $CH_2Cl_2$/EtOH | | | 301 |
| 3-(2-*t*-BuOOCNH-3-Ph-Propanoyloxy)-5-Ph-7-Cl | | | | | |
| Isomer A | 205 | | | | 478 |
| Isomer B | ~175d | | | | 478 |
| 3-(4-*t*-BuOOCNH-Butanoyloxy)-5-Ph-7-Cl | 118–120 | $Et_2O$/Petr ether | | | 478 |
| | Amorphous | | | | 339 |
| 3-(4-*t*-BuOOCNH-Butanoyloxy)-5-(2-$ClC_6H_4$)-7-$NO_2$ | 182–184 | EtOAc | | | 478 |
| 3-(5-*t*-BuOOCNH-Pentanoyloxy)-5-Ph-7-Cl | 117 | | | | 478 |
| 3-HOOC-5-Ph-7-Cl | | | | | |
| Dipotassium salt | | | | ir, uv | 104 |
| Potassium salt | | | | ir, uv | 104 |
| 3-HOOC-5-(2-$ClC_6H_4$)-7-Cl | | | | | |
| Dipotassium salt·$2H_2O$ | | EtOH | | | 108 |
| 3-(2-HOOC-Ethoxy)-5-Ph-7-Cl | 240–241 | MeCN | 77 | ir, pmr | 357 |
| 3-$HOOCCH_2NHCO$-5-Ph-7-Cl | 242d | EtOH | | | 301 |
| 3-$HOOCCH_2O$-5-Ph-7-Cl | 120 | MeCN | | | 227 |
| | 218–220d | MeCN | 65 | ir, pmr | 357 |
| 3-(3-HOOC-Propanoyloxy)-5-Ph-7-$H_2N$ | >190d | THF/$CH_2Cl_2$/Hexane | | | 33b |
| 3-(3-HOOC-Propanoyloxy)-5-Ph-7-Br | | | | | |
| (+)-Enantiomer | 157 | EtOAc/Hexane | | [α] | 499 |

| | | | | | |
|---|---|---|---|---|---|
| 3-(3-HOOC-Propanoyloxy)-5-Ph-7-Cl | 152–153 | EtOH/$H_2O$ | 74 | | 337 |
| Hydrate | 110–112 | EtOH/$H_2O$ | | | 227 |
| Pyridinium salt | 139–141 | EtOAc | | | 227 |
| (+)-Enantiomer | 178 | | 80 | [α] | 499 |
| | 158–159 | EtOAc/Hexane | | [α] | 499 |
| (−)-Ephedrine salt | 159–160 | | | [α] | 499, 499 |
| 3-(3-HOOC-Propanoyloxy)-5-Ph-7-$NO_2$ | 187–190d | $CH_2Cl_2$/Hexane | | | 33b |
| (+)-Enantiomer·$2H_2O$ | 136 | EtOH/$H_2O$ | | [α] | 499 |
| 3-(3-HOOC-Propanoyloxy)-5-(2-$ClC_6H_4$)-7-$H_2N$ | 130 | THF/$CH_2Cl_2$/Hexane | | | 33b |
| 3-(3-HOOC-Propanoyloxy)-5-(2-$ClC_6H_4$)-7-$NO_2$ | 172–180d | THF/$CH_2Cl_2$/Heptane | | | 33b |
| 3-(3-HOOC-Propoxy)-5-Ph-7-Cl | 226–227 | MeCN | 70 | ir, pmr | 357 |
| 3-R-Benzyl-5-Ph-7-Cl | 108–110 | | 82 | [α], pmr | 40 |
| 3-S-Benzyl-5-Ph-7-Cl | 108–109 | | 82 | [α], pmr | 40 |
| 3-[(Benzyl-NHCO)methoxy]-5-Ph-7-Cl | 187–188 | | | pmr | 357 |
| 3-Benzyl(Me)N-5-Ph-7-Cl | 193–194 | EtOH | 44; 67 | | 369, 348 |
| 3-(Benzyloxy)CO-5-Ph-7-Cl | 220 | EtOH | | | 301 |
| 4-Oxide | 193–195d | EtOH | | | 108 |
| 3-[2-(Benzyloxy)CO(Me)N-ethoxy]CO-5-Ph-7-Cl | | | | | |
| 4-Oxide | 161–162 | $CH_2Cl_2$/Hexane | | | 108 |
| 3-BuNHCOO-5-Ph-7Cl | 214–215 | | | | 336 |
| 3-(Butanoyloxy)-5-Ph-7-Cl | 182–186 | EtOH | 50 | | 337 |
| 3-(Carboxy)$CH_2$5-Ph-7-Cl | | | | | |
| Hydrochloride | 277–283 | $H_2O$/HCl | | | 80 |
| Dipotassium salt | 271–272d | EtOH | | | 80 |
| 3-[(N-HOOC$CH_2$-N-Me-Amino)acetoxy]-5-(2-$ClC_6H_4$)-7-Cl | >170d | | | pmr | 333 |
| Sodium salt·2 $H_2O$ | | | | | |
| 3-(3-Carboxy)propanoyloxy-5-Ph-7-$NO_2$ | 187–190d | $CH_2Cl_2$/Hexane | 64 | ir, pmr, uv | 8 |
| (+)-Enaniomer | 156 | | | [α] | 499 |
| (−)-Ephedrine salt | | | | [α] | 499 |
| 3-(3-Carboxy)propanoyloxy-5-Ph-7-$NH_2$ | | | 48 | pmr, uv | 8 |
| 3-(3-Carboxy)propanoyloxy-5-(2-$ClC_6H_4$)-7-$NO_2$ | 172–180d | $CH_2Cl_2$/THF/Heptane | 67 | ir, uv | 8 |
| 3-(3-Carboxy)propanoyloxy-5-(2-$ClC_6H_4$)-7-$NH_2$ | | | 70 | ir | 8 |

TABLE VII-1. —(contd.)

| Substituent | mp (°C) or; [bp (°C/torr)] | Solvent of Crystallization | Yield (%) | Spectra | Refs. |
|---|---|---|---|---|---|
| 3-Cl-5-Ph-7-Cl | 179d | | | | 263 |
| | 120–122 | PhH | 95 | pmr | 356, 334 |
| | 139–140d | PhH | 95 | pmr | 357 |
| Hydrochloride | 151–153 | | | | 263 |
| 4-Oxide | 210–211d | THF/Hexane | 87 | ir | 143 |
| 3-Cl-5-Ph-7-$F_3C$ | | | | | |
| 4-Oxide | 222–223d | $(CH_2OMe)_2$/Hexane | | | 143 |
| 3-Cl-5-(2-$ClC_6H_4$)-7-Cl | 133–138 | | 98 | | 334 |
| | 196–199d | | 97 | pmr | 356 |
| 3-Cl-5-(2-$ClC_6H_4$)-7-$NO_2$ | 208–210d | THF/Hexane | | | 302 |
| 3-(Cl-Acetoxy)-5-Ph-7-Cl | 230–231 | EtOH | 75 | | 262, 263 |
| | 217–218 | PhH | 85 | | 337 |
| 3-(Cl-Acetoxy)-5-(2-$ClC_6H_4$)-7-Cl | 232–234 | $CH_2Cl_2/Et_2O$ | 75 | pmr | 265, 328 |
| 3-($Cl_2$-Acetoxy)-5-Ph-7-Cl | 210–217 | PhH | 82 | | 337 |
| 3-($Cl_3$-Acetoxy)-5-Ph-7-Cl | 246–248 | $C_6H_6$/Petr ether | 50 | | 334 |
| 3-Cl-5-Ph-7-$NO_2$ | | | | | |
| 4-Oxide | 215–216 | THF/Hexane | | | 143 |
| 3-($Cl_3CONHCO$)-5-Ph-7-Cl | 231–233 | MeCN | | ir, ms, pmr | 313 |
| 3-(2-Cl-Ethoxy)-5-Ph-7-Cl | 189–191 | $i$-$Pr_2O/Et_2O$ | 35 | ir, pmr | 260 |
| 3-(2-Cl-Ethoxy)-5-Ph-7-$NO_2$ | 197–198d | $EtOAc/Et_2O$ | 38 | | 260 |
| 3-(2-Cl-Ethoxy)CO-5-Ph-7-Cl | 206–207 | $CH_2Cl_2/Et_2O$ | | | 108 |
| 4-Oxide | 188–190d | $CH_2Cl_2$/EtOAc | | | 108 |
| 3-(4-$ClC_6H_4$)CHN-5-Ph-7-Cl | 180–182 | | 40 | | 202 |
| 3-(2,6-$Cl_2C_6H_3$)NH-5-Ph-7-Cl | 227–229 | Acetone | | | 33b |
| 3-{[2-(4-Cl-Phenoxy)acetoxy ethyl]-$(Me)_2$-ammonium}-5-(2-$ClC_6H_4$)-7-Cl | | | | | |
| Chloride | 198–201 | Acetone | 85 | | 356 |
| 3-(2-Cl-Propanoyloxy)-5-(2-$ClC_6H_4$)-7-Cl | 204–206 | EtOH | 53 | | 337b |
| 3-(3-Cl-Propoxy)-5-(2-$ClC_6H_4$)-7-$NO_2$ | 215d | $CH_2Cl_2$/EtOH | | | 301 |

| | | | | | |
|---|---|---|---|---|---|
| 3-(3-Cl-Propoxy)CO-5-Ph-7-Cl | 196–198 | $CH_2Cl_2$ | | | 301 |
| 3-CN-5-Ph-7-Cl | 241–244d | | 57 | ir, pmr | 260 |
| 3-CN-5-Ph-7-$NO_2$ | 261d | $CHCl_3$ | 66 | ir, pmr | 260 |
| 3-(2-CN-Ethyl)-5-(2-$ClC_6H_4$)-7-$NO_2$ | 255–257d | $CH_2Cl_2$/Petr ether | | | 301 |
| 3-(2-CN-Ethyl)-5-(2-$FC_6H_4$)-7-Cl | 167–168 | $Et_2O$ | | | 301 |
| 3-CN(HO)CH-5-(2-$ClC_6H_4$)-7-$NO_2$ | 216d | $CH_2Cl_2$ | | | 301 |
| 3-(Cyclohexyl)NHCOO-5-Ph-7-Cl | 231–233 | Dioxane/EtOH | | | 336 |
| 3-(Cyclopentyl)NH-5-(2-$ClC_6H_4$)-7-$NO_2$ | 185–187 | $CH_2Cl_2$/EtOH | | | 301 |
| 3-(Cyclopentyl)MeN-5-(2-$ClC_6H_4$)-7-$NO_2$ | 196–198d | $CH_2Cl_2$/EtOH | | | 301 |
| 3-(Cyclopropyl)NH-5-(2-$ClC_6H_4$)-7-$NO_2$ | 148–150 | EtOH | | | 301 |
| 3-{3-[10,11-Dihydro-5*H*-dibenzo[*a,d*]cyclohepta-trien-5-ylidene]propyl}MeNH-5-Ph-7-Cl | 194–196 | EtOAc/Hexane | | | 33b |
| 3-{3-(10,11-Dihydro-5*H*-Dibenzo[*a,d*]cyclohept-5-ylidene)propyl-$Me_2$-ammonium}-5-(2-$ClC_6H_4$)-7-Cl Chloride | 168–170 | EtOAc | 70 | | 356 |
| 3-(4,5-Dihydrothiazol-2-yl)thio-5-Ph-7-Cl | 213–215 | DMF | | | 227 |
| 3-Et-5-(2-$ClC_6H_4$)-7-$NO_2$ | 242–243 | $CH_2Cl_2/Et_2O$ | | | 231 |
| 3-Et-5-(2-$FC_6H_4$)-7-$NO_2$ | 260 | $CH_2Cl_2$ | | | 231 |
| 3-(3-Et-Imidazolium)-5-Ph-7-Cl Chloride | 268–272d | Acetone | 67 | | 356 |
| 3-(3-Et-Imidazolium)-5-(2-$ClC_6H_4$)-7-Cl Chloride | 256–260d | Acetone | 93 | | 356 |
| 3-EtNH-5-Ph-7-Cl | 199–200 | MeCN | | ir, pmr, uv | 215b |
| 3-EtNH-5-(2-$ClC_6H_4$)-7-$NO_2$ Hydrochloride | 192–195d | MeOH/Acetone | | | 302 |
| 3-(2-Et-Butanoyloxy)-5-Ph-7-Cl | 171–172 | EtOH | 68 | | 326 |
| 3-(2-Et-Hexanoyloxy)-5-Ph-7-Cl | 131–134 | EtOH | 83 | | 326 |
| 3-EtO-5-Ph-7-Cl | 214–216 | $MeCN/H_2O$ | | | 223 |
| 3-EtO-5-Ph-7-Br | 225–227 | MeCN | | | 225 |
| (+)-Enantiomer | 171 | | | [α] | 366 |
| 3-EtO-5-Ph-7-Cl | 22–225 | $CH_2Cl_2/Et_2O$ | 44 | uv | 190 |
| (+)-Enantiomer | 194 | EtOH | | [α] | 366 |
| 3-EtO-5-Ph-7-$NO_2$ (+)-Enantiomer | 242 | $Dioxane/H_2O$ | | [α] | 366 |

TABLE VII-1. —(contd.)

| Substituent | mp (°C) or; [bp (°C/torr)] | Solvent of Crystallization | Yield (%) | Spectra | Refs. |
|---|---|---|---|---|---|
| 3-EtO-5-(2-$ClC_6H_4$)-7-$NO_2$ | 253–255d | EtOH | | | 301 |
| 3-$(EtO)_2$CH-5-(2-$ClC_6H_4$)-7-$NO_2$ | 212d | $Et_2O$ | | | 301 |
| 3-EtO(HO)CH-5-Ph-7-Cl | 183–197 | $CH_2Cl_2/Et_2O$ | | | 152c |
| 3-EtO(HO)CH-5-(2-Cl-$C_6H_4$)-7-$NO_2$ | 225–226 | $Et_2O$/Hexane | | | 301 |
| 3-EtOOC-5-Cyclohexyl-7-Cl | 208 | EtOAc | 40 | | 105 |
| 3-(EtOOC)$CH_2$-5-Ph-7-Cl | 187–189 | MeCN | | pmr | 80 |
| 3-EtOOC-5-Ph-7-$H_2N$ | 305d | DMF/EtOH | 90 | | 105 |
| 3-EtOOC-5-Ph-7-Cl | 232–239 | EtOH | 60 | | 39 |
| | 244 | EtOAc | 74 | | 105 |
| 4-Oxide | 159–161d | PhH/$CH_2Cl_2$ | | | 108 |
| 3-EtOOC-5-Ph-7-F | 225–230d | MeCN | | | 302 |
| 3-EtOOC-5-Ph-7-Me | 260 | | 25 | | 105 |
| | 229–231 | MeCN | | | 302 |
| 3-EtOOC-5-Ph-7-$NO_2$ | 237–238d | AcOH | 66 | | 104 |
| 4-Oxide | 172–176d | EtOH/$Et_2O$ | | | 108 |
| | 271 | EtOAc | 55 | | 105 |
| 3-EtOOC-5-(2-$ClC_6H_4$)-7-$H_2N$ | 235–238 | $CH_2Cl_2$/EtOH | | | 301 |
| 3-EtOOC-5-(2-$ClC_6H_4$)-7-Cl | 138–140 | $Et_2O$ | | | 82 |
| EtOH Solvate | 128–130d | EtOH | | | 108 |
| 4-Oxide | 200–201d | $CH_2Cl_2$/Hexane | | | 108 |
| 3-EtOOC-5-(2-$ClC_6H_4$)-7-$NO_2$·EtOH | 224 | EtOH | | | 301 |
| 4-Oxide | 213 | $CH_2Cl_2/Et_2O$ | | | 301 |
| 3-EtOOC-5-(4-$ClC_6H_4$)-7-Cl | 189–192 | MeCN | | | 302 |
| 3-EtOOC-5-(2-$FC_6H_4$)-7-Cl | 200 | $Et_2O$ | | | 82 |
| | 193–194d | | | | 145 |
| 3-EtOOC-5-(2-$FC_6H_4$)-7-$NO_2$ | 201d | | | | 145 |
| 3-EtOOC-5-(4-$MeC_6H_4$)-7-Me | 215–220d | EtOH | | | 302 |
| 3-EtOOC-5-(3-$O_2NC_6H_4$)-7-$NO_2$ | | | | | |
| 4-Oxide | 190–193d | PhH/Hexane | | | 108 |

| | | | | |
|---|---|---|---|---|
| 3-EtOOC-5-(2-Pyridyl)-7-Br | 224–225 | EtOAc | | | 145 |
| 3-[1,2-$(EtOOC)_2$-Ethyl]-5-Ph-7-Cl | 140–145 | $Et_2O$/Petr ether | | | 30 |
| 3-$EtOOCCH_2NHCO$-5-Ph-7-Cl | 235–245 | MeOH | | | 152c |
| 3-$(EtO)_2OP$-5-(2-$ClC_6H_4$)-7-Cl | 172–174 | MeCN | 74 | ir, pmr | 243, 369 |
| 3-$Et_2N$-5-Ph-7-Cl | 175–177 | EtOH | 50 | | |
| 3-$Et_2N$-5-Ph-7-$NO_2$ | 180–182d | EtOH | | | 302 |
| 3-$Et_2N$-5-(2-$ClC_6H_4$)-7-$NO_2$ | 195–197 | EtOH | | | 302 |
| Methiodide | 181–184d | EtOH/$Et_2O$ | | | 302 |
| 3-($Et_2NCO$)$CH_2$-5-Ph-7-Cl | 256–258 | EtOH | | | 80 |
| 3-(3-$Et_2NCO$-1-Pyridinium)-5-Ph-7-Cl | | | | | |
| Chloride | 200–203d | Acetone | 70 | | 356 |
| 3-(3-$Et_2NCO$-1-Pyridinium)-5-(2-$ClC_6H_4$)-7-Cl | | | | | |
| Chloride | 182–185d | Acetone | 86 | | 356 |
| 3-(2-$Et_2N$-Ethoxy)CO-5-Ph-7-Cl | | | | | |
| Ethobromide | 194–196d | $CH_2Cl_2$/EtOAc | | | 147 |
| 3-(2-$Et_2N$-Ethyl)NH-5-Ph-7-Cl | 185–186 | EtOH | 41 | | 348 |
| 3-(2-$Et_2N$-Ethyl)NH-5-(2-$ClC_6H_4$)-7-$NO_2$ | 180d | EtOAc | | | 301 |
| 3-(2-$Et_2N$-Ethyl)NHCO-5-Ph-7-Cl | 220 | EtOAc | 90 | | 105 |
| 3-(2-$Et_2N$-Ethyl)NHCO-5-Ph-7-Cl | 169–170 | Acetone/Petr ether | | | 336 |
| 3-(2-$Et_2N$-Ethyl)S-5-Ph-7-Cl | | | | | |
| Hydrochloride | 260d | | | | 367 |
| 3-(2-$Et_2N$-1-Me-Ethoxy)-CO-5-Ph-7-Cl | | | | | |
| Hydrochloride | 225d | $CH_2Cl_2$/Acetone | | | 147 |
| 3-(3-$Et_2N$-Propyloxy)CO-5-Ph-7-Cl | 149–150d | Cyclohexane | | | 147 |
| Hydrochloride | 204–206d | $CH_2Cl_2$/Acetone | | | 147 |
| Ethobromide | 155–160 | PhH | | | 147 |
| Methobromide | 170–174d | $CH_2Cl_2$/EtOAc | | | 147 |
| 4-Oxide, hydrochloride | 183d | | | | 147 |
| 3-(3-$Et_2N$-Propyloxy)CO-5-Ph-7-$NO_2$ | | | | | |
| Hydrochloride | 165d | $CH_2Cl_2$/Acetone | | | 147 |
| 3-(3-$Et_2N$-Propyloxy)CO-5-(2-$ClC_6H_4$)-7-Cl | | | | | |
| Hydrochloride | 193–198d | $CH_2Cl_2$/EtOAc | | | 147 |
| 3-(3-$Et_2N$-Propyl)NHCO-5-Ph-7-Cl | 216–218 | $CH_2Cl_2$/EtOAc | | | 301 |
| 3-Et(Me)N-5-(2-$ClC_6H_4$)-7-$NO_2$ | 213–214d | $CH_2Cl_2$ | | | 301 |

TABLE VII-1. —(contd.)

| Substituent | mp (°C) or; [bp (°C/torr)] | Solvent of Crystallization | Yield (%) | Spectra | Refs. |
|---|---|---|---|---|---|
| 3-F-5-Ph-7-Br | 207–209d | PhH/Hexane | 88 | ir, F-nmr, pmr | 266, 267 |
| 3-F-5-Ph-7-Cl | 190–192d | | 82 | F-nmr, pmr | 266, 267 |
| 4-Oxide | 207–208d | $H_2O$ | 93 | ir, pmr, F-nmr | 110 |
| 3-F-5-Ph-7-$NO_2$ | 174–175 | PhH/Heptane | 83 | F-nmr, pmr | 266, 267 |
| 3-F-5-(2-$ClC_6H_4$)-7-Cl | 210–211d | | | F-nmr, pmr | 266, 267 |
| 3-F-5-(2-$FC_6H_4$)-7-Br | 195–197d | PhH/Hexane | | F-nmr, pmr | 266, 267 |
| 3-F-5-(2-$FC_6H_4$)-7-Cl | 206–207d | PhH/Heptane | 67 | F-nmr, pmr | 266, 267 |
| 3-$F_3CCOO$-5-Ph-7-Br | 181–183 | | 99 | F-nmr, pmr | 266, 267 |
| 3-$F_3CCOO$-5-(2-$FC_6H_4$)-7-Br | 175–177d | | 80 | F-nmr, pmr | 266, 267 |
| 3-(2-Furyl)$CH_2NH$-5-Ph-7-Cl | 186–188 | EtOH | 10 | | 348 |
| 3-$H_2NNH$-5-Ph-7-Cl | 153–157 | MeCN | 74 | | 249 |
| 3-$H_2NNH$-5-(2-$ClC_6H_4$)-7-Cl | 137–142 | | 71 | | 249 |
| 3-$H_2NNHCO$-5-Ph-7-Cl | 215–220 | EtOH/$Et_2O$ | | | 152c |
| 3-HO-5-$C_6D_5$-7-Cl | 194–195 | | 78 | | 97 |
| 3-HO-5-Ph-7-Ac | 209–210 | Acetone | | | 2b |
| | 194–195 | EtOH | 90 | | 44 |
| 3-HO-5-Ph-7-AcNH | 215–216 | MeCN | | | 1b |
| 3-HO-5-Ph-7-$H_2N$ | >320 | MeCN | 17 | | 252 |
| 3-HO-5-Ph-7-$N_3$ | 188–190d | THF/Hexane | 65 | | 252 |
| 3-HO-5-Ph-7-Br | 202–203 | EtOH | 70 | | 202 |
| | 190–192 | | 85 | pmr | 266, 267 |
| 3-HO-5-Ph-7-Cl | 205–207 | EtOH | 81; 66 | | 199, 260 |
| Sodium salt | 196–210d | EtOH | 50 | | 148 |
| 4-Oxide | 174–175d | MeCN | | ir | 206 |
| 3-HO-5-Ph-7-$F_3C$ | 190–191d | EtOH | | | 1b |
| 3-HO-5-Ph-7-(1-HO-Ethyl) | 172–174 | Acetone/Hexane | | | 33b |
| 3-HO-5-Ph-7-I | 208–210d | EtOH | | | 371 |
| 3-HO-5-Ph-7-MeS | 195–197d | THF/Hexane | | | 1b |
| 3-HO-5-Ph-7-$NO_2$ | 211d | | 68 | | 260 |

| | | | | | |
|---|---|---|---|---|---|
| 3-HO-5-(4-$H_2NC_6H_4$)-7-Cl | >350 | THF/MeOH | 50 | ir, ms, pmr | 8 |
| 3-HO-5-(2-$ClC_6H_4$)-7-AcNH | 187–190d | MeOH/$H_2O$ | | | 134 |
| 3-HO-5-(2-$ClC_6H_4$)-7-$H_2N$ | 340d | THF/MeOH | | | 152c |
| 3-HO-5-(2-$ClC_6H_4$)-7-Cl | 160–162 | | | | 263 |
| 3-HO-5-(2-$ClC_6H_4$)-7-$NO_2$ · 0.5 Acetone | 159–160 | Acetone | 30 | ir | 8 |
| 3-HO-5-(2-$FC_6H_4$)-7-Ac · 0.5 THF | 153–155 | THF/Petr ether | | | 2b |
| 3-HO-5-(2-$FC_6H_4$)-7-AcNH · 0.5 $H_2O$ | 185–187 | | | | 2b |
| 3-HO-5-(2-$FC_6H_4$)-7-$H_2N$ | >300 | EtOH/$Et_2O$ | | | 152c |
| 3-HO-5-(2-$FC_6H_4$)-7-Br | 196–198 | EtOH | 62 | F-nmr, pmr | 266 |
| 3-HO-5-(2-$FC_6H_3$)-7-Cl | 197–200 | THF/Hexane | 96 | | 233 |
| 3-HO-5-(2-$FC_6H_4$)-7-(1-HO-Ethyl) | 125–140 | $CH_2Cl_2$ | | | 33b |
| 3-HO-5-(2-$FC_6H_4$)-7-$NO_2$ | 193–195 | $CH_2Cl_2$/$Et_2O$ | | | 152c |
| 3-HO-5-(2,6-$F_2C_6H_4$)-7-Cl | 197–200 | THF/$CHCl_3$ | | | 152c |
| 3-HO-5-(4-$HOC_6H_4$)-7-Cl | 178–179 | Dioxane/$H_2O$ | 67 | ir, pmr | 148 |
| 3-HO-5-(2-Pyridyl)-7-Br | 197–198 | MeCN | 33 | | 148 |
| 3-HONH-5-Ph-7-Cl | 178–180d | | 58 | ir, pmr | 260 |
| 3-(HO)PhCH-5-Ph-7-Cl | 225 | | 13 | ir | 304 |
| 3-(HO)PhCH-5-(2-$ClC_6H_4$)-7-Cl | 181–183 | EtOAc | | | 192 |
| 3-S-(4-HO-Benzyl)-5-Ph-7-Cl | 143–146 | | | | 72c |
| 3-(4-HO-Butoxy)-5-Ph-7-Cl | 169–170 | MeCN | 60 | ir, pmr | 357 |
| 3-(2-HO-Ethoxy)-5-Ph-7-Cl | 210–212 | EtOH | 75 | | 334 |
| | 216–217 | Acetone | 97 | ir, pmr | 357, 260 |
| 3-(2-HO-Ethoxy)-5-Ph-7-$NO_2$ | 222–224d | MeOH | 34 | ir, pmr | 260 |
| 3-(2-HO-Ethoxy)-5-(2-$ClC_6H_4$)-7-Cl | 225–226 | | | | 334 |
| 3-S-(1-HO-Ethyl)-5-Ph-7-Cl | 118–121 | | 80 | [α], pmr | 40 |
| 3-(2-HO-Ethyl)NH-5-(2-$ClC_6H_4$)-7-$NO_2$ | 208–210d | $CH_2Cl_2$/EtOH | | | 301 |
| 3-(2-HO-Ethyl)$_2$N-5-(2-$ClC_6H_4$)-7-$NO_2$ · $CH_2Cl_2$ | 118–120d | $CH_2Cl_2$ | | | 302 |
| 3-(2-HO-Ethyl)NHCO-5-Ph-7-Cl | 265–268d | $CH_2Cl_2$ | | | 301 |
| 3-(2-HO-Ethyl)NHCOO-5-Ph-7-Cl | 168–170 | EtOH | | | 336 |
| 3-(2-HO-Ethyl)MeNH-5-(2-$ClC_6H_4$)-7-$NO_2$ | 202–203d | $CH_2Cl_2$/EtOH | | | 301 |
| 3-[1-(2-HO-Ethyl)-1-morpholinium]-5-Ph-7-Cl Chloride | 169–172d | Acetone | 50 | | 356 |
| 3-[1-(2-HO-Ethyl)-1-morpholinium]-5-(2-$ClC_6H_4$) 7-Cl · Chloride | 168–170d | Acetone | 70 | | 356 |

TABLE VII-1. —(contd.)

| Substituent | mp (°C) or; [bp (°C/torr)] | Solvent of Crystallization | Yield (%) | Spectra | Refs. |
|---|---|---|---|---|---|
| 3-[2-(4-(2-HO-Ethyl)piperazin-l-yl) acetoxyl]-5-(2-$ClC_6H_4$)-7-Cl | | | | | |
| Dihydrochloride · 2.5 $H_2O$ | 195–202d | $EtOH/Et_2O$ | 46 | pmr | 265, 328 |
| 3-(2-HO-Ethyl)S-5-(2-$ClC_6H_4$)-7-$NO_2$ · MeOH | 126–128d | MeOH | | | 302 |
| 3-(Hydroxyimino)methyl-5-(2-$ClC_6H_4$)-7-$NO_2$ | 215–220 | $MeOH/H_2O$ | | | 301 |
| 3-$HOCH_2$-5-Ph-7-Cl | 201–202 | $MeOH/Et_2O$ | | | |
| 3-$HOCH_2$-5-(2-$ClC_6H_4$)-7-AcNH | 255–260d | EtOAc | | | 301 |
| 3-$HOCH_2$-5-(2-$ClC_6H_4$)-7-$H_2N$ | 245–250d | $EtOAc/Et_2O$ | | | 301 |
| 3-$HOCH_2$-5-(2-$ClC_6H_4$)-7-$NO_2$ · EtOH | 260–265 | $EtOH/CH_2Cl_2$/ | | | 301 |
| | | Hexane/Petr ether | | | 91a |
| 3-$HOCH_2$-5-(2-$FC_6H_4$)-7-$H_2N$ | 265–267 | EtOAc | | | 301 |
| 3-$HOCH_2$-5-(2-$FC_6H_4$)-7-$NO_2$ · EtOH | 220–224 | $CH_2Cl_2$/EtOH | | | 301 |
| 3-HO(MeO)CH-5-Ph-7-$NO_2$ | 140d | MeOH | | | 301 |
| 3-HO(MeO) CH-5-(2-$ClC_6H_4$)-7-$NO_2$ | 225 | MeOH/Petr ether | | | 301 |
| 3-[4-$HOC_6H_4$]$CH_2$-5-Ph-7-Cl | 151–153 | PhH | 42 | | 2 |
| 3-*S*-(4-$HOC_6H_4$)$CH_2$-5-Ph-7-Cl | 139–141 | $Et_2O$/Cyclohexane | 76 | [α], pmr | 40 |
| 3-(3-HO-Propoxy)-5-Ph-7-Cl | 190–191 | MeCN | 60 | ir, pmr | 357 |
| 3-[2,3-$(OH)_2$-Propoxy]-5-Ph-7-Cl | 166–168 | | | | 334 |
| 3-(2-HO-Prop-2-yl)-5-Ph-7-Cl | 193–196d | EtOH | | | 134 |
| | 179–180 | | | | 367 |
| 3-HS-5-Ph-7-Cl · 0.5 EtOH | 140–141 | EtOH | | | 227 |
| 3-(Imidazol-1-yl)-5-Ph-7-Cl | 249 | | | | 478 |
| 3-*S*-(3-Indolyl)$CH_2$-5-Ph-7-Cl | 149–151 | $Et_2O$ | 79 | [α], pmr | 40 |
| · Ether · Acetone | 150–152 | Acetone | | pmr | 40 |
| 3-(3-Indolyl)methyl-5-Ph-7-(1-carboxy-1-ethyl) | 244–250 | | 82 | pmr | 20 |
| 3-(Menthyloxy)acetoxy-5-(2-$ClC_6H_4$)-7-Cl | > 100d | | | pmr, [α] | 338 |
| 3-Me-5-Cyclohexyl-7-Cl | 140; 159 | Hexane | 17 | | 18 |
| 3-Me-5-Ph-7-Br | 225 | | 22 | | 100 |

| | | | | | |
|---|---|---|---|---|---|
| 3-Me-5-Ph-7-(1-Carboxy-1-ethyl) | 135–137 | | 86 | pmr | 20 |
| 3-Me-5-Ph-7-Cl | 220–221 | PhH/Petr ether | 75 | | 2 |
| Hydrochloride | 294–295 | EtOH | | | 34 |
| 4-Oxide | 268d | | | | 34 |
| 3-*R*-Me-5-Ph-7-Cl | 200–203 | | | [α] | 40 |
| 3-*S*-Me-5-Ph-7-Cl | 200–203 | Acetone/$H_2O$ | 90 | [α], pmr | 40 |
| 3-*S*-Me-5-Ph-7-$F_3C$ | 87–90 | Acetone/$H_2O$ | | [α] | 81 |
| 3-Me-5-Ph-7-$NO_2$ | 221–222 | PhH/Hexane | | | 136 |
| (+)-Enantiomer | 144–146 | EtOH | | | 301 |
| 3-*S*-Me-5-Ph-7-$NO_2$ | 96–98 | Acetone/$H_2O$ | | [α] | 81 |
| 3-Me-5-(2-$ClC_6H_4$)-7-AcNH | 272–274 | EtOH/Ligroin | | | 301 |
| 3-Me-5-(2-$ClC_6H_4$)-7-$H_2N$ | 229–230 | EtOH/Ligroin | | | 301 |
| | 283–285 | $CH_2Cl_2$/MeOH | | | 301 |
| 3-*R*-Me-5-(2-$ClC_6H_4$)-7-Cl | | | | | |
| Hydrochloride | 260–265d | EtOH | | | 134 |
| 3-*S*-Me-5-(2-$ClC_6H_4$)-7-Cl | 172–173 | Acetone/$H_3O$ | | | 112 |
| Hydrochloride | 265–268d | EtOH | | | 134 |
| 3-*S*-Me-5-(2-$ClC_6H_4$)-7-F · 0.5 $Et_2O$ | 73–75d | $Et_2O$ | | | 134 |
| Hydrochloride · 0.25 $H_2O$ | 230–235d | EtOH | | | 134 |
| 3-Me-5-(2-$ClC_6H_4$)-7-$NO_2$ | 193–196 | $Et_2O$/Petr ether | | | 301 |
| (+)-Enantiomer | 193–196 | $CH_2Cl_2$/Hexane | | | 301 |
| (−)-Enantiomer | 228–229 | $Et_2O$/Hexane | | | 301 |
| 4-Oxide | 300d | $CH_2Cl_2$/EtOH | | | 301 |
| 3-*S*-Me-5-(2-$ClC_6H_4$)-7-$NO_2$ | 198–200 | $Et_2O$ | | [α] | 79 |
| 3-Me-5-(3-$ClC_6H_4$)-7-$NO_2$ | 211–213 | MeOH | | | 301 |
| 3-Me-5-(2-$FC_6H_4$)-7-Ac | 203 | $CH_2Cl_2$/Cyclohexane | | | 251 |
| 3-Me-5-(2-$FC_6H_4$)-7-$H_2N$ | 270 | | | | 251 |
| 3-Me-5-(2-$FC_6H_4$)-7-Cl | 188–191 | $CH_2Cl_2$/$Et_2O$/Petr ether | | | 421 |
| (+)-Enantiomer | 162–164 | $Et_2O$ | | | 301 |
| (−)-Enantiomer | 158–162 | $Et_2O$/Petr ether | | | 301 |
| 3-Me-5-(2-$FC_6H_4$)-7-CN | 216 | Cyclohexane | | | 251 |
| 3-Me-5-(2-$FC_6H_4$)-7-I | 211 | Cyclohexane | | | 251 |
| 3-Me-5-(2-$FC_6H_4$)-7-$NO_2$ | 230–234 | EtOAc/Petr ether | | | 231 |
| 3-*S*-Me-5-(2-$FC_6H_4$)-7-$NO_2$ | 130–140 | $Et_2O$/Petr ether | | [α] | 79, 231 |

TABLE VII-1. —(contd.)

| Substituent | mp (°C) or; [bp (°C/torr)] | Solvent of Crystallization | Yield (%) | Spectra | Refs. |
|---|---|---|---|---|---|
| 3-*R*-Me-5-(2-$FC_6H_4$)-7-$NO_2$ | 130–140 | $Et_2O$/Hexane | | | 231 |
| 3-Me-5-(2-Pyridyl)-7-Br | 228–229 | Acetone | 30 | | 3 |
| 3-($Me_2$-Acetoxy)-5-Ph-7-Cl | 223–225 | *i*-PrOH | 79 | | 326 |
| 3-[2-Me-3-(2-Br-Ethyl)-4-$NO_2$-1-imidazolium]-5-Ph-7-Cl | | | | | |
| Chloride | 209–211 | Acetone/MeOH | 70 | | 356 |
| 3-[2-Me-(2-Br-Ethyl)-4-$NO_2$-1-imidazolium]-5-(2-$ClC_6H_4$)-7-Cl | | | | | |
| Chloride | 234–237 | Acetone | 70 | | 356 |
| 3-[(2,2-$Me_2$-1,3-Dioxolan-4-yl)methoxy]-5-Ph-7-Cl | 194–196 | | 80 | | 334 |
| | 210–211 | | | | 334 |
| 3-(3-Me-1-Imidazolium)-5-Ph-7-Cl | | | | | |
| Chloride | 215–218 | Acetone | 70 | | 356 |
| 3-(3-Me-1-Imidazolium)-5-(2-$ClC_6H_4$)-7-Cl | | | | | |
| Chloride | 260–263 | Acetone | 78 | | 356 |
| 3-(1-Me-Hydrazino)-5-(2-$ClC_6H_4$)-7-$NO_2$ | 184d | $CH_2Cl_2$ | | | 301 |
| 3-(4-$MeC_6H_4SO_3$)$CH_2$-5-Ph-7-Cl | 189–190 | Dioxane/$H_2O$ | | | 38b |
| 3-(4-Me-Piperazin-1-yl)-7-Cl | 223–225 | | 13 | | 149 |
| 3-MeNH-5-Ph-7-Cl | 227–230 | | | | 358 |
| | 197–200 | MeOH/$Et_2O$ | 99 | ir, pmr | 260 |
| | 182–183 | EtOH | 15 | | 348 |
| 3-MeNH-5-(2-$ClC_6H_4$)-7-$NO_2$ | 196–198d | $CH_2Cl_2$/EtOH | | | 301 |
| 3-$MeNHCH_2$-5-(2-$ClC_6H_4$)-7-$NO_2$ | 163 | $CH_2Cl_2$/$Et_2O$/Hexane | | | 301 |
| 3-Me(2-MeO-Ethyl)N-5-(2-$ClC_6H_4$)-7-$NO_2$ | 194–195 | EtOAc/Petr ether | | | 301 |
| 3-(4-Me-Piperazin-1-yl)-5-(2-$ClC_6H_4$)-7-$NO_2$ · 0.5 THF | 198–200d | THF/Hexane | | | 302 |
| 3-(4-Me-Piperazin-1-yl)-COO-5-Ph-7-Cl | 233–234 | DMF/EtOAc | | | 336 |
| 3-[2-(4-Me-Piperazin-1-yl)acetoxy]-5-(2-$ClC_6H_4$)-7-Cl | | | | | |
| Hydrochloride · $H_2O$ | 233–234 | MeOH/$Et_2O$ | 82 | pmr | 265, 328 |
| Dihydrochloride · $2H_2O$ | 202–204d | MeOH/$Et_2O$ | 33 | pmr | 265, 328 |
| Dihydrochloride · $H_2O$ | 218–220 | MeOH/$Et_2O$ | 42 | | 265 |

| | | | | | |
|---|---|---|---|---|---|
| Methanesulfonate | >160d | MeOH/$Et_2O$ | 80 | pmr | 265, 328 |
| Maleate · 0.5 $H_2O$ | >125d | MeOH/$Et_2O$ | 51 | pmr | 265, 328 |
| 3-[2-(4-Me-Piperzin-1-yl)-ethoxy]CO-5-Ph-7-Cl | | | | | |
| Dihydrochloride | 204–206d | $CH_2Cl_2$/MeOH/EtOAc | | | 147 |
| 3-Me(Ph)N-5-Ph-7-Cl | 232–233 | EtOH | 33; 80 | | 369, 348 |
| 3-(2-Me-3-Ph-Propanoyloxy)-5-(2-$ClC_6H_4$)-7-Cl | 200–206 | MeOH | 44 | | 337b |
| 3-(2-Me-Propanoyloxy)-5-Ph-7-Cl | 197–198 | EtOH | 81 | | 326 |
| 3-(2,2-$Me_2$-Propanoyloxy)-5-(2-$ClC_6H_4$)-7-Cl | 198–200 | EtOH | 87 | | 329 |
| 3-(1,1-$Me_2$-2-Propyn-1-yl)NH-5-(2-$ClC_6H_4$)-7-$NO_2$ | 204–206 | EtOAc/Petr ether | | | 301 |
| 3-MeNHCO-5-Ph-7-Cl | 294 | EtOH | 90 | | 105 |
| 3-MeNHCOCONH-5-Ph-7-Cl | 307–313 | $CH_2Cl_2$/MeOH | 20 | ms, pmr, uv | 190 |
| 3-(2-MeNH-Ethoxy)CO-5-Ph-7-Cl | 230–231 | | | | 336 |
| 4-Oxide, hydrobromide | 195–200d | Acetone/EtOH | | | 147 |
| 3-$Me_2$N-5-Ph-7-Cl | 217–218d | $CH_2Cl_2$ | 79 | | 149 |
| | 210–212 | Hexane | 41 | | 369 |
| 3-$Me_2$N-5-Ph-7-Me | 220–222 | | 88 | | 149 |
| 3-$Me_2$N-5-(2-$ClC_6H_4$)-7-$NO_2$ | | | | | |
| Hydrochloride · 0.5 Acetone | 248–250 | MeOH/Acetone | | | 302 |
| 3-$Me_2$N-(2-Pyridyl)-7-Br | >210d | $CH_2Cl_2$ | | | 251b |
| 3-$Me_2$NCO-5-Ph-7-Cl | 297 | | | | 105 |
| 3-$Me_2$NCOO-5-Ph-7-Cl | 234–235 | | | | 336 |
| 3-(2-$Me_2$N-Ethoxy)CO-5-Ph-7-Cl | 62 | $CH_2Cl_2$/Petr ether | | | 147 |
| Hydrochloride | 206–208d | $CH_2Cl_2$/Acetone | | | 147 |
| Methobromide | 194–196 | $CH_2Cl_2$/EtOAc | | | 301 |
| 4-Oxide, hydrochloride | 155d | Acetone/EtOAc | | | 147 |
| 3-(2-$Me_2$N-Ethyl)NH-5-Ph-7-Cl | 182–183 | EtOH | | | 227 |
| 3-(2-$Me_2$N-Ethyl)NHCO-5-Ph-7-Cl | 240–242 | EtOAc | | | 301 |
| 3-[2-(2-$Me_2$N-Ethylmethylamino)acetoxy]-5-(2-$ClC_6H_4$)-7-Cl | | | | | |
| Dihydrochloride · $2H_2O$ | 221–223d | EtOH/$Et_2O$ | 20 | pmr | 265, 328 |
| 3-(2-$Me_2$N-Ethyl)S-5-Ph-7-Cl | | | | | |
| Hydrochloride | 255–260 | EtOH | | | 227 |
| 3-$Me_2$$NCH_2$-5-Ph-7-$NO_2$ | 191–193d | MeOH/$Et_2O$ | | | 301 |

TABLE VII-1. —(contd.)

| Substituent | mp (°C) or; [bp (°C/torr)] | Solvent of Crystallization | Yield (%) | Spectra | Refs. |
|---|---|---|---|---|---|
| 3-$Me_2NCH_2$-5-(2-$ClC_6H_4$)-7-$NO_2$ | 166d | EtOH/$Et_2O$ | | | 301 |
| 3-(2-$Me_2N$-2,2-$Me_2$-Ethoxy)CO-5-Ph-7-Cl | | | | | |
| Hydrochloride | 176–178 | $CH_2Cl_2$/EtOAc | | | 147 |
| 3-[(4-Me-Piperazin-1-yl)carbonyl]$CH_2$-5-Ph-7-Cl | 254–255 | MeOH | | | 80 |
| 3-(2-Me-Prop-1-yl)-5-Ph-7-Cl | 213–214 | $CHCl_3$/Hexane | 15 | | 2 |
| *S*-Enantiomer | 160–161 | Hexane | 39 | [α] | 109 |
| 3-MeO-5-Ph-7-Cl | 230–233 | MeCN | | | 101 |
| | 258–260 | MeOH/$Et_2O$ | 87 | | 260, 227 |
| (+)-Entantiomer | 136 | EtOH | | [α] | 366 |
| 3-MeO-5-Ph-7-$NO_2$ | 270d | | 21 | ir | 260 |
| | 255–259d | MeCN/MeOH | | | 302 |
| 3-MeO-5-(2-$ClC_6H_4$)-7-$NO_2$ | 220–225 | THF/Hexane | | | 302 |
| 3-(2-MeO-Ethyl)NH-5-(2-$ClC_6H_4$)-7-$NO_2$ | 182d | EtOH | | | 301 |
| 3-(2-MeO-Ethoxy)-5-(2-$ClC_6H_4$)-7-$NO_2$ | 217–219 | $CH_2Cl_2$/EtOH | | | 301 |
| 3-$(MeO)_2$CH-5-(2-$ClC_6H_4$)-7-$NO_2$ | 205–206 | $Et_2O$ | | | 301 |
| 3-(MeO)Ph-Acetoxy-5-(2-$ClC_6H_4$)-7-Cl | 163–164 | | | pmr, [α] | 338 |
| Mixture of diastereoisomers | >116 | $Et_2O$ | | pmr | 338 |
| 3-[3,4,5-$(MeO)_3C_6H_2$]CONH-5-Ph-7-Cl | 267 | EtOH | 63 | ir | 202 |
| 3-MeOOC-5-Ph-7-Ac | 195–196 | PhH | | | 32 |
| 3-MeOOC-5-Ph-7-Cl | 217–219 | $CH_2Cl_2$/MeOH | 76 | | 39 |
| 4-Oxide | 182–184d | MeOH | | | 108 |
| 3-MeOOC-5-Ph-7-$NO_2$ | 245d | EtOH | | | 145 |
| | 226 | MeOH | 47 | | 105 |
| 3-MeOOC-5-(2-$ClC_6H_4$)-7-Cl | 216–219 | MeOH/$Et_2O$ | 70 | pmr | 39 |
| 3-MeOOC-5-(2-$FC_6H_4$)-7-I | 230–236d | $CH_2Cl_2$/$Et_2O$ | | | 77 |
| 3-MeOOC-5-(2-$FC_6H_4$)-7-$NO_2$ | 220–222 | EtOAc/MeOH | | | 108 |
| 4-Oxide | 185–186d | EtOAc | | | 108 |
| 3-MeOOCNH-5-Ph-7-Cl | 209d | *i*-PrOH | | | 488 |
| 3-MeOOCO-5-(2-$ClC_6H_4$)-7-Cl | 223–225 | $CH_2Cl_2$ | | pmr | 265 |

| | | | | | |
|---|---|---|---|---|---|
| 3-(3-MeOOC-Propanoyloxy)-5-Ph-7-Cl | 145–148 | EtOH | 69 | | 326 |
| 3-(3-MeOOC-Pyridinium)-5-(2-$ClC_6H_4$)-7-Cl | | | | | |
| Chloride | 216–219 | Acetone | 70 | | 356 |
| 3-$MeOCH_2$-5-Ph-7-Cl | 166–167 | $Et_2O$ | 10 | | 2 |
| 3-(4-$MeOC_6H_4$)-5-Ph-7-Cl | 237–238 | | | | 86b |
| 3-$(MeO)_2OP$-5-(2-$ClC_6H_4$)-7-Cl | 248d | MeCN | 73 | ir, pmr | 243 |
| 3-MeS-5-(2-$ClC_6H_4$)-7-$NO_2$ | 244–246d | $CH_2Cl_2/Et_2O$ | | | 301 |
| 3-(2-MeS-Ethyl)-5-Ph-7-Cl | 179–180 | EtOH | 7 | | 2 |
| | 184 | EtOAc | 50 | | 105 |
| 3-(2-MeS-Ethyl)-5-(2-$ClC_6H_4$)-7-$NO_2$ | 148–150 | $Et_2O$/Hexane | | | 301 |
| 3-Me(O)S-5-(2-$ClC_6H_4$)-7-$NO_2$ | 234–236 | $CH_2Cl_2$/MeOH | | | 301 |
| 3-Morpholino-5-Ph-7-Cl | 226–228 | EtOH | 60; 42 | | 149, 369 |
| | 211–213 | | | | 358 |
| 3-Morpholino-5-(2-$ClC_6H_4$)-7-$NO_2$ | 217–219d | EtOH | | | 302 |
| 3-(Morpholino)acetoxy-5-Ph-7-Cl | | | | | |
| Hydrochloride | 223–224 | EtOH | | | 263 |
| | 228–229 | EtOH | | | 262 |
| 3-(Morpholino)acetoxy-5-(2-$ClC_6H_4$)-7-Cl | | | | | |
| Hydrochloride · $H_2O$ | 255–257d | | 40 | pmr | 265 |
| 3-[4-(Morpholino)butoxy]CO-5-Ph-7-Cl | 128–130d | EtOAc | | | 147 |
| Hydrochloride | 160–165 | $CH_2Cl_2$/MeOH | | | |
| | | EtOAc | | | 147 |
| 3-(Morpholino)CO-5-(2-$ClC_6H_4$)-7-$NO_2$ | 270–272 | | | | 301 |
| 3-(Morpholino)CO-5-(2-$FC_6H_4$)-7-$NO_2$ | 176 | $Et_2O$ | | | 301 |
| 3-[2-(Morpholino)ethoxy]CO-5-Ph-7-Cl · 0.4cyclohexane | 100–109d | Cyclohexane | | | 147 |
| Hydrochloride | 174–178d | Acetone | | | 147 |
| 3-[2-(Morpholino)ethyl]NHCO-5-Ph-7-Cl | 254–256 | EtOAc | | | 301 |
| 3-[6-(Morpholino)hexoxy]CO-5-Ph-7-Cl | 147–150 | Acetone/Petr ether | | | 147 |
| Hydrochloride | 210–212d | $CH_2Cl_2$/MeOH/EtOAc | | | 147 |
| 3-(Morpholino)$CH_2$-5-(2-$ClC_6H_4$)-7-$NO_2$ | 194–195 | $CH_2Cl_2$/Petr ether | | | 301 |
| 3-[3-(Morpholino)propoxy]CO-5-Ph-7-Cl | | | | | |
| Hydrochloride | 170–172d | $CH_2Cl_2$/Acetone | | | 147 |
| 3-[3-(Morpholino)propoxy]CO-5-(2-$ClC_6H_4$)-7-$H_2N$ | 215d | $Et_2O$ | | | 301 |

TABLE VII-1. —(contd.)

| Substituent | mp (°C) or; [bp (°C/torr)] | Solvent of Crystallization | Yield (%) | Spectra | Refs. |
|---|---|---|---|---|---|
| 3-[3-(Morpholino)propoxy]CO-5-(2-Cl$C_6H_4$)-7-$NO_2$ | 203–204d | PhH | | | 147 |
| Hydrochloride | 188–190d | | | | |
| | 229–232d | $CH_2Cl_2$ | | | 147 |
| Methobromide | 140–150d | PhH/Acetone | | | 147 |
| 3-[3-(Morpholino)propoxy]CO-5-(2-F$C_6H_4$)-7-Cl | 170–172d | $Et_2O$ | | | 147 |
| 3-(4-$O_2NC_6H_4$)CHN-5-Ph-7-Cl | 156–157 | | 25 | | 202 |
| 3-(3-Oxobutyl)-5-(2-Cl$C_6H_4$)-7-$NO_2$ | 185 | EtOAc/Petr ether | | | 301 |
| 3-(2-Oxopyrrolidin-1-yl)acetoxy-5-Ph-7-Cl | 229–230 | | | | 478 |
| 3-(2-Oxopyrrolidin-1-yl)acetoxy-5-(2-Cl$C_6H_4$)-7-$NO_2$ · THF | 142 | THF | | | 478 |
| 3-(1-Ph-Eth-1-yl)NHCOO-5-(2-Cl$C_6H_4$)-7-Cl | 193–196 | $Et_2O$ | | [α] | 338 |
| 3-PhCHN-5-Ph-7-Cl | 185–187 | EtOH | 29 | ir | 202 |
| 3,5-$Ph_2$-7-(1-Carboxy-1-ethyl) | 151–155 | | 64 | pmr | 20 |
| 3,5-$Ph_2$-7-Cl | 269–270 | DMF | 52 | | 2 |
| | 279d | PhMe | | | 35 |
| 3-Ph-5-(2-Cl$C_6H_4$)-7-$NO_2$ | 220–225d | THF/Hexane | | | 302 |
| 3-(2-Ph-Ethyl)NH-5-Ph-7-Cl | 181–183 | EtOH | 20 | | 348 |
| 3-(3-Ph-Propanoyloxy)-5-(2-Cl$C_6H_4$)-7-Cl | 175–178 | EtOH | 44 | | 337b |
| 3-PhOCOO-5-Ph-7-Cl | 162–164 | Dioxane/Ligroin | | | 336 |
| 3-[2-(Piperazin-1-yl)ethoxy]CO-5-Ph-Cl Dihydrochloride | 216–217d | Acetone/$H_2O$ | | | 147 |
| 3-[2-(Piperazin-1-yl)ethyl]NHCO-5-Ph-7-Cl Dihydrochloride · 0.5MeOH · 0.5$H_2O$ | 244–248 | $CH_2Cl_2$/MeOH | | | 301 |
| 3-(Piperidin-1-yl)-5-Ph-7-Cl | 229–230 | EtOH | 72; 58 | | 149, 348 |
| | 260–262 | EtOH | 45 | | 369 |
| 3-(Piperidin-1-yl)-5-Ph-7-$NO_2$ | 205–207d | | 57 | ir | 260 |
| 3-(Piperidin-1-yl)COO-5-Ph-7-Cl | 224–226 | EtOAc | | | 336 |
| 3-[2-(Piperidino)ethoxy]CO-5-Ph-7-Cl | 152–153d | $CH_2Cl_2$/Acetone | | | 147 |
| Dihydrochloride | 172–173d | $CH_2Cl_2$/Acetone | | | 147 |

| | | | | | |
|---|---|---|---|---|---|
| 3-[2-Piperidino-1-Me-ethoxy]CO-5-Ph-7-Cl | | | | | |
| Hydrochloride | 190d | Acetone/EtOAc | | | 147 |
| 3-Propanoyloxy-5-Ph-7-Cl | 228–230 | EtOH | 87 | | 326 |
| 3-*i*-Pr-5-Ph-7-Br | 234 | | 8 | | 100 |
| 3-*i*-Pr-5-Ph-(1-Carboxy-1-ethyl) | 130–135 | | 55 | pmr | 20 |
| 3-*i*-Pr-5-Ph-7-Cl | 226–227 | $Et_2O$/Petr ether | 9 | | 2 |
| 3-S-*i*-Pr-5-Ph-7-Cl | 192–194 | $CH_2Cl_2$/Petr ether | 87 | [α], pmr | 40 |
| 3-Pr-5-(2-Cl$C_6H_4$)-7-$NO_2$ | 228–230 | $CH_2Cl_2$/Petr ether | | | 301 |
| 3-*i*-Pr-5-(2-F$C_6H_4$)-7-Cl | | | | | |
| (+)-Enantiomer | 190 | $Et_2O$ | | | 301 |
| 3-*i*-Pr-5-(2-F$C_6H_4$)-7-$NO_2$ | 208–210 | $Et_2O$/Petr ether | | | 301 |
| 3-Pr-5-(2-F$C_6H_4$)-7-$NO_2$ | 245–246 | $CH_2Cl_2$/EtOH. | | | 231 |
| 3-PrNH-5-(2-Cl$C_6H_4$)-7-$NO_2$ | 178–180 | EtOAc | | | 301 |
| 3-*i*-PrNH-5-(2-Cl$C_6H_4$)-7-$NO_2$ | 183–185 | MeCN | | | 302 |
| 3-$Pr_2$N-5-(2-Cl$C_6H_4$)-7-$NO_2$ | 204–205 | $CH_2Cl_2$/EtOH | | | 301 |
| 3-PrO-5-(2-Cl$C_6H_4$)-7-$NO_2$ | 240–245 | $CH_2Cl_2$/EtOH | | | 301 |
| 3-*i*-PrO-5-(2-Cl$C_6H_4$)-7-$NO_2$ | 270–275d | $CH_2Cl_2$/EtOH | | | 301 |
| 3-*i*-PrOOC-5-Ph-7-Cl | 235–238d | | | | 302 |
| 3-(2-Pr-Pentanoyloxy)-5-(2-Cl$C_6H_4$)-7-Cl | 191–193 | MeCN | | | 326 |
| 3-(2-Pr-Pentanoyloxy)-5-(2-F$C_6H_4$)-7-Cl | 147–149 | MeCN/$H_2O$ | | | 326 |
| 3-(2-Propyn-1-yl)NH-5-(2-Cl$C_6H_4$)-7-$NO_2$ | 212–214d | EtOAc/Hexane | | | 301 |
| 3-(1-Pyridinium)-5-Ph-7-Cl | | | | | |
| Chloride | 250–251 | MeCN | | | 227 |
| | 234–236d | $H_2O$ | 80 | | 356 |
| Iodide | 240–246d | MeOH | | | 67b |
| 3-(1-Pyridinium)-5-(2-Cl$C_6H_4$)-7-Cl | | | | | |
| Chloride | 234–237d | Acetone | 90 | | 356 |
| 3-(Pyrrolidin-1-yl)-5-Ph-7-Cl | 193–195 | EtOH | 20 | | 348 |
| 3-(Pyrrolidin-1-yl)-5-(2-Cl$C_6H_4$)-7-$NO_2$ | | | | | |
| Hydrochloride | 238–240 | MeOH/Acetone | | | 302 |
| 3-(1,2,3,4-Tetrahydroisoquinolin-2-yl)CO-5-Ph-7-Cl | 241d | MeOH/$CH_2Cl_2$/*i*-$Pr_2O$ | | | 488 |
| 3-Thioacetamino-5-Ph-7-Cl | 284 | PhH | 48 | ir | 202 |
| 3-(Thiazolidin-3-yl)-5-(2-Cl$C_6H_4$)-7-$NO_2$ | 237–239d | $CH_2Cl_2$ | | | 301 |

TABLE VII-1. —(contd.)

| Substituent | mp (°C) or; [bp (°C/torr)] | Solvent of Crystallization | Yield (%) | Spectra | Refs. |
|---|---|---|---|---|---|
| ***3,5,8-Trisubstituted*** | | | | | |
| 3-$EtO_2C$-5-Ph-8-Cl | 214–217d | MeCN | | | 302 |
| ***3,5,9-Trisubstituted*** | | | | | |
| 3,5,9-$Me_3$ | 153–154 | Cyclohexane | | | 477 |
| ***5,6,7-Trisubstituted*** | | | | | |
| 5-Ph-6-Cl-7-$H_2N$ | 300–303 | MeCN | | | 33b |
| 5-(2-$FC_6H_4$)-6-Cl-7-$H_2N$ | 310–320d | $CH_2Cl_2$/EtOH | | | 301 |
| ***5,7,8-Trisubstituted*** | | | | | |
| 5-Me-7,8-$(MeO)_2$ | | | | | |
| 4-Oxide | 259–260 | DMF/$Et_2O$ | | | 2b |
| 5-Ph-7-$H_2NSO_2$-8-Cl | 291–294 | Acetone/MeOH | | | 67b |
| 5-Ph-7-Br-8-MeO | 260–261 | PhH/Hexane | 60 | | 2 |
| 5-Ph-7,8-$Cl_2$ | 242–244 | EtOAc | | | 475 |
| 5-Ph-7,8-$Me_2$ | 255–256 | MeOH | 75 | | 2 |
| 4-Oxide | 234–235 | $CH_2Cl_2$/Petr ether | | | 91a |
| 5-Ph-7-$NO_2$-8-Me | 218–219 | PhH | | | 2b |
| 5-(2-$ClC_6H_4$)-7-F-8-Me | 192–193 | EtOH/Hexane | | | 134 |
| 5-(2-$ClC_6H_4$)-7,8-$Me_2$ | 259–260 | $CH_2Cl_2$/Petr ether | 10 | | 2 |
| 5-(2-$FC_6H_4$)-7-Br-8-Cl | 216–217 | $CH_2Cl_2$/Cyclohexane | | | 257 |
| 5-(2-$FC_6H_4$)-7,8-Cl | 238–239 | $CH_2Cl_2$/EtOH | | | 257 |
| 5-(2-$FC_6H_4$)-7-Cl-8-$H_2N$ | 262–264 | | | | 257 |
| 5-(2-$FC_6H_4$)-7-Cl-8-($H_2N$-Acetamino) | 270–272 | $CH_2Cl_2$/THF | | | 257 |
| 5-(2-$FC_6H_4$)-7-Cl-8-(3-HOOC-Propionyl)NH · 0.5 $H_2O$ | 270–275 | DMSO/$H_2O$ | | | 257 |
| 5-(2-$FC_6H_4$)-7-Cl-8-Me | 237–239 | $CH_2Cl_2$/EtOH | | | 257 |
| 5-(2-$FC_6H_4$)-7-Cl-8-$NO_2$ | 150–155 | $CH_2Cl_2$/Cyclohexane | | | 257 |
| 5-(2-$FC_6H_4$)-7-CN-8-Cl | 217–218 | $CH_2Cl_2$/Cyclohexane | | | 257 |

| | | | | | |
|---|---|---|---|---|---|
| 5-(2-$FC_6H_4$)-7-Me-8-Cl | 255–259 | $CH_2Cl_2$/EtOH | | | 257 |
| 5-(2-$F_6H_4$)-7-$NO_2$-8-Cl | 248–249 | $CH_2Cl_2$/Cyclohexane | | | 257 |
| 5-(4-$O_2NC_6H_4$)-7,8-$Me_2$ | | | | | |
| 4-Oxide | 254–255 | | | | 2b |
| 5-(2-Ph-Ethyl)-7,8-$(MeO)_2$ | 182–183 | MeCN | | | 2b |
| 5-{2-[3,4-$(MeO)_2C_6H_3$]ethyl}-7,8-$(MeO)_2$ | 161–162 | EtOH | | | 2b |
| ***5,7,9-Trisubstituted*** | | | | | |
| 5-Ph-7,9-$Br_2$ | 195–197 | EtOH | | | 67b |
| 5-Ph-7,9-$Cl_2$ | 207–208 | Acetone | 40 | | 2 |
| 5-Ph-7-Cl-9-HO | 274–276 | PhH/EtOH | | pmr | 67a |
| 4-Oxide | 236–238 | PhH/MeOH | 70 | | 67a |
| 5-Ph-7-Cl-9-(2-HO-3-*i*-PrNH-Propoxy) | | | | | |
| Oxalate | 217–218 | MeOH | | | 492 |
| 5-Ph-7-Cl-9-I | 173–175 | EtOAc | | | 493 |
| 5-Ph-7-Cl-9-Me | 186–189 | EtOH | | | 371 |
| 5-Ph-7-Cl-9-MeS | 189–191 | MeCN | | | 7 |
| 5-Ph-7,9-$I_2$ | 209–210 | THF/$Et_2O$ | | | 371 |
| 5-Ph-7,9-$Me_2$ | 210–211 | Acetone | | | 91 |
| 5-Ph-7-$NO_2$-9-Me | 202–204 | $Et_2O$ | | | 2b |
| 5-(2-$ClC_6H_4$)-7,9-$Br_2$ | 216–217 | EtOAc | | | 475 |
| 5-(2-$ClC_6H_4$)-7,9-$Cl_2$ | 201–202 | $CH_2Cl_2$/$Et_2O$ | | | 134 |
| 5-(2-$ClC_6H_4$)-7-Cl-9-Br | 199–200 | | | | 494 |
| 5-(2-$ClC_6H_4$)-7-Cl-9-HOOC | 260d | $CHCl_3$/EtOAc | | | 486 |
| 5-(2-$ClC_6H_4$)-7-Cl-9-MeOOC | 213–215 | EtOAc | | | 486 |
| 5-(2-$FC_6H_4$)-7-$H_2N$-9-Cl | 200 | $CH_2Cl_2$/$Et_2O$ | | | 301 |
| 5-(2-$FC_6H_4$)-7,9-$Br_2$ | 206–207 | EtOAc | | | 475 |
| 5-(2-$FC_6H_4$)-7-Cl-9-HOOC | 260d | MeOH/EtOAc | | | 486 |
| 5-(2-$FC_6H_4$)-7-Cl-9-I | 211–213 | EtOAc/$CHCl_3$ | | | 493 |
| 5-(2-$FC_6H_4$)-7-(Cyclopentyl)CONH-9-Cl | 275 | EtOAc/MeOH | | | 301 |
| 5-(2-$FC_6H_4$)-7,9-$I_2$ | 231–233 | THF/Petr ether | | | 192 |
| 5-(2-$FC_6H_4$)-7-$NO_2$-9-Cl | 179 | EtOAc | | | 301 |
| 5-(3-$O_2NC_6H_4$)-7-Cl-9-$NO_2$ | 164–167 | MeOH/$CH_2Cl_2$ | 20 | pmr | 152a |
| 5-(2-Pyridyl)-7,9-$Br_2$ | 239–241 | BuOH | | | 67b |

TABLE VII-1. —(contd.)

| Substituent | mp (°C) or; [bp (°C/torr)] | Solvent of Crystallization | Yield (%) | Spectra | Refs. |
|---|---|---|---|---|---|
| ***Tetrasubstituted*** | | | | | |
| ***1,3,3,5-Tetrasubstituted*** | | | | | |
| 1-Me-3-AcO-3-EtOOC-5-Ph- | 154–156 | EtOH | | | 108 |
| 1-Me-3-EtOOC-3-EtOOCNHCOO-5-Ph-7-Cl | 165–168 | $CH_2Cl_2$/Hexane | | | 302 |
| 1-Me-3-EtOOC-3-HO-5-Ph | 171–173 | EtOH | | | 108 |
| ***1,3,5,7-Tetrasubstituted*** | | | | | |
| 1-Ac-3-AcO-5-Ph-7-Ac | 150–153 | EtOH | 37 | | 44 |
| 1-Ac-3-AcO-5-Ph-7-Cl | 170–172 | Cyclohexane | 70 | | 309 |
| 1-Ac-3-AcO-5-Ph-7-$NO_2$ | 178–180 | PhH/Hexane | | | 33b |
| 1-Ac-3-AcO-5-(2-$ClC_6H_4$)-7-Cl | 171–173 | | | | 308 |
| 1-Ac-3-AcO-5-(2-$ClC_6H_4$)-7-$NO_2$ | 206–208 | $CH_2Cl_2$/MeOH | | | 108 |
| 1-Ac-3-AcO-5-(2-$FC_6H_4$)-7-AcNH | 218–224 | $CH_2Cl_2$ | | | 301 |
| 1-Ac-3-AcO-5-(2,6-$F_2C_6H_3$)-7-Cl | 158–162 | $CH_2Cl_2$/Hexane | | | 152c |
| 1,3-$(AcO)_2$-5-Ph-7-Cl | 196–198 | THF/Petr ether | 80 | ir, ms, pmr, uv | 189 |
| 1,3-$(AcO)_2$-5-(2-Pyridyl)-7-Br | 190–191 | PhH/Hexane | 20 | ir, pmr | 148 |
| 1-(2-AcO-Ethyl)-3-AcO-5-Ph-7-Cl | 155–156 | *i*-PrOH | 78 | | 365 |
| 1-(2-AcO-Ethyl)-3-AcO-5-Ph-7-$NO_2$ | 173–174 | MeOH | 77 | | 365 |
| 1-(2-AcO-Ethyl)-3-AcO-5-(2-$ClC_6H_4$)-7-Cl | 162–164 | *i*-PrOH | 92 | | 365 |
| 1-(2-AcO-Ethyl)-3-AcO-5-(2-$F_6H_4$)-7-Cl | 129–130 | *i*-PrOH | 78 | | 365, 282 |
| 1-(3-AcO-Propyl)-3-AcO-5-Ph-Cl | 152–153 | MeOH | 47 | | 365 |
| 1-(2-AcO-Ethyl)-3-EtO-5-Ph-7-Cl | 112–114 | Acetone/Petr ether | 48 | | 365 |
| 1-Allyl-3-AcO-5-Ph-7-Cl | 155–156 | $CH_2Cl_2$/$Et_2O$/Petr ether | | | 2b |
| 1-Allyl-3-EtOOC-5-(2-$ClC_6H_4$)-7-Cl | 196 | $Et_2O$ | | | 275 |
| 1-Allyl-3-F-5-Ph-7-Cl | 138–140 | Heptane | | F-nmr, pmr | 266 |
| 1-Allyl-3-HO-5-Ph-7-Cl | 149–151 | $Et_2O$ | | | 2b |
| 1-Benzoyl-3-benzoyloxy-5-(2-$ClC_6H_4$)-7-$NO_2$ | 172–182d | THF/Hexane | | | 302 |
| 1-*t*-Bu-3-Me-5-Ph-7-Cl | 144–146 | $Et_2O$/Pentane | | | 391b |

| | | | | | |
|---|---|---|---|---|---|
| 1-*t*-Bu-3-Me-5-(2-$ClC_6H_4$)-7-$H_2N$ | | | | | |
| Picrate | 202–204d | EtOH/$H_2O$ | | | 391b |
| 1-*t*-Bu-3-Me-5-(2-$ClC_6H_4$)-7-$NO_2$ | 190–191 | Acetone/Hexane | | | 391b |
| 1-*t*-Bu-5-Me-5-(2-$ClC_6H_4$)-7-Phthalimide | 123–125d | Acetone/Hexane | | | 391b |
| 1-Cl-3-Me-5-Ph-7-Cl | 185d | PhH | 58 | | 224 |
| 1-(Cl-Acetyl)-3-AcO-5-Ph-7-Cl | 195–197 | | | | 308 |
| 1-(2-Cl-Ethyl)-3-EtO-5-Ph-7-Cl | 209–210 | EtOAc | 68 | | 365 |
| 1-(2-Cl-1-MeO-Ethyl)-3-AcO-5-Ph-7-Cl | 193–195 | EtOH | | | 72 |
| | | $CH_2Cl_2$/Hexane | 65 | | 143 |
| 1-(2-Cl-1-MeO-Ethyl)-3-EtOOC-5-Ph-7-Cl | 157–159 | EtOH | | | 278 |
| 1-(2-Cl-MeO-Ethyl)-3-HO-5-Ph-7-Cl | 191–193 | EtOAc | | | 72 |
| | 205–206 | EtOH | | | 199 |
| 1-(2-CN-Ethyl)-3-(2-CN-ethoxy)-5-Ph-7-Cl | 215–218 | MeCN | | | 280 |
| 1-(2-CN-Ethyl)-3-(2-CN-ethoxy)-5-(2-$ClC_6H_4$)-7-Cl | 210–213 | MeCN | | | 280 |
| 1-(2-CN-Ethyl)-3-(2-CN-ethoxy)-5-(2-$FC_6H_4$)-7-Cl | 204–207 | MeCN | | | 280 |
| 1-(2-CN-Ethyl)-3-HO-5-Ph-7-Cl | 192–194 | $CHCl_3$ | | | 280 |
| 1-(2-CN-Ethyl)-3-HO-5-Ph-7-$NO_2$ | 185–188 | | | | 280 |
| 1-(2-CN-Ethyl)-3-HO-5-(2-$ClC_6H_4$)-7-Cl | 198–202 | EtOAc | | | 280 |
| 1-(2-CN-Ethyl)-3-HO-5-(2-$ClC_6H_4$)-7-$NO_2$ | 191–194 | | | | 280 |
| 1-(2-CN-Ethyl)-3-HO-5-(2-$FC_6H_4$)-7-Cl | 190–193 | EtOH | | | 280 |
| 1-(2-CN-Ethyl)-3-HO-5-(2-$FC_6H_4$)-7-$NO_2$ | 183–186 | | | | 280 |
| 1-[2-(N-CN-N-Et-amino)ethyl]-3-AcO-5-(2-$FC_6H_4$)-7-Cl | 94–101 | MeOH/$Et_2O$ | | | 152c |
| 1-(Cyclopropyl)$CH_2$-3-AcO-5-Ph-7-Cl | 195–197 | *i*-PrOH | | | 232 |
| 1-(Cyclopropyl)$CH_2$-3-HO-5-Ph-7-Cl | 159–161 | EtOH | 96 | | 232 |
| | 272–275 | $CH_2Cl_2$/$Et_2O$ | | | 301 |
| 1-Et-3-F-5-Ph-7-Cl | 156–158 | Heptane | 57 | pmr | 266 |
| 1-Et-3-Me-5-(2-$ClC_6H_4$)-7-$H_2N$ | | | | | |
| *S*(+)-Enantiomer | 209 | | | | 495 |
| 1-Et-3-Me-5-(2-$ClC_6H_4$)-7-$NO_2$ | | | | | |
| *S*(+)-Enantiomer | 141–142 | | | | 495 |
| 1-EtNHCO-3-F-5-Ph-7-Cl | 110–112 | Cyclohexane | | F-nmr, pmr | 266 |
| 1,3-$(EtOOC)_2$-5-Ph-7-Cl | 147–150d | $CH_2Cl_2$/$Et_2O$ | | | 108 |
| 1-EtOOC-3-EtOOCNH-5-Ph-7-Cl | 190–192 | MeCN | | | 227 |

TABLE VII-1. —(contd.)

| Substituent | mp (°C) or; [bp (°C/torr)] | Solvent of Crystallization | Yield (%) | Spectra | Refs. |
|---|---|---|---|---|---|
| 1-EtOOCCH$_2$-3-Me-5-(2-ClC$_6$H$_4$)-7-NO$_2$ | | | | | |
| (+)-Enantiomer | 155–157 | Et$_2$O/Hexane | | | 301 |
| 1-(2-Et$_2$N-Ethyl)-3-AcO-5-Ph-7-Cl | 132–133 | EtOH/H$_2$O | 71 | | 365 |
| 1-(2-Et$_2$N-Ethyl)-3-AcO-5-(2-ClC$_6$H$_4$)-7-Cl | 144–146 | CH$_2$Cl$_2$/Hexane | 70 | pmr | 238, 365 |
| 1-(2-Et$_2$N-Ethyl)-3-AcO-5-(2-FC$_6$H$_4$)-7-Cl | | | | | |
| Hydrochloride | 214–218 | Acetone/Et$_2$O | 84 | | 233 |
| 1-(2-Et$_2$N-Ethyl)-3-(3-HOOC-propanoyloxy)-5-(2-FC$_6$H$_4$)-7-Cl | Oil | | | | 152c |
| Succinate | 101–102 | Acetone/Et$_2$O | | | 152c |
| 1-(2-Et$_2$N-Ethyl)-3-HO-5-Ph-7-Cl | 115–117 | Hexane | 88 | | 365 |
| 1-(2-Et$_2$N-Ethyl)-3-HO-5-(2-ClC$_6$H$_4$)-7-Cl | 138–140 | Et$_2$O/Pentane | 12 | pmr | 288 |
| | 128–129 | *i*-PrOH | 60 | | 365 |
| 1-(2-Et$_2$N-Ethyl)-3-HO-5-(2-FC$_6$H$_4$)-7-Cl | 118–121 | Et$_2$O/Petr ether | | | 233 |
| Hydrochloride | 196–203 | Acetone/Et$_2$O | 73 | | 233 |
| 1-(2-Et$_2$N-Ethyl)-3-HOCH$_2$-5-(2-ClC$_6$H$_4$)-7-Br | Amorphous | | | | 264 |
| 1-(2-Et$_2$N-Ethyl)-3-Me-5-(2-ClC$_6$H$_4$)-7-NO$_2$ | | | | | |
| (+)-Enantiomer | 148 | Et$_2$O/Petr ether | | | 301 |
| 1-(2-Et$_2$N-Ethyl)-3-MeO-2-(2-ClC$_6$H$_4$)-7-Cl | 123–125 | Cyclohexane/Et$_2$O | 68 | pmr | 288 |
| 1-(2-Et$_2$N-Ethyl)-3-EtOOC-5-Ph-7-Cl | 92–95 | EtOH | | | 147 |
| 1-(2-Et$_2$N-Ethyl)-3-EtOOC-5-(2-FC$_6$H$_4$)-7-Cl | | | | ir, ms, pmr | 145 |
| 1-(2-Et$_2$N-Ethyl)-3-(3-Et$_2$N-propyloxy)CO-5-Ph-Cl | | | | | |
| Dihydrochloride | 200–204 | | | | 147 |
| 1-F$_3$CO-3-F$_3$COO-5-Ph-7-Cl | 159–161 | (F$_3$CCO)$_2$O | | | 33b |
| 1-F$_3$CCH$_2$-3-AcO-5-Ph-7-Cl | 193–194 | Acetone/Hexane | 75 | | 69 |
| 1-F$_3$CCH$_2$-3-AcO-5-(2-FC$_6$H$_4$)-7-Cl | 156–159 | CH$_2$Cl$_2$/Hexane | | | 69 |
| 1-F$_3$CCH$_2$-3-HO-5-Ph-7-Cl | 186–187 | EtOH/Hexane | 75 | | 69 |
| 1-F$_3$CCH$_2$-3-HOCH$_2$-5-(2-ClC$_6$H$_4$)-7-Br | Amorphous | | | | 264 |
| 1,3-(HO)$_2$-5-Ph-7-Cl | 176–178d | EtOH/H$_2$O | 23 | ir, ms, pmr, uv | 189 |

| | | | | |
|---|---|---|---|---|
| 1-(2-HO-Ethyl)-3-AcO-5-Ph-7-Cl | 203–205 | Acetone/$H_2O$ | 15 | 365 |
| 1-(2-HO-Ethyl)-3-EtO-5-Ph-7-Cl | 213–215 | EtOH | 63 | 365 |
| 1-(2-HO-Ethyl)-3-EtOOC-5-Ph-7-Cl | 168–170 | EtOH | | 147 |
| 1-(2-HO-Ethyl)-3-(3-$Et_2N$-propoxy)CO-5-Ph-7-Cl | 118–120 | $Et_2O$/Petr ether | | 147 |
| 1-(2-HO-Ethyl)-3-HO-5-Ph-7-Cl | 158–160 | MeOAc/Hexane | 85 | 365 |
| 1-(2-HO-Ethyl)-3-HO-5-Ph-7-$NO_2$ | 178d | PrOH | 45 | 365 |
| 1-(2-HO-Ethyl)-3-HO-5-(2-$ClC_6H_4$)-7-Cl | 171–173 | *i*-PrOH | 73 | 365 |
| 1-(2-HO-Ethyl)-3-HO-5-(2-$FC_6H_4$)-7-Cl | 138–140 | $CH_2Cl_2$/$Et_2O$ | 81 | 282, 365 |
| 1-(2-HO-Ethyl)-3-$Me_2NCO$-$CH_2$-5-Ph-7-Cl | 163–165 | $CH_2Cl_2$/$Et_2O$ | | 302 |
| 1-(3-HO-Propyl)-3-HO-5-Ph-7-Cl | 151–153 | EtOH/$Et_2O$ | 63 | 365 |
| 1-Me-3-Ac-5-Ph-7-Cl | 148 | Pentane | 12 | 299 |
| Methylhydrazone | 174–176 | $Et_2O$/Petr ether | | 301 |
| 1-Me-3-(4-AcNH-Butanoyloxy)-5-Ph-7-Cl | 131–132 | MeCN/*i*-$Pr_2O$ | | 339 |
| 1-Me-3-(4-AcNH-Butanoyloxy)-5-(2-$ClC_6H_4$)-7-Cl | Amorphous | | | 339 |
| 1-Me-3-Ac-(EtOOC)CH-5-(Cyclohexen-1-yl)-7-Cl | 191 | EtOH | | 72c |
| 1-Me-3-$AcCH_2$-5-(Cyclohexen-1-yl)-7-Cl | 163–165 | MeOH | | 72c |
| 1-Me-3-AcO-5-Ph-7-AcNH | >245 | EtOH/$Et_2O$ | | 251c |
| 1-Me-3-AcO-5-Ph-7-Cl | 262–263 | | | 263 |
| 1-Me-3-AcO-5-Ph-7-$F_3C$ | 189–190 | Acetone/Hexane | | 1b |
| 1-Me-3-AcO-5-Ph-7-MeS | 209–212d | $CH_2Cl_2$/$Et_2O$/Hexane | | 192 |
| 1-Me-3-AcO-5-Ph-7-$NO_2$ | 248–249 | THF/Hexane | | 1b |
| 1-Me-3-AcO-5-(2-$ClC_6H_4$)-7-Cl | 219–221 | $CH_2Cl_2$/$Et_2O$ | | 247 |
| 1-Me-3-AcO-5-(2-$ClC_6H_4$)-7-$NO_2$ | 234–236 | THF/Hexane | | 302 |
| 1-Me-3-AcO-5-(4-$ClC_6H_4$)-7-Cl | 249–253 | Acetone | | 152c |
| 1-Me-3-AcO-5-(2-$FC_6H_4$)-7-Ac | 145–147 | $Et_2O$ | | 2b |
| 1-Me-3-AcO-5-(2-$FC_6H_4$)-7-AcNH | 280 | $Ac_2O$/$Et_2O$ | | 301 |
| 1-Me-3-AcO-5-(2-$FC_6H_4$)-7-(2-AcO-Ethyl)NHCONH | 183–186 | EtOAc | | 301 |
| 1-Me-3-AcO-5-(2-$FC_6H_4$)-7-$H_2N$ | 259–260 | $CH_2Cl_2$/$Et_2O$ | | 2b |
| 1-Me-3-AcO-5-(2-$FC_6H_4$)-7-Cl | 252–253 | $CH_2Cl_2$/$Et_2O$/Petr ether | | 152c |
| 1-Me-3-AcO-5-(2-$FC_6H_4$)-7-I | 239–241d | $CH_2Cl_2$/Hexane | | 134 |
| 1-Me-3-AcO-5-(2-$FC_6H_4$)-7-$NO_2$ | 219–220 | | | 2b |
| 1-Me-3-AcO-5-(2-Pyridyl)-7-Br | 259–265d | MeOH | | 152c |
| 1′-Oxide | 235–245 | | | |
| | >350 | $CH_2Cl_2$/MeOH | | 152c |

TABLE VII-1. —(contd.)

| Substituent | mp (°C) or; [bp (°C/torr)] | Solvent of Crystallization | Yield (%) | Spectra | Refs. |
|---|---|---|---|---|---|
| 1-Me-3-AcOCH$_2$-5-(2-ClC$_6$H$_4$)-7-Br | 121–123 | | | | 264 |
| 1-Me-3-AcOCH$_2$-5-(2-ClC$_6$H$_4$)-7-Cl | Amorphous | | | | 264 |
| 1-Me-3-AcOCH$_2$-5-(2-FC$_6$H$_4$)-7-H$_2$N | 205–207 | Et$_2$O/Petr ether | | | 301 |
| 1-Me-3-AcOCH$_2$-5-(2-FC$_6$H$_4$)-7-Cl | 168–170 | | | | 264 |
| 1-Me-3-AcOCH$_2$-5-(2-FC$_6$H$_4$)-7-(2-HO-Ethyl)NHCONH | 175–177 | EtOAc/Et$_2$O | | | 301 |
| 1-Me-3-Allyl-5-Ph-7-Cl | | | | | |
| 4-Oxide | 183–185 | MeOH | 70 | ir, pmr, uv | 298 |
| 1-Me-3-(Allyl)OOC-5-Ph-7-Cl | 171–172 | PhH/Hexane | | | 490 |
| 1-Me-3-H$_2$NCO-5-Ph-7-Cl | 242–247d | MeOH | 98 | ir, pmr | 226 |
| 1-Me-3-H$_2$NCONHCO-5-Ph-7-Cl | 235–240d | MeOH | | | 302 |
| 1-Me-3-H$_2$NCOO-5-Ph-7-Cl | 216–218 | Dioxane/Ligroin | | | 336 |
| 1-Me-3-(3-H$_2$NCO-1-Pyridinium)-5-Ph-7-Cl | | | | | |
| Chloride | 205–207d | Acetone | | | 334 |
| 1-Me-3-(4-H$_2$NCO-1-Pyridinium)-5-Ph-7-Cl | | | | | |
| Chloride | 273–276d | Acetone | 80 | | 356 |
| 1-Me-3-N$_3$-5-Ph-7-Cl | 151–153 | EtOH | | | 33b |
| 1-Me-3-N$_3$CO-5-Ph-7-Cl | 120d | MeOH | | | 488 |
| 1-Me-3-Benzoyl-5-Ph-7-Cl | 199–200 | CHCl$_3$/Hexane | 20 | | 299 |
| 1-Me-3-(Benzoyloxy)CH$_2$-5-(2-FC$_6$H$_4$)-7-Cl | 120–123 | | | | 264 |
| 1-Me-3-Benzyl-5-Ph-7-Cl | 151–152 | Hexane | 53 | | 299 |
| 4-Oxide | 180–182 | MeOH | 59 | pmr | 297 |
| 1-Me-3-S-Benzyl-5-Ph-7-Cl | 135–137 | Petr ether | 85–90 | [α], pmr | 40 |
| 1-Me-3-(4-Br-Butyl)-5-Ph-7-Cl | | | | | |
| 4-Oxide | 146–147 | CH$_2$Cl$_2$/Petr ether | | | 134 |
| 1-Me-3-(5-Br-Pentyl)-5-Ph-7-Cl | | | | | |
| 4-Oxide | 163–165 | CH$_2$Cl$_2$/Et$_2$O | | | 134 |
| 1-Me-3-(3-Br-Propyloxy)-5-Ph-7-Cl | 164–165 | EtCOMe | | | 1b |
| 1-Me-3-(*trans*-But-2-en-1-yl)-5-Ph-7-Cl | | | | | |
| 4-Oxide | 164–167 | MeOH | 30 | pmr | 298 |

| | | | | | |
|---|---|---|---|---|---|
| 1-Me-3-Bu-5-Ph-7-Cl | 159–160 | $Et_2O$/Pentane | 22 | | 299 |
| 1-Me-3-(3-*t*-BuOOC-Propyl)-NHCOO-5-Ph-7-Cl | 157–159 | | | | 478 |
| 1-Me-3-(3-Br-Propoxy)-5-Ph-7-Cl | 153–154 | MeCOEt | | | 324 |
| 1-Me-3-(3-HOOC-Propanoyloxy)-5-(2-$FC_6H_4$)-7-$NO_2$ | 162–165 | $CH_2Cl_2/Et_2O$ | | | 152c |
| 1-Me-3-(3-HOOC-Propanoyloxy)$CH_2$-5-(2-$ClC_6H_4$)-7-Br | 175–177d | | | | 264 |
| 1-Me-3-(3-HOOC-*trans*-Propenoyloxy)$CH_2$-5-(2-$ClC_6H_4$)-7-Br | | | | | |
| (2-HO-Ethyl)$Me_2$-ammonium salt | 133–135 | | | | 264 |
| 1-Me-3-Cl-5-Ph-7-Cl | 98–100 | $Et_2O$ | 98 | pmr | 334, 356 |
| | 108–110 | $Et_2O$ | | ir, uv | 215b |
| 1-Me-3-Cl-5-(2-$ClC_6H_4$)-7-Cl | 217–219 | MeCN | 90 | pmr | 265 |
| 1-Me-3-Cl-5-(2-$ClC_6H_4$)-7-$NO_2$ | 240–245d | $CH_2Cl_2$ | | | 302 |
| 1-Me-3-Cl-5-(2-$FC_6H_4$)-7-Cl | 163–166 | $CH_2Cl_2$/Hexane | | | 134 |
| 1-Me-3-(Cl-Acetoxy)-5-(2-$ClC_6H_4$)-7-Cl | 172–174 | $Et_2O$ | 83 | pmr | 265, 328 |
| 1-Me-3-($Cl_3$-Acetoxy)-5-Ph-7-Cl | 194–196 | PhH/Petr ether | 60 | | 334 |
| 1-Me-3-(4-Cl-Benzoyloxy)-$CH_2$-5-(2-$FC_6H_4$)-7-Cl | 160–163 | | | | 264 |
| 1-Me-3-(4-Cl-Benzyl)-5-Ph-7-Cl | 167–168 | Hexane | 60 | | 299 |
| 1-Me-3-(4-$ClC_6H_4$)-5-Ph-7-Cl | 200–201 | | | | 86b |
| 1-Me-3-CN-5-Ph-7-Cl | 209–211 | EtOH | 76 | ir, pmr | 226 |
| 1-Me-3-(2-CN-Ethyl)-5-Ph-7-Cl | | | | | |
| 4-Oxide | 238–240 | $CH_2Cl_2/Et_2O$ | | | 134 |
| 1-Me-3-(2-CN-Ethyl)-5-Ph-7-Cl | 140–144 | $Et_2O$/Petr ether | | | 301 |
| 1-Me-3-[3-(Cyclohexyl)-NH-propoxy]CO-5-Ph-7-Cl | 109–111 | $Et_2O$ | | | 302 |
| 1-Me-3-[3-(10,11-Dihydro-5*H*-dibenzo[*a,d*]cyclohepten-5-ylidene)propyl]MeNCOO-5-Ph-7-Cl | 80–100 | $Et_2O$/Hexane | | | 33b |
| 1-Me-3-(2-$Et_2N$-Ethoxy)CO-5-Ph-7-Cl | 103–105 | EtOAc/Hexane | | | 302 |
| 1-Me-3-F-5-Ph-7-Br | | | | F-nmr, pmr | 266 |
| 1-Me-3-F-5-Ph-7-Cl | 145–147 | EtOH | 98 | pmr, F-nmr | 110 |
| | 138–140 | Heptane | 90 | pmr, F-nmr | 267 |
| | 138–140 | Heptane | 90 | F-nmr, pmr | 266 |
| 4-Oxide | 167–169 | MeCOEt | 75 | ir, pmr, F-nmr | |
| | 190–192d | $CH_2Cl_2$/EtOH | 85 | $^{13}C$-nmr | 110 |

TABLE VII-1. —(contd.)

| Substituent | mp (°C) or; [bp (°C/torr)] | Solvent of Crystallization | Yield (%) | Spectra | Refs. |
|---|---|---|---|---|---|
| 1-Me-3-F-5-Ph-7-$NO_2$ | | | 60 | F-nmr, pmr | 266 |
| 1-Me-3-F-5-(2-$ClC_6H_4$)-7-Cl | 204–205 | PhH/Hexane | | F-nmr, pmr | 266 |
| 1-Me-3-F-5-(2-$FC_6H_4$)-7-Br | 127–130 | Heptane | 50 | F-nmr, pmr | 266 |
| 1-Me-3-F-5-(2-$FC_6H_4$)-7-Cl | 91–95 | | | pmr | 266 |
| 1-Me-3-$F_3CCOO$-5-Ph-7-Cl | 209–211 | $CH_2Cl_2$/Pentane | | | 33b |
| 1-Me-3-Et-5-Ph-7-Cl | 180–181 | $CH_2Cl_2$/Petr ether | | | |
| 1-Me-3-Et-5-(2-$ClC_6H_4$)-7-$H_2N$ | 198–199 | EtOAc/$Et_2O$ | | | 231 |
| 1-Me-3-Et-5-(2-$ClC_6H_4$)-7-$NO_2$ | 161–162 | EtOAc/$Et_2O$ | | | 231 |
| 1-Me-3-Et-5-(2-$FC_6H_4$)-7-$H_2N$ | 185–186 | EtOAc | | | 231 |
| 1-Me-3-Et-5-(2-$ClC_6H_4$)-7-(2-HO-Ethyl)NHCONH | 177–180d | Acetone/$Et_2O$ | | | 301 |
| 1-Me-3-Et-5-(2-$FC_6H_4$)-7-$NO_2$ | 175–176 | EtOAc/Petr ether | | | 231 |
| 1-Me-3-(3-Et-1-Imidazolium)-5-Ph-7-Cl | | | | | |
| Chloride | 230–233 | Acetone | 60 | | 356 |
| | | ether | 60 | | 299 |
| 1-Me-3-(3-$Et_2NCO$-1-Pyridinium)-5-Ph-7-Cl | | | | | |
| Chloride | 169–172 | Acetone | 70 | | 356 |
| 1-Me-3-(3-$Et_2N$-Propoxy)-5-Ph-7-Cl | 153–154 | MeCOEt | | | 324 |
| 1-Me-3-(3-$Et_2N$-Propyloxy)CO-5-Ph-7-Cl | | | | | |
| Hydrochloride | 204–206d | $CH_2Cl_2$/Acetone | | | 147 |
| 1-Me-3-(2,4-Dinitrocyclohexadien-1-yl)-5-Ph-7-Cl | | | | | |
| $Me_4N$-salt | | | 76 | ir, $^{13}C$-nmr, pmr, uv | 303 |
| 1,3-$Me_2$-5-Ph-7-Cl | 110–112 | $CHCl_3$/Hexane | 75 | | 299 |
| 4-Oxide | 185–188 | EtOH | 73 | ir, pmr, uv | 297 |
| 1,3-R-$Me_2$-5-Ph-7-Cl | 48–50 | Petr ether | 85–90 | [α] | 40 |
| 1,3-S-$Me_2$-5-Ph-7-Cl | 47–50 | Petr ether | 85–90 | [α], pmr | 40 |

| | | | | |
|---|---|---|---|---|
| 1,3-$Me_2$-5-(2-$ClC_6H_4$)-7-$H_2N$ | 180 | EtOAc/Petr ether | | 80b |
| *R*(+)-Enantiomer | 218–220 | EtOH | | 301 |
| *S*(−)-Enantiomer | 219–222 | EtOH/$Et_2O$ | | 301 |
| 1,3-$Me_2$-5-(2-$ClC_6H_4$)-7-Cl | 130 | $Et_2O$/Ligroin | | 301 |
| 1,3-$Me_2$-5-(2-$ClC_6H_4$)-7-(2-HO-Ethyl)NHCONH | 233–237 | | | 301 |
| 1,3-$Me_2$-5-(2-$ClC_6H_4$)-7-(4-$MeC_6H_4SO_2$)NHCONH | 196–200 | Acetone/$Et_2O$ | | 301 |
| 1,3-$Me_2$-5-(2-$ClC_6H_4$)-7-$NO_2$ | 170–174 | $CH_2Cl_2$/EtOH | | 80b |
| (+)-Enantiomer | 149 | $Et_2O$/EtOH/Hexane | | 301 |
| (−)-Enantiomer | 152–154 | $Et_2O$ | | 301 |
| 1,3-$Me_2$-5-(2,4-$Cl_2C_6H_3$)-7-Cl | 158–159 | $Et_2O$ | | 134 |
| 1,3-$Me_2$-5-(2-$FC_6H_4$)-7-Ac | 65 | Cyclohexene | | 251 |
| 1,3-$Me_2$-5-(2-$FC_6H_4$)-7-$H_2N$ | 178–180 | EtOAc | | 231 |
| *R*(−)-Enantiomer·$Et_2O$ | 95–110d | $Et_2O$ | [α] | 231 |
| *S*(+)-Enantiomer·$Et_2O$ | 100–110d | $Et_2O$ | [α] | 231 |
| 1,3-$Me_2$-5-(2-$FC_6H_4$)-7-(1-$H_2N$-Ethyl) | 48–50 | | | 251 |
| 1,3-$Me_2$-5-(2-$FC_6H_4$)-7-Cl | 122–127 | MeOH | | 421 |
| 1,3-$Me_2$-5-(2-$FC_6C_4$)-7-(1-HO-Ethyl) | 69 | | | 251b |
| 1,3-$Me_2$-5-(2-$FC_6H_4$)-7-(2-HO-Ethyl)NHCONH | 193–198 | Acetone/$CH_2Cl_2$ | | 301 |
| *R*(−)-Enantiomer | >130 | EtOAc/Hexane | | 301 |
| *S*(+)-Enantiomer | >113d | EtOAc/Hexane | | 301 |
| 1,3-$Me_2$-5-(2-$FC_6H_4$)-7-(1-Hydroxyimino)ethyl | 101–103 | | | 251 |
| 1,3-$Me_2$-5-(2-$FC_6H_4$)-7-MeNHCONH | 109–111 | $Et_2O$/Hexane | | 231 |
| 1,3-$Me_2$-5-(2-$FC_6H_4$)-7-(4-$MeC_6H_4SO_2$)NHCONH | 229–230d | Acetone/$Et_2O$ | | 301 |
| 1,3-$Me_2$-5-(2-$FC_6H_4$)-7-MeS | 137–139 | $Et_2O$ | | 192 |
| 1,3-$Me_2$-5-(2,6-$F_2C_6H_3$)-7-$NO_2$ | 186–195 | $CH_2Cl_2$/Petr ether | | 152c |
| 1,3-$Me_2$-5-Ph-7-(2-Me-1,3-Dioxolan-2-yl) | 137–139 | Hexane | | 33b |
| 1,3-$Me_2$-5-(2-$FC_6H_4$)-7-$NO_2$ | 154–156 | EtOAc/Petr ether | | 231 |
| | 230–234 | EtOAc/$Et_2O$ | | 301 |
| *R*(−)-Enantiomer | 120 | $Et_2O$/Hexane | [α] | 231 |
| *S*(+)-Enantiomer | 118–126d | $Et_2O$/Petr ether | [α] | 231 |
| 1,3-$Me_2$-5-(3-$O_2NC_6H_4$)-7-$NO_2$ | 247–253 | Acetone | | 152c |
| 1,3-$Me_2$-5-(2-Pyrimidyl)-7-Cl | 198–200 | EtOAc | | 1b |
| 1-Me-3-EtNH-5-Ph-7-Cl | 214–215 | MeCN | ir, ms, pmr, uv | 215b |

TABLE VII-1. —(contd.)

| Substituent | mp (°C) or; [bp (°C/torr)] | Solvent of Crystallization | Yield (%) | Spectra | Refs. |
|---|---|---|---|---|---|
| 1-Me-3-(2-$Et_2$N-Ethyl)-5-Ph-7-Cl | | | | | |
| 4-Oxide | 122–123 | $CH_2Cl_2$/MeOH/Petr ether | | | 134 |
| Hydrochloride | 218–220d | MeOH/$Et_2O$/Acetone | | | 134 |
| 1-Me-3-EtO-5-Ph-7-Cl | 222–224 | $CH_2Cl_2$/$Et_2O$ | 28 | | 215 |
| 1-Me-3-EtO(HO)CH-5-Ph-7-Cl | 148–150 | EtOH | | | 152c |
| 1-Me-3-$(EtO)_2$CH-5-(2-Cl$C_6H_4$)-7-Br | 122–124 | $Et_2O$ | | | 314 |
| 1-Me-3-$(EtO)_2$CH-5-(2-F$C_6H_4$)-7-(2-HO-Ethyl)-NHCONH | Amorphous | | | | 301 |
| 1-Me-3-EtOOC-5-Ph-7-Cl | 196–199 | EtOH | 73 | | 39 |
| 4-Oxide | 184–185 | EtOH | | | 108 |
| 1-Me-3-$(EtOOC)_2$CH-5-(Cyclohexen-1-yl)-7-Cl | 165–166 | EtOH | | | 72c |
| 1-Me-3-$(EtO)_2$OP-5-Ph-7-Cl | 163–166 | EtOAc/Hexane | 55 | ir, pmr | 243 |
| 1-Me-3-(2-Furyl)(HO)CH-5-Ph-7-Cl | 213–215 | $CH_2Cl_2$/Petr ether | | | 192 |
| 1-Me-3-$H_2$NNHCO-5-Ph-7-Cl | 208d | *i*-PrOH | | | 488 |
| 1-Me-3-HO-5-Ph-7-AcNH | >240 | EtOH | | | 251c |
| 1-Me-3-HO-5-Ph-7-Cl | 125–126 | $Et_2O$/Pentane | 51, 69 | | 215 |
| 1-Me-3-HO-5-Ph-7-$F_3C$ | 183–185 | | | | 241 |
| 1-Me-3-HO-5-Ph-7-MeS | 158–161 | $CH_2Cl_2$/Petr ether | | | 192 |
| 1-Me-3-HO-5-Ph-7-$NO_2$ | 156–157 | EtOAc/$Et_2O$ | 56 | | 260 |
| | 201–202 | | | | 241 |
| 1-Me-3-HO-5-(2-Cl$C_6H_4$)-7-Cl | 209–211 | THF/EtOH | 81 | | 265 |
| | 192–194 | | | | 263 |
| 1-Me-3-HO-5-(2-Cl$C_6H_4$)-7-$NO_2$ | 225–230d | $MeOCH_2CH_2OH$/$Et_2O$ | | | 301 |
| 1-Me-3-HO-5-(4-Cl$C_6H_4$)-7-Cl | 195–197 | THF/Hexane | | | 152c |
| 1-Me-3-HO-5-(2-F$C_6H_4$)-7-Ac | 188–189 | $CH_2Cl_2$/Petr ether | | | 2b |
| 1-Me-3-HO-5-(2-F$C_6H_4$)-7-AcNH·0.5$Et_2O$ | 158–160 | Acetone/$Et_2O$ | | | 2b |
| 1-Me-3-HO-5-(2-F$C_6H_4$)-7-$H_2N$ | 210–212 | $CH_2Cl_2$/Petr ether | | | 152c |
| 1-Me-3-HO-5-(2-F$C_6H_4$)-7-Cl | 159–161 | $Et_2O$ | | | 152c |
| 1-Me-3-HO-5-(2-F$C_6H_4$)-7-(2-HO-Ethyl)NHCONH | 205–207 | EtOH | | | 301 |

| | | | | |
|---|---|---|---|---|
| 1-Me-3-HO-5-(2-$FC_6H_4$)-7-I | 188–190d | | | 108 |
| 1-Me-3-HO-5-(2-$FC_6H_4$)-7-$NO_2$ | 216–217 | | | 2b |
| 1-Me-3-HO-5-(2-Pyridyl)-7-Br | | | | |
| 1′-Oxide | 189–192 | MeOH/$Et_2O$ | | 152c |
| 1-Me-3-HO-5-(2-Thiazolyl)-7-Cl | 224–225d | THF/Hexane | | 192 |
| 1-Me-3-[3,4-$(HO)_2$-Benzyl]-5-Ph-7-Cl | 209–211d | $CHCl_3$/Heptane | | 192 |
| 1-Me-3-(4-HO-Benzyl)-5-Ph-7-Cl | 217–218 | MeOH | | 61 |
| 1-Me-3-(2-HO-But-2-yl)-5-(2-$FC_6H_4$)-7-$NO_2$ | 185–190 | EtOH | | 496 |
| 1-Me-3-(1-HO-Cyclohexan-1-yl)-5-Ph-7-Cl | 214–215 | $CHCl_3$/Hexane | 75 | 299 |
| 1-Me-3-(2-HO-Ethoxy)-5-Ph-7-Cl | 217–219 | $CHCl_3$/Petr ether | 85 | 334 |
| 1-Me-3-$HOCH_2$-5-Ph-7-Cl | | | | |
| Hydrochloride | 214–217d | | | 264 |
| 1-Me-3-$HOCH_2$-5-(2-$ClC_6H_4$)-7-Br | 198–201 | | | 264 |
| 1-Me-3-$HOCH_2$-5-(2-$ClC_6H_4$)-7-Cl | 182–184 | $Et_2O$/Petr ether | | 264 |
| Hydrochloride | 176d | | | 264 |
| 1-Me-3-$HOCH_2$-5-(2-$ClC_6H_4$)-7-F | 202–204 | | | 264 |
| 1-Me-3-$HOCH_2$-5-(2-$ClC_6H_4$)-7-I | 192–196d | | | 264 |
| 1-Me-3-$HOCH_2$-5-(2-$ClC_6H_4$)-7-$NO_2$ | 229–231 | | | 264 |
| 1-Me-3-$HOCH_2$-5-(2-$FC_6H_4$)-7-Br | 152–155 | | | 264 |
| 1-Me-3-$HOCH_2$-5-(2-$FC_6H_4$)-7-Cl | 141–143 | PhMe | | 264 |
| Hydrochloride | 202–204d | | | 264 |
| 1-Me-3-$HOCH_2$-5-(2-$FC_6H_4$)-7-(2-HO-Ethyl)-NHCONH | 200–204 | Acetone/$Et_2O$ | | 301 |
| 1-Me-3-[$(HO)Me_2C$]-5-Ph-7-Cl | 165–168 | $CHCl_3$/Hexane | 16 | 299 |
| 1-Me-3-[(HO)MePhC]-5-Ph-7-Cl | 216–219 | Hexane/$CH_2Cl_2$ | 22 | 299 |
| 1-Me-3-[(HO)PhCH]-5-Ph-7-Cl | 170–171 | $CHCl_3$/Hexane | 44 | 299 |
| Isomer A | 172–175 | $CH_2Cl_2$/Heptane | | 192 |
| Isomer B | 218–220 | $CH_2Cl_2$/Heptane | | 192 |
| 1-Me-3-[$(HO)Ph_2C$]-5-Ph-7-Cl | 201–204 | $CHCl_3$/Hexane | 60 | 299 |
| 1-Me-3-(3-HO-Propoxy)CO-5-Ph-7-Cl | 146–148 | PhH/Hexane | | 302 |
| 1-Me-3-[2,3-$(HO)_2$-Propoxy]-5-Ph-7-Cl | 199–202 | $Et_2O$/Petr ether | 98 | 334 |
| 1-Me-3-[(HO)(2-Pyridyl)CH]-5-Ph-7-Cl | 190–192 | $CH_2Cl_2$/$Et_2O$/Petr ether | | 192 |
| 1-Me-3-(Imidazol-1-yl)COO-5-Ph-7-Cl | 195–197 | | | 478 |

TABLE VII-1. —(contd.)

| Substituent | mp (°C) or; [bp (°C/torr)] | Solvent of Crystallization | Yield (%) | Spectra | Refs. |
|---|---|---|---|---|---|
| 1-Me-3-(3-Me-But-2-en-1-yl)-5-Ph-7-Cl | | | | | |
| 4-Oxide | 149–151 | $CH_2Cl_2$/Petr ether | | | 134 |
| 1-Me-3-[2-Me-3-(2-Br-Ethyl)-4-$NO_2$-1-imidazolium]-5-Ph-7-Cl | | | | | |
| Chloride | 199–202d | Acetone/MeOH | 60 | | 356 |
| 1-Me-3-[1-Me-1,3-Dihydro-5-(2-$FC_6H_4$)-7-Cl-2-oxo-2*H*-1,4-benzodiazepin-3-yl]-5-(2-$FC_6H_4$)-7-Cl | 336–338 | $CH_2Cl_2$/MeOH/ether | | | 152c |
| 1-Me-3-(3-Me-1-Imidazolium)-5-Ph-7-Cl | | | | | |
| Chloride | 209–212 | Acetone | 80 | | 356 |
| 1-Me-3-MeNH-5-Ph-7-$NO_2$ | 153–157 | EtOAc/*i*-$Pr_2O$ | 56 | ir, pmr | 260 |
| 1-Me-3-MeNHCOO-5-Ph-7-Cl | 227–229 | Dioxane/Ligroin | | | 336 |
| 1-Me-3-(1-Me-1-Aziridinium)-5-Ph-7-Cl | | | | | |
| Chloride | 200–202 | | | | 334 |
| 1-Me-3-(3-Me-But-2-en-1-yl)-5-Ph-7-Cl | | | | | |
| 4-Oxide | 149–151 | MeOH | 57 | pmr | 298 |
| 1-Me-3-[(2,2-$Me_2$-1,3-Dioxolan-4-yl)methoxy]-5-Ph-7-Cl | 146–148 | $Et_2O$/Petr ether | 75 | | 334 |
| 1-Me-3-(1-Me-1-Morpholinium)-5-Ph-7-Cl | | | | | |
| Chloride | 237–239d | Acetone | 67 | | 356 |
| | 223–225d | | | | 334 |
| 1-Me-3-(1-Me-2-Et-5-[2-HO-Ethyl]-1-morpholinium)-5-Ph-7-Cl | | | | | |
| Chloride | 198–201 | | | | 334 |
| 1-Me-3-(1-Me-3-Et-1-Morpholinium)-5-Ph-7-$NO_2$ | | | | | |
| Chloride | 235–237d | | | | 334 |
| 1-Me-3-[2-(4-Me-Piperazin-1-yl)acetoxy]-5-(2-$ClC_6H_4$)-7-Cl | | | | | |
| Hydrochloride·$H_2O$ | 270–271 | MeOH/$Et_2O$ | 71 | pmr | 265, 328 |
| Dihydrochloride·1.5$H_2O$ | $>$210d | | 36 | pmr | 265, 328 |

| | | | | | |
|---|---|---|---|---|---|
| Methanesulfonate | 238–241 | $MeOH/Et_2O$ | 85 | pmr | 328 |
| 1-Me-3-(4-Me-Piperazin-1-yl)COO-5-Ph-7-Cl | 185–187 | | | | 336 |
| 1-Me-3-(1-Me-1-Piperidinium)-5-(2-$ClC_6H_4$)-7-MeO | | | | | |
| Chloride | 218–220d | | | | 334 |
| 1-Me-3-(1,2-$Me_2$-Pyrimidinium)-5-Ph-7-Cl | | | | | |
| Chloride | 230–232 | | | | 334 |
| 1-Me-3-(2,2-$Me_2$-Propanoyloxy)-5-Ph-7-Cl | 199–201 | EtOH | 88 | | 329 |
| 1-Me-3-$Me_2N$-5-Ph-7-Cl | 144–145 | $CH_2Cl_2$/Hexane | 78 | ir, pmr | 149 |
| 1-Me-3-$Me_2NCOO$-5-Ph-7-Cl | 174–176 | EtOAc | | | 335 |
| 1-Me-3-(2-$Me_2N$-Ethoxy)CO-5-Ph-7-Cl | 131–132 | EtOAc/Pentane | | | 302 |
| 1-Me-3-$Me_2NCH_2$-5-(2-$ClC_6H_4$)-7-$NO_2$ | 143–144 | $Et_2O$/Petr ether | | | 301 |
| 1-Me-3-MeO-5-Ph-7-Cl | 146–147 | $Et_2O$ | | | 2b |
| 1-Me-3-$(MeO)_2CH$-5-(2-$ClC_6H_4$)-7-Br | 125–127 | | | | 314 |
| 1-Me-3-MeOOC-5-Ph-7-Cl | 224–226 | $CH_2Cl_2$/MeOH | 91 | | 39 |
| 1-Me-3-MeOOC-5-(2-$ClC_6H_4$)-7-Cl | 200–201 | MeOH | | | 108 |
| 1-Me-3-MeOOC-5-(2-$FC_6H_4$)-7-I | 135–139 | $CH_2Cl_2$/Hexane | | | 60 |
| 1-Me-3-(2-MeOOC-Ethyl)-5-Ph-7-Cl | | | | | |
| 4-Oxide | 179–182 | $CH_2Cl_2/Et_2O$ | 34 | pmr | 298 |
| 1-Me-3-MeOOCO-5-(2-$ClC_6H_4$)-7-Cl | 193–195 | EtOAc | 64 | | 265 |
| 1-Me-3-(4-MeO-Benzyl)-5-Ph-7-Cl | 121–122 | $Et_2O$/Pentane | | | 299 |
| 1-Me-3-(4-$MeOC_6H_4$)-5-Ph-7-Cl | 177–178 | | | | 86b |
| 1-Me-3-[3,4,5-$(MeO)_3$-Benzoyloxy]-5-Ph-7-Cl | 266–230 | Acetone | | | 2b |
| 1-Me-3-[3,4-$(MeO)_2$-Benzyl]-5-Ph-7-Cl | 155–157 | $CH_2Cl_2$/Petr ether | | | 192 |
| 1-Me-3-[3,4-$(MeO)_2$-$\alpha$-HO-Benzyl]-5-Ph-7-Cl | 227–230d | $CH_2Cl_2$/Petr ether | | | 192 |
| 1-Me-3-(MeO)(HO)OP-5-Ph-7-Cl | 171d | MeCN | 36 | ir, pmr | 243 |
| 1-Me-3-$(MeO)_2OP$-5-Ph-7-Cl | 185–187 | EtOAc/Hexane | 61 | ir, pmr | 243 |
| 1-Me-3-$MeO_2S$-5-Ph-7-Cl | 250–255 | $CHCl_3$/EtOH | | | 152c |
| 1-Me-3-$MeO_2S$-5-(2-$FC_6H_4$)-7-Cl | 224–228 | $CH_2Cl_2/Et_2O$ | | | 343 |
| 1-Me-3-(3-$MeSO_3$-Propoxy)CO-5-Ph-7-Cl | 138–139 | PhH/Pentane | | | 302 |
| 1-Me-3-(Morpholino)acetoxy-5-(2-$ClC_6H_4$)-7-Cl | | | | | |
| Hydrochloride·$H_2O$ | 237–238 | | 61 | pmr | 265 |
| 1-Me-3-(Morpholino)COO-5-Ph-7-Cl | 200–202 | | | | 336 |
| 1-Me-3-[2-(Morpholino)ethoxy]CO-5-Ph-7-Cl | 167–169 | EtOAc/Pentane | | | 302 |

TABLE VII-1. —(contd.)

| Substituent | mp (°C) or; [bp (°C/torr)] | Solvent of Crystallization | Yield (%) | Spectra | Refs. |
|---|---|---|---|---|---|
| 1-Me-3-[2-(Morpholino)ethoxy]CO-5-(2-$ClC_6H_4$)-7-$NO_2$ | 160–162 | $CH_2Cl_2$/EtOAc | | | 147 |
| Hydrochloride | 172–175d | $Et_2O$ | | | 147 |
| 1-Me-3-$(Morpholino)_2PO_2$-5-Ph-7-Cl | 232–234 | Acetone/Hexane | | | 33b |
| 1-Me-3-[3-(Morpholino)propoxy]CO-5-(2-$ClC_6H_4$)-7-$NO_2$ | 160–162 | $CH_2Cl_2$/EtOAc | | | 301 |
| 1-Me-3-(3-Oxo-but-1-yl)-5-Ph-7-Cl | | | | | |
| 4-Oxide | 179–182 | MeOH | | | 134 |
| 1-Me-3-(Piperidin-1-yl)COO-5-Ph-7-Cl | 201–203 | | | | 336 |
| 1-Me-3-PhOOC-5-Ph-7-Cl | 211–212d | $CH_2Cl_2$/Petr ether | | | 490 |
| 1-Me-3-PhOCOO-5-Ph-7-Cl | 176–178 | Dioxane/EtOH | | | 336 |
| 1-Me-3-Pr-5-(2-$FC_6H_4$)-7-$H_2N$ | 112–116 | $Et_2O$ | | | 231 |
| 1-Me-3-Pr-5-(2-$FC_6H_4$)-7-(2-HO-Ethyl)NHCONH | 105–110 | EtOAc/Hexane | | | 231 |
| 1-Me-3-Pr-5-(2-$FC_6H_4$)-7-$NO_2$ | 128–130 | $Et_2O$/Petr ether | | | 231 |
| 1-Me-3-*i*-Pr-5-(2-$FC_6H_4$)-7-$H_2N$ | 213–215 | $Et_2O$ | | | 301 |
| 1-Me-3-*i*-Pr-5-(2-$FC_6H_4$)-7-(2-HO-Ethyl)NHCONH | 125–130d | Acetone/$Et_2O$ | | | 301 |
| 1-Me-3-*i*-Pr-5-(2-$FC_6H_4$)-7-$NO_2$ | 136–138 | | | | 301 |
| 1-Me-3-(1-Pyridinium)-5-Ph-7-Cl | | | | | |
| Chloride | 229–231 | Acetone | 80 | | 356, 334 |
| 1-Me-3-(Pyrrolidin-1-yl)COO-5-Ph-7-Cl | 210–212 | | | | 336 |
| 1-Me-3-*i*-PrO-5-Ph-7-Cl | 210–212 | $Et_2O$/Pentane | 65 | ir, pmr, uv | 215 |
| 1-Me-3,5-$Ph_2$-7-Cl | 223–224 | PhH/MeOH | | | 61 |
| 1-Me-3-$Ph_2PO_4$-5-Ph-7-Cl | 184–186 | EtOAc/$Et_2O$ | | | 33b |
| 1-Me-3-(2-Pr-Pentanoyloxy)-5-Ph-7-Cl | 134–136 | *i*-PrOH | | | 326 |
| 1-MeNHCO-3-AcO-5-Ph-7-Cl | 170–171 | EtCOMe | 82 | | 311 |
| 1-MeNHCO-3-EtOOC-5-(2-$ClC_6H_4$)-7-Cl | 155d | Cyclohexane | | | 275 |
| 1-MeNHCO-3-EtOOC-5-(2-$FC_6H_4$)-7-Cl | 169 | $Et_2O$ | | | 275 |
| 1-MeNHCO-3-F-5-Ph-7-Cl | 224–225 | EtOH | | F-nmr, pmr | 266, 267 |
| 1-MeNHCO-3-MeNHCOO-5-Ph-7-Cl | 174–175 | *i*-PrOH | 70 | | 311 |

| | | | | | |
|---|---|---|---|---|---|
| 1-MeNHCOCH$_2$-3-AcO-5-Ph-7-Cl | 189–191 | Acetone/Et$_2$O | | | 293 |
| 1-MeNHCOCH$_2$-3-HO-5-Ph-7-Cl | 140–145 | CH$_2$Cl$_2$/Hexane | | | 293 |
| 1-(2-Me$_2$N-Ethyl)-3-AcO-5-Ph-7-Cl | 187–188 | CH$_2$Cl$_2$/Petr ether | | | 254 |
| 1-(2-Me$_2$N-Ethyl)-3-Cl-5-(2-ClC$_6$H$_4$)-7-Cl | 158–160 | Et$_2$O/Pentane | 2.4 | pmr | 288 |
| 1-(2-Me$_2$N-Ethyl)-3-EtOOCO-5-(2-ClC$_6$H$_4$)-7-Cl | 119–121 | Et$_2$O/Pentane | 15 | pmr | 288 |
| 1-(2-Me$_2$N-Ethyl)-3-HO-5-Ph-7-Cl | 144–146 | EtOH | | | 2b |
| 1-(2-Me$_2$N-Ethyl)-3-MeO-5-(2-ClC$_6$H$_4$)-7-Cl | 136–137 | Et$_2$O/Pentane | 24 | pmr | 288 |
| 1-(3-Me$_2$N-Propyl)-3-MeO-5-(2-ClC$_6$H$_4$)-7-Cl | | | | | |
| Hydrochloride·H$_2$O | 115–130d | Et$_2$O | 16 | | 288 |
| 1-(2-MeO-Ethyl)-3-HO-5-Ph-7-Cl | 160–161 | EtOAc | | | 280 |
| 1-MeOCH$_2$-3-AcO-5-Ph-7-Cl | 131–133 | Et$_2$O | | | 72 |
| 1-MeOCH$_2$-3-AcO-5-Ph-7-NO$_2$ | 163–165 | EtOAc/Hexane | | | 72 |
| 1-MeOCH$_2$-3-allyl-5-Ph-7-Cl | | | | | |
| 4-Oxide | 137–139 | EtOH | 50 | pmr | 298 |
| 1-MeOCH$_2$-3-[(7-Cl-1,3-Dihydro-1-MeOCH$_2$-2-oxo-2*H*-1,4-benzodiazepin-3-yl)Methyl]-5-Ph-7-Cl | 255–260 | CH$_2$Cl$_2$/MeOH | | | 108 |
| 1-MeOCH$_2$-3-EtOOC-5-Ph-7-H$_2$N | 218–220 | EtOH | | | 108 |
| 1-MeOCH$_2$-3-EtOOC-5-Ph-7-Cl | 161–164 | CH$_2$Cl$_2$/Et$_2$O | | | 72 |
| 1-MeOCH$_2$-3-EtOOC-5-Ph-7-NO$_2$ | 156–159 | CH$_2$Cl$_2$/Hexane | | | 108 |
| 1-MeOCH$_2$-3-(3-Et$_2$N-Propyloxy)CO-5-Ph-7-Cl | | | | | |
| Hydrochloride | 103–105d | CH$_2$Cl$_2$/Acetone | | | 147 |
| 1-MeOCH$_2$-3-HO-5-Ph-7-H$_2$N | 165–167 | EtOH/Petr ether | | | 72c |
| 1-MeOCH$_2$-3-HO-5-Ph-7-Cl | 138–139 | EtOH | | | 72 |
| 1-MeOCH$_2$-3-HO-5-Ph-7-NO$_2$ | 160–162 | CH$_2$Cl$_2$/Et$_2$O | | | 72 |
| 1-MeOCH$_2$-3-Me-5-Ph-7-(benzyloxy)CO(Me)N | 95–98 | CH$_2$Cl$_2$/Et$_2$O/Petr ether | | | 108 |
| 1-MeOCH$_2$-3-Me-5-Ph-7-Cl | | | | | |
| 4-Oxide | 133–136 | EtOH | 55 | pmr | 277 |
| 1-MeOCH$_2$-3-Me-5-Ph-7-Me(MeO)N | 148–150 | MeOH/H$_2$O | 15 | pmr, uv | 245 |
| 1-MeOCH$_2$-3-Me-5-(2-ClC$_6$H$_4$)-7-NO$_2$ | | | | | |
| (+)-Enantiomer | 112–115 | Et$_2$O/Hexane | | | 301 |
| 1-MeOCH$_2$-3-(3-Morpholinopropyloxy)CO-5-Ph-7-Cl | | | | | |
| Hydrochloride | 190d | Acetone/Et$_2$O | | | 147 |
| Methobromide | 168d | PhH/Et$_2$O | | | 147 |

TABLE VII-1. —(contd.)

| Substituent | mp (°C) or; [bp (°C/torr)] | Solvent of Crystallization | Yield (%) | Spectra | Refs. |
|---|---|---|---|---|---|
| 1-$MeOCH_2$-3-(3-Morpholinopropyloxy)CO-5-(2-$ClC_6H_4$)-7-$NO_2$ | 128–130 | $Et_2O$/EtOAc | | | 147 |
| Hydrochloride | 153d | $Et_2O/CH_2Cl_2$ | | | 147 |
| 1-[2-MeS(O)-Ethyl]-3-AcO-5-Ph-7-Cl | 128–130 | *i*-$Pr_2O$ | | | 234 |
| 1-[2-MeS(O)-Ethyl]-3-HO-5-Ph-7-Cl | 95–97 | EtOAc | | | 234 |
| 1-[2-$MeSO_2$-Ethyl]-3-AcO-5-Ph-7-Cl | 103–105 | Hexane | | | 234 |
| 1-[2-$MeSO_2$-Ethyl]-3-HO-5-Ph-7-Cl | 90–92 | | | | 234 |
| 1-Ph-3-Me-5-(2-$ClC_6H_4$)-7-$NO_2$ | 233–235 | $CH_2Cl_2$/MeOH | | | 391b |
| 1-(2-PhO-Ethyl)-3-AcO-5-(2-$FC_6H_4$)-7-Cl | Oil | | | ir | 234 |
| 1-(2-PhO-Ethyl)-3-HO-5-(2-$FC_6H_4$)-7-Cl | 180–181 | | | | 234 |
| 1-(Propyn-3-yl)-3-EtOOC-5-(2-$ClC_6H_4$)-7-Cl | 164 | $Et_2O$ | | | 275 |
| 1-(Propyn-3-yl)-3-EtOOC-5-(2-$FC_6H_4$)-7-Cl | 173 | EtOH | | | 275 |
| 1-[2-(Pyrrolidin-1-yl)ethyl]-3-EtOOC-5-Ph-7-Cl | 126–128 | EtOH | | | 302 |
| 1-(2-Tetrahydropyran-2-yloxy)ethyl-3-AcO-5-(2-$FC_6H_4$)-7-Cl | Oil | | | ir | 234 |
| 1-(2-Tetrahydropyran-2-yloxy)ethyl-3-HO-5-(2-$FC_6H_4$)-7-Cl | Oil | | | ir | 234 |
| ***1,5,6,7-Tetrasubstituted*** | | | | | |
| 1-Me-5-(2-$ClC_6H_4$)-6-Cl-7-$H_2N$ | 215–216 | EtOAc/$CH_2Cl_2$ | | | 301 |
| 1-Me-5-(2-$ClC_6H_4$)-6-Cl-7-(2-HO-Ethyl)NHCONH | 156–160 | Acetone/$Et_2O$ | | | 301 |
| 1-Me-5-(2-$ClC_6H_4$)-6-Cl-7-MeNHCONH | 224–228d | EtOAc/Petr ether | | | 301 |
| 1-Me-5-(2-$FC_6H_4$)-6-Br-7-$H_2N$ | 252–254 | EtOAc/$CH_2Cl_2$ | | | 301 |
| 1-Me-5-(2-$FC_6H_4$)-6-Br-7-(2-HO-Ethyl)NHCONH | 155–160d | Acetone/$Et_2O$ | | | 301 |
| 1-Me-5-(2-$FC_6H_4$)-6-Cl-7-AcNH | 175–176 | EtOH/$Et_2O$ | | | 301 |
| | 159–162 | $Et_2O$/Petr ether | | | 301 |
| 1-Me-5-(2-$FC_6H_4$)-6-Cl-7-(Allyl)OOCNH | 138–140 | $Et_2O$/Petr ether | | | 301 |
| 1-Me-5-(2-$FC_6H_4$)-6-Cl-7-$H_2N$ | 240 | EtOAc/$Et_2O$/Hexane | | | 231 |
| 1-Me-5-(2-$FC_6H_6$)-6-Cl-7-$H_2NCONH$ | 234 | $CH_2Cl_2/Et_2O$ | | | 301 |

| | | | |
|---|---|---|---|
| 1-Me-5-(2-$FC_6H_4$)-6-Cl-7-$H_2N$(NH)CNH | 153–155 | EtOAc | 301 |
| 1-Me-5-(2-$FC_6H_4$)-6-Cl-7-(Benzyl)NHCONH | 185–187 | EtOAc/$Et_2O$ | 301 |
| 1-Me-5-(2-$FC_6H_4$)-6-Cl-7-(Benzyloxy)CONH·$Et_2O$ | 70–80d | $Et_2O$/Petr ether | 301 |
| 1-Me-5-(2-$FC_6H_4$)-6-Cl-7-(Benzoloxy)CO(Me)N | 145–150 | $Et_2O$/Hexane | 301 |
| 1-Me-5-(2-$FC_6H_4$)-6-Cl-7-(Benzyloxy)CSNH | 170–174 | $Et_2O$/Hexane | 301 |
| 1-Me-5-(2-$FC_6H_4$)-6-Cl-7-BuNHCONH | 196–198 | Acetone/$Et_2O$ | 231 |
| 1-Me-5-(2-$FC_6H_4$)-6-Cl-7-*t*-BuNHCONH | 206–208d | $Et_2O$/Hexane | 231 |
| 1-Me-5-(2-$FC_6H_4$)-6-Cl-7-BuOOCNH | 160–162 | $Et_2O$/Hexane | 301 |
| 1-Me-5-(2-$FC_6H_4$)-6-Cl-7-(But-2-yloxy)CONH | 174–176d | $Et_2O$/Petr ether | 301 |
| (+)-Enantiomer | 160–162 | $Et_2O$/Petr ether | 301 |
| (−)-Enantiomer | 162–164 | $Et_2O$/Petr ether | 301 |
| 1-Me-5-(2-$FC_6H_4$)-6-Cl-7-(But-2-yl)SOCNH | 194–198d | $CH_2Cl_2$/$Et_2O$/Petr ether | 301 |
| 1-Me-5-(2-$FC_6H_4$)-6-Cl-7-(2-But-2-yl)SCSNH | 168–172 | $Et_2O$/Petr ether | 301 |
| 1-Me-5-(2-$FC_6H_4$)-6-Cl-7-(4-Cl-Benzyl)OOCNH | 134–138 | $Et_2O$ | 301 |
| 1-Me-5-(2-$FC_6H_4$)-6-Cl-7-[6-Cl-1,3-Dihydro-5-(2-$FC_6H_4$)-2-oxo-2*H*-1,4-benzodiazepin-7-yl]-NHCONH | 276–277 | EtOAc | 301 |
| 1-Me-5-(2-$FC_6H_4$)-6-Cl-7-(5-Cl-2-$MeOC_6H_3$)NHCONH | 270–274 | EtOAc | 301 |
| 1-Me-5-(2-$FC_6H_4$)-6-Cl-7-(2-$ClC_6H_4$)NHCONH | 220–222 | EtOAc/$Et_2O$ | 301 |
| 1-Me-5-(2-$FC_6H_4$)-6-Cl-7-(4-Cl-$C_6H_4$)NHCONH | 285 | $CH_2Cl_2$/EtOH | 231 |
| 1-Me-5-(2-$FC_6H_4$)-6-Cl-7-(4-$ClC_6H_4$)OOCNH | 206–210 | $Et_2O$/Petr ether | 301 |
| 1-Me-5-(2-$FC_6H_4$)-6-Cl-7-(Cyclopentyl)CONH | 224–228 | $CH_2Cl_2$/$Et_2O$ | 301 |
| 1-Me-5-(2-$FC_6H_4$)-6-Cl-7-[(Cyclopentyl)CO]$_2$N | 170–174 | $Et_2O$/Petr ether | 301 |
| 1-Me-5-(2-$FC_6H_4$)-6-Cl-7-EtNHCONH | 155–160d | EtOAc/$Et_2O$ | 231 |
| 1-Me-5-(2-$FC_6H_4$)-6-Cl-7-$Et_2$NCONH | 190–191 | EtOAc/$Et_2O$ | 231 |
| 1-Me-5-(2-$FC_6H_4$)-6-Cl-7-Et(Me)NCONH | 184–186 | EtOAc/$Et_2O$ | 231 |
| 1-Me-5-(2-$FC_6H_4$)-6-Cl-7-EtOCSNH | 147 | $Et_2O$/$CH_2Cl_2$/Petr ether | 301 |
| 1-Me-5-(2-$FC_6H_4$)-6-Cl-7-EtSCSNH | 178–182d | $Et_2O$/Petr ether | 301 |
| 1-Me-5-(2-$FC_6H_4$)-6-Cl-7-(Hexyl)NHCONH | 175–176 | EtOAc | 301 |
| 1-Me-5-(2-$FC_6H_4$)-6-Cl-7-(2-HO-Ethyl)NHCONH | 150–153 | EtOH/$Et_2O$ | 231 |
| 1-Me-5-(2-$FC_6H_4$)-6-Cl-7-(2-HO-Ethyl)(Me)NCONH | 218–220 | Acetone | 231 |
| 1-Me-5-(2-$FC_6H_4$)-6-Cl-7-[4-(2-HO-Ethyl)piperazin-1-yl]CONH | 173 | EtOAc/$Et_2O$ | 301 |

TABLE VII-1. —(contd.)

| Substituent | mp (°C) or; [bp (°C/torr)] | Solvent of Crystallization | Yield (%) | Spectra | Refs. |
|---|---|---|---|---|---|
| 1-Me-5-(2-$FC_6H_4$)-6-Cl-7-(2-HS-Ethyl)NHCONH | 168–170 | EtOAc/$Et_2O$ | | | 301 |
| 1-Me-5-(2-$FC_6H_4$)-6-Cl-7-(1-Iminoeth-1-yl)NH | 198–200d | EtOAc/$Et_2O$ | | | 301 |
| 1-Me-5-(2-$FC_6H_4$)-6-Cl-7-MeNH | 204–206 | $Et_2O$ | | | 301 |
| 1-Me-5-(2-$FC_6H_4$)-6-Cl-7-MeNHCONH | 145–160d | EtOAc/$Et_2O$ | | | 231 |
| 1-Me-5-(2-$FC_6H_4$)-6-Cl-7-MeNHCSNH | 194–195 | EtOAc/$Et_2O$ | | | 301 |
| 1-Me-5-(2-$FC_6H_4$)-6-Cl-7-$Me_2$NCONH | 174–175 | $Et_2O$ | | | 301 |
| 1-Me-5-(2-$FC_6H_4$)-6-Cl-7-(3-Me-Butyl)NHCONH | 230 | EtOAc/$Et_2O$ | | | 301 |
| 1-Me-5-(2-$FC_6H_4$)-6-Cl-7-(4-$MeC_6H_4SO_2$)NHCONH | 186 | $CH_2Cl_2$/$Et_2O$ | | | 301 |
| 1-Me-5-(2-$FC_6H_4$)-6-Cl-7-(4-Me-Piperazin-1-yl)CONH | 154d | EtOAc/$Et_2O$/Hexane | | | 231 |
| 1-Me-5-(2-$FC_6H_4$)-6-Cl-7-(2-MeO-Ethoxy)CONH | 148–150 | $Et_2O$/Petr ether | | | 301 |
| 1-Me-5-(2-$FC_6H_4$)-6-Cl-7-[2-(2-MeO-Ethoxy)ethoxy]-CONH | 106–110 | EtOAc/$Et_2O$/Petr ether | | | 301 |
| 1-Me-5-(2-$FC_6H_4$)-6-Cl-7-(Morpholin-1-yl)CONH | 126 | EtOAc/$Et_2O$ | | | 231 |
| 1-Me-5-(2-$FC_6H_4$)-6-Cl-7-(4-$NO_2$-Benzyl)OOCNH | 170–174 | $CH_2Cl_2$/$Et_2O$ | | | 301 |
| 1-Me-5-(2-$FC_6H_4$)-6-Cl-7-(2-Oxoimidazilidin-1-yl) | 148 | $CH_2Cl_2$/Hexane | | | 301 |
| 1-Me-5-(2-$FC_6H_4$)-6-Cl-7-(Pentyl)NHCONH | 208 | EtOAc/$Et_2O$ | | | 301 |
| 1-Me-5-(2-$FC_6H_4$)-6-Cl-7-PhOOCNH | 184–190d | $Et_2O$/Petr ether | | | 301 |
| 1-Me-5-(2-$FC_6H_4$)-6-Cl-7-(Piperazin-1-yl)CONH | Amorphous | | | | 301 |
| 1-Me-5-(2-$FC_6H_4$)-6-Cl-7-(Piperidin-1-yl)CONH | 195 | EtOAc/$Et_2O$ | | | 301 |
| 1-Me-5-(2-$FC_6H_4$)-6-Cl-7-*i*-PrNHCONH | 130–140 | EtOAc/Hexane | | | 231 |
| 1-Me-5-(2-$FC_6H_4$)-6-Cl-7-*i*-PrOOCNH | 162–166d | $Et_2O$/Petr ether | | | 301 |
| 1-Me-5-(2-$FC_6H_4$)-6-Cl-7-SCN | 165 | $Et_2O$ | | | 301 |
| 1-Me-5-(2-$FC_6H_4$)-6-HO-7-$H_2N$ | 195–197d | $CH_2Cl_2$/EtOH | | | 301 |
| 1-Me-5-(2-$FC_6H_4$)-6-MeNHCOO-7-$H_2N$ | 222d | EtOAc | | | 301 |
| 1-Me-5-(2-$FC_6H_4$)-6-MeNHCOO-7-MeNHCONH | 166–170 | $CH_2Cl_2$/EtOAc | | | 301 |
| 1-Me-5-(2-$FC_6H_4$)-6-MeO-7-$H_2N$ | 234 | $CH_2Cl_2$/EtOAc | | | 301 |
| 1-Me-5-(2-$FC_6H_4$)-6-MeO-7-(2-HO-Ethyl)NHCONH | 166–168d | $CH_2Cl_2$/$Et_2O$ | | | 301 |
| 1-$MeOCH_2$-5-Ph-6-HO-7-$H_2N$ | 204–206 | $CH_2Cl_2$/MeOH/$Et_2O$ | | pmr, uv | 245 |

| | | | | |
|---|---|---|---|---|
| ***1,5,6,8-Tetrasubstituted*** | | | | |
| 1-Me-5-(2-$FC_6H_4$)-6,8-$Cl_2$ | 193 | $Et_2O$ | | 301 |
| ***1,5,7,8-Tetrasubstituted*** | | | | |
| 1,5-$Me_2$-7,8-$(MeO)_2$ | 160–163 | $CH_2Cl_2$/Hexane | | 152c |
| 1-Me-5-Ph-7,8-$Cl_2$ | 185–186 | MeOH | | 475 |
| 1-Me-5-(2-$FC_6H_4$)-7-Cl-8-$H_2N$ | 215–218 | $CH_2Cl_2$/Cyclohexane | | 257 |
| ***1,5,7,9-Tetrasubstituted*** | | | | |
| 1-(2-$Et_2N$-Ethyl)-5-Ph-7,9-$Cl_2$ Dihydrochloride | 208–211d | MeOH/$Et_2O$ | | 302 |
| 1-Me-5-EtO-7-Cl-9-$NO_2$ | 100–102 | $CHCl_3$/Pentane | 53 | 195 |
| 1-Me-5-Ph-7,9-$Cl_2$ | 143–145 | EtOH | | 496 |
| 1-Me-5-Ph-7,9-$Me_2$ | 108–109 | $Et_2O$/Petr ether | | 2b |
| 1-Me-5-(2-$ClC_6H_4$)-7,9-$Br_2$ | 142–143 | EtOAc | | 475 |
| 1-Me-5-(2-$FC_6H_4$)-7-$H_2N$-9-Cl | 210 | EtOAc/$CH_2Cl_2$ | | 301 |
| 1-Me-5-(2-$FC_6H_4$)-7,9-$Br_2$ | 156–157 | EtOAc | | 475 |
| 1-Me-5-(2-$FC_6H_4$)-7,9-$I_2$ | 173–175 | EtOH | | 192 |
| 1-Me-5-(2-$FC_6H_4$)-7-$NO_2$-9-Cl | 184 | EtOAc | | 301 |
| 1-Me-5-(2-Pyridyl)-7,9-$Br_2$ | 183–185 | EtOH | | 67b |
| ***3,3,5,7-Tetrasubstituted*** | | | | |
| 3-AcO-3-(2-Cl-Ethoxy)CO-5-Ph-7-Cl | 192–195 | MeOH | | 108 |
| 3-Ac-3-$(EtO)_2$OP-5-(2-$ClC_6H_4$)-7-Cl | 186–188 | EtOAc/Hexane | | |
| 3-AcO-3-Me-5-Ph-7-Cl | 179–180 | EtOH | | 325 |
| 3-AcO-3-Me-5-(2-$ClC_6H_4$)-7-$NO_2$ | 206–210 | $CH_2Cl_2$/Petr ether | | 301 |
| 3-CN-3-Me-5-Ph-7-Cl | 230–234d | EtOH | | 302 |
| 3,3-(2-CN-Ethyl)2-5-(2-$FC_6H_4$)-7-Cl | 188–189 | $Et_2O$ | | 301 |
| 3-Et-3,5-$Ph_2$-7-Cl | 219–221 | Acetone/Hexane | | 1b |
| 3-$Et_2N$-3-Me-5-Ph-7-Cl | 178–179 | | | 325 |
| 3-EtO-3-EtOOC-5-Ph-7-Cl·0.5 EtOH | 164–166 | EtOH/$H_2O$ | | 101 |
| 3-EtO-3-Me-5-Ph-7-Cl | 167–168 | Hexane | | 1b |
| 3-EtOOC-3-HO-5-Ph-7-Cl | 180–182 | MeCN | | 101 |
| | 190–193 | $CH_2Cl_2$/EtOH | | 242 |

TABLE VII-1. —(contd.)

| Substituent | mp (°C) or; [bp (°C/torr)] | Solvent of Crystallization | Yield (%) | Spectra | Refs. |
|---|---|---|---|---|---|
| 3-EtOOC-3-MeO-5-Ph-7-Cl | 168–170 | EtOH/$H_2O$ | | | 101 |
| Hydrobromide | 180–181 | EtOAc | | | 101 |
| 3,3-$Me_2$-5-(2-$ClC_6H_4$)-7-$H_2N$ | 214 | $Et_2O$/Hexane | | | 301 |
| 3,3-$Me_2$-5-(2-$ClC_6H_4$)-7-(2-HO-Ethyl)NHCONH | 255 | EtOAc | | | 301 |
| 3,3-$Me_2$-5-(2-$ClC_6H_4$)-7-$NO_2$ | 242 | $CH_2Cl_2$/Hexane | | | 114 |
| 3,3-$Me_2$-5-(2-$FC_6H_4$)-7-Ac | 206 | | | | 251 |
| 3,3-$Me_2$-5-(2-$FC_6H_4$)-7-$H_2N$ | 229–231 | EtOAc | | | 251 |
| 3,3-$Me_2$-5-(2-$FC_6H_4$)-7-CN | 212 | Cyclohexane | | | 251 |
| 3,3-$Me_2$-5-(2-$FC_6H_4$)-7-Diazonium | | | | | |
| Tetrafluoroborate | 203 | | | | 251 |
| 3,3-$Me_2$-5-(2-$FC_6H_4$)-7-(1-HO-Ethyl) | 205 | | | | 251b |
| 3,3-$Me_2$-5-(2-$FC_6H_4$)-7-I | 202 | Cyclohexane | | | 251 |
| 3,3-$Me_2$-5-(2-$FC_6H_4$)-7-$NO_2$ | 241 | $Et_2O$ | | | 114 |
| 3-Me-3-MeO-5-Ph-7-Cl | 177–178 | Hexane | | | 1b |
| 3-Me-3-MeO-5-(2-$ClC_6H_4$)-7-$NO_2$ | 233–234 | EtOH | | | 301 |
| 3,3-Tetramethylene-5-Ph-7-Cl | 238–240 | EtOH | | | 35 |
| 3,3-(*spiro*-Adamant-2-yl)-5-Ph-7-Br | 305–307 | | | | 115 |
| 3,3-(*spiro*-Adamant-2-yl)-5-Ph-7-Cl | 300–302 | EtOH | 55 | | 115 |
| 3,3-(*spiro*-Adamant-2-yl)-5-Ph-7-Me | 278–280 | | | | 115 |
| ***3,5,6,7-Tetrasubstituted*** | | | | | |
| 3-Me-5-(2-$ClC_6H_4$)-6-Br-7-$H_2N$ | | | | | |
| *S*(−)-Enantiomer | 231–232 | | | | 495 |
| 3-Me-5-(2-$ClC_6H_4$)-6-Br-7-(2-HO-Ethyl)NHCONH | | | | | |
| *S*(+)-Enantiomer | 100–101 | | | | 495 |
| 3-Me-5-(2-$ClC_6H_4$)-6-Br-7-(3-HO-Propyl)NHCONH | | | | | |
| *S*(+)-Enantiomer | 160 | | | | 495 |
| 3-Me-5-(2-$ClC_6H_4$)-6-Br-7-(Morpholino)CONH | | | | | |
| *S*(+)-Enantiomer | 120d | | | | 495 |

| | | | | | |
|---|---|---|---|---|---|
| 3-Me-5-(2-$ClC_6H_4$)-6-Cl-7-$H_2N \cdot MeOH$ | 248–250 | $CH_2Cl_2/MeOH$ | | | 301 |
| 3-Me-5-(2-$ClC_6H_4$)-6-Cl-7-$NO_2$ | 253–254 | $CH_2Cl_2/MeOH$ | | | 301 |
| ***3,5,7,9-Tetrasubstituted*** | | | | | |
| 3-AcO-5-Ph-7,9-$Br_2$ | 236–240 | MeOH | | | 152c |
| 3-AcO-5-(2-Pyridyl)-7,9-$Br_2$ | 215–218d | PhH | | | 225 |
| 3-EtOOC-5-Ph-7,9-$Cl_2$ | 182–184d | MeCN | | | 302 |
| 3-Me-5-(2-$ClC_6H_4$)-7-$NO_2$-9-Cl | 213–214 | $EtOH/CH_2Cl_2$ $Et_2O$ | | | 301 |
| ***5,6,7,8-Tetrasubstituted*** | | | | | |
| 5-Ph-6,7,8-$(MeO)_3$ | 188–189 | $Et_2O$ | | | 486 |
| 4-Oxide | 199–200 | EtOAc | | | 486 |
| 5-Ph-6,8-$(MeO)_2$-7-HO | 239–240 | EtOAc | | | 486 |
| ***5,7,8,9-Tetrasubstituted*** | | | | | |
| 5-Cyclopropyl-7,8-$Me_2$-9-CN | 192–194 | Acetone | 69 | | 29 |
| 5,7,8-$Me_3$-9-CN | 223–225 | Acetone | 91 | ir, pmr | 29 |
| 5-Me-7,8,9-$(MeO)_3$ | 222–224 | MeOH | | | 486 |
| 5-Ph-7,8-$Me_2$-9-CN | 214–215 | Acetone | 82 | | 29 |
| 5-Ph-7,8,9-$(MeO)_3$ | 196–197 | $EtOAc/Et_2O$ | | | 486 |
| 4-Oxide | 203–204 | $CH_2Cl_2/EtOAc$ | | | 486 |
| 5-(2-$FC_6H_4$)-7,8-$Me_2$-9-CN | 229–231 | $CH_2Cl_2/EtOH$ | 75 | | 29 |
| **Pentasubstituted** | | | | | |
| ***1,3,3,5,7-Pentasubstituted*** | | | | | |
| 1-Bu-3,3-$Me_2$-5-(2-$ClC_6H_4$)-7-$H_2N$ | 155–156 | $Et_2O$/Hexane | | | 301 |
| 1-Bu-3,3-$Me_2$-5-(2-$ClC_6H_4$)-7-(2-HO-Ethyl)NHCONH | 163–164 | Acetone/$Et_2O$ | | | 301 |
| 1-Bu-3,3-$Me_2$-5-(2-$ClC_6H_4$)-7-$NO_2$ | 129 | $Et_2O$/Hexane | | | 230 |
| 1-Et-3,3-$Me_2$-5-(2-$ClC_6H_4$)-7-$H_2N$ | 178 | $Et_2O$ | | | 230 |
| 1-Et-3,3-$Me_2$-5-(2-$ClC_6H_4$)-7-(2-HO-Ethyl)NHCONH | 160–162 | Acetone/$Et_2O$ | | | 301 |
| 1-Et-3,3-$Me_2$-5-(2-$ClC_6H_4$)-7-$NO_2$ | 174 | $Et_2O$ | | | 230 |
| 1-(2-$Et_2N$-Ethyl)-3,3-$Me_2$-5-(2-$ClC_6H_4$)-7-$H_2N$ | 70 | $EtOH/H_2O$ | | | 230 |

TABLE VII-1. —(contd.)

| Substituent | mp (°C) or; [bp (°C/torr)] | Solvent of Crystallization | Yield (%) | Spectra | Refs. |
|---|---|---|---|---|---|
| 1-(2-$Et_2$N-Ethyl)-3,3-$Me_2$-5-(2-$ClC_6H_4$)-7-[5-(2-$ClC_6H_4$)-1-(2-$Et_2$N-Ethyl)-1,3-dihydro-3,3-$Me_2$-2-oxo-2*H*-1,4-benzodiazepin-7-yl]NHCONH | 155–175d | EtOH/$H_2O$ | | | 301 |
| 1-(2-$Et_2$N-Ethyl)-3,3-$Me_2$-5-(2-$ClC_6H_4$)-7-(2-HO-ethyl)NHCONH | 165–170 | Acetone/$Et_2O$ | | | 301 |
| 1-(2-$Et_2$N-Ethyl)-3,3-$Me_2$-5-(2-$ClC_6H_4$)-7-$NO_2$ | 109 | $Et_2O$/Hexane | | | 230 |
| 1-(2-HO-Ethyl)-3,3-$Me_2$-5-(2-$ClC_6H_4$)-7-$H_2N$ | 192–194 | EtOAc/$Et_2O$ | | | 230 |
| 1-(2-HO-Ethyl)-3,3-$Me_2$-5-(2-$ClC_6H_4$)-7-(2-HO-ethyl)NHCONH | 175–176 | EtOH/EtOAc | | | 230 |
| 1-(2-HO-Ethyl)-3,3-$Me_2$-5-(2-$ClC_6H_4$)-7-$NO_2$ | 124–125 | $Et_2O$/Hexane | | | 230 |
| 1,3-$Me_2$-3-AcO-5-Ph-7-Cl | 150–152 | $CH_2Cl_2$/Hexane | 66 | pmr, uv | 277 |
| 1,3-$Me_2$-3-CN-5-Ph-7-Cl | 205–207 | EtOH | | | 302 |
| 1,3-$Me_2$-3-HO-5-Ph-7-Cl | 156–158d | $CH_2Cl_2$/Hexane | 79 | ir, pmr, uv | 277 |
| 1,3,3-$Me_3$-5-Ph-7-Cl | 125–128 | $Et_2O$/Petr ether | | | 152c |
| 1,3,3-$Me_3$-5-(2-$ClC_6H_4$)-7-$H_2N$ | 177–178 | $Et_2O$/Hexane | | | 230 |
| 1,3,3-$Me_3$-5-(2-$ClC_6H_4$)-7-$NO_2$ | 125–127 | $CH_2Cl_2$/EtOAc | | | 230 |
| 1,3,3-$Me_3$-5-(2-$ClC_6H_4$)-7-(2-HO-Ethyl)NHCONH | 85–110 | | | | 230 |
| 1,3,3-$Me_3$-5-(2-$ClC_6H_4$)-7-(4-$MeC_6H_4SO_2$)NHCONH Hydrochloride | 180–185 | Acetone | | | 301 |
| 1,3,3-$Me_3$-5-(2-$ClC_6H_4$)-7-(Pyrrolidin-1-yl)CONH | 248–250 | EtOAc/$Et_2O$ | | | 230 |
| 1,3,3-$Me_3$-5-(2-$FC_6H_4$)-7-Ac | 59 | Cyclohexane | | | 251 |
| 1,3,3-$Me_3$-5-(2-$FC_6H_4$)-7-(1-AcO-Ethyl) | Oil | | | | 251 |
| 1,3,3-$Me_3$-5-(2-$FC_6H_4$)-7-$H_2N$ | 190–191 | $Et_2O$ | | | 230 |
| 1,3,3-$Me_3$-5-(2-$FC_6H_4$)-7-(1-$H_2$N-Ethyl) | 50–51 | | | | 251 |
| 1,3,3-$Me_3$-5-(2-$FC_6H_4$)-7-(2-$H_2$N-2-Me-propanoyl)NH | 250 | EtOAc/$Et_2O$ | | | 230 |
| 1,3,3-$Me_3$-5-(2-$FC_6H_4$)-7-(1-BuNHCOO-Ethyl) | 54–56 | | | | 251b |
| 1,3,3-$Me_3$-5-(2-$FC_6H_4$)-7-(1-HO-Ethyl) | 68 | | | | 251 |
| 1,3,3-$Me_3$-5-(2-$FC_6H_4$)-7-(2-HO-Ethyl)NHCONH | 148–150 | EtOAc/Hexane | | | 301 |

| | | | | | |
|---|---|---|---|---|---|
| 1,3,3-$Me_3$-5-(2-$FC_6H_4$)-7-[1-(2-HO-Ethylaminocarbonylamino)ethyl] | 83 | | | | 251 |
| 1,3,3-$Me_3$-5-(2-$FC_6H_4$)-7-(1-Hydroxyiminoethyl | 199 | $Et_2O$ | | | 251 |
| 1,3,3-$Me_3$-5-(2-$FC_6H_4$)-7-(4,4-$Me_2$-2,5-Dioxoimidazolidin-1-yl) | 286–287 | EtOAc/$Et_2O$ | | | 230 |
| 1,3,3-$Me_3$-5-(2-$FC_6H_4$)-7-(4-$MeC_6H_4SO_2$)NHCONH Hydrochloride·$Et_2O$ | 208–214 | Acetone/$Et_2O$ | | | 301 |
| 1,3,3-$Me_3$-5-(2-$FC_6H_4$)-7-$NO_2$ | 128–129 | PhH/Hexane/Petr ether | | | 230 |
| 1,3-$Me_3$-3-$MeO_2S$-5-(2-$FC_6H_4$)-7-Cl | 226–229 | $CH_2Cl_2$/Petr ether | | | 343 |
| 1-Me-3-AcO-3-EtOOC-5-Ph-7-Cl | 159–161 | $CH_2Cl_2$/EtOH | | | 108 |
| 1-Me-3-Ac-3-$(EtO)_2OP$-5-Ph-7-Cl | 186–188 | | | | |
| 1-Me-3-AcNH-3-EtOOC-5-Ph-7-Cl | 213–215 | EtOH | | | 302 |
| 1-Me-3-$H_2N$-3-EtOOC-5-Ph-7-Cl | 147–149 | $CH_2Cl_2$/$Et_2O$/ | | | |
| 1-Me-3-$H_2N$-3-MeOOC-5-Ph-7-Cl | 169–172 | MeOH/$Et_2O$/Hexane | | | 302 |
| 1-Me-3-$H_2NCO$-3-EtOOC-5-Ph-7-Cl | 200–204d | | 93 | ir, pmr | 226 |
| 1-Me-3-$H_2NCO$-3-Me-5-Ph-7-Cl | 236–238 | *i*-PrOH | 72 | ir, pmr | 226 |
| 1-Me-3-$H_2NCO$-3-MeO-5-Ph-7-Cl | 235–238 | EtOH | 96 | ir, pmr | 226 |
| 1-Me-3-$H_2NCO$-3-MeOOC-5-Ph-7-Cl | 232–234 | MeCN | | | 302 |
| 1-Me-3-$H_2NCONHCO$-3-EtOOC-5-Ph-7-Cl | 198–199 | EtOH | | | 302 |
| 1-Me-3,3-$Bu_2$-5-Ph-7-Cl | 137–138 | $CH_2Cl_2$/Hexane | 10 | | 299 |
| 1-Me-3-HOOC-3-HO-5-Ph-7-Cl Sodium salt | 220–225d | Acetone/$H_2O$ | | | 242 |
| 1-Me-3-CN-3-$CNCH_2$-5-Ph-7-Cl | 181–183 | EtOH | | | 302 |
| 1-Me-3-CN-3-EtO-5-Ph-7-Cl | 165–167 | EtOH | | | 302 |
| 1-Me-3-CN-3-EtOOC-5-Ph-7-Cl | 146–148 | | 68 | ir, pmr | 226 |
| 1-Me-3-CN-3-$EtOOCCH_2$-5-Ph-7-Cl | 172–174 | EtOH | | | 302 |
| 1-Me-3-CN-3-Me-5-Ph-7-Cl | 204–206 | EtOH | 77 | ir, pmr | 226 |
| 1-Me-3-CN-3-MeO-5-Ph-7-Cl | 172–176d | | 78 | ir, pmr | 226 |
| 1-Me-3,3-(2-CN-Ethyl)$_2$-5-Ph-7-Cl | 194–196 | $Et_2O$ | | | 301 |
| 1-Me-3,3-$Et_2$-5-Ph-7-Cl | 133–136 | $Et_2O$/Petr ether | | | 152c |
| 1-Me-3-EtO-3-EtOOC-5-Ph-7-Cl | 142–143 | $Et_2O$/Hexane | | | 302 |
| 1-Me-3-EtOOC-3-HO-5-Ph-7-Cl | 174–176 | EtOH/$Et_2O$ | | | 242 |
| 1-Me-3-HO-3-MeOOC-5-Ph-7-Cl | 208–211d | $CH_2Cl_2$/MeOH | | | 108 |
| 1-$MeOCH_2$-3-EtOOC-3-HO-5-Ph-7-Cl | 187–189 | EtOH | | | 72 |

TABLE VII-1. —(contd.)

| Substituent | mp (°C) or; [bp (°C/torr)] | Solvent of Crystallization | Yield (%) | Spectra | Refs. |
|---|---|---|---|---|---|
| 1-$MeOCH_2$-3-AcO-3-Me-5-Ph-7-Cl | 154–157 | $CH_2Cl_2/Et_2O$/Hexane | 57 | ir, pmr | 277 |
| 1-Me-3-(3-$Et_2N$-Propyloxy)CO-3-HO-5-Ph-7-Cl | | | | | |
| Hydrochloride | 195–196d | $CH_2Cl_2$/EtOAc | | | 147 |
| 1-*i*-Pr-3,3-$Me_2$-5-(2-$ClC_6H_4$)-7-$H_2N$ | 169–170 | $Et_2O$/Hexane | | | 230 |
| 1-*i*-Pr-3,3-$Me_2$-5-(2-$ClC_6H_4$)-7-(2-HO-Ethyl)NHCONH | 243–244 | $CH_2Cl_2/ClCH_2CH_2Cl$ | | | 230 |
| 1-*i*-Pr-3,3-$Me_2$-5-(2-$ClC_6H_4$)-7-$NO_2$ | 197–198 | $Et_2O$/Hexane | | | 230 |
| ***1,3,5,6,7-Pentasubstituted*** | | | | | |
| 1-Et-3-Me-5-(2-$ClC_6H_4$)-6-Br-7-$H_2N$ | | | | | |
| *S*(−)-Enantiomer | 206–207 | | | | 495 |
| 1-Et-3-Me-5-(2-$ClC_6H_4$)-6-Br-7-(2-HO-Ethyl)-NHCONH | | | | | |
| *S*(+)-Enantiomer | 75 | | | | 495 |
| 1-Et-3-Me-5-(2-$ClC_6H_4$)-6-Br-7-(3-HO-Propyl)-NHCONH | | | | | |
| *S*(+)-Enantiomer | 118–120 | | | | 495 |
| 1-Et-3-Me-5-(2-$ClC_4H_4$)-6-Br-7-(Morpholino)CONH | | | | | |
| *S*(+)-Enantiomer | 115d | | | | 495 |
| 1-Me-3-Et-5-(2-$ClC_6H_4$)-6-Br-7-$H_2N$ | 251d | $CH_2Cl_2/Et_2O$/Petr ether | | | 231 |
| 1-Me-3-Et-5-(2-$ClC_6H_4$)-6-Br-7-(2-HO-Ethyl)-NHCONH | 236–237 | Acetone | | | 231 |
| 1-Me-3-Et-5-(2-$ClC_6H_4$)-6-Cl-7-(2-HO-Ethyl)-NHCONH | 234–235 | Acetone/$CH_2Cl_2$ | | | 301 |
| 1-Me-3-Et-5-(2-$FC_6H_4$)-6-Br-7-$H_2N$ | 248 | $Et_2O$ | | | 231 |
| 1-Me-3-Et-5-(2-$FC_6H_4$)-6-Br-7-(2-HO-Ethyl)-NHCONH | 218–222 | Acetone/$Et_2O$ | | | 231 |
| 1-Me-3-Et-5-(2-$FC_6H_4$)-6-Cl-7-$H_2N$ | 250 | EtOH | | | 231 |
| 1-Me-3-Et-5-(2-$FC_6H_4$)-6-Cl-7-(2-HO-Ethyl)-NHCONH | 243–244 | Acetone/$Et_2O$ | | | 231 |

| | | | |
|---|---|---|---|
| 1,3-$Me_2$-5-(2-$ClC_6H_4$)-6-Br-7-$H_2N$ | 254–256d | EtOAc/$Et_2O$ | 231 |
| *R*(+)-Enantiomer | 241–242d | MeCN | 301 |
| *S*(−)-Enantiomer | 241–243d | EtOH | 301 |
| 1,3-$Me_2$-5-(2-$ClC_6H_4$)-6-Br-7-(2-HO-Ethyl)- | | | |
| NHCONH | 204–206d | Acetone/$Et_2O$ | 231 |
| *R*(−)-Enantiomer | 224–226 | Acetone/$Et_2O$ | 301 |
| *S*(+)-Enantiomer | 224–226 | Acetone/$Et_2O$ | 301 |
| 1,3-$Me_2$-5-(2-$ClC_6H_4$)-6-Br-7-(3-HO-Propyl)- | | | |
| NHCONH | | | |
| *S*(+)-Enantiomer | 155 | | 301 |
| 1,3-$Me_2$-5-(2-$ClC_6H_4$)-6-Br-7-(2-HS-Ethyl)- | | | |
| NHCONH·$H_2O$·0.5 *i*-PrOH | 200 | *i*-PrOH | 301 |
| 1,3-$Me_2$-5-(2-$ClC_6H_4$)-6-Br-7-(4-$MeC_6H_4SO_2$)- | | | |
| NHCONH | 166–175d | Acetone/$Et_2O$ | 301 |
| 1,3-$Me_2$-5-(2-$ClC_6H_4$)-6-Br-7-(2-MeS-Ethyl)- | | | |
| NHCONH | 210 | EtOAc | 301 |
| 1,3-$Me_2$-5-(2-$ClC_6H_4$)-Br-7-(2-Me(O)S-Ethyl)- | | | |
| NHCONH | 215 | EtOAc | 301 |
| 1,3-$Me_2$-5-(2-$ClC_6H_4$)-6-Br-7-(2-$MeSO_2$-Ethyl)- | | | |
| NHCONH | 144d | EtOAc | 301 |
| 1,3-$Me_2$-5-(2-$ClC_6H_4$)-6-Br-7-(Morpholino)CONH | | | |
| *S*(+)-Enantiomer | 189–190 | | 301 |
| 1,3-$Me_2$-5-(2-$ClC_6H_4$)-6-Br-7-(2-$Ph_3CS$-Ethyl)- | | | |
| NHCONH | 240 | EtOH | 301 |
| 1,3-$Me_2$-5-(2-$ClC_6H_4$)-6-Cl-7-$H_2N$ | 240–242d | EtOAc | 231 |
| 1,3-$Me_2$-5-(2-$ClC_6H_4$)-6-Cl-7-(2-HO-Ethyl)NHCONH | 233–236d | Acetone | 231 |
| 1,3-$Me_2$-5-(2-$ClC_6H_4$)-6-Cl-7-(4-$MeC_6H_4SO_2$)- | | | |
| NHCONH | 168–174d | Acetone/$Et_2O$ | 301 |
| 1,3-$Me_2$-5-(2-$FC_6H_4$)-6-Br-7-$H_2N$ | 255d | EtOAc/$Et_2O$ | 231 |
| 1,3-$Me_2$-5-(2-$FC_6H_4$)-6-Br-7-(2-HO-Ethyl)- | | | |
| NHCONH | 210–212 | Acetone/$Et_2O$ | 231 |
| 1,3-$Me_2$5-(2-$FC_6H_4$)-6-Br-7-(4-$MeC_6H_4SO_2$)- | | | |
| NHCONH·Acetone | 155–160d | Acetone/$Et_2O$ | 301 |

TABLE VII-1. —(contd.)

| Substituent | mp (°C) or; [bp (°C/torr)] | Solvent of Crystallization | Yield (%) | Spectra | Refs. |
|---|---|---|---|---|---|
| 1,3-$Me_2$-5-(2-$FC_6H_4$)-6-Cl-7-$H_2N$ | | | | | |
| *R*(−)-Enantiomer | 224–226 | $Et_2O$ | [α] | | 231 |
| *S*(+)-Enantiomer | 222–226 | EtOAc/$Et_2O$ | [α] | | 231 |
| 1,3-$Me_2$-5-(2-$FC_6H_4$)-6-Cl-7-(2-HO-Ethyl)-NHCONH | | | | | |
| *R*(−)-Enantiomer | > 100d | EtOAc/Hexane | [α] | | 231 |
| *S*(+)-Enantiomer | 147d | EtOAc/Hexane | [α] | | 231 |
| 1,3-$Me_2$-5-(2-$FC_6H_4$)-6-Cl-7-MeNHCONH | 178–184d | $CH_2Cl_2$/EtOAc | | | 301 |
| 1,3-$Me_2$-5-(2-$FC_6H_4$)-6-Cl-7-(4-$MeC_6H_4SO_2$)-NHCONH | 163–170d | Acetone/$Et_2O$ | | | 301 |
| ***3,3,5,7,9-Pentasubstituted*** | | | | | |
| 3,3-$Me_2$-5-(2-$ClC_6H_4$)-7-$NO_2$-9-Cl | 105 | $Et_2O$/Hexane | | | 230 |
| 3,3-$Me_2$-5-(2-$FC_5H_4$)-7-$H_2N$-9-Cl | 195–197 | EtOAc/Hexane | | | 301 |
| 3,3-$Me_2$-5-(2-$FC_5H_4$)-7,9-Cl | 124–125 | Hexane/Petr ether | | | 301 |
| 3,3-$Me_2$-5-(2-$FC_6H_4$)-7-$NO_2$-9-Br | 164 | $Et_2O$/Hexane | | | 230 |
| 3,3-$Me_2$-5-(2-$FC_6H_4$)-7-$NO_2$-9-Cl | 174–175 | | | | 230 |
| | 168–169 | $Et_2O$/Petr ether | | | 301 |
| ***3,5,6,7,8-Pentasubstituted*** | | | | | |
| 3-Me-5-(2-$ClC_6H_4$)-6,8-$Cl_2$-7-$H_2N$ | 273 | EtOAc/MeOH | | | 301 |
| ***3,5,7,8,9-Pentasubstituted*** | | | | | |
| 3-AcO-5-Ph-7,8,9-$(MeO)_3$ | 213–214 | EtOAc | | | 486 |
| 4-HO-5-Ph-7,8,9-$(MeO)_3$ | 192–193 | EtOAc | | | 486 |
| ***5,6,7,8,9-Pentasubstituted*** | | | | | |
| 5-Ph-6,7,8,9-$Cl_4$ | 262–265 | AcOH | | | 497 |

| Substituents | mp (°C) | Solvent | Ref. |
|---|---|---|---|
| ***1,5,6,7,8-Pentasubstituted*** | | | |
| 1-Me-5-Ph-6,8-$Cl_2$-7-$H_2N$ | 228 | $CH_2Cl_2$/EtOAc | 80b |
| | 205 | $CH_2Cl_2$/EtOAC | 251c |
| 1-Me-5-Ph-6,7,8-$(MeO)_3$ | 129–130 | Cyclohexane | 486 |
| 1-Me-5-Ph-6,8-$(MeO)_2$-7-HO | 198–199 | MeOH | 486 |
| 1-Me-5-(2-$ClC_6H_4$)-6,8-$Cl_2$-7-$H_2N$ | 186–187 | $CH_2Cl_2$/Hexane | 80b |
| 1-Me-5-(2-$FC_6H_4$)-6,8-$Br_2$-7-$H_2N$ | 213–216 | $CH_2Cl_2$/EtOAc | 301 |
| 1-Me-5-(2-$FC_6H_4$)-6,8-$Cl_2$-7-AcNH | 198–200 | $Et_2O$ | 251b |
| 1-Me-5-(2-$FC_6H_4$)-6,8-$Cl_2$-7-$H_2N$ | 234 | $CH_2Cl_2$/EtOAc | 80b |
| 1-Me-5-(2-$FC_6H_4$)-6,8-$Cl_2$-7-[4-(2-Cl-Ethyl)piperazin-1-yl]CONH | 130–133 | | 251b |
| 1-Me-5-(2-$FC_6H_4$)-6,8-$Cl_2$-7-$F_3CCONH$ | 238–240 | $Et_2O$ | 80b |
| 1-Me-5-(2-$FC_6H_4$)-6,8-$Cl_2$-7-(Formyl)NH | 213 | $Et_2O$ | 251b |
| 1-Me-5-(2-$FC_6H_4$)-6,8-$Cl_2$-7-(2-HO-Ethyl)NHCONH | 242–246d | Acetone | 301 |
| 1-Me-5-(2-$FC_6H_4$)-6,8-$Cl_2$-7-MeNHCONH | 213–214 | $Et_2O$ | 80b |
| 1-Me-5-(2-$FC_6H_4$)-6,8-$Cl_2$-7-$Me_2NCONH$ | 145–147 | $Et_2O$ | 80b |
| 1-Me-5-(2-$FC_6H_4$)-6,8-$Cl_2$-7-(4-Me-Piperazin-1-yl)CONH | 193–195 | $Et_2O$ | 251b |
| 1-Me-5-(2-$FC_6H_4$)-6,8-$Cl_2$-7-[4-(2-HO-Ethyl)piperazin-1-yl]CONH | 232–233 | $CH_2Cl_2$/Hexane | 80b |
| 1-Me-5-(2-$FC_6H_4$)-6,8-$Cl_2$-7-[2-Morpholino-ethylamino]CONH | 174–176 | $Et_2O$ | 80b |
| 1-Me-5-(2-$FC_6H_4$)-6,8-$Cl_2$-7-(Pyrrolidin-1-yl)CONH | 166–167 | $Et_2O$ | 251b |
| ***1,5,6,7,9-Pentasubstituted*** | | | |
| 1-Me-5-(2-$FC_6H_4$)-6,9-$Cl_2$-7-$H_2N$ | 208–209 | $CH_2Cl_2$/$Et_2O$ | 301 |
| ***1,5,6,8,9-Pentasubstituted*** | | | |
| 1-Me-5-(2-$FC_6H_4$)-6,8,9-$Cl_3$ | 190–192 | $CH_2Cl_2$/$Et_2O$ | 301 |
| ***1,5,7,8,9-Pentasubstituted*** | | | |
| 1,5-$Me_2$-7,8,9-$(MeO)_3$ | 138–139 | EtOAc | 486 |
| 1-Me-5-Ph-7,8,9-$(MeO)_3$ | 109–110 | $Et_2O$/Cyclohexane | 486 |

TABLE VII-1. —(contd.)

| Substituent | mp (°C) or; [bp (°C/torr)] | Solvent of Crystallization | Yield (%) | Spectra | Refs. |
|---|---|---|---|---|---|
| ***Hexasubstituted*** | | | | | |
| ***1,3,3,5,6,7-Hexasubstituted*** | | | | | |
| 1,3,3-$Me_3$-5-(2-$ClC_6H_4$)-6-Br-7-$H_2N$ | 188–189 | EtOAc/$Et_2O$ | | | 230 |
| 1,3,3-$Me_3$-5-(2-$ClC_6H_4$)-6-Br-7-(2-HO-Ethyl)-NHCONH | 214–218 | Acetone/$Et_2O$ | | | 230 |
| 1,3,3-$Me_3$-5-(2-$ClC_6H_4$)-6-Cl-7-$H_2N$ | 195–196 | EtOAc/$Et_2O$ Hexane | | | 230 |
| 1,3,3-$Me_3$-5-(2-$ClC_6H_4$)-7-$H_2N$-9-Cl | 254 | $Et_2O$/Hexane | | | 230 |
| 1,3,3-$Me_3$-5-(2-$ClC_6H_4$)-6-Cl-7-(2-HO-Ethyl)-NHCONH | 222d | Acetone/$Et_2O$ | | | 301 |
| 1,3,3-$Me_3$-5-(2-$ClC_6H_4$)-6-Cl-7-(4-$MeC_6H_4SO_2$)-NHCONH | 158–160d | Acetone/$Et_2O$ | | | 301 |
| 1,3,3-$Me_3$-5-(2-$FC_6H_4$)-6-Br-7-$H_2N$ | 217–220 | EtOAc/$Et_2O$ | | | 230 |
| 1,3,3-$Me_3$-5-(2-$FC_6H_4$)-6-Br-7-(2-HO-Ethyl)-NHCONH | 202 | EtOH/EtOAc/Hexane | | | 230 |
| 1,3,3-$Me_3$-5-(2-$FC_6H_4$)-6-Cl-7-$H_2N$ | 192–193 | $Et_2O$/Petr ether | | | 230 |
| 1,3,3-$Me_3$-5-(2-$FC_6H_4$)-6-Cl-7-(2-HO-Ethyl)-NHCONH | 146–147 | Acetone/$Et_2O$ | | | 230 |
| 1,3,3-$Me_3$-5-(2-$FC_6H_4$)-6-Cl-7-MeNHCONH | 235 | $Et_2O$ | | | 301 |
| ***1,3,3,5,7,8-Hexasubstituted*** | | | | | |
| 1,3,3-$Me_3$-5-(2-$FC_6H_4$)-7-$H_2N$-8-Cl | 206–209 | $Et_2O$/Hexane | | | 301 |
| 1,3,3-$Me_3$-5-(2-$FC_6H_4$)-7-$H_2N$-8-Me | 214–215 | $CH_2Cl_2$/$Et_2O$/Hexane | | | 301 |
| 1,3,3-$Me_3$-5-(2-$FC_6H_4$)-7-$H_2N$-8-(Morpholino)$SO_2CH_2$ | 276 | EtOAc/Petr ether | | | 301 |
| 1,3,3-$Me_3$-5-(2-$FC_6H_4$)-7-HO-8-Me | 219–220 | $Et_2O$/Hexane | | | 301 |
| 1,3,3-$Me_3$-5-(2-$FC_6H_4$)-7-(2-HO-Ethyl)NHCONH-8-Me | | EtOAc | | | 301 |

| | | | |
|---|---|---|---|
| 1,3,3-$Me_3$-5-(2-$FC_6H_4$)-7-(2-HO-Ethyl)NHCONH-8-morpholino)$SO_2CH_2$ | 230–231 | EtOAc/$Et_2O$ | |
| | | Acetone | 301 |
| 1,3,3-$Me_3$-5-(2-$FC_6H_4$)-7-*i*-PrNH-8-Cl | 199–200 | $Et_2O$/Hexane | 301 |
| 1,3,3-$Me_3$-5-(2-$FC_6H_4$)-7-$NO_2$-8-Me | 210–212 | $Et_2O$ | 301 |
| 1,3,3-$Me_3$-5-(2-$FC_6H_4$)-7-$NO_2$-8-$Me_2NSO_2CH_2$ | 209–210 | $Et_2O$ | 301 |
| 1,3,3-$Me_3$-5-(2-$FC_6H_4$)-7-$NO_2$-8-(Morpholino)-$SO_2CH_2$ | 213–214 | EtOAc | 301 |
| 1,3,3-$Me_3$-5-(2-$FC_6H_4$)-7-$NO_2$-8-(Morpholino)-$SO_2CHCl$ | 178–179 | $Et_2O$/Hexane/ Petr ether | 301 |
| ***1,3,3,5,7,9-Hexasubstituted*** | | | |
| 1-Bu-3,3-$Me_2$-5-(2-$FC_6H_4$)-7-$H_2N$-9-Cl | 151 | $Et_2O$/Hexane | 230 |
| 1-Bu-3,3-$Me_2$-5-(2-$FC_6H_4$)·7-$NO_2$-9-Cl | 125–127 | Hexane/ether | 230 |
| 1-Et-3,3-$Me_2$-5-(2-$FC_4H_4$)-7-$H_2N$-9-Cl | 156 | $Et_2O$/Hexane | 230 |
| 1-Et-3,3-$Me_2$-5-(2-$FC_6H_4$)7-$NO_2$-9-Cl | 117–120 | $Et_2O$/Hexane | 230 |
| 1-(2-$Et_2N$-Ethyl)-3,3-$Me_2$-5-(2-$FC_6H_4$)-7-$H_2N$-9-Cl | 180–183 | EtOAc/$Et_2O$ | 301 |
| 1-(2-$Et_2N$-Ethyl)-3,3-$Me_2$-5-(2-$FC_6H_4$)-7-$NO_2$-9-Cl | 111–112 | Hexane | 301 |
| 1-(2-HO-Ethyl)-3,3-$Me_2$-5-(2-$FC_6H_4$)-7-$H_2N$-9-Cl | 188–189 | $Et_2O$ | 301 |
| 1-(2-HO-Ethyl)-3,3-$Me_2$-5-(2-$FC_6H_4$)-7-$NO_2$-9-Cl | 145–146 | $Et_2O$/Hexane | 301 |
| 1,3,3-$Me_3$-5-(2-$ClC_6H_4$)-7-$NO_2$-9-Cl | 121 | $Et_2O$/Hexane | 230 |
| 1,3,3-$Me_3$-5-(2-$FC_6H_4$)-7-$H_2N$-9-Br | 228 | $CH_2Cl_2$/Petr ether | 230 |
| 1,3,3-$Me_3$-5-(2-$FC_6H_4$)-7-$H_2N$-9-Cl | 227 | $Et_2O$/Petr ether | 301 |
| 1,3,3-$Me_2$-5-(2-$FC_6H_4$)-7-(Benzyloxy)CONH-9-Cl | 212 | $Et_2O$ | 230 |
| 1,3,3-$Me_3$-5-(2-$FC_6H_4$)-7-(Benzyloxy)CON(Et)-9-Cl | 125–127 | Hexane | 230 |
| 1,3,3-$Me_3$5-(2-$FC_6H_4$)-7,9-$Cl_2$ | 100–101 | Hexane | 301 |
| 1,3,3-$Me_3$-5-(2-$FC_6H_4$)-7-BuNH-9-Cl | 113–115 | Hexane | 230 |
| 1,3,3-$Me_3$-5-(2-$FC_6H_4$)-7-EtNH-9-Cl | 150–151 | $Et_2O$/Petr ether | 230 |
| 1,3,3-$Me_3$-5-(2-$FC_6H_4$)-7-MeNH-9-Cl | 168–169 | $Et_2O$/Hexane | 301 |
| 1,3,3-$Me_3$-5-(2-$FC_6H_4$)-7-$Me_2N$-9-Cl | 167 | Hexane | 301 |
| 1,3,3-$Me_3$-5-(2-$FC_6H_4$)-7-$NO_2$-9-Br | 180 | $Et_2O$/Petr ether | 230 |
| 1,3,3-$Me_3$-5-(2-$FC_6H_4$)-7-$H_2N$-9-Cl | 237–238 | *i*-PrOH | 230 |
| 1,3,3-$Me_3$-5-(2-$FC_6H_4$)-7-$NO_2$-9-Cl | 146–147 | $CH_2Cl_2$/*i*-PrOH | 230 |
| 1,3,3-$Me_3$-5-(2-$FC_6H_4$)-7-$Me_2N$-9-Cl | 227 | $Et_2O$/Petr ether | 230 |

TABLE VII-1. —(contd.)

| Substituent | mp (°C) or; [bp (°C/torr)] | Solvent of Crystallization | Yield (%) | Spectra | Refs. |
|---|---|---|---|---|---|
| 1,3,3-$Me_3$-5-(2-$FC_6H_4$)-7-(Pyrrolidin-1-yl)CONH-9-Cl | 218–219 | EtOAc/$Et_2O$ | | | 301 |
| 1-Pr-3,3-$Me_2$-5-(2-$FC_6H_4$)-7-$H_2N$-9-Cl | 178–179 | EtOAc/$Et_2O$/Hexane | | | 230 |
| 1-Pr-3,3-$Me_2$-5-(2-$FC_6H_4$)-7-$NO_2$-9-Cl | 142–143 | $Et_2O$/Hexane | | | 230 |
| 1-*i*-Pr-3,3-$Me_2$-5-(2-$FC_6H_4$)-7-$H_2N$-9-Cl | 248–249 | $Et_2O$ | | | 301 |
| 1-*i*-Pr-3,3-$Me_2$-5-(2-$FC_6H_4$)-7-$NO_2$-9-Cl | 168 | $Et_2O$/Petr ether | | | 301 |
| ***1,3,5,6,7,8-Hexasubstituted*** | | | | | |
| 1-Me-3-Et-5-(2-$FC_6H_4$)-6,8-$Cl_2$-7-$H_2N$ | 197–198 | EtOH/$Et_2O$ | | | |
| 1-Me-3-Et-5-(2-$FC_6H_4$)-6,8-$Cl_2$-7-(2-HO-Ethyl)-NHCONH | 213–215 | Acetone/$Et_2O$ Cyclohexane/$CH_2Cl_2$ | | | 301 |
| 1,3-$Me_2$-5-(2-$ClC_6H_4$)-6,8-$Br_2$-7-$H_2N$ | 212 | | | | 301 |
| *S*(−)-Enantiomer·EtOH | 114–115d | EtOH | | | 301 |
| 1,3-$Me_2$-5-(2-$ClC_6H_4$)-6,8-$Br_2$-7-(2-Cl-Ethyl)NHCONH | 180–200d | EtOAc/$Et_2O$ | | | 301 |
| 1,3-$Me_2$-5-(2-$ClC_6H_4$)-6,8-$Br_2$-7-(2-HO-Ethyl)-NHCONH | 222–225 | EtOAc | | | 301 |
| 1,3-$Me_2$-5-(2-$ClC_6H_4$)-6,8-$Br_2$-7-(2-HS-Ethyl)NHCONH | 195–200d | EtOAc/$CH_2Cl_2$/$Et_2O$ | | | 301 |
| 1,3-$Me_2$-5-(2-$ClC_6H_4$)-6,8-$Br_2$-7-(2-MeS-Ethyl)-NHCONH | 234 | EtOAc/$Et_2O$ | | | 301 |
| 1,3-$Me_2$-5-(2-$ClC_6H_4$)-6,8-$Br_2$-7-(2-Me(O)S-Ethyl)-NHCONH | 226 | Acetone | | | 301 |
| 1,3-$Me_2$-5-(2-$ClC_6H_4$)-6,8-$Br_2$-7-(2-$MeSO_2$-Ethyl)-NHCONH | 259 | $Et_2O$/Cyclohexane | | | 301 |
| 1,3-$Me_2$-5-(2-$ClC_6H_4$)-6,8-$Cl_2$-7-$H_2N$ | 186–187 | $CH_2Cl_2$ | | | 301 |
| 1,3-$Me_2$-5-(2-$FC_6H_4$)-6,8-$Cl_2$-7-$H_2N$ | | | | | 301 |
| *R*(−)-Enantiomer | 165 | $Et_2O$ | | | 301 |
| *S*(+)-Enantiomer | 165–166 | $Et_2O$ | | | 301 |

| | | | | | |
|---|---|---|---|---|---|
| ***1,3,5,7,8,9-Hexasubstituted*** | | | | | |
| 1-Ac-3-AcO-5-Ph-7,8,9-$(MeO)_3$ | 156–157 | EtOAc | | | 486 |
| ***1,5,6,7,8,9-Hexasubstituted*** | | | | | |
| 1-Me-5-(2-$FC_6H_4$)-6,8,9-$Cl_3$-7-$H_2N$ | 150 | $Et_2O$ | | | 301 |
| ***3,3,5,6,7,8-Hexasubstituted*** | | | | | |
| 3,3-$Me_2$-5-(2-$FC_6H_4$)-6,8-$Br_2$-7-$H_2N$ | 255d | | | | 251b |
| ***Heptasubstituted*** | | | | | |
| 1,3,3-$Me_3$-5-(2-$ClC_6H_4$)-6,8-$Br_2$-7-$H_2N$ | 110 | $CH_2Cl_2$/Hexane | | | 230 |
| 1,3,3-$Me_3$-5-(2-$ClC_6H_4$)-6,8-$Cl_2$-7-$H_2N$ | 105 | Cyclohexane/$CH_2Cl_2$ | | | 230 |
| 1,3,3-$Me_3$-5-(2-$FC_6H_4$)-6,8-$Br_2$-7-$H_2N$ | 202 | | | | 230 |
| 1,3,3-$Me_3$-5-(2-$FC_6H_4$)-6,8-$Cl_2$-7-$H_2N$ | 208 | Cyclohexane/$CH_2Cl_2$ | | | 230 |
| 1,3,3-$Me_3$-5-(2-$FC_6H_4$)-6,9-$Cl_2$-7-$H_2N$ | 190–192 | $Et_2O$/Hexane | | | 230 |
| 1,3,3-$Me_3$-5-(2-$FC_6H_4$)-7-$H_2N$-8,9-$Cl_2$ | 169–172 | $Et_2O$ | | | 230 |
| 1,3-$Me_2$-5-(2-$ClC_6H_4$)-6,8,9-$Cl_3$-7-$H_2N$ | 271 | $Et_2O$ | | | 251b |
| *S*(−)-Enantiomer | 227 | $Et_2O$ | | | 251b |
| ***Octasubstituted*** | | | | | |
| 1,3,3-$Me_3$-5-(2-$FC_6H_4$)-6,8,9-$Cl_3$-7-$H_2N$ | 165–166 | $Et_2O$/Hexane | | | 230 |

*3-Methylene-1,3-dihydro-1,4-benzodiazepin-2(2H)-ones*

| | | | | | |
|---|---|---|---|---|---|
| ***$R_1$, $R_2$; other*** | | | | | |
| H, H; 5-Ph-7-Cl | | | | | |
| H, H; 5-(2-$ClC_6H_4$)-7-Cl | 200–202d | MeCN | 63 | pmr | 243 |
| H, H; 1-Me-5-(2-$ClC_6H_4$)-7-Cl | 163–165 | MeCN | 52 | ir, pmr | 243 |
| AcO, H; 1-Me-5-(2-$ClC_6H_4$)-7-Br | 155–157 | MeOH | | | 314 |
| $H_2N$, H; 5-Ph-7-Cl | 244–247 | | | | 314 |
| $H_2N$, H; 5-(2-$ClC_6H_4$)-7-Br | 216–218d | | | | 315 |

TABLE VII-1. —(contd.)

| Substituent | mp (°C) or; [bp (°C/torr)] | Solvent of Crystallization | Yield (%) | Spectra | Refs. |
|---|---|---|---|---|---|
| $H_2N$, H; 5-(2-$ClC_6H_4$)-7-Cl | 220–222 | | | | 315 |
| $H_2N$, H; 5-(2-$ClC_6H_4$)-7-F | 173–174 | | | | 315 |
| $H_2N$, H; 5-(2-$ClC_6H_4$)-7-I | 167–172d | | | | 315 |
| $H_2N$, H; 5-(2-$FC_6H_4$)-7-Cl | 275–280 | | | | 315 |
| $H_2N$, H; 1-Me-5-Ph-7-Cl | Amorphous | | | | 315 |
| $H_2N$, H; 1-Me-5-(2-$ClC_6H_4$)-7-Br | 186–189 | | | | 315 |
| $H_2N$, H; 1-Me-5-(2-$ClC_6H_4$)-7-Cl | 201–204 | | | | 315 |
| $H_2N$, H; 1-Me-5-(2-$ClC_6H_4$)-7-F | 105–110 | | | | 315 |
| $H_2N$, H; 1-Me-5-(2-$ClC_6H_4$)-7-I | 130d | | | | 315 |
| $H_2N$, H; 1-Me-5-(2-$ClC_6H_4$)-7-$NO_2$ | 140–143d | | | | 315 |
| $(Allyl)_2N$, H; 5-Ph-7-Cl | 173–175 | | | | 315 |
| $(Allyl)_2N$, H; 5-Ph-7-$NO_2$ | 194–195 | | | | 315 |
| $(Allyl)_2N$, H; 5-(2-$ClC_6H_4$)-7-Cl | 165–167 | | | | 315 |
| $(Allyl)_2N$, H; 5-(2-$ClC_6H_4$)-7-$NO_2$ | 179–181 | | | | 315 |
| $(Allyl)_2N$, H; 5-(2-$FC_6H_4$)-7-$NO_2$ | 201–202 | | | | 315 |
| $(Allyl)_2N$, H; 1-Me-5-(2-$ClC_6H_4$)-7-Cl | Amorphous | | | | 315 |
| $(Allyl)_2N$, H; 1-Me-5-(2-$ClC_6H_4$)-7-$NO_2$ | 55–60 | | | | 315 |
| BuNH, H; 5-(2-$ClC_6H_4$)-7-Cl | 173 | | | | 315 |
| BuNH, H; 5-(2-$ClC_6H_4$)-7-$NO_2$ | 222–224 | | | | 315 |
| *t*-BuNH, H; 5-(2-$ClC_6H_4$)-7-Cl | 230 | | | | 315 |
| BuNH, H; 1-Me-5-Ph-7-Cl | Amorphous | | | | 315 |
| BuNH, H; 1-Me-5-(2-$ClC_6H_4$)-7-Br | 142–145 | | | | 315 |
| BuNH, H; 1-Me-5-(2-$ClC_6H_4$)-7-Cl | 110–113 | | | | 315 |
| BuNH, H; 1-Me-5-(2-$ClC_6H_4$)-7-$NO_2$ | 90d | | | | 315 |
| *t*-BuNH, H; 1-Me-5-(2-$ClC_6H_4$)-7-Cl | 175–177 | | | | 315 |
| $Bu_2N$, H; 5-Ph-7-Cl | 138–140 | | | | 315 |
| $Bu_2N$, H; 5-(2-$ClC_6H_4$)-7-Cl | 211–213 | | | | 315 |
| $Bu_2N$, H; 1-Me-5-Ph-7-Cl | 95–97 | | | | 315 |
| $Bu_2N$, H; 1-Me-5-(2-$ClC_6H_4$)-7-Cl | Amorphous | | | | 315 |

| | | | | |
|---|---|---|---|---|
| Cyclohexyl(Me)N, H; 5-Ph-7-Cl | 247–250 | | | 315 |
| Cyclohexyl(Me)N, H; 5-(2-ClC$_6$H$_4$)-7-Cl | 197–200 | | | 315 |
| Cyclohexyl(Me)N, H; 5-(2-ClC$_6$H$_4$)-7-NO$_2$ | 209–211 | | | 315 |
| Cyclohexyl(Me)N, H; 1-Me-5-Ph-7-Cl | 167–169 | | | 315 |
| Cyclohexyl(Me)N, H; 1-Me-5-(2-ClC$_6$H$_4$)-7-Cl | 170–172 | | | 315 |
| Cyclohexyl(Me)N, H; 1-Me-5-(2-ClC$_6$H$_4$)-7-NO$_2$ | 107–109 | | | 315 |
| (Cyclopropyl)NH, H; 1-Me-5-(2-ClC$_6$H$_4$)-7-Br | 136–138d | | | 315 |
| EtNH, H; 5-(2-ClC$_6$H$_4$)-7-Br | 227–229 | | | 315 |
| EtNH, H; 5-(2-ClC$_6$H$_4$)-7-Cl | 213–215 | EtOH | | 315 |
| EtNH, H; 5-(2-ClC$_6$H$_4$)-7-F | 138–141 | | | 315 |
| EtNH, H; 5-(2-ClC$_6$H$_4$)-7-I | 212–214 | | | 315 |
| EtNH, H; 5-(2-ClC$_6$H$_4$)-7-NO$_2$ | 260–261 | | | 315 |
| EtNH, H; 5-(2-FC$_6$H$_4$)-7-NO$_2$ | 240–241 | | | 315 |
| EtNH, H; 1-Me-5-(2-ClC$_6$H$_4$)-7-Br | 173–175 | | | 315 |
| EtNH, H; 1-Me-5-(2-ClC$_6$H$_4$)-7-Cl | 122 | | | 315 |
| EtNH, H; 1-Me-5-(2-ClC$_6$H$_4$)-7-I | 202–206 | | | 315 |
| EtNH, H; 1-Me-5-(2-ClC$_6$H$_4$)-7-NO$_2$ | 159–161 | | | 315 |
| EtNH, H; 1-Me-5-(2-FC$_6$H$_4$)-7-Cl | 151–154 | | | 315 |
| EtNH, H; 1-Me-5-(2-FC$_6$H$_4$)-7-NO$_2$ | 175–176 | | | 315 |
| Et$_2$N, H; 5-Ph-7-Cl | 236–238 | | | 315 |
| Et$_2$N, H; 5-(2-ClC$_6$H$_4$)-7-Cl | 204–206 | | | 315 |
| Et$_2$N, H; 5-(2-ClC$_6$H$_4$)-7-NO$_2$ | 208–209 | | | 315 |
| Et$_2$N, H; 1-Et-5-(2-ClC$_6$H$_4$)-7-Cl | Amorphous | | ir | 315 |
| Et$_2$N, H; 1-(2-Me$_2$N-Ethyl)-5-(2-ClC$_6$H$_4$)-7-NO$_2$ | 103 | | | 315 |
| Et$_2$N, H; 1-Me-5-Ph-7-Cr | 192–194 | | | 315 |
| Et$_2$N, H; 1-Me-5-(2-ClC$_6$H$_4$)-7-Br | 131–133 | | | 315 |
| Et$_2$N, H; 1-Me-5-(2-ClC$_6$H$_4$)-7-Cl | 135–137 | | | 315 |
| Et$_2$N, H; 1-Me-5-(2-ClC$_6$H$_4$)-7-NO$_2$ | 148–150 | | | 315 |
| (2-Et$_2$N-Ethyl)NH, H; 5-(2-ClC$_6$H$_4$)-7-Cl | Amorphous | | | 315 |
| (2-Et$_2$N-Ethyl)NH, H; 5-(2-ClC$_6$H$_4$)-7-NO$_2$ | 164–165 | | | 315 |
| (2-Et$_2$N-Ethyl)NH, H; 1-Me-5-Ph-7-Cl | Amorphous | | | 315 |
| (2-Et$_2$N-Ethyl)NH, H; 1-Me-5-(2-ClC$_6$H$_4$)-7-Cl | Amorphous | | | 315 |
| EtMeN, H; 1-Me-5-Ph-7-Cl | 98–101 | | | 315 |
| EtMeN, H; 1-Me-5-(2-ClC$_6$H$_4$)-7-Cl | 88–91 | | | 315 |

TABLE VII-1. —(contd.)

| Substituent | mp (°C) or; [bp (°C/torr)] | Solvent of Crystallization | Yield (%) | Spectra | Refs. |
|---|---|---|---|---|---|
| EtO, H; 1-Me-5-(2-ClC$_6$H$_4$)-7-Br | 175–176 | Et$_2$O | | | 314 |
| (EtOOCCH$_2$)NH, H; 1-Me-5-(2-ClC$_6$H$_4$)-7-Br | 175–177 | | | | 315 |
| 2-Furyl, H; 5-Ph-7-Cl | 275–277 | | 52 | ir | 304 |
| 2-Furyl, H; 1-Me-5-Ph-7-Cl | 169–171 | Et$_2$O/Petr ether | | | 192 |
| (2-Furyl)NH, H; 1-Me-5-(2-ClC$_6$H$_4$)-7-Br | 81–83 | | | | 315 |
| Hexahydroazepino, H; 5-Ph-7-Cl | 248–250 | | | | 315 |
| Hexahydroazepino, H; 5-(2-ClC$_6$H$_4$)-7-Cl | 238–239 | | | | 315 |
| Hexahydroazepino, H; 5-(2-ClC$_6$H$_4$)-7-NO$_2$ | 217–218 | | | | 315 |
| Hexahydroazepino, H; 1-Me-5-Ph-7-Cl | 196–198 | | | | 315 |
| Hexahydroazepino, H; 1-Me-5-(2-ClC$_6$H$_4$)-7-Br | 140–142 | | | | 315 |
| Hexahydroazepino, H; 1-Me-5-(2-ClC$_6$H$_4$)-7-Cl | Amorphous | | | | |
| Hexahydroazepino, H; 1-Me-5-(2-ClC$_6$H$_4$)-7-NO$_2$ | 98–105 | | | | 315 |
| HO, H; 5-Ph-7-Cl | 155–157 | | | | 314 |
| HO, H; 5-(2-ClC$_6$H$_4$)-7-Br | | | | | |
| Hydrochloride | 205d | | | | 314 |
| HO, H; 5-(2-ClC$_6$H$_4$)-7-Cl | 117–123d | | | | 314 |
| HO, H; 5-(2-ClC$_6$H$_4$)-7-F | | | | | |
| Hydrochloride | $>$120d | | | | 314 |
| HO, H; 5-(2-ClC$_6$H$_4$)-7-NO$_2$ | 213–219d | | | | 314 |
| HO, H; 1-(2-Et$_2$N-Ethyl)5-(2-ClC$_6$H$_4$)-7-Br | | | | | |
| Sodium salt | 155–165d | | | | 314 |
| HO, H; 1-(2,2,2-F$_3$-Ethyl)5-(2-ClC$_6$H$_4$)-7-Br | | | | | |
| Hydrochloride | 175–177d | | | | 314 |
| HO, H; 1-Me-5-(2-ClC$_6$H$_4$)-7-Br | 137–139d | | | | 314 |
| Hydrochloride | 185–187d | THF/Et$_2$O | | | 314 |
| Sodium salt | $>$260d | | | | 314 |
| HO, H; 1-Me-5-(2-ClC$_6$H$_4$)-7-Cl | 117–123d | | | | 314 |
| Sodium salt | 183–184d | | | | 314 |

| | | | |
|---|---|---|---|
| HO, H; 1-Me-5-(2-$ClC_6H_4$)-7-I | 115–125d | | 314 |
| Sodium salt·$H_2O$ | 287–294d | | 314 |
| HO, H; 1-Me-5-(2-$ClC_6H_4$)-7-$NO_2$ | 149–151d | | 314 |
| Sodium salt | 215–217d | | 314 |
| HO, H; 1-Me-5-(2-$FC_6H_4$)-7-Br | | | |
| Sodium salt | 180–190d | | 314 |
| HO, H; 1-Me-5-(2-$FC_6H_4$)-7-Cl | | | |
| Sodium salt | >220d | | 314 |
| (2-HO-Ethyl)NH, H; 5-Ph-7-Cl | 228–230 | | 315 |
| (2-HO-Ethyl)NH, H; 5-(2-$ClC_6H_4$)-7-Cl | 231–233 | | 315 |
| (2-HO-Ethyl)NH, H; 1-Me-(2-$ClC_6H_4$)-7-Cl | 155–157 | | 315 |
| (2-HO-Ethyl)NH, H; 1-Me-5-Ph-7-Cl | Amorphous | | 315 |
| MeO, H; 1-Me-5-(2-$ClC_6H_4$)-7-Br | 203–205 | | 314 |
| MeNH, H; 5-Ph-7-Cl | 210 | | 315 |
| MeNH, H; 5-Ph-7-$NO_2$ | 247–249 | DMF/MeOH | 152c |
| MeNH, H; 5-(2-$ClC_6H_4$)-7-Br | 157–158 | | 315 |
| MeNH, H; 5-(2-$ClC_6H_4$)-7-Cl | 225 | | 315 |
| MeNH, H; 1-Me-5-Ph-7-Cl | 201–202 | | 315 |
| MeNH, H; 1-Me-5-(2-$ClC_6H_4$)-7-Cl | 155–158 | | 315 |
| $Me_2N$, H; 5-Ph-7-Cl | 237–240 | | 315 |
| $Me_2N$, H; 5-Ph-7-$NO_2$ | 260–264d | | 315 |
| $Me_2N$, H; 5-(2-$ClC_6H_4$)-7-Br | 220–223 | | 315 |
| $Me_2N$, H; 5-(2-$ClC_6H_4$)-7-Cl | 239–241 | MeOH | 315 |
| $Me_2N$, H; 5-(2-$ClC_6H_4$)-7-F | 210–211 | | 315 |
| $Me_2N$, H; 2-(2-$ClC_6H_4$)-7-I | 227–230 | | 315 |
| $Me_2N$, H; 5-(2-$ClC_6H_4$)-7-$NO_2$ | 241d | | 315 |
| $Me_2N$, H; 5-(2-$FC_6H_4$)-7-$NO_2$ | 243–244d | | 315 |
| $Me_2N$, H; 1-(Cyclopropyl)methyl-5-(2-$ClC_6H_4$)-7-Cl | 188–190 | | 315 |
| $Me_2N$, H; 1-(2,2,2-$F_3$-Ethyl)-5-(2-$ClC_6H_4$)-7-Br | 216–217 | | 315 |
| $Me_2N$, H; 1-Me-5-Ph-7-Cl | 204–205 | | 315 |
| $Me_2N$, H; 1-Me-5-Ph-7-$NO_2$ | 228–230 | | 315 |
| $Me_2N$, H; 1-Me-5-(2-$ClC_6H_4$)-7-Br | 198–199 | | 315 |
| $Me_2N$, H; 1-Me-5-(2-$ClC_6H_4$)-7-Cl | 202–203 | *i*-PrOH | 315 |
| $Me_2N$, H; 1-Me-5-(2-$ClC_6H_4$)-7-F | 177–179 | | 315 |

TABLE VII-1. —(contd.)

| Substituent | mp (°C) or; [bp (°C/torr)] | Solvent of Crystallization | Yield (%) | Spectra | Refs. |
|---|---|---|---|---|---|
| $Me_2N$, H; 1-Me-5-(2-$ClC_6H_4$)-7-I | 182–184 | | | | 315 |
| $Me_2N$, H; 1-Me-5-(2-$ClC_6H_4$)-7-$NO_2$ | 182–183 | | | | 315 |
| $Me_2N$, H; 1-Me-5-(2-$FC_6H_4$)-7-Br | 172–173 | | | | 315 |
| $Me_2N$, H; 1-Me-5-(2-$FC_6H_4$)-7-Cl | 157–159 | | | | 315 |
| $Me_2N$, H; 1-Me-5-(2-$FC_6H_4$)-7-$NO_2$ | 193–195 | | | | 315 |
| $Me_2N$, H;1-(2-$Me_2N$-Ethyl)-5-(2-$ClC_6H_4$)-7-Cl | Amorphous | | | | 315 |
| Me(2-$Me_2N$-ethyl)N, H; 5-Ph-7-$NO_2$ | 200–202 | | | | 315 |
| Me(2-$Me_2N$-Ethyl)N, H; 5-(2-$ClC_6H_4$)-7-Cl | 202–205 | | | | 315 |
| Me(2-$Me_2N$-Ethyl)N, H;1-Me-5-(2-$ClC_6H_4$)-7-Cl | 130–132 | | | | 315 |
| Me(Morpholinocarbonyl)methylamino, H; 1-Me-5-(2-$ClC_6H_4$)-7-Br | 200–202 | THF | | | 315 |
| Me(Ph)N, H; 1-Me-5-(2-$ClC_6H_4$)-7-Cl | 197–199 | | | | 315 |
| Me(Ph)N, H; 1-Me-5-(2-$ClC_6H_4$)-7-$NO_2$ | 183–184 | | | | 315 |
| 4-Me-Piperazino, H; 5-(2-$ClC_6H_4$)-7-Cl | 194–196 | | | | 315 |
| 4-Me-Piperazino, H; 1-Me-5-Ph-7-Cl | 141–143 | | | | 315 |
| 4-Me-Piperazino, H; 1-Me-5-(2-$ClC_6H_4$)-7-Cl | Amorphous | | | | 315 |
| (2-Me-Prop-1-en-1-yl), H; 1-Me-5-Ph-7-Cl | 163–166 | $CH_2Cl_2$/Petr ether | 45 | ir, pmr, uv | 298 |
| Morpholino, H; 5-Ph-7-Cl | 251–252 | | | | 315 |
| Morpholino, H; 5-Ph-7-$NO_2$ | 243–245d | | | | 315 |
| Morpholino, H; 5-(2-$ClC_6H_4$)-7-Br | 156–185d | | | | 315 |
| Morpholino, H; 5-(2-$ClC_6H_4$)-7-Cl | 209–211 | | | | 315 |
| Morpholino, H; 5-(2-$ClC_6H_4$)-7-F | 243–246d | | | | 315 |
| Morpholino, H; 5-(2-$ClC_6H_4$)-7-I | 218–221 | | | | 315 |
| Morpholino, H; 5-(2-$ClC_6H_4$)-7-$NO_2$ | 246–248 | | | | 315 |
| Morpholino, H; 5-(2-$FC_6H_4$)-7-$NO_2$ | 246–249d | | | | 315 |
| Morpholino, H; 1-(cyclopropyl)$CH_2$-5-(2-$ClC_6H_4$)-7-Cl | Amorphous | | | | 315 |
| Morpholino, H; 1-Me-5-Ph-7-Cl | 169–171 | | | | 315 |
| Morpholino, H; 1-Me-5-Ph-7-$NO_2$ | 211–212 | | | | 315 |
| Morpholino, H; 1-Me-5-(2-$ClC_6H_4$)-7-Br | 148–153 | | | | 315 |

| | | | | | |
|---|---|---|---|---|---|
| Morpholino, H; 1-Me-5-(2-$ClC_6H_4$)-7-Cl | 158–160 | | | | 315 |
| Morpholino, H; 1-Me-5-(2-$ClC_6H_4$)-7-F | 110–115 | | | | 315 |
| Morpholino, H; 1-Me-5-(2-$ClC_6H_4$)-7-I | 193–195 | | | | 315 |
| Morpholino, H; 1-Me-5-(2-$ClC_6H_4$)-7-$NO_2$ | 102–105 | | | | 315 |
| Morpholino, H; 1-Me-5-(2-$FC_6H_4$)-7-Cl | 172–174 | | | | 315 |
| Morpholino, H; 1-Me-5-(2-$FC_6H_4$)-7-$NO_2$ | 187–188 | | | | 315 |
| Ph, H; 5-Ph-7-Cl | 235 | | 40 | ir, ms, uv | 304 |
| Ph, H; 5-Ph-7-Me | 210 | | 55 | ir | 304 |
| Ph, H; 1-Me-5-Ph-7-Cl | 189–191 | MeOH | 67 | | 298, 299 |
| Ph, H; 5-(2-$ClC_6H_4$)-7-Cl | 230–232 | $CH_2Cl_2$/Hexane | | | 192 |
| 4-AcO-3-$MeOC_6H_3$, H; 5-Ph-7-Cl | 198–200 | | 43 | | 304 |
| 4-$ClC_6H_4$, H; 5-Ph-7-Cl | 220–222 | | 36 | ir | 304 |
| 4-$Me_2NC_6H_4$, H; 5-Ph-7-Cl | 248–249 | | 48 | ir | 304 |
| 3,4-$(MeO)_2C_6H_3$, H; 1-Me-5-Ph-7-Cl | 125–128 | EtOH | | | 192 |
| 2-$NO_2C_6H_4$, H; 5-Ph-7-Cl | 260–261 | | 53 | | 304 |
| 4-$NO_2C_6H_4$, H; 5-Ph-7-Cl | 284 | | 42 | ir | 304 |
| 4-$iPrC_6H_4$, H; 5-Ph-7-Cl | 222–223 | | 40 | ir | 304 |
| Piperidino, H; 5-Ph-7-Cl | 242–243 | | | | 315 |
| Piperidino, H; 5-(2-$ClC_6H_4$)-7-Cl | 245–248d | | | | 315 |
| Piperidino, H; 5-(2-$ClC_6H_4$)-7-$NO_2$ | 235–236 | | | | 315 |
| Piperidino, H; 1-(2-$Me_2N$-ethyl)-5-(2-$ClC_6H_4$)-7-$NO_2$ | 103–106 | | | | 315 |
| Piperidino, H; 1-Me-5-Ph-7-Cl | 158d | | | | 315 |
| Piperidino, H; 1-Me-5-(2-$ClC_6H_6$)-7-Cl | Amorphous | | | | 315 |
| Piperidino, H; 1-Me-5-(2-$ClC_6H_4$)-7-$NO_2$ | 139–141 | | | | 315 |
| Piperidino, H; 1-Me-5-(2-$ClC_6H_4$)-7-Br | 176–178 | | | | 315 |
| PrNH, H; 5-(2-$ClC_6H_4$)-7-Cl | 186 | | | | 315 |
| PrNH, H; 1-Me-5-(2-$ClC_6H_4$)-7-Cl | 146–147 | | | | 315 |
| $Pr_2N$, H; 5-Ph-7-Cl | 200–201 | | | | 315 |
| $Pr_2N$, H; 5-(2-$ClC_6H_4$)-7-Cl | 226–227 | | | | 315 |
| $Pr_2N$, H; 5-(2-$ClC_6H_4$)-7-$NO_2$ | 241–242 | | | | 315 |
| $Pr_2N$, H; 1-Me-5-Ph-7-Cl | 134–135 | | | | 315 |
| $Pr_2N$, H; 1-Me-5-(2-$ClC_6H_4$)-7-Cl | 130–132 | | | | 315 |
| $Pr_2N$, H; 1-Me-5-(2-$ClC_6H_4$)-7-$NO_2$ | 71–75 | | | | 315 |
| *i*-PrNH, H; 5-(2-$ClC_6H_4$)-7-Cl | 188–191 | | | | 315 |

TABLE VII-1. —(contd.)

| Substituent | mp (°C) or; [bp (°C/torr)] | Solvent of Crystallization | Yield (%) | Spectra | Refs. |
|---|---|---|---|---|---|
| *i*-$Pr_2N$, H; $iPr_2NH$; 5-Ph-7-Cl | 248–250 | | | | 315 |
| *i*-$Pr_2N$, H; 5-(2-$ClC_6H_4$)-7-Cl | 222–224 | | | | 315 |
| *i*-$Pr_2N$, H; 5-(2-$ClC_6H_4$)-7-$NO_2$ | 211–213 | | | | 315 |
| *i*-$Pr_2N$, H; 1-Me-5-Ph-7-Cl | Amorphous | | | | 315 |
| *i*-$Pr_2N$, H; 1-Me-5-(2-$ClC_6H_4$)-7-Cl | 147–149 | | | | 315 |
| *i*-$Pr_2N$, H; 1-Me-5-(2-$ClC_6H_4$)-7-$NO_2$ | 156–158 | | | | 315 |
| 3-Pyridyl, H; 1-Me-5-Ph-7-Cl | 149–151 | $Et_2O$/Petr ether | | | 192 |
| Pyrrolidino, H; 5-Ph-7-Cl | 252–254 | | | | 315 |
| Pyrrolidino, H; 5-(2-$ClC_6H_4$)-7-Cl | 247–249 | | | | 315 |
| Pyrrolidino, H; 5-(2-$ClC_6H_4$)-7-$NO_2$ | 251–252 | | | | 315 |
| Pyrrolidino, H; 1-(2-$Me_2N$-ethyl)-5-(2-$ClC_6H_4$)-7-$NO_2$ | 209–210 | | | | 315 |
| Pyrrolidino, H; 1-Me-5-Ph-7-Cl | 184–186 | | | | 315 |
| Pyrrolidino, H; 1-Me-5-(2-$ClC_6H_4$)-7-Cl | 181–183 | | | | 315 |
| Pyrrolidino, H; 1-Me-5-(2-$ClC_6H_4$)-7-$NO_2$ | 184–185 | | | | 315 |
| Pyrrolidino, H; 1-(2-$Me_2N$-ethyl)-5-Ph-7-Cl | 139–141 | | | | 315 |
| Pyrrol-2-yl, H; 5-Ph-7-Cl | 250–252 | | 47 | | 304 |
| Pyrrol-2-yl, H; 1-Me-5-Ph-7-Cl | 223–226d | $CHCl_3$/Petr ether | | | 192 |
| Thien-2-yl, H; 5-Ph-7-Cl | 252–254 | | 40 | ir | 304 |
| Thiomorpholine, H; 1-Me-5-(2-$ClC_6H_4$)-7-Cl | Amorphous | | | ir | 315 |
| *S*-Oxide | 185–191d | | | | 315 |

*1,3-Dihydro-1,4-benzodiazepin-2(2H)-thiones*

| | | | | |
|---|---|---|---|---|
| ***Monosubstituted*** | | | | |
| 5-Ph | 256–257 | EtOH | 40 | 100, 255 |
| | 234–235 | $CH_2Cl_2$/EtOH | 34 | 416 |
| 5-(2-$ClC_6H_4$) | 228–229 | EtOH | 78 | 419 |
| ***Disubstituted*** | | | | |
| 1-Me-5-(4-$ClC_6H_4$) | 181–182 | MeOH | | 255 |
| 1-Me-5-(2-$F_3CC_6H_4$) | 133–136 | EtOH | | 255 |
| 1-($MeOOCCH_2$)-5-(2-$ClC_6H_4$) | 166–167 | MeOH | | 301 |
| ***5,6-Disubstituted*** | | | | |
| 5-Ph-6-Cl | 221–223 | $CHCl_3$/Hexane | | 478 |
| ***5,7-Disubstituted*** | | | | |
| 5-Me-7-Cl | 205–206 | EtOAc | | 417 |
| 5-Ph-7-Br | 255–256 | EtOH | 57 | 100, 255 |
| 5-Ph-7-Cl | 244–246 | EtOH | 53 | 100, 255 |
| | 238–239 | | 72 | 418 |
| 5-Ph-7-Et | 215–217 | MeCN | | 33b |
| 5-Ph-7-$F_3C$ | 223d | $CH_2Cl_2$/EtOH | 64 | 24 |
| | 228–229 | $CH_2Cl_2$/EtOH | | 416 |
| 5-Ph-7-I | 234–236 | THF/Hexane | | 33b |
| 5-Ph-7-Me | 260–261 | EtOH | | 255 |
| 5-Ph-7-$Me_2N$ | 250–253 | EtOH | | 255 |
| 5-Ph-7-Mes | 222–224 | EtOH | | 255 |
| | 210–212d | *i*-PrOH | 25 | 416 |
| 5-Ph-7-$NO_2$ | 209–214 | EtOH | | 255 |
| | 211–212d | $CH_2Cl_2$/EtOH | 57 | 416 |
| 5-(2-$ClC_6H_4$)-7-Cl | 251–253 | EtOH | | 255 |
| | 242–244 | | 78 | 418 |
| 5-(2-$ClC_6H_4$)-7-$NO_2$ | 219–221d | EtOH | 66 | 416 |
| 5-(2-$ClC_6H_4$)-7-Me | 245–247 | $CH_2Cl_2$/EtOH | | 255 |
| 5-(2-$FC_6H_4$)-7-Cl | 229–232 | EtOH | 73 | 255, 416 |
| 5-(2-$FC_6H_4$)-7-Et | 210–215d | EtOH | | 33b |

TABLE VII-1. —(contd.)

| Substituent | mp (°C) or; [bp (°C/torr)] | Solvent of Crystallization | Yield (%) | Spectra | Refs. |
|---|---|---|---|---|---|
| 5-(2-$FC_6H_4$)-7-I | 226–229 | MeCN | | | 33b |
| 5-(2,6-$F_2C_6H_3$)-7-Cl | 222–224 | EtOH/$H_2O$ | 74 | | 24, 416 |
| 5-(2-$MeOC_6H_4$)-7-Cl | 222–224 | EtOH | | | 255 |
| 5-(3-$MeOC_6H_4$)-7-Cl | 225–236d | MeOH | 80 | | 313 |
| 5-(2-Pyridyl)-7-Br | 245–246d | EtOH | | uv | 417 |
| ***Trisubstituted*** | | | | | |
| ***1,5,7-Trisubstituted*** | | | | | |
| 1-(2-$Et_2N$-Ethyl)-5-(2-$FC_6H_4$)-7-Cl | 98–100 | MeOH | | | 415 |
| Hydrochloride | 223–225 | MeOH/$Et_2O$ | | | 415 |
| 1-$F_3CCH_2$-5-Ph-7-Cl | 169–170 | $CH_2Cl_2$/ | | | 422a |
| 1-$F_3CCH_2$-5-(2-$FC_6H_4$)-7-Cl | 137–139 | $CH_2Cl_2$/Hexane | | | 422a |
| 1-Me-5-Ph-7-Cl | 162–164 | EtOH | | | 255 |
| 1-Me-5-Ph-7-$Me_2N$ | 185–187 | EtOH | | | 255 |
| 1-Me-5-(2-$ClC_6H_4$)-7-Cl | 160–163 | PhH/Hexane | | | 255 |
| 1-Me-5-(2-$ClC_6H_4$)-7-$NO_2$ | 204–206 | MeOH | | | 415 |
| 1-Me-5-(2-$FC_6H_4$)-7-$H_2N$ | 255–257d | $CH_2Cl_2$ | | | 301 |
| 1-Me-5-(2-$FC_6H_4$)-7-Cl | 144–146 | EtOH | | | 255c |
| 1-Me-5-(2-$FC_6H_4$)-7-MeNHCONH | 153–160d | EtOAc/$Et_2O$ | | | 301 |
| 1-Me-5-(2-$FC_6H_4$)-7-$NO_2$ | 220–226d | $CH_2Cl_2$/$Et_2O$ | | | 301 |
| 1-(2-CN-Ethyl)-5-(2-$FC_6H_4$)-7-Cl | 144–146 | $Et_2O$/Petr ether | | | 301 |
| 1-(2-$Me_2N$-Ethyl)-5-(2-$ClC_6H_4$)-7-Cl | 128–130 | EtOH | 67 | ir, pmr | 179 |
| 1-($MeOOCCH_2$)-5-Ph-7-Cl | 188–189 | $Et_2O$ | | | 301 |
| 1-($MeOOCCH_2$)-5-(2-Cl-$C_6H_4$)-7-Cl | 193–194 | MeOH | | | 301 |

| | | | | | |
|---|---|---|---|---|---|
| ***3,5,7-Trisubstituted*** | | | | | |
| 3-(Benzyloxy-COO)-5-Ph-7-Cl | 113d | | | | 425 |
| 3-Me-5-(2-$ClC_6H_4$)-7-$NO_2$ | 260 | $CH_2Cl_2$/EtOH | | | 301 |
| (+)-Enantiomer | 258–260 | $CH_2Cl_2$/Petr ether | | [α] | 79 |
| (−)-Enantiomer | 250 | $CH_2Cl_2$/EtOH/Hexane | | [α] | 301 |
| 3-Me-5-(2-$FC_6H_4$)-7-$NO_2$ | | | | | |
| (+)-Enantiomer | 250d | | | [α] | 79 |
| 3-(2-Me-1-Propyl)-5-Ph-7-Cl | 170–171 | EtOH | 43 | | 109 |
| ***Tetrasubstituted*** | | | | | |
| ***1,3,5,7-Tetrasubstituted*** | | | | | |
| 1,3-$Me_2$-5-(2-$FC_6H_4$)-7-Cl | 104–109 | $Et_2O$/Petr ether | 58 | | 343 |
| 1-Me-3-(2-CN-Ethyl)-5-Ph-7-Cl | 186–188 | $Et_2O$/Petr ether | | | 301 |
| ***1,5,6,7-Tetrasubstituted*** | | | | | |
| 1-Me-5-(2-$FC_6H_4$)-6-Cl-7-$H_2N$ | 250d | EtOAc/$Et_2O$ | | | 301 |
| 1-Me-5-(2-$FC_6H_4$)-6-Cl-7-MeNHCONH | 165–170d | EtOAc/$Et_2O$ | | | 301 |

*1,5-Dihydro-1,4-benzodiazepin-2(2H)-ones*

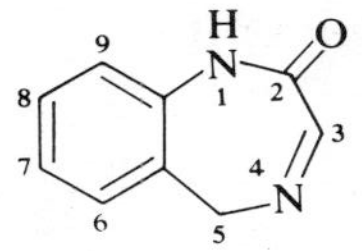

| | | | | | |
|---|---|---|---|---|---|
| ***Disubstituted*** | | | | | |
| ***5,7-Disubstituted*** | | | | | |
| 5-Ph-7-Ac | 200–201 | $CH_2Cl_2$/Petr ether | | | 32 |
| 5-Ph-7-Cl | 202–210 | $CH_2Cl_2$/Hexane | 35 | | 170 |
| | 215–217 | | | ir, pmr, uv | 455 |
| 5-(2-$ClC_6H_4$)-7-Cl | 248–250 | | | | 455 |

TABLE VII-1. —(contd.)

| Substituent | mp (°C) or; [bp (°C/torr)] | Solvent of Crystallization | Yield (%) | Spectra | Refs. |
|---|---|---|---|---|---|
| ***Trisubstituted*** | | | | | |
| ***1,5,7-Trisubstituted*** | | | | | |
| 1-Me-5-(2-$FC_6H_4$)-7-I | 191–195 | $CH_2Cl_2$/Hexane | | | 141, 175 |
| ***3,5,7-Trisubstituted*** | | | | | |
| 3-MeNH-5-Ph-7-Cl | 197–199 | EtOAc/*i*-$Pr_2O$ | 31 | ir, pmr | 260 |
| 3-MeO-5-Ph-7-Cl | 249–251d | MeOH | | ir, pmr | 260 |
| 3-MeO-5-(2-$ClC_6H_4$)-7-Cl | 282–284 | EtOH | 75 | pmr | 457 |
| ***5,5,7-Trisubstituted*** | | | | | |
| 5-MeO-5-Ph-7-Cl | 188 | *i*-$Pr_2O$ | 68 | ir, pmr | 260 |
| 5-MeO-5-Ph-7-$NO_2$ | 205d | $CH_2Cl_2$/$Et_2O$ | 92 | ir | 260 |
| ***Tetrasubstituted*** | | | | | |
| ***1,3,5,7-Tetrasubstituted*** | | | | | |
| 1-(2-$Et_2N$-Ethyl)-3-MeO-5-(2-$ClC_6H_4$)-7-Cl | 148–150 | Cyclohexane | 46 | pmr | 457 |
| 1-Me-3-MeNH-5-Ph-7-$NO_2$ | 211–213 | EtOAc | 84 | ir, pmr | 260 |
| 1-Me-3-$Me_2N$-5-Ph-7-Cl | 193–194 | $CH_2Cl_2$/Hexane | 88 | ir, pmr | 149 |
| 1-Me-3-MeO-5-(2-$ClC_6H_4$)-7-Cl | 179–180 | Cyclohexane | | ir, pmr | 457 |
| ***1,5,5,7-Tetrasubstituted*** | | | | | |
| 1-Me-5-MeO-5-Ph-7-Cl | 146–147 | $Et_2O$ | 61 | ir | 260, 456 |
| 1-Me-5-MeO-5-Ph-7-$NO_2$ | 163–164 | $CH_2Cl_2$/*i*-$Pr_2O$ | 74 | ir, pmr | 260, 456 |

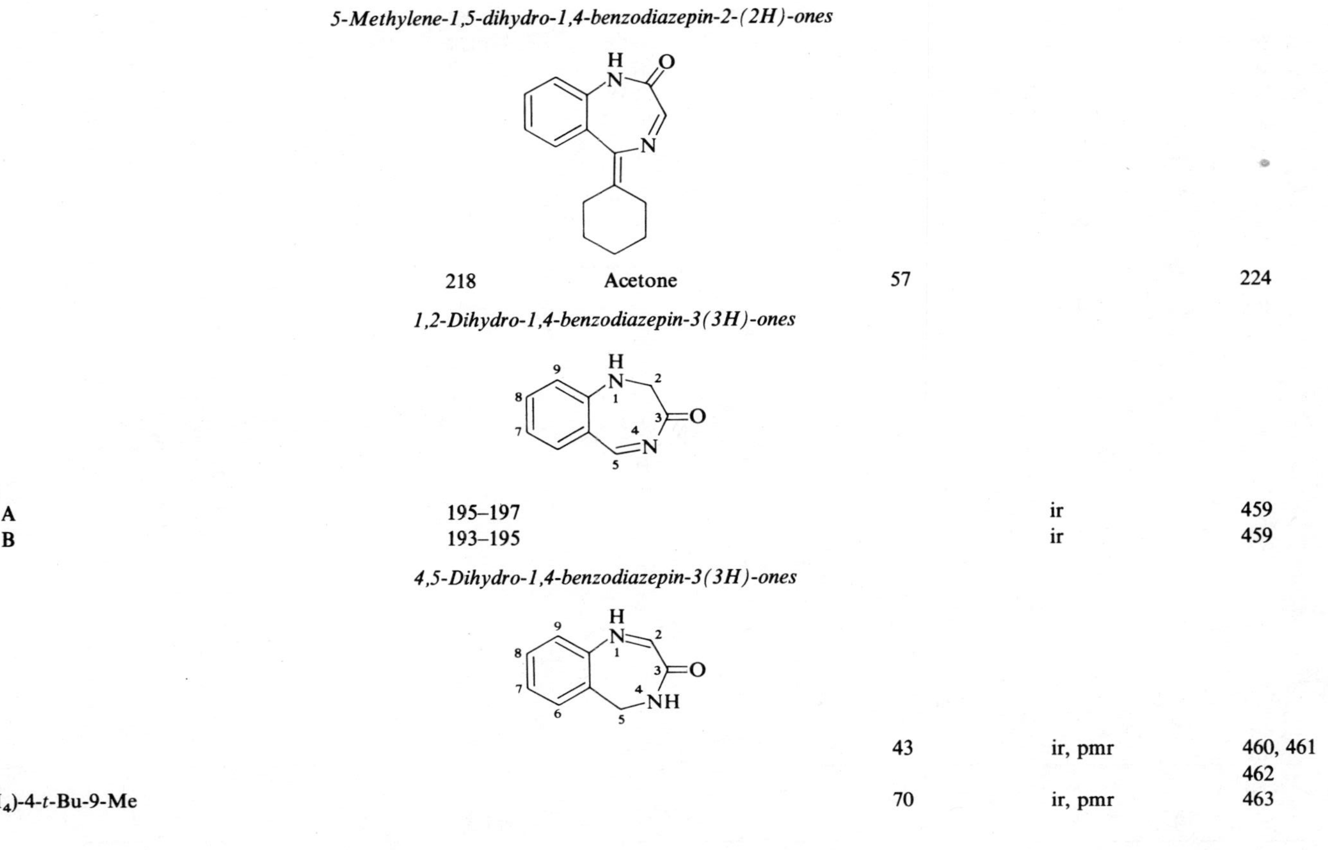

| | | | | | |
|---|---|---|---|---|---|
| *5-Methylene-1,5-dihydro-1,4-benzodiazepin-2-(2H)-ones* | | | | | |
| None | 218 | Acetone | 57 | | 224 |
| *1,2-Dihydro-1,4-benzodiazepin-3(3H)-ones* | | | | | |
| 2,2-$Me_2$-5-Ph | | | | | |
| Conformer A | 195–197 | | | ir | 459 |
| Conformer B | 193–195 | | | ir | 459 |
| *4,5-Dihydro-1,4-benzodiazepin-3(3H)-ones* | | | | | |
| None | | | 43 | ir, pmr | 460, 461 |
| 2-Et-5-Ph | | | | | 462 |
| 2-(4-$NO_2C_6H_4$)-4-*t*-Bu-9-Me | | | 70 | ir, pmr | 463 |

TABLE VII-1. —(contd.)

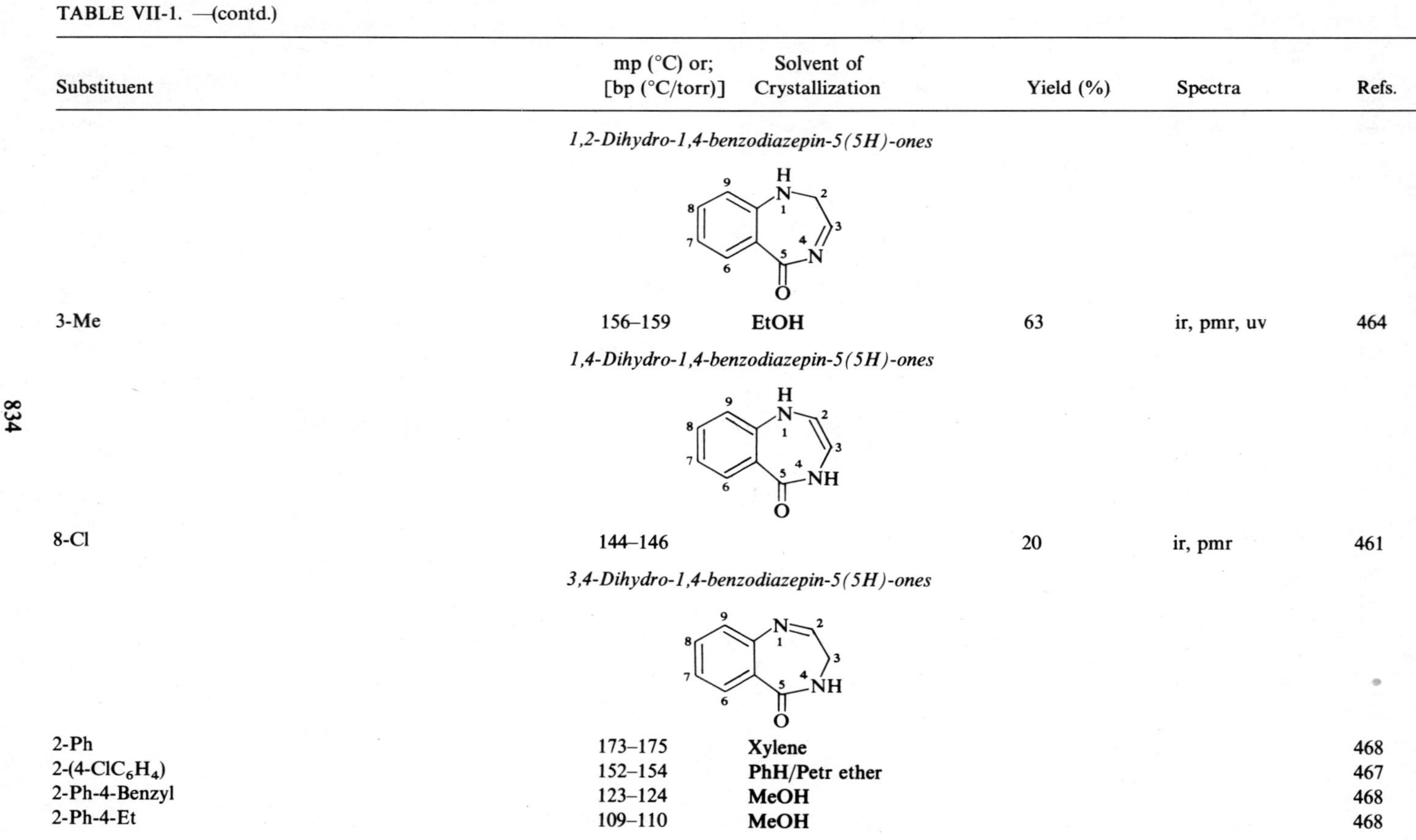

| Substituent | mp (°C) or; [bp (°C/torr)] | Solvent of Crystallization | Yield (%) | Spectra | Refs. |
|---|---|---|---|---|---|
| | *1,2-Dihydro-1,4-benzodiazepin-5(5H)-ones* | | | | |
| 3-Me | 156–159 | EtOH | 63 | ir, pmr, uv | 464 |
| | *1,4-Dihydro-1,4-benzodiazepin-5(5H)-ones* | | | | |
| 8-Cl | 144–146 | | 20 | ir, pmr | 461 |
| | *3,4-Dihydro-1,4-benzodiazepin-5(5H)-ones* | | | | |
| 2-Ph | 173–175 | Xylene | | | 468 |
| 2-(4-$ClC_6H_4$) | 152–154 | PhH/Petr ether | | | 467 |
| 2-Ph-4-Benzyl | 123–124 | MeOH | | | 468 |
| 2-Ph-4-Et | 109–110 | MeOH | | | 468 |

| | | | | | |
|---|---|---|---|---|---|
| 2-Ph-4-Me | 133–135 | MeOH | 90 | | 468 |
| 2-Ph-4-*i*-Pr | 138–139 | MeOH | | | 468 |
| 2-(2-Ph-Ethyl)NH-4-Me | 150–151 | EtOAc | | | 471 |
| 2-MeNH-7-Cl | 246–249d | MeOH | | | 469 |
| 2-Me(NO)N-7-Cl | 193–195 | $Et_2O$ | | | 134 |
| 2-MeOOCC(NOH)-7-Cl | 189–191 | THF/Hexane | | | 134 |
| 2-Ph-7-Cl | 160–162 | Xylene | | ir | 465 |
| 2-(4-$ClC_6H_4$)-4-Me | 186–187 | MeOH | | | 468 |
| 2-(4-$MeC_6H_4$)-4-Me | 172–174 | MeOH | | | 468 |
| 2-$H_2N$-3-Ph-4-Me | 282–286 | EtOH/$Et_2O$ | | | 471 |
| 2-(2-$Me_2N$-Ethyl)NH-3-Ph-4-Me Dihydrochloride·0.5$H_2O$ | 246–250d | | | | 471 |
| 2-MeS-3-Ph-4-Me | 130–132 | | | | 471 |
| 2-(2-Ph-Ethyl)NH-3-Ph-4-Me | 173–175 | EtOAc/$Et_2O$ | | | 471 |
| 2-Benzylamino-4-Me-7-Cl | 158–160 | EtOAc | | | 471 |
| 2-(2,6-$Cl_2C_6H_3$)NH-4-Me-7-Cl | 247–250 | *i*-PrOH/MeOH | | | 471 |
| 2-[2-(3,4-[Benzyloxy]$_2$-$C_6H_3$)ethyl]amino-4-Me-7-Cl Hydrochloride | 158–162 | EtOH/$Et_2O$ | | | 471 |
| 2-[2-(3,4-[HO]$_2C_6H_3$)-Ethyl]amino-4-Me-7-Cl Hydrochloride·0.5MeOH | 179–180d | EtOH/$Et_2O$ | | | 471 |
| 2-[2-(Indol-3-yl)ethyl]amino-4-Me-7-Cl | 220–222 | MeCOEt/$Et_2O$ | | | 471 |
| 2-[2-(3,4-[MeO]$_2C_6H_3$)Ethyl]amino-4-Me-7-Cl | 177–179 | EtOH/$Et_2O$ | | | 471 |
| 2-MeNH-4-(Morpholino)2-PO-7-Cl | 236–238 | MeCN | 80 | ir, ms, pmr, uv | 344 |
| 2-(3-$Me_2N$-Propyl)NH-4-Me-7-Cl Dihydrochloride | 261–264d | MeOH | | | 471 |
| 2-MeS-4-Me-7-Cl | 124–127 | MeOH/$H_2O$ | | | 471 |
| 2-Ph-3,4-$Me_2$ | 106–108 | MeOH | | | 468 |
| 2-Ph-4-Me-7-Br | 144–146 | MeOH | | | 468 |
| 2-Ph-4-Me-7-Cl | 156–157 | MeOH | | | 468 |
| 2-Ph-4,7-$Me_2$ | 165–166 | MeOH | | | 468 |
| 2-Ph-4-Me-8-Br | 145–146 | MeOH | | | 468 |
| 2-Ph-4-Me-8-Cl | 133–134 | MeOH | | | 468 |

TABLE VII-1. —(contd.)

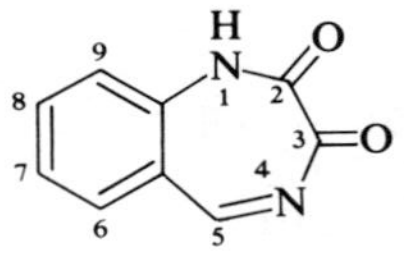

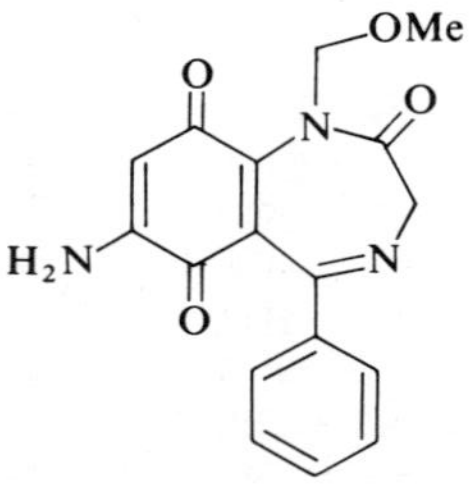

| Substituent | mp (°C) or; [bp (°C/torr)] | Solvent of Crystallization | Yield (%) | Spectra | Refs. |
|---|---|---|---|---|---|
| 2-(2-Ph-Ethyl)NH-4-Me-7-Cl | 196–198 | EtOH/$Et_2O$ | | | 471 |
| 2-$H_2$N-3-Ph- | | | | | |
| 2-Ph-4-Me-7,8-$(MeO)_2$ | 157–159 | MeOH | | | 468 |
| *1H-1,4-Benzodiazepin-2,3(2H,3H)-diones* | | | | | |
| 5-Ph-7-Cl | 248–251d | $CH_2Cl_2$/Petr ether | 57 | | 157, 158 |
| 5-(2-$ClC_6H_4$)-7-Cl | 258d | MeCN | 43 | ir, pmr | 243 |
| 1-Me-5-Ph-7-Cl | 169–174 | $CH_2Cl_2$/Hexane | 45 | | 157, 158 |
| 1-Me-5-(2-$ClC_6H_4$)-7-Cl | 204–206 | EtOAc | 21 | ir, pmr | 243 |
| *1,4-Benzodiazepinetriones* | | | | | |
| | 219–220d | $CH_2Cl_2$/MeOH/$Et_2O$ | 17 | pmr, uv | 245a |

# 11. REFERENCES


1. (a) A. Stempel, I. Douvan, and L. H. Sternbach, *J. Org. Chem.*, **33**, 2963 (1968). (b) A. Stempel, unpublished data, Hoffmann-La Roche, Nutley, NJ.
2. (a) L. H. Sternbach, R. Ian Fryer, W. Metlesics, E. Reeder, G. Sach, G. Saucy, and A. Stempel, *J. Org. Chem.*, **27**, 3788 (1962). (b) L. H. Sternbach, unpublished data, Hoffmann-La Roche, Nutley, NJ.
3. R. Ian Fryer, R. A. Schmidt, and L. H. Sternbach, *J. Pharm. Sci.*, **53**, 264 (1964).
4. G. Saucy and L. H. Sternbach, *Helv. Chim. Acta*, **45**, 2226 (1962).
5. L. H. Sternbach, G. Saucy, F. A. Smith, M. Mueller, and J. Lee, *Helv. Chim. Acta*, **46**, 1720 (1963).
6. L. H. Sternbach and G. Saucy, U.S. Patent 3,341,592, September 1967.
7. O. Keller, N. Steiger, and L. H. Sternbach, U.S. Patent 3,121,103, February 1964.
8. J. V. Earley, R. Ian Fryer, and R. Y. Ning, *J. Pharm. Sci.*, **68**, 845 (1979).
9. R. V. Davis and R. Ian Fryer, U.S. Patent 4,083,948, April 1978.
10. S. A. Andronati, A. V. Bogat-skii, G. N. Gordiichuk, Z. I. Zhilina, and L. M. Yagupol'skii, *Chem. Heterocycl. Comp.*, **11**, 230 (1975).
11. R. Y. Ning and L. H. Sternbach: (a) U.S. Patent 3,686,308, August 1972, (b) U.S. Patent 3,823,166, July 1974.
12. A. Focella and A. I. Rachlin, U.S. Patent 3,335,181, August 1967.
13. (a) R. Kalish, T. V. Steppe, and A. Walser, *J. Med. Chem.*, **18**, 222 (1975), (b) R. Kalish, unpublished data, Hoffmann-La Roche, Nutley, NJ.
14. N. W. Gilman and L. H. Sternbach, *J. Heterocycl. Chem.*, **8**, 297 (1971).
15. G. Saucy, F. A. Smith, and L. H. Sternbach, U.S. Patent 3,329,701, July 1967.
16. L. H. Sternbach and A. Stempel, U.S. Patent 3,336,295, August 1967.
17. W. J. Houlihan, U.S. Patent 3,609,146, September 1971.
18. French Patent 1,391,752, February 1965 (Etablissements Clin-Byla, France).
19. A. Bauer, K.-H. Weber, P. Danneberg, and K. Minck, U.S. Patent 3,786,051, January 1974.
20. M. Zinic, D. Kolbah, N. Blasevic, F. Kajfez, and V. Sunjic, *J. Heterocycl. Chem.*, **14**, 1225 (1977).
21. Yu. S. Shabarov, T. P. Surikova, and S. S. Mochalov, *Chem. Heterocycl. Compd.*, **10**, 498 (1974). [*Khim. Geterotsikl. Soedin.*, 572 (1974)].
22. Belg. Patent 619,101, December 1962 (Hoffmann-La Roche & Co. AG, Switzerland).
23. T. L. Lemke, J. B. Hester, and A. D. Rudzik, *J. Pharm. Sci.*, **61**, 275 (1972).
24. (a) J. B. Hester, Jr., U.S. Patent 3,701,782, October 1972. (b) J. B. Hester and A. R. Hanze, U.S. Patents 3,933,794, January 1976; 3,917,627, November 1975; 3,927,016, December 1975.
25. E. Wolf, H. Kohl, and G. Haertfelder, U.S. Patent 3,849,559, November 1974.
26. G. F. Field and L. H. Sternbach: (a) Swiss Patent 561,706, May 1975, (b) Swiss Patent 562,222, April 1975.
27. Ger. Offen. 2,223,648, November 1972 (Hoffmann-La Roche & Co. AG, Switzerland).
28. G. A. Archer, R. I. Kalish, R. Y. Ning, B. C. Sluboski, A. Stempel, T. V. Steppe, and L. H. Sternbach, *J. Med. Chem.*, **20**, 1312 (1977).
29. A. A. Fatmi, N. A. Vaidya, W. B. Iturrian, and C. DeWitt Blanton, Jr., *J. Med. Chem.*, **27**, 772 (1984).
30. G. M. Clarke, J. B. Lee, F. J. Swinbourne, and B. Williamson, *J. Chem. Res. (S)*, 398 (1980).
31. A. Ziggiotti, G. Riva, and F. Mauri, U.S. Patent 3,852,270, December 1974.
32. Belg. Patent 777,098, June 1972 (Hoffmann-La Roche & Co. AG, Switzerland).
33. (a) R. Y. Ning and L. H. Sternbach, U.S. Patent 3,682,892, August 1972. (b) R. Y. Ning, unpublished data, Hoffmann-La Roche, Nutley, NJ.
34. S. C. Bell and S. J. Childress, U.S. Patent 3,714,145, January 1973.
35. S. C. Bell, T. S. Sulkowski, C, Gochman, and S. J. Childress, *J. Org. Chem.*, **27**, 562 (1962).
36. T. Sugasawa, M. Adachi, T. Toyoda, and K. Sasakura, *J. Heterocycl. Chem.*, **16**, 445 (1979).

37. H. Moriyama, H. Yamamoto, S. Inaba, H. Nagata, T. Tamaki, and T. Hirohashi, U.S. Patent 3,524,848, August 1970.
38. (a) M. E. Derieg, R. M. Schweininger, and R. Ian Fryer, *J. Org. Chem.*, **34**, 179 (1969). (b) M. E. Derieg, unpublished data, Hoffmann-La Roche, Nutley, NJ.
39. A. Walser, A. Szente, and J. Hellerbach, *J. Org. Chem.*, **38**, 449 (1973).
40. V. Sunjic, F. Kajfez, I. Stromar, N. Blasevic, and D. Kolbah, *J. Heterocycl. Chem.*, **10**, 591 (1973).
41. Neth. Patent 7,207,637, December 1972 (Sumitomo Chemical Co., Ltd., Japan).
42. R. Ian Fryer, W. Leimgruber, and E. J. Trybulski, *J. Med. Chem.*, **25**, 1050 (1982).
43. N. W. Gilman, P. Levitan, and L. H. Sternbach, *J. Org. Chem.*, **38**, 373 (1973).
44. Belg. Patent 778,191, May 1972 (Grindstedvaerket, Denmark).
45. Neth. Patent 7,117,351, June 1972 (Sumitomo Chemical Co., Ltd., Japan).
46. J. Ackrell, E. Galeazzi, J. M. Muchowski, and L. Tokes, *Can. J. Chem.*, **57**, 2696 (1979).
47. Ger. Offen. 2,504,937, August 1975 (Specta International B.V., Netherlands).
48. N. Blasevic and F. Kajfez, *J. Heterocycl. Chem.*, **7**, 1173 (1970).
49. Z. Vejdelek, M. Rajsner, E. Svatek, J. Holubek, and M. Protiva, *Collect. Czech. Chem. Commun.*, **44**, 3604 (1979).
50. G. O. Chase, unpublished data, Hoffmann-La Roche, Nutley, NJ.
51. G. M. Clarke, J. B. Lee, F. J. Swinbourne, and B. Williamson, *J. Chem. Res. (S)*, 399 (1980); 400 (1980).
52. G. O. Chase, U.S. Patent 3,886,141, May 1975.
53. G. O. Chase, U.S. Patent 3,996,209, December 1976.
54. W. Schlesinger, U.S. Patent 4,155,904, May 1979.
55. Belg. Patent 871,280, February 1979 (KRKA, Yugoslavia).
56. S. C. Bell, R. J. McCaully, and S. J. Childress, *J. Heterocycl. Chem.*, **4**, 647 (1967).
57. S. C. Bell, U.S. Patent 3,401,200, September 1968.
58. S. C. Bell, U.S. Patent 3,458,501, July 1969.
59. R. J. McCaully: (a) U.S. Patent 3,763,171, October 1973, (b) U.S. Patent 3,896,170, July 1975.
60. S. C. Bell, U.S. Patent 3,445,458, May 1969.
61. Neth. Patent 6,500,446, July 1965 (Delmar Chemicals Ltd., Canada).
62. F. H. McMillan and I. Pattison, U.S. Patent 3,304,313, February 1967.
63. S. P. Hsi, *J. Labelled Compds.*, **10**, 389 (1974).
64. C. Podesva and K. Vagi, U.S. Patent 3,842,094, October 1974.
65. Z. Vejdelek, M. Rajsner, A. Dlabac, M. Ryska, J. Holubek, E. Svatek, and M. Protiva, *Collect. Czech. Chem. Commun.*, **45**, 3593 (1980).
66. French Patent 2,253,501, July 1975 (Synthelabo S.A. France).
67. (a) E. E. Garcia, L. E. Benjamin, R. Ian Fryer, and L. H. Sternbach, *J. Med. Chem.*, **15**, 986 (1972). (b) E. E. Garcia, unpublished data, Hoffmann-La Roche, Nutley, NJ.
68. R. I. Fryer and L. H. Sternbach, U.S. Patent 3,691,157, September 1972.
69. M. Steinman, J. G. Topliss, R. Alekel, Y.-S. Wong, and E. E. York, *J. Med. Chem.*, **16**, 1354 (1973).
70. Belg. Patent 772,825, September 1971 (Lab. Pharmedical S.A., Luxembourg).
71. Belg. Patent 794,536, July 1973 (Fujisawa Pharmaceutical Co., Japan).
72. (a) J. Hellerbach and A. Walser, U.S. Patent 3,784,542, January 1974. (b) J. Hellerbach and A. Walser, U.S. Patent 3,886,214, May 1975. (c) J. Hellerbach, unpublished data, Hoffmann-La Roche & Co. AG, Basle, Switzerland.
73. T. Kishimoto, M. Matsuo, and I. Ueda, *Chem. Pharm. Bull.*, **30**, 1477 (1982).
74. I. Mikami, Y. Hara, and Y. Usui, *J. Takeda Res. Lab.*, **31**, 11 (1972).
75. A. Stempel and F. W. Landgraf, *J. Org. Chem.*, **27**, 4675 (1962).
76. A. Stempel, U.S. Patent 3,202,699, August 1965.
77. G. F. Field and L. H. Sternbach, Swiss Patent 561,705, May 1975.
78. Zs. Tagyey, G. Maksay, and L. Otvos, *J. Labelled Compd.*, **16**, 377 (1978).
79. A. Szente, U.S. Patent 4,031,078, June 1977.

80. Q. Branca, A. E. Fischli, and A. Szente: (a) U.S. Patent 4,361,511, November 1982, (b) European Patent A1 0,033,973, August 1981.
81. F. Kajfez, N. Blasevic, and V. Sunjic, U.S. Patent 3,998,811, December 1976.
82. H. Demarne and A. Hallot, European Patent A1 0,022,710, January 1981.
83. H. H. Kaegi and W. Burger, *J. Labelled Compd.*, **19**, 975 (1982).
84. J. Hellerbach, A. Walser, H. Bretschneider, and W. Rudolph, U.S. Patent 3,772,271, November 1973.
85. K. Ishizumi, K. Mori, S. Inaba, and H. Yamamoto, U.S. Patent 3,991,048.
86. H. Yamamoto, S. Inaba, T. Hirohashi, M. Yamamoto, K. Ishizumi, M. Akatsu, I. Maruyama, Y. Kume, K. Mori, and T. Izumi: (a) U.S. Patent 3,812,101, May 1974, (b) U.S. Patent 3,778,433, December 1973, (c) U.S. Patent 3,867,372, February 1975, (d) U.S. Patent 4,010,154, March 1977, (e) U.S. Patent 3,989,829, November 1976.
87. Ger. Offen. 2,166,472, February 1974 (Sumitomo Chemical Co., Ltd., Japan).
88. Belg. Patent 764,034, March 1971 (Sumitomo Chemical Co., Ltd., Japan).
89. M. Akatsu, Y. Kume, T. Hirohashi, K. Ishizumi, M. Yamamoto, I. Maruyama, K. Mori, T. Izumi, S. Inaba, and H. Yamamoto, U.S. Patent 3,910,889, October 1975.
90. G. N. Walker, *J. Org. Chem.*, **27**, 1929 (1962).
91. (a) E. Reeder and L. H. Sternbach, U.S. Patent 3,371,085, February 1968, (b) E. Reeder and L. H. Sternbach, U.S. Patent 3,402,171, September 1968, (c) E. Reeder and L. H. Sternbach, unpublished data, Hoffmann-La Roche, Nutley, NJ.
92. R. Littell and D. S. Allen, Jr., *J. Med. Chem.*, **8**, 892 (1965).
93. O. Keller, N. Steiger, and L. H. Sternbach, U.S. Patent 3,121,075, February 1964.
94. (a) L. Berger and L. H. Sternbach, U.S. Patent 3,179,656, April 1965, (b) L. Berger and L. H. Sternbach, U.S. Patent 3,268,586, August 1966, (c) L. Berger, unpublished data, Hoffmann-La Roche, Nutley, NJ.
95. Belg. Patent 629,352, October 1963 (Hoffmann-La Roche & Co AG, Switzerland).
96. R. I. Fryer, L. H. Sternbach, and J. V. Earley, U.S. Patent 3,520,877, July 1970.
97. A. F. Fentiman, Jr. and R. L. Foltz, *J. Labelled Compd.*, **13**, 579 (1977).
98. R. B. Moffett, U.S. Patent 3,609,145, September 1971.
99. G. F. Field and L. H. Sternbach, Swiss Patent 561,703, May 1975.
100. A. V. Bogatskii, S. A. Andronati, V. P. Gul'tyai, Yu. I. Vikhylaev, A. F. Galatin, Z. I. Zhilina, and T. A. Klygul', *J. Gen. Chem., USSR*, **41**, 1364 (1971).
101. R. J. McCaully, U.S. Patent 3,678,043, July 1972.
102. E. Manghisi and A. Salimbeni, *Bull. Chim. Farm.*, **113**, 642 (1974).
103. F. Camps, J. Cartells, and J. Pi, *Ann. Quim.*, **70**, 848 (1974).
104. J. Schmitt, P. Comoy, M. Suquet, G. Callet, J. Le Meur, T. Clim, M. Brunaud, J. Mercier, J. Salle, and G. Siou, *Chim. Ther.*, 239 (1969).
105. J. Schmitt, U.S. Patent 3,966,793, June 1976.
106. Neth. Patent 7,500,364, April 1975 (CM Industries S.A., France).
107. Neth. Patent 6,507,637, December 1965 (Etablissement Clin-Byla S.A., France).
108. A. Walser and J. Hellerbach, unpublished data, Hoffmann-La Roche & Co. AG, Basel, Switzerland.
109. K. Meguro, H. Tawada, H. Miyano, Y. Sato, and Y. Kuwada, *Chem. Pharm. Bull.*, **21**, 2382 (1973).
110. F. Krausz, U.S. Patent 4,051,127, September 1977.
111. H. Boemches and H. Meyer, Ger. Offen. 2,252,378, May 1973.
112. F. Kajfez and V. Sunjic, U.S. Patent 4,010,153, May 1977.
113. Neth. Patent 7,117,350, June 1972 (Sumitomo Chemical Co., Ltd., Japan).
114. Q. Branca, A. E. Fischli, and A. Szente, European Patent A1 0,033,974, August 1981.
115. A. V. Bogat-skii, S. A. Andronati, Z. I. Zhilina, S. D. Isaev, and A. G. Yurchenko, *Chem. Heterocycl. Compd.*, **13**, 694 (1977).
116. S. Inaba, M. Akatsu, T. Hirohashi, and H. Yamamoto, *Chem. Pharm. Bull.*, **24**, 1076 (1976).
117. H. Yamamoto, S. Inaba, T. Hirohashi, and K. Ishizumi, *Chem. Ber.*, **101**, 4245 (1968).

118. S. Inaba, K. Ishizumi, T. Okamoto, and H. Yamamoto, *Chem. Pharm. Bull.*, **23**, 3279 (1975).
119. K. Ishizumi, S. Inaba, and H. Yamamoto, *Chem. Pharm. Bull.*, **20**, 592 (1972).
120. S. Inaba, K. Ishizumi, K. Mori, and H. Yamamoto, *Chem. Pharm. Bull.*, **19**, 722 (1971).
121. S. Inaba, K. Ishizumi, and H. Yamamoto, *Chem. Pharm. Bull.*, **19**, 263 (1971).
122. S. Inaba, T. Hirohashi, and H. Yamamoto, *Chem. Pharm. Bull.*, **17**, 1263 (1969).
123. G. N. Walker, A. R. Engle, and R. J. Kempton, *J. Org. Chem.*, **37**, 3755 (1972).
124. H. Yamamoto, S. Inaba, T. Hirohashi, M. Akatsu, I. Maruyama, and T. Izumi, U.S. Patent 3,781,299, December 1973.
125. H. Yamamoto, S. Inaba, T. Okamoto, T. Hirohashi, K. Ishizumi, M. Yamamoto, I. Maruyama, K. Mori, and T. Kobayashi: (a) U.S. Patent 3,632,574, January 1972, (b) U.S. Patent 3,362,573, January 1972, (c) U.S. Patent 3,770,767, November 1973, (d) U.S. Patent 3,723,464, March 1973, (e) U.S. Patent 3,652,551, March 1972, (f) U.S. Patent 3,925,364, December 1975, (g) U.S. Patent 3,828,027, August 1974, (h) U.S. Patent 3,922,264, November 1975, (i) U.S. Patent 3,631,029, December 1971.
126. H. Yamamoto, S. Inaba, T. Hirohashi, K. Ishizumi, I. Maruyama, and K. Mori, U.S. Patent 3,632,805, January 1972.
127. Ger. Offen. 2,166,473, February 1974 (Sumitomo Chemical Co., Ltd., Japan).
128. East Ger. Patent 91,045 July 1972 (Sumitomo Chemical Co., Ltd., Japan).
129. H. Yamamoto, S. Inaba, T. Hirohashi, M. Akatsu, I. Maruyama, and T. Izumi, U.S. Patent 3,634,402, January 1972.
130. H. Yamamoto, S. Inaba, T. Hirohashi, M. Akatsu, I. Maruyama, and T. Izumi, Ger. Offen. 1,966,206, January 1972.
131. K. Ishizumi, K. Mori, S. Inaba, and H. Yamamoto, *Chem. Pharm. Bull.*, **21**, 1027 (1973).
132. H. Yamamoto, S. Inaba, T. Okamoto, T. Hirohashi, K. Ishizumi, M. Yamamoto, I. Maruyama, K. Mori, T. Kobayashi, and T. Izumi, U.S. Patent 3,702,323, November 1972.
133. K. Ishizumi, K. Mori, S. Inaba, and H. Yamamoto, U.S. Patent 3,812,102, May 1974.
134. A. Walser, unpublished data, Hoffmann-La Roche, Nutley, NJ.
135. L. H. Sternbach and E. Reeder, *J. Org. Chem.*, **26**, 4936 (1961).
136. L. H. Sternbach, R. Ian Fryer, O. Keller, W. Metlesics, G. Sach, and N. Steiger, *J. Med. Chem.*, **6**, 261 (1963).
137. Neth. Patent 7,204,248, October 1972 (Hoffmann-La Roche & Co. AG, Switzerland).
138. G. F. Field and L. H. Sternbach, U.S. Patent 3,594,365, July 1971.
139. G. F. Field and L. H. Sternbach, U.S. Patent 3,625,959, December 1971.
140. Neth. Patent 6,512,614, March 1966 (Hoffmann-La Roche & Co. AG, Switzerland).
141. G. F. Field and L. H. Sternbach, Swiss Patent 561,189, April 1975.
142. W. J. Middleton, U.S. Patent 4,182,760, January 1980.
143. A. Stempel, E. Reeder, and L. H. Sternbach, *J. Org. Chem.*, **30**, 4267 (1965).
144. W. J. Middleton, U.S. Patent 4,141,895, February 1979.
145. J. Hellerbach, A. Szente, and A. Walser, U.S. Patent 3,657,223, April 1972.
146. G. F. Field and L. H. Sternbach, Swiss Patent 562,221, May 1975.
147. J. Hellerbach, A. Szente, and A. Walser, U.S. Patent 3,671,518, June 1972.
148. A. Stempel, I. Douvan, E. Reeder, and L. H. Sternbach, *J. Org. Chem.*, **32**, 2417 (1967).
149. H. Natsugari, K. Meguro, and Y. Kuwada, *Chem. Pharm. Bull.*, **27**, 2084 (1979).
150. (a) Neth. Patent 7,110,495, February 1972 (Richter Gedeon Vegyeszeti Gyar RT, Hungary). (b) G. F. Field and L. H. Sternbach, Swiss Patent 561,704, March 1975. (c) R. Jaunin and J. Hellerbach, Ger. Offen. 2,150,075, April 1972. (d) R. Jaunin and J. Hellerbach, Swiss Patent 566,324, July 1975.
151. R. I. Fryer, G. A. Archer, B. Brust, W. Zally, and L. H. Sternbach, *J. Org. Chem.*, **30**, 1308 (1965).
152. (a) R. I. Fryer, J. V. Earley, and L. H. Sternbach, *J. Org. Chem.*, **30**, 521 (1965). (b) R. I. Fryer and L. H. Sternbach, U.S. Patent 3,322,753, May 1967. (c) R. I. Fryer and J. V. Earley, unpublished data, Hoffmann-La Roche, Nutley, NJ.
153. S. J. Childress, S. C. Bell, and T. S. Sulkowski, U.S. Patent 3,321,522, May 1967.
154. R. I. Fryer and L. H. Sternbach, U.S. Patent 3,567,710, March 1971.

155. S. Inaba, T. Okamoto, T. Hirohashi, K. Ishizumi, M. Yamamoto, I. Murayama, K. Mori, T. Kobayashi, and H. Yamamoto, U.S. Patent 3,666,643, May 1972.
156. K. Ishizumi, K. Mori, S. Inaba, and, H. Yamamoto, *Chem. Pharm. Bull.*, **23**, 2169 (1975).
157. (a) A. M. Felix, J. V. Earley, R. I. Fryer, and L. H. Sternbach, *J. Heterocycl. Chem.*, **5**, 731 (1968). (b) A. M. Felix, unpublished data, Hoffmann-La Roche, Nutley, NJ.
158. A. M. Felix, R. I. Fryer, and L. H. Sternbach, U.S. Patent 3,546,212, December 1970.
159. Neth. Patent 7,508,659, February 1976 (Sumitomo Chemical Co., Ltd., Japan).
160. P. A. Wehrli, R. I. Fryer, and L. H. Sternbach, U.S. Patent 3,553,206, January 1971.
161. K.-H. Wuensch, H. Dettmann, and S. Schoenberg, *Chem. Ber.*, **102**, 3891 (1969).
162. M. E. Derieg, R. I. Fryer, and L. H. Sternbach, U.S. Patent 3,651,046, March 1972.
163. I. Nakatsuka, K. Kawahara, T. Kamada, F. Shono, and A. Yoshitake, *J. Labelled Compd.*, **13**, 453 (1977).
164. K. Meguro and Y. Kuwada, *Yakugaku Zasshi*, **93**, 1263 (1973).
165. K. Meguro, H. Tawada, Y. Kuwada, and T. Masuda, U.S. Patent 3,692,772, September 1972.
166. H. Liepmann, W. Milkowski, and H. Zeugner, *Eur. J. Med. Chem. Chim. Ther.*, **11**, 501 (1976).
167. W. Milkowski, R. Hueschens, and H. Kuchenbecker, *J. Heterocycl. Chem.*, **17**, 373 (1980).
168. W. Milkowski, S. Funke, R. Hueschens, H. G. Liepmann, W. Stuehmer, and H. Zeugner, Ger. Offen. 2,221,536, November 1973.
169. L. H. Sternbach and R. Y. Ning, U.S. Patent 3,873,525, March 1975.
170. R. I. Fryer, and D. Winter, and L. H. Sternbach, *J. Heterocycl. Chem.*, **4**, 355 (1967).
171. R. I. Fryer and L. H. Sternbach, U.S. Patent 3,706,734, December 1972.
172. R. I. Fryer and L. H. Sternbach, U.S. Patent 3,625,957, December 1971.
173. G. F. Field and L. H. Sternbach, Swiss Patent 562,224, May 1975.
174. G. F. Field and L. H. Sternbach, Swiss Patent 561,190, April 1975.
175. G. F. Field and L. H. Sternbach, Swiss Patent 562,223, May 1975.
176. (a) P. Nedenskov and Mandrup, *Acta Chem. Scand.*, **31**, 701 (1977). (b) Neth. Patent 6,608,039, December 1966 (Grindstedvaerket, Denmark).
177. S. A. Christensen, Br. Patent 1,247,554, September 1971.
178. A. Walser, R. I. Fryer, L. H. Sternbach, and M. A. Archer, *J. Heterocycl. Chem.*, **11**, 619 (1974).
179. M. Matsuo, K. Taniguchi, and I. Ueda, *Chem. Pharm. Bull.*, **30**, 1141 (1982).
180. A. Walser and R. I. Fryer, *J. Heterocycl. Chem.*, **20**, 551 (1983).
181. K. Meguro, H. Tawada, and Y. Kuwada, *Yakugaku Zasshi*, **93**, 1253 (1973).
182. H. Natsugari, K. Meguro, and Y. Kuwada, *Chem. Pharm. Bull.*, **27**, 2589 (1979).
183. L. H. Sternbach, E. Reeder, A. Stempel, and A. I. Rachlin, *J. Org. Chem.*, **29**, 332 (1964).
184. S. C. Bell, C. Gochman, and S. J. Childress, *J. Org. Chem.*, **28**, 3010 (1963).
185. K. Meguro and Y. Kuwada, *Chem. Pharm. Bull.*, **21**, 2375 (1973).
186. (a) K. Meguro and Y. Kuwada, *Tetrahedron Lett.*, 4039 (1970). (b) G. A. Archer and L. H. Sternbach, *J. Org. Chem.*, **29**, 231 (1964).
187. H. Tawada, H. Natsugari, K. Meguro, and Y. Kuwada, U.S. Patent 4,102,881, July 1978.
188. Neth. Patent 6,412,484, May 1965 (Hoffmann-La Roche & Co. AG, Switzerland).
189. R. Y. Ning, R. I. Fryer, and B. C. Sluboski, *J. Org. Chem.*, **42**, 3301 (1977).
190. R. I. Fryer, J. V. Earley, and J. F. Blount, *J. Org. Chem.*, **42**, 2212 (1977).
191. T. Masuda, Y. Usui, Y. Hara, and T. Komatsu, U.S. Patent 3,644,339, February 1972.
192. E. Broger, unpublished data, Hoffmann-La Roche, Nutley, NJ.
193. M. Ogata and H. Matsumoto, *Chem. Ind.*, (London), 1067 (1976).
194. M. Gates, *J. Org. Chem.*, **45**, 1675 (1980).
195. J. H. Gogerty, R. G. Griot, D. Habeck, L. C. Iorio, and W. J. Houlihan, *J. Med. Chem.*, **20**, 952 (1977).
196. R. G. Griot, U.S. Patent 3,414,563, December 1968.
197. P. C. Wade, B. R. Vogt, and M. S. Puar, *Tetrahedron Lett.*, 1699 (1977).
198. (a) B. R. Vogt, P. C. Wade, and M. S. Puar, *Tetrahedron Lett.*, 1931 (1976). (b) P. C. Wade, B. R. Vogt, B. Toeplitz, and M. S. Puar, *J. Org. Chem.*, **44**, 88 (1979). (c) P. C. Wade and B. R. Vogt, U.S. Patent 4,022,765, May 1977. (d) B. R. Vogt and P. C. Wade, U.S. Patent 3,895,005, July

1975. (e) B. R. Vogt and P. C. Wade, U.S. Patent 3,894,011, July 1975. (f) P. C. Wade and B. R. Vogt, U.S. Patent 3,898,212, August 1975. (g) B. R. Vogt and P. C. Wade, U.S. Patent 3,898,211, August 1975.
199. S. C. Bell, R. J. McCaully, and S. J. Childress, *J. Org. Chem.*, **33**, 216 (1968).
200. S. C. Bell, R. J. McCaully, and S. J. Childress, *Tetrahedron Lett.*, 2889 (1965).
201. S. C. Bell, S. J. Childress, U.S. Patent 3,344,136, September 1967.
202. Z. I. Zhilina, A. V. Bogat-Skii, S. A. Andronati, and N. I. Danilina, *Chem. Heterocycl. Compd.*, **15**, 447 (1979).
203. S. C. Bell, U.S. Patent 3,652,634, March 1972.
204. Neth. Patent 7,502,283, September 1975 (Hoffmann-La Roche & Co AG, Switzerland).
205. S. C. Bell, U.S. Patent 3,257,382, June 1966.
206. R. J. McCaully, U.S. Patent 3,875,213, April 1975; 3,873,604, March 1975.
207. R. J. McCaully, U.S. Patent 3,926,952, December 1975.
208. H. Tawada, H. Natsugari, K. Meguro, and Y. Kuwada, U.S. Patent 3,746,701, July 1973.
209. R. I. Fryer and L. H. Sternbach, U.S. Patent 3,376,290, April 1968.
210. J. Bergman, A. Brynolf, and B. Elman, *Heterocycles*, **20**, 2141 (1983).
211. J. Bergman, A. Brynolf, and B. Elman, *Heterocycles*, **21**, 511 (1984).
212. Y. Ban and M. Nagai, Swiss Patent 513,886, October 1971.
213. F. Kajfez and N. Blasevic, Swiss Patent 547,293, February 1974.
214. Neth. Patent 7,402,020, August 1974 (Sumitomo Chemical Co., Ltd., Japan).
215. (a) R. Y. Ning, W. Y. Chen, and L. H. Sternbach, *J. Org. Chem.*, **38**, 4206 (1973). (b) R. Y. Ning, W. Y. Chen, and L. H. Sternbach, *J. Org. Chem.*, **36**, 1064 (1971).
216. L. H. Sternbach, B. A. Koechlin, and E. Reeder, *J. Org. Chem.*, **27**, 4671 (1962).
217. G. F. Field and L. H. Sternbach, *J. Org. Chem.*, **33**, 4438 (1968).
218. R. Y. Ning, G. F. Field, and L. H. Sternbach, *J. Heterocycl. Chem.*, **7**, 475 (1970).
219. J. V. Earley, R. I. Fryer, R. Y. Ning, and L. H. Sternbach, U.S. Patent 3,681,341, August 1972.
220. Neth. Patent 7,111,035, February 1972 (Chugai Seiyaku, Japan).
221. Neth. Patent 7,117,532, June 1972 (Sankyo Co., Ltd., Japan).
222. J. Schmitt, P. Comoy, M. Suquet, J. Boitard, J. Le Meur, J.-J. Basselier, M. Brunaud, and J. Salle, *Chim. Ther.*, **2**, 254 (1967).
223. R. I. Fryer, E. E. Garcia, and L. H. Sternbach, U.S. Patent 3,371,083, February 1968.
224. Neth. Patent 6,600,095, July 1966 (S.A. Etablissement Clin-Byla, France).
225. R. I. Fryer, E. E. Garcia, and L. H. Sternbach, U.S. Patent 3,371,084, February 1968.
226. R. Jaunin and W. Arnold, *Helv. Chim. Acta*, **56**, 2569 (1973).
227. S. C. Bell, R. J. McCaully, C. Gochman, S. J. Childress, and M. I. Gluckman, *J. Med. Chem.*, **11**, 457 (1968).
228. E. Poetsch, J. Uhl, D. Marx, W. Strehlow, H. Mueller-Calgan, and G. Dolce, U.S. Patent 4,232,016, November 1980.
229. A. Ziggiotti, F. Mauri, and G. Riva, U.S. Patent 3,852,271, December 1974.
230. Q. Branca, A. E. Fischli, and A. Szente, U.S. Patent 4,377,522, March 1983.
231. A. E. Fischli and A. Szente: (a) U.S. Patent 4,294,758, October 1981, (b) U.S. Patent 4,299,767, November 1981.
232. R. M. Novack, U.S. Patent 3,549,623, December 1970.
233. J. V. Earley, R. I. Fryer, D. Winter, and L. H. Sternbach, *J. Med. Chem.*, **11**, 774 (1968).
234. Belg. Patent 786,951, November 1972 (Sumitomo Chemical Co., Ltd., Japan).
235. L. H. Schlager, *Tetrahedron Lett.*, 4519 (1970).
236. J. Hellerbach, H. Hoffmann, and G. Zanetti, U.S. Patent 3,963,777, June 1976.
237. Belg. Patent 805,054, March 1974 (Hoffmann-La Roche & Co. AG, Switzerland).
238. A. Walser and R. I. Fryer, *J. Med. Chem.*, **17**, 1228 (1974).
239. P. Nedenskov, U.S. Patent 3,970,645, July 1976.
240. W. Sadee, K.-H. Beyer, and J. Schwandt, Third International Congress of Heterocyclic Chemistry, Japan, 1971; *Abstracts*.

241. H. Yamamoto, S. Inaba, H. Wada, S. Ogino, and F. Kishimoto, U.S. Patent 3,806,418, April 1974.
242. J. Hellerbach and A. Walser, U.S. Patent 3,763,144, October 1972.
243. (a) J. H. Sellstedt, *J. Org. Chem.*, **40**, 1508 (1975), (b) J. H. Sellstedt, U.S. Patent 3,915,961, October 1975.
244. R. Y. Ning and L. H. Sternbach, U.S. Patent 3,781,353, December 1973.
245. A. Walser, G. Zenchoff, and R. I. Fryer, *J. Med. Chem.*, **19**, 1378 (1976).
246. R. I. Fryer and A. Walser: (a) U.S. Patent 3,989,681, November 1976, (b) U.S. Patent 3,865,815, February 1975.
247. W. Metlesics, unpublished data, Hoffmann-La Roche, Nutley, NJ.
248. J. Kollonitsch and G. A. Doldouras, U.S. Patent 3,859,276, January 1975.
249. Belg. Patent 880,419, April 1980 (KRKA, Farmacevtika Kemija, Yugoslavia).
250. W. Metlesics, R. F. Tavares, and L. H. Sternbach, *J. Org. Chem.*, **30**, 1311 (1965).
251. (a) Q. Branca, A. E. Fischli, and A. Szente, U.S. Patent 4,327,026, April 1982, (b) Q. Branca, unpublished data, Hoffmann-La Roche & Co. AG, Basel, Switzerland, (c) A. E. Fischli, unpublished data, Hoffmann-La Roche & Co. AG, Basel, Switzerland.
252. (a) R. Y. Ning, L. H. Sternbach, W. Pool, and L. O. Randall, *J. Med. Chem.*, **16**, 879 (1973). (b) R. Y. Ning and L. H. Sternbach, U.S. Patent 3,644,334, February 1972.
253. G. F. Field and L. H. Sternbach, Swiss Patent 561,191, March 1975.
254. G. A. Archer, R. I. Fryer, and L. H. Sternbach, U.S. Patent 3,299,053, January 1967.
255. (a) G. A. Archer and L. H. Sternbach, U.S. Patent 3,678,036, July 1972. (b) G. A. Archer and L. H. Sternbach, U.S. Patent 3,422,091, January 1969. (c) G. A. Archer, unpublished data, Hoffmann-La Roche, Nutley, NJ.
256. R. I. Fryer, B. Brust, and L. H. Sternbach, *J. Chem. Soc.*, 4977 (1963).
257. D. L. Coffen, unpublished data, Hoffmann-La Roche, Nutley, NJ.
258. H. Hoffmann, unpublished data, Hoffmann-La Roche & Co. AG, Basel, Switzerland.
259. M. Gerecke, unpublished data, Hoffmann-La Roche & Co. AG, Basel, Switzerland.
260. M. Ogata, H. Matsumoto, and K. Hirose, *J. Med. Chem.*, **20**, 776 (1977).
261. A. Gagneux, R. Heckendron, and R. Meier, U.S. Patent 3,852,300, December 1974.
262. S. C. Bell and S. J. Childress, *J. Org. Chem.*, **27**, 1691 (1962).
263. S. C. Bell, U.S. Patent 3,296,249, January 1967.
264. H. Pieper, G. Krueger, J. Keck, and K.-R. Noll, Ger. Offen. 2,311,714, September 1974.
265. G. Nudelman, R. J. McCaully, and S. C. Bell, *J. Pharm. Sci.*, **63**, 1880 (1974).
266. E. M. Bingham and W. J. Middleton, U.S. Patent 4,246,270, January 1981.
267. E. M. Bingham and W. J. Middleton, U.S. Patent 4,120,856, October 1978.
268. T. Okamoto, T. Akase, T. Izumi, M. Akatsu, Y. Kume, S. Inaba, and H. Yamamoto, U.S. Patent 3,832,344, April 1974.
269. M. Maziere, J.-M. Godot, G. Berger, Ch. Prenant, and D. Comar, *J. Radioanal. Chem.*, **56**, 229 (1980).
270. M. E. Derieg, R. I. Fryer, and L. H. Sternbach, U.S. Patent 3,534,021, October 1970.
271. J. Hester, Jr., Ger. Offen. 2,302,525, August 1973.
272. Neth. Patent 6,412,300, April 1965 (Hoffmann-La Roche & Co. AG, Switzerland).
273. Belg. Patent 803,315, December 1973 (L. Zambeletti S.A., Italy).
274. E. Reeder and L. H. Sternbach, French Patent 1,343,085, October 1963.
275. H. Demarne and A. Hallot, U.S. Patent 4,235,897, November 1980.
276. Belg. Patent 766,098, June 1971 (Sumitomo Chemical Co., Ltd., Japan).
277. A. Walser, G. Silverman, J. Blount, R. I. Fryer, and L. H. Sternbach, *J. Org. Chem.*, **36**, 1465 (1971).
278. R. I. Fryer and A. Walser, U.S. Patent 3,796,722, March 1974.
279. Neth. Patent 6,510,539, February 1966 (Hoffmann-La Roche & Co. AG, Switzerland).
280. L. H. Schlager, Ger. Offen, 2,950,235, June 1980 (Gerot-Pharmazeutica GmbH, Austria).
281. H. Najer and P. M. J. Manoury, Ger. Offen. 3,008,852, September 1980.

282. G. F. Tomagnone, M. V. Torrielli, and F. de Marchi, *J. Pharm. Pharmacol*, **26**, 567 (1974).
283. R. I. Fryer and L. H. Sternbach, U.S. Patent 3,819,602, June 1974.
284. G. A. Archer and L. H. Sternbach, U.S. Patent 3,391,138, July 1968.
285. M. Akatsu, Y. Kume, T. Hirohashi, S. Inaba, H. Yamamoto, and H. Sato, U.S. Patent 3,906,003, September 1975.
286. H. Yamamoto, S. Inaba, T. Hirohashi, M. Yamamoto, K. Ishizumi, M. Akatsu, I. Maruyama, K. Mori, Y. Kume, and T. Izumi, U.S. Patent 4,079,053, March 1978.
287. L. H. Sternbach, G. A. Archer, J. V. Earley, R. I. Fryer, E. Reeder, N. Wasyliw, L. O. Randall, and R. Banziger, *J. Med. Chem.*, **8**, 815 (1965).
288. Br. Patent 1,346,176, February 1974 (American Home Products Corp., New York).
289. H. Yamamoto, S. Kitagawa, S. Inaba, S. Sakai, T. Hirohashi, I. Maruyama, M. Akatsu, and T. Izumi, U.S. Patent 3,641,002, February 1972.
290. W. J. Welstead, Jr., R. F. Boswell, Jr., U.S. Patent 4,151,285, April 1979.
291. M. E. Derieg, R. I. Fryer, R. M. Schweininger, and L. H. Sternbach, *J. Med. Chem.*, **11**, 912 (1968).
292. J. Szmuskovicz, U.S. Patents 3,901,881, August 1975; 3,882,112, May 1975; 3,818,003, June 1974; 3,933,816; January 1976.
293. G. A. Archer and L. H. Sternbach, U.S. Patent 3,236,838, February 1966.
294. J. V. Earley, R. I. Fryer, and L. H. Sternbach, U.S. Patent 3,703,510, November 1972.
295. Belg. Patent 775,256, November 1971 (A.H. Robins Co. Inc., Richmond, VA).
296. R. B. Moffet, U.S. Patent 4,016,165, April 1977.
297. A. Walser, G. Silverman, R. I. Fryer, L. H. Sternbach, and J. Hellerbach, *J. Org. Chem.*, **36**, 1248 (1971).
298. A. Walser, G. Silverman, and R. I. Fryer, *J. Org. Chem.*, **38**, 3502 (1973).
299. B. E. Reitter, Y. P. Sachdeva, and J. F. Wolfe, *J. Org. Chem.*, **46**, 3945 (1981).
300. R. I. Fryer, J. V. Earley, and L. H. Sternbach, *J. Org. Chem.*, **34**, 649 (1969).
301. A. Szente, unpublished data, Hoffmann-La Roche & Co. AG, Basel, Switzerland.
302. R. Jaunin, unpublished data, Hoffmann-La Roche & Co. AG, Basel, Switzerland.
303. K.-A. Kovar and B. Biegert, *Arch. Pharm.*, **309**, 522 (1976).
304. A. V. Bogat-skii, S. A. Andronati, Z. I. Zhilina, O. V. Kobzareva, P. A. Sharbatyan, R. Yu. Ivanova, and T. K. Chumachenko, *J. Gen. Chem., USSR*, **45**, 384 (1975).
305. M. E. Derieg, J. V. Earley, R. I. Fryer, and L. H. Sternbach, U.S. Patent 3,965,151, June 1976; 3,905,956, September 1975; 3,997,591, December 1976; 4,049,666, September 1977; 4,049,667, September 1977; 4,017,531, April 1977; 4,017,532, April 1977.
306. M. E. Derieg, J. V. Earley, R. I. Fryer, R. J. Lopresti, R. M. Schweiniger, L. H. Sternbach, and H. Wharton, *Tetrahedron* **27**, 2591 (1971).
307. R. Tachikawa, H. Takagi, T. Kamioka, M. Fukunaga, Y. Kawano, and T. Miyadera, U.S. Patent 3,696,094, October 1972.
308. S. C. Bell, U.S. Patent 3,514,445, May 1970.
309. S. C. Bell and P. H. L. Wei, *J. Org. Chem.*, **30**, 3576 (1965).
310. R. B. Moffett, U.S. Patent 3,822,259, July 1974.
311. R. B. Moffett, and A. D. Rudzik, *J. Med. Chem.*, **15**, 1079 (1972).
312. R. B. Moffett, U.S. Patent 3,718,646, February 1973.
313. R. B. Moffett and B. V. Kamdar, *J. Heterocycl. Chem.*, **16**, 793 (1979).
314. H. Pieper, G. Krueger, J. Keck, and K. Noll, Ger. Offen. 2,304,095, August 1974.
315. (a) H. Pieper, G. Krueger, J. Keck, K.-R. Noll, and J. Kahling, U.S. Patent 3,872,090, March 1975. (b) E. Ger. Patent 106,172, June 1974 (Dr. K. Thomae, GmbH).
316. S. Kobayashi, *Tetrahedron Lett.*, 967 (1974).
317. S. Kobayashi, *Bull Chem. Soc. Japan*, **48**, 302 (1975).
318. J. Szmuszkovicz, C. G. Chidester, D. J. Duchamp, F. A. MacKellar, and G. Slomp, *Tetrahedron Lett.*, 3665 (1971).
319. V. Hach, Ger. Offen. 2,512,092, September 1975.
320. J. Hellerbach and A. Szente, U.S. Patent 3,696,095, October 1972.
321. D. L. Coffen, J. P. DeNoble, E. L. Evans, G. F. Field, R. I. Fryer, D. A. Katonak, B. J. Mandel, L. H. Sternbach, and W. J. Zally, *J. Org. Chem.*, **39**, 167 (1974).

322. T. E. Gunda and C. Enebaeck, *Acta Chem. Scand.*, **37**, 75 (1983).
323. R. Jaunin, W. E. Oberhaensli, and J. Hellerbach, *Helv. Chim. Acta*, **55**, 2975 (1972).
324. L. H. Sternbach and A. Stempel, U.S. Patent 3,450,695, June 1969.
325. S. C. Bell: (a) U.S. Patent 3,401,159, September 1968, (b) U.S. Patent 3,418,315, December 1968.
326. L. H. Schlager U.S. Patent 4,261,987, April 1981.
327. A. Nudelman and R. J. McCaully, U.S. Patent 3,825,533, July 1974.
328. (a) R. J. McCaully and A. Nudelman, U.S. Patent 3,886,276, May 1975. (b) Neth. Patent 7,204,228, October 1972 (American Home Products Corp. New York).
329. Germ. Patent 2,519,329, April 1975 (Laboratories Ferrer S.L., Spain).
330. R. B. Moffett, *J. Org. Chem.*, **39**, 568 (1974).
331. E. F. Williams, K. C. Rice, S. M. Paul, and P. Skolnick, *J. Neurochem.*, **35**, 591 (1980).
332. K. C. Rice, E. Williams, and P. Skolnick, 178th Meeting of the American Chemical Society, September 1979; *Abstracts.*
333. A. Nudelman and R. J. McCaully, U.S. Patent 3,903,276, September 1975.
334. F. Kajfez, T. Kovac, and V. Sunjic, U.S. Patent 3,852,274, December 1974.
335. Belg. Patent 828,222, August 1975 (Siphar S.A., Switzerland).
336. G. Ferrari and C. Casagrande, U.S. Patent 3,799,920, March 1974.
337. (a) G. Maksay, Z. Tegyey, and L. Oetvoes, *Z. Physiol. Chem.*, **359**, 879 (1978). (b) E. Simon-Trompler, G. Maksay, I. Lukovits, J. Volford, and L. Oetvoes, *Arzneim.-Forsch.*, **32**, 102 (1982).
338. A. Nudelman and R. J. McCaully, U.S. Patent 3,801,568, April 1974.
339. A. Baglioni, European Patent application A1, 0,122,889, October 1984.
340. A. Nudelman and R. J. McCaully, U.S. Patent 3,860,581, January 1975.
341. R. I. Fryer and L. H. Sternbach, *J. Org. Chem.*, **30**, 524 (1965).
342. R. I. Fryer, B. Brust, J. V. Earley, and L. H. Sternbach, *J. Chem. Soc., C*, 366 (1967).
343. R. I. Fryer, D. L. Coffen, J. V. Earley, and A. Walser, *J. Heterocycl. Chem.*, **10**, 473 (1973).
344. R. Y. Ning, R. I. Fryer, P. B. Madan, and B. C. Sluboski, *J. Org. Chem.*, **41**, 2720 (1976).
345. A. Walser, U.S. Patent 4,118,386, October 1978.
346. R. Y. Ning, R. I. Fryer, P. B. Madan, and B. Sluboski, *J. Org. Chem.*, **41**, 2724 (1976).
347. A. Nussbaum, unpublished data, Hoffmann-La Roche, Nutley NJ.
348. M. R. Del Giudice, F. Gatta, C. Pandolfi, and G. Settimj, *Farmaco, Ed. Sci.*, **37**, 343 (1982).
349. S. C. Bell and S. J. Childress, *J. Org. Chem.*, **29**, 506 (1964).
350. J. Volke, M. M. Ellaithy, and O. Manousek, *Talanta*, **25**, 209 (1978).
351. B. Maupas and M. B. Fleury, *Analysis*, **10**, 187 (1982).
352. Belg. Patent 823,615, April 1975 (Asta-Werke AG, Germany).
353. G. Scheffler, F. Bourseaux, N. Brock, and J. Pohl, Ger. Offen. 2,947,076 June 1981.
354. G. Scheffler, F. Bourseaux, N. Brock, and J. Pohl, Ger. Offen. 2,947,075 June 1981.
355. J. H. Sellstedt, U.S. Patent 3,882,101, May 1975.
356. T. Kovac, F. Kajfez, V. Sunjic, and M. Oklobdzija, *J. Med. Chem.*, **17**, 766 (1974).
357. W. A. Khan and P. Singh, *Org. Prep. Proc. Int.*, **10**, 105 (1978).
358. S. C. Bell, U.S. Patent 3,198,789, August 1965.
359. H. Scholl, P. Laufer, G. Kloster, and G. Stoecklin, *J. Labelled Compd.*, **19**, 1294 (1982).
360. R. Y. Ning and L. H. Sternbach, U.S. Patent 3,801,569, April 1974.
361. M. Nakano, N. Inotsume, N. Kohri, and T. Arita, *Int. J. Pharm.*, **3**, 195 (1979).
362. N. Inotsume and M. Nakano, *Int. J. Pharm.*, **6**, 147 (1980).
363. J. B. Hester, Jr., D. J. Duchamp, and C. G. Chidester, *Tetrahedron Lett.*, 1609 (1971).
364. J. Gasparic, J. Zimak, P. Sedmera, Z. Breberova, and J. Volke, *Collect. Czech. Chem. Commun.*, **44**, 2243 (1979).
365. G. F. Tamagnone, R. De Maria, and F. De Marchi, *Arzneim.-Forschung.*, **25**, 720 (1975).
366. G. Jommi, F. Mauri, and G. Riva, U.S. Patent 3,689,478, September 1972.
367. Br. Patent 1,035,918, July 1966 (J. Wyeth & Brother Ltd., England).
368. R. I. Fryer, J. V. Earley, and L. H. Sternbach, *J. Org. Chem.*, **32**, 3798 (1967).
369. F. Gatta, M. R. Del Giudice, and G. Settimj, *Synthesis*, 718 (1979).
370. F. G. Kathawala, U.S. Patent 3,869,450, March 1975.
371. G. F. Field, unpublished data, Hoffmann-La Roche, Nutley, NJ.

372. G. F. Field, R. Y. Ning, and L. H. Sternbach, U.S. Patent 3,591,581, July 1971.
373. H. J. Roth and M. Adomeit, *Arch. Pharmaz.*, **306**, 889 (1973).
374. (a) P. J. G. Cornelissen and G. M. J. Beijersbergen van Henegouwen, *Pharm. Wkbl.*, **3**, 800 (1981). (b) P. J. G. Cornelissen and G. M. J. Beijersbergen van Henegouwen, *Photochem. Photobiol.*, **30**, 337 (1979).
375. P. J. G. Cornelissen and G. M. J. Beijersbergen van Henegouwen, and K. W. Gerritsma, *Int. J. Pharm.*, **1**, 173 (1978).
376. R. I. Fryer, J. V. Earley, and L. H. Sternbach, *J. Am. Chem. Soc.*, **88**, 3173 (1966).
377. J. P. Freeman, D. J. Duchamp, C. G. Chidester, G. Slomp, J. Szmuszkovicz, and M. Raban, *J. Am. Chem. Soc.*, **104**, 1380 (1982).
378. A. V. Bogat-skii, S. A. Andronati, L. V. Komogortseva, and Z. I. Zhilina, *J. Gen. Chem., USSR*, **46**, 1828 (1976).
379. C. Preti and G. Tosi, *Transition Met. Chem.*, **3**, 246 (1978).
380. M. Gajewska, E. Lugowska, M. Ciszewska-Jedrasik, and M. Rzedowski, *Chem. Anal.*, **26**, 517 (1981).
381. J. H. Mendez, C. G. Perez, and M. I. G. Martin, *An. Quim., B*, **80**, 390 (1984).
382. J. A. Real and J. Borras, *Synth. React. Inorg. Met.-Org. Chem.*, **14**, 843 (1984).
383. J. A. Real and J. Borras, *Synth. React. Inorg. Met.-Org. Chem.*, **14**, 857 (1984).
384. J. A. Groves and W. F. Smith, *Spectrochim. Acta*, **35**, 603 (1979).
385. G. Snatzke, A. Konowal, A. Sabljic, N. Blasevic, and V. Sunjic, *Croat. Chem. Acta*, **55**, 435 (1982).
386. P. Linscheid and J. M. Lehn, *Bull. Soc. Chim. Fr.*, 992 (1967).
387. W. Bley, P. Nuhn, and G. Benndorf, *Arch. Pharm.*, **301**, 444 (1968).
388. M. Sarrazin, M. Bourdeaux-Pontier, C. Briand, and E. J. Vincent, *Org. Magn. Resonance*, **7**, 89 (1975).
389. M. Raban, E. H. Carlson, J. Szmuszkovicz, G, Slomp, C. G. Chidester, and D. J. Duchamp, *Tetrahedron Lett.*, 139 (1975).
390. L. Cazaux, C. Vidal, and M. Pasdeloup, *Org. Magn. Resonance*, **21**, 190 (1983).
391. N. W. Gilman, P. Rosen, J. V. Earley. C. Cook and L. Todaro, *J. Am. Chem. Soc.*, **112**, 3969 (1990).
391b. N. W. Gilman, Unpublished data, Hoffman-La Roche Inc., Nutley, NJ.
392. W. Sadee, H.-J. Schwandt, and K.-H. Beyer, *Arch. Pharm.*, **306**, 751 (1973).
393. V. Sunjic, A. Lisini, A. Sega, T. Kovac, and F. Kajfez, *J. Heterocycl. Chem.*, **16**, 757 (1979).
394. H.-H. Paul, H. Sapper, W. Lohmann, and H.-O. Kalinowski, *Org. Magn. Resonance*, **19**, 49 (1982).
395. R. Haran and J. P. Tuchagues, *J. Heterocycl. Chem.*, **17**, 1483 (1980).
396. A. Patra, A. K. Mukhopadhyay, A. K. Mitra, and A. K. Acharyya, *Org. Magn. Resonance*, **15**, 99 (1981).
397. K.-A. Kovar, D. Linden, and E. Breitmaier, *Arch. Pharm.*, **314**, 186 (1981).
398. H.-H. Paul, H. Sapper, W. Lohmann, and H. O. Kalinowski, *Org. Magn. Resonance*, **21**, 319 (1983).
399. B. Unterhalt, *Arch. Pharm.*, **314**, 733 (1981).
400. T. A. Scahill and S. L. Smith, *Org. Magn. Resonance*, **21**, 621 (1983).
401. F. M. Vane and W. Benz, *Org. Mass Spectrom.*, **14**, 233 (1979).
402. A. Camerman and N. Camerman, *J. Am. Chem. Soc.*, **94**, 268 (1972).
403. L. H. Sternbach, F. D. Sancilio, and J. F. Blount, *J. Med. Chem.*, **17**, 374 (1974).
404. G. Gilli, V. Bertolasi, M. Sacerdoti, and P. A. Borea, *Acta Crystallogr.*, **B33**, 2664 (1977).
405. G. Gilli, V. Bertolasi, M. Sacerdoti, and P. A. Borea, *Acta Crystallogr.*, **B34**, 2826 (1978).
406. M. Sikirica and I. Vickovic, *Cryst. Struct. Commun.*, **11**, 1293 (1982).
407. A. A. Karapetyan, V. G. Andrianov, Y. T. Struchkov, A. V. Bogatskii, S. A. Andronati, and T. I. Korotenko, *Bioorg Khim.*, **5**, 1684 (1979).
408. G. Bandoli and D. A. Clemente, *J. Chem. Soc., Perkin Trans. II*, 413 (1976).
409. Z. Galdecki and M. L. Glowka, *Acta Crystallogr.*, **B36**, 3044 (1980).
410. P. Chananont, T. A. Hamor, and I. L. Martin, *Cryst. Struct. Commun.*, **8**, 393 (1979).
411. P. Chananont, T. A. Hamor, and I. L. Martin, *Acta Crystallogr.*, **B37**, 1371 (1981).

412. R. F. Dunphy and H. Lynton, *Can. J. Chem.*, **49**, 3401 (1971).
413. G. Brachtel and M. Jansen, *Cryst. Struct. Commun.*, **10**, 669 (1981).
414. P. Chananont, T. A. Hamor, and I. L. Martin, *Acta Crystallogr.*, **B36**, 2115 (1980).
415. J. V. Earley, R. I. Fryer, and A. Walser, U.S. Patent 3,869,448, March 1975.
416. J. B. Hester, Jr., A. D. Rudzik, and B. V. Kamdar, *J. Med. Chem.*, **14**, 1078 (1971).
417. R. B. Moffet, U.S. Patent 3,846,443, November 1974.
418. R. S. P. Hsi and T. D. Johnson, *J. Labelled Compd. Radiopharm.*, **12**, 613 (1976).
419. J. B. Hester, Jr., U.S. Patent 4,018,788, April 1977.
420. J. B. Hester, Jr., U.S. Patent 3,741,957, June 1973.
421. J. V. Earley, R. I. Fryer, and A. Walser, U.S. Patent 3,836,521, September 1974.
422. M. Steinman: (a) U.S. Patent 3,920,818, November 1975, (b) U.S. Patent 3,845,039, October 1974.
423. (a) R. B. Moffett, Ger. Offen 2,252,079, May 1973, (b) J. B. Hester, Jr., Ger. Offen. 2,220,623, November 1972.
424. J. B. Hester, Jr., U.S. Patent 3,674,777, July 1972.
425. M. Foersch and H. Gerhards, European Patent A1 0,041,242 December 1981.
426. J. B. Hester, Jr., U.S. Patent 3,649,617, March 1972.
427. I. Ueda and M. Matsuo, U.S. Patent 4,094,870, June 1978.
428. J. B. Hester, Jr., U.S. Patent 3,897,446, July 1975.
429. M. Steinman, U.S. Patent 3,856,787, December 1974.
430. Belg. Patent 798,677, August 1973 (Centre d'Etudes pour l'Industrie Pharmaceutique, France).
431. J.-P. Maffrand, G. Ferrand, and F. Eloy, *Eur. J. Med. Chem.*, **9**, 539 (1974).
432. French Patent 2,244,525, April 1975 (Centre d'Etudes pour l'Industrie Pharmaceutique, France).
433. R. B. Moffett, and A. D. Rudzik, *J. Med. Chem.*, **16**, 1256 (1973).
434. R. B. Moffett, U.S. Patent 3,847,935, November 1974.
435. M. Gall, U.S. Patent 3,910,946, October 1975.
436. M. Gall and B. V. Kamdar *J. Org. Chem.*, **46**, 1575 (1981).
437. J. B. Taylor and D. R. Harrison, U.S. Patent 4,185,102, January 1980.
438. J. B. Taylor and D. R. Harrison, U.S. Patent 4,134,976, January 1979.
439. I. R. Ager, G. W. Danswan, D. R. Harrison, D. P. Kay, P. D. Kennewell, and J. B. Taylor, *J. Med. Chem.*, **20**, 1035 (1977).
440. M. Gall, U.S. Patent 3,992,393, November 1976.
441. M. Gall, U.S. Patent 3,763,179, October 1973.
442. J. P. Maffrand, G. Ferrand, and E. F. Eloy, *Tetrahedron Lett.*, 3449 (1973).
443. Belg. Patent 634,438, January 1964 (Hoffmann-La Roche & Co. AG, Switzerland).
444. J. B. Hester, Jr., and A. D. Rudzik, *J. Med. Chem.*, **17**, 293 (1974).
445. J. B. Hester, Jr., U.S. Patent 3,857,854, December 1974.
446. H.-G. Schecker and G. Zinner, *Arch. Pharm.*, **313**, 926 (1980).
447. J. B. Hester, Jr., U.S. Patent 4,082,761, April 1978.
448. J. B. Hester, Jr., U.S. Patent 3,734,922, May 1973.
449. R. B. Moffett, U.S. Patent 3,743,652, July 1973.
450. J. B. Hester, Jr., U.S. Patent 3,995,043, November 1976.
451. J. Szmuskovicz, U.S. Patent 3,856,802, December 1974.
452. Neth. Patent 7,206,300, November 1972 (Upjohn Co., Kalamazoo, MI).
453. Neth. Patent 7,205,705, October 1972 (Upjohn Co., Kalamazoo, MI).
454. J. B. Hester, Jr., U.S. Patent 3,886,174, May 1975.
455. S. C. Bell, R. J. McCaully, and S. J. Childress, *J. Med. Chem.*, **11**, 172 (1968).
456. M. Ogata and H. Matsumoto, U.S. Patent 4,041,026, August 1977.
457. R. J. McCaully and A. Nudelman, U.S. Patent 3,803,129, April 1974.
458. R. J. McCaully, U.S. Patent 3,446,800, May 1969.
459. J. Bergman and A. Brynolf, *Heterocycles*, **20**, 2145 (1983).
460. F. Hollywood, E. F. V. Scriven, H. Suschitzky, and D. R. Thomas, *J. Chem. Soc., Chem. Commun.*, 806 (1978).

461. F. Hollywood, Z, U. Khan, E. F. V. Scriven, R. K. Smalley, H. Suschitzky, and D. R. Thomas, *J. Chem. Soc., Perkin Trans. I*, 431 (1982).
462. J. Bergman, A. Brynolf, and B. Elman, *Heterocycles*, **20**, 2141 (1983).
463. J. A. Deyrup and J. C. Gill, *Tetrahedron Lett.*, 4845 (1973).
464. G. F. Field, W. J. Zally, and L. H. Sternbach, *J. Org. Chem.*, **36**, 777 (1971).
465. A. A. Santilli and T. S. Osdene, *J. Org. Chem.*, **29**, 1998 (1964).
466. A. A. Santilli and T. S. Osdene, *J. Org. Chem.*, **30**, 2100 (1965).
467. A. A. Santilli and T. S. Osdene, U.S. Patent 3,336,300, August 1967.
468. K. H. Weber, *Arch. Pharm.*, **302**, 584 (1969).
469. J. V. Earley, R. I. Fryer, and L. H. Sternbach, U.S. Patent 3,644,335, February 1972.
470. G. F. Field, L. H. Sternbach, and W. J. Zally: (a) U.S. Patent 3,624,073, November, 1971, (b) U.S. Patent 3,678,038, July 1972.
471. R. J. Mohrbacher and P. P. Grous: (a) U.S. Patent 4,022,767, May 1977, (b) U.S. Patent 4,031,079, June 1977, (c) U.S. Patent 4,020,055, April 1977, (d) U.S. Patent 4,002,610, January 1977.
472. S. Gaertner, *Justus Liebigs Ann. Chem.*, **332**, 226 (1904).
473. N. W. Gilman, J. F. Blount, and R. I. Fryer, *J. Org. Chem.*, **41**, 737 (1976).
474. C.-M. Liu, unpublished data, Hoffmann-La Roche, Nutley, NJ.
475. C. W. Perry, unpublished data, Hoffmann-La Roche, Nutley, NJ.
476. H. Lehr, unpublished data, Hoffmann-La Roche, Nutley, NJ.
477. J. M. Osbond, unpublished data, Roche Products Ltd., Welwyn, England.
478. W. Hunkeler, unpublished data, Hoffmann-La Roche, Basel, Switzerland.
479. E. Wenis, unpublished data, Hoffmann-La Roche, Nutley, NJ.
480. H. Shimizu, unpublished data, Nippon Roche K.K., Tokyo.
481. W. Voegtli, unpublished data, Hoffmann-La Roche & Co., AG, Basel, Switzerland.
482. K. E. Fahrenholtz, unpublished data, Hoffmann-La Roche, Nutley, NJ.
483. R. W. Lambert, unpublished data, Roche Products Ltd., Welwyn, England.
484. R. F. Lauer, unpublished data, Hoffmann-La Roche, Nutley, NJ.
485. E. Kyburz, unpublished data, Hoffmann-La Roche & Co., AG, Basel, Switzerland.
486. G. Zanetti, unpublished data, Hoffmann-La Roche & Co., AG, Basel, Switzerland.
487. R. A. LeMahieu, unpublished data, Hoffmann-La Roche, Nutley, NJ.
488. H. Ramuz, unpublished data, Hoffmann-La Roche & Co., AG, Basel, Switzerland.
489. U. Koelliker, unpublished data, Hoffmann-La Roche & Co., AG, Basel, Switzerland.
490. W. Aschwanden, unpublished data, Hoffmann-La Roche & Co., AG, Basel, Switzerland.
491. H. Pauling, unpublished data, Hoffmann-La Roche & Co., AG, Basel, Switzerland.
492. P. J. Machin, unpublished data, Roche Products Ltd., Welwyn, England.
493. A. A. Liebman, unpublished data, Hoffmann-La Roche, Nutley, NJ.
494. R. Schwob, unpublished data, Hoffmann-La Roche & Co., AG, Basel, Switzerland.
495. J. M. Cassal, unpublished data, Hoffmann-La Roche & Co., AG, Basel, Switzerland.
496. W. Zwahlen, unpublished data, Hoffmann-La Roche & Co., AG, Basel, Switzerland.
497. B. Pecherer, unpublished data, Hoffmann-La Roche, Nutley, NJ.
498. A. V. Bogat-skii, Yu. I. Vikhlyaev, S. A. Andronati, T. A. Klygul, and Z. I. Zhilina, *Chem. Heterocycl. Compd.*, **9**, 1413 (1973).
499. (a) G. Jommi, G. Riva, F. Mauri, and L. Mauri, U.S. Patent 3,654,267, April 1972. (b) G. Jommi, G. Riva, and F. Mauri, U.S. Patent 3,798,212, March 1974.
500. R. I. Fryer, B. Brust, J. Earley, and L. H. Sternbach, *J. Med. Chem.*, **7**, 386 (1964).
501. J. F. Blount, R. I. Fryer, N. W. Gilman, and L. J. Todaro, *Molec. Pharm.*, **24**, 425 (1983).

# CHAPTER VIII

# Tetrahydro- and Polyhydro-1,4-Benzodiazepines

**A. Walser**

*Chemical Research Department,*
*Hoffmann-La Roche Inc.,*
*Nutley, New Jersey*

and

**R. Ian Fryer**

*Department of Chemistry, Rutgers,*
*State University of New Jersey,*
*Newark, New Jersey*

## INTRODUCTION

This chapter describes the chemistry of tetra-, octa-, and decahydro-1,4-benzodiazepines, as well as the carbonyl and thiocarybonyl derivatives, such as 2-ones, 3-ones, 5-ones, 2,3-diones, 2,5-diones, and 3,5-diones. The octa- and decahydro compounds discussed in Section 9 could also be considered as cyclohexa[*e*][1,4]diazepines. The related carbocyclic system cyclopenta[*e*][1,4]diazepine ring system is reviewed in Chapter IX, Section 1.

## 1. 2,3,4,5-TETRAHYDRO-1*H*-1,4-BENZODIAZEPINES

### 1.1. Synthesis

#### *1.1.1. By Reduction*

Most tetrahydro-1,4-benzodiazepines were prepared by reduction of compounds with higher oxidation levels. Thus, the parent compound **2** ($R_1$, $R_2$, $R_3$, $R_4$, X = H) was obtained by reduction of the corresponding 2,5-dione **1** with lithium aluminum hydride (Eq. 1).[1,2] The same compound was also prepared by cleavage of the 1-benzyl derivative **2** ($R_1$ = $PhCH_2$; $R_2$, $R_3$, $R_4$, X = H), which was accessible by reduction of the appropriate 2,5-dione with lithium aluminum hydride.[3] Several 1-, 3-, and 4-substituted tetrahydro-1,4-benzodiazepines were analogously obtained from the substituted 2,5-diones.[2,3–7]

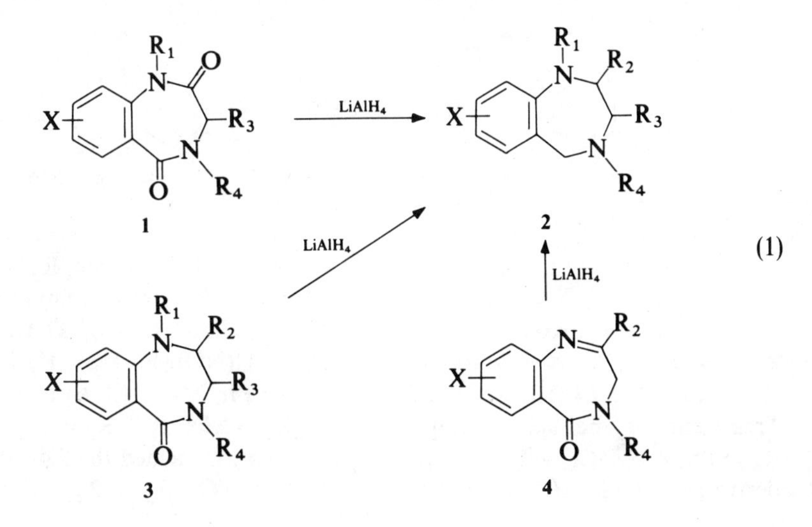

Similar reductions were also carried out on the 5-ones **3**[8,9] and the 5-ones **4**.[10,11]

5-Substituted tetrahydro derivatives **8** were synthesized by reduction of **5** and **6**.[12–15] The transformation of **6** to **8** worked well with lithium aluminum hydride in the presence of aluminum chloride[13] or with diborane in boiling tetrahydrofuran (Eq. 2).[14,15] Treatment of **6** with a combination of sodium borohydride and boron trifluoride in refluxing tetrahydrofuran also gave good yields of the totally reduced products **8**.[15] The 5-phenyl derivatives **8** ($R_1$, $R_2$ = H; $R_3$ = Ph; X = H, 7-Cl) were obtained by reduction of the 3-ones **7** (X = H, Cl) with lithium aluminum hydride.[16]

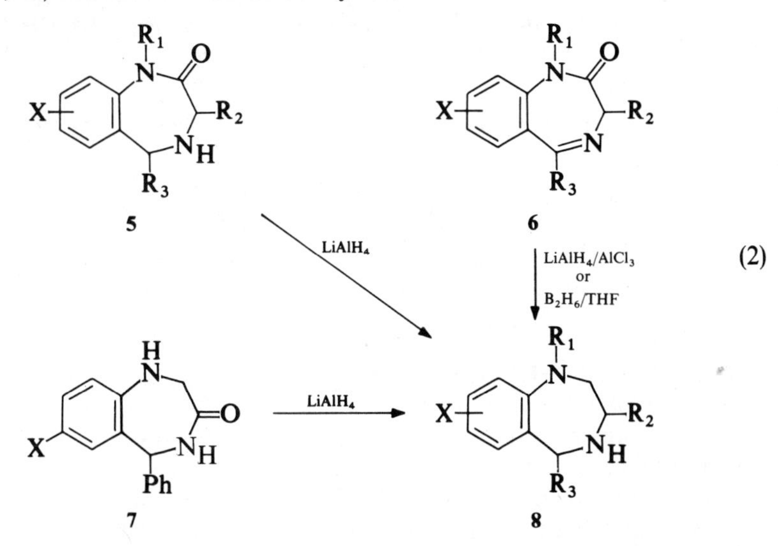


(2)

The pentasubstituted derivative **10** was prepared by treatment of the 2,3-dione **9** with lithium aluminum hydride in refluxing tetrahydrofuran (Eq. 3).[17]

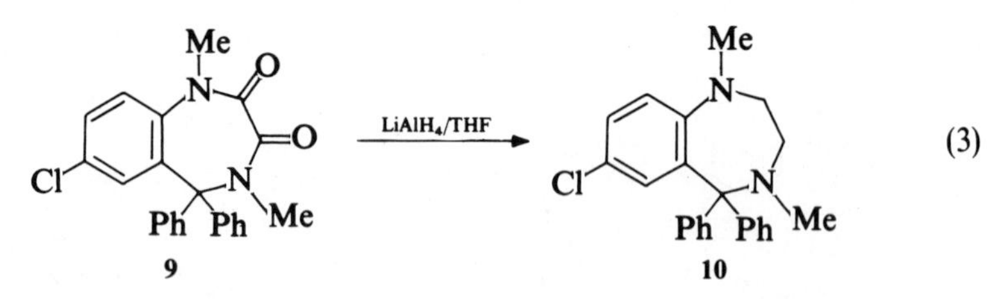


(3)

The reduction of the nitrones **11** and **12**, where ($R_1$ = H, Me; $R_3$ = H, Me; $R_4$ = Ph; $R_5$ = H; X = 7-Cl)[20,21] and **13**[18,22] with lithium aluminum hydride afforded the 4-hydroxy compounds **14** (Eq. 4). Reduction of **11** ($R_2$ = Me, $R_3$ = H, $R_4$ = Ph, X = 7-Cl)[18] and **11** [$R_2$ = N(NO)Me, $R_3$ = H, $R_4$ = Ph, X = 7-Cl] gave **14** ($R_1$, $R_2$, $R_3$, $R_5$ = H; $R_4$ = Ph; X = 7-Cl) as a product.[19] Treatment of the spiro compound **12** [$R_1$ = Me; $R_3$, $R_5$ = —$(CH_2)_4$—; $R_4$ = Ph; X = 7-Cl] with lithium aluminum hydride yielded the 2,4-dihydroxy derivative **14** [$R_1$ = Me; $R_2$ = OH; $R_3$, $R_5$ = —$(CH_2)_4$—; $R_4$ = Ph; X = 7-

Cl].[17] The sterically hindered carbonyl function at the 2-position was reduced only to the carbinolamine.

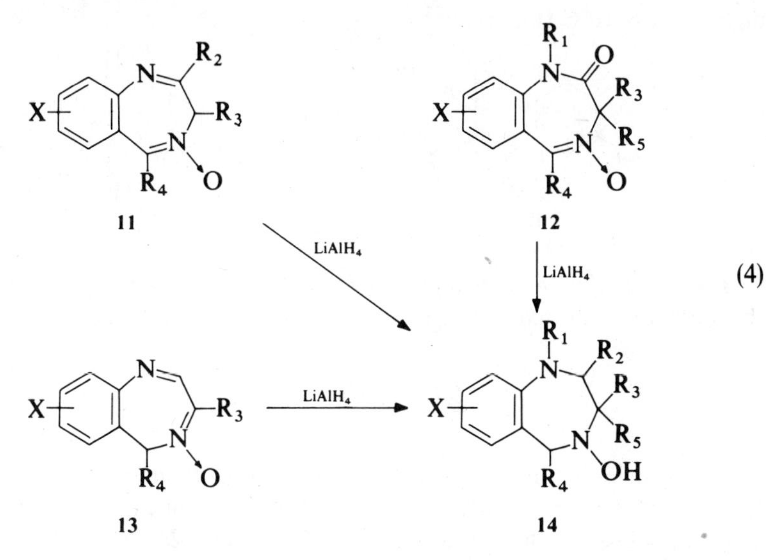


The nitrone **15** could be reduced by lithium aluminum hydride to the 4-hydroxy compound **16** with retention of the acetyl group. Under more vigorous conditions the acetyl group could be reduced to give **16** (R = H) (Eq. 5).[23] A commonly applied method for the preparation of the tetrahydro-1*H*-1,4-benzodiazepines **18** involved the reduction of the imine bond in **17**. Zinc and acetic acid,[24] iron and hydrochloric acid,[25] and catalytic hydrogenation using platinum[26] have been successfully used for this transformation. Hydrogenation of the 3-hydroxy derivative **17** ($R_1$, $R_2$ = H; $R_3$ = OH; $R_4$ = Ph; X = Cl) over platinum as a catalyst led to the tetrahydro compound **18** ($R_1$, $R_2$, $R_3$ = H; $R_4$ = Ph; X = Cl).[27]

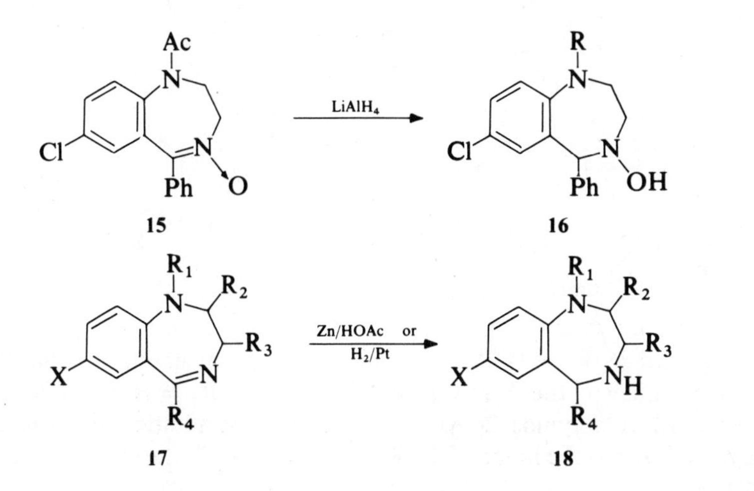

The 4-hydroxy compounds **20** (R = H, Me) were products of the reduction of the aziridinoquinazolines **19** (R = H, Me, $CH_2Cl$) with sodium borohydride in diglyme at 5–10°C (Eq. 6).[18,22]

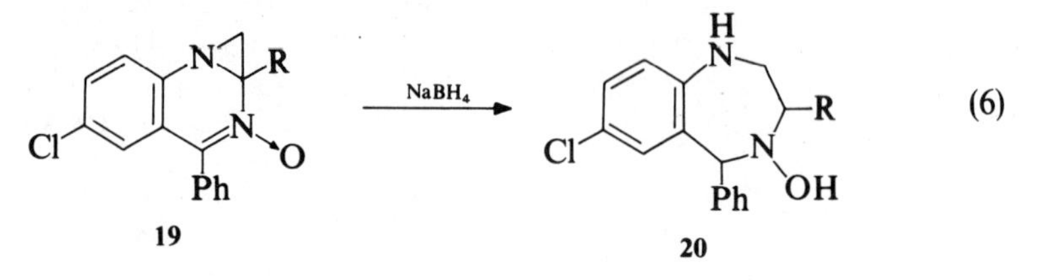


The quaternary salts **21** ($R_2$ = alkyl) could be reduced by hydride reagents to the 4-alkyl-substituted tetrahydrobenzodiazepines **22** (Eq. 7).[28]

### *1.1.2. Other Syntheses*

The 4-acyliminium intermediate **21** ($R_2 = R_4CO$), generated by addition of acyl chlorides to the imine, were reacted with such nucleophiles as hydroxide and alkoxides to form the 4-acyl derivatives **24**.[17,29] Compounds **24** ($R_4$ = MeO, benzyloxy, EtO, Me; $R_5$ = H, Me, Et) were thus obtained by treatment of the imines **23** with alkoxycarbonyl chloride or acetyl chloride in the presence of aqueous or alcoholic potassium carbonate.

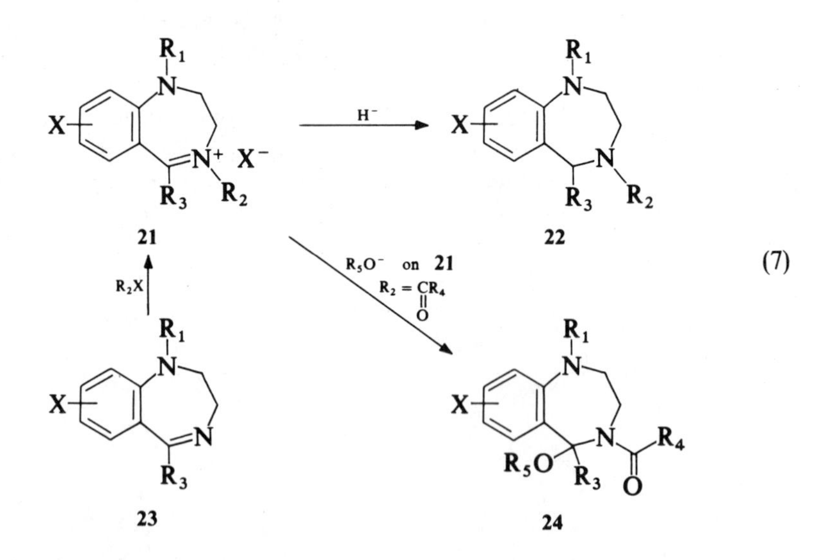


Compound **26** ($R_1$ = H, $R_2$ = Ph) was accessible by addition of phenylmagnesium bromide to the 5-unsubstituted imine **25** ($R_1$ = H),[30] while the 5,5-disubstituted compounds **26** were synthesized by addition of Grignard or lithium reagents to the imine of **25** ($R_1$ = Ph) (Eq. 8).[29] The reagents that could successfully react with the imine **25** ($R_1$ = Ph) included phenyllithium, butyllithium, and 3-$Me_2$N-propylmagnesium chloride.

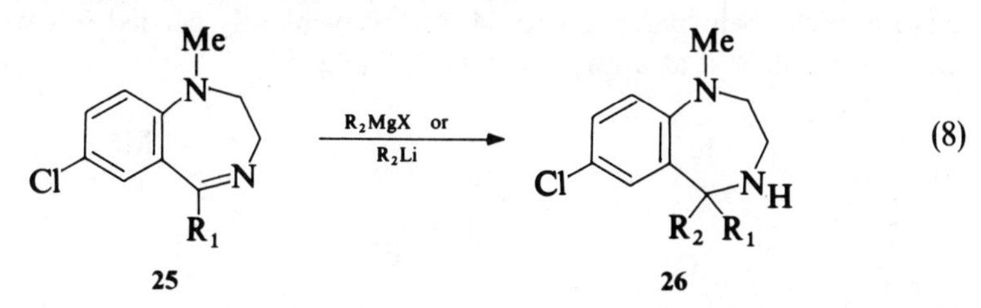


The 4-formyl derivatives **28** (R = H, Ph) were synthesized by reaction of the *N*-(2-aminoethyl)aniline **27** with formaldehyde and benzaldehyde, respectively, in formic acid (Eq. 9).[30,31] The tropanyl analogs **29** were similarly transformed into the benzodiazepines **30** by heating with formaldehyde in formic acid.[32]

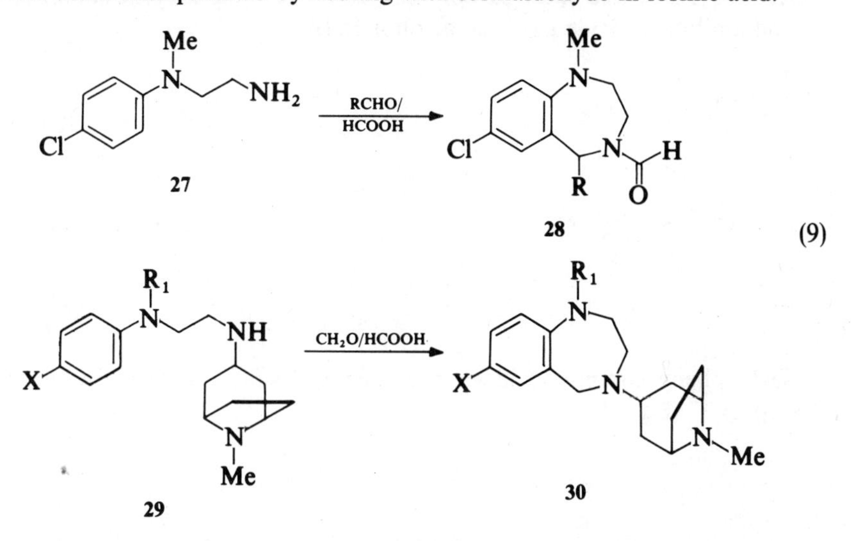


The 1,4-ditosylate **32** [$R_1 = R_2$ = (4-$MeC_6H_4$)$SO_2$, $R_3$ = X = H] was prepared by a double alkylation of the aniline **31** [$R_1 = R_2$ = (4-$MeC_6H_4$)$SO_2$, $R_3$ = X = H] with 1,2-dibromoethane and sodium butoxide (Eq. 10).[33] Removal of the tosyl groups by hydrolysis with sulfuric acid led in high yield to the parent compound. A related alkylation of the diamine **31** ($R_1$ = Me, $R_2$ = H, $R_3$ = Ph, X = Cl) with 1,2-dibromoethane at reflux temperature led to the benzodiazepine **32** ($R_1$ = Me, $R_2$ = H, $R_3$ = Ph, X = Cl).[34]

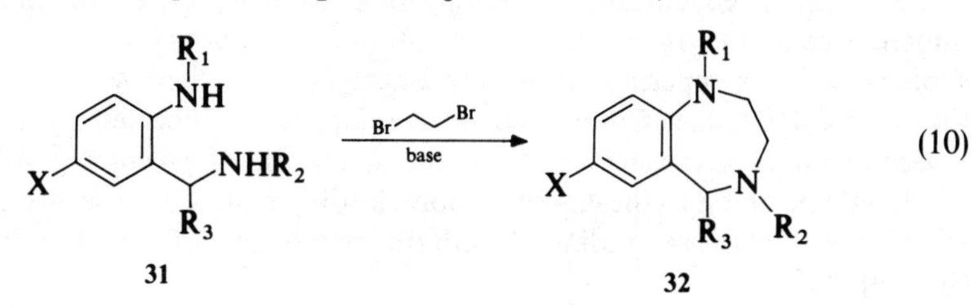

Intramolecular trapping of the carbonium ion by an amide nitrogen was found to be a successful way of obtaining tetrahydrobenzodiazepine **34**. The ring closure of the benzhydrol **33** to **34** was conveniently carried out with concentrated sulfuric acid at approximately 0°C (Eq. 11).[17]

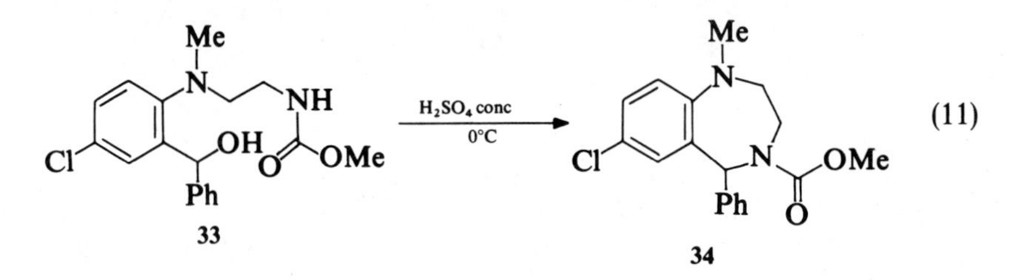


(11)

Reductive cleavage of the carbon–oxygen bond in the tricyclic compound **35** by sodium borohydride gave the alcohol **36** (Eq. 12).[35]

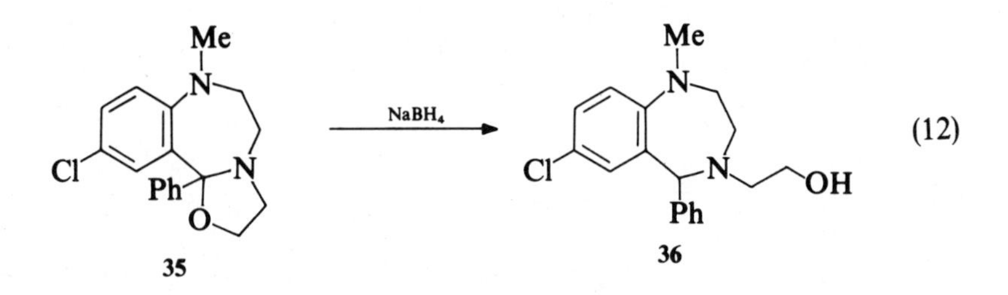


(12)

The 1-phenyl compound **38** was prepared by desulfurization of the tetracyclic diazepine **37** (Eq. 13).[33]

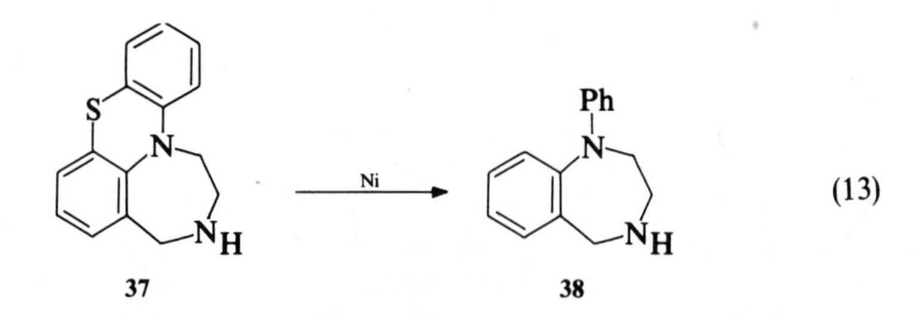


(13)

Reaction of the 2-chloromethylquinazoline 3-oxide **39** with the anion of *N,N*-dimethylacetamide, generated with lithium diisopropylamide in tetrahydrofuran at low temperature, gave the 2-acetylidene derivative **40** in addition to the expected nitrone. Compound **40** was apparently formed by addition of a second mole of the acetamide anion to the initial product.[17] Addition of nitromethane anion to the 4,5-imine bond leading to **42** was observed during the reaction of the nitrosoamidine **41** with nitromethane and potassium *t*-butoxide (Eq. 14).[36]

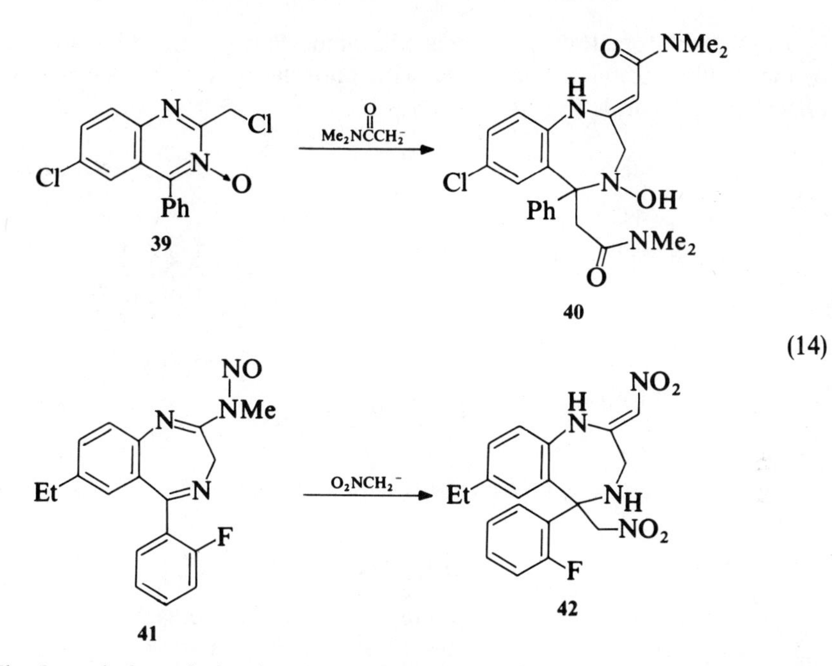


The 2-methylene derivative **44** resulted from the reduction of the oxime **43** (R = CN) with zinc and acetic acid followed by a reaction with dimethylformamide dimethylacetal. Compound **45** was similarly prepared by the reduction of its corresponding 4,5-azomethine precursor with zinc and acetic acid.[17] The tetrahydro derivative **46** resulted from the treatment of the oxime **43** (R = COOEt) with sodium borohydride (Eq. 15).[17]

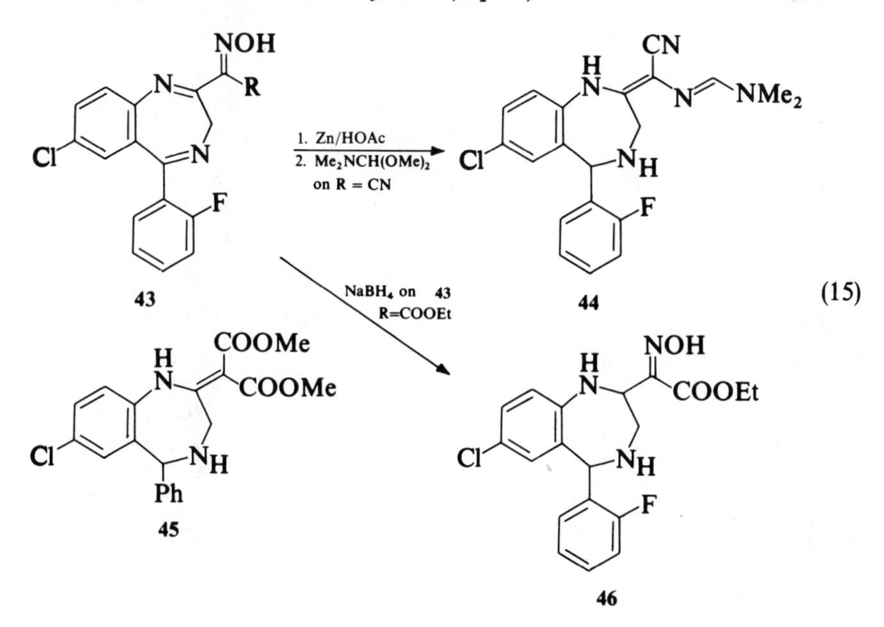

The 5-dichloromethylene-1,4-benzodiazepine **48** was formed by the reaction of the dichloroacetyl derivative **47** with phosphorus oxychloride and phosphorus pentachloride (Eq. 16).[17]

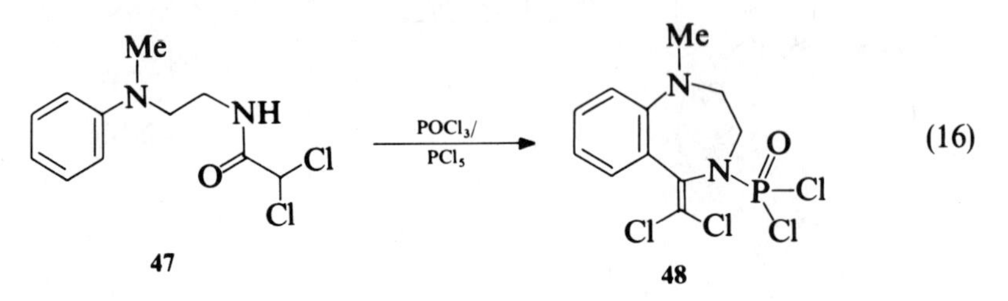


(16)

## 1.2. Reactions

### *1.2.1. Reactions with Electrophiles*

#### 1.2.1.1. Oxidation

The oxidation of the 2,3,4,5-tetrahydro-1*H*-1,4-benzodiazepines **18** to the 2,3-dihydro-1*H*-1,4-benzodiazepines **17** and of the 4-hydroxy analogs to the corresponding nitrones was discussed in Chapter VI, Section 1.1.7. Introduction of a carbonyl group at the 2-position of tetrahydro derivatives will be discussed in this chapter, in Section 2.1.

Oxidation of the 1-methyl-5,5-diphenyl compound **49** with chromium trioxide led to the 1-formyl analog **50** (Eq. 17).[29]

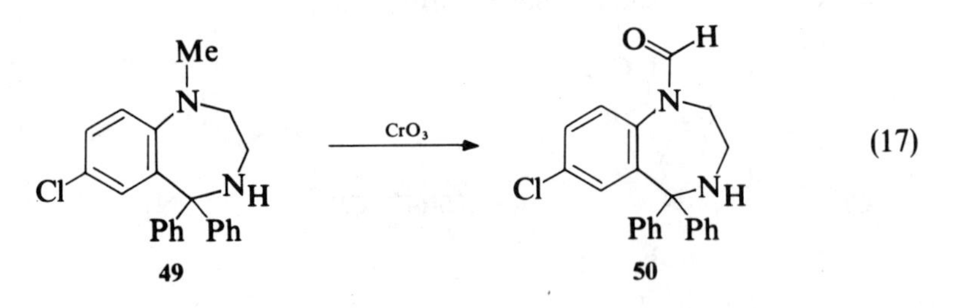


(17)

#### 1.2.1.2. Reactions with Nitrogen Electrophiles

The 4-acetyl compounds **51** (X = H, Cl) were nitrosated at the 1-position to give compounds **52** (Eq. 18).[37]

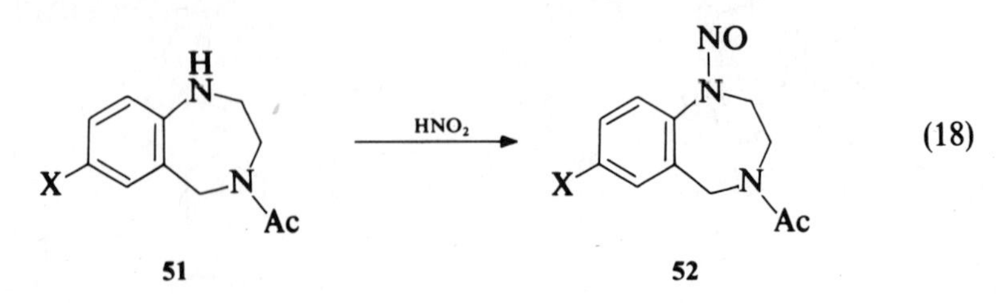


(18)

### 1.2.1.3. Reaction with Carbon Electrophiles

A. Alkylation. A basic side chain was attached by alkylation at the 4-position of **53** (X = Cl, $NO_2$) to yield the 4-dialkylaminoalkyl derivatives **54** (Eq. 19).[38]

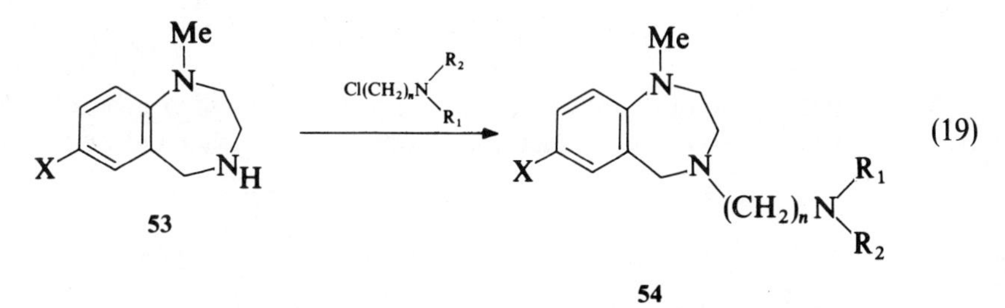


Arylation of the 4-tosyl derivative **55** with iodobenzene in the presence of copper and potassium carbonate yielded the 1-Ph-4-tosyl compound **56** (R = 4-$MeC_6H_4SO_2$). Subsequent hydrolysis yielded the 4-hydrogen analog (Eq. 20).[33]

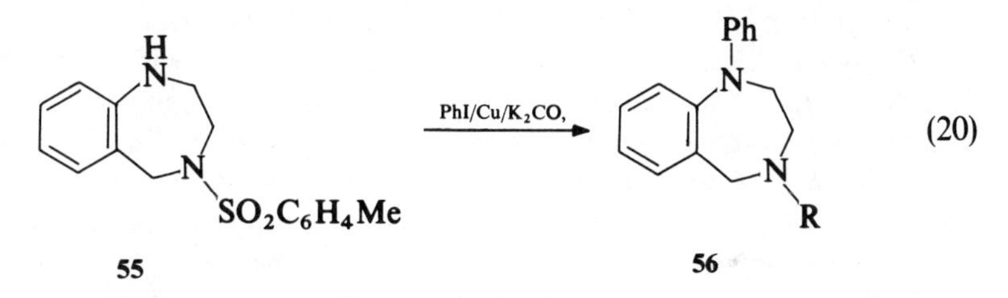


Quaternizations of the 4-position nitrogen were carried out on several 4-alkyl derivatives.[6] The 4-tropanyl compounds **30** ($R_1$ = Me, Et; X = H), when heated with methyl iodide, formed the diquaternary salts **57** (Eq. 21).[32] The monomethiodide was also isolated with the tropanyl nitrogen quaternized.

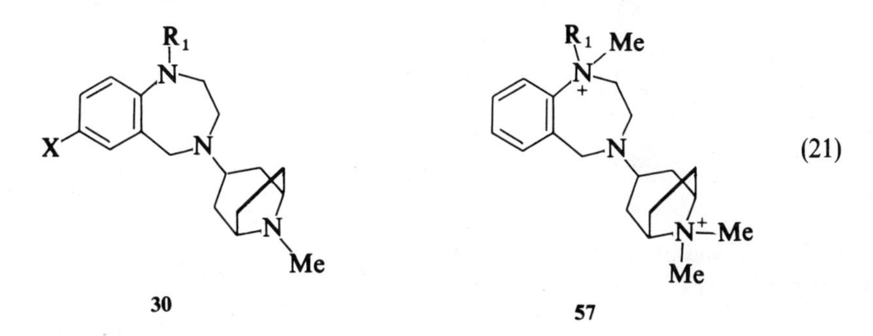


B. Reaction with Aldehydes and Ketones. Reaction of the parent compound **58** ($R_1$ = H) with formaldehyde or benzaldehyde yielded the 1,4-bridged compounds **59** ($R_1$ = H; $R_2$ = H, Ph).[39] Treatment of the 5-phenyl analog **58** ($R_1$ = Ph) with formaldehyde in refluxing benzene gave similarly 92% of **59** ($R_1$ = Ph, $R_2$ = H) (Eq. 22).[26]

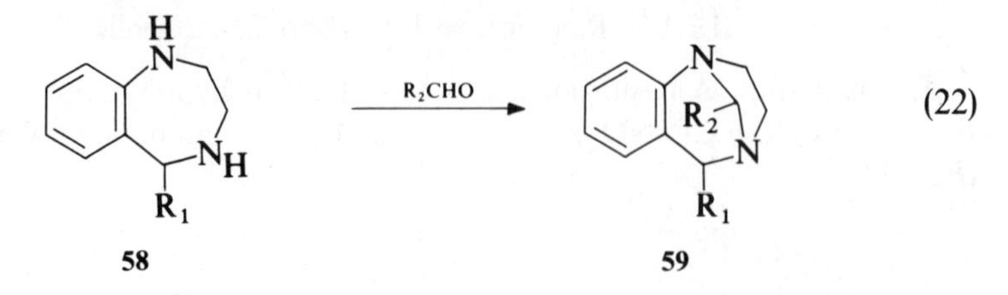


(22)

The 1-amino compound **60** was converted to the tetrahydrocarbazole **61** by reaction with cyclohexanone under conditions of the Fischer indole synthesis (Eq. 23).[37]

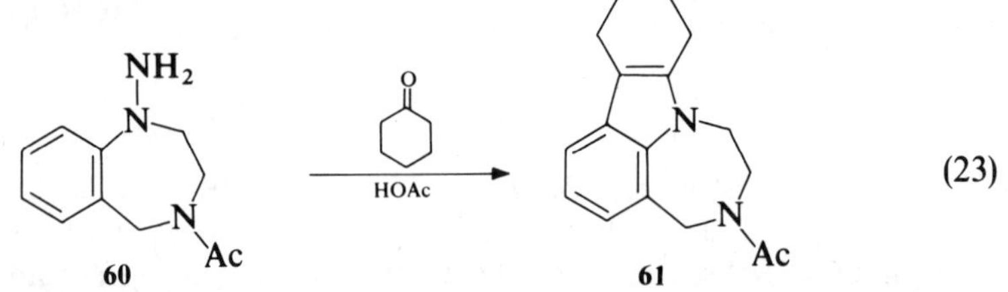


(23)

C. Acylation. Double acylations at the 1- and 4-positions were carried out on the parent 2,3,4,5-tetrahydro-1*H*-1,4-benzodiazepine with acetic anhydride at 100 °C.[6] Selective acetylation of the more basic nitrogen in the 4-position was possible with acetic anhydride in refluxing ether in the presence of triethylamine.[37] Compounds substituted at the 1-position were acylated at the 4-position using standard procedures.[15,40] Compounds bearing a substitutent at the 4-position were likewise acylated at the 1-position.[2,5] Introduction of the guanidino functionality at the 4-position, to yield **64**, was achieved by transfer of the guanidino moiety from the pyrazole **63** to the parent compound **62** ($R_1 = X = H$), as well as to the 1-methyl and 7-chloro analogs of **62** by heating at 190–200 °C (Eq. 24).[1]

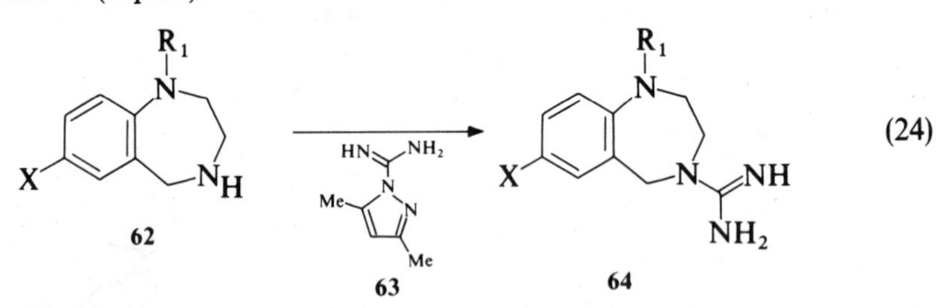


(24)

The guanidine **66** was prepared by reaction of the 4-phenyl derivative **65** with cyanamide (Eq. 25).[4]

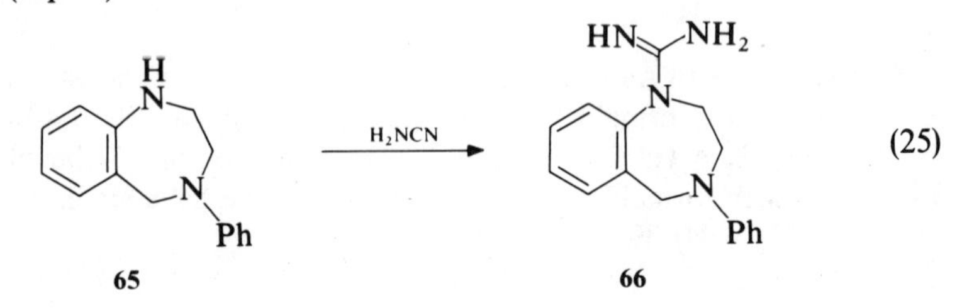


(25)

#### 1.2.1.4. Sulfonation

The parent tetrahydrobenzodiazepine **67** ($R_1 = R_2 = H$) was selectively tosylated at the 4-position nitrogen to yield **68** ($R_1 = R_2 = H$, $R_3 = 4$-$MeC_6H_4SO_2$) (Eq. 26).[33] Other tosylations were carried out on **67** ($R_1$ = Me, $R_2$ = Ph, X = Cl)[31] and on **67** ($R_1$ = Me, $R_2$ = H, X = Cl).[30] Reaction of the latter compound with methanesulfonyl chloride led similarly to the mesylate **68** ($R_1 = R_3$ = Me, $R_2$ = H, X = Cl) (Eq. 26).[30]

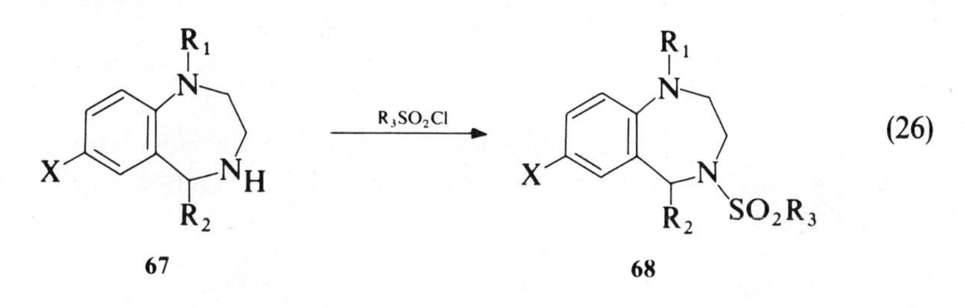


### *1.2.2. Reactions with Nucleophiles*

#### 1.2.2.1. Reduction

Lithium aluminum hydride was used to reduce the 1,4-diacetyl derivative **69** to the 1,4-diethyl analog **70** (Eq. 27).[6] The 1-nitroso compound **52** (X = H) was reduced to the 1-amino derivative **60** by zinc and acetic acid.[37]

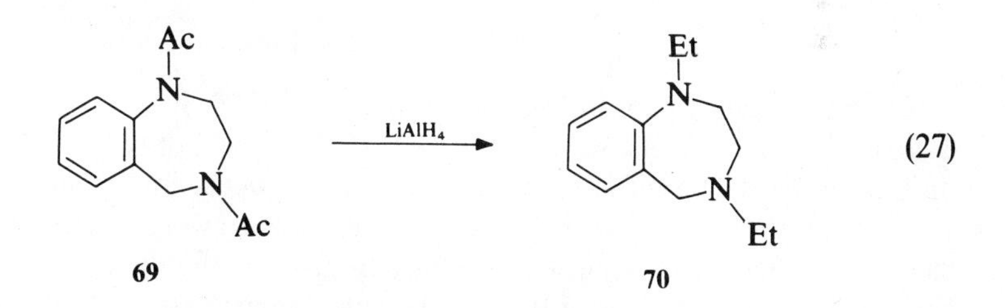


#### 1.2.2.2. Hydrolytic Reactions

Tosyl groups attached at the 1 and 4-positions were cleaved by treatment with 90% sulfuric acid at 110°C.[33] Selective hydrolysis of the 4-acetoxy moiety of **71** afforded the hydroxyamine **72** (Eq. 28).[20]

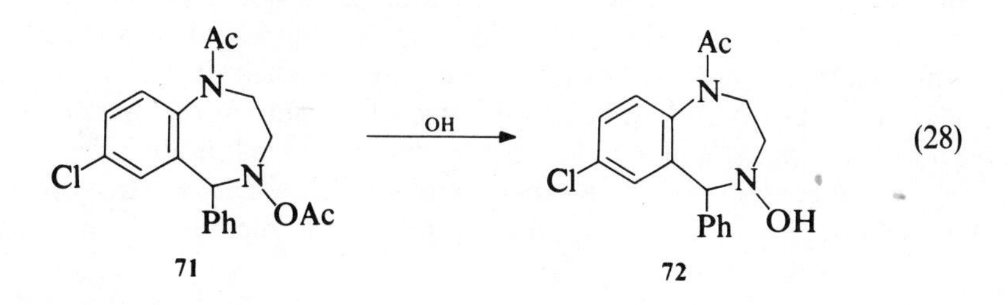

Hydrolytic ring opening of compounds **24** ($R_1$ = Me; $R_3$ = Ph; $R_4$ = Me, MeO, EtO, benzyloxy; X = 7-Cl) afforded the corresponding benzophenones **73** (Eq. 29).[29]

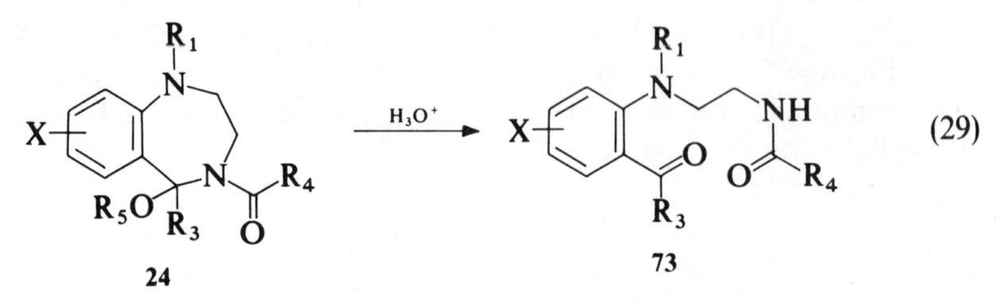


Treatment of the bismethiodide **57** with the hydroxide form of the ion-exchange resin IRA-400 was reported to yield the vinylamine **74**, characterized as a dipicrate (Eq. 30).[32]

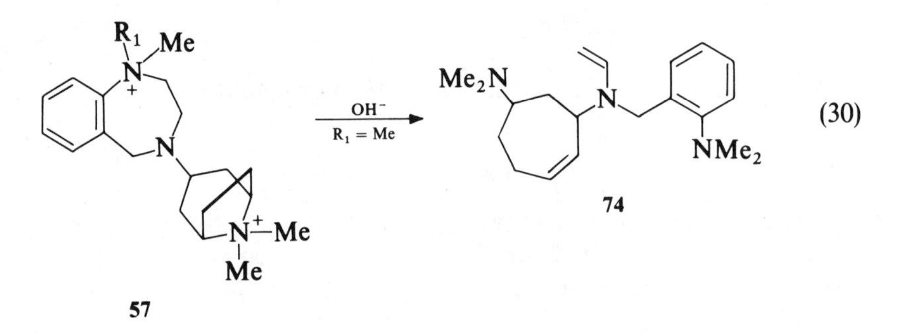


### 1.2.2.3. Other Reactions

Abstraction of a proton from the 5-position of **75**, in which $R_3$ represents a leaving group, leads to the formation of the 2,3-dihydro-1*H*-1,4-benzodiazepines **76**. This synthesis of **76** is discussed in detail in Chapter VI, Section 1.1.8. When the 4-hydroxy compounds **75** ($R_3$ = HO) were treated with thionyl chloride, the tetrahydroquinoxalines **80** were formed.[20] When phosphorus oxychloride was used in place of thionyl chloride, the formation of the quinoxaline became the main reaction. An electron-rich nitrogen at the 1-position of the benzodiazepine **75** is essential for this ring contraction to proceed, since treatment of the 1-acetyl derivative **75** ($R_1$ = Ac, $R_2$ = H, $R_3$ = HO, X = 7-Cl) with phosphorus oxychloride gave only the dehydration product **76**. The mechanism of this ring contraction was formulated as shown in Eq. 31. The intermediacy of the ions **78** or **79** was supported by isolation of the 4-benzyltetrahydroquinoxaline from a reductive workup. Phenyl isocyanate could also effect the ring contraction as well as the dehydration steps. When **75** ($R_1$ = Me, $R_2$ = H, $R_3$ = HO, X = 7-Cl) was heated in toluene to reflux in the presence of excess phenyl isocyanate, the urea **81** was the major isolated product.[20]

The oxidation of 7-chloro-1-methyl-5-phenyl-2,3,4,5-tetrahydro-1*H*-1,4-benzodiazepine to the corresponding imine was reported to proceed in di-

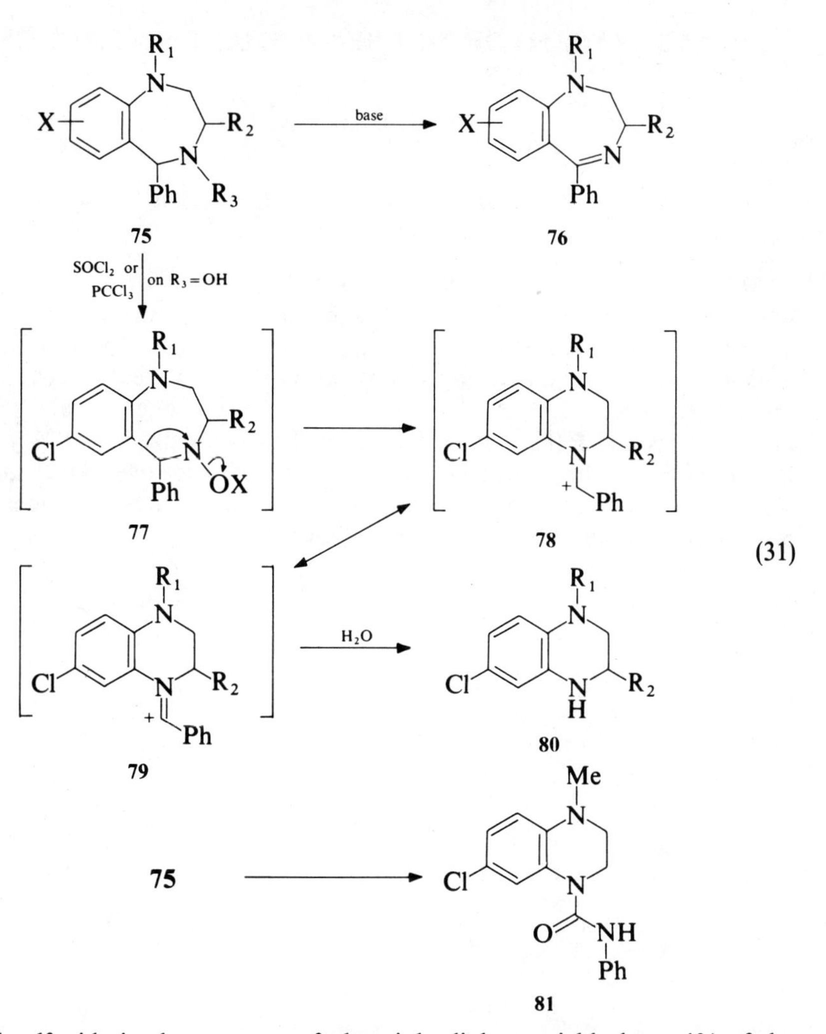


methyl sulfoxide in the presence of ultraviolet light to yield about 4% of the imine.[40]

The amidine **44** was thermally cyclized to the imidazobenzodiazepine **82** by heating in acetic acid (Eq. 32).[17]

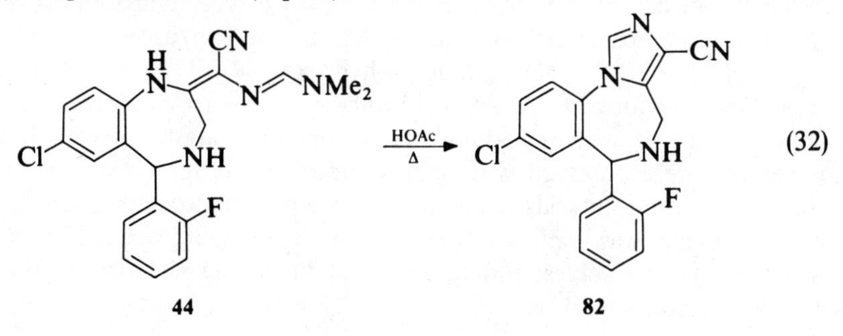

# 2. 1,3,4,5-TETRAHYDRO-1,4-BENZODIAZEPIN-2(2*H*)-ONES

## 2.1. Synthesis

### 2.1.1. *By Reduction and Oxidation*

Most tetrahydrobenzodiazepin-2-ones were prepared by reduction of the imine bond in the dihydrocompounds **6** (Eq. 33). Reagents of choice were zinc and acetic acid[41,42] and hydrogen over platinum catalyst.[43–56] Catalytic hydrogenation over palladium on carbon was also reported to effect reduction of the imine **6** ($R_1 = R_2 = X = H$, $R_3 = Ph$) to the corresponding amine **5**.[57,58] The parent compound **5** ($R_1 = R_2 = R_3 = X = H$) was obtained in 60% yield by hydrogenation of the corresponding imine **6** or its 4-oxide over Raney nickel.[59] This catalyst generally left the imine untouched. The reduction of the imine in 5-substituted compounds **6** was rarely observed.[60]

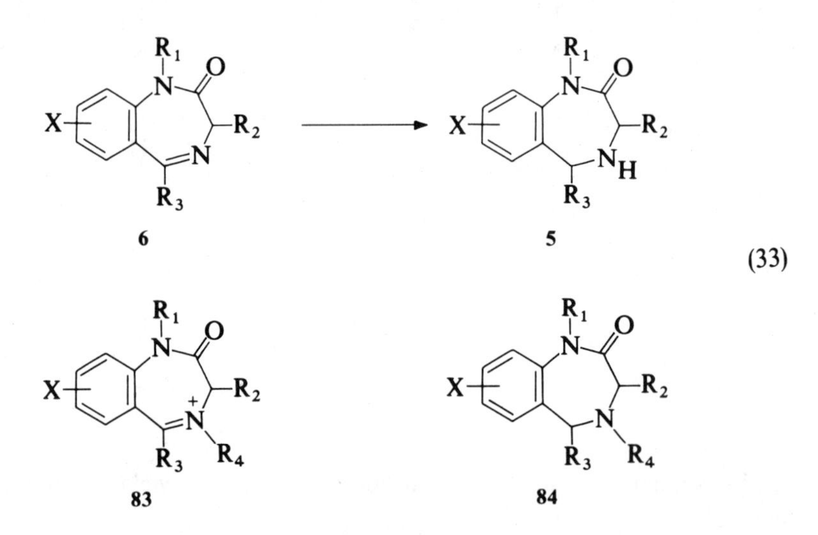


The 4-oxides of **6** were reduced to the amines **5** in quantitative yield by zinc and acetic acid.[41,61,62] Reduction of **83** ($R_4 = O$), with hydrogen in acetic acid solution, using platinum as catalyst, led to the 4-hydroxy compounds **84** ($R_4 = OH$).[41,47,49,51] The parent 4-hydroxy **84** ($R_1 = R_2 = R_3 = X = H$, $R_4 = OH$) was formed in 15% yield during the hydrogenation of the corresponding 4-oxide with palladium on carbon as catalyst.[59] High yield imine reductions were observed with sodium borohydride in acetic acid[63] or with sodium cyanoborohydride in methanol containing hydrochloric acid.[60] Aluminum amalgam was reported to reduce **6** ($R_1 = H$, $R_2 = COOEt$, $R_3 = Ph$, $X = 7\text{-Cl}$) to the corresponding amine of undefined stereochemistry.[64] The 4-hydroxy compound **84** ($R_1 = Me$, $R_2 = H$, $R_3 = 2\text{-}FC_6H_4$, $R_4 = OH$,

X = 7-I) was prepared by treatment of the appropriate nitrone with sodium borohydride in ethanol–tetrahydrofuran at 50–60°C.[65] This reagent allowed also the conversion of the quaternary salts **83** to the 4-alkyl derivatives **84** ($R_4$ = alkyl).[54,58,66]

Catalytic hydrogenation of the oxazolobenzodiazepines **85** and **86** (R = Me) over platinum catalyst led to the 4-(2-hydroxyethyl) derivative **87** (R = Me) (Eq. 34).[67] The reduction of the carbon–oxygen bond of the oxazole **86** (R = H) to yield **87** (R = H) was also possible with sodium borohydride.[35]

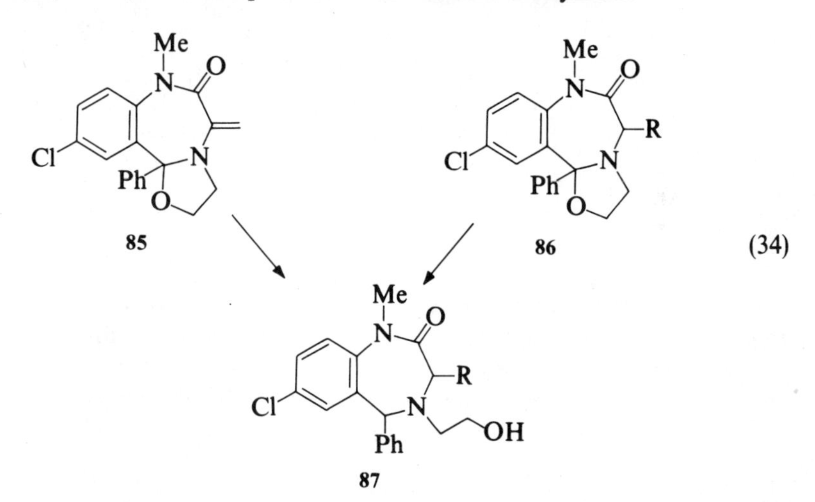


As demonstrated in Eq. 35, the 1,5-dihydro tautomer **88** could also be reduced to the amine **89** by hydrogen over platinum catalyst.[68]

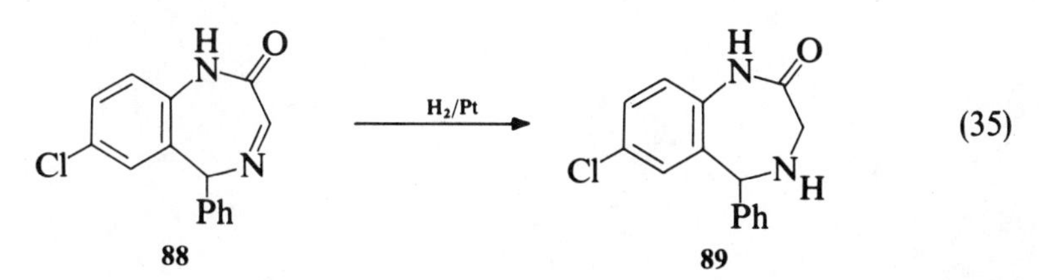


A reductive alkylation of the imines **90** (X = H, Cl, $NHCH_2SMe$) was observed, when these compounds were treated with Raney nickel in refluxing ethanol. The 7-chloro compound was thus converted to the 4-ethyl derivative **91** (X = H) by concomitant dehalogenation. The methylthio function of

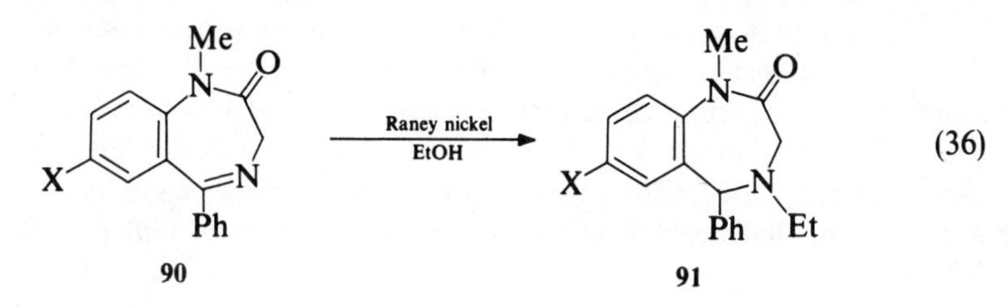

**90** (X = $MeSCH_2NH$) was cleaved under these conditions and led to **91** (X = MeNH) (Eq. 36).[37]

The patent literature describes the direct conversion of the carbobenzoxy compounds **92** to the tetrahydro-2-ones **93** by catalyic hydrogenation over palladium on carbon in the presence of acetic or hydrochloric acid. The 7-amino derivative **93** ($R_1$ = H, X = $NH_2$) was also formed, in 84% yield, by this process (Eq. 37).[69]

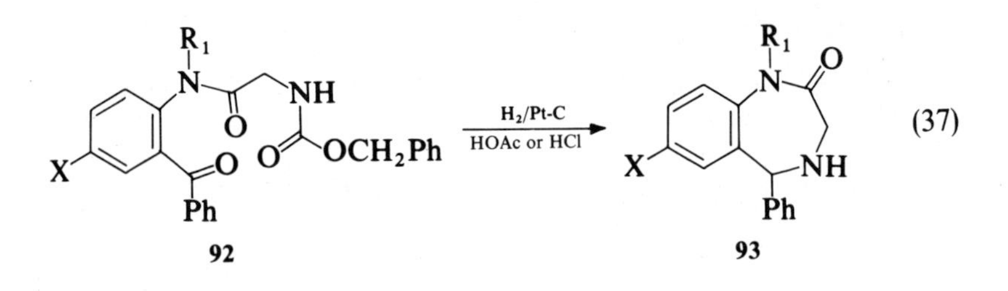


A few syntheses of tetrahydrobenzodiazepin-2-ones by oxidation were also reported. The 4-acetyl derivative **94** (R = Ac) was oxidized to the corresponding 2-one **95** by potassium permanganate in approximately 95% yield (Eq. 38).[70] Introduction of a 2-carbonyl function into the tetrahydro compound **94** (R = H) was achieved in 15% yield by using chromium trioxide in sulfuric acid.[52,71]

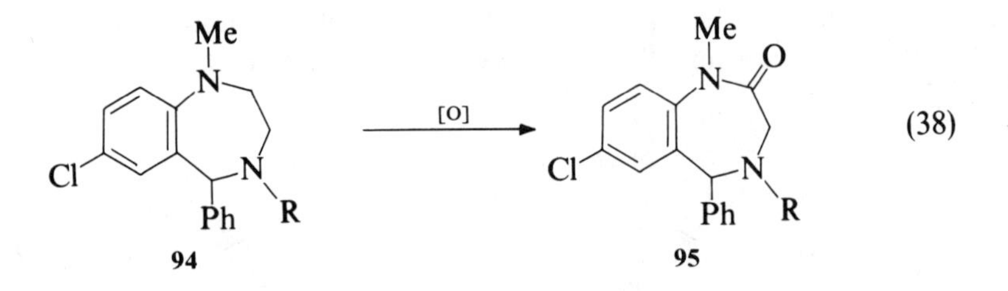


### 2.1.2. *By Additions to the 4,5-Double Bond*

5,5-Disubstituted compounds **97** were obtained by addition of a variety of nucleophiles to the quaternary salts **96** ($R_2$ = alkyl). Nucleophiles added to **96** include ammonia, amines,[72] alkoxides,[72] and Grignard reagents[72,73] (Eq. 39). The intermediate acyliminium ions **96** ($R_2$ = acyl), generated by addition of acyl chlorides on the appropriate imine, also added nucleophiles such as alkoxides.[17,29] Compounds **97** ($R_1$ = H, Me; $R_2$ = COOMe, COOEt; $R_3$ = Ph, 2-pyridyl; $R_4$ = OEt, OMe) were formed when the corresponding imines were stirred in methylene chloride with alkyl chloroformate and solid potassium carbonate.

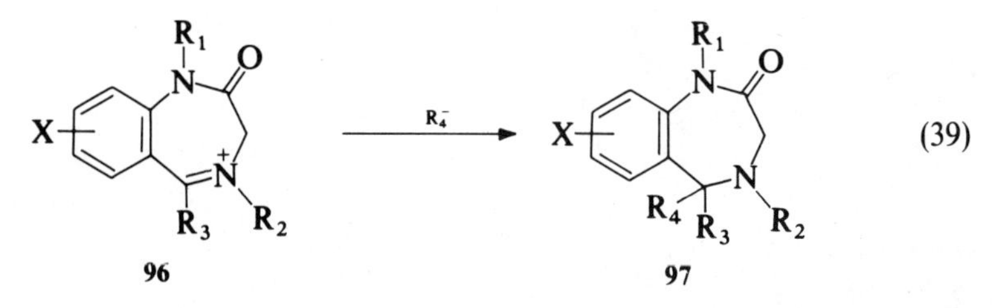


The formation of the 4-nitro compound **97** ($R_1$ = Cl, $R_2$ = $NO_2$, $R_3$ = Ph, $R_4$ = AcO, X = 7-Cl) by the reaction of the corresponding imine with nitric acid in acetic anhydride may be rationalized by addition of acetate anion to the intermediate 4-nitroiminium species **96** ($R_2$ = $NO_2$).[74] The reaction of diazepam (**98**) with acetylene dicarboxylate in methanol led to **99** (Eq. 40).[75] This transformation represents another example of the successive addition of electrophile and nucleophile to the imine bond.

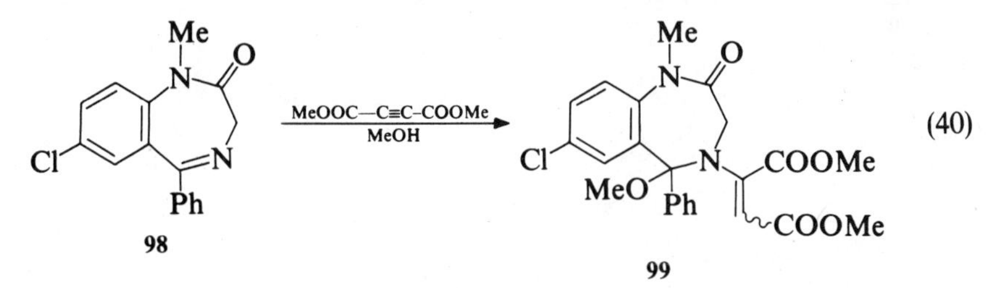


The 4-hydroxybenzodiazepin-2-ones **101** resulted from the addition of Grignard reagents to the nitrone **100** (Eq. 41). The Grignard reagents used were phenylmagnesium bromide[75,76] and methylmagnesium bromide.[29]

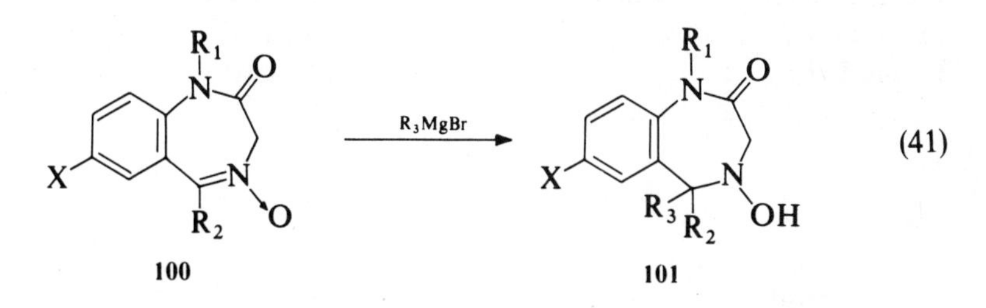


### 2.1.3. *By Ring Expansion*

A Beckmann rearrangement of the oximes of tetrahydroquinolin-4-ones **102** provided another method for the preparation of benzodiazepin-2-ones **103**.[56,77,78] The Schmidt reaction on the quinolone **102** gave higher yields of the product **103**.[56,78] Compound **103** (R = H, X = Cl) was obtained in 48% yield by treating the quinolone **102** (R = H, X = Cl) with excess sodium azide in concentrated sulfuric acid (Eq. 42).[78]

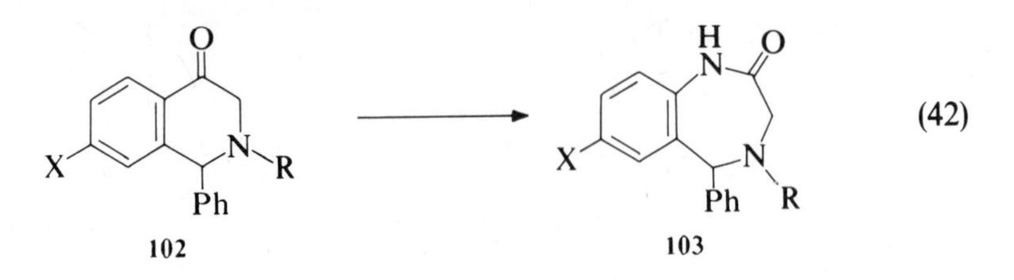


### 2.1.4. *Other Syntheses*

According to a Russian patent,[79] the benzhydrols **104** may be converted in high yields to the tetrahydrobenzodiazepin-2-ones by first treating the benzhydrole with trifluoroacetic anhydride and triethylamine at low temperature and subsequently treating with glycine methyl ester hydrochloride and triethylamine in refluxing chloroform (Eq. 43).

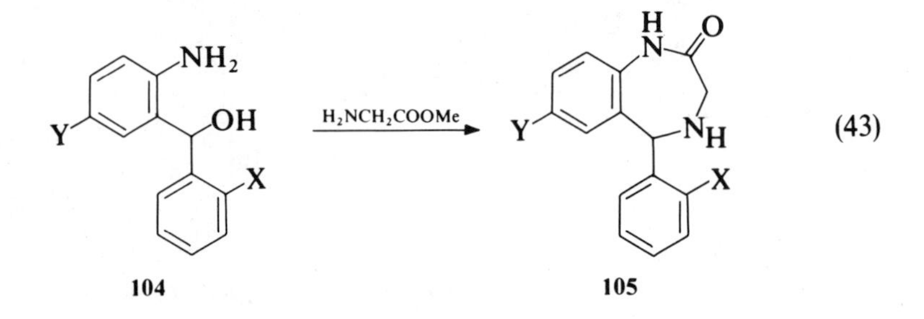


The method mentioned above may be considered a variation of the synthesis, previously reported, which involved preparation of the glycine derivative **106** followed by ring closure and by dehydration to form **89**. The ring closure could be effected by heating in xylene and azeotropically removing water,[71,80] or by using phosphorus pentachloride (Eq. 44).[81]

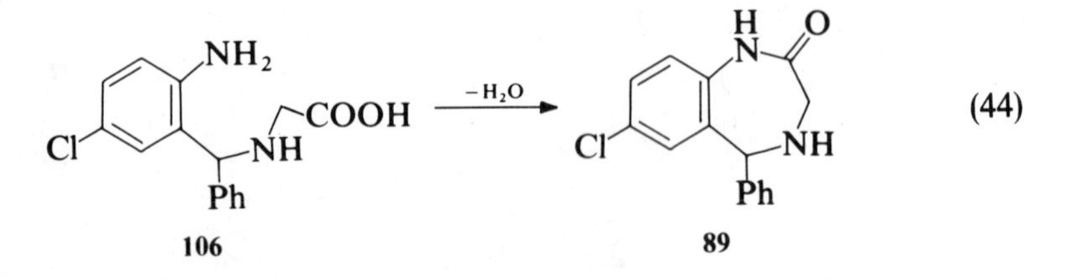


Generating the carbonium ion of the benzhydrol **107** led to ring closure through an intramolecular trapping by the nitrogen atom of the glycine moiety. Thus **107** ($R_1 = Et_2NCH_2CH_2$, $R_2 = H$, $R_3 = 2\text{-}FC_6H_4$) was converted to the benzodiazepine **108** by treatment with hydrogen bromide in acetic acid (Eq. 45).[82] The 4-acyl derivatives **108** ($R_2$ = Ac, COOEt) were obtained in high yield when the corresponding benzhydrols were stirred in cold concentrated sulfuric acid.[17]

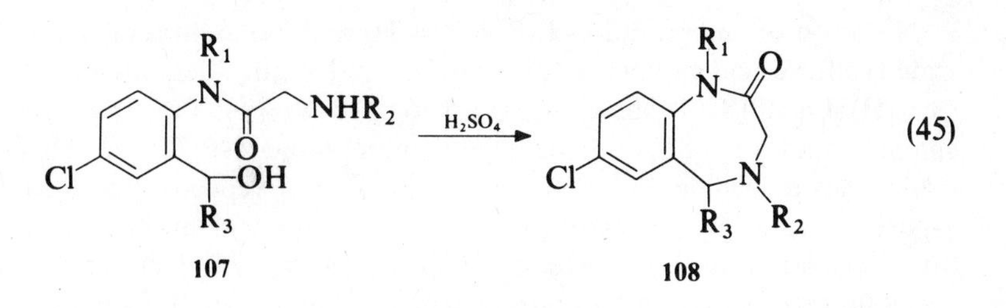


## 2.2. Reactions of 1,3,4,5-Tetrahydro-1,4-benzodiazepin-2(2*H*)-ones

### 2.2.1. *Reactions with Electrophiles*

#### 2.2.1.1. Halogenation

The 4-bromo derivatives **110** (X = Br) are most likely intermediates in the oxidation of **109** to the corresponding imines.[83,84] The 4-bromo compound **110** [$R_1$ = Me, $R_2$ = 3,4-$(MeO)_2$-benzyl, X = Br] was isolated and characterized by Broger.[85]

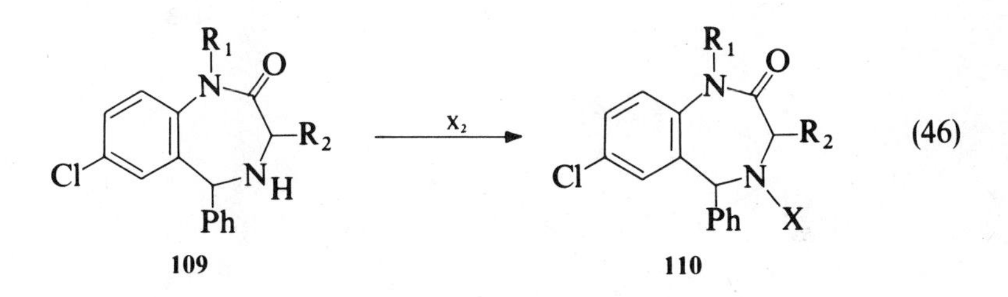


#### 2.2.1.2. Oxidation

The oxidation of 1,3,4,5-tetrahydro-1,4-benzodiazepin-2(2*H*)-ones to the corresponding 1,3-dihydro and 1,5-dihydro derivatives was discussed in Chapter VII, Section 1.1.5.

#### 2.2.1.3. Reactions with Nitrogen Electrophiles

The 4-nitroso compounds **111** were prepared by reaction of the tetrahydrobenzodiazepin-2-ones **93** with nitrous acid (Eq. 47).[86]

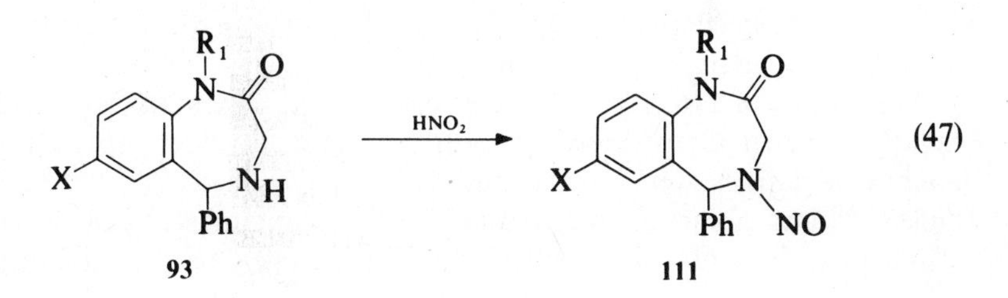

Nitration of compound **112** (R = X = H) with potassium nitrate in concentrated sulfuric acid gave a mixture of the 4′- and 3′-nitro derivatives, **113** and **115** (X = H) (Eq. 48).[52,87] The 2′-fluoro analog **112** (R = H, X = F) afforded, under similar conditions, 81% of the 5′-nitro compound **115** (X = F).[87] The 4-hydroxybenzodiazepine **112** (R = OH, X = H) was reported to nitrate at the 7-position when treated with nitric acid in acetic acid–methylene chloride at 0°C.[76] Nitration of the same compound with nitric acid in sulfuric acid at 0°C was found to lead directly to nitrazepam, by elimination of water and nitration at the 7-position.[76] The 7-chloro derivative of **112** (R = OH, X = F) was nitrated in the para position to the fluorine, yielding **114**.[28]

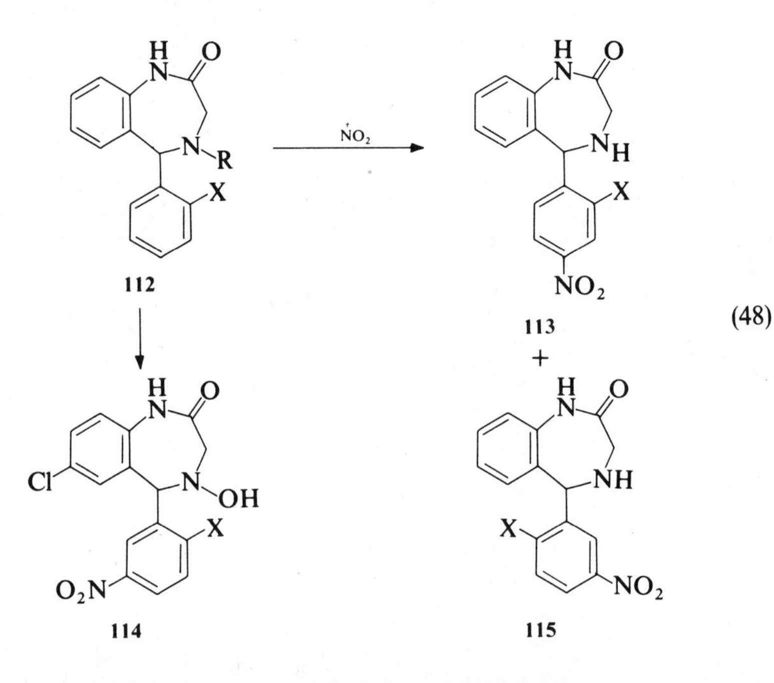


2.2.1.4. Reactions with Carbon Electrophiles

A. Alkylation. Compounds **116** were selectively alkylated at the 1-position by generating the anion with a strong base such as sodium methoxide or sodium hydride and reacting this anion with the alkylating agent at temperatures ranging from −10 to 25°C. Alkylations at the 1-position were thus carried out with methyl iodide[43] and with 4-bromobutanoic acid ethyl ester (Eq. 49).[88] 4-Alkyl and 4-acyl derivatives **116** (R = alkyl, acyl) were readily alkylated at the 1-position under similar conditions.[28,55,84]

The anions of **116** ($R_2$ = H) were reacted with excess alkylating agents at slightly elevated temperature to form the 1,4-dialkyl derivatives **118**. Such double alkylations were carried out with methyl iodide,[43,45,46,49] allyl bromide,[43,49] and *N*-methylbromoacetamide.[89] Compounds **116** ($R_2$ = H) could also be selectively alkylated at the 4-position nitrogen. Successful monoalkyla-

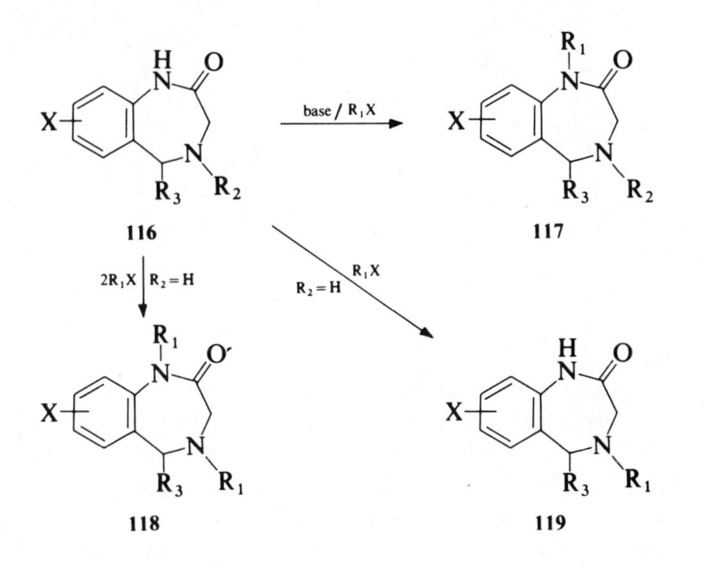


(49)

tions at the 4-position were reported with methyl iodide,[43,55] benzyl chloride,[56] and ethyl bromoacetate[35] to give the 4-alkyl derivatives **119**. Compounds bearing a substituent at the 1-position were similarly reacted with methyl iodide,[29,43,54,55] dimethyl sulfate,[29] ethyl iodide,[43] allyl bromide,[29,43] propargyl bromide,[29] *N*-methylbromoacetamide,[54] and 1-chloro-2-(diethylamino)-ethane[44,55] to give the corresponding 4-alkyl derivatives.

Ethyl acrylate, added to the 4-position nitrogen in a Michael fashion and yielded the corresponding 4-(2-ethoxycarbonyl)ethyl derivative.[35] Quaternizations of the 4-methyl compounds[49] with methyl iodide was reported to give the 4,4-dimethyldiazepinium salts. However, for **117** ($R_1$ = 2-$Et_2N$-ethyl, $R_2$ = Me, $R_3$ = 2-$FC_6H_4$, X = 7-Cl), quaternization yielded the quaternary salt on the nitrogen in the side chain.[55]

B. Reactions with Aldehydes and Ketones. The 4-amino compound **120** was reacted with a variety of substituted benzaldehydes to form the imines **121** (Eq. 50).[86]

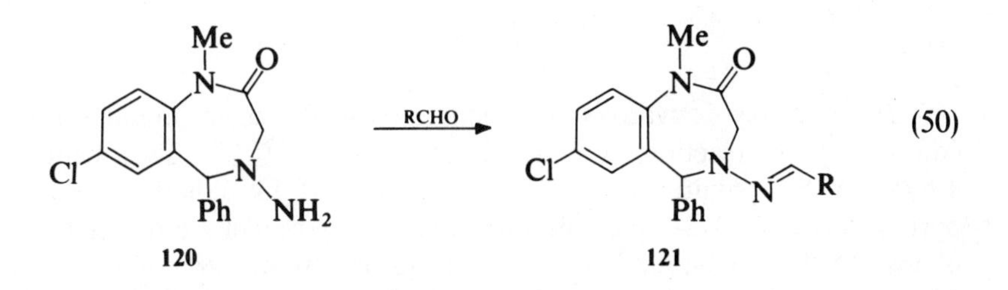


(50)

Condensation of the diamine **122** (X = Br, Cl) with benzaldehyde and acetone led to the tetracyclic quinazolinobenzodiazepines **123** (Eq. 51).[53]

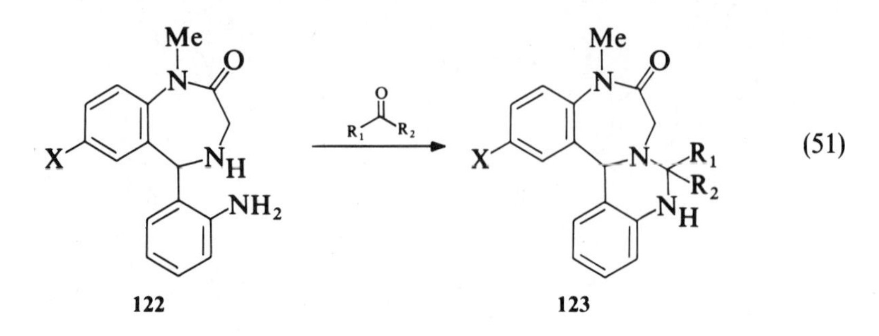


C. Acylation. Acylations of the 4-position nitrogen were carried out with a variety of reagents under standard conditions. They included acid anhydrides,[29,60b,90] acid chlorides,[29,90] phosgene,[90] chloroformates,[83,84,90] and isocyanates.[60b,90] The 4-formyl derivatives were obtained by reaction of **124** with ethyl formate.[90] Addition of phosgene to **124** resulted in the formation of the 4-chlorocarbonyl compounds **125** ($R_3$ = Cl) (Eq. 52).[90] The urea **126** was obtained in 66% yield from the reaction of **124** ($R_1$ = Me, $R_2$ = Ph, X = 7-Cl) with phosgene.[90] The 4-aminocarbonyl derivatives **125** ($R_3$ = $NH_2$) were prepared by means of potassium cyanate and acetic acid.[63,90] Amino acids were attached the 4-position of **124** by the mixed anhydride method, followed by deprotection of the amino function.[73,90]

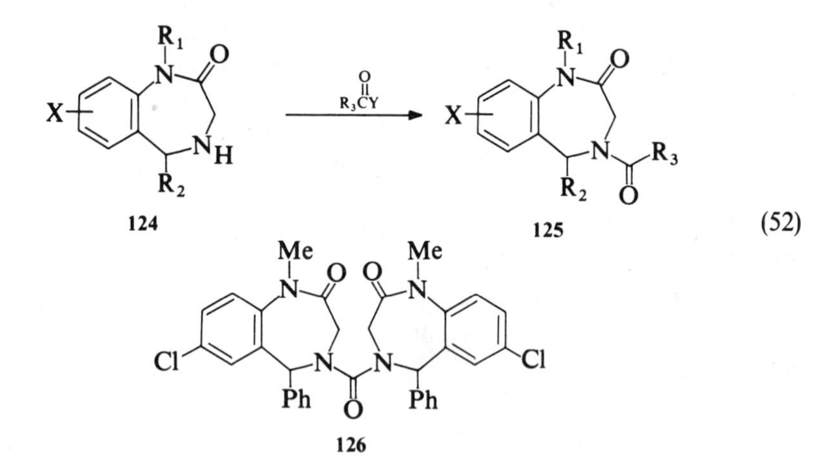


The 4-hydroxy derivatives were converted to the corresponding 4-acetoxy compounds by reaction with acetic anhydride.[41,47,91,92] Treatment of the 4-hydroxy compound with phenylisocyanate and 1,4-dimethylpiperazine or pyridine in refluxing 2-acetoxybutane led to elimination of water with formation of the 4,5-double bond.[75] The same transformation was noticed when the 4-hydroxy compounds were reacted with dicyclohexylcarbodiimide in pyridine[76] or in refluxing toluene.[41] Acylations of the 4-amino group in **120** were also reported.[86]

An intramolecular acylation was observed with the ester **127**, which upon heating in quinoline, cyclized in 54% yield to the pyrrolobenzodiazepine **128** of undefined stereochemistry (Eq. 53).[62]

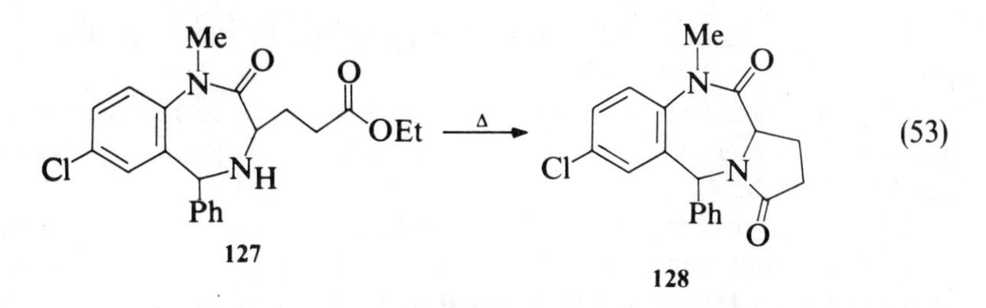


Reaction of the diamines **122** (X = Br, Cl) with carbonyldiimidazole afforded **129**.[53] Compound **130** (R = H) resulted from the reaction of **122** (X = Cl) with triethyl orthoformate, while the methyl analog **130** (R = Me) was formed by cyclization of the 2′-acetylamino derivative (Eq. 54).[53]

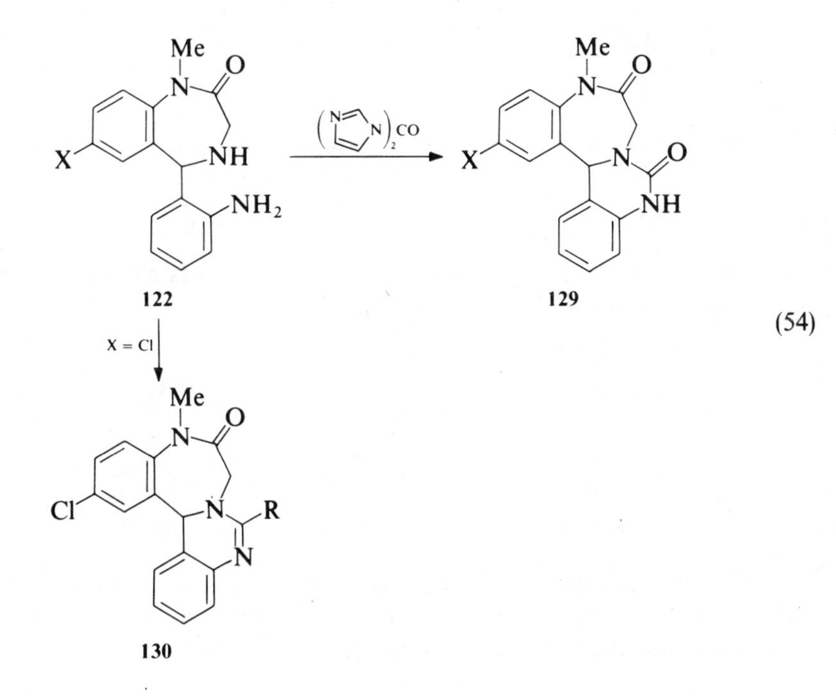


### 2.2.1.5. Sulfonation

Reaction of the 4-position nitrogen with *p*-toluenesulfonyl chloride[41,47,68,78,82,93] and with methanesulfonyl chloride[68,94] led to the 4-sulfonylated benzodiazepines. Treatment of the 4-hydroxy compound with thionyl chloride in chloroform yielded 34% of the dehydration product.[76] This dehydration was also effected by phosphorus oxychloride in pyridine.[65]

### *2.2.2. Reactions with Nucleophiles*

#### 2.2.2.1. Reduction

The reduction of the 2-carbonyl group to the hydrocarbon level was discussed in Section 1.1.1 above.

The 4-nitroso derivatives **111** were reduced to the corresponding 4-amino derivatives by zinc and acetic acid.[86] The 7-nitro-4-nitroso analog was similarly converted to the 4,7-diamino compound.[86]

The 1-chloro compound **131** was reduced by methylamine in methylene chloride at room temperature to the lactam **132** (Eq. 55).[74] The chloronium ion was apparently transferred from **131** to the methylamine.

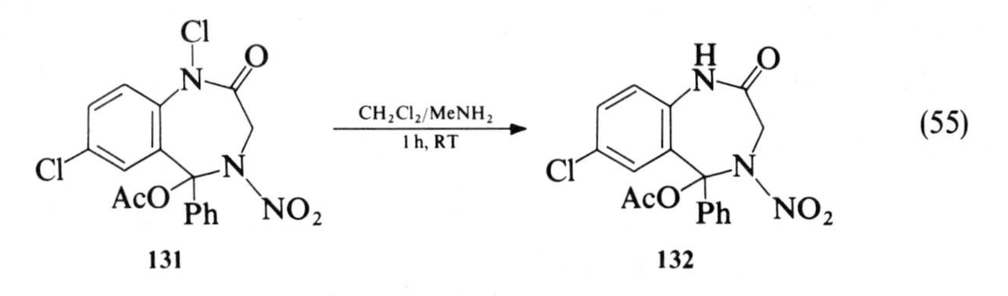


Hydrogenolysis of the 4-benzyl derivative was carried out over palladium catalyst in the presence of acetic and hydrochloric acids.[56,78]

#### 2.2.2.2. Reactions with Oxygen Nucleophiles

The hydrolysis of **89** to the ring-opened acid **106** was effected by barium hydroxide in boiling methanol–water (Eq. 56).[51,80]

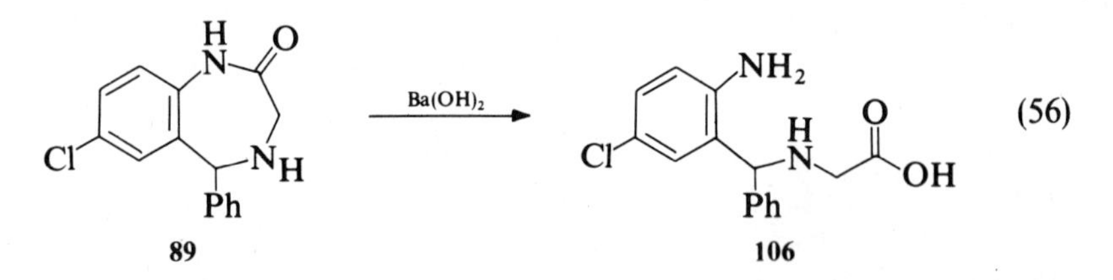


The 4-nitro derivatives **133** were hydrolyzed by water to the benzophenones **134**.[74] The 4-acyl compounds **135** underwent the same type of hydrolytic ring opening to yield the benzophenones **136** (Eq. 57).[17,29]

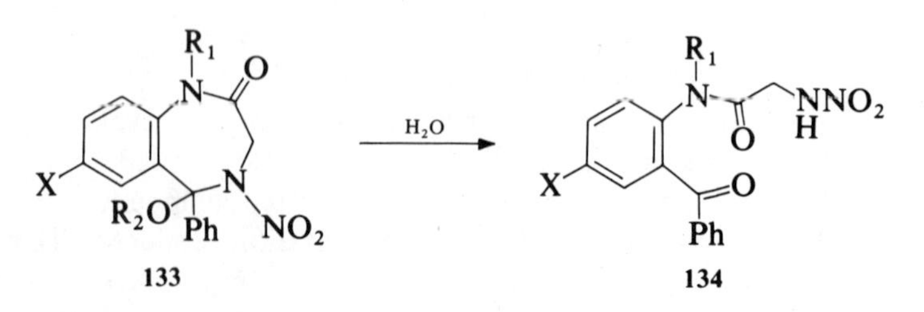

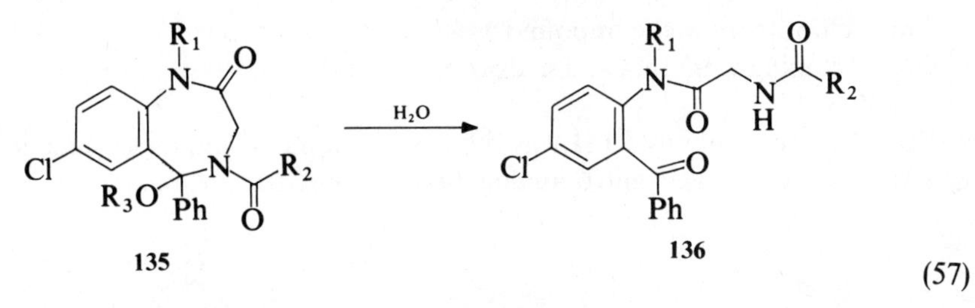


(57)

Other reported hydrolytic transformations include removal of a 4-acetyl group[70] or conversion of side chain ester functions attached at the 1- and 4-positions of the corresponding acid[88] or alcohol.[35]

### 2.2.2.3. Reactions with Nitrogen Nucleophiles

Compound **89** was converted to the amidine **137** by reaction with methylamine and titanium tetrachloride (Eq. 58).[95]

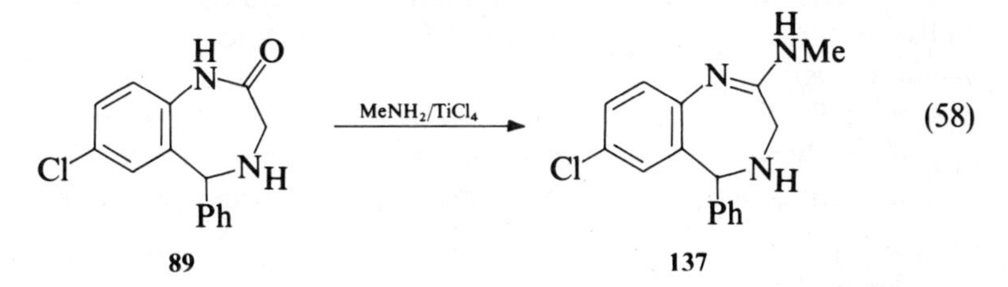


(58)

The 4-nitro compounds **133** reacted with ammonia and other amines in a mixture of methanol and methylene chloride at room temperature to form the 3-amino benzodiazepines **138** (Eq. 59).[74] Reaction of the same compounds with triethylamine resulted in deprotonation at the 3-position followed by elimination of the 4-nitro group to give compounds **139**, which were converted to the 3-amino derivatives **138** by reaction with amines.[74]

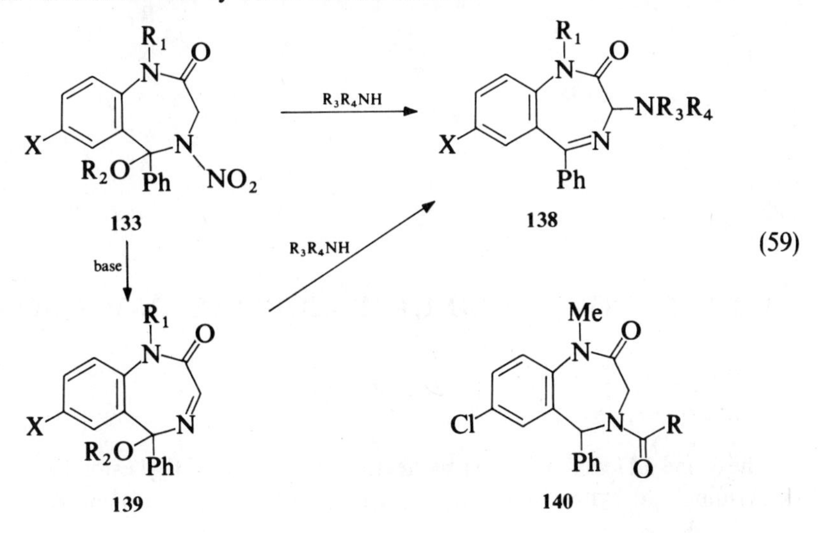


(59)

The 4-chlorocarbonyl compound **140** (R = Cl) was reacted with a variety of amines, including ammonia and hydrazine, to yield the corresponding ureas **140** ($R = NR_1R_2$).[90]

The 7-amino compound **141** was subjected to a nitro Sandmeyer reaction to give the corresponding 7-nitro analog **142** (Eq. 60).[83]

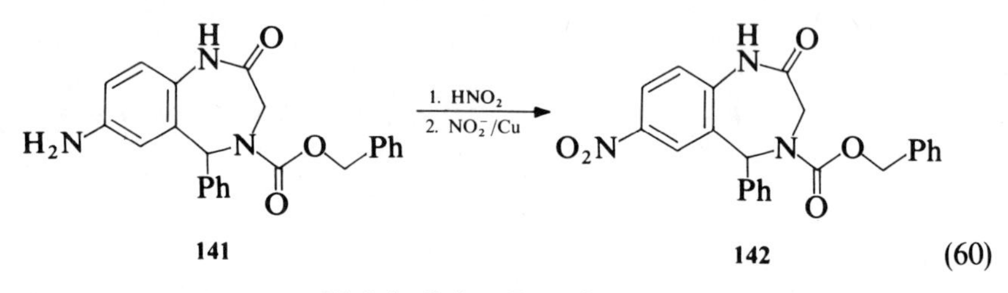


(60)

### 2.2.2.4. Other Reactions

Elimination of the elements HY from the 4-substituted compounds **143** led to a mixture of the tautomers **144** and **145**. Such eliminations were carried out with 4-sulfonyl derivatives[41,68,82,93,94] using a variety of strong bases, with 4-acetoxy compounds,[41,91,92] 4-nitro derivatives,[74,96] and 4-hydroxy compounds (Eq. 61).[75] The 4-hydroxy compounds were, for this purpose, converted in situ to acyl derivatives with phenyl isocyanate[75] or dicyclohexylcarbodiimide.[41]

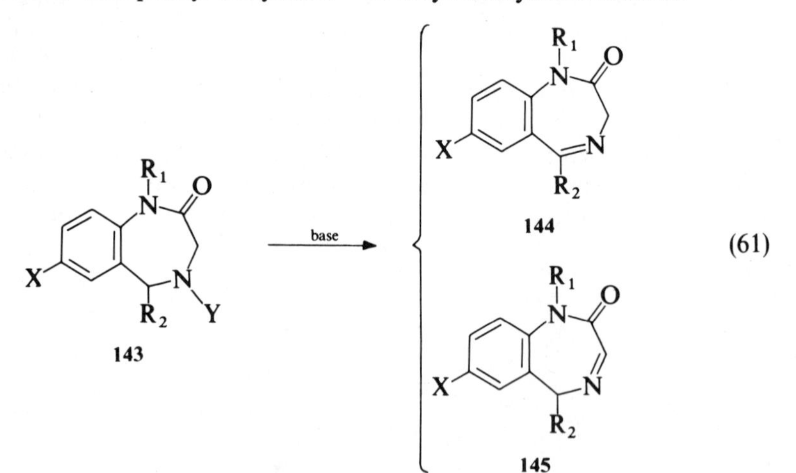


(61)

## 3. 1,2,4,5-TETRAHYDRO-1,4-BENZODIAZEPIN-3(3*H*)-ONES

### 3.1. Synthesis

The 3-one **147** was prepared by heating to reflux the ethyl ester **146** (R = Et) in pyridine and hydrochloric acid.[16] The corresponding acid **146** (R = H) was also converted to **147**, in this case by boiling in xylene and azeotropically removing water (Eq. 62).[16]

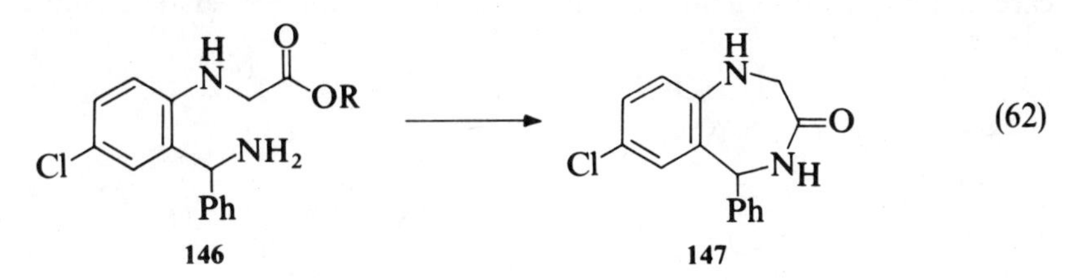


Compound **148** was isolated from the mixture resulting from the oxidation of the tetrahydrobenzodiazepine **49** with chromic acid (Eq. 63).[29]

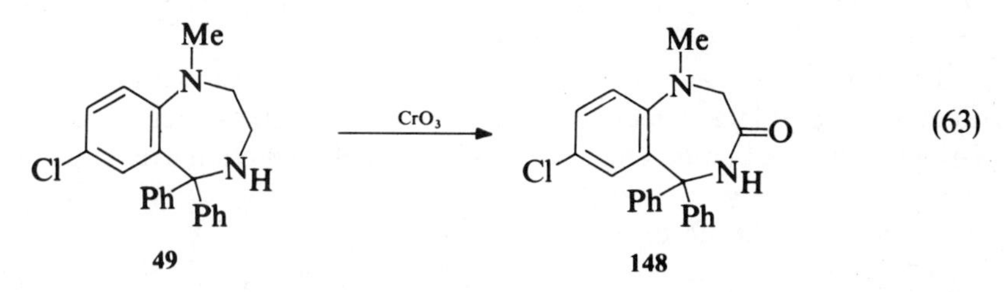


Bergman and coworkers[97] reported the formation of the 3-one **150** by the reaction of the α-chloro amide **149** with phenylmagnesium bromide (Eq. 64). The authors rationalized the formation of **150** by invoking the intermediates **151** and **152**. Intramolecular attack of the imine anion on the carbonyl group of the aziridinone **151** would generate **152**, which could then undergo a hydride transfer to form a 1,2-imine bond, which could be susceptible to another addition of phenylmagnesium bromide to give **150** as a product. Enolization of the 3-one may protect it from further reaction with the Grignard reagent.

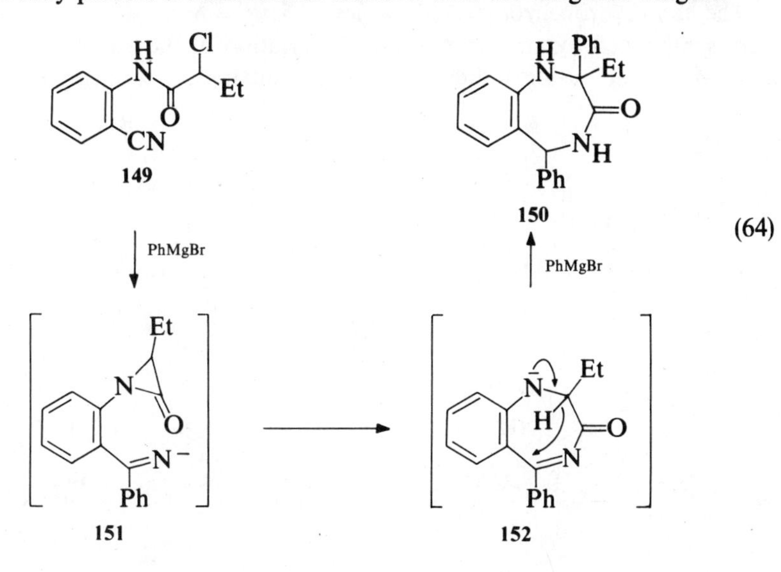

The structure **153** was assigned to one of the products obtained from the treatment of diazepam (**98**), with phenyllithium (Eq. 65).[29] It is possible that the carbonyl group was created by air oxidation of an intermediate enamine.

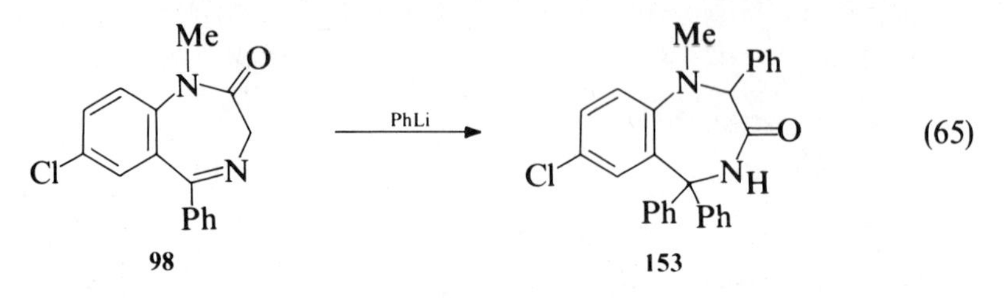


### 3.2. Reactions

The reduction of the carbonyl group of **147** to the hydrocarbon by lithium aluminum hydride was mentioned in Section 1.1. Removal of the chlorine at 7-position from **147** was achieved by catalytic hydrogenation over palladium on carbon in the presence of potassium acetate.[16] Alkylation of **147** with dimethyl sulfate afforded the 1,4-dimethyl derivative.[89]

## 4. 1,2,3,4-TETRAHYDRO-1,4-BENZODIAZEPIN-5(5*H*)-ONES

### 4.1. Synthesis

#### *4.1.1. From Anthranilic Acids*

The tetrahydrobenzodiazepin-5-ones **155** (R = Me, benzyl; X = H, 7-Cl) were synthesized by an intramolecular alkylation of the anthranilamides **154** (R = Me, benzyl) using sodium hydride in refluxing benzene (Eq. 66).[3,8,98,99]

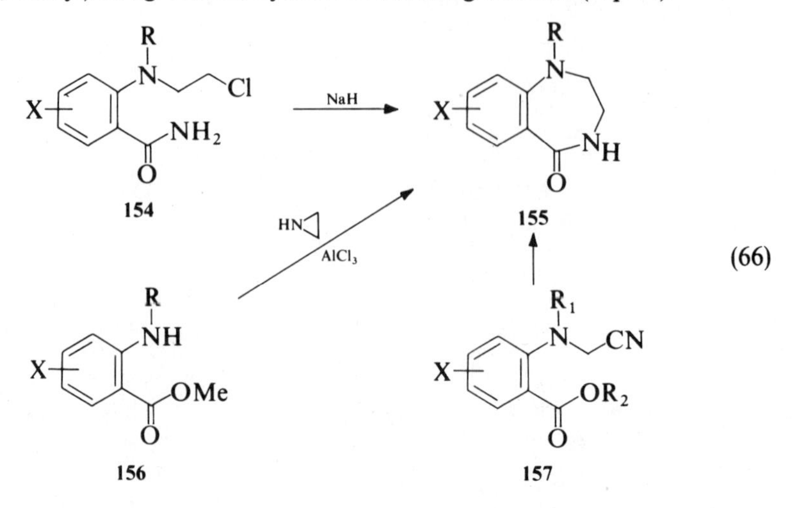

Several 1-substituted compounds **155** were prepared in 50–70% yield by reaction of the anthranilate **156** with ethylenimine and aluminum chloride in benzene.[100]

Reduction of the nitrile **157** ($R_2$ = Me) followed by ring closure provided another route to compounds **155**, in particular to 1-aryl-substituted analogs.[101] Reduction of the carboxylic acid **157** ($R_1$ = Ph, $R_2$ = H, X = 4-$NO_2$) with Raney nickel and hydrazine led to the amino acid, which was cyclized by methoxide in boiling methanol to yield the 1-phenyl-8-aminobenzodiazepine.[102]

### *4.1.2. By Ring Expansion*

The ring expansion of variously substituted tetrahydroquinazolin-4-ones was extensively studied by Field and coworkers.[103] Treatment of **158** (R = H) with potassium *t*-butoxide in methanol at room temperature gave an almost quantitative yield of the 3-methoxybenzodiazepin-5-one **159** (Eq. 67). The 3-methyl analog **158** (R = Me) rearranged to the 3-methylene derivative **160** when it was reacted with potassium *t*-butoxide in tetrahydrofuran. The exocyclic double bond was hydrogenated over platinum to afford the 3-methyl compound **161**.

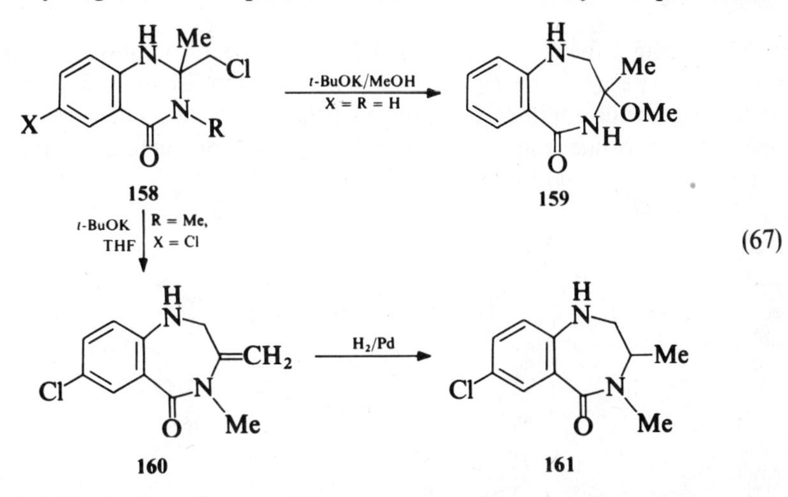


(67)

The 1-methylquinazolinone **162** reacted with potassium *t*-butoxide in tetrahydrofuran at room temperature to give 48% of the 2-methylene-benzodiazepin-5-one **163**, which was similarly hydrogenated to the 2-methyl compound (Eq. 68).[103]

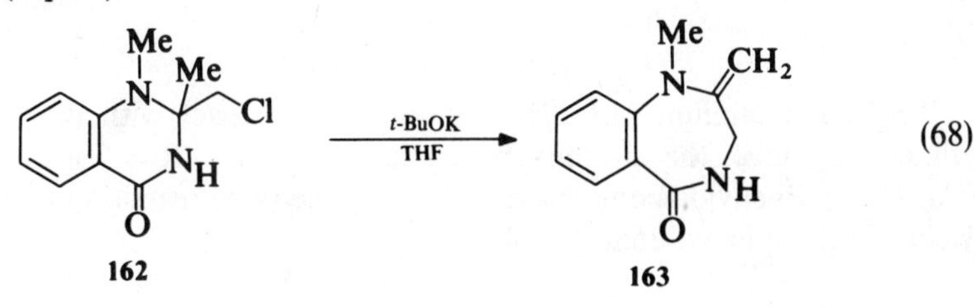


(68)

Refluxing the dichloromethylquinazolinone **164** in methanol–methoxide led, in high yield, to the 2,3-dimethoxybenzodiazepine **165**, while the quinazolinone **166** was rearranged by potassium *t*-butoxide in tetrahydrofuran to the 3-chloromethylene derivative **167** (Eq. 69). Further transformations of these compounds are described in Section 4.2.

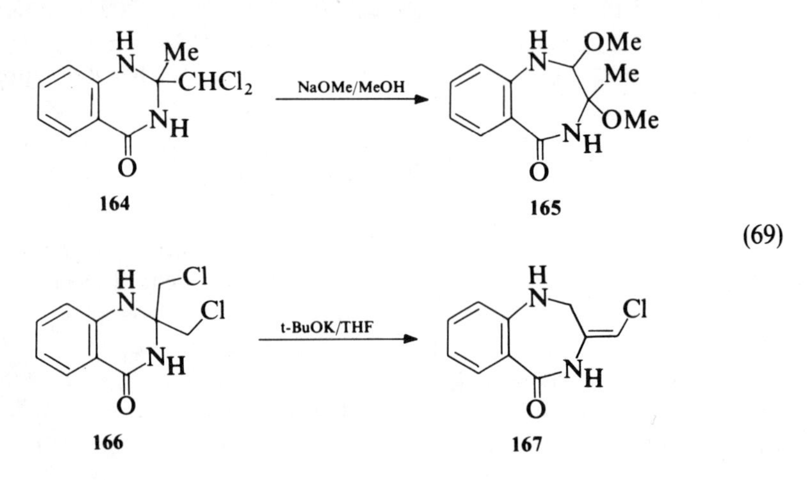


(69)

These ring expansions were explained by the formation of intermediate aziridines. The aziridine **169** (R = Ph) was actually isolated, and its reaction with various nucleophiles was studied (Eq. 70).[104] The nucleophiles used to open the aziridine ring were hydride, ethanol or methanol, acetate, and ethanethiol or phenylthiol.[104]

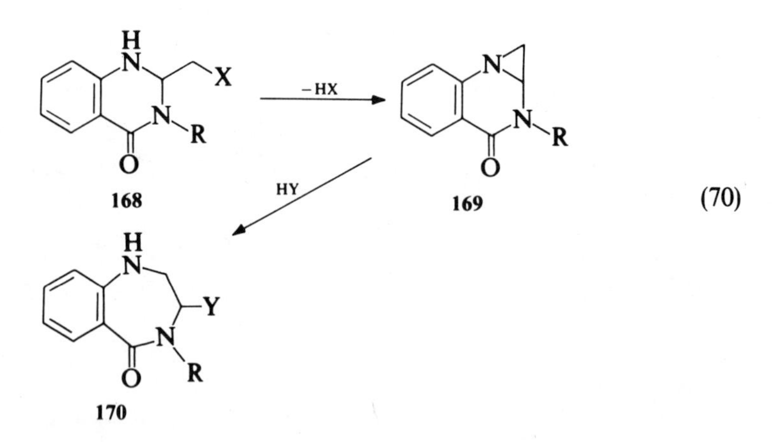


(70)

The quinazolinium salt **171** was reported to react with diazomethane in ethanol to give the 3-ethoxybenzodiazepine **172** (R = H) in 67% yield (Eq. 71).[105] Phenyldiazomethane underwent the same reaction but gave a lower yield of the 2-phenyl analog **172** (R = Ph).

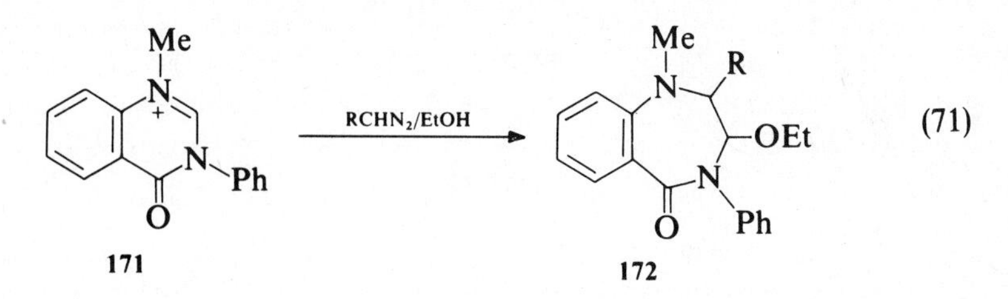


The Schmidt reaction on tetrahydroquinolin-4-ones represents another ring expansion path to benzodiazepin-5-ones. This reaction was first studied by Itterah and Mann,[106] on **173** ($R_1$ = Me, Ph). These authors assigned the structure of a 1,5-benzodiazepine **175** to the products. These products were later shown by Wuensch and coworkers[107] to be 1,4-benzodiazepin-5-ones **174**. Misiti and coworkers[9] investigated this reaction in detail and reported that it leads to a mixture of 1,2,3,4-tetrahydro-1,4-benzodiazepin-5(5*H*)-ones (**174**) and 1,2,3,5-tetrahydro-1,5-benzodiazepin-4(4*H*)-ones (**175**). The ratio of these two products was dependent on the substituent on the quinoline nitrogen. Thus, a phenyl substituent at the 1-position of the quinoline favored alkyl migration to predominantly form the 1,4-diazepines **174**.

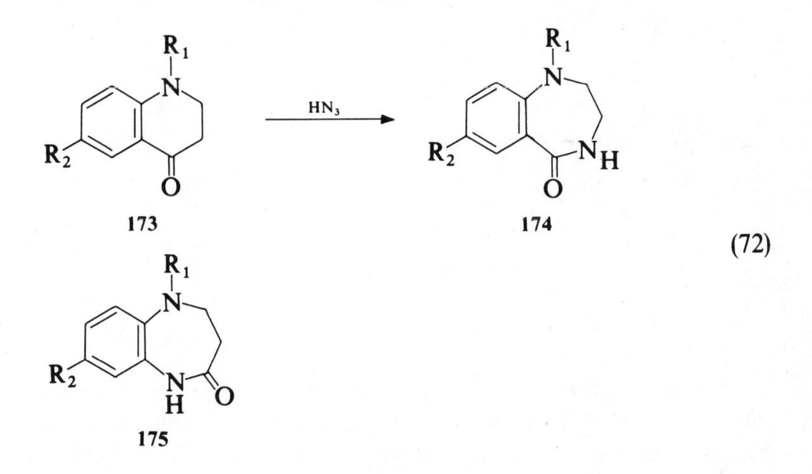


The ring expansion reaction of Eq. 72 was used to prepare a variety of 1-phenyl derivatives with substituents in the para position.[108]

### 4.1.3. By Reduction

The 1,2-imine bond in **176** was reduced catalytically with platinum and hydrogen[10,109] or by lithium aluminum hydride in tetrahydrofuran (Eq. 73).[11,110] Compound **178** was similarly reduced by catalytic hydrogenation over platinum to the 3-methyl derivative **179**.[103]

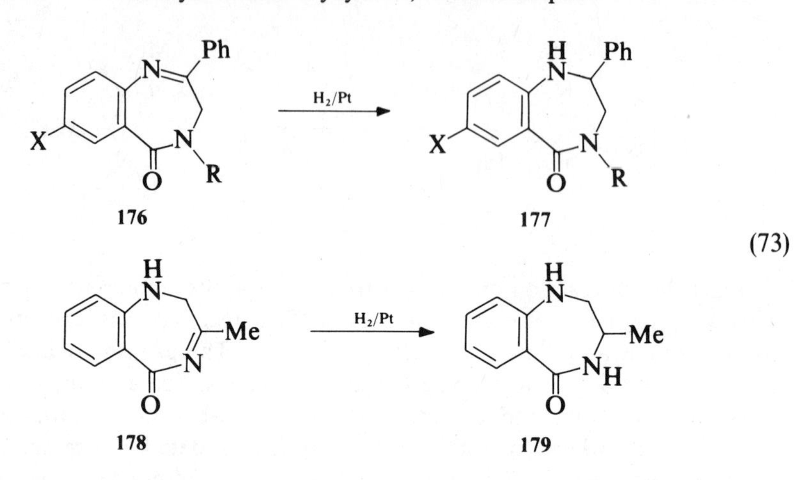


## 4.1.4. Other Syntheses

Reaction of the isocyanates **180** (X = H, Cl) with aluminum chloride in dichlorobenzene at 150°C afforded the 1-acylbenzodiazepin-5-ones **181** (X = H, Cl) in 30% yield (Eq. 74).[100]

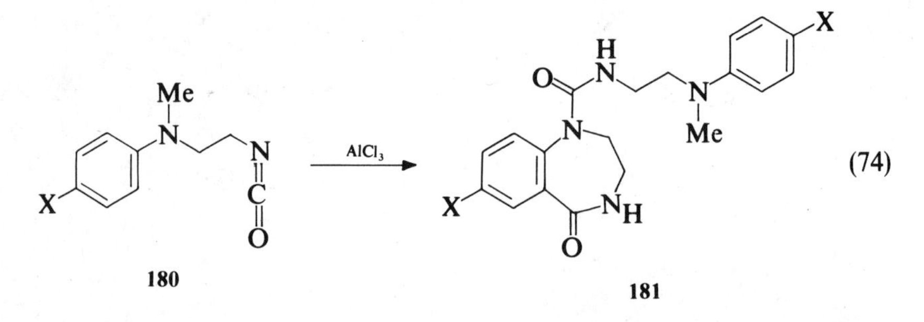


Benzodiazepin-5-ones **183** with exocyclic methylene groups in the 2-position were obtained by reaction of **182** bearing a leaving group at the 2-position with carbanions (Eq. 75). Useful leaving groups were the *N*-nitrosomethylamino functionality,[17] as well as the chloride and the phosphoryloxy groups.[111] The carbanions that reacted with **182** include those of nitromethane and dimethyl malonate.

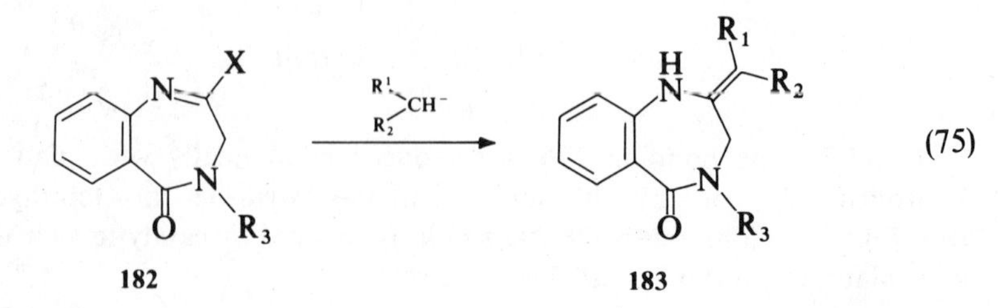

## 4.2. Reactions

### 4.2.1. *Reactions with Electrophiles*

Nitrosation of the acetylidene derivatives **184** with sodium nitrite in acetic acid led to the oximes **185** (Eq. 76).[17,111] The oximes were intermediates for the synthesis of imidazo[1,5-*a*][1,4]benzodiazepines.

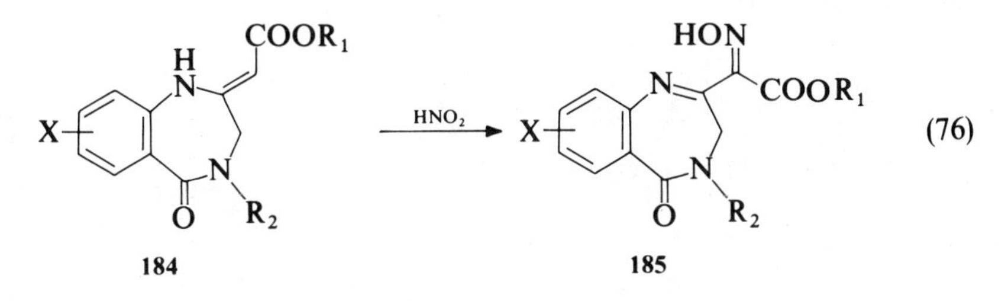


The 8-amino derivative **186** was diazotized and the resulting diazonium salt was reacted in situ with various nucleophiles.[101,102] During the Sandmeyer reactions with bromide and nitrite, bromination and nitration were observed to occur concomitantly at the para position of the 1-phenyl substituent.[101] Compound **186** was thus converted to the substituted analogs **187** (X = Br, $NO_2$) (Eq. 77).

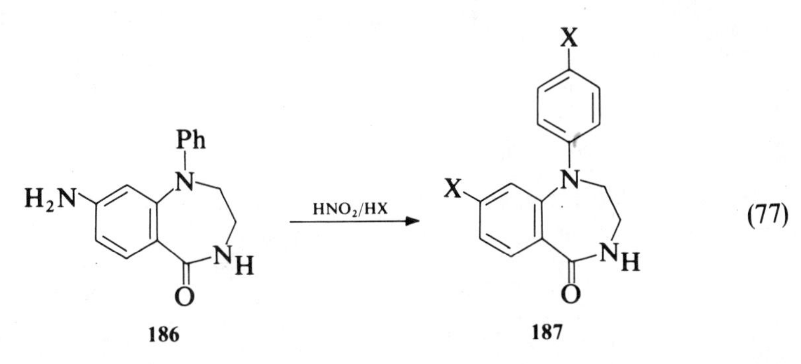


Methylation of the 1-position was carried out reductively with formaldehyde and hydrogen over palladium in carbon.[105] The 1-phenyl-8-chlorobenzo-diazepin-5-one was methylated at the 4-position by treatment with methyl iodide and sodium methoxide in tetrahydrofuran.[101,102]

The conversion of the 5-one to 5-ethoxy-2,3-dihydro-1*H*-1,4-benzo-diazepines by reaction with Meerwein salt is discussed in Chapter VI.

1,2,3,4-Tetrahydro-1,4-benzodiazepin-5(5*H*)-ones were acylated at the 1-position,[112] at the 4-position,[38,101] and simultaneously at both nitrogens.[10] The reagents used include acetic anhydride, propionic anhydride, and benzoyl chloride.

### *4.2.2. Reactions with Nucleophiles*

#### 4.2.2.1. Reduction

The reduction of the 5-carbonyl function to the hydrocarbon level was discussed in Section 1.1, which dealt with the synthesis of 2,3,4,5-tetrahydro-1*H*-1,4-benzodiazepines. The 2,3-dimethoxy compound **165** was reduced by lithium aluminum hydride to the 3-methyl analog **179** with retention of the carbonyl group (Eq. 78).[103]

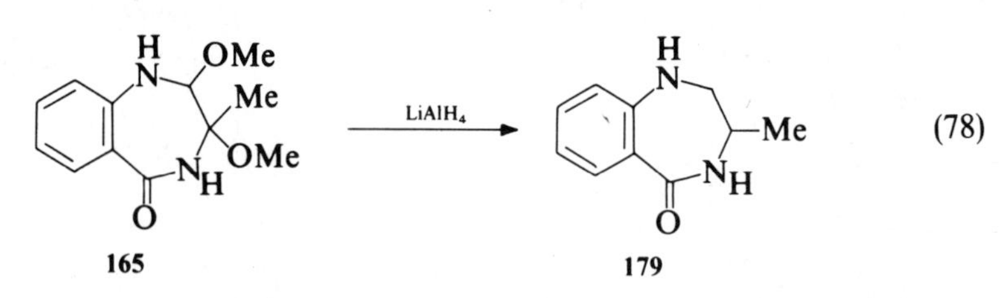


Reduction of the 3-chloromethylene derivative **167** with sodium borohydride yielded **179**.[103] When tetramethylammonium borohydride in methanol was used as the reducing agent, the chloromethyl functionality survived and yielded **188** (Eq. 79).

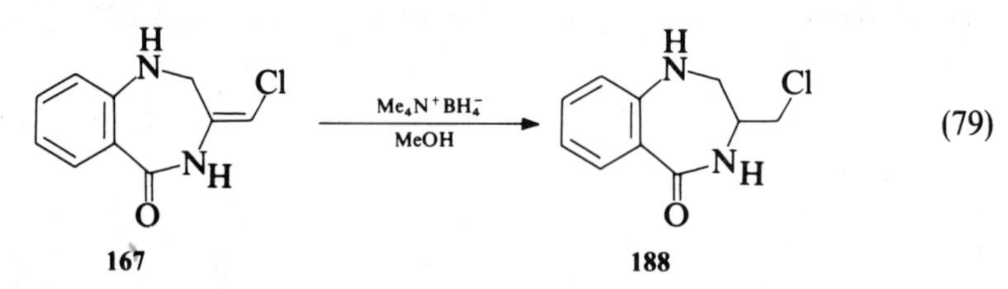


Stannous chloride in acetic acid selectively reduced the 7-nitro group of **189** and gave 59% of the 7-amino analog **190** (Eq. 80).[108]

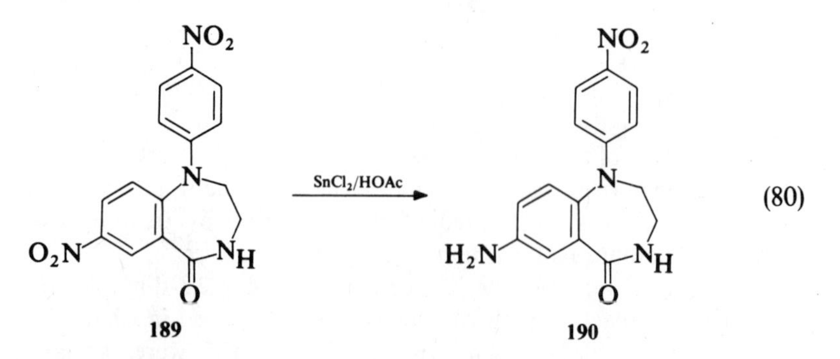


The 2-nitromethylene derivative **191** was reduced catalytically with Raney nickel to the intermediate 2-aminomethyl compound **192**, which was further reacted with triethyl orthoacetate to yield the imidazoline **193** (Eq. 81).[17]

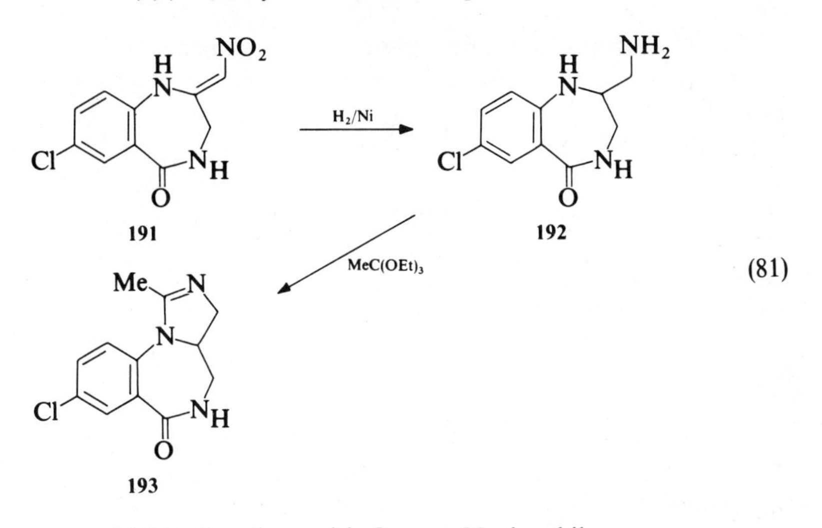


### 4.2.2.2. Reactions with Oxygen Nucleophiles

The amide bond in the parent compound **194** was cleaved by hydroxide to give the ring-opened amino acid **195** (Eq. 82).[107]

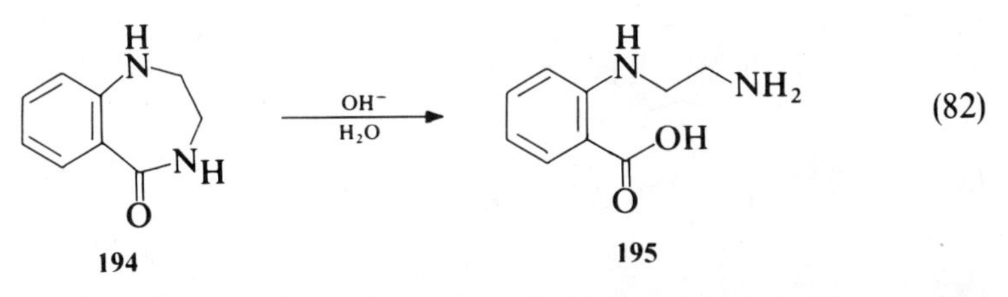


Acid hydrolysis of the 3-methoxy compound **159** led to the methyl ketone **196** (R = X = H).[103] Compound **196** (R = Me, X = Cl) was the hydrolysis product of the 3-methylene-4-methyl derivative **160**, while **197** resulted from the hydrolytic cleavage of the 1-methyl-2-methylene analog **163** (Eq. 83).[103]

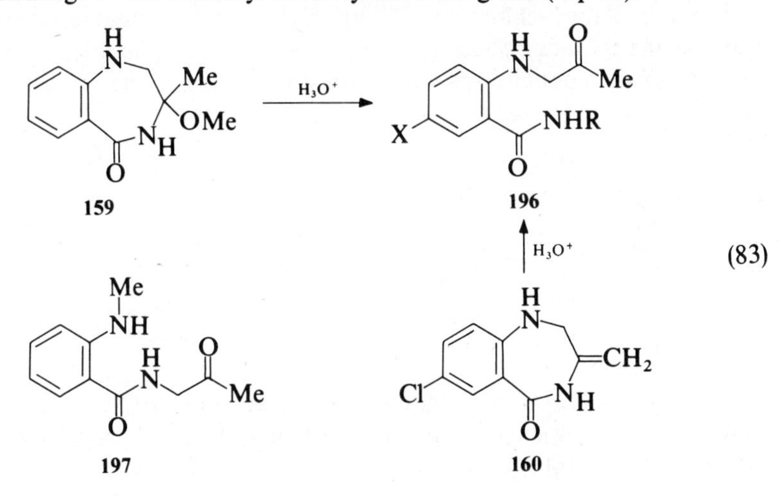

The malonylidene derivatives **198** were transformed into the corresponding acetylidene compounds **184** by treatment with hydroxide in refluxing methanol (Eq. 84).[17,111]

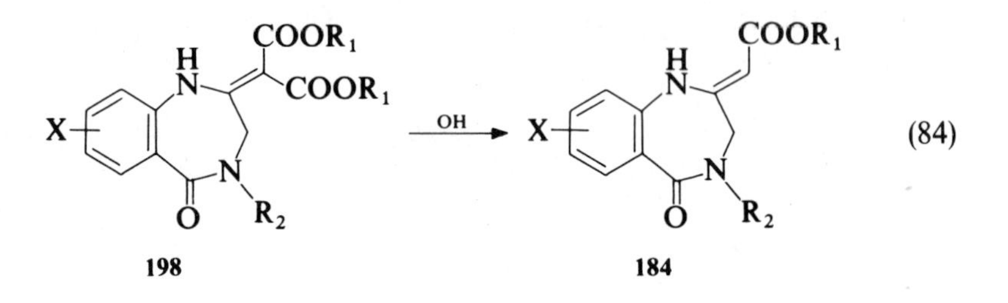


Ethanolysis in the presence of hydrochloric acid removed the acetyl group from a 1-(4-acetylaminophenyl)-substituted compound to give the corresponding amine.[108]

The 2-methoxy group in **165** was displaced by the acetoxy moiety by treatment with sodium acetate in acetic acid (Eq. 85).[103] Compound **199** was thus obtained in 40% yield. The stereochemistry remains unknown.

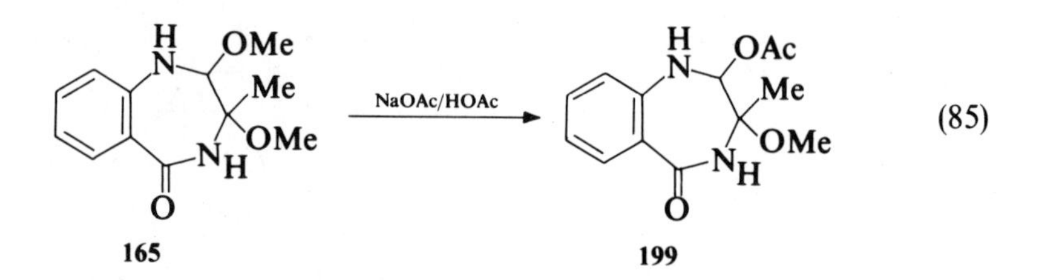


4.2.2.3. Reactions with Nitrogen Nucleophiles

Reaction of the 3-chloromethyl derivative **188** with amines resulted in substitution of the chloride by such groups as piperidine, morpholine, and butylamine (Eq. 86).[103] Compounds **200** were formed in yields ranging from 70 to 80%.

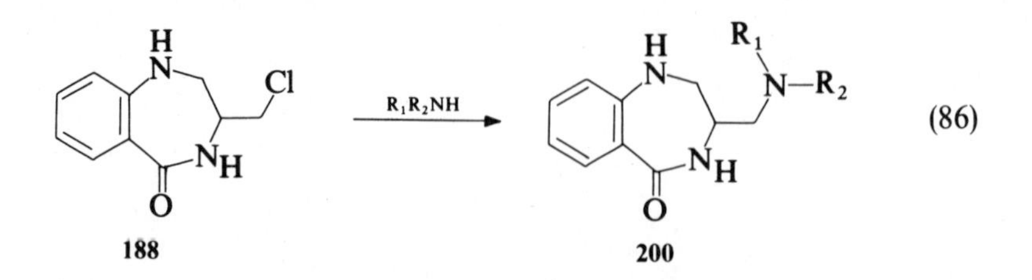


The conversion of the 5-ones **201** to the amidines **202** was carried out by reaction with phosphorus pentachloride followed by treatment with an amine (Eq. 87).[101] This transformation was also described in Chapter VI.

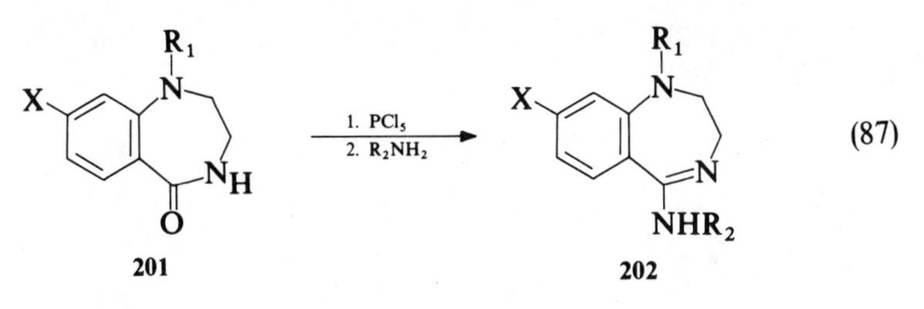


(87)

4.2.2.4. Other Reactions

The conversion of the 5-ones to the 5-thiones by means of phosphorus pentasulfide will be discussed in Section 8.1.

# 5. 4,5-DIHYDRO-1*H*-1,4-BENZODIAZEPIN-2,3(2*H*,3*H*)-DIONES

## 5.1. Synthesis

Most of the 5-phenyl derivatives **204** ($R_2$ = Ph, substituted phenyl) were prepared by protic rearrangement of the 3-hydroxy compounds **203**. This transformation was achieved in high yields by treatment of the 3-hydroxybenzodiazepines with a strong base such as hydroxide or alkoxide (Eq. 88).[113–118]

Compound **204** ($R_1$ = H, $R_2$ = Ph, X = 7-Cl) was formed in low yield by nitrosation of the 3-amino derivative **205**.[119] Chromic acid oxidation of the pyrrolobenzodiazepine **206** also gave this 2,3-dione.[62]

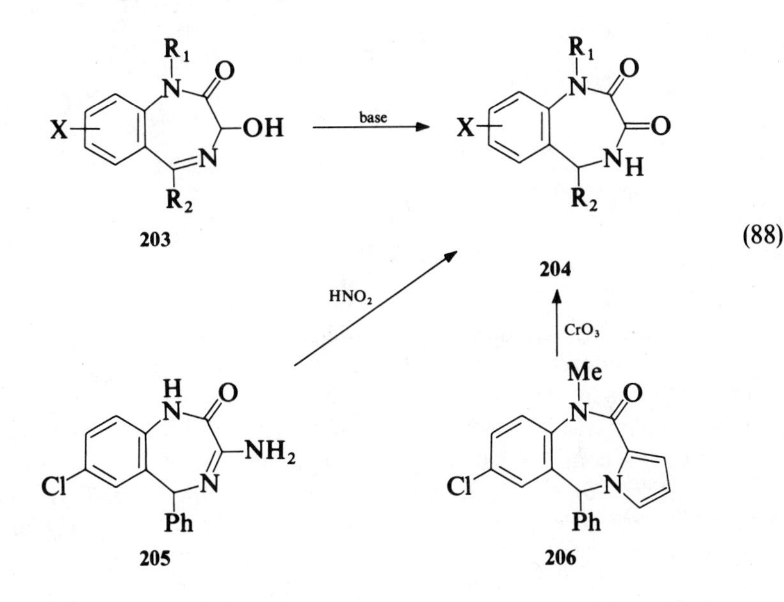


(88)

Compound **9** was prepared by reaction of the diamine **207** with oxalyl chloride (Eq. 89).[17]

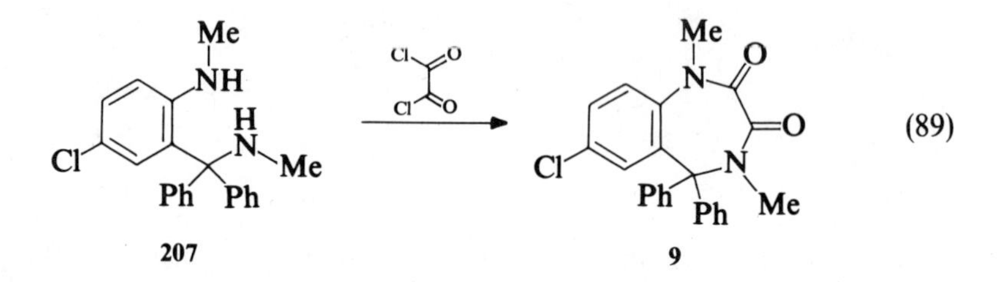


Treatment of the 3-cyano derivatives **208** (R = H, Me) with oxygen in the presence of sodium ethoxide led to the 5-cyano-2,3-diones **209**, apparently via a 3-hydroperoxide (Eq. 90).[120]

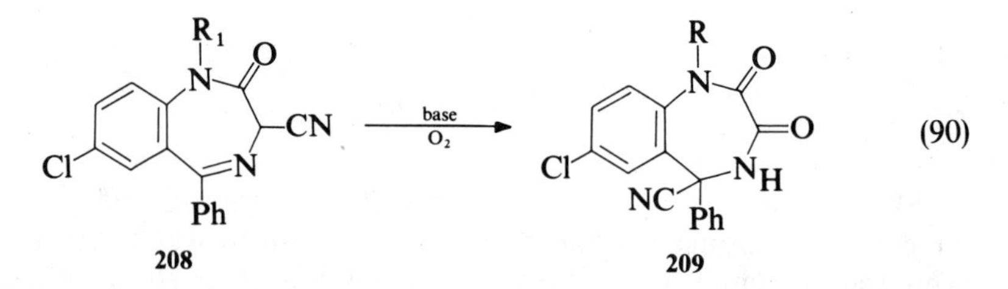


The 5-methyl derivative **211** was isolated from the reaction of the 4-oxide **210** with methylmagnesium iodide (Eq. 91).[29] Its formation may involve an oxidation, possibly by air.

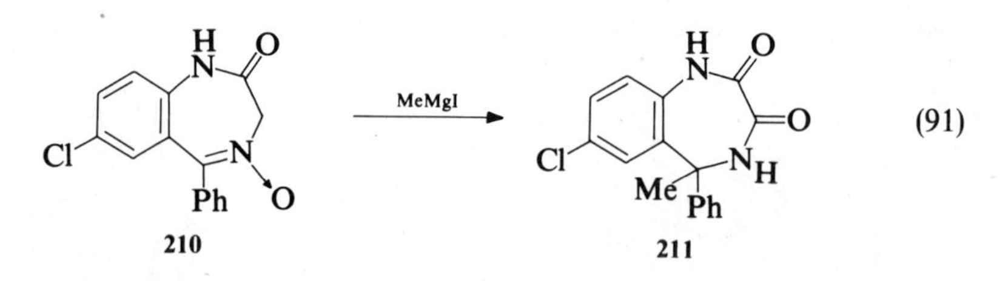


## 5.2. Reactions

Oxidation of the 2,3-dione **212** (R = Me) with ruthenium tetroxide yielded 25% of the quinazolinone **213** (Eq. 92).[121] Methylation of **212** (R = H) with dimethyl sulfate in aqueous ethanol, in the presence of hydroxide, afforded the 1-methyl analog **212** (R = Me).[113] The 1,4-dimethyl derivative **214** was obtained under the same conditions with an excess of dimethyl sulfate.[113] Compound **209** (R = Me) (Eq. 90) was methylated at the 4-position by methyl iodide and a base.[120] The same compound was also benzoylated by means of benzoyl chloride.[120]

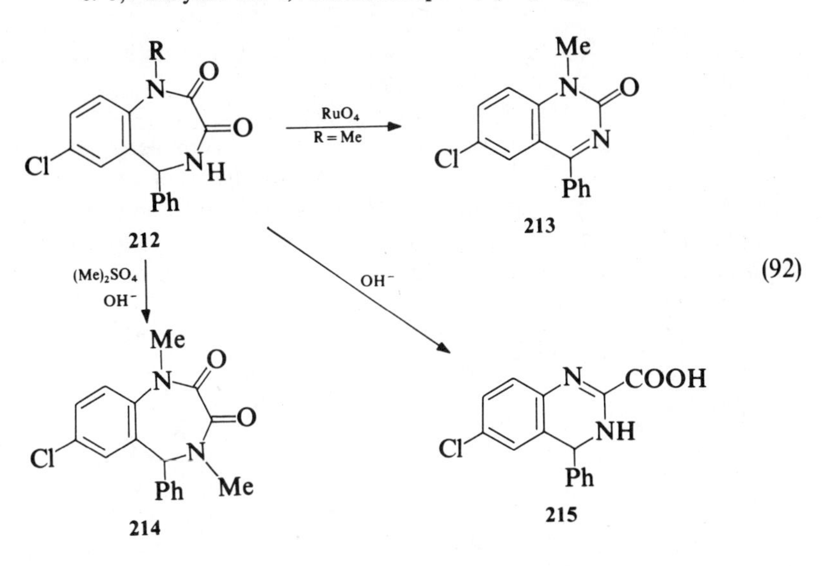


Treatment of the 2,3-dione **212** (R = H) with sodium hydroxide in boiling ethanol led to the dihydroquinazoline **215**.[113]

## 6. 3,4-DIHYDRO-1*H*-1,4-BENZODIAZEPIN-2,5(2*H*,5*H*)-DIONES

### 6.1. Synthesis

#### *6.1.1. From Anthranilic Acids*

Most syntheses of benzodiazepin-2,5-diones **217** start from derivatives of anthranilic acids and are built up by sequential formation of the two amide bonds. The precursors **216** are converted to **217** by elimination of $R_4OH$. Compounds **216** were prepared by a variety of procedures. The most convenient way was by reduction of the appropriate 2-nitrobenzoyl compound, which led to **216** ($R_1$ = H).[2,5,122–127] When **216** ($R_1$ = H; $R_4$ = H, Me, Et) was prepared by reduction of the appropriate nitro compound, cyclization to the benzodiazepin-2,5-dione occurred during the hydrogenation step.[2,123,127] This observation was also made during the reduction of 2-nitrohippuric acid with Raney nickel, which constituted the first synthesis of a 1,4-benzodiazepine.[122] Reduction of the nitro group with iron in a mixture of ethanol, acetic acid, and water also led in high yield to the benzodiazepines.[5] The esters **216** ($R_4$ = Et) could be cyclized in several ways: thermally,[127] by treatment with methoxide in methanol,[126] or by heating to reflux in methanol in the presence of piperidine.[124] The acids **216** ($R_4$ = H) were cyclized neat or by heating in solvents such as water, xylene, or ethylene glycol.[128]

Compounds **216** are intermediates in the reaction of the isatoic anhydrides **218** with amino acids or their esters, which directly leads to the benzodiazepin-2,5-diones.[1,2,5,6,111,112,129–131] The condensation of sarcosine with various isatoic anhydrides was carried out in dimethyl sulfoxide at approximately 100°C,[111,112] while the reaction with glycine ethyl ester hydrochloride was performed in refluxing pyridine.[1,129,130] In a two-step modification, the reaction of **218** ($R_1$ = Me, X = 6-Cl) with glycine in aqueous triethylamine followed by heating to reflux in acetic acid, was reported to give a 92% yield of the benzodiazepin-2,5-dione.[131]

Heating the piperidine **219** to reflux in acetic acid also afforded the parent benzodiazepin-2,5-dione **217** (Eq. 93).[132,133]

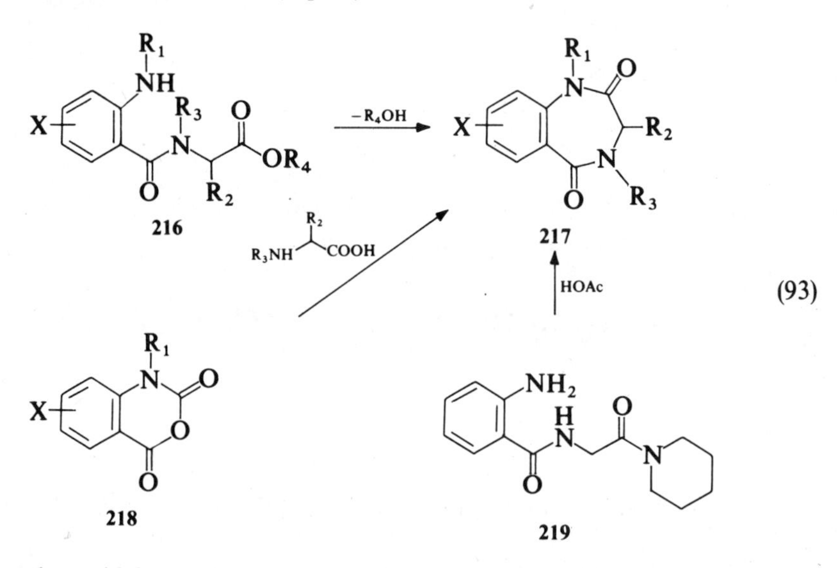


Another widely used method for the preparation of benzodiazepine-2,5-diones is the reaction of the haloacylated anthranilic acid esters **220** with amines (Eq. 94).[7,129,132–135] It is likely that the amine is first displacing the halide Y in **220** to give the intermediate **221**, which in situ forms the benzodiazepine. Syntheses, in which the 3,4-bond was formed by an alkylation reaction in the last step, were also reported. The haloacylated anthranilamides **222** were treated with a strong base, such as methoxide, to give **217** (Eq. 94).[7,136,137] This method was used to prepare the 4-methoxy[136] derivatives and the 4-phosphorylated compound.[137]

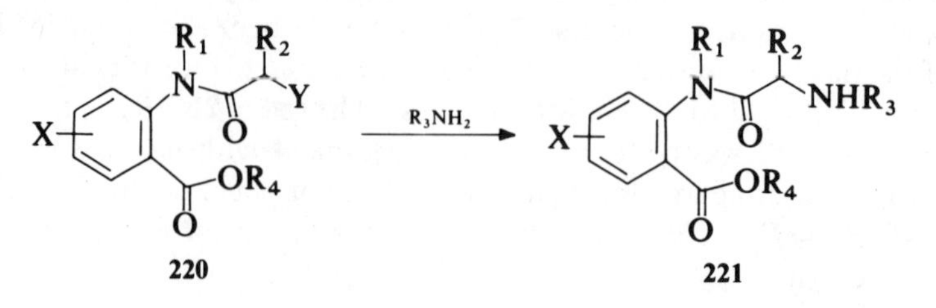

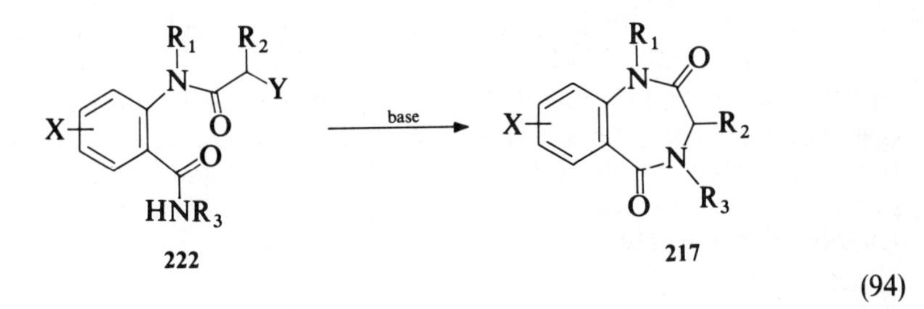


(94)

Intramolecular alkylation by an epoxide was used for the synthesis of **224**. The epoxide **223**, obtained from the corresponding olefin, was reacted with potassium *t*-butoxide in *t*-butanol to give the benzodiazepine **224** (Eq. 95).[125,138] The stereochemistry of this compound is based on the trans opening of the epoxide, but the orientation of the 3-position substituent has not been determined.

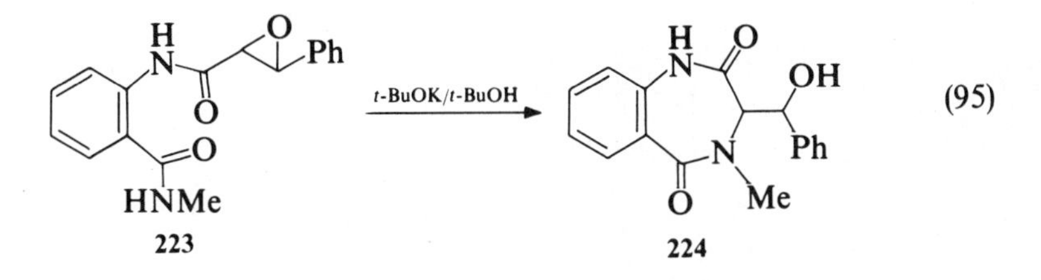


(95)

## 6.1.2. By Ring Expansion

Quinazolines of structure **225** (X = Cl, Br; R = aryl) were found to rearrange to the benzodiazepin-2,5-diones **226** in the presence of hydroxide. This ring expansion may proceed in a fashion similar to that for the rearrangement of the 2-chloromethylquinazoline-3-oxides; i.e., hydroxide adding to the 2-position. This would result in the opening of the 2,3-bond to give the anion of the haloacetylated anthranilamide which, as shown in Eq. 94, may be converted to the benzodiazepin-2,5-diones by an intramolecular alkylation. When **225** (X = Cl, R = 2-$MeC_6H_4$) was reacted with sodium carbonate in water–dioxane at 70°C for 2 hours, 63% of the corresponding product **226** was formed.[139] The closely related 4-phenyl compound **226** (R = Ph) was reported to be a result of the reaction of the bromide **225** (X = Br, R = Ph) with dimethyl sulfoxide and base (Eq. 96).[140]

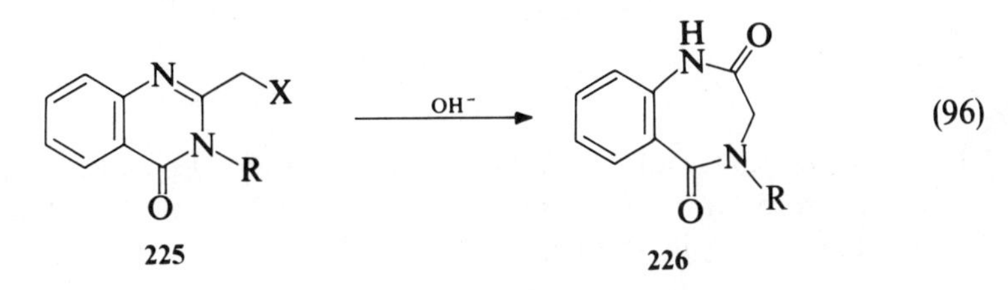


(96)

Podesva and coworkers[141] described the rearrangement of the quinoline **227** to the benzodiazepin-2,5-dione **228** by treatment with a catalytic amount of Triton B in refluxing methanol. Similarly, the same compound was formed by reacting the 3-phenyl-3-aminoquinoline **229**, which was proposed to be a possible intermediate in the conversion of **227** to **228** (Eq. 97). Interestingly, the benzodiazepine **228** forms in much better yield from **227** than from the postulated intermediate **229**.

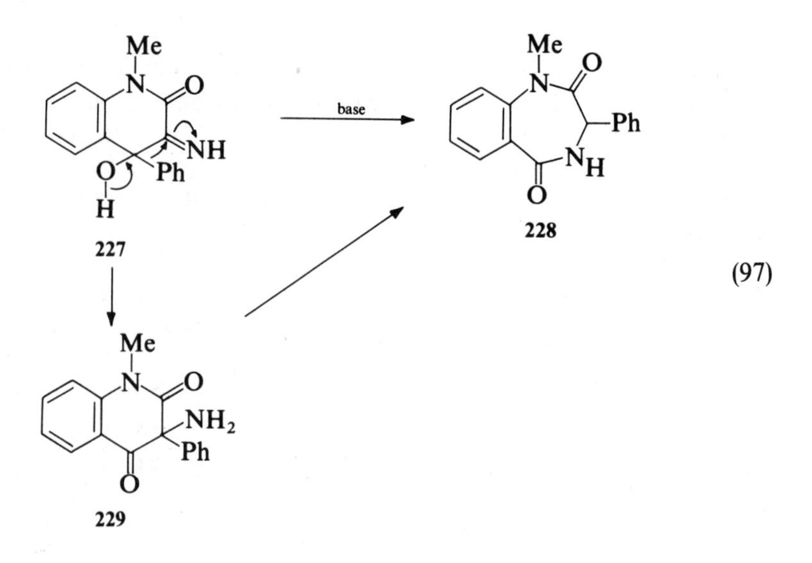


### 6.1.3. Other Syntheses

Palladium-catalyzed carbonylation of the aryl bromides **230** represents another useful method for the preparation of the 2,5-diones **231** (Eq. 98). These reactions were carried out with **230** ($R_1$ = Me, $CH_2OMe$; $R_2$ = Ac, Me, benzyl; X = H, Cl) and palladium acetate, triphenyl phosphine, tributyl amine, and 4 atm of carbon monoxide in HMPA at 110–120°C for 40–48 hours.[142,143] The yields of the benzodiazepines obtained varied from between 40 to 50%.

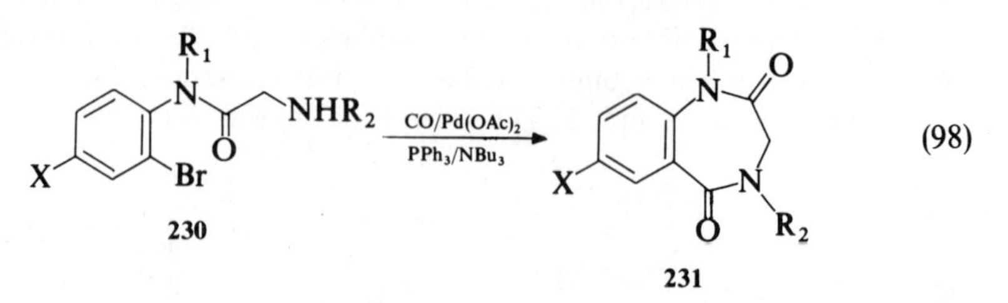


Benzodiazepin-2,5-diones **233** (R = Ph, Me) were also isolated in low yields from the ozonization of the corresponding indoles **232** ($R_1$ = Ph, $R_2$ = Me; $R_1$ = Me, $R_2$ = H), in acetic acid (Eq. 99).[144]

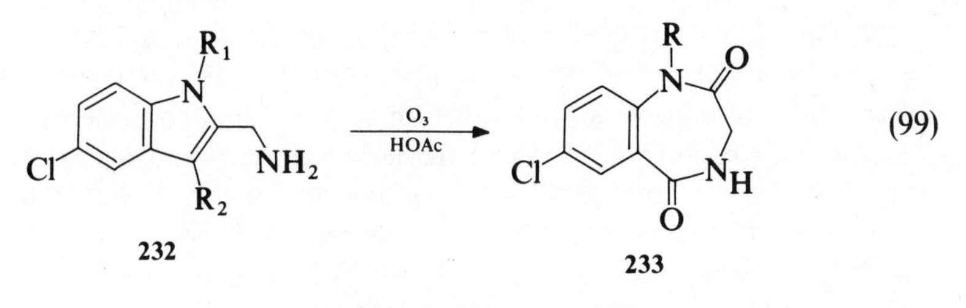


## 6.2. Reactions

### 6.2.1. *Reactions with Electrophiles*

Reaction of the 1-acetyl derivative **234** with sodium hypochlorite in aqueous dioxane led to the 1-chloro derivative **235** (Eq. 100).[123,143]

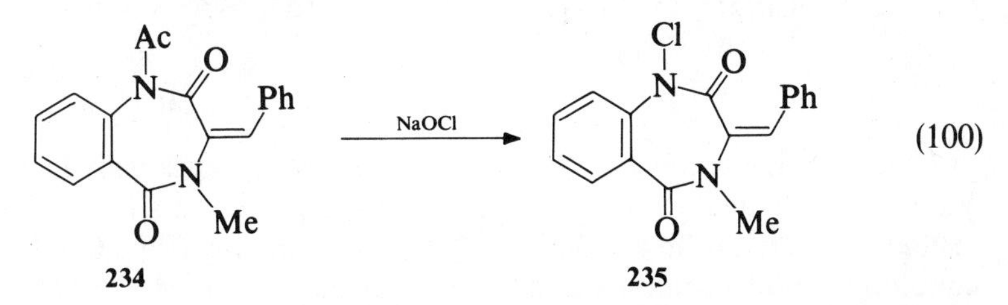


The 7-amino group was converted to the 7-diazo functionality by reaction with nitrous acid.[145] Nitration of the 1-methyl compound **236** (X = H) with sodium nitrate in sulfuric acid yielded the 7-nitro derivative **236** (X = $NO_2$). The 7-chloro compound **236** (X = Cl), under the same conditions, nitrated at the 9-position to give **237** (Eq. 101).[129]

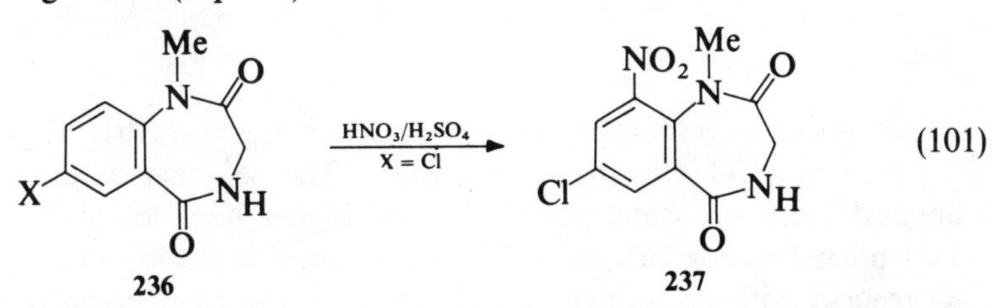


Oxidation of the double bond in **238** was effected by *m*-chloroperoxybenzoic acid in methylene chloride for 14–27 days to give 10–37% of the epoxides **239** (R = H, OAc) (Eq. 102).[138,143]

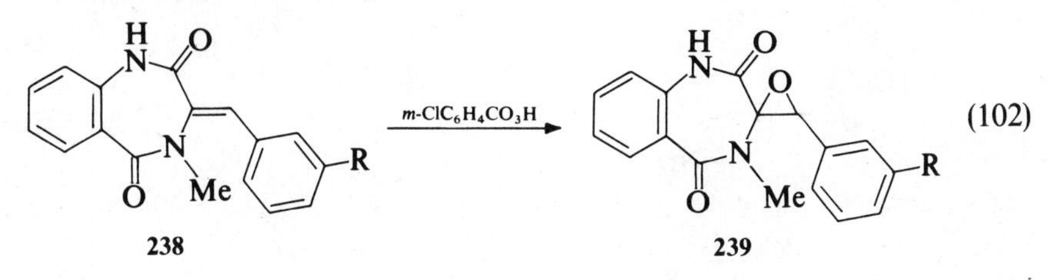

Diazomethane methylated the 1-position of the 4-methyl derivative to give the 1,4-dimethyl analog.[123] Compounds **239** (R = H, OH) were likewise alkylated by this reagent at the 1-position with simultaneous methylation of the phenolic hydroxide.[123,146] Double alkylation at the 1- and 4-positions was also carried out with methyl iodide and sodium methoxide.[133] Reaction of 1-alkyl derivatives with Meerwein salts led to the imino ethers **241** (X = OMe, OEt) (Eq. 103).[135] Compound **241** (X = Cl) was formed by the reaction of the lactam **240** with phosphorus pentachloride in refluxing chloroform. The displacement of the chloride in **241** (X = Cl) and of the alkoxide in the imino ethers **241** (X = OMe, OEt) was discussed in Chapter VII.

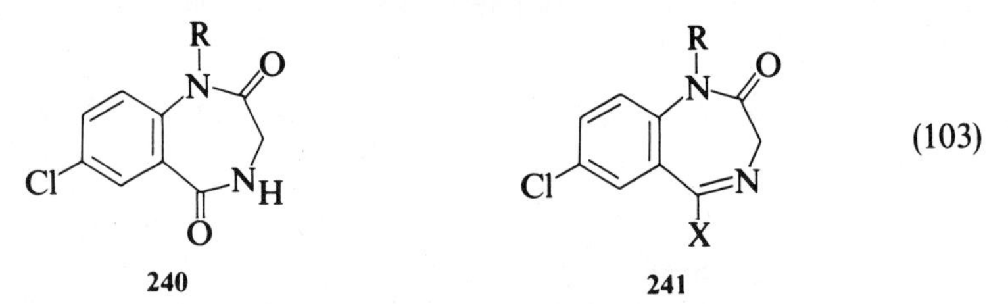


Condensation of the 4-methyl derivative **242** with benzaldehyde by means of sodium acetate and acetic anhydride at 150 °C for 3 hours led to a mixture of stereoisomers with the geometry of **238** predominating (Eq. 104). Under these conditions, the 1-position was also partially acetylated.[123,138,143]

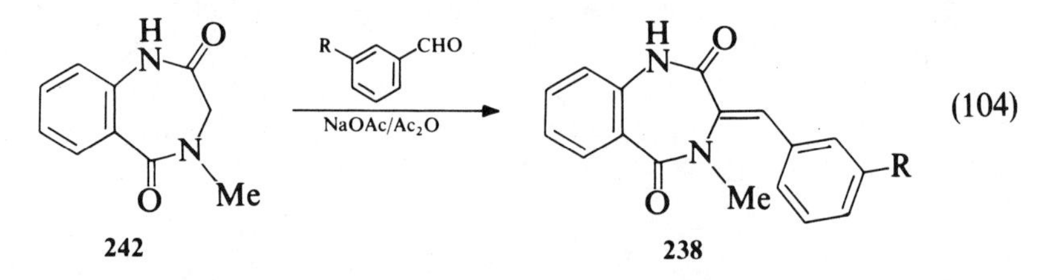


High yield acylations were reported for the 1-methyl compound **240** (R = Me) to yield the 4-acyl derivatives. The reagents used include acetic anhydride, trifluoroacetic anhydride, and higher homologous anhydrides.[131] The 1-phenyl analog **243** was similarly acetylated at the 4-position. Compound **243** reacted with propionyl chloride in the presence of pyridine to give the 4-propionyl derivative, while the 4-formyl compound **244** resulted from the treatment of **243** with dimethylformamide and phosphorus pentachloride (Eq. 105).[147]

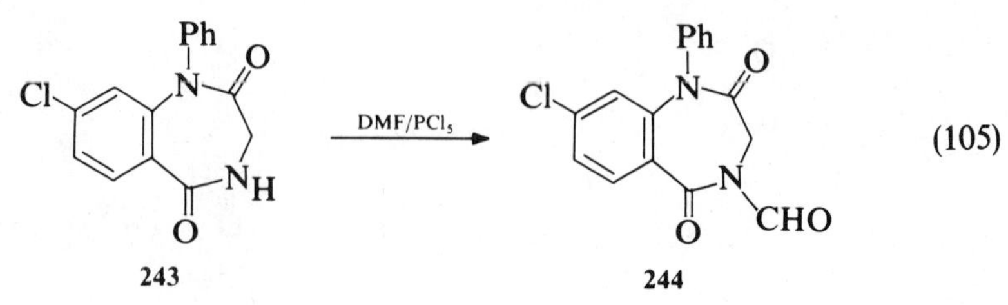

The hydroxy function of **224** (Eq. 95) was acetylated under standard conditions.[138]

The anions of the 4-substituted 2,5-diones **245** were phosphorylated by diethyl chlorophosphate or diphenyl chlorophosphate to give the imino phosphates **246**. Compounds **246** were reacted in situ with nucleophiles, in particular carbanions such as malonate anion or the anion of isocyanoacetates, to form **247** and **248**, respectively (Eq. 106).[111,112]

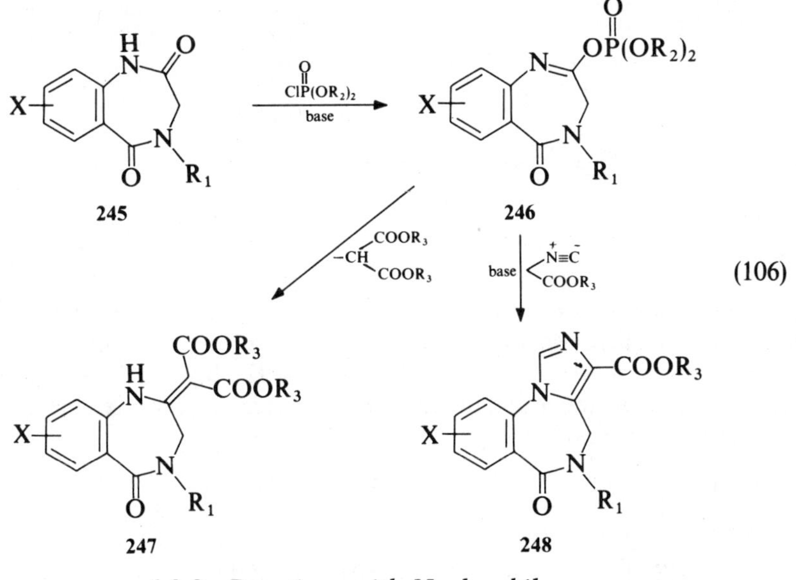


### 6.2.2. *Reactions with Nucleophiles*

The reductions of the carbonyl groups of the 2,5-diones were discussed in Section 1.1 of this chapter.[1,2,5-7,132,133] The 1-chloro derivative **235** (Eq. 100) was reduced by potassium iodide to the corresponding 1-*H* compound.[123,143] Catalytic hydrogenation of the double bond in **238** (R = H) was achieved over a platinum catalyst.[143] A 1-benzyl group was submitted to hydrogenolysis with palladium catalyst in acetic acid.[132] Reduction of a 7-nitro group to the 7-amino function was achieved by hydrogenation over the same catalyst.[145]

Hydrolysis of the parent 2,5-dione with 70% sulfuric acid at 140°C for 15 minutes led to anthranilic acid.[132] Treatment of the 4-methyl analog with 2*N* hydrochloric acid at reflux for 3 hours effected the partial cleavage of either amide bond. The reaction led to sarcosine and anthranilic acid as well.[123] The epoxides **239** (R = H, OH), on the other hand, underwent rearrangement under mild hydrolytic conditions. Thus **239** (R = H) yielded the quinolone **249** (R = H), methylamine, and carbon dioxide when treated with 2*N* hydrochloric acid at 87 °C for 3 hours.[123,146] The oxidative degradation of **239** (R = H) with hydrogen peroxide in acetic acid gave the quinazolinone **250** in 30% yield (Eq. 107).[146]

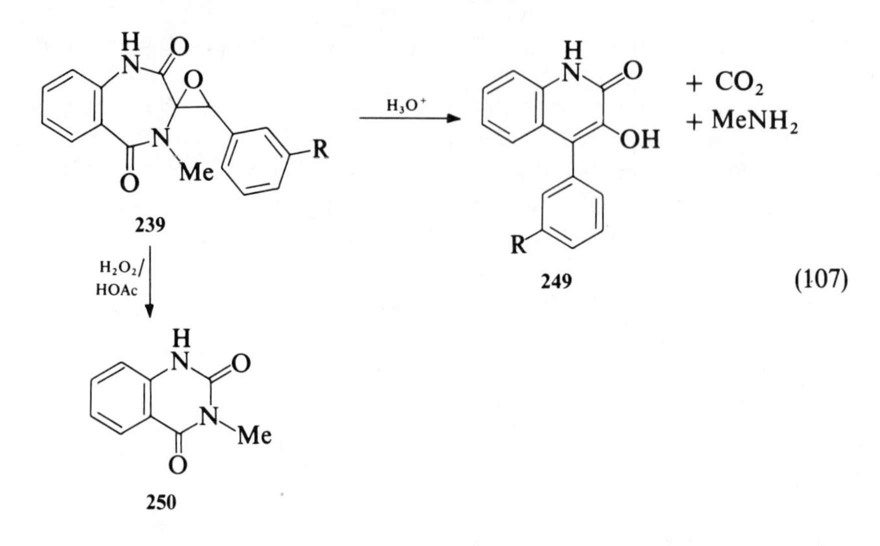


(107)

Acid hydrolysis allowed the removal of the 1-methoxymethyl moiety from the 1-methoxymethyl-4-methyl compound to give the 4-methyl 2,5-dione **242**.[143]

The conversion of the 2,5-diones **245** ($R_1$ = H, alkyl) to the amidines **251** by reaction with an amine and titanium tetrachloride was discussed in Chapter VII, Section 8. (Eq. 108).[95,137,148,149]

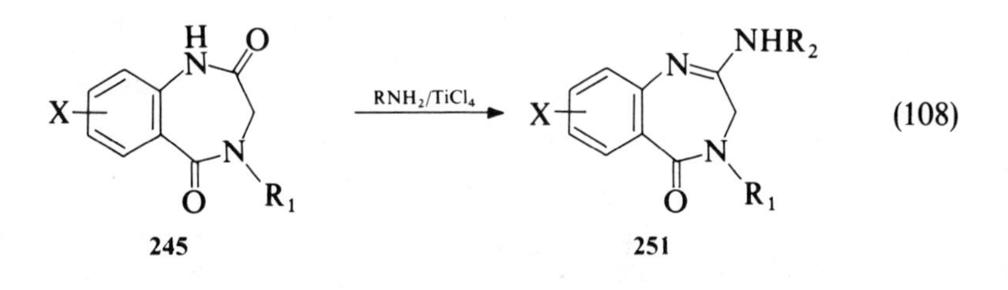


(108)

Reaction of the diazonium salt **252** with piperidine led to the triazene **253** (Eq. 109).[145]

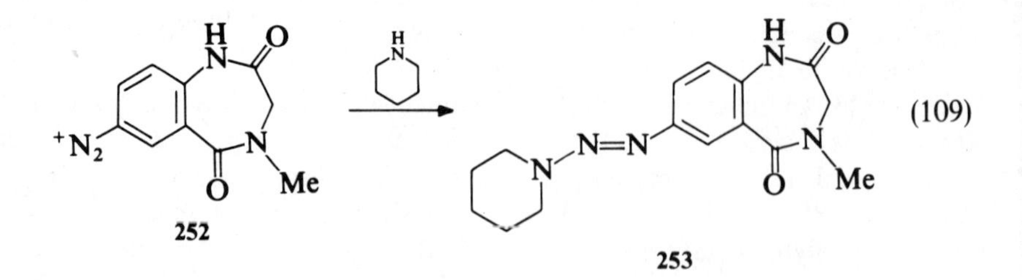


(109)

The 4-acyl derivatives **254** (R = Me, $F_3C$) reacted with phenylmagnesium halides to form the benzophenones **255**. Hydrolysis of **255** gave diazepam as a product (Eq. 110).[131]

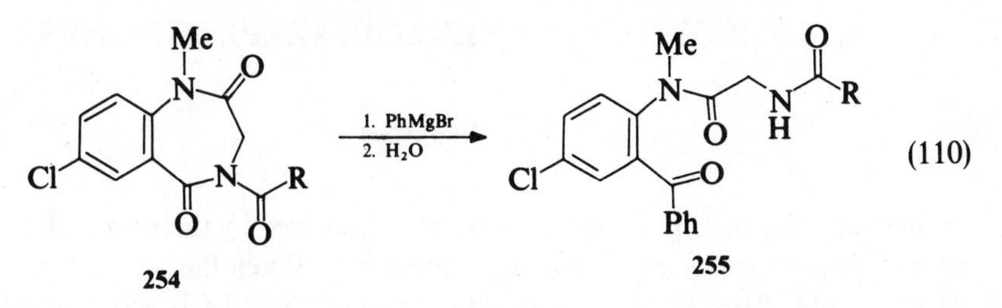


(110)

Thermal elimination of acetic acid converted the acetate **256** to the olefin **257**, yielding only one stereoisomer (Eq. 111).[125,138]

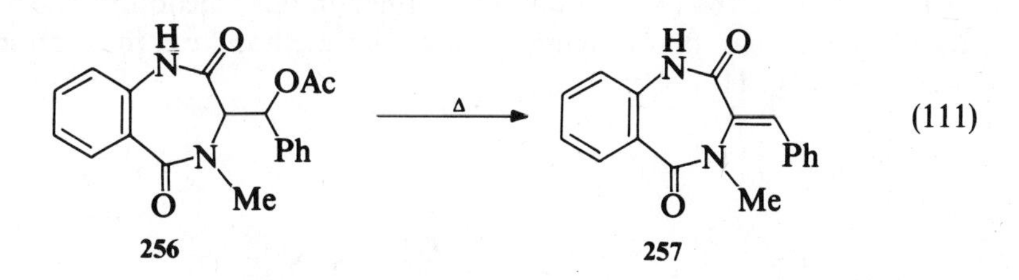


(111)

## 7. 2,4-DIHYDRO-1*H*-1,4-BENZODIAZEPIN-3,5(3*H*,5*H*)-DIONES

The 4-substituted 3,5-diones **259** were synthesized by dehydration of the carboxylic acids **258** using dicyclohexylcarbodiimide.[150] The esters corresponding to the acids **258** were obtained by alkylation of the anthranilamide with bromoacetates (Eq. 112).

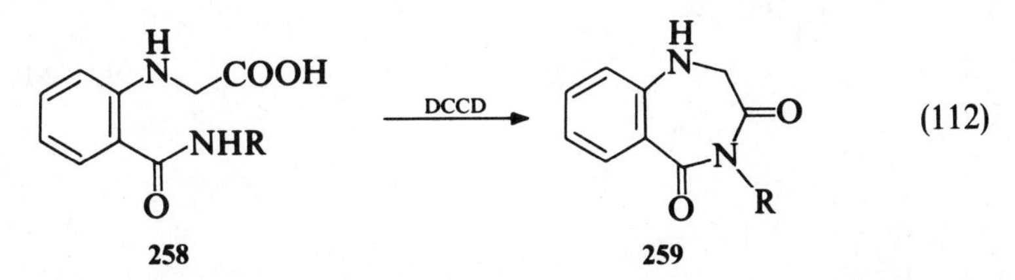


(112)

The 2-acetylidene derivatives **261** were prepared in moderate yields by treatment of the diesters **260** (X = H, Cl) with sodium methoxide in boiling methanol.[151] The compounds **260** were obtained in high yield by addition of anthranilamide to acetylene dicarboxylate (Eq. 113).

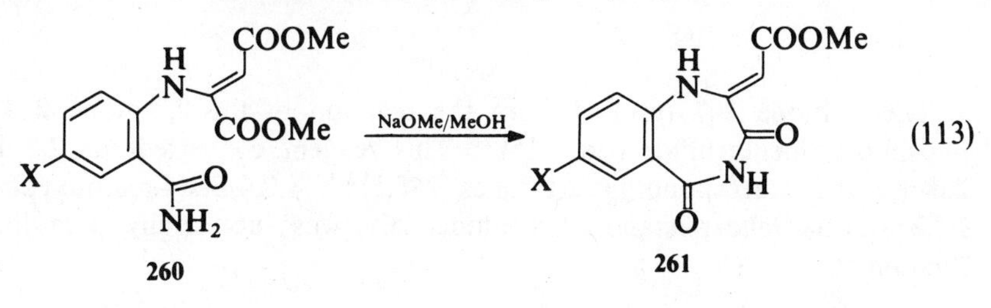


(113)

# 8. TETRAHYDRO-1,4-BENZODIAZEPINTHIONES

## 8.1. Synthesis

Several benzodiazepin-2-thiones **262** were prepared by thiation of the 2-ones **84** with phosphorus pentasulfide in pyridine.[28,66] When the 4-acetyl derivative **84** ($R_1$ = Me, $R_2$ = H, $R_3$ = 2-$FC_6H_4$, $R_4$ = Ac, X = 7-Cl) was reacted with these reagents, the acetyl group was simultaneously converted to the thioacetyl moiety.[28] The 3-one **263** was similarly converted to the 3-thione **264** (X = $H_2$).[28] The 2,3-dithione **264** (X = S) was also formed in this reaction.[28] The 5-thiones **266** (R = Me, benzyl) were synthesized by this method from the corresponding 5-ones **265** (Eq. 114).[3,8,152]

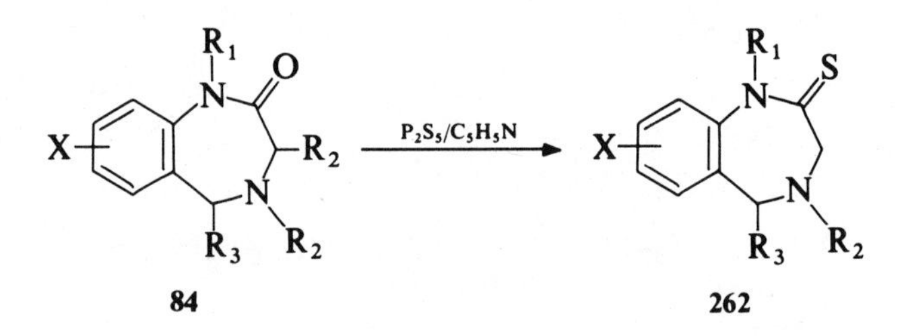


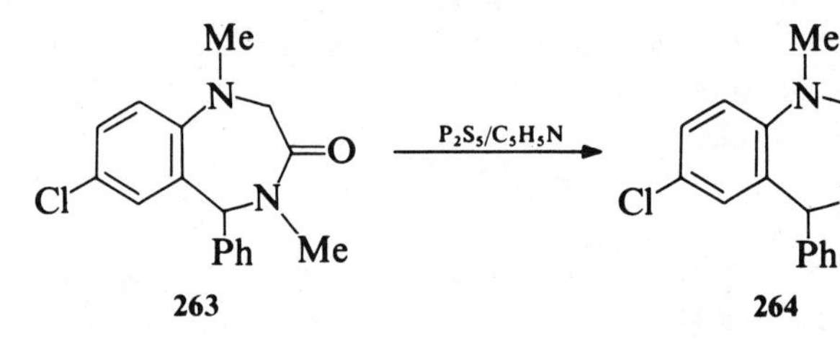


(114)

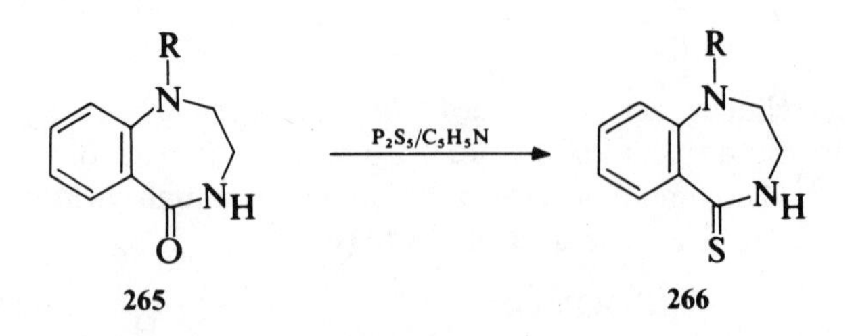


The 3-thione **267** resulted from the reaction of the 2,3-dione **214** with phosphorus pentasulfide (Eq. 115).[28] This reagent converted the 2,5-diones **268** to the corresponding 2-thiones **270**.[112,149] 2,4-Bis(4-methoxyphenyl)-1,3,2,4-dithiadiphosphetane-2,4-disulfide, **269**, was successfully used for this thiation.[112]

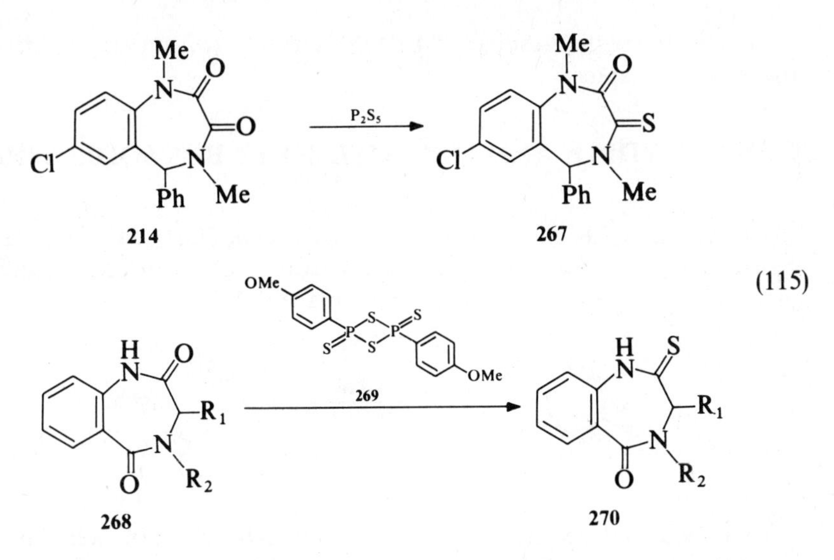


The 2,5-dithione **272** was obtained in 75% yield by reaction of the iminophosphate **271** with hydrogen sulfide and triethylamine in tetrahydrofuran (Eq. 116).[153]

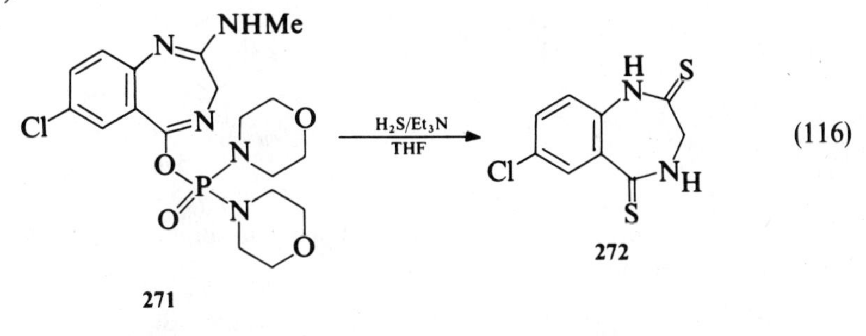


### 8.2. Reactions

The 2-thione **262** ($R_1$ = Me, $R_2 = R_4 = X$ = H, $R_3$ = Ph) was methylated and acetylated at the 4-position.[28] Compound **262** ($R_1 = R_2$ = H, $R_4$ = Me, $R_3$ = 2-Cl$C_6H_4$, X = 7-Cl) and the 2-thiones **270** were alkylated at sulfur to give the iminothioethers **273** and **274**, respectively.[66,149]

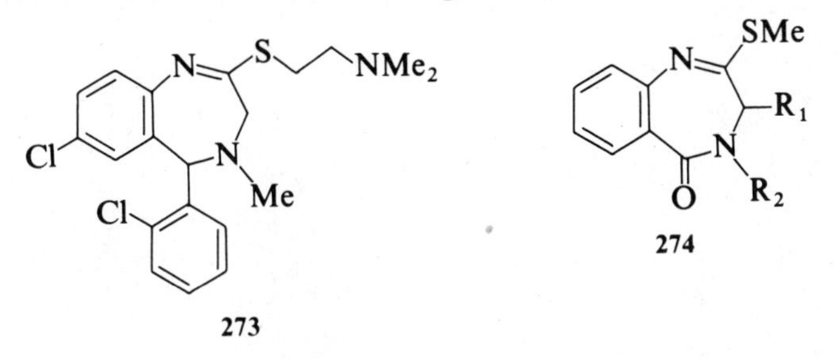

The reaction of the 2-thiones **270** with amines to give 2-amino derivatives was discussed in Chapter VII.[149]

## 9. OCTAHYDRO- AND DECAHYDRO-1,4-BENZODIAZEPINES

The 2,4,5,5*a*,6,7,8,9-octahydro-1,4-benzodiazepin-3(3*H*)-one **276** was synthesized by reacting the cyclohexanone derivative **275** with bromoacetyl bromide and subsequently with liquid ammonia (Eq. 117).[154]

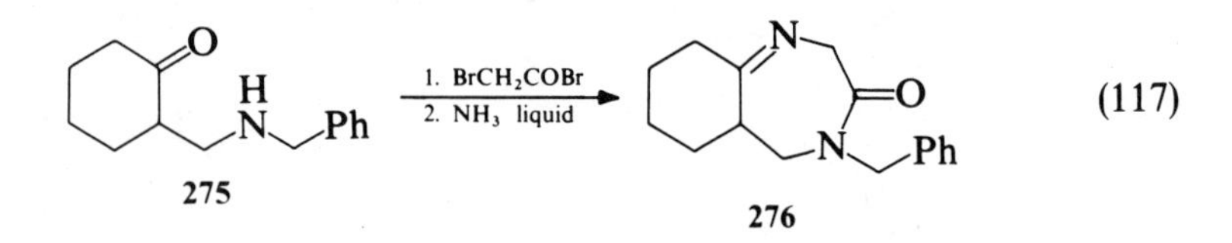


Two isomeric decahydro-1,4-benzodiazepin-2-ones **278** were prepared through ring closure of cyclohexylamines **277** by boiling in acetic acid (Eq. 118).[154]

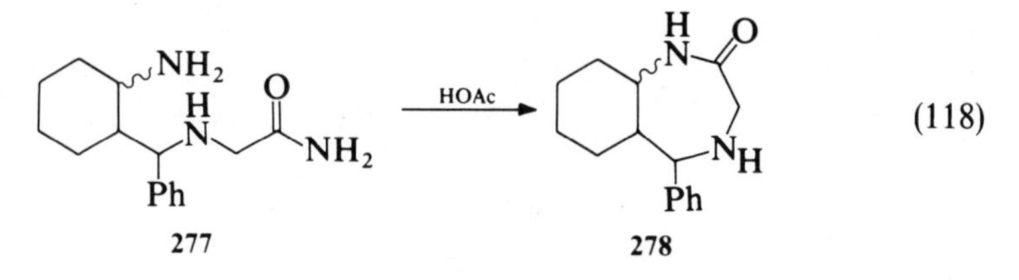


The decahydro-2,5-dione **280** resulted from a similar ring closure of the cyclohexanecarboxylic acid **279** (Eq. 119).[155]

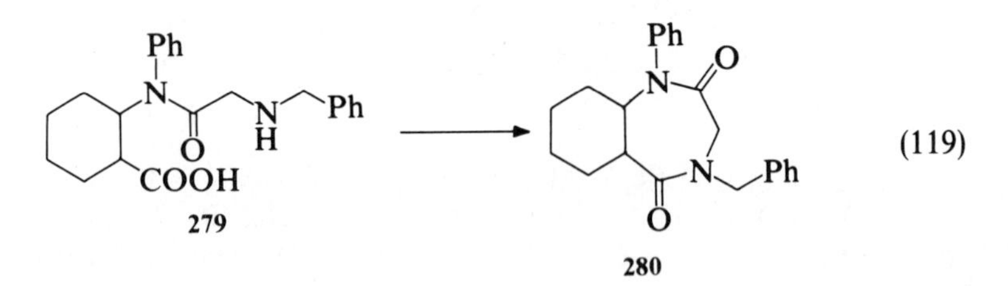

# 10. TABLE OF COMPOUNDS

TABLE VIII.1. TETRAHYDRO- AND POLYHYDRO-1,4-BENZODIAZEPINES

| Substituent | mp (°C) or; [bp (°C)/(torr)] | Solvent of Crystallization | Yield (%) | Spectra | Refs. |
|---|---|---|---|---|---|
| | *2,3,4,5-Tetrahydro-1H-1,4-benzodiazepines* 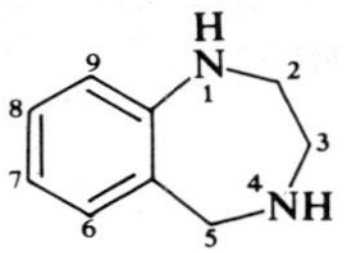 | | | | |
| None | 93–96 | $Et_2O$ | 90 | | 1, 2, 33 |
| Dihydrochloride | 246–248 | MeOH | 97 | | 5 |
| ***Monosubstituted*** | | | | | |
| 1-Ac | [115–120/0.05] | | | ir, pmr | 9 |
| 1-Benzyl dihydrochloride·$H_2O$ | 134–136 | EtOH/Acetone | | | 3, 8 |
| 1-Me | [60–70/0.005] | | | ir, pmr | 9 |
| Dihydrochloride | 188–190 | MeOH/Acetone | | ir | 3, 8 |
| 1-Ph | [155/2 mm] | | | uv | 33 |
| Maleate | 132 | | | uv | 33 |
| 4-Ac | 84–86 | $Et_2O$ | | | 37 |
| 4-Allyl | [84–85/0.13] | | | | 5 |
| Dihydrochloride | 190–199d | MeOH | 82 | | 2, 5 |
| 4-$H_2N$(iminomethyl) nitrate | 235–237 | $H_2O$ | | | 2 |
| 4-(Cyclopropyl)$CH_2$ dihydrochloride | 210–215 | EtOH | 99 | | 2, 5 |
| 4-Me dihydrochloride | 210–215 | MeOH | 92 | | 5 |
| Methiodide | 201–203 | EtOH | | | 6 |
| 4-(4-$MeC_6H_4$)$SO_2$ | 110 | | | | 33 |
| | 115–117 | $CH_2Cl_2/Et_2O$ | | | 28 |

TABLE VIII.1. —(contd.)

| Substituent | mp (°C) or; [bp (°C)/(torr)] | Solvent of Crystallization | Yield (%) | Spectra | Refs. |
|---|---|---|---|---|---|
| 4-Ph | 103–105 | EtOH | 62 | | 4 |
| 4-(2-Ph-Ethyl) | 80–83 | Hexane | 71 | | 2, 5 |
| 5-Ph | 82–83 | Hexane | 58 | pmr | 26 |
| 5-(2-Thiazolyl) maleate | 163–165 | *i*-PrOH/$Et_2O$ | | | 119b |
| 7-Cl | 95–98 | $Et_2O$ | | | 2 |
| ***Disubstituted*** | | | | | |
| ***1,4-Disubstituted*** | | | | | |
| 1,4-$Ac_2$ | 119–120 | EtOH/$Et_2O$ | | | 6 |
| 1-Ac-4-Allyl hydrochloride | 230–231 | EtOH/$Et_2O$ | 84 | | 2, 5 |
| 1-Ac-4-(2-Ph-Ethyl) hydrochloride | 240–241 | *i*-PrOH | 75 | | 2, 5 |
| 1-$H_2N$(Iminomethyl)-4-Ph hydrochloride | 249–251 | | 31 | | 4 |
| 1-Benzoyl-4-Allyl hydrochloride | 219–221d | EtOH/$Et_2O$ | 64 | | 2, 5 |
| 1-Butanoyl-4-allyl hydrochloride | 185–188 | EtOH/$Et_2O$ | 61 | | 2, 5 |
| 1,4-$Et_2$ methiodide | 174–176 | EtOH | | | 6 |
| 1-Et-4-(3-Tropanyl) | [171–175/0.2] | | | | 32 |
| Methiodide | 227–229 | EtOH | | | 32 |
| Bismethiodide | 266–267d | $H_2O$ | | | 32 |
| 1,4-$Me_2$ methiodide | 155–156 | EtOH | | | 6 |
| 1-Me-4-$H_2N$(Iminomethyl) nitrate | 210–213d | $H_2O$ | | | 2 |
| 1-Me-4-Ph | [160–162/0.3] | | 57 | | 7 |
| 1-Me-4-(2-$MeOC_5H_4$) | 78–79 | Petr ether | 64 | | 7 |
| 1-Me-4-(4-$MeOC_5H_4$) | 64–65 | Petr ether | 75 | | 7 |
| 1-Me-4-[3,4,5-$(MeO)_3C_6H_2$] | 113–114 | Petr ether | 62 | | 7 |
| 1-Me-4-(3-Tropanyl) | [180–182/0.7] | | | | 32 |
| Bismethiodide | 253–254 | $H_2O$ | | | 32 |
| 1,4-[(4-$MeC_6H_4$)$SO_2$]$_2$ | 155 | | | | 33 |

| | | | | |
|---|---|---|---|---|
| 1-Propanoyl-4-allyl hydrochloride | 237–239 | EtOH | 63 | 2, 5 |
| 1-Propanoyl-4-(cyclopropyl)$CH_2$ hydrochloride | 226–227 | MeOH | 72 | 2, 5 |
| 1-Propanoyl-4-Me hydrochloride | 229–232 | MeOH | 74 | 5 |
| 1-Propanoyl-4-(2-Ph-ethyl) hydrochloride | 214–216 | *i*-PrOH | 77 | 2, 5 |
| ***1,5-Disubstituted*** | | | | |
| 1-Ac-5-Ph | 128–129 | Cyclohexane | 63 | 26 |
| 1-Et-5-Ph | [155–163/0.3] | | 48 | 26 |
| 1-Me-5-Ph hydrochloride | 238–244d | MeOH/*i*-PrOH | | 119b |
| ***1,7-Disubstituted*** | | | | |
| 1-Me-7-Cl hydrochloride | 260–261 | MeOH | | 30 |
| 1-Me-7-$NO_2$ | 106–107 | Acetone/Petı ether | | 38 |
| ***2,4-Disubstituted*** | | | | |
| 2-Ph-4-Me dihydrochloride | 238d | EtOH/$Et_2O$ | 50 | 10 |
| ***2,7-Disubstituted*** | | | | |
| 2-Ph-7-Cl | 60–68 | MeOH | 25 | 11 |
| ***3,4-Disubstituted*** | | | | |
| 3-Me-4-Allyl | [90–93/0.2] | | | 5 |
| ***3,8-Disubstituted*** | | | | |
| 3-(3-Indolyl)$CH_2$-8-Cl, (*S*)-enantiomer Dihydrochloride | 205d | EtOH/$Et_2O$ | | 156 |
| ***4,5-Disubstituted*** | | | | |
| 4-Me-5-(3-Indolyl) maleate | 157–160 | MeOH | | 119b |
| ***4,7-Disubstituted*** | | | | |
| 4-Ac-7-Cl | 95–96 | $Et_2O$ | | 37 |
| 4-Allyl-7-benzyloxy | 77–79 | | 65 | 2, 5 |
| 4-Allyl-7-HO | 97–111 | EtOAc | | 5 |
| 4-Benzyl-7-MeO | 43–49 | $Et_2O$/Petr ether | | 28 |

TABLE VIII.1. —(contd.)

| Substituent | mp (°C) or; [bp (°C)/(torr)] | Solvent of Crystallization | Yield (%) | Spectra | Refs. |
|---|---|---|---|---|---|
| ***4,8-Disubstituted*** | | | | | |
| 4-Allyl-8-MeO dihydrochloride | 182–186 | MeOH | 95 | | 5 |
| 4-$H_2$N(Iminomethyl)-7-Cl nitrate | 270–272 | $H_2O$ | 52 | | 2 |
| Dihydrochloride | 183–186 | EtOH | | | 5 |
| 4-HO-5-Ph | 142–144 | $CH_2Cl_2$/Petr ether | | | 17 |
| ***5,7-Disubstituted*** | | | | | |
| 5-Ph-7-Cl hydrochloride | 259–261 | MeOH/Acetone | 72 | | 12, 14, 16 |
| 5-Ph-7-MeS | 101–103 | Hexane | | | 89 |
| 5-Ph-7-$NO_2$ | 216–217 | EtOH | | | 14 |
| 5-(2-Cl$C_6H_4$)-7-Cl hydrochloride | 275–280d | MeOH/$Et_2O$ | | | 23d |
| 5-(2-Cl$C_6H_4$)-7-$NO_2$ | 138–140 | $CH_2Cl_2$/Hexane | | | 29 |
| 5-(3-F$C_6H_4$)-7-Cl hydrochloride | 283–285 | MeOH/$Et_2O$ | | | 28 |
| ***Trisubstituted*** | | | | | |
| ***1,3,4-Trisubstituted*** | | | | | |
| 1-Propanoyl-3-Me-4-allyl hydrochloride | 200–201 | EtOH/$Et_2O$ | 52 | | 2,5 |
| ***1,4,7-Trisubstituted*** | | | | | |
| 1-Ac-4-Allyl-7-Cl hydrochloride | 235–236d | EtOH | 57 | | 2, 5 |
| 1,4-$Me_2$-7-$NO_2$ | 57–58 | $Et_2O$ | | | 38 |
| 1-Me-4-(2-$Et_2$N-Ethyl)-7-Cl dihydrochloride | 239–240 | MeOH/$Et_2O$ | | | 38 |
| 1-Me-4-(2-$Et_2$N-Ethyl)-7-$NO_2$ dihydrochloride | 235–236 | MeOH/$Et_2O$ | | | 38 |
| 1-Me-4-Formyl-7-Cl | 90–91 | Hexane | | | 30 |
| 1-Me-4-Formyl-7-$NO_2$ | 154–156 | Acetone | | | 38 |
| 1-Me-4-(3-$Me_2$N-Propyl)-7-Cl dihydrochloride | 273–274 | MeOH | | | 38 |
| 1-Me-4-(4-MeO$C_6H_4$)-7-Cl | 87–88 | Petr ether | 58 | | 7 |
| 1-Me-4-(4-Me$C_6H_4$)$SO_2$-7-Cl | 138–139 | PhH | | | 30 |
| 1-Me-4-Me$SO_2$-7-Cl | 119–120 | $CH_2Cl_2$/Hexane | | | 30 |

| | | | | |
|---|---|---|---|---|
| 1-Me-4-(2-Morpholinoethyl)-7-Cl dihydrochloride | 288–289 | $EtOH/H_2O$ | | 38 |
| 1-Me-4-(2-Piperidinoethyl)-7-Cl dihydrochloride | 273–275d | $EtOH/H_2O$/Acetone | | 38 |
| 1-Me-4-(3-Tropanyl)-7-Me | [180–183/0.1] | | | 32 |
| 1-Propanoyl-4-allyl-7-HO hydrochloride | 245–246 | MeOH | 43 | 2,5 |
| ***1,4,8-Trisubstituted*** | | | | |
| 1-Propanoyl-4-allyl-8-MeO hydrochloride | 210–212 | MeOH | 41 | 2,5 |
| ***1,5,5-Trisubstituted*** | | | | |
| 1-Me-5,5-$Ph_2$ | 115 | MeOH | | 29 |
| ***1,5,7-Trisubstituted*** | | | | |
| 1-(3-$H_2$N-Propyl)-5-Ph-7-Cl | 122–125 | Cyclohexane/Hexane | | 35 |
| 1-Me-5-Ph-7-(2-HOOC-Benzoyl) | 242–250d | | | 28 |
| 1-Me-5-Ph-7-Cl | 66–68 | Pentane | | 12, 14 |
| Hydrochloride | 257–259d | $Et_2O$ | | 12, 13 |
| Picrate | 202–204 | EtOH | | 12 |
| 1-Me-5-Ph-7-$NO_2$ | 96–99 | *i*-PrOH | | 14 |
| Hydrochloride | 295–297d | EtOH | | 14 |
| 1-Me-5-(2-$H_2NC_6H_4$)-7-Cl | 147–148 | $Et_2O$/Hexane | | 35 |
| 1-Me-5-(2-$ClC_6H_4$)-7-Cl | 133–135 | PhH/Hexane | | 23d |
| 1-Me-5-(2-$FC_6H_4$)-7-Cl | 90–91 | $Et_2O$/Hexane | | 17 |
| Hydrochloride | 172–173 | $MeOH/Et_2O$ | | 89 |
| 1-Me-5-(Thiazol-2-yl)-7-$NO_2$ | 103–105 | $CH_2Cl_2/Et_2O$ | | 25 |
| Maleate | 177–178 | $MeOH/Et_2O$ | | 25 |
| 1-Propyl-5-(2-$FC_6H_4$)-7-Cl | 82–83 | Petr ether | | 35 |
| ***2,5,7-Trisubstituted*** | | | | |
| 2-$H_2NCH_2$-5-(2-$FC_6H_4$)-7-Cl | 127–128 | $Et_2O$ | 27 | 24 |
| 2-[*t*-BuOOC(HON)C]-5-(2-$FC_6H_4$)-7-Cl | 204–205d | $Et_2O$ | | 17 |
| 2-$MeOOCCH_2$-5-(2-$FC_6H_4$)-7-Cl | 155–158 | $Et_2O$ | | 28 |

TABLE VIII.1. —(contd.)

| Substituent | mp (°C) or; [bp (°C)/(torr)] | Solvent of Crystallization | Yield (%) | Spectra | Refs. |
|---|---|---|---|---|---|
| 2-(4-$MeC_6H_4SO_2$)$NHCH_2$-5-(2-$FC_6H_4$)-7-Cl | | | | | |
| Isomer A | 196–198 | MeOH | 60 | ir, ms, pmr | 24 |
| Isomer B | 180–182 | MeOH | 40 | ir, pmr | 24 |
| ***4,5,7-Trisubstituted*** | | | | | |
| 4-Benzyl-5-Ph-7-Cl | 103–106 | $Et_2O$/Hexane | | | 28 |
| 4-Benzyl-5-Ph-7-MeO | 43–49 | $Et_2O$/Petr ether | 35 | | 56 |
| 4-HO-5-Ph-7-Cl | 165–167 | PhH/Hexane | 65 | | 18, 22 |
| | 170–172 | *i*-PrOH | 80 | | 19, 21 |
| 4-HO-5-(2-$ClC_6H_4$)-7-Cl | 153–154 | $Et_2O$/Hexane | | | 17 |
| 4-Me-5-Ph-7-Cl hydrochloride | 155–175 | Acetone/$Et_2O$ | | | 28 |
| ***Tetrasubstituted*** | | | | | |
| ***1,3,5,7-Tetrasubstituted*** | | | | | |
| 1,3-$Me_2$-5-Ph-7-Cl hydrochloride | 261–263d | MeOH/$Et_2O$ | | | 17 |
| ***1,4,5,5-Tetrasubstituted*** | | | | | |
| 1-Me-4-(Benzyloxy)CO-5-MeO-5-Ph | 110–113 | $Et_2O$/Petr ether | | | 29 |
| ***1,4,5,7-Tetrasubstituted*** | | | | | |
| 1-Ac-4-AcO-5-Ph-7-Cl | 134–136 | $Et_2O$/Hexane | 68 | ir, uv | 20 |
| 1-Ac-4-HO-5-Ph-7-Cl | 160–162 | EtOAc/Hexane | 77 | | 20 |
| 1-(3-AcNH-Propyl)-4-Ac-5-Ph-7-Cl | 246–248 | Hexane | | | 35 |
| 1,4-$Et_2$-5-(2-$FC_6H_4$)-7-Cl hydrochloride | 189–196 | Acetone/$Et_2O$ | | | 28 |
| 1,4-$Me_2$-5-Ph-7-Cl | 79–80 | EtOH | | | 28 |
| 1,4-$Me_2$-5-(2-$FC_6H_4$)-7-Cl | 68–70 | MeOH | | | 28 |
| 1-Me-4-Ac-5-Ph-7-Cl | 99–103 | Pentane | 11 | | 15 |
| | 106–108 | $Et_2O$/Petr ether | | | 31 |
| 1-Me-4-Ac-5-(2-Pyrimidyl)-7-AcNH | 228–229 | PhH | | | 89 |

| | | | | |
|---|---|---|---|---|
| 1-Me-4-Benzoyl-5-Ph-7-Cl | 156–157 | MePh/Cyclohexane | | 157 |
| 1-Me-4-[7-Cl-2,3-Dihydro-5-Ph-2*H*-1,4-benzodiazepin-2-yl]-5-Ph-7-Cl | 281–285 | THF | 23 | 15 |
| 1-Me-4-Et-5-(2-$FC_6H_4$)-7-Cl hydrochloride | 182–185d | MeOH/$Et_2O$ | | 28 |
| 1-Me-4-($Et_2N$-Acetyl)-5-Ph-7-Cl dihydrochloride | 240–242 | MeOH/$Et_2O$ | | 17 |
| 1-Me-4-(2-$Et_2N$-Ethyl)-5-Ph-7-Cl | [172/0.05] | | | 17 |
| 1-Me-4-Formyl-5-Ph-7-Cl | 120–123 | *i*-PrOH | 87 | 40 |
| | 121–122 | Hexane | | 31 |
| 1-Me-4-Formyl-5-(2-$FC_6H_4$)-7-Cl | 100–101 | PhH/Hexane | | 89 |
| 1-Me-4-Formyl-5-[2-(diformyl)$NC_6H_4$]-7-Cl | 201–204 | EtOH | | 35 |
| 1-Me-4-HO-5-Ph-7-Cl | 20–143 | MeOH | 52 | 20 |
| 1-Me-4-HO-5-(2-$ClC_6H_4$)-7-Cl | 158–159 | $Et_2O$/Hexane | | 17 |
| 1-Me-4-(2-HO-Ethyl)-5-Ph-7-Cl hydrochloride·$H_2O$ | 134–136 | EtOH/$Et_2O$ | | 35 |
| 1-Me-4-(MeNH-Acetyl)-5-Ph-7-Cl dihydrochloride | 165–180d | EtOH/$Et_2O$ | | 17 |
| 1-Me-4-(2-MeNH-Ethyl)-5-Ph-7-Cl hydrochloride | 222–224 | EtOH/$Et_2O$ | | 17 |
| 1-Me-4-MeOOC-5-Ph-7-Cl | 111–114 | $Et_2O$/Hexane | | 17 |
| 1-Me-4-(4-$MeC_6H_4$)$SO_2$-5-Ph-7-Cl | 127–130 | $CH_2Cl_2$/Hexane | | 31 |
| ***1,4,5,5-Tetrasubstituted*** | | | | |
| 1,4,5-$Me_3$-5-Ph | 88–89 | $Et_2O$/Hexane | | 73 |
| ***1,5,5,7-Tetrasubstituted*** | | | | |
| 1-Formyl-5,5-$Ph_2$-7-Cl | 154–155 | $Et_2O$/Petr ether | 29 | |
| 1-Me-5-Bu-5-Ph-7-Cl | 89–92 | MeOH | | 29 |
| 1-Me-5-(3-$Me_2N$-Propyl)-5-Ph-7-Cl dihydrochloride | 205–210d | Acetone/EtOH | | 29 |
| 1-Me-5,5-$Ph_2$-7-Cl hydrochloride·0.3 MeOH | 260–261d | MeOH/$Et_2O$ | | 29 |
| ***2,3,5,7-Tetrasubstituted*** | | | | |
| 2,3-$Me_2$-5-Ph-7-Cl | 129–131 | MeOH | | 18d |
| ***2,4,5,7-Tetrasubstituted*** | | | | |
| 2-Me-4-HO-5-Ph-7-Cl | 168–173 | EtOH | | 18a–c |
| 2-$O_2NCH_2$-4-HO-5-(2-$FC_6H_4$)-7-Cl | 202–203d | EtOAc/Hexane | | 18d |

TABLE VIII.1. —(contd.)

| Substituent | mp (°C) or; [bp (°C)/(torr)] | Solvent of Crystallization | Yield (%) | Spectra | Refs. |
|---|---|---|---|---|---|
| ***3,4,5,7-Tetrasubstituted*** | | | | | |
| 3-Me-4-HO-5-Ph-7-Cl | | | | | |
| Isomer A | 204–210d | EtOH | | | 18a, c |
| Isomer B | 185–200d | EtOAc | | | 18d |
| 3-Me-4-HO-5-Ph-7-$NO_2$ | 222–228d | THF/$H_2O$ | | | 18a–c |
| ***Pentasubstituted*** | | | | | |
| 1,3-$Me_2$-4-HO-5-Ph-7-Cl | 135–136 | $Et_2O$/Petr ether | 75 | pmr, uv | 20 |
| 1-Me-3-Benzyl-4-HO- | | | | | |
| 1-Me-4-Ac-5-EtO-5-Ph-7-Cl | 168–174 | MeOH | | | 29 |
| 5-Ph-7-Cl | 140–143 | $CH_2Cl_2$/Hexane | 81 | pmr | 20 |
| 1-Me-4-(Benzyloxy)CO-5-EtO-5-Ph-7-Cl | 198–202 | $CH_2Cl_2$/EtOH | | | 29 |
| 1-Me-4-(Benzyloxy)CO-5-EtO-5-(2-$FC_4H_4$)-7-Cl | 170d | $CH_2Cl_2$/EtOH | | | 29 |
| 1-Me-4-(Benzyloxy)CO-5-HO-5-Ph-7-Cl | 115–116 | $Et_2O$/Heptane | | | 29 |
| 1-Me-4-(Benzyloxy)CO-5-MeO-5-Ph-7-Cl | 175–180 | MeOH | | | 29 |
| 1-Me-4-(EtOOC-Acetyl)-5,5-$Ph_2$-7-Cl | 128–129 | EtOH | | | 29 |
| 1,4-$Me_2$-5,5-$Ph_2$-7-Cl | 153–155 | MeOH | | | 17 |
| 1-Me-4-EtSOC-5-EtO-5-Ph-7-Cl | 160–162 | MeOH | | | 17 |
| 1-Me-4-MeOOC-5-EtO-5-Ph-7-Cl | 127–130 | EtOH/$H_2O$ | | | 29 |
| 1-Me-4-MeOOC-5-HO-5-Ph-7-Cl | 116–117 | $Et_2O$/Heptane | | | 29 |
| 1-Me-4-MeOOC-5-MeO-5-Ph-7-Cl | 163–165 | MeOH | | | 17 |
| 1-Me-4-Propenoyl-5,5-$Ph_2$-7-Cl | 176–178 | Petr ether | | | 29 |
| 2,2-$Me_2$-4-HO-5-Ph-7-Cl | 165–167 | EtOH | | | 18d |

*2-Methylene-1,3,4,5-tetrahydro-2H-1,4-benzodiazepines*

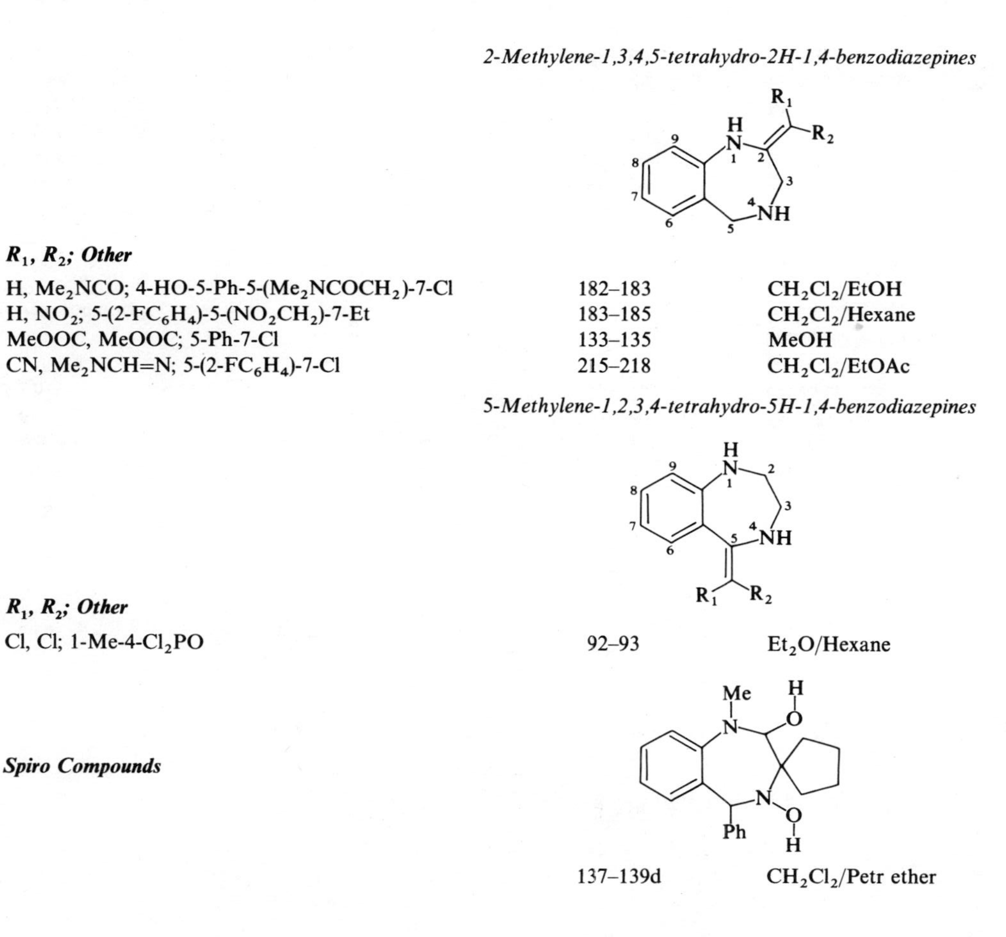

| ***$R_1$, $R_2$; Other*** | | | |
|---|---|---|---|
| H, $Me_2NCO$; 4-HO-5-Ph-5-($Me_2NCOCH_2$)-7-Cl | 182–183 | $CH_2Cl_2$/EtOH | 17 |
| H, $NO_2$; 5-(2-$FC_6H_4$)-5-($NO_2CH_2$)-7-Et | 183–185 | $CH_2Cl_2$/Hexane | 36 |
| MeOOC, MeOOC; 5-Ph-7-Cl | 133–135 | MeOH | 17 |
| CN, $Me_2NCH$=N; 5-(2-$FC_6H_4$)-7-Cl | 215–218 | $CH_2Cl_2$/EtOAc | 17 |

*5-Methylene-1,2,3,4-tetrahydro-5H-1,4-benzodiazepines*

| ***$R_1$, $R_2$; Other*** | | | |
|---|---|---|---|
| Cl, Cl; 1-Me-4-$Cl_2$PO | 92–93 | $Et_2O$/Hexane | 17 |

***Spiro Compounds***

| | | | |
|---|---|---|---|
| | 137–139d | $CH_2Cl_2$/Petr ether | 17 |

TABLE VIII.1. —(contd.)

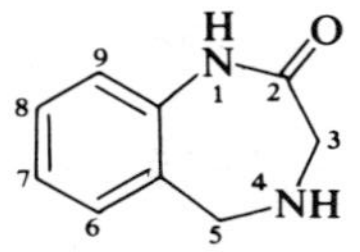

| Substituent | mp (°C) or; [bp (°C)/(torr)] | Solvent of Crystallization | Yield (%) | Spectra | Refs. |
|---|---|---|---|---|---|
| *1,3,4,5-Tetrahydro-1,4-benzodiazepin-2(2H)-ones* | | | | | |
| None | 151–154 | MeCN | 67 | ir, pmr | 59 |
| ***Monosubstituted*** | | | | | |
| 4-HO | 207–208 | MeCN | 15 | ir, ms, pmr | 59 |
| 5-Benzyl | 169–171 | $CH_2Cl_2/Et_2O$ | | | 73 |
| 5-Me | 139–140 | PhH/Hexane | | | 38 |
| 5-Ph | 145–146 | Cyclohexane | 50 | | 58 |
| | 146–147 | 2-PrOH | 80 | | 69 |
| | 147–148 | Acetone/Petr ether | 84 | | 43 |
| Hydrochloride | 247 | MeCN | | | 57 |
| 5-(2-$ClC_6H_4$) | 187–189 | EtOAc | | | 49 |
| 5-(4-$ClC_6H_4$) | 192–195 | $CH_2Cl_2$/Hexane | 98 | | 43, 52 |
| 5-(2-$FC_6H_4$) | 162–163 | $CH_2Cl_2$ | 92 | | 43, 49 |
| 5-(2-F-4-$ClC_6H_3$) | 178–179 | Acetone/Hexane | | | 52 |
| 5-(2-F-5-$NO_2C_6H_3$) | 188–189 | Acetone/Petr ether | 81 | | 87 |
| 5-(3-$NO_2C_6H_4$) | 158–160 | Acetone/Hexane | 23 | | 87 |
| Hydrochloride | 285–300 | | | | 52 |
| 5-(4-$NO_2C_6H_4$) | 235–237 | MeOH | 48 | | 87 |
| 5-(2-Thiazolyl) | 144–145 | Acetone/Hexane | | | 119b |
| 7-Cl | 188–190 | MeOH | | | 28 |

| | | | | |
|---|---|---|---|---|
| ***Disubstituted*** | | | | |
| ***1,5-Disubstituted*** | | | | |
| 1-Me-5-Ph | 141–143 | $Et_2O$ | | 38 |
| 1-Me-5-(2-$ClC_6H_4$) | 177–180 | $Et_2O$ | | 49 |
| 1-Me-5-(2-$FC_6H_4$) | 123–124 | Acetone/Petr ether | | 28 |
| ***4,5-Disubstituted*** | | | | |
| 4-$H_2NCO$-5-Ph | 227–228 | EtOH | 74 | 90 |
| 4-Benzyl-5-Ph | 156–159 | $CH_2Cl_2$/$Et_2O$/ Petr ether | | 28 |
| 4-BuNHCO-5-Ph | 202–204 | MeCN | 88 | 90 |
| 4-HO-5-Ph | 200–202 | | | 76 |
| 4,5-$Me_2$ | 123–124 | $CH_2Cl_2$/Hexane | | 73 |
| 4-Me-5-Ph methiodide | 190–191 | EtOH | | 49 |
| Methochloride | 199–201 | MeOH/$Et_2O$ | | 49 |
| 4-Me-5-(2-$FC_6H_4$) | 123–125 | $CH_2Cl_2$ | 50 | 43 |
| 4-MeNHCO-5-Ph | 140–144 | MeCN | 96 | 90 |
| 4-MeOOC-5-Ph | 215–218 | $CH_2Cl_2$/$Et_2O$/Hexane | | 73 |
| 4-PhNHCO-5-Ph | 227–228 | MeCN | 82 | 90 |
| ***4,7-Disubstituted*** | | | | |
| 4-Benzyl-7-MeO | 145–148 | $CH_2Cl_2$/Petr ether | | 28 |
| ***5,7-Disubstituted*** | | | | |
| 5-Ph-7-Ac | 184–186 | MeCN | | 47 |
| 5-Ph-7-AcNH | 148–154d | EtOAc | | 29 |
| 5-Ph-7-$H_2N$ | 182–184 | PhH | 85 | 69 |
| 5-Ph-7-$H_2NSO_2$ | 155–158 | DMF/$H_2O$ | | 35 |
| 5-Ph-7-Br | 191–192 | DMF/$H_2O$ | 79 | 43, 49 |
| 5-Ph-7-Cl | 184–186 | DMF/$H_2O$ | 87 | 51 |
| 5-Ph-7-$F_3C$ | 152–153 | MeOH | 72 | 43 |
| 5-Ph-7-(1-HO-1-Ethyl) | 159–161 | $Et_2O$/Pentane | | 48 |
| 5-Ph-7-MeO | 150–153 | Acetone/Hexane | 87 | 56 |
| 5-Ph-7-Me | 174–176 | EtOH | 50 | 43, 49 |

TABLE VIII.1. —(contd.)

| Substituent | mp (°C) or; [bp (°C)/(torr)] | Solvent of Crystallization | Yield (%) | Spectra | Refs. |
|---|---|---|---|---|---|
| 5-Ph-7-$Me_2N$ | 174–176 | EtOH | | | 50 |
| 5-Ph-7-(2-Me-1,3-Dioxolan-2-yl) | 181–182 | MeCN | | | 47 |
| 5-Ph-7-MeS | 151–153 | EtOH | | | 46 |
| 5-Ph-7-$NO_2$ | 232–234 | $CH_2Cl_2$ | | | 83 |
| 5-(4-$BrC_6H_4$)-7-Cl | 180–184 | $CH_2Cl_2$/Hexane | | | 28 |
| 5-(2-$ClC_6H_4$)-7-Cl | 235–237 | AcOH/$H_2O$ | 40 | | 43, 49 |
| 5-(2,4-$ClC_6H_3$)-7-Cl | 208–210 | Acetone | 75 | | 43 |
| 5-(4-$ClC_6H_4$)-7-F | 177–178 | PhH | 49 | | 43 |
| 5-(2-$FC_6H_4$)-7-AcNH | 150–155d | $CH_2Cl_2$/EtOAc | | | 29 |
| 5-(2-$FC_6H_4$)-7-Br | 224–225 | MeOH | 87 | | 43, 49 |
| 5-(2-$FC_6H_4$)-7-Cl | 214–215 | Acetone | 90 | | 43, 49 |
| 5-(2-$FC_6H_4$)-7-(Cyclopentyl)CONH | 230–232 | $CH_2Cl_2$/Hexane/EtOH | | | 29 |
| 5-(2-$FC_6H_4$)-7-F | 174–177 | Acetone/Hexane | 96 | | 43 |
| 5-(2,6-$F_2C_6H_3$)-7-Cl | 243–245 | $CHCl_3$/EtOH | | | 28 |
| 5-(2-$Me_2NC_6H_4$)-7-Cl | 160–168d | $Et_2O$ | | | 28 |
| 5-(2-$MeC_6H_4$)-7-Cl | 248–249 | DMF/$H_2O$ | 91 | | 43, 49 |
| 5-(2-Pyridyl)-7-Br | 192–194 | EtOH | | | 158 |
| ***Trisubstituted*** | | | | | |
| ***1,4,5-Trisubstituted*** | | | | | |
| 1,4-$Me_2$-5-Ph | 115–116 | $CH_2Cl_2$ | 32 | | 43 |
| 1,4-$Me_2$-5-(2-$ClC_6H_4$) hydrochloride | 240–241 | Acetone/$Et_2O$ | | | 49 |
| 1,4-$Me_2$-5-(4-$ClC_6H_4$) | 132–133 | $Et_2O$/Petr ether | 76 | | 43 |
| 1,4-$Me_2$-5-(2-$FC_6H_4$) | 119–122 | Hexane | 59 | | 43 |
| 1-Me-4-(2S-$H_2N$-3-Ph-Propanoyl)-5-Ph | 146–147 | $Et_2O$ | | | 73 |
| 1-Me-4-(2R-$H_2N$-3-Ph-Propanoyl)-5-Ph | 143–144 | $CH_2Cl_2$/$Et_2O$/Hexane | | | 73 |
| 1-Me-4-[2-(Benzyloxy)CONH-3-Ph-propanoyl]-5-Ph | 178–179 | EtOH/$Et_2O$ | | | 73 |
| 1-Me-4-Et-5-Ph | 115–120 | $Et_2O$/Hexane | | | 17 |
| 1-Me-4-MeOOC-5-Ph | 136–138 | $Et_2O$ | | | 73 |

| | | | | | |
|---|---|---|---|---|---|
| ***1,5,7-Trisubstituted*** | | | | | |
| 1-(2-$H_2NCOCH_2O$-Ethyl)-5-(2-$FC_6H_4$)-7-Cl | 168–170 | EtOH | | | 84 |
| 1-(4-HOOC-Butyl)-5-(2-$FC_6H_4$)-7-Cl | 149–155 | MeOH | | | 88 |
| 1-(3-Cl-Propyl)-5-Ph-7-Cl hydrochloride | 215–220 | MeOH | | | 35 |
| 1-Et-5-(2-$FC_6H_4$)-7-Cl | 165–167 | MeOH | 87 | | 43 |
| 1-Et-5-(2-$FC_6H_4$)-7-[1-(2-HO-Ethyl)amino-1-ethyl] | 49–50 | | | | 60 |
| 1-Et-5-(2-$FC_6H_4$)-7-(1-HO-Imino-1-ethyl) | 237–239 | $Et_2O$ | | | 60 |
| 1-Et-5-(2-$FC_6H_4$)-7-(1-MeNH-Ethyl) | 140–142 | | | | 60b |
| 1-Et-5-(2-$FC_6H_4$)-7-(1-$Me_2N$-1-Ethyl) dihydrochloride | 160 | | | | 60 |
| 1-Et-5-(2-$FC_6H_4$)-7-(1-Pyrrolidin-1-yl)ethyl | Oil | | | | 60b |
| 1-(4-EtOOC-Butyl)-5-(2-$FC_6H_4$)-7-Cl hydrochloride | 162–164 | EtOH/$Et_2O$ | | | 88 |
| 1-(2-EtNH-Ethyl)-5-(2-$FC_6H_4$)-7-Cl dihydrochloride | 235–240 | MeOH/$Et_2O$ | | | 28 |
| 1-(4-$Et_2N$-Butyl)-5-(2-$FC_6H_4$)-7-Cl dihydrochloride | 230–238d | MeOH/$Et_2O$ | | | 55 |
| 1-(2-$Et_2N$-Ethyl)-5-(2-$FC_6H_4$)-7-Cl | 98–99 | $Et_2O$/Petr ether | 72 | | 44, 55 |
| Dihydrochloride | 215–225 | EtOH/$Et_2O$ | | | 82 |
| | 175–190 | MeOH/$Et_2O$ | 72 | | 44 |
| 1-(2-$Et_2N$-Ethyl)-5-(2-pyridyl)-7-Br dihydrobromide | 214–220d | MeOH/$Et_2O$ | | | 55 |
| 1-(3-$Et_2N$-Propyl)-5-(2-$FC_6H_4$)-7-Cl dihydrochloride | 237–242d | MeOH/$Et_2O$ | 70 | | 44, 55 |
| 1-Me-5-Ph-7-AcNH | 131–133 | EtOH/$Et_2O$ | | | 60c |
| 1-Me-5-Ph-7-Cl | 144–145 | $Et_2O$ | 76 | | 43, 52 |
| | 110–112 | | | | 70 |
| Hydrochloride | 266–269d | EtOH/$H_2O$ | | | 70 |
| 2-$^{14}C$ | | | | | 42 |
| (−)-Enantiomer | | | | | |
| 2-$^{14}C$ | | | | [α] | 42 |
| (+)-Enantiomer | 116 | | | [α] | 42, 166 |
| Camphor sulfonate | 293 | | 70 | [α] | 166 |
| 1-Me-5-Ph-7-CN | 134–135 | $Et_2O$ | | | 38 |
| 1-Me-5-Ph-7-$Me_2N$ | 152–154 | EtOH | | | 50 |
| 1-Me-5-(2-$AcNHC_6H_4$)-7-Cl | 203–205 | EtOH | 88 | | 53 |
| 1-Me-5-(2-$H_2NC_6H_4$)-7-Br | 205–208 | $CH_2Cl_2$/Hexane | | | 53 |

TABLE VIII.1. —(contd.)

| Substituent | mp (°C) or; [bp (°C)/(torr)] | Solvent of Crystallization | Yield (%) | Spectra | Refs. |
|---|---|---|---|---|---|
| 1-Me-5-(2-$H_2NC_6H_4$)-7-Cl | 203–205 | $CH_2Cl_2$/Hexane | | | 28 |
| 1-Me-5-(2-$ClC_6H_4$)-7-Cl | 169–172 | $CH_2Cl_2/Et_2O$ | 51 | | 43 |
| 1-Me-5-(2,4-$Cl_2C_6H_3$)-7-Cl | 169–172 | Acetone/Hexane | 69 | | 43 |
| 1-Me-5-(4-$ClC_6H_4$)-7-Cl | 134–135 | $Et_2O$ | 63 | | 43 |
| 1-Me-5-(4-$ClC_6H_4$)-7-F hydrochloride | 201–205 | MeOH/$Et_2O$ | 92 | | 43 |
| 1-Me-5-(2-$FC_6H_4$)-7-AcNH·0.5 EtOAc | 130–133 | EtOAc | | | 29 |
| 1-Me-5-(2-$FC_6H_4$)-7-AcN(Me) | 140–141 | EtOAc | | | 29 |
| 1-Me-5-(2-$FC_6H_4$)-7-$H_2N$ | 218 | $CH_2Cl_2$ | | | 60b |
| 1-Me-5-(2-$FC_6H_4$)-7-Cl | 135–136 | Acetone/Petr ether | | | 55 |
| 1-Me-5-(2-$FC_6H_4$)-7-F hydrochloride | 219–239 | MeOH | 35 | | 43 |
| 1-Me-5-(2-$FC_6H_4$)-7-I | 171–173 | $CH_2Cl_2$/Hexane | | | 61 |
| 1-Me-5-(2-$FC_6H_4$)-7-$BuNHCONHCH_2$ | 172 | EtOAc | | | 60 |
| 1-Me-5-(2-$FC_6H_4$)-7-MeNHCONH | 120–140 | EtOAc/$Et_2O$ | | | 29 |
| 1-Me-5-(2,6-$F_2C_6H_3$)-7-Cl | 150–155 | $CH_2Cl_2$/MeOH | | | 28 |
| 1-$MeNHCOCH_2$-5-Ph-7-Cl | 179–181 | EtOH/$H_2O$ | | | 54 |
| Hydrochloride | 299–302 | MeOH/Acetone | | | 54 |
| 1-(3-MeNH-Propyl)-5-(2-$FC_6H_4$)-7-Cl dihydrochloride·$H_2O$ | 193–196d | MeOH/$Et_2O$ | 73 | | 44, 55 |
| 1-(3-$Me_2N$-Propyl)-5-Ph-7-Cl dihydrochloride | 273–278d | EtOH/$H_2O$ | | | 35 |
| 1-(3-$Me_2N$-Propyl)-5-(2-$FC_6H_4$)-7-Cl dihydrochloride | 186–200d | MeOH/$Et_2O$ | 85 | | 44, 55 |
| 1-(3-$Me_2N$-Propyl)-5-(2-Pyridyl)-7-Br dihydrochloride | 243–248d | MeOH/$Et_2O$ | | | 158 |
| 1-$MeOCH_2$-5-Ph-7-Cl | 136–138 | EtOH | | | 41 |
| ***3,5,7-Trisubstituted*** | | | | | |
| 3-(Benzyloxy)CO-5-Ph-7-Cl | 150–155 | EtOH | | | 29 |
| 3-EtOOC-5-Ph-7-Cl | 172–174 | | | | |
| | 180–182 | MeCN | | | 64 |
| 3-(3-$Et_2N$-Propyloxy)CO-5-Ph-7-Cl | 104–106 | $Et_2O$/Petr ether | | | 29 |
| 3-$HOCH_2$-5-Ph-7-Cl | 135–136 | $CH_2Cl_2$/Hexane | | | 35 |

| | | | | |
|---|---|---|---|---|
| ***4,5,5,-Trisubstituted*** | | | | |
| 4,5,5-$Me_3$ | 185–186 | $CH_2Cl_2$/Hexane | | 73 |
| ***4,5,7-Trisubstituted*** | | | | |
| 4-Ac-5-Ph-7-AcNH | > 230 | MeOH/$Et_2O$ | | 60c |
| 4-Ac-5-Ph-7-Cl | 232–233 | Acetone | | 38 |
| 4-Ac-5-Ph-7-$NO_2$ | 258–260 | MeCN | 98 | 90 |
| 4-Ac-5-(4-$BrC_6H_4$)-7-Cl | 231–236 | $CH_2Cl_2$/MeOH | | 28 |
| 4-Ac-5-(2-$FC_6H_4$)-7-AcNH·0.5 EtOAc | 270–274 | EtOAc | | 29 |
| 4-Ac-5-(2-$FC_6H_4$)-7-Cl | 202–203 | Acetone/Hexane | | 28 |
| 4-AcO-5-Ph-7-Ac | 187–188 | $Ac_2O$ | | 47 |
| 4-AcO-5-Ph-7-Cl | 193–195 | $Ac_2O$ | 40 | 92 |
| 4-AcO-5-(2-$ClC_6H_4$)-7-Cl | 193–195 | $Ac_2O$ | | 92 |
| 4-Allyl-5-(2-$FC_6H_4$)-7-(cyclopentyl)CONH | 274–276 | $CH_2Cl_2$/$Et_2O$ | | 29 |
| 4-$H_2N$-5-Ph-7-$H_2N$ | 218–221 | MeCN | 56 | 86 |
| 4-$H_2NCO$-5-Ph-7-Cl | 242–245 | EtOH | 98 | 90 |
| 4-$H_2NCO$-5-Ph-7-$NO_2$ | 239–241 | EtOH | 73 | 90 |
| 4-Benzyl-5-Ph-7-Cl | 197–204 | $CH_2Cl_2$/$Et_2O$/Petr ether | 55 | 78, 56 |
| 4-Benzyl-5-Ph-7-MeO | 145–148 | $CH_2Cl_2$/Petr ether | 81 | 78, 56 |
| 4-Benzyloxy(CO)-5-Ph-7-$H_2N$ | 165–168 | EtOH | 90 | 83 |
| 4-Benzyloxy(CO)-5-Ph-7-$NO_2$ | 179–181 | *i*-PrOH | | 83 |
| 4-Benzyloxy(CO)-5-(2-$FC_6H_4$)-7-Cl | 169–170 | | | 84 |
| 4-(Br-Acetyl)-5-(2-$FC_6H_4$)-7-Cl | 232–234 | Acetone | | 28 |
| 4-*t*-BuOOC-5-Ph-7-$NO_2$ | 138–139 | EtOH | 65 | 90 |
| 4-BuNHCO-5-Ph-7-Cl | 226–229 | EtOH | 91 | 90 |
| 4-BuNHCO-5-Ph-7-$NO_2$ | 204–208 | MeCN | 88 | 90 |
| 4-(2-HOOC-Ethyl)-5-Ph-7-Cl | 218–219 | EtOH/$H_2O$ | | 35 |
| 4-$HOOCCH_2$-5-Ph-7-Cl | 217–220 | DMF/$H_2O$ | | 35 |
| 4-(4-Cl-Benzyloxy)CO-5-Ph-7-$NO_2$ | 203–205 | EtOH | 67 | 90 |
| 4-ClCO-5-Ph-7-Cl | 181–184 | PhH | 64 | 90 |
| 4-Cyclohexanoyl-5-Ph-7-Cl | 181–183 | EtOH | 70 | 90 |
| 4-Cyclopentanoyl-5-Ph-7-$NO_2$ | 248–251d | EtOH | 80 | 90 |
| 4-Et-5-(2-$FC_6H_4$)-7-Cl | 154–156 | MeOH/$H_2O$ | | 28 |
| 4-EtOOC-5-Ph-7-$NO_2$ | 130–132 | EtOH | 75 | 90 |

TABLE VIII.1. —(contd.)

| Substituent | mp (°C) or; [bp (°C)/(torr)] | Solvent of Crystallization | Yield (%) | Spectra | Refs. |
|---|---|---|---|---|---|
| 4-(2-EtOOC-Ethyl)-5-Ph-7-Cl | 146–148 | $CH_2Cl_2$/Hexane | | | 35 |
| 4-EtOOC$CH_2$-5-Ph-7-Cl | 185–186 | EtOH | | | 35 |
| 4-EtNHCO-5-Ph-7-$NO_2$ | 185–189 | EtOH | 72 | | 90 |
| 4-Formyl-5-Ph-7-$H_2N$ | 239–241 | EtOH | | | 90 |
| 4-Formyl-5-Ph-7-$NO_2$ | 258–261 | Acetone | 90 | | 90 |
| 4-HO-5-Ph-7-Ac | 201–203 | $CH_2Cl_2$ | | | 47 |
| 4-HO-5-Ph-7-Cl | 215–216 | AcOH | 60 | | 51 |
| 4-HO-5-Ph-7-$NO_2$ | 219–220d | Acetone | 77 | | 75 |
| 4-HO-5-(2-Cl$C_6H_4$)-7-Cl | 203–205 | MeCN | | | 92 |
| 4-HO-5-(2-F$C_6H_4$)-7-Cl | 235–253 | MeOH | | | 28 |
| 4-HO-5-(2-F-5-$O_2NC_6H_3$)-7-Cl | 240–265 | MeOH | | | 28 |
| 4-(Menthyloxy)CO-5-Ph-7-Cl | 220 | | 57 | | 90 |
| 4-Me-5-Ph-7-AcNH | 160–164 | EtOAc | | | 29 |
| 4-Me-5-Ph-7-Cl | 206–208 | EtOH | 38 | | 43, 58 |
| Methiodide | 231–240 | MeOH/$Et_2O$ | | | 49 |
| 4-Me-5-Ph-7-MeO | 214–215 | $CH_2Cl_2$ | 20 | | 56, 77 |
| 4-Me-5-Ph-7-$NO_2$ | 182–185 | MeOH | | | 28 |
| 4-Me-5-(2-Cl$C_6H_4$)-7-Cl | 232–235 | EtOH/Acetone | 60 | ir, pmr | 66 |
| 4-Me-5-(2-F$C_6H_4$)-7-NHAc | 160–170 | EtOAc | | | 29 |
| 4-Me-5-(2-F$C_6H_4$)-7-Cl | 185–186 | $CH_2Cl_2$/Petr ether | | | 49 |
| 4-Me-5-(2-Pyridyl)-7-Br | 211–213 | EtOH | | | 158 |
| 4-MeNHCO-5-Ph-7-Cl | 235–238 | EtOH | 88 | | 90 |
| 4-MeNHCO-5-Ph-7-$NO_2$ | 179–180 | EtOH | 97 | | 90 |
| 4-MeOOC-5-Ph-7-Cl | 173–175 | MeOH/$Et_2O$ | | | 17 |
| 4-(4-Me$C_6H_4$)$SO_2$-5-Ph-7-Ac | 265–266 | MeCN | | | 47 |
| 4-(4-Me$C_6H_4$)$SO_2$-5-Ph-7-Br | 177–179 | $CH_2Cl_2$/EtOH | | | 17 |
| 4-(4-Me$C_6H_4$)$SO_2$-5-Ph-7-Cl | 246–252 | $CHCl_3$/EtOH | 80 | | 68, 92 |
| 4-(4-Me$C_6H_4$)$SO_2$-5-Ph-7-$NO_2$ | 245–247 | $CH_2Cl_2$/$Et_2O$ | | | 17 |

| | | | | | |
|---|---|---|---|---|---|
| 4-(4-Me$C_6H_4$)$SO_2$-5-(2-F$C_6H_4$)-7-Cl | 242–243 | $CH_2Cl_2$/Petr ether | | | 31 |
| 4-Me$SO_2$-5-Ph-7-Cl | 203–206 | $CHCl_3$/EtOH | 86 | | 68 |
| 4-(1-Naphthyl)NHCO-5-Ph-7-$NO_2$ | 194–196 | EtOAc | 95 | | 90 |
| 4-NO-5-Ph-7-$NO_2$ | 211–212 | EtOH | 80 | | 86 |
| 4-PhNHCO-5-Ph-7-Cl | 232–237 | MeCN | 82 | | 90 |
| 4-PhNHCO-5-Ph-7-$NO_2$ | 222–224 | THF/Hexane | 63 | | 90 |
| 4-(1-Ph-1-Ethoxy)CO-5-Ph-7-Cl | 170–172 | EtOAc/$Et_2O$ | 78 | | 90 |
| 4-Propenoyl-5-Ph-7-$NO_2$ | 225–227 | EtOH | 72 | | 90 |
| 4-(2-Propyn-1-yl)-5-(2-F$C_6H_4$)-7-(cyclopentyl)CONH | 230–234 | $CH_2Cl_2$/$Et_2O$ | | | 29 |
| ***5,7,8-Trisubstituted*** | | | | | |
| 5-Ph-7-$H_2NSO_2$-8-Cl | 237–240 | THF/$Et_2O$ | | | 28 |
| ***5,7,9-Trisubstituted*** | | | | | |
| 5-(2-Pyridyl)-7,9-$Br_2$ | 199–203 | EtOH | | | 158 |
| ***Tetrasubstituted*** | | | | | |
| ***1,3,5,7-Tetrasubstituted*** | | | | | |
| 1,3-$Me_2$-5-Ph-7-Cl | | | | | |
| 3*S*,5*S*-Enantiomer | 102–103 | EtOH | 95 | [α] | 63 |
| 3*R*,5*R*-Enantiomer | 102–104 | EtOH | 86 | [α] | 63 |
| 1-Me-3-EtOOC-5-Ph-7-Cl | | | | | |
| Isomer A | 150–152 | $Et_2O$/Hexane | | | 17 |
| Isomer B | 153–154 | $Et_2O$/Hexane | | | 17 |
| 1-Me-3-[3,4-$(MeO)_2$-Benzyl]-5-Ph-7-Cl | | | | | |
| Isomer A | 173–175 | $CHCl_3$/Petr ether | | | 85 |
| 1-Me-3-[3,4-$(MeO)_2C_6H_3$](HO)CH-5-Ph-7-Cl | | | | | |
| hydrochloride | 219–221d | MeOH/$Et_2O$ | | | 85 |
| Isomer B | 150–155 | $CH_2Cl_2$/Petr ether | | | 85 |
| 1-Me-3-(2-MeOOC-1-Ethyl)-5-Ph-7-Cl | 177–180 | $CH_2Cl_2$/$Et_2O$ | 66 | ir, pmr, uv | 62 |
| ***1,4,5,5-Tetrasubstituted*** | | | | | |
| 1,4,5-$Me_3$-5-Ph | 120–121 | $Et_2O$/Hexane | | | 73 |

TABLE VIII.1. —(contd.)

| Substituent | mp (°C) or; [bp (°C)/(torr)] | Solvent of Crystallization | Yield (%) | Spectra | Refs. |
|---|---|---|---|---|---|
| ***1,4,5,7-Tetrasubstituted*** | | | | | |
| 1,4-$Ac_2$-5-Ph-7-Cl | 185–186 | $CH_2Cl_2$ | | | 38 |
| 1,4-$(Allyl)_2$-5-Ph-7-Cl hydrochloride | 190–191 | $CH_2Cl_2/Et_2O$ | 43 | | 43, 49 |
| 1-(4-HOOC-Butyl)-4-Me-5-(2-$FC_6H_4$)-7-Cl | 167–174 | MeOH | | | 88 |
| 1,4-$Et_2$-5-(2-$FC_6H_4$)-7-Cl | 92–93 | Hexane | 47 | | 43 |
| 1-Et-4-Ac-5-(2-$FC_6H_4$)-7-Cl | 180–181 | MeOH | | | 28 |
| 1-Et-4-Me-5-(2-$FC_6H_4$)-7-Cl | 132–133 | Acetone/Hexane | 57 | | 43 |
| 1-(4-EtOOC-Butyl)-4-Me-5-(2-$FC_6H_4$)-7-Cl hydrochloride | 158–167 | $EtOH/Et_2O$ | | | 88 |
| 1-(2-$Et_2$N-Ethyl)-4-Me-5-(2-$FC_6H_4$)-7-Cl | 83–85 | Petr ether | 78 | | 44, 55 |
| Methiodide | 185–190 | Acetone/ether | | | 55 |
| 1-(2-$Et_2$N-Ethyl)-4-(4-$MeC_6H_4$)$SO_2$-5-(2-$FC_6H_4$)-7-Cl | 179–182 | MeOH | | | 82 |
| 1,4-$Me_2$-5-Ph-7-AcNH | 218–219 | $EtOH/Et_2O$ | | | 60c |
| 1,4-$Me_2$-5-Ph-7-Br | 166–172 | $MeOH/Et_2O$ | 20 | | 43, 49 |
| 1,4-$Me_2$-5-Ph-7-Cl | 90–91 | Hexane | 31 | | 43, 49 |
| 1,4-$Me_2$-5-Ph-7-$F_3C$ | 77–79 | Hexane | 48 | | 43, 45 |
| 1,4-$Me_2$-5-Ph-7-Me | 71–73 | Hexane | 15 | | 43, 49 |
| Methiodide | 160–161d | $MeOH/Et_2O$ | | | 49 |
| 1,4-$Me_2$-5-Ph-7-MeS | 96–98 | $Et_2O$/Hexane | | | 46 |
| 1,4-$Me_2$-5-Ph-7-MeS(O) | 160–161 | Acetone/Hexane | | | 46 |
| 1,4-$Me_2$-5-(2-$ClC_6H_4$)-7-Br | 134–135 | $Et_2O$ | | | 49 |
| 1,4-$Me_2$-5-(2-$ClC_6H_4$)-7-Cl hydrochloride | 240–241 | Acetone/$Et_2O$ | 50 | | 43 |
| 1,4-$Me_2$-5-(2,4-$Cl_2C_6H_3$)-7-Cl | 150–153 | Hexane | 30 | | 43 |
| | | | | X-ray | 159 |
| 1,4-$Me_2$-5-(4-$ClC_6H_4$)-7-F | 109–112 | Hexane | 60 | | 43 |
| 1,4-$Me_2$-5-(2-$FC_6H_4$)-7-AcNH | 242–244 | EtOAc | | | 29 |
| 1,4-$Me_2$-5-(2-$FC_6H_4$)-7-Br | 134–135 | $Et_2O$ | 41 | | 43 |
| 1,4-$Me_2$-5-(2-$FC_6H_4$)-7-Cl | 124–125 | $Et_2O$ | | | 49 |

| | | | | | |
|---|---|---|---|---|---|
| 1,4-$Me_2$-5-(2-$FC_6H_4$)-7-F | 118–120 | Hexane | 84 | | 43 |
| 1,4-$Me_2$-5-(2-$FC_6H_4$)-7-(Me)AcN | 130–135 | $Et_2O$/Petr ether | | | 29 |
| 1,4-$Me_2$-5-(2-$FC_6H_4$)-7-MeNHCONH | 170–176d | EtOAc/$Et_2O$ | | | 29 |
| 1,4-$Me_2$-5-(2,6-$F_2C_6H_3$)-7-Cl | 183–187 | $Et_2O$/Petr ether | | | 28 |
| 1,4-$Me_2$-5-(2-$Me_2NC_6H_4$)-7-Cl dihydrochloride | 152–163 | $CH_2Cl_2$/Hexane | | | 28 |
| 1,4-$Me_2$-5-(2-$MeC_6H_4$)-7-Cl hydrochloride | 197–215 | MeOH/$Et_2O$ | 66 | | 43, 49 |
| 1-Me-4-Ac-5-Ph-7-NHAc | > 240 | EtOH/$Et_2O$ | | | 60c |
| 1-Me-4-Ac-5-Ph-7-Cl | 188–192 | $Et_2O$ | 76 | ir | 40 |
| | 177–182 | $Et_2O$ | 95 | | 70 |
| | 201–203 | EtOH | 74 | | 90 |
| 1-Me-4-Ac-5-(2-$FC_6H_4$)-7-AcNH·0.25 EtOAc | 256 | EtOAc | | | 29 |
| 1-Me-4-Ac-5-(2-$FC_6H_4$)-7-Cl | 206–207 | Acetone/Hexane | | | 28 |
| 1-Me-4-Ac-5-(2-$FC_6H_4$)-7-(Me)AcN | 200–201 | EtOAc | | | 29 |
| 1-Me-4-AcO-5-(2-$FC_6H_4$)-7-I | 172–174 | MeOH | | | 91 |
| 1-Me-4-(2-AcO-Ethyl)-5-Ph-7-Cl | 128–130 | $CH_2Cl_2$/Hexane | | | 35 |
| 1-Me-4-$H_2N$-5-Ph-7-Cl | 147–148 | *i*-PrOH | 59 | | 86 |
| 1-Me-4-AcNH-5-Ph-7-Cl | 198–206 | MeCN | 74 | | 86 |
| 1-Me-4-($H_2N$-Acetyl)-5-Ph-7-Cl | 158–160 | PhH | 79 | | 90 |
| 1-Me-4-($H_2N$-Acetyl)NH-5-Ph-7-Cl | 173–175 | EtOAc | 76 | | 86 |
| 1-Me-4-(2S-$H_2N$-3-Ph-Propanoyl)-5-Ph-7-Cl | 179–182 | $CH_2Cl_2$/$Et_2o$ | | | 73 |
| 1-Me-4-(2R-$H_2N$-3-Ph-Propanoyl)-5-Ph-7-Cl | 172–174 | | | | 73 |
| 1-Me-4-Allyl-5-Ph-7-Cl | 108–109 | Hexane | 57 | | 43, 49 |
| 1-Me-4-$H_2NCO$-5-Ph-7-Cl | 217–219 | EtOH | 95 | | 90 |
| (+)-Enantiomer | 215–217 | EtOH | 87 | [α] | 90 |
| | | | | X-ray | 160 |
| (−)-Enantiomer | 214–271 | EtOH | 87 | [α] | 90 |
| 1-Me-4-Benzyl-5-Ph-7-Cl | 151–154 | MeOH | | | 78 |
| 1-Me-4-(Benzyloxycarbonylamino)acetyl-5-Ph-7-Cl | 158–160 | EtOH | 90 | | 90 |
| 1-Me-4-[(Benzyloxycarbonylamino)acetyl]NH-5-Ph-7-Cl | 184–190 | EtOH | 72 | | 86 |
| 1-Me-4-*t*-BuOOC-5-Ph-7-Cl | 177–178 | EtOH | 90 | | 90 |
| 1-Me-4-BuNHCO-5-(2-$FC_6H_4$)-7-BuNHCONH | 172 | EtOAc | | | 60b |
| 1-Me-4-(2-HOOC-Benzoyl)-5-Ph-7-Cl | 280–282 | EtOH | | | 89 |
| 1-Me-4-(Cl-Acetyl)-5-Ph-7-Cl | 197–200 | Acetone | 81 | | 90 |

TABLE VIII.1. —(contd.)

| Substituent | mp (°C) or; [bp (°C)/(torr)] | Solvent of Crystallization | Yield (%) | Spectra | Refs. |
|---|---|---|---|---|---|
| 1-Me-4-(Cl-Acetyl)NH-5-Ph-7-Cl | 227–229 | MeCN | 74 | | 86 |
| 1-Me-4-(2-Cl-Benzoyl)-5-Ph-7-Cl | 250–253 | MeCN | 88 | | 90 |
| 1-Me-4-(4-Cl-Benzoyl)-5-Ph-7-Cl | 238–240 | MeCN | 96 | | 90 |
| 1-Me-4-(2-$ClC_6H_4$-Methylene)amino-5-Ph-7-Cl | 185–187 | EtOH | 91 | | 86 |
| 1-Me-4-(4-Cl-Benzyloxy)CO-5-Ph-7-Cl | 153–154 | EtOH | 82 | | 90 |
| 1-Me-4-ClCO-5-Ph-7-Cl | 186–187 | PhH | 91 | | 90 |
| (−)-Enantiomer | 201–203 | Acetone | 67 | [α] | 90 |
| 1-Me-4-Cyclohexanoyl-5-Ph-7-Cl | 189–191 | EtOH | 74 | | 90 |
| 1-Me-4-Cyclopentanoyl-5-Ph-7-Cl | 229–232d | EtOH | 80 | | 90 |
| 1-Me-4-Et-5-$H_2NCO$-7-Br | 190–192 | $CH_2Cl_2/Et_2O$ | | | 85 |
| 1-Me-4-Et-5-Ph-7-Cl | 111–114 | $Et_2O$ | | | 28 |
| 1-Me-4-Et-5-Ph-7-MeNH | 168–170 | $CH_2Cl_2/Et_2O$/Hexane | | | 17 |
| 1-Me-4-Et-5-(2-$FC_6H_4$)-7-Cl | 115–116 | Hexane | | | 28 |
| 1-Me-4-EtOOC-5-Ph-7-Cl | 173–174 | EtOH | 92 | | 90 |
| 1-Me-4-($Et_2N$-Acetyl)-5-Ph-7-Cl | 129–130 | EtOH | | | 17 |
| 1-Me-4-(2-$Et_2N$-Ethyl)-5-(2-$FC_6H_4$)-7-Cl | | | | | |
| hydrochloride | 186–193 | Acetone/$Et_2O$ | 30 | | 44, 55 |
| 1-Me-4-Formyl-5-Ph-7-Cl | 162–164 | *i*-PrOH | 47 | ir | 40 |
| | 167–169 | EtOH | 68 | | 90 |
| 1-Me-4-$H_2NNHCO$-5-Ph-7-Cl | 187–189 | PhH/$CHCl_3$ | 87 | | 90 |
| 1-Me-4-HO-5-Ph-7-Cl | | | | | |
| Isomer A | 190 | PhH | 20 | | 75 |
| Isomer B | 209 | PhH/Petr ether | 33 | | 75 |
| | 212–218 | MeOH | | | 17 |
| 1-Me-4-HO-5-(2-$FC_6H_4$)-7-I | 216–220 | $CH_2Cl_2$/Hexane | | | 65 |
| 1-Me-4-(2-HO-Ethyl)-5-Ph-7-Cl | 135–136 | $Et_2O$ | | | 35 |
| 1-Me-4-(4-$HOC_6H_4$-Methylene)amino-5-Ph-7-Cl | 135–137 | PhH | 89 | | 86 |
| 1-Me-4-(2-Me-Propanoyl)-5-(2-$FC_6H_4$)-7-Cl | 203–204 | EtOAc | | | 60c |
| 1-Me-4-MeOOC-5-Ph-7-Cl | 155–157 | MeOH | | | 17 |

| | | | | |
|---|---|---|---|---|
| 1-Me-4-MeOOC-5-(2-$FC_6H_4$)-7-MeNHCONH | 240–244d | EtOAc/$Et_2O$ | | 29 |
| 1-Me-4-(Menthyloxy)CO-5-Ph-7-Cl | 144–145 | EtOAc/Hexane | | 90 |
| 1-Me-4-(Menthyloxy)CO-5-Ph-7-$NO_2$ | 206–207 | EtOH | 39 | 90 |
| 1-Me-4-(4-$MeC_6H_4$)$SO_2$-5-Ph-7-Cl | 260–262 | $CH_2Cl_2$/$Et_2O$ | | 31 |
| 1-Me-4-(4-$MeC_6H_4$)$SO_2$-5-(2-$FC_6H_4$)-7-I | 255–258 | $CH_2Cl_2$/$Et_2O$ | | 93 |
| 1-Me-4-(4-Me-Piperidino)CO-5-Ph-7-Cl | 240–242 | MeCN | 95 | 90 |
| 1-Me-4-(Morpholino)CO-5-Ph-7-Cl | 184–185 | EtOH | 89 | 90 |
| 1-Me-4-(1-Naphthyl)NHCO-5-Ph-7-Cl | 271–273 | DMF | 72 | 90 |
| 1-Me-4-(2-$O_2NC_6H_4$-Methylene)amino-5-Ph-7-Cl | 187–188 | EtOH | 90 | 86 |
| 1-Me-4-(3-$O_2NC_6H_4$-Methylene)amino-5-Ph-7-Cl | 224–226 | EtOH | 93 | 86 |
| 1-Me-4-(4-$O_2NC_6H_4$-Methylene)amino-5-Ph-7-Cl | 202–204 | EtOH | 88 | 86 |
| 1-Me-4-NO-5-Ph-7-Cl | 180–182 | EtOH | 88 | 86 |
| 1-Me-4-(Ph-Acetyl)-5-Ph-7-Cl | 228–231 | EtOH | 94 | 90 |
| 1-Me-4-PhNHCO-5(2-$FC_6H_4$)-7-PhNHCONH | Amorphous | | | 60b |
| 1-Me-4-(1-Ph-Ethoxy)-CO-5-Ph-7-Cl | 163–165 | EtOH | 75 | 90 |
| 1-Me-4-(N-Ph-Methylene)amino-5-Ph-7-Cl | 157–158 | EtOH | 80 | 86 |
| 1-Me-4-(3-Ph-Propanoyl)-5-Ph-7-Cl | 169–171 | $CH_2Cl_2$/$Et_2O$ | | 161 |
| 1-Me-4-(Piperidino)CO-5-Ph-7-Cl | 136–137 | EtOH | 87 | 90 |
| 1-Me-4-Propenoyl-5-Ph-7-Cl | 169–171 | EtOH | 85 | 90 |
| 1,4-$(MeNHCOCH_2)_2$-5-Ph-7-Cl | 177–179 | Acetone | | 54 |
|   Hydrochloride | 208–210 | MeOH/$Et_2O$ | | 54 |
| 1,4-$(MeNHCOCH_2)_2$-5-Ph-7-Cl | 177–179 | Acetone | | 89 |
| 1-$MeNHCOCH_2$-4-Me-5-Ph-7-Cl | 155–157 | $CH_2Cl_2$/Hexane | | 54 |
|   Hydrochloride | 231–233 | MeOH/$Et_2O$ | | 54 |
| 1-$MeOCH_2$-4-AcO-5-Ph-7-Cl | 103–106 | MeOH | | 41 |
| 1-$MeOCH_2$-4-HO-5-Ph-7-Cl | 173–175 | EtOH | | 41 |
| 1-$MeOCH_2$-4-(4-$MeC_6H_4$)$SO_3$-5-Ph-7-Cl | 216–218 | EtOAc | | 17 |
| ***1,5,6,7-Tetrasubstituted*** | | | | |
| 1-Me-5-(2-$FC_6H_4$)-6-Cl-7-$H_2N$ | 230–234 | $CH_2Cl$ | | 29 |
| 1-Me-5-(2-$FC_6H_4$)-6-Cl-7-(Cyclopentyl)CONH | 190–192 | $CH_2Cl_2$/$Et_2O$/Hexane | | 29 |
| ***4,5,5,7-Tetrasubstituted*** | | | | |
| 4-(Benzyloxy)CO-5-EtO-5-Ph-7-Cl | 210–212 | $Et_2O$/Petr ether | | 29 |
| 4-EtOOC-5-EtO-5-Ph-7-Cl | 230–232 | $CH_2Cl_2$/EtOH | | 17 |

TABLE VIII.1. —(contd.)

| Substituent | mp (°C) or; [bp (°C)/(torr)] | Solvent of Crystallization | Yield (%) | Spectra | Refs. |
|---|---|---|---|---|---|
| 4-EtOOC-5-EtO-5-(2-Pyridyl)-7-Br | 236 | $Et_2O$ | | | 29 |
| 4-HO-5-Me-5-Ph-7-Cl | 211–213d | MeOH | | | 29 |
| 4,5-$Me_2$-5-Ph-7-Cl | 178–181 | MeOH | | ir | 72 |
| 4-Me-5-{-[3-Azabicyclo(3,2,2)non-3-yl]propyn-1-yl}-5-Ph-7-Cl | 202–203 | | | ir, pmr | 72 |
| 4-Me-5-CN-5-Ph-7-Cl | 196–198d | *i*-PrOH | 85 | ir, pmr | 72 |
| 4-Me-5-MeO-5-Ph-7-Cl | 201–203 | MeOH | | ir, pmr | 72 |
| 4-Me-5-MeNH-5-Ph-7-Cl | 193–195 | | | ir | 72 |
| 4-Me-5-Ph-5-(3-Tetrahydropyranyloxy-propyn-1-yl)-7-Cl | 161–162 | MeOH | | ir, pmr | 72 |
| 4-MeOOC-5-MeO-5-Ph-7-Cl | 227–230 | EtOAc | | | 17 |
| 4-$NO_2$-5-AcO-5-Ph-7-Cl | 138–142d | EtOAc/$Et_2O$ | 94 | ir, pmr | 74 |
| 4-$NO_2$-5-AcO-5-Ph-7-$NO_2$ | 174–175d | $CH_2Cl_2$/EtOAc | 75 | ir, pmr | 74 |
| 4-$NO_2$-5-MeO-5-Ph-7-Cl | 208–210d | MeOH | 81 | ir, pmr | 74 |
| 4-$NO_2$-5-MeO-5-Ph-7-$NO_2$ | 208d | $CH_2Cl_2$/MeOH | 88 | ir | 74 |
| ***5,6,7,8-Tetrasubstituted*** | | | | | |
| 5-Ph-6,7,8-$(MeO)_3$ | 191–192 | $Et_2O$ | | | 162 |
| ***5,7,8,9-Tetrasubstituted*** | | | | | |
| 5-Me-7,8,9-$(MeO)_3$ | | | | | |
| Acetate | 130–132 | EtOAc | | | 162 |
| 5-Ph-7,8,9-$(MeO)_3$ | 114–115 | EtOAc | | | 162 |
| ***Pentasubstituted*** | | | | | |
| 1,3-$Me_2$-4-$H_2$NCO-5-Ph-7-Cl | 241–242 | EtOH | | | 63 |
| 3*S*,5*S*-Enantiomer | 208 | EtOH | 81 | [α] | 63 |
| 3*R*,5*R*-Enantiomer | 207–208 | EtOH | | [α] | 63 |
| 1,3-$Me_2$-4-(2-HO-Ethyl)-5-Ph-7-Cl | 186–189 | | 15 | | 67 |

| | | | | | |
|---|---|---|---|---|---|
| 1-Me-3-[3,4-$(MeO)_2C_6H_4$]-(HO)CH-4-Br-5-Ph-7-Cl | 176–178 | $Et_2O$/Petr ether | | | 85 |
| 1-Cl-4-$NO_2$-5-AcO-5-Ph-7-Cl | 132d | $Ac_2O/Et_2O$ | 76 | ir | 74 |
| 1,4,5-$Me_3$-5-Ph-7-Cl | 149–150 | MeOH | | ir, pmr | 72 |
| 1,4-$Me_2$-5-$H_2N$-5-Ph-7-Cl | 185–186 | | | ir, ms | 72 |
| 1,4-$Me_2$-5-$H_2N$-5-Ph-7-$NO_2$ | 214–216 | | | ir | 72 |
| 1,4-$Me_2$-5-$H_2N$-5-(2-$FC_6H_4$)7-Cl | 186–188 | | | ir | 72 |
| 1,4-$Me_2$-5-(Benzyl)NH-5-Ph-7-Cl | 170–172 | | | ir | 72 |
| 1,4-$Me_2$-5-Bu-5-Ph-7-Cl | 137–140 | MeOH | | ir, pmr | 72 |
| 1,4-$Me_2$-5-BuNH-5-Ph-7-Cl | 116–118 | | | ir | 72 |
| 1,4-$Me_2$-5-CN-5-Ph-7-Cl ethanolate | 134–135 | EtOH | 95 | ir, pmr | 72 |
| 1.4-$Me_2$-5-CN-5-Ph-7-$NO_2$ | 164–166 | 2-PrOH | 92 | ir, pmr | 72 |
| 1,4-$Me_2$-5-CN-5-(2-F-$C_6H_4$)-7-Cl | 165–166 | EtOH | 79 | ir, pmr | 72 |
| 1,4-$Me_2$-5-Et-5-Ph-7-Cl | 139–141 | MeOH | | ir, pmr | 72 |
| 1,4-$Me_2$-5-EtO-5-Ph-7-Cl | 100–102 | MeOH | | ir, pmr | 72 |
| 1,4-$Me_2$-5-EtNH-5-Ph-7-Cl | 144–145 | | | | 72 |
| 1,4-$Me_2$-5-EtNH-5-(2-$FC_6H_4$)-7-Cl | 173–175 | | | | 72 |
| 1,4-$Me_2$-5-(2-$Et_2N$-Ethyl)NH-5-Ph-7-Cl | 162–164 | | | ir | 72 |
| 1,4-$Me_2$-5-$H_2NNH$-5-Ph-7-Cl | Oil | | | ir | 72 |
| 1,4-$Me_2$-5-(2-HO-Ethyl)NH-5-Ph-7-Cl | Oil | | | ir | 72 |
| 1,4-$Me_2$-5-MeO-5-Ph-7-Cl | 107–108 | MeOH | | ir, pmr | 72 |
| 1,4-$Me_2$-5-MeO-5-(2-$FC_6H_4$)-7-Cl | 91–93 | EtOH | | ir, pmr | 72 |
| 1,4-$Me_2$-5-MeNH-5-Ph-7-Cl | 196–198 | | | ir, ms | 72 |
| 1,4-$Me_2$-5-MeNH-5-Ph-7-$NO_2$ | 238–240 | | | ir | 72 |
| 1,4-$Me_2$-5-MeNH-5-(2-$FC_6H_4$)-7-Cl | 202–203 | | | ir | 72 |
| 1-Me-4-EtOOC-5-EtO-5-Ph-7-Cl | 207–208 | PhH/EtOH | | | 17 |
| 1-Me-4-EtOOC-5-EtO-5-(2-Pyridyl)-7-Br | 220 | EtOAc | | | 29 |
| 1-Me-4-[1,2-$(MeOOC)_2$-Ethenyl]-5-MeO-5-Ph-7-Cl | 162–163d | MeCN | | | 119b |
| 1-Me-4-$NO_2$-5-AcO-5-Ph-7-Cl | 158–159d | $CH_2Cl_2$/EtOAc | 84 | ir | 74 |
| 1-Me-4-$NO_2$-5-AcO-5-Ph-7-$NO_2$ | 146–147d | $CH_2Cl_2$/EtOAc | 77 | ir | 74 |
| 1-Me-4-$NO_2$-5-MeO-5-Ph-7-Cl | 198–199d | EtOAc/$i$-$Pr_2O$ | 67 | ir, pmr | 74 |
| 1-Me-4-$NO_2$-5-MeO-5-Ph-7-$NO_2$ | 202–204d | EtOAc/$i$-$Pr_2O$ | 59 | ir, pmr | 74 |
| 1,4-$Me_2$-5-(2-$FC_6H_4$)-6-Cl-7-(Cyclopentyl)CONH | 218–220 | EtOAc | | | 29 |
| 1,4-$Me_2$-5-(2-$FC_6H_4$)-6-Cl-7-(Cyclopentyl)CON(Me) | 195–200 | EtOAc/Petr ether | | | 29 |
| 1-Me-5-Ph-6,7,8-$(MeO)_3$ hydrochloride | 238d | MeOH/EtOAc | | | 162 |

TABLE VIII.1. —(contd.)

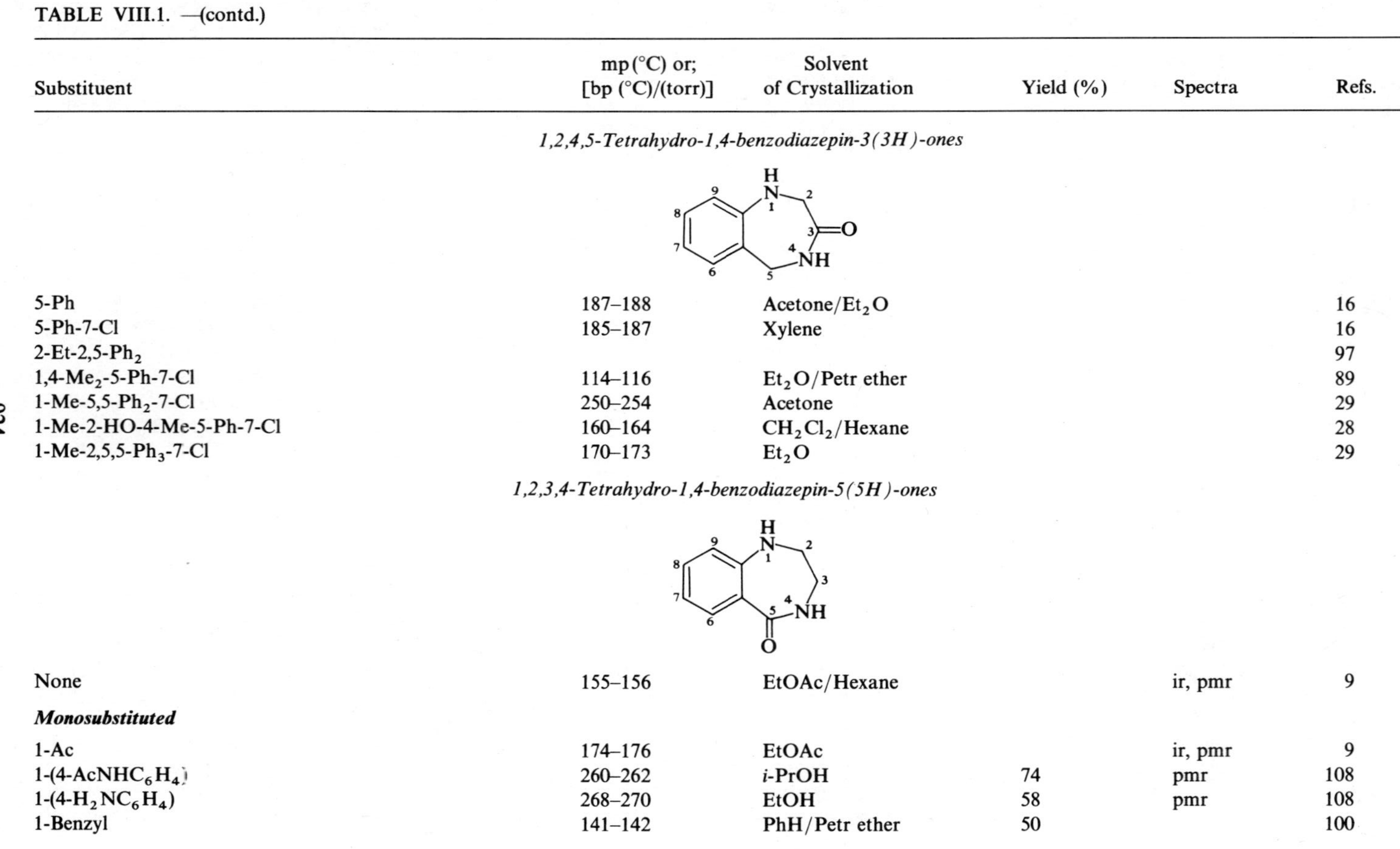

| Substituent | mp (°C) or; [bp (°C)/(torr)] | Solvent of Crystallization | Yield (%) | Spectra | Refs. |
|---|---|---|---|---|---|
| *1,2,4,5-Tetrahydro-1,4-benzodiazepin-3(3H)-ones* | | | | | |
| 5-Ph | 187–188 | Acetone/$Et_2O$ | | | 16 |
| 5-Ph-7-Cl | 185–187 | Xylene | | | 16 |
| 2-Et-2,5-$Ph_2$ | | | | | 97 |
| 1,4-$Me_2$-5-Ph-7-Cl | 114–116 | $Et_2O$/Petr ether | | | 89 |
| 1-Me-5,5-$Ph_2$-7-Cl | 250–254 | Acetone | | | 29 |
| 1-Me-2-HO-4-Me-5-Ph-7-Cl | 160–164 | $CH_2Cl_2$/Hexane | | | 28 |
| 1-Me-2,5,5-$Ph_3$-7-Cl | 170–173 | $Et_2O$ | | | 29 |
| *1,2,3,4-Tetrahydro-1,4-benzodiazepin-5(5H)-ones* | | | | | |
| None | 155–156 | EtOAc/Hexane | | ir, pmr | 9 |
| ***Monosubstituted*** | | | | | |
| 1-Ac | 174–176 | EtOAc | | ir, pmr | 9 |
| 1-(4-$AcNHC_6H_4$) | 260–262 | *i*-PrOH | 74 | pmr | 108 |
| 1-(4-$H_2NC_6H_4$) | 268–270 | EtOH | 58 | pmr | 108 |
| 1-Benzyl | 141–142 | PhH/Petr ether | 50 | | 100 |

| | | | | | |
|---|---|---|---|---|---|
| 1-Cyclohexyl | 110 | PhH/Petr ether | 35 | | 100 |
| 1-Me | 176 | PhH | 68 | | 100 |
| | 169–170 | PhH | | ir, pmr | 9 |
| 1-[2-(N-Me-N-Ph-Ethylamino)]NHCO | 135–136 | MeOH | 30 | ir, pmr | 100 |
| 1-Ph | 227 | EtOH | 50 | | 100 |
| 3-$BuNHCH_2$ | 145–147 | EtOAc | 70 | ir | 103 |
| 3-$ClCH_2$ | 179–181 | EtOH | 85 | ir, uv | 103 |
| 3-Me | 214–216 | EtOH | 62 | ir, uv | 103 |
| 3-(Morpholino)$CH_2$ | 151–152 | EtOAc | 70 | ir | 103 |
| 3-(Piperidino)$CH_2$ | 175–177 | EtOH | 80 | ir | 103 |
| 4-Me | 185–188 | | | | 112 |
| 4-Ph | 227–230 | $Et_2O$/Petr ether | | | 23d |
| | 223–224 | EtOH | 50 | pmr, uv | 104 |
| 7-Cl | 150–151 | EtOH | | ir, pmr | 9 |
| 8-MeO | 192–194 | EtOAc | | ir, pmr | 9 |
| | | | | ir, pmr | 9 |
| ***Disubstituted*** | | | | | |
| ***1,2-Disubstituted*** | | | | | |
| 1,2-$Me_2$ | 170–172 | PhH/Hexane | 45 | ir, pmr, uv | 103 |
| ***1,4-Disubstituted*** | | | | | |
| 1-Me-4-Ac | 104–106 | EtOH | | | 18d |
| 1-Ph-4-(2-$Et_2$N-Ethyl) | 84–85 | Hexane | | | 108 |
| Methiodide | 190–192 | Acetone | | | 108 |
| 1-Ph-4-(2-$Me_2$N-Ethyl) | 124–126 | Hexane | 75 | | 108 |
| Methiodide | 256–258 | Acetone | | | 108 |
| 1-Ph-4-(1-$Me_2$N-Prop-2-yl) | 107–108 | Hexane | | | 108 |
| Methiodide | 224–226 | Acetone | | | 108 |
| 1-Ph-4-(3-$Me_2$N-Propyl) | 125–127 | Petr ether | | | 108 |
| Methiodide | 258–259 | Acetone | | | 108 |
| 1-Ph-4-(2-Piperidinoethyl) | 118–120 | Petr ether | | | 108 |
| Methiodide | 190–192 | Acetone | | | 108 |

TABLE VIII.1. —(contd.)

| Substituent | mp (°C) or; [bp (°C)/(torr)] | Solvent of Crystallization | Yield (%) | Spectra | Refs. |
|---|---|---|---|---|---|
| 1-Ph-4-(2-Pyrrolidinoethyl) | 138–140 | Petr ether | | | 108 |
| Methiodide | 176–178 | Acetone | | | 108 |
| 1-Propanoyl-4-Me | 173–175 | EtOAc | | | 112 |
| ***1,7-Disubstituted*** | | | | | |
| 1-Ac-7-Cl | 204–206 | EtOAc | | ir, pmr | 9 |
| 1-[2-(N-4-ClC$_6$H$_4$-N-Me-Amino)ethyl]NHCO-7-Cl | 179–180 | MeOH | 31 | ir, pmr | 100 |
| 1-Cyclohexyl-7-Cl | 167–168 | MeOH | 41 | ir, pmr | 100 |
| 1-Me-7-Cl | 188–190 | EtOH/THF | 42 | | 100 |
| 1-(4-O$_2$NC$_6$H$_4$)-7-NH$_2$ | 243–245 | | 59 | pmr | 108 |
| 1-(4-O$_2$NC$_6$H$_4$)-7-NO$_2$ | 184–186 | MeOH/*i*-PrOH | 69 | pmr | 108 |
| | | | | ir, pmr | 9 |
| ***1,8-Disubstituted*** | | | | | |
| 1-Ac-8-MeO | 205–206 | EtOAc | | ir, pmr | 9 |
| 1-Me-8-MeO | 161–162 | EtOAc | | ir, pmr | 9 |
| 1-Cyclohexyl-8-Cl | 184–185 | PhH | 41 | ir, pmr | 100 |
| 1-Ph-8-H$_2$N | 217–218 | MeOH | 87 | ir, pmr | 101 |
| 1-Ph-8-Br | 218–220 | MeOH | 34 | | 101 |
| 1-Ph-8-Cl | 182 | PhH | 20 | | 100 |
| | 179–181 | MeOH | 78 | | 101 |
| 1-Ph-8-F | 173–175 | MeOH | 43 | | 101 |
| 1-Ph-8-F$_3$C | 231 | MeOH | 90 | ir, ms, pmr | 101 |
| 1-Ph-8-HO | 163–164d | MeOH | 50 | | 101 |
| 1-Ph-8-NO$_2$ | 211–216 | | | | 102 |
| 1-(4-BrC$_6$H$_4$)-8-Br | 262–264 | EtOAc | 29 | ms | 101 |
| 1-(2-ClC$_6$H$_4$)-8-Cl | 216–217 | MeOH | 42 | | 101 |
| 1-(4-O$_2$NC$_6$H$_4$)-8-NO$_2$ | 229–231 | | 10 | pmr | 101 |

| | | | | | |
|---|---|---|---|---|---|
| ***2,4-Disubstituted*** | | | | | |
| 2-Ph-4-Me | 202–203 | MeOH | 63 | | 10 |
| ***2,7-Disubstituted*** | | | | | |
| 2-Ph-7-Cl | 170–173 | EtOAc | 33 | ir | 11, 109 |
| ***3,3-Disubstituted*** | | | | | |
| 3-MeO-3-Me | 165–168d | MeOH | 90 | ir, uv | 103 |
| ***3,4-Disubstituted*** | | | | | |
| 3-AcO-4-Ph | 162–164d | PhH | 74 | pmr, uv | 104 |
| 3-EtO-4-Ph | 163–165 | EtOH | 85 | | 104 |
| 3-EtS-4-Ph | 192–193 | EtOH | 19 | pmr, uv | 104 |
| 3-MeO-4-Ph | 227–231d | MeOH | 67 | pmr, uv | 104 |
| 3,4-$Me_2$ | 190–192 | EtOH | 61 | ir, pmr | 103 |
| 3-PhS-4-Ph | 183–185 | EtOH | 60 | pmr, uv | 104 |
| ***3,7-Disubtituted*** | | | | | |
| 3-Me-7-Cl | 212–214 | *i*-PrOH | | | 18d |
| ***4,7-Disubstituted*** | | | | | |
| 4-Ph-7-Cl | 240–244 | DMF/$H_2O$ | | | 23d |
| ***Trisubstituted*** | | | | | |
| ***1,2,4-Trisubstituted*** | | | | | |
| 1-Ac-2-Ph-4-Me | 140–142 | EtOH | | | 10 |
| ***1,3,4-Trisubstituted*** | | | | | |
| 1-Me-3-EtO-4-Ph | 89–90 | *i*-$Pr_2O$ | 67 | pmr, uv | 105 |
| ***1,4,7-Trisubstituted*** | | | | | |
| 2-Me-4-Ac-7-Cl | 142–145 | EtOH | | | 18d |
| ***1,4,8-Trisubstituted*** | | | | | |
| 1-Ph-4-Ac-8-Cl | 107–108 | EtOAc/$Et_2O$ | 78 | | 101 |
| 1-Ph-4-Ac-8-$F_3C$ | 97–99 | *i*-$Pr_2O$ | 93 | ir, pmr | 101 |

TABLE VIII.1. —(contd.)

| Substituent | mp (°C) or; [bp (°C)/(torr)] | Solvent of Crystallization | Yield (%) | Spectra | Refs. |
|---|---|---|---|---|---|
| 1-Ph-4-Benzoyl-8-$F_3C$ | 100–102 | *i*-$Pr_2O$ | 87 | | 101 |
| 1-Ph-4-Me-8-Cl | 126–127 | | 90 | ir, pmr | 101 |
| ***1,7,9-Trisubstituted*** | | | | | |
| 1-Me-7,9-$Cl_2$ | 97–99 | EtOH/$H_2O$ | 37 | ir, pmr | 100 |
| ***2,3,3-Trisubstituted*** | | | | | |
| 2-AcO-3-Me-3-MeO | 180–183 | EtOAc | 40 | ir, pmr, uv | 103 |
| 2,3-$(MeO)_2$-3-Me | 153–156d | MeOH | 80 | ir, pmr, uv | 103 |
| ***3,4,7-Trisubstituted*** | | | | | |
| 3-AcO-4-Ph-7-Cl | 170–173d | $CH_2Cl_2$/Petr ether | | | 23d |
| 3-Benzylthio-4-Ph-7-Cl | 217–219 | $CH_2Cl_2$/PhH | | | 23d |
| 3-EtO-4-Ph-7-Cl | 176–177 | EtOH | 76 | | 104 |
| 3-EtO-4-Ph-7-$NO_2$ | 234–236 | THF | 19 | | 104 |
| 3-EtO-4-(2-$ClC_6H_4$)-7-Cl | 132–133 | *i*-PrOH | 28 | | 104 |
| 3-EtO-4-(2-$MeC_6H_4$)-7-Cl | 174–176 | EtOH | 88 | | 104 |
| 3-EtO-4-(4-$MeC_6H_4$)-7-MeO | 125–128 | *c*-PrOH | 67 | | 104 |
| 3-EtO-4-(2,4-$Me_2C_6H_3$)-7-Cl | 176–179 | EtOH | 76 | | 104 |
| 3,4-$Me_2$-7-Cl | 190–192 | EtOAc/Hexane | | | 18d |
| 3-MeO-4-Ph-7-Cl | 227–229 | $CH_2Cl_2$/MeOH | | | 23d |
| ***Tetrasubstituted*** | | | | | |
| ***1,2,3,4-Tetrasubstituted*** | | | | | |
| 1-Me-2,4-$Ph_2$-3-EtO | 184–186d | EtOH | 30 | pmr, uv | 105 |
| ***2,3,3,7-Tetrasubstituted*** | | | | | |
| 2,3-$(MeO)_2$-3-Me-7-Cl | 189–192 | THF | | | 18d |

| Substituents | mp (°C) | Solvent | Yield (%) | Spectra | Ref. |
|---|---|---|---|---|---|
| ***Pentasubstituted*** | | | | | |
| 2-MeO-3-Benzyl-4,8-$Me_2$-9-HO | 122–123 | MeOH/$Et_2O$ | | | 163 |

*2-Methylene-1,2,3,4-tetrahydro-1,4-benzodiazepin-5(5H)-ones*

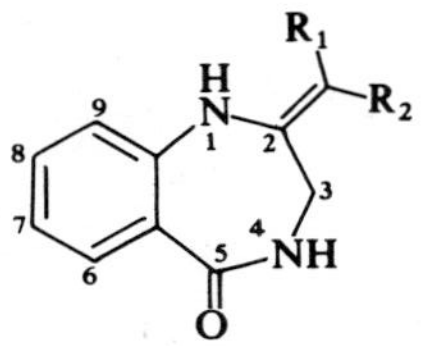

| $R_1$, $R_2$; *Other* | mp (°C) | Solvent | Yield (%) | Spectra | Ref. |
|---|---|---|---|---|---|
| H, H; 1-Me | 127–129 | PhH/Hexane | 48 | ir, pmr, uv | 103 |
| H, EtOOC; 4-Me | 149–150 | | | | 112 |
| H, MeOOC; 7-Cl | 284–287 | THF | | | 17 |
| H, $NO_2$; 7-Cl | 241–244d | DMF | | | 17 |
| EtOOC, EtOOC; 4-Benzyl | 141–142 | MeOH | | | 112 |
| EtOOC, EtOOC; 4-Me | 139 | Hexane | | | 112 |
| EtOOC, EtOOC; 4-[3,4-$(MeO)_2$-Benzyl] | 133–134 | EtOH | | | 112 |
| MeOOC, MeOOC; 7-Cl | 300–303 | DMF | | | 17 |

*3-Methylene-1,2,3,4-tetrahydro-1,4-benzodiazepin-5(5H)-ones*

| $R_1$, $R_2$; *Other* | mp (°C) | Solvent | Yield (%) | Spectra | Ref. |
|---|---|---|---|---|---|
| H, H; 4-Me-7-Cl | 162–165 | *i*-PrOH | 70 | ir, pmr, uv | 103 |
| H, Cl | 107–108d | EtOAc | 60 | ir, pmr | 103 |

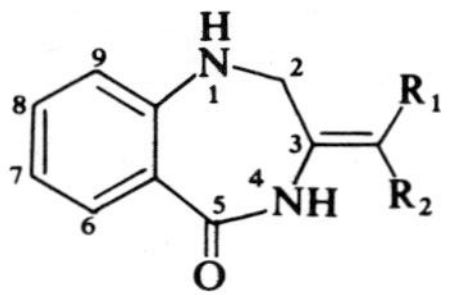

TABLE VIII.1. —(contd.)

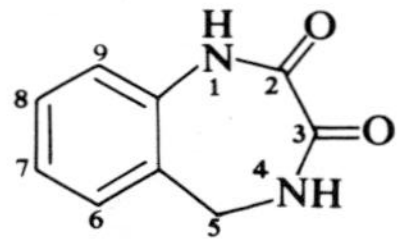

| Substituent | mp (°C) or; [bp (°C/torr)] | Solvent of Crystallization | Yield (%) | Spectra | Refs. |
|---|---|---|---|---|---|
| *4,5-Dihydro-1H-1,4-benzodiazepin-2,3(2H,3H)-diones* | | | | | |
| ***5,7-Disubstituted*** | | | | | |
| 5-Ph-7-Cl | 297–298 | EtOH | | | 113 |
| 5-(2-$ClC_6H_4$)-7-Cl | 330–331 | | | pmr | 114 |
| 5-(2-Pyridyl)-7-Br | 269–272d | DMF/$H_2O$ | | ir, pmr | 118 |
| ***1,5,7-Trisubstituted*** | | | | | |
| 1-(2-$Et_2N$-Ethyl)-5-(2-$ClC_6H_4$)-7-Cl · 0.5 $H_2O$ | 157–159 | $CH_2Cl_2$/Pentane | 62 | pmr | 115 |
| 1-(2-$Et_2N$-Ethyl)-5-(2-$FC_6H_4$)-7-Cl | 169–171 | Acetone/Petr ether | 65 | | 117 |
| 1-(2-HO-Ethyl)-5-Ph-7-Cl | 203–205 | EtOH | 15 | | 116 |
| 1-(2-HO-Ethyl)-5-(2-$FC_6H_4$)-7-Cl | 184–187 | $CH_2Cl_2$/$Et_2O$/EtOH | | | |
| 1-Me-5-Ph-7-Cl | 224–225 | | | | 113 |
| 1-Me-5-(2-$ClC_6H_4$)-7-Cl | 239–241 | | | pmr | 114 |
| 1-Me-5-(4-$ClC_6H_4$)-7-Cl | 264–267d | THF/Hexane | | | 28 |
| 1-Me-5-(2-$FC_6H_4$)-7-Cl | 233–235 | $CH_2Cl_2$/$Et_2O$ | | | 17 |
| 1-Me-$NHCOCH_2$-5-Ph-7-Cl | 230–232 | Acetone | | | 89 |
| 1-(2-$Me_2N$-Ethyl)-5-Ph-7-Cl | 194–195 | $CH_2Cl_2$/Petr ether | | | 38 |
| ***5,5,7-Trisubstituted*** | | | | | |
| 5-CN-5-Ph-7-Cl | 192–195d | MeCN | | | 120 |
| 5-Me-5-Ph-7-Cl | 210–213d | MeOH | | | 29 |

| | | | | | |
|---|---|---|---|---|---|
| ***Tetrasubstituted*** | | | | | |
| 1,4-$Me_2$-5-Ph-7-Cl | 268–270 | $CH_2Cl_2$/Hexane | | | 28 |
| 1-Me-5-CN-5-Ph-7-Cl | 195–200d | MeCN | | | 120 |
| ***Pentasubstituted*** | | | | | |
| 1,4-$Me_2$-5-CN-3-Ph-7-Cl | 234–236 | EtOH | | | 120 |
| 1,4-$Me_2$-5,5-$Ph_2$-7-Cl | 276–278 | PhH/$CH_2Cl_2$ | | | 17 |
| 1-Me-4-Benzoyl-5-CN-5-Ph-7-Cl | 224–226 | MeCN | | | 120 |

*3,4-Dihydro-1H-1,4-benzodiazepin-2,5(2H,5H)-diones*

| | | | | | |
|---|---|---|---|---|---|
| None | 327–328 | AcOH/$H_2O$ | | ir, uv | 132 |
| ***Monosubstituted*** | | | | | |
| 1-Benzyl | 189–190 | $CH_2Cl_2$/$Et_2O$ | | ir, uv | 132 |
| 1-Me | 194–197 | Acetone | 52 | | 130, 133 |
| 1-(4-$NO_2$-Benzyl) | 262–264 | MeOH | | | 133 |
| 1-Ph | 221–224 | $CH_2Cl_2$/$Et_2O$ | 40 | ir, uv | 134 |
| 3-Benzyl | 270–272 | Acetone | 38 | ir, ms, pmr | 124 |
| | 275–278 | DMF/$H_2O$ | 44 | | 128 |
| 3-Me | 320–321 | MeOH | | | 133 |
| | 331–333 | | 46 | | 130 |
| 3-(2-Me-Propyl) | 252 | DMF/$H_2O$ | 55 | | 128 |
| 3-*i*-Pr | 266 | DMF/$H_2O$ | 64 | | 128 |
| 4-Benzyl | 172–175 | | | | 112 |
| 4-Me | 248–252 | MeOH | 84 | ir, ms, pmr | 133, 143 |
| 4-[2,4-$(MeO)_2$-Benzyl] | 151–152 | EtOAc | | | 111 |

TABLE VIII.1. —(contd.)

| Substituent | mp (°C) or; [bp (°C)/(torr)] | Solvent of Crystallization | Yield (%) | Spectra | Refs. |
|---|---|---|---|---|---|
| 4-Ph | 203–204 | | 52 | ir, pmr | 140 |
| | 199–200 | MeCN | 88 | ir | 127 |
| 4-(2-Benzoyl-4-$ClC_6H_3$) | 246–247 | 1,3-Propandiol | | | 164 |
| 4-(4-$ClC_6H_4$) | 192–194 | MeCN | | | 164 |
| 4-(4-$HOC_6H_4$) | 304–306 | $DMF/H_2O$ | | | 164 |
| 4-(2-$MeC_6H_4$) | 249–250 | EtOH | 63 | ms | 139 |
| 4-*i*-Pr | 196–200 | | | | 112 |
| 4-(2-Pyridyl) | 242–243 | MeCN | | | 164 |
| 7-Cl | 325–328d | | 72 | | 131 |
| ***Disubstituted*** | | | | | |
| ***1,3-Disubstituted*** | | | | | |
| 1-Benzyl-3-Me | 209–233 | Acetone/$Et_2O$ | | | 133 |
| 1-(Cyclohexyl)$CH_2$-3-Me | 184–186 | Acetone/$Et_2O$ | | | 133 |
| 1,3-$Me_2$ | 253–255 | MeOH | | | 133 |
| ***1,4-Disubstituted*** | | | | | |
| 1-Benzyl-4-Bu | 113–116 | $Et_2O$/Hexane | | | 133 |
| 1-Benzyl-4-Me | 150–151 | Acetone/$Et_2O$ | | | 133 |
| 1-Benzyl-4-MeO | 147 | PhMe | 58 | | 136 |
| 1-(4-Cl-Benzyl)-4-MeO | 132–134 | PhMe | 54 | ir, ms, pmr | 136 |
| 1-(4-$ClC_6H_4$)-4-MeO | 197–199 | PhMe | 56 | | 136 |
| 1-(4-$NO_2$-Benzyl)-4-Me | 199–200 | MeOH | | | 133 |
| 1,4-$Me_2$ | 148–151 | Acetone | | | 133 |
| 1-$MeOCH_2$-4-Me | Oil | | 41 | ir, ms, pmr | 143 |
| ***1,7-Disubstituted*** | | | | | |
| 1-Benzyl-7-Cl | 202–203 | Acetone/$Et_2O$ | | | 133 |
| 1-Et-7-Cl | 146–147 | EtOAc/EtOH | 55 | | 135, 129 |

| | | | | | |
|---|---|---|---|---|---|
| 1,7-$Me_2$ | 170 | Acetone/EtOH | 50 | | 129 |
| 1-Me-7-Cl | 171–173 | $CH_2Cl_2/Et_2O$ | | | 133 |
| | 178–179 | MeOH | | | 135 |
| | 177–179 | $Et_2O$ | 92 | | 131 |
| 1-Me-7-$NO_2$ | 271–272 | DMF | 71 | | 129 |
| 1-Ph-7-Cl | 199–201 | *i*-PrOH/*i*-$Pr_2O$ | 28 | | 144 |
| | 203–205 | EtOAc | 35 | | 129 |
| ***1,8-Disubstituted*** | | | | | |
| 1-Me-8-Cl | 210–211 | Acetone/$Et_2O$ | 40 | | 129, 133 |
| ***3,4-Disubstituted*** | | | | | |
| 3-Benzyl-4-Me | 100–103 | Acetone/Hexane | | ir, ms, pmr | 143 |
| (*S*)-Enantiomer | 95–98 | | 45 | [α] | 125 |
| 3-(AcOCHPh)-4-Me | 233–235 | | | | 138 |
| 3-(HOCHPh)-4-Me | 195–197 | | 80 | | 138, 125 |
| 3-Me-4-[3,4-$(MeO)_2$-Benzyl] | | | | | |
| (*S*)-Enantiomer | 140–144 | EtOH | | | 112 |
| (*R*)-Enantiomer | 136–137 | PhMe | | | 112 |
| 3-Ph-4-MeO | 236 | PhMe | 52 | | 136 |
| | 242–244 | DMF/$H_2O$ | | | 137 |
| ***3,8-Disubstituted*** | | | | | |
| 3-(Indol-3-yl)$CH_2$-8-Cl | | | | | |
| *S*-Enantiomer | 255–257 | EtOH | | | 156 |
| ***4,6-Disubstituted*** | | | | | |
| 4-Me-6-Br | 230–232 | MeOH/$Et_2O$ | | | 165 |
| 4-Me-6-F | 214–217d | | | | 111 |
| | 245–247 | DMSO/$H_2O$ | | | 165 |
| 4-Me-6-Cl | 237–238 | EtOH | | | 111 |
| 4,6-$Me_2$ | 200–202 | EtOH | | | 111 |
| 4-Me-6-MeO | 196–198 | | | | 112 |
| 4-Me-6-$NO_2$ | 287–289 | | | | 145 |

TABLE VIII.1. —(contd.)

| Substituent | mp (°C) or; [bp (°C)/(torr)] | Solvent of Crystallization | Yield (%) | Spectra | Refs. |
|---|---|---|---|---|---|
| ***4,7-Disubstituted*** | | | | | |
| 4-Benzyl-7-Cl | 183–184 | | | | 112 |
| 4-Me-7-$H_2N$ | 266d | | | | 145 |
| 4-Me-7-Br | 260–261 | MeOH | | | 111 |
| 4-Me-7-Cl | 253–255 | Acetone/$Et_2O$ | | | 133 |
| 4-Me-7-Diazonium hexaflurophosphate | 215d | | | | 145 |
| | 259–262 | | 47 | | 130 |
| 4-Me-7-F | 262–263 | DMSO/$H_2O$ | | | 111 |
| 4-Me-7-$F_3C$ | 203–206 | EtOH | | | 111 |
| 4-Me-7-MeO | 208–209 | EtOAc | | | 112 |
| 4-Me-7-(Piperidino)azo | 194–194d | EtOAc/Petr ether | | | 145 |
| 4-[2,4-$(MeO)_2$-benzyl]-7-F | 190–192 | DMSO/$H_2O$ | | | 111 |
| 4-$(Morpholino)_2$OP-7-Cl | 216–218 | EtOH | 65 | ir, pmr, uv | 137 |
| 4-Ph-7-Cl | 197–199 | MeOH/$Et_2O$ | | | 23d |
| ***4,8-Disubstituted*** | | | | | |
| 4-Et-8-Cl | 212–214 | EtOH | 75 | | 126 |
| 4-Me-8-Cl | 282–285 | | | | 112 |
| 4-Me-8-MeO | 216–218 | | | | 112 |
| ***4,9-Disubstituted*** | | | | | |
| 4-Me-9-Cl | 183–185 | EtOAc | | | 112 |
| 4-Me-9-MeO | 147–148 | EtOAc/Hexane | | | 112 |
| ***8,9-Disubstituted*** | | | | | |
| 8-Me-9-HO | 292–294d | MeOH | | | 163 |
| ***Trisubstituted*** | | | | | |
| ***1,3,4-Trisubstituted*** | | | | | |
| 1-Benzyl-3,4-$Me_2$ | 137–139 | $Et_2O$ | | | 133 |
| 1-Benzyl-3-Ph-4-MeO | 183 | PhMe | 60 | | 136 |

| | | | | | |
|---|---|---|---|---|---|
| 1,3,4-$Me_3$ | 137–139 | Acetone | | | 132b |
| ***1,3,7-Trisubstituted*** | | | | | |
| 1,3-$Me_2$-7-Cl | 210–211 | EtOAc/EtOH | 44 | | 131 |
| 1-Me-3-Ph-7-Cl | 234–235 | MeOH | 51 | | 141 |
| ***1,4,7-Trisubstituted*** | | | | | |
| 1,4-$Me_2$-7-$H_2N$ | 197–198 | MeOH | | | 145 |
| 1,4-$Me_2$-7-Cl | 182–183 | Acetone/$Et_2O$ | | | 133 |
| 1,4-$Me_2$-7-$NO_2$ | 217–218 | | | | 145 |
| 1-Me-4-Ac-7-Cl | 207–209 | Acetone | 48 | ir, ms, pmr | 143 |
| | | $Ac_2O$ | 95 | ir, ms, pmr | 131 |
| 1-Me-4-Benzyl-7-Cl | 135–137 | Acetone/Hexane | 42 | ir, ms, pmr | 143 |
| 1-Me-4-Butanoyl-7-Cl | 136–137 | EtOAc/Cyclohexane | 95 | | 131 |
| 1-Me-4-$F_3CCO$-7-Cl | 176–178 | $(F_3CCO)_2O$ | | pmr | 131 |
| 1-Me-4-MeO-7-$NO_2$ | 183–185 | MeOH | 51 | ir, ms, pmr | 136 |
| 1-Me-4-(2-Me-Propanoyl)-7-Cl | 111–112 | EtOAc/Cyclohexane | 90 | pmr | 131 |
| 1-Me-Propanoyl-7-Cl | 158–159 | MeOH | 96 | pmr | 131 |
| ***1,4,8-Trisubstituted*** | | | | | |
| 1-(4-Cl-Benzyl)-4-MeO-8-Cl | 95–98 | Cyclohexane | 57 | | 136 |
| 1,4-$Me_2$-8-Cl | 198–200 | Acetone/$Et_2O$ | | | 133 |
| 1-Ph-4-Ac-8-Cl | 137–138 | EtOAc/*i*-$Pr_2O$ | 84 | | 147 |
| 1-Ph-4-Ac-8-$NO_2$ | 204–205 | | | | 147 |
| 1-Ph-4-(4-$H_2N$-Benzoyl)-8-Cl | 210–212 | | | | 147 |
| 1-Ph-4-Benzoyl-8-Cl | 210–211 | | | | 147 |
| 1-Ph-4-Benzoyl-8-$NO_2$ | 205–206 | | | | 147 |
| 1-Ph-4-(Benzyloxy)CO-8-Cl | 138–140 | | | | 147 |
| 1-Ph-4-Butanoyl-8-$NO_2$ | 183–184 | | | | 147 |
| 1-Ph-4-(Butoxy)CO-8-Cl | 96–99 | | | | 147 |
| 1-Ph-4-(Cl-Acetyl)-8-Cl | 189–191 | | | | 147 |
| 1-Ph-4-(Cl-Acetyl)-8-$NO_2$ | 236–238 | | | | 147 |
| 1-Ph-4-Cyclohexanoyl-8-Cl | 185–186 | | | | 147 |

TABLE VIII.1. —(contd.)

| Substituent | mp (°C) or; [bp (°C)/(torr)] | Solvent of Crystallization | Yield (%) | Spectra | Refs. |
|---|---|---|---|---|---|
| 1-Ph-4-Cyclohexanoyl-8-$NO_2$ | 193–194 | | | | 147 |
| 1-Ph-4-EtNHCO-8-Cl | 196–198d | | | | 147 |
| 1-Ph-4-EtOOC-8-Cl | 189–191 | | | | 147 |
| 1-Ph-4-Formyl-8-Cl | 196–198 | | 64 | | 147 |
| 1-Ph-4-Formyl-8-$NO_2$ | 183–185 | | | | 147 |
| 1-Ph-4-(I-Acetyl)-8-Cl | 141–142 | | | | 147 |
| 1-Ph-4-(I-Acetyl)-8-$NO_2$ | 189–190 | | | | 147 |
| 1-Ph-4-MeO-8-Cl | 165–170 | Ligroin | 53 | | 136 |
| 1-Ph-4-($Me_2$-Acetyl)-8-$NO_2$ | 167–168 | | | | 147 |
| 1-Ph-4-(3-Me-Butanoyl)-8-Cl | 149–150 | | | | 147 |
| 1-Ph-4-(4-$MeC_6H_4SO_2$)-8-Cl | 186–187 | | | | 147 |
| 1-Ph-4-(2-Me-Propanoyl)-8-Cl | 158–160 | | | | 147 |
| 1-Ph-4-(2-Me-Propanoyl)-8-$NO_2$ | 215–216 | | | | 147 |
| 1-Ph-4-PhOOC-8-Cl | 197–198 | | | | 147 |
| 1-Ph-4-PhNHCO-8-Cl | 206–208 | | | | 147 |
| 1-Ph-4-(3-Ph-Propenoyl)-8-$NO_2$ | 237–238 | | | | 147 |
| 1-Ph-4-Propanoyl-8-Cl | 185–186 | | | | 147 |
| 1-Ph-4-Propanoyl-8-$NO_2$ | 191–192 | $CH_2Cl_2/Et_2O$ | 74 | | 147 |
| 1-(2-$ClC_6H_4$)-4-Ac-8-Cl | 207–208 | | | | 147 |
| 1-(2-$ClC_6H_4$)-4-Benzoyl-8-Cl | 191–192 | | | | 147 |
| 1-(2-$FC_6H_4$)-4-Ac-8-Cl | 162–163 | | | | 147 |
| ***1,7,9-Trisubstituted*** | | | | | |
| 1-Me-7-Cl-9-$NO_2$ | 237–238 | DMF | 56 | | 129 |
| ***3,8,9-Trisubstituted*** | | | | | |
| 3-(*S*)-Benzyl-8-Me-9-benzyloxy | 240–241 | $CH_2Cl_2$/MeOH | | [α] | 163 |
| 3-(*S*)-Benzyl-8-Me-9-HO | 305–306 | MeOH/$Et_2O$ | | [α] | 163 |
| 3-(3-HO-Benzyl)-8-Me-9-benzyloxy | 256–258 | MeOH | | | 163 |

| | | | | |
|---|---|---|---|---|
| 3-(3-HO-Benzyl)-8-Me-9-HO | 272–275 | Acetone | | 163 |
| 3-(*S*)-(4-HO-Benzyl)-8-Me-9-benzyloxy | 273–239 | $CH_2Cl_2/Et_2O$ | | 163 |
| 3-(*S*)-(4-HO-Benzyl)-8-Me-9-HO | 325–335 | Acetone | | 163 |
| 3-(*S*)-[3,4-$(HO)_2$-Benzyl]8-Me-9-benzyloxy | 244–246 | Acetone/$Et_2O$ | | 163 |
| 3-(*S*)-[3,4-$(HO)_2$-Benzyl]8-Me-9-HO | 155–157 | MeOH/$Et_2O$ | | 163 |
| ***4,8,9-Trisubstituted*** | | | | |
| 4,8-$Me_2$-9-Benzyloxy | 238–239 | $CH_2Cl_2/Et_2O$ | | 163 |
| 4,8-$Me_2$-9-HO | 239–241 | Acetone | | 163 |
| ***7,8,9-Trisubstituted*** | | | | |
| 7,8,9-$(MeO)_3$ | 194–196 | EtOAc | | 162 |
| ***Tetrasubstituted*** | | | | |
| ***1,3,4,7-Tetrasubstituted*** | | | | |
| 1,3-$Me_3$-4-Ac-7-Cl | 165–166 | $CH_2Cl_2$/MeOH | pmr | 131 |
| ***3,4,8,9-Tetrasubstituted*** | | | | |
| 3-Benzyl-4,8-$Me_2$-9-HO | 196–197 | $CH_2Cl_2/Et_2O$ | | 163 |
| ***Pentasubstituted*** | | | | |
| 1,4-$Me_2$-7,8,9-$(MeO)_3$ | 141–143 | $Et_2O$ | | 162 |

*3-Methylene-3,4-dihydro-1H-1,4-benzodiazepin-2,5(2H,5H)-diones*

| | | | | |
|---|---|---|---|---|
| ***R, $R_2$; Other*** | | | | |
| H, Ph; 1-Ac-4-Me | 177–179 | MeOH | pmr, uv | 123 |
| H, Ph; 1-Cl-4-Me | 129–130 | $CHCl_3$/Petr ether | uv | 123 |
| H, Ph; 4-Me | 208–209 | | pmr | 138 |
| | 207–208 | $CHCl_3$/Petr ether | ir, pmr, uv | 123 |

TABLE VIII.1. —(contd.)

| Substituent | mp (°C) or; [bp (°C)/(torr)] | Solvent of Crystallization | Yield (%) | Spectra | Refs. |
|---|---|---|---|---|---|
| H, 3-$AcOC_6H_4$; 4-Me | Amorphous | | 10 | ir, ms, pmr | 143 |
| H, 3-$AcOC_6H_4$; 1-Ac-4-Me | 180–182 | EtOAc/Hexane | 24 | ir, ms, pmr | 143 |
| Ph, H; 4-Me | 185 | | | ir, pmr, uv | 123 |
| ***Spiro Compounds*** | | | | | |
| ***R; Other*** | | | | | |
| Ph; 4-Me | 179–180 | $CH_2Cl_2/Et_2O$ | | [α] | 133 |
| | 183–184 | | | ir, pmr | 146 |
| Racemic | 193–195 | $Et_2O$/Hexane | | ir, pmr, uv | 123, 138 |
| Ph; 1,4-$Me_2$ | 206 | | | ir, pmr | 146 |
| 3-$AcOC_6H_4$; 4-Me | 193–195 | | 24 | ir, ms, pmr | 143 |
| 3-$HOC_6H_4$; 4-Me | 210–211 | EtOAc/PhH | | [α], pmr, uv | 123 |
| Racemic | 209–211 | Acetone/Hexane | | ir, ms, pmr | 143 |
| 3-$MeOC_6H_4$; 1,4-$Me_2$ | 167–169 | | | ir, ms | 123, 146 |

*2,4-Dihydro-1H-1,4-benzodiazepin-3,5(3H,5H)-diones*

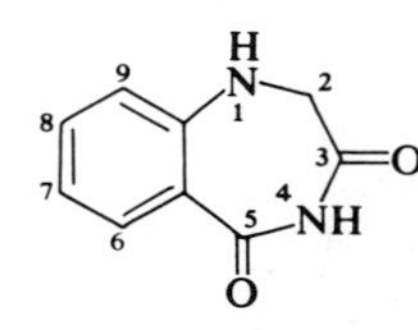

| | | | | | |
|---|---|---|---|---|---|
| 4-Benzyl | 140–142 | *i*-PrOH | 60 | ir | 150 |
| 4-(4-$MeC_6H_4$) | 193–194 | *i*-PrOH | 70 | ir | 150 |
| 4-Ph | 198–199 | *i*-PrOH | | ir, pmr | 150 |

*2-Methylene-2,4-dihydro-1H-1,4-benzodiazepin-3,5(3H,5H)-diones*

| ***$R_1$, $R_2$; Other*** | | | | | |
|---|---|---|---|---|---|
| H, MeOOC; | 234–236 | MeOH | | | 151 |
| H, MeOOC; 7-Cl | 291–292 | | | pmr | 151 |

*Tetrahydro-1,4-benzodiazepinthiones*

| ***1,3,4,5-Tetrahydro-1,4-benzodiazepin-2(2H)-thiones*** | | | | | |
|---|---|---|---|---|---|
| 1-Me-5-Ph | 117–125 | Hexane | | | 28 |
| 1,4-$Me_2$-5-Ph | 141–145 | MeOH | | | 28 |
| 1-Me-4-Ac-5-Ph | 225–227 | MeOH | | | 28 |
| 1-Me-5-(2-$FC_6H_4$)-7-Cl | 128–132 | MeOH | | | 28 |
| 4-Me-5-(2-$ClC_6H_4$)-7-Cl | 157–160 | PhH | 50 | ir, pmr | 66 |
| 1,4-$Me_2$-5-(2-$FC_6H_4$)-7-Cl | 138–140 | $CH_2Cl_2$/MeOH | | | 28 |
| 1-Me-4-Thioacetyl-5-(2-$FC_6H_4$)-7-Cl | 183–187 | $CH_2Cl_2$/Petr ether | | | 28 |

TABLE VIII.1. —(contd.)

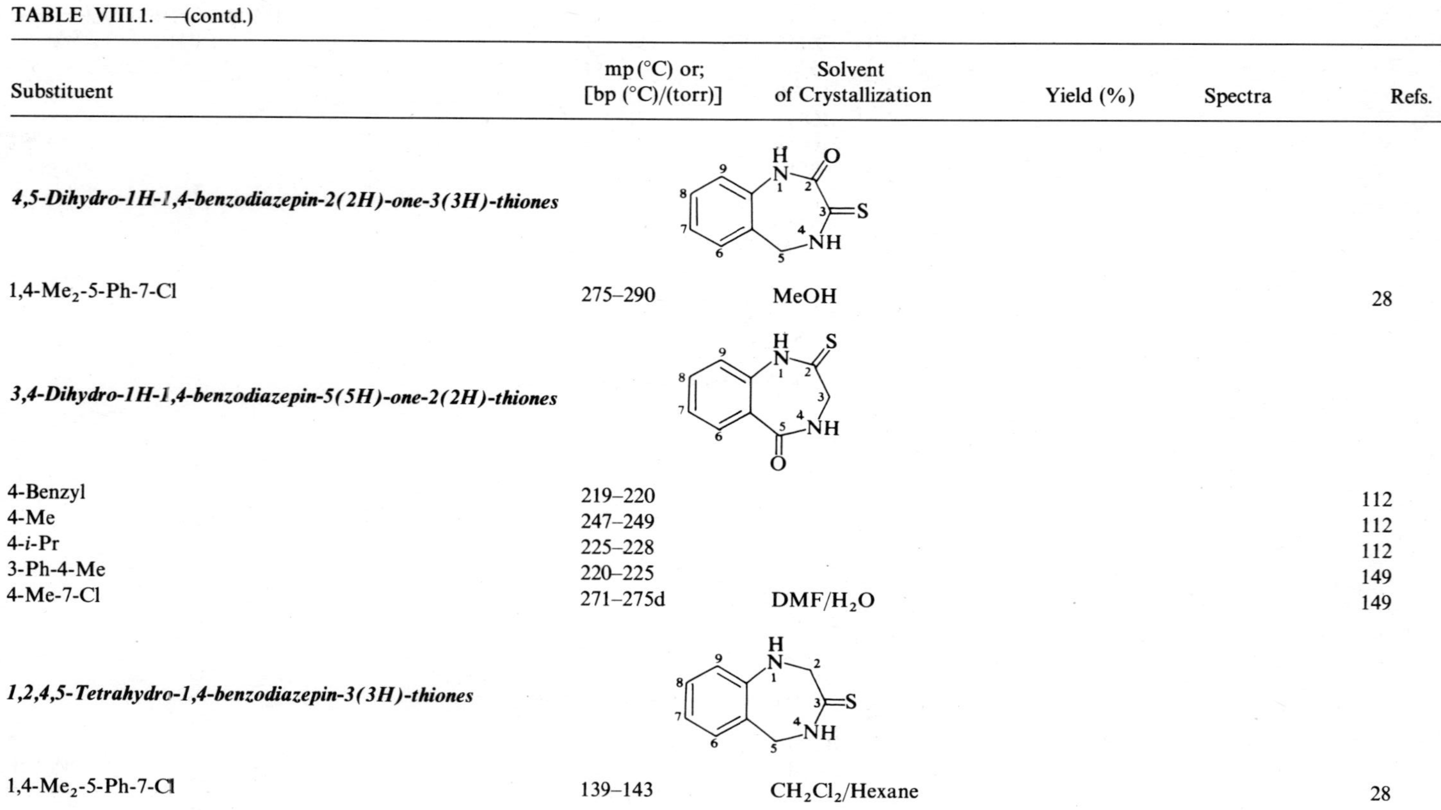

| Substituent | mp (°C) or; [bp (°C)/(torr)] | Solvent of Crystallization | Yield (%) | Spectra | Refs. |
|---|---|---|---|---|---|
| ***4,5-Dihydro-1H-1,4-benzodiazepin-2(2H)-one-3(3H)-thiones*** | | | | | |
| 1,4-$Me_2$-5-Ph-7-Cl | 275–290 | MeOH | | | 28 |
| ***3,4-Dihydro-1H-1,4-benzodiazepin-5(5H)-one-2(2H)-thiones*** | | | | | |
| 4-Benzyl | 219–220 | | | | 112 |
| 4-Me | 247–249 | | | | 112 |
| 4-*i*-Pr | 225–228 | | | | 112 |
| 3-Ph-4-Me | 220–225 | | | | 149 |
| 4-Me-7-Cl | 271–275d | DMF/$H_2O$ | | | 149 |
| ***1,2,4,5-Tetrahydro-1,4-benzodiazepin-3(3H)-thiones*** | | | | | |
| 1,4-$Me_2$-5-Ph-7-Cl | 139–143 | $CH_2Cl_2$/Hexane | | | 28 |

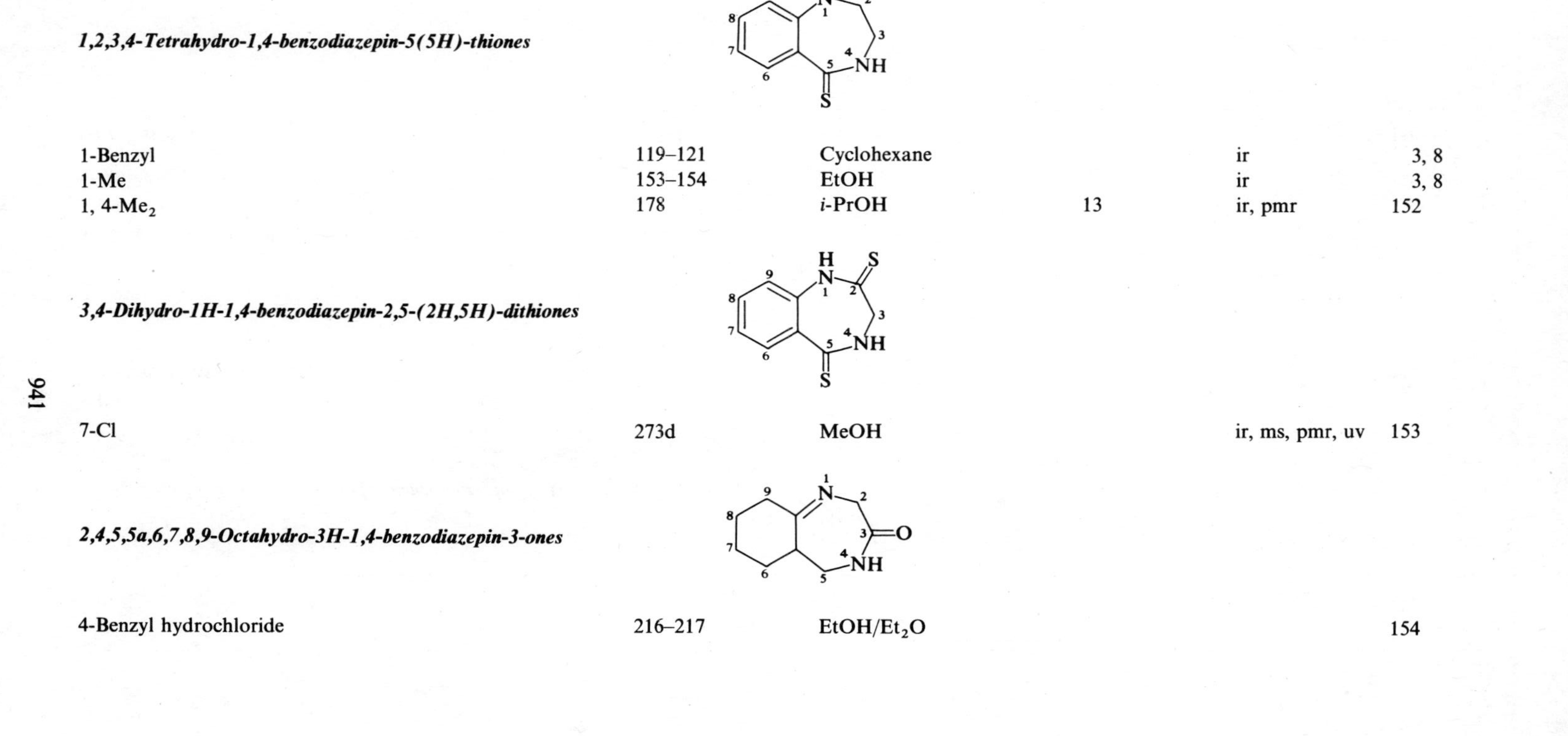

| Substituent | mp (°C) | Solvent | Yield (%) | Spectra | Ref. |
|---|---|---|---|---|---|
| ***1,2,3,4-Tetrahydro-1,4-benzodiazepin-5(5H)-thiones*** | | | | | |
| 1-Benzyl | 119–121 | Cyclohexane | | ir | 3, 8 |
| 1-Me | 153–154 | EtOH | | ir | 3, 8 |
| 1, 4-$Me_2$ | 178 | *i*-PrOH | 13 | ir, pmr | 152 |
| ***3,4-Dihydro-1H-1,4-benzodiazepin-2,5-(2H,5H)-dithiones*** | | | | | |
| 7-Cl | 273d | MeOH | | ir, ms, pmr, uv | 153 |
| ***2,4,5,5a,6,7,8,9-Octahydro-3H-1,4-benzodiazepin-3-ones*** | | | | | |
| 4-Benzyl hydrochloride | 216–217 | EtOH/$Et_2O$ | | | 154 |

TABLE VIII.1. —(contd.)

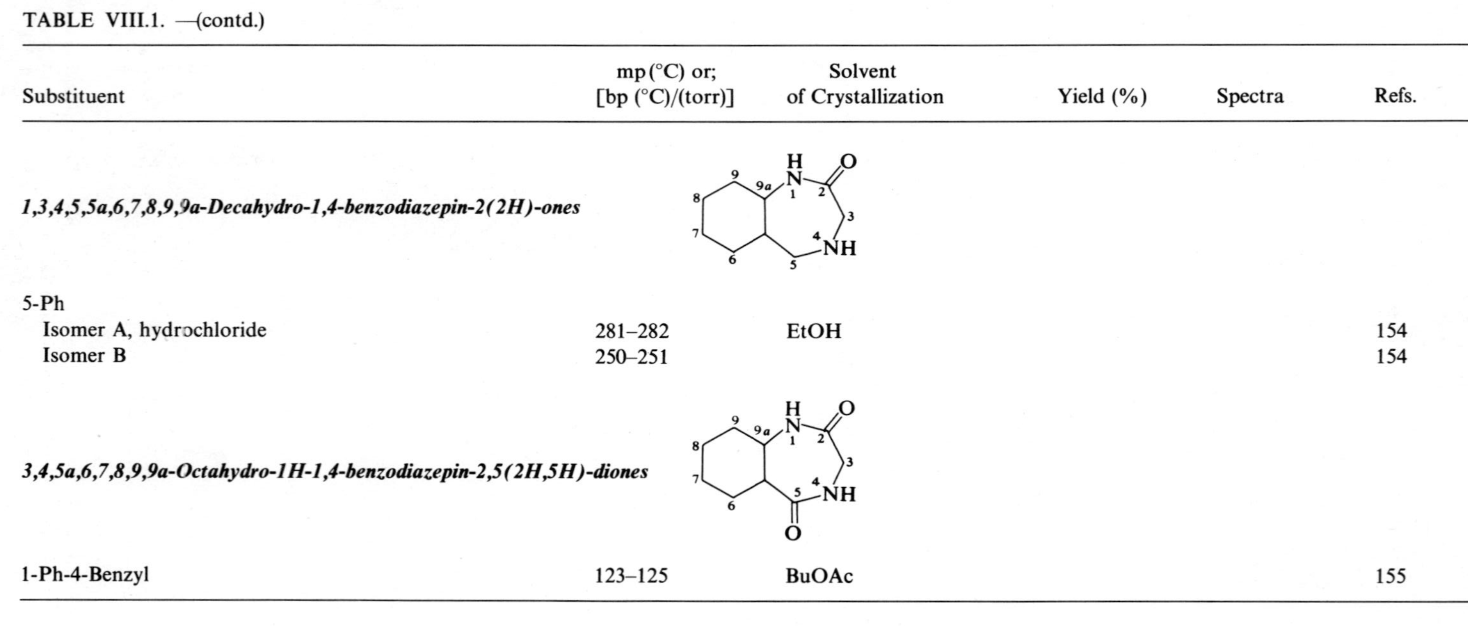

| Substituent | mp (°C) or; [bp (°C)/(torr)] | Solvent of Crystallization | Yield (%) | Spectra | Refs. |
|---|---|---|---|---|---|
| ***1,3,4,5,5a,6,7,8,9,9a-Decahydro-1,4-benzodiazepin-2(2H)-ones*** | | | | | |
| 5-Ph | | | | | |
| Isomer A, hydrochloride | 281–282 | EtOH | | | 154 |
| Isomer B | 250–251 | | | | 154 |
| ***3,4,5a,6,7,8,9,9a-Octahydro-1H-1,4-benzodiazepin-2,5(2H,5H)-diones*** | | | | | |
| 1-Ph-4-Benzyl | 123–125 | BuOAc | | | 155 |

# 11. REFERENCES


1. D. H. Kim, U.S. Patent 3,925,361, December 1975.
2. P. M. Carabateas and L. S. Harris, *J. Med. Chem.*, **9**, 6 (1966).
3. A. A. Santilli and T. S. Osdene, *J. Org. Chem.*, **31**, 4268 (1966).
4. J. Krapcho, U.S. Patent 3,869,447, March 1975.
5. P. M. Carabateas, U.S. Patent 3,384,635, May 1968.
6. D. H. Kim, U.S. Patent 3,904,603, September 1975.
7. C.-M. Lee, *J. Hetercycl. Chem.*, **1**, 235 (1964).
8. A. A. Santilli and T. S. Osdene, U.S. Patent 3,457,258, July 1969.
9. D. Misiti, F. Gatta, and R. Landi-Vittory, *J. Heterocycl. Chem.*, **8**, 231 (1971).
10. K. H. Weber, *Arch. Pharmaz.*, **302**, 584 (1969).
11. A. A. Santilli and T. S. Osdene, *J. Org. Chem.*, **29**, 1998 (1964).
12. L. H. Sternbach, E. Reeder, and G. A. Archer, *J. Org. Chem.*, **28**, 2456 (1963).
13. Neth. Patent 7,207,500, December 1972 (Sumitomo Chemical Co. Ltd., Japan).
14. K. Ishizumi, K. Mori, T. Okamoto, T. Akase, T. Izumi, M. Akatsu, Y. Kume, S. Inaba, and H. Yamamoto, U.S. Patent 4,044,003, August 1977.
15. K. Ishizumi, S. Inaba, and H. Yamamoto, *J. Org. Chem.*, **37**, 4111 (1972).
16. G. A. Archer and L. H. Sternbach: (a) U.S. Patent 3,531,467, September 1970, (b) U.S. Patent 3,317,518, March 1967.
17. A. Walser, unpublished data, Hoffmann-La Roche, Nutley, NJ.
18. (a) G. F. Field and L. H. Sternbach, U.S. Patent 3,625,959, December 1971. (b) G. F. Field and L. H. Sternbach, U.S. Patent 3,594,365, July 1971. (c) G. F. Field and L. H. Sternbach, U.S. Patent 3,594,364, July 1971. (d) G. F. Field, unpublished data, Hoffmann-La Roche, Nutley, NJ.
19. A. Walser and R. I. Fryer, *J. Org. Chem.*, **40**, 153 (1975).
20. A. Walser, G. Silverman, R. I. Fryer, L. H. Sternbach, and J. Hellerbach, *J. Org. Chem.*, **36**, 1248 (1971).
21. T. S. Sulkowski and S. J. Childress, *J. Org. Chem.*, **28**, 2150 (1963).
22. G. F. Field, W. J. Zally, and L. H. Sternbach, *J. Am. Chem. Soc.*, **89**, 332 (1967).
23. (a) W. Metlesics, G. Silverman, and L. H. Sternbach, *J. Org. Chem.*, **28**, 2459 (1963). (b) W. Metlesics and L. H. Sternbach, U.S. Patent 3,644,336, February 1972. (c) W. Metlesics and L. H. Sternbach, U.S. Patent 3,498,973, March 1970. (d) W. Metlesics, unpublished data, Hoffmann-La Roche, Nutley, NJ.
24. R. I. Fryer, J. Blount, E. Reeder, E. J. Trybulski, and A. Walser, *J. Org. Chem.*, **43**, 4480 (1978).
25. E. Broger, G. F. Field, and L. H. Sternbach, Ger. Offen. 2,223,648, November 1972.
26. F. Gatta and S. Chiavarelli, *Farmaco, Ed. Sci.*, **32**, 33 (1977).
27. W. Metlesics, G. Silverman, and L. H. Sternbach, *J. Org. Chem.*, **29**, 1621 (1964).
28. R. I. Fryer and J. V. Earley, unpublished data, Hoffmann-La Roche, Nutley, NJ.
29. A. Szente, unpublished data, Hoffmann-La Roche & Co., AG, Basel, Switzerland.
30. G. A. Archer and L. H. Sternbach: (a) U.S. Patent 3,671,517, June 1972, (b) Swiss Patent 511,866, August 1971.
31. R. I. Fryer and L. H. Sternbach, U.S. Patent 3,706,734, December 1972.
32. S. Archer, T. R. Lewis, M. J. Unser, J. O. Hoppe, and H. Lape, *J. Am. Chem. Soc.*, **79**, 5783 (1957).
33. T. Ichii, *J. Pharm. Soc. Japan*, **82**, 999 (1962).
34. G. A. Archer and L. H. Sternbach, U.S. Patent 3,553,199, January 1971.
35. M. E. Derieg, unpublished data, Hoffmann-La Roche, Nutley, NJ.
36. R. Kalish, unpublished data, Hoffmann-La Roche, Nutley, NJ.
37. D. H. Kim, U.S. Patent 3,914,250, October 1975.
38. L. H. Sternbach and E. Reeder, unpublished data, Hoffmann-La Roche, Nutley, NJ.
39. S. Shiotani and K. Mitsuhashi, *Yakugaku Zasshi*, **84**, 656 (1964).
40. K. Ishizumi, K. Mori, S. Inaba, and H. Yamamoto, *Chem. Pharm. Bull.*, **23**, 2169 (1975).

41. J. Hellerbach and A. Walser, U.S. Patent 3,784,542, January 1974.
42. Zs. Tegyey, G. Maksay, and L. Otvos, *J. Labelled Compd.*, **16**, 377 (1978).
43. R. I. Fryer, B. Brust, J. Earley, and L. H. Sternbach, *J. Med. Chem.*, **7**, 386 (1964).
44. L. H. Sternbach, G. A. Archer, J. V. Earley, R. I. Fryer, E. Reeder, N. Wasyliw, L. O. Randall, and R. Benziger, *J. Med. Chem.*, **8**, 815 (1965).
45. L. H. Sternbach and G. Saucy, U.S. Patent 3,341,592, September 1967.
46. O. Keller, N. Steiger, and L. H. Sternbach, U.S. Patent 3,121,103, February 1964.
47. Belg. Patent 777,096, June 1972 (Hoffmann-La Roche & Co. AG, Switzerland).
48. R. Y. Ning and L. H. Sternbach, U.S. Patent 3,682,892, August 1972.
49. E. Reeder and L. H. Sternbach, U.S. Patent 3,371,085, February 1968.
50. Belg. Patent 629,352, October 1963 (Hoffmann-La Roche & Co. AG, Switzerland).
51. L. H. Sternbach and E. Reeder, *J. Org. Chem.*, **26**, 4936 (1961).
52. R. I. Fryer and L. H. Sternbach, U.S. Patent 3,322,753, May 1967.
53. M. E. Derieg, R. I. Fryer, and L. H. Sternbach, U.S. Patent 3,651,046, March 1972.
54. G. A. Archer and L. H. Sternbach, U.S. Patent 3,236,838, February 1966.
55. G. A. Archer, R. I. Fryer, E. Reeder, and L. H. Sternbach, U.S. Patent 3,299,053, January 1967.
56. R. I. Fryer, J. V. Earley, E. Evans, J. Schneider, and L. H. Sternbach, *J. Org. Chem.*, **35**, 2455 (1970).
57. S. C. Bell and S. J. Childress, U.S. Patent 3,714,145, January 1973.
58. S. C. Bell and T. S. Sulkowski, C. Gochman, and S. J. Childress, *J. Org. Chem.*, **27**, 562 (1962).
59. A. Stempel, I. Douvan, and L. H. Sternbach, *J. Org. Chem.*, **33**, 2963 (1968).
60. (a) Q. Branca, A. E. Fischli, and A. Szente, U.S. Patent 4,327,026, April 1982. (b) Q. Branca, unpublished data, Hoffmann-La Roche & Co. AG, Basel, Switzerland. (c) A. E. Fischli, unpublished data, Hoffmann-La Roche & Co. AG, Basel, Switzerland.
61. G. F. Field and L. H. Sternbach, Swiss Patent 561,704, March 1975.
62. A. Walser, G. Silverman, and R. I. Fryer, *J. Org. Chem.*, **38**, 3502 (1973).
63. J. Roechricht, L. Kisfaludy, M. Kajtar, E. Palosi, and L. Szporny, U.S. Patent 4,329,341, May 1982.
64. R. J. McCaully, U.S. Patent 3,410,844, November 1968.
65. G. F. Field and L. H. Sternbach, Swiss Patent 561,189, March 1975.
66. M. Matsuo, K. Taniguchi, and I. Ueda, *Chem. Pharm. Bull.*, **30**, 1141 (1982).
67. A. Terada, Y. Yabe, T. Miyadera, and R. Tachikawa, presentation at the Third International Congress of Heterocyclic Chemistry, August 1971, Sendai, Japan.
68. R. I. Fryer, D. Winter, and L. H. Sternbach, *J. Heterocycl. Chem.*, **4**, 355 (1967).
69. Neth. Patent 7,110,356, February 1972 (Richter Gedeon Vegyeszeti Gyar, Hungary).
70. Neth. Patent 7,311,309, February 1974 (Sumitomo Chemical Co, Japan).
71. R. I. Fryer, G. A. Archer, B. Brust, W. Zally, and L. H. Sternbach, *J. Org. Chem.*, **30**, 1308 (1965).
72. J. Katsube, Y. Takashima, T. Hirohashi, K. Ishizumi, M. Akatsu, K. Mori, I. Katsuki, Y. Kume, H. Sato, S. Inaba, and H. Yamamoto, U.S. Patent 3,864,330, February 1975.
73. P. M. Mueller and A. Fischli, unpublished data, Hoffmann-La Roche & Co. AG, Basel, Switzerland.
74. M. Ogata, H. Matsumoto, and K. Hirose, *J. Med. Chem.*, **20**, 776 (1977).
75. P. Nedenskov and M. Mandrup, *Acta Chem. Scand.*, **31**, 701 (1977).
76. S. A. Christensen, Br. Patent 1,247,554, September 1971.
77. R. I. Fryer and L. H. Sternbach, U.S. Patent 3,501,474, March 1970.
78. J. V. Earley, R. I. Fryer, and L. H. Sternbach, U.S. Patent 3,551,415, December 1970.
79. O. P. Rudenko, T. L. Karaseva and A. I. Lisitsyna, Russ. Patent 1,051,081-A, October 1983, Derwent No. 84-170270/27.
80. G. A. Archer and L. H. Sternbach, U.S. Patent 3,370,091, February 1968.
81. S. J. Childress, S. C. Bell, and T. S. Sulkowski, U.S. Patent 3,321,522, May 1967.
82. R. I. Fryer and L. H. Sternbach, U.S. Patent 3,567,710, March 1971.
83. Neth. Patent 7,110,495, February 1972 (Richter Gedeon Vegyeszeti Gyar, Hungary).
84. R. Jaunin and J. Hellerbach, Ger. Offen. 2,150,075, April 1972.

85. E. Broger, unpublished data, Hoffmann-La Roche, Nutley, NJ.
86. L. Kisfaludy, J. Roehricht, L. Uroegdi, S. Szeberenyi, E. Palosi, and L. Szporny, U.S. Patent 4,021,421, May 1977.
87. R. I. Fryer, J. V. Earley, and L. H. Sternbach, *J. Org. Chem.*, **30**, 521 (1965).
88. J. V. Earley, R. I. Fryer, and L. H. Sternbach, U.S. Patent 3,703,510, November 1972.
89. G. A. Archer, unpublished data, Hoffmann-La Roche, Nutley, NJ.
90. J. Roehricht, L. Kisfaludy, L. Urogdi, E. Palosi, S. Szeberenyi, and L. Szporny, U.S. Patents 4,045,433, August 1977, and 4,342,755, August 1982.
91. G. F. Field and L. H. Sternbach, Swiss Patent 561,190, March 1975.
92. S. C. Bell, R. J. McCaully, and S. J. Childress, *J. Med. Chem.*, **11**, 172 (1968).
93. G. F. Field and L. H. Sternbach, Swiss Patent 562,224, April 1975.
94. R. I. Fryer and L. H. Sternbach, U.S. Patent 3,625,957, December 1971.
95. J. V. Earley, R. I. Fryer, and L. H. Sternbach, U.S. Patent 3,644,335, February 1972.
96. M. Ogata and H. Matsumoto, U.S. Patent 4,041,026, August 1977.
97. J. Bergman, A. Brynolf, and Bjoern Elman, *Heterocycles*, **20**, 2141 (1983).
98. J. B. Hester, U.S. Patent 3,896,109, January 1975.
99. J. B. Hester, U.S. Patent 3,714,178, January 1973.
100. C. Corral, R. Madronero, and S. Vega, *J. Heterocycl. Chem.*, **14**, 99 (1977).
101. A. Bauer and K.-H. Weber, *Z. Naturforsch.*, **29b**, 670 (1974).
102. A. Bauer and K.-H. Weber, Ger. Offen. 2,165,310, July 1973.
103. G. F. Field, W. J. Zally, and L. H. Sternbach, *J. Org. Chem.*, **36**, 777 (1971).
104. Y. Yamada, T. Oine, and I. Inoue, *Chem. Pharm. Bull.*, **22**, 601 (1974).
105. Y. Yamada, T. Oine, and I. Inoue, *Bull. Chem. Soc. Japan*, **47**, 339 (1974).
106. P. I. Itterah and F. G. Mann, *J. Chem. Soc.*, 471 (1958).
107. K.-H. Wuensch, K.-H. Stahnke, and P. Gomoll, *Z. Chem.*, **10**, 219 (1970).
108. C. Bagolini, P. de Witt, L. Pacifici, and M. T. Ramacci, *J. Med. Chem.*, **21**, 476 (1978).
109. A. A. Santilli and T. S. Osdene, *J. Org. Chem.*, **30**, 2100 (1965).
110. A. A. Santilli and T. S. Osdene, U.S. Patent 3,336,300, August 1967.
111. (a) M. Gerecke, W. Haefely, W. Hunkeler, E. Kyburz, H. Moehler, L. Pieri, and P. Polc, U.S. Patent 4,363,762, December 1982. (b) W. Hunkeler and E. Kyburz, U.S. Patent 4,352,818, October 1982.
112. W. Hunkeler, unpublished data, Hoffmann-La Roche & Co. AG, Basel, Switzerland.
113. S. C. Bell and S. J. Childress, *J. Org. Chem.*, **27**, 1691 (1962).
114. A. Nudelman, R. J. McCaully, and S. C. Bell, *J. Pharm. Sci.*, **63**, 1880 (1974).
115. Brit. Patent 1,346,176, February 1974 (American Home Products Corp., New York).
116. G. F. Tamagnone, R. De Maria, and F. De Marchi, *Arznemit. Forsch.*, **25**, 720 (1975).
117. J. V. Earley, R. I. Fryer, D. Winter, and L. H. Sternbach, *J. Med. Chem.*, **11**, 774 (1968).
118. A. Stempel, I. Douvan, E. Reeder, and L. H. Sternbach, *J. Org. Chem.*, **32**, 2417 (1967).
119. (a) R. Y. Ning, W. Y. Chen, and L. H. Sternbach, *J. Org. Chem.*, **36**, 1064 (1971). (b) R. Y. Ning, unpublished data, Hoffmann-La Roche, Nutley, NJ.
120. R. Jaunin, unpublished data, Hoffmann-La Roche & Co. AG, Basel, Switzerland.
121. A. M. Felix, J. V. Earley, R. I. Fryer, and L. H. Sternbach, *J. Heterocycl. Chem.*, **5**, 731 (1968).
122. K. Miyatake and S. Kaga, *J. Pharm. Soc. Japan*, **72**, 1160 (1952).
123. P. K. Martin, H. Rapoport, H. W. Smith, and J. L. Wong, *J. Org. Chem.*, **34**, 1359 (1969).
124. R. P. Rhee and J. D. White, *J. Org. Chem.*, **42**, 3650 (1977).
125. J. Framm, L. Nover, A. El Azzouny, H. Richter, K. Winter, S. Werner, and M. Luckner, *Eur. J. Biochem.*, **37**, 78 (1973).
126. K.-H. Weber, A. Bauer, and K.-H. Hauptmann, *Justus Liebigs Ann. Chem.*, **756**, 128 (1972).
127. J. Krapcho and C. F. Turk, *J. Med. Chem.*, **9**, 191 (1966).
128. E. Hoffmann and B. Jagnicinski, *J. Heterocycl. Chem.*, **3**, 348 (1966).
129. J. H. Gogerty, R. G. Griot, D. Habeck, L. C. Iorio, and W. J. Houlihan, *J. Med. Chem.*, **20**, 952 (1977).
130. D. H. Kim, *J. Heterocycl. Chem.*, **12**, 1323 (1975).

131. M. Gates, *J. Org. Chem.*, **45**, 1675 (1980).
132. (a) M. Uskokovic, J. Iacobelli, and W. Wenner, *J. Org. Chem.*, **27**, 3606 (1962). (b) M. Uskokovic, unpublished data, Hoffmann-La Roche, Nutley, NJ.
133. M. R. Uskokovic and W. Wenner, U.S. Patent 3,261,828, July 1966.
134. J. Iacobelli, M. Uskokovic, and W. Wenner, *J. Heterocycl. Chem.*, **2**, 323 (1965).
135. R. G. Griot, U.S. Patent 3,414,563, December 1968.
136. E. Wolf and H. Kohl, *Justus Liebigs Ann. Chem.*, 1245 (1975).
137. R. Y. Ning, R. I. Fryer, P. B. Madan, and B. C. Sluboski, *J. Org. Chem.*, **41**, 2720 (1976).
138. H. Smith, P. Wegfahrt, and H. Rapoport, *J. Am. Chem. Soc.*, **90**, 1668 (1968).
139. C. Bogentoft, O. Ericsson, and B. Danielsson, *Acta Pharm. Suec.*, **11**, 59 (1974).
140. M. Z. Kirmani and K. Sethi, *Tetrahedron Lett.*, 2917 (1979).
141. C. Podesva, K. Vagi, and C. Solomon, *Can. J. Chem.*, **46**, 2263 (1968).
142. M. Mori, M. Ishikura, T. Ikeda, and Y. Ban, *Heterocycles*, **16**, 1491 (1981).
143. M. Ishikura, M. Mori, T. Ikeda, M. Terashima, and Y. Ban, *J. Org. Chem.*, **47**, 2456 (1982).
144. S. Inaba, M. Akatsu, T. Hirohashi, and H. Yamamoto, *Chem. Pharm. Bull.*, **24**, 1076 (1976).
145. E. Kyburz, unpublished data, Hoffmann-La Roche & Co. AG, Basel, Switzerland.
146. Y. S. Mohammed and M. Luckner, *Tetrahedron Lett.*, 1953 (1963).
147. A. Bauer, K.-H. Weber, P. Danneberg, and F. J. Kuhn, Ger. Offen. 2,257,171, May 1974.
148. G. F. Field, L. H. Sternbach, and W. J. Zally: (a) U.S. Patent 3,624,073, November 1971. (b) U.S. Patent 3,678,038, July 1972.
149. R. J. Mohrbacher and P. P. Grous, (a) U.S. Patent 4,022,767, May 1977, (b) U.S. Patent 4,031,079, June 1977, (c) U.S. Patent 4,020,055, April 1977, (d) U.S. Patent 4,002,610, January 1977.
150. G. Stavropoulos and D. Theodoropoulos, *J. Heterocycl. Chem.*, **14**, 1139 (1977).
151. N. D. Heindel and T. F. Lemke, *J. Heterocycl. Chem.*, **3**, 389 (1966).
152. C. Corral, R. Madronero, and S. Vega, *J. Heterocycl. Chem.*, **14**, 985 (1977).
153. R. Y. Ning, R. I. Fryer, P. B. Madan, and B. C. Sluboski, *J. Org. Chem.*, **41**, 2724 (1976).
154. J. T. Plati and H. A. Albrecht, unpublished data, Hoffmann-La Roche, Nutley, NJ.
155. O. Schnider, unpublished data, Hoffmann-La Roche & Co. AG, Basel, Switzerland.
156. A. Focella, unpublished data, Hoffmann-La Roche, Nutley, NJ.
157. J. Maricq, unpublished data, Hoffmann-La Roche, Nutley, NJ.
158. E. Garcia, unpublished data, Hoffmann-La Roche, Nutley, NJ.
159. J. Karle and I. L. Karle, *J. Am. Chem. Soc.*, **89**, 804 (1967).
160. M. Czugler and A. Kalman, *Tetrahedron Lett.*, 917 (1977).
161. E. M. Wilka, unpublished data, Hoffmann-La Roche & Co. AG, Basel, Switzerland.
162. G. Zanetti, unpublished data, Hoffmann-La Roche & Co. AG, Basel, Switzerland.
163. W. Leimgruber, unpublished data, Hoffmann-La Roche, Nutley, NJ.
164. A. I. Rachlin, unpublished data, Hoffmann-La Roche, Nutley, NJ.
165. R. Joos, unpublished data, Hoffmann-La Roche & Co. AG, Basel, Switzerland.
166. Neth. Patent 7,110,498, February 1972 (Richter Gedeon Vegyeszeti Gyar, Hungary).

CHAPTER IX

# Hetero Ring[*e*][1,4]Diazepines

**A. Walser**

*Chemical Research Department,*
*Hoffmann-La Roche Inc.,*
*Nutley, New Jersey*

and

**R. Ian Fryer**

*Department of Chemistry, Rutgers,*
*State University of New Jersey,*
*Newark, New Jersey*

# INTRODUCTION

This chapter deals with the chemistry of 1,4-diazepines having a heterocycle fused to the *e*-bond, represented by the general structure **1**. Also included in this chapter is a discussion of the carbocyclic cyclopenta[*e*][1,4]diazepine ring system. The related cyclohexa[*e*][1,4]diazepines have been reviewed as perhydro derivatives of the benzodiazepine ring system (Chapter VIII, Section 9). The various ring systems reported in the literature will be discussed in alphabetical order. Unpublished data generated in the laboratories of Hoffmann-La Roche in Nutley, New Jersey, in Basel, Switzerland, and in Welwyn Garden City, United Kingdom are included in the review.

# 1. CYCLOPENTA[*e*][1,4]DIAZEPINES

While the parent ring system **2** appears to be still unknown, derivatives with a higher degree of saturation were prepared by Broger[1] in our laboratories. The

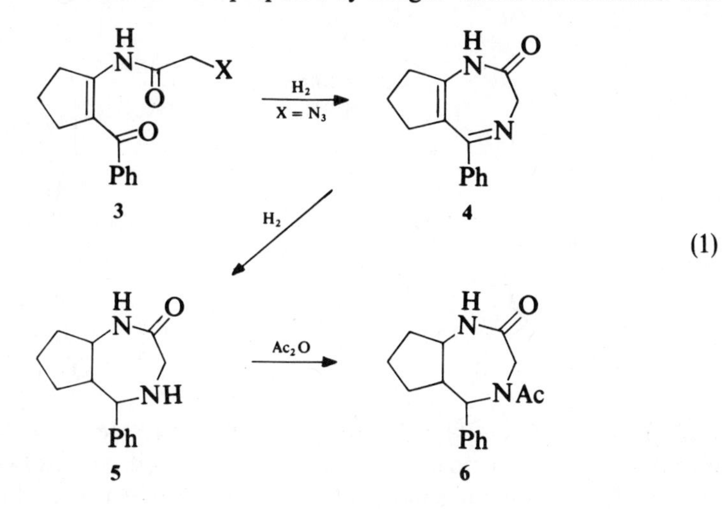

bromoacetylated enamine **3** (X = Br), derived from 2-benzoylcyclopenantone was converted to the corresponding azide **3** ($X = N_3$), hydrogenation of which led to the diazepine **4** (Eq. 1). Further hydrogenation of **4** afforded the saturated derivative **5** of undetermined stereochemistry. Acetylation of **5** with acetic anhydride gave **6**, again of unknown stereochemistry.

## 2. FURO[*e*][1,4]DIAZEPINES

Of the three possible furo[*e*][1,4]diazepines **7**–**9**, derivatives of the furo-[2,3-*e*][1,4]diazepine **7** with a higher degree of saturation have been reported in the literature.

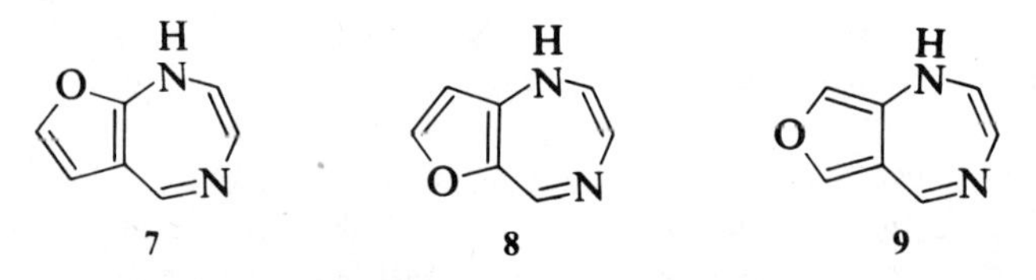


Reaction of the nitrile **10** with ethylenediamine in water at room temperature gave 30% yield of the diazepine **11** ($R_1 = R_3 = Me$, $R_2 = H$) (Eq. 2).[2] This compound and a few analogs were later obtained by treatment of the amino esters **12** with ethylenediamine in refluxing ethanol.[3] Acetylation of **11** ($R_1 = R_2 = R_3 = H$) in boiling acetic anhydride afforded the 1,4-diacetyl derivative **13**.

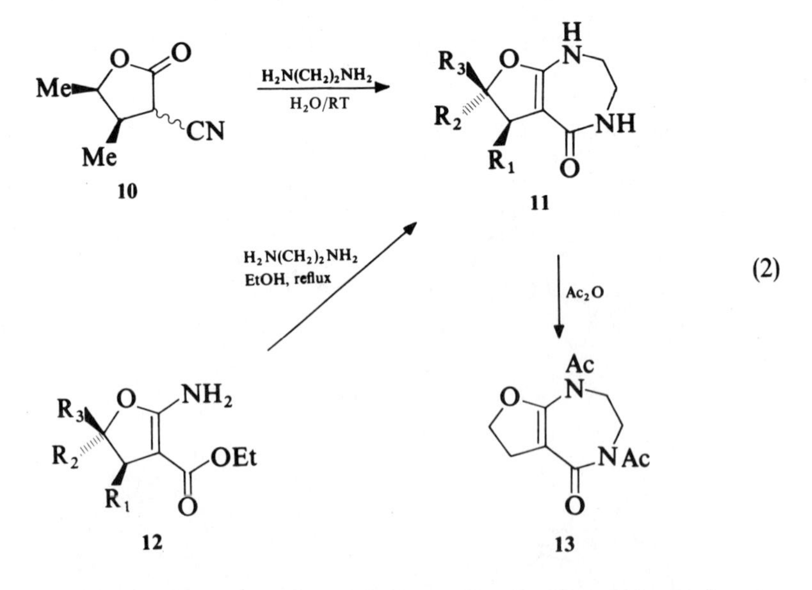


Perhydrofuro[2,3-*e*][1,4]diazepines **16** ($R_1$ = H, Me; $R_2$ = Me, Bu) were more recently[4a] prepared by condensation of the dihydrofurans **14** ($R_1$ = H, Me) with the 1,3-dialkylimidazolidines **15** ($R_2$ = Me, Bu) by means of trifluoroacetic acid (Eq. 3). Compound **16** ($R_1$ = H; $R_2$ = Me) was obtained as the cis-fused isomer, while a mixture of cis and trans isomers was generally formed.

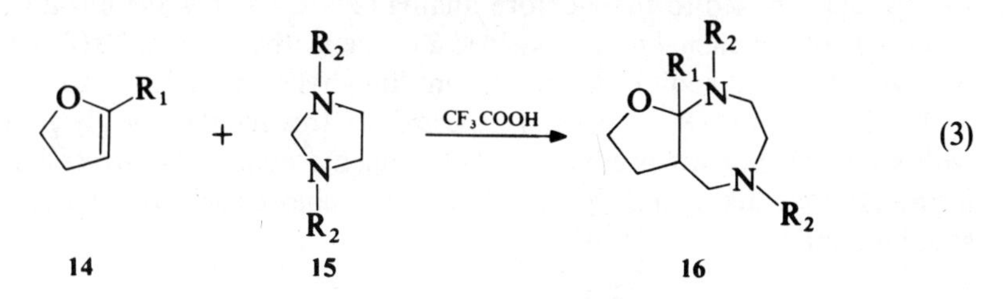


# 3. IMIDAZO[4,5-*e*][1,4]DIAZEPINES

Derivatives of the parent ring **17**, in particular compounds with an intact aromatic imidazole ring, were synthesized by applying the methods developed for the preparation of 1,4-benzodiazepines. Thus the aminoimidazole **18** (X = O) was reacted with bromoacetyl bromide to form the amide **19**. Treatment of this compound with ammonia in 1,2-dichloroethane gave the diazepine **20** (R = Me) in 57% yield (Eq. 4).[5]

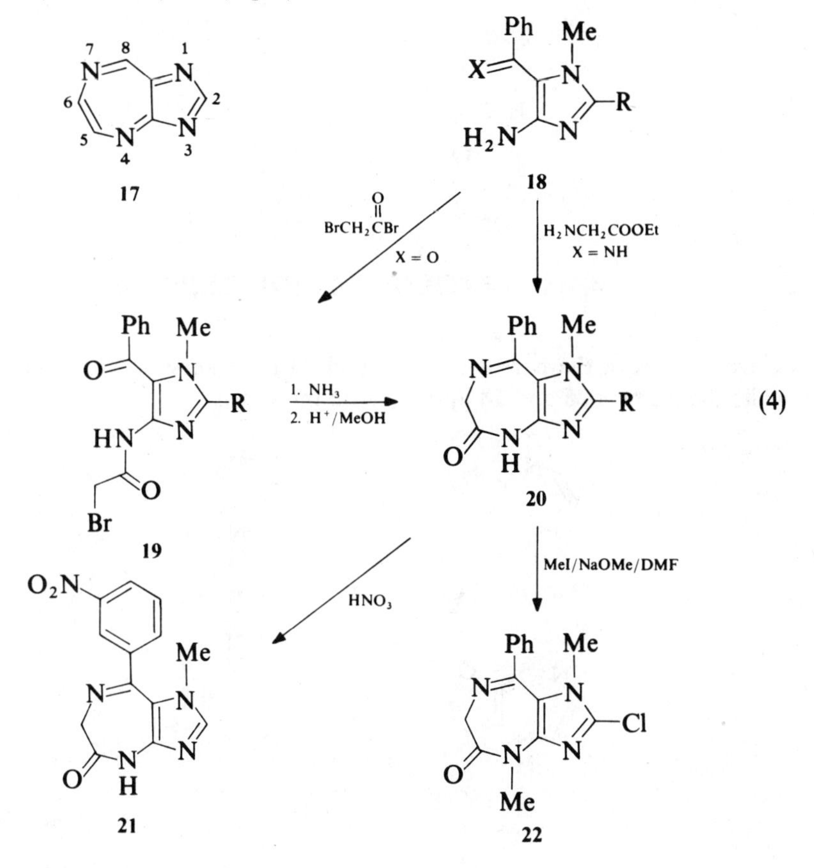

Compound **20** (R = H) was obtained by reaction of the imine **18** (R = H, X = NH) with glycine ethyl ester hydrochloride.[6] Chlorination of **20** with sulfuryl chloride led to the 2-chloro analog (**20**: R = Cl), while nitration at the 3-position of the phenyl moiety yielded **21**. The methylation of **20** (R = Cl) with methyl iodide and sodium methoxide in dimethylformamide led to **22**.[6]

The 5,8-dione **24** (R = H) was synthesized by treatment of the glycine ester **23** with sodium ethoxide in ethanol.[7] The 7-phenyl analog was accessible by an intramolecular alkylation, by reaction of the chloroacetyl derivative **25** with sodium methoxide (Eq. 5).[7]

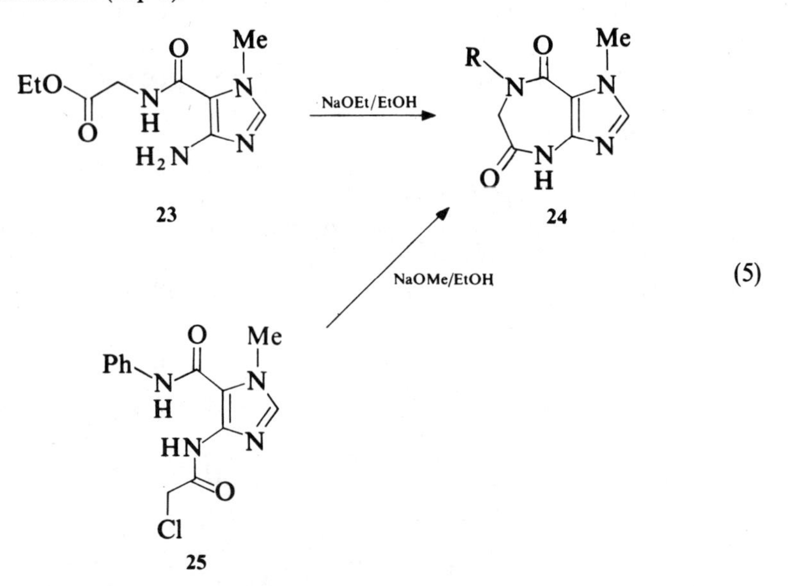


## 4. ISOXAZOLO[e][1,4]DIAZEPINES

Derivatives of the parent rings **27** and **29** have appeared in the literature, while the isomers **26** and **28** appear to be unknown.

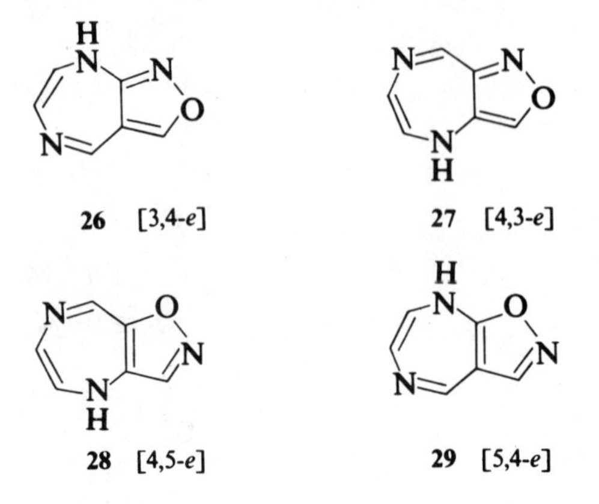

### 4.1. Isoxazolo[4,3-*e*][1,4]diazepines

The 8-phenyl-substituted compounds **34** were recently patented and synthesized as shown in Eq. 6.[8] The chlorooximes **30** (X = F, Cl) were reacted with nitroacetone in the presence of triethylamine to form the isoxazoles **32**. The nitro group was reduced to the amino function by stannous chloride, and the amine was bromoacetylated to give **33**. The standard reaction with ammonia followed by a thermally effected ring closure in methanol and acetic acid, led to the diazepine **34** (R = H). Alkylations of **34** were carried out with methyl iodide, 2-diethylaminoethyl chloride, and propargyl bromide.

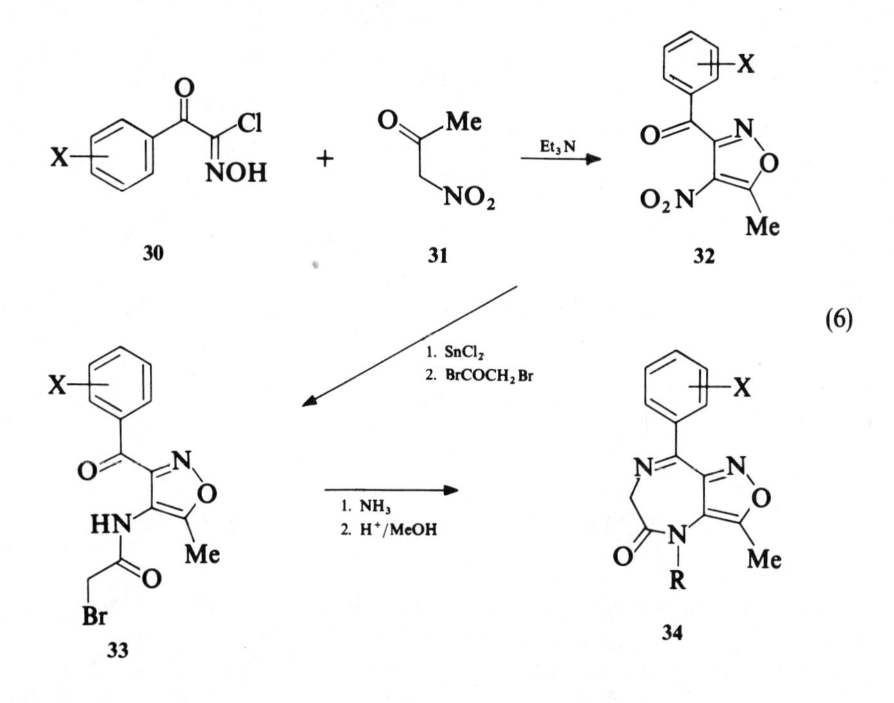


### 4.2. Isoxazolo[5,4-*e*][1,4]diazepines

The synthesis of the 3-methyl-4-phenyl derivative **38** was reported by Jaunin.[9] The protected amino nitrile **35** was reacted with phenylmagnesium bromide to yield, after acid hydrolysis, the amino ketone **36**. This intermediate was converted to the diazepine **38** (R = H) via the bromoacetyl compound **37**. Hexamethylenetetramine was used for the conversion of **37** to **38** (Eq. 7). When **38** (R = H) was subjected to diazomethane, methylation occurred predominantly on the nitrogen to yield **38** (R = Me). The 7-methoxy compound **39** was isolated in 8% yield as a by-product of this reaction.

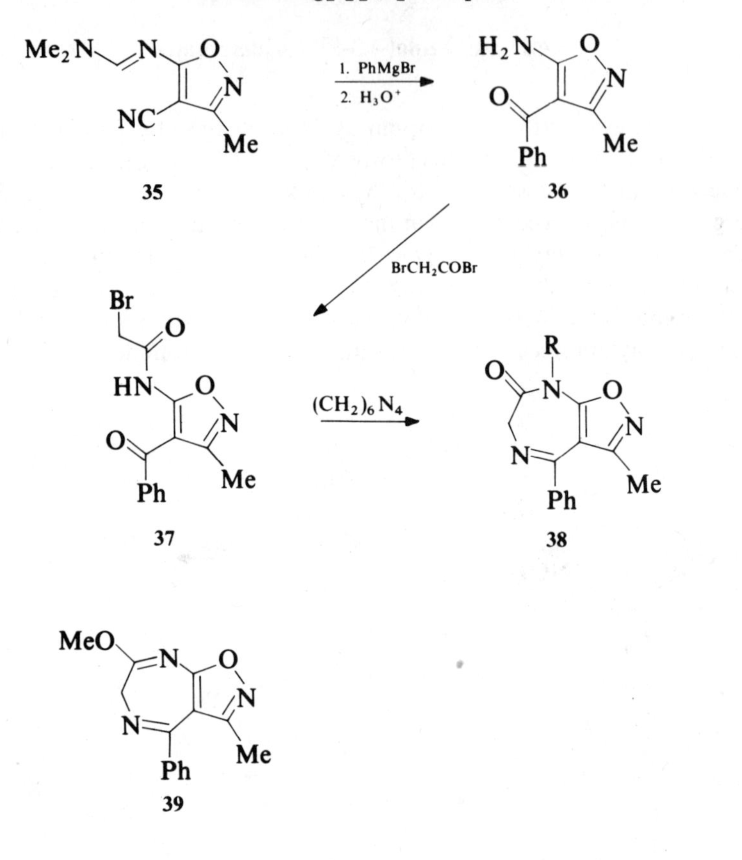


## 5. ISOTHIAZOLO[3,4-*e*][1,4]DIAZEPINES

A representative of this ring system, which is one of four possible isothiazolo[*e*][1,4]diazepines, was prepared by Fryer and coworkers.[10] The protected nitrile **40** was reacted with 2-thiazolyl lithium to form an intermediate imine, which after hydrolysis led to the amino ketone **41**. The conversion to the diazepine was carried out in a standard fashion by bromoacetylation of **41** to form **42** which upon treatment with ammonia, yielded the diazepine **43** (Eq. 8).

## 6. OXAZOLO[*e*][1,4]DIAZEPINES

This ring system has received little attention. Of the two possible parent rings **44** and **45**, only one derivative with the oxazolo[5,4-*e*][1,4]diazepine structure was described in the literature. The 4-thione **47** resulted from the reaction of the dithio ester **46** with ethylenediamine (Eq. 9).[11]

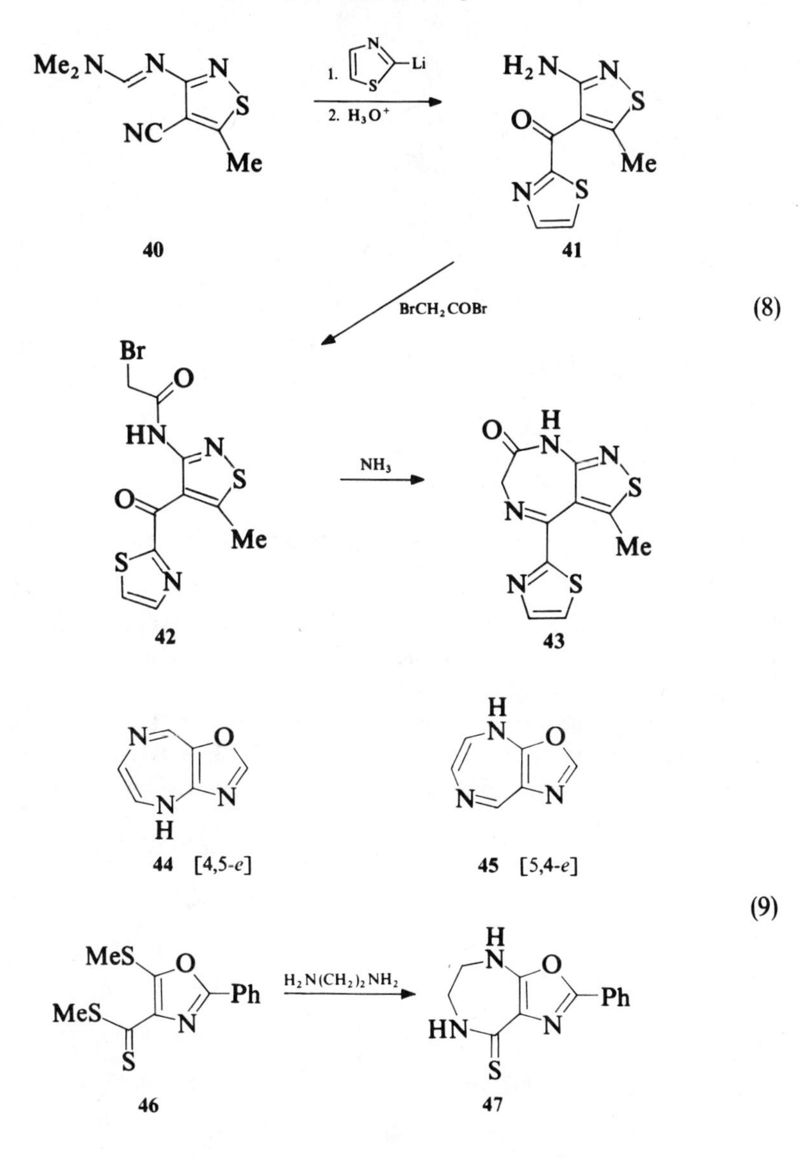


## 7. PYRANO[2,3-*e*][1,4]DIAZEPINES

The perhydro derivatives **49** of the parent ring **50** were synthesized by condensation of dihydropyran with 1,3-dialkylimidazolidines in the presence of trifluoroacetic acid.[4] The products **49** were obtained as a mixture of cis/trans isomers, with the trans ring fusion predominating. Treatment of **49** with lithium aluminum hydride led to the diazepines **51**. Representatives of the other three possible pyrano[*e*][1,4]diazepines appear to be unknown.

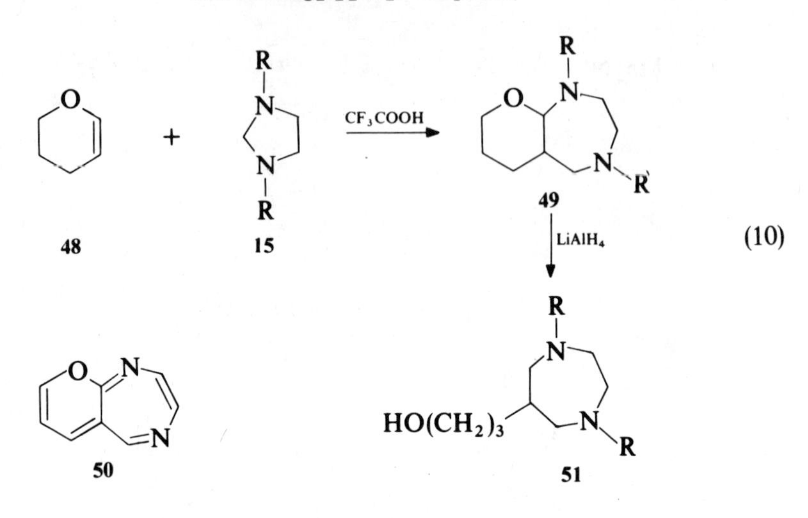


## 8. PYRAZINO[2,3-*e*][1,4]DIAZEPINES

The synthesis of the 2-ones **55** (X = H, Cl; R = H, Me) has been described in the patent literature.[12] The amino ketone **54** was accessible by reaction of the nitrile **52** or the pyrrolidine derivative **53** with phenylmagnesium bromide (Eq. 11). The transformation of **54** to the diazepines **55** (R = H) was carried out by standard methods. Compound **55** (X = Cl, R = H) was methylated at the 1-position by means of dimethyl sulfate and potassium ethoxide.

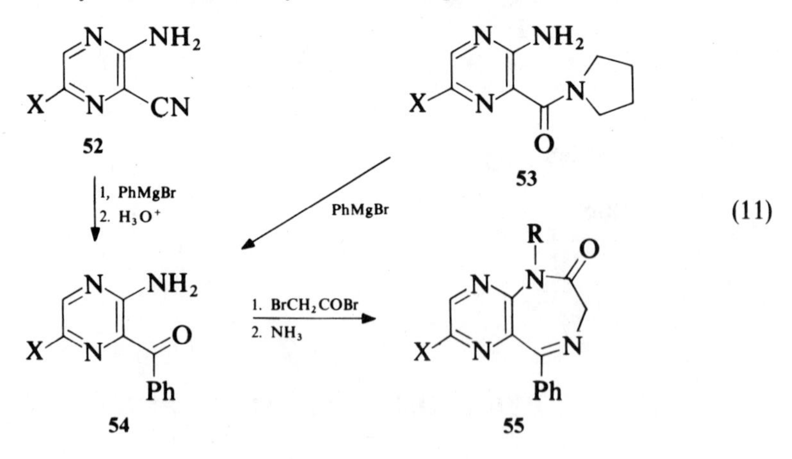


## 9. PYRAZOLO[*e*][1,4]DIAZEPINES

The parent rings **56** and **57** remain unknown, but many derivatives with a higher degree of saturation have been described for both ring systems. In all these compounds the aromaticity of the pyrazole ring has been retained.

## 9.1. Pyrazolo[3,4-e][1,4]diazepines

### 9.1.1. 1,6-Dihydropyrazolo[3,4-e][1,4]diazepines

#### 9.1.1.1. Synthesis

The 1,3,4-trisubstituted compound **59** was obtained by oxidation of the corresponding tetrahydro derivative **58** ($R_1 = R_2 = Me$, $R_3 = 2\text{-}ClC_6H_4$, $X = H_2$) with manganese dioxide (Eq. 12).[13] Compounds **60** bearing a methylthio group at the 7-position were accessible by methylation of the 7-thiones **58** ($X = S$) with methyl iodide.[14] The 7-amino derivatives **61** were synthesized by reaction of the 7-thiones **58** ($X = S$) or the 7-methylthio compounds **60** with a primary amine[15] or by amination of the 7-ones **58** ($X = O$) with a primary amine and titanium tetrachloride (Eq. 12).[10]

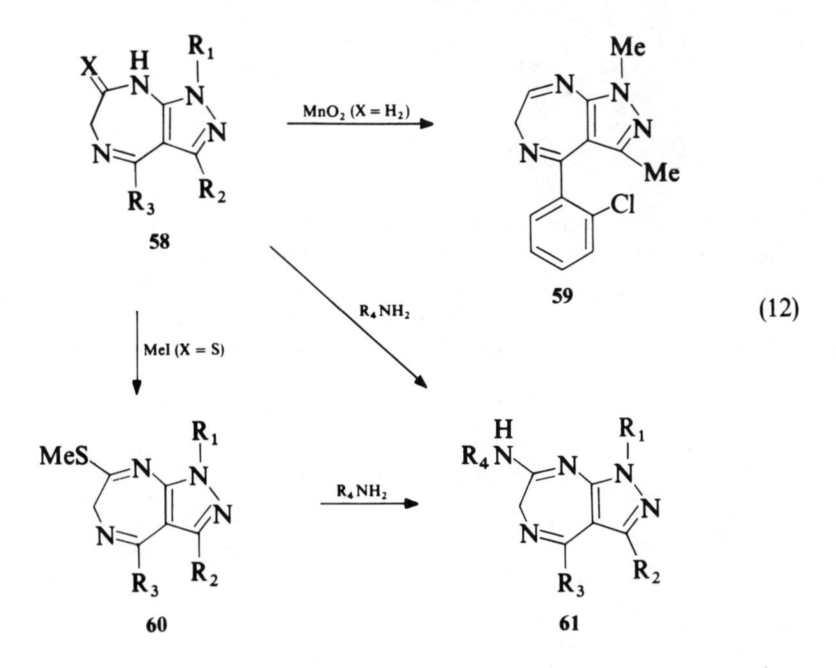


(12)

The oxime **63** was prepared from the malonylidene derivative **62** via a hydrolysis, decarboxylation, and nitrosation of the intermediate acetylidene compound (Eq. 13).[16]

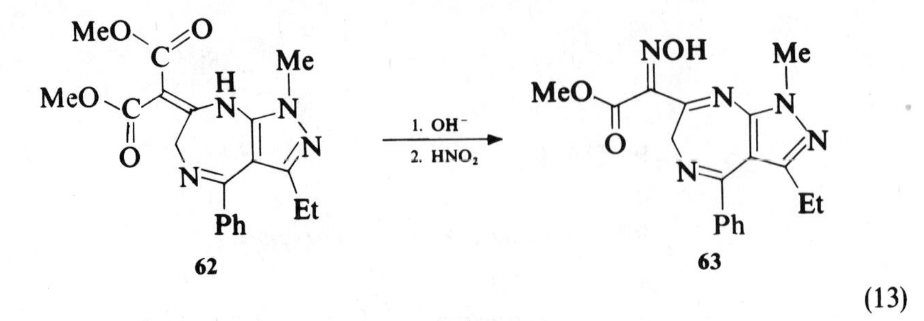


(13)

### 9.1.1.2. Reactions

Nitrosation of the amidine **64** with nitrosyl chloride in the presence of pyridine led to the nitrosoamidine **65** in high yield.[16] This compound reacted with the anion of dimethyl malonate to give the malonylidene derivative **62** (Eq. 14).

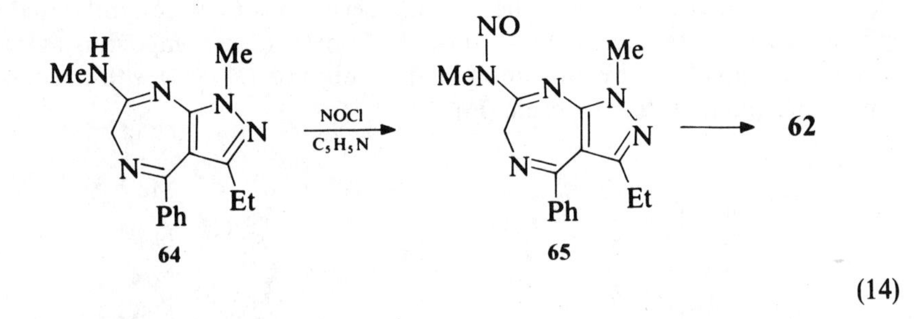


(14)

Reaction of the 7-methylthio compound **66** with acetylhydrazine afforded the triazolodiazepine **67** (Eq. 15).[14] The imidazopyrazolodiazepines **69** (R = H, Me; X = H, F, Cl) were obtained by treatment of the acetylenes **68** with sulfuric acid in the presence of mercuric oxide.[15] The amidines **68** were accessible by reaction of the 7-methylthio compounds **66** or the 7-thiones with the acetylenic amines in refluxing ethanol.

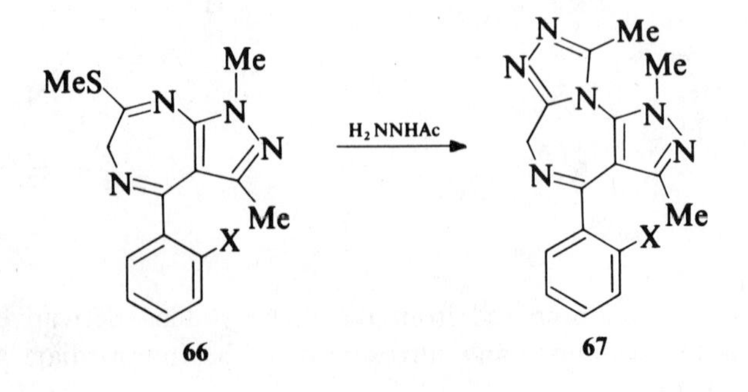

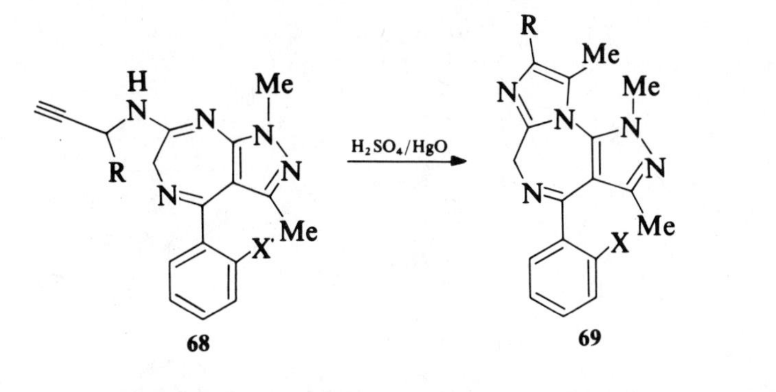


(15)

The imidazo[1,5-*a*]pyrazolo[4,3-*f*][1,4]diazepine **71** was synthesized via the oxime **63**. Reduction of this oxime with zinc dust and acetic acid in methylene chloride gave presumably the intermediate enamine **70**, which was reacted in situ with triethyl orthoacetate to yield **71** (Eq. 16).[16]

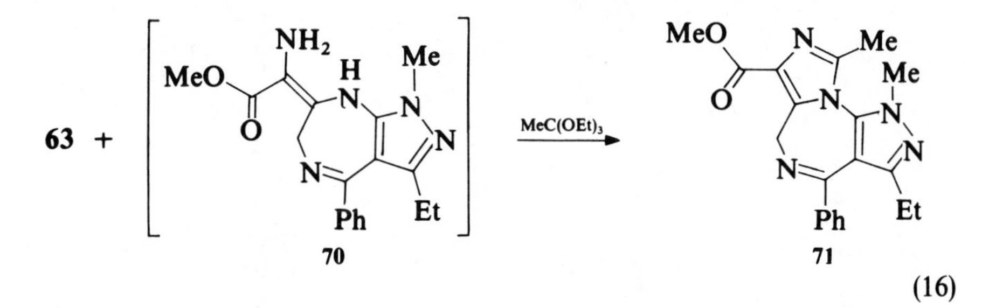


(16)

Addition of hydrogen cyanide to the 7,8-imine bond of **59** led to the 7-cyano derivative **72** (Eq. 17).[13]

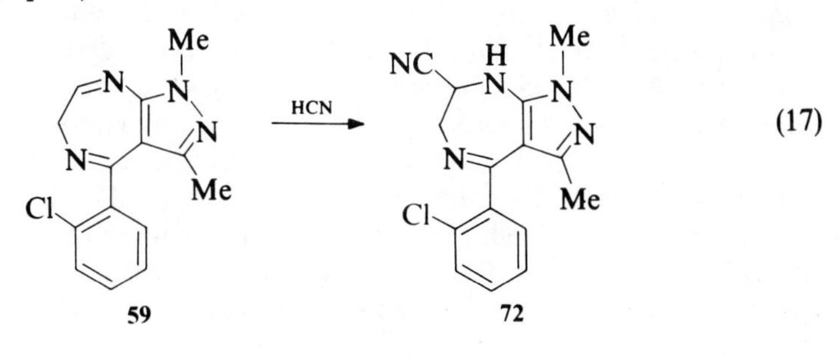


(17)

## 9.1.2. *1,6,7,8-Tetrahydropyrazolo[3,4-e][1,4]diazepines*

### 9.1.2.1. Synthesis

A variety of 1,6,7,8-tetrahydro compounds **75** were synthesized by reaction of the 5-chloropyrazoles **73** with the diamine **74** (Eq. 18).[13,17] The best results were obtained by refluxing the chloropyrazole with the diamine and by partial distillation of the diamine. Yields of diazepines varied from 25 to 78%.

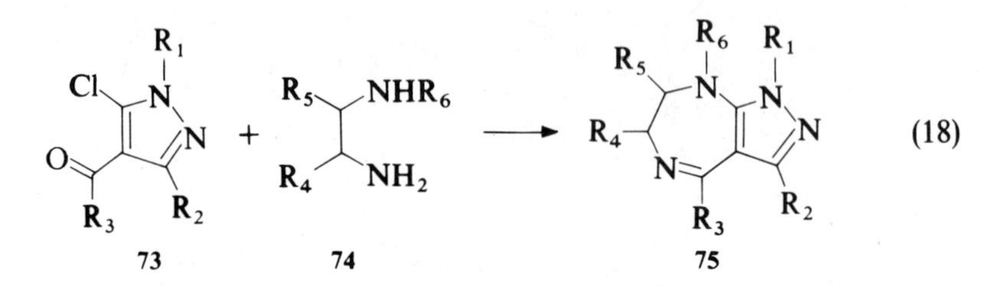


As exemplified by the preparation of **77** (Eq. 19), 1,6,7,8-tetrahydro derivatives may also be synthesized by reduction of the 7-ones, such as **76**, with aluminum hydride.[13]

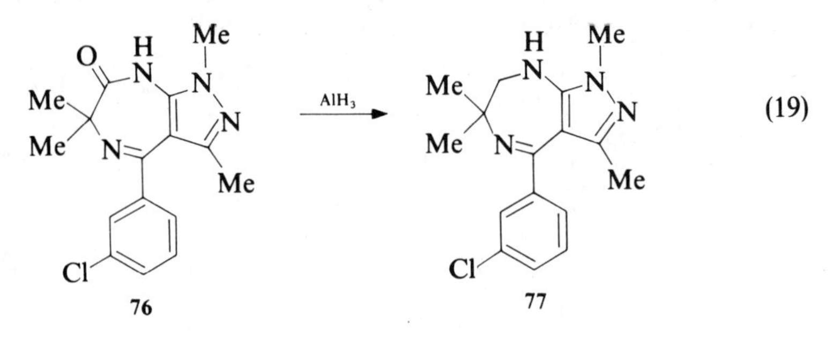


### 9.1.2.2. Reactions

The 4-phenyl compound **78** (X = R = H) was brominated in the meta position of the phenyl moiety with bromine and silver sulfate in concentrated sulfuric acid.[17] Oxidation of the 8-trifluoroacetyl derivative **78** (X = Cl, R = $CF_3CO$) with *m*-chloroperoxybenzoic acid afforded the 5-oxide **79** (R = $CF_3CO$). The trifluoroacetyl group was readily removed by alkaline hydrolysis to yield **79** (R = H) (Eq. 20).[13] Nitration of **78** (X = R = H) with potassium nitrate in concentrated sulfuric acid occurred at the 3-position of the 4-phenyl group, giving **80**. The nitro group was reduced to the amino function by hydrogenation over Raney nickel, and this amine was diazotized. The diazonium salt was reacted with azide and cyanide to yield **81** (X = $N_3$) and **81** (X = CN).

Nitrosation of **78** (X = Cl, R = H) led to the 8-nitroso analog **78** (X = Cl, R = NO). Acylations at the 8-position were carried out with acetic and trifluoroacetic anhydrides. Alkylations at this nitrogen were performed with 3-dimethylaminopropyl chloride and 1-(2-chloroethyl)pyrrolidine in dimethylformamide using sodium hydride as base. Reduction of 8-acyl derivatives by aluminum hydride, as shown by the conversion of the 8-acetyl compound **78** (X = Cl, R = Ac) to the corresponding 8-ethyl derivative, provided another method for the preparation of 8-alkyl compounds.

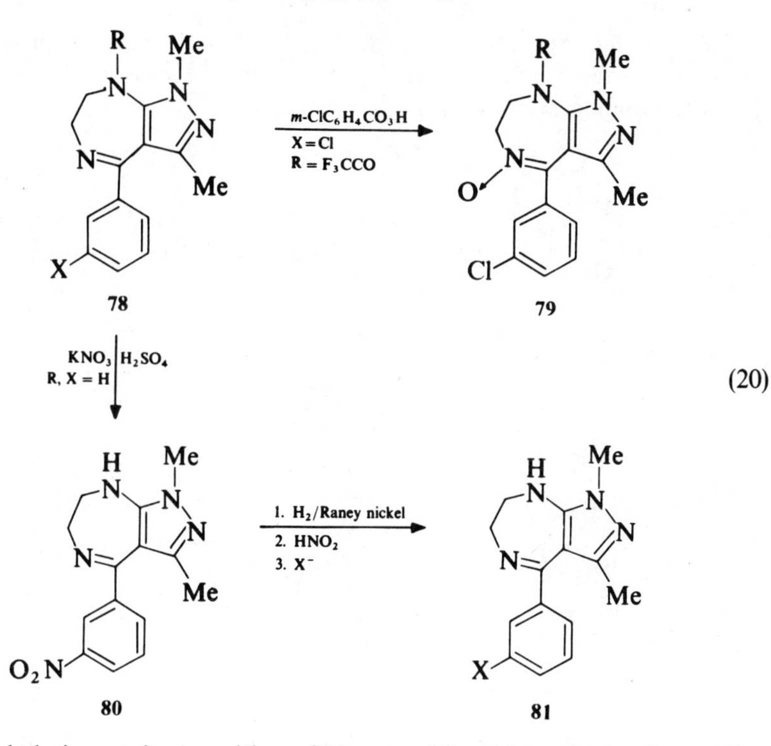


Demethylation at the 1-position of **82** was achieved by refluxing in pyridine hydrochloride for 6 hours. Compound **83** was thus obtained in 40% yield (Eq. 21).[13]

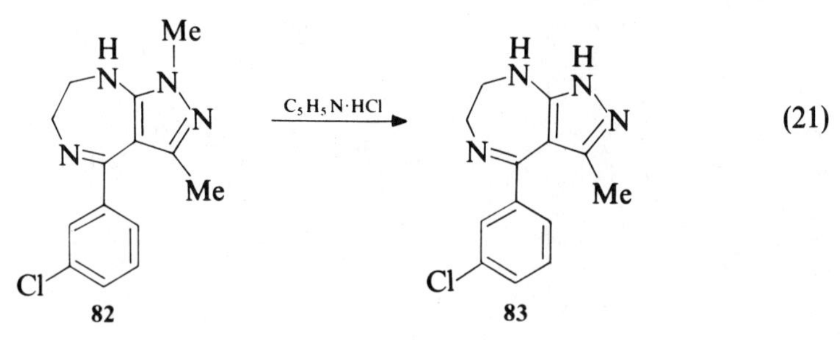


### 9.1.3. *1,6,7,8-Tetrahydropyrazolo[3,4-e][1,4]diazepin-7-ones*

#### 9.1.3.1. Synthesis

The 4-aroyl-5-amino-pyrazoles **84** were converted to the haloacetamides **85**. The halogen Y was then displaced by ammonia, azide, or phthalimide anion. The azide was reduced to the amine catalytically or by treatment with triphenylphosphine, while the phthalimido group was cleaved by standard hydrazinolysis. The aminoacetamides **85** ($X = NH_2$) were not characterized but were cyclized in situ to the diazepinones **86** (Eq. 22).[18,19] The azide method was also applied to convert the bromide **85** ($R_1 = R_2 = R_5 = Me$, $R_3 = 3\text{-Cl}$, $R_4 = H$) to the corresponding diazepinone **86**.[13]

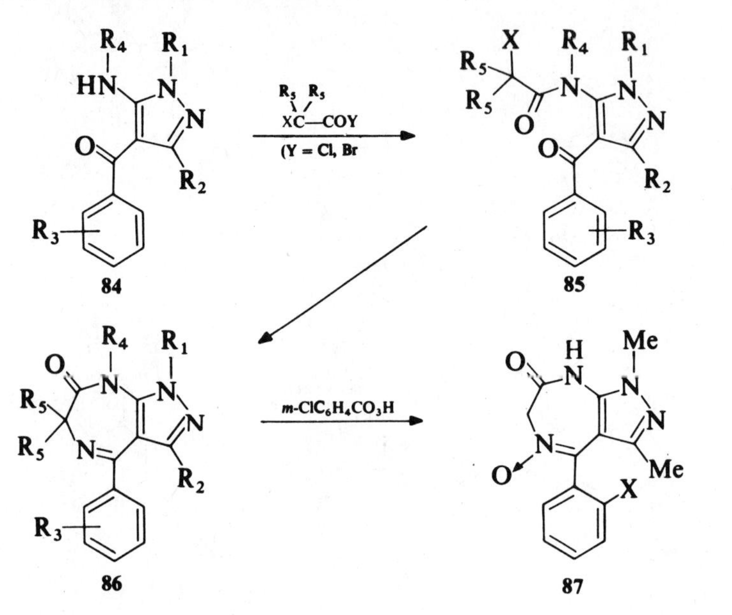


(22)

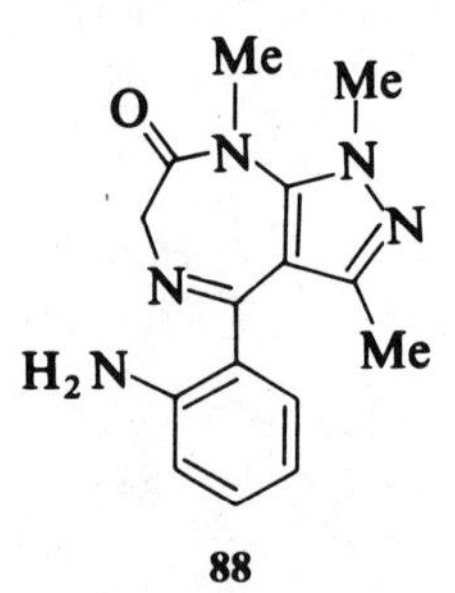


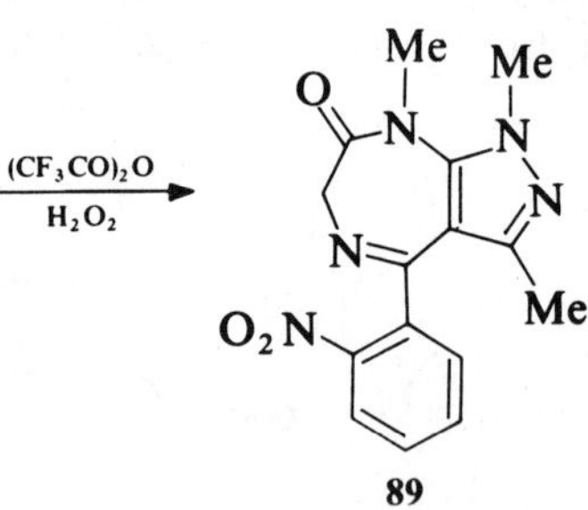


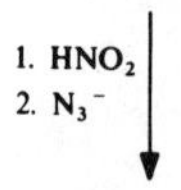


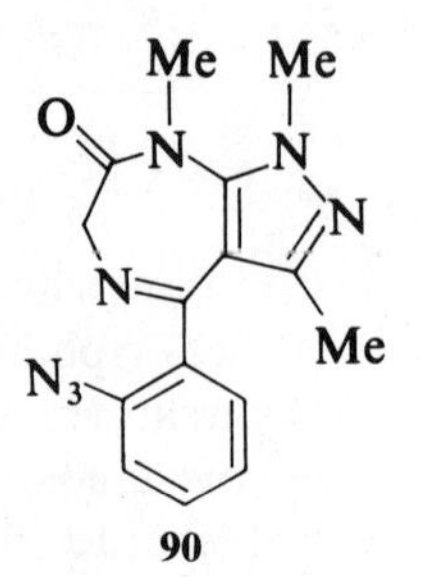


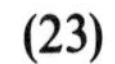

The 5-oxides **87** (X = H, F) were prepared by oxidation of the appropriate compounds **86** with *m*-chloroperoxybenzoic acid in acetic acid.[18]

### 9.1.3.2. Reactions

Oxidation of the 2′-amino group of **88** with trifluoroperoxyacetic acid led to the nitro derivative **89** (Eq. 23).[18] The azide **90** resulted from diazotization of the amine **88** followed by displacement of the diazonium group by azide.[18]

Alkylations of **86** ($R_4$ = H) to the corresponding 8-alkyl analogs **86** ($R_4$ = alkyl) were carried out with methyl iodide, ethyl iodide, and allyl bromide in dimethylformamide using sodium hydride for deprotonation.[10,18–20] The *N*-oxides **87** were reacted with acetic anhydride to give the 6-acetoxy compounds **91**, which, upon hydrolysis with aqueous sodium hydroxide in methanol, gave the corresponding 6-hydroxy derivatives **92** (Eq. 24).[18]

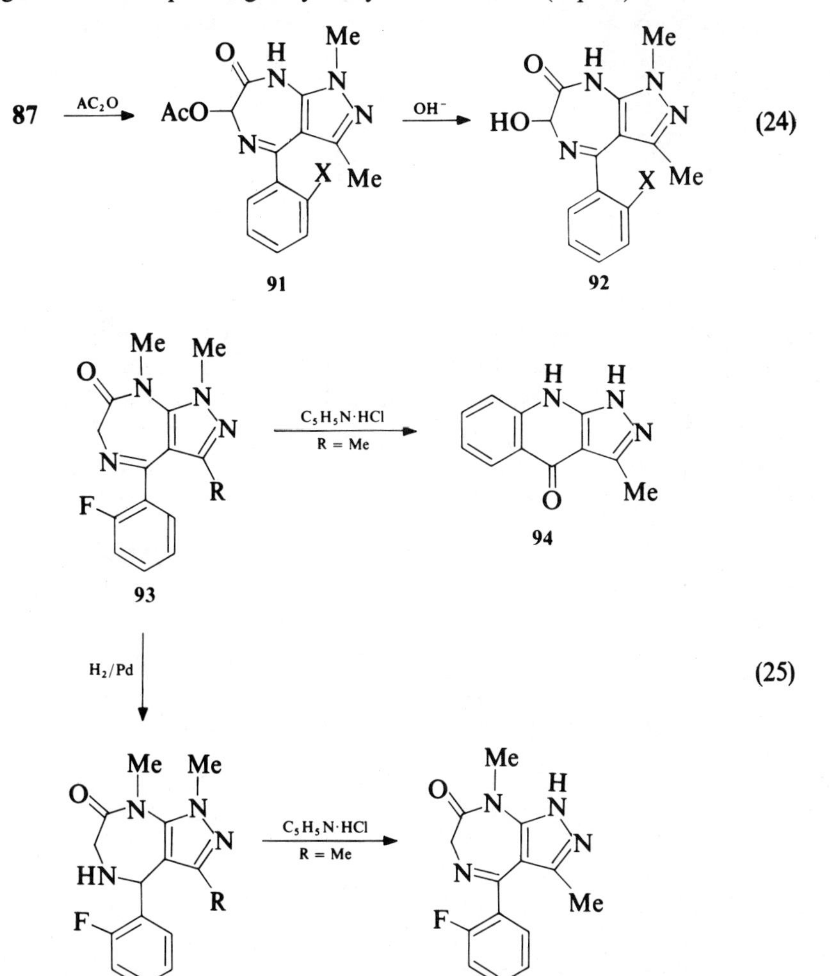

Reduction of the imine bond in **93** (R = Me, Et) was achieved by catalytic hydrogenation over palladium catalyst, giving **95** in high yields (Eq. 25).[18,21] While an attempt to remove the 1-methyl group from **93** (R = Me) led to the pyrazoloquinolone **94**, the demethylation of the reduced imine **95** (R = Me) in refluxing pyridine hydrochloride was succcssful and gave directly the desired product **96**, apparently by simultaneous oxidation of the amine to the imine.[21]

The hydrolytic opening of the imine bond of **97** to the amino ketone **98** was studied in detail (Eq. 26).[22] At pH 3.7, for example, the equilibrium mixture was determined to contain 18% of the open-chain **98**.

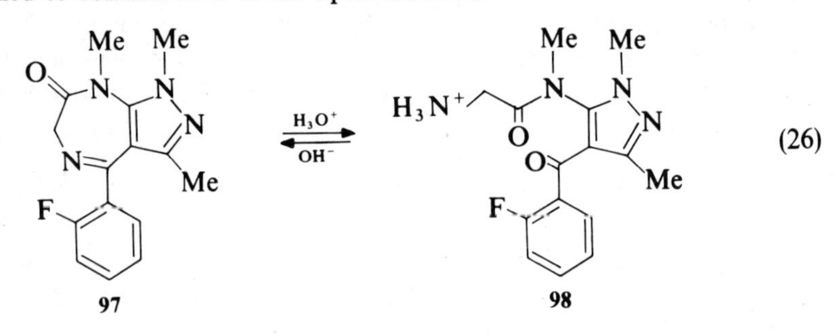


(26)

### 9.1.4. *1,4,5,6,7,8-Hexahydropyrazolo[3,4-e][1,4]diazepines*

The hexahydro derivative **100** (R = Et) was obtained by reduction of the 8-acetyl compound **99** (X = $H_2$ R = Ac) with aluminum hydride (Eq. 27).[13] Reoxidation to the imine **99** (X = $H_2$, R = Et) was achieved in low yield by treatment with manganese dioxide. The carbonyl compound **99** (X = O, R = H) was reduced by aluminum hydride to the hexahydro analog **100** (R = H),[17] which was reoxidized to the tetrahydro compound **99** (X = $H_2$, R = H) by means of diethyl azodicarboxylate.

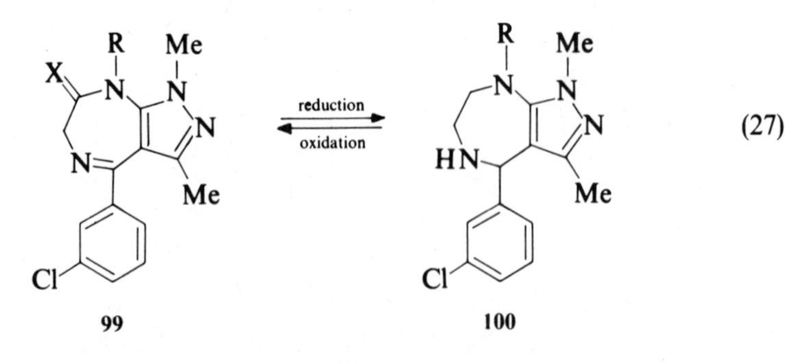


(27)

## 9.2. Pyrazolo[4,3-e][1,4]diazepines

The fully unsaturated parent ring system **57** remains unknown. However, derivatives with a higher level of saturation and, in particular, compounds that retain the aromaticity of the pyrazole ring, have been described.

### *9.2.1. 1,6-Dihydropyrazolo[4,3e][1,4]diazepines*

The 5-methylthio derivatives **102** and the amidine **103**, apparently the only representatives of this class described in the literature, were prepared by alkylation of the thione **101** with dimethyl sulfate and sodium hydroxide in methanol,[23] followed by displacement of the methylthio group by methylamine (Eq. 28).[23a]

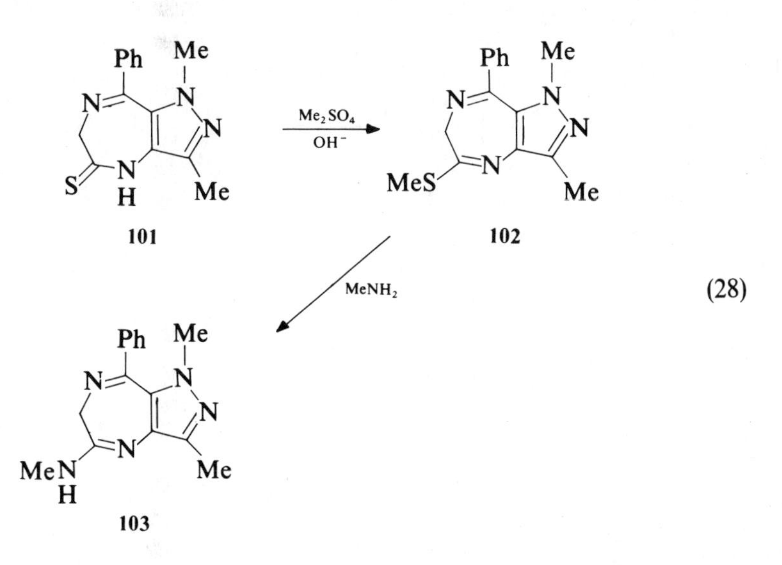


(28)

### *9.2.2. 1,4,5,6-Tetrahydropyrazolo[4,3-e][1,4]diazepin-5-ones*

#### 9.2.2.1. Synthesis

The 1,3-dialkyl derivatives **106** were synthesized by the standard methods from the aminopyrazoles **104b**. These aminopyrazoles were obtained by reduction of the corresponding nitropyrazoles, which were derived from the nitropyrazole carboxylic acid **104a** by Friedel–Crafts arylation[24–26] or by reaction of the corresponding nitrile with aryllithium reagents (Eq. 29).[24,27] The amines **104b** could be converted in one step to the diazepinones **106** by reaction with glycine ethyl ester hydrochloride in refluxing pyridine,[24] or via the bromoacetate **105** (X = Br). The bromide in **105** (X = Br) was displaced by azide, and the azido compound **105** (X = $N_3$) was reduced catalytically over palladium on carbon. The resulting amine was heated in toluene in the presence of acetic acid to complete the cyclization to the diazepinone.[24] Compound **105** (X = benzyloxycarbonylamino) was formed by reaction of **104b** with benzyloxycarbonyl glycine and dicyclohexylcarbodiimide and was cleaved by hydrogen bromide in acetic acid to give the dihydrobromide of the amine **105** (X = $NH_2$).[24]

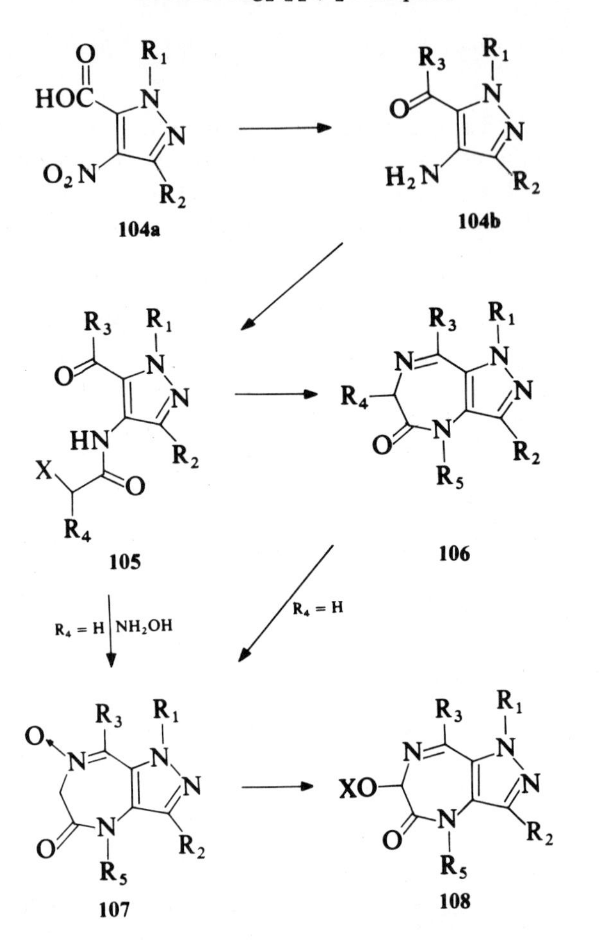


The 7-oxides **107** were accessible by reaction of the bromides **105** (X = Br) with hydroxylamine or by oxidation of the imines **106** with *m*-chloroperoxybenzoic acid.[28,29] 6-Acetoxy compounds **108** (X = Ac) were prepared by heating the 7-oxides **107** with acetic anhydride.[29] Alkaline hydrolysis of these acetoxy derivatives led to the corresponding 6-hydroxy analogs **108** (X = H).

The thiation of the lactam **106** ($R_1 = R_2$ = Me, $R_3$ = Ph, $R_4 = R_5$ = H) to the thione **101** was performed with phosphorous pentasulfide in pyridine.[23]

### 9.2.2.2. Reactions

Bromination of the 8-phenyl compound **109** with bromine and silver sulfate in concentrated sulfuric acid afforded the 3-bromophenyl derivative **110** (X = Br) (Eq. 30).[24] The 3-nitrophenyl analog was similarly prepared by nitration in the same medium.[24]

Alkylations of **106** ($R_5$ = H) or the *N*-oxides **107** ($R_5$ = H) were carried out in the standard fashion using sodium hydride and the alkyl halide in dimethylformamide.[24,26,28] The reagents employed include methyl iodide, ethyl iodide,

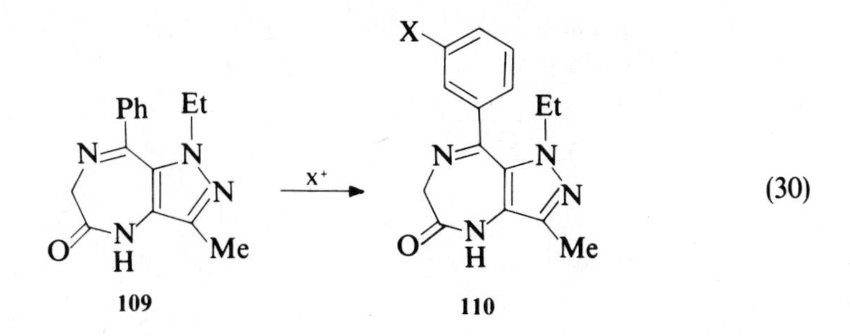


3-dimethylaminopropyl chloride, 2-diethylaminoethyl chloride, and 2,2,2-trifluoroethyl bromide and propargyl bromide. Reaction of **111** (R = Me) with fluoromethyl sulfate followed by hydrolytic workup led to the open-chain methylated amine **112**, characterized as a hydrochloride (Eq. 31).[30] Quaternization with methyl iodide in acetonitrile at 100°C afforded the methiodide salt **114**. The equilibria of the diazepinones **111** (R = H, Me) with the corresponding ring-opened hydrolysis products **113** were investigated.[30] At a given pH, the 4-Me analog **111** (R = Me) was ring opened to a larger degree than the 4-H compound.

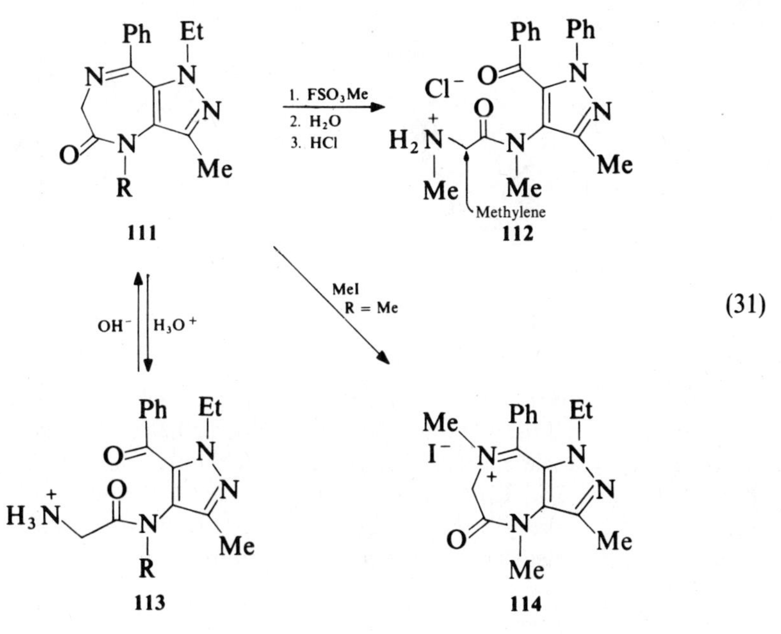


### 9.2.3. *2,4,5,6-Tetrahydropyrazolo[4,3-e][1,4]diazepin-5-ones*

Compounds **116** ($R_4$ = H) were prepared by reaction of the aminopyrazoles **115** with glycine ethyl ester in boiling pyridine (Eq. 32).[24] The aminopyrazoles **115** were accessible via the nitropyrazoles mentioned in Section 9.2.2.1. The substituent at the 4-position was introduced by alkylating **116** ($R_4$ = H), using the method described for the preparation of **116** ($R_4$ = Me, Et).[24]

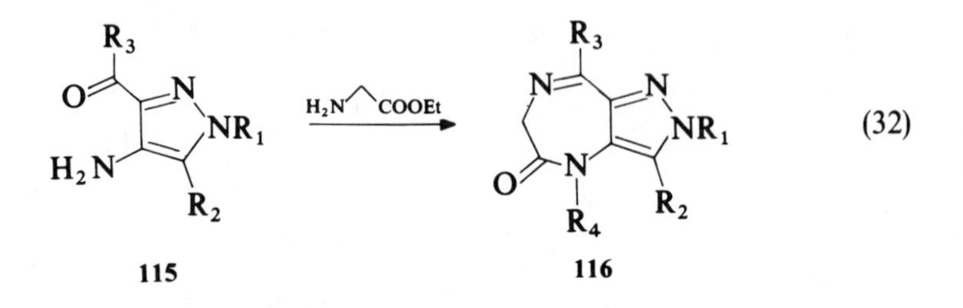


### 9.2.4. *1,6,7,8-Tetrahydropyrazolo[4,3-e][1,4]diazepin-8-ones*

Baraldi and coworkers[31] recently synthesized the 5-phenyl derivatives **118** (X = H, Me, Cl, Br, Ph) by reductive cyclization of the nitro compounds **117**, using iron powder in a refluxing mixture of water and 2-methoxyethanol (Eq. 33).

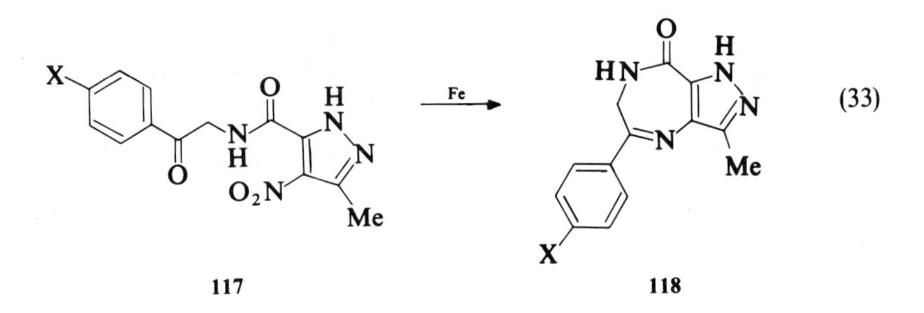


The nitro amides **117** were obtained by treating the diketopiperazine **120** with the appropriate α-amino ketone. The diketopiperazine resulted from the treatment of the pyrazole carboxylic acid **119** with thionyl chloride (Eq. 34). The protected amino acid **121** was alternatively reacted with the amino ketone by means of dicyclohexylcarbodiimide to yield **122**, which cyclized to **118** (X = $NO_2$) upon cleavage of the protecting group.

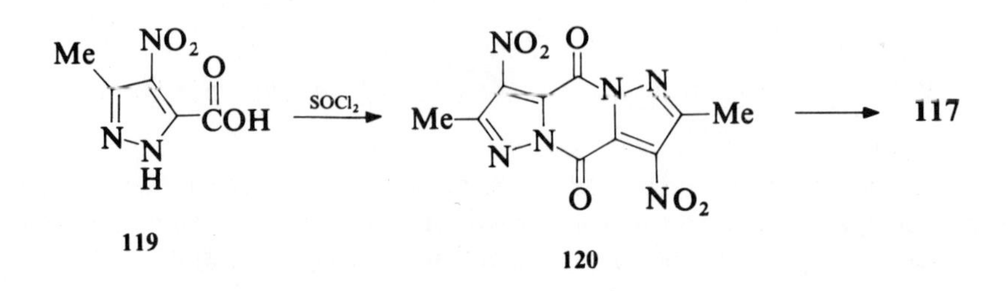

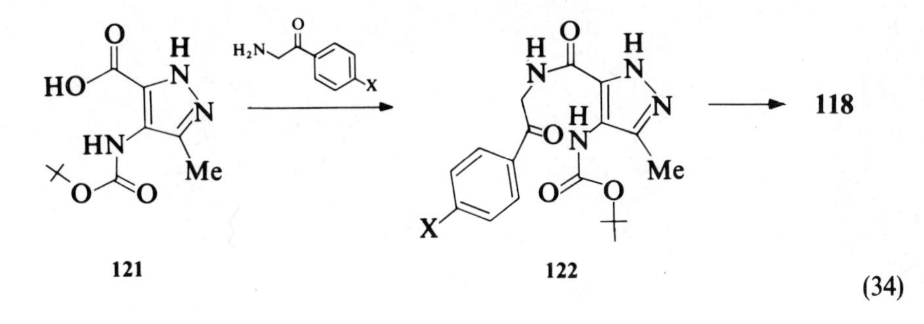


(34)

### 9.2.5. *1,4,5,6,7,8-Hexahydropyrazolo[4,3-e][1,4]diazepin-5-ones*

The hexahydro derivatives **124** were prepared by catalytic hydrogenation of the imine bond in **123**, employing palladium on carbon as the catalyst of choice (Eq. 35).[32] Reaction of the amines **124** with formaldehyde and formic acid led to the 7-methyl analogs **125**.

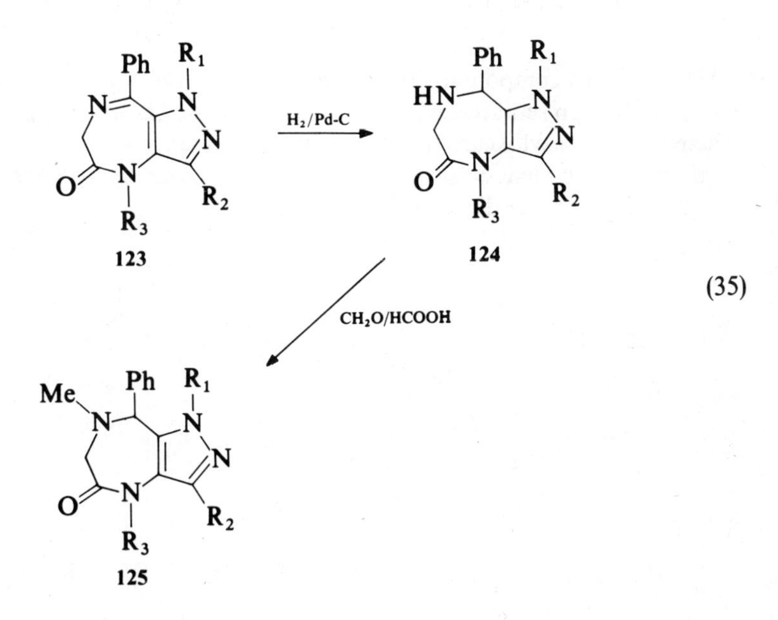


(35)

## 10. PYRIDO[*e*][1,4]DIAZEPINES

Representatives of all four possible pyrido[*e*][1,4]diazepines, the 1*H* tautomers of which are shown, (**126**–**129**), have been reported in the literature. The [3,2-*e*] isomers have received most attention, because of their interesting pharmacological properties.

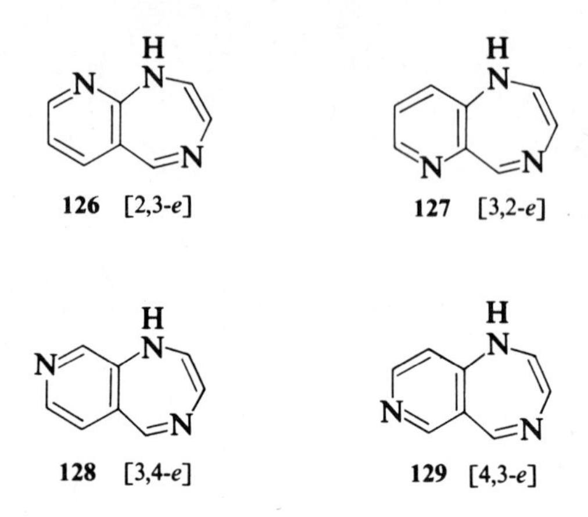


## 10.1. Pyrido[2,3-*e*][1,4]diazepines

### *10.1.1. 1,3-Dihydropyrido[2,3-e][1,4]diazepin-2(2H)-ones*

The 5-phenyl compound **131** was prepared by Littell and Allen[33] in 1965, and it is the only representative with this oxidation state. The aminopyridine **130** was reacted with benzyloxcarbonylglycine in the presence of dicyclohexylcarbodi-imide, and subsequent cleavage of the protecting group by hydrogen bromide in acetic acid led to the diazepinone **131** (Eq. 36).

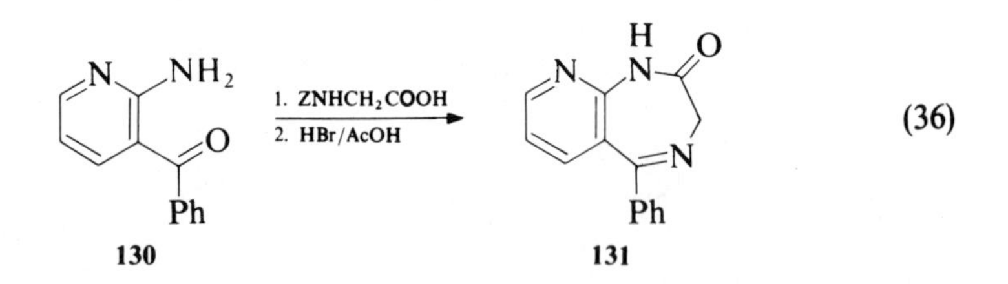


(36)

### *10.1.2. 2,3,4,5-Tetrahydro-1H-Pyrido[2,3-e][1,4]diazepines*

The 8-methyl compounds **133** were prepared by desulfurization of the 5-thiones **132** (X = H, Br) by Raney nickel in ethanol (Eq. 37).[34] The 1-phenyl derivative **135** (R = H) was obtained by reduction of the 3-one **134** with lithium aluminum hydride.[35] Compound **135** (R = H) was methylated at the 4-position to give **135** (R = Me) and acylated with acetic anhydride or 4-chloro-benzoyl chloride to yield the 4-acyl derivatives **135** (R = Ac, 4-$ClC_6H_4CO$), respectively.[35]

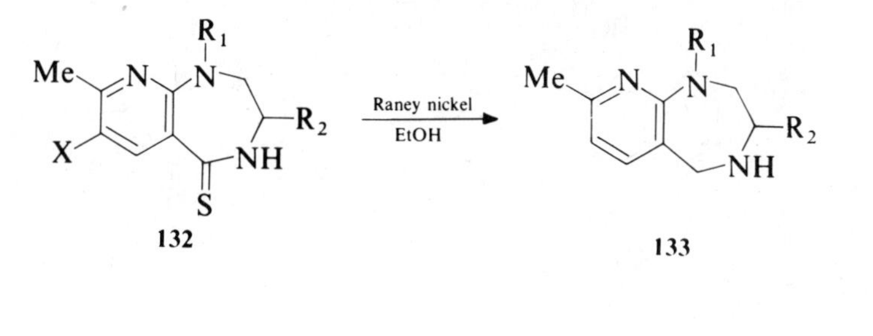


(37)

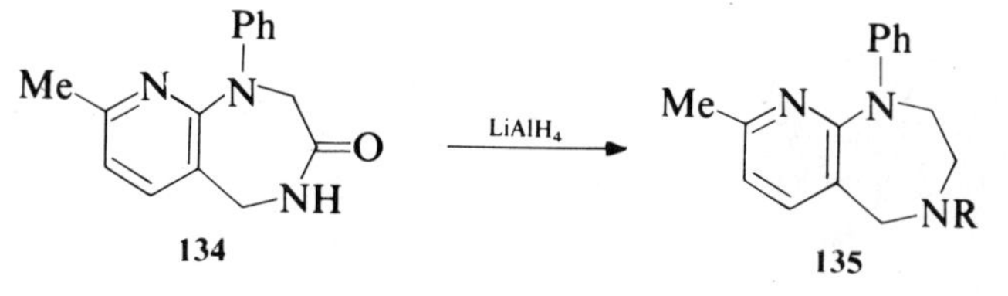


### *10.1.3. 1,2,4,5-Tetrahydropyrido[2,3-e][1,4]diazepin-3(3H)-ones*

Compound **134**, the only representative of this class, was synthesized by alkylation of **136** with ethyl bromoacetate to give **137** (Eq. 38), which was, in turn, subjected to catalytic hydrogenation and ring closure to the diazepinone **134**. The ring closure and hydrogenation processes occurred simultaneously.[35]

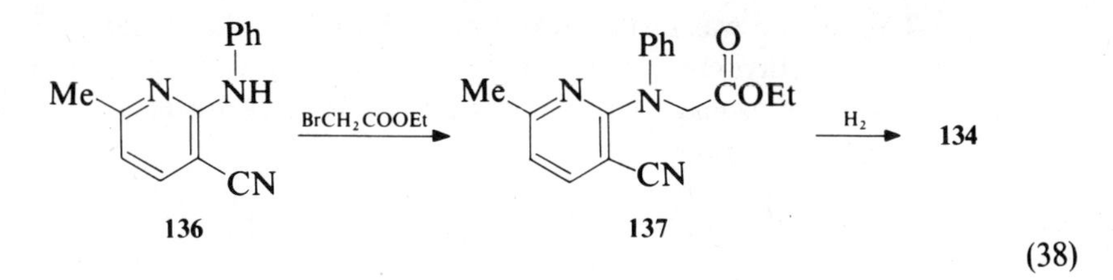


(38)

### *10.1.4. 1,2,3,4-Tetrahydropyrido[2,3-e][1,4]diazepin-5(5H)-ones*

Several substituted 8-methyl derivatives **139** were accessible in high yields by performing the Schmidt reaction on the tetrahydronaphthyridines **138** (Eq. 39).[34,36] Compound **139** ($R_1 = R_2 = X = H$) was also prepared by ring closure of the amino ester **140**.[35] The conversion of the lactams **139** to the corresponding 5-thiones **132** was carried out with phosphorus pentasulfide in refluxing pyridine.[34,36]

Acetylation of **139** ($R_1 = R_2 = X = H$) with acetyl chloride led to the 1-acetyl derivative.[34] The 7-nitro compound **139** ($R_1 = R_2 = H$, $X = NO_2$) was prepared by nitration of the 7-H analog.[34]

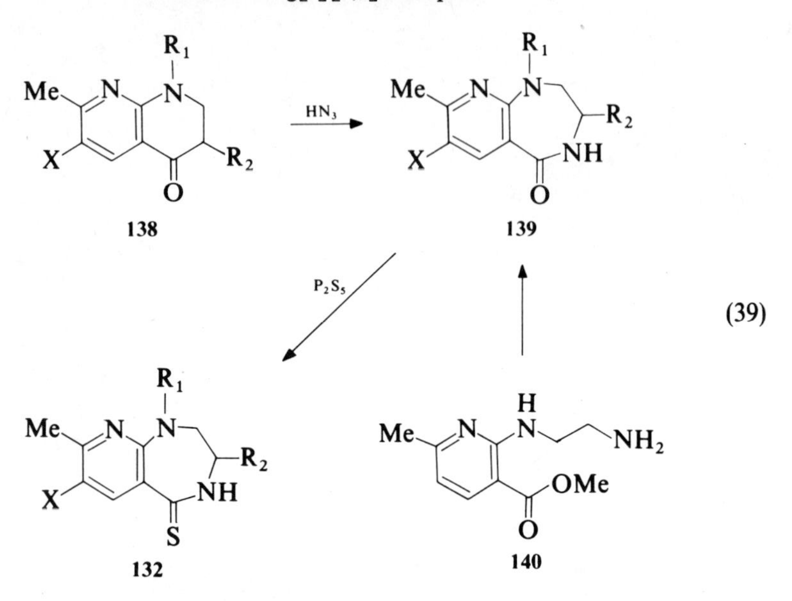


*10.1.5. 3,4,-Dihydro-1H-pyrido[2,3-e][1,4]diazepin-2,5(2H,5H)-diones*

The 4-methyl compound **142** was prepared by reaction of the oxazinone **141** with sarcosine ethyl ester in dimethylformamide and triethylamine at 80°C (Eq. 40).[37] This lactam was further transformed into the imidazodiazepine **143** by reaction with diethyl chlorophosphate and subsequent condensation of the intermediate iminophosphate with the anion of ethyl isocyanoacetate.

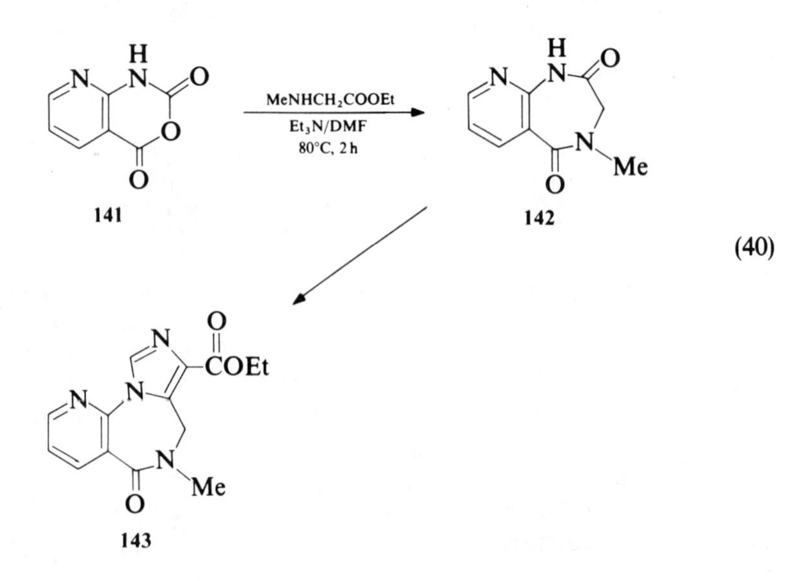

### *10.1.6. 1,2,3,4,5a,6,7,8-Octahydropyrido[2,3-e][1,4]diazepin-5(5H)-ones*

The octahydro derivatives **145** (R = H, Et) were obtained when the tetrahydropyridines **144** (R = H, Et) were heated with ethylenediamine in boiling ethanol (Eq. 41).[38,39]

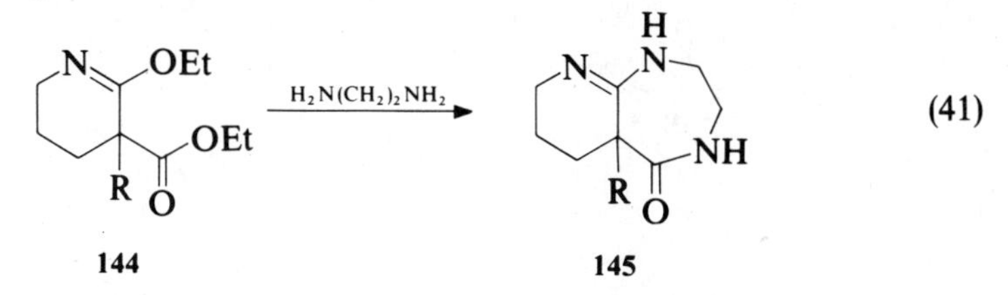


### *10.1.7. 1,2,3,4,6,7,8,9-Octahydropyrido[2,3-e][1,4]diazepin-5(5H)-ones*

The 9-alkyl derivatives **147** (R = Me, Et) were synthesized by reaction of the tetrahydropyridines **146** (R = Me, Et) with ethylenediamine (Eq. 42).[39,40]

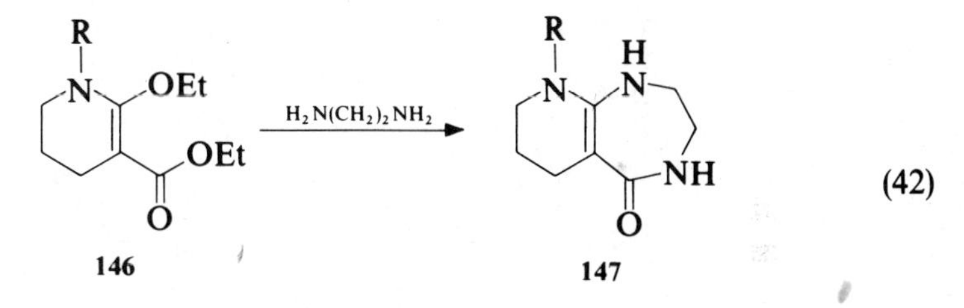


## 10.2. Pyrido[3,2-*e*][1,4]diazepines

Many of the reactions carried out with the 1,4-benzodiazepines were applied to this ring system, but, in particular to the 7-chloro-substituted genus.

### *10.2.1. 3H-Pyrido[3,2-e][1,4]diazepines*

In analogy to the benzodiazepines, reaction of the pyridopyrimidine **148** with methylamine resulted in ring expansion to the diazepine **149**.[41] The 4-desoxy analog of **149** was also prepared by amination of the lactam **150** with methylamine and titanium tetrachloride (Eq. 43).[41] The 2-hydrazino compounds **152** were accessible by reaction of the nitrosoamidines **151** with hydrazine. The nitrosoamidines were formed by nitrosation of the corresponding amidines. The hydrazines **152** were converted to the tetrazolodiazepines **153** by nitrous acid and to several triazolodiazepines **154** (R = H, OEt) and **155** by reaction with orthoesters and carbonyldiimidazole.

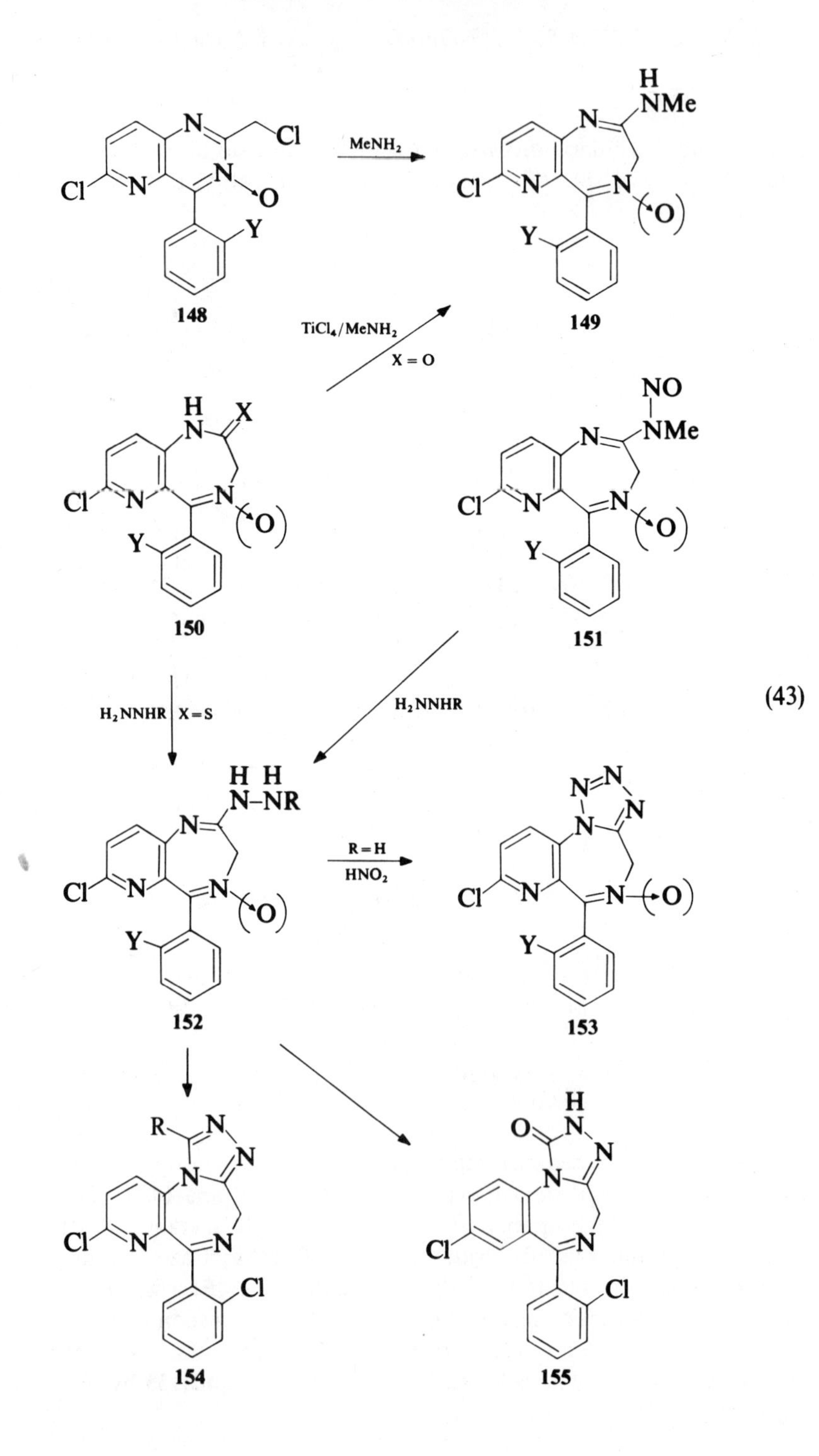
MeNH₂
TiCl₄/MeNH₂
X = O
H₂NNHR
X=S
H₂NNHR
R=H
HNO₂
148
149
150
151
152
153
154
155

(43)

### *10.2.2. 2,3-Dihydro-1H-Pyrido[3,2-e][1,4]diazepines*

The 2-hydroxy compound **157** (R = OH) was prepared by reduction of the 2-one **156** with sodium diethoxyaluminum hydride at 0°C in tetrahydrofuran (Eq. 44).[42] Treatment of the 2-hydroxy derivative with ethanolic hydrogen chloride led to the 2-ethoxy analog **157** (R = OEt). Lithium aluminum hydride in tetrahydrofuran reduced the carbonyl group to a hydrocarbon, yielding **157** (R = H).[43]

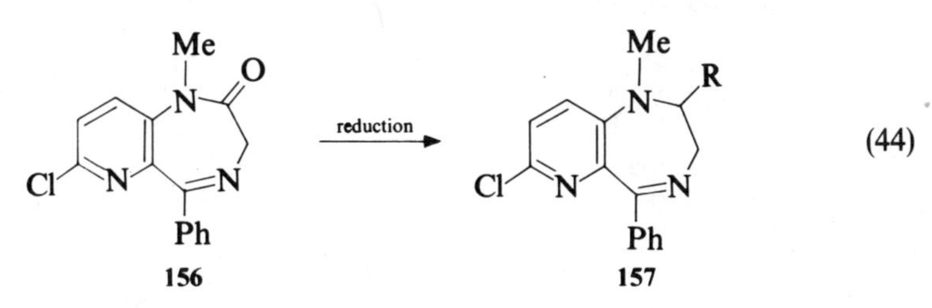


(44)

### *10.2.3. 1,3-Dihydropyrido[3,2-e][1,4]diazepin-2(2H)-ones*

#### 10.2.3.1. Synthesis

The 5-phenyl compound **159** was synthesized by Littell and Allen[33] by coupling 3-amino-2-benzoylpyridine **158** with benzyloxycarbonyl glycine. Subsequent cleavage of the protecting group, followed by cyclization, yielded **159** (Eq. 45).

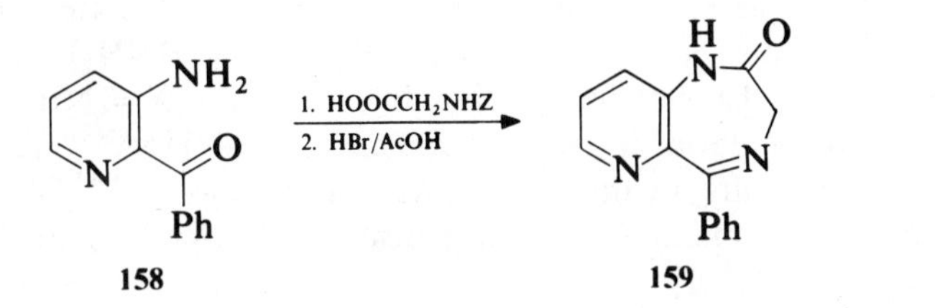


(45)

Synthesis of 7-substituted analogs of **159** was achieved by using the 2,6-dichloro-3-nitropyridine **160** as the starting material. Compound **160** was reacted with phenylacetonitriles and base to give **161**. These intermediates were then oxidized by alkaline hydrogen peroxide to the ketones **162** (Eq. 46). Since chloride was partially displaced by hydroxide during this oxidation, the 6-hydroxypyridine had to be reconverted to the 6-chloro compound by treatment with a mixture of phosphorus pentachloride, phosphorus oxychloride, and phosphorus trichloride. The nitro group was reduced by catalytic hydrogenation over Raney nickel to the amines **163**.[43,44] Displacement of the chloride in **162** (X = Cl) by various secondary amines led to the 6-amino pyridines **162** (X = $NR_1R_2$).[44]

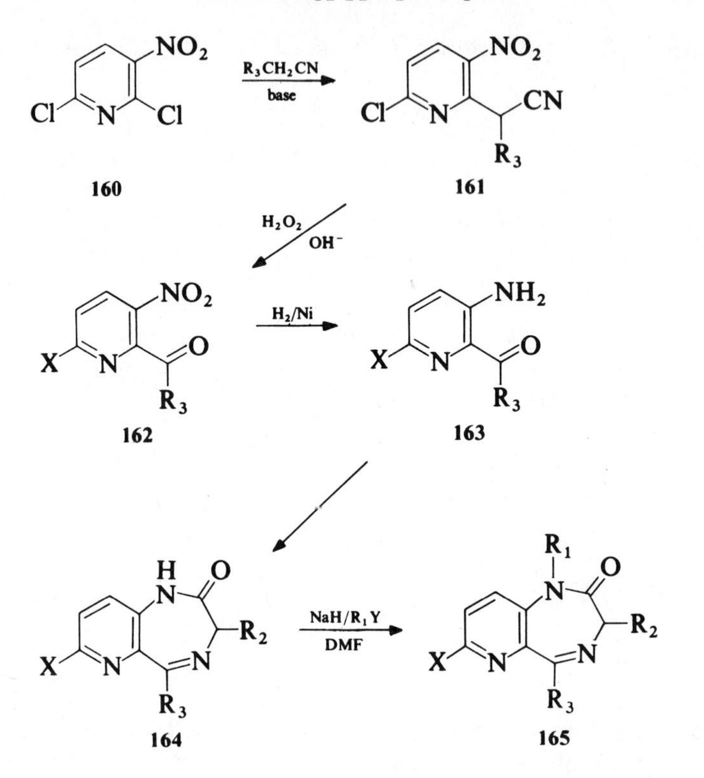


The aminopyridines **163** were then transformed into the diazepinones **164** by standard methods, such as by bromoacetylation followed by treatment with ammonia or by the addition of a benzyloxycarbonyl-protected amino acid.[43]

The 7-amino derivative **164** ($R_3$ = Ph, $R_2$ = H, X = $NH_2$) was prepared by condensation of the diaminopyridine **163** ($R_3$ = Ph, X = $NH_2$) with glycine ethyl ester hydrochloride in an imidazole melt at 110–115°C.[44] The preparation of the 7-chloro analog by this method was patented as well.[45] The 4-oxides **150** (X = O) were also obtained by ring expansion of the pyridopyrimidines **148** with hydroxide (Eq. 43).[43,46]

### 10.2.3.2. Reactions

Oxidation of **164** ($R_2$ = H, $R_3$ = Ph, X = Cl) with *m*-chloroperoxybenzoic acid led to the corresponding 4-oxide.[43]

A variety of substituents were introduced at the 1-position of **164** ($R_3$ = Ph, substituted Ph) by the standard alkylation using sodium hydride and the appropriate alkyl halide ($R_1$Y) in dimethylformamide. Moieties ($R_1$) attached in this fashion include methyl, propyl, butyl, allyl, cyclopropylmethyl, 2-dimethylaminoethyl, 2-hydroxyethyl, 2-morpholinoethyl, and 2-piperidinoethyl.[43,46] Similar alkylations were carried out with chloroacetone, chloroacetonitrile, ethyl bromoacetate, α-chloroacetophenones, 3-bromopropionic acid, and 4-chlorobutyronitrile.[47,48] A 3-hydroxy derivative, **165** ($R_1$ = Me,

$R_2$ = OH, $R_3$ = 2-$ClC_6H_4$), was converted to the 3-methoxy analog by reaction with methyl iodide and sodium hydride in dimethylformamide.[43] The dimethylamino group at the 7-position of **164** ($R_2$ = H, $R_3$ = Ph, X = $Me_2N$) was quaternized by reaction with methyl iodide.[44]

Acylations at the 1-position of **164** were performed with acetic anhydride under reflux[43] and with a variety of isocyanates.[49] The 4-oxides were rearranged in the usual fashion to the 3-acetoxy derivatives by treatment with acetic anhydride.[43,46,48] The 3-hydroxy compound **164** ($R_2$ = OH, $R_3$ = 2-$ClC_6H_4$, X = Cl) was acylated by succinic anhydride to the 3-hemisuccinate.[43]

The chloride at the 7-position of **164** ($R_2$ = H, $R_3$ = Ph, 2-halophenyl; X = Cl) was displaced by a variety of amines, such as dimethylamine, ethanolamine, morpholine, pyrrolidine, and *N*-methylpiperazine.[44]

Compounds **166** (X = H, Cl) were converted to the thioamides **167** by treatment with hydrogen sulfide in methanolic ammonia (Eq. 47).[48]

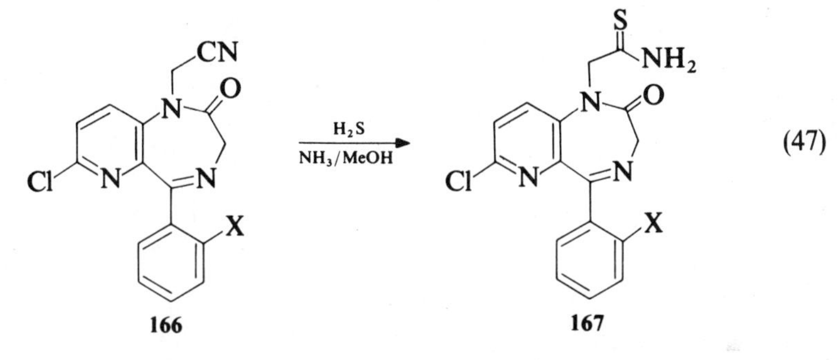


A few of the 2-ones were transformed into the corresponding thiones by the standard reaction with phosphorus pentasulfide in pyridine.[43] Reaction of the 2-thione **150** (X = S, Y = H) with acetylhydrazine in dioxane afforded the 2-hydrazino derivative **152** (R = Ac, Y = H) (Eq. 43).

### *10.2.4. 3,4-Dihydro-1H-pyrido[3,2-e][1,4]diazepin-2,5(2H,5H)-diones*

The 4-methyl compound **169** was synthesized by Hunkeler and Kyburz[37] by reaction of the oxazinone **168** with sarcosine ethyl ester (Eq. 48).

This dione (**169**) was converted to the corresponding imidazodiazepine-3-carboxylic acid ethyl ester as shown for compound **143** depicted in Eq. 40.

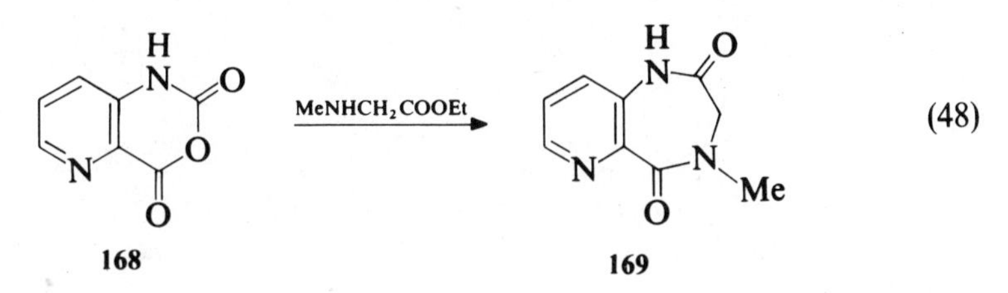

## 10.3. Pyrido[3,4-*e*][1,4]diazepines

The only representative of this ring system described in the literature was prepared by Littell and Allen.[33] They converted the 3-amino-4-benzoylpyridine to the diazepinone **172** by the benzyloxycarbonyl glycine method. The benzoylpyridine **171** resulted from addition of phenylmagnesium bromide to the oxazinone **170** (Eq. 49).

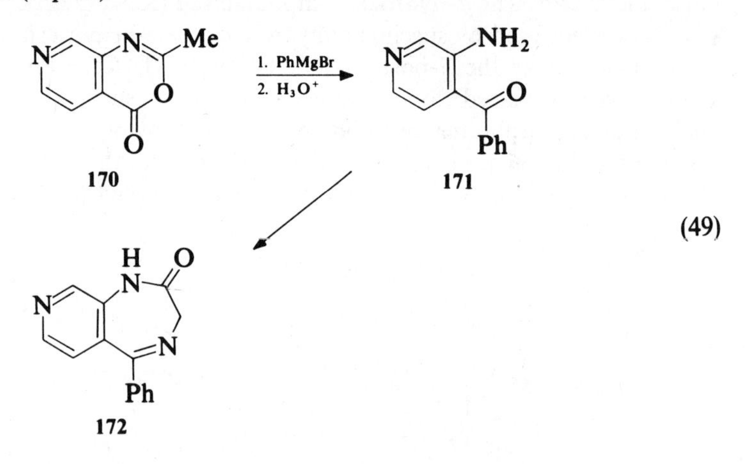

## 10.4. Pyrido[4,3-*e*][1,4]diazepines

The 5-phenyl derivative **176** was synthesized, according to the procedures discussed above for the positional isomers, from the oxazinone **173** (R = Me), via the aminobenzoylpyridine **174** (Eq. 50).[33] The related oxazinedione **173** (R = OH) was used for the preparation of the 2,5-dione **175**, which was further

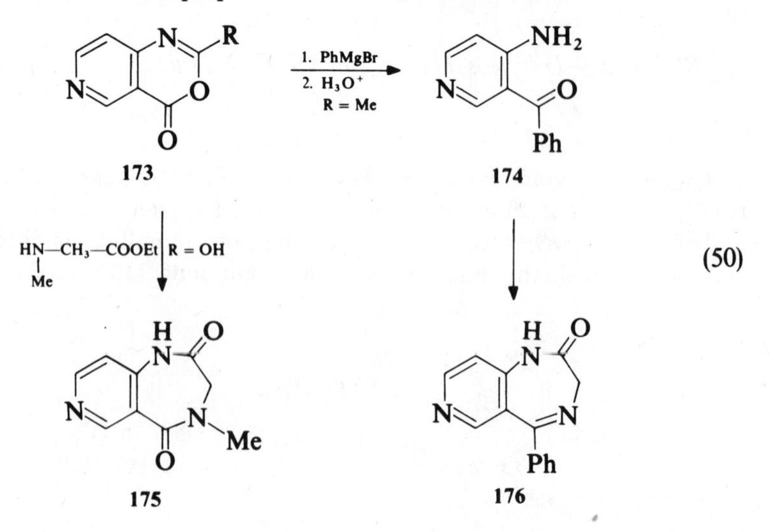

reacted to form the imidazodiazepine in the same fashion as the two positional isomers previously discussed.[37]

The 1,3,4,5-tetrahydro-2-one **178** was one of four products formed in 15% yield from the reaction of bromopyridine **177** with potassium amide in liquid ammonia (Eq. 51).[50] A possible mechanism for the formation of **178** is given by an addition of ammonia to the intermediate aryne, followed by a ring closure.

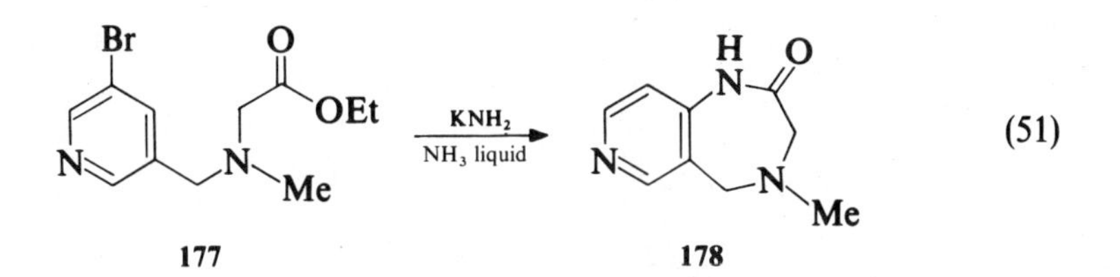


## 11. PYRIMIDO[*e*][1,4]DIAZEPINES

### 11.1. Pyrimido[4,5-*e*][1,4]diazepines

None of the many possible tautomers of the parent ring system has been reported in the literature, although compounds with a higher degree of saturation and with an aromatic pyrimidine ring are known.

The structure 6,9-dihydro-5*H*-pyrimido[4,5-*e*][1,4]diazepine (**182**: R = OH) was assigned to the product resulting from the treatment of thiamine **179** with hydroxide (Eq. 52).[51,52] The cyclization is believed to proceed via the ring-open intermediate **180** by loss of hydrogen sulphide. The reaction of thiamine anhydride **181** with 4-substituted phenylthiols was reported to lead to the diazepines **182** (R = $SC_6H_4X$ with X = H, Me, Br).[53] The diazepine was hydrolyzed at pH 8-9 and at 100°C to give the open aminoketone **183**. Reaction of **182** with 2 equivalents of 5-hydroxy-3-mercaptopentan-2-one formed **184** with stereochemistry unassigned.[51]

Kim and Santilli[54] prepared the 6,7,8,9-tetrahydro-5-one **186** by reaction of the chloropyrimidine **185** with *N,N'*-dimethylethylenediamine (Eq. 53).

Koch[35] synthesised 5,6,8,9-tetrahydro-7-one **188** by catalytic hydrogenation of the nitrile **187** with concomitant ring closure (Eq. 54).

Jaunin[9b] prepared several 4,8-diones **191** by ring closure of the imines **190**, which resulted from the reaction of the ketones **189** with glycine ethyl ester. Compounds with $R_1$ = H, Me, benzyl, 2-dimethylaminoethyl and $R_2$ = Ph, Me were thus obtained. The 5-phenyl compounds were further alkylated at the 9-position by methyl iodide and 2-dimethylaminoethyl chloride using standard procedures (Eq. 55).[9b]

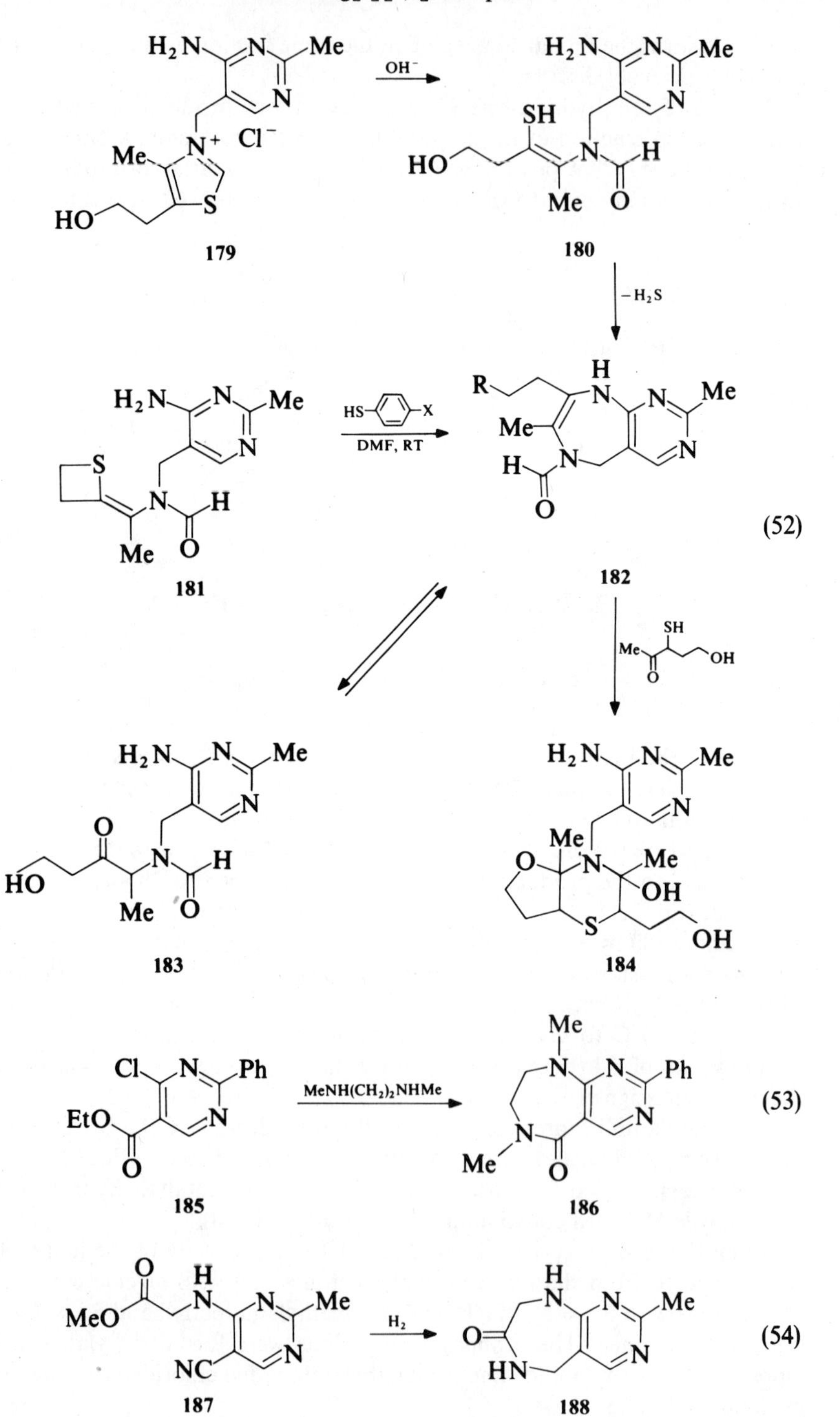

OH⁻
−H₂S
HS—C₆H₄—X
DMF, RT
MeNH(CH₂)₂NHMe
H₂
179
180
181
182
183
184
185
186
187
188
(52)
(53)
(54)

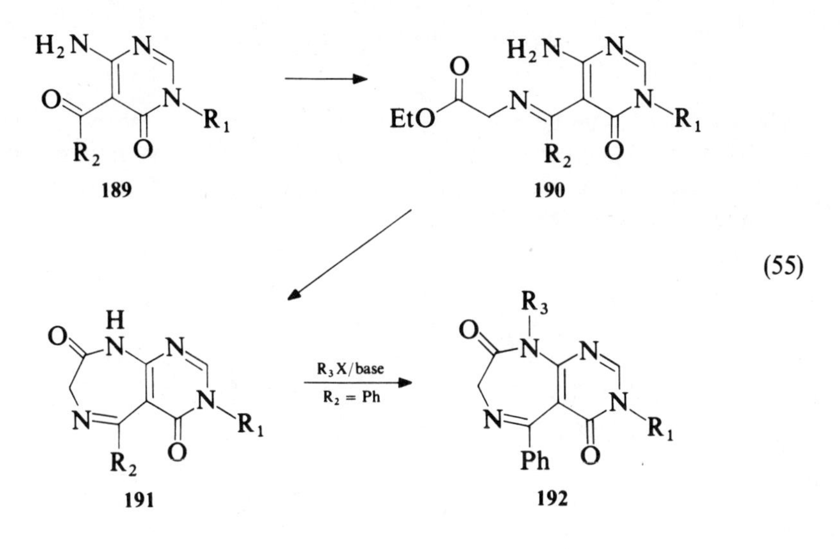


(55)

### 11.2. Pyrimido[5,4-*e*][1,4]diazepines

The only representatives of this heterocyclic system were described by Santilli and Scotese.[55] These investigators reacted the oxadiazinones **193** with ethylenediamine in boiling methanol (Eq. 56). Compounds **193** resulted from the addition of dimethyl acetylenedicarboxylate to the appropriate amidoxime and subsequent cyclization. Diazepines **194** (R = 4-$ClC_6H_4$, morpholinocarbonylmethyl, piperidinocarbonylmethyl) were reported.

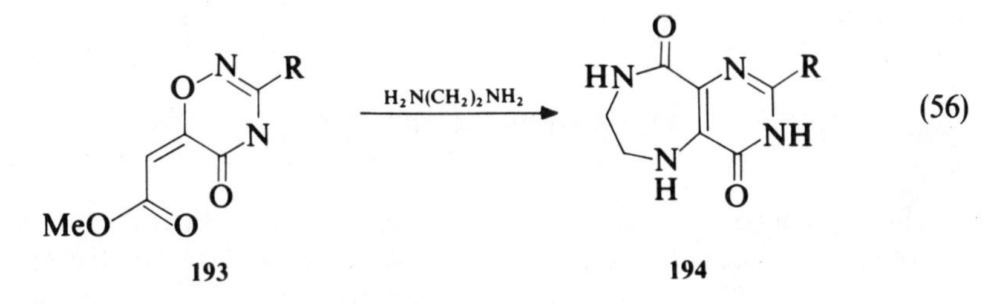


(56)

## 12. PYRROLO[*e*][1,4]DIAZEPINES

Partially saturated representatives of all three positional isomers have been synthesized.

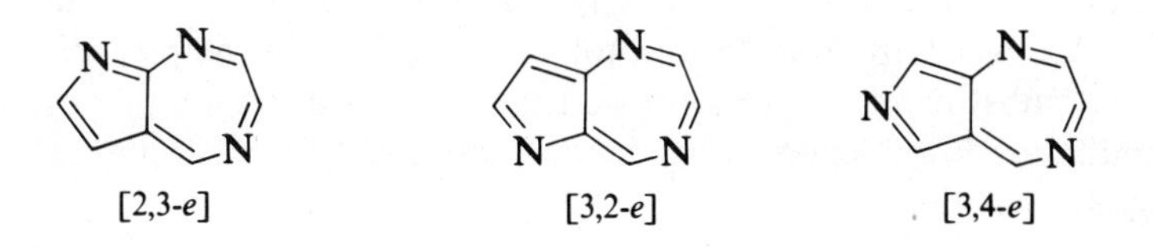

### 12.1. Pyrrolo[2,3-*e*][1,4]diazepines

The octahydro-5-one **196** was obtained by refluxing the ester **195** with ethylenediamine in ethanol for 16 hours (Eq. 57).[38] The question of stereochemistry was not addressed, but the trans configuration may be assumed to be thermodynamically preferred.

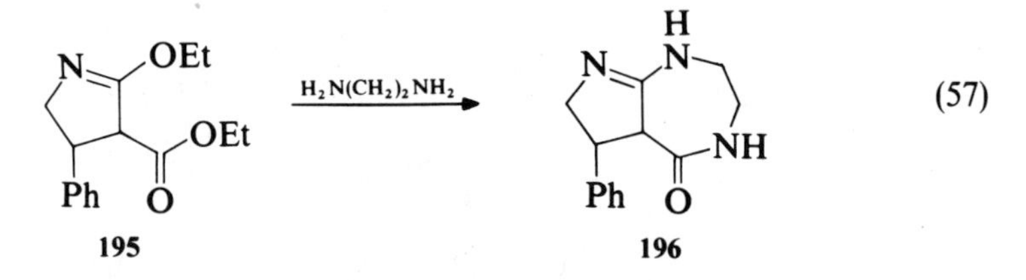


### 12.2. Pyrrolo[3,2-*e*][1,4]diazepines

Garcia[56] synthesized the 5-phenyl-2-ones **198** (R = CN, COOEt, $CONH_2$) from the corresponding aminopyrroles **197** by the now standard method of bromoacetylation and subsequent cyclization reaction with ammonia (Eq. 58).

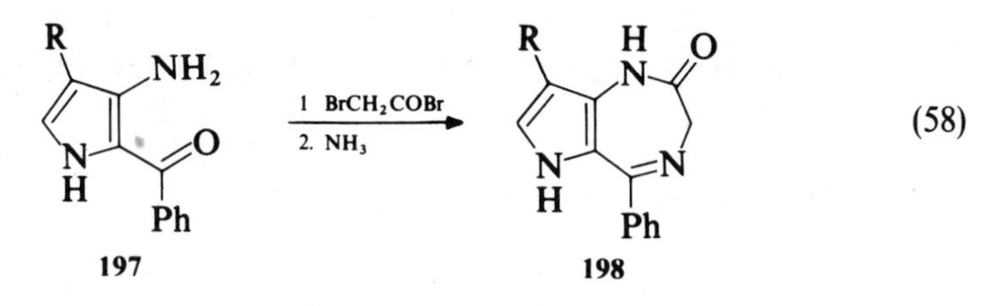


### 12.3. Pyrrolo[3,4-*e*][1,4]diazepines

This class of compounds has received much attention because of the interesting pharmacological properties of such 5-phenyl-2-ones as premazepam **202** ($R_1$ = Ph, $R_2$ = $R_3$ = Me). Fontanella and co-workers[57,58] prepared several substituted 2-ones **202** starting with the enamines **199**, which were cyclized to the unstable aminopyrroles **200** by treatment with sodium ethoxide (Eq. 59). The aminopyrroles were acylated with bromoacetyl chloride, bromide or phthalimidoacetyl chloride to give the amides **201** (X = Cl, Br, phthalimido). Hydrazinolysis of the phthalimido group led to the corresponding amine, which was thermally cyclized to the diazepine. We[59] prepared the same amines from the bromide via the azide **201** (X = $N_3$) by hydrogenation of the latter. Ring closure took place upon heating in toluene or xylene in the presence of acetic acid. The substituent $R_4$ was introduced by alkylation of the pyrrole **201** (X = phthalimido or azido) using methyl iodide and potassium carbonate in boiling methyl ethyl ketone.

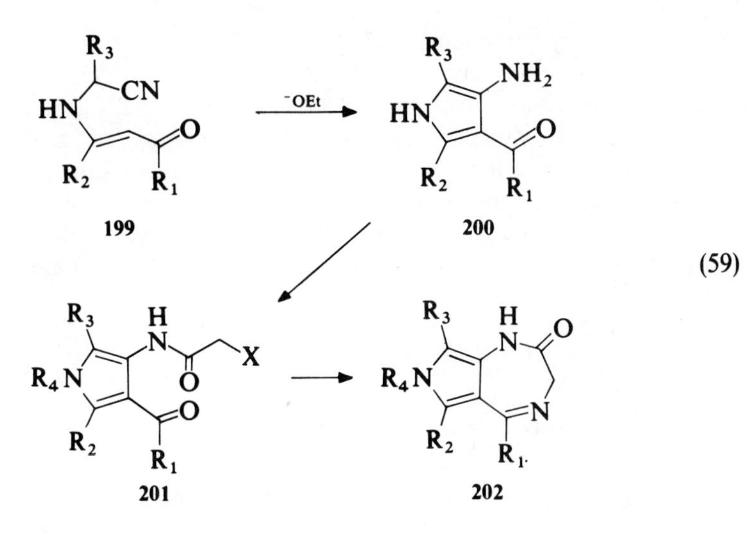


The 4-oxides of **202** were similarly prepared by reacting the iodoacetates **201** (X = I) with hydroxylamine.[60]

Halogenation or nitration of premazepam (**203**) afforded the 8-chloro, bromo, or nitro derivatives **204** (E = Cl, Br, $NO_2$) (Eq. 60).[61]

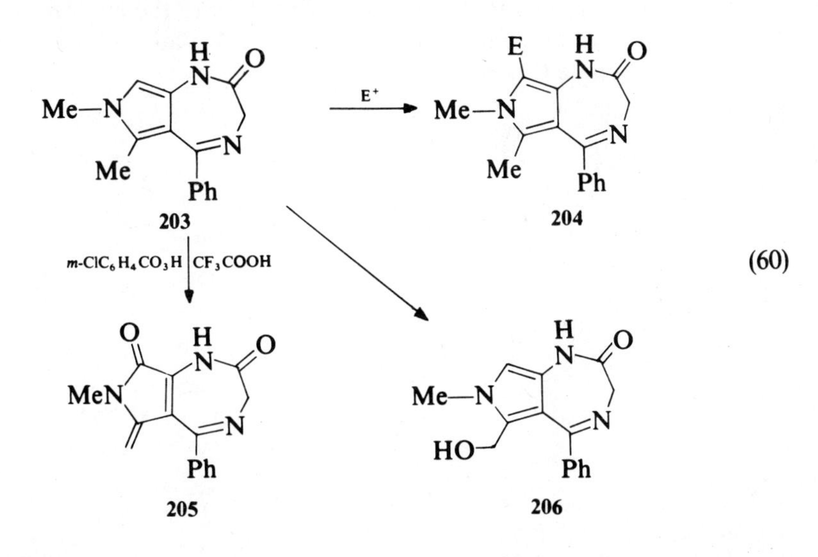


Treatment of the same compound with *m*-chloroperoxybenzoic acid in trifluoroacetic acid led to the methylenepyrrolinone **205**.[59] The hydroxymethyl derivative **206** was reported to be a metabolite of premazepam,[62] but was only characterized spectroscopically.[63]

Substituents such as methyl and ethyl groups were introduced at the 1-position of **202** by alkylation with an alkyl iodide in liquid ammonia containing sodium amide.[57,58]

The 3-hydroxy compounds **207** ($R_2$ = H) (Eq. 61) were obtained by hydrolysis of the 3-acetoxy analogs **207** ($R_2 = COCH_3$), which resulted in turn from the Polonovski rearrangement of the 4-oxides with acetic anhydride.[60] The 3-hydroxy group was acylated by a variety of acid chlorides.[60] The 3-alkoxy compounds **207** ($R_2$ = Me, Et) and the 3-amino derivative **208** [X = $NH(CH_2)_3COOEt$] were synthesized by converting the 3-hydroxy compounds[207] ($R_2$ = H) to the chloride **208** (X = Cl) by means of thionyl chloride and then displacing the chloride either by alkoxide or by an amine.[60]

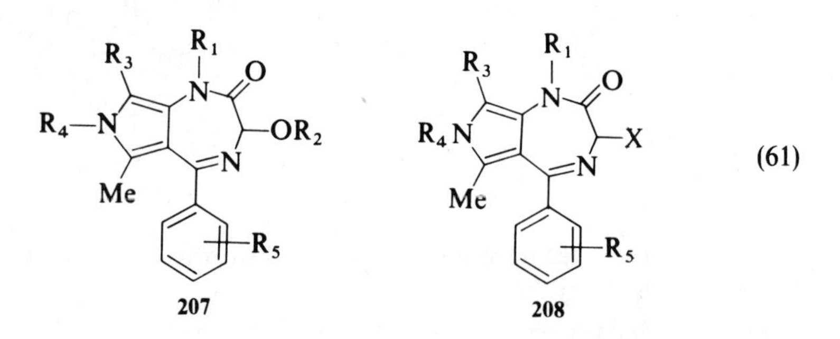


The lactams **209** (X = H, Cl) were transformed into the imidazodiazepines **210** by conversion to the iminophosphate and condensation with the anion of ethyl isocyanoacetate (Eq. 62).[59]

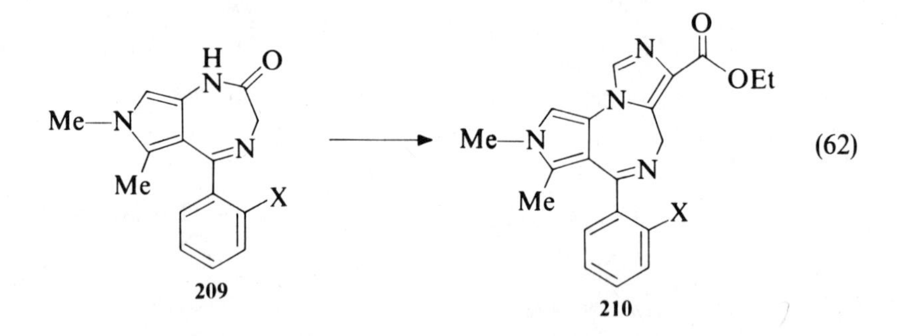


The 6,8-diones **212** (R = H, Ph) were prepared in 77% yield by reacting the maleimides **211** with ethylenediamine (Eq. 63).[64]

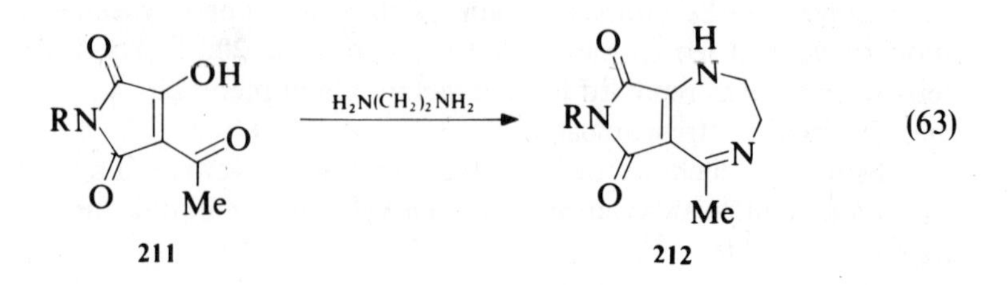

# 13. [1,2,5]-THIADIAZOLO[3,4-*e*][1,4]DIAZEPINES

Compound **214**, the only representative of this heterocycle, was prepared by Ning[65] by the bromoacetylation of 3-amino-4-benzoylthiadiazole (**213**) and subsequent reaction with ammonia (Eq. 64).

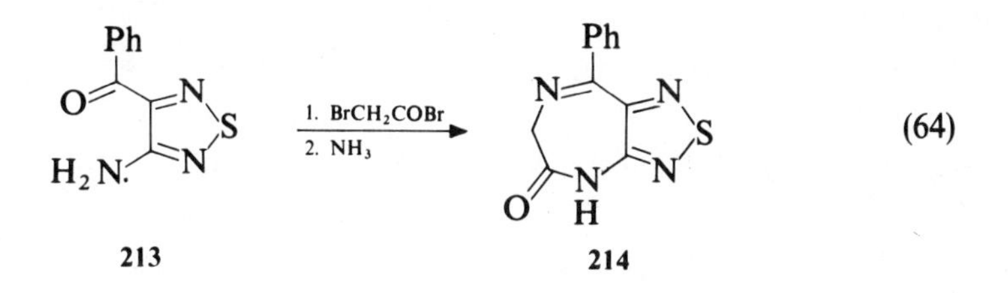


# 14. THIAZOLO[*e*][1,4]DIAZEPINES

Compounds belonging to either of the two possible thiazolo[*e*][1,4]diazepines illustrated were reported in the literature.

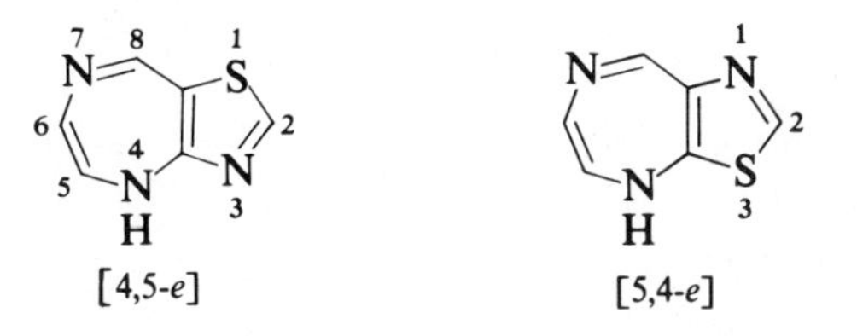


## 14.1. Thiazolo[4,5-*e*][1,4]diazepines

The oxathiolium salts **215** (X = H, Cl; $R_1$ = $Me_2N$, piperidino, morpholino, 4-$ClC_6H_4$) reacted with cyanamide in the presence of base to form the aminothiazoles **216** ($R_2$ = H) (Eq. 65).[66] Methylation of these compounds afforded the methylamino derivatives **216** ($R_2$ = Me). Among the several methods investigated, the condensation of the benzyloxycarbonyl-protected amino acid with **216** by means of thionyl chloride, followed by deprotection with hydrogen bromide in acetic acid and acid-catalyzed cyclization, proved to be best suited for the synthesis of the diazepinones **217**. Compound **217** ($R_1$ = morpholino, $R_2$ = $R_3$ = X = H) was converted to the thione **218** by phosphorus pentasulfide and pyridine in dichloromethane. Reaction of the thione with acetylhydrazine led to **219**, which was converted to the triazolo compound **220** by heating in acetic acid.[66]

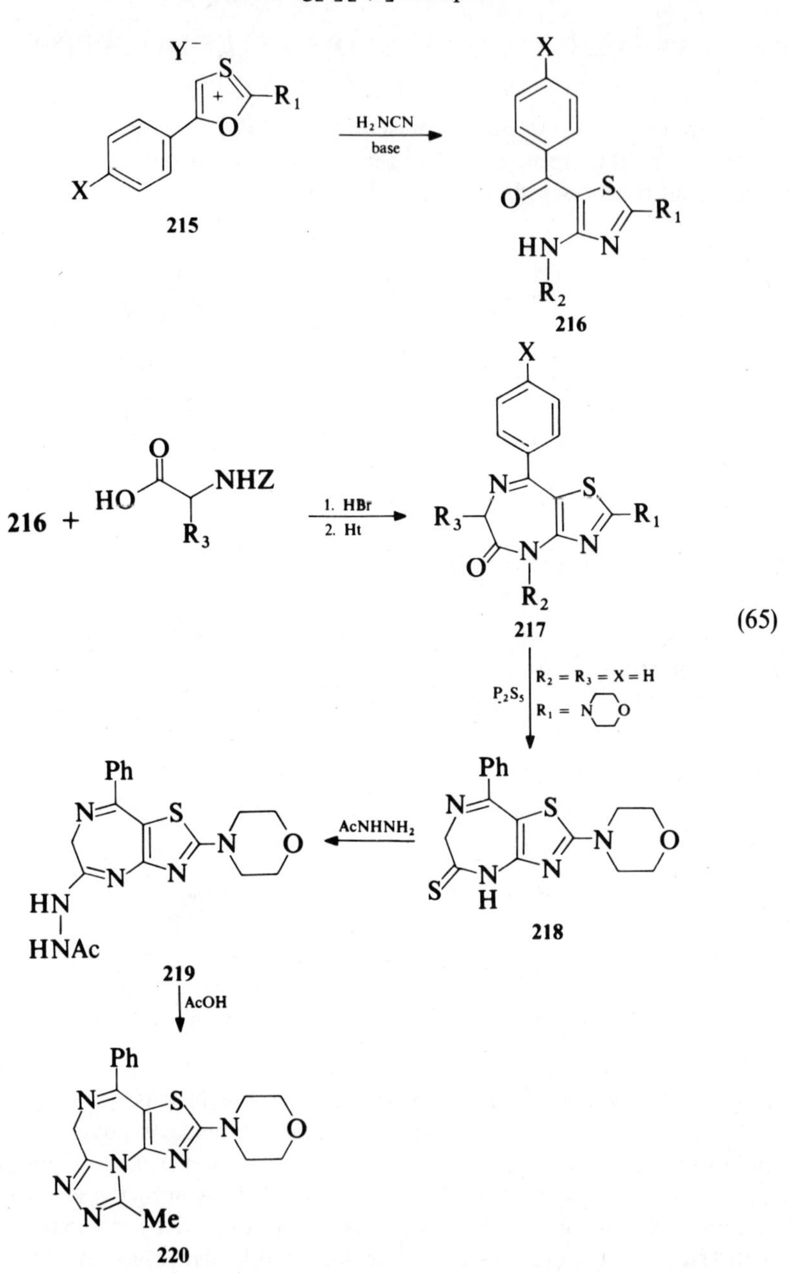


The thiazoline-2-thiones **221** (R = Me, Et) were obtained by reacting benzoylacetonitrile with methyl or ethyl isothiocyanate and sulfur in dimethylformamide in the presence of triethylamine.[67] The amino ketones **221** were transformed into the diazepinones **222** (R = Me, Et) by chloroacetylation and reaction with liquid ammonia (Eq. 66).[67]

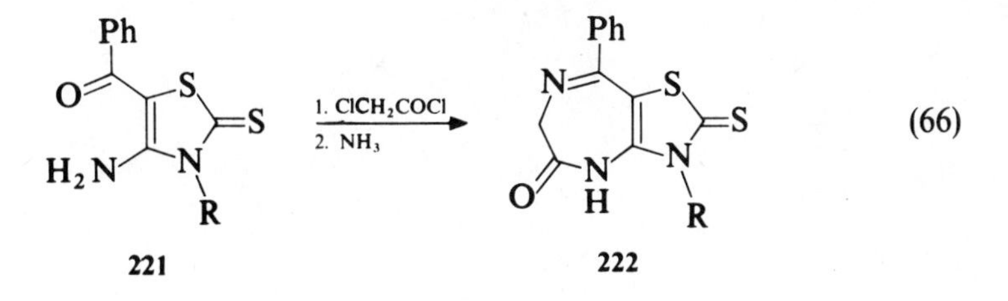


(66)

## 14.2. Thiazolo[5,4-*e*][1,4]diazepines

The 8-phenyl derivative **224** ($R_1$ = Ph, $R_2$ = H) was synthesized by Szente,[6] who reacted the imine **223** ($R_1$ = Ph) with bromoacetyl bromide in the presence of sodium bicarbonate (Eq. 67). The imine **223** resulted from the addition of phenyllithium to the corresponding nitrile. Fryer and Earley prepared the 2-thiazolyl analog **224** ($R_1$ = 2-thiazolyl) by the same method.[10] Methylation of **224** ($R_1$ = Ph, $R_2$ = H) with dimethyl sulfate and potassium carbonate afforded the 4-methyl derivative **224** ($R_2$ = Me), which was nitrated by nitric acid in concentrated sulfuric acid to yield **225**.[6]

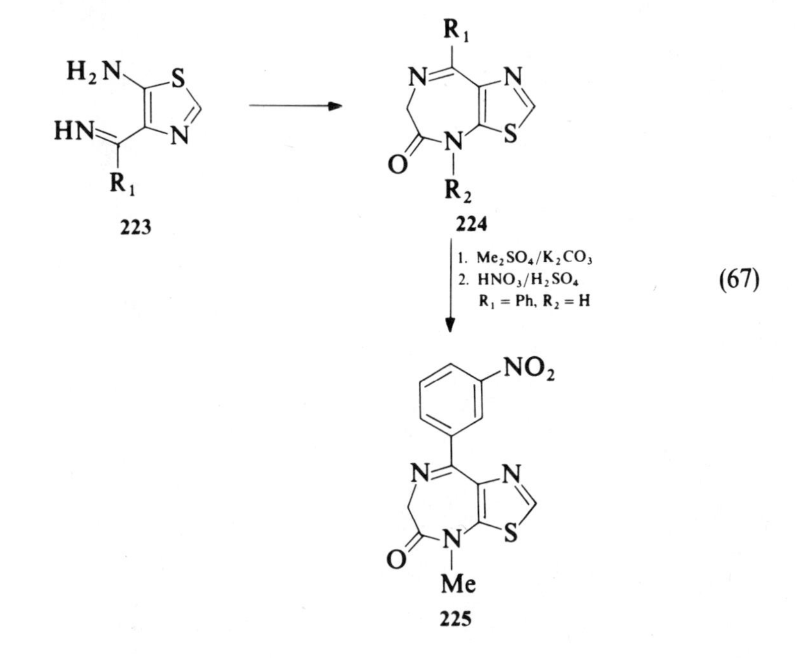


(67)

# 15. THIENO[*e*][1,4]DIAZEPINES

Representatives of all three positional isomers have been reported in the literature.

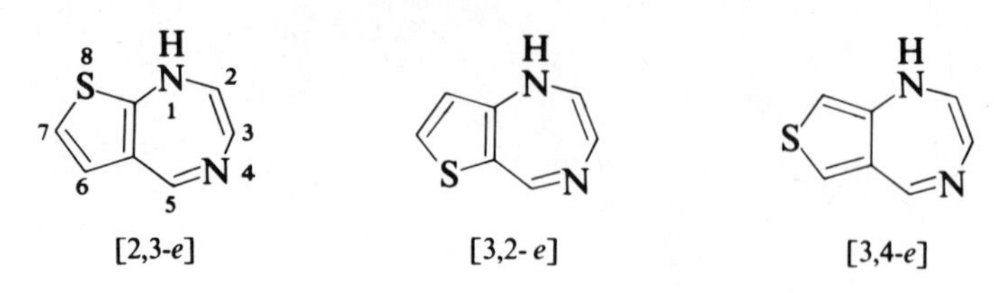


## 15.1. Thieno[2,3-*e*][1,4]diazepines

### *15.1.1. 3H-Thieno[2,3e][1,4]diazepines*

The 2-amino-3*H*-thieno[2,3-*e*] [1,4]diazepines **227** were prepared either by reacting the 2-ones **226** (X = O) with titanium tetrachloride and an amine,[68–71] or by treatment of the 2-thiones **226** (X = S) with an amine.[72] The 2-thiomethyl derivative **228** resulted from methylation of the appropriate 2-thione. Compound **227** [$R_1 = 2\text{-ClC}_6H_4$, $R_2 = R_4 = H$, $R_3 = Et$, $R_5 = CH_2CH(OEt)_2$)] was converted to the imidazothienodiazepine **229** (Eq. 68).[72,73]

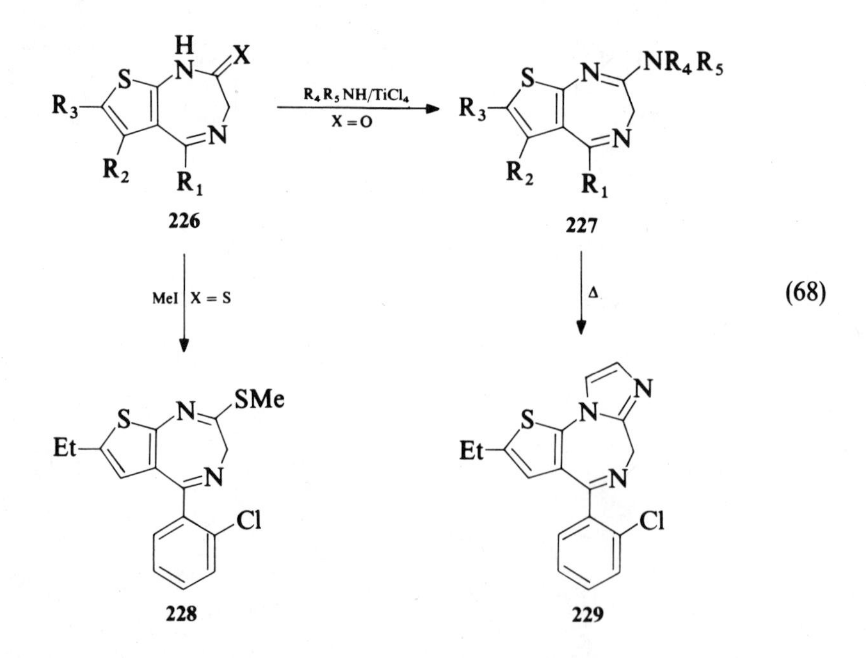


(68)

The nitrosoamidines **230** (X = H, Cl) accessible by nitrosation of the corresponding amidines, were reacted with the anions of nitromethane and dimethyl malonate to give the 2-methylene derivatives **231** ($R_1 = NO_2$, $R_2 = H$) and **231** ($R_1 = R_2 = COOMe$), respectively (Eq. 69).[16] The latter was hydrolyzed, decarboxylated, and nitrosated in acetic acid to yield the oxime **232**. Compound **232** was further converted to the imidazo compound **233** (R = H, Me) by hydrogenation over Raney nickel and subsequent condensation with triethyl orthoformate or acetate.[16]

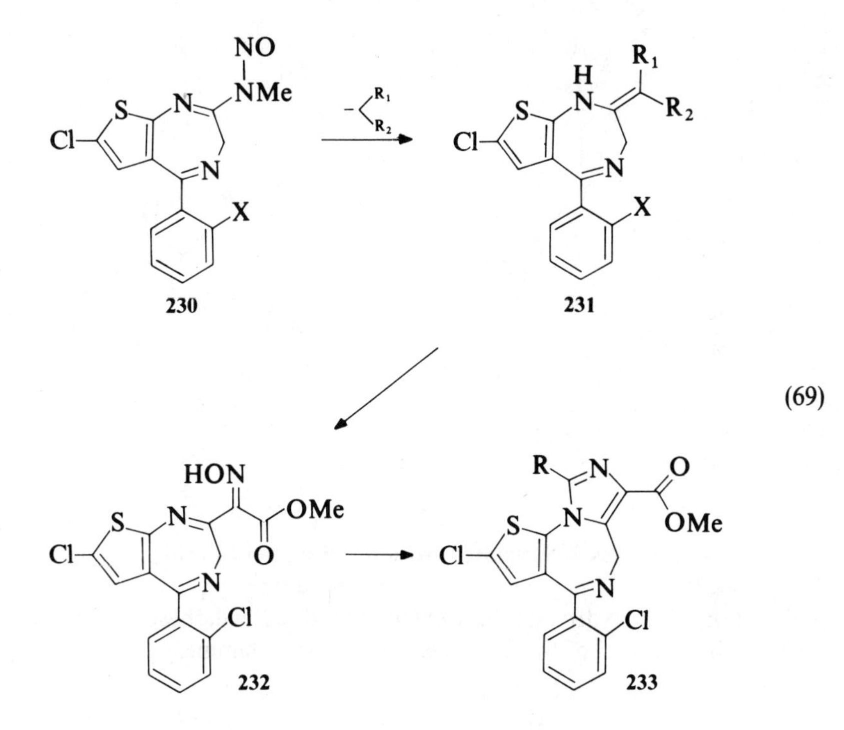


The 2-hydrazino compounds **234** ($R_4$ = H) were predominantly synthesized by treatment of the 2-thiones **226** (X = S) with hydrazine.[68,69,74–76] Displacement of a 2-amino group by a 2-hydrazino functionality was reported to succeed by a catalyzed reaction with *N*-methylimidazole.[69] 2-Hydrazino compounds have been demonstrated to form 1,2-disubstituted hydrazines upon boiling in ethanol–acetic acid.[68] Compounds **234** ($R_4$ = acyl) were similarly obtained from the reaction of 2-thiones with acylhydrazines or by acylation of the 2-hydrazino derivatives.[68,69,74,77,78] Many of these 2-hydrazino compounds **234** were converted to the triazolothienodiazepines **235**, by one of two routes: (a) the reaction with orthoesters, imidates, or amidines when $R_4$ = H[68,69,74,76] or (b) cyclization with dehydration ($R_4 = COR_5$) (Eq. 70).[68,69,74,77–79] The ethylcarbamate **234** ($R_1$ = 2-$ClC_6H_4$, $R_2$ = H, $R_3$ = Br, $R_4$ = COOEt) was cyclized to the triazolo analog **235** ($R_5$ = EtO) by heating in xylene in the presence of silica gel.[78] Phosgene in the presence of triethylamine converted the hydrazine **234** ($R_1$ = 2-$ClC_6H_4$, $R_2 = R_4$ = H, $R_3$ = Et) to the triazolone **237**,[75] while the tetrazolo derivative **236** was formed by nitrosation of the same hydrazine.

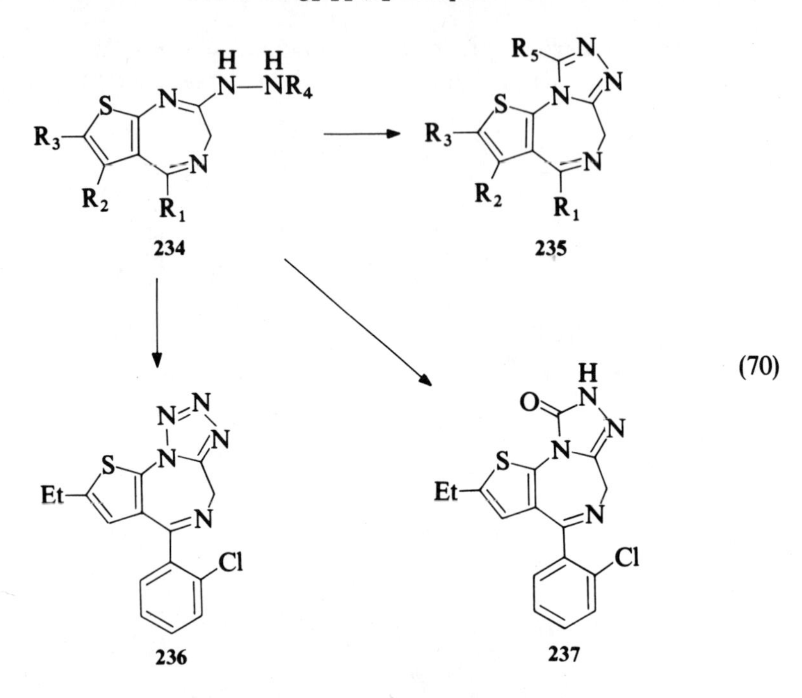


The oxadiazolones **239** resulted from the treatment of the 2-hydroxyamines **238** (R = Me, Et) with phosgene and triethylamine (Eq. 71).[73] Although the preparation of the hydroxyamines **238** was not described, these products may be obtained by reaction of the 2-thiones with hydroxylamine.

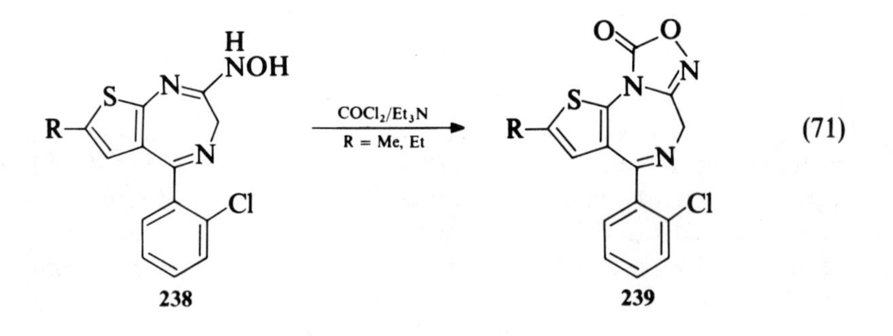


### *15.1.2. 2,3-Dihydro-1H-thieno*[*2,3-e*][*1,4*]*diazepines*

The 5-phenyl derivatives **241** were obtained in high yields by reduction of the 2-ones **240** with the sodium aluminium hydride reagent at room temperature (Eq. 72).[80,81] These compounds were reacted with a variety of isocyanates, aminocarbonyl chlorides, and thioisocyanates to form the 1-aminocarbonyl derivatives **242** (Y = O) or the thio analogs **242** (Y = S).

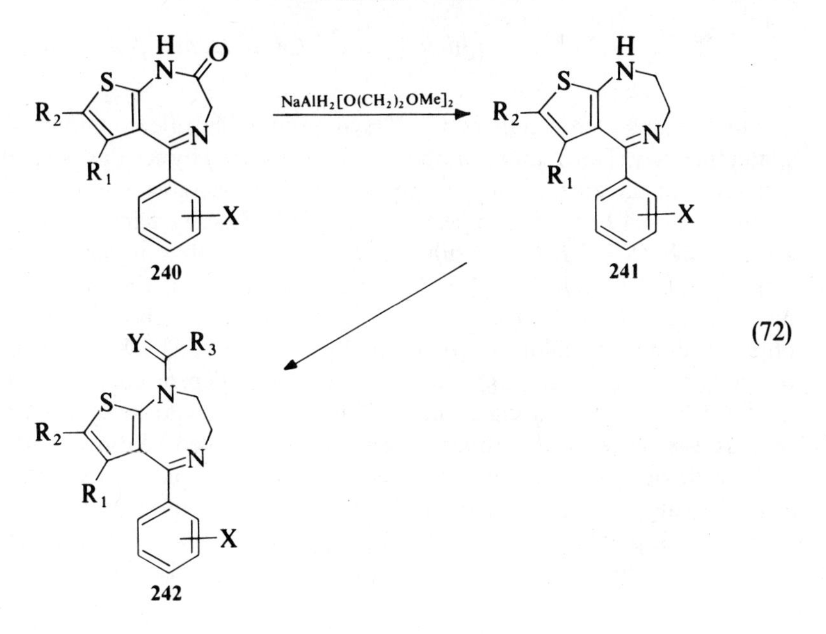


Lithium aluminum hydride reduced the 2-nitromethylene compound **243** to the 2-aminomethyl derivative **244**. Compound **244** was converted in low yield to the imidazothienodiazepine **245** by reaction with triethyl orthoacetate and subsequent oxidation of the intermediate imidazoline to the imidazole by activated manganese dioxide (Eq. 73).[16]

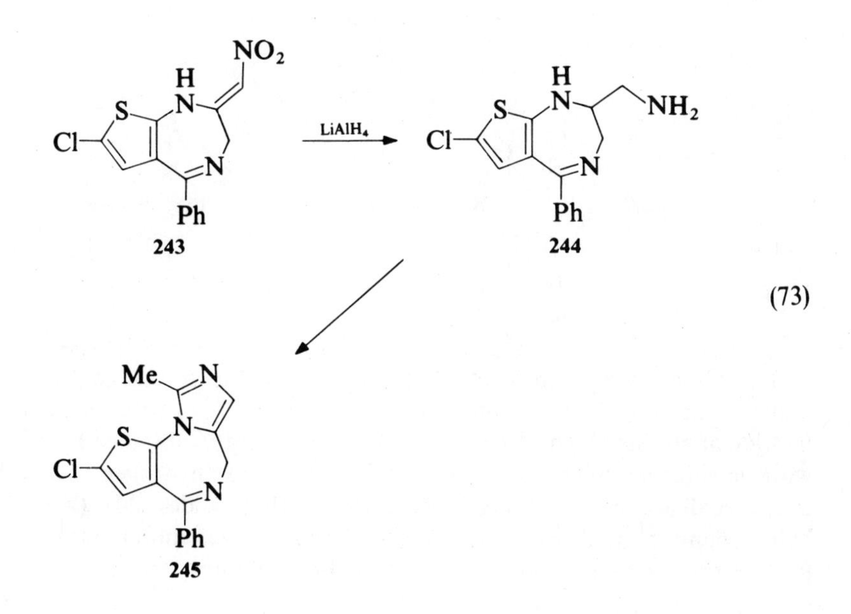

### *15.1.3. 1,3-Dihydrothieno[2,3-e][1,4]diazepin-2(2H)-ones*

Three teams of researchers[82,85,86] synthesized the title compounds almost simultaneously. The 2-aminothiophenes **246**, prepared by several methods, were converted to the diazepinones by the standard procedures worked out for the synthesis of 1,4-benzodiazepines. The methods described include haloacetylation to **247** (X = Br, I) and subsequent reaction with ammonia[82–85] hydrogenation of the azide **247** ($X = N_3$),[82] and coupling of the amine **246** with *N*-protected amino acids, such as benzyloxycarbonyl glycine or phthalimidoacetyl chloride to form the amides **247** (X = benzyloxycarbonylamino; phthalimido), which were then deprotected by standard methods.[82,86] The resulting amines **247** ($X = NH_2$) could be cyclized to compounds **248** by several methods. The methods included heating in ethanol or with acetic acid and pyridine in benzene with azeotropic removal of water,[86] and refluxing in ethanol in the presence of formic acid or heating in acetic, isobutyric, or pivalic acid (Eq. 74).[87] The application of a polymeric acrylic acid resin in a mixture of dioxane and ethylene glycol was also described.[88]

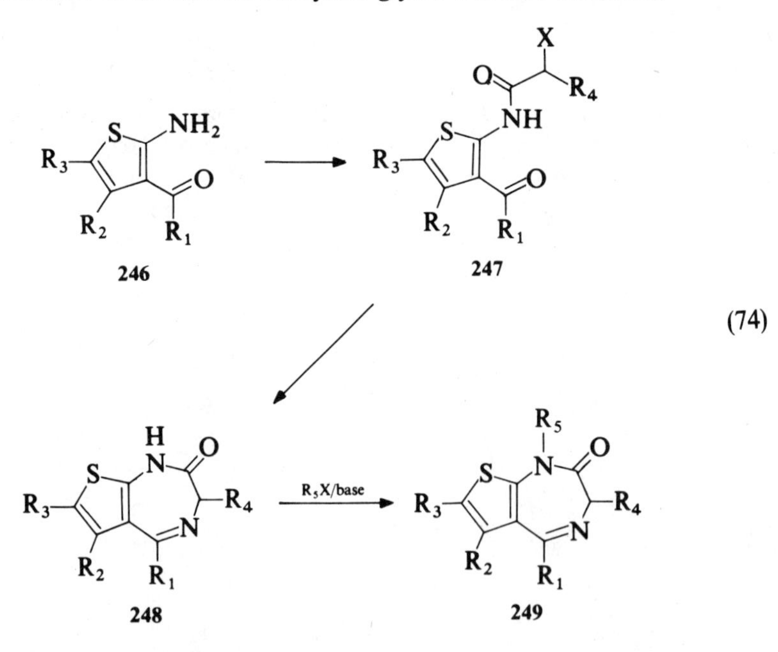


The thienodiazepinones **248** ($R_4$ = H) were also accessible by reaction of the aminothiophenes **246** with glycine ethyl ester[89,90] or with the Leuch's anhydride of glycine.[91] The 4-oxides of **248** were generally obtained by oxidation with *m*-chloroperoxybenzoic acid,[85,92,93] or with hydrogen peroxide in acetic acid–acetic anhydride.[94] Reaction of the iodides **246** (X = I) with hydroxylamine and ring closure under acid catalysis was another viable method for the preparation of the 4-oxides of **248**.[82] In turn, the 4-oxides of **248**

underwent the usual rearrangement to the 3-acetoxy derivatives **248** ($R_4$ = OAc) upon heating in acetic anhydride.[82,94] Hydrolysis of the acetoxy group afforded the corresponding 3-hydroxy analogs.[82,94] A variety of substituents were introduced at the 1-position by alkylation. The $R_5$ residues of **249** thus attached include the methyl, ethyl, isopropyl, butyl, benzyl, propargyl, allyl, cyclopropylmethyl, methoxymethyl, 2-hydroxyethyl, and 2-diethylaminoethyl groups.[82,85–87,89,92,94–99] A methylaminocarbonyl group was attached at the 1-position of **248** ($R_1$ = 2-$ClC_6H_4$, $R_2$ = $R_4$ = H, $R_3$ = Et) by reaction with methyl isocyanate to give the corresponding derivative **249** with $R_5$ = CONHMe.[100]

Electrophilic reagents attacked **250** ($R_4$ = H) at the 7-position. Thus, halogenation of these thienodiazepines with sulfuryl chloride in acetic acid or with bromine in chloroform led to the 7-halo derivatives **251** (X = Cl, Br).[83–85,88,92,97] The 7-nitro analogs were similarly obtained by nitration in concentrated sulfuric acid.[83,85,92,97] Iodine in the presence of mercuric oxide allowed the iodination of **250** ($R_4$ = H) to give the 7-iodo analogs **251** (X = I).[97] If the 7-position in **250** was occupied, nitration would occur at the 6-position, leading to **252** (X = $NO_2$).[88,95] Reaction of **250** ($R_1$ = $R_2$ = H, $R_3$ = 2-$ClC_6H_4$, $R_4$ = Et) with chloramine or 2,4-dinitrophenoxyamine and sodium hydride in dimethylformamide gave in high yield the 1-amino derivative **253** (Eq. 75).[100]

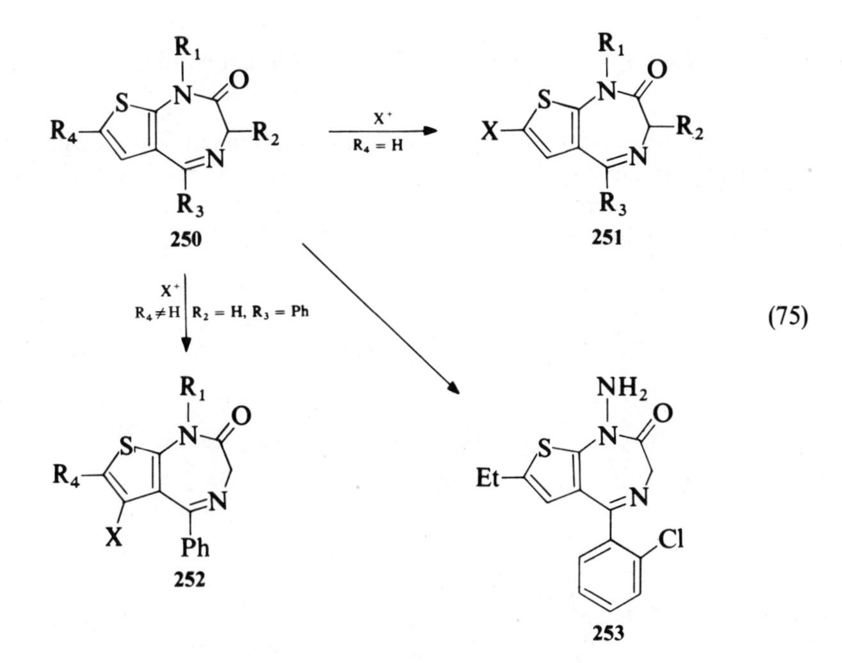


The hydrolytic ring opening of etizolam (**254**) to the amino ketone **255** was investigated and compared with that of triazolobenzo- and triazolothienodiazepines.[101] Etizolam was found to be little affected by treatment at 37°C for

100 minutes in 0.1$N$ hydrochloric acid. The p$K_a$ of this compound was given as 4.11.[101]

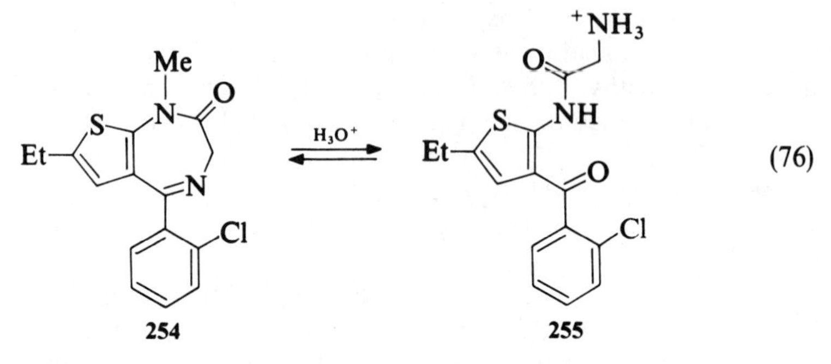


(76)

Nmr spectroscopy of thienodiazepinones **250** ($R_1$ = H, Me; $R_2$ = H; $R_3$ = Et, Ph, 2-halophenyl; $R_4$ = H, Cl, $NO_2$) and their 4-oxides revealed that the chemical shift of the proton in the 6-position depended on the bulkiness of the ortho substituent on the 5-phenyl ring. The degree of shielding was determined not only by the out-of-plane orientation of the phenyl ring but also by the ring system.[93] With 5-(2-fluorophenyl)-substituted compounds, a long-range coupling of the fluorine with the proton at the 6-position was observed. The value of this coupling constant depended on the spatial relation of the two nuclei.[102]

### *15.1.4. 3,4-Dihydrothieno[2,3-e][1,4]diazepin-5(5H)-ones*

The 2-amino derivative **259** was prepared by treatment of the nitrile **258** with sodium methoxide (Eq. 77). The 2-aminothiophene **258** resulted from the reaction of dimeric mercaptoacetaldehyde **256** with the cyanoacetamide **257** in refluxing ethanol containing triethylamine (Eq. 77).[103]

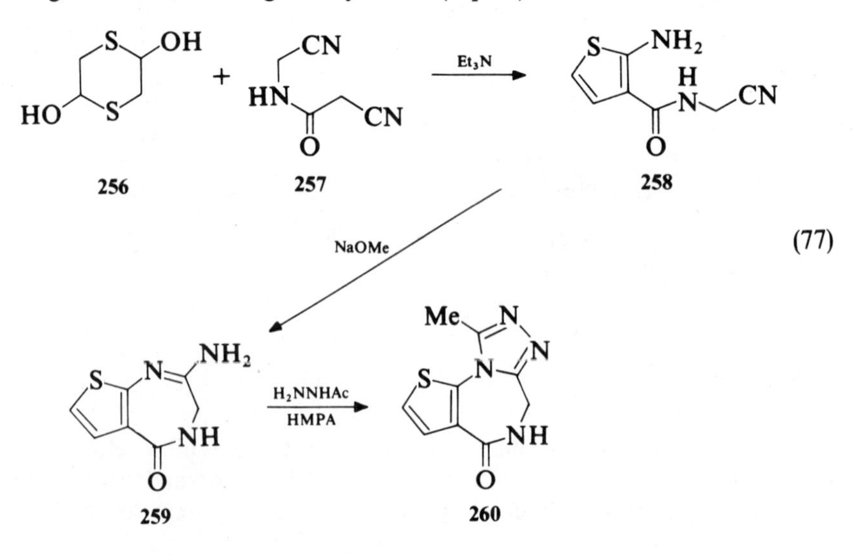


(77)

The diazepine **259** was converted to the triazolo derivative **260** by heating with acetylhydrazine in hexamethylphosphoric triamide for 2 hours at 140°C.[103]

## 15.2. Thieno[3,2-*e*][1,4]diazepines

### *15.2.1. 3H-Thieno[3,2-e][1,4]diazepines*

The 2-methylamino compound **262** (R = H) was obtained by amination of the lactam **261** with methylamine and titanium tetrachloride. It was nitrosated to form the nitrosoamidine **262** (R = NO), which was reacted with the anion of nitromethane to yield the 2-nitromethylene derivative **263** (Eq. 78). The 2-aminomethyl derivative **264** was obtained from the catalytic hydrogenation of the nitro compound **263** over Raney nickel and was characterized as a maleate salt.[16] Compound **264** was further converted to an imidazothienodiazepine by condensation with triethyl orthoacetate and subsequent oxidation with activated manganese dioxide.

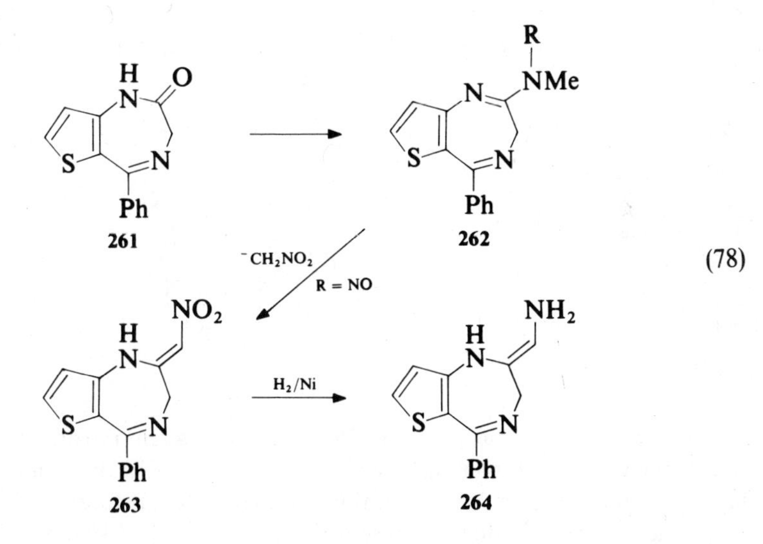


(78)

### *15.2.2. 1,3-Dihydrothieno[3,2-e][1,4]diazepin-2(2H)-ones*

The 2-ones **271** were synthesized by haloacetylation of the 3-aminothiophenes **269** to give **270** (X = Cl, Br, I). Displacement of the halide by ammonia, followed by ring closure, afforded **271** as a product.[66,104–108] The required aminothiophenes were synthesized by the three methods shown in Eq. 79. Compounds **269** ($R_1$ = Ph; $R_2$ = H, $CF_3$; $R_3$ = H) were obtained by treatment of the oxazinone **265** with aryl Grignard reagents and subsequent hydrolysis.[104,105] Friedel–Crafts acylation of the thiophenes **266** ($R_2$ = H, Cl) provided another access to the aminothiophenes.[106,108] The 3-amino-4-cyanothiophenes **269** ($R_1$ = Ph; $R_2$ = morpholino, piperidino; $R_3$ = CN) were derived from the 1,3-oxathiolium ions **267**. Reaction of these ions with malononitrile in the presence of triethylamine may proceed via the intermediate **268**.[66]

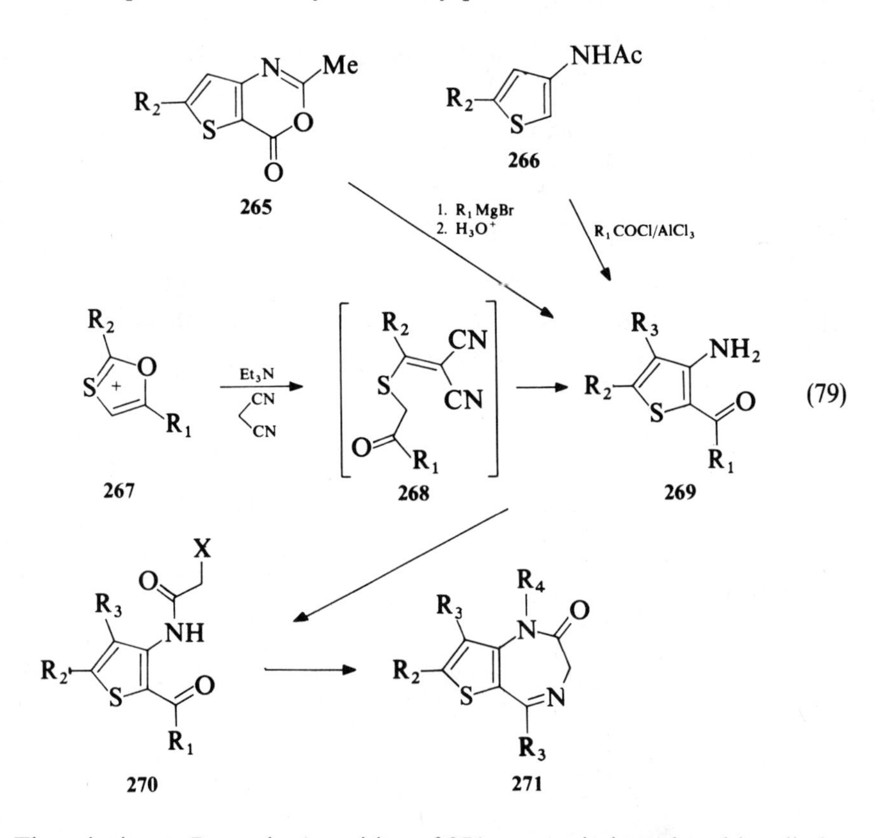


The substituent $R_4$ at the 1-position of **271** was again introduced by alkylation of the parent compound ($R_4$ = H).[104,106,107,109] The possibility of alkylating the amino group of the aminothiophene **269** before establishing the diazepine ring also existed. Because of the propensity of *N*-methylated acetamides to undergo a Smiles-type rearrangement during ammonolysis or hydrazinolysis, the diazepine ring was best formed by the benzyloxycarbonyl glycine method.[66] The 4-oxides of **271** can be obtained by the standard oxidation with *m*-chloroperoxybenzoic acid, as demonstrated for **261**.[104] Nitration of **261** with nitric acid in concentrated sulfuric acid at 0–5°C led to the 8-nitro derivative **272** (Eq. 80). The location of the nitro group was established by hydrolysis of **272** to the ketone **274** and removal of the amino group from **274** by diazotization and treatment with hypophosphoric acid. The resulting 2-benzoyl-4-nitrothiophene was compared with an authentic sample.[109] Nitration under more vigorous conditions led to the dinitro compound **273**. The position of the nitro group in the phenyl ring was proved by oxidative degradation to 3-nitrobenzoic acid by permanganate.

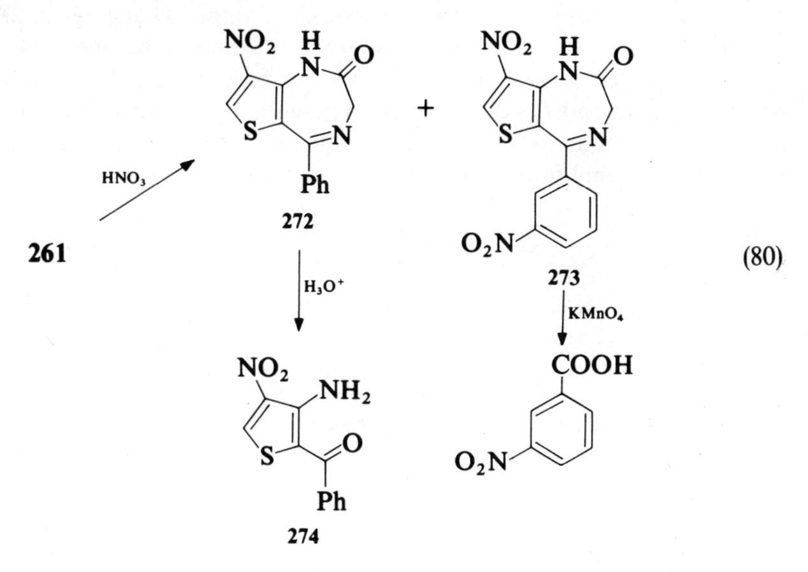


(80)

### *15.2.3. 3,4-Dihydro-1H-thieno[3,2-e][1,4]diazepin-2,5(2H,5H)-diones*

The 4-methyl compound **276** was prepared by heating the oxazinone **275** with sarcosine ethyl ester in dimethyl sulfoxide for 1.5 hours at 110°C (Eq. 81).[110] This compound was further transformed into the imidazothienodiazepine as shown in Eq. 84 for the [3,4-*e*]-fused isomer.

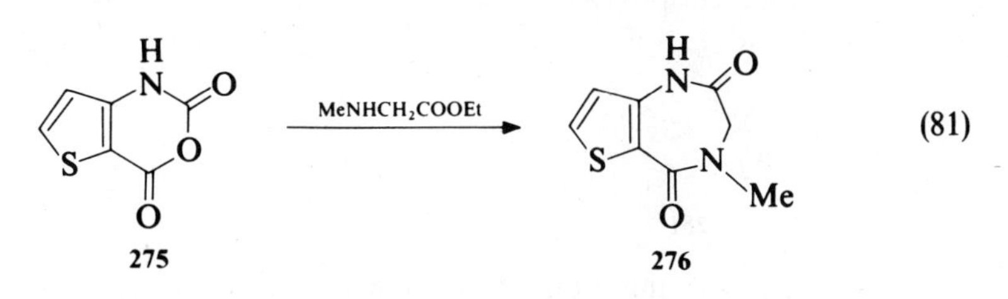


(81)

## 15.3. Thieno[3,4-*e*][1,4]diazepines

### *15.3.1. 1,3-Dihydrothieno[3,4-e][1,4]diazepin-2(2H)-ones*

The key to the synthesis of the diazepines **281** was again the preparation of the aminothiophenes **280**. Hromatka and coworkers obtained **280** ($R_1$ = Ph, 2-$F_3CC_6H_4$, $R_2 = R_3$ = H) by reaction of the oxazinone **277** with the aryl

Grignard reagent (Eq. 82).[111–113] The 3-amino-4-benzoylthiophene **280** ($R_1$ = Ph, $R_2$ = $R_3$ = H) was also accessible by Friedel–Crafts acylation, by reacting the acid chloride **278** with aluminum chloride in benzene.[111] This method was later applied to prepare several 2,5-dimethyl analogs of **280** from the thiophene **279**.[114] The conversion of the amino ketones **280** to the diazepinones was carried out in the standard fashion by haloacetylation and subsequent ammonolysis of the halide, preferentially the iodide. Ring closures of intermediate aminoacetamides occurred upon heating in pyridine[114] or refluxing in ethanol in the presence of pivalic acid.[112,113]

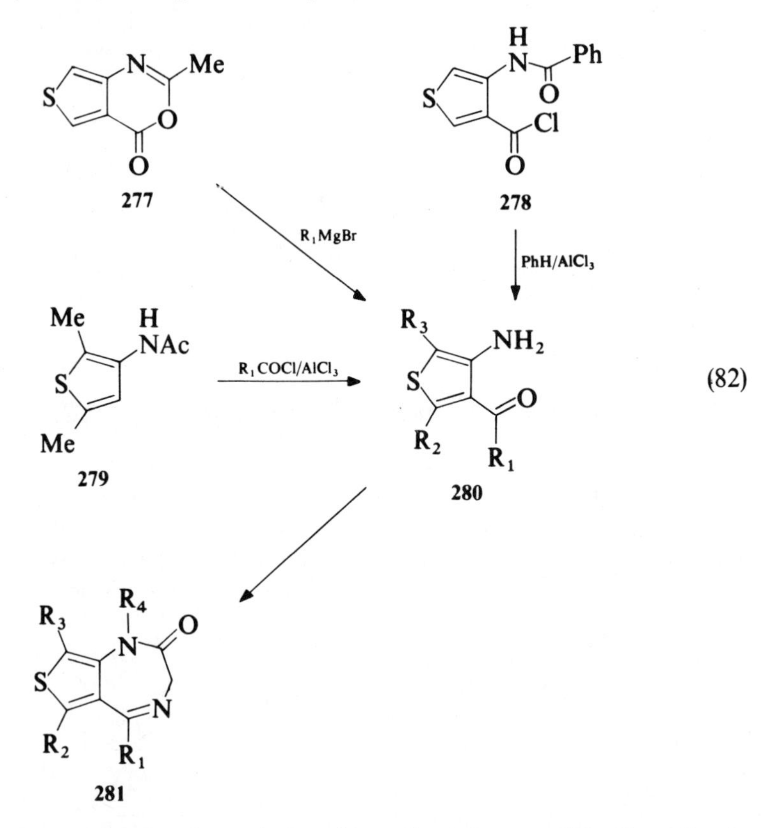


(82)

Chlorination and nitration of the thienodiazepinones **282** (R = Ph, 2-$F_3CC_6H_4$) were shown to occur at the 8-position, leading to **283** (X = Cl, $NO_2$) (Eq. 83).[112,115]

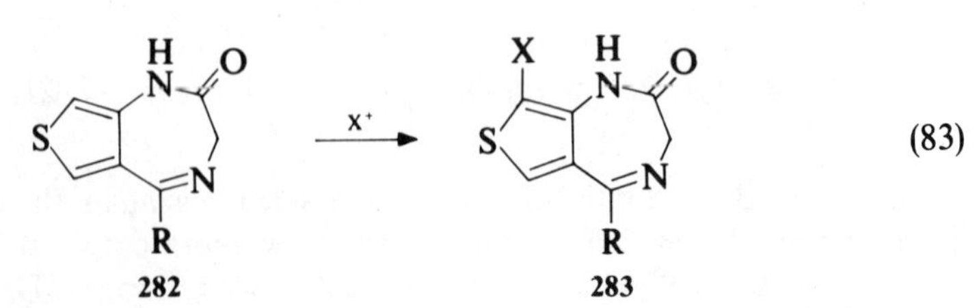


(83)

### *15.3.2. 3,4-Dihydro-1H-thieno[3,4-e][1,4]diazepine-2,5(2H,5H)-diones*

Compound **285**, the only representative of this class of compounds reported in the patent literature,[110] was prepared by thermal cyclization of the methylamino derivative **284** (X = NHMe) (Eq. 84). This amine was obtained by treatment of the iodide (X = I) with methylamine. The dione **285** was transformed into the imidazo derivative **287** via the intermediate iminophosphate **286** by means of ethyl isocyanoacetate.[110]

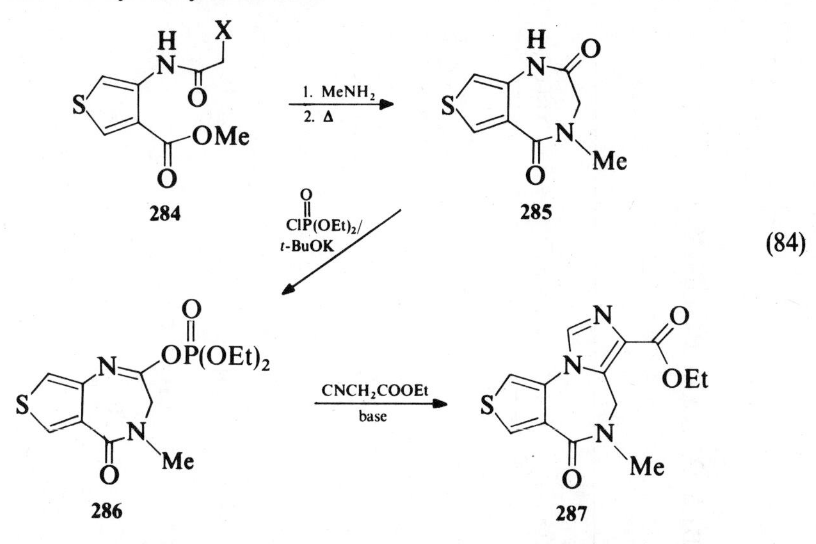


(84)

## 16. [1,2,4]TRIAZINO[5,6-*e*][1,4]DIAZEPINES

The 9-one **290** (Eq. 85) appears to be only derivative of the parent ring **288** that is described in the literature.[116] This triazinodiazepine was obtained by reaction of the triazine **289** with symmetrical *N,N′*-dimethylethylenediamine.

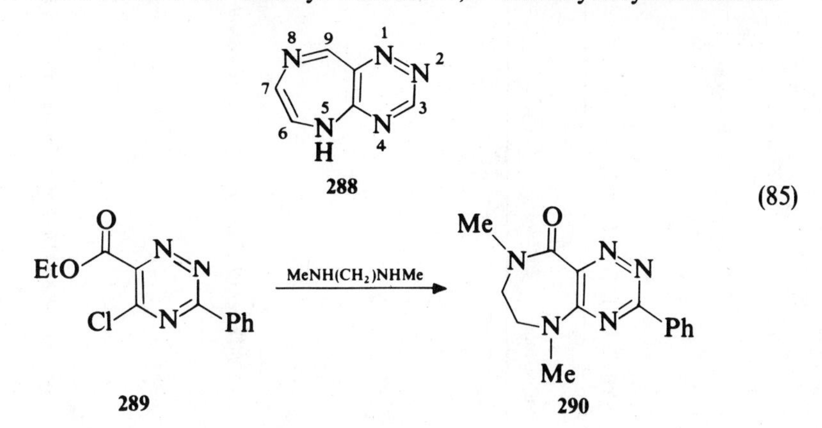


(85)

# 17. TABLE OF COMPOUNDS

TABLE IX-1. [*e*]-FUSED[1,4]DIAZEPINES

| Substituent | mp (°C) or; [bp (°C)/(torr)] | Solvent of Crystallization | Yield (%) | Spectra | Refs. |
|---|---|---|---|---|---|
| | 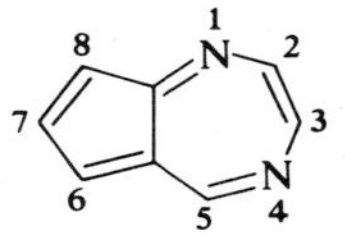 | *Cyclopenta[e][1,4]diazepines* | | | |
| ***3,6,7,8-Tetrahydrocyclopenta[e][1,4]diazepin-2(1H)-ones*** | 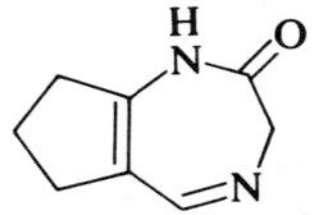 | | | | |
| 5-Ph | 195–197 | $CH_2Cl_2/Et_2O$ | | | 1 |
| ***3,4,5,5a,6,7,8,8a-Octahydrocyclopenta[e][1,4]diazepin-2(1H)-ones*** | 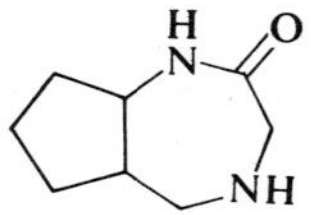 | | | | |
| 5-Ph | 201–203 | MeOH | | | 1 |
| 4-Ac-5-Ph | 194–196 | $CH_2Cl_2/Et_2O$ | | | 1 |

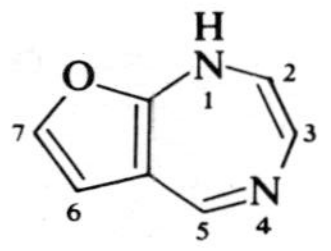

***1,2,3,4,6,7-Hexahydrofuro[2,3-e][1,4]diazepin-5(5H)-ones***

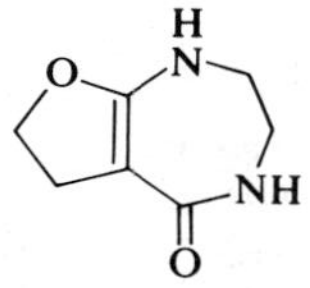

| Substituent | mp [bp/mm] | Solvent | Yield (%) | Spectra | Ref. |
|---|---|---|---|---|---|
| None | 155–158 | EtOH | 45 | pmr, uv | 3 |
| 7-Me | 146–149 | EtOH | 69 | pmr, uv | 3 |
| 1,4-$Ac_2$ | 120–125 | EtOH/Petr ether | | | 3 |
| 6,7-$Me_2$ | 166–170 | EtOH | 60 | pmr, uv | 3 |
| | 168–170 | EtOH | 31 | pmr, uv | 2 |
| 7,7-$Me_2$ | 206–210 | EtOH | 63 | pmr, uv | 3 |

***2,3,4,5,5a,6,7,8a-Octahydro-1H-furo[2,3-e][1,4]diazepines***

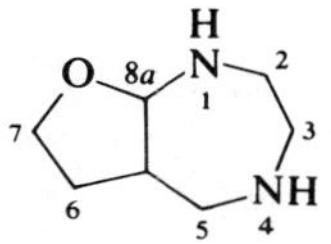

| Substituent | mp [bp/mm] | Solvent | Yield (%) | Spectra | Ref. |
|---|---|---|---|---|---|
| 1,4-$Bu_2$, *cis/trans* | [82–84/0.003] | | 35 | ms, pmr | 4 |
| 1,4-$Me_2$, *cis* | [110–111/28] | | 41 | ms, pmr | 4 |
| 1,4, 8*a*-$Me_3$ *cis/trans* | [115–120/15] | | 20 | pmr | 4 |

TABLE IX-1. —(contd.)

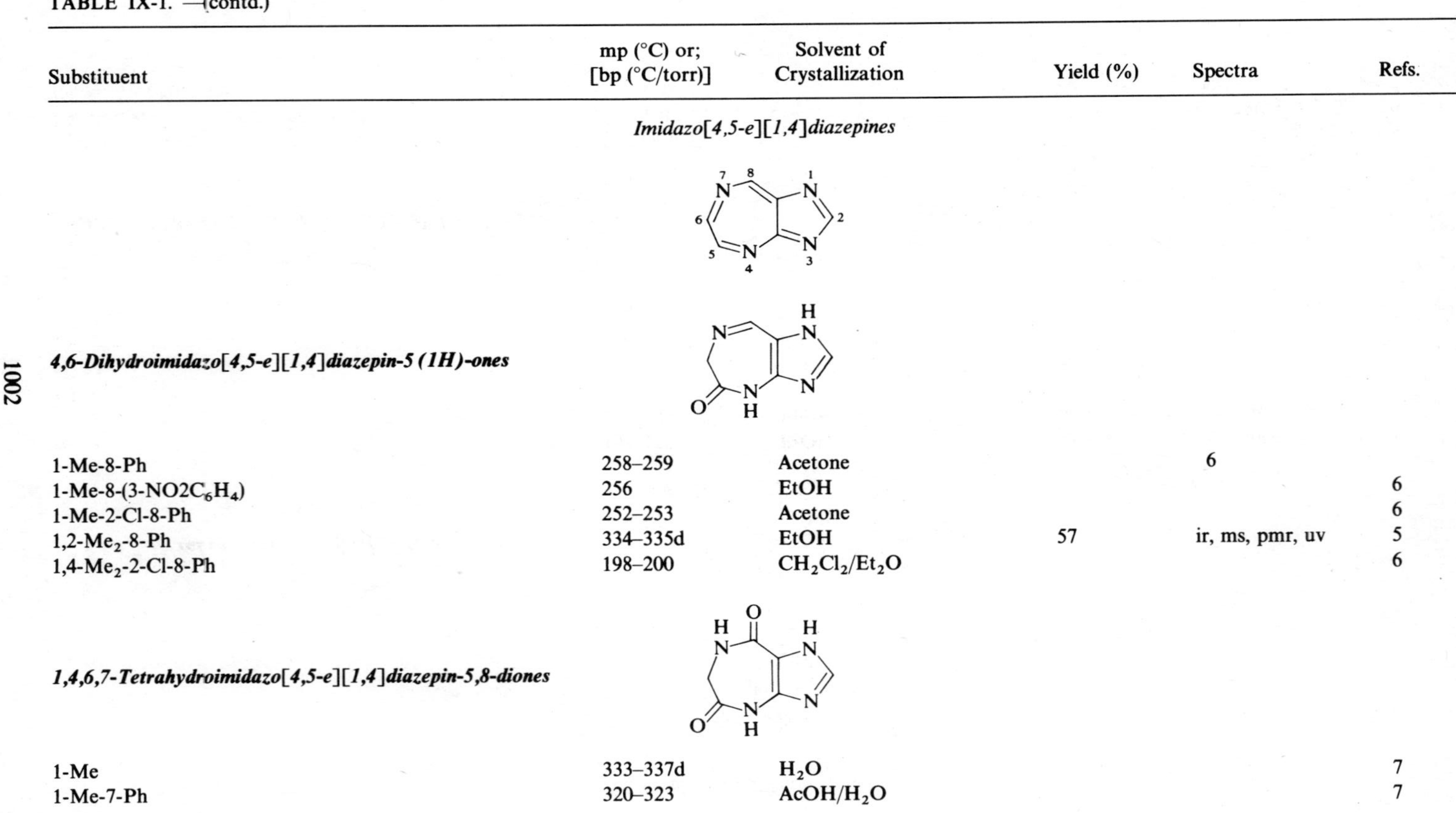

| Substituent | mp (°C) or; [bp (°C/torr)] | Solvent of Crystallization | Yield (%) | Spectra | Refs. |
|---|---|---|---|---|---|
| *Imidazo[4,5-e][1,4]diazepines* | | | | | |
| ***4,6-Dihydroimidazo[4,5-e][1,4]diazepin-5 (1H)-ones*** | | | | | |
| 1-Me-8-Ph | 258–259 | Acetone | | 6 | |
| 1-Me-8-(3-NO2$C_6H_4$) | 256 | EtOH | | | 6 |
| 1-Me-2-Cl-8-Ph | 252–253 | Acetone | | | 6 |
| 1,2-$Me_2$-8-Ph | 334–335d | EtOH | 57 | ir, ms, pmr, uv | 5 |
| 1,4-$Me_2$-2-Cl-8-Ph | 198–200 | $CH_2Cl_2/Et_2O$ | | | 6 |
| ***1,4,6,7-Tetrahydroimidazo[4,5-e][1,4]diazepin-5,8-diones*** | | | | | |
| 1-Me | 333–337d | $H_2O$ | | | 7 |
| 1-Me-7-Ph | 320–323 | $AcOH/H_2O$ | | | 7 |

*Isoxazolo[4,3-e][1,4]diazepines*

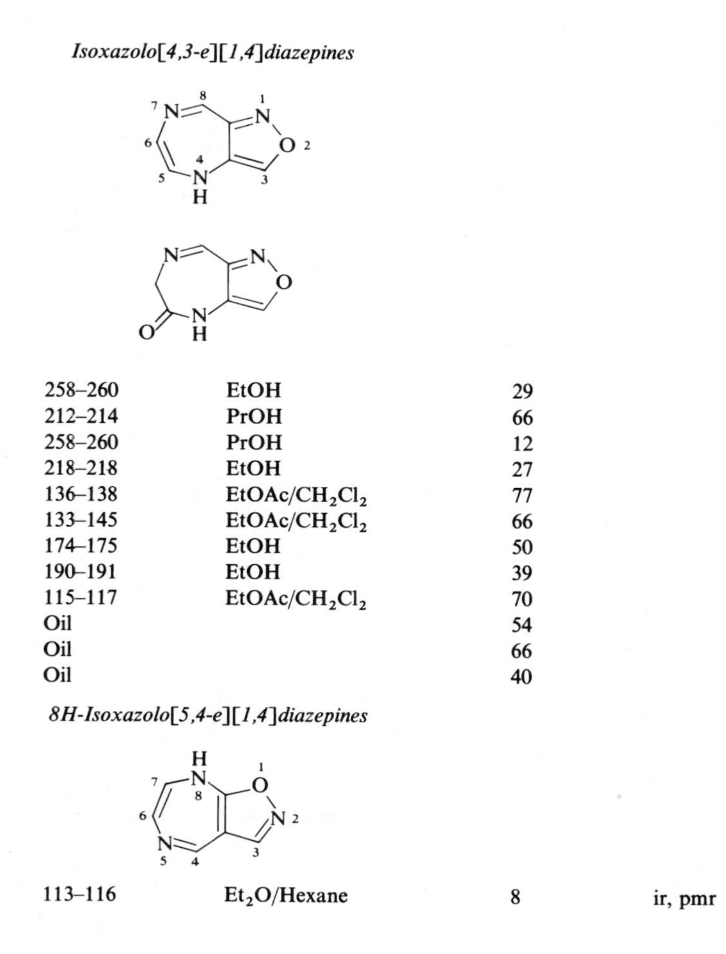

***4,6-Dihydroisoxazolo[4,3-e][1,4]diazepin-5(5H)-ones***

| | | | | | |
|---|---|---|---|---|---|
| 3-Me-8-(2-$ClC_6H_4$) | 258–260 | EtOH | 29 | | 8 |
| 3-Me-8-(2-$FC_6H_4$) | 212–214 | PrOH | 66 | | 8 |
| 3-Me-8-(2,6-$F_2C_6H_3$) | 258–260 | PrOH | 12 | | 8 |
| 3-Me-8-(2-$F_3CC_6H_4$) | 218–218 | EtOH | 27 | | 8 |
| 3,4-$Me_2$-8-(2-$ClC_6H_4$) | 136–138 | EtOAc/$CH_2Cl_2$ | 77 | | 8 |
| 3,4-$Me_2$-8-(4-$ClC_6H_4$) | 133–145 | EtOAc/$CH_2Cl_2$ | 66 | | 8 |
| 3,4-$Me_2$-8-(2-$FC_6H_4$) | 174–175 | EtOH | 50 | | 8 |
| 3,4-$Me_2$-8-(2,6-$F_2C_6H_3$) | 190–191 | EtOH | 39 | | 8 |
| 3,4-$Me_2$-8-(2-$F_3CC_6H_4$) | 115–117 | EtOAc/$CH_2Cl_2$ | 70 | | 8 |
| 3-Me-4-(2-$Et_2$N-Ethyl)-8-(2-$FC_6H_4$) | Oil | | 54 | | 8 |
| 3-Me-4-(2-$Et_2$N-Ethyl)-8-(2-$F_3CC_6H_4$) | Oil | | 66 | | 8 |
| 3-Me-4-(2-Propyn-1-yl)-8-(2-$F_3CC_6H_4$) | Oil | | 40 | | 8 |

*8H-Isoxazolo[5,4-e][1,4]diazepines*

| | | | | | |
|---|---|---|---|---|---|
| 3-Me-4-Ph-7-MeO | 113–116 | $Et_2O$/Hexane | 8 | ir, pmr | 9 |

TABLE IX-1. —(contd.)

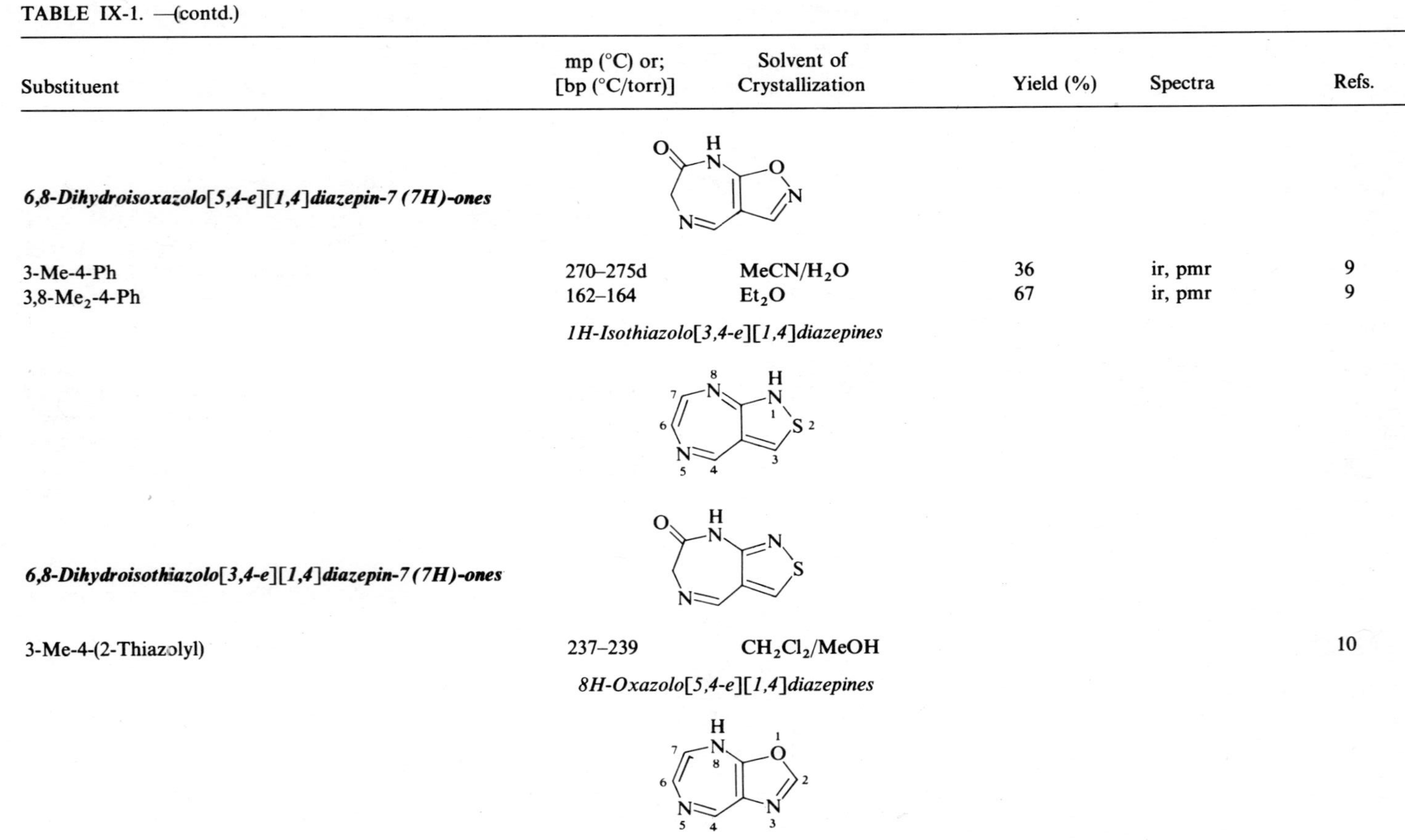

| Substituent | mp (°C) or; [bp (°C/torr)] | Solvent of Crystallization | Yield (%) | Spectra | Refs. |
|---|---|---|---|---|---|
| ***6,8-Dihydroisoxazolo[5,4-e][1,4]diazepin-7(7H)-ones*** | | | | | |
| 3-Me-4-Ph | 270–275d | $MeCN/H_2O$ | 36 | ir, pmr | 9 |
| 3,8-$Me_2$-4-Ph | 162–164 | $Et_2O$ | 67 | ir, pmr | 9 |
| *1H-Isothiazolo[3,4-e][1,4]diazepines* | | | | | |
| ***6,8-Dihydroisothiazolo[3,4-e][1,4]diazepin-7(7H)-ones*** | | | | | |
| 3-Me-4-(2-Thiazolyl) | 237–239 | $CH_2Cl_2$/MeOH | | | 10 |
| *8H-Oxazolo[5,4-e][1,4]diazepines* | | | | | |

| Substituent | mp/[bp] | Solvent | Yield | Spectra | Ref. |
|---|---|---|---|---|---|
| ***5,6,7,8-Tetrahydrooxazolo[5,4-e][1,4]diazepin-4(4H)-thiones*** | | | | | |
| 2-Ph | 192 | EtOH/$H_2O$ | | ir, pmr, uv | 11 |
| *Pyrano[2,3-e][1,4]diazepines* | | | | | |
| ***1,2,3,4,5,5a, 6,7,8,9a-Decahydropyrano[2,3-e][1,4]diazepines*** | | | | | |
| 1,4-$Me_2$, *cis/trans* | [115–120/15] | | 54 | ms, pmr | 4 |
| 1,4-$(Benzyl)_2$, *cis/trans* | [176–183/0.006] | | 43 | pmr | 4 |
| 1,4-$Bu_2$, *cis/trans* | [88–92/0.004] | | 42 | pmr | 4 |
| *1H-Pyrazino[2,3-e][1,4]diazepines* | | | | | |
| ***1,3-Dihydropyrazino[2,3-e][1,4]diazepin-2(2H)-ones*** | | | | | |
| 5-Ph | 213–215d | $CH_2Cl_2$/EtOAc | | | 12 |
| 5-Ph-7-Cl | 223–226d | $CH_2Cl_2$/EtOAc | | | 12 |
| 1-Me-5-Ph-7-Cl | 105–107 | $Et_2O$/Hexane | | | 12 |

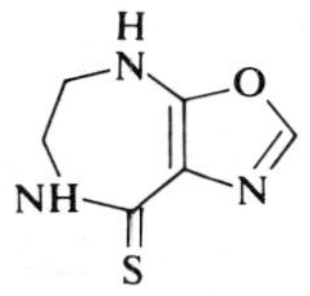

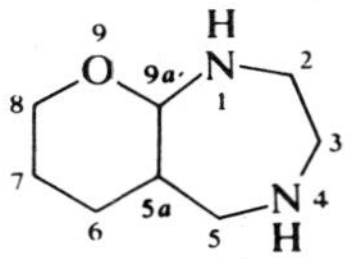

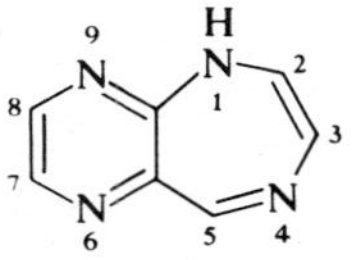

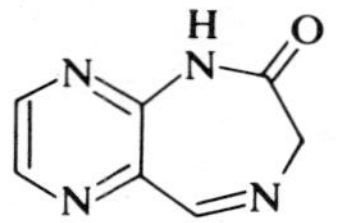

TABLE IX-1. —(contd.)

| Substituent | mp (°C) or; [bp (°C/torr)] | Solvent of Crystallization | Yield (%) | Spectra | Refs. |
|---|---|---|---|---|---|
| | *Pyrazolo[3,4-e][1,4]diazepines* | | | | |
| ***1,6-Dihydropyrazolo[3,4-e][1,4]diazepines*** | | | | | |
| 1,3-$Me_2$-4-(3-$ClC_6H_4$) | 105–107 | $Et_2O$ | 63 | | 13 |
| 1-Me-4-Ph-7-EtNH | 187–189 | EtOAc | | | 117 |
| 1,3-$Me_2$-4-Ph-7-(1-Butyn-3-yl)NH | 193–195d | EtOH | | | 15 |
| 1,3-$Me_2$-4-Ph-7-MeS | 117–118 | | | | 14 |
| 1,3-$Me_2$-4-Ph-7-(2-Propyn-1-yl)NH | 193–195 | $Et_2O$ | | | 15 |
| 1,3-$Me_2$4-(2-$ClC_6H_4$)-7-MeS | Oil | | | | 14 |
| 1,3-$Me_2$-4-(2-$ClC_5H_4$)-7-(2-Propyn-1-yl)NH | 176–178d | | | | 15 |
| 1,3-$Me_2$-4-(2-$FC_6H_4$)-7-MeS | 110–112 | $Et_2O$ | | | 14 |
| 1,3-$Me_2$-4-(2-$FC_6H_4$)-7-(2-Propyn-1-yl)NH | 202–204d | $Et_2O$ | | | 15 |
| 1-Me-3-Et-4-Ph-7-MeNH | 218–220 | $CH_2Cl_2/Et_2O$ | 81 | | 16 |
| 1-Me-3-Et-4-Ph-7-Me(NO)N | 120–122 | $CH_2Cl_2/Et_2O$/Hexane | 82 | | 16 |
| 1-Me-3-Et-4-Ph-7-MeOOC(HON)C | 225–227 | $CH_2Cl_2/Et_2O$ | 35 | | 16 |
| ***1,6,7,8-Tetrahydropyrazolo[3,4-e][1,4]diazepines*** | | | | | |
| ***Disubstituted*** | | | | | |
| 3-Me-4-(3-$ClC_6H_4$) | 200–203 | MeCN | 40 | | 13 |

| Substituents | mp (°C) | Solvent | Yield (%) | Refs. |
|---|---|---|---|---|
| ***Trisubstituted*** | | | | |
| 1-Et-3-Me-4-(3-$F_3CC_6H_4$) | 184–185 | MeCN | 40 | 13 |
| 1,3-$Me_2$-4-Ph | 140–141 | $CHCl_3$/Isooctane | | 17 |
| Hydrochloride | 300 | *i*-PrOH | 40 | 13 |
| 1,3-$Me_2$-4-(3-$BrC_6H_4$) | 153–155 | Acetone | 62 | 13 |
| 1,3-$Me_2$-4-(2-$ClC_6H_4$) | 209–212 | PhMe | 40 | 13 |
| 1,3-$Me_2$-4-(3-$ClC_6H_4$) | 186–188 | MeCN | 50–70 | 13 |
| Hydrochloride | 279–281 | *i*-PrOH/THF | | 13 |
| 5-Oxide | 225–228 | *i*-PrOH | 65 | 13 |
| 1,3-$Me_2$-4-(4-$ClC_6H_4$) | 184–186 | EtOAc/Petr ether | 53 | 13 |
| 1,3-$Me_2$-4-(3,4-$Cl_2C_6H_3$) hydrochloride·0.5 $H_2O$ | 150 | *i*-PrOH | 52 | 13 |
| 1,3-$Me_2$-4-(3-$CNC_6H_4$) | 183–185 | Acetone | 25 | 13 |
| 1,3-$Me_2$-4-(3-$FC_6H_4$) hydrate | 114–116 | Acetone | 37 | 13 |
| Hemihydrate | 95–98 | $Et_2O$/Petr ether | 22 | 13, 17 |
| 1,3-$Me_2$-4-(3-$H_2NC_6H_4$) dihydrochloride | 296 | *i*-PrOH | 89 | 13 |
| 1,3-$Me_2$-4-(3-$HOC_6H_4$) dihydrochloride·*i*PrOH | 115–118 | *i*-PrOH | 26 | 13 |
| 1,3-$Me_2$-4-(2-Me-3-$ClC_6H_3$) | 204–206 | Acetone | 52 | 13 |
| 1,3-$Me_2$-4-(3-$MeC_6H_4$) hydrochloride | 244–246 | *i*-PrOH | 78 | 13 |
| 1,3-$Me_2$-4-(4-$MeC_6H_4$) | 227–229 | EtOAc/Petr ether | 51 | 13 |
| 1,3-$Me_2$-4-(2,5-$Me_2C_6H_3$) hydrochloride | 197–200 | *i*-PrOH | 17 | 13 |
| 1,3-$Me_2$-4-(3,5-$Me_2C_6H_3$) | 221–223 | EtOAc/Petr ether | 35 | 13 |
| 1,3-$Me_2$-4-(4-$MeOC_6H_4$) | 198–200 | EtOAc/Petr ether | 57 | 13 |
| 1,3-$Me_2$-4-[3,5-$(MeO)_2C_6H_3$] | 145–147 | EtOAc/Petr ether | 11 | 13 |
| 1,3-$Me_2$-4-(3-$N_3C_6H_4$) | 129–131 | Acetone | 38 | 13 |
| 1,3-$Me_2$-4-(3-$NO_2C_6H_4$) hydrochloride | 242–245 | *i*-PrOH | 41 | 13 |
| 1-Me-3-Et-4-(3-$BrC_6H_4$) hydrochloride | 289–291 | *i*-PrOH | 50 | 13 |
| 1-Me-3-Et-4-(3-$FC_6H_4$) | 143–145 | EtOAc/Petr ether | 52 | 13 |
| 1-Me-3-Et-4-(3-$MeC_6H_4$) oxalate | 177–180 | MeCN | 41 | 13 |
| ***Tetrasubstituted*** | | | | |
| 1-Et-3-Me-4-(2-$ClC_6H_4$)-8-Me hydrochloride | 274–275 | *i*-PrOH | 30 | 13 |
| 1,3,6-$Me_3$-4-(3-$ClC_6H_4$) | 96–98 | EtOAc/Petr ether | 37 | 13 |
| 1,3,8-$Me_3$-4-Ph hydrochloride | 260 | *i*-PrOH | 42 | 13 |

TABLE IX-1. —(contd.)

| Substituent | mp (°C) or; [bp (°C/torr)] | Solvent of Crystallization | Yield (%) | Spectra | Refs. |
|---|---|---|---|---|---|
| 1,3,8-$Me_3$-4-(2-$ClC_6H_4$) dihydrochloride | 255 | *i*-PrOH | 35 | | 13 |
| 1,3,8-$Me_3$-4-(3-$ClC_6H_4$) dihydrochloride | 170–172 | *i*-PrOH | 30 | | 13 |
| 1,3,8-$Me_3$-4-(4-$HOC_6H_4$) dihydrochloride·$H_2O$ | 195–197 | *i*-PrOH | 36 | | 13 |
| 1,3-$Me_2$-4-Ph-8-(3-$Me_2$N-Propyl) | 79–81 | EtOAc/Petr ether | 35 | | 13 |
| 1,3-$Me_2$-4-(2-$ClC_6H_4$)-8-(3-$Me_2$N-Propyl) | 50–53 | Petr ether | 44 | | 13 |
| 1,3-$Me_2$-4-(2-$ClC_6H_4$)-7-CN·$H_2O$ | 140–142 | MeCN | 72 | | 13 |
| 1,3-$Me_2$-4-(3-$ClC_6H_4$)-8-Ac hydrochloride | 267–268 | *i*-PrOH | 85 | | 13 |
| 1,3-$Me_2$-4-(3-$ClC_6H_4$)-8-Et dihydrochloride | 175 | *i*-PrOH | 12 | | 13 |
| 1,3-$Me_2$-4-(3-$ClC_6H_4$)-8-$F_3$CCO | 183–185 | Petr ether | | | 13 |
| 1,3-$Me_2$-4-(3-$ClC_6H_4$)-8-(3-$Me_2$N-Propyl) | 75–77 | Petr ether | 20 | | 13 |
| 1,3-$Me_2$-4-(3-$ClC_6H_4$)-8-NO | 118–120 | EtOAc/Petr ether | 28 | | 13 |
| 1,3-$Me_2$-4-(3-$ClC_6H_4$)-8-(2-Pyrrolidinoethyl)-dihydrochloride | 265 | *i*-PrOH | 69 | | 13 |
| 1,3-$Me_2$-4-(3-$FC_6H_4$)-8-(3-$Me_2$N-Propyl) | 80–82 | Petr ether | 30 | | 13 |
| 1,3-$Me_2$-4-(3-$MeC_6H_4$)-8-(3-$Me_2$N-Propyl) | 120–124 | Petr ether | 65 | | 13 |
| 1,3-$Me_2$-4-(4-$MeC_6H_4$)-8-(3-$Me_2$N-Propyl) | 99–101 | Petr ether | 35 | | 13 |
| 1-Me-3-Et-4-(3-$FC_6H_4$)-8-(3-$Me_2$N-Propyl) ·0.5 $H_2O$ | 73–75 | Petr ether | 55 | | 13 |
| ***Pentasubstituted*** | | | | | |
| 1,3,6,6-$Me_4$-4-(3-$ClC_6H_4$) | 155–157 | EtOAc/Petr ether | 57 | | 13 |
| 1,3,6,6-$Me_4$-4-(3-$MeC_6H_4$)·0.5 $H_2O$ | 128–130 | EtOAc/Petr ether | 20 | | 13 |
| 1,3,6,7-$Me_4$-4-Ph·0.5$H_2O$ | 101–102 | Acetone | 35 | | 13 |

***7-Methylene-1,6,7,8-tetrahydropyrazolo[3,4-e][1,4]diazepines***

$R_1$, $R_2$, H, N, H, N, N, N

| | | | | |
|---|---|---|---|---|
| ***R*$_1$*, R*$_2$*, Other*** | | | | |
| COOMe, COOMe; 1-Me-3-Et-4-PH | 145–148 | MeOH | 24 | 16 |
| ***1,6,7,8-Tetrahydropyrazolo-[3,4-e][1,4]diazepin-7-ones*** | | | | |
| ***Disubstituted*** | | | | |
| 1-Me-4-Ph | 218–220 | EtOH/THF | | 117 |
| Hydrochloride·$H_2O$ | 285 | *i*-PrOH/$Et_2O$ | 20 | 18 |
| 1-Ph-4-(2-Thiazolyl) | 237–240 | $CH_2Cl_2$/Hexane | | 10 |
| ***Trisubstituted*** | | | | |
| 1-Cyclohexyl-3-Me-4-(2-$ClC_6H_4$) | 203–205 | MeOH | 65 | 18 |
| 1-Bu-3-Me-4-(2-$ClC_6H_4$) | 183–185 | $Et_2O$ | 40 | 18 |
| 1-Bu-3-Me-4-(3-$ClC_6H_4$) | 137–139 | $Et_2O$ | 65 | 18 |
| 1-Et-3-Me-4-(2-$ClC_6H_4$) | 246–248 | PhMe | 30 | 18 |
| 1-Et-3-Me-4-(2-$FC_6H_4$) | 250–252 | PhMe | 17 | 18 |
| 1,3-$Me_2$-4-Ph | 265–267 | PhMe | 65 | 18 |
| 1,3-$Me_2$-4-(2-$ClC_6H_4$) | 245–248 | PhMe | 84 | 18 |
| 1,3-$Me_2$-4-(3-$ClC_6H_4$) | 255–257 | Acetone | 95 | 18 |
| 1,3-$Me_2$-4-(2-$FC_6H_4$) | 235–237 | PhMe | 40 | 18 |
| Hydrobromide | 295d | Acetone/AcOH | | 19 |
| 1,3-$Me_2$-4-(2-Thienyl) | 228–230 | PhMe | 4 | 18 |
| 1,8-$Me_2$-4-Ph | 163–165 | THF/Petr ether | | 117 |
| 1-Me-3-Bu-4-(2-$ClC_6H_4$) | 191–193 | $Et_2O$ | 41 | 18 |
| 1-Me-3-Cl-4-Ph | 253–255 | MeCN | 44 | 18 |
| 1-Me-3-Et-4-Ph | 268–270 | MeOH | 35 | 18 |
| 1-Me-3-Et-4-(2-$ClC_6H_4$) | 258–260 | PhMe | 18 | 18 |
| 1-Me-3-Et-4-(2-$FC_6H_4$) | 253–255 | Acetone | 81 | 18 |
| 1-Me-3-Et-4-(3-$FC_6H_4$) | 263–265 | Acetone | 19 | 18 |
| 1-Me-3-Et-4-(4-$FC_6H_4$) | 247–250 | Acetone | 33 | 18 |
| 1-Me-3-Pr-4-(2-$ClC_6H_4$) | 179–180 | Acetone | 48 | 18 |

TABLE IX-1. —(contd.)

| Substituent | mp (°C) or; [bp (°C/torr)] | Solvent of Crystallization | Yield (%) | Spectra | Refs. |
|---|---|---|---|---|---|
| 1-Me-3-*i*-Pr-4-(2-$ClC_6H_4$) | 223–224 | PhMe | 29 | | 18 |
| 1,4-$Ph_2$-3-Me | 255–257 | Acetone | 30 | | 18 |
| 1-Ph-3-Me-4-(2-$ClC_6H_4$) | 229–231 | PhMe | 12 | | 18 |
| 1-Ph-4-(2-Thiazolyl)-8-Me | 213–223 | $CH_2Cl_2$/Hexane | | | 10 |
| 1-Pr-3-Me-4-(2-$ClC_6H_4$) | 171–173 | $Et_2O$ | 40 | | 18 |
| 1-*i*-Pr-3-Me-4-(2-$ClC_6H_4$) | 253–255 | EtOAc | 20 | | 18 |
| ***Tetrasubstituted*** | | | | | |
| 1-Bu-3,8-$Me_2$-4-(2-$ClC_6H_4$) | Amorphous | $Et_2O$ | 95 | | 18 |
| 1-Cyclohexyl-3,8-$Me_2$-4-(2-$ClC_6H_4$) | Amorphous | $Et_2O$ | 60 | | 18 |
| 1-Et-3,8-$Me_2$-4-(2-$ClC_6H_4$) | 105–108 | $Et_2O$ | 20 | | 18 |
| 1-Et-3,8-$Me_2$-4-(2-$FC_6H_4$) | 74–77 | Hexane | 70 | | 18 |
| 1-Et-3-Me-4-(2-$ClC_6H_4$)-8-Allyl | Amorphous | $Et_2O$ | 50 | | 18 |
| 1,3,8-$Me_3$-4-Ph | 183–185 | PhMe | 34 | | 18 |
| Hydrochloride | 180d | *i*-PrOH/HCl | 90 | | 18 |
| 5-Oxide | 232–235 | Acetone | 73 | | 18 |
| 1,3,8-$Me_3$-4-(2-$BrC_6H_4$) | 169–171 | $Et_2O$ | 80 | | 18 |
| 1,3,8-$Me_3$-4-(2-$ClC_6H_4$) hydrochloride | 265d | *i*-PrOH | 90 | | 18 |
| 1,3,8-$Me_3$-4-(2-$FC_6H_4$) | 183–185 | MeOH/$Et_2O$ | 86 | | 18 |
| Hydrochloride | 248d | *i*-PrOH | | | 18 |
| 5-Oxide | 182–184 | *i*-PrOH/$Et_2O$ | 85 | | 18 |
| 1,3,8-$Me_3$-4-(2,6-$F_2C_6H_3$) | 247–250 | *i*-PrOH/THF | 81 | | 18 |
| 1,3,8-$Me_3$-4-(2-$H_2NC_6H_4$) | 160 | $Et_2O$ | 90 | | 18 |
| 1,3,8-$Me_3$-4-(2-$N_3C_6H_4$) | 137–139 | EtOAc/Petr ether | 40 | | 18 |
| 1,3,8-$Me_3$-4-(2-$NO_2C_6H4$) | 184–186 | MeCN | 11 | | 18 |
| 1,3-$Me_2$-4-(2-$FC_6H_4$)-8-Et | 133–135 | $Et_2O$ | 27 | | 18 |
| 1,8-$Me_2$-3-Bu-4-(2-$ClC_6H_4$) | 96–98 | $Et_2O$ | 42 | | 18 |
| 1,8-$Me_2$-3-Et-4-Ph | 193–195 | $Et_2O$ | 79 | | 18 |
| 1,8-$Me_2$-3-$Et_4$-(2-$ClC_6H_4$) | 115–117 | $Et_2O$ | 70 | | 18 |

| | | | | |
|---|---|---|---|---|
| 1,8-$Me_2$-3-Et-4-(2-$FC_6H_4$) | 165–168 | $Et_2O$ | 50 | 18 |
| 1,8-$Me_2$-3-Et-4-(3-$FC_6H_4$) | 163–165 | $Et_2O$ | 50 | 18 |
| 1,8-$Me_2$-3-Et-4-(4-$FC_6H_4$) | 216–218 | PhMe | 40 | 18 |
| 1,8-$Me_2$-3-MeO-4-(2-$FC_6H_4$) | 212–214 | $CHCl_3/Et_2O$ | 80 | 18 |
| 1,8-$Me_2$-3-Pr-4-(2-$ClC_6H_4$) | 120–122 | $Et_2O$/Petr ether | 20 | 18 |
| 1,8-$Me_2$-3-*i*-Pr-4-(2-$ClC_6H_4$) | 168–170 | $Et_2O$ | 56 | 18 |
| 1-Ph-4-(2-Thiazolyl)-6,8-$Me_2$ | 168–170 | $CH_2Cl_2$/Hexane | | 10 |
| 1-Pr-3,8-$Me_2$-4-(2-$ClC_6H_4$) | Amorphous | $Et_2O$ | 60 | 18 |
| ***Pentasubstituted*** | | | | |
| 1,3,6,6-$Me_4$-4-(3-$ClC_6H_4$) | 210 | Acetone | | 13 |
| 1,3,8-$Me_3$-4-Ph-6-AcO | 218–220 | EtOAc | 83 | 18 |
| 1,3,8-$Me_3$-4-Ph-6-HO | 243–245 | $MeOH/H_2O$ | 56 | 18 |
| 1,3,8-$Me_3$-4-(2-$FC_6H_4$)-6-AcO | 203–205 | PhH/Petr ether | 81 | 18 |
| 1,3,8-$Me_3$-4-(2-$FC_6H_4$)-6-HO | 250–252 | $MeOH/H_2O$ | 90 | 18 |
| ***1,6,7,8-Tetrahydropyrazolo[3,4-e][1,4]diazepin-7-thiones*** | | | | |
| 1,3-$Me_2$-4-Ph | 268–269 | | | 14 |
| 1,3-$Me_2$-4-(2-$ClC_6H_4$) | 243–245 | | | 14 |
| 1,3-$Me_2$-4-(2-$FC_6H_4$) | 243–245 | | | 14 |
| ***1,4,5,6,7,8-Hexahydropyrazolo[3,4-e][1,4]diazepines*** | | | | |
| 1,3-$Me_2$-4-(3-$ClC_6H_4$) dihydrochloride | 255d | *i*-$PrOH/Et_2O$ | | 17 |
| 1,3-$Me_2$-4-(3-$ClC_6H_4$)-8-Et dihydrochloride | 190d | *i*-PrOH/EtOAc | | 17 |

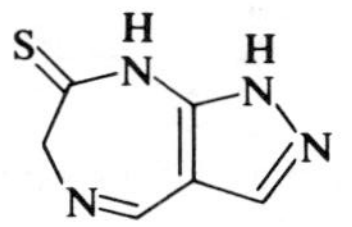

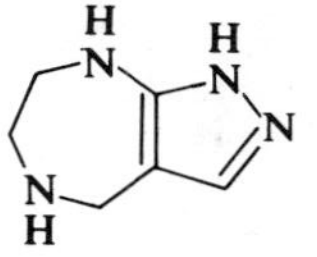

TABLE IX-1. —(contd.)

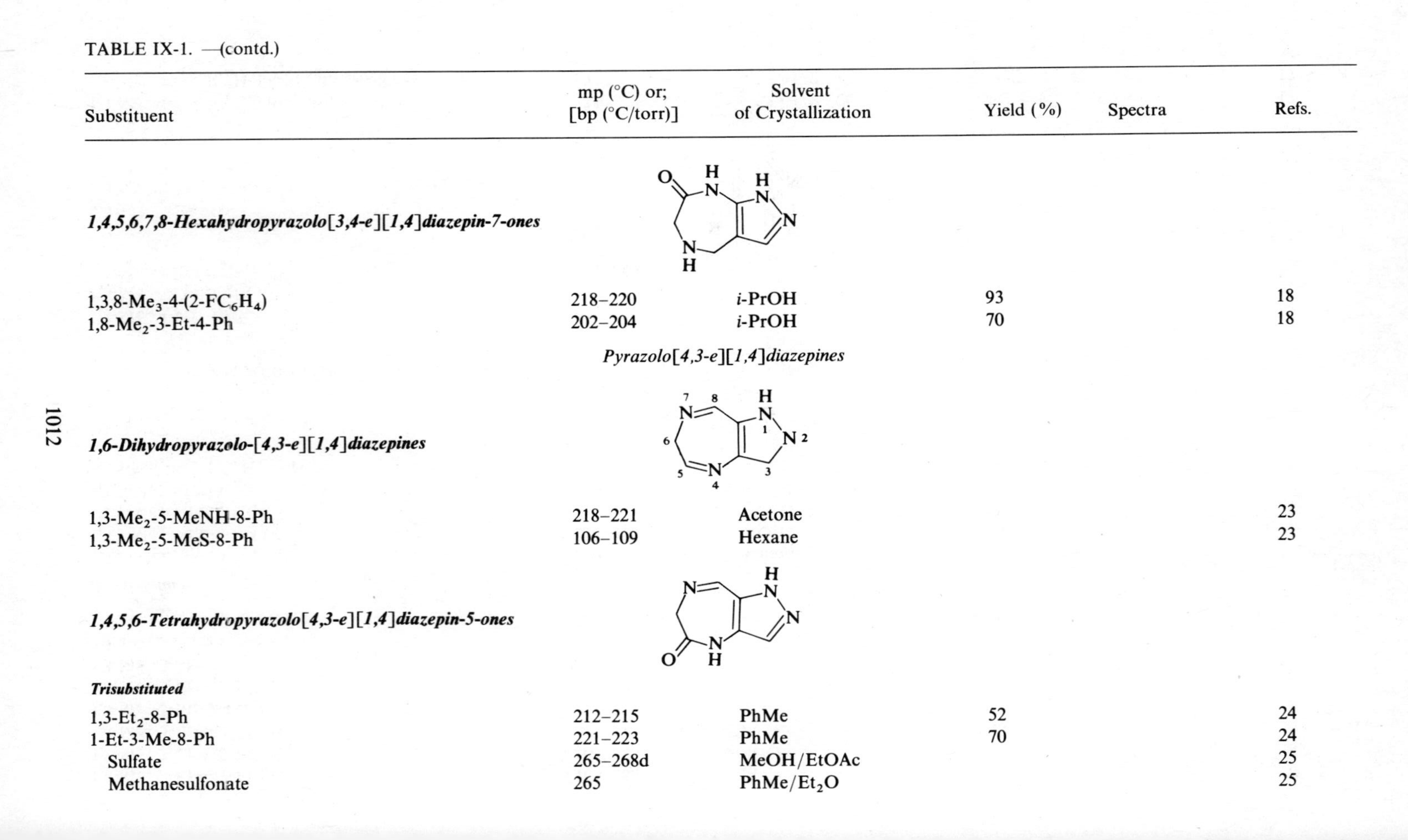

| Substituent | mp (°C) or; [bp (°C/torr)] | Solvent of Crystallization | Yield (%) | Spectra | Refs. |
|---|---|---|---|---|---|
| ***1,4,5,6,7,8-Hexahydropyrazolo[3,4-e][1,4]diazepin-7-ones*** | | | | | |
| 1,3,8-$Me_3$-4-(2-$FC_6H_4$) | 218–220 | *i*-PrOH | 93 | | 18 |
| 1,8-$Me_2$-3-Et-4-Ph | 202–204 | *i*-PrOH | 70 | | 18 |
| | *Pyrazolo[4,3-e][1,4]diazepines* | | | | |
| ***1,6-Dihydropyrazolo-[4,3-e][1,4]diazepines*** | | | | | |
| 1,3-$Me_2$-5-MeNH-8-Ph | 218–221 | Acetone | | | 23 |
| 1,3-$Me_2$-5-MeS-8-Ph | 106–109 | Hexane | | | 23 |
| ***1,4,5,6-Tetrahydropyrazolo[4,3-e][1,4]diazepin-5-ones*** | | | | | |
| ***Trisubstituted*** | | | | | |
| 1,3-$Et_2$-8-Ph | 212–215 | PhMe | 52 | | 24 |
| 1-Et-3-Me-8-Ph | 221–223 | PhMe | 70 | | 24 |
| Sulfate | 265–268d | MeOH/EtOAc | | | 25 |
| Methanesulfonate | 265 | PhMe/$Et_2O$ | | | 25 |

| | | | | | |
|---|---|---|---|---|---|
| Sodium salt·$H_2O$ | > 300d | DMF/$Et_2O$ | | | 25 |
| 7-Oxide | 198–200 | EtOH | | | 28 |
| 1-Et-3-Me-8-(3-$BrC_6H_4$) | 27–228 | EtOAc | 71 | | 24 |
| 1-Et-3-Me-8-(3-$ClC_6H_4$) | 224–225 | Xylene | 30 | | 24 |
| 1-Et-3-Me-8-(4-$ClC_6H_4$) | 240–242 | PhMe | 51 | | 24 |
| 1-Et-3-Me-8-(5-Cl-2-Thienyl) | 189–192 | EtOH | 15 | | 24 |
| 1-Et-3-Me-8-(4-$FC_6H_4$) | 198–200 | MeCN | 69 | | 24 |
| 1-Et-3-Me-8-(2-$F_3CC_6H_4$) | 195–197 | PhMe | 49 | | 24 |
| 1-Et-3-Me-8-(4-$MeC_6H_4$) | 199–201 | PhMe | 37 | | 24 |
| 1-Et-3-Me-8-(4-$MeOC_6H_4$) | 198–200 | PhMe | 36 | | 24 |
| 1-Et-3-Me-8-(3-$NO_2C_6H_4$) | 155–157 | MeOH | 88 | | 24 |
| 1-Et-3-Me-8-(2-Thienyl) | 205–206 | EtOH | 10 | | 24 |
| 7-Oxide | 226–228 | EtOH | | | 27 |
| 1,3-$Me_2$-8-Ph | 267–270 | PhMe | 70 | | 24 |
| Hydrochloride | 295 | $CHCl_3$/MeOH/$Et_2O$ | | | 25 |
| 7-Oxide | 242–243 | MeCN | | | 28 |
| 1,3-$Me_2$-8-(4-$FC_6H_4$) | 218–222 | PhMe | 24 | | 24 |
| 1,3-$Me_2$-8-(2-Thienyl) | 246–247 | EtOH | 19 | | 24 |
| 1-Me-3-Et-8-Ph | 236–239 | PhMe | 50 | | 24 |
| 1-Pr-3-Me-8-Ph | 168–172 | *i*-PrOH | 71 | | 24 |
| 1-*i*-Pr-3-Me-8-Ph | 207–210 | PhMe | 34 | | 24 |
| Hydrobromide | 305d | AcOH/HBr | | | 25 |
| ***Tetrasubstituted*** | | | | | |
| 1-Et-3,4-$Me_2$-8-Ph | 99–102 | Cyclohexane | 72 | | 24 |
| 7-Oxide | 145–147 | EtOAc | | | 28 |
| 7-Methiodide | 230d | MeCN/$Et_2O$ | 79 | uv | 30 |
| 1-Et-3,4-$Me_2$-8-(2-Thienyl)-7-oxide | 189–192 | *i*-PrOH | | | 27 |
| 1-Et-3,6-$Me_2$-8-Ph | 223–226 | PhMe | 43 | | 24 |
| 1,3-$Et_2$-4-Me-8-Ph | 79–81 | $Et_2O$ | 54 | | 24 |
| 1-Et-3-Me-4-(2-$Et_2N$-Ethyl)-8-Ph hydrobromide | 188–193 | *i*-PrOH/$Et_2O$ | 48 | | 24 |
| 1-Et-3-Me-4-(3-$Me_2N$-Propyl)-8-Ph hydrobromide | 178–180 | *i*-PrOH/$Et_2O$ | 10 | | 24 |
| 1-Et-3-Me-6-AcO-8-Ph | 201–203 | EtOAc | | | 29 |
| 1-Et-3-Me-6-HO-8-Ph | 217–219 | MeCN/$Et_2O$ | | | 29 |

TABLE IX-1. —(contd.)

| Substituent | mp (°C) or; [bp (°C/torr)] | Solvent of Crystallization | Yield (%) | Spectra | Refs. |
|---|---|---|---|---|---|
| 1,4-$Et_2$-3-Me-8-Ph | 91–93 | Cyclohexane | 68 | | 24 |
| 1,3,4-$Me_3$-8-Ph hydrochloride·$H_2O$ | 213–215 | *i*-PrOH | 37 | | 24 |
| 7-Oxide | 150–152 | *i*-PrOH/Pentane | | | 28 |
| 1,3-$Me_2$-4-$F_3CCH_2$-8-Ph | 186–187 | PhMe | 50 | | 24 |
| 1,3-$Me_2$-4-(Propyn-3-yl)-8-Ph | 128–130 | $Et_2O$/Petr ether | | | 26 |
| 1,3-$Me_2$-6-AcO-8-Ph | 235–241 | | | | 29 |
| 1,3-$Me_2$-6-HO-8-Ph | 246–248 | DMF | | | 29 |
| 1,4-$Me_2$-3-Et-8-Ph | 91–93 | $Et_2O$ | 32 | | 24 |
| 1-Pr-3,4-$Me_2$-8-Ph | 82–84 | PhH/Petr ether | 82 | | 24 |
| 1-*i*-Pr-3,4-$Me_2$-8-Ph | 145–147 | PhMe | 20 | | 24 |
| ***Pentasubstituted*** | | | | | |
| 1-Et-3,4-$Me_2$-6-AcO-8-Ph | 173–175 | $EtOH/Et_2O$ | | | 29 |
| 1-Et-3,4-$Me_2$-6-HO-8-Ph | 188–190 | MeCN | | | 29 |
| 1,3,4-$Me_3$-6-AcO-8-Ph | 180–182 | PhMe/Pentane | | | 29 |
| 1,3,4-$Me_3$-6-HO-8-Ph | 211–213 | $MeCN/Et_2O$ | | | 29 |
| ***1,6,7,8-Tetrahydropyrazolo[4,3-e][1,4]diazepin-8-ones*** | | | | | |
| 3-Me-5-Ph | 249–251 | $DMF/H_2O$ | 53 | | 31 |
| 3-Me-5-(4-$BrC_6H_4$) | 266–268 | $DMF/H_2O$ | 53 | | 31 |
| 3-Me-5-(4-$ClC_6H_4$) | 258–260 | $DMF/H_2O$ | 48 | | 31 |
| 3-Me-5-(4-$MeOC_6H_4$) | 234–236 | $DMF/H_2O$ | 61 | | 31 |
| 3-Me-5-(4-$NO_2C_6H_4$) | > 310 | $DMF/H_2O$ | 56 | | 31 |
| 3-Me-5-(4-$PhC_6H_4$) | 281–283 | $DMF/H_2O$ | 49 | | 31 |

| | | | | |
|---|---|---|---|---|
| ***2,4,5,6-Tetrahydropyrazolo[4,3-e][1,4]diazepin-5-ones*** | | | | |
| ***Trisubstituted*** | | | | |
| 2-Et-3-Me-8-Ph | 213–215 | EtOH | 45 | 24 |
| 2,3-$Me_2$-8-Ph | 265–267 | EtOh | 55 | 24 |
| 2,3-$Me_2$-8-(5-Cl-1,3-$Me_2$-Pyrazol-4-yl) | 253–255 | PhMe | 53 | 24 |
| 2-*i*-Pr-3-Me-8-Ph | 210–212 | MeOH | 30 | 24 |
| ***Tetrasubstituted*** | | | | |
| 2-Et-3,4-$Me_2$-8-Ph | 130–132 | PhH/Petr ether | 44 | 24 |
| 2,3,4-$Me_3$-8-Ph | 170–173 | $CHCl_3$/Petr ether | 50 | 24 |
| 2,3-$Me_2$-4-Et-8-Ph | 154–156 | EtOAc/Petr ether | 71 | 24 |
| ***1,4,5,6,7,8-Hexahydropyrazolo-[4,3-e][1,4]diazepin-5-ones*** | | | | |
| ***Trisubstituted*** | | | | |
| 1-Et-3-Me-8-Ph | 185–187 | EtOH | | 32 |
| 1,3-$Me_2$-8-Ph | 186–188 | EtOH | | 32 |
| 1-Pr-3-Me-8-Ph | 176–178 | EtOH/Pentane | | 32 |
| ***Tetrasubstituted*** | | | | |
| 1-Et-3,7-$Me_2$-8-Ph | 176–179 | EtOH | | 32 |
| 1,3,4-$Me_3$-8-Ph hydrochloride | 248–250 | EtOH/$Et_2O$ | | 32 |
| 1,3,7-$Me_3$-8-Ph | 179–181 | EtOH | | 32 |
| 1-Pr-3,4-$Me_2$-8-Ph hydrochloride | 222–225 | EtOH/$Et_2O$ | | 32 |
| 1-Pr-3,7-$Me_2$-8-Ph | 166–168 | MeCN | | 32 |

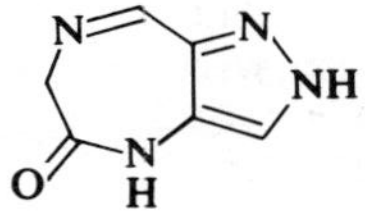

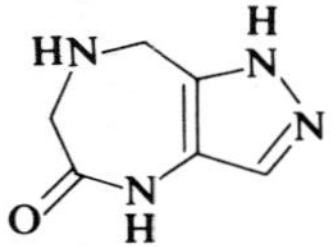

TABLE IX-1. —(contd.)

| Substituent | mp (°C) or; [bp (°C/torr)] | Solvent of Crystallization | Yield (%) | Spectra | Refs. |
|---|---|---|---|---|---|
| ***Pentasubstituted*** | | | | | |
| 1,3,4,7-$Me_4$-8-Ph hydrochloride | 229–232 | $EtOH/Et_2O$ | | | 32 |
| 1-Pr-3,4,7-$Me_3$-8-Ph | 97–99 | PhH/Pentane | | | 32 |
| | *Pyrido[2,3-e][1,4]diazepines* | | | | |
| ***1,3-Dihydropyrido[2,3-e][1,4]diazepin-2(2H)-ones*** | | | | | |
| 5-Ph | 203–205 | $CHCl_3$/Hexane | 45 | | 33 |
| ***2,3,4,5-Tetrahydro-1H-pyrido[2,3-e][1,4]diazepines*** | | | | | |
| 8-Me | 103–105 | PhH/Ligroin | 83 | pmr | 34 |
| 1-Et-8-Me picrate | 182–184 | AcOH | 85 | pmr | 36 |
| 1,8-$Me_2$ picrate | 211–212 | AcOH | 85 | pmr | 36 |
| 1-Ph-8-Me oxalate | 175–179d | $MeOH/Et_2O$ | | | 35 |
| 1-Ph-4-Ac-8-Me | 129–131 | $CH_2Cl_2/Et_2O$ | | | 35 |
| 1-Ph-4-(4-Cl-Benzoyl)-8-Me | 145–146 | $CH_2Cl_2/Et_2O$ | | | 35 |
| 1-Ph-4,8-$Me_2$ hydrochloride | 260d | $MeOH/Et_2O$ | | | 35 |
| 3-Benzyl-8-Me | 72–74 | Ligroin | 76 | pmr | 36 |

*1,2,4,5-Tetrahydropyrido[2,3-e][1,4]diazepin-3(3H)-ones*

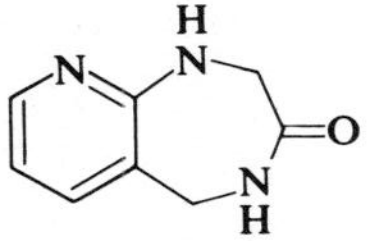

| | | | | | |
|---|---|---|---|---|---|
| 1-Ph-8-Me | 157–158 | $CH_2Cl_2/Et_2O$ | | | 35 |
| 1-Ph-4-(2-$Me_2$N-Ethyl)-8-Me hydrochloride | 225–226d | $MeOH/Et_2O$ | | | 35 |
| 1-Ph-4-(3-$Me_2$N-Propyl)-8-Me hydrochloride | 213–215d | $MeOH/Et_2O$ | | | 35 |

*1,2,3,4-Tetrahydropyrido[2,3-e][1,4]diazepin-5(5H)-ones*

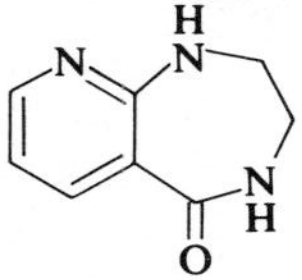

| | | | | | |
|---|---|---|---|---|---|
| 8-Me | 270–272 | MeOH | 79 | pmr | 34 |
| | 272–274 | $H_2O$ | | | 35 |
| Hydrochloride | 320–321d | MeOH | | | 35 |
| 1-Ac-8-Me | 190–191 | PhH | 57 | pmr | 34 |
| 1-Et-8-Me | 150–152 | PhH | 79 | pmr | 36 |
| 1,8-$Me_2$ | 166–169 | PhH | 78 | pmr | 36 |
| 3-Benzyl-8-Me | 227–230 | EtOH | 64 | pmr | 36 |
| 4-Allyl-8-Me | 85–86 | PhH/Ligroin | 69 | pmr | 36 |
| 4-Benzyl-8-Me | 105–107 | PhH/Ligroin | 80 | pmr | 36 |
| 4-Et-8-Me | 109–112 | PhH/Petr ether | 69 | pmr | 36 |
| 4,8-$Me_2$ | 186–188 | PhH | 92 | pmr | 36 |
| 7-Br-8-Me | 302–305d | DMF | 73 | pmr | 34 |
| 7-$NO_2$-8-Me | > 320 | DMF | 72 | pmr | 34 |
| 4-Benzyl-7-Br-8-Me | 154–156 | EtOAc | 33 | pmr | 36 |
| 4-Et-7-Br-8-Me | 135–137 | PhH/Ligroin | 45 | pmr | 36 |
| 4-Me-7-Br-8-Me | 221–223 | MeOH | 66 | pmr | 36 |

TABLE IX-1. —(contd.)

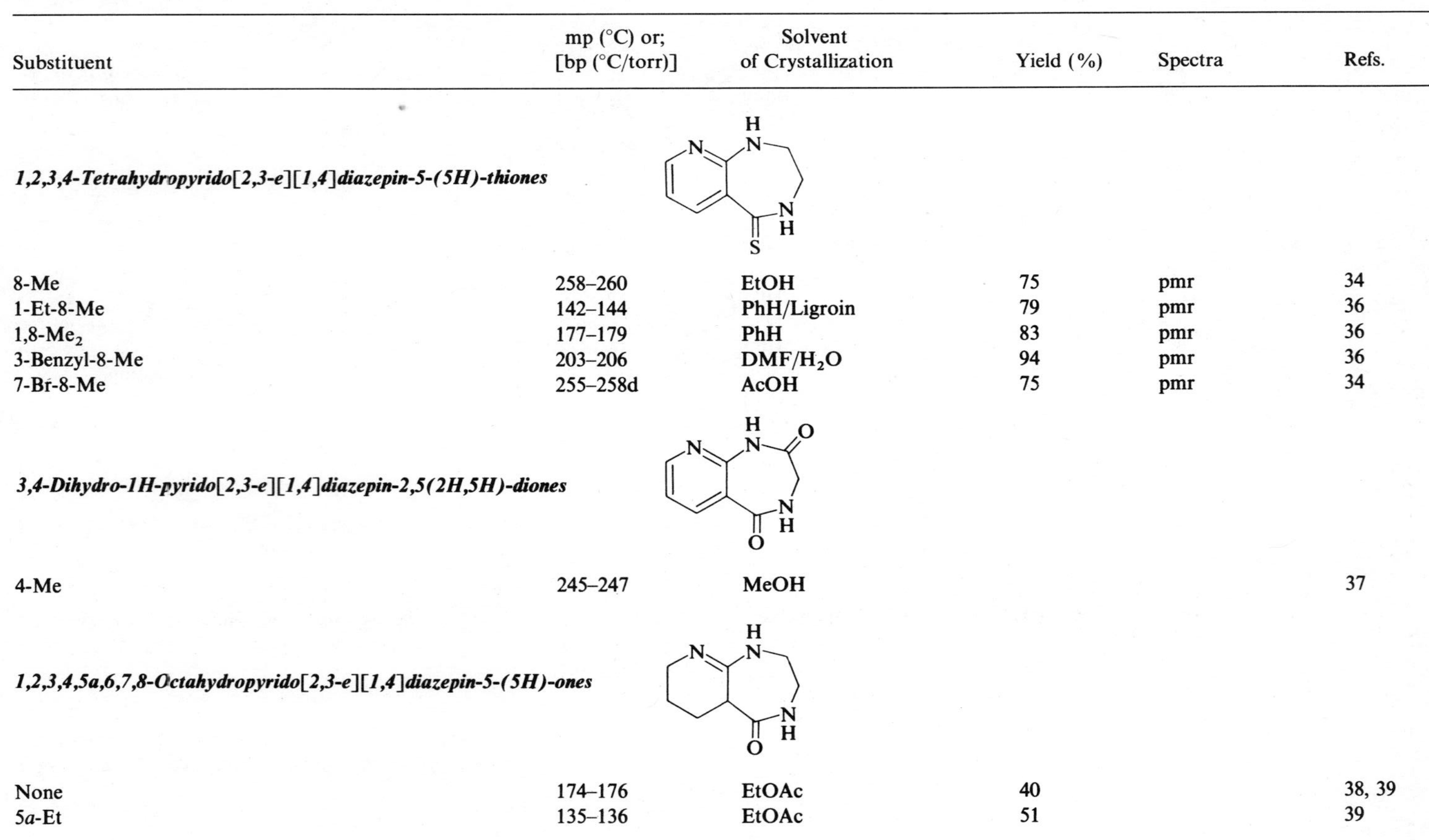

| Substituent | mp (°C) or; [bp (°C/torr)] | Solvent of Crystallization | Yield (%) | Spectra | Refs. |
|---|---|---|---|---|---|
| ***1,2,3,4-Tetrahydropyrido[2,3-e][1,4]diazepin-5-(5H)-thiones*** | | | | | |
| 8-Me | 258–260 | EtOH | 75 | pmr | 34 |
| 1-Et-8-Me | 142–144 | PhH/Ligroin | 79 | pmr | 36 |
| 1,8-$Me_2$ | 177–179 | PhH | 83 | pmr | 36 |
| 3-Benzyl-8-Me | 203–206 | $DMF/H_2O$ | 94 | pmr | 36 |
| 7-Br-8-Me | 255–258d | AcOH | 75 | pmr | 34 |
| ***3,4-Dihydro-1H-pyrido[2,3-e][1,4]diazepin-2,5(2H,5H)-diones*** | | | | | |
| 4-Me | 245–247 | MeOH | | | 37 |
| ***1,2,3,4,5a,6,7,8-Octahydropyrido[2,3-e][1,4]diazepin-5-(5H)-ones*** | | | | | |
| None | 174–176 | EtOAc | 40 | | 38, 39 |
| 5*a*-Et | 135–136 | EtOAc | 51 | | 39 |

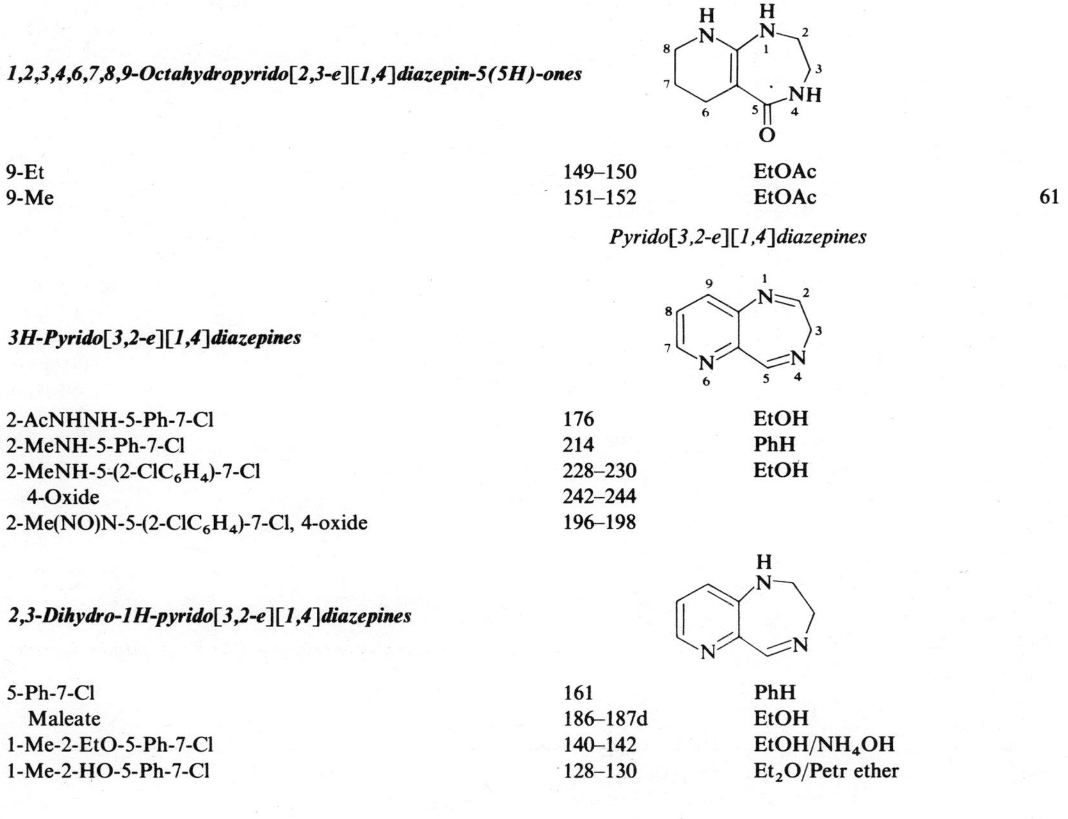

| | | | | |
|---|---|---|---|---|
| ***1,2,3,4,6,7,8,9-Octahydropyrido[2,3-e][1,4]diazepin-5(5H)-ones*** | | | | |
| 9-Et | 149–150 | EtOAc | | 40 |
| 9-Me | 151–152 | EtOAc | 61 | 40 |
| | *Pyrido[3,2-e][1,4]diazepines* | | | |
| ***3H-Pyrido[3,2-e][1,4]diazepines*** | | | | |
| 2-AcNHNH-5-Ph-7-Cl | 176 | EtOH | | 43 |
| 2-MeNH-5-Ph-7-Cl | 214 | PhH | | 43 |
| 2-MeNH-5-(2-$ClC_6H_4$)-7-Cl | 228–230 | EtOH | | 41 |
| 4-Oxide | 242–244 | | | 41 |
| 2-Me(NO)N-5-(2-$ClC_6H_4$)-7-Cl, 4-oxide | 196–198 | | | 41 |
| ***2,3-Dihydro-1H-pyrido[3,2-e][1,4]diazepines*** | | | | |
| 5-Ph-7-Cl | 161 | PhH | | 43 |
| Maleate | 186–187d | EtOH | | 43 |
| 1-Me-2-EtO-5-Ph-7-Cl | 140–142 | EtOH/$NH_4OH$ | | 42 |
| 1-Me-2-HO-5-Ph-7-Cl | 128–130 | $Et_2O$/Petr ether | | 42 |

TABLE IX-1. —(contd.)

| Substituent | mp (°C) or; [bp (°C/torr)] | Solvent of Crystallization | Yield (%) | Spectra | Refs. |
|---|---|---|---|---|---|
| *1,3-Dihydropyrido[3,2-e][1,4]diazepin-2-(2H)-ones* | 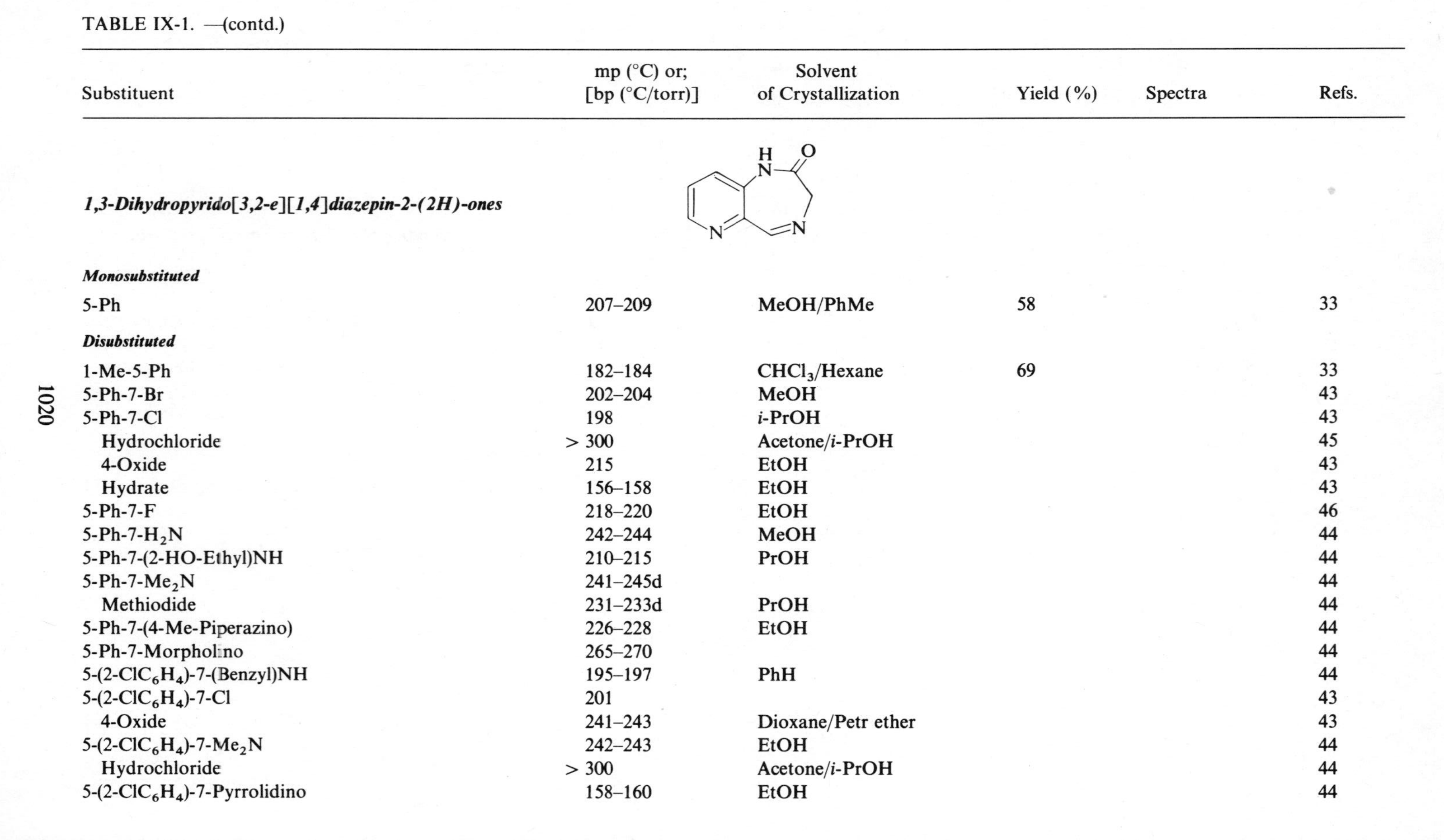 | | | | |
| ***Monosubstituted*** | | | | | |
| 5-Ph | 207–209 | MeOH/PhMe | 58 | | 33 |
| ***Disubstituted*** | | | | | |
| 1-Me-5-Ph | 182–184 | $CHCl_3$/Hexane | 69 | | 33 |
| 5-Ph-7-Br | 202–204 | MeOH | | | 43 |
| 5-Ph-7-Cl | 198 | *i*-PrOH | | | 43 |
| Hydrochloride | > 300 | Acetone/*i*-PrOH | | | 45 |
| 4-Oxide | 215 | EtOH | | | 43 |
| Hydrate | 156–158 | EtOH | | | 43 |
| 5-Ph-7-F | 218–220 | EtOH | | | 46 |
| 5-Ph-7-$H_2N$ | 242–244 | MeOH | | | 44 |
| 5-Ph-7-(2-HO-Ethyl)NH | 210–215 | PrOH | | | 44 |
| 5-Ph-7-$Me_2N$ | 241–245d | | | | 44 |
| Methiodide | 231–233d | PrOH | | | 44 |
| 5-Ph-7-(4-Me-Piperazino) | 226–228 | EtOH | | | 44 |
| 5-Ph-7-Morpholino | 265–270 | | | | 44 |
| 5-(2-$ClC_6H_4$)-7-(Benzyl)NH | 195–197 | PhH | | | 44 |
| 5-(2-$ClC_6H_4$)-7-Cl | 201 | | | | 43 |
| 4-Oxide | 241–243 | Dioxane/Petr ether | | | 43 |
| 5-(2-$ClC_6H_4$)-7-$Me_2N$ | 242–243 | EtOH | | | 44 |
| Hydrochloride | > 300 | Acetone/*i*-PrOH | | | 44 |
| 5-(2-$ClC_6H_4$)-7-Pyrrolidino | 158–160 | EtOH | | | 44 |

| | | | |
|---|---|---|---|
| 5-(2,5-$Cl_2C_6H_3$)-7-Cl | 240 | | 43 |
| 5-(2-$FC_6H_4$)-7-Cl | 195–196 | PrOH | 43 |
|   4-Oxide | 239 | Acetone/MeOH | 46 |
| 5-(2-$FC_6H_4$)-7-Morpholino | 227–240 | EtOH/Petr ether | 44 |
| ***Trisubstituted*** | | | |
| 1-Ac-5-Ph-7-Cl | 256–260 | DMSO | 43 |
| 1-$AcCH_2$-5-Ph-7-Cl | 176 | EtOH | 47 |
| 1-$AcCH_2$-5-(2-$ClC_6H_4$)-7-Cl, 4-oxide | 112–114 | EtOH | 48 |
| 1-Allyl-5-Ph-7-Cl | 94 | | 43 |
| 1-Allyl-5-(2-$ClC_6H_4$)-7-Cl hydrochloride | 200–202d | Acetone/*i*-PrOH | 43 |
|   4-Oxide | 220 | DMF/EtOH | 43 |
| 1-Allyl-5-(2-$ClC_6H_4$)-7-$Me_2N$ | 113–115 | | 44 |
| 1-(Allyl)NHCO-5-Ph-7-Cl | 137–139 | EtOH | 49 |
| 1-(Allyl)NHCO-5-(2-$FC_6H_4$)-7-Cl | 138–140 | EtOH | 49 |
| 1-(Benzoyl)$CH_2$-5-(2-$ClC_6H_4$)-7-Cl | 161–163 | | 47 |
|   4-Oxide | 240 | DMF/EtOH | 47 |
| 1-Bu-5-Ph-7-Cl | 108–110 | | 43 |
| 1-(4-Cl-Benzoyl)$CH_2$-5-Ph-7-Cl | 216–218 | DMF/MeOH | 47 |
| 1-(Cyano)$CH_2$-5-Ph-7-Cl | 222–224 | Acetone | 47 |
| 1-(Cyano)$CH_2$-5-(2-$ClC_6H_4$)-7-Cl | 176 | DMF/EtOH | 48 |
|   4-Oxide | 220 | DMF/EtOH | 48 |
| 1-(3-Cyanopropyl)-5-Ph-7-Cl | 170–174 | $CHCl_3$/Petr ether | 47 |
| 1-(Cyclohexyl)NHCO-5-Ph-7-Cl | 252 | DMF/MeOH | 49 |
| 1-(Cyclopropyl)$CH_2$-5-Ph-7-Cl hydrochloride | 180–188 | *i*-PrOH/$Et_2O$ | 43 |
| 1-EtNHCO-5-Ph-7-Cl | 127 | *i*-PrOH | 49 |
| 1-EtNHCO-5-(2-$ClC_6H_4$)-7-Cl, 4-oxide | 129 | Petr ether | 49 |
| 1-EtNHCO-5-(2-$FC_6H_4$)-7-Cl | 119–121 | THF | 49 |
| 1-$EtOOCCH_2$-5-Ph-7-Cl | 184–186 | EtOAc | 47 |
| 1-$H_2NCSCH_2$-5-Ph-7-Cl | 216d | DMF/EtOH | 48 |
| 1-$H_2NCSCH_2$-5-(2-$ClC_6H_4$)-7-Cl | 204 | DMF/EtOH | 48 |
| 1-(2-HO-Ethyl)-5-(2-$FC_6H_4$)-7-Cl | 154–156 | EtOH | 43 |
| 1-(2-HOOC-Ethyl)-5-Ph-7-Cl | 198–202 | $CHCl_3$/Petr ether | 47 |
| 1-Me-5-Ph-7-Br | 148–150 | EtOH | 43 |

TABLE IX-1. —(contd.)

| Substituent | mp (°C) or; [bp (°C/torr)] | Solvent of Crystallization | Yield (%) | Spectra | Refs. |
|---|---|---|---|---|---|
| 1-Me-5-Ph-7-Cl | 154 | PhH/Petr ether | | | 43 |
| 1-Me-5-(2-$ClC_6H_4$)-7-Cl | 204–206d | EtOH | | | 43 |
| 4-Oxide | 231 | EtOH/Petr ether | | | 43 |
| 1-Me-5-(2-$ClC_6H_4$)-7-$Me_2N$ | 158–162 | MeOH | | | 44 |
| 1-Me-5-(2-$FC_6H_4$)-7-Cl | 139 | EtOH | | | 43 |
| 1-(2-$Me_2N$-Ethyl)-5-Ph-7-Cl | 154 | PhH/Petr ether | | | 43 |
| 1-(2-$Me_2N$-Ethyl)-5-(2-$ClC_6H_4$)-7-$Me_2N$ | 119–120 | $CH_2Cl_2$/Hexane | | | 44 |
| 1-(2-Morpholinoethyl)-5-Ph-7-Cl | 162–164 | EtOH | | | 43 |
| 1-(2-Piperidinoethyl)-5-Ph-7-Cl | 136–137 | EtOH | | | 43 |
| 1-Pr-5-Ph-7-Cl | 139–142 | MeOH | | | 43 |
| 3-AcO-5-(2-$ClC_6H_4$)-7-Cl | 243 | $Ac_2O$/AcOH | | | 43 |
| 3-Benzyl-5-Ph-7-Cl | 234 | | | | 43 |
| 3-(3-HOOC-Propanoyloxy)-5-(2-$ClC_6H_4$)-7-Cl | 170–171 | EtOH/$H_2O$ | | | 43 |
| 3-HO-5-Ph-7-Cl | 177 | EtOH | | | 43 |
| 3-HO-5-(2-$ClC_6H_4$)-7-Cl | 200–202 | EtOH | | | 43 |
| 3-HO-5-(2-$FC_6H_4$)-7-Cl | 177–179 | PrOH | | | 43 |
| 3-Me-5-Ph-7-Cl | 182 | PhH/Petr ether | | | 43 |
| *S*-Enantiomer | 113–116 | PhH/Petr ether | | | 43 |
| 3-*i*-Pr-5-Ph-7-Cl | 225–226 | PhH/Petr ether | | | 43 |
| ***Tetrasubstituted*** | | | | | |
| 1-Ac-3-AcO-5-(2-$ClC_6H_4$)-7-Cl | 203–207 | MeOH | | | 43 |
| 1-$AcCH_2$-3-HO-5-(2-$ClC_6H_4$)-7-Cl | 235–237 | EtOH | | | 48 |
| 1-Allyl-3-AcO-5-2-Cl | | | | | |
| 1-(Cyano)$CH_2$-3-AcO-5-(2-$ClC_6H_4$)-7-Cl | 248 | $Ac_2O$/AcOH | | | 48 |
| 1-(Cyano)$CH_2$-3-HO-5-(2-$ClC_6H_4$)-7-Cl | 178–180 | $CHCl_3$/$Et_2O$ | | | 48 |
| 1-EtNHCO-3-HO-5-(2-$ClC_6H_4$)-7-Cl | 159–163d | | | | 49 |
| 1-Me-3-AcO-5-(2-$ClC_6H_4$)-7-Cl | 178–179 | $Ac_2O$/AcOH | | | 43 |

| | | | | |
|---|---|---|---|---|
| 1,3-$Me_2$-5-Ph-7-Cl | 132–134 | PhH/Petr ether | | 43 |
| *S*-Enantiomer | 143–144 | PhH/Petr ether | | 43 |
| 1-Me-3-HO-5-(2-$ClC_6H_4$)-7-Cl | 247–250 | MeOH | | 43 |
| 1-Me-3-MeO-5-(2-$ClC_6H_4$)-7-Cl | 247–249 | EtOH | | 43 |

***1,3-Dihydropyrido[3,2-e][1,4]diazepin-2-(2H)-thiones***

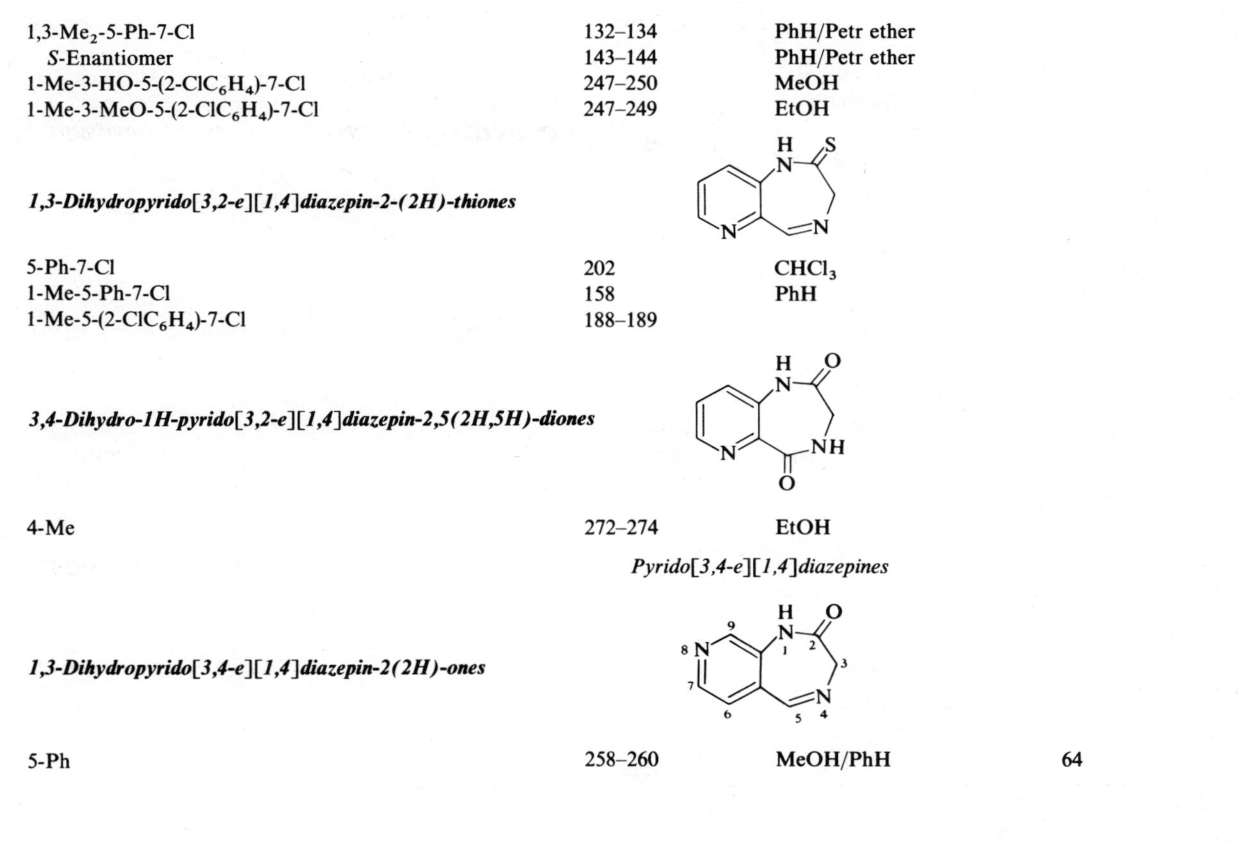

| | | | | |
|---|---|---|---|---|
| 5-Ph-7-Cl | 202 | $CHCl_3$ | | 43 |
| 1-Me-5-Ph-7-Cl | 158 | PhH | | 43 |
| 1-Me-5-(2-$ClC_6H_4$)-7-Cl | 188–189 | | | 43 |

***3,4-Dihydro-1H-pyrido[3,2-e][1,4]diazepin-2,5(2H,5H)-diones***

| | | | | |
|---|---|---|---|---|
| 4-Me | 272–274 | EtOH | | 37 |

*Pyrido[3,4-e][1,4]diazepines*

***1,3-Dihydropyrido[3,4-e][1,4]diazepin-2(2H)-ones***

| | | | | |
|---|---|---|---|---|
| 5-Ph | 258–260 | MeOH/PhH | 64 | 33 |

TABLE IX-1. —(contd.)

| Substituent | mp (°C) or; [bp (°C/torr)] | Solvent of Crystallization | Yield (%) | Spectra | Refs. |
|---|---|---|---|---|---|
| | *Pyrido[4,3-e][1,4]diazepines* | | | | |
| ***1,3-Dihydropyrido[4,3-e][1,4]diazepin-2-(2H)-ones*** | 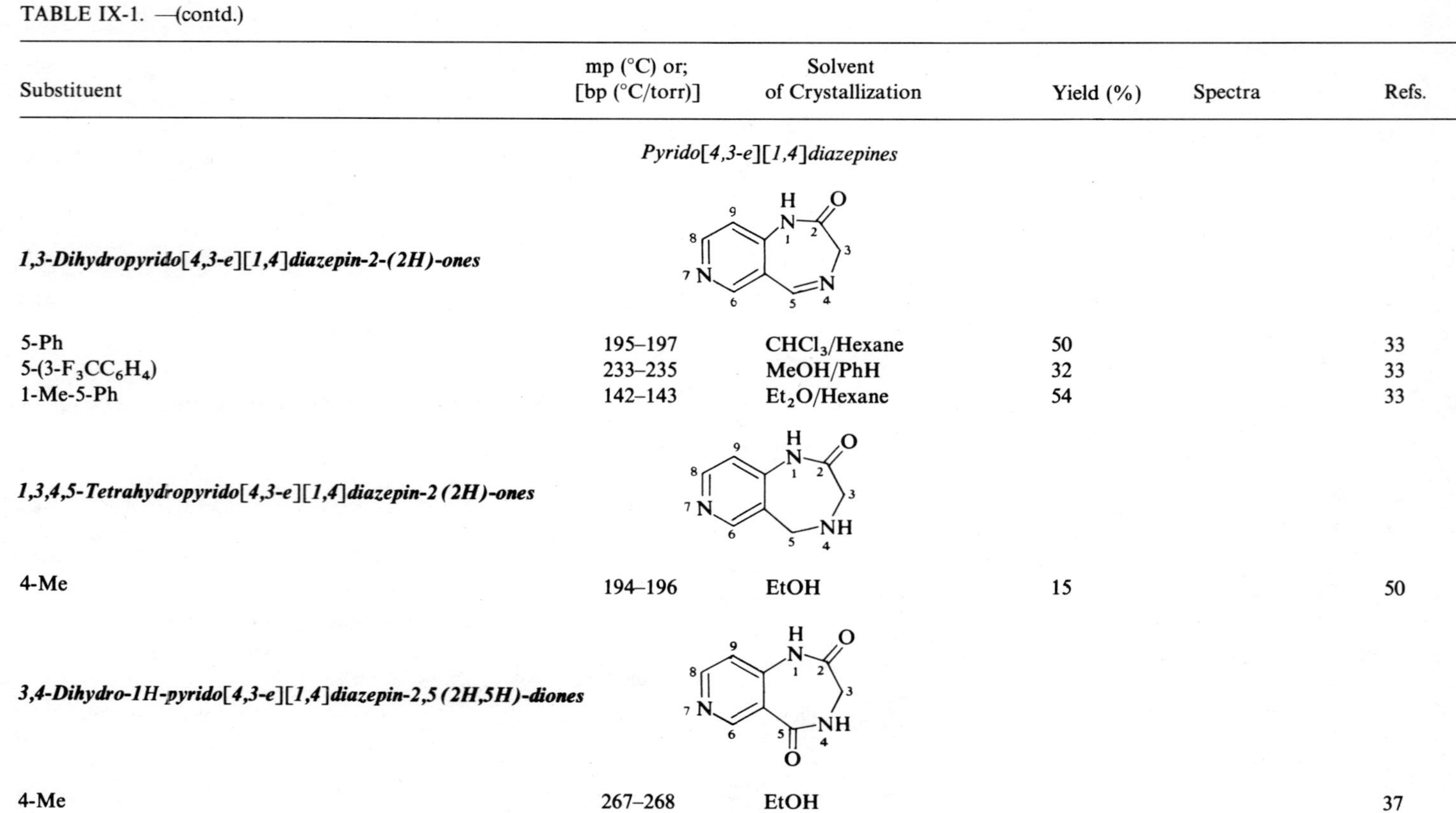 | | | | |
| 5-Ph | 195–197 | $CHCl_3$/Hexane | 50 | | 33 |
| 5-(3-$F_3CC_6H_4$) | 233–235 | MeOH/PhH | 32 | | 33 |
| 1-Me-5-Ph | 142–143 | $Et_2O$/Hexane | 54 | | 33 |
| ***1,3,4,5-Tetrahydropyrido[4,3-e][1,4]diazepin-2(2H)-ones*** | | | | | |
| 4-Me | 194–196 | EtOH | 15 | | 50 |
| ***3,4-Dihydro-1H-pyrido[4,3-e][1,4]diazepin-2,5(2H,5H)-diones*** | | | | | |
| 4-Me | 267–268 | EtOH | | | 37 |

*1H-Pyrimido[4,5-e][1,4]diazepines*

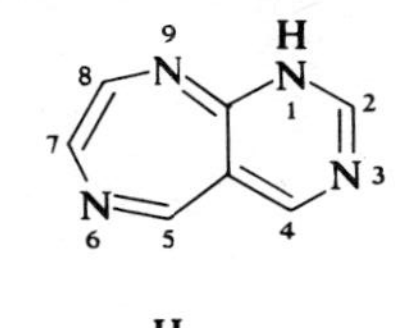


***6,9-Dihydro-5H-pyrimido[4,5-e][1,4]diazepines***

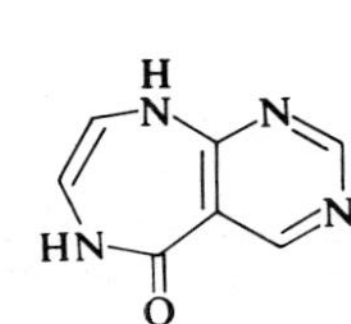


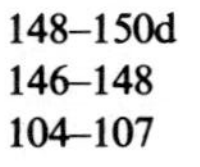

| | | | | | |
|---|---|---|---|---|---|
| 2,7-$Me_2$-8-(2-HO-Ethyl) | | | | | 51 |
| 2,7-$Me_2$-6-Formyl-8-(2-HO-ethyl) | | | | | 51, 52 |
| 2,7-$Me_2$-6-Formyl-8-(2-PhS-ethyl) | 148–150d | | 34 | pmr, uv, p*K* | 53 |
| 2,7-$Me_2$-6-Formyl-8-[2-(4-$BrC_6H_4$)S-ethyl] | 146–148 | | 13 | p*K* | 53 |
| 2,7-$Me_2$-6-Formyl-8-[2-(4-$MeC_6H_4$)S-ethyl] | 104–107 | | 27 | p*K* | 53 |

***6,7,8,9-Tetrahydropyrimido[4,5-e][1,4]diazepin-5(5H)-ones***

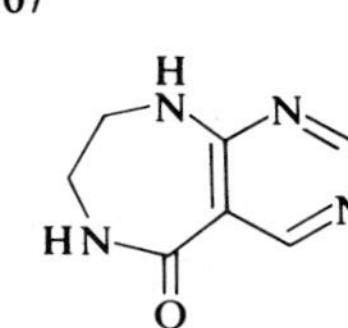


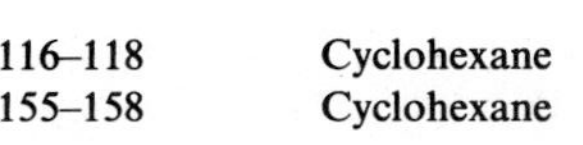

| | | | | | |
|---|---|---|---|---|---|
| 2-Ph-6,9-$Et_2$ | 116–118 | Cyclohexane | | | 54 |
| 2-Ph-6,9-$Me_2$ | 155–158 | Cyclohexane | | pmr | 54 |

***5,6,8,9-Tetrahydropyrimido[4,5-e][1,4]diazepin-7(7H)-ones***

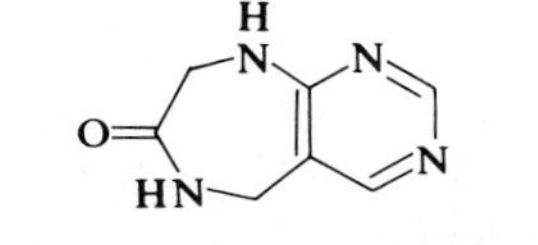


| | | | | | |
|---|---|---|---|---|---|
| 2-Me | 274–275 | MeOH/$Et_2O$ | | | 35 |

TABLE IX-1. —(contd.)

| Substituent | mp (°C) or; [bp (°C/torr)] | Solvent of Crystallization | Yield (%) | Spectra | Refs. |
|---|---|---|---|---|---|
| ***3,4,7,9-Tetrahydropyrimido[4,5-e][1,4]diazepin-4,8 (8H)-diones*** 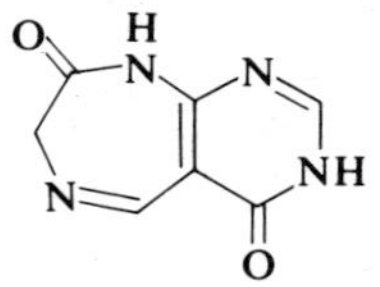 | | | | | |
| 5-Ph | 270d | EtOH | | | 9b |
| 3,5-$Me_2$ | 245–250d | EtOH | | | 9b |
| 3-Benzyl-5-Ph | 246–248 | EtOH | | | 9b |
| 3-Me-5-Ph | 259–261 | MeCN | | | 9b |
| 3-(2-$Me_2$N-Ethyl)-5-Ph | 226–228 | MeCN | | | 9b |
| 3-Benzyl-5-Ph-9-Me | 132–134 | EtOH | | | 9b |
| 3,9-$Me_2$-5-Ph | 154–156 | MeCN/$Et_2O$/Hexane | | | 9b |
| 3,9-(2-$Me_2$N-Ethyl)2-5-Ph | 160–162 | EtOAc | | | 9b |
| ***3,4,5,6,7,8-Hexahydropyrimido[5,4-e][1,4]diazepin-4,9 (9H)-diones*** 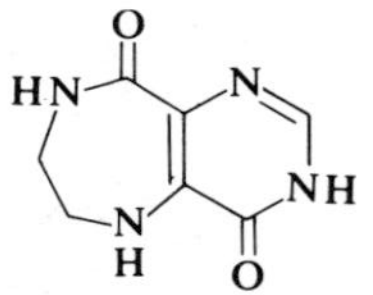 | | | | | |
| 2-(4-$ClC_6H_4$) | 264–266 | EtOH | | ir | 55 |
| 2-(Morpholino)$COCH_2$ | 234–236 | EtOH | | | 55 |
| 2-(Piperidino)$COCH_2$ | 234–236 | EtOH | | | 55 |

*Pyrrolo[2,3-e][1,4]diazepines*

***1,2,3,4,5,5a,6,7-Octahydropyrrolo[2,3-e][1,4]diazepin-5-ones***

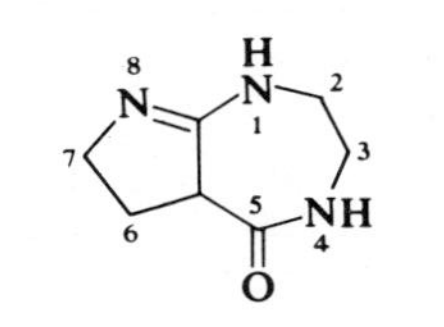


| | | | | |
|---|---|---|---|---|
| 6-Ph | 175–176 | Acetone | | 38 |

*Pyrrolo[3,2-e][1,4]diazepines*

***1,2,3,6-Tetrahydropyrrolo[3,2-e][1,4]diazepin-2-ones***

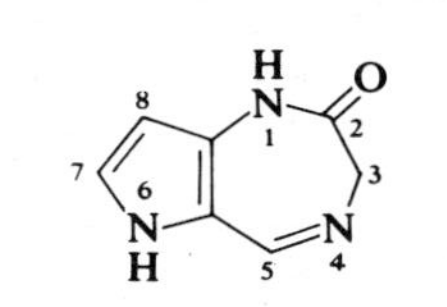


| | | | | |
|---|---|---|---|---|
| 5-Ph-8-CN | 315–318d | DMF/*i*-PrOH | | 56 |
| 5-Ph-8-EtOOC | 211–212 | $CH_2Cl_2$/MeOH | | 56 |
| 5-Ph-8-$H_2$NCO | 200–203d | DMF/$Et_2O$ | | 56 |

*Pyrrolo[3,4-e][1,4]diazepines*

***1,2,3,7-Tetrahydropyrrolo[3,4-e][1,4]diazepin-2-ones***

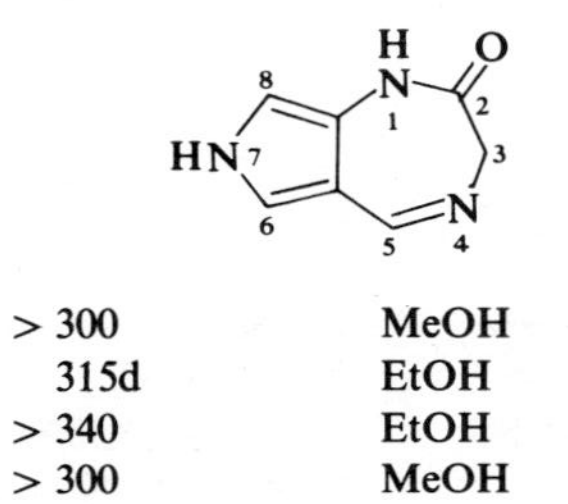


| | | | | |
|---|---|---|---|---|
| 5-Ph-6-Me | > 300 | MeOH | 64 | 57 |
| 5-(2-$ClC_6H_4$)-6-Me | 315d | EtOH | 37 | 57 |
| 5-(2-$FC_6H_4$)-6-Me | > 340 | EtOH | 45 | 57 |
| 5-(4-$MeOC_6H_4$)-6-Me | > 300 | MeOH | 35 | 57 |

TABLE IX-1. —(contd.)

| Substituent | mp (°C) or; [bp (°C/torr)] | Solvent of Crystallization | Yield (%) | Spectra | Refs. |
|---|---|---|---|---|---|
| ***Trisubstituted*** | | | | | |
| 5-Ph-6,7-$Me_2$ | 270d | MeOH | 67 | | 57 |
| 4-Oxide | 255–258 | $H_2O$ | | | 60 |
| 5-Ph-6-Me-8-Et | 290d | *i*-PrOH | 52 | | 57 |
| 5-Ph-6-Me-7-Bu | 152–153 | EtOAc | 36 | | 57 |
| 5-Ph-6-Me-7-Et | 229–231 | EtOAc | 42 | | 57 |
| 5-Ph-6-Me-7-(2-Me-Prop-1-yl) | 191–193 | EtOAc | 41 | | 57 |
| 5-Ph-6-Me-7-Pr | 163–165 | EtOAc | 33 | | 57 |
| 5,8-$Ph_2$-6-Me | 310–312 | EtOH | 43 | ir, pmr | 57 |
| ***Tetrasubstituted*** | | | | | |
| 1-Et-5-Ph-6,7-$Me_2$ | 136–138 | $CHCl_3$/EtOH | 47 | ir, pmr | 57 |
| 1,6,7-$Me_3$-5-Ph | 237–239 | *i*-PrOH | 62 | | 57 |
| 1,6,7-$Me_3$-5-(2-$ClC_6H_4$), 4-oxide | 198–200 | *i*-PrOH | | | 60 |
| 3-AcO-5-Ph-6,7-$Me_2$ | 255–256 | $Ac_2O$ | | | 60 |
| 3-Butanoyloxy-5-Ph-6,7-$Me_2$ | 198–200 | *i*-PrOH | | | 60 |
| 3-(4-*t*-Bu-Benzoyloxy)-5-Ph-6,7-$Me_2$ | 235–237 | *i*-PrOH | | | 60 |
| 3-(Cyclohexyl)COO-5-Ph-6,7-$Me_2$ | 209–210 | *i*-PrOH | | | 60 |
| 3-(Decanoyloxy)-5-Ph-6,7-$Me_2$ | 187–188 | EtOH | | | 60 |
| 3{3-[N-(3-EtOOC-Propyl)aminocarbonyl] propanoyloxy}-5-Ph-6,7-$Me_2$ hydrochloride | 125–127 | Acetone/$Et_2O$ | | | 60 |
| 3-(4-F-Benzoyloxy)-5-Ph-6,7-$Me_2$ | 211–213 | EtOH | | | 60 |
| 3-HO-5-Ph-6,7-$Me_2$ | 243d | EtOH | | | 60 |
| 3-(3-HOOC-Propanoyloxy)-5-Ph-6,7-$Me_2$ | 153–155 | EtOH | | | 60 |
| 3-(4-Phthalimidobutanoyloxy)-5-Ph-6,7-$Me_2$ | 210–212 | EtOH | | | 60 |
| 5-Ph-6,7-$Me_2$-8-Br | 200–202 | MeOH | | | 61 |
| 5-Ph-6,7-$Me_2$-8-Cl | 195d | EtOH | | | 61 |

| | | | | | |
|---|---|---|---|---|---|
| 5-Ph-6,7-$Me_2$-8-Et | 254–256 | MeOH | 32 | | 57 |
| 5-Ph-6,7-$Me_2$-8-$NO_2$ | 232–234 | MeOH | | | 61 |
| ***Pentasubstituted*** | | | | | |
| 1,6,7-$Me_3$-3-AcO-5-(2-$ClC_6H_4$) | 202–204 | EtOH | | | 60 |
| 1,6,7-$Me_3$-3-Benzoyloxy-5-(2-$ClC_6H_4$) | 228–229 | EtOH | | | 60 |
| 1,6,7-$Me_3$-3-EtO-5-(2-$ClC_6H_4$) hydrochloride | 110–111 | $Et_2O$ | | | 60 |
| 1,6,7-$Me_3$-3-(3-EtOOC-Propyl)NH-5-(2-Cl)$C_6H_4$ | 105–107 | *i*-PrOH/$Et_2O$ | | | 60 |
| 1,6,7-$Me_3$-3-HO-5-(2-$ClC_6H_4$) | 182–183 | EtOAc | | | 60 |
| 1,6,7-$Me_3$-3-(3-HOOC-Propanoyloxy)-5-(2-$ClC_6H_4$ | 173–175 | *i*-PrOH | | | 60 |
| 1,6,7-$Me_3$-3-(2-Me-Benzoyloxy)-5-(2-$ClC_6H_4$) | 171–173 | EtOH | | | 60 |
| 1,6,7-$Me_2$-3-(2-Me-Propanoyloxy)-5-(2-$ClC_6H_4$) | 173–175 | EtOH | | | 60 |
| 1,6,7-$Me_3$-3-(2,2-$Me_2$-Propanoyloxy)-5-(2-$ClC_6H_4$) | 221–222 | EtOH | | | 60 |
| 1,6,7-$Me_3$-3-MeO-5-(2-$ClC_6H_4$) hydrochloride·$H_2O$ | 101–103 | $Et_2O$ | | | 60 |
| 1,6,7-$Me_3$-3-(4-MeO-Benzoyloxy)-5-(2-$ClC_6H_4$) | 138–140 | EtOH | | | 60 |
| 1,6,7-$Me_3$-3-(4-Ph-Benzoyloxy)-5-(2-$ClC_6H_4$) | 173–175 | $CHCl_3$/MeOH | | | 60 |
| 1,6,7-$Me_3$-5,8-$Ph_2$ | 232–234 | MeOH | 45 | | 57 |
| ***1,2,3,7-Tetrahydropyrrolo[3,4-e][1,4]diazepin-6,8 (6H,8H)-diones*** | | | | | |
| 5-Me | 262–264 | MeOH | 77 | ir, pmr | 64 |
| 5-Me-7-Ph | 220–221 | EtOH | 77 | ir, pmr | 64 |
| ***6-Methylene-1,3,6,7-tetrahydropyrrolo[3,4-e][1,4]diazepin-2,8(2H,8H)-diones*** | | | | | |
| 5-Ph-7-Me | 215–217 | EtOAc/Hexane | | | 59 |

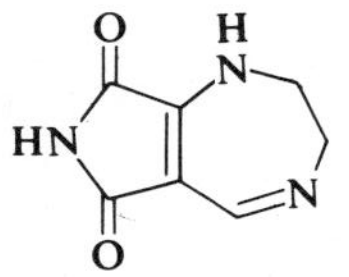

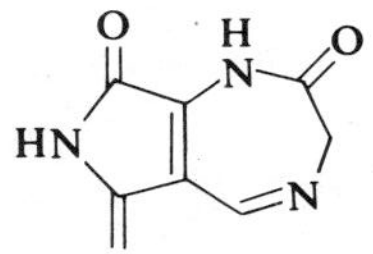

TABLE IX-1. —(contd.)

| Substituent | mp (°C) or; [bp (°C/torr)] | Solvent of Crystallization | Yield (%) | Spectra | Refs. |
|---|---|---|---|---|---|
| | *1H-Thiadiazolo[3,4-e][1,4]diazepines* | | | | |
| **4,6-Dihydrothiadiazolo[3,4-e][1,4]diazepin-5(5H)-thiones** | | | | | |
| 8-Ph | 178–180 | Acetone/Hexane | | | 65 |
| **6H-Thiazolo[4,5-e][1,4]diazepines** | | | | | |
| 2-Morpholino-5-AcNHNH-8-Ph | 226–228 | Acetone | 44 | pmr | 66 |
| **4,6-Dihydrothiazolo[4,5-e][1,4]diazepin-5(5H)-ones** | | | | | |
| 2-(4-$ClC_6H_4$) | 194–195 | EtOAc | 77 | | 66 |
| 2-$Me_2N$-8-Ph | 226–227 | EtOAc | 17 | | 66 |

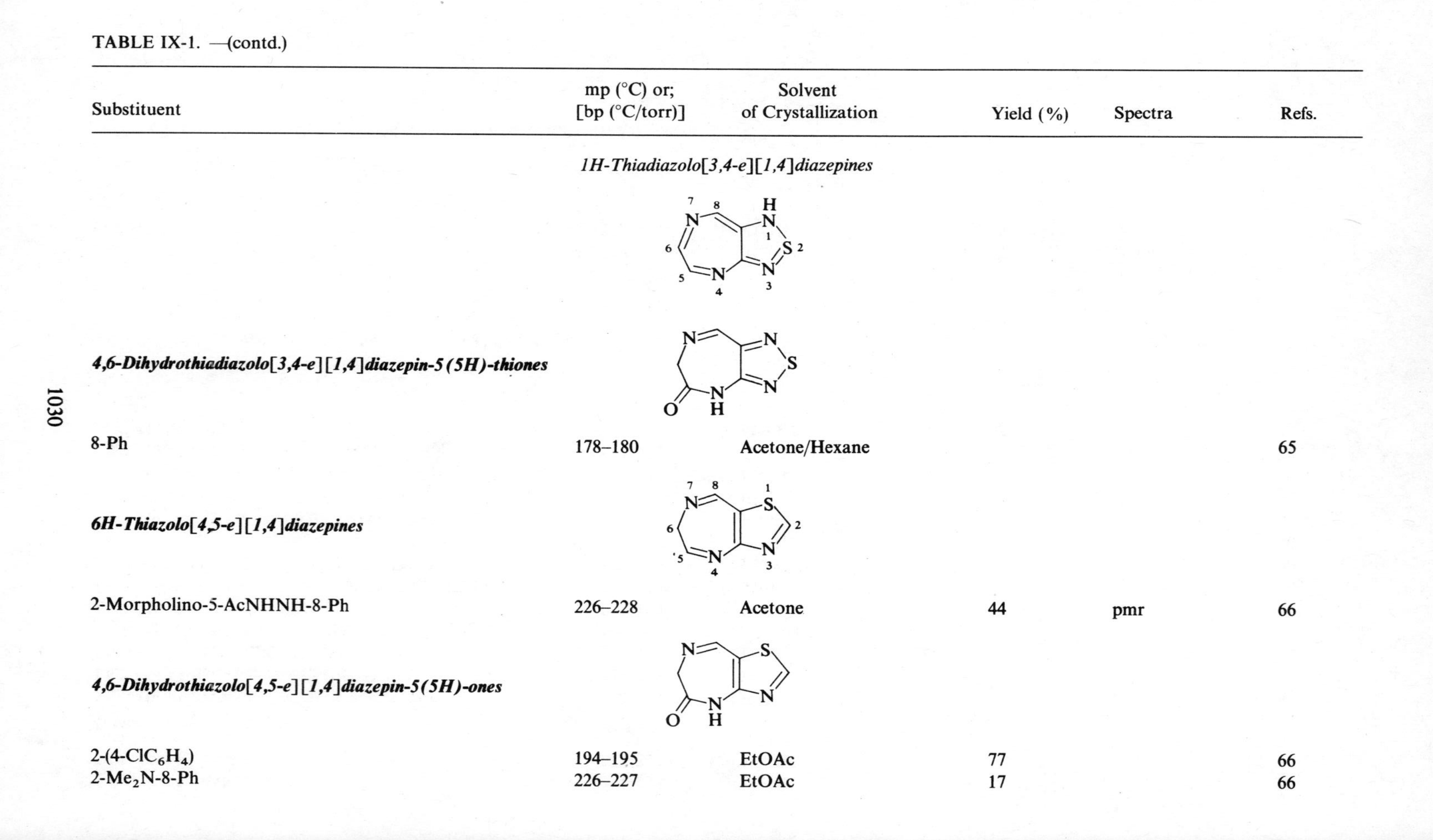

| | | | | | |
|---|---|---|---|---|---|
| 2-Morpholino-8-Ph | 200–201 | EtOAc | 30 | $^{13}C$ -nmr | 66 |
| 2-Piperidino-8-Ph | 191–193 | EtOAc | 29 | pmr | 66 |
| 2-$Me_2N$-4-Me-8-Ph | 138–139 | EtOAc | 22 | | 66 |
| 2-$Me_2N$-6-Benzyl-8-Ph | 181–182 | EtOAc | 66 | | 66 |
| 2-$Me_2N$-6-Me-8-Ph | 204–207 | EtOAc | 19 | | 66 |
| 2-Morpholino-4-Me-8-(4-$ClC_6H_4$) | 222–223 | EtOAc | 37 | | 66 |
| 2-Piperidino-4-Me-8-Ph | 145–147 | EtOAc | 37 | | 66 |

***4,6-Dihydrothiazolo[4,5-e][1,4]diazepin-5(5H)-thiones***

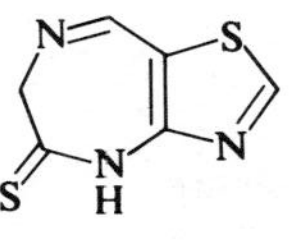

| | | | | | |
|---|---|---|---|---|---|
| 2-Morpholino-8-Ph | | | | | 66 |

*4H-Thiazolo[5,4-e][1,4]diazepines*

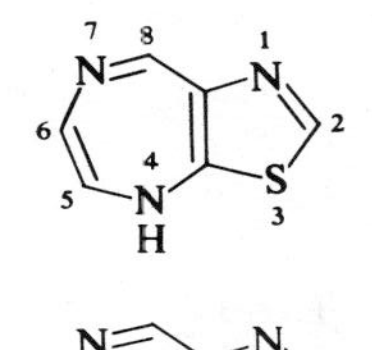

***4,6-Dihydrothiazolo[5,4-e][1,4]diazepin-5(5H)-ones***

| | | | | | |
|---|---|---|---|---|---|
| 8-Ph | 258–260 | $CH_2Cl_2$/EtOH | | | 6 |
| 8-(2-Thiazolyl) | 183–185d | $CH_2Cl_2$/MeOH | | | 10 |
| 4-Me-8-Ph | 120 | $Et_2O$/Petr ether | | | 6 |
| 3-Me-8-(3-$O_2NC_6H_4$) | 186 | $CH_2Cl_2$/EtOH | | | 6 |

TABLE IX-1. —(contd.)

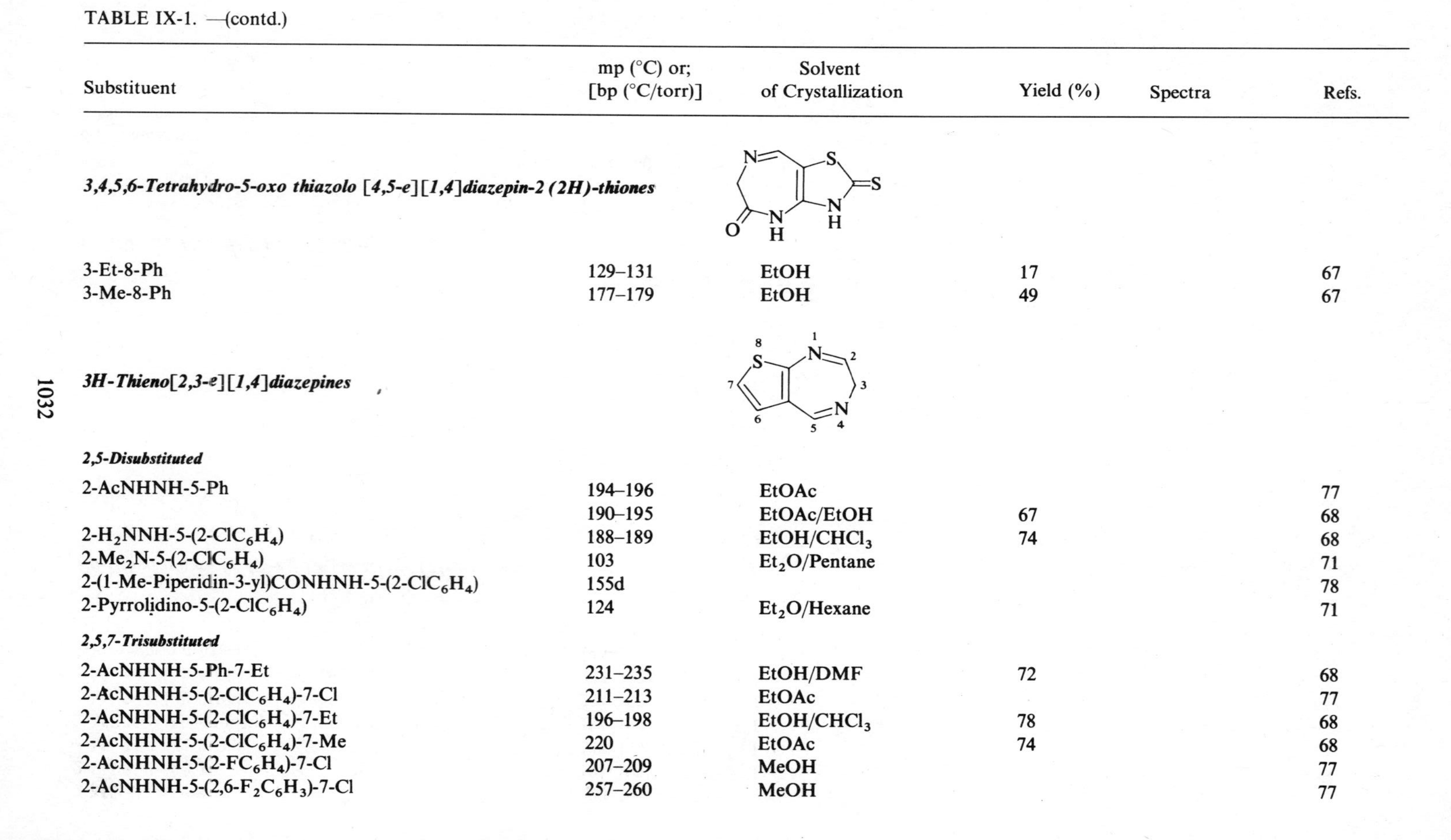

| Substituent | mp (°C) or; [bp (°C/torr)] | Solvent of Crystallization | Yield (%) | Spectra | Refs. |
|---|---|---|---|---|---|
| ***3,4,5,6-Tetrahydro-5-oxo thiazolo [4,5-e][1,4]diazepin-2 (2H)-thiones*** | | | | | |
| 3-Et-8-Ph | 129–131 | EtOH | 17 | | 67 |
| 3-Me-8-Ph | 177–179 | EtOH | 49 | | 67 |
| ***3H-Thieno[2,3-e][1,4]diazepines*** | | | | | |
| ***2,5-Disubstituted*** | | | | | |
| 2-AcNHNH-5-Ph | 194–196 | EtOAc | | | 77 |
| | 190–195 | EtOAc/EtOH | 67 | | 68 |
| 2-$H_2$NNH-5-(2-Cl$C_6H_4$) | 188–189 | EtOH/$CHCl_3$ | 74 | | 68 |
| 2-$Me_2$N-5-(2-Cl$C_6H_4$) | 103 | $Et_2$O/Pentane | | | 71 |
| 2-(1-Me-Piperidin-3-yl)CONHNH-5-(2-Cl$C_6H_4$) | 155d | | | | 78 |
| 2-Pyrrolidino-5-(2-Cl$C_6H_4$) | 124 | $Et_2$O/Hexane | | | 71 |
| ***2,5,7-Trisubstituted*** | | | | | |
| 2-AcNHNH-5-Ph-7-Et | 231–235 | EtOH/DMF | 72 | | 68 |
| 2-AcNHNH-5-(2-Cl$C_6H_4$)-7-Cl | 211–213 | EtOAc | | | 77 |
| 2-AcNHNH-5-(2-Cl$C_6H_4$)-7-Et | 196–198 | EtOH/$CHCl_3$ | 78 | | 68 |
| 2-AcNHNH-5-(2-Cl$C_6H_4$)-7-Me | 220 | EtOAc | 74 | | 68 |
| 2-AcNHNH-5-(2-F$C_6H_4$)-7-Cl | 207–209 | MeOH | | | 77 |
| 2-AcNHNH-5-(2,6-$F_2C_6H_3$)-7-Cl | 257–260 | MeOH | | | 77 |

| | | | | |
|---|---|---|---|---|
| 2-AcNHNH-5-(2-Pyridyl)-7-Cl | 193–195 | MeOH | | 77 |
| 2-(Benzoyl)NHNH-5-(2-ClC$_6$H$_4$)-7-Et | 228–230 | EtOH | 64 | 68 |
| 2-BuNH-5-(2-ClC$_6$H$_4$)-7-Et | 165–168 | | | 70 |
| 2-(Cyclohexyl)CONHNH-5-(2-BrC$_6$H$_4$)-7-Br | 220d | | | 78 |
| 2-(Cyclohexyl)CONHNH-5-(2-ClC$_6$H$_4$)-7-Br | 140d | | | 78 |
| 2-Cyclohexyl)CONHNH-5-(2-ClC$_6$H$_4$)-7-Cl | 177d | | | 78 |
| 2-(Cyclohexyl)CONHNH-5-(2-ClC$_6$H$_4$)-7-Et | 203 | | | 78 |
| 2-(Cyclopentyl)CONHNH-5-(2-ClC$_6$H$_4$)-7-Br | 110d | | | 78 |
| 2-(Cyclopropyl)CONHNH-5-(2-ClC$_6$H$_4$)-7-Br | 236d | | | 78 |
| 2-EtNH-5-(2-ClC$_6$H$_4$)-7-Et | 188–190 | | | 70 |
| Oxalate | 186–187 | | | 70 |
| 2-EtOOC(Me)NNH-5-(2-ClC$_6$H$_4$)-7-Et | 172–174 | Ligroin | | 75 |
| 2-EtOOCNHNH-5-(2-ClC$_6$H$_4$)-7-Br | 236 | BuOH | | 78 |
| 2-[2,2-(EtO)$_2$-Ethyl]NH-5-(2-ClC$_6$H$_4$)-7-Et | 143–146 | EtOH | | 72 |
| 2-H$_2$N-5-(2-ClC$_6$H$_4$)-7-Et | 247–248 | CHCl$_3$/EtOH | | 70 |
| 2-H$_2$NNH-5-Ph-7-Cl | 204–207 | MeOH | | 77 |
| 2-H$_2$NNH-5-Ph-7-Et | 195–196 | EtOH/CHCl$_3$ | 75 | 68 |
| 2-H$_2$NNH-5-(2-BrC$_6$H$_4$)-7-Et | 189–190 | EtOH/DMF | 90 | 68 |
| 2-H$_2$NNH-5-(2-ClC$_6$H$_4$)-7-Br | 300d | THF/Et$_2$O | 80 | 74 |
| 2-H$_2$NNH-5-(2-ClC$_6$H$_4$)-7-Et | 214–216 | EtOH/DMF | 92 | 68 |
| 2-H$_2$NNH-5-(2-ClC$_6$H$_4$)-7-I | 145d | Et$_2$O | | 76 |
| 2-H$_2$NNH-5-(2-ClC$_6$H$_4$)-7-Me | 216–218 | EtOH/DMF | 73 | 68 |
| 2-H$_2$NNH-5-(2-FC$_6$H$_4$)-7-Et | 203–204 | EtOH | 58 | 68 |
| 2-H$_2$NNH-5-(4-FC$_6$H$_4$)-7-Et | 218–219 | MeOH/CHCl$_3$ | 53 | 68 |
| 2-H$_2$NNH-5-(2-MeC$_6$H$_4$)-7-Et | 205–207 | EtOH/DMF | 84 | 68 |
| 2-H$_2$NNH-5-(2-MeOC$_6$H$_4$)-7-Et | 207–208 | EtOH/CHCl$_3$ | 62 | 68 |
| 2-H$_2$NNH-5-(4-MeOC$_6$H$_4$)-7-Et | 184–186 | EtOH/CHCl$_3$ | 76 | 68 |
| 2-H$_2$NNH-5-(2-Pyridyl)-7-Et | 164–165 | MeOH | 62 | 68 |
| 2-H$_2$NNH-5-(3-Pyridyl)-7-Et | 182–183 | EtOH | 75 | 68 |
| 2-MeNH-5-Ph-7-Cl | 247–250 | CH$_2$Cl$_2$ | 68 | 16 |
| 2-MeNH-5-Ph-7-Et | 243–245 | | | 70 |
| 2-MeNH-5-(2-ClC$_6$H$_4$)-7-Cl | 259–262 | THF/Hexane | 81 | 16 |
| 2-MeNH-5-(2-ClC$_6$H$_4$)-7-Et | 216–217 | CHCl$_3$/EtOH | | 70 |
| Hydrochloride | 232–234 | | | 70 |

TABLE IX-1. —(contd.)

| Substituent | mp (°C) or; [bp (°C)/(torr)] | Solvent of Crystallization | Yield (%) | Spectra | Refs. |
|---|---|---|---|---|---|
| 2-MeNH-5-(2-ClC$_6$H$_4$)-7-Me | 241–243 | | | | 70 |
| 2-MeNHCONHNH-5-(2-ClC$_6$H$_4$)-7-Et | 213–215 | CHCl$_3$/Ligroin | | | 75 |
| 2-Me$_2$N-5-Ph-7-Et | 141–143 | | | | 70 |
| 2-Me$_2$N-5-(2-ClC$_6$H$_4$)-7-Et Hydrochloride | 249–250d | EtOH/EtOAc | | | 70 |
| 2-Me(NO)N-5-Ph-7-Cl | 111–113 | Et$_2$O/Hexane | 80 | | 16 |
| 2-Me(NO)N-5-(2-ClC$_6$H$_4$)-7-Cl | 104–107 | Et$_2$O/Petr ether | 92 | | 16 |
| 2-(MeO-Acetyl)NHNH-5-(2-NO$_2$C$_6$H$_4$)-7-Cl | 198–200 | MeOH | | | 77 |
| 2-MeOOC(HON)C-5-(2-ClC$_6$H$_4$)-7-Cl | 242–245d | THF/MeOH | 23 | | 16 |
| 2-(4-Me-Piperazino)-5-(2-ClC$_6$H$_4$)-7-Et dimaleate | 156–158 | | | | 70 |
| 2-(1-Me-Piperidin-3-yl)CONHNH-5-(2-ClC$_6$H$_4$)-7-Br | 217–218d | | | | 78 |
| 2-(1-Me-Piperidin-3-yl)CONHNH-5-(2-ClC$_6$H$_4$)-7-Et | 165–168d | | | | 78 |
| 2-Morpholino-5-(2-ClC$_6$H$_4$)-7-Et dihydrochloride | 167–17 | | | | 70 |
| 2-Piperidino-5-(2-ClC$_6$H$_4$)-7-Et dihydrochloride | 190–192 | EtOH/EtOAc | | | 70 |
| 2-(Piperidin-4-yl)CONHNH-5-(2-ClC$_6$H$_4$)-7-Br | 196d | | | | 78 |
| 2-(Propanoyl)NHNH-5-(2-ClC$_6$H$_4$)-7-Et | 187 | EtOH/Ligroin | 68 | | 68 |
| 2-PrNH-5-(2-ClC$_6$H$_4$)-7-Et | 174–176 | | | | 70 |
| 2-(Pyridin-2-yl)CONHNH-5-(2-ClC$_6$H$_4$)-7-Br | 217d | | | | 78 |
| 2-Pyrrolidino-5-(2-ClC$_6$H$_4$)-7-Et | 140–142 | | | | 70 |
| 2-(Pyrrolidin-2-yl)CONHNH-5-(2-ClC$_6$H$_4$)-7-Br | 215–220d | | | | 78 |
| 2-(Tetrahydrofuran-2-yl)CONHNH-5-(2-BrC$_6$H$_4$)-7-Br | 172d | | | | 78 |
| 2-(Tetrahydrofuran-2-yl)CONHNH-5-(2-ClC$_6$H$_4$)-7-Br | 199–200 | EtOH | 60 | | 78 |
| 2-(Tetrahydrofuran-3-yl)CONHNH-5-(2-ClC$_6$H$_4$)-7-Br | 212d | | | | 78 |
| 2-(Tetrahydropyran-2-yl)CONHNH-5-(2-ClC$_6$H$_4$)-7-Br | 185 | EtOH | | | 78 |
| 2-(Tetrahydropyran-3-yl)CONHNH-5-(2-ClC$_6$H$_4$)-7-Br | 200d | | | | 78 |
| 2-(Tetrahydropyran-3-yl)CONHNH-5-(2-ClC$_6$H$_4$)-7-Et | 197d | | | | 78 |
| 2-(Tetrahydrothien-2-yl)CONHNH-5-(2-ClC$_6$H$_4$)-7-Br | 197–198d | | | | 78 |
| 2-(Tetrahydrothiopyran-2-yl)CONHNH-5-(2-ClC$_6$H$_4$)-7-Br | 213d | | | | 78 |

| | | | | |
|---|---|---|---|---|
| 2-(Tetrahydrothiopyran-4-yl)CONHNH-5-(2-$ClC_6H_4$)-7-Br | 204–206 | | | 78 |
| 2-(Thien-2-yl)CONHNH-5-(2-$ClC_6H_4$)-7-Br | 215–218d | | | 78 |
| ***Tetrasubstituted*** | | | | |
| 2-$H_2NNH$-5-Ph-6,7-$Me_2$ | 225–227 | EtOH/$CHCl_3$ | 84 | 68 |
| 2-$H_2NNH$-5-(2-$ClC_6H_4$)-6,7-$Me_2$ | 220–222 | EtOH/$CHCl_3$ | 83 | 68 |
| 2-MeNH-5-Ph-6,7-$Me_2$ | 275–278 | | | 70 |
| ***2,3-Dihydro-1H-thieno[2,3-e][1,4]diazepines*** | | | | |
| ***Disubstituted*** | | | | |
| 5-Ph-7-Et | 190–191 | | | 80 |
| 5-(2-$ClC_6H_4$)-7-Et | 178–179 | $CH_2Cl_2$/Petr ether | 90 | 80 |
| 5-(2-$ClC_6H_4$)-7-Me | 205–206 | | | 80 |
| 5-(2-$MeC_6H_4$)-7-Et | 185–186 | | | 80 |
| 5-(2-$MeOC_6H_4$)-7-Et | 177–178 | | | 80 |
| ***Trisubstituted*** | | | | |
| 1-BuNHCO-5-(2-$ClC_6H_4$)-7-Et hydrochloride | 201 | | | 80 |
| 1-(2-$ClC_6H_4$)NHCO-5-(2-$ClC_6H_4$)-7-Et | 128–130 | | | 80 |
| 1-$Et_2NCO$-5-(2-$ClC_6H_4$)-7-Et hydrochloride | 224–226d | MeOH/EtOAc | | 81 |
| 1-(2-$Et_2N$-Ethyl)NHCO-5-(2-$ClC_6H_4$)-7-Et dihydrochloride·$H_2O$ | 150–151d | MeOH/EtOAc | | 81 |
| 1-Me-5-(2-$ClC_6H_4$)-7-Et hydrochloride·$H_2O$ | 247–248 | | | 80 |
| 1-MeNHCO-5-(2-$ClC_6H_4$)-7-Et | 137–138 | $CHCl_3$/Petr ether | 83 | 80 |
| 1-MeNHCO-5-(2-$MeC_6H_4$)-7-Et hydrochloride | 207–208 | | | 80 |
| 1-MeNHCO-5-(2-$MeOC_6H_4$)-7-Et hydrochloride | 213–214 | | | 80 |
| 1-MeNHCS-5-(2-$ClC_6H_4$)-7-Et hydrochloride | 155–156 | MeOH | 67 | 80 |
| 1-$Me_2NCO$-5-Ph-7-Et hydrochloride | 237–238d | EtOH | | 81 |
| 1-$Me_2NCO$-5-(2-$BrC_6H_4$)-7-Et hydrochloride | 242–243d | EtOH | | 81 |

TABLE IX-1. —(contd.)

| Substituent | mp (°C) or; [bp (°C)/(torr)] | Solvent of Crystallization | Yield (%) | Spectra | Refs. |
|---|---|---|---|---|---|
| 1-$Me_2NCO$-5-(2-$ClC_6H_4$)-7-Cl | 132–133 | EtOH/Ligroin | | | 81 |
| 1-$Me_2NCO$-5-(2-$ClC_6H_4$)-7-Et hydrochloride | 240–241d | EtOAc/MeOH | | | 81 |
| 1-$Me_2NCO$-5-(2-$ClC_6H_4$)-7-Me hydrochloride | 238–239d | Acetone/$H_2O$ | | | 81 |
| 1-$Me_2NCO$-5-(4-$ClC_6H_4$)-7-Et hydrochloride | 208–210d | EtOH | | | 81 |
| 1-$Me_2NCO$-5-(3,4-$Cl_2C_6H_3$)-7-Et hydrochloride | 219–221d | EtOH | | | 81 |
| 1-$Me_2NCO$-5-(4-$FC_6H_4$)-7-Et hydrochloride | 216–218d | EtOH | | | 81 |
| 1-$Me_2NCO$-5-(2-$MeC_6H_4$)-7-Et hydrochloride | 225–226d | MeOH/EtOAc | | | 81 |
| 1-$Me_2NCO$-5-(4-$MeC_6H_4$)-7-Et hydrochloride | 216–218d | EtOH | | | 81 |
| 1-$Me_2NCO$-5-(2-$MeOC_6H_4$)-7-Et hydrochloride | 219–220d | MeOH/EtOAc | | | 81 |
| 1-$Me_2NCO$-5-(4-$MeOC_6H_4$)-7-Et hydrochloride | 204–206d | EtOH | | | 81 |
| 1-$Me_2NCO$-5-(2-Pyridyl)-7-Et dihydrochloride | 183–192d | EtOH | | | 81 |
| 1-(2-$Me_2N$-Ethyl)NHCO-5-(2-$ClC_6H_4$)-7-Et dihydrochloride·$H_2O$ | 160–162d | MeOH/EtOAc | | | 81 |
| 1-(4-Me-Piperazino)CO-5-(2-$ClC_6H_4$)-7-Et·0.5$H_2O$ | 248–250d | Acetone/$H_2O$ | | | 81 |
| 1-(3-$Me_2N$-Propyl)NHCO-5-(2-$ClC_6H_4$)-7-Et dihydrochloride | 223–224d | MeOH/EtOAc | | | 81 |
| 1-(Morpholino)CO-5-(2-$ClC_6H_4$)-7-Et hydrochloride | 234–235d | MeOH/EtOAc | | | 81 |
| 1-(3-Morpholinopropyl)NHCO-5-(2-$ClC_6H_4$)-7-Et dihydrochloride | 225–238d | | | | 81 |
| 1-(Piperidino)CO-5-(2-$ClC_6H_4$)-7-Et hydrochloride | 237–238d | MeOH/EtOAc | | | 81 |
| 1-(2-Piperidinoethyl)NHCO-5-(2-$ClC_6H_4$)-7-Et dihydrochloride·0.5$H_2O$ | 215–216d | EtOH | | | 81 |
| 1-(Pyrrolidino)CO-5-(2-$ClC_6H_4$)-7-Et hydrochloride | 236–237d | MeOH/Acetone | | | 81 |
| 2-$H_2NCH_2$-5-Ph-7-Cl dimaleate | 176–178 | EtOAc/EtOH | 28 | | 16 |
| 5-Ph-6,7-$Me_2$ | 203–205 | | | | 80 |
| ***Tetrasubstituted*** | | | | | |
| 1-MeNHCO-5-Ph-6,7-$Me_2$ | 225–227 | | | | 80 |
| 1-$Me_2NCO$-5-(2-$ClC_6H_4$)-6,7-$Me_2$ hydrochloride | 246d | MeOH/EtOAc | | | 81 |

| Compound | mp (°C) | Solvent | Yield (%) | Spectra | Ref. |
|---|---|---|---|---|---|
| ***2-Methylene-1,3-dihydro-2H-thieno[2,3-e][1,4]diazepines*** | | | | | |
| ***$R_1$, $R_2$; Other*** | | | | | |
| H, $NO_2$; 5-Ph-7-Cl | 164–165 | EtOAc/Hexane | 68 | | 16 |
| MeOOC, MeOOC; 5-(2-$ClC_6H_4$)-7-Cl | 158–160 | EtOH | 70 | | 16 |
| ***1,3-Dihydrothieno[2,3-e][1,4]diazepin-2(2H)-ones*** | | | | | |
| ***Monosubstituted*** | | | | | |
| 5-Et | 137–140 | $Et_2O$ | 17 | ir, pmr, uv | 93 |
| 5-Me | 200d | Acetone | | | 6 |
| 5-Ph | 203 | EtOAc | 75 | | 85 |
| 4-Oxide | 260–263d | EtOH | | | 85 |
| 5-(2-$ClC_6H_4$) | 222–225 | EtOH | | | 97 |
| 5-(2,3-$Cl_2C_6H_3$) | 201–203 | PhH | 63 | | 87 |
| 5(2-$FC_6H_4$) | 196–199 | Dioxane | 89 | | 87 |
| | 201–202 | EtOH | 75 | ir, pmr, uv | 93 |
| 5-(2,6-$F_2C_6H_3$) | 235–237 | Dioxane | 42 | | 87 |
| 5-(2-$F_3CC_6H_4$) | 205–207 | EtOH | 52 | | 87 |
| 5-(2-$MeC_6H_4$) | 213–215 | PhH | 80 | | 87 |
| 5-(2-Me-3-$NO_2C_6H_3$) | 217–220 | MeOH | 45 | | 87 |
| 5-(2-$MeOC_6H_4$) | 270–271 | Dioxane | 56 | | 87 |
| 5-(2-$MeSO_2C_6H_4$) | 258–260 | EtOH | 45 | | 87 |
| 5-(2-$NO_2C_6H_4$) | 255–257 | $CH_2Cl_2$ | 53 | | 88 |
| 5-(3-$NO_2C_6H_4$) | 223–225 | EtOH | | | 95 |
| 5-(2-Pyridyl) | 263–266d | EtOH | | | 84 |
| 5-(3-Me-2-Pyridyl) | 272–274d | EtOH | | | 84 |

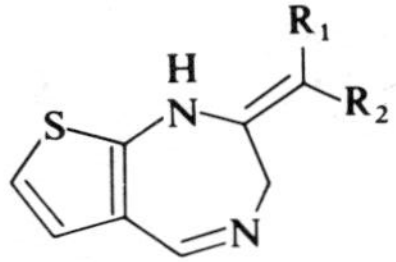

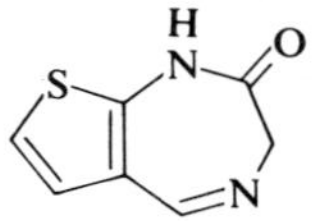

TABLE IX-1. —(contd.)

| Substituent | mp(°C) or; [bp (°C)/(torr)] | Solvent of Crystallization | Yield (%) | Spectra | Refs. |
|---|---|---|---|---|---|
| ***1,5-Disubstituted*** | | | | | |
| 1,5-$Me_2$ | 116–120 | $Et_2O$ | | | 6 |
| 1-Me-5-Et | 82–83 | $Et_2O$ | 18 | | 93 |
| 4-Oxide | 169–171 | $EtOH/Et_2O$ | 47 | ir, pmr, uv | 93 |
| 1-Me-5-Ph | 136–138 | EtOH | | | 85 |
| | 134–135 | EtOH | 86 | | 93 |
| 4-Oxide | 158–159 | $EtOH/CCl_4$ | 67 | ir, pmr, uv | 93 |
| 1-Me-5-(2-$BrC_6H_4$) | 109–113 | EtOH | 32 | ir, pmr, uv | 93 |
| 4-Oxide | 162–164 | $EtOH/CCl_4$ | 61 | ir, pmr, uv | 93 |
| 1-Me-5-(2-$ClC_6H_4$) | 109–111 | $Et_2O$ | 56 | ir, pmr, uv | 93 |
| 4-Oxide | 163–164 | $EtOH/CCl_4$ | 73 | ir, pmr, uv | 93 |
| 1-Me-5-(4-$ClC_6H_4$) | 139–141 | EtOH | 76 | ir, pmr | 93 |
| 1-Me-5-(2,3-$Cl_2C_6H_3$) | 148–150 | EtOH | 87 | | 87 |
| 1-Me-5-(2-$FC_6H_4$) | 113–116 | Cyclohexane | 71 | | 87 |
| | 115–117 | EtOH | 84 | ir, pmr, uv | 93 |
| 4-Oxide·EtOH | 81–85 | EtOH | 83 | ir, pmr, uv | 93 |
| 1-Me-5-(2,6-$F_2C_6H_3$) | 99–100 | $Et_2O$ | 77 | | 87 |
| | 97–100 | $Et_2O$ | 55 | ir, pmr, uv | 102 |
| 4-Oxide | 104–107 | EtOH | 55 | ir, pmr, uv | 102 |
| 1-Me-5-(2-$F_3CC_6H_4$) | 110–113 | MeOH | 89 | | 87 |
| 1-Me-5-(2-$IC_6H_4$) | 136–138 | EtOH | 48 | ir, pmr, uv | 93 |
| 4-Oxide | 161d | $EtOH/CCl_4$ | 18 | ir, pmr, uv | 93 |
| 1-Me-5-(2-$MeC_6H_4$) | 123–124 | EtOH | 63 | | 87 |
| | 125–127 | EtOH | 74 | ir, pmr, uv | 93 |
| 4-oxide | 149–151 | EtOH | 55 | ir, pmr, uv | 93 |
| 1-Me-5-(2-Me-3-$NO_2C_6H_3$) | 130–132 | EtOH | 72 | | 87 |
| 1-Me-5-(2-$MeOC_6H_4$) | 129–130 | Cyclohexane | 53 | | 87 |
| 1-Me-5-(2-$NO_2C_6H_4$) | 149–151 | MeOH | 83 | | 88 |

| | | | | |
|---|---|---|---|---|
| 1-Me-5-(3-$NO_2C_6H_4$) | 196–198 | EtOH | | 95 |
| 1-Me-5-(2-Pyridyl) | 142–144 | MeOH | | 84 |
| 1-Me-5-(3-Me-2-Pyridyl) | 137–140 | Cyclohexane | | 84 |
| ***3,5-Disubstituted*** | | | | |
| 3-EtOOC-5-Ph | 122 | $Et_2O$/Petr ether | | 6 |
| 3-Me-5-Ph | 189–190 | $Et_2O$ | | 6 |
| ***5,6-Disubstituted*** | | | | |
| 5,6-$Ph_2$ | 251–253 | MeCN | 23 | 82 |
| 5-Ph-6-Cyclopropyl | 160 | | | 98 |
| Hydrochloride | 264 | Acetone/$Et_2O$ | | 98 |
| 5-Ph-6-Me | 248–250 | MeOH | 60 | 82 |
| | 253–254d | $CH_2Cl_2$/EtOH | | 83 |
| 5-Ph-6-*i*-Pr | 230–232 | EtOH/Hexane | | 86 |
| 5-(2-$NO_2C_6H_4$)-6-Et | 220–222 | EtOH | 52 | 88 |
| 5-(2-$NO_2C_6H_4$)-6-Me | 245–248 | EtOH | 62 | 88 |
| 5-(2-Thienyl)-6-Me | 211–213 | MeOH | 30 | 82 |
| 5-(2-Thienyl)-7-$NO_2$ | 260d | Acetone | | 77b |
| ***5,7-Disubstituted*** | | | | |
| 5,7-$Ph_2$ | 230–232 | MeCN | 40 | 82 |
| 5-Ph-7-Bu | 163–164 | *i*-PrOH | | 86 |
| 5-Ph-7-Cl | 250–252d | EtOH | | 85 |
| 5-Ph-7-Et | 194–195 | PhMe | 70 | 86 |
| 5-Ph-7-Me | 213–215 | EtOH/PhH | | 92, 83 |
| | 207–210 | | | 86 |
| 5-Ph-7-MeOOC | 246–248 | EtOH | | 96 |
| 5-Ph-7-$NO_2$ | 266–268d | EtOH | | 85 |
| 5-(2-$BrC_6H_4$)-7-Et | 208–209 | EtOH/Hexane | 79 | 86 |
| (5-(2-$ClC_6H_4$)-7-Br | 248d | EtOH | | 97 |
| 5-(2-$ClC_6H_4$)-7-Cl | 237d | EtOH | | 97 |
| 5-(2-$ClC_6H_4$)-7-Et | 204–206 | PhMe | 81 | 86 |
| 4-Oxide | 225d | | | 94 |

TABLE IX-1. —(contd.)

| Substituent | mp (°C) or; [bp (°C/torr)] | Solvent of Crystallization | Yield (%) | Spectra | Refs. |
|---|---|---|---|---|---|
| 5-(2-$ClC_6H_4$)-7-I | 214–216 | Dioxane | | | 97 |
| | 222–224 | EtOH | 87 | | 76 |
| 5-(2-$ClC_6H_4$)-7-Me | 212–213 | EtOH/Hexane | 74 | | 86 |
| 5-(2-$ClC_6H_4$)-7-$NO_2$ | 269d | Dioxane | | | 97 |
| 5-(2-$ClC_6H_4$)-7-*i*-Pr | 243–246d | EtOH | 84 | | 86 |
| 5-(2,3-$Cl_2C_6H_3$)-7-Cl | 253–255 | EtOH | 50 | | 87 |
| 5-(2,3-$Cl_2C_6H_3$)-7-$NO_2$ | 253–255 | PhH | 92 | | 87 |
| 5-(2-$FC_6H_4$)-7-Cl | 256–259 | EtOH | 77 | | 87 |
| 5-(2-$FC_6H_4$)-7-Et | 178–180 | EtOH/Hexane | 82 | | 86 |
| 5-(2-$FC_6H_4$)-7-I | 212–214 | EtOH | 60 | | 87 |
| 5-(2-$FC_6H_4$)-7-$NO_2$ | 258–259 | EtOH | 89 | | 87 |
| 5-(2,3-$F_2C_6H_3$)-7-Cl | 245–247 | EtOH | 52 | | 87 |
| 5-(2,6-$F_2C_6H_3$)-7-Cl | 245–247 | EtOH | | | 97 |
| 5-(2,6-$F_2C_6H_3$)-7-$NO_2$ | 268–269 | Dioxane | | | 97 |
| 5-(2-$F_3CC_6H_4$)-7-Cl | 278–281 | EtOH | 75 | | 87 |
| 5-(2-$MeC_6H_4$)-7-Cl | 237–241 | EtOH | 62 | | 87 |
| 5-(2-$MeC_6H_4$)-7-Et | 182–183 | PhMe | 76 | | 86 |
| 5-(2-Me-3-$NO_2C_6H_3$)-7-Cl | 275–280 | EtOH | 45 | | 87 |
| 5-(2-Me-3-$NO_2C_6H_3$)-7-$NO_2$ | 238–240 | Dioxane | 57 | | 87 |
| 5-(2-$MeOC_6H_4$)-7-Cl | 250–252 | EtOH | 41 | | 87 |
| 5-(2-$MeOC_6H_4$)-7-Et | 168–169 | PhMe | 64 | | 86 |
| 5-(2-$MeSO_2C_6H_4$)-7-Cl | 270–272 | EtOH | 74 | | 87 |
| 5-(2-$MeSO_2C_6H_4$)-7-$NO_2$ | 268–270 | EtOH | 80 | | 87 |
| 5-(2-$NO_2C_6H_4$)-7-Ac | 260–262 | PhH | 21 | | 88 |
| 5-(2-$NO_2C_6H_4$)-7-Cl | 243–246 | $CH_2Cl_2$ | 28 | | 88 |
| | 253–254 | $CH_2Cl_2$ | | | 97 |
| 5-(2-$NO_2C_6H_4$)-7-Et | 190–192 | EtOH | 46 | | 88 |
| 5-(2-$NO_2C_6H_4$)-7-Me | 225–227 | EtOH | 34 | | 88 |

| | | | | | |
|---|---|---|---|---|---|
| 5-(2-$NO_2C_6H_4$)-7-$NO_2$ | 267–269 | Dioxane | 80 | | 88 |
| 5-(3-$NO_2C_6H_4$)-7-Cl | 232–234d | EtOH | | | 95 |
| 5-(4-$NO_2C_6H_4$)-7-Cl | 267 | $CH_2Cl_2$ | | | 77b |
| 5-(2-Pyridyl)-7-Cl | 250–252d | EtOH | | | 77 |
| 5-(2-Pyridyl)-7-Et | 228–234d | EtOH | 48 | | 89 |
| 5-(3-Me-2-Pyridyl)-7-Cl | 237–239d | Dioxane | | | 84 |
| ***1,5,6-Trisubstituted*** | | | | | |
| 1-Allyl-5-Ph-6-Me | 116–118 | Cyclohexane | 74 | | 82 |
| 1-Et-5-Ph-6-Me | 158–160 | MeOH | 80 | | 82 |
| 1-Me-5-Ph-6-Cyclopropyl hydrochloride | 238 | Acetone/$Et_2O$ | | | 98 |
| 1-Me-5-Ph-6-Me | 202–203 | MeOH | 83 | | 82 |
| | 205–206 | EtOH | 80 | | 83 |
| 1-Me-5-(2-$NO_2C_6H_4$)-6-Et | 167–169 | EtOH | 77 | | 88 |
| 1-Me-5-(2-$NO_2C_6H_4$)-6-Me | 204–206 | EtOH | 67 | | 88 |
| ***1,5,7-Trisubstituted*** | | | | | |
| 1-Allyl-5-(2-$NO_2C_6H_4$)-7-Cl | 93–95 | Cyclohexane | | | 97 |
| 1-(Cyclopropyl)$CH_2$-5-Ph-7-Cl | 95–97 | Cyclohexane | | | 85 |
| 1-(Cyclopropyl)$CH_2$-5-(2-$NO_2C_6H_4$)-7-Cl | 91–93 | MeOH | | | 97 |
| 1-Et-5-(2-$NO_2C_6H_4$)-7-Cl | 167–168 | MeOH | | | 97 |
| 1-(2-$Et_2N$-Ethyl)-5-(2-$NO_2C_6H_4$)-7-Cl | 110–112 | MeOH | | | 97 |
| 1-$H_2N$-5-(2-$ClC_6H_4$)-7-Et | 153–154 | EtOH | 90 | | 100 |
| 1-$H_2N$-5-(2-$NO_2C_6H_4$)-7-Cl | 157–159 | EtOH | | | 77b |
| 1-(2-HO-Ethyl)-5-(2-$NO_2C_6H_4$)-7-Cl | 143–145 | $Et_2O$ | | | 97 |
| 1,5-$Me_2$-7-$NO_2$ | 170d | $CH_2Cl_2/Et_2O$ | | | 6 |
| 1-Me-5-Et-7-Cl | 76–78 | $Et_2O$ | 16 | ir, pmr | 93 |
| 4-Oxide | 171–172 | $Et_2O$ | 25 | ir, pmr | 93 |
| 1-Me-5-Ph-7-Cl | 120–122 | EtOH | | | 85 |
| | 117–119 | EtOH | 80 | | 93 |
| 4-Oxide | 194–195 | EtOH/$CCl_4$ | 40 | ir, pmr, uv | 93 |
| 1-Me-5-Ph-7-Et | 103–104 | EtOH | 64 | | 86 |
| 1-Me-5-Ph-7-Me | 97–100 | $CH_2Cl_2$ | 85 | | 83 |
| Hydrochloride | 234–235d | EtOH/EtOAc | 70 | | 86 |

TABLE IX-1. —(contd.)

| Substituent | mp (°C) or; [bp (°C/torr)] | Solvent of Crystallization | Yield (%) | Spectra | Refs. |
|---|---|---|---|---|---|
| 1-Me-5-Ph-7-MeOOC | 147–149 | MeOH | | | 96 |
| 1-Me-5-Ph-7-$NO_2$ | 195–197 | EtOH | | | 85 |
| 1-Me-5-(2-$BrC_6H_4$)-7-Cl | 86–89 | $Et_2O$ | 49 | ir, pmr, uv | 93 |
| 1-Me-5-(2-$BrC_6H_4$)-7-Et | 100–102 | EtOH/Hexane | 80 | | 86 |
| 1-Me-5-(2-$ClC_6H_4$)-7-Cl | 83–85 | Cyclohexane | | | 97 |
| | | | 28 | ir, pmr, uv | 93 |
| 4-Oxide | 95–97 | $CCl_4$ | 54 | ir, pmr, uv | 93 |
| 1-Me-5-(2-$ClC_6H_4$)-7-Et | 105–106 | Hexane | 88 | | 86 |
| 4-Oxide | 117–118 | PhH/Hexane | 95 | | 94 |
| 1-Me-5-(2-$ClC_6H_4$)-7-I | 117–118 | $Et_2O$ | | | 97 |
| 1-Me-5-(2-$ClC_6H_4$)-7-Me hydrochloride | 232–234d | EtOH/EtOAc | 68 | | 86 |
| 1-Me-5-(2-$ClC_6H_4$)-7-$NO_2$ | 162–165 | EtOH/PhH | | | 97 |
| 1-Me-5-(4-$ClC_6H_4$)-7-Cl | 212–214 | EtOH/$CHCl_3$ | 83 | ir, pmr, uv | 93 |
| 1-Me-5-(2,3-$Cl_2C_6H_3$)-7-Cl | 100–105 | $Et_2O$ | 70 | | 87 |
| 1-Me-5-(2-$FC_6H_4$)-7-Cl | 97–98 | Cyclohexane | 65 | | 87 |
| | 101–103 | *i*-PrOAc | 73 | ir, pmr, uv | 93 |
| 4-Oxide | 153–154 | $Et_2O$/Hexane | 83 | ir, pmr, uv | 93 |
| 1-Me-5-(2-$FC_6H_4$)-7-Et hydrochloride | 203–204d | Acetone/EtOH | 68 | | 86 |
| 1-Me-5-(2-$FC_6H_4$)-7-$NO_2$ | 174–177 | PhH | 83 | | 87 |
| 1-Me-5-(2-$F_3CC_5H_4$)-7-Cl | 100–102 | Cyclohexane | 65 | | 87 |
| 1-Me-5-(2-$F_3CC_5H_4$)-7-$NO_2$ | 224–225 | Dioxane | | | 97 |
| 1-Me-5-(2-$IC_6H_4$)-7-Cl hydrochloride | 217–221 | EtOH | 32 | ir, pmr | 93 |
| 1-Me-5-(2-$MeC_6H_4$)-7-Cl | 81–82 | EtOH | 65 | | 87 |
| 1-Me-5-(2-Me-3-$NO_2C_6H_3$)-7-Cl | 128–130 | EtOH | 44 | | 87 |
| 1-Me-5-(2-Me-3-$NO_3C_6H_3$)-7-$NO_2$ | 171–173 | Dioxane | 52 | | 87 |
| 1-Me-5-(2-$NO_2C_6H_4$)-7-Cl | 145–147 | MeOH | 69 | | 88 |
| 1-Me-5-(2-$NO_2C_6H_4$)-7-Et | 126–127 | EtOH | 80 | | 88 |
| 1-Me-5-(2-$NO_2C_6H_4$)-7-Me | 106–108 | $Et_2O$ | 81 | | 88 |
| 1-Me-5-(2-$NO_2C_6H_4$)-7-$NO_2$ | 143–145 | MeOH | 78 | | 88 |

| | | | | |
|---|---|---|---|---|
| 1-Me-5-(3-$NO_2C_6H_4$)-7-Cl | 201–203 | EtOH | | 95 |
| 1-Me-5-(2-Pyridyl)-7-Cl | 250–252d | EtOH | | 84 |
| 1-Me-5-(2-Pyridyl)-7-Et | 120–121 | | 31 | 89 |
| 1-Me-5-(3-Me-2-Pyridyl)-7-Cl | 129–131 | Cyclohexane | | 84 |
| 1-Me-5-(2-Thienyl)-7-Cl | 171–173 | EtOH | | 77b |
| 1-MeNHCO-5-(2-$ClC_6H_4$)-7-Et | 126–127 | EtOH | | 100 |
| 1-$MeOCH_2$-5-(2-$NO_2C_6H_4$)-7-Cl | 147–149 | EtOH | | 97 |
| 1-*i*-Pr-5-(2-$NO_2C_6H_4$)-7-Cl | 126–128 | MeOH | | 97 |
| ***3,5,7-Trisubstituted*** | | | | |
| 3-AcO-5-(2-$ClC_6H_4$)-7-Et | 219–220 | | 81 | 94 |
| 3-HO-5-(2-$ClC_6H_4$)-7-Et | 153–154 | EtOH | 97 | 94 |
| Sodium salt | 239–240d | EtOH/$H_2O$ | | 94 |
| 3-Me-5-Ph-7-$NO_2$ | 251 | $CH_2Cl_2$/EtOH | | 6 |
| ***5,6,7-Trisubstituted*** | | | | |
| 5-Ph-6,7-$Cl_2$ | 226–228d | PhH | | 99 |
| 5-Ph-6-Et-7-Me | 218–220 | MeOH | 40 | 82 |
| 5-Ph-6,7-$Me_2$ | 236–237 | DMF/$H_2O$ | 75 | 86 |
| | 242–244 | EtOH | | 83 |
| 5-Ph-6-Me-7-Br | 225–230d | $CH_2Cl_2$ | | 83 |
| 5-Ph-6-Me-7-Cl | 233–236 | EtOH | | 83 |
| 5-Ph-6-Me-7-$NO_2$ | 235–260d | EtOH | | 83 |
| 5-Ph-6-Me-7-*i*-Pr | 207–208 | EtOH | 70 | 86 |
| 5-Ph-6-$NO_2$-7-Cl | 245–250d | EtOH | | 95 |
| 5-Ph-6-Pr-7-Et | 193–195 | MeCN | 32 | 82 |
| 5-(2-$ClC_6H_4$)-6,7-$Me_2$ | 251 | EtOH/PhMe | 72 | 86 |
| 4-Oxide | 150–151 | | | 94 |
| 5-(4-$ClC_6H_4$-6,7-$Me_2$ | 243–245d | EtOH/Hexane | 80 | 86 |
| Hydrobromide | 262–263 | | | 86b |
| 5-(3-$F_3CC_6H_4$)-6,7-$Me_2$ | 220–222 | EtOH/Hexane | 75 | 86 |
| 5-(2-$NO_2C_6H_4$)-6-Et-7-Cl | 218–221 | EtOH | 53 | 88 |
| 5-(2-$NO_2C_6H_4$)-6-Et-7-$NO_2$ | 250–255 | MeOH | 75 | 88 |
| 5-(2-$NO_2C_6H_4$)-6,7-$Me_2$ | 202–204 | MeOH | 42 | 88 |

TABLE IX-1. —(contd.)

| Substituent | mp (°C) or; [bp (°C/torr)] | Solvent of Crystallization | Yield (%) | Spectra | Refs. |
|---|---|---|---|---|---|
| 5-(2-$NO_2C_6H_4$)-6-Me-7-Cl | 215–217 | EtOH | 28 | | 88 |
| 5-(2-$NO_2C_6H_4$)-6-Me-7-$NO_2$ | 273 | EtOH | 56 | | 88 |
| 5-(2-$NO_2C_6H_4$)-6-$NO_2$-7-Et | 219–221 | PhH | | | 77b |
| 5-(2-$NO_2C_6H_4$)-6-$NO_2$-7-Me | 261 | EtOH | 85 | | 88 |
| ***1,3,5,7-Tetrasubstituted*** | | | | | |
| 1-Me-3-AcO-5-(2-$ClC_6H_4$)-7-Et | Oil | | | | 94 |
| 1-Me-3-AcO-5-(2-$ClC_6H_4$)-7-Me | 201–202 | | | | 94 |
| 1-Me-3-HO-5-(2-$ClC_6H_4$)-7-Et | 142–143 | | | | 94 |
| 1-Me-3-HO-5-(2-$ClC_6H_4$)-7-Me | 168–169 | | | | 94 |
| ***1,5,6,7-Tetrasubstituted*** | | | | | |
| 1-Me-5-Ph-6, 7-$Cl_2$ | 147–149 | MeOH | | | 99 |
| 1-Me-5-Ph-6, 7-$Me_2$ | 121 | EtOH/Hexane | 70 | | 86 |
| | 137–140 | Cyclohexane | 90 | | 83 |
| 1-Me-5-Ph-6-Me-7-Br | 193–194 | EtOH | | | 83 |
| 1-Me-5-Ph-6-Me-7-Cl | 139–142 | EtOH | | | 83 |
| 1-Me-5-Ph-6-Me-7-$NO_2$ | 199–201 | $CCl_4$ | | | 83 |
| 1-Me-5-Ph-6-$NO_2$-7-Cl | 179–181 | PhH | | | 95 |
| 1-Me-5-(2-$ClC_6H_4$)-6,7-$Me_2$ hydrochloride | 235–236 | EtOH | 83 | | 86 |
| 4-Oxide | 180–181 | | | | 94 |
| 1-Me-5-(4-$ClC_6H_4$)-6,7-$Me_2$ | 182–184 | EtOH/Hexane | 76 | | 86 |
| 1-Me-5-(2-$NO_2C_6H_4$)-6-Et-7-Cl | 207–210 | EtOH | 82 | | 88 |
| 1-Me-5-(2-$NO_2C_6H_4$)-6-Et-7-$NO_2$ | 217–219 | Dioxane | 71 | | 88 |
| 1-Me-5-(2-$NO_2C_6H_4$)-6,7-$Me_2$ | 149–151 | EtOH | 82 | | 88 |
| 1-Me-5-(2-$NO_2C_6H_4$)-6-Me-7-Cl | 189–190 | EtOH | 58 | | 88 |
| 1-Me-5-(2-$NO_2C_6H_4$)-6-Me-7-$NO_2$ | 169–171 | EtOH | 81 | | 88 |
| 1-Me-5-(2-$NO_2C_6H_4$)-6-$NO_2$-7-Et | 164–166 | PhH | 85 | | 88 |
| 1-Me-5-(2-$NO_2C_6H_4$)-6-$NO_2$-7-Me | 193–195 | EtOH | 89 | | 88 |

| | | | | |
|---|---|---|---|---|
| ***Pentasubstituted*** | | | | |
| 1-Me-3-AcO-5-(2-$ClC_6H_4$)-6,7-$Me_2$ | 111–112 | | | 94 |
| 1-Me-3-HO-5-(2-$ClC_6H_4$)-6,7-$Me_2$ | 183–184d | | | 94 |
| ***1,3-Dihydrothieno[2,3-e][1,4]diazepin-2(2H)-thiones*** | | | | |
| ***Monosubstituted*** | | | | |
| 5-Ph | 210 | $MeOH/CHCl_3$ | 52 | 68 |
| 5-(2-$ClC_6H_4$) | 206–208 | $MeOH/CHCl_3$ | 85 | 68 |
| | 221–223 | $CH_2Cl_2$ | | 77 |
| 5-(2-$NO_2C_6H_4$) | 207–209 | MeOH | | 77 |
| ***Disubstituted*** | | | | |
| 5-Ph-7-Cl | 223–225d | MeOH | | 77 |
| 5-Ph-7-Et | 202–203 | EtOH | 58 | 68 |
| 5-(2-$BrC_6H_4$)-7-Et | 194–195 | $MeOH/CHCl_3$ | 71 | 68 |
| 5-(2-$ClC_6H_4$)-7-Br | 214d | $CH_2Cl_2$ | 50 | 74 |
| 5-(2-$ClC_6H_4$)-7-Cl | 223–225d | MeOH | | 77 |
| 5-(2-$ClC_6H_4$)-7-I | 202 | $Et_2O$ | 92 | 76 |
| 5-(2-$ClC_6H_4$)-7-Et | 198–199 | $EtOH/CHCl_3$ | 84 | 68 |
| 5-(2-$ClC_6H_4$)-7-Me | 218–219 | $EtOH/CHCl_3$ | 64 | 68 |
| 5-(4-$ClC_6H_4$)-7-Et | 230 | $EtOH/CHCl_3$ | 76 | 68 |
| 5-(2-$FC_6H_4$)-7-Et | 188–189 | EtOH | 43 | 68 |
| 5-(4-$FC_6H_4$)-7-Et | 225–227 | $EtOH/CHCl_3$ | 63 | 68 |
| 5-(2-$MeC_6H_4$)-7-Et | 223–225 | EtOH | 48 | 68 |
| 5-(2-$MeOC_6H_4$)-7-Et | 189–190 | $EtOH/CHCl_3$ | 68 | 68 |
| 5-(4-$MeOC_6H_4$)-7-Et | 202–203 | MeOH | 81 | 68 |
| 5-(2-$NO_2C_6H_4$)-7-Cl | 213–215 | MeOH | | 77 |
| 5-(2-Pyridyl)-7-Et | 199–200 | $CHCl_3$ | 44 | 68 |
| 5-(3-Pyridyl)-7-Et | 223–225 | EtOH | 75 | 68 |

TABLE IX-1. —(contd.)

| Substituent | mp (°C) or; [bp (°C/torr)] | Solvent of Crystallization | Yield (%) | Spectra | Refs. |
|---|---|---|---|---|---|
| ***Trisubstituted*** | | | | | |
| 5-Ph-6,7-$Me_2$ | 232–233 | EtOH | 62 | | 68 |
| 5-(2-$ClC_6H_4$)-6,7-$Me_2$ | 220–221 | EtOH | 65 | | 68 |
| ***3,4-Dihydrothieno[2,3-e][1,4]diazepin-5(5H)-ones*** | 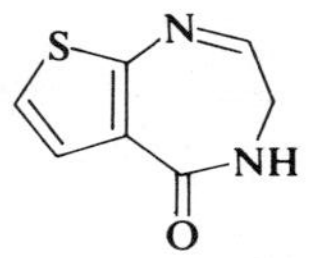 | | | | |
| 2-$H_2N$ | 254–255 | EtOH | 73 | pmr | 103 |
| | *Thieno[3,2-e][1,4]diazepines* | | | | |
| ***3H-Thieno[3,2-e][1,4]diazepines*** | 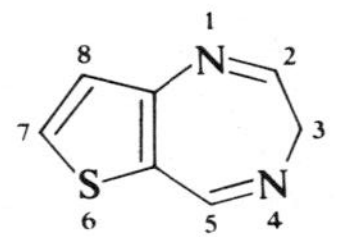 | | | | |
| 2-MeNH-5-Ph | 227–229 | $CH_2Cl_2$ | 92 | | 16 |
| 2-Me(NO)N-5-Ph | 158–160 | $Et_2O$/Hexane | 91 | | 16 |
| ***2,3-Dihydro-1H-thieno[3,2-e][1,4]diazepines*** | 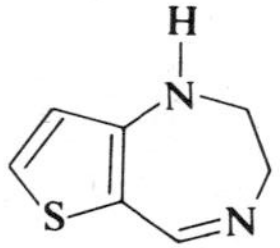 | | | | |
| 2-$H_2NCH_2$-5-Ph dimaleate | 187–189 | *i*-PrOH/MeOH | 49 | | 16 |

| | | | | | |
|---|---|---|---|---|---|
| ***2-Methylene-1,3-dihydro-2H-thieno[3,2-e][1,4]diazepines*** | | | | | |
| ***$R_1$, $R_2$; Other*** | | | | | |
| H, $NO_2$; 5-Ph | 163–164 | MeOH | 45 | | 16 |
| ***1,3-Dihydrothieno[3,2-e][1,4]diazepin-2(2H)-ones*** | | | | | |
| ***Monosubstituted*** | | | | | |
| 5-Me | 170 | | | ir, pmr | 108 |
| 5-Ph | 205–206 | EtOH | | | 104 |
| 4-Oxide | 266–270d | EtOH | | | 104 |
| ***Disubstituted*** | | | | | |
| 1-(Cyclopropyl)$CH_2$-5-Ph | 142–145 | EtOH | | | 104 |
| 1-Me-5-Ph | 159–161 | EtOH | | | 104 |
| 5-Ph-7-Cl | 220–221 | PhH | | | 106 |
| 5-Ph-8-Cl | 200–203 | EtOH | | | 107 |
| 5-Ph-8-Me | 233–238 | EtOH | | | 77b |
| 5-Ph-8-$NO_2$ | 215d | EtOH | | | 109 |
| 5-(2-$ClC_6H_4$)-7-Cl | 220 | EtOH | 58 | | 106 |
| 5-(2-$ClC_6H_4$)-7-$F_3C$ | 227–228 | PhH | 60 | | 105 |
| 5-(3-$NO_2C_6H_4$)-8-$NO_2$ | 230d | EtOH | 60 | | 109 |
| ***Trisubstituted*** | | | | | |
| 1-Me-5-Ph-8-Cl | 147–149 | EtOH | | | 107 |
| 1-Me-5-Ph-8-Me | 152–154 | Cyclohexane | | | 77b |
| 1-Me-5-Ph-8-$NO_2$ | 183–186 | EtOH | 50 | | 109 |
| 1-Me-5-(2-$ClC_6H_4$)-7-Cl | 127–128 | Cyclohexane | 62 | | 106 |

TABLE IX-1. —(contd.)

| Substituent | mp (°C) or; [bp (°C/torr)] | Solvent of Crystallization | Yield (%) | Spectra | Refs. |
|---|---|---|---|---|---|
| 5-Ph-7,8-$Cl_2$ | 226–227 | PhH | | | 107 |
| 5-Ph-7-Morpholino-8-CN | 235–236 | $CHCl_3$ | | ir, pmr, uv | 66 |
| ***Tetrasubstituted*** | | | | | |
| 1-Me-5-Ph-7,8-$Cl_2$ | 122 | EtOH | | | 107 |
| 1-Me-5-Ph-7-Morpholino-8-CN | 208–210 | EtOAc | | pmr | 66 |
| 1-Me-5-Ph-7-Piperidino-8-CN | 210–212 | EtOAc | 81 | pmr | 66 |
| ***3,4-Dihydro-1H-thieno[3,2-e][1,4]diazepin-2,5(2H,5H)-diones*** | 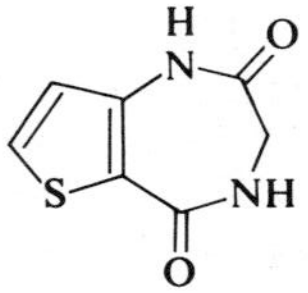 | | | | |
| 4-Me | 270–272 | DMSO/$H_2O$ | | | 110 |
| *Thieno[3,4-e][1,4]diazepines* | | | | | |
| ***1,3-Dihydrothieno[3,4-e][1,4]diazepin-2(2H)-ones*** | 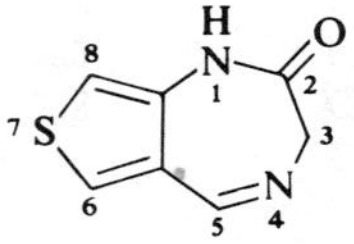 | | | | |
| ***Monosubstituted*** | | | | | |
| 5-Ph | 177–178 | PhH | 72 | | 111 |
| 5-(2-$F_3CC_6H_4$) | 186–187 | EtOH | 70 | | 112 |
| ***Disubstituted*** | | | | | |
| 1-Me-5-Ph | [130/0.003] | | | | |
| Picrate | 215–218d | EtOH | | | 111 |

| | | | | | |
|---|---|---|---|---|---|
| 5-Ph-8-Cl | 209–210d | EtOH | | | 115 |
| 5-Ph-8-$NO_2$ | 195–197d | EtOH | | | 115 |
| 5-(2-$ClC_6H_4$)-6-$F_3C$ | 190–191 | Cyclohexane | 50 | | 113 |
| 5-(2-$F_3CC_6H_4$)-8-Cl | 215–217 | EtOH | 75 | | 112 |
| 5-(2-$F_3CC_6H_4$)-8-$NO_2$ | 167–169d | EtOH | 60 | | 112 |
| ***Trisubstituted*** | | | | | |
| 1-Me-5-Ph-8-Cl | 150 | Cyclohexane | | | 115 |
| 5,6,8-$Me_3$ | 226–228 | $CHCl_3$/Petr ether | 74 | | 114 |
| 5-Ph-6,8-$Me_2$ | 271–272 | *i*-PrOH | 65 | | 114 |
| 5-(3-$ClC_6H_4$)-6,8-$Me_2$ | 221–222 | DMF/EtOH | 63 | | 114 |
| 5-(4-$ClC_6H_4$)-6,8-$Me_2$ | 263–264 | DMF/EtOH | 56 | | 114 |
| 5-(3-$MeC_6H_4$)-6,8-$Me_2$ | 210–212 | DMF | 54 | | 114 |
| ***3,4-Dihydro-1H-thieno[3,4-e][1,4]diazepin-2,5(2H,5H)-diones*** | | | | | |
| 4-Me | 263–265 | DMF/$Et_2O$ | | | 110 |
| | *1,2,4-Triazino[5,6-e][1,4]diazepines* | | | | |
| ***5,6,7,8-Tetrahydrotriazino[5,6-e][1,4]diazepin-9(9H)-ones*** | | | | | |
| 3-Ph-5,8-$Me_2$ | 268 | EtOH | 34 | ir, pmr, uv | 116 |

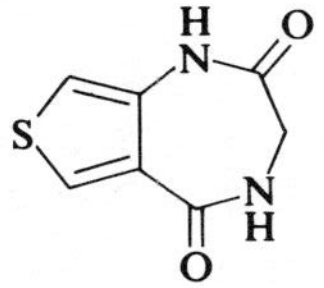

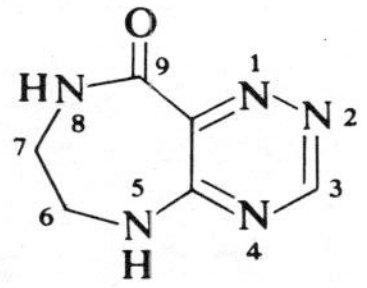

## 18. REFERENCES


1. E. Broger, unpublished data, Hoffmann-La Roche, Nutley, NJ.
2. P. L. Pacini and R. G. Ghirardelli, *J. Org. Chem.*, **31**, 4133 (1966).
3. H. Wamhoff, C. Materne, and F. Knoll, *Chem. Ber.*, **105**, 753 (1972).
4. (a) H. Griengl, G. Prischl, and A. Bleikolm, *Justus Liebigs Ann. Chem.*, 400 (1979). (b) H. Griengl and A. Bleikolm, Ger. Offen. 2,609,601, September 1976.
5. A. Edenhofer, *Helv. Chim. Acta*, **58**, 2192 (1975).
6. A. Szente, unpublished data, Hoffmann-La Roche & Co. AG, Basel, Switzerland.
7. B. P. Tong, unpublished data, Roche Products Ltd., Welwyn, England.
8. J. J. Tegeler and J. Diamond, U.S. Patent 4,514,410, April 1985.
9. (a) R. Jaunin, *Helv. Chim. Acta*, **57**, 1934 (1974). (b) R. Jaunin, unpublished data, Hoffmann-La Roche & Co. AG, Basel, Switzerland.
10. R. I. Fryer and J. V. Earley, unpublished data, Hoffmann-La Roche, Nutley, NJ.
11. R. F. C. Brown, I. D. Rae, J. S. Shannon, S. Sternhell, and J. M. Swan, *Aust. J. Chem.*, **19**, 503 (1966).
12. G. F. Field, L. H. Sternbach, and A. Walser, U.S. Patent 3,880,840, April 1975.
13. H. A. DeWald, S. Lobbestael, and B. P. H. Poschel, *J. Med. Chem.*, **24**, 982 (1981).
14. D. E. Butler, U.S. Patent 3,770,762, November 1973.
15. D. E. Butler, U.S. Patent 4,075,408, February 1978.
16. R. I. Fryer, J. V. Earley, and A. Walser, *J. Heterocycl. Chem.*, **15**, 619 (1978).
17. H. A. DeWald and S. J. Lobbestael, U.S. Patent 3,823,157, July 1974.
18. H. A. DeWald, S. Lobbestael, and D. E. Butler, *J. Med. Chem.*, **20**, 1562 (1977).
19. H. A. DeWald and D. E. Butler, U.S. Patent 3,558,605, January 1971.
20. H. A. DeWald and D. E. Butler, Ger. Offen. 2,023,453, October 1975.
21. H. A. DeWald, *J. Heterocycl. Chem.*, **11**, 1061 (1974).
22. W.-H. Hong and D. H. Szulczewski, *J. Pharm. Sci.*, **70**, 691 (1981).
23. L. R. Swett: (a) U.S. Patent 3,657,271, April 1972, (b) Br. Patent 1,357,978, June 1974.
24. H. A. DeWald, I. C. Nordin, I. J. L'Italien, and R. F. Parcell, *J. Med. Chem.*, **16**, 1346 (1973).
25. H. A. DeWald, U.S. Patent 3,557,095, January 1971.
26. L. R. Swett, U.S. Patent 3,764,688, October 1973.
27. Y. J. L'Italien and I. C. Nordin, U.S. Patent 3,553,209, January 1971.
28. I. C. Nordin, U.S. Patent 3,553,210, January 1971.
29. I. C. Nordin, U.S. Patent 3,553,207, January 1971.
30. W. H. Hong, C. Johnston, and D. Szulczewski, *J. Pharm. Sci.*, **66**, 1703 (1977).
31. P. G. Baraldi, S. Manfredini, V. Periotto, D. Simoni, M. Guarneri, and P. A. Borea, *J. Med. Chem.*, **28**, 683 (1985).
32. I. C. Nordin, U.S. Patent 3,700,657, October 1972.
33. R. Littell and D. S. Allen, *J. Med. Chem.*, **8**, 722 (1965).
34. S. Carboni, A. Da Settimo, D. Bertini, P. L. Ferrarini, O. Livi, and I. Tonetti, *Farmaco, Ed. Sci.*, **30**, 237 (1975).
35. W. Koch, unpublished data, Hoffmann-La Roche & Co. AG, Basel, Switzerland.
36. S. Carboni, A. Da Settimo, D. Bertini, P. L. Ferrarini, O. Livi, and I. Tonetti, *Farmaco, Ed. Sci.*, **31**, 322 (1976).
37. W. Hunkeler and E. Kyburz, U.S. Patent 4,362,732, December 1982.
38. M. K. Eberle and W. J. Houlihan, U.S. Patent 3,682,897, August 1972.
39. B. M. Pyatin and R. G. Glushkov, *Khim. Farm. Zh.*, **3**, 26 (1969).
40. B. M. Pyatin and R. G. Glushkov, *Khim. Farm. Zh.*, **3**, 13 (1969).
41. W. von Bebenburg, N. Schulmeyer, and V. Jakovlev, U.S. Patents 4,110,455, August 1978, 4,207,322, June 1980.
42. W. von Bebenburg and H. Offermanns, U.S. Patent 3,917,629, November 1975.
43. W. von Bebenburg and H. Offermanns, U.S. Patent 4,008,223, February 1977.
44. W. von Bebenburg and H. Offermanns, U.S. Patent 4,009,271, February 1977.

45. W. von Bebenburg and N. Schulmeyer, Ger. Offen. 2,428,469, August 1976.
46. Belg. Patent 821,016, March 1975 (Degussa, Germany).
47. W. von Bebenburg and H. Offermanns, U.S. Patent 3,900,466, August 1975.
48. W. von Bebenburg and H. Offermanns, U.S. Patent 3,972,873, August 1976.
49. W. von Bebenburg and H. Offermanns, U.S. Patent 3,920,633, November 1975.
50. I. Ahmed, G. W. H. Cheeseman, and P. Jaques, *Tetrahedron*, **35**, 1145 (1979).
51. C. Kawasaki and H. Yokoyama, *Vitamins* (Japan), **41**, 190 (1970).
52. C. Kawasaki, H. Yokoyama, G. Kurata, T. Sakai, and T. Miyahara, *Vitamins* (Japan), **37**, 165 (1968).
53. A. Takamizawa, K. Hirai, and T. Ishaba, *Tetrahedron Lett*., 437 (1970).
54. (a) D. H. Kim and A. A. Santilli, *J. Med. Chem*., **12**, 1121 (1969). (b) D. H. Kim and A. A. Santilli, U.S. Patent 3,535,310, October 1970.
55. A. A. Santilli and A. C. Scotese, *J. Heterocycl. Chem*., **16**, 213 (1979).
56. E. Garcia, unpublished data, Hoffmann-La Roche, Nutley, NJ.
57. L. Fontanella, L. Mariani, G. Tarzia, and N. Corsico, *Chim. Ther*., **11**, 217 (1976).
58. L. Fontanella, L. Mariani, G. Tarzia, U.S. Patent 4,022,766, May 1977.
59. A. Walser, unpublished data, Hoffmann-La Roche, Nutley, NJ.
60. L. Mariani and G. Tarzia, European Patent 0,102,602 Al, March 1984.
61. L. Mariani and G. Tarzia, Ger. Offen. 3,221,400 Al, December 1982.
62. B. Vitiello, G. Buniva, A. Bernareggi, A. Assandri, A. Perazzi, L. M. Fuccella, and R. Palumbo, *Int. J. Clin. Pharm*., **22**, 273 (1984).
63. A. Assandri, D. Baroni, P. Ferrari, A. Perazzi, A. Ripamonte, G. Tuan, L. F. Zerilli, *Drug Met. Disp*., **12**, 257 (1984).
64. Y. Sakamoto and T. Kurihara, *Yakugaku Zasshi*, **99**, 818 (1979).
65. R. Y. Ning, unpublished data, Hoffmann-La Roche, Nutley, NJ.
66. K. Hirai, H. Sugimoto, and T. Ishiba, *J. Org. Chem*., **45**, 253 (1980).
67. K. A. Maier and O. Hromatka, *Monatsh. Chem*., **102**, 1010 (1971).
68. T. Tahara, K. Araki, M. Shiroki, H. Matsuo, and T. Munakata, *Arzneim.-Forsch*., **28**, 1153 (1978).
69. M. Nakanishi, T. Tahara, K. Araki, and M. Shiroki, U.S. Patent 3,904,641, September 1975.
70. M. Nakanishi, T. Tahara, K. Araki, and M. Shiroki, U.S. Patent 3,828,039, August 1974.
71. Q. Branca, unpublished data, Hoffmann-La Roche & Co. AG, Basel, Switzerland.
72. T. Tahara, H. Matsuki, K. Araki, and M. Shiroki, U.S. Patent 3,952,006, April 1976.
73. M. Nakanishi, T. Tahara, K. Araki, and M. Shiroki, U.S. Patent 3,920,679, November 1975.
74. K. -H. Weber, A. Bauer, P. Danneberg, and J. Kuhn, U.S. Patent 4,094,984, June 1978.
75. M. Nakanishi, K. Araki, T. Tahara, and M. Shiroki, U.S. Patent 3,965,111, June 1976.
76. K. -H. Weber, A. Langbein, E. Lehr, K. Boeke, and F. J. Kuhn, Ger. Offen. 2,701,752, July 1978.
77. (a) J. Hellerbach, P. Zeller, D. Binder, and O. Hromatka, U.S. Patent 4,155,913, May 1979. (b) O. Hromatka and J. Hellerbach, unpublished data, Hoffmann-La Roche & Co. AG, Basel, Switzerland.
78. K.-H. Weber, A. Bauer, P. Danneberg, and F. J. Kuhn, U.S. Patent 4,199,588, April 1980.
79. J. Hellerbach, P. Zeller, D. Binder, and O. Hromatka, Ger. Offen. 2,405,682, August 1974.
80. M. Nakanishi, K. Araki, T. Tahara, and M. Shiroki, U.S. Patent 3,840,558, October 1974.
81. M. Nakanishi, K. Araki, T. Tahara, and M. Shiroki, U.S. Patent 4,010,184, March 1977.
82. F. J. Tinney, J. Sanchez, and J. A. Nogas, *J. Med. Chem*., **17**, 624 (1974).
83. O. Hromatka, D. Binder, C. R. Noe, P. Stanetty, and W. Veit, *Monatsh. Chem*., **104**, 715 (1973).
84. O. Hromatka, D. Binder, P. Stanetty, and G. Marischler, *Monatsh. Chem*., **107**, 233 (1976).
85. O. Hromatka, and D. Binder, *Monatsh. Chem*., **104**, 704 (1973).
86. (a) M. Nakanishi, T. Tahara, K. Araki, M. Shiroki, T. Tsumagari, and Y. Takigawa, *J. Med. Chem*., **16**, 214 (1973). (b) M. Nakanishi, K. Araki, T. Tahara, and M. Shiroki, U.S. Patent 3,849,405, November 1974.
87. D. Binder, O. Hromatka, C. R. Noe, F. Hillebrand, and W. Veit, *Arch. Pharm*., **313**, 587 (1980).
88. D. Binder, O. Hromatka, C. R. Noe, Y. A. Bara, M. Feifel, G. Habison, and F. Leierer, *Arch. Pharm*., **313**, 636 (1980).

89. M. Nakanishi, M. Shiroki, T. Tahara, and K. Araki, U.S. Patent 3,806,512, April 1974.
90. Belg. Patent 771,951, September 1971 (Yoshitomi Pharm. Ind., Ltd., Japan).
91. Belg. Patent 811,726, June 1974 (Lab. Made S.A., Spain).
92. O. Hromatka and D. Binder, U.S. Patent 3,669,959, June 1972.
93. T. Hirohashi, S. Inaba, and H. Yamamoto, *Bull. Chem. Soc. Japan*, **48**, 147 (1975).
94. M. Nakanishi, M. Shiroki, T. Tahara, and K. Araki, U.S. Patent 3,859,275, January 1975.
95. O. Hromatka, D. Binder, and P. Stanetty, *Monatsh. Chem.*, **104**, 709 (1973).
96. O. Hromatka, D. Binder, and W. Veit, *Monatsh. Chem.*, **104**, 973 (1973).
97. O. Hromatka and D. Binder, U.S. Patent 3,872,089, March 1975.
98. J.-C. Cognacq, U.S. Patent 4,156,009, May 1979.
99. O. Hromatka, D. Binder, and P. Stanetty, *Monatsh. Chem.*, **104**, 920 (1973).
100. M. Nakanishi, M. Shiroki, T. Tahara, and K. Araki, U.S. Patent 3,887,543, June 1975.
101. M. Inotsume and M. Nakano, *Chem. Pharm. Bull.*, **28**, 2536 (1980).
102. T. Hirohashi, S. Inaba, and H. Yamamoto, *Bull. Chem. Soc. Japan*, **48**, 974 (1975).
103. K.-H. Weber, and H. Daniel, *Justus Liebigs Ann. Chem.*, 328 (1979).
104. O. Hromatka, and D. Binder, *Monatsh. Chem.*, **104**, 1343 (1973).
105. O. Hromatka, D. Binder, and K. Eichinger, *Monatsh. Chem.*, **105**, 138 (1974).
106. O. Hromatka, D. Binder, and G. Pixner, *Monatsh. Chem.*, **106**, 1103 (1975).
107. O. Hromatka, D. Binder, and G. Pixner, *Monatsh. Chem.*, **104**, 1348 (1973).
108. G. Ah-Kow, C. Paulmier, and P. Pastour, *Bull. Soc. Chim.*, 151 (1976).
109. O. Hromatka, and D. Binder, *Monatsh. Chem.*, **104**, 1105 (1973).
110. M. Gerecke, W. Haefeli, W. Hunkeler, E. Kyburz, H. Moehler, L. Pieri, and P. Polc, U.S. Patent 4,316,839, February 1982.
111. O. Hromatka, D. Binder, and K. Eichinger, *Monatsh. Chem.*, **104**, 1513 (1973).
112. O. Hromatka, D. Binder, and K. Eichinger, *Monatsh. Chem.*, **105**, 123 (1974).
113. O. Hromatka, D. Binder, and K. Eichinger, *Monatsh. Chem.*, **105**, 135 (1974).
114. (a) K. Grohe and H. Heitzer, *Justus Liebigs Ann. Chem.*, 1947 (1977). (b) K. Grohe, F. Hoffmeister, and W. Wuttke, Ger. Offen. 2,047,013, March 1972.
115. O. Hromatka, D. Binder, and K. Eichinger, *Monatsh. Chem.*, **104**, 1599 (1973).
116. M. Brugger, H. Wamhoff, and F. Korte, *Justus Liebigs Ann. Chem.*, **758**, 173 (1972).
117. G. F. Field, unpublished data, Hoffmann-La Roche, Nutley, NJ.

# Author Index

Page citations followed by numbers in parentheses indicate that an author's work is discussed on that page without his name being given in the text. The parenthetical numbers are used to designate the appropriate literature references cited on any given page. Page numbers giving full literature references at the end of each chapter are italicized and are to be used with the citations that immediately preceed them. Page numbers that appear in italics at the beginning of each chapter indicate full literature references for previous review articles. In a general way the italicized numbers may also be used to indicate chapter separations. Unpublished data from the files of the Hoffmann-La Roche company reported in Chapter VII, were in general, not included in this index.

# Subject Index

Page numbers followed by the letter "t", indicate the beginning page for the appropriate table of compounds. References to "Name" reactions and rearrangements within the text are exemplary only and were not used consistently. Each entry under "Generic Names" is followed by chemical nomenclature, in order to avoid possible confusion. References to spectral data listed in the tables are not indexed. Due to the volume of tabular material for dihydro-1,4-benzodiazepin-2-ones (Chapter VII), indexing of this material has been subdivided according to the number of ring substituents for ease of reference. This index should be used in conjunction with the individual chapter indices and does not duplicate information provided therein.